The Geological Evolution of the Eastern Mediterranean

edited by

J. E. Dixon
and
A. H. F. Robertson

Department of Geology, University of Edinburgh,
West Mains Road, Edinburgh, EH9 3JW, Scotland

1984

Published for
The Geological Society
by Blackwell Scientific Publications
Oxford London Edinburgh
Boston Palo Alto Melbourne

Published by

Blackwell Scientific Publications
Osney Mead, Oxford OX2 0EL
8 John Street, London WC1N 2ES
9 Forrest Road, Edinburgh EH1 2QH
52 Beacon Street, Boston, Massachusetts 02108, USA
706 Cowper Street, Palo Alto California 94301, USA
99 Barry Street, Carlton, Victoria 3053, Australia

First published 1984

Printed in Great Britain
at The Pitman Press, Bath

DISTRIBUTORS

USA and Canada
Blackwell Scientific Publications Inc
PO Box 50009, Palo Alto
California 94303

Australia
Blackwell Scientific Book Distributors
31 Advantage Road, Highett
Victoria 3190

British Library Cataloguing in Publication Data

The Geological evolution of the Eastern Mediterranean.—(Special publications of the Geological Society, ISSN 0305–8719)
1. Geology—Near East
I. Dixon, J.E. II. Robertson, A.H.F.
III. Series
551.091822 QE319.N4

ISBN 0–632–01144–0

Preface

In the years since the 1977 Aegean conferences in Athens and Izmir there has been an explosion of new information on the Eastern Mediterranean, particularly on the Turkish area. Some three years ago we sensed that the time might be ripe for a meeting to consider all aspects of geology relevant to the tectonic evolution of the Eastern Mediterranean. We sounded out opinion in most of the European laboratories involved and received enthusiastic support for the idea. The boundaries of the area to be discussed were fixed early on: the Apennines to the west, the Carpathians and Caucasus to the north, the Zagros to the east and North Africa to the south. The meeting concentrated on the Late Palaeozoic to Recent evolution, as few coherent tracts of older rocks exist in the area.

The conference was held in Edinburgh from the 28 to 30 September 1982 and was attended by 220 scientists from 13 countries, 88 papers were read and 22 given in poster form. Sixty one contributions are published in this volume.

One of the key starting points of Eastern Mediterranean geology is deceptively easy to state. Several global reconstructions of the continents for Permian time indicate that a substantial tract of ocean, the 'Tethys', existed between Africa, and a Eurasian landmass to the north. Since little or none of this oceanic crust remains, the history of the Tethys from the Late Palaeozoic onwards must involve destruction of this ocean area. Perhaps surprisingly, however, the geology of the Eastern Mediterranean land areas is not dominated by this process but by the products of Mesozoic rifting and the formation and destruction of new oceanic basins. From this it is evident that closure of the Palaeozoic Tethyan ocean can not simply have involved continuing subduction culminating in continental collision in the Neogene. Tectonic events must have involved comprehensive re-organizations of plate boundaries so that the distribution of continent and ocean is almost bound to have been complicated and frequently changing.

The book is organized by age of events discussed and by area within this framework, and it is in five sections. Following the editors' introductory chapter the first is concerned with early Mesozoic events linked to ocean closure and so by implication deals with the fate of the Palaeozoic Tethys or *Palaeotethys*. Later sections deal with events following the birth and growth of wholly Mesozoic ocean basins of the *Neotethys*. This division follows the interpretation of the contributors and is not inherently clear-cut. We use the term Palaeotethys to imply an ocean basin already in existence at the end of the Palaeozoic, but one that may have continued in existence in the Mesozoic and Tertiary. We argue, for example, in our own introductory chapter that the destruction of Palaeotethys in this sense was a long-drawn-out process lasting until Tertiary times, and even that early Mesozoic compressional events in the area, though related to Palaeotethyan subduction, could also have involved the formation of short-lived Neotethyan ocean basins. While questions of this kind dominate the book as a whole, much of the research documented here is also concerned with fundamental questions of continent–ocean dynamics. How do continental margins evolve from the early rift stage? Where are ophiolites formed and how are they emplaced? How do blueschist belts come to be created and exhumed? What is the relationship between consumption of oceanic crust and the construction of volcanic arcs? With its great geological diversity and relative accessibility the Eastern Mediterranean is an exceptionally valuable field laboratory, producing results complementary, for example, to those of the Deep Sea Drilling Project or marine geophysical exploration.

The reader will find disparity in the depth of knowledge and the coverage of different areas and processes. He may be disappointed that the jig-saw puzzle is not yet fully solved, but he may in return come to appreciate how many of the pieces are turning out to have the same pattern on them, and he may catch occasional glimpses of parts of the picture. He will also find amid the tentative models and the many expressions of uncertainty, flat contradictions and diametrically opposed interpretations. Perhaps these simply reflect the state of earth science a decade and a half after the plate-tectonic revolution. They are, none-the-less, a stimulus for us to question the validity of the ground rules. Do palaeomagnetic inclinations faithfully record palaeo-latitudes? Are ophiolites really remnants of oceanic crust, from marginal or any other kind of basin? Does calc-alkaline magmatism mean active subduction? Must blueschists and melanges also be linked to subduction? Whatever one's doubts, the philosophy behind a book like this is clear enough. Directing the contributors to the wider implications of their work should help progress towards greater overall understanding which must come from the dynamic interaction of large-scale thinking and small-scale research.

ACKNOWLEDGEMENTS:

Esso Exploration Inc. provided generous financial support for which we are particularly grateful. Financial support was also received from BP.

We would like to thank Professor G. Y. Craig of the University of Edinburgh, Department of Geology, for making available to us the full range of departmental facilities both in initially organizing the conference, then editing this book. Our sincere thanks go to our secretaries Mrs M. Wright and Mrs H. Hooker for their unfailing help, particularly with retyping edited manuscripts. Mrs D. Baty assisted with photography and, together with Mrs F. Tullis, with drafting also.

Finally we wish to record out thanks to Mr N. Palmer of Blackwell Scientific Publications for his friendly advice and assistance at all stages.

Contents

4. NEOTETHYS: GREECE AND THE BALKANS

5. NEOGENE

Introduction: aspects of the geological evolution of the Eastern Mediterranean

A. H. F. Robertson & J. E. Dixon

SUMMARY: We review extensively the evidence and arguments bearing on the nature of Palaeotethys in relation to the age of formation, location and multiplicity of Neotethyan strands and their fate. We conclude that Palaeotethys did not die early but was only finally subducted northwards in the Tertiary along the Vardar-Intra Pontide-East Anatolian suture. Neotethyan strands must have opened into it at all times. The Adriatic promontory remained attached to Africa but rotated anti-clockwise in the mid-Tertiary. The Pontides are considered to be Eurasian and the Cimmerides are viewed as a 'collage terrain' formed along an oblique-convergence margin. The south Aegean, Greek and Turkish micro-continental blocks were rifted-off Gondwana in the Triassic but formation of braided Neotethyan oceanic crustal strands was essentially confined to mid-Jurassic in the Hellenides and to the Cretaceous in Turkey.

We propose a new model of ophiolite genesis by asymmetrical spreading-ridge collapse in an attempt to explain both arc-like ophiolite chemistry prior to major volcanic arc edifice construction, and the synchroneity of sub-ophiolite metamorphic sole formation with Atlantic opening phases.

Jurassic dispersal of Hellenide blocks had little effect in the unexpanded Turkish mosaic, but northwards Cretaceous opening of Tauride Neotethyan strands caused oblique collision deformation in the Pelagonian zone and unresolvable complexity in the Aegean. Late Cretaceous and Tertiary arc-volcanism was related in part to continuing Palaeotethyan subduction, and in part to Neotethyan destruction initiated after ridge-collapse. Diachronous collisions ensued from the Late Cretaceous onwards but significant oceanic tracts must have persisted at least to Mid-Tertiary to satisfy Africa-Eurasia separation constraints determined from Atlantic anomaly fitting. Our favoured plate evolutionary model is presented in 7 sketch-maps.

The problems posed

From the 60 papers included in this book it is clear that the Eastern Mediterranean is an extremely complicated area which has only recently become well enough known to attempt overall tectonic synthesis (see Smith & Woodcock 1982, for a recent summary of views, Fig. 1). An outline solution emerged well over a decade ago when increased palaeomagnetic data for the Africa and Eurasian continents suggested that in Permo-Triassic time the two super-continents forming Pangaea were separated by a wedge-shaped westward-narrowing gulf, the Tethys (Wegener 1924; Bullard *et al.* 1965; Smith & Briden 1977). Then, as the magnetic anomalies in the Atlantic ocean were dated and correlated it became clear that Africa and Eurasia must have undergone successive phases of relative shear and compression during Mesozoic and Tertiary times. An outline solution to Tethyan evolution must thus have involved closure of both an essentially Palaeozoic ocean (*Palaeotethys*) as well as younger ones formed during the Mesozoic and even the Tertiary (*Neotethys*).

A number of fundamental questions arise if we look closer at the problem. What was the Late Palaeozoic continental configuration in detail? Did the Palaeozoic ocean close northwards or southwards, or in both directions? Did the Palaeotethys remain open and perhaps actively spreading, and so continue to influence plate boundaries into the Mesozoic and Tertiary, or was it by then already long dead, consumed by subduction? Are we to seek a solution in which many of the continental fragments have ended up not far removed from where they were in the Late Palaeozoic relative to Africa or Eurasia, or should we, for example, seek an origin for some components of the Tethys completely outside the present East Mediterranean area? How great a role has strike-slip played in reversibly or irreversibly displacing microcontinental blocks?

Hopefully palaeomagnetism and the Atlantic spreading record can provide a framework, but then a successful Tethyan synthesis depends on correctly interpreting and integrating a mass of regional geological data of often widely differing scope, detail and reliability, with the very concept of plate subduction meaning that most if not all of the record of ocean floor spreading has been destroyed without a trace.

A hoped-for consensus has not really emerged from the contributions in this book despite many being wide-ranging syntheses.

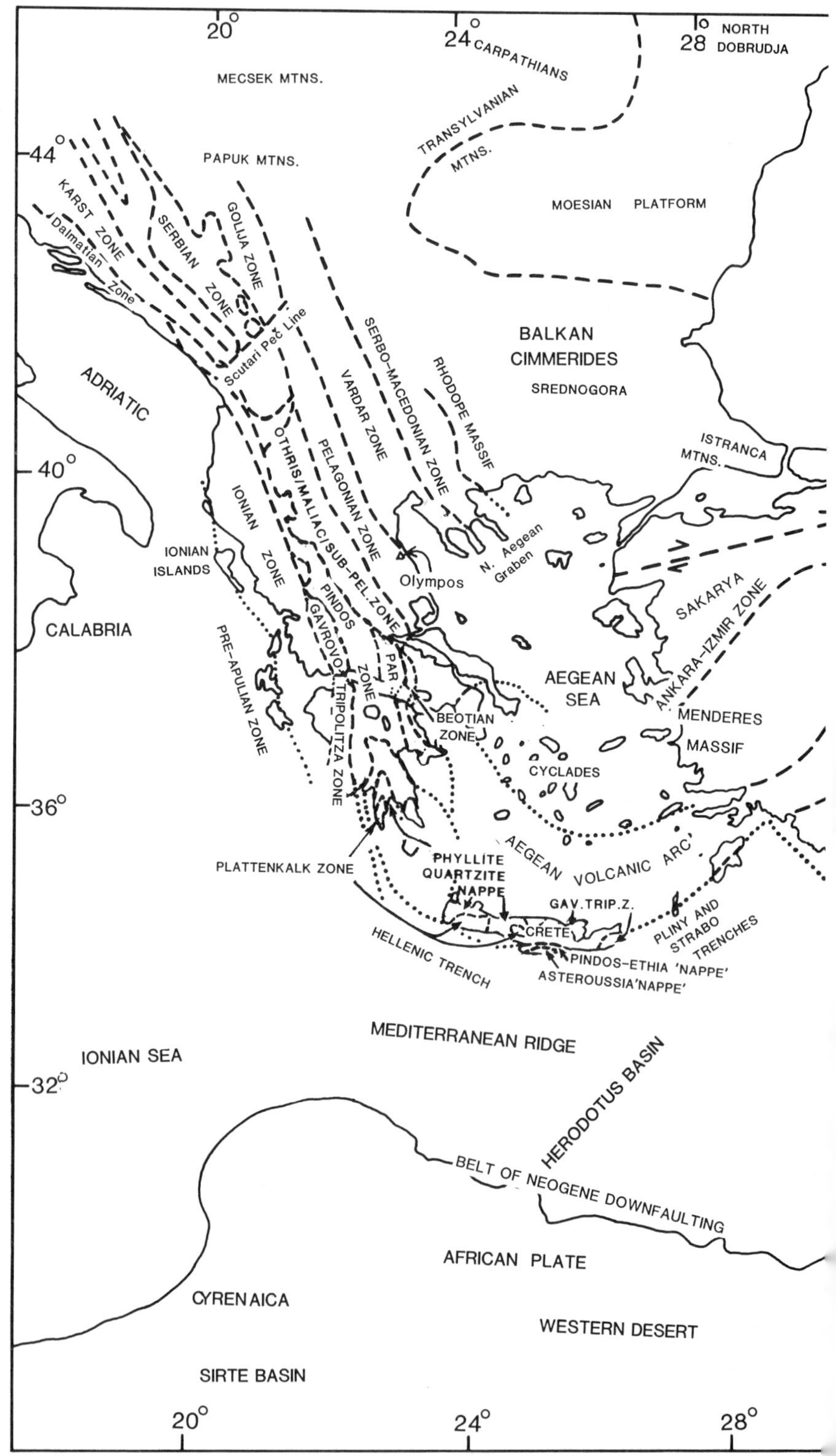

FIG. 1. Outline sketch of the Eastern Mediterranean showing many of the important tectonic units mentioned in the text.

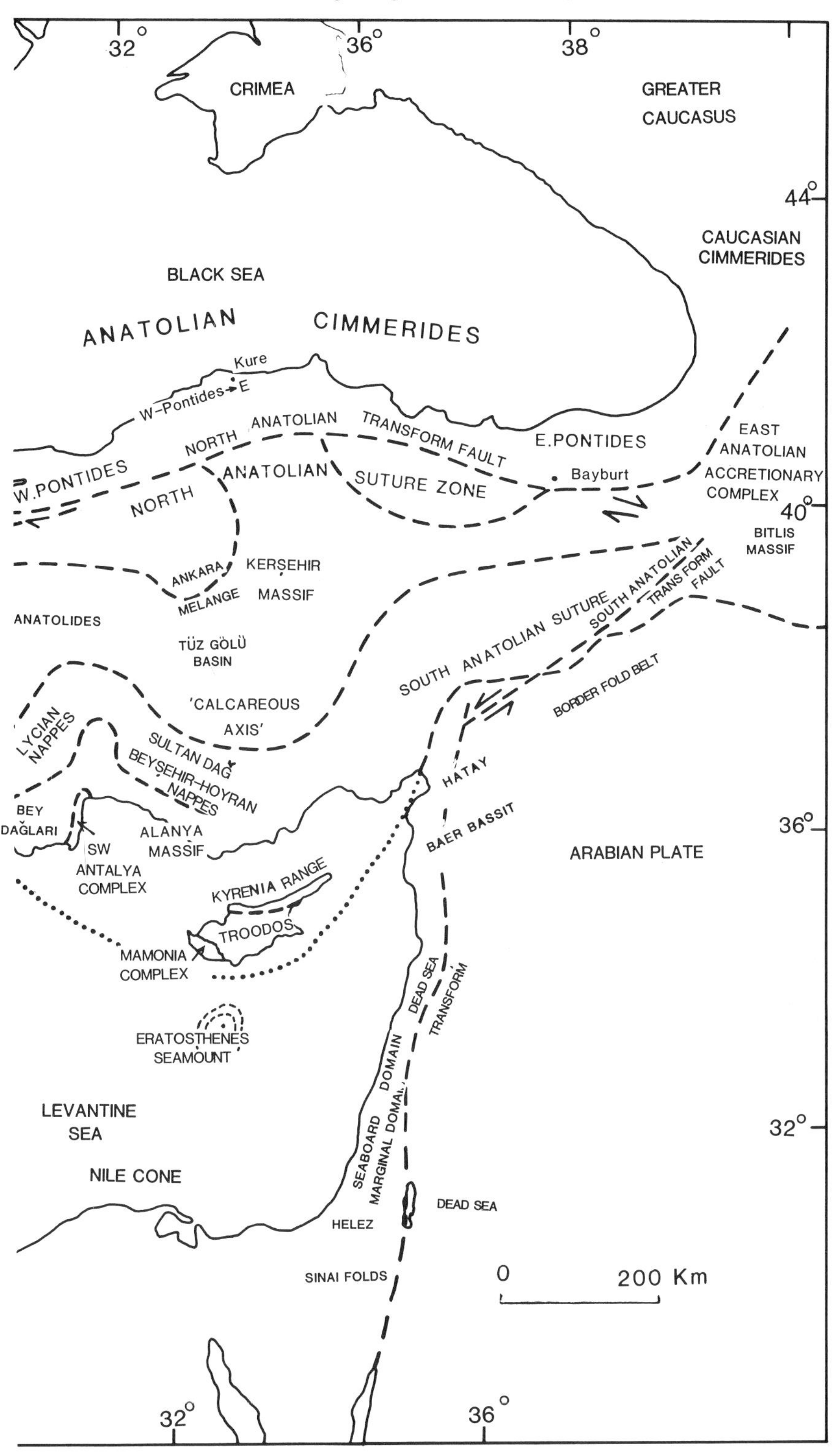

32°
36°
38°
CRIMEA
GREATER CAUCASUS
44°
CAUCASIAN CIMMERIDES
BLACK SEA
ANATOLIAN CIMMERIDES
Kure
W-Pontides E
NORTH ANATOLIAN TRANSFORM FAULT
E.PONTIDES
W.PONTIDES
NORTH ANATOLIAN SUTURE ZONE
Bayburt
EAST ANATOLIAN ACCRETIONARY COMPLEX
40°
BITLIS MASSIF
ANKARA MELANGE
KERŞEHIR MASSIF
SOUTH ANATOLIAN TRANS FORM FAULT
SOUTH ANATOLIAN SUTURE
ANATOLIDES
TÜZ GÖLÜ BASIN
'CALCAREOUS AXIS'
BORDER FOLD BELT
LYCIAN NAPPES
SULTAN DAĞ
BEYŞEHIR-HOYRAN NAPPES
HATAY
BEY DAĞLARI
ALANYA MASSIF
SW ANTALYA COMPLEX
BAER BASSIT
36°
ARABIAN PLATE
KYRENIA RANGE
TROODOS
MAMONIA COMPLEX
DEAD SEA TRANSFORM
ERATOSTHENES SEAMOUNT
DOMAIN
SEABOARD
MARGINAL DOMAIN
LEVANTINE SEA
32°
NILE CONE
DEAD SEA
HELEZ
SINAI FOLDS
0 200 Km
32°
36°

Nevertheless, on many of the critical issues, for example the number of major Neotethyan strands that were created, the debate reduces to a clear but limited choice of alternatives. Our aim is to clarify arguments in the chain of decisions, offer our own preferred options and so end up with what we hope is a consistent set of component conclusions for an overall solution. In our concluding section we look at the constraints imposed by plate geometry—how to disperse and reassemble the pieces of the jigsaw through the Mesozoic and Tertiary to comply with our set of conclusions and, perhaps most importantly, how to fit the entire puzzle within the Eurasia-Africa convergence frame.

We have attempted to find the simplest solution to each component problem, which is generally good science, but is liable to lead to an unwarranted simplification of the geological record. Conversely, a genuine solution is liable to be so complicated that it may well not be understood and accepted by workers in other areas. Tectonic modelling is founded on the individual's faith that solutions are there to be found.

For brevity, where possible, we cite individual review contributions in the book rather than original sources. We are aware of major gaps in our own knowledge, particularly of eastern Yugoslavia and Bulgaria, areas where important Mesozoic sutures certainly exist, but which we have not fully incorporated into our models.

The shape of Pangaea

Previous syntheses of the Eastern Mediterranean have tended to take for granted Wegener's (1924) reconstruction of Pangaea which placed NW Africa against SE North America, with the well known Tethyan gulf in Permo-Carboniferous and Triassic time. While there is, indeed, a consensus that the Tethyan gulf did exist by Early Jurassic time (Smith *et al.* 1981; Irving 1977; see also Hallam 1980; Fig. 2), controversy persists as to the stability of this configuration prior to the Jurassic. Palaeopoles for Eurasia, North America and Africa imply that Gondwana should be placed further north than in Wegener's famous reconstruction and this can readily be achieved by placing the NW part of South America against the SE Atlantic margin of North America (recent review of Irving 1982); an acceptable alternative for the Permian places the NW coast of South America along the Gulf coast of North America (Peinado *et al.* 1982).

The transition from any Permo-Carboniferous Pangaean configuration to the Early Juras-

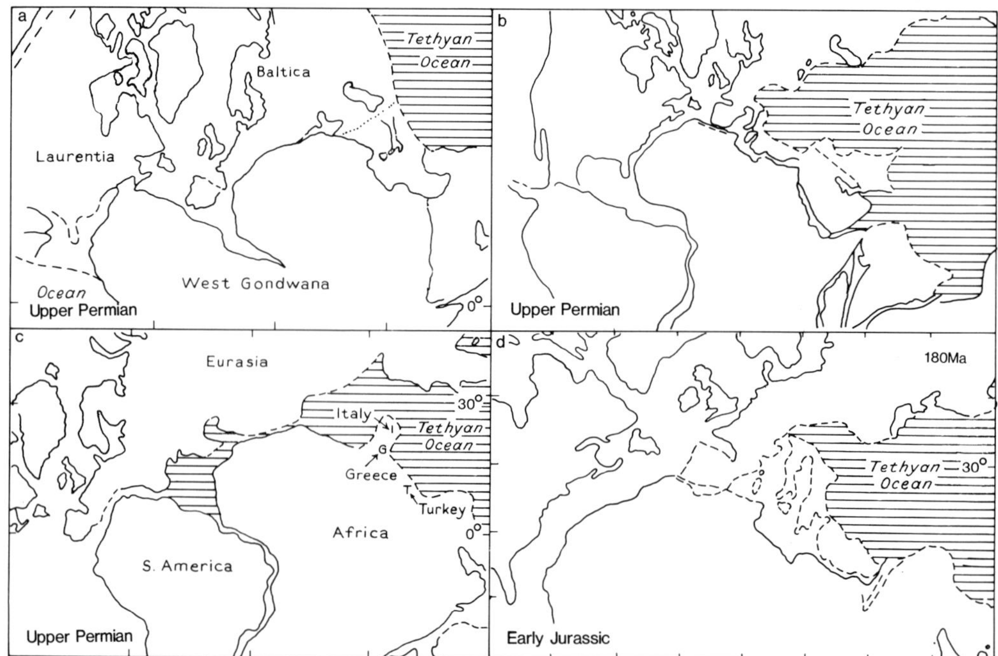

FIG. 2. Alternative reconstructions of Pangaea in the Late Permian and Early Jurassic. a, Late Permian modified after Irving (1982); b, Late Permian, modified after Smith and Briden (1977); c, Late Permian, modified after Smith & Woodcock (1981); d, Early Jurassic, modified after Smith & Woodcock (1981).

sic one must involve dextral translation of Eurasia relative to Africa; proposals range from about 500 km to as much as 9000 km (Irving 1982). Smith *et al.* (1981) rotate Gondwana about 7000 km eastward relative to Laurasia, with South America initially against SW Mediterranean Europe and NW Africa against central southern Eurasia (see also Smith & Woodcock 1982). Both a relatively small (ca. 500 km, Van der Voo & French 1974) and a very large (Smith *et al.* 1981) rotation require a significant oceanic separation in the Permian between North and South America and Eurasia, and between South America and Africa, for which there is still little evidence (e.g. see Ogg *et al.*, 1984, for synthesis of early Atlantic evolution). Notably, Smith *et al.*'s (1981) reconstruction very greatly reduces the width of the Upper Permian Palaeotethys (Fig. 2), while an acceptable alternative using 3500 km of clockwise rotation (Irving 1977) potentially removes it altogether leaving an essentially N–S western margin to the super-ocean, Panthalassa.

When did any clockwise rotation of Eurasia relative to Africa take place? Hurley, in Smith & Woodcock (1982) proposes a transform zone passing through the northern edge of Gondwana. Recently determined palaeopoles for the Triassic and Jurassic of North America (Colorado Plateau) suggest that much of the clockwise rotation may have occurred as late as the Middle Triassic and again in the Early Jurassic (Steiner 1983, see below). Notably, the rotation appears to have been relative to an effectively stable position of Africa.

The general point is thus that far from being an established fact, the very former existence of a major Palaeotethyan ocean in the East Mediterranean area, distinct from the super-ocean, Panthalassa, is a controversial hypothesis based on still scanty palaeomagnetic data of arguable reliability. Any synthesis which does not accommodate major clockwise rotation of Eurasia-North America relative to Africa in the Permian to Early Jurassic interval is unlikely to be fully correct. According to the timing of dextral translation of Eurasia relative to Africa, there could thus have been a wide Upper Permian Palaeotethys in the Eastern Mediterranean extending into the Western Mediterranean, a narrow one or even no ocean at all.

We adopt a configuration for the Triassic to Mid-Jurassic interval in our concluding section which maintains a wide ocean connection between the East and West Tethyan areas during the period of dextral, and subsequent sinistral, shear of Eurasia relative to Africa. This does not exclude the possibility of an earlier, Permian, configuration with the ocean connection largely eliminated. The crucial constraints are the amount of pre-Atlantic-opening convergence and divergence between West Africa and North America that the geology will allow and the range of possible pure-slip transform paths that can pass between the two and still be consistent with the palaeomagnetic data. After a long period of controversy on the significance of palaeomagnetic data from the Italian area (especially Umbria; see recent reviews by Vandenberg & Zijderveld 1982, and Wensink 1981), there is now a growing, but still by no means complete, consensus that the Adriatic was indeed a major peninsula of Gondwana until the post-Eocene when a 25° anticlockwise rotation (of at least Umbria) is needed to bring the palaeopoles into conformity with those of stable Africa. This conclusion is now strengthened by **Márton's*** new palaeomagnetic data which help confirm that the Italian peninsula south of the Po Basin, and the Istria peninsula of Yugoslavia together with some adjacent areas formed part of Africa in the Mesozoic. If, as we favour, the Adriatic area formed a major promontory of Gondwana (Argand, 1924) then the major strike-slip faults on which Africa-Eurasia shear occurred would probably have been constrained to run north of it, possibly along an active southern Eurasian margin.

The fossil evidence

Faunal and floral distribution can, in theory, provide an independent check of the former existence of Tethyan oceans postulated from palaeomagnetic data. In an essentially east-west trending orogenic belt the fossil evidence has the advantage of potentially reflecting geographical separation as well as palaeolatitude. It is however increasingly being realized that biogeography is influenced by a whole range of factors including temperature, bathymetry, facies, eustatic sea level changes, and ocean currents, among others. Facies differed strongly along the African and Eurasian shores of the Mesozoic Tethys (Bernoulli & Jenkyns 1974) and some early generalizations about provinciality are now questionable.

(i) Palaeotethys

Does the fossil evidence help with the Palaeotethyan problem? Plant distributions are potentially useful during the Late Palaeozoic time of pronounced provinciality (reviewed by Truswell 1981). Early Permian coals in Turkey

* References to papers in this volume are in **bold** type

(Kayseri region) are reported to contain genera and species in common with SE China. On Ziegler's (1981) paleogeographical maps, for example, China is placed as a block contiguous with the eastern edge of Laurasia. Megafossil floras on the Gondwana margin of SE Turkey (Hazro inlier) contain both Gondwanian and Cathaysian elements (Wagner 1962; Akyol 1975), which is significant as Cathaysian forms are not reported from Europe and the USSR (Truswell 1981), implying, in turn, that migrations along the southern shores of the Tethyan ocean were unable to extend as far as Eurasia. **Kerey** points out that the Upper Carboniferous floras of the Pontides (Black Sea coast) are more comparable with those of Eurasia than with Gondwana.

The limited fossil evidence is quite consistent with the existence of a Palaeotethyan ocean separating Africa from Laurasia. Taken with **Kerey's** sedimentological evidence of provenance of the Upper Carboniferous facies of the Black Sea coast from a major landmass to the north, the palaeontological evidence, however, provides no support for the Pontides having formed part of Gondwana until the Mid Triassic as in **Şengör *et al.*'s** model (Fig. 3). It appears very probable that the Pontides had an Eurasian affinity, while the Tauride-Anatolide platform areas to the south clearly formed part of Gondwana. We will consider later whether they were then attached to North African or somewhere off the eastern margin of Gondwana as in **Lauer's** hypothesis.

(ii) Neotethys

In the Mesozoic the faunal evidence centres around the provinciality of ammonites and of benthos, particularly bivalves, brachiopods, foraminifera and calcareous algae. The distributions of pelagic organisms must be viewed with caution: for example, periods of environmental stability combined with topographic barriers (Dommergues 1982) (e.g. carbonate platforms) tended to promote faunal differentiation while phases of ocean floor spreading and increased faunal differentiation may in fact have favoured increased mixing and dispersal of faunas (Enay & Mangold 1982). However, the existence of distinctive boreal and Tethyan ammonite faunas generally, has long been known (Arkell 1956), particularly for the Middle Liassic which was a time of marked provinciality (Cèczy 1973; Enay 1976). There is a marked affinity between south Tethyan ammonite faunas and those of India and Madagascar which favours an Early Jurassic oceanic connection, or at least a seaway along the north margin of Gondwana isolated from the Eurasian margin (Thierry 1982).

The distributions of benthic foraminifera and calcareous algae link the Adriatic promontory with Gondwana and identify them as distinct from the south margin of Eurasia (e.g. parts of the Balkans, Yugoslavia, and Hungary; Pélissié *et al.* 1982; Schroeder *et al.* 1974), thus providing no support for the concept of an oceanic separation between southern Italy and Gondwana (c.f. Biju-Duval *et al.* 1977). This conclusion appears also now to be supported by the palaeomagnetic data and facies continuity.

Support for the former existence of some form of a Mesozoic Tethyan ocean comes from the distribution of benthos, mostly in the western Tethys, including Liassic rhynchonellids (Ager, 1967), Pliensbachian brachiopods (Vörös, 1977) and upper Liassic *Lithiotis* fauna, which includes large bivalves, brachiopods, foraminifera and algae typical of the south Tethys but hardly known on the Eurasian margin (Broglio-Loriga & Neri, 1976).

In addition, **Hirsch** summarizes the still scanty faunal data for Turkey and Greece in the Mesozoic. The Mesozoic brachiopod faunas of the 'Apulian' area of Greece are apparently distinctively south-Tethyan, as are the Late Palaeozoic conodont, the Mesozoic vertebrate, and the ammonite faunas of the Turkish Tauride carbonate platforms ('Tauride Calcareous Axis'). Tantalizingly, **Ricou *et al.*** mention a distinctive ammonite fauna in one part of the Tauride carbonate platform (Şeydişehir area), but few details are given.

The faunal evidence is thus consistent with a Mesozoic oceanic basin lying between Gondwana and Eurasia, but has little bearing on whether that ocean had persisted from the Palaeozoic (Palaeotethys), or whether it was newly created in the Mesozoic and if so, when. One must note that, especially in the western Tethys, a wide area of carbonate platforms (i.e. like the modern Bahamas) divided by deep rifted marine basins could have been a more effective barrier to faunal dispersal than any open oceanic separation.

Palaeo-oceanography

One other related line of evidence is the palaeo-oceanography of Triassic and Jurassic radiolaria. As reviewed by Jenkyns & Winterer (1982) (see also Ricou & Marcoux 1980) there is now a growing view that the radiolarites of Upper Triassic age in the eastern Tethys, and those of the Upper Jurassic (Tithonian) in the Tethys generally, reflect periods of accentuated productivity related to upwelling in rifts or small ocean basins marked by a high carbonate compensation depth.

In the western Tethys, where there is little evidence of a significant Mesozoic oceanic separation of Gondwana from Eurasia until the Upper Jurassic (see below), extensive non-calcareous radiolarites appear with opening of Upper Jurassic ocean basins which accompanied intial spreading of the central North Atlantic (e.g. Hsü 1977). The western Tethyan radiolarites could then have formed in small ocean basins of a scale analogous, for example, to the modern of Gulf of California (e.g. Kelts 1981). Here it should be noted that, except for the Upper Jurassic (Tithonian) radiolarites, and the Upper Cretaceous pelagic carbonates, the pelagic facies of the Eastern and Western Tethys are really quite different (Ogg *et al.* 1984). The implication is that either the two areas were so far separated in the Mesozoic that they experienced quite different palaeo-oceanographical controls (e.g. deep currents, upwelling) or they were closer together but effectively isolated from deep water exchange. The notable feature is that in the southerly ('external') basins of the Eastern Tethys (e.g. Antalya Complex, Turkey, see below) non-calcareous radiolarian facies persisted for as long as Mid Triassic to Mid Cretaceous. If the radiolarites generally are *correctly* interpreted as really having formed in narrow basins, then throughout the Mesozoic these southerly 'external' basins could have remained relatively small without major deep water exchange with the open oceanic areas, particularly the Atlantic. If on the other hand, relatively wide southerly ('external') basins really did exist in the Eastern Mediterranean, as in the models of Şengör & Yilmaz (1981), **Smith & Spray** and Pe-Piper (1982) then the concept of a rift and narrow ocean basin origin for the Tethyan radiolarites must be quite wrong. Below we will see that, in fact, geologically, there is little need for wide 'southerly' Early Jurassic and Triassic oceans in the East Mediterranean.

Quest for the Palaeotethys

Since the Alpine-Mediterranean Tethyan terrains are mainly composed of Mesozoic and Tertiary rocks it was natural to look further north for the original location of an older ocean, Palaeotethys (Smith 1971; Dewey *et al.* 1973; Argyriadis 1975; Hsü 1977). There are now two radically different published conceptions of the fate of the Palaeotethys, which it is now useful to compare, particularly in view of different implications for the opening of the Mesozoic Tethys, or Neotethys. In one of the two models the Palaeotethys was destroyed by southward subduction under a northward-migrating strip of Gondwana, while in the other it was consumed by northward subduction under Eurasia over a period extending from Late Palaeozoic to Neogene.

(i) Southerly subduction model

For the first time in this volume, **Şengör *et al.*** assemble a huge body of previously not easily accessible information on a region extending from central Europe, through the Black Sea to the Caucasus. As summarized in Fig. 3, in their view the Palaeotethys, or Palaeo-Tethys as they prefer to call it, opened after the Hercynian orogeny, then was subducted southwards under the southern margin of the Pangaean embayment during Mid-Triassic to Early Jurassic time. During the Triassic the southward consumption of Palaeotethyan crust was accompanied by opening of a strand of the Neotethys, south of an elongate E-W trending 'Cimmerian continent' which was rifted off Gondwana and then migrated northwards towards Eurasia. Closer to the trench a marginal basin (Karakaya basin) had already opened and it too travelled north with the 'Cimmerian continent'. From the Upper Triassic to the Early Jurassic the Eurasian margin collided with the 'Cimmerian continent' creating the 'Cimmerides'. The 'Karakaya marginal basin' closed and the compression culminated in complete allochthoneity of the Pontides (Şengör *et al.* 1980, 1982). In the Liassic, the 'Cimmerian continent' was then rifted not far south of the Palaeotethyan suture isolating a narrow sliver to the north which continued to undergo 'Tibetan-type' collisional orogeny and separating the Sakarya and Kirşehir blocks as independent entities. From the Early Jurassic onwards the Palaeotethyan suture zone had become a stable platform adjacent to the northern passive margin of the northern strand of the Neotethys. This model thus not only derives the Pontides from Gondwana, but opens a Triassic ocean basin between the Taurides and the African margin, then another, bifurcating one further north from the Early Jurassic onwards.

(ii) Northerly subduction model

A radically different hypothesis has been developed by Soviet workers (Adamia *et al.* 1981) further east in the Caucasus and also applied by them to the Eastern Mediterranean (Fig. 4). The Pontides and their extension into the Caucasus are considered to reflect long-lived northward subduction, active episodically

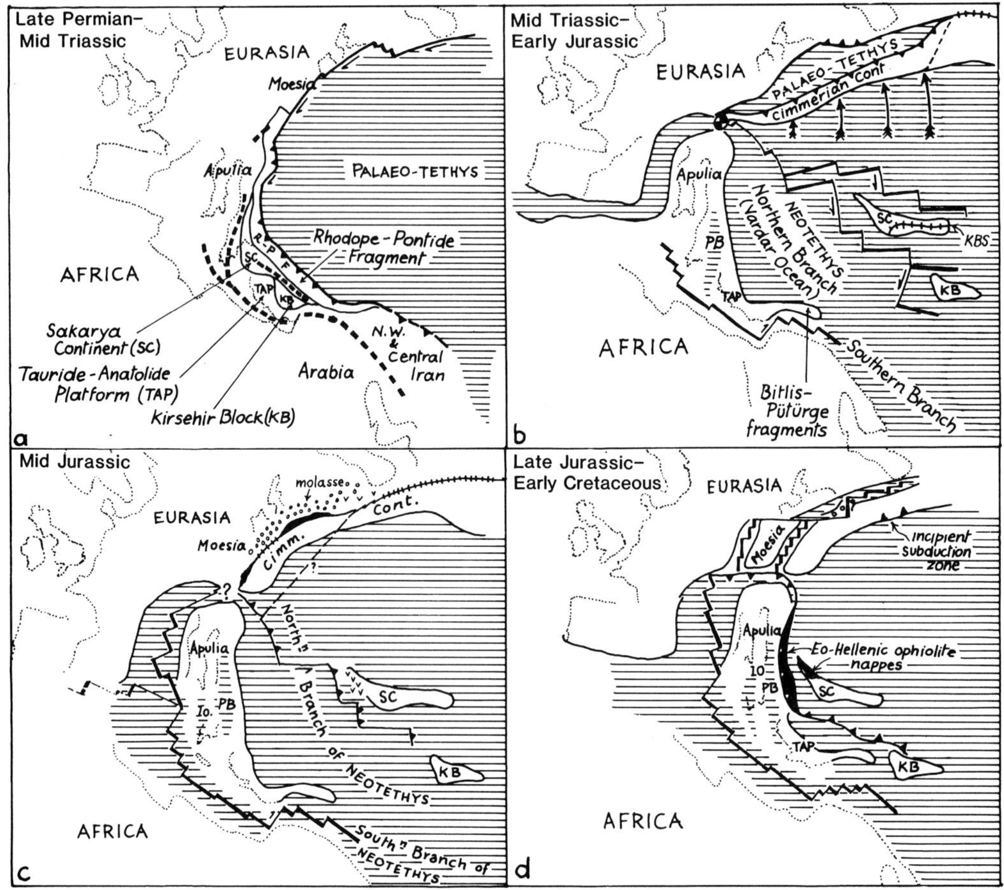

FIG. 3. Concept of the geological evolution of the Palaeotethys according to **Şengör** ***et al.*** (simplified). PB = Pindos-Budva basin; IO = Ionian zone; SC = Sakarya continent; KBS = Karakaya Basin suture. a, a wide Late Permian-Mid Triassic Palaeotethys undergoes southward subduction under 'Gondwana-land'; b, Mid-Triassic to Early Jurassic, a Cimmerian continent is rifted off 'Gondwana-land' and sweeps northwards opening a 'Northern Branch of the Neotethys'. Note the inferred pole of rotation; C, Mid-Jurassic, the Cimmerian continent has collided with the Eurasian margin; the 'Southern Neotethyan Branch' opens; intra-oceanic subduction of the Northern Neotethys is initiated. d, Late Jurassic to Early Cretaceous. Subduction of the Northern Neotethyan Branch leads to ophiolite emplacement in Greece and establishes the mechanism for future Tauride ophiolite obduction. The 'Southern Neotethyan Branch' remains unaffected at this stage. Compare this reconstruction with that of Şengör & Yilmaz (1981) (see Fig. 17).

throughout later Palaeozoic, Mesozoic and Tertiary time. The ophiolitic and related deep-sea sediments of the Minor Caucasus are correlated with the North Anatolian suture zone as the master Tethyan suture. They interpret the 'Palaeo-Tethyan' suture further north in the Greater Caucasus instead as essentially a small Mesozoic marginal ocean basin which they believe was separated from their main arc by the Transcaucasus median mass. The major platform south of the Minor Caucasus (Armenian-Nakhichevan subplatform) is identified as part of the Iranian structural block. In their model the Iranian block was located in a southerly Gondwanan position until Late Triassic to Early Jurassic when it migrated rapidly northwards, eliminating with it most of the pre-existing Palaeozoic oceanic crust, but leaving Palaeotethys more or less intact in the Eastern Mediterranean area (Fig. 4).

Renewed speading in the vicinity of the master subduction zone is then needed to create a Jurassic Neotethyan ocean basin which was later episodically subducted northwards. In Adamia *et al.*'s (1981) model Turkey remained part of Gondwana until the Late Cretaceous,

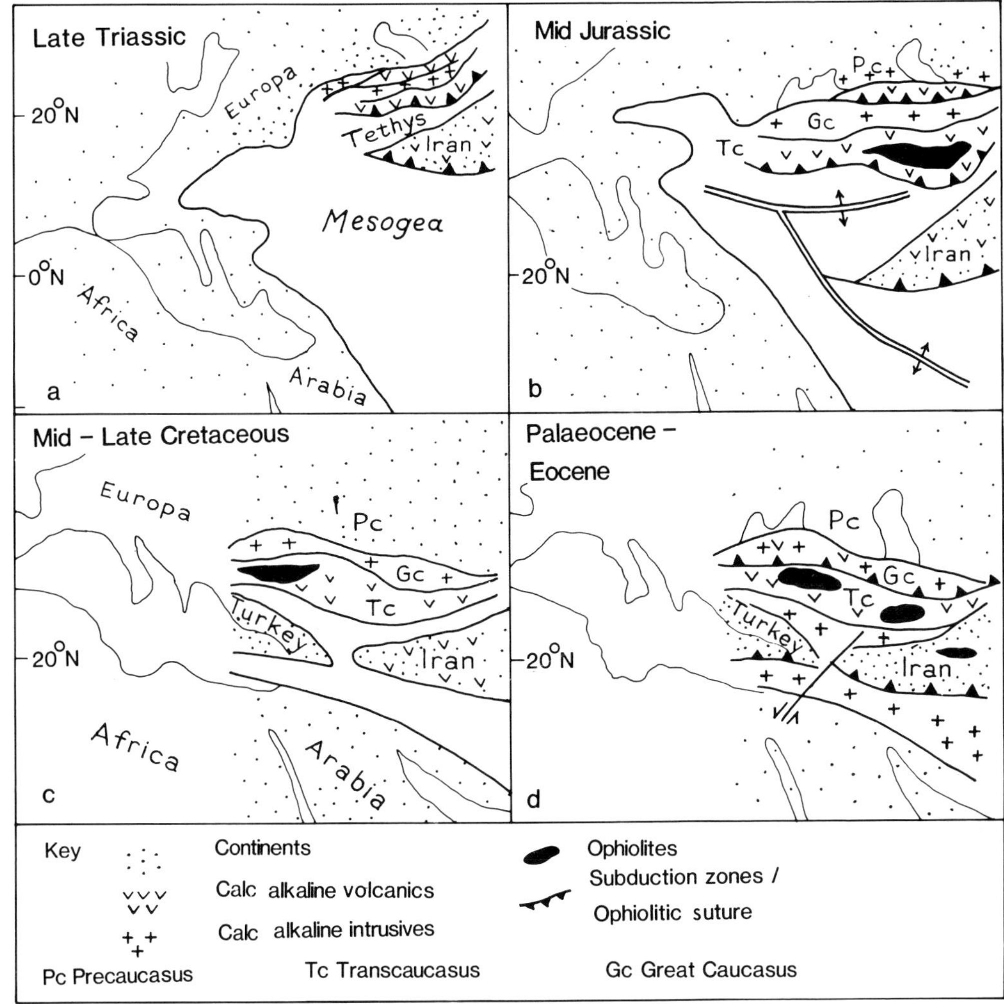

FIG. 4. Concept of the fate of the Palaeotethys according to Adamia *et al.* (1981). Contrast with that of **Şengör *et al.*** (Fig. 3). Late Triassic, a wide southern Tethyan Mesogea already exists. The Palaeotethys is consumed by northward subduction; b, Mid-Jurassic, a northern Neotethyan branch re-opens, splitting-off Iran; C, Mid-Late Cretaceous. A southern Neotethyan branch south of Turkey opens; d, Palaeocene-Eocene. The Tethys is sutured and strongly influenced by strike-slip faulting.

when a separate southern arm of the Neo-Tethys opened.

(iii) Discussion

Both models do have elements in common in that both postulate northward drift of Gondwanan elements early in the Mesozoic (Iran for Adamia *et al.* 1981; the whole 'Cimmerian continent' for **Şengör *et al.***). Both models also postulate almost complete elimination of Palaeozoic oceanic crust by early in the Mesozoic. Although detailed east-west correlations are still contentious (Şengör *et al.* 1982), it seems unlikely that the two areas did genuinely have consistently opposing subduction polarities from the Late Palaeozoic onwards.

In fact the proponents of both models are so far able to produce remarkably little firm evidence of the existence of definite Late Palaeozoic oceanic crust in either of the two alternative suture zones. Within the 'Anatolian Cimmerides', **Şengör *et al.*** identify a major oceanic element of the Palaeotethys as the Strandja nappe in eastern Bulgaria. This is a 'geosynclinal' assemblage including the Diabase-Phyllitoid Complex, an assemblage of metamorphosed sediments, basic igneous rocks, plus Besshi-type stratiform massive sulphide ores, together tentatively dated as Devonian to Triassic. The equivalent unit identified in North Turkey, the Kure nappe is an assemblage of clastic sediments, olistostrome melange and all the elements of an ophiolitic suite, again including massive sulphides, which were deformed and metamorphosed in Early to Mid-Jurassic time. From the descriptions it is by no means clear that the meta-sediments need have originated as a Palaeotethyan passive margin, nor that the meta-igneous rocks really amount to definite Palaeozoic oceanic crust, as opposed to a relatively young (Triassic) crust formed in a small ocean basin, or in rifts, possibly more on the scale of the modern Gulf of California than to a wide truly Palaeotethyan ocean. **Şengör *et al.*'s** 'Karakaya marginal basin', which is inferred to have opened further south, includes Late Carboniferous to Late Triassic olistoliths, cut by felsic intrusives, and dismembered ophiolitic rocks, all unconformably overlain by Late Triassic sediments. Where this is now in tectonic contact to the south with Mesozoic elements of the Neotethys (the Ankara melange, **Norman;** see below) any Palaeozoic oceanic crust *there* could also be interpreted in terms of a northward subduction model (see below), although further west the Karakaya suture appears to cut through the Sakarya zone independent of Neotethyan sutures.

The interpretation of calc-alkaline volcanism related to Palaeotethyan subduction is particularly critical. Adamia *et al.* (1981) refer to chemical evidence (e.g. K_2O-depth relationships) in favour of northerly subduction polarity but the validity of this evidence in often altered or metamorphosed 'arc' rocks of varying age in a number of different tectonic units is open to question. **Şengör *et al.*** relate the huge Kirkareli nappe of the Strandja and Northern Dobrudja areas to Triassic subduction, but here again there is as yet little definite evidence of polarity. How this arc extended eastward into the Pontides is also unclear as this particular nappe is not exposed.

It is also appropriate to note here that west of our main area of interest, in Northern Italy, Marcucci & Passerini (1981) describe a continuity of magmatism and tectonism between the Hercynian orogeny 'proper' and the formation of ophiolites in the Upper Jurassic. Also, during the Late Palaeozoic to early Mesozoic the southern (Gondwanan) margin in the broad sense was far from passive either, as shown by the pre-Upper Triassic uplift, and thrusting, documented in the Taurides by **Monod** and the successive Late Palaeozoic and Mesozoic angular unconformities in south-east coastal Turkey described by **Demirtaşlı.**

Our general impression from the available information is thus that it is more likely that the Pontides formed part of Eurasia than Gondwana by the Late Permian to Early Triassic. Other parts of **Şengör *et al.*'s** Cimmerian continent further east may well be of Gondwanan origin. The alternative which we prefer is that the Eurasian margin acted as a long-lived active (East Pacific-type) margin, and as such was affected by major strike-slip translation of units, and the opening and closing of pull-apart or marginal basins, and oblique rifts like the modern Gulf of California. It seems more likely that any early Mesozoic collisional deformation is somehow to be explained more in the context of a long-lived obliquely convergent margin than by the opening and subsequent closure of an Atlantic-type ocean basin. This continuously evolving 'collage' model for the Eurasian margin is adopted in our concluding section.

Şengör *et al.*'s model can be further tested by examining the nature of the rifting which is inferred to have detached parts of the 'Cimmerian continent' from Gondwana. We now consider this.

One of the most surprising features of **Şengör *et al.*'s** model is the requirement that a thin sliver of crust rifted-off the northern edge of the 'Cimmerian continent' prior to final closure of the Palaeotethys to become the northern passive margin of a northern strand of the Neotethys after Liassic time. The continuing calc-alkaline volcanism on this sliver has to be explained as post-collisional ('Tibetan') volcanism, while in the alternative model the volcanism is easily explained by episodic northward subduction along the master Tethyan subduction zone (North Anatolian suture). In Şengör and Yilmaz's (1981) (Fig. 17) model the whole of the Taurides and the Pontides formed part of a wider but still elongate 'Cimmerian' continent which rifted-off Gondwana from the Mid-Triassic onwards to open a southern Neotethyan strand. By contrast, in line with **Lauer's** palaeomagnetic data, **Şengör *et al.*** now show the Tauride platform units close up against Gondwana for much of the Mesozoic, thus restricting further the 'Cimmerian' continent to a narrow sliver consisting in Turkey of little more than the Pontides. On the other hand Adamia *et al.*'s model of successive accretion to the Eurasian margin over a long period from the Late Palaeozoic onwards, even if valid for the Caucasus, is not exactly applicable to the Eastern Mediterranean where there is little evidence of collisions along the North Anatolian suture during Early Jurassic to Upper Cretaceous time. The model also has difficulty in explaining why a new Neotethyan strand opened *after* Iran has reached the Eurasian margin by early in the Mesozoic. This re-opening is more akin to the Early Jurassic opening of a separate northern Neotethyan strand as in **Şengör *et al.*'s** model (see also Şengör & Yilmaz, 1981). One possibility is that the Arabian margin essentially shielded the North Anatolian suture zone from more continuous accretion from the south related to northward subduction, as in the Caucasus.

Rifting of southerly units

Extensive Triassic rifting is one of the most characteristic features of the Neotethyan terrains extending throughout the Eastern Mediterranean and into adjacent areas (Bernoulli & Jenkyns 1974). Overall, major rifting (but not necessarily spreading) becomes progressively younger westwards from Upper Permian in the Oman Mountains (Glennie *et al.* 1973), Lower Triassic in the south Turkish area, Mid-Triassic (Anisian/Ladinian) in Greece, the Dinarides and Italy, to Lower Jurassic (Liassic) in the Atlantic Tethys (e.g. Iberia) (see Channel *et al.* 1979). The best developed Triassic rift assemblages are located in relatively southerly ('external') locations within the Neotethyan fold belt. We will first discuss the evidence for Triassic rifting and passive margin development in these southerly units, then see if comparable rifting has taken place in the northerly ('internal') zones. A very important question is to assess the timing of first true ocean floor spreading after major rifting. For example, did Triassic rifting mark the time of initial ocean floor spreading, or did the rifts persist for some considerable time before the appearance of significant areas of oceanic crust? We return later to the separate question of whether all the allochthonous rift units could have been derived from a single zone or whether southerly ('external') and northerly ('internal') zones must have existed.

(i) Levant

The Levant is the only part of the stable margin of Gondwana now well documented by deep drilling. **Garfunkel & Derin** show that the differential vertical movements date back to the Upper Permian, with uplift of a buried Hercynian tectonic high, the 'Geanticline of Helez', which is described in detail by **Gvirtzman & Weissbrod.** Rifting proper took place in three pulses: Early Triassic (Late Anisian to Ladinian), Late Triassic (Carnian-Norian) and early Liassic. Subsidence was intially rapid (50–100 m/Ma) then slowed (Fig. 5). After taking account of sediment-loading and the effects of sea-level fluctuations, **Garfunkel & Derin** favour a model of pulsed crustal extension, rather than merely progressive cooling after an Upper Permian thermal event. A previously little-known Liassic rift event accounted for a further 2–3 km of differential motion accompanied by important volcanism, then a major transgression took place from Upper Jurassic (Bathonian) onwards.

There is, however, wide disagreement as to the significance of the rifting. **Garfunkel & Derin** review the existing geophysical evidence that the crust thins progressively westwards under the coastal plain of Israel with strong NW-SE trending basement magnetic and gravity anomalies suggesting that the continent-ocean boundary could be located at the base of the present continental slope. In support of the hypothesis of the Levant as a surviving Mesozoic passive continental margin, **Gvirtzman & Weissbrod** point out that the NW limb of their Hercynian 'Geanticline of Helez' is missing, apparently because it has been transposed to somewhere in the Taurides. Robertson & Woodcock (1980) and **Gvirtzman & Weissbrod** suggest that the location was the SW segment of the Antalya Complex west of Antalya Bay, but this now seems less likely as there the Ordovician to Upper Cretaceous successions are locally preserved intact (Robertson & Woodcock 1982), without the major hiatus present in the Helez geanticline.

A very different conclusion is reached by **Druckman** who interprets the successions in the Helez 1A deep borehole, not as a Mesozoic passive margin but in terms of a failed intracontinental rift. In his view the Upper Triassic (Carnian) and Liassic successions ought to be deep marine in a continental margin model, whereas they are shallow marine or subaerial. Also, gravity and magnetic anomalies, hitherto interpreted as rift-related mafic igneous rocks, were shown to be metamorphic basement by the deep drilling. **Druckman's** belief is that faulting had terminated by the mid-Jurassic followed by gradual subsidence. This view is consistent with **Ricou *et al.*'s** and **Hirsch's** assessment of the Levant as part of the Gondwana platform located well south of the Tethyan suture.

(ii) Southern Turkey and Syria

The record of rifting is also clear in the allochthonous Antalya Complex of SW Turkey (Fig. 6), where once again, the crustal extension was pulsed, first in the Early Triassic (Scythian-Anisian), then in the Late Triassic (Carnian-Norian) (Brunn *et al.* 1970; Marcoux 1974; Robertson & Woodcock 1982). Early Triassic block-faulting and differential subsidence of a stable epeiric sea area was accompanied by localized volcanism of Pietre verde-type, which is well exposed in the Beyşehır-Hoyran nappes (Monod 1977), followed by more quiescent mid-Triassic pelagic carbonate and radiolarite deposition, prior to renewed major subsidence and volcanism in the Upper Triassic (Carnian-Norian) (Juteau 1975; Robertson & Woodcock 1981; this volume). In the SW Antalya Complex segment, these mafic alkalic extrusives exceed 1000 m in unbroken succession, considerably greater than normal oceanic crustal thicknesses.

Whitechurch *et al.* review occurrences of rift-related mafic volcanics throughout the Taurides and areas to the south. Very similar Late Triassic mafic volcanics are present in the Mamonia Complex of SW Cyprus (Lapierre 1975; Swarbrick 1980) and Baër-Bassit area of N Syria (Parrot 1977; **Delaune-Mayere; Delaloye & Wagner**). Deformed mafic alkalic lavas are also associated with pelagic sediment structurally underlying the Beyşehir-Hoyran and Pozantı-Karsantı ophiolites. By contrast, in the Lycian nappes, alkalic basalts in 'coloured melange' are inferred to be of Cretaceous age. Petrologically, these mafic extrusives range from tholeiites (Baër-Bassit), to alkalic basalts and trachytes (e.g. Antalya Complex). Plots of 'immobile' trace elements of Upper Triassic basalts from the Mamonia Complex (Pearce 1975) and from both Baër-Bassit and Hatay (**Delaloye & Wagner**) point to a 'within plate' genesis, typical of continents, continental margins and oceanic islands (e.g. Pearce 1980; see below).

Robertson & Woodcock discuss evidence in favour of 'braided' rifting in the Antalya Complex (Fig. 6) to produce a series of horsts, which persisted as carbonate build-ups along the continental margin throughout much of the Mesozoic. In the SW segment of the complex, the Upper Triassic mafic volcanism was followed by localized continued input of quartzose turbidites and *Halobia*-bearing hemi-pelagic limestones, passing into non-calcareous successions of radiolarian sediments which continue to the mid-Cretaceous. This is consistent with subsidence and slow net accumulation beneath a shallow carbonate compensation depth adjacent to a progressively subsiding passive carbonate margin.

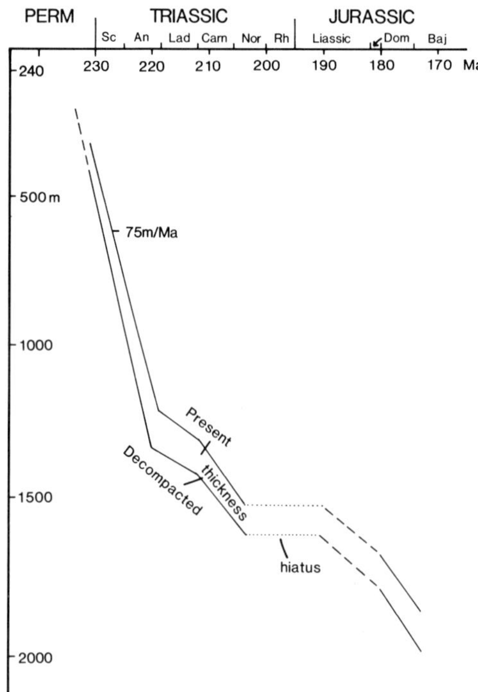

FIG. 5. Inferred subsidence history of the Levant Margin, after Garfunkel & Derin.

(iii) Greece and Yugoslavia

In Greece and Yugoslavia (Fig. 7) the southerly ('external') units are taken to be those located west of the Pelagonian-Golija zone (recent review: Dercourt and others, 1980). In Greece the evidence of rifting is clearest in the Othris area (Sub-Pelagonian or Maliac zone) where the pattern follows that in SW Turkey with the disintegration of a continental platform and onset of deep water sedimentation coupled with alkalic volcanism, (Smith *et al.* 1975, 1979; see **Smith & Spray** for literature review). Other scattered occurrences of Middle Triassic volcanism in the Greek southern units are reviewed by Pe-Piper (1982). The westernmost and stratigraphically lowest examples in the Pindos zone are in sequences transitional to the Gavrovo-Tripolitza carbonate platform (Caron 1975; Fleury 1976) and may be rift-related. Other possible rift-related extrusives in the Argolis Peninsula (Peloponnesus) comprise Lower Triassic acidic volcanics, while mafic Triassic lavas occur in the HP-LT Phyllite unit in the Peloponnese lying structurally between sequences in the Ionian Zone and the Gavrovo-Tripolitza zone (see **Papanikolaou** & **Hall *et al.***). Mafic lavas of Permo-Triassic age from the equivalent Phyllite-Quartzite nappe in Crete

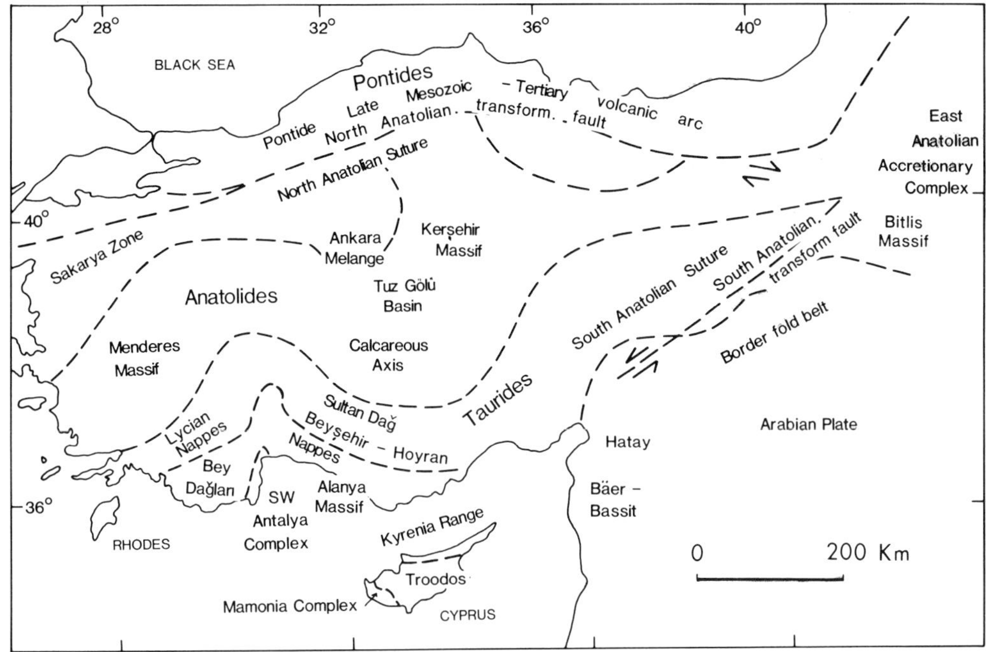

FIG. 6. Outline map of Turkey showing the main tectonic units dealt with in the text.

are described by Seidel *et al.* (1891) and include alkalic basalts, andesites and latites of generally calc-alkaline affinities similar to those known from Calabria. Pe-Piper (1982) argues from their geochemistry that these lavas, and possibly the more clearly rift-related ones also, were generated above an active subduction zone and are not the products of 'normal' intra-continental rifts. This paradox is more acute in the Dolomites (Pisa *et al.* 1979) and the Dinarides (Bébien *et al.* 1978). Middle Triassic volcanism is extensive including basaltic to keratophyric volcanics and tuffs, and in places granitoids (e.g. Northern Montenegro: Djordjević *et al.* 1982). The mafic igneous rocks can be clearly related to the break-up and subsidence of carbonate platforms, but by many criteria are calc-alkaline in character. The Dinaride volcanics ('porphyrite-radiolarite formation'; Bébien *et al.* 1978; Aubouin *et al.* 1970) which are of uppermost Anisian to Landinian in age, occur in all the zones, including both platformal and basinal, located west of the Vardar zone. Evidence of possible rifting within the Vardar zone is discussed in relation to the other 'internal' zones below.

Chemistry

Confirmation of a rift origin of the various Triassic mafic extrusives has been sought in studies of the mineralogy and chemistry of the lavas. Some Triassic lavas *appear* on this basis to be typical undersaturated products. These include Othris, Greece (Hynes 1974), the Mamonia Complex, Cyprus (Lapierre 1975) and Baër-Bassit, Syria (Parrot 1977). On the other hand in parts of Greece, in the Dinarides and in the Dolomites more evolved rocks are quite abundant including basaltic andesites, andesites and rhyolites, with a distinct calc-alkaline character. This by itself is insufficient to rule out a rift origin or to establish a subduction zone origin, as is well established by the Pliocene to Recent volcanic history of the Rio Grande rift in the SW USA, where besides tholeiitic and alkalic basalts, there are extensive basaltic andesites, latites and voluminous rhyolites and ignimbrites (e.g. Riecker 1979; Basaltic Volcanism Study Project, BVSP, 1981). Without going into detail, it would appear that discrimination between rift and subduction-related origins on the basis of mineralogy and major element chemistry *alone* is probably not possible. Thus, the conclusion of Bébien *et al.* (1978), using these criteria, that the Triassic volcanics of the Dinarides are calcalkaline, and thus subduction-related, could well be premature.

A more reliable discrimination could theoretically be achieved in the few cases where

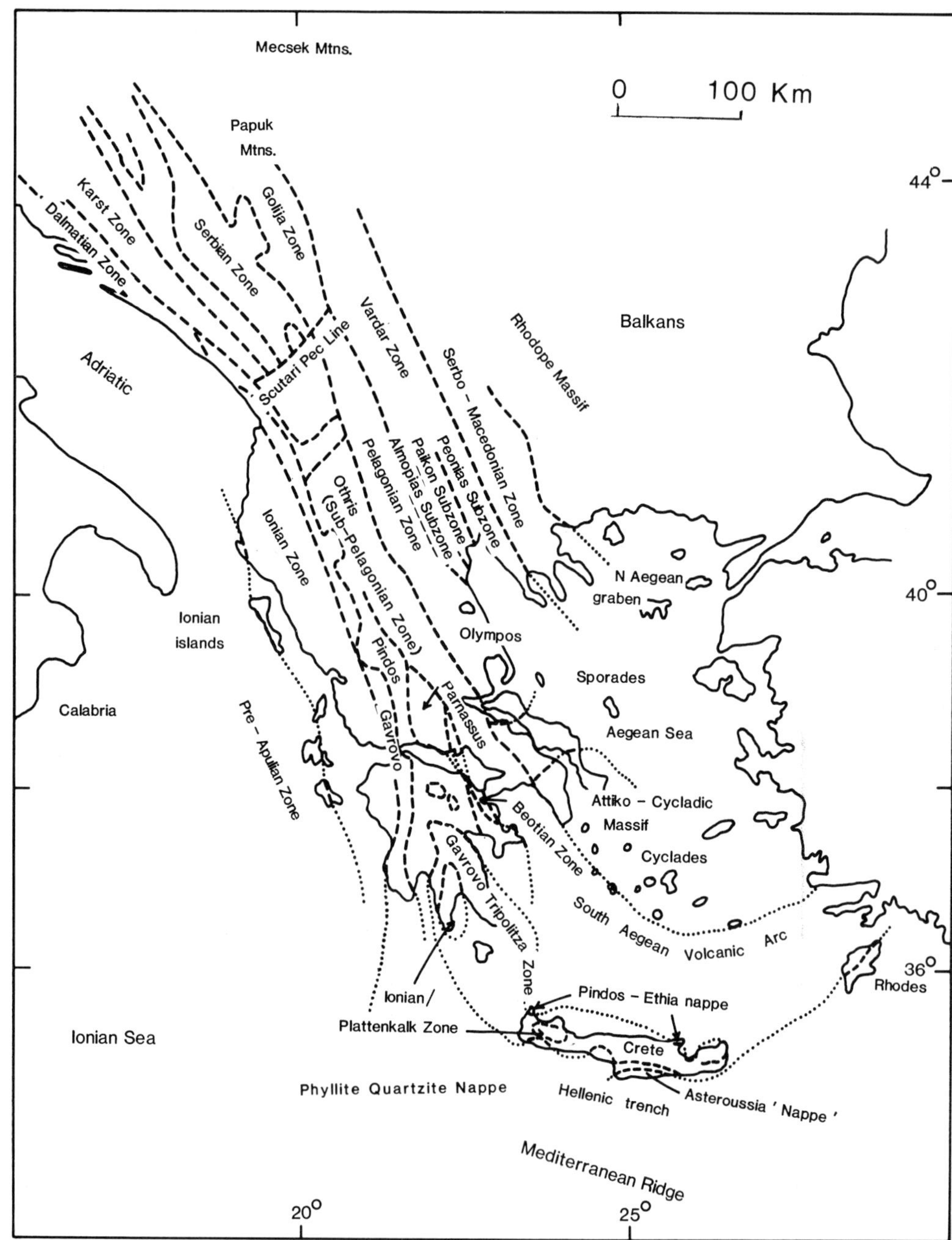

FIG. 7. Outline map of Greece showing the main tectonic units dealt with in the text.

extensive trace element data are available (e.g. Pearce 1980), although Pearce himself advocates caution as the chemistry of rift lavas could be significantly contaminated by continental crust. In some cases, for example, the Mamonia Complex (Pearce 1975) and Baër-Bassit (**Delaloye & Wagner**), the trace elements are fully consistent with an intra-continental rift origin unrelated to subduction. On the other hand Pe-Piper (1982), working in Greece, and Pisa *et al.* (1979) in the Dolomites point to trace element abundances with a clear above-subduction zone signature. Indeed, some of the Greek samples show MORB-normalized plots with a marked depletion in Ta, Zr, Ti and Y and enrichment in Rb, Ba and K. Others, however, lie within the envelope for Rio Grande rift basalts despite matching 'transitional volcanic arc basalt' patterns. It does seem to us that at a time when little trace element data are available for definite intra-continental rifts elsewhere, an above-subduction-zone origin for any of the Triassic Tethyan lavas is not yet at all well established, particularly as other aspects of the geology are quite consistent with an intra-continental rift origin. For example, although Şengör & Yilmaz (1981) suggested that the Triassic Tauride Pietre Verde basalts could be related to back-arc extension over a southward subducting Palaeotethys, **Şengör *et al.*** now observe that these rifts would be up to 1500 km from the inferred trench, much further than in present, or definite past, marginal basins (e.g. Rocas Verdes, Chile: Bruhn & Dalziel 1975).

Discussion

Quite apart from the question discussed above of whether or not the Triassic rifts were ultimately related to active subduction zones, there has in the past been considerable debate as to the timing of first ocean floor spreading in each of the areas concerned. Very different conclusions have been reached from different standpoints.

First, **Garfunkel & Derin's** assessment that ocean floor spreading in the Levant began after the Liassic is based on comparisons with other non-emplaced passive margins where major faulting along the margin ended at the time of final continental break-up when passive subsidence normally begins. On the other hand, these authors infer that *Triassic* spreading did not take place in the Levant since major faulting and volcanism continued into the Liassic (see also Klang & Folkman 1982). The second approach has been to attempt to infer the history of the margins from the remaining deformed successions in allochthonous units like the Antalya Complex. The continuing record of Upper Jurassic to Lower Cretaceous alkaline volcanism in the Antalya Complex (**Robertson & Woodcock; Yılmaz**) and of Lower Cretaceous per-alkaline volcanism of Baër-Bassit (**Delaune-Mayere**), plus the evidence of extensive hydrothermal manganese accumulation (Robertson 1981), can, again by analogy with other passive margins, be taken to indicate that the margins remained in an active slow rift phase long after the Triassic. This could suggest that significant spreading might not have taken place until well into the Cretaceous. A similar problem is met with in the Pindos zone, Greece, where chemically unusual 'hybrid' lavas of Late Jurassic to Early Cretaceous age occur within successions related to Triassic rifting (**Pe-Piper & Piper**). The difficulty here is that although volcanism normally ceases on passive margins after final break-up, at least in some cases, alkaline volcanism can persist (e.g. Hold with Hope region, E Greenland, Upton *et al.* 1980). A third approach has been to infer the timing of spreading in relation to tectonic hypotheses for the belt as a whole. Thus, **Şengör *et al.*'s** (see also Şengör & Yilmaz 1981) hypothesis that the Palaeotethys was subducted southwards under Gondwana in the Triassic and simultaneously a continental sliver of Gondwana migrated northwards, requires a Neotethyan ocean to open behind this sliver of width comparable with Palaeotethys, since Africa and Eurasia remain widely separated until the Cretaceous. In view of the uncertain history of Palaeotethys, as discussed above, not too much can however be put on this. Conversely, **Druckman's & Hirsch's** assessment that the Levant was all part of Gondwana far from the main Neotethyan margin implies to them that the rifting in the Levant was still-born and never progressed to ocean spreading. Below we will conclude that the allochthonous southerly ('external') units (e.g. Antalya Complex) provide independent evidence of southerly Neotethyan strands located not far north of the Levant rift zone.

One other line of evidence that could theoretically be helpful in assessing the timing of first ocean floor speading is the relative rate of subsidence of adjacent platform units. **Garfunkel & Derin** calculate subsidence rates of 50–100 metres/Ma for the Mid, Upper and Early Jurassic rifting events (Fig. 5), which in their view, preceded final break-up and ocean floor spreading. The limited available data for the Taurides suggest that subsidence rates of the Tauride carbonate platforms as a whole were relatively low in the Upper Triassic (ca <25 metres/Ma), but increased from the Liassic onwards reaching a maximum in the Cenomanian (Poisson 1977). **Smith & Spray** note that carbonate platforms adjacent to their 'Triassic Othris ocean' subsided 2 km in less than 35 Ma and more generally throughout the Adriatic area the carbonate platforms subsided at >100 metres/Ma in the Upper Triassic, then rates progressively decreased to a few metres per Ma by the end of the Mesozoic (D'Argenio 1976; Channel *et al.* 1979).

It is thus clear that while rifting is very well documented in the southern ('external') units, the exact timing of final decoupling is not at all easy to establish and may leave little trace on the adjacent margins. In view of the fact that in the Levant rapid subsidence almost certainly preceded initial speading and in the Western Tethys there is little, if any, evidence of ocean genesis prior to the Upper Jurassic, we conclude that rapid subsidence does indeed correspond to rifting but need not in itself *require* spreading in adjacent areas either then or later. Even in cases where the lava chemistry can be matched with typical intra-continental rifts this is in itself no guarantee of progression to the ocean crust stage. For example, the SW Antalya Complex Triassic rift lavas exceed 1000 m in unbroken succession (Juteau 1975) without any known transition laterally or vertically suggestive of immediately adjacent ocean floor spreading. There is in any case as yet insufficient evidence that intra-continental rift lavas can be unambiguously characterized chemically.

Those in favour of major Triassic spreading can always argue that Triassic ocean basins were later subducted without trace. At least in the case of the Antalya Complex this is less likely as the later tectonic emplacement involved a major component of strike-slip which has entrained ocean basins remnants along the margin, rather than simply subducting them (Woodcock & Robertson 1982), yet no Jurassic ocean crust is known despite extensive study of the area. In the southerly ('external') Greek area ocean crust is known to have been created prior to the Upper Jurassic (see below), but in southern Turkey, SW Cyprus and Baër-Bassit, the field relations are quite consistent with Triassic rifting to form a broad deep trough floored by alkalic lavas and overlain by radiolarites, which persisted from the Upper Triassic through the Jurassic and possibly into the Cretaceous before much ocean floor need have been created. Our main conclusion here is thus that evidence of rifting need not indicate spreading. Whatever was the case we now go on to see if similar Triassic rifting took place in the northerly ('internal') units. This is particularly important as according to some hypotheses (e.g. **Ricou *et al.***) all the allochthonous rift-related rocks, as well as the ophiolites, were derived from a single northerly rift zone of Late Triassic age, while in Şengör & Yilmaz's (1981) model major rifting took place in the Pontides in the Early Jurassic.

Possible rifting of northerly units

The relevant areas in Turkey, the North Anatolian suture, and in Greece and Yugoslavia, the Vardar zone and the Pelagonian zone—should throw light on the nature of the margins, particularly whether they were active or passive early in the Mesozoic and if any possible rift history matches that of the southern ('external') units.

(i) Turkey

Interpretation of the critical successions in the North Anatolian suture zone (Fig. 6) is complicated by later deformation and metamorphism (**Okay**), but in essence appear to include the following: southwest of the Izmir-Ankara suture (Afyon zone) Precambrian basement of the Menderes massif is overlain by 1500 metres of meta-clastics and recrystallized limestones of Carboniferous to Upper Permian age which were deformed during the 'Cimmerian' orogeny of **Şengör *et al.*** The metamorphic rocks are then unconformably overlain by shallow water clastics, passing into the Anatolide platform carbonates which persisted until the Maastrichtian, when ophiolitic allochthons were thrust southwards from the Izmir-Ankara suture zone. Servais (1982) stresses the existence locally of Triassic fragmental volcanics and lava flows interbedded with clastics (e.g. Porsuk). North of the North Anatolian suture (Sakarya zone of Şengör & Yilmaz 1981), the pre-Liassic metamorphic basement is overlain by Mesozoic to Tertiary clastics and shallow water limestones, then by Upper Cretaceous volcanic arc rocks. In the Eskişehır area Servais (1982) reconstructs a metamorphic basement passing first into latest Triassic conglomerates in proximal successions (Günyarık Formation), then into Middle and Upper Liassic shales and clastics, in turn overlain by the Middle and Upper Jurassic condensed neritic Bilecik limestones, 300–800 m thick. The Lower Cretaceous is extremely condensed or missing altogether, followed by Upper Cretaceous flysch, then volcanics related to the Pontide arc.

(ii) Greece

In the Vardar zone in Greece (Fig. 7) the Triassic history is generally obscured in the west by deformation and metamorphic events in the Lower Cretaceous and Tertiary. In north and central Euboea the basic volcanics are succeeded by Upper Triassic and Jurassic neritic carbonates (Katsikatsos 1977). Further north in Yugoslavia, (Rampnoux 1970, p. 950–953) describes a greenschist facies succession from the western Vardar zone from beneath the ophiolite nappe, in which a Lower and Middle Carboniferous clastic sequence is overlain apparently conformably by transgressive Upper Permian-Lower Triassic calc-schists, massive

Anisian dolomites, 1–2 m of porphyritic tuff and then Ladinian *Halobia*-bearing cherty limestones. This sqeuence is consistent with Ladinian subsidence in the more 'external' units, but may indicate that an earlier basin was present.

On the eastern side of the Vardar zone in Greece the western margin of the Serbo-Macedonian zone appears to be a structurally complex generally east-dipping succession of Hercynian basement slices, ?Permian arkosic clastics and rhyolitic volcanics, Triassic carbonates, ?Lower Jurassic flysch, ?Jurassic intermediate volcanics and hypabyssal intrusives (Chortiatis series), and ophiolites continuous with the main Vardarian belt (Mercier 1968; Kockel *et al.* 1977; see also **Dixon & Dimitriadis**). This stack is overlain locally by unconformable Upper Jurassic conglomerates, the Doubcon molasse. Despite the apparently fault-controlled deposition of the ?Permian clastics in the easternmost part of this sequence the evidence for a Triassic or Jurassic rift-origin for this margin is not very strong, especially as the platform break-up stage recognized further west in Triassic carbonate sequences is not clearly seen.

Discussion

In Greece and Turkey it has generally been assumed that Triassic intracontinental rifting took place in the northerly units just as in the southerly ('external') zones. Although still poorly known, the comparable areas in Turkey did not experience the major phase of Upper Jurassic-Lower Cretaceous thrusting and metamorphism which affected Greece and Yugoslavia (see below). Despite this simpler situation, opinions again vary. Bergougnan (1975), Fourquin (1975) and Bergougnan & Forquin (1980) argued that the Hercynian orogen rifted in the Early Triassic, giving rise to distinctive north and south Tethyan ammonite faunas by the Liassic (Enay 1976). For **Ricou *et al.*** this was the basin from which all the south Tauride Triassic rift-related assemblages were later derived (e.g. Antalya Complex). On the other hand, Şengör & Yilmaz (1981) believe that both margins belonged to the 'Cimmerian' continent until the Liassic, when prior to final closure of the Palaeotethys, a northern stand of the Neotethys rifted and went on to form the main northern arm of the Neotethys. In their view, two ocean basins formed in the west, the 'Inner Tauride ocean' to the south and the 'Intra-Pontide ocean' to the north. A third radical alternative is proposed by **Norman** (see also Tekeli 1981) who argues that the Mesozoic components of the Ankara melange (e.g. limestone-block melange and ophiolitic melange) represent a subduction complex, or accretionary wedge, formed by northward subduction of Tethyan ocean crust at least from the Early Jurassic onwards. To explain the 'mega-debris-flow' character of much of the melange he infers a major oblique component to the subduction. Servais (1982) also raises the possibility that the Palaeotethys might have been still open with the North Anatolian suture persisting as an active margin from the Late Palaeozoic into the Mesozoic.

One other aspect is that **Monod** points out that any interpretation must take account of regional compression involving uplift, erosion, clastic deposition (Çayır Formation), thrusting (e.g. Beyşehir), and possibly metamorphism and intrusion, which affected much of the Anatolide-Tauride platform prior to and during Upper Triassic time. A significant sedimentary hiatus is now known to extend as far west as the Olympus window in eastern central Greece (Schmitt, 1983). Surprisingly, the compression of the Anatolide Platform is at least partly contemporaneous with the extension and rifting of the southern Neotethyan units (e.g. Antalya Complex). Indeed, the uplift and erosion could well have helped generate much of the siliciclastic material associated with the Upper Triassic rifting there.

We conclude then that there is very little evidence of Triassic successions in the North Anatolide platform which would favour a close correlation with those of the southern Tauride area (c.f. Ricou). It also seems unlikely that the Pontide margin was really a typical passive margin from either Upper Triassic (c.f. Ricou) or from Liassic to Mid-Cretaceous times (c.f. Şengör & Yilmaz 1981). The continuing calc-alkaline volcanism in the Pontides (**Akinci**), numerous disconformities (Fourquin 1975) and absence of an Atlantic-type margin subsidence history are all consistent with a more active margin, affected by shear tectonics and/or subduction. We believe that the Paleotethys remained open and the North Anatolian suture zone continued to operate episodically as an active margin. This in turn raises the problem of the origin of the 'Cimmerian' orogeny which could not then have resulted from collision and suturing of a 'Cimmerian continent' with a south-dipping Palaeotethyan subduction zone to the north, but instead would represent one in a series of events (e.g. collision with intra-oceanic subduction zones or volcanic edifices) along something more akin to a Pacific-type margin. The exact reason for the pre-Upper Triassic deformation of the Tauride-Anatolide platform remains obscure; the timing was too *early* for closure of the Palaeotethys from the

Upper Triassic to Mid-Jurassic as postulated by **Şengör *et al.***, while in the 'open-ocean' model the Anatolide-Tauride platform was essentially still attached to Gondwana in the Upper Triassic and was thus presumably far from Eurasia. One possibility is that the deformation related to major crustal shear as Eurasia and much of Palaeotethys were transported dextrally (see above) relative to Africa, prior to establishing a Tethyan gulf geography in the Early Jurassic.

In summary, our main conclusions concerning these northerly ('internal') successions are first, that they have not been shown to have resulted from Triassic, or even Jurassic, intracontinental rifting, and secondly, that they cannot be matched with the Triassic rift-related sequences in the southerly ('external') units. We now go on to look at the record of ophiolites associated with both the southerly and the northerly zones.

Genesis of Neotethyan oceanic crust

Assuming that the ophiolites *do* represent some form of oceanic crust (Fig. 8), the key questions are when and in what exact tectonic setting did they form. First we look at the age and chemistry of the ophiolites in the Greek and Turkish areas in relation to the relative motion of Africa and Eurasia, then we discuss the alternative models of ocean crust genesis. We will return in a later section to the difficult question of the number of root-zones involved. Comprehensive reviews of the petrology and field relations of Greek and Turkish ophiolites have been published by Bébien *et al.* (1980) and Juteau (1980), and for Cyprus, Gass (1980).

(i) Age of the ophiolites

In contrast to the Triassic rift lavas, the ophiolites, proper, have so far proved difficult to date mostly because of their dismembered and often deeply eroded state, their alteration and low potassium contents.

Spray *et al.* review the existing and new radiometric age data which now support genesis of the ophiolites in both the western (Pindos zone) and eastern (Vardar zone) Greek belts and extensions into Yugoslavia essentially in the late Middle Jurassic (174 Ma, Bathonian, Othris and Pindos). In the Taurides the ophiolites which have so far been dated are all Cretaceous (Thiuzat *et al.* 1981). The plutonic ophiolitic rocks in the Antalya Complex yield a wide range of ages tending to cluster in the Upper Cretaceous to the Tertiary boundary, but showing a dispersion back towards the Mid-Cretaceous (**Yilmaz**). The Troodos Massif, Baër-Bassit and the Pozantı-Kersantı are also dated as Upper Cretaceous ranging from 87–75 Ma (Santonian-Campanian), as are some of the Guleman ophiolitic rocks and related slices of mafic extrusives (Killan Group) in the Maden Complex area of SE Anatolia, 65 ± 3 Ma (**Aktaş & Robertson**). Elsewhere (e.g. South Aegean area), the formation age of ophiolites remains very poorly constrained. An important point stressed by **Whitechurch *et al.*** is that, after allowing for later rotations (e.g. Troodos), all the Upper Cretaceous ophiolites appear to have been created at essentially east-west spreading axes, offset by north-south transform faults.

(ii) The chemical evidence

There is now growing evidence from analysis of the 'immobile' trace elements that ophiolites in the Eastern Mediterranean show considerable chemical variation, often differing significantly from MORB (Mid-ocean-ridge-basalt; e.g. Capedri *et al.* 1980, 1982; Noiret *et al.* 1981). Relative enrichment in large-ion-lithophile elements (LIL, e.g. K, Rb, Ba) and depletion in high-field-strength elements (HFS, e.g. Nb, Ta, Zr) is generally taken as an indicator of generation above subduction zones. Pearce *et al.* (1984) further argue that highly magnesian boninitic lavas (e.g. Cameron *et al.* 1980) arise by partial melting of hydrated oceanic lithosphere, while more typical island arc tholeiites are generated where hydrated asthenosphere is the main magma source. Pearce *et al.* (1984) claim to be able to recognize supra-subduction zone (SSZ) character superimposed on a normal MORB or within-plate magma type signature in hybrid cases.

Using these criteria, for **Smith & Spray** the Pindos ophiolite is calc-alkaline, while for Pearce *et al.* (1984) it is of hybrid type showing features of both MORB and the boninites, as found in the Mariana fore-arc region. **Smith & Spray** refer to the Vourinos ophiolite as transitional MORB/ocean island/continental tholeiite, whereas for Pearce *et al.* (1984) Vourinos is boninitic. There is agreement that Othris is MORB-like, while according to **Smith & Spray** the Guevgueli magmatic complex of the Vardar Zone has a tholeiitic to calc-alkaline affinity. According to Pearce *et al.* (1984) the Cretaceous ophiolites further east (Troodos, Baër-Bassit, Hatay) show consistent signatures indicating an above-subduction zone genesis (see also **Delaloye & Wagner**) as do the gabbros of the Guleman ophiolite and mafic Upper Cretaceous mafic extrusives of the Maden Com-

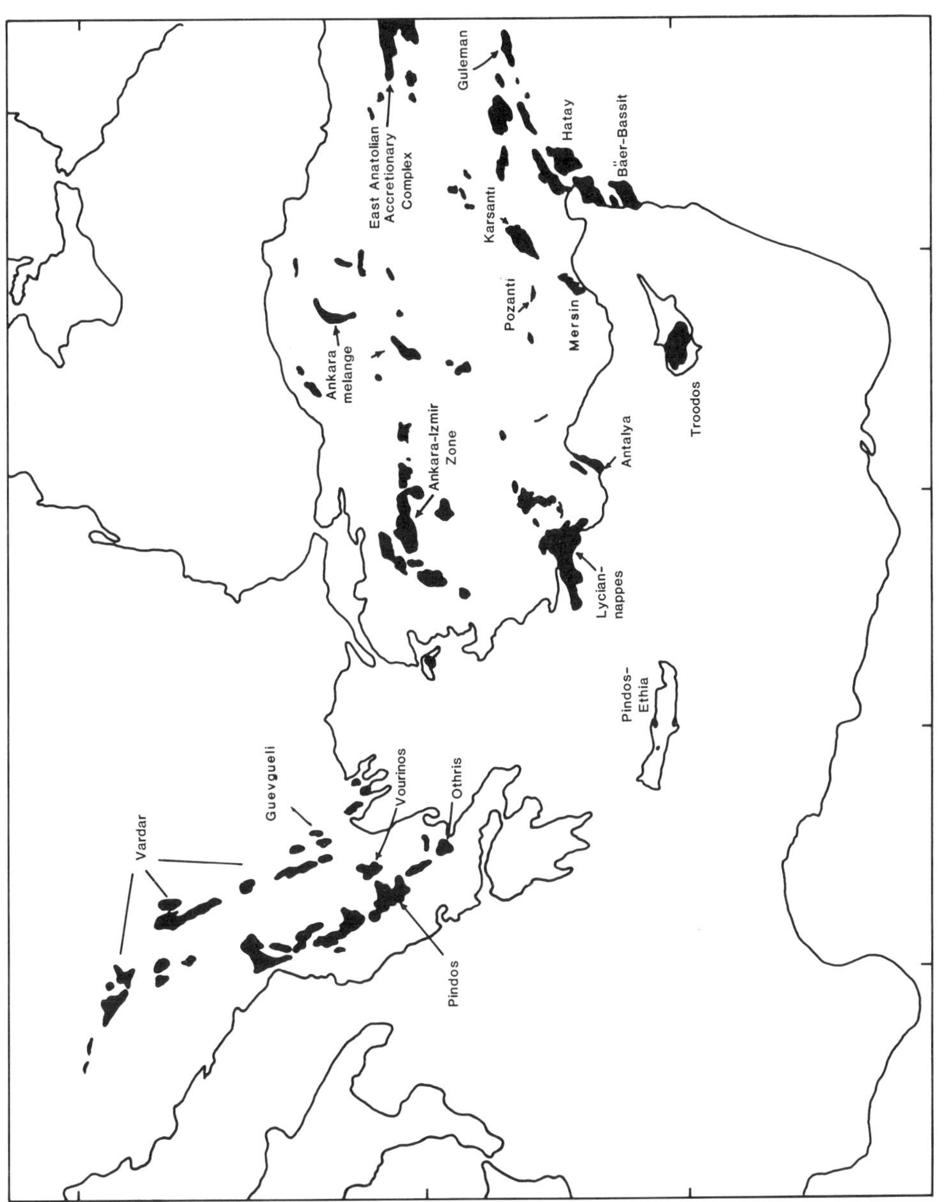

Fig. 8. Location of the main ophiolites in the Eastern Mediterranean.

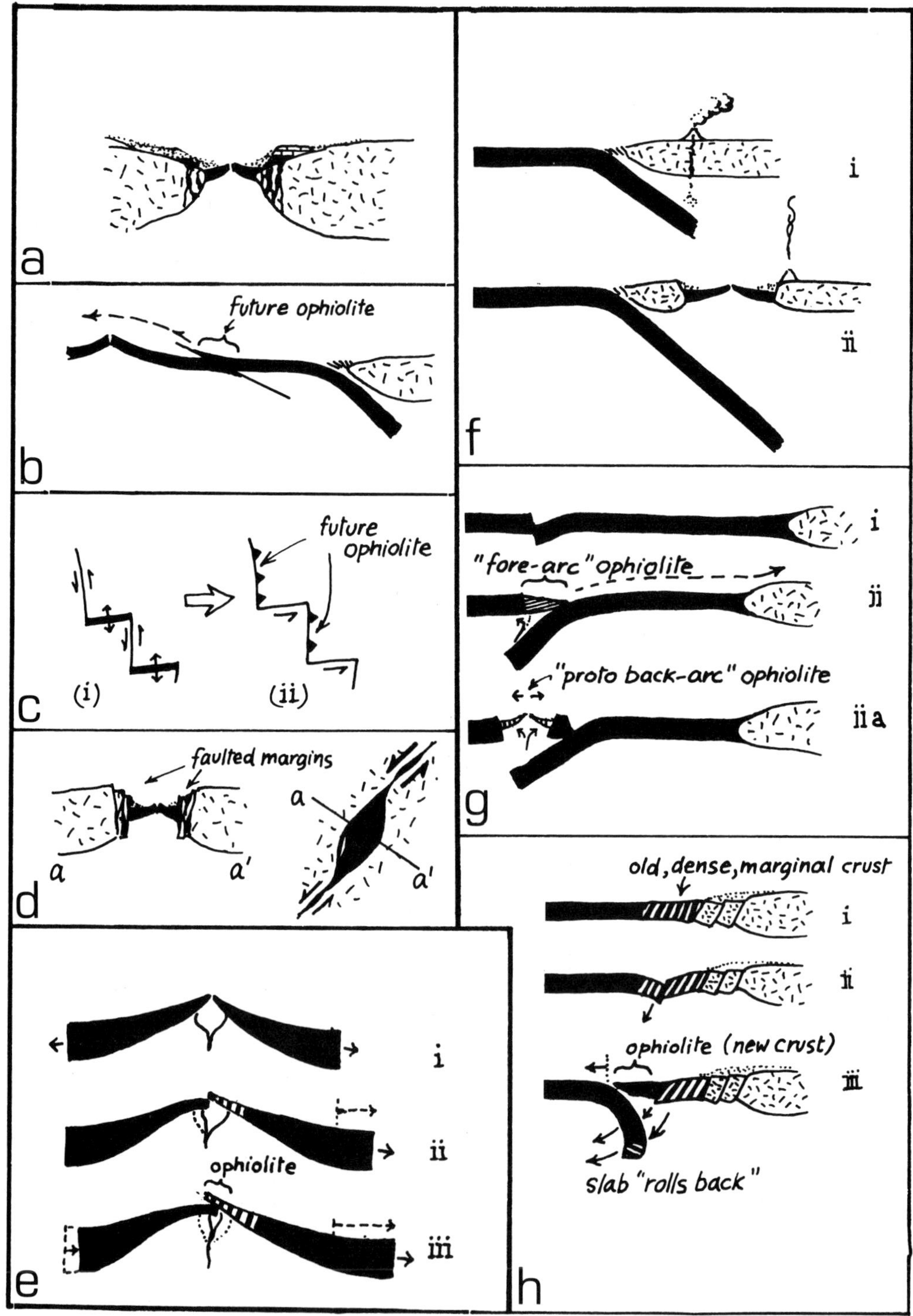

FIG. 9. Alternative settings for ophiolite generation proposed for the Eastern Mediterranean Tethyan region. (a-d, 'Normal' oceanic crust created at mid-ocean ridges or rifts. e, 'Abnormal' mid-ocean ridge crust created by asymmetrical ridge-collapse as proposed in this paper. f-h, 'Above subduction-zone' crust formed in various settings). a, Narrow 'Red Sea' type rifts, e.g. Robertson & Woodcock (1981). b, Young oceanic crust approaching a subduction zone e.g. **Whitechurch *et al.***

plex (Killan Group) in eastern Anatolia described by **Aktaş & Robertson**. The chemical variation correlates with differences in the plutonic stratigraphy, particularly the relative abundance of harzburgite versus lherzolites and the abundance of podiform chromites and trondjemites (Pearce *et al.* 1984). Whitechurch *et al.* contrast the plutonic sequence of the Troodos & Hatay with those ophiolites such as Baër-Bassit, Antalya, and those within the Lycian nappes and the Pozantı-Karsantı ophiolite in which harzburgites are more abundant and cumulates are much thicker (Fig. 8).

Taken at face value the chemistry and petrology would tend to indicate that the Greek Jurassic ophiolites formed either in a range of tectonic settings or even all in back-arc basins since some back-arc basins show normal MORB character (Saunders *et al.* 1980), while the relatively few Cretaceous one analyzed so far (Troodos, Hatay, Baër-Bassit, Guleman) all show the signatures of above subduction zone genesis. The implications are discussed towards the end of this chapter where we present a model for ophiolite genesis at asymmetrically-collapsing ridges which could help explain some of this diversity.

(iii) Relationship to Atlantic opening

It has long been appreciated that the relative motion of Africa and Eurasia has exerted an important control on Mesozoic and Tertiary Tethyan evolution, especially as the north margin of Gondwana was highly indented with Arabia and the Adriatic as major promontories (e.g. Smith 1971; Hsü 1982). Earlier correlations of Atlantic magnetic anomalies (Pitman & Talwani 1972) led to inferred trajectories which involved sinistral motion of Africa relative to Eurasia from the initial Atlantic opening (then estimated at ca. 180 Ma), then dextral motion from around 100 to about 40 Ma, followed by north-south convergence (Dewey *et al.* 1973). More recently, correlations by Tapponier (1977) and Olivet (1982) indicate a smooth, more arcuate, first sinistral then convergent path of Africa relative to Eurasia and this has now been substantiated by Livermore & Smith (1984). The results of DSDP Site 534 and dating of the M-Series magnetic anomalies (Sheridan *et al.* 1982; Ogg *et al.* 1984) now further suggest that spreading in the North Atlantic was initiated from late Bathonian to early Callovian (the projected age of the Blake Spur anomaly, 174–169 Ma), and was followed by a pulse of relatively fast spreading (half-rate ca. 3.4 cm/a) in Callovian to Kimmeridgian time. Recent palaeomagnetic results also suggest that relative motion between Africa and Laurasia was occurring in the Triassic to Early Jurassic prior to Atlantic opening (Steiner 1983). Given that other palaeomagnetic data (see below) suggests that the Adriatic was indeed a major northward promontory of Gondwana, Steiner's results indicate that dextral shear between Africa and Eurasia in the Late Triassic to early Upper Jurassic could have favoured *opening* of a roughly north-south trending ocean basin in the external zones of mainland Greece and Yugoslavia. The opening of the central Atlantic in the Bathonian and associated rapid relative E-W motion of Africa and Eurasia appears to coincide with ophiolitic sole formation in these Hellenide and Dinarde ophiolites, which we interpret as recording the *collapse* of the N-S ridge systems. The association of Atlantic and Tethyan events is strengthened still further by the coincidence of Tauride ophiolite sole formation in the mid-Cretaceous, and the initiation of north Atlantic opening and rapid relative N-S motion (in our area), at around 108 to 90 Ma. In our concluding section we show how these episodes of acclerated motion, following the period of early Mesozoic shear, can help to account for many of the dominant features of East Mediterranean geology.

Discussion

Although any interpretation of Tethyan ophiolites must use the evidence of their slicing and emplacement histories, it is useful to review first the range of tectonic settings that have been proposed for their generation (as summarized in Fig. 9). These fall into two main

following Nicolas & Le Pichon, 1980. c, Normal oceanic crust subject to motion direction change: transforms become subduction zones, e.g. **Şengör *et al.*** d, 'Pull-apart' basins along irregular intra-continental transforms e.g. Bébien *et al.* 1980. e, Asymmetrical ridge-collapse (see this chapter and Fig. 22). f, Intra-continental back-arc basins e.g. Guevgueli ophiolite, Péonias zone, NE Greece. Bébien & Mercier, 1977. g, Fore-arc or 'proto-back-arc' ophiolite mechanism proposed by Pearce *et al.* 1981, for Oman. (i) Initiation of sea-ward intra-oceanic subduction. (ii) Initial rupture leads to 'spreading' in the future fore-arc region. (iia) (alternative) Spreading occurs behind a 'toe' of older crust but pre-dates volcanic arc formation. h, 'Roll-back' model of **Smith & Spray** (i) & (ii) Subduction continent-wards initiated in old, dense oceanic lithosphere. (iii) If asthenosphere flow is right to left, slab will 'roll back', the trench will retreat and new 'above-subduction-zone' type crust will form in the fore-arc region. Emplacement subsequently is by 'roll-to', see Fig. 10.

categories: settings unrelated to subduction (Red Sea-type oceans, 'leaky' transforms, etc., Fig. 9, a-d) and a variety of above-subduction-zone settings ranging from fore-arc to distant back-arc and from intra-oceanic to intra-continental Fig. 9, f-h. Subduction-related settings are increasingly put forward to explain apparent geochemical divergence from normal MORB and apparent short time intervals between creation and initial slicing. Useful literature reviews may be found in **Smith & Spray**, **Whitechurch *et al.***, Channel *et al.* (1979), Bébien *et al.* (1980) and Smith & Woodcock (1982).

As a general observation, there seems no compelling reason to suppose that all Tethyan ophiolites were created in the same tectonic setting, despite their having many features in common. It may also be extremely difficult to distinguish between a slice of 'normal' ocean crust and an intra-oceanic MORB-type marginal basin ophiolite, if no substantial oceanic volcanic arc with evolved products was ever constructed. The continental margin sequences in each case could be identical.

The evidence that many of the ophiolites differ significantly from typical mid-ocean ridge basalts has led to increasingly more *ad hoc* non-actualistic solutions (e.g. Hall 1982) to the extent that the wheel has almost come full-circle with pre-plate tectonic ideas (e.g. Brunn 1956) of ophiolite genesis coming back into favour (e.g. Passerini 1981). From a recent comprehensive review of the Greek ophiolites, Bébien *et al.* (1980) conclude in favour of genesis of most in a series of small sub-parallel intra-continental rifts, or 'leaky' transform faults (Fig. 9d) as a result of diapiric mantle upwelling without significant ocean floor spreading. Such explanations however appear to overlook the palaeomagnetic and fossil evidence confirming a significant oceanic separation of Eurasia and Africa throughout the Mesozoic. If as Bébien *et al.* (1980) imply, the Greek ophiolites, both external and internal, formed in small basins, then some major hitherto undiscovered oceanic suture must exist to the northeast, presumably within the Serbo-Macedonian zone. Instead we firmly align ourselves with those like Pamić (1983) and Şengör and Yilmaz (1981) who argue in favour of a significant oceanic separation through Yugoslavia, Greece and Turkey during much of the Mesozoic at least. We look first at what can be inferred from the field relations then consider the implications of the chemistry.

The preserved *margin successions* of the southern ('external') units, for example, Othris, Pindos and the Antalya Complex, are quite consistent with genesis in Red Sea-type basins which retained their passive margins from the Triassic right up to the Upper Jurassic to Lower Cretaceous in Greece, and to the Upper Cretaceous in Turkey and Baër-Bassit. Consistent with this a variety of tectonic settings have been postulated including failed rifts, narrow 'ocean basins' floored by continental crust, or leaky intracontinental transform faults. Although attractive, the 'leaky' transform fault notion is yet to be supported by much specific evidence, for example, from fabrics in the ophiolites produced by shearing, as seen in the Antalya ophiolite (**Reuber**). Pamić (1983) points out that the Dinaride ophiolites are more reasonably interpreted as being *transected* by transform faults like the Scutari-Peč line which now separates ophiolitic sub-provinces, rather than *created* along them. Pearce *et al.* (1984) point out that oceanic fracture zones are characterized by specific type of alkalic mid-ocean ridge basalts, and thus contrast with the wide chemical variation in the Greek southerly ('external') ophiolites. Also it should be recalled that intra-continental transform faults are rarely straight, but curve, leading to alternating phases of crustal extension and compression along individual strands. If, as suggested by Bébien *et al.* (1980), many of the Greek ophiolites, both 'internal' and 'external', formed in 'leaky' transform faults then great local variations in the timing of genesis, slicing and emplacement would be expected but this is apparently not observed. Instead, the *field relations* of the southern ('external') basins (e.g. Pindos, Othris and possibly Vourinos) are consistent with genesis in small Red Sea-type basins, or even a wider ocean, while the Guevgueli ophiolite in the northerly 'internal' Vardar zone shows many features in common with subduction-related intra-continental marginal basins (Fig. 9f), particularly in the abundance of silicic and fragmental extrusives and the intrusion of the Upper Jurassic Fanos granite (Bébien & Mercier 1977). On the other hand the very complex relationships of the basic and ultrabasic units further SW in the Almopias zone, including serpentinitic tectonic melange (Mercier 1968) could instead be consistent with a range of possible settings, including intra-oceanic arc rocks incorporated in an accretionary complex.

While a good case, based on field relations, can be made for genesis of some of the northerly ('internal') ophiolites above subduction zones, either intra-continental or intra-oceanic, the similarities of the chemistry of some of the southerly ('external') ophiolites with those of volcanic arcs (e.g. Pindos, Capedri *et al.* 1980) has suggested (e.g. **Smith & Spray**) that they too were generated above subduction zones. For example, it has been proposed that both the Greek and the more easterly Baër-Bassit and Hatay ophiolites formed above subduction zones dipping essentially southwards from northern Tethyan strands (e.g. Miyashiro 1977; **Delaloye & Wagner**). A major problem here is the absence of volcanic arcs of appropriate ages in either the Greek Pelagonian zone or the Turkish Tauride carbonate platforms.

An alternative has been to propose genesis of some of the ophiolites as intra-oceanic marginal basins formed above subduction zones as in the present SW Pacific. The chief snag here is that modern marginal basins appear to form by splitting the frontal portion of older arcs, thus isolating a substantial remnant arc (Karig 1974), yet in the case of at least the Greek and Turkish southerly ('external') basins no such older arc rocks units are known; the margin successions instead remained essentially passive during the time of ophiolite genesis (e.g. Antalya Complex). A possible exception here could be the Eastern Anatolian ophiolites in which the fossil and radiometric ages appear to overlap with those of adjacent volcanic arc units (**Michard *et al.***, **Aktaş & Robertson**). A further difficulty with a classical intra-oceanic marginal basin origin is the absence of the expected preserved slabs of older oceanic crust stranded in the arc-trench gap. These have to be explained away by fortuitous non-exposure or some 'subduction erosion' mechanism. The notion that the lower part of the Troodos lava pile (Lower Pillow Lavas) formed at an ocean ridge, while the higher lavas (Upper Pillow Lavas) were created in an incipient marginal basin (Pearce 1975) is opposed by the field relations (e.g. Robertson & Woodcock 1980) and the petrology (e.g. Smewing *et al.* 1976) which all point to genesis essentially within a *single* tectonic setting. Pearce *et al.* (1984) now believe that the whole of the Troodos has an above subduction zone signature.

In an attempt to circumvent these serious problems, Pearce *et al.* (1984) and **Smith & Spray** ingeniously propose that ophiolites in general can form during the initial stages of subduction when sinking slabs of oceanic crust 'roll-back' leading to a brief period of crustal extension at, or just beyond, the trench (i.e. 'pre-arc' or 'fore-arc' spreading) (Figs 9g & h). If the original oceanic basin was quite narrow then all the older crust could be consumed before any significant arc was constructed. As first applied to the Oman ophiolite (Fig. 9g) (Pearce *et al.* 1981) it was proposed that subduction took place well within the ocean and *away* from the Arabian passive margin located to the SE. Continued subduction then consumed the remaining older oceanic crust leading to collision and ophiolite emplacement of the 'pre-' or 'fore-arc' ophiolite onto the Arabian margin before any substantial intra-oceanic volcanic arc was constructed. In marked contrast, **Smith & Spray** propose that in the Greek and Yugoslavian southern 'external' ophiolites subduction was initiated *towards* and close to an adjacent passive margin (e.g. Othris) with 'roll-back' occurring towards the ocean (Fig. 9h). The effect was thus to preserve a strip of older Triassic marginal crust beside the oceanic crust newly created in the Upper Jurassic by 'roll-back'. Pearce *et al.*'s (1980, 1984) models explain the emplacement of the Oman ophiolite onto the former Arabian passive margin to the SW, while **Smith & Spray's** model emplaces the ophiolite over the, by then, active western Pelagonian margin leaving open ocean and the Apulian passive margin to the SW. Conceivably both mechanisms could have operated in the different circumstances. The 'roll-back' model however lacks confirmation from modern active margins so far (e.g. SW Pacific). As pointed out earlier there is little evidence that the Triassic mafic crust (e.g. Agrilia lavas of Othris) really is marginal oceanic crust, as opposed to intra-continental rift lavas. We now go on to consider the related problems of the origin of the ophiolite 'sole' rocks and the number of root-zones, then in a later section see if an alternative mechanism can reconcile some of the various apparently conflicting indicators of ophiolite genesis.

Formation of ophiolitic 'soles'

Bound up with the formation of the Eastern Mediterranean ophiolites is the question of the origin of the underlying ophiolitic 'soles' or 'inverted metamorphic aureoles' (Woodcock & Robertson 1977). These are dominantly mafic in composition but lie beneath tectonized ultramafics and show highest 'grade' adjacent to their upper contacts. There is now general agreement that the 'soles' formed by slicing or 'double-stacking' of still-hot oceanic crust, often long before first tectonic emplacement onto adjacent margins. The extensive recent literature is summarized by **Spray *et al.***

Several models have been proposed involving slicing of hot mafic crust as illustrated in Fig. 10. One model (10c) explores further implications of roll-back genesis of the ophiolites (see above, **Spray *et al.***). Another (10a) envisages that young crust in wide ocean basins is hard to subduct and thus may be sliced near a subduc-

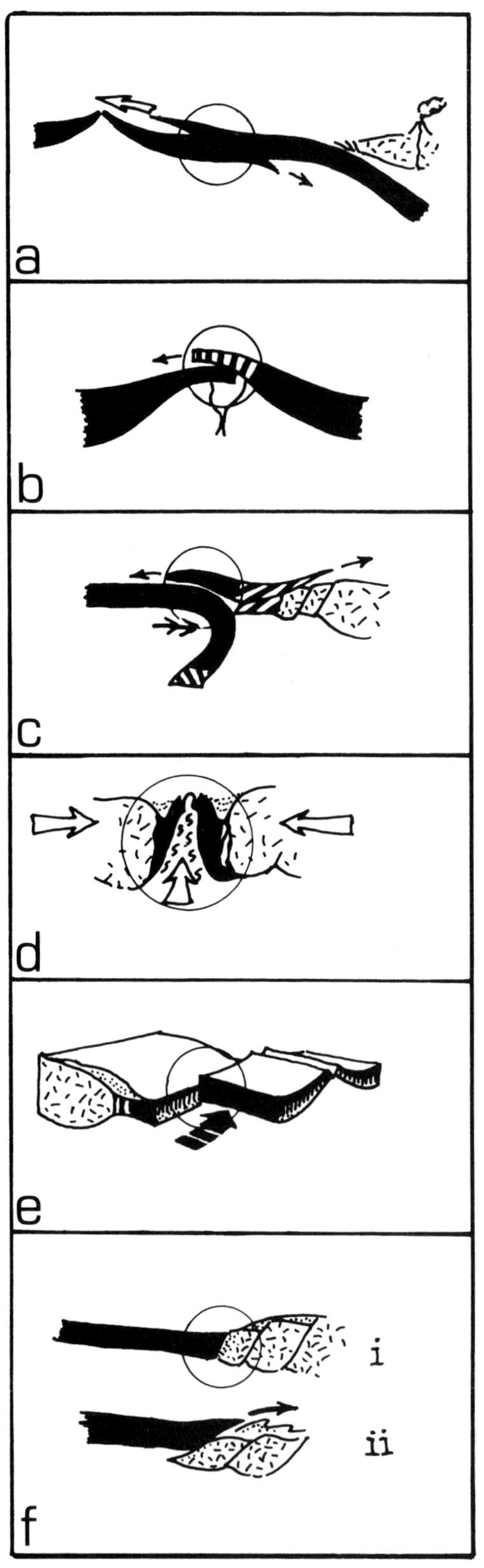

tion zone (Nicholas & Le Pichon 1980; **Whitechurch *et al.***, Fourquin *et al.* 1982). A third suggestion (10d) is that metamorphism took place during creation of ophiolites by diapiric uprise of hot magma in narrow intra-continental rift zones (Bébien *et al.* 1980; Hall 1982). Finally, a possible origin of ophiolitic-metamorphic rocks related to transform faults (10e) has been suggested (Robertson & Woodcock 1980; Spray & Roddick, 1981). Late shearing during emplacement, superimposed on a continent-ocean boundary aureole has also been invoked, (10f), (Hall, 1982).

The new radiometric age data quoted by **Spray *et al.*** now suggest that at least the ophiolites of the western ('exernal') Greek belt were sliced at a time essentially indistinguishable from the age of their genesis at a spreading axis or at most a few Ma after. **Smith & Spray** further point out that the slicing and ophiolite formation is close to the time when the central North Atlantic started to open, put by them at Bathonian, 173 Ma, slightly earlier than the latest Bathonian-basal Callovian age, 169 Ma, inferred by Ogg *et al.* (1984). In their 'pre-arc spreading' model (see above) the ophiolites were first created by 'roll-back' as subduction began, then, soon after, they were sliced at or near the ridge by 'roll-forward', as convergence continued leading to emplacement onto the margins in the Upper Jurassic for the eastern ophiolites, and the Lower Cretaceous for the western ones.

Whitechurch *et al.* point out an apparent progression from west to east in the ages of the Tauride ophiolitic soles. The ages are reported to be around 104 Ma in the Lycian nappes, 94 Ma in the Pozantı-Karsantı ophiolite and 88 Ma in Baër-Bassit. This apparent age prog-

FIG. 10. Alternative mechanisms proposed for ophiolite amphibolitic sole generation. Site of sole formation is ringed. a, Low-angle shearing, parallel to an adjacent subduction zone, in young lithosphere e.g. **Whitechurch *et al.*** as 9b. b, At an overlapped ridge crest in the asymmetrical ridge-collapse model proposed here (see Fig. 22). c, During 'roll-to' following 'roll-back' (**Smith & Spray**) see 9h. Fore-arc ophiolite is thrust out over older crust. d, As a result of diapiric or forceful up-welling of upper mantle and basaltic crust in a pull-apart rift-zone, Bébien *et al.* 1980. e, As a result of transform-activated juxtaposition of an active ridge segment against cold, marginal oceanic crust, Robertson & Woodcock 1982. f, Emplacement-related shearing superimposed on an earlier thermal aureole generated at the continent-ocean boundary, Hall 1982.

ression is complicated by the Antalya Complex, where Yilmaz reports amphibolites of both 122 ma from the coastal Tekirova zone), and 73 and 78 Ma from the Gödene zone further west. The Hatay ophiolite lacks a basal aureole while no base to the Troodos is exposed. **Whitechurch *et al.*** & **Ricou *et al.*** feel that this reported age progression is in keeping with an origin of all the Tauride ophiolites and the Troodos in a single Neotethyan ocean basin (see below). Their model, following Nicolas & Le Pichon's (1980) theoretical analysis is that the ophiolites were decoupled along a shallowly-dipping plane of maximum resolved shear-stress, which in young lithosphere approaching the down-bend of a subduction zone, can lie at shallow depths and intersect the surface. Shear can then occur some distance from both the trench and the spreading axis, with the plane of detachment dipping towards the trench. **Whitechurch *et al.*** suggest that as the young Tethyan crust attempted to subduct under the Pontide active margin to the north (North Anatolian suture zone), thrusting took place progressively further to the south-east, while spreading continued. Thus the Troodos and Hatay ophiolites were created well after slicing of the Lycian and Antalya ophiolites crust further west in the same basin. This model *assumes* only one suture for all the Tauric ophiolites, which, as discussed below, may well not be tenable. There is also little evidence of modern young oceanic crust failing to be subducted in its entirety. Also the model *requires* that continental collision of the Arabian and Eurasian margin took place in the latest Cretaceous to emplace the sliced ophiolites assembled near the Pontide active margin to the north, back southwards onto the Arabian margin. This poses acute problems, particularly for the single ocean basin model, unless the Pontides themselves were separate from, and well to the south of the Eurasian margin, as Eurasia and Africa were still a long way apart at this time (Livermore & Smith 1984).

It has also been argued that strike-slip faulting can play an important role in the genesis of ophiolite-related metamorphic rocks. In SW Cyprus Swarbrick (1980) showed that the amphibolitic rocks (e.g. Ayia Varvara massif) formed sub-vertical screens sandwiched between Troodos ultramafic rocks to the east and Triassic rift-related mafic lavas and sediments to the west. In the Antalya Complex amphibolites occur tectonically-bounded in isolated steeply-dipping screens (**Whitechurch *et al.***). The whole setting of both genesis (**Reuber**) and emplacement (Woodcock & Robertson 1982) of the Antalya ophiolite is now attributed to successive events along a strike-slip controlled Mesozoic margin. As discussed by Spray & Roddick (1981), it is thus possible to envisage high temperature metamorphism occurring as one ridge segment moves past the other, as well as during ocean closure when still-hot oceanic crust is faulted against a transform-type passive margin (Fig. 10e), thus explaining why marginal volcanics and sediments hae also been metamorphosed (e.g. boundary between the Antalya Complex Tekirova zone and the Gödene zone, Robertson & Woodcock 1982).

It is thus probable that rather than having a unique origin, ophiolitic 'sole' rocks can in fact form in a range of tectonic settings which may involve subduction and transform faulting. The key requirement is that hot crust must be faulted against older, cooler mafic crust. We suggest in our penultimate section that this most readily occurs at ridge crests, (Figs 10b & 22) in response to plate-motion changes initiated elsewhere, and that features of both ophiolite genesis and tectonic slicing can thereby be explained as related events.

How many ophiolitic root-zones?

We now come to what is probably the most controversial single aspect of Turkish and Greek Mesozoic geology, the number and location of ophiolitic root-zones. Proposals range from suggestions that all the Turkish and Greek ophiolites, at least, originated in a single basin (Ricou 1971; Bernoulli & Laubscher 1972; Aubouin *et al.* 1976), to ones in which virtually every ophiolite body is considered to be rooted almost *in situ* in a separate basin. We consider Turkey first where the situation is less complicated as major deformation did not take place there until latest Cretaceous time.

(i) Turkish area

In the Turkish area as shown in Fig. 11 **Ricou *et al.*** (see also Marcoux & Ricou 1979) forcefully argue in favour of derivation of all the allochthonous ophiolite-related-units, including the Troodos, from a single root-zone in North Anatolia. This is in total contradiction to reconstructions which invoke the existence of at least a separate southern strand of the Neotethys (Dumont *et al.* 1972; Biju-Duval *et al.* 1977; Robertson & Woodcock 1980) and those which infer a complex palaeogeography involving a braided pattern of oceanic strands and microcontinental slivers (Şengör & Yilmaz 1981).

The single ocean basin model of **Ricou *et al.*** certainly has its attractions in that it promises

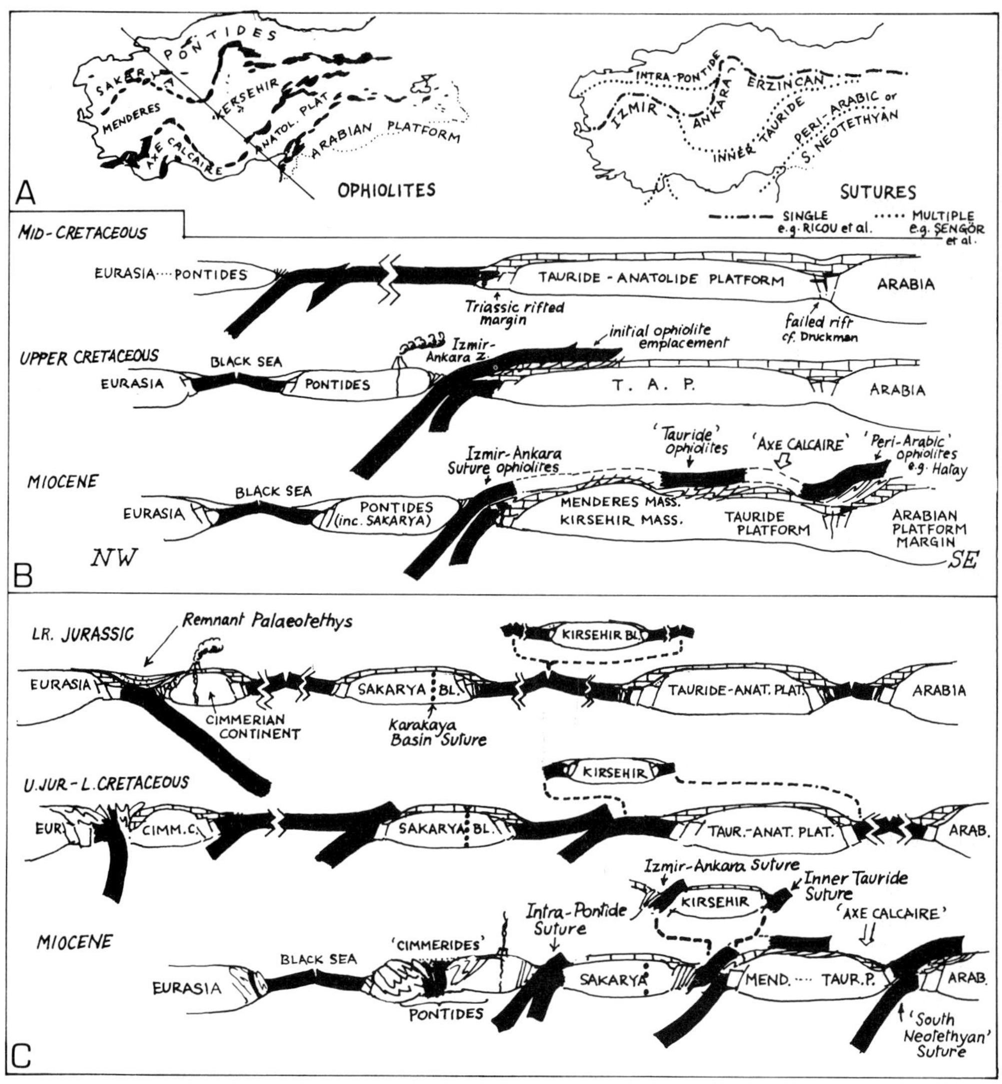

FIG. 11. Ophiolite root-zone interpretations for Turkey: two alternative 'end-members'. *A*. The principal ophiolite belts, 'massifs', and postulated sutures in B and C. Approximate line of sections shown *B*. Cartoon showing the main features of the single-northern-root-zone model of **Ricou *et al.*** and **Whitechurch *et al.*** Turkey as far north as the Izmir-Ankara-Erzincan ophiolite belt is considered part of the northern margin of Africa/Arabia. A major ophiolite nappe emplaced initially in the Upper Cretaceous is disrupted and transported south in stages in Tertiary thrusting events to yield three ophiolite belts (see text). *C*. Multiple-root-zone-model. Synthetic cross-section extending the pre-Cretaceous model of **Sengor *et al.*** to the Tertiary, after Şengör & Yilmaz, 1981. The Sakarya, Kirşehir, Tauride-Anatolide and Cimmerian blocks were rifted off Gondwana-land by Palaeotethyan southward subduction (see Fig. 3 for plan view and note the incomplete picture given in cross-section). Each rift acted as an ophiolite source once the oceanic stage was reached. Four Neotethyan sutures result, which merge in a complex region in eastern Anatolia. The Inner Tauride Suture joins the Izmir-Ankara zone to the west. The 'Tauride' ophiolites, particularly those in the Lycian nappes are the most remote from their source suture. The multiple-suture model proposed later in this chapter adopts the separate blocks shown here but maintains a *northward-subducting* Palaeotethys between Sakarya and Kirşehir blocks, and the Pontides, until the Mid Tertiary. Timing and sense of subduction elsewhere also differ.

a unified view of the complicated Turkish geology. In this model, after the Hercynian orogeny, rifting took place in the Triassic to form a single Mesozoic ocean basin with the Tauride carbonate platforms ('Calcareous axis') then forming part of the passive northern margin of Gondwana. The ocean then began to close by northward subduction under the Pontides from the Mid-Cretaceous onwards (possibly also with southward subduction as specified by Model 1 of **Michard *et al.***), leading first to rapid collapse of platforms to pelagic depths, followed by deposition of ophiolitic olistostromes, then gravity emplacement of ophiolitic nappes. After successive stages of southward transport over the Arabian platform the nappes reached their final positions by Late Eocene to Late Miocene time.

We now look at the evidence in more detail. It has been generally believed that the Lycian nappes were derived from north of the Menderes massif, and it is now accepted that the unbroken Early Cretaceous to Recent sedimentary cover of the Bey Dağları carbonate platform unit precludes simple overthrusting of the Antalya 'nappes' from the northeast (**Poisson**; see discussion in **Hayward**). Over a wide area in the latest Cretaceous ophiolitic nappes were emplaced southwards over the northern edge of the Tauride carbonate platforms ('Calcareous axis') as the Bozkır nappes of Şengör & Yilmaz 1981. For **Ricou *et al.*** ophiolites were also thrust from north of the Kirşehir massif (part of the Anatolides of Şengör & Yilmaz 1981). On the other hand **Görür *et al.*** point out that the structural history of the Kirşehir massif contrasts with that of the Menderes massif sufficiently to propose that the Tauride carbonate platforms were separated from the Kirşehir massif (Anatolides) by an 'Inner Tauride ocean', (see Fig. 11c) from which, in their model, the Bozkır nappes (including the Hoyran-Beyşehir and Pozantı-Karsantı ophiolites) were derived.

Eastwards again, the Anatolide-Tauride platform splits into two arms represented to the north by the Munzur Dağ carbonate platform unit and to the south by the Pütürge-Bitlis metamorphic massifs. In **Ricou *et al.*'s** model all the allochthonous units were derived from north of the Munzur Dağ, while for Şengör & Yilmaz (1981) oceanic crust existed on both sides of the Pütürge-Bitlis metamorphic massifs.

To explain why southerly located units (e.g. Troodos and Antalya complexes) were involved in Upper Cretaceous tectonics while some others to the north either remained undeformed, or experienced mainly Late Eocene transport (e.g. Beyşehir-Hoyran nappes), Ricou *et al.* (1979) have proposed a major Tertiary reorganization by which some units emplaced in the Upper Cretaceous (e.g. Antalya and Mamonia 'nappes') were translated to a more southerly position by a combination of strike-slip faulting and dextral shear. The Troodos and Antalya ophiolites are considered to have been located northwest of the Menderes metamorphic massif, then to have been transported by dextral strike-slip and rotation through the Isparta angle to reach their present positions. In support of the 'internal' hypothesis **Ricou *et al.*** tabulate common elements on both sides of the Tauride 'Calcareous axis', while **Whitechurch *et al.*** conclude that the apparently common origin of all the Tauride ophiolites and those of the Troodos, Baër-Bassit and Hatay at originally roughly E-W spreading axes offset by N-S transform faults, favour an origin in a single basin unrelated to the present shape of the Tauride 'Calcareous axis'.

The central question is whether or not the Antalya Complex and the Troodos massif were once located north of the Tauride 'Calcareous axis', together with the Hoyran-Beyşehir nappes and the Pozantı-Karsantı ophiolites. A similar problem exists with the Hatay and Baër-Bassit ophiolites (Parrot 1977). Şengör & Yilmaz (1981; see also Monod 1977) list five localities where successions on the Tauride carbonate platforms extend intact into Palaeocene-Early Eocene time. This in turn raises the possibility that the ophiolitic allochthons could have been initially assembled north of the Tauride 'axis' in the latest Cretaceous then thrust southwards in the Upper Eocene phase of major regional compression. A number of lines of evidence appear to rule this out at least in the vicinity of the Isparta angle. First, **Waldron** sets out important new structural evidence of major, early, NE-directed thrusting in the NE segment of the Antalya Complex (E of Lake Eğridır) which almost certainly took place in latest Cretaceous time. The thrust stack was later deformed by a SE-directed phase of compression which, regionally, is of Late Eocene age. Secondly, the SW segment of the Antalya Complex was not finally emplaced by easterly-directed thrusting (Woodcock & Robertson 1982; **Hayward**) until the Late Miocene (Poisson 1977), yet in the Isparta area, near the apex of the Isparta angle, the Antalya Complex is unconformably overlain by flysch of Lower and Middle Miocene age which is in turn locally overthrust by the Lycian nappes from the west. Recently Poisson *et al.*, 1983, report that in the heart of the Isparta angle ophiolite-derived flysch of Upper Oligocene and Early Miocene age unconformably overlies emplaced Antalya

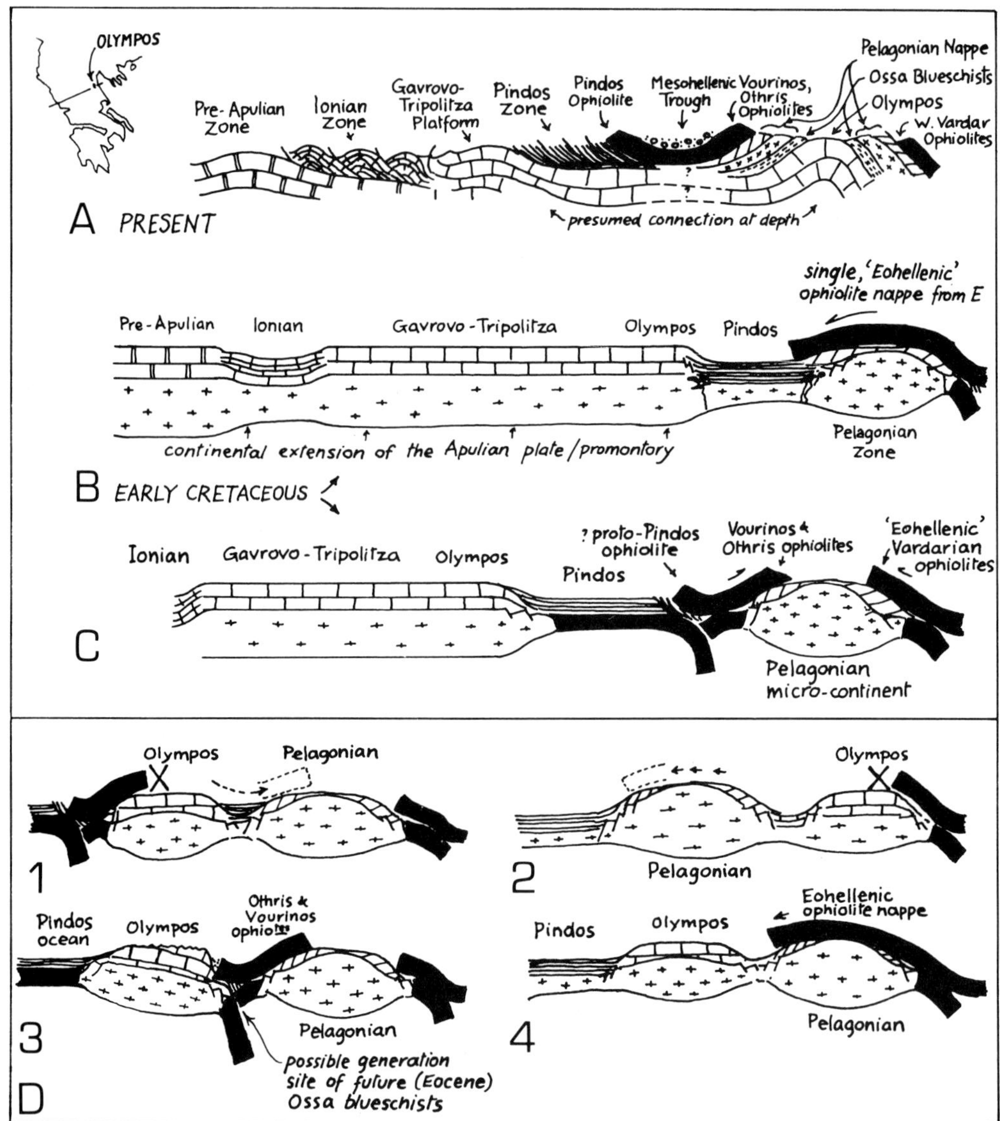

Fig. 12. Principal alternative models for ophiolite emplacement in northern mainland Greece. *A*. Present structural section across the northern mainland (schematic), after Jacobshagen, 1977. The Olympos carbonate sequence is assumed to be a continuation of the Gavrovo-Tripolitza carbonate platform and the Pindos and Vourinos ophiolites are assumed to be continuous beneath the Mesohellenic molasse trough. *B. Single*, Vardarian, root-zone model for ophiolites emplaced onto the Pelagonian zone and its carbonate cover in the Late Jurassic-Early Cretaceous ('Eohellenic') event. After Jacobshagen, 1977. *C. Double* (Pindos & Vardar) root-zone model favoured by Ferrière and Vergely 1976, Smith *et al.* 1979, Bébien *et al.* 1980, and others. Vourinos is attributed variously to E or W derivation. The Pindos-Othris ophiolite emplacement mechanism illustrated is based on the **Smith & Spray** roll-back/roll-to principle (see Figs 9g and 10c): other authors may favour alternative methods from the same western source. See Fig. 18 for the subsequent eastward Tertiary thrusting models. *D*. Alternative possibilities for an *internal* Olympos carbonate-bank, east of the Pindos basin, and their implications for ophiolite emplacement in the Late Jurassic-Early Cretaceous. Each section assumes that the Olympos platform cannot have been overthrust by an ophiolite in the Triassic to Palaeocene interval. Along-strike continuity of the Olympos platform is here assumed but may not be real (see text). *1* and *3*—Oceanic Pindos basin as in *C*. *2* and *4*—Stretched-continental Pindos basin and single, eastern ophiolite root zone, as in *B*. *1* and *2* appear to be impossible: the Olympos sequence bars the way to west Pelagonian ophiolites in *1* and

Complex units. Southward translation of ophiolitic allochthons would thus have to be complete by the mid-Oligocene at the latest well prior to final thrusting of the Antalya Complex in the Miocene further south.

When considering the possibility of complex multi-stage southward thrusting, it should not be forgotten that throughout this time the northern and eastern margins of the Troodos massif remained undeformed, accumulating a continuous pelagic sediment cover from the time of its genesis at an Upper Cretaceous spreading axis through the Campanian, Maastrichtian and into the Lower Tertiary. This culminated in Oligocene-Miocene gentle emergence without significant later deformation (Robertson & Hudson 1974; Robertson 1978). Is it really conceivable that these parts of Troodos underwent more than 400 km of tectonic transport in at least two major separate phases (supposedly latest Cretaceous and Late Eocene) without in any way disturbing the quiet unbroken pelagic accumulation? We think not. It is also worth mentioning the contrast with the Semail ophiolite nappe, Oman, which is *known* to have been emplaced onto the Arabian margin in the Upper Cretaceous (Glennie *et al.* 1973), and where pelagic sedimentation above the ophiolitic extrusives ended abruptly, followed by overthrusting of various allochthonous units of continental margin affinities (Woodcock & Robertson 1982).

We therefore conclude that, contrary to the hypothesis of **Ricou *et al.*** the southern ophiolite massifs (Troodos, Antalya, Baër-Bassit and Hatay) were not thrust over the Tauride carbonate platform units to the north. However, since the objections to **Ricou *et al.*'s** single ocean basin model are based on outcrop geometry and field relations rather than any theoretical problems with the still attractive, simple concept of a single basin, we must see if juxtaposition of sub-parallel ophiolite belts could have been achieved in any other way, for example, by strike-slip faulting, before discounting a unique origin for the various Turkish ophiolites altogether.

(ii) Greek area

The situation is further complicated in the Greek area by successive major phases of deformation and metamorphism in Upper Jurassic-Lower Cretaceous and Tertiary time. Also, there are few long-lived intact carbonate platform successions such as occur in the Taurides to constrain the timing of overthrusting of ophiolitic nappes. There is general agreement that the extreme one-basin model applied to Turkey by **Ricou *et al.*** cannot work in Greece because a passive margin was maintained along the eastern margin of the Gavrovo-Tripolitza platform into the Tertiary. The Gavrovo-Pindos margin could thus not have been the sole westward margin to a Mesozoic Tethys ocean in the same way as the Taurides are thought by **Ricou *et al.*** to lie south of the Tethys in Turkey.

More potentially feasible options, illustrated in Fig. 12, involve emplacement of either all the ophiolites from an eastern Vardar zone (internal origin') or some of them from a separate westerly ('external') basin located within the Pindos zone (Sub-Pelagonian, or Maliac zone) and the remainder from the Vardar.

The alternative view (Fig. 12c) that ophiolites were also rooted within a separate westerly ('external') Pindos ocean basin was based largely on evidence from Othris (Smith *et al.* 1975). Key points in the argument are the presence of a westwards-facing passive margin (Price 1976) and the sense of movement of thrust-related structures (Smith & Woodcock 1976b; see **Smith & Spray** for summary). On structural criteria Naylor & Harle (1975) determined that the Vourinos ophiolite was also emplaced eastwards, while Barton (1976) described structures in the critical Olympos 'window' (see below) indicating tectonic transport in the Tertiary towards the NE. Much of the polarity evidence for Othris now seems applicable also to the Kastoria ophiolite further north (Moundrakis 1982). Some authors at the time of the Athens Aegean Colloquium in 1977 were advocating simultaneously an external origin for the Othris ophiolite and an internal one for the more northerly Vourinos and Pindos ophiolites (e.g. Ferrière & Vergely 1976; Vergely 1977), while many of the contributors to that colloquium were still clearly favouring the unique 'internal' origin (e.g. Aubouin *et al.* 1977; Jacobshagen

all Pelagonian ophiolites in *2*. *2* also requires subsequent Pelagonian nappe transport *eastwards* to generate the present stack (*A*). *4* is internally consistent but does not provide an obvious mechanism for generating a Tertiary blueschist sheet between the Olympos sequence and the Pelagonian nappe. It ignores the evidence for westward emplacement of the Othris ophiolite (**Smith & Spray**). *3* seems the least unacceptable. The west Pelagonian ophiolites would be generated in a narrow oceanic tract separate from the source of the Pindos thrust sheets. The hiatus in carbonate deposition recognized by Schmitt, 1983, could be related to aborted subduction.

1977; cf. Smith 1977) for all of them. By the time of the 1982 Edinburgh conference multiple origins seemed to have been more widely accepted, if only implicitly, (e.g. **Smith & Spray; Moudrakis; Bonneau; Hall** *et al.*; **Papanikolaou; Monod** and Bébien *et al.* 1980). We now look at some of the critical evidence in slightly more detail.

The status of the Olympos carbonate platform succession seen now in a tectonic window through the Pelagonian zone gneisses has always been crucial. Until recently it was considered to show an unbroken succession from Triassic to Eocene and from its close affinities with the Gavrovo-Tripolitza platform sequences on the western side of the Pindos basin, was considered an eastern extension of this platform. On this basis in either the single (Vardar), root-zone conception, or the double Pindos plus Vardar root zones model, all Late Jurassic and Cretaceous thrusting events of ophiolites onto the Pelagonian zone from either direction took place well to the east of the Pindos basin and did not affect the Olympos sequence. In the Tertiary, the Pelagonian zone together with its ophiolitic nappe cover was then transported eastwards onto the Gavrovo platform, later to be revealed *en fenêtre* as the Olympos sequence. With Olympos as an *external* sequence the arguments about root zones then revolve around the polarity of marginal successions bordering the Pelagonian zone and the significance of structural data as indicators of tectonic transport. For example Ferrière & Vergely (1976) accepted the Smith *et al.* (1975) polarity data for Othris but followed Vergely (1976) in attributing the eastward-vergent structures under the Vourinos ophiolite to 'retro-charriage' in the Early Cretaceous rather than as evidence of the 'primary' emplacement direction, which they considered to be from the Vardar zone to the east (Fig. 12). There is now more structural evidence to link the Vourinos and Pindos ophiolites. Ross & Zimmerman (1982) show that fabrics in ultramafics associated with initial obduction are consistent with transport to the ENE thus implying a Pindos basin origin for both, as with the Othris ophiolite.

Barton's (1976) conclusions remain fundamentally opposed to this family of models as they invoke Tertiary *eastward* transport of the Pelagonian zone onto a platform which in the conventional view was always to the west of it.

The situation is now complicated further by the recent observations of Schmitt (1983) that a hiatus exists in the Olympos sequence from Upper Triassic to Upper Cretaceous, setting it apart from the Gavrovo platform and linking it to Pelagonian carbonate successions further south. By implication the Olympos platform was thus on the 'active' eastern side of the Pindos basin rather than on the western 'inactive' side during the Early Cretaceous. To which side of the Pelagonian zone itself it lay is an open question. What seems clear from the lack of any appropriate debris is that no ophiolitic nappes can have crossed over it in the Late Jurassic or Early Cretaceous, in turn raising the question of the continuity or lack of it of both carbonate platforms and ophiolitic nappes along the length of the Pelagonian zone. In the simplest model of a single generally applicable cross-section, an 'internal' Olympos platform *west* of the Pelagonian zone bars the way to Pindos-rooted ophiolites, just as an Olympos platform east of the Pelagonian zone bars the way to Vardarian ophiolite (Fig. 12d, 1–4). As the balance of the evidence now favours two root zones, at least for the Othris and Almopias ophiolites, an 'internal' Olympos sequence requires a more complex model with *discontinuous* carbonate banks which ophiolitic nappes could skirt around or a model in which *present* across-strike juxtapositions (e.g. of Pindos, Vourinos and Olympos) are accidental artifacts of strike-slip tectonics from the Late Jurassic onwards. Both factors could well be important. The Pindos and Vourinos ophiolites could thus have been initially emplaced from different sides of the Pelagonian zone in a region where the Olympos carbonate barrier did not exist, with the Olympos 'block' being transported along the margin into its present position after the emplacement event(s).

Our conclusion is that the Olympos sequence, if external, (i.e. Gavrovo-Tripolitza), does not affect the root-zone argument and that, if internal, the possible tectonic configurations are so many and varied that no postulated root zone for any *individual* ophiolite can yet be ruled out solely because of its presence. The best documented case for an ophiolite root-zone is for an external (Pindos) origin for the Othris ophiolite of Smith *et al.* (1975). If all the Hellenide and Dinaride ophiolites were derived from a single *external* basin in the same way, the inverse of the classical French school mode, then an internal Olympos platform on the western Pelagonian margin would have to have special properties as a block around which all ophiolites 'flowed' as they moved east. Alternatively, it could have lain always to the east of the Pelagonian margin, posing insurmountable problems for **Smith & Spray** who have no eastern side to the Pelagonian zone until after ophiolite slicing in the Jurassic, and posing equally awkward problems for the conventional model of westward stacking of nappes in the Tertiary as it now lies beneath the Pelagonian zone. We will return to the Tertiary nappe stacking problem later.

On the root-zone question in Greece a minimum of two basins seem to be indicated by the available evidence, one 'external' (Pindos), one 'internal' (Vardar). Both are 'Neotethyan' from the age of the oceanic crust involved but this does not preclude the possibility of Mesozoic spreading on a ridge system in a Palaeotethyan basin, giving rise to the Vardarian ophiolites, a possibility developed further in our concluding section.

We are less familiar with the evidence from Yugoslavia, but we feel nevertheless, along with **Smith & Spray**, that the extensions of the Othris and Vourinos ophiolites into the Serbian zone were not rooted east of the Pelagonian-Golija zones but instead formed within the Pindos-Budva ocean basin (c.f. Pamić 1983). Overall, it appears that the extent of thrust-stacking is much greater in Greece and Yugoslavia than in Turkey, possibly because of more intense compression and narrower units. For example, there is no real exposed equivalent in Greece of the relatively autochthonous Turkish carbonate platforms.

We now go on to discuss whether this juxtaposition of parallel ophiolite belts in both Turkey and Greece could have resulted from major strike-slip faulting.

How suspect are the Neotethyan terrains?

In the last few years it has come to be realized, from palaeomagnetic and palaeontological evidence, that colossal lateral displacements, in the order of hundreds to thousands of kilometres have taken place within orogenic belts, leaving few clues for the field geologist to decipher (e.g. Pacific margin of USA, Coney *et al.* 1980). The overall E-W trend of the Tethyan belt makes it unsuitable for palaeomagnetic detection of major strike-slip displacements, but now **Smith & Spray** propose lateral offsets of hundreds of kilometers in the Greek area, while **Lauer** infers displacements in Turkey on the scale of thousands of kilometres.

(i) Smith & Spray's hypothesis

Smith & Spray suggest that the western and eastern ophiolite belts in Greece and Yugoslavia were once parts of a laterally continuous Neotethyan marginal ocean basin formed by 'roll-back' above a subduction zone in the Late Jurassic. After initial tectonic slicing to form high temperatures subophiolite aureoles, continuing left-lateral motion of Eurasia with respect to Africa then detached a strip of the Eurasian continental margin, represented for them by the Serbo-Macedonian/Pelagonian zones together with adjacent oceanic crust and entrained both southeastwards by ca 700 km, as shown in Fig. 13. This motion created a parallel

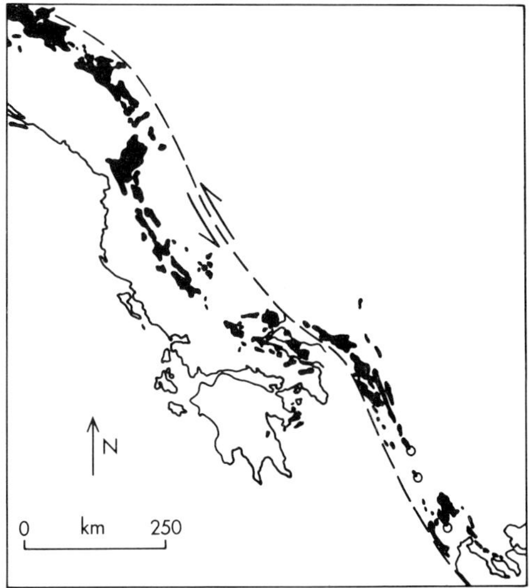

FIG. 13. Sketch-map to illustrate the concept of duplication of the Greek and Yugoslavian ophiolite belts by strike-slip faulting in the Upper Jurassic, as favoured by **Smith & Spray**.

pair of continental margin-ocean segments each with a 'pre-sliced' ophiolite nappe offshore. Shortly after, still in the Jurassic, the eastern Vardar ophiolites were emplaced, followed in the Lower Cretaceous by the western Othris/Pindos ones. The model predicts that the western margins of the Pelagonian and Serbo-Macedonian zones should have essentially identical histories up to the Early Jurassic since they were along strike from each other. As discussed earlier the Triassic successions in the Serbo-Macedonian zone are poorly known due to deformation and metamorphism but do not obviously match those of the eastern Pelagonian margin (e.g. Othris). On the other hand the eastern margin of the Pelagonian zone should show no matching Triassic rifting history to the western side as it is bounded by a Late Jurassic intra-continental transform fault in this model. These predictions may be borne out by field data but there are a number of more directly puzzling features. Why did strike slip-motion intervene so precisely between the time of initial slicing and continental margin emplacement? Why, in the emplacement phase, were Vardarian ophiolites thrust SW onto the Pelagonian 'strike-slip' margin whereas the ophiolites on the other, western, side were thrust the other way, NE, onto the Pelagonian passive margin somewhat later, in the Early Cretaceous? Why is there apparently much more evidence of Upper Jurassic plutonism and Upper Cretaceous and Tertiary volcanism and plutonism in the Vardar and Serbo-Macedonian zones than in the Pelagonian and sub-Pelagonian zones and their extension to the north. As we discussed earlier, the status of the Olympos carbonate platform as an 'internal' block poses severe problems for this model.

In **Smith & Spray**'s model the Vardar ocean was but a stranded remnant ocean which ceased to exist when the ophiolites were emplaced onto its margins, yet in both northern mainland Greece and its probable eastward extension into Turkey active margin conditions appear to have persisted into the Tertiary. On the other hand the Pindos basin shows little evidence of having accommodated the *full* width of the Neotethys after the Early Cretaceous until closure in the Eocene. If the Pindos basin was not thousands of kilometres wide, another as yet unrecognized major basin is required to separate Africa and Eurasia in the Lower Cretaceous. These 'space-filling' considerations are discussed further in our concluding section. No evidence has yet come to light of a major intra-continental transform fault somewhere in Yugoslavia which must mark the site of detachment of the Pelagonian zone in the Upper Jurassic. Thus we conclude that, although stimulating, there is, as yet, little positive evidence in favour of **Smith & Spray**'s hypothesis. If strike-slip translation of ophiolite *did* however play an important role in duplicating ophiolite belts then the model could also be applicable to Turkey. We show in our concluding section that plate kinematic arguments point to a major strike-slip zone existing along the Vardar zone in the Early Cretaceous, but with a *dextral* sense, opposite to that invoked by **Smith & Spray.** We now consider **Lauer's** evidence for displacement on an altogether larger scale based on palaeomagnetic evidence.

(ii) Lauer's hypothesis

Lauer's (1982) hypothesis is that Turkey can be considered as three blocks, A the Pontides, B the Western Taurides and C, the Eastern Taurides, which moved independently until continental collision in the Neogene (Fig. 14). There are no relevant data from autochthonous units of Block B, but the hypothesis *requires* that blocks A and C were located near or south of the equator in the Triassic and Jurassic. For blocks A and C the declination and inclination data for the allochthonous units do not differ significantly from the associated relatively autochthonous ones, showing that, not unexpectedly, none of these units are relatively allochthonous on a palaeomagnetic scale. Accepting this general result, the data from the *allochthonous* units in Block B (Western Taurides) imply an equatorial origin for this area also. This would, in turn, suggest that much of Greece also originated off the eastern Arabian margin, or that the two areas were once separated by thousands of kilometres.

Lauer's hypothesis meets with some immediate general geological support. The boundary between blocks A and C (Pontides and Eastern Taurides) corresponds with the North Anatolian suture zone, while that between the Blocks

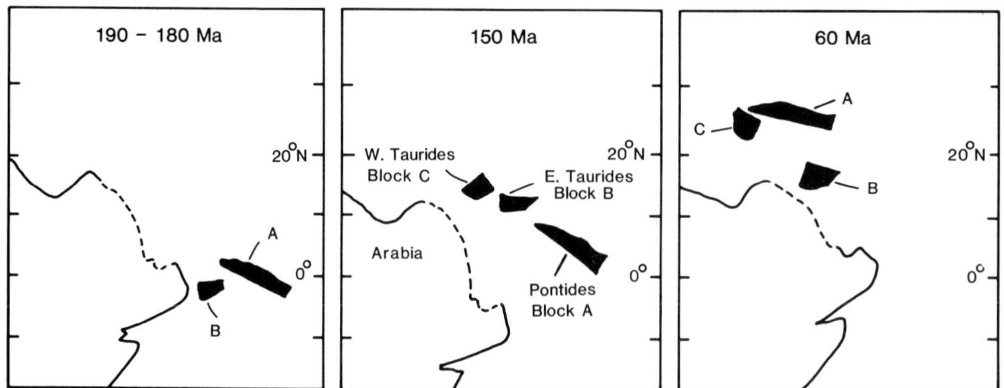

Fig. 14. Lauer's hypothesis of division of Turkey into 3 separate blocks, A, B, C, which drifted northwestwards from an initial position off southeastern Arabia. Various specific geological problems are discussed in the text.

B and C is represented by the Isparta angle. Some oceanic separation across the Isparta angle, at least in the Late Cretaceous, is suggested by **Waldron's** palinspastic restoration of the NE segment of the Antalya Complex. Also, the implied large-scale shear between the blocks would be consistent, at least superficially, with the evidence of long-lived strike-slip faulting involved in both the genesis (**Reuber**) and emplacement (Woodcock & Robertson 1982) of the Antalya Complex.

As a way of theoretically explaining the duplication of ophiolites north and south of the Tauride 'Calcareous axis' each of **Lauer's** three separate blocks A, B and C could have been rifted off adjacent stretches of the Eastern Arabian margin in the Triassic, possibly with block A, the Pontides, being the most southerly, and Block C, the West Taurides, being the most northerly. These three units could then have been transported northward never far from the Arabian margin, consistent with **Hirsch's** faunal evidence, and then juxtaposed prior to emplacement of the ophiolites in latest Cretaceous time to produce parallel ophiolite belts.

In evaluating **Lauer's** hypothesis, the first question is how accurate the palaeo-inclination data really are in relation to the position of Africa. For example **Kissel *et al.*** obtained unreasonably low inclinations for Tertiary sediments of the Aegean arc, and dip corrections for both subaerial and submarine lavas are notoriously difficult. Illustrative of the problems which can arise, earlier palaeomagnetic data for the Lebanon suggested that it also had operated as an independent micro-plate located near the equator during Late Mesozoic to Cenozoic time, despite the lack of confirming geological evidence (Gregor *et al.* 1974). More recently, the discrepancies in declinations relative to the African palaeopoles have been attributed to block-rotation along faults connected with the northward extension of the Dead Sea transform fault system into Lebanon (Freund & Tarling 1978; see also **Quennell**). The problem there of unexpectedly low palaeo-inclinations, however, remains.

How firm then are the actual inclination data? Taking Block A, the Pontides, first, Lauer (1981) relies heavily on the Permian inclination data of Gregor & Zijderveld (1964) and the Jurassic data of Van der Voo (1968) supplemented by a new Upper Triassic site (No. 60 Bayla) and one of Upper Jurassic age (Site 18, Görede 111). As noted by **Lauer**, the Permian data are open to alternative interpretations (normal or reversed polarity?), while the ?Lower Jurassic result from Bayburt was not considered reliable by Van der Voo (1968) due to unstable remanence. Much more palaeomagnetic work is clearly needed to assess the Eurasian versus Gondwanan affinities of the Pontides. From a thorough analysis of 191 samples of Upper Jurassic limestones (Bilecik limestones) in the western Pontides, (Evans *et al.* 1983) obtained a mean palaeo-latitude of about 19°N and 92°E declination, pointing to ca 90° clockwise rotation of the area since the Late Jurassic. Thus a southerly origin of the Block A is not yet confirmed, particularly, as discussed earlier, the fossil evidence points to an Eurasian affinity for the Pontides.

For the Tauride blocks B and C, there are many more new palaeomagnetic data and **Lauer's** hypothesis does not appear to contradict the palaeontological evidence (**Hirsch**). The allochthonous units of the Antalya Complex have yielded a wide spread of inclinations equivalent to palaeolatitudes mostly ranging from about 10°N to 10°S. Near-equatorial palaeolatitudes are also obtained from an Upper Triassic relatively autochthonous site (No. 20) and allochthonous sites (57 and 59) in Block C, the eastern Taurides. The scanty available palaeomagnetic data from the Othris ophiolite also points to a near-equatorial palaeo-latitude in the early Mesozoic (**Smith & Spray**).

Where then, exactly, was Africa situated at this time? Although additional palaeopoles are now being published, Irving & Irving's (1982) summary of African Apparent Polar Wander (APW) (Fig. 15) shows that during the latest Carboniferous and Early Permian there was little APW, then by the latest Triassic to Early

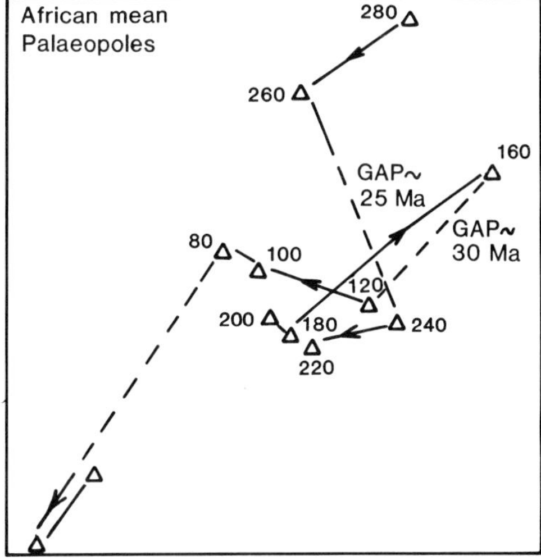

FIG. 15. African mean palaeopoles as compiled by Irving & Irving (1982). Note the major gaps still persisting in the record.

Jurassic the palaeopole migrated to a new stable position. This transition may well have taken place in the Permian but there is a 20 Ma gap in the record from 200–180 Ma. The relative position of Africa is thus not precisely constrained during the critical Upper Triassic to Early Jurassic period. The present North African margin is placed at less than 10°N on the Smith *et al.* (1981) palaeo-continental maps. The critical northward bulge of Arabia which at least the Turkish blocks A and C would have passed in **Lauer's** hypothesis is placed by Lauer (1981; **Lauer**) at ca 16°N, only very slightly further north than shown by Smith *et al.* (1981).

Until a more rigorous statistical analysis of the possible cumulative errors in the inclination data has been made one has to provisionally conclude that the apparent 10–20° difference in inclinations between both the eastern and western Taurides (Blocks B and C) and stable Africa does not yet preclude these areas from having been located against North Africa rather than south-east Arabia. The need for a several thousand kilometre translation, with all its accompanying geological problems, has not yet been convincingly demonstrated. For example, the model implies that the independent blocks should be bordered by essentially Palaeozoic continental margins which are, as yet, unknown; there is little evidence of the long history of transform faulting on this scale around the margins of all of the blocks—the transform faults in the ophiolites are north-south, rather than east-west—nor is there any record of the vast oceanic expanse (e.g. accretionary wedges, arc-trench units) between Greece and the Adriatic peninsula of Africa which this hypothesis would also suggest (assuming low-latitude palaeo-inclination for Greece also). Doubt must also remain as to whether the two Tauride blocks, B and C, really moved independently in the Mesozoic, since, for example, facies patterns on either side of the Isparta angle show little evidence of a significant separation (Gutnic *et al.* 1979) until possibly in the Late Cretaceous (**Waldron**). Setting aside **Lauer's** hypothesis removes obstacles to correlations between the geology of the Levant and the Taurides (**Hirsch, Garfunkel & Derin**) and does not exclude traditional correlations across the Aegean.

(iii) Possible strike-slip in Turkey

Even if Lauer's hypothesis is not confirmed at present, it is worth briefly considering the possibility that less drastic strike-slip faulting could have juxtaposed the ophiolite belts, as postulated by **Smith & Spray** for the Greek and Yugoslavian areas. There is little doubt that strike-slip faulting did indeed play an important role in Turkey. For example, Hall (1976) pointed out that successions in the Bitlis massif in SE Anatolia can not be matched in detail with those on the adjacent Arabian margin (Hazro inlier). **Aktaş & Robertson** model the Eocene Karadere Formation, also in SE Anatolia, in terms of oblique northwards subduction analogous to the modern Andaman Sea, while **Norman** interprets the Ankara melange in terms of a long history of oblique northward subduction in the Mesozoic from the Early Jurassic onwards. In the Antalya Complex both the formation (**Reuber**) and tectonic emplacement (**Robertson & Woodcock**) are now interpreted in terms of transform faulting.

For the ophiolite belts to have been duplicated, however, the offset would have had to have taken place during a brief period in the Upper Cretaceous after formation of the youngest ophiolites (Troodos, Baër-Bassit), but prior to their emplacement onto adjacent carbonate margins in the latest Cretaceous (Maastrichtian). Any later juxtaposition would have been along major intra-continental transform faults which are not known. From the Mid-Cretaceous onwards the relative motion of Africa and Eurasia was convergent, without any driving mechanism for very large scale strike-slip in the area (e.g. Livermore & Smith 1984, Fig. 16). Known transform faults in the ophiolite were orientated essentially north-south regionally (e.g. Antalya Complex, **Reuber, Whitechurch *et al.***) with no evidence of E-W displacement. Strike-slip duplication of the ophiolite derived from a single ocean basin with only one southern (Arabian) margin also implies that the southern edge of the Tauride carbonate platforms should be a major intra-continental transform fault, yet passive margin facies are exposed, as in the Alanya Massif (**Okay & Özgul**) north of Cyprus.

We are therefore led to conclude that the present distribution of sub-parallel ophiolite belts in Anatolia cannot easily be explained either by overthrusting or by some judicious application of strike-slip faulting. This returns us to the concept of genesis and duplication of the Upper Cretaceous Neotethyan ophiolites in several inter-connected basins separated by continental slivers. We look now at the possible arrangements in more detail.

Ophiolite emplacement: open margin or continental collision?

We have already discussed the timing of major ophiolite emplacement events in both the

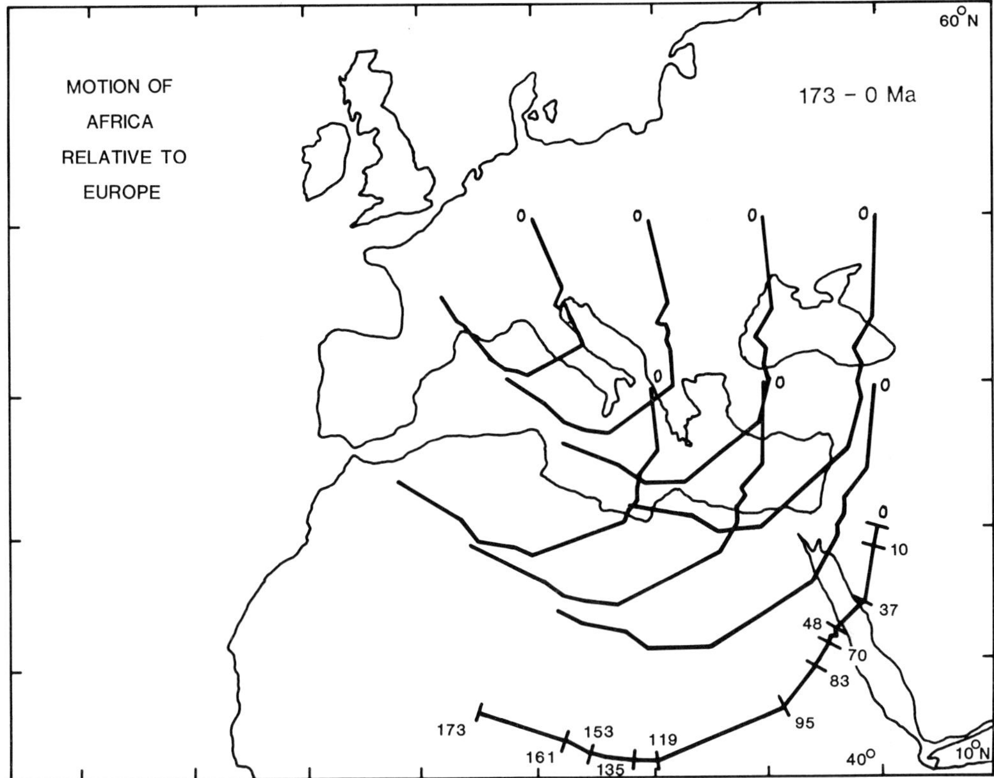

FIG. 16. The motion of Africa relative to Eurasia deduced from the Atlantic opening history by Livermore & Smith (1984). Motion trajectories illustrated are of points on a rectilinear grid fixed to Africa. Note the relatively smooth curves in the East Mediterranean region, contrasting with some earlier interpretations, and the periods of accelerated motion around 170 Ma and 100 Ma.

Greek and Turkish areas. The key question now is whether any of the ophiolites were driven onto adjacent margins by continental collisions or by some other processes acting along open, essentially Andean-type, continental margins.

(i) Greek and Yugoslavian Upper Jurassic-Lower Cretaceous emplacement mechanisms

In the Greek area the dilemma is first encountered in the Upper Jurassic. Along with **Smith & Spray** and others, as discussed earlier, we favour emplacement of ophiolites onto the Pelagonian margin both from the east in the Upper Jurassic (Vardar ophiolites, Evvia, Argolis), and from the west in the Early Cretaceous (e.g. Othris, Vourinos, Pindos, Kastoria). It is still unclear if Vardarian ophiolites were emplaced in a similar way some distance *over* the south-western margin of the Serbo-Macedonian zone or whether they represent some form of subduction-complex accreted along the European margin at this time (see below). In Crete ophiolites were also formed and emplaced in the Upper Jurassic, associated with a range of crystalline rocks (Kreuzer *et al.* 1982), but mechanisms remain obscure.

In the case of the Pindos zone in mainland Greece we *know* that the ophiolite emplacement was not driven by continental collision in the Upper Jurassic-Lower Cretaceous since passive margin conditions persisted into the Tertiary, as seen for example in the Parnassus and Gavrovo-Tripolitza carbonate platform units (Dercourt *et al.* 1973, 1976; Celet 1977; Fleury 1980). Siliciclastics were mostly derived from west of the Pindos zone throughout the Mesozoic (Piper and Pe-Piper 1980; 1982). As a mechanism of emplacement of the Othris ophiolite along an open active margin, Smith & Woodcock (1976a) suggested uplift and gravity-sliding related to serpentinization of a strip of marginal oceanic crust above a subduction zone. In **Smith & Spray's** current model the emplacement is attributed to 'roll-to' during steady-state subduction (Fig. 10c). Evidence of active subduction along the western margin of the Pelagonian zone is non-existent in Greece but present in Yugoslavia in the form of extensive blueschists in the 'Zone of Western Macedonia' adjacent to the Golija zone (Grubić 1980), and in scattered calc-alkaline magmatism within the Golija-Pelagonian zone discussed further below. The Olympos window suggests that the present Pelagonian nappe in Greece could be an attenuated allochthonous remnant of the original Pelagonian microcontinental sliver and that any evidence of active subduction may be obscured. The lack of this evidence in Greece does not preclude subduction being the ultimate driving force for NE-ward ophiolite emplacement.

In contrast to the Pindos zone it has generally been assumed that the ophiolite emplacement and major thrusting and metamorphism along the eastern margin of the Pelagonian zone (e.g. Pelagonian and Eohellenic nappes of the Sporades, **Jacobshagen *et al.***), resulted from closure of the Vardar ocean and collision with the Serbo-Macedonian zone to the north. **Jacobshagen *et al.*** tentatively derive the Skyros unit in the Sporades from a 'Vardar ocean' which *re-opened* in the Late Cretaceous, although Altherr *et al.* (1982b) suggest that this ocean never closed. Once again, the wide separation of Eurasia and Africa until Tertiary times requires at least one major oceanic tract to remain open through the Cretaceous. We will argue that a long-lived Vardar ocean presents the fewest associated problems. From the Early Cretaceous onward, the dominant sense of movement between Eurasia and Africa is *NS* in our part of Tethys (i.e. *parallel* to the Pelagonian and Vardar zones), making strike-slip tectonics an important feature during elimination of this oceanic tract. We will also argue that oblique convergence of micro-continental blocks of Gondwanan origin with the Pelagonian zone may have been instrumental in emplacing ophiolitic rocks from the Vardar zone *sensu lato* in the Aegean area, as for example in the Sporades and Pelion as well as in Evvia and Argolis (e.g. Diabase-Chert Unit and Dhidhimi-Trapezona Unit, Baumgartner & Bernoulli 1976; 1982). This process, combined with the accretion of intra-oceanic-emplaced ophiolites as the 'Vardar Ocean' closed obliquely, could account for much of the complexity. Below we summarize some of the evidence for continuing active subduction on both sides of the Pelagonian zone until the Tertiary.

(ii) Upper Cretaceous Turkish ophiolite emplacement

The problem here is to decide whether or not the ophiolite emplacement onto the north margins of the Tauride-Anatolide platform ('Bozkır nappe' of Şengör & Yilmaz 1981 Fig. 17) and the Arabian margin (e.g. Hatay & Baër-Bassit ophiolites) corresponded to closure of either the northerly or the southerly strands of the Neotethys which must have both existed as concluded above. We refer again to the Africa-Eurasia separation constraint which precludes Upper Cretaceous collision for *all* the ophiolitic sutures together (Fig. 16). One at least must remain open. In Turkey it was traditionally assumed that the ophiolite emplacement resulted from continental collision followed by subsidence of the suture zones to form the Lower Tertiary flysch troughs of central Anatolia (Ketin 1966; Brinkmann 1966; see **Görür *et al.***). Now influenced largely by the continuing record of Tertiary calc-alkaline volcanism in the Pontides (**Akinci**) and sediment provenance, **Görür *et al.*** believe that, except in western Anatolia, two branches of the northern Neotethys, (a main northern Neotethyan strand and their 'Inner Tauride ocean') remained open and continued to subduct northward along an 'Andean'-margin until continental collision in the Upper Eocene (Fig. 17). They re-interpret the central Anatolian flysch basins (Haymana and 'Tuz Gölü') as fore-arc basins active from latest Cretaceous to Upper Eocene then subject to complex rotations and collisions, which finally fortuitously preserved these basins effectively undeformed in a strain-free 'hole' in central Anatolia. Similarly, **Aktaş & Robertson** propose a model for SE Turkey in which the southern strand of the Neotethys remained open and continued to accrete material (Killan Group) above a northward-dipping subduction zone in the Palaeocene-Eocene, followed by subsidence and flysch deposition within a fore-arc basin in the Palaeocene-Eocene. Upper Cretaceous ophiolite emplacement was followed by resumed passive margin deposition along the Arabian margin until the Miocene when the initial stages of continental collision were marked by subsidence and flysch accumulation (Lice Formation), then final southward nappe emplacement. Leaving two open ocean tracts between Eurasia and Arabia would of course satisfy the large scale plate constraints.

To test **Görür *et al.*'s** 'open ocean' model for the central Turkish flysch basins one looks for evidence of Tertiary accreted material and organized arc volcanism. The Ankara melange, which locally depositionally underlies part of the Tuz Gölü basin (Bala area), is indeed interpreted by **Norman** as an accretionary wedge, but he relates it to the oblique northward, rather than *eastward*, subduction of Neotethyan oceanic crust, without any known Tertiary accreted ophiolitic rocks. In **Görür *et al.*'s** hypothesis the Ankara melange should ideally comprise two parts, one related to northward subduction under the Pontide active margin, the other to eastward subduction under

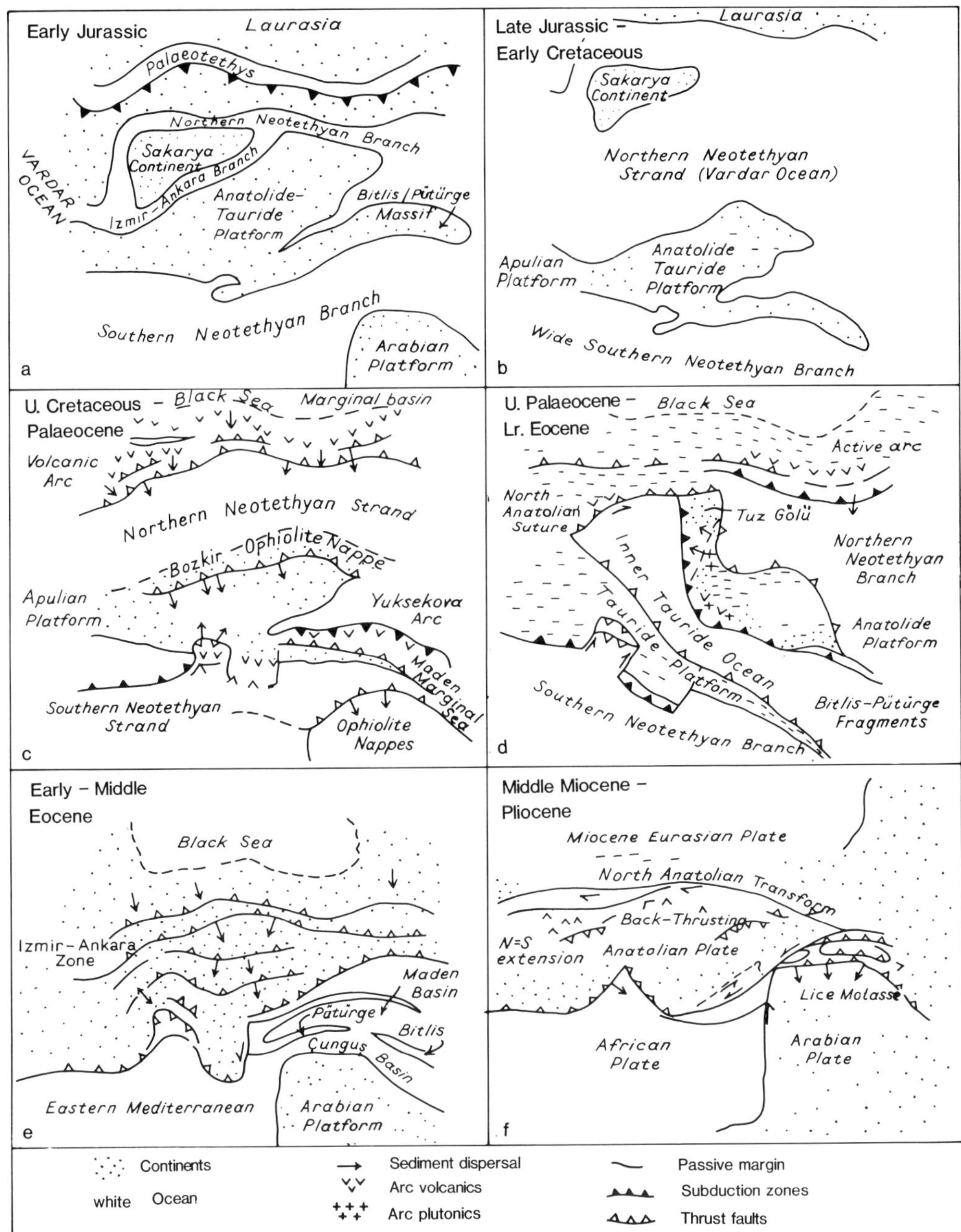

FIG. 17. Plate tectonic reconstruction of Turkey in the Mesozoic and Tertiary. a, b, c, e, f, simplified after Şengör & Yilmaz (1981); d, contrast with **Görür *et al.*** Şengör & Yilmaz's (1981) model involved both a wide southern and also a wide northern branch of the Neotethys which largely closed by early Tertiary while, according to **Görür *et al.*** a distinct 'Inner Tauride ocean' persisted into Eocene time.

the Kirşehir massif, but, as yet, there is little or no field evidence of this. The eastern margin of the Tuz Gölü basin (Bala sub-basin) is cut by the major Palaeocene Barnadağ granitic pluton, and Lower Tertiary calc-alkaline volcanics are known further east on the Kirşehir massif, but it is by no means clear yet that this is part of an organized volcanic arc related to steady-state eastward subduction. Indeed, extensive Lower Tertiary volcanics are reported from deep wells in the Tuz Gölü which is unexpected in a typical fore-arc basin. A further problem is that the units exposed south of the Haymana basin comprise recrystallized platform carbonates (Dereköy Formation: Ünalan *et al.* 1976) rather than any remains of an accretionary wedge as the fore-arc model would predict. On the other hand if major continental collision did indeed take place in the Late Cretaceous, it is unclear why rapid subsidence of the 'collisional' areas should have immediately taken place to form deep flysch basins as early as the latest Cretaceous (e.g. Haymana Formation). Similarly, the Upper Cretaceous to Middle Eocene flysch basin of Turkish Thrace (Gökcen & Gökcen 1982) could have formed as a fore-arc basin related to northward subduction of a northern Neotethyan strand, *if* the Biga peninsula to the south, which was a sediment source area, had already been incorporated into the European active margin in the Lower Tertiary, possibly as a westward extension of the Pontide active margin.

On balance, then, although problems of interpretating local field relations still exist, we favour the view that *both* northern and southern strands of the Neotethys remained broadly open from the Mesozoic until well into the Tertiary. The latest Cretaceous ophiolite emplacement would thus have taken place along open Andean margins. If the Upper Cretaceous ophiolites were generated above active subduction zones, or as suggested in a subsequent section, along spreading ridges immediately prior to their collapse, then subsequent intra-oceanic subduction could have led eventually to collisions with adjacent passive margins. The fact that all these collisions took place within a relatively short period in the Late Cretaceous (Maastrichitan) suggests that the oceans were not very wide or that, fortuitously, any intra-oceanic subduction zones were created at approximately the same distance from the adjacent margins in several basins. Specifically, the 'Bozkır' ophiolites would record the collision of an intra-oceanic subduction zone in the northern Neotethyan strand (the Inner Tauride Ocean) with the Tauride carbonate margin to the south. The Eastern Anatolian ophiolites Baër-Bassit and Hatay, which formed in the southern Neotethyan strand were emplaced over the Arabian margin in the Maastrichtian by the same mechanism. A possibility worth considering is that the Baër-Bassit ophiolite could have been emplaced eastwards from ocean crust in the northern part of the present Levant Sea area rather than from the north over major carbonate platform units (Jebel Aqraa). However, the latest Cretaceous tectonic emplacement did not involve the Troodos massif, which continued to undergo unbroken pelagic deposition along its northern and eastern margins (e.g. Robertson & Hudson 1974).

Summarizing, in both Greece and Turkey our view is that both the northern and southern strands of the Neotethys remained open from the Mesozoic into the Tertiary and that both the Late Jurassic-Early Cretaceous emplacement event in Greece and the Late Cretaceous ones in Turkey took place along open 'Andean' margins for the most part, with localized oblique microcontinental block-collision being perhaps important in the Aegean region.

Subduction and arc volcanism

There is now an increasing record of calc-alkaline volcanism which could be related to subduction of the various Neotethyan strands in the Eastern Mediterranean. The problem, however, is to decide when calc-alkaline magmatism indicates steady-state subduction, as opposed to other processes which could have taken place later. These include the rejuvenation of older only partly subducted oceanic slabs (e.g. Model 1 of **Michard *et al.***) or general deep level crustal anatexis following pervasive continental collision. Above, we concluded that at least in Turkey both northerly and southerly Neotethyan strands remained open into the Tertiary, following major ophiolite emplacement, but micro-continental collisions could have begun the Upper Cretaceous.

(i) Hellenides and Dinarides

(a) Serbo-Macedonian Margin, Eastern Vardar Zone in Greece and Yugoslavia

Unequivocal evidence for active subduction beneath the Pelagonian/Golija margin and beneath the Serbo-Macedonian zone to the east is not easy to accumulate (see Smith & Moores 1974). The clearest indication is the sequence of granitic intrusions closely associated with ophiolites which stretch north-westwards from Guevgueli in the Eastern Vardar Zone on the Greece-Yugoslavia border—the granites of

Fanos, Furka, Stip and Karadagh (Bébien & Mercier 1977). The Fanos granite, which cuts the Guevgueli ophiolite, has been recently dated again as Late Jurassic (148 ± 3 Ma) from a K-Ar determination on biotite (**Spray *et al.***). Other granitic bodies in the Serbo-Macedonian zone in Greece (Monopigadhon: 180 Ma, Ricou 1965; Lachanas: 140 Ma, Marakis 1970) may belong to the same phase. The interpretation of the Guevgueli ophiolite as marginal basin crust telescoped in the Upper Jurassic and then cut by continuing calc-alkaline magmatic intrusions related to eastward subduction seems the most reasonable. The Fruška Gora HP-LT metamorphic rocks, which project from the Pannonian basin more or less along strike from the Serbo-Macedonian/Vardar border may be related to subduction on this margin but their age is not known (see Grubić 1980 p. 326). The duration of subduction-related magmatic activity on this margin does not seem to be well established. Grubić (1980) quotes the Bujanovac granite pluton in the Serbo-Macedonian zone as dated by U-Pb and other methods at 234 Ma (Ladinian), which may indicate pre-Upper Jurassic activity. Mercier (1966) records Early Cretaceous 'trachyandesites' from the Païkon zone (which includes Guevgueli) and Late Cretaceous spilites of unknown affinity from the same zone, but it is not until the Tertiary that references to calc-alkaline volcanism in the Vardar and western Serbo-Macedonian zones become more prominent. Middle and Late Eocene volcanics are known from the Peonias zone of Greece (eastern Vardar) and its continuation in Yugoslavia, as well as from the Almopias zone in the western Vardar where activity is again found in the Pliocene (Mercier 1966). Within the Serbo-Macedonian zone in Yugoslavia 'alpine' granites (e.g. Jastrebac) with 'Tertiary andesites and dacites' are recorded by Grubić, (1980, p. 302). The situation is complicated in the more internal part of the Serbo-Macedonian zone in Yugoslavia by proximity to the Danube trough, which is regarded as having closed in the Aptian and Albian by subduction of its floor westwards under the Serbo-Macedonian zone. This gave rise to extensive calc-alkaline vulcanism in the Timok zone (a trough within the Serbo-Macedonian massif) in eastern Serbia through much of the Late Cretaceous, (Grubić, op. cit. p. 329). It seems that eastward subduction-related magmatic activity may well have diminished in the Cretaceous, possibly through the change in convergence patterns introduced by the elimination of the Danube basin between the Serbo-Macedonian massif and the Rhodope. The significance of the Danube basin is difficult to gauge and it may be that active subduction of 'Vardarian' ocean crust largely ceased after Upper Jurassic ophiolite emplacement and as Africa-Eurasia convergence rates slowed in the Early Cretaceous. We examine this and other possibilities further in our concluding section.

(b) Pelagonian Margin: Greece and Yugoslavia

Evidence is lacking of Upper Jurassic or Cretaceous magmatism that might be related to eastward subduction of the Pindos ocean floor beneath the Greek or Yugoslavian Pelagonian zones or the Golija zone further north. In Greece this may reflect the fact that the Pelagonian 'zone' is a narrow, wholly allochthonous nappe complex which may not be representative of the original autochthonous Pelagonian micro-continental sliver. It may always have been too narrow, and the angle of subduction too shallow, for magmatism to occur beneath the Pelagonian zone itself. Atlernatively, Pindos ocean crust may not have been extensive enough or the dip again steep enough for an arc to develop until much later. Grubić (op. cit.) records that Tertiary andesites and 'granitoids' occur in the Golija zone, the northward continuation of the zone in Yugoslavia.

Aegean Area

In the south Aegean, ophiolites emplaced in the Upper Jurassic on Crete and Gavdos are associated with calc-alkaline intrusives and high grade metamorphic rocks (Kreuzer *et al.* 1982). There is also evidence of LP/HT metamorphic rocks, including meta-ophiolites and related granitoids of Late Cretaceous age in central Crete, and on the smaller islands of Anaphi, Nikouria, and Donoussa (Reinecke *et al.* 1982). On Crete, the granitoids form part of the composite Asteroussia nappe (see Fig. 19) of Pelagonian affinity (**Bonneau**) which is associated with the ultramafic rocks and micro-gabbros (the Upper Jurassic 'Serpentinite-amphibolite association' of Creutzberg & Seidel 1975) and was last metamorphosed in the Upper Cretaceous (75 Ma; Seidel *et al.* 1981). **Bonneau** recognizes equivalents of his Asteroussia nappe in the Cyclades (the Vari unit of Syros, and on the islet of Nikouria, Dürr *et al.* 1978). In Crete the Asteroussia rocks form the highest levels of the nappe pile, structurally overlying an assemblage traditionally correlated with the Pindos zone of mainland Greece (Fig. 19, see below) and which **Hall *et al.*** suggest could be a major far-travelled olistostrome. It seems that the nappe-pile that is now Crete is recording a tectonic history more akin to Turkey rather than mainland Greece in the Upper Cretaceous and yet has the characteristic Hellenide-Dinaride Upper Jurassic ophiolite event also represented.

However, as we hope to show, the potential for tectonic complexity in the south Aegean area is enormous from the Jurassic onwards and attempts at 'Hellenidic' or 'Tauridic' interpretations are probably unrealistic.

There remain considerable problems even with the more obvious Tertiary subduction-related magmatic and metamorphic events in this area in identifying the strands of Neotethys responsible. **Fytikas *et al.*** argue for a progressive southward migration of Hellenic-arc-related activity from Thrace in the Eocene-Oligocene to the present arc initiated at around 3 Ma. However, Altherr *et al.* (1982a) attribute the belt of Oligocene granitic intrusions across the South Aegean and the HP-LT metamorphism of the Phyllite-Quartzite and Plattenkalk series of Crete to NE-ward subduction of a southern Neotethyan strand in the Oligocene to Miocene interval. It is difficult to envisage a configuration for a subducting slab that could account for both the Thrace and the south Aegean activity at the same time. It may be more realistic to accept the possibility of other oceanic tracts linked to the Vardar ocean and possibly to the Intra-Pontide ocean to the east which could have been eliminated in the northern Aegean by northward subduction, perhaps even with post-collisional magmatism continuing for a few million years. Indeed, Papavassiliou and Sideris (1982) attribute the Tertiary lavas of Western Thrace to north-eastwards subduction of Vardarian oceanic crust.

2. Turkish area

The situation is generally more straightforward in Turkey (see Fig. 6) where the major Upper Jurassic-Lower Cretaceous magmatic events are not known. There is now general agreement that extensive Upper Cretaceous calc-alkaline volcanism in the Pontides can be related to northwards subduction of Neotethyan oceanic crust (Şengör & Yilmaz 1981; **Akinci**) possibly leading in the latest Cretaceous to Early Tertiary to opening of the Black Sea as a marginal basin (Letouzey *et al.* 1977, Fig. 17). In the Eastern Pontides, the calc-alkaline rocks are divided into the Lower Basic Series (195–88 Ma), the Dacitic Series (88–65 Ma), the Upper Basic Series (65–37 Ma), Tertiary granitoids (37–22.5 Ma) and the Neogene volcanics (**Akinci**). In this area the volcanism began well prior to the Upper Cretaceous and persisted without a break into the Tertiary. In the western Pontides ('Sakarya continent' of Şengör & Yilmaz, 1981) extensive calc-alkaline volcanism began in the Turonian (ca 90 Ma, **Okay**) but this block may have been some distance to the south of the Eurasian margin at this time (see below).

A major Upper Cretaceous-Early Tertiary volcanic arc complex is also known along the south margins of the Tauride platforms in SE Anatolia. This is the Eliziğ-Palu nappe of **Aktaş & Robertson** and the Upper Cretaceous arc of **Michard *et al.*** which comprises mafic to acid extrusives and granitic intrusives dated as Coniacian to Maastrichtian (ca 88 Ma-65 Ma). In various areas, the intrusives can be shown to cut older ophiolitic and platform rocks (**Michard *et al.*; Aktaş & Robertson**). The entire assemblage could contain elements of both intra-continental and intra-oceanic volcanic arcs related to the northward subduction of a southern strand of the Neotethys.

In Eastern Anatolia, the calc-alkaline igneous activity appears to stop abruptly in the latest Cretaceous, although it resumes more locally in the Eocene (**Aktaş & Robertson**). Further west there is an important record of thick Upper Cretaceous to Oligocene andesites associated with a narrow ophiolite belt along the northern margin of the local Tauride platform, the Bolkar Dağ, and further north a dismembered ophiolitic complex (Alihoca) occurs with calc-alkaline volcanics and thick flysch-type sediments (Ciftehan) of Upper Cretaceous Palaeocene age together with dioritic and granitoid intrusives (reviewed by Juteau 1980), which together can probably be related to the northward subduction of a Neotethyan oceanic strand located immediately north of the Tauride platforms ('Inner Tauride ocean' of **Görür *et al.***).

South of the Taurides, in the Kyrenia Range of North Cyprus mafic volcanics associated with deformed Tauride platform units began in the Maastrichtian and show calc-alkaline trends based on major element chemistry (Baroz 1980). Subsequent more evolved Palaeocene lavas reported as calc-alkaline, include shoshonites. Believing, along with **Ricou *et al.*** that the Troodos was emplaced over the Arabian margin in the Upper Cretaceous, Baroz (1980) thought that the 'calc-alkaline' rocks of the Kyrenia Range were post-collisional and related to transcurrent motion. This contrasts with the view of Robertson & Woodcock (1980), which we favour here, that they can be related to continued northward subduction of a southern strand of the Neotethys in the present Eastern Mediterranean area.

If we take the general volume of calc-alkaline products again as an indication of the extent and duration of subduction of Neotethyan oceanic crust, then it can be inferred that the northern branch of the Neotethys, in which calc-alkaline igneous activity extended through the Upper Cretaceous and into the Tertiary, was probably considerably wider than the southern Neotethyan strand, where the volcanism is volumetrically smaller, more variable, and less persistent.

Collisions large and small

The Late Eocene to Miocene was dominated by a series of diachronous continental collisions of varying scale and intensity eventually leaving only remnants of the southern Neotethyan strand south of Crete. With onset of continental collisional tectonics comes the problem of the origins of the various Late Cretaceous and Tertiary blueschists in the East Mediterranean, and the formation of the Aegean arc (Brunn 1976).

(i) Turkish area

Okay suggests that the NW part of the Anatolide platform first collided with the Sakarya block to the north in the Upper Cretaceous (Cenomanian) leading to the formation of blueschists in, for example, the Eskişehır region which were then uplifted and eroded to provide material for the Haymana basin within 25 Ma of this event (Campanian). The blueschist protolith was a thick volcano-sedimentary unit which **Okay** suggests was part of the former northern passive margin of the Anatolide platform. There are a number of problems with this interpretation. **Okay** implies that the Sakarya block to the north was part of the Pontide active margin and so attached to Eurasia at this stage. This not only neglects the possibility of an open 'Intra-pontide suture' (Şengör & Yilmaz 1981, Fig. 17) north of the Sakarya block but would require an enormously wide Neotethyan ocean south of the western Tauride-Anatolide block to accommodate all the remaining excess Africa-Eurasia separation at this time. There are similar difficulties even if **Okay's** Cretaceous collision occurs between two blocks remote from either the African or Eurasian margin, but the combined block still collides with Eurasia in the Palaeocene along the Intra-pontide suture, as envisaged by Şengör & Yilmaz (1981, Fig. 17). Nevertheless, Servais *et al.* (1982) also support a Late Cretaceous collisional mechanism for the blueschists in this belt and identity Pontide (i.e. Sakarya) elements affected by HP-LT metamorphism, as well as Tauride units, to support their case. Welding the western Taurides and the Sakarya block to Eurasia by the Palaeocene would, incidentally, be completely incompatible with a single-ophiolite-root-zone model which envisages the Tauride platform as part of the African margin and relatively autochthonous. Africa and Eurasia are simply too far apart in the Palaeocene for all Neotethyan oceans to have been eliminated by then (Livermore & Smith, 1984, Fig. 16, and see Figs 23a-f below). We cannot yet resolve the paradox presented by early collisions between blocks and would still favour an accretionary prism—active open margin interpretation for Cretaceous blueschist generation in the Izmir-Ankara Zone, if the field evidence will sustain it. Collision would then only occur in the Eocene when the Lycian nappes were driven southward over the internally deforming Menderes massif. We will return to the 'gap-filling' problem in our concluding section.

If an 'intra-oceanic' Cretaceous collision is unavoidable it may well be similar to that which appears to have generated the newly discovered blueschists in the Alanya massif on the Mediterranean coast of Turkey, north of Cyprus (**Okay & Özgül**). There, HP-LT rocks in the Sogüzü nappe are sandwiched between a low-grade Mesozoic platform-carbonate upper nappe and a greenschist-facies meta-clastic and marble-bearing lower nappe. All three units were thrust together in post-Maastrichtian times over Antalya Complex rocks now exposed in the Alanya window. The Sugözü blueschists could relate to the elimination in the late Mesozoic of a small Neotethyan strand between the main eastern Tauride block to the north and an 'Alanya micro-continent' to the south. Southward thrusting was complete by the early Middle Eocene (**Okay & Özgül**). In addition, scattered blueschist occurrences associated with the emplacement of the ophiolitic 'Bozkır nappes' (Şengör & Yilmaz 1981) occur along the northern margin of the Tauride platform, for example in the Konya region, the Bünyan-Pinarbaşı area of the Kayseri region and north of the Bolkar Dağ (review by Juteau 1980). They can be related to destruction of oceanic crust between the Tauride and Anatolide platform units—the 'Inner Tauride ocean' (**Görür *et al.***).

During the Upper Eocene and Oligocene more general collision of the Tauride-Anatolide platform with the Pontide margin was the main force driving southward emplacement of the major allochthons, including the Hoyran-Beyşehir, Hadim and Lycian nappes. After possible initial collision in the Upper Cretaceous, the Lycian nappes were apparently driven southwards over the Menderes massif in the ?latest Palaeocene and Eocene (Graciansky 1972; Akkok 1982; **Poisson**) giving rise to slicing and progressive metamorphism of complex poorly understood units below, which include meta-carbonate, meta-clastics and serpentinite. Further east in Hoyran-Beyşehir and Hadim nappes were thrust southwards, the deformation front reaching as far south as the Kyrenia range in North Cyprus which experienced southward thrusting in the mid-Late Eocene (Baroz 1980). At least in Eastern Anatolia, after a brief period of ophiolite emplacement in the latest Cretaceous, the northern margin of Arabia remained passive until the Miocene when it subsided rapidly and was overlain by a thick flysch wedge (Lice and Çunguş flysch) which can be related to the advance of the adjacent northern active margin of the southern Neotethyan strand (**Aktaş & Robertson**). It can be reasonably inferred that the southern Neotethyan strand remained open until the Miocene at least in Eastern Anatolia.

(ii) Greek area

In the Greek area (Figs 7 & 12), the situation is apparently considerably more complex and even more controversial. Major problems are the extent of allochthoneity of mainland Greece, whether or not it was grossly telescoped by the Mid-Tertiary, and how to correlate events with those which affected the South Aegean and Turkey.

In mainland Greece, west of the Vardar zone and north of the island of Evvia, the present pattern of 'isopic zones' is largely the result of progressive westward-stacking of nappes, documented by the general westward-younging of the onset of flysch sedimentation: from Maastrichtian in the Pindos zone ('Premier flysch'), Upper Eocene on the eastern side of the Gavrovo-Tripolitza carbonate platform (and in the Olympos window), Oligocene on its western side and in the Ionian zone. There is general agreement that ophiolites had been emplaced onto the Pelegonian zone in the Lower Cretaceous following Upper Jurassic slicing (e.g. Aubouin 1977; Bébien *et al.* 1980; Jacobshagen 1977), giving rise to a 'typical' Pelagonian sequence of pre-Mesozoic basement (e.g. Hercynian granite as in Pieria: Yarwood & Dixon 1977), metamorphosed and deformed in the Lower Cretaceous along with its Triassic-Jurassic carbonate cover, and already with emplaced ophiolites, and a transgressive cover of Middle or Upper Cretaceous age (Celet & Ferrière 1978; see Fig. 18).

While it is generally agreed that the westward telescoping is the result of new or continued collision between a linear Pelagonian basement block and the presumed Apulian/African margin west of the Ionian zone, the amount of shortening involved and particularly the role played by subduction of oceanic crust are still in dispute and critically dependent on the model adopted for the Late Jurassic—Early Cretaceous events and the palaeogeographic attribution of the Olympos window carbonate-platform sequence.

The problems are compounded by the fact that the Pelagonian zone loses its role as an autochthon after ophiolite emplacement as Celet & Ferrière (1978) have pointed out. Pelagonian sequences in mainland Greece are clearly allochthonous and in the Olympos area rest on a composite blueschist nappe (Ossa Series, e.g. Derycke & Godfriaux 1976) including ophiolites and Palaeogene sediments. These in turn overlie the Olympos platform carbonate sequence and its Eocene flysch cover. The simplifying assumption is that the Pelagonian nappes have come to rest in a complex stack somewhere not far removed from their original location even if they are now interleaved with blueschist, basinal and ophiolitic sequences. This simplification carries the implication that major displacements along strike might have occurred within the Pelagonian zone, or that it was fragmented by across-strike transform-related faulting without either kind of displacement being detected in the reassembled 'zone' (Celet *et al.* 1982).

If the Olympos carbonates did form the internal edge of the Gavrovo-Tripolitza platform, then the overlying ophiolitic blueschist nappe and Pelagonian gneisses are most readily explained by eastward Tertiary subduction on the western side of the Pelagonian zone, within a Pindos ocean basin (see **Papanikolaou**), leading to collision (Fig. 18a). This would in turn be consistent with models for the earlier emplacement of all or part of the ophiolites onto the Pelagonian zone from the west (e.g. **Smith & Spray**; Smith 1977; Bébien *et al.* 1980), or even with models which emplace all the ophiolites from the east (Vardar) but leave a substantial Pindos ocean basin to the west (e.g. Aubouin, 1977). Models which relegate the Pindos trough to a deep basin floored by continental crust have difficulty in accounting for the Ossa ophiolitic blueschists except by intra-continental subduction of a previously emplaced ophiolite nappe (Jacobshagen 1977; Fig. 18b). The blueschists of the 'zone of Western Macedonia' (Grubić 1980), which we noted earlier as evidence for subduction of the Pindos ocean could be a relatively autochthonous equivalent

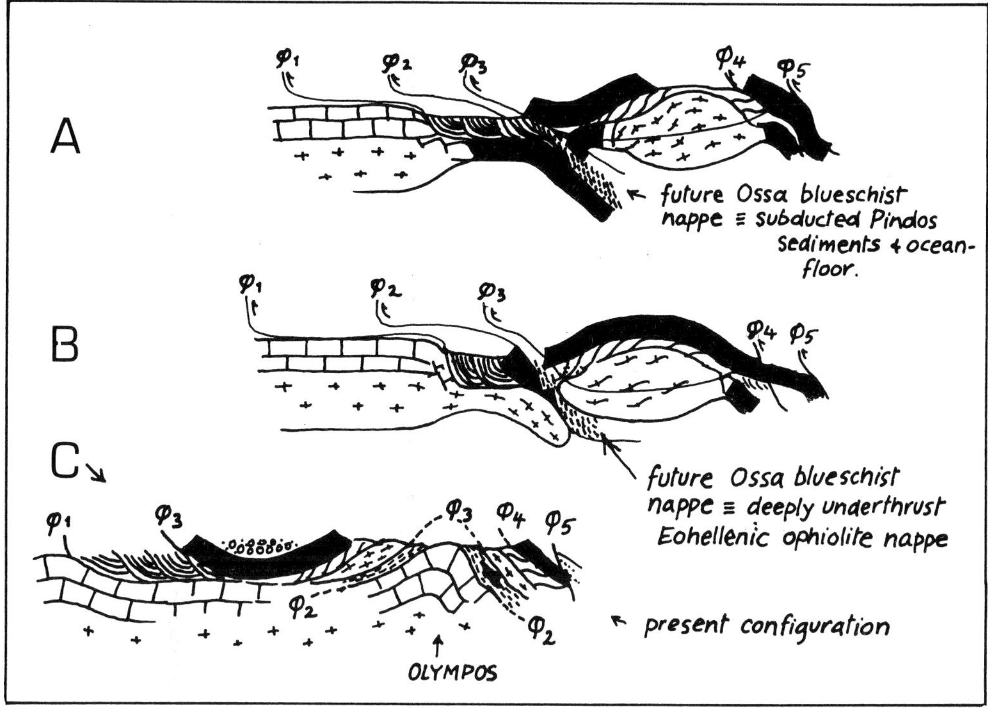

FIG. 18. Schematic cross-section through northern Mainland Greece showing the onset of Eocene/Oligocene nappe-stacking, following the models in Fig. 12 q.v. *A*: Double root-zone model as *12C*. Continued eastward subduction of the Pindos ocean generated blueschists beneath the W Pelagonian margin. The Pindos basinal sheets are expelled completely. The Pelagonian nappe complete with ophiolite cover is thrust westwards emplacing the Pindos ophiolite and sandwiching Ossa blueschists above Olympos, e.g. **Papanikolaou**. *B*: Single root-zone model as *12B*. Ossa blueschists have to be generated by complex underthrusting of an emplaced Eohellenic ophiolite nappe. (If Olympos is an *internal* carbonate bank, Fig. *12D-3* provides a means of generating the observed structural succession).

of the Ossa series and would repay a detailed isotopic dating study.

However first emplaced, the Pindos ophiolite nappe was finally rethrust westwards onto Pindos zone sequences in the Upper Eocene. **Kemp & McCaig** summarize the evidence bearing on these two emplacement stages of the Pindos ophiolite and argue in favour of initial eastwards emplacement as do Ross & Zimmerman (1982).

We discussed earlier the problems raised by the new results from the Olympos sequence (Schmitt 1983) for simple ophiolite-emplacement models which require any 'event' to occur along the length of the Pelagonian zone. If the Olympos sequence is more closely related to internal (Pelagonian) platform sequences or to Tauride platform successions (**Monod**), then the structural position of the Ossa series blueschists is also a problem. If we assume that the Olympos sequence was deposited on a micro-continental block separate from the Pelagonian sliver and *west* of it, Pindos-derived ophiolites (i.e. Othris and perhaps Vourinos too) could have been generated from between the Olympos block and the Pelagonian margin. This possibility is illustrated in Fig. 12D3. The blueschists would have originated from the elimination of this remaining oceanic crust in the Eocene. Clastic sediments from the developing orogen may well have bypassed the Olympos block and passed into the Pindos basin proper in the Upper Cretaceous. Eocene telescoping then emplaced the Pelagonian nappe over the Olympos sequence, which was by then jammed in the subduction zone, and gave rise to the re-emplacement westwards of the Pindos ophiolitic nappe. As we observed earlier quite realistic extra degrees of freedom are introduced into the models if carbonate-topped micro-continental blocks were not continuous along the margin and furthermore could have been displaced along-strike from their position in the Jurassic. Ophiolites could then have been emplaced round them, and blueschists formed on segments of the margin between them before being subsequently emplaced over them. It should be recalled that, for example, in Turkey the Antalya nappes were first 'emplaced' varying distances over the eastern edge of the Bey Dağları in the latest Cretaceous, while pelagic deposition continued further west, long prior to final major overthrusting in the Late Miocene (e.g. **Robertson & Woodcock**).

The picture is further complicated by the record of the metamorphism and emplacement of units structurally overlying the Pelagonian zone to the east. Here, as discussed earlier, it is notable that the Upper Jurassic to Miocene intrusive and extrusive igneous activity in the Vardar and Serbo-Macedonian zones in Greece and Yugoslavia would be consistent with continuing subduction after the initial phase of ophiolite emplacement. Flysch sedimentation appears to have set in throughout the Vardar zone as now exposed, by the Eocene (Mercier 1966; Rampnoux 1970), but thrusting continued into the Lower Miocene, raising the possibility of hidden sutures. A number of blueschist occurrences are now known in the Vardar zone in Yugoslavia (Majer & Mason 1983) but no details of age or tectonic setting are known to us. In the Almopias zone in Greece, HP-LT assemblages occur in meta-ophiolites structurally above the Pelagonian zone (Braud 1967). Mercier (1966) records glaucophane from elsewhere in the Almopias zone in Tertiary flyschoid sediments. If the meta-ophiolitic rocks were indeed pre-Late Cretaceous in age of metamorphism, this relationship would then be similar to that described by Walbrecher (1977) and **Jacobshagen & Wallbrecher** in Pelion and the Sporades which lie on the southern extension of the Almopias sub-zone in Greece. Here they argue that an already metamorphosed HP-LT nappe including ophiolite material was emplaced over 'Pelagonian' meta-clastics in the Late Jurassic/Early Cretaceous (Eohellenic) phase prior to transgressive Upper Cretaceous sealing of thrust contacts. This Pelagonian *relative* authochthon would appear to be the equivalent of the highest level Pelagonian greenschist/epidote amphibolite nappe found in northern Pelion (Ferrière 1977) and the Olympos Ossa region. Pelagonian 'basement' rocks on the mainland are thus now sandwiched between Tertiary ophiolitic blueschists below (Ossa series: Derycke & Godfriaux 1977; Godfriaux & Pichon 1979) and both Late Jurassic—Early Cretaceous blueschists, and Tertiary blueschists, above, in different segments. The conclusion that subduction zones bounded both sides of the Pelagonian zone at some stage in its evolution thus seems inescapable.

(iii) Central and Southern Aegean

In the central and southern Aegean the key problem is usually seen as one of disentangling events related to the final closure of the northern Neotethyan Vardar ocean from those connected with closure of one or more the southern Neotethyan strands north and south of Crete. We argue later that oceanic strands definable outside the Aegean area may well have been merged, branched and undergone an inherently more complex evolution within it, than in the Tauride and Hellenide-Dinaride belts on either side. There is however a general continuity of Pelagonian-type sequences at high structural levels in the nappe pile along the eastern margin of mainland Greece (Celet & Ferrière 1978). In Evvia, meta-sedimentary successions including metabasites, tectonically overlain by serpentinite and transgressively overlain by Upper Cretaceous limestones, (Katsikatsos *et al.* 1976), now structurally overlie marbles terminating in Upper Cretaceous flysch (Almyropotamos series). The imbricate thrust-stack includes the major blueschist units (300 and 800 m thick) which Bavay *et al.* (1980) dated as Eocene (45–50 Ma), but older Mid-Late Cretaceous ages (78–120 Ma) events are also apparent from $^{39}Ar/^{40}Ar$ age spectra. Taken with the evidence of Upper Cretaceous granitoids in Crete (Asteroussia nappe, **Bonneau**) and in the nappe sequence on Anafi (Reinecke *et al.* 1982) it seems possible that the southern part of the Pelagonian zone *sensu lato* was involved in Upper Cretaceous tectonics comparable to those which affected the Tauride-Anatolide platform.

The structural succession in the Cyclades is complex and attempts at a unified nappe interpretation in a region with so much sea must inevitably be treated with reserve. The succession is discussed at length by **Bonneau** q.v. The 'lower unit' (Dürr *et al.* 1978; Henjes-Kunst & Kreuzer 1982) is made up of thrust sheets of pre-Mesozoic basement (e.g. Ios), Mesozoic carbonates, clastics, volcanics and ophiolitic melange (e.g. Syros, **Ridley-a**; Samos, **Okrusch**), all metamorphosed or re-metamorphosed under blueschist conditions and yielding K-Ar ages around 41 Ma. The relict ages of around 300 Ma obtained by Henjes-Kunst & Kreuzer (1982) in Ios from the 'basement' strongly suggest affinities with the Pelagonian zone of Mainland Greece (cf. **Moundrakis's** review of U/Pb ages) and meta-sedimentary successions as a whole are clearly indicative of continental or shallow-marine deposits rather than typical hemi-pelagic trench sequences (e.g. Syros: **Ridley-a**).

Ridley (1982) points out that the highest grade blueschists are presently exposed in a NE-trending line from Milos through Sifnos, Syros to Tinos, perpendicular to the syn-metamorphic underthrusting direction to the NW deduced from an analysis of Syros fabrics. Lower grade blueschists of the same age are exposed as relicts in the SE of Naxos around the later 25 Ma thermal dome, and in Ios (Van der Maar 1980), where pre-Mesozoic basement is exposed. If these outcrops are still roughly in their Eocene configuration the simplest explanation is that a NE-trending NW-dipping subduction zone was finally jammed in the Eocene by a southern Aegean block of Pelagonian-type continental crust and its Mesozoic cover, perhaps even complete with an Upper Jurassic emplaced ophiolite melange high in the sequence as on Syros.

Ridley (1982), modelling the thermal evolution of Syros blueschists showed that uplift from peak metamorphic conditions at highest pressure is quasi-isothermal for a significantly long period. This suggests, but does not prove, that K-Ar dates, particularly from micas with blocking temperatures lower than the peak temperatures achieved, may be cooling ages rather than crystallization ages and that peak conditions may have occurred as much as 25 Ma earlier in the Late Cretaceous. This possibility strengthens the suggestion of **Ridley-a** that the Cycladic blueschist belt is linked to Late Cretaceous-Early Tertiary convergence of the Menderes massif and Sakarya block documented by **Okay**. One perplexing feature of Cycladic geology is the significance of the remnants of largely unmetamorphosed Mesozoic and Tertiary sequences which occur mostly around the edges of the metamorphic islands (Dürr *et al.* 1978; **Bonneau**, **Papanikolaou** & **Ridley-b** for further discussion). While acknowledging that many of these remnants are clearly from higher-level nappes **Ridley** makes the important point, supported indirectly by **Jackson & McKenzie**, that many of the low-angle contacts are more probably the flat segments of listric normal faults related to Aegean extension, and that the timing and direction of true nappe emplacement may not be deducible from the structures or present altitude of these contacts. Some of the younger Tertiary clastic sedimentary 'nappe remnants' may be derived from a high-level autochthonous cover.

Summarizing, the Attic-Cycladic complex can be related to collision and final closure of a number of complexly orientated northern Neotethyan strands, followed by long distance southward thrusting over the southward extension of the Pelagonian platform from mainland Greece onto the Aegean to form a vast nappe pile. Detailed palaeogeographical reconstructions incorporating the high-level, relatively unmetamorphosed nappe remnants are inevitably speculative and we refer the reader to the brave attempts of **Bonneau** & **Papanikolaou.**

Crete

While we take the pessimistic view that the whole Aegean area may well be inherently uninterpretable because of its location at the Tauride-Hellenide 'junction', and the amount of sea, the best hope for enlightenment must come from Crete, the largest landmass. Crete is a Tertiary nappe pile (Fig. 19) and is included in our review of collisional events even though there is little clear evidence that a *collision* actually piled up the nappes. There are recognizable but fragmentary 'Pelagonian' components at the highest structural level but no relatively autochthonous platformal sequence that might represent an opposing continental block at the lowest level. Furthermore, it now seems well established from the palaeomagnetic work of Laj *et al.* (1982) and **Kissel *et al.*** that Crete, in contrast to the Peloponnesos and the Ionian islands, has not rotated significantly since the end of the Miocene, but that the arc was apparently essentially linear at around the Oligocene-Miocene boundary. It is also generally agreed that oceanic crust existed to the south of Crete at least until Recent times, to provide subductable material for the Hellenic arc. In the central part of the Cretan nappe pile two major units are generally accepted as directly correlateable with Hellenide equivalents—a Gavrovo-Tripolitza Mesozoic carbonate platform sequence and an overlying basinal Pindos-type nappe complex. The first problem is thus to account for the absence of an Apulian or African margin sequence at the base of, or south of Crete. This palaeogeographic problem is compounded by the identification of the lowest Cretan nappe, the Plattenkalk, which can be traced into the Peloponnese (Thiebault 1977) with the intra-platformal basin of the Ionian zone in mainland Greece, which apparently developed within the Gavrovo-Tripolitza platform in the Upper Jurassic. **Poisson** recognizes the Ionian/Plattenkalk zone in the Kitzilca korak gol trough in the Lycian nappes, but neither the Gavrovo-Tripolitza carbonate sequences, nor clearly equivalent Pindos rocks

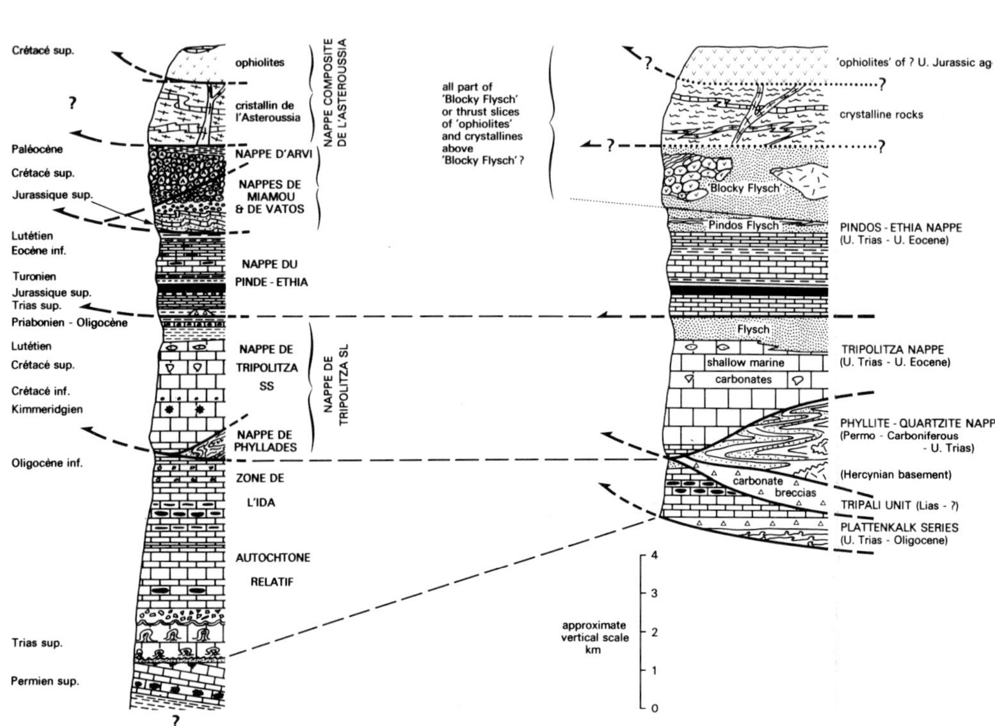

FIG. 19. Structural successions in Crete after **Hall *et al.*** & Bonneau *et al.* (1977), reproduced from **Hall *et al.*** by kind permission.

occur there in higher nappes, and the relative autochthon in western Turkey is the Bey Dağları carbonate platform which cannot be traced west into Crete (for a detailed discussion of correlations in the SE Aegean see Bernoulli *et al.* 1974). It seems clear that the Mesozoic palaeogeography must have changed quite markedly from mainland Greece through into Western Turkey yet still preserve the identity of the Ionian/Plattenkalk zone as a *locally* linear region where Late Jurassic subsidence of a carbonate-platform occurred. In Crete and the Peloponnese the palaeogeographical interpretation bears directly upon the problem of the Phyllite-Quartzite nappe discussed by **Hall *et al.*** & **Bonneau** (=the Arna unit of Papanikolaou), a unit containing Permo-Triassic rift-related volcanics (Seidel *et al.* 1982) and Hercynian metamorphic basement slices lying between the Gavrovo-Tripolitza nappe and the Plattenkalk. This unit is not traceable into Turkey or northern mainland Greece. Both it and the underlying Plattenkalk have undergone HP/LT metamorphism. The Phyllite-Quartzite unit is interpreted variously as the former basement of the structurally overlying Tripolitza nappe (**Bonneau** & **Hall *et al.***), as a Permo-Triassic rift related to opening of a southern Neotethyan strand (Seidel *et al.* 1982, Pe-Piper 1982), or as a far-travelled nappe from the Cycladic area which reached its present position by complicated re-thrusting (**Papanikolaou**). Disagreement also persists on the significance of the Tripali unit, composed of Liassic carbonate rudites intercalated between the Phyllite-Quartzite unit and the Tripolitza nappe, which for Creuzburg & Seidel (1975) and Seidel *et al.* (1982) and **Hall *et al.*** represents an important palaeogeographic unit, but for **Bonneau** is merely tectonic melange caught between two major nappes.

Disagreements remain as to the variation in metamorphic grade through the nappe pile. **Hall *et al.*** suspect that the structurally underlying Plattenkalk unit may be of lower grade than the acknowledged high pressure assemblages of the Phyllite-Quartzite unit, although Seidel *et al.* (1982) quote mineral assemblages in Plattenkalk meta-bauxites (pure magnesiocarpholites with co-existing pyrophyllite and diaspore) which also occur in the Phyllite-Quartzite unit and which they take to be of high pressure origin from recent experimental data. On the one hand, Seidel *et al.* (1982) state that the Tripolitza nappe above is essentially unmetamorphosed, while **Bonneau** points out that in the absence of diagnostic mineral assemblages in the carbonate rocks, the evidence of colour alteration of conodonts from at least two localities in Crete (Dikti Mtns. and Psilorita Mtns.) indicates temperatures of 300°–400°.

A marked metamorphic grade discontinuity between the Phyllite-Quartzite unit and the Tripolitza unit leads to considerable difficulty in interpreting intersliced units which appear to be lithologically transitional. This applies particularly to the Ravdoucha beds in Western Crete and the Tyros beds in the Peloponnese which are akin to the Phyllite-Quartzite unit but appear to have suffered only low-grade metamorphism. Seidel *et al.* (1982) now link the HP/LT metamorphism of the Phyllite-Quartzite unit and the Plattenkalk, (which must both be of post-Oligocene but pre-Miocene age), to the Oligocene-Miocene HT Barrovian belt and accompanying granites in the Cyclades (e.g. Naxos) as a paired metamorphic belt related to north-eastward subduction of Neotethyan oceanic crust south of Crete. This leads to problems. First, as stressed by **Hall *et al.*** & **Bonneau**, none of the HP/LT protoliths can be shown to be of definitely deep water or oceanic origin, secondly there remains the generally accepted lack of a southern block to provide a collision and thirdly, the apparently transitional lithologies between the Phyllite-Quartzite and Tripolitza nappes are hard to reconcile with an abrupt change from HP/LT to essentially unmetamorphosed assemblages.

Potential controversies also stem from **Hall *et al.*'s** division of the Cretan nappe pile into two quite different parts separated by the base of the Phyllite-Quartzite unit. According to them the upper nappe was folded on N-S trending axes during eastward compression in the Eocene. The Plattenkalk unit and supposedly more proximal Tripali unit then represent part of the southern passive margin of 'Apulia' which was transported into the area along an east-west transform fault, then thrust beneath the upper nappe unit in the Early Oligocene producing the later E-W (D2) structures. An immediate problem is that in western Crete, Greiling & Skala (1977) were not able to separate the widely varying orientations of flattened isoclinal folds into two distinct fold phases from a statistical analysis. **Ridley-a** shows that on Syros (Cyclades) variable fold hinge orientation has resulted from wrench-shear superimposed on thrust-sense shear and a similar explanation may apply. A further difficulty, as pointed out above, is that **Poisson** correlates the Plattenkalk unit with an *intra-carbonate platform* basin in SW Turkey (Kizilica-corak gol facies). The similarity sugests that some other unexposed platformal carbonate unit, possibly a westward extension of the Bey Dağları in SW Turkey, may exist beneath Crete or offshore areas to the south. The Plattenkalk facies do appear to differ strikingly from the non-calcareous Mesozoic radiolarian facies of, for example, the Pindos and Antalya units, which in our view record genuine continental margin successions.

Faced with so many apparent contradictions, the palaeogeography of Crete remains obscure. But for the apparently major metamorphic discontinuity above the Phyllite-Quartzite unit the most simple explanation, in line with **Bonneau**'s view, would be that the whole of Crete is a vast nappe pile which is sufficiently thick to have developed HP/LT assemblages at its base during southward transport from Eocene to Early Miocene time. On the other hand if, as appears likely, there is a major metamorphic discontinuity, then the implication is that a separate small southern Neotethyan strand may have existed within the Plattenkalk unit, separating the Tripolitza carbonate platform unit to the north from the Plattenkalk unit to the south, which was subducted northwards. The main objection to this is the lack of pelagic or ophiolitic components in the Phyllite-Quartzite unit. We suggest that some form of intracontinental abortive subduction may have occurred along a line of attenuated continental crust below the Triassic Phyllite-Quartzite rift zone, driven by the early stages of subduction proper of the main southern strand of Neotethys south of Crete. The Plattenkalk in the south Aegean would then be an early Mesozoic platformal region lying between two rifts, one abortive to the north (the Phyllite-Quartzite rift), one successfully developing into an ocean in the Late Jurassic-Early Cretaceous to the south and causing, in the process, subsidence of the intervening strip of platform, as shown in Fig. 20. This successful rift propagated east and north-west into the Bey Dağları platform and the Apulian Gavrovo-Tripolitza platforms respectively as linear areas of subsidence directly correlatable with the Plattenkalk in Crete.

More controversy surrounds the higher nappes in Crete. **Hall *et al.*** following the work of Hussin & Tee (1982), argue that Pindos facies trends run north-south across the island from palaeocurrent and slump-fold analysis. In our concluding section we offer implicit support for this idea in that it is virtually impossible to preserve either straight or continuous oceanic tracts in a region like the South Aegean, lying between two evolving orogenic belts. A

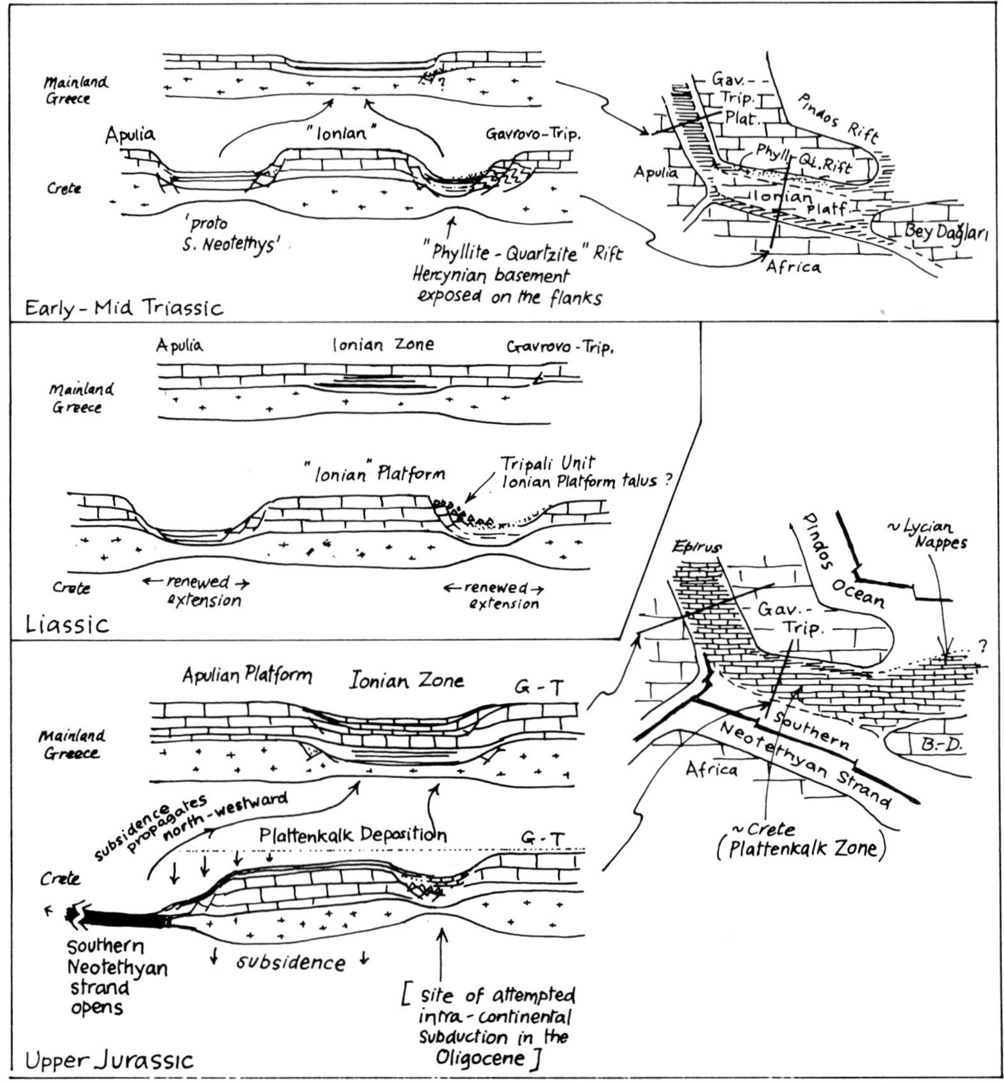

FIG. 20. Schematic palaeogeographic maps and sections to explain the relationships between the Ionian zone, Plattenkalk, Phyllite-Quartzite Nappe and Kizilca-corak gol trough, from Mainland Greece, through Crete into W Turkey.

straightforward rift-rift-rift triple-junction can also provide a general mechanism for the ultimate telescoping of nappes derived from one rift arm onto nappes formed on another. In Fig. 20 note the Pindos and Phyllite-Quartzite rift orientations. Such considerations can also perhaps account for the structural discontinuities noted earlier.

The Neotectonic phases: the last oceanic vestiges

The post-Miocene Neotectonic phase has been dominated by the later stages of pervasive collision of Africa and Eurasia, which has tightened up and rotated the pre-existing mosaic of continental blocks, and has been coupled with subduction of the last remnants of Neotethyan oceanic crust (McKenzie 1972; Fig. 21). Space permits only a few general points to be made. The overall pattern was established by the collision (Tapponnier 1977) of the Adriatic and Arabian promontories with Eurasia. In SE Turkey this collision drove much of Anatolia westwards towards the Aegean along major transform faults. (Dewey & Şengör 1979; Şengör & Canitez 1982). **Barka & Hancock** now establish, from detailed lithological and structural

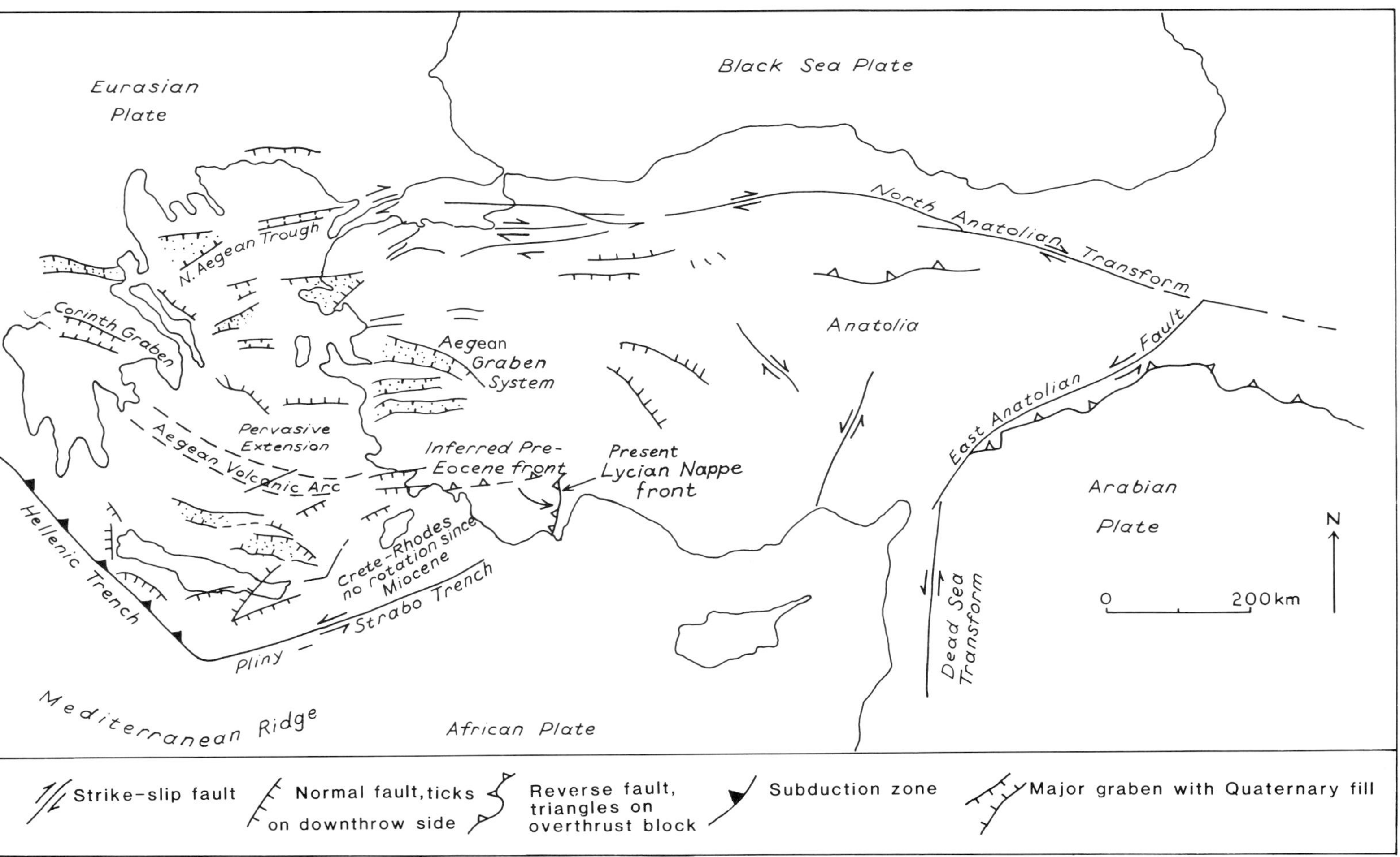

FIG. 21. Main Neotectonic elements in the Eastern Mediterranean, modified after **Barka & Hancock**.

correlations, that 25 km of right-lateral motion took place along the North Anatolian transform fault since the Late Miocene. This transform fault is resolved westwards into the Aegean graben system, of which the northern part is discussed in detail by **Lyberis**.

At the same time up to several hundred kilometres width of Neotethyan oceanic crust in the Ionian Sea was subducted north-eastwards under Crete forming the Hellenic trench (Le Pichon & Angelier 1979; Angelier *et al.* 1982), while further east the more V-shaped Pliny and Strabo trenches mark a transform plate boundary. We suggest in our final section that much of this crust may have been created in the Tertiary during the anti-clockwise rotation of the Apulian-Adriatic promontory. There is now growing evidence that the Hellenic trench cannot be dismissed as an intra-continental shear-zone, or the result of subduction of continental crust (c.f. Channel *et al.* 1979; Horvath & Berckhemer 1982). As summarized by Angelier *et al.* (1982), the Hellenic arc shows many features in common with the SW Pacific area, including a well-defined underthrusting slab, volcanism and back-arc extension. **Fytikas *et al.*** review the growth of the South Aegean volcanic arc from Early Miocene onwards, while recent geochemical studies have failed to detect any evidence of continental crustal contamination in the lavas at least of Santorini (Briqueu *et al.* 1982). Proponents of derivation of all the ophiolitic allochthons from a single ocean basin located much further north (e.g. **Ricou *et al.***) have yet to satisfactorily explain the existence of this southerly oceanic crust in the Neogene. In line with a subduction model, recent work suggests that the Mediterranean ridge, which Stride *et al.* (1977) had already established as a major zone of compression extending 100–150 km south of Crete, could in fact be the surface expression of an accretionary wedge, or subduction complex (Ryan *et al.* 1982), suggesting, in turn, that the Hellenic trench is now a fore-arc basin rather than an active slip-zone. According to Giese *et al.* (1982), however, the main seismically active zone would intersect the sea floor ca 50 km south of Crete. A further problem is that there is no unambiguous evidence that oceanic crust still exists south of Crete, raising the possibility that collision with the African continental margin may already be under way, with the Mediterranean ridge as the southward-propagating thrust-front.

The history of back-arc extension in the Aegean has now been carefully documented and modelled. Estimates of overall crustal extension however vary greatly from 30–50% by Le Pichon & Angelier (1979) to up to three times this amount by McKenzie (1982). In this volume **Le Pichon *et al.*** present a detailed thermal analysis of the North Aegean graben system. One of the main points is that the stretching model which fits the observations must involve very significant lateral heat conduction which will effectively prevent the asthenosphere/lithosphere boundary from rising to the point where a back-arc basin transforms into an active spreading centre. Back-arc spreading has indeed not taken place in the Aegean, despite northward subduction, if **Fytikas *et al.*** are correct, from the Oligocene onwards. A major drawback in modelling the crustal extension history of the Aegean in the past has been the enigmatic evidence that major phases of simultaneous compression took place at 12–9 Ma, 7–5 Ma and 3–1 Ma (Mercier 1976; Mercier *et al.* 1979). Now **Ridley-b** shows that on Syros (Cyclades) low-angle tectonic contacts interpretable as 'Neogene' thrusts (e.g. Papanikolaou 1980) are in fact the toes of listric normal faults produced by pervasive crustal extension. If **Ridley's** evidence is applicable to the South Aegean generally, as he suggests, the 'compressional phases' may on the contrary be the times of maximum crustal extension with 'roll-back' of the Aegean plate towards Africa.

Despite much research and a voluminous literature, the status of the crust further east in the Levant Sea has still not finally been resolved (see Giese *et al.* 1982, for a recent review). The possibility still exists that the crust south of Cyprus could either be thinned continental crust (i.e. African continental margin), or oceanic crust with a thick sediment blanket. A recent view expressed by Makris (in Giese *et al.* 1982) is that full continental crust exists from mainland Turkey as far south as the Eratosthenes seamount, but beyond that the SE corner of the Levant Sea is still Mesozoic ocean crust. Heat flow data (reviewed by Giese *et al.* 1982) do seem to oppose an origin as Tertiary oceanic crust as in the Western Mediterranean (e.g. Ligurian basin). Any interpretation must also take account of the rotation of Cyprus away from SW Turkey since its formation (reviewed by **Lauer**), which could have opened transtensional basins or leaky-transform faults now deeply buried under a thick later Tertiary sediment cover.

An origin of the Levant Sea as part of a southern Neotethyan oceanic strand is consistent with **Garfunkel & Derin's** view of the Levant as a north-south rifted margin of Arabia. A major problem here, however, is that, from a careful analysis of the available (often unpublished) borehole and seismic data, **Sestini** can find no trace of a rifted margin under the North African margin at least as far north as the Herodotus basin, and, in his view, possibly as far as southern Turkey. This is, of course, quite consistent with **Ricou *et al.*'s** hypothesis that all the Tauride allochthons travelled far over the Arabian platform, with the Levant merely as a Triassic intra-continental rift (**Druckman**). Resolution of these contradictions might lie in the suggestion that the rifted African continental margin runs from under the Nile cone, for which there are no data, north-eastwards past the Eratosthenes seamount and then into SW Turkey. This, though, is hardly consistent with the evidence of Mesozoic continental margin-type facies exposed in the Alanya window along the Mediterranean coast north of Cyprus (**Okay & Özgül**). A further problem is that if much of the Levant Sea is still oceanic, why then has Anatolia been translated westwards towards the Aegean rather than 'bulging' southwards, and why is there no obvious eastward extension of the Hellenic trench-arc system? The most plausible explanation is that northward subduction of a southern Neotethyan strand continued through the Tertiary, with the Miocene flysch of Kythrea (N. Cyprus) and Adana (mainland Turkey) as fore-arc basins in front of related intra-continental volcanic arcs sited in central Anatolia (e.g. Hasan Dağ; Innocenti *et al.* 1982). During this time Cyprus including the Troodos Massif was essentially stranded in a fore-arc setting, being rotated towards its present position by the Mid Tertiary (Shelton & Gass 1980). Then, in the Pleistocene drastic updoming of the Troodos Massif (Robertson 1978) could reflect the initial underthrusting of thinned African continental margin crust, implying that little fully oceanic crust still persists in the East Mediterranean sea south of Cyprus.

Once into the realm of general continental collision, a whole series of geological processes come into play as yet only dimly perceived. **Fabricius** discusses the enigmatic progressive subsidence of the Eastern Mediterranean sea areas at an average rate of 1 metre/Ma since the Miocene, accompanied by episodic uplift of marginal areas. In his view the famous Messinian evaporities could have formed in shallow basins prior to this subsidence. Despite all the nappe empilement, mysteriously the East Mediterranean sea areas are now mass-deficient, which is perhaps most plausibly explained by large-scale asthenosphere flow away from collisional areas (mantle creep'), coupled with localized mantle diapirism (e.g. Wezel 1981). The effects are considered by **Fabricius** to be sufficiently characteristic to establish a 'Mediterranean-type' of collisional orogeny.

Model of oceanic crustal genesis and slicing

Before concluding with an outline of our conception of the evolution of the Eastern Mediterranean we introduce a new model for the genesis and slicing of the ophiolites which we hope may help to reconcile the conflicting interpretations stemming from field and igneous geochemical studies.

In the earlier discussion we saw that major Triassic intra-continental rifting took place in the southerly ('external') areas of the Dinarides, Hellenides and Taurides, but that further north the margin may already have been open, more akin, for example, to the California borderland of western USA. We also argued that rifting and subsidence should not be equated with onset of spreading which was probably delayed, until sometime in the Jurassic in Greece and possibly not until the Cretaceous in Turkey. But for the ophiolite trace element chemistry, which points to diverse tectonic settings, the margin succession would be quite compatible with formation in Red Sea-type basins in most cases. We then saw that neither thrusting nor strike-slip faulting can realistically be used to derive all the ophiolites from a single basin and that in both Greece and Turkey a number of inter-connected basins must have existed. From the record of arc volcanism we argued that both the northern (wider) and southern (narrower) strands of the Neotethys remained open into the Tertiary after major ophiolite emplacement in the Late Jurassic (Greece) and in the Late Cretaceous (Turkey).

The realization that the units preserved with the ophiolites provide no support for the former existence of major marginal basins formed behind conventional volcanic arcs (i.e. absence of older adjacent volcanic arc) now favours less actualistic models involving some form of 'pre-arc' spreading with 'roll-back' of an older oceanic plate which then leads to crustal extension and short-lived spreading to form a marginal basin adjacent to the original ocean-continent boundary. With later 'roll-to' this crust is then emplaced over the older oceanic edge creating the sub-ophiolite metamorphic sole-rocks. There are a number of problems with this model. First, there is no proof that older (Triassic) oceanic crust actually existed. Secondly, the intra-oceanic subduction has to be initiated through thick old marginal oceanic crust, yet no preserved transition to young oceanic crust has yet been reported (e.g. Upper Jurassic dykes through Triassic oceanic crust) although such material should eventually be preserved in the emplaced arc-trench gap. Thirdly, the absence of associated arc volcanism is all the harder to explain if ocean floor spreading commenced in the Triassic. Fourthly, although **Smith & Spray** point out that protoliths in the sub-ophiolitic sole rocks range from alkalic basalts to MORB, Searle & Malpas (1982) show that in the case of Oman, where the sole rocks are now well documented, the high temperature amphibolites differ significantly from both the rift-related Triassic lavas (Haybi unit) and from the Upper Cretaceous Semail nappe, being considerably closer to MORB than either of the two other units. On the other hand, the protoliths of the structurally underlying greenschist facies metamorphic rocks are indeed the marginal igneous rocks and sediments (A. E. S. Kemp & A. H. F. Robertson, unpublished data), but these may owe more to frictional heat generated during the final emplacement of the Semail nappe than to the earlier overthrusting of hot oceanic crust. 'Sole' rocks can thus not necessarily be taken to be preferentially metamorphosed and preserved marginal units.

As a radical alternative, illustrated in Fig. 22, we propose that the 'anomalous' extrusives *and* the sub-ophiolite 'sole' rocks all formed when spreading ridges collapsed following a regional change from extension to compression. We suggest that during active extension the asthenosphere upwelling beneath the ridge represents a localized convective system which cannot necessarily respond rapidly to sudden changes in motion of the lithosphere above. Normally it will remain under the ridge axis. However, if the plate on one side attempts to move *towards* the ridge-axis because of a major change in boundaries elsewhere, such as the separation of Africa from North America, the accreting edge of the 'pushed' plate may initially slow down relative to the ridge axis and eventually spreading on that side will cease. The leading edge of the plate will cool and subside faster than the opposite leading edge which will *continue to spread* away from the ridge axis. If 'one-sided' convergence continues, the 'dead' leading edge may begin to over-ride, or curl over and drop through the zone of asthenospheric upwelling under the ridge axis, while *net* plate separation is still occurring. This continuing separation will be by one-sided accretion above what is essentially an incipient subduction zone. As the relative motion between the two plates finally drops to zero and becomes convergent the leading edge of the upper plate now made of abnormal oceanic crust will actively over-ride the lower plate at the modified spreading centre itself, producing a hot amphibolitic aureole of the same age as the upper plate above. In this model the lithosphere in the 'pushed' plate is moving *towards* the ridge relative to the asthenosphere, so that once subsidence and subduction starts the leading edge will tend to roll-back just as envisaged by **Smith & Spray** as it will tend to be anchored in the asthenosphere.

The effect of the introduction of the 'dead' plate edge into the uprising asthenosphere, through the magma chamber, will be to dehydrate it rapidly and even to partially remelt it, giving rise to magmas with a high ratio of large-ion-lithophile-elements relative to high-field-strength-elements, as in conventional island arc rocks. During the intense shearing, extension and vertical tectonics above the 'rolling-back' lower plate, extrusion of the magnesian andesites, or boninites can take place, reflecting hydrous melting of relatively depleted mantle source material (e.g. Cameron *et al.* 1980). As the subducting wedge thickens resistance to underthrusting increases and the latest lavas are derived from a successively depleted mantle source (e.g. Troodos Upper Pillow Lavas, Smewing & Potts 1976), before spreading is finally extinguished.

Discussion

Both the Upper Jurassic and Upper Cretaceous Greek and Turkish ophiolites appear during a time of changed relative plate motion. In the case of Greece this was related to opening of the central North Atlantic and in the case of the Cyprus and Turkish ophiolites to northward convergence of Africa and Eurasia from Mid-Cretaceous (108 Ma) onwards (Livermore & Smith, 1984, Fig. 16). In the Greek areas ridge collapse was possibly sudden, allowing only a brief period for continued spreading following ridge collapse, hence the preserved trace element signatures ranging from MORB, alkalic MORB to boninitic and volcanic arc rocks. The model predicts that the underthrust plate will be the one that converges on the asthenospheric upwelling. Which of the two this is, will depend on the relative motion of lithosphere and asthenosphere on a wider, possibly global, scale and in particular on the location of downgoing slabs which probably tend to anchor the plates (Chase 1978; Tucholke & Ludwig 1982). It is not yet clear therefore whether Africa/Apulia or the Pelagonian microcontinent would have been the upper plate in a Pindos ridge collapse initiated by Africa-North America divergence.

In contrast to the inferred rapid change in

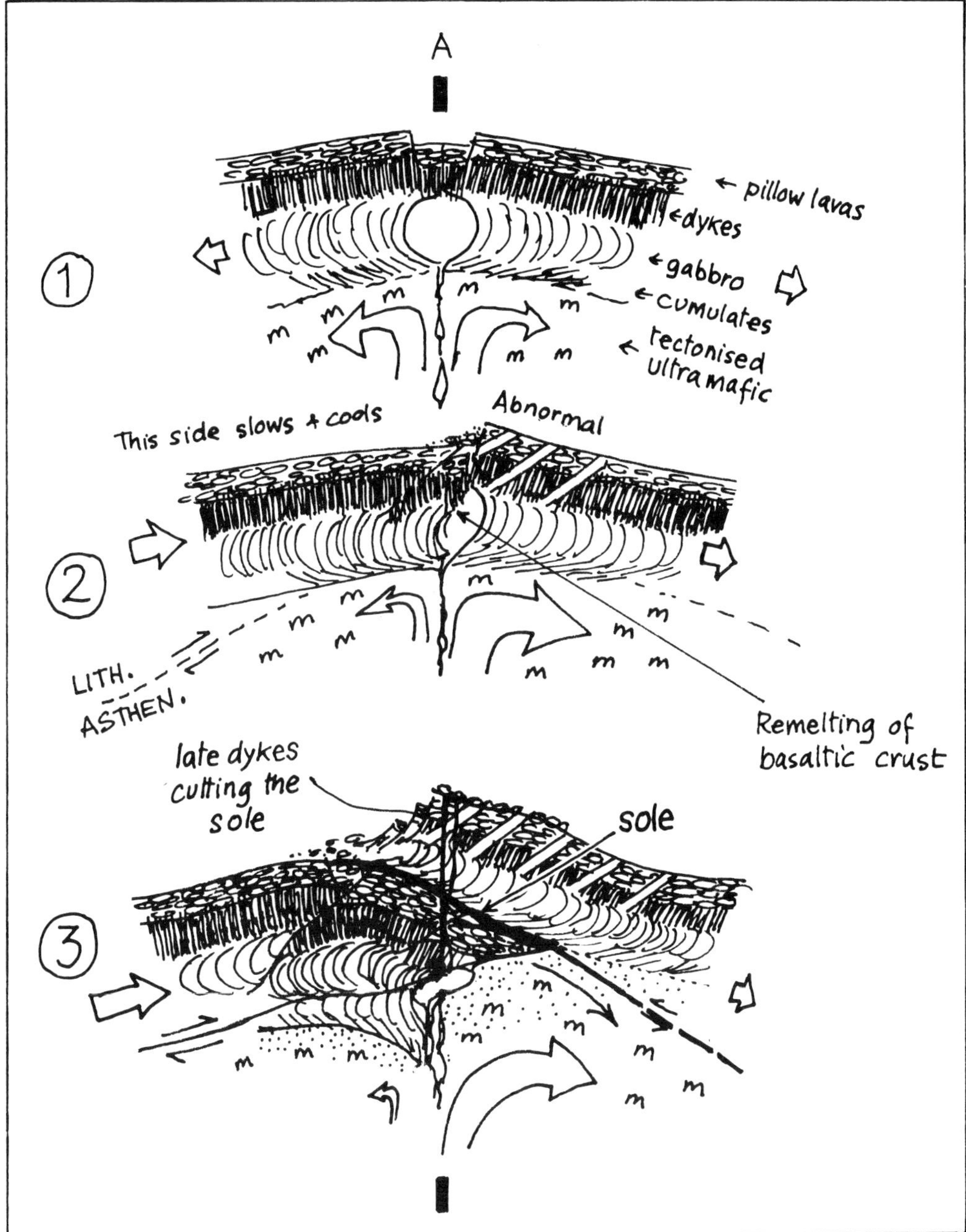

FIG. 22. Sketches to illustrate the concept of asymmetrical ridge collapse and the generation of ophiolites and their metamorphic soles. 1. Normal spreading to L & R relative to an asthenospheric upwelling inertial frame. 2. L-Plate 'pushed' from L. Spreading continues asymmetrically. Nose of L-Plate subsides and invades active melt area and generates LIL-enriched anomalous basalt in R-plate. 3. Motion of L-Plate now to R relative to 'inertial frame'; active underthrusting may begin once R-plate ceases to spread faster than L-plate is converging. Sole forms but may be cut by thoroughly 'calc-alkaline' dykes.

motion in the Greek area, the onset of compression in the Turkish area in the Mid-Late Cretaceous was more gradual which may have allowed a longer period for spreading above an underthrust ridge segment, thus creating the consistent above-subduction zone signatures of the Troodos, Baër-Bassit and Hatay ophiolites. If valid, this model implies that the trace element chemistry has little, if any, bearing on the original tectonic setting of genesis of the oceanic crust, but is instead an indication of the events involved in the collapse and final phase of anomalous spreading of particular ridge segments.

In a more regional context, the model would imply that, in the Upper Jurassic, the onset of sinistral motion of Eurasia with respect to Africa resulted in the collapse of spreading ridges located both in the Vardar zone and in the Pindos zone. Subsequent intra-oceanic subduction then led to a collision of the Pelagonian margin with a trench and the emplacement of ophiolites. In the case of the Pindos zone, any pre-existing ocean basin was probably too narrow to promote calc-alkaline volcanism prior to ophiolite emplacement, or the angle of slab dip too shallow to give rise to volcanism within the narrow Pelagonian microcontinent itself.

In the Late Cretaceous the problem is to slice and create oceanic crust in *both* northern and southern strands of the Neotethys. Here it is possible that after initial collapse and 'roll-back' of one underthrust ridge segment, continued northward motion of Africa relative to Eurasia then resulted in cessation of spreading and thickening of the axial oceanic crust sufficiently to impede further subduction. The next weakest link would then have been either an adjacent offset spreading segment in the same basin, or a spreading axis in an adjacent interconnected basin.

The timing of the critical transitions from asymmetrical spreading, to one-sided creation with 'roll-back', to actual convergence and sole formation will depend on the positions of the poles and the rates of relative motion of the two plates and the local asthenosphere. One would in general expect the transitions to progress systematically with time along a ridge system rather than be synchronous everywhere. The net result would be that over a period of time ridges progressively subside both within and between small ocean basins, explaining why, for example, the Antalya Complex was formed and sliced prior to formation of the Troodos and Baër-Bassit ophiolites further east along an offset ridge segment within the *same* basin.

It seems unlikely that collapsed ridges always went on immediately to form subduction zones since this would create widespread major intra-oceanic volcanic arcs, yet, as discussed earlier, in Turkey intra-continental volcanic arcs were constructed along both the north and the south margins of Neotethyan strands in the Upper Cretaceous. In the case of the Semail nappe, Oman intra-oceanic subduction has been postulated to explain the later collision and emplacement of the Semail nappe over the Arabian margin (Pearce *et al.* 1980), yet a separate subduction zone appears to have been active northwards under the Makran at least from the Early Cretaceous onwards (McCall & Kidd 1981), implying that several northeastward dipping subduction zones existed in the Upper Cretaceous. It may be that the Makran slab effectively anchored the oceanward side of the ridge allowing the continent-ward side to underthrust it. Similarly in the case of SE Turkey two north dipping subduction zones are apparently needed both to create a volcanic arc on the Bitlis-Pütürge continental units to the north and to emplace ophiolites southwards

---

FIGS. 23a-f. General key.
All cartoons show motion relative to a fixed Africa, with Arabia restored to a pre-Red Sea position. Continental blocks are shown idealized and un-stretched. The Adriatic-Apulian promontory has been restored to a more N-S position by being rotated 25° clockwise about a pole near its NW end. This choice of pole allows the Rhodope promontory of Eurasia to make an oblique rather than a head-on collision with it in the Upper Cretaceous. Thrust-barb symbols imply a subduction zone complete with calc-alkaline magmatism.

FIGS. 23a. Mid-Triassic. 240 Ma.
The positions of Eurasia and Gondwana are those for 245 Ma *and* 173 Ma. The vector AA′ gives the approximate magnitude and direction of the displacement of the Eurasian plate to the *NE*, in the Lower to Mid Triassic interval, and back to the *SW* in the succeeding Upper Triassic to Mid Jurassic interval, that will account for the separate N. American and African A.P.W. paths at this time and also keep the NW African Atlantic margin as a transform zone. Coupling of Palaeotethys to Eurasia during the NE-ward phase generates the Gondwanan margin braided rift net-work. Palaeotethyan spreading, subduction, and/or strike-slip zones if they existed, will have added components to the relative motion across the north and south margins. The configuration of continental fragments along the northern, 'E Pacific-type' margin is conjectural.

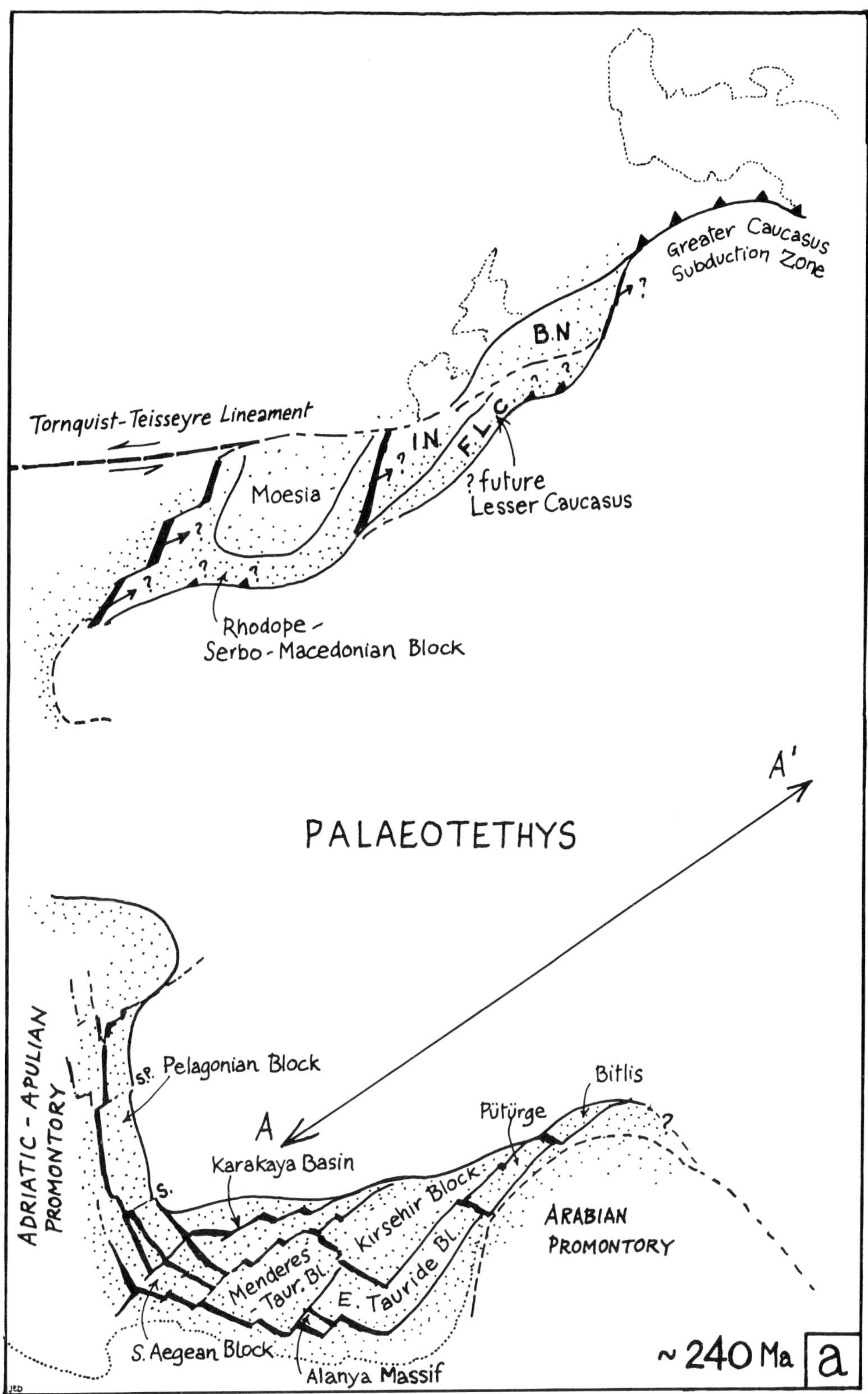
Greater Caucasus Subduction Zone
B.N
Tornquist-Teisseyre Lineament
I.N.
F.L.C.
Moesia
?future Lesser Caucasus
Rhodope-Serbo-Macedonian Block
A'
PALAEOTETHYS
ADRIATIC-APULIAN PROMONTORY
S.P.
Pelagonian Block
Bitlis
Pütürge
A
Karakaya Basin
S.
Kirsehir Block
ARABIAN PROMONTORY
Menderes Taur. Bl.
E. Tauride Bl.
S. Aegean Block
Alanya Massif
~240 Ma
a

onto the Arabian margin in the Maastrichtian.

Summarizing, this ridge collapse model differs significantly from other 'roll-back' models (e.g. **Smith & Spray**), in that *both* slicing to form ophiolitic soles *and* continued spreading to form chemically anomalous lavas can *all* be related to ongoing events in the progressive collapse of spreading axes in inter-connected ocean basins. This removes the need to form separate marginal basins related to steady-state spreading above subduction zones, and also implies that the stratigraphy of the associated margin units may be a better guide to the tectonic setting of ophiolite genesis than the trace element chemistry of the ophiolites themselves. Finally, it predicts that in general the highest grade sole rocks will be chemically closer to normal MORB than the ophiolite itself.

Conclusions: towards a plate-tectonic solution

With the abundance of controversies and unsolved problems, the fragmentary nature of the record and the daunting complexity of the possible configurations of plates, microplates and micro-continental blocks a realistic solution is still hardly feasible. (For a summary of models published up to 1977 see Brunn *et al.* 1977). However a framework is now available in the relative motion history of Africa and Eurasia (Livermore & Smith, 1984, Fig. 16) and in so far as we have favoured particular solutions to major geological questions it seems appropriate to see if a mutually consistent pattern of plate boundaries can be found. One possible sequence is displayed in a series of cartoons, Figs 23a-f. The three most important conclusions incorporated in it are: (i) Palaeotethys remained open until the Early Tertiary, (ii) the Adriatic 'promontory', including the Apulian platform, remained an undisturbed part of Africa until sometime in the Tertiary when it rotated clockwise to its present position (e.g. Channel *et al.* 1979), and (iii) the constituent blocks of Turkey were derived from essentially within the Mediterranean basin.

The cartoons demonstrate an important general result, that there was insufficient continental crust, even when stretched, to bridge the Africa-Eurasia gap *anywhere* until the Upper Cretaceous/Palaeocene, when the two projecting blocks of the Apulian platform and Rhodope-Moesia came into 'glancing' contact. Even at 37 Ma (Oligocene) the distance between the Crimea and Sinai was 2400 km or still 60% greater than the present separation of 1500 km. This compares with a Late Jurassic separation of some 3800 km, a figure not significantly reduced until mid-Cretaceous time. These long-lasting extended separations impose constraints on more local reconstructions. At least one major oceanic tract has to remain open until mid-Tertiary time in the 'circuit': Arabian platform/Tauride block/Sakarya continent/Pontides/Eurasia, even though each pair of blocks has been interpreted by various authors as sutured by pre-Eocene times. Similarly, any cross-section through northern Greece for the Cretaceous-Tertiary boundary (65 Ma) which purports to show both Africa-Apulia and Rhodope-Moesia-Eurasia 'basements' at each side, ought to acknowledge that nearly 1000 km of post-65 Ma strike-slip displacement has yet to be accommodated somewhere in the section, probably along with a significant amount of oceanic opening.

In constructing the diagrams we follow the approach of Smith (1971) in attributing the motion of crustal blocks between the two major plates to direct coupling to one or other plate as far as possible. This seems to lead to sensible results at times of major changes in plate boundary configurations, but it is clear that independent motion of substantial plates away from Africa must dominate the geological evolution at other times. We have not incorporated any stretching into the individual blocks which thus travel in their final re-compressed form. Stretching will lead to earlier collision and is obviously a refinement for future models.

Ocean basins have been kept of restricted width except when long-distance travel is required. The Gulf of California and Red Sea are both thoroughly oceanic in the centre and yet only 250 km wide coast to coast. Conversely, the modest subduction rate of 5 cm/a will destroy 1000 km of ocean in 20 Ma, or the duration of the Albian and Aptian together.

Late Permian to Mid Triassic, Fig. 23a

It appears likely that the Permian to Mid Triassic interval saw a major change in the shape of Pangaea with substantial dextral (i.e. west to east) motion of Eurasia and North America relative to Africa, creating Palaeotethys, an eastward widening arm of the super ocean Panthalassa. This Palaeotethyan ocean crust was undergoing subduction under an active Eurasian margin of Pacific-type with a long history of accretion and strike-slip faulting, subduction and marginal basin formation. In our view this probably remained the case from Iran westwards until the arrival of Gondwanan fragments and promontories in the latest Cretaceous and Tertiary.

To the south the northern edge of Gondwana was active in the sense that it was undergoing extension and volcanism from the Late Permian, possibly in a kind of California borderland situation. Whether or not southward subduction of Palaeotethyan crust took place at this time remains unclear but it may well not be a necessary requirement to explain the chemistry of the rift-related igneous rocks.

We adopt a clear-cut assignment of continental blocks to one margin or to the other. To the north the Moesian block and the Rhodope and Pontide fragments are essentially Eurasian and underwent periodic episodes of strike-slip and subduction-driven displacement and compression. The Sakarya block is placed on the northern edge of Gondwana and partly disassembled along the Karakaya suture, largely because it appears easier to accommodate the later collisional history of the reassembled block this way. More uncertainty surrounds the Serbo-Macedonian massif which we assign to Eurasia, thereby implying that the ocean basin remnants now found in the complex of Mesozoic belts in eastern Serbia (Grubić 1981) are all Neotethyan rather than Palaeotethyan. The Vardar zone on this assumption is the site of Palaeotethyan consumption. However, an almost equally acceptable alternative would derive the Serbo-Macedonian massif by Early Mesozoic (perhaps even Late Palaeozoic) rifting from the Adriatic-Apulian promontory, perhaps from a northern extension of the Pelagonian block, and take it in stages across the gap to Eurasia. An arm of Palaeotethys would in this alternative be eliminated along the Getic suture in the Lower Cretaceous. This is our sole concession to **Şengör *et al.*'s** derivation of the 'Cimmerian continent' from Gondwana. The eastern margin of the Vardar ocean, if it was ever passive, appears to have been active early in the Jurassic, with subduction to the east beneath the Serbo-Macedonian massif. For much of the Jurassic and Cretaceous an unknown amount of ocean lay between the massif and the Rhodope, and is now represented by oceanic remnants in the Danube, Luznića and Timok zones. Since the location of the Serbo-Macedonian massif remains uncertain for long periods the question of its source is relatively less crucial.

Mid Triassic to Mid Jurassic, Fig. 23b

New palaeomagnetic data (Steiner 1983) reveal a significant discrepancy between the palaeopoles for North America and Africa from the Early to Mid Triassic up to the time of opening of the central Atlantic in the early Late Jurassic (173 Ma). We assume that the displacements responsible for the discrepancy have to be accommodated as strike-slip motion along the future site of Atlantic opening, without any major separation or convergence there. If this is so, approximate stereographic calculations suggest that a pair of finite rotations about poles which are not coincident and so can duplicate the 'dog-leg' excursion in the palaeopole of North America relative to Africa, can at the same time yield small circle traces in the crucial Africa-North America contact zone which are nearly superimposed. This analysis yields large displacement vectors for Eurasia relative to Africa of some 2500 km for points on the Gondwanan margin, directed towards the north-east for the Early to Mid Triassic, with reversal back along a similar track to the pre-Atlantic opening configuration in the Early Jurassic.

We see this major dextral shear of Eurasia coupled to Palaeotethys as the driving force for the widespread extension of the northern Gondwana margin from the Late Permian to Middle Triassic. As a result, a series of intracontinental rifts propagaged westwards from south-east Arabia, reaching the Levant by the Early Triassic and Greece by the Middle Triassic. We assume that the majority of these remained deep rifts floored by stretched continental crust which underwent non-calcareous radiolarian deposition in a series of only partly connected gulfs and bays narrower than the modern Gulf of California. We have placed the Sakarya continent (see **Şengör *et al.***) as the most northerly continental sliver of Gondwana and suggest that the closure of the Karakaya basin within it might be related to oblique convergence and strike-slip processes associated with the establishment of active spreading on the north-south Pindos rift and eastward motion of the Pelagonian block. At a later stage the re-united Sakarya continent moved northwards once the way was clear after collapse of the Pindos ocean ridge system. Because all the southern continental fragments are located in a major Gondwanan margin embayment between the Adriatic and Arabian promontories the sense and direction of shear motion favoured the establishment of active spreading on the north-south aligned rifts in the Greek-Yugoslavian area and kept the Turkish fragments largely together. The main shear displacement was accommodated within Palaeotethys and along its northern margin, creating a Rhodope-Pontide collage of Late Triassic-Early Jurassic age. Periodic propagation of shear into the network of braided Turkish rifts may be the cause of local thrusting, uplift and coarse clastic

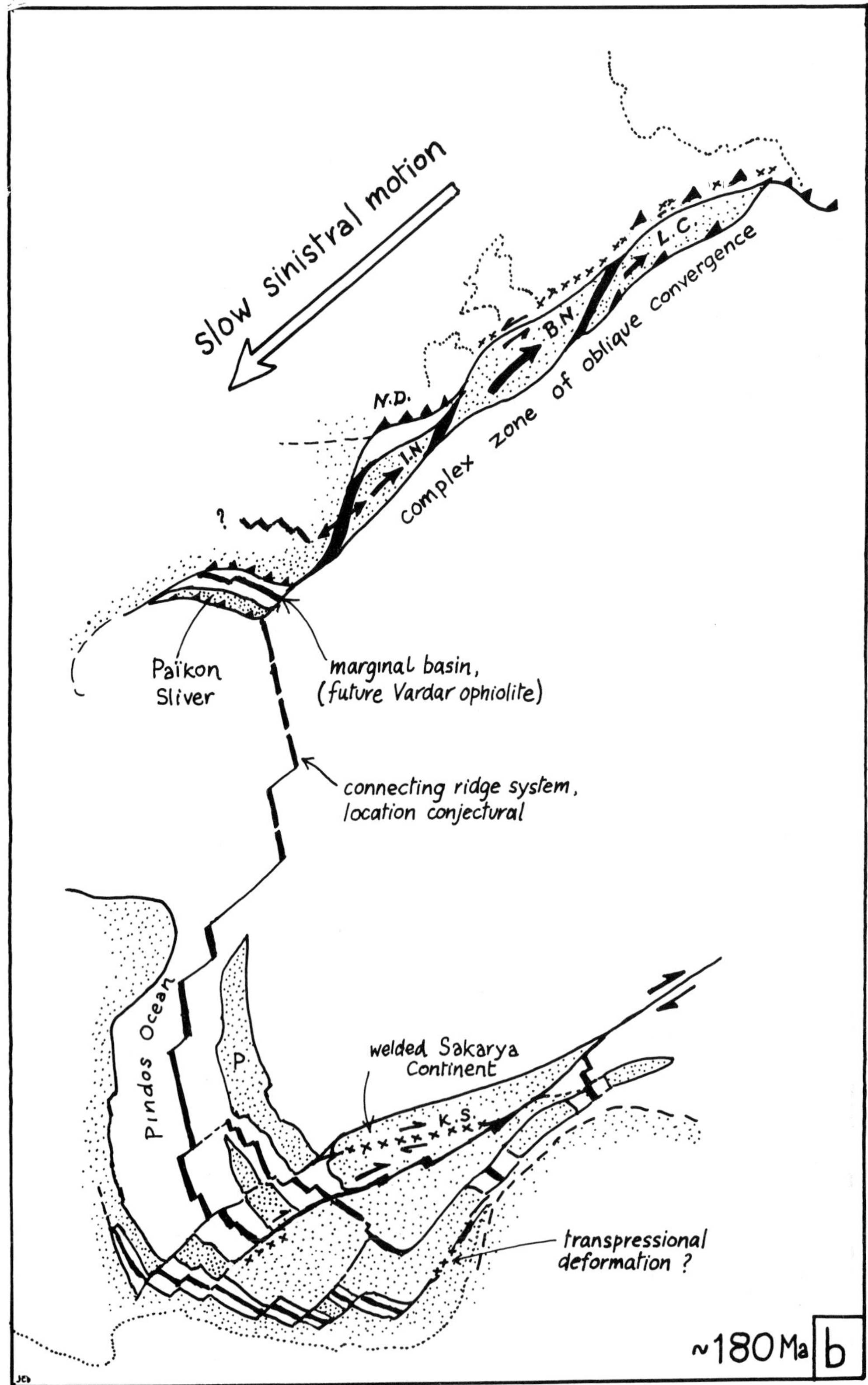
Slow sinistral motion
N.D.
I.N.
B.N.
L.C.
complex zone of oblique convergence
?
Païkon Sliver
marginal basin, (future Vardar ophiolite)
connecting ridge system, location conjectural
Pindos Ocean
P
welded Sakarya Continent
K.S.
transpressional deformation ?
~180 Ma
b

sedimentation documented by **Monod & Demirtaşlı**. The Arabian promontory prevented easy eastwards escape for Turkish blocks at this stage.

The Pelagonian microcontinent may itself have been broken up along the transform zones that bounded subsidiary extensional basins like the Boetian zone (Clément 1977) that in some instances left isolated continental blocks capped by carbonate build-ups such as Parnassus. These transforms are preserved, as Dercourt (1970) suggested, as discontinuities like the Scutari-Peč and Sperchios lines, defined by the effects of the original offsets, not the offsets themselves which suffered later collisional realignment.

It appears inevitable that a split occurred between the Pelagonian and Menderes blocks during the Early Jurassic to accommodate Greek but not Turkish oceanic opening. The Pelagonian zone and possibly a significant part of future Aegean 'basement' may even have been displaced north-eastwards to a position north of the Menderes-Tauride block.

At some stage in the Early Jurassic the sense of shear reversed between Africa and Eurasia (Steiner 1983). By this time spreading had become established in the Pindos ocean and by implication on a spreading ridge system connecting the Pindos centre with the Eurasian margin plate boundary to the north. This connecting system may have passed out into Palaeotethys as a new intra-oceanic rift as in Fig. 23b or may have continued as a northern prolongation of the Pindos ocean separating the Serbo-Macedonian massif from the Apulian platform. In either case the component of motion eastwards away from the Adriatic promontory generated by one (or more) N-S ridge systems, *plus* the now sinistral sense of shear between the major plates would have produced profound changes in the tectonic status of the Eurasian margin. Accretion and active subduction would now have dominated the segments more nearly perpendicular to the resolved direction of oceanic convergence.

We attribute the formation of the ophiolite chain of the eastern Vardar zone (e.g. Guevgueli) to the opening of a marginal basin in the Middle Jurassic within the continental crust of the active western Serbo-Macedonian margin. Whether this process occurred when the Serbo-Macedonian massif was itself separated from Eurasia by a Palaeotethyan remnant in the Danube basin is not clear.

Upper Jurassic, Fig. 23c

A significant change in the direction of motion of Eurasia relative to Africa occurred when the central Atlantic began to open at 173 Ma (Livermore & Smith, 1984, Fig. 16), from south-westwards convergence to a more westerly motion. More importantly the *rate* of motion in the initial period of 10 Ma or so, was very much faster than at any subsequent time up to the opening of the North Atlantic in the Cretaceous for all points attached to a Eurasian frame in the future eastern Mediterranean region. Steiner's (1983) results indicate that Africa and Eurasia were probably already in relative motion at the time of opening. If this existing motion was parallel to both the initial split and so to the early anomalies, it may not be detectable by fitting operations and so could add a further component to the early Upper Jurassic motion in the Tethyan region. This sudden burst of sinistral east-west motion had several effects. In the western Tethys, beyond the Adriatic promontory, a series of interconnected ocean basins of Red Sea-type were activated (e.g. Liguria). On the eastern side, the north-south spreading centres in the Pindos basin, between the Pelagonian zone and the Eurasian margin, and the Paikon-Serbo-Macedonian marginal basin were all thrown into compression. The process of progressive asymmetrical spreading and ridge collapse, described earlier, ensued, generating oceanic crust with supra-subduction-zone geochemistry and sub-ophiolite metamorphic soles at around 174 Ma. This process extinguished the north-south ridges and created emplaceable

FIG. 23b. Mid-Jurassic. 180 Ma.
The earlier NE-ward extensional motion established the N-S trending Hellenide-Dinaride rifts as ocean basins (Pindos). S. Turkey remains largely 'unexpanded' in the shadow of the Arabian promontory. Dextral shear propagating intermittently into the Turkish 'mosaic' during the Late Triassic-Early Jurassic causes local transpressional deformation in the Taurides (see **Monod & Demirtaşlı**). The most northerly E-W continental sliver of Gondwana is welded by dextral shear along the Karakaya suture in Late Triassic to form the Sakarya continent. At least three major plates now exist implying a connecting spreading system from Neotethyan Pindos ocean, through Palaeotethys, to the northern complex margin. The Danube basin between the Serbo-Macedonian and Rhodope Massifs may be open at this time. The Istanbul Nappe (IN), Bayburt Nappe (BN) and Lesser Caucasus (LC) blocks are emplaced NE-wards by strike-slip processes in the Late Triassic-Early Jurassic, to form the Eurasian margin 'collage zone'.

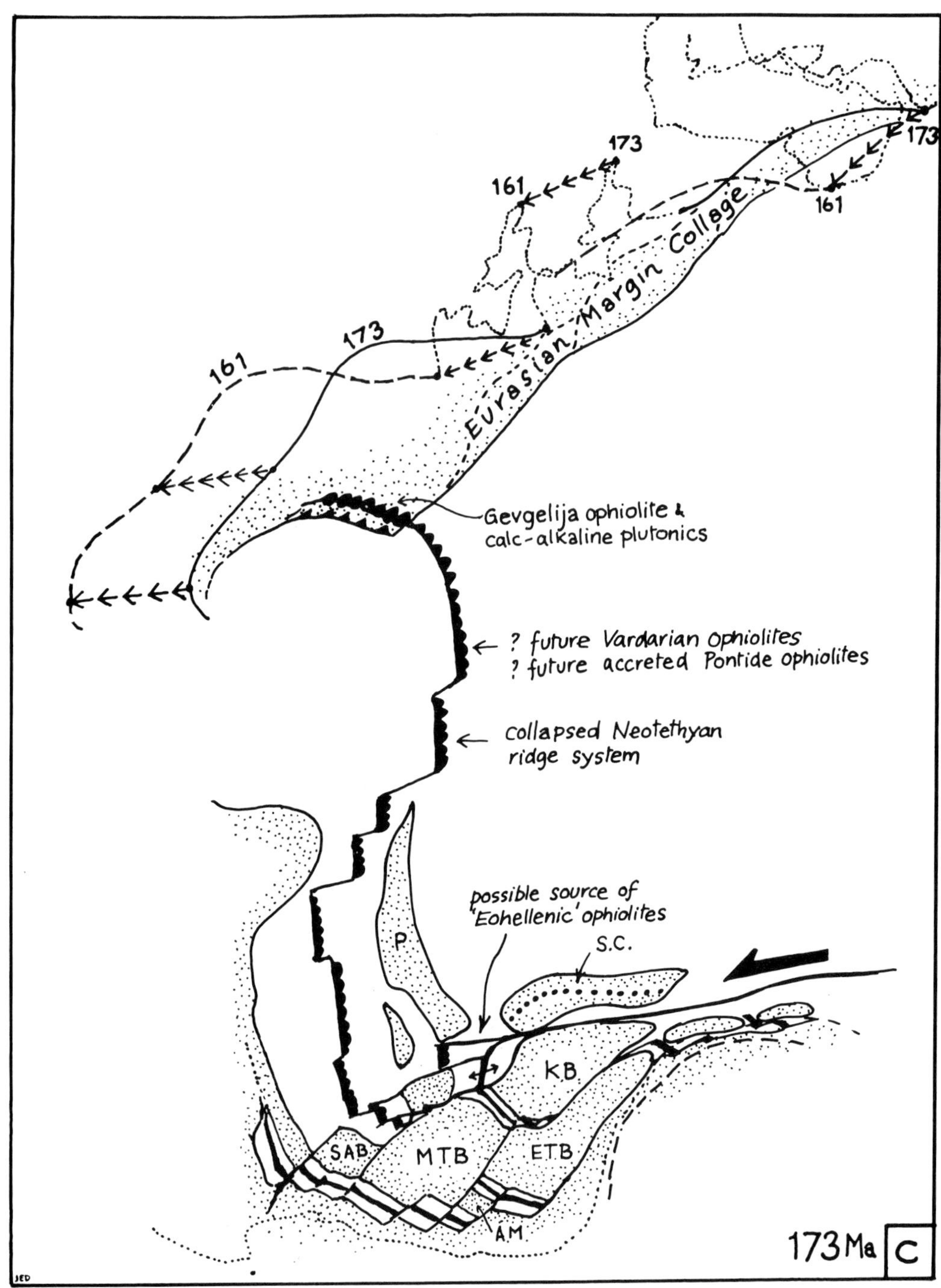

FIG. 23c. Late Mid-Jurassic (Bathonian). 173 Ma.
Relative to Africa, Eurasia shifts rapidly westwards as the central Atlantic begins to open. The position of the boundary of 'stable' Europe at 161 Ma (Oxfordian) is shown dashed as an indication of the scale and rate of this motion. The Pindos-Budva ocean ridge and linking ridges to the north, including the Gevgelija marginal basin, are driven into compression and collapse at the crest generating emplaceable ophiolites with 173 Ma soles. An E-W transform bounds the Turkish Gondwanan mosaic to the south along the future Izmir-Ankara zone. Pull-apart basins along this zone provide future Eo-Hellenic ophiolites for the eastern Pelagonian zone margin. Rifts within the Turkish mosaic and south of the South Aegean block remain undisturbed. The Sakarya continent (SC) is transported passively westwards with the Pelagonian microcontinent.

ophiolites. What is less clear is whether the transport of these ophiolitic bodies to the adjacent continental margins was always necessarily the result of the collapsed ridges evolving into full subduction zones with the same polarity. In the case of the eastern Vardar ophiolites the polarity may be immaterial if the entire marginal basin was telescoped.

A general east-over-west sense of collapse and consequent eastward subduction would be in general accord with the ultimate sense of ophiolite nappe vergence, as well as with the evidence from Yugoslavia of extensive blueschists in the western part of the Golija (Pelagonian) zone and scattered calc-alkaline magmatism within the Pelagonian zone (Grubić 1981). However, to emplace Pindic ophiolites from the ridge axis eastwards onto the Pelagonian zone in Greece in the Lower Cretaceous requires an additional mechanism such as gravity-sliding, strike-slip and transform-offset interaction or the initiation of a subduction zone at the Pelagonian margin. These other mechanisms may have been rendered more effective if the ocean basin became, or always was, extremely asymmetrical so that the collapsed ridge lay close to the eastern side of the basin, an option developed by **Smith & Spray**.

It appears likely that emplaceable intra-oceanic ophiolites and intra-oceanic subduction zones developed from the collapsed ridges linking the southern and northern plate boundary systems, were both created at this time. Both would have been eventually swept into the gap between the Pelagonian and Serbo-Macedonian massifs, to become the complex of ophiolitic remnants and calc-alkaline volcanics found in the Almopias subzone of the Vardar zone (Mercier 1966) or into the future Intra-Pontide suture zone. Older Palaeotethyan ophiolitic material may also be present in both.

The Turkish microplate mosaic remained tucked into the North African embayment effectively decoupled from the Hellenides and Dinarides by a transform zone. The elusive Sakarya continent may however already have 'escaped' to lie north of this zone and so be subject to westward transport with Palaeotethys, and its escape may well explain the subsidence of its margins to pelagic depths in the Late Jurassic (Şengör & Yilmaz 1981). The occurrence of Upper Jurassic ophiolite remnants in the Asteroussia nappe of eastern Crete (see **Bonneau**) suggests that they were generated in the north-south 'Hellenide regime', perhaps at its southern limit, lending support in a general way to **Hall *et al.*'s** contention that Cretan facies belts followed a Hellenide trend in the early Mesozoic at least.

Early to Mid-Cretaceous, Fig. 23d

The rate of sinistral motion declined steadily until the mid-Cretaceous. Towards the end of this period a change to north-south convergence occurred with another spectacular acceleration coinciding with the period around 108 Ma when the north Atlantic began to open. With the extinction of the Hellenide and Dinaride, and possibly Getic, north-south ridge systems in the Upper Jurassic, the emplacement in the Lower Cretaceous of Pelagonian and 'Vardarian' ophiolites, and the progressive reduction in Africa-Eurasia relative motion, Palaeotethys ceased moving actively towards the Adriatic promontory. The Arabian margin and the Pelagonian zone no longer impeded opening of the east-west basins in the Turkish area and the older Triassic rifts were reactivated converting the attenuated Tauride-Anatolide platform into a series of microplates. A major and possibly gradual shift in the motion of Palaeotethys relative to Africa took place from the Late Jurassic to Mid Cretaceous, from westwards through northwestwards to northerly as the contribution from the east-west Turkish ridges increased and the motion of the Pelagonian zone westwards declined.

Relative to Eurasia, consumption of Palaeotethyan crust along the Pontide margin probably became more and more orthogonal, extending the zone of consumption westwards from the Caucasus. Net rates of convergence were perhaps still quite low but subject to an unknown extra contribution from any Palaeotethyan ridges still active. The transition from oblique convergence to strike-slip motion along the eastern margin of the Pelagonian zone is recorded in the widespread Late Jurassic-Early Cretaceous deformation and metamorphism (the 'Eohellenic' phase) in the Greek part of the zone. Examples are the southerly-directed syn-metamorphic thrusting in Pieria dated at 130 Ma (Yarwood & Dixon 1977) and the pre-Upper Cretaceous longitudinal shear zone ('zone de broyage') in the western part of the Almopias described by Braud (1967). The deformation along this margin may have been enhanced by any irregular projections at transform offsets and perhaps also by oblique collision with the Sakarya continent as it migrated northwards. Again, a complex, probably shifting boundary passing through the Aegean coupled and decoupled the northward motion of dispersing Turkish fragments from the relatively quiescent 'external' zones of Greece and the Dinarides.

Several separate Neotethyan ocean strands developed. A southern strand south of Turkey

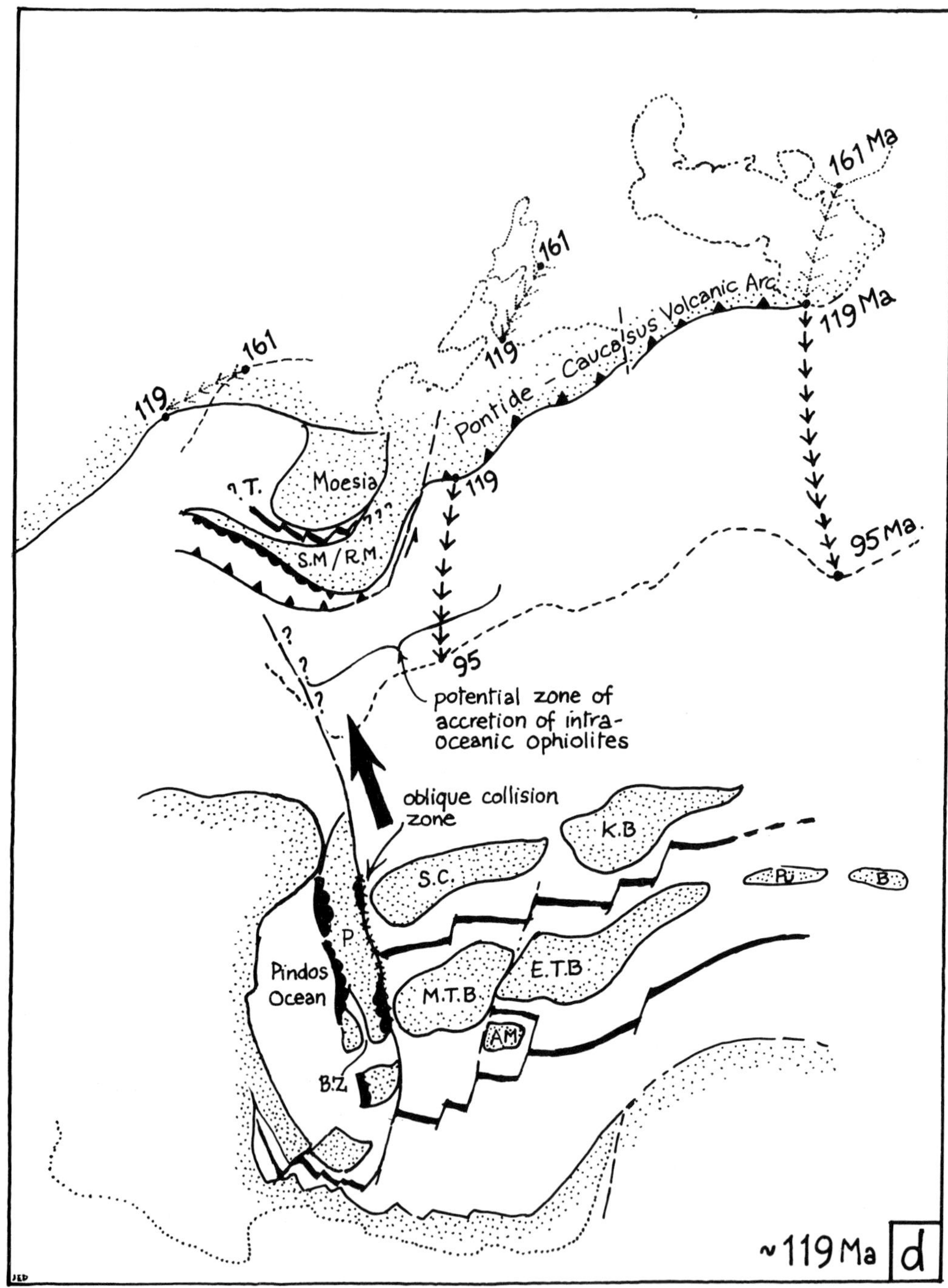

FIG. 23d. Early Cretaceous (The position of the Pontide margin is shown for 119 Ma—Barremian/Aptian—with the motion in the preceding 42 Ma and succeeding 24 Ma, shown by the arrows). Collapse of N-S spreading ridges and continued westward motion of Palaeotethys with the Pelagonian and Sakarya blocks led to emplacement of ophiolites on the western Pelagonian margin in the latest Jurassic-Early Cretaceous. E-W rifts in the Turkish mosaic began actively spreading; the E-W leaky transform south of the Sakarya and Kirşehir blocks also became an active spreading zone. The Sakarya and Menderes-Tauride blocks travelled northwards, deforming the eastward Pelagonian margin and emplacing ophiolitic melange on to it. The Pindos ocean became an inactive basin

connected with a strand south of Crete and the Peloponnese. This strand then terminated as a true ocean basin at the Adriatic-Apulian platform but was probably traceable northwards as the intra-platformal Ionian zone basin. In the Crete-Peloponnese area we suggest the Ionian zone became differentiated as the northern margin of a deeper east-west basinal area (Plattenkalk) which subsided as the Neotethyan rift still further south widened in the Upper Jurassic and Lower Cretaceous. To the north the future components of the higher structural units in Crete belonged to the fundamentally Hellenide north-south regime. To the east **Poisson's** recognition of Ionian zone sequences in the Lycian nappes suggests that an intra-platformal arm of the zone analogous to that in mainland Greece separated an eastward-narrowing Gavrovo-Tripolitza platform from the eastward-widening Bey Dağları platform while a southern branch continued out into the strand of Neotethys south of the Bey Dağları.

Further east the southern strand connected with northern strands NE-wards through the present Isparta angle, and the Kirşehir block to the north-east was separated from the Taurides by one or more strands.

We have identified the Pelagonian zone, and particularly its eastern margin, as forming a boundary through most of the Cretaceous between the relatively inactive 'external' Hellenides and the northward moving Turkish microplate complex to the east. The corollary of this is that the southern Neotethyan strand south of Crete did not open into a significantly wide basin, because if it had, the external Hellenide zones would have been translated northwards relative to the Apulian platform, for which there is no evidence. Enough Neotethyan oceanic crust must be created south-west of Crete at some stage to fuel the Tertiary to Recent Aegean arc. We identify the anticlockwise rotation of the Adriatic-Apulian promontory as the mechanism for generating crust in the mid-Tertiary, but, additional crust could have been created during the Cretaceous if the Hellenides moved rather further north along a cryptic 'Apulian-margin transform'. These adjustments do not substantially affect the main feature of the Cretaceous configurations, the northward expansion of Turkey on active east-west ridge systems.

Looking ahead to the collapse of these ridges and the emplacement of ophiolites in the Upper Cretaceous, in Turkey we may infer from the absence of ophiolites of this age in Crete and the Aegean that the active transform boundary delimiting east-west ridges lay at the eastern edge of the 'proto-Aegean'. This is also consistent with a discontinuity in the spreading history of the southern Neotethyan stand between an active eastern end and a subdued western end. Upper Cretaceous ophiolite nappes may perhaps be buried beneath the Mediterranean ridge south of Crete. There is still great scope for tectonic complexity in this Aegean boundary region during the Cretaceous: an example may be the puzzling Cretaceous acid plutonics found at a high structural level on Anafi in the southern Cyclades (Reinecke *et al.* 1982) which could imply localized subduction of a small ocean or crustal thickening followed by transtensional rifting and basic magmatism.

Upper Cretaceous, Fig. 23e

Between 119 Ma and 95 Ma a major change in the motion of Eurasia relative to Africa occurs, from slow sinistral shear to rapid convergence, for all points on a Eurasian reference frame lying in our area of interest. This coincides with the opening of the north Atlantic at around 108 Ma (Tucholke & Ludwig 1982). The spreading ridges separating the southern Turkish blocks from each other and from Arabia, and active ridge segments in the remaining part of Palaeotethys all came under compression and underwent the process of asymmetrical spreading decline, collapse and the generation of an upper plate ophiolite with supra-subduction zone characteristics, that had befallen the north-south Hellenide ridges in the Upper Jurassic. The timing of ridge collapse and active ridge overlap to generate a dateable sole, varied progressively along the ridge segments from west to east (see **Whitechurch *et al.***). This is to be expected in a general way from our ridge-collapse model. The process of asymmetrical spreading rate decline will start at the same time, at last along a single oceanic strand, but

---

but was connected to the main active southern Neotethyan strand and thence to Palaeotethys. In the north a progressive change in Africa-Eurasia motion to convergent, plus the spreading in the south, led to enhanced subduction and associated magmatism spreading from E to W. Intra-oceanic ophiolites and island-arcs created in the Upper Jurassic were accreted along the Eurasian margin. Continuing subduction under the Serbo-Macedonian massif may have initiated the Timok rift zone (T) but its SE-ward status and continuation is uncertain.

The accelerated convergence from 108 Ma on results from N Atlantic opening and causes collapse of the E-W ridges.

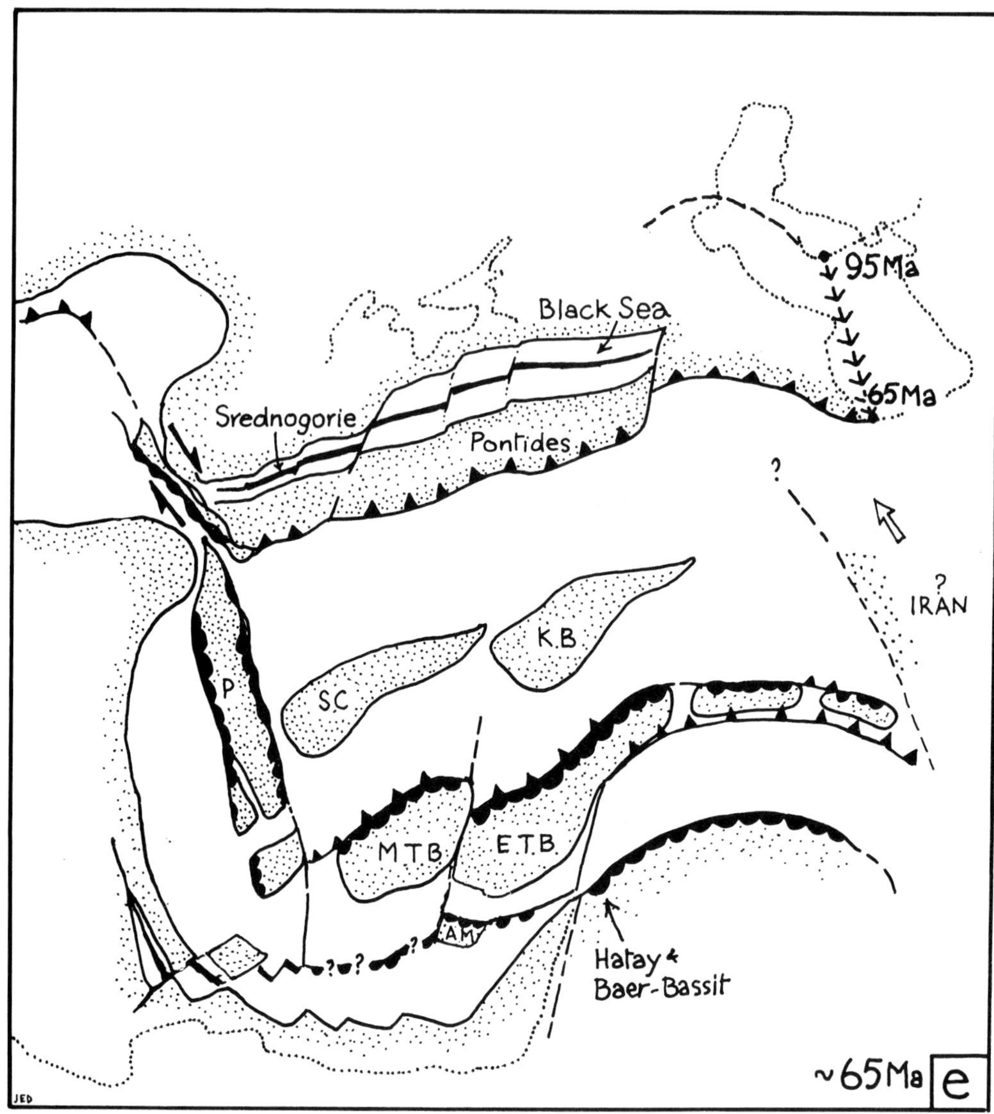

Fig. 23e. Maastrichtain. 65 Ma.
Collapse of E-W ridges in the period 108–85 Ma, following the opening of the N Atlantic, led to Maastrichtian emplacement of ophiolites southwards onto adjacent continental margins and the development of northward-dipping subduction zones. The Pindos basin and most of the future Aegean remained unaffected. Earlier ridge collapse in the main Palaeotethyan-Neotethyan basins caused oblique convergence along the Pelagonian margin to cease and to go into reverse, leading to widespread subsidence. In the north, the Srednogorie and Black Sea basins may have been initiated behind the Pontide arc following the drop in net convergence rates. An independently advancing Iran plate kept the Caucasus are active and compressed. Impingement of the Eurasian margin began at the northern end of the Apulian promontory.

the length of time needed to bring one side of the ridge system to a stop and then induce active overlap and a dateable sole, will be a function of the difference between the velocity of the ridge relative to a hot-spot frame before the change, and the velocity it needs to have after the change to stay active and avoid collapse. This parameter must vary systematically with distance for a two plate system but the detailed timing of sole formation in a complex braided rift system could be esentially unpredictable.

Subduction zones were initiated adjacent to the Kirşehir block and to the south of the

Tauride platform, with N-S segments acting as transform margins (e.g. Antalya complex). Some may have begun at the ridge-collapse zones themselves as intra-oceanic arcs leading by the Maastrichtian to collision of the southern passive margins with the trenches and emplacement of ophiolites.

By the end of the Cretaceous the northern bulge of the Adriatic-Apulian promontory was probably close enough to Eurasia to induce oblique collisions in the western Rhodope-eastern Serbo Macedonian region, closing basins opened since the Late Jurassic (e.g. Timok zone, Grubić 1981). In the south-facing part of the Rhodope, the Srednogorie basin opened as a marginal sea. It was perhaps an eastern arm of the Black Sea, which itself may have opened at this time behind the Pontide arc (Letouzey *et al.* 1977). Unless the Pelagonian zone had moved a long way north through the Cretaceous, which we have discounted, there was still no part of Eurasia close to the future Greek segment of the Hellenides west of the Vardar zone.

The location of the Turkish microcontinental blocks at the end of the Cretaceous is an area of critical uncertainty. Our ridge-collapse model for ophiolite generation carries with it the consequence that all basins so affected become inactive and can henceforth only get narrower unless spreading somehow restarts. If all the Neotethyan strands in Turkey were affected in this way, the microcontinental blocks south of the Pontides, including the Sakaraya continent, can only have converged on Eurasia after about 85 Ma, at a rate *equal to*, or *slower than*, the convergence rate of Africa and Eurasia. They were not then being pushed away from Africa by an ridge-system. The fact of Upper Cretaceous continental-margin emplacement implies a narrowing of more than one southern Neotethyan strand and motion *towards* Africa of the blocks involved.

In the initial mid-Cretaceous period of accelerated convergence the subduction rate under the northern Pontide arc increased dramatically (see **Akinci**) but towards the end of the Cretaceous it may have actually declined as the component of extra northward motion from the Neotethyan ridges to the south fell to zero and then reversed.

One deduction from this is that a block which collides with Eurasia early, (Upper Cretaceous to Palaeocene) has to be already close to Eurasia by the Mid-Cretaceous. We think the Sakarya continent may have been such a case. It then becomes extremely difficult to accept that the Menderes-Tauride block was also in a position to collide with the Sakarya continent in the Upper Cretaceous as suggested by **Okay** from blueschist evidence. Such a configuration would place the Menderes massif adjacent to northern Yugoslavian external zones with drastic consequences for the continuity of belts across the Aegean. The southern Neotethyan strand that would have separated this block from Arabia would then have been enormously wide making southward emplacement of ridge-generated ophiolites onto the Arabian margin by the Maastrichtian virtually impossible, with only Africa-Eurasia motion to close this southern basin. Decoupling the eastern and western Taurides through the Isparta angle and allowing hundreds of kilometers of relative motion would increase the range of options but seems unjustifiable at this early stage of our 'balanced cross-section' approach to Turkish plate evolution.

We prefer to leave a substantial oceanic tract through the Izmir-Ankara-Erzincan zone in the earliest Tertiary, which would be Neotethyan west of Ankara and Palaeotethyan in origin to the east, since we derived the Sakarya continent from Gondwanaland and the Pontides from Eurasia.

Oceanic tracts could also have remained after ophiolite emplacement on southern margins, in all the ophiolite-generating Neotethyan strands.

Early Tertiary

A complex pattern of progressive convergence and collision now ensued as Africa and Eurasia continued their rapid approach. Major early Tertiary flysch basins developed in 'fore-arc' areas such as the Hazar Group basin in SE Anatolia, the Tuz Gölü basin in central Anatolia and in Turkish Thrace. Intermittent calc-alkaline magmatism occurred round the Eurasian active margins from the Vardar zone to Caucasus. The Sakarya continent was accreted to Eurasia and was then well to the north of its present position, relative to the Pelagonian zone on the other side of the Aegean. The southward movement of this corner of Eurasia down the Pelagonian zone caused oblique collision and progressive compression of the external Hellenide zones. The Eocene Aegean blueschists may record the north-westward subduction of an oceanic extension of the Izmir-Ankara zone separating the Pelagonian north Aegean and Menderes massifs and the reuniting of these blocks.

At some stage the major impingement of the Adriatic-Apulian promontory with Eurasia caused it to begin its anticlockwise rotation. The south-western Neotethyan rifts became active spreading zones, swinging the southern

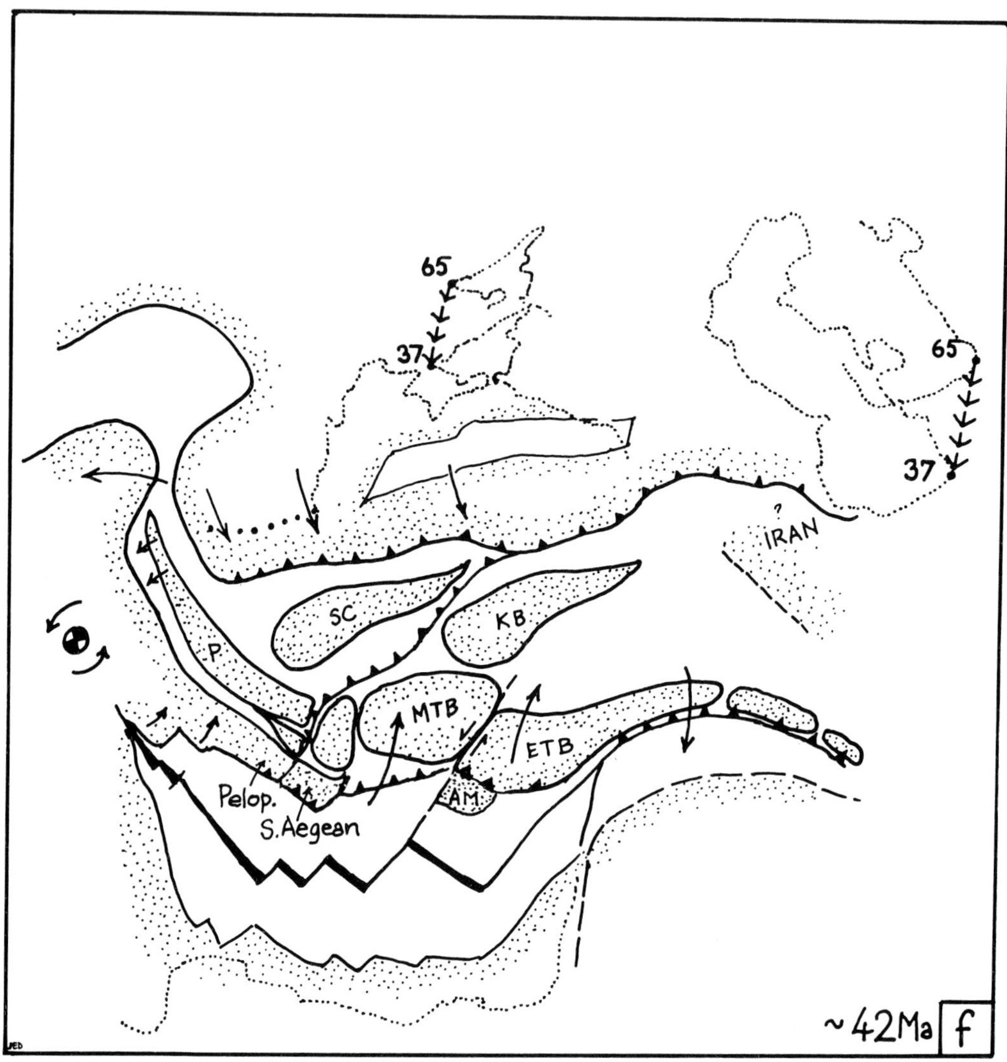

FIG. 23f. Late Eocene. 42 Ma.
The Pontide 'front' is shown at its 37 Ma (Lower Oligocene) position, with its previous Maastrichtian position indicated. The state of microplate collision to the south is shown for Mid to Late Eocene times. The advance of Eurasia closes the Pindos basin and initiates the rotation of the Adriatic promontory anti-clockwise about a pole lying within it. The south-westernmost Neotethyan rifts now spread actively and propagate eastwards swinging the Peloponnesan and S Aegean blocks northwards to create a complex E-W proto-Hellenic arc. The Neotethyan remnant lying between the S end of the Pelagonian block and the N Aegean block is destroyed by NW-ward subduction creating the Aegean blueschist belt. Further E the Alanya Massif (AM), collides with the E Tauride Block, which rotates clockwise. The Sakarya and Kirşehir blocks are relatively independent and can be positioned to suit local geological or palaeomagnetic constraints. The Srednogorie basin within the Rhodope closes, but the Black Sea remains open. The largest remaining oceanic tract is a complex mix of Palaeotethyan and Neotethyan crust lying between Arabia and the E Pontides, the future E Anatolian accretionary complex of Şengör & Yilmaz (1981).

Hellenide and south Aegean blocks northwards and creating a nearly E-W 'proto-Hellenic arc'. The Oligocene HP-LT metamorphism in Crete (Seidel *et al.* 1982) presumably reflects a transformation from passive to active for the northern margin of this Neotethyan strand (the Plattenkalk subsided platform).

From the Late Eocene to the Miocene a series of collisions occurred differing in timing and geometry until the northern Neotethyan and Palaeotethyan strands closed completely leaving only a remnant southern strand south of Turkey, Crete and the Peloponnese. By this stage we can no longer distinguish plate interaction effects from the complex pattern of local nappe movements, uplift and subsidence which remain in many cases very poorly understood.

ACKNOWLEDGEMENTS: In this writing this review we have drawn on our experiences of the Eastern Mediterranean area over a number of years, but we would also like to acknowledge our debt to our fellow contributors, whose unpublished results, available to us through the editing of the book, have greatly augmented our understanding of the area and its problems. We are particularly grateful to Alan Smith and R. A. Livermore for access to their Africa-Eurasia motion results in advance of publication. We would particularly like to thank our secretaries, Mrs M. Wright and Mrs H. Hooker for all their constant efforts in typing and re-typing the manuscript. Mrs D. Baty and Mrs F. Tullis helped draft diagrams for us.

References

ADAMIA, S. A., CHKHOTUA, MB., KEKELIA, M., LORDKIPANIDZE, M., SHAVISHVILI, I. & ZACHARIADZE, G. 1981. Tectonics of the Caucasus and adjoining regions: implications for the evaluation of the Tethys ocean. *J. Struct. Geol.* **3,** 437–447.

AGER, D. V. 1967. Some Mesozoic brachiopods in the Tethys region. *In:* ADAMS, C. G. & AGER, D. V. (eds). Aspects of Tethyan Biogeography. *System. Assoc. Publ.* **7,** 135–151.

AKKÖK, R. 1982. The Menderes Massif: an intracrustal shear zone beneath the Lycian nappes. *In:* DIXON, J. E. & ROBERTSON A. H. F. (eds). *The Geological Evolution of the Eastern Mediterranean*, Abstracts, Edinburgh, 1982, p. 7.

AKYOL, E. 1975. Palynologie du Permien inférieur de Sariz (Kayseri) et de Pamucak (Antalya-Turquie) et contamination jurassique observée, due aux ruisseaux 'Pamucak' et 'Goynuk', *Pollen and Spores*, **17,** 141–179.

ALTHERR, R. KREUZER, H., WENDT, I., LENZ, H., WAGNER, G. A., KELLER, J., HARRE, J., HÖHNDORF, A. 1982(a) A Late Oligocene/Early Miocene high temperature belt in the Attic-Cycladic crystalline complex (SE Pelagonian, Greece), *Geol. Jb.* **E23,** 97–164.

——, SEIDEL, E. & KREUZER, H. 1982b Petrological and geochronological constraints on geodynamic models for the Hellenides, *In:* DIXON, J. E. & ROBERTSON, A. H. F. (eds), *The Geological Evolution of the Eastern Mediterranean* Abstracts, Edinburgh, 1982, p. 10.

ANGELIER, J., LYBERIS, N., LE PICHON, X., BARRIER, E., HUCHON, P. 1982. The tectonic development of the Hellenic arc and the sea of Crete: a synthesis. *Tectonophysics*, **86,** 159–196.

ARGAND, E. 1924. La tectonique de l'Asie. Proc. Int. Geol. Congr. **13,** 171–372.

ARGYRIADIS, I. 1975. Mésogé permienne, chaîne hercynienne et cassure téthysienne. *Bull. Soc. géol. Fr.* **17,** 56–59.

ARKELL, W. J. 1956. *Jurassic Geology of the World.* Oliver and Boyd, Edinburgh and London, 806 pp.

AUBOUIN, J., BONNEAU, M., CELET, P., CHARVET, J., CLEMENT, B., DEGARDIN, J. M., DERCOURT, J., FERRIÈRE, J., FLEURY, J. J., GUERNET, C., MAILLOT, H., MANIA, J. H., MANSY, J. L. TERRY, J., THIEBAULT, P., TSOFLIAS, P. & VERRIEX, J. J. 1970. Contribution à la géologie des Hellénides: le Gavrovo, le Pinde et la zone ophiolitique subpélagonienne. *Ann. Soc. géol. Nord*, **90,** 277–306.

——, BONNEAU, M., DAVIDSON, J., LEBOULENGER, P., MATESCO, S. & ZAMBETAKIS, A. 1976. Esquisse structurale de l'Arc égéen externe: des Dinarides aux Taurides. *Bull. Soc. géol. Fr.* **18,** 327–336.

——, LE PICHON, X., WINTERER, E. & BONNEAU, M. 1977. Les Hellénides dans l'optique de la tectonique des plaques. *6th Coll. Geol. Aegean region (Athens)*, **3,** 1333–1354.

BARTON, C. M. 1976. The tectonic vector and emplacement age of an allochthonous basement slice in the Olympos area, NE Greece. *Bull. Soc. géol. Fr.* **18,** 253–58.

BAROZ, F. 1980. Volcanism and continent-island arc collision in the Pentadactylos Range, Cyprus. *In:* PANAYIOTOU, A. (ed.) *Ophiolites*, Proc. Int. Ophiolite Symp., Nicosia, Cyprus, 1979, 73–86.

BASALTIC VOLCANISM STUDY PROJECT 1981. *Basaltic Volcanism on the Terrestrial Planets.* Pergamon Press Inc., NY, 1286 pp.

BAVAY, D., BAVAY, P., MALUSKI, H., VERGÉLY, P. & KATSIKATSOS, G. 1980. Datation par la méthode $^{40}Ar/^{39}Ar$ de minéraux de métamorphisme de haute pression en Eubée du Sud (Grèce). Corrélations avec les évènements tectonométamorphique des Hellénides internes. *C.r. Acad. Sci. Paris*, **290,** (Series D), 1051–1054.

BAUMGARTNER, P. O. & BERNOULLI, D. 1976. Stratigraphy and radiolarian fauna in a Late Jurassic-Early Cretaceous section near Achladi (Evvoia, Eastern Greece). *Eclogae Geol. Helv.* **69,** 601–626.

——, & BERNOULLI, D. 1982. Late Mesozoic and Tertiary nappe structure of the Argolis peninsula (Peloponnesus, Greece). *In:* DIXON, J. E. & ROBERTSON, A. H. F. (eds). *The Geological Evolution of the Eastern Mediterranean.* Abstracts, Edinburgh, 1982, p. 16.

Bébien, J. & Mercier, J. L. 1977. Le cadre structural de l'association ophiolites-migmatites-granites de Guévguéli (Macédoine, Grèce): une croûte de bassin interarc? *Bull. Soc. géol. Fr.* **19,** 927–34.

——, Blanchet, R., Cadet, J. P., Charvet, J., Chorowicz, J., Lapierre, H. & Rampnoux, J. P. 1978. Le volcanisme triasique des Dinarides en Yougoslavie: sa place dans l'évolution géotectonique. *Tectonophysics*, **47,** 159–176.

——, Ohnenstetter, D., Ohnenstetter, M. & Vergély, P. 1980. Diversity of Greek ophiolites: birth of ocean basins in transcurrent systems. *Ofioliti*, **2,** 129–197.

Bergougnan, H. 1975. Relations entre les édifices pontique et taurique dans les nord-est de L'Anatolie. *Bull. Soc. géol. Fr.* **17,** 1045–1057.

——, Fourquin, C. 1980. Un ensemble d'éléments communs à une marge active alpine des Carpathes méridionales à l'Iran central: le domaine iranobalkanique. *Bull. Soc. géol. Fr.* **12,** 61–83.

——, & Fourquin, C. 1982. Remnants of a pre-Late Jurassic ocean in northern Turkey: fragments of Permian-Triassic Palaeo-Tethys? Discussion and reply by Şengör, A. M. C., Yilmaz, Y. & Ketin, I. *Bull. geol. Soc. Am.* **93,** 929–936.

Bernoulli, D. & Laubscher, H. 1972. The palinspastic problem of the Hellenides. *Eclog. geol. Helv.* **65,** 107–118.

——, Graciansky, P. C. de & Monod, O. 1974. The extension of the Lycian Nappes (South-West Turkey) into the South-East Aegean Islands. *Eclog. geol. Helv.* **67,** 39–90.

——, & Jenkyns, H. C. 1974. Alpine, Mediterranean and North Atlantic Mesozoic facies in relation to the early evolution of the Tethys. *In:* Dott, R. H. & Shaver, R. H. (eds). *Symposium on modern and ancient geosynclinal sedimentation. Spec. Publ. Soc. Econ. Paleontol. Mineral. Tulsa*, **19,** 129–160.

Biju-Duval, B., Dercourt, J. & Le Pichon, X. 1977. From the Tethys ocean to the Mediterranean seas: a plate-tectonic model of the evolution of the western Alpine system. *In;* Biju-Duval, B. & Montadert, L. (eds). *Structural history of the Mediterranean basins.* Editions Technip., Paris, 143–164.

Bonneau, M., Angélier, J. & Epting, M. 1977. Réunion extraordinaire de la Société géologique de France en Crète. *Bull. Soc. géol. Fr.* **19,** 87–102.

Braud, J. 1967. *Stratigraphie, tectonique, métamorphisme et ophiolites, dans le Vermion septentrional (Macédoine, Grèce).* Thèse, 3ème cycle, Univ. Paris Sud., Orsay, 213 pp.

Brinkmann, R. 1966. Geotektonische Gliederung von Westanatolian. *Neues Jahrb. Geol. Palaontol., Monatshefte*, **10,** 603–618.

Briqueu, L., Lancelot, J. R., Tatsumoto, M., Coffran, D. & Vilminôt, J. C. 1982. Sr, Nd and Hf isotopic constraints on magma genesis in the Aegean island arc. *In;* Dixon, J. E. & Robertson, A. H. F. (eds). *The Geological Evolution of the Eastern Mediterranean* Abstracts, Edinburgh, 1982, p. 18.

Broglio-Loriga, C. & Neri, C. 1976. Aspetti paleobiologici e paleogeografici della facies a 'Lithiotis' (Givrese inf.). *Riv. Ital. Paleontol. Stratigr.* **82,** 651–706.

Brunn, J. H. 1956. Contribution a l'étude géologique du Pinde septentrional et d'une partie de la Macédoine occidentale. *Ann. géol. Pays. Hell.* **7,** 358 pp.

—— 1976. L'arc concave Zagro-Taurique et les arcs convexes Taurique et Egéen: collision des arcs induits. *Bull. Soc. géol. Fr.* **18,** 553–567.

——, Graciansky, P. C. de, Gutnic, M., Juteau, T., Lefèvre, R., Marcoux, J., Monod, O. & Poisson, A. 1970. Structures majeures et corrélations stratigraphiques dans les Taurides occidentales. *Bull. Soc. géol. Fr.* **12,** 515–556.

——, Clément, B. & Dercourt, J. 1977. Histoire des recherches géologiques dans ies Hellénides. *6th Colloq. Aegean Geol. Athens*, 1977, **1,** 21–60.

Bruhn, R. L. & Dalziel, I. W. D. 1975. Destruction of the early Cretaceous marginal basin in the Andes of Tierra del Fuego. *In:* Talwani, M. & Pitman, III, W. C. (eds). *Island arcs, Deep Sea Trenches and Back-arc Basins. Am. Geophys. Un., M. Ewing Ser.* **1,** 395–405.

Bullard, E. C., Everett, J. E. & Smith, A. G. 1965. Fit of continents around the Atlantic. *In:* Blackett, P. M. S., Bullard, E. C. & Runcorn, S. K. (eds). *A Symposium on Continental Drift. Phil. Trans. Roy. Soc. Lond., Ser. A*, **258,** 41–75.

Cameron, W. E., Nisbet, E. G. & Dietrich, V. J. 1980. Petrographic dissimilarities between ophiolitic and ocean-floor basalts. *In:* Panayiotou, A. (ed.). *Ophiolites. Proc. Int. Ophiolite Symp., Nicosia, Cyprus, 1979*, 182–193.

Capedri, S., Venturelli, G., Bocchi, G., Dostal, J., Garuti, G. & Rossi, A. 1980. The geochemistry and petrogenesis of an ophiolitic sequence from Pindos, Greece. *Contrib. Mineral. Petrol.* **74,** 189–200.

Caron, D. 1975. *Sur la géologie du Pinde méridional: les Monts Lakmon (Epire, Greece). La série des radiolarites.* Thèse, 3ème cycle, Univ. P. et M. Curie, Paris, 119 pp.

Céczy, B. 1973. The origin of the Jurassic faunal provinces and the Mediterranean plate tectonics. *Ann. Univ. Sci. Budap. Rolando Eotvos Nominatae, Sect. Geol.* **16,** 99–114.

Celet, P. 1977. Les bourdures de la zone du Parnasse (Grèce). Evolution paleogéographique au Mesozoique et charactères structuraux. *6th Coll. Aegean Geol. Athens, 1977*, 725–741.

—— & Ferrière, J. 1978. Les Hellénides internes: Le Pélagonien. *Eclogae geol. Helv.*, **73,** 467–495.

——, Clément, B. & Ferrière, J. 1982. Les éléments structuraux transverses en Grèce Moyene. *In:* Dixon, J. E. & Robertson, A. H. F. (eds). *The Geological Evolution of the Eastern Mediterranean.* Abstracts, Edinburgh, 1982, p. 23.

Channel, J. E. T., D'Argenio, B. & Horvath, F. 1979. Adria, the African promontory, in Mesozoic Mediterranean palaeogeography. *Earth Sci. Rev.* **15,** 213–292.

CHASE, C. G. 1978. Extension behind island arcs and motions relative to hot spots. *J. Geophys. Res.* **83,** 5385–5387.

CLÉMENT, B. 1977. Relations structurales entre la Zone du Parnasse et la Zone Pélagonienne en Béotie (Grèce continentale). *6th Coll. Aegean Geol. Athens, 1977*, 237–251.

CONEY, P. J., JONES, D. L. & MONGER, J. W. H. 1980. Cordilleran suspect terrains. *Nature*, **288,** 329–333.

CREUTZBERG, N. & SEIDEL, E. 1975. Zum Stand der Geologie des Präneogens auf Kreta. *Neues Jahrb. Geol. Palaeontol. Abhandlungen*, **149,** 363–383.

D'ARGENIO, B. 1976. Le Piattaforme carbonatiche Peridriatiche: una rassegna di problemi nel quadro geodinamico Mesozoico dell'area Mediterranean. *Mem. Soc. Geol. Ital.* **13,** 1–28.

DERCOURT, J. 1970. L'expansion océanique actuelle et fossile; ses implications géotectoniques. *Bull. Soc. géol. Fr.* **12,** 267–317.

——, FLAMENT, J. M., FLEURY, J. J. & MEILLEZ, F. 1973. Stratigraphie des couches situées sous les radiolarites de la zone de Pindos-Olonos (Grèce): Le Trias supérieur et la jurassique inférieur. *Ann. géol. Pays Hell.* **25,** 397–406.

——, DE WEVER, P. & FLEURY, J. J. 1976. Données sur la style tectonique de la nappe de Tripolitza septentrional (Grèce). *Bull. Soc. géol. Fr.* **18,** 317–326.

——, and others 1980. Les zones externes de l'édifice Hellénique. *In:* DERCOURT, J. *et al.* (eds). *Géologie des pays européens*. Bordas and 26th International Geological Congress, Paris, France (Dunod), 361–371.

DERYCKE, F. & GODFRIAUX, I. 1976. Métamorphismes 'schistes bleus et schistes verts' dans l'Ossa et le Bas-Olympe (Thessalie, Grèce). *Bull. Soc. géol. Fr.* **18,** 252.

—— & GODFRIAUX, I. 1977. A cross section in the Olympus area, Thessaly, Greece. *Proc. 6th Coll. Aegean Geol.* **1,** 353–354.

DEWEY, J. F., PITMAN, W. C. III, RYAN, W. B. F. & BONNIN, J. 1973. Plate tectonics and the evolution of the Alpine System. *Bull. geol. Soc. Am. 84,* 3137–3180.

—— & ŞENGÖR, A. M. C. 1979. Aegean and surrounding regions: complex multi-plate and continuum tectonics in a convergent zone. *Bull. geol. Soc. Am.* **90,** 82–91.

DJORDJEVIĆ, V., KARAMATA, S. & PAMIĆ, J. 1982. The Triassic rifting phase and development of oceanic basins in the Dinarides. *In;* DIXON, J. E. & ROBERTSON, A. H. F. (eds). *The Geological Evolution of the Eastern Mediterranean.* Abstracts, Edinburgh, 1982, p. 34.

DOMMERGUES, J. L. 1982. Le provincialisme des Ammonites nord-ouest européennes au Lias moyen. Une crise faunique sous controle paleogéographique. *Bull. Soc. géol. Fr.* **24,** 1047–1053.

DUMONT, J. F., GUTNIC, M., MARCOUX, J., MONOD, O. & POISSON, A. 1972. Le Trias des Taurides occidentales (Turquie). Définition du bassin pamphylien: Un nouveau domaine à ophiolites à la marge externe de la chaîne taurique. *Zeit. Deutsch. Geol. Ges.* **123,** 385–409.

DÜRR, ST., ALTHERR, R., KELLER, J., OKRUSCH, M. & SEIDEL, E. 1978. The median Aegean crystalline belt: stratigraphy, structure, metamorphism, magmatism. *In:* CLOSS, H., ROEDER, D. & SCHMIDT, K. (eds). *Alps, Apennines, Hellenides. Inter-Union Comm. Geodyn. Sci. Rep.* **38,** 455–477.

ENAY, R. 1976. Faunes anatoliennes (*Ammonitina*, Jurassique) et domaines biogéographiques nord et sud téthysienne. *Bull. Soc. géol. Fr.* **18,** 533–541.

—— & MANGOLD, C. 1982. Dynamique biogéographique et évolution des faunes d'Ammonites au jurassique, *Bull. Soc. géol. Fr.*, **24,** 1025–1047.

EVANS, I., HALL, S. A., CARMAN, M. F., SENALP, M. & COSKUN, S. 1982. A palaeomagnetic study of the Bilecik limestone (Jurassic), Northwestern Anatolia. *Earth planet. Sci. Lett.* **61,** 399–411.

FERRIÈRE, J. 1977. Le secteur méridional du 'massif métamorphique de Thessalie'. Le massif du Pélion et ses environs. *6th Coll. Aegean Geol., Athens*, **2,** 291–309.

—— & VERGELY, P. 1976. A propos des structures tectoniques et microtectoniques observées dans les nappes anté-Crétacé supérieur d'Othrys centrale (Grèce continentale): Conséquences. *C.R. Acad. Sci. Paris*, **283D,** 1003–1006.

FLEURY, J. J. 1976. Unité paleogéographique originale sous le front de la nappe du Pinde-Olonos: l'Unité du Megdovas (Grèce continentale). *C.R. Acad. Sci. Paris*, **282D,** 25–28.

—— 1980. Evolution d'une platforme et d'une bassin dans leur cadre alpin: les zones Gavrovo-Tripolitza et du Pinde-Olonos. *Soc. géol. du Nord, Spec. Publ.* **4,** 651 pp.

FOURQUIN, C. 1975. L'Anatolie du Nord-Ouest, marge méridionale du continent européen, historie paleogéographique tectonique et magmatique durant le Secondaire et Tertiaire. *Bull. Soc. géol. Fr.* **17,** 1058–1070.

——, WHITECHURCH, H. & JUTEAU, T. 1982. Intra-oceanic events in the Tauric ophiolites and magmatic activity in the Pontic active margin: correlations and implications. *In:* DIXON, J. E. & ROBERTSON, A. H. F. (eds). *The Geological Evolution of the Eastern Mediterranean.* Abstracts, Edinburgh, 1982, p. 42.

FREUND, R. & TARLING, D. H. 1978. Preliminary Mesozoic palaeomagnetic results from Israel and inference for a microplate structure in the Lebanon. *Tectonophysics*, **60,** 189–205.

GASS, I. G. 1980. The Troodos massif, Cyprus: its role in the unravelling of the ophiolite problem and its significance in the understanding of constructive margin processes. *In:* PANAYIOTOU, A. (ed.). *Ophiolites. Proc. Int. Ophiolite Symp., Nicosia, Cyprus, 1979*, 23–36.

GIESE, P., NICOLICH, R. & REUTTER, K. J. 1982. Explosion seismic crustal studies in the Alpine-Mediterranean region and their implications to tectonic processes. *In: Alpine-Mediterranean*

Geodynamics, Amer. Geophys. Un., Geodynamics Series, **7,** 39–75.

GLENNIE, K. W., BOEUF, M. G. A., HUGHES-CLARKE, M. W., MOODY-STUART, M., PILAAR, W. F. H. & REINHARDT, B. M. 1973. Late Cretaceous nappes in the Oman Mountains and their geologic evolution. *Bull. Am. Ass. Petrol. Geol.* **57,** 5–27.

GODFRIAUX, I. & PICHON, J. F. 1979. Sur l'importance des évènements tectoniques et métamorphiques d'age tertiaire en Thessalie septentrionale (Olympe, Ossa, Flambouron). *Ann. Soc. géol. Nord*, **29,** 367–376.

GÖKÇEN, S. L. & GÖKÇEN, N. S. 1982. Geological evolution of Alpine Turkish Thrace. *In:* DIXON, J. E. & ROBERTSON, A. H. F. (eds). *The Geological Evolution of the Eastern Mediterranean.* Abstracts, Edinburgh, 1982, p. 44.

GRACIANSKY, P. C. DE 1972. *Recherches géologiques dans le Taurus Lycien.* Thèse, Université Paris-Sud Orsay no. 896.

GREILING, R. & SKALA, W. 1977. The petrofabrics of the Phyllite-Quartzite Series of western Crete as an example for the pre-Neogene structures of the Cretan arc. *Proc. 6th Coll. Aegean Geol. Athens, 1977*, **1,** 97–102.

GREGOR, C. B. & ZIJDERVELD, J. D. A. 1964. Paleomagnetic and Alpine tectonics of Eurasia, Part 1. The magnetism of some Permian red sandstones from northwestern Turkey. *Tectonophysics*, **1,** 189–306.

——, MERTZMAN, S., NAIRN, A. E. M. & NEGENDANK, J. 1974. The palaeomagnetism of some Mesozoic and Cenozoic volcanic rocks from the Lebanon. *Tectonophysics*, **21,** 375–395.

GRUBIĆ, A. 1980. Yugoslavie. *In:* DERCOURT, J. *et al.* (eds). *Géologie des pays Européens.* Bordas and 26th International Geological Congress, Paris, France (Dunod), 291–342.

GUTNIC, M., MONOD, O., POISSON, A. & DUMONT, J.-F. 1979. Geologie des Taurides occidentales (Turquie). *Mem. Soc. géol. Fr., nouv. sér. 58*, 112 pp.

HALL, R. 1976. Ophiolite emplacement and evolution of the Taurus suture zone, SE Turkey. *Bull. geol. Soc. Am.* **87,** 1078–1088.

—— 1982. Ophiolites and passive continental margins. *Ofioliti*, **7,** 279–299.

HALLAM, A. 1980. A reassessment of the fit of Pangaea components and the time of their initial breakup. *In:* STRANGWAY, D. W. (ed.). *The Continental Crust and Its Mineral Deposits. Geol. Assoc. Can. Spec. Pap.* **20,** 375–387.

HENJES-KUNST, F. & KREUZER, H. 1982. Isotopic dating of pre-Alpidic rocks from the island of Ios (Cyclades, Greece). *Contrib. Mineral. Petrol.* **80,** 245–253.

HORVATH, F. & BERCKHEMER, H. 1982. Mediterranean back arc basins. *In: Alpine-Mediterranean Geodynamics. Am. Geophys. Un., Geodynamics Series*, **7,** 141–175.

HSÜ, K. J. 1977. Tectonic evolution of the Mediterranean basins. *In:* NAIRN, A. E. M., KANES, W. H. & STEHLI, F. G. (eds). *The Ocean Basins and Margins, 4A, The Eastern Mediterranean.* Plenum, New York, 29–75.

——, 1982. Editor's introduction: Alpine Mediterranean Geodynamics: Past, Present and Future. *In: Alpine-Mediterranean Geodynamics. Amer. Geophys. Un., Geodynamics Series*, **7,** 7–17.

HUSSIN, A. & TEE, A. T. 1982. Significance of sedimentary facies of the Pindos nappe in southern Crete. *In:* DIXON, J. E. & ROBERTSON, A. H. F. (eds). *The Geological Evolution of the Eastern Mediterranean*, Abstracts, Edinburgh, 1982, p. 53.

HYNES, A. 1974. Igneous activity at the birth of an ocean basin in Eastern Greece. *Can. J. Earth Sci.* **11,** 842–853.

INNOCENTI, F., MANETTI, P., MAZZUOLI, R., PASQUARÉ, G. & VILLARI, L. 1982. Anatolia and north-western Iran. *In:* THORPE, R. S. (ed.). *Andesites. Orogenic andesites and related rocks.* John Wiley & Sons, London & New York, 328–349.

IRVING, E. 1977. Drift of the major continental blocks since the Devonian. *Nature*, **270,** 307–309.

—— 1982. Fragmentation and assembly of the continents, Mid-Carboniferous to present. *Geophys. Surv.* **5,** 299–333.

—— & IRVING, G. A. 1982. Apparent polar wander paths Carboniferous through Cenozoic and the assembly of Gondwana. *Geophys. Surv.* **5,** 141–189.

JACOBSHAGEN, V. 1977. Structure and geotectonic evolution of the Hellenides. *6th Coll. Aegean Geol., Athens*, 1355–1367.

JENKYNS, H. C. & WINTERER, E. L. 1982. Paleooceanography of Mesozoic ribbon radiolarites. *Earth planet. Sci. Lett.* **60,** 351–375.

JUTEAU, T. 1975. Les ophiolites des Nappes d'Antalya (Taurides occidentales, Turquie). Pétrologie d'un fragment de l'ancienne croûte océanique téthysienne. *Sciences de la Terre, Nancy, Mém.* **32,** 692 pp.

JUTEAU, T. 1980. Ophiolites of Turkey. *Ofioliti, Spec. Issue, Tethyan Ophiolites*, **2,** 199–237.

KARIG, D. E. 1974. Evolutions of arc systems in the Western Pacific. *Ann. Rev. Earth Planet. Sci.* **2,** 51–75.

KATSKIKATSOS, G. 1977. La structure tectonique d'Attique et de l'ile d'Eubée. *6th Coll. Aegean Geol., Athens, 1977*, 211–228.

——, MERCIER, J.-L. & VERGÉLY, P. 1976. La fenêtre d'Attique-Cyclades et les fenêtres métamorphiques des Hellénides internes (Grèce). *C.r. Acad. Sci. Paris (Ser. D)*, **283,** 1613–1616.

KAUFFMANN, G. KOCKEL, F. & MOLLAT, H. 1976. Notes on the stratigraphic and palaeogeographic position of the Svoula Formation in the innermost zone of the Hellenides. *Bull. Soc. géol. Fr.* **18,** 225–230.

KELTS, K. 1981. A comparison of some aspects of sedimentation and translational tectonics from the Gulf of California and the Mesozoic Tethys, Northern Penninic margin. *Eclog. geol. Helvetiae*, **74,** 317–338.

KETIN, I. 1966. Tectonic units of Anatolia (Asia Minor). *Turk. Min. Res. Expl. Inst. Bull.* **66,** 23–34.

KLANG, A. & FOLKMAN, Y. 1982. Seismic interpretation of the Eastern Mediterranean off Israel. *In:*

DIXON, J. E. & ROBERTSON, A. H. F. (eds). *The Geological Evolution of the Eastern Mediterranean*, Abstracts, Edinburgh, 1982, p. 62.

KOCKEL, F., MOLLAT, H. & WALTHER, H. W. 1977. *Erlauterungen zur geologischen Karte de Chalkidhiki und angrenzender Gebiete 1:100,000 (Nord-Griechenland)*. Bundesanst. fur Geowiss. u. Rohstoffe, Hannover, 119 pp.

KREUZER, H., SEIDEL, E., KOEPKE, J. & ALTHERR, R. 1982. Ophiolites and associated crystalline rocks in the Aegean arc—a geodynamic puzzle. *In:* DIXON, J. E. & ROBERTSON, A. H. F. (eds). *The Geological Evolution of the Eastern Mediterranean*, Abstracts, Edinburgh, 1982, p. 64.

LAJ, C., JAMET, M., SOREL, D. & VALENTE, J. P. 1982. First paleomagnetic results from Mio-Pliocene series of the Hellenic sedimentary arc. *In:* LE PICHON, X., AUGUSTITHIS, S. S. & MASCLE, J. (eds). *Geodynamics of the Hellenic Arc and Trench. Tectonophysics*, **86,** 45–67.

LAPIERRE, H. 1975. Les formations sédimentaires et eruptives des nappes de Mamonia et leur relation avec le massif du Troodos (Chypre occidentale). *Mem. Soc. géol. Fr.* **123,** 322 pp.

LAUER, J.-P. 1981. *L'évolution géodynamique de la Turquie et de la Chypre deduite de l'étude paléomagnétique.* Unpubl. Doctorât-Es-Sciences thèse, University of Strasbourg, France, 292 pp.

LE PICHON, X. & ANGELIER, J. 1979. The Hellenic arc and trench system: a key to the neotectonic evolution of the Eastern Mediterranean area. *Tectonophysics*, **60,** 1–42.

LETOUZEY, J., BIJU-DUVAL, B., DORKEL, A., GONNARD, R., KRISTCHEV, K., MONTADERT, L. & SUNGURLU, O. 1977. The Black Sea: a marginal basin. Geophysical and geological data. *In;* BIJU-DUVAL, B. & MONTADERT, L. (eds). *Stuctural History of the Mediterranean Basins.* Editions Technip., Paris, 363–376.

LIVERMORE, R. A. & SMITH, A. G. 1984. Some boundary conditions for the evolution of the Mediterranean region. *In:* NATO A.R.I., *Erice, Sicily, Volume on Mediterranean tectonics.*

MAJER, V. & MASON, R. 1983. High-pressure metamorphism between the Pelagonian Massif and the Vardar Ophiolite Belt, Yugoslavia. *Mineral. Mag.* **47,** 139–142.

MARAKIS, G. 1970. Geochronology studies of some granites from Macedonia. *Ann. géol. Pays. Hell.* **21,** 121–152.

MARCOUX, M. J. 1974. Alpine type Triassic of the upper Antalya nappe (western Taurides, Turkey). *In:* ZAPFE, H. (ed.) *Die Stratigraphie der alpin-mediterranean Trias*, Wien, 145–146.

——, & RICOU, L.-E. 1979. Classification des ophiolites et radiolarites alpino-mediterranéenes d'après leur contexte paléogéographique et structural. Implications sur leur signification géodynamique. *Bull. Soc. géol. Fr.* **21,** 643–652.

MARCUCCI, M. & PASSERINI, P. 1981. Northern Apennine ophiolites, Hercynian and Alpine Orogenesis. *Ofioliti*, **6,** 255–261.

MCCALL, G. J. H. & KIDD, R. G. W. 1981. The Makran, southeastern Iran: the anatomy of a convergent plate margin active from Cretaceous to present. *In:* LEGGETT, J. K. (ed.). *Trench-Forearc Geology. Geol. Soc. Lond. Spec. Publ.* **10,** 387–401.

MCKENZIE, D. P. 1972. Active tectonics of the Mediterranean region. *Geophys. J.R. astron. Soc.* **30,** 109–185.

—— 1982. Active tectonics of the Alpine-Himalayan belt: the Aegean Sea and surrounding regions. *Geophys. J.R. astron. Soc.* **55,** 217–254.

MERCIER, J. 1966. Paléogéographie, orogenèse, métamorphisme et magmatisme des zones internes des Hellénides en Macédoine (Grèce): vue d'ensemble. *Bull. Soc. géol. France*, **8,** 1020–1049.

—— 1968. Etude géologique des zones internes des Hellénides en Macédoine centrale (Grece). *Ann. géol. Pays Hell.* **20,** 1–792.

—— 1976. La néotectonique—ses methods et ses buts. Un exemple: l'arc egéen (Mediterranée orientale). *Rev. Géog. phys. Géol. dyn.* **18,** 323–346.

MERCIER, J.-L., DELIBASSIS, N., GAUTHIER, A., JARRIGE, J., LEMEILLE, F., PHILIP, H., SEBRIER, M. & SOREL, D. 1979. La néotectonique de l'arc Egéen. *Rev. Géogr. phys. Géol. dyn.* **21,** 67–92.

MIYASHIRO, A. 1977. Subduction-zone ophiolites and island-arc ophiolites. *In;* SAXENA, X. & BHATTACHARJI, S. (eds). *Energetics of Geological Processes.* Springer-Verlag, Berlin, 188–213.

MONOD, O. 1977. *Recherches géologiques dans le Taurus occidental au Sud de Beyşehir (Turquie).* Thesis (unpubl.), Univ. Paris-Sud, Orsay, France, 511 pp.

MOUNDRAKIS, D. 1982. The Kastoria ophiolites on the western edge of the internal Hellenides (Greece). *Ofioliti*, **7,** 397–307.

NAYLOR, M. A. & HARLE, T. J. 1975. Palaeogeographical significance of rocks and structures beneath the Vourinos ophiolite, Northern Greece. *J. geol. Soc. Lond.* **132,** 667–677.

NICOLAS, A. & LE PICHON, X. 1980. Thrusting of young lithosphere in subduction zones with special reference to structures in ophiolitic peridotites. *Earth planet. Sci. Lett.* **46,** 397–406.

NOIRET, G., MONTIGNY, R. & ALLÈGRE, C. J. 1981. Is the Vourinos complex an island arc ophiolite? *Earth planet. Sci. Lett.* **56,** 375–386.

OGG, J. G., ROBERTSON, A. H. F. & JANSA, L. F. 1984. Jurassic sedimentation on Site 534 (Western North Atlantic) and of the Atlantic-Tethys seaway. Initial Reports of the Deep Sea Drilling Project for Leg 76.

OLIVET, J.-J. 1982. Cinématique des plaques et paléogéographie: une revue. *Bull. Soc. géol. Fr.* **24,** 875–892.

PAMIĆ, J. 1983. Considerations on the boundary between lherzolitic and harzburgite sub-provinces in the Dinarides and Northern Hellenides. *Ofioliti*, **8,** 153–164.

PAPAVASSILIOU, C. T. & SIDERIS, C. 1982. Geochemistry and mineralogy of Tertiary lavas of Sappai-Ferrai area (W. Thrace), Greece. Implications on their origin. *In:* DIXON, J. E. & ROBERTSON, A. H. F. (eds). *The Geological Evolution of the Eastern Mediterranean*, Abstracts, Edinburgh, 1982, p. 85.

PAPANIKOLAOU, D. J. 1980. Contribution to the geology of the Aegean Sea: The island of Paros. *Ann. géol. Pays Hell.* **30,** 65–96.

PARROT, J.-F. 1977. Assemblage ophiolitique du Baër-Bassit et termes éffusifs du volcano-sédimentaire. Pétrologie d'un fragment de la croûte océanique téthysienne, charriée sur la plateforme syrienne. *Trav. et Doc., Série géol., O.R.S.T.O.M.*, **6,** 97–126.

PASSERINI, P. 1981. What are ophiolites? Speculations on Tethyan ophiolites in a convective earth. *Ofioliti*, **6,** 261–269.

PEARCE, J. A. 1975. Basalt geochemistry used to investigate past tectonic environments on Cyprus. *Tectonophysics*, **25,** 41–75.

—— 1980. Geochemical evidence for the genesis and eruptive setting of lavas from Tethyan ophiolites. *In:* PANAYIOTOU, A. (ed.) *Ophiolites.* Proc. Int. Ophiolite Symp., Nicosia, Cyprus, 261–272.

——, ALABASTER, T., SHELTON, A. W. & SEARLE, M. P. 1981. The Oman ophiolite as a Cretaceous arc-basin complex: evidence and implications. *Phil. Trans. R. Soc. Lond.* **300,** 299–317.

——, LIPPARD, S. J. & ROBERTS, S. 1984. Characteristics and tectonic significance of supra-subduction zone ophiolites. *Spec. Publ. geol. Soc. Lond.*

PEINADO, J., VAN DER VOO, R. & SCOTESE, C. 1982. A re-evaluation of Pangea reconstructions. *EOS*, **63,** 307.

PÉLISSIÉ, T., PEYBERNÈS, B. & REY, J. 1982. Tectonique des plaques et paléobiogéographie des grands Foraminifères benthiques et des Algues calcaires du Dogger a l'Albien sur le pourtour de la Mesogée. *Bull. Soc. géol. Fr.* **24,** 1069–1077.

PE-PIPER, G. 1982. Geochemistry, tectonic setting and metamorphism of mid-Triassic volcanic rocks of Greece. *Tectonophysics*, **85,** 253–272.

PIPER, D. W. J. & PE-PIPER, G. 1980. Was there a western (external) souce of terrigenous sediment for the Pindos zone of the Peloponnese (Greece)? *N. Jb. Geol. Paläont., Monats.*, **2,** 107–115.

—— & PE-PIPER, G. 1982. Depositional setting of Pindos zone rocks of the Peloponnese, Greece. *In:* DIXON, J. E. & ROBERTSON, A. H. F. (eds). *The Geological Evolution of the Eastern Mediterranean* Abstracts, Edinburgh, p. 87.

PISA, G., CASTELLARIN, A., LACCHINI, F., ROSSI, P. L., SIMBOLI, G. BOSELLINI, A. & SOMMAVILLA, E. 1979. Middle Triassic magmatism in the Southern Alps, I: a review of general data in the dolomites. *Riv. ital. Paleont. Stratigr.* **85,** 1093–1110.

PITMAN, W. C. & TALWANI, M. 1972. Sea floor spreading in the North Atlantic. *Bull. geol. Soc. Am.* **83,** 619–646.

POISSON, A. 1977. *Recherches géologiques dans les Taurides occidentales, Turquie.* Thesis (unpubl.), Univ. Paris-Sud, Orsay, France, 790 pp.

——, AKAY, E., CRAVETTE, J., MULLER, C. & UYSALL, S. 1983. Données nouvelles sur la chronologie de mise en place des nappes d'Antalya au centre de l'angle d'Isparta (Taurides occidentales, Turquie), *C.r. Acad. Sci. Paris*, **296,** 923–926.

PRICE, I. 1976. Carbonate sedimentology in a pre-Upper Cretaceous continental margin sequence, Othris, Greece. *Bull. Soc. géol. Fr.* **7,** 18: 273–279.

RAMPNOUX, J.-P. 1970. Regards sur les Dinarides internes yougoslaves (Serbie-Monténégro oriental): stratigraphie, évolution paléogéographique, magmatisme. *Bull. Soc. géol. Fr.* **12,** 948–966.

REINECKE, T., ALTHERR, R., HARTUNG, B., HATZIPANAGIORGIOU, K., KREUZER, H., HARRE, W., KLEIN, H., KELLER, J., GEENEN, E. & BÖGER, H. 1982. Remnants of a Late Cretaceous high temperature belt on the island of Anafi (Cyclades, Greece). *N. Jb. Mineral., Abh.* **145,** 157–182.

RICOU, L.-E. 1965. *Contribution a l'étude géologique de la bordure sud-ouest du Massif Serbo-Macédonien aux environs de Salonique.* Thèse, 3ème cycle, Univ. de Paris, 121 pp.

—— 1971. Le croissant ophiolitique peri-arabe: une ceinture de nappes mises en place au Cretacé supérieur. *Rev. Géogr. phys. Géol. dynam.* **13,** 327–349.

——, MARCOUX, J. & POISSON, A. 1979. L'allochtonie des Bey Dağları orientaux. Reconstruction palinspastique des Taurides occidentales. *Bull. Soc. géol. Fr.* **21,** 125–133.

——, & MARCOUX, J. 1980. Organisation générale et rôle structural des radiolarites et ophiolites du système alpino-méditerranéen. *Bull. Soc. géol. Fr.* **22,** 1–14.

RIDLEY, J. R. 1982. *Tectonic style, strain history, and fabric development in a blueschist terrain, Syros, Greece.* Ph.D. thesis, unpubl., Edinburgh University, 283 pp.

RIECKER, R. E. 1979. *Rio Grande Rift: tectonics and magmatism.* Amer. Geophys. Union, Washington D.C., 438 pp.

ROBERTSON, A. H. F. 1978. Tertiary uplift history of the Troodos Massif, Cyprus. *Bull. geol. Soc. Am.* **88,** 1763–1772.

—— 1981. Metallogenesis on Mesozoic passive continental margin, Antalya Complex, SW Turkey. *Earth planet. Sci. Lett.* **54,** 323–345.

—— & HUDSON, J. D. 1974. Pelagic sediments in the Cretaceous and Tertiary history of the Troodos Massif, Cyprus. *In:* HSÜ, K. J. & JENKYNS, H. C. (eds). *Pelagic sediments: On land and under the sea. Spec. Publ. Internat. Assoc., Sedimentol.* **1,** 403–436.

—— & WOODCOCK, N. H. 1980. Tectonic setting of the Troodos Massif in the East Mediterranean. *In:* PANAYIOTOU, A. (ed.) *Ophiolites. Proc. Int. Ophiolite Symp., Nicosia, Cyprus, 1979*, 36–49.

—— & WOODCOCK, N. H. 1981. Gödene Zone, Antalya Complex, SE Turkey: volcanism and sedimentation of Mesozoic marginal ocean crust. *Geol. Rdsch.* **70,** 1177–1214.

—— & WOODCOCK, N. H. 1982. Sedimentary history of the south-western segment of the Mesozoic-Tertiary Antalya continental margin, south-western Turkey. *Eclogae geol. Helv.* **75,** 517–562.

ROESLER, G. 1978. Relics of non-metamorphic sediments on Central Aegean islands. *In:* CLOSS, H., ROEDER, D. & SCHMIDT, K. (eds). *Alps, Apennines, Hellenides. Inter-Union Comm. Geodyn. Sci. Rep., No. 38*, 480–482.

ROSS, J. V. & ZIMMERMAN, J. 1982. The Pindos Ophiolite Complex, Northern Greece: evolution and tectonic significance of the tectonic suite. *In:* DIXON, J. E. & ROBERTSON, A. H. F. (eds). *The Geological Evolution of the Eastern Mediterranean*, Abstracts, Edinburgh, 1982, p. 95.

RYAN, B. F., KASTENS, K. A. & CITA, M. B. 1982. Geological evidence concerning compressional tectonics in the Eastern Mediterranean. *Tectonophysics*, **86**, 213–243.

SAUNDERS, A. D., TARNEY, J., MARSH, N. G. & WOOD, D. A. 1980. Ophiolites as ocean crust or marginal basin crust: a geochemical approach. *In:* PANAYIOTOU, A. (ed.). *Ophiolites, Proc. Intl. Ophiolite Symp., Nicosia, Cyprus, 1979*, 193–204.

SCHROEDER, R., CHERCHI, A., GUELLAL, S. & VILA, J. M. 1974. Biozonation par les grands Foraminifères du Jurassique supérieur et du Cretacé inférieur et moyen des séries néritiques en Algérie NE. Considérations paléobiogéographiques. *6th Coll. Afr. Micropaleontol., Tunis*, 1–8.

SCHMITT, A. 1983. *Nouvelle contributions a l'étude géologique des Pieria, de l'Olympe et de l'Ossa (Grèce du Nord)*. Thèse, Fac. Polytech. de Mons (Belgique) (2 vols.), 385 pp.

SEARLE, M. P. & MALPAS, J. 1982. Petrochemistry of sub-ophiolite metamorphic and related rocks in the Oman Mountains. *J. geol. Soc. Lond.* **139,** 235–249.

SEIDEL, E., OKRUSCH, M., KREUZER, H., RASCHKA, H. & HARRE, W. 1981. Eoalpine metamorphism in the uppermost unit of the Cretan nappe system petrology and geochronology. *Contrib. Mineral. Petrol.* **72,** 351–361.

——, KREUZER, H. & HARRE, W. 1982. A Late Oligocene/Early Miocene high pressure belt in the external Hellenides. *Geol. Jb.* **E23,** 165–206.

ŞENGÖR, A. M. C., YILMAZ, Y. & KETIN, İ. 1980. Remnants of a pre-late Jurassic ocean in northern Turkey: fragments of Permian-Triassic Palaeo-Tethys? *Bull. geol. Soc. Am.* **91,** 599–609.

——, —— & —— 1982. Reply. *Bull. geol. Soc. Am.*, **93,** 932–936.

—— & MONOD, O. 1980. Océans sialique et collisions continentales. *C.r. Acad. Sci., Paris*, **290,** 1459–1462.

—— & YILMAZ, Y. 1981. Tethyan evolution of Turkey: a plate tectonic approach. *Tectonophysics*, **75,** 181–241.

—— & CANITEZ, N. 1982. The North Anatolian Fault. *In: Alpine-Mediterranean Geodynamics, Am. Geophys. Un., Geodynamics Series*, **7,** 205–216.

SERVAIS, M. 1982. *Collision et suture téthysienne en Anatolie centrale: Etude structurale et métamorphique (HP-BT) de la zone Nord Kutahya.* Unpubl. Thèse, 3ème cycle, Univ. Paris-Sud, Orsay, France, 349 pp.

SERVAIS, M., KIENAST, J.-R. & MONOD, O. 1982. High pressure belt resulting from continental collision in NW Anatolia. *In:* DIXON, J. E. & ROBERTSON, A. H. F. (eds). *The Geological Evolution of the Eastern Mediterranean*, Abstracts, Edinburgh, 1982, p. 100.

SHELTON, A. W. & GASS, I. G. 1980. Rotation of the Cyprus microplate. *In:* PANAYIOTOU, A. (ed.). *Ophiolites. Proc. Int. Symp. Ophiolites, Nicosia, Cyprus, 1979*, 61–66.

SHERIDAN, R. E. and others 1982. Early history of the Atlantic Ocean and gas hydrates on the Blake Outer Ridge: Results of the Deep Sea Drilling Project, Leg 76. *Bull. geol. Soc. Am.* **93,** 876–885.

SMEWING, J. D., SIMONIAN, K. O. & GASS, I. G. 1975. Metabasalts from the Troodos Massif, Cyprus: genetic implication deduced from petrology and trace element composition. *Contr. Miner. Petr.* **51,** 49–64.

—— & POTTS, P. J. 1976. Rare-earth abundance in basalts and metabasites from the Troodos Massif, Cyprus. *Contr. Mineral. Petrol.* **57,** 245–258.

SMITH, A. G. 1971. Alpine deformation and the oceanic areas of the Tethys, Mediterranean and Atlantic. *Bull. geol. Soc. Am.* **82,** 2039–2070.

—— 1977. Othris, Pindos and Vourinos ophiolites and the Pelagonian Zone. *Proc. 6th Coll. Aegean Geol., Athens, 1977*, 1369–1374.

—— & BRIDEN, J. C. 1977. *Mesozoic-Cenozoic Palaeocontinental Maps.* Cambridge University Press, 63 pp.

—— & MOORES, E. M. 1974. The Hellenides. *In:* SPENCER, A. M. (ed.). *Mesozoic and Cenozoic orogenic belts. Geol. Soc. Lond. Spec. Publ.*, **4,** 159–186.

—— & WOODCOCK, N. H. 1976a. Emplacement model for some 'Tethyan' ophiolites. *Geology*, **4,** 653–656.

—— & —— 1976b. The earliest Mesozoic structures in the Othris region, Eastern Central Greece. *Bull. Soc. géol. Fr.* **18,** 245–251.

——, —— & NAYLOR, M. A. 1979. The structural evolution of a Mesozoic continental margin, Othris mountains, Greece. *J. geol. Soc. Lond.* **136,** 589–603.

——, HURLEY, A. M. & BRIDEN, J. C. 1981. *Phanerozoic Paleocontinental Maps.* Cambridge University Press, Cambridge, 102 pp.

——, HYNES, A. J., MENZIES, M., NISBET, E. G., PRICE, I., WELLAND, M. J. P. & FERRIÈRE, J. 1975. The stratigraphy of the Othris mountains, eastern central Greece: a deformed Mesozoic continental margin sequence. *Eclog. geol. Helv.* **68,** 463–481.

—— & WOODCOCK, N. H. 1982. Tectonic syntheses of the Alpine-Mediterranean region: A review. *In: Alpine-Mediterranean Geodynamics. Am. Geophys. Un., Geodynamics Series*, **7,** 15–39.

SPRAY, J. G. & RODDICK, J. C. 1981. Evidence for Upper Cretaceous transform metamorphism in West Cyprus. *Earth planet. Sci. Lett.* **55,** 273–291.

STEINER, M. B. 1983. Mesozoic apparent polar wander and plate motion of North America. *In:* REYNOLDS, N. W. & DOLLY, E. D. (eds). *Mesozoic paleogeography of West-Central United States. Soc. Econ. Paleont. Mineral. Spec. Publ.*,

Rocky Mountain Sect., Denver, Colorado, 1983, 1–11.

Stride, A. H., Belderson, R. H. & Kenyon, N. H. 1977. Evolving miogeanticlines of the East Mediterranean (Hellenic, Calabrian and Cyprus Outer Ridges) *Phil. Trans. R. Soc. Lond.* **284,** 255–285.

Swarbrick, R. E. 1980. The Mamonia Complex of SW Cyprus: A Mesozoic continental margin and its relationship with the Troodos Complex. *In:* Panayiotou, A. (ed.) *Ophiolites. Proc. Int. Oph. Symp. Cyprus, 1979*, 86–92.

Tapponnier, P. 1977. Evolution tectonique du Système Alpin en Méditerranée: poinçonnement et écrasement rigide-plastique. *Bull. Soc. géol. Fr.* **19,** 437–460.

Tekeli, O. 1981. Subduction complex of pre-Jurassic age, northern Anatolia, Turkey. *Geology*, **9,** 68–72.

Thiebault, F. 1977. Stratigraphie de la série des calcschistes et marbres ('plattenkalk') en fenêtre dans les massifs du Taygète et du Parnon (Peloponnèse-Grèce). *Proc. 6th Coll. Aegean Geol. Athens 1977*, **2,** 691–701.

—— 1981. Modèle d'évolution géodynamique d'une portion des Hellénides externes (Peloponnèse méridional, Grèce) de l'Eocène à l'époque actuelle. *C.R. Acad. Sci. Paris*, **292,** 1491–1496.

Thierry, J. 1982. Tethys, Mesogée et Atlantique au Jurassique: quelques reflexions basées sur les faunes d'Ammonites. *Bull. Soc. géol. Fr.* **24,** 1053–1069.

Thuizat, R., Whitechurch, H., Montigny, R. & Juteau, T. 1981. K-Ar dating of some infra-ophiolitic metamorphic soles from the Eastern Mediterranean: new evidence for oceanic thrustings before obduction. *Earth planet. Sci. Lett.* **52,** 302–310.

Truswell, E. M. 1981. Pre Cenozoic palynology and continental movements. *In:* McElhinny, M. W. & Valencio, D. A. (eds). *Paleoreconstruction of the Continents. Am. Geophys. Un. Geodynamics Series,* **2,** 13–27.

Tucholke, B. E. & Ludwig, W. J. 1982. Structure and origin of the J-anomaly ridge, western North Atlantic Ocean. *J. geophys. Res.* **87,** 9389–9407.

Ünalan, G., Yüksel., V., Tekeli, T., Gönenc, O., Seyirt, Z. & Hüseyin, S. 1976. Haymana-Polatlı yöresinin (güney batı Ankara) Üst Kretase-Alt Tersiyer stratigrafisi ve paleocografik evrimi. *Türk. Jeol. Kur. Bülteni*, **19,** 159–176.

Upton, B. G. J., Emeleus, C. H. & Hald, N. 1980. Tertiary volcanism in northern E. Greenland: Gauss Halvø and Hold with Hope. *J. geol. Soc. Lond.* **137,** 491–508.

Van den Berg, J. & Zijderveld, J. D. A. 1982. Paleomagnetism in the Mediterranean. *In: Alpine-Mediterranean Geodynamics. Am. Geophys. Un. Geodynamics Series*, **7,** 83–113.

Van der Maar, P. 1980. The geology and petrology of Ios, Cyclades, Greece. *Ann. géol. Pays Hell.* **30,** 206–224.

Van der Voo, R. 1968. Paleomagnetism and the Alpine tectonics of Eurasia, Part 4, Jurassic, Cretaceous and Eocene pole positions from NE Turkey. *Tectonophysics*, **6,** 251–269.

—— & French, R. B. 1974. Apparent polar wandering for the Atlantic bordering continents: Late Carboniferous to Eocene. *Earth Sci. Rev.* **10,** 99–119.

Vergely, P. 1976. Chevauchement vers l'ouest et rétrocharriage vers l'est des ophiolites: deux phases tectoniques au cours du Jurassique supérieure—Eocretacé dans les Hellénides Internes. *Bull. soc. géol. Fr.* **18,** 231–244.

Vergely, P. 1977. Ophiolites et phases tectoniques superposées dans les Hellénides. *6th Coll. Aegean Geol., Athens, 1977*, 1293–1302.

Vörös, A. 1977. Provinciality of the Mediterranean Lower Jurassic brachiopod fauna: causes and plate-tectonic implications. *Palaeogeogr., Palaeoclimatol., Palaeoecol.*, **21,** 1–16.

Wagner, R. H. 1962. On a mixed Cathaysia and Gondwana flora from SE Anatolia (Turkey). *C.R. 4ème Congrés pour l'avancement des études stratigraphie et de géologie du Carbonifère, Heerlen, 1958*, **3,** 745–752.

Wallbrecher, E. 1977. Nappe units of the Southern Pelion Peninsula and their origins. *Proc. 6th Coll. Aegean Geol. Athens, 1977*, 281–290.

Waldron, J. W. H. 1982. *Mesozoic sedimentary and tectonic evolution of the northeast Antalya Complex, Eğrıdır, SW Turkey.* Unpubl. Ph.D. thesis, University of Edinburgh, 239 pp.

Wegener, A. 1924. *The Origin of Continents and Oceans.* Methuen, London, 212 pp. (English translation of 3rd German edition of 1922).

Wensink, H. 1981. Pre-Cenozoic palaeomagnetism Europe and the Middle East. *In:* McElhinny, M. W. & Valencio, D. A. (eds) *Paleoreconstruction of the Continents. Am. Geophys. Un. Geodynamics Ser.* **2,** 151–159.

Wezel, F. C. 1981. The structure of the Calabri-Sicilian arc: result of a post-orogenic intra-plate deformation. *In:* Leggett, J. K. (ed.). *Trench-Forearc Geology. Geol. Soc. Lond. Spec. Publ.* **10,** 345–347.

Woodcock, N. H. & Robertson, A. H. F. 1977. Origins of some ophiolite-related metamorphic rocks of the 'Tethyan' belt. *Geology*, **5,** 373–376.

—— & —— 1982. Wrench and thrust tectonics along a Mesozoic-Cenozoic continental margin: Antalya Complex, SW Turkey. *J. geol. Soc. Lond.* **139,** 147–163.

Yarwood, G. & Dixon, J. E. 1977. Lower Cretaceous and younger thrusting in the Pelagonian rocks of the high Pieria, Greece. *Proc. 6th Coll. Aegean Geol. Athens, 1977*, 269–280.

Ziegler, A. M. 1981. Palaeozoic palaeogeography. *In:* McElhinny, M. W. & Velencio, D. A. (eds) *Paleoreconstruction of the continents. Am. Geophys. Un. Geodynamics, Ser.* **2,** 31–39.

1. PALAEOTETHYS

Among the fundamental, 'boundary condition' questions in East Mediterranean geology are: How large was Palaeotethys? How many strands did it have and where were they at the beginning of the Mesozoic? Ophiolites of the right age are not known from the south Tethyan area so the natural tendency has been to seek evidence for the fate of Palaeotethys in the north. As we discussed in the introductory chapter, if Palaeotethys was large but was destroyed completely, early in the Mesozoic, then a Neotethyan ocean would have to have grown rapidly to balance its destruction. A different Early Triassic fit of Africa to Eurasia, with a much narrower Palaeotethyan embayment could diminish the scale of both Palaeotethyan destruction and balancing Neotethyan construction until major re-organization ensued, once Atlantic opening began.

The first two contributors to this section (Fig. 1), **ŞENGÖR** ***et al.*** and **MONOD** attempt to trace the fate of Palaeotethys in Turkey by looking for evidence of early Mesozoic subduction or crustal shortening attributable to collision. Late Triassic–Early Jurassic shortening events, so goes the logic, must relate to the destruction of an ocean which existed before the Triassic. **MONOD** working in the Taurides, south of the Palaeotethyan sutures identified by **ŞENGÖR** ***et al.*** collects together evidence of regional Upper Triassic–Lower Jurassic hiatuses terminated by distinctive red, polymict subaerial clastics which occur from Chios to Eastern Anatolia. In addition thrusts involving sediments in the Beyşehir area are sealed by early Middle Jurassic shallow water limestones. Whether these uplift, relief-generating and thrusting events must be the consequence of collision or could be related to continuing subduction at a distinct trench is perhaps an open question.

ŞENGÖR ***et al.***'s contribution is much more specific. They champion the case here, following on from earlier papers, for a major northern Palaeotethyan ocean basin which closed, virtually completely, by southward subduction before Middle Jurassic times. The essence of their concept is that a continental sliver, the 'Cimmerian continent', comprising a substantial part of Turkey and part of Iran, was rifted off Gondwanaland in the Late Triassic–Early Jurassic and moved northwards, eliminating Palaeotethys north of it as it went, and opening an early Neotethyan ocean behind it to the south. **ŞENGÖR** ***et al.*** describe the fate of the 'main trunk' of Palaeotethys in four segments from the Balkans to the Caucasus. The segment in northern Turkey contains a subsidiary Palaeotethyan suture representing a minor southern 'trunk' to Palaeotethys created as a marginal basin behind the subduction zone, before the Cimmerian continent was detached. **ŞENGÖR** ***et al.*** go on to chart the fragmentation of the Cimmerian continent by Early Jurassic rifting as Palaeotethys was being eliminated. They discuss the evolution of the Eastern Mediterranean sector of Palaeotethys in relation to parallel events further east and, in its final stages, to the opening of the Atlantic in the west.

KEREY focuses on the sedimentology of one part of **ŞENGÖR** ***et al.***'s Cimmerian continent. Along the western Black Sea coast Visean platform carbonates are overlain unconformably by 3500 m of Upper Carboniferous clastics and then by further Permian clastics. The Upper Carboniferous sediments were deposited in a vast deltaic complex and derived from the north (from a region now rifted or strike-slipped away to form the Black Sea) under the varying influences of faulting, sea-level fluctuations and climatic changes. **KEREY** stresses the close match between his Upper Carboniferous successions and those of Laurasia, which poses an unresolved problem since in the **ŞENGÖR** ***et al.*** model **KEREY**'s area was then part of Gondwana.

Also relevant to the configuration of the

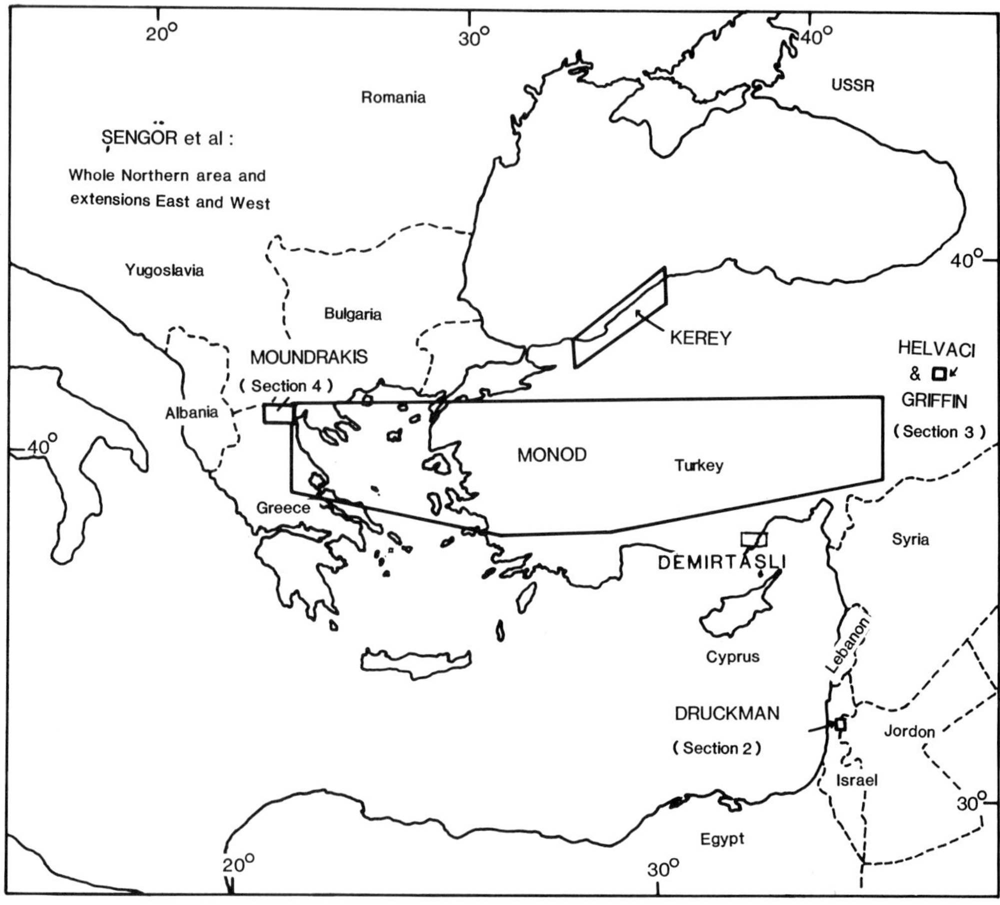

FIG. 1. The Palaeotethyan areas discussed by different authors in this section.

Palaeotethys, **DEMIRTAŞLI** outlines the structure and stratigraphy of an important little-studied area along the S.E. Mediterranean Turkish coast. Four zones are recognized each with contrasting stratigraphical development from the Early Palaeozoic onwards. Important deformation apparently took place in pre-Late Permian, Triassic and Late Cretaceous to Early Tertiary times. Strong evidence is produced for derivation of the northerly (internal) zones from the north, although the driving mechanism of the earlier deformations remains obscure.

Aspects of Palaeotethys are touched on in other papers later in the book. **DRUCKMAN** in section 2 describes the Palaeozoic succession encountered in the Helez 1A deep borehole under the coastal plain of Israel. **HELVACI & GRIFFIN** have identified Lower and Upper Palaeozoic components in the Avnik area of the Bitlis Massif in Eastern Anatolia by isotopic dating techniques. Similar evidence for Carboniferous and older granites is included in **MOUDRAKIS**' account of part of the northern Pelagonian zone in Greece.

Tectonics of the Mediterranean Cimmerides: nature and evolution of the western termination of Palaeo-Tethys

A. M. C. Şengör, Y. Yılmaz & O. Sungurlu

SUMMARY: The Cimmeride orogenic system was produced during the closure of Palaeo-Tethys, the original, triangular embayment of the Permo-Triassic Pangaea, and its immediate dependencies, and includes a multi-strand suture network that extends from the eastern Carpathians to the Pacific shores of Asia. Its westernmost segment, the Mediterranean Cimmerides, extends from the eastern Carpathians and the North Dobrudja to the eastern end of the Greater Caucasus thereby enclosing the Black Sea. It may be subdivided into three longitudinal sections, *viz.* the Balkan/Carpathian Cimmerides, the Anatolian Cimmerides (including the North Dobrudja), and the Caucasian Cimmerides. The Balkan/Carpathian Cimmerides reach from the eastern Dacides to the island of Thásos and are characterized by a diachronous Lias-Dogger (? locally later) collision that developed from north to south along a semi-cryptic suture. Owing to extreme later overprinting, neither the original orogenic polarity nor the fingerprints of the Palaeo-Tethyan suture are everywhere recognizable in them. The Anatolian Cimmerides stretch between Thásos and the Artvin region and are represented by a 'double-orogen'. The main trunk in the north consists of five main north-vergent allochthons resting on the southern periphery of the Moesian platform and probably on those parts of the Scythian platform now separated from it by the later opening of the Black Sea. The lowermost allochthons represent remnants of Palaeo-Tethyan oceanic material with ages from possibly Devonian to the earliest Dogger, whereas the higher ones are pieces of the former Cimmerian continent. All of these allochthons were emplaced during the latest Lias-earliest Dogger. The subordinate S trunk of the Anatolian Cimmerides is also north-vergent and is the product of the Late Triassic collapse of the Karakaya marginal basin of Palaeo-Tethys, the effects of which cover most of Turkey south of the north trunk. The Caucasian Cimmerides are south-vergent and consist almost entirely of the Greater Caucasus. Palaeo-Tethyan closure along the Caucasian Cimmerides took place during the ?latest Triassic. The northern foreland of the Mediterranean Cimmerides was the site of compressional deformation during the Late Triassic and Early Jurassic, whereas to the south, the medial Triassic opening of the eastern Mediterranean and the Liassic rifting of the northern branch of Neo-Tethys indicate the presence of dominantly extensional tectonics that in part may have been related to back-arc activity. Around the African promontory, these Palaeo-Tethys-related rifting events interfered with one another, and with the Atlantic-opening-related rifting phases to generate a hitherto problematic, complex, protracted, and polyphase rifting history, so characteristic of the central Mediterranean Tethyan orogens. The most outstanding problems of the Mediterranean Cimmerides are the structural/palaeogeographic relations between their constituent tectonic units and the precise delineation of the areal extent of the Cimmeride orogen.

The large, westward-narrowing, triangular embayment of the Permo-Triassic Pangaea, which extended from the E as an enormous gulf (Wilson 1963, Bullard *et al.*, 1965, Ziegler *et al.*, 1979, Smith *et al.*, 1981), has been widely presumed to have been destroyed as a result of subduction and collision during post-Triassic times (Smith 1971, Dewey *et al.*, 1973, Biju-Duval *et al.*, 1977, Hsü 1977, *Ad Hoc* Panel 1979). However, the details of this collision with Laurasia and demise of this oceanic embayment, now commonly referred to as Palaeo-Tethys (after Stöcklin's 1974 usage: Laubscher & Bernoulli 1977, Şengör 1979), have been much less clear than the history of the Tethyan oceans that opened mostly during and after the Triassic by the disruption of the northern margin of the Palaeozoic Gondwana-Land (e.g., Smith 1973), and that are now collectively called Neo-Tethys (Laubscher & Bernoulli 1977, Şengör 1979). Recently, Şengör (1979, 1981) and Şengör *et al.* (1980) have argued that the disappearance of Palaeo-Tethys and the coeval generation of Neo-Tethys were consequences of the rifting, from north Gondwana-Land, of a narrow continental strip or archipelago, called the Cimmerian Continent by Şengör (1979), that swept across the original Pangaean embayment by rotating anticlockwise about a pole located somewhere in the present Carpathian region during the Early Triassic to medial Jurassic interval (Fig. 1). The large,

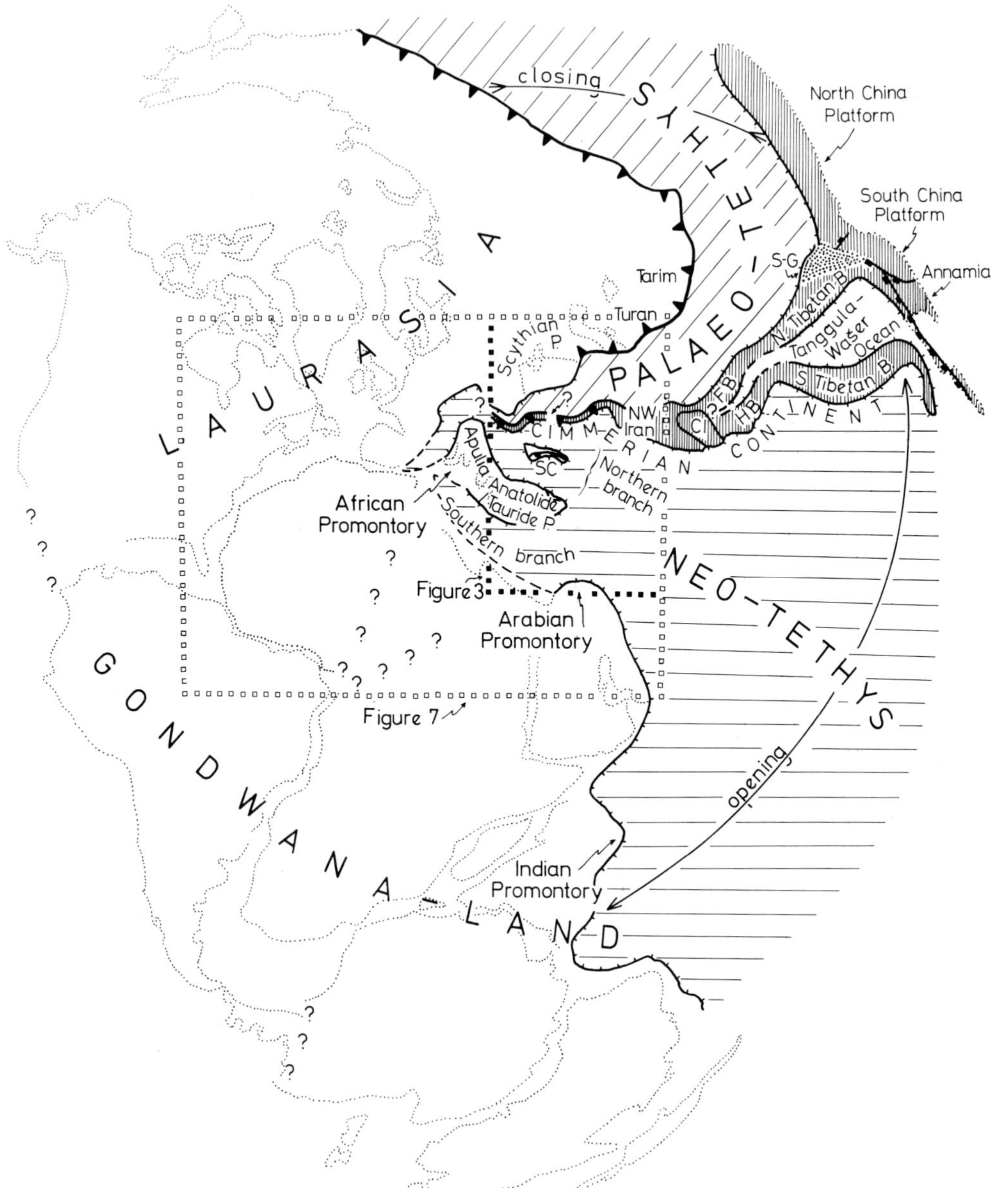

FIG. 1. Liassic reconstruction of Pangaea showing the locations of and structural relations between Laurasia, Palaeo-Tethys, Cimmerian continent, Neo-Tethys and Gondwana-Land. Intra-continental structures within Pangaea are not shown. Base map is from Smith *et al.* (1981) with modifications according to De Wit (1977) and Pindell & Dewey (1982). Except for showing the possible geometry of the Tanggula-Waser ocean very schematically, the shape of the Cimmerian continent has not been subjected to any palinspastic correction, which may be the source of several inconsistencies in this reconstruction, such as the open ocean N of Iran. Area surrounded by black squares represents regions depicted in Figs 2 & 3, whereas that surrounded by white squares represents those shown in Fig. 7.

SC, Sakarya continent; FB, Farah Block; HB, Helmand Block.

multi-branched, collisional orogen that is the product of the closure of Palaeo-Tethys and that now extends from the eastern Carpathians and the North Dobrudja to the Asiatic shores of the Pacific ocean is called the *Cimmerides* to contrast them with the products of the demise of the Neo-Tethyan oceans, termed the *Alpides*.

The Cimmerides are divided into four major longitudinal segments on the basis of tectonic style and history of development (Şengör in press). These are, from west to east, the Mediterranean Cimmerides, the Southwest Asian or Ghaznian Cimmerides, the Chinese or Sino-Cimmerides, and the Southeast Asian or Indochinese Cimmerides. Of these, the Mediterranean Cimmerides, the subject of this paper, are the least understood.

For a number of reasons, the Mediterranean Cimmerides may be regarded as the most important of the four subdivisions with respect to the tectonic evolution of Palaeo-Tethys in its Pangaean framework. First, they include regions that once formed the western terminus of the original Permo-Triassic embayment of Pangaea. Their geology must therefore contain evidence regarding the nature of that termination. Secondly, and related closely to the first point, the solution to the long-standing problem of the relations of the Hercynides to Palaeo-Tethys (Dewey & Burke 1973) probably resides in the geology of the Mediterranean Cimmerides, because they are constructed largely on a Hercynian basement. Thirdly, the development of the circum-Mediterranean Alpides has long been viewed as a by-product of the multi-phase opening of the Atlantic ocean (Smith 1971, Dewey *et al.*, 1973, Biju-Duval *et al.*, 1977, Tapponnier 1977, Channel *et al.*, 1979, Burchfiel 1980). However, the opening of the Atlantic is of little help in interpreting the pre-Jurassic evolution of the Mediterranean area in general. It seems that we should look to Palaeo-Tethyan evolution, as recorded in the geology of the Mediterranean Cimmerides, for explanations of pre-Jurassic and some Jurassic-Cretaceous phenomena in the western Tethyan chains. The key to the understanding of the marked dissimilarity, particularly during the early stages of their history, between the western and eastern parts of the circum-Mediterranean Alpides appears to lie in the appreciation of the significance of Palaeo-Tethyan evolution. Fourthly, the area of complex block (*scholle*) tectonics in central Europe (Lotze 1974) is located between the North Sea graben complexes and the pole of rotation of the Cimmerian continent (Şengör 1979, Ziegler 1982). Whether and to what extent the rotation of the Cimmerian continent affected the early Mesozoic tectonics of central Europe is an important question from the viewpoint of the relations between the effects of Atlantic and Tethyan tectonics in central and northwestern Europe (cf. Ziegler 1982). Finally, from a more general standpoint, the Mediterranean Cimmerides provide a fossil example of orogenic evolution near the pole of closure of an ocean. Such a geometry poses interesting tectonic and palaeogeographic problems that have been among the chief factors contributing to the late recognition of Palaeo-Tethyan evolution in the Mediterranean area and that are shared by the western end of the Alpides in the western Mediterranean (e.g. Maxwell 1970).

This paper is a progress report. We cannot yet hope to provide fully-developed solutions to any of the problems listed in the foregoing paragraph mainly because there are still critical areas within the Mediterranean Cimmerides that are extremely poorly known, such as the Serbo-Macedonian Massif in the Balkan/Carpathian Cimmerides or the Tokat Massif in the Anatolian Cimmerides. Here we first present a summary of new fieldwork in Turkey on Palaeo-Tethyan problems in the context of a review of new data from the Mediterranean Cimmerides in general. We then propose a new evolutionary model for the tectonic development of the Mediterranean Cimmerides in terms of four palaeo-tectonic reconstructions. Finally, we make suggestions as to how some existing problems might be solved and more importantly, identify problems that remain.

Extent and subdivisions of the Mediterranean Cimmerides

The Mediterranean Cimmerides extend from the northernmost tip of the Rhodope-Pontide fragment (Burchfiel 1980, Şengör & Yılmaz 1981) in the eastern Carpathians in the west to the eastern end of the Greater Caucasus, north of the Khoura depression, in the east (Figs 2 & 3). Although outcrop is discontinuous, they form a continuous orogenic belt along the south shore of the Black Sea, whereas in the north, present continuity between the North Dobrudja and the Crimea is not documented. Apart from numerous younger interruptions on various scales, ranging from simple faults to large basins such as that of Riou, the unity of the Mediterranean Cimmerides, was disrupted by the Late Cretaceous to the Eocene opening of the Black Sea marginal basin behind the S-facing Pontide magmatic arc (Letouzey *et al.*, 1977, Şengör & Yılmaz 1981) which split the

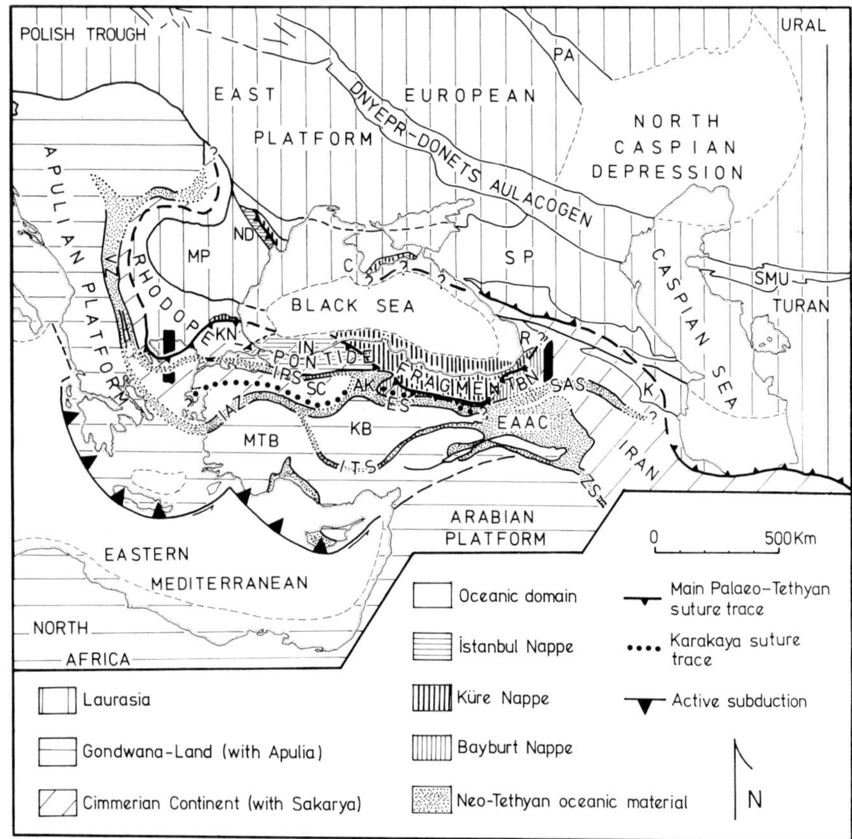

Fig. 2. Semi-schematic tectonic map showing the Tethyan palaeotectonic elements in the area of the eastern Mediterranean and surrounding regions, compiled from Şengör & Yılmaz (1981), Şengör *et al.* (1982), and Khain & Leonov (1975).

PA, Pachelma Aulacogen; SP Scythian Platform; SMU, South-Mangyshlak/Üst Yurt uplift; C, Crimea; ND, North Dobrudja; MP, Moesian Platform; SC, Sakarya continent; MTB, Menderes-Taurus Block; KB, Kırşehir block (MTB + KB = Anatolide/Tauride platform, see Fig. 1); VZ, Vardar Zone; IPS, Intra-Pontide suture; IAZ, İzmir-Ankara Zone; AK, Ankara Knot; ES, Erzincan suture; SAS, Sevan-Akera suture; EAAC, East Anatolian accretionary complex; ZS, Zagros suture. KN, IN, and BN are the Kırklareli, Istanbul, and Bayburt nappes respectively. R, Riou; K, Khoura depressions. Two heavy, short, black, vertical lines depict the segment divisions of the Mediterranean Cimmerides.

orogen along its strike (Figs 2 & 3) and now hides considerable portions of it beneath its waves.

On the basis of tectonic style and history of development, Mediterranean Cimmerides may be subdivided into three segments of unequal length. These are, from west to east the Balkan/Carpathian Cimmerides, the Anatolian Cimmerides (including the North Dobrudja), and the Caucasian Cimmerides (Figs 2 & 3). Figures 3, 4 & 5 form the basis of our descriptions of each segment. We begin with the Anatolian Cimmerides, where later overprinting disrupted the Cimmeride orogen much less than in the 2 adjoining segments, and from where we have the largest amount of data.

Anatolian Cimmerides

The Anatolian Cimmerides lie on both sides of the Black Sea. In the S, they extend as scattered inliers surrounded by younger cover from about the meridian of the island of Thásos to the Artvin area (Figs 2 & 3). Their southern boundary is extremely ill-defined because of intensive later reworking and extensive younger cover in Turkey. From a limited number of outcrops it appears clear, however, that even the autochthonous rocks of the Taurus range (Geyik Dağı unit of Özgül 1976) in southern Turkey were affected by Late Triassic compressional deformation (Fig. 3), although its significance is not yet fully clear (see below). Along the N shores of the Black Sea, the Anatolian Cimmerides include the North Dobrudja and the Yayla Range in the Crimea (Figs 2 & 3). In Turkey, their type locality, they consist of two parallel, east to west trending orogens. The one containing the main Palaeo-Tethyan suture is here termed the 'main trunk' and described first: it lies N of the smaller orogen of Karakaya, which grew out of a marginal basin of Palaeo-Tethys.

Main trunk of the Anatolian Cimmerides

All along its strike, the main trunk is believed to have a north-vergent triple-decker structure, much like the Eastern Alps, consisting of autochthonous Laurasian foreland units at the base, allochthonous Palaeo-Tethyan oceanic rocks in the middle, and allochthonous units belonging to the Cimmerian continent on the top. The complete sandwich is seen only in the Strandja Mountains and in the North Dubrudja (Fig. 3, locs. 3 & 13), whereas only the two upper units are exposed in Northern Turkey and the two lower units in the Crimea. The topmost unit of the main trunk consists of three nappes apparently succeeding one another along the strike. The main trunk may be conveniently described in terms of three longitudinal compartments that are co-extensive with the highest nappes (from east to west the Kırklareli, İstanbul and Bayburt Nappes: Figs 2 & 3) and whose relationships with one another are unclear owing to younger cover overlying their mutual contacts and later structural disruption obliterating original relationships (Fig. 3).

Western compartment

The western compartment of the main trunk is formed from the eastern Rhodopian nappes (Gocev 1979) and the Strandja Mountains (Fig. 3, locs. 3 & 13). They comprise, from top to bottom, the Kırklareli Nappe, the Strandja Nappe (Chatalov 1977, 1979), and the autochthonous units, with respect to Cimmeride evolution, of the Rhodope Massif and the Srednogora (Gocev 1979) (Fig. 3). Hsü *et al.* (1977) have shown that the Balkanides *sensu lato* developed by rifting from, and subsequent accretion to, the Moesian platform of the Rhodope Massif at least twice during the Tithonian to the early Cainozoic interval (also see Aiello *et al.*, 1977). Because the evolution of the Cimmerides in this area had already ended by Tithonian times with one possible exception in Northern Greece (see below), we entirely disregard the Balkanide evolution, which introudced the separation of the Rhodope Massif from the Moesian platform, and call the lowermost tectonic unit of the western compartment the autochthonous 'Rhodope-Moesian platform'.

During the Triassic, the Rhodope-Moesian platform to the immediate north and northwest of the Strandja Mountains was the site of quiet platform sedimentation characterized mainly by basal arkosic clastics unconformably overlying Palaeozoic felsic intrusives, and passing up into a km-thick succession of dolomites and fine-grained clastics (Chatalov 1979). This sequence is a typical example of what Janishevski (1946) termed the Balkanide-type Triassic. At the top of the Balkanide-type Triassic of the Strandja Mountains is a 'varicoloured terrigenous-carbonate formation' composed of laminated dolomicrites, micritic limestones, argillites, and limestone conglomerates (Chatalov 1979). Chatalov (1979) relates the argillites and the limestone conglomerates containing Upper Triassic clasts to the onset of orogenic activity creating structural highs (dry land areas) and lows forming basins with abnormal salinity. The Balkanide-type Triassic carries an unconformable Early to medial Jurassic cover that is tectonically overlain by the Strandja Nappe along thrust faults (Chatalov 1979).

The north-vergent Strandja Nappe is exposed only in the northeastern part of the Strandja Mountains (Fig. 3). Chatalov (1977) has shown that the Triassic rocks of the Strandja Nappe differ markedly from those of the 'autochthon', i.e. from the Balkanide-type Triassic and indicate a 'geosynclinal' environment. This new facies type of the Triassic in Bulgaria has been called by Chatalov (1977, 1979) the 'Strandja-type' Triassic and consists mainly of the flyschoid Lipacka Formation and the conodont-bearing Kondolovo Limestone (Fig. 4, col. 13; Chatalov 1977). The Lipacka Formation is a dark coloured, rhythmically bedded, terrigenous-calcareous sequence, possibly of turbiditic orgin. Although Chatalov (1979) prefers to reserve judgement on the exact environment of deposition and mechanism of transport of the Lipacka Formation, he nevertheless favours a continental slope/rise setting for its deposition. The Strandja diabase-phyllitoid complex consisting dominantly of greyish black argillites and shales, grey-green chlorite-sericite-calcschists, limestones, spilitic diabases, keratophyres, and albitophyres with associated stratiform copper-pyrite and copper-polymetallic ores in the Gramatikovo ore field (Chatalov 1978) tectonically overlies the Lipacka Formation

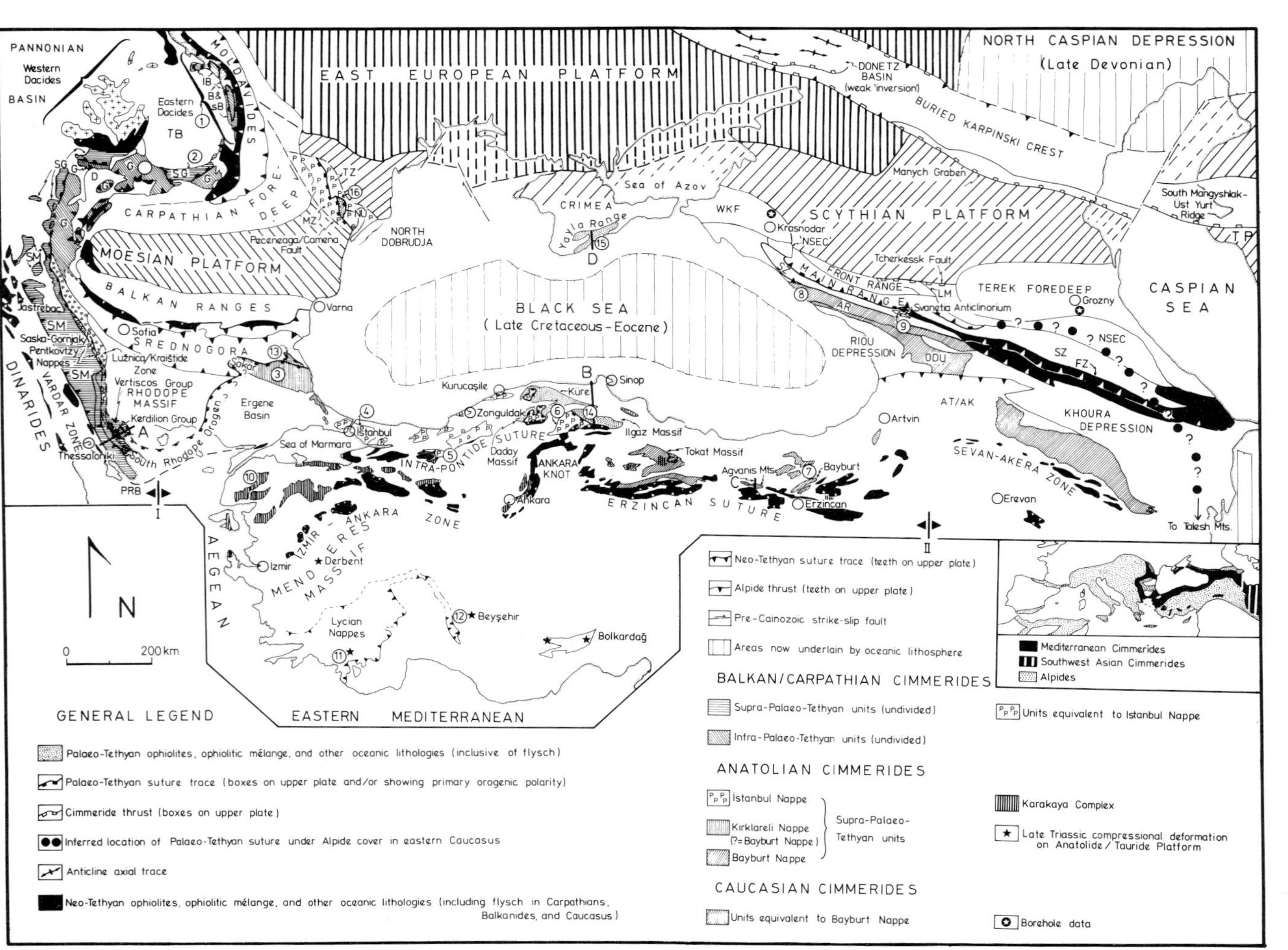
PANNONIAN BASIN
Western Dacides
Eastern Dacides
TB
MOLDAVIDES
EAST EUROPEAN PLATFORM
NORTH CASPIAN DEPRESSION (Late Devonian)
DONETZ BASIN (weak 'inversion)
BURIED KARPINSKI CREST
South Mangyshlak-Üst Yurt Ridge
Manych Graben
CARPATHIAN FORE-DEEP
Peceneaga/Camena Fault
NORTH DOBRUDJA
MOESIAN PLATFORM
BALKAN RANGES
Varna
Sofia
SREDNOGORA
Lužnica/Kraištide Zone
Vertiscos Group
RHODOPE MASSIF
Kerdilion Group
Saska-Gornjak Pentkovtzy Nappes
Jastrebac
DINARIDES
VARDAR ZONE
Thessaloniki
South Rhodope Orogen
PRB
Ergene Basin
Sea of Marmara
Istanbul
Sea of Azov
CRIMEA
Yayla Range
WKF
Krasnodar
NSEC
SCYTHIAN PLATFORM
Tcherkessk Fault
FRONT RANGE
MAIN RANGE
Svanetia Anticlinorium
TEREK FOREDEEP
Grozny
CASPIAN SEA
BLACK SEA (Late Cretaceous-Eocene)
Kurucaşile
Zonguldak
Kure
Sinop
RIOU DEPRESSION
DDU
AT/AK
Artvin
KHOURA DEPRESSION
SZ FZ
INTRA-PONTIDE SUTURE
Daday Massif
Ilgaz Massif
Tokat Massif
ANKARA KNOT
Ankara
Agvanis Mts
Bayburt
Erzincan
ERZINCAN SUTURE
SEVAN-AKERA ZONE
Erevan
To Talesh Mts.
IZMIR-ANKARA ZONE
Izmir
Derbent
MENDERES MASSIF
AEGEAN
Lycian Nappes
Beyşehir
Bolkardağ
EASTERN MEDITERRANEAN
N
0 200 km
Mediterranean Cimmerides
Southwest Asian Cimmerides
Alpides
GENERAL LEGEND
Palaeo-Tethyan ophiolites, ophiolitic mélange, and other oceanic lithologies (inclusive of flysch)
Palaeo-Tethyan suture trace (boxes on upper plate and/or showing primary orogenic polarity)
Cimmeride thrust (boxes on upper plate)
Inferred location of Palaeo-Tethyan suture under Alpide cover in eastern Caucasus
Anticline axial trace
Neo-Tethyan ophiolites, ophiolitic mélange, and other oceanic lithologies (including flysch in Carpathians, Balkanides, and Caucasus)
Neo-Tethyan suture trace (teeth on upper plate)
Alpide thrust (teeth on upper plate)
Pre-Cainozoic strike-slip fault
Areas now underlain by oceanic lithosphere
BALKAN/CARPATHIAN CIMMERIDES
Supra-Palaeo-Tethyan units (undivided)
Infra-Palaeo-Tethyan units (undivided)
ANATOLIAN CIMMERIDES
Istanbul Nappe
Kirklareli Nappe (?=Bayburt Nappe)
Bayburt Nappe
Supra-Palaeo-Tethyan units
CAUCASIAN CIMMERIDES
Units equivalent to Bayburt Nappe
Units equivalent to Istanbul Nappe
Karakaya Complex
Late Triassic compressional deformation on Anatolide/Tauride Platform
Borehole data

(Chatalov 1979). The Gramatikovo massive sulphide occurrence is very similar to Besshi-type deposits and clearly indicates an oceanic environment (early arc development: Mitchell & Bell 1973; sea-floor spreading followed by tectonic interleaving of oceanic basalts with continental margin sediments: Sillitoe 1972). We view the diabase-phyllitoid complex of the Strandja Nappe as simply the scraped-off top of a former ocean floor. Its rather long age span, possibly from the Devonian to the Triassic (Kalvacheva & Chatalov 1974) and the extremely deformed internal structure (Chatalov 1979) suggest such an origin. If our correlation of the Strandja Nappe with the Küre Nappe in north Turkey (Fig. 2) is correct (see below), the existence in the Küre Nappe of Cyprus-type massive sulphides (Güner 1980) associated with the Küre ophiolites supports our inference.

Thrust from the south onto both the Strandja Nappe and the Rhodope-Moesian platform is the enormous body of the Kırklareli Nappe comprising nearly the whole of the Turkish part of the Strandja Mountains (Pamir & Baykal 1947, Aydın 1974) and the Sakar region in Bulgaria (Gocev 1979) (Fig. 3, loc. 3). The Kırklareli Nappe consists (Fig. 4, col. 3), at the base, of a polyphase deformed and metamorphosed basement (the Koruköy 'Formation' of Y. Aydın pers. comm. 1982) intruded by the Kırklareli plutonic complex of granites, granodiorites, tonalites, quartz-diorites, and gabbro-diorites (Aykol 1979) during the Late Permian (Rb/Sr whole-rock age of 245 Ma: Aydın 1974). Meta-conglomerates, meta-sandstones, garnet- and staurolite-bearing mica-schists, quartz-amphibolites, and calcite and dolomite marbles form a metamorphosed sedimentary and volcanic sequence laid down onto the older metamorphic and igneous basement. A dominantly pelitic and carbonate succession forms the Jurassic rocks of the Kırklareli Nappe (Aydın 1974). Its later Mesozoic evolution is related to the development of the Balkanides (Hsü *et al.*, 1977) and the Vardar-Intra-Pontide ocean (Şengör & Yılmaz, 1981) and lies outside the scope of this paper.

The only data now available to constrain the date of emplacement of the Kırklareli and the Strandja Nappes onto the Rhodope-Moesian platform are provided by the youngest sediments in the nappes and the oldest sediments sealing the thrust contacts. The youngest sediments beneath the Strandja Nappe are of medial Jurassic age (Chatalov 1978, 1979), whereas the youngest rocks in that nappe are probably no younger than the Late Triassic (Chatalov 1979). The youngest event recognizable in the Kırklareli Nappe before the onset of the Balkanide-related phenomena is the latest Jurassic rejuvenation of the Kırklareli plutonic complex (144 Ma on biotite: Aydın 1974). The oldest rocks unconformably covering the thrusts bounding the Kırklareli and the Strandja nappes are of Cenomanian age (Aykol 1979).

From the existing data we draw the following picture for the evolution of the western compartment of the main trunk of the Anatolian Cimmerides. South-dipping (present geographic directions) subduction of ocean floor in the latest Palaeozoic beneath the Kırklareli Nappe caused the diabase-phyllitoid complex to accumulate beneath and in front of the nappe, while giving rise to arc magmatism in the main body of the nappe as represented by the Kırklareli plutonic complex. The same probably continued through the Triassic. Towards the end of the Trias and during the early

FIG. 3. Simplified tectonic map of the Mediterranean Cimmerides and their 'foreland' drawn after the sources cited in text and our own field work. In Turkey all areas younger than Lias are shown in white to emphasize the *outcrop distribution* of the Cimmerides. The northernmost Neo-Tethyan ophiolitic outcrops, indicating the main Neo-Tethyan suture in Turkey are shown to emphasize that most of the known Cimmeride outcrops are now located N of that suture and in the Sakarya continent. Areas of Late Triassic compressional deformation on the Anatolide/Tauride platform probably also belong to the Cimmeride orogen, but because Neo-Tethyan sutures are now located between them and the main body of the Cimmeride orogen, no definitive statements can be made. Because of extreme Alpide tectonism it has not been possible to isolate Cimmeride outcrops in the Balkan/Carpathian and Caucasian Cimmerides.

IB, infra-Bucovinian Bretila unit; B & sB, Bucovinian and sub-Bucovinian units respectively; G, Getics; SG, Supra-Getics; SM, Serbo-Macedonian Massif; PRB, Peri-Rhodopian Belt; TZ, Tulcea; MZ, Macin zones; NU, Niculitel unit; AT/AK, Adzharia-Trialeti/Artvin-Karabakh zone; AR, Abkhasia-Racha zone; SZ, Slate Zone; FZ, Flysch Zone; LM, Laba-Malka zone; NSEC, northern Slope of eastern Greater Caucasus; WKF, west Kubanian foredeep; I and II, segment boundaries of the Mediterranean Cimmerides.

Inset shows the areal extent of the Mediterranean Cimmerides in the Alpine System. A, B, C and D indicate the locations of the geological cross-sections displayed in Fig. 5.

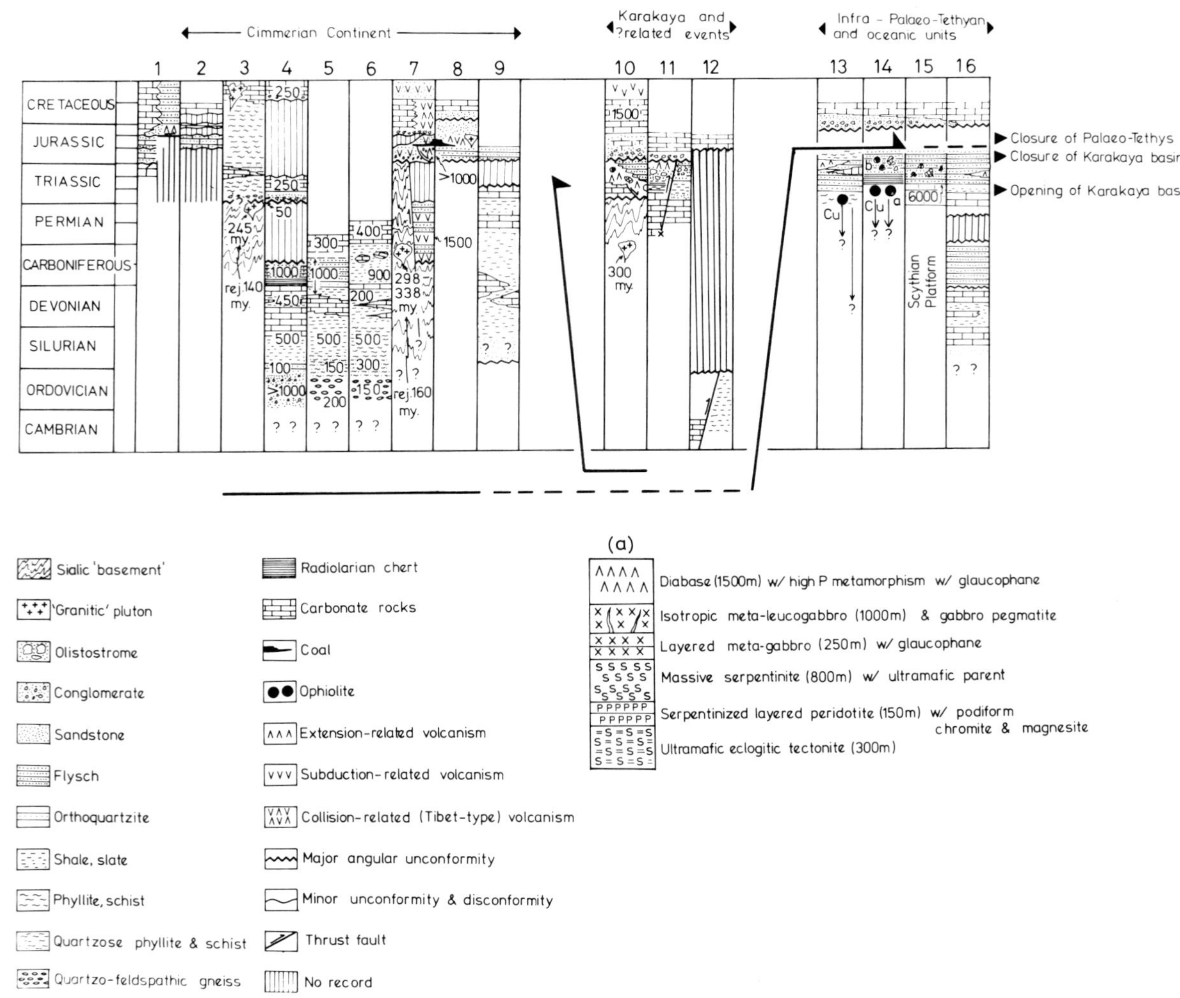
Cimmerian Continent
Karakaya and ?related events
Infra - Palaeo-Tethyan and oceanic units
1
2
3
4
5
6
7
8
9
10
11
12
13
14
15
16
CRETACEOUS
JURASSIC
TRIASSIC
PERMIAN
CARBONIFEROUS
DEVONIAN
SILURIAN
ORDOVICIAN
CAMBRIAN
Closure of Palaeo-Tethys
Closure of Karakaya basin
Opening of Karakaya basin
Scythian Platform
(a)
Diabase (1500m) w/ high P metamorphism w/ glaucophane
Isotropic meta-leucogabbro (1000m) & gabbro pegmatite
Layered meta-gabbro (250m) w/ glaucophane
Massive serpentinite (800m) w/ ultramafic parent
Serpentinized layered peridotite (150m) w/ podiform chromite & magnesite
Ultramafic eclogitic tectonite (300m)
Sialic 'basement'
'Granitic' pluton
Olistostrome
Conglomerate
Sandstone
Flysch
Orthoquartzite
Shale, slate
Phyllite, schist
Quartzose phyllite & schist
Quartzo-feldspathic gneiss
Radiolarian chert
Carbonate rocks
Coal
Ophiolite
Extension-related volcanism
Subduction-related volcanism
Collision-related (Tibet-type) volcanism
Major angular unconformity
Minor unconformity & disconformity
Thrust fault
No record

Lias, sedimentation ceased in the Strandja Nappe, either because the accretionary wedge it represents had been thickened by continuous underplating through accretion, and surfaced, or because it may have been already overridden by the Kırklareli Nappe. The cessation of volcanism on the Kırklareli Nappe during the Late Triassic–early Liassic might indicate the latter. At the same time a peripheral bulge formed on the southern edge of the Rhodope-Moesian platform north of the Strandja mountains, possibly as a consequence of beginning thrust-loading by the toe of the accretionary complex of the Strandja Nappe, and perhaps analogous to the area of Miocene deformation in Cyrenaica, now topographically high and seismically active, in front of the eastern Mediterranean ridge (Burollet & Rouvier, 1971). This peripheral bulge and the associated deformation is expressed by the Late Triassic conglomerates and the basin and ridge topography along the southern edge of the Rhodope-Moesian platform, unconformably covered by Jurassic sediments. Finally, during the Middle to Late Jurassic the ocean finally closed. The resetting of isotopic clocks in the Kırklareli plutonic complex probably reflects post-collisional heating of the Kırklareli Nappe, much like the post-collisional heating of the Lower Austro-alpine units around the SE part of the Tauern window in the eastern Alps (Oberhauser 1980a).

Central compartment

As seen in Fig. 3, the Kırklareli Nappe plunges east-southeastwards under the Eocene to Pliocene sediments of Thrace. Some 12 km to the east of the easternmost outcrops of the Kırklareli Nappe, are the westernmost exposures of the İstanbul Nappe, whose relations to the former are consequently totally unknown. However, the easternmost outcrop of the Kırklareli Nappe exhibits, near Çatalca, strong, polyphase deformation and metamorphism, characterized by the development of phyllites, meta-quartzites, chlorite-sericite-mica schists, marbles and calcschists (Akartuna 1953), folded around north-east vergent, overturned F_2 axial planes (Şengör, unpub. data). The ages of these rocks in the Çatalca region are unknown, but Dr. Y. Aydın includes them in his Fatmakaya Formation of Triassic age on the basis of lithologic similarity to dated lithologies in the Strandja Mountains (Y. Aydın, pers. comm. 1982). Because the Çatalca lithologies bear a striking resemblance to those of the Sakar Triassic (Gocev 1979), not only individually, but also as an association, we here adopt Y. Aydın's age assignment. In sharp contrast to the rocks of the Kırklareli Nappe, those belonging to the İstanbul Nappe west of İstanbul (Fig. 3) show no metamorphism and their deformation is much weaker than that of the Kırklareli Nappe. On the basis of similar arguments Tollmann (1965, 1968) argued, following a hint of Kober (1931), that the presently exposed Strandja Mountains (i.e. the Kırklareli + the Strandja Nappes—Tollmann naturally could not know about these recently discovered nappes) might be a tectonic window over which a higher, but now eroded nappe might have lain. Tollmann (1965) thought that the Palaeozoic rocks of İstanbul might have been a part of that hypothetical higher nappe (*Stanbuldecke* in Fig. 1 of Tollmann 1965). We agree with this interpretation of Tollmann (1965), with the exception of his erroneous correlation of the İstanbul Nappe with the younger units of the Balkanides. The lack of any kind of structural work in the Kırklareli Nappe and its completely obscured contacts with its neighbouring units in the east make it impossible to formulate a more precise hypothesis.

The İstanbul Nappe extends from the western tracts of the province of İstanbul to the Ilgaz (or Kargı) Massif (Fig. 3). Its massive body thins towards the northeast (?north), where its numerous klippen rest on the rocks of the Küre Nappe without any trace of intervening rocks belonging to the Kırklareli Nappe, roughly between the meridians of Zonguldak and Sinop (Fig. 3). It seems clear that even if the Kırklareli Nappe did extend westwards under the İstanbul Nappe, it clearly must end west of Zonguldak. The southern limit of the İstanbul Nappe is now formed by the Intra-Pontide suture (Fig. 3, Şengör & Yılmaz 1981), although before the opening of the Intra-Pontide ocean during the Jurassic, it was probably

Fig. 4. Highly simplified stratigraphic columns from selected localities in the Mediterranean Cimmerides. Emphasis is on those units whose geology was not discussed in Şengör *et al.* (1980) and which best illustrate the triple-decker structure of the Mediterranean Cimmerides. Karakaya units are thrust onto the highest allochthons of the Anatolian Cimmerides' main trunk, forming the highest, fourth unit. Separate column (a) shows the ophiolite pseudostratigraphy of the 'ordered' Elekdağ ophiolite nappe (see Fig. 5B). Letter b in column 14 refers to table I. Sources are cited in text.

more extensive toward the south. Its eastern termination is partly erosional but it is partly cut-off by the *retrocharriage* of Neo-Tethyan ophiolitic mélange (Figs 3 & 5b). As seen in Fig. 4 (cols. 4, 5 & 6), the main bulk of the İstanbul Nappe is made up of a fairly complete Palaeozoic succession known as the 'Palaeozoic of İstanbul' (Şengör & Yılmaz 1981), which consists of a SSW-facing, Late Silurian–Late Devonian Atlantic-type continental margin (mainly carbonates and shales, Fig. 4) that prograded across an Ordovician–early Silurian 'graben facies' (arkoses, shales, quartzites, Fig. 4) (Abdüsselâmoğlu 1977). Cambrian rocks overlying a gneissic basement are reported only from the Safranbolu region (Arpat *et al.*, 1978), where they lie in thrust contact on the rocks of the Küre Nappe, whereas probable Precambrian basement, composed of quartzo-feldspathic gneisses is reported from the Bolu Massif north of locality 5 in Fig. 3 (Prof. İ. Ketin, unpub. data). This ?Precambrian to Late Devonian succession was strongly deformed during the Early to middle Carboniferous with a northeasterly vergence accompanied by Early Carboniferous flysch (Kaya 1971, Abdüsselamoğlu 1977) and Late Carboniferous molasse, and intruded by Permian granites. An interesting and, from the viewpoint of Palaeo-Tethan evolution, critical feature of the Carboniferous molasse is its content of *northerly-derived* Westphalian andesitic and porphyritic clasts indicating probably coeval andesitic volcanism north of the present Zonguldak area (E. Kerey, pers. comm. 1982). The Carboniferous deformation of the İstanbul Nappe was related to the closure of an 'Hercynian' ocean located to its south and unrelated to Palaeo-Tethys (İ.T.Ü. group unpub. data). In large parts of the İstanbul Nappe, the Permian and the Scythian are represented by continental redbeds. Scythian andesitic volcanism is known from Hereke east of İstanbul (our observations), whereas Kaya and Lys (1980) have reported 'basic' Scythian volcanism from north of İstanbul. Around İstanbul, the rest of the Triassic up to and including the Carnian developed in Alpine-type Triassic facies and is unconformably overlain by Late Cretaceous carbonates. Although there is no Jurassic around İstanbul, towards the east, where the İstanbul Nappe thins markedly, the Permo-Triassic continental redbeds are unconformably overlain by sandstones, shales and purple conglomerates of ?Liassic-Dogger age (e.g. the Himmetpaşa Formation: MTA 1974). These Jurassic rocks differ widely from the partly coeval deep-sea lithologies of the Küre Nappe. Both, however, are unconformably covered by Late Jurassic–Early Cretaceous rocks (see below). The post-Jurassic evolution of the İstanbul Nappe is not relevant to the evolution of the Cimmerides and that is why we do not extend our description to later rocks, at the time of the deposition of which, the İstanbul Nappe had long lost its individuality.

In the rectangle defined by Kurucaşile-Sinop-Daday Massif-Ilgaz Massif (Fig. 3), klippen of the İstanbul Nappe overlie deformed and metamorphosed lithologies of the Küre Nappe. Recently, the geology of the Küre Nappe and its relationship with the İstanbul Nappe have been the subject of a number of detailed investigations (Yılmaz 1979, Güner 1980, M. Aydın, pers. comm. 1982) on which we base our brief summary.

Much like the Strandja Nappe of the western compartment of the main trunk of the Anatolian Cimmerides, the main bulk of the Küre Nappe consists of a flyschoid association here known as the Akgöl Formation (M. Aydın, pers, comm. 1982) made up of dark greyish-black, largely non-fossiliferous, bituminous, detrital chromite-bearing shales (Güner 1980), siltstones, and sandstones. Locally, olistostromal horizons contain blocks of recrystallized limestones, and nearly all the members of an ophiolitic association (abyssal tholeiites, diabase, gabbro peridotite, serpentinite: Güner 1980) (Table 1). Southwards, in the direction of the Ilgaz Massif and structurally downwards in Akgöl outcrop areas, metamorphism increases from anchizonal to greenschist facies. Undistrubed meta-ophiolitic slabs, such as that of Elekdağ (Figs 4a & 5b), and glaucophane-bearing greenschist and blueschist-facies metavolcanics are tectonically intercalated into the metamorphic parts of the Akgöl Formation. Aydın *et al.* (in prep.) have collected fossils ranging in age from the medial Triassic to the Early Jurassic from thin micritic horizons within the Akgöl Formation that are probably comparable with the Kondolovo Limestone of the Strandja Nappe (see above). On the basis of regional considerations, Güner (1980) assigns, as did Ketin (in MTA 1961), a Permian age to the bituminous shales of the Akgöl Formation. In the Küre area, Cyprus-type massive sulphide deposits are associated with the ophiolitic volcanics. The simplified cross-section in Fig. 5b shows the internal structure of the Küre Nappe and its relationships with the İstanbul Nappe. Structures such as the Çangal Dağı anticline (Şengör *et al.*, 1980) and the large strike-slip fault to its south are related to Alpide evolution and do not concern Cimmeride tectonics. As this cross-section reveals, most of the Cimmeride structures in the Küre Nappe are mesoscopic in scale and indicate predomi-

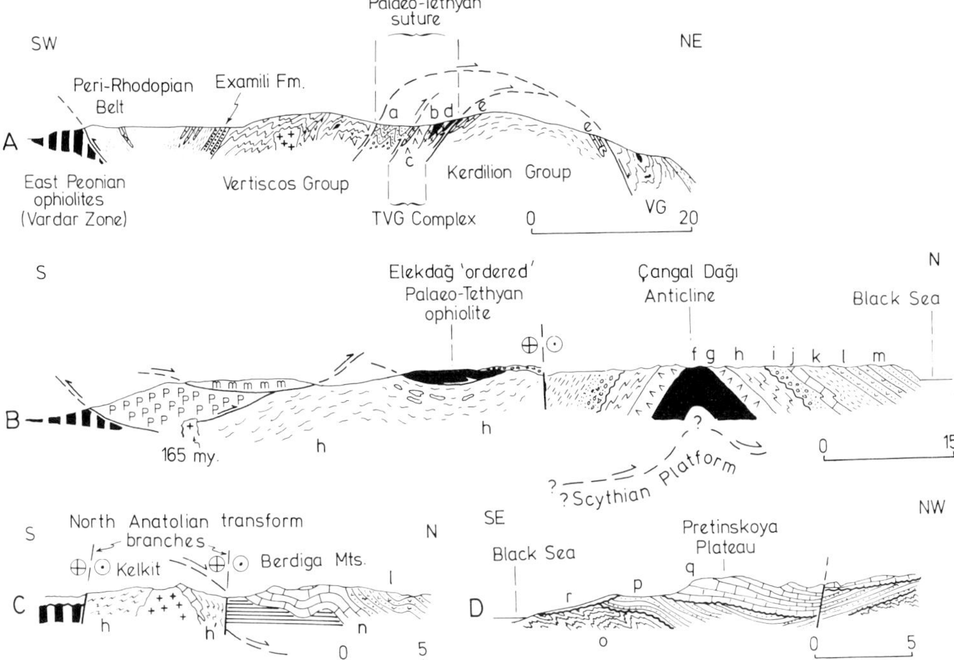

Fig. 5. Simplified geological cross-sections from selected areas in the Mediterranean Cimmerides (for locations see Fig. 3). A is from an unpublished manuscript kindly provided by Dr S. Dimitriadis. B is the result of our own field work since 1979, C is from Şengör *et al.* (1980), and D is from Kotanski (1978).

Key to small lettering: a, metaflysch; garnetiferous phyllite, quartzite, b, metagabbro; c, metavolcanics; d, metaultramafics; e, marble; f, Küre ophiolite; g, metavolcanics and metadiabases; h, Phyllites and slates (Liassic 'Flysch'); i, Malm basal conglomerates (Bürnük Formation); j, Upper Jurassic–Lower Cretaceous limestones; k, Lower Cretaceous black shales; 1, Upper Cretaceous volcanogenic flysch; m, Palaeocene-Eocene flysch; n, Upper Jurassic–Lower Cretaceous reefal carbonates (equivalents of unit j); o, Tauridian series; p, Middle Jurassic sandstones, shales and volcanoclastics; q, Oxfordian and later sediments; r, Neogene cover.

Notice the north-vergent mesoscopic structures in h and the sharp unconformity between h and i. The rocks covering the Elekdağ Nappe are of Cretaceous age. Unit h in C forms the Agvianis Mountain.

Table 1. *List showing the dominant block types found in the Akgöl flysch and its metamorphic equivalents.*

Unit name	Lithologic Content
Metapelite (Slate)	Phyllitic quartz mica-schist in greenschist facies; of sedimentary origin; with foliation
Metabasite	(1) Metalava } (2) Volcanoclastic } Greenschist → Glaucophane-bearing greenschist → Blueschist facies
Serpentinite	Partially or completely altered to listvenite
Amphibolite	Of metabasite origin; garnet-bearing
Recrystallized limestone marble	
Spilite } Red chert }	No metamorphism

nant northward vergence. Above the Akgöl Formation and truncating its structures along a sharp angular unconformity are latest Dogger to Malm fluviatile clastics of the Bürnük Formation passing upwards into shallow marine limestones of latest Jurassic–Early Cretaceous age. The youngest age reported from the Akgöl Formation (early Lias) and the oldest age known from the Bürnük molasse (latest Dogger) provide a stratigraphic bracket for the timing of deformation of the Küre Nappe. The timing of the juxtaposition of the Küre Nappe and the İstanbul Nappe is provided by the age of the youngest sediments beneath the İstanbul Nappe (Lias) and the istopic ages of 165 ± 6 Ma reported from the dioritic and quartz monzonitic plutons cutting its basal thrust (Yılmaz 1979). As indicated above, the Late Jurassic–Early Cretaceous neo-autochthon of the central compartment of the main trunk covers both the İstanbul and the Küre nappes.

As Şengör *et al.* (1980) have pointed out, the Küre Nappe is the lowest presently exposed tectonic unit in northern Turkey and therefore its 'nappe' character is based on circumstantial evidence. In addition to those observations reported from north Turkey to support the nappe character of the Küre unit by Şengör *et al.* (1980), the following arguments based on the geology of the Crimea also suggest that the Küre unit may be a nappe resting on the southern edge of the Scythian platform. We assume that the Late Cretaceous to Eocene opening of the Black Sea involved no significant strike-slip motion along its long axis (Letouzey *et al.*, 1977, Şengör and Yılmaz 1981). Therefore, if the geography of the Black Sea region is restored to its pre-opening geometry, the Crimea would abut against the regions now lying between Kurucaşile and Sinop in northern Turkey. In the Crimea, the Yayla Range (Fig. 3) is partly made up of the Tauridian flysch sequences (Kotanski 1978, Ager 1980, Sandulescu 1980) (Fig. 4, col. 15), which have a Late Triassic to Liassic age span, and which rest an older, northerly-derived, shallow shelf deposits of the Scythian platform. A peculiar characteristic of the dominantly argillaceous Tauridian series is the presence in it of large exotic early Permian limestone blocks, Triassic blocks, and blocks of pillowed basalt (Sandulescu 1980). The Tauridian flysch succession was deformed, with a southerly vergence (Figs. 5D & 6), during the Lias–earliest Dogger interval, because Bajocian and Bathonian sandstones, shales, and associated andesitic volcanics and tuffs (Ager 1980) cover the deformed Tauridian rocks above a sharp unconformity (Fig. 5D). The Gurzuf pluton was intruded

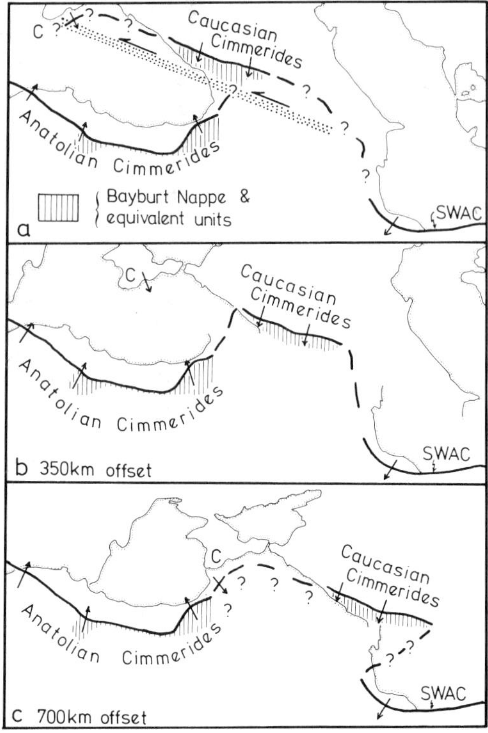

FIG. 6. Maps showing the present geometry of the nature of the junction between the Anatolian Cimmerides and the Caucasian Cimmerides (a) and two possible interpretations of its origin (b & c). Our choice is shown in (b). Small arrows indicate the predominant vergence (in the sense originally introduced by Stille). Discussion in text.

during the Dogger, much like its north Turkish counterparts.

As seen from the foregoing description and from Fig. 4, the similarities between the Crimea and the Küre Nappe of the central compartment in terms of stratigraphic development and the timing of events are striking, as was noted by Petrascheck (1960). The only contrast between the two areas lies in the opposing vergences of Cimmeride structures. Disregarding for the time being this contrast, to which we shall return later, we propose to correlate the Akgöl Formation and the Tauridian flysch. Whereas the Akgöl Formation probably formed on oceanic crust, the Tauridian series developed on the continental crust of the Scythian platform. It seems that in the circum-Black Sea regions, during the latest Triassic, Palaeo-Tethys had been reduced to a remnant ocean basin because of the Late Triassic collision of the Iranian segment of the Cimmerian continent with Turan farther E (Şengör, in press) (Fig. 7b). At that time, the Akgöl Formation was probably forming a large flysch/mélange wedge in front of the S-dipping subduction zone of Palaeo-Tethys, much like the Strandja Nappe and probably as its eastern continuation. As the toe of this wedge began impinging onto the southern border of the Scythian platform, it must have begun shedding flysch onto the previously quiet platform regions. From that time onwards (± Late Triassic) the Tauridian and the Akgöl lithologies developed together on the back and in front of the older Akgöl accretionary wedge.

Taking into account the continued convergence of the Cimmerian continent and the Scythian platform, we hypothesize that the southern extremities of the Scythian platform must underlie at least the northern parts of the Küre rocks as schematically shown with a query in Fig. 5B, perhaps in the same manner as the Rhodope-Moesian platform underlies the Strandja Nappe. Or, alternatively, the Küre 'nappe' may still have an oceanic substratum. Only seismic refraction and/or deep seismic reflection studies can choose between these two models.

Eastern Compartment

The eastern compartment of the main trunk of the Anatolian Cimmerides consists of the eastern Pontide tectonic unit of Ketin (1966) comprising regions lying north of the Erzincan suture and east of the Ilgaz Massif (Fig. 3). Vast areas of the eastern Pontides are hidden beneath an extensive volcanic and volcanoclastic blanket making it difficult to study their pre-Late Cretaceous geology. Fig. 3 shows the distribution of Liassic and older rocks in the eastern Pontides and their small areal extent is striking.

Because no new data have been gathered to change the general picture presented by Şengör *et al.* (1980) on the Palaeo-Tethyan evolution of the eastern Pontides since the publication of that paper, we here refrain from going into details and refer the interested reader to Şengör *et al.* (1980, 1982).

Two tectonic units characterize the pre-Malm geology of the eastern Pontides, much as for the pre-Malm geology of the central compartment. The lower unit, the Küre Nappe, is common to both the central and the eastern compartments. It corresponds with the 'oceanic assemblage' of Şengör *et al.* (1980). The higher unit is the 'continental assemblage' of Şengör *et al.* (1980), but, contrary to their correlation, it is distinct from the 'continental assemblage' of the central compartment, which is the İstanbul Nappe, and apparently forms an independent tectonic unit. For that reason, it is termed the Bayburt Nappe. Unfortunately, the *retrocharriage* of the Neo-Tethyan ophiolitic mélange north of the Ankara Knot intervenes between the İstanbul and the Bayburt Nappes completely obscuring their mutual relationships (Fig. 3). Rocks belonging to both the Küre and the Bayburt nappes in the eastern compartment appear as erosional inliers within the Late Cretaceous-Eocene volcanic and volcaniclastic cover and nowhere are their mutual contacts exposed. Şengör *et al.* (1980) listed a number of observations suggesting that the Bayburt Nappe is an allochthonous slab resting on the rocks of the Küre Nappe. Except for their erroneous correlation of the continental assemblages of the central and the eastern compartments, the observations itemized by Şengör *et al.* (1980) as well as the nappe interpretation of the Bayburt unit remain valid.

One area in the eastern Pontides that Şengör *et al.* (1980) did not consider is the poorly-known Tokat Massif. Since Blumenthal's (1950) pioneering work, only isolated studies have been undertaken on small areas in the Tokat Massif (e.g., Alp 1972, Seymen 1975). Recently, Tekeli (1981) included the whole of the Tokat Massif in the Karakaya orogenic zone (his North Antolian Belt). This interpretation is attractive because the Tokat Massif has a transgressive Liassic cover that is identical to that of the Sakarya Continent as well as to the Bayburt Lias and covers a basement extremely similar to the Karakaya Complex complete with limestone knockers (see below). However, A. I. Okay (pers. comm. 1982) has suggested that a part of that basement, devoid both of the transgressive Liassic cover and the exotic limestone knockers, may well represent parts of the Küre Nappe and correlate directly with the Küre lithologies in the Agvanis Mountains (Fig. 5C) (Şengör *et al.*, 1980). Thus it is possible that in the Tokat Massif both the Karakaya and the Küre lithologies are represented. This interpretation implies the possible union of the main Palaeo-Tethyan ocean with its marginal basin between the continental blocks represented by the İstanbul and the Bayburt nappes and further implies their original separation. Because this question bears directly on the along-strike continuity vs. discontinuity of the Cimmerian continent, the geology of the Tokat Massif appears to be of cardinal importance for our understanding of the Palaeo-Tethyan palaeogeography in the Anatolian Cimmerides.

North Dobrudja

The North Dobrudja (Fig. 3) is an isolated

fragment of the Anatolian Cimmerides' main trunk placed between the Scythian and the Moesian platforms in pre-Cenomanian times probably by strike-slip faulting (Burchfiel 1976, 1980, also see below). It consists of two main units, the Macin and the Tulcea zones separated by a major zone of NE-vergent thrusting locally marked by the Niculitel unit of oceanic origin (Fig. 3) (Ianovici *et al.*, 1961). The Macin zone is a broad anticlinorium whose core is made up of mesozonal amphibolites, schists, and quartzites. These rocks are overlain by weakly metamorphosed, interbedded quartzite and phyllite forming the Boelugea series. Ages of these rocks are unknown and estimates range from Precambrian to Ordovician (Burchfiel 1976). Above these are weakly metamorphosed to unmetamorphosed argillaceous shales, phyllites, limestones, and quartzites containing Silurian fossils in the lower horizons, followed by Devonian sandy quartzites, shales, and limestones (Burchfiel 1976). Lying unconformably on older rocks is the ?latest Devonian to Carboniferous Carapelit flysch. The Macin zone was deformed during the deposition of the Carapelit flysch and intruded by granites (Burchfiel 1976). It has only a thin and incomplete Triassic succession covered unconformably by younger rocks.

The basement of the Tulcea zone is identical to that of the Macin zone and exposed in the cores of several northwest-trending anticlines (Burchfiel 1976). This basement is unconformably overlain by an Alpine-type Triassic succession giving way upwards to the latest Triassic-early Liassic Nalbantian flysch (Fig. 4, col. 16) (Sandulescu 1978, 1980), which has often been compared with that of the Tauridian series (e.g. Stille 1953, Sandulescu 1980).

The Niculitel 'ophiolite' (Savu 1980) was obducted from the southwest onto the Tulcea zone during the Late Triassic and is rooted in the boundary between the Macin and the Tulcea zones. It consists of spilitic rocks including basalt flows, amygdaloidal basalts, and variolitic basalts that are commonly pillowed. They are intercalated with horizons of volcanic tuff, calcite-cemented mafic breccias and grey or red Triassic limestones. Small bodies or sills of gabbro-dolerites and spilitic dolerites are associated with the volcanic carapace of the Niculitel 'ophiolite'. Savu (1980) has shown that the mafic association of Niculitel was generated from the differentiation of a tholeiitic parent magma. He interprets these rocks as having formed seamounts on the oceanic floor of what Herz & Savu (1974) termed the Siret ocean. We adopt this interpretation, although we disagree with Herz & Savu's (1974) interpretation of the tectonic evolution of this oceanic realm that once separated the Macin zone from the Tulcea zone (see below). Now, the Niculitel unit is found along the Luncavita-Consul suture zone, a northeast-convex thrust front bringing the ophiolites of the Niculitel unit onto the Tulcean foreland (see map in Ianovici *et al.*, 1961).

The pre-Cenomanian location of the North Dobrudja has long been a problem (cf. Burchfiel 1980), because it resembles none of the units that now surround it. We have been struck by its similarity, with respect to both its Palaeozoic assemblage and the thin, lacunar development of its Alpine-type Triassic, to the İstanbul Nappe, particularly to areas just east of İstanbul (Fig. 4, cols. 4 & 16). Not only do the shelf sediments of Siluro-Devonian age and the Carboniferous flysch resemble closely their counterparts in the Palaeozoic sediments in the Gebze area east of İstanbul (Abdüsselâmoğlu 1977), but the sequence of deformational and magmatic events is closely similar to those known from the İstanbul Nappe. In both areas, Alpine-type Triassic successions occur, which are thin and incomplete (Özdemir *et al.*, 1973, Abdüsselâmoğlu 1977, Ianovici *et al.*, 1961, Sandulescu 1978), spanning only the Scythian-Carnian interval. On the İstanbul Nappe the Carnian lithologies are shallower marine than older Triassic rocks, except the continental Lower Scythian. Locally they contain terrestrial rocks and coal, all truncated above by an angular unconformity. In the North Dobrudja, by contrast, the deposition of the Nalbantian flysch set in during the Late Triassic. From these similarities, we place the North Dobrudja, at least the Macine zone, adjacent to the Kocaeli peninsula (east of İstanbul) during the Palaeozoic to Middle Jurassic interval in such a way that it would be possible to strike-slip it into its present location during the Late Jurassic–Early Cretaceous interval (see below).

Karakaya Orogen

The smaller, northward-vergent, southern trunk of the Anatolian Cimmerides extends from the northern tip of the Karaburun Peninsula (Brinkmann *et al.*, 1972) and the island of Lesbos (Hecht 1972) to the Tokat Massif (Tekeli 1981) and perhaps further east to Erzincan (Bergougnan & Parrot 1981) (Figs 2 & 3). Tekeli (1981) has recently reviewed the suture zone of the southern trunk of the Anatolian Cimmerides, called the 'Karakaya Orogen' after the Karakaya Complex (Karakaya Formation of Bingöl *et al.*, 1973, and Karakaya Group of Bingöl 1978), which makes up its suture material. The Karakaya Complex (Figs. 3 & 4, loc. 10) consists dominantly of metamorphic rocks in blueschist, greenschist, and amphibolite facies, and a deformed ophiolitic mélange. The metamorphic rocks represent deformed and metamorphosed oceanic crust and upper mantle, whereas the mélange rocks originated in diverse environments in abyssal plain, trench, and arc-trench gap regions. The mélange consists of a greywacke-slate (flysch with minor conglomerate, radiolarite, and pelagic limestone interlayers) and, locally, a spilite-basalt matrix containing blocks of limestone and marble of late Palaeozoic–early Mesozoic age. The ages of such limestone 'knockers' range from the Late Carboniferous to the Late Triassic (with a local Silurian development of ?interlayered limestones reported from the Tokat Massif (Alp 1972) which may be also blocks). Local felsic intrusives cut and/or are deformed with the Karakaya complex. Bingöl (1978) and Tekeli (1981) have argued that the Karakaya Complex records the opening and subsequent closure by south-dipping subduction of an oceanic area, which Şengör & Yılmaz (1981) interpreted to have been a marginal basin of Palaeo-Tethys, because of its short life-span, limited along-strike continuity, and location behind the main, S-dipping subduction zone of Palaeo-Tethys. South of the central compartment of the main trunk of the Anatolian Cimmerides, the Karakaya back-arc basin opened by disrupting the pre-Mesozoic basement of the Sakarya Continent (Fig. 2), whereas in the Tokat Massif, it is not clear what split to accommodate the Karakaya Complex. In Figs 2 & 3, we tentatively indicate that the original basement in the Tokat Massif might have been the Bayburt Nappe on the basis of the similarity of the transgressive Liassic cover on both (F. Alkaya, pers. comm., 1982).

As in other marginal orogens of the main Cimmeride trunk (e.g. the Waser and the Tanggula/Sittang Valley/Mytkyina orogens) data for the age of opening of the Karakaya oceanic area are sparse. The best evidence comes from the Bilecik-Söğüt area (Karakaya outcrops northwest of loc. 5 in Fig. 3), where a Permian neritic carbonate platform began rupturing during the Late Permian–Early Triassic to produce basalt flows (Yılmaz 1981). From the ophiolitic rocks of the Erzincan area (Fig. 3) Bergougnan & Parrot (1981) report a K/Ar age of at least 256 Ma obtained from a plagiogranite associated with the ophiolitic massif of Keşiş Dağ. O. Tekeli (pers. comm. 1980) had earlier indicated the possibility of the Karakaya Complex's extension into the Erzincan area, mainly from his reinterpretation of Tatar's (1973) map and the existence there of pre-Liassic, Karakaya-like rock associations (Ketin 1961). Because Bergougnan & Parrot (1981) give no details of the age determination we are unable to express an opinion on its reliability. It may support Tekeli's view and provide new evidence for his interpretation of the opening age of the Karakaya ocean as being at least Carboniferous (Tekeli 1981). The closure of the Karakaya back-arc basin is rather tightly constrained by the Late Triassic age of its youngest deformed lithologies and the latest Triassic-Liassic age of the oldest unconformable sediments (Tekeli 1981). Tekeli (1981) also pointed out that closure-related deformation and metamorphism may have terminated diachronously from the Late Triassic in the west to the late Liassic in the east, as shown by the ages of the oldest unconformable cover sequences, perhaps as a consequence of the diachronous closure from east to west of the Karakaya back-arc basin.

Because of the scattered outcrops and extreme structural and metamorphic overprinting, the southern limit of the Karakaya orogen is difficult to define. Moreover, we have little data to establish how much, if any, lateral movement occurred along the main Neo-Tethyan suture which cuts across Anatolia (İzmir-Ankara zone/Erzincan suture, Figs 2 & 3) and consequently whether any correlation of events between the Pontides and the rest of Turkey is to be expected. Whatever may be the answer to these questions, there is evidence for compressional deformation during the Late Triassic in the northern part of the Menderes Massif, in the western Lycian Nappes, in the autochthonous carbonate platform of the Taurides (the Geyik Dağı unit of Özgül 1976) near Beyşehir, and in the Bolkardağ region (Fig. 3). In the Menderes Massif, near Derbent, Akkök (1982) has shown the presence of strong folding and development of associated foliation accompanied by metamorphism and intrusion of the Dede Dağı granite, during the Permian to Late Cretaceous interval. He has argued that these events could most reasonably be interpreted as being related to the closure of the Karakaya basin to the north and therefore that they must be of Late Triassic age. Akkök (1982) also pointed out the striking similarity of the Permo-Carboniferous rocks on both sides of the İzmir-Ankara zone to argue against major sideways motion on this suture since the late Palaeozoic. The timing of early Mesozoic compressional deformation is much better constrained in the western Lycian Nappes. In the Gözyeri Tepe and the Haticeana Dağı (Fig. 3, loc. 11 and Fig. 4, col. 11), the Permo-Triassic Darıyeri and Ahat series had been structurally juxtaposed by thrusting and were subsequently covered by latest Triassic sediments containing clasts of both the Ahat and the Darıyeri rocks. Unfortunately, the later deformation and the cover make it impossible to deduce the primary vergence of the associated structures. In the Beyşehir area (Fig. 3, loc. 12 and Fig. 4, col. 12), Akay (1981) ascribed a Late Triassic age (Rhaetian) to the thrust juxtaposition of Cambrian rocks and the Cambro-Ordovician Seydişehir schists, where the thrust contacts are covered by late Liassic–early Dogger carbonates. Akay's (1981) argument is based on the observation that no compressional deformation of Palaeozoic age is known anywhere in the western and central Taurus Mountains (Özgül 1976). In the Bolkardağ region (Fig. 3) one of us (O.S.) observed the Permo-Triassic rocks imbricated beneath an unconformable, conglomeratic Lias.

Whatever the cause may have been, Late Triassic compressional deformation seems to have been widespread on the Anatolide/Tauride platform at a time when extension was going on in the eastern Mediterranean (Şengör & Yılmaz 1981). Although this phenomenon has been recognized very recently, and the available data are of reconnaissance type, we believe that it has wide-ranging implications not only for the delineation of the areal extent of the Cimmerides in Turkey, but also for its possible influence on the later evolution of the Anatolide/Tauride platform (e.g. the Triassic 'palaeohigh' of the Sultandağ region: Gutnic *et al.*, 1979).

Balkan/Carpathian Cimmerides

The northern Aegean Sea interrupts the outcrop continuity of the Cimmeride orogen south of the Rhodope Massif (Fig. 3). Şengör *et al.* (1980) and Şengör & Yılmaz (1981) correlated the main trunk of the Anatolian Cimmerides with the peri-Rhodopian belt of Kauffmann *et al.* (1976). However, recently S. Dimitriadis (written comm. 1982) has shown that the main Palaeo-Tethyan suture may be located further to the north-east than the peri-Rhodopian belt, within the eastern margin of the Serbo-Macedonian Massif, and that the peri-Rhodopian belt probably continues into the innermost zones of what Şengör *et al.* (1980) termed the intra-Pontide suture, in accordance with Gocev's (1979) earlier suggestion. Dimitriadis' interpretation makes much more sense in the face of the available data from the Strandja Mountains (see above) and from eastern Serbia (see below) than that of Şengör *et al.* (1980) and Şengör & Yılmaz (1981), as it eliminates the difficulty of bringing the Palaeo-Tethyan suture across the east Rhodopian nappes (Gocev 1979). Figs. 3 and 5A portray the present structure of the proposed Palaeo-Tethyan suture in northern Greece after an unpublished manuscript kindly provided by S. Dimitriadis. There, Dimitriadis has shown that the Vertiscos and the Kerdilion metamorphic complexes of the Serbo-Macedonian Massif, made up mainly of gneisses, subordinate amphibolites, marbles, and granitic intrusions are separated by a major zone of northeasterly-directed thrusting marked by the metamorphosed mafic volcanics, gabbros, and ultramafics collectively making up what he calls the TVG (Therma-Volvi-Gomati) Complex. He interprets the TVG Complex as representing the remnants of an oceanic lithosphere, perhaps including some ensimatic island arc lithologies. Preliminary work by Dimitriadis and Dixon (1982) on the Volvi part of the TVG complex appears to substantiate Dimitriadis' view and supports the interpretation of at least the Volvi occurrence as an ophiolite, now much disrupted and metamorphosed. Although Dimitriadis had originally thought that the main deformation of the TVG Complex and the thrust juxtaposition of the Vertiscos and the Kerdilion groups had taken place during the Middle Jurassic, on the basis of the earlier regional syntheses of Kockel *et al.* (1971, 1972, 1977), a re-evaluation by Dimitriadis and Dixon (1982) (and this volume) of the isotopic age of 104 Ma from the Volvi ophiolite obtained by Kreuzer *et al.* (1968), suggests that the deformation of the TVG Complex may have been delayed until as late as the mid-Cretaceous. Interpretation of this isotopic age is difficult, because all along the proposed Palaeo-Tethyan suture from NW Turkey and south Bulgaria to Yugoslavia (Fig. 3, see also below), major deformation seems to have occurred during Middle to Late Jurassic. Until more data become available we follow Dimitriadis' original age assignment because of its conformity to the overall regional picture.

Although Tertiary and Quaternary sediments interrupt the outcrop continuity northwards from the TVG Complex, north of the Greek frontier, the Pentkovtzy Nappes are overthrust toward the east, onto the Tithonian flysch sequences of the Luznica/Kraistide zone (Fig. 3). The Pentkovtzy Nappes contain rocks that span in age an interval from the Ordovician to the Carboniferous (Sandulescu 1980). Siliceous sediments and recrystallized limestones probably belong to the Silurian, whereas the Devonian is made up mainly of shales, carbonates, and terrigenous turbidites. The Carboniferous is represented by flysch, whereas the limited amount of Permian rocks are all of continental origin (redbeds and conglomerates) (Karagjuleva *et al.*, 1974). We view the Pentkovtzy Nappes as the westernmost equivalents of the Istanbul Nappe of the main trunk of the Anatolian Cimmerides, although we can say little about what their primary relations might have been. The Pentkovtzy Nappes are thrust eastwards onto the Luznica/Kraistide zone which is characterized by a metamorphic diabase-phyllitoid association of unknown age and intruded by granitic rocks of also uncertain ages. These rocks are overlain (?tectonically) by Triassic-Jurassic carbonates, which are capped by Tithonian flysch (Karagjuleva *et al.*, 1974, Sandulescu 1980). After the Hercynian deformation, two important episodes of Mesozoic deformation affected the Pentkovtzy Nappes/Luznica/Kraistide zones: an 'old Cimmerian' and a mid-Cretaceous one. In the old literature, the mid-Cretaceous deformation has been referred to as the 'Austrian orogenic phase' and can now be readily related to the closure of the ocean located between the Rhodope-Pontide fragment and the Moesian platform during the Albian (Burchfiel 1980). In fact this 'Meso-Cretaceous' deformation is mainly confined, in the Pentkovtzy Nappes/Luznica/Kraistide zones, to their central and northern parts (Karagjuleva *et al.* 1974), which are adjacent to the circum-Moesian oceanic area. What are called the 'old Cimmerian' movements in the Luznica/Kraistide zone in the older literature actually span a time interval from the Late Triassic to the Late Jurassic and the evidence for them is mainly stratigraphic in nature. Structures associated with these events are difficult to recognize owing to later overprinting, but Karagjuleva *et al.* (1974) have reported Late Triassic-Jurassic anticlines with complicated geometries from the east Serbian and the Kraistide segments of the Luznica/Kraistide zone, whereas in the Banat section 'the nature of the Old Cimmerian structures has not been cleared up, owing to the limited outcrops and considerable later reworking' (Karagjuleva *et al.*, 1974, p. 333). We place the Palaeo-Tethyan suture along the boundary of the Pentkovtzy Nappes and the Luznica/Kraistide zone, mainly because of our correlation of the former with the Istanbul Nappe, and because the Tithonian 'flysch' of the latter resemble a foredeep fill. The diabase-phyllitoid association in this area may mark the remnants of Palaeo-Tethyan oceanic lithologies. Northwards, in Yugoslavia, the regions of Veliki Jastrebac and Mali Jastrebac, just north of and along the strike from the point where the Pentkovtzy Nappes end (Fig. 3), two age groups of structures have been distinguished: 'Cimmerian' and Late Cretaceous-Palaeogene. The 'Cimmerian' structures are represented by a major zone of north-east-vergent thrusting whereby crystalline schists are pushed over the '*schistes lustrés*' of Jurassic age (Grubic 1980), which may represent Palaeo-Tethyan oceanic sediments.

Except for the later discredited (see Stille 1953) work of Preda (1940), the continuation into the Carpathians of the Cimmerides has received very little attention. The possible location of the Palaeo-Tethyan suture in the Carpathians has not yet been discussed in the literature. Palaeogeographically, such authors as Dewey *et al.* (1973), Gocev (1976), and Laubscher & Bernoulli (1977) discussed the westernmost extension of Palaeo-Tethys, but no suture zone beyond Yugoslavia has yet been proposed.

No well-defined ophiolite complexes, nor any well-developed active continental margin assemblages have yet been reported from within Burchfiel's (1980) Rhodopian fragment north of Greece. On the map of the ophiolitic belts of the Mediterranean (Dietrich 1979), several outcrops of 'ultramafic and mafic rocks' of uncertain origin are shown, and on the basis of experience in Greece (Dimitriadis & Dixon 1982) and Turkey (Şengör *et al.*, 1980) they may very well turn out to be dismembered ophiolites. Those located within the Danubian terrain (Fig. 3) consist of serpentinized dunites, harzburgites, wherlites, and gabbros, but are of pre-late Carboniferous age as shown by the ophiolitic clasts in the Late Carboniferous conglomerates, for example, of the Svinita zone (Codarcea *et al.*, 1961). Others, with less certain ages, lie close to the Getic/Supra-Getic boundary within the Supra-Getic Nappes (Fig. 3). As seen in Fig. 3, this boundary is exactly on strike with the boundary between the Pentkovtzy Nappes and the Luznica/Kraistide zones and the Jastrebac in Yugoslavia. Several very tenuous lines of indirect evidence suggest that the Getic/Supra-Getic boundary and its equivalents further north may represent what has remained of the Palaeo-Tethyan suture in the south and east Carpathians. First of all, intra-Jurassic and Triassic unconformities are widespread in both the Getic and Supra-Getic successions (Fig. 4. Col. 1). A little Triassic is only locally present in the Supra-Getics and its development is very incomplete. The Upper Triassic and the Lias are entirely missing and the Dogger begins with sandy limestones, the succession giving way upwards to more and more quiet-water environment sediments (Sandulescu 1975). In the Getics, the Upper Triassic is also absent and the Lias is generally developed in typical coal-bearing Gresten facies, locally volcanic. This facies is indicative of a continental environment and commonly overlies deformed rocks (Sandulescu 1975). Farther north in the eastern Dacides (Fig. 3), the Bucovinian and the Sub-Bucovinian Nappes are considered to be the equivalents of the Supra-Getic Nappes, whereas the 'infra-Bucovinian' Bretila unit is regarded as the equivalent of the Getic Nappes (Sandulescu 1975, 1980) (Fig. 3). As in the Supra-Getics and the Getics, both the Sub-Bucovinian and the Bretila units show increased tectonic activity during the Lias, as evidenced by the development of the Gresten facies resting directly on pre-Triassic crystalline basement and stratigraphic breaks spanning an interval from mid-Triassic to mid-Jurassic.

Between the Sub-Bucovinian and the Bretila units, the rootless Argestru slice consists of metamorphosed sandy-shaly units, mainly represented by phyllitic schists. Rare basaltic volcanics are associated with these (Sandulescu 1975). On the basis of the existence of pieces of crystalline schists and purported limnic sediments, a Carboniferous age has been assigned to the contents of the Argestru slice (Sandulescu & Bercia 1974), although there is no direct evidence. The tectonic equivalent of the Argestru slice, the Poleanca slice, consists of volcanic Permian and Triassic black bituminous schists. By analogy, the Argestru rocks may also be younger, i.e. Permo-Triassic. We tentatively interpret the Argestru and the Poleanca slices as what has remained of Palaeo-Tethyan oceanic and active margin assemblages, mainly from their similarities both in age and in lithology to the Akgöl pelites and volcanics (see above), and on their present structural location between the Sub-Bucovinian and the Bretila units. We believe that they mark the location of the Palaeo-Tethyan suture in the eastern Dacides, north of which the suture does not seem to extend. In the southern Carpathians, the tectonic equivalents of the Argestru and the Poleanca slices may be represented by the ultramafics sporadically present along the lower boundary of the Supra-Getic Nappes.

Owing to extreme later tectonism, it has not been possible to decipher the geometry and kinematic evolution of the Cimmeride structures north of Yugoslavia. In these regions, the present large-scale structure is almost entirely dictated by the Alpide evolution beginning in the Early Cretaceous. In most of the areas that we describe, Cimmeride structures also overprinted, and are commonly confused with, the older Hercynian structures, particularly in the metamorphic terrains where there is no biostratigraphic control and little isotopic age-dating. In most of these areas, the separation of the Hercynides from the Cimmerides, and the Cimmerides from the Alpides is far from being complete; in some of them this question has not yet even been addressed. Because of such shortcomings, in the Balkan/Carpathian Cimmerides we have been able to distinguish only three tectonic units on the basis of their *present* position with respect to the proposed Palaeo-Tethyan suture. The units lying structurally above the Palaeo-Tethyan suture material are simply termed the 'supra-Palaeo-Tethyan units', whereas those underlying it constitute the 'infra-Palaeo-Tethyan' units (Fig. 3). These two units and the suture itself define the first-order tectonic units of the Balkan/Carpathian Cimmerides. They do not, however, give any information as to what the geometry of the *original* Cimmeride orogen here may have been. For example, we have as yet no information to constrain the primary orogenic polarity in the Balkan/Carpathian Cimmerides. Of their evolutionary history, all we can now tell is that Palaeo-Tethys probably shut terminally during the latest Triassic or Liassic times between the eastern Dacides and Yugoslavia, and that collision was probably a late Dogger–early Malm affair further south in the Grecian part. An important factor contributing to the difficulties of recognition of the Cimmeride suture in the northern part of the Balkan/Carpathian Cimmerides is the extremely feeble development of its fingerprints such as subduction-related magmatism and subduction complex generation. This is probably a consequence of orogenic evolution very close to the closure pole of Palaeo-Tethys, where plate motion rates must have been very slow.

Caucasian Cimmerides

Together with the North Dobrudja and the Crimea, the Caucasian Cimmerides as a whole represent regions in which the importance of early to middle Mesozoic orogenic deformation in the Mediterranean area was first recognized (Suess 1909). They are distinguished from the Anatolian Cimmerides by their original south-facing orogenic polarity and pre-Liassic closure of the Palaeo-Tethyan ocean along them. In contrast to such structural differences, the facies realms of the main trunk of the Anatolian Cimmerides continue directly into the Caucasian Cimmerides, where they are only partly exposed. The Karakaya orogen does not seem to continue further east than Erzincan (Fig. 3).

In the Caucasus area, the northernmost unit

that is directly correlatable with the Anatolian Cimmerides is the Abkhasia-Racha zone (Kain 1975), which correlates with the Bayburt Nappe (Fig. 3). The exposed rocks of this zone consist of two sequences separated by an angular unconformity (Khain & Milanovsky 1963, Khain 1975, Nalivkin 1976). The lower sequence consists dominantly of clastic sediments of Liassic age that are partly in flysch facies (Aalenian). Above these is the Bajocian 'prophyrite series' and, locally, a regressive sequence including coal of Bathonian age. This sequence, except for the 'porphyrite series', is extremely similar to the Jurassic sequence of Bayburt (Ketin 1951); unlike Bayburt, however, rocks older than the Liassic clastics are unknown from the Abkhasia-Racha zone. These 'lower' rocks of the Abkhasia-Racha zone are unconformably overlain, particularly in its western part by lagoonal carbonates of Malm age, and by those of the Cretaceous, which are similar to those of the Bayburt Nappe. Granitoids of Dogger age have intruded the central part of the Abkhasia-Racha zone (Nalivkin 1976).

The Abkhasia-Racha zone has the structure of a doubly-plunging anticlinorium with an axial trace paralleling the Main Range. Owing to this structure the unconformable Malm and Cretaceous sediments are exposed only along the periphery of the anticlinorium which makes it difficult to study the effects of pre-Malm deformation because of later tectonism. However, particularly at the north-western end of the anticlinorium, along the southern flank of the Bzybskiy Mountains, sharply folded Aalenian clastics and the Bajocian 'porphyrite series' are seen to be covered unconformably by Malm sediments (Nalivkin 1976). Farther north, in the southern limb of the flysch synclinorium, in Svanetia, there is a small anticlinal outcrop of a very slightly metamorphosed Palaeozoic succession (Figs 3, loc. 9 4, col. 9) which consists of conformable, almost entirely volcanic-free, carbonate-terrigenous sediments spanning ages from early Devonian or latest Silurian to latest Permian or earliest Triassic (Khain 1975, and pers. comm. 1980). This sequence of rocks differs widely from the Palaeozoic successions known from the more northerly parts of the Caucasus. Noteworthy is the complete lack of any sign of post-Silurian orogenic activity until the later Triassic, when the Svanetia anticline was unconformably covered by Early Jurassic clastics. Although the Palaeozoic–Early Triassic succession in the Svanetia anticline bears no relation to the Palaeozoic rocks of the areas north of it, it resembles closely those of the southern Transcaucasus, which is one of the five regions in the whole world for which it has been cliamed that marine sedimentation was continuous from the Late Permian into the Early Triassic (Kummel & Teichert 1970; also see Bonnet & Bonnet 1947). The Svanetia stratigraphy is also very similar to that of central Iran (Berberian & King 1981), which is continuous with the Transcaucasus. Stöcklin (1974), Alavi (1979), Şengör (1979) and Berberian & King (1981) have reviewed evidence showing that central Iran was an integral part of Gondwana-Land until the Early Triassic when it rifted from it as a part of Şengör's (1979) Cimmerian continent and later collided with Laurasia before the deposition of the Rhaeto-Liassic Shemshak Formation which covers the Palaeo-Tethyan suture zone in northern Iran. We therefore view the Palaeozoic succession of the Svanetia anticline as a part of the Cimmerian continent, which collided with the western Main Range representing the southern, active margin in Laurasia. Although Peive *et al.* (1976) also placed the Palaeo-Tethyan suture in the Svanetia area, the complete absence of any ophiolitic rocks makes it difficult to determine exactly the location of the suture. Stratigraphical arguments leave no room for doubt, however, about its location roughly between the Main Range and the Svanetia anticline.

Because pre-Jurassic rocks are virtually absent in the eastern part of the Greater Caucasus, it is not possible to trace the Palaeo-Tethyan suture farther east and so connect it in the Talesh Mountains, with the westernmost outcrops of the main Palaeo-Tethyan suture of the Southwest Asian Cimmerides. However, in one of the frontal anticlines south of Grozny (Fig. 3), intensely deformed Upper Permian was penetrated by drilling beneath comparatively thin and unconformable Jurassic strata. This zone of deformation extends in a belt striking E-W from Svanetia (Khain 1975). We suggest that this east-west band of pre-Jurassic tectonism may mark the trace of the Palaeo-Tethyan suture in the Caucasus, now completely hidden under Alpide thrusts and post-Triassic sedimentary cover. Significantly, its strike is slightly discordant to that of the Alpide structures (Fig. 3), supporting the interpretation that it is an independent structure from the Alpides. From Grozny eastwards, the suture must somehow turn southward to connect with the Talesh Mountains, although the abrupt southerly turn depicted in Fig. 3 is probably much accentuated by the later opening of the South Caspian oceanic basin and the rifting of the Khoura depression.

That the shape as depicted in Figs 2 & 3 of the present suture trace of Palaeo-Tethys in and around the Caucasian Cimmerides is the result of deformation post-dating the closure of Palaeo-Tethys is clearly seen in Figs 2 & 3, because the Bayburt nappe and the Palaeo-Tethyan suture in the eastern Pontides overlap with the Abkhasia-Racha zone and the suture in the Caucasus, in a 350 km wide band, as shown schematically in Fig. 6a. Figure 6 shows two possible interpretations to circumvent this problem, both of which involve post-suturing left-lateral displacement along a narrow shear zone paralleling the long axis of the Main Range of the Greater Caucasus (Fig. 6). Figure 6b displaces the Greater Caucasus and the north shore of the Black Sea, right-laterally for 350 km. This reconstruction solves the overlap problem and produces a much smoother trace of the Palaeo-Tethyan suture. Figure 6c explores the possibility of extending the Caucasian Cimmerides all the way to the Crimea through the Krasnodar area (Fig. 3), where drilling has revealed a Triassic/Jurassic assemblage very similar to that of the Crimea (Khain and Milanovsky 1963), and so include all the southward-vergent Cimmeride structures in the circum-Black Sea area. The reconstruction shown in Fig. 6c is produced by displacing right-laterally the Greater Caucasus and the north shore of the Black Sea along the same shear zone as in Fig. 6b until the north- and south-vergences no longer overlap. This produces the picture of two orogenic belts of opposing vergence meeting along a rather sharp boundary, as in the case of the Alps and the Apennines. However, the reconstruction shown in Fig. 6c not only produces suture overlap problems in the area of the present Southern Caspian, but also puts the Crimea much farther away from the North Dobrudja and from the central compartment of the main trunk of the Anatolian Cimmerides than their stratigraphic similarities would warrant. We therefore discard the hypothesis represented by Fig. 6c and adopt the one depicted in Fig. 6b. Opposing vergences are not a unique phenomenon for an orogen and are mimicked by the craton-vergent overthrusts of the Idaho–Wyoming thrust belt of the Cordillera and the west-vergent Gros Ventre and Windriver basement thrust, which oppose one another around the Hoback basin (Dorr *et al.*, 1977). We believe that a similar situation may have occurred north of the Black Sea as we shall discuss below. The reader should bear in mind, however, that the Figs 6b and 6c were drawn on the entirely unrealistic assumption that the areas in question have maintained torsional rigidity since the proposed strike-slip movement. Our sketches are drawn only to illustrate the problem and to hint at a possible solution and expressedly *not* as final answers.

Northern 'foreland' of the Mediterranean Cimmerides

The northern 'foreland' is a lump term for all the cratonic areas lying north of the Mediterranean Cimmerides that were not directly involved in their orogenic evolution. In this context it is a misnomer, because these cratonic areas are foreland only to the Anatolian Cimmerides, whereas with respect to the Caucasian Cimmerides they constitute a hinterland. Because we do not know the original orogenic polarity of the Balkan/Carpathian Cimmerides, its relation to them is uncertain. For brevity we employ the term 'foreland', but put it between inverted commas.

The dominating structure of the northern 'foreland' of the Mediterranean Cimmerides is the enormous, partially 'inverted' aulacogen of the Dnyepr-Donets (Fig. 2). Burke and Dewey (1973), Burke (1977), and Milanovsky (1981) have reviewed the origin and the Palaeozoic evolution of this structure. That compressional deformation affected the Dnyepr-Donets aulacogen, especially the Donets basin, during the Mesozoic was first noticed by Borisyak (1903), who indicated the existence of particularly strong compressional movements during the Early Jurassic, which had re-deformed rocks already once folded during the Late Palaeozoic. Borisyak (1903) reported dips as much as 80° from the Liassic to Lower Bajocian sediments that were unconformably overlain by younger strata. Later, von Bubnoff (1931) termed this deformation the 'Donets phase'. More recently, Konashov (1980) showed that compressional deformation in the Donets basin had already set in during the Late Triassic as evidenced by the emergence of a series of anticlinal crests affecting the distribution of Late Triassic sediments. Widespread pre-Toarcian erosion was followed by incipient transgression; during the Toarcian-Aalenian interval the entire eastern part of the Dnyepr-Donets aulacogen (the buried Karpinski crest) continued to fold (Konashov 1980). Later during the Jurassic and during the Cretaceous (related perhaps to Alpide compression) the Donets basin was pushed, *en bloc*, along the Manytch Graben 'synclinorium' and fault system (Krasny 1961), which we view simply as a compressed foredeep in front of the block uplift of the Donets basin, towards the south. This south-vèrgent thrusting affected the whole of the circum-Azov Sea regions and was probably also responsible for the south-vergent folding of the Tauridian series in the Crimea.

Some of the Mesozoic folding of the Pachelma aulacogen (Milanovksy 1981) may also be related to the foreland deformation area of the Mediterranean Cimmerides.

A model for the tectonic evolution of the Mediterranean Cimmerides

In this section we present an evolutionary model of the tectonic development of the Mediterranean Cimmerides in terms of plate reconstructions for four selected time frames from the latest Permian to the Early Cretaceous. It is clear that for a satisfactory understanding to be achieved, the Mediterranean Cimmerides should be considered in the broader context of the evolution of the western part of the Tethyan orogenic system. For that reason we have included in our reconstructions all regions from eastern North America to the meridian of the Urals. The basic data used for these reconstructions cannot be fully documented in this paper, although the following publications would give the reader a good review of much of the data we used: Berberian and King (1981), Biju-Duval *et al.* (1977), Boillot and Capdevila (1979), Bourrouilh & Grosline (1979), Burchfiel (1980), Channel *et al.* (1979), Cohen (1980), Dewey *et al.* (1973), Kelts (1981), Lowry (1980), Nairn *et al.* (1977a, 1977b), Oberhauser (1980b), Ogniben *et al.* (1975), Sclater *et al.* (1977), Şengör & Yılmaz (1981), and Ziegler (1982). The latest Permian–Middle Triassic Pangaea (Fig. 7a) is a somewhat modified combination of the Sclater *et al.* (1977) fit of the Atlantic-bordering continents and the Mediterranean assembly of Smith *et al.* (1981). Our plate boundary reconstructions are not satisfactory at every point on the four maps, mainly because of the lack of constraints and because of possibly erroneous initial assumptions. We made no attempts to 'smooth out' such inconsistencies, because we feel that they may lead to the detection of weak points in our model and their subsequent correction.

The tectonic evolution of the circum-Mediterranean Tethyan orogenic belts, of which the Mediterranean Cimmerides from a subset (Fig. 3, inset), proceeded during the latest Permian to the Early Cretaceous interval under the influence of two main factors: the progressive contraction and the final elimination of Palaeo-Tethys, and the rifting and subsequent opening by sea-floor spreading of the central Atlantic ocean. Because the latter does not seem to have affected the evolution of the Cimmerides, we shall not dwell on it except briefly when discussing the implications of Palaeo-Tethyan evolution for the early development of the central Mediterranean Alpides.

Late Permian to Middle Triassic Events

Figure 7a shows our reconstruction of the Late Permian to Middle Triassic geometry of the west Tethyan regions. During this time interval, initial separation of Gondwana-Land from Laurasia along the future central Atlantic and the Caribbean/central American region began, possibly involving a complex series of exensional and strike-slip motion between the two megacontinents (Swanson 1982), although the Mediterranean area was apparently not yet feeling the effects of this separation. In the western and central Mediterranean regions, as far east as Turkey, a continental to shallow water environment prevailed (Dewey *et al.*, 1973, Biju-Duval *et al.*, 1977, Laubscher & Bernoulli 1977, Ziegler *et al.*, 1979, Ziegler 1982); by Middle Triassic times, regions east of the Corso-Sardinian block and Tunisia, all the way to the margins of Palaeo-Tethys, were dominated by Alpine-type Triassic carbonate and locally evaporitic and terrigenous (mainly Carnian) deposition. The Triassic carbonates do not form a wedge continuously thickening towards the margin of Palaeo-Tethys, as is frequently depicted (e.g. Laubscher & Bernoulli, 1977, Fig. 4), but after having reached a certain maximum away from the terrestrial environments of central Europe and the western Mediterranean, they thin again towards the margin of Palaeo-Tethys. Wherever Alpine-type Triassic is seen near the main Palaeo-Tethyan suture (e.g., the Macin zone, and near Istanbul) in the Mediterranean Cimmerides, it is much thinner than its counterparts in such areas as the Northern Calcareous Alps and its development is much less complete. Near the Palaeo-Tethyan suture, it is commonly volcanic-bearing.

Major tectonic activity at this time was concentrated along the margins of Palaeo-Tethys. In northern Turkey, Greece, and southern Bulgaria a west- and southwest-dipping subduction zone had been consuming Palaeo-Tethyan ocean floor probably since the latest Carboniferous. The Kırklareli and the Gümüşhane (Yılmaz 1973; see Şengör *et al.*, 1980) granodioritic and granitic intrusive complexes may be related to this subduction. Permo-Carboniferous andesitic volcanism, accompanied by extensive silicic tuff deposition on the back of the Bayburt Nappe, Westphalian andesitic volcan-

GREENLAND
Dnyepr-Donetz
Basin
TURAN
Polish Trough
EURASIA
ROCKALL
BANK
MOESIA
NORTH
AMERICA
TV
RHODOPE
GRAND
BANKS
Bay of Fundy
IBERIA
CS
APULIA
PALAEO-TETHYS
B
Newark
Connecticut
Haha
BT
KC
LNg
PONTIDE
IB
PB
SC
FRAGMENT
Palaeo-Tethys-
Pangaea Pole
KB
TAP
Casamance
AFRICA
NORTH-WESTERN
&
CENTRAL
IRAN
ARABIA
?Propagating Neo-Tethyan ridge
a

ism on the Istanbul Nappe, and the Scythian andesites and basalts again of the İstanbul Nappe, and perhaps some of the Permian volcanics known in the Balkan/Carpathian Cimmerides are here viewed as manifestations of the main Palaeo-Tethyan subduction zone. For Permian time only, Stämpfli (1978) painted a similar picture for northern Iran. Recently, Pisa *et al.* (1979), Castellarin *et al.* (1979) and Castellarin & Rossi (1981) postulated that the widespread Ladinian 'Pietra Verde' volcanism of the north and east parts of the Apulian-Anatolide/Tauride platform, which includes ignimbrites, rhyolitic, rhydacitic, andesitic, and basaltic lava flows and associated pyroclastic deposits may have been related to Palaeo-Tethyan subduction. Most outcrops of the 'Pietra Verde'-type volcanics now lie at distances of 200 to 700 km from the nearest Palaeo-Tethyan suture and these distances should at least be doubled to obtain the Triassic distances from the postulated Palaeo-Tethyan subduction zone. Although arc-type volcanics do occur as much as 1500 km from the nearest coeval subduction zone in the W United States (e.g. Snyder *et al.*, 1976), that distance must be nearly halved to allow for post-subduction basin-and-range extension. Because present-day arcs are never as wide as the 'Pietra Verde' volcanic area seems to have been, and because the 'Pietra Verde' volcanism is seen to have been associated with widespread normal faulting, we reject the subduction-related arc model of Castellarin *et al.* (1979) and Castellarin & Rossi (1981). Instead, the association of a predominantly bimodal volcanism with diffuse extensional tectonics and only very localized and subordinate compression (Castellarin *et al.* 1979) invites comparison with the Basin-and-Range region. We adopt here the Basin-and Range analogy as a working hypothesis for the 'Pietra Verde' volcanism and incoporate the small areas of normal subduction-related volcanism, much as the small Cascade volcanic arc lies north of the enormous area of the Basin-and-Range proper. We follow basically Livaccari's (1979) card-deck shear model for the origin of the Basin-and-Range and believe that the 'Pietra Verde' basin-and-range was probably also related to strike-slip faulting along much of the northern and western boundary of Palaeo-Tethys near its western termination (Fig. 7a; also see Dewey *et al.*, 1973, Fig. 9). The Triassic basin, in which the lithologies of the Transylvanian Nappes were deposited (Sandulescu 1975, 1980), is here viewed as a rift related to the same strike-slip motion, in which extension went far enough to produce local oceanic crust, perhaps much like the modern Gulf of California (Kelts 1981), and not, as hitherto assumed, as an early-opening segment of the northern branch of Neo-Tethys. It is true, however, that once the Liassic opening of the latter began along the Vardar-Mureş-Pieniny Klippen Belt zones (Burchfiel 1980), it seems to have utilized the pre-existing Translylvanian basin in Transylvania (not to be confused with the Neogene Transylvanian basin: Royden *et al.*, 1982).

While dominantly strike-slip and subduction tectonics characterized the margins of Palaeo-Tethys, and the associated basin-and-range-type activity occupied the future area of the central Mediterranean orogens, in south and south-east Turkey and in Cyprus, Carnian-Norian alkaline mafic volcanism, normal faulting, and a transition from neritic to pelagic sedimentation heralded the opening of the southern branch of Neo-Tethys (Şengör & Yılmaz 1981). This opening was suggested by Şengör (1979) and Şengör *et al.* (1980) to have been, at least locally (e.g. in the cross-section of Erzincan), a back-arc basin-opening event, related to the subduction of Palaeo-Tethyan

---

FIG. 7a. Latest Permian to medial Triassic reconstruction of the Mediterranean area and surrounding regions. Thin line with dots along it represents the present-day coast-lines for reference. Simple thin line denotes edge of continent. Discontinuous thin line shows block boundaries inferred mainly from geological data. Heavy lines with half arrows are transform faults, double heavy lines are oceanic spreading centres, heavy lines with black triangles are subduction zones with triangles on the upper plate. Discontinuous lines with hachures indicate rifts. Closely spaced horizontal lines in Northeast Apulia and West Turkey are areas of widespread 'Pietra Verde' volcanism. v's indicate subduction-related volcanism, whereas upside-down v's represent rift volcanics.

Key to lettering: B, Balearic/Pre-Betic block; BT, Betic block; CS, Corso-Sardinian block; KC, Kabylia-Calabria block; IB, Sclafani-Imerese basin; LNg, Lagonegro basin; PB, Pindos-Budva basin; TV, Transylvanian basin; SC, future Sakarya continent; KB, Karakaya back-arc basin; TAP, future Anatolide/Tauride platform. Sources and discussion in text.

The Palaeo-Tethys/Pangaea pole concerns only regions shown on this map and not the Tethyan domain in its entirety.

ocean floor. This process was probably also helped by the westerly propagation of a Neo-Tethyan ridge coming from the nascent Zagros ocean at the time (Fig. 7a). Piper & Piper (1982) have recently suggested, following Capedri *et al.* (1980), that the Pindos-Budva trough (Fig. 7a) may also have opened as a marginal basin, but they believe the associated magmatic arc to have lain west of the present position of the basin, because from the Jurassic to the Early Tertiary the major sediment supply was from a crystalline region in the west with associated volcanics. Piper & Piper (1982) state that 'no clear evidence of an eastern margin to the Pindos zone is seen in the Peloponnese' (p. 87). We believe that this absence of evidence is only natural, because the eastern margin of the Pindos-Budva trough was later overriden by west-vergent thrusts. In contrast to Piper & Piper (1982), we place the associated magmatic arc of the Pindos-Budva marginal basin to the north-east, in the area of the active margin of Palaeo-Tethys. No suitable oceanic area existed during the early Mesozoic to the west of the Pindos-Budva trough, whose subduction could have created an arc. That coeval mafic magmatism existed on the western side of the Pindos-Budva marginal basin is also compatible with our view and probably resembled the location of the Changpaishan volcano in Manchuria and the unnamed Holocene volcanoes in Korea which are located on the 'continentward' side of the Japan Sea marginal basin.

Along with the Pindos-Budva trough, several deep pelagic basins with mafic volcanism disrupted the continuity of the Apulian neritic platform during the Early to Middle Triassic. The main ones are the Lagonegro and the Sclafani/Imerese basins of southern Italy and Sicily. These basins seem to be the products of the attempts of the propagating Neo-Tethyan ridge of the southern branch of Neo-Tethys to tear further into the central Mediterranean area. Numerous, mainly north-south trending graben complexes of the Southern Alps and the Northern Calcareous Alps are probably also the manifestations of the same phenomenon and have nothing to do with the later opening of the Alpine ocean, which cut across them. A detailed analysis of these basins is outside the scope of this paper.

Another marginal basin related to the Palaeo-Tethyan subduction was that of Karakaya which began opening during the latest Permian (?Carboniferous).

By Permian times, the central and north-west European regions had been strongly deformed as a consequence of the terminal Hercynian collision and its foreland effects (Dewey & Burke 1973, Şengör 1976, Arthaud & Matte 1977) that had produced a number of impactogens on the foreland such as the Oslo graben (Ziegler 1982) and intra-montane basins paralleling the trends of earlier orogenic structures south of the northern Hercynian suture (Boigk & Schöneich 1970) during the latest Carboniferous-earliest Permian times. The Early Permian was a time of active rifting in the future North Atlantic/Arctic area, but the rest of central and north-west Europe, with the exception of the earlier established impactogens, was relatively quiescent (Ziegler 1982). Intense rifting activity began in these areas during the Triassic, which Ziegler (1982) attributes to beginning separation of Laurasia from Gondwanaland along a broad swath of rift terrains occupying much of westernmost and north-west Europe and the western Mediterranean, although there was not much direct relation between the rifts related to the future Atlantic region and those in the Mediterranean area, which were mainly in the eastern and central Mediterranean. As indicated above, our view is that the Mediterranean Triassic rifts were unrelated to the Atlantic/north-west European systems and evolved mainly under the direct influence of the propagating Neo-Tethyan ridge systems and, therefore, indirectly under the influence of the beginning rotation of the Cimmerian continent.

The tectonic picture presented by the Permo-Triassic geology of the Mediterranean region is one of a continental domain in the west passing gradually into a broad carbonate shelf, perforated during the Triassic, by a number of deep pelagic basins with local mafic volcanism in the hinterland of an arc complex. This passed northwards into a transform-type continental margin of Palaeo-Tethys as displayed in Fig. 7a. A portion of this shelf rifted behind the Palaeo-Tethyan arc, to give rise to a number of oceanic basins, the southernmost of which began developing into the southern branch of Neo-Tethys, perhaps because it accommodated the main Neo-Tethyan spreading centre(s) entering the Mediterranean area from the east. The other marginal basins either did not evolve into major oceanic tracts and collapsed later during the Neo-Tethyan evolution, as did the Pindos-Budva trough, or closed during the Palaeo-Tethyan evolution to be incorporated into the Cimmeride orogenic edifice, as did the Karakaya basin.

An analogue for nearly the whole of the latest Permian–Middle Triassic tectonics of the west Tethyan regions is today represented by the Indonesian area. The Sumatra/Java subduction zone swings northward into a transform orientation in the Bay of Bengal, where the

Andaman basin may represent another analogue of the Triassic Transylvanian basin. The Sunda shelf is the site of extensive carbonate deposition. It is perforated by a number of deep basins (Hamilton 1974) and contains isolated, 'intra-plate' volcanism as in the case of the Veteran and the Cu-Lao Re island volcanoes. The Flores back-arc basin (Hamilton 1974) may be comparable to the Karakaya back-arc basin. The active tectonics of the whole of south-east Asia is complex as subduction-related tectonism interferes with intracontinental tectonism influenced by the ongoing India/Eurasia convergence (Peltzer *et al.*, 1982). A similarly complex picture seems to have existed in the Mediterranean area during Permo-Triassic times, with the existence also of a basin-and-range type domain in its central part. Palaeo-Tethys-dominated tectonics clearly had an important role in shaping that picture, perhaps with a little additional help from the rift regions of north-west-Europe.

Late Triassic–Early Jurassic events

During this time interval, rifting of all the major Neo-Tethyan oceans began in the Mediterranean domain as seen in Fig. 7b. In the eastern Mediterranean, the Early to Middle Triassic rift zone probably developed into an actively spreading ridge, to judge from the cessation of extensional tectonic activity along its margins and development of quiet shelf-rise-slope triplets along the margins of the Sinai and off the coast of Israel (Friedman *et al.*, 1971).

Within Turkey and in the island of Lesbos, the Karakaya marginal basin closed during the latest Triassic and the earliest Jurassic. Very shortly thereafter the northern branch of Neo-Tethys, now represented by the Vardar-Intra Pontide-İzmir/Ankara-Erzincan sutures and their along-strike continuations, began opening, perhaps at the same time as the opening of the poorly-known Inner Tauride suture (Şengör & Yılmaz 1981, Oktay 1982). We here follow the interpretation of Şengör *et al.* (1980) and Şengör & Yılmaz (1981) in regarding this new rifting event as renewed back-arc activity over the still active Palaeo-Tethyan subduction zone. This double opening not only separated an independent Kırşehir block (Görür *et al.* this volume) from the rest of the Anatolide/Tauride platform, but reduced the Cimmerian continent to the areas now represented by the Supra-Palaeo-Tethyan units in the Balkan/Carpathian Cimmerides and the Kırklareli, İstanbul, and the Bayburt Nappes in the Anatolian Cimmerides. Because the mutual tectonic and palaeogeographic relations of these nappe systems are largely unknown in northern Turkey, we have little control on the geometry of the Cimmerian continent in the Anatolian Cimmerides. Even if the present discontinuity between the İstanbul and the Bayburt nappes reflects an original gap in the along-strike extension of the Cimmerian continent, it can not have been very large, from the similarity of the structural events and sedimentary facies in both of these nappes during the Dogger to Neocomian interval.

The Palaeo-Tethyan ocean finally closed in north Iran along the suture zone of the South-west Asian Cimmerides during the Late Triassic (pre-Rhaetian: Berberian & King, 1981), whereas collision began during the Lias in the northern parts of the Balkan/Carpathian Cimmerides. These collision events reduced Palaeo-Tethys in the Mediterranean Cimmerides to a remnant ocean basin, in which enormous thicknesses of flysch were deposited, locally formed into large accretionary flysch/mélange wedges, much like the present eastern Mediterranean ridge (Le Pichon, in press). It was in this environment that much of the Akgöl, Lipacka, Nalbantian, and the Tauridian lithologies developed (Fig. 7b). Locally, deep-sea deposition continued into the earliest Dogger, albeit under the influence of powerful clastic supply, as shown by the epi-ophiolitic slate-diabase association of the Artvin area (Şengör *et al.*, 1980).

Although data from the Svanetia anticline and the Abkhasia-Racha zone in the Greater Caucasus point to the terminal elimination of Palaeo-Tethys in the Caucasian Cimmerides during the latest Triassic, the slate-diabase association of Lias-Dogger age in the Slate Zone of the southern slope of the Greater Caucasus (Figs 3 & 7b; see Khain 1975) implies an incipient rifting event of the same age. We interpret this localized rifting, which occurred so shortly after the closure of the Palaeo-Tethyan ocean, as being possibly due to sideways escape of a continental fragment into the still-open oceanic area in the west, as shown in Fig. 7b. The Tscherkessk fault, which was active during the Jurassic (Khain & Milanovsky 1963), may have been one of the boundaries of the sideways moving block, and the slate-diabase rift may have opened as a result of the imperfections of the escape tectonics, analogous to the rifting of the Saros Graben at the west end of the North Anatolian transform fault (Dewey & Şengör 1979). The suggested left-lateral movement of the Greater Caucasus may at least partly be due to this hypothetical escape tectonics.

The pivoting of the Cimmerian continent around a pole somewhere in the Carpathian

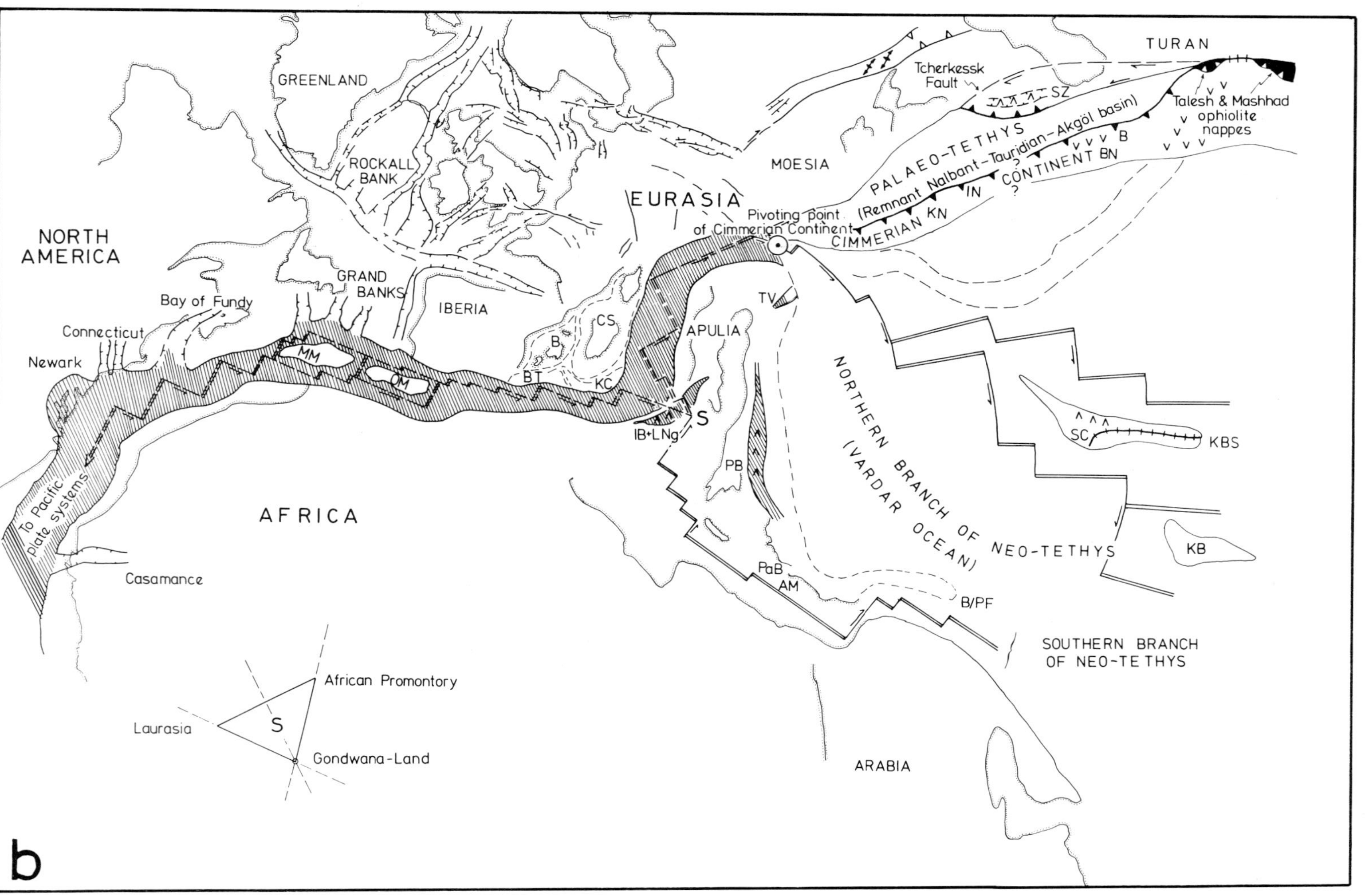
GREENLAND
ROCKALL BANK
TURAN
Tcherkessk Fault
SZ
Talesh & Mashhad ophiolite nappes
MOESIA
PALAEO-TETHYS
(Remnant Nalbant-Tauridian-Akgöl basin)
B
CONTINENT BN
IN
CIMMERIAN KN
EURASIA
Pivoting point of Cimmerian Continent
NORTH AMERICA
GRAND BANKS
IBERIA
Bay of Fundy
Connecticut
Newark
CS
B
BT
KC
MM
OM
APULIA
TV
S
IB+LNg
PB
NORTHERN BRANCH OF NEO-TETHYS
(VARDAR OCEAN)
SC
KBS
KB
To Pacific plate systems
AFRICA
Casamance
PaB
AM
B/PF
SOUTHERN BRANCH OF NEO-TETHYS
African Promontory
Laurasia
S
Gondwana-Land
ARABIA
b

area as shown in Fig. 7b is indicated by the termination of the Palaeo-Tethyan suture within the Rhodope-Pontide fragment and by the wrapping of the latter onto itself as implied by the identical basements of the Macin and the Tulcea zones of the North Dobrudja. While the Cimmerian continent was rotating anticlockwise during the Lias, the accumulation of about 1000 m of Early Jurassic sediments in the Polish Trough was accompanied by considerable tectonic activity that locally involved shoulder uplift along its margins (Ziegler 1982). This deformation may be indicative of the coupling of the west end of the Cimmerian continent to Europe, which may have led to across-the-rotation-pole extension.

We have no direct information to constrain the geometry of the constructive plate margins behind the Cimmerian continent, except that they must have been compatible with the proposed pivoting of the Cimmerian continent. As a working hypothesis we have adopted the view that the pivoting of the Cimmerian continent may have been a consequence of it overriding oceanic lithosphere whose age increased eastward along the strike of the subduction zone. Dewey (1980) has argued that in such a case the orientations of ridge segments responsible for the opening of a marginal basin must rotate from nearly perpendicular to the associated trench near the pivoting point to nearly parallel to it away from the hinge. When such a scheme is applied to the northern branch of Neo-Tethys, it is seen that the ridge orientations change from nearly north-south near the pole of rotation of the Cimmerian continent to east-west farther away. As Dewey (1980) has pointed out, such kinematics violate rigid plate tectonics. In a highly speculative vein, it might be argued that the initial separation between the İstanbul and the Bayburt Nappes may have been due to such non-rigid kinematics. The marginal basin interpretation for the opening of the northern branch of Neo-Tethys carries an important implication that even this Neo-Tethyan opening may have been independent of Atlantic-related tectonics. In fact, the meeting of the two plate boundary systems at the eastern end of the Alpine 'ocean' creates an extremely involved plate boundary configuration that may eventually have led to the tearing away from the rest of Europe of the Moesian platform.

The Late Triassic–Early Jurassic geology of the Mediterranean Cimmerides was dominated by the continued growth of flysch/mélange wedges or molasse deposition, depending on where terminal collision had already occurred between the Cimmerian continent and Laurasia, and island arc magmatism where subduction remained active as in the Bayburt area (Şengör *et al.*, 1980).

The Mediterranean regions outside the Cimmerides were dominated by widespread extension, induced in the east by Palaeo-Tethys-related back-arc activity and in the west by the advanced intra-continental extension in the Central Atlantic. Diffuse extension reaching into the Mediterranean area along intra-continental transform systems led to disintegration of the existing carbonate platforms. As shown in Fig. 7b, the left-lateral transform systems entered the Mediterranean area mainly through the High Atlas, the Saharan Atlas south of the Oran and Moroccan Mesetas (Fig. 7b), and the Mauretanian-Massylian flysch trough (Laubscher & Bernoulli 1977) depicted by the single, dashed ridge/transform system extending between S Iberia and Sicily in Fig. 7b. In the Alps and the Apennines, the diffuse extension later led to the generation of oceanic crust and mantle, whereas further west extension never went far enough to generate real oceanic lithosphere. As Fig. 7b clearly illustrates, the Atlantic-related plate boundary systems were responsible for the opening of all the Mesozoic oceanic and quasi-oceanic areas in the Mediterranean area as far east as the eastern Alps. From the Carpathians westwards, it was Palaeo-Tethyan tectonics that governed the Triassic/Jurassic evolution of the Neo-Tethyan oceans. These two systems of plate boundaries met at two main triple junctions, one near Sicily (**S** in Fig. 7b) and the other near the pivoting point of the Cimmerian continent.

Fig. 7b. The latest Triassic–Early Jurassic reconstruction of the Mediterranean area and surrounding regions. Symbols and lettering same as Fig. 7a.

PaB; Pamphylian basin; AM, Alanya Massif; KBS, Karakaya suture; B/PF, Bitlis/Pötürge fragment; KB, kırşehir block; MM and OM, Moroccan and Oran Mesetas respectively. Obliquely ruled areas are regions of intracontinental extension. The triangle in the southwest corner shows the stability condition of the triple junction S. Notice that 'Tethyan' oceans as far east as the eastern Alps were in reality parts of the Atlantic ocean and had no relation to the Tethyan realm. Sources and discussion in text.

Middle Jurassic events

Fig. 7c displays the geometry of the Mediterranean area and its surroundings as it may have appeared 165 ma ago, when Palaeo-Tethys terminally shut in all segments of the Mediterranean Cimmerides with the possible exception of the S part of the Balkan/Carpathian Cimmerides (Dimitriadis & Dixon 1982, see above), and when ocean-floor generation began in the Central Atlantic/Apennine/Alpine ocean (Sclater *et al.*, 1977). Terminal closure and the final welding of the Cimmerian continent to the southern margin of Laurasia is evidenced by extensive molasse deposition in the Crimea, in the Caucasus, and partly in the central compartment of the main trunk of the Anatolian Cimmerides (Bürnük and the upper parts of the Himmetpaşa Formations), and by quartz-diorite and granodiorite intrusions sealing the thrust contacts that juxtaposed Laurasian and Cimmerian continent rocks and Palaeo-Tethyan lithologies. Throughout the Caucasus, Bajocian/Bathonian granitic magmatism (Khain 1975) may be indicative of crustal thickening and melting as a consequence of ongoing intra-continental convergence, which is suggested by Middle Jurassic angular unconformities (Nalivkin 1976). Curiously, the slate-diabase zone continued its extensional activity amidst this intense collisional tectonics, in the manner of the Gulf of Saros graben of the N Aegean.

In the Balkan/Carpathian Cimmerides, the Liassic molasse sedimentation continued in the northern parts and during the Dogger gradually gave way to quieter, shallow water sedimentation. By contrast, in the southern parts, flysch sedimentation and possibly oceanic conditions continued into the Malm or even into the early Cretaceous, if the re-interpretation by Dimitriadis & Dixon (1982) of ages from the Volvi complex of NE Greece is valid.

By the end of the Middle Jurassic, Palaeo-Tethyan tectonics had no remaining influence in the Mediterranean area, except locally within the Cimmeride orogen. That is why we do not discuss the rest of Mediterranean tectonics, which was independent of the Palaeo-Tethyan evolution.

Late Jurassic–Early Cretaceous events

Except for the Late Jurassic/Neocomian 'Tibet-type' magmatism in the Caucasus area (Şengör *et al.*, 1980), no trace of Cimmeride orogenic events remained anywhere in the west Tethyan domain except for the ruins of the orogenic edifice, on which molasse deposition during the Malm was widespread, particularly in northern Turkey (Fig. 7d). By early Cretaceous times even that edifice had been levelled down to sea-level as shown by the deposition of a thick Lower Cretaceous carbonate prism and a clastic continental apron along the south-facing Atlantic-type continental margin of the northern shore of the northern branch of Neo-Tethys. It was probably during this time interval that the North Dobrudja parted company with the Rhodope-Pontide fragment, and, coevally with the generation of the circum-Moesian oceans, was transferred to its present location along the latter's north-east margin by means of a proto-Peceneaga-Camena fault (Fig. 7d). Thus, by the Cretaceous, the earth's cemetry of fossil orogens had already been enriched by the addition of the corpse of the first child of the maternal Tethys, the Cimmerides.

Conclusions

In the foregoing paragraphs we have summarized the present state of knowledge on the Cimmeride orogenic system in the circum-Black Sea regions and interpreted it in terms of the progressive contraction and final elimination of the Permo-Triassic oceanic embayment of Pangaea, Palaeo-Tethys. For many years, the recognition of the Palaeo-Tethyan suture(s) in the circum-Black Sea area has been a problem for the students of Mediterranean tectonics, mainly because of the paucity of knowledge on ophiolite complexes and convergent plate-margin petro-tectonic assemblages such as fossil magmatic arcs and subduction complexes related to oceanic areas that existed during the late Palaeozoic and that subsequently vanished during the Mesozoic. After intensive detailed search began in the field for possible remnants of Palaeo-Tethyan ocean-floor and active margins around the Black Sea areas following a suggestion of Hsü (1977), it became apparent at once that the reported lack of ophiolites representing the remnants of Palaeo-Tethyan floor was largely due to inadequate or misleading descriptions of these rocks in the literature so hampering their recognition as ophiolites, as the studies of Şengör *et al.* (1980) and Tekeli (1981) in Turkey, and possibly those of Dimitriadis (writ. comm. 1982) and Dimitriadis & Dixon (1982) in Greece have shown.

In only 2 years, mapping by a number of groups has revealed much larger outcrop areas of Palaeo-Tethyan oceanic lithologies than Şengör *et al.* (1980) were able display on their maps. Many occurrences recorded previously as 'basalts', 'amphibolites' or just 'basic rocks'

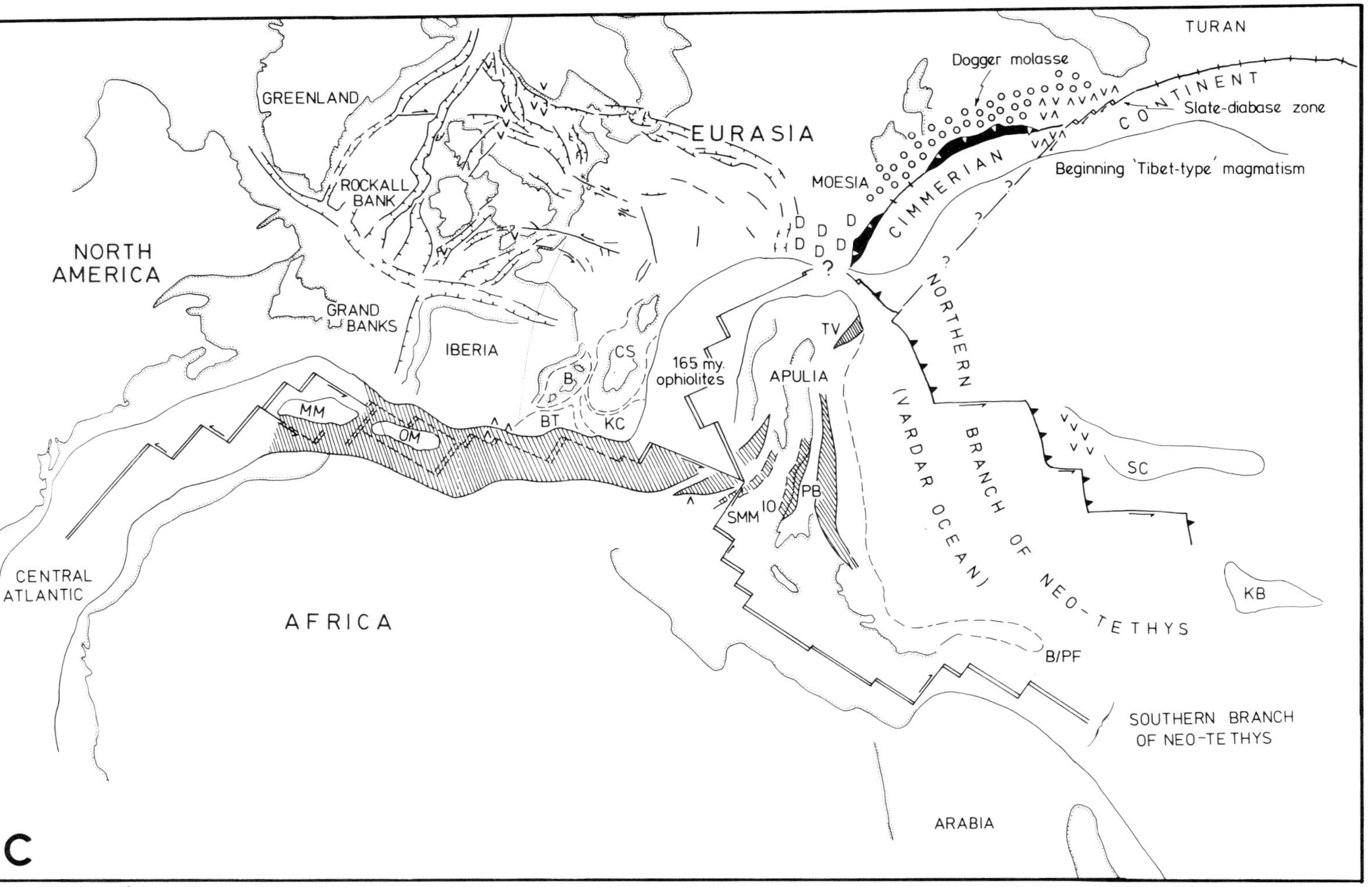

FIG. 7c. Medial Jurassic reconstruction of the Mediterranean area and surrounding regions. Symbols and lettering same as in Figs 7a and 7b. SMM, Sicano-Molise-Marsica basin complex; and IO, Ionian zone. Positions of Africa, Europe and North America are after Sclater *et al*. (1977). Sources and discussion in text.

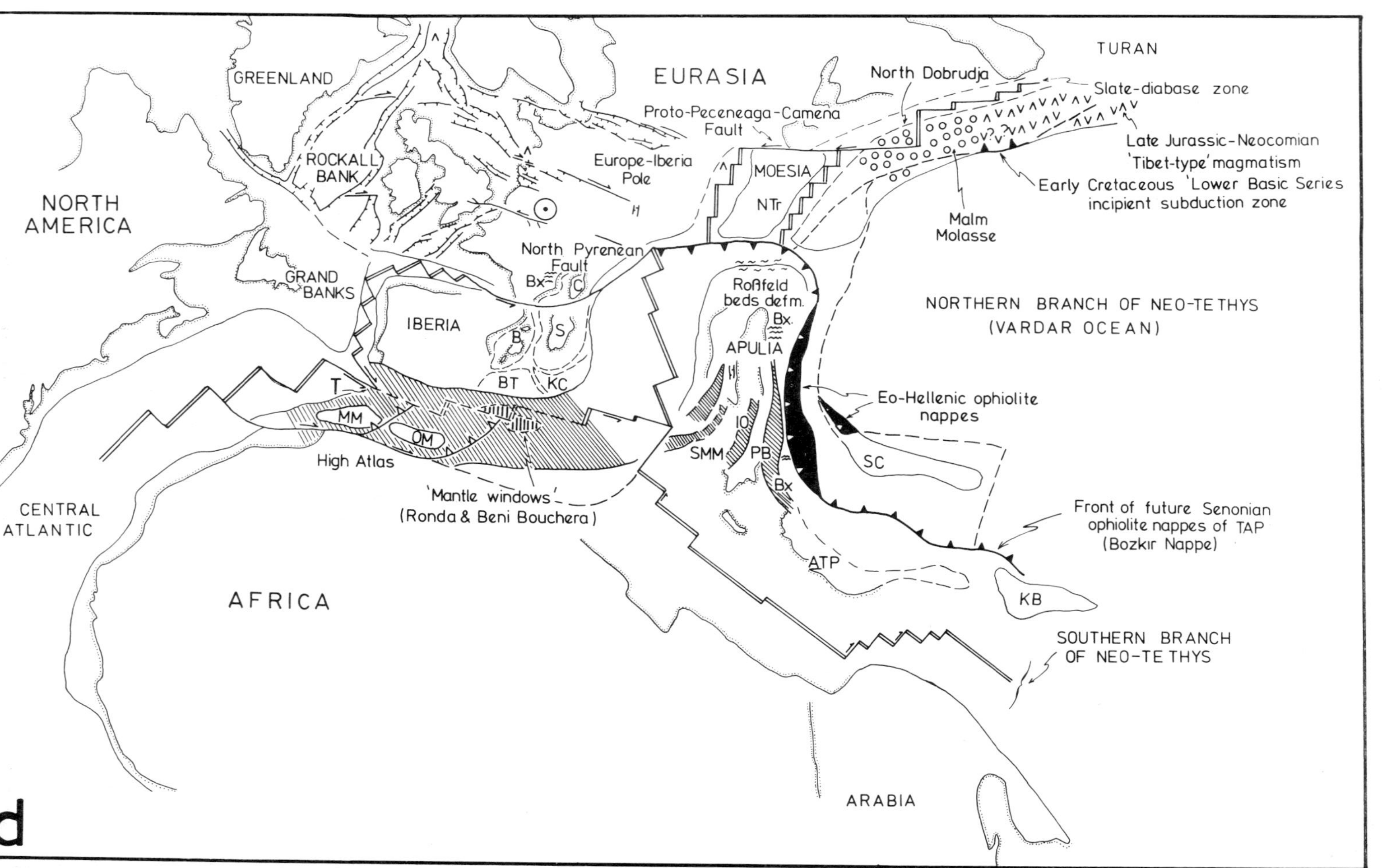

FIG. 7d. The Late Jurassic to Early Cretaceous reconstruction of the Mediterranean area and surrounding regions. Lettering and symbols are as in Figs 7a, b, and c. Positions of Africa, Europe, and North America are after Sclater *et al.* (1977). Discussion in text.

turned out to be fairly well-developed, but now-disrupted, ophiolites, whereas large areas of 'old metamorphic basement' proved to be metamorphosed late Palaeozoic–early Mesozoic flysh/mélange complexes related to Palaeo-Tethyan evolution. Ophiolites and associated oceanic sediments are much more common within the Rhodope-Pontide fragment and in the Sakarya continent than hitherto believed and, doubtlessly, future research will increase the number of known occurrences.

A more serious problem is the cryptic nature of the Palaeo-Tethyan suture along considerable stretches of the Mediterranean Cimmerides. In the Caucasus, for example, not only the large crystalline nappe of the Main Range probably structurally overlies parts of the Palaeo-Tethyan suture, but the post-Triassic sediments of the southern foothills of the Main Range completely hide it. In the eastern Pontides, it is the vast Late Cretaceous-Eocene volcanic and volcaniclastic blanket that covers not only the Palaeo-Tethyan suture, but nearly the entire bulk of the Cimmeride orogen. Experience in northern Turkey has revealed that the only way to deal with the problem of discontinuous outcrop and to delineate the first-order tectonic units of the Cimmeride orogen is terrain-mapping.

A number of Alpide and Hercynian sutures also lie very close to the suture trace of Palaeo-Tethys (see map in Şengör *et al.* 1982). Intense deformation associated with each suturing episode not only obliterates older sutures, but also locally erases the traces of its own products, as shown by the discontinuous ophiolite outcrops along the Indus-Yarlung-Zangbo suture in the Himalaya (Gansser 1980), whereas metamorphism frequently destroys biostratigraphic age information and confuses isotopic ages. Experience in the Cimmerides shows that in such extraordinarily complex areas field mapping alone can easily lead the geologist astray, and a careful, iterative combination of the existing field data with large-scale regional modelling is a necessity to be able to make sense out of individual mapping results.

The recognition of the Palaeo-Tethyan suture from its traditional 'fingerprints' becomes increasingly more difficult westwards along the strike of the Mediterranean Cimmerides, i.e. as the western termination of Palaeo-Tethys is approached. In the Caucasus, even though the suture itself is hidden by thrusts and younger sediment cover, its approximate location is readily discernible from the very disparate stratigraphic evolution of its two sides. Farther west, although palaeogeographic differences between the two sides of Palaeo-Tethys are not remarkable, the development of subduction-accretion complexes and magmatic arcs in the Anatolian Cimmerides clearly indicate where the suture is located. By contrast, neither palaeogeography, nor development of petrotectonic facies offer much help to locate the Palaeo-Tethyan suture in the Balkan/Carpathian Cimmerides. Progressive deterioration of suture indicators along the strike of a mountain belt generally indicates that in that direction the original oceanic area had its terminus. As can be seen in Fig. 7, the Mediterranean Cimmerides provide an excellent example of such a situation.

If the oceanic 'terminus' coincides with the pole of closure of the ocean and only one plate boundary is involved, closure on one side of the pole must lead to opening on the opposite side. Although the pivoting point of the Cimmerian continent *was* connected, since the Lias, to the dominantly transform plate boundary of the Alpine ocean, it seems that now and then the rotating Cimmerian continent may have been coupled to the European lithosphere to induce extension in it as shown by the Liassic tectonism of the Polish Trough. Unfortunately, the geometry and the possible factors influencing the development of the rift and strike-slip tectonics of central and north-west Europe are so complex that it does not now seem possible to isolate with confidence the effects of the rotation of the Cimmerian continent on the coeval intra-continental deformation in Europe.

Argyriadis (1975, 1978) used the occurrence of Hercynian orogeny as a criterion to delimit Laurasia to the south against Gondwanaland, in which he was followed by Ziegler, 1982. However, zones from which Hercynian orogeny is known nearly completely encircle the western termination of Palaeo-Tethys and are now seen on both sides of its suture, particularly in the Balkan/Carpathian and the Anatolian Cimmerides as we have seen. The story of the development of the Hercynian orogens south and north of the Palaeo-Tethyan suture in the Mediterranean Cimmerides cannot be told in this paper, but the terminal collisions that formed those orogens appear to have been responsible for initiating subduction along the margin of Palaeo-Tethys. Thus, although the evolution of the Palaeo-Tethys-bordering Hercynian orogens were causally unrelated to Palaeo-Tethys, they may have triggered its destruction.

Palaeo-Tethyan studies in the Mediterranean region are only just beginning. Despite considerable progress, some of the most fundamental questions posed by Şengör (1979) are still awaiting answers, such as the problem of the continuity of the Cimmerian continent along its strike. Unless little-known areas such as the Tokat Massif are mapped and areas of discontinuous outcrop are explored with geophysical methods, we cannot achieve a satisfactory understanding of the present structure of the Cimmeride orogen in the circum-Black Sea areas. Its evolution poses even grander problems that can only be addressed by a cosmopolitan group of researchers including palaeomagnetists, structural geologists, metamorphic and igneous petrologists, sedimentologists, and palaeontologists.

At present the areas that need the most urgent attention are the eastern Dacides, the southern Carpathians, the Serbo-Macedonian Massif, and the Tokat Massif. In all these areas the main problem is still the exact location of the Palaeo-Tethyan suture.

ACKNOWLEDGMENTS: We are grateful to a very large number of friends and colleagues in Turkey, too many to be listed individually here, for making their data available to us, for valuable discussions and friendly criticisms. Prof. İhsan Ketin has put at our disposal his enormous reservoir of knowledge on the geology of Turkey and surrounding regions during all stages of our work. Without Ketin's active support and encouragement, this paper would have never been presented at Edinburgh and would have never been written. We are grateful to Drs Dixon and Robertson, the conveners of the Edinburgh meeting, for having invited this contribution and for their very generous and friendly help in general. Drs G. Chatalov and P. Gocev generously supplied us with literature on the Bulgarian Strandja. Dr. Gocev also very kindly translated for us some of his work written in Bulgarian. Prof. N. Pantic has been our main guide to the Yugoslavian geological literature, much of which he provided for us himself. Drs M. Sandulescu and H. Savu supplied us with literature on the Romanian Carpathians and the Dobrudja. Prof. V. E. Khain has provided some of the Caucasian geological literature for us and has been a very helpful discussion partner. Prof. S. Erinç kindly translated some of the Soviet literature for us. Dr S. Dimitriadis deserves a very special note of thanks for having given us his unpublished data on the Grecian part of the Serbo-Macedonian Massif. Without his assistance our story of the Balkan/Carpathian Cimmerides would have been much less complete. We owe the initial stimulus for this research to Prof. K. J. Hsü when he visited Turkey in 1978. Prof. B. C. Burchfiel has contributed to our work substantially by discussions. For similar reasons we are grateful to Professors Kevin Burke and John Dewey.

References

ABDÜSSELÂMOĞLU, M. Ş. 1977. *The Palaeozoic and Mesozoic in the Gebze region—explanatory text and excursion guidebook, 4th Colloquium on the Geology of the Aegean Region*, Exc. 4. İ.T.Ü. Maden Fak., İstanbul, 16 pp.

AD HOC PANEL. 1979. *Continental margins, geological and geophysical research needs and problems*. The Nat. Res. Council of the Nat. Acad. Sci., Washington, D.C., 302 pp.

AGER, D. V. 1980. *The Geology of Europe*. John Wiley, New York, 535 pp.

AIELLO, E., BARTOLINI, C., BOCCALETTI, M., GOCEV, P., KARAGJULEVA, J., KOSTADINOV, V. & MANETTI, P. 1977. Sedimentary features of the Srednogorie zone (Bulgaria): an Upper Cretaceous intra-arc basin. *Sed. Geol.* **19,** 36–68.

AKARTUNA, M. 1953. Çatalca-Karacaköy bölgesinin jeolojisi. *Ist. Univ. Fen Fak. Monogr. (Tabiî İlimler Kısmi)* **13,** 88 pp.

AKAY, E. 1981. Beyşehir yöresinde (Orta Toroslar) olası Alt Kimmeriyen dağoluşumu izleri. *Türk. Jeol. Kur. Bülteni* **24,** 25–29.

AKKÖK, R. 1982. *Menderes Masifi'nin Jeolojisi Paneli*. TMMOB Jeoloji Mühendisleri Odası & Türkiye Jeoloji Kurumu, Ankara, 39–42.

ALAVI, M. 1979. The Virani ophiolite complex and surrounding rocks. *Geol. Rundsch.* **68,** 331–341.

ALP, D. 1972. Amasya yöresinin jeolojisi. *İst. Üniv. Fen Fak. Monogr. (Tabiî Ilimler Kısmı)* **22,** 101 pp.

ARGYRIADIS, I. 1975. Mésogée Permienne, chaîne hercynienne et cassure téthysienne. *Bull. Soc. géol. Fr.* 17(7) 56–67.

—— 1978. *Le Permien Alpino-Méditerranéen à la charnière entre l'Hercynien et l'Alpin*. Thèse de Docteur ès sciences, Univ. Paris-Sud, Orsay **1,** 302 pp., **2,** 190 pp.

ARPAT, E., TÜTÜNCÜ, K., UYSAL, Ş. & GÖĞER, E. 1978. Safranbolu yöresinde Kambriyen-Devoniyen istifi. *Türk. Jeol. Kur. 32. Bilimsel ve Teknik Kurultayı, Bildiri Özetleri*, 67–68.

ARTHAUD, F. & MATTE, P. 1977. Late Paleozoic strike-slip faulting in southern Europe and northern Africa: Result of a right-lateral shear zone between the Appalachians and the Urals. *Bull. geol. Soc. Am.* **88,** 1305–1320.

AYDIN, Y. 1974. *Etude pétrographique et géochimique de la partie centrale du Massif d'Istranca (Turquie)*. Thèse de Docteur Ingénieur, Université de Nancy, **1,** 131 pp.

AYKOL, A. 1979. *Kırklareli-Demirköy sokulumu'nun petroloji ve jeokimyası*. Habilitation Thesis, İ.T.Ü. Maden Fak., İstanbul, 204 pp.

BERBERIAN, M. & KING, G. C. P. 1981. Towards a palaeogeography and tectonic evolution of Iran. *Can. J. Earth Sci.* **18,** 210–265.

BERGOUGNAN, H. & PARROT, J.-F. 1981. Le carrefour ophiolitique d'Erzincan, Néo et Paléotéthys. *Cah. O.R.S.T.O.M., sér. Géol.*, **11,** 165–188.

BIJU-DUVAL, B., DECOURT, J. & LE PICHON, X. 1977. From the Tethys ocean to the Mediterranean seas: A plate tectonic model for the evolution of the western Alpine System. *In:* Biju-Duval, B. Montadert, L. (eds), *Structural History of the Mediterranean Basins*, 143–164. Editions Technip, Paris.

Bingöl, E. 1978. Explanatory notes to the metamorphic map of Turkey. *In:* Zwart, H. J. (ed) *Metamorphic Map of Europe 1:2,500,000 Explanatory Text*, 148–154, Unesco, Leiden.

——, Akyürek, B. & Korkmazer, B. 1973. Geology of the Biga Peninsul and some characteristics of the Karakaya Formation. *In: Congress of Earth Sciences on the Occasion of the 50th Anniversary of the Turkish Republic*, 71–77, Mineral Research & Exploration Institute, Ankara.

Blumenthal, M. 1950. Beitraege zur Geologie des Landschaften am Mittleren und Unteren Yeşil Irmak ((Tokat, Amasya, Havza, Erbaa, Niskar). *Veröff. Inst. Lagerstaettenforsch. Türkei*, Ser. D, **4**, 153 pp.

Boigk, H. & Schöneich, H. 1970. Die Tiefenlage der Permbasis im nördlichen Teil des Oberrheingrabens. *In:* Illies, J. H. & Mueller, S., eds., *Graben Problems*, E. Schweizerbart'sche Verlagsbuchhandlung, Stuttgart, 45–55.

Boillot, G. & Capdevila, R. 1977. The Pyrenees: subduction and collision? *Earth planet. Sci. Lett.* **35**, 151–160.

Bonnet, P. & Bonnet, N. M. 1947. Description géologique de la Transcaucasie méridionale. *Mém. Soc. géol. Fr.*, nouv. sér. **25**, 292 pp.

Borisyak, A. 1903. Ueber die Tektonik des Donez-Höhenzuges in seinen nordwestlichen Ausläufern. *Centralbl. Min. Geo. Palaeont.*, **1**, 644–649.

Bourrouilh, R. & Gorsline, D. S. 1979. Pre-Triassic fit and alpine tectonics of continental blocks in the western Mediterranean. *Bull. geol. Soc. Am.* **90(I)**, 1074–1083.

Brinkmann, R., Flügel, E., Jacobshagen, V., Lechner, H., Rendel, B. & Trick, P. 1972. Trias, Jura und Unterkreide der Halbinsel Karaburun (West-Anatolien). *Geologica et Palaeontologica* **6**, 139–150.

von Bubnoff, S. 1931. *Grundprobleme der Geologie*. Gebr. Borntraeger, Berlin, 237 pp.

Bullard, E. C., Everett, J. E. & Smith, A. G. 1965. The fit of the continents around the Atlantic. *Royal Soc. London Phil. Trans.* A258, 41–51.

Burchfiel, B. C. 1976. Geology of Romania. *Geol. Soc. Am. Spec. Pap.* 158, 82 pp.

—— 1980. Eastern European Alpine System and the Carpathian orocline as an example of collision tectonics. *Tectonophysics* **63**, 31–61.

Burke, K. 1977. Aulacogens and continental breakup. *Ann. Rev. Earth planet. Sci.* **5**, 371–396.

—— & Dewey, J. F. 1973. Plume-generated triple junctions: key indicators in applying plate tectonics to old rocks. *Jour. Geol.* **81**, 406–433.

Burollet, P. F. & Rouvier, H. 1971. La Tunisie. *In: Tectonique de l'Afrique*, 91–100, Unesco, Paris.

Capedri, S., Venturelli, G., Bocchi, G., Dostal, J., Garuti, G. & Rossi, A. 1980. The geochemistry and petrogenesis of an ophiolitic sequence from Pindos, Greece. *Contrib. Mineral. Petrol.* **71**, 1–12.

Castellarin, A., Lucchini, F., Rossi, P. L., Simboli, G., Bosellini, A. & Sommavilla, E. 1979. Middle Triassic magmatism in the Southern Alps, II: A geodynamic model. *Riv. ital. Paleont. (Stratigr.)* **85**, 1111–1124.

—— & Rossi, P. M. 1981. The Southern Alps: an aborted Middle Triassic mountain chain? *Eclog. geol. Helvet.* **74**, 313–316.

Channel, J. E. T., D'Argenio, B. & Horvath, F. 1979. Adria, the African promontory, in Mesozoic Mediterranean palaeogeography. *Earth Sci. Rev.* **15**, 213–292.

Chatalov, G. 1977. A new facial type of Triassic in the Strandja Mountain. *C.r. Acad. bulgare Sci.* **30**, 1605–1608.

—— 1978. Triassic rocks around Bosna Pea, (Strandja Mountain). *C.r. Acad. bulgare Sci.* **31**, 1163–1166.

—— 1979. Two facies types of Triassic in Strandza Mountain, SE Bulgaria. *Riv. İtal. Paleont. (Stratigr.)* **85**, 1029–1046.

Codarcea, A., Raileanu, G., Pavelescu, L., Gherasi, N., Nastaseanu, S., Bercia, I. & Mercus, D. 1961. *Aperçu sur la structure géologique des Carpates Méridionales éntre le Danube et l'Olt.* V ème Congr. Assoc. géol. Carpato-Balkanique, Guide des Excursions, C.-Carpates Méridionales, Bucarest, 131 pp.

Cohen, C. 1980. Plate tectonic model for the Oligo-Miocene evolution of the Western Mediterranean. *Tectonophysics* **68**, 283–311.

Dewey, J. F. 1980. Episodicity, sequence, and style at convergent plate boundaries. *Geol. Assoc. Canada Spec. Pap.* **20**, 553–573.

—— & Burke, K. 1973. Tibetan, Variscan and Precambrian basement reactivation: products of continental collision. *Jour. Geol.* **81**, 683–692.

——, Pitman, W. C. III, Ryan, W. B. F. & Bonnin, J. 1973. Plate tectonics and the evolution of the Alpine System: *Bull. geol. Soc. Am.* **84**, 3137–3180.

—— & Şengör, A. M. C. 1979. Aegean and surrounding regions: complex multi-plate and continuum tectonics in a convergent zone *Bull. geol. Soc. Am.* **90(I)**, 82–91.

De Wit, M. J. 1977. The evolution of the Scotia arc as a key to the reconstruction of southwestern Gondwanaland. *Tectonophysics* **37**, 53–81.

Dietrich, V. J. 1979. Ophiolitic belts of the central Mediterranean. *Internat. Atlas of Ophiolites, Geol. Soc. Am. Map Chart Ser.* MC-33, sheet 2.

Dimitriadis, S. & Dixon, J. E. 1982. Metamorphosed ophiolitic rocks from the Serbo-Macedonian Zone, Lake Volvi, Greece. *The Geological Evolution of the Eastern Mediterranean, Abstracts*, **32**, Edinburgh.

Dorr, J. A., Spearing, D. R. & Steidtmann, J. R. 1977. Deformation and deposition between a foreland uplift and an impinging thrust belt: Hoback basin, Wyoming. *Geol. Soc. Am. Spec. Pap.* **177**, 82 pp.

Enay, R. 1976. Faunes anatoliennes (Ammonitina, Jurassique) et domaines biogéographiques nord et sud téthysienne. *Bull. Soc. géol. Fr.* **18(7),** 533–541.

FRIEDMAN, G. M., BARZEL, A. & DERIN, B. 1971. Paleoenvironments of the Jurassic in the coastal belt of northern and central Israel and their significance in the search for petroleum reservoirs. *Geol. Surv. Israel, Rep.* OD/1/81.

GANSSER, A. 1980. The significance of the Himalayan suture zone. *Tectonophysics 62*, 37–52.

GOCEV, P. 1976. L'évolution géotectonique du mégabloc bulgare pendant le Trias et le Jurassique. *Bull. Soc. géol. Fr.* **18(7),** 209–216.

—— 1979. The place of Strandza in the Alpine structure of the Balkan Peninsula. *Rev. Bul. Geol. Soc.* **40,** 27–46 (in Bulgarian with English summary).

GRUBIC, A. 1980. Yougoslavie. *In: Géologie des Pays Européens (Espagne, Grèce, Italie, Portugal, Yougoslavie)*, 287–342, Dunod, Paris.

GÜNER, M. 1980. Küre civarının masif sülfit yatakları ve jeolojisi, Pontidler (Kuzey Türkiye). *Maden Tetkik ve Arama Enst. Der.* 93/94, 65–109.

GUTNIC, M., MONOD, O., POISSON, A. & DUMONT, J.-F. 1979. Géologie des Taurides occidentales (Turquie). *Mém. Soc. géol. Fr.*, nouv. sér. **58,** 112 pp.

HAMILTON, W. 1974. Sedimentary basins of the Indonesian Region. *U.S. Geol. Survey.*

HECHT, J. 1972. Zur Geologie von Südost-Lesbos (Griechenland). *Zeitschr. deutsch. geol. Ges.* **123,** 423–432.

HERZ, N. & SAVU, H. 1974. Plate tectonics history of Romania. *Bull. geol. Soc. Am.* **85,** 1429–1440.

HSÜ, K. J. 1977. Tectonic evolution of the Mediterranean basins. *In:* NAIRN, A. E. M., KANES, W. H. & STEHLI, F. G. (eds.) *The Ocean Basins and Margins, 4A, The Eastern Mediterranean,* 29–75, Plenum, New York.

——, NACHEV, I. K. & VUCHEV, V. T. 1977. Geologic evolution of Bulgaria in light of plate tectonics. *Tectonophysics* **40,** 245–256.

IANOVICI, V., GIUŞCA, D., MUTINAC, V., MIRAUTA, O. & CHIRIAC, M., 1961. *Aperçu général sur la géologie de la Dobrogea.* Véme Congr. Assoc. géol. Carpato-Balkanique, Guide des Excursions, D.-Dobrogea, Bucarest 92 pp.

JANISHEVSKI, A. 1946. Short overview of the geology of the Strandja Mountain in SE Bulgaria. *In:* COHEN, E. R., DIMITROV, T. & KAMENOFF, B. (eds) *Geology of Bulgaria* **4,** 380–389, Sofia (In Bulgarian).

KALVACHEVA, R. K. & CHATALOV, G. 1974. Palynomorphen aus den phyllitoiden Tonschiefern des Strandza-Gebirges. *C.r. Acad. bulgare Sci.* **27,** 1419–1422.

KARAGJULEVA, J., KOSTADINOV, V. & ZAGORCEV, I. 1974. Tectonic characteristic of the Kraistides. *In:* MAHEL, M. (ed.) *Tectonics of the Carpathian/Balkan Regions*, 332–340, Geological Institute of Dionyz Stur, Bratislava.

KAUFFMANN, G., KOCKEL, F. & MOLLAT, H. 1976. Notes on the stratigraphic and palaeogeographic position of the Svoula Formation in the innermost zone of the Hellenides. *Bull. Soc. géol. Fr.* **18(7),** 225–230.

KAYA, O. 1971. Istanbul'un Karbonifer stratigrafisi. *Türk. Jeol. Kur. Bült.* **14,** 143–199.

—— & LYS, M. 1980. İstanbul Boğazının batı yakasında yeni bir Triyas bulgusu. *Maden Tetkik ve Arama Enst. Der.* **93/94,** 20–26.

KELTS, K. 1981. A comparison of some aspects of sedimentation and translational tectonics from the Gulf of California and the Mesozoic Tethys, Northern Penninic Margin. *Eclog. geol. Helvet.* **74,** 317–338.

KETİN, İ. 1951. Über die Geologie der Gegend von Bayburt in Nordost Anatolien. *Rev. Fac. Sci. Univ. İstanbul* B16, 113–127.

—— 1961. Erzincan-Tercan bölgesinin jeolojisi ve maden yatakları hakkında kısa not. *Maden Mecmuası* **2,** 3–9.

—— 1966. Tectonic units of Anatolia (Asia Minor). *Min. Res. Expl. Inst. Bull.* **66,** 23–34.

KHAIN, V. E. 1975. Structure and main stages in the tectonomagmatic development of the Caucasus: An attempt at geodynamic interpretation. *Am. Jour. Sci.* 275–A, 131–156.

—— & LEONOV, Y. 1975. *Carte tectonique de l'Europe et des régions avoisinantes*, 1:10,000,000. Dir. Gén. Géodésie et de Cartogr., Conseil des Ministres de l'URSS (published 1979).

—— & MILANOVSKY, E.-E. 1963. Structure tectonique du Caucase d'après données modernes. *In: Livre à la Mémoire du Professeur Paul Fallot* **2,** 663–703, Mém. hors sér. Soc. géol. Fr., Paris.

KOBER, L. 1931. *Das Alpine Europa.* Gebr. Borntraeger, Berlin, 310 pp.

KOCKEL, F., MOLLAT, H. & WALTHER, H. W. 1971. Geologie des Serbo-Mazedonischen Massivs und seines mesozoischen Rahmens (Nordgriechenland) *Geol. Jb.* **89,** 529–551.

——, —— & —— 1972. New Facts and ideas on the innermost zones of the Hellenides (a comprehensive view). *Zeitschr. deutsch. geol. Ges.* **123,** 349–352.

——, —— & —— 1977. *Erläuterungen zur geologischen Karte der Chalkidhiki und angrenzender Gebiete, 1:100,000 (Nordgriechenland).* Bundesstalt für Geowissenschaften und Rohstoffe, Hannover, 119 pp.

KONASHOV, V. C. 1980. Expression of early Kimmerian folding phase in the Donetz basin. *Geotectonics* **14,** 268–272.

KOTANSKI, Z. 1978. The Caucasus, Crimea and their foreland (Scythian platform). The Black Sea and Caspian Sea. *In:* LEMOINE, M. (ed.) *Geological Atlas of Alpine Europe and Adjoining Alpine Areas*, 545–576, Elsevier, Amsterdam.

KRASNY, L. I. 1960. *Structure géologique de l'U.R.S.S., v. III, Tectonique.* CNRS, Paris, 502 pp.

KUMMEL, B. & TEICHERT, C. 1970. Stratigraphy and paleontology of the Permian-Triassic boundary beds, Salt Range and Trans-Indus Ranges, West Pakistan. *In:* KUMMEL, B. & TEICHERT, C. (eds) *Stratigraphic Boundary Problems*, 1–151, The University of Kansas Press, Lawrence.

LAUBSCHER, H.-P. & BERNOULLI, D. 1977. Mediterranean and Tethys. *In:* NAIRN, A. E. M., KANES,

W. H. & STEHLI, F. G. (eds) *The Ocean Basins and Margins, 4A, The Eastern Mediterranean*, 1–28, Plenum, New York.

LE PICHON, X. 1983. Land-locked oceanic basins and continental collision: the eastern Mediterranean as a case example. *In:* HSÜ, K. J., ed., *Symposium on Mountain-Building*. Academic Press, London.

LETOUZEY, J., BIJU-DUVAL, B., DORKEL, A., GONNARD, R., KRISTCHEV, K., MONTADERT, L. & SUNGURLU, O. 1977. The Black Sea: a marginal basin. Geophysical and geological data. *In:* BIJU-DUVAL, B. & MONTADERT, L. (eds), *Structural History of the Mediterranean Basins*, 363–376. Editions Technip, Paris.

LIVACCARI, R. 1979. Late Cenozoic tectonic evolution of the western United States. *Geology* **7**, 72–75.

LOTZE, F. 1974. *Geologie Mitteleuropas*, E. Schweizerbart'sche Verlagsbuchhandlung, Stuttgart, 491 pp.

LOWRIE, W. 1980. A palaeomagnetic overview of the Alpine system. *In: Colloque 5, Géologie des Chaînes Alpines issues de la Téthys*, 26e Congr. Géol. Int., Paris 316–330.

MAXWELL, J. C. 1970. The Mediterranean, ophiolites, and continental drift. *In:* JOHNSON, H. & BENNETT, L. S. (eds), *The Megatectonics of Continents and Oceans*, 167–193, Rutgers University Press, New Brunswick.

MILANOVSKY, E.-E. 1981. Aulacogens of ancient platforms: problems of their origin and tectonic development. *Tectonophysics* **73,** 213–248.

MITCHELL, A. H. G. & BELL, J. D. 1973. Island arc evolution and related mineral deposits. *J. Geol.* **81,** 381–405.

MTA, 1961. *Geological Map of Turkey, 1:500,000, The Sinop Sheet.* Maden Tetkik ve Arama Enst., Ankara.

—— 1974. *Geologic Map of the Cide-Kuruçaşile Region, 1:50,000.* Maden Tetkik ve Arama Enstitüsü, Ankara.

NAIRN, A. E. M., KANES, W. H. & STEHLI, F. G. (eds) 1977a. *The Ocean Basins and Margins*, 4A, *The Eastern Mediterranean*. Plenum, New York, 503 pp.

——, —— & —— (eds). 1977b. The Ocean *Basins and Margins*, 4B, *The Western Mediterranean*, Plenum, New York, 447 pp.

NALIVKIN, V. (ed). 1976. Geological Map of the Caucasus, 1:500,000.

OBERHAUSER, R. 1980a. Das Altalpidikum. *In:* OBERHAUSER, R. (ed.), *Der Geologische Aufbau Österreichs*, 35–48, Springer-Verlag, Wien.

OBERHAUSER, R. (ed.) 1980b. *Der Geologische Aufbau Österreichs.* Springer-Verlag, Wien, 700 pp.

OGNIBEN, L., PAROTTO, M., PRATURLON, A. (eds) 1975. *Structural Model of Italy.* Consiglio Nazionale delle Ricerce, Roma 502 pp. 4 sheets.

OKTAY, F. Y. 1982. Ulukışla ve çevresinin stratigrafisi ve jeolojik evrimi. *Türk. Jeol. Kur. Bülteni* **25,** 15–25.

ÖZDEMİR, Ü., TALAY, G. & YURTSEVER, A. 1973. Biostratigraphy of the rocks from Kocaeli Peninsula. *In: Congress of Earth Sciences on the Occasion of the 50th Anniversary of the Turkish Republic*, 115–130, Mineral Research & Exploration Institute, Ankara.

ÖZGÜL, N. 1976. Torosların bazı temel jeoloji özellikleri. *Türk. Jeol. Kur. Bült.* **19,** 65–78.

PAMİR, H. N. & BAYKAL, F. 1947. Le Massif de Stranca. *Türk. Jeol Kur. Bült.* **1,** 26–43.

PEIVE, A. V., YANSHIN, A. L., ZONENSHAIN, L. P., KNIPPER, A. L., MARKOV, M. S., MOSSAKOVSKY, A. A., PERFILIEV, A. S., PUSCHAROVSKY, Y. M., SHLEZINGER, A. E. & SHTREIS, N. A. 1976. *Fundamental Principles in the compilation of the tectonic map of northern Eurasia.* Mimeographed text, 106 pp.

PELTZER, G., TAPPONNIER, P. & COBBOLD, P. 1982, Les grands décrochements de l'Est Asiatique: évolution dans le temps et comparaison avec un modèle expérimental. *C.r. hébd. Acad. Sci. Paris* **294,** 1341–1348.

PETRASCHECK, W. E. 1960. Über ostmediterrane Gebirgszusammenhänge. *Abh. dtsch. Akad. Wiss. Berlin*, K1. III **1,** 9–18.

PINDELL, J. & DEWEY, J. F. 1982. Permo-Triassic reconstruction of western Pangaea and the evolution of the Gulf of Mexico/Caribbean region. *Tectonics* **1,** 179–212.

PIPER, D. J. W. & PIPER, G. 1982. Depositional setting of Pindos zone rocks of the Pelponnese, Greece. *The Geological Evolution of the Eastern Mediterranean, Abstracts*, **87,** Edinburgh, U.K.

PISA, G., CASTELLARIN, A., LUCCHINI, F., ROSSI, P. L., SIMBOLI, G., BOSELLINI, A. & SOMMAVILLA, E. 1979. Middle Triassic magmatism in the Southern Alps, I: a review of general data in the dolomites. *Riv. ital. Paleont. (Stratigr.)* **85,** 1093–1110.

PREDA, D. M. 1940. Sur la présence d'une tectonique cimmérienne dans les Carpates orientales. *C.r. Inst. géol. Rom.* **24,** 68–77.

ROYDEN, L. H., HORVATH, F. & BURCHFIEL, B. C. 1982. Transform faulting, extension and subduction in the Carpathian Pannonian region. *Bull. geol. Soc. Am.* **93,** 717–725.

SANDULESCU, M. 1975. Essai de synthèse structurale des Carpates. *Bull. Soc. géol. Fr.* **17(7),** 299–358.

—— 1978. The Moesic platform and the North Dobrogean Orogene. In: LEMOINE, M. (ed), *Geological Atlas of Alpine Europe and Adjoining Alpine Areas*, 427–442, Elsevier, Amsterdam.

—— 1980. Analyse géotectonique des chaînes alpines situées autour de la Mer Noire occidentale. *Ann. Inst. Géol. Géophys.* **56,** 5–54.

—— & BERCIA, I. 1974 The East Carpathians: The Crystalline-Mesozoic zone. *In:* MAHEL, M. (ed), *Tectonics of the Carpathian/Balkan Regions*, 240–253, Geological Institute of Dionyz Stur, Bratislava.

SAVU, H. 1980. Genesis of the Alpine cycle ophiolites from Romania and their associated calc-alkaline volcanics. *Ann. Inst. Géol. Géophys.* **56,** 55–77.

SCLATER, J. G., HELLINGER, S. & TAPSCOTT, C. 1977. The palaeobathymetry of the Atlantic ocean from the Jurassic to the present. *Jour. Geol.* **85,** 509–552.

ŞENGÖR, A. M. C. 1976. Collision of irregular con-

tinental margins: implications for foreland deformation of Alpine-type orogens. *Geology* **4,** 79–82.
—— 1979. Mid-Mesozoic closure of Permo-Triassic Tethys and its implications. *Nature London* **279,** 590–593.
—— 1981. The evolution of Palaeo-Tethys in the Tibetan segment of the Alpides. *In: Proc. Symp. Qinghai-Xizang (Tibet) Plateau* **1,** 51–56, Science Press, Beijing.
—— Evolution of the Cimmeride orogenic system and its implications for the tectonics of Eurasia: Products of the closure of Palaeo-Tethys. *Geol. Soc. America Spec. Pap.* (in press).
—— & Yilmaz, Y. 1981. Tethyan evolution of Turkey: a plate tectonic approach. *Tectonophysics* **75,** 181–241.
——, —— & Ketin, İ. 1980. Remnants of a pre-late Jurassic ocean in northern Turkey: fragments of Permian-Triassic Palaeo-Tethys? *Bull. geol. Soc. Am.* **91(I),** 599–609.
——, —— & —— 1982. Reply. *Bull. geol. Soc. Am.* **93,** 932–936.
Seymen, İ. 1975. *Kelkit vadisi kesiminde Kuzey Anadolu Fay Zonunun tektonik özelliği.* İ.T.Ü. Maden Fak. Yay., İstanbul, 192 pp.
Sillitoe, R. H. 1972. Relation of metal provinces in western America to subduction of oceanic lithosphere. *Bull. geol. Soc. Am.* **83,** 813–818.
Snyder, W. S., Dickinson, W. R. & Silberman, M. L. 1976. Tectonic implications of space-time patterns of Cenozoic magmatism in the Western United States. *Earth planet. Sci. Lett.* **32,** 91–106.
Smith, A. G. 1971. Alpine deformation and the oceanic areas of the Tethys, Mediterranean and Atlantic. *Bull. geol. Soc. Am.* **82,** 2039–2070.
—— 1973. The so-called Tethyan ophiolites. *In:* Tarling, D. H. & Runcorn, S. K. (eds), *Implications of Continental Drift to the Earth Sciences* **2,** 977–986, Academic Press, London.
——, Hurley, A. M. & Briden, J C. 1981. *Phanerozoic Palaeocontinental Maps.* Cambridge University Press, Cambridge, 102 pp.
Stille, H. 1953. Der geotektonische Werdegang der Karpaten. *Beih. geol. Jb.* **8,** 239 pp.
Stöcklin, J. 1974. Possible ancient continental margins in Iran. *In:* Burk, C. A. & Drake, C. L. (eds), The Geology of Continental Margins, 873–887, Springer-Verlag, Berlin.
Stämpfli, G. M. 1978. *Etude géologique générale de l'Elburz oriental au S de Gonbad-e-Qabus, Iran N-E.* Thèse de Docteur ès Sciences, no. 1868, Univ. Genève, 315 pp.
Suess, E. 1909. *Das Antlitz der Erde*, 3/II. Tempsky, Wien, 789 pp.
Swanson, M. T. 1982. Preliminary model for an early transform history in central Atlantic rifting. *Geology* **10,** 317–320.
Tapponnier, P. 1977. Evolution tectonique du Système Alpin en Méditerranée: poinçonnement et écrasement rigid-plastique. *Bull. Soc. géol. Fr.* **19(7),** 437–460.
Tatar, Y. 1973. Ophiolites around Conur village to the southeast of Refahiye. *In: Congress of Earth Sciences on the Occasion of the 50th Anniversary of the Turkish Republic*, 437–446, Mineral Research & Exploration Institute, Ankara.
Tekeli, O. 1981. Subduction complex of pre-Jurassic age, northern Anatolia, Turkey. *Geology* **9,** 68–72.
Tollmann, A. 1965. Das Strandscha-Fenster, ein neues Fenster der Metamorphiden im alpinen Nordstamm des Balkans. *N. Jb. Geol. Paläont., Mh.* **4,** 234–248.
—— 1968. Der Deckenbau im mediterranen Orogen mit besonderer Berücksichtigung des Balkangebirges. *Bull. geol. Inst., ser. Geotectonics, Stratigraphy and Lythology* **17,** 53–60.
Wilson, J. T. 1963. Continental drift. *Scientific American*, April 1963, 2–16.
Yilmaz, O. 1979. *Daday-Devrekâni Masifi Kuzeydoğu Kesimi Metamorfitleri.* Habilitation Thesis, Hacettepe Üniversitesi, Ankara 134 pp.
Yilmaz, Y. 1973. emplacement of the Gümüşhane Granite. *In: Congress of Earth Sciences on the Occasion of the 50th Anniversary of the Turkish Republic*, 489–492, Mineral Research & Exploration Institute, Ankara.
—— 1981. Sakarya Kıtası güney kenarının evrimi *Istanbul Yerbilimleri* **1,** 33–52.
Ziegler, P. A. 1982. *Geological Atlas of Western and Central Europe.* Shell Internationale Petroleum Maatschappij, The Hague, 130 pp.
Ziegler, A. M., Scotese, C. R., McKerrow, W. S., Johnson, M. E. & Bambach, R. K. 1979. Palaeozoic paleogeography. *Ann. Rev. Earth Planet. Sci.* **7,** 473–502.

A. M. C. Şengör, İ.T.U. Maden Fakültesi, Jeoloji Mühendisliği Bölümü, Teşvikiye İstanbul, Turkey.
Y. Yilmaz, İstanbul Üniversitesi, Mühendislik Fakültesi, Jeoloji Mühendisliği Bölümü, Vezneciler, İstanbul, Turkey.
O. Sungurlu, T.P.A.O. Arama Grubu, Müdafaa Cad. 22, P.K. 209, Ankara, Turkey.

Evidence for a Late Triassic–Early Jurassic orogenic event in the Taurides

O. Monod & E. Akay

SUMMARY: A simple view of the Tethys during most of the Mesozoic emphasizes the contrast between the active northern European margin and the passive Southern Gondwanan margin which is considered to remain undeformed until the end of the Cretaceous. However, evidence from various places in the Taurus chain demonstrates the existence of a pre-Jurassic unconformity and a thrusting event. Indirect evidence is also provided by coarse, red oligomict clastic sediments (Çayır Formation) of Late Triassic to Lower Liassic age which are widespread on the older Tauride carbonate platform. A review of the data reveals a wide distribution of these sediments in southern Turkey, and their possible extension into the Hellenides (Chios). Finally, the evidence of Late Triassic to Early Jurassic deformation is considered in the light of recent syntheses of the evolution of the 'Palaeotethys'.

In the Middle Eastern sector of the Alpine chain, a simplified view of the Mesozoic Tethys ('Neotethys') emphasizes the contrasts between its two margins: to the north, the European (Pontic) active margin exhibits extensive magmatic activity in Lower Jurassic and in Upper Cretaceous to Oligocene times, whereas to the south, the Taurus chain is considered to have been the passive margin of Gondwanaland, without magmatism or orogeny until the Late Cretaceous–Eocene period. This simple view is apparently confirmed by most of the Mesozoic successions, recording the opening of Neotethys during the Middle to Late Triassic through to its final closure after the Late Cretaceous.

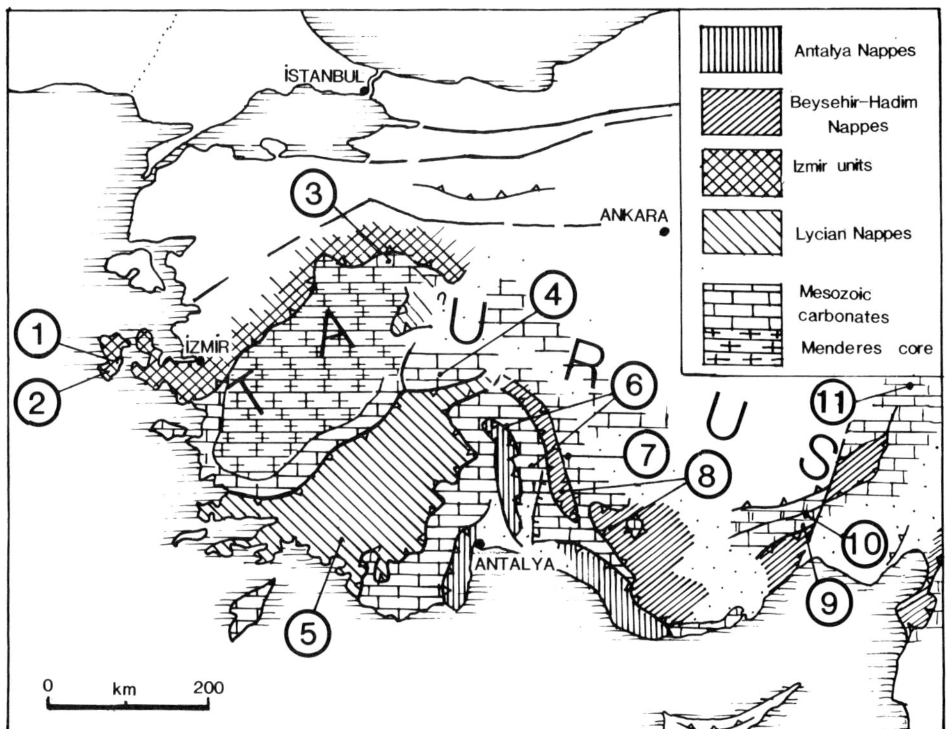

FIG. 1. Outline sketch of the main tectonic units of S.W. Turkey showing locations of successions discussed in the text.

However, observations in the Taurus range show that certain orogenic movements did occur along the Gondwanan margin as early as Lower Jurassic times thus modifying the picture of a quiet passive margin. This evidence includes unconformities, coarse detrital sediments and a definite case of pre-Jurassic thrusting.

These southern margin movements were not as important as those in the Pontides during the same period, but although volcanism is absent, the wide extent of the other evidence and its synchroneity point to an important regional event that has hitherto not been fully appreciated.

Here we review the various localities in the Taurides (Fig. 1) where these movements have been detected and then we put forward a tentative interpretation.

The sedimentary evidence: extension of the Çayır formation

From west to east in the Taurides, we have selected several sections (Fig. 2) which all display an important horizon of clastic sediments of Upper Triassic to Middle Liassic age. That these sequences are now situated in different tectonic units is not important, as only their origin within a carbonate platform is relevant to our discussion.

Chios

On Chios (Fig. 2, localities 1 and 2) two important units are present. In the autochthonous series, very thick Upper Triassic carbonates contain *Conchodus infraliassicus* Stoppani, algae and corals and are directly overlain by sandstones and quartz-bearing conglomerates up to 50 m thick, in turn overlain by dolomites and limestones of probable Liassic age (Besenecker *et al.*, 1968). In contrast, the allochthonous unit of Chios, for which no specific name has been given, is characterized by a major stratigraphic gap from Late Permian to Liassic. The first horizon to transgress onto the Permian limestone was a coarse red polymict conglomerate which was followed by sandstones and siltstones, 10–20 m thick, grading

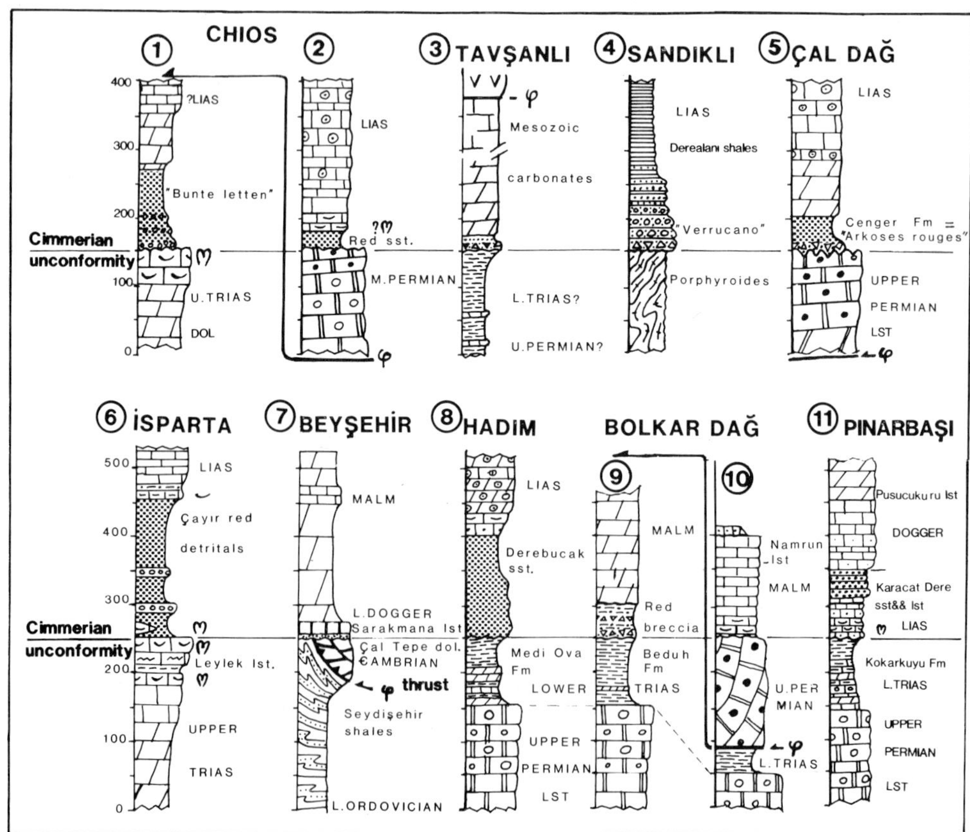

Fig. 2. Generalized sedimentary logs of successions shown in Fig. 1.

into thickly bedded limestones of 'Dachstein' type with large megalodonts and algal limestones with *Paleodasycladus mediterraneus* Pia of Middle to Upper Liassic age.

Chios shows two closely associated but different types of red clastic sediments. In the autochthonous unit a gradual emergence is responsible for the conformable deposition of the red sandstones on the Late Triassic carbonates. In the allochthonous unit a clear angular unconformity with a large time gap separates the Late Permian limestones from the overlying red detrital sediments of Liassic age. This distinction also applies in the Taurus as discussed below.

Lycian Taurus

In the Lycian Taurus (Fig. 2, locality 5), north of Fethiye, Graciansky (1972) described several sedimentary nappe units thrust over Lower Miocene flysch. In the Haticeana unit a conspicuous red arkosic horizon ('les arkoses rouges') is clearly unconformable on various Permian formations but passes upwards transitionally into Liassic dolomites and algal limestones. This horizon, now called the Cenger Formation (Turkish Petroleum Company report, 1981) consists of coarse red conglomerates, arkosic sandstones and red silts, and has been recently dated as Late Triassic by a sparse Dipnoid fauna at the base of the unit (Monod *et al.*, 1983). There is a striking similarity with the allochthonous unit of Chios. Graciansky (*op. cit.*) also described two different Palaeozoic units (Karadağ and Tekedere) which are in fact tectonically superposed, with the Tekedere unit always above the Karadağ unit. Since the overlying Cenger Formation unconformably overlies the Tekedere unit, the tectonic superposition of the two Palaeozoic units may have occurred in Late Triassic times. However, because of Eocene and Miocene thrusting, their present position south of the Menderes Massif cannot be regarded as original and they would have been located north of the Massif in the Mesozoic.

Tavşanlı region

South of Tavşanlı (Fig. 2, locality 3), along the road to Örencik, a well exposed section of the northern part of the Menderes Massif shows, at the base, alternating red and yellow shales and siltstones, with occasional limestone beds more than 200 m thick, of unknown age with Lower Triassic as the most probable age, from facies similarities. Near Akça village this formation is followed by coarse red polymict conglomerates and calcareous sandstones up to 20 m thick passing up into thick-bedded dolomites and recrystallized limestones of Mesozoic aspect, the whole overlain tectonically by an ophiolitic nappe, as seen 10 km south of Tavşanlı. Although not directly dated, this section clearly shows a hiatus in the probable Upper Triassic–Liassic interval, with the intercalation of the red clastic sediments on the relatively authochthonous platform.

Western Taurus

Originally defined east of Isparta (Gutnic *et al.*, 1979) in the autochthonous platform unit of the Western Taurus, the Çayır formation has since been recognized throughout the range as far as Silifke. This formation consists of coarse polygenic conglomerates, red siliceous sandstones and siltstones, gradually passing up into calcareous sandstones and algal limestones containing *Paleodasycladus* sp. and *Orbitopsella praecursor* of Middle Liassic age. At the base, the Çayir formation conformably overlies thick-bedded dolomites and limestones with megalodonts and *Triasina hantkeni* of Upper Norian age (Vegh-Neubrandt *et al.*, 1976). Within the conglomerates, a wide range of lithologies is found, most of the pebbles being quartzites and siliceous sandstones of presumably Palaeozoic age; some however are dark limestones with algae and fusulinids of Upper Permian age; others are red or yellow radiolarian cherts of unknown age, probably Triassic. Occasional plant stems are found in the sandstones. Marine limestone intercalations at the base, composed of high-energy oolitic grainstones, as well as other limestones of a sandy beach facies with lamellibranchs, higher in the formation, indicate a marine origin for most of the Çayır Formation which is here interpreted as deltaic. The overall reddish colour is attributed to weathering of the source rocks. As we shall see, a denuded source-area has been identified (Dumont *et al.*, 1972) and is precisely where the pre-Jurassic tectonism took place.

Hadim nappe

Recognized as early as 1944 by M. Blumenthal, the Hadim Nappe (Fig. 2, locality 8) is now considered as a proximal para-autochthonous unit although its true displacement may have exceeded 100 km (Monod, 1977). Conformably overlying the Late Permian Mizzia limestones are the yellow to pink marly limestones, oolitic grainstones and blue marls of the Medi Ova Formation, with a sparse molluscan fauna of Lower Triassic age. In sharp contrast, the overlying formation is a thick coarse-grained conglomerate, the Derebucak Formation, passing into red sandstones and silts which contain not only material from the underlying Palaeozoic rocks but also red cherts, quartzites and even rhyolites of unknown origin. At the top, clastic facies are transitional into algal limestones and dolomites of Upper Liassic age, containing *Haurania* and *Gutnicella* (Gutnic *et al.*, 1979).

The Hadim Nappe has been recognized over a large area by Özgul (1976) who describes conglomerates of Upper Triassic age more than 500 m thick, south of Hadim, which are unconformable on Lower Triassic and Upper Permian limestones.

Bolkar Dağ

South of Bolkar Dağ (Fig. 2, locality 9) the relatively autochthonous series lying under the ophiolites west of Namrun also show a sharp break in the Triassic to Liassic interval. The deep gorge of Cehenem Dere is cut through Upper Permian limestones overlain by typical Lower Triassic red marls of the Beduh Formation. Above, coarse calcareous breccias with reworked Permian limestone fragments underlie the thick Jurassic dolomites and limestones of the autochthonous series (Turkish Petroleum Company report, 1981).

Eastern Taurus

Fifty kilometers east of Kayseri, in the Pınarbaşı region (Fig. 2, locality 11), Altıner (1981) describes the Palaeozoic and Mesozoic stratigraphy of a thick platform series (Aygörmez unit) which bears a close resemblance to the Hadim Nappe units. The Kokarkuyu Formation of Lower Triassic age, composed of marly limestones, dolomites and red marls, rests conformably on Upper Permian algal limestones. A clear marine transgression above is confirmed by a richly fossiliferous sandy limestone unit with lamellibranchs, megalodonts, gastropods and *Orbitopsella primaeva* (Henson), the last of Middle Liassic age. This unit is followed by a unit of pink to red siliceous sandstones alternating with yellow sandy limestone horizons also containing *Orbitopsella*. Both units are included in the Karacat Dere Formation by Altıner (*op cit.*). The overlying oolitic limestones and dolomites are of Dogger age and extend uninterrupted to the Upper Cretaceous. In spite of the lithological differences, particularly the lack of conglomerates, the same gap separates the Lower Triassic formations from the overlying Liassic clastics in the Eastern Taurus as in the Western Taurus, thus supporting a tentative correlation between the Karacat Dere Formation and the clastics of the Çayır Formation.

Summarising the sedimentary evidence, the Taurus range contains platform units spread over 800 km that are characterized by an important break within Mesozoic carbonate successions. The timing of this break is tightly constrained to the interval from latest Triassic to Middle/Upper Liassic. This event is unique in the sense that no similar polygenetic detrital units are known before or after it. The derived material clearly indicates that a variety of lithologies from different depositional environments were subject to erosion. These include Palaeozoic quartzites, Permian limestones, radiolarian cherts, and acid volcanics. This strongly suggests a mechanism related to orogenic deformation. However, no important pre-Jurassic deformation event below the Çayır clastic facies has hitherto been established. Below, three localities are described which clearly demonstrate the reality of the 'Cimmerian' deformation event.

The tectonic evidence

Eocene and Miocene tectonics have often obscured earlier deformation which can consequently only be seen in restricted areas. The most significant of these is near Beyşehir; two others are in the Lycian Taurus and in the Bolkar Mountains.

Beyşehir area

Early Cimmerian movements have been established in the Beyşehir area (Fig. 2, locality 7; Fig. 3; Akay, 1981) and suspected but not proved in the nearby Seydişehir area (Monod, 1977). In the Beyşehir area, Figs 3, 4, 5 clearly show that the Seydişehir shales of Arenig age are tectonically overlain by coarse grained dolomites and fossiliferous limestones of Cambrian age (Çal Tepe Formation). The tectonic contract is visible in several places especially along the track from Beyşehir to Ağılönü where it is marked by several meters of mylonitised shales and quartzites beneath the Çal Tepe dolomite. The critical evidence is that both the Seydişehir shales and the Cambrian dolomites are transgressively overlain by the Sarakmana limestone (Akay, 1981) of Lower Dogger age (Monod 1977; Gutnic *et al.*, 1979) thus demonstrating the pre-Jurassic age of the overthrust. An older age, Palaeozoic for example, is ruled out by the presence of Middle to Upper Triassic rocks below the same thrust in the Seydişehir region

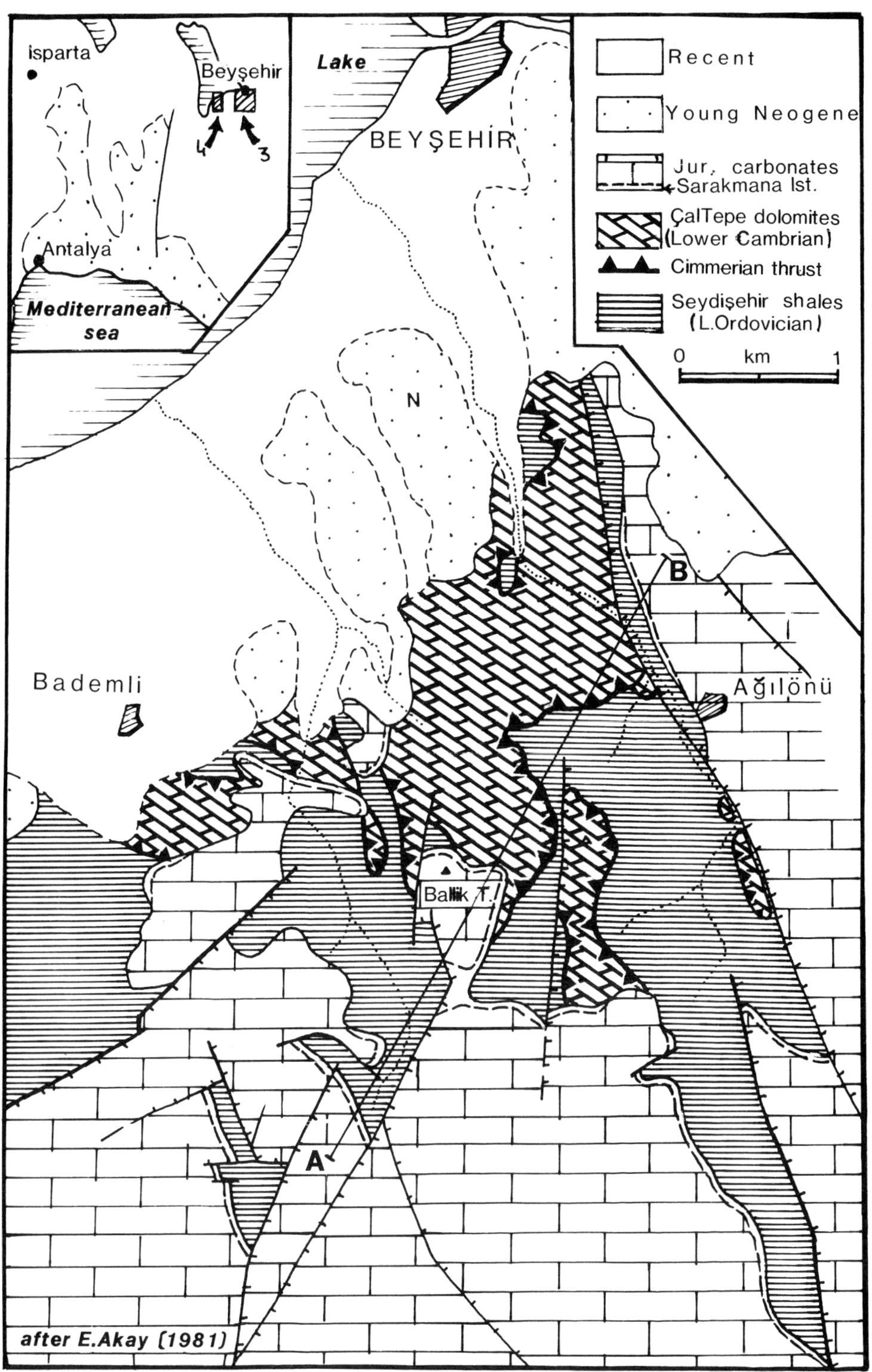

FIG. 3. Simplified geological map of an area S. of Beyşehir, S.W. Turkey. Location inset.

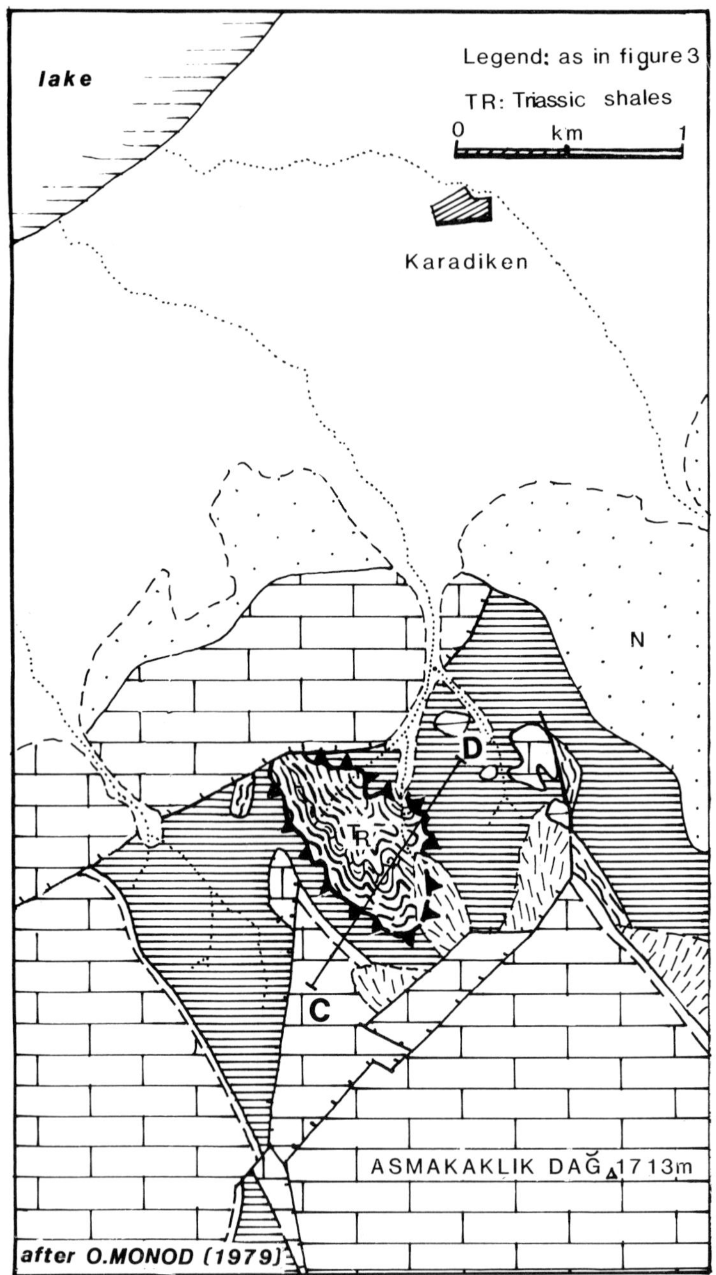

FIG. 4. Simplified geological map of an area S.E. of Lake Beyşehir, S.W. Turkey, location is inset in Fig. 3.

(Monod 1977), and near the shores of Lake Beyşehir, north of Asmakaklik Dağ (Figs 4 and 5).

Elsewhere in the Taurus

In spite of the limited nature of the area where Cimmerian movements can actually be demonstrated, a wide region must be considered in order to appreciate their significance. In this respect, the presence of a major unconformity beneath Jurassic limestones in the Hadim (Özgül 1976) and in the Tufanbeyli region (Özgul 1973) should be regarded as due to these tectonic events.

Another locality of interest is south of Afyon,

BALLIK TEPE SECTION

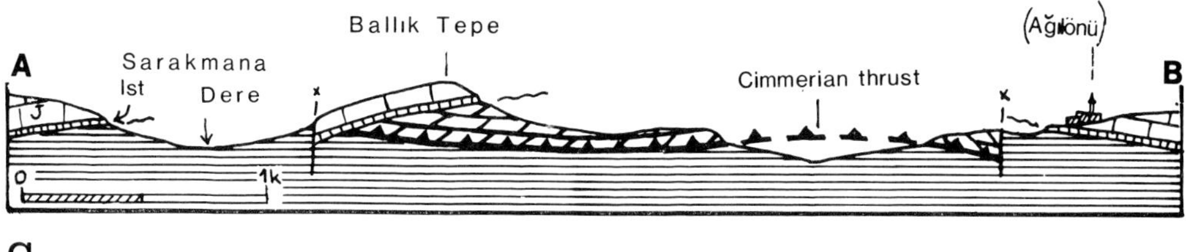

ASMAKAKLIK SECTION

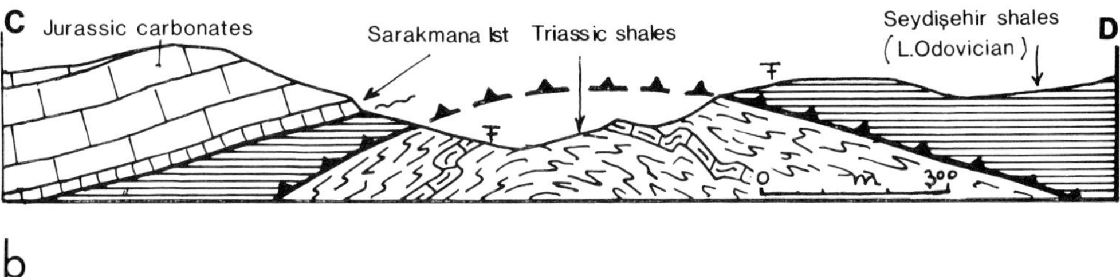

FIG. 5. Cross-section across the area shown in Figs 3 and 4 S. and S.W. of Beyşehir.

near Sandıklı, where a major pre-Upper Liassic unconformity has been established (Gutnic *et al.*, 1979, Fig. 1, locality 4). The basement rocks are highly deformed meta-rhyolites ('porphyroids of Sandıklı') and schists of greenschist facies. Lateral sedimentary facies variations have introduced fusulinid-bearing limestone lenses with *Pseudoschwagerina* sp. and algae determined by M. Lys as of Lower Permian age. These are found near the village of Umrali. Directly above the porphyroids of Sandıklı lies a thick clastic formation, the Derealani shales, of Upper Liassic to Dogger age which commences with a characterstic red siliceous conglomerate, 10 to 100 m thick, of typical 'Verrucano' facies (Parejas 1943). The unconformity is marked and the underlying metamorphism can therefore be dated as between Lower Permian and Upper Liassic, i.e. within the age range of the Cimmerian movements.

Lycian Taurus

As already mentioned, the tectonic superposition of the two Palaeozoic Tekedere and Karadağ units could well have been a pre-Liassic tectonic event (Fig. 7). The Palaeozoic succession in the Karadağ unit is poorly known but certainly contains Middle Carboniferous limestones and Permian sediments, and is transgressively overlain by Ladinian quartzites and limestones which are succeeded by black shales of the Carnian Belenkavak Formation. No younger Mesozoic rocks occur in the succession which is then invariably interrupted by the tectonically emplaced Permian dolomites and limestones of the overlying Tekedere unit. This upper unit is in turn unconformably overlain by the transgressive Cenger Formation, composed of red conglomerates and sandstones of Late Triassic age. Although this formation has not been observed in direct contact with the Karadağ unit, we consider the juxtaposition by thrusting of the two Palaeozoic units to reflect significant pre-Liassic tectonism (Monod *et al.*, 1983).

If the Lycian Nappes are restored to the north of the Menderes Massif this pattern strongly resembles that described long ago from near Balya Maden, where Upper Permian limestones are thrust over Triassic shales, beneath a transgressive Liassic cover (Aygen 1956, Radelli, 1971).

Bolkar Dağ

South of the central Bolkar Dağ (Fig. 6) two platform units are recognized under the ophiolitic nappes. The lowest unit (Özgül 1976) comprises Permian limestones and Lower Triassic

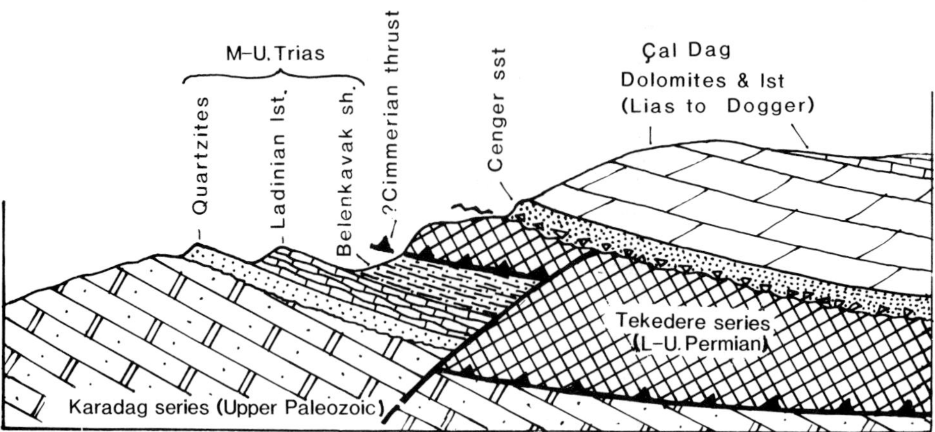

FIG. 6. Generalized cross-section across part of the Lycian Nappes, S.W. Turkey. See text for explanation.

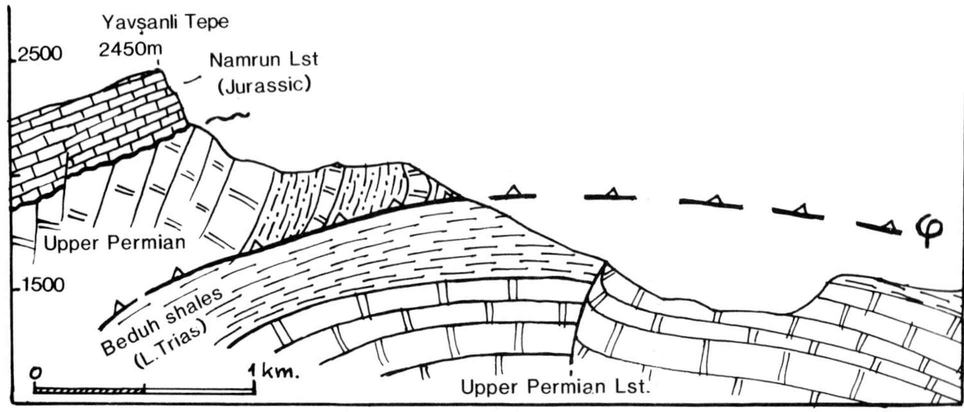

FIG. 7. Generalized cross-section across the South Boekar Dağ, W. Turkey, see text for explanation.

shales, overlain by red breccias containing reworked fossils including fusulinids, but without derived cherts as in the other instances. Above lie Upper Jurassic limestones and dolomites passing up into thick Cretaceous carbonates.

This unit in turn is overlain tectonically by the 'Namrun Unit' (Turkish Petroleum Company report, 1981) which contains evidence of a strong pre-Jurassic deformation event. Its lowest part consists of steeply dipping fossiliferous Upper Permian limestones. These are overlain by almost underformed black micrites over 100 m thick, the Namrun limestones, which contain a poor Jurassic microfauna. These black limestones were formerly considered as Permian by Blumenthal (1956). No clastic sediments are present at the base of this discordant sequence.

Thus the Bolkar Dağ shows a structural succession similar to that of Chios, with a lower unit containing Lower Jurassic red clastics and an upper unit exhibiting a sharp discordance below Jurassic limestones.

Conclusion

This brief review of several critical areas in the Western Taurus points to repeated instances of pre-Jurassic discordances and folding and large volumes of coarse polymict clastics ranging in age from latest Triassic to Upper Liassic. The wide distribution, overall similarity over great distances and the narrow age range of these phenomena strongly suggest a common tectonic origin. The clearest evidence of such tectonism is found in the Beyşehir area where Cambrian dolomites are thrust on Ordovician shales. These thrusts imply that more important related events occurred further north, as the red polygenetic material of the Çayır Formation cannot have been derived solely from the nearby platform units.

What were these tectonic events? At present, answers can only be tentative. Firstly, the deformed areas must have been along the Gondwana margin for no such deformation is known within the platform to the south, nor within the allochthonous units of oceanic origin derived from the central Tethys realm to the north (see Gutnic *et al.*, 1979) where Triassic to Liassic pelagic limestone successions are continuous. The style of deformation is still obscure, but may include metamorphism and large scale thrusting, the most far-travelled nappes being preserved in the Beyşehir area. The timing of deformation is much better established as Upper Triassic to Upper Liassic, and may well have varied locally.

In their attempt at an initial synthesis, Şengör & Yılmaz (1981) considered that a 'Cimmerian continent' of southern, Gondwanan, origin moved slowly northwards from Lower Triassic times onwards thus opening the Neotethys. As a consequence of Palaeotethyan subduction, the Cimmerian continent was split by the opening of a narrow marginal basin (the 'Karakaya' basin) which was then closed as early as the Liassic, thus leaving traces on nearby platforms of an early orogenic event.

Another approach is to consider the motion of Gondwanaland relative to Eurasia. If, as Şengör (1979) proposed, part of Gondwanaland was rifted away as early as the Triassic to form the Cimmerian continent, the collision of this fragment with Eurasia, presumably during the Liassic, should have suddenly changed the rate of opening of the newly created Neotethys. Before collision the opening of the Neotethys was unrestricted being controlled by the rate of subduction of Palaeotethys beneath the European margin. The Cimmerian collision would have modified the relative motion of the two 'super-plates', Gondwanaland and Eurasia, with respect to the young and narrow Neotethys and caused deformation on its margins. This may be more easily recognizable now on the former passive Gondwanaland margin than on the opposite active Pontic one.

This interpretation could also be valid for other parts of the Gondwanaland margin in the Middle East. Indeed several successions in south-east Turkey, Iran and Iraq include siliceous clastics which accumulated on the edge of the stable Arabian platform specifically between Upper Triassic and Middle Liassic times (Fontaine 1981, Ricou 1976). These facies can now be understood in terms of the Cimmerian deformation which affected the Gondwana margin.

ACKNOWLEDGEMENTS: This work has been supported by CNRS (ERA N°804) (Paris), by MTA Institute (Temel Arastirma Subesi) and by TPAO (Arma Grubu) (Ankara). We would like to thank B. Purser and J. E. Dixon for their help with the manuscript.

References

AKAY, E. 1981. Probable early Cimmerian orogenic phase in Beyşehir area, Central Taurus. *Bull. geol. Soc. Turkey*, **24**, 25–9 (in Turkish).

ALTINER, D. 1981. *Recherches stratigraphiques et micropaléontologiques dans le Taurus Occidental au NW de Pinarbaşi (Turquie)*. Thèse Univ. Genève, 450 pp.

AYGEN, T. 1956. Balya Maden Bolgesi jeolojisinin incelenmesi. *M.T.A. Publ.*, **D11,** 95 pp.

BESENECKER, H., DÜRR, ST., HERGET, G., JACOBSHAGEN, V., KAUFFMANN, G., LUDTKE, G., ROTH, W. & TIETZE, K. 1968. Geologie von Chios (Agäis). *Geologia et Paleontologia*, **2,** 121–50.

BLUMENTHAL, M. 1944. Schichtfolge und Bau der Taurusketten im Hinterland von Bozkir. *Rev. Fac. Univ. Istanbul*, **Ser B, 9,** 1–36.

—— 1951. Recherches géologiques dans le Taurus Occidental dans l'arrière pays d'Alanya. *M.T.A. Inst. Publ.*, **5,** 1–134.

DUMONT, J.-F., GUTNIC, M., MARCOUX, J., MONOD, O. & POISSON, A. 1972. Le Trias des Taurides occidentales. *Z. deutsch. Geol. Ges.*, **Bd 123,** 385–409.

FONTAINE, J.-M. 1981. *La plateforme arabe et sa marge passive au Mésozoique: l'exemple de Hazro (S.E. Turquie)*. Thèse 3°Cycle Université Paris Sud, Orsay, Ed. Technip, 270 pp.

GRACIANSKY, P. C. DE 1972. *Recherches géologiques dans le Taurus Lycien*. Thèse d'état Université Paris Sud, Orsay, 762 pp.

GUTNIC, M., MONOD, O., POISSON, A. & DUMONT, J.-F. 1979. Géologie des Taurides Occidentales. *Mém. Soc. géol. France*, **137,** 112 pp.

Lexique stratigraphique international, **3,** 10a, Iraq. Ed. C.N.R.S.

MONOD, O., MEŞHUR, M., MARTIN, M. & LYS, M. 1983. Découverte de Dipneustes triasiques dans la formation de Cenger ('Arkoses rouges') du Taurus Lycien (Turquie Occidentale). *Géobios*, **16,** 161–68.

MONOD, O. 1977. *Recherches géologiques dans le Taurus occidental au sud de Beyşehir* (*Turquie*). Thèse d'état Université Paris Sud, Orsay, 450 pp.

—— 1979. *Carte géologique du Taurus Occidental au Sud de Beyşehir et Notice explicative*, 55 pp., 1 Carte 1/100000°, Publ. C.N.R.S. Paris.

ÖZGÜL, N. 1976. Some geological aspects of the Taurus orogenic belt-Turkey. *Turk. Jeol. Kurumu Bull.* **119,** 65–78, in Turkish with English abstract.

—— METIN, S. GÖĞER, E., BINGÖL, I., BAYDAR, O. & ERDOĞAN, B. 1973. Cambrian-Tertiary rocks in the Tufanbeyli region, Eastern Taurus. *Bull. Soc. Geol. Turkey*, **16,** 82–100, in Turkish with English abstract.

PAREJAS, E. 1943. Le substratum ancien du Taurus Occidental au Sud d'Afyon Karashissar (Anatolie). *C.R. Soc. Sc. Phys. Hist. Nat. Genève*, **60,** 110–14.

RADELLI, L. 1970. La nappe de Balya, la zone des plis égéens et l'extension de la zone du Vardar en Turquie occidentale. *Geol. Alpine*, **46,** 169–75.

RICOU, L.-E. 1976. Evolution structurale des Zagrides, la région clef de Nieriz (Zagros iranien). *Mem. Soc. geol. France*, **125,** 1–140.

ŞENGÖR, A. M. C. 1979. Mid-Mesozoic closure of Permo-Triassic Tethys and its implications. *Nature*, **279,** 590–3.

——, YILMAZ, Y. & KETIN; I. 1980. Remnants of a pre-Late Jurassic ocean in northern Turkey: Fragments of Permian-Triassic Palaeo-Tethys. *Bull. Geol. Soc. Am.* **91,** 599–609.

STÖCKLIN, J. 1968. Structural history and tectonics of Iran: a review. *Bull. Am. Assoc. Petroleum Geol.* **7,** 1229–58.

TURKISH PETROLEUM COMPANY REPORT (TPAO) 1981. *Géologie de la région de Yaylapinar* (*Namrun*), 15 pp., unpublished.

VEGH-NEUBRANDT, E., DUMONT, J.-F., GUTNIC, M., MARCOUX, J., MONOD, O. & POISSON, A. 1976. Megalodontidae du Trias supérieur dans la chaîne taurique (Turquie méridionale). *Geobios*, **9,** 199–222.

O. MONOD,
E. AKAY,

† OLIVIER MONOD, Laboratoire de Géologie Historique, Bat 504, Orsay University 91405F.
†† ERGÜN AKAY, Temel Arastirma Sb. MTA Institute Ankara (Turkey).

Note add in proof

Similarly, a recent re-interpretation of the carbonate succession of Mt Olympus, Greece, by A. Schmitt (1982, 1983) has shown that a major angular unconformity separated the Upper Triassic carbonates at the base of the mountain from the overlying turbiditic limestones of Upper Cretaceous age. The importance of a compressional episode followed by extension involving only the Triassic, and not the overlying Cretaceous limestones, is stressed by the author who ascribes an 'inner' (northerly) position for the Olympus platform within the Hellenic orogenic area. A tentative correlation may thus be made with the inner part of the Taurides where a similar deformation episode is recognized.

References

A. SCHMITT & VRIELYNK, B. (1981) Sur la présence de Trias moyen carbonaté à la base de la série autochtone de l'Olympe (Thessalie) Grèce. *C.r. Acad. Sc. Paris*, **292,** 1485–1490.

A. SCHMITT (1983) *Nouvelle contributions à l'étude géologique des Pietia, de l'Olympe et de l'Ossa (Grèce du Nord).* Thesis of the Faculté Polytechnique de Mons (Belgique) 2 vol 385p. (Annales de la Société Géologique du Nord).

Facies and tectonic setting of the Upper Carboniferous rocks of Northwestern Turkey

I. E. Kerey

SUMMARY: In various outcrops along the Black Sea coast, a thick Dinantian shallow water carbonate succession is overlain by four formations of highest Visean to highest Westphalian age. The succession changes upwards from generally shallow marine, to non-marine giving way in the upper levels of flood-plain, lacustrine and fluviatile deposits. All the present outcrops are believed to have belonged to a single large basin in which material was derived from a former Black Sea continent to the north. In contrast to some recent suggestions, the basin was located near the southern edge of the Laurasian plate.

A model is discussed in which uplift of the area north of the modern North Anatolian Fault is attributed to dextral strike-slip faulting in approximately E-W and NNE-SSW directions. Although evidence of strike-slip faulting has been obscured by later tectonic events, in general the North Turkish Upper Carboniferous sedimentary rocks were deformed during a late Hercynian phase of large-scale wrench-faulting caused by oblique collision of the Gondwana and Laurasian plates after Carboniferous time, which may have also displayed the former Black Sea continent eastwards.

Geological setting

This paper gives a general synthesis of the palaeoenvironmental and palaeogeographical evolution of the Upper Carboniferous rocks of northwestern Turkey, which occur in a number of outcrops near the Black Sea coast (Fig. 1). Four main regions have been examined, the Zonguldak, Amasra, Pelitova, and Azdavay areas.

It was found possible to assign the Upper Carboniferous rocks to four lithostratigraphic formations, which have been sub-divided where appropriate. These lithostratigraphic units are: the Alacaağzı Formation (Uppermost Visean-Upper Namurian), the Kozlu Formation (Upper Namurian-Westphalian A), the Karadon Formation (Westphalian B-C), and the Kızıllı Formation (Westphalian D). These formations of Namurian to Westphalian age comprise approximately 3500 m of clastics overlying Visean platform carbonates and are unconformably, or structurally, succeeded by younger rocks, mainly Permian red sandstone and Cretaceous limestone and flysch.

Kerey (1982) summarised the earlier stratigraphy and previous work. Kerey *et al.* (in press) also provided an outline stratigraphy and palaeobotanical data for Northwestern Turkey, which is not repeated here.

Paleogeographical evolution

During the Visean, the Carboniferous sea appears to have transgressed northwards over the northern part of Turkey (Brinkmann, 1971), producing platform carbonates. Eventually, the entire region formed part of an extensive shallow epicontinental sea. A few areas may have remained emergent and provided local sources for small amounts of clastic sediment, represented by occasional clastic intercalations in the Visean limestone succession. During the Late Visean and Early Namurian there was increased tectonic activity throughout the Black Sea landmass (Brinkmann's 1974 Pontian Massif) leading to renewed clastic input and formation of a regressive sequence. The Black Sea landmass was uplifted, and coarse sediments were shed southwards into the shallow sea, rapidly forming a series of coarsening-upwards deltaic sequences. Rapid subsidence coupled with recurrently shifting delta-lobes produced a vertical stacking of delta complexes and caused lateral facies changes. The western section (Zonguldak) was formed by the gradual southwards progradation of a river-dominated delta while the Amasra succession, to the east, was formed by progradation of a wave-dominated delta, with associated shoreline environments displaying more evidence of longshore transport. The compositional uniformity of the sands contained in both deltaic sequences indicates that they were supplied by the same fluvial input and thus probably formed parts of one major delta system, which was at least 100 km wide (Fig. 2). Towards the end of the Namurian the nature of the Alacaağzı Formation indicates that fluvial cycles became well established in the delta-plain environment. Marine processes were limited and sedimentation was mainly controlled by reworking in the vicinity of the river mouth. Progradation was terminated by

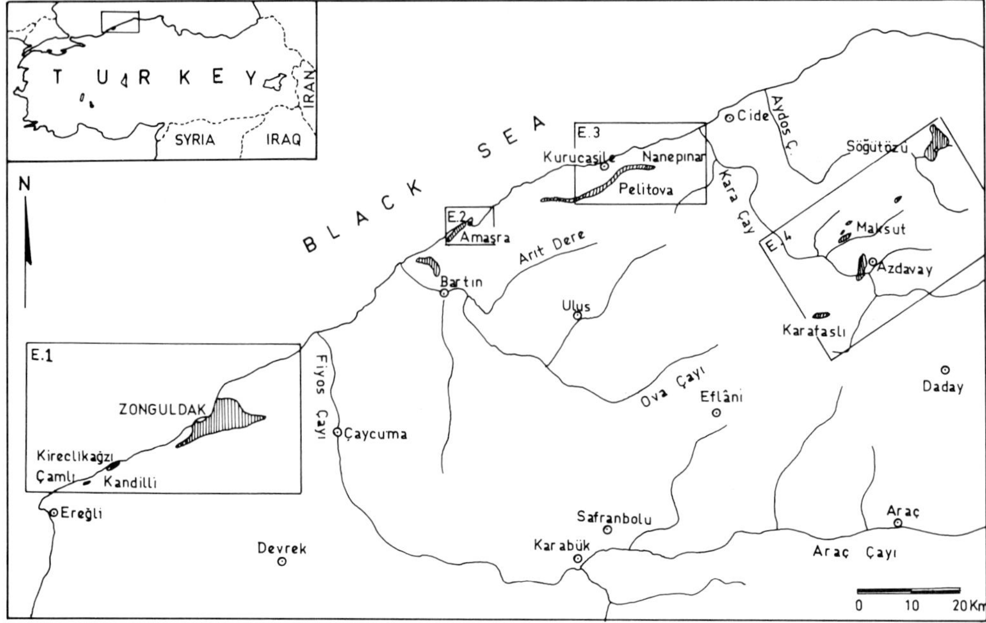

FIG. 1. Distribution of Upper Carboniferous outcrops in north-western Turkey.

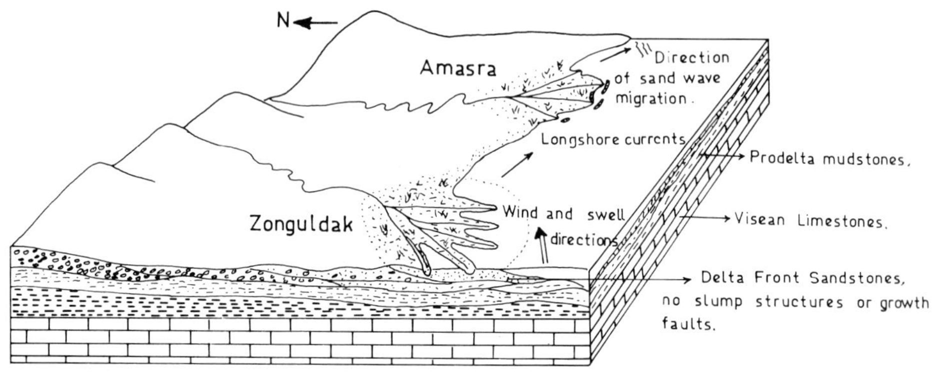

FIG. 2. Schematic block-diagram illustrating the general palaeogeographical relationships for the Alacaağzı Formation.

delta abandonment, in some cases producing thin beds of fossiliferous siltstones, mudstones and rootlet beds with, or without, coals.

Abandonment was followed by a new fluvial input, which brought pebbly material and coarse sands southwards into the area. Faunal evidence suggests that by the end of the Alacaağzı deposition the deltas may have discharged into a brackish rather than a fully marine basin. Later still, in the Namurian, the Kılıç unit of the Kozlu Formation records establishment of a delta-plain environment. Westphalian A time was marked by widespread accumulation of coals, which can be correlated throughout the Zonguldak area (Fig. 3), and are well represented by the Dilaver unit of the Kozlu Formation. This formed on the floodplains of rapidly migrating high-sinuosity streams. The meandering streams responsible for depositing these fining-upwards Kozlu sequences flowed southwards across areas that previously formed parts of a delta-plain and rapid basin subsidence probably controlled the general topographic slopes in the area.

During Westphalian B and C times (Karadon Formation), the appearance of sandy low-sinuosity and braided rivers was followed by coarse pebbly braided streams and finally again

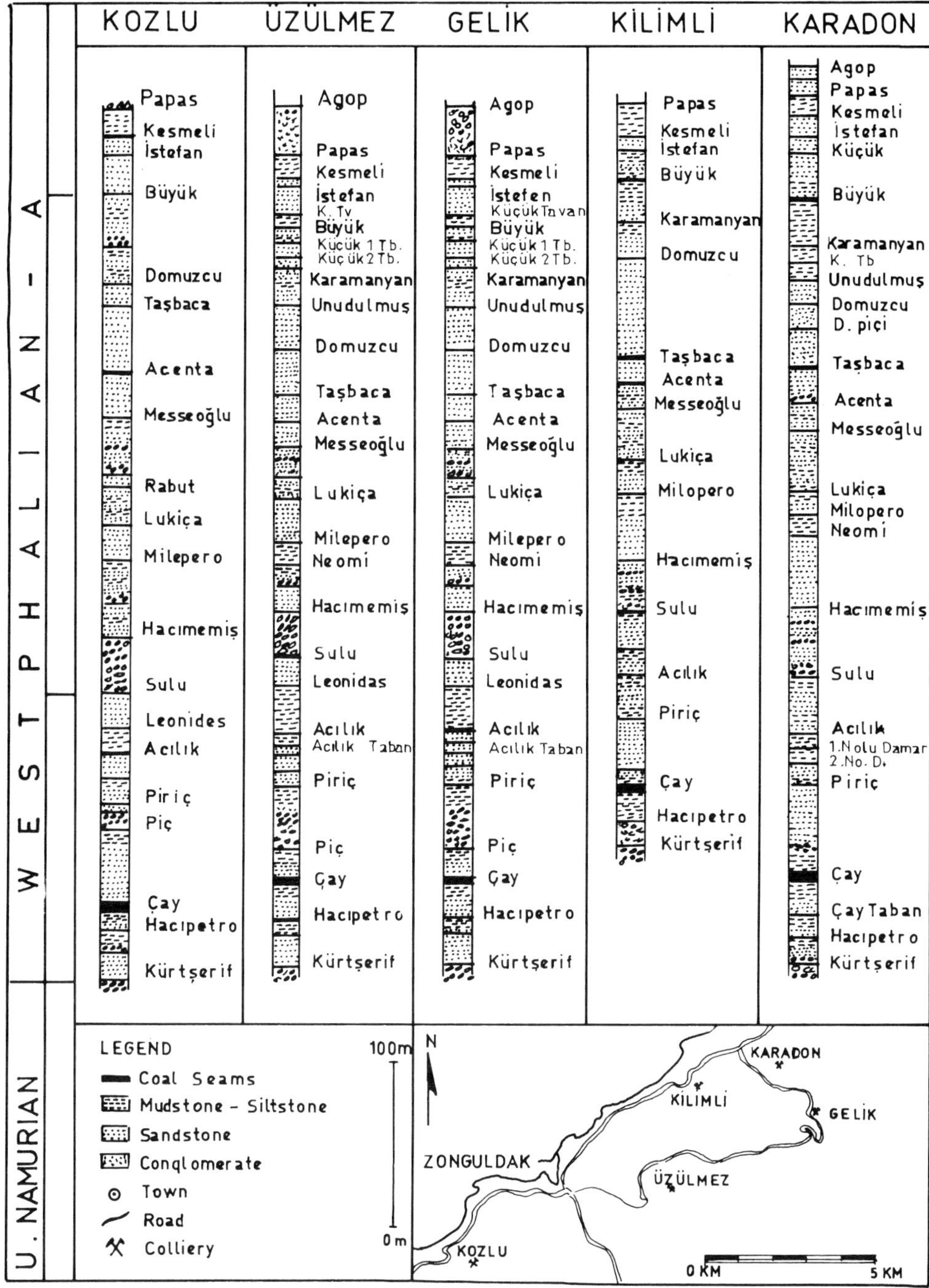

FIG. 3. Correlation of coal seams in the Kozlu Formation of the Zonduldak area.

by low-sinuosity streams. Systematic lateral migration of the braided stream channels over broad, low-gradient fans ('humid fans') resulted in the deposition of laterally persistent sheet sandstones. The generation and preservation of thick sequences of conglomerate and coarse sandstone in the Karadon Formation requires a substantial hinterland relief, attributed to activity on basin-border faults. This process of episodic uplift caused a slight change in the braided-stream pattern which induced streams to flow in a southwesterly direction. This episode of tectonic activity culminated in the Late Westphalian B with the deposition of several extremely coarse and laterally persistent conglomerates. Sandy low sinuosity and braided-stream complexes, with conspicuous fine-grained sequences and thin coal beds were developed in the Westphalian C time interval, suggesting a temporary pause in the tectonic activity. A similar situation has been recorded in the coal-bearing Upper Cretaceous strata of Central Alberta in Canada by Jerzykiewicz & Mclean (1978).

During Early Westphalian D time (Kızıllı Formation), the northwestern Turkish basin was characterized by a complex of meandering streams with wide floodplains on which vegetation flourished. These meander-belt deposits pass upstream into sediments formed in low-sinuosity river channels, and ultimately are succeeded by possible anastomosing stream sequences. The general features of the Kızıllı Formation sequences resemble those of Schäfer's (1981) freshwater molasse basin in the Permo-carboniferous of Saar-Nahe. Schäfer has suggested that the prime sedimentary control was increasing aridity. As a result, sedimentation in the Saar-Nahe region became more alluvial, combined with tectonic activity and magmatic intrusions. Similarly, the maturity and well rounded nature of the grains in the red sandstones of the Kızıllı Formation suggests a high degree of reworking in an intermontane alluvial plain environment, possibly with associated wind action, which may also indicate a climate change from a semi-arid to an arid environment. Possible evidence for an earlier climatic change may be provided by the occurrence, within the upper part of the Karadon Formation, of thick units of refractory clay ('schieferton'), which are attributed to prolonged leaching of palaeosols. This suggests a change from a hot humid climate to a semi-arid climate during Westphalian C time.

Regional significance

Because of a lack of information from other parts of Turkey, it is not possible to provide here a precise palaeoenvironmental picture of the northwestern Turkish Upper Carboniferous successions in a regional tectonic setting. However, since Turkey is part of the Alpine-Himalayan mobile belt, there have been several attempts to explain the geological features in terms of plate tectonics (e.g. Ketin, 1966; Brinkmann, 1971; Şengör *et al.*, 1980; Şengör & Yılmaz 1981; Dewey & Burke, 1973). All these attempts have involved considerations of gross stratigraphic and structural data but none has been based on detailed sedimentological studies. Also, most of these syntheses have been concerned with the Alpine orogenic belt and the Tethys ocean, with little treatment of the Variscan (Hercynian) orogenic belt.

Demirtaşlı (1981) summarized the Palaeozoic stratigraphy and Variscan events in the Taurus belt of southern Turkey. Similar rocks of Bashkirian to Moscovian age may be correlated with the Karlık Formation of the Kasımlar area (Dumont & Kerey, 1975), which is a regressive sequence of red sandstones and conglomerates.

Bingöl (1982), on the other hand, suggests that the region was rifted during the Middle Devonian or probably earlier. The resultant oceanic crust was subducted to the north along the Tokat-Ankara-İzmir line before the Carboniferous (Fig. 4, Bingöl, 1982). The area just north of the modern North Anatolian Fault, starting from the Late Devonian, was uplifted from south to north accompanied by right-lateral strike-slip faulting in E-W and NE-SW directions (Bingöl 1982).

According to Wagner (pers. comm. 1982), the northwest Turkish Carboniferous outcrops probably formed part of a single large basin of sedimentation on the southern margin of the Laurasian plate. Although southern Turkey belonged to Gondwanaland, the boundary between Laurasia and Gondwanaland remains controversial.

Kerey (1982) suggested that Middle and Upper Carboniferous sequences in northwestern Turkey display close stratigraphical and sedimentological similarities with coeval successions in northern Europe, including the presence of abundant coal seams and comparable floras.

Recently Şengör & Yılmaz (1981) have suggested that during the Permian the entire present area of Turkey formed part of the northern margin of Gondwanaland. However, the floral and faunal evidence presented in Kerey *et al.* (1982), together with the stratigraphical and sedimentological findings, demonstrate that, at least in Upper Carboniferous time, northern Turkey was not part of Gondwanaland but is more plausibly regarded as part of the Laurasian Carboniferous assemblage (e.g. Smith *et al.*, 1981). Similarly, Bergougnan and Fourquin (1980) have suggested that the North Anatolian belt (Pontides) belonged to the 'Eurasiatic continent' facing the Tethyan ocean during the Mesozoic.

In recent years work on Variscan plate tectonics has shown that the sedimentary and structural features of this orogen are not readily accounted for by conventional plate-tectonic models involving rectilinear continental colli-

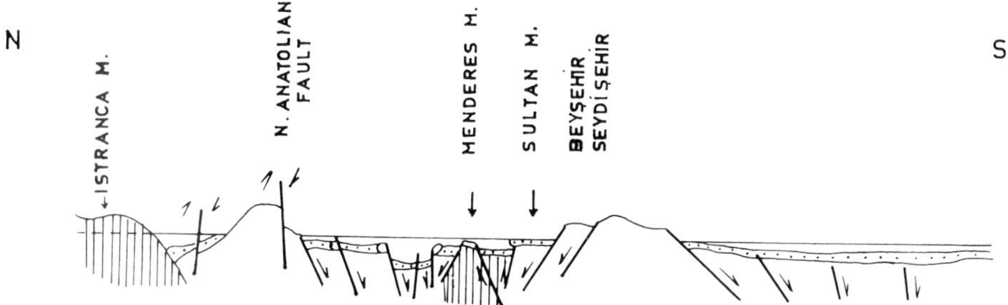

FIG. 4. Inferred pre-Carboniferous rifting of northern Turkey (after Bingöl, 1982).

sions. For example, Lefort & Van der Voo (1981) suggested a kinematic model for the oblique collision and suturing of Gondwanaland and Laurasia in the Carboniferous, based on the general distribution of major strike-slip faults in western Europe, northern Africa, and eastern North America. Scotese *et al.* (1979) and Rau & Tongiorgi (1981) also suggested continental collision between Africa and North Europe as the prime cause of the formation of Gondwanaland and they prefer to exclude the hypothesis of an infra-Palaeozoic oceanic proto-Tethys in southwestern Europe, in accordance with Flugel's (1975) suggestion for Turkey and with the Lower Palaeozoic maps of Scotese *et al.* (1979).

Finally, as indicated above, the deposition of the Late Carboniferous successions in the north-west Turkish basin was probably fault-controlled. While the fault motions responsible for the supply of sediment were essentially vertical, this does not exclude the possibility that the major controlling fractures underwent strike-slip motion as in some modern basins, including the San Andreas fault zone, and many others (Ballance & Reading, 1980). Also, the uplift and deformation which followed Late Carboniferous sedimentation, would have prevented identification of any lateral displacement of the basin margins. Heward & Reading (1980) used the present separation of alluvial fan and associated sequences from their source areas as evidence for lateral displacement of the Early Stephanian basinal successions in Cantabria, northern Spain. The difference is that the Spanish Carboniferous successions are much thicker than those in north-west Turkey; only the Karadon Formation is comparable. Since the sedimentology of the north-western Turkish basin shows close similarity with the north-west European Upper Carboniferous successions, this may indicate a similar geotectonic history, involving a significant amount of strike-slip faulting.

Conclusions

The four formations recognized in this study apparently constitute a continuous succession ranging from highest Visean to highest Westphalian D in age. This succession overlies a Dinantian carbonate sequence. The basal part of the succession is generally marine in the lower part, becoming non-marine in the middle part. The upper part was laid down on a flood-plain and shows lacustrine, fluviatile and humid fan deposits.

The stratigraphical, sedimentological and faunal data indicate that the various small, structurally-separated, outcrops of Upper Carboniferous rocks in northwestern Turkey were originally deposited in a single large basin, with a northerly source. This basin was formed on the southern margin of the Laurasian plate.

An area just north of the North Anatolian fault started to be uplifted from south to north possibly in Late Devonian time, with possible synchronous right-lateral strike-slip faulting in approximately E-W and NNE-SSW directions. The uplift and deformation which followed Late Carboniferous sedimentation has however obscured any original lateral displacement of the basin margins. Deformation of the Carboniferous successions relates to the initial stage of a late Hercynian phase of large scale wrench-faulting caused by post-Carboniferous oblique collision of the Eurasian and Gondwana plates (Lefort & Van der Voo 1981). Strike-slip faulting may also explain the eastward displacement of the northern Black Sea continent, which was the source of the Late Carboniferous clastics.

References

Bergougnan, H. & Fourquin, C. 1980. The Pontids. *In: The Alpine Middle East between the Aegean and Oman Traverses.* Int. geol. Congr. Paris, **115,** 128–130.

Ballance, P. F. & Reading, H. G. 1980. Sedimentation is oblique-slip mobile zones: an introduction. *In:* Ballance, P. F. & Reading, H. G. (eds.) *Spec. Publ. Int. Ass. Sediment.*, **4,** 1–5.

Bingöl, E. 1982. Probable geotectonic evolution of Turkey between Precambrian and Jurassic. *Bull. geol. Soc. Turkey* (in press).

Brinkmann, R. 1971. The geology of Western Anatolia. *In:* Campbell, A. S. (ed.), *Geology and History of Turkey*, Tripoli 171–90.

—— 1974. Geologic relations between Black Sea and Anotolia. *In:* Degens, E. T. & Ross, D. A. (ed.) *The Black Sea, geology, chemistry and biology.* Mem. Am. Assoc. Petrol. Geol. **20,** 63–76.

Demirtaşli, E. 1981. Summary of the Palaeozoic stratigraphy and Variscan events in the Taurus Belt. *In:* Karamata, S. & Sassi, F. P. (ed.) *IGGP. No. 5 Newsletter*, **3,** 44–57.

Dewey, J. F. & Burke, K. C. A. 1973. Tibetan, Variscan and Precambrian basement reactivation: products of continental collision *J. Geol.* **81,** 683–692.

Dumont, J. F. & Kerey, I. E. 1975. Eğridir gölü günüyinin temel jèolojik etüdü. *Bull. geol. Soc. Turkey*, **18,** 163–174.

Flugel, H. W. 1975. Einige Probleme des Variszikums von Neo-Europa. *Geol. Rundschau*, **64,** 1–12.

Heward, A. P. & Reading, H. G. 1980. Deposition associated with a Hercynian to late Hercynian continental strike-slip system Cantabria Mountains, northern Spain. *In:* Ballance, P. F. & Reading, H. G. (ed.) *Sedimentation in Oblique slip Mobile Zones.* Spec. Publ. Int. Ass. Sediment., **4,** 105–125.

Jerzykiewicz, T. & McLean, J. R. 1980. Lithostratigraphical and sedimentological framework of coal-bearing Upper Cretaceous and Lower Tertiary strata, Coal Valley area, Central Alberta Foothills. *Geol. Survey. Canada, Paper* **79,** 47.

Kerey, I. E. 1982. *Stratigraphical and Sedimentological Studies of Upper Carboniferous rocks in northwestern Turkey.* Keele University Unpublished Ph.D Thesis, U.K. 232 p.

——, Kelling, G. & Wagner, R. H. 1982. *An outline stratigraphy and palaeobotanical records from the Middle Carboniferous rocks of Northwestern Turkey.* M.T.A. Special Publication of the Coal Conference held in Zonguldak, 1978. (In press).

Ketin, I. 1966. Tectonic units of Anatolia (Asia Minor). *Bull., M.T.A.* **66,** 23–34.

Lefort, J. P. & Van der Voo, R. D. 1981. A kinematic model for the collision and complete suturing between Gondwanaland and Laurasia in the Carboniferous. *J. Geol.*, **89,** 537–550.

Rau, A. & Tongiorgi, M. 1981. Some problems regarding the Paleozoic palaeogeography in Mediterranean Western Europe. *J. Geol.* **89,** 663–673.

Schäfer, A. 1981. Permocarboniferous Saar-Nahe Basin S.W. Germany. Modern and Ancient Fluvial Systems: Sedimentology and Processes. Abstracts of a meeting held at Keele University, U.K. 107 p.

Scotese, C. R. *et al.*, 1979. Paleozoic base maps. *J. Geol.*, **87,** 217–277.

Smith, A. G., Hurley, A. M. & Briden, J. C. 1981. *Phanerozoic Palaeocontinental Maps.* Cambridge University Press, 102 p.

Şengör, A. M. C., Yilmaz, Y. & Ketin, I. 1980. Remnants of a pre-late Jurassic ocean in northern Turkey: Fragments of Permian-Triassic Palaeo-Tethys. *Bull. geol. Soc. Am.*, **91,** 599–609.

——, —— 1981. Tethyan evolution of Turkey: a plate tectonic approach. *Tectonophysics*, **75,** 181–241.

I. E. Kerey, Department of Geology, The University of Fırat, Elazığ, Turkey.

Stratigraphic evidence of Variscan and early Alpine tectonics in Southern Turkey

E. Demirtaşlı

SUMMARY: The area between Silifke and Anamur forms the central Taurus belt, in which four geotectonic zones are recognized: the Southern, Intermediate and Northern zones and the Hadim nappe. Stratigraphic successions in each zone show considerable differences from the early Palaeozoic onwards. For example, a Late Triassic non-marine clastic basin dominated the Intermediate and Northern zones, while a marine carbonate basin existed to the south.

The contrasting stratigraphic development of each of the three main tectonic zones studied, particularly reflects Early Permian (Variscan) and Late Triassic ('Eocimmerian') orogenic events. Overthrusting of the Northern zones over the Intermediate zone took place before a Late Permian transgression. Later overthrusting of the Intermediate zone over the Southern zone occurred during Middle to Upper Triassic times producing large-scale nappes. The Variscan and early Alpine structures were then unconformably overlain by platform carbonates of Early Jurassic to Late Cretaceous age. Middle and late Alpine movements are also observed in the Intermediate and Northern zones. Ophiolitic melange was thrust over Late Cretaceous–Palaeocene flysch and olistostrome while platformal carbonate deposition continued in the Southern zone, confirming that an ocean basin existed to the north. Orogenic movements ended before the Early Miocene, followed by block-faulting to produce graben (e.g. Çavuşlar basin). Formations of Middle and Late Miocene age unconformably overlying all other units are post-tectonic.

In the Tauride mountains of southern Turkey intense Alpine tectonics have largely obscured the record of important earlier tectonic events of Hercynian and early Mesozoic age. While previously the Tauride mountains were thought to have formed part of the stable northern margin of Gondwana in the Palaeozoic to Late Mesozoic (Blumenthal, 1955, 1960) there is now growing evidence of marked regional variation (Dumont & Kerey, 1975; Dumont, 1976), with some areas undergoing undisturbed deposition from Ordovician to Upper Cretaceous (e.g. Tahtahlı Dağ unit of the Antalya Complex, Robertson and Woodcock, 1982; Şenel, 1983) while others experienced important deformation and thrusting in Early Permian and Late Triassic times. Monod & Şengör *et al.* (both this volume) point out that the early Mesozoic 'Eocimmerian' movements affect much of the Tauride area. In this paper a detailed account is given of the stratigraphy and tectonic evolution of a critical area located in the central part of the Tauride mountain belt along the Turkish Mediterranean coast between Silifke and Anamur, where both the Hercynian and 'Eocimmerian' tectonics can be distinguished from later Alpine overprinting.

Geological setting

In the area studied (Figs 1 & 2), the following tectono-stratigraphic units have been recognized: (1) The autochthonous Southern zone. (2) The Intermediate zone (lower allochthon). (3) The Northern zone (middle allochthon), and (4) The Hadım nappe (upper allochthon; the Aladağ unit of Özgül, 1976) (Figs 3 & 4). The stratigraphic sequences of each of these

FIG. 1. Location map of the area studied along the SE Mediterranean coast of Turkey.

units show considerable differences (Fig. 5) indicating a contrasting tectonic development of each unit. The Northern zone was emplaced over the Intermediate zone before the Late Permian. The pre-Late Permian formations of both units were tightly folded and faulted as a result of the Variscan orogeny. Late Permian limestone (Kırtıldağı Formation) then unconformably overlies the older formations of both the other zones. After a short-lived Late Permian marine transgression, both zones were

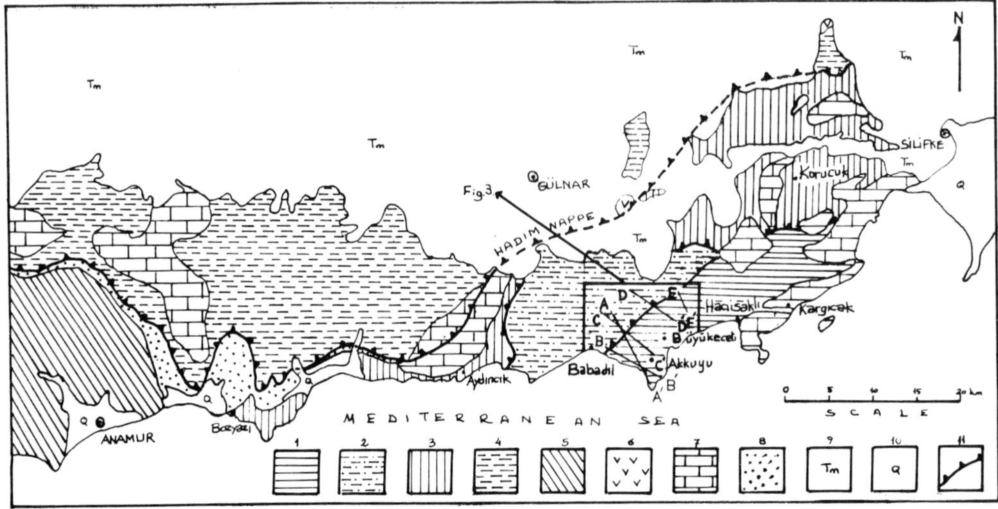

FIG. 2. Outline geological map of the major tectonic units along the Mediterranean coast of Southern Turkey, lists of sections shown in Fig. 5. 1. Southern zone; 2. Intermediate zone; 3. Northern zone; 4. Aladağ unit (Hadim nappe); 5. Alanya massif; 6. Ophiolites; 7. Jurassic–Cretaceous platform carbonates; 8. Palaeocene olistostrome; 9. Marine Miocene; 10. Quaternary deposits; 11. Thrust.

uplifted in Early and Middle Triassic time. Then during Late Triassic time, continental and lagoonal facies were deposited in small fault-controlled depressions.

In contrast to the Intermediate and the Northern zones, the relatively autochthonous Southern zone experienced only local uplift, block-faulting and non-deposition of Carboniferous to Lower Permian rock units during the Variscan orogeny. Marine Triassic deposition prevailed in the Southern zone in contrast to further north. During the Late Triassic the Intermediate zone was emplaced over the Southern zone along the Büyükeceli overthrust which trends NE–SW for over 20 km (Figs 2, 3 & 4) giving rise to a series of klippen and windows in the north-eastern part of the area mapped (Fig. 3) which can be attributed to the early Alpine 'Eocimmerian' orogeny. Outside the area mapped the Büyükeceli overthrust can be seen to be covered by Jurassic–Cretaceous facies (Tokmar Formation) (Fig. 2). The transgression was most extensive during the Late Cretaceous with the result that Late Cretaceous to Palaeocene platform carbonates (Hayvandağı Formation) transgressively overlie all the older formations in each of the tectonic zones (Figs 3, 4 & 5).

The effects of 'middle' Alpine (Late Eocene) orogeny have also been observed in the Intermediate and Northern zones. Flysch and olistostrome melange of Late Cretaceous to Early Tertiary age unconformably overlies a Jurassic to Cretaceous succession and then was overthrust by the ophiolitic melange and Palaeozoic rocks of the Aladağ unit (upper allochthon) together with the Hadım nappe. A similar olistostrome facies with conglomeratic interbeds unconformably overlying both the Alanya metamorphics and the Northern zone (middle allochthon) occurs in the western part of the area beneath the Hadım nappe (Fig. 2). Ophiolitic rocks observed in the area are all associated with the upper allochthon.

Major movements ceased before the Early Miocene. Block faulting then produced depressions, for example the Çavuşlar basin, in the north where conglomerates, sandstones and non-marine limestones were deposited (Tepeköy Formation) in the Early Miocene. Marine limestones and marls of Middle and Late Miocene age (Silifke Formation) overlie all the other units and are considered to be post-tectonic. The stratigraphy of the main formations in each of the tectonic zones is now described (Fig. 5).

Stratigraphy

Autochthonous Southern zone

Cambrian

The following formations are recognized in the relatively autochthonous Southern zone (Fig. 6). The Cambrian *Ovacık Işıklı Formation* is composed of medium- to thick-bedded, grey crystalline dolomitic limestone containing conodonts of Middle to Upper Cambrian age (Fig.

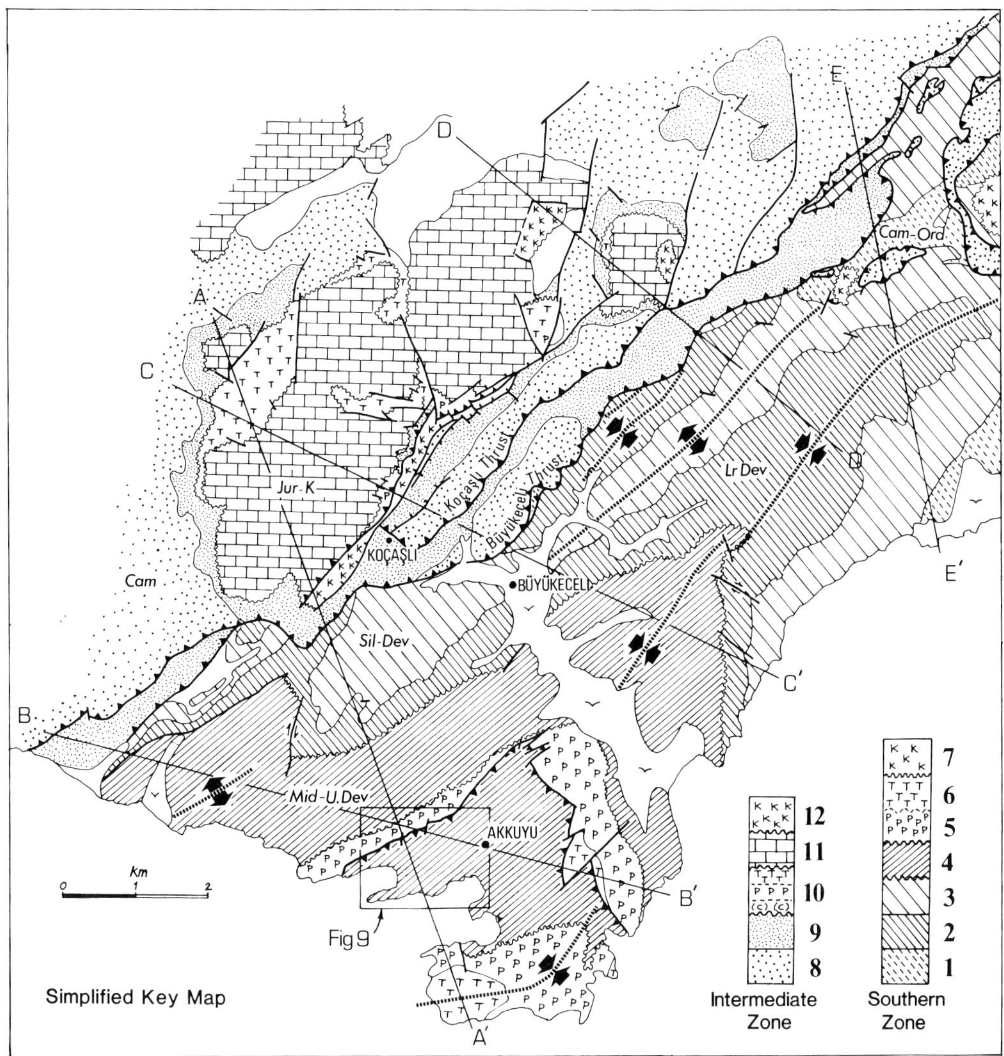

FIG. 3. Simplified geological map of the area round Büyükeceli, based on 1:25 000 scale mapping of sheets *Silifke-P31d₁* and *-P30C₂*, showing the major structural and stratigraphic elements. The *Intermediate zone* to the north-west lies structurally above the relatively autochthonous *Southern zone*, along the *Büyükeceli thrust* and associated *Koçaşlı thrust*. Stratigraphic units have been grouped to emphasis the main angular unconformities, shown by wavy contacts. The axial traces of major NE-trending anticlines and synclines in the Southern zone are also shown.
Key (U/C) = angular unconformity: *Southern zone*. 1. Cambrian to Ordovician *Ovacik Fm*.; 2. Silurian to Lr. Devonian *Eğripınar, Hırmanlı* and *Karayar Fms*.; 3. Lr. Devonian *Siğircik* (U/C); 4. Mid to U. Devonian *Büyükeceli* and *Akdere Fms*. (U/C); 5. Permian *Kırtıldağı Fm*.; 6. disconformable Triassic *Kuşyuvasıtepe Fm*. (U/C); 7. Cretaceous–Palaeocene *Hayvandağı Fm*.
Intermediate Zone. 8. Cambrian *Sipahili Fm*.; 9. U. Cambrian to Devonian *Babadıl Gp*. (U/C); 10: Permian *Kırtıldağı Fm*. and disconformably overlying Triassic *Murtcukuru Fm*. Carboniferous *Korukuk Fm*. occurs locally at the base (U/C); 11. Jurassic *Yanişli, Dibekli* and *Tokmar Fms*. (see text for discussion of intra-Jurassic unconformities) (U/C); 12. Cretaceous–Palaeocene *Hayvandağı Fm*. Post tectonic Miocene and Quarternary deposits are left unornamented. A–A′, etc.: cross-sections, see Fig. 4.

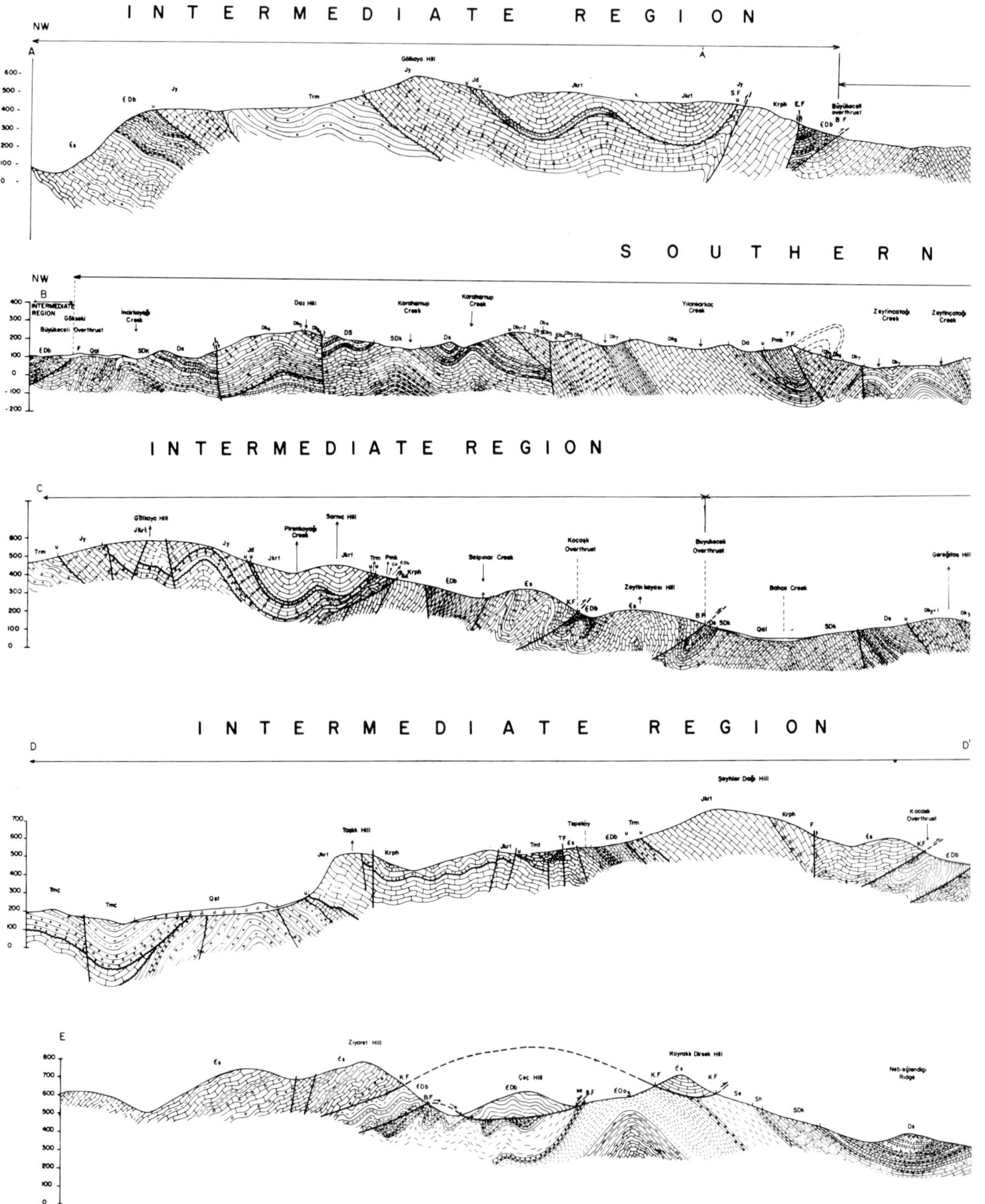

FIG. 4. Geological cross-sections of the area mapped (see Figs 2 & 3 for locations).

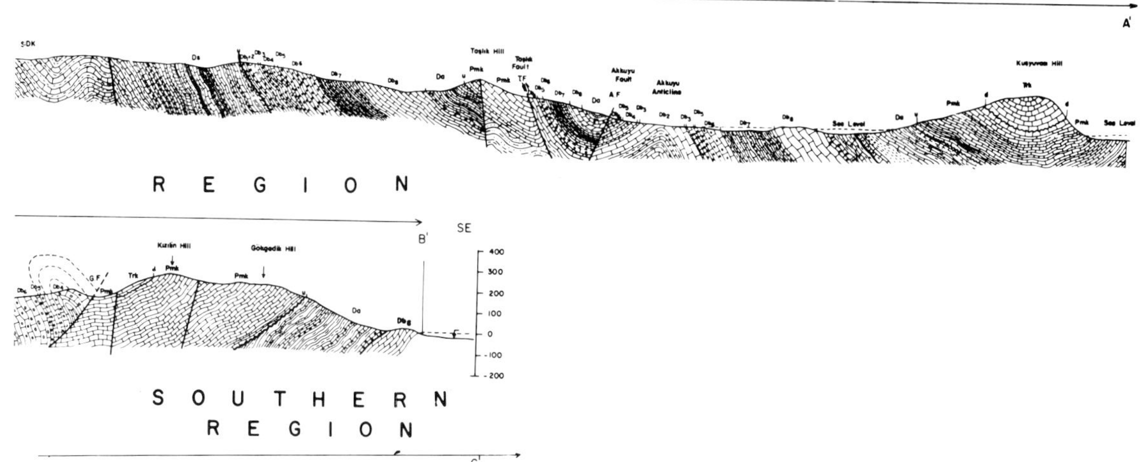

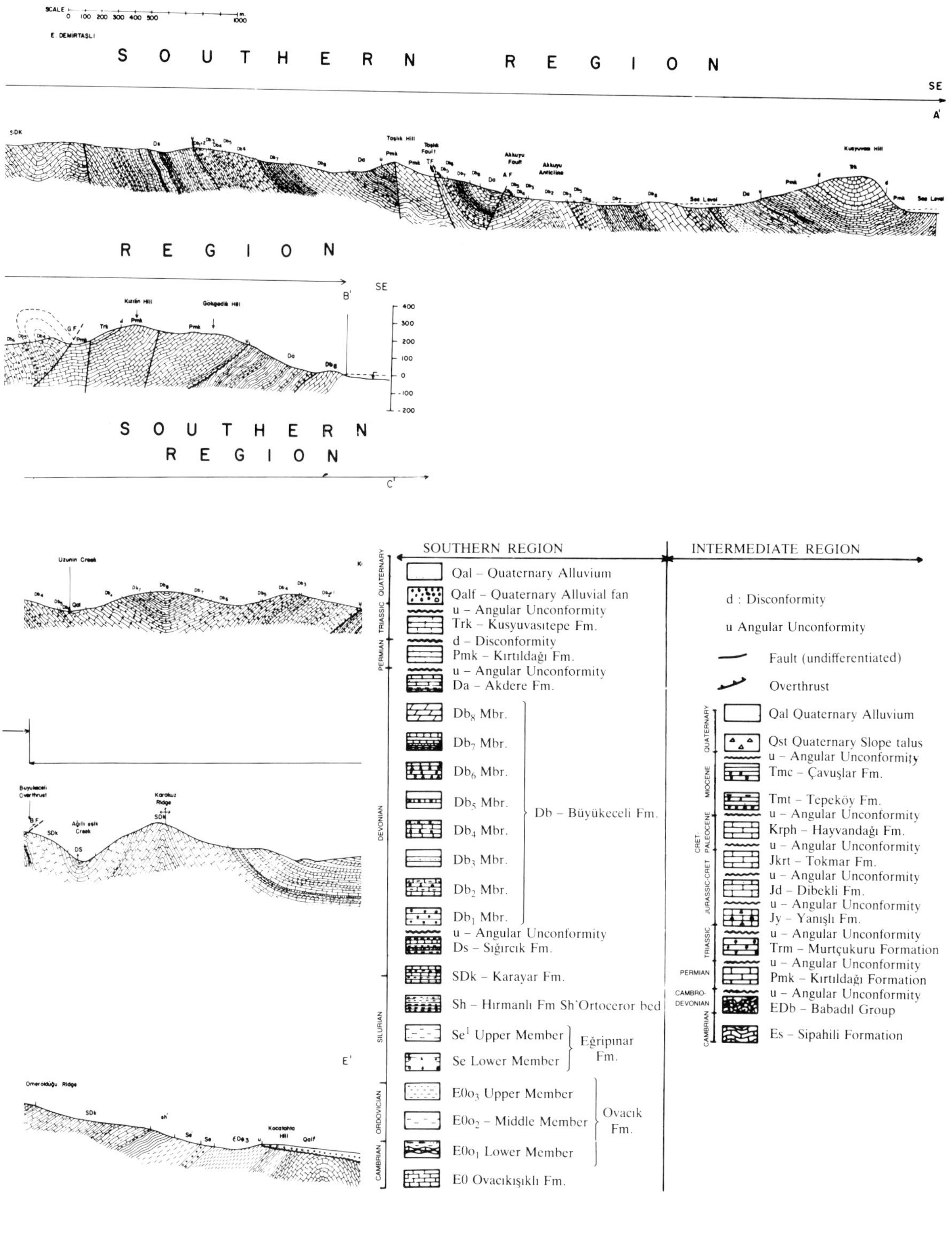

REGION

SE B'

Kızılin Hill Gökgedik Hill

400 300 200 100 0 -100 -200

SOUTHERN REGION

C'

Uzunin Creek

Büyükeceli Overthrust

Karakuz Ridge

Ağıllı eşik Creek

E'

Omeroldugu Ridge

Kocatahta Hill

SOUTHERN REGION		INTERMEDIATE REGION	
QUATERNARY	Qal – Quaternary Alluvium		d : Disconformity
	Qalf – Quaternary Alluvial fan		u Angular Unconformity
	u – Angular Unconformity		Fault (undifferentiated)
TRIASSIC	Trk – Kusyuvasıtepe Fm.		Overthrust
	d – Disconformity	QUATERNARY	Qal Quaternary Alluvium
PERMIAN	Pmk – Kırtıldağı Fm.		Qst Quaternary Slope talus
	u – Angular Unconformity		u – Angular Unconformity
	Da – Akdere Fm.	MIOCENE	Tmc – Çavuşlar Fm.
DEVONIAN	Db_8 Mbr. (Db – Büyükeceli Fm.)		Tmt – Tepeköy Fm.
	Db_7 Mbr.		u – Angular Unconformity
	Db_6 Mbr.	CRET-PALEOCENE	Krph – Hayvandağı Fm.
	Db_5 Mbr.		u – Angular Unconformity
	Db_4 Mbr.	JURASSIC-CRET	Jkrt – Tokmar Fm.
	Db_3 Mbr.		u – Angular Unconformity
	Db_2 Mbr.		Jd – Dibekli Fm.
	Db_1 Mbr.		u – Angular Unconformity
	u – Angular Unconformity		Jy – Yanışlı Fm.
	Ds – Sığırcık Fm.	TRIASSIC	u – Angular Unconformity
	SDk – Karayar Fm.		Trm – Murtçukuru Formation
SILURIAN	Sh – Hırmanlı Fm Sh'Ortoceror bed		u – Angular Unconformity
	Se^1 Upper Member (Eğripınar Fm.)	PERMIAN	Pmk – Kırtıldağı Formation
	Se Lower Member (Eğripınar Fm.)		u – Angular Unconformity
ORDOVICIAN	E0o$_3$ Upper Member (Ovacık Fm.)	CAMBRO-DEVONIAN	EDb – Babadıl Group
	E0o$_2$ – Middle Member (Ovacık Fm.)	CAMBRIAN	Es – Sipahili Formation
CAMBRIAN	E0o$_1$ Lower Member (Ovacık Fm.)		
	E0 Ovacıkışıklı Fm.		

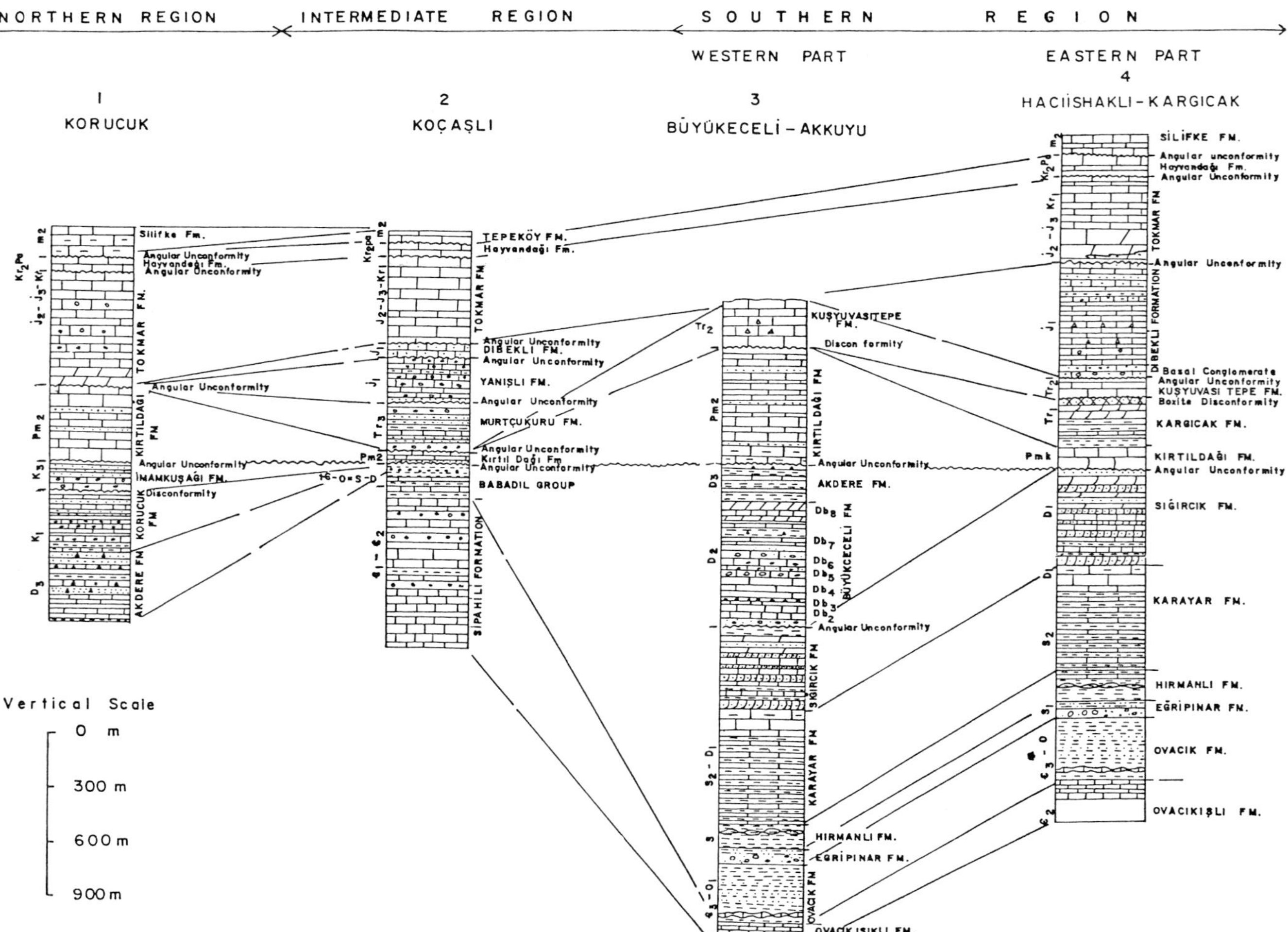

FIG. 5. Correlated composite successions of the main tectonic units in the area studied; No. 1, see Fig. 2; nos 2, 3 & 4 (see Fig. 3).

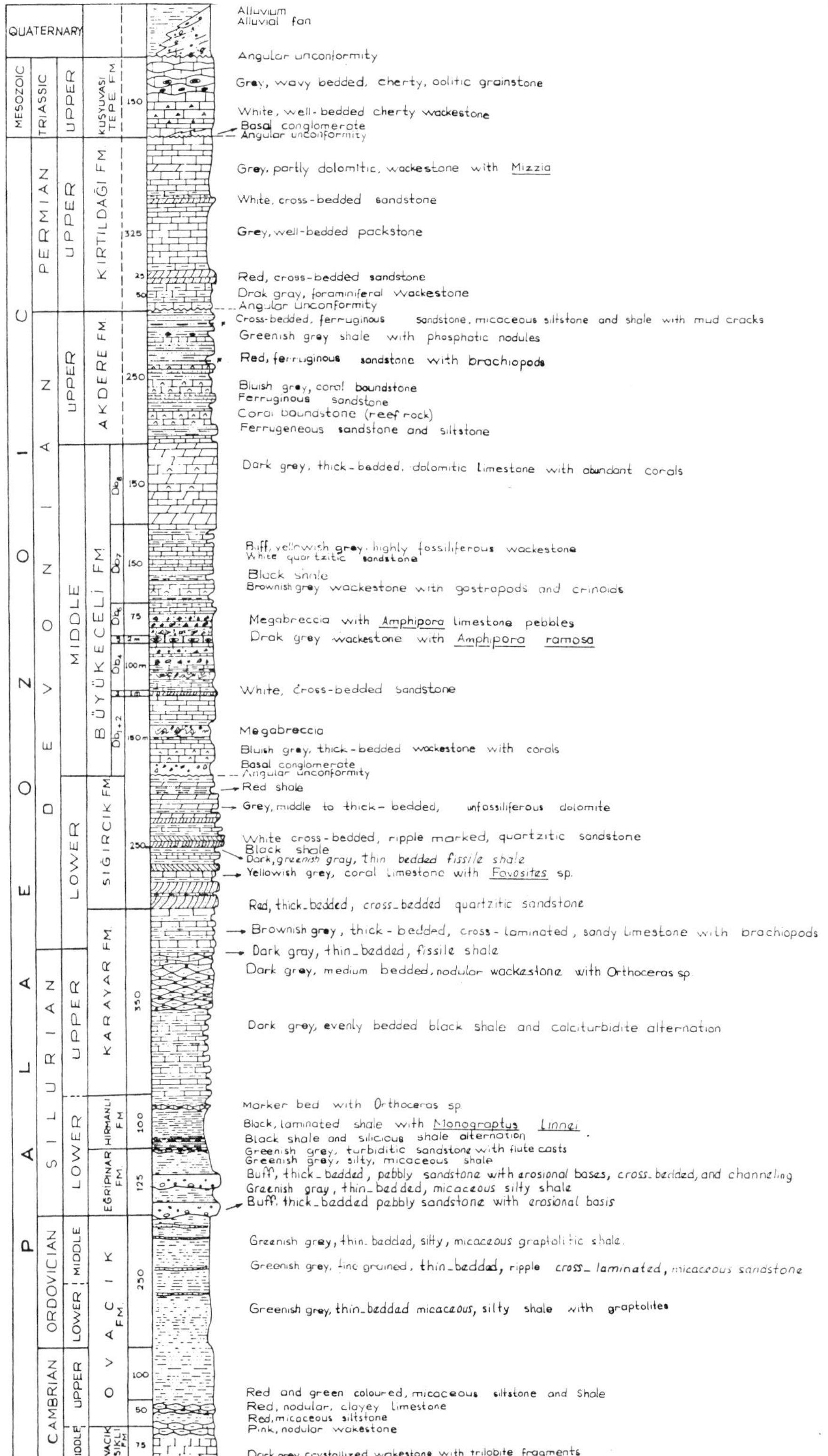

FIG. 6. Composite succession of the relatively autochthonous Southern zone (see Fig. 3).

6) and is then conformably overlain by the Ovacǐk Formation of Late Cambrian to Ordovician age and may be correlated within the Koruk Formation of south-eastern Turkey (Tuna, 1973). This unit appears to have been deposited under stable shelf conditions.

Cambrian–Ordovician

The lower part of the *Ovacık Formation* is composed of red to buff siltstone containing abundant small inarticulate brachiopods of Cambrian Ordovician age. This siltstone is followed by 400 m of monotonous olive-green to grey silty shale. Ordovician graptolites, including *Didymograptus* sp. and *Tetragraptus* sp., occur in the overlying shales.

The Ovacık Formation is conformably overlain by the Eğripınar Formation (Early Silurian) and can be correlated with the Sosink and Bedinan Formations of southeastern Turkey (Tuna 1973). The Ovacık Formation was also deposited on an open marine shelf, which became less tectonically stable with time as indicated by an increasing number of turbiditic and conglomeratic intercalations (Fig. 6).

Silurian

The lower part of the conformably overlying *Eğripınar Formation*, which is ca. 300 m thick, is composed of thick-bedded, yellow to buff, arkosic sandstone and conglomerate containing well-rounded quartz pebbles. A sandstone–siltstone member is recognized in the upper part of the formation. The age of the Eğripınar Formation is Early Silurian from miospores. The existence of three thick sandstone intercalations with erosional bases, and towards the top of the formation, turbiditic sandstones with well-developed sole-structures all point to a form of proximal turbiditic deposition.

The conformably overlying *Hırmanlı Formation*, ca. 100 m thick, is mainly composed of very thin bedded laminated siliceous black shale, often pyritous, containing the following graptolites of Early Silurian age: *Monograptus* cf. *M. spiralis* Geinitz, *Monograptus* cf. *M. scitulum* Lapworth, *Climacograptus* cf., *C. scalaris* Hisinger, *Rastrites* sp. The high organic content, regularity of lamination without bioturbation, and the siliceous nature of the shales all suggest deposition in a deep restricted basin analogous to many other Lower Palaeozoic black shale deposits.

Late Silurian–Early Devonian

The conformably overlying *Karayar Formation* is composed of medium- to thick-bedded, dark limestone with regular thin-bedded black shale alternations. At the base a 10–20 m thick limestone member is almost entirely composed of orthocone shells which together with miospores, confirm a Late Silurian age. The 600 m upper part of the formation is dominated by thick-bedded limestone with Early Devonian brachiopods and corals. In the lower part of the formation the limestones are interpreted as calciturbidites deposited on a carbonate slope, while those higher in the formation appear to be regressive shelf deposits, including corals.

Early Devonian

The again conformably overlying *Sığırcık Formation* is composed of limestone, quartzite with shaly alternations near the base and quartzite–dolomite alternations towards the top. The quartzite is white to yellowish and cross-bedded, while the limestone is thick-bedded, dark grey to buff, with crinoids, corals (e.g. *Favosites* sp.) and brachiopods of Early Devonian age. In this formation abundant herringbone cross-bedding together with *in situ* colonial corals point to conditions rapidly alternating between shallow marine and coastal. Overlying unfossiliferous fine-grained red clastics with dolomitic interbeds thus appear to be non-marine deposits.

Middle Devonian

The *Büyükeceli Formation*, which follows unconformably, is mainly light to dark grey, medium- to thick-bedded, partly dolomitic limestone. Eight members are distinguished, with frequent occurrences of intra-formational conglomerates and mega-breccias. The thickness of the formation gradually decreases to the east. Fossils such as *Calceola sandalina* and *Amphipora ramosa* indicate a Middle Devonian age (Fig. 6).

The Büyükeceli Formation shows evidence of having been deposited on a carbonate platform. The existence of slump structures and mega-breccias points to unstable tectonic conditions during deposition, except in the two upper members.

Late Devonian

The conformably overlying *Akdere Formation* is composed of alternations of limestone, quartzitic sandstone, siltstone and shale. The limestone is dark grey, partly sandy, dolomitic and highly fossiliferous (Fig. 6). The quartzitic sandstone is white to pink and is intercalated with reddish-brown to buff siltstone and olive-grey shale. Minor oolitic haematite exists in the lower part of the formation and phosphatic nodules occur in the upper part. Interbedded reefal limestone contains fossils such as *Disphyllum caespitosum* and *Hexagonaria* sp. indicating a Late Devonian age. Associated clastic intercalations contain numerous brachiopods especially spirifers.

This facies is clearly indicative of reefal deposition affected by a series of regressive clastic incursions which become more abundant upwards, culminating in subaerial exposure marked by mud-cracked horizons.

Permian

In the Akkuyu area (Figs 3 & 9) the *Kırtıldağı Formation* unconformably overlies the Akdere Formation starting with 20 m of dark grey, medium- to thick-bedded lime-mudstone and foraminiferal wackestone which in turn is conformably overlain by 20 m of pink cross-bedded quartzite (Fig. 6). Fossils collected from the lower limestone unit are of early Late Permian (Murgapian) age (e.g. *Pachyphlaina* sp., *Staffella* sp., *Pseudovermiporella* sp., *Eopolydioxedina alghenensis*). Above comes highly fossiliferous grey well-bedded foraminiferal wackestone with packstone intercalations rich in brachiopods, corals, gastropods and foraminifera, indicating a Late Permian (Djulfian) age. *Mizzia velebitana, Permocalculus* sp., *Geinitzina* sp., *Gymnocodium* sp., *Agathammina* sp., *Hemigordius* sp.). Overlying pink, cross-bedded quartzitic sandstone then grades into dark-grey thick-bedded partly dolomitic wackestone with an increasing number of dolomite intercalations and stromatolites towards the top of the formation.

To the east and south of the Akkuyu area the Kırtıldagı Formation is disconformably overlain by the Kusyuvasıtepe Formation of Middle to Late Triassic age (see below, Fig. 6). To the west and north of Akkuyu the Kırtıldağı Formation is overthrust by the Büyükeceli Formation along the Taşlik fault (Figs 4 & 9). The Kırtıldağı Formation unconformably overlies the Akdere Formation, except north of Taşlik Hill where it is underlain by the quartzitic sandstone member of the Ovacık Formation (Fig. 9, see above), which may indicate pre-Late Permian faulting.

As generally throughout Turkey the Permian fauna confirm shallow marine stable shelf-type deposition.

Triassic

The unconformably overlying *Kargıcak Formation*, up to 250 m thick, crops out only in the eastern part of the Southern zone, in Haciishakli area outside the area mapped (Figs 2 & 7), where it begins with yellowish grey to maroon varicoloured marls with interbeds of bluish grey thin-bedded partly oolitic grainstone and wackestone. The oolitic grainstone interbeds include a dwarf gastropod, *Naticella* sp. The microfauna obtained from the limestones and marls (e.g. *Ammodiscus* sp., *Meandrospira* sp., and *Cyclogyra* sp.) indicate a Scythian (Early Triassic) age for the lower part of the formation. Varicoloured marls with thin limestone interbeds are then conformably overlain by grey, thick-bedded dolomitic limestones and dolomites with locally a thin bauxite at the top of the succession in which fossils generally indicate a Middle Triassic age (*Citaella iulia* Silva, *Agathammina* sp., *Diplotremina* sp., *Triasina* sp., *Aulotortus* sp.). The Kargıcak Formation can be correlated with the Katarası Formation in the Eastern Taurus (Demirtaşlı 1978). A gradual increase in fine-grained clastics and local conglomeratic intercalations points to deepening, less stable tectonic conditions.

The disconformably overlying Kuşyuvasıtepe Formation, ca. 150 m thick, is mainly composed of white, thick-bedded, very fine-grained mudstone and wackestone with occasional chert nodules (Figs 5 & 6). A local thin basal conglomerate at Kızılin hill (Figs 3 & 9) consists of white, well-bedded cherty wackestone overlain by grey wavy-bedded, partly nodular, oolitic grainstone with abundant crinoids. The existence of a shallow marine fauna and oolites confirms that the Kuşyuvasıtepe Formation was deposited on an open carbonate shelf. This is the youngest Mesozoic unit cropping out in the Akkuyu region and is unconformably overlain by Quaternary deposits. However, east of Hacıishaklı village the Kuşyuvasıtepe Formation is disconformably overlain by the Dibekli Formation (see below) (Fig. 7). Fossils in the Kuşyuvasıtepe Formation indicating a Middle–Late Triassic age include *Endothyra* cf. *Kupferi*, *Endothyronella wirzi*, *Trochammina* cf. *jaunensis, Ammobaculites* cf. *radstattensis, Haplofragmella inflata*, *Involutina eomesozoica*, *Involutina* cf. *gaschei*, *Diplopora* sp., *Variostoma* sp.

Jurassic

The *Dibekli Formation*, up to 30 m thick, starts with a basal conglomerate which is overlain by dark-grey well-bedded lime-mudstones and wackestones. Slope-breccias and intra-formational conglomerates are common at the base. Towards the top of the formation oolitic and oncolitic grainstones and packstones intercalated with oolitic ironstones and iron-stained quartzitic sandstones, are distinguished as the Işıklıkızıtepe Member (Fig. 6). The Dibekli Formation occurs largely in the eastern part of the southern zone, where it unconformably overlies either the Kargıcak or the Kuşyuvasıtepe Formations. It is absent in the Akkuyu

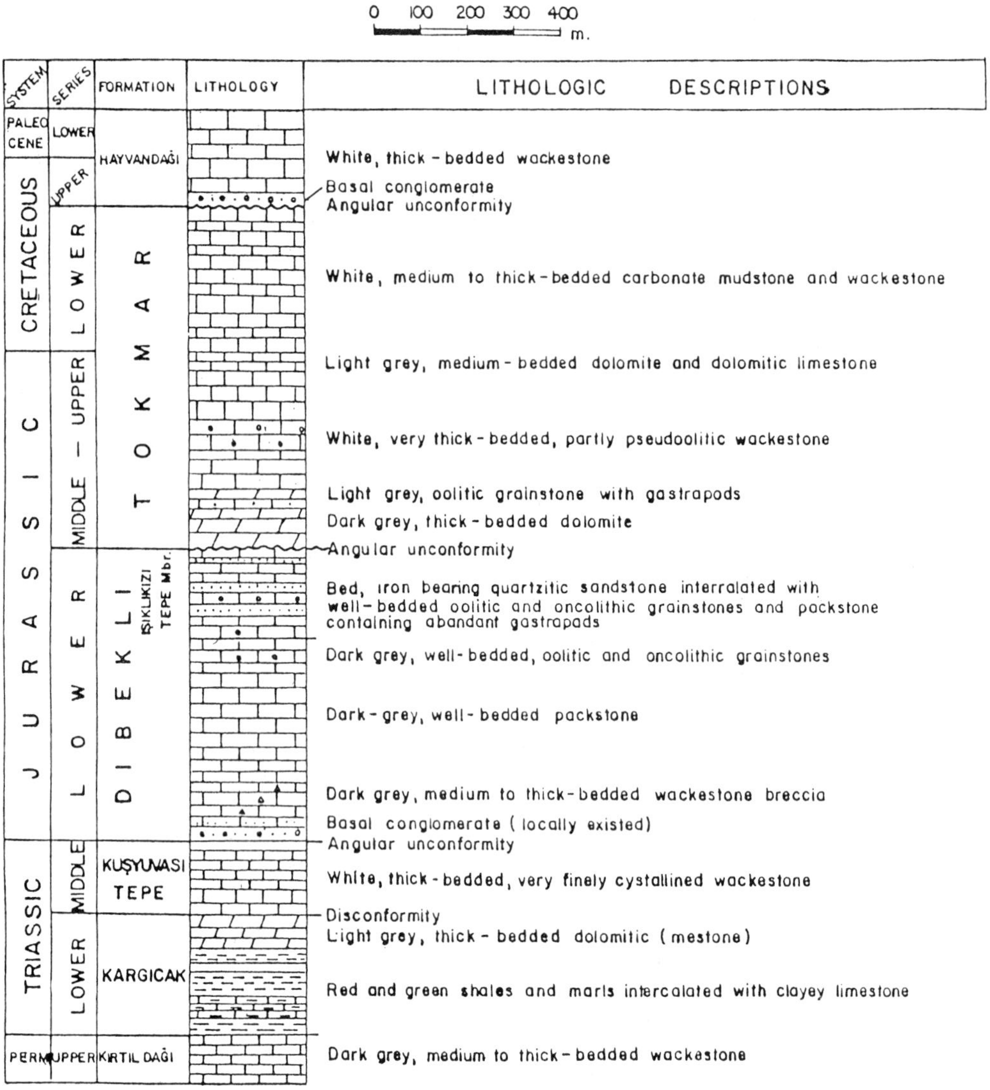

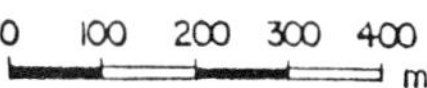

SYSTEM	SERIES	FORMATION	LITHOLOGY	LITHOLOGIC DESCRIPTIONS
PALEOCENE	LOWER	HAYVANDAĞI		White, thick-bedded wackestone
CRETACEOUS	UPPER			Basal conglomerate Angular unconformity
CRETACEOUS	LOWER	TOKMAR		White, medium to thick-bedded carbonate mudstone and wackestone
JURASSIC	MIDDLE – UPPER	TOKMAR		Light grey, medium-bedded dolomite and dolomitic limestone White, very thick-bedded, partly pseudoolitic wackestone Light grey, oolitic grainstone with gastrapods Dark grey, thick-bedded dolomite Angular unconformity
JURASSIC	LOWER	DIBEKLI (IŞIKLIKIZI TEPE Mbr.)		Bed, iron bearing quartzitic sandstone interralated with well-bedded oolitic and oncolithic grainstones and packstone containing abandant gastrapods
JURASSIC	LOWER	DIBEKLI		Dark grey, well-bedded, oolitic and oncolithic grainstones Dark-grey, well-bedded packstone Dark grey, medium to thick-bedded wackestone breccia Basal conglomerate (locally existed) Angular unconformity
TRIASSIC	MIDDLE	KUŞYUVASI TEPE		White, thick-bedded, very finely cystallined wackestone Disconformity
TRIASSIC	LOWER	KARGICAK		Light grey, thick-bedded dolomitic (mestone) Red and green shales and marls intercalated with clayey limestone
PERM	UPPER	KIRTIL DAĞI		Dark grey, medium to thick-bedded wackestone

FIG. 7. Composite succession of the eastward extension of the relatively autochthonous Southern zone between Hacıishakli and Kargıcak (see Fig. 2).

region possibly because of uplift and erosion after the Triassic. The Dibekli Formation may partly correlate with the Yanışlı Formation of the Intermediate zone. In the type section microfossils from the Dibekli Formation indicating a Liassic age are as follows: *Mayncina termieri*, *Palaeodasycladus medius*, *Thaumatoporella* sp., *Ammobaculites* sp., *Orbitopsella praecursor*, *Lituosepta recoarensis*, *Textrataxis* cf. *conica*, *Haurania amiji*, *Pseudocyclamina* sp., *Reophax* sp., *Pseudocyclamina lituus*, *Trocholina* sp., *Haplophragminium* sp., *Pseudopfenderina* sp. The reappearance of mega-breccias and intra-formational conglomerates suggest that the Dibekli Formation were deposited near the edge of a carbonate platform, possibly resulting from tectonic dissection into horsts and graben. The succession becomes more generally regressive upwards with numerous oolitic and oncolitic horizons.

Above this the up to 700 m thick unconformably overlying *Tokmar Formation* begins with light-grey, thick- to medium-bedded dolomites and dolomitic limestones which are in turn overlain by white well-bedded carbonate mudstone and wackestones (Fig. 7). A very thick-bedded to massive, white, partly pseudo-oolitic biomicrite bed overlies the dolomitic lower part of the Tokmar Formation forming a widespread marker horizon in each of the tectonic zones (Figs 5 & 7) except in the Akkuyu area where it probably has been eroded or not deposited. Microfossils from this formation indicate a Middle–Late Jurassic to Lower Cretaceous age: *Pfenderina neocomiensis*, *Kurnubia palestiniensis*, *Salpingoporella annulata*, *Valvulina lugeoni*, *Macroporella sellii*, *Kurnubia jurassica*, *Everticyclammina* cf. *greigi*, *Trocholina* cf. *elongata*, *Cuneolina* sp., *Actinoporella* sp., *Pseudochrysolidina* sp., *Hensonella* sp.

The Tokmar Formation may be correlated with the upper part of the Köroğultepe Formation of the Eastern Taurus (Demirtaşlı 1978). The fully marine fauna and facies point to accumulation again in a carbonate shelf environment.

Cretaceous–Palaeocene

The 200 m thick *Hayvandağı Formation* is mainly composed of white medium- to thick-bedded wackestone (Fig. 7) with a local basal conglomerate and transgresses all the rock units older than Late Cretaceous, including those of Lower Palaeozoic age. Fossils obtained from the base of the Hayvandağı Formation indicate a Late Cretaceous age: *Globotruncana arca*, *G. conica, Dicyclina schlumb.*, *Nezzazata simplex*, *Lepidorbitoides* sp., *Siderolites* sp. From the upper part of the formation fossils indicate a Palaeocene age: *Gavelinella* sp., *Miscellenea* sp. The environment of deposition does not appear to have differed significantly from the underlying Jurassic–Cretaceous Tokmar Formation.

Intermediate Zone

The *Intermediate* zone which has been thrust southwards over the Southern zone includes a thick Lower to Middle Cambrian sequence composed of limestones, shales and intraformational conglomerates, the *Sipahili Formation*, in turn conformably overlain by a thin Palaeozoic succession ranging from Late Cambrian to Late Devonian (*Babadıl Group*) which will not be described separately as it is difficult to differentiate in the field (Figs 5 & 7). The Late Cambrian successions are represented by red quartzitic sandstone and siltstone similar to the Ovacık Formation (see above). The Silurian is represented by graptolitic shales which have their counterparts in the coastal area as the Ovacık and Hırmanlı Formations. The Devonian is represented by limestone, shale and quartzite alternations which are similar to the Karayar and Sığırcık Formations. Brachiopods and corals indicate a Late Devonian age for the top of the Babadıl Group. Possible disconformities may exist between units in this group. Locally, very thin Lower Carboniferous limestone and quartzite (Korucuk Formation) also occur (Fig. 8). Late Permian limestone (Kırtıldağı Formation) unconformably overlies the Babadıl Group and is in turn unconformably overlain by red sandstone, shale and conglomerate of Late Triassic age (Murtçukuru Formation) (Fig. 8). The Dibekli Formation (Lower Jurassic) and the Tokmar Formation (Upper Jurassic–Lower Cretaceous) unconformably overlies the Sipahili Formation and the Babadıl Group (Figs 5 & 8).

In the central part of the Intermediate zone (at Tepeköy), conglomerates were deposited in a molassic basin which developed within both the Intermediate and Northern zones during Early Miocene time and in which fresh-water and lagoonal, partly coal-bearing sediments (Çavuşlar Formation), were deposited. This formation is unconformably overlain by flat-lying, marine limestone of Middle Miocene age (Silifke Formation) which was subjected only to small-scale normal faulting (Gökten 1976; Gedik *et al.* 1979).

Northern zone

The Northern zone has also been thrust southwards, this time over the Intermediate zone (Figs 2, 3 & 4). The overthrusting *could* have occurred before the Late Permian transgression (Kırtıldağı Formation), since it unconformably overlies the older formations of these two tectonic zones, however the Variscan movements have been obscured by later Alpine thrusting.

The Palaeozoic of the Northern zone is comparable with the Southern zone until the Carboniferous which is much more extensively developed in the Northern zone (Fig. 5). The best exposures of the Carboniferous successions in the Northern zone occur around Korucuk and Imamuşağı villages northwest of Silifke. There the Late Permian is represented by limestone with quartzite intercalations, unconformably overlain by Jurassic limestones, generally without any Triassic (Fig. 5). Jurassic–Cretaceous platform carbonates are then overlain by a Late Cretaceous olistostrome over which ophiolites were thrust in Late Cretaceous–Early Palaeocene time. In contrast, carbonate deposition continued on the platform in the Southern zone where the Hayvandağı limestone of Cretaceous–Palaeocene age was deposited. This is therefore strong evidence that the ophiolitic nappes were derived from the north.

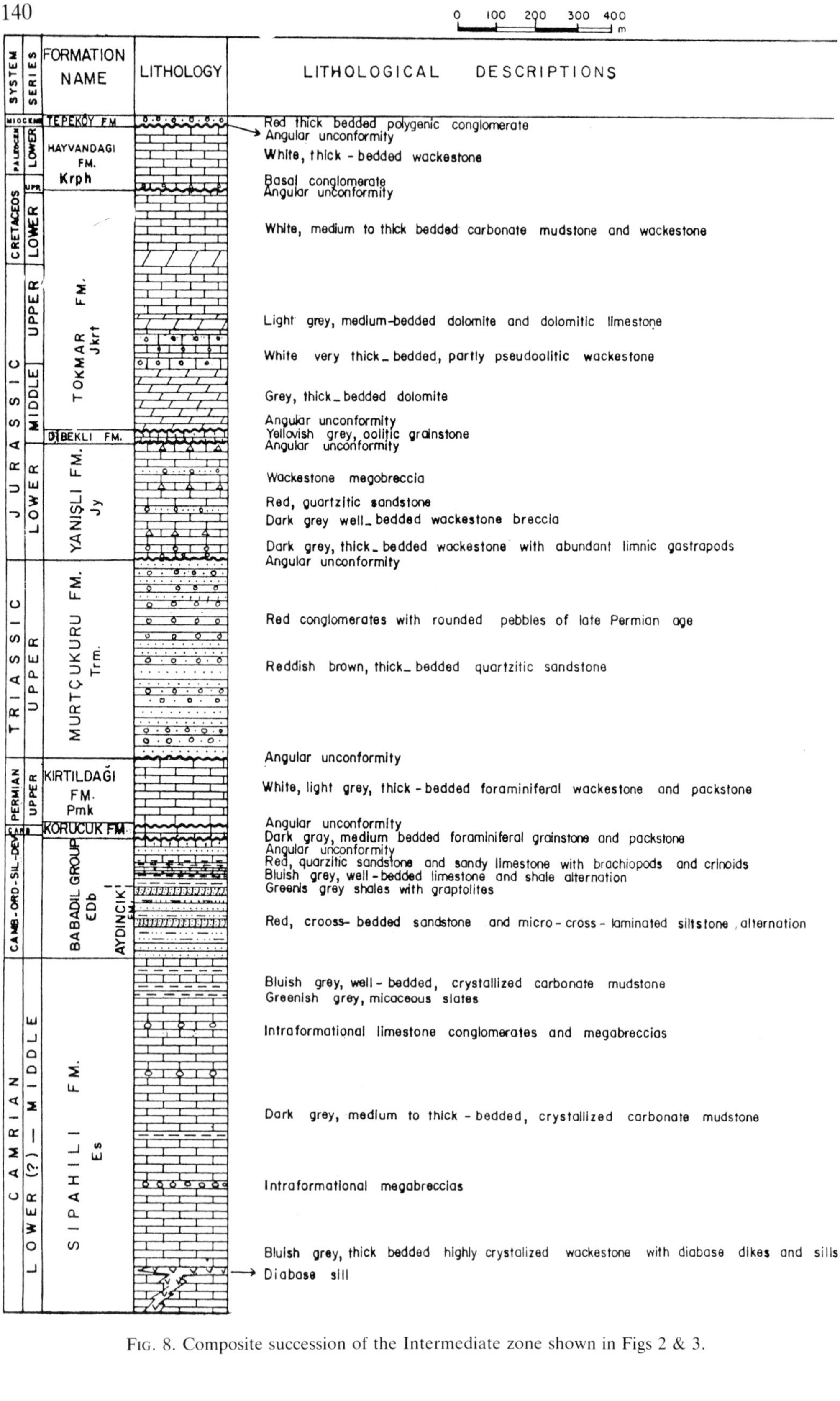

0 100 200 300 400 m

SYSTEM	SERIES	FORMATION NAME	LITHOLOGY	LITHOLOGICAL DESCRIPTIONS
MIOCENE		TEPEKÖY FM.		Red thick bedded polygenic conglomerate
PALEOCENE	LOWER	HAYVANDAGI FM. Krph		Angular unconformity White, thick - bedded wackestone Basal conglomerate
CRETACEOUS	UPR.			Angular unconformity
CRETACEOUS	LOWER	TOKMAR FM. Jkrt		White, medium to thick bedded carbonate mudstone and wackestone
JURASSIC	UPPER	TOKMAR FM. Jkrt		Light grey, medium-bedded dolomite and dolomitic limestone White very thick_ bedded, partly pseudoolitic wackestone
JURASSIC	MIDDLE	TOKMAR FM. Jkrt		Grey, thick_ bedded dolomite Angular unconformity
JURASSIC	MIDDLE	DİBEKLİ FM.		Yellowish grey, oolitic grainstone Angular unconformity
JURASSIC	LOWER	YANIŞLI FM. Jy		Wackestone megabreccia Red, quartzitic sandstone Dark grey well_ bedded wackestone breccia Dark grey, thick_ bedded wackestone with abundant limnic gastropods Angular unconformity
TRIASSIC	UPPER	MURTÇUKURU FM. Trm.		Red conglomerates with rounded pebbles of late Permian age Reddish brown, thick_ bedded quartzitic sandstone Angular unconformity
PERMIAN	UPPER	KIRTILDAĞI FM. Pmk		White, light grey, thick - bedded foraminiferal wackestone and packstone Angular unconformity
CARB.		KORUCUK FM.		Dark gray, medium bedded foraminiferal grainstone and packstone Angular unconformity
CAMB.-ORD.-SIL.-DEV.		BABADIL GROUP ЄDb; AYDINCIK FM.		Red, quarzitic sandstone and sandy limestone with brachiopods and crinoids Bluish grey, well - bedded limestone and shale alternation Greenis grey shales with graptolites Red, crooss- bedded sandstone and micro - cross - laminated siltstone , alternation
CAMBRIAN	LOWER (?) — MIDDLE	SIPAHILI FM. Єs		Bluish grey, well - bedded, crystallized carbonate mudstone Greenish grey, micaceous slates Intraformational limestone conglomerates and megabreccias Dark grey, medium to thick - bedded, crystallized carbonate mudstone Intraformational megabreccias Bluish grey, thick bedded highly crystalized wackestone with diabase dikes and sills Diabase sill

FIG. 8. Composite succession of the Intermediate zone shown in Figs 2 & 3.

Stratigraphic breaks

The differences between each of the three tectonic zones are further illustrated by the record of both angular unconformities and disconformities.

First there is a *pre-Middle Devonian angular unconformity* which is observed at the base of the Büyükeceli Formation in the Southern zone (Figs 3 & 5). Following this, a *pre-Middle Carboniferous disconformity* exists at the base of the Carboniferous Imamuşaği Formation which crops out only in the Northern zone (Fig. 5). A possible pre-Carboniferous unconformity may also exist between the Cambrian to Devonian Babadıl Group and the Early Carboniferous Koruçuk Formation which crops out north of Koçaslı village in the Intermediate zone (Figs 3, 5 & 8).

An important *pre-Late Permian angular unconformity* exists at the base of the Upper Permian Kırtıldaği Formation in all the tectonic zones (Figs 3 & 5). The Cambrian–Ordovician Ovacık Formation crops out (near Taşlik hill) beneath the Upper Permian Kırtıldaği Formation indicating that Variscan faulting took place before transgression of the Kırtıldaği Formation. In the Intermediate zone the Kırtıldaği Formation is considerably thinner and is unconformably underlain by the Korucuk Formation, the Babadıl Group and the Sipahili Formation, together of Cambrian to Early Carboniferous age (Fig. 8). In the Northern zone it unconformably overlies the Imamuşaği Formation (Fig. 5), or different units of the Korucuk and Akdere Formations, confirming that the Variscan orogeny affected particularly the north of the area.

In addition a *pre-Late Triassic disconformity* has been observed between the Middle Triassic Kuşyuvasıtepe and the Permian Kırtıldaği Formation (Fig. 6). A basal conglomerate locally exists at the base of the Kuşyuvasıtepe Formation on Kızılın hill, east of Akkuyu. At Kargıcak 20 km east of Akkuyu a 2 m thick bauxite bed separates the Kuşyuvasıtepe and the Kargıcak Formations indicating uplift and subaerial exposure during the Triassic (Fig. 7). In the Intermediate zone at Yanıslı village an angular unconformity exists at the base of the Upper Triassic Murtcukuru Formation where it is underlain by the Cambrian–Devonian Babadıl Group (Figs 3, 5 & 8).

A *pre-Early Jurassic angular unconformity* exists at the base of the Lower Jurassic Dibekli Formation which crops out east of Hacıishaklı in the Southern zone (Figs 3 & 7). The Dibekli Formation is transgressive over the Lower Devonian Sıgırcık, the Lower Triassic Kargıcak and the Middle Triassic Kuşyuvasıtepe Formations (Fig. 5). The Lower Jurassic Dibekli Formation is also transgressive onto the Intermediate zone where it is unconformably underlain by the Lower Jurassic Yanışlı Formation, the Upper Triassic Murtcukuru Formation and the Cambrian–Devonian Babadıl Group (Fig. 8).

A *pre-Late Jurassic angular unconformity* also exists at the base of the Tokmar Formation which crops out in each of the geotectonic zones (Fig. 5), reflecting an 'Eocimmerian' orogenic event. In the southern zone the Middle–Upper Jurassic Tokmar Formation is unconformably underlain by the Lower Jurassic Dibekli Formation, whereas in the Intermediate zone it unconformably overlies the Upper Triassic Murtcukuru and the Lower Jurassic Yanışlı Formations. In the Northern zone the Tokmar Formation unconformably overlies the Permian Kırtıldağı Formation reflecting 'Eocimmerian' erosion, or non-deposition of Triassic rocks (Fig. 5).

There is a *pre-Late Cretaceous–Palaeocene angular unconformity* at the base of the Upper Cretaceous Hayvandağı Formation, which unconformably overlies different Palaeozoic formations in the Southern zone, and is unconformably underlain by the Tokmar Formation in the Intermediate zone (Figs 3 & 5). In the Northern zone the Late Cretaceous–Palaeocene olistostrome unconformably overlies the Akdere Formation and in turn has been overthrust by ophiolitic melange.

Pre–Early Miocene angular unconformities have also been observed at the base of the Tepeköy Formation and the Çavuşlar Formation (Fig. 5) and are assumed to represent the same discontinuity since the age of both formations is Early Miocene.

The *pre–Middle–Miocene angular unconformity* is the most major in each of the geotectonic zones (Fig. 5). The Silifke Formation of Middle Miocene age unconformably overlies all the older rock units and is assumed to be post-tectonic (Fig. 5).

Structure

Folds

In the Southern zone the main fold axes trend approximately NE–SW. North of Akkuyu lies the major Büyükeceli anticline/syncline pair. In the centre of the Büyükeceli syncline is a piercement structure associated with two conjugate faults, the Taşlık and Gökgedik thrusts, between which is exposed the earliest major

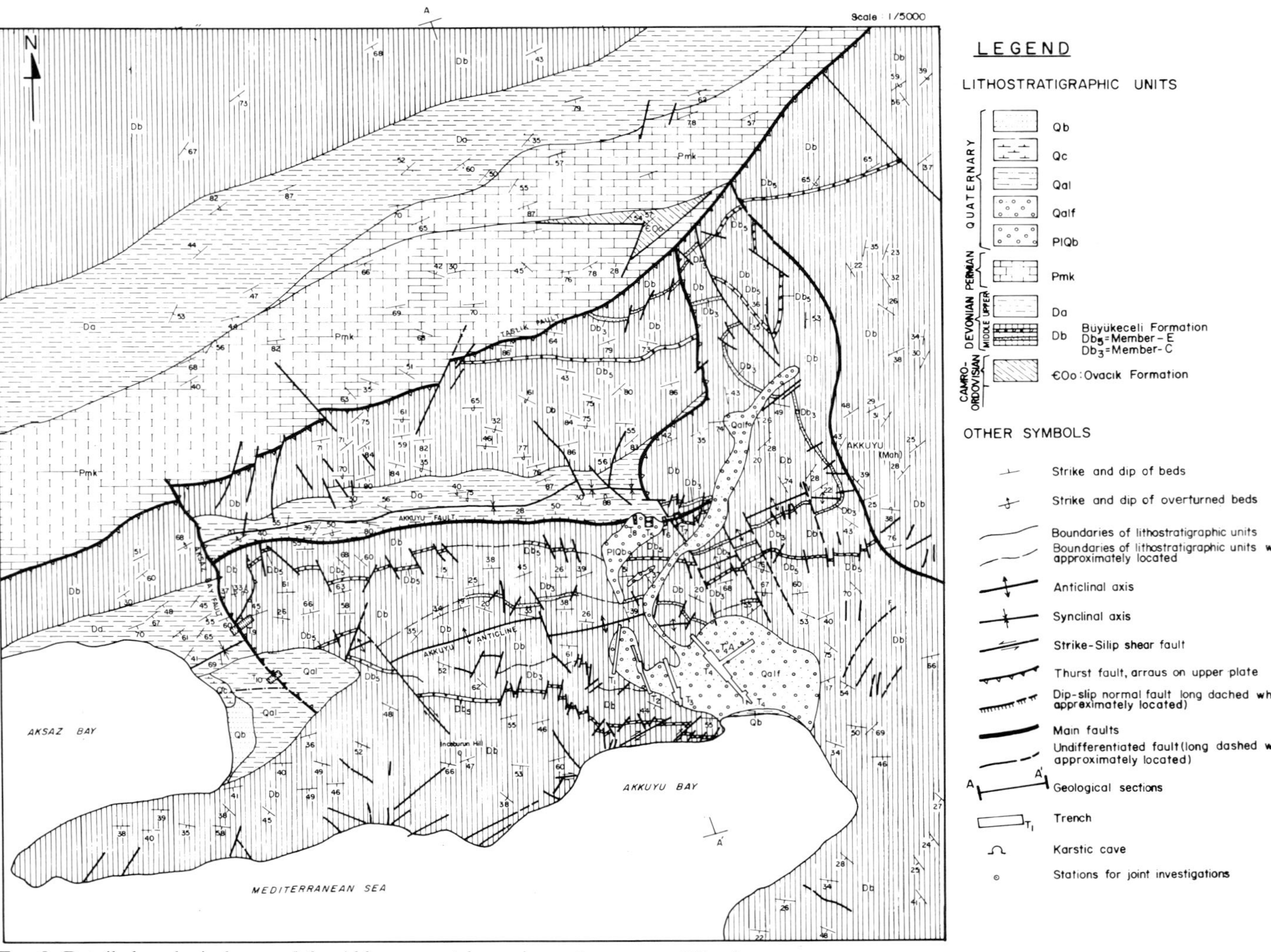

FIG. 9. Detailed geological map of the Akkuyu coastal area in southern part of Fig. 3.

folds in the area, the ENE–WSW Akkuyu anticline (Figs 3, 4 & 9), which has been displaced by strike-slip faults. Analysis of associated minor structures shows that this major fold plunged 10° towards 250°. Dispersal of fold axes is attributed to the superimposition of early Alpine ('Eocimmerian') deformation on the earlier Variscan pattern. A series of isoclinal folds overturned towards the south, in the northern part of the area, are also attributed to 'Eocimmerian' thrusting (e.g. on the Büyükeceli thrust, Figs 3 & 4).

Folds in the Intermediate zone between the Büyükeceli and Koçaşlı thrusts are strongly overturned towards the south. Fold axes are approximately parallel to the outcrop of the thrusts (NE–SW). A nappe composed of the Sipahili Formation, thrust from the north, is now represented by klippen resting on the Babadıl Group and the autochthonous Hırmanlı and Karayar Formations (Figs 2, 3 & 4).

In contrast to the NE–SW trending fold axes in the Babadıl Group and Sipahili Formation, axes in the Tokmar and Hayvandağı Formations trend approximately NNE–SSW and are assumed to reflect later folding during the Tertiary phase of the Alpine orogeny.

Evidence of Variscan folding in the Northern zone is seen north of Dedeler village where gentle dipping, open-folded Kırtıldağı Formation rocks unconformably overlie tightly folded and faulted lower members of the Korucuk Formation and/or the Akdere Formation. The Imamuşağı Formation and the upper members of the Korucuk Formation (Visean and Late Tournasian) were truncated due to Variscan uplift and erosion before deposition of the Kırtıldağı Formation (Fig. 5).

Faults

Various types of faults have been mapped in the area.

Thrusts

The Büyükeceli thrust is the most persistent and important structure and can be traced for 20 km in a NE–SW direction dipping 40° to the northwest (Figs 3 & 4). Along the Büyükeceli thrust the Sipahili Formation and the Babadıl Group of the Intermediate zone were thrust over the Sığırcık and the Karayar Formations of the Southern zone. The Büyükeceli and the Koçaşlı thrusts are unconformably sealed by the Tokmar Formation of Jurassic–Cretaceous age at Payamgediği (16 km NE of Büyükeceli) and outside the area mapped in detail (see Fig. 2), indicating a pre-Jurassic age of thrusting.

Another major thrust is the *Koçaşlı thrust* which runs parallel to the Büyükeceli thrust, and along which the Sipahili Formation moved over the Babadil Group (Figs 3 & 4). Klippen of the Koçaşlı thrust-sheet consisting of Sipahili Formation rocks lie tectonically above the Ovacik and the Hırmanlı Formations. Both the Koçaşlı and the Büyükeceli thrusts are sealed by the Tokmar Formation at Payamgediği but further southwest are covered by the Hayvandağı Formation. It is believed that both the Büyükeceli and Koçaşlı thrusts are related to the same episode of pre-Late Jurassic thrusting.

Also of note is the *Sarnıc thrust* which is exposed on the main road between Koçaşlı and Gülnar, 2 km north of Koçaslı, taking its name from Sarnıc hill where the Kırtıldağı Formation (Permian) is thrust over the Hayvandağı Formation (Upper Cretaceous). This thrust trends NE–SW until 2 km south of Tepeköy where it is terminated by the NNE–SSW-trending Tepeköy fault of pre-Early Miocene age.

The *Taşlik* and *Gökgedik* thrusts or reverse faults occur to the NW and NE of Akkuyu respectively and are thought to be a conjugate pair of reverse faults developed during the 'Eocimmerian' orogenic phase (Figs 3 & 9). They emplace members of the Devonian Büyükeceli Formation to the NW (Taşlık) and to the NE (Gökgedik), over the Permian Kırtıldağı Formation and, in the case of the Gökgedik structure, over the Triassic Kuşyuvasıtepe Formation. The Taşlık thrust is well exposed along the road from Büyükeceli to Akkuyu. The Gökgedik thrust is best exposed on the coast in a bay 1 km south of Gökgedik hill.

Discussion and conclusions

During Pre-Cambrian and Early Palaeozoic times the entire Taurus belt formed part of the Arabian continent. Intra-cratonic block-faulting and rifting produced thick intraformational conglomerates during Early–Middle Cambrian time in the Intermediate zone. General regression took place during the Late Cambrian in the entire Taurus belt and red conglomerates, quartzites and siltstones were deposited. A widespread marine transgression during the uppermost Cambrian and Lower Ordovician led to deposition of open marine mudstones and minor turbidites (Ovacık Formation). Uplift and renewed erosion of the source areas to the south in the Late Ordovician to Early Silurian resulted in deposition of arkosic conglomerates and sandstones forming the proximal turbidites of the Eğripınar Formation. Graptolitic black shales were deposited in a deep euxinic basin during the Early Sılurian while calci-turbidites and black shales accumulated on a carbonate slope during the Late Silurian (Eğripınar and Hırmanlı Formations). The Early Devonian was characterized by regression with tidal-flat-type deposition comprising cross-bedded red sandstones intercalated with shales and sandy limestones (Sığırcık Formation).

Continued uplift of the southern zone during Early Devonian time resulted in subaerial exposure and continental-type deposition with a regional angular unconformity between the Lower and Middle Devonian formations. Syn-sedimentary faulting and deposition of mega-breccias of the Middle Devonian in the Akkuyu area and intercalations of basaltic volcanics with mega-breccias of the same age in the Sariz-Tufanbeyli area of the Eastern Taurus belt (Demirtaşlı 1981) are strong evidence of some form of rifting at this time. Vertical movements were again active in Late Devonian, Carboniferous and Early Permian times in each of the three tectonic zones. Carboniferous sediments were not deposited in the southern zone due to local uplift, whereas thick Carboniferous was deposited in the Northern zone. The Northern zone was then thrust over the Intermediate zone prior to the Late Permian.

Major crustal extension and rifting of the Tauride area then took place during Middle and Late Triassic time. During the Late Triassic an intracratonic, molasse basin developed in the Intermediate zone and the Northern zone whereas a marine Triassic basin existed in the Southern zone. Then further compression and thrusting took place related to the 'Eocimmerian' orogeny prior to a general Early Jurassic transgression.

The ophiolites now located north of the Central Taurus relatively autochthonous Jurassic–Cretaceous carbonate platform are thought to have formed in an oceanic basin located between the Inner Taurus belt (Bolkar Mountains) and the Central Taurus. In the Karaman and Ermenek areas south of Konya marine Triassic rocks including basic extrusives, pelagic limestones and radiolarites, indicate opening of this Neotethyan ocean (Koçyiğit, 1976). Obduction of the ophiolites from this oceanic basin took place during Late Cretaceous to Palaeocene time from the north. In contrast in the southern zone, stable continental shelf conditions prevailed during Late Cretaceous–Palaeocene time (neritic Hayvandağı Limestone) showing that ophiolites could not have been emplaced from the south before this time.

During Late Eocene–Oligocene time strong horizontal movements affected the western Taurus, seen in the emplacement of the Hadım and Beyşehir–Hoyran nappes during the 'Middle Alpine (Pyrenean) Orogeny'. A basin north of the Intermediate zone filled with turbiditic sandstones and conglomerates early in the Miocene. Syn-sedimentary deformation affected the already deposited Miocene sediments (Çavuşlar Formation). Stable platform conditions then resumed in the region during the Middle to Upper Miocene. During uppermost Miocene (Messinian) a general regression took place in the East Mediterranean and the Taurus Mountains were uplifted towards their present elevation.

ACKNOWLEDGEMENTS: The author thanks the Turkish Electricity Authority (TEK) for general support of this work. Fossil determinations were provided by MTA palaeontologists, Z. Dağer, M. Serdaroğlu, E. Çatal, F. Armağan and B. Sözeri. Useful discussions both in the field and laboratory took place with, in particular, Prof. I. Ketin, Prof. S. Abdüsselamoğlu, Dr I. Seymen, Dr N. Görür, Dr R. Akkok and Dr W. Harsch. A. H. F. Robertson and J. E. Dixon helped produce the final manuscript.

References

BLUMENTHAL, M. 1955. *Geologische untersuchungen in den Kosten Ketten des Sudanatolischen Taurus zwischen Silifke und Anamur.* Unpublished MTA Report No. 2823, Anakara.

——, 1960. Le système structural du Taurus sudanatolien. In: *Livre à la mémoire du Prof. P. Fallot, Mém h. ser. Soc. géol. Fr.* **2,** 611–662.

BRUNN, J. H., DUMONT, J. H., DE GRACIANSKY, P. C., GUTNIC, M., JUTEAU, T., MARCOUX, J., MONOD, O. & POISSON, A. 1971. Outline of the Geology of the Western Taurides. In: A. S. CAMPBELL, ed., *Geology and History of Turkey*, Petroleum Exploration Society of Libya, Tripoli, 225–255.

DEMIRTAŞLI, E. 1978. Carboniferous of the area between Pinarbasi and Sariz. In: *Guide Book of Field Excursions on the Carboniferous Stratigraphy of Turkey*, Spec. Publ., MTA, Ankara, 25–29.

——, ÇATAL, E., DIL, N., KIRAĞLI, C. & SALANCI, A. 1978. Carboniferous of the Silifke area. In: *Guide Book of Field Excursions on the Carboniferous Stratigraphy in Turkey*, Spec. Publ., MTA, Ankara, 31–37.

——, 1980. Correlation of Palaeozoic stratigraphy of

Turkey, Iran and Pakistan. In: IBRAHIM SHAH, S. M. & QUENNELL, A. M. (eds.), *Stratigraphic correlation of Turkey, Iran and Pakistan*, Vol. 1, Overseas Development Administration, London.

——, 1981. Summary of the Palaeozoic stratigraphy and Variscan events in the Taurus Belt. Newsletter, I.G.C.P. Project No. 5, *Correlation of Variscan and pre-Variscan events in the Alpine Mediterranean Belt*, **3,** 44–57.

DUMONT, J.-F. & KEREY, E. 1975. L'accident de Kirkkavak: un décrochement NS à la limite du Taurus occidental et de la depression du Köprüçay. *Bull. géol. Soc. Turkey*, **18,** 59–62.

—— 1976. *Etudes geologiques dans les Taurides Occidental; Les formations paléozoiques et mésozoiques de la couple de Karacahisar (Province d'Isparte, Turquie)*. Thesis, Université de Paris Sud, 213pp.

GEDIK, A., BIRGILI, Ş., YILMAZ, H. & YOLDAŞ, R. 1979. Mut-Ermenek-Silifke yöresinin jeolojisi ve petrol olanaklari. *Türkiye İeol. Kur. Büll.* **22,** 7–26.

GÖKTEN, E. 1976. Silifke yöresinin temel kaya birimleri ve Miyosen stratigrafisi, *Turkish İeol. Kur. Büll.* **19,** 117–126.

KOÇYIĞIT, A. 1976. Karaman-Ermenek (Konya) bölgesinde ofiyolitli melanj ne diğer olusuk-lar. *Türkiye İeol. Kur. Büll.* **22,** 103–115.

ÖZGÜL, N. 1976. Toroslarin bau temel jeologi özellikleri (Some geological aspects of the Taurus orogenic belt, Turkey). *Türkiye İeol. Kür. Büll.* **19,** 65–78.

ROBERTSON, A. H. F. & WOODCOCK, N. H. 1982. Sedimentary history of the south-western segment of the Mesozoic–Tertiary Antalya continental margin, south-western Turkey. *Eclogae geol. Helv.* **75,** 517–562.

ŞENEL, M., SERDAROĞLU, M., KENGIL, R., ÜNVERDI, M. & GÖZLER, M. Z. 1983. Teke Toroslari Güneydoğusurun Jeolojisi. *MTA Bull.* **95/96,** 13–43.

AUTHOR'S ADDRESS: M.T.A. Enstitüsü, Jeoloji Dairesi, Ankara, Turkey.

2. NEOTETHYS: Levant and North African offshore

It is appropriate to begin the discussion of Neotethyan evolution with the Levant and North African offshore areas, which are the only parts of the jig-saw agreed by everyone to have remained relatively fixed, as parts of the stable margin of Gondwana. Off North Africa and in Israel, the continental margin sequences described in this section (Fig. 1) are all essentially in place, and the objectives of the contributors are to infer indirectly from the stratigraphic record and the history of faulting and subsidence, the location and timing of Neotethyan ocean formation beyond these areas to the north and west. In the Hatay and Baër-Bassit area of north western Syria, the evidence is more direct since an Upper Cretaceous ophiolite has been emplaced onto a stacked sequence of continental margin and pelagic sediments described by **DELAUNE-MAYERE**.

With or without an ophiolite, the key Neotethyan question remains: when does rifting and subsidence imply that an adjacent ocean basin has formed and when is it attributable merely to crustal attenuation without ocean-floor generation? **SESTINI**, reviewing seismic and borehole data from west of the Nile Cone, sees a remarkably stable subsiding shelf, albeit with faults, small half-grabens, basins and highs, but with no indications of a nearby Mesozoic continent/ocean boundary at all, at least as far out as the Herodotus Basin in the central East Mediterranean. Pelagic deposition became widespread in the Mesozoic with a major eustatic transgression in the Upper Cretaceous. The North African off-shore area experienced only mild subsidence, tilt reversal and faulting in the latest Cretaceous at a time when major nappe transport was occurring in the Taurides. The most marked subsidence was linked to the outbuilding of the Nile cone in the Neogene.

In Israel, the two opposing inferences from a generally accepted clear history of rifting are both expressed in this section. **GVIRTZMAN & WEISBROD** and **GARFUNKEL & DERIN** both conclude that rifting reflects Mesozoic ocean-floor generation along a presenting N–S offshore line. The movement of the inferred block away from the Israel margin (to ?Turkey) has to be accommodated outside the area of **SESTINI**'s stable shelf. **DRUCKMAN** describes and interprets the sequence in the Helez 1A Deep borehole on the coastal plain of Israel, which penetrated the Precambrian, and arrives at the alternative view, that although Middle Triassic clastics were eroded from uplifted faulted areas, the fact that the exposed basement was metamorphic rather than mafic igneous, and the shallow rather than deep marine nature of Late Triassic and Early Jurassic facies both argue against a nearby N–S continental margin in the early Mesozoic. **GVIRTZMAN & WEISBROD** show that the Helez 1A borehole, in which virtually the entire Palaeozoic succession is missing, is located near the crest of a major Hercynian geanticline, the 'Geanticline of Helez', towards the crest of which, regionally, progressively older Palaeozoic units are cut out. In the 500 km wide axial region, Permian and Triassic rocks rest directly on a Precambrian/Cambrian metamorphic basement.

The geanticline is thought to have developed in two phases, pre-Carboniferous and Pre-Permian. The authors point to the existence of other major geanticlinal structures located around the periphery of the Hercynian orogenic belt. The western limb of the 'Geanticline of Helez' is now missing, and the authors believe it was rifted-off the Levant continental margin and is possibly now located in the Tauride Mountains.

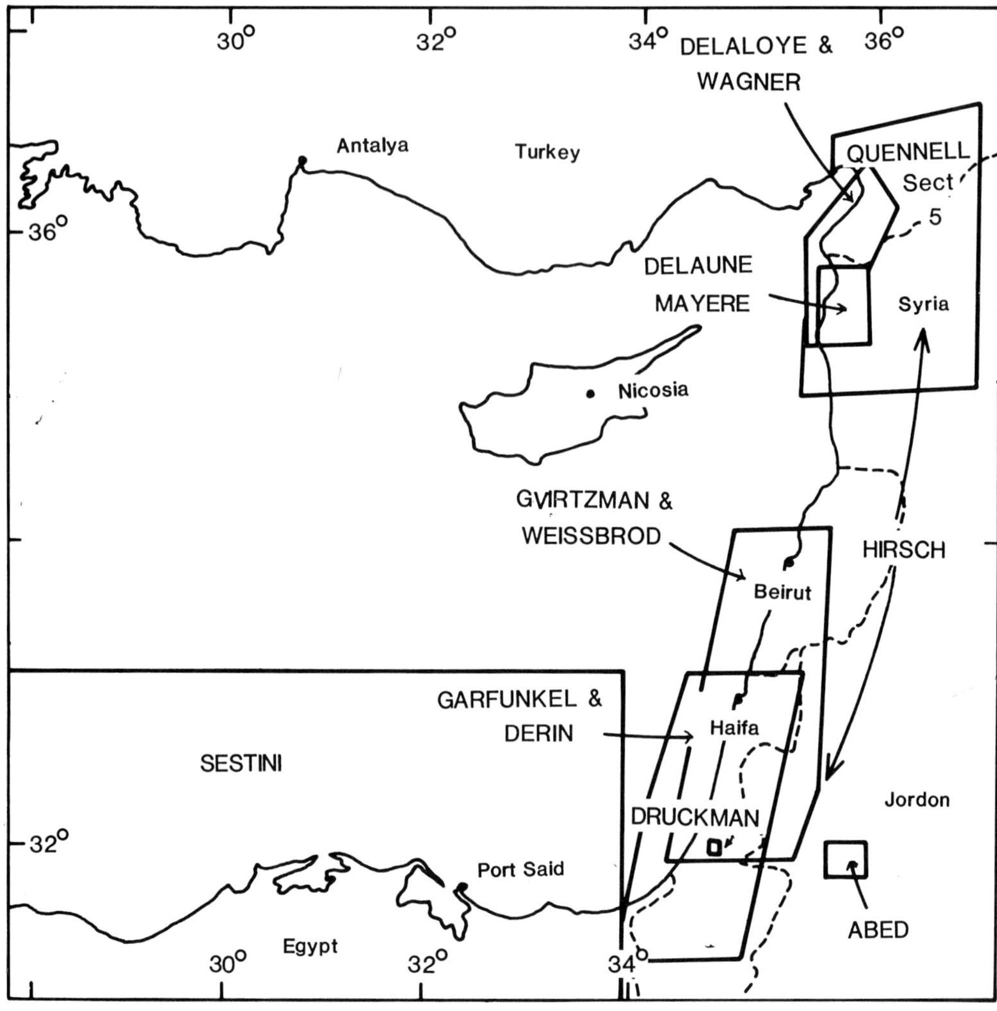

FIG. 1. The areas of the Levant discussed by various authors in this section.

A thorough analysis of the Levant also in terms of a Mesozoic continental margin is given by **GARFUNKEL & DERIN**. Subsidence of the coastal areas of Israel was first evident in the Late Permian then pulses of rapid subsidence took place in the Lower Triassic (Scythian-Anisian), Upper Triassic (Carnian-Norian) and Early Jurassic (early Liassic). The subsidence rates of 50–100 m/Ma are typical of rifted and faulted areas. By Late Liassic time faulting had ceased, followed by passive subsidence and the construction of a Mesozoic carbonate platform. Geophysical data confirm that, as expected for a passive continental margin, the crust thins towards the Eastern Mediterranean Sea, and there are strong NW–SE trending magnetic and gravity anomalies in the basement. **GARFUNKEL & DERIN** see the rifting of the Levant as a long prelude to ocean floor spreading related to opening of the Atlantic in the Upper Jurassic. The left-lateral shear inferred between the Eurasian and African plates during early Atlantic opening suggests that the Levant opened by E–W spreading possibly bounded by a transform margin to the south.

The record of a major Upper Cretaceous transgression over the Levant is summarized by **ABED** from work in Jordan. During the Cenomanian an island appeared and was cut by channels, the author believes, related to a period of syn-sedimentary tectonism.

HIRSCH takes the opposite view to **GVIRTZMANN & WEISBROD** and **GARFUNKEL & DERIN**, arguing from a compilation of comparative stratigraphic and faunal-province data that during the Mesozoic, the Levant from Sinai northwards was contiguous with the Tauride mountains and that no suture can have existed between them. This is based mainly on brachiopod and bivalve associations which separate the north and south Tethyan areas into faunal provinces. The brachiopods of the Greek 'Apulian' area relate to Gondwana and an important ammonite fauna in the eastern Taurides ('Pisidien') belongs to the south-Tethyan or 'Ethiopian' fauna.

The Hatay and Baër-Bassit massifs of North Syria and the Turkish border area are the subject of the two final papers in the section. **DELAUNE-MAYERE** concentrates on the pelagic sedimentary rocks; **DELALOYE & WAGNER** on the geochemistry and ages of the mafic igneous rocks. The rifting and subsidence history is recorded in the passive margin sequences subsequently emplaced over the Arabian margin in Late Cretaceous time. Alkali basalts, deep water micritic limestones and minor redeposited quartzose sandstones are related by **DELAUNE-MAYERE** to Upper Triassic rifting, followed by siliceous pelagic sediment accumulation in the Mesozoic. An important phase of deep-water per-alkaline volcanism occurred in the Upper Jurassic. The author points out the similarities with successions in SW Cyprus (Mamonia Complex) and SW Turkey (Antalya Complex) which she has elsewhere related to former Mesozoic passive margins. On top of the stack are the Cretaceous ophiolitic nappes of Hatay and Baër-Bassit which **DELALOYE & WAGNER** show to have formed at 73–99 Ma, and after a negligible interval to have been sliced while still hot (86–95 Ma), and then finally emplaced in the Campanian-Maastrichtian.

The Cretaceous ophiolitic rocks show marginal basin or arc-like chemistry in contrast to the within-plate characteristics of the Triassic alkalic lavas. **DELALOYE & WAGNER** conclude that both ophiolites formed in a single basin produced by east–west shear related to the opening of the Atlantic.

These two papers form an apt prelude to the Turkish section which follows. Those who equate Triassic rifting with continental margin formation and conclude that the Cretaceous ophiolite formed as a continuation of this process in the same basin, are in opposition to **HIRSCH** in this section and, for example, to **RICOU *et al.*** in the next, who would derive the ophiolitic nappe from a single ocean basin much further north beyond the Tauride carbonate platform or 'axe calcaire'. This section also highlights the plate-kinematic problems which remain for proponents of Mesozoic continental margin formation: how to link Hatay, Baër-Bassit, off-shore Israel and, if **DELAUNE-MAYERE** is correct, Antalya and SW Cyprus in a coherent evolving plate-boundary network in the Mesozoic.

Evolution of a Mesozoic passive continental margin: Baër-Bassit (NW Syria)

M. Delaune-Mayere

SUMMARY: The Baër-Bassit area comprises imbricate sheets of ophiolities and pelagic sedimentary rocks thrust eastward onto the Arabian platform.

Deep-water sediments were deposited from Late Triassic to Mid-Cretaceous time (Cenomanian/Turonian). Triassic sediments consist of marls and Halobia-bearing limestones. From Jurassic to Cenomanian the deposits were siliceous mudstones, siltstones and cherts.

Comparisons with the pelagic sediments of the Mamonia Complex (Cyprus) and the Antalya Complex (Turkey) show that the three areas in the East Mediterranean had a similar evolution from Upper Triassic to Lower Cretaceous time. In the Upper Triassic there was continental rifting and formation of oceanic crust in a rift similar to the Red Sea. The opening was more marked in the west than in the east. This was followed by subsidence and sedimentation of siliceous rocks from Jurassic to Lower Cretaceous. Per-alkaline volcanism took place in the Baër-Bassit area at this time. Sedimentation stopped in Hauterivian time in Cyprus, in the Cenomanian/Turonian in Syria, and during Late Cretaceous in Antalya. Sedimentation related to closure of the oceanic basin still remains to be defined.

The Baër-Bassit area of north-west Syria comprises sheets of ophiolite and radiolarites thrust from north-west to south-east onto the northern margin of the Arabian platform during Upper Maastrichtian time (Parrot 1977). These sheets are part of the 'peri-arabic crescent' adjacent to the platform, which extends as far as Oman (Ricou 1971) and they are also a link with the ophiolitic and radiolarite sheets of the Mamonia Complex, south-west Cyprus (Lapierre & Parrot 1972).

Structurally, the sedimentary and ophiolitic units are all bounded by tectonic contacts and are allochthonous with respect to the underlying Upper Cretaceous limestones of the Arabian platform (Fig. 1). Post-Maastrichtian tectonism has further dismembered the sedimentary units which now exist as scattered outcrops (Fig. 2).

There are changes in the sedimentary successions from east to west. To the east the deposits are thicker and more clastic, mostly sandstones and calcarenites with redeposited platform carbonate material; to the west sequences are more distal with monotonous mudstones and cherts (Delaune-Mayere *et al.*, 1977).

Stratigraphy

The stratigraphy of the sedimentary formations was established by Dubertret (1953) and Kazmin & Kulakov (1968), then modified by Delaune-Mayere & Saint-Marc (1979, 1980). The deep water sediments (Fig. 3) were deposited from the Late Triassic to the Mid-Cretaceous time (Cenomanian/Turonian) and consist of two sequences:

Upper Triassic Halobia limestones

No units older than Late Triassic are seen in the Baër-Bassit pelagic successions. The oldest deposits consist of marls and limestones with abundant fragments of *Halobia* shells and calcite-replaced Radiolaria. Thin beds of carbonaceous sandstones with plant material are interbedded in the lower part of this succession. These sandstones are quite similar to the Vlambouros sandstones of the Mamonia Complex, Cyprus (Lapierre 1975; Ealey & Knox 1975; Robertson & Woodcock 1979) and the Upper Triassic sandstones of the Antalya Complex, Turkey (Alakır, Çay unit of Marcoux 1976 or the Kumluca Zone of Robertson & Woodcock 1981a,b,c; this volume).

Upwards in Baër-Bassit, the sequence passes into calcilutite with replacement chert and intercalated red ribbon radiolarite. Calcarenites interbedded with the clastic sequence, mostly marls and clays, have been dated as Carnian/Norian by presence of *Trocholina sp.*, *Trochammina aff. jaunensis* Bronnimann and Page, *Semiinvoluta clari* Kristan. *Frondicularia c. woodwardi* Howch.

Rare weathered mafic lava sequences are associated with these limestones, either in small lenses interbedded at the top of the carbonate sequence, or surrounding horsts formed by white, pink or red massive limestones (Beit

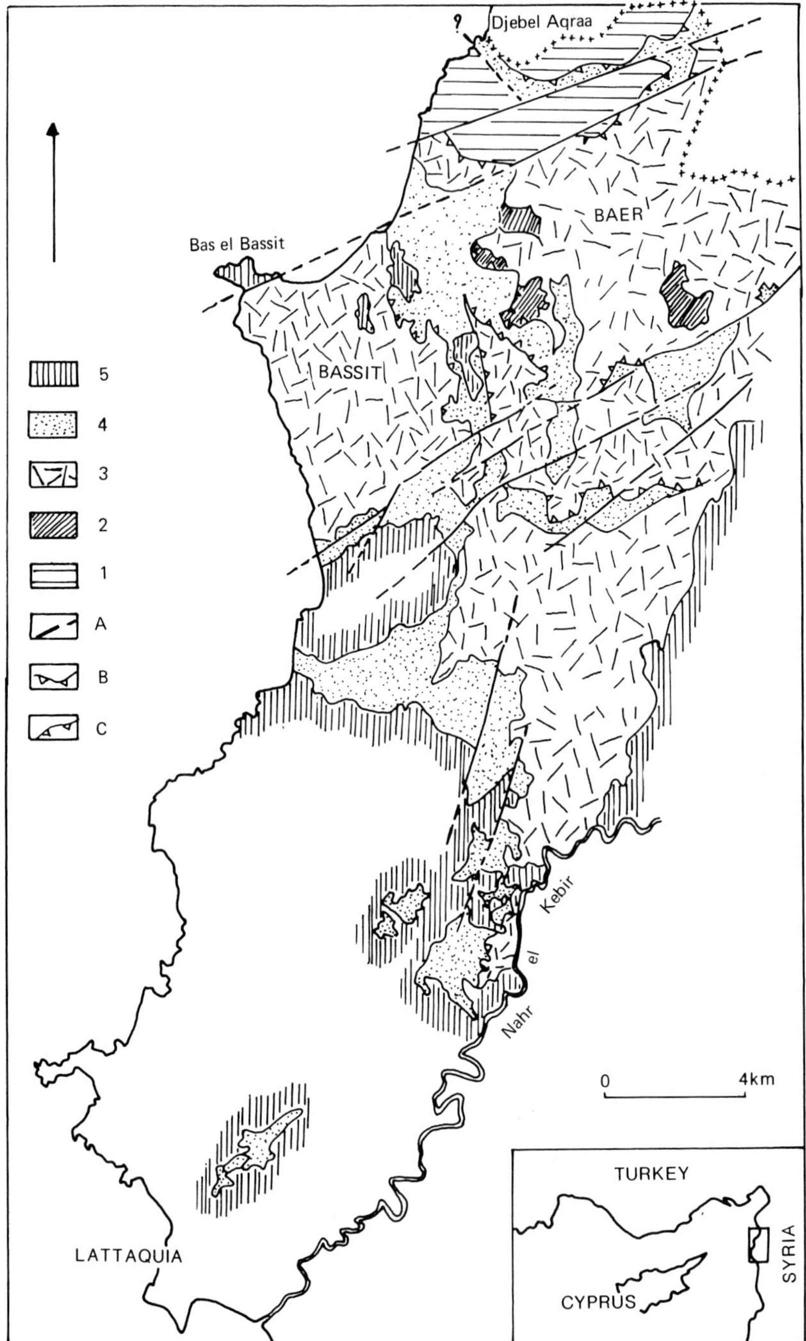

FIG. 1. Schematic geological map of ophiolitic and pelagic sediment sheets in the Baër-Bassit area, Syria. Symbols: 1. Arabian platform limestones of the Djebel Aqraa; 2. Amphibolitic sole rocks; 3. Ophiolitic assemblage; 4. Pelagic sediment sheets; 5. Upper-Maastrichtian transgressive units, C. Tectonic contacts between platform limestones and ophiolites, B. Tectonic contacts between ophiolites and pelagic sediments, A. Later thrust faults.

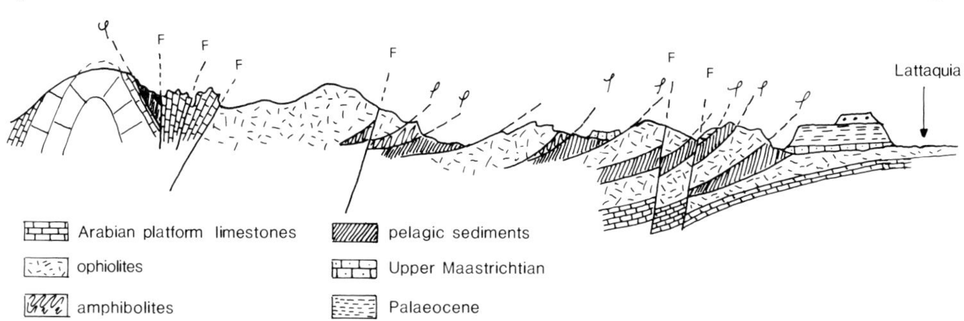

FIG. 2. Structural cross-section through the Baër-Bassit area to show the relationships of the various tectonic units in Fig. 1.

Ouéli Hassane facies, 'B.O.H.' on Fig. 3). These limestones are similar to the Petra tou Romiou reef limestones in Cyprus (Lapierre & Rocci 1970; Swarbrick & Robertson 1980). The horst-graben topography and the occurrence of tholeiitic volcanism with alkaline trends are interpreted as the result of rifting at the end of Triassic time (Parrot 1977).

Jurassic-Cretaceous siliceous deposits

Halobia-limestones pass abruptly into siliceous sediments composed of siltstones in the east, and cherts in the west which extend to Mid-Cretaceous (Cenomanian-Turonian). These deposits vary lithologically from red and green siltstones and mudstones, to cherts and calcareous siltstones. Calcareous deposits appear only in Cenomanian time in the upper levels of the succession in which calcilutites are often partly replaced by microcrystalline quartz.

The Jurassic age of the units overlying the Triassic limestones has not been confirmed in the Baër-Bassit area, but a Cretaceous age is shown by substantial intercalations of rudite- to lutite-grade re-deposited limestones which contain benthic foraminifera and platform-derived carbonates.

From Berriasian to Hauterivian time important peralkaline volcanism began with the successive production of basanites, lamprophyres, trachytes and phonolites (Kazmin & Kulakov, *op. cit.*; Parrot 1974, 1977). The sediments associated with the volcanics are monotonous red ribbon cherts (Delaune-Mayere 1978). Until the Cenomanian-Turonian, sedimentation remained siliceous in the whole Baër-Bassit area. Beds of calcirudite in the chert units yielded the following ages: Valanginian-Hauterivian (*Acruliaminina longa* (Tappan), *Flabellamina, Bacinella irregularis* Radoicic),

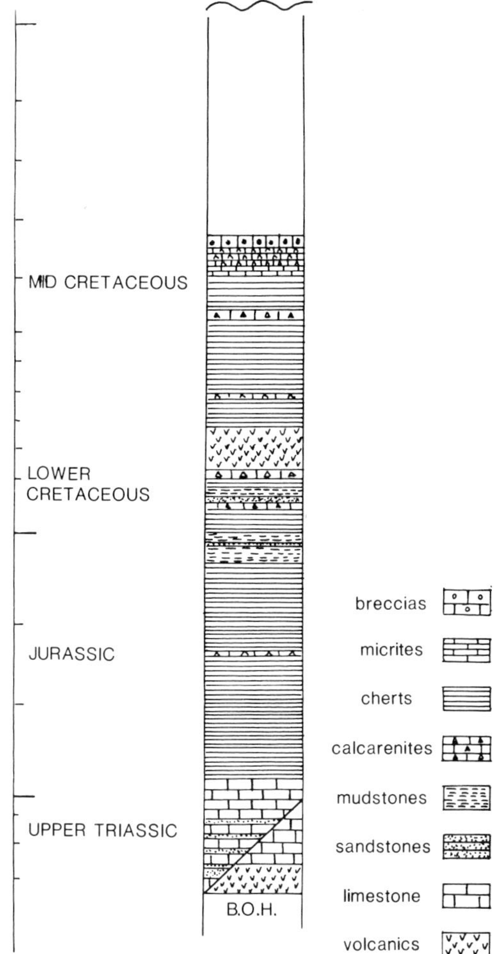

FIG. 3. Generalized stratigraphic succession of pelagic sediments in the Baër-Bassit area not drawn to scale.

Barremian-Lower Aptian (*Orbitolinidae, Choffatella decipiens* Schlumberger).

The upper levels of the pelagic sediments are calcareous siltstones and calcilutites with a pelagic micro-fauna: *Pithonella ovalis, Pithonella sphaerica, Hedbergella sp., Gavelinellidae.* Cenomanian time was characterised by calcarenites with *Praealveolina cretacea* (d'Archiac), *Pseudolituonella reicheli* Marie, *Pithonella sp., Hedbergella sp., Heterohelix sp.* Calcirudites and breccias are the latest known deposits in the basin.

Tectonic units in the eastern part of Baër-Bassit show stratigraphic sequences with quartzitic sandstones and calcareous tubidites towards the base and breccias dated as Albian-Mid-Cenomanian at the top. These facies represent an intermediate depositional area originally located between the siliceous pelagic deposits and a platform. They constitute the 'Kepir formation' of Delaune-Mayere & Parrot (1976) and Delaune-Mayere & Saint-Marc (*op. cit.*). *Orbitolina* with arenaceous tests typical of the 'Kepir formation' are also present in calcarenites of the more pelagic western tectonic units and this confirms a relationship between the two formations.

Mineralogy

Upper Triassic

The clastic intercalations in the Halobia limestones including marls and calcareous mudstones contain calcite, dolomite (sometimes up to 15% by volume), clays in the form of illite, kaolinite and mixed-clays, and quartz. In the sandstone beds quartz predominates over minor plagioclase and goethite. Also present are small quantities of the heavy minerals zircon, tourmaline, rutile and anatase, together with pyroxene and apatite from weathered volcanics and barite and pyrite probably derived from eroded hydrothermal deposits elsewhere in the basin. The calcilutites contain more than 90% of calcite and traces of smectites. The deposits close to the volcanics are often dolomitized (up to 45% dolomite) and contain haematite, attapulgite and Mn-hydroxides.

Jurassic-Cretaceous

The siliceous sediments of Jurassic to Cretaceous age contain, besides quartz, significant amounts of other minerals generated by subaerial and submarine erosion, and by hydrothermal processes.

Quartz is the main mineral, dominating the cherts and radiolarites and forming between 9 and 70% of the mudstones and siltstones. It was mostly derived from biogenic and chemical silica. Detrital input was abundant during some periods, for example, the Early Cretaceous. The fine-grained sediments contain an illite and smectite clay assemblage, thought to be partly of volcanic origin especially in mudstones close to the Early Cretaceous volcanics. Well-crystallized kaolinite is always present and can reach 30% in some horizons. Carbonate silts were derived from the erosion of adjacent shallow water limestones and their redeposition into the oceanic basin, particularly in Upper Jurassic and Cenomanian times.

Minerals formed in an oceanic environment are present in small amounts; some are derived by weathering of volcanic rocks (e.g. haematite, albite, pyroxenes, apatite), while some are hydrothermal deposits. Deposits of hydrothermal origin are present in mudstones of Early Cretaceous age. The most typical assemblage is an association of goethite and Mn-hydroxides with 5 cm thick black beds rich in silica and siderite. Dolomite, barite, magnetite and smectite are also present in small weathered outcrops. Thin horizons of pale yellow jarosite are interbedded with the mudstones. A similar mineral association is known from modern oceanic metalliferous deposits (Soler *et al.* 1982).

Geochemistry

Preliminary data allow comparisons to be made with the geochemistry of the sediments of the Antalya Complex, South-west Turkey, described by Robertson (1980).

Upper Triassic

Only two samples, a red calcareous siltstone interbedded with Halobia limestones and a radiolarite close to the mafic volcanics are relatively enriched in Fe and Mn. All the other samples analysed are marls and calcilutites which on the Al-Fe-Mn diagram (Fig. 4) do not show relatively high concentrations of Fe or Mn (Table 1).

Jurassic-Lower Cretaceous

The post-Triassic sediments are enriched in Fe, as seen from the ternary Al-Fe-Mn diagram and Table 2. There is a progression from chemically normal pelagic sediments to highly Fe-enriched sediments (Fig. 5). The high concentrations in Fe are interpreted as the result of

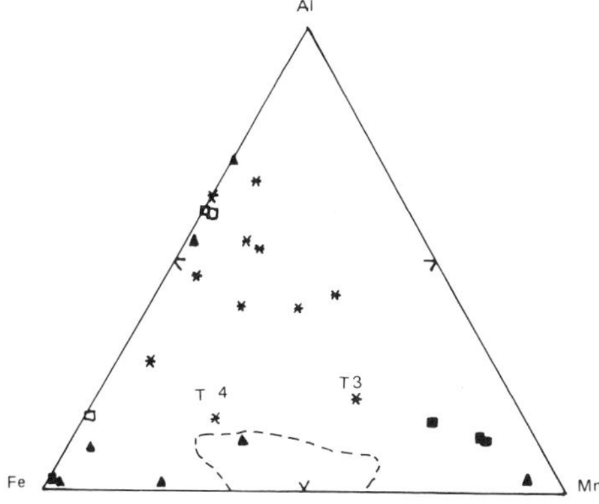

FIG. 4. Triangular plot of Baër-Bassit and Antalya sediments of Upper Triassic age.

Symbols:
* Baër-Bassit, passive margin
□ Antalya, passive margin
▲ Antalya, interlava sediments
■ Antalya, basal supra-lavas
◌ Antalya, Cyprus intra- and supra-lava sediments
} from Robertson 1981
$T_3 - T_4$: see Table 1

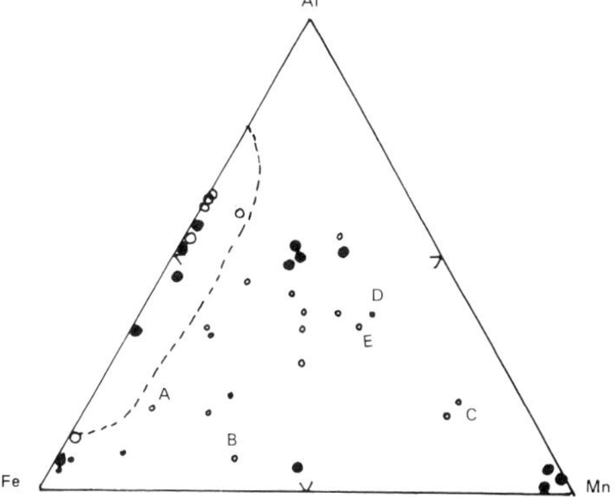

FIG. 5. Triangular plot of Baër-Bassit and Antalya sediments of Jurassic and Lower Cretaceous age.

Symbols:
◌ Baër-Bassit, passive margin
○ Antalya, passive margin
● Antalya, supra-lava, Jurassic
} from Robertson 1981
A to E: see Table 2.

TABLE 1. *Analyses of major and trace elements in Upper Triassic sediments of Baër-Bassit. T1, marls; T2, Halobia-bearing limestones; T3, radiolarite close to volcanics; T4, red calcareous siltstone. Major elements in percent. Trace elements in ppm*

	T_1	T_2	T_3	T_4
SiO_2	6.84	1.26	93.27	62.3
Al_2O_3	1.86	0.06	1.94	1.81
Fe_2O_3	0.89	0.06	0.78	0.86
MnO	0.05	0.11	1.07	0.05
MgO	1.42	0.44	0.42	0.48
CaO	48.27	54.70	0.33	18.35
Na_2O	0.06	0.05	0.14	0.03
K_2O	0.39	0.04	0.27	0.29
TiO_2	0.09	0.01	0.02	0.06
P_2O_5	0.05	0.05	nd.	0.05
H_2O^+	40.08	43.22	1.76	15.71
Total	100.00	100.00	100.00	100.00
Ba	69	7	144	56
Co	<10	<10	17	<10
Cr	<10	<10	184	130
Cu	<3	<3	176	39
Mo	<3	<3	<3	<3
Ni	26	52	737	500
Pb	<15	<15	nd.	26
Sr	2500	340	266	88
V	<5	<5	<10	11
Zr	380	380	nd.	99

hydrothermal leaching of the coeval volcanic rocks. Many authors, including Bischoff & Dickson (1975), Hajash & Archer (1980) and Seyfried & Mottl (1982) have shown that hydrothermal leaching of oceanic basalts is a source of Fe, Mn and some trace elements. The large amount of Fe, (11.37% in basanite, and 11.53% in monchiquite) in the lower part of the peralkaline volcanics (Parrot 1977) could have been the source of Fe-enrichment in the Lower Cretaceous sediments of Baër-Bassit (Delaune-Mayere, in press). The metalliferous deposits are black or ochre-coloured concretions composed of Mn-hydroxides and goethite, which occur mostly in the mudstones, associated with siderite, jarosite and barite. There was a sharp segregation of the ferruginous and the manganiferous components during diagenesis (Fig. 5), with Mn-hydroxides concentrated in small lenses or beds intercalated with mudstones and cherts and also as a cement in the sandstones and radiolarites (c.f. Lynn & Bonatti 1965).

Mid Cretaceous

Sediments of this age contain lower amounts of Fe than those described above (Fig. 6). In the western part of Baër-Bassit, at the top of the chert successions, close to calcarenites dated as Cenomanian, two kinds of metalliferous deposits have been observed. First cherts with high concentrations of Mn and some trace elements, as in umbers (Robertson & Hudson 1973; Elderfield *et al.* 1972; Parrot & Delaune-Mayere 1974); secondly, black or red coloured

TABLE 2. *Analyses of major and trace elements in Fe/Mn enriched sediments. A, mudstone with a high Fe content (Early Cretaceous). B, brown goethite concretion in mudstone (Early Cretaceous). C, Mn-concretion. D, E, radiolarite cemented by Mn-hydroxides (D, eastern part; E, western part of Baër-Bassit). F, G, Mn-enriched cherts, Mid-Cretaceous. H, average of analyses of Baër-Bassit umbers. Comparison of Early Cretaceous Mn-Concentrations: B.B (Baër Bassit), Chy. (Mamonia Complex), Ant. (Antalya Complex, in Robertson 1981): major elements in percent, trace elements in ppm*

	A	B	C	D	E	F	G	H	B-B	Chy.	Ant.
SiO_2	42.33	8.03	47.92	92.72	93.77	53.29	57.48	21.34	47.92	39.06	24.37
Al_2O_3	16.58	4.74	7.64	1.89	1.84	4.20	1.72	6.00	7.64	6.35	2.12
Fe_2O_3	19.45	35.99	3.68	1.21	0.46	13.37	9.97	43.41	3.68	6.62	1.12
MnO	5.14	18.20	19.78	1.09	0.67	18.50	20.73	7.12	19.78	25.95	67.33
MgO	1.16	2.04	2.25	0.31	0.36	0.56	0.60	2.87	2.25	1.33	1.17
CaO	1.50	4.66	2.42	0.25	0.61	0.62	1.23	4.14	2.42	1.33	0.36
Na_2O	0.10	0.06	0.27	0.08	0.08	0.09	0.25	0.64	0.27	0.36	0.07
K_2O	1.09	0.27	1.28	0.36	0.26	0.65	0.36	0.75	1.28	1.04	3.31
TiO_2	0.88	0.21	0.43	0.10	0.87	0.41	0.11	0.33	0.43	0.38	0.16
P_2O_5	0.36	0.14	0.35	0.05	0.05	0.24	0.17	nd.	0.35	0.05	0.36
H_2O^+	11.42	25.66	13.97	1.94	1.82	8.07	7.38	13.46	13.52	17.52	
Total	100.00	100.00	100.00	100.00	100.00	100.00	100.00		100.00	100.00	97.38
Ba	94	420	330	120	13	760	<20	1028	330	59	561
Co	30	11	65	14	<10	79	217	111	65	33	454
Cr	180	47	100	180	190	60	291	67	100	28	84
Cu	51	9	390	120	26	67	350	788	390	940	905
Mo	<3	nd.	nd.	<3	<3	<3	192	38	nd.	<	549
Ni	250	25	150	550	590	140	923	366	150	68	130
Pb	550	nd.	nd.	19	37	45	24	153	nd.	42	108
Sr	140	1200	370	30	28	420		506	370	100	888
V	150	27	150	17	22	43	79	818	150	120	343
Zr	150	320	110	<50	<50	90	<50	112	110	85	259

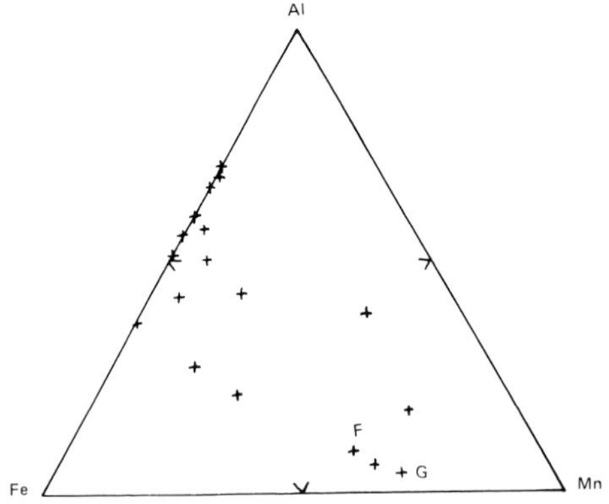

FIG. 6. Triangular plot of Baër-Bassit sediments from the Mid-Cretaceous passive margin.
F, G: see Table 2.

clasts which are re-worked in calcarenites and can be related to coeval volcanism in the basin.

Discussion

The sedimentary sequence described above displays the following features: A transition from shallow water deposition of marls and plant-rich sandstones to deeper-water Halobia limestones in Upper Triassic time. The volcanic rocks associated with these sediments resulted from extensional rifting in the area. The siliceous pelagic rocks are interpreted as deposits on a passive margin existing from Jurassic to Mid-Cretaceous time. This succession is quite similar to the pelagic deposits of similar age exposed in other areas of the East Mediterranean, particularly the Mamonia Complex S.W. Cyprus, and the Antalya Complex, SW Turkey. There are marked structural and lithological similarities between the Baër-Bassit and Mamonia areas (Lapierre and Parrot 1972; Delaune-Mayere & Saint-Marc, *op. cit.*). Green quartzitic sandstones are interbedded with mudstones in the eastern part of Baër-Bassit as in the Akamas sandstones of the Episkopi Formation of the Mamonia Complex (Swarbrick & Robertson *op. cit.*). Similarly, the Early Cretaceous volcanism in Baër-Bassit may be represented in the Mamonia Complex as hydrothermal deposits (jarosite) and as manganiferous concentrations (Table 2).

Although the comparison between the Baër-Bassit and the Antalya Complex pelagic deposits is less clear, recent sedimentological and geochemical studies (Robertson & Woodcock

1981a, & b; Robertson 1981; this volume) reveal great similarities in the tectonic evolution of all three areas.

The successive stages of evolution of the margin were:

Upper Triassic

In Antalya continental rifting occurred in Upper Anisian – Lower Ladinian time, characterized by horst-graben tectonics and appearance of tuffs (Marcoux 1976; Monod 1977; Argyriadis *et al.* 1980). A more important second stage of rifting in Carnian to Norian times is observed in the three areas. This is characterized by major subsidence and the deposition of clastic sediments. These are the Telektaş Tepe Formation and the Karabük Formation in Antalya (Robertson & Woodcock 1981a), and the Vlambouros Formation in Mamonia (Swarbrick & Robertson *op. cit.*).

Above come deeper water deposits, the Halobia-limestones, in a more tectonically stable environment. These limestones were deposited in deep water adjacent to carbonate platforms and reefs, by re-deposition as blocks, calcarenites and calcilutites (Robertson & Woodcock 1981c). The opening of the basin was accompanied by important mafic volcanism exceeding 600 m in the Gödene Zone (Kara Dere) of the Antalya Complex. Sedimentary intercalations include pink limestone and dark brown mudstones rich in Fe/Mn hydroxides (Robertson & Woodcock 1981c). In the Mamonia Complex this volcanism is represented by the Phasoula Formation which is up to 250 m thick (Swarbrick & Robertson *op. cit.*). As in Antalya, pink and grey limestones with manganiferous concentrations are associated with these lavas. In the Baër-Bassit area the Triassic terrigenous sandstones are very reduced relative to the Halobia limestones. Volcanics are restricted to scattered outcrops without known hydrothermal deposits.

It seems therefore that the oceanic opening occurred earlier in the western part of the East Mediterranean (Upper Anisian/Lower Ladinian in Antalya) than in the east in Baër-Bassit. Rift tectonism was less marked in the east as there are only thin deposits of detrital material.

Jurassic-Cretaceous

The sedimentation pattern changed abruptly in the Jurassic. The margins subsided below the carbonate compensation depth in all three areas and the period spanning the Jurassic to the Lower (or Mid) Cretaceous saw the accumulation of siliceous sediments.

In Baër-Bassit peralkaline volcanism was accompanied by hydrothermal activity leading to a high Fe/Al ratio in the sediments. In the Mamonia Complex Mn enrichment in some beds and in concretions is attributed to a hydrothermal origin (Swarbrick & Robertson, *op. cit.*). In the Antalya Complex, Fe-Mn enriched horizons are interpreted, as in the Mamonia Complex, as being related to renewed genesis of the ocean floor in the area (Robertson 1981). In the Mamonia Complex volcanics of alkaline character (lamprophyres) have been found in the same position intercalated with mudstones as in Baër-Bassit.

In all three areas the Early Cretaceous is marked by renewed influx of detrital material, as terrigenous mudstones and sandstones of the Akamas Member of the Mamonia Complex, and similar sediments found locally in the Antalya Complex (Robertson & Woodcock 1981c). In the Baër-Bassit region there was also re-deposition of carbonate material as calcarenites associated with Calpionnellid-bearing limestones.

In Mid- and Upper Cretaceous time each area exhibits a more specific evolution. In Baër-Bassit sedimentation in the intermediate ('Kepir formation') and oceanic areas had ended by Cenomanian to Turonian time; coeval volcanism at this time is implied by Mn-enriched horizons. In the Mamonia Complex among the last sediments are the Akamas sandstones, dated as Hauterivian (Ealey & Knox *op. cit.*). In the Antalya Complex the sedimentation extends into the Upper Cretaceous in both the Karabük and the Taskaya Tepe Formations (Kumluca Zone) and the Doğu Group of the more oceanic Gödene Zone (Robertson and Woodcock op. cit.; see also Yılmaz, this volume).

Conclusion

Regional comparisons of the pelagic sediments in three areas of the East Mediterranean show a similar evolution from Upper Triassic to Lower Cretaceous time. This involved continental rifting and the initiation of a rift in Mid-Triassic time (Antalya), then in the Upper Triassic (Antalya, Cyprus and Baër-Bassit) a rift zone like the Red Sea formed. More evidence for this opening is seen in the west than in the east.

This was followed by subsidence and siliceous sedimentation from Jurassic to Lower Cretaceous time. The deep water sedimentation also reflects periods of renewed uplift when sandstones and calcareous turbidites were re-deposited. Peralkaline volcanics were erupted in the east and metal-rich horizons observed in the west could be related to this event.

The final stages of evolution of the continental margin from the Mid-Cretaceous to the later closing of the basin remain to be defined.

References

ARGYRIADIS, I., DE GRACIANSKY, P C., MARCOUX, J. & RICOU, L E. 1980. The opening of the Mesozoic Tethys between Eurasia and Arabia-Africa. *In*: Geology of the Alpine chains born of the Tethys, *Mem. BRGM*, **115**, 119–214.

BISCHOFF, J. L. & DICKSON, F. W. 1975. Seawater-basalt interaction at 200°C and 500 bars: implications for origin of sea-floor heavy-metal deposits and regulation of sea water chemistry. *Earth and Planet. Sci. Letters*, **25**, 385–97.

DELAUNE-MAYERE, M. 1978. Cherts mésozoiques du bassin téthysien oriental: minéralogie et géochimie des sédiments siliceux du secteur de Tamima^{h} (NW Syrien). *Cah. O.R.S.T.O.M., ser. Géol.* **10**, 191–202.

DELAUNE-MAYERE, M. 1983. Polarités géochimiques et paléogéographie des séries volcano-sédimentaires pélitiques du NW Syrien au Crétacé basal. *Cah. O.R.S.T.O.M. sér. Géol.* **13**, (in press).

DELAUNE-MAYERE, M. & PARROT, J. F. 1976. Evolution du mésozoique de la marge continentale méridionale du bassin téthysien oriental d'après l'étude des séries sédimentaires de la région ophiolitique du nordouest syrien. *Cah. O.R.S.T.O.M., ser. Géol.* **8**, 173–84.

DELAUNE-MAYERE, M., MARCOUX, J., PARROT, J. F. & POISSON, A. 1977. Modèle d'évolution mésozoique de la paléo-marge téthysienne au niveau des nappes radiolaritiques et ophiolitiques du Taurus Lycien, d'Antalya et du Baër-Bassit. Intern. Symp. on Struct. Hist. of the Mediterranean Basins, Split, 1976. Technip (ed.), 79–94.

DELAUNE-MAYERE, M. & SAINT-MARC, P. 1979/80. Données stratigraphiques nouvelles sur les sédiments océaniques mésozoiques associés aux nappes ophiolitiques du Baër-Bassit (NW syrien). *Cah. O.R.S.T.O.M., ser. Géol.* **11**, 151–64.

DUBERTRET, L. 1953. Géologie des roches vertes du nordouest de la Syrie et du Hatay. *Notes et Mém. Moy. orient*, **6**, 179.

EALEY, P J. & KNOX, G. J. 1975. The pre-Tertiary rocks of SW Cyprus. *Geol. en Mijnb.* **54**, 85–100.

ELDERFIELD, H., GASS, I. G., HAMMOND, A. & BEAR, L M. 1972. The origin of ferromanganese sediments associated with the Troodos massif of Cyprus. *Sedimentology*, **19**, 1–19.

HAJASH, A. & ARCHER, P. 1980. Experimental seawater/basalt interactions: effects of cooling. *Contr. Mineral. Petrol.* **75**, 1–13.

KAZMIN, V. G. & KULAKOV, V. V. 1968. Geological map of Syria. Report on the Geological Survey. Techno. Export, Moscou, 124.

LAPIERRE, H. 1975. Les formations sédimentaires et éruptives des nappes de Mamonia et leurs relations avec le massif du Troodos (Chypre occidentale). *Mém. Soc. géol. Fr.* **123.**

LAPIERRE, H. & ROCCI, G. 1970. Un bel exemple d'association cogénétique laves-radiolarites-calcaires: la formation triasqiue de Petra tou Romiou (Chypre). *C.R. Acad. Sci. Paris*, **268**, 2637–40.

LAPIERRE, H. & PARROT, J. F. 1972. Identité géologique des régions de Paphos (Chypre) et du Baër-Bassit (Syrie). *C.R. Acad. Sci. Paris*, **274**, 1999–2002.

LYNN, D. C. & BONATTI, E. 1965. Mobility of manganese in the diagnenesis of deep-sea sediments. *Mar. Geol.* **3**, 457–74.

MARCOUX, J. 1976. Les séries triasiques des nappes à radiolarites et ophiolites d'Antalya (Turquie): homologie et signification probable (résumé). *Bull. Soc. géol. Fr.* **7**, 511–2.

MONOD, O. 1977. *Recherches géologiques dans le Taurus occidental du sud de Beyşehir, Turquie.* Thèse no 1839. Fac. Sci. Univ. Paris Sud (Orsay). 442 pp.

PARROT, J. F. 1974. Le secteur de Tamima^{h} (Tourkmannli); étude d'une séquence volcano-sédimentaire de la région ophiolitique du Baër-Bassit (NW de la Syrie. *Cah.O.R.S.T.O.M., ser. Géol.***6**, 127–46.

PARROT, J. F. 1977. Assemblage ophiolitique du Baër-Bassit et termes éffusifs du volcano-sédimentaire. Pétrologie d'un fragment de la croûte océanique téthysienne charriée sur la plate-forme syrienne. *Trav. et Doc. O.R.S.T.O.M.* **72**, 333.

PARROT, J. F. & DELAUNE-MAYERE, M. 1974. Les terres d'ombre du Bassit (NW syrien). Comparison avec les termes similaires du Troodos (Chypre). *Cah. O.R.S.T.O.M. ser. Géol.* **6**, 147–60.

RICOU, L E. 1971. Le croissant ophiolitique péri-arabe, une ceinture de nappes mises en place au Crétacé supérieur. *Rev. Géogr. phys. Géol. dyn.* **13**, 327–49.

ROBERTSON, A. H. F. 1980. Metallogenesis on a Mesozoic passive continental margin, Antalya Complex, southwest Turkey. *Earth planet. Sci. Letters*, **54**, 323–45.

ROBERTSON, A. H. F. & HUDSON, J. D. 1973. Cyprus umbers: chemical precipitates on a Tethyan ocean ridge, *Earth planet. Sci. Lett*, **28**, 385–94.

ROBERTSON, A. H. F. & WOODCOCK, N. H. 1979. Mamonia Complex, southwest Cyrprus: evolution and emplacement of a Mesozoic continental margin. *Bull. geol. Soc. Amer.* **90**, 651–65.

ROBERTSON, A. H. F. & WOODCOCK, N. H. 1981a. Bilelyeri Group, Antalya Complex: deposition on a Mesozoic passive continental margin, southwest Turkey. *Sedimentology*, **28**, 381–99.

ROBERTSON, A. H. F. & WOODCOCK, N. H. 1981b. Alakır Çay Group, Antalya Complex, SW Turkey: a deformed mesozoic carbonate margin. *Sedim. Geol.* **30**, 95–131.

ROBERTSON A. H. F. & WOODCOCK, N. H. 1981c. Gödene Zone, Antalya Complex: volcanism and sedimentation along a Mesozoic continental margin, SW Turkey. *Geol. Rundschau*, **70,** 1177–211.

ROBERTSON, A. H. F. & WOODCOCK, N. H. 1982. Sedimentary history of the south-western segment of the Mesozoic-Tertiary Antalya continental margin, south-western Turkey. *Eclog. geol. Helv.* **75,** 517–62.

SEYFRIED, W. E. & MOTTL, M. J. 1982. Hydrothermal alteration of basalt by seawater under seawater dominated conditions. *Geochim. Cosmoch. Acta*, **46,** 985–1002.

SOLER, E., BERNARD, A. J. & NESTEROFF, W. D. Activité hydrothermale sulfurée de dorsale: comparaison entre dépôts contemporains (EPR, 21°N) et dépôts anciens (Chypre). *Oceanol. Acta*, **5,** 105–20

SWARBRICK, R. E. & ROBERTSON, A. H. F. 1980. Revised stratigraphy of the Mesozoic rocks of southern Cyprus. *Geol. Mag.* **117,** 547–63.

M. DELAUNE-MAYERE, ORSTOM-70, 74 route d'Aulnay-93140 Bondy, France.

Tectonic and sedimentary history of the NE African margin (Egypt—Libya)

G. Sestini

SUMMARY: The present structure and thickness variations of the Mesozoic-Cenozoic sediments of the Mediterranean coastal belt of Egypt (west of Sinai) and of North Cyrenaica, indicate that the African Plate at this latitude was part of an unstable zone characterized by a number of fault-controlled marginal depocentres. Triassic to Middle Cretaceous facies are almost everywhere shallow marine, nearshore, deltaic and continental, with no indication of a paleobasin margin. Deeper-water sedimentation occurred in parts, from Senonian onwards. During the Campanian-Eocene phase of plate collision, overthrusting and folding, the African margin was raised in the north and depressed southwards, and mild folding was superimposed on the block structures. Widespread Oligo-Miocene faulting and differential vertical movements shifted the focus of deposition to the Nile Delta region and the Gulf of Suez, while transcurrent movements to NNE (Aqaba-Dead Sea faults) created local compression that magnified and skewed Eocene folds. It is concluded that the onshore coastal belt of NE Egypt-Cyrenaica belongs to a relatively southerly part of the African plate margin.

Introduction

The SE margin of the Mediterranean Sea, from Northern Cyrenaica to the Nile Delta, is characterized by a narrow continental shelf (15–50 km) and a steep faulted continental slope which in Egypt has a rectilinear WNW diection (Fig. 1). The coast and the continental shelf margin are offset rather abruptly to the NNW at several places, especially at the Gulf of Sallum and the Gulf of Bomba.

The African coast lies as much as 300 km south of the folded arc which forms the Mediterranean Ridge in the east, where the sea floor is occupied by the vast Nile cone and by the Herodotus Basin (Ross & Uchupi 1977). To the west the distance is as little as 70 km where Crete and Northern Cyrenaica are only separated by the Herodotus Trough, a graben filled by sediments (Sancho *et al.* 1973). The rest of the sea floor of the Levant Basin is underlain by 3–5 km of Late Miocene-Pliocene sediments. It is generally agreed that the main physiographic and tectonic units are of geologically recent origin, having resulted from a combination of Late Miocene and Pliocene subsidence, progradation of the Nile in the Pleistocene, faulting and halokinesis in the Levant Platform, and folding and slumping on the Mediterranean Ridge (Finetti 1976, Mulder *et al.* 1975. Ross & Uchupi 1977, Said 1981, Stanley & Maldonado 1977).

According to some studies (Finetti & Morelli 1973, Malovitskiy *et al.* 1975, Moskalenko 1966) below these units lies a Mesozoic-Cenozoic sequence that is part of the African platform. The crust underneath appears to be of modified continental, or semi-continental type (Lort 1977, Morelli 1978), with the exception of possible oceanic crust in the SE corner of the Mediterranean (Ginzburg & Gvirtzman 1979). Woodside (1977) has proposed an extension of the African crust as far as the Taurus, and according to Biju-Duval & Montadert (1977) as much as 280 km of African Plate have disappeared under the Aegean Arc. However, virtually nothing is known about the stratigraphy, paleogeography and tectonics of the subacqueous portion of the Mediterranean African margin.

The purpose of this paper is therefore to see what indirect evidence can be obtained from the Mesozoic-Cenozoic geology of the coastal belt of Egypt and Cyrenaica (East of the Sirte basin), during the Neo-Tethys (South Tethys) history. The region shows a surprising tectonic complexity. It is part of the 'Unstable Shelf' of Said (1962), a zone characterized by deeper Pre-Cambrian basement showing greater tectonic mobility, and by thicker Palaeozoic-Cenozoic sediments than the 'Stable Shelf' to the south, which borders the Nubian Shield (Fig. 2).

Main features of the 'Unstable Shelf'

The surface geology is simple (Desio 1968, El Shazly *et al.* 1975, 1976; Kleinsmede & Van Der Berg 1968; Said 1962; Thiele *et al.* 1970). It is dominated by tabular Middle Miocene limestones, with Miocene clastics exposed at the north margin of the Qattara Depression and on the north slope of the Cyrenaica Plateau.

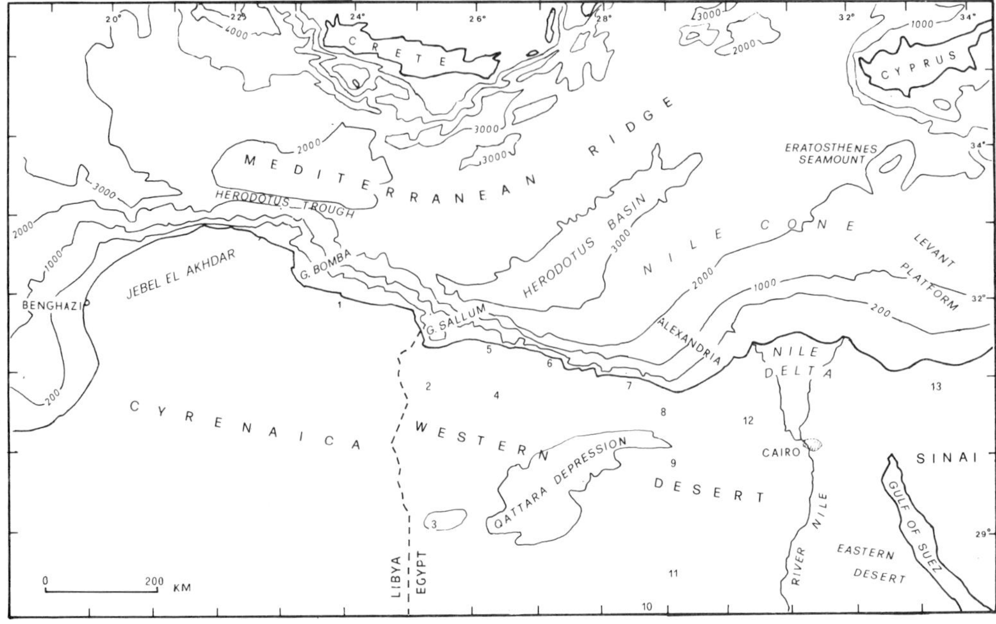

FIG. 1. Index map of the region described, with localities mentioned in the text. 1. Tobruk, 2. Faghur, 3. Siwa, 4. Umbarka, 5. Mamura, 6. Marsa Matruh, 7. Dabaa, 8. Alamein, 9. Rabat High, 10. Farafra, 11. Baharia, 12. Natrun, 13. Jebel Maghara.

Deformed older rocks (Cenomanian-Eocene) outcrop only in the inlier of Jebel El Akhdar.

The subsurface geology is known from oil exploration with some 70 wells drilled in a belt 50 km south of the coast, 25 within 20 km of the coast and only one offshore. Relatively little of this information has been released. Some data are however available in several publications, which allow an outline of the stratigraphy, structure and tectonic evolution to be drawn. The following summary and the illustrations (Figs 2 & 4) have been gleaned from the following sources: Amin (1961), Bayoumi & El Gamili (1970), Conant & Goudarzi (1967), Colley (1963), El Etr & Moustafa (1980), Ezzat & Dia El Din (1974), Garfunkel & Bartov (1977), Goudarzi (1980) Kamel *et al.* (1978), Khalil & El Mofty (1972), Khalil *et al.* (1976), Meshref (1982), Meshref *et al.* (1980), Metwalli & Abdel Hady 1973), Neev (1977), Neev *et al.* (1976), Riad (1977), Riad *et al.* (1980), Röhlich (1980), Ross & Uchupi 1977), Said (1962, 1981), Salem (1976), Sigaev (1969).

The 'Unstable Shelf' is characterized by the following structural elements (Fig. 2, 3).

1. Subsidence depocenters

Some of these have been continuously active from the Jurassic (possible earlier) to the Miocene (Abu Gharadig, Shushan, Razzak-Rabat 'basins'), from the Early Cretaceous to the Eocene (Guindi Basin), or from the Late Cretaceous to the Miocene (Sirte Basin). Others have been inverted, like the Northern Basin, which subsided strongly during the Jurassic and Early Cretaceous, but was then partly inverted in the Senonian and Eocene. The North Sinai-Natrun Jurassic-Early Cretaceous depocenter was inverted in the Late Cretaceous and the Palaeozoic basin of NE Libya-NW Egypt (Faghur region, Said 1962, El Sweifi 1975) was inverted in the Jurassic, although parts subsided again in the Cretaceous and Eocene.

The cumulative thickness of Mesozoic sediments in the coastal depocenters (Northern, Razzak-Rabat and North Sinai 'basins') appears to be in excess of 5000 m. The more internal depocenters may have sediments in the order of 4500 m (Abu Gharadig) to 3000 m (Sirte).

2. Positive structures

There are several types:

(a) Basement-Palaeozoic ridges that remained high until Late Cretaceous to Eocene times (e.g. Siwa—Alamein Ridge).

(b) Basement-Palaeozoic highs under the

present coast, that acted passively during the Jurassic-Early Cretaceous, as part of the North Basin subsidence, but rose again from Coniacian to Middle Eocene times.

(c) Basement-Palaeozoic ridges, or blocks, that have been inverted, like the Umbarka and Jaghbub highs, then have subsided since the Aptian and Cenomanian, and the Rabat and parts of the Natrun High, which subsided since the Paleogene.

(d) Late Cretaceous-Tertiary uplifts near the coast, the most prominent of which are Jebel El Akhdar, the North Sinai folds and the Mamura-Matruh High. These rose over above previously subsided areas. Others, the Dabaa High for example, are rejuvenated older blocks.

The main trend of these subsurface features changes from W to E. East of 25°E faulting is mainly NW–SE to NNW–SSE and E–W. Between 25°E and 29°E the trend is either E–W or ESE–ENE, or a combination of WNW and ENE (e.g. the margins of Abu Gharadig and Shushan basins, Ezzat & Dia El Din 1974). East of 29°E the dominant orientation is NE and ENE, the so-called 'Syrian Arc' folds (surface and subsurface), which range from broad anticlines like Farafra and Baharia (Soliman *et al.* 1970), to narrower, asymmetrical folds between Cairo and Alexandria and in North Sinai. All these have SE asymmetry and even overthrusting (Jebel Maghara, Al Far 1966).

Stages of tectonic activity

Stratigraphic and structural relations indicate the following periods of tectonic activity in the coastal regions.

(1) Between the Late Palaeozoic and Triassic;

(2) Between the Jurassic and Cretaceous.

(3) Within the Cretaceous in the more southern areas, towards the zone of linear uplifts, and on the northern blocks in three phases; (a) Late Aptian, (b) between Early and Late Cenomanian, (c) at the end of the Turonian, or in the Coniacian (c.f. Sinai, Lewy 1975).

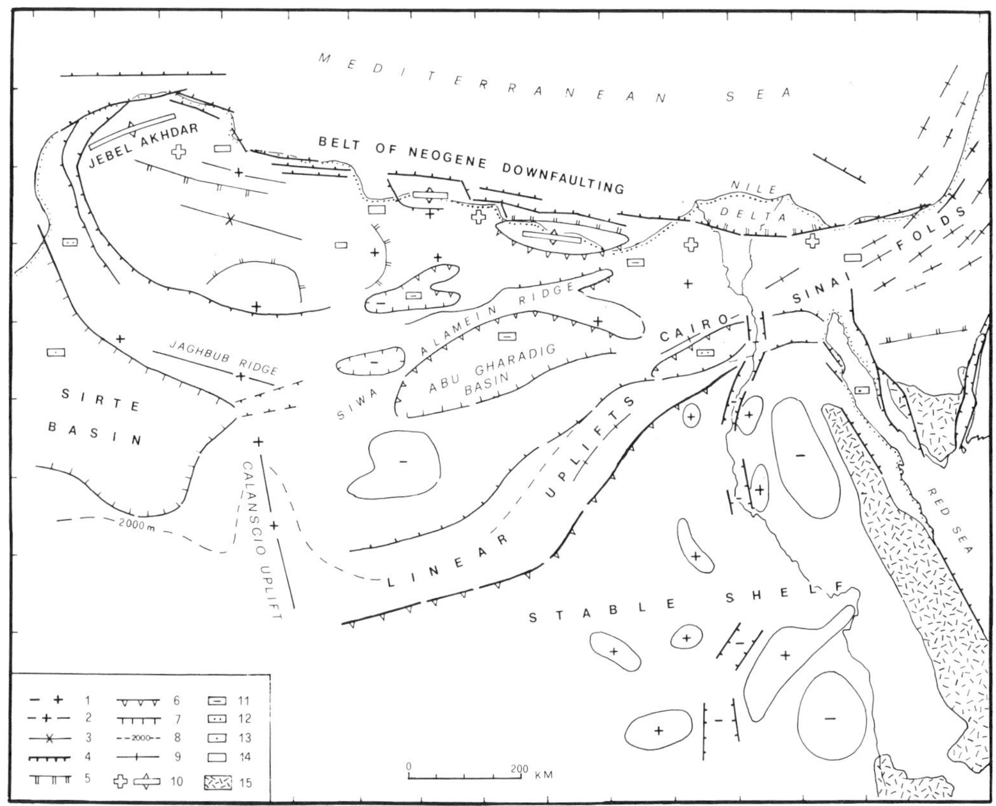

FIG. 2. Tectonic elements of the 'Unstable Shelf.' 1. Basement lows|highs, 2. Basement ridges|blocks, 3. Basement troughs, 4. Main faults, 5. Flexures, 6. Main faulted basin margins, 7. Basin margins (faulted or not), 8. Depth of Basement, 9. Folds, 10. Late Cretaceous-Tertiary uplifts, 11. Jurassic-Early Miocene depocenters, 12. Early Cretaceous-Eocene depocenters, 13. Tertiary depocenters, 14. Inverted depocenters, 15. Exposed PreCambrian Basement.

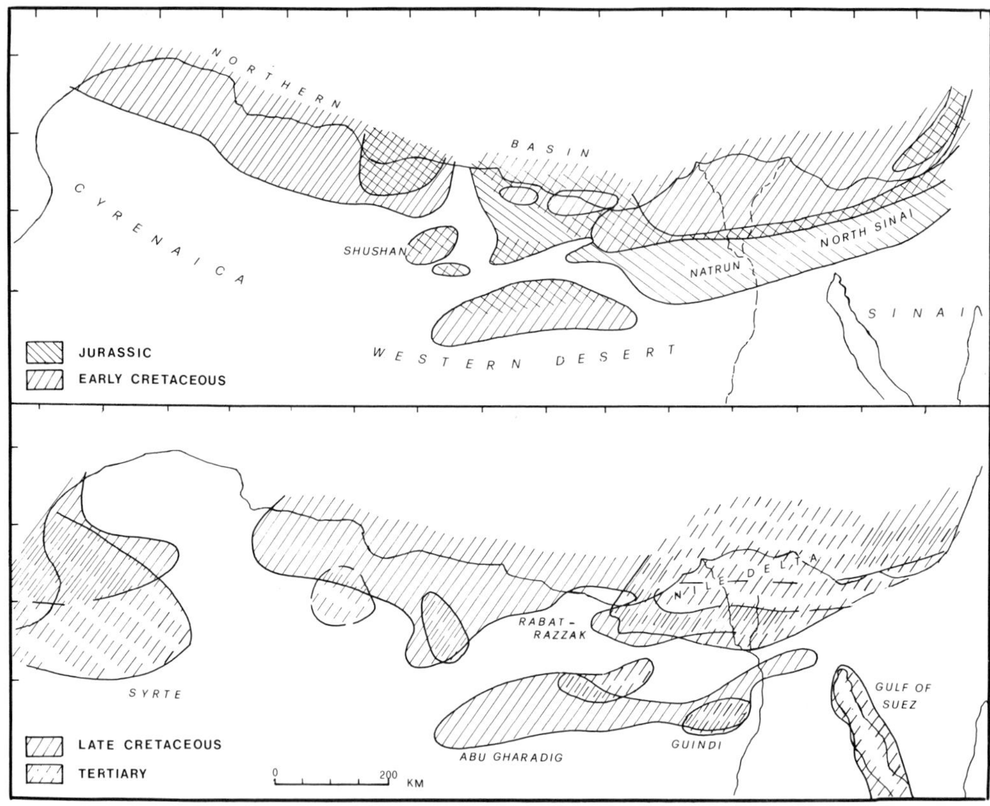

FIG. 3. Areas of major subsidence; compiled from references cited in the text.

(4) From Campanian to Early Tertiary, with a widespread unconformity between Maastrichtian and Early Eocene except in the deeper parts of the basins, and between the Middle Eocene and Late Eocene-Oligocene on the major folds and highs.

(5) Uplift of the Western Desert after the Middle Miocene, accompanied by minor warping along NW axes (El Shazly *et al.* 1976), and regional tilting towards the east, with the formation of the present Nile Delta Basin (Hantar 1975, Rizzini *et al.* 1978; Said 1981; Salem 1974), and large scale downfaulting towards the Mediterranean.

The dominant pre-Tertiary structural style of the Western Desert, west of the Nile River, and of the Sirte Basin is normal faulting, essentially a horst-graben structure (Fig. 4). In Egypt, most basins are half-graben, tilted towards the north, with a thickening of Jurassic-Cretaceous sediments towards bounding faults to the north. The basins are not simple synclinal lows, but are themselves fragmented into a large number of smaller tilted blocks (Fig. 4).

Various systems of younger (Tertiary) tensional faults cut across all the elements mentioned, as well as as across the 'Stable' and 'Unstable shelves'. The main ones are the N–S Nile Graben (Said 1981), the NW–SE trending Gulf of Suez (Garfunkel & Bartov 1977) and the Red Sea rifts (Garson & Krs 1976). A second system, the E-W to WNW belt of Neogene downfaulting that coincides with the coast in NE Cyrenaica runs along the continental slope in NE Egypt and crosses the northern part of the Nile Delta, where it overlies a deeper faulted flexure zone (Barber 1980, Rizzini *et al.* 1978, Said 1981). This belt is accompanied by an erosional escarpment, which is buried under Plio-Pleistocene sediments east of 28°E but roughly coincides with the upper continental slope in the west.

Satellite imagery studies by El Shazly *et al.* (1975, 1976) have also shown the existence of a large number of more superficial fractures that in the Western Desert are orientated predominantly NW–SE. The main faults over 100 km in length, appear to have undergone dextral

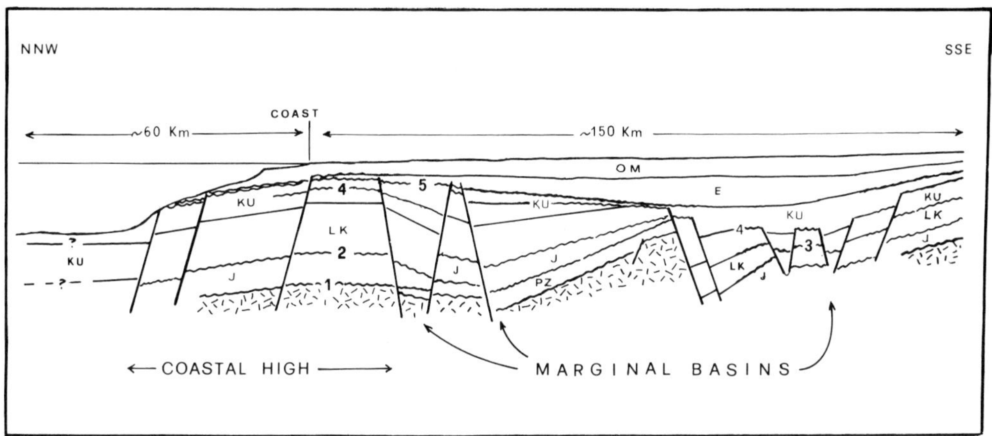

FIG. 4. Ideal structural cross-section of northern Western Desert, Egypt (constructed mainly from data in: Aaland & Hassan 1972, El Gezeery *et al.* 1972, Ezzat & Dia El Din 1974, Metwalli & Abdel Hady 1973). Unconformities: 1. Basement-Palaeozoic, or Jurassic. 2. Jurassic-Cretaceous. 3. Aptian-Albian, or within Cenomanian. 4. Late Turonian/Coniacian. 5. Senonian-Eocene.

movement in relatively recent times (Pliocene-Pleistocene?), and are possibly responsible for the repeated offsets of the coast. In the Nile Delta—Cairo-Suez—North Sinai region, fault orientations are more varied , including common E–W and ENE sets (El Shazly *et al.* 1974, 1975). The combination of NW–SE and NE–SW fractures has been generally interpreted as a conjugate set due to compression from NNE (Bentz & Gutman 1977, Garfunkel & Bartov 1977, Youssef 1968).

Coastal stratigraphy and palaeogeography

The preliminary sedimentary mega-units of Said (1962) were a Palaeozoic-Early Cretaceous *Lower Clastic Division*; a Cenomanian-Eocene *Middle Calcareous Division*; and a post-Eocene *Upper Clastic Division*. These can now be refined into at least five Mesozoic-Tertiary sedimentary cycles, based on major transgressions and regressions, most of which are related to global sea level variations (Vail *et al.* 1977), and on major unconformities (Fig. 5).

Triassic sediments are poorly known west of the Nile Delta. Perhaps they are generally absent (e.g. Middle Jurassic overlies Carboniferous in the Mamura well (Said 1962), or they could be part of the Permo-Jurassic, clastic-continental Eghei Group (wells South of 30°30′ N, Aaland & Hassan 1972). In North Sinai and the Negev, Triassic rocks are of shallow marine, lagoonal and intertidal facies, with some evaporites (Druckman 1974). Off the Egyptian-Cyrenaica coast seismic data show no evidence of Triassic (or Hettangian) evaporites, or salt structures.

Jurassic sediments are better known (Barakat 1976, Khalid 1975). In the east (Natrun-North Sinai 'basin'), the Middle Jurassic Khatatba formation is predominantly lagoonal (Jebel Maghara: Goldberg & Friedman, 1976). The coastal Western Desert sequence appears to be more open marine (Khalid 1975, Prior 1976), but there are no signs of an abrupt carbonate-platform margin as in Israel (Bein & Gvirtzman 1977). The Oxfordian-Kimmeridgian Masajid carbonates are a transgressive shallow marine facies in Egypt. In northern Cyrenaica the Late Jurassic is largely clastic and nearshore.

The *Early Cretaceous sequence* is very sandy in the east, with continental or deltaic intercalations, including thin coals and occasional dolomite and anhydrite beds (Abdine & Deibis 1972, Metwalli & Abdel Hadi 1973; Soliman & Amer 1972). Limestones become more and more frequent towards the NW. In the Mamura-Sallum area shallow marine carbonates with common grainstones, but no reefs, are predominant while sands are very reduced. An exception is the thick (>1000 m) shale sequence of the Matruh well (Abdine & Deibis 1972, Said 1962), which has been interpreted as a basinal, delta-front facies (Prior, 1974). In northern Cyrenaica, there are shallow marine sandstone-limestone-shale alternations; however, the sand percentage quickly increases southwards into a continental 'Nubian' facies (Barr, 1968).

The *Aptian Alamein Dolomite* represents lagoonal to supratidal environments, with uniform dolosparites, over large parts of the Western Desert (Hamed 1972). North of 31°N there are apparently areas where the facies is less dolomitic, and even includes grainstones (Metwalli & Abdel Hadi, 1975).

The *Albian Kharita Fm* is very sandy and largely deltaic, or reashore marine, from the east to 28°E. In the Marsa Matruh region (26°30'E to 28°E), and also further west, it is entirely shallow marine, with several limestone intercalations. A carbonate bank must have existed in the Mamura area, as the Albian in the Mamura well is entirely represented by limestones and dolomite (Said 1962).

The *Cenomanian transgression* extended these shallow marine environments much further south (Fig. 6). Nevertheless, depths at the present coast appear to have remained moderate as shown by oolitic and bioclastic intercalations in the Baharia Formation between Dabaa and Matruh. The *Late Cenomanian-Turonian Abu Roash Fm* is almost wholly carbonate. The lower part is predominantly dolomitic east of Matruh, and even contains a regionally extensive anhydrite intercalation. The upper part is mainly lime mudstone, partly dolomitic. Oolitic and other grainstone facies appear only on the coastal highs, under the Turonian-Coniacian unconformity. In northern Cyrenaica, on the other hand, there are Cenomanian and Turonian dark grey deep water shales near the present coast, grading southwards to progressively shallower water sands and limestones (Barr 1968).

A substantial change of sedimentary environment took place after the Turonian-Coniacian unconformtiy. Uniform, open marine chalky limestone deposition extended over large parts of NE Egypt (Fig. 6) and persisted from *Santonian* to *Maastrichtian*. Facies were shallower and sandier in the south; but, middle to deep neritic in the north (Bartov & Steinitz 1977, Said 1962), and even bathyal in North Cyrenaica (Barr 1968). In the north, however, the rise of coastal structures and differential subsidence caused considerable variations of thickness, if not of facies.

This pattern continued in *the Palaeogene*, with an extension of marine conditions further south into Upper Egypt and the Sirte Basin. Palaeocene deposits were mainly mudstones, but in the north they were reduced or totally eliminated by erosion during the Early Eocene (Jebel El Akhdar, Roehlich 1980), or Middle Eocene (Said 1962). The Eocene rocks bear the evidence of this syn-sedimentary tectonism, not only in thickness variations, but also in facies changes. Thinner shallower-water cherty bioclastic limestones (Mokattam) or dolomitic siliceous limestones (Apollonia) were deposited on, or around highs, with chalky micritic limestones in the basins (Salem 1976).

During the *Late Eocene-Oligocene*, neritic carbonate deposition persisted over the more positive areas from Sallum to Jebel El Akhdar, while in the eastern Western Desert sedimentation was characterized by thick, open marine calcareous shales (Dabaa), followed, in the Early Miocene, by a delta-front turbiditic depocenter (Moghra formation) just west of the Nile Delta (Salem 1976). In the *Middle Miocene* the shallow water carbonates of the Marmarica Formation (Jaghbub and Regima Formations in north Cyrenaica) transgressed as far as Alexandria. Clastic deposition continued only in the Nile Delta region (Rizzini *et al.* 1978, Said 1981, Soliman & Faris 1963).

The coastal sequence west of Alexandria practically terminates with the Marmarica limestones. There are no Late Miocene sediments, (and scanty Pliocene clastics) either onshore, offshore under the continental shelf, or on the steep continental slope. East of Dabaa, however, the Marmarica Formation is covered offshore by an increasing thickness of Plio-Pleistocene shales and sands of Nile derivation. The Nile Delta area and the Gulf of Suez were the only active foci of deposition from Tortonian onwards, with the deltaic regression of the Nile starting essentially no earlier than Late Pliocene (Said, 1981).

Discussion

Facies and nature of depocenters

Available information on Mesozoic facies suggests that several of the depocenters mentioned were not 'basins' in a physiographic sense, with deeper water deposition. Also, those further south relative to the present coast (e.g. Shushan, Abu Gharadig, Guindi, Sirte) were situated in transitional, marine nearshore to continental facies belts from Jurassic to at least Late Albian. There are however indications that a number of stratigraphic units in some depocenters tended to be more shaley, possibly due to the north-bounding ridge barriers.

The situation changed in the Senonian and the Cenozoic, when depocenters also became areas of deeper water deposition, for example the Early Miocene in the eastern Western Desert, and the Early-Middle Miocene in the Nile Delta and Gulf of Suez.

Could the 'Northern Basin' depocenter be regarded as an ancient continental margin prism? Several formations appear to become thicker and more marine towards north, but the coastal wells bear evidence only of shallow-water, carbonate and clastic environments, with little sign of slope or basin facies, while seismic information suggests that the main carbonate

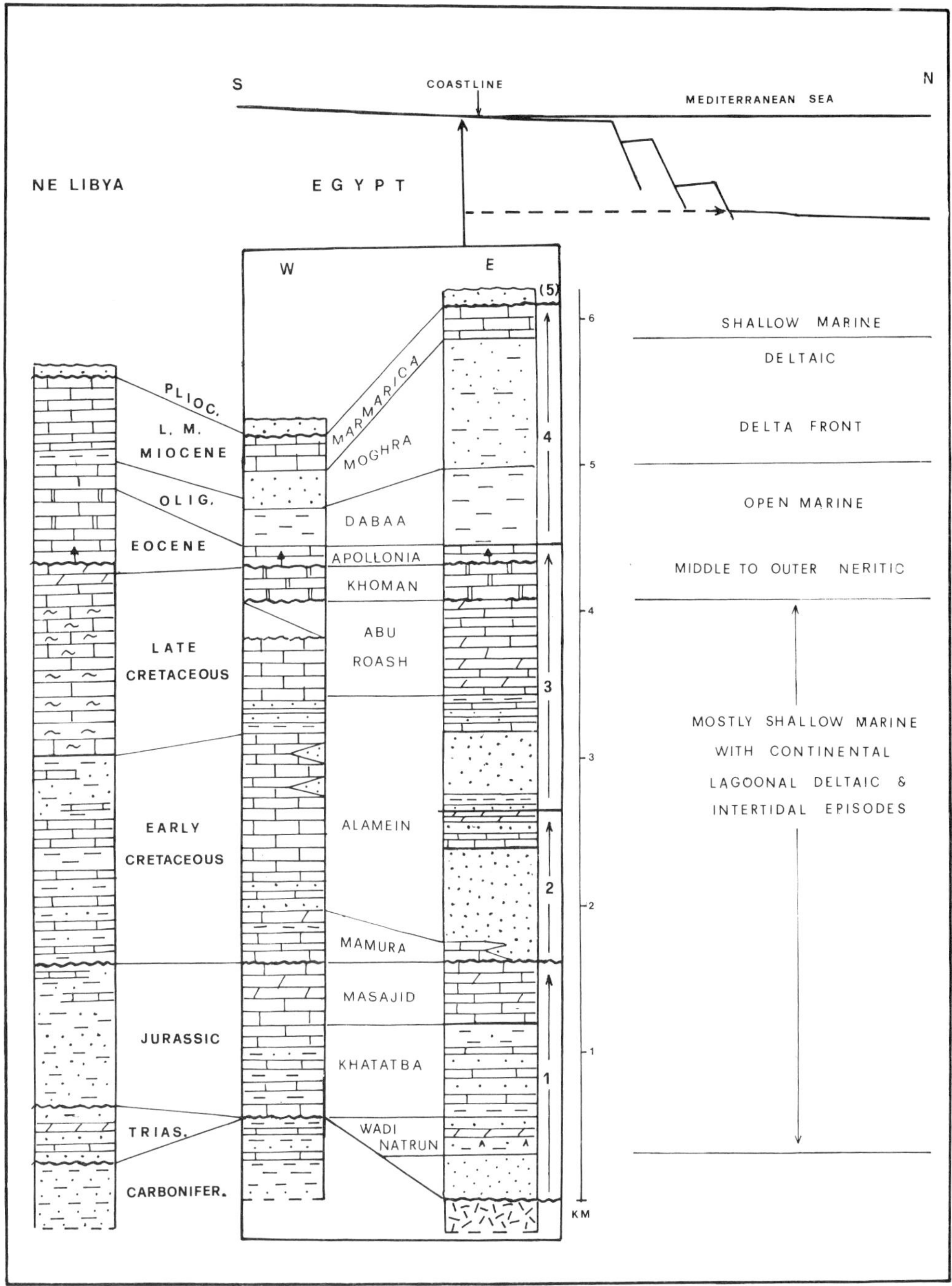

FIG. 5. Summary of the coastal sequence of Egypt and north Cyrenaica. (Based on: Abdine & Deibis 1972, Barr 1968, Conant & Goudarzi, Deibis 1976, Goudarzi 1980, Metwalli & Abdelhady 1973, Said 1962). Sedimentary cycles: 1. (Permo?) Triassic-Late Jurassic. 2. Early Cretaceous-Aptian. 3. Albian-Early-Middle Eocene. 4. Late Eocene-Middle Miocene; 5. Late Miocene to Recent (Nile Delta and Sirte Embayment)—half cycle.

units (Fig. 5) continue at least to the downfaulted zone, and to the southern part of the Herodotus Basin. Further north, unequivocal seismic data are lacking. Data in Malovitsky *et al.* (1975) and in Lort (1977) suggest a total of 11 km of sediments under the Herodotus Basin (? including Palaeozoic). Taking into account a thickness of post-Eocene sediments of at least 4 km including Plio-Pleistocene clastics, Messinian salt and Oligo-Miocene pelagic marls (Mulder *et al.* 1975, Ross & Uchupi 1977), the remaining 7 km of Palaeozoic-Eocene rocks must be essentially the same thickness as at the coast. The 'slope' must have been located yet further north.

Towards the NNW, between Crete and Cyrenaica, the 'Northern Basin' may have closed. A profile in Moskalenko (1966) showed only 2.5 km of 'Mesozoic rocks' over crystalline basement, a thickness less than under Jebel El Akhdar (3–4 km, Goudarzi 1980, Fig. 5). In Crete, the mainly carbonatic, Triassic-Early Eocene sequences of the autochthon and of the overlying Tripolitsa and Pindus nappes, are also thin (2–2.5 km) and indicate facies belonging to the African, or the Apulian Plate margin. (Cadet *et al.* 1980)

The location of the Mesozoic 'basin' margin to the east is an unsolved problem. Israeli studies have shown that the SE corner of the Mediterranean, and even north-central Negev inland, had been a site of persistent subsidence (e.g. >5.5 km of Triassic-Jurassic, Hirsch, this volume), and that a NNE facies and tectonic hinge-line, situated near the present coast, was active during the Jurassic and Early Cretaceous, with platform carbonates to the east, and slope sediments to the west (Bein & Gvirtzman 1977).

Did this hinge-line turn westwards in North Sinai to continue across the Nile Delta? No data have been published from wells drilled at the coast and offshore in North Sinai, and none of the score of wells in the northern Nile Delta has even reached the Late Eocene. The platform carbonates of the South Delta Block plunge deeply under the Tertiary of the 'Faulted Flexure' (Barber 1980, Said 1982), and no guess can be made of a facies change. The 'flexure' could be a purely tectonic Miocene feature, or faulting could have developed there because of an underlying palaeofacies hinge-line. However, west of the Nile delta, downfaulting of the same Mesozoic carbonates does not seem to coincide with a facies change. Either the palaeoslope turned NW at the East side of the Nile delta, or the margin was no longer a sharp hinge-line, but a gradual 'slope', as in the model proposed by Ben Avraham (1978). Another hypothesis is that a Jurassic-Early Cretaceous oceanic opening extended in a NW direction from North Sinai-Negev to Antalya, in Turkey, persisting at least until the Senonian (Robertson & Woodcock 1979).

Shorelines and depositional strike

The Jurassic to Cenomanian shorelines appear to have run fairly close to the present coast (Fig. 6), with a WNW trend from Cyrenaica to the Gulf of Suez. Evidence of a Triassic shoreline is limited to the Negev-Nile Delta area. The effect of the Cenomanian transgression was to increase the SE trend, with the creation of an 'ancestral Nile basin' gulf. During the Tertiary the depositional strike was reversed to the SW, the likely consequence of the Sirte Basin subsidence and of the post-Eocene uplift of Sinai and the Eastern Desert. The author tends to agree with Garfunkel & Bartov (1977) that there is little evidence of a Gulf of Suez embayment extending as far back as the Palaeozoic (c.f. maps in Said, 1962).

It is clear, therefore, that prior to the Late Cretaceous-Eocene pre-rift doming of the Arabo-Nubian Shield, the Oligo-Pliocene sinistral transcurrent movement of the Aqaba-Dead Sea faults, totalling 110 km (Freund *et al.* 1970), and the Oligocene-Late Miocene uplift of South Sinai-Eastern Desert, the WNW depositional strike continued undeviated east of Sinai, along the north Arabian margin. It could be suggested that the present N75°W Jurassic-Early Cretaceous trend parallels the original South Tethys margin.

Tectonic and depositional history

1. Triassic to Early Cretaceous phase

Due to lack of Triassic rocks west of North Sinai-Negev, little can be said about the early rifting stage on the Egyptian-Libyan margin of the South Tethys. It may be that the present structural framework of faulted ridges and basins developed mainly during the Jurassic over a still-unknown pattern of Permo-Triassic blocks. The block-tilting process appears to have been continuous, with accelerated phases in the Late Jurassic (particularly on, and south of the Siwa-Alamein Ridge), and in the late Aptian and middle Cenomanian (ref. Fig. 4). This suggests a stretching of the crust in the present N–S direction, accompanied by uplift in the south. The period from the Triassic to the Early Cretaceous was therefore one of passive plate margin development, the present NE African coastal zone being a rather marginal belt of a gradually widening marine basin situated further north.

The extent of the possible influence of Precambrian and Palaeozoic tectonic trends on Jurassic structures cannot be established. Published attempts at picturing Palaeozoic palaeogeography and isopachs have suggested broad N–S elements (El Shazly 1977, Said 1962, Soliman & El Fetouh 1970). N–S highs do exist on the basement surface (e.g. Calanscio, Umbarka-Mamura) but:

(a) the present configuration of the basement surface (maps in Goudarzi 1980, Khalil *et al.* 1976, El Shazly 1977, based on gravity data) is the net result of all tectonic movements, which mostly have E–W and ENE trends (similar trends are shown in the basement configuration derived from analysis of magnetic anomalies, Meshref 1982, Meshref *et al.* 1980);

(b) subsurface well data are too scattered to produce anything but a speculative and subjective picture of pre-Mesozoic structure. Nevertheless, it is possible that most of the area between Cairo and the Shushan Basin was high, a geanticline with N–S axis, at least in the Late Palaeozoic.

2. *Late Cretaceous to Middle Eocene phase*

This important phase of tectonic activity was characterized by three principal events:

(a) The uplift of sections of the African margin in the vicinity of the present coast in the Coniacian or Early Santonian. Coniacian beds were not deposited, or were largely eroded together with a variable thickness of Turonian carbonates. Jebel El Akhdar was folded in the Santonian (Roehlich 1980). In coastal Egypt (Western Desert) folding is not evident, but block tilting could have occurred, as in the Abu Gharadig Basin (Ezzat & Dia El Din 1974).

(b) The gradual formation since the Campanian of broad folds which are generally independent of the trends of the earlier structures (i.e. ENE versus WNW to E–W), though some are conformable as drapes over major fault blocks.

According to Said (1962) the growth of these folds occurred in various stages up to the Middle Eocene, but at different times in different places. On Jebel El Akhdar, Campanian-

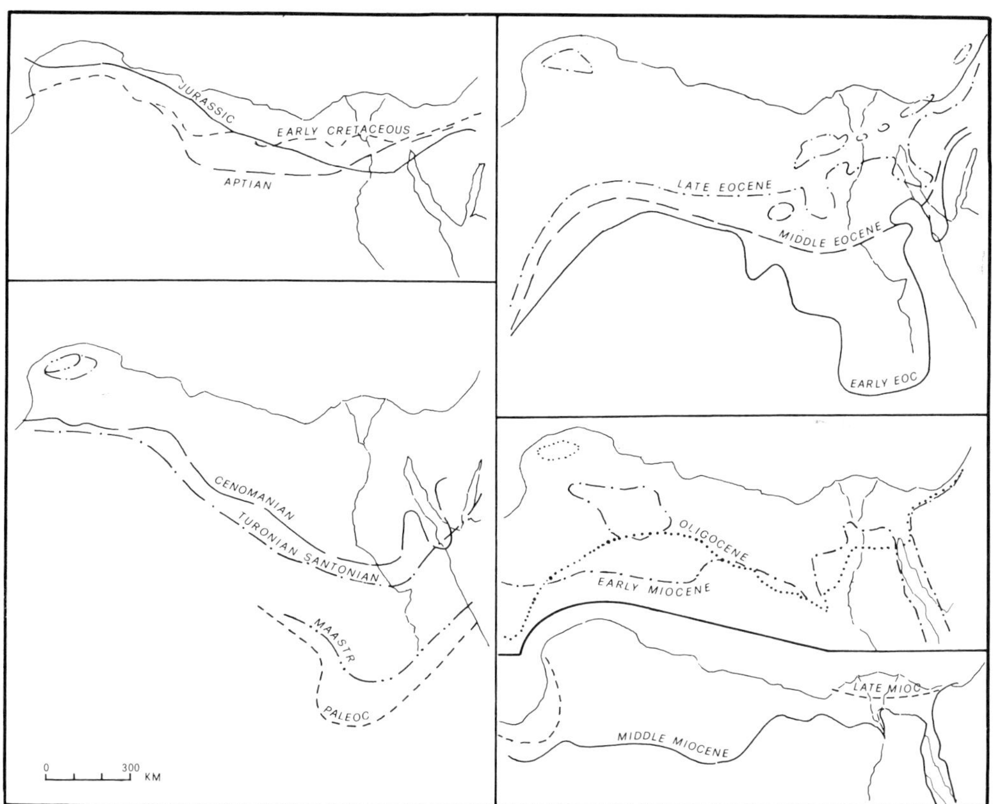

FIG. 6. Shoreline trends from Jurassic to Late Miocene (a compilation from many sources, ref. bibliography).

Eocene sediments were mildly folded, then again uplifted and truncated in the Early Eocene. Since the Senonian was a time of general subsidence characterized by open to deep neritic chalks, folding must have developed in a submarine environment.

(c) The accentuation of the coastal uplift in the Early-Middle Eocene, with the emersion and erosion of several structures, and an increased subsidence in the internal depocenters, the two accentuating the regional tilt to the south.

The first two events may be viewed as grossly contemporaneous with the stages of oceanic closure in the South Tethys, and the collision of the Eurasian and Afro-Arabian Plates that caused overthrusting or gravity gliding, of ophiolites onto carbonate platforms between Campanian and Maastrichtian (Gass & Masson Smith 1963: Ricou 1971; Stoneley 1981). More compression followed in the Eocene, with the folding of the internal Aegean zone and of the Tauric arcs (Cadet *et al.* 1980, Brunn 1976).

The Senonian-Eocene folds continue offshore: NE of Jebel El Akhdar (ref. profile in Sancho *et al.* 1973; gravity data in Morelli *et al.* 1975), under the Nile Delta (Salem 1976), and offshore North Sinai (Ginzburg *et al.* 1975, Neev 1975, Neev *et al.* 1976). Under the Eastern Mediterranean the (?) Late Mesozoic-Eocene lithified sediments are said to be gently folded (Malovitsky *et al.* 1975, Mulder *et al.* 1975). The northward extension of the coastal uplift is more problematic. It could be that the fairly similar thickness of the Mesozoic under the Herodotus Basin and under coastal Egypt results from their both having been relatively high areas with a consequent considerable reduction of the Senonian-Paleocene interval. The evidence includes thinner Paleocene-Eocene with hiatuses in the Pelagian Block, East of Tunisia (Burollet *et al.* 1977) and the Late Cretaceous-Eocene tectonic inversion in the Zagros of Lurestan, relative to the 'internal' Gurpi-Pabdeh basin (Koop & Stoneley 1982).

Regarding the Late Cretaceous-Early Eocene transgression into southern Egypt, and the Paleocene-Eocene tilt to the south, it could be suggested that the migration of subsidence to the internal depocenters, the 'Zone of Linear Uplifts' and the 'Stable Shelf,' was the combined effect of the uplift of the plate margin resulting from collision, and the doming of the Arabo-Nubian shield that preceded the Red Sea rifting. The margin of the African Plate responded differently to these stresses according to the different composition of its parts:

(1) A more rigid belt by the coast, with buried crystalline blocks, mainly overlain by platform carbonates.

(2) A more mobile internal belt resulting from the small-block-fragmentation in the half-graben of the internal depocenters, overlain by thick transitional facies sediments.

(3) A southern belt, characterized by a rapid transition from deep to shallow basement, and by a thinner cover of sands and shales, at the border of a more rigid area with a mainly plutonic basement, lacking a Palaeozoic to Early Cretaceous cover.

A deep zone of weakness may indeed run along the latter margin (the Cairo-Baharia Line of Said 1962; Baharia Fault, in Nagy *et al.* 1974). The idea that this NE structure is aligned with others, like the Pelusium Line of Neev (1976), or the Sinai fold belt, is certainly attractive in a global perspective. Its continuity, however, would require confirmation at least of its more superficial effects in critical areas. In the Suez Canal to Cairo region, for instance, such evidence is lacking (El Shazly *et al.* 1974; and other unpublished data).

The Senonian-Eocene folding may be attributed to the compression being transmitted through the unconsolidated calcareous oozes, while the underlying blocks were re-adjusted. A certain amount of rotation to the south of some blocks could have been the initial cause of the present fold asymmetry.

3. Post-Eocene phase

The dominant deformation style of the post-collision period on the African margin was differential vertical movements. Active subsidence (limited to NE Egypt and the Sirte basin) and hinterland uplift operated within a framework of fractures created, on one hand, by the opening of the Red Sea and the NNE-ward movement of Arabia, and on the other, by the gradual sinking of the Mediterranean basin, in association with the early stages of African margin subduction.

The only 'response' to compressional events in the north (e.g. Oligocene-Middle Miocene nappe emplacement in Crete, Oligocene overthrusting in Rhodes, L.–M. Miocene emplacement of Lycian nappes, Late Miocene folding of Kythrea Flysch, Cyprus, Cadet *et al.* 1980) was mild warping in the Western Desert and northern Cyrenaica at the end of the Middle Miocene.

NE African events in the Oligocene to Early Miocene included: ocean spreading in the northern Red Sea (Garson & Krs 1976), the first phase of lateral movement of the Aqaba-Dead Sea faults (Freund *et al.* 1970), the shaping of the Clysmic (Suez) Gulf, with extensive block-faulting (Garfunkel & Bartov 1977), and basaltic intrusions and lava flows (El Shazly

1977, Said 1962). Major differential movements occurred in the Middle to Late Miocene. Starting in the Langhian, there was significant uplift and faulting of the NE African and Levantine margin (Gvirtzman and Buchbinder 1977) and epeirogenic uplift in the Western Desert and North Cyrenaica, but, at the same time (Langhian-Serravallian) deeper water deposition in the Nile delta (Rizzini *et al.* 1978), widening and subsidence of the Clysmic Basin, and the formation of the Cairo graben, which caused the diversion of the palaeo-Nile to the present delta region (Said 1981).

The coastal belt of faulting from the Nile delta to Cyrenaica probably started at this time, together with the initial sinking of the Herodotus Trough and Basin. A marked escarpment began to form, which was then deeply incised in various places (e.g. Beersheva 'canyon', Neev & Ben Avraham 1977; the Nile graben, and buried valleys at the NE margin of the Sirte Basin, Barr & Walker, 1973). It is envisaged that it was mainly this pre-Messinian uplift that was responsible for the cutting of the marginal channels, later to be filled by Messinian as well as Pliocene deposits (Gvirtzman & Buchbinder 1977), rather than entirely Messinian erosion related to a 'deep desiccated Mediterranean' (c.f. Barber 1980, 1981, Hsü *et al.* 1973, Ryan 1978).

In the northern Nile Delta, the thick, coarse clastic fluvial-deltaic Qawasim Formation of Tortonian-Messinian age (Rizzini *et al.* 1978; Zaghlul *et al.* 1977) witnesses both the active subsidence of the basin, and the uplift and stripping of the Eastern Desert uplands down to the crystalline basement.

The later Neogene was characterized by important extensional movements and by considerable subsidence in the Eastern Mediterranean and Gulf of Suez-Red Sea. Plio-Pleistocene subsidence in the Nile delta led to the accumulation of over 2500 m of clastics. There was a gradual shift in the location of the depocenter, and a change of trend: from N–S, onshore, in the Pliocene, to E–W, under the continental shelf, in the Early Pleistocene to NW–SE, under the western continental slope and rise, in the Late Pleistocene (Ross & Uchupi 1977, Said 1981, Stanley & Maldonado 1977).

Plio-Pleistocene deformation is generally attributed to an increased rate of convergence of the African and Eurasian plates. Direct frontal compression was limited to the Mediterranean Ridge and Crete (Angelier 1978, Finetti 1976), while the rest of the African Plate margin was subject to downfaulting (e.g. Herodotus Trough, Levant Platform). More localized compression has been related to transcurrent movement; in North Sinai, with NNE direction (Dead Sea rift, and presumed 'Pelusium Line'), possibly responsible for halokinesis in the Levant platform, offshore (Ross & Uchupi 1977, Neev 1975), and for the accentuation of the Eocene fold asymmetry inland.

Other transcurrent movements, with NW trend, have been suggested in the Levant Basin, e.g. between Cyprus and the Eratosthenes seamount, possibly associated with the general westward movement of the south margin of the Turkish Plate (Dewey & Şengör 1979, Robertson & Woodcock 1980). Also NW-trending shears appear to be common in the Western Desert (El Shazly *et al.* 1975, 1976; Meshref *et al.* 1980; Riad 1977), and may be so in North Cyrenaica as well, to judge from the numerous NW and WNW faults. It is suggested that the NW shearing, west of Sinai, resulted from the northward movement of Africa towards Europe since middle Eocene times, and a certain clockwise rotation since Late Miocene, associated with subduction under Crete in the Pleistocene.

General comparisons

The Mesozoic structural pattern of NE Africa (Fig. 4) is reminiscent of the block faulting fragmentation of the Atlantic and African margins during the Triassic and Early Jurassic (Bernoulli & Jenkyns 1974; Argyriadis *et al.* 1980). The 'Unstable Shelf' of Egypt, however, did not subside rapidly in the Early Jurassic, so that horsts and fault-bound promontories did not become seamounts with condensed deposition later smothered by pelagic facies, as for instance occurred in NW Sicily (Catalano & D'Argenio 1982). On the contrary, in the zone of the present Egyptian-Libyan coasts, deposition remained shallow marine to transitional, even where Jurassic to Early Cretaceous subsidence became strong. It is only in the 'Ionian' sequence of the Pindic Nappe of Crete that a change of facies occurs from shallow carbonate shelf in the Triassic, to deep basin in the Liassic, while the more external (southern) 'pre-Apulian' sequence of the Ida autochthon and of the Tripolitza Nappe is entirely in platform facies. Thus it should be fair to assume that the Mesozoic under the Mediterranean Ridge is represented by carbonate platform facies and belongs to the African Plate, like the Mesozoic under the Ionian Sea north of the Sirte Gulf (Burollet *et al.* 1977), and that the Jurassic and later basin development was confined to the north of Crete and did not extend south.

References

AALAND, A. J. & HASSAN, A. D. 1972. Hydrocarbon potential of the Abu Gharadig Basin in Western Desert, Arab Republic of Egypt. *Eighth Arab Petrol. Congr.*, Paper 81 (B-3).

ABDEL RAZZIK, I. M. 1972. Comparative studies on the Upper Cretaceous-Early Paleogene sediments of the Red Sea Coast, Nile Valley and Western Desert, Egypt. *Eighth Arab Petrol. Congr.*, Algiers, III, Paper 71 (B-3).

ABDINE, A. A. 1974. Oil and gas discoveries in the Western Desert of Egypt. *Proceed. E.G.P.C. Fourth Explo. Seminar*, Cairo.

ABDINE, A. A. & DEIBIS, S. 1972. Lower Cretaceous-Aptian sediments and their oil prospects in the northern Western Desert, Egypt. *Eighth Arab Petrol. Cong.*, Algiers, Paper 74 (B-3).

AL FAR, D. M. 1966. Geology and coal deposits of Gebel El Maghara, northern Sinai. *Geol. Survey, Min. Res. Dept.*, Paper 37, Cairo.

AMIN, M. S., 1961. Subsurface features and oil prospects of the Western Desert, Egypt, U.A.R. *Third Arab Petrol. Congr.*, Alexandria.

ANGELIER, J. 1978. Tectonic evolution of the Hellenic Arc since Late Miocene. *Tectonophysics*, **49,** 23–36.

ARGYRIADIS, I., DE GRACIANSKY, P. C., MARCOUX, J. & RICOU, L.E. 1980. The opening of the Mesozoic Tethys between Eurasia and Arabia-Africa. *Mem. B.R.G.M.*, **115,** 199–214.

AWAD, G. H. & FAWZI, I. M. A. 1956. The Cenomanian transgression over Egypt. *Bull. Inst. Desert Egypte*, **6,** 169–184.

BARAKAT, M. G. 1970. A stratigraphical review of the Jurassic formations in Egypt and their oil potentialities. *Seventh Arab Petrol. Cong. Kuwait*, II, Paper 58 (B-3).

BARBER, P. M. 1980. Paleogeographic evolution of the Proto-Nile Delta during the Messinian salinity crisis. *Géologie Medit.*, **7,** 13–18.

BARBER, P. M 1981. Messinian subaerial erosion of the Proto-Nile Delta. *Mar. Geol.*, **44,** 353–272.

BARR, F. T. 1968. Upper Cretaceous stratigraphy of Jebel El Akhdar, Northern Cyrenaica. In: BARR, F. T. (ed.) *Geology and Archaeology of Northern Cyrenaica, Libya.* Petrol. Explo. Soc. Libya, 10th anl. Field Conf., 131–142.

BARR, F. T. & WALKER, B. R. 1973. Late Tertiary channel system in northern Libya and its implication on Mediterranean sea level changes. *Init. Rept. Deep Sea Drill. Proj.*, XIII, pt. 2, 1244–1250.

BARTOV, R. & STEINITZ, G. 1977. The Judea and Mt. Scopus groups in Negev and Sinai with trend surface analysis of the thickness data. *Isr. J. Earth Sci.*, **26,**

BAYOUMI, A. A. & EL GAMILI, M. M. 1970. A geophysical study on the Fayoum-Ryan area with refrence to its subsurface structures. Seventh *Arab Petrol. Congr., Kuwait, II*, Papers 35 (B-2).

BEIN, A. & GVIRTZMAN, G. 1977. A Mesozoic fossil edge of the Arabian Plate along the Levant coastline and its bearing on the evolution of the Eastern Mediterranean. In: BIJU-DUVAL B. & MONTADERT, L. (eds) *Structural History of Mediterranean Basins.* Editions Technip, Paris, 95–110.

BEN AVRAHAM, Z., SHOHAM, Y. & GINZBURG, A. 1976. Magnetic anomalies in the Eastern Mediterranean and the tectonic setting of the Heratosthenes Seamount. *Geophys. J. Roy Astron. Soc.*, **45,** 105–123.

BEN AVRAHAM, Z. 1978. The structure and tectonic setting of the Levant continental margin, Eastern Mediterranean. *Tectonophysics*, **44,** 313–331.

BENTZ, F. P. & GUTMAN, S. J. 1977. Landsat data contributions to hydrocarbon exploration in foreign regions. *Prof. Pap. U.S. geol. Surv.*, **1015,** 83–92.

BERNOULLI, D. & JENKYNS, H. C. 1974. Alpine, Mediterranean and central Atlantic Mesozoic facies in relation to the early evolution of the Tethys. In: DOTT, R. H. JR & SHAVER, R. H. (eds) *Modern and ancient geosynclinal sedimentation.* Spec. Publ. S.E.P.M., **19,** 129–160.

BIJU-DUVAL B. & MONTADERT L. 1977. Introduction to the structural history of the Mediterranean Basins. In: BIJU-DUVAL B. & MONTADERT L. (eds) *Structural History of Mediterranean Basins.* Editions Technip, Paris, 1–12.

BUROLLET, P. F., MUGNIOT, J. M. & SWEENEY, P. 1977. The geology of the Pelagian Block: the margins and basins off southern Tunisia and Tripolitania. In: NAIRN, A. E. M., STEHLI, F. G. & KANES, W. H. (eds). *The Ocean Basins and Margins, the Western Mediterranean*, Plenum.

BRUNN, J. H. 1976. L'Arc concave Zagro-Taurique et les arcs convexes Tauriques et Egeéns. *Bull. Soc. géol. France*, **18,** 553–567.

CADET, J. P., BONNEAU, M., CHARVET, J., DURR, S., ELTER, P., FERRIERE, J., SCANDONE, P. & THIEBAULT, F. 1980. Les chaines de la Méditerranée moyenne et orientale. *Mem. B.R.G.M.*, **115,** 98–118.

COLLEY, B. B. 1963. Libya: petroleum geology and development. *6th World Petrol. Congr., Frankfurt*, Sect. 1. Paper 43, 10 pp.

CONANT, L. C. & GOUDARZI, G. H. 1967. Stratigraphic and tectonic framework of Libya. *Bull. Am. Assoc. Petrol. Geol.*, **51,** 719–730.

DEIBIS, S. 1976. Oil potential of the Upper Cretaceous sediments of Northern Western Desert, Egypt. Proceed. *Fifth E.G.P.C. Exploration Seminar*, Cairo.

DESIO, A. 1968. History of geological exploration in Cyrenaica. In: F. T. BARR (ed.) *Geology and Archaeology of Northern Cyrenaica*, Petrol. Explo. Soc. Libya, 10th anl. Field Conf., 79–114.

DEWEY, J. F. & ŞENGÖR, A. M. C. 1979. Aegean and surrounding regions: complex multi-plate and continuum tectonics in a convergent zone. *Bull. Geol. Soc. Amer.*, **90,** 89–92.

DIA EL DIN, M. 1974. Stratigraphic and structural studies of Abu Gharadig oil and gas field. *Proceed. Fourth E.G.P.C. Explo. Seminar*, Cairo.

DRUCKMAN, Y. 1974. Triassic paleogeography of

southern Israel and the Sinai Peninsula. *Wien Symp. Strat. Alpine-Medit. Triassic.*

El Etr, H. A. & Moustafa, A. R. 1980. Delineation of regional lineation pattern of central-Western Desert of Egypt, with particular emphasis on the Baharia region. In: Salem, M. I. & Busrewil M. T. (eds). *The Geology of Libya*, III, 933–954. Academic Press.

El Gezeery, M. N., Mohsen, S. M. & Farid, M. 1972. Sedimentary basins of Egypt and their petroleum prospects. *Eighth Arab Petrol. Congr.*, Algiers, Paper 83 (B-3).

El Shazly, E. M. 1977. Geology of the Egyptian region. In: Nairn A. E. M., Stehli, F. G. & Kanes W. H. (eds). *The Oceans Basins and Margins*,4A,*The Eastern Mediterranean*,379–344.

El Shazly, E. M., Abel Hady, A. M., El Ghawaby, A. M. & El Kassas, I. A. 1974. Geology of Sinai Peninsula from ERTS-1 Satellite images. *Remote Sensing Res. Project*, Acad. Sci. Res. Techn., Cairo.

El Shazly, E. M., Abdel Hady, A. M., El Ghawaby, A. M., El Kassas, I. A., Khawasik, S. M., El Shazly, M. M. & Sanad S. 1975. Geologic interpretation of landsat satellite images for West Nile Delta area, Egypt. *Remote Sensing Res. Project*, Cairo, 34 pp.

El Shazly, E. M., Abdel Hady, M A., El Ghawaby, M. A., Khawasik, S. M., El Shazly, M. M. 1976. Geologic interpretation of Landsat Satellite images for the Qattara Depression area, Egypt. *OSU—Remote Sensing Center*, Cairo, 54 pp.

El Sweifi, A. 1975. Subsurface Paleozoic stratigraphy of Siwa-Faghur area, Western Desert, Egypt. *Ninth Arab Petrol. Congr.*, Dubai, Paper 119 (B-3).

Ezzat, M. R. & Dia El Din, M. 1974. Oil and gas discoveries in the Western Desert, Egypt (Abu Gharadig and Razzak fields). *Proceed. Fifth E.G.P.C. Exploration Seminar*, Cairo, 16 pp.

Finetti, I. 1976. Mediterranean Ridge: a young submerged chain associated with the Hellenic arc. *Boll. Geofis. teor. appl.*, **19, 69,** 31–65.

Finetti, I. & Morelli, C. 1973. Geophysical exploration of the Mediterranean Sea. *Boll. Geofis. teor. app.*, **15, 60,** 263–340.

Freund, R., Zak, T. & Garfunkel, Z., Goldberg, M., Weissbrod, T. & Derin, B. 1970. The shear along the Dead Sea rift. *Phil. Trans. roy. Soc. London*, A 267, 107–130.

Garfunkel, Z. & Bartov, J. 1977. Tectonics of the Suez Rift. *Bull. geol. Surv, Isr.*, **71,** 1–44.

Garson, M. S. & Krs, M. 1976. Geophysical and geological evidence of the relationship of the Red Sea transverse tectonics to ancient fractures. *Bull. geol. Soc. Am.*, **87,** 169–181.

Gass, I. G. & Masson Smith, D. 1963. The geology and gravity anomalies of the Troodos Massif, Cyprus. *Phil. Trans. roy. Soc. London*, A **255,** 417–467.

Ginzburg, A., Cohen, S. S., Hay-Roe, H. & Rossenzweig, A. 1975. Geology of the Mediterranean shelf of Israel. *Bull. Am. Assoc. Petrol. Geol.*, **59,** 2142–2160.

—— & Gvirtzman, G. 1979. Changes in the crust and in the sedimentary cover across the transition from the Arabian Platform to the Mediterranean Basin: evidence from seismic refraction and sedimentary studies in Israel and Sinai. *Sediment. Geol.*, **23,** 19–36.

Goldberg, M. & Friedman, G. M. 1974. Paleoenvironment and paleogeographic evolution of the Jurassic system in southern Israel. *Bull. geol. Surv. Israel*, **61,** 44 pp.

Goudarzi, G. H. 1980. Structure of Libya. In: M. J. Salem & M. T. Busrewil (eds). *The Geology of Libya, III*, Academic Press, 879–892.

Gvirtzman, G. & Buchbinder, B. 1977. The dessication events in the Eastern Mediterranean during Messinian times, as compared with other dessication events in basins around the Mediterranean. In: Biju-Duval, B. & Montadert, L. (eds). *Structural History of the Mediterranean Basins* Editions Technip, Paris, 411–420.

Hamed, A. R. A. 1972. Environmental interpretation of the Aptian carbonates of the Western Desert, Egypt. *Eighth Arab Petrol. Congr.*, Algiers, Paper 79 (B-3).

Hantar, G. 1975. Contributions to the origin of the Nile Delta. *Ninth Arab Petrol. Congr.*, Dubai, Paper 117 (B-3).

Hsü, K. J., Ryan, W. B. F. & Cita, B. B. 1973. Late Miocene desiccation of the Mediterranean. *Nature*, **242,** 239–243.

Issawi, B. 1972. Review of Upper Cretaceous-Lower Tertiary stratigraphy in central and southern Egypt. *Bull. Am. Assoc. Petrol. Geol.*, **56,** 1448–1463.

Kamel, H., Awad, B. & Ahmed, I. 1978. Geotectonic investigations of Shaltut–Burg El Arab Area, A.R.E. from gravity measurements. *Tenth Arab Petrol. Congr.*, Dubai, Paper 28 (B-3).

Khalid, A. D. 1975. Jurassic prospects in Western Desert, Egypt. *Ninth Arab Petrol. Congr.*, Dubai, Paper 100 (B-3).

Khalil, N., Omar, K. Z. & El Demerdash, A. 1976. Evaluation of open areas. *Proceed. Fifth E.G.P.C. Exploration Seminar*, 44 pp.

—— & El Mofty, S. 1972. Some geophysical anomalies in the Western Desert, Egypt. Their geological significance and oil prospects. *Eighth Arab Petrol. Conqr.*, Algiers, II, Paper 42 (B-2).

Kleinsmede, W. F. J. & Van Der Berg, N. J., 1968. Surface geology of the Jabal Al Akhdar, Northern Cyrenaica, Libya. In: Barr, F. T. (ed.). *Geology and Archeology of Northern Cyrenaica, Libya.* Petrol. Explo. Soc. Libya, 10th anl. Field Conf., 115–124.

Koop, W. J. & Stoneley, R. 1982. Subsidence history of the Middle East Zagros Basin, Permian to Recent. *Phil. Trans. roy. Soc.*, A **305,** 149–168.

Kostandi, A. B. 1963. Eocene facies maps and tectonic interpretation in Western Desert, U.A.R. *Rev. Inst. franc. Pétrol.*, **18,** 17–29.

Lewy, Z. 1975. The geological history of southern Israel and Sinai During the Coniacian. *Isr. J. Earth. Sci.*, **24,** 19–43.

Lort, J. M., 1977. Geophysics of the Mediterranean Sea basins. In: Nairn, A. E. M., Stehli, F. G. &

KANES, W. H. (eds). *The Oceans Basins and Margins*, 4a *The Eastern Mediterranean*, 151–213.

MALOVISKIY, Y. A. P., EMELYANOV, E. M., KAZAKOV, O. V., MOSKAELENKO, V. N., OSIPOV, G. V., SHIMKUS, K. M., CHUMAKOV, I. S. 1975. Geological structure of the Mediterranean Sea floor (based on geological-geophysical data). *Mar. Geol.* **18,** 231–261.

MARZOUK, I. 1970. Rock stratigraphy and oil potentialities of the Oligocene and Miocene in the Western Desert, U.A.R. *Seventh Arab Petrol. Congr.*, Kuwait, Paper 54 (B-3).

MESHREF, W. M. 1982. Regional structural setting of Northern Egypt. *Proceed. Sixth E.G.P.C. Explo. Seminar*, Cairo, 11 pp.

——, ABDEL BAKI, S. H., ABDEL HADY, H. M. & SOLIMAN, S. A. 1980. Magnetic trend analysis of the Arabian-Nubian Shield and its tectonic implications. *Annls. geol. Surv. Egypt*, **10,** 939–953.

METWALLI, M. H. & ABDEL HADY, Y. E. 1973. Stratigraphic setting, lithofacies and tectonic analysis of the subsurface sedimentary succession in the Alamein oilfield, northern Western Desert. Egypt. *Bull. Inst. Desert Egypte*, **2,** 23–48.

—— & —— 1975. Petrographic characteristics of oilbearing rocks in Alamein oilfield, significance in source-reservoir relations in northern Western Desert, Egypt. *Bull. Am. Assoc. Petrol. Geol.*, 59, 510–523.

MORELLI, C. 1978. Eastern Mediterranean: geophysical results and implications. *Tectonophysics*, **46,** 333–346.

MORELLI, C., PISANI, M. & GANTAR, C. 1975. Geophysical studies in the Aegean Sea and Eastern Mediterranean. *Boll. Geofis. teor. appl.*, **13, 66,** 127–167.

MOSKALENKO, V. 1966. New data on the structure of the sedimentary strata and basement in the Levant Sea. *Oceanology*, **6,** 828–836.

MULDER, C. J., LEHNER, P. & ALLEN, D. C. K. 1975. Structural evolution of the Neogene salt basins in the East Mediterranean and the Red Sea. *Geol. Mijnb.*, **54,** 208–221.

NAGY, R. M., CHOUMA, M. A., ROGERS, J. J. W. 1976. A crustal suture and lineament in North Africa. *Tectonophysics*, **31,** T67–T72.

NEEV, D. 1975. Tectonic evolution of the Middle East and the Levantine Basin. (Easternmost Mediterranean). *Geology*, **3,** 683–686.

—— 1976. The Pelusium Line, a major transcontinental shear. *Tectonophysics*, **38,** T1–T8.

——, ALMAGOR, G., ARAD, A., GINZBURG, A. & HALL, J. K. 1976. Geology of the Southeastern Mediterranean Sea. *Rept. Isr. Geol. Surv.*, MG/73/5.

—— & BEN AVRAHAM, Z. 1977. The Levantine countries: the Israeli coastal region. In: NAIRN, A. E. M., STEHLI, P. F. & KANES, W. H. (eds). *The Ocean Basins and Margins*, 4a *The Eastern Mediterranean*, 355–377.

OMARA, S. & OUDA, K. 1972. A lithostratigraphic revision of the Oligocene-Miocene succession in the northern Western Desert, Egypt. *Eighth Arab Petrol. Congr.*, Algiers, Paper 94 (B-3).

PAPAZACHOS, B. C. & COMNINAKIS, P. E. 1978. Deep structure and tectonics of the Eastern Mediterranean. *Tectonophysics*, **46,** 285–296.

PRIOR, S. W. 1976. Matruh Basin: possible failed arm of Mesozoic crustal rift. *Proceed. Fifth E.G.P.C. Exploration Seminar*, Cairo, 11 pp.

RÖHLICH, P. 1980. Tectonic development of Jebel El Akhdar, Libya. In: SALEM, M. I. & BURREWIL, M. T. (eds). *The Geology of Libya*, III, 879–982, Academic Press.

RIAD, S. 1977. Shear zones in North Egypt interpreted from gravity data. *Geophysics*, 42, 1207–1214.

——, EL ETR, H. A. & MOHAMMED, M. A. 1980. Gravity-tectonic trend analysis in Siwa—Al Jaghbub region. In: SALEM, M. I. & BUSREWIL, M. T. (eds). *The Geology of Libya*, III, Academic Press, 979–992.

RICOU, L.-E. 1971. Le Croissant ophiolitique periarabe, une ceinture de nappes mises en places au Cretacé supérieur. *Rev. Géogr. phys. Géol. dyn.*, **13,** 327–349.

RIZZINI, A., VEZZANI, F., COCOCCETTA, V. & MIIAD, G. 1978. Stratigraphy and sedimentology of Neogene-Quaternary section in the Nile Delta area. *Mar. Geol.*, **27,** 327–348.

ROBERTSON, A. H. F., & WOODCOCK, N. H. 1980. Tectonic setting of the Troodos in the Eastern Mediterranean. In: *Ophiolites*, A. PANAYIOTOU (ed.) 36–49.

ROSS, D. A. & UCHUPI, E. 1977. The structure and sedimentary history of the SE Mediterranean Sea. *Bull. Am. Ass. Petr. Geol.*, **61,** 879–902.

RYAN, W. F. B. 1978. Messinian badlands on the SE margin of the Mediterranean Sea. *Mar. Geol.*, **27,** 339–363.

SAID, R. 1962. *Geology of Egypt.* Elsevier, Amsterdam, 377 pp.

—— 1981. *The River Nile.* Springer Verlag.

SALEM, R. 1976. Evolution of Eocene-Miocene sedimentation patterns in Northern Egypt. *Bull. Am. Ass. Peol. Geol.* **60,** 34–64.

SANCHO, J., LETOUZEY, J., BIJU-DUVAL, B., COURRIER, I. P., MONTADERT, L. & WINNOCK, E., 1973. New data on the structure of the Eastern Mediterranean basin from seismic reflection data. *Earth planet. Sci. Lett.*, **18,** 189–204.

SIGAEV, N. A. 1959. The main tectonic features of Egypt. *Geol. Surv. Egypt*, Paper 39, 25 pp.

SOLIMAN, S. M. & EL BADRY, O. 1970. Nature of Cretaceous sedimentation in Western Desert, Egypt. *Bull. Am. Ass. Petrol. Geol.*, **54,** 2349–2370.

—— & EL FETOUH, M. A. 1970. Carboniferous of Egypt- isopach and lithofacies maps. *Bull. Am. Ass. Petr. Geol.*, **54,** 1918–1930.

—— & FARIS, M. I. 1963. General geologic setting of the Nile Delta Province, and its evaluation for petroleum prospecting. *Fourth Arab Petrol. Congr.*, Paper 31.

——, FARIS, M. I. & EL BADRY, O. 1970. Lithostratigraphy of the Cretaceous formations in the Baharia Oasis, Western Desert, Egypt. *Seventh Arab Petrol. Congr*, Kuwait, II, Paper 59 (B-3).

——, FARIS, M. I. & HASSAN, M. F. 1965. Geological

setting of the Gulf of Suez during the Eocene period. *Fifth Arab Petrol. Congr.*, Cairo, Paper 30 (B-3).

—— & AMER, K. M. 1972. Cretaceous sedimentation in Eastern Desert, Egypt and evaluation for petroleum prospecting. *Eighth Arab Petrol. Congr.*, Algiers, *III*, Paper 77 (B-3).

STANLEY, D. J. & MALDONADO, A. 1977. Nile Cone: Late Quaternary stratigraphy and sediment dispersal. *Nature*, **266**, 129–135.

STONELEY, R. 1980. The geology of Kuh-e-Dalnashin area of Southern Iran, and its bearing on the evolution of the Southern Tethys. *J. geol. Soc.*, **138**, 509–526.

THIELE, J., GRAMANN, F. & KLEINSORGE, H. 1970. Zur Geologie zwischen der Nordrand der ostlichen Kattara-Senke und der Mittelmeer-Kuste, Aegypten, Westliche Wuste. *Geol. Jahrb.*, **88**, 321–346.

VAIL, P. R., MITCHUM, R. M. JR. & THOMPSON, S. 1977. Global cycles and relative changes of sea level. In: PAYTON, C. E. (ed.). *Seismic Stratigraphy—Applications to hydrocarbon exploration*, Mem. Am. Ass. Petr. Geol., **26**, 183–189.

WOODSIDE, J. M. 1977. Tectonic elements and crust of the Eastern Mediterranean Sea. *Mar. geophys. Res.*, **3**, 317–354.

—— & BOWIN, C. 1970. Gravity anomalies and inferred crustal structure in the Eastern Mediterranean Sea. *Bull. geol. Soc. Am.*, **81**, 1107–1122.

YOUSSEF, M. I. 1968. Structural pattern of Egypt and its interpretation. *Bull. Am. Ass. Petrol. Geol.*, **52**, 601–614.

ZAGHLUL, Z. M., TAHA, A. A., HEGAB, O. & EL FAWAL, F. 1977. The Neogene-Quaternary sedimentary basins of the Nile Delta. *Egypt. J. Geol.*, **21**, 1–19.

G. SESTINI, Via Giambologna 8, Florence, Italy.

The Hercynian Geanticline of Helez and the Late Palaeozoic history of the Levant

Gdaliahu Gvirtzman & Tuvia Weissbrod

SUMMARY: The presence of a large structure of Late Palaeozoic age has been deduced from analysis of Palaeozoic columnar sections from the Levant countries, Israel, Jordan, Egypt, Saudi Arabia and Syria. This structure, which was uplifted and truncated in two phases, pre-Carboniferous and pre-Permian, is here termed the Geanticline of Helez and is related to Hercynian movements. Permian to Triassic sediments overlie the Precambrian basement at its crest, whereas on the flanks of the structure, Palaeozoic sediments from Cambrian to Carboniferous in age wedge out gradually towards the culmination. Similar Hercynian geanticlines are known from the Hazro-Singar area near the Syrian-Turkish border, from Ennedi- Al Awaynat near the Libyan-Egyptian-Sudanese border, as well as elsewhere in North Africa. The culmination of the structure, some 500 km in diameter, extends over a large area of Israel, Sinai and Jordan. The flanks of the geanticline, originally dome-shaped, could be identified only in three directions. The missing northwestern flank, dipping towards the southeastern corner of the Mediterranean Sea, was probably removed during rifting and spreading from Late Triassic to Late Cretaceous times and may be located today in the Antalya area or elsewhere in Turkey. Its original location is today occupied by the Mesozoic basaltic ocean-floor of the Neo-Tethys, which is assumed to underlie the Levantine Basin of the Mediterranean.

Analysis of the information and working assumptions

Reconstruction of the Palaeozoic history of the Levant is hindered by lack of adequate information. Scattered outcrops are distributed around the core of the Arabo-Nubian massif, whereas the bulk of the Palaeozoic sediments are buried and have been penetrated only in a few deep boreholes.

Other difficulties in its reconstruction are the uneven areal distribution of data points, frequent stratigraphical gaps within the Palaeozoic sequence, lack of age-diagnostic fossils, inconsistency of available data from the different countries and misleading correlations between the various locations.

Recent publications on the stratigraphy of deep boreholes from Syria (Yankauskas & Talli, 1978) and those of the Palaeozoic of the Negev in Israel and of Sinai (Weissbrod, 1969a, 1969b, 1970a, 1970b, 1976, 1981), together with Weissbrod's (1981) revised stratigraphy and correlations of the Palaeozoic of the Eastern Desert of Egypt, northwestern Saudi–Arabia and Jordan, overcame some of these difficulties. Nevertheless, lithostratigraphic correlation is limited to the southeastern part of the Levant. The only correlatable intervals over the Levant area as a whole are time-stratigraphic units, relating to the main systems of the Palaeozoic.

Such correlation, which considers the presence or absence of time-stratigraphic units in the various Palaeozoic sections, has made it possible to present a preliminary palaeostructural compilation. From the onlapping and wedge-out of Palaeozoic sediments, deduced from the above criteria, it appears that in Israel, northern Sinai and northern Jordan, there is a large Late Palaeozoic structure. The present paper attempts to reconstruct this structure, designated the *Hercynian Geanticline of Helez*.

The available information from Palaeozoic sections in the Levant area was grouped into 18 stations (Fig. 1), some of which are single boreholes, while the others represent typical sequences of certain regions. The detailed stratigraphic information for each station is shown in the Appendix. The stratigraphy is further summarized in Fig. 2.

Some of the units are absent due to Late Palaeozoic erosion. Others are missing due to later erosion, mainly during the Early Cretaceous, Tertiary and Quaternary. For the purpose of the present study the Palaeozoic section of every station is reconstructed *as it appeared at the end of the Palaeozoic times*. Sediments which were removed during subsequent erosional processes were considered as present.

In some of the sections the distinction between the Permian and the Triassic is not clear. This sequence may be regarded as a single cycle in the Middle East and North Africa where its sediments unconformably overlie vast areas that were truncated before the Permian. This conclusion is based on the regional studies of Demaison (1965), Sander (1970), Assereto & Benelli (1971), Bosellini & Hsü (1973),

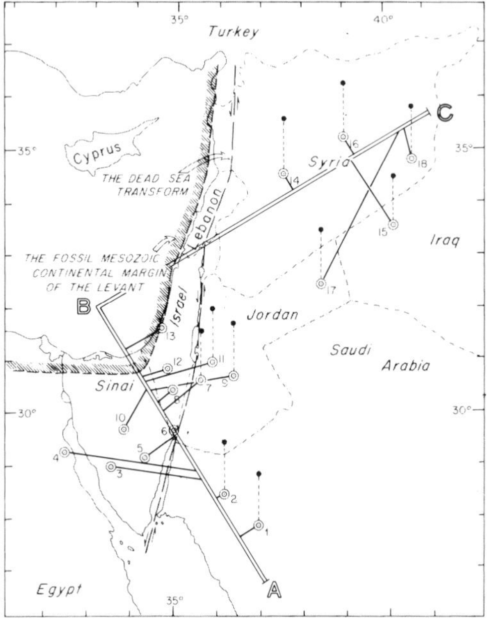

FIG. 1. Location map of 18 selected stations of columnar sections of the Palaeozoic in the Levant countries. The stations located on the eastern side of the Dead Sea Transform have been displaced some 105 kms southwards in order to match their original position prior to the movements along the transform (Freund *et al.*, 1970). The stations are listed in Fig. 2. The cross-section A-B-C is shown in Fig. 7. Stations 1–18 were projected onto the cross-section at various angles in order to illustrate their position in accordance with stratigraphic contacts shown in the contours (Figs 4, 5 & 6), and in the cross-section (Fig. 7).

STATIONS	TABUQ-JA'UF, NW SAUDI ARABIA	MUDAWWARA, S. JORDAN	UM-BOGMA - G.FOKA, SW SINAI	WADI ARABA - ABU DARAG, E. EGYPT	GEBEL DHALAL, E. SINAI	TIMNA-ELAT, S. NEGEV	ZERQA MA'IN, C. JORDAN	HAMEISHAR 1-SINAF 1, C. NEGEV	SAFRA 1, N. JORDAN	NAHL-ABU HAMTH, C. SINAI	SUWEILA 1, N. JORDAN	N. NEGEV BOREHOLES	HELEZ DEEP 1A, C. ISRAEL	KHANASSER 1, NW SYRIA	SWAB 1, E. SYRIA	ABBA 1, N. SYRIA	TANF 1, SE SYRIA	MARKHADA 101, NE SYRIA
	1	2	3	4	5	6	7	8	9	10	11	12	13	14	15	16	17	18
PERMIAN-TRIASSIC	+(1)	+(1)	+	+	+(1)	+(1)	+	+	+	+	+	+	+	+	+(1)	+	+(1)	+
CARBONIFEROUS	+(1)	+(1)	+	+	?	?	–	–	–	–	–	–	–	+	+	+	+	+
DEVONIAN	+	–	–	NE	–	–	–	–	–	–	–	–	–	–	–	–	–	–
SILURIAN	+	+	–	?	–	–	–	–	–	–	–	–	–	–	–	–	+	+
ORDOVICIAN	+	+	–	?	–	–	–	–	–	–	–	–	–	+	+	+	+	+
CAMBRIAN	+	+	+	?	+	+	+	+	+	–	–	–	–	+	+	NP	NP	NP
PRECAMBRIAN	+	+	+	?	+	+	NE	+	+	+	+	+	+	NP	NP	NP	NP	NP
REFERENCES	4 8	2 7 8	6 7 8	1 8	8	6 8	8	5 8	8	7 8	8	5 7 8	3	9	9	9	9	9

FIG. 2. Stratigraphic data for 18 Palaeozoic stations shown in Fig. 1. The presence or absence of time-stratigraphic units is based on information summarized in the Appendix. References: 1—Abdallah & Adindani, 1965; 2—Bender, 1968; 3—Druckman & Kashai, 1981; 4—Powers, 1968; 5—Weissbrod, 1969a; 6—Weissbrod, 1969b; 7—Weissbrod, 1976; 8—Weissbrod, 1981; 9—Yankauskas & Talli, 1978.
(1) = eroded during the Early Cretaceous; NE = not exposed; NP = not penetrated.

Druckman *et al.* (1975), and Weissbrod (1976). Consequently, the Permian and the Triassic are included in the present study as one unit.

Early Cretaceous erosion removed the Permian-Triassic section from large areas. In the proposed reconstruction it is assumed that the entire area studied was covered by Permian-Triassic sediments and therefore, the Permian-Triassic is considered as *present* in all the sections (Fig. 2). This concept, *a priori*, is based on a detailed reconstruction of the Permian by Weissbrod (1976) and of the Triassic by Druckman *et al.* (1975).

The stratigraphical table (Fig. 2) includes six chronostratigraphic systems: *Cambrian*, *Ordovician, Silurian, Devonian, Carboniferous* and *Permian-Triassic.* Only at a few stations are there clearly differentiated series, stages and sub-stages, such as the subdivision of the Ordovician. Since these subdivisions were not determined for most of the stations, they were disregarded. Even if at a given station only part of a system is present, the system was considered to be present.

The numbers of the stations in Fig. 2 are arranged according to their geographical order, from south to north. It appears that full sections are found in the extreme south and in the extreme north only. The sections of the central part, namely stations 10, 11, 12 and 13, where the Permian-Triassic directly overlies Precambrian basement, show the biggest gap.

Stratigraphical results

The stratigraphic data (Fig. 2) can be grouped into six combinations as illustrated in Fig. 3. These six combinations are expressed as map units in Fig. 4. The most typical case, unit A, is a full Palaeozoic sequence, with no system missing, but not precluding minor stratigraphic gaps. This unit is only represented by station 1 in northwestern Saudi-Arabia. In unit B the Devonian is missing and Carboniferous overlies the Silurian. This is the case for station 2 in southern Jordan as well as stations 17 and 18 in northern Syria. In unit C the Silurian is also missing; in unit D the Ordovician is missing and

MAP UNIT →	A	B	C	D	E	F	SUBCROP GEOLOGICAL MAP
PERMIAN-TRIASSIC	+	+	+	+	+	+	
CARBONIFEROUS	+	+	+	+	—	—	1
DEVONIAN	+	—	—	—	—	—	2
SILURIAN	+	+	—	—	—	—	
ORDOVICIAN	+	+	+	—	—	—	
CAMBRIAN	+	+	+	+	+	—	
PRECAMBRIAN	+	+	+	+	+	+	
STATION NOS.	1	2 17 18	14 15 16	3 4?	5? 6? 7 8 9	10 11 12 13	

FIG. 3. Classification of the Palaeozoic sections of Fig. 2 into units according to the presence or absence of time-stratigraphic sequences. The sections are grouped into six units A to F, from a complete section in A to a section with the largest internal stratigraphic gap in F. The units are mapped in Fig. 4. 1—A pre-Permian surface, shown in map of Fig. 5 and as surface B in the cross-section of Fig. 7. 2—A pre-Carboniferous surface, shown in map of Fig. 6 and as surface A in the cross section of Fig. 7.

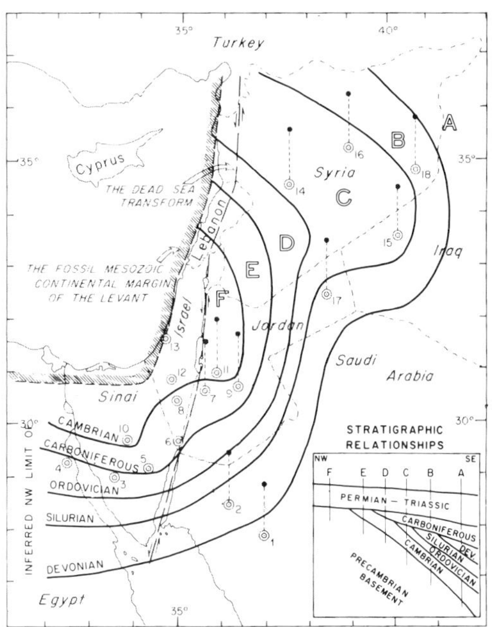

FIG. 4. A geological map showing the onlapping stratigraphic relationship on the Geanticline of Helez. The map units A through F are defined in Fig. 3 and in the diagram at the lower right corner.

in unit E the Carboniferous also disappears. In unit F, where the Permian-Triassic directly overlies the Precambrian, almost the entire Palaeozoic sequence is missing. The classification of station 4 (Wadi Araba-Abu Darag, eastern Egypt) is questionable since the base of the section is not exposed. There is also doubt as to stations 5 and 6 (eastern Sinai, Elat and Timna) since the assumption that the Carboniferous was absent is not conclusive (see the remarks on station 5 in the Appendix). However, as mentioned above, even if the Carboniferous was *present* there before the Early Cretaceous erosion, thus classifying stations 5 and 6 as unit D, contours of the maps of Figs 4 & 5 might be shifted to a very limited extent but this will not affect the overall palaeostructural picture.

The six units, A through F, are mapped in Fig. 4. Their relationships show onlaps and wedging-out towards a structural high in which truncations and a maximum stratigraphic gap occur. The culmination of the high is represented by station 13 (Helez Deep 1A borehole) in the southern Coastal Plain of Israel. This high is reconstructed as a palaeogeographic structure from Permian to Triassic times. From the map (Fig. 4) can be plotted the maximum areal distribution and the northwesternmost limit of the sediments of every system of the Palaeozoic, towards the culmination of the structure.

The boundary between units A and B on the map is also the reconstructed northwestern limit of the Devonian. The boundary between B and C is the northwestern limit of the Silurian and so on. These stratigraphic relationships indicate that the Cambrian, Ordovician, Silurian and Devonian might conformably overlie each other. An exception to this pattern is the Carboniferous, which unconformably overlies an erosional surface, under which the Cambrian is the northwesternmost system, and the Devonian the most southeastern. The reconstructed maximum extension of the Carboniferous is the boundary between units D and E. The area of unit F, approximately 500 km in diameter, forms the crest of the structure, and is covered only by the Permian-Triassic. It is also possible that at Helez Deep 1A (station 13), which is

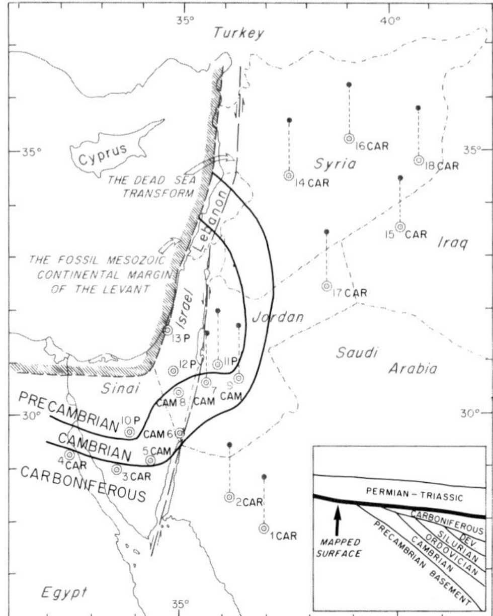

FIG. 5. Subcrop geological map reconstructing the base of the Permian-Triassic in the area of the Geanticline of Helez. The map shows the outcrops which were exposed after the pre-Permian uplift. The units underlying the Permian-Triassic at every station are indicated.

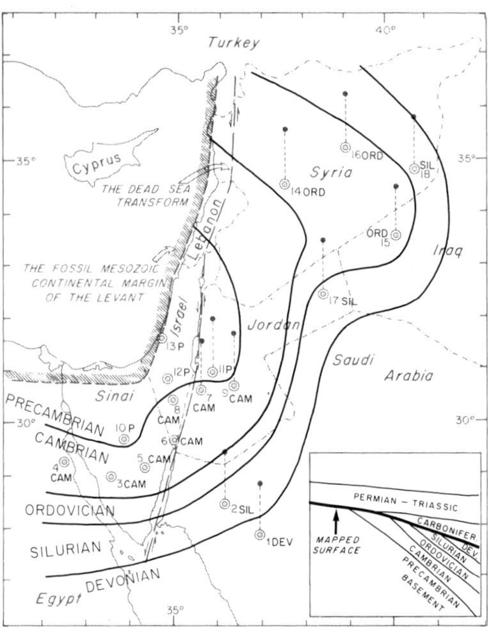

FIG. 6. Subcrop geological map reconstructing the base of the Carboniferous in the area of the Geanticline of Helez. The map shows the outcrops which were exposed after the pre-Carboniferous uplift. The units underlying the Carboniferous at every station are indicated.

probably located at the centre of the culmination, only Middle Triassic covers the Precambrian basement (station 13 in the Appendix).

The exposed geology of the structure prior to the Permian, which is derived from data given in Fig. 4, is shown in Fig. 5). This map indicates that the Carboniferous covered most of the northern Arabian Shield. The structure is delineated by a belt of Cambrian outcrops around an exposed Precambrian core.

The exposed geology of the structure prior to the Carboniferous, likewise derived from Fig. 4, is shown in Fig. 6. Around the exposed Precambrian basement of the core of the structure, concentric belts of outcrops delineate the shape and the dimensions of the structure. The Cambrian belt is the inner and the Silurian belt the outer. The Devonian was probably distributed further to the southwest.

The cross-section A-B-C in Fig. 7 illustrates the onlaps and wedge-outs of the Palaeozoic sediments around the structure. The various stations (1–18) were projected onto the cross-section at different angles (Fig. 1), in order to retain the regular wedging-out arrangement of the units which onlap the core of the structure in accordance with the contours of Fig. 4. The southeastern flank dipping towards Saudi–Arabia (A-B) is relatively well documented. The northeastern flank, dipping towards Syria (B-C), is less well documented. However, the symmetrical shape of both flanks seems to confirm the deduced stratigraphic relationships of the Syrian flanks, in spite of the fact that these are based only on data from 5 boreholes, none of which penetrated the basement.

Interpretation and discussion

The Hercynian Geanticline of Helez

The surface of the unconformity below the Carboniferous (Fig. 6 and surface A of cross-section in Fig. 7) was formed as a result of an uplift of the Helez area prior to the Carboniferous. The onlapping stratigraphic relationship of this surface may be interpreted in two ways:

(1) A long continuous uplift with changing rates during the Palaeozoic, from the Cambrian onward. As a result of these processes there were repeated periods of non-deposition during

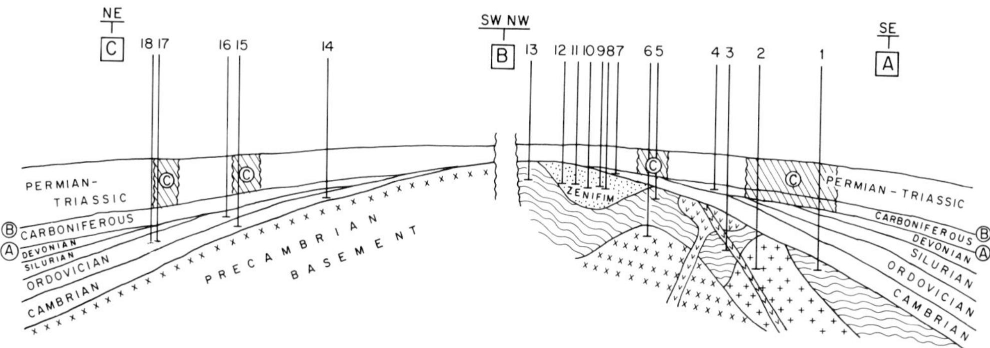

FIG. 7. Cross-section A-B-C across the Geanticline of Helez. The direction of the cross-section is shown in Fig. 1. Stations 1 to 18 were projected onto the cross-section line at various angles, in accordance with the stratigraphic order, the onlap relationships, and the contours of the maps of Figs 4, 5 & 6. The illustrated details of the basement represent a generalized conceptual model deduced from outcrops and boreholes, in which regional metamorphism (wavy lines) was followed by two phases of granitic intrusions (+ and × graphic symbols). The latter was followed by extrusions of acid volcanics and by dykes (v graphic symbols). Finally, the basement was topped by arkosic sedimentation of the Zenifim Formation (dots).
A—The surface at the base of the Carboniferous mapped in Fig. 6.
B—The surface at the base of the Permian mapped in Fig. 5.
C—Reconstructed parts of the sequence, which were removed by erosion during the Early Cretaceous. The criteria used for the reconstruction are discussed in the text.

the time span from the Cambrian to the Devonian and hence sedimentary wedge-outs resulted.

(2) A short and a strong pulse of uplift, prior to the Carboniferous. The result was intensive erosion and truncation of sediments ranging in age from the Cambrian to Devonian, as well as of the top of the Precambrian. It is reasonable to assume that before this uplift there was no high around the Helez area. Normal and regular sedimentation could have taken place from the Cambrian to the Devonian.

A successive unconformity surface below the Permian (Fig. 5 and surface B of the cross-section in Fig. 7) was also formed as a result of an uplift prior to the Permian. The interpretations mentioned above may also be applied to the pre-Permian event.

The second alternative, namely pre-Carboniferous and pre-Permian phases of uplifts and erosion, appears preferable due the following considerations:

(1) Palaeocurrent measurements of fluviatile Palaeozoic sediments indicate a general north to northwest trend. This is true for the *Cambrian* of southern Israel (Karcz & Key, 1966; Karcz, Weiler & Key, 1971), of southern Jordan (Selley, 1972) and of southwestern Sinai (Soliman & El-Fetouh, 1969); also for the *Ordovician* of southern Jordan (Selley, 1972) and for the *Carboniferous* of southwestern Sinai (Soliman & El-Fetouh, 1969). From this, it may be deduced that a base level of erosion rather than an emergent structure, was situated in the north or northwest. Only relatively short phases of uplift could change the northward direction of the drainage system, which probably prevailed during most of the Palaeozoic times.

(2) The concentric outcrop belts shown in Fig. 6 are organized according to their stratigraphical order, from the inner to the outer belt, which also represents a change from older to younger units. This order could better be explained as a result of normal sedimentation of Cambrian to Devonian sediments over a non-active area which was followed by a single truncation event. If their concentric arrangement is a result of wedge-outs and non-sedimentation due to continuous tectonic activity during deposition, why should the zero-lines of the successive wedge-outs follow a chronological order?

(3) A solution might be deduced by analogy with similar structures in the Middle East and in North Africa. The timing of the pre-Carboniferous and the pre-Permian movements fits with those of the Hercynian orogeny which took place elsewhere from the end of the Devonian to Early Permian. It is highly probable that the Helez structure was active during the Hercynian movements.

It is therefore proposed that the Helez structure developed as a result of two successive pulses. Pre-existing sediments, which were probably deposited over a non-active area, were truncated during a pre-Carboniferous uplift. The Carboniferous strata, which also covered part or the entire area, were truncated during a second pre-Permian pulse.

In summary, on the assumption that the Helez structure is a Hercynian element it is named the '*Hercynian Geanticline of Helez*'. A similar geanticline, close to Helez, extends across the Turkish-Syrian boundary and is known as the Hazro structure in Turkey (Rigo de Righi & Cortesini, 1964) and as the Singar structure in northeastern Syria (Metwalli *et al.*, 1974); both belong to the same high, which is similar in dimensions to that of Helez. Another geanticline, also close to Helez, is the Hercynian Ennedi-Al Awaynat uplift, which extends across the Libyan-Egyptian border, close to where it meets the Sudanese border (Bellini & Massa, 1980). Sander (1970) shows Hercynian geanticlines in North Africa as well. The characteristics common to these geanticlines are their relatively large dimensions, their stratigraphic relationships, including onlaps of Palaeozoic units over eroded central cores, and their regional cover of Permian to Triassic sediments overlying the truncated highs. The end products of the Hercynian orogeny in Europe, where strong deformation, regional metamorphism and erosion, occurred, are the European 'massifs'. The end product of the Hercynian movements in the African and Arabian cratons are the 'geanticlines' which underwent only epeirogenic uplift, erosion and onlapping sedimentation. The exposed Precambrian core in Helez is about 500 km in diameter (Figs. 5 and 6), bigger than any other structure in the area.

The northwest extension of the Helez Geanticline

A sedimentary and a palaeostructural belt which was designated the 'hinge line' or the 'hinge belt' (Gvirtzman & Klang, 1972), extends along the present-day coastline of the Levant. This line is shown in Figs. 1, 4, 5 and 6 and is designated '*The fossil Mesozoic continental margin of the Levant*'. This belt was interpreted later as the transition from a thick continental crust in the east to a thin oceanic crust in the west (Ginzburg & Gvirtzman, 1979). Along this line changes in Jurassic and especially Cretaceous sediments were observed from platform and shelf-edge environments of deposition in the east to continental margin environments of deposition in the west (Bein & Weiler, 1976; Bein & Gvirtzman, 1977). This belt was interpreted as the fossil edge of the Arabian Platform, in which the transition between the Arabian craton and the Tethys ocean took place during the time span from Late Triassic to Late Cretaceous (Bein & Gvirtzman, 1977). It was also proposed that the Levant margin was formed as a result of Late Triassic rifting and initial Early Jurassic spreading, which continued up to the Late Cretaceous (Bein & Gvirtzman, 1977; Ginzburg & Gvirtzman, 1979).

The eastern basin of the Mediterranean has developed as an ocean-floor of the 'Neo-Tethys', between the Levant margin and a margin in Turkey (Bein & Gvirtzman, 1977). Robertson & Woodcock (1980) pointed out the similarity between the Levant and the Antalya region in Turkey and proposed the location and the matching of these lines prior to the opening of what they designate as the 'Troodos Ocean'.

The Hercynian Geanticline of Helez predates the rifting, opening and spreading of the Neo-Tethys of the Levantine Basin. The northwestern part of the dome-shaped Helez Geanticline has not been recognized. The crust in the area of the Helez Geanticline is of typical continental composition. Quartz-porphyry and schists which were penetrated in the Helez Deep 1A borehole (station 13) indicates a cratonic-continental history. The thickness of the crust from the Moho to the surface, estimated by seismic refraction in the Negev, is about 42 km, of which 40 km are a crystalline crust (Ginzburg & Gvirtzman, 1979). The thickness of the crust at the southeastern corner of the Mediterranean, according to seismic refraction measurements in the El-Arish area (northern Sinai, at the western part of the fossil hinge-belt) is assumed to be 20 km, 10 km of which are a crystalline crust. The reduced thickness of the crystalline crust towards the hinge-belt, was interpreted by Ginzburg & Gvirtzman (1979) as indicating a transition from a continental crust below the Negev to an oceanic-basaltic crust beneath El-Arish.

It is therefore useless to look for the missing part of the Helez Geanticline in the Mediterranean off Helez. It is highly probable that a continental fragment of Helez was detached from its original position in the process of rifting and spreading, and that it reached southern Turkey. Younger ocean-floor was accreted in its original place and then was covered by a thick pile of sediments. It should be emphasized however, that the location of the fragment of the Helez Geanticline in Turkey can only be pinpointed after a closer comparison between the Palaeozoic sections in the Levant and those in Turkey, either in Antalya, as suggested by

Robertson & Woodcock, 1980, or in Alanya or Amanos.

Conclusions

Despite the limited information on the Palaeozoic stratigraphy of the Levant countries, a preliminary outline of the Late Palaeozoic history can be suggested.

(1) The presence of a Late Palaeozoic structure, which is designated here the *Hercynian Geanticline of Helez*, can be deduced from an analysis of 18 sections from the Palaeozoic of the Eastern Desert of Egypt, Sinai, northwestern Saudi-Arabia, Jordan, the Negev and the Coastal Plain of Israel and from Syria.

(2) The crest of the dome-shaped Helez Geanticline was truncated during a first pre-Carboniferous phase and in a second pre-Permian phase. In the area of the culmination, some 500 km in diameter, Permian-Triassic sediments overlie the Precambrian basement. Palaeozoic sediments, Cambrian to Carboniferous in age, wedge out from the flanks to the culmination.

(3) The Permian to Triassic sediments unconformably overlie the truncated geanticline. Similar geanticlines were formed during the Hercynian movements in the Hazro-Singar area extending across the Turkish-Syrian border, in the Ennedi-El Awaynat area extending across the Libyan-Egyptian-Sudanese border, as well as in other regions in North Africa.

(4) The culmination of the Helez Geanticline covers the areas of central and northern Israel, the northern Negev, northern Sinai and northern Jordan. The flanks dip towards the southwest to the Eastern Desert of Egypt, towards the southeast to south Sinai, southern Jordan and Saudi-Arabia, and towards the northeast, to Syria.

(5) The northwestern flank of the geanticline was probably removed from its original position by rifting and opening of the Neo-Tethys Ocean, during Late Triassic to Late Cretaceous times and may have reached southern Turkey. Its former position is probably occupied today by basaltic ocean-floor of the eastern basin of the Mediterranean.

Acknowledgements. This paper is a result of project 28977 of the Geological Survey of Israel. The authors wish to express their thanks to Drs. A. H. F. Robertson, G. Sestini & J. R. Evans, for their useful comments; to Dr. H. Sokolin for a translation of Russian literature; to Mrs. B. Katz, for preliminary editing; to Ms. R. Gordon for typing the manuscript; to Mrs. D. Ashkenazi & Mrs. C. Alafi for their technical assistance and to Mr. S. Levy, Mrs. N. Shragai and Mrs. C. Hadar for drafting the figures.

References

Abdallah, A. M. & Adindani, A. 1965. Stratigraphy of Upper Palaeozoic rocks, western side of the Gulf of Suez. *Egypt. Geol. Survey. & Min. Res. Dept.* **25,** 18 pp.

Assereto, R. & Benelli, F. 1971. Sedimentology of the pre-Cenomanian formations of Jebel Gharian, Libya. *In:* Gray, C. (ed.). *Symposium on the Geology of Libya,* Faculty of Science, University of Libya. 37–83.

Bein, A. & Gvirtzman, G. 1977. A Mesozoic fossil edge of the Arabian Plate along the Levant coastline and its bearing on the evolution of the eastern Mediterranean. *In:* Biju-Duval, B. & Montadert, L. (eds). *International Symposium on the Structural History of the Mediterranean Basins. Split (Yugoslavia).* Editions Technip Paris. 95–110.

—— & Weiler, Y. 1976. The Cretaceous Talme Yafe Formation: A contour current shaped sedimentary prism of calcareous detritus at the continental margin of the Arabian Craton. *Sedimentology,* **23,** 512–32.

Bellini, E. & Massa, D. 1980. A stratigraphic contribution to the Palaeozoic of the southern basins of Libya. *In:* Salem, M. J. & Bousrewil, T. M. (eds). *The Geology of Libya,* **1,** Academic Press, London, 3–56.

Bender, F. 1968. *Geologie von Jordanien.* Gebruder Bornträger, Berlin, 230 pp.

Bosellini, A. & Hsü, K. J. 1973. Mediterranean plate tectonics and Triassic palaeogeography. *Nature, London,* **244,** 144–46.

Daniel, E. J. 1963. *Lexique stratigraphique international.* **3,** Asie, Fasc. 10c1, Liban, Syrie, Jordanie. Centre Nation. Rech. Sci. Paris, 159–289.

Demaison, G. J. 1965. The Triassic salt in the Algerian Sahara. *In: Salt Basins around Africa.* The Institute of Petroleum, London, 91–100.

Derin, B. 1979. Helez Deep 1A stratigraphic log. *Isr. Petrol. Inst. Rep.* 1/79, 1 p.

Druckman, Y. & Kashai, E. 1981. The Helez Deep and Devora 2A boreholes and their implication to oil prospects in pre-Jurassic strata in Israel. *Geol. Surv. Isr. Rep.* OD/1/81, 23 pp.

——, Gvirtzman, G. & Kashai, E. 1975. Distribution and environment of deposition of Upper Triassic on the northern margins of the Arabian Shield and around the Mediterranean Sea. *Proc. 9th Internat. Congr. Sedimentol. Nice. 1975.* Theme **5(1),** 183–92.

Freund, R., Garfunkel, Z., Zak, I., Goldberg, M., Weissbrod, T. & Derin, B. 1970. The shear along the Dead Sea Rift. *Phil. Trans. Roy. Soc. London,* A267, 107–30.

GINZBURG, A. & GVIRTZMAN, G. 1979. Changes in the crust and in the sedimentary cover across the transition from the Arabian Platform to the Mediterranean Basin: evidence from seismic refraction and sedimentary studies in Israel and in Sinai. *Sediment. Geol.*, **23,** 19–36.

GVIRTZMAN, G. & KLANG, A. 1972. A structural and depositional hinge-line along the Coastal Plain of Israel evidenced by magneto-tellurics: *Isr. Geol. Surv. Bull.* **55,** 18 pp.

KARCZ, I. & KEY, C. A. 1966. Note on the pre-Palaeozoic morphology of the basement in the Timna area (southern Israel). *Isr. J. Earth-Sci.* **15,** 47–56.

——, WEILER, Y. & KEY, C. A. 1971. Lithology and environments of deposition of the Amudei Shelomo Sandstone ('NSP') in Nahal Shani, Elat. *Isr.J. Earth-Sci.* **20,** 119–24.

METWALLI, M. H., PHILIP, G. & MOUSSLY, M. M. 1974. Petroleum-bearing formations in northeastern Syria and northern Iraq. *Bull. Am. Assoc. Petrol. Geol.* **58,** 1781–96.

PONIKAROV, V. P., KAZMIN, V. G., MIKHAILOV, I. A., RAZVALIAYEV, A. V., KRASHENINNIKOV, V. A. KOZLOV, V. V., SOULIDI-KONDRATIYEV, E. D. & FARADZHEV, V. A. 1966. *The Geological Map of Syria, 1:100,000, Explanatory Notes.* Syrian-Arab Republic, Ministry of Industry Damascus, 111 pp.

POWERS, R. W. 1968. *Lexique stratigraphique international.* 3 Asie, Fasc. 10b1-Arabi Saoudite. Centre Nation. Rech. Sci. Paris, 172 pp.

RIGO DE RIGHI, M. & CORTESINI, A. 1964. Gravity tectonics in foothills structure belt of southwest Turkey. *Bull. Am. Assoc. Petr. Geol.* **48,** 1911–37.

ROBERTSON, A. H. F. & WOODCOCK, N. H. 1980. Tectonic setting of the Troodos in the east Mediterranean. *In:* PANAYIOTOU, A. (ed.). *Ophiolites.* Proc. Int. Ophiolite Symp. Cyprus, 1979, 36–49.

SANDER, N. J. 1970. Structural evolution of the Mediterranean region during the Mesozoic era. *In:* ALVAREZ, W. & KLAUS, H. A. (eds) *Geology and History of Sicily.* Venice, Petroleum Exploration Society, Libya. 43–132.

SELLEY, R. C. 1972. Diagnosis of marine and non-marine environments from the Cambro-Ordovician sandstones of Jordan. *J. geol. Soc. Lond.* **128,** 135–50.

SOLIMAN, S. M. & EL-FETOUH, M. A. 1969. Petrology of the Carboniferous sandstones in west central Sinai. *J.U.A.R.* **13,** 61–143.

WEISSBROD, T. 1969a. The Palaeozoic of Israel and adjacent countries. Part 1: The subsurface Palaeozoic stratigraphy of southern Israel. *Isr. Geol. Surv. Bull.* **47,** 35 pp.

—— 1969b. The Palaeozoic of Israel and adjacent countries, Part 2: The Palaeozoic outcrops in southwestern Sinai and their correlation with those of southern Israel. *Isr. Geol. Surv. Bull.* **48,** 32 pp.

—— 1970a. The stratigraphy of the Nubian Sandstone in southern Israel. *Isr. Geol. Surv. Rep.* OD/2/70, 22 pp.

—— 1970b. 'Nubian Sandstone': Discussion. *Bull. Am. Assoc. Petrol. Geol.* **54,** 526–29.

—— 1976. The Permian in the Near East. *In:* FALKE, H. (ed.), *The Continental Permian in Central, West and South Europe.* D. Reidel Publishing Company, Dordrecht—Holland. 200–14.

—— 1981. The Palaeozoic of Israel and adjacent countries (lithostratigraphic study). *Isr. Geol. Surv. Rep. M.P. 600/81*, 276 pp. (in Hebrew, with English abstract).

WETZEL, R. & MORTON, D. M. 1959. Contribution a la geologie de la Transjordanie. *Notes et Mémoires sur le Moyen-Orient*, **7,** 95–191.

WOLFART, R. W. 1967. *Geologie von Syrien und dem Libanon.* Gebruder Bornträger, Berlin-Nikolassee. 326 pp.

YANKAUSKAS, T. V. & TALLI, S. 1978. The Palaeozoic of the Arab Republic of Syria (ARS). *Proceedings of the USSR Academy of Sciences, Geological Series*, **ii,** 92–7 (in Russian).

Appendix

Palaeozoic sections in the Levant, stratigraphic data and remarks

(See also Figs 1 and 2)

STATION 1—TABUQ—JAUF, NW SAUDI ARABIA

The sequence: Precambrian crystalline basement/Cambrian Saq Formation/Ordovician Saq and Tabuq formations/Silurian Tabuq Formation/Devonian Tabuq and Jauf formations/Lower Cretaceous Wasia Formation (Powers, 1968; see also Weissbrod, 1981).

Remarks: The Carboniferous was penetrated in the ST-8 borehole in north An-Nafud area (Berwath Formation, Lower—? Middle Carboniferous, Powers, 1968). Weissbrod (1981) suggests that this formation overlies the Devonian Jauf Formation and underlies the Permian Khuff Formation. It is assumed that the Carboniferous as well as the Permian was deposited over a wide area in Saudi-Arabia but was eroded in some places during the Early Cretaceous.

STATION 2—MUDAWWARA—WADI ARABA, S JORDAN

The sequence: Precambrian Saramuj Conglomerate and other sediments/Cambrian basal conglomerate, bedded arkose sandstones, dolomite-limestone-shale, massive brownish weathered sandstones/Ordovician massive whitish weathered sandstones, bedded brownish weathered sandstones, graptolite sandstones, *Sabellarifex* sandstones, *Conularia* sandstones/Silurian nautiloidea sandstones, worm burrows sandstones and red brown argillaceous sandstones/Lower Cretaceous massive white sandstones, varicoloured sandstones, (Bender, 1968; see also Weissbrod, 1981).

Remarks: Although Carboniferous outcrops were

not mapped in Jordan by Quennell and by Bender, R. C. Mitchell found in the Ras en-Naqeb area, S. Jordan, an outcrop of 300 m of Lower-?Middle Carboniferous with diagnostic fossils (in Weissbrod, 1981). Though this finding was not confirmed elsewhere, Weissbrod suggests that if the Carboniferous is in fact present, it might have been protected from the Lower Cretaceous erosion, as in some other locations, due to its low structural position. The Permian-Triassic was probably deposited in this area and was eroded during the Lower Cretaceous.

STATION 3—UM BOGMA—GEBEL FOKA, SW SINAI

The sequence: Precambrian crystalline basement/ Cambrian Amudei-Shelomo, Timna, Shehoret and Netafim formations/Carboniferous Um Bogma and Abu Thora formations/Permo-Triassic Budra Formation/Lower Cretaceous Amir and Hatira formations (Weissbrod, 1969b, 1981).

Remarks: According to Weissbrod (1981) the Netafim Formation is younger than Lower Cambrian and older than Lower Ordovician. However, a Cambrian age is much more probable than lowermost Ordovician. In the present paper, a Cambrian age was attributed to the Netafim Formation (see also stations 5, 6).

STATION 4—WADI ARABA—ABU DARAG, E EGYPT

The sequence: Carboniferous Rod el Hamal, Abu Darag and Aheimar formations/Permo-Triassic Qiseib Formation/Lower Cretaceous Malha Formation (Abdallah and Adindani, 1965; see also Weissbrod, 1981).

Remarks: The base is not exposed.

STATION 5—GEBEL DHALAL, E SINAI

The sequence: Precambrian crystalline basement/ Cambrian Netafim Formation/Lower Cretaceous Amir and Hatira formations. (Weissbrod, 1981).

Remarks: See the remarks on the Netafim Formation at Station 3. Contrary to the assumption of the presence of Carboniferous at stations 1 and 2, it is not assumed that the Carboniferous was deposited at this station, or at station 6, before the Permian. No Carboniferous was found in the vicinity of East Sinai and the Timna-Elat area; however, the absence of the Carboniferous at stations 5 and 6 is not conclusive.

STATION 6—TIMNA—ELAT, S NEGEV

The sequence: Precambrian crystalline basement/ Cambrian Amudei-Shelomo, Timna, Shehoret and Netafim formations/Lower Cretaceous Amir and Hatira formations (Weissbrod, 1969b, 1981).

Remarks: See the remarks on the Netafim Formation at station 3. See the remarks on the absence of Carboniferous at station 5.

STATION 7—ZERQA MAIN, CENTRAL JORDAN

The sequence: Cambrian Burj and Qunaya formations/Permian—Triassic Humrat Main Deltaic Formation (Wetzel and Morton, 1959; Weissbrod, 1976, 1981).

Remarks: The base is not exposed.

STATION 8—HAMEISHAR 1—SINAF 1 BOREHOLES, CENTRAL NEGEV

The sequence: Precambrian Zenifim Formation/ Cambrian Amudei-Shelomo, Timna and Shehoret formations (in Hameisher 1 the Cambrian is not differentiated)/Permian-Triassic Negev and Ramon groups (missing in Sinaf 1)/Lower Cretaceous Hatira Formation (Weissbrod 1969a, 1981).

STATION 9—SAFRA 1, N JORDAN

The sequence: Precambrian Zenifim Formation/ Cambrian Yam Suf Group/Permian—Triassic Negev and Ramon groups/Lower Cretaceous Hatira Formation (Weissbrod, 1981).

Remarks: The stratigraphy of Bender (1968) was revised by Weissbrod, (1981) using electric-log correlation with Hameishar 1 borehole in Israel.

STATION 10—NAHL 1 AND ABU HAMTH 1 BOREHOLES, CENTRAL SINAI

The sequence: Precambrian Zenifim Formation/ Permian Yamin and Arqov formations/Permian-Triassic Budra Formation (Weissbrod, 1976, 1981).

STATION 11—SUWEILAH 1, N JORDAN

The sequence: Precambrian Zenifim Formation/ Permian—Triassic Negev and Ramon groups/Lower Cretaceous Hatira Formation (Weissbrod, 1981).

Remarks: The stratigraphy of Bender (1968) was revised by Weissbrod (1981), using electric-log correlation with Lot 1 borehole in Israel.

STATION 12—NORTHERN NEGEV BOREHOLES

The sequence: Precambrian Zenifim Formation/ Permian—Triassic Negev and Ramon groups/Jurassic. (Weissbrod, 1969a, 1976, 1981).

STATION 13—HELEZ DEEP 1A, CENTRAL ISRAEL

The sequence: Precambrian schist/Jurassic dyke or sill of quartz-porphyry (180 Ma by K-Ar)/? Permian-Triassic Karmia Shale/Middle Triassic (Anisian) Erez Conglomerate and Mohilla Formation/ Jurassic. (Druckman and Kashai, 1981, absolute age: Steinitz, personal communication).

Remarks: Derin (1979) suggests some different lithostratigraphic boundaries but the ages are not different. There are no fossils in the Karmia Shale. The Erez Conglomerate contains Permian limestone fragments (Derin, 1979) which indicate the presence of Permian not very far from Helez.

STATION 14—KHANASSER 1, NW SYRIA

The sequence: Cambrian sandstone and limestone/ Ordovician Khanasser, Swab and Afendi formations/ Carboniferous sandstones/? Permian/Mesozoic (Yankauskas & Talli, 1978).

Remarks: In the Baflioun 1 borehole, 100 km northwest of Khanasser 1, an Ordovician age was attributed by Wolfart (1967) to a sequence which, in his opinion, resembles the Ordovician of Abba 1 (Station 16). However, Ponikarov *et al.* (1966) suggested a Late Proterozoic (Upper Riphaean) age for this sequence. In the model presented here, the sequence in Baflioun 1 is considered to be similar to the sequence in Khanasser 1. This assumption does not contradict any available information.

STATION 15—SWAB 1, E SYRIA

The sequence: Cambrian sandstone/Ordovician Khanasser, Swab and Afendi formations/Carboniferous sandstone/Lower Cretaceous (Yankauskas & Talli, 1978).

Remarks: Permian-Triassic sediments were probably removed during the Early Cretaceous erosion.

STATION 16—ABBA 1, N SYRIA

The sequence: Ordovician Swab and Afendi formations/Carboniferous sandstone/? Permian/Triassic (Yankauskas & Talli, 1978).

Remarks: The Cambrian was recorded at 10 stations. At 4 other stations, (10, 11, 12, 13) where the Cambrian is missing, the overlying Ordovician, Silurian, Devonian and Carboniferous are also missing. It is inferred that this is due to pre-Permian erosion. It is therefore assumed that when the Ordovician is present, it also overlies the Cambrian in the same location. Hence it is believed that in stations 16, 17 and 18 the Cambrian, though not penetrated, is present (see Fig. 2).

STATION 17—TANF 1, SE SYRIA

The sequence: Ordovician Khanasser, Swab and Afendi formations/Silurian Tanf Formation/Carboniferous sandstone/Lower Cretaceous (Yankauskas & Talli, 1978).

Remarks: Permian—Triassic sediments were probably removed during the Early Cretaceous erosion. See also the remarks for station 16.

STATION 18—MARKHADA 101, SE SYRIA

The sequence: Ordovician Swab and Afendi formations/Silurian Tanf Formation/Carboniferous sandstone and shale/Mesozoic (Yankauskas & Talli, 1978).

Remarks: See the remarks of station 16.

In the Qamishlya 1 borehole, 120 km north of Markhada 101, Ordovician and Silurian sediments were penetrated (Wolfart, 1967; Ponikarov *et al.*, 1966). In the El-Bouab 1 borehole, 30 km north of the Markhada 101, Wolfart (1967) reported a sequence, whose facies resembles the Silurian of Qamishlya 1 borehole. Ponikarov *et al.* (1966) also attributed a Silurian age to the bottom-sequence of El-Bouab 1. In the model presented here the sequence of the Qamishlya 1 and El-Bouab 1 boreholes is considered to be similar to the sequence of Markhada 101. This assumption does not contradict any known information.

GENERAL STRATIGRAPHICAL COMMENTS

A. *The List of Boreholes from Syria:* The boreholes *Qamishlya 1, Baflioun 1,* and *El-Bouab 1* were not included in this list above, since their Palaeozoic stratigraphy (Daniel, 1963; Wolfart, 1967) is not well established. The relevant information from these boreholes is included in the remarks for stations 14 and 18, situated nearby. The borehole *Dola'a 1* and *Doubayat 1* were not included in the same lists, since they bottomed in the Carboniferous. Ponikarov *et al.* (1966), however, raised the possibility that the lowermost part of these boreholes may be of Devonian age. This possibility is not supported by Daniel (1963), Wolfart (1967) or Yankauskas & Talli (1978). In any case, the relevant data for these boreholes would not have affected the contours on the maps of figures 4, 5 and 6.

B. *Presence or absence of the Carboniferous:* At 7 of the 18 stations, the Carboniferous is present today. In another 7 cases, there is no doubt that the Carboniferous was *absent before the Permian* since the Permian-Triassic overlies units older than Carboniferous. At the remaining 4 stations, the Carboniferous and the Permian-Triassic are absent today. The Permian-Triassic was assumed to be present in all cases. It is not clear whether the Carboniferous was present at these 4 stations before the Early Cretaceous removal. At stations 1 and 2, the Carboniferous was assumed to be present since Carboniferous was found in nearby locations. At stations 5 and 6, it is assumed, although doubtful, that the Carboniferous was removed before the Permian-Triassic. This again is based on information from nearby areas. It should be emphasized that even if the assumption is changed for stations 5 and 6 and the Carboniferous is considered as *present*, the original picture (Figs. 4, 5 & 6) would be changed only in a minor way.

GDALIAHU GVIRTZMAN and TUVIA WEISSBROD, Geological Survey of Israel, 30, Malkhei Yisrael Street, Jerusalem 95 501, Israel.

Permian-early Mesozoic tectonism and continental margin formation in Israel and its implications for the history of the Eastern Mediterranean

Z. Garfunkel & B. Derin

SUMMARY: The Early Mesozoic tectonic reorganization of the Tethys region also affected the Levant area. Here two processes were superimposed: (a) The pattern of long-wavelength vertical motions of the Arabo-Nubian platform changed in the Late Permian when Israel and nearby areas began to subside. (b) Rifting occurred in this subsiding area, probably in several phases: Late Anisian (and Ladinian?), Carnian-Norian, and Liassic. Differential movements reached 2–3 km and magmatism occurred in the Liassic and perhaps also in the Triassic. The tectonism was strongly felt up to 50 km landward of the present coast. The crust was thinned and modified under the present continental margin and in the main rifts. Passive margin conditions were established in Late Liassic times over the previously faulted area. Later in the Jurassic a carbonate shelf was constructed along this subsiding margin; its basinward edge was 1–1.5 km high, which shows that the adjacent SE Mediterranean was already a deep sea. Hence this basin and its passive margin are considered to have been shaped by the Early Mesozoic rifting which is recognized in Israel. The rifting process probably signifies oblique separation of the Tauride block from the Levant part of Gondwana.

In Early Mesozoic times the tectonic and palaeogeographic conditions changed fundamentally in the Tethyan realm: fragmentation of the Pangea super-continent began, motions of the major lithospheric plates changed, and new continental margins formed (Smith 1971; Dewey *et al.* 1973; Bernoulli & Lemoine 1980). It has been proposed that the Eastern Mediterranean basin also originated from Early Mesozoic rifting (Monod *et al.* 1974; Hsü 1977; Bein & Gvirtzman 1977; Biju-Duval & Dercourt 1980). Recent deep drilling in Israel has revealed evidence of eventful Early Mesozoic tectonism. Because of its nature and timing, this activity is here interpreted as a rifting process which produced the present passive margin of the SE Mediterranean basin. In addition, the pattern of the long-wavelength vertical motions within the neighbouring stable Arabo-Nubian platform was reorganized in Permian times. Combined, these events produced the Mesozoic tectonic pattern of the region.

The purpose of the present work is to summarize this chapter of earth movements in the Levant. We also attempt to place the local data in the regional context of Early Mesozoic continental breakup in the Tethyan region.

General setting

The crystalline basement of the Levant was formed in the Pan-African orogeny of Late Precambrian age (Engel *et al.* 1979; Bielski 1982). It is exposed around the Red Sea, and in the subsurface it extends at least as far as the Mediterranean coast where it was reached in the Helez Deep well (Fig. 1; Steinitz *et al.* 1981). In the subsurface of the Negev a successor basin developed and was filled with several kilometres of immature clastics and some volcanics (Weissbrod 1969).

Later on, in the Early Cambrian, the region became a stable platform and was covered by a veneer of mature sediments. The platformal history can be divided into two distinct stages which are characterized by different patterns of vertical movements. During the first stage, up to the Permian, the region was still a part of the Gondwana continent and its vertical movements were not related to the present margins of the Arabo-Nubian platform. During this first stage two major sedimentary series were deposited, separated by an unconformity, probably of Devonian age (Fig. 1). The Permian transition between the two main stages of the platformal history is marked by a major erosive unconformity whose subcrop map (Fig. 1A) reveals several long-wavelength undulations. In the main lows, in the Syrian Desert and in the Gulf of Suez region, more than 1000 m of Carboniferous sediments were preserved, whereas on the main highs in the Negev region and in central Arabia Precambrian rocks were exposed.

During the second, Late Permian-Mesozoic, stage of the platformal history, the present peripheral parts of the platform subsided strongly relative to its interior, and on them

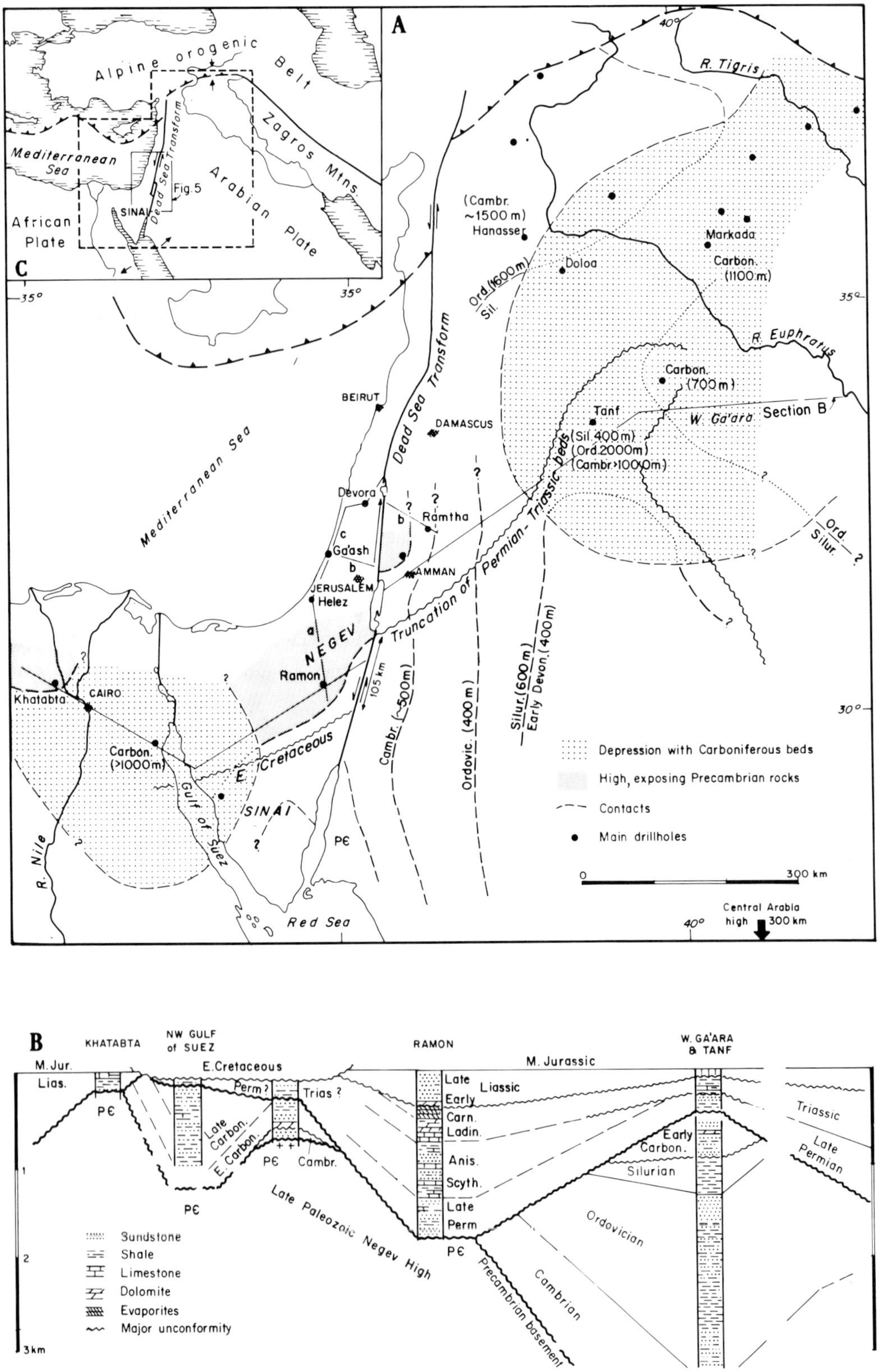

A
C
B
Alpine orogenic Belt
Zagros Mtns.
Arabian Plate
African Plate
Mediterranean Sea
Dead Sea Transform
Fig.5
SINAI
R. Tigris
R. Euphrates
(Cambr. ~1500 m)
Hanasser
Doloa
Markada
Carbon. (1100 m)
Ord.(600m)
Sil.
Carbon. (700 m)
Tanf
W. Ga'ara
Section B
(Sil.400m)
(Ord.2000m)
(Cambr.>1000m)
BEIRUT
DAMASCUS
Mediterranean Sea
Devora
Ramtha
Ga'ash
JERUSALEM
AMMAN
Helez
NEGEV
Ramon
Truncation of Permian-Triassic beds
105 km
Khatabta
CAIRO
Carbon. (>1000m)
E. Cretaceous
Gulf of Suez
SINAI
R. Nile
Red Sea
PЄ
Cambr. (~500m)
Ordovic. (400m)
Silur.(600m)
Early Devon.(400m)
Ord.
Silur.
Depression with Carboniferous beds
High, exposing Precambrian rocks
Contacts
Main drillholes
0
300 km
Central Arabia high 300 km
35°
40°
30°
KHATABTA
NW GULF of SUEZ
RAMON
W. GA'ARA & TANF
M.Jur.
Lias.
E.Cretaceous
M. Jurassic
Perm ?
Trias ?
Late
Early
Liassic
Carn.
Ladin.
Anis.
Scyth.
Late Perm
Late Carbon.
E. Carbon.
Cambr.
Early Carbon.
Silurian
Ordovician
Cambrian
Triassic
Late Permian
Late Paleozoic Negev High
Precambrian basement
Sandstone
Shale
Limestone
Dolomite
Evaporites
Major unconformity
1
2
3 km

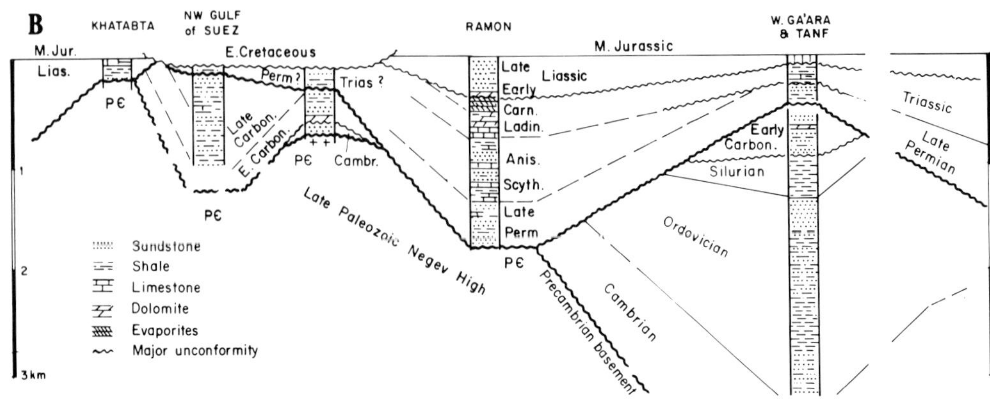

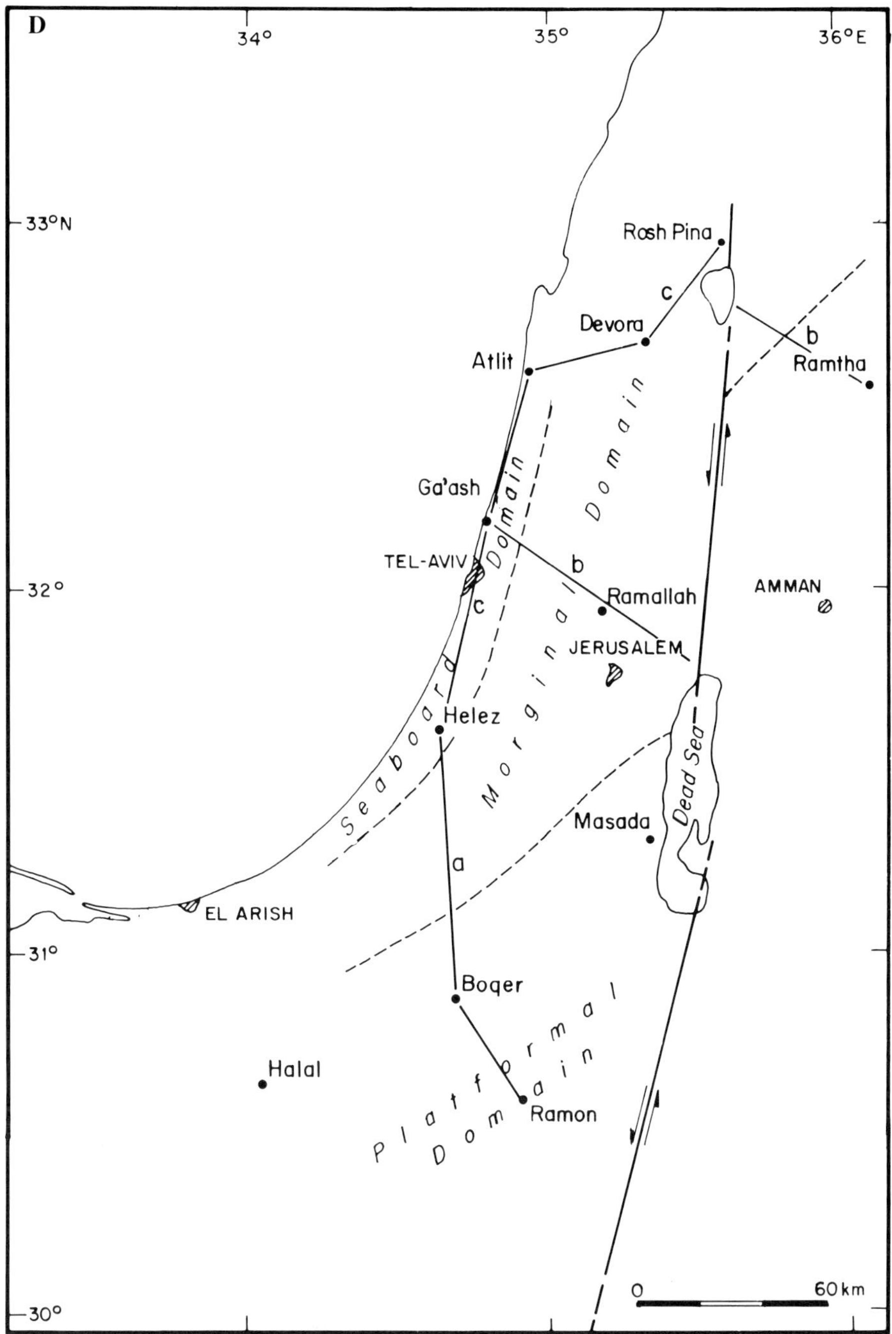

FIG. 1. The transition from the Palaeozoic to the Mesozoic regimes of the Arabo–Nubian platform.

(A) Subcrop map of the pre Late-Permian unconformity, showing the areas of Late Palaeozoic subsidence and uplift.

(B) Generalized cross-section (location shown in A) showing distrbution of pre-Middle Jurassic sediments. Data from Abdallah and el-Adindani 1965; Bender 1968; Murris 1980; Powers *et al.* 1966; Said 1962; Van Bellen *et al.* 1959; Yankauskas and Talli 1978.

(C) Location map and present plate configuration.

(D) Domains of sedimentation.

thick wedges of predominantly marine sediments were deposited in three domains of sedimentation (Fig. 1D): (1) Inland is the *platformal domain*, where the sections are rich in clastics, mainly sand, but the amount of clastics decreases towards the periphery of the platform while the amount of carbonate increases. (2) The *marginal domain* comprises the most subsided part of the platform, where thick sections consisting predominantly of carbonates accumulated on shallow-water shelves. On the side of the platform sloping towards the Mediterranean the transition from the platformal to the marginal domain is located 100–150 km inland from the present coast (Picard 1959; Garfunkel 1978), but it is much farther inland on the E flank of the Arabian peninsula (Murris 1980). During the main transgressions, sheets of carbonates were deposited far inland over the platform. (3) The *seaboard domain*, revealed by deep drilling under the coastal area and offshore from Israel, is characterized by important and abrupt facies and thickness changes and by unconformities. These record the tectonism which shaped this continental margin (see below), and the transition it underwent through continental slope and deep basin stages since the Middle Jurassic (Derin 1974; Bein & Gvirtzman 1977). Under the seaboard domain the crust thins and perhaps also becomes more mafic (Ginzburg & Folkman 1980; Folkman 1976), so that it is intermediate between the normal continental crust of the platform, and the semi-oceanic crust of the Eastern Mediterranean basin. A similar domain probably also existed along the E margin of the Arabo-Nubian platform (Murris 1980).

This pattern was modified after Late Cretaceous times when plate convergence and thrusting began in the Alpine orogenic belt bordering the Arabo-Nubian platform. The E margin of the platform was obliterated by the thrusts of the Zagros chain, but the Mediterranean margin was only mildly tectonized and its original features are still recognizable. The orogenic activity and plate convergence still continue, but in mid-Cenozoic times the Arabo-Nubian platform was rifted and broken into separate plates (Fig. 1C). Along one of the new plate boundaries, the Dead Sea transform, left-lateral slip of about 105 km took place (Quennell 1959; Freund *et al.* 1970).

Permian–Liassic earth movements in Israel and nearby areas

Knowledge of the Permian to Liassic history of this region comes principally from drill-holes, augmented by scarce outcrop data. Therefore, the main tool used to unravel the history and scope of earth movements is analysis of sediment thicknesses and of facies changes. In this way the overall picture is revealed even though the data come from widely spaced points. After the Late Palaeozoic uplift phase, the region was influenced by the super-position of: (a) regional long-wavelength vertical motions of the Arabo-Nubian platform, which acted since the Late Permian, and (b) more local but intense continental margin tectonism that probably operated in several pulses, from Late Anisian to Late Liassic times.

Late Palaeozoic uplift

The buried Negev High (Fig. 1) formed in the Permian during the transition between the two stages of Arabo-Nubian platform evolution, probably as a result of transient heating of the lithosphere. This is suggested by the finding by Kohn & Eyal (1981) that the fission-track ages of apatite and sphene in Precambrian basement rocks are reset to 245–215 Ma, i.e. to Late Permian or Triassic ages (van Eysinga 1978; Odin 1981). The thickness of sediments in the region (Fig. 1) shows that the reset samples were never covered by more than 1.5–2.5 km of Palaeozoic sediments. The fission-track age of sphene is reset only on heating to ca. 300°C (Naeser 1979). For the basement to reach 300°C, or even 200°C, under 1.5–2.5 km of sediments, heat flow must have been much higher than that of areas with a Late Precambrian basement. Subsequent cooling of the Negev High would then cause subsidence and sedimentation.

Late Permian–Liassic sediments of the platformal and marginal domain

The first phase of renewed sedimentation was of Late Permian to Late Triassic age. Its main features were summarized by Zak (1957), Druckman (1974), Picard & Flexer (1974), Weissbrod (1969, 1976), Bender (1968), Bandel & Khouri (1981) and in the palaeontological works of Parnes (1975), Hirsch (1976), and Derin & Gerry (1981). Sedimentation continued in the Jurassic (Goldberg & Friedman 1974; Derin 1974; Bandel 1981). In the platformal domain the section consists of vertical alternations and lateral interfingering of clastic-rich and clastic-poor units (Figs 2, 3) which record regressive and transgressive pulses. In the marginal domain the section is quite uniform and consists primarily of carbonates alternating with shales. Here the lithologic types and

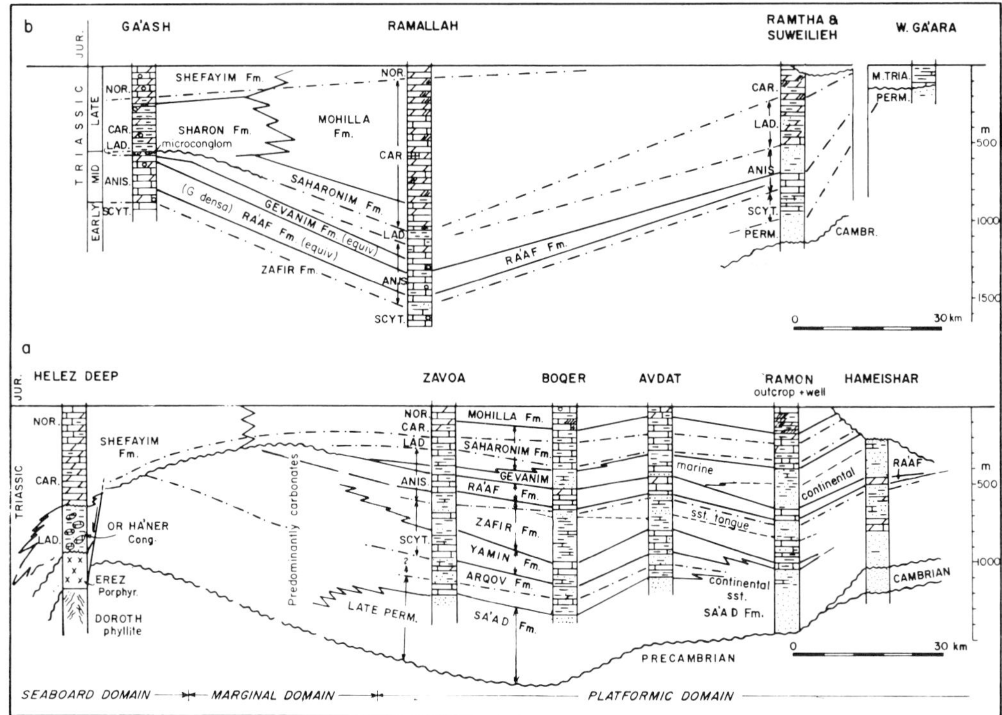

FIG. 2. Late Permian–Triassic sequences in Israel. For locations see Fig. 1 (see also Fig. 5). Lithologic symbols as in Fig. 1B. After Druckman (1974), Picard & Flexer (1974), Bandel & Khouri (1981), Derin & Gerry (1981), and unpublished reports by Derin & Gerry. The mainly carbonatic units of the platformic domain were deposited in normal shallow seas during transgressions when influx of clastics was reduced. The units of mixed lithologies were deposited in tidal flat and deltaic environments. The continental sandy units are mainly of fluviatile origin and formed during the most pronounced regressions. The sequences of the marginal domain formed on shallow-water shelves. *Glomospira densa* is a guide fossil of Middle Anisian Age (Derin & Gerry, 1981).

faunal assemblages mostly resemble those of the platformal domain, indicating that shallow water conditions were generally also maintained in the marginal domain.

Until the Late Triassic there are no major stratigraphic breaks, but short lacunas are probable (Hirsch 1976). The Permian–Triassic boundary occurs in a continental unit, where a hiatus is likely. The main Triassic transgression occurred in Late Anisian–Ladinian times. Later in the Ladinian, hypersaline conditions developed in both the platformal and marginal domains, leading in the Carnian to deposition of shallow water evaporites, mainly dolomite and anhydrite (Mohilla Formation). The youngest Triassic beds record more normal conditions and are probably of Norian age. The subsequent unconformity spans 15–10 Ma, when limited erosion and slight angular truncation occurred. Only subdued relief, locally karstic, was produced, and the surface was covered by a layer of varicoloured residual clays, locally pisolitic and leached (Mish'hor Formation). The first Liassic transgression (Ardon Formation) was quite widespread and led to the occasional development of restricted and evaporitic conditions (Fig. 3). The exact age of this transgression is uncertain, but is probably of Early Liassic age, in view of the great thickness of the sediments below well dated Domerian (Pliensbachian) beds (Qeren Member and equivalents, Derin 1974; Fig. 3); some fossils from the Ramon outcrop may allow even a Hettangian age (Parnes 1980). The main Jurassic transgression began in Bathonian times, when predominantly carbonate sedimentation again extended far into the platformal domain.

The distribution of clastics and general facies trends in the Negev and east of the Dead Sea (Druckman 1974; Goldberg & Friedman 1974; Bandel 1981) show that a permanent continental area existed in the S and SE. In these directions the section is truncated by Early Cretaceous erosion (Figs 1A, 2, 3). However,

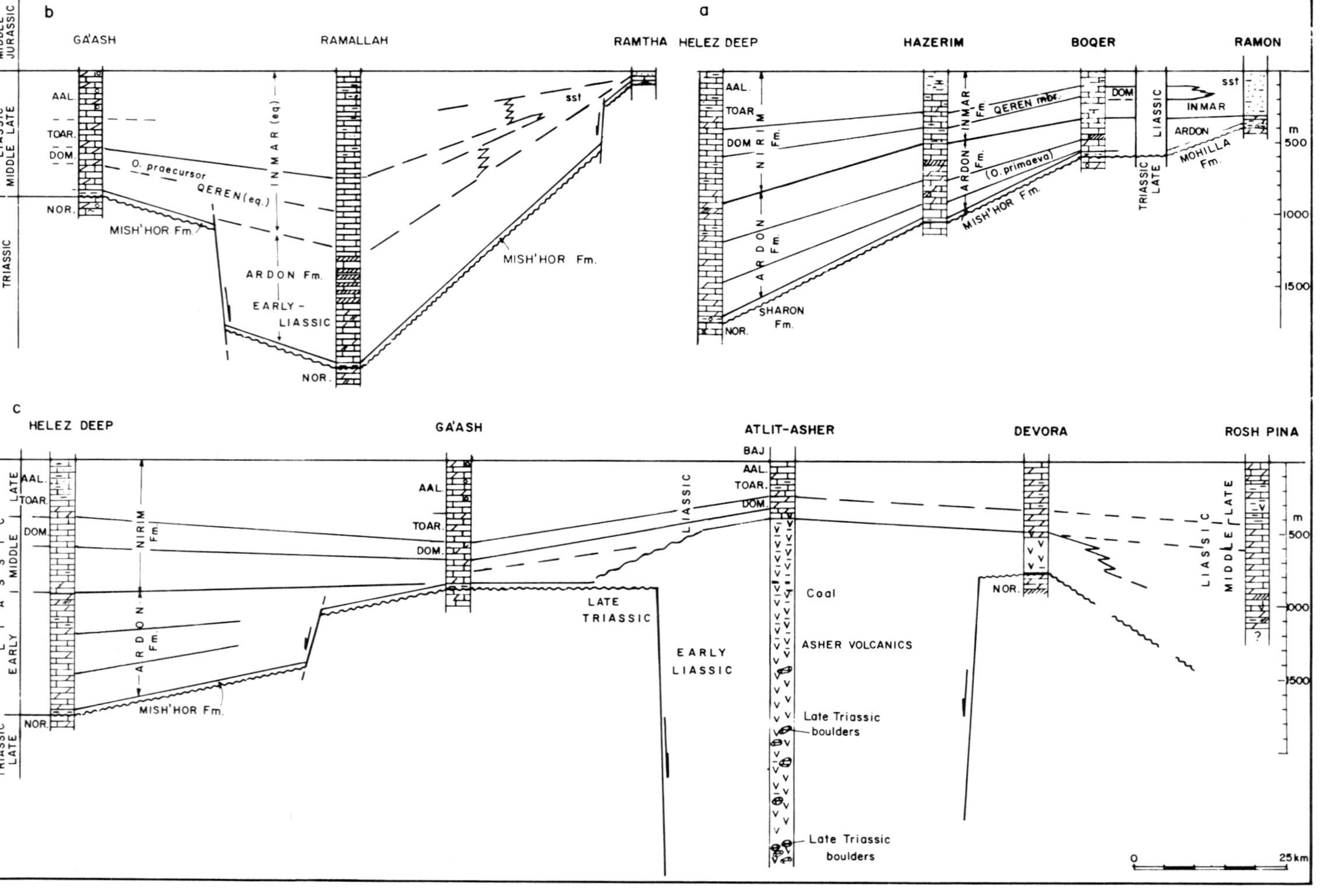

FIG. 3. Liassic cross-sections in Israel. For locations see Fig. 2A (see also Fig. 5), lithologic symbols as in Fig. 1B. After Derin (1974), and unpublished reports. Significance of lithologic units as explained in Fig. 2. *Orbitopsella praecursor* is a guide fossil for Qeren Member times (Domerian), and *Orbitopsella primaeva*—for the Lower Ardon Formation times (Derin, 1979).

as the truncated sections are thick and without signs of rapid pinching-out, it is likely that originally they extended several hundred kilometres beyond their present limits. A continental area, with sandy sedimentation, also existed in central Sinai during Late Permian–Triassic times, according to Druckman (1974). Still further west, in N Egypt, Permo-Triassic sediments are absent, and Jurassic (Late Liassic) beds lie directly on pre-Permian rocks (Fig. 1; Said 1962; Barakat 1970). Thus the thick Late Permian–Triassic sections occupy an embayment bulging into the platform in the Negev (Fig. 1B), whereas thick Late Liassic and younger Jurassic sequences extend westward along the present Mediterranean margin of the Arabo–Nubian platform.

History of subsidence of the platformal domain

The thick Late Permian to Late Liassic sections, reaching 2.5 km in the Negev, record rapid subsidence and sedimentation. In fact it is here, in Scythian–Anisian times, that the fastest sedimentation rate attained during the Mesozoic occurred, reaching an average of ca. 75 m/Ma, even before correction for post-depositional compaction (Fig. 4). This is also fast in comparison with other cratonic basins (Schwab 1976). In the entire studied region higher sedimentation rates occurred only where continental margin tectonism was strong (see below).

In the present case sediment thicknesses essentially record subsidence relative to sea level, as most of the section consists of very shallow marine sediments. However, for a better insight it is necessary to isolate the various factors whose combination determines the original sediment thickness. These are (e.g. Steckler & Watts 1978): (1) the tectonic subsidence; (2) eustatic changes of sea level; (3) the isostatic effect of loading by sediments. It is easy to see that removing the isostatic effect, i.e. replacing the sediments by *air*, leaves a residual subsidence which is the sum of the first two effects, provided the sediments always extended exactly to sea level. This condition is closely approached in the platformal domain, as indicated by the nature of the sediments. Hence the residual subsidence in the Negev can be estimated (Fig. 4), and from this the tectonic subsidence is obtained by subtracting a eustatic sea-level component, using for example the estimates of Vail *et al.* (1977). The results show that at first the tectonic subsidence was rapid, but it then slowed from the Ladinian on. The tectonic subsidence could have been proportional to $\sqrt{}$time (Fig. 4, model A), implying

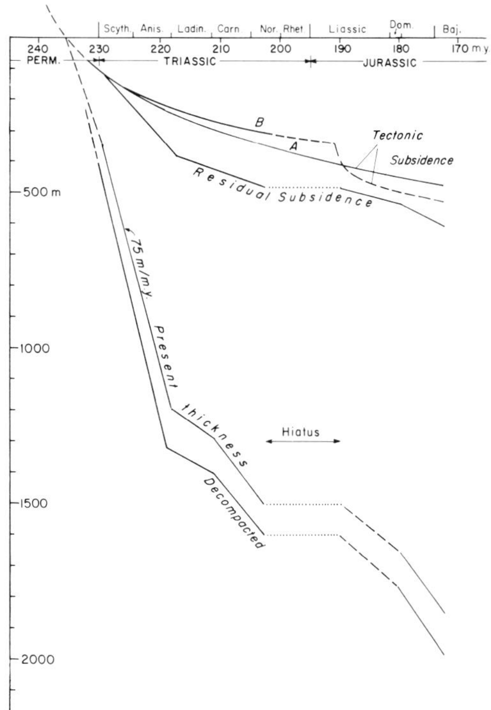

FIG. 4. Late Permian to Liassic subsidence in the central Negev. For discussion see text. Time scale after van Eysinga, 1978. Decompaction calculated with the method of Perrier & Quiblier, 1974.

that it resulted essentially from cooling of the lithosphere since the Late Permian. In this model the Late Triassic–earliest Jurassic unconformity is interpreted as resulting from an eustatic oscillation of the sea level. The alternative model B in Fig. 4 is prompted by the occurrence of an important Early Liassic transgression in the studied region, and the absence of a transgression of this age on the E side of the Arabo–Nubian platform (Powers *et al.* 1965; Setudehnia 1978). The Early Liassic is also not a time of important transgression on a worldwide scale (Hallam 1981). Therefore, we favour model B in which the tectonic subsidence of the Negev was accelerated in the Early Liassic. This is probably a weak echo of the very fast Early Liassic subsidence of the adjacent marginal domain.

The succession in the platformal domain thins gradually to the SE (Figs 2, 3), which shows that the area was tilted as a whole towards the present Mediterranean coast. This analysis thus actually records the history of tilting of the platformal domain.

Continental margin tectonism

Late Anisian earth movements (Fig. 5A).

Until late in the Anisian (Ra'af Formation times, Fig. 2) local tectonism is not recognizable in the studied area. Beds of this and of older ages encountered in Ga'ash-2, Ramallah and Devora-2A wells resemble the coeval beds of the northern Negev in thickness, lithology and faunal assemblages, suggesting a rather uniform regional subsidence. From Late Anisian times, however, intense differential movements occurred in the seaboard and marginal domains, and were also felt weakly in the platformal domain.

In the northern Negev the general trend of regional tilting and thickening of sediments to the NW was temporarily reversed in the Late Anisian. As a result the Gevanim Formation (Fig. 2) thins by 150–200 m over a distance of about 40 km perpendicular to the facies belts strike. If this trend can be extrapolated to the NW, then deep parts of the section closer to the present coast should have been exposed in a structural high (Figs 2, 5A). Further north, in Ga'ash-2 well, the Late Anisian–Ladinian section is reduced compared with those in the Ramallah and Devora-2A wells located further inland (Fig. 2). Moreover, at Ga'ash-2 the older Anisian carbonate sedimentation was interrupted by the formation of a micro-conglomerate that was followed by shale and sand deposition. These features are taken as evidence for an unconformity and formation of a structural high close to the present coast.

The most important Late Anisian tectonism is recorded in the Helez Deep well. Here the ca. 350 m thick Or-Haner Conglomerate (Fig. 2) consists of dolomite and limestone clasts which contain fossils of Late Permian to Anisian age (Ra'af Formation times), but not of younger ages, and is followed closely by Carnian beds. Extrapolation of the thickness trends of the Negev shows that the carbonate-rich section from which the clasts were derived was more than 0.5 km, perhaps 1 km thick. Moreover, the clasts are poorly sorted, and range up to boulder size, which indicates derivation from a nearby source with high relief. These observations are taken as evidence for Late Anisian to Ladinian faulting, with a displacement of 0.5–1.0 km, that created a prominent fault-scarp near the Helez drillhole, probably to the east of it. A similar interpretation was proposed also by Druckman & Kashai (1981) and Druckman (this volume). This high was most likely the continuation of the feature recognized in the Negev, and probably continued toward the stuctural high at Ga'ash (Fig. 5A). In detail the local structure at Helez is not clear, however. The main constraint is the proximity of the Precambrian basement to the conglomerate, though the latter does not contain any clastic material derived from the basement. Possibly they were juxtaposed by faulting after deposition of the conglomerate, and the initial faulting did not in fact expose the basement. This implies that the drill-hole actually crosses a major fault. Alternatively, the conglomerate could have covered very low-lying exposures of the basement which were hardly eroded.

The great amplitude of the tectonism inferred from the Helez area, and the widespread occurrence of coeval differential movements, are taken as evidence for an important tectonic event which we interpret as the *beginning of rifting*.

Carnian and Norian earth movements (Fig. 5B)

The great variations in thickness and facies of the sediments of this time interval prove widespread and strong syn-sedimentary tectonism. This led to major subsidence of the marginal and seaboard domains as well as to important differential movements that extended into the platformal domain. These movements produced silled basins in which the evaporitic Mohilla Formation was deposited (Figs 2, 5B).

The strongest subsidence produced a basin in which the Ramallah well penetrated about 1200 m of mostly Carnian with some Norian strata. The thicknesses of older beds in the two wells show that rapid subsidence began close to the Ladinian–Carnian boundary. West of this interior basin was a shallower sill, penetrated by Ga'ash-2. Here the lower part of the Carnian sequence (Sharon Formation, Fig. 2) consists of alternating shale, very fine grained dolomite, and some siltstone and sandstone, in which palynomorphs are the only fossils. The clastics may have been derived from a still higher part of the sill. The overlying beds (Shefayim Formation), which range up to Norian in age, also differ from the coeval dolomitic-evaporitic part of the Mohilla Formation. Their fauna and lithology indicate a high energy, back-reef or shallow lagoonal environment with oolitic shoals, resembling Alpine Triassic facies (Derin & Gerry 1981). Similar Norian limestones are found as blocks in Liassic volcanics in Atlit-1 (Fig. 3). These occurrences probably signify the proximity of a reefal belt which separated the interior evaporitic basin from the open sea in the west. At Helez the Carnian section is as thick as at Ga'ash-2, but more dolomitic. Therefore, a shallow sill west of Helez is post-

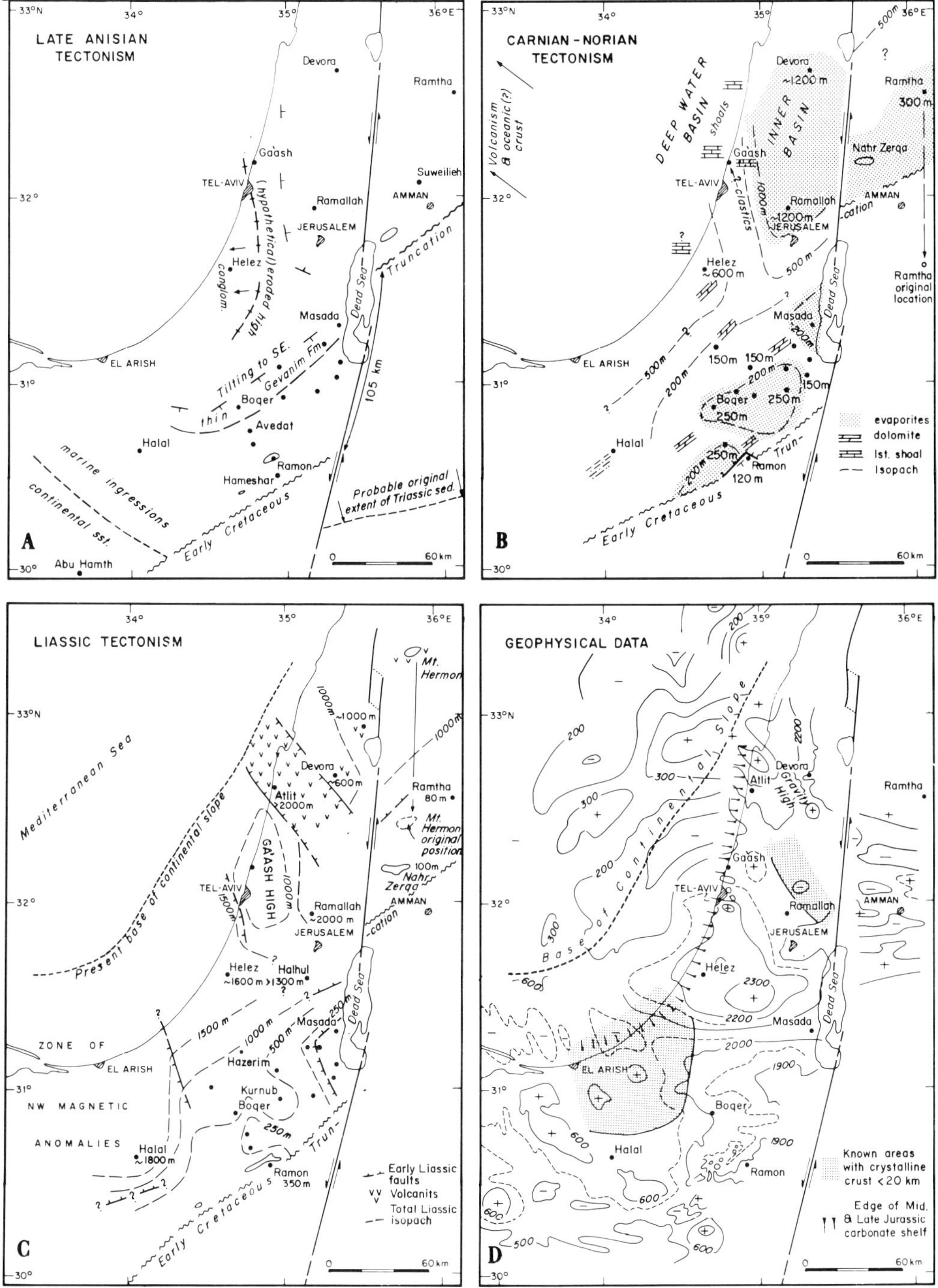

FIG. 5. The Early Mesozoic tectonic elements in Israel and geophysical data. For details see text. Sources of information (wells and outcrops) shown.

(A) Known Late Anisian tectonic features.

(B) Known Carnian–Norian tectonic features. The important thickness changes are probably fault controlled.

(C) Known Liassic (mainly early Liassic) tectonic features. This figure is to be compared with D.

(D) Geophysical data: Magnetic anomalies after Folkman & Assael, 1981, & Hatcher *et al.* 1981. Crustal structure after Ginzburg & Folkman, 1980; shown only where data exist (not shown at sea).

ulated. The Carnian rates of overall subsidence and of differential movement in the seaboard and marginal domains were high, reaching average values of about 100 m/Ma and 50 m/Ma respectively, in terms of present sediment thicknesses. Such rates are exceptional for platformal basins, but are usual for movements of fault blocks in rifted areas and in continental margins (Schwab 1976).

In contrast with the marginal and seaboard domains, the Carnian subsidence of the platformal domain slowed down. This domain was also affected by differential movements that are recorded as thickness variations of the Mohilla Formation (Fig. 5B) which is well exposed in the Ramon Valley (Zak 1957; Garfunkel 1964) and also known from the subsurface (Druckman 1974). As these variations are accompanied by transitions from thin dolomitic sequences to thick evaporite-bearing sequences, syn-sedimentary movements are inferred. In the Ramon outcrop these changes are seen to be fault-controlled. Therefore, the thickness variations in the subsurface are also assumed to be controlled by faulting and by mild tilting of blocks. As the Ladinian units are very uniform in the Negev, it seems that the Carnian movements signify a distinct tectonic pulse, and are not the continuation of the Late Anisian activity.

Liassic earth movements (Fig. 5C)

The basal Liassic marine beds are of nearly the same age in the platformal domain as in the more strongly subsided Helez and Ramallah areas, according to a local biozonation (Gerry, pers. communication). Hence, the Early Liassic sea inundated a rather flat area, which implies a reduction in tectonism after the Late Triassic. Renewed tectonism later in the Early Liassic (mainly Ardon Formation times, Fig. 3) produced differential movements reaching an amplitude of 2 km, that were accompanied by important volcanism. In the Late Liassic, however, local movements virtually ceased, and regional subsidence, increasing towards the Mediterranean, became the dominant feature.

In the Early Liassic, a rift-like trough was formed in the north. Its fill, drilled at Atlit-1, consists of more than 2 km of basic volcanics, becoming acid downward, interbedded with subordinate sediments that include some coal beds (Fig. 3). Blocks of Norian limestone found among the volcanics indicate nearby relief, probably a fault-scarp. The NW trend of the structure is inferred from geophysical data (see below). Further inland the fill of the trough consists of sediments. In Ramallah, at the margin of the trough, the drill-hole crossed ca. 1 km of shallow-water carbonates, mainly dolomite, and some anhydrite. NE of the Atlit trough (Fig. 5C) was a less actively subsiding area where the Devora-2A drillhole crossed a thin volcanic sequence, while SW of the trough was the Ga'ash High where Early Liassic rocks are absent (Fig. 3). These structures formed before the Domerian, when they were all buried by a uniform blanket of shallow-water limestones. Other strongly subsiding regions included the Helez area and in particular, NE Sinai (Druckman 1977). More than 1 km of carbonates of a restricted environment, with some evaporites, were found in the Halal and Helez drillholes. In the latter, a body of quartz-porphyry located beneath Triassic beds, yielded an Early Liassic age (Steinitz *et al.* 1981), implying that the body was intrusive. However, its very fine-grained matrix casts some doubt on this conclusion. The platformal domain also subsided during the Early Liassic, but much less than the other areas. The amplitude of local movements in this domain hardly reached 200 m, as for example between the Masada block and the adjacent Kurnub low which were probably separated by a fault (Fig. 5C).

Though the pattern of the Early Liassic structures resembles the Carnian structures, there are significant differences. For instance, the Masada block (Fig. 5) cuts across the Carnian structure and the depositional history in Devora-2A was also different in the two periods. Such cases demonstrate the independent nature of the Early Liassic tectonism.

The nature of the earth movements had changed markedly by the Domerian (Pliensbachian) (Qeren member times; Fig. 3): local differential movements practically stopped, and the entire area was covered by uniform shallow-water marine beds which pass into continental sandstones in the platformal domain. In contrast to the earlier movement pattern, general subsidence and tilting towards the present coast, expressed by sediment thicknesses, now also occurred further to the west. Therefore the wedge of Late Liassic and younger Jurassic sediments extends from Israel through Sinai into northern Egypt (Barakat 1970), where Permian–Triassic beds are absent. The new regime of vertical movements signifies the end of rifting in the studied area, and the production of a stable subsiding passive margin. Along this margin a bank of shallow-water carbonates was constructed. Its steep edge, reaching a height of ca. 1.5–2 km by the end of the Jurassic, marked the transition to the deep Eastern Mediterranean basin (Fig. 5D) (Derin 1974; Cohen 1976).

Relation to crustal structure (Fig. 5D)

The interpretation of the Early Mesozoic activity as rifting and extensional tectonics implies that it changed the crustal structure of the studied area. In fact, the published seismic refraction, aeromagnetic and gravity studies (Folkman 1976; Ginzburg & Folkman 1980; Folkman & Assael 1981) show a general thinning of the crust towards the Mediterranean, which is here interpreted as a result of the Early Mesozoic rifting. In addition, more local geophysical features can be related to the Early Mesozoic structures discussed above.

Seismic refraction reveals two areas with a thinner than normal crust and an exceptionally thick sedimentary cover. Remarkably, these areas coincide with the areas of maximal Early Liassic (and in part also Carnian) subsidence (compare Fig. 5D with 5B, 5C). It is inferred that in these areas the crust was thinned by the Early Mesozoic extension. Accordingly, the extent of thin crust can be used to delineate the Ga'ash High and the trough NE of it. There is no direct evidence of the mechanism of crustal thinning, but by analogy with, for example, the Suez rift and the Bay of Biscay (Garfunkel & Bartov 1977; Montadert *et al.* 1979), it is assumed that the crust was attenuated by listric-normal faults with associated tilting of blocks. Deep intrusions may have also been important, especially in the Atlit trough.

The strongly subsided areas also have other distinct geophysical characteristics: the Atlit trough is marked by NW-trending strong positive magnetic and gravity anomalies which most likely originate from its volcanic fill and possibly also from deeper intrusions (Folkman 1976; Ben Avraham & Hall 1977). These anomalies reveal the shape of the trough on land, and also show that it continues some distance into the sea (Fig. 5). The strongly depressed area in NE Sinai is marked by low-amplitude magnetic anomalies with a distinct NW trend, which differ from the magnetic anomalies in the adjacent areas. These anomalies may originate from the basement, or from Early Mesozoic features, but the important point is that they define a distinct crustal unit which extends from well inland to as far as the continental slope (Fig. 5D). These observations show that in the studied area there is no abrupt change in crustal structure at the present coast-line. An abrupt crustal boundary can only exist beyond the base of the continental slope, where a large positive magnetic anomaly is located (Folkman 1976).

Only the E-W magnetic gradient south of Jerusalem (Fig. 5D), which is associated with a gravity gradient (Folkman, 1976), is not related to an identifiable Early Mesozoic structure. Neither does it follow any important change in crustal thickness (Ginzburg & Folkman 1980). Moreover, E of the Dead Sea transform, where Early Mesozoic tectonism was hardly felt, E-W trending magnetic anomalies are prominent, and seem to have been offset by the transform (Hatcher *et al.* 1981). Therefore, the feature south of Jerusalem most likely originates from a lithologic change in the Precambrian basement.

Discussion

In the foregoing description the prominent Early Mesozoic thickness and facies changes, subsidence, and magmatism in the Levant region were interpreted in terms of syn-sedimentary extensional tectonics and rifting. This fits well with the local data, and is strongly supported by the observation that similar activity happened contemporaneously throughout the Tethys area (Scandone 1975; Bernoulli & Lemoine 1980; Argyriadis *et al.* 1980). On this regional scale the Triassic and Liassic rifting and volcanism are a long prelude to the separation of Gondwana from Laurasia and the opening of the central Atlantic Ocean between them, which became important only towards the Middle Jurassic. In reconstructions of the Triassic Tethys the Tauride block is usually placed opposite the Levant and Egypt (Fig. 6), following Smith (1971) and Dewey *et al.* (1973). This is reasonable, because the 'calcareous axis' (i.e. the autochthon) of the Taurides is a fragment of Gondwana: it has a Precambrian basement and a Palaeozoic sedimentary cover similar to the Arabo-Nubian platform (Ricou *et al.* 1975; Gutnic *et al.* 1979). The precise original position of this block is poorly constrained, but it may be noted that in its western part (Anamas Mountains) Carboniferous beds overlie the basement (Gutnic *et al.* 1979). In this it resembles western Sinai and northern Egypt (Fig. 1), but no other place. These areas could therefore have been originally opposite one another (Fig. 6), implying that the Tauride block must have rotated after its detachment, or that it was much deformed internally. Information about the detachment of this block comes from the allochthonous complexes of Antalya (S. Turkey), Mamonia (Cyprus) and Baër-Bassit (NW Syria) which were probably derived from the basin that opened when the Tauride block separated from the Arabo-Nubian platform (Monod *et al.* 1974; Robertson & Woodcock 1980).

Within this framework, the significance of the Early Mesozoic events in the studied area can be evaluated. Here strong subsidence took

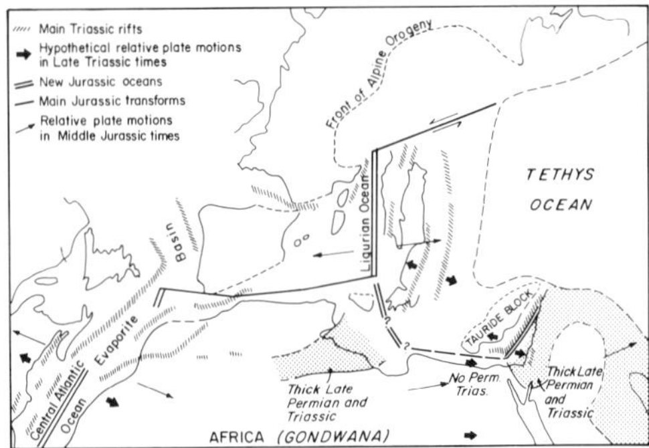

FIG. 6. Triassic and Jurassic rifting in the Tethys. According to Scandone, 1965; Van Houten 1977; Jansa *et al.* 1980; Bernoulli & Lemoine, 1980.

place 15–20 Ma before the onset of rifting in Late Anisian times. Still earlier this area had been uplifted in the Late Palaeozoic, when the pattern of vertical movement of the entire Arabo–Nubian platform changed. A direct relation between these events is difficult to define, but the Late Palaeozoic uplift (and heating, perhaps by a hot spot) could have weakend the lithosphere and thus determined the site of its future failure.

The evidence from the Helez Deep borehole, and the presence of Anisian–Ladinian volcanics and breccias in the Antalya Complex (Delaune-Mayere *et al.* 1977, Robertson & Woodcock 1982) show that important faulting occurred in the Middle Triassic. Submarine rifts may have formed at that time, but the similarity of the Ladinian faunas of the Levant and the Tauride block (Hirsch 1976) argues against much separation between them. Separation became important, in the Late Triassic, according to evidence from the Antalya, Mamonia, and Baër-Bassit allochthonous complexes. These contain Carnian-Norian deep-water sediments and submarine basic volcanics, all overlain by condensed Jurassic and Cretaceous deep-water deposits (Monod *et al.* 1974; Delaune-Mayere *et al.* 1977; Robertson & Woodcock 1980, 1982), implying that a deep basin, perhaps with an oceanic-like crust, was opened in the Late Triassic between the Levant and the Tauride block. The (now) southern margin of the latter was thus shaped in the Late Triassic. The contemporaneous activity in Israel is thus just a marginal manifestation of this important rifting event. In this regard the absence of similar features in northern Egypt calls for special attention. If the Tauride block were detached from there, one would expect formation of a passive subsiding margin in Egypt, as in other places. A possible explanation is that the motion between the Tauride block and Egypt was essentially strike-slip in a roughly E-W direction (Fig. 6).

This history fits the regional context (e.g. Bernoulli & Lemoine 1980): Middle Triassic rifting and volcanism occurred as far west as the Apulian block, but as in the Eastern Mediterranean area it did not lead to plate fragmentation, and many structures were short-lived. Late Triassic rifting was more widespread, extending as far west as North America, and in the Atlantic Ocean it produced wide evaporite basins (Van Houten 1977; Jansa *et al.* 1980). The pattern of the structures is compatible with left-lateral shear between Gondwana and Laurasia. Such a motion on a much larger scale, and perhaps in a slightly different direction, led in the Jurassic, to the opening of the Central Atlantic, and the smaller Ligurian oceans (Fig. 6). Throughout, rifts would be dominantly N-S, and strike-slip faults E-W, as inferred for the Eastern Mediterranean.

It is of note that the Early Liassic activity occurred at all in the Levant area. If the Late Triassic rift had continued to widen by sea-floor spreading, activity would be expected to have become confined to its centre, and to have ceased at its margins. The Early Liassic phase of tectonism and magmatism in the Levant implies that the Late Triassic rifting stopped, or slowed, and that the centre of renewed activity shifted eastward. This phase ended abruptly with the formation of a subsiding passive margin stretching from the Levant to Egypt, and the creation, presumably, of a widening basin between the Tauride block and the Arabo–Nubian platform.

The separation of the Tauride block was evidently oblique and not pure transform motion as first envisaged by Dewey *et al.* (1973) (Fig. 6). In addition, the geophysical data (Fig. 5D) show that the continent-ocean boundary is at least at the base of the present continental slope, and possibly much further out to sea. Rifting may well have occurred episodically along several axes, stranding continental blocks in the Eastern Mediterranean. One such may be the Eratosthenes seamount which has the magnetization (Ben Avraham *et al.* 1976) expected from an Early Mesozoic palaeopole position (Morel and Irving 1980). The seamount could be founded on Triassic or Liassic volcanics.

These observations support the interpretation of the SE Mediterranean as a relic of a Mesozoic basin (e.g. Hsü 1977; Biju-Duval & Dercourt 1980). The main relevant points are: (a) Important early Mesozoic rifting, magmatism and crustal thinning occurred in the area. (b) By Late Jurassic times a deep basin bordered by a carbonate platform with a high slope existed off Israel. (c) The Tauride block seems to have been detached from this platform in the Early Mesozoic. The remaining unanswered question is the extent of the Jurassic basin, and the amount of Early Mesozoic sea-floor spreading and stretching of older continental crust. If the interpretation of the Eratosthenes seamount is correct, it implies that the ca. 200 km wide Levant basin between the coast and this seamount was essentially shaped in the Early Mesozoic. It may be underlain in part by strongly stretched continental crust and by volcanics formed during the rifting process. Cretaceous sea-floor spreading may have occurred only N and NW of this basin.

ACKNOWLEDGEMENTS. We are very grateful to Y. Langotzky (Oil Exploration (Investments) Ltd.) for pushing us to start this study, and to E. Barak & J. Fisher (Israel National Oil Co.) for permission to publish subsurface data. F. Hirsch, L. E. Ricou, Z. Reches, E. Sass & A. Starinsky reviewed the paper and offered helpful comments.

References

ABDALLAH, A. M. & EL-ADINDANI, A. 1965. Stratigraphy of Upper Paleozoic rocks, western side of Gulf of Suez. *UAR Geol. Surv., Bull.* **25,** 18 pp.

ARGYRIADIS, I., DE GRACIANSKY, P. C., MARCOUX, J. & RICOU, L. E. 1980. The opening of the Mesozoic Tethys between Eurasia and Arabia-Africa. *Bur. Res. Géol. Min. Mem.* **115,** 199–214.

BANDEL, K. 1981. New Stratigraphical and Structural Evidence for lateral dislocation in the Jordan Rift Valley connected with a description of the Jurassic Rock column in Jordan. *N. Jb. Geol. Paläont. Abh.* **164,** 271–308.

—— & KHOURI, H. 1981. Lithostratigraphy of the Triassic in Jordan. *Facies*, **4,** 1–26.

BARAKAT, M. G. 1970. A stratigraphical review of the Jurassic Formations in Egypt and their oil potentialities, *7th Arab. Petrol. Congr.* **2,** paper 58 (B-3), 14 pp.

BEIN, A. & GVIRTZMAN, G. 1977. A Mesozoic fossil edge of the Arabian plate along the Levant coastline and its bearing on the evolution of the Eastern Mediterranean. *In:* BIJU-DUVAL, B. & MONTADERT, L. (eds.) *Structural history of the Mediterranean basins,* Editions Technip (Paris) 95–110.

BEN AVRAHAM, Z. & HALL, J. K. 1977. Geophysical survey of Mount Carmel Structure and its extension into the Eastern Mediterranean. *J. Geophys. Res.* **82,** 793–802.

BEN AVRAHAM, Z., SHOHAM, X. & GINZBURG, A. 1976. Magnetic anomalies in the Eastern Mediterranean and the Tectonic Setting of the Erasthenes Seamount. *Geophys. J. R. astr. Soc.* **45,** 105–123.

BENDER, F. 1968. *Geologie von Jordanien.* Gebrüder Borntraeger, Berlin, 230 pp.

BERNOULLI, D. & LEMOINE, M. 1980. Birth and early evolution of the Tethys: the overall situation. *Bur. Res. Géol. Min. Mem.* **115,** 168–179.

BIELSKI, M. 1982. *Stages in the evolution of the Arabian-Nubian massif in Sinai.* Thesis Ph.D., Hebrew Univ., Jerusalem, Israel, 155 pp. (in Hebrew, English abs.).

BIJU-DUVAL, B. & DERCOURT, J. 1980. Les bassins de la Méditerranée orientale représentent-ils les restes d'un domain océanique, la Mésogée, ouvert au Mésozoique et distinct de la Téthys? *Bull. Soc. géol Fr.* (7), **22,** 43–60.

COHEN, Z. 1976. Early Cretaceous buried canyon: Influence on accumulation of hydrocarbons in Helez Oil Field. *Bull. Amer. Assoc. Petroleum Geol.* **60,** 108–114.

DELAUNE-MAYERE, M., MARCOUX, J., PARROT, J. F. & POISSON, A. 1976. Modèle d'évolution mésozoïque de la paléomarge téthysienne au niveau de nappes radiolaritiques et ophiolitiques du Taurus Lycien, d'Antalya et du Baër-Bassit. *In:* BIJU-DUVAL, B. & MONTADERT, L. (eds.) *Structural history of the Mediterranean basins*, Editions Technip (Paris) 79–94.

DERIN, B. 1974. *The Jurassic of central and northern Israel.* Thesis Ph.D., Hebrew Univ., Jerusalem, Israel. 152 pp. (in Hebrew, English summary).

—— & GERRY, E. 1981. Late Permian–Late Triassic stratigraphy in Israel and its significance to oil exploration. *Isr. Geol. Soc., Proc. Symposium on 'Oil Exploration in Israel'*, 9–10.

DEWEY, J. F., PITMAN, C. C. III, RYAN, W. B. F. & BONNIN, J. 1973. Plate tectonics and the evolution of the Alpine system. *Geol. Soc. Amer. Bull.* **84,** 3137–3180.

DRUCKMAN, Y. 1974. The stratigraphy of the Triassic sequence in southern Israel. *Geol. Surv. Isr., Bull.* **64,** 62 pp.

—— 1977. Differential subsidence during the deposition of the Lower Jurassic Ardon Formation in Western Jordan, Southern Israel and Northern Sinai. *Isr. J. Earth Sci.* **26,** 45–54.

—— & KASHAI, E. 1981. The Helez Deep and Devora 2A boreholes, and their implications to oil prospects in Israel. *Geol. Surv. Isr. Report* OD/1/81, 23 pp.

ENGEL, A. E. J., DIXON, T. H. & STERN, R. J. 1980. Late Precambrian evolution of Afro-Arabian crust from ocean arc to craton. *Geol. Soc. Amer. Bull.* **91,** 699–706.

FOLKMAN, Y. 1976. *Magnetic and gravity investigations of the crustal structure in Israel.* Thesis Ph.D., Tel Aviv Univ., Israel, 203 pp. (in Hebrew).

—— & ASSAEL, R. 1981. *Magnetic map of the south-eastern Mediterranean Sea*, scale 1 : 250 000. The Survey of Israel.

FREUND, R., GARFUNKEL, Z., ZAK, I., GOLDBERG, M., WEISSBROD, T. & DERIN, B. 1970. The shear along the Dead Sea rift. *Phil. Trans. R. Soc. London*, **A267,** 107–130.

GARFUNKEL, Z. 1964. *Tectonic problems along the Ramon lineament.* Thesis M.Sc., Hebrew Univ., Jerusalem, Israel. 68 pp. (in Hebrew).

—— 1978. The Negev—Regional Synthesis of Sedimentary basins. *10th Int. Congr. Sedimentology,* Jerusalem, Excursion A2, 35–110.

—— & BARTOV, Y. 1977. Tectonics of the Suez Rift. *Geol. Surv. Isr. Bull.* **71,** 44 pp.

GINZBURG, A. & FOLKMAN, Y. 1980. The crustal structure between the Dead Sea Rift and the Mediterranean Sea. *Earth Planet. Sci. Lett.* **51,** 181–188.

GOLDBERG, M. & FRIEDMAN, G. M. 1974. Paleoenvironments and paleogeographic evolution of the Jurassic system in southern Israel. *Geol. Surv. Isr. Bull.* **61,** 44 pp.

GUTNIC, M., MONOD, O., POISSON, A. & DUMONT, J. F. 1979. Géologie des Taurides Occidentales (Turquie) *Soc. géol. Fr., Mém.* **58,** 112 pp.

HALLAM, A. 1981. A revised sea-level curve for the early Jurassic. *J. Geol. Soc. Lond.,* **138,** 735–743.

HATCHER, A. D. JR., ZIETZ, I., REGAN, R. D. & ABU-AJAMIEH, M. 1981. Sinistral strike-slip motion on the Dead Sea rift: Confirmation from new magnetic data. *Geology,* **9,** 458–462.

HIRSCH, F. 1976. Sur l'origine des particularismes de la faune du Trias et du Jurassique de la plateforme africano-arabe. *Bull. Soc. géol. Fr.* (7), **18,** 543–552.

HSÜ, J. K. 1977. Tectonic evolution of the Mediterranean basins. *In:* NAIRN, E. M., KANES, W. H. & STEHLI, F. G. (eds.) *The Ocean Basins and Margins*, **4A,** 29–76, Plenum Press.

JANSA, L. F., BUJAK, J. P. & WILLIAMS, G. L. 1980. Upper Triassic salt deposits of the western North Atlantic. *Can. J. Earth Sci.*, **17,** 547–559.

KOHN, B. P. & EYAL, M. 1981. Late Paleozoic–Late Tertiary evolution of the Sinai crystalline basement: Evidence from fission track dating of apatites and spherens. *Abs. ECOG VII*, Jerusalem.

MONOD, O., MARCOUX, J., POISSON, A. & DUMONT, J. F. 1974. Le domaine d'Antalya, témoin de la fracturation de la plate-forme africaine au cours du Trias. *Bull. Soc. géol. Fr.* (7), **16,** 116–125.

MONTADERT, L., ROBERTS, D. G., DE CHARPAL, O. & GUENNOC, P. 1979. Rifting and subsidence of the northern continental margin of the Bay of Biscay. *Init. Rep. DSDP*, **48,** 1025–1060. U.S. Government Printing Office, Washington D.C.

MOREL, P. & IRVING, E. 1981. Paleomagnetism and the evolution of Pangea. *J. Geophys. Res.* **86,** 1858–1872.

MURRIS, R. J. 1980. Middle East: stratigraphic evolution and oil Habitat. *Amer. Assoc. Petrol. Geol., Bull.* **64,** 597–618.

NAESER, C. W. 1979. Fission-Track Dating and Geologic Annealing of Fission Tracks. *In:* JAEGER, E. & HUNZIKER, J. C. (eds.) *Lectures in Isotope Geology*, 154–169. Springer Verlag.

PARNES, A. 1975. Middle Triassic ammonite biostratigraphy in Israel. *Geol. Surv. Isr., Bull.* **66,** 44 pp.

—— 1980. Lower Jurassic (Liassic) Invertebrates from Makhtesh Ramon (Negev, Southern Israel). *Isr. J. Earth Sci.* **29,** 107–113.

PERRIER, R. & QUIBLIER, J. 1974. Thickness changes in sedimentary layers during compaction; methods for quantitative evaluation. *Amer. Assoc. Petrol. Geol., Bull.* **58,** 507–520.

PICARD, L. 1959. Geology and oil exploration of Israel. *Bull. Res. Council Isr.* **G-8,** 1–30.

—— & FLEXER, A. 1974. Studies on the stratigraphy of Israel: The Triassic. *Isr. Inst. Petrol.*, Tel Aviv, Israel, 62 pp.

POWERS, R. W., RAMIREZ, L. F., REDMOND, C. P. & ELBERG, E. J. JR. 1966. Geology of the Arabian Peninsula: sedimentary geology of Saudi Arabia. *U.S. Geol. Surv. Prof. Paper*, **560-D,** 147 pp.

QUENNELL, A. M. 1959. Tectonics of the Dead Sea Rift. *Int. Geol. Congr. 20 (Mexico), Assoc. Serv. Geol. Afr.*, 385–405.

ODIN, G. S. 1981. Isotopic dating of Mesozoic rocks and the Numerical Time Scale. *Abs. Proc. ECOG VII*, Jerusalem.

RICOU, L. E., ARGYRIADIS, I. & MARCOUX, J. 1975. L'axe calcaire du Taurus, un alignement de fenêtres arabo-africaines sous les nappes radiolaritiques ophiolitiques et métamorphiques. *Bull. Soc. géol. Fr.* (7), **17,** 1024–1044.

ROBERTSON, A. H. F. & WOODCOCK, N. H. 1980. Tectonic setting of the Troodos massif in the east Mediterranean. *In:* PANAIOTOU, A. (ed.) *Ophiolites*, Proc. Int. Oph. Symp. Cyprus, 1979, 36–49.

——, —— 1982. Sedimentary history of the southwestern segment of the Mesozoic-Tertiary Antalya continental margin, south-western Turkey. *Eclogae geol. Helv.* **75,** 517–562.

SAID, R. 1962. *The Geology of Egypt.* Elsevier, 377 pp.

SCANDONE, P. 1975. Triassic seaways and the Jurassic Tethys Ocean in the central Mediterranean area. *Nature*, **256,** 117–120.

SCHWAB, F. L. 1976. Modern and ancient sedimentary basins: Comparative accumulation rates. *Geology*, **4,** 723–727.

SETUDEHNIA, A. 1978. The Mesozoic sequence in south-west Iran and adjacent areas. *J. Petrol. Geol.* **1,** 57–82.

SMITH, A. G. 1970. Alpine deformation and the oceanic areas of the Tethys, Mediterranean and Atlantic. *Geol. Soc. Amer. Bull.*, **82,** 2070–2089.

STECKLER, M. S. & WATTS, A. B. 1978. Subsidence of the Atlantic-type continental margin off New York. *Earth Planet. Sci. Lett.* **41,** 1–13.

STEINITZ, G., BIELSKI, M. & STARINSKY, A. 1981. The extension of the Arabian shield beneath the coastal plain of Israel in light of Rb-Sr age determinations. *Abs. ECOG VII*, Jerusalem.

VAIL, R. P., MITCHUM, R. M., THOMPSON, S., TODD, R. G., SANGREE, J. B., WIDMIER, J. M., BUBB, J. N. & HATTFIELD, W. G. 1977. Seismic stratigraphy and global changes of sea level. *Mem. Amer. Assoc. Petrol. Geol.* **26,** 49–50.

Van Bellen, R. C., Dunnington, H. V., Wetzel, R. & Morton, D. M. 1959. *Lexique Strat. Int.* **3,** *Asie*, pt. **10a,** *Iraq*. 333 pp.

Van Eysinga, F. W. B. 1978. *Geological time scale*, 3rd ed. Elsevier.

Van Houten, F. B. 1977. Triassic–Liassic deposits of Morocco and Eastern North America: Comparison. *Amer. Assoc. Petrol. Geol., Bull.* **61,** 79–99.

Weissbrod, T. 1969. The Paleozoic of Israel and adjacent countries, part I: The subsurface Paleozoic stratigraphy of southern Israel. *Geol. Surv. Isr., Bull.* **47,** 35 pp.

—— 1976. The Permian in the Near East. *In:* Falke, H. (ed.) *The Continental Permian in Central, West and South Europe, D. Reidel,* 200–214.

Yankauskas, T. V. & Talli, S. 1978. The Paleozoic of the Syrian Arabic Republic. *Izv. AN SSSR, Ser. Geol.*, **11,** 92–97. (In Russian).

Zak, I. 1957. *The Triassic in Makhtesh Ramon.* Thesis M.Sc., Hebrew Univ., Jerusalem, Israel, 98 pp. (In Hebrew).

Z. Garfunkel, Inst. Earth Sciences, Hebrew University, Jerusalem, Israel.
B. Derin, The Israel National Oil Co., Tel Aviv, Israel.

Evidence for Early-Middle Triassic faulting and possible rifting from the Helez Deep Borehole in the coastal plain of Israel

Yehezkeel Druckman

SUMMARY: A 600 m thick conglomerate, the Erez conglomerate, was penetrated in the Helez Deep borehole, drilled in the Coastal Plain of Israel. This conglomerate is underlain by a 40 m thick sandy shale of probable Early Triassic age, and is overlain by 190 m of Late Triassic peritidal dolomites. This sedimentary sequence is underlain by metamorphic Precambrian basement.

The Erez conglomerate is polymictic, consisting of various types of carbonate fragments, of ages varying from Upper Permian to Anisian. The fragments are angular to subrounded, poorly sorted, ranging in diameter from fractions of a millimetre to several decimetres. The matrix consists of black argillaceous micrite, yielding palynomorphs of Scythian-Anisian age.

It is suggested that the Erez Conglomerate is analogous to the thick, well-developed accumulation of polymictic conglomerates along the margins of the Miocene Suez and Plio-Pleistocene Dead Sea rifts, and thus indicating a major faulting phase and possibly an initial stage of rifting during Early-Middle Triassic times. Although an Upper Palaeozoic updoming prior to rifting may be considered, it is not a necessary postulate to explain the borehole data.

The nature and origin of the Eastern Mediterranean basin, as well as the N-S trending coast line, have attracted much research. As a result, numerous hypotheses have been proposed during the last decade, most of them based on geophysical data, because of lack of direct geological observation. Few boreholes in the Eastern Mediterranean have penetrated beyond the Tertiary and almost none have penetrated the entire Cretaceous. Thus the Helez Deep borehole, which is located only some 11 km from the Mediterranean shoreline (Fig. 1) and which penetrated the entire sedimentary section down to the crystalline Precambrian basement is of particular interest. It provides basic geological information that may aid in the evaluation of some of the ideas proposed concerning the early history of the eastern Mediterranean.

The Pre-Jurassic platform sedimentary sequence in Israel

The Precambrian basement in Israel consists of both metamorphic and igneous rocks overlain by a thick Late Precambrian pile of arkoses (Zenifim Formation) more than 2000 m thick (Weissbrod, 1969). The entire Palaeozoic section is missing, except for the Permian which directly overlies the Infracambrian arkoses (Weissbrod, 1969; Gvirtzman & Weissbrod, this volume). At the southernmost tip of Israel, relics of a Cambrian sequence occur and consist of fluvial and shallow marine sediments.

The Permian section consists of a few hundred metres of clastic and carbonate rocks (Weissbrod, 1969, 1976). The Sa'ad Formation, Lower Permian in age, consists of quartz arenites, interpreted as being distal deposits of fluviatile systems, alternating with carbonaceous shales which were deposited in marshes or lagoons. The Arqov Formation of Upper Permian age consists of alternating sandstone, shale and carbonate sequences believed to have accumulated in nearshore marine environments (Weissbrod, 1981). This rock sequence gradually thickens and becomes more marine in character towards the north and northwest. Its nature in central and northern Israel is unknown, as it is overlain there by several kilometres of Mesozoic rocks.

The overlying Triassic section is more than 1000 m thick in southern Israel and exceeds 2000 m in northern Israel (Druckman, 1974a). While the Triassic in southern Israel consists of alternating clastic, carbonate and evaporitic sections, reflecting transgressions and regressions of the Tethys, in the north an almost continuous shallow marine sequence of carbonates and evaporites was recorded (Druckman & Kashai, 1981).

The Negev (southern Israel) Triassic section can be viewed as a series of transgressive-regressive cycles (Druckman, 1974b), the first of which (Scythian-Lower Anisian) includes the Yamin and Zafir Formations (Fig. 2), which consist of shallow marine carbonates and sand-bodies, nearshore and lagoonal mudstones and sandstones, and possibly deltaic complexes.

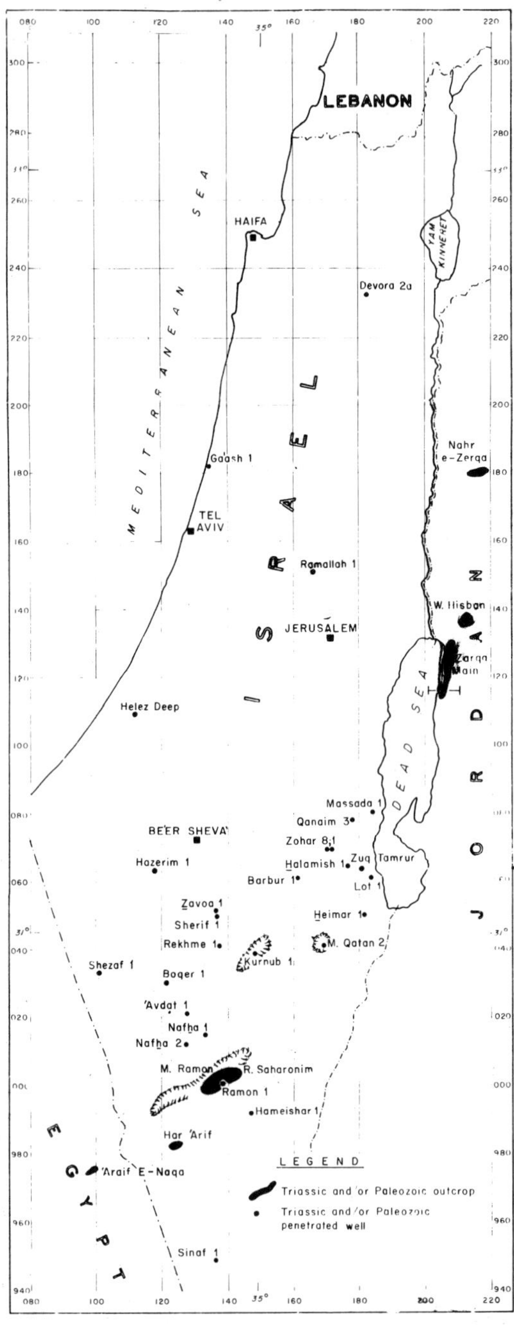

FIG. 1. Location map of Palaeozoic and/or Triassic sections in outcrops and boreholes.

The second cycle (early Upper Anisian) includes the Ra'af Formation and Gevanim Formation and consists of shallow marine carbonate (Ra'af Fm.), and fluvial, fluvio-deltaic and nearshore tidal sandstones, siltstones and mudstones (Gevanim Fm.). The third cycle (late Upper Anisian-Carnian) is the most extensive, both in duration and area; it includes the Saharonim and Mohilla Formations. The sedimentary sequence of this cycle consists of shallow marine carbonates, tidal flats and lagoonal carbonates and evaporites (Druckman, 1976). A fourth cycle, terminating the Triassic sedimentary sequence may have taken place during late Carnian-Norian times. However, insufficient palaeontological data at present hinder definition of the duration of this cycle. It begins with shallow marine carbonates, a few tens of metres thick, and ends with a regional emergence of a low relief that has exposed the entire Triassic terrain to karst solution and lateritization in Late Triassic-Early Jurassic times (Druckman, 1974a; Druckman, Hirsch & Weissbrod, 1982).

The Pre-Jurassic section in the Helez Deep Borehole

Unlike the previously described Palaeozoic and Triassic sections, the Helez section is unexpected, and different from the widely distributed, well known, documented sections from all over Israel (Druckman, 1974; Druckman & Kashai, 1981). Its stratigraphy is given below (Fig. 2), following Druckman & Kashai (1981).

The Dorot Schist (interval: 5946–6086 m T.D.) (Derin, 1979)

The Dorot Schist is a dark grey muscovite-chlorite schist, which has undergone several phases of deformation. In addition to muscovite, chlorite and ore minerals, it consists of quartz and plagioclase. This rock-type crystallized under low grade metamorphism of the greenschist facies as deduced from its mineralogy (determination by O. Amit, 1980, personal communication).

Age

Radiometric dating (Rb-Sr) of the Dorot Schist yielded a Precambrian to Early Cambrian age. Whole rock dating revealed 636 ± 107 million years Ma. (Steinitz, 1981). Similar ages have been reported from igneous and metamorphic rocks of the Arabo-Nubian Massif.

Gevim Quartz Porphyry (interval: 5769–5946 m)

The Gevim Quartz-Porphyry is reddish, pink or light grey in colour. It is composed of

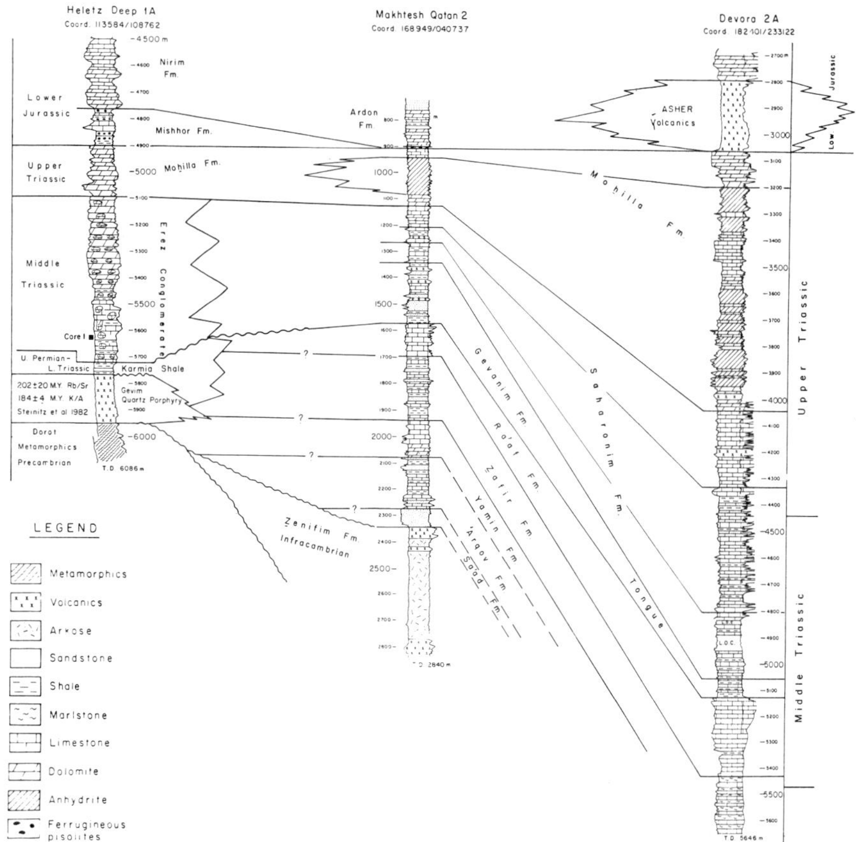

FIG. 2. Stratigraphic correlation of typical pre-Jurassic sections in southern and northern Israel. Note the completely different sequence, including a 600 m thick conglomerate of early Middle Triassic age, penetrated in the Helez Deep borehole (after Druckman & Kashai, 1981).

feldspar and quartz phenocrysts up to several millimetres in size set in a very fine crystalline matrix. Abundant pyrite and calcite crystals are scattered throughout the rock. In core 2 (interval: 5828–5831 m) a 15 cm thick section was found, consisting of dark brown quartz-porphyry with scattered clasts of quartz-porphyry and green clay up to 1 cm in size.

Age

Radiometric dating of the Gevim Quartz Porphyry yielded a Late Triassic-Early Triassic age (Rb-Sr 202 ± 20 Ma; K-Ar: 184 ± 4 Ma, G. Steinitz, personal communication).

Karmia Shale (interval: 5726–5768 m)

This 43 m thick unit consists of dark grey calcareous shale, often silty and sandy. This shale is intercalated by several metres of fine to medium-grained quartzitic sandstones, angular to subangular and well cemented. Pressure solution contacts among grains are common; some calcite cement is also often present.

Age

Upper Permian to Lower Triassic. This age was defined from palynomorphs (Cousminer, written communication, 1981).

The Erez Conglomerate (interval: 5090–5726 m)

The Erez Conglomerate consists of angular to subrounded packed carbonate fragments ranging in size from 0.5 m to a fraction of a millimetre (Figs. 3–6). The conglomerate is polymictic and comprises various carbonate types: light grey lumpy micrites with abundant stylolites; oosparites, consisting of well-sorted spherical oolites with concentric rims a few hundred microns in diameter; various types of bioclastic limestones, reddish ostracod and gastropod sparites, grey pelecypod sparites,

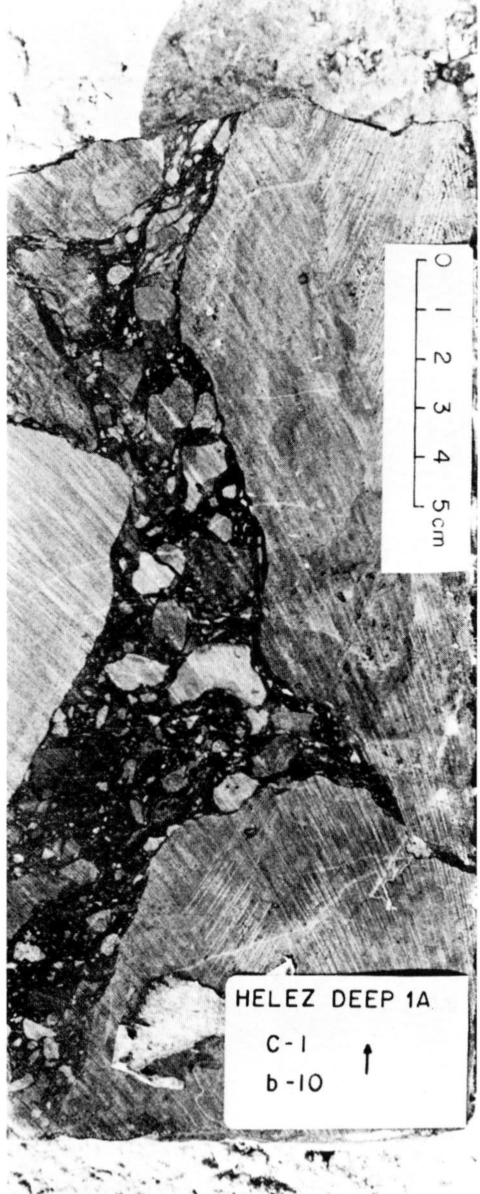

FIG. 3. Photograph of slabbed core of the Erez Conglomerate. Decimeter-sized fragments of dense, stylolitized micrite; interfragmental voids filled with carbonate fragments of various lithologies cemented by dark argillaceous carbonate mud. (Core 1, box 10).

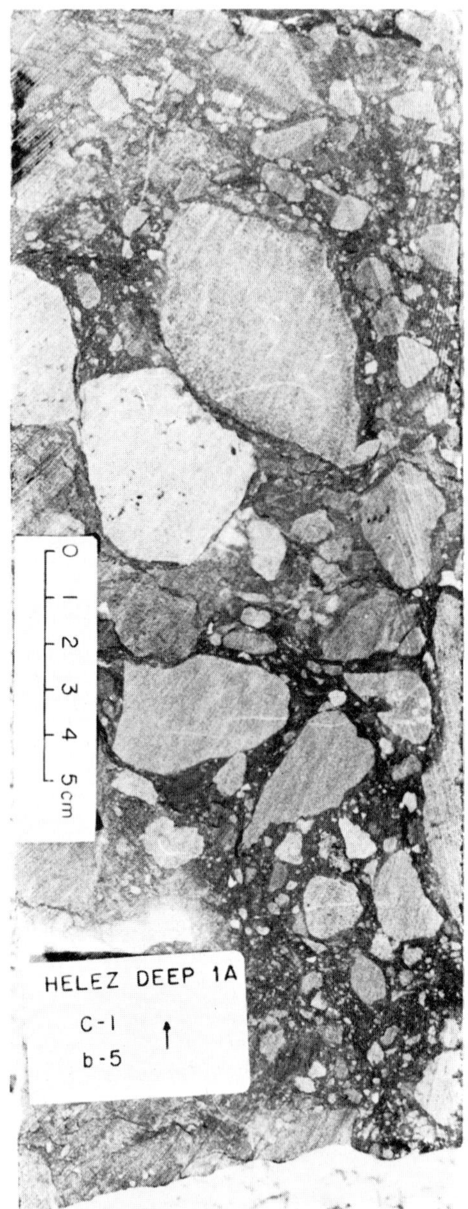

FIG. 4. Photograph of slabbed core of the Erez Conglomerate. Bimodal carbonate fragments of various lithologies tightly cemented by dark argilaceous carbonate mud. (Core 1, box 5).

foraminiferal micrites, laminated silty micrites; and black limy, silty shale. Above 5450 m, the conglomerate becomes monomictic and the fragments are a dense mosaic of xenotopic dolomite crystals, a few tens of microns in size. Peldolomicrites are also common. A few occurrences of recrystallized oodolomicrites, biodolomicrites with bivalves, gastropods and algal debris were also recorded.

The matrix in the lower part of the Erez Conglomerate (below 5450 m) consists of an argillaceous micrite that is occasionally silty.

FIG. 5. Photography of slabbed core of the Erez Conglomerate. Note calcite veins (lower centre of photograph), not extending beyond fragment into the matrix and adjacent fragments, indicating a significant burial history prior to erosion and deposition within the conglomerate. (Core 1, box 5).

FIG. 6. Photography of slabbed core of the Erez Conglomerate. Note graded fine carbonate particles. The graded laminae are disrupted and distorted by a fragment which fell into semi-consolidated matrix sediment. (Core 1, box 18).

Since the fragments are closely packed, the amount of matrix is small. An exception was found in core 1A, box 18, where a 10 cm thick layer of graded and folded matrix was found (Fig. 6). There, the matrix consists of laminated micritic fragments ranging from a few microns to several hundred microns, in several recurring graded cycles. Above 5450 m the matrix consists of a reddish-brownish ferruginous argillaceous, silty dolomicrite. Core 1 recovered a 25 cm long piece of fractured dolomicrite. The fractures are filled with argillaceous material and prismatic quartz crystals, growing perpendicularly to the fracture walls. In addition to this core, fractures were observed in several thin sections of cuttings, again filled with argillaceous and ferruginous material. A few horizons, 7–10 m thick, of reddish brown silty shale are intercalated with the Erez Conglomerate, roughly separating the lower and upper parts.

The upper boundary and the continuity of the Erez Conglomerate are questionable. The difficulty in defining the boundary arises from the fact that only two cores (cores 1, 1a) were taken from this interval, one of which (core 1) recovered only 25 cm. Electric logs did not help in this respect. Therefore, the description of the fragmented nature of the rocks is based on thin sections of cutting chips. Several conglomerate intervals were observed, the highest of which was found at 5090 m. It is therefore suggested that the interval between 5090–5373 m consists of alternating layers of conglomerates and dolomite layers, whereas the interval below 5373 m consists of continuous conglomerates intercalated by reddish brown silty shale at the interval 5440–5483 m. The dolomitic interlayers in the upper part consist of dolomitized pelmicrite, algal and mollusc debris and micron- to decimicron-sized hipidiotopic dolomite mosaic. It is thus suggested that the Erez Conglomerate accumulated in a shallow marine environment.

Age

The age of the Erez Conglomerate is Middle Triassic. The latest age obtained from fauna in the conglomerate fragments is Anisian (*Glomospira densa, Meandrospira* cf. *dinarica*). Palynomorphs obtained from the matrix also indicate an Anisian age, whereas the fragments contain palynomorphs ranging from Permian to Anisian (Cousminer, written communication, 1981; Derin, 1979).

The Mohilla Formation (4900–5090 m)

The Mohilla Formation in the Helez Deep borehole is composed of dolomites, usually dolomitized muds and pelmicrites. Some oomicrites and algal and molluscan debris were found in the sequence, a few samples of xenotopic and hipidiotopic dolomites, several tens to a hundred microns in diameter, were also observed.

This 190 m thick section correlates with the Mohilla Formation in the Negev.

Age

Carnian—based on palynomorph assemblages (Cousminer, written communication, 1981).

The Mishhor Formation (4760–4900 m)

The Mishhor Formation consists of three distinct units: an upper (4760–4812 m) and a lower unit (4850–4900 m), consisting of red brown claystones with ferruginous pisolites and some intercalations of limestone layers, separated by a thick limestone middle unit (4812–4850 m), consisting of pelmicrites, some oomicrites and some intrabiosparites. In a few of the samples ferruginous vadose pisolites are observed; in others, tiny ferruginous clay particles are found. It is thus inferred that subareal exposure and lateritization took place in the Helez area as it did in southern and central Israel.

Age

Although the regional emergence and lateritization, which is well known in the entire Negev and Jordan, is still not precisely dated, it occurred somewhere in the time span between the late Carnian and early Liassic.

Discussion

The findings in the Helez Deep borehole, reveal a completely different geological history from that of the rest of Israel, as known from well and outcrop data elsewhere, and may bear significantly on the geological history of the Eastern Mediterranean. Two possible interpretations may be advanced to explain the data obtained in the Helez Deep borehole.

(1) Updoming and faulting model (Fig. 7)

The absence of Permian rocks overlying the metamorphic Precambrian basement complex or the Infracambrian arkose Zenfim Formation, which is the common case in southern Israel, implies non-deposition or truncation of the Permian in the Helez area. Since nowhere in Israel has there been any indication of a major erosion phase between the Permian and the Triassic, non-deposition seems more applicable. The well-established pre-Permian unconformity, documented by Weissbrod (1969, 1976, 1981) all over southern Israel, and the

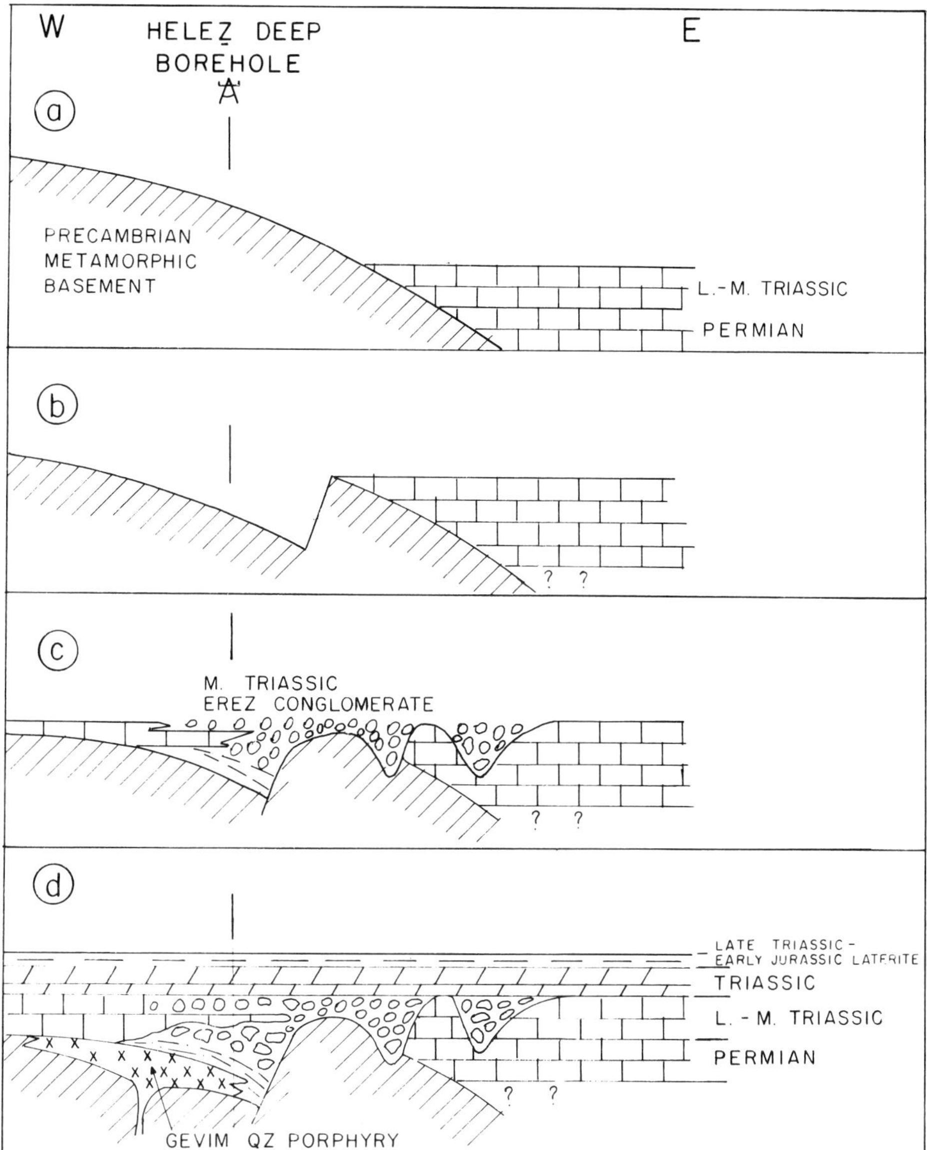

Fig. 7. Model of the Triassic geologic history of the Helez area. Interpretation (1):

(a) Updoming during Upper Palaeozoic times and onlap of Permian, Lower and Middle Triassic platform sediments.

(b) Faulting and possible rifting.

(c) Continuous faulting, erosion of platform sediments and accumulation of the Erez Conglomerate in erosional channels and graben.

(d) Termination of faulting phase, subsidence and deposition of the Upper Triassic Mohilla Formation dolomites; Late Triassic-Early Jurassic regional emergence and lateritization, intrusion of the Gevim Quartz Porphyry.

absence of the Permian in Helez, may indicate onlap relations of the Triassic over an updomed basement (Gvitzman & Weissbrod, this volume). The time of this postulated undoming may have been during the Permian; however, on the basis of the present limited data, this cannot be confirmed. The occurrence of the Erez Conglomerate in the Helez well and nowhere else in Israel, definitely indicates erosion and accumulation over high relief gradients, most probably over a fault escarpment.

Along the eastern margins of the Gulf of Suez, polymictic carbonate conglomerates a few hundred metres thick occur (The Abu Alaqa Group). These conglomerates are interlayered with shallow marine limestones similar to those found in the Erez Conglomerate. Garfunkel & Bartov (1978) interpreted the Abu Alaqa conglomerate as representing the main Middle Miocene rifting phase of the Suez Rift.

Numerous canyons cut the border faults of the Dead Sea Rift, fanning out into deltas. These fan-deltas consist of polymictic carbonate conglomerates in proximal areas, which pass into interlayers of conglomerates and varved lake sediments and finally into varved lake sediments in distal areas (Sneh, 1979). These conglomerates are several hundred metres thick in the proximal parts of the fan-deltas.

By analogy with these well-documented conglomerates along the Suez and Dead Sea Rifts, it is suggested that the Erez Conglomerate too, reflects a Triassic rifting event. The dimensions and orientation of this feature cannot be discerned without additional well data. From the age of the matrix and youngest fragments it is concluded that the faulting and possible rifting did not take place earlier than Anisian times (early Middle Triassic). As the overlying Mohilla Formation is of Carnian age, it is clear that the faulting did not extend into the early Upper Triassic. If the proposed rifting did take place in the Helez area, it was not followed by further spreading, as evidenced by the shoaling nature of the overlying Mohilla Formation dolomites, and even more so by the regional emergence and lateritization event of the Mishhar Formation, directly overlying the Mohilla Formation.

An East-West trending pre-Late Jurassic fossil continental margin under central Israel was suggested by Folkman & Bein (1978) on the basis of geophysical data. Their model seems to be incompatible with the Helez Deep borehole data presented here. The predicted basaltic composition of the basement underneath the sedimentary column was not found. Instead, a Precambrian metamorphic basement similar to those occurring in the well-exposed Arabo-Nubian massif (Steinitz *et al.*, 1981) was penetrated in the well. The shallow marine nature of both the Triassic and Jurassic sediments, and in particular the Mishhor laterite found in central and northern Israel, exclude the possibility of oceanic sedimentation in these areas, as would have resulted according to Folkman and Bein's model (1978).

(2) Faulting model

This interpretation (Fig. 8) differs from the previously suggested model, but still fits the borehole data, without requiring an updoming phase of the Precambrian basement over which the Permian and Triassic sediments onlapped. It assumes normal sedimentation of Permian and Lower Triassic, as known from other parts of Israel, as well as faulting and possible rifting, starting in early Middle Triassic and terminating prior to early Upper Triassic, as in the previously described model. In the updoming model, however, the borehole is located on the down-faulted block only, whereas in the present model, it penetrates part of the down-faulted block, the fault plane itself and continues in the upthrown block. A similar model was proposed by Freund (1981). In so locating the borehole, the 43 m of the Karmia Shale can be explained as being part of a much thicker Lower Triassic clastic section, and the missing Permian as being replaced by the Gevim Quartz Porphyry intrusion. However, the thickness of the Permian is expected to be greater than the 177 m thick intrusion; therefore, one would expect some Permian sediments, either above or below, or both above and below the intrusive body (Fig. 8). On the other hand, the first model fails to explain the absence of metamorphic basement fragments within the Erez Conglomerate. Keeping in mind that both models are based on data obtained from one single borehole, it is almost impossible to favour one model over the other.

Conclusions

(1) Continental Precambrian crust similar to that known from the Arabian Massif underlies the southern Coastal Plain of Israel.

(2) A pre-Triassic updoming event may have occurred, although the Helez Deep data can also be explained without it.

(3) Faulting and possible rifting took place in the area of the southern Coastal Plain of Israel, during Middle Triassic times; however, it was not followed by further spreading.

(4) Faulting and possible rifting activity ceased prior to the Upper Triassic and was not resumed throughout at least the entire Lower Mesozoic.

ACKNOWLEDGEMENTS: I acknowledge with thanks the Oil Exploration (Investments) Ltd., for permission to use and publish the borehole data. I wish to express my appreciation to my colleagues, Dr E. Kashai and

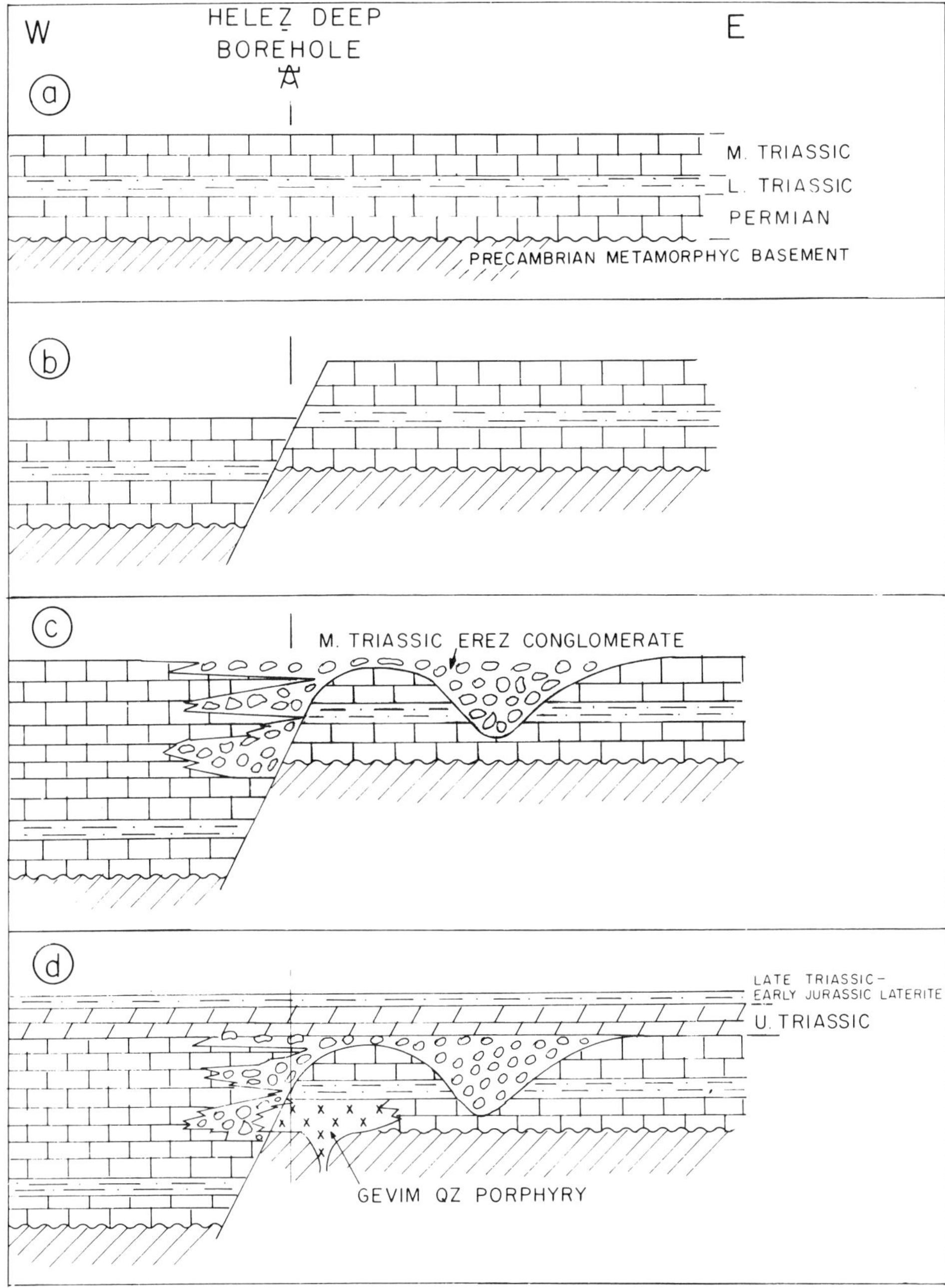

FIG. 8. Model of the Triassic geological history of the Helez area. Interpretation (2):

(a) Deposition of Permian, Lower and Middle Triassic sediments overlying Precambrian basement.

(b) Faulting and possible rifting.

(c) Continuous faulting, erosion of platform sediments and accumulation of the Erez Conglomerate in erosion channels and graben.

(d) Termination of faulting phase, subsidence and deposition of the Upper Triassic-Mohilla Formation dolomites; Late Triassic-Early Jurassic regional emergence and lateritization; intrusion of the Gevim Quartz Prophyry.

This interpretation is also compatible with the borehole data without an updoming phase as suggested in Fig. 7.

L. Fleischer, for their very helpful suggestions and to Dr A. Poisson, for his useful review of the manuscript. Many thanks are also due to Mrs B. Katz & C. Alafi for linguistic editing and typing. The study was carried out as part of project No. 28332 of the Geological Survey of Israel.

References

DERIN, B. 1979. Helez Deep 1A. Stratigraphic Log, scale 1:2000. *Israel Inst. Petrol.* Report No. 1/79.

DRUCKMAN, Y. 1974a. The stratigraphy of the Triassic sequence in southern Israel *Isr. Geol. Surv. Bull.* **64,** 76 pp.

—— 1974b. Triassic palaeogeography of southern Israel and Sinai Peninsula: Proc. Inter. Symp. of the Alpine-Mediterranean Triassic stratigraphy, Vienna, May 1973, *Erdewiss. Komm. Osterr. Akad. Wiss.* **2,** 79–86.

—— 1976. The Triassic in southern Israel and Sinai: A sedimentological model of marginal, epicontinental, marine environments. *Geol. Surv. Israel Rep. OD/1/76*, 188 pp. (in Hebriew, English summary).

—— & KASHAI, E. 1981. The Helez Deep and Devora 2A boreholes and their implication to oil prospects in pre-Jurassic strata in Israel. *Isr. Geol. Surv.* Report *OD/1/81*, 23 pp.

——, HIRSCH F. & WEISSBROD, T. 1982. The Triassic of the southern margin of the Tethys in the Levant and its correlation across the Jordan Rift Valley. *Geologische Rundschau*, **71,** 1–7.

FOLKMAN, Y. & BEIN, A. 1978. Geophysical evidence for a pre-Late Jurassic fossil continental margin oriented East-West under Central Israel. *Earth planet. Sci. Lett.* **39,** 335–40.

FREUND, R. 1981. Drot 1. Recommendation for drilling. *Oil Explor.* (*Invest.*) *Ltd.* Rep. **81/49,** 1–26.

GRUNKEL, Z. & BARTOV, Y. 1977. The tectonics of the Suez Rift. *Isr. Geol. Surv. Bull.* **71,** 1–44.

SNEH, A. 1979. Late Pleistocene fan-deltas along the Dead Sea Rift. *J. sediment. Petrol.* **49,** 541–52.

STEINITZ, G., BIELSKI, M. & STARINSKY, A. 1981. The extension of the Arabian shield beneath the Coastal Plain of Israel in light of Rb-Sr age determinations. *Ecoq VII.* 7th European Colloq. of Geochronology, Cosmochronology and Isotope Geology, Jerusalem.

WEISSBROD, T. 1969. The Paleozoic of Israel and adjacent countries: Part 1. The subsurface Palaeozoic stratigraphy of southern Israel. *Isr. Geol. Surv. Bull.* **47,** 23 pp.

—— 1976. The Permian in the Near East. *In:* H. FALKE (ed.). *The Continental Permian in Central, West and South Europe.* Reidel Publ. Co., Dordrecht, 200–14.

—— 1981. The Palaeozoic of Israel and adjacent countries (Lithostratigraphic Study). *Isr. Geol. Surv. Rep. M.P. 600/81*, 1–256 pp.

YEHEZKEEL DRUCKMAN, Geological Survey of Israel, 30, Malkhei Israel St. Jerusalem 95501, Israel.

Emergence of Wadi Mujib (Central Jordan) during Lower Cenomanian time and its regional tectonic implications

Abdulkader M. Abed

SUMMARY: Detailed sedimentological investigations of Wadi Mujib (Central Jordan), east of the Dead Sea, show that the regional contact of Lower Cretaceous fluviatile deposits on Upper Cretaceous carbonates ushered in a shallow marine environment all over Jordan. Later, an emergent area, in Wadi Mujib only, passes upwards and laterally eastwards into a supratidal environment, followed again by shallow marine conditions. The emergence is marked by a major erosional unconformity with wadi deposits, caliche and karstification. Wadi Mujib is believed to have been an island during Lower Cenomanian time and is thought to be an example of vertical tectonics in the area prior to Tertiary rifting of the Dead Sea to the west.

As part of the Levant, Jordan underwent major marine transgression during Upper Cretaceous times. This transgression, which also affected the area to the south-east, penetrated far into northern Saudi Arabia, and is possibly the largest transgression in the history of the area (Bender 1968, 1974; Burdon 1959; Wetzel & Morton 1959). At this time Jordan was a continental shelf adjacent to the Upper Cretaceous sea. The aim here is to outline an example of possible syn-sedimentary vertical tectonics in Lower Cenomanian time.

Geological setting

The area investigated lies along the course of Wadi Mujib, a few kilometers west of Mujib bridge (Fig. 1). The section studied is part of the 'Nodular Limestone Member' (Bender 1974), which is of Cenomanian age (Table 1). Other wadis north and south of Wadi Mujib were also investigated, from north Amman down to Ras en Naqb in the south (Fig. 1).

Field observations

A 60 m thick section was measured along the course of Wadi Mujib (Fig. 2). The rocks are whitish nodular limestones interbedded with brownish-yellowish dolomites. The section includes a major unconformity. The facies from bottom to top are:

Burrowed biomicrite

This is a white limestone of two types: an original, white relatively soft biomicrite, and a brownish, relatively hard fine grained dolomite, filling burrows. The nodular character of the limestone is believed to result from burrowing (Abed & Schneider 1980) in a subtidal environment (Abed & Schneider 1979, Fig. 2).

Brownish-yellowish fine grained dolomites

Interbeds with the previous facies range from 50 to 100 cm thick and show no sedimentary structures (Fig. 2). These dolomites are believed to be of early diagenetic origin.

A major erosional unconformity (Fig. 3)

The unconformity surface shows scouring and karstification especially of the burrowed biomicrite facies. This surface continues for more than 1 km westwards where it disappears under slope debris.

Conglomerate facies (Fig. 3)

Above the unconformity surface there is a 15 m thick conglomerate horizon containing clasts of brownish-yellowish dolomite up to 30 cm in diameter. They decrease in size upwards and pass into poorly bedded fine grained dolomites. Lamination is hardly visible near the top of these dolomites. Examination of pebbles in the conglomerate showed them to be reworked burrow-infills and fragments of the underlying two facies. However, larger boulders of the limestone can also be seen. A channel, or wadi, trending NW–SE, can be recognized in this area, indicating that the conglomerates can be considered to be channel-fill deposits.

Primary dolomites

Towards the top of the measured section a thinly bedded limestone and primary dolomite horizon is present. Careful examination reveals several laminated horizons with mudcracks. Small cavities are believed to be pseudomorphs after gypsum. Fossils are absent. This facies is interpreted as having been deposited in a supratidal environment (Friedman & Saunders

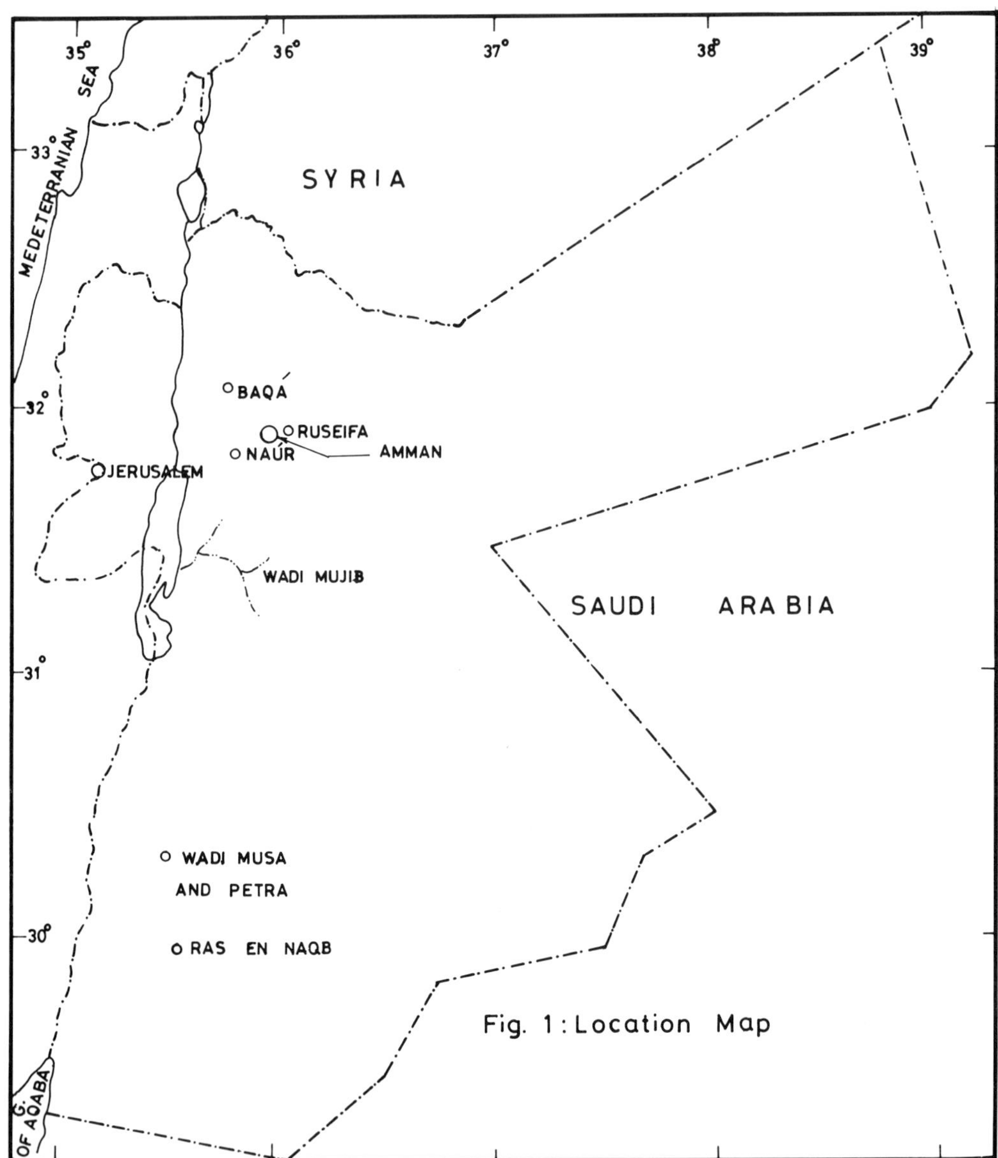

FIG. 1. Sketch map showing the location of Wadi Mujib in Jordan.

1978). Fig. 2 shows that the facies change laterally westwards into the upper part of the conglomerate facies (Fig. 2).

Burrowed biomicrite

The top of the measured section is again a burrowed white biomicrite facies but the beds are more massive.

There is thus evidence of almost total emergence and subaerial erosion of Wadi Mujib in Lower Cenomanian time.

Discussion and conclusions

It is clear that two environments existed in a small area: a supratidal environment (primary dolomite facies), changing laterally into a continental environment undergoing subaerial erosion (the channelized conglomerates). These sediments were both underlain and overlain by subtidal to intertidal marine deposits (the burrowed micrites).

Twenty metres below the base of the section

Period	Epoch	Formation (Various Authors)	Units Bender (1968)	Wetzel & Morton (1959)
Upper Cretaceous	Turonian	Wadi Sir	Massive Limestone	Judea Limestone
	Cenomanian	Shueib	Echinoidal Limestone	
		Hummar		
		Fuheis	Nodular Limestone	
		Naúr		
Lower Cretaceous		Kurnub (Hathira) Sandstones		

TABLE 1. *Stratigraphy of the 'Nodular Limestone Member' in Jordan.*

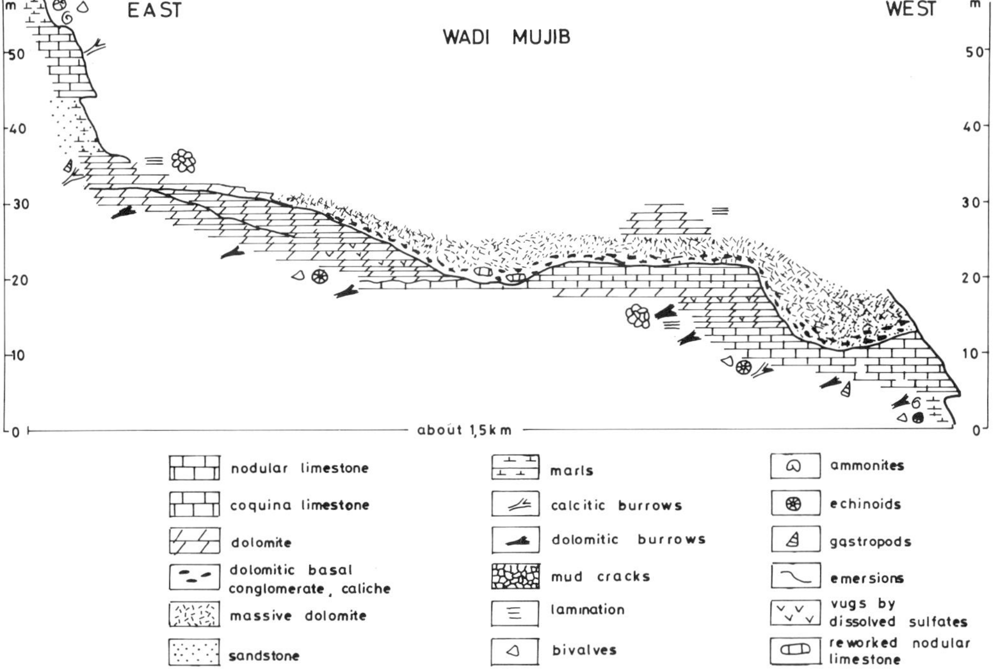

FIG. 2. Cross-section taken from east to west along Wadi Mujib to show the various facies discussed in the text.

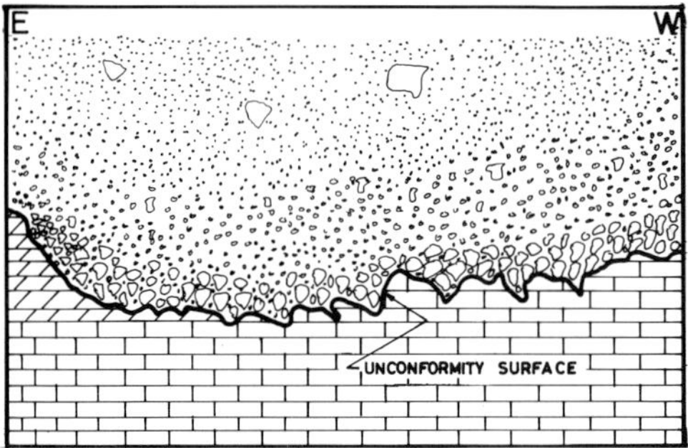

Fig. 3. Sketch of the limestone and dolomite facies above and below the major erosional unconformity in Wadi Mujib. Note the irregular unconformity and the fining-upwards nature of the conglomerate.

studied is the contact between the Lower Cretaceous fluviatile Kurnub sandstones and marine Upper Cretaceous limestones (Abed 1982). This regionally important contact can be easily traced through Jordan, Palestine, Syria and Lebanon (Abed 1982, and literature therein).

It thus seems that Jordan, and adjacent areas, was emergent during Lower Cretaceous times when the fluviatile Kurnub sandstone was deposited. During the Lower Cenomanian there was then a major marine transgression. Jordan, including Wadi Mujib, become a shallow marine environment, ranging from subtidal to intertidal (Abed & Schneider 1979) and the 20 m of normal marine limestones and marls were deposited below the sections studied. Later, still in Lower Cenomanian time, an island formed in the area on which subaerial erosion and channeling took place, while around the edges supratidal deposits accumulated. The simplest explanation is syn-sedimentary uplift of this part of Central Jordan during Lower Cenomanian time, followed by subsidence giving rise to a thick sequence of Upper Cretaceous marine limestone, marl, chert and phosphate with various fossils (Bender 1974; Wetzel & Morton 1959).

For comparison, several wadis north and south of Wadi Mujib were examined, including Baqa' and Naur in the north, and Wadi Musa—Petra and Ras en Naqb in the south. None of the sections examined show an unconformity or other evidence of emergence. Even the Ras en Naqb section, which is far to the south and lay near the 'Nubian sandstone continent', does not show sedimentary structures indicating emergence (Abed & Schneider 1979). The Wadi Mujib event is thus local compared with the regional Lower Cretaceous-Upper Cretaceous facies trends discussed above.

In conclusion, the topography of the older rocks coupled with syn-sedimentary uplift then subsidence explain the sedimentology of the Mujib area. Eustatic regional sea level fluctuations are thought to have been important because of the lack of evidence of emergence at the same stratigraphic level elsewhere in Jordan. There is no obvious evidence that the Wadi Mujib area was already a topographic high prior to Lower Cenomanian time. On the other hand, Mikbel & Bandel (personal communication) mention that there are syn-sedimentary tectonic features in the Maastrichtian Phosphorite Member, southeast of Ruseifa.

References

Abed, A. M. 1982. Depositional environments of the early Cretaceous Kurnub (Hathira) sandstones, north Jordan. *Sedimentary Geology* **31,** 267–279.

Abed, A. M. & Schneider, W. 1979. Palaeogeography of the so-called nodular limestones (Cenomanian), Jordan. *Proc. 1st Geol. Cong. Middle East (GEOCOME-1),* Ankara, 206–221.

Abed, A. M. & Schneider, W. 1980. A general aspect in the genesis of nodular limestones documented by the Upper Cretaceous limestone of Jordan. *Sedimentary Geology,* **26,** 329–335.

Bender, F. 1968. *Geologie von Jordanien*, Bornträger, Stuttgart. 230 pp.

Bender, F. 1974. *Geology of Jordan.* Gebruder Borntraeger, Berlin. 196 pp.

Burdon, D. J. 1959. Handbook of the geology of Jordan. Government of the Hashemite Kingdom of Jordan. 82 pp.

Friedman, G. M. & Saunders, J. E. 1978. *Principles in sedimentology*, Wiley, N.Y. 792 pp.

Masri, M. R. 1963. Unpublished report on the Amman-Zerqa area. Central Water Authority, Jordan.

Wetzel, R. & Morton, D. M. 1959. Contribution à la géologie de la Transjordanie. *Notes Mém. Moyen-Orient* **71,** 95–191.

Abdulkader M. Abed, Department of Geology and Mineralogy, the University of Jordan, Amman, Jordan.

The Arabian sub-plate during the Mesozoic

F. Hirsch

SUMMARY: The Mesozoic facies, distribution, palaeo-zoo-geography and tectonic history of the Levantine and Tauride-Zagride autochthonous and para-autochthonous units, suggest that these areas were part of the Arabian sub-plate and never lay far apart. This would tend to invalidate the hypothesis of an oceanic basin (Mesogea) in the present Eastern Mediterranean during the Mesozoic.

Intensive research by many authors during the last decade allows us to draw a relatively detailed picture of the development of Triassic, Jurassic and Cretaceous deposits in the Levantine region, Southern Turkey and the Arabian peninsula (Fig. 1). The bio- and lithofacial similarities of these deposits and the synchroneity of major tectonic events affecting these areas has been recently summarized by Ricou & Marcoux (1980). Lacunae due to regional uplifts and the presence of continental deposits in both the Levant and in southern Turkey provide further evidence for the reconstruction of the tectonic evolution and plate configuration of the Eastern Mediterranean.

Mesozoic tectonic and sedimentary evolution

Palaeozoic substratum

On the Arabian subplate the Cambrian is followed, in places, by a huge lacunae (Monod 1977), that may extend into the Permian, Triassic or even Jurassic. The Ordovician is present in the Pisidian Taurus and in Jordan, the Carboniferous in Sinai, the Lower Permian in the subsurface of Southern Israel.

In the Levant, the Mesozoic is underlain by Upper Permian (Djulfian), whereas in the Pisidian Taurus, Middle Triassic or Middle Jurassic rests on Ordovician metamorphics. In the Levant, the Mesozoic overlap took place not earlier than during the Upper Scythian, thus leaving a lacuna from Lower to Middle Scythian.

Triassic

The Lower Triassic is developed in a typical Alpine-Dinaride Werfen-type facies. This facies is found in Israel, Jordan, Arabia and in the Bey Dağları in South Turkey. The Werfen facies persisted in the Levant into the Lower Anisian (cycle I). Mesozoic Tethyan rifting (Neotethys) during the Middle Triassic (Marcoux 1978) modified the Werfenian palaeogeography inherited from the Late Permian. During the Middle Triassic the differentiation between facies of the southern Tethys shelf (Gondwanian shelf-platform) and the Eutethyan pelagic sea-way become well defined. The former consisted of a Muschelkalk-type lithofacies (Saharonim Formation) with sephardic faunal elements while the latter is characterized by Hallstatt-type lithofacies with cosmopolitan Tethyan faunal elements. The maximum extent of the sephardic realm (cycle III) occurred during the Ladinian and Early Carnian, characterized by *Israelites, Gevanites, Iberites* and endemic forms of *Protrachyceras* as well as conodonts (*Pseudofurnishius*). It extended at that time from the Iberian peninsula through to the Arabian peninsula, including North Africa, the Levant and the Pisidian Taurides (Druckman *et al.* 1982).

From the Middle Carnian, evaporites become a dominant facies in the Levant. A more Alpinotype Hauptdolomit-like facies of Late Carnian-Norian age appears in boreholes in the coastal plain of Israel. According to Ager *et al.* (1980), Norian brachiopods from the Tauride authochthonous units also have close affinities with 'Tethyan' forms of South Europe.

Jurassic

The Triassic-Liassic boundary is marked by a distinct sedimentary break between the Carnian or Norian and the Early Pliensbachian. This break is characterized by laterites (flint-bearing clays and iron pisolites) and palaeosoils of early Liassic age (Mishhor Formation). The same abrupt lithological change, indicative of a hiatus, occurs in Arabia where the late Liassic (Toarcian) Marrat Formation rests on the Upper Triassic–Liassic Minjur Formation with its sands and lateritic soil intercalations. Lower Jurassic iron-pisolites 'Fasulia demir' are observed in the Pisidian region of South Turkey, resting on Triassic limestone or Ordovician schists, overlain by Early to Mid-Jurassic *Orbi-*

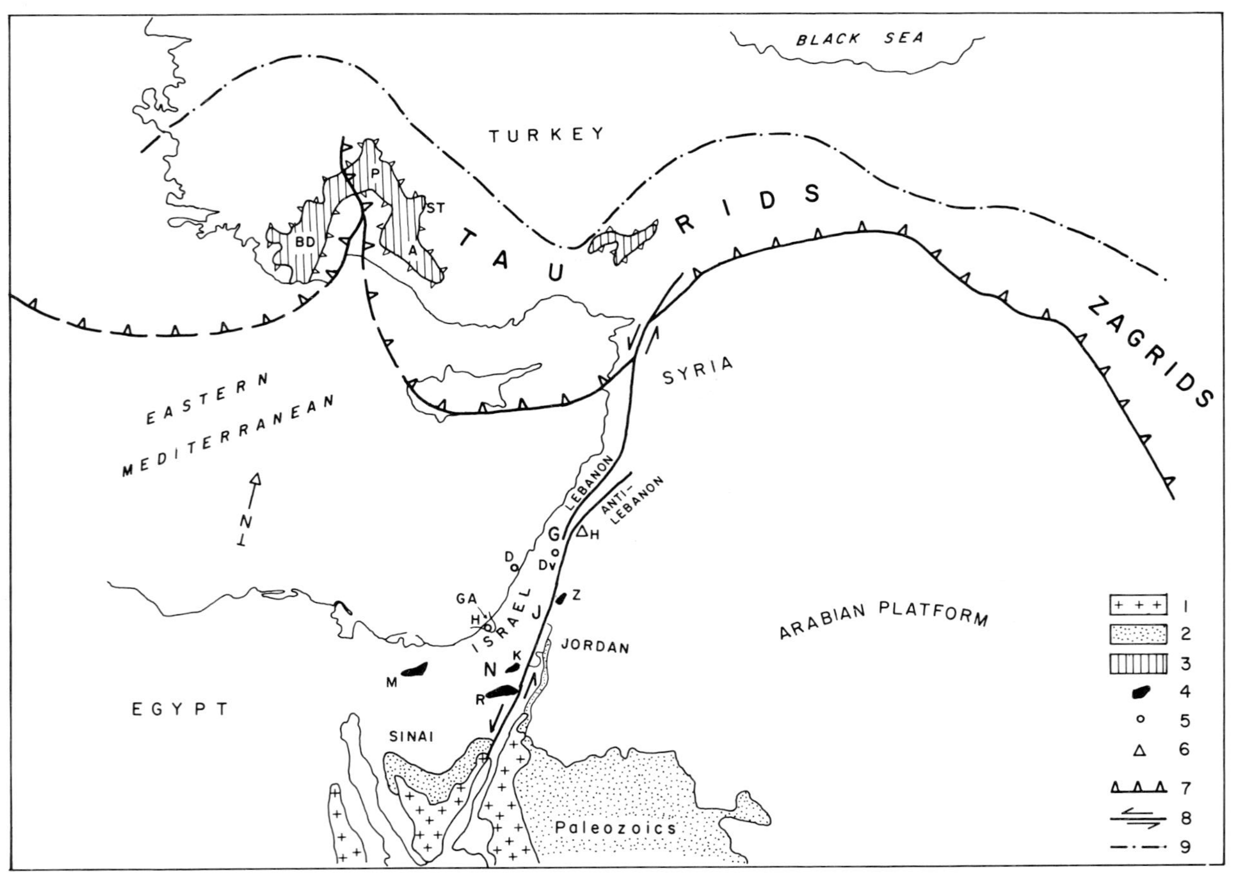

Fig. 1. Location map. 1, Precambrian massif; 2, Palaeozoic outcrops; 3, Tauride autochthonous units; 4, Significant Early Mesozoic outcrops mentioned in text; 5, Bore-hole; 6, Mountain-top; 7, Thrust; 8, Strike-slip fault; and 9, Margin of Arabian Sub-plate.

D, Delta borehole; H, Helez borehole; Dv, Devorah borehole; R, Ramon outcrop (Tr-J); K, Kurnub-Hamaktesh Hagadol outcrop (J); M, Gebel Maghara (N. Sinai) (J); Z, Zarqa-River (Yabbok) (Tr-J); H, Mt. Hermon; GA, Gevar-Am trough; G, Galilee; J, Judea; N, Negev; A, Akseki; ST, Seydişehir-Tarasci; BD, Bey Dağları; and P, Pisidian Taurus.

topsella and *Haurania*-bearing sandstones and limestones (Fig. 2). Lateritic or bauxite deposits of the same age are found in Greece, Yugoslavia and Sicily (Goldberg & Friedman 1974).

The late Lower to early Middle Jurassic marine sediments (Cycle I, Fig. 2) of central and northern Israel (Nirim Formation) are an up to 1500 m thick sequence of platform-limestones and dolomites with lagoonal anhydrites. The equivalent horizon in southern Israel (Ramon) and Jordan (Nahr Zerqa) is still semi-continental. The age of the marine sequence is defined as Pliensbachian and Toarcian by *Orbitopsella* and calcareous nannoplankton. This facies is interbedded towards the south with deltaic and paralic sandstones with thin coal intercalations (Cycles Ia, Ib, Fig. 2).

Towards Lebanon, Syria and the Tauride autochthon of the Pisidian region near Akseki, the same facies can be identified, but, with decreasing thickness (Monod 1977). The same facies characterizes the Bajocian, and is well defined by ammonites and the calcareous algae *Selliporella donzelli*, reaching some 700 m in Central Israel (Cycle Ic, Fig. 2).

The Lower Bathonian is characterized by paralic facies in Southern Israel which terminates cycle I. The Bathonian becomes more and more carbonaceous in central Israel and the Anti-Lebanon (Hermon) where it reaches up to 800 m in thickness. Well defined by ammonites in the semiclastic facies of the Northern Sinai (Gebel Maghara) and Arabia, the Bathonian shales are followed by up to 180 m of Callovian carbonates (Cycle II).

In Sinai, southern Israel (Hamakhtesh Hagadol) and in Central Arabia (Gebel Tuwaiq) the Callovian is characterized by ammonites of the Erythrean province (Gill & Tintant 1974). Towards the north and west the Callovian may be absent. In central Israel and Southern Anti-Lebanon (Hermon), Oxfordian shales (Kidod Formation) transgressively overlap various truncated levels of the Callovian. In the Negev and Lebanon, bedded limestones (Beersheba Formation) represent the Oxfordian–Early Kimmeridgian. Shales containing allodapic limestone fragments are found in the Delta borehole in offshore Israel (Fig. 1, 'D'). The limestone fragments are derived from the massive reefoid build-ups (Niram Formation), that occur in patches along the present Levantine coast.

The Lower Oxfordian transgression (Cycle III) is followed by a slow regressive trend that ends in the oolitic facies of the Late Oxfordian–Early Kimmeridgian, accompanied by volcanic tuffs in Galilee. The reef-like structures (Niram Formation) and the adjacent basinal sediments (Delta Formation) suggest the formation of a N-S graben along the actual Levant coastal region (Fig. 2). The question remains open as to whether the Gevar Am Shales in the subsurface of the coastal plain of Israel may start as early as uppermost Jurassic (Tithonian), as suggested by the finding of a fragment belonging to the ammonite subfamily Virgatitinae (Raab 1962).

Late Jurassic uplift accentuated the picture of a N-S graben along the present Levant coast, causing on the one hand, possibly continuous sedimentation from the Tithonian into the Neocomian in the Gevar-Am trough, in contrast to intensive denudation of the emerged Levantine block to the east from Tithonian to Valanginian time.

In southern Turkey, up to 200 m of bituminous shales with ammonites (Akkuyu) range from Middle Oxfordian into the Hauterivian. They overlie transgressively the Middle Jurassic and pass laterally into *Cladocoropsis*- & *Clypeina*-bearing limestones.

The long Early Jurassic emergence of the entire African-Arabian shelf, followed by the essentially shallow marine character of the Levantine Jurassic seas, facilitated the differentiation of faunal realms within the Jurassic Tethys. One can clearly define southern Tethyan Ethiopian ammonite, bivalve and brachiopod associations that characterize the African-Arabian platform. This was recognized by Arkell (1956). Enay (1975) has reported northern and southern Tethyan associations in Turkey, putting the rich Akkuyu ammonite faunas into the southern Tethyan or Ethiopian realm.

During the Lower Callovian, the influx into Western Europe of boreal waters from the newly born northern Atlantic turned the North African, Levantine, Arabian and Erythrean (Somalian) regions into an ecological shelter in which warm-water faunas kept their Bathonian character for a while and continued their evolution. This Erythrean Province (Gill & Tintant 1974; Hirsch 1979) was limited to a narrow belt of the African-Arabian shelf. In Turkey, Ager *et al.* (1980) have found that late Liassic brachipods show affinities with extra-alpine European forms. In northern Sinai and southern Israel, Hirsch (1975) reported cosmopolitan 'European' bivalves together with the endemic forms that characterize the Ethiopian Province on the African-Arabian platform. However, according to Vörös (1980) most brachiopod assemblages found in areas belonging to fragments of the Apulian plate belong to this Mediterranean province.

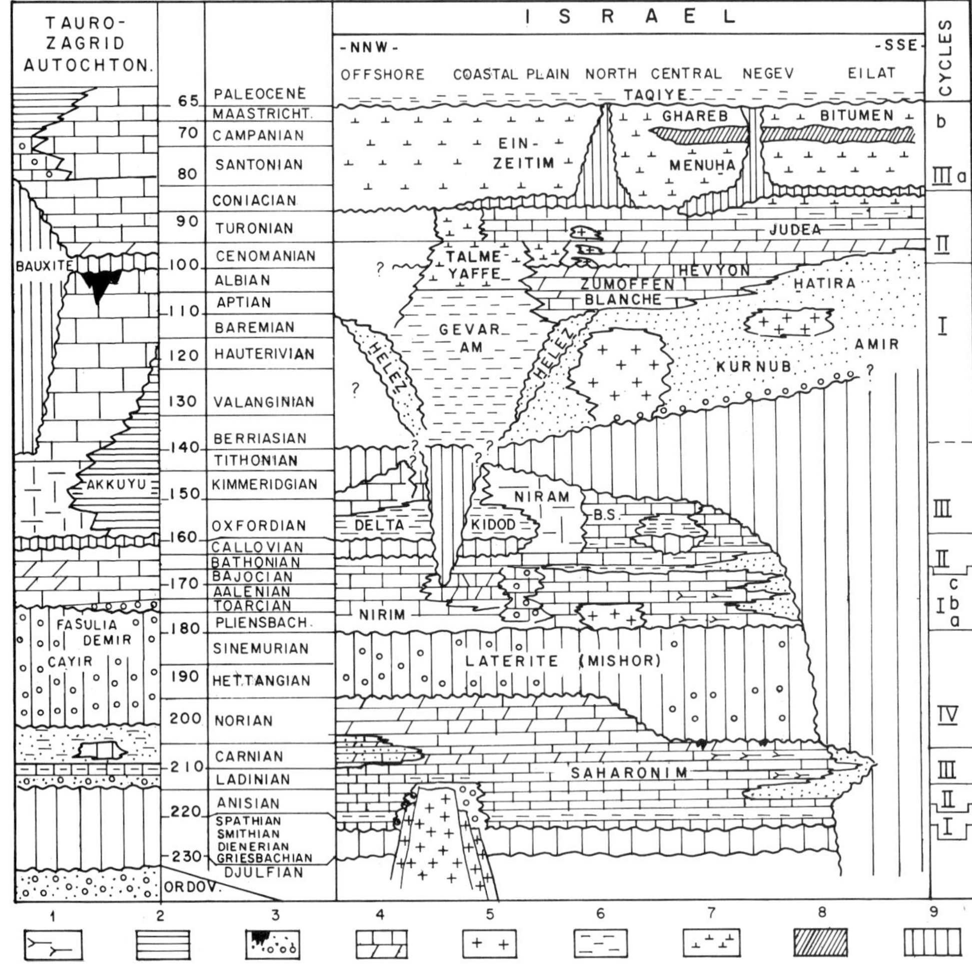

Fig. 2. Schematic cross-section of the Mesozoic of Israel, correlated with the Tauride-Zagride autochthonous units of southern Turkey.

1, Gypsum and anhydrite; 2, Pelagic limestone and shale; 3, Sandstone, conglomerates, pisolites, laterites and bauxites; 4, Carbonate, limestones and dolomites; 5, Intrusive, basement and eruptives; 6, Marls and shales; 7, Chalk; 8, Chert and porcellanite; and 9, Lacunae; B.S., Beersheba formation.

The Late Jurassic–Early Cretaceous erosion, which took place primarily due to tilting of the Arabian platform, has progressively removed, from NW to SE, the entire Jurassic, Triassic and marine Upper Palaeozoic, bringing, in the Southern Negev, the Lower Cretaceous Kurnub Sandstone in direct stratigraphic contact with Lower Palaeozoic Nubian Sandstones.

Evidence for the Late Jurassic instability is also noted in the Zagro-Tauride autochthonous terrains where, with the exception of the continuous sedimentation within the Akkuyu basin, there are huge lacunae between the Jurassic and the Middle Cretaceous (Ricou 1979).

Cretaceous

The Cretaceous of the Levantine region as a whole can be subdivided into a Neocomian-Turonian lower part and a Senonian upper part, separated by the pre-Senonian lacuna. The lower part consists of three groups with different facies. The Kurnub Group comprises mainly dune sandstones (Amir Formation), deltaic sandstones, alluvial fans and longshore sandstones (Hathira Formation) from Sinai far into Lebanon. Intensive volcanic activity accompanied the Neocomian in the Negev, Galilee and Lebanon. Carbonates occur not earlier than the Aptian (Falaise de Blanche). The Kurnub Group passes laterally and vertically from NW to SE into the Judea Group which consists of limestones with abundant biostromes and dolomites (Albian-Turonian). Both groups pass laterally to the west into the Gevar-Am-Talme Yafe Group consisting of shales and pelagic chalks. A slight regression marks the end of the Albian (Top Hevyon Formation, Cycle I) followed by Cenomanian and Turonian limestones and dolomites which are accompanied by volcanics in Galilee, Cycle II). Both, the Kurnub sandstones and Gevar Am shales prograde unconformably, from NW to SE over, Kimmeridgian to Palaeozoic deposits. The base of the Kurnub sediments is diachronous and varies from Valanginian to Hauterivian in age, becoming younger towards the S.E. (Fig. 2). Coniacian deposits are recorded only in south Israel and Sinai (Lewy 1975).

Following Coniacian–Early Santonian folding, the Senonian (Cycle III) comprise (a) Santonian chalks (Menuha Formation) ending in a Campanian regression with shallow chert and phosphates and (b) Maastrichtian bituminous chalks (Ghareb Formation). Towards the northwest continuous pelagic chalk sedimentation occurs (Ein Zeitim Formation). The late Maastrichtian uplift is sealed by a Palaeocene palaeosoil and shales (Taqiye Formation).

The Cretaceous of the autochthonous units of the Turkish-Syrian border region is characterized by Aptian/Albian to Turonian carbonates, resting unconformably on Jurassic and overlain unconformably by Campanian/Maastrichtian marly limestones (Scaglia) passing into flysch-type sediments. In the Pisidian Taurus, Lower Cretaceous neritic limestones occur, and are covered by bauxites. The gap represented by the bauxites may range from Albian to Turonian (Monod 1977) corresponding roughly to the regression of the first cycle in the Levant. Late Cretaceous neritic limestones contain rudists.

In the Peri-Arabian border-zone the end of the Senonian corresponds to the tectonic emplacement of allochthonous ophiolites, radiolarites and metamorphics of the 'Peri-Arabian ophiolite belt' (Ricou 1971).

Palaeogeography of the African-Arabian plate

During the Triassic, the African-Arabian craton was bordered by a shallow sea with decreasing clastic deposition through time (Scythian to Carnian) and from south to north in the Levantine region. Steady subsidence allowed deposition of an average of 1000 m of Triassic sandstones, shales and carbonates in the Negev. Towards the north, in a borehole in Galilee (Devorah-1) over 2500 m of carbonates and evaporites occur. In Syria, Arabia and Southern Turkey, the thickness of the Triassic is strongly reduced. It must be noted that the Ladinian (Saharonim, Tarascim sections) show an extremely constant thickness of about 150 m all over the area discussed.

Middle Triassic block-faulting caused a horst-like structure in the area of the coastal plain of Israel (Druckman this volume). The extension of the Jurassic seas on the African-Arabian platform did not much exceed that of the Triassic, but its palaeogeographic differentiation was accentuated by differential subsidence and vertical movements. The African-Arabian craton was still bordered by a belt of clastics of Lower and Middle Jurassic age. In a trough crossing central Israel from west to east (the Judean Embayment), the Jurassic consists of 4000 m of carbonates and minor shales. Towards the north, on the Syrian horst that includes Lebanon, Syria and the South Turkey Pisidian region, the thickness becomes progressively reduced to less than 1000 m. The Oxfordian Delta-Kidod Shales that lie in a belt following roughly the present Levantine coastline for 300 km from North Sinai to Central Israel, extend at least 100 km east into the Judean Embayment (Hermon).

The Pisidian Jurassic strongly recalls the Levantine sequence with Liassic clastics and carbonates, Bajocian-Bathonian limestones, Oxfordian-Kimmeridgian carbonates and shales (Akkuyu Kormation). Monod (1977) put forward the possibility of infra-Liassic uplift on the northern edge of the African-Arabian platform from which the Liassic clastics may have been derived.

The Cretaceous pre-Coniacian palaeogeography is dominated by an approximately N-S graben (Gevar Am-Talme Yafe Formations) separating an offshore western province from an onshore eastern province (Rosenfeld and Raab, 1980). This eastern province remained a shallow marine platform with rudist biohermal and biostromal structures, whereas the western province seems to have had a more pelagic character. The Senonian palaeogeography of the Levantine region was controlled by the Coniacian-Santonian folding of the Syrian arch.

The Tauride autochthon units are dominated during the Cretaceous by shallow platforms which were partly subaerial with bauxites; whereas during the Senonian the northern edge of platform limestones with rudists was bordered by marly limestone and flysch sedimentation associated with the Maastrichtian and Palaeocene thrust-fronts.

The area extending from the Sinai Peninsula in the south to the Pisidian Taurus to the north formed a single sedimentary basin with up to 6–7 km of epicontinental shelf deposits. Clastic intercalations are increasingly important both towards the Arabo-Nubian craton in the south, and towards the inferred uplifted platform edge in the north. An intracratonic north to south graben system, with conglomerates, turbidites and deep-water shales, subparallel to the present Levant coast was active from the Middle Triassic, onwards.

The description of the present Eastern Mediterranean as a relic of an oceanic seaway (Mesogea, or Tethys 3) separating the Levantine region from its northern opposite margin in the Pisidian Taurus and Bey Dağları at any time during the Mesozoic seems highly improbable. On the contrary, it seems more likely that from Late Palaeozoic to Late Cretaceous, the Eastern Mediterranean remained part of the African-Arabian and Apulian plate-system with a thinned continental crust. The true Neo-Tethys was located north of the Arabian subplate and originated with rifting in the Middle Triassic followed by oceanic spreading during the Jurassic and Cretaceous.

ACKNOWLEDGEMENTS: I am thankful to Dr. Y. Druckman, and Dr. D. Neev of the Geological Survey of Israel, and to Prof. L. Picard of the Israel Academy of Science, for fruitful discussions. I am greatly indebted to Ellis Owen (British Museum, Natural History, London) for assistance with the writing, and to Prof. Z. Garfunkel of the Hebrew University of Jerusalem, and to Dr. A. M. Quennell (Bristol), Dr. B. Z. Begin and R. Bogoch of the Geological Survey of Israel, for critical reading of the manuscript, as well as Dr. O. Monod of the University of Paris for kindly reviewing the material on the Tauride-Zagrid autochthonous units.

References

ADAMIA, S., BERGOUGNAN, H., FOURQUIN, C., HABHIPOUR, A., LORKIPANIDZE, M., OZGUL, N., RICOU, L. E. & ZAKARIADZE, G. 1980. The Alpine Middle East between the Aegean and the Oman traverses. *Mém. BRGM.* **115,** 122–136.

AGER, D. V., GUTNIC, M., JUTEAU, TH. & MONOD, O. 1980. New Early Mesozoic brachiopods from Southern Turkey. *Bull. Min. Research & Explor. Inst.* **91,** 61–75.

D'ARGENIO, B., HORVATH, F. & CHANNELL, J. E. T. 1980. Palaeotectonic evolution of Adria, the African promontory. *Mém. BRGM.* **115,** 331–351.

ARGYRIADIS, I. 1975. Mesogée permienne, chaine hercynienne et cassure téthysienne. *Bull. Soc. géol., France* **17,** 56–67.

——, DE GRACIANSKY, P. C., MARCOUX, J. & RICOU, L. E. 1980. The opening of the Mesozoic Tethys between Eurasia and Arabia-Africa. *Mém. BRGM.* **115,** 199–214.

ARKELL, W. J. 1956. *Jurassic Geology of the World.* Oliver & Boyd, Edinburgh.

BANDEL, K. 1981. New stratigraphical and structural evidence for lateral dislocation in the Jordan Rift connected with a description of the Jurassic rock column in Jordan. *N. Jb. Geol. Palaont. Abh.*

—— & KHOURY, H. 1981. Lithostratigraphy of the Triassic in Jordan. *Facies*, **4,** 1–26.

BARTOV, Y., LEWY, Z., STEINITZ, G. & ZAK, I. 1980. Mesozoic and Tertiary stratigraphy, palaeogeography and structural history of the Gebel Areif en Naqa Area, Eastern Sinai. *Isr. J. Earth Sci.* **29,** 114–139.

BEIN, A. & GVIRTZMAN, G. 1976. A Mesozoic fossil edge of the Arabian plate along the Levant coastline and its bearing on the evolution of the Eastern Mediterranean. *In*: BIJU-DUVAL, B. & MONTADERT, L. (eds) *Histoire Structurale des Bassins Mediterranéens*, 95–109. Technip, Paris.

—— & WEILER, Y. 1976. The Cretaceous Talme Yafe Formation: A contour current shaped sedimentary prism of calcareous detritus at the continental margin of the Arabian craton. *Sedimentology*, **23,** 511–532.

BERNOUILLI, D. & LEMOINE, M. 1980. Birth and early evolution of the Tethys: the overall situation. *Mém. BRGM.* **115,** 168–179.

COUSMINER, H. L. & CONWAY, B. 1980. Initial results and potential of an Israeli palynology program. *Geol. Survey Israel, Current Research.* 57–61.

DERCOURT, J. 1972. The Canadian Cordillera, the Hellenides, and the sea floor Speading Theory. *Can. J. Earth Sci.* **9,** 709–743.

DUMONT, J. F., GUTNIC, M., MARCOUX, J., MONOD, O. & POISSON, A. 1972. Le Trias des Taurides occidentales (Turquie). Définition du bassin pamphylien: un nouveau domaine à ophiolites à la marge externe de la chaîne taurique. *Z. Deutsche geol. Ges.* **123,** 385–409.

DRUCKMAN, Y., HIRSCH, F. & WEISSBROD, T. 1982. The Triassic of the Southern Tethys Margin in the Levant and its correlation across the Jordan Rift Valley. *Geol. Rundschau* **71,** 919–936.

ENAY, R. 1972. Paléobiogéographie des Ammonites de Jurassique terminal (Tithonique, Volgien, Porthandien) en mobilité continentale. *Géobios* **5,** 4, 355–407.

—— 1980. Evolutions et relations paléobiogéographiques de la Téthys mésozoique et cénozoique. *Mém. BRGM.* **115,** 276–283.

GILL, G. A. & TINTANT, H. 1975. Les ammonites calloviennes du Sud d'Israel—Stratigraphie et relations paléogéographiques. *Soc. Géol. France C.r.* **4,** 103–106.

GOĆER, D. M. 1976. L'évolution géotectonique du Mégabloc bulgare pendant le Trias et le Jurassique. *Bull. Soc. Géol. France*, **18,** 209–216.

GOLDBERG, M. & NRIEDMAN, G. M. 1974. Paleoenvironment and paleogeographic evolution of the Jurassic System in Southern Israel. *Bull. Israel Geol. Surv.* **61,** 1–44.

HIRSCH, F. 1972. Middle Triassic conodonts from Israel, Southern France and Spain. *Mitt. Ges. Geol. Bergbaustud.* **21.**

—— 1976. Sur l'origine des particularismes de la faune du Trias et du Jurassique de la plateforme africano-arabe. *Bull. Soc. géol. France.* **18,** 543–552.

—— 1977. Essai de correlation biostratigraphique des niveaux méso-et néotriasiques de faciès 'Muschelkalk' du domaine sépharde. *Cuadernos Geologia Iberica.* **4,** 511–526.

KAMEN-KAYE, M. 1976. Mediterranean Permian Tethys. *Am. Ass. Petr. Geol. Bull.* **60,** 623–626.

LETOUZEY, J. & TREMOLIERES, P. 1980. Paleo-stress fields around the Mediterranean since the Mesozoic derived from microtectonics: comparisons with plate tectonic data. *BRGM. Mém.* **115,** 261–273.

LEWY, Z. 1975. The geological history of Southern Israel and Sinai during the Coniacian. *Isr. J. Earth-Sci.* **24,** 1-2, 19–43.

—— & RABB, M. 1976. Mid Cretaceous stratigraphy of the Middle East. *Mus. d'Hist. Nat. Nice. Ann.* **4,** 321–322.

MARCOUX, J. 1974. 'Alpine type' Triassic of the Upper Antalya Nappe (Western Taurids, Turkey). *Schriftenreihe Erdwiss-Komm. Oesterr. Akad. Wiss.* **2,** 145–146.

—— 1978. *A scenario for the birth of a new oceanic realm: the Alpine Neotethys.* 10th Cong. Int. Sedimentology, Jerusalem.

MONOD, O. 1977. *Recherches géologiques dans le Taurus Occidental au Sud de Beyşehir (Türquie).* Thèse; Univ. Paris Sud; No. 1839; p. 442.

MOSHKOVITZ, S. & EHRLICH, A. 1980. Late Jurassic calcareous nannoplankton in Israel's offshore and onland areas. *Geol. Surv. Isr. Current Research, 65–72.*

POISSON, A. 1977. *Recherches géologiques dans le Taurus occidental.* Tome 2: Synthèse. Thèse. Univ. Paris-Sud. 550–795.

RAAB, M. 1962. Jurassic–Early Cretaceous Ammonites from the southern coastal plain, Israel. *Bull. Geol. Surv. Isr.* **34,** 24–30.

RICOU, L. E. 1971. Le Croissant ophiolitique péri-arabe, une ceinture de nappes—mise en place au Crétace Superieur. *Rev. Géographie physique et Géol. dynamique*, **4,** 327–349.

—— ARGYRIADIS, I. & MARCOUX, J. 1975. L'axe calcaire du Taurus, un alignement de fenêtres arabo-africaines sous des nappes radiolaritiques, ophiolithiques et métamorphiques. *Bull. Soc. géol. France*, **17,** 1024–1044.

—— & MARCOUX, J. 1980. Organisation générale et rôle structural des radiolarites et ophiolites le long du système Alpino-Mediterranéen. *Bull. Soc. Géol. France*, **22,** 1–14.

——, —— & POISSON, A. 1979. L'allochtonie des Bey Dăglari oreintaux. Reconstruction palinspastique des Taurides occidentales. *Bull. Soc. géol. France*, **21,** 125–133.

ROSENFELD, A. & RAAB, M. 1980. Lower Cretaceous ostracods from Israel. *Geol. Surv. Israel Current Research*, 62–65.

SANDER, N. J. 1970. Structural evolution of the Mediterranean region during the Mesozoic Era. Geology and history of Sicily. *In*: ALVAREZ & GOHRBARDT (eds) Ann. Field Conference No. 12. *Petrol. Explor. Soc. Libya (ESSL)* **291,** 100–189.

SCANDONE, P. 1975. Triassic seaways and the Jurassic Tethys ocean in the central Mediterranean area. *Nature* **256,** 117–119.

VÖRÖS, S. 1980. Lower and Middle Jurassic brachiopod provinces in the western Tethys. *Bull. Hungarian Geol. Soc.* **110,** 395–416.

F. HIRSCH, Geological Survey of Israel, Malkhei Yisrael Street 30, 95501 Jerusalem, Israel.

Ophiolites and volcanic activity near the western edge of the Arabian plate

Michel Delaloye & Jean-Jacques Wagner

SUMMARY: The geology of the Eastern Mediterranean is governed partly by the collision of the Arabian plate with the Anatolian block and partly by the Dead Sea Rift.

The oceanic elements which have been formed between these two blocks are still visible as obducted fragments represented amongst others by the two ophiolitic massifs of Hatay and Baër-Bassit.

The opening of the Atlantic ocean give rise, in the Eastern Mediterranean, to complex relative movements which can partly be resolved into a west to east shear and a south to north compression. The shear pre-dates the compression and took place between the Triassic and middle Cretaceous. It is responsible for the rifting and the volcanism in Syria, as well as the formation of a basin capable of producing ophiolites such as Hatay and Baër-Bassit.

The compressive phase began in the middle Cretaceous and resulted in the detachment and the emplacement of the ophiolites on the Arabian continental margin. The timing of this phase can be determined from the age of the Baër-Bassit amphibolitic sole. Traces of this compressive episode are seen as late as the Neogene.

The volcanic evolution of the border zone of the northwestern section of the Afro-Arabian plate (later part of a separate Arabian plate) can be correlated with the kinematics of the zone deduced from relative movements between Europe and Africa. These movements, which have been worked out from magnetic lineations (Patriat *et al.* 1982) in the North Atlantic and from continental palaeopoles (Irving, 1981), suggest from the opening of the North Atlantic that two principal stages can be distinguished. The first stage was an Upper Triassic-Lower Cretaceous east to west shear movement while the second was a north to south compression, which continues to the present day.

The aim of the present study is to give a more detailed understanding of the Eastern Mediterranean by using trace element geochemistry and isotopic dating. In this context we will discuss the ophiolitic areas of the Hatay (Parrot 1973; Çoğulu *et al.* 1975; Delaloye *et al.* 1977, 1980; Tinkler *et al.* 1981; Selçuk 1981) and of the Baër-Bassit (Delaune-Mayere *et al.* 1977; Parrot 1977a and b) (Fig. 1). The Hatay and Baër-Bassit ophiolitic massifs are included in the southern Turkey fold belt. They belong to the peri-Arabian ophiolitic crescent and represent a link between the ophiolitic massifs of Oman, Neyriz and Kermanshah in the East and Troodos (Cyprus) in the West. The Hatay is a complete ophiolitic sequence including tectonites, cumulates, sheeted dyke complex and lavas. The Baër-Bassit also shows the various members of an ophiolite but they are much more dismembered. Volcaniclastic sequences are strongly developed attesting to the presence of a graben system in the neighbourhood. Further investigation of volcanism along the Dead Sea Rift system is also included in this work.

The Kızıl Dağ massif (Fig. 2), 25 km wide and 45 km long, is composed of a typical ophiolite sequence, up to 6 km thick, composed of serpentinized peridotites and harzburgites, layered gabbros, a sheeted dyke complex and pillow lavas. The ophiolites have been subjected to zeolite facies metamorphism already in an oceanic environment because the sediments which lie stratigraphically above and tectonically below are not metamorphosed. The ophiolites occur as a thick thrust sheet which lies tectonically on Lower Cretaceous limestones of Albian to Aptian age. The oldest sediments covering the ophiolites are Upper Cretaceous limestones of Upper Maastrichtian age. At their base they rest discordantly on the ophiolites and are composed of an extensive conglomerate, 2–5 m thick, which contains abundant ophiolitic detritus.

The Baër-Bassit ophiolitic complex (Fig. 2) consists of harzburgitic tectonites, peridotitic cumulates, layered gabbros, diabase dykes and two layers of pillow lavas. One of these is tholeiitic related to the sheeted complex and the other hypertholeïitic and lies unconformably above the two earlier formations. The lower contacts of the Baër unit are marked by the presence of an amphibolitic sole. We observe in the area a tangential tectonism of intra-Maastrichtian age corresponding to the emplacement of the basic—ultrabasic and re-

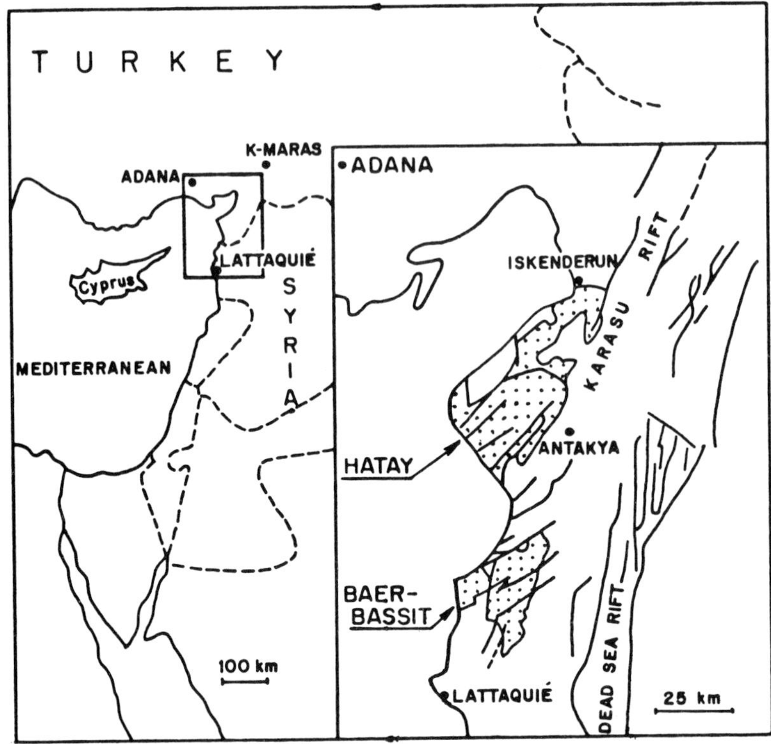

FIG. 1. Location map and insert showing the relations between the Dead Sea Rift and the Kara Su Rift.

lated rocks onto the Arabian margin. During the thrusting, the ophiolitic sheets pushed volcano-sedimentary series of Upper Triassic, Jurassic and Lower Cretaceous age in front of them. The Triassic lavas are tholeïitic with an alkaline tendency, in flows or pillow form. The Jurassic and Cretaceous volcanic horizons consist of tephrites, basanites and lamprophyres, all in pillow form as well as numerous phonolite flows.

Geochemical character of the Hatay and Baër-Bassit ophiolites

Major element geochemistry is not by itself sufficient to fingerprint the environment in which the ophiolites of Hatay and Baër-Bassit were formed. The greatest problem is probably the remobilization and hydrothermal transport of elements that took place during the zeolite to greenschist facies metamorphism to which the rocks were subjected (Delaloye *et al.* 1980).

In contrast many trace elements (mainly those of high field strengths) are little affected by such alteration and metamorphic processes. They are therefore useful for determining the environment of formation of ophiolitic sequences. Analytical data are given in Table 1. Diagrams, such as those devised by Pearce & Cann (1973), can be used to identify the eruptive setting of the volcanic rocks.

The plot of trace elements normalized to an average MORB composition (Fig. 3) (Pearce 1980) preclude the possibility that the lavas from Hatay and Baër-Bassit belong to typical oceanic crust environment. Volcanic arc basalts or marginal basin basalts remain the most probable sources.

The Ti-Zr-Y diagram (Fig. 4) constructed to discriminate between island arc lavas, ocean floor basalts and within-plate lavas shows data points shifted toward the Y corner but still close to the fields A and B. This shift can be best explained by the fact that the asthenosphere which melted to produce the Hatay lavas was already depleted in trace elements. As the depletion is strongly different for Ti and Zr compared to Y the observed trend is as expected.

The Ti vs Zr plot (Fig. 5) shows that both Hatay and Baër-Bassit ophiolites have a volca-

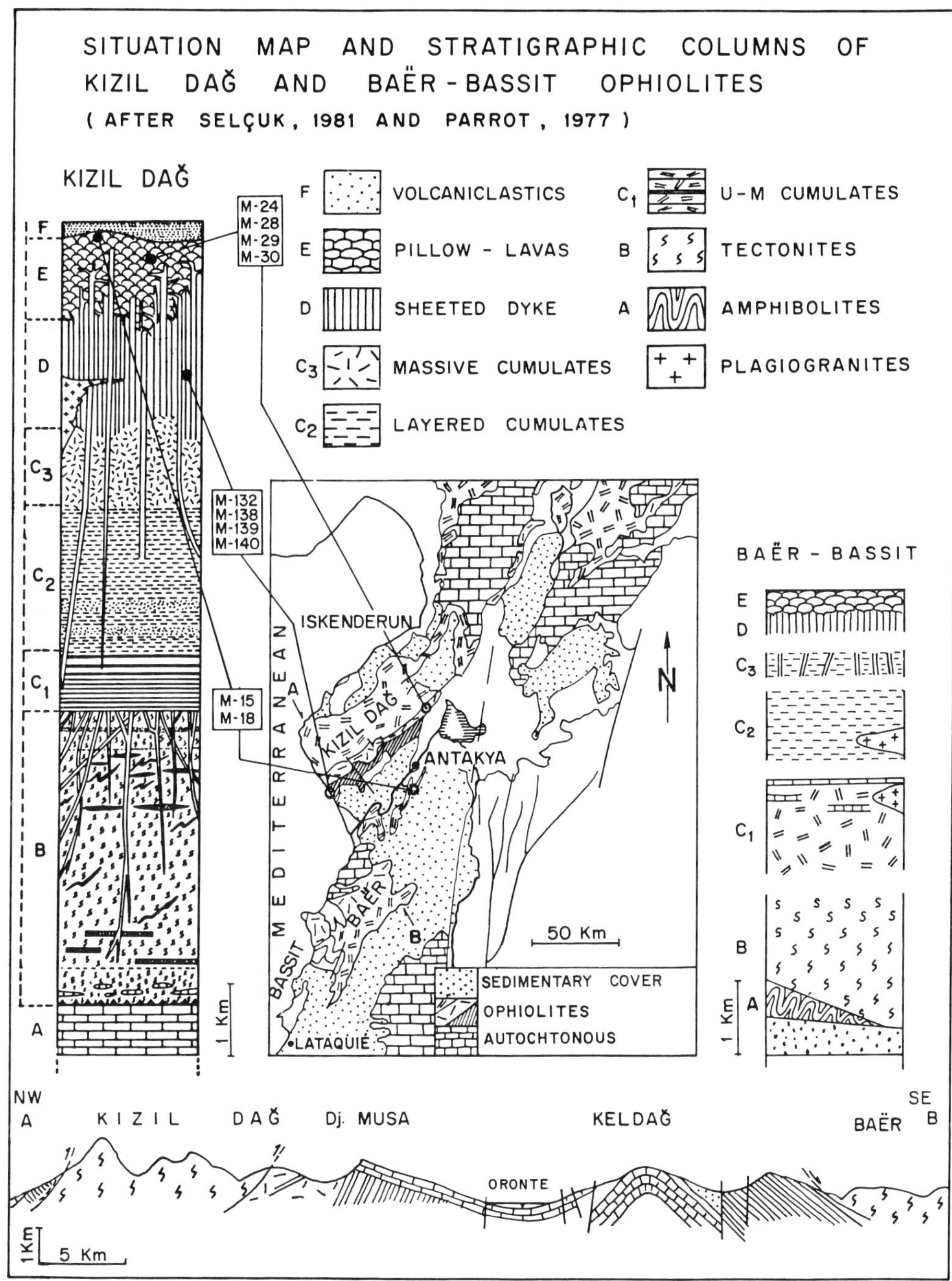

FIG. 2. Geological map and schematic cross-section through the Hatay and Baër-Bassit ophiolites also showing sample locations.

nic arc affinity but the volcanic elements constituting the upper sequence of the volcaniclastic sediments associated with the Baër-Bassit ophiolites have a typical within-plate lava composition. Large volcanic flows associated with pillow lavas occur near Merdavan (80 km northeast of Antakya) on the southeastern rim of the Kara Su Rift. These volcanics belong also to the within-plate lava group.

In conclusion, the use of trace elements to

TABLE 1. *Trace element concentrations for Hatay and Baër-Bassit rocks discussed in the present paper.*

	HATAY											KURD-DAĞ		BAËR-BASSIT							
	MORB	TATAKÖPRÜ PILLOW-LAVAS				SILPIUS PILLOW-L.		ÇEVLIK - Sea Shore Sheeted dyke complex				MERDAVAN BASALT		after Parrot et al. 1974 OPHIOLITE				JURA-CRET. VOLCANICL. ELEMENTS			
		M-24	M-28	M-29	M-30	M-15	M-18	M-132	M-138	M-139	M-140	M-21	M-23	dykes	sheeted dykes	lower P.L.	upper P.L.	Tephrites (5 samples)	Lamprophyres (6)	Monchiquites (5)	Trachytes (4)
SYMBOLS ON FIGURES		+	+	+	+	⊙	⊙	○	○	○	○	×	×	●	●	●	●	△			
Sr	120	100	60	105	325	48	43	102	107	101	137	149	147	(3 s.)	(5 s.)	(13 s.)	(11 s.)				
K	1219	2845	812	5609	2845	650	569	1381	3414	3983	813	3983	3886								
Rb	2	24	7	15	12	10	10	9	8	7	10	17	15								
Ba	20	66	11	28	28	26	44	27	55	44	55	87	92								
Th	0.2	2	4	3	1	2	2		1	4	5	Tr.	Tr.								
Ce	10					10	32					22									
P	524	175	262	306	436	305	262	523	436	2575	524	787	699								
Zr	90	23	20	18	17	23	24	32	39	21	35	83	100	87	94	100	43	305	505	463	675
Hf	2.4	6	3	4	8	4	2	5	8	2	4	14	11								
Sm	3.3					0,7	2,5					3,4									
Ti	9000	2280	2220	2340	4020	2100	2100	4320	4320	3300	4560	8400	8850	0,58%	0,68	1,17	0,39	3.16	3.0	3.2	1.24
Y	30	17	20	13	21	14	14	20	21	18	20	23	26					39	34	53	16
Yb	3.4					1,3	2,1					2,4									
Cr	250	68	90	143	123	476	515	236	105	184	84	224	238								

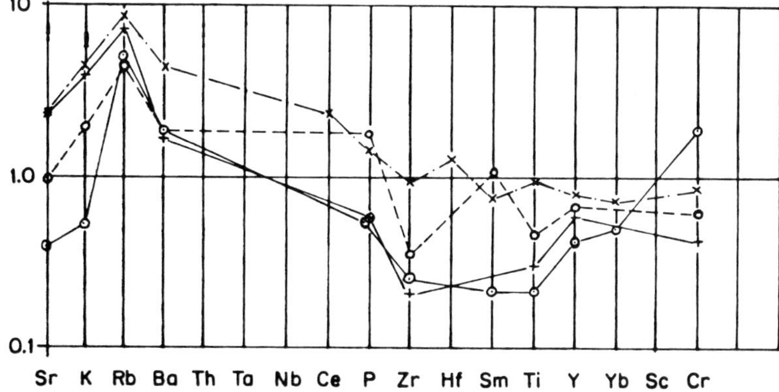

FIG. 3. Trace element geochemical pattern of lavas from the Hatay area.
+ Pillow lavas from Tataköprü (4 samples)
⊙ Pillow lavas from Silpius (2 samples)
○ Sheeted dyke complex on the shore (4 samples)
× Non-ophiolitic sequence of Merdavan (2 samples)

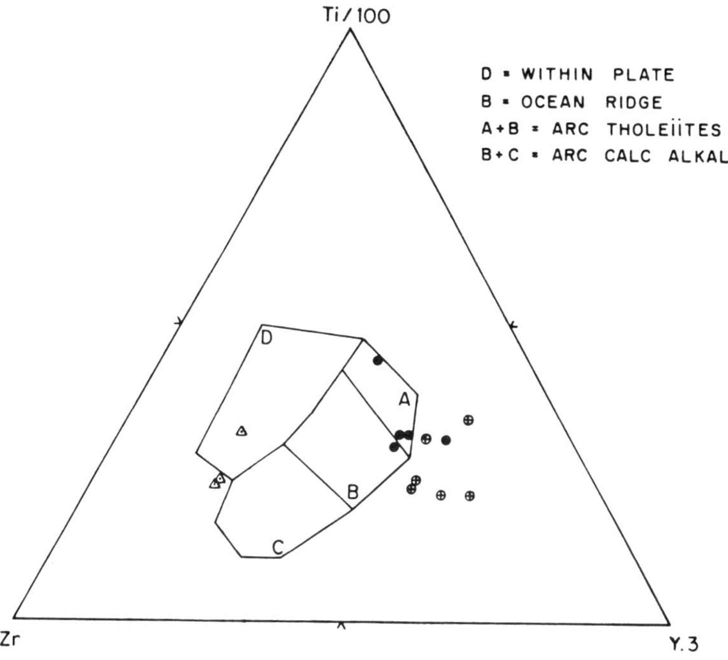

FIG. 4. Ti–Zr–Y diagram for the Hatay ophiolite (⊕), the Baër-Bassit ophiolite (●) and the Baër-Bassit Jurassic to Cretaceous volcaniclastics (△).

characterize the volcanic setting of ophiolitic sequences near the northwestern border of the Arabian plate indicates a possible marginal basin origin. Petrological arguments, the most important of which is the absence of typical andesitic lavas, suggest that neither Hatay nor Baër-Bassit belong to a typical volcanic arc: a marginal basin origin is therefore most likely.

Isotopic ages of the Hatay and Baër-Bassit areas

Knowledge of the ages of formation and emplacement of these massifs provides important constraints on the development of accurate models of the Mesozoic geological evolution of the Eastern Mediterranean region. For inter-

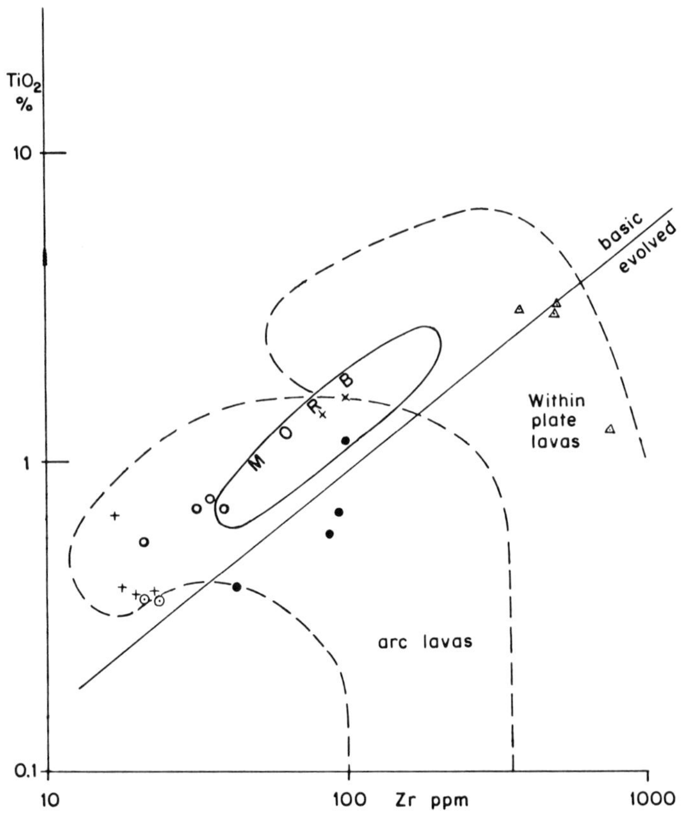

FIG. 5. TiO_2-Zr diagram for Hatay pillow lavas (+, ⊙), Hatay sheeted dyke complex diabases (○), non-ophiolitic pillow lavas from Merdavan (×) and Baër-Bassit dykes and pillow-lavas (●).

pretation purposes we use the time scale published by Odin (1982).

Several rock formations have been dated by the K-Ar method (Delaloye *et al.* 1977, 1980). The oldest volcanic rocks in these areas are in volcaniclastic sediments of Baër-Bassit (Parrot 1973) where Triassic and Jurassic volcanic units have been recognized. A determination of the biotite from the monchiquite outcrop near Turkmannli gave an age of 150 Ma. This fits with the range of ages deduced from palaeontological studies (Delaune-Mayere *et al.* 1976). From the ophiolite, we have dated pillow lavas, dykes and gabbros (Coğulu *et al.* 1975; Delaloye *et al.* 1977, 1980). Due to intense weathering, pillow-lavas gave unreliable values. The dykes have been subjected to only weak greenschist facies metamorphism on the ocean floor (Sarp & Tanner 1977); there is no evidence of post-emplacement metamorphism. In this study only the least metamorphosed dykes from an extensive collection of samples were dated. With one exception, the dates obtained range from 73 to 99 Ma and indicate that the complex is mid-Cretaceous or older. Secondary green amphiboles separated from gabbro samples gave dates which are on average older than those from the dykes so that a Late Jurassic age cannot be ruled out. Owing to their low K content the dates of these samples are not considered to be very reliable.

Also dated were amphiboles from metamorphic rocks. The metamorphic sole at Baër-Bassit (high grade amphibolitic facies, de Souza, oral communication) gives an age of 85–95 Ma in accord with the ages given by Thuizat *et al.* (1981). If one considers the sole to have been formed by detachment and overriding of the oceanic crust (Coleman 1981), and if one accepts the hypothesis that the spreading axis is a zone of weakness (Boudier *et al.* 1982) and a possible site of the detachment, then the ophiolitic material associated with the amphibolitic sole represents the youngest part of the lithosphere. As a consequence, isotopic data obtained in this type of environment represent an upper limit for the formation age of the ophiolite. It is then quite possible that the ophiolites of Hatay and Baër-Bassit represent the youngest elements of a ridge already functioning during Jurassic times. K-Ar ages on gabbros and diabases in Hatay and Baër-Bassit indicating Jurassic and Early Cretaceous ocean floor activity are good evidence (Delaloye *et al.* 1980).

Tectonic emplacement of ophiolite

Dubertret (1955) first described the emplacement of the basic-ultrabasic body of the Kızıl Dağ. Subsequent studies (Vuagnat & Çoğulu 1967; Aslaner 1973; Çoğulu 1973; Tinkler *et al.* 1981) were carried out on this subject. Selçuk (1981) studied in detail the stratigraphy of the underlying and overlying sediments.

In the northern part of the massif (south of the village of Kömürçukuru), in a small window, the ultramafic unit rests tectonically on Lower Cretaceous limestones, ranging in age from Albian to Aptian. This unit shows the basal thrust along which the ophiolite was emplaced (Aslaner 1973; Dubertret 1955). The ultramafic rocks, in the vicinity of the contact, contain large tectonic slices of limestone 10–20 m thick, and 20–40 m long, the youngest blocks being Campanian (Upper Cretaceous) and the oldest ones being possibly Triassic. The lithologies of these limestone wedges are similar to carbonate sequences found in neighbouring parts of Syria (Al-Maleh 1976). It is therefore thought that the limestone wedges were originally part of carbonate sequences of the Arabian continental platform which were later detached and included in the ultrabasics during emplacement of the ophiolites.

The NW-SE emplacement of the ophiolites is thought to be late Campanian to Maastrichtian because:

(1) along the tectonic contact between the ophiolites and the Albian/Aptian limestones there are wedges of Campanian limestones included in the serpentinites, and,

(2) the Maastrichtian sediments which lie on the ophiolites are the earliest to contain ophiolitic detritus and they are associated with other Upper Cretaceous sediments which show continental affinities (Dubertret 1955). Further north in the Amanos, elements of the ophiolite rest on Palaeozoic formations of Arabian block affinity (Dean & Krummenacher 1961). In Hatay, one sees also the cover of the massive Maastrichtian limestones with basal conglomerate resting on pillow-lavas. In other places sedimentary transgressions occur as late as Mid-Miocene (Tinkler *et al.*, 1981).

The tectonic emplacement of this section of the ocean floor onto the Arabian platform took place as a cold rigid slab during Maestrichtian or immediately pre-Maestrichtian times. From the age of the amphibolitic sole a period of almost 20–25 Ma elapsed between the initial (pre-Campanian) detachment of the ophiolitic sheets and their final emplacement to their present positions. This has also been observed by Thuizat *et al.* (1981) in ophiolites from Pindos and the Taurides.

Discussion and conclusions

Using the model of Patriat *et al.* (1982) based on the magnetic lineations of the Atlantic ocean, the relative movement between Africa and Eurasia can be resolved into three main phases (Fig. 6). In the southeastern Mediterranean these are:

(1) a West to East translation of about 1200 km between 180 and 110 Ma,

(2) the translation continues but changes its direction to ENE to attain 800 km in our area between 110 and 85 Ma,

(3) a true compressive movement oriented more or less south to north as Africa rotates anticlockwise around an axis situated near the NW part of Morocco. This last phase was active from 85 Ma ago up to the present day. Such large scale movements should have profoundly affected the geology of the Eastern Mediterranean in particular its volcanism. The phases (1) and partly (2) of translation have induced rifting of the northern margin of Gondwana. This rifting is recognised by the Carnian-Norian volcanism found in some units of the volcaniclastics associated with the Baër-Bassit ophiolite. The intra-plate volcanism continued stepwise as attested to by Middle-Jurassic activity in the Anti-Liban (170 Ma) near Qastal Djendal, by Upper-Jurassic activity recognised in the volcaniclastics of the Baër-Bassit (150 Ma) and also by Lower-Cretaceous flows in the Djebel Ansarieh in northern Syria (100 Ma). This global movement favoured the formation of the Neo-Tethysian oceanic system. The ophiolites of Hatay and Baër-Bassit are elements of the southern part of this system.

From our discussion on the isotopic ages of ophiolitic material, we favour oceanic ridge activity in the studied area at least from the Late-Jurassic. The trace element geochemistry of these ophiolites suggests an origin in a marginal basin.

The studies of the Dead Sea Rift system have demonstrated the existence of pull-apart basins induced by lateral movements (Quennell 1959). Could an extrapolation of this phenomenon to the large scale African west to east lateral

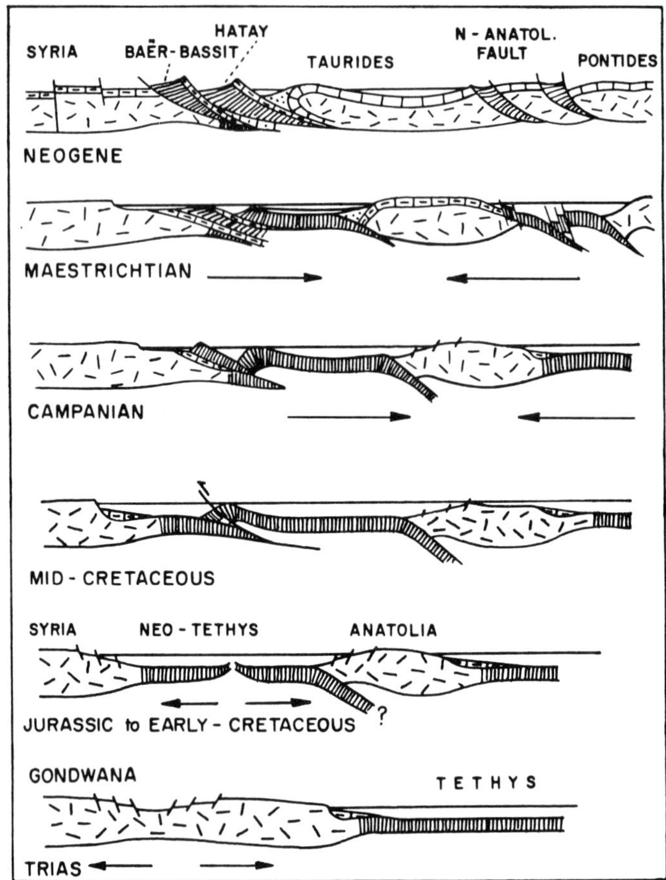

FIG. 6. Reconstruction of the tectonic history of the Levant between Trias and Neogene along a NW-SE section passing through Ankara-Adana-Antakya-Damascus.

translation also be invoked to describe a mechanism which give birth to oceanic crust having the geochemical character of a marginal basin defined by the Pearce diagrams? This is at present being investigated.

The youngest sediments (Selçuk 1981) discovered under the thrust-plane of the ophiolite of the Kızıl Dağ are dated palaeontologically as Cenomanian to Santonian (95 to 83 Ma). It is remarkable that the ages obtained on the amphibolitic sole of the Baër-Bassit (Thuizat *et al.*, 1981 and unpublished data by ourselves) are so close to the interruption of sedimentation on the Arabian margin.

The oldest sediments found above the pillow-lavas and the associated volcaniclastics are of Middle to Upper Maastrichtian age (69 to 75 Ma). The sedimentary gap is therefore of 14 to 30 Ma. The closure of the Neo-Tethys and the obduction of oceanic material onto the Arabian shelf must have occurred during the same time span. Patriat *et al.* propose a south to north movement of 900 to 1200 km of the Arabian plate towards Eurasia in order to close the Neo-Tethyan ocean. Both estimations of time and distance would necessitate a speed of closure ranging between 3 and 4 cm/year which is compatible with the dynamics of such systems.

The opening of the Red Sea in the Lower Neogene led to the formation of a separate Arabian plate and induced more compressive movements. The widespread effect of this opening is suggested by several disconformities in the sedimentary record as well as by very active Neogene volcanism associated with the Dead Sea Rift and particularly well developed in Syria.

ACKNOWLEDGEMENTS: We would like to thank J. Pearce & M. Vuagnat for their constructive criticism of the manuscript. We acknowledge comments from the reviewers J. Cann & T. Juteau. We thank Michèle Senn for technical assitance in chemical analysis, Jacqueline Berthoud and Helen McKeating for typing. The authors gratefully acknowledge financial support from the Swiss National Science Foundation (grant 2.077–81), the director of Maden Tetkik ve Arama in Ankara for assistance and cooperation. M. Mouti, M. Otaki & K. Al-Malch from Damascus University are thanked for assistance and fruitful dicussions during fieldwork in Syria.

References

AL-MALEH, K. 1976. *Etude stratigraphique, pétrographique, sédimentologique et géochimique du Crétacé du NW syrien (Kurd Dagh et environs D'Aafrine)*. Thesis, Ph D. (unpubl.), University of Paris VI, 620 pp.

ASLANER, M. 1973. Iskenderun-Kırıkhan Bölgesindeki Ofiolitlerin Jeoloji ve Petrografisi. *Maden Tetkik ve Arama Inst. Bull.* **150,** 78 pp.

BOUDIER, F., NICOLAS, A. & BOUCHEZ, J. L. 1982. Kinematics of oceanic thrusting and subduction from basal sections of ophiolites. *Nature*, **296,** 825–828.

ÇOĞULU, E. 1973. New data on the petrology of the Kızıl Dağ massif (Hatay, Turkey). *Proc. of Congress of Earth Sciences '50th Anniversary of the Turkish Republic'*, 409–423.

——, DELALOYE, M., VUAGNAT, M. & WAGNER, J.-J. 1975. Some geochemical, geochronological and petrophysical data on the ophiolitic Massif from the Kızıl Dağ, Hatay, Turkey. *C.r. Séances Soc. Phys. Hist. nat. Genève*, **10,** 141–150.

COLEMAN, R. G. 1981. Tectonic setting for ophiolite obduction in Oman. *J. geophys. Res.* **86,** 2497–2508.

DEAN, W. T. & KRUMMENACHER, R. 1961. Cambrian trilobites from the Amanos Mountains, Turkey. *Palaeontology, London*, **4,** 71–81.

DELALOYE, M., VUAGNAT, M. & WAGNER, J.-J. 1977. K-Ar ages from the Kızıl Dağ ophiolitic complex (Hatay, Turkey) and their interpretation. *In:* BIJU-DUVAL, B. & MONTADERT L. (eds). *The Structural History of the Mediterranean Basins*, Editions Technip, Paris, 73–77.

——, DE SOUZA, H., WAGNER, J.-J. & HEDLEY, I. 1980. Isotopic ages of ophiolites from the Eastern Mediterranean. *In:* PANAYIOTOU A. (ed.). *Ophiolites, Proc. Int. Oph. Symp. Cyprus*, 292–295.

DELAUNE-MAYERE, M., MARCOU, J., PARROT, J. F. & POISSON, A. 1976. Modèle d'évolution mésozoïque de la paléo-marge téthysienne au niveau des nappes radiolaritiques et ophiolitiques du Taurus Lycien d'Antalya et du Baër-Bassit. *In:* BIJU-DUVAL, B. & MONTADERT, L. (eds). *The Structural History of the Mediterranean Basins*, Editions Technip, Paris, 79–94.

DUBERTRET, L. 1955. Géologie des roches vertes du nord-ouest de la Syrie et du Hatay (Turquie). *Notes Mém. Moyen-Orient*, **6,** 227 pp.

IRVING, R. 1981. Phanerozoic continental drift. *Phys. Earth planet. Inter.* **24,** 197–204.

ODIN, G. S. 1982. The Phanerozoïc time scale revisited. *Episodes* 1982/3, 3–9.

PARROT, J.-F. 1973. Pétrologie de la coupe du Djebel Moussa: Massif basique-ultrabasique du Kızıl Dağ (Hatay, Turquie). *Sci. Terre, Nancy*, **28,** 43–172.

—— 1977. Ophiolites du Nord Ouest Syrien et évolution de la croûte océanique téthysienne au cours du mésozoïque. *Tectonophysics*, **41,** 251–268.

—— 1977. Assemblage ophiolitique du Baër-Bassit et termes effusifs du volcano-sédimentaire. Pétrologie d'un fragment de la croûte océanique téthysienne, charriée sur la plateforme syrienne. *Trav. et Doc. ORSTOM, Série géol.* **6,** 97–126.

PATRIAT, P., SEGOUFIN, J., SCHLICH, R., GOSLIN, J., AUZINDE, J.-M., BEUZART, P., BONNIN, J. & OLIVET, J.-L. 1982. Les mouvements relatifs de l'Inde, de l'Afrique et de l'Eurasie. *Bull. Soc. géol. France*, **24,** 363–373.

PEARCE, J. A. & CANN, J. R. 1973. Tectonic setting of basic volcanic rocks determined using trace element analysis. *Earth planet. Sci. Lett.* **19,** 290–300.

—— 1980. Geochemical evidence for the genesis and eruptive setting of lavas from Tethyan ophiolites. *In:* PANAYIOTOU, A. (ed.) *Ophiolites. Proc. Int. Oph. Symp, Cyprus*, 261–272.

QUENNELL, A. M. 1959. *Tectonics of the Dead Sea Rift.* Int. Geol Congress Mexico. Assoc. Serv. Geol. Afr. 385–405.

SARP, H. & TANNER, M. 1977. Mise en évidence dans le sheeted complex du Kızıl Dağ (Hatay) d'un métamorphisme (Abstract). *Sixth Colloquium on Geology of the Aegean Region, Ege University, İzmir*, **89.**

SELÇUK, J. 1981. *Etude géologique de la partie méridionale du Hatay (Turquie)*. Thesis No 1997; PhD. (publ.). University of Geneva, 116 pp.

TINKLER, C., WAGNER, J.-J., DELALOYE, M. & SELÇUK, H. 1981. Tectonic History of the Hatay Ophiolites (South Turkey) and their relation with the Dead Sea Rift. *Tectonophysics*, **72,** 23–41.

THIUZAT, R., WHITECHURCH, H., MONTIGNY, R. & JUTEAU, T. 1981. K-Ar dating of some infra-ophiolitic metamorphic soles from the Eastern Mediterranean: new evidence for oceanic thrustings before obduction. *Earth planet. Sci. Lett.* **52,** 302–310.

VUAGNAT, M. & ÇOĞULU, E. 1967. Quelques réflexions sur le massif basique-ultrabasique du Kızıl Dağ, Hatay, Turquie. *C.r. Séances Soc. Phys. Hist. nat., Genève*, **2,** 210–216.

MICHEL DELALOYE & JEAN-JACQUES WAGNER, Earth Sciences Section, The University, CH-1211 Geneva 4.

3. NEOTETHYS: Turkey

The papers in this section are all concerned with aspects of the Mesozoic evolution of Turkey, which has seen an upsurge in international research interest in recent years. The dominant themes are the processes of *initiation*, *growth* and *destruction* of Mesozoic ocean basins documented in the development of rift-related sedimentary and volcanic sequences and in the internal character and 'organisation' of Mesozoic ophiolites, and the ocean *closure* indicators—the calc-alkaline volcanic belts, blueschist zones and nappe emplacement sequences involving ophiolites.

If regional plate-kinematic synthesis is the ultimate aim then an obvious precursor stage is the building of a network of reasonably well understood transects where ocean creation and destruction are recorded, coupled with more summary attention to variations along-strike and with time in relevant linear belts. This is the stage represented by the papers in this section. Mapping and detailed interpretation of the many vast tracts that lie between the discontinuous ophiolite belts will probably continue to take a minor role compared to 'suture documentation', and models will remain largely two-dimensional until the areal density of information is much greater.

Figures 1 & 2 show the areas covered by contributors to this section. The scale of the map should be noted. There are two papers not so directly ophiolite or continental-margin oriented. The paper by **POISSON** would be a good starting point for anyone seeking a sample of the stratigraphic complexities and interpretation problems in one tract of nappes—the Lycian Nappes of SW Turkey. The other exception is the paper by **LAUER** reporting palaeomagnetic data bearing on the number and location of Turkey's component blocks during the Mesozoic. It is a fair reflection of the gap between what are presently realistic interpretation objectives on the ground and what might actually have happened, that **LAUER**'s far-reaching inferences of large-scale independent block-motion are only discussed in one of the other contributions, **RICOU *et al.*** (see also the introductory chapter). One important reason is that he concludes that no significant displacement is detectable between allochthonous sequences and the local autochthon with which they are associated so justifying for the moment local two-dimensional interpretations.

Returning to the main theme of the majority of the contributions, the reader should recognize that a fundamental and resilient dichotomy of interpretation exists among the contributors, which will probably take a future Eastern Mediterranean conference to resolve. On the one hand **RICOU *et al.*** argue that so remarkable are the parallels and systematic variations in the histories of rifting, sea-floor spreading, and emplacement in the two discontinuous ophiolite belts north and south of the Tauride carbonate axis, that the belts must have a common origin in a *single* basin, in this case a northern one. The opposing view, implicit for example in the model of **ROBERTSON & WOODCOCK** for the Antalya area is that parallel evolution in different belts reflects the parallel generation and destruction of *more than one* Neotethyan ocean basin, with common timing and common stratigraphic evolution being a natural consequence of the common crustal structure, large-scale plate-configuration, climate, etc., in the region as a whole. These opposing views are discussed at length in the introductory chapter. The reader seeking first-hand statements should perhaps start with **RICOU *et al.*** for the basic philosophy and support it with **WHITECHURCH *et al.*** in which ophiolites in the two belts are reviewed in some detail. **ROBERTSON & WOODCOCK** review the history of the models proposed for the Antalya area and focus specifically on areas and field relations which constitute critical tests of the **RICOU *et al.*** concept.

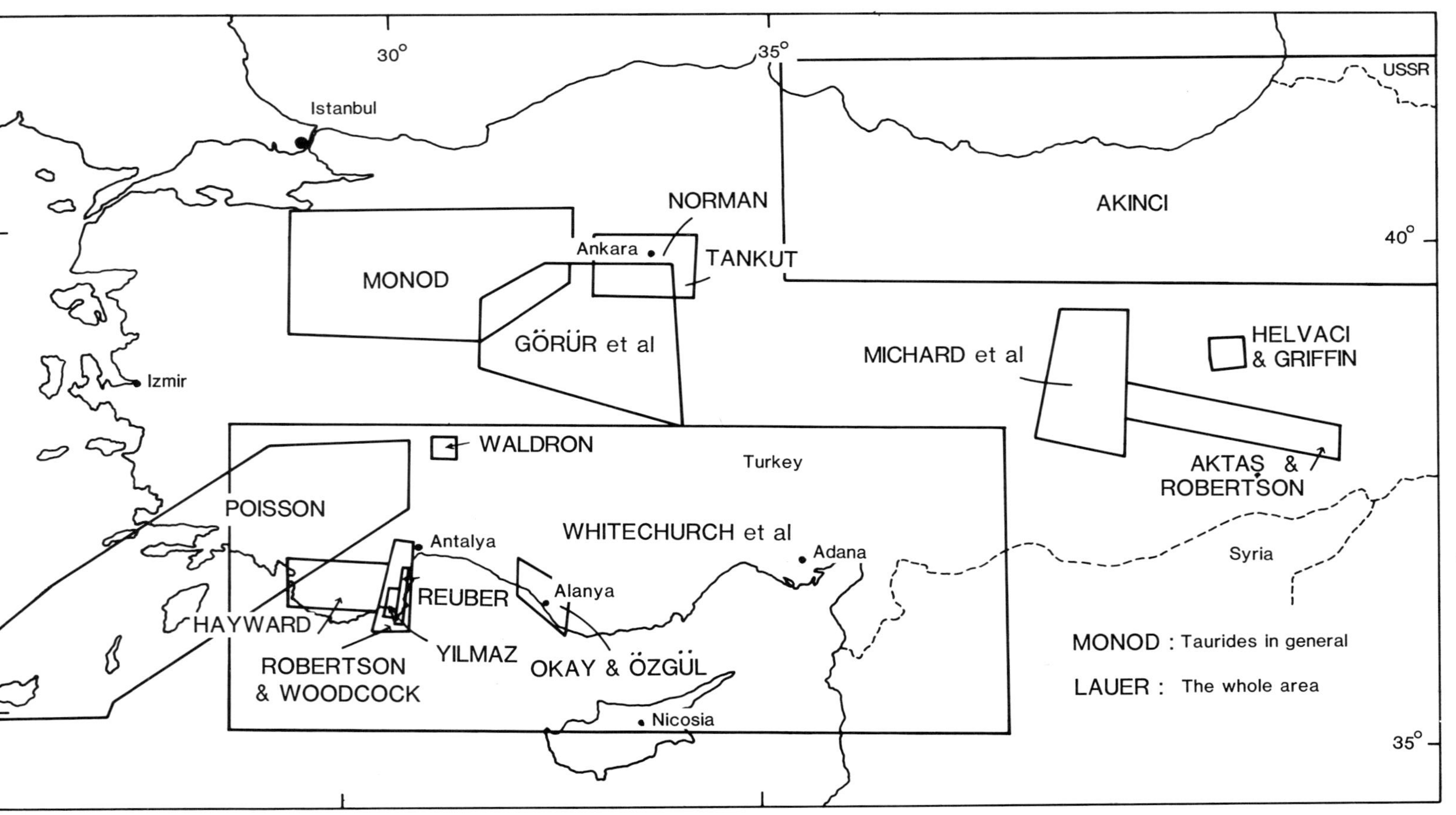

FIG. 1. The areas of Turkey discussed by various authors in this section.

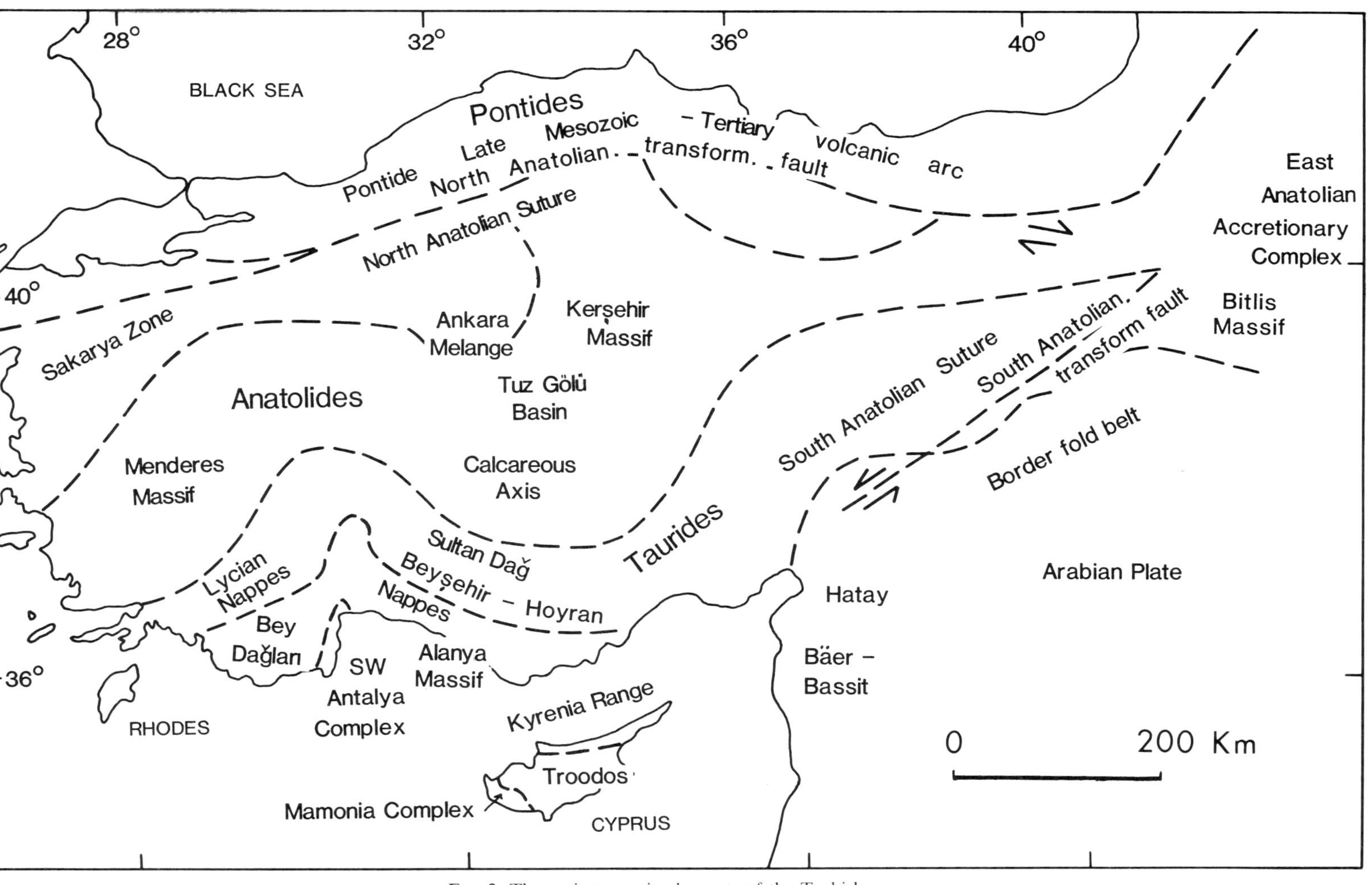

FIG. 2. The main tectonic elements of the Turkish area.

The two opposing views are discussed further in the papers by **MICHARD *et al.*** and **AKTAŞ & ROBERTSON** which consider segments of the southern suture belt in Eastern Anatolia, and these two contributions well illustrate the enormous diversity in structure, stratigraphy, magmatism and the timing of major events to be found even along what is apparently a 'single' belt, so great is the scale of Turkish geology.

The papers are arranged geographically, which means that the information they contain on the often synchronous phases of rifting, ocean growth, subduction and collision is rather scattered. Here we provide a brief guide to where contributions under these headings may be found.

Most of the discussion on rifted and passive margin development in the existing literature has centred on the Antalya–Isparta area of SW Turkey. **POISSON**, working in the Lycian nappes, to the west correlates stratigraphic sequences as far as the Aegean Sea in the far west. He draws attention to the Kızılıca–Corak gol trough which originated by Early Jurassic faulting with deposition of pelagic carbonates throughout much of the Mesozoic to Early Tertiary. He points out that this trough could well be an extension of the Ionian Zone as the Aegean arc runs from western Greece, through Crete and Rhodes (see **BONNEAU**, Section 4 and the introductory chapter). The appearance of ophiolite-derived clastics in the trough as late as Upper Eocene fits with an origin south of the Menderes massif, rather than north as has been assumed previously for the Lycian nappes. Since the ophiolites themselves in the Lycian nappes were not apparently rooted in this trough, **POISSON** still prefers to derive them from a basin north of the Menderes massif. This has profound implications for the scale of nappe transport and the curvature of the Aegean arc, which **POISSON** then goes on to discuss.

The Tauride 'Calcareous Axis' is the critical zone either across which ophiolite nappes were transported from the north or, alternatively, which separated northern and southern basins, according to one's view in the great debate. **WALDRON**, describing a key area near Eğridir in the Isparta angle, on the edge of this axis, uses detailed mapping, facies analysis and balanced cross-sections to reconstruct the Mesozoic palaeogeography. The result is a series of irregularly shaped carbonate banks separated by deep basins with pelagic sediments, similar to the modern Bahamas. Latest Cretaceous thrusting to the north or east is clear from thrust duplex geometry, followed by SW or W-directed refolding and thrusting in the north-east of the area, probably in the Eocene. The structural analysis lends no support to the case for ophiolite nappe transport from a northern basin as in **RICOU *et al.***'s concept.

ROBERTSON & WOODCOCK review the sedimentary and structural evidence for the rifting and passive margin development (and subsequent tectonic emplacement) of the segment of the Antalya allochthon, located on the SW limb of the Isparta angle. The competing models share a common view of the Triassic rifting history, but still contentious issues are the former existence of off-margin carbonate banks, the timing and significance of mafic volcanism and ophiolite genesis and the extent and timing of strike-slip faulting. **ROBERTSON & WOODCOCK** outline their conception of the Mesozoic palaeogeography which involves progressive subsidence of a braided Triassic rift during much of the Jurassic followed by genesis of ophiolites in a new strike-slip-controlled tectonic regime. Strike-slip faulting along a N–S margin dominated the active margin phase from latest Cretaceous to Mid-Tertiary.

Evidence from the volcanics and sediments related to Triassic rifting in the various allochthonous units, including the Lycian Nappes, the Beyşehır-Hoyran Nappes (Turkey), Baër-Bassit (N. Syria) and the Mamonia 'nappes' (SW Cyprus) is summarized by **RICOU *et al.*** and **WHITECHURCH *et al.*** Ricou states that the Mid-Triassic units of the 'Calcareous Axis' show a northwards slope to basin transition.

Ophiolites

WHITECHURCH *et al.* and **RICOU *et al.*** review between them much new data on the age, petrology and 'organization' of the Tauride ophiolites and those of Antalya (Tekirova), Cyprus and Hatay/Baër-Bassit. The internal fabric of the Tekirova ophiolite and its age of creation and slicing are discussed by **REUBER** and **YILMAZ** respectively. **DELALOYE & WAGNER**'s geochemical and age data on Hatay and Baër-Bassit, presented in the previous section, augment this information.

On age of creation, firm data are still sparse where the sedimentary cover is missing, as in the 'Tauride-type' of eroded, dismembered ophiolite with thick cumulate sequences: the Antalya (Tekirova), Lycian, Baër-Bassit and Pozantı-Kersantı ophiolites (**RICOU/WHITECHURCH**). Isotopic ages from plutonics in the Antalya (Tekirova) (**YILMAZ**) and Baër-Bassit (**DELALOYE & WAGNER**) ophiolites are *Upper Cretaceous*, in accord with the sedimentary ages from the thinner intact, cumulate-poor 'Troodos-type' ophiolites of Hatay and Troodos itself. **RICOU/WHITECHURCH** using in part the detailed fabric analysis of **REUBER** argue that sheeted-dyke orientations and internal structure in those ophiolites so far studied, are all consistent with original E–W spreading axes cut by N–S transform faults.

This consistency of creation age and orientation extends to the 'slicing' ages considered below and constitutes the basis of the **RICOU/WHITECHURCH *et al.*** single-basin argument.

Destructive margin processes

From the Upper Cretaceous onwards there is abundant evidence from many parts of Turkey of destructive margin processes until continental collision in the Miocene. Looking at the two southern ophiolite belts first, **WHITECHURCH *et al.*** and **RICOU *et al.*** consider that intra-oceanic slicing and the generation of amphibolite soles beneath the ophiolites (other than Troodos and Hatay) are the first evidence of compressional tectonics. A progression from west to east in the ages of the soles, from 104 Ma in the Lycian ophiolites to 88 Ma for Baër-Bassit in the east, is attributed to the northward motion of African affecting a single ocean basin.

There is general agreement that ophiolites were emplaced onto the Arabian margin in the Upper Cretaceous (e.g. see **DELAUNE-MAYERE**, section 2). **RICOU *et al.*** see this as the result of complete closure of the single northern Neotethyan basin, with expulsion of the ophiolites, following the precursor intra-oceanic slicing. Further collision-related nappe-telescoping then followed in the Tertiary. As the carbonate platform in the Antalya area, and the unseen basement of Troodos, are envisaged as the continuation of the African margin west of Hatay, the emplacement of these ophiolites is considered to be the same event.

The alternative two-ocean-basin concept admits the possibility of ophiolite emplacement from a southern basin south onto the Arabian margin, as well as north onto the southern edge of the 'calcareous axis' or an intervening arc. The papers of **ROBERTSON & WOODCOCK** on Antalya, **MICHARD *et al.*** on the Malatya transect of the Arabian margin suture, **AKTAŞ & ROBERTSON** on the adjacent Maden area, east of Malatya, **HAYWARD** west, and **WALDRON** north of Antalya all deal with these basic questions of transport polarity in the southern belt of ophiolites.

ROBERTSON & WOODCOCK's Antalya review, as noted earlier, presents a case for Upper Cretaceous strike-slip-dominated emplacement of the passive margin sequences and Upper Cretaceous Tekirova ophiolite, from the south and east onto the adjacent Bey Dağları platform. **REUBER**'s independent documentation of N–S transform fabrics in the Tekirova ophiolite is of note in this respect. Further north near Eğridir **WALDRON**'s first and major Upper Cretaceous thrust phase is directed towards the NE quadrant. **HAYWARD**, describing the incoming of ophiolitic debris in continuous Miocene sedimentary sequences on the Bey Dağları autochthon, documents the final mutual approach from opposite directions of the Lycian and Antalya ophiolitic nappes but sees no evidence for an earlier emplacement event across the platform. **ROBERTSON & WOODCOCK** discuss at length how **RICOU *et al.***'s model attempts to explain these observations.

East of Hatay and Baër-Bassit, along the suture zone, the structural succession is markedly different and complex and many interpretation problems remain. The two papers by **MICHARD *et al.*** and **AKTAŞ & ROBERTSON** which deal with areas with several structural units in common produce between them three fundamentally different alternative models. The major Upper Cretaceous volcanic arc north of the suture and clear evidence of ophiolite generation south of this arc are important common components. **MICHARD *et al.***'s one-basin model has to invoke a subduction-generated back-arc basin to generate this southern Elaziğ ophiolite and both north- and south-dipping subduction to explain the present structural sequence. Their alternative two-basin model shares with **AKTAŞ & ROBERTSON**'s favoured alternative interpretation, the conclusion that the southern basin remained in existence into Tertiary times. However, **AKTAŞ & ROBERTSON** conclude that northward subduction generated the arc, that an Upper Cretaceous ophiolite was emplaced northwards over the fore-arc complex and southwards elsewhere onto the Arabian margin, and that further northward subduction led to imbrication of the remaining basin and ultimately collision.

HELVACI & GRIFFIN illuminate the structural metamorphic and magmatic history of one component in the Eastern Anatolia structural stack—the Bitlis massif, which forms the uppermost unit in the east and in part the basement on which the Upper Cretaceous volcanic arc was constructed. They present new radiometric age data and argue that in the Avnic area a Lower Ordovician–Devonian meta-volcanic unit was intruded by granitoids and then covered unconformably by Carboniferous to ?Permian shales and limestones, which were deformed and metamorphosed in the Cretaceous (90 Ma) and Eocene (40 Ma).

Still problematic is the significance of important new discoveries by **OKAY & ÖZGÜL** of a blueschist nappe in the Alanya massif along the Turkish Mediterranean coast. Eclogites and blueschists occur in one of three nappes thrust over the Antalya Complex, in which a fine Ordovician to Late Mesozoic succession is exposed in the Alanya window.

In contrast to the North Turkish blueschists described by **OKAY**, those in the Alanya massif show evidence of two distinct metamorphic events prior to exposure and transgression in the Middle Eocene.

For the remaining papers, the coverage of destructive margin processes switches further north to the North Anatolian suture zone south of the Pontides. In these areas, clear evidence of a previous rifting phase is often obscure and there is also evidence of active margin processes earlier than the Upper Cretaceous.

NORMAN and **TANKUT** summarize the evidence of accretionary sedimentation, tectonics and magmatism along the south Pontide active margin. The famous ca. 100 km wide Ankara melange consist of 7 parallel belts, including a 'metamorphic belt' melange, a 'limestone block' melange and an ophiolitic melange. **NORMAN** views the Ankara melange as the result of subduction under a continental block then located to the east (Sakarya or Cimmerian continent of **ŞENGÖR** ***et al.***) possibly from as early as mid-Jurassic onwards. He attributes major debris-flow units to oblique subduction, uplift of trench sediments and flow roughly at right angles to the direction of subduction. **TANKUT** summarizes the limited data on the field relations, mineralogy and geochemistry of mafic igneous rocks in the Ankara melange. Only rare basic or ultramafic rocks are known from the 'metamorphic block' and the 'limestone block' melange. All the mafic rocks are altered but initial data from 'immobile' trace elements points to a low-Ti type chemistry typical of many of the Eastern Mediterranean Upper Cretaceous ophiolites.

OKAY discusses the major developments of blueschists located in the Western Pontides in a regional depression between the 'Sakarya zone' of the Pontides and the carbonate cover of the Menderes metamorphic massif to the south. The HP/LT belt is tectonically overlain by a major undated, little-metamorphosed peridotite nappe. **OKAY** shows that both the blueschist nappe and the structurally underlying volcano–sedimentary unit were derived from similar protoliths, including abundant pyroclastics, ranging in age from at least Late Jurassic to Upper Cretaceous. Influenced by a downward transition of the blueschist nappe into a basal marble unit, **OKAY** suggests that the blueschists originated as the northern passive margin of the 'Anatolide-Tauride platform' which was subducted from the Turonian, followed by continental collision. The blueschists were exposed as early as Upper Campanian and problems inherent in this remarkably rapid uplift are discussed.

The possible complexities involving closure of a multi-strand Mesozoic Tethys are discussed by **GÖRÜR** ***et al.*** The contrasting deposition in the Haymana and Tuz Gölü basins of central Anatolia is summarized, from isopachyte maps. After ophiolite emplacement, flysch troughs formed then filled in Palaeocene and Eocene time. **GÖRÜR** ***et al.*** unfold a scenario in which by the late Maastrichtian northward subduction had pinned the tip of the Kerşehir micro-continent against the Pontide active margin to the north, while subduction continued during the early Tertiary to the north west and along the western margin of the Kirşehir massif. By the late Eocene, closure was complete and a 90° rotation preserved the Haymana and Tuz Gölü basins in a strain-free area between the colliding continental blocks.

The extension of the Ionian trough into southwestern Turkey

A. Poisson

SUMMARY: During Mesozoic time the Ionian trough in Greece shows a relatively complex and distinctive geodynamic evolution compared to the adjacent platforms. A similar evolution is found in several series of the Western Taurides in Turkey. The existence of a Mesozoic intra-Tauric trough (previously named 'Kizilca-Çorak göl trough'), is established and interpreted as an homologue of the Ionian trough.

This situation, external with respect to the Helleno-Tauric belt, is used as a basis for a reconstruction of the general evolution of the eastern part of the Aegean 'curvature' that seems to have been initiated during the mid-Miocene (after Langhian but prior to Tortonian). The recent (neotectonic) evolution of the arc started after the Tortonian, possibly as late as the Lower Pliocene.

In the Helleno-Tauric belt most authors recognize many similarities between the external zones, especially between the Apulian and Bey Dağları platforms (references in Aubouin *et al.* 1976, Brunn *et al.* 1976). These platforms have been considered as the northern border of the African plate (Brunn *et al.* 1976), or as the southern border of an Apulian-Anatolian plate (Biju-Duval *et al.* 1977). The recognition of an intra-Tauric trough in the middle of the western Taurides (Brunn *et al.*, 1976), named the Kizilca-Çorak göl trough (Poisson & Sarp 1977; Poisson 1977) permits a more precise correlation. This trough can be interpeted as homologous to the Ionian trough. The proposed prolongation of the Kizilca trough into the Isparta angle again raises the problem of the birth of this angle, which is not purely tectonic in origin, and the related problem of the original location of the Antalya Nappes. Models explaining the genesis of the curved structure of the Aegean and Tauric arcs consider the curvature as ancient and do not take into account Plio-Quaternary movements in the Aegean domain (Brunn 1976, Tapponier 1977, Aubouin *et al.* 1976, Ricou 1980). The importance of these recent movements seems well established and generally accepted (McKenzie 1978; Mercier *et al.* 1979; Le Pichon & Angelier 1979). Recent palaeomagnetic studies (Laj *et al.* 1982 and this volume), also seem to prove that the extension is more important than previously calculated by McKenzie (1978), and by Le Pichon & Angelier (1979). Laj *et al.* (1982) deduced a recent age for the main part of the curvature. At the same time, Sorel & Cushing (1982) demonstrate the importance of nappe displacements during the Langhian phase in the Ionian Zone.

The Ionian trough in northwestern Greece

To characterize the Ionian trough in northwestern Greece (Fig. 1, 3) it is necessary to consider more than one section. The detailed work presented jointly by l'Institut de Géologie et Recherches du Sous-Sol-Athènes and l'Institut Français du Pétrole-Paris (1966), has shown a great diversity of lithological successions between the central part of the zone and its borders during the Lias and Dogger. These variations contrast with the homogeneity of the series in the adjacent carbonate platforms. (Apulian platform in the west and Gavrovo platform in the east).

During the Upper Lias a significant palaeogeographic differentiation occurred: elongate highs emerged separated by small basins with hemipelagic carbonate sediments (ammonitico rosso, 'schistes à Posidonies', 'calcaires à filaments'). This new palaeogeography was related to an early tectonism of the Pantokrator limestones which initiated the Ionian trough and clearly differentiated it from the adjacent platforms. The emergent parts were progressively submerged and covered by sediments during the Dogger and Malm. The most important appearance of pelagic sediments occurred at the Jurassic-Cretaceous boundary, and during the Lower Cretaceous widespread pelagic sedimentation occurred as the Vigla limestones. These cherty micritic limestones ('siliceous limestones' or 'calcaires à zones siliceuses') are typical facies of the Ionian Zone in the external Aegean arc extending as far as the islands of Crete and Rhodes.

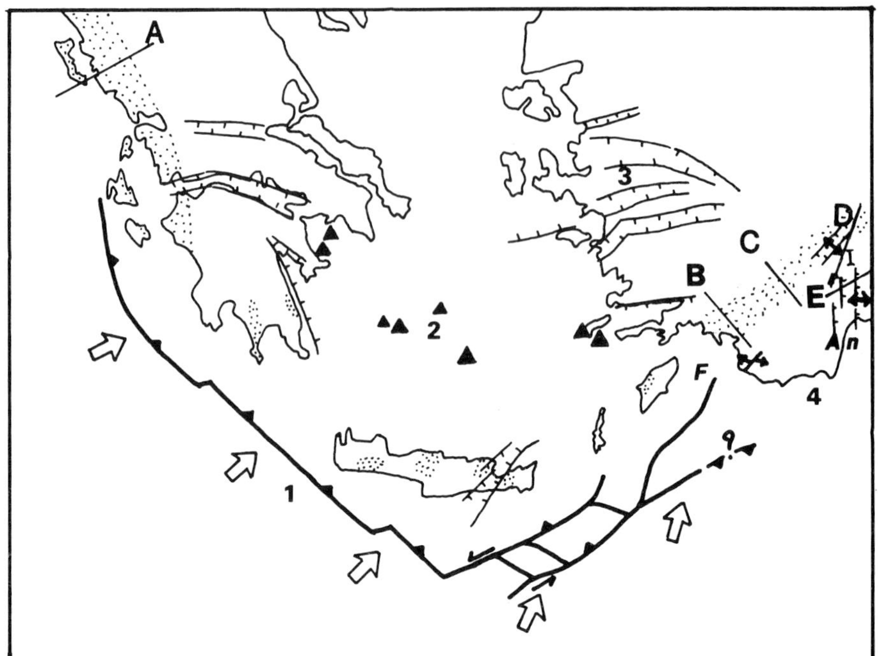

FIG. 1. Present configuration of the Aegean arc. 1, Hellenic trenches with indication of the relative motion between Africa plate and Aegean arc; 2, Internal volcanic arc; 3, Grabens and main direction of normal faulting; 4, Normal faulting in the regions of Antalya (An), Isparta (I), and Fethiye (F); G, Göcek windows. A, B, C, D, E, locality of the sections in Fig. 3. Dotted areas: area of the Ionian and Kizilca outcrops.

The external intra-Tauric trough (southwestern Taurides-Turkey)

Compared to the Ionian series, several units of the Lycian nappes show a similar sedimentary development during Mesozoic times. They have been grouped in a single palaeogeographic zone named 'Kizilca-Çorak göl Zone' (Poisson & Sarp 1977; Poisson 1977).

The main characteristics of this zone are the following:

(1) It was situated in the middle of the Tauric platform and became a trough during late Jurassic times.

(2) It is presently allochthonous, due to an important late thrusting phase during the Middle Miocene.

(3) Its original position is thus subject to discussion.

Two possibilities may be considered:

(1) the original location either was near the northern margin of the Menderes Massif, between the Menderes and the Manisa Massifs;

(2) or it was situated near the southern margin of the Menderes Massif, between the Menderes and Bey Dağları Massifs. In both cases the Kizilca Zone was flanked by carbonate platforms. Fig. 2 summarizes the major series of Western Anatolia along a NW-SE cross-section between the Tethys suture (Izmir-Ankara Zone), and the Bey Dağları platform. Between the Manisa and Bey Dağları sections we observe a gradual increase in detrital material. Coarse ophiolitic sediments are restricted to the north during the Maastrichtian (Manisa), while the southern border of the Menderes Massif is covered by finer sand and marls interbedded with calciturbidites (Maastrichtian), and by coarse detrital sediments (Palaeocene-Lower Eocene). A marly flysch with calciturbidites is present at the top of the Köyceğiz section considered here as a part of the Kizilca Zone (Çamova flysch, de Gracianksy 1972, Bernoulli *et al.* 1974). Ophiolitic detrital sediments appear in this trough as late as the Lower Eocene (Kizilca) or Upper Eocene (Gökceovacik). These facts seem in better agreement with the hypothesis of a south-Menderes origin for the Kizilca zone than with derivation from further north.

Figure 3 shows various reconstructions of sections of the Kizilca trough prior to Tertiary

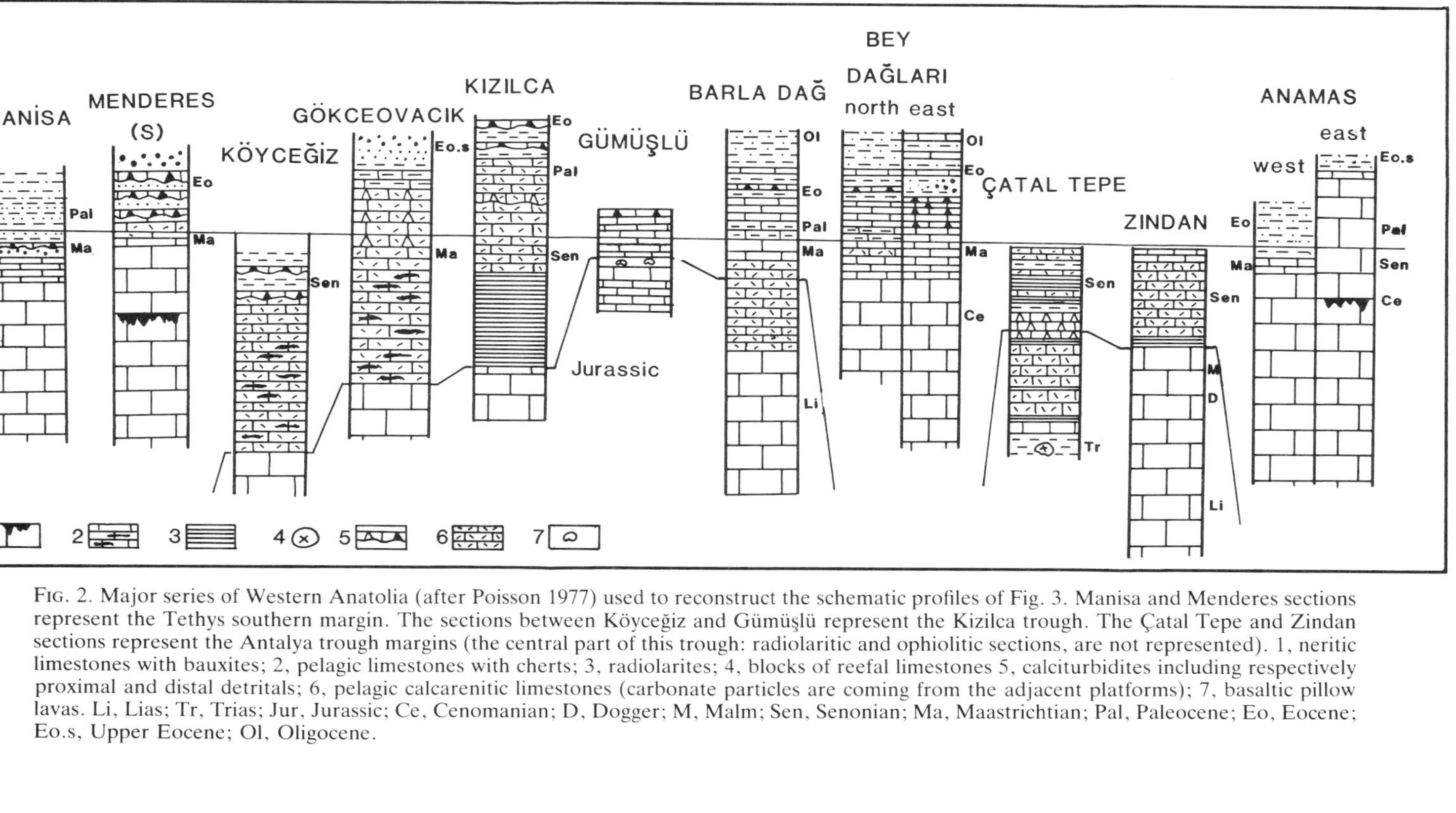

FIG. 2. Major series of Western Anatolia (after Poisson 1977) used to reconstruct the schematic profiles of Fig. 3. Manisa and Menderes sections represent the Tethys southern margin. The sections between Köyceğiz and Gümüşlü represent the Kizilca trough. The Çatal Tepe and Zindan sections represent the Antalya trough margins (the central part of this trough: radiolaritic and ophiolitic sections, are not represented). 1, neritic limestones with bauxites; 2, pelagic limestones with cherts; 3, radiolarites; 4, blocks of reefal limestones 5, calciturbidites including respectively proximal and distal detritals; 6, pelagic calcarenitic limestones (carbonate particles are coming from the adjacent platforms); 7, basaltic pillow lavas. Li, Lias; Tr, Trias; Jur, Jurassic; Ce, Cenomanian; D, Dogger; M, Malm; Sen, Senonian; Ma, Maastrichtian; Pal, Paleocene; Eo, Eocene; Eo.s, Upper Eocene; Ol, Oligocene.

times (approximately latest Cretaceous). The examples are taken from different parts of the Lycian Nappes between the region of Göcek and the Isparta angle.

Connection between the troughs of Kizilca and Antalya

The problem of the genesis of the Isparta angle has been a mater of discussion for several years particularly in relation to the origin of the Antalya Nappes. Different hypotheses have been proposed concerning the original source of these Nappes. Dumont *et al.* (1972) proposed a southern origin from the 'Pamphylian basin'. After that there were different proposals in favour of a northern origin (Ricou *et al.* 1974, 1975, 1979, Monod 1976, Dumont *et al.* 1980a), as well as for a southern origin (Biju-Duval *et al.* 1977, Robertson & Woodcock 1980, 1981). It is not the intention to discuss these different hypotheses. Nevertheless the detailed work carried out during the last few years on the carbonate platforms as well as in the Neogene basins, leads to the conclusion that the initial hypothesis proposed by Dumont *et al.* (1972) is still acceptable, and is now more precisely documented than 10 years ago. The new data particularly concern the Mesozoic and Tertiary sediments. One major fact that should be noted is the possibility of reconstructing a slope basin to the Bey Dağları platform to the east and of the Anamas-Seydişehir platform to the west. The Çatal Tepe unit represents the transitional unit between the Bey Dağları and the Antalya trough (Poisson 1977, Robertson & Woodcock 1981). The central part of this basin was occupied by various radiolaritic series and by oceanic crust. On the opposite edge of the basin, the margin comprises the units of Zindan (Gutnic 1977, Dumont *et al.* 1980a), Karacahisar (Dumont 1976) and probably Kemer (as defined by Marcoux in Delaune-Mayère *et al.* 1977). Across this eastern edge of the basin one can observe the progressive appearance of pelagic facies and detrital sediments during Upper Cretaceous and Paleogene times from the southwestern units (Zindan-Karacahisar) to the northeastern one (Eastern Anamas-Seydişehir) (Dumont *et al.* 1980b). This implies a western location for the source of the detritals: the Antalya Nappes. Waldron (1981, and this volume) arrives at the same conclusion. The Antalya Nappes and the Bey Dağları autochton south of Isparta have an unconformable sedimentary cover of Lower to Middle Miocene age (Gutnic & Poisson 1970, Poisson 1977). The same cover was recently discovered in the middle of the Antalya Neogene basin in several places. This implies a different age and consequently a different process for the emplacement of the Lycian and Antalya nappes this being in complete disagreement with the hypothesis of Ricou *et al.* (1980 and this volume). Because of this fact, plus the probable connection between the Kizilca and Antalya troughs, it seems likely that the Antalya Nappes have a southern origin.

The initiation of the Antalya trough by rifting during Middle to Upper Triassic times is well documented (Marcoux 1978), but according to this author the trough was located to the north of the Tauric platform (Argyriadis *et al.* 1980). The trough was later enlarged during the Jurassic and Cretaceous by distensional faulting (Waldron 1981, this volume; Robertson & Woodcock 1982, this volume).

A connection between the troughs of Antalya and Kizilca through the Barla Dağ region is possible (north of the Isparta angle). As early as Upper Triassic times the Barla Dağ, Antalya and Anamas basins were connected (Kasimlar shales and Çayir clastics), this connection persisting during the Mesozoic (Fig. 3). On the opposite side of the basin (in a western direction) during the Palaeogene, the presence of the same flysch unit (Yavuz type, Poisson 1977) in the Barla Dağ, Bey Dağları and Kizilca basins indicates that they were connected (Fig. 3) (Dumont *et al.* 1980b). Consequently, it is logical to consider that the connection between the Kizilca and Antalya troughs existed. In this trough rifting was initiated during middle-Upper Triassic times in the east (Antalya) and progressively reached the western side: oceanic crust appearing only in the eastern part, perhaps as late as the middle Cretaceous (Thuizat *et al.* 1980).

Comparison of the Ionian and Kizilca zones

These zones have the following aspects in common:

(1) They occupy the same position in the external part of the belt, between two stable carbonate platforms.

(2) They were initiated as troughs during the Jurassic (Triassic for Antalya) and became pelagic synchronously at the Jurassic-Cretaceous boundary.

(3) The pre-pelagic sequence is characterized by the variability of the sedimentary series related to the beginning of the extensional movements (Lias to Malm).

As there is a wide consensus concerning the continuity of the Ionian zone from the northwest of continental Greece to the island of

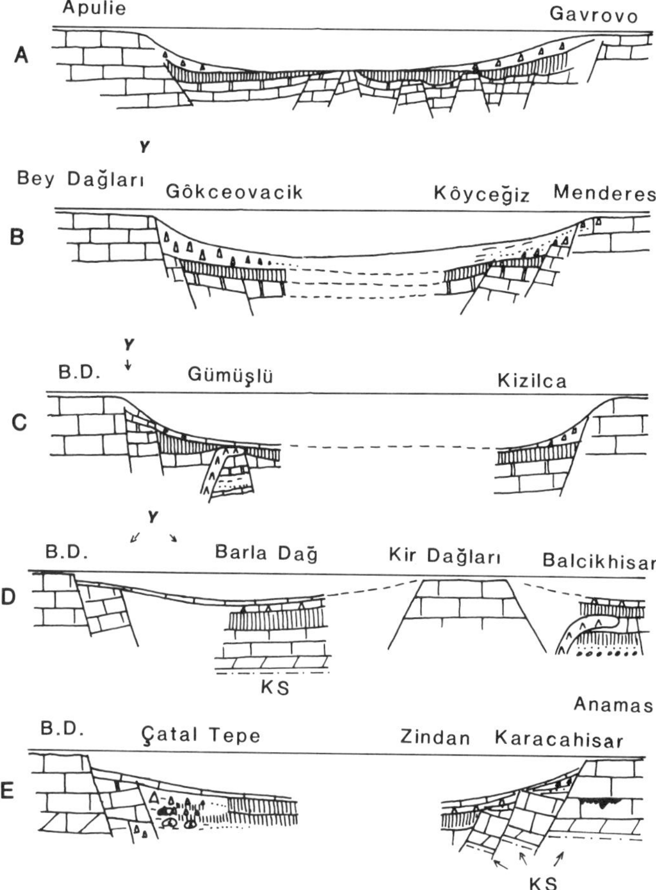

FIG. 3. Reconstructed sections at the end of the Cretaceous. A, Ionian trough; B, C and D, Kizilca trough; E, Antalya trough.

See Fig. 2 for detailed logs. KS: Kasimlar shales (connection between Barla Dağ and Karacahisar-Anamas-Seydişehir during Late Triassic). Y, future depositional emplacement of the Yavuz flysch (connection between Barla Dağ and Bey Dağları during Paleogene).

Rhodes (through the Peloponnesus and Crete), it seems justifiable to consider the Kizilca zone as its direct prolongation or, more probably as an homologous zone in the Western Taurides.

Geodynamic evolution of the eastern part of the Aegean curvature

The shortening of the external zones in the Western Taurides (Fig. 3)

The Kizilca zone, as defined above, is used as a guide to reconstruct the geodynamic evolution of the Western Taurides. The zone was thrust onto the adjacent platforms (mainly towards the south), during Tertiary compressive phases. One can attempt to reconstruct the initial width of the trough by unfolding the Lycian Nappe pile.

A good section is present in the southwestern part of the Taurides between the Göcek windows and the Bey Dağları (Fig. 4) where the following units have been recorded: Yavuz (Tertiary flysch); Gökceovacik; Köyceğiz, attributed to Kizilca zone; coloured melange and ophiolites. A north-Menderes origin was first assigned to these units (de Graciansky 1972), but a south-Menderes origin was postulated later (Brunn *et al.* 1976) for the lower three:

A first step in the estimation of the tectonic shortening is to consider the distance between the front line of the nappes along the Bey Dağları, and the Göcek windows (where the Bey Dağları re-appears), which is 80 km. The 3 units surrounding the windows outcrop

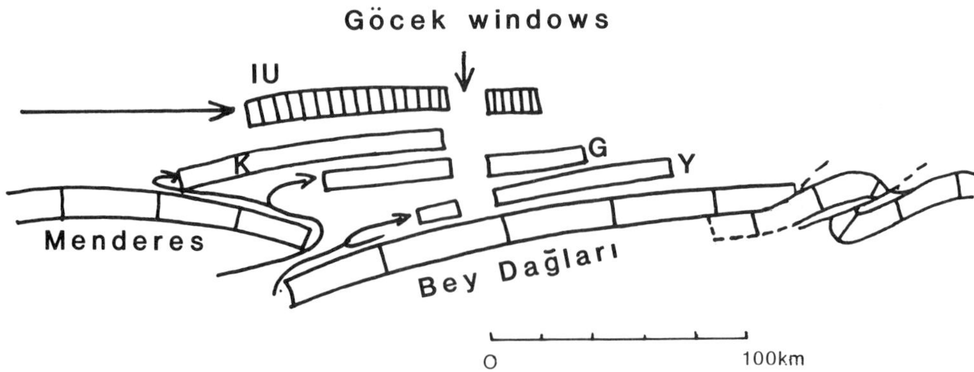

FIG. 4. The pile of the Lycian nappes in the region of Göcek. Y, Yavuz unit; G, Gökceovacik unit; K, Köyceğiz unit; IU, internal units (peridotites, coloured melange). Interpretation of the initial position of the Kizilca zone between the Menderes and Bey Dağları platforms.

continuously for distances of 70 to 100 km. If we replace these units behind the windows to the northwest, we may calculate a minimum of 290 km of northwestern shortening. If we add an internal shortening of about 25% (obtained by graphic unfolding along cross-sections) we have an aggregate 375 km of northwestern shortening. In reality the shortening is probably greater as numerous small thrusts affect the Lycian Nappes. At the other end of the Kizilca zone, in the region of the Isparta angle, the estimates of shortening are less precise. Nevertheless, the established proximal connection between Barla Dağ and Bey Dağları (Dumont *et al.* 1980b), during Eocene time (Yavuz flysch), is in good agreement with only minor shortening across this section. In conclusion, it appears that the shortening increases from the Isparta angle towards the southwest.

From the recent Aegean arc to the Middle Miocene Ionian-Lycian arc

Present day situation

Figure 1 shows the present situation (Late Pliocene to Quaternary), after Le Pichon & Angelier (1979, 1982), simplified and completed for the Eastern part (Mascle *et al.* 1982, Angelier *et al.*, 1981, Dumont *et al.* 1979, and own data for Antalya).

Late Miocene situation

The importance of the Plio-Quaternary extension of the Aegean region has been recognized by many authors (McKenzie 1978, Mercier *et al.* 1979, Le Pichon & Angelier 1979). The beginning of the Neotectonic evolution is dated as Upper Miocene by Le Pichon & Angelier (1979). Recent palaeomagnetic studies (Laj *et al.* 1982, and this volume) show that there was no rotational movement in the Ionian islands between 11.5 and about 5 Ma. These authors give an age of 5 ± 1 Ma as the beginning of the Neotectonic episode. Fig. 5 shows the configuration of the region at the end of the Miocene (coastlines after Le Pichon & Angelier 1979). The compressional events are important at both extremities of the curvature. In the Antalya region the east to west direction of compression is completely different from the Middle Miocene direction. This must be related to a complete change in the geodynamic evolution of the Taurus belt probably connected with change in the relative motion of the plates along the Levant fault system (Letouzey & Tremolières 1980).

Middle Miocene situation

The Kizilca trough was completely closed by the Middle Miocene compressive phase (Lycian phase—after or during late Langhian). At the same time the Ionian trough was also strongly deformed (Sorel & Cushing 1982). As discussed above, the total amount of nappe motion, from the Isparta angle, over the Tauric platform, increases to the southwest. The maximum translation (according to Aubouin *et at.* 1976), was probably localized in the Crete area. The front line of the Nappes delineate the curvature of a Middle Miocene arc, named the 'Ionian-Lycian arc', distinct from the recent 'Aegean arc', with a less important curvature.

The Aegean region during Eocene time

The reconstruction of the Aegean region during Eocene time is more difficult and problematic. For the Western Taurides, Fig. 6 shows

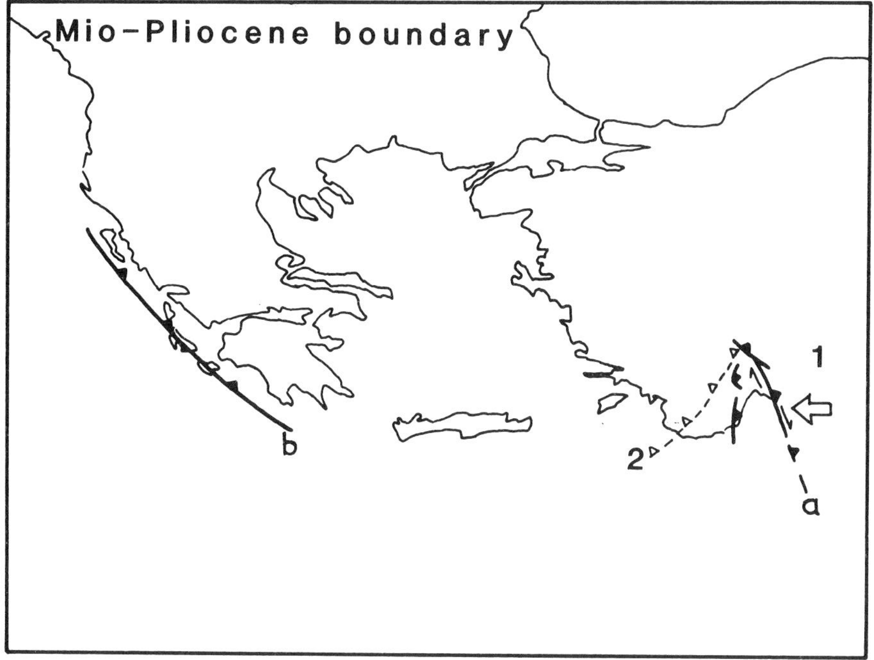

FIG. 5. Helleno-Tauric belt at the Mio-Pliocene boundary. 1, compressional movements; a: before Lower Pliocene; b: during Lower Pliocene; 2, front line of the former Ionian-Lycian arc.

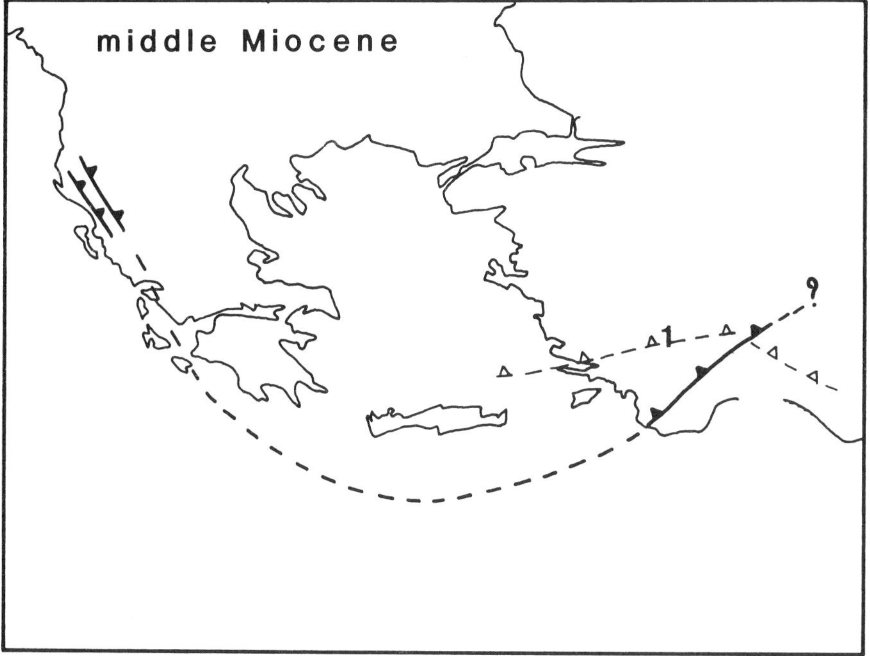

FIG. 6. The Ionian-Lycian arc (Middle Miocene). 1, front line of the former Eocene thrusts.

the results obtained by unfolding the Middle Miocene structures by means of rotation of the nappe fronts around a pole situated in the region of Isparta. If we extrapolate this to the direction of the Aegean, the curvature of the Middle Miocene arc disappears. Under these conditions, it seems that the Ionian-Lycian arc was the first curved structure in the Aegean region. Laj *et al.* (1982, and this volume), using another method arrived at the same result at the western end of the arc. Further studies are necessary in order to establish the angle of rotation at both ends of the curvature.

Conclusions

The connection proposed here between the Kizilca and Antalya troughs suggests that the Antalya basin is the western end of a southern branch of the Tethys ocean, or at least an appendix of it. Towards the east its continuation should be found in Cyprus and Hatay. In the opposite direction the ophiolitic crust did not extend further than the Isparta area, and disappears in the 'Isparta angle'.

The evolution of the Aegean and Tauric curvatures is characterized by a succession of movements from Eocene to recent. This evolution is discontinuous and not synchronous in the two curvatures. The Aegean arc seems to have been initiated during the Middle Miocene and its curvature increased progressively from the Lower Pliocene. The Tauric arc was initiated during the Late Eocene and its curvature was achieved during the Late Miocene. In the area of Antalya the Pliocene and Quaternary periods are characterized by normal extensional faulting without any known compressional event.

In the Helleno-Tauric belt, the Eocene-Oligocene tectonic phase seems to have been the most extensive. Translations occurred from internal to external zones. During the Middle Miocene phase these translations took place in the same direction but were restricted to the Aegean province: the Ionian-Lycian arc was initiated while the former Tauric arc (Eocene) was not deformed.

At the end of the Miocene the compressional directions changed from N-S to E-W and the Alpine-type character of the Helleno-Tauric belt was established; the Neotectonic period had thus begun (Late Miocene Tauric arc).

References

Angelier, J., Dumont, J. F., Karamanderesi, H., Poisson, A., Şimsek, Ş. & Uysal, Ş. 1981. Analysis of fault mechanisms and expansion of southwestern Anatolia since Late Miocene. *Tectonophysics*, **75**, T_1-T_9.

——, Lyberis, N., Le Pichon, X. Barrier, E. & Huchon, P. 1982. The tectonic development of the Hellenic Arc and the Sea of Crete: a synthesis. *Tectonophysics*, **86**, 159–96.

Argyriadis, I., Graciansky, P. C. de, Marcoux, J. & Ricou, L. E. 1980. The opening of the Mesozoic Tethys between Eurasia and Arabia-Africa. *26th Intern. Geol. Cong., Paris, Coll. C5, Mém. B.R.G.M.*, **115**, 199–214.

Aubouin, J., Bonneau, M., Davidson, J., Leboulenger, P., Matesco, S. & Zambetakis, A. 1976. Esquisse structurale de l'arc égéen externe; des Dinarides aux Taurides. *Bull. Soc. géol. France*, **18**, 327–36.

Bernoulli, D., Graciansky, P. C. de & Monod, O. 1974. The extension of the Lycian Nappes (SW Turkey) into the SE Aegean Islands. *Eclog. Geol. Helvetiae*, **67**, 39–90.

Biju-Duval, B., Dercourt, J. & Le Pichon, X. 1977. From the Tethys ocean to the Mediterranean sea: a plate tectonic model of the evolution of the western alpine system. *In:* B. Biju-Duval & L. Montadert (ed.) *Histoire Structurale des Bassins Méditerranéens*. Technip Paris, 143–64.

Brunn, J. H. 1976. L'arc concave zagro-taurique et les arcs convexes taurique et égéen: collision et arcs induits. *Bull. Soc. géol. France*, **18**, 553–67.

——, Argyriadis, I. Ricou, L. E., Poisson, A., Marcoux, J. & Graciansky, P. C. De. 1976. Eléments majeurs de liaison entre Taurides et Hellénides. *Bull. Soc. géol. France*, **18**, 481–97.

Delaune-Mayere, M., Marcoux, J., Parrot, J. F. & Poisson, A. 1977. Modèle d'évolution mésozoique de la paléo-marge téthysienne au niveau des nappes radiolaritiques et ophiolitiques du Taurus lycien, d'Antalya et du Baer-Bassit. *In:* B. Biju-Duval & L. Montadert (ed.) *Histoire Structurale des Bassins Méditerranéens*. Technip Paris, 79–94.

Dumont, J. F. 1976. *Etudes géologiques dans les Taurides occidentales: les formations paléozoiques et mésozoiques de la coupole de Karacahisar (Isparta, Turquie)*. Thése, 3è Cycle (non publiée), Univ. Paris-Sud Orsay, France, 213 pp.

——, Gutnic M., Marcoux, J., Monod, O. & Poisson, A. 1972. Le Trias des Taurides occidentales (Turquie). Définition du bassin pamphylien: un nouveau domaine à ophiolites à la marge externe de la chaîne taurique. *Zeits. Deutsch. Geol. Gesell.* **123**, 385–409.

——, Poisson, A. & Şahinci, A. 1979. Sur l'existence de coulissements sénestres récents à l'extrémité orientale de l'arc égéen (SW de la Turquie). *C.r. Acad. Sc. Paris*, **289**, 261–4.

——, Uysal, Ş. & Monod, O. 1980a. La série de

Zindan:un élément de liaison entre plate-forme et bassin à l'est d'Isparta (Taurides occidentales, Turquie). *Bull. Soc. géol. France*, **22**, 225–32.
——, Uysal, Ş. & Poisson, A. 1980b. *Les Plates-formes des taurides Occidentales*. Rapport (inédit), M.T.A. No 0823, Ankara, 53 pp.
Graciansky, P. C. De 1972. *Recherches géologiques dans le Taurus lycien occidental*. Thèse Doctorat Etat (inédite), Univ. Paris-Sud Orsay France, 762 pp.
Gutnic, M. 1977. *Géologie du Taurus Pisidien au N d'Isparta (Turquie)*. Trav. Lab. Géol. Historique, Univ. Paris-Sud Orsay (inédit), 130 pp.
—— & Poisson, A. 1970. Un dispositif remarquable des chaînes tauriques dans le Sud de la courbure d'Isparta. *C.r. Acad. Sc. Paris*, **270**, 672–5.
Institut de Geologie et de Recherches du Sous-Sol Athenes et Institut Francais du Petrole. 1966. *Etude Géologique de l'Epire*. Ed. Technip Paris.
Jamet, M. 1982. *Etude néotectonique de Corfou et étude paléomagnétique des sédiments néogènes des îles de Corfou, Cephalonie et Zanthe*. Thèse 3è cycle (inédite) Univ. Paris-Sud Orsay, 150 pp.
Laj, C., Jamet, M., Sorel, D. & Valente, J. P. 1982. First paleomagnetic results from Mio-Pliocene series of the Hellenic sedimentary arc. *Tectonophysics*, **86**, 45–68.
Le Pichon, X. & Angelier, J. 1979. The Hellenic arc and trench system: a key to the neotectonic evolution of the Eastern Mediterranean area. *Tectonophysics*, **60**, 1–42.
Letouzey, J. & Tremolieres, P. 1980. Paleo-stress fields around the Mediterranean since the Mesozoic derived from microtectonics: comparisons with plate-tectonic data. *26th Int. Geol. Cong. Paris, Mém. B.R.G.M.*, **115**, 261–73.
Marcoux, J. 1978. A scenario for the birth of a new oceanic realm: the alpine Neotethys. *Xth Int. Sedim. Cong., Jerusalem*, 419–20.
Mascle, J., Jongsma, D., Campredon, R., Dercourt, J., Glacon, G., Lecleach, A., Lyberis, N., Malod, J. A. & Mitropoulos, D. 1982. The Hellenic margin from Eastern Crete to Rhodes: preliminary results. *Tectonophysics*, **86**, 133–47.
McKenzie, D. 1978. Active tectonics of the Alpine-Himalayan belt: the Aegean Sea and surrounding regions. *Geophys. J. R. astron. Soc.* **55**, 217–54.
Mercier, J. L., Delibasis, N., Gauthier, A., Jarrige, J. J., Lemeille, F., Philip, H., Sebrier, M. & Sorel, D. 1979. La néotectonique de l'Arc Egéen. *Rev. Géol. dyn. Géogr. phys.* **21**, 67–92.
Monod, O. 1976. La courbure d'Isparta:une mosaique de blocs autochtones surmontés de nappes composites à la jonction de l'arc hellénique et de l'arc taurique. *Bull. Soc. géol. France*, **18**, 521–31.
Poisson, A. 1977. *Recherches géologiques dans les Taurides occidentales*. Thèse Doctorat Etat (inédite), Univ. Paris-Sud Orsay, No 1902, 795 pp.
—— & Akay, E. 1981. Miocene transgression forms in the Gökbük-Çatallar basin (Western Taurus Range, Turkey). *T.C. Petrol. Isleri Genel Müdürlüğü Dergisi, Ankara*, **25**, 217–21.
—— & Sarp, H. 1977. La Zone de Kızılca-Çorak göl. Un exemple de sillon inta-plate-forme à la marge externe du massif du Menderes. *Abs. 6th Coll. Aegean Geol., Izmir*.
——, Cravatte, J., Muller, C., Akay, E. & Uysal, Ş. 1983. Le Bassin du Köprü Çay (Golfe d'Antalya), et la chronologie de mise en place des nappes d'Antalya. *C.r. Acad. Sc. Paris*, **296**, 933–26.
Ricou, L. E. 1980. La tectonique de coin et la genèse de l'Arc Egéen. *Rev. Géol. dyn. Géogr. phys.* **22**, 147–56.
——, Argyriadis, I. & Lefevre, R. 1974. Proposition d'une origine interne pour les nappes d'Antalya et le massif d'Alanya (Taurides occidentales). *Bull. Soc. géol. France*, **16**, 107–11.
——, Argyriadis, I. & Marcoux, J. 1975. L'axe calcaire du Taurus, un alignement de fenêtres arabo-africaines sous les nappes à matériel radiolaritique, ophiolitique et métamorphique. *Bull. Soc. géol. France*, **17**, 1024–44.
——, Marcoux, J. & Poisson, A. 1979. L'allochtonie des Bey Dağları orientaux. Reconstruction palinspastique des Taurides Occidentales. *Bull. Soc. géol. France*, **21**, 125–33.
Robertson, A. H. F. & Woodcock, N. H. 1980. Tectonic setting of the Troodos massif in the East Mediterranean. *In:* Panayiotou A. (ed.) *Ophiolites, Proc. Int. Oph. Symp. Cyprus* 1979, 36–49.
Robertson, A. H. F. & Woodcock, N. H. 1981. Bilelyeri Group, Antalya complex: deposition on a Mesozoic passive continental margin, SW Turkey. *Sedimentology*, **28**, 381–99.
—— 1982. Sedimentary history of the southwestern segment of the Mesozoic-Tertiary Antalya continental margin, southwest Turkey. *Eclog. geol. Helv*, **75**, 517–562.
Sorel, D. & Cushing, M. 1982. Mise en évidence d'un charriage de couverture dans la zone ionienne en Grèce occidentale: la nappe d'Akarnanie-Levkas. *C.r. Acad. Sc. Paris*, **294**, 675–78.
Tapponnier, P. 1977. Evolution tectonique du système alpin en Méditerranée, poinçonnement et écrasement rigid plastique. *Bull. Soc. géol. France*, **19**, 437–60.
Thuizat, R., Whitechurch, H., Montigny, R. & Juteau, T. 1980. K-Ar dating of some infra-ophiolitic metamorphic soles from the Eastern Mediterranean: new evidence for oceanic thrustings before obduction. *Earth planet. Sci. Letter*, **52**, 302–10.
Waldron, J. W. F. 1981. *Mesozoic sedimentary and tectonic evolution of the Northeast Antalya Complex, Eğridir, SW Turkey*. Ph.D. thesis (unpubl.), Univ. Edinburgh, U.K., 312 pp.

A. Poisson, Laboratoire de Géologie Historique, Université de Paris-Sud, Bâtiment 504, 91405 Orsay, France.

The SW segment of the Antalya Complex, Turkey as a Mesozoic-Tertiary Tethyan continental margin

A. H. F. Robertson & N. H. Woodcock

SUMMARY: The SW segment of the Antalya Complex is an assemblage of Mesozoic carbonate platform, margin and ophiolitic rocks which record the formation and tectonic emplacement of a small Mesozoic ocean basin.

The pros and cons of three alternative models in the literature are outlined. The original model was of a linear continental margin of a southerly Mesozoic ocean basin (Dumont *et al.* 1972), emplaced in latest Cretaceous and Miocene time as three major nappes. Ricou *et al.* (1975) reinterpreted these nappes as being rooted in a unique northerly Tethys. In the light of knowledge of modern ocean basins, Robertson and Woodcock then developed the concept of a palaeogeographically complex small southern ocean basin in which strike-slip faulting played an important role in both formation (Reuber, this volume) and emplacement (Woodcock & Robertson 1977).

We discuss here a number of important and controversial topics, including the timing of volcanism and ocean crust genesis, the origin of mountain-sized shallow water limestone masses in the allochthon, and the role of strike-slip versus nappe tectonics in the Antalya Complex. An updated version of our palaeogeographical model views the Antalya Complex as an early Mesozoic (Triassic-Jurassic) rift expanded in the Cretaceous by formation of a small ocean basin strongly influenced by strike-slip faulting during both the opening and closing stages.

Vigorous controversy has existed for some years now over the origins of the Antalya Complex, SW Turkey, an assemblage of Mesozoic platform, margin and ophiolitic rocks. Interpretations founded on classical Alpine nappe tectonics are being steadily modified to incorporate knowledge of modern ocean basins and margins.

Following early work (Lefevre 1967, Colin 1962), the first integrated interpretations of the area resulted from regional mapping by a French team (reviewed by Brunn *et al.* 1970, 1971), and particularly from detailed work by Juteau (1975) on the igneous rocks, and by Marcoux (e.g. 1970, 1974) on the stratigraphy (Fig. 1). Using this as a framework, we embarked on a study (1976–1979) of the sedimentology and structure of the SW Antalya area (Fig. 1; Woodcock & Robertson 1977, 1981, 1982, 1984; Robertson & Woodcock 1981a, b, c, 1982; Robertson 1981). Other results from regional mapping (Şenel, 1980), palaeontology and radiometric age dating (Yılmaz, 1981, this volume) and the structure of the ophiolite (Reuber 1982, this volume) are now available.

The aim here is to review three alternative models for the Antalya Complex, then to discuss specific controversial aspects in more detail. Finally, we will update our concept of the palaeogeographical evolution. A parallel study of the NE segment of the Antalya Complex is reported by Waldron (1981, this volume).

Outline Geology

The structural units recognized by different workers are summarized in Fig. 2. Yılmaz (this volume) reviews the available fossil and radiometric age data. In brief we have divided the SW Antalya Complex into the following tectonic zones (Fig. 3):

(1) *Bey Dağları Zone*: a relatively autochthonous Mesozoic carbonate platform unit which became pelagic by mid-Late Cretaceous time (Poisson 1967, 1976, 1978), and which passes passing conformably up into Miocene ophiolite-derived clastics (Hayward & Robertson 1982; Hayward, this volume).

(2) *Kumluca Zone*: a north-south trending imbricate thrust unit of Upper Triassic to Upper Cretaceous quartzose clastics, hemi-pelagic and pelagic sediments of passive margin affinities (Robertson & Woodcock 1981a, b).

(3) *Gödene Zone*: a complexly deformed, mostly steeply dipping zone of Upper-Triassic alkalic mafic extrusives and associated Late Triassic to Upper Cretaceous deep sea sediments, plus intrusive ophiolitic rocks, mountain-sized shallow water carbonate massifs, and minor metamorphic rocks (Juteau 1968, 1975; Delaune-Mayere *et al.* 1977; Robertson & Woodcock 1981c; Yılmaz 1981).

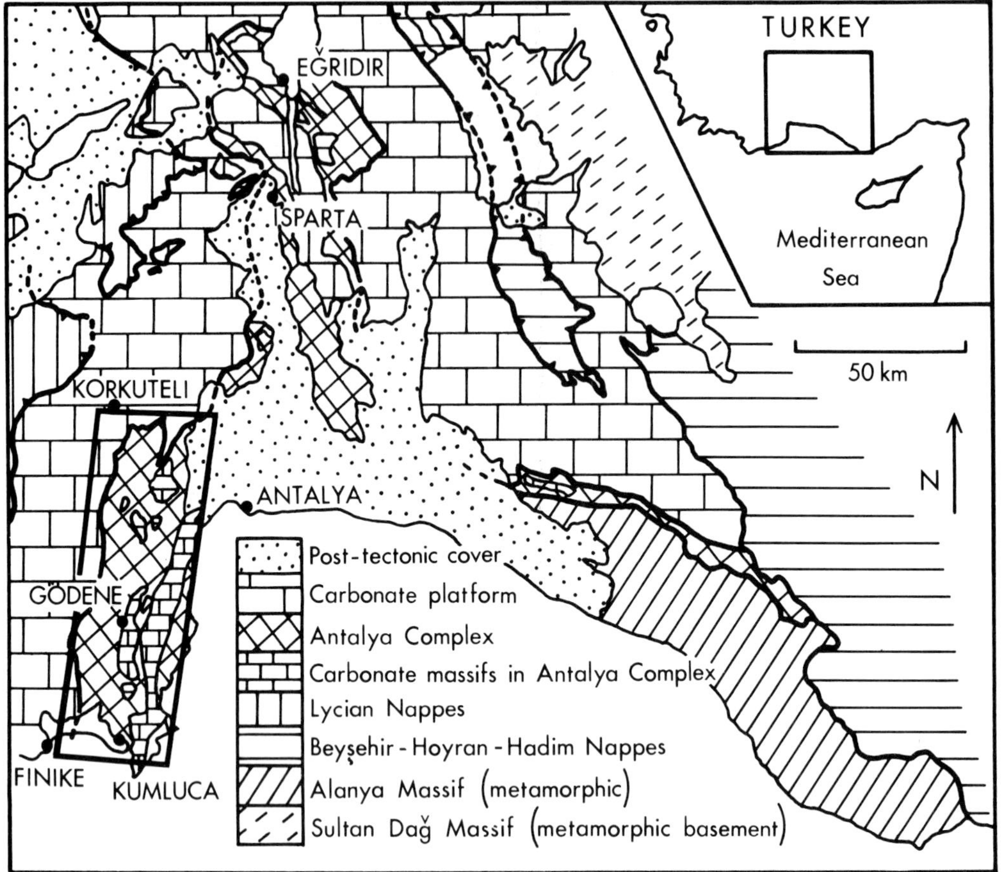

FIG. 1. Outline geology of the Isparta angle. The SW segment of the Antalya Complex is the area in the box (left lower). The NE Antalya discussed by Waldron (this volume) is located east of Eğridir.

	Tectonic Units					
Marcoux (in Delaune-Mayere et al. 1976)	Upper Antalya Nappe					
	Bey Dağları	Lower Nappe	Middle Antalya Nappe			Ophiolite Nappe
Yilmaz (1981)	Bey Dağları Massif	Kumluca Complex	Alakır Çay Melange sheet 1	sheet 2	sheet 3	Tekirova ophiolite
Woodcock & Robertson (1977)	Bey Dağları Zone	Kumluca Zone	Gödene Zone		Kemer Zone	Tekirova Zone

FIG. 2. The alternative tectonic units recognized by various workers.

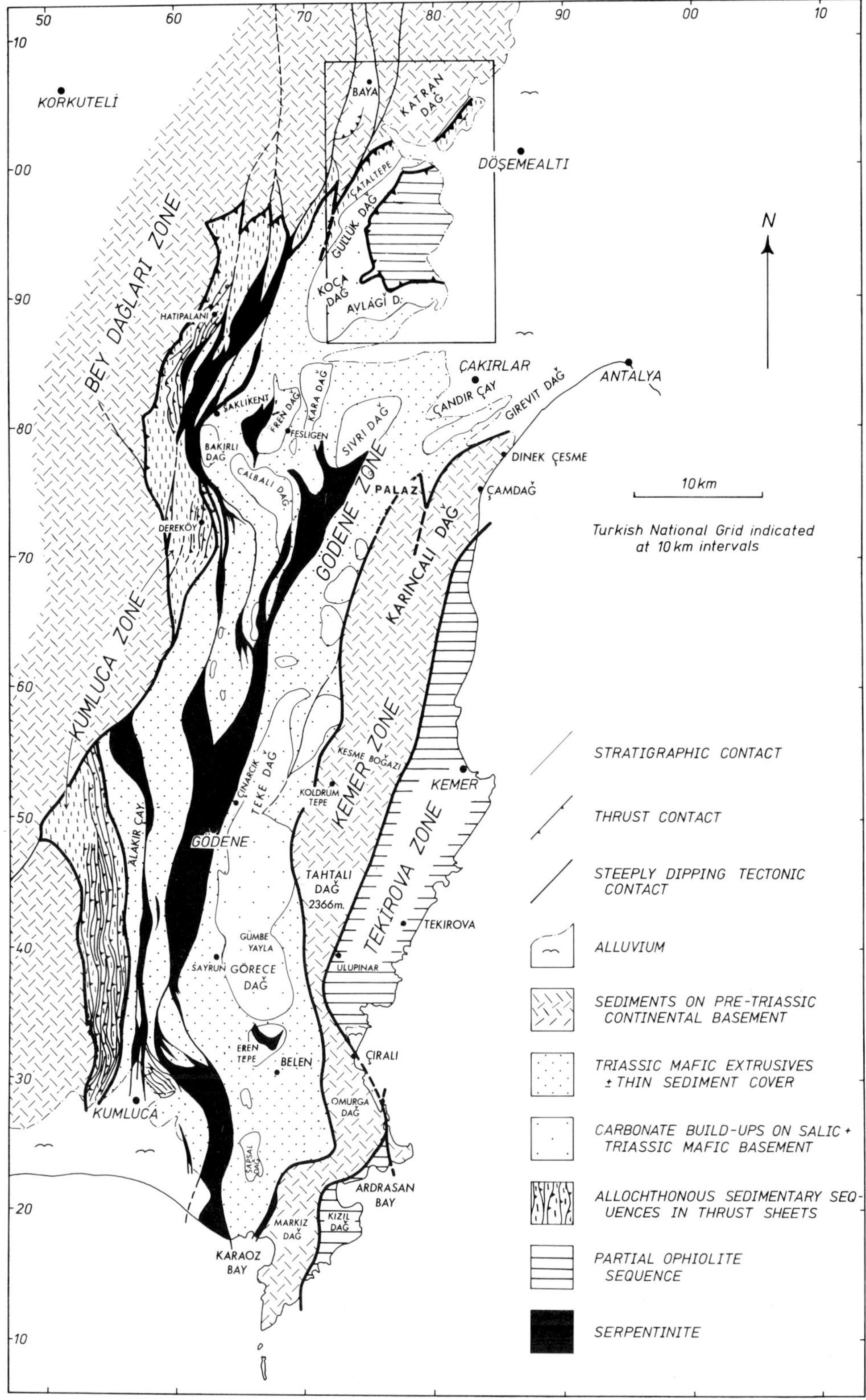

FIG. 3. Simplified geological map to show the main lithologies and tectonic zonation of the SW Antalya Complex. Compiled from 1:100,000 and 1:25,000 mapping.

(4) *Kemer Zone*: a zone dominated by again mostly steeply dipping north-south trending masses of Ordovician to Upper Cretaceous sedimentary rocks, plus smaller volumes of mafic lavas and basinal sediments similar to the Gödene Zone (Robertson & Woodcock 1982).
(5) *Tekirova Zone*: a major ophiolite complex exposed along the present coast and dated by Thuizat & Montigny, & Thuizat *et al.* (1981), and Yılmaz (this volume) as mostly Cretaceous (Juteau 1970, Juteau *et al.* 1977). Only intrusive parts of the ophiolite are preserved.

Summary logs of the chief sedimentary successions in each of the zones are given in Fig. 4.

Model 1: The 'externally'—(or southerly)—derived nappe model

In this model of Dumont *et al.* (1972) a Mesozoic continent-ocean transition was thrust a substantial distance northwards (>100 km) over the 'Tauride limestone axis' as three major nappes, first emplaced in latest Cretaceous time (Figs 5b, 6). This southerly basin was conceived by Dumont *et al.* (1972) as the 'external' Pamphylian basin. Monod (1976) envisaged the Tauride limestones as a mosaic of independent carbonate blocks overthrust by major ophiolitic and limestone nappes in latest Cretaceous time, Marcoux & Ricou (1979) have revived the Pamphylian basin as a Triassic rift located within the Arabian margin to the south but separate from a northern basin which they still considered was the 'internal' root zone of the ophiolitic nappes. In this model and its derivative, the Bey Dağları zone is not viewed as having been originally at all close to the actual margin of the southerly basin from which the three nappes (equivalent to the Kumluca, Gödene, Kemer and Tekirova zones) were derived (Fig. 2).

Pros

(1) The model conforms to the traditional view of the Antalya area as grossly allochthonous Alpine-type nappes which were emplaced in latest Cretaceous and Miocene time.
(2) All the edge-facies of the carbonate platform arc now in thc allochthon (c.g. Çatal Tepe Unit of the 'Lower Antalya Nappe'; Keşme Boğazı sequence of the Kemer Zone). The oldest relatively *in situ* exposed successions in the Bey Dağları are of Early Cretaceous age and are mostly back-reef facies which do not constrain the location of the Mesozoic platform edge. In this model the carbonate platform extends far southwards under the Antalya allochthon. This can not be contradicted from evidence of the exposed relatively autochthonous units.
(3) The model fits the structural data given by Waldron (this volume) that at least the north-east segment of the Antalya Complex was emplaced northwards in latest Cretaceous time. As thrusting in the south-west Antalya segment was mostly from east to west, the structure of this area has little bearing on whether the root zone was originally to the north or to the south.
(4) The model is consistent with seismic evidence that much of the Eastern Mediterranean is now underlain by continental crust (Sestini, this volume). No continental margin separating the Taurides from Africa has yet been confirmed.

Cons

(1) In the south-west Antalya area lithologies within the 'nappes' can be considered as facies variants along a single passive margin. Thus, the 'Lower Antalya Nappe' forming the Çatal Tepe unit of Poisson (1978) is consistent with more proximal facies of the 'Middle Antalya Nappe' exposed further east. Also, carbonate massifs of the 'Upper Antalya Nappe' within the allochthon are reported to show depositional contact with underlying Upper Triassic mafic lavas and sediments of the 'Middle Antalya Nappe' (Robertson & Woodcock 1982). The establishment of such spatial relationships between units removes the need for any large scale internal shortening within the allochthon and indeed makes nappes very difficult to define.
(2) There is no easy explanation of the structurally high position of the limestones of the 'upper nappe' (e.g. Tahtalı Dağ), which have to be variously explained as part of the southerly 'African' margin (Dumont *et al.* 1972), the west facing margin of the Isparta angle (Monod 1976, 1977, or complexly re-thrust slices of carbonate platform (Poisson 1978).
(3) If the nappes were thrust onto the carbonate platform from the south during the latest Cretaceous, the southern part

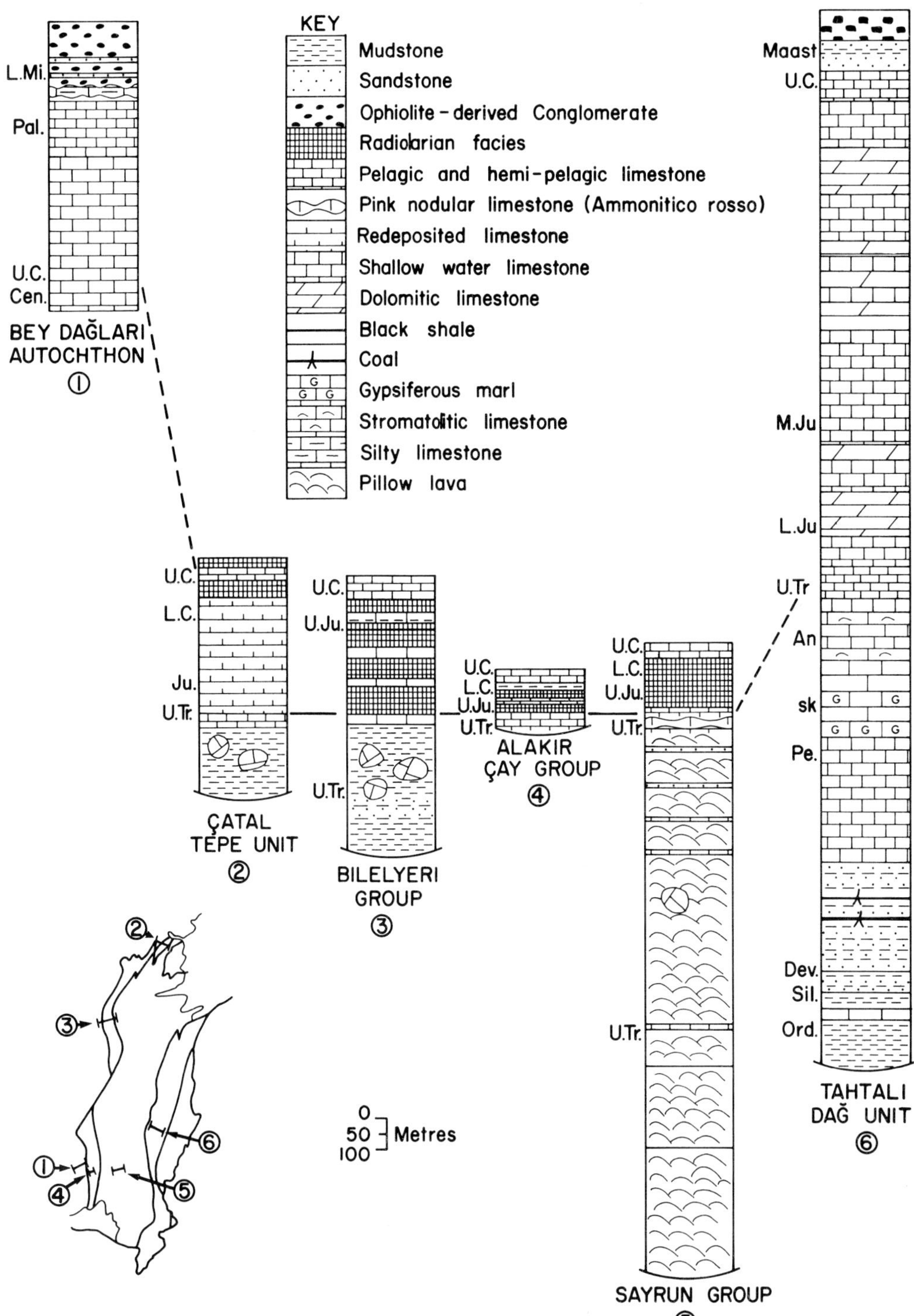

FIG. 4. Composite sedimentary logs summarizing the main sedimentary succession in each tectonic zone. More detailed sedimentary logs are given in our earlier papers. Location of composite logs are shown (bottom left).

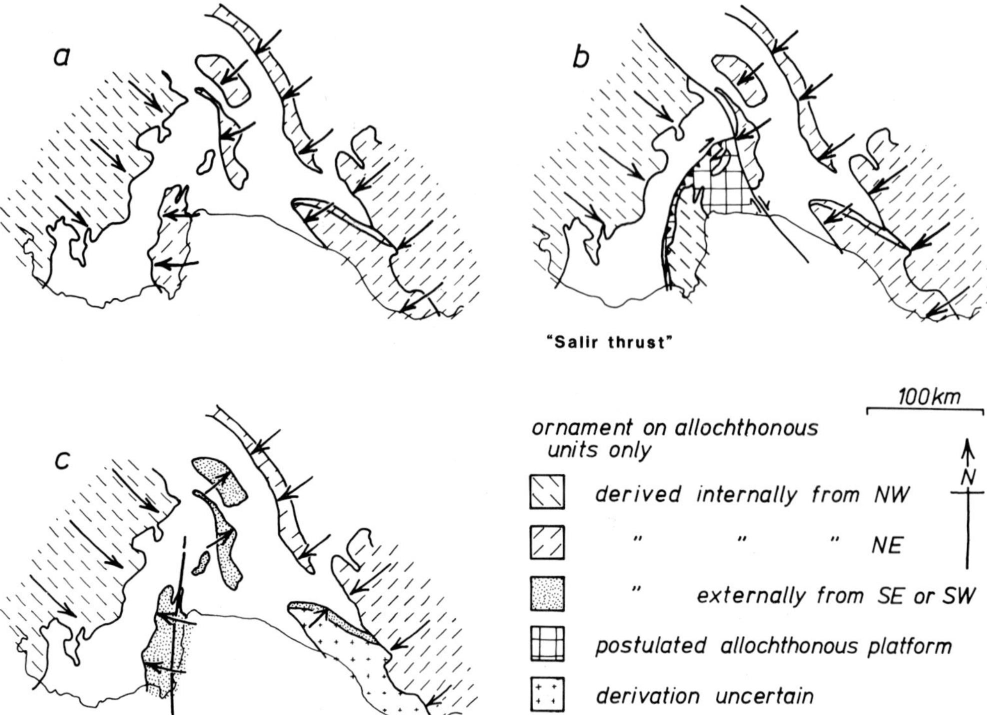

FIG. 5. Illustration of the concept 'external' (southerly) and 'internal' (northerly) nappe emplacement of the Antalya complex as postulated by French workers (Brunn, 1974) a, the 'internal' origin of Ricou *et al.* (1975). All the nappes root in a unique northerly Mesozoic Tethys; b, Ricou *et al's.* (1979) modified internal model. The Antalya allochthon was emplaced over a carbonate platform in latest Cretaceous ('Eastern Bey Dağları'), then transported by dextral shear southwards and emplaced over another carbonate platform unit in Miocene time ('Western Bey Dağları'). In c, the 'external' model the Antalya allochthon is thrust from a southerly Mesozoic ocean basin (Dumont *et al.* 1972). At issue here is the relative importance of thrust versus strike-slip tectonics.

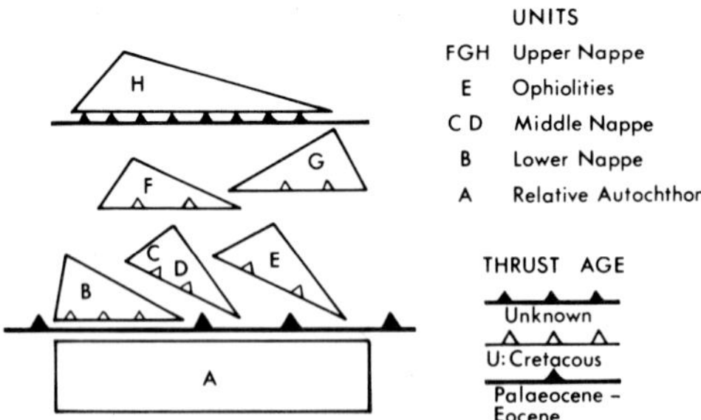

FIG. 6. Stacking order of 'nappes' inferred by Delaune-Mayere *et al.* (1977).

of the Bey Dağları should have been the first to have been effected. Once ophiolitic nappes are ramped from oceanic basins onto continental platforms, erosion products are spread widely and this in fact did happen during the subsequent Miocene final emplacement (Hayward, this volume). By contrast, in the southern Bey Dağları pelagic deposition continued from the Late Cretaceous through to Palaeocene time without a break (Poisson 1978; Hayward & Robertson 1982); this tends to oppose long-distance nappe transport over the platform in latest Cretaceous time.

(4) Long distance emplacement of a major limestone nappe in the latest Cretaceous is also hard to square with a dominant strike-slip mode of both genesis and emplacement of the SW Antalya Complex. The Antalya ophiolite was both formed along a sinistral fracture zone (Reuber, this volume), then emplaced from latest Cretaceous onwards along strike-slip faults (Woodcock & Robertson 1982), leaving little scope for a separate major phase of Upper Cretaceous low-angle nappe emplacement.

(5) There is little evidence of long distance westward Miocene thrusting. Along the north-east margin of the Bey Dağları Zone relatively autochthonous Eocene redeposited limestones are carbonate slope facies which are most simply interpreted as being derived from the Bey Dağları margin at this time. Ophiolite-derived clastics, including reworked Oligocene benthonic foraminifera, which imply the imminent approach of ophiolitic and continental-margin nappes, first appear in the Early Miocene along the south-east Bey Dağları margin. About 20 km to the west the sequence is complete from Lower to Upper Miocene thus limiting Miocene westward thrusting to ca. 10 km (Hayward & Robertson 1982, Hayward, this volume).

Model 2: The 'internal' or northerly-derived ophiolite nappe model

Stemming from his early work on the 'Croissant ophiolitique peri-Arabe', from the Levant to Oman, Ricou *et al.* (1974, 1975, and this volume) developed the hypothesis that all the major Turkish ophiolites were rooted in a single Mesozoic ocean basin sited well north of the Mesozoic carbonate platforms of central Anatolia ('Tauride limestone axis', Fig. 5a). The circum-Mediterranean ophiolitic rocks and associated carbonate margin and pelagic units are seen as far-travelled nappes, emplaced first southwards on to the Arabian margin in Upper Cretaceous time, then thrust progressively further south until Neogene time. In this model the whole of the East Mediterranean would always have been underlain by continental crust, essentially the northern extension of the African and Arabian plates. After Dumont *et al.* (1972) and Poisson (1978) confirmed that the sedimentary successions of the southern part of the Bey Dağları carbonate platform continued unbroken into the Miocene with no record of any Cretaceous thrusting, Ricou *et al.* (1979) devised an ingenious scheme whereby the Bey Dağları massif had been split into two separate parts, the 'Eastern' and the 'Western Bey Dağları'. The ophiolites and associated units were considered to have been emplaced onto the 'Eastern Bey Dağları' in Upper Cretaceous time (Fig. 5b), when it was still far to the north of its present position. The Antalya allochthon, plus a portion of the carbonate platform structurally beneath (Eastern Bey Dağları), was then transported south by dextral shear, followed by a final phase of westward emplacement along the 'Salir thrust' (Fig. 5b). In the south the Antalya allochthon supposedly overthrust the Eastern Bey Dağları totally so that the critical evidence of two 'separate' Bey Dağları units can only be seen in the poorly exposed mountainous northern area.

Pros

(1) These are set out in full by Ricou (this volume):
The similarity of sedimentary and igneous rocks in both the 'internal' and 'external' zones suggests that all could have formed in a single palaeogeographically simple Atlantic-type basin, without need to postulate the complexities of an area like the Caribbean.

(2) The inferred east-west orientation of spreading axes in both 'internal' and 'external' zones, followed not long after by 'intra-oceanic slicing' then later by emplacement onto margins all favour derivation from a single oceanic basin.

(3) The fossil evidence opposes any great separation of at least the Pisidian (eastern) Taurides from the Arabian margin in the Mesozoic (Hirsch, this volume).

Cons

(1) The main arguments put forward in support of this model ignore the fact that similar facies may originate in similar depositional environments but in different geographical locations. Triassic alkalic lavas, carbonate edge facies and radiolarites, are for example, ubiquitous throughout the Mesozoic Tethyan region during Late Triassic time and were clearly not necessarily formed in a single connected rift. Also, the Upper Cretaceous emplacement coincided with worldwide reorganization of plate boundaries and of basins throughout the Tethyan area, ranging from the West Mediterranean to the Himalayas. In no sense need all these roughly synchronous events have occurred in a single ocean basin.

(2) The contrast between the supposed 'Eastern' and 'Western Bey Dağları units' in the north should according to Ricou be a very major thrust separating units which formed far apart, yet similar platform carbonates of similar age are exposed on both sides without evidence of major thrusting (Poisson 1978). Pelagic deposition continued on the 'Eastern

Bey Dağları' (and in the 'Western Bey Dağları') from latest Cretaceous through to Palaeocene, opposing the concept of Upper Cretaceous emplacement of the Antalya allochthon on to this part of carbonate platform at this time. Also, similar Eocene carbonate edge facies exist both north and south of the 'Salir thrust', suggesting that during Eocene time the Bey Dağları formed a continuous east-facing carbonate edge (Hayward & Robertson 1982) rather than being split into two units spaced far apart.

(3) In the critical north-east Antalya Complex area Waldron (this volume) finds clear evidence of northerly thrusting of probably latest Cretaceous age, rather than the southward displacement required by Ricou's model.

(4) South of Isparta the Antalya Complex is unconformably overlain by Lower and Middle Miocene sedimentary cover rocks and in the core of the Isparta angle Poisson (this volume) now reports a similar relationship. This precludes thrusting of the Antalya Complex through the Isparta angle to reach its present location in the south-western area only by Middle to Late Miocene time.

Model 3: The thrust- and wrench-emplaced southern basin margin model

In previous papers (*op. cit.*) we have put forward a model of a southerly basin with a palaeogeographically complex margin, generated by 'braided' Triassic rifting which led to microcontinental slivers being stranded offshore, followed by formation of a small ocean basin later in the Mesozoic. Emplacement during latest Cretaceous to Miocene was by a combination of wrench and thrust tectonics. There was no major overthrusting of the southern part of the Bey Dağları relative autochthon until Miocene time (Robertson & Woodcock 1982).

Pros

(1) The Upper Triassic to Upper Cretaceous sedimentary facies of the Kumluca Zone, including the Çatal Tepe unit, can be interpreted as the eastern margin of the relatively autochthonous Bey Dağları carbonate platform (see also Poisson, this volume).

(2) Identification of Upper Triassic to mid-Cretaceous carbonate build-ups overlying slivers of pre-rift continental basement allows the present structural complexity to be easily explained by a varied palaeogeography, as in the modern Caribbean.

(3) The high-angle tectonics is exlained by latest Cretaceous to Miocene predominant strike-slip faulting without need for complex polyphase low-angle thrusting.

(4) The strike-slip model is strengthened both by the recognition by Reuber (1982 and this volume) of formation of the Tekirova ophiolite along a sinistral oceanic transform fault, and by the existence of Tertiary ophiolite-derived clastics shed into strike-slip controlled basins during the later stages of tectonic emplacement (Robertson & Woodcock 1980a).

Less well-constrained features of the model

(1) The timing of ocean floor genesis remains to be clarified. Spreading could have begun in Late Triassic time or could have been delayed until Late Jurassic to Early Cretaceous or even later (see also Whitechurch *et al.* this volume). The possibility could theoretically exist that the off-margin carbonate build-ups all originally belonged to a linear carbonate margin but were later sliced-off and entrained by strike-slip faulting to reach their present position. No detailed facies analysis of the individual carbonate massifs has yet been undertaken.

(2) Some of the steep faults could conceivably be listric or over-steepened thrusts perhaps formed in several phases from Upper Cretaceous to Miocene time.

(3) The distance of strike-slip motion during emplacement and the sense of offset is hard to determine. The offset margins cannot be matched; sea is to the south and the lineament disappears under the Lycian Nappes to the north.

Comparison of the three models

There is some consensus between the models, despite the fundamental conflict between Model 2 (Ricou) and Models 1 and 3. Models 1 and 3 differ from each other in the contrasting *styles* and *scales* of original palaeogeography and of emplacement mechanisms, but both are similar in that a southerly ('external') ocean basin is involved, distinct from a northern branch of the Mesozoic Tethys (Şengör & Yılmaz 1981).

(1) The Lower to Middle Triassic carbonates, evaporites and siliceous pelagic sediments record the initial stages of continental rifting (e.g. Delaune-Mayere *et al.* 1977).

(2) The Late Triassic rifting (Carnian-Norian) was accompanied by extrusion of large volumes of mafic alkalic extrusives (Juteau 1975), together with deposition of terrigenous and calcareous clastics, and pelagic sediments (Robertson & Woodcock 1981c). Upper Triassic tholeiitic extrusives have not been reported.

(3) During Late Cretaceous time the relatively autochthonous Bey Dağları carbonate massif was tectonically juxtaposed with the thrust-imbricated distal part of a Mesozoic passive margin (Robertson & Woodcock 1981a, b). More proximal slope facies of the same margin are recorded in the Upper Triassic to Cretaceous reef-derived calcirudites of the Çatal Tepe unit.

(4) The serpentinites, and other ultramafic rocks, gabbros and diabase of both the Tekirova and the Gödene Zone are remnants of essentially Upper Cretaceous oceanic crust distinct in origin from the Triassic mafic alkalic extrusives (but see Yılmaz, this volume). They represent some form of Mesozoic oceanic crust (Biju-Duval *et al.* 1977, Argyriadis *et al.* 1980).

(5) The entire Antalya allochthon was finally emplaced by westward thrusting over the relatively autochthonous Bey Dağları in Early to Late Miocene time.

Controversial topics

We go on to discuss specific topics which are still controversial.

Nature and age of volcanism

Marcoux (*op. cit.*) considers the preserved volcanics in the Antalya allochthon to be exclusively Upper Triassic. Yılmaz (1981, this volume) confirms a Late Carnian to late Middle Norian age for the mafic alkalic volcanism based on fossils. Yılmaz confirms fossils of Late Jurassic to Early Cretaceous age from between alkalic flows, and Thuizat *et al.* (1978, 1981) gave a Late Cretaceous (Santonian) K-Ar age for an alkaline sill in shales (81 ± 3 Ma). In the Saklikent area (Fig. 3) we have observed mafic pillow lavas interbedded with pelagic chalks, and elsewhere pelagic chalks of Late Cretaceous age locally directly overlie pillow lavas (Süğütcuma). A sample of radiolarian chert within sheared mafic lava near Dereköy (Fig. 2) yielded radiolaria of Late Jurassic to Early Cretaceous age (E. A. Pessagno, personal communication, 1980).

Above the Triassic rift lavas, unbroken pelagic successions span Upper Triassic to locally Upper Cretaceous age. None of these successions have been seen by us to contain post-Triassic volcanics, although intercalated manganese ores in the Late Jurassic to Early Cretaceous interval do imply renewed volcanism somewhere in the basin (Robertson 1980). For this reason it seems that Cretaceous volcanics in the Antalya Complex (Gödene Zone) were probably formed in more axial parts of the basin and were then tectonically juxtaposed with the former rift margins in which major volcanism had ceased by latest Triassic time.

Problems of nappe tectonics

Although major nappes have been invoked by French workers supporting both the 'internal' (northerly) and the 'external' (southerly) origins of the Antalya allochthon (Fig. 6), there has been little consideration of how low-angle thrusting could have produced the present structural complexity of the Antalya allochthon, which is by common consent now dominated by steep structures. The stacking order of Delaune-Mayere *et al.* (1977) could theoretically have been achieved in two quite different ways (Fig. 7).

By rethrusting of the carbonate platform

This model assumes a palaeogeographically simple continent-ocean transition. In Late Cretaceous time the margin facies and the ophiolitic rocks were thrust as low-angle sheets a substantial distance over the adjacent relatively autochthonous carbonate platform (Fig. 7). This would have been similar to the emplacement of the Semail Nappe in Oman (Glennie *et al.* 1973: Fig. 7a). Later, possibly in Miocene time, the entire stack would have been re-thrust so that the relative autochthon at the base was thrust up to high structural level as the 'Upper Antalya Nappe' (Fig. 6). However the resulting geometry does not match the Antalya Complex. Contrary to expectation, the 'Upper Antalya Nappe' limestones are rarely seen 're-stacked' above ophiolite, the implied highest

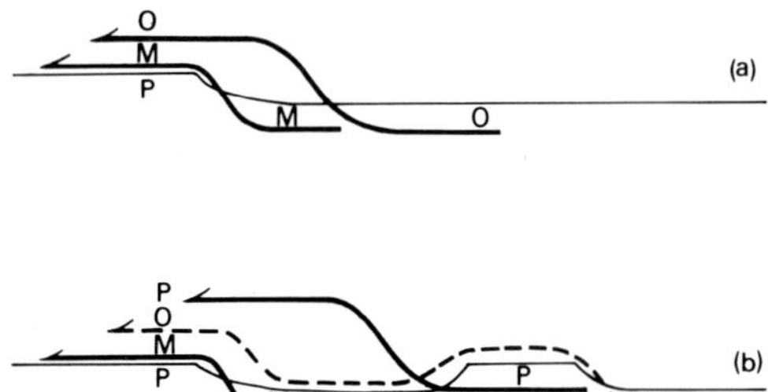

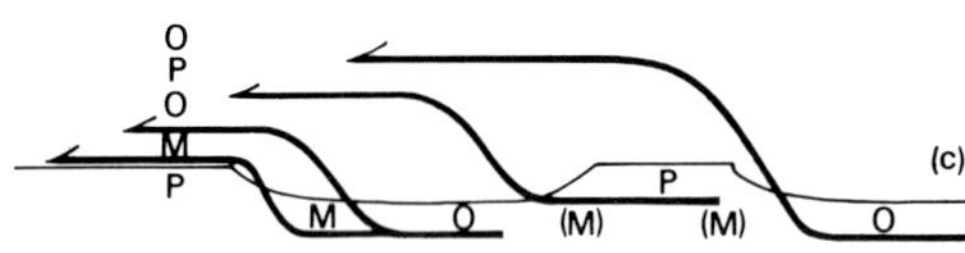

P = platforms
M = margins including mafic basement
O = ophiolite

FIG. 7. Attempts to match the structural complexity of the Antalya allochthon by thrusting *without* strike-slip. a, the continental margin is emplaced over the carbonate platform (U. Cretaceous), then faulted, or re-thrust (?Miocene) carrying platform as the 'Upper Antalya Nappe' (Poisson, 1978; see Fig. 6); b, This illustrates Marcoux and Ricou's (1979) suggestion of a separate rift and ophiolite basin with platform limestones between (Kemer Zone). During the latest Cretaceous, ophiolites were supposedly emplaced over the Kemer platform and into the older Triassic rift (broken line). Later (?Miocene) rethrusting could have intercalated rift and ophiolitic lithologies in the Gödene Zone; c, The Gödene Zone forms as an ophiolitic basin separate from the coastal ophiolites (Tekirova Zone), then is stacked in latest Cretaceous and Miocene (? 'obduction melange' of Yilmaz, 1981). None of these thrust schemes matches the true complexity. Notably for b, and c, Reuber (this volume) maps the coastal ophiolite under the Kemer Zone.

nappe unit after Late Cretaceous thrusting, but are commonly in contact with Upper Triassic mafic lavas and their sedimentary cover. Restacking implies gross duplication of an original thrust-stack, but this is not seen, particularly along the contact between the limestones of the Kemer Zone ('Upper Antalya Nappe') and the adjacent Tekirova Zone. Along the contact the entire Kumluca and Gödene Zones should be repeated but are not.

By contrast, in the NE segment of the Antalya Complex (Waldron, this volume), there is no structural evidence of systematic re-thrusting (e.g. cross-cutting thrusts and refolded folds). Particularly, the regularly imbricated Kumluca Zone shows no signs of the polyphase deformation expected close to the sole of a supposedly sequentially deformed major nappe pile (Woodcock & Robertson 1982).

By thrusting involving two basins

The second possible interpretation stems from Marcoux & Ricou's (1979) proposal that the Upper Triassic lavas and sediments formed in a southerly rift in the Arabian continental platform. In this view, the ophiolite was generated separately in a northerly basin and the two were brought together by thrusting in the latest Cretaceous time. Located somewhere between these two basins were the Kemer limestones. As shown in Fig. 7b the present tectonic configuration could theoretically have formed by thrusting of the ophiolite over the Kemer limestones into a separate rift. Re-thrusting could have then interleaved ophiolite with rift lithologies and left the Kemer limestones as the 'Upper Antalya Nappe' (Fig. 7b). This model is consistent with the existence of ophiolite-derived olistostrome melange in the highest levels of the Kemer limestones stratigraphy (e.g. Belen Yayla, Robertson and Woodcock 1982), and the existence of an ophiolite klippen above the Kemer limestones in the north-east of the area (Fig. 2). The chief problem is that Reuber's (this volume) mapping of the Kemer-Tekirova Zone contact shows that in the south the ophiolite structurally underlies, rather than overlies, the Kemer limestones (Ardrasan-Omurga Dağ, Fig. 2). It seems most unlikely that re-thrusting produced the anastomosing laterally continuous steeply dipping serpentinite strands characteristic of the Gödene Zone without imposing any regular stacking order.

Summarizing, scrutiny of the two alternative published nappe models show that neither matches the internal complexity of the Antalya Complex. Postulating a dominant strike-slip regime immediately solves most of these problems.

The Antalya Complex is not an 'obduction melange'

Yılmaz (1981 & Yılmaz *et al.* 1981) introduced a new dimension into the Antalya debate by reinterpreting much of the allochthon as an Upper Cretaceous 'obduction melange' (Figs 7c, 8). This includes the area of our Gödene

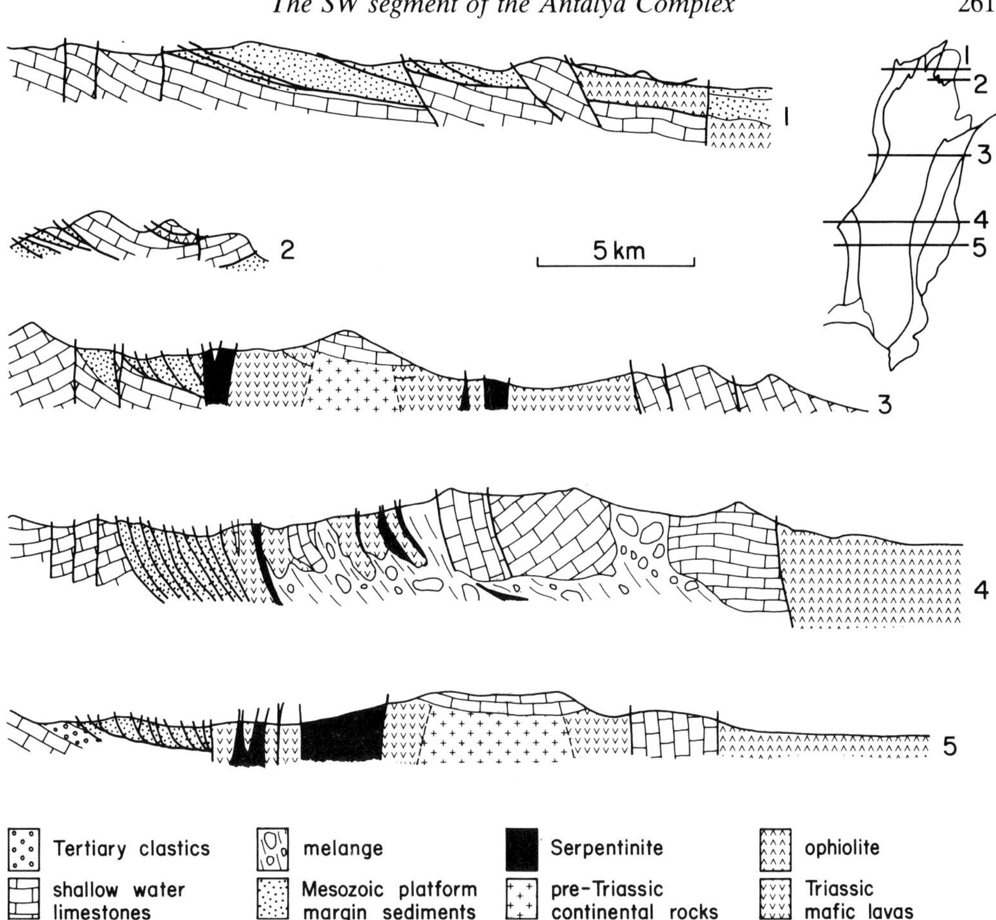

FIG. 8. Published cross-sections illustrating interpretations of the SW Antalya Complex. Same horizontal scale but vertical scales vary. Locations shown on outline tectonic zonation (top right). 1, 2 from Poisson (1978) showing his concept of the 'Upper Antalya Nappe' as upfaulted platform basement slices previously overthrust by the Antalya allochthon; 3 and 5 from Robertson and Woodcock, (1981c) showing from W to E the imbricated carbonate margin facies, subvertical screens of serpentinite, continental slivers and carbonate build-ups within the Triassic rift lavas, 4, from Yılmaz (1981) showing her concept of the Antalya allochthon as a vast 'obduction melange'. Cross-sections illustrating Marcoux and Juteau's (1982) interpretation have not been published.

and Kemer Zones, and units of the French 'Middle' and 'Upper Antalya Nappe'. Yılmaz draws a sharp distinction between the regularly stacked 'Kumluca complex' (our Kumluca Zone) which she interprets as a thrust passive margin, and a major zone of 'obduction melange' to the east. Three melange 'sheets' are recognized, each comprising clast-to mountain-sized detached blocks in a matrix ranging from serpentinite to clay and basalt. From west to east: Sheet 1 is described as melange with blocks of basalt, *Halobia*-limestone, Jurassic cherty limestone and Cenomanian marl in a matrix of basalt and clay. Sheet 2 contains, in addition, ultramafic rocks in a matrix of serpentinite and clay. Sheet 3, to the east, the structurally highest, comprises mountain-sized masses of Ordovician to Upper Cretaceous limestones with a clastic matrix. In striking contrast with the French nappe interpretation, Yılmaz (*op. cit.*) shows the limestone massifs as rootless blocks in a mega-melange (Fig. 8).

While we agree that parts of the Antalya Complex, particularly in the Gödene Zone, are melange, several points strongly oppose this 'obduction melange' hypothesis. First, no sedimentary or tectonic matrix can be identified around the major limestone massifs. These are relatively internally coherent or thrust-deformed. The major limestone massifs are either in sharp tectonic contact with other units (e.g. west edge of Bakirlı Dağ, Fig. 3) or else stratigraphically overlie pre-rift Palaeozoic successions (Kemer limestones, Fig. 8). Secondly, the melange sheets have little lateral continuity; the more ophiolitic 'sheet 2' well developed in the Gödene area (Fig. 3) thins to narrow subvertical ophiolite screens north and south.

Thirdly, we find no support for the inferred distinction between the Kumluca imbricate 'complex' and the adjacent 'obduction melange'. In several cases we have observed facies transitions across the Kumluca-Gödene Zone boundary (Robertson & Woodcock 1981c). Within the southern area of the Gödene Zone adjacent to the Kumluca Zone we have also mapped imbricated successions almost identical to the adjacent Kumluca Zone (Fig. 9). The chief difference is that, whereas no basement to the Kumluca Zone is seen, the imbricated lozenges in the Gödene Zone in places overlie Upper Triassic mafic lavas with normal contact. A similar imbricate wedge of 'Kumluca Zone-type' sediments is also mapped in the north-east area (Çandır Çay, Robertson & Woodcock 1980c).

The Kumluca Zone and the Gödene Zone thus did not form in different areas subjected to different emplacement. Instead the lavas and deep sea sediments of the Gödene Zone formed adjacent to the distal carbonate margin of the Kumluca Zone. In the latest Cretaceous both zones experienced a wrench- and thrust-regime, although competency of the rocks in the adjacent zones differed. Sediments in the southwest of the Gödene Zone were tectonically imbricated just as in the adjacent Kumluca Zone. Mapping shows that these imbricated successions originated as deep sea sediments ponded in hollows in the Triassic lava surface. During early deformation the lava basement shortened while more competent sediments in hollows deformed into partly rooted thrust sheets.

In our view, melange in the South-West Antalya area is of three main types. First, during the initial deformation still-unconsolidated Upper Cretaceous marls and pelagic chalks slid off underlying more consolidated strata to form disorganized incompetent sheets, best seen near the Gödene Zone-Kumluca Zone contact in the south-west. Secondly, with continued shortening of the Upper Triassic lavas more competent sheets of deep water sedimentary rocks were detached as a series of gently inclined rootless sheets without matrix. Such sheets are particularly well-exposed in the southern part of the Gödene Zone. Thirdly, in contrast to these superficial processes, shearing along strike-slip faults has created zones of tectonic melange, best seen along the northwest Gödene Zone margin (e.g. east of Hatıpalanı). This is 'coloured melange' with clasts of lava, deep sea sediments and shallow water limestones in a serpentinite matrix.

Summarizing, the Antalya Complex is far from being an Upper Cretaceous 'obduction melange' and in fact varies in coherence from relatively intact to thrust-deformed. It is clearly cut by tectonic melange generated by pervasive strike-slip faulting. There is no evidence that the Antalya Complex was 'obducted' over the southern part of the Bey Dağları carbonate autochthon until Miocene time. The Antalya Complex differs strongly, for example, from the Oman nappes, where 'melanges' beneath the Semail ophiolitic nappe, are mostly olistostromes free of ophiolite-derived material (Glennie *et al.* 1973; Searle 1980, Woodcock & Robertson 1984). Serpentinite melange is restricted to local zones below the Semail nappe where 'basal serpentinite' was injected into regularly imbricated passive continental margin lithologies.

Role of strike-slip faulting

The role of strike-slip faulting in the genesis and emplacement of the Antalya Complex has been emphasized by Reuber's (this volume) work on fabrics in the coastal ophiolites. This points to genesis in the Upper Cretaceous at a broadly east-west spreading axis offset by a north-south oceanic fracture zone (Reuber *et al.* 1982). Emplacement of the Tekirova ophiolite was accompanied by extensive brittle shearing and deposition of ophiolite-derived mega-breccias with shallow water Maastrichtian fossils (Robertson & Woodcock 1982).

Although more dismembered, the ophiolitic rocks of the Gödene Zone are petrologically similar to those of the coastal ophiolite (Tekirova Zone). Reuber (*op. cit.*) interprets the absence of well defined layered cumulate sequence (Juteau & Whitechurch 1979) in both the westward part of the Tekirova Zone and in the Gödene Zone as an indication of formation close to a transform fault. Other indications of strike-slip faulting in the genesis of the Gödene Zone ophiolite are numerous pegmatitic segregations, zones of intense hydrothermal alteration in massive gabbros, and irregularly orientated dykes which cut other parts of the ophiolite (Juteau 1975, Yılmaz 1978). It is to be expected that along oceanic transform faults, where seawater has easy access, magma chambers can be suppressed, while hydrous

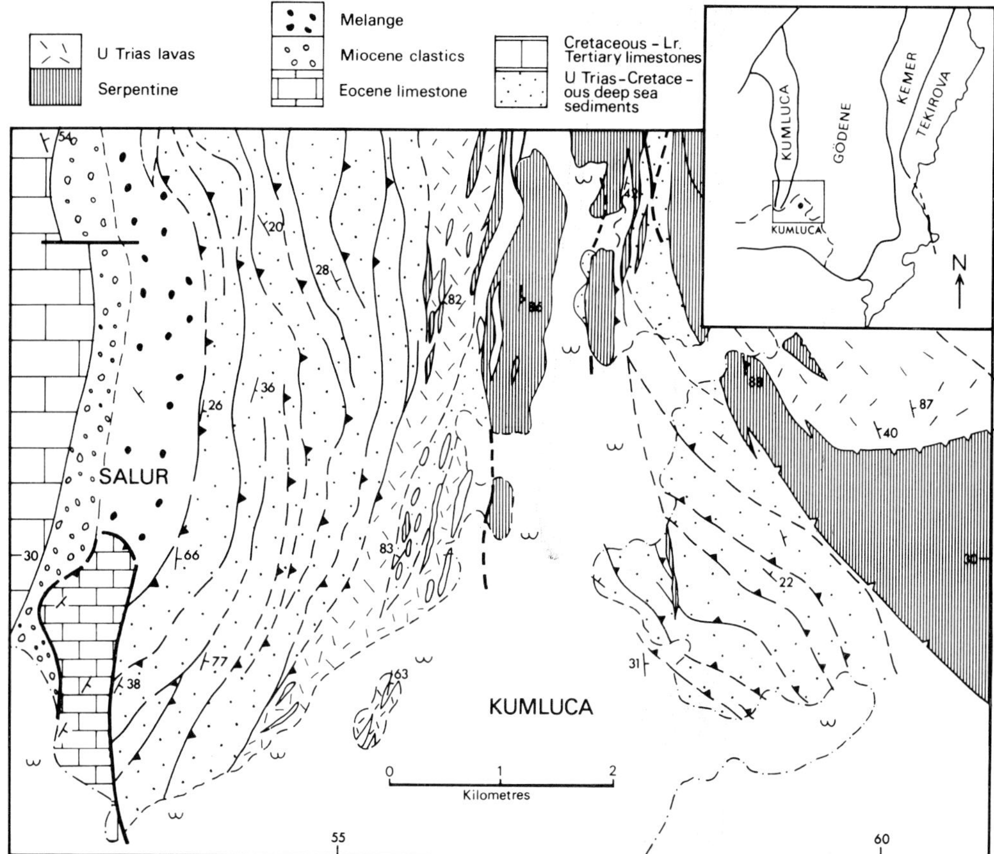

FIG. 9. Map of an area near the contact of the Kumluca Zone and the Gödene Zone to show the similarity of tectonic styles. In both zones similar Mesozoic deep sea sediments are tectonically imbricated, but the Gödene Zone was later cut by sub-vertical sheared serpentinite protruded along strike-slip faults. This pattern opposes the idea of the Gödene Zone as an 'obduction melange' (Yilmaz, 1981). Map modified after Woodcock & Robertson, (1982).

mineral phases are abundant and dykes could be variably intruded, for example, during the passage of offset spreading axes.

In the Gödene Zone the pattern of braided anastomosing strands of vertical sheared serpentinite can now be confirmed as the result of continued sinistral strike-slip faulting during the emplacement of the Antalya Complex. Any doubt that the serpentinite strands are indeed vertical should be removed by reiterating the definite observation that the sheared serpentinites map out over a rugged terrain without deflection and show parallel-sided near vertical tectonic contacts with other lithologies (Fig. 9). In places serpentinite strands cut obliquely across gently dipping imbricate sheets showing that the serpentinite emplacement *followed* initial thrust deformation. Several major lava massifs which are only gently inclined (Çalbalı Dağ. Fig. 3) are truncated on both sides by steeply dipping serpentinite sheets.

Further indications of wrench tectonics are the subaerially deposited ophiolite-derived clastics found as discontinuous lenses within the Gödene Zone (e.g. Çınarcık breccias; Robertson & Woodcock 1980a). These are mostly scree breccias shed from steep ephemeral fault scarps. Facies shed from major nappes advancing over land are quite different. For example the south-western USA Keystone Thrust sheet shed mature alluvial fan facies which were progressively overthrust as the nappe advanced (Johnson 1981). Indeed the final phases of thrusting of both the Lycian Nappes and the Antalya Complex in the Miocene involved deposition of just such mature ophiolite-derived alluvial facies. By contrast highly immature breccias shed from ephemeral scarps are a feature of deformed pull-apart basins in strike-slip systems, for example, along the San Andreas Fault, or the Alpine Fault of New Zealand (see papers in Ballance & Reading 1980).

Summarizing, both the Tertiary ophiolite-derived clastics in the Gödene Zone and their more spectacular Maastrichtian counterparts along the Kemer-Tekirova tectonic contact can be attributed to strike-slip genesis, and are definitely not typical of the deposits shed by major nappes advancing over land.

The evident importance of strike-slip faulting in the Antalya Complex raises the question of the extent to which its whole history could be related to alternating phases of compression and extension along a long-lived strike-slip system, for example, similar to the Western USA active margin. In the geologically similar Hatay and Baër-Bassit areas, near the Syrian-Turkish border, Delaloye & Wagner (this volume) and Garfunkel & Derin (this volume) refer to the oblique Atlantic opening during Late Triassic and Early Jurassic and infer major sinistral strike-slip during initiation of a small Mesozoic ocean basin in the East Mediterranean. Yılmaz (1981) proposed, from radiometric data, that the Gödene Zone ophiolites formed in a small trans-tensional basin which was then emplaced during the Maastrichtian before, or during, genesis of the coastal Tekirova ophiolite in a separate ocean basin.

There is, however, some evidence that strike-slip faulting only became dominant later in the genesis of the South-West Antalya Complex. Studies of modern strike-slip passive margins show little basement tilting or magmatism (Scrutton 1979). By contrast, the Antalya area in the Late Triassic can be closely compared, for example, with the post-Miocene Gulf of Suez (Garfunkel & Bartov 1977), an area of rifting without significant strike-slip offset. The Antalya Upper Triassic to Upper Jurassic-Lower Cretaceous basinal successions are mostly deep water pelagics without the clastic intercalations expected of an actively strike-slip deforming area like the modern Gulf of Elat (Ben-Avraham *et al.* 1979).

Local clastic intercalations, hydrothermally precipitated manganese and mafic volcanics appear in the Upper Jurassic to Lower Cretaceous interval. Oblique spreading to produce a small strike-slip controlled ocean basin could have begin at this time or later. Evidence from the Gulf of Elat suggests that later deformation of previously rifted areas involves little renewed uplift, tilting, or faulting away from master strike-slip faults bounding the active basin (Ben-Avraham *et al.* 1979). General comparisons show that strike-slip passive margins (Scrutton 1979) remain topographically high without strong subsidence until collapse occurs outside the transform offset zone. This could help explain why shallow water limestone deposition continued on the adjacent margins until mid-Late Cretaceous during or some time after the inferred onset of spreading.

Role of off-margin carbonate build-ups

We have previously argued that many of the mountain-sized limestone massifs within the Gödene Zone, are not parts of the formerly continuous 'Upper Antalya Nappe' (e.g. Bakirlı Dağ, Çalbalı Dağ, Fig. 3), but instead are carbonate build-ups constructed on basement horsts *within* the Triassic rift zone (Robertson & Woodcock 1981c). Mapping suggested to Kalafatçioğlu (1973); Şenel (1980); Robertson & Woodcock (1980c) and Yılmaz (1980) that the limestones were autochthonous or para-autochthonous with respect to adjacent Upper Triassic rift extrusives and sedimentary cover, but for Marcoux & Ricou (1979), Marcoux and Juteau (1982) and Ricou & Marcoux, (1980) they are grossly allochthonous nappes transported many hundreds of kilometres from the north.

It is not disputed by Marcoux (1976) that some of the small-mountain-sized reefal limestones in the Gödene Zone are Upper Triassic carbonate build-ups. Marcoux & Juteau (1982) also concede that some of the Upper Triassic lava successions pass with *normal contact* into condensed ammonitico rosso facies which accumulated on seamounts (Dumont *et al.* 1972). In some cases contacts at the base of the limestone massifs are indeed tectonic. The north-east area was particularly affected by thrusting (Kalafatçioğlu 1973 e.g. Sivridağ, Karadağ, Eren Dağ, Fig. 3). Clear tectonic contacts are also produced where limestone massifs are cut by steep serpentinite strands (e.g. west margin of Bakirlı Dağ, Fig. 2) which we relate to strike-slip faulting.

Unfortunately, except where fault-offsets can be specifically matched, the distance of offset along any particular shear-plane is always to some extent a matter of interpretation. For this reason, although we believe that relative displacement along the base of the carbonate massifs was, at most, minor, we point to the gross stratigraphy as the best indication of an original relatively *in situ* origin. Away from the carbonate massifs we have identified basinal successions of Upper Triassic mafic lavas overlain by pink ammonitico rosso facies (e.g.

Bakirlı Dağ, Çalbalı Dağ, Fig. 2). Above come thick successions of shallow water limestones of Jurassic and Early Cretaceous age. Is it merely coincidental that the shallow water limestones apparently conformably overlie what by consent were *already* Upper Triassic seamounts, while shallow water limestones are *not* found thrust over the basinal facies in the same way? Instead the basinal facies, mostly radiolarian cherts in the Gödene Zone, are typically in tectonic contact with serpentinites and other ophiolitic rocks. The obvious explanation is that the carbonate build-ups were originally horsts within the Triassic rift zone. Such basement highs are to be expected in rifts produced by listric normal faulting (e.g. see Ridley this volume). The existence of such intra-rift highs greatly helps to explain the structural complexity of the Gödene Zone which, as discussed above, cannot be reconciled with thrust emplacement alone. No complicated multi-phase thrusting is then needed to produce the present outcrop pattern which was instead initiated by Mesozoic palaeogeography.

The Kemer limestone: microcontinent rather than 'suspect terrain'

The succession in the Kemer Zone (Teke-Tahtalı Dağ Zone of Yılmaz, this volume) comprises Ordovician to Permian pre-rift sedimentary 'basement', passing conformably into Triassic to Cretaceous mostly shallow water carbonate facies, then terminating locally in ophiolite-derived olistostrome melange of Maastrichtian age (Gümbe Yala, Fig. 3). In the north the main Kemer limestone units are steeply east-dipping to near vertical, repeated by thrusts with mostly westward displacement. Macroscopic west-facing folds are present, particularly close to the Gödene Zone contact. To the east the Kemer limestones are in moderate to steeply west-dipping tectonic contact with the Tekirova ophiolite, and in the north with Upper Triassic lavas and sediments similar to the Gödene Zone.

Although much of the Kemer-Gödene Zone contact is concealed by Neogene limestone screes, locally in the north-east (near Palaz, Fig. 3) Upper Triassic limestone facies depositionally overlying mafic pillow lavas match those in the adjacent Kemer Zone, suggesting an original lateral facies transition once existed between the two zones. Further south major limestone massifs (e.g. Tahtahlı Dağ) are bordered by Upper Triassic quartzose turbidites and redeposited limestones which we interpreted as possible slope-basin facies (Robertson & Woodcock 1982).

On the other hand, over most of its length the Gödene-Kemer Zone contact is clearly tectonic, and the smaller limestone massifs of the south-east area (e.g. Omurgadağ, Fig. 3) are in tectonic contact with Upper Triassic lavas and sediments similar to the Gödene Zone. Locally in the south (3.5 km SW of Ardrasan) the Gödene-Kemer zone contact is marked by a ca. 70 m thick intercalation of up to greenschist-facies pelites, psammites, marbles and metalavas. The metamorphism is attributed to localised shear-heating during emplacement *unrelated* to the ophiolite. This raises the possibility that the Kemer Zone could be more grossly allochthonous than the tectonic zones to the west.

We have already stressed the problem with emplacement of the Kemer limestones as a major nappe between two zones (Tekirova & Kemer) which experienced pervasive strike-slip tectonics from Maastrichtian to Neogene time. An alternative, is that the Kemer Zone could have been sliced-off a southward continuation of the Bey Dağları carbonate platform then entrained along strike-slip faults, and so be a small-scale analogue of a 'suspect terrain' of the Western USA, (Coney *et al.* 1980). The sinistral offset inferred for both genesis and emplacement, suggests any such tectonic displacement of slivers along the margin would have been northwards. This process should have duplicated parts of the continent-ocean transition in a recognizable way (e.g. shelf, slope, rise facies). If strike-slip faults were to have cut progressively deeper into a margin, more proximal units could have ended up furthest east.

Could the Kemer Zone have been sliced-off and entrained in this way? The most easterly Kemer limestones, exposed in Kesme Boğazı section (Fig. 3), are dominated by limestone rudites and breccias, interpreted as proximal carbonate slope facies, reported by Demirtaşlı (pers comm. 1982) to be mostly Cretaceous in age. These slope facies do have features in common with the Çatal Tepe unit in Kumluca Zone, itself interpreted as thrust edge facies of the Bey Dağları relative autochthon.

However there are serious problems with this model. First, for parts of a carbonate margin to have been sliced-off and entrained along the margin, some part of the Bey Dağları margin, presumably to the south, must have been involved in major latest Cretaceous tectonics, yet, as discussed above, pelagic sedimentation continued along the eastern margin of the Bey Dağları in to the Tertiary without a break. Secondly, major slices of a carbonate platform edge should be easily recognizable in the Antalya allochthon. This is particularly true of the thick proximal edge facies, equivalent for example, to the Sumeini Group in Oman (Glennie *et al.* 1974), which consists of thick carbonate conglomerates of a type seen only in the Antalya allochthon, as the Çatal Tepe unit and interpreted as the emplaced edge of the formerly adjacent Bey Dağları carbonate platform. Similar proximal edge facies are not present elsewhere in the Antalya allochthon. Thirdly, a relatively proximal slice of a carbonate margin would have been underlain by pre-rift Palaeozoic lithologies, as indeed seen in the largest carbonate massif of the Kemer Zone (Kezme Boğazı), yet many of the smaller carbonate massifs to the south-east (e.g. Omurgadağ) are depositionally underlain by, or in tectonic contact with, Upper Triassic mafic lavas and sediments, suggesting an origin *within* the rift rather than as slices derived from well within a carbonate platform.

Summarizing, although strike translation of units could play a role in both the genesis and emplacement of the Antalya Complex it seems unlikely that the off-margin carbonate massifs in the Kemer and Gödene Zones could have been strike-slipped off a single linear parent carbonate margin. Instead, we believe the Kemer Zone originated as a very large off-margin carbonate build-up floored by Palaeozoic basement rocks, plus several smaller build-ups now located further west in the Gödene Zone. The various slivers of Upper Triassic mafic lavas and sediments east of the Kemer Zone (sandstones, cherts, *Halobia* limestone) formed in an eastward continuation of the Triassic rift. During the Cretaceous the eastern edge of the Kemer Zone could have accommodated a north-south strike-slip fault which created a steep fault scarp along which limestones, breccias and conglomerates accumulated (Kesme Boğazı). At this stage the Kemer margin topography would have been comparable with the modern strike-slip controlled Gulf of Elat (Ben-Avraham *et al.* 1979). Later, oceanic crust formed in the mid-Late Cretaceous at a spreading axis offset by sinistral transform faults.

An important question is whether oceanic crust was created within the Gödene Zone separate from the origin of the Tekirova ophiolite as suggested by Yılmaz (1981), or if both formed in a single oceanic basin. As noted above both are lithologically similar and, on available radiometric age data, formed in Late Cretaceous time (Yılmaz, this volume) near an oceanic fracture zone (Reuber, this volume). If, as seems more likely, all the SW Antalya ophiolites formed in a single basin, it is possible, as shown in Fig. 10, that during final oblique separation to form ocean crust, the older Triassic off-margin highs were 'strung' out along the strike-slip lineament. During subsequent oblique closure oceanic crust could have been intercalated as ophiolite slivers between the main carbonate margin and the former off-margin carbonate build-ups.

By latest Cretaceous the small ocean basin was closing and impinging on the SE Antalya margin with sinistral strike-slip (Fig. 10b). Oblique convergence thrust ophiolite under the Kemer Zone in the south-east. Transpression was concentrated on master-faults near the Tekirova-Kemer Zone contact and resulted in westward thrusting of the Kemer limestones, low grade metamorphism of lavas and basinal Mesozoic sediments and, within the ophiolite, caused intense shearing, brecciation, uplift and genesis of chaotic mega-breccias close to the Kemer Zone contact.

Palaeogeographical summary

The Antalya Complex originated as a 'braided' Triassic intra-continental rift, comparable with the Neogene Gulf of Suez (Fig. 11a). Crustal extension was probably achieved by large-scale listric faulting generating both deep pelagic basins into which thick piles of mafic alkalic lavas were erupted, and basement highs forming seamounts capped by carbonate complexes. To the west the Bey Dağları carbonate platform became defined. After the end of volcanism the area subsided progressively under thermal control. Reefs on the main margin and on the off-margin highs grew apace. During Early and Mid Jurassic time the area formed an elongate sediment-starved gulf, possibly ca 150 km across and several kilometres deep, comparable, for example, with the Neogene Red Sea prior to ocean spreading.

Probably in Late Jurassic to Early Cretaceous time a regime dominated by north-south strike-slip faulting was initiated (Fig. 11). This was marked by fresh volcanism, hydrothermal manganese accumulation, redeposition of shelf quartzose sandstone into pelagic basins and high productivity of radiolaria as circulation and upwelling increased. From this time onwards the basin progressively widened becoming comparable in scale with the Gulf of California by the Mid Cretaceous.

During the mid to Late Cretaceous the southern Neo-Tethys small ocean basin opened. The edge of a major basement high within the rift accommodated a submarine strike-slip controlled fault scarp during the final crustal separation. During strike-slip opening the former off-margin highs may have been 'strung' out

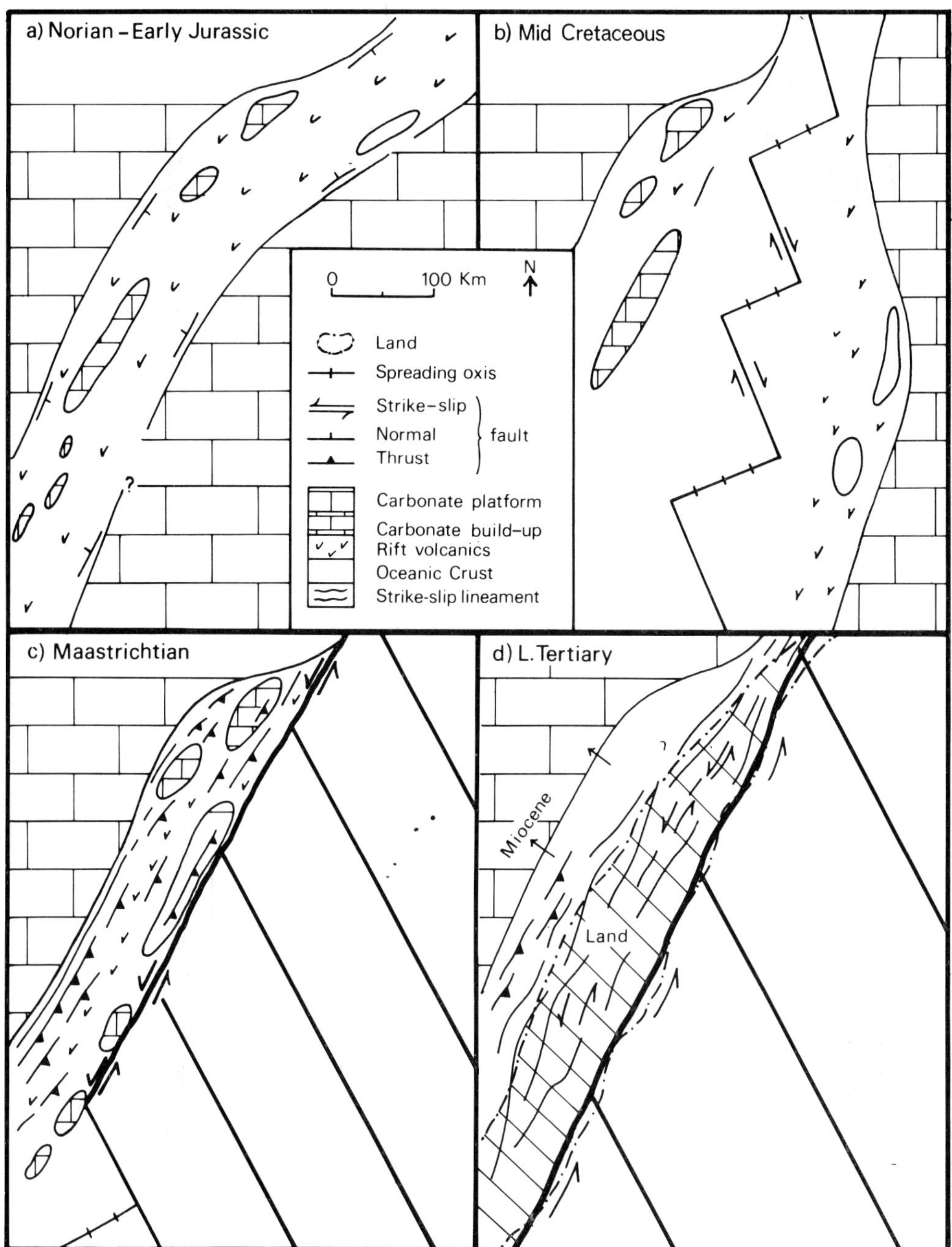

FIG. 10. Schematic evolution of the Antalya continental margin. a, Upper Triassic (Norian) to Early Jurassic. An elongate gulf, formed by continental rifting, is floored by mafic extrusives and deep sea sediments; carbonate build-ups cap basement highs, b, In the Upper Jurassic to Lower Cretaceous interval oblique spreading begins, offset by sinistral transform faults; c, By Maastrichtian the small ocean basin is closing obliquely compressing and shearing the adjacent Antalya margin; d, During the Lower Tertiary an elongate island is pervasively deformed by strike-slip faulting and serpentinite protrusion. In the Miocene the whole deformed area is thrust over the carbonate platform and the small ocean basin is virtually closed.

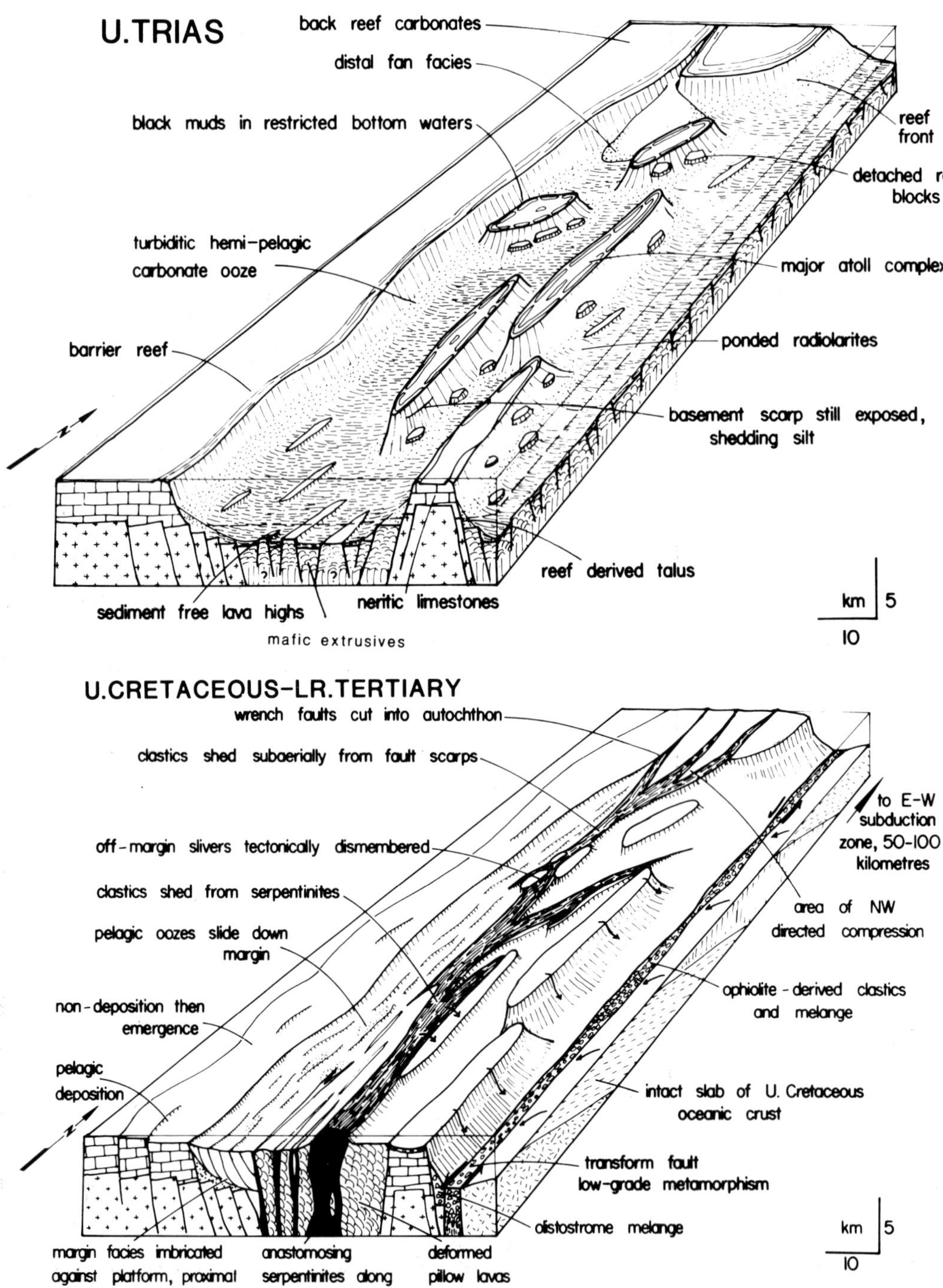

FIG. 11. Block diagrams showing the SW Antalya palaeogeography during Upper Triassic and Latest Cretaceous to Early Tertiary time. a, shows the pattern of off-margin carbonate build-ups on basement highs formed in a wide extensional rift. Strike slip faulting played little role until onset of oblique ocean floor spreading possibly in Upper Jurassic to Lower Cretaceous time; b, by Latest Cretaceous to Early Tertiary the carbonate margin was stacked against, but not over, the carbonate platform. The Tekirova coastal ophiolite was emplaced by sinistral strike-slip which also effected the ophiolites further west where anastomosing strands of serpentinites were protruded. Note: these diagrams illustrate the *style* of the margin and are *not* intended to be accurate palinspastic reconstructions.

along the margin. Oceanic crust formed at essentially east-west spreading axes offset by sinistral transform faults (Reuber, this volume, Fig. 10). By latest Cretaceous the basin was closing northwards. The Antalya margin acted as a sinistral strike-slip fault zone. During closure, slivers of oceanic crust could have been entrained between the carbonate platform (Bey Dağları Zone) and the former off-margin slivers to the east (Gödene and Kemer Zones). Close to the former continent-ocean boundary (Kemer-Tekirova ophiolite contact) the ophiolite was intensely sheared, brecciated, uplifted and eroded to form shallow water megabreccias of Maastrichtian age. Low grade metamorphism took place along key strike slip faults. Further west the edge facies of the Kumluca Zone were imbricated (Kumluca Zone). The most proximal slope facies are now only exposed locally in the north (Çatal Tepe unit).

Strike-slip faulting continued either continuously or episodically from the Upper Cretaceous throughout Palaeogene into Neogene time (Fig. 10b). Previously imbricated zones were cut by sub-vertical anastomosing strands of sheared serpentinite interpreted as low temperature protrusions up strike-slip faults. Ophiolite-derived clastics accumulated in short-lived pull-apart basins. During Early to Mid Miocene time the deformed Antalya allochthon was thrust westwards over the edge of the carbonate platform (Bey Dağları Zone) shedding large volumes of first turbiditic, then alluvial fan clastics (Hayward, this volume). Only later in the Neogene did deformation end followed by drastic uplift of limestone massifs (Kemer and Gödene Zones), possibly isostatically controlled. By this time the southern Neo-Tethys had probably completely closed explaining why continental crust is now apparently present beneath the Antalya Bay area. It follows therefore that the whole of the Antalya Complex is allochthonous with respect to the original parent carbonate platform and basement, although primary facies relationships can still be determined within the allochthon.

Finally, it should not be thought at all surprising that the history of the south-west segment of the Antalya Complex, which is dominated by wrench and thrust tectonics, should differ markedly from that of the north-east segment of the Antalya Complex which shows larger scale polyphase thrusting as described by Waldron, this volume. Both areas document originally complex Mesozoic palaeogeography with carbonate build-ups of various scales with intervening pelagic basins. Such variety in structural styles is exactly as expected from knowledge of modern passive and active continental margins.

ACKNOWLEDGEMENTS: Fieldwork was supported by the British Natural Environmental Research Council (NERC) and the Turkish Geological Survey (MTA). Our thoughts on the Antalya Complex benefited from discussion with E. Demirtaşlı, J. Dixon, A. Hayward, O. Monod, A. Poisson, A. M. C. Şengör, J. Waldron and P. Yılmaz. The manuscript was typed by Mrs H. A. Hooker.

References

ARGYRIADIS, I., DE GRACIANSKY, P. C., MARCOUX, J. & RICOU, L.-E. 1980. The opening of the Mesozoic Tethys between Eurasia and Arabia-Africa. *Proc. 26th Int. Geol. Cong. Paris*, c5, 199–214.

BALLANCE, P. F. & READING, H. G. (eds.) Sedimentation in oblique-slip mobile zones. *Internat. Assoc. of Sedimentol. Spec. Publ.*, **4,** 265 pp.

BEN-AVRAHAM, Z., ALMAGOR, G., & GARFUNKEL, Z. 1979. Sediments and the structure of the Gulf of Elat (Aqaba)—Northern Red Sea. *Sedimentary Geology*, **23,** 239–261.

BIJU-DUVAL, B., DERCOURT, J. & LE PICHON, X. 1977. From the Tethys Ocean to the Mediterranean Sea: A plate tectonic model of the evolution of the western Alpine system. In: BIJU-DUVAL, B. & MONTADERT, L. (eds.) *Structural History of the Mediterranean Basins*, Editions Technip. Paris, 143–164.

BRUNN, J. H. 1974. Le problème de l'origine des nappes et leurs translations dans les Taurides occidentales. *Bull. Soc. géol. Fr.* **16,** 101–106.

——, GRACIANSKY, P. C. DE, GUTNIC, M., JUTEAU, T., LEFEVRE, R., MARCOUX, J. MONOD, O. & POISSON, A. 1970. Structures majeures correlations stratigraphiques dans les Taurides occidentales. *Bull. Soc. géol. Fr.* **12,** 515–551.

——, DUMONT, J. F., GRACIANSKY, P. C., DE, GUTNIC, M., JUTEAU, T., MARCOUX, J., MONOD, O. & POISSON, A. 1971. Outline of the geology of the western Taurides. In: CAMPBELL, A. S. (ed.) *Geology and History of Turkey*. Petroleum Exploration Society of Libya. Tripoli. 225–255.

COLIN, H. J. 1962. Geologische Untersuchungen im Raume Fethiye-Antalya-Kaş-Finike. *Bull. Miner. Res. Explor. Inst. Turkey*, **59,** 19–61.

CONEY, P. J., JONES, D. L., MONGER, J. W. H. 1980. Cordilleran suspect terrains, *Nature*, **288,** 329–333.

DELAUNE-MAYERE, M., MARCOUX, J., PARROT, J.-F. & POISSON, A. 1977. Modèle d'évolution Mésozoïque de la paleómarge Téthysienne au niveau des nappes radiolaritiques et ophiolitiques du Taurus Lycien, d'Antalya et du Baër-Bassit. In: BIJU-DUVAL, B. & MONTADERT, L. (eds.) *Structural History of the Mediterranean Basins*, Editions Technip, Paris, 79–94.

DUMONT, J.-F., GUTNIC, M., MARCOUX, J., MONOD, O. & POISSON, A. 1972. Le Trias des Taurides occidentales (Turquie). Definition du bassin pamphylien: Un nouveau domaine à ophiolites à la marge externe de la chaîne taurique. *Zeit. Deutsch. Geol. Ges.* **123,** 385–409.

GARFUNKEL, Z. & BARTOV, T. 1977. The tectonics of the Suez Rift. *Bull. geol. Surv. Israel.* **71,** 44 pp.

GLENNIE, K. W., BOEUF, M. G. A., HUGHES CLARKE, M. W., MOODY-STUART, M., PILAAR, W. F. H., REINHARDT, B. M. 1973. Late Cretaceous nappes in the Oman Mountains and their geologic evolution. *Bull. Am. Ass. Petrol. Geol.* **57,** 5–27.

HAYWARD, A. B. 1982. *Tertiary ophiolite-related sedimentation in S.W. Turkey.* Unpublished University of Edinburgh Ph.D thesis, Scotland, 420 pp.

—— & ROBERTSON, A. H. F. 1982. Direction of ophiolite emplacement inferred from Tertiary and Cretaceous sediments of an adjacent autochthon; the Bey Dăgları, S.W. Turkey. *Bull. geol. Soc. Amer.* **93,** 68–75.

JOHNSON, M. R. W. 1981. The erosion factor in the emplacement of the Keystone thrust sheet (South East Nevada) across a land surface, *Geol. Mag.*, **118,** 501–507.

JUTEAU, T. 1968. Commentaire de la carte géologique des ophiolites de la région de Kumluca (Taurus lycien, Turquie méridionale). *Bull. Miner. Res. Explor. Inst. Turkey*, **70,** 70–91.

——, 1970. Pétrogenèse des ophiolites des Nappes d'Antalya (Taurus Lycien Oriental). Leur liaison avec une phase d'éxpansion océanique active au Trias supérieur. *Mém. Sciences Terre*, **15,** 265–288.

—— 1975. Les ophiolites des nappes d'Antalya (Taurides occidentales, Turquie). *Mem. Sciences Terre*, **32,** 692 pp.

——, NICHOLAS, A., DUBESSY, J., FRUCHARD, J. C. & BUCHEZ, J. L. 1977. Structural relationships in the Antalya ophiolite complex, Turkey: Possible model for an oceanic ridge. *Bull. geol. Soc. Am.* **88,** 1740–1748.

—— & WHITECHURCH, H. 1979. The magmatic cumulates of Antalya (Turkey): Evidence of multiple intrusions in an ophiolitic magma chamber. In: *Panayiotou, A.* (ed.) *Ophiolites.* Proc. Int. Oph. Symp. Cyprus 1979. 377–391.

KALAFACIOĢLU, A. 1973. Geology of the western part of Antalya Bay. *Bull. Miner. Res. Explor. Inst. Turkey*, **81,** 31–84.

LEFEVRE, R. 1967. Un nouvel élément de la géologie du Taurus Lycien: les nappes d'Antalya (Turquie). *C.r. Acad. Sci. Paris*, **265,** 1365–1368.

MARCOUX, J. 1970. Age carnien des termes effusives du cortège ophiolites des Nappes d'Antalya (Taurus Lycien oriental, Turquie). *C.r. Acad. Sci. Paris*, **271,** 285–267.

—— 1974. Alpin type Triassic of the upper Antalya nappe (western Taurides, Turkey). In: ZAPFE, H. (ed.) *Die Stratigraphie der alpin-mediterranean Trias*, Wien, 145–146.

—— 1976. Les séries triasiques des nappes à radiolarites et ophiolites d'Antalya (Turquie): homologies et signification probable. *Bull. Soc. géol. Fr.* **18,** 511–2.

—— & RICOU, L.-E. 1979. Classification des ophiolites et radiolarites alpino-mediteranéenes d'après leur contexte paléogéographique et structural. Implications sur leur signification géodynamique. *Bull. Soc. géol. France*, **21,** 643–652.

—— & JUTEAU, T. 1982. Alkaline upper Triassic submarine volcanism and associated sediments in the Antalya nappes, Turkey. In: J. E. DIXON & A H. F. ROBERTSON (eds). Abstracts, *Geological evolution of the East Mediterranean*, Edinburgh, U.K. September 1982, p. 70.

MONOD, O. 1976. La 'courbure d'Isparta': une mosaïque de blocs autochtones surmontés de nappes composites à la jonction de l'arc hellénique et de l'arc taurique. *Bull. Soc. géol. Fr.* **18,** 521–532.

—— 1977. *Recherches géologiques dans le Taurus occidental au sud de Beyşehir (Turquie).* Thèse de Docteur Es Sciences, Univ. Paris-Sud. Orsay, 442 pp.

——, MARCOUX, J., POISSON, A., & DUMONT, J. F. 1975. Le domaine d'Antalya, témoin de la fracturation de la plateforme africaine au cours du Trias. *Bull. Soc. géol. Fr.* **16,** 116–127.

POISSON, A. 1967. Données nouvelles sur le Cretacé superieur et le Tertiaire du Taurus au NW d'Antalya (region de Korkuteli Turquie). *C.r. Adac. Sci., Paris.* **264,** 218–221.

—— 1976. Présence d'un Trias supérieur de facies récifal dans le Taurus lycien au NW Antalya (Turquie). *C.r. Acad. Sci. Paris*, **264,** 2443–2446.

—— 1978. *Recherches géologiques dans les Taurides occidentales (Turquie).* Thèse de Docteur Es Sciences, Université de Paris-Sud. 795 pp.

REUBER, I. 1982. *Générations successives de filons grenus dans le complexe ophiolitique d'Antalya (Turquie): origine, évolution et mécanisme d'injection des liquides.* Thèse de 3 ème cycle, Strasbourg University, 246 pp.

——, WHITECHURCH, H. & CARON, J. M. 1982. Setting of gabbro dikelets in an ophiolite complex (Antalya, Turkey) by hydraulic fracturing: downward injections of residual liquids. *Nature*, **296,** 141–143.

RICOU, L.-E. ARGYRIADIS, I. & LEFEVRE, R. 1974. Proposition d'une origine interne pour les nappes d'Antalya et le massif d'Antalya (Taurides occidentales, Turquie). *Bull. Soc. géol. Fr.* **16,** 107–111.

——, ARGYRIADIS, I. & MARCOUX, J. 1975. L'axe calcaire du Taurus, un alignement de fenêtres arabo-africaines sous des nappes radiolaritiques, ophiolitiques et métamorphiques. *Bull. Soc. géol. Fr.* **17,** 1024–1044.

——, MARCOUX, J. & POISSON, A. 1979. L'allochtonie des Bey Dağları orientaux. Reconstruction palinspastique des Taurides occidentales. *Bull. Soc. géol. Fr.* **21,** 125–33.

RICOU, L.-E. & MARCOUX, J. 1980. Organisation générale et rôle structural des radiolarites et ophiolites du système alpino-méditerranéen. *Bull. Soc. géol. Fr.*, **22,** 1–14.

ROBERTSON, A. H. F. 1981. Metallogenesis on a Mesozoic passive continental margin, Antalya Complex, S.W. Turkey. *Earth planet. Sci. Lett.* **54,** 323–345.

—— & WOODCOCK, N. H. 1980a. Strike-slip related sedimentation in the Antalya Complex, S.W. Turkey. *Spec. Publ. Int. Assoc. Sedimentol.* **4,** 127–145.

—— & WOODCOCK, N. H. 1980b. Tectonic setting of the Troodos Massif in the East Mediterranean. *Proc. Int. Ophiolite Symp. Cyprus 1979*, 36–49.

—— & WOODCOCK, N. H. 1981a. Bilelyeri Group, Antalya Complex: deposition on a Mesozoic continental margin in S.W. Turkey. *Sedimentology*, **28,** 381–401.

—— & WOODCOCK, N. H. 1981b. Alakır, Çay Group, Antalya Complex, S.W. Turkey: A deformed Mesozoic carbonate margin. *Sediment. Geol.* **30,** 95–131.

—— & WOODCOCK, N. H. 1981c. Gödene Zone, Antalya Complex, S.W. Turkey: volcanism and sedimentation on Mesozoic marginal ocean crust. *Geol. Rdsch.* **70,** 1177–1214.

—— & WOODCOCK, N. H. 1982. Sedimentary history of the south-western segment of the Mesozoic-Tertiary Antalya continental margin, south-western Turkey. *Eclogae geol. Helv.* **75,** 517–562.

SCRUTTON, R. A. 1979. On sheared passive margins. *Tectonophysics*, **59,** 293–305.

SEARLE, M. P. 1980. *The metamorphic sheet and underlying volcanic rocks beneath the Semail ophiolite in the Northern Oman Mountains of Arabia.* Unpublished Open University Ph.D. thesis, 213 pp.

ŞENEL, Y. M. 1980. *Teke Torosları Güneydoğusunun Jeologisi, Finike-Kumluca-Kemer (Antalya).* Unpublished report, M.T.A. Ankara, Turkey, 106 pp.

ŞENGÖR, A. M. C. & YILMAZ, Y. 1981. Tethyan evolution of Turkey: A plate tectonic approach. *Tectonophysics*, **75,** 181–241.

THUIZAT, R., MONTIGNY, R., ÇAKIR, U. & JUTEAU, T. 1978. K-Ar Investigations on two Turkish ophiolites. *In:* ZARTMAN, R. E. (ed.) Short Pap. 24th Int. Conf. Geochr. Cosmoschr. Isot. Geol. Geol. Surv. Open-File Rep. 78–701, 430–32.

——, WHITECHURCH, H., MONTIGNY, R. & JUTEAU, T. 1981. K-Ar Dating of some infra-ophiolitic metamorphic soles from the Eastern Mediterranean: new evidence for oceanic thrustings before obduction. *Earth planet. Sci. Lett.* **52,** 302–10.

WALDRON, J. W. H. 1981. *Mesozoic sedimentary and tectonic evolution of the northeast Antalya Complex, Eğridir, S.W. Turkey.* Unpublished University of Edinburgh Ph.D. thesis, 239 pp.

WOODCOCK, N. H. & ROBERTSON, A. H. F. 1977. Imbricate thrust belt tectonics and sedimentation as a guide to emplacement of part of the Antalya Complex, S.W. Turkey. *Abst. 6th Coll. Aegean Geol. Izmir* 1977, p. 98.

—— & ROBERTSON, A. H. F. 1981. Wrench-related thrusting in Turkey. *In:* MCCLAY, K. R. & PRICE, N. J. (eds). *Thrust and Nappe Tectonics, Spec. Publ. Geol. Soc. Lond.* 359–362.

—— & ROBERTSON, A. H. F. 1982. Wrench and thrust tectonics along a Mesozoic-Cenozoic continental margin: Antalya Complex, S. W. Turkey. *J. geol. Soc. Lond.* **139,** 147–163.

—— & ROBERTSON, 1984. The structural variety in Tethyan ophiolite terrains. *Spec. Publ. Geol. Soc. Lond.*, **13**.

YILMAZ, P. O. 1978. *Alakır Çay Unit of the Antalya Complex (Turkey): an example of ocean floor obduction.* Unpublished M.S. thesis, Bryn Mawr College, U.S.A., 91 pp.

—— 1981. *Geology of the Antalya Complex, S.W. Turkey.* Unpublished Ph.D. thesis, Univ. Texas. Univ. Microfilms Int. Michigan, 194 pp.

——, MAXWELL, J. C. & MUEHLBERGER, W. R. 1981. Antalya kompleksinin yapişal evrimi ve doğu Akdeniz' deki yeri. *Yerbilimleri*, **7,** 119–127.

Structural history of the Antalya Complex in the 'Isparta angle', Southwest Turkey

J. W. F. Waldron

SUMMARY: The Isparta angle is the north-pointing cusp where the arc of the Hellenides meets that of the Taurides in southwestern Turkey. The Antalya Complex is an assemblage of allochthonous deep- to shallow-water Mesozoic sedimentary rocks and ophiolite fragments present at the centre of the angle, surrounded by massive platform carbonate units, also of Mesozoic age.

Structural mapping of the folded and faulted Antalya Complex in the Eğridir region at the centre of the Isparta angle indicates at least two major episodes of deformation. An earlier phase, probably latest Cretaceous in age, resulted in the thrusting of deep-water sediments over shelf-edge units which originally lay to the northeast. Subsequent southwest-vergent folding and thrusting in the Tertiary era locally reversed the earlier stacking order. Regularly imbricated thrust complexes were converted into chaotic mélange-like terrains.

The sedimentary cover was shortened by at least 72% as a cumulative result of these events. Palinspastic reconstruction of the Antalya Complex sedimentary basin indicates a wide original separation of the carbonate platform units on the east and west limbs of the Isparta angle, with at least one smaller platform between. These units were therefore probably separate carbonate banks, analogous to the present-day Bahama Banks.

The Hellenide and Tauride orogens are both southward-concave arcuate belts; they meet in southwestern Turkey in a north-pointing cusp, variously termed the Courbure d'Isparta (Blumenthal 1963), the Taurus orocline, or the Isparta angle. Fig. 1 shows the main tectonic units in this area. The Lycian and Hadim nappe systems represent the eastern and western extremities of the Hellenide and Tauride orogens respectively. Both are emplaced above relatively autochthonous units dominated by Mesozoic 'platform' carbonate sequences. The Antalya Complex (roughly requivalent to the Antalya Nappes of Lefèvre, 1967) is a third group of allochthonous rocks, which is confined to the Isparta angle, although closely similar rocks occur in the Mamonia Complex of southwestern Cyprus (Lapierre 1975, Ealey & Knox 1975, Robertson & Woodcock 1979, Swarbrick 1980). The Complex occupies a key position between the east and west limbs of the Isparta angle; its structures and relationships with surrounding units are the main subject of this paper. Most of the conclusions are based on field mapping in the north of the Complex, to the southeast of Lake Eğridir (Fig. 1). Complementary studies in the southwest of the Complex are described by Robertson & Woodcock (1981a, 1981b, 1981c; Woodcock & Robertson 1982).

Regional geology of the Isparta Angle

Antalya Complex and Alanya Massif

The Antalya Complex occupies an irregular belt through the centre of the Isparta angle, separating the major carbonate massifs of the two limbs (Fig. 1).

In the Antalya Complex a great variety of basinal Mesozoic sedimentary facies have been distinguished, including turbiditic sandstones and shales, pelagic limestones, radiolarian cherts, redeposited limestones, and ophiolite-derived sandstones (Brunn *et al.*, 1971, Marcoux *in* Delaune Mayère *et al.* 1978, Allasinaz *et al.* 1974, Monod 1977, 1978, Dumont 1976, Poisson 1977, Gutnic *et al.*, 1979, Robertson & Woodcock 1981a, 1981b, 1981c). Shallow-water carbonates also occur, resting on Ordovician or Triassic sandstones, mudstones and limestones; certain of the carbonate massifs shown as part of the Complex in Fig. 1 have been excluded from the Antalya Complex by previous authors (e.g. Brunn *et al.*, 1971). In addition, a Cretaceous ophiolite suite is represented by dolerites, gabbros, dunites, and peridotites (Juteau 1975, Thiuzat & Montigny 1979). Minor occurrences of metamorphic rocks have also been described (Juteau 1975, Woodcock & Robertson 1977b, Waldron 1981).

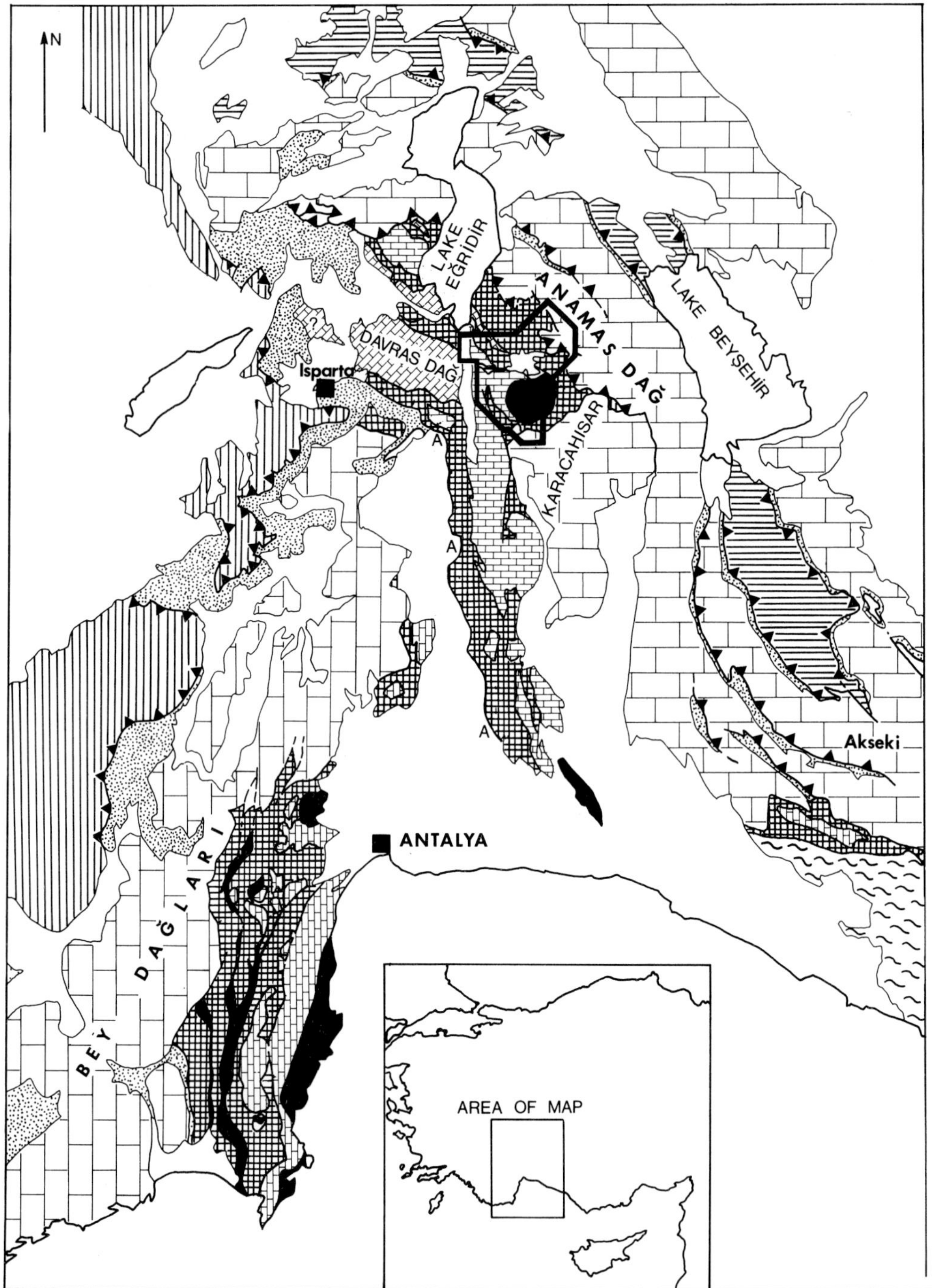

FIG. 1. Tectonic map of the Isparta angle, after Brunn *et al.* (1971) with modifications by Woodcock & Robertson (1982), Gutnic *et al.* (1979), and Waldron (1981). Box encloses the area of detailed mapping (Fig. 4). Units with shallow-water Mesozoic carbonate sequences (listed under Antalya Complex and Tauride Platform Units) have been distinguished on the basis of tectonic position not stratigraphic sequence. Karacahisar region may in fact be a separate unit from Anamas Dağ.

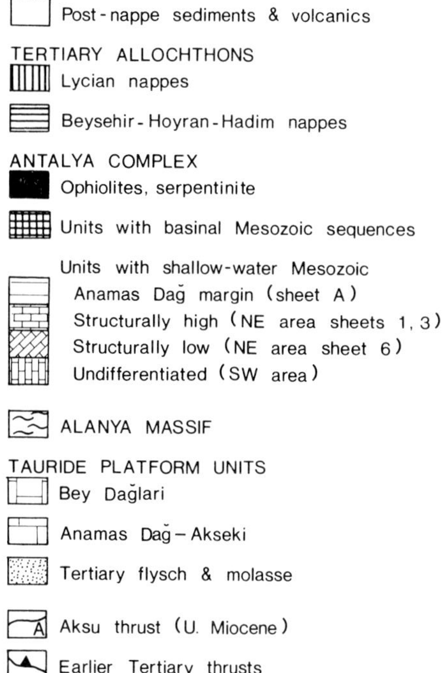

The Antalya Complex was originally named the Antalya Nappes (Lefèvre 1967). Brunn *et al.* (1971) mapped three nappes, but these were identified mainly by facies correlations, without demonstration of structural continuity. According to Woodcock & Robertson (1982), on the other hand, the structure of these rock units west of Antalya is dominated by steep strike-slip faults. The term Antalya Complex (Woodcock & Robertson 1977a) is therefore used in this paper, in order to avoid implications of structural style.

The Alanya massif (Fig. 1) is an area of metamorphic rocks which lies to the southwest of the Antalya Complex on the east limb of the Isparta angle. According to Argyriadis (1974), the massif is an allochthonous sheet thrust over the Antalya Complex, which re-appears in a tectonic window to the east of the area of Fig. 1.

Lycian (Teke) Taurus

The region west and southwest of the Antalya Complex (Fig. 1) is known as the Lycian or Teke Taurus. Its structures are generally accepted to represent an extension of the Hellenides, although the precise correlation is disputed (Brunn *et al.* 1976, Bernoulli *et al.* 1974, Özgül & Arpat (1973). The region is dominated by the Lycian Nappes, an edifice of allochthonous sheets that were transported from the northwest towards the southeast in a number of phases during the early Tertiary era, ending with emplacement into their present position in Miocene times. Within the Lycian Nappes, Poisson (1977) and Graciansky (1972) have described a number of distinct stratigraphic sequences of Mesozoic to early Tertiary age. Delaune-Mayère *et al.* (1977) interpret these sequences as different parts of a Mesozoic continental margin, which was subsequently telescoped during late Cretaceous and early Tertiary orogeny. So-called 'ophiolitic' units, consisting of slivers of periodotite and diabase, are intercalated between the nappes; the western part of the edifice is capped by the 'Peridotite Nappe', consisting mainly of harzburgites cut by pyroxenite and dolerite dykes (Graciansky 1972).

The Lycian Nappes rest on a 'relative autochthon' consisting of Jurassic to Cretaceous shallow-marine carbonates of the Bey Dağları massif. These are overlain by Tertiary detrital sediments derived both from the advancing nappes to the northwest and from the Antalya Complex, which was thrust onto the eastern edge of the same autochthon (Hayward & Robertson 1982 and Hayward, this volume).

Taurus Occidental

The ranges forming the eastern limb of the Isparta angle represent the western extremity of the Taurus proper, and are known as the *Taurus Occidental.* As in the Lycian Taurus, a series of nappes here rests on an 'autochthon' dominated by Mesozoic shallow water carbonates constituting the massifs of Anamas Dağ & Akseki (Fig. 1). Erosion has reached a deeper level than in the Bey Dağları, revealing Triassic sandstones and shales resting unconformably on a Palaeozoic basement (Monod, 1977). The 'autochthon' here actually consists of a series of parautochthonous slices, thrust towards the southeast. Resting upon these slices are the Beyşehir-Hoyran-Hadim Nappes (Brunn *et al.* 1971, Monod 1977). Like the Lycian Nappes, these contain an ophiolitic unit, together with a variety of Upper Palaeozoic to Late Cretaceous sediments. These nappes were emplaced from the northeast during Eocene time. They are preserved partly as an enormous klippe along the axis of a gentle synform; their roots presumably lie to the north or northeast in central Anatolia.

Structure of the Northeastern Antalya Complex

The Northeastern Antalya Complex (south and east of Lake Eğridir, Fig. 1) lies at the centre of

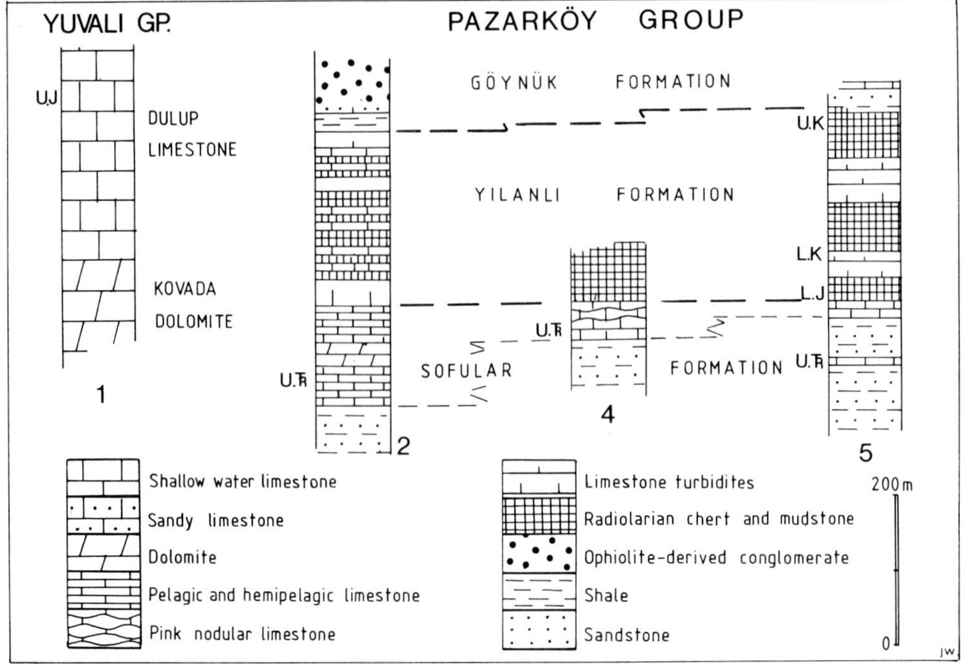

FIG. 2. Stratigraphic sequences typical of thrust sheets 1, 2, 4 and 5 in the southwest of the mapped area, illustrating stratigraphic units characteristic of shelf, slope, and basin.

the Isparta angle, between the Anamas Dağ and Davras Dağ carbonate massifs. The Complex in this region is subdivided by faults into numerous tectonic slices, as shown in Figs 3, 5, 6 and 7.

The lithostratigraphic units recognized in these slices are assigned to two groups (Fig. 2). Formal definitions will be proposed elsewhere. The *Yuvalı Group* consists entirely of Mesozoic platform carbonates of shallow-water origin. The *Pazarköy Group* comprises a variety of platform-edge, slope, and basin facies (Fig. 2). The most extensive are turbiditic sandstones, shales, and hemipelagic *Halobia*-bearing limestones of Late Triassic to Early Jurassic age (Sofular Formation). These are generally overlain by Jurassic and Cretaceous bedded radiolarian cherts and mudstones with interbeds of shelf-derived carbonate material (Yılanlı Formation). Late Cretaceous ophiolite-derived mudstones, sandstones and conglomerates are assigned to the Göynük Formation. Isolated slices of altered tholeiitic basalt and radiolarite (Havutlu Lava Fm.) are the same age as part of the Yılanlı Formation; these lavas and a tectonically bounded massif of harzburgite in the south of the area studied (Kızıl Dağ massif: Fig. 5 and Juteau 1975) are probably parts of an ophiolite suite, representing ancient oceanic lithosphere. Along the NE edge of the Complex, slices of oolitic, redeposited, and pelagic limestones represent platform-edge environments in the Late Triassic to Late Cretaceous interval. They are lateral equivalents of the more basinal pelagic and hemipelagic sediments.

'Platform' and 'basin' sequences equivalent to the Yuvalı and Pazarköy Groups are recognized throughout the Antalya Complex (Brunn *et al.* 1971, Allasinaz *et al.* 1974, Akbulut 1977, Marcoux *in* Delaune-Mayère *et al.* 1977, Monod 1977, Poisson 1977, Gutnic *et al.* 1979, Robertson & Woodcock 1981a, 1981b, 1981c, 1982).

Minor structures

Folds

These are the most conspicuous structures in many outcrops in the Eğridir region especially in the thinner bedded turbidite and radiolarian chert units. A few of these folds can be shown to be slumps truncated by erosion. The slump folds tend to be disharmonic and have strongly thickened hinges in coarser beds. The majority of folds show no evidence of having formed at or near the sediment surface, and affect considerable thicknesses of strata. Fold shape is

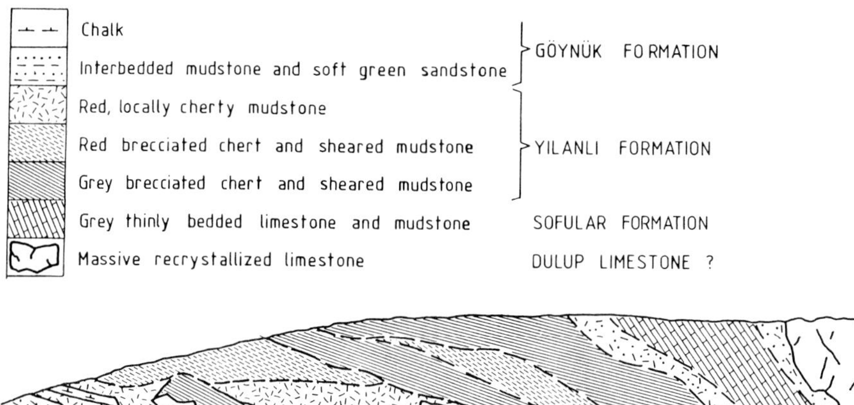

FIG. 3. Field sketch of a megabreccia outcrop in a road cutting 5 km south of southern tip of Lake Eğridir, showing the characteristic form of slices.

controlled largely by lithology. Competent layers (chert, sandstone, limestone) showing class 1C geometry (classification of Ramsay 1967) may be distinguished from incompetent layers (generally shale) with class 3 geometry. Chevron folds occur when competence contrasts are most marked (as in bedded cherts) and shale layers are thin.

A number of outcrops show folds in superimposed relationships. Owing to the scarcity of axial planar fabrics and the wide scatter of fold orientations, folding sequences are difficult to disentangle. In most cases one or more early sets of structures (designated the f_1 group) are refolded by later (f_2) folds with axial surfaces dipping north and northeast. The early folds themselves are extremely varied, but where a single set is present, they are generally asymmetrical and have northward vergence in upright sequences.

Faults

Minor faults are often seen in the hinges of folds, and elsewhere, where they are usually marked by zones of shaley gouge or recemented fault breccia. The major fault planes revealed by mapping are seldom well exposed.

Cleavage

This is only locally developed, and is never penetrative at outcrop scale. Cleavage is most apparent in fine-grained limestones, where it is stylolytic, resembling that described by Alvarez *et al* (1978) in Italian pelagic limestones.

Joints and veins

These are ubiquitous, and are particularly abundant in cherts, where they cause beds to distintegrate at outcrop into small (1 cm) cubes. The orientation of joints has not been investigated systematically. Fractures showing appreciable dilation are usually filled with calcite, though early generations of fractures in chert have chalcedony, quartz, or analcite fills.

Mélange and megabreccia

Many outcrops in the Antalya Complex show a variety of rock-types, juxtaposed as fragments from a few millimetres to tens of metres across. The term *mélange* is restricted to outcrops where a fine-grained matrix produced by mixing is present between the blocks (Greenly 1919, Hsü 1968, Naylor 1978). *Megabreccia* describes outcrops in which the soft sheared material surrounding the more competent lithologies has not been mixed, but consists entirely of large (1–50 m) slices of Antalya Complex lithologies (Fig. 3). Even in outcrop, these two alternatives may be difficult to distinguish. Mixing may be indicated by rare exotic fragments in apparently homogeneous shales. In poorly exposed areas the single mapping category 'mélange/megabreccia' is used.

All the main sedimentary lithologies of the Antalya Complex may be recognized as blocks in mélange and megabreccia terrains, but the larger slices are generally derived from adjacent structural units. Shales, impure limestones, and ophiolite-derived sandstones of the Late Cretaceous Göynük Formation are significantly over-represented and typically show complete disruption of stratification. This suggests that the formation was very soft and possibly still wet at the time of deformation.

Mélange and megabreccia were probably produced during thrusting of the Complex (see

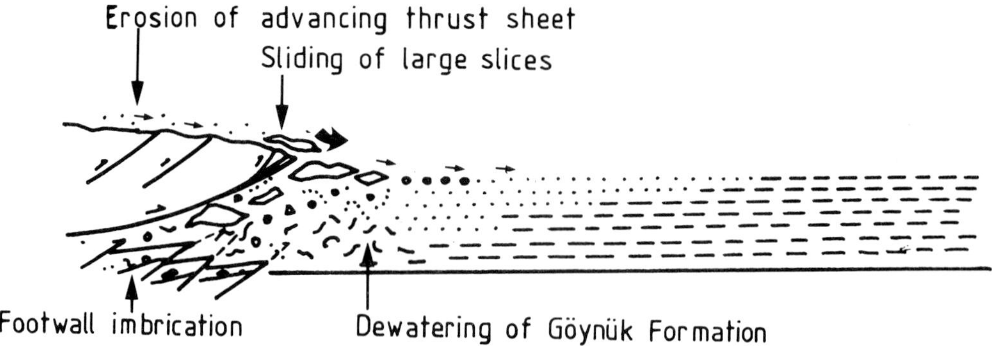

FIG. 4. Diagram to show possible processes involved in the creation of mélange-megabreccia terrains at the toe of an advancing thrust sheet. No scale implied.

below: major structures). Syntectonic sediments of the Göynük Formation were deposited immediately ahead of advancing thrust sheets and preferentially incorporated into mélange. Fig. 4 indicates the processes which may have operated; the relative importance of 'tectonic' and 'sedimentary' processes is uncertain.

Major structures

The megascopic structure of the northern Antalya Complex is dominated by tectonic contacts, the majority of which are at low angles to the stratification in the units they separate; they are here termed *thrusts*, even though some have been rotated into near-vertical present-day orientations. Two types of thrusts are distinguished. *First-order thrusts* are mainly near-horizontal at the present day; they can be traced for long (10 km+) distances and separate units with strikingly different stratigraphic sequences. *Second-order thrusts* are often steeper in dip and separate units with similar stratigraphies. Tectonic units bounded by first-order thrusts are termed *thrust-sheets* (or simply *sheets*) even when sub-divided by second order thrusts. Units bounded by second-order thrusts or immersed in mélange or megabreccia are termed *slices*.

The extent of the main tectonic units is shown in Fig. 5 and in cross-sections (Fig. 6). A composite schematic cross-section from southwest to northeast across the area is shown in Fig. 7.

In the southern part of the region (south of the Yılanlı-Pazarköy Fault: Fig. 4) a stack of generally south-dipping thrust sheets is exposed; these have been numbered from 1 to 6, starting at the structurally highest. Sheet 6 has previously been regarded as an autochthonous unit (Brunn *et al.* 1971), but mapping of its northeastern boundary indicates a southwest-dipping surface at a low angle to bedding, indicating that it too is allochthonous with respect to the units exposed further northeast. Sheets 1, 3 and 6 form rugged ridges and plateaus consisting of Yuvalı Group limestone and dolomites. They have relatively simple structures characterized by open folding. Sheet 3 is sharply lenticular in cross-section. Sheet 2 consists of a number of slices elongated northwest to southeast, within which softer, more easily eroded slope sediments are exposed. Its internal structure is shown in Fig. 6 (section AA′). Thrust sheets 4 and 5 consist of deeper-water slope and basin sediments. They are also divided into slices separated by second-order thrusts dipping south and southwest (Fig. 6, section BB′).

A larger lenticular thrust-bounded body of harzburgite (Eğridir Kızıl Dağ Peridotite) is intercalated between sheets 2 and 4 in the southern part of the area (Figs. 5, 6 & 7). This and an adjacent slice of lava are probably fragments of an ophiolite suite, representing ancient oceanic crust and upper mantle.

Thrust sheets 1 to 6 display a relatively orderly structure of imbricated slices, although areas of mélange and megabreccia are present within some sheets. Sequences are generally upright, and the relationships of first and second-order thrusts in 'duplexes' (Dahlstrom 1970) indicate overthrusting generally to the north or east (Figs 6 & 7).

In contrast, the northern part of the mapped area has a more chaotic structure with numerous overturned regions and extensive megabreccia terrains. The lowest structural unit is an 'imbricated unit' which resembles thrust sheets 4 and 5 in consisting of a stack of slices of basinal sediment striking northwest to south-

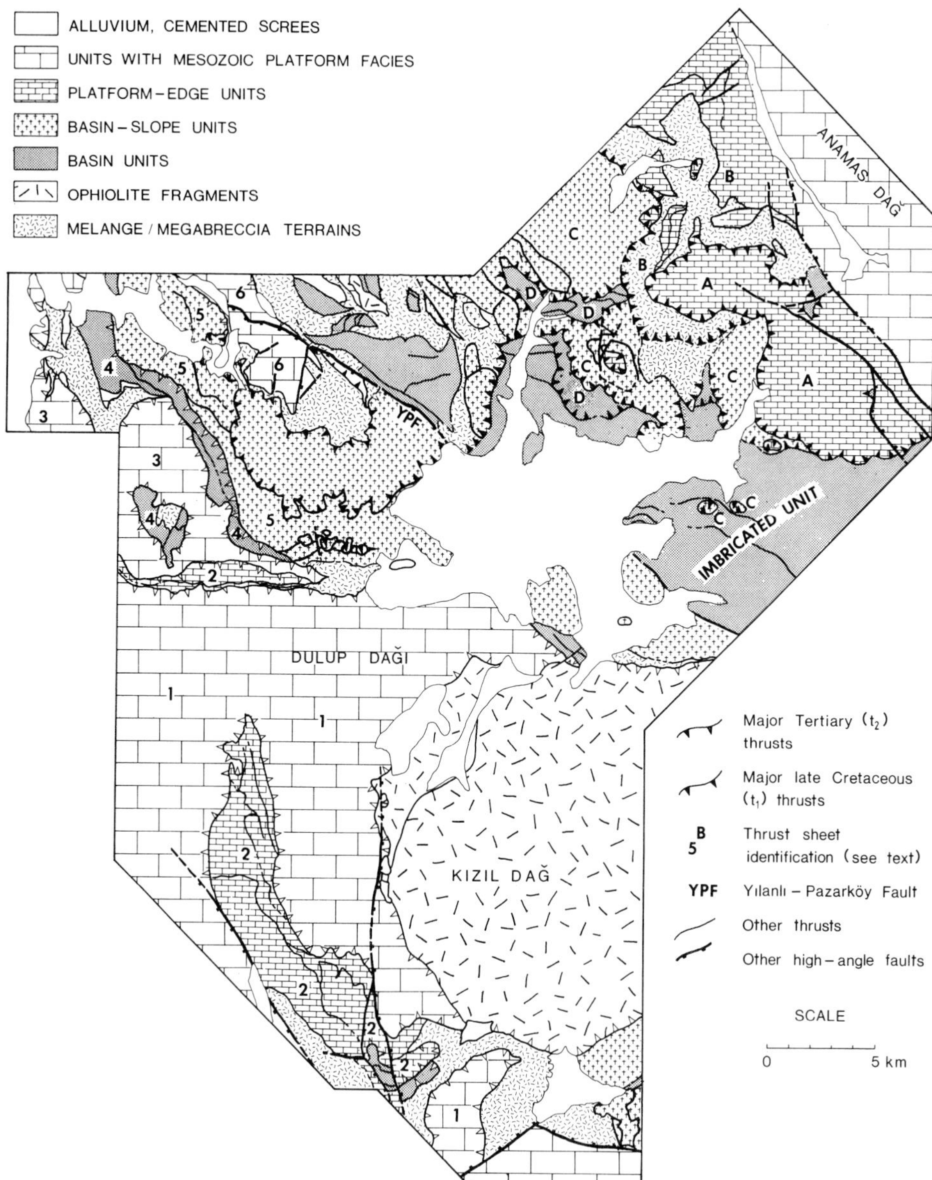

FIG. 5. Map showing tectonic units present to SE of Lake Eğridir. Southeast-transported thrust sheets of inferred Tertiary age are lettered A-D. Earlier thrust sheets are numbered 1–5. Some units (e.g. ophiolite fragments) are neither lettered nor numbered. Boundary of sheet C in the region of 'YPF' is uncertain. Location shown in Fig. 1.

east and younging southwest. These are overlain and truncated by three lenticular thrust sheets (B, C and D), each consisting of many slices. Megascopic folds within the slices face generally southwest (Fig. 6 CC′). The sediments of sheets C and D are basinal, but sheet B shows platform edge facies. The highest sheet in this part of the area is designated sheet A, which rests in turn from northwest to southeast on sheets B, C, and 'imbricated unit'. Unlike the underlying sheets it has a relatively simple structure without major overturning of sequences. Sheet A consists of massive Triassic lavas and reef carbonates in faulted contact with spectacularly exposed Jurassic to Cretaceous platform-edge sediments (Akpınar

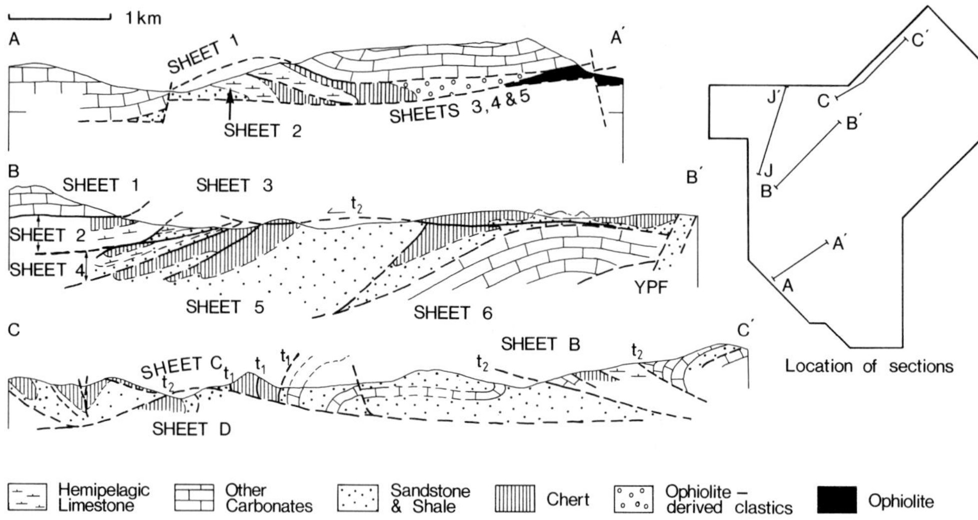

FIG. 6. Cross-sections without vertical exaggeration. The location map represents the area of Fig. 5. AA′ shows stacked slices in sheets 1 and 2, produced during episode t_1. BB′ shows imbricated t_1 slices, truncated by relatively minor t_2 thrust. CC′ from the NE or the area shows major t_2 thrusts and SW facing folds. Note that faults within t_2 thrust sheets are interpreted as folded t_1 thrusts. YPF, Yılanlı-Pazarköy fault.

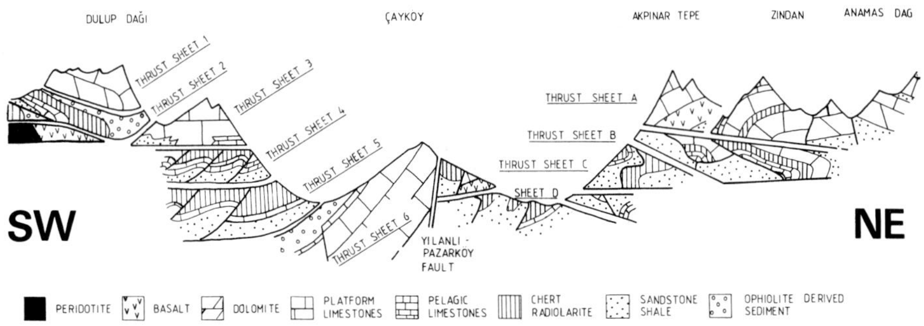

FIG. 7. Cartoon showing tectonic units present to southeast of Lake Eğridir. This diagram is an attempt to show all the major units in the area of Fig. 5. In fact not all the units may be seen on any one cross section. Internal structure of sheets is shown very schematically. Thrust sheets 1 to 6 were probably formed in a latest Cretaceous episode (t_1). The northeast boundary of sheet 6 is affected by the late Yılanlı-Pazarköy Fault, but was originally a thrust contact. Thrust sheets B to D and the underlying imbricated unit were remobilized by Tertiary thrusting (t_2). Sheet A is interpreted as the overthrust margin of the Anamas Dağ platform. It shows no evidence of earlier thrusting and was probably not affected by t_1. Not to scale.

Tepe Limestone & Bucak Lava: Juteau 1975; Zindan series: Dumont *et al.* 1980). At the northeast edge of the Complex, Late Triassic dolomites of sheet A are indistinguishable from those of the adjacent Anamas Dağ massif. The underlying sandstones and overlying algal limestones are also closely similar in the two units. Sheet A is therefore interpreted as the southwest edge of the Anamas Dağ massif, which has been thrust over the other Antalya Complex units.

Structural history

The structure of the Northeastern Antalya Complex is consistent with a two-phase history of thrusting. An early northeast-directed phase (t_1) is responsible for the stack of sheets (1–5) seen in the southern part of the area. A later phase (t_2) of thrusting in almost the opposite direction affects the northeast part of the area, remobilizing and overturning earlier slices as shown diagramatically in Fig. 8d. Phase t_2 prob-

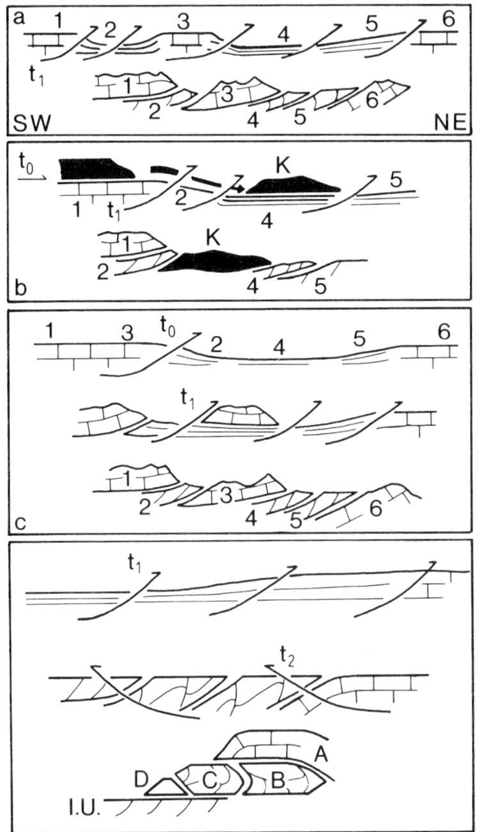

FIG. 8. Qualitative stacking models.

(a) Probable original arrangement and present configuration of sheets 1–5.

(b) Model to account for the position of the Kızıl Dağ peridotite massif (K) between sheets 2 and 4 in a region where sheet 3 is absent. The initial ophiolite emplacement event (t_0) contrasts with later purely thin-skinned thrusting (t_1).

(c) Alternative model for the initial configuration of sheets 1–5.

(d) Model for the origin of sheets A-D and imbricated unit (I.U.).

ably corresponds to f_2 folding. Some or all of the f_1 folds were produced during the early thrust phase.

Because only a recent unconformable cover is seen, the ages of these thrust events are poorly constrained. The youngest rocks deformed are Late Cretaceous (Maastrichtian); this is therefore the lower limit for t_1. However, the composition of sandstones and conglomerates of the Late Cretaceous Göynük Formation indicates syn-tectonic deposition, and its structural style suggests deformation of wet sediment. The early thrust episode is therefore believed to have started in Maastrichtian time, but possible continued into earliest Tertiary. The youngest autochthonous rocks underlying the Complex in the Karacahisar region to the south (Fig. 1) are Maastrichtian sediments including ophiolite and radiolarite detritus (Dumont 1976). This suggestst that episode t_1 caused the emplacement of the complex onto this massif.

There is little evidence in the mapped area of the age of t_2. However, southwest-vergent thrusting and folding is widespread in the Anamas Dağ and Akseki regions to the east and southeast (Monod 1977, Gutnic 1977), where it is dated as Eocene. This deformation would be expected to affect the Antalya Complex. Event t_2 is therefore assumed to have occurred during Eocene time.

Evidence of later deformational events is seen locally, as at Ağılköy where thrust sheets 3, 4 and 5 are folded into near-vertical orientation by a N-S trending fold with east-dipping axial surface (Fig. 9a). The orientation of this structure is oblique to both t_1 and t_2 structures. It is perhaps related to the 'Aksu phase' (Poisson 1977) of late Miocene westward thrusting which strongly affects the Sütçüler region further south (Akbulut 1977).

A number of high-angle faults, generally with north to south strike, occur south and east of Eğridir. The most conspicuous of these define an alluvium-filled graben to the south of Lake Eğridir; these faults have probably been recently active, suggesting a regime of east to west extension at the present-day.

Palinspastic reconstructions

The style of thrusting in the Northeastern Antalya Complex may be described as *thin-skinned*, since the Mesozoic sedimentary cover has been largely removed from its original basement, which is never seen. There is no evidence in surrounding areas for any large region of tectonic denudation, from which the Antalya Complex sheets have been derived by gravity sliding. Thrusting of the sedimentary cover in the Antalya Complex is therefore probably a response to a corresponding shortening of the underlying basement, though the exact position and nature of this shortening are unknown. Analogy with modern continental margins suggests that the Complex in the Mesozoic was underlain by thinned continental crust. Thickening by telescoping during orogenesis would then have led to décollement of cover sequences.

There is no evidence for cross-cutting of major t_1 structures in thrust sheets 1 to 6; their

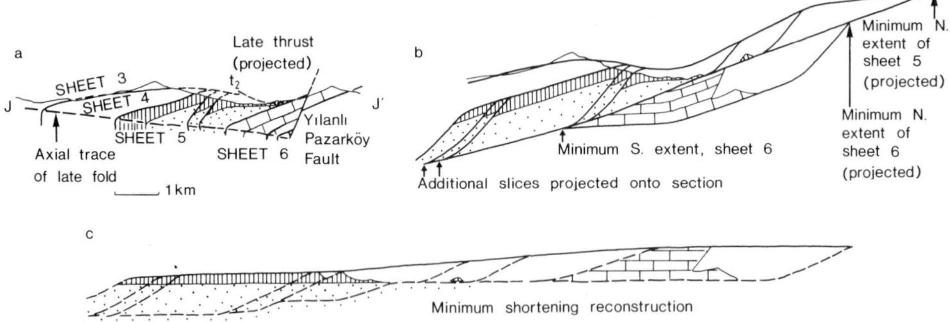

FIG. 9. Minimum shortening for thrust sheets 5 & 6. Line of section JJ′ shown on Fig. 6.
(a) Present day configuration (detail not shown in sheets 3 & 4).
(b) Effects of late (t_2) thrust and late fold removed.
(c) Effects of early (t_1) thrusting removed.
Reconstructions are not unique, but are constructed to show minimum shortening consistent with the outcrops. Shaded parts of units are those preserved at the present day. Symbols as Fig. 6.

stacking order probably reflects their original geographical configuration, as shown in Fig. 8a. In this reconstruction, the shallow-water carbonates of thrust sheets 1, 3 and 6 represent distinct carbonate banks, separated by deeper-water areas.

The Kızıl Dağ peridotite massif is an anomalous unit of 'basement' within a stack of thin-skinned thrust slices; it is believed to be oceanic upper mantle material of deep origin (Juteau 1975). The associated lavas probably represent the upper levels of oceanic crust, but the expected intermediate parts of an ophiolite suite (gabbro, dyke complex) are missing in the northern part of the Antalya Complex. The presence of these lithologies as clasts in syntectonic sediments of thrust sheet 2 suggests that an ophiolite suite underwent emplacement and erosion before the thrusting of sheets 1 to 6. The now-isolated Kizil Dağ massif may be a block derived from this ophiolite as shown in Fig. 8b. An analogous origin may be postulated for thrust sheet 3, as shown in Fig. 8c, which represents an alternative model to Fig. 8a.

The northern part of the mapped area is even more difficult to reconstruct because of the intensity of the second episode of thrusting. Several initial arrangements are consistent with the structual evidence. However, sheets A, B and C show a regular succession of facies from shelf to basin during the Mesozoic. This suggests a stacking model such as that shown in Fig. 8d.

Fig. 9 shows a more quantitative attempt at reconstruction for sheets 5 and 6 using the 'balanced' cross-section technique of Dahlstrom (1969), and based on a number of simplifying assumptions. The cross-sections are shown without vertical exaggeration, and are drawn perpendicular to the mean fold axis direction for sheet 5, this being the inferred direction of thrusting (Elliot & Johnson 1980). The minimum extent of eroded and subsurface portions of the sheets are derived by projection from outcrop away from the line of section. The cross-sectional area of slices in the present-day and reconstructed sections are kept constant; this is equivalent to assuming plane strain. Finally, no stratal shortening is assumed other than that seen at thrusts and megascopic folds. While none of these assumptions can strictly be justified, their net result is that Fig. 9 is a *minimum-shortening* reconstruction for the two units; the real shortening was probably considerably greater than that shown, suggesting an even wider separation of platforms.

Similar reconstruction for the other major thrust sheets lead to the conclusion that the Antalya Complex east of Lake Eğridir has suffered shortening by *at least 72% of its original across-strike width*.

Correlation with adjacent areas

Many controversies in Alpine-Mediterranean geology result from uniformity of shallow-water carbonate platform facies throughout the Mesozoic Tethys ocean. Fig. 1 shows the proposed correlation of platform units in the central Isparta angle based on the results outlined above. The following features should be noted:

(1) The southeast edge of the Anamas Dağ platform massif is shown as thrust *over* the other Antalya Complex units as thrust sheet A. This sheet includes Triassic Lavas and reefs,

as well as platform-edge carbonates (Zindan sequence: Dumont *et al.*, 1980). West of Lake Eğridir, the Barla Dağ carbonate massif shows an analogous structural relationship with the Antalya Complex.

(2) The platform limestones here assigned to thrust sheet 6 are correlated, following Brunn *et al.* (1971) and Gutnic *et al.* (1979) with the more extensive Davras Dağ and Eğridir massifs. In contrast with earlier interpretations, however, these are regarded as an allochthonous thrust sheet, intercalated between sheets of deeper-water sediments, and thought to represent an originally isolated carbonate platform. They are therefore logically regarded as part of the Antalya Complex along with the structurally higher platform carbonates of thrust sheet 1.

(3) The Karacahisar massif to the southeast of the Antalya Complex in the area studied is regarded by Dumont (1976) as part of the Tauride platform, continuous with the Anamas Dağ. Nevertheless, a small slice of redeposited and cherty limestone (Série de Camova) is described by Dumont tectonically intercalated between the two massifs. This raises the possibility that these two platforms may also have been separated by a basinal area, which would neatly explain the absence of early Jurassic sandy facies (Çayır Formation) from the Karacahisar massif, the stratigraphy of which is otherwise comparable to the Anamas Dağ (Gutnic *et al.*, 1979). Continuity of this unit with thrust sheet 6 and the Davras Dağ massif is conceivable.

(4) The Davras Dağ massif is bounded to the southwest by the Late Miocene northeast dipping Aksu thrust zone (Akbulut 1977, Poisson 1977) shown in Fig. 1. Because the displacement on this thrust is not known, the original position and orientation of the units to the southwest cannot be unequivocally determined. These units include the Bey Dağları platform massif, which forms the relatively autochthonous backbone of the western limb of the Isparta angle, and the structurally overlying parts of the Antalya Complex (Fig. 1).

Original location of the Antalya Complex

Ever since its original identification by Lefèvre (1967), controversy has raged over the original location of the Antalya Complex (Brunn *et al.*, 1971). All previous interpretations have assumed initial continuity of the Mesozoic carbonate massifs (Bey Dağlari, Karacahısar, Barla Dağ, Anamas Dağ) onto which the Antalya Complex was emplaced in late Cretaceous and early Tertiary times. Discussion has therefore become polarized between those proposing an origin to the north of this 'Taurus Limestone Axis' (Ricou *et al.* 1974, 1979), and those who favour an origin broadly to the south (Dumont *et al.* 1972a & b, Robertson & Woodcock 1980).

A radical alternative to the hypothesis of a continuous Tauride platform is here proposed. Structural evidence in this study clearly implies a late Cretaceous northeastward thrusting of the Antalya Complex against the margin of the Anamas Dağ. The original location of the complex must therefore be sought generally to the southwest of this massif. On the other hand, transport of the Complex *over* the Bey Dağları platform, which at present lies to its southwest, cannot have occurred, even allowing for a partially allochthonous Bey Dağları as envisaged by Ricou *et al.* (1979), since sedimentation in the central Bey Dağları continued into the Upper Miocene (Hayward & Robertson 1982). By this time the Antalya Complex was already in place and receiving an unconformable sedimentary cover (Poisson 1977). Structures and facies in the southwestern Antalya Complex (Woodcock & Robertson 1977, Robertson & Woodcock 1981a, 1982 and this volume), and in clastic sediments of the adjacent Bey Dağları massif (Hayward & Robertson 1982, Hayward, this volume) confirm that this part of the Complex was thrust westward.

The simplest remaining alternative is therefore that the Antalya Complex originated in a basin which lay *between* the Anamas Dağ and Bey Dağları massifs. Arguments based on facies similarities for continuity of the massifs Barla Dağ-Davras Dağ-Bey Dağları (Fig. 1) are not compelling. Gutnic *et al.* (1979) have documented important contrasts, particularly between the Davras Dağ platform carbonates and the platform-edge facies of Barla Dağ, here interpreted as the margin of the Anamas Dağ platform.

Figure 10 shows a schematic palinspastic map of the norhteastern part of the Complex, based on the assumption of minimum shortening consistent with the structural evidence; original dimensions may well have been much larger than shown in this reconstruction. Thrust sheets have been unstacked as shown in Figs 8a, b and d. The direction of thrusting has where possible been estimated separately for each sheet using mean fold axis orientation; errors in these estimates would lead to different relative positions of platforms but a broadly similar overall picture. At least a narrow oceanic zone must have been present by Cretaceous time to explain the presence of ophiolite fragments; it may well have been wider than that shown in

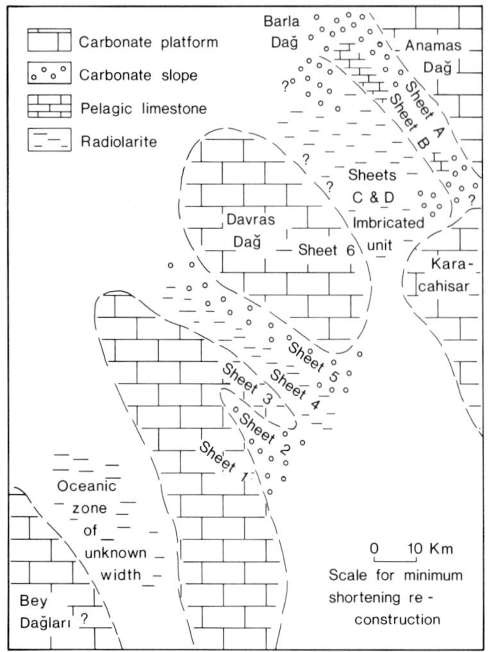

FIG. 10. Schematic U. Cretaceous (Campanian-Senonian) palinspastic lithofacies map for units in the Northwestern Antalya Complex and adjacent areas. Units have been unstacked without rotation and assuming minimal translation for consistency with the outcrop. Actual size of banks and/or basins may have been much larger.

Fig. 10. This oceanic area was probably continuous to the southeast with that represented in the Troodos ophiolite of Cyprus, and possible communicated to the northwest with oceanic areas represented by ophiolites of the Lycian nappes and Hellenides.

The carbonate platform represented by the Bey Dağları massif must have lain broadly to the south or west of this oceanic area. The overall scale and configuration of carbonate platforms and basins was closely comparable to the present day Bahama Banks, which probably began their evolution around the same time, but escaped Alpine orogenic deformation.

Conclusion—the Isparta Angle

The Isparta angle shows no evidence to suggest major oroclinal bending of an originally straight mountain chain. Instead, it originated as two regions of continental crust, on which were developed carbonate platforms, separated by a zone of Cretaceous oceanic crust. Major late Cretaceous changes in the relative motion of Africa and Europe (Smith 1971, Pitman & Talwani 1972) led to emplacement of the Antalya Complex between the Anamas Dağ and Bey Dağları. Emplacement involved mainly thrust tectonics along the margin of the Anamas Dağ, but along the Bey Dağları margin the Complex displays important strike slip faults (Woodcock & Robertson 1982).

Following emplacement of the Antalya Complex, convergence resulting from the Paleocene to Miocene northward movement of Africa was taken up along the northern margins of the new-joined platform massifs, resulting in the emplacement of the Lycian and Beysehir-Hoyran-Hadim nappe systems southward across their relative autochthons. Remobilized parts of the Antalya Complex are perhaps present in the Lycian nappes as ophiolitic mélange units described by Poisson (1977). Finally, thrusting took place in Miocene time along a narrow belt to the south of the Davras Dağ (the Aksu phase of Poisson, 1977), representing the last stage of relative movement between the Anamas Dağ and Bey Dağları blocks, leaving the Isparta angle essentially in its present-day configuration.

ACKNOWLEDGMENTS: I am grateful for field support provided by Erdoğan Demirtaşlı and Mineral Research and Exploration Institute (MTA), Ankara, without whose assistance this research would not have been possible. Orhan Akgül provided considerable assistance in the field. For discussions, I thank A. H. F. Robertson, E. Demirtaşlı, J. F. Dumont, O. Monod, A. Poisson, N. H. Woodcock, J. E. Dixon and A. B. Hayward. This work was supported by an NERC studentship, held at Edinburgh University.

References

AKBULUT, A. 1977. *Étude Géologique d'une partie du Taurus occidentale au sud d'Eğridir (Turquie).* Thèse de Docteur 3 ème cycle, Université de Paris Sud, Orsay, France, 164 pp.

ALLASINAZ, A., GUTNIC, M. & POISSON, A. 1974. La formation de L'Isparta Cay: calcaires à Halobies, grès à plantes, et radiolarites d'âge Carnien(?)-Norien (Taurides—Région d'Isparta—Turquie). *Schr. Erdwiss. Komm. Oster. Akad.* **2,** 11–21.

ALVAREZ, W., ENGELDER, T. & GEISER, P. A. 1978. Classification of solution cleavage in pelagic limestones. *Geology* **6,** 263–266.

ARGYRIADIS, I. 1974. Le Paléozoique supérieur du massif d'Alanya, Turquie méridionale). Déscription, corrélations et position structurale. *Bull. Soc. géol. Fr.* **16,** 112–115.

Bernoulli, D., Graciansky, P. C. de & Monod, O. 1974. The extension of the Lycian Nappes (S. W. Turkey) into the southeastern Aegean islands. *Eclog. geol. Helv.* **67,** 39–90.

Blumenthal, M. 1963. Le système structural du Taurus sud-Anatolien *in* Livre à la mémoire du Professeur Paul Fallot. *Mém. hors-série Soc. géol. Fr.* 611–662.

Brunn, J. H., Dumont, J. F., Graciansky, P. C. de, Gutnic, M., Juteau, T., Marcoux, J., Monod, O. & Poisson, A. 1971. Outline of the geology of the western Taurids. *In:* Campbell A. S. (ed.) *Geology and History of Turkey.* Petrol. Explor. Soc. Libya. Tripoli, 225–252.

——, Argyriadis, I., Marcoux, J., Monod, O., Poisson, A. & Ricou, L. E. 1973. Arguments pour et contre l'origine méridionale des nappes à ophiolites d'Antalya. Geological congress: 50th Anniversary of Turkish Republic. M.T.A. & D.S.I. Ankara 58–73.

——, ——, Ricou, L. E., Poisson, A., Marcoux, J. & Gracianksy, P. C. de 1976. Eléments majeurs de liaison entre Taurides et Hellénides. *Bull. Soc. géol. Fr.* **18,** 481–497.

Dahlstrom, C. D. A. 1960. Balanced cross sections. *Can. J. Earth Sci.* **6,** 743–757.

—— 1970. Structural Geology in the eastern-margin of the Canadian Rocky Mountains. *Bull. Can. Pet. Geol.* **18,** 332–406.

Delaune-Mayère, M., Marcoux, J., Parrot, J.-F. & Poisson, A. 1977. Modèle d'évolution mésozoique de la paléo-marge tethysienne au niveau des nappes radiolaritiques et ophiolitiques du Taurus lycien, d'Antalya et du Baër-Bassit. *In:* Biju-Duval B. & Montadert, L. (eds.) *Structural History of the Mediterranean Basins.* Technip Paris, 79–94.

Dumont, J. F. 1976. *Etudes géologiques dans les Taurides Occidentales: Les formations paléozoiques et mésozoiques de la coupole de Karacahisar (Province d'Isparta, Turquie).* Thesis. Paris-Sud, Orsay, France, 213 pp.

——, Gutnic, M., Marcoux, J., Monod, O. & Poisson, A. 1972a. Le Trias des Taurides occidentales (Turquie). Définition du bassin pamphylien: un nouveau domaine à ophiolites à la marge externe de la chaîn Taurique. *Z. Deutsch. geol. Ges.* **123,** 385–409.

——, ——, ——, —— & —— 1972b. Essai de reconstitution d'un bassin triasique à ophiolites à la marge externe des Taurides: le bassin pamphylien. *C. r. Somm. Soc. géol. Fr.* **2,** 73–74.

——, Uysal, Ş. & Monod, O. 1980. La série de Zindan: un élément de liaison entre plate-forme et bassin a l'est d'Isparta (Taurides occidentales, Turquie). *Bull. Soc. géol. Fr.* **22,** 225–232.

Ealey, P. J. & Knox, G. J. 1975. The pre-Tertiary rocks of S.W. Cyprus. *Geol. Mijnbouw* **54,** 85–100.

Elliott, D. & Johnson, M. R. W. 1980. Structural evolution in the northern part of the Moine thrust belt, NW Scotland. *Trans. R. Soc. Edinburgh Earth Sciences* **71,** 69–96.

Graciansky, P. C. de. 1972. *Recherches géologiques dans le Taurus lycien occidental.* Thesis. Paris-Sud, Orsay, France.

Greenly, E. 1919. *The Geology of Anglesey.* Mem. geol. Surv. G.B. 980 p.

Gutnic, M. 1977. *Géologie du Taurus pisidien au Nord d'Isparta (Turquie).* Principaux résultats extraits des notes de M. Gutnic entre 1964 et 1971 per O. Monod. Publ. Faculté des Sciences, Univ. Paris-Sud, Orsay. 130 p.

——, Monod, O., Poisson, A. & Dumont, J. F. 1979. Géologie des Taurides occidentales (Turquie). *Mém. Soc. géol. Fr.* **58,** 1–112.

Hayward, A. B. & Robertson, A. H. F. 1982. Clastic sedimentation related to the Tertiary emplacement of the Antalya ophiolite, S.W. Turkey. *Bull. geol. Soc. Amer.* **93,** 68–75.

Hsü, K. J. 1968. Principles of mélanges and their bearing on the Franciscan-Knoxville paradox. *Bull. geol. Soc. Am.* **79,** 1063–74.

Juteau, T. 1975. Les ophiolites des nappes d'Antalya (Taurides occidentales, Turquie). *Mém. Sci. Terre Nancy* **32,**

Lapierre, H. 1975. Les formations sédimentaires et eruptives des nappes de Mamonia et leurs relations avec le Massif de Troodos. *Mém. Soc. géol. Fr.* **123,** 132 p.

Lefèvre, R. 1967. Un nouvel élément de la géologie du Taurus lycien: les nappes d'Antalya (Turquie). *C. r. Séances Acad. Sci. Paris* **265,** 1365–1368.

Monod, O. 1977. *Recherches géologiques dans le Taurus occidental au sud de Beyşehir (Turquie).* Thèse de Docteur-ès Sciences, Universite de Paris-Sud, Orsay, France, 442 pp.

—— 1978. Güzelsü Akseki bölgesindeki Antalya Napları üzerine açıklama (Orta Batı Toroslar-Türkiye)—Precisions upon the Antalya Nappes in the region of Güzelsü-Akseki (Western Taurus, Turkey). *Bull. geol. Soc. Turkey* **21,** 27–29.

Naylor, M. A. 1978. *A geological study of some olistostromes and melanges.* Thesis. Ph.D. (unpubl.) Cambridge University, U.K. 257 pp.

Özgül, N. & Arpat, E. 1973. Structural units of the Taurus orogenic belt and their continuation in neighbouring regions. *Bull. geol. Soc. Greece* **10,** 156–164.

Pitman, W C. & Talwani, M. 1972. Sea-floor spreading in the north Atlantic. *Bull. geol. Soc. Am.* **83,** 619–649.

Poisson, A. 1977. *Recherches géologiques dans les Taurus occidentales (Turquie).* Thèse de Docteur ès Sciences, Université de Paris-Sud, Orsay, France. 2 vols. 795 pp.

Ramsay, J. G. 1967. *Folding and Fracturing of Rocks.* McGraw Hill. New York.

Ricou, L. E., Argyriadis, I. & Lefèvre, R. 1974. Proposition d'une origine interne pour les nappes d'Antalya et le massif d'Alanya (Taurides occidentales, Turquie). *Bull. Soc. géol. Fr.* **16,** 107–111.

Ricou, L E., Marcoux, J. & Poisson, A. 1979. L'Allochthonie des Bey Dağları orientaux. Reconstruction palinspastique des Taurides occidentales. *Bull. Soc. géol. Fr.* **21,** 125–134.

ROBERTSON, A. H. F. & WOODCOCK, N. H. 1979. Mamonia Complex, southwest Cyrprus: Evolution and emplacement of a Mesozoic continental margin. *Bull. geol. Soc. Am.* Part I, **90,** 651–665.
—— & —— 1980. Tectonic setting of the Troodos Massif in the east Mediterranean. *In:* PANAYIOTOU, A. (ed.) *Ophiolites: Proceedings International Ophiolite Symposium Cyprus 1979.* Geological Survey Department, Cyprus, 36–49.
—— & —— 1981a. Alakir Çay Group, Antalya Complex, S.W. Turkey: deposition on a Mesozoic passive carbonate margin. *Sediment. Geol.* **30,** 95–131.
—— & —— 1981b. Bilelyeri Group, Antalya Complex, S.W. Turkey: deposition on a Mesozoic passive continental margin. *Sedimentology* **28,** 381–399.
—— & —— 1981c. Gödene zone, Antalya Complex, S.W. Turkey: volcanism and sedimentation on Mesozoic marginal oceanic crust. *Geol. Rdsch.* **70,** 1177–1214.
—— & —— 1982. Sedimentary history of the Southwestern segment of the Mesozoic-Tertiary Antalya continental margin, southwestern Turkey, *Eclogae geol. Helv.* **75,** 517–562.
SMITH, A. G. 1971. Alpine deformation and the oceanic areas of the Tethys, Mediterranean and Atlantic. *Bull. geol. Soc. Amer.* **82,** 2039–2070.
SWARBRICK R. E. 1980. The Mamonia Complex, S.W. Cyprus; a Mesozoic continental margin and its relationship to the Troodos Complex. *In:* PANAYIOTOU, A. (ed.) *Ophiolites: Proceedings International Ophiolite Symposium Cyprus 1979.* Geological Survey Department, Cyprus, 86–92.
THIUZAT, R. & MONTIGNY, R. 1979. *K-Ar Geochronology of Three Turkish Ophiolites.* Abstracts, International Ophiolite Symposium. Geological Survey Department, Cyprus. 80–81.
WALDRON, J. W. F. 1981. *Mesozoic sedimentary and tectonic evolution of the northeast Antalya Complex, Eğridir, S.W. Turkey.* Thesis. Ph.D. (unpubl.), Edinburgh University, U.K.
WOODCOCK, N. H. & ROBERTSON, A. H. F. 1977a. Imbricate thrust belt tectonics and sedimentation as a guide to emplacement of part of the Antalya Complex, S.W. Turkey. *Abstracts 6th Colloquium on the Geology of the Aegean Region, Izmir, Turkey.* p. 98.
—— & —— 1977b. Origins of some ophiolite-related metamorphic rocks of the 'Tethyan' belt. *Geology* **5,** 373–6.
—— & —— 1982. Imbricate thrusting and wrench tectonics in the Antalya Complex, S.W. Turkey. *J. geol. Soc. Lond.* **139,** 147–165.

J. W. F. WALDRON, Department of Geology, Saint Mary's University, Halifax, Nova Scotia, Canada, B3H 3C3.

Miocene clastic sedimentation related to the emplacement of the Lycian Nappes and the Antalya Complex, S.W. Turkey

A. B. Hayward

SUMMARY: The western Tauride Mountains of S.W. Turkey comprise a central relatively autochthonous carbonate platform unit, the Tauride autochthon bordered by two allochthonous units, the Lycian Nappes to the west and the Antalya Complex to the east. Sequences of Miocene clastic sediments up to 1000 m thick that were derived from both the allochthons document the timing and direction of their emplacement onto the carbonate platform.

Along the western margin of the Miocene basin intial emplacement of the Lycian Nappes, from the northwest, in the Lower Miocene, was coupled with rapid subsidence of the previously stable carbonate platform. The resulting basin was ca 170 km across. Fan-deltas were derived from the leading edge of the nappes and passed basinwards into a series of small submarine fans. At the same time rapid uplift of the central parts of the carbonate-platform, along the margin of the basin opposite the nappe pile, led to a thick wedge of carbonate-derived clastics being shed northwestwards into the basin. Varying sedimentation rates and migrating facies belts in the overlying Middle and Upper Miocene basin-fill document the progressive emplacement of the Lycian Nappes ca 100 km from the northwest over the basin margin.

Along the eastern margin of the basin, palaeocurrent analysis and downslope facies-transitions in the Miocene sediments show that the Antalya Complex was emplaced from the east but only advanced a short distance beyond the eastern margin of the basin.

The Lycian Nappes and the Antalya Complex approached the basin from opposite directions and so their respective ophiolite units must have originated in separate ocean basins on either side of the Bey Dağlari and the Susuz Dağ carbonate platform.

Palinspastic reconstructions of the 'Neo-Tethys' ocean in the Mesozoic and Tertiary are particularly dependent on knowledge of the direction and timing of emplacement of tectonically transported allochthonous units. Diametrically opposed solutions are put forward, particularly for the Hellenides (e.g. Barton, 1975 and discussion) and the Taurides (Dumont *et al.*, 1972a, b; Brunn *et al.*, 1973; Ricou *et al.*, 1974, 1979; Brunn, 1976; Robertson and Woodcock, 1981a; Woodcock & Robertson, 1981b). Most of the arguments are based on regional lithostratigraphical correlations, comparisons and detailed structural analysis of the allochthonous units.

In this study an alternative approach is adopted, the determination of emplacement direction from detailed facies analysis of *in situ* sedimentary sequences in autochthonous blocks adjacent to, and underlying, the allochthons. These deposits, which were mostly eroded from the allochthonous units, record the timing and direction of emplacement in sequences which have not subsequently been much deformed (Hayward & Robertson, 1982).

In southwestern Turkey (Fig. 1) the debate is as to whether the Mesozoic rocks of the Antalya allochthon have been transported far from the north 'internally', over a contemporaneous carbonate platform, or alternatively whether the rocks were rooted 'externally', that is to the south in the area of the present Mediterranean Sea. The local carbonate platform authochthon (Fig. 1) passes up into a thick sequence of Miocene clastic sediments, which yield critical data on the timing, mechanism and direction of emplacement.

Geological setting

The Tauride Mountains form an extension of the Alpine orogenic belt into southwestern Turkey (Fig. 1). The Western Taurides (Fig. 1) define an arcuate belt divided into two limbs on either side of Antalya Bay.

The Lycian Taurus (Fig. 1), in the west, is considered to be an extension of the Hellenide orogenic belt of Greece (Brunn *et al.*, 1976; Özgül and Arpat, 1973). On a regional scale the Lycian Taurus consists of a central relatively autochthonous unit, the 'Tauride autochthon' (Brunn *et al.*, 1970, 1971; Dumont *et al.*, 1972a, b), either side of which lie two allochthonous ophiolite units, the Lycian Nappes to the northwest (Fig. 1) and the Antalya Complex to the east.

The Lycian Nappes comprise a regionally

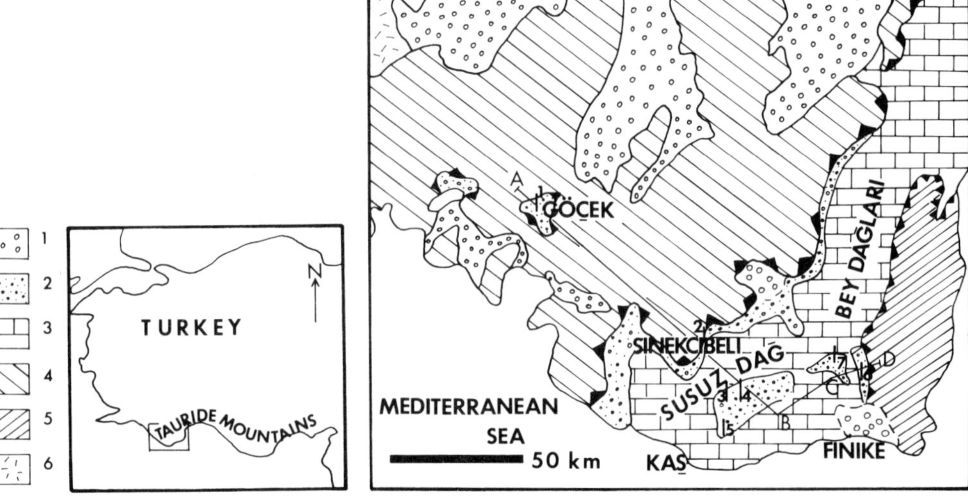

FIG. 1a. Simplified geological map of the Susuz Dağ, Bey Dağları and adjoining areas showing distribution of Miocene clastic sediments (dotted). Letters refer to sedimentological sections in Fig. 2. Line A-D shows location of structural cross-section in Fig. 1b. Key. 1-Pliocene-Recent: 2-Miocene clastic sediments; 3-Tauride carbonate platform (Liassic-Miocene); 4-Lycian Nappes; 5-Antalya Complex; 6-Menderes massif metamorphic complex.

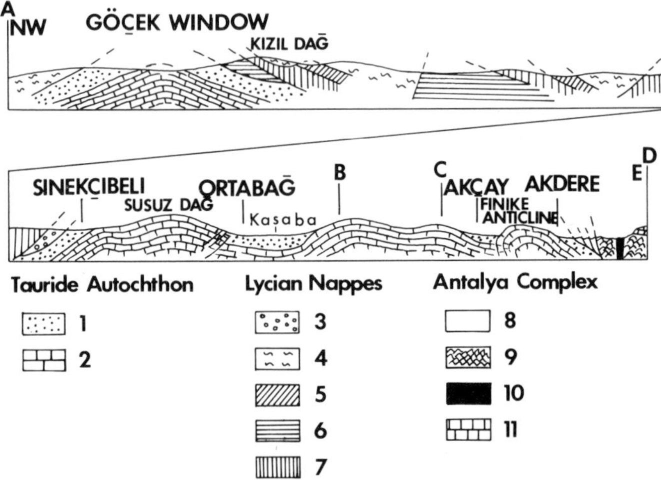

FIG. 1b. Structural cross-section along line A-D on Fig. 1a.

KEY
1-Miocene clastic sediments
2-Tauride carbonate platform (Liassic-Miocene)
3-Eocene flysch
4-Peridotite
5-Dolerite
6-Radiolarian chert
7-Limestone
8-Pelagic limestone, radiolarian chert (Triassic-Jurassic)
9-Basaltic lavas
10-Serpentinite
11-Limestone (Trias-Jur.)

extensive series of allochthonous sheets (ca 120 × 150 km) transported from the northwest towards the southeast in a number of phases during the early Tertiary era. Mesozoic sedimentary facies within the Lycian Nappes range from shallow water platform carbonates through redeposited slope sequences, to pelagic limestones and cherts (Brunn *et al.*, 1970, 1971, Graciansky 1972). Early Tertiary sediments comprise a flysch seqeunce of Eocene age (Poisson, 1977).

The whole assemblage has been interpreted by Delaune-Mayere *et al.* (1974) as different parts of a Mesozoic continental margin which was subsequently emplaced during Late Cretaceous and early Tertiary orogenic events. Ophiolitic units in the Lycian Nappe consist of slices of peridotite and diabase intercalated with the nappes.

The Antalya Complex (Woodcock & Robertson, 1977a), formerly the Antalya Nappes (Lefevre, 1967; Brunn *et al.*, 1971; Graciansky, 1972; Poisson, 1977), comprises a wide variety of Mesozoic sedimentary facies, including turbiditic sandstones, pelagic limestones, radiolarites, redeposited limestones and ophiolite-derived sandstones.

Massive shallow water limestones overlie Ordovician to Permian sandstones, mudstones and limestones (Brunn *et al.*, 1971; Marcoux (in Delaune-Mayere *et al.*, 1977); Allasinaz *et al.*, 1974; Kalafatçıoğlu, 1973; Monod, 1977a, 1978; Dumont, 1976a; Robertson & Woodcock, 1981c, 1982). The ophiolite suite is represented by pillow lavas, dolerites, gabbros, and periodotites (Juteau, 1975). Minor occurrences of metamorphic rocks are also known (Juteau, 1975; Woodcock & Robertson, 1977b).

On the basis of sedimentary correlation (Fig. 1), Brunn *et al.* (1971) distinguished three nappes, the Lower, Middle and Upper Antalya Nappes. However, no structural continuity can be demonstrated between the various isolated tectonic slices. Robertson & Woodcock (1981a, b, c) recently reinterpreted some of the *critical* thrust contacts of Brunn *et al.* (1971) as stratigraphic contacts and recognised five N-S trending zones within the Complex. From west to east these tectonic zones record the transition from the Mesozoic carbonate platform (Bey Dağları and Susuz Dağ, Fig. 1) across Mesozoic continental margin sediments, into oceanic crust formed during the initial stages of continental rifting (Woodcock & Robertson, 1982). Zones further east are tectonically displaced with respect to the western zones. They consist of carbonate-platform and basement lithologies and portions of Late Cretaceous ocean crust (Juteau *et al.*, 1977; Robertson & Woodcock, 1982, Yılmaz, this volume; Reuber, this volume).

Taken as a whole the Antalya Complex records the initiation, construction and later tectonic disruption of part of the continental margin of a small Mesozoic-Cainozoic oceanic basin (Robertson & Woodcock, this volume).

The central unit of the Lycian Taurus comprises the Taurus autochthon, a regionally extensive autochthonous unit of shallow water limestones belonging to a carbonate platform ranging in age from Liassic to Lower Miocene (Aquitanian) with several non-sequences in the Lower Tertiary (Poisson, 1977; Dumont *et al.*, 1972b). This unit which forms the limestone massifs of the Susuz Dağ and Bey Dağları has been the subject of extensive biostratigraphical and sedimentological studies (Poisson 1967a, b, 1974b, 1977, Poisson & Poignant 1974, Graciansky *et al.*, 1974, Gutnic & Poisson 1974).

By comparison, thick sequences of up to 1000 m of Miocene clastic sediments which unconformably overlie the carbonate platform have been relatively little studied (Poisson 1977, Önalon 1980, Poisson & Akay 1981). Detailed facies analysis of the Miocene sediments demonstrates derivation from both ophiolitic allochthons and also documents the timing and direction of emplacement of both the ophiolite units onto the carbonate platform.

Miocene clastic sediments

The Miocene clastic basin shows evidence of derivation from both an eastern and western margin. Sediments of the eastern area (Bey Dağları, Fig. 1) were derived from the east and record the emplacement of the Antalya Complex, while those of the western area (Susuz Dağ, Fig. 1) are related to the emplacement of the Lycian Nappes, generally from the northwest.

The detailed stratigraphy of the Miocene clastic sediments is presented elsewhere (Hayward, in press b); the principal palaeontological data on which the stratigraphy is based are outlined in Table 1.

Eastern margin

Sediments related to the emplacement of the Antalya Complex are well exposed both sides of the Finike anticline (Fig. 1).

Lower Miocene

In the east, close to the tectonic contact with the Antalya Complex, the Lower Miocene

TABLE 1. *Stratigraphy of the Miocene clastic sediments in the Bey Dağları and Susuz Dağ area (after Hayward, in press b) showing principal palaeotological beterminations.*

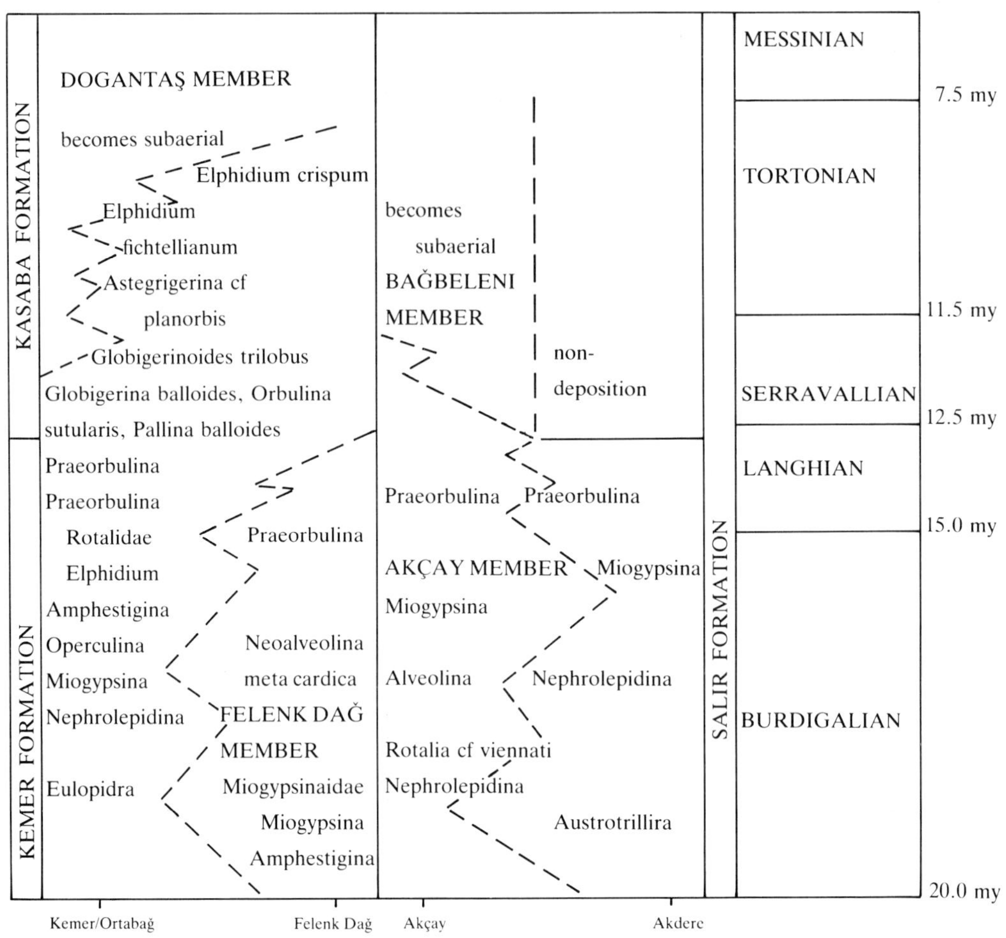

1 This study
2 Pisoni 1967
3 Önalon 1980

(Burdigalian-Langhian) sequence comprises 200 m of conglomerate and sandstone interbedded with mudstone and pelagic chalk (Fig. 2 section 7). The clastic sediments lie with slight angular discordance on either green-grey calcareous marl of Oligocene age or on limestone of Maastrichtian to Palaeocene age. The conglomerate and sandstone form fining-upward cycles between 5 m and 10 m thick. The conglomerate is both matrix- and clast-supported. Well-rounded clasts include basalt, gabbro, dolerite, variably serpentinized ultramafic rocks, radiolarian chert, limestone and quartzose sandstone.

In marked contrast, contemporaneous sediments in the Akçay syncline, 8 km to the west (Figs 1 & 2 section 6), are composed of 350 m of ophiolite-derived sandstone interbedded with siltstone, mudstone and pelagic chalk (Fig. 2). The sandstones are markedly channelized and form a series of fining-upward cycles. Palaeocurrent analysis of both these areas based on flute-marks, clast-imbrication and the orientation of slumps shows that sediment transport was from the east and north-east (Fig. 2) (Hayward & Robertson, 1982).

The well-rounded nature of most of the clasts suggests reworking in a high energy area, either shallow marine or fluviatile, prior to redeposition by a combination of turbidity currents and mass-flow processes. The coarsest deposits are seen in the east, the sediments becoming finer

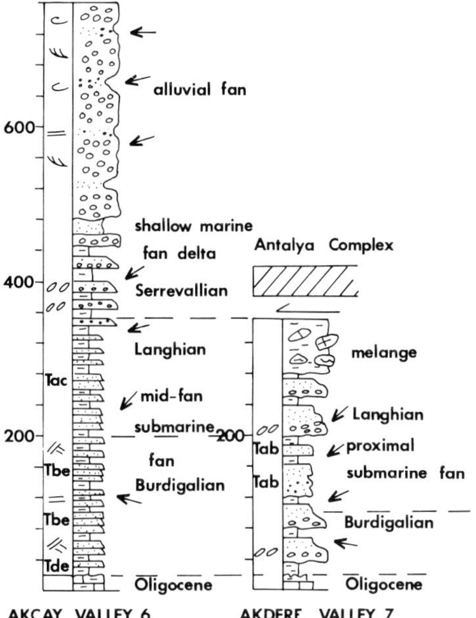

FIG. 2. Sedimentological sections illustrating facies relationships in the eastern half of the Miocene basin. These sequences were derived from the Antalya Complex. Note the contrast between the proximal sequences close to the Antalya Complex (7) and more westerly distal sequences (6). For location of sections see Fig. 1a.

grained westward. This is confirmed by palaeocurrent evidence of supply from the east and northeast.

Sedimentary facies suggest a fan-delta to submarine-fan system. In the model shown in Fig. 3, based also on additional data (Hayward, 1982, Hayward & Robertson, 1982), the topographically elevated Antalya Complex lay adjacent to a marine basin. Coarse-grained well rounded alluvium formed a series of steep-fronted fan-deltas which prograded directly into deep water as a series of anastomizing sediment lobes. These passed downslope into a series of small submarine fans. The inner fan area consisted of amalgamated conglomerate-sandstone units formed in a low relief, braided, broadly channelled area. Interbedded over-bank sediments consisted of thin sandstone, mudstone and chalk.

To the west, the mid-fan environment consisted of amalgamated sandstone units deposited in shallow distributary channels and sandstone packets deposited in a mid-fan depositional lobe area (Fig. 3).

Middle-Upper Miocene

In the Akçay area, the submarine ophiolite derived sequence passes up into ca 250 m of poorly stratified conglomerate with subordinate ophiolitic sandstone and minor mudstone and calcrete (Fig. 2 section 6). Sedimentary structures comprise tabular and trough-cross-

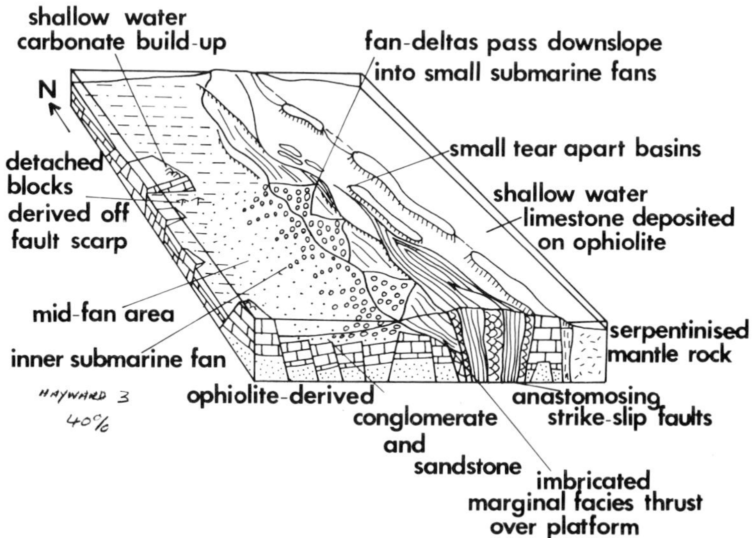

FIG. 3. *Lower Miocene emplacement of the Antalya Complex.* Fan-deltas derived from the elevated Antalya Complex pass basinwards (eastwards) into a series of small submarine fans. Strike-slip faulting in the Antalya Complex results in a series of small tear-apart basins in which coarse subaerial clastic sediments were deposited.

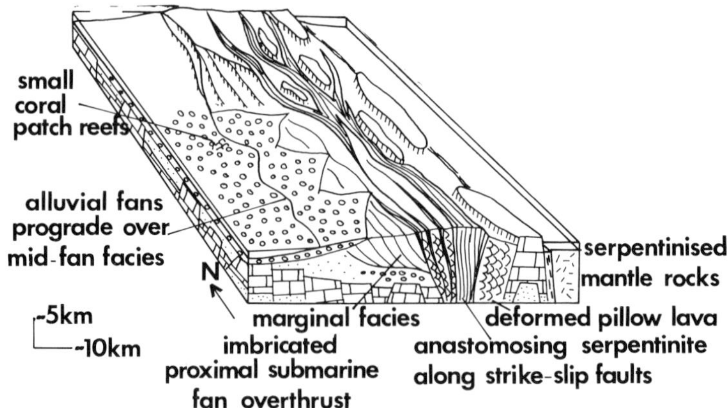

FIG. 4. *Upper Miocene emplacement of the Antalya Complex.* Lower Miocene proximal submarine-fan sediments are overthrust. Fan deltas derived from the Antalya Complex prograde into central areas of the Bey Dağları autochthon.

bedding and well developed imbrication. The presence of calcrete and rootlet horizons indicates subaerial exposure. Clast types in the conglomerate are well rounded dolerite, gabbro, chert and ultramafic cumulates, with subordinate serpentinite and basalt. Sandstones comprise poorly sorted grains of dolerite, basalt, serpentinite, chert, quartz and feldspar are poorly cemented by carbonate. Palaeocurrent data from clast-imbrication and cross-bedding show that supply was dominantly from the east and north-east (Fig. 2).

In the east, along the tectonic contact with the structurally overlying Antalya allochthon, the submarine clastic sequence passes abruptly upward into a tectonic melange c.a. 70 m thick. The melange contains blocks of all the Antalya Complex lithologies in a matrix of ophiolitic mudstone. The thick clast-supported conglomerates seen in the Akçay valley to the west are not present.

The clast-supported conglomerates and associated sediments in the west represent deposition on a stream-flow-dominated alluvial fan (Fig. 4) of Middle to Upper Miocene age.

Palaeocurrents clearly indicate derivation from the east (Hayward & Robertson, 1982). The exposed sequence terminated sometimes in the Middle to Upper Miocene, although an absence of fossils precludes accurate dating of the upper parts of the sequence.

Sedimentation related to the emplacement of the Antalya Complex

Initial tectonic deformation of the Antalya Complex took place in the Late Cretaceous (Woodcock & Robertson, 1982), but this did not impinge on the Bey Dağları authochthon at this time, except possibly in the far north, where olistostrome melange is present (Poisson, 1977; Hayward & Robertson, 1982). Along the eastern flank of the Bey Dağları pelagic carbonate deposition continued through the Palaeocene. By Eocene times a marked north-south trending depositional hinge line had become established close to the present eastern outcrop area of the Bey Dağları, dividing a shallow carbonate-depositing area to the west from a deeper water basinal area to the east (Haryward & Robertson, 1982).

By Early Miocene times the Antalya Complex formed an elevated landmass to the east of the already subsided eastern margin of the Bey Dağları carbonate platform (Fig. 3). As the Complex was emplaced the carbonate platform continued to subside possibly as a result of flexural-loading associated with thrusting. The Antalya Complex was rapidly eroded, rivers transported conglomerates to fan-deltas which in turn fed a series of small submarine fans. Palaeocurrent orientations and grain-size variations are consistent with a broadly ENE-WSW palaeoslope. Irregular subsidence and faulting in the underlying carbonate platform resulted in considerable volumes of carbonate clastics, and fault-derived detached blocks being shed into the basin (Fig. 3). In the north of the Akdere valley, submarine faulting and local submarine impingement of the Antalya Complex against the platform is indicated by a very thick (ca 250 m) sequence of debris-flows with large olistoliths (up to 5 × 10 m) derived from the front of the Antalya Complex as it advanced (Fig. 3).

Throughout this period the Antalya Complex is inferred to have been affected by extensive strike-slip faulting resulting in small tear-apart basins that are now represented by deformed intercalations of subaerially deposited ophiolite-derived conglomerates and sandstones within the Antalya Complex (Robertson & Woodcock, 1980a) (Fig. 3).

The submarine-fan phase of sedimentation was abruptly terminated in mid-Miocene times by westward thrusting of the Antalya Complex to near its present position. In the east, the Antalya Complex over-rode the earlier Miocene submarine-fan sequence, forming a basal tectonic melange and preventing further deposition. Slivers of marginal, Eocene, Bey Dağları limestones were stripped-off and entrained along the sole thrust of the Antalya allochthon (Hayward & Robertson, 1982). To the west alluvial fans prograded over the autochthon, where coarse-grained ophiolite-derived deposition continued until Upper Miocene time (Fig. 4).

Western margin

Sediments related to the emplacement of the Lycian Nappes are well exposed in the Kasaba syncline, beneath the present-day thrust-front of the Lycian Nappes (Fig. 1) and within a tectonic window in the nappe pile at Göçek (Fig. 1). Generalised sedimentological sections at various localities (Fig. 5) illustrate the sedimentary facies and allow the sedimentary sequence to be related to progressive subsidence and nappe emplacement (Fig. 5). The sedimentological models developed are based on more data than can be presented here (Hayward, 1982a, b, c, in press a).

Lower to Middle Miocene

In the area of Göçek (Fig. 1), a tectonic window in the nappe pile reveals an autochthonous carbonate platform sequence overlain by Miocene clastic sediments of Burdigalian age (Fig. 5.1). Shallow-water Aquitanian limestones are overlain by 75 m of interbedded stratified conglomerate, sandstone and mudstone of Lower to Upper Burdigalian age. The conglomerates are composed of well rounded clasts of dolerite, basalt, peridotite, pyroxenite, radiolarian chert and limestone. Disorientated coral blocks are also present. The sandstones contain an abundant shallow marine fauna of gastropods and bivalves.

The fauna, coral blocks and well rounded stratified conglomerate indicate deposition in a shallow marine environment. This sequence represents deposition on the submarine part of a series of fan-deltas derived from the front of the Lycian Nappes.

The overlying sequence comprises approximately 20 m of melange which in turn is overthrust by the Lycian Nappes. The melange contains large blocks of all lithologies from the overlying nappe pile, along with chaotically slumped and deformed sandstone and conglomerate from the underlying sedimentary sequence.

Beneath the present-day thrust-front of the Lycian Nappes in the area around Sinekçibeli (Fig. 1), Aquitanian shallow water algal limestones of a stable carbonate platform pass upwards into a thin-bedded, turbidite sandstone-mudstone sequence of Burdigalian age (Fig. 5.2). The transition occurs over approximately 70 m and is marked by a gradual upward increase in terrigenous material and decrease in carbonate content and associated shallow water faunas including gastropods and bivalves. Redepositied algal calcarenites decrease in thickness and abundance upwards, reflecting the termination of shallow-water carbonate deposition both on local basement highs and around the margins of the basin.

This transition is overlain by a coarsening-upward sequence (Fig. 5.2) which spans Burdigalian to Langhian and is in turn overthrust by the Lycian Nappes. Thin bedded turbiditic sandstones and mudstones pass upwards into thick bedded turbiditic sandstones and then into massive conglomerates. Slump horizons indicate a NW to SE palaeoslope and palaeocurrent measurements from flute- and groove-marks are consistent with this (Fig. 5).

The transition from medium to coarse turbiditic sandstone and mudstone to conglomerate is abrupt. Upwards, beds of massive and stratified conglomerate are interbedded with very coarse grained sandstone and mudstone (Fig. 5b). The conglomerates are composed of well rounded clasts of peridotite, diabase, pyroxenite, basalt, red radiolarite and limestone. Disorientated coral blocks are also present. Green-grey sandstone are structureless or rarely plane-laminated. Massive silty dark green mudstones contain an abundant shallow marine fauna of gastropods and bivalves.

The more proximal sedimentary sequence outlined above, as exposed in the Sinekçibeli area, is not continuous across the Susuz Dağ anticline (Fig. 1). However, identical ages, similar sedimentary facies and palaeocurrent dispersal patterns indicate the clastic sequence exposed in the Kasaba syncline can be correlated with the sequence to the northwest and represents the development of more basinal sedimentary facies.

In the Kemer area a similar transition marks the initiation of clastic sedimentation. Carbonate platform limestones of Aquitanian age pass

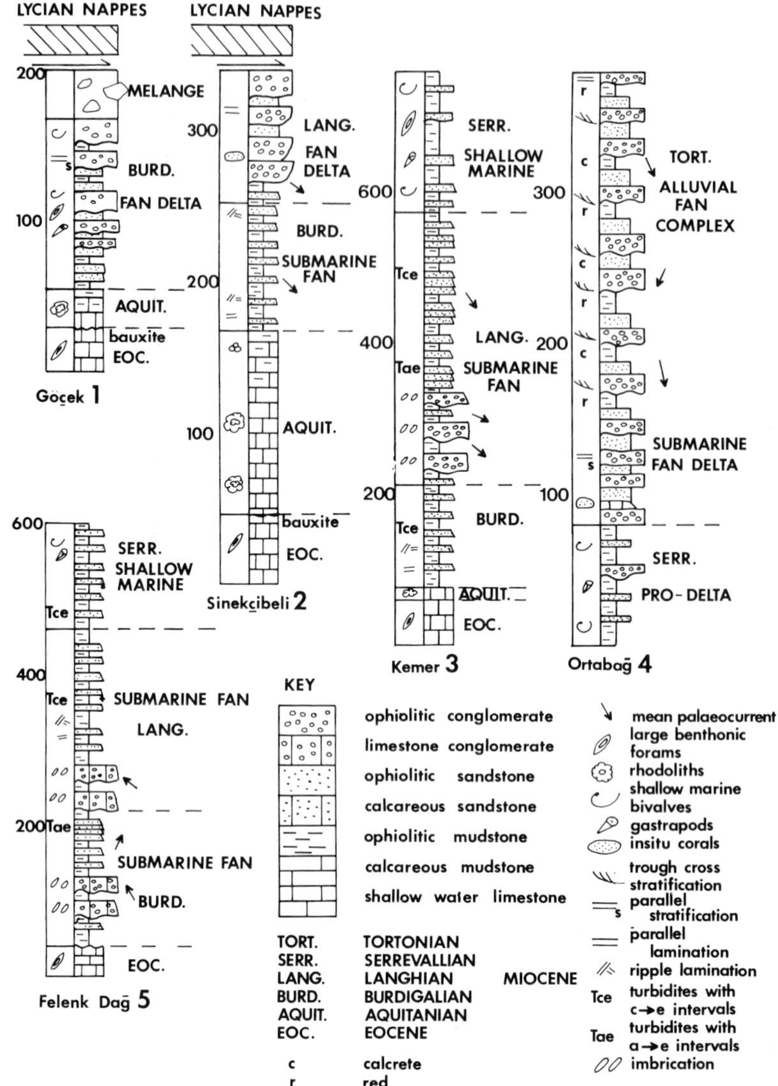

FIG. 5. Sedimentological sections in the western half of the Miocene basin (Susuz Dağ). Sediments of this area are related to emplacement of the Lycian Nappes. For location of sections see Fig. 1.

upwards into calcareous marls and then into a distal turbiditic sandstone and mudstone sequence of Burdigalian age (Fig. 5c). This is overlain by a sequence of turbiditic sandstones and mudstones interbedded with a number of conglomerate-sandstone horizons which spans Burdigalian to Langhian. Amalgamated units are between 10 m and 25 m thick. Beds of dark green to buff conglomerates are between 0.3–3.5 m thick. Poorly sorted, they comprise moderately to well rounded clasts up to 0.6 m in diameter of diabase, peridotite, basalt, harzburgite, red radiolarian chert and limestone. Sandstones are thick bedded, graded or structureless with erosional bases and flat or truncated tops.

Both conglomerates and sandstones show no evidence of deposition by traction currents and textures in the conglomerate indicate deposition by a variety of sediment-gravity-flows.

In individual exposures the amalgamated conglomerate-sandstone units appear to be sheets. However, when traced laterally they are lenticular over several kilometres across palaeoslope and interfinger with inter-bedded sandstone, mudstone and palagic chalk horizons. They are interpreted as large submarine fan channels with a general NW-SE orientation.

Sedimentary model

The Lower Miocene sequence exposed at both Sinekçibeli and Kemer can be synthesised into the following sedimentological model (Fig. 6). Palaeocurrent dispersal patterns, the general decrease in clast-size and the increasingly more basinal aspects of the sedimentary sequence from northwest to southeast are all consistent with a general northwest to southeast palaeoslope.

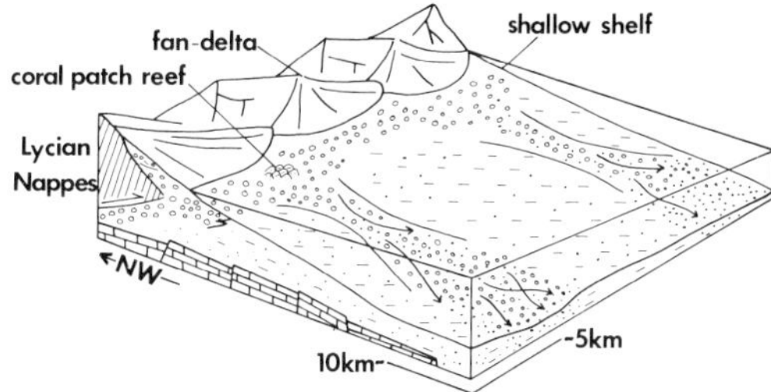

FIG. 6. Lower-Middle Miocene sedimentological model for emplacement of the Lycian Nappes. Fan-deltas derived off the leading edge of the nappe pile pass basinwards to the southeast into large sumbarine fan channels.

Proximal sequences at Sinekcibeli show an overall coarsening-upward tendency, in which turbiditic sandstones are overlain by a thick conglomerate unit. Following rapid subsidence at the end of the Aquitanian, this sequence represents overall shallowing-upwards caused by infilling of the basin. Within the conglomerates a marine fauna of bivalves, gastropods and disorientated coral-blocks indicate deposition in a shallow marine environment. This sequence is interpreted as a series of fan-deltas derived from the leading edge of the advancing nappes (Fig. 6).

The formerly immediately down-slope sequence has been removed by erosion. Distal facies-equivalents of the shallow-marine sequence are large channels of a submarine fan. Within the channels conglomerate-sandstone units were deposited by a variety of sediment-gravity-flows in a complexly braided-channel system.

Felenk Dağ

In contrast to the sediments derived from the Lycian Nappes as described above, the Lower Miocene sediments in the Felenk Dağ area (Figs 1 and 5.5) comprise limestone conglomerate and calcarenite. Eocene shallow-water nummulitic limestone is overlain by grey-green calcerous mudstone which pass upwards into a sequence of thin-bedded turbiditic calcarenite and calcareous mudstone. These in turn are overlain by up to 750 m of redeposited limestone conglomerate and turbiditic calcarenite (Fig. 5e).

The conglomerate and calcarenite comprise a complete admixture of carbonate lithoclasts and carbonate bioclastic debris. The conglomerate is composed of moderately to well rounded lithoclasts of calcilutite, recrystallized calcarenite and more subordinate generally poorly rounded bioclastic material.

Many of the limestone clasts in the conglomerate can be correlated palaeontologically and lithologically with Eocene rock units exposed in the subjacent carbonate platform, indicating that substantial areas of the platform were uplifted and subject to erosion at this time, while in areas to the north (e.g. Kemer, Sinekçibeli) the carbonate platform was undergoing rapid subsidence (see discussion below).

The high sphericity and roundness of many of the conglomerate clasts is indicative of subaerial erosion with transport through a high energy shallow marine or fluvial environment prior to redeposition. The sedimentary sequence in this area is characterized by: (i) an absence of shallow water indicators; (ii) broadly undirectional palaeocurrents consistently from the south and southwest; (iii) abundant hemipelagic chalk horizons; (iv) channelled conglomerates and calcarenites deposited by a variety of sediment-gravity flows. All these features suggest deposition in a submarine fan environment.

In general terms, the sediments of the Felenk Dağ area represent the introduction of carbonate terrigenous material derived from an area of uplifted carbonate platform to the south and southwest of Kasaba, into the dominantly ophiolite-derived basin-fill (Figs 9a and 6). Sedimentary facies-associations suggest deposition is one, or a series of, submarine fans fed by fan-deltas.

Middle-Upper Miocene

Middle to Upper Miocene sequences, in the western (Susuz Dağ) area are preserved only along the northern margin of the Kasaba syncline (Figs 1 and 5.4). In this area redeposited ophiolite-derived conglomerates and sandstones of Lower Miocene age (above) are overlain by an overall shallowing- and fining-upward sequence. The lower 200 m of the sequence is characterized by: (i) an upward increase in the mudstone to sandstone ratio, (ii) turbiditic sandstone beds passing up into massive graded units with evidence of wave-reworked tops, and (iii) the introduction of a shallow-marine fauna.

This general fining-upward sequence marks the end of the initial emplacement of the Lycian Nappes, relief in the source area having been lowered by erosion. Shallowing upwards is related both to the progressive infill of the sedimentary basin, and probably also to a lowering of sea level.

The shallow marine regressive-upwards sequence is overlain by a thick succession (ca 350 m) of conglomerates and sandstones of Tortonian age. This represents the second major incursion of coarse terrigenous clastic sediments along the northwestern margin of the sedimentary basin.

Petrographically, the sediments consist of an admixture of rock types. Clasts in the conglomerate consist of moderate to well rounded limestone, diabase, gabbro, chert and subordinate serpentinite and basalt. The sandstones are generally poorly sorted; serpentinite, chert, dolerite, basalt and subordinate quartz and feldspar are cemented by carbonate. Cross-stratification in conglomerates and sandstones in most of this sequence indicates that palaeocurrents flowed dominantly north to south.

The sedimentary sequence can be broadly subdivided into sediments deposited in either a continental or a marine environment. Within the continental sequence two facies-associations are recognized. In proximal areas poorly sorted reddened conglomerate is interbedded with subordinate sandstone and red and green mudstone. The conglomerate, which makes up more than 80% of this association, is dominantly clast-supported although rare matrix-supported conglomerate also occurs. Downslope interbedded conglomerate and sandstone occur as well defined fining-upward units 15–20 m thick. Trough-cross-stratified conglomerate at the base of a unit passes laterally and vertically into massive conglomerate of similar grain size. The conglomerates are overlain by coarse to medium grained parallel- and cross-laminated red sandstones, red to green mudstones and calcrete.

The marine facies-association, laterally equivalent to the continental facies-association, comprises interbedded conglomerate, sandstone and mudstone with an abudant marine fauna and small coral patch-reefs.

Evidence of a generally north to south palaeoslope is indicated by: (i) the general decrease in clast size from north to south; (ii) the overall palaeocurrent trend which is uniformly to the south for the continental sequence; (iii) the increasing marine influence seen in the sediments from north to south. Facies associations and downslope transitions indicate deposition on an alluvival fan, passing through a fluvial braid-plain into a shallow sea (Hayward, 1983).

Sedimentation related to the emplacement of the Lycian Nappes

Palaeocurrents and downslope facies-transitions clearly indicate that the ophiolite-derived sediments of the western margin were derived from the Lycian Nappes to the northwest. Initial emplacement of the Lycian Nappes in Burdigalian times was to the northwest of Göçek. Fan-deltas derived from the leading edge of the nappe pile passed basinward into redeposited sandstones and mudstones.

The subsidence of the carbonate platform and subsequent basin formation are closely related to the initial emplacement event.

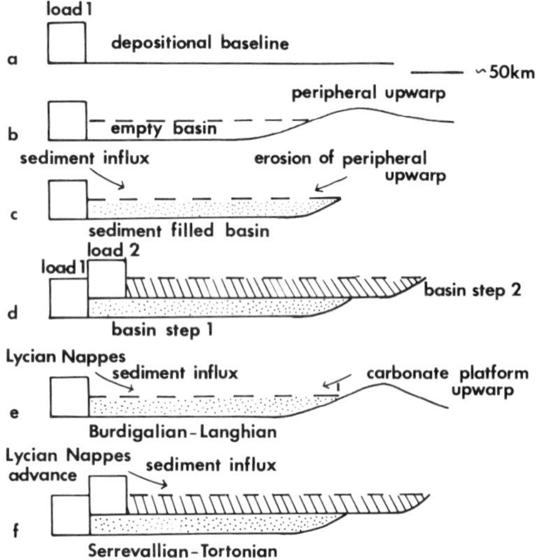

FIG. 7. Model for basin development (after Beaumont 1981).
(a and b) Passive loading of the lithosphere results in downward flexure and the development of a coupled trough in front of the nappe pile and a flexural upwarp along the opposite margin of the basin.
(c) Overthrusting of the nappe pile results in overstepping of the peripheral upwarp.
(d and e) Application of this model to the Lycian Nappes (see text for details).

Beaumont (1981) has recently proposed a model whereby sedimentary basins subside in response to the lateral tranfer of a rock mass over an adjacent part of the lithosphere. The resulting downward flexure produces a coupled trough in front of the fold- and thrust-belt in which sediments accumulate (Fig. 7).

One aspect of the flexural-loading model for basin formation is the flexural upwarp predicted along the margin of the basin opposite the thrust belt. This is applicable to the present area. Initial emplacement of the Lycian Nappes and loading of the carbonate platform to the northwest of Göçek developed a sedimentary basin ca 170 km across. Over most of the carbonate platform the Lower Miocene (Burdigalian-Langhian) was a period of rapid subsidence (Fig. 8): in the Kemer area ca 750 m of sediment accumulated in 5 Ma at an average rate of 15 cm/1000 yrs. In the area south and southeast of Kaş the carbonate platform was rapidly uplifted and a thick wedge of limestone conglomerates and calcarenites was shed northwards into the basin (Figs 9a, b).

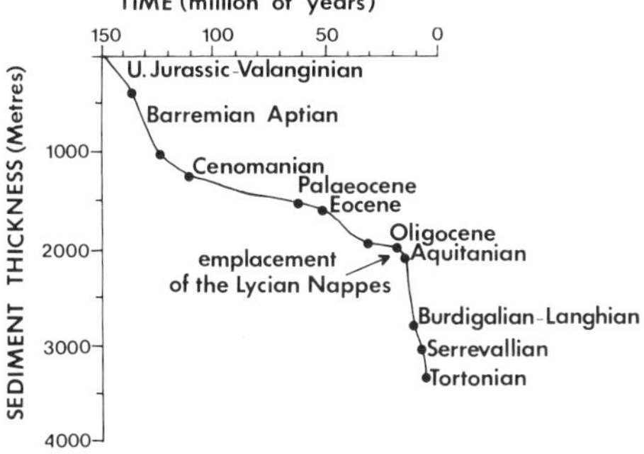

FIG. 8. Sediment accumulation curve for the Susuz Dağ (Sinekçibeli area). Note the rapid increase in rate of sedimentation following emplacement of the Lycian Nappes.

Continued southeastward-thrusting of the Lycian Nappes terminated deposition in the Göçek area in Upper Burdigalian time (Fig. 9a, b). In the area of Sinekçibeli deposition continued into the Langhian. Fan-deltas were derived from the front of the nappe pile and passed basinwards into a series of small submarine-fans (Fig. 9b). In Upper Langhian times the sedimentary sequence was truncated by continued south-eastward thrusting of the nappes. The Göçek window lies approximately 70 km to the west of Sinekçibeli indicating that in a period of approximately 3 Ma the Lycian Nappes were translated ca 70 km at an average rate of 2.5 cm/yr (Fig. 10).

During Serrevallian time subsidence slowed and coarse terrigenous clastic sediment input was minimal. This period may represent a pause in nappe emplacement. A final phase of southeastward thrusting in Tortonian times is marked by rapid subsidence and a coarse grained clastic wedge of sediment which prograded into a shallow sea.

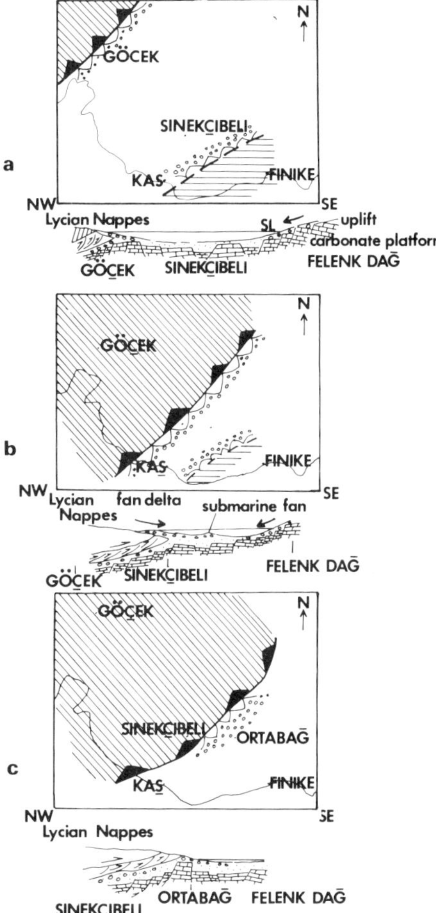

FIG. 9. Model for progressive nappe emplacement and basin formation.
(a) Burdigalian: Initial emplacement of the Lycian Nappes to the west of Göçek, produces loading of the underlying platform and results in a sedimentary basin ca 170 km across. Ophiolite-derived clastics are shed off the front of the nappe pile. Carbonate clastics are derived from a peripheral bulge in the carbonate platform opposite the nappe pile.
(b) Langhian: continued southwestward thrusting of the Lycian Nappes terminates deposition in the Göçek area. Fan-deltas derived off the nappe front pass basinwards to the S.E. into a series of submarine fans. Carbonate clastics, derived from the peripheral bulge, continued to be shed northwestwards.
(c) Tortonian: Final stage of emplacement. The peripheral bulge is overstepped and alluvial-fans prograde into the central areas of the carbonate platform.

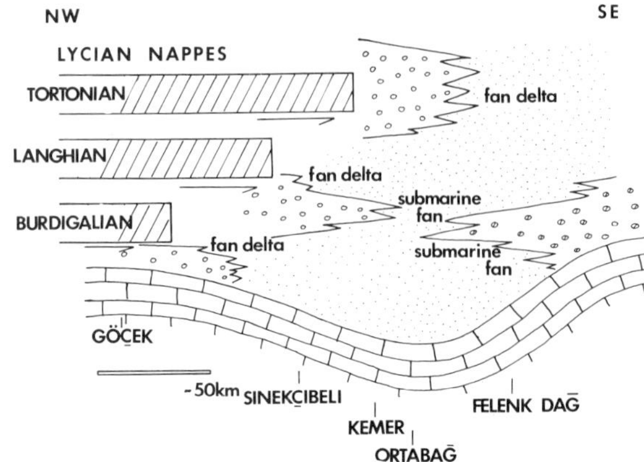

FIG. 10. Schematic cross-section illustrating development of the sedimentary basin in front of the Lycian nappes, progressive nappe emplacement and variations in sedimentary facies within the basin.

Conclusions: Implications for the regional geology

Following the original identification of the Antalya Complex as a series of thrust sheets, the original location of the Antalya Complex ocean basin has been the subject of continued debate.

Initial work assumed that the Complex formed a series of far travelled tectonic klippen related to the other major ophiolitic nappe-units of western Turkey, the Lycian Nappes and the Beyşehir-Hoyran-Hadim Nappes which were thrust southward over the autochthonous carbonate platform sequences during the Tertiary. Subsequent stratigraphic studies of Brunn *et al.* (1970, 1971) demonstrated continuous sequences up to Eocene times in northern areas of the carbonate platform, which precluded emplacement prior to this time whereas parts of the Antalya Complex are known to have been emplaced in Late Cretaceous to Palaeocene times. In the light of this evidence Dumont *et al.* (1972a, b) postulated a southern 'external' origin for the Complex, suggesting an original location south of the relatively autochthonous carbonate platform, in the 'Pamphyllian basin' which lay between the Tauride platform and the African continental margin separate from the main Tethys ocean.

In recent years the 'internalist' hypothesis has again found favour with a number of geologists (Ricou *et al.*, 1974, 1975, 1979; Ricou & Marcoux, 1980; Monod, 1976b; Dumont, 1976b; Dumont *et al.*, 1980; Ricou *et al.*, this volume). The most recent model proposed by Ricou *et al.* (1979) involves large areas of the Tauride carbonate platform, hitherto thought as relatively authochthonous or para-autochthonous, being regarded instead as far-travelled allochthonous sheets. In this model the Bey Dağları is made up of two separate tectonic units, a 'Western Bey Dağları' and an 'Eastern Bey Dağları', supposedly separated by a major north-south thrust which runs into the Miocene basal thrust of the Antalya Complex along the S.E. flank of the Bey Dağları.

The sediments of the *in situ* terrigenous clastic cover of the Bey Dağları and the Susuz Dağ carbonate platform are unequivocal evidence of a south-eastern 'external' origin for the Antalya Complex (Hayward & Robertson, 1982). Sediments of the eastern margin of the Bey Dağları show a clear derivation from the east and northeast from the direction of the Antalya Complex.

In contrast, the Miocene clastic sediments of the Susuz Dağ area were clearly derived from the northwest and document the progressive emplacement in several stages of the Lycian Nappes from the northwest to the southeast onto the central parts of the carbonate platform during the Miocene (Fig. 10). Critically, the sedimentary sequence in the Kasaba syncline is continuous into the Upper Miocene (Tortonian) (Figs. 5d and 10), wheres the Antalya Complex and Bey Dağları north of Antalya have an unconformable cover up to Middle and Upper Miocene age. The Antalya Complex cannot therefore have been emplaced over the western Bey Dağları during Middle Miocene time as required by the 'internalist' hypothesis of Ricou *et al.* (1979). The two ophiolite allochthons must have originated in separate, distinct ocean basins either side of the Bey Dağları and Susuz Dağ carbonate platform. Similar detailed facies analysis could well help solve comparable problems elsewhere in the Eastern Mediterranean.

ACKNOWLEDGMENTS: This research was carried out during the tenure of a N.E.R.C. Studentship at the University of Edinburgh. The Turkish Geological Survey (M.T.A.) provided logistical support. In particular, I thank E. Demirtaşlı and K. Tanner for their support in Turkey. Alastair Robertson provided help and encouragement during my time at Edinburgh. André Poisson, Alastair Robertson, Bryan Lovell and John O'Leary made constructive comments on an earlier version of this manuscript. I thank Maureen Fulton for drafting some of the diagrams.

References

ALLASINAZ, A., GUTNIC, M. & POISSON, A. 1974. La formation de l'Isparta Cay: Calcaires à Halobies, Grès à plantes, et Radiolarites d'age Carnien (?)—Norien (Taurides—Region d'Isparta—Turquie). *Schr. Erdwiss. Komm. Oster. Akad.* **2,** 11–21.

BARTON, C. M. 1975. Mount Olympus, Greece: new light on an old window. *Q.J. Geol. Soc. Lond.* **131,** 389–396.

BEAUMONT, C. 1981. Foreland basins. *Geophys. J.R. astr. Soc.* **65,** 291–329.

BRUNN, J. H., GRACIANSKY, P. C. DE, GUTNIC, M., JUTEAU, T., LEFEVRE, R., MARCOUX, J., MONOD, O. & POISSON, A. 1970. Structures majeurs et corrélations stratigraphiques dans les Taurides occidentales. *Bull. Soc. géol. Fr.* **12,** 515–551.

——, DUMONT, J. F., GRACIANSKY, P. C. DE, GUTNIC, M., JUTEAU, T., MARCOUX, J., MONOD, O. & POISSON, A. 1971. Outline of the geology of the western Taurides. *In* CAMPBELL, A. S. (ed.), *Geology and History of Turkey.* Petroleum Exploration Society of Libya, Tripoli. 225–257.

——, ARGYRIADIS, I., MARCOUX, J., MONOD, O., POISSON, A. & RICOU, L. E. 1973. Arguments pour et contre l'origine méridionale des nappes a ophiolites d'Antalya. *Geological congress: 50th Anniversary of Turkish Republic.* M.T.A. & D.S.I. Ankara. 58–73.

——, ——, RICOU, L. E., POISSON, A., MARCOUX, J. & GRACIANSKY, P. C. DE 1976. Elements majeurs de liaison entre Taurides et Hellénides. *Bull. Soc. géol. Fr.* **18,** 481–497.

DELAUNE-MAYERE, M., MARCOUX, J., PARROT, J. F. & POISSON, A. 1977. Modèle d'évolution mésozoique de la paléomarge Tethysienne au niveau des nappes radiolaritiques et ophiolitiques du Taurus lycien, d'Antalya et du Baër-Bassit. *In* BIJU-DUVAL, B. & MONTADERT, L. (eds), *Structural History of the Mediterranean Basins.* Editions Technip, Paris, 79–94.

DUMONT, J. F. 1976. *Etudes géologiques dans les Taurides Occidentales: Les formations paléozoiques et mésozoiques de la coupole de Karacahisar (Province d'Isparta, Turquie).* Thesis: Univ. Paris-Sud, Orsay. 213 pp.

——, GUTNIC, M., MARCOUX, M., MONOD, O. & POISSON, A. 1972a. Le Trias des Taurides Occidentales (Turquie). Definition du bassin pamphylien: Un nouveau domaine a ophiolithes à la marge externe de la chaîne taurique. Z. *Deutsch geol. Ges.*, **123,** 385–409.

——, ——, ——, —— & —— 1972b. Essai de reconstitution d'un bassin triasique à ophiolites à la marge externe des Taurides: la bassin pamphylien. *C.r. Somm. Soc. géol. Fr.* **2,** 73–74.

——, UYSAL, S. & MONOD, O. 1980. La série de Zindan: un élément de liaison entre plate-forme et bassin a l'est d'Isparta (Taurides occidentales, Turquie). *Bull. Soc. géol. Fr.* **22,** 225–232.

GRACIANSKY, P. C. DE 1972. *Recherches géologiques dans le Taurus Lycien Occidental.* Thèse, Université de Paris-Sud, 571 pp.

——, LORENZ, C. & MAGNE, J. 1970. Sur les étapes de la transgression du Miocène inférieur observée dans les fenêtres de Göçek (Sud-Quest de la Turquie). *Bull. Soc. géol. Fr.* **12,** 557–564.

GUTNIC, M. & POISSON, A. 1970. Un dispositif remarquable des chaînes tauriques dans le sud de la courbure d'Isparta. *C.r. Séances Acad. Sci. Paris* **270,** 672–675.

HAYWARD, A. B. 1982a. *Tertiary ophiolite-related sedimentation in S.W. Turkey.* Unpubl. Ph.D. thesis, Univ. of Edinburgh, 420 pp.

—— 1982b. Coral reefs in a clastic sedimentary environment: fossil (Miocene, S.W. Turkey) and modern (Recent Red Sea) analogues. *Coral Reefs*, **1,** 109–114.

——1983. Coastal alluvial fans of Miocene age in S.W. Turkey. *In:* COLLINSON, J. D. & LEWIN, J. (eds). *Modern and Ancient Fluvial Systems. Sedimentology.* Spec. Publ. **6,** 323–336.

—— Hemipelagic chalks in a clastic submarine fan sequence. *In:* STOW, D. A. V. & PIPER, D. J. W. (eds). *Fine Grained Sediments.* Spec. Publ. Geol. Soc. Lond., in press a.

—— Stratigraphy of the Miocene clastic sediments of the Bey Dağları and Susuz Dağ, Tauride autochthon S.W. Turkey. *Bull. geol. Soc. Turkey.* In press b.

—— & ROBERTSON, A. H. F. 1982. Direction of ophiolite emplacement inferred from Cretaceous and Tertiary sediments of an adjacent autochthon, the Bey Dağları, S. W. Turkey. *Bull. geol. Soc. Am.* **93,** 68–75.

HSÜ, K. J. 1973. The desiccated deep-basin model for the Messinian events. *In* DROOGER, C. W. (ed.). *Messinian Events in the Mediterranean.* North Holland, Amsterdam, 60–67.

JUTEAU, T. 1975. Les ophiolites des nappes d'Antalya (Taurides) occidentales (Turquie). *Mém. Sci. Terre Nancy* **32.**

——, NICOLAS, A., DUBESSY, J., FRUCHARD, J. C. & BOUCHEZ, J. L. 1977. The Antalya ophiolite complex Western Taurides (Turkey): a structural model for an oceanic ridge. *Bull. geol. Soc. Am.* **88,** 1740–1748.

KALAFATÇIOĞLU, A. 1973. Geology of the western part of Antalya Bay. *Bull. Miner. Res. Explor. Inst. Turkey* **81**, 31–84.

LEFEVRE, 1967. Un nouvel élément de la géologie du Taurus lycien: les nappes d'Antalya (Turquie). *C.r. Séances Acad. Sci. Paris* **265**, 1365–1368.

MONOD, O. 1976. La«courbure d'Isparta»: une mosaîque de blocs autochtones surmontés de nappes composites à la jonction de l'arc hellénique et de l'arc taurique. *Bull. Soc. géol. Fr.* **18**, 521–531.

—— 1977. *Recherches géologiques dans le Taurus occidentales au sud de Beyşehir Turquie*). Thesis, Univ. Paris-Sud, Orsay.

—— 1978. Güzelsu Akseki Bölgesindeki Antalya Naplari uzerine aciklama (Orta Bati Toroslar—Turkiye). [Precisions upon the Antalya Nappes in the region of Guzelsu-Akseki (Western Taurus, Turkey)]. *Bull. geol. Soc. Turkey* **21**, 27–29.

ÖNALON, M. 1980. Elmali-Kaş (Antalya) arasirdaki Bölgerin jeolojisi. *Istanbul Universitesi, Fen Fakultesi Monografileri, Sayi* **29**, 140 p.

ÖZGUL, N. & ARPAT, E. 1973. Structural units of the Taurus orogenic belt and their continuation in neighbouring regions. *Bull. geol. Soc. Greece* **10**, 156–164.

POISSON, A. 1967a. Données nouvelles sur le Crétacé et le Tertiaire du Taurus occidental au Nord-Ouest d'Antalya (région du Korkuteli, Turquie). *C.r. Séances Acad. Sci. Paris* **264**, 218–221.

—— 1967b. Présence d'un Trias supérieur de faciès récifal dans le Taurus lycien au Nord-Quest d'Antalya (Turquie). *C.r. Séances Acad. Sci. Paris* **264**, 2443–2446.

—— 1974. Présence de Jurassique et de Crétacé inférieur à faciès de type plateforme dans l'autochtone lycien près d'Antalya (Bey Dağları) Turquie. *C.r. Séances Acad. Sci. Paris* **278**, 835–838.

—— 1977. *Recherches géologiques dans les Taurides occidentales* (*Turquie*). Thesis: Univ. Paris-Sud, Orsay, 705 pp.

—— & AKKAY, E. 1981. Miocene transgression forms in the Gökbuk-Çatallar basin (western Taurus range Turkey). *T.C. Petrol Isleri Genel Mudurlugu Dergisi*, **25**, 217–221.

—— & POIGNANT, A. F. 1974. La formation de Karabayir base de la transgression miocène dans la région de Korkuteli (Antalya-Turquie). *Lithothamnium pseudoramossissimum* nouvelle espèce d'Algue rouge de la formation de Karabayir. *Bull. Miner. Res. Explor. Inst. Turkey* **82**, 67–71.

RICOU, L. E., ARGYRIADIS, I. & LEFEVRE, R. 1974. Proposition d'une origine interne pour les nappes d'Antalya et le massif d'Alanya (Taurides occidentales, Turquie). *Bull. Soc. géol. Fr.* **16**, 107–111.

——, ARGYRIADIS, I. & MARCOUX, J. 1975. L'axe calcaire du Taurus, un alignement de fenêtres arabo-africaines sous des nappes radiolaritiques, ophiolitiques et métamorphiques. *Bull. Soc. géol. Fr.* **17**, 1024–1043.

——, MARCOUX, J. & POISSON, A. 1979. L'allochtonie des Bey Dağları orientaux. Reconstruction palinspastique des Taurides occidentales. *Bull. Soc. géol. Fr.* **21**, 125–134.

—— & —— 1980. Organisation générale et rôle structurale des radiolarites et ophiolites le long du système alpine-méditerranéan. *Bull. Soc. géol. Fr.* **22**, 1–14.

ROBERTSON, A. H. F. & WOODCOCK, N. H. 1980. Strike-slip related sedimentation in the Antalya Complex, S.W. Turkey. *In*: BALANCE, P. F. & READING, H. G. (eds) *Sedimentation in Oblique-Slip Mobile Belts. Spec. Publ. Int. Assoc. Sedimentol.* **4**, 127–145.

—— & —— 1981a. Alakır, Çay Group, Antalya Complex, S.W. Turkey: deposition on a Mesozoic passive carbonate margin. *Sediment. Geol.* **30**, 95–131.

—— & —— 1981b. Bilelyeri Group, Antalya Complex, S.W. Turkey: deposition on a Mesozoic passive continental margin. *Sedimentology* **28**, 381–399.

—— & —— 1981c. Gödene zone, Antalya Complex, S.W. Turkey: volcanism and sedimentation on Mesozoic marginal oceanic crust. *Geol. Rdsch.*

VAIL, P. R., MITCHUM, R. M. & THOMPSON, S. 1977. Global cycles of relative changes in sea level. *In:* PAYTON, C. E. (ed). *Seismic Stratigraphy—applications to hydrocarbon exploration.* Am. Assoc. Pet. Geol. Mem., **26**, 83–97.

WOODCOCK, N. H. & ROBERTSON, A. H. F. 1977. Imbricate thrust belt tectonics and sedimentation as a guide to the emplacement of part of the Antalya Complex, S.W. Turkey. Abstracts, *6th Colloquium Geology Aegean Region, Izmir, Turkey*. p. 98.

—— & —— 1982. Imbricate thrusting and wrench tectonics in the Antalya Complex, S.W. Turkey. *J. geol. Soc. Lond.* **139**, 147–56.

A. B. HAYWARD, British Petroleum plc, West Britannic House (a), Moor Lane, London EC 2, England. *Formerly:* Grant Institute of Geology, University of Edinburgh, EH9 3JW, Scotland.

Role of the Eastern Mediterranean ophiolites (Turkey, Syria, Cyprus) in the history of the Neo-Tethys

H. Whitechurch, T. Juteau and R. Montigny

SUMMARY: The internal organization and structure of the Taurus, Troodos, Hatay and Baër-Bassit ophiolite massifs are used, along with K-Ar geochronological data for the different parts of the assemblages, to outline stages in the evolution of the Neo-Tethys ocean during Mesozoic times. The authors distinguish:

(1) an intra-continental stage, during Middle and Upper Triassic, when alkaline submarine basalts were erupted during rifting of the Arabian platform;

(2) a Lower Cretaceous to Campanian phase, marked by active spreading for at least 40 Ma, along east-west trending palaeo-spreading axes, offset by north-south transform fracture zones. During that period, the oldest oceanic lithosphere (Tauride ophiolites and the Baër-Bassit complex) began to be sliced up with the development of metamorphic soles (104 to 88 Ma), later intruded by swarms of tholeiitic diabase dykes of Campanian age (ca. 75 Ma). At that time, the latest spreading zones, now represented by the Troodos and Hatay ophiolites, were still active in the Neo-Tethys; and

(3) during the Maastrichtian (70–65 Ma), all ophiolitic massifs were thrust over the Arabo-African continental margin from a single Neo-Tethys oceanic basin.

Today ophiolite complexes are widely considered to be fragments of ancient oceanic lithosphere formed at constructive margins. As a consequence, any piece of ophiolite must record an oceanic history, starting with formation of oceanic lithosphere in a geotectonic context which may vary from, for example, a spreading plate boundary, inter-arc basin, or marginal sea, and ending with final thrusting over a continental margin or an island arc. Between these two events, the ocean floor is affected by various magmatic, tectonic and metamorphic processes which result from spreading, ageing, transform faulting, hot-spot activity, intra-oceanic slicing, and other events likely to occur in an oceanic domain.

The aim of this paper is to combine structural, petrological and geochronological data on ophiolite complexes of the East Mediterranean, which includes the Tauride belt, Cyprus, and northern Syria, to identify the main oceanic events in the Neo-Tethys oceanic lithosphere during the Mesozoic, and thus to help constrain general models for the geotectonic evolution of the Eastern Mediterranean.

Geological setting and structural organization

The ophiolites referred to in this paper include the massifs of the Taurus Range (S. Turkey), the Hatay massif (S.E. Turkey), the Baër-Bassit complex (NW Syria), and the Troodos Massif (Cyprus). Fig. 1 shows the distribution of these massifs in simplified geological setting.

Until recently the state of knowledge of these ophiolites was very unequal which discouraged comparisons. Now all these massifs have been mapped, mostly at a scale of 1/25,000, and their internal organization and structure are well known. Adequate petrological data are now available. The reader is referred to Gass (1980) for the Troodos ophiolite, to Parrot (1977b) for Baër-Bassit, to Tinkler *et al.* (1980) for Hatay and to Juteau (1979, 1980) for the Taurus ophiolites. We do not describe the massifs again here, but review salient features which contribute to a better understanding of their oceanic evolution. This should also take into account (from west to east) the massifs of Lycia (de Graciansky 1972), Antalya, Beyşehir-Hoyran, Mersin, Pozantı-Karsantı (Juteau 1979), also Pinarbaşı, Gürün and Divriği, and the North Anatolian ophiolite belt. With the exception of Antalya, these other ophiolites have not been discussed much before.

Geological setting

Most of the ophiolites considered in this paper are clearly allochthonous, forming nappes thrust over the Arabo-African platform carbonates, either seen in tectonic windows as the 'Taurus Limestone Axis' of Ricou *et al.* (1975) (Taurus ophiolites), or seen as the direct continuation of the Arabian continental margin (Hatay, Baër-Bassit). The age of first tectonic emplacement has been dated everywhere as Upper Cretaceous (Campanian to Palaeocene; Ricou 1971; Ricou *et al.* 1975). In the so-called 'Western Taurides' (Brunn *et al.* 1971) Tertiary

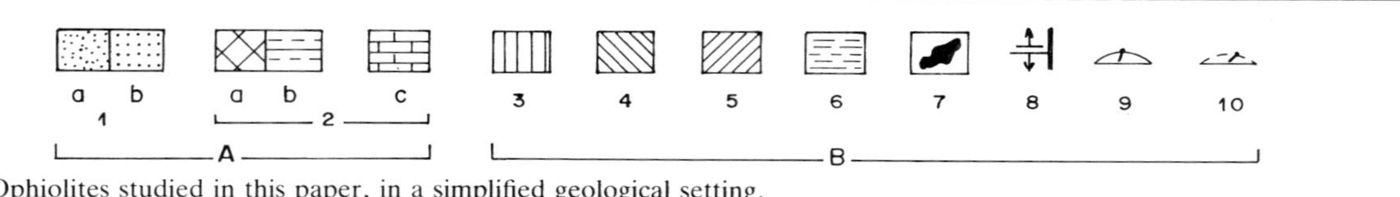

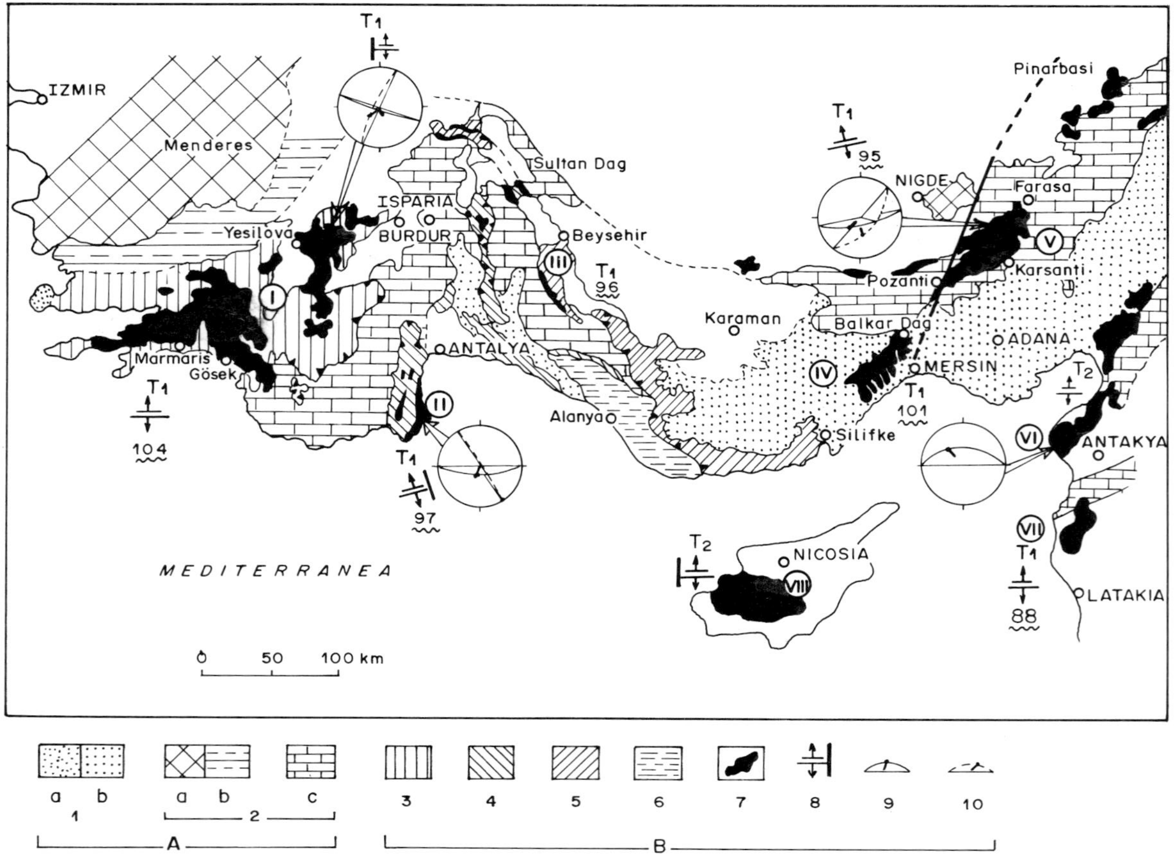

Fig. 1. Ophiolites studied in this paper, in a simplified geological setting.

(A) *Autochthonous formations*: (1) post-tectonic formations, (a) Pliocene; (b) Miocene. (2) Relatively autochthonous units of the Tauride nappes, (a) crystalline massifs of Menderes and Nigde; (b) metamorphic cover of Menderes massif; (c) 'Taurus Limestone Axis'.

(B) *Allochthonous formations*: (3) Lycian nappes; (4) nappes of Antalya; (5) nappes of Beyşehir-Hoyran; (6) massif of Antalya; (7) ophiolites (I, Lycian ophiolites; II, Antalya; III, Beyşehir-Hoyran; IV, Mersin; V, Pozantı-Karsantı; VI, Hatay; VII, Baër-Bassit; VIII, Troodos ophiolite); (8) inferred orientation of the palaeo-spreading zones and associated fracture zones. The stereograms give the present mean attitude of the (S_1) and (S_2) foliations in the tectonites, with their associated mineral lineations (see text). Projection on the upper hemisphere (approximate number of measurements: Yesilova: 150; Antalya: 1,000: Pozantı: 300; Hatay: 30); (9) (S_1) and association lineation; (10) S_2 and associated lineation. T_1, T_2, Taurus-type and Troodos-type ophiolites respectively (see Fig. 2). 97 Ma: Age of the main metamorphic crystallization of the metamorphic sole.

orogenic movements reactivated the nappe piles emplaced during the Upper Cretaceous in two main tectonic phases, Eocene-Oligocene and intra-Miocene. These phases are well known in the Lycian, Antalya and Beyşehir-Hoyran nappes (Brunn *et al.* 1970, 1971, 1976).

Nobody has seriously contested the allochthoneity of the following ophiolites: the Lycian Nappes (de Graciansky 1972; Brunn *et al.* 1970, 1971, 1976), the Beyşehir-Hoyran nappes (Brunn *et al.* 1970; Monod 1977), the Mersin and Pozantı-Karsantı complexes (Metz 1956; Ricou *et al.* 1975; Çakir *et al.* 1978; Juteau 1979, 1980), and the Hatay and Baër-Bassit massifs (Dubertret 1939; Ricou 1971; Parrot, 1973, 1977a, 1977b; Parrot & Whitechurch 1978).

As the base of Troodos is not exposed, geophysical and geological data have been interpreted differently. Gass and co-workers proposed that the Troodos Massif is autochthonous (Gass & Masson-Smith 1963; Gass 1968, 1980) whereas Biju-Duval *et al.* (1976) considered it as an allochthonous massif. Gass's model did, however, propose under-thrusting of African lithosphere, either continental or oceanic (see Gass 1980, his Fig. 6b and c) under the Troodos ophiolite. Robertson and Woodcock (1980) pointed out that this model is consistent with tectonic obduction of the Troodos over the African continental margin.

The Antalya ophiolites are also controversial with regard to their allochthoneity. French geologists have claimed that a major nappe pile exists (Lefèvre 1967; Brunn *et al.* 1970, 1971, 1976; Delaune-Mayere *et al.* 1976) and interpretations of original tectonic setting range from either proximal, the 'Pamphylian basin' of Dumont *et al.* 1972 (Monod *et al.* 1974) to far northern (Brunn 1974; Ricou *et al.* 1974, 1975, 1979; Ricou & Marcoux 1980). More recently, Robertson and Woodcock (1980a and b, 1982, this volume; Woodcock & Robertson 1982) pointed out the importance of strike-slip movements during the evolution and emplacement of their Antalya Complex. They developed a wrench and thrust tectonic model where the thrusting component is minimized and original palaeogeographical continuity can still be recognized within various parts of the Antalya allochthon. In this view the present tectonic zonation reflects the initial Mesozoic palaeogeographic organization to some extent. This conception is in disagreement with Marcoux's interpretations (Marcoux 1976a, Marcoux, in Delaune-Mayere *et al.* 1976) of a grossly allochthonous nappe pile. Our view is that the supporters of 'autochthonous' relationships for the Troodos and the Antalya ophiolites can not escape the need for some overthrusting in their models. Like other ophiolites around the world (Coleman 1977), the Eastern Mediterranean ophiolites have been obducted, in this case over the Arabo-African platform carbonates. Most have been detached from their roots as true tectonic nappes. We admit however that both the Troodos and the Antalya ophiolites could still be attached to subjacent mantle, as in Gass and Masson-Smith's model (*op. cit.*), and also that wrench-faulting movements could have played an important part during the emplacement of the Antalya ophiolites, as suggested by Robertson & Woodcock (*op. cit.*).

Internal organization, structures and petrology

We distinguish two main groups of ophiolites, each with specific characteristics. The first group includes the Taurus and Baër-Bassit ophiolites which have a rather complex structural organization, whereas the second group, including the Troodos and Hatay ophiolites, exhibit a very simple structure. We call them 'Taurus-type' and 'Troodos-type' ophiolites respectively (Fig. 2).

Taurus-type ophiolites

This group includes the ophiolites of the whole of the Taurus belt and the Baër-Bassit complex of N. Syria. Composite successions and a detailed account have been given by Juteau (1979, 1980). Fig. 2a shows the typical organization of these complexes, which are composed of the following lithological units, from bottom to top:

(1) A basal metamorphic aureole composed of slices of polyhase-folded metamorphic rocks, which are mainly amphibolites with sedimentary protoliths, as well as subordinate quartzites, mica-schists and marbles. The sub-ophiolitic position of these metamorphic slices is very clear in the Lycian nappes (de Graciansky 1972), in the Beyşehir-Hoyran nappes (Monod 1977; Juteau 1979), in the Mersin ophiolites (Juteau 1979), in the Pozantı-Karsantı massif (Çakir 1978; Çakir *et al.* 1978; Juteau 1979, 1980) and in the Baër-Bassit complex (Whitechurch 1977; Parrot and Whitechurch 1978). The situation is unclear in the Antalya complex, where the ophiolite-related metamorphic rocks occur in the Alakır Çay valley as vertical blocks and slivers in serpentinite-filled vertical fault zones (Juteau 1975; Robertson & Woodcock 1980). In previous papers it was concluded (Juteau 1979, 1980; Thuizat *et al.* 1981) that a sub-ophiolitic position was probable by analogy with other

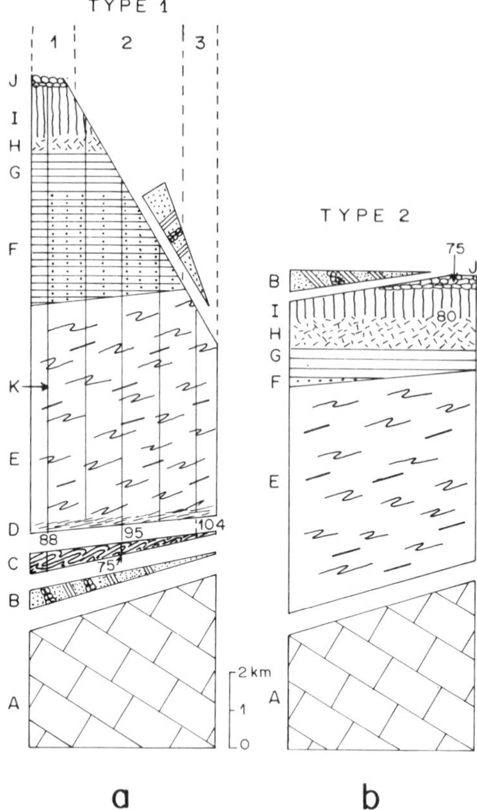

FIG. 2. Composite successions of the structural organization and petrographical units of (a) the Taurus-type and (b) Troodos-type ophiolites. The Taurus-type column shows the progressive variation of the present erosion level of the ophiolites from NW to SE of the studied area, relative to the ages of the metamorphic soles (ages, in Ma).

A, Autochthonous Mesozoic platform carbonates (Arabo-African continental margin). Not visible under the Troodos and Antalya ophiolites. B, Slices of mainly Upper Triassic alkali basalts (pillow-lavas) and associated sediments, below or over the Cretaceous ophiolites. C, Infra-ophiolitic metamorphic rocks, mainly amphibolites (ages in Ma).

Ophiolite sequence: D, Harzburgitic tectonites affected by a porphyroclastic mylonitic foliation (S_2), linked to the intra-oceanic slicing; E, Harzburgitic tectonites affected by a granoblastic foliation (S_1) due to the asthenospheric plastic flow during spreading; F, Layered ultramafic cumulates; G, Layered gabbroic cumulates; H, Isotropic gabbros, quartz diorites, plagiogranites; I, Sheeted dyke complex; J, Tholeiitic pillow-lavas; and K, Isolated tholeiitic diabase dykes intruding the whole ophiolitic pile, and sometimes the sub-ophiolitic metamorphic rocks. One of them intersects the contact between amphibolites and harzburgites (Pozantı-Karsantı).

Erosion level of the Taurus-type ophiolites: 1, Baër-Bassit ophiolites; 2, Antalya Mersin, Pozantı-Karsantı ophiolites; 3, Marmaris, Yesilova, Beyşehir-Hoyran ophiolites (see Fig. 1).

Taurus ophiolites. A careful re-examination of the outcrops, described in Juteau (1975), and a comparison with similar outcrops described in west Cyprus by Spray & Roddick (1981) lead us to admit that these metamorphic slivers could alternatively represent the products of transform fault metamorphism, or at least record, as suggested by Robertson & Woodcock (1980b, 1982), some degree of frictional heating along transpressive segments of strike-slip faults. This would be in good agreement with the recent discovery of important mylonitic wrench shear zones in the Tekirova coastal ophiolite complex (Reuber 1982a and b, and this volume), interpreted as having been formed along an active transform fault zone.

Most of the authors agree that the protoliths of these metamorphic rocks have an oceanic or continental margin origin, having been derived mostly from oceanic basalts, cherts and pelagic limestones. They probably represent metamorphosed oceanic crust and associated sediments (Williams & Smyth 1973; Whitechurch & Parrot 1974; Woodcock & Robertson 1977; Çakir *et al.* 1978; Parrot 1977a and b; Whitechurch 1977; Parrot & Whitechurch 1978; Malpas 1979; Spray & Roddick 1981). Folding of these rocks is polyphase with a first schistosity contemporaneous with the main metamorphic recrystallization. The age of the metamorphism has been determined on separated minerals by the K-Ar method in all the massifs (Thuizat *et al.* 1981). It ranges between 104 ± 4 Ma (Albian) for the Lycian sub-ophiolitic rocks to 88 ± 2 Ma (Santonian) for the Baër-Bassit ones; intermediate ages cluster around 95 Ma (Cenomanian) in the metamorphic soles of the Beyşehir-Hoyran nappes, Mersin and Pozantı-Karsantı complexes, and also in the Alakır Çay amphibolite slivers of the Antalya ophiolite. Comparison between hornblende and biotite ages in the same amphibolite sample from Pozantı-Karsantı suggests a rapid cooling with a brief metamorphic episode, at most over a few million years (Thuizat *et al.* 1981).

(2) A huge mass of harzburgitic tectonites is well preserved in all the massifs, representing 60 to 90% of the whole ophiolitic mass in

volume. These rocks exhibit the imprints of two main mantle processes: plastic flow at high temperatures, and partial melting (Juteau 1975; Juteau *et al.* 1977). They are foliated, lineated and folded, with structures and fabrics indicating high temperatures, around 1200°C, and low stresses during deformation. Partial melting is shown by lenses of clinopyroxenites and gabbros surrounded by aureoles of dunitic harzburgites. In the Antalya and the Pozantı-Karsantı massifs, where huge cumulate magma chambers are still preserved above the tectonites, the harzburgites are of residual mantle type, with clinopyroxene almost absent, and a dense net of gabbroic veins probably representing conduits which collected the melt near the base of the magma chambers (Juteau *et al.* 1977; Çakir 1977). In the Lycian ophiolites, where cumulates are not visible, the harzburgites are sub-lherzolitic (2–3% clinopyroxene on average, locally up to 5%) and hence these ultramafics are richer in Al_2O_3 and CaO, and do not exhibit the gabbroic net veins (de Graciansky 1972; Juteau 1980). These rocks are less depleted, and were probably deeper tectonites than the harzburgites which immediately underlie the cumulates of the Antalya and Pozantı-Karsantı massifs.

Recent studies of the Pozantı-Karsantı and Antalya massifs show that a second foliation, characterized by mylonitic textures developed under lower temperatures, around 800°C, and higher stress conditions overprints the first one. In the Pozantı-Karsantı massif, this second foliation is obviously linked to the formation of the sub-ophiolitic metamorphic sole, as it is located in the harzburgites for several hundreds of meters above the contact, and is parallel to the schistosity of the metamorphic rocks (Çakir 1978; Çakir *et al.* 1978). In the Antalya massif, the orientation of the second foliation is quite different: it develops in mylonitic shear zones intersecting the first foliation at a high angle, and bears a sub-horizontal lineation formed during a ductile strike-slip movement, probably due to transform faulting processes (Reuber 1982a and b, and this volume).

(3) Undeformed units of ultramafic to mafic layered cumulates, over 3–5 km thick are well developed in the Antalya, Pozantı-Karsantı and Baër-Bassit complexes, and can be recognized in small tectonic slices in other complexes (see Juteau 1980). The detailed stratigraphy and petrology of the cumulate sequence have been studied by Juteau (1975) and Juteau and Whitechurch (1980) in Antalya, by Bingöl (1978) and Çakir (1978) in the Pozantı-Karsantı massif, and by Parrot (1977a) in the Baër-Bassit complex. In these three examples, two main features characterize the cumulate sequence:

(a) the ratio of ultramafic to mafic rocks is unusually high, from 1/1 to 3/1; and

(b) these cumulates have a high MgO content and show a total absence of cryptic variations in chemistry of the main mineral phases which are olivine, clinopyroxene, orthopyroxene and plagioclase.

Parrot (1977a) and Juteau & Whitechurch (1980) interpret these features as resulting from continuous replenishment of a steady-state magma chamber by influxes of Mg-rich primitive magma. Moreover, the thickness of the layered cumulates (4–5 km), and the thinness of the overlying massive gabbros suggest their crystallization in magma chambers which were wider than they were deep (Dewey & Kidd 1977; Pallister & Hopson 1981). Such magma chambers are likely to be found under rather fast-spreading zones with a total separation rate of greater than 5 cm/year (Nisbet & Fowler 1978).

(4) Massive isotropic gabbros, quartz diorites and subordinate plagiogranites, which form the host-rocks of a dyke complex are visible in the Kemer-Tekirova area of the Antalya ophiolite only. In this case, the dyke complex is composed of tholeiitic diabase and quartz diabase dykes, with plagiogranite veinlets and late basaltic dykelets.

The rocks of this unit are the only ones suitable for K-Ar datings, and thus considerable recent efforts have been made to determine the primary cooling ages of the Taurus-type ophiolites, particularly the Antalya complex (Thuizat *et al.* 1978, 1979; Yılmaz and Maxwell 1981; Montigny *et al.* in press), and the Pozantı-Karsantı massif (Thuizat *et al.* 1978). About 50 age determinations are now available for the Antalya ophiolites, most of them on separated minerals. Fifteen are presented by Yılmaz (this volume) and show a wide scatter of apparent ages from 54 Ma to 130 Ma. 35 new dates determined by Montigny *et al.* (in press) are from separated amphiboles, plagioclases and chlorites from hornblende gabbros, diabases and plagiogranites from the dyke complex of Kemer (Antalya) and its host-rocks, and from gabbro dykelets and plagiogranite veins intruding the underlying tectonites and cumulates. They also display a wide range of apparent ages from 35 to 130 Ma, with however a cluster of ages between 70 and 80 Ma. Hornblende gabbros and plagiogranites from the Pozantı-Karsantı ophiolite yield ages clustering around 85 Ma (Thuizat *et al.* 1978). These data are difficult to interpret, and will be discussed below.

(5) Tholeiitic pillow-lavas are only preserved

in the Baër-Bassit complex (Parrot 1977a) as a thin (150 m) unit of lower pillow-lavas, directly linked to the dyke complex, and discordant Mg-rich hyper-tholeiitic unit of upper pillow-lavas showing komatiitic affinities. Metalliferous sediments (umbers) were deposited on the upper pillow lavas as in the Troodos Massif.

(6) Swarms of isolated tholeiitic and fine-grained diabase dykes, 1–10 m thick, exhibit fine chilled margins and intersect tectonites and cumulates of all the Taurus-type massifs without exception. There are many thousands of these dykes which are a peculiar and distinctive feature of the belt. The mineralogy and chemistry of these dykes show an extremely uniform composition as low-K tholeiites with very weak evolutionary trends. Although some porphyritic lithologies have been described by Sarp (1976) from the northern Lycian nappes, these dykes are usually totally aphyric. K-Ar dating of the diabases on separated amphiboles and plagioclases yields ages clustering between 70 Ma and 75 Ma, i.e. Campanian (Thuizat *et al.* 1978).

Some of these dykes also intersect the sub-peridotite metamorphic slices. This is the case in the Lycian Nappes (Sarp 1976) and in the Pozantı-Karsantı ophiolites (Çakir 1978; Çakir *et al.* 1978). In the latter massif, one of these dykes even intersects a flat tectonic contact between amphibolites and overlying harzburgites at right angle, and hence post-dates their final tectonic emplacement.

Occurring separately either below or above the Taurus-type ophiolites are tectonic slices of volcanic and sedimentary rocks. These units are composed mainly of basaltic pillow-lavas and associated terrigenous to pelagic sediments which were first attributed to the ophiolite suite (Dubertret 1955; Kazmin & Kulakov 1968; Juteau 1968, 1970). It was later found that these TiO_2-rich lavas, rich in olivine, clinopyroxene and plagioclase phenocrysts belonged to an alkalic trend and could not be co-genetic with the tholeiitic ophiolites (de Graciansky 1972; Juteau *et al.* 1973; Juteau 1975, 1979, 1980; Montigny 1975; Parrot 1974, 1977a).

The petrology of these lavas has been studied unequally in the different massifs and their age, where known, is rather variable. In the Lycian complex, they constitute the 'diabase nappe' of de Graciansky (1972), and appear immediately under the peridotites or the metamorphic slices as a 'coloured melange' in which submarine alkalic basalts predominate, with a trachytic differentiation trend. The associated sediments are mainly radiolarites and pelagic limestones, of partly Cretaceous age (Barremian-Aptian and Lower Senonian) (de Gracianksy 1972), but this does not preclude older ages (? Upper Triassic), for other lavas in the melange. The 'diabases nappe' rests tectonically on a Maastrichtian to Palaeocene 'wildflysch' containing exotic blocks of all sizes, including alkalic lava from the 'diabases nappe'. Also, basaltic flows of pillow-lavas are interbedded with oolitic limestones of Dogger-Malm age in the Domuz Dağ, at the eastern front of the Lycian Nappes, and in the neighbouring unit massive basaltic flows are interbedded with pelagic *Globotruncana*-bearing limestones of Upper Cretaceous age (Poisson 1968, 1977). In the Yesilova area, in the northern Lycian nappes, Sarp (1976) describes slices of very deformed alkalic basalts, and pillow-lavas associated with radiolarites which rest tectonically on peridotites.

In the Antalya nappes, thick piles of pillow-lavas and massive flows are splendidly exposed and more than 1 km thick on the eastern flank of the Alakır Çay valley, interbedded with *Halobia* and ammonite-bearing limestones of Upper Triassic age (Marcoux 1970, 1976a; Guerin-Franiatte & Juteau 1970). The pillows are of TiO_2-rich alkalic basalts, hawaiites and mugearites (Juteau & Marcoux 1973; Juteau 1975). It can be demonstrated that the submarine lavas were emplaced by numerous fissural eruptions occurring in a horst-graben topography (Juteau & Marcoux 1973; Robertson & Woodcock 1981). In the Beyşehir-Hoyran nappes, severely deformed alkali-basaltic pillow-lavas, associated with radiolarites, Halobia-bearing pelagic limestones of Upper Triassic age as well as schists, in places tectonically overlie peridotite (Monod 1977). The same formations occur as tectonic slices between the infra-ophiolitic metamorphic soles and the autochthonous platform carbonates in the Mersin nappes (Juteau 1979) and in the Pozantı-Karsantı opiolites, where the alkalic lavas are of Upper Triassic age (Çatakli *et al.* 1977). Finally, the Baër-Bassit nappes include volcanic and sedimentary units intercalated at the base of the Baër peridotites and the metamorphic soles. The sediments are of Upper Triassic to Upper Cretaceous age, and include two volcanic episodes. The first one consists of Upper Triassic tholeiitic pillow-lavas showing alkalic tendencies, the second one of the Upper Jurassic age is made of undersaturated alkalic lavas, mostly basanites and phonolites (Parrot 1974a and b; Parrot & Vatin-Perignon 1974; Parrot 1977a).

A puzzling fact is that the upper parts of the tholeiitic complexes in the Taurus Range have disappeared systematically. The Lycian ophiolites, and those of Yesilova and Beyşehir-Hoyran are eroded down to the tectonites. The most complete section is that of Antalya where massive gabbros and part of the dyke complex are preserved. In the case of this massif, parts of the dyke complex are seen as slabs and blocks within a Maastrichtian mega-breccia overlying the ophiolites and interpreted by Reuber (1982) and Robertson & Woodcock (1982) as related to an active NS transform fault. As only one block of tholeiitic pillow-lava has been recognized in this breccia we believe the ophiolitic extrusives were probably eroded before obduction. At Pozantı-Karsantı, the upper massive gabbros overlying the cumulates are visible, but without a dyke complex. Tholeiitic pillow-lavas are not preserved anywhere in the ophiolitic complexes of the Taurus Range. The absence of those lavas must have resulted from submarine erosion as tholeiitic basaltic fragments or pebbles in abundant Tertiary detrital sediments of the Taurides are lacking. The Baër-Bassit assemblage of NW Syria is the only Taurus-type massif where the tholeiitic volcanics are preserved. Thus, there is a generally increased depth of erosion of the ophiolites going from west to east. The older the age of deformation from the radiometric dates, the deeper the erosional level now is (Fig. 2a).

Troodos-type ophiolites

These ophiolites, represented by the Troodos and the Hatay massifs (Fig. 2b), exhibit a complete ophiolitic sequence from tectonites to pillow-lavas. There are great differences from the Taurus-type ophiolites: first their cumulate sequences are thinner compared to the Taurus-type, 1 km maximum compared to 3 to 5 km. In the Hatay massif the magmatic cumulates occur as mappable lenses on a kilometre scale; two of them have been recently reported (Whitechurch *et al.* 1981). Smewing (1975) and Smewing *et al.* (1975) describe small cumulate chambers in the Troodos Massif (see also Gass 1980). In contrast to the Taurus-type ophiolites, the cumulate sequences of the Hatay and Troodos ophiolites contain very few ultramafic rocks, these being mainly restricted to a few metres of troctolites at the base. Layered gabbros with some troctolites are the main lithology of the whole cumulate sequence. Another characteristic is that massive gabbros are as abundant as layered ones, indicating, following Dewey & Kidd (1977), that the magma chambers were as wide as they are high. Finally, the geochemistry of those cumulates is different from the Taurus-type ophiolites; they are richer in Fe (Parrot & Ricou 1976) and exhibit a cryptic enrichment of Fe in the Fe-Mg minerals from bottom to the top (Parrot & Verdoni 1976; Whitechurch, unpublished data). These small chambers of kilometre size were filled by more evolved liquids and this could be characteristic of slow-spreading (1–2 cm/year) zones.

Another difference is that these Troodos-type ophiolites do not contain isolated cross-cutting diabase dykes, and metamorphic soles are also probably absent. Unfortunately, the base of the Troodos Massif is not exposed. However, a metamorphic sole is not present at the base of the Hatay ophiolite. This is separated from the platform carbonates of the Amanos Mountains by an olistostrome containing ophiolitic and carbonate blocks within a serpentine matrix (Selçuk 1981; Tinkler *et al.* 1981).

The formation age of the Troodos igneous rocks is generally given as Campanian, between 85 Ma and 75 Ma (Mantis 1970; Vine *et al.* 1973). Desmet *et al.* (1978) obtained a K-Ar isochron age of 79 Ma for the sheeted dyke complex, and Delaloye & Desmet (1979) reported an isochron age of 75 Ma for the pillow-lavas, thus confirming previous age determinations. Delaloye *et al.* (1980) confirm these dates and add two gabbro ages of 80 Ma. The Hatay complex seems to be more difficult to date since the pillow-lavas given unexpectedly young ages (44–75 Ma, Delaloye *et al.* 1977), relative to the dyke complex which gives an isochron age of 94 Ma and, recently, several isochron ages scattered between 150 and 110 Ma (Delaloye *et al.* 1980). The authors themselves point out that their ^{40}Ar-^{40}K isochron 'can not be considered significant' because of insufficient measurements, and so we think these apparent ages must be taken with caution. The conclusion is that Troodos at least can safely be considered to be the youngest remaining ophiolite in the Eastern Mediterranean area.

Also in the Taurus massifs and in the Baër-Bassit area, sliced allochthonous units of Upper Triassic alkalic volcanics and associated terrigenous to pelagic sediments of Upper Triassic to Upper Cretaceous age are tectonically juxtaposed with the Troodos ophiolite as the Mamonia Nappes (Lapierre 1975), or the Mamonia Complex (Robertson & Woodcock 1979; Swarbrick & Robertson 1980; Swarbrick 1980). Great similarities with the Baër-Bassit and Alakır Çay volcano-sedimentary series have been pointed out by several authors (Lapierre & Parrot 1972; Juteau *et al.* 1973; Delaune-Mayere *et al.* 1977; Robertson & Woodcock 1979; Rocci *et al.* 1980; Delaune-Mayere, this volume).

Orientation of the spreading axes of the Neo-Tethys

Do the ophiolites record the orientation of the spreading axes in the Neo-Tethys? Two main criteria can be used to restore the direction of palaeo-spreading axes in ophiolitic complexes:

(1) The safest indication is without doubt the average direction of the sheeted dyke complex. Where well exposed, chilling statistics indicate the side of the ridge the ophiolite came from (Kidd & Cann 1974; Kidd 1977), although the validity of the method has been contested (e.g. Gass 1980).

(2) The plane perpendicular to the average high temperature mineral lineation in the harzburgitic tectonites can also be used; this lineation, following the model proposed by Nicolas & Poirier (1976), indicates the flow direction of the asthenospheric upper mantle beneath the spreading zones and is supposed to be at a high-angle to the ridge direction. Systematic measurements of the shear sense deduced from the mineral fabrics can also help to indicate the origin of the ophiolite block relative to the spreading axis.

The first criterion was used on the Troodos and Hatay ophiolites where the upper parts of the oceanic section are well preserved. The second has been used in the Taurus ophiolites where the upper parts have been eroded, and the tectonites are well exposed.

Using both the criteria, six ophiolitic massifs give surprisingly consistent results (Fig. 1). They all indicate, without exception, an inferred east-west trending spreading axis.

The average direction of the sheeted dyke complex of the Troodos massif has been pointed out by many authors (Wilson 1959; Bear 1960; Gass 1960; Kidd & Cann 1974; Kidd 1977) to be north-south. Taking into account the anticlockwise rotation of this massif (Moores & Vine 1971; Lauer & Barry 1976; Shelton & Gass 1980), probably during the Miocene (Shelton & Gass 1980), the dykes trended east-west in the Upper Cretaceous. The same east-west direction has been found in the sheeted dyke complex of the Hatay massif (Parrot 1973; Tinkler *et al.* 1980) and Baër-Bassit (Parrot 1977a). The chilled margin statistics, in the case of the Troodos, indicate that the ophiolite belonged to the southern flank of the spreading zone (Kidd & Cann 1974) and this would also be the case for the Hatay complex using the same criteria (Erendil, unpubl. data), thus strengthening comparison between the Hatay and Troodos massifs.

The average pole of the lineation in the harzburgitic tectonites, calculated from a great number of measurements from cliffs on the Antalya coast, indicates a N 70 E, spreading axis (Juteau *et al.* 1977). The same direction has been found by the same method in the Pozantı-Karsantı massif, more than 500 km to the east (Çakir 1978; Juteau 1980). To the west, measurements made recently in the northern Lycian Nappes in the Yesilova area also indicate an east-west spreading axis (Çapan, Juteau & Reuber, unpubl. data). Finally, this method has recently been applied to the tectonites of the Hatay massif (Whitechurch & Tekeli, unpubl. data), where the average mineral lineation is north-south, confirming the east-west spreading axis already indicated by the sheeted dyke complex. Also, measurements of the shear sense indicate that the Marmaris, Antalya and Pozantı-Karsantı massifs originated on the northern flank of the spreading zone, whereas the Hatay massif would have originated on the southern flank, confirming the indications from the dyke complex.

These data fit with the discovery of palaeotransform faults in several ophiolitic massifs; the famous Arakapas fault zone, in the southern part of the Troodos complex, is at right angles to the sheeted dyke complex and would have an original north-south direction (Simonian & Gass 1978; Spray & Roddick 1981). More recently evidence of transform mylonite shear zones at N 150 E have been discovered in the Antalya massif (Reuber 1982a and b, and this volume). This orientation fits well with the orientation of slabs of dyke complex rocks within the mega-breccia found in the same massif. This shear zone would have had a topographic expression where the breccia was deposited. The important north-south fault zone of Doğanbaba in the eastern Lycian Nappes described by Sarp (1976) also has the characteristics of a palaeotransform fault, including rotation of dykes, characteristic internal structure and the development of amphibolites with sub-horizontal lineations.

Thus, over more than one thousand kilometres, Cretaceous fragments of the Neo-Tethys lithosphere exhibit very consistent internal structures, indicating that they originated at east-west trending spreading axes, offset by north-south trending transform fault zones.

This consistency of internal structures between ophiolitic massifs separated by hundreds of kilometres also indicates that their tectonic emplacement and later Tertiary deformation did not result in significant rotation of the ophiolite blocks. Only the Troodos Massif is known to have rotated 90° anticlockwise during the Miocene after its Upper Cretaceous emplacement. Palaeomagnetic data obtained on the Taurus ophiolites and the Troodos Massif by Lauer (1981, 1982 and this volume) confirm these conclusions.

Oceanic history of the Neo-Tethys ocean

The data presented above, though incomplete and often fragmentary, show that the ophiolites record important magmatic, tectonic, metamorphic and sedimentary events during their oceanic evolution. We now discuss the chronology of these events.

Upper Triassic rifting history

The oldest magmatic events recorded in most of the studied areas are the Upper Triassic alkalic volcanic episodes, associated with terrigenous clastic to hemipelagic sediments. These submarine alkalic lavas belong to a vast alkalic volcanic province of Upper Triassic age, known all across the Eastern Mediterranean from the Pindos massif in Greece (Terry 1972; Dumont *et al.* 1972; Juteau *et al.* 1973; Lapierre & Rocci 1976; Marcoux 1976b) to the Baër-Bassit area in Syria, through the Mamonia complex in Cyprus and the whole Taurus belt. Several authors have suggested that this association of submarine alkalic lavas and both terrigenous and pelagic sediments, may witness rifting and early ocean-floor spreading in an intra-cratonic sea (Juteau 1970; Dumont *et al.* 1972; Juteau *et al.* 1973; Lapierre & Rocci 1976; Marcoux 1976b; Delaune-Mayere *et al.* 1977; Robertson & Woodcock 1980a, 1981a, b, c, 1982).

Unfortunately the precise nature of the crustal basement is unknown, since these units are clearly allochthonous. We want to point out, for example, that on the eastern flank of the Alakır Çay valley (Antalya), in the Gödene Zone of Robertson & Woodcock 1981, the Triassic volcanic sedimentary units lie tectonically on the underlying peridotites and tholeiitic gabbros of Cretaceous age and so can not thus represent an original basement. It is often argued that these Upper Triassic submarine lavas could represent some kind of 'Red Sea stage'. Recent studies, however, show that the basalts from the Red Sea bottom are typical oceanic tholeiites (Matveenkov *et al.* 1982; Eissen 1982; Juteau *et al.* 1983), and that transitional to alkaline basalts are found only on the continental margins or in some islands of the Red Sea (Coleman *et al.* 1975; Gass *et al.* 1973). Hence our present opinion is that the Upper Triassic volcanics probably represent a 'pre-Red Sea' stage, with an attenuated continental crust. They correspond without doubt to a phase of rifting of the Arabian platform (Delaune-Mayere *et al.* 1976), but the existence of a true oceanic lithosphere in Triassic time remains highly speculative. Woodcock & Robertson (1982, this volume) propose that the Antalya complex, analogous to the Mesozoic of Belize in the Caribbean, underwent a two-phase evolution, Triassic rifting followed by genesis later in the Mesozoic of a small ocean basin in which strike-slip faulting played an important part both in the genesis and the emplacement. During the rifting micro-continental slivers were rifted-off the main margin.

Jurassic (?) to Cretaceous formation of Neo-Tethys oceanic lithosphere

There is a total lack of information concerning the Jurassic and Early Cretaceous ocean genesis, although the platform sediments of the marginal units show a switch to deep water pelagic carbonate deposition consistent with rapid subsidence (Delaune-Mayere *et al.* 1976; Robertson & Woodcock 1981a, b, c, 1982).

In the Taurus belt, the primary ages of the ophiolites are still uncertain, but a Jurassic age can not be excluded. For the moment, the most reliable ages obtained by K-Ar dating are those of the metamorphic soles. As this metamorphic event affected the overlying ophiolites, at least in the Pozantı-Karsantı massif, the ophiolites can not be younger than their metamorphic soles. By analogy with the Semail nappe in Oman, where U-Pb ages on zircons from plagiogranites are only about 10 Ma older than the basal metamorphic sheets (Ghent & Stout 1981), we propose a similar time span for the Taurus-type ophiolites, i.e. a minimum age of 98 Ma for the Baër-Bassit, 114 Ma for the Lycian ophiolites, and around 100–105 Ma for the other Tauric ophiolites.

If the apparent ages obtained on hornblende gabbros or plagiogranites of the plutonic suite are true cooling ages for the ophiolites this would lead to obvious contradictions. In the Pozantı-Karsantı massif for example, the hornblende gabbros and plagiogranites yield an apparent age of about 85 Ma, whereas the metamorphic soles give 94 Ma. The ophiolite would be younger than its metamorphic sole, which is nonsense, and so we feel that apparent cooling ages of the plutonic suite should be used with great care until confirmed by other methods, for example U-Pb dating of separated zircons from plagiogranites, as presently in progress. The available age data however have been used to support formation of the Taurus ophiolites in one or more small oceanic basins during Late Cretaceous time (Yilmaz & Maxwell 1981; Robertson & Woodcock 1981, and this volume).

The age of the Hatay ophiolite is difficult to interpret. Delaloye *et al.* (1980) give very different ages for pillow-lavas and for the plutonic suite. We are sceptical about the Late Jurassic age (150–110 Ma) they give for the dyke complex and most of the gabbros analysed, because of the scatter of data and the absence of consideration of the possibility of a radiogenic Ar excess in the dykes. The close analogy of Hatay with Troodos, whose Campanian age is now well established, suggests that both ophiolites could have a similar age, but we recognize that the question remains open. Whatever the case, the Campanian Troodos ophiolite is the youngest fragment of the Neo-Tethys oceanic lithosphere, and probably represents the latest ocean floor spreading in the Mesozoic East Mediterranean.

Summarizing, after the Upper Triassic intracontinental rifting episode, we have no record of any ocean-floor spreading in the area before late in the Lower Cretaceous (Aptian), basing our age estimations of the Taurus ophiolites on their metamorphic soles. Assuming that, as in Oman, the 'Taurus type' ophiolites are about 10 Ma older than their metamorphic sole, we get a minimum time span of 40 Ma for ocean floor spreading recorded in the different ophiolites, from the oldest Taurus ophiolite (Lycian nappes, c. 115 Ma) to the younger Troodos ophiolite (c. 75 Ma).

A simple calculation indicates that during that time span, about 1600 km of oceanic lithosphere could have been created: 1200 km created by the 'Tauric' spreading centres during the first 20 Ma with a rather fast spreading rate (say 6 cm/year) corresponding to the dimensions of the cumulates chambers (see above), and 400 km created by the slower Troodos (and Hatay?) spreading centre during the following 20 Ma. Though highly speculative, this gives an idea of the order of magnitude of the oceanic lithosphere which could have been created during these 40 Ma.

Now we must explain the absence of Jurassic and Lower Cretaceous ophiolites. Two interpretations are possible. In the first, no sea floor spreading occurred between Late Triassic and late Lower Cretaceous. In this case after a long quiescence, the aborted Upper Triassic rift was reactivated and evolved towards a Red Sea type ocean during Upper Cretaceous. In the second interpretation, ocean floor spreading began earlier, perhaps as soon as the Lower Jurassic and did not stop until Campanian period. In this case the missing oceanic lithosphere would have disappeared by subduction, and only some of the youngest parts of the Neo-Tethys oceanic lithosphere would have been preserved and tectonically juxtaposed with the older Upper Triassic rocks. We feel that the field data support the second interpretation, but in the present state of knowledge the first possibility can not be discounted.

Chronology of intra-oceanic events

Three main types of oceanic events affected the ophiolites discussed here: transform faulting, intra-oceanic slicing and diabase dyke intrusions.

Transform faulting

This is recorded in several ophiolite massifs, in the Arakapas fault zone of the Troodos Massif, in the Tekirova complex (Antalya) (Reuber, this volume) and in the Yesilova area, Lycian ophiolites (Fig. 1). We agree with Woodcock and Robertson (1982) that the amphibolites and associated greenschists slivers occurring in the Alakır Çay valley could have been generated by strike-slip faulting. The 95 Ma age of these amphibolites (Thuizat *et al.* 1981) implies earlier oceanic crust formation.

Intra-oceanic slicing

This corresponds to formation of the metamorphic soles of the Taurus-type ophiolites. As we have seen the age of the main metamorphic episode recorded in these soles varies from Albian to Santonian (104–88 Ma), and is thus distinctly older than the time of the first emplacement of the ophiolites onto the continental margin, around 65 Ma. The time span between the two events is ca. 25–30 ma, confirming that the metamorphic soles were generated during the oceanic evolution of the Tauric-type ophiolites prior to initial obduction.

Thuizat *et al.* (1981) also found the same age distinction in the somewhat older Pindos ophiolites (Greece), in agreement with Spray & Roddick (1980). Very similar ages were also obtained in the metamophric sole of the Semail Nappe (Oman) (Ghent & Stout 1981; Searle & Malpas 1980). These data strengthen the idea that the main recrystallization event in the metamorphic slices occurred in aborted subduction zones during periods of intense slicing of oceanic lithopshere, as suggested by Parrot & Whitechurch (1978) and recently discussed by Nicolas & Le Pichon (1980).

Another point of interest is that the metamorphic soles become younger from west (Lycian nappes, 104 Ma) to east (Baër-Bassit, 88 Ma, Fig. 1). This is in good agreement with the dates obtained further west in Greece, where the metamorphic soles are even older, of

Middle Jurassic age, and also further east in Oman, where their age is similar to the Baër-Bassit metamorphics. This eastward age migration of the oceanic slicing in the Neo-Tethys during the Cretaceous is probably the result of the north-eastward drift of the Arabo-African continent during that period, and reflects the progressive eastward migration of the beginning of compressive movements in the Neo-Tethys. The present erosion level of the Taurus-type ophiolites shows a parallel evolution (Fig. 2); the older the slicing, the deeper the erosion level. We suggest that this could be partly due to submarine erosion beginning early, soon after the intraoceanic slicing. Such processes have been recently well documented on the Gorringe Bank, an old piece of the Atlantic oceanic floor which has been sliced and uplifted along the Azores-Gibraltar plate boundary (Auzende *et al.* 1978; La Gabrielle & Auzende 1982; La Gabrielle *et al.* 1981).

Diabase dyke intrusions

This a distinctive feature of the Taurus-type ophiolites. The low-K tholeiitic diabase dykes intersect tectonites and cumulates in each of these massifs as swarms of isolated dykes, but never form dyke complexes. Some of them also intersect the sub-ophiolite metamorphic rocks and in the Pozantı-Karsantı massif one of them even intersects the tectonic contact between the amphibolites and the overlying harzburgites. K-Ar dating of these dykes in several massifs gives a Campanian age. Since the autochthonous platform limestones under the ophiolite nappes do not contain diabase dykes, they must have been injected before obduction. On the other hand, as the dykes are fine grained, with chilled margins and cut tectonites, cumulates and metamorphic soles, the time of their injection must have been late relative to accretion. They were injected into an already cold oceanic crust after oceanic slicing had occurred. We conclude that this important magmatic event took place at the end of the oceanic genesis of the Neo-Tethys, after intra-oceanic slicing had taken place, during a period of intense fracturing and extension of a rather cold ocean floor. At the same time the Troodos spreading zone was active, implying that the Taurus-type ophiolites represent older ocean floor than the Troodos-type ones.

East Mediterranean ophiolites: one or several oceanic basins?

On a broader scale, we are faced with the difficult problem of the margins to the East Mediterranean ophiolites. The North-Anatolian ophiolite belt, with its high-pressure blueschists and calc-alkaline intrusions, obviously represents an important suture zone marking closure of a Mesozoic ocean (Fourquin 1975; Bergougnan 1975; Juteau 1979, 1980; Şengör & Yilmaz 1981; Okay, this volume), but it is the only one?

Interpreting the 'Taurus Limestone Axis' as an integral part of the Arabo-African continental margin, now exposed as tectonic windows below or behind the Taurus ophiolite nappes, Ricou *et al.* (1975) proposed that the root of the radiolaritic-ophiolitic nappes was north of the Arabo-African continental block, hence north of the 'Taurus Limestone Axis'. As some windows through to this axis, such as the Munzur Dağ are located immediately south of the North Anatolian ophiolitic suture (Bergougnan 1975), they concluded that the 'periarabic' radiolaritic-ophiolitic nappes were thrust southwards from the North-Anatolian suture. In this concept, there was only one ophiolitic suture for all the Eastern Mediterranean ophiolites which was located in North Anatolia.

The main difficulty with this hypothesis is the large horizontal displacements implied for the southernmost massifs: Hatay, Baër-Bassit and Troodos are separated by hundreds of kilometers from the North Anatolian suture. This difficulty would disappear if a second suture zone existed closer to the southern ophiolite belt. Juteau (1980) has proposed that the northern margin of the Taurus range, for instance north of the Bolkar Dağ (Alihoca complex) is a possible candidate.

Alternatively, models involving less allochthoneity have been developed for the East Mediterranean ophiolites. Brinkmann, (1972), 1976 p. 122–126), for instance, proposed that the present Turkish ophiolite belts represent ancient rifts due to crustal extension and vertical uprise of ultramafic diapirs. This hypothesis can not, of course, accommodate the complex tectonic history of the ophiolite belts, or explain the development of high-pressure blueschist metamorphism in the North Anatolian ophiolite belt. Gass and Masson-Smith (1963) first proposed that the Troodos ophiolite is in effect autochthonous, and several authors have suggested the existence of a southern oceanic basin so needing only short distance displacements to account for the Antalya, Troodos, Hatay and Baër-Bassit ophiolites ('Pamphylian basin' of Dumont *et al.* 1972; 'Troodos ocean' of Robertson & Woodcock 1980; 'Mesoge' of Biju-Duval *et al.* 1977; Southern Neo-Tethys of Şengör & Yilmaz 1981). During the last few years, the Antalya area has become a testing ground for these two opposite conceptions, Robertson and Woodcock (e.g., this volume), Waldron (this volume) and Hayward (this volume) favouring derivation from a southerly oceanic basin with less displacement than Marcoux and Ricou (e.g. Ricou *et al.*, this volume), who argue for long-distance transport from a single northern ocean basin.

We think that the remarkable consistency of internal structures of the East Mediterranean ophiolites especially the common east-west orientation of their inferred palaeo-spreading axes, is a strong argument in favour of a single oceanic domain. Most of these complexes are associated with similar Upper Triassic rift-related rocks and all of them were obducted at the same time late in the Upper Cretaceous. The progressive eastward migration of the age of the metamorphic soles in the Taurus-type ophiolites, regardless of their present position with respect to the Taurus Limestone Axis, also favours a common origin. Consequently, we are reluctant to dissociate the origin of the Antalya, Cyprus and Hatay ophiolites from the more internal Lycian and Beyşehir-Hoyran nappe complexes, as proposed by Robertson & Woodcock (1980). Our point of view is that the Taurus-type ophiolites were generated some time before the 'Troodos-type' ones in the same oceanic basin, probably at the same spreading axes after a change in spreading rate.

It is difficult to choose between a 'super Troodos ocean' branching from the northern Neo-Tethys ocean somewhere to the east, and an ultra-northern origin for all the ophiolites in the North Anatolian suture. Palaeomagnetic data obtained by Lauer (1981) on the ophiolites of Antalya, Pozantı-Karsantı and Troodos are fairly concordant: they indicate a common initial position for these massifs, close to a 17°N palaeo-latitude. This is an argument for a common origin in the same basin. Linear interpolations deduced from Lauer's data concerning the Bey Dağları platform sediments indicate that, at the same time, this part of the Taurus Limestone Axis lay close to 12°N. Lauer's data thus suggest that the Antalya, Pozantı-Karsantı and Troodos ophiolites were initially located in a same oceanic basin north of the Taurus Limestone Axis.

Summarizing, we consider that the single basin hypothesis is the more likely to explain the data available for the East Mediterranean ophiolites and their associated allochthonous slices.

Conclusions

Most of the oceanic lithosphere created in the Neo-Tethys ocean during Mesozoic time has probably disappeared by subduction under the Pontides or some other fossil subduction zone. The ophiolite complexes examined in this paper represent fragments probably of the youngest parts of the original oceanic lithosphere.

Our main conclusions are:

(1) Some of the ophiolites (the Taurus-type) are somewhat older than others (the Troodos-type). When the Troodos (and Hatay?) spreading centres were still active, between ca. 85 and 75 Ma, the Taurus-type ophiolite were already cold, sliced (metamorphic soles crystallized between 104 and 88 Ma), or affected by transform faulting (94 Ma for the amphibolites of Antalya). Also, the extrusion of the Troodos pillow-lavas during the Campanian is synchronous with the intrusion of numerous isolated diabase dykes in the already sliced Taurus-type ophiolites. We feel that in spite of great difficulty in dating the primary cooling ages of the ophiolites, the available data support this distinction. Particularly, we are confident that the K-Ar ages obtained on the metamorphic amphibolites and associated rocks are reliable and do constrain the primary cooling ages of the Taurus-type ophiolites to be somewhat older than their metamorphic soles.

(2) If there are good reasons for a chronological distinction between these two groups, we see no compelling reason to dissociate their origin. On the contrary, their remarkable consistency of internal structures is a strong argument for a common origin in the same ocean basin. Palaeomagnetic data are rather in favour of a northern origin of the whole set of ophiolite nappes with respect to the Taurus Limestone Axis, but many more data are needed for a convincing demonstration. Alternatively, a southern branch of the Neo-Tethys ocean ('Troodos ocean' or 'Mesogea') could also have existed, but the precise location of its northern active margins remains unclear (Michard *et al.*, this volume).

(3) This common ocean had an east-west spreading zone offset by north-south fracture zones, as already proposed by Robertson and Woodcock (1980) for their 'Troodos ocean'. Admitting points (1) and (2), we get a minimum time span of 40 Ma of recorded Upper Cretaceous seafloor spreading, during which about 1600 km of oceanic lithosphere could have been

created. The Neo-Tethys ocean however almost certainly never reached that width, because of slicing and subduction.

ACKNOWLEDGEMENTS: We wish to thank the General Direction of the M. T. A. Enstitüsü, Ankara, and Professor G. Ataman from Haccettepe University, Ankara, for their constant help during our fieldwork in Turkey between 1976 and 1981. Many thanks also are due to our Turkish colleagues U. Çapan, E. Yazgan, O. Tekeli, and to our former students F. Bingöl, U. Çakir, J. Dubessy, A. Blanco-Sanchez, P. Malley and I. Reuber. This work was financially supported by the French I.N.A.G. (Institut National d'Astronomie et de Géophysique, A.T.P. 'Géodynamique 1'). We wish to express our gratitude to the reviewers of our initial manuscript Prof. I. G. Gass and P. Yilmaz, and to the editors; A. H. F. Robertson not only helped us to produce more readable English, but also compelled us to a more rigorous presentation of our data and interpretations.

References

AUZENDE, J. M., OLIVET, J. L., CHARVET, J., LE LANN, A., LE PICHON, X., MONTEIRO, J. H., NICOLAS, A. & RIBEIRO, A. 1977. Sampling and observation of oceanic mantle and crust on Gorringe Bank. *Nature, London*, **273,** 45–49.

BEAR, L. M. 1960. The geology and mineral resources of the Akaki-Lythrodonda area Cyprus. *Geol. Surv. Dept. Mem.* **3,** 122 pp.

BERGOUGNAN, H. 1975. Relations entre les édifices pontique et taurique dans le Nord-Est de l'Anatolie. *Bull. Soc. géol. Fr.* **17,** 1045–57.

BIJU-DUVAL, B., LAPIERRE, H. & LETOUZEY, J. 1976. Is the Troodos Massif (Cyprus) allochthonous? *Bull. Soc. géol. Fr.* **18,** 1347–56.

——, DERCOURT, J. & LE PICHON, X. 1977. From the Tethys ocean to the Mediterranean seas: a plate tectonic model of the evolution of the western alpine System. *In*: BIJU-DUVAL, B. & MONTADERT, L. (eds), *Intern. Symp. on the structural history of the Mediterranean basins, Split, Oct. 1976*, Ed. Technip, Paris, 143–64.

BINGÖL, A. F. 1978. *Pétrologie du massif ophiolitique de Pozanti-Karsanti (Taurus cilicien, Turquie); étude de la partie orientale.* Thesis 3d cycle (unpubl.), Louis Pasteur Univ., Strasbourg, France, 227 pp.

BRINKMANN, R. 1972. Mesozoic troughs and crustal structure in Anatolia. *Bull. geol Soc. Am.* **83,** 819–26.

—— 1976. *Geology of Turkey.* Elsevier, New-York, 158 pp.

BRUNN, J. H. 1974. Le problème de l'origine des nappes et de leurs translations dans les Taurides occidentales. *Bull. Soc. géol. Fr.* **16,** 101–06.

——, GRACIANSKY, P. C. (DE), GUTNIC, M., JUTEAU, T., LEFÈVRE, R., MARCOUX, J., MONOD, O. & POISSON, A. 1970. Structures majeures et corrélations stratigraphiques dans les taurides occidentales. *Bull. Soc. géol. Fr.* **12,** 515–66.

——, DUMONT, J. F, GRACIANSKY, P. C. (DE), GUTNIC, M., JUTEAU, T., MARCOUX, J., MONOD, O. & POISSON, A. 1971. Outline of the Geology of the western Taurids, *In*: A. S. CAMPBELL (eds), *Geology and History of Turkey*, Petroleum Exploration Society of Libya, Tripoli, 225–55.

——, ARGYRIADIS, I., RICOU, L. E., POISSON, A., MARCOUX, J. & GRACIANKSY, P. C. (DE). 1976. Eléments majeurs de liaison entre Taurides et Hellénides. *Bull. Soc. géol. Fr.* **18,** 481–97.

ÇAKIR, Ü. 1978. *Pétrologie du massif ophiolitique de Pozantı-Karsantı (Taurus cilcien, Turquie): étude de la partie centrale.* Thesis 3d cycle (unpubl.), Louis Pasteur Univ., Strasbourg, France, 251 pp.

——, JUTEAU, T. & WHITECHURCH, H. 1978. Nouvelles preuves de l'écaillage intraocéanique précoce des ophiolites téthysiennes: les roches métamorphiques infra-péridotitiques du massif de Pozanti-Karsanti (Turquie). *Bull. Soc. géol. Fr.* **20,** 61–70.

CATAKLI, A., ROCCI, G. & LAPIERRE H. 1977. Nouvelles donnés pétrographiques et structurales sur l'assemblage ophiolitique et les roches associées de la partie occidentale du massif de Pozanti (Turquie). *5th Annual Meeting Earth Sciences, Soc. geol. France, Rennes, Abstr.*, **139.**

COLEMAN, R. G. 1977. *Ophiolites.* Springer-Verlag, New York, 229 pp.

——, FLECK, R. J., HEDGE, C. E., & GHENT, E. D. 1975. The volcanic rocks of south-west Saudi Arabia and the opening of the Red Sea. *U.S. Geol. Surv. Saudi Arabian Project Report*, **194,** 60 pp.

DELALOYE, M., VUAGNAT, M. & WAGNER, J. J. 1977. K-Ar ages from the Kizil Dag ophiolitic complex (Hatay, Turkey) and their interpretation. *In*: BIJU-DUVAL, B. & MONTADERT, L. eds, *Intern. Symp. on the Structural History of the Mediterranean Basins, Split, Oct. 1976*, Technip, Paris, 73–78.

—— & DESMET, A. 1979. Nouvelles données radiométriques sur les pillow lavas du Troodos. *C.R. Acad. Sci. Paris*, **288,** 461–64.

——, SOUZA, H. (DE), WAGNER, J. J. & HEDLEY, I. 1980. Isotopic ages on ophiolites from the eastern Mediterranean. *In*: PANAYIOTOU. A. (ed.). *Ophiolites,* Proc. Int. Oph. Symp. Cyprus 1979, 80–81.

DELAUNE-MAYERE, M., MARCOUX, J., PARROT, J. F. & POISSON, A. 1977. Modèle d'évolution mésozoïque de la paléomarge téthysienne au niveau des nappes radiolaritiques et ophiolitiques du Taurus lycien, d'Antalya et du Baër-Bassit. *In*: BIJU-DUVAL, B. & MONTADERT, L., (eds.), *Intern. Symp. on the Structural History of the Mediterranean Basins, Split, Oct. 1976*, Technip, Paris, 79–94.

DESMET, A., LAPIERRE, H., ROCCI, G., GAGNY, C. C., PARROT, J. F. & DELALOYE, M. 1978. Constitution and significance of the Troodos sheeted complex. *Nature, London,* **273,** 527–530.

DEWEY, J. F. & KIDD, W. S. F. 1977. Geometry of plate accretion. *Bull. Soc. geol. Am.* **88,** 960–68.

Dubertret, L. 1939. Sur la genèse et l'âge des roches vertes syriennes. *C.R. Acad. Sci. Paris*, **209,** 763 pp.

—— 1955. Géologie des roches vertes du NW de la Syrie et du Hatay (Turquie). *Notes et Mém. Moy. Orient.* **6,** 13–179.

Dumont, J. F., Gutnic, M., Marcoux, J., Monod, O. & Poisson, A. 1972. Le Trias des Taurides occidentales (Turquie). Définition du bassin pamphylien: un nouveau domaine à ophiolites à la marge externe de la chaîne taurique. *Z. Deutsch. Geol. Ges.* **123,** 385–409.

Eissen, J. P. 1982. *Pétrologie comparée de basaltes de différentes segments de zones d'accrétion océaniques à taux d'accrétion variés (Mer Rouge, Atlantique, Pacifique).* Thesis 3d cycle (unpubl.), Louis Pasteur Univ. Strasbourg, France, 202 pp.

Fourquin, C. 1975. L'Antolie du Nord-Ouest, marge méridionale du continent européen, histoire paléogéographique, tectonique et magmatique durant le Secondaire et le Tertiaire. *Bull. Soc. géol. Fr.* **17,** 1065–69.

Gass, I. G. 1960. The geology and mineral resources of the Dhali area. *Cyprus Geol. Surv. Dept. Mem.* **4,** 116 pp.

—— 1968. Is the Troodos massif of Cyprus a fragment of Mesozoic ocean floor? *Nature, London*, **220,** 39–42.

—— 1980. The Troodos massif: its role in the unravelling of the ophiolite problem and its significance in the understanding of constructive plate margin processes. *In*: Panayiotou. A. (ed.). *Ophiolites*, Proc. Int. Oph. Symp. Cyprus 1979, 23–35.

—— & Masson-Smith, D. 1963. The geology and gravity anomalies of the Troodos massif, Cyprus. *Roy. Soc. London Phil. Trans.* **A255,** 417–67.

——, Mallick, D. I. J. & Cox, K. G. 1973. Volcanic islands of the Red Sea. *J. geol. Soc. Lond.*, **129,** 275–310.

Ghent, E. D. & Stout, M. Z. 1981. Metamorphism at the base of the Samail Ophiolite, Southeastern Oman Mountains. *J. geophys. Res.* **88,** 2557–71.

Graciansky, P. C. (de) 1972. *Recherches géologiques dans le Taurus lycien.* Thesis (unpubl.), Paris XI-Orsay Univ., France, 571 pp.

Guérin-Franiatte, S. & Juteau, T. 1970. Découverte de blocs calcaires à Ammonites et Halobies triasiques dans les pillow-lavas de Sayrun, Province d'Antalya (Turquie). *C.R. Acad. Sci. Paris*, **270,** 2897–99.

Juteau, T. 1968. Commentaire de la carte géologique des ophiolite de la région de Kumluca (Taurus lycien, Turquie méridionale): cadre structural, modes de gisement et description des principaux faciès du cortège ophiolitique. *M.T.A. Enst. Bull. Ankara*, **70,** 70–91.

—— 1970. Pétrogenèse des ophiolites des Nappes d'Antalya (Taurus lycien oriental, Turquie). Leur liaison avec une phase d'expansion océanique active au Trias supérieur. *Sciences de la Terre, Nancy*, **15,** 265–88.

—— 1975. Les ophiolites des Nappes d'Antalya (Taurides occidentales, Turquie). Pétrologie d'un fragment de l'ancienne croûte océanique téthysienne. *Sciences de la Terre, Nancy, Mém.* **32,** 692 pp.

—— 1979. Ophiolites des Taurides: Essai sur leur histoire océanique. *Revue de Géologie dynamique et de Géographie physique*, **21,** 191–14.

—— 1980. Ophiolites of Turkey. *Ofioliti*, spec. issue: *Tethyan ophiolites*, G. Rocci (ed.), **2,** 199–37.

—— & Marcoux, J. 1973. Un exemple du volcanisme sous-marin au Trias supérieur: Le stratovolcan du Karadere-Çalbalı Dağ. Etude des manifestations éruptives et de leur contexte sédimentaire. *1st Annual Meeting Earth Sciences, Soc. geol. France, Paris, Abstr. volume*, 238.

—— & Whitechurch, H. 1980. The magmatic cumulates of Antalya (Turkey): Evidence of multiple intrusions in an ophiolitic magma chamber. *In*: Panayiotou. A. (ed.). *Ophiolites*, Proc. Int. Oph. Symp. Cyprus 1979, 377–91.

——, Lapierre, H., Nicolas, A., Parrot, J. F., Ricou, L. E., Rocci, G. & Rollet, M. 1973. Idées actuelles sur la constitution, l'origine et l'évolution des assemblages ophiolitiques mésogéens. *Bull. Soc. géol. Fr.* **15,** 478–93.

——, Nicolas, A., Dubessy, J., Fruchard, J. C. & Bouchez, J. L. 1977. The Antalya ophiolite complex (Western Taurides, Turkey): a structural model for an oceanic ridge. *Bull. geol. Soc. Am.* **88,** 1740–48.

——, Eissen, J. P., Monin, A. S., Zonenshain, L. P., Sorokhtin, O. G., Matveenkov, V. & Almukhamedov, I. 1983. Structure et pétrologie du rift axial de la Mer Rouge vers 18°Nord. *In*: *Rifts et Fossés Anciens, Bull. Centre Rech. Explor. Prod. Elf-Aquitaine*, Paris, C.N.R.S. Colloquium, Marseille 1982, 217–31.

Kazmin, V. G. & Kulakov, V. V. 1968. *Geological map of Syria.* Report on the geological survey. Techno. Export, Moscou, 124 pp.

Kidd, R. G. W. 1977. A model for the process of formation of the upper oceanic crust. *Geophys. J. Roy. astr. Soc.* **50,** 149–83.

—— & Cann, J. R. 1974. Chilling statistics indicate an ocean-floor spreading origin for the Troodos Complex, Cyprus. *Earth planet. Sci. Lett.* **24,** 151–55.

LaGabrille, Y., & Auzende, J. M. 1982. Active *in situ* disaggregation of oceanic crust and mantle on Gorringe Bank: analogy with ophiolitic massifs. *Nature, London*, **297,** 490–93.

——, ——, Cornen, G., Juteau, T., Lensch, G., Mevel, C., Nicolas, A., Prichard, H., Ribeiro, A., & Vanney, J. R. 1981. Sable, cailloutis et blocs de gabbros en domaine sous-marin profond. Evidence d'un démantèlement et d'un transfert actif d'après des observations par submersible sur le banc de Gorringe (S-W Portugal). *C.R. Acad. Sci., Paris*, **293,** 827–32.

Lapierre, H. 1975. Les formations sédimentaires et éruptives des nappes de Mamonia et leur relation avec le massif du Troodos (Chypre occidentale). *Mem. Soc. géol. France*, **123,** 322 pp.

—— & Parrot, J. F. 1972. Identité géologique des régions de Paphos (Chypre) et du Baër-Bassit (Syrie). *C.R. Acad. Sci., Paris*, **274,** 1999–2002.

—— & Rocci, G. 1976. Le volcanisme alcalin du sud-ouest de Chypre et le problème de l'ouverture des régions téthysiennes au Trias. *Tectonophysics*, **30**, 299–313.

Lauer, J. P. 1981. *L'évolution géodynamique de la Turquie et de Chypre déduite de l'étude paléomagnétique.* Thesis (unpubl.), Louis Pasteur Univ., Strasbourg, France, 292 pp.

—— & Barry, P. 1979. Etude paléomagnétique des ophiolites du Troodos (Chypre). *C.R. Acad. Sci., Paris*, **289**, 977–80.

Lefèvre, R. 1967. Un nouvel élément dans la géologie du Taurus lycien: les nappes d'Antalya (Turquie). *C.R. Acad. Sci., Paris*, **265**, 1365–68.

Malpas, J. 1979. The dynamothermal aureole of the Bay of Islands ophiolite suite. *Can. J. Earth. Sci.* **16**, 2086–101.

Mantis, M. 1970. Upper Cretaceous—Tertiary Foraminiferal zones in Cyprus. *Cyprus. Res. Centre* **111**, 227–41.

Marcoux, J. 1970. Age carnien de termes effusifs du cortège ophiolitique des nappes d'Antalya (Taurus lycien occidental. Turquie). *C.R. Acad. Sci. Fr.* **271**, 285–87.

—— 1976a. Les séries des nappes à radiolarites et ophiolites d'Antalya (Turquie): homologies et signification probable. *Bull. Soc. géol. Fr.* **18**, 511–12.

—— 1976b. La fracturation de la plate-forme scythienne et les nappes initiaux du développement de la Téthys alpine en Méditerranée. *4th Annual Meeting Earth Sciences, Soc. geol. France, Paris, Abstr. volume*, 285 pp.

Metz, K. 1956. Ein Beitrag zur Kennits des gebirgshauer an Aladag und Karanfildag und ihres Westrandes (Kilikisher Taurus). *M.T.A. Enst. Bull.*, Ankara, no. 48, 68–78.

Monod, O. 1977. *Recherches géologiques dans le Taurus occidental au Sud de Beyşehir (Turquie).* Thesis (unpubl.), Paris XI-Orsay Univ., France, 511 pp.

——, Marcoux, J., Poisson, A. & Dumont, J. F. 1974. Le domainc d'Analya, témoin de la fracturation de la plate-forme africaine au cours du Trias. *Bull. Soc. géol. France*, **16**, 116–25.

Montigny, R. 1975. Géochimie comparée des cortèges de roches océaniques et ophiolitiques. Problèmes de leur genèse. thesis (unpubl.), Paris VII, Univ., France, 288 pp.

Montigny, R., Whitechurch, H., Reuber, I., Thuizat, R. & Juteau, T. K-Ar investigation of the Antalya ophiolites. Geological implications. *Sciences Géologiques, Strasbourg*, (in press).

Moores, E. M. & Vine, F. J. 1971. Troodos Massif, Cyprus and other ophiolites as oceanic crust: evaluation and implications. *Roy. Soc. London Philos. Trans.* **268**, 443–66.

Nicolas, A. & Poirier, J. P. 1976. *Crystalline Plasticity and Solid State Flow in Metamorphic Rocks.* Wiley, London, 444 pp.

—— & Le Pichon, X. 1980. Thrusting of young lithosphere in subduction zones, with special reference to structures in ophiolitic peridotites. *Earth Planet. Sci. Lett.* **46**, 397–06.

Nisbet, E. G. & Fowler, C. M. 1978. The Mid-Atlantic Ridge at 37 and 45°N: some geophysical and petrological constraints. *Geophys. J.R. astr. Soc.* **54**, 631–60.

Pallister, J. S. & Hopson, C. A. 1981. Samail Ophiolite Plutonic Suite: Field Relations, Phase Variation, Cryptic Variation and Layering, and a Model of a Spreading Ridge Magma Chamber. *J. Geophys. Res.* **86**, 2593–644.

Parrot, J. F. 1973. Pétrologie de la Coupe Djebel Moussa Massif Basique-Ultrabasique du Kizil Dağ (Hatay, Turquie). *Sciences de la Terre, Nancy*, **18**, 143–172.

—— 1974a. Les Différentes manifestations effusives de la région ophiolitique du Baër-Bassit (Nord-ouest de la Syrie): comparaison pétrographique et géochimique. *C.R. Acad. Sci., Paris*, **279**, 627–30.

—— 1974b. L'assemblage ophiolitique du Baër-Bassit (Nord-ouest de la Syrie): Etude pétrogaphique et géochimique du complexe filonien, des laves en coussins qui lui sont associées, et d'une partie des formations effusives du volcano-sédimentaire. *Cah. ORSTOM, sér, Géol.* **6**, 97–126.

—— 1977a. Assemblage ophiolitique du Baër-Bassit et termes effusifs du volcano-sédimentaire. Pétrologie d'un fragment de la croûte téthysienne charriée sur la plate-forme syrienne. *Trav. et Doc. ORSTOM Paris*, **72**, 333 pp.

—— 1977b. Ophiolites du NW syrien et évolution de la croûte océanique téthysienne au cours du Mésozoïque. *Tectonophysics*, **41**, 251–69.

—— & Ricou, L. E. 1976. Evolution des assemblages ophiolitiques au cours de l'expansion océanique *Cah. ORSTOM Paris, sér. Géol.* **8**, 49–68.

—— & Vatin-Perignon, N. 1974. Répartition de quelques éléments en trace dans les différentes roches effusives de la région ophiolitique du nord-ouest syrien. *Cah. ORSTOM Paris, sér. Géol.* **6**, 185–26.

—— & Verdoni, P. A. 1976. Conditions de formations de deux assemblages ophiolitiques méditerranéens (Pinde et Hatay) d'après l'étude des minéaux constitutifs. *Cah. ORSTOM Paris, sér. Géol.* **8**, 69–94.

—— & Whitechurch, H. 1978. Subductions antérieures au charriage nord-sud de la croûte téthysienne: facteur de métamorphisme de séries sédimentaires et volcaniques liées aux assemblages ophiolitiques syro-turcs, en schistes verts et amphibolites. *Rev. Géogr. phys. Géol. dyn.* **20**, 153–170.

Poisson, A. 1968. L'unité inférieure du Domuz Dăg (Taurus lycien, Turquie), série sédimentaire avec intercalation de coulées sous-marines en coussins. *M.T.A. Enst. Bull.*, Ankara, no. 70, 100–105.

—— 1977. *Recherches géologiques dans les Taurides occidentales, Turquie.* Thesis (unpubl.), Paris XI-Orsay Univ., France, 790 pp.

Reuber, I. 1982a. *Pétrologie du cortège filonien aux tectonites et cumulats dans quelques massifs. ophiolitiques tauriques.* Thesis 3d cycle (unpubl.), Louis Pasteur Univ., Strasbourg, France, 246 pp.

—— 1982b. Mylonitic shear zones of millimetric to kilometric scale related to a transform fault in an ophiolitic complex (Antalya, Turkey). *Int. conf. on Planar and Linear fabrics of deformed rocks, Mitt. ETH Zurich*, 239a, 233–236.

Ricou, L. E. 1971. Le croissant ophiolitique péri-arabe, une ceinture de nappes mises en place au Crétacé superieur. *Rev. Geogr. phys. Geol. Dynam.* **18,** 327–349.

—— & Marcoux, J. 1980. Organisation générale et rôle structural des radiolarites et ophiolites le long du système alpino-méditerranéen. *Bull. Soc. géol. France* **22,** 1–14.

——, Argyriadis, I. & Lefèbre, R. 1974. Proposition d'une origine interne pour les nappes d'Antalya et le massif d'Alanya (Taurides occidentales, Turquie). Bull. Soc. géol. Fr. **16,** 107–11.

——, —— & Marcoux, J. 1975. L'axe calcaire du Taurus, un alignement de fenêtres arabo-africaines sous des nappes radiolaritiques, ophiolitiques et métamorphiques. *Bull. Soc. géol. Fr.* **17,** 1024–43.

——, Marcoux, J. & Poisson, A. 1979. L'allochtonie des Bey Dağları orientaux. Reconstruction palinspastique des Taurides occidentales. *Bull. Soc. géol. Fr.* **21,** 125–133.

Robertson, A. H. F. & Woodcock, N. H. 1979. The Mamonia Complex, southwest Cyprus; the evolution and emplacement of a Mesozoic continental margin. *Bull. geol. Soc. Am.*, **90,** 651–665.

—— & —— 1980a. Strike-slip related sedimentation in the Antalya Complex, SW Turkey, *Spec. Publ. Int. Assoc. Sedimentol.* **4,** 127–145.

—— & —— 1980b. Tectonic setting of the Troodos massif in the east Mediterranean. *Proc. Int. Ophiolite Symp. Cyprus* 1979, 36–49.

—— & —— 1981a. Gödene Zone, Antalya Complex continental margin, SW Turkey: volcanism and sedimentation along a Mesozoic marginal ocean crust. Geol. Rdsch. 70, 1177–1214.

—— & —— 1981b. Bilelyeri Group, Antalya Complex: deposition on a Mesozoic passive continental margin in SW Turkey. Sedimentology, **28,** 381–399.

—— & —— 1981c. Alakır Çay Group, Antalya Complex, SW Turkey. A deformed Mesozoic carbonate margin. Sediment. Geol. 30, 95–131.

—— & —— Sedimentary history of the south-western segment of the Mesozoic-Tertiary Antalya continental margins, south-western Turkey. *Eclogae geol. Helv*, **75,** 517–562.

Rocci, G., Baroz, F., Bebien, J., Desmet, A., Lapierre, H., Ohnenstetter, D., Ohnenstetter, M. & Parrot, J. F. 1980. The Mediterranean ophiolites and their related Mesozoic volcano-sedimentary sequences. *In*: Panayiotou A. (ed.) *Ophiolites*, Proc. Int. Oph. Symp. Cyprus 1979, 273–286.

Sarp, H. 1976. *Etude géologique et minéralogique de la région située au Nord-Ouest de Yesilova (Burdur, Turquie)*. Thesis (unpubl.) Geneve, Univ., 373 pp.

Searle, M. P. & Malpas, J. 1980. Structure and metamorphism of rocks beneath the Semail ophiolite of Oman and their significance in ophiolite obduction. *Trans. R. Soc. Edinburgh*, **71,** 247–62.

Selçuk, H. 1981. *Etude géologique de la partie méridionale du Hatay (Turquie)*. Thesis (unpubl.) Genève, Univ., 116 pp.

Şengör, A. M. C. & Yilmaz, Y. 1981. Tethyan evolution of Turkey: a plate tectonic approach. *Tectonophysics*, **75,** 181–241.

Shelton, A. W. & Gass, I. G. 1980. Rotation of the Cyprus microplate. *In*: Panayiotou, A. (ed.) *Ophiolites*. Proc. Int. Oph. Symp. Cyprus 1979, 61–65.

Simonian, K. O. & Gass, I. G. 1978. The Arakapas fault belt, Cyprus: a fossil transform fault. *Geol. Soc. Amer. Bull.* **89,** 1220–30.

Smewing, J. D. 1975. *Metamorphism of the Troodos massif, Cyprus*. Thesis (unpubl.) Open Univ., 267 pp.

——, Simonian, K. O. & Gass, I. G. 1975. Metabasalts from the Troodos massif, Cyprus genetic implication deduced from petrography and trace element geochemistry. *Contr. Miner. Petr.* **51,** 49–64.

Spray, J. G. & Roddick, J. C. 1981. Evidence for Upper Cetaceous transform fault metamorphism in West Cyprus. *Earth planet. Sci. Lett.* **55,** 273–91.

Swarbrick, R. E. 1980. The Mamonia Complex of SW Cyprus: A Mesozoic continental margin and its relationship with the Troodos Complex. *In*: Panayiotou, A. (ed.) *Ophiolites*. Proc. Int. Oph. Symp. Cyprus 1979, 86–92.

Terry, S. 1972. Sur l'âge triasique de laves associées à la nappe ophiolitique du Pinde septentrional (Epire et Macédoine, Grèce). *C.R. somm. Soc. géol. Fr.* 384–86.

Thuizat, R. & Montigny, R. 1979. K-Ar geochronology of three Turkish ophiolites. *Intern. Ophiolite Symposium, Nicosia, Cyprus*, 80–81 (abstract).

——, ——, Çakir, U. & Juteau, T. 1978. K-Ar investigations on two Turkish ophiolites. *Short Pap. Fourth Intern. Conf. Geochr. Cosmoschr. Isot. Geol.* (ed.) Zartman, R. E. Geol. Surv. Open-File Rep., 78–701, 430–432.

——, Whitechurch, H., Montigny, R. & Juteau, T. 1981. K-Ar dating of some infraophiolitic metamorphic soles from the eastern Mediterranean: new evidences for oceanic thrustings before obduction. *Earth planet. Sci. Lett.* **52,** 302–10.

Tinkler, Ch., Wagner, J. J., Delaloye, M. & Selçuk, H. 1981. Tectonic history of Hatay ophiolite (S. Turkey) and their relation with the Dead Sea rift. *Tectonophysics*, **72,** 23–41.

Vine, F. J., Poster, C. K. & Gass, I. G. 1973. Aeromagnetic survey of the Troodos igneous massif, Cyprus. *Nature, London*, **244,** 34–38.

Whitechurch, H. 1977. *Les roches métamorphiques infrapéridotitiques du Baër-Bassit (NW Syrien), témoins de l'écaillage intraocéanique téthysien. Etude pétrologique et structurale.* Thesis (unpubl.), Nancy, Univ., France, 194 pp.

—— & Parrot, J. F. 1974. Les écailles métamorphiques infrapéridotitiques du Baër-Bassit (Nord-Ouest de la Syrie). *Cah. ORSTOM. Paris, sér. Géol.* **7,** 173–184.

——, TEKELI, G. & ERENDIL, U. 1981. Pecularity of the Hatay ophiolite in the western tauric ophiolitic belt (Turkey). *Terra Cognita Spec. issue, 1st E.U.G. Meeting, Strasbourg 1981*, A34 (abstract).

WILLIAMS, H. & SMYTH, W. R. 1973. Metamorphic aureoles beneath ophiolite suites and alpine peridotites: tectonic implications with west Newfoundland examples. *Amer. J. Sci,* **273,** 594–621.

WILSON, R. A. M. 1959. The geology of the Xeros-Troodos area. *Cyprus Geol. Surv. Dept. Mem.* no. 1, 184 pp.

WOODCOCK, M. H. & ROBERTSON, A. H. F. 1977. Origin of some ophiolite-related metamorphic rocks of the "Tethyan Belt". *Geology*, **5,** 373–76.

—— & —— 1982. Wrench and thrust tectonics along a Mesozoic-Cenozoic continental margin: Antalya Complex, SW Turkey. *J. geol. Soc. London*, **139,** 147–163.

YILMAZ, P. O. & MAXWELL, J. C. 1981. K-Ar investigations from the Antalya Complex ophiolites, SW Turkey. *Abst. Ophiolites and Actualism Conf., Florence 1981*, Ofioliti 6, 49.

HUBERT WHITECHURCH & THIERRY JUTEAU, Laboratorie de Minéralogie et Pétrographie, 1, rue Blessig, 67084 Strasbourg Cedex, France.

RAYMOND MONTIGNY, Laboratoire de Géochimie, Institut de Physqiue du Globe, 5, rue René Descartes, 67084 Strasbourg Cedex, France.

Mylonitic ductile shear zones within tectonites and cumulates as evidence for an oceanic transform fault in the Antalya ophiolite, S.W. Turkey

Ingrid Reuber

SUMMARY: In the coastal ophiolite belt of Antalya, especially in the massifs of Adrasan and Cıralı-Tekirova, tectonites and cumulates have preserved their primary structures. Tectonites display a first, high temperature-low deviatoric stress deformation S1, linked to the diapiric uplift of the asthenosphere below an east-west orientated spreading centre. A second mylonitic deformation produced under high deviatoric stress and lower temperature conditions produced shear zones of millimetric to kilometric scale throughout the tectonites and even in the cumulates. Their mean orientation (N 150 E) is near to perpendicular to the ridge axis, and the associated lineation indicates horizontal, essentially sinistral strike-slip displacement. This deformation must have taken place during the final crystallisation of the cumulates, as indicated by its relationship to intrusive gabbroic 'dykelets' representing residual liquids of the cumulates. The S2 deformation is attributed to a transform fault situated SW of the coastal complexes. Later tectonic events include thrusting and wrench faulting, again sinistral, in a north-south trending direction.

Introduction and geological setting

The Antalya ophiolite is situated near the southern coast of Turkey, south of Antalya (Fig. 1). It occurs in association with pelagic volcano-sedimentary series of Triassic age and Mesozoic platform carbonates. Most authors agree that the alkaline Triassic volcanics indicate an early rifting phase (see references below), distinct from the ophiolitic sequence (harzburgitic tectonites, ultrabasic and basic cumulates, diabase dykes) of probably Upper Cretaceous age (Yılmaz this vol., Montigny *et al.* in press).

Yılmaz (this volume) and Robertson & Woodcock (this volume) provide an extensive discussion of the different tectonic interpretations of the Antalya Complex. Briefly, there are two schools: Brunn (1974), Monod (1976), Ricou *et al* (1974, 1975), Delaune-Mayere *et al.* (1977) recognize three Antalyan nappes: the Tahtalı Dağ carbonates (upper nappe), which tectonically overlie the pelagic volcano-sedimentary series and partial ophiolitic sequences of the median (Alakır Çay) nappe. These two, together with the lower nappe, were thrust onto the eastern margin of the autochtonous Bey Dağları unit during late Maastrichtian to early Paleocene times. However, Robertson & Woodcock (1980, 1981 a,b,c; 1982) and Woodcock & Robertson (1982) explain the present assemblage essentially as a passive Mesozoic margin of a south-Tethyan oceanic basin, subjected to north–south trending, sinistral, wrench tectonics in latest Cretaceous/early Paleocene to Miocene times. They propose a subdivision of the Antalya Complex into five zones parallel to these strike-slip movements (Fig. 2). The two main occurrences of ophiolites (the Gödene zone in the Alakır Çay valley and the coastal Tekirova zone) are separated by the shallow water carbonates of the Kemer zone (Robertson & Woodcock *op. cit*) cf. the Tahtalı Dağ nappe of Delaune-Mayere *et al.* (1977). The contacts between these shallow water carbonates and the pelagic–ophiolitic sequences are of a tectonic nature and their difference in paleogeography requires considerable differential displacements. These contacts are frequently north–south trending faults, dipping steeply eastwards, but subhorizontal superposition can often be observed, especially in the northern part of the Alakır Çay valley.

In contrast to the Alakır Çay, in which the ophiolitic rocks are essentially represented by serpentinite slices (Juteau 1968, Robertson & Woodcock *op. cit.*), in the coastal belt a nearly complete sequence from the tectonites up to a sheeted dyke complex has largely preserved its original structures (Juteau 1970, 1974). On the basis of statistical measurements of the foliation and the lineation within the tectonites, as well as the layering in the cumulates, Juteau *et al.* (1977) conclude that the coastal ophiolite represents oceanic crust south of a N 70 E trending spreading centre. This azimuth is the direction perpendicular to the lineation obtained after rotation of the foliation plane to the horizontal. Remapping of such internal structures and the recognition of a sequence of different deformation events has enabled us to propose a model for the oceanic setting of the Antalya coastal ophiolite which is the subject of this paper.

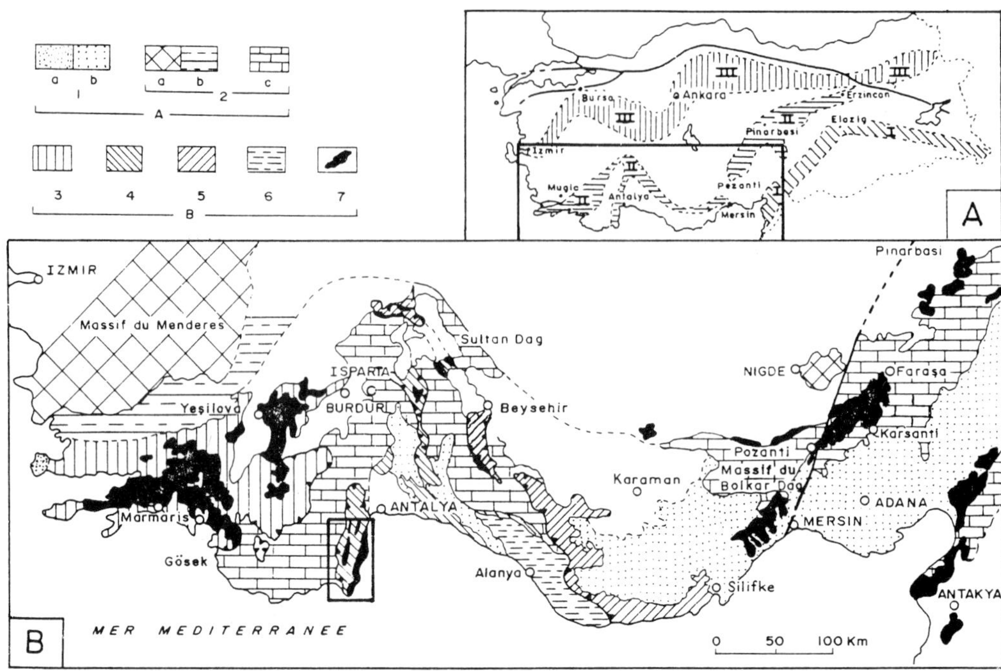

FIG. 1. A. Location of the study area south of Antalya, belonging to the median ophiolitic belt in Turkey, after Juteau (1979). I- southern (or peri-arabic) ophiolitic belt, II- median (or tauric) ophiolitic belt, III-northern ophiolitic belt. B. Structural setting of the ophiolites within the Tauric range after Juteau (1979). The inset indicates the area of Fig. 2. Al-post-tectonic formations: (a) Pliocene, (b) Miocene, A2-autochtonous substratum: (a) crystalline massifs of Menderes and Niğde, (b) metamorphic cover of the Menderes massif, (c) continuous carbonate series of the 'Tauric limestone axis' (Ricou *et al.* 1975), B3-Lycian nappes, B4-Antalyan nappes, B5-nappes of Hoyran-Beyşehir and Hadim, B6-Alanya massif, B7-ophiolites.

Thrust and wrench fault tectonics

The coastal ophiolite complex is disrupted by faults (Fig. 2) into the tectonite massif of Adrasan in the south, the mixed tectonite-cumulate massif of Cıralı–Tekirova in the central part and several smaller tectonite massifs, as well as outcrops of gabbro and more or less brecciated diabase dykes around Kemer. These brittle emplacement tectonics have to be analysed, before one can unravel the ophiolite's earlier ductile structures.

In the northern part of the coastal belt, high-angle strike-slip faults trend mainly north–south (N 15 E), and are especially frequent at the contact between the Kemer limestones and the Tekirova zone. Further south, the importance of a NW–SE direction increases, as the number and outcrop length of the north–south faults decreases (Fig. 2). Throughout the region, short faults of minor importance, trending around east–west (N 105 E ± 20°) can be found. They are especially well exposed along the coastal outcrops of the Cıralı–Tekirova massif (Fig. 4A). Here detailed analysis of the intersection relationships of all faults and of their relative displacements, reveals that they belong to a principal north–south trending, sinistral strike-slip system (Fig. 3), which is confirmed by the pattern of lineaments (Fig. 2). As this strike-slip deformation is less intense in the southern parts, the angles between different synthetic faults are greater and low-angle contacts of the Kemer carbonates ('Tahtalı Dağ nappe') upon the harzburgites can be observed (southern limit of the Cıralı–Tekirova massif and northern limit of the Adrasan massif, Fig. 4). These thrust contacts predate the high-angle wrench faults. Minor reverse faults within the ophiolitic sequence also predate high-angle faults (Fig. 4A). The observed thrust contacts are certainly related to the nappe emplacement in latest Cretaceous to early Tertiary time (see above), whereas a Miocene age seems reasonable for the high-angle faults. In some cases the Neogene cover is affected by these high-angle faults, in others it is not (or only weakly so),

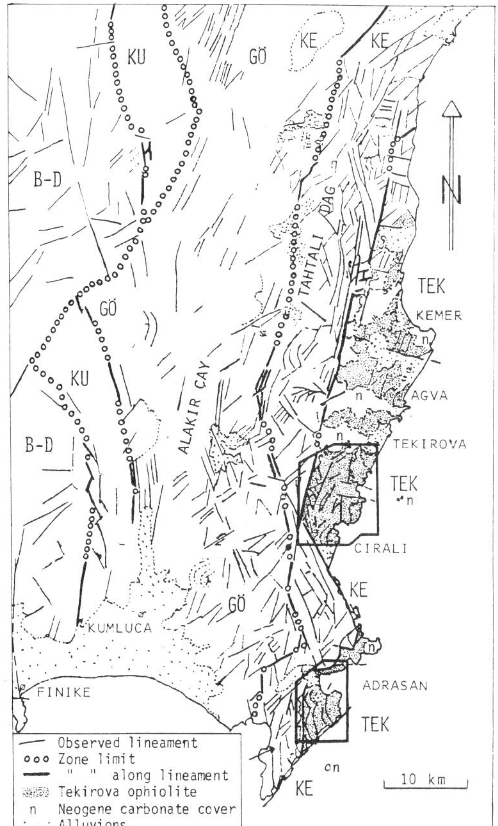

Fig. 2. Map of lineaments visible in satellite images; most correspond to steeply dipping faults. Superposed the principal zones after Robertson and Woodcock (*op. cit.*): B-D-Bey Dağları, KU-Kumluca zone, GÖ-Gödene zone, KE-Kemer zone, TEK-Tekirova zone. The insets show the areas of the detailed maps (Figs 4 & 7).

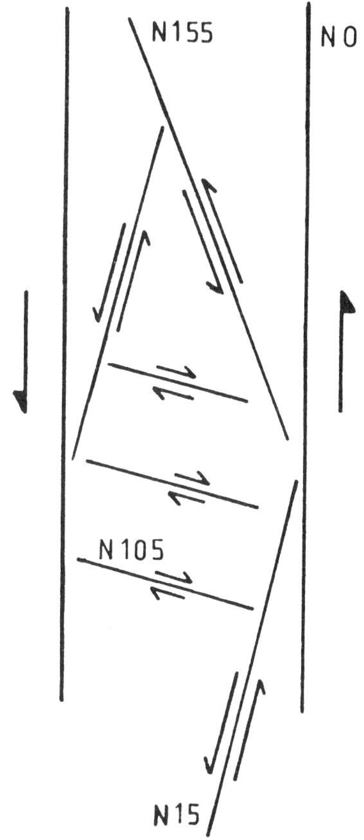

Fig. 3. Mean orientations of secondary faults associated with a north–south trending wrench-fault system, according to Riedel (1929), Tchalenko (1970), and Ruhland (1973).

suggesting that the strike-slip movements have been active over a long period of time.

The most important faults are accompanied by a penetrative schistosity in the adjacent serpentinized harzburgite. The projection of the lineation, defined by boudinaged orthopyroxenes, onto the shear plane probably indicates the direction of shear and thus the direction of displacement along the fault. The thrust contact with overlying carbonates in the north of the Adrasan massif and high angle faults around Kemer are the main occurrences of such local schistosities. They certainly indicate quite significant total displacement in a horizontal north–south direction (a thrust movement towards N 170 E in the first, and a strike-slip, N 15 E, movement in the latter case).

In spite of these complex tectonics, the two main massifs, Adrasan and Cıralı–Tekirova, have preserved their primary structures quite well. Moreover, the consistency of orientations of corresponding internal structures from one massif to the other suggests that they both belong to one unit, offset by strike-slip faults without rotation of the individual blocks.

Distribution of the petrographic facies

A petrographic succession can be observed from NE to SW within the cumulates, as well as within the tectonites, corresponding to a sequence from top to bottom of the ophiolite (Fig. 4):

Cumulates (cf. Juteau & Whitechurch 1980) The uppermost, isotropic, amphibolitic gabbros are exposed at the NE edge of the Cıralı–Tekirova massif, just south of the Tekirova

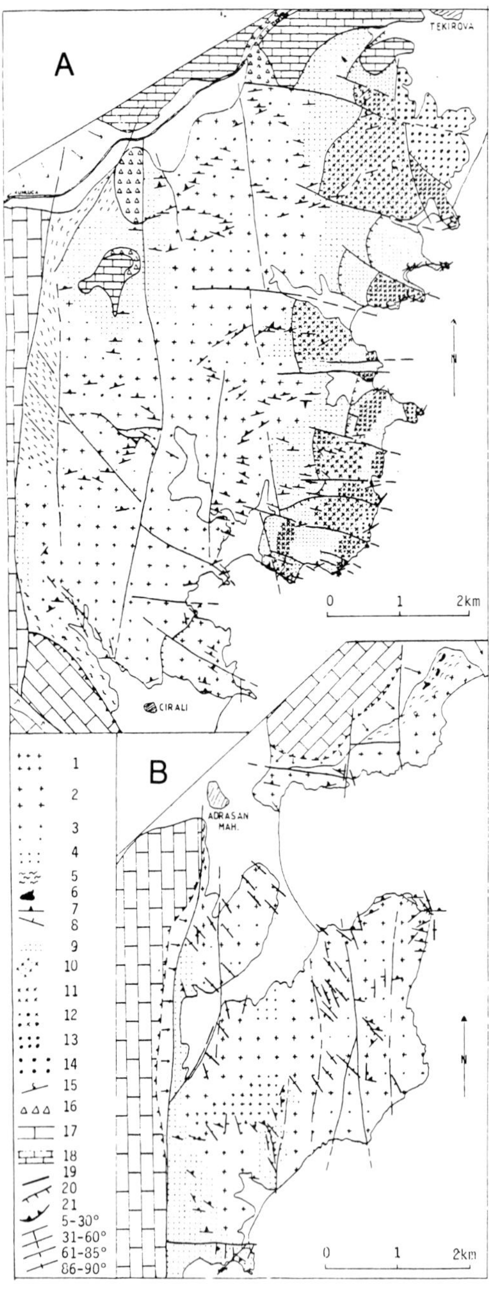

FIG. 4. Maps of Çıralı–Tekirova (A) and Adrasan massif (B), showing the distribution of petrographic facies and the orientations of foliations S1 and S2. 1-lherzolithic harzburgites; 2-massive harzburgites; 3-'layered' harzburgites; 4-dunites belonging to the tectonites; 5-faulted serpentinites near to the major contacts; 6-plagiogranites intrusive in the tectonites; 7-foliation S1; 8-foliation S2; 9-cumulitic dunites; 10-wehrlites; 11-proxenites; 12-plagioclase-bearing wehrlites; 13-layered gabbros; 14-isotropic gabbros; 15-layering S0; 16-breccias derived from cumulates and diabase dykes; 17-Mesozoic platform carbonates; 18-Neogene limestone; 19-fault; 20-minor thrusts within the ophiolite complex; 21-thrust of the platform carbonates onto the ophiolite series.

coastal plain; Layered gabbros are found on the coastal cliff and are frequently in tectonic contact with underlying dunitic cumulates; Plagioclase-bearing wehrlites occur on top of and also within the alternating series of wehrlite and dunite; the basal part of the cumulates is composed of undeformed dunites. Spinel chemistry and the occurrence of greenish clinopyroxene layers within the first 50 m from the cumulate/tectonite contact help distinguish them from tectonized dunites; the dunite-dunite contact between tectonites and cumulates is a favoured site of fault displacement, due to their higher ductility. However, this E to SE dipping reverse fault is parallel to the primary contact and of minor importance, since the rocks in contact still represent the highest levels of tectonites and the lowest levels of cumulates respectively.

Tectonites The uppermost dunites, the most depleted relics of the upper mantle, grade into layered harzburgite, containing dunitic bodies whose size and frequency diminish away from the contact with the cumulates. Towards the southern parts of the Çıralı–Tekirova massif the abundance of dunitic layers (50% of the total rock in the north) decreases, and the facies grades into massive harzburgites containing about 25% orthopyroxene. The same massive harzburgites are the most abundant facies in the Adrasan massif. However, the orthopyroxene content increases southwards, and locally some clinopyroxene-bearing harzburgites can be found (about 40% orthopyroxene and less than 1% clinopyroxene).

Injection of small-scale dykelets occurred during an early stage of facies differentiation and a similar distribution of different petrographical types can be found in them along a vertical section through the ophiolite (Reuber 1982). These early dykelets vary from rare coarse grained websterite or gabbro (lowest tectonites) to orthopyroxenite dykelets (widely occurring in all intermediate levels) to frequent dunitic 'layering' (upper tectonites). This differentiation can be explained by different percentages of partial melting of the host rock and partial crystallization of the upward moving magma. Similarly the different dykelets in the cumulates, ranging from clinopyroxenite to leucocratic gabbro, represent different stages in the evolution of the residual liquid from the adjacent cumulate host rock. Injection of the most differentiated liquids was on a larger scale, as these leucocratic gabbro dykelets occur in all cumulate levels and in the upper tectonites. In the latter they clearly postdate all the early dykelets as defined above.

The high temperature deformation S1

All the tectonites have been affected by deformation at high temperature (1000–1200°C) and low deviatoric stress (200 bars). This can be deduced from the activated glide-system of kinked olivine porphyroblasts (Fig. 5, Raleigh 1968, Nicolas & Poirier 1976). Spinel grains have also been deformed during this phase, S1 (Fig. 6). The intensity of the deformation varies from one outcrop to the next, especially with regard to the dispersion of elongated spinel or orthopyroxene grains defining the mineral lineation L1. The orientation of the foliation S1 is relatively constant, with strike direction N 70 E to N 100 E and steep northerly dip throughout all the outcrops of the coastal ophiolite belt. The same orientation has been occasionally observed for the S1 foliation in harzburgites of the Alakır Çay valley. The associated lineation L1 is generally sub-parallel to the dip of the corresponding foliation plane, thus plunging steeply northwards (Figs 7 & 8A). Some deviations from this general trend can be explained as weak reorientation during the following deformation phase S2.

The shear sense has been determined in the field by statistical observation of lamellar orthopyroxenes (Darot & Boudier 1975, Reuber *et al.* 1982) and confirmed for some examples by olivine petrofabric studies. The results clearly show a predominance of 'inverse' shear sense (477 : 82 in total, Figs 7 & 8A) indicating a southward and upward displacement of each upper and northern slice with respect to the underlying ones.

The mylonitic deformation S2

Shear zones are found at all scales, millimetric to kilometric, throughout the tectonites and even within the cumulates. The mylonitic granulation of olivine indicates high strain rates at high deviatoric stress (about 2 kb) and at relatively low temperature (about 800°C, Nicolas & Le Pichon 1980). The strike directions and dip of the S2 foliation are grouped around N 145 E, 80 NE (Fig. 8B). Deviations from this orientation are more noticeable where the in-

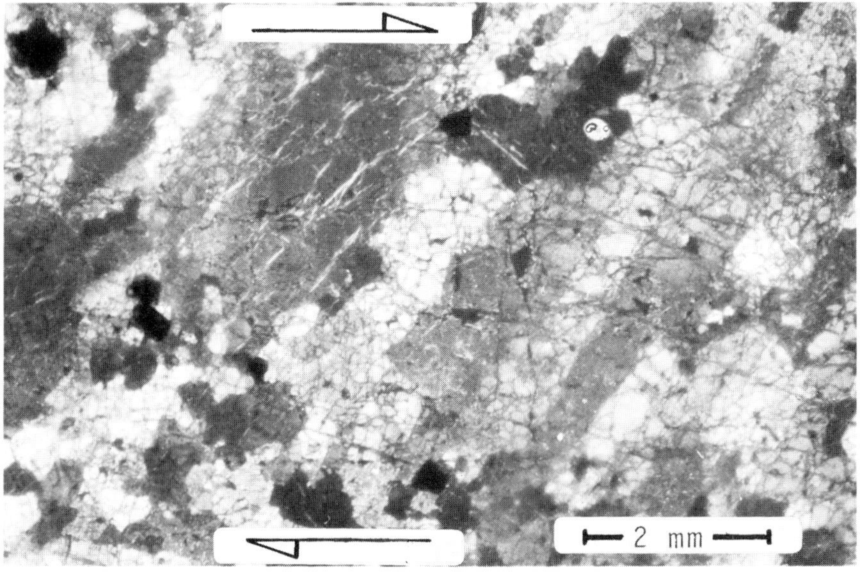

Fig. 5. Kink bands in a large olivine grain, typical of the high temperature deformation S1, in which the (0k1) [100] glide system ('pencil glide') of the olivine has been activated. Thin section from the Adrasan massif, south of the main shear belt, cross-polarized light.

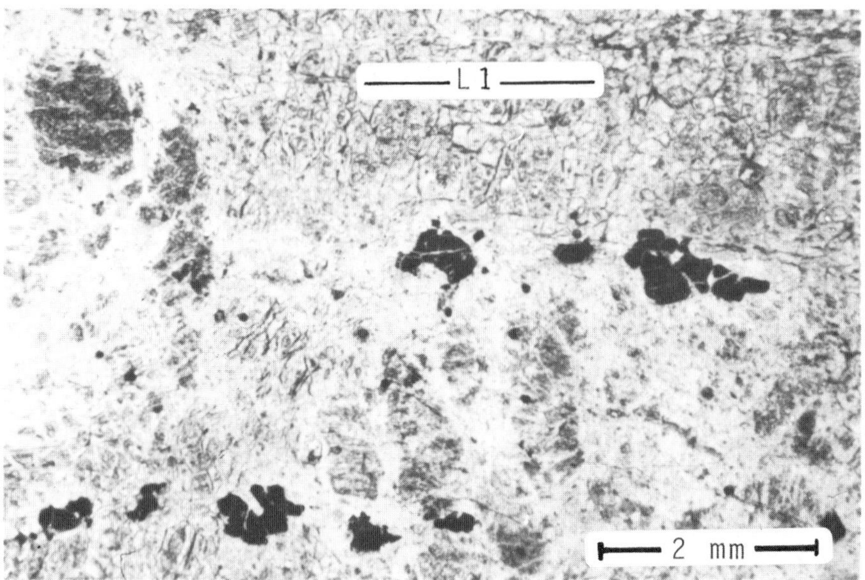

FIG. 6. Plastically elongated and boudinaged chromite grains, typical of intense high temperature deformation S1, in highly serpentinized harzburgite. Thin section from 'layered' harzburgites of Cıralı–Tekirova massif, plane-polarized light.

tensity of the S2 deformation is weaker, or in the extremely small guided shear zones (see below). Under the same restrictions the orientations of the lineation L2, in the intensely deformed parts, are well grouped around horizontal to low plunges towards the NW (Fig. 8B). The mean angle between S1 and S2 foliations is about 60°, but it may reach 90° locally.

S2 shear zones in tectonites

The most important shear belt is the 2 km wide central part of the Adrasan massif displaying porphyroclastic to mylonitic harzburgite. The foliation S2 is vertical, or nearly so, and strikes around N 150 E (values from N 130 E to N 170 E occur, Fig. 7). The lineation is essentially indicated by lamellar orthopyroxenes, as spinel grains within an olivine matrix are not deformed during this phase (due to the considerably lower temperatures Fig. 9). The lineation L2 is generally horizontal to gently northwards-plunging (Fig. 8B). The shear sense has again been determined by statistical observation of the lamellar orthopyroxenes and confirmed by olivine petrofabric study. It generally indicates sinistral shear sense, except for one main lens at the northern margin of the shear belt, where it is dextral (Fig. 7B).

Smaller shear zones occur further to the north, the most important, 0.5 km wide, is situated at the southern end of the Cıralı–Tekirova massif. The mylonitization is here more intense, but directions of foliation and lineation are identical to those observed in the main shear zone of Adrasan. Overall, equal numbers of orthopyroxenes are sheared dextrally as sinistrally. This is also the case on the scale of some individual outcrops, whereas in others simple shear in one direction or the other could be determined (Fig. 7A). However, for this zone the overall pattern of deformation appears to be heterogenous pure shear.

Horizontal lineations on foliation planes with an orientation intermediate between the S1 and S2 mean orientations (N 100 E to N 120 E) occur further north in the Cıralı–Tekirova tectonites. They indicate dextral as well as sinistral simple shear and probably result from weak overprinting of S1 structures by the S2 deformation.

A mylonitic fabric is particularly well developed locally in very small shear bands that occur south of the Adrasan main shear belt. Here dunitic bands, several centimetres to a decimetre wide, formed as a result of complete recrystallization of olivine, and tectonic segregation of pyroxene and spinel grains towards the centre of these bands. At the margins of the bands kinked olivine porphyroblasts, enclosed within some less deformed lenses, exhibit the relict S1 fabric (Fig. 9).

The orientation of the S1 and the S2 fabrics, as well as the associated shear senses, are

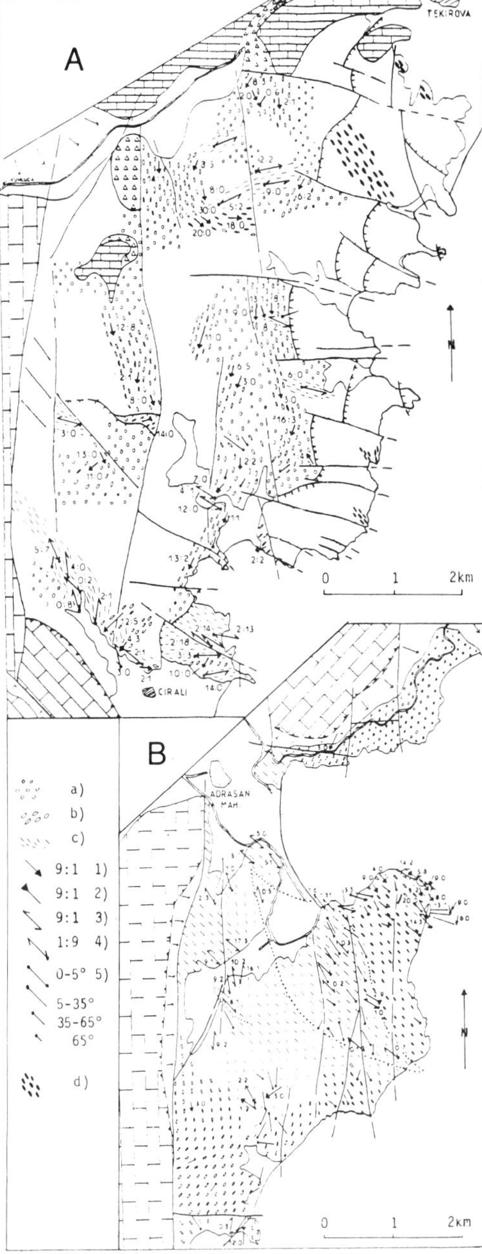

FIG. 7. Maps of the Cıralı-Tekirova (A) and the Adrasan massif (B), showing the lineations L1 and L2, a qualitative estimation of the shear intensity and the shear sense obtained from statistical observation of lamellar orthopyroxene. a-lineation L1 weakly or not visible; b-lineation L1 well developed; c-lineation L2; 1-northward dipping lineation L1 associated with inverse shear of 9 orthopyroxenes (displacement of the upper and northern part towards the south and upwards in respect to the underlying unit), and one indicating normal shear; 2-southward dipping lineation L1 associated with 9 orthopyroxenes indicating normal shear and one inverse (here the normal shear corresponds to the displacement of the upper part towards the south in respect to the underlying one); 3-horizontal (0–5°) lineation L2 indicating sinistral shear (9 orthopyroxenes against one dextral one); 4-northward dipping lineation L2 indicating dextral shear (9 dextral orthopyroxenes against one sinistral one); 5-indications of dip; d-location of S2 shear within the cumulates.

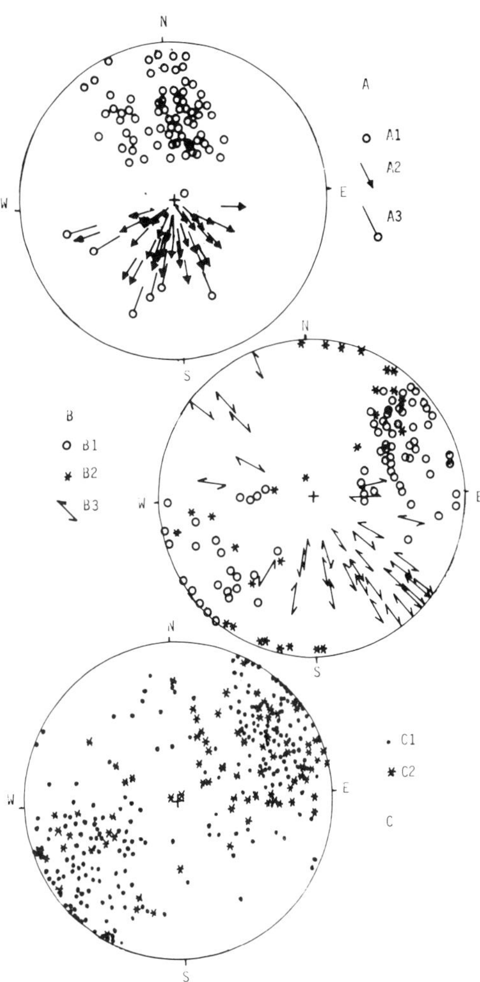

FIG. 8. Stereographic projection (equal-angle, upper hemisphere) of poles to foliation planes and lineations for deformations S1 (A) and S2 (B); and of the poles to leucocratic gabbro dykelets (C). A1-foliation S1; A2-lineation L1 with inverse shear sense; A3-lineation L1, undetermined shear sense; B1-foliation S2 in tectonites; B2-foliation S2 in cumulates; B3-lineation L2 and shear sense; C1-leucocratic gabbro dykelets in the tectonites and C2-in the cumulates, those at a high angle to foliation S2 are commonly undeformed.

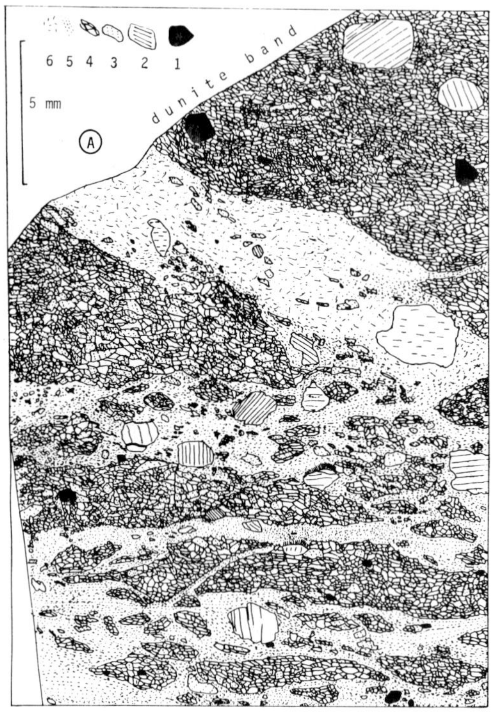

FIG. 9. Mylonitic microcrystalline shear zones (lower part) with porphyroblastic lenses and a dunite band (upper part). 1-spinel; 2-orthopyroxene; 3-olivine porphyroclasts; 4-olivine neoblasts; 5-mylonitic olivine; 6-actinolite. Drawing of thin section from Adrasan, south of the main shear belt.

clearly different (Figs 9 & 10). Moreover, the petrofabric diagrams (Fig. 10) show that the deformation mechanisms of the olivine were different during the different deformation phases. The unimodal textures of olivine grains in the completely S2-dependent fabric of the dunite bands show the weak preferred orientation typical of olivine mylonites, in which low temperature glide in the [001] direction has been active (Fig. 10A). The fabric of the relict lenses (Fig. 10B) shows, in addition, a second texture which corresponds to the orientation of kinked olivine porphyroblasts. Their [100] axes are near to the lineation (the weak obliquity indicates sinistral shear) the general pattern is that of the high temperature (0kl) [100] 'pencil' glide system (Raleigh 1968, Nicolas & Vialon 1980).

S2 shear zones in cumulates

Mylonitic shear zones in the cumulates are less common and narrower than those in the tectonites but they have been found throughout the cumulate units. For example in the isotropic amphibolitic gabbros centimetre—to metre—sized shear zones are frequent, orientated around N 150 E, and displaying amphiboles recrystallized in the S2 foliation planes. A large part of the northernmost wehrlites (around Geren Tepe) show a subvertical serpentinite schistosity trending N 130 E to N 150 E. A brecciated zone, re-cemented by leucocratic gabbro and bordered by flaser-gabbro, and a metre-wide highly mylonitized band (Fig. 11, Juteau *et al.* 1977) have been observed in the layered gabbros. They indicate localized intense shear which induced the dynamic recrystallization of calcic plagioclase.

S2 shear zones related to leucocratic gabbro dykelets

Leucocratic gabbro dykelets are the latest generation of the small scale, coarse grained dykelets. Their evolved chemistry, shown by whole rock and by ortho- and clino-pyroxene compositions, indicates that they represent residual liquids of the cumulates, injected in all directions, especially downwards, into cumulates and upper tectonites (Reuber *et al.* 1982). The mechanism of injection is hydraulic fracturing under isotropic to tensile stress conditions as shown by the dyke structures. The liquid represents the most evolved phase of the cumulates and so this injection took place during the final stage of the cumulates' crystallization. The orientation of these dykelets varies with time, as shown by the intersection of different generations. The older ones are generally underformed, whereas the younger ones are deformed and parallel to the S2 foliation (Fig. 8C).

These leucogabbro dykelets are frequently the site of narrow, but intense zones of mylonitization. The internal foliation is generally parallel or weakly oblique to the dikelets themselves. It is also parallel to the S2 foliation of the immediately adjacent host rock and to the mean orientation of S2 foliation; the lineation is also subhorizontal. The shear sense is indicated by bending of ortho- and clinopyroxenes and by lamellar orthopyroxenes; both dextral and sinistral senses have been recorded.

In some places the deformation is concentrated wholly within, or even just in part of the dykelet (most often the centre, Fig. 12), indicating that the ductility of the dykelet, or part of it, was much higher than that of the host rock, which, in these cases, is completely devoid of S2 deformation. On the other hand, a local abundance of such dykelets has been

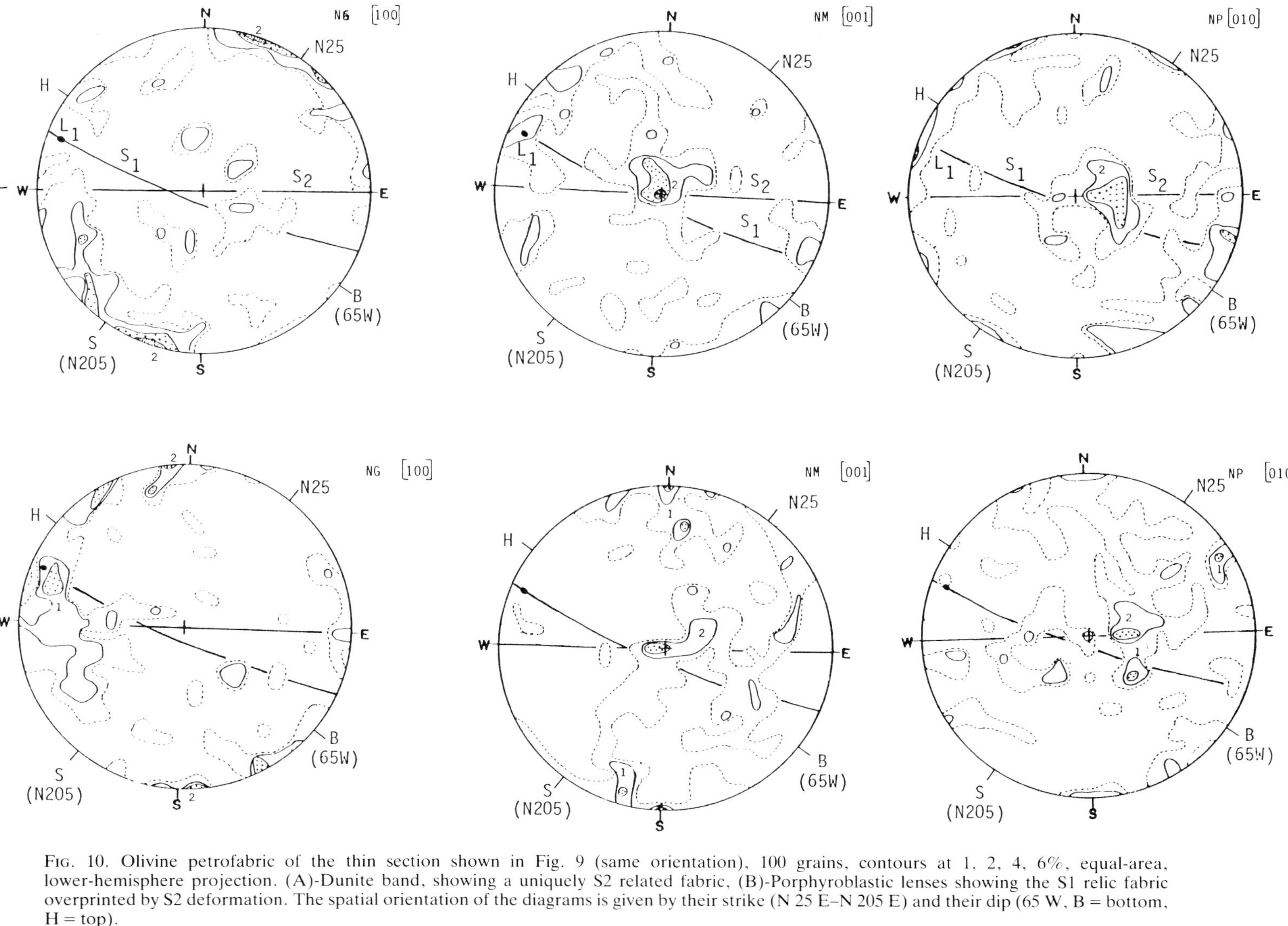

FIG. 10. Olivine petrofabric of the thin section shown in Fig. 9 (same orientation), 100 grains, contours at 1, 2, 4, 6%, equal-area, lower-hemisphere projection. (A)-Dunite band, showing a uniquely S2 related fabric, (B)-Porphyroblastic lenses showing the S1 relic fabric overprinted by S2 deformation. The spatial orientation of the diagrams is given by their strike (N 25 E–N 205 E) and their dip (65 W, B = bottom, H = top).

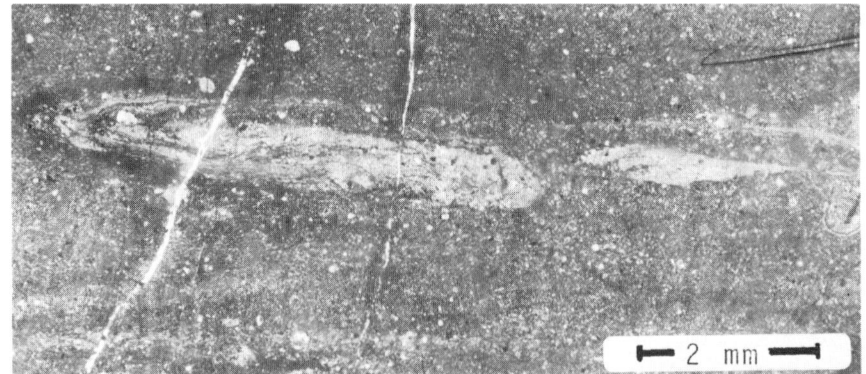

Fig. 11. Intensely mylonitized gabbro showing pyroxene fragments in a microcrystalline plagioclase matrix, thin section from Cıralı–Tekirova massif, cross-polarized light.

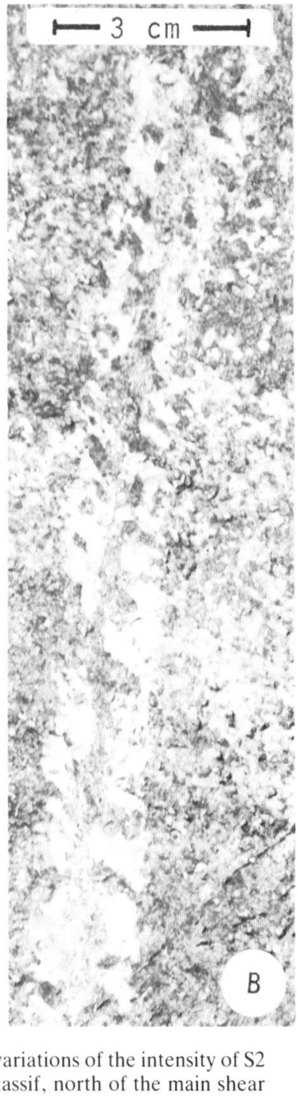

Fig. 12. Partially deformed leucocratic gabbro dykelets: (A) Multiple variations of the intensity of S2 deformation within a dykelet in S1 deformed harzburgite; Adrasan massif, north of the main shear belt, outcrop photo. (B) Mylonitic deformation of the central part of the dykelet within non deformed layered gabbros, Cıralı–Tekirova massif, outcrop photo.

observed in association with a metre wide local shear zone (Fig. 13), as if the presence of the dykelets increased the ductility of the whole zone.

Both observations point to coeval dykelet emplacement and S2 deformation. The ductility of the still not fully crystallized dykelet is so much greater than that of its host rock, that all the local stress is dissipated within the dykelet, or even in a tiny zone within it, corresponding to the least crystallized central part. On the other hand, the intrusion of numerous dykelets may heat the host rock in such a way that ductile shear becomes possible and a wider shear zone will develop. This contemporaneity does not necessarily indicate a genetic link between S2 deformation and dykelet intrusion, as dykelets of the same type may occasionally be undeformed. In addition the emplacement mechanism has been interpreted as hydraulic fracturing. Such dykelets have never been observed to form 'en echelon' arrays, which would be typical for shear emplacement.

Timing of the S2 shear event

The contemporaneity of the S2 deformation with the emplacement of the leucocratic gabbro dykelets, indicates that these ductile shear zones have developed at a very early stage of ophiolite genesis, i.e. during the final stage of crystallization of the cumulate magmatic chamber. S2 clearly postdates the S1 deformation, but could still have occurred very near to, the accretion axis. More precisely, for a reasonable half-width of the cumulate magma chamber of less than 10 km, (with final crystallization at about 15 km from the ridge axis, Reuber 1982) and a relatively rapid spreading rate of 6–10 cm/ which is consistent with the type and size of the cumulate magmatic chamber (Juteau & Whitechurch 1980), the final crystallization would have taken place 1 to 1.5 Ma after the central protrusion for each vertical section considered. This time span is smaller than the analytical and statistical error of the radiometric dating. The available data (Yılmaz this vol., Montigny *et al.* in press) indicate that these early oceanic events of ophiolite genesis took place during middle Upper Cretaceous time. However, some geological problems remain unsolved, and these dates may correspond to some yet unknown event later than the ophiolite genesis.

Interpretation of the geometric relationships of the internal structures—model for the oceanic accretion zone influenced by transform movements

A fairly consistent model can be constructed from the structural elements in their present orientation, without considering rotation during emplacement. In contrast to the model of Juteau *et al.* (1977) no palaeohorizontal is now proposed, as neither layering of the cumulates nor the S1 foliation of the tectonites needs to have been horizontal (cf. models of Dewey & Kidd 1977, Girardeau & Nicolas 1981). Also the position of the palaeo-Moho need not be horizontal in the proximity of the S2 shear zones (see below). However, a rotation of 30° around an east–west trending horizontal axis brings the data into the best internal coherence. This rotation changes the position of most of the structures very little except in the case of the S1 foliation, lineation and associated shear sense (Fig. 14). The position of these now corresponds to the classic model of a convective cell as the driving force for flow in the upper mantle (Fig. 15). The S1 deformation phase is linked to that convective mantle flow and the consequent uprise of asthenosphere below a spreading centre. In the upper levels of the upper mantle, corresponding to the pile of observed tectonites, plastic flow only appears to take place near to the ridge axis. Thus the S1 foliation is 'fossilized' in a high-angle position, dipping towards the ridge axis. The differential shear is thought to have taken place where the upward movement bends to a horizontal one, still near to the axis, just before the S1 structures became 'fossilized'. The accretion zone, parallel to the strike of S1 foliation and perpendicular to the direction of the lineation L1, would be situated south of the studied complex (Fig. 15).

The S2 deformation represents oceanic strike-slip movements. The position of the S2 shear zones at a high angle (up to subperpendicular) with respect to the ridge axis, indicates that they cross-cut the accretion zone. The absence of cumulates west of the studied complex and the general increase of S2 deformation towards the WSW strongly suggest that this deformation was linked to an oceanic transform fault situated west of the studied complex (Fig. 15). Moreover, these shear zones limit the cumulate magma chamber longitudinally. Thus its boundary bends round progressively from parallel to the S2 shear zones towards its 'normal' ridge orientation (parallel to and gently

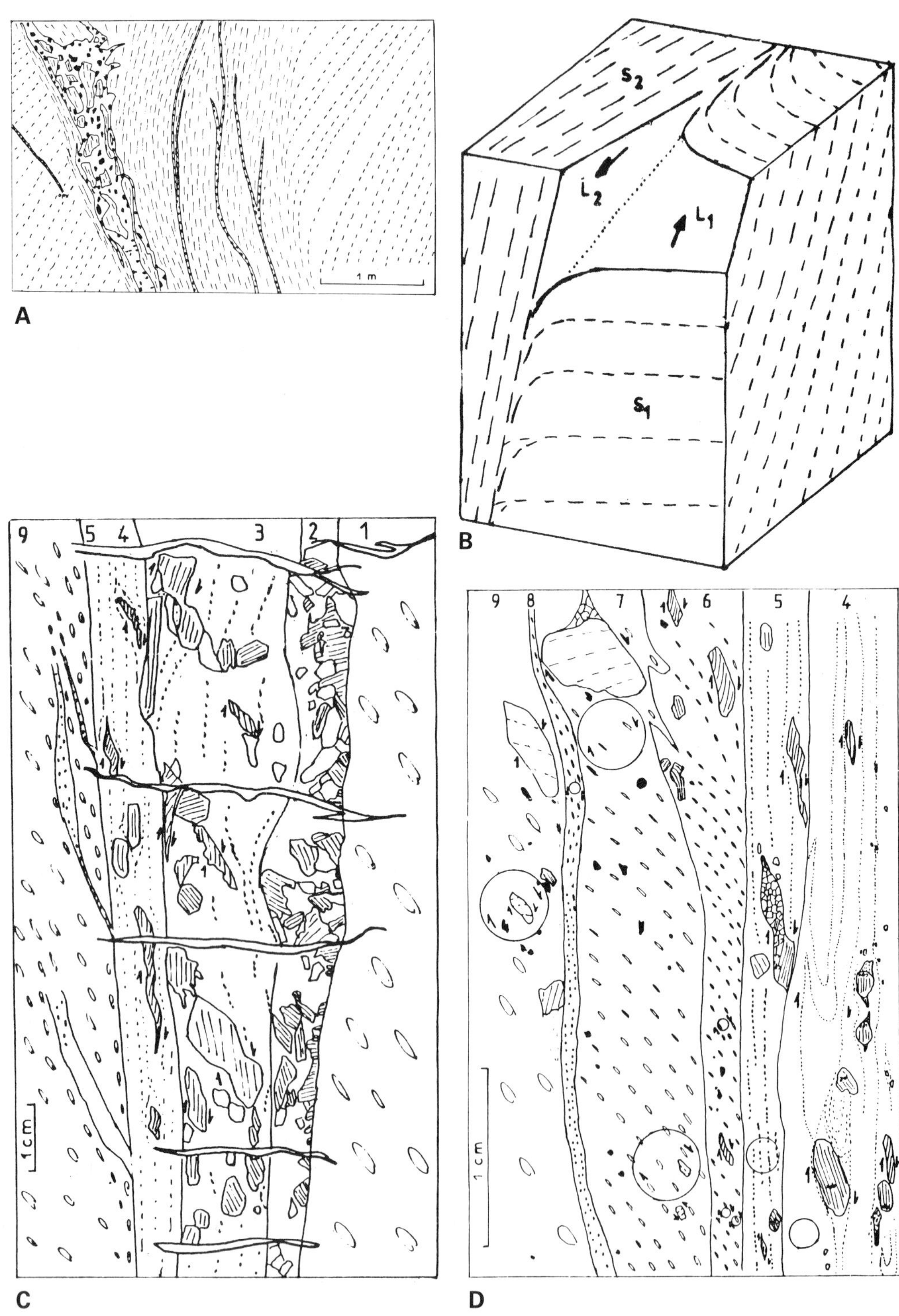
1 m
A
S2
L2
L1
S1
B
9
5
4
3
2
1
1 cm
C
9
8
7
6
5
4
1 cm
D

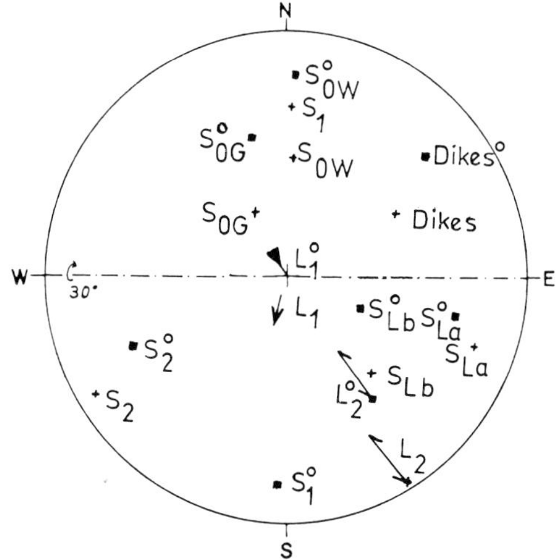

FIG. 14. Equal-angle, upper-hemisphere projection of the average poles of the main structures in their present position (small crosses) and in their supposed initial position (black squares and ° indexes). S1, L1, S2, L2 are foliations and associated lineations as described in the text. S_{OG}-layering in the gabbros; S_{OW}-layering in the wehrlites; S_L-major lithological contacts (a-close to the tectonites, b-further east, inside the cumulate sequence); Dykes-huge dyke slabs in the breccia.

dipping towards the ridge axis). A similar bending can actually be observed in the major lithological contact within the cumulates (Fig. 4A, 14): the contacts between dunite and wehrlite are more strongly influenced by the S2 shear zones than those between pyroxenite and gabbro in the upper levels of the cumulates.

As the lineation L2 is gently dipping northward, the essentially sinistral movements also have the effect of slightly raising the block immediately to the WSW of each shear zone relative to the opposing one. This is especially true of the main shear zone of Adrasan and explains why the facies of deepest origin crop out SW of this zone. The magma chamber thus has a transverse 'static' wall composed of 'hot' mantle, similar to its underlying moving floor. Thus it is not surprising to find exactly the same succession of facies within the cumulates from west to east as from bottom to top.

The obliquity of the S2 foliation with respect to a direction perpendicular to S1 should result in extension and subsidence of the block located northeast of the intersection of transform structures and the accretion zone, i.e. the studied complex. In fact the abundance of slumping and syn-magmatic faulting, as well as the relatively pronounced dip of the cumulate layering (Figs 4A & 14), do indicate tectonic instability during crystallization. Moreover, *in situ* brecciation of the diabase dykes results in huge dyke slabs, which cluster in orientation around N 150 E. Their orientation is parallel to the S2 shear zones and their eastward dip indicates movement towards the east, i.e. towards the slightly subsiding block corresponding to the study area (Fig. 15).

The Antalya coastal ophiolite thus represents oceanic lithosphere, generated at a normal spreading centre, intersected by strike-slip shear zones. These were probably related to a sinistral transform fault, located further west, possibly in the Alakır Çay ophiolite.

Conclusions

The proposed interpretation is in good agreement with the general framework of ophiolites in the eastern Mediterranean. Whitechurch *et al.* (this vol.) conclude that east–west orientated accretion zones existed, offset by generally north–south trending transform faults, in all the Tauric ophiolites.

More precisely, the ductile shear zones described in this paper add one more strike-slip event to the long history of north–south trending wrench tectonics of the Antalya complex, reported from Lake Cretaceous to Miocene times (Robertson & Woodcock 1980, 1982, Woodcock & Robertson 1982), and possibly related to a north–south trending sinistral transform system. The ductile shear described here however, results from deformation in a deeper seated environment than the faulting of the Gödene zone, which appears to have taken place under much colder conditions. This could

FIG. 13. A two metre-wide shear zone affecting harzburgitic host rock at a site of abundant leucocratic gabbro dykelets. The intensity of S2 deformation varies within the host rock and the gabbro dykelets. (A) Sketch of the outcrop (steeply northward dipping plane). (B) Block diagram to illustrate the geometric relations between the foliations S1 and S2 and the corresponding lineations; arrows indicate the displacement of the missing block. (C) Detail of a gabbro dykelet at the outcrop showing a non-deformed border (2) and increasing deformation (3) to (5); arrows show shear sense in a subhorizontal plane. (D) Sketch of thin section of the same dykelet showing mylonitized zones (4) and (5) in the dykelet and decreasing intensity of the mylonitic deformation in the harzburgitic host rock, (6) to (9). (8) is a 'branching off' little dykelet; arrows indicate shear sense as in (C).

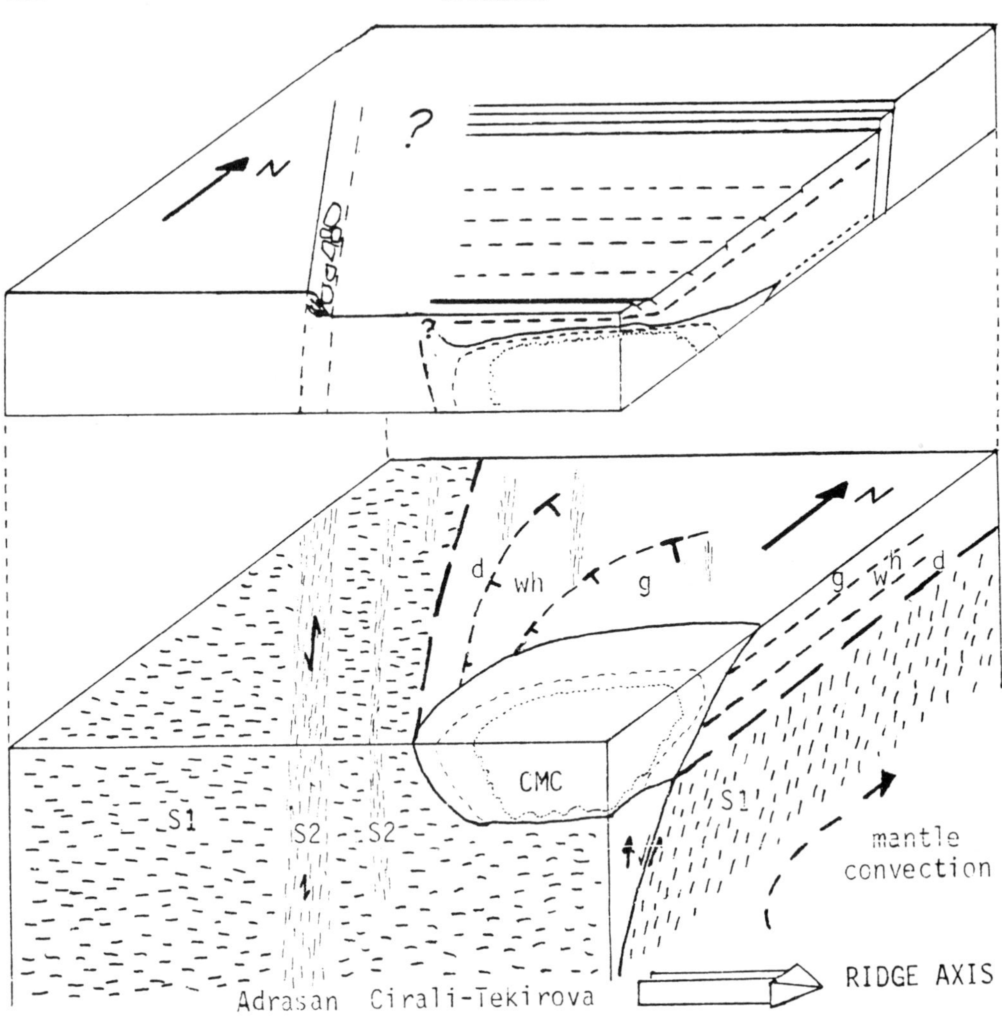

FIG. 15. Schematic representation of the palaeo-zone of accretion of the Antalya ophiolite in its original genetic context, as reconstructed after the structural data described in this paper. The upper part shows the submarine cliff, supposed to be responsible for the brecciation of the dyke-complex. The lower part represents a section through the magma chamber, corresponding approximately to the mean actual level of erosion. The southwestern limit of the cumulate magmatic chamber (CMC) is due to the transform shear zones (S2). Consequently the tectonite/cumulate contact and major lithological contacts within the cumulates (d-dunites, wh-wehrlites, g-gabbros) migrate from parallel to the S2 shear zones towards a normal flat position.

be explained by a more pervasive circulation of water in the main transform fault zone, if it already existed at the time of the ductile shear event. This would be consistent with the total absence of cumulates, as the shear zones of the coastal belt operate as longitudinal limit of the cumulate magmatic chamber (Fig. 15).

ACKNOWLEDGEMENTS: This paper summarizes part of the results of my Ph.D. thesis, carried out at the University Louis Pasteur, Strasbourg, in the research group of T. Juteau. I am very grateful for the stimulating discussions and advice of T. Juteau and H. Whitechurch, Strasbourg, as well as A. Nicolas and F. Boudier, Nantes. I also thank A. Siddans (Strasbourg) for revising the English.

References

BRUNN, J. H. 1974. Le problème de l'origine des nappes et de leurs translations dans les Taurides occidentales. *Bull. Soc. géol. France*. **16**, 101–106.

DAROT, M. & BOUDIER, F. 1975. Mineral lineations in deformed peridotites: kinematic meaning. *Petrologie*., **1**, 225–236.

DELAUNE-MAYERE, M., MARCOUX, J., PARROT, J. F. & POISSON, A. 1977. Modèle d'évolution Mésozoïque de la paléomarge Téthysienne au niveau des nappes radiolaritiques et ophiolitiques du Taurus Lycien, d'Antalya et du Baër-Bassit. *In*: BIJU-DUVAL, B. & MONTADERT, L. (ed.): *Structural history of the Mediterranean basins*, Technip, Paris, 79–94.

DEWEY, J. F. & KIDD, W. S. F. 1977. Geometry of plate accretion. *Bull. geol. Soc. Amer*. **88**, 960–968.

GIRARDEAU, J. & NICOLAS, A. 1981. The structures of two ophiolite massifs, Bay-of-Islands, Newfoundland: a model for the oceanic crust and upper mantle. *Tectonophysics*., **77**, 1–34.

JUTEAU, T. 1968. Commentaire de la carte géologique des ophiolites de la région de Kumluca (Taurus Lycien, Turquie méridionale): cadre structural, modes de gisement et description des principaux faciès du cortège ophiolitique. *M.T.A. Enst. Bull. Ankara*., **70**, 70–91.

—— 1970. Pétrogenèse des ophiolites des nappes d'Antalya, leur liaison avec une phase d'expansion ocèanique active au Trias supérieur. *Sci. Terre*., **15**, 265–306.

—— 1975. Les ophiolites des nappes d'Antalya (Taurides occidentales, Turquie). Pétrologie d'un fragment de l'ancienne croûte océanique tethysienne. *Mém. Sci. Terre. Nancy*., **32**, 692 p.

—— 1979. Ophiolites des Taurides: Essai sur leur historie océanique. *Rev. Géogr. phys. Géol. dyn*., **21**, 191–214.

——, NICOLAS, A., DUBESSY, J., FRUCHARD, J. C. & BOUCHEZ, J. L. 1977. Structural relationships in the Antalya ophiolite complex, Turkey: possible model for an oceanic ridge. *Bull. géol. Soc. Amer*., **88**, 1740–1748.

—— & WHITECHURCH, H. 1980. The magmatic cumulates of Antalya (Turkey): evidence of multiple intrusions in ophiolitic magma chamber. *In*: 'Ophiolites' *Proc. Int. Oph. Symposium, Cyprus, 1979*, 337–391.

MONOD, O. 1976. La 'courbure d'Isparta' une mosaïque de blocs autochtones surmontés de nappes composites à la jonction de l'arc hellénique et de l'arc taurique. *Bull. Soc. géol. France*, **18**, 521–532.

MONTIGNY, R., THUIZAT, R., REUBER, I., WHITECHURCH, H. & JUTEAU, T. K-Ar investigation of the Antalya ophiolites. Geological implications. *Sci. Terre. Strasbourg*., (in press).

NICOLAS, A. & POIRIER, J. P. 1976. *Crystalline plasticity and solid state flow in metamorphic rocks*, J. Wiley, London, 444 p.

—— & LE PICHON, X. 1980. Thrusting of young lithosphere in subduction zones with special reference to structures in ophiolitic peridotites. *Earth and Planet. Sc. Lett*., **46**, 397–406.

—— & VIALON, P. 1980. Les mécanismes de déformation ductile dans les roches. *Mém. h. Sér., Soc. Geol. France*., **10**, 127–139.

RALEIGH, C. B. 1968. Mechanism of plastic deformation of olivine. *J. geophys. Res*., **73**, 5391–5406.

REUBER, I. 1982. Générations successives de filons grenus dans le complexe ophiolitique d'Antalya (Turquie): origine, évolution, et mécanisme d'injection des liquides. *Thesis Strasbourg*., 246 p.

——, WHITECHURCH, H. & CARON, J. M. 1982. Setting of gabbro dikeletes in an ophiolitic complex (Antalya, Turkey) by hydraulic fracturing: downward injection of residual liquids. *Nature*, **296**, 141–143.

——, MICHARD, A., CHALOUAN, A., JUTEAU, T. & JERMOUNI, B. 1982. Structure and emplacement of the alpine-type peridotites from Beni-Bousera, Rif Marocco: a polyphase tectonic interpretation. *Tectonophysics*., **82**, 231–251.

RICOU, L. E., ARGYRIADIS, I. & LEFEVRE, R. 1974. Proposition d'une origine interne pour les nappes d'Antalya et le massif d'Alanya (Taurides occidentales, Turquie). *Bull. Soc. géol. France*, **16**, 107–111.

——, —— & MARCOUX, J. 1975. L'axe calcaire du Taurus, un alignemant de fenêtres arabo-africaines sous des nappes radiolaritiques, ophiolitiques et métamorphiques. *Bull. Soc. géol. France*, **17**, 1024–1044.

RIEDEL, W. 1929. Zur Mechanik geologischer Brucherscheinungen. *Centrabl. f. Min. Geol. Paläont*., 1929 B, 354–368.

ROBERTSON, A. H. F. & WOODCOCK, N. H. 1980. Strike-slip related sedimentation in the Antalya Complex, SW Turkey. *Spec. Publ. int. Ass. Sediment*., **4**, 127–145.

—— & —— 1981a. Alakır Çay Group, Antalya Complex, SW Turkey: a deformed Mesozoic Carbonate margin. *Sedimentary Geology*., **30**, 95–131.

—— & —— 1981b. Bileyeri Group, Antalya Complex: deposition on a Mesozoic passive continental margin, south-west Turkey. *Sedimentology*., **28**, 381–399.

—— & —— 1981c. Gödene Zone, Antalya Complex: volcanism and sedimentation along a mesozoic continental margin, SW Turkey. *Geol. Rdsch*., **70**, 1177–1214.

—— & —— 1982. Sedimentary history of the south-western segment of the Mesozoic-Tertiary Antalya continental margin, south-western Turkey. *Eclogae. geol. Helv*., **75**, 517–562.

RUHLAND, M. 1973. Méthode d'étude de la fracturation naturelle des roches associées à divers modèles structuraux. *Bull. Sci. Geol. Strasbourg*., **26**, 91–113.

TCHALENKO, J. S. 1970. Similarities between shear zones of different magnitudes. *Bull. geol. Scc. Amer*. **81**, 1625–1640.

Woodcock, N. H. & Robertson, A. H. F. 1982. Wrench and thrust tectonics along a Mesozoic-Cenozoic continental margin: Antalya Complex SW Turkey. *J. geol. Soc.*, **139**, 147–163.

Ingrid Reuber, Laboratoire de Pétrologie, Université Louis Pasteur, 1, rue Blessig, 67084 Strasbourg-Cedex, France.

Fossil and K-Ar data for the age of the Antalya complex, S W Turkey

Pınar O. Yılmaz

SUMMARY: The Antalya Complex of southwestern Turkey contains, from west to east, the autochthonous Mesozoic-Cenozoic platform carbonates of the Bey Dağları massif and three allochthonous units: the Kumluca complex of imbricated Mesozoic continental-margin sedimentary rocks; units of the Late Cretaceous Alakır Çay ophiolite sequence and the Late Cretaceous Tekirova partial ophiolite sequence; and the Teke-Tahtalı Dağ unit of Palaeozoic-Mesozoic sedimentary rocks originating on continental crust. The allochthonous units are overlain by conglomerates containing Maastrichtian clasts, ophiolitic olistostromes and locally by pelagic marls bearing Lower Miocene foraminifera. Samples from the Alakır Çay and Tekirova ophiolites have been isotopically dated. These ages cluster in the latest Cretaceous.

Fossils from pelagic sediments interbedded with pillowed basalt flows yielded ages of Late Triassic-Jurassic and Late Jurassic-Early Cretaceous, indicating intermittent volcanic activity through Mesozoic time. It is believed that the ophiolites of the Alakır Çay and Tekirova unit were formed in the same basin but were emplaced during separate events.

This paper provides additional K-Ar radiometric age determinations of the ophiolite rocks of the Antalya Complex. The Antalya Complex lies along the Cyprus-Hatay ophiolitic belt of the Eastern Mediterranean. It is also part of the southwest Tauride mountains of the Alpine system which continue westward to the Hellenides and eastward to the Zagros. The Alpine-Himalayan system is characterized throughout by slivers of ophiolites which are remnants of the Mesozoic-Cenozoic Tethys oceanic realm.

The Antalya Complex in southwest Turkey (Fig. 1) consists of: allochthonous units of imbricated Mesozoic continental margin rocks; a Late Cretaceous ophiolitic unit; tectonic slices of Palaeozoic-Mesozoic platform rocks; a Late Cretaceous partial ophiolite sequence; and the autochthonous Mesozoic-Cenozoic continental platform of Bey Dağları.

The region has been studied by many geologists of various nationalities for 15 years yet there is no accepted solution for the complex geology of the Antalya Complex. The author spent three field seasons in the Antalya area doing detailed geological mapping, reconnaissance and field collation of available geologic maps, and collecting samples for dating and analysis. Contributions of other workers in the area (Baykal & Kalafatçıoğlu 1973; Juteau 1968, 1975; Woodcock & Robertson 1977, 1981, 1982; Robertson & Woodcock 1980a, 1981a, 1981b, 1981c, 1982; Şenel 1980; Yılmaz 1978) have been compiled and referenced and an evolutionary model has been developed for the geology and tectonics of the Antalya Complex (Yılmaz 1981, Yılmaz and Maxwell 1982, 1983, see also Robertson & Woodcock, this volume). Samples were submitted for K-Ar analysis (potassium content determined at Mobil Research & Development Corp.; argon content at FM consultants Ltd) and palaeontological study (University of Texas, Min. Research & Explor. Inst. Turkey, Mobil Expl. & Prod. Serv. Inc.). These additional radiometric dates supplemented by stratigraphic data will provide additional constraints for the timing of tectonic events and help in reconstructing an evolutionary model for the area. All ages are quoted as Ma.

Previous work

The Antalya Complex (after Woodcock & Robertson 1977) was originally termed the Antalya Nappes by Lefèvre (1967) and 'Antalya Birliği' (unit) by Ozgül (1976). Previous mapping in this area had displayed north-trending structures (Colin 1962; Juteau 1968, 1970, 1977, 1979; Juteau *et al.*, 1973, 1977; Brunn *et al.*, 1970, 1971; Baykal & Kalafatçıoğlu 1973; Dumont *et al.*, 1972; Poisson 1977, 1978; Marcoux 1976, 1977; Yılmaz 1978, 1981; Yılmaz *et al.*, 1981). The dominant north-trending tectonic grain of the Complex is due to major faults which juxtapose components of the Complex with contrasting lithology, age and structure. Lefèvre and later workers (ref. op. cit.) considered the area to be an allochthonous assemblage dominated by Mesozoic ophiolitic and related sedimentary rocks, emplaced along low-angle faults (lower, middle and upper nappes; ref. op. cit.; Table 2). It was postulated that the nappes were transported eastward over the Bey Dağları autochthonous carbonate platform.

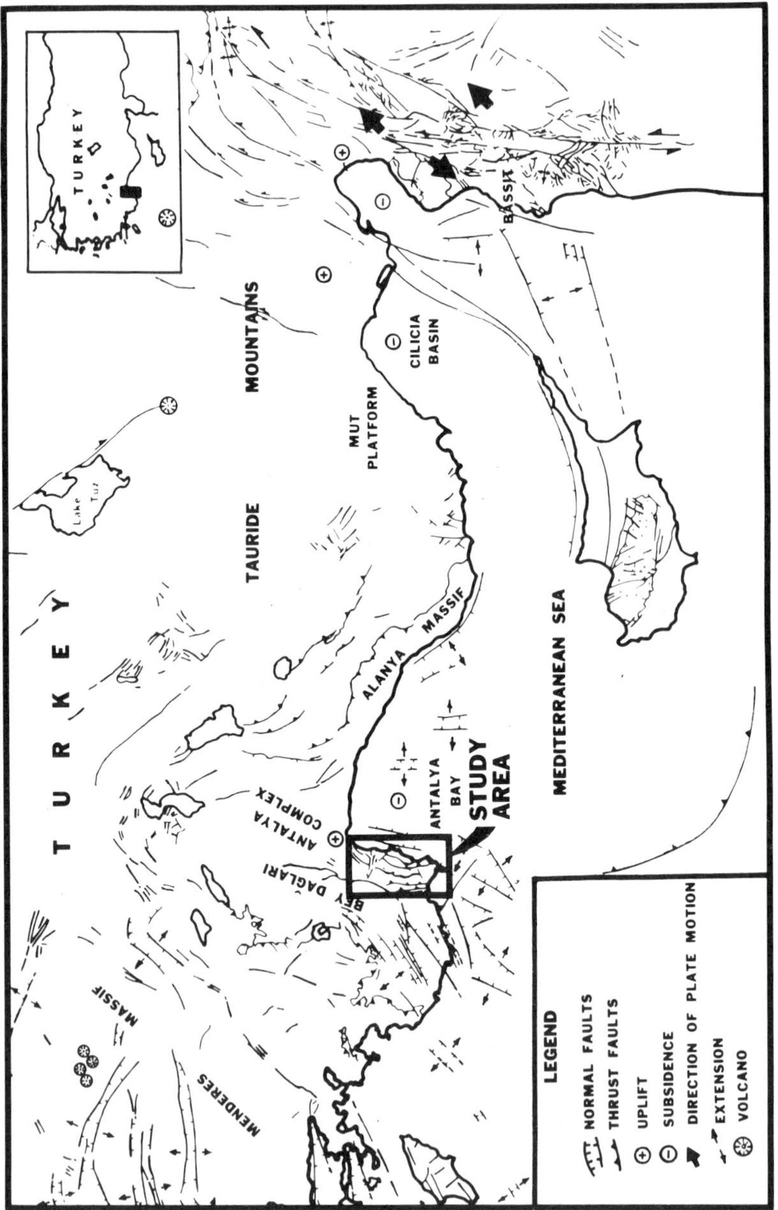

FIG. 1. Index map. Locality of study area in southern Turkey. Present-day tectonics shown.

TABLE 1. *K-Ar analytical data for Tekirova and Alakır Çay units.*

Sample #	Locality	Rock type	Mineral	%K	%Ar^{40} rad	v/m	Calculated age* (my)
POY-2A	Kozaklı Tepe-Coastal Tekirova	Amphibolite	Hornblende	0.048	8.7	$(2.50 \pm 0.22)10^{-4}$	130.4 ± 11.5
					7.6	$(2.21 \pm 0.21)10^{-4}$	114.4 ± 11.0
POY-3	Arap Yalısı Coastal Tekirova	Diabase	Hornblende	0.050	8.9	$(1.95 \pm 0.17)10^{-4}$	98.0 ± 8.7
					7.1	$(1.83 \pm 0.18)10^{-4}$	92.1 ± 9.2
POY-51	Tekerlek Tepe-Coastal Tekirova	Quartz diabase	Whole rock	0.158	15.9	$(4.40 \pm 0.16)10^{-4}$	70.4 ± 3.9
					15.1	$(4.17 \pm 0.16)10^{-4}$	66.8 ± 3.8
TT-1	Tekerlek Tepe-Coastal Tekirova	Quartz hornblende diabase	Whole rock	0.116	77.3	$(3.49 \pm 0.08)10^{-4}$	91.0 ± 1.8
					77.8	$(3.59 \pm 0.08)10^{-4}$	92.3 ± 1.8
					77.5	$(3.55 \pm 0.08)10^{-4}$	92.4 ± 1.8
GB-1	Göynük Çeşme-Tekirova	Quartz hornblende gabbro	Whole rock	0.183	58.6	$(4.90 \pm 0.07)10^{-4}$	67.6 ± 1.9
					52.8	$(4.86 \pm 0.06)10^{-4}$	67.2 ± 1.8
					55.4	$(4.84 \pm 0.06)10^{-4}$	66.8 ± 1.8
GB-2	Göynük Çeşme-Tekirova	Quartz hornblende gabbro	Plagioclase	0.297	69.9	$(7.88 \pm 0.10)10^{-4}$	68.1 ± 0.9
					69.2	$(7.64 \pm 0.11)10^{-4}$	65.2 ± 1.0
					58.4	$(8.04 \pm 0.12)10^{-4}$	68.5 ± 1.1
GB-3	Göynük Çeşme-Tekirova	Quartz hornblende diabase	Whole rock	0.126	80.8	$(4.03 \pm 0.08)10^{-4}$	84.4 ± 2.2
					56.8	$(4.01 \pm 0.08)10^{-4}$	84.0 ± 2.2
					78.0	$(4.08 \pm 0.08)10^{-4}$	85.5 ± 2.2
POY-117B	Dedeler mah.-Kemer	Gabbro	Plagioclase	0.070	8.8	$(1.56 \pm 0.10)10^{-4}$	67.8 ± 4.8
					9.0	$(1.58 \pm 0.13)10^{-4}$	68.6 ± 6.0
POY-117C	Dedeler mah.-Kemer	Gabbro	Plagioclase	0.026	4.0	$(5.13 \pm 0.77)10^{-4}$	49.5 ± 9.9
					3.6	$(5.05 \pm 0.76)10^{-4}$	48.7 ± 9.7
POY-118C	Aldağlı Boğazı-Kemer	Diabase	Hornblende	0.386	26.0	$(8.24 \pm 0.35)10^{-4}$	54.1 ± 3.2
					27.3	$(8.68 \pm 0.33)10^{-4}$	56.9 ± 3.4
POY-64	Karkadındömek Tepe-Alakır Çay	Gabbro	Whole Rock	0.032	5.3	$(7.09 \pm 1.35)10^{-5}$	55.5 ± 10.9
					5.1	$(7.93 \pm 1.19)10^{-5}$	61.8 ± 9.8
K-4	Karkadındömek Tepe-Alakır Çay	Quartz-bearing hornblende gabbro	Plagioclase	0.212	50.5	$(5.01 \pm 0.08)10^{-4}$	59.8 ± 1.7
					37.8	$(4.90 \pm 0.08)10^{-4}$	58.5 ± 1.7
					49.2	$(5.05 \pm 0.08)10^{-4}$	60.3 ± 1.7
8-D	Sarsaldömek Tepe-Alakır Çay	Plagiogranite	Whole rock	0.214	61.5	$(5.63 \pm 0.08)10^{-4}$	66.5 ± 2.1
					60.8	$(5.36 \pm 0.08)10^{-4}$	63.5 ± 2.0
					60.2	$(5.45 \pm 0.08)10^{-4}$	64.5 ± 2.2
POY-14	Akkayalar-Alakır Çay	Amphibolite	Hornblende	0.793	70.0	$(1.89 \pm 0.03)10^{-3}$	72.4 ± 1.1
					69.1	$(1.96 \pm 0.03)10^{-3}$	75.0 ± 1.1
POY-59	Zincirkıracak Dere-Alakır Çay	Amphibolite	Hornblende	0.529	68.3	$(1.34 \pm 0.03)10^{-3}$	76.8 ± 1.7
					69.0	$(1.40 \pm 0.03)10^{-3}$	80.2 ± 1.7

* $\lambda_\varepsilon = 0.581 \times 10^{-10} yr^{-1}$, $\lambda_\beta = 4.962 \times 10^{-10} yr^{-1}$, K^{40}/K total $= 1.167 \times 10^{-4}$ atom/atom. Estimated error of all %K Values is 1%.
v/m = Volume of radiogenic Ar^{40} in mm^3/gram. $\sigma = 2$.
Argon analyses by FM Consultants Ltd., Cambridge, England.
Potassium analyses by Mobil Research and Development Corp.

Later workers accepted the nappe structure for this area (Delaune-Mayere *et al.*, 1977; Juteau 1968, 1975, 1977; Marcoux 1976, 1977; Monod *et al.*, 1974; Ricou *et al.*, 1974) but have debated the place of nappe origin and the direction and distance of tectonic transport. Brunn *et al.* (1977), Gutnic *et al.* (1979) and Ricou *et al.* (1974, 1975) support an internal origin for the Antalya Nappes in a Mesozoic ocean basin located to the north. Since the sedimentation was continuous on the autochthon from Jurassic to Miocene (Poisson 1978), the internal origin requires either great distances of transport (250 km, Ricou *et al.*, 1974) or that part of the Bey Dağları platform be allochthonous (Ricou *et al.*, 1979). An external origin for the Complex in a Mesozoic ocean basin located to the south is proposed by Dumont *et al.* (1972), Brunn (1974), Monod *et al.* (1974) and Monod (1976). Recent works by Delaune-Mayere *et al.* (1977) and Ricou *et al.* (1979) favour the internal origin, while Robertson and Woodcock (1980a) propose emplacement from an external, southern basin. Palaeogeographic reconstructions by Şengör & Yılmaz (1981) propose emplacement of the

Antalya Nappes from an external southern basin which is a remnant of the Neo-Tethys ocean.

Robertson & Woodcock (1980a, 1980b) and Woodcock & Robertson (1981a, b, c), have postulated that strike-slip components are present on the north-trending faults. They explain the Antalya Complex as an originally intact part of the Mesozoic continent-ocean boundary which has been tectonically disrupted by wrench-faulting during the Late Cretaceous to Tertiary. Part of their field evidence is the presence of elongate anastomosing strands of different rock types bounded by presumed strike-slip faults.

The southern part of the Antalya Complex is divided into five north-trending units based on differences in constituent rocks, tectonic style and age (Table 2). The most recent subdivision of the area is that of Yılmaz (1981) which is after Robertson & Woodcock (1980a). In this paper, the named units of the Antalya Complex from west to east are: the autochthonous Mesozoic-Cenozoic platform carbonates of the Bey Dağları massif and four allochthonous units: the Kumluca unit of imbricate Mesozoic continental-margin sedimentary rocks; the ophiolitic units of Alakır Çay (the equivalent of Robertson & Woodcock's Gödene Zone); the Teke-Tahtalı Dağ unit of Palaeozoic and Mesozoic shelf sedimentary rocks (Table 2) (Kemer Zone of Robertson & Woodcock); and the Tekirova unit, a partial ophiolite sequence of Late Cretaceous age.

Geology

The following account gives brief summaries of the geology of the Antalya Complex units. Field and laboratory studies by the writer were concentrated in the southern portion of the Antalya Complex.

Bey Dağları Massif

The Bey Dağları massif is a northeast-trending anticlinorium with Mesozoic and early Tertiary carbonates overlain by late Tertiary detrital clastics. This massif has been described in detail by Poisson (1978) and Hayward (this volume).

Kumluca unit

The Kumluca unit is an elongated, north-trending, imbricate thrust-complex stacked on the eastern flank of the Bey Dağları massif. Sedimentary rocks involved are mostly Triassic and Jurassic in age, overlain by Cretaceous rocks in the north and east. This unit has been mapped by Woodcock & Robertson (1981b). The northern Kumluca unit is described as the lower nappe (Brunn *et al.*, 1971; Dumont *et al.*, 1972), while there is no equivalent nappe unit for the southern part of the Kumluca unit (Table 2).

Alakır Çay unit

The Alakır Çay unit has been designated as the Middle Nappe, transported over large distances (Brunn *et al.*, 1971; Dumont *et al.*, 1972; Ricou *et al.*, 1974, 1975, 1979; Marcoux 1976, 1977; Juteau 1975; Table 2). It is also described as rocks of a deformed Mesozoic carbonate margin representing outer-margin facies which lapped over adjacent mafic crust (Robertson & Woodcock 1981b; Table 2). The Alakır Çay ophiolitic unit consists of pelagic sedimentary and igneous rocks ranging in age from Upper Triassic to Upper Cretaceous. It is overlain by coarse clastics and ophiolitic olistostromes with Maastrichtian-fauna-bearing clasts. Upper Triassic alkaline lavas interbedded with pelagic sediments are widespread. The recent discovery of alkaline lavas of Jurassic-Early Cretaceous age (Yılmaz 1981) in the Alakır Çay unit has increased the possibility of an oceanic protolith existing from Late Triassic to Late Cretaceous time. This is contrary to earlier assumptions that the Triassic alkaline lavas are rift-related and not part of the oceanic crust (Robertson & Woodcock 1980b, 1981b).

The author's structural mapping, the results of the data compilation and the laboratory work together suggest that the Alakır Çay unit includes rocks of oceanic and carbonate margin origin, assembled by tectonic and gravitational mechanisms into multiple slices interlayered with melange during Latest Cretaceous-Palaeocene time.

Teke-Tahtalı Dağ unit

The Teke-Tahtalı Dağ unit is interpreted in the literature as the Upper Nappe which has been transported large distances (Brunn *et al.*, 1970, 1971; Dumont *et al.*, 1972; Ricou *et al.*, 1974, 1975, 1979; Marcoux 1977; Gutnic *et al.*, 1979; Table 2). It is alternatively interpreted as a sequence of rocks which are part of carbonate buildups on pre-Triassic basement slivers later emplaced in tectonic slices with other basin rocks on the adjacent platform (Robertson & Woodcock 1980a, 1980b, 1980c, 1982; Table 2). This unit consists of Upper Triassic-Cretaceous shallow water carbonates overlying Ordovician-

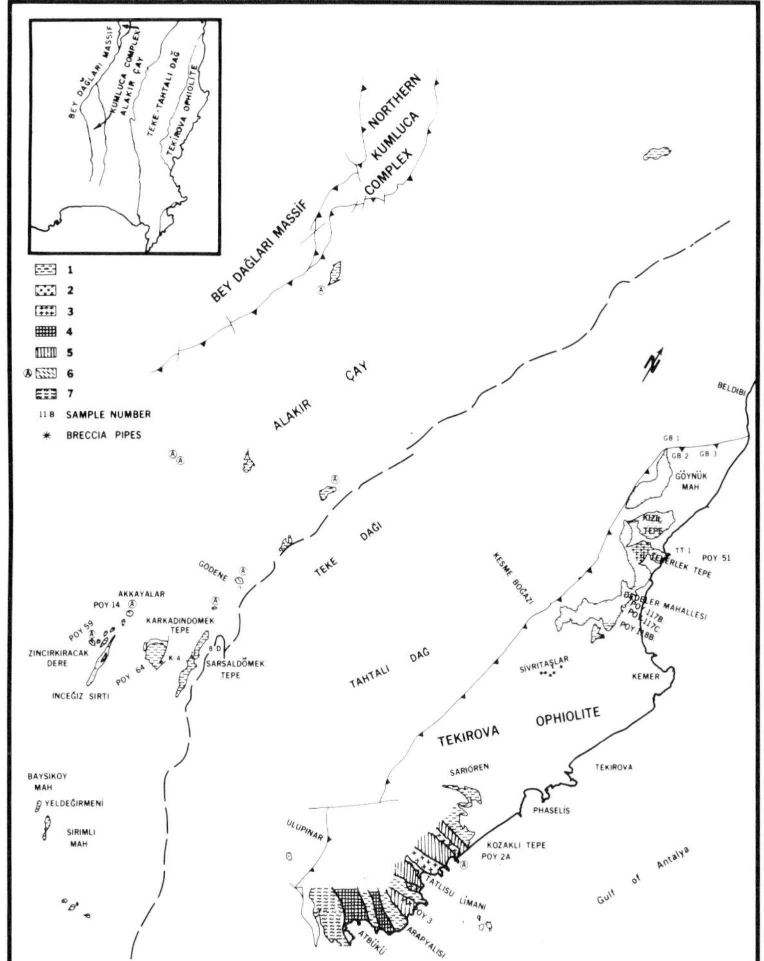

FIG. 2. Location map for chemical and K-Ar dating samples. Geologic units of the Antalya complex are also shown. Legend: 1-Gabbro; 2-Diabase; 3-Pyroxenite; 4-Dunite-Pyroxenite interlayered; 5-Massive Dunite; 6-Amphibolite; 7-Harzburgite.

Permian sandstone-shale-evaporite-limestone sequences.

Tekirova ophiolite

This unit consists of a partial ophiolite sequence and overlying sedimentary rocks. The ophiolite sequence present consists of a basal section at Adrasan and in the Çıralı-Tekirova area (Fig. 2) which is made up of: serpentinized ultramafic rocks at the base; harzburgite with diabase dikes; cumulate ultramafics (pyroxenite, dunite with chromite, websterite, and diallagite); a cumulate-gabbro complex with diabase dikes, pegmatitic gabbro dikes, magma-chamber feeder dikes, and breccia pipes. North of Kemer (Fig. 2), the ophiolite sequence consists of cumulate gabbro with diabase dikes, pegmatitic gabbro dikes and a locally well-developed diabase sheeted-dike complex. This partial ophiolite is interpreted to be of Late Cretaceous age (Table 1). The petrology and chemistry of these rocks have been studied in detail by Juteau (1975), and Juteau and Whitechurch (1979; see also Reuber, this volume).

Petrology and K-Ar radiometric dates from the Alakır Çay and Tekirova units

The ophiolites of the Alakır Çay and Tekirova units are very similar petrologically. The major differences between the two are the completeness of the sequence and amount of tectonic

TABLE 2. *Comparison between the Antalya Complex units from published data and this study.*

Juteau, 1975 Juteau & Whitechurch, 1979 Delaune-Mayere *et al.*, 1977	Robertson & Woodcock, 1980a; 1981a, b, c; 1982 this volume	Yılmaz, 1981 and this study
BEY DAĞLARI PLATFORM Upper Mesozoic-Tertiary carbonate platform.	BEY DAĞLARI ZONE Upper Mesozoic-Tertiary carbonate platform.	BEY DAĞLARI MASSIF Upper Mesozoic-Tertiary carbonate platform.
LOWER ANTALYA NAPPE Çatal Tepe Fm.: Mesozoic shelf and pelagic sedimentary rocks.	KUMLUCA ZONE (1) Northern part: Bilelyeri/Dereköy Group sediments (Late Triassic-Upper Cretaceous). (2) Southern part: Alakır Çay Group sediments (Late Triassic-Late Cretaceous).	KUMLUCA UNIT Imbricated Mesozoic pelagic sediments.
MIDDLE ANTALYA NAPPE Alakır Çay Fm. Late Triassic ophiolite and Late Triassic-Upper Cretaceous pelagic sedimentary rocks. Late Cretaceous ophiolite at Adrasan-Tekirova with no genetic relationship to the Alakır Çay ophiolite.	GÖDENE ZONE Tectonized Late Cretaceous ophiolite and Alakır Çay Group sediments. Upper Triassic alkaline basalts and pelagic sediments formed on mafic crust.	ALAKIR ÇAY UNIT Late Cretaceous tectonically imbricated ophiolitic melange. Upper Triassic-Jurassic alkaline basalts with pelagic sediments and Upper Jurassic-Early Cretacous alkaline basalt with pelagic sediments formed on oceanic-type crust. Late Cretaceous ophiolite which consists of serpentinite, ultramafic cumulates, gabbro & diabase.
UPPER ANTALYA NAPPE Kemer Fm. Palaeozoic-Mesozoic shelf/platform sedimentary rocks	KEMER ZONE Palaeozoic-Mesozoic sediments and carbonate build-ups on continental basement.	TEKE-TAHTALI DAĞ UNIT Palaeozoic-Mesozoic sedimentary sequences formed on continental crust.
	TEKIROVA ZONE Late Cretaceous partial ophiolite and Late Cretaceous ophiolite derived sedimentary cover.	TEKIROVA OPHIOLITE Late Cretaceous partial ophiolite with basal tectonites and ultramafic/mafic cumulate rocks overlain unconformably by conglomerates with Maastrichtian-age fauna bearing clasts.

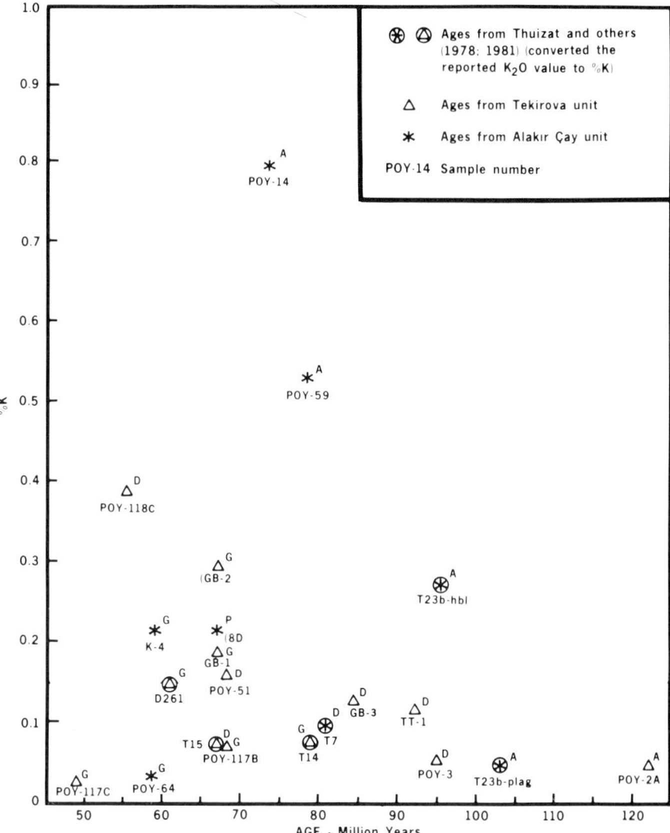

FIG. 3. Plot of age determined in relation to % K. A = amphibolite, G = gabbro, D = diabase, P = plagiogranite.

disruption. Only amphibolite, gabbro, diabase and plagiogranite rocks have been analysed and one similarity in composition of these rocks is the low potassium content, which is characteristic of ophiolites.

The K-Ar method is a powerful chronological tool for the whole range of geologic time with greatest precision being in the Mesozoic-Cenozoic (Schaeffer & Zahringer 1966). In this paper, the term K-Ar 'date' indicates the numerical result and the K-Ar 'age' indicates the geologic event.

Samples were chosen carefully in the field and further selected petrographically. Analytical technique for potassium measurement was the combination of flame-photometry and precipitation described by Burke *et al.* (1969). Table 1 gives the average of potassium measurements for each sample.

One of the most important features of potassium-argon dating is availability of minerals with different thermal diffusion properties for argon (Dalrymple & Lanphere 1969). Thus, the dating of two or more minerals from a rock can yield information on more than one geologic event or on rates of cooling (Hamilton & Farquhar 1968). In this study potassic minerals were dated when they could be separated, otherwise the whole-rock was dated. Hornblende has the highest argon closure temperature whereas plagioclase has some sites which can exchange argon at very much lower temperature and so can yield ages significantly lower. Whole rock ages of amphibolitic gabbros will be expected to lie somewhere between amphibole and plagioclase mineral ages.

Samples low in potassium are susceptible to errors in the K-Ar dating process, mostly because of the difficulty of accurately determining the low content of radiogenic argon produced as well as the consequent larger error caused by any excess radiogenic argon, at the time of crystallization. Some error may be introduced in the potassium determination itself for very low concentrations. The present analytical techniques are not refined enough to allow evaluation of the dates calculated from radiogenic argon contents of less than 10% of total argon (F. McDowell, pers. comm. 1982). Five samples contained less than 1% ^{40}Ar (Table 1). These sample dates are not used in the final geological interpretation of the Antalya Complex.

Figure 3 is a plot of the K-Ar ages determined in relation to per cent potassium of samples. The sample locations are shown in Fig. 2. The values include all of the results reported here and published values (Thuizat *et al.*, 1978, 1981; Yılmaz & Maxwell 1982).

Alakır Çay unit

The igneous and metamorphic rocks in the Alakır Çay unit are of six kinds: amphibolites, ultramafics, gabbros, plagiogranites, diabases and basalts. The ultramafics and gabbros appear to be associated and gradational contacts are observed, with the abundance of plagioclase increasing upwards.

In this study two amphibolites (POY-14 and POY-59), and three magmatic rocks, two gabbros (POY-64 and K4) and a plagiogranite (8D) have been dated. One of the gabbros (POY-64) had a ^{40}Ar content below the acceptable level although it gave a similar final result to the other gabbro.

Amphibolites

The two samples were from metamorphic soles to thick serpentinite thrust sheets and are interpreted as having formed during the detachment of an ophiolite slice from its substrate or at some later stage during emplacement when it was thrust over or against rocks of basaltic composition. They are composed of blue-green hornblendes in a fine grained matrix of pyroxenes and plagioclases. The pyroxenes are altered to hornblende around the rims. The two samples give similar ages of 73.7 ± 2.5 and 78.5 ± 3.4 (Late Cretaceous). The 'slicing' ages if real must be younger than the formation ages of both the overlying ophiolite and the underlying mafic rocks, but not necessarily younger than the cooling age of the overlying ophiolite since it may have been very hot at the time of slicing-up.

Gabbros

The two gabbro samples K4 (Plagioclase date) and POY-64 are from the Karkadindomek Tepe mass east of the ultramafic mass with the sole amphibolite and give ages of 59.5 ± 2.6 and 58.7 ± 13 though the latter has a large attached error because of its low ^{40}Ar content and is technically outside the acceptable limit.

Plagiogranite

The plagiogranite (8D) comes from an ophiolitic mass still further to the east, Sarsaldomek Tepe, and yielded a whole-rock age of 64.8 ± 3.8. The ages of these 'magmatic' rocks are interpreted as being cooling ages following the hydration of the original high-temperature igneous assemblages. This is most simply interpreted as being close to the age of formation rather than related to later emplacement events which are only likely to have affected rocks in the immediate vicinity of the thrusts, particularly as these rocks are undeformed and contain hornblende rather than low grade greenschist assemblages.

The implication is that the Alakır Çay unit contains ophiolitic slices with a range of formation and emplacement ages—one formed pre-75 Ma and emplaced at 75 Ma and one or perhaps two, formed at about 65 Ma and emplaced at the same time or later.

The Upper Cretaceous ages complement the fossil ages reviewed in detail below which confirm that the igneous activity also occurred in the Upper Triassic and also show that it occurred in the Upper Jurassic-Lower Cretaceous interval.

Tekirova ophiolite

Pillow-basalts and associated sediments are apparently absent from the Tekirova unit and therefore fossil ages are not available. T. Juteau (pers. comm. 1981) reports Upper Triassic pillow lava outcrops from the Tekirova unit but they are in tectonic contact with the ultramafics in Çiralı. The only stratigraphic constraints are given by the overlying clastics with ophiolitic clasts bearing Maastrichtian fauna and local hemipelagic marls of Miocene age. The Alakır Çay unit is overlain by similar clastics with Maastrichtian age clasts.

In this study six samples met the analytical criteria for acceptability, all gabbros or diabases with hornblende, plagioclase ± quartz (Table 1). The whole rock ages are 91.9 ± 2.3 (TT-1), 84.7 ± 3.0 (GB-3), 68.6 ± 5.7 (POY-51), 67.2 ± 2.3 (GB-1). Mineral ages were obtained from a plagioclase from GB-2, 67.1 ± 3.0 and hornblende from POY-118C, 55.5 ± 4.7. These ages are in broad agreement with the results of Thuizat *et al.* (1978) who obtained mineral ages of 79 ± 2 (plag in 92-diorite), 67 ± 2 (hornblende from sheeted diabase in the same quartz diorite) and 61 ± 2 (plag from gabbro). The second date was from the same dyke complex as the sample POY-51 of this study, dated at 68.6 ± 5.7.

These ages present considerable interpretation problems in detail, even if it is clear that loosely they represent Upper Cretaceous ophiolite generation. Thuizat *et al.* (1978) interpreted the younger age of 61 ± 2 from plagioclase, as representing a distinct post-emplacement 'event'. There is however no supporting textural evidence of a distinct metamorphic or deformational event. It is not clear whether the older ages obtained in this study 91.9 ± 2.3 84.7 ± 3.0 represent genuinely older ophiolite generation or whether the spread of ages as a whole is a result of complex and long-lasting processes of potassium and argon diffusion in hydrating mafic plutonics. It must not be forgotten that the rocks sampled are all thoroughly 'altered' by normal igneous geochemical standards.

It does seem apparent that the plutonic ophiolitic (gabbroic and doleritic) bodies of both the Tekirova and Alakır Çay units cannot readily be distinguished radiometrically and both are essentially Upper Cretaceous in age of formation and emplacement. How many separate creation and emplacement events are represented is not clear in either unit. Dyke emplacement may possibly have continued even after emplacement if the Palaeocene age from POY-118C is real.

Thuizat *et al.* (1978) interpreted the age discrepancy between quartz diorite host at 79 ± 2 and sheeted diabase within it at 67 ± 2 to be indicative of a real difference in age of intrusion, but quote the results of geochemical work suggesting that the two rock types are co-genetic.

Tekirova amphibolite ages

Sample POY-2A in Table 1 is an amphibolite inferred to be from a metamorphic sole which yields a hornblende age of 122.2 ± 19.7. The sample gave less than the accepted yield of ^{40}Ar and the date should perhaps be ignored. However, Thuizat *et al.* (1981) obtained hornblende and plagioclase ages of 97 ± 4 and 105 ± 7 respectively from a ?Tekirova garnet amphibolite which they interpret as the initial intra-oceanic obduction event. They point to a close convergence of amphibolite ages from Tauride ophiolites at around 95 ma. This conclusion contradicts their previous interpretation of the younger quartz-diorite age of 79 ± 2 in the Antalya Complex (Thuizat 1978) as being the age of genesis of the cumulates and tectonites, i.e. the ophiolite creation age. If the amphibolite age of 122.2 ± 19.7 is at all reliable, then this study confirms the same paradox—the amphibolite sole ages are older than the gabbro and dolerite ages, both in the Tekirova and in the Alakır Çay, yet the gabbros and dolerites show no evidence of a major metamorphic event post-obduction.

Sedimentary rock ages

Within the Alakır Çay melange unit, the sedimentary rocks associated with the ophiolitic elements provide fossil ages to complement the K-Ar dates. Both sets of dates should be used as time constraints on tectonic events in the evolution of the Antalya Complex.

The sedimentary rocks associated with the ophiolite units include pelagic sediments interbedded with pillowed basalt flows. The age of these sediments are interpreted as the age of volcanic activity. These include Upper Carnian to Upper Middle Norian pelagic limestones with calcareous nannofossils, foraminifera and radiolaria dated by E. A. Pessagno (pers. comm. 1980). Upper Triassic volcanic activity is widespread. The Sayrun volcanic massif is a 1.4 km thick pile of pillowed basalt flows interbedded with Upper Triassic Halobia-bearing and ammonitico rosso facies limestones (Juteau 1975; Marcoux 1976; Robertson & Woodcock 1980b, 1981b; Yılmaz 1981).

Upper Triassic volcanic activity is well documented in the Antalya area. However, the pelagic sediments interbedded with alkaline pillow-basalt flows locally yield younger ages. These have been reported in the Ülupınar/Kara Tepe area by Yılmaz (1981) and are Upper Jurassic to Lower Cretaceous. The location of these sections are shown in Fig. 4. Fossils identified in these pelagic units by Pessagno and by Mobil Expl. & Prod. Serv. Inc., Stratigraphic Laboratories indicate ages ranging from Kimmeridgian-Tithonian through Lower Hauterivian. These fossils include radiolaria, foraminifera, calcareous nannofossils and palynomorphs (Appendix I, fossil identification lists; Yılmaz 1981).

The sedimentary section overlying the Upper Triassic pillow-basalt flows contains pelagic limestone and radiolarite units which yield ages ranging from Upper Triassic to Middle Cretaceous (Yılmaz 1981; Robertson & Woodcock 1981a, b, c). The occurrence of pillow-basalt flows interbedded with pelagic sediments of Kimmeridgian-Tithonian to Lower Hauterivian

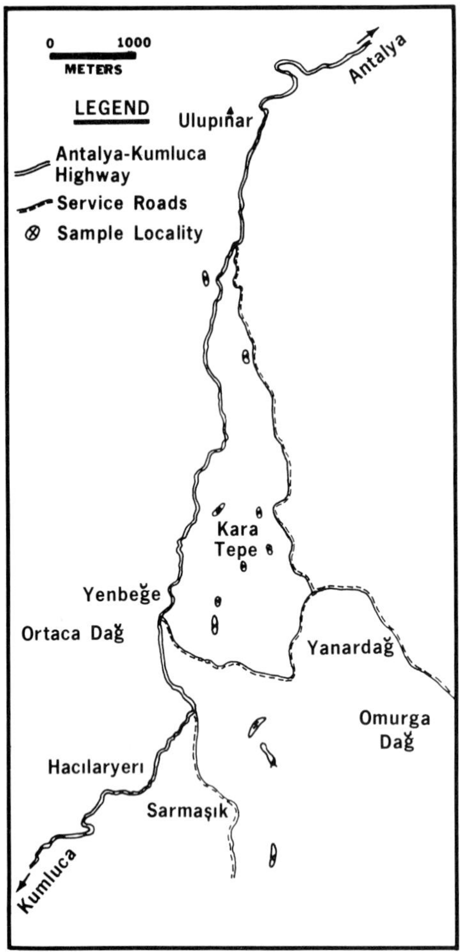

FIG. 4. Location map for pillowed basalt flows of Jurassic–Early Cretaceous age.

age indicates local contemporaneous volcanic activity. The relationship of this activity to the Upper Triassic volcanic sequences and to the Upper Cretaceous mafic and ultramafic rocks is not known. The radiometric dates in the Tekirova suggest ocean crust formation perhaps as old as 92 Ma (Cenomanian), narrowing the gap from the youngest pillow-basalt activity (Hauterivian) 130 Ma to some 40 Ma. The data presented here are not inconsistent with oceanic crust generation from the Triassic on.

Clastics with characteristics of rudites, mudflows, olistostromes and alluvial-fans unconformably overlie the ophiolitic units (Robertson & Woodcock 1982). They are exposed mostly along the Antalya-Kumluca highway. These clastics unconformably overlie parts of the Alakır Çay, Teke-Tahtalı Dağ and Tekirova units. The age of these clastics are deduced from the included fossil-bearing clasts. They range from middle Maastrichtian 69 Ma to Albian-Aptian. Radiolaria, foraminifera and palynomorph fossils have been identified (Yılmaz & Maxwell, 1983). These clastic sediments cannot be any older than middle Maastrichtian, but they could be a lot younger.

A local outcrop of highly deformed pelagic marls and shales occurs at Kemer (Fig. 2). These sediments are presumed to lie with an unconformable contact on the serpentinized ultramafic rocks SW of Kemer, but the contact is not exposed. These sediments are late Early to early Middle Miocene based on calcareous nanno-fossils and foraminifera fossil dates. These sediments were deposited between the time of ophiolite obduction and emplacement of the Antalya rocks on the Bey Dağları platform.

Discussion of fossil and K-Ar dates

This study presents laboratory results for K-Ar age analyses on ophiolitic rocks and their interpretation in terms of evolution of the Antalya Complex. With fossil age constraints, the K-Ar dates indicate the following sequence of events.

The igneous activity first started in Carnian-Norian with the extrusion of basalt flows. This lasted intermittently until Early Cretaceous, as indicated by fossil dates from the sediments interbedded with basalt flows. Oceanic crust formation was possibly continuous from 92–65 Ma judging from diabase and gabbro dates. At least one ophiolite detachment or emplacement event occurred during Campanian times (75 Ma). The ophiolite sequence was uplifted and exposed to erosion after late Maastrichtian, as indicated by fossil ages of clasts within ophiolite-rich clastics. A small scale thermal event possibly resulted in some argon loss giving apparent ages of some gabbros as Early Palaeocene (60 Ma).

There is no evidence of any igneous activity younger than Late Cretaceous. Any tectonic event after the Late Cretaceous-Palaeocene (e.g. the final emplacement of ophiolitic units against the Eocene-limestone units of the Bey Dağları) was not a thermal event sufficient to affect the earlier radiometric ages.

Conclusions

The Antalya Complex consists of: the allochthonous Kumluca complex of imbricated Mesozoic continental margin sediments; the Alakır Çay unit of Late Triassic to Late Cretaceous age incorporating a tectonically imbricated ophiolite sequence; the Teke-Tahtalı Dağ unit of Palaeozoic-Mesozoic sedimentary rocks; the Tekirova unit, a partial ophiolite sequence of Late Cretaceous age; and the authochthonous Bey Dăgları Mesozoic-early Tertiary carbonate platform.

Fossil and K-Ar radiometric age determinations have been utilized to model the evolution and emplacement of the ophiolitic units of the Antalya Complex. The youngest stratigraphic constraint is Middle Miocene hemipelagic marls which are considered to locally overlie the ophiolite sequence. Clastics including ophiolite debris are post-middle Maastrichtian, based on clasts of middle Maastrichtian rocks. The ophiolite sequence is overthrust against the Kumluca unit in which the sediments range from Late Triassic to Upper Cretaceous, and the Upper Cretaceous to Eocene carbonates of the Bey Dağları autochthon.

The age relationships of different parts of the ophiolite sequence are as follows: Late Triassic to Early Cretaceous, possibly intermittent, volcanic activity (fossiliferous sediments between pillowed basalt flows); Campanian amphibolites found in mylonitized serpentinites at soles of thrust sheets; Late Cretaceous time of crystallization of gabbroic rocks; a possible early Palaeocene thermal event which reset some gabbro ages; and Cenomanian to Palaeocene? diabase activity.

The results indicate the following sequence of events: Late Triassic—fracturing of continental crust, volcanics of alkaline basalt composition; Late Triassic to Early Cretaceous—formation of a small oceanic basin, general subsidence, deposition of pelagic sediments, radiolarites and local volcanic activity; Campanian—emplacement of ophiolites, major thrusting starts, diabase intrusion; Maastrichtian—continued ocean floor generation, compression, formation of local ophiolitic melange, emplacement of ophiolite sequence and melange, clastics with ophiolite debris deposited; early Palaeocene—further emplacement of ophiolite units, local diabase intrusions, clastic deposition; Palaeocene–early Miocene—erosion and clastic deposition; mid-Miocene—subsidence, hemipelagic marls deposited locally; late Miocene—final emplacement of allochthonous units on the Bey Dağları carbonate platform.

ACKNOWLEDGMENTS: The author would like to thank Wm. A. Belden, The University of Texas-Austin and Mobil Exploration and Producing Services Inc. (MEPSI) for funding this study. Logistic support in the field was provided by the Mineral Research and Exploration Institute of Turkey (MTA). K-Ar analyses were provided by R. E. Denison and fossil dates by E. Pessagno, K. Young, MEPSI-Stratigraphy Laboratories and MTA (Mualla Serdaroğlu). John C. Maxwell guided this study throughout, and the author expresses particular gratitude to him for his help and encouragement. R. E. Denison, F. McDowell, A. M. C. Şengör, A. H. F. Robertson, N. Özgül, A. W. Bally, T. Juteau, R. E. Cobb, and M. Delaloye provided helpful comments. J. E. Dixon helped produce a substantially improved revision of the manuscript.

References

ARGYRIADIS, I., DE GRACIANSKY, P. C., MARCOUX, J. & RICOU, L.-E. 1980. The opening of the Mesozoic Tethys between Eurasia and Arabian-Africa. *Proc. 26th Int. Geol. Cong. Paris*, c5, 199–214.

BAYKAL, F. & KALAFATÇIOĞLU, A. 1973. New geological observations in the area west of Antalya Bay. *Bull. Miner. Res. Explor. Inst. Turkey*, **80,** 33–42.

BRUNN, J. H. 1974. Le problème de l'origine des nappes et de leurs translations dans les Taurides occidentales. *Bull. Soc. géol. Fr.* **12,** 515–56.

——, DE GRACIANSKY, P. CH., GUTNIC, M., JUTEAU, T., LEFEVRE, R., MARCOUX, J., MONOD, O. & POISSON, A. 1970. Structures majeures et corrélations stratigraphiques dans les Taurides occidentales. *Bull. Soc. geol. Fr.* **12,** 515–56.

——, DUMONT, J. F., DE GRACIANSKY, P. C., GUTNIC, M., JUTEAU, T., MARCOUX, J., MONOD, O. & POISSON, A. 1971. Outline of the geology of the western Taurides. *In:* CAMPBELL, A. S. (ed.). *Geology and History of Turkey*, Petroleum Exploration Society of Libya, Tripoli, 225–55.

BURKE, W. H., OTTO, J. B. & DENISON, R. E. 1969. Potassium-Argon dating of basaltic rocks. *J. geophys. Res.* **74,** 1082–6.

COLIN, H. J. 1962. Geologische Untersuchungen im Raume Fethiye-Antalya-Kaş-Finike. *Bull. Miner. Res. Explor. Inst. Turkey*, **59,** 19–61.

DALRYMPLE, G. B. & LANPHERE, M. A. 1969. *Potassium-Argon Dating.* W. H. Freeman & Co. San Francisco, 258 pp.

DELAUNE-MAYERE, M., MARCOUX, J., PARROT, J.-F. & POISSON, A. 1977. Modèle d'évolution Mésozoïque de la paléomarge Téthysienne au niveau des nappes radiolaritiques et ophiolitiques du Taurus Lycien, d'Antalya et du Baër-Bassit. *In:* BIJU-DUVAL., B. & MONTADERT, L. (eds.). *Structural History of the Mediterranean Basins.* Editions, Technip, Paris, 79–94.

DUMONT, J.-F., GUTNIC, M., MARCOUX, J. MONOD, O. & POISSON, A. 1972. Le Trias des Taurides occidentales (Turquie). Définition du bassin pamphylien: Un nouveau domaine à ophiolites à la marge éxterne de la chaîne taurique. *Zeit. Deutsch. Geol. Ges.* **123,** 385–409.

De Gracianksy, P. C. De, Lorenz, C. & Magne, J. 1970. Sur les étapes de la transgression du Miocene inférieur observée dans les fenêtres de Göçek (Sud-Ouest de la Turquie). *Bull. Soc. géol. Fr.* **17**, 557–64.

Gutnic, M., Monod, O., Poisson, A. & Dumont, J.-F. 1979. Géologie des Taurides occidentales (Turquie). *Mem. Soc. géol. Fr.* **137**, 112 pp.

Hamilton, E. I. & Farquhar, R. M. 1968. *Radiometric Dating for Geologists.* Interscience Publ. London, 506 pp.

Juteau, T. 1968. Commentaire de la carte géologique des ophiolites de la région de Kumluca (Taurus lycien, Turquie méridionale). *Bull. Miner. Res. Explor. Inst. Turkey,* **70,** 70–91.

——, 1970. Pétrogenèse des ophiolites des Nappes d'Antalya (Taurus Lycien Oriental). Leur liaison avec une phase d'éxpansion océanique active au Trias supérieur. *Mém. Sciences Terre,* **15,** 265–88.

——, 1975. Les Ophiolites des nappes d'Antalya (Taurides occidentales, Turquie) *Mém. Sciences Terre,* **32,** 692.

—— 1977. The ophiolitic complex of Antalya, *In:* Güvenç, T., Marcoux, J., Demirtaşlı, E. & Özgül, N. (eds.). *Western Taurus Excursion Geological Guidebook.* 6th Coll. Aegean Geol. Izmir 1977, 22 pp.

—— 1979. Ophiolites des Taurides: Essai sur leur historie océanique. *Rev. Géog. phys. Géol. dyn.* **21,** 191–214.

——, Lapierre, H., Nicolas, A., Parrot, J. F., Ricou, L. E., Rocci, G. & Rollet, M. 1973. Idées actuelles sur la constitution, l'origine et l'évolution des assemblages ophiolitiques mésogéens: *Bull. Soc. géol. Fr.* **15,** 478–93.

——, Nicolas, A., Dubessy, J., Fruchard, J. C. & Bouchez, J. L. 1977. The Antalya Complex (Western Taurides, Turkey): a structural model for an oceanic ridge. *Bull. geol. Soc. Am.* **88,** 1740–8.

——, & Whitechurch, H. 1979. The magmatic cumulates of Antalya (Turkey): Evidence of multiple intrusions in an ophiolitic magma chamber. *In:* Panayiotou, A. (ed.). *Ophiolites.* Proc. Int. Oph. Symp. Cyprus 1979, 377–391.

Lefèvre, R. 1967. Un nouvel élément de la géologie du Taurus Lycien: les nappes d'Antalya (Turquie). *C.r. Séances. Acad. Sci. Paris,* **D265,** 1365–8.

Marcoux, J. 1976. Les séries triasiques des nappes à radiolarites et ophiolites d'Antalya (Turquie): homologies et signification probable. *Bull. Soc. géol. Fr.* **18,** 511–2.

—— 1977. Geological sections of the Antalya region. *In:* Güvenç, T., Marcoux, J., Demirtaşlı, E. & Özgül, N. (eds.). *Western Taurus Excursion Geological Guidebook.* 6th Coll. Aegean Geol., Izmir 1977.

Monod, O. 1976. La 'courbure d'Isparta': une mosaïque de blocs authochtones surmontés de nappes composites à la jonction de l'arc hellénique et de l'arc taurique. *Bull. Soc. géol. Fr.* **18,** 521–32.

——, Marcoux, J., Poisson, A. & Dumont, J.-F. 1974. Le domaine d'Antalya, témoin de la fracturation de la plateforme africaine au cours du Trias. *Bull. Soc. géol. Fr.* **16,** 116–27.

Özgül, N. 1976. Torosların bazı jeolojik özellikleri. *Türkiye Jeol. Kur. Bul.* **19,** 65–78.

Poisson, A. 1977. The massif of Bey Dağları. *In:* Güvenç, T., Marcoux, J., Demirtaşlı, E. & Özgül, N. (eds.), *Western Taurus Excursion Geological Guidebook.* 6th Coll. Aegean Geol., Izmir 1977.

——, 1978. *Recherches géologiques dans les Taurides occidentales (Turquie).* Thèse de Docteur ès Sciences, Université de Paris-Sud, 815 pp.

Ricou, L.-E., Argyriadis, I. & Lefèvre, R. 1974. Proposition d'une origine interne pour les nappes d'Antalya et le massif d'Alanya (Taurides occidentales, Turquie). *Bull. Soc. géol. Fr.* **16,** 107–11.

——, —— & Marcoux, J. 1975. L'axe calcaire du Taurus, un alignement de fenêtres arabo-africaines sous des nappes radiolaritiques, ophiolitiques et métamorphiques. *Bull. Soc. géol. Fr.* **17,** 1024–44.

——, Marcoux, J. & Poisson, A. 1979. L'allochtonie des Bey Dağları orientaux. Reconstruction palinspastique des Taurides occidentales. *Bull. Soc. géol. Fr.* **21,** 125–33.

Robertson, A. H. F. & Woodcock, N. H. 1980a. Strike-slip related sedimentation in the Antalya Complex, SW Turkey. *Spec. Publ. Int. Assoc. Sediment.* **4,** 127–45.

—— & —— 1980b. Tectonic setting of the Troodos Massif in the East Mediterranean. *In:* Panayiotou, A. (ed.). *Ophiolites,* Proc. Int. Oph. Symp. Cyprus 1979, 36–49.

—— & —— 1981a. Bilelyeri Group, Antalya Complex: deposition on a Mesozoic passive continental margin in SW Turkey. *Sedimentology,* **28,** 381–99.

—— & —— 1981b. Gödene Zone, Antalya Complex continental margin, SW Turkey: volcanism and sedimentation along a Mesozoic marginal ocean crust. *Geol. Rdsch.* **70,** 1177–214.

—— & —— 1981c. Alakır Çay Group, Antalya Complex, SW Turkey. A deformed Mesozoic carbonate margin. *Sediment. Geol.* **30,** 95–131.

—— & —— 1982. Sedimentary history of the Southwestern segment of the Mesozoic-Tertiary Antalya continental margin, south-western Turkey. *Eclogae geol. Helv.,* **75,** 517–62.

Schaeffer, O. A. & Zahringer, J. 1966. Potassium Argon Dating. Springer-Verlag. New York, 234 pp.

Şenel, M. 1980. *Teke Torosları Güneydoğusunun Jeolojisi, Finike-Kumluca-Kemer (Antalya).* Unpublished report, M.T.A. Ankara, Turkey, 106 pp.

Şengör, A. M. C. & Yılmaz, Y. 1981. Tethyan Evolution of Turkey: A plate tectonic approach. *Tectonophysics.* **75,** 181–241.

Thuizat, R., Montigny, R., Çakir, U. & Juteau, T. 1978. K-Ar Investigations on two Turkish ophiolites. *In:* Zartman, R. E. (ed.), *Short Pap. 4th Int. Conf. Geochr. Cosmoschr. Isot. Geol. Geol. Surv. Open-File Rep. 78–701,* 430–32.

——, Whitechurch, H., Montigny, R. & Juteau, T. 1981. K-Ar dating of some infra-ophiolitic metamorphic soles from the Eastern Mediterranean: new evidence for oceanic thrustings before obduction. *Earth planet. Sci. Lett.* **52,** 302–10.

Woodcock, N. H. & Robertson, A. H. F. 1977. Imbricate thrust belt tectonics and sedimentation as a guide to emplacement of part of the Antalya Complex, SW Turkey. *Abs. 6th Coll. Aegean Geol., Izmir 1977*, 98.

—— & —— 1981. Wrench-related thrusting along a Mesozoic-Cenozoic continental margin: Antalya Complex, SW Turkey. *In:* McClay, K. R. & Price, N. J. (eds.). *Thrust and Nappe Tectonics*, Spec. Publ. geol. Soc. Lond, **9,** 359–62.

—— & —— 1982. Wrench and thrust tectonics along a Mesozoic-Cenozoic continental margin: Antalya Complex, SW Turkey. *J. geol. Soc. Lond.*, **139,** 147–65.

Yilmaz, P. O. 1978. *Alakır Çay Unit of the Antalya Complex (Turkey): an example of ocean floor obduction.* Thesis, M.S. (unpubl.), Bryn Mawr College, U.S.A., 91 pp.

—— 1981. *Geology of the Antalya Complex, SW Turkey.* Dissertation, PhD, Univ. Texas. Univ. Microfilms Int. Michigan, 268 pp.

—— & Maxwell, J. C. 1983. An example of an obduction melange: The Alakır Çay unit, Antalya Complex, SW Turkey. *Spec. Publ. geol. Soc. Am.* (in press).

—— & —— 1982. K-Ar Investigations from the Antalya Complex ophiolites, SW Turkey. *Ofioliti*, **2/3,** 527–38.

——, Maxwell, J. C. & Muehlberger, W. R. 1981. Antalya kompleksinin yapısal evrimi ve doğu Akdeniz' deki yeri. *Yerbilimleri*, **7,** 119–27.

Pinar O. Yilmaz, Mobil Oil Corporation, P.O. Box 5444, Denver, Colorado, 80217, U.S.A.

The Mesozoic organization of the Taurides: one or several ocean basins?

L. E. Ricou, J. Marcoux and H. Whitechurch

SUMMARY: The Taurides are an Alpine chain composed of both continental and oceanic units including ophiolitic massifs organized in three longitudinal belts. The Taurides are also an area of nappes, where overthrusting occurred successively from the Upper Cretaceous on and greatly modified the pre-tectonic organization. A single or a multiple ocean basin system have both been postulated.

It is shown that the nappes composed of slope-basin material, ophiolites and their metamorphic soles vary from place to place in the Taurides. They exhibit a coherent organization which attests to a common origin with transport as a whole entity over the presently underlying platform units. The latter can not be continental slivers within the basin. The inferred large-scale allochthoneity is supported also by palaeontological and palaeomagnetic data, and by a consideration of the mechanisms of ocean-basin opening and obduction.

We concluded that the Tauride ophiolites, the associated slope-basin and horst deposits constitute a unique set of nappes derived from a single northern oceanic basin bordered to the south by a large epi-continental platform. This basin was located north of the platform units which now constitute the Calcareous Axis of the Taurides.

The Taurides are an Alpine chain composed of both continental and oceanic units, the latter represented by numerous ophiolitic nappes roughly organized into three longitudinal belts. Overthrusting has occurred successively in phases since the Upper Cretaceous and has greatly modified the pre-tectonic organization. It is consequently not surprising that this pre-tectonic distribution of continental and oceanic areas has been the subject of contradictory interpretations.

Some authors assume that tectonic displacement did not drastically modify the original organization and consequently tend to reconstruct several oceanic basins separated by microcontinents (Dumont *et al.* 1972; Biju Duval *et al.* 1977; Robertson & Woodcock 1981; Şengör & Yılmaz 1981). Other authors assume the contrary, that tectonic displacements played a major role, and tend consequently to consider that most of the ophiolites came from a single major ocean basin bordering a large epicontinental platform (Ricou *et al.* 1974, 1975; Brunn *et al.* 1976).

The data are now sufficiently numerous and precise to evaluate the different interpretations of the Tauride part of the chain, including the central and southern ophiolitic belts and surrounding terrains (Fig. 1) in the Eastern Mediterranean.

The tectonic units which constitute the chain can be classified into three groups according to their character during the Mesozoic: platform units, slope-basin units, and ophiolitic units. The ophiolites rest tectonically on the platform units and the intercalated slope-basin units. This is commonly accepted as due to obduction in which material was transported from the ocean to the continental margin, and on to the epi-continental platform. Study of the slope-basin material reveals a very important phase of Triassic crustal extension which includes normal faulting, horst and graben formation, basin foundering and submarine volcanism, related to the birth of an oceanic domain during early Mesozoic time (Marcoux 1978; Marcoux & Ricou 1979: Argyriadis *et al.* 1980; Robertson & Woodcock 1981). A second epoch of foundering, shown by a switch to pelagic deposition over the platforms, occurred during the Turonian to Senonian, prior to a major obduction event. Metamorphic events are dated in places as Upper Cretaceous or Tertiary and appear to be linked to tectonic events. The metamorphic massifs seem to be mostly of platform type (Özgul, in Adamia *et al.* 1980), or are too poorly known to be integrated into the discussion.

Special mention must be made of the Malatya-Elazığ ophiolites which do not obey the general Tauride rule for ophiolites, as they are not associated with Triassic slope-basin material. They appear to record the existence of an Upper Cretaceous basin, as discussed in Michard *et al.* (this volume).

In its present day position, the platform material throughout the Taurides constitutes the basal units of the tectonic pile and crops out in two belts: the edge of the Arabian platform to the south-east and the Taurus Calcareous

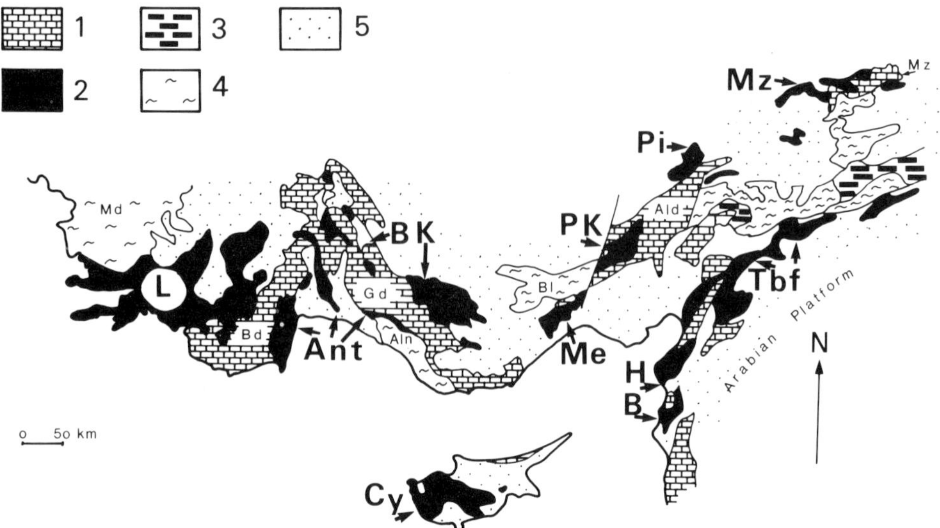

FIG. 1. Geological sketch-map of Taurides.
(1) Platform units: Bd, Bey Dağları; Gd, Geyik Dağ; Ald, Aladağ; and Md, Munzur Dağ constitute the Calcareous Axis of the Taurus.
(2) Ophiolites and slope-basin units: L, Lycian Nappes; Ant, Antalya; B, Beyşehir; K, Karaman; Cy, Cyprus; Me, Mersin; H, Hatay; B, Bäer Bassit; PK, Posantı, Karsantı; Pi, Pinarbaşı; Mz, Munzur; and Tbf, Turkish border folds.
(3) Malatya-Eliazığ ophiolites and associated rocks.
(4) Metamorphic massifs: Md, Menderes; Aln, Alanya; and Bl, Bolkar Dağ
(5) Post tectonic formations.

Axis along the axial part of the chain (Fig. 1). The two major belts of ophiolites crop out north (or over) and south of the Calcareous Axis.

The key issue is the rôle of this Calcareous Axis during the Mesozoic. First, could it have been a continental sliver separating two oceanic basins formed in the Triassic, the Tethys to the north and the Mesogea to the south, as proposed by Biju Duval *et al.* 1977? This opinion is supported by Robertson & Woodcock (1981; 1982, this volume), and Yılmaz & Maxwell (1981, and Yılmaz this volume). Yılmaz *et al.* (1981) even distinguish sub-basins in the Antalya part of the Mesogea, while Şengör & Yılmaz (1981) consider the Malatya ophiolites as a third basin initiated in the Triassic. Alternatively, the Calcareous Axis could have been an integral part of the Arabian-African continent, now exposed as tectonic windows below and behind ophiolitic nappes derived from a single Tethyan basin to the north. This is the interpretation proposed by Ricou *et al.* (1975), and Ricou & Marcoux (1980).

We intend to evaluate the different hypotheses from various standpoints. Any analysis of the problem must attempt to establish whether the Calcareous Axis was a Mesozoic extension of the Arabian platform, or was a separate platform; it must also consider if the slope-basin and ophiolitic nappes possess a similar internal organization thus indicating a common origin for them all, or alternatively, if each of the two belts shows a distinct kind of internal organization indicating a separate origin. Palaeontological data from faunal provinces can be used for the same purposes. Palaeomagnetic data should also be able to show if the allochthonous terrains were originally sited north or south of the Calcareous Axis. Analysis of the tectonic evolution in space and time ought to show if the Calcareous Axis can be considered as a set of windows or as a unit which was never overthrust by the nappes. Finally, one must appeal to general plate-tectonic considerations to decide if coeval rifting, ophiolite genesis, obduction, gravity sliding, nappe emplacement and metamorphism in multiple basins, is more likely to have occurred than a single set of events in a single basin.

The units

The similarities of both ophiolitic and slope-basin units on both flanks of the Calcareous

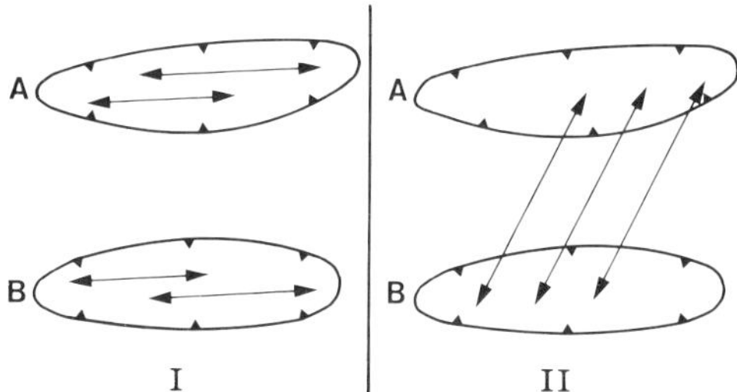

FIG. 2. Close correlation (arrows) to be expected if the terrains cropping out in A and B have a distinct (I) or common (II) origin.

Axis, and of the platform units of the Calcareous Axis with the Arabian platform are the starting point of the nappe-window interpretation. Comprehensive recent studies show various significant local variations in the tectonic organization. Detailed correlations along the present belts of outcrop are to be expected if these formed in different basins; on the contrary, if the correlations can be extended from one belt to the other this would indicate a common origin (Fig. 2).

Platform units

The Mesozoic history of both the Arabian platform and the Calcareous Axis shows the same main features. There was Triassic uplift and erosion with extension and foundering of the slope-basin units. Progressive transgression during the Jurassic was probably related to sea floor spreading. Turonian to Senonian subsidence preceded the emplacement of the ophiolitic nappes. Local features include diachronism of the Jurassic transgression, occurrences of lavas or of low-energy facies, and the diachronism of Upper Cretaceous subsidence. Unfortunately, these variations can not help us to solve our problem. As far as we know they are not simply organized and can not be correlated precisely along a single belt, or from one belt to the other. Nevertheless, unpublished field observations, by J. Marcoux and O. Tekeli and M.T.A. geologists show a local transition from the platform facies of the Calcareous Axis (Aladağ Massif) to pelagic facies; the Middle Triassic level of the upper units of the Aladağ Massif shows an increasing content of slope-basin material, in the form of calcareous and terrigeneous turbidites with reworking of submarine volcanic clasts, red nodular limestones and cherts. Such a transition has not been documented from the Arabian platform.

These data show that the Arabian platform and the Calcareous Axis platform were in the same geodynamic environment during the Mesozoic, with a transition towards open sea to the north. They are not sufficient to establish if the two platforms were continuous or detached, or to show when the Eastern Mediterranean basin came into being.

Slope-basin units

These units accompany the ophiolitic nappes over the whole of the Taurides. The best known outcrops are situated both north (Lycian Nappes, Beyşehir, Karaman, Pınarbaşı, Munzur) and south (Antalya, Cyprus, W Syria, Turkish border folds) of the Calcareous Axis. The general similarities between northern and southern outcrops of the Western Taurides have been stressed since 1974 (Ricou *et al.* 1974). Delaune-Mayere *et al.* (1977) and Robertson & Woodcock (1980) established precise correlations between Antalya, Cyprus and NW Syria. Special mention must be made of recent works on the Karaman nappes (Gökdeniz 1981), the Pınarbaşı (Altiner 1981) and Divriği (Çapan 1980) areas, all situated north of the Calcareous Axis. Specific data on the Western Taurides are given in Monod (1977), Poisson (1977) and Gutnic *et al.* (1979).

Numerous contrasting facies of different ages are known from these units. In order to compare the different outcrops over the Taurides we have prepared a set of tables indicating occurrences and age of the main facies (Fig. 3). Absence of an entry in the tables could simply be due to lack of data. The outcrops are ordered from west to east; those situated north of the Calcareous Axis are hatched while those situated to the south are left blank.

The tables show clearly that no individual feature or group of features is specifically restricted to either the northern or the southern belt. On the contrary, the closest correlations are apparent between the Antalya and the Beyşehir-Karaman outcrops which are situated near each other but on opposite flanks of the Calcareous Axis. Features common to both include mass-flow deposits, allodapic breccias, turbiditic terrigenous sandstones, pelagic and hemipelagic limestones, and volcanics.

It is possible to correlate not only different facies but complete stratigraphic sequences. In the Karaman-Beyşehir area, the Oyuklu Dağ sequence of Hallstatt-type Ammonitico Rosso of Carnian age overlain by Mesozoic shallow water carbonates, the Bayirköy-Ishaniye sequence of Ladinian Pietraverde tuffites with a thin Mesozoic cover of pelagic limestones and cherts and the Norian turbiditic sandstones (dated by Doubinger, Strasbourg, pers. comm., 1982)

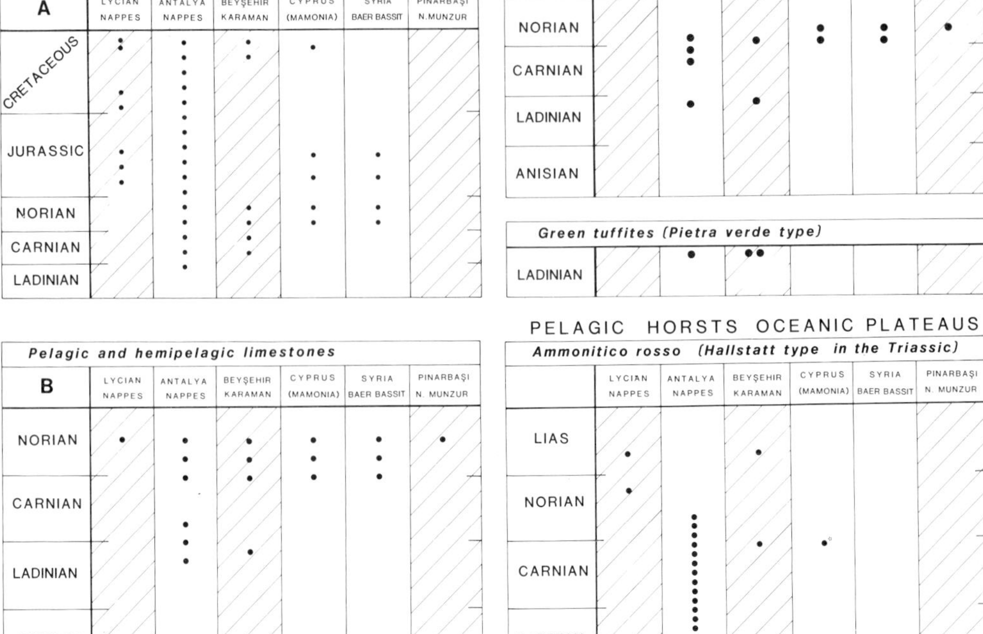

FIG. 3. Tabulated occurrences of the major rock-types of the slope-basin units. Outcrops situated north of the Calcareous Axis are hatched; those situated south of it are left blank.

which tectonically underlie the Bayirköy-Ishaniye sequence are similar, respectively, to the Bakırlı Dağ, Kemer and Alakır, Çay sequences of the Antalya area. Only the Çatal Tepe sequence of Antalya has no exact equivalent in the Karaman area.

The sequences corresponding to pelagic horsts are particularly interesting because rapid changes in facies with time give them a specific character. Plotting the succession of facies with time (Fig. 4) reveals distinct types of sequences. Some are known from both the Lycian and the Beyşehir-Karaman areas, others from both Beyşehir-Karaman and the Antalya areas. They show particular links with these three western outcrops situated on both flanks of the Calcareous Axis (Fig. 1). Taking account of all types of facies the closest correlations are established between Antalya and Karaman, specially between their closest parts (Güselsu-Demirtaş sector and Karaman Ermenek sector) which are located near each other but separated by the Calcareous Axis. This shows that close correlations depend on proximity and are independent of location in the northern as distinct from the southern belt of outcrop. We cannot establish

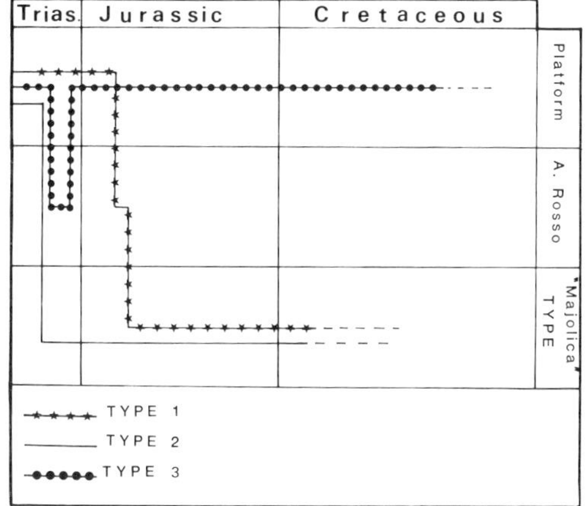

FIG. 4. Types of pelagic horst sequences and their location.
(1) Lycian nappes (Domuz Dağ) and Beyşehir-Karaman (Boyalı Tepe);
(2) Antalya (Kemer gorge) and Beyşehir-Karaman (Huğlu);
(3) Antalya (Bakırlı Dağ) and Beyşehir-Karaman (Oyuklu Dağ).

any distinction between north and south; on the contrary, detailed correlations extend from north to south (Fig. 2, II), favouring the hypothesis of a common origin for both the northern and the southern nappes.

The ophiolites

Detailed comparisons of the numerous ophiolitic masses of the Taurides have not so far been taken into account. Although knowledge of these ophiolites is uneven, recent syntheses (Juteau 1979, 1980; Çapan 1980; Whitechurch *et al.* this volume) enable us to integrate the ophiolitic data into the discussion.

Most of the ophiolites of the Taurides are tectonically associated with slope and basin units deposited during the Triassic extension of the margin. Consequently, those ophiolites came from, or over, a basin which opened during the Triassic. They differ from the Malatya-Elazığ ophiolites which are intruded by Senonian calc-alkaline plutons and associated with calc-alkaline volcanics and volcano-detritic flysch of the same age, without any Triassic slope basin units (Yazgan *et al.* 1982). These ophiolites appear to have formed in a Senonian back-arc context (Michard *et al.* this volume) and will not be discussed here as they are not part of the pre-Senonian Mesozoic organization and so form an important exception.

The major type of Tauride ophiolite crops out partly north of the Calcareous Axis (Lycia; Beyşehir-Karaman; Pınarbası; Divriği; Munzur), over the Axis itself (Pozantı-Karsantı) and south of it (Antalya; Mersin; Troodos; Hatay; Baër-Bassit and Turkish Border Folds). Again, their consistent association with Triassic slope-basin material is a unifying feature. However, examination of the differences between the massifs shows that there are systematic variations along WSW-ENE zones in the Taurus ophiolites.

The first variation appears in the erosion level. Erosion has been:

(1) severe in the Lycian ophiolites where only tectonites are preserved;

(2) moderate in the Antalya, Beyşehir-Karaman, Mersin, Pozantı-Kersantı, Pınarbaşı and Divriği ophiolites where tectonics and cumulates are preserved, while most of the lavas have been eroded; and

(3) negligible in the Troodos, Hatay, and Baër-Basit ophiolites, where the whole pile including the upper lavas is preserved.

The stratigraphy and petrological characteristics of the cumulates, where preserved (i.e. excluding the Lycian ophiolites) shows a similar systematic west to east variation. It appears that the Antalya, Behşehir-Karaman, Mersin, Pozantı-Karsantı ophiolites probably originated at a fast- to moderate-spreading axis, the Troodos and Hatay ones from a slow-spreading ridge, while further south the Baër-Bassit ophiolites again indicate a fast- to moderate-spreading ridge (Whitechurch *et al.* this volume).

Analysis of the ages of pre-obduction oceanic slicing, as obtained from the metamorphic soles of most of the massifs (Parrot & Whitechurch 1978; Thuizat *et al.* 1981) shows first that their distribution in time is unimodal, indicating a common history, as does the common 75 Ma age obtained from isolated tholeiitic dykes cutting the ophiolites and their metamorphic soles. Secondly, metamorphic sole ages vary systematically from west to east: the oldest are encountered under the Lycian ophiolites and the youngest below the Baër-Bassit ophiolite, with a regular age decrease between; i.e. the oceanic slicing migrated from NNW to SSE.

Determination of the primary ages of the ophiolites is still in progress, but two points are already established. The ophiolites affected by oceanic slicing are everywhere older than the appropriate slicing age and consequently the Lycian ophiolites appear to be the oldest. The Troodos and Hatay ophiolites appear to be the youngest; their ages of formation are respectively 80 Ma (Mantis 1970; Delaloye & Desmet 1979) and 90 Ma (Delaloye *et al.* 1980) though neither metamorphic soles nor isolated tholeiitic dykes are present. Here again we find the same regional pattern, the youngest spreading axis before obduction being the WSW-ENE Troodos-Hatay line.

Tectonic analysis of internal structures has so far been restricted to two massifs, Antalya (Juteau *et al.* 1977) and Posantı-Karsantı (Çakir 1978). These data indicate that both massifs have been created in a similar position, namely on the northern flank of a spreading zone orientated N 70°E. This is consistent with the orientation of the dyke complexes of the Hatay and the Troodos Massifs, if the anticlockwise rotation indicated by the palaeomagnetic data from Cyprus is allowed for (Vine *et al.* 1973; Parrot 1973; Shelton & Gass 1980).

There is thus a common spreading-axis trend of N 70°E in all the Tauride ophiolites. This is oblique to the Calcareous Axis and the same trend is found both south (Antalya, Mersin) and north (Beyşehir-Karaman, Pozantı-Karsantı) of the Axis. It is consequently not possible for the Calcareous Axis to have been a pre-obduction continental sliver separating two different oceanic basins. On the contrary, the similar organization present in all shows that these ophiolites constitute a single set of nappes derived from the same oceanic basin in which the spreading rate steadily decreased, and in which the last spreading activity is represented by the Troodos-Hatay line. This single ocean basin was subject to oceanic slicing, migrating from north to south prior to obduction.

Conclusions from the local field-data

The study of local variations in slope-basin sediments and the ophiolites permits close correlations to be made between the different outcrops. These correlations do not apply exclusively along the northern or southern outcrop belts nor do the two belts show the systematic differences that would be expected if they were derived from two different basins (Fig. 2, I). On the contrary, correlations link outcrops in the southern belt with those of the northern belt, as would be expected if they were derived from the same basin (Fig. 2, II). The slope-basin and the ophiolitic nappes thus appear to constitute a single set of nappes thrust over the Calcareous Axis and not different nappes derived from different basins separated by platform slivers such as the Calcareous Axis.

Palaeontological data

Studies in the Western Taurides enable us to compare the Calcareous Axis in Seydişehir and the slope-basin units in Antalya, in faunal province terms.

Following the work of Asseretto & Monod (1974) subsequent authors (Hirsch 1977; Beltan *et al.* 1979; Nicora 1980; Gedik 1977, 1981; Krystyn 1982 unpubl. data) who have studied the fauna of Seydişehir (conodonts, ammonites, vertebrates) have stressed its affinities with that of the Arabian platform (Hirsch, this volume). Conodonts of both areas are attributed to the same Mediterranean province by Gedik (1981), and to the Sephardic province by Hirsch (1977). Ammonites of Seydişehir attributed to the Tethyan province by Asseretto & Monod (1974) are now considered by Krystyn (pers. comm. 1982) to be new forms unknown in the Tethyan realm. Krystyn emphasizes that on the contrary the Antalya ammonites of Ladinian age are fully Tethyan and quite similar to those from the classical Hallstatt sections (e.g. Feuerkogel in Austria). Conodonts from Antalya are also typically Tethyan, according to Kristan-Tollmann and Krystyn (1975). This indicates a direct link between the Calcareous Axis and the Arabian platform during Triassic time. The Antalya fauna thus appears to favour the hypothesis of large-scale allochthoneity. Nevertheless, we accept that these considerations are not sufficient alone. They are however reinforced by other data discussed here.

Palaeomagnetic data

The important work of Lauer (1981 and this volume), which includes a discussion of previous data for the whole of Turkey, has led its author to propose that the Taurides have moved in a direction along the strike of the chain from a position near the southeastern end of the Arabian peninsula in Triassic time to their present location. This hypothesis depends greatly on the latitude assigned to Arabia during the Mesozoic, a point which will not be discussed here. However, if such great longitudinal displacements took place at a pre-obduction stage, the palaeolatitudes obtained for Triassic and Jurassic times cannot help to solve our problem. We can however still use the Cretaceous data to check if the southern allochthonous terrains were situated north or south of the Calcareous Axis immediately before obduction. To do this we must compare the palaeolatitude of the Troodos Massif with that of the Calcareous Axis before thrusting. Lauer (1981) gives measurements on the Upper Cretaceous ophiolites of Antalya, Troodos and Pozantı-Karsantı which all indicate a position for these massifs around 17°N. This common position means that they must have come from the same basin, either north or south of the Calcareous Axis. For the Calcareous Axis to the north-east of Antalya Bay, i.e. to the north of the present post-obduction position of the Antalya and Troodos massifs, we have no Upper Cretaceous data and so must interpolate between the 150–160 Ma and the 50–60 Ma palaeolatitudes. A linear interpolation gives a position about 12°N. More realistic interpolations could be proposed using the assumption that the displacement in palaeolatitude was of the same type as the displacement of the Arabian peninsula estimated by different authors. They do not lead to a more northerly position.

It consequently appears that the Troodos and Antalya massifs were situated to the north of the Calcareous Axis before obduction. They cannot represent an ocean basin sited to the south of it and were, on the contrary, obducted over it from north to south as proposed in the single basin hypothesis.

Tectonic data

Major obduction occurred during the Late Cretaceous and was followed by successive overthrusting phases during the Teriary. In the eastern and central Taurides, Tertiary tectonics did not drastically change the north to south organization which resulted from the Late Cretaceous events. In contrast, the Western Taurides have been subjected to important reorganization during the Tertiary, the effects of which must be deciphered in order to reconstruct the Late Cretaceous situation.

In the Western Taurides, the Western Bey Dağları, which was unaffected by tectonism before the Miocene, underlies allochthonous units affected by earlier events including the emplacement of the ophiolites and slope-basin units of the Antalya and Lycian nappes. The Western Bey Dağları is interpreted as a Miocene half-window below nappes transported from the inner parts of the chain. To the north and north-east, the Menderes Massif and the Geyik Dağ units were unaffected by tectonism before the Lower Tertiary and underlie allochthonous units affected by earlier events—the Lycian ophiolites and slope-basin units lying above the Menderes, and the Antalya and Behsehir-Karaman ophiolites and slope-basin units above the Geyik Dağ. The Menderes and Geyik Dağ are interpreted as 'Eo-Oligocene' windows through nappes derived from the inner part of the chain. This interpretation led to a step-by-step palinspastic reconstruction (Ricou *et al.* 1979) which accounts for all known data. The early Tertiary stage shows that the Late Cretaceous obduction event had covered the whole of the Taurides east of a line running from the north-eastern flank of the Menderes to the western approaches of Cyprus. An objection to this reconstruction has recently been made by Hayward and Robertson (1982) who report evidence that ophiolites were located not far east of the Bey Dağları during the Tertiary. However, Ricou *et al.* (1979) drew the limit of the Cretaceous obduction as immediately east of the Bey Dağları.

This post-obduction configuration must be used as a test of the two hypotheses; either it is the result of a single obduction from the north over the Calcareous Axis extending as far south as Troodos and north-western Syria, or it follows multiple obduction from at least two basins, i.e. southward obduction from the Tethys over the Calcareous Axis, southward obduction from the Mesogea over the Arabian Platform, and most probably a northward obduction from the Mesogea over the Calcareous Axis, if the latter is not to be considered as a window, an assumption which would otherwise lead us back directly to the first hypothesis. The total amount of overthrusting is broadly the same for the two hypotheses though probably greater for the two-basins version, as some overlap is to be expected between converging thrusts over the Calcareous Axis.

The observed modes of tectonic emplacement appear to be similar over the whole area:

(1) Remnants of tectonic amphibolites testify to an episode of pre-obduction syn-metamorphic slicing;

(2) Slices of slope and basin material testify to transport over a continental margin;

(3) The whole set of nappes rests flat on the uppermost levels of the autochthonous sequence which are locally olistostromic.

Prior to nappe emplacement, the autochthonous platforms foundered as indicated by pelagic deposits, followed by flysch, olistostrome deposits and nappe emplacement. These features show that the final emplacement was by gravity tectonics, as emphasized by various authors since Rigo de Righi & Cortesini (1964). The similarity of tectonic style over the whole area leads to the following conclusion: if there were multiple obductions, they were so similar that they can not be distinguished from a single one.

In order to defend the two-basins hypothesis, platform areas showing continuous sedimentation during the tectonic events have been searched for, as they could represent areas never overthrust by nappes and consequently separating two areas of obduction. A case was first presented by the Kyrenia Range situated north of the Troodos Massif (Biju-Duval *et al.* 1976), but the ages of the units show that a break in sedimentation accompanied by olistostrome deposition was coeval with near-by nappe emplacement (Brunn *et al.* 1977; Baroz 1979). More recently, the Sariz section (near Tufanbeyli) has been quoted as showing continuous sedimentation during the critical period (Şengör & Yılmaz 1981). Unfortunately the section has not been described, nor have its relations been made clear with the near-by sections, which include ophiolitic nappes, olistostromes, and detrital sediments of the same critical age. Even if true, this locality would not constitute a line never passed by nappes, but only a single point which remained unaffected by nappes which extend over an area over 500 × 200 Km. This is insufficient to preclude the hypothesis of a unique source, especially in the context of gravity emplacement with the volume of allochthonous material varying greatly from place to place.

The timing of tectonic events is also important. As already mentioned, the age of intra-oceanic pre-obduction slicing shows a progressive and apparently continuous variation from NW to SE, and the absence of age discontinuities does not fit with the hypothesis of two basins separated by the Calcareous Axis. The age of arrival of nappes on the platform is precisely established as intra-Maastrichtian on the southern front; in the internal part, this age is less precisely known but the different sets of possible age brackets all encompass the Maastrichtian. In the extreme north, the Munzur Dağ is said to have received ophiolitic material as early as the Campanian (Özgül, in Adamia *et al.* 1980). None of these data can sustain the multiple basins and multiple obductions hypothesis. On the contrary they fit very well with the hypothesis of a single obduction from the north, as they show progressive migration of the tectonic stacking in the oceanic basin from NW to SE, followed by north to south gravity emplacement over the platform.

None of these observations can directly demonstrate that the multiple-basin hypothesis is wrong. This type of hypothesis cannot in fact be disproved as it has no requirement other than the presence of different ophiolitic outcrops, each of which could be supposed to hide a suture zone. The defender can always explain the similarities of style by duplication of tectonic circumstances and can always propose a tectonic history capable of fitting the observed series of events, but this does not constitute proof. Conversely, the single obduction hypothesis could have been disproved as it implies that there should be no significant area without traces of nappe transport along the supposed obduction path. These conditions are satisfactorily fulfilled in an area longer than 500 Km along strike, thus the single basin hypothesis is significantly strengthened.

The mechanisms

Defenders of the multiple-basin hypothesis must explain why the different basins show exactly the same behaviour at the same time. Along the transect from Konya to Alanya, for instance, the Antalya and Karaman basins would have to have had in common: the same Triassic extension phase ending in the Jurassic, creation of the same detailed stratigraphic sequences, the same renewal of oceanic activity creating Cretaceous ophiolites, the same oceanic compression inducing oceanic slicing, the same obduction towards an 'axis of symmetry', the same blueschist metamorphism in Konya-Bolkar and Alanya and the same 'Eo-Oligocene' thrusting involving platform units. Such identical events in two different basis over a 100 Ma period would be unique in the world. Furthermore, a third similar basin would have to be envisaged south of the Antalya platform to account for the Troodos and Mamonia units.

On the contrary, any similarity is a positive argument for the single basin and large allochthoneity hypothesis, which is also enhanced by local features. The situation recalls the early debates on the Glarus double fold or the origin of the Calcareous Alps, which were resolved by the acceptance of large-scale allochthoneity.

A great difficulty for the multiple-basin-hypothesis is to explain coeval obduction. Actualism and earth history show that obduction is an exceptional way to resolve ocean-continent convergence, subduction being far more frequent and easier. Obduction appears to be a catastrophic event occurring in response to rare geodynamic conditions. This stands against the multiple-basin model which implies at least two coeval southward-obductions from both the Tethys and Mesogea and even coeval northward obduction from the Mesogea over the Calcareous Axis. How likely would it be that two or three of these catastrophic events would occur at the same time in independent basins in the same area?

Moreover, the obduction mechanism as it appears in the Taurides implies two steps in the tectonic process; first, movement up-slope from the oceanic basin over the continental margin, then gliding down-slope over the platform as indicated by the olistostromic and gravity-driven character of the final emplacement. The multiple-basin model implies three coeval up-slope movements against gravity, which seems unrealistic. The single-basin-model implies simply a concentration of tectonic 'effort' on a single margin followed by gliding southwards as far as the slope existed.

On the latter point, a control is given by the morphology of the platform which was subject to differential subsidence during the Upper Cretaceous. The greatest subsidence is documented in the Senonian south-Tauride basin situated in the Turkish Border Fold Zone and now partly hidden below the Bitlis thrust. It shows a thickness of 800 m of pelagic and detrital Senonian sediments (Sayındere and Kastel Formations; Perincek 1979). This basin was not filled by the nappe emplacement as flysch sedimentation continued above the ophiolites during the Maastrichtian and Palaeocene (Hazar Formation). This southern basin clearly appears to have acted as an 'attraction zone' for the gravity nappes; the deepest part of the platform broadly coincides with the southern limit of the nappes. These data strongly sustain the single-basin model which implies a single set of nappes from the north. On the other hand, the multiple-basins hypothesis implies the convergence of gravity nappes toward the Calcareous Axis, a place where the prethrust foundering of the platform was not specially important and which cannot consequently be presented as a specific line of 'attraction'.

Conclusions

Ten years ago significant progress in our understanding of the geology of the Middle East came with recognition of the allochthonous character of the ophiolites and associated rocks. The full implications of this allochthoneity, i.e. the possibility of large displacements drastically changing the relative position of the different components of the chain, were not then realized. Nevertheless this possibility is of fundamental importance as it is not reasonable to suppose that each present outcrop of ophiolites represents a past oceanic basin.

Confirming the systematic allochthoneity of the Tauride ophiolites and slope-basin material and assessing their similarities and tectonic histories has led to the hypothesis that they all came from the same northern areas over the Taurus Calcareous Axis which now forms a multiple tectonic window. Some of the geologists involved have accepted the interpretation, as similar situations are frequently encountered in mountain chains affected by nappe tectonics. Opponents appear to consider that small, yet multiple and convergent, displacements are more likely to occur than single major ones. They believe that the fit of the data with our model could be a misleading coincidence. In fact, the data had at one time only limited precision and one could then state that similar basins and platforms evolving in the same context were difficult to distinguish with the data then available.

That era is over. Accumulating data have continued to stress the close similarities between the different ophiolites and slope-basin material, similarities which cannot now be treated as a misleading coincidence. Also, the distinctions that we can make between the different outcrops lead to two major conclusions: the Malatya-Elazığ ophiolites do not obey the general Tauride rule and must be treated separately. The rest, the great majority, show local variations from place to place which only fit a single-basin model and not models involving the Calcareous Axis as a continental sliver between two or more basins. The slope-basin material shows the closest similarities between the neighbouring Antalya and Karaman nappes situated on opposite flanks of the Axis. This can not be explained if they came from two distinct basins. This is exactly what must be expected if they came from the same basin and more precisely from neighbouring parts of the same basin. Even more demonstrative are the various data obtained on the ophiolitic material. They all show an internal organization consistent with spreading along the N 70°E direction which has nothing to do with the irregular orientation of the Calcareous Axis. This indicates that they belong to a single set of nappes and that the disposition of the underlying windows which constitute the Axis relates to some other process. The palaeontological and palaeomagnetic data independently enhance the single-basin model. A consideration of possible mechanisms shows the single-basin model to be more satisfying. Improvements in our knowledge now lead us to interpret the Calcareous Axis as a set of windows and to locate the pre-obduction Triassic-formed basin and ophiolites to the north. Opponents' models were built on the assumption that small displacements are more likely than large ones but nappe tectonics, continental drift and plate tectonics have successively demonstrated the mobility of the Earth's lithosphere and repeatedly taught us that mobilistic models are more adequate than authochthonist to explain our mobile Earth.

ACKNOWLEDGEMENTS: This is a contribution of Laboratoire Associé au CNRS 215, Geologie Structurale Paris and Equipe de Recherche Associée au CNRS 887, Géologie et Pétrographies Structurales, Strasbourg. Field work facilities were made available by the MTA, Turkey.

References

ADAMIA, S., BERGOUGNAN, H., FOURQUIN, C., HAGHIPOUR, A., LORDKIPANIDZE, M., ÖZGÜL, N., RICOU, L. E. & ZAKARIADZE, G. 1980. The Alpine Middle East between the Aegean and the Oman traverses. *26ᵉ Cong. Géol. Internat. Paris 1980, Colloque C 5 & Mém. B.R.G.M.* **115,** 122–136.

ALTINER, D. 1981. *Recherches stratigraphiques et micropaléontologiques dans le Taurus Oriental au NW de Pınarbaşı (Turquie).* Thèse n° 2005, Univ. de Genève, section Sciences de la Terre, 450 p.

ARGYRIADIS, I., GRACIANSKY, DE, P.C., MARCOUX, J. & RICOU, L. E. 1980. The opening of the Mesozoic Tethys between Eurasia and Arabia Africa. *26ᵉ Cong. Géol. Internat. Paris 1980, Colloque C 5 & Mém. B.R.G.M.* **115,** 199–214.

ASSERETO, R. & MONOD, O. 1974. Les formations triasiques du Taurus occidental à Seydişehir (Turquie méridionale). Stratigraphie et interprétation sédimentologique. *Riv. Italiana di Pal. e Strat., Mem. XIV Contributi stratigrafici e paleogeografici sul Mesozoico della tetide*', 159–191.

BAROZ, F. 1979. *Etude géologique dans le Pentadaktylos e la Mesaoria (Chypre septentrionale).* Thèse Université Nancy 1. **1,** la série stratigraphique, **2,** les structures, la synthèse.

BELTAN, L., JANIVIER, P., MONOD, O. & WESTPHAL, F. 1979. A new marine fish and placodont reptile fauna of Ladinian age from Southwestern Turkey. *N. Jb. Geol. Paläont. Mh.* **5,** 257–267.

BIJU-DUVAL, B., LAPIERRE, H. & LETOUZEY, J. 1976. Is the Troodos massif allochthonous? *Bull. Soc. géol. Fr.* **18,** 1347–1356.

——, DERCOURT, J. & LE PICHON, X. 1977. From the Tethys ocean to the Mediterranean seas: a plate tectonic model of the evolution of the western alpine system. *In*: BIJU-DUVAL B. & MONTADERT, L. (eds), *Structural history of the Mediterranean Basins,* Editions Technip, Paris, 143–164.

BRUNN, J. H., ARGYRIADIS, I., RICOU, L. E., POISSON, A., MARCOUX, J. & DE GRACIANSKY, P. C. 1976. Eléments majeurs de liasion entre Taurides et Hellénides. *Bull. Soc. géol. Fr.* **18,** 481–497 & *Coll. intern. CNRS, Paris,* **244,** 285–301.

——, ARGYRIADIS, I., MARCOUX, J. & RICOU, L. E. 1977. Commentaires sur la note: 'Is the Troodos massif allochthonous?' présentée par B. Biju-Duval, H. Lapierre et J. Letouzey. Discussion d'une origine nord- ou sud-taurique. *C.r. somm. Soc. géol. Fr., fasc.* **6,** 344–345.

ÇAKIR, Ü. 1978. *Pétrologie du massif ophiolitique de Pozantı-Karsantı (Taurus Cilicien, Turquie): Etude de la partie centrale.* Thèse, Univ. Louis Pasteur Strasbourg, 251 p.

ÇAPAN, U. 1980. *Toros kuşağıofiolit masiflerinin (Marmaris, Mersin, Pozantı, Pınarbaşı ve Divriği) iç yapilari, Petroloji ve petrokimyalarina yaklaşimlar.* Doc. thesis, Hacettepe Un., 406 p.

DELALOYE, M. & DESMET, A. 1979. Nouvelles données radiométriques sur les pillow lavas du Troodos (Chypre). *C.R. Acad. Sc. Paris,* **288,** D, 461–464.

——, DE SOUZA, H., WAGNER, J. J. & HEDLEY, I. 1980. Istopic ages on ophiolites from the eastern Mediterranean. *In*: *Ophiolites* A. PANAYIOTOU, Ed. Proc. Int. Oph. Symp, Cyprus, 1979, 292–295.

DELAUNE-MAYERE, M., MARCOUX, J., PARROT, J.-F. & POISSON, A. 1977. Modèles d'évolution mésozoïque de la paléomarge téthysienne au niveau des nappes radiolaritiques et ophiolitiques du Taurus Lycien, d'Antalya et du Baer-Bassit. *In*: B. BIJU-DUVAL, & L. MONTADERT, (eds), *Structural history of the Mediterranean Basins.* Editions Technip, Paris, 79–94.

DUMONT, J.-F., GUTNIC, M., MARCOUX, J., MONOD, O. & POISSON, A. 1972. Le Trias des Taurides occidentales (Turquie). Définition du bassin pamphylien: un nouveau domaine à ophiolites à la marge externe de la chaîne taurique. *C.R. Somm. Soc. géol. Fr., fasc,* **2,** 73.

GEDIK, I. 1977. Conodont biostratigraphy in the Middle Taurus. *Bull. Geol. Soc. Turkey,* **20,** 35–48.

—— 1981. Conodont provinces in the Triassic of Turkey and their tectonic-palaeogeographic significance. *Black Sea Technical University Earth Sciences Bulletin, Geology,* **1,** 1, 1–14 (in Turkish with Engish abstract).

GÖKDENIZ, S. 1981. *Recherches géologiques dans les Taurides occidentales entre Karaman et Ermenek, Turquie. Les séries à Tuffites vertes triasiques.* Thèse n° 3006. Univ. Paris-Sud, Centre d'Orsay. 202 p.

GUTNIC, M., MONOD, O., POISSON, A. & DUMONT, J. F. 1979. Géologie des Taurides occidentales (Turquie) *Mém. Soc. géol. Fr., Nouv. Sér.* **58,** 137, 108 p.

HAYWARD, A. B. & ROBERTSON, A. H. F. 1982. Direction of ophiolite emplacement inferred from Cretaceous and Tertiary sediments of an adjacent autochthon, the Bey Dağları, southwest Turkey. *Bull. geol. Soc. Amer.* **93,** 68–75.

HIRSCH, F. 1977. Essai de correlation biostratigraphique des niveaux méso et neotriasiques de faciès 'Muschelkalk' du domaine sepharde. *Cuadernos Geologia Iberica,* **4,** 511–526.

JUTEAU, T. 1979. Ophiolites des Taurides: essai sur leur histoire océanique. *Rev. Géol. dyn & Géogr. phys.,* **21,** 191–214.

—— 1980. Ophiolites of Turkey. *Ofioliti, spec. issue: 'Tethyan ophiolites', G. Rocci (ed.)* **2,** 199–237.

——, NICOLAS, A., DUBESSY, J., FRUCHARD, J. C. & BOUCHEZ, J. L. 1977. Structural relationships in the Antalya ophiolite complex, Turkey: possible model for an oceanic ridge. *Bull. geol. Soc. Amer.* **88,** 1740–1748.

KRISTAN-TOLLMANN, E. & KRYSTYN, L. 1975. Die Mikrofauna der ladinisch-karnischen Hallstätter Kalke von Saklibeli (Taurus Gebirge, Turkei) I. *Sitzungsberichten der Österr. Akad. Wiss. Math.-nat. Kl., Abt. I,* **184,** 8. bis 10. Heft, 259–340.

LAUER, J. P. 1981. *L'évolution géodynamique de la Turquie et de Chypre déduite de l'étude paléomagnétique.* Thèse Inst. Phys. du Globe de l'Univ. Louis Pasteur Strasbourg. 291 pp.

MANTIS, M. 1970. Upper Cretaceous Tertiary foraminiferal zones in Cyprus. *Cyprus Research Centre,* **4,** 227–241.

MARCOUX, J. 1978. A scenario for the birth of a new oceanic realm: the alpine Neotethys. *10th Congrès Internat. Sédimentologie, Jerusalem 9–14 Juillet 1978,* 419–420.

—— & RICOU, L. E. 1979. Classification des ophiolites et radiolarites alpino-méditerranéennes d'après leur contexte paléogéographique et structural. Implications sur leur signification géodynamique. *Bull. Soc. géol. Fr.* **21,** 643–652.

MONOD, O. 1977. *Recherches géologiques dans le Taurus occidental au Sud de Beyşehir (Turquie).* Thèse, Univ. Paris Sud Orsay. 442 pp.

NICORA, A. 1980. *Pseudofurnishius murcianus* Van Den Boogaard in the upper Triassic of Southern Alps and Turkey. *Riv. Ital. Paleont. e Strati.* **86,** 769–778.

PARROT, J. F. 1973. Pétrologie de la coupe du Djebel Moussa, massif basique-ultrabasique du Kızıl Dağ (Hatay, Turquie). *Sciences de Terre, Nancy,* **18,** 143–172.

—— & WHITECHURCH, H. 1978. Subductions antérieures au charriage Nord-Sud de la croûte téthysienne: facteur de métamorphisme de séries sédimentaires et volcaniques liées aux assemblages ophiolitiques syroturcs, en schistes verts et amphibolites. *Rev. Géol. dyn. & Géogr. phys.* **20,** 153–170.

PERINCEK, D. 1979. *Interrelations of the Arab and Anatolian Plates.* Guide Book excursion B First Geol. Congr. Middle East, Ankara, 34 pp.

POISSON, A. 1977. *Recherches géologiques dans les Taurides Occidentales (Turquie).* Thèse n° 1902, Univ. Paris Sud Orsay, Tome 1: Description des séries, Tome 2: Synthèse. 795 pp.

RICOU, L. E., ARGYRIADIS, I. & LEFEVRE, R. 1974. Proposition d'une origine interne pour les nappes d'Antalya et le massif d'Alanya (Taurides occidentales, Turquie). *Bull. Soc. géol. Fr.* **16,** 107–111.

——, ARGYRIADIS, I. & MARCOUX, J. 1975. L'Axe Calcaire du Taurus, un alignement de fenêtres arabo-africaines sous des nappes radiolaritiques, ophiolitiques et métamorphiques. *Bull. Soc. géol. Fr.* **17,** 1024–1043.

——, MARCOUX, J. & POISSON, A. 1979. L'allochtonie des Bey Dağları orientaux. Reconstruction palinspastique des Taurides occidentales. *Bull. Soc. géol. Fr.* **21,** 125–133.

—— & —— 1980. Organisation générale et rôle structural des radiolarites et ophiolites le long du système alpino-méditerranéen. *Bull. Soc. géol. Fr.* **22,** 1–14.

RIGO DE RIGHI, M. & CORTESINI, A. 1964. Gravity tectonics in foothills structure belt of the southeast Turkey. *Bull. Am. Assoc. Petrol. Geol.* **48,** 1911–1937.

ROBERTSON, A. H. F. & WOODCOCK, N. H. 1980. Tectonic setting of the Troodos massif in the East Mediterranean *In*: 'Ophiolites' A. PANAYIOTOU (ed.) Proc. Int. Oph. Symp. Cyprus 1979, 36–49.

—— & —— 1981. Gödene Zone, Antalya Complex, SW Turkey: volcanism and sedimentation on Mesozoic marginal ocean crust. *Geol. Rdsch.* **70,** 1177–1214.

SENGÖR, A. M. C. & YILMAZ Y. 1981. Tethyan evolution of Turkey: a plate tectonic approach. *Tectonophysics,* **75,** 181–241.

SHELTON, A. W. & GASS, I. G. 1980. Rotation of the Cyprus microplate. *In*: 'Ophiolites' A. PANAYIOTOU (ed.). Int. Oph. Symp. Cyprus 1979, 61–65.

THUIZAT, R., WHITECHURCH, H., MONTIGNY, R. & JUTEAU, T. 1981. K. Ar. Dating of some infra-ophiolitic metamorphic soles from the Eastern Mediterranean: new evidence for oceanic thrustings before obduction. *Earth planet. Science Lett.* **52,** 302–310.

VINE, F. J., POSTER, C. K. & GASS, I. G. 1973. Aeromagnetic survey of the Troodos igneous massif, Cyprus. *Nature Phys. Sci.* **244,** 34–38.

YILMAZ, P. O. & MAXWELL, J. C. 1981. K-Ar. Investigations from the Antalya complex ophiolites, S.W. Turkey. *Ofioliti,* **6,** 49–50.

——, —— & MUEHLENBERGER, W. R. 1982. Structural evolution of the Antalya complex (SW Turkey) within the Eastern Mediterranean framework. *Proceedings, International Symposium on the Hellenic Arc and Trench, Athens, 8–10 April 1981.* pp. 419–445.

YAZGAN, E., MICHARD, A., WHITECHURCH, H. & MONTIGNY, R. 1982. Le Taurus de Malatya (Turquie orientale), élément de la suture sud-téthysienne. *Bull. Soc. géol. Fr.* **25,** 59–69.

L. E. RICOU, LA 215/Labo. Géologie Structurale, Univ. P. & M. Curie, 4 Place Jussieu, Tour 26, 75230 Paris Cedex 05, France.

J. MARCOUX, Labo. Sciences de la Terre, Univ. de Reims, B.P. 347, 51062 Reims Cedex France & LA 215.

H. WHITECHURCH, Institut de Géologie, Labo. de Petrologie, 1 rue Blessig, 67084 Strasbourg, France.

Tauric subduction (Malatya-Elazığ provinces) and its bearing on tectonics of the Tethyan realm in Turkey

A. Michard, H. Whitechurch, L. E. Ricou, R. Montigny & E. Yazgan

SUMMARY: In the Eastern Taurus, The Elazığ nappes shown an ophiolitic association overlain by andesites and intruded by calc-alkalic granites, both of Late Cretaceous age. The basin was compressed in the Late Cretaceous between two metamorphic massifs each with a sialic basement and platform-type Permian to Mesozoic cover. This history indicates Late Cretaceous subduction under the Taurides and subsequent compression of the basin and arc.

Attempts to place this Tauric subduction in the geodynamic history of Turkey lead to conflicting interpretations and two alternative models are presented.

The first involves a single Tethyan ocean subducting northwards below the Pontides and southwards below the Taurides. The latter led to the Late Cretaceous opening of back-arc basins which split the formerly continuous Tauric-Arabian platform. Southward subduction ended when the ridge reached the trench, leading to compression of the Elazığ back-arc basin, southward obduction and closure of the ocean. The residual, Upper Cretaceous marginal basin controlled the subsequent Tertiary development of the area.

The second model involves a northern Tethyan ocean and a southern Mesogean ocean, both subducting northwards. Subduction of the southern ocean generated calc-alkaline magmatism and deformation of the leading edge of the Tauric blocks. The Elazığ basin closed as a result, having earlier formed either as a subduction-related marginal basin or as a pre-existing extensional basin. Southward obduction took place from both oceans. After closure of the Tethyan ocean, Tertiary development was controlled by reactivation of Mesogean subduction.

The main feature of the Mesozoic history of the Alpine Middle East is the sharp contrast between its northern and southern parts. The Pontides, Caucasus and Central Iran, constituting a North-Tethyan domain, acted as an active margin at the southern edge of the Eurasian continent during the whole period. The Taurides and Zagrides, constituting a South-Tethyan domain, acted as a passive margin at the northern edge of the African-Arabian continent until the Upper Cretaceous, when the now classic 'peri-Arabian' obduction took place (Rigo de Righi & Cortesini 1964; Ricou 1971; Juteau 1979; Adamia *et al.* 1980).

It now appears that the southern margin became active some time before the Maastrichtian obduction events. An important subduction complex of Late Cretaceous age has been described in the Eastern Taurus (Fig. 1) stretching from Yükeskova to Malatya (Perincek 1979; Şengör & Yilmaz 1981; Yazgan 1981; Yazgan *et al.* 1983; Özkaya 1982a, b). In this area the syn-metamorphic deformation of the Pütürge Massif and northern-most nappes has also been dated as Late Cretaceous and inferred to be structurally linked to subduction (Yazgan *et al.* 1982).

This Tauric eo-Alpine subduction, and the obducted ophiolites of the area, have to be explained within the framework of Middle East plate tectonic reconstructions. Various models have been proposed, involving one, two or more oceans between the Pontic and Arabian margins (Smith 1971; Dewey *et al.* 1973; Ricou *et al.* 1975; Biju-Duval *et al.* 1977; Şengör & Yilmaz 1981; Ozkaya 1982 a, b). In this paper we discuss interpretations of the Tauric subduction event based on the well-documented Malatya transect. We consider two models: a 'one-ocean-' and a 'two-ocean-' reconstruction. We also discuss the late-Alpine tectonics of this same transect and the associated calc-alkaline, subduction-related 'Maden' volcanism (Perincek 1979).

Structural and stratigraphic setting of the Malatya transect ophiolites

The Malatya Taurus data are from Yazgan *et al* (1983). The main component of the Elazığ Nappe, which constitutes the axial part of the belt, is the Upper Cretaceous calc-alkaline magmatic complex (Figs 2 & 3). This complex comprises intrusives, dated as from 85 to 76 Ma (all quoted radiometric ages are from unpublished data of Montigny *et al.*). These intrusives are associated laterally with extrusives overlain by Campanian-Maastrichtian volcaniclastic flysch. In northern parts of the nappe (Şişman-

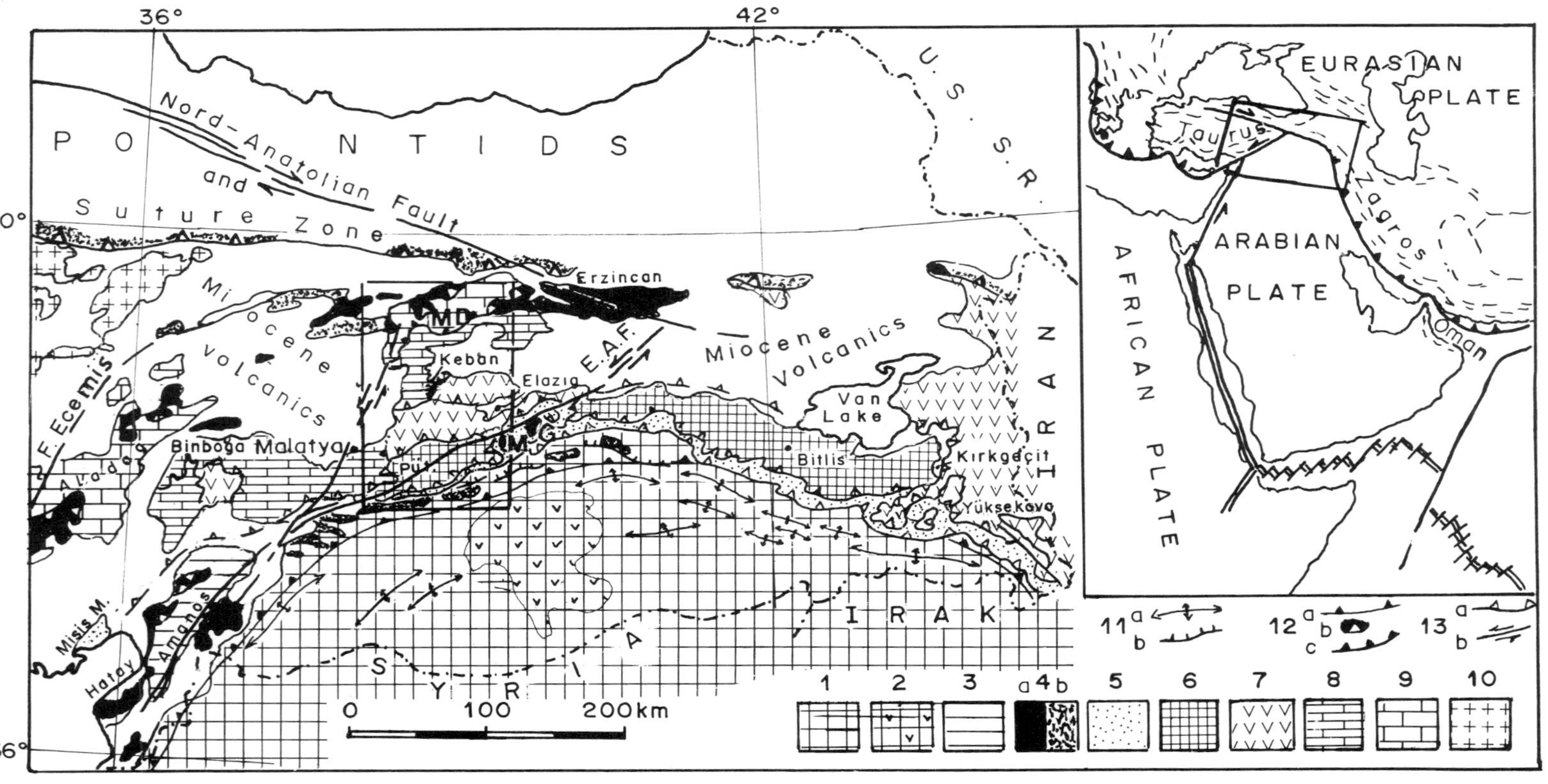

FIG. 1. The Malatya Tauric range in the Tethyan suture zone: (A) At the scale of the Middle-East plates. (B) At the scale of Eastern Turkey. From Yazgan & others 1982 (inset = Fig. 2).

1, Arabian platform; 2, *Idem*, overlain by Plio-Quaternary basalts; 3, Amanos massif; 4, Ophiolites (a) and ophiolitic melanges (b); 5, Guleman-Maden and Çüngüs nappes; 6, Pütürge and Bitlis nappes; 7, Elazığ-Yüksekova nappes; 8, Keban and Malatya nappes and Binboga massif; 9, Tauric Calcareous axis; 10, Kirşehir massif. *Blank*: Miocene-Pliocene volcano-sedimentary cover of the Tauric and Anatolian platform (units 7 to 10), North-Anatolian suture zone and Pontids.

E.A.F., East-Anatolian fault; G, Guleman; M, Maden; MD, Munzur Dağ; Püt, Pütürge.

Baskil units) roof-pendants of marble, similar to the Permian-to-Mesozoic limestones of the Keban nappe immediately to the north indicate that this magmatism affected the southern margin of the Keban platform. The same intrusives are known from immediately south of the Şişman-Baskil units, cutting the Ispendere-Karga Dağ sheared ophiolitic units. These ophiolites are overlain by an andesitic, volcaniclastic flysch dated as Coniacian (lowermost micrites) to Maastrichtian. The whole Elazığ nappe is affected by a southward-verging, pre-late Maastrichtian polyphase deformation. The deepest shear-zones formed under conditions of lower greenschist-facies metamorphism, dated locally at 75 ± 3 Ma. However the earliest shear zones are cross-cut by the granitoids, which in turn are foliated and occasionally folded, as is the flysch itself.

The Malatya-Keban nappe overthrust the Elazığ nappe from the north. An Upper Maastrichtian neritic cover overlies the contact. The Malatya-Keban marbles are typical carbonate platform facies dated from Permian to Mesozoic (Özgül 1981). Their continental basement can be observed locally (Perincek & Özkaya 1981; Yazgan *et al.* 1983). The limestones were affected by a Late Cretaceous upper greenschist facies metamorphism and by contemporaneous, northward-verging or upright folds. Small alkaline granitic bodies are seen near Keban, dated at 78 Ma.

Ricou *et al.* (1975) proposed that the Keban marbles were thrust from north to south on to the Munzur Dağ limestones, together with the ophiolites of this area. However, Özgül (1981) and two of us (A.M., E.Y.) observed that the Keban marbles are, in fact, overlain by Upper Cretaceous flysch, converted locally into a tectonic melange with radiolarites, basalts and greywacke slices. The flysch and melange are in turn tectonically overlain by the Munzur Dağ unit.

The Munzur Dağ includes a non-metamorphic slab of Mesozoic limestones and a blanket of ophiolites and Triassic to Cretaceous slope-and-basin units, obducted on to the limestones during the Campanian (Özgul, *in* Adamia *et al.* 1980). This is the northernmost unit of the transect, which is bounded to the north by the North-Anatolian suture zone.

South of the calc-alkaline magmatic axis lies the metamorphic massif of Pütürge, which is overthrust from the north by the Elazığ nappe. This massif is mainly composed of gneisses and mica-schists, whose last syn-tectonic recrystallization is dated at from 80 to 70 Ma. It is unconformably overlain by Eocene calc-alkaline volcanic and sedimentary series rocks of the Maden Formation that pre-date the southward thrusting of the Elaziğ nappe and the Pütürge Massif. The later is the equivalent of the Bitlis Massif (Baştug 1976; Perincek 1979; Özkaya 1982b).

Oliogocene to Miocene thrusting carried the Tauric nappes over a complex of South Tauric thrust sheets and slices composed of two units: the *Guleman* ophiolites and their volcanic and sedimentary cover (Late Maastrichtian-Paleocene Hazar flysch, Eocene calc-alkaline Maden Formation); and the Tertiary *Çungus* flysch which includes Eocene olistoliths of Maden-facies rocks (Yazgan 1981; Perincek 1979). These units all rest upon the 'Foothill Structure Belt' (Rigo de Righi & Cortesini 1964) along the margin of the Arabian platform. This border belt is characterized by: (1) deep water sediments of Campanian-Maastrichtian age, deposited immediately prior to gravity-emplacement of the Koçalı ophiolites and related flysch units, during the Maastrichtian; (2) proximal flysch of Upper Miocene age 'Lice Formation) deposited just before the final phase of nappe emplacement; (3) folding of the whole sedimentary and tectonic pile during the Mio-Pliocene, parallel to the Tauric front.

Along the Malatya transect, ophiolites occur at four structural levels, (Figs 2 & 3) not three as claimed by Özkaya (1982b). The *Munzur* ophiolites (the Bozkır ophiolites of Şengör & Yilmaz 1981) and the *Kocalı* ophiolites are petrographically similar, and their manner of emplacement, by Late Cretaceous gravity gliding on to unmetamorphosed limestones, and their ages, are the same. The younger associated sediments are dated as Late Creaceous (Bergougnan 1975; Özkaya 1982b), but older pelagic or slope deposits are Triassic in the Munzur complex (Bergougnan 1975) and Late Jurassic in the Koçalı (Perincek 1979). In the western equivalent of the latter ophiolite, the Bäer-Bassit, Triassic slope-and-basin units are also known (Delaune-Mayere *et al.* 1977; this volume). The Munzur and Koçalı ophiolites were thus formed in a Triassic basin (Whitechurch *et al.* this volume). In contrast, the *Elazığ* ophiolite, which comprises the ophiolitic basement of the southern part of the Elazığ nappe (e.g. Ispendere and Karga Dağ sections) is devoid of associated pre-Late Cretaceous sediments. It probably originated by post-Triassic, perhaps even Late Cretaceous rifting, since it is overlain by the same Late Cretaceous volcaniclastic cover as the Permo-Triassic Keban marbles to the north. Its tectonic style is also very different from that of Munzur and Koçalı ophiolites, as it suffered the same Late Cretaceous deformation and

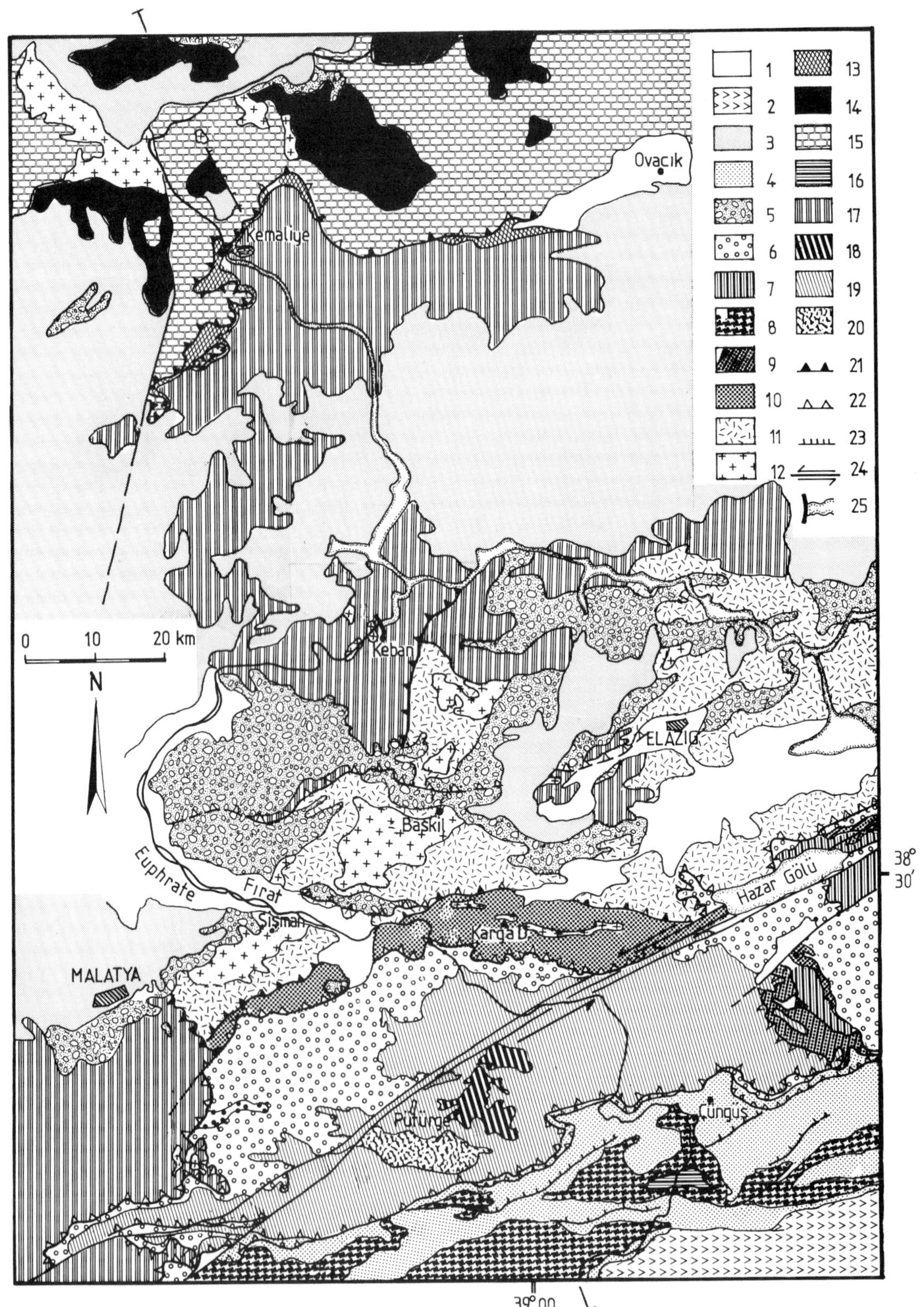
Ovacık
Kemaliye
Keban
ELAZIĞ
Başkil
Euphrate
Fırat
Şişman
Karga D.
Hazar Gölü
MALATYA
Pütürge
Çüngüş
0 10 20 km
N
38°
30′
39°00
1
2
3
4
5
6
7
8
9
10
11
12
13
14
15
16
17
18
19
20
21
22
23
24
25

calc-alkaline intrusive events as the Keban-Malatya and Pütürge belts on either side. The fourth ophiolitic complex, the *Guleman* ophiolite, which is still undated (but see Aktaş & Robertson, this volume), occurs in two different tectonic positions: within the South Tauric thrust sheets, below the Pütürge metamorphics and above the Çüngüs flysch (Perincek 1979; Yazgan *et al.* 1983), and in thrust sheets overlying the Bitlis metamorphics, associated with Late Cretaceous calcalkaline volcaniclastics (Özkaya 1978; Perincek 1979). The upper sheets should be correlatable with the Elazığ ophiolite. The lower ones could correlate with either the Elazığ or the Koçalı ophiolites. Below the Bitlis massif in the Baykam-Mutki area (Hall 1976, 1980), two different ophiolitic melanges are thrust on to the Foothills Structure Belt, but only the lowermost is comparable to the Koçalı complex. Further east, below the Şemdinli metamorphics, which are equivalent to the Bitlis massif, the Karadăg ophiolites are overlain by Campanian-Maastrichtian volcanic and sedimentary rocks of Yuksekova type (Özkaya 1982a).

Regional setting of Tauric ophiolites

Other relevant studies concern the plate-tectonic settings in which Tauric subduction took place. Özkaya (1982a, b) emphasized that some ophiolites of Eastern Turkey may represent an Upper Cretaceous back-arc or leaky-transform system and not a basin formed by Triassic extension. In our opinion, this certainly applies to the Elazığ ophiolites, which are clearly distinct from those of Koçalı and Munzur.

Throughout the Taurides, ophiolites associated with rock associations indicative of Triassic extension show a common structural trend of N 70°E representing the direction of spreading in the Antalya, Posantı-Kersantı, Hatay and Troodos ophiolites (taking late rotation into account). These ophiolites were subject to pre-obduction shearing, which migrated from north to south between 104 Ma and 88 Ma. Spreading was still active and the southern ophiolites (Troodos, Hatay), which are the youngest, formed around 80 Ma. Tholeiitic dykes were still being intruded at 75 Ma, prior to obduction (Whitechurch *et al.* this volume).

Palaeomagnetic data presented by Lauer (1982 and this volume) led him to assume important longitudinal displacements in the Mesozoic.

The Taurides show no indication of subduction before the Upper Cretaceous but Jurassic and Upper Cretaceous subduction occurred below the Pontides (Adamia *et al.* 1980). The intervening ocean was consquently not symmetrical when the Tauric subduction started.

The Mesozoic organisation of the Taurides is the subject of conflicting interpretations, involving either one or several oceanic basins initiated by Triassic extension between the Pon-

Fɪɢ. 2. Geological sketch map of the Malatya-Elazığ Taurus. The lower part (South of Baskil) is drawn after recent detailed mapping (E. Y., in progress; Perincek 1979), the upper part after the geological map of Turkey, 1:500 000, modified after the personal observations of E.Y. & A.M.

1, Alluvium; 2, Plio-Quaternary basalts; 3, Miocene volcanics and sediments, Pliocene continental deposits (Tauric neo-platform); 4, Upper Maastrichtian to Miocene, Arabian platform facies (Midyat carbonates, Lice proximal flysch). Also Tertiary Çüngüs flysch (the thin thrust sheet just below Pütürge metamorphics); 5, Upper Maastrichtian-Paleocene (Baskil units, Malatya nappe), Eocene and Lower Oligocene, Tauric neoplatform facies (Seske carbonates and clastics, Kirkgeçit marls). Includes some Eocene volcanics at Gambaş tepe (Ispendere unit); 6, Lower and Middle Eocene, Maden calc-alkaline volcanic and sedimentary formation (Guleman and Pütürge nappes); 7, Upper Maastrichtian-Paleocene flysch (Hazar-Simaki formation, Guleman nappes); 8, Ophiolitic mélange (Koçalı) with associated Cretaceous flysch (Karadut) obducted on (1.16) before (1.4); 9, Guleman ophiolites; 10, Elazığ ophiolites, Ispendere and Karga Dağ units, with some Coniacian-Maastrichtian andesitic flysch cover (Ispendere: this village is located 20 km east of Malatya); 11, Senonian arc complex: calc-alkaline hypovolcanics; volcanics and volcaniclastics (basalts, andesites, rhyodacites, agglomerates . . .) and Late Campanian volcaniclastic flysch cover; 12, Senonian arc complex (continuation): Granitoids, dated as Coniacian-Santonian (K/Ar), sometimes foliated (Karga Dağ); 13, Kemaliye mélange: Late Cretaceous greywackes including pillow basalts, radiolarites and limestones in tectonic lenses; 14, Munzur (Bozkır) ophiolites, Late Cretaceous obduction upon the Munzur limestones; 15, Late Triassic to Late Cretaceous Munzur Dağ limestones (Taurus Calcareous Axis); 16, Permian to Senonian limestones, autochthonous Arabian platform; 17, Metamorphic limestones and calcschistes, mostly Permian and Triassic; Malatya, Elazığ and Keban nappes; 18, Corresponding marbles and calcschistes in the Pütürge nappe; 19, Pütürge polymetamorphic schists; 20, Pütürge mylonitic orthogneiss; 21, Late Cretaceous (Early Maastrichtian) thrust; 22, Tertiary (Late Oligocene to Pliocene) thrust; 23, Late Miocene to Pliocene reverse faults; 24, East Anatolian transcurrent faults; 25, Keban dam.

N.B.: The Eo-alpine (Senonian) folded, metamorphic and/or magmatic units are numbered 10 to 12 and 17 to 20 in the legend.

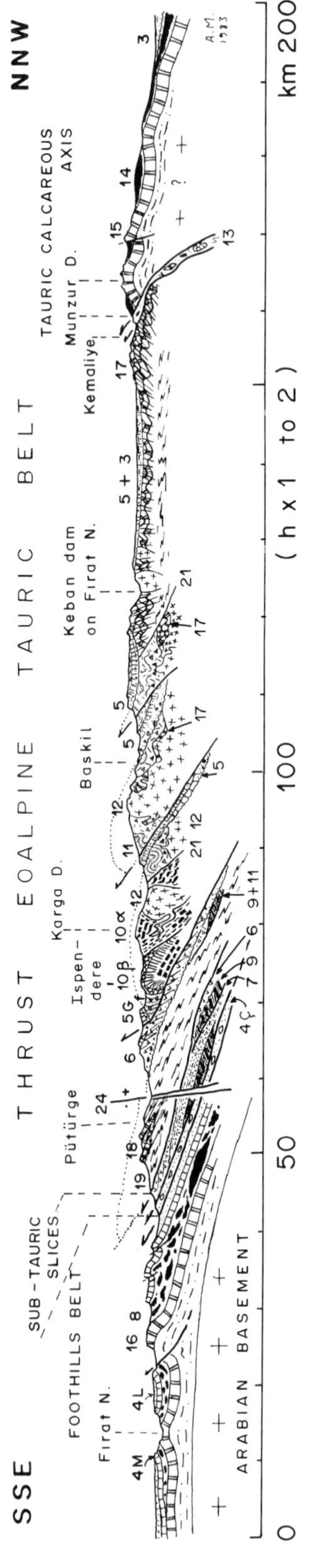

Fig. 3. The Malatya Taurus transect, generalized from 3 cross-sections by Yazgan & others (1983), except the Munzur area, from Bergougnan (1975), with new observations (A.M., E.Y.).

C, Çüngüs; G, Gambaş Tepe; L, Lice; M, Midyat; α, sheared gabbros and diorites, with some wherlite; β, pillow basalts overlying dyke complex and gabbros; D, Dağ (mountain); N, Nehri (river). Lithostratigraphical index as on Fig. 2.

tides and Arabia. Following Ricou *et al.* (1975), Biju-Duval *et al.* (1977), Şengör & Yilmaz (1981), Ozkaya (1982b), the problem is discussed further in this volume by Ricou *et al.* Robertson & Woodcock, Waldron, Hayward, Juteau *et al.*, Yilmaz, Aktaş & Robertson. Since in our view the data presently available do not permit a definitive solution of this problem, we discuss two alternative models.

Model 1: One oceanic basin subducting northwards and southwards

This model (Fig. 4) follows the hypothesis of Ricou *et al.* (this volume) that a single Tethyan ocean existed between the Pontides and the Tauric-Arabian platform.

At 100 Ma the northern margin of this ocean had already been affected by Jurassic subduction and was now the site of resumed, Upper Cretaceous, subduction below the Pontides. The southern margin is known from slope-basin units which occur at the base of the obducted Tauric ophiolites and was still evidently a passive margin originally formed by Triassic extension and most probably bordered by early Mesozoic oceanic crust (Fig. 4a). The ocean was thus asymmetrical: old and dense lithosphere was present to the south while younger, less dense lithosphere (Fig. 4b) was being subducted to the north. Tectonic shearing of the oceanic lithosphere occurred between 104 Ma and 88 Ma and migrated southwards (Thuizat *et al.* 1981; Whitechurch *et al.* this volume), probably resulting from the relative difficulty of subducting young lithosphere below the Pontides. The asymmetry of the ocean may well have led to a southward shift of the spreading ridge as shown in Fig. 4, during the Late Cretaceous.

At 90 Ma the southern platform was becoming increasingly unstable, as documented by pelagic deposits on both the Arabian platform and on the future 'Calcareous Axis' (Ricou *et al.* this volume). This is interpreted as the beginning of southward-dipping subduction. Plate kinematics shows that Africa and Eurasia were then converging (Patriat *et al.* 1982) and it is probable that the relatively inefficient northward-dipping subduction zone was not capable of keeping pace with the supply of lithosphere from both plate convergence and Tethyan spreading, and instead subduction of more dense southern lithosphere was initiated towards the south. (Fig. 4a).

By 80 Ma the southward subduction had led to opening of the Elazığ back-arc basin, with its ophiolitic basement and calc-alkaline magmatism. Tectonic shearing in the ocean ceased, probably indicating that the two subduction zones were able to accommodate the excess lithosphere resulting from convergence. The ridge now spread at a decreasing rate and produced the last ophiolites (Troodos, Hatay, Fig. 4d). To account for the fact that the obducted ophiolites mainly represent the northern half of an ocean, we propose that the old southern lithosphere had already been consumed by southward subduction. The spreading ridge and adjacent low-density lithosphere then arrived at the trench and obduction followed.

At 75 Ma, the last oceanic magmatism is documented by tholeiitic dykes cutting the ophiolites. The 'compressional crisis' occurred in both Pontides and Taurides, with the Elazığ basin crushed in a complex way, as granitic intrusions persisted during compression. The oceanic lithosphere began to be obducted on to the Taurides, as documented from the Munzur area.

By 65 Ma, obduction was ending and the collision stage followed. Gravity-driven ophiolitic nappes were emplaced at this time. The shape of the basin into which they arrived can be inferred from the subsidence pattern which controlled the Upper Cretaceous pelagic sedimentation on the previously stable platform. The front of the gravity nappes reached the South Tauric basin where subsidence was at a maximum (Ricou *et al.* this volume). The basin persisted while to the north the area from Munzur Dağ to Pütürge was emergent, uplift probably assisting southward nappe transport.

Palaeogeographic data for the Early Tertiary shows that the Tethyan ocean was probably completely closed in Turkey, but still existed along the Zagros where collision occurred only during the Late Tertiary (Adamia *et al.* 1980). The Pontides and Taurides then constituted an Anatolian mass bordered by two areas of thinned lithosphere, the Black Sea to the north and the South Tauric basin to the south. This Upper Cretaceous basin linked the Eastern Mediterranean and the Zagros remnant of the Tethyan ocean. This weak zone then controlled the subsequent evolution of the area.

At 50–40 Ma, magmatism reappeared in both the Pontides (South Pontic volcanic belt) and the Taurides (for example the Maden volcanics on the northern flank of the South Tauric basin). This could have been due to the residual influence of the two subducted slabs, or alternatively could have been caused by thermal adjustment of previously deformed lithosphere.

By 20 Ma, the Oligocene-Miocene regime was established, characterized by tectonic shortening of the South Tauric basin and associ-

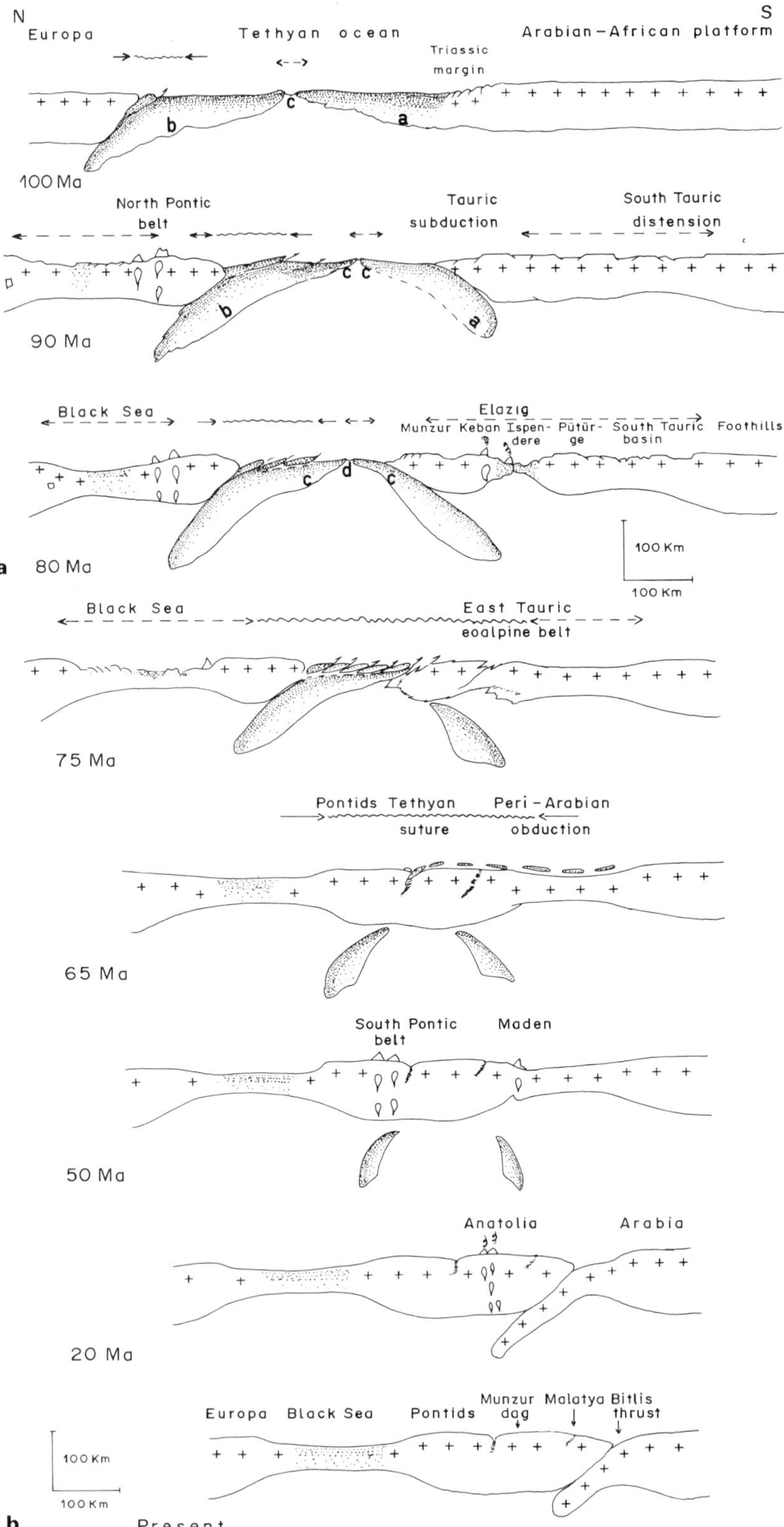
N
S
Europa
Tethyan ocean
Arabian–African platform
Triassic margin
b
c
a
100 Ma
North Pontic belt
Tauric subduction
South Tauric distension
b
c c
a
90 Ma
Black Sea
Elazig
Munzur Keban Ispen-dere Pütür-ge South Tauric basin Foothills
c
d
c
100 Km
100 Km
a
80 Ma
Black Sea
East Tauric eoalpine belt
75 Ma
Pontids Tethyan suture
Peri-Arabian obduction
65 Ma
South Pontic belt
Maden
50 Ma
Anatolia
Arabia
20 Ma
Europa
Black Sea
Pontids
Munzur dag
Malatya
Bitlis thrust
100 Km
100 Km
b
Present

ated volcanism on the Anatolian massif. Consumption of lithosphere took place along the new weak zone linked to on going plate convergence.

At the present time, the Zagros and South Tauric basins have reached the collision stage, while their western extension in the Aegean area is still undergoing subduction.

Model 2: Two northward subducting oceanic basins

This second model (Fig. 5) follows Biju-Duval *et al.* 1977) who postulate the existence of two oceanic basins between Eurasia and Africa during the Mesozoic. One was north of the Taurus 'Calcareous Axis' or Anatolian platform, called 'Tethys', the other was south of the same continental platform, between its southernmost units, the Pütürge and Bitlis massifs, and the present Arabian platform, and is called 'Mesogea'. These two oceans are equivalent, after the Triassic, to the northern and southern branches of Neo-Tethys (Şengör & Yilmaz 1981).

At 100 Ma, the northern ocean (Tethys) is envisaged as being similar to the single ocean in the preceding model, though possibly narrower. Mesogea is also assumed to have had a complicated and asymmetrical structure. It opened as early as the Triassic, as indicated by slope-and-basin units obducted together with the Baër-Bassit ophiolites, and also by the Triassic pillow lavas associated with some of the southern Bitlis slices (Perincek, 1980); it was still opening in Early Cretaceous time. The intervening platform is presumed to have been a fragment of Gondwana, covered by Permo-Triassic limestones predating or accompanying rifting and identical to those of Arabia, which was then subject to extension perhaps as early as the Middle Jurassic as suggested by basaltic pillows in the West Taurus Middle Jurassic. The co-existence of extension and continental convergence may be explained by a sinistral transcurrent regime, which would agree with the palaeomagnetic data of Lauer (1982; this volume).

The Tauric subduction is inferred to have begun at 90 Ma, since the oldest calc-alkaline intrusions are dated at 85 ± 2 Ma. Three to five million years are thought to be enough for subduction at a moderate rate to induce arc magmatism. The extension of the Elazığ basin was possibly increased at first by this process. A variant of the model would be to consider the Elazığ basin as a marginal basin, the opening of which would have been linked to the initiation of subduction (see the one-ocean model). However, the Upper Cretaceous arc magmatism would then be anomalous in position and age with respect to the supposed back-arc basin. In island arc systems back-arc extension generally seems to post-date the initiation of the arc magmatism (Toksöz & Hsui 1982; Barker, 1982). In contrast, field relations in the Elazığ nappe show that calc-alkaline magmatism post-dates the generation of ophiolites and even the early stages of their deformation.

On the 80 Ma sketch (Fig. 5), the observed field relations are accounted for by accelerated convergence of Africa and Eurasia. This initiated deformation in the Elazığ ophiolites and the Pütürge continental block, caused Mesogean subduction to progress to the magmatic stage and caused the last pre-obduction thrusting in the Tethyan and Mesogean oceans. Spreading ceased in the northern ocean but was still active in the southern one to produce the Troodos and Hatay ophiolites.

During the Late Campanian (75 Ma sketch, Fig. 5), subduction, and related magmatism and tectonics continued. Progressive syn-metamorphic shearing and folding deformed the southern part of the Anatolia-Taurus block (i.e. the Keban-Malatya, Elazığ and Pütürge nappes) at high temperature. The high geothermal gradient is related to the effect of the underlying subducted Mesogean slab. The asymmetrical structure of this Eo-alpine Tauric belt is consistent with the supposed northward-dipping subduction. The major southern part of the belt is a pile of southeast-verging sheared units, probably parallel to the subduction plane; the northern and less metamorphosed part of the belt (Keban) shows slightly reclined folds verging north-westwards, towards the colder passive parts of the platform (Munzur). A Late Cretaceous flysch basin probably separated the Munzur area from the Keban area at that time, and only closed during Tertiary thrusting

FIG. 4. A plate tectonic reconstruction for the Malatya Taurus transect: the 'one-ocean' model.

(a) The earliest stages of the Tauric subduction at 100 Ma (late Early Cretaceous), 90 Ma (Cenomanian) and 80 Ma (Early Campanian). Note that intra-oceanic slicing terminated some time before this latter sketch.

(b) The latest stages of the Tauric subduction and the collisional stages. Eoalpine folding (75 Ma: Late Campanian), suturing and obduction slipping (65 Ma: Late Maastrichtian); fossil slabs sinking (50 Ma: Eocene); subduction of thinned continental lithosphere (20 Ma: Miocene) and finally continental collision.

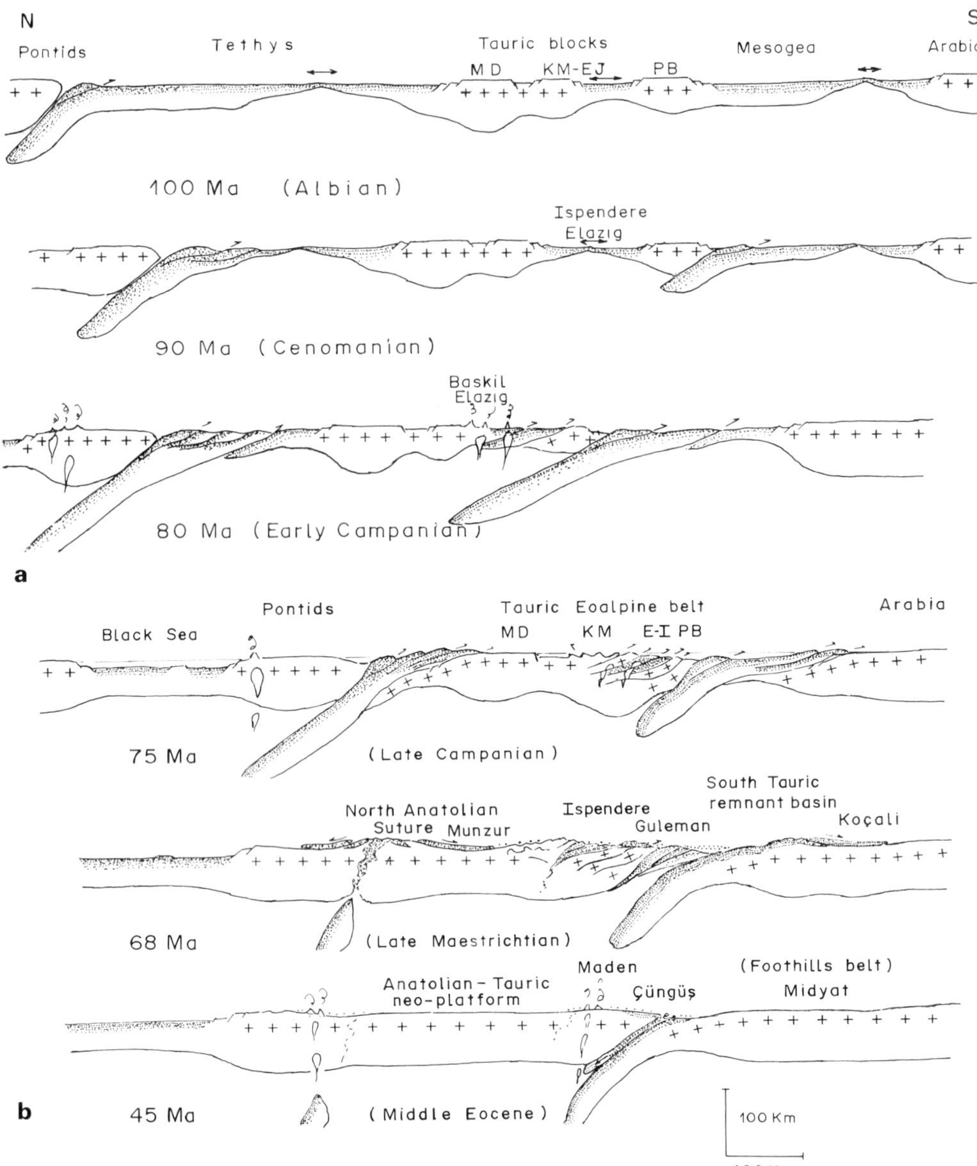

Fig. 5. An alternative reconstruction for the East Tauric belt: a 'two-ocean' model.
(a) From extension to subduction in the Tauric intermediate blocks.
(b) From Eoalpine folding to the closure of Mesogea. The latest stages (20 and 0 Ma) would be similar to those of Fig. 4, except that the Miocene volcanism would be explained by fossil slab sinking or by thermal adjustment of the deformed lithosphere.

(Kemaliye-Ovacik thrust). The Pütürge Massif is shown on the sketch to be overlain by the Elazığ ophiolites. This hypothesis accounts for the metamorphism of the Permian-to-Mesozoic cover of the Pütürge massif and for the ophiolite blocks included in its transgressive Eocene cover. In addition, some pre-Eocene ophiolitic klippen associated with Late Cretaceous volcanism are seen on the Bitlis Massif (Perincek 1980). A slice of rocks between the Pütürge massif and the subduction plane is inferred to have disappeared by tectonic erosion during Late Cretaceous to Late Eocene times, along with most of the trench deposits (45 Ma, Fig. 5).

During the late Campanian, closure of the Tethyan ocean would have been almost complete, while the Mesogean ocean would have remained slightly open, a hypothesis which explains the Maden Eocene magmatism by subduction. On the southern border of both oceans, sheared slices of ophiolites and sediments were stacked on to continental crust prior to gravity-sliding.

On the 68 Ma sketch (Late Maastrichtian, Fig. 5), the late obduction events are now shown. Ophiolite slid towards lower, underformed parts of the continental blocks, especially the 'Calcareous Axis' and the Foothills Belt. On the other hand the deformed basement of the Upper Cretaceous arc was uplifted, as shown by conglomerates at the base of the late Maastrichtian neritic, transgressive cover of the Elazığ nappe, which includes folded Campano-Maastrichtian volcaniclastic flysch. We can not say if this uplift pre-dates or post-dates transport of the ophiolites. The post-tectonic and post-obduction cover is everywhere of the same neritic facies, except in the South Tauric units where the Guleman ophiolites are unconformably overlain by a Late Maastrichtian-Paleocene proximal flysch. This flysch would thus represent a fore-arc deposit, the Guleman ophiolites being interpreted either as obducted Elazığ slices or as sheared Mesogean elements accreted on to the Tauric mass.

On the 45 Ma sketch, the Eocene calc-alkaline magmatism is explained in two different ways. In the Pontides, volcanism can be attributed to a fossil subducting slab (as in Fig. 4), while in the East Taurides, it results from the end stages of Mesogean subduction, a situation similar to that which still prevails in the Oman-Makran transect (Boudier & Michard 1981).

After the closure of Mesogea, continental collision led to Miocene and Plio-Quaternary suturing. In this model, Miocene volcanism is attributed to a sinking fossil slab and/or thermal adjustment of the deformed lithosphere.

Discussion and conclusions

In addition to Pontic subduction we are forced to conclude that Tauric subduction also occurred, although the direction of dip of the slab is open to discussion: to the south according to Şengör and Yilmaz (*op. cit.*) as in our first model; to the north in our second model; both to the north and to the south according to Özkaya (1982b). As already stressed by Şengör & Yilmaz (1981) and by Özkaya (1982a) having to account for this Tauric subduction implies that the Late Cretaceous development of Turkey is more complicated than previously assumed.

Our data on the Baskil-Keban area do not fit with the formation of an accretionary prism and ensimatic Upper Cretaceous arc north of the Malatya-Keban-Bitlis block, as advocated by Şengör & Yilmaz (*op. cit.*). Transgression of the calc-alkaline volcaniclastic Maden Formation on the Pütürge metamorphics is not consistent with the genesis of the Maden Formation in a deep marginal sea opened above a subduction zone north of the Bitlis block.

Özkaya's latest model (1982b) visualises the Pütürge-Bitlis massifs as a Tertiary para-autochthonous slice of the Arabian platform. This assumption does not explain the Upper Cretaceous deformation of Pütürge, Elazığ and Keban zones. We also disagree with the assumption that the Tauric ophiolites apart from the North Anatolian ones, originated in two leaky transform faults on both sides of the Keban plate. Subducting both these narrow oceanic basins below the same narrow Keban plate, with opposite subduction dips would hardly have permitted a single arc to be constructed on the southern margin of this block.

In fact the Tauric Upper Cretaceous arc is located both on the southern margin of the Keban block and on the northern margin of the small Elazığ oceanic basin, the southern margin of which is the Pütürge-Bitlis block. In our models, the opening of the Elazığ basin is considered to be due to back-arc extension or to rifting, either normal or oblique. Upper Cretaceous magmatism is known westwards as far as the Kyrenia Range, North Cyprus (Baroz 1979). No equivalent is known in the Zagros Range to the east. The subduction would seem to have been parallel to and have the same areal extent as the 'Calcareous Axis' at the end of the Cretaceous. Both phenomena seem to be restricted to this peculiar segment of the Middle East Alpine chain. Were they limited at both ends by transverse faults, as suggested by Ricou (1980) on other grounds? Neogene faulting and sedimentary cover in the Van area obscure the critical outcrops.

The first model considers the 'Calcareous Axis' as simply the northern prolongation of the Arabian platform during Mesozoic time. It explains their common Upper Cretaceous extension as due to their similar position above the south-dipping Tauric subduction zone. This extension results in their final separation by opening of the Elazığ and South Tauric basins. Southward subduction can also explain why obducted ophiolites appear to be only the axial and northern parts of an oceanic basin, if it is assumed that the southern part has been subducted before obduction. The model explains why the ophiolites and associated basinal units are similar on both the 'Calcareous Axis' and the Arabian platform, which are themselves also similar. It explains the similar internal organization of the various ophiolites, which show decreasing spreading rates and southward migration of oceanic shearing. A surprising feature of the model is the large displacement of the obducted slabs, ca. 300 km, over an uneven basement. Nappes had to pass over the Eo-Alpine fold-belt after closure of the Elazığ basin, which did not occur before the Maastrichtian, and before the uplift of the fold-belt in the Late Maastrichtian, which in turn possibly helped further gliding southwards.

Our second, two-ocean model, postulates a north-dipping Tauric subduction zone, similar to the Pontic one. The palaeogeographic and tectonic reconstruction is more complicated. For example, the early Lage Cretaceous extension of the Tauric platform is considered to be an ?oblique rifting process, pre-dating subduction, timing which is consistent with known arc geodynamics, as already noted. Ophiolitic thrust translation is correspondingly reduced. The timing problem for the crossing of the Eo-Alpine belt by the ophiolites is thereby avoided. The obduction process takes place on cold, down-going continental margins, as in Oman (Boudier & Michard *et al.* 1982). No change in the dip of subduction is needed between Upper Cretaceous and Miocene times, nor is a fossil subducted slab required to explain Eocene magmatism, at least in the Taurus. The asymmetrical opposite vergence of the northern and southern parts of the Eo-Alpine belt is accounted for by synthetic, high temperature shearing above the subduction plane and antithetic reclined folds on the colder fringe of the back-arc trough, a general pattern which is frequently observed in subduction margins (e.g. Japan: Sato 1979; Alaska: Hudson & Plafker 1982). The similarities between the 'Calcareous Axis'—Northern ophiolites and Arabian platform—Southern ophiolites doublets are however not so readily explained as in the first model.

In conclusion, we emphasise that the authors are not in complete agreement as to which model is the best. L.E.R. and H.W. prefer the first for its ability to explain the local data and for its good fit with their general interpretation of the Taurides (Ricou *et al.* this volume; Whitechurch *et al.* this volume). A.M. and E.Y. prefer the second from several points of view but fully acknowledge the weak points of both models. A.M. wishes to emphasize that the importance of Mesozoic and Tertiary strike-slip displacements, slightly oblique to the length of the Taurides, needs to be carefully explored; a lateral accretion during Late Maastrichtian to Eocene time would explain the repetition of the same doublet (calcareous platform and obduced ophiolites) twice on the same transect. This hypothesis needs further development and probably more field data.

ACKNOWLEDGEMENTS: This work was carried out with the support of the Équipe de Recherche Associée au C.N.R.S. no 887 Géologie et Pétrologie structurales, Strasbourg and the Laboratoire Associé au C.N.R.S. no 215, Géologie Structurale, Paris. Fieldwork facilities have been supplied by the Maden Tetkik ve Arama Enstitüsü, and financial support by the Institut National d'Astronomie et Géophysique (A.T.P. Géodynamique I). The manuscript was reviewed by N.H. Woodcock and G. Aktaş.

References

ADAMIA, S., BERGOUGNAN, H., FOURQUIN, C., HAGHIPOUR, A., LORDKIPANIDZE, M., ÖZGÜL, N., RICOU, L. E. & ZAKARIADZE, G. 1980. The alpine Middle East between the Aegean and the Oman traverses. Public. 26e Congr. Géol. Int. Paris (Mém. BRGM no 115) Colloque C5, 122–136.

BARKER, P. 1982. Back-arc geodynamics (News and views). *Nature* **298,** 119–120.

BAROZ, F. 1979. *Etude géologique dans le Pentadaktylos et la Mesaoria* (*Chypre septentrionale*). Unpubl. thesis, Nancy **I,** 365 pp.

BAŞTUG, C. 1976. Bitlis napinin stratigrafisi ve Güneydogu Anadolu sütür sonurun evrini. *Yeryuvari ve Insan*, **3,** 55–61.

BERGOUGNAN, H. 1975. Relations entre les édifices pontique et taurique dans le Nord-Est de l'Anatolie. *Bull. Soc. géol. Fr.* **17,** 1045–1057.

BIJU-DUVAL, B., DERCOURT, J. & LE PICHON, X. 1977. From the Tethys ocean to the Mediterranean seas: a plate tectonic model of the evolution of the western alpine System. In: BIJU-DUVAL, B. & MONTADERT, L. (eds). *Structural History of the Mediterranean Basins.* Editions Technip, Paris, 143–164.

BOUDIER, F. & MICHARD, A. 1981. Oman ophiolites. The quiet obduction of oceanic crust. *Terra cognita*, **1,** 109–118.

BRAUD, J. & RICOU, L. E. 1975. Eléments de continuité entre le Zagros et la Turquie du Sud-Est. *Bull. Soc. géol. Fr.* **17,** *1015–1023.*

DELAUNE-MAYERE, MARCOUX, J., PARROT, J. F. & POISSON, A. 1977. Modèle d'évolution mésozoïque de la paléo-marge téthysienne au niveau des nappes radiolaritiques et ophiolitiques du Taurus Lycien, d'Antalya et du Bäer-Bassit. In: BIJU-DUVAL B. & MONTADERT, L. (eds). *Structural history of the Mediterranean Basins.* Editions Technip, Paris, 79–94.

DEWEY, J. F., PITMAN, W. C. III, RYAN, W. B. F. &

BONNIN, J. 1973. Plate tectonics and the evolution of the alpine System. *Bull. geol. Soc. Am.* **84,** 3137–3180.

HALL, R. 1976. Ophiolite emplacement and the evolution of the Taurus suture Zone, southeastern Turkey. *Bull. geol. Soc. Am.* **87,** 1978–1988.

—— 1980. Unmixing a melange: the petrology and history of a disrupted and metamorphosed ophiolite, SE Turkey. *J. geol. Soc. Lond., 137,* 195–206.

HUDSON, T. & PLAFKER, G. 1982. Paleogene metamorphism of an accretionary flysch terrane, eastern Gulf of Alaská. *Bull. geol. Soc. Am.* **93,** 1280–1290.

JUTEAU, T. 1979. Ophiolites des Taurides: essai sur leur histoire océanique. *Rev. Géol. dyn. Géogr. phys.* **21,** 191–214.

LAUER, J. P. 1982. *L'évolution géodynamique de la Turquie et de Chypre, déduite de l'étude paléomagnétique.* Unpublished thesis Univ. Louis Pasteur Strasbourg, 300 pp.

ÖZGÜL, N. 1981. Munzur Dagkrinun geolojisi. Internal Report, M. T.A. no 6995.

ÖZKAYA, I. 1978. Stratigraphy of the Ergani-Maden Region. *Türk. Geol. Kurumu. Bült. Turkey,* **21,** 41–67.

—— 1982a. Marginal basin ophiolites at Oramar and Karadağ, SE Turkey. *J. geol. Soc. Lond.* **139,** 203–210.

—— 1982b. Upper Cretaceous plate rupture and development of leaky transcurrent fault ophiolites in southeast Turkey. *Tectonophysics,* **88,** 103–106.

PARROT, J. F. & WHITECHURCH, H. 1978. Subductions antérieures au charriage Nord-Sud de la croûte téthysienne: facteur de métamorphisme des séries sédimentaires et volcaniques liées aux assemblages ophiolitiques syro-turcs, en schistes verts et amphibolites. *Rev. Géogr. phys. Géol. dynam.* **20,** 153–170.

PATRIAT, PH., SÉGOUFIN, J., SCHLICH, R., GOSLIN, J., AUZENDE, J. M., BEUZART, P., BONNIN, J. & OLIVET, J. L. 1982. Les mouvements relatifs de l'Inde, de l'Afrique et de l'Eurasie. *Bull. Soc. géol. Fr.* **24,** 363–372.

PERINCEK, D. 1979. Interrelations of the arab and anatolian plates. 'Guide Book excursion B' First Geol. Congr. Middle East. Ankara, 34 pp.

—— 1980. Bitlis metamorfiterlinde volkanitli Triyas. *Türk. Geol. Kurumu Bült.,* **23,** 201–211.

—— & ÖZKAYA, I. 1981. Arabistan levnasi kusey kenari tektonik evrimi. *Yerbilimleri* **8,** 91–101.

RICOU, L. E. 1971. Le croissant ophiolitique périarabe: une ceinture de nappes mises en place au Crétacé supérieur. *Rev. Géogr. phys. Géol. dynam.* **13,** 327–349.

—— 1980. La tectonique de coin et la genèse de l'arc égéen. *Rev. Géol. dyn. Géogr. phys.* **22,** 147–155.

——, ARGYRIADIS, I. & MARCOUX, J. 1975. L'axe calcaire du Taurus, un alignement de fenêtres arabo-africaines sous des nappes radiolaritiques, ophiolitiques et métamorphiques. *Bull. Soc. géol. Fr.* **17,** 1024–1040.

RIGO DE RIGHI, M. & CORTESINI, A. 1964. Gravity tectonics in the foothills structure belt of southeast Turkey. *Am. Assoc. Pet. Geol. Bull.* **48,** 1911–1937.

SATO, T. 1979. La structure du Japon. *Rev. Géol. dyn. Géogr. phys.,* **21,** 161–179.

ŞENGÖR, A. M. C. & YILMAZ, Y. 1981. Tethyan evolution of Turkey: a plate tectonic approach. *Tectonophysics,* **75,** 181–241.

SMITH, A. G. 1971. Alpine deformation and the oceanic areas of Tethys, Mediterranean and Atlantic. *Bull. geol. Soc. Am.* **82,** 2039–2070.

THUIZAT, R., WHITECHURCH, H., MONTIGNY, R. & JUTEAU, T. 1981. K-Ar dating of some infra-ophiolitic metamorphic soles from the Eastern Mediterranean: new evidences for oceanic thrustings before obduction. *Earth planet. Sci. Lett,* **52,** 302–310.

TOKSÖZ, M. N. & HSUI, J. K. 1978. Numerical studies of back-arc convection and the formation of marginal basins. *Tectonophysics,* **50,** 177–196.

YAZGAN, E. 1981. Dogu Toroslarda etkin bir paleokita kenari etüdü (üst kretaseorta eosen) Malatya-Elazig, dogu Anadolu. *Yerbilimleri,* **7,** 83–104.

——, MICHARD, A., WHITECHURCH, H. & MONTIGNY, R. 1983. Le Taurus de Malatya (Turquie orientale), élément de la structure sud-téthysienne. *Bull. Soc. géol. Fr.* **25,** 59–69.

A. MICHARD, & H. WHITECHURCH, Université L. Pasteur, Laboratoire de Géologie Structurale et Laboratoire de Pétrographie, 1 rue Blessig, 67084 Strasbourg Cédex, France.

L. E. RICOU, Laboratorie de Géologie Structurale, Université P. et M. Curie, 4 place Jussieu, 75230 Paris Cédex 05, France.

R. MONTIGNY, Institut de Physique du Globe, 5 rue Descartes, 67084 Strasbourg Cédex, France.

E. YAZGAN, M.T.A. Enstitüsü, Jeol. Dairesi Eskişehir Yolu, Ankara, Turquie.

The Maden Complex, SE Turkey: evolution of a Neotethyan active margin

G. Aktaş & A. H. F. Robertson

SUMMARY: In this paper we describe and interpret a critical segment of a Tethyan suture zone in SE Turkey dominated by a major nappe pile which, from the structural top downwards, consists of the following units: first, the *Elazığ-Palu nappe*, a major Upper Cretaceous volcanic arc complex and cover related to subduction of a southern strand of the Mesozoic Tethys. The arc rocks are regionally underlain by metamorphic rocks of the Bitlis-Pütürge nappe, but in the area studied this is absent and the next unit is the *Hazar-Guleman nappe*, which consists of a sliced ophiolitic assemblage of presumed Upper Cretaceous age unconformably overlain by a relatively undeformed succession of Palaeogene red beds (Ceffan Formation), flysch (Simaki Formation) and both pelagic and redeposited neritic limestones (Gehroz Formation). This nappe is cut by the *South Anatolian transform fault*. Below it the *Kıllan Imbricate Unit* consists of imbricated Upper Cretaceous ophiolitic and other mafic extrusives (Killan Group) associated with, and locally unconformably overlain by, Palaeocene and Eocene rocks ranging from deep sea sediments to sedimentary melange and detached blocks, with minor volcanics. Towards the sole-thrust the Killan Imbricate Unit is mostly tectonic melange. Below, are the autochthonous and para-autochthonous successions of the *Arabian foreland*, including ophiolitic olistostrome of latest Cretaceous age (Koçali Complex). Our Hazar-Guleman nappe and the Killan Imbricate Unit together constitute the *Maden Complex*. We also report on a distinctive assemblage of Middle Eocene mafic extrusives and interbedded deep water sediments further east along the suture zone (*Karadere Formation*), and briefly summarise implications for economic sulphide mineralization in the area.

We interpret the suture zone in terms of formation of a Mesozoic ocean basin followed by northward subduction. The Karadere Formation further east would in one possible model have formed by rifting of the fore-arc, possibly related to oblique convergence in the Middle Eocene. In an alternative model the Neo-Tethys had essentially closed by the latest Cretaceous facilitating ophiolite emplacement over both the northern active margin and the former passive southern margin. In this case the Killan Imbricate Unit would be related to tightening of the suture and the Karadere Formation to renewed rifting controlled by strike-slip faulting. In both models continental collision was well advanced by the Miocene with flysch and sedimentary melange deposition (Lice and Çüngüş Formations), followed by translation along the South Anatolian transform fault.

The region discussed is a segment of the major suture zone in SE Turkey separating the Arabian platform to the south from discontinuous metamorphic massifs representing one or more former micro-continents to the north (Fig. 1). These massifs are overlain, and intruded by, calc-alkaline igneous rocks of the Upper Cretaceous *Elaziğ-Palu nappe* volcanic arc complex and its sedimentary cover, which can be traced for some 200 km parallel to the suture, although partly obscured by later sediments and volcanics. The Elazığ-Palu arc complex is generally agreed to have formed as a direct result of subduction of a strand of the Mesozoic Tethys, or Neotethys, although the size of the ocean basin and the polarity of subduction are still in dispute (e.g. Perincek & Özkaya 1981; Michard *et al.* this volume; Ricou *et al.* this volume). Along much of the suture, particularly E and W of the Maden area (Fig. 2), the metamorphic massifs are either in direct tectonic contact with the Arabian foreland sediments beneath, or are separated from them only by thin slices of highly deformed mafic volcanics and sediments of uncertain significance. The Maden area in SE Turkey (Fig. 2) is of crucial importance because, in contrast to adjacent areas, the metamorphic massifs are virtually absent there and a complex stack of north-dipping thrust sheets of volcanic and sedimentary rocks is exposed between the basal thrust of the nappe representing the Elazığ-Palu arc complex and the sole-thrust overlying the Arabian foreland below. This intervening thrust-stack, previously termed the *Maden Complex*, essentially consists of Upper Cretaceous ophiolitic rocks and Lower Tertiary sedimentary and some volcanic rocks which were assembled during the Upper Cretaceous and Lower Tertiary and finally emplaced southwards over the Arabian margin in the Miocene. Further east, in an equivalent structural position to the Maden Complex, is a more intact succession of Eocene mafic lavas and deep sea

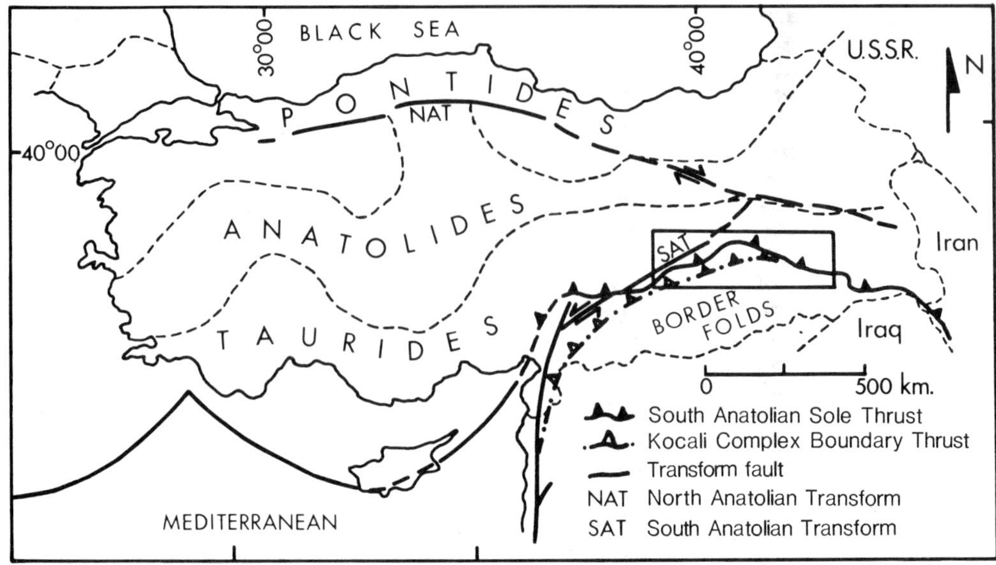

Fig. 1. Sketch map of the major tectonic units of Turkey. Inset: the area discussed here.

sediments, here termed the *Karadere Formation* (Fig. 2).

Our main objective in this paper is to infer the history of closure of a strand of the Mesozoic Tethys from Late Cretaceous to Miocene time in terms of active margin processes. We will conclude in favour of at least a major part of the Maden Complex being a form of accretionary wedge with emplaced ophiolites and a basin complex, all related to overall northward subduction during Late Cretaceous and Lower Tertiary time. Localized pull-apart basins of Middle Eocene age floored by mafic crust are attributed to strike-slip faulting during oblique convergence. We also draw attention to the implications for the genesis of major economic sulphide ore bodies within Upper Cretaceous and Middle Eocene mafic lava units.

We have concentrated our study on two areas: the *western area* around Maden itself, where the full thrust-stack is developed, and the *eastern area*, around Karadere and Mizik where more intact successions of Eocene volcanics and deep-sea sediments are present (Karadere Formation, Fig. 2). The western area is discussed first.

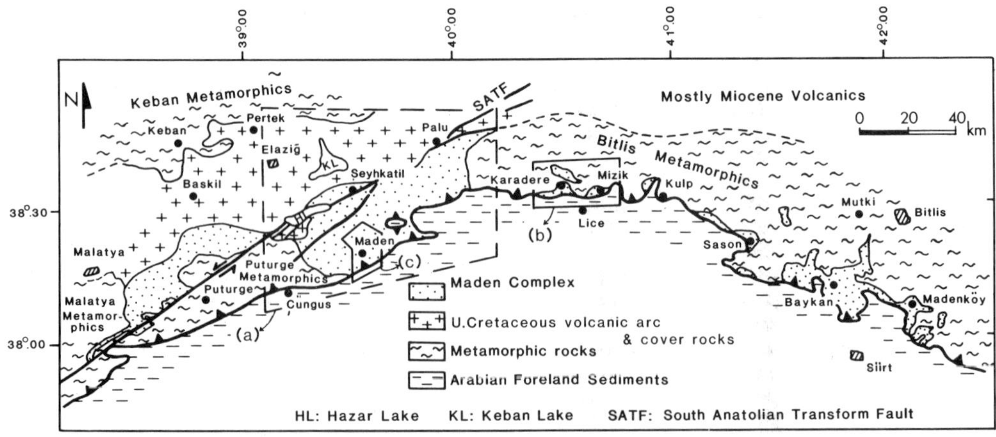

Fig. 2. Sketch map of the major structural units in SE Turkey. Insets: (a) the *western area* studied here (see Fig. 3), (b) the *eastern area* (see Fig. 12a), (c) the area around Maden town mapped in more detail (see Fig. 6a).

The Western Maden Complex area

Structural Succession

The structural succession in the *western area* of the Maden Complex (Fig. 3) can be divided into 5 units from the top downwards:

(1) *The Elazığ-Palu nappe.* This is composed of Upper Cretaceous mafic extrusives and pyroclastics, mostly calc-alkaline, with minor sediments, all of Upper Cretaceous age.

(2) *The Bitlis-Pütürge metamorphic nappe.* This major unit extends several hundred kms along the suture zone but is reduced to small slices and detached blocks in the type-area of the Maden Complex studied here.

(3) *The Hazar-Guleman nappe.* This consists of a variably deformed and partly eroded ophiolite of probably Upper Cretaceous age unconformably overlain by a Palaeogene sedimentary succession with minor volcanics.

(4) *The Killan Imbricate unit.* This consists of tectonic slices with a total structural thickness in places exceeding 13 km. The individual thrust-bounded packets consist of two main components, the Upper Cretaceous Killan Group, comprising mafic extrusives, and mafic and ultramafic plutonics, and the Maden Group, composed of sediments ranging from hemi-pelagic mudstones, to turbidites, sedimentary melange and pelagic carbonates with minor volcanics. The lowest structural levels of the Killan Imbricate Unit are *tectonic melange*, containing material mostly from the Killan Imbricate unit.

(5) *The Arabian foreland succession.* To the south of the area studied, this consists of Albian to Upper Cretaceous sediments of passive margin affinities overlain by ophiolite-bearing sedimentary melange, then by Palaeogene to Miocene clastics and shallow water carbonates showing considerable lateral and vertical facies variation.

Our work mainly on structural units 3 and 4 (Fig. 3) has been augmented by results of earlier studies and by detailed stratigraphical and sedimentological work, new micro-palaeontological results provided by E. A. Pessagno & G. C. Adams, as well as petrography and geochemistry of the igneous rocks, and new radiometric age data. It should be noted that the structural succession and stratigraphical terms used in this type area of the Maden Complex differ somewhat from those of the other workers (Table 1). Previously all the ophiolite rocks in our western area have been named the 'Guleman ophiolite', assumed to be of Upper Cretaceous age. However we distinguish the ophiolitic rocks of the structurally higher Hazar-Guleman nappe from those of the Killan Imbricate Unit. We now summarize the main features of structual units 1, 2 and 5 which we have not studied in detail ourselves but which must be taken account of in any tectonic interpretation of the area.

Elazıg-Palu nappe

Regionally, the Elazığ-Palu nappe consists of an assemblage of calc-alkaline intrusive and extrusive igneous rocks of Upper Cretacous age (e.g. Perincek 1979; Yazgan 1981). In the Malatya area the Upper Cretaceous arc rocks are reported to locally depositionally overlie metamorphic rocks (Elazığ-Keban highway; Perincek and Özkaya 1981), but more generally this contact appears to be tectonic. On the other hand Perincek (1981) described a depositional contact with Guleman ophiolitic rocks of probable Upper Cretaceous age in the Palu-Elazığ area (Fig. 2) and in this volume Michard *et al.* state that in the Malatya region their Elazığ nappe and the adjacent Palaeozoic Keban metamorphics are intruded by calc-alkaline rocks of the Upper Cretaceous arc complex. In the area studied the base of the Elazığ-Palu arc complex is marked by a tectonic contact with the underlying Hazar-Guleman nappe, while to the north-east the arc rocks have been thrust southward over the Bitlis metamorphic massif (Fig. 3; Tuna 1979; Perincek 1979, 1980).

Where well exposed as in the Malatya area (Fig. 2), the arc complex consits of both extrusives (andesites, silicic ash-flows, and volcaniclastics) and intrusives (diorite, granodiorite, monzonite and gabbro) with intercalated sediments (Yazgan 1981; Perincek 1979; Michard *et al.* this volume). In the area studied here the arc complex crops out north of Hazar Lake (Figs 2 & 3), where the succession consists mostly of massive and pillowed andesite and basaltic andesite lava flows with minor hyaloclastites and flow-breccias. It may be concluded that regionally the complex consists of an Upper Cretaceous (Coniacian-Maastrichtian) volcanic arc possibly constructed on both continental crust and an ophiolitic basement. That this calc-alkaline volcanism had ended by latest Cretaceous is shown by an unconformable sedimentary cover reported in some areas (e.g. Palu-Elazığ, Fig. 3).

Bitlis-Pütürge metamorphic nappe

In the areas studied, the Bitlis metamorphic nappe structurally overlies the eastern area of the Maden Complex, and after a break in outcrop reappears further west as the Pütürge

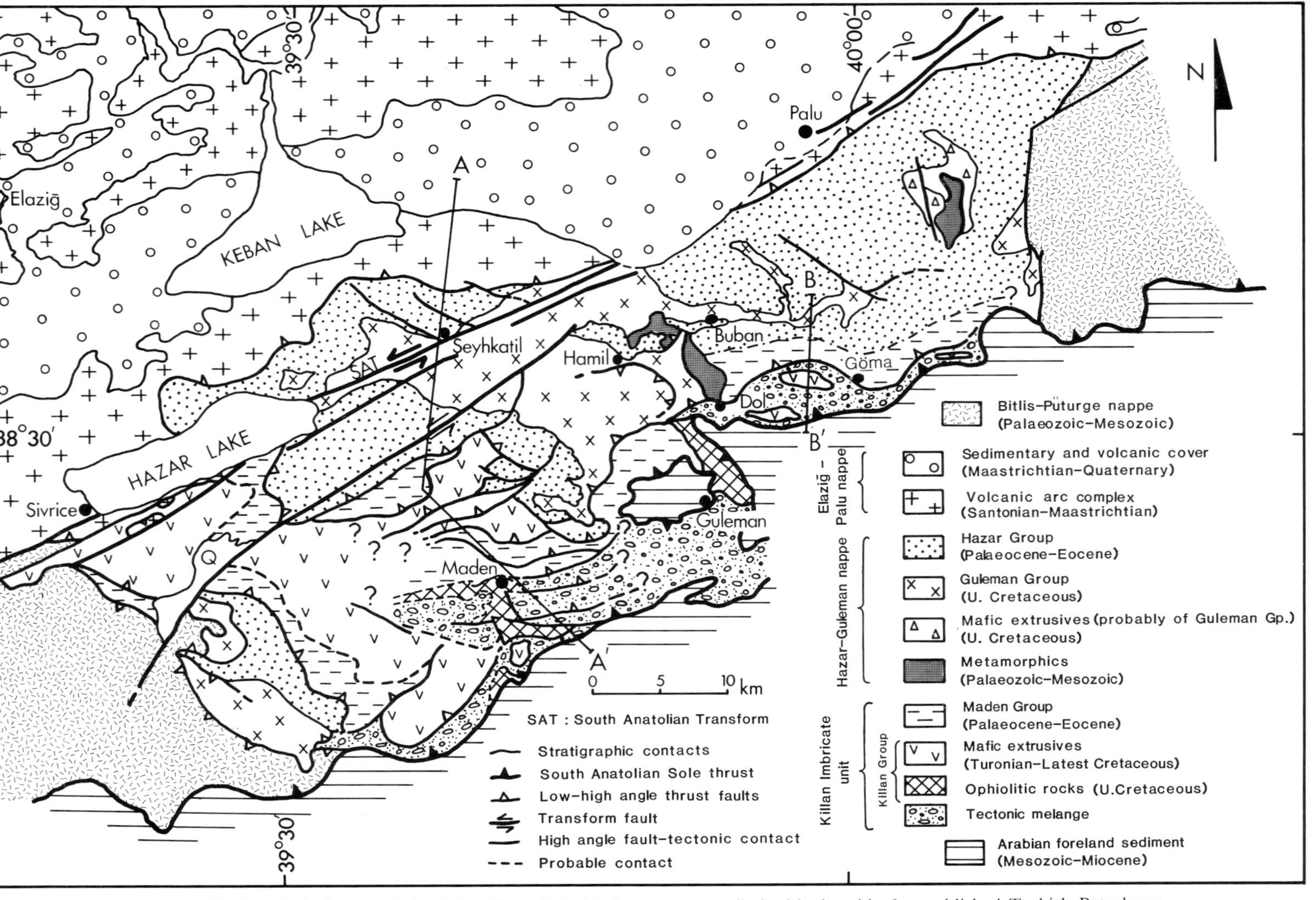

FIG. 3. Simplified geological map of the Lake Hazar-Palu-Maden area, compiled with the aid of unpublished Turkish Petroleum Corporation maps.

TABLE 1. *Tectono-stratigraphic classification of the* western *and* eastern *areas, based on this study and Perincek (1979, 1981)*

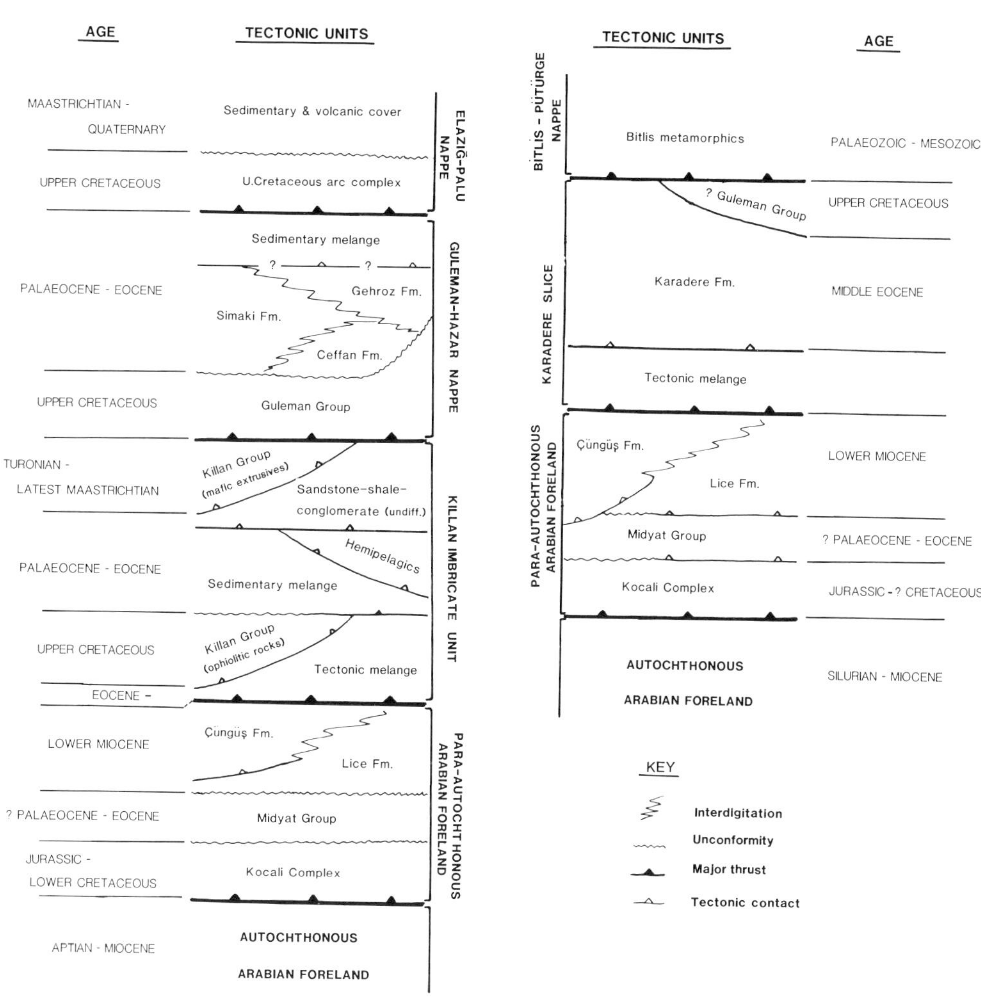

metamorphic nappe (Figs 2 & 3). The Bitlis metamorphic rocks are mostly greenschist facies calc-schists, mica-schist and meta-igneous rocks. The west, the Pütürge metamorphic massif is reported to consist mostly of gneiss and schist, with a latest Late Cretaceous syntectonic metamorphism of 80–70 Ma (Michard *et al.* this volume). Hempton (1982) suggests that the Pütürge nappe could have originated as a Mesozoic continental margin.

The new radiometric work reported by Helvaci & Griffin in this volume on the Avnik (Bingöl) area of the western Bitlis Massif confirms the presence of felsic metavolcanics, granitoids, mica-schists and Permian marbles (Tolun 1953), now dated as both Lower-Palaeozoic (Ordovician to Devonian) and Upper Palaeozoic (Carboniferous-Permian), affected by Late Cretaceous, and to a lesser extent by Eocene tectonics. Upper Triassic metamorphosed mafic volcanics, meta-cherts and marbles in the Bitlis Massif (e.g. near Dol, Perincek, 1981) are comparable to Triassic rocks related to rifting elsewhere in the Taurides (Göncüoğlu & Turhan 1982). Overall the Bitlis-Pütürge nappe thus consists of Pre-Triassic continental basement which was rifted in the Late Triassic, as the precursor to development of a continental margin. This was juxtaposed with ophiolites and metamorphosed during the Late Cretaceous, then uplifted, eroded and finally thrust southwards over the Arabian margin in Late Miocene time. The Bitlis and Pütürge metamorphic massifs may be joined and the outcrop of the nappe interrupted by thrusting, or conceivably two separate crustal slivers could always have existed.

Arabian foreland successions

Regionally, the Palaeozoic-Mesozoic succession is well exposed south of the eastern area (Fig. 2) in the Hazro inlier, a major east-west trending 'geanticline', in which deeply eroded Palaeozoic rocks are unconformably overlain by Cretaceous limestones (Rigo de Righi & Cortesini 1964; Perincek 1981). The Palaeozoic succession includes two cycles of mostly clastic rocks of Cambrian to Upper Permian age, unconformably overlain by a carbonate-evaporite succession which was slightly deformed in Jurassic to Lower Cretaceous time. This is overlain by Upper Cretaceous shallow marine carbonates and dolomites. The area north of the Hazro 'geanticline' subsided strongly during Campanian time (Karaboğaz Formation) with deposition of up to 600 m of calcareous shales and greywackes of the Kastel Formation, culminating in regional emplacement of the Besni olistostrome (Kocali Complex of Perincek, 1979) and major ophiolites. Ophiolites emplaced at this time include the Çermik, Oramar, Karadağ and others extending from Hatay in the west along the 'croissant ophiolitique peri-Arabe' of Ricou (1971) to the east (Rigo di Righi & Cortesini 1964; Çoğulu 1975, Perincek 1979, Özkaya 1982) as far as Oman (Glennie *et al.* 1973).

Ophiolite emplacement was followed, again regionally, by uplift and deposition of up to 1000 m of alluvial fan, deltaic, and shallow marine facies (Antak and Germav Formations; Rigo de Righi & Cortesini 1964; Perincek 1979; Baştuğ 1980). The overlying mostly shallow water carbonates with clastic intercalations (Midyat Group) record stable shelf conditions in the Middle and Upper Eocene (Baştuğ, 1980). A major transgression in the Early Miocene was then accompanied by faulting. Highs were capped by patch-reefs, while thick successions of chalky turbidites (up to 1250 m) accumulated in the lows (Lice Formation; Perincek 1979, 1981; Baştuğ 1980, 1982). Stratigraphically higher and coeval units towards the north include clastic and carbonate rocks with large detached blocks mostly derived from the structurally overlying Maden Complex (Çüngüş Formation).

We now consider the two units in the *western area* which we have studied in greater detail, the Hazar-Guleman nappe and the Killan Imbricate Unit.

The Hazar-Guleman Nappe

Guleman Group

The bulk of this nappe consists of ophiolitic rocks of assumed Upper Cretaceous age (Guleman Group) which are unconformably overlain by the Palaeogene sediments of the Hazar Group (see below). The basal serpentinized ultramafics (2 km E of Hamil village, Fig. 3) are tectonically interleaved with slices of greenschist to amphibolite facies rocks up to 40 m in structural thickness which may have formed as a high temperature metamorphic aureole. The type area of the Guleman ophiolite consists of sliced, partly serpentinised dunite, harzburgite, lherzolite, banded gabbro, and microgabbro, but no true sheeted complex or basalts and sediments have been reported (Erdoğan 1977; Özkan 1982).

In our principal area of study further west (Fig. 3) the Guleman ophiolitic rocks are very deformed internally and consist of mafic and ultramafic plutonics well exposed immediately north and south of the depression created by the South Anatolian transform fault.

Hazar Group

Structure

In the area studied the Hazar Group unconformably overlies the deformed and eroded basement of the Guleman ophiolitic rocks but is itself relatively underformed although cut by the South Anatolian transform fault (Fig. 3). To the north, shearing increases towards the thrust contact with the Elazığ-Palu volcanic arc rocks and there are deformed slices of Eocene andesitic extrusives (2–40 m thick) in the upper stratigraphical levels of the Hazar Group. Close to the transform fault (near Şeyhkatil village, Fig. 3), the rocks of the Hazar Group are cut by at least three major NW-SE trending high-angle reverse faults which dip northwards and many indicate compression along this segment of the transform fault (trans-pression). Regionally, the South Anatolian transform fault forms a depression 1–10 km wide; Hazar Lake has been interpreted as a classic pull-apart basin (Hempton & Dunne, 1982).

South of the South Anatolian transform fault sedimentary rocks of the Hazar Group are folded into megascopic open asymmetrical folds mostly verging southwards. In the Maden gorge (Fig. 4) the contact with the Killan Imbricate Unit is a high-angle north-dipping reverse fault, but elsewhere it takes the form of major monoclinal structure, termed the Putyan monocline (Figs 5a, b). This monocline possesses short, gently dipping, often faulted upper limbs and long, steeply dipping to inverted southern limbs. For example, near Putyan (Fig. 6) the Hazar group rocks swing southwards from nearly horizontal to steeply dipping and are then truncated southwards by a moderately north-dipping tectonic contact which brings in Upper Cretaceous mafic rocks of the Killan Imbricate Unit (see below).

Stratigraphy and age

There are three formations in the Hazar Group (Table 1), which, in generally ascending order, are the brilliant red clastics of the Ceffan Formation, the greyish and brownish flysch-like sediments of the Simaki Formation and the pelagic and redeposited limestones with minor volcanics of the Gehroz Formation. The Ceffan Formation is unfossiliferous. The Simaki and the Gehroz Formations yield the planktonic foraminifera *Globorotalia* (*morozovella*) sp., *Discocyclina* sp., *Distichoplax biserialis* (G. C. Adams, pers. comm. 1982) of Upper Palaeocene and Eocene age. Elsewhere, in the Malatya and Palu areas, the possible equivalents of the Hazar Group are reported to be Maastrichtian to Middle Eocene (Perincek 1979; Yazgan 1981). These three formations are distinctive mapping units which in the type area (Şeyhkatil, Fig. 3) follow each other conformably, but elsewhere, (North of Buban or Putyan, Figs. 3 & 6) wedges of the Ceffan Formation interfinger with the Simaki Formation. Also, in some areas the Gehroz Formation is laterally equivalent to both the Ceffan and the Simaki Formations. The individual outcrops are not sufficiently continuous to allow a three dimensional facies analysis.

Ceffan Formation: sub-aerial clastics

In the area studied the Ceffan Formation consists of distinctive 'red beds' from zero to

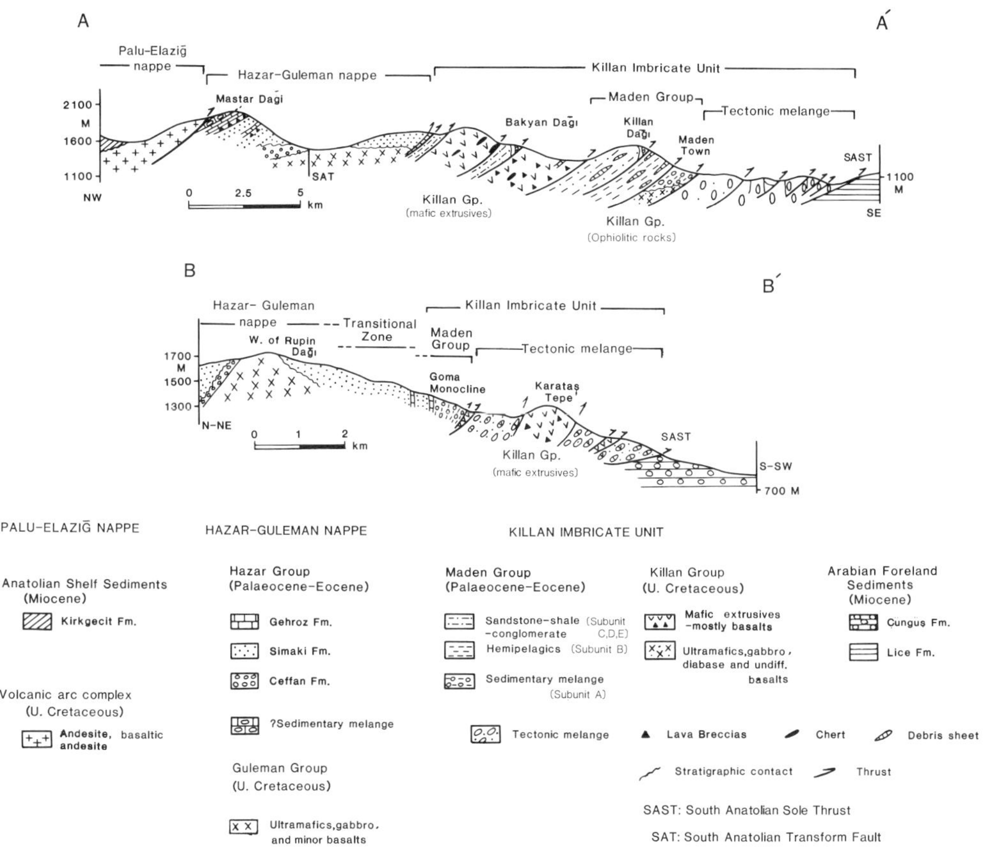

FIG. 4. Geological cross-sections along the lines A-A′ and B-B′ shown in Fig. 3.

320 m thick. Where present the formation overlies a rugged basement of eroded ophiolitic rocks of the Guleman Group. The type succession, measured near Şeyhkatil (Fig. 3), fines upwards overall from ophiolite-derived clastics to shallow marine coastal deposits. As shown in Fig. 5.1, the basal 30 m of the formation consists of almost massive clast-supported conglomerates with well-rounded cobbles and boulders with small volumes of interstitial red mudstone. The conglomerates consist mainly of clasts of basalt, chert, pelagic sediments, granodiorite, serpentinite and rare metamorphic rocks. Above comes a 130 m unit (Fig. 5.2) of several-metre-thick fining-upward cycles composed of trough-cross-bedded conglomerate and sandstone, interbedded with red siltstone and minor mudstone. Each cycle typically ends with mud-cracked caliche horizons (Fig. 5.2). There follows ca. 95 m of generally finer grained, medium to thinly bedded, lenticular, cross-bedded sandstone, also siltstone and mudstone, often pebbly, with abundant caliche. The highest unit, transitional to the Simaki Formation (Fig. 5.3), consists of ca. 60 m of alternations of greyish green and reddish well-sorted conglomerate, tabular cross-bedded sandstone, ripple cross-laminated siltstone and red, purple and greenish grey mudstone. The redder horizons generally contain calcrete while the greener beds possess a marine fauna, including encrusting calcareous algae, shells and echinoderm fragments. Petrographically, the Ceffan Formation was mainly derived from ophiolitic to acidic intrusive and extrusive rocks.

Although still easily recognisable by the bright red colour, most Ceffan facies elsewhere are finer grained than the conglomerates of the Şeyhkatil type-succession. Further south, (e.g. on the north flank of the Putyan monocline, (Figs 6a & 6b), successions consist of shallow marine fine-grained conglomerate, sandstone, and cross-bedded and rippled sandstone and mudstone with shallow marine fauna. Locally (e.g. Buban area, Fig. 3), the Ceffan Formation interdigitates with the Simaki Formation where it consists of red sandstone, siltstone, and mudstone and caliche. Several sandstone horizons up to 2 m thick contain concentrations of heavy mineral grains, particularly chrome-spinel and magnetite derived by erosion of the nearby ophiolitic and intermediate rocks. Locally (Buban area, Fig. 3), Ceffan sequences contain small volumes of tuffaceous sediment and thin mafic lava flows.

Interpretation: semi-arid peneplanation

The brilliant red colour, the complete absence of marine fossils except in the highest levels transitional to the Simaki Formation, plus the abundance of caliche all point to subaerial deposition. Prior to accumulation, an intrusive ophiolitic basement had been exposed adjacent to sources of both intrusive and extrusive (intermediate to acidic) igneous rocks typical of the Elazığ-Palu volcanic arc complex. The ophiolitic rocks could have been derived locally but the nearest intrusive arc rocks are now exposed south of Lake Hazar (Fig. 3). The rarity of metamorphic rocks clasts is consistent with the now virtual absence of the metamorphic nappe to the south.

The general distribution of the Ceffan Formation is suggestive of an E-W trending basin deepening to the south. The red conglomerates (e.g. Şeyhkatil succession) were laid down as small alluvial fans shed from a rugged basement. With time the relief diminished due to some combination of peneplanation and subsidence. The fining-upwards cycles of sandstone and caliche were then deposited as channelized braided stream and flood-plain deposits, while the ubiquitous red colour and abundance of caliche point to a hot semi-arid climate. By some time in the Palaeocene the area was being transgressed, presumably from the south, and for a time the coastline oscillated, giving rise to relatively high-energy coastal plain deposits, then with accelerated subsidence, the first sediments of the Simaki Formation began to accumulate. We thus infer that the Ceffan Formation records a period of Early Tertiary subaerial erosion of the ophiolitic and magmatic arc rocks which had been tectonically assembled in the area by the latest Cretaceous.

Simaki Formation: flysch basin

The Simaki Formation, which in our area is dated Palaeocene and Eocene, either conformably overlies the Ceffan Formation or unconformably overlies the ophiolitic rocks of the Guleman Group; to the south the formation is in tectonic contact with the rocks of the Killan Imbricate Unit (Figs 3, 4 & 6a).

The Simaki succession is well exposed and complete ca. 2.5 km northwest of Şeyhkatil village (Fig. 3), where it is over 500 m thick (Fig. 5.4).

Above the coastal facies at the top of the Ceffan Formation (Fig. 5.4) there is a ca. 260 m thick succession of greyish-green mostly medium bedded soft-weathering calcareous sandstone, siltstone and interbedded mudstone. The sandstone shows grading, micro-cross-lamination, plane-lamination and other features of the Bouma turbidite divisions. The

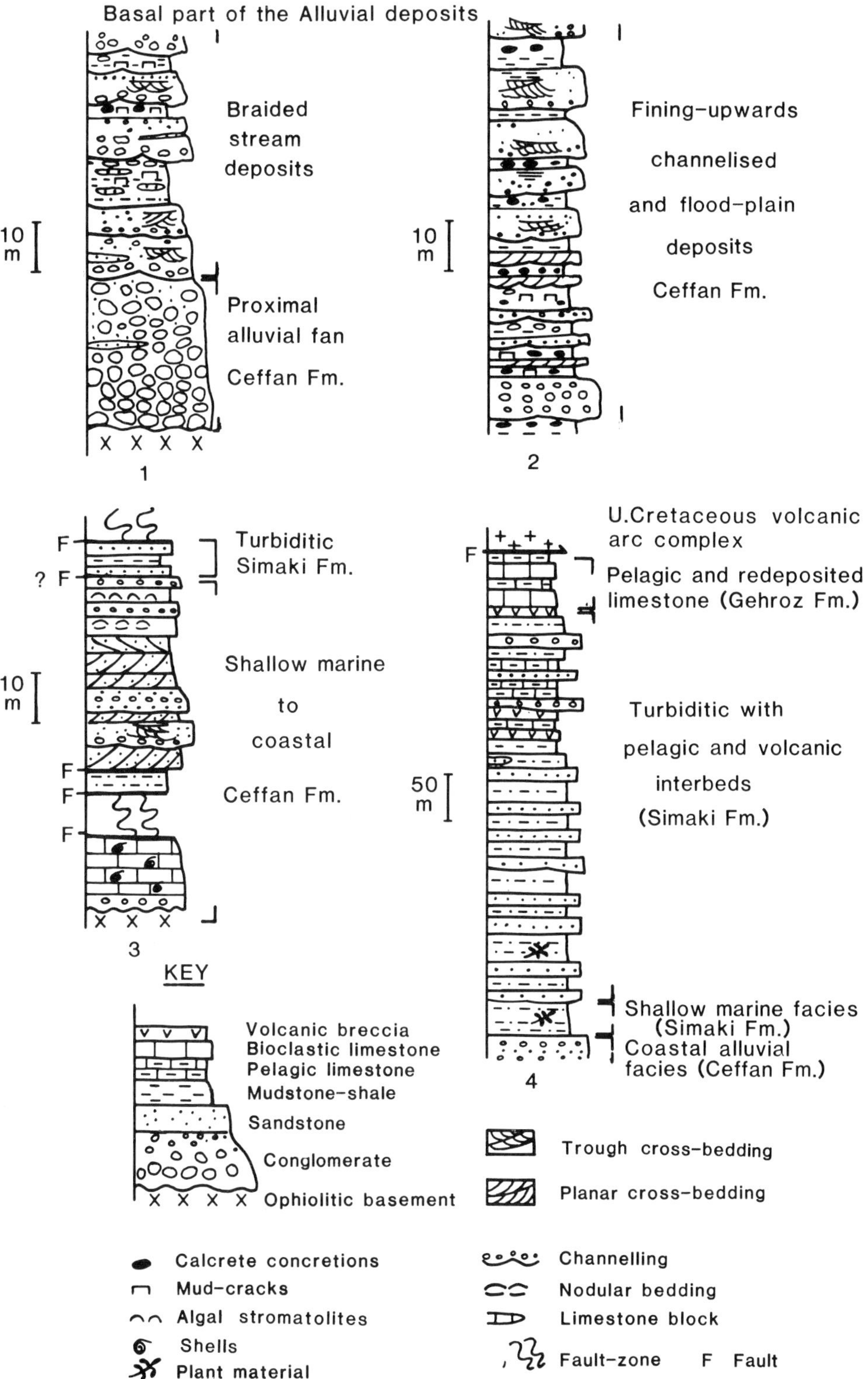

Fig. 5. Representative generalized sedimentary logs of units in the Hazar Group near Şeyhkatil village (Fig. 3). Logs 1, 2 and 3 are mainly of the Ceffan Formation; Log 4 includes both of the Simaki and the Gehoroz Formations.

turbidites first appear a few tens of metres above the Ceffan Formation. Interbeds of bioclastic limestone and minor matrix-supported calcirudites up to 0.8 m thick then appear higher in the succession. The highest levels, up to 165 m thick (Fig. 5.4), consist of finer grained sandstone, siltstone, and mudstone, with an increasing proportion of pink and white pelagic limestone, and rare laterally discontinuous horizons of mafic lava-breccia up to 4 m thick.

Microscopy shows that the typical sandstone of the Simaki Formation consist of variable mixtures of the following components: sedimentary (neritic and pelagic limestone, the latter whith planktonic foraminifera and radiolaria), metamorphic (polycrystalline quartz and muscovite), acid intrusive (plutonic quartz), mafic intrusive (ferromagnesian minerals and plagioclase) and mafic extrusive (basalt).

The Simaki Formation shows considerable thickness variation. Locally (e.g. near Hamil Fig. 3) it is condensed to ca. 30 m, where it locally unconformably overlies Guleman ophiolitic rocks, interdigitating with red clastics of the Ceffan Formation and with limestones of the Gehroz Formation. To the south, along the crest of the Putyan monocline (Fig. 6a), Guleman ophiolitic rocks are unconformably overlain by ca. 45 m of bioturbated pebbly calcareous sandstone with a rich fauna including gastropods, molluscs, and echinoderms. Upwards the succession passes into ca. 240 m of the fine-grained flysch typical of the upper part of the Simaki Formation in the Şeyhkatil area.

Interpretation: a subsiding basin

The Simaki Formation is interpreted as deposits of a deepening basin located north of the Killan Imbricate Unit during Palaeocene and Eocene time (Fig. 11, see below). After a long period of relative tectonic stability and subaerial exposure, the Simaki basin was initiated by rapid subsidence and block-faulting. The main basin trended roughly east-west (present coordinates), presumably with a now concealed shoreline to the north; to the south the Simaki basin was separated from the area of deposition of the Killan Imbricate Unit by discontinuous topographic highs undergoing shallow water carbonate deposition. Some areas (e.g. north of Buban and north of the Putyan monocline) remained near sea level interdigitating with subaerial facies of the Ceffan Formation and shallow marine sediments of the Gehroz Formation. Elsewhere, (e.g. near Simaki village) subsidence was gradual allowing shallow water deposition to continue. The southern long limb of the Goma monocline (Figs 3, 4b) appears to have been a depositional hinge with thick turbiditic deposition to the south. Small volumes of andesitic lavas, lava breccias and hyaloclastites were erupted throughout the period of subsidence.

Gehroz Formation: shallow water and pelagic carbonates

The Gehroz Formation, also dated Palaeocene and Eocene, either conformably overlies the Simaki Formation (Fig. 5.4), interdigitates with the Simaki and Ceffan Formations, or locally directly overlies Guleman ophiolitic rocks (Fig. 6a). Where best exposed on the Mastar Dağı (3 km NW of Şeyhkatil village, Figs. 3 and 4a), the formation consists of up to 80 m of limestones which overlie the Simaki Formation with a transitional contact. The succession consists of alternations of pink and grey pelagic and bioclastic limestones with occasional volcanic breccias of calc-alkaline composition (Table 2). Upwards, the succession is increasingly sheared, and then tectonically overlain first by melange containing blocks of Upper Cretaceous age, followed by the Upper Cretaceous Elazığ-Palu arc complex (Table 1, Fig. 4).

Petrographically, the pelagic material in the Gehroz limestones is foraminiferal micrite with Radiolaria replaced by calcite. The calcarenite comprises shallow water carbonate allochems, including calcareous algae, coral, echinoderm- and shell-material, intraclasts and benthonic foraminifera.

The Gehroz Formation again shows marked regional variation. In some areas (west of Putyan, Fig. 6a), Gehroz limestones up to 110 m thick are almost entirely hemi-pelagic, while elsewhere (e.g. north of Buban and north of the Putyan and Göma monoclines, Fig. 3), the formation comprises up to 80 m of shallow water limestones which rest on Simaki or Ceffan facies, or locally, directly on ophiolitic rocks of the Guleman Group.

Interpretation: marginal and basinal carbonate deposition

The Gehroz Formation accumulated in both basinal and coastal areas, and possibly on isolated topographic highs (Fig. 11, see below). The Simaki flysch-basin, as seen at Seyhkatil, is overlain by pelagic carbonates interbedded with bioclastic material, which is all re-deposited rather than *in situ* carbonate build-ups. The thin bedded lenticular and poorly sorted nature favours a local source, possibly from a carbonate shoreline to the south or to the north, now overthrust by the Upper Cretaceous volcanic arc rocks. Clastic input into the basin greatly diminished later, either because the basin was by then largely filled, or clastic deposition switched elsewhere, or possibly because regional transgression of source areas had taken place. By this time calc-alkaline volcanism was quite extensive.

In the south marine conditions were established early and persisted locally throughout the Palaeocene and the Eocene. During Simaki flysch deposition major topographic highs depositing shallow water carbonates came into existence between the Simaki flysch basin to the north and the areas of much deeper water accumulation now in the Killan Imbricate Unit to the south (Fig. 11, see below). From these highs large volumes of shallow water carbonate material were shed southwards as major wedges of sedimentary melange and detached blocks now in the Killan Imbricate Unit. The topographic highs were probably discontinuous and pelagic carbonates were deposited in intervening deeper water areas starved of terrigenous clastic input (e.g. west of Putyan, Fig. 6a).

The Hazar Group thus records, overall, an initially rugged subaerial landmass undergoing gradual peneplanation, followed by rapid differential subsidence to form at least one major E-W trending flysch basin bordered by carbonate build-ups. This basin then progressively filled with both pelagic and redeposited limestones, becoming increasingly volcanically active with calc-alkaline extrusions.

The Killan Imbricate Unit

Strikingly different deposition is recorded in the structurally underlying Killan Imbricate Unit which consists of imbricate slices of Upper Cretaceous ophiolitic rocks and mafic extrusives (Killan Group) and mostly sedimentary rocks (Maden Group) of Upper Palaeocene to Eocene age, as well as extensive tectonic melange.

Structure

The imbricate structure is most clear in the type area, the Maden river gorge, where the total structural thickness is estimated to be up to 13 km (Figs 6a, b). The contact with the structurally overlying Hazar-Guleman nappe is a major reverse fault in the Maden gorge, but elsewhere in the same structural position the Putyan and Goma monoclines are developed (Figs 4, 6). Beneath this the structurally upper part of the Killan Imbricate Unit in Maden Gorge is dominated by moderately northward-dipping slices of Upper Cretaceous mafic lavas intercalated with sedimentary rocks. Some of the individual mafic sheets which can be traced up to 6 km laterally are internally sheared. Below, the structurally intermediate part of the Killan Imbricate Unit is dominated by the sedimentary rocks of the Maden Group, which are increasingly deformed downwards by a combination of slumping and penetrative deformation. Lithologies include extensive slump-sheets up to hundreds of metres long and tens of metres thick and detached blocks up 20 m in diameter. Cleavage is locally well developed particularly close to the tectonic contacts between individual thrust packets. Frequent younging direction reversal indicates the existence of numerous isoclinal often recumbent folds, which are mostly south-facing and locally possess axial planar cleavage. The lower part of the Killan Imbricate Unit above the sole-thrust ('South Anatolian Sole Thrust', Figs 4, 6) consists of a tectonic melange including both Maden Group and Killan Group rocks. Near the sole-thrust, serpentinite becomes abundant as a matrix. Individual thrust packets in the Killan Imbricate Unit can often be traced for several km laterally, usually terminating in major dip-slip or wrench-faults (e.g. near Simaki village, Fig. 6a). On a larger scale the lateral contacts with the Pütürge and Bitlis metamorphic massif to the west and east (Fig. 3) appear to be complicated wrench-fault zones.

The Killan Group: ophiolitic rocks and mafic extrusives

The lower structural levels of the Killan Imbricate Unit contain sliced ophiolitic rocks including gabbro, ultramafics and mafic extrusives, while higher in the stack the Killan Group is restricted to mafic extrusives.

Ophiolitic rocks

In the structurally lower part of the Killan Imbricate Unit the ophiolitic rocks are in thrust sheets up to 1 km thick, composed of serpentinized dunite, peridotite, gabbro, diabase and both tholeiites and basaltic andesites. The contacts between the lithologies appear always to be tectonic. The underlying tectonic melange also contains over 40 m long bodies of trondjemite and lenses of radiolarian chert in which E. A. Pessagno (pers. comm. 1982) identified *Pseudoaulophacus* sp. of upper Turonian to upper Campanian age, and blocks possibly derived from a sheeted dyke complex. This assemblage can thus reasonably be interpreted as an Upper Cretaceous dismembered ophiolite.

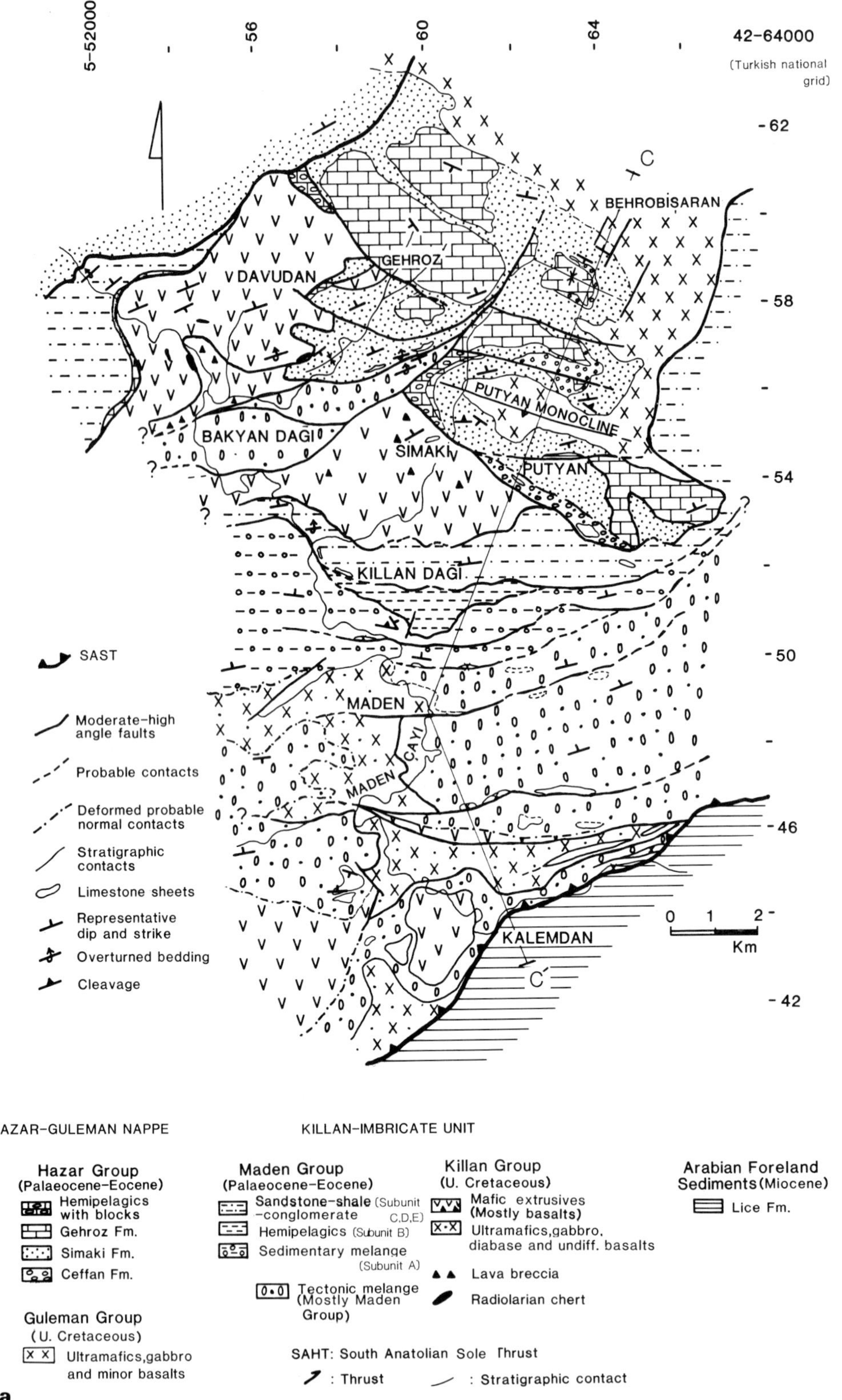

FIG. 6. (a) Geological map of the Maden area (see Fig. 2 for location).

Mafic extrusives

Tectonic slices of mafic extrusives are present at several different structural levels of the Killan Imbricate Unit. In addition to the ophiolitic assemblage low in the thrust-stack, two major thrust slices dominate the upper levels of the Killan Imbricate Unit and there are also numerous basaltic slices in the tectonic melange in the Göma area further east (Figs 3, 4, 6a).

The upper of these two major tectonic slices (eg. SW of Davudan village Fig. 6a) is mostly pillow basalt sometimes vesicular, with subordinate hyaloclastite and radiolarian chert intercalations. The lavas are mostly aphyric or clinopyroxene-phyric. Definite sedimentary interbeds of red ribbon radiolarian chert yield Radiolaria of late Turonian and Santonian age (E. A. Pessagno, pers. comm. 1982) including *Turris smewingi, Turris* sp., *Pseudoaulophacus* sp., *Dictyomitra formosa* and *Archaeodictyomitra* sp.

The lower tectonic slice (e.g. around Simaki village) consists mostly of thick sequences of hyaloclastite, flow breccia and pillow basalt, which is locally vesicular. Also present are massive basalts which are aphyric or clinopyroxene-phyric, and minor basaltic andesites which show similarities to those of the upper slice. Petrographically, most of these mafic extrusives have counterparts in the dismembered ophiolite in the lower part of the Killan Imbricate Unit.

Further east in the Göma area (Fig. 3) are slices of pillowed and massive basalt with minor flow-breccias, several kilometres long and several hundred metres thick. These basalts usually possess a sub-ophitic texture with fresh clinopyroxene and plagioclase and contain occasional ultramafic inclusions. These basalts are tectonically intercalated with red radiolarian cherts again dated as Upper Cretaceous based on *Patellula* sp. aff. *P. verteroensis* Pessagno, *Dictyomitra* sp., *Alievium gallowayi* White of Turonian to lower Campanian age (E. A. Pessagno, pers. comm. 1982).

In this study M. McIntyre (pers. comm. 1983) has used the K/Ar method to date mafic lavas from the Göma region and west of Maden Gorge. The reliable ages range from 64 ± 2.7 to 65 ± 3 Ma, within the Upper Cretaceous to the Palaeocene boundary.

Chemistry of mafic extrusives

On the basis of the analyses (Table 2) two suites of mafic extrusives can be distinguished. The first type, or Killan-type are mainly tholeiites and basaltic andesites and include the ophiolite-related extrusives low in the Killan Imbricate Unit, most of the basalts from the upper volcanic slice, and from the few available analyses, the mafic extrusives of the Hazar-Guleman nappe (Guleman Group). Gabbros from both the Killan Imbricate Unit and the Hazar-Guleman nappe are of this type (Table 2). The second suite includes the basalts of the Göma area and some of those in the lower tectonic slice.

The tholeiites and basaltic andesites of Killan-type have high abundances of large-ion lithophile elements (LIL, e.g. K, Rb, Sr, Ba) and show depletion of high-field-strength elements (HFS, e.g. Ti, P, Zr, Nb) relative to MORB (Fig. 7). Together with the gabbros, all these rocks possess a volcanic arc tholeiitic character (c.f. Ringwood 1977; Saunders *et al.* 1980; Saunders & Weaver 1980). However, on the TiO_2 vs. Zr and Cr vs. Y diagram of Pearce (1982) most of these rocks plot in the MORB and island arc tholeiitic field (IAT; Figs 8 and 9). The ratios of La/Nb, Ba/Zr and the Cr abundances however approximate to those of MORB (Table 2). Basalts of this type are generally attributed to an above-subduction zone genesis (Gill 1976; Weaver *et al.* 1980; Pearce, 1982; Alabaster *et al.* 1982).

The second type (Göma-type) of extrusives are relatively enriched in all incompatible elements relative to the first type and to MORB, except for Ti, Y and Sc (Fig. 7). This fact and

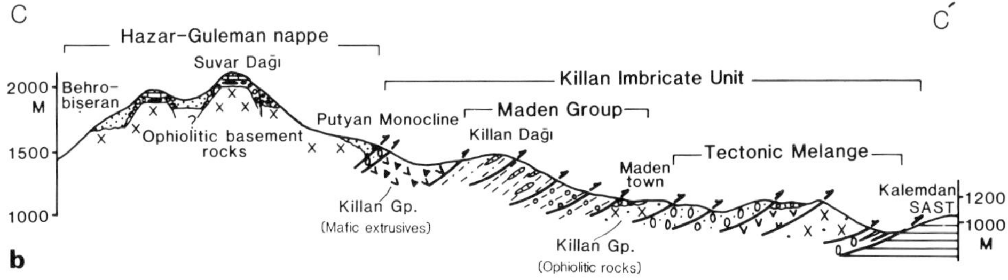

(b) Cross-section along the line C-C′ shown in Fig. 6a (Key as in Fig. 6a).

TABLE 2. *Representative analysis of igneous rocks from the* western *and* eastern *areas. Western area* (1–9 Upper Cretaceous, 10 Eocene). 1, 2 Killan-type basalts; 3, 4 gabbros of Guleman Group; 5 gabbro of Killan Group; 6, 7 Göma-type basalts; 8 andesite of Killan Group; 9 andesite of Elazığ-Palu arc complex; 10 andesite of Hazar Group. *Eastern area* (Karadere Formation, Middle Eocene); 11, 12 high-Al basalts; 13, 14 high-Ti basalts; 15 tholeiitic basalt (Karadere area); 16 tholeiitic basalt (Mizik area); 17: andesite

	WESTERN AREA										*EASTERN AREA*						
	Upper Cretaceous									Eocene	Middle Eocene						
	R9	57R	0135	0132	EM22	G5	G24	E25	R22	S62	K302	M58	K301	M57	1K6	L433	K316
	1	2	3	4	5	6	7	8	9	10	11	12	13	14	15	16	17
SiO_2(%)	48.93	49.04	48.97	45.52	47.02	48.68	49.80	56.25	56.48	57.59	48.17	47.77	50.17	52.11	48.55	50.95	56.47
Al_2O_3	15.94	14.99	16.84	16.28	16.26	18.23	15.18	16.70	15.40	17.13	20.14	20.26	17.42	15.88	16.73	15.51	18.02
Fe_2O_3	8.71	8.17	9.50	8.77	4.80	9.11	8.19	5.44	7.58	8.71	6.19	5.60	8.98	10.37	7.93	9.03	6.51
MgO	9.60	6.80	8.56	6.41	10.55	6.48	6.41	3.29	3.15	4.58	4.94	4.33	7.05	1.23	4.89	7.41	4.35
CaO	9.22	10.71	10.73	18.76	15.22	7.67	7.67	7.25	10.96	6.26	10.27	11.35	8.31	6.70	10.90	8.75	1.15
Na_2O	3.12	4.04	1.95	—	—	4.47	5.07	5.54	2.94	3.67	3.88	4.44	4.44	6.91	4.60	4.54	7.65
K_2O	0.87	0.54	0.06	0.01	2.47	0.24	1.16	0.25	0.68	0.06	0.88	0.47	0.25	0.68	0.37	0.34	1.43
TiO_2	0.74	1.05	0.93	0.44	0.15	1.30	1.28	0.55	0.69	0.54	1.24	1.01	1.90	2.54	1.47	1.46	1.16
MnO	0.15	0.15	0.13	0.12	0.11	0.15	0.13	0.10	0.17	0.12	0.12	0.11	0.16	0.18	0.13	0.15	0.11
P_2O_5	0.08	0.13	0.08	0.02	—	0.26	0.24	0.19	0.13	0.05	0.30	0.17	0.38	0.52	0.27	0.18	0.30
Total	97.34	95.60	97.73	96.15	96.49	96.59	95.15	95.56	98.17	98.71	96.13	95.51	99.06	97.07	95.83	97.83	97.42
LOI	3.17	4.05	2.16	3.33	4.07	3.95	4.80	4.96	2.48	2.68	3.77	4.64	2.65	3.05	4.17	6.67	2.67
Ni(ppm)	76	88	88	164	123	58	105	28	29	11	70	68	30	36	142	77	27
Cr	278	221	254	567	770	135	253	45	55	30	123	194	131	37	263	229	65
V	198	291	247	320	142	326	241	182	446	223	146	145	250	326	205	250	109
Sc	37	41	53	69	51	35	28	15	31	36	21	22	36	30	30	37	18
Cu	88	35	2	11	7	53	5	24	72	44	36	31	49	23	49	59	24
Zn	64	75	12	10	17	71	61	53	57	83	38	39	67	102	64	64	57
Sr	215	306	177	247	88	517	386	1216	336	316	494	443	304	222	352	223	121
Rb	14	7	3	4	52	6	10	5	15	1	11	8	2	20	7	2	11
Zr	52	71	66	8	3	90	148	101	52	128	164	113	228	335	184	123	361
Nb	1	6	1	1	2	10	6	4	3	20	14	8	12	22	9	4	28
Ba	82	42	26	39	86	111	98	129	80	46	104	71	65	103	35	31	137
Pb	1	9	6	6	5	4	8	26	8	11	9	10	9	8	6	8	9
Th	—	4	8	2	—	7	16	6	2	4	—	—	—	—	12	—	—
La	4	4	0	—	—	7	9	21	8	17	10	4	11	20	8	4	17
Ce	3	9	4	—	—	19	21	47	15	26	23	13	28	45	19	10	42
Nd	4	7	5	—	—	12	13	22	10	12	14	9	17	28	14	9	22
Y	23	24	21	11	6	23	26	19	17	26	23	19	37	49	29	30	38
Zr/Nb	52	12	66	8	1.5	9	25	25	17	6	12	15	20	15	21	30	13
Ba/Zr	1.6	0.6	0.4	4.7	28.6	1.2	0.7	1.3	1.5	0.4	0.6	0.6	0.3	0.3	0.2	0.3	0.4
La/Nb	4	0.7	—	—	—	0.7	1.5	5.2	2.7	0.9	0.8	0.6	0.9	0.9	0.9	0.9	0.6

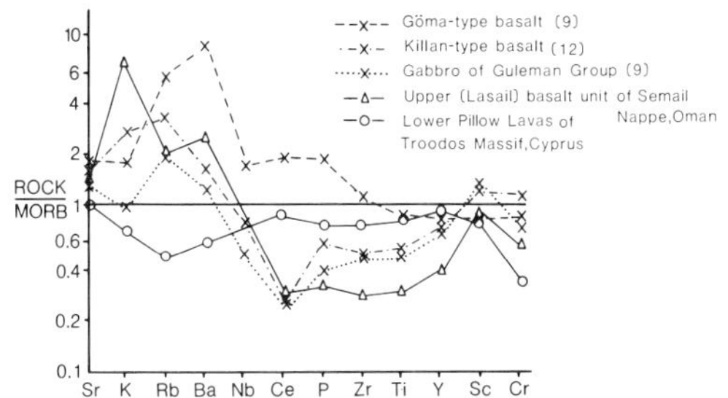

FIG. 7. Average MORB—normalized trace element patterns for Upper Cretaceous basalts of the Killan Group and gabbros of the Guleman Group (*western area*). Number of samples shown in brackets. Included for comparison are the Upper (Lasail) basalt unit of the Semail Nappe, Oman (Alabaster *et al.* 1982) and the Lower Pillow Lavas, Troodos Massif, Cyprus (Pearce 1980). MORB values are from Pearce (1982).

the ratios of Zr/Nb, Ba/Zr and La/Nb (Table 1) suggest they are more enriched than N-type MORB. On the Pearce (1982) diagrams (Figs 8 and 9) these basalts plot in the MORB and within-plate (WPB) fields and are compatible with a seamount origin (Wood *et al.* 1979; Alabaster *et al.* 1982).

Compared to other Tethyan ophiolites, the MORB-normalized plots (Fig. 7) show that the Killan-type extrusives are intermediate in composition between the lower units of both the Semail nappe (Geotimes unit), Oman, and the Troodos Lower Pillow Lavas, and two of the

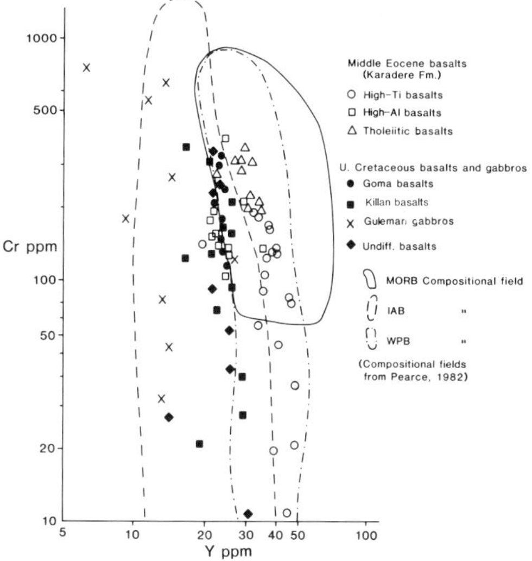

FIG. 8. Cr-Y discrimination diagram for basalts and gabbros. In the *western area*, the Killan-type basalts (Upper Cretaceous) plot in the IAT (island arc tholeiite) field towards the MORB (mid-ocean ridge basalt) field while the Göma-type basalts (Upper Cretaceous) plot between the MORB and IAT fields. Gabbros of the Guleman Group also plot in the IAT field. Basalts of the Karadere Formation (Middle Eocene) in the *eastern area* occupy the WPB field (within plate basalts) overlapping with the MORB and IAT fields.

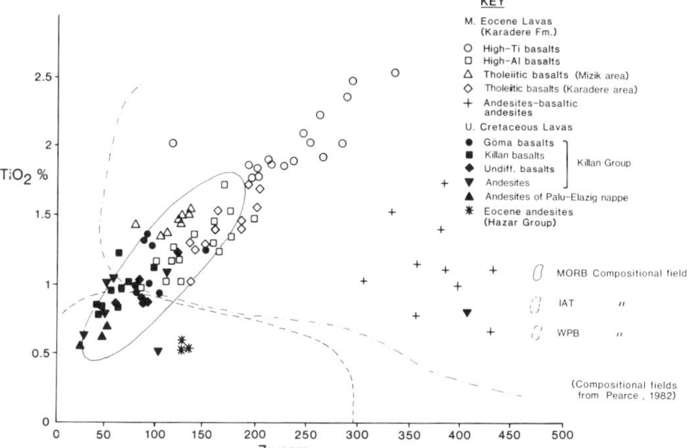

FIG. 9. TiO_2 versus Zr plot for both Upper Cretaceous and Tertiary basalts and andesites; compositional fields of Pearce, 1982. The basalts and andesites of the middle Eocene Karadere Formation (*eastern area*) plot in the WPB field and partly in the MORB field. Basalts and andesites of Upper Cretaceous age (*western area*) plot mainly in the MORB and IAT field. Andesites of the Hazar Group (Eocene, *western area*) plot in the IAT field.

upper units, the Troodos Upper Pillow Lavas and the Upper (Lasail) unit of the Semail nappe (Pearce 1980; Alabaster *et al.* 1982). The Göma-type basalt is akin to the 'enriched' Salahi basalt unit present locally at the highest levels of the extrusive succession in the Semail nappe (Alabaster *et al.* 1982).

Thus we may reasonably conclude that the mafic extrusives and the ophiolitic rocks of both the Killan Imbricate Unit and the Hazar Guleman nappe originated in a single area of Upper Cretaceous oceanic crust which, following conventional reasoning, was probably related to a subduction zone. If, as discussed below, the Killan Imbricate Unit is some form of subduction complex, then the 'enriched' extrusives could be preferentially accreted seamounts.

The Maden Group: sediments and minor volcanics

Sedimentary rocks and minor volcanics predominate in the mid part of the Killan Imbricate Unit, but also exist between mafic extrusive sheets (Killan Group) and in the tectonic melange (Fig. 6a, b). The succession of the Putyan and the Göma monoclines are critical as they form a link with the structurally overlying sedimentary rocks of the Hazar-Guleman nappe.

The internal structure and stratigraphy of the individual thrust packets in the mid part of the Killan Imbricate Unit is complex and laterally variable, but we have been albe to recognise five main sub-units in ascending structural order (Fig. 10):

Sub-Unit A

As shown in Fig. 10a, deformed ophiolitic rocks of the Killan Group, including pillow lavas, are locally unconformably overlain by more gently dipping fine grained mudstones and sedimentary melange. In strong contrast to the basement of the Hazar Group (Hazar-Guleman nappe), there are no coarse grained clastics or any obvious erosive products of the ophiolite. Instead the basal sediment is reddish brown and green siliceous and pebbly mudstone, ranging from thinly to thickly bedded (up to 2 m thick) and generally massive. Upwards, the mudstone becomes more silty with an increasing number of sub-rounded detached blocks of bioclastic limestone and tuffaceous extrusives. The higher part of the succession consists of ca. 160 m of siltstone, medium bedded calciturbidites, and both matrix- and clast-supported conglomerate with extensive slump-folding. There are also detached blocks of limestone up to 20 m in diameter, and many smaller blocks of mafic volcanics.

Sub-unit B (Fig. 10b)

This sub-unit, over 250 m thick, is mostly composed of hemi-pelagic calcareous mudstone and siltstones, in places pebbly, with a few lenses of pelagic limestone and intercalations of vesicular lava. Higher, there are intercalations of limestone conglomerate, mass-flow deposits containing quartzose clasts, as well as detached limestone blocks and several limestone debris-sheets ca. 25 m thick and over 150 m long. The conglomerates are both clast- and matrix-supported; individual pebbles are mostly rounded.

Sub-unit C (Fig. 10c)

Sub-unit C consists of alternations up to 200 m thick, of hemipelagic siltstone, mudstone and limestone with intercalations of fine grained sandstone, occasional limestone conglomerate, and thin horizons of vesicular lava and tuffaceous sediment. Fine grained sandstones coarsen upwards with a few grain-flow intervals.

Sub-unit D (Fig. 10d)

This is dominated by mudstones, often pebbly, with subordinate pelagic limestone and channelised debris-flows up to 15 m thick repeated several times over 350 m.

Sub-unit E (Fig. 10e)

This is by far the most homogeneous, composed of fine to medium grained, thin to medium bedded calcareous sandstones and mudstones which are mostly turbidites totalling over 260 m. The basal ca. 100 m contains tectonic inclusions of mafic lava over 50 m long.

Depositional link with the Hazar Group

In the Maden gorge area, Sub-unit E is tectonically overlain by mafic extrusives of the Killan Group, discussed above, with deformed sedimentary intercalations totalling 1200 m thick (partly melange in Fig. 6a), which are mostly flysch-like as in Sub-unit E with some debris-flow deposits. These are in turn, in tectonic contact with Hazar Group rocks. An original sedimentary transition can however be inferred between the Killan and the Hazar Group rocks further east on the southern limb of the Göma monocline (Figs 3, 4b). Traced northwards there, typical Upper Cretaceous melange of the Killan Imbricate Unit is separated from the southern limb of the Göma monocline by a moderately northward-dipping strike-fault. Near the fault, the succession on the Göma monocline is at first composed of inverted calciturbidites, passing northwards (generally stratigraphically downwards) into ca. 1000 m of

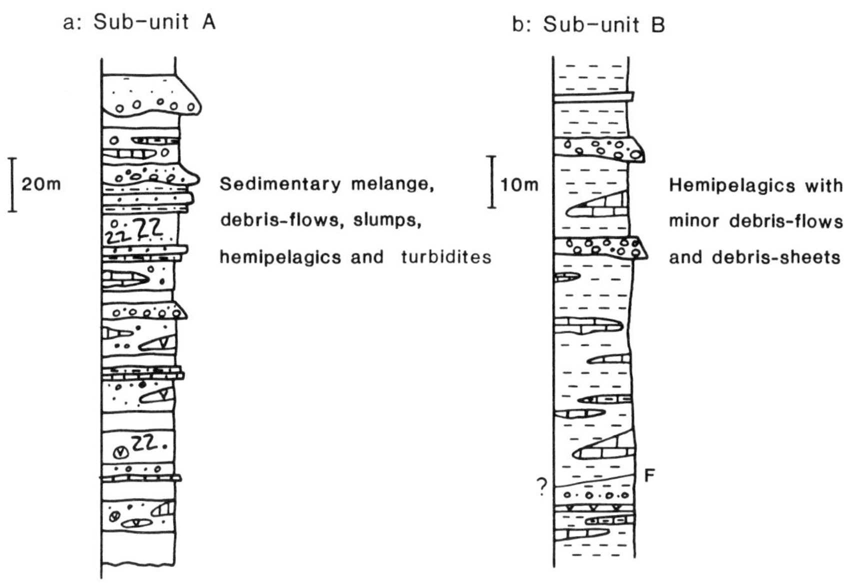

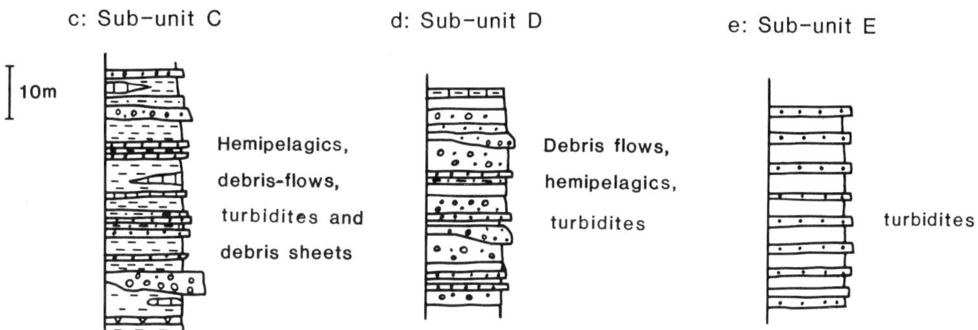

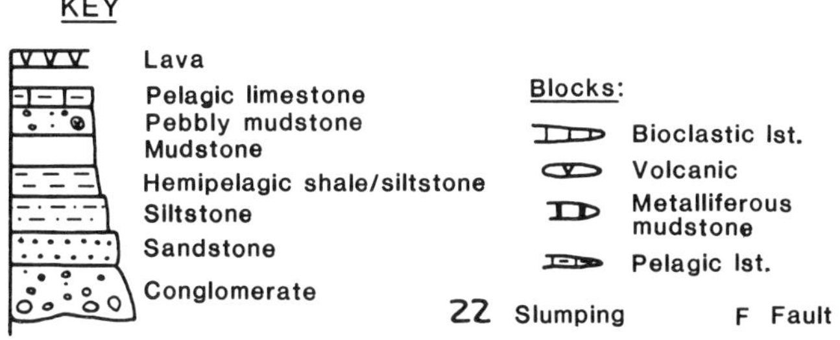

FIG. 10. Representative sedimentary logs of the Maden Group in the *western area*. Sub-units A and B measured just N. of Maden town in sedimentary melange and hemipelagic units. Sub-units C, D and E were measured in the undifferentiated Maden Group (sandstone-shale-conglomerate unit), in the Maden Gorge, just NW of the Killan Dağı (Fig. 5a).

limestone conglomerates, mass-flow deposits, detached blocks of neritic limestones and several debris sheets, individually ca. 25 m thick and over 150 m long. The conglomerates are both clast- and matrix-supported; clasts are generally well-rounded and include quartzose material in a mudstone matrix. The redeposited lithologies are typical of the Gehroz Formation which may still be seen *in situ* 5 km to the north (e.g. north of Buban, Fig. 3). Northwards, on the right-way-up limb of the Göma monocline the succession gives way to ca. 250 m of hemi-pelagic calcareous mudstones and siltstones, in places pebbly, with a few lenses of pelagic limestones and intercalations of vesicular lava. The conclusion is thus that the Göma monocline originated as a steep depositional hinge down which material typical of the Hazar Group was redeposited into the deeper water areas of accumulation of the Maden Group sedimentary rocks.

Interpretation: a south-facing active margin

The sediments of the Maden Group accumulated on steep continuously unstable slopes as shown by the great abundance of debris-flows, slump-sheets, detached blocks and sedimentary melange (Fig. 11). The south margin of the Hazar-Guleman block to the north acted as a steep south-facing depositional hinge (Göma monocline) with redeposition of shallow water limestone on a vast scale.

Two models of Maden Group deposition can theoretically be considered. In the first a subsiding basin floored by previously sliced Upper Cretaceous ophiolites, possibly assembled in the arc-trench gap, subsided, accumulated deep sea and redeposited shallow water sediments and some volcanics, and then was later deformed as a conventional thrust belt related to southward emplacement over the Arabian margin during, or prior to, the Miocene. In this model the Maden Group could conceivably have accumulated in a Tertiary marginal basin related to a southward-dipping subduction zone, as in the model of Perincek & Özkaya (1981) and Şengör & Yılmaz (1981). The second model would involve northward underthrusting of ophiolitic basement throughout Maden Group deposition. In this case the Killan Imbricate Unit could be a type of accretionary wedge related to northward subduction in the Palaeocene and Eocene. The ophiolitic slices would then be accreted oceanic crust of some type, while the Maden Group sediments should be akin to trench and trench-slope deposits.

The first model of a Palaeogene subsiding basin, thrust-stacked, can be ruled out. No proximal-distal relationship from the margin to the centre of a coherent basin can be inferred after allowing for the thrusting. Indeed in the Maden Gorge the most 'distal' turbidites of Sub-unit E (structurally highest) would originally have been located closest to the major source area of the Hazar-Guleman block to the north. On the other hand no southerly source is known, for example, for the sedimentary melange in Sub-units A and B (structurally lowest) which accumulated further south. Some of the longer successions coarsen upwards from deep water pelagic or hemipelagic sediments, the opposite to that expected in a subsiding basin, as seen in the Hazar Group. Instead, all the successions reflect accumulation at the base of steep tectonically active slopes.

Many of these problems disappear if the Maden Group is interpreted as an accretionary wedge formed above a northward-dipping subduction zone (Fig. 11). The Maden Group can be compared with a range of trench-slope (Sub-units A, B and C), trench-floor (Sub-unit D) and basin-plain (Sub-unit E) settings analogous to other active margins related to subduction, and recognised, for example, on Kodiak Island in Alaska (Nilsen & Moore 1979), Nias Island in Indonesia (Moore *et al.* 1980) and the Mendocino coast of N. California (Bachman, 1981). The ophiolitic rocks and mafic extrusives of the Killan Group would then be accreted older oceanic material, possibly including volcanic edifices. In this case the Killan Imbricate Unit can be viewed as subduction complex with the Hazar Group as a fore-arc basin to the north floored by a previously emplaced ophiolite (see discussion section) with a discontinuous fore-arc high in between. The Göma and Putyan monoclines would have formed by the underthrusting during subduction. In this model the arc would then comprise the Eocene calc-alkaline volcanics of the Malatya-Elazığ area (Yazgan 1981; Michard *et al.* this volume) and the more minor calc-alkaline extrusives of the Hazar Group.

Although attractive, some problems still remain in this model. First, the fossil evidence is as yet insufficient to confirm a north to south age transition in the Maden Group sediments as required by the accretionary hypothesis; secondly, the deep-sea Maden Group sediments (sub-unit A) unconformably overlie, at least locally, ophiolitic rocks of the Killan Group, which must have been at least partly sliced prior to deposition. This raises the possibility that some form of deformed ophiolitic basement could have existed prior to slicing, although it should be remembered that ophiolites could have been sliced near the ridge

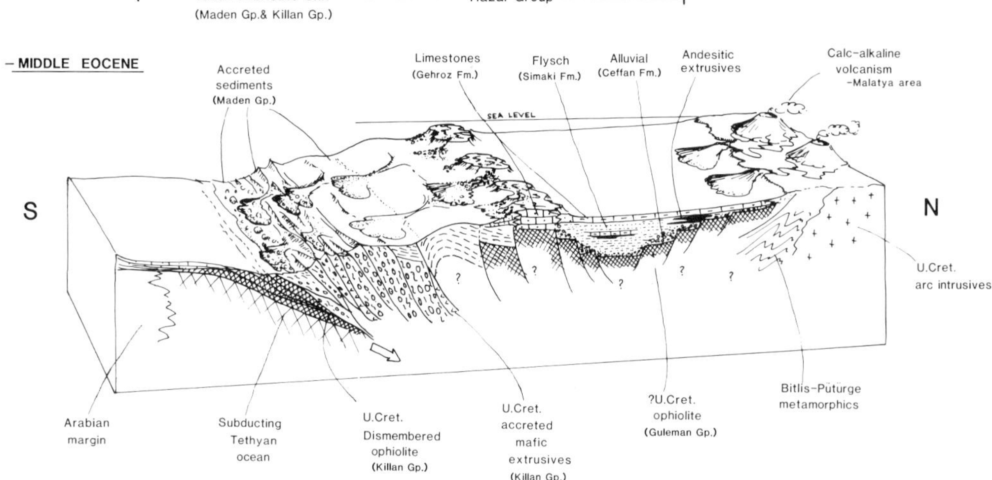

Fig. 11. Proposed depositional and tectonic setting of the Hazar Group in a subsiding fore-arc basin, and the Killan Imbricate Unit (including the Maden Group and Killan Group) as an accretionary prism. Emplaced ophiolitic rocks of the Guleman Group form the basement to the Hazar Group. Where ophiolitic rocks are absent the Hazar Group overlies metamorphics of the Pütürge and Bitlis massifs and possibly older fore-arc rocks.

(e.g. Spray, this volume) or have formed along a fracture zone where plutonic rocks were exposed on the sea floor (e.g. Mid-Atlantic ridge). Thirdly, the individual thrust-packets rarely if ever repeat complete transitions from ocean, to trench, then trench-slope deposits as conventionally expected in sequentially accreted slices in a steady-state subduction complex. Instead each thrust-sheet tends to be significantly different.

At this stage we conclude that the Maden Group is best explained by some form of accretionary process involving underthrusting of ophiolitic basement, and sediment accumulation on an active margin. How this fits into the wider picture is discussed later.

The Eastern Maden Complex area

Any regional interpretation of the Maden Complex must also take account of the contrasting *eastern area* (Figs 2 & 12), where both the Hazar-Guleman nappe and the Killan Imbricate Unit are absent and the Maden Complex is represented only by sliced mafic extrusives and sediments, termed the Karadere Formation (Fig. 12). The Karadere Formation is tectonically overlain by the Bitlis metamorphic nappe and local slivers of ophiolitic rocks dated by the K/Ar method as Upper Cretaceous (M. McIntyre, pers comm. 1983). The formation is then tectonically underlain by the para-autochthonous Arabian foreland successions (Fig. 12). Pelagic chalks at various levels in the Karadere Formation have yielded planktonic foraminifera, including *Trunco-rotaloides* of Middle Eocene age (G. C. Adams, pers comm. 1982).

The Karadere Formation crops out in two main areas, Karadere and Mizik, which are separated by a klippe of Bitlis metamorphic rocks and offset by a wrench-fault (Fig. 12). The two areas probably originally constituted an intact lava-sediment succession with three intergradational units (Fig. 13). The lowest unit, locally underlain by tectonic melange, consists of 10–20 m of discontinuous reddish brown siliceous to calcareous silty mudstone which is extremely deformed. Above this, there is a ca. 450 m thick unit (Figs 12b, 13) composed of alternating pillowed and massive lavas with minor lava breccia, hyaloclastite, mudstone, siltstone, sandstone and pelagic limestone. The upper levels of this unit are sheared and include wedges of melange up to several metres thick. Where the lavas are particularly well exposed (e.g. north of Palahin Tepe and south of Ahmet Dağı, Fig. 12), individual sedimentary horizons, up to several metres thick, can often be traced for several hundred metres laterally. These are clearly sedimentary interbeds rather than tectonic intercalations.

The stratigraphically highest unit of the Karadere Formation, (best exposed around Kahkik village, Fig. 12) consists of more than 600 m of

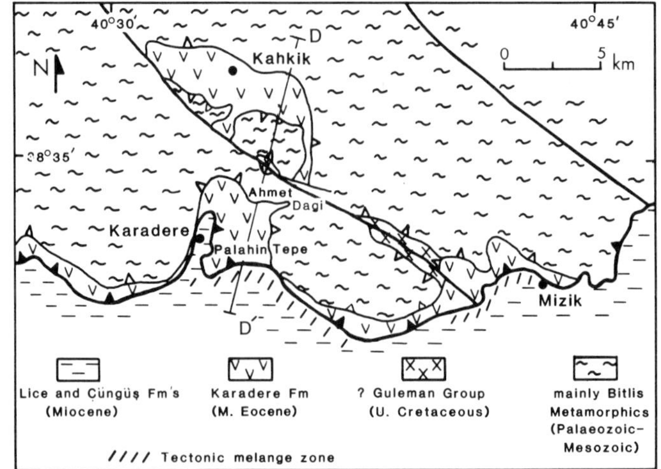

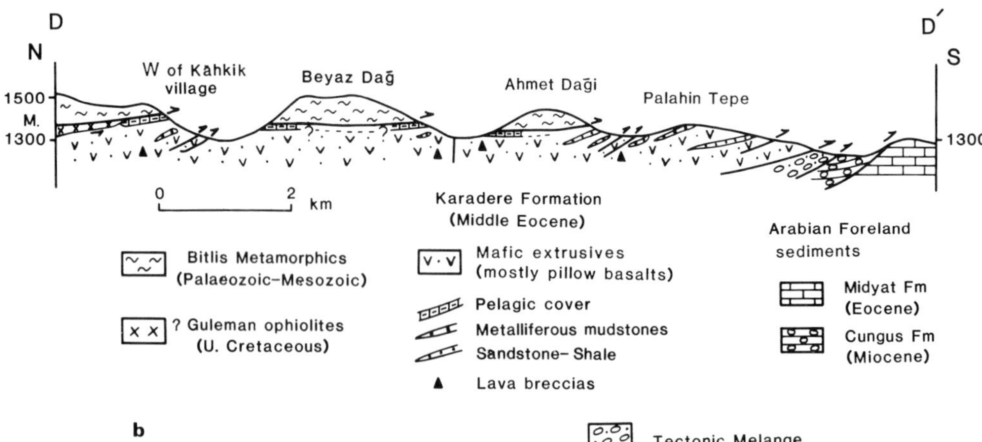

FIG. 12.(a) Simplified geological map of the major tectonic and stratigraphic units in the *eastern area* (see Fig. 2 for location).
(b) Cross-section along the line D-D′ shown in 12a.

both pillowed and mafic lavas, again with pelagic material interstitially and as thin interbeds, but clastics are much less abundant than lower in the succession. The highest lavas exposed are overlain depositionally by up to 10 m of homogeneous pink pelagic chalk which can be traced laterally up to 4 km. The entire formation is then overthrust by Bitlis metamorphic rocks to the north.

The Karadere Formation again crops out about 20 km to the east, near Mizik (Fig. 12), where pelagic and metalliferous mudstones tectonically overlie the Cüngüş Formation with a discontinuous zone of tectonic melange between. This unit is overlain by an up to 350 m thick lava-sediment unit which includes thin (up to 10 m thick) turbiditic sandstones and hyaloclastite intervals especially in the lower levels. Between the Karadere Formation and the structurally overlying Bitlis nappe are tectonic slivers of ophiolitic rocks not seen at Karadere.

Chemistry of Karadere Formation lavas

Analysis (Table 2) reveals a wide spectrum from transitional tholeiites to enriched basalts with a consistent antithetic relationship between Al_2O_3 and TiO_2. This allows subdivision into: (i) a high-Ti basalt type (TiO_2 1.7–2.4%) and (ii) a high-Al type (Al_2O_3 18.0–22.1%) and (iii) a tholeiitic basalt type. The typical MORB-normalised trends of these three types are shown in Fig. 14, although a continuous spectrum does exist between them.

The high-Ti basaltic type is enriched in all incompatible elements relative to MORB, and resembles Azores alkalic basalts (Fig. 14). The

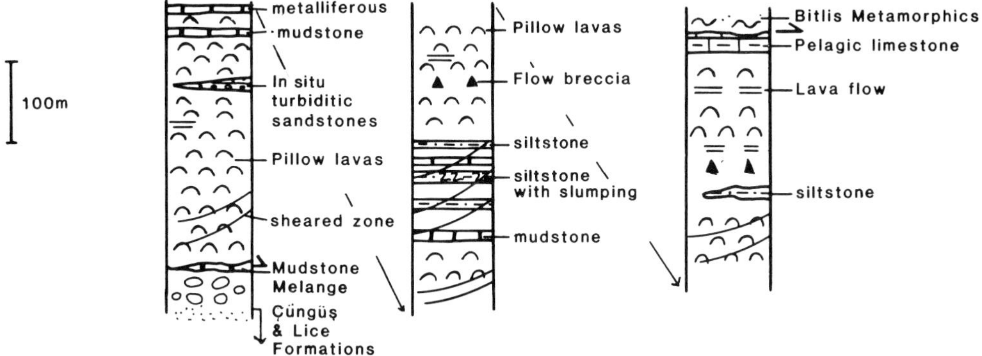

FIG. 13. Generalised stratigraphic column of the Karadere Formation in the *eastern area.*

high-Al basalts are less enriched in most incompatible elements relative to MORB, with the exception of Ti and Y. The tholeiitic type is enriched in the LIL elements, Zr, Nb and P but is otherwise similar to MORB.

The high-Ti enriched basaltic type is comparable with many within-plate or 'hot-spot' basalts (Wood *et al.* 1979; Saunders *et al.* 1980). The high Al_2O_3 content of the high-Al type basalts along with the high values of Ba, Sr, K and Rb relative to some HFS elements (e.g. Ti, Y) are a calc-alkaline feature (Ringwood 1977; Hawkesworth *et al.* 1977; Tarney *et al.* 1977), although these basalts are also enriched in some other HFS elements (Nb, P, Zr). The tholeiitic basalts are more similar to MORB than the other two types. On the TiO_2 vs. Zr diagram and the Cr vs. Y discriminant diagrams (Figs 8 & 9; Pearce 1982), the high-Ti type basalts consistently plot in the WPB field, while the high-Al and tholeiitic types plot in the MORB and WPB fields; some of the high-Al type basalts also plot close to the IAT field. The basaltic andesites and andesites of the Karadere Formation are also enriched in incompatible elements (Table 2), as are certain 'non-orogenic' andesites (Saunders & Weaver 1980).

Clearly the petrogenesis of the Karadere Formation must be complicated. The chemistry differs both from the Upper Cretaceous mafic rocks of the western area (Killan and Guleman Groups) many of which show a volcanic arc to MORB character and from other Eocene extrusives within the suture zone (e.g. Malatya area) which are unambiguously calc-alkaline (Yazgan, 1981). Although not diagnostic, the chemistry is consistent with the hybrid tectonic setting proposed below for the Karadere Formation as a pull-apart basin in a general fore-arc region.

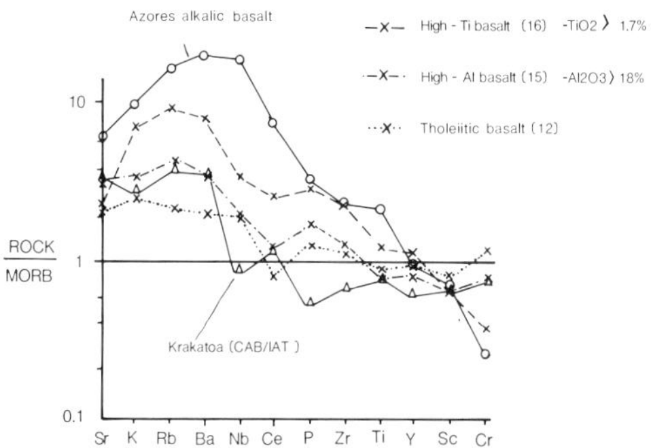

FIG. 14. Average MORB-normalized trace element patterns of basalts of the Karadere Foramation (Middle Eocene). Number of samples shown in brackets. The Azores alkalic and Krakatoa calc-alkalic trends are included for comparison (Pearce, 1982). MORB values are those of Pearce (1982).

Interpretation: Middle Eocene rifted basin

It might be anticipated that the Karadere Formation could be explained by some form of accretion of oceanic crustal slices as in the Killan Imbricate Unit, which in this case would have to be Middle Eocene rather than Upper Cretaceous. This hypothesis is however virtually ruled out by the presence of intact successions, rather than imbricate slices, the absence of any exotic material (e.g. sedimentary melange), the interbeds of quartzose clastics throughout the lava pile, and the presence of a pelagic chalk cover preserved in some successions. Instead, it is much more plausible that the Karadere Formation formed as a small extensional basin located adjacent to the Bitlis metamorphic massif to the north. The large volumes of fragmental mafic extrusives particularly lower in the succession point to eruption on an unstable relatively tectonically active sea-floor. The interbedded turbidites with abundant metamorphic quartz would have been derived from the adjacent Bitlis metamorphic terrain. Shallow water carbonate material in the calciturbidites was presumably derived from concealed carbonate shorelines to the north and possibly to the south. At least one 40 m thick sheet of bioclastic shallow water limestone which structurally underlies the Karadere Formation (SW of Mizik village, Fig. 12a) could be a remnant of a southern margin Later extrusion of pillowed and massive lava took place on a more even seafloor with diminished clastic input. Eruption was continuously in deep water with pelagic chalk intercalations throughout the succession. After volcanism ended, the mafic crust was blanketed by pelagic chalks marking a period of tectonic quiescence still within the Middle Eocene. The regional implications of this extensional basin origin are discussed later.

Baykan and Siirt-Madenköy areas for comparison

Before further discussion we briefly report on an area ca. 170 km east of the type area of the Karadere Formation (Fig. 2). For the first 100 km east of the Karadere and Mizik outcrops the Maden Complex is either tectonically thinned or cut out totally. About 120 km further east in the Baykan area, regional mapping (Turkish Petroleum geologists, Ozkaya 1974, Hall 1976) has revealed rocks similar to the Killan Imbricate Unit and the Hazar-Guleman nappe.

Further east again, in the Siirt-Madenköy area, we have observed an assemblage virtually identical to the Karadere Formation (Fig. 2). According to mapping by Yıldırım & Alyamaç (1976), the Maden Complex there is sandwiched between the Bitlis nappe and the Miocene foreland sediments and comprises mafic lavas, lava breccias, pelagic carbonates, conglomerates, diabase intrusives and cupriferous sulphides. Again there are long intact successions of pillowed and some massive lavas, flow breccias, and hyaloclastites with interbedded pink pelagic limestones, dated as Middle Eocene (G. C. Adams, pers. com. 1982). The lavas are chemically very similar to the Karadere Formation (Table 2); once again there is a several metre-thick laterally continuous pelagic blanket over the mafic crust.

Discussion: alternative tectonic models

We now discuss the emplacement of the Maden Complex in its regional tectonic setting. Michard *et al.* (this volume) discuss two alternative models for the evolution of the adjacent Malatya-Elaziğ area. In their first model a single Mesozoic Tethys ocean is assumed and the Upper Cretaceous arc complex is seen as the product of southward subduction in latest Cretaceous leading to continental collision, then further southwards thrusting over the Arabian foreland in the Late Miocene. In their second model a separate southern Mesozoic strand of the Mesozoic Tethys existed and is assumed to have closed northwards, creating first the Upper Cretaceous 'Elazığ' arc and then later, Eocene calc-alkaline volcanism ('Maden Formation') prior to Miocene continental collision. In our view the widespread existence of ophiolitic rocks on both the northern active margin and the southern passive Arabian margin strongly favours the former existence of oceanic crust between the Arabian margin and micro-continental slivers represented by the Bitlis and Pütürge metamorphic nappes to the north.

Polarity of subduction

What was the polarity of the subduction zone which created the Upper Cretaceous Elazığ-Palu arc complex? Perincek & Özkaya (1981) and Şengör & Yılmaz (1981) proposed that the Maden Complex originated as an Eocene marginal basin behind a south-dipping subduction zone, but our work now shows that a major part of the Maden Complex, previously thought to be largely Tertiary (Killan Group), is in fact Upper Cretaceous. There is no evidence of a rifted southern margin of a 'Maden basin' (c.f. Baştuğ 1980) nor of any underlying Tertiary mafic crust in the western area. Özkaya (1982) proposed the further east the Oramar and Kara Dağ ophiolites exposed near the Iranian border were formed in a Late Mesozoic marginal basin behind a south-dipping subduction zone, then emplaced onto the Arabian margin in Campanian-Maastrichtian, but the evidence of subduction polarity is far from unambiguous and in any case need not relate directly to the Maden Complex in our area. Upper Cretaceous calc-alkaline arc volcanism has not been reported from any part of the *in situ* Arabian margin from the Eastern Mediterranean: (e.g. Delaloye & Wagner, this volume) to Oman (Glennie *et al.* 1973), although admittedly southward subduction under isolated micro-continents might not have affected the Arabian margin. Most models for the adjacent areas have however postulated northward subduction of the Mesozoic Tethys (e.g. Eastern Mediterranean, Robertson & Woodcock 1980; Oman, Pearce *et al.* 1982; southern Iran, McCall &

Kidd 1982; Berberian *et al.* 1982) and we believe this was also the case in SE Turkey.

Genesis of the ophiolites

Accepting this, what then was the tectonic setting of the Upper Cretaceous ophiolitic rocks and mafic extrusives of the Maden Complex? The chemical similarities suggest genesis of all the mafic extrusive and ophiolitic rocks of the Killan Imbricate Unit and of the Hazar-Guleman nappe in a single area of oceanic crust. The 'immobile' trace elements, although not clearly diagnostic of a specific tectonic setting, by current convention do point to an above-subduction-zone genesis, with some lavas more akin to MORB. Here it must be significant that the available ages of the ophiolites overlap with those of the Elazığ-Palu arc complex (and equivalents). The Santonian and younger radiometric ages for granitoids given by Michard *et al.* (this volume) could suggest subduction in the area as early as mid-Cretaceous, while the ophiolitic rocks, from our work, range from Santonian to the Cretaceous-Tertiary boundary. Since at least in our area the Upper Cretaceous arc rocks and the ophiolites form different lithological and structural units a genesis actually within the arc system (e.g. arc roots or fore-arc limb) seems unlikely. Also, the Tertiary sediments of the Maden Group are hard to reconcile with an origin entirely within an embryonic island arc as proposed by Erdoğan (1977). On the other hand the Upper Cretaceous ophiolites could have formed above intra-oceanic subduction zones, followed by tectonic juxtaposition with the Upper Cretaceous volcanic arc rocks. Similar problems of interpretation exist with the much larger more intact ophiolites elsewhere in the Neotethys including the Semail nappe, Oman (Pearce *et al.* 1982) and the Troodos massif, Cyprus (Pearce 1980), and the problem of their origin to some extent still remains.

Fore-arc organisation

Assuming the ophiolites do represent oceanic crust formed at some type of Upper Cretaceous spreading axis, the next problem is how similar ophiolitic rocks came to exist both structurally beneath, above, and between the Bitlis and Pütürge metamorphic nappes. Michard *et al.* (this volume) suggest that their 'Ispendere' ophiolite in the Malatya-Elaziğ area formed as a Late Mesozoic ensialic marginal basin. The main problem here is that, on Michard *et al.*'s data, the ophiolite appears to predate the Upper Cretaceous volcanic arc, which is unexpected. (cf. Rocas Verdes, Chile; Bruhn & Dalziel 1975). A second possibility is that the Bitlis and Pütürge massifs were originally separate micro-continental slivers in the Late Mesozoic, floating in Upper Cretaceous oceanic crust. However the chemical similarities of all the ophiolitic rocks (i.e. island arc tholeiitic-MORB chemistry) offers no obvious support for genesis in separate basins north and south of the Pütürge-Bitlis nappe. A third possibility, which on the whole we favour is that major sheets of ophiolitic crust were thrust northward from the suture over the arc trench gap onto the continental crust in the latest Cretaceous. This model of emplacement would be analogous, for example, to that proposed for the Othris ophiolite, Greece, in the Upper Jurassic to Lower Cretaceous interval (see Smith & Spray, this volume, and references therein). In this case the ophiolitic nappe would have been eroded from above the Bitlis and Pütürge massifs prior to the reported formation of an unconformable cover of Palaeocene-Eocene sediments and volcanics (Hempton 1982; Michard *et al.* this volume). Alternatively, some major unknown structural organization possibly related to strike-slip faulting either before, or after, Miocene emplacement over the Arabian margin could conceivably have taken place.

An open or closed ocean in the Lower Tertiary?

The next important question is whether the southern strand of the Neotethys was still open in the Palaeogene, or essentially closed by the latest Cretaceous, as this strongly influences the interpretation of the Killan Imbricate Unit and the Karadere Formation. If open ocean persisted the Killan Imbricate Unit and the Hazar Group can be reasonably interpreted as an accretionary wedge and adjacent fore-arc basin related to northward subduction. The Karadere Formation then has to be interpreted as a Middle Eocene rifted basin in the fore-arc area. If the later stages of subduction were markedly oblique, subduction could have been resolved in the fore-arc area into some regions undergoing calc-alkaline volcanism and other areas where extension produced pull-apart basins floored by mafic crust. A rifted fore-arc setting would be consistent with the hybrid chemistry of the mafic extrusives (see above). A helpful analogy would be the Andaman Sea which is floored by a series of small extensional basins offset by faults in a general fore-arc setting, behind obliquely subducting crust of the Indian Ocean (Curray *et al.* 1979; Karig *et al.* 1980). One problem for this 'open-ocean' model is the absence of any preserved accretionary wedge or outer-arc high between the Karadere Formation and the structurally underlying Arabian margin. However, the Miocene Çüngüş Formation structurally underneath the Karadere Formation indicates massive erosion of the Maden Complex. Also, prior to the Middle Eocene this part of the margin could have been undergoing 'subduction erosion', or removal of material by strike-slip faulting. Maden Complex rocks could also be concealed by final southward overthrusting.

If, however, the oceanic crust had essentially all been consumed in the latest Cretaceous, with ophiolite emplacement onto both margins, then the Killan Imbricate Unit and the Middle Eocene Karadere Formation would have to be explained in terms of further crustal convergence and strike-slip faulting during progressive crustal thickening of the suture zone. In this model the ophiolite slices of the Killan Imbricate Unit were already assembled but remained below sea-level in the latest Cretaceous. Then, with slow renewed convergence in the Palaeocene-Eocene the sediments of the Maden Group would have accumulated on a south-facing active margin and slices of ophiolitic basement rocks would have been incorporated into the accretionary wedge. In this model the Middle Eocene Karadere basin of the eastern area would have formed as a short-lived pull-apart basin probably related to major strike-slip faulting. Any Palaeogene calc-alkaline volcanism, as in the Hazar Group (mainly in the Malatya area) would then be of post-collisional type, rather than related to continuing subduction. This could be consistent with the widespread but minor calc-alkaline volcanism in the Hazar group and the mafic volcanism in the Killan Imbricate Unit, which is not restricted to a well defined laterally continuous Paleogene volcanic arc. If modern active margins are anything to go by the actual tectonic history of the Neotethyan suture zone in this area was probably very complicated. For example, even if the ocean remained open in the Palaeogene, as we suspect, a strong possibility is that the ophiolitic rocks of the Killan Imbricate Unit were assembled in the fore-arc area in the latest Cretaceous, then shortened further in the Tertiary, accompanied by active margin sedimentation to form an 'accretionary wedge', while the remaining Mesozoic Neotethyan crust proper was subducted without trace.

Implications for the mineralization

Major economically important cupriferro sulphide orebodies include Ergani in the western area, and Siirt-Madenköy in the east. The field relations and origins of both these sulphides have been extensively studed by Sirel (1952), Griffith *et al.* (1972), Bamba (1976), Takishima (1975) Erdoğan (1977), and İleri *et al.* (1976). Virtually every known mode of sulphide genesis has been proposed, ranging from sedimentary and volcanic exhalative to several types of syngenetic origin, with or without structural control.

In newly opened pits at Anayatak mine (Ergani) we have observed that the mineralization occurs in mafic pillow lavas of the Upper Cretaceous ophiolitic rocks of the Killan Group. There is a sharp break between the mineralized ophiolitic rocks and the overlying red mudstones and other clastics of the Palaeocene-Eocene Maden Group which are unmineralized. Many of the 'mudstones' of earlier workers are very weathered and chloritized mafic lavas. İleri *et al.* (1976) report Upper Cretaceous planktonic foraminifera from mineralized sediments in the area. In our view the tholeiitic chemistry of the unmineralized equivalent of the basaltic hosts-rocks, the high temperature ore mineral assemblages (Çağatay, 1977) and the massive ore with sedimentary structures (Sirel 1952) all point to an origin as a Cyprus-type orebody. The mineralization would thus have formed in a Late Cretaceous ophiolite which was then incorporated into the Killan Imbricate Unit in Palaeocene-Eocene time.

Mineralization in our *eastern* Maden outcrop area (Karadere-Mizik) is restricted to very small sulphide bodies and minor magnetite mineralization close to the Bitlis nappe. Large volumes of hydrothermal oxide-sediments in the lavas could be dispersion halos from larger sulphide orebodies in the area not now exposed. Very much larger orebodies at Siirt-Madenköy would have formed in the inferred Middle Eocene extensional basin floored by mafic crust. Such a tectonic setting is consistent with the known range of modern oceanic sulphide mineralization at spreading centres in open oceans (e.g. Pacific), Red sea-type basins and small ocean basins controlled by strike-slip faults (e.g. Gulf of California, Ballance & Reading 1980).

Conclusions: tectonic evolution

The overall inferred tectonic evolution can be summarized as follows:

(1) Early Mesozoic

A southern strand of the Neotethys was rifted-off the Arabian margin not necessarily adjacent to its present position. This history could well be akin to that of the Antalya

Complex in SW Turkey (Robertson & Woodcock, this volume). How much oceanic crust was created in the Mesozoic is unknown, but by the Late Cretaceous the Bitlis and Pütürge metamorphic massifs are considered to have been part of one or more micro-continental slivers surrounded by oceanic crust.

(2) Late Cretaceous

North-dipping subduction was initiated under the Bitlis and Pütürge continental margin, possibly as early as Mid-Cretaceous, creating an Upper Cretaceous volcanic arc (Fig. 15a). The ophiolites formed in some unspecified tectonic setting, possibly above an intra-ocean subduction zone. In the latest Cretaceous ophiolites were emplaced northwards (Fig. 15b), possibly over the arc-trench gap, as large slices, and also southwards by gravity-sliding onto the formerly passive Arabian margin.

(3) Palaeocene-Eocene

There are then two main alternatives. In the first case (Fig. 15 ci) the ocean remained open but for some reason subduction ceased until the Palaeocene-Eocene when there was renewed calc-alkaline volcanism. In the other case (Fig. 15, cii), the Mesozoic ocean crust had been essentially eliminated by the latest Cretaceous, halting subduction and facilitating ophiolite emplacement. In both cases emplaced ophiolitic and volcanic arc rocks were deeply eroded and then peneplaned in a semi-arid environment on a northern active margin during a period of relative tectonic quiescence (Ceffan Formation). Very unstable tectonic conditions ensued and the former ophiolitic and arc terrain subsided rapidly to form one or more E-W trending deep marine basins (Figs. 15, ci and cii) which progressively filled, first with flysch (Simaki Formation), then with both pelagic and redeposited limestones (Gehroz Formation) accompanied by minor calc-alkaline volcanism. The margin to the south was activated, leading to sliding of large volumes of shallow water limestones into deep water, and this was followed by underthrusting and successive accretion of ophiolitic rocks and sediments of the Killan Imbricate Unit. In the 'open ocean' model preferred the Killan Imbricate Unit could be a true subduction complex, or accretionary wedge, formed above a northward-dipping subduction zone; the Hazar group would then be a classic fore-arc basin. The alternative 'closed ocean model' involves regional crustal shortening of a previously assembled ophiolitic basement leading to formation of the Killan Imbricate Unit. In both models the contrasting Middle Eocene mafic extrusives of the Karadere Formation and equivalents elsewhere in the suture zone are attributed to formation of trans-tensional or 'pull-apart' basins in which important sulphide mineralization took place.

(5) Miocene

By early Miocene continued compression had taken up any 'slack' remaining from the original Mesozoic crustal stretching of the Arabian margin and the Bitlis and Pütürge metamorphic massifs, and the Arabian margin began to exhibit collisional tectonics (Fig. 15d). The foreland was faulted with carbonate build-ups on highs and flysch deposition in lows (Lice Formation). The outer edge of the advancing Maden allochthon disintegrated to form sedimentary melange (Çüngüs Formation), then was overthrust. With further compression the Tertiary Arabian margin successions

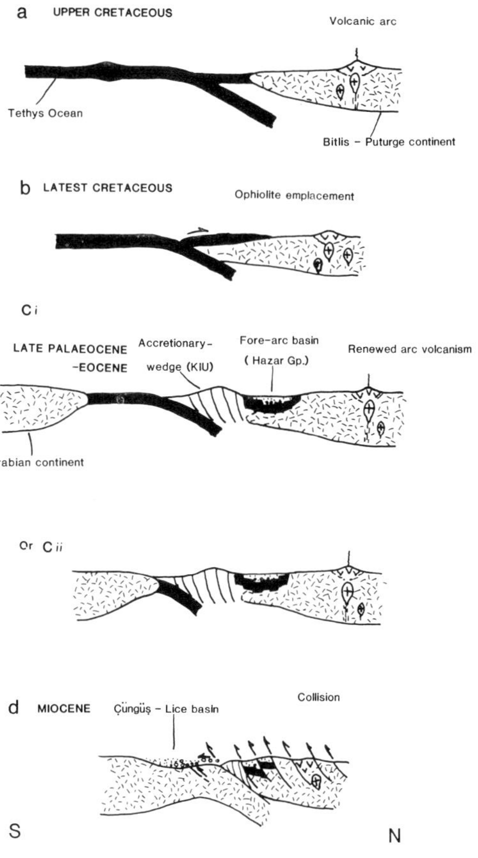

FIG. 15. Plate tectonic model of the Upper Cretaceous to later Tertiary evolution of the Malatya-Hazar-Maden area, in SE Anatolia (see text for discussion). KIU, Killan Imbricate Unit.

peeled-off and were shunted southwards to their present position. In the Pliocene the depocentre on the Arabian platform shifted southwards, filled and then was deformed.

(6) Quaternary

By Late Pliocene times continental collision had advanced to the stage when further shortening could only be achieved by strike-slip faulting along the South Anatolian transform fault. Along a considerable distance this fault must have followed the Lower Tertiary fore-arc basin since similar facies are now present on both sides.

ACKNOWLEDGEMENTS: This work was carried out during the tenure of a M.T.A. Enstitüsü (Turkey) scholarship to G.A. A.H.F.R's fieldwork was assisted by the Travel and Research Fund of the University of Edinburgh. For helpful discussion in Turkey we thank D. Perincek, O. Balkaş, O. Sungurlu, E. Tuna, G. Kurt, F. Alyamaç, Y. Özkan and in Edinburgh, J. Dixon, A. Kemp, R. Gill, A. Hayward, M. Johnson, M. Fisk & D. Winter. G. Fitton, D. James & P. Hill assisted with X.R.F. and microprobe analysis. Microfossils were determined by E. A. Pessagno (radiolaria) & G. C. Adams (foraminifera). The radiometric ages were determined by M. McIntyre of the Scottish Universities Reactor Centre, East Kilbride. The manuscript was reviewed by J. Dixon, A. Kemp, J. Leggett & M. Fisk. We are most grateful to J. Dixon for suggesting ways to improve presentation. Mrs D. Baty assisted with drafting and photography. Mrs H. Hooker & Mrs M. Wright patiently and efficiently typed both the initial draft and the final version of the manuscript.

References

ALABASTER, T., PEARCE, J. A. & MALPAS, J. 1982. The volcanic stratigraphy and petrogenesis of the Oman ophiolite complex. *Contrib. Mineral Petrol.* **81,** 168–183.

BACHMAN, S. B. 1981. The coastal Belt of the Franciscan: youngest phase of northern Californian subduction. In: LEGGETT, J. K. (ed.). *Trench-forearc Geology. Spec. Publ. Geol. Soc. Lond.*, **10,** 401–417.

BALLANCE, P. F. & READING, H. R. 1980. Sedimentation in oblique-slip mobile zones: an introduction. In: BALANCE, P. F. & READING, H. R. (eds.). *Sedimentation in oblique-slip mobile zones. Spec. Pub. int. Ass. Sediment.* **4,** 265 pp.

BAMBA, T. 1976. The ophiolite of Ergani-Maden ara, in SE Anatolia and associated copper mine. *Bull. MTA*, **86,** 35–48 (in Turkish).

BAŞTUĞ, M. C. 1980. *Sedimentation, deformation and melange emplacement in the Lice basin, Dicle-Karabeğan area, Southeast Turkey*: Unpubl. Ph.D. thesis, Middle East Technical University, Ankara, Turkey.

—— 1982. Late Alpine depositional and deformational history of the northern margin of the Arabian plate. In: J. E. DIXON & A. H. F. ROBERTSON (eds.) *Abstracts of meeting, the Geological Evolution of the Eastern Mediterranean*, Edinburgh, Sept. 1982, p. 15.

BERBERIAN, F., MUIR, I. D., PANKHURST, R. J. & BERBERIAN, M. 1982. Late Cretaceous and early Miocene Andean-type plutonic activity in northern Makran and Central Iran. *J. geol. Soc. London.* **139,** 605–614.

BRUHN, R. L. & DALZIEL, I. W. D. 1977. Destruction of the early Cretaceous marginal basin in the Andes of Tierra del Fuego. In: TALWANI, M. & PITMAN III, W. C. (eds.). *Island Arcs, Deep Sea Trenches and Back-arc Basins. Am. geophys. Un., M. Ewing Ser.* **1,** 395–405.

ÇAĞATAY, A. 1977. Geological and mineralogical study of SE Anatolian copper mines with implications for their genesis. *Bull. MTA.* **89,** 46–68 (in Turkish).

ÇOĞULU, E. 1975. New findings for the genesis of Hatay-Kizildağ massif. *MTA Spec. Publ., Earth Sciences Congress of 50th Anniversary*, 409–423 (in Turkish).

CURRAY, J. R., MOORE, D. G., LAWYER, L. A., EMMEL, F. J., RAITT, R. W., HENRY, M. & KIECKHEFER, 1979. Tectonics of the Andaman Sea and Burma. In: WATKINS, J. S., MONTADERT, L. & DICKERSON, P. W. (eds.). *Geological and geophysical investigations of continental margins. Mem. Am. Ass Petrol. Geol.* **29,** 189–198.

ERDOĞAN, B. 1977. *Geology, geochemistry and genesis of the sulphide deposits of the Ergani-Maden region. Southeast Turkey.* Unpubl. Ph.D. thesis, University of New Brunswick, Canada, 228 pp.

GILL, J. B. 1976. Composition and age of Lau Basin and Ridge volcanic rocks: implications for evolution of an inter-arc basin and remnant arc. *Bull. geol. Soc. Am.* **87,** 1384–1395.

GLENNIE, K. W., BOEUF, M. G. A., HUGHES-CLARKE, M. W., MOODY-STUART, M., PILAAR, W. F. H. & REINHART, B. M. 1973. Late Cretaceous nappes in Oman mountains and their geologic evolution. *Bull. Am. Assoc. Petrol. Geol.* **57,** 5–27.

GÖNCÜOĞLU, M. C. & TURHAN, N. 1981. New age determinations in the Bitlis metamorphics. *MTA Bull.* **95/96,** 44–49 (in Turkish).

GRIFFITHS, W. R., ALBERS, J. P. & ÖNER, O. 1972. Massive sulfide copper deposits of the Ergani-Maden area, SE Turkey. *Econ. Geol.* **67,** 701–713.

HALL, R. 1976 Ophiolite emplacement and the evolution of the Taurus suture zone, South-east Turkey. *Bull. geol. Soc. Am.* **87,** 1078–1088.

HAWKESWORTH, C. J., O'NIONS, R. K., PANKHURST, R. J., HAMILTON, P. J. & EVENSON, N. M. 1977. A geochemical study of island-arc and back-arc tholeiites from the Scotia Sea. *Earth planet. Sci. Lett.* **36,** 253–262.

HEMPTON, M. R. 1982. *Structure of the Northern margin of the Bitlis suture zone near Sivrice, southeastern Turkey.* Unpbl. PhD. thesis, State Univ. of New York, Albany, U.S.A.

—— & DUNNE, L. A. 1982. Sedimentation in the Lake Hazar strike-slip basin. SE Turkey. In: *Abstracts of 11th Congress, Internat. Assoc. of Sedimentol. McMaster University, Hamilton, Canada, Sept. 1982,* p. 39.

ILERI, S., SALANCI, B., BITEM, M. & DOĞAN, R. 1976. Ergani (Maden) copper mine and plate tectonics. *Bull. geol. Soc. Turkey.* **19,** 133–144 (in Turkish).

KARIG, D. E., LAWRENCE, M. B., MOORE, G. F. & CURRAY, J. R. 1980. Structural framework of the fore-arc basin, SW Sumatra. *J. geol. Soc. Lond.* **137,** 77–91.

MCCALL, G. J. H. & KIDD, R. G. W. 1982. The Makran, southern Iran: the anatomy of a convergent plate margin active from Cretaceous to present. In: LEGGETT, J. K. (ed.) *Trench Fore-arc Geology. Spec. Publ. geol. Soc. Lond.* **10,** 387–401.

MOORE, G. F., BILLMAN, H. G., HEHANUSSA, P. E. & KARIG, D. E. 1980. Sedimentology and paleobathymetry of Neogene trench-slope deposits, Nias island, Indonesia. *J. Geol.* **88,** 161–180.

NILSEN, T. H. & MOORE, G. W. 1979. Reconnaissance study of Upper Cretaceous to Miocene stratigraphic units and sedimentary facies, Kodiak and adjacent islands, Alaska. *Prof. Pap. U.S. geol. Surv.* **1093,** 1–34.

ÖZKAN, Y. Z. 1982. *Geology and petrology of the Guleman ophiolite.* Unpubl. PhD. thesis, University of Istanbul. Turkey (in Turkish).

ÖZKAYA, I. 1974. Stratigraphy of Sason and Baykan area, SE Anatolia. *Bull. geol. Soc. Turkey.* **17,** 51–72. (in Turkish).

—— 1982. Marginal basin ophiolites of Oramar and Karadağ, SE Turkey. *J. geol. Soc. Lond.* **139,** 203–211.

PEARCE, J. A. 1980. Geochemical evidence for the genesis and eruptive setting of lavas from Tethyan ophiolites. In: PANAYIOTOU, A. (ed.). *Ophiolites, Proc. Int. Ophiolite Symp., Cyprus 1979. Geol. Surv. Cyprus, Nicosia,* 261–271.

——, ALABASTER, T., SHELTON, A. W. & SEARLE, M. P. 1981. The Oman ophiolite as a Cretaceous arc-basin complex: evidence and implications. *Phil. Trans. R. Soc. London,* **A300,** 299–317.

—— 1982. Trace element characteristics of lavas from destructive plate boundaries. In: THORPE, R. S. (ed.). *Andesites,* John Wiley & Sons 525–547.

PERINCEK, D. 1979. The geology of Hazro- Korudağ-Çüngüş-Maden-Ergani-Hazar-Elaziğ-Malatya area. *Spec. Publ. geol. Soc. Turkey.* 33 pp.

—— 1980. Volcanics of Triassic age in Bitlis metamorphic rocks. *Bull. geol. Soc. Turkey,* **23,** 201–211. (in Turkish).

—— 1981. Sedimentation on the Arabian shelf under the control of tectonic activity in Taurid belt. *Proc. 5th Petrol. Congr., Turkey,* 77–93. (in Turkish).

—— & ÖZKAYA, I. 1981. Tectonic evolution of the northern margin of Arabian plate. *Yerbilimleri, Bull. Inst. Earth Sci. Hacettepe Univ.* **8,** 91–101. (in Turkish).

RICOU, L. E. 1971. Le Croissant ophiolitique periarabe, une ceinture de nappes mises en place au cretacé supérieur. *Rev. Géogr. phys. Géol. dyn.* **13,** 327–349.

RIGO DE RIGHI, M. & CORTESINI, A. 1964. Gravity tectonics in foothills structure belt of SE Turkey. *Bull. Am. Assoc. Petrol. Geol.* **48,** 1911–1937.

RINGWOOD, A. E. 1977. Petrogenesis in island arc, systems. In: TALWANI, M. & PITMAN III, W. C. (eds). *Island Arcs, Deep Sea Trenches and Back-arc Basins, Am. geophys. Un. Monographs, M. Ewing Series,* **1,** 311–324.

ROBERTSON, A. H. F. & WOODCOCK, N. H. 1980. Tectonic setting of the Troodos Massif in the East Mediterranean. In: PANAYIOTOU, A. (ed.) *Ophiolites. Proc. Int. Ophiolite Symp., Cyprus., 1979. Geol. Surv. Cyprus, Nicosia,* 36–49.

SAUNDERS, A. D. & WEAVER, S. D. 1980. Transverse geochemical variations across the Antarctic Peninsula: implications for the genesis of calc-alkaline magmas. *Earth planet. Sci. Lett.* **46,** 344–60.

——, ——, MARSH, N. G. & WOOD, D. A. 1980. Ophiolites as ocean crust or marginal basin crust: a geochemical approach. In: PANAYIOTOU, A. (ed.). *Proc. Int. Ophiolite Symp., Cyprus, 1979. Geol. Surv. Cyprus, Nicosia,* 193–214.

ŞENGÖR, A. M. C. & YILMAZ, Y. 1981. Tethyan evolution of Turkey: a plate tectonic approach. *Tectonophysics,* **75,** 181–241.

SIREL, M. A. 1952. Die Kupferlagerstaette Ergani-Maden in der Turkei, *N. Jb. Miner. Abh.* **80,** Abt. A, 36–100.

TAKASHIMA, K. 1975. Geology of cupriferous pyrite deposits in ophiolite series between Ergani and Siirt-Madenköy, SE Anatolia, Turkey. *MTA unpubl. report.*

TARNEY, J., SAUNDERS, A. D. & WEAVER, S. D. 1977. Geochemistry of volcanic rocks of the island arcs and marginal basins of the Scotia Sea region. In: TALWANI, M. & PITMANN III, W. C. (eds.). *Island arcs, deep sea trenches and back arc basins. Am. geophys. Un. Monographs, M. Ewing Ser.* **1,** 395–405.

TUNA, E. 1979. Geology of Elazığ-Pertek area. Turkish Petroleum unpublished report 39 pp. (in Turkish).

TOLUN, N. 1953. Contribution a l'étude géologique des environs du sud et du sud-ouest du lac de Van. *MTA Bull.* **44–45,** 77–114.

WEAVER, S. D., SAUNDERS, A. D., PANKHURST, R. J. & TARNEY, J. 1979. A geochemical study of magmatism associated with the initial stages of back-arc spreading: the Quaternary volcanics of Bransfield Strait, South Shetland Islands. *Contrib. Mineral. Petrol.* **68,** 151–169.

WOOD, D. A., TARNEY, J., VARET, J. SAUNDERS, A. D., BOUGAULT, H., JORON, J.-L. TREUIL, M. & CANN, J. R. 1979. Geochemistry of basalts drilled in the North Atlantic by IPOD Leg 49: Implications for mantle heterogeneity. *Earth planet. Sci. Lett.* **42,** 77–97.

YAZGAN, E. 1981. A study of active continental paleomargin in the Eastern Taurids (Upper Cretaceous-Middle Eocene) Malatya-Elaziğ (Eastern Turkey). *Yerbilimleri, Bull. Inst. Earth Sci. Hacettepe Univ.* **7,** 83–104 (in Turkish).

YILDIRIM, R. & ALYAMAÇ, F. 1976. Geology of Siirt-Madenköy copper mineralisation. MTA unpublished report (in Turkish).

G. AKTAŞ & A. H. F. ROBERTSON, Department of Geology, University of Edinburgh, West Mains Road, Edinburgh, EH9 3JW, Scotland.

Rb-Sr geochronology of the Bitlis Massif, Avnik (Bingöl) area, S.E. Turkey

Cahit Helvaci & William L. Griffin

SUMMARY: In the Avnik area, felsic metavolcanics are interbedded with banded and massive apatite-rich iron ores, and are intruded by the Avnik and Yayla granitoids. These rocks are unconformably overlain with depositional contact by micaschists and (?) Permian marbles, which were folded and metamorphosed during the Alpine orogeny. The metavolcanics and the granitoids are extensively feldspathized and silicified.

The Yayla granite gives a poorly defined Rb-Sr age of 347 ± 52 Ma (IR = 0.7217 ± 80) but the albitized Avnik body gives only a scatter of data suggesting an age of 250–425 Ma. Feldspathized metavolcanics from a 100-metre section define an age of 91 ± 9 Ma. One sample of the Avnik granitoid gives an amphibole-rock-feldspar age of 71 ± 28 Ma and another gives a biotite-rock age of 41 ± 1 Ma; a micaschist gives a chlorite-muscovite age of 38 ± 2 Ma.

An isochron age of 454 ± 13 Ma (IR = 0.7105) for metavolcanics in the Cacas area can be extracted from the data of Yilmaz *et al.* (1981). We interpret this as the age of eruption, and suggest a similar age for the Avnik metavolcanics. Our 90 Ma (Eoalpine) data on these metavolcanics requires metamorphic resetting of whole-rock Rb-Sr systems over tens of metres, and probably records the time of pervasive feldspathization and silicification. The data for the Yayla granite are interpreted as the age of intrusion; the high IR indicates an origin by anatexis of older crust. The late-Alpine mineral ages probably reflect recrystallization during folding and thrusting.

The Avnik (Bingöl) area lies in the western part of the Bitlis massif, a large area of greenschist to amphibolite-facies metamorphic rocks in the Eastern Taurus fold-belt of S.E. Turkey (Fig. 1). Geological and palaeontological evidence (Altinli 1966), and geochronological evidence (Yılmaz 1971) suggest that these rocks were deposited, deformed and metamorphosed in Palaeozoic time. The southern boundary of the Bitlis Massif separates the Anatolian and the Arabian plates along the SE-Antolian thrust fault. The metamorphic rocks of the Bitlis Massif are thrust southward over an ophiolite-flysch complex, which is in turn thrust southward over sedimentary rocks of the Arabian foreland (Altınlı 1966, Ketin 1966, Yılmaz 1971, Hall & Mason 1972, Aykulu & Evans 1974, Hall 1976, Genç 1977).

In the Avnik area, the metamorphic rocks are subdivided into a Lower and an Upper unit (Fig. 2). The Lower Unit includes two metamorphosed granitic bodies with recognizably intrusive contacts. These contacts are truncated at an unconformity surface which forms the base of the meta-sedimentary Upper Unit.

We report Rb-Sr age determinations on metavolcanics, granitoids and granites within the Lower Unit, and on the micaschists from the Upper Unit from the Avnik region. We have also attempted to reinterpret previous Rb-Sr studies of the Bitlis Massif in the light of our investigations.

Lithostratigraphy

Lower Unit

The Lower Unit (Erdoğan *et al.* 1981) consists of gneisses, amphibolites, meta-volcanics, meta-tuffs and meta-agglomerates, with porphyritic and spherulitic textures that are often well preserved. They are interbedded with banded and massive apatite-rich iron ores and are intruded by the Avnik granitoid and the Yayla granite. The total thickness of the units is approximately 2800 m.

The *quartzo-feldspathic gneiss*, the lowest rock observed in the Avnik sequence, consists principally of strained and recrystallized quartz and feldspar with variable amounts of amphibole, muscovite and secondary chlorite after biotite and amphibole. The groundmass contains polygonized quartz megacrysts (phenocrysts/porphyroblasts). The chemical composition and petrographic evidence suggest that these gneisses are strongly foliated and recrystallized felsic meta-volcanics. The gneisses alternate with amphibolite and are migmatized along the contact with the granitoid.

The amphibolites consist of amphibole (actinolite, rarely hornblende and crossite), rare diopside, albite, epidote, apatite, biotite, muscovite, quartz, and minor talc, chlorite, calcite, sphene, and opaques (magnetite, haematite). Field relationships and structures

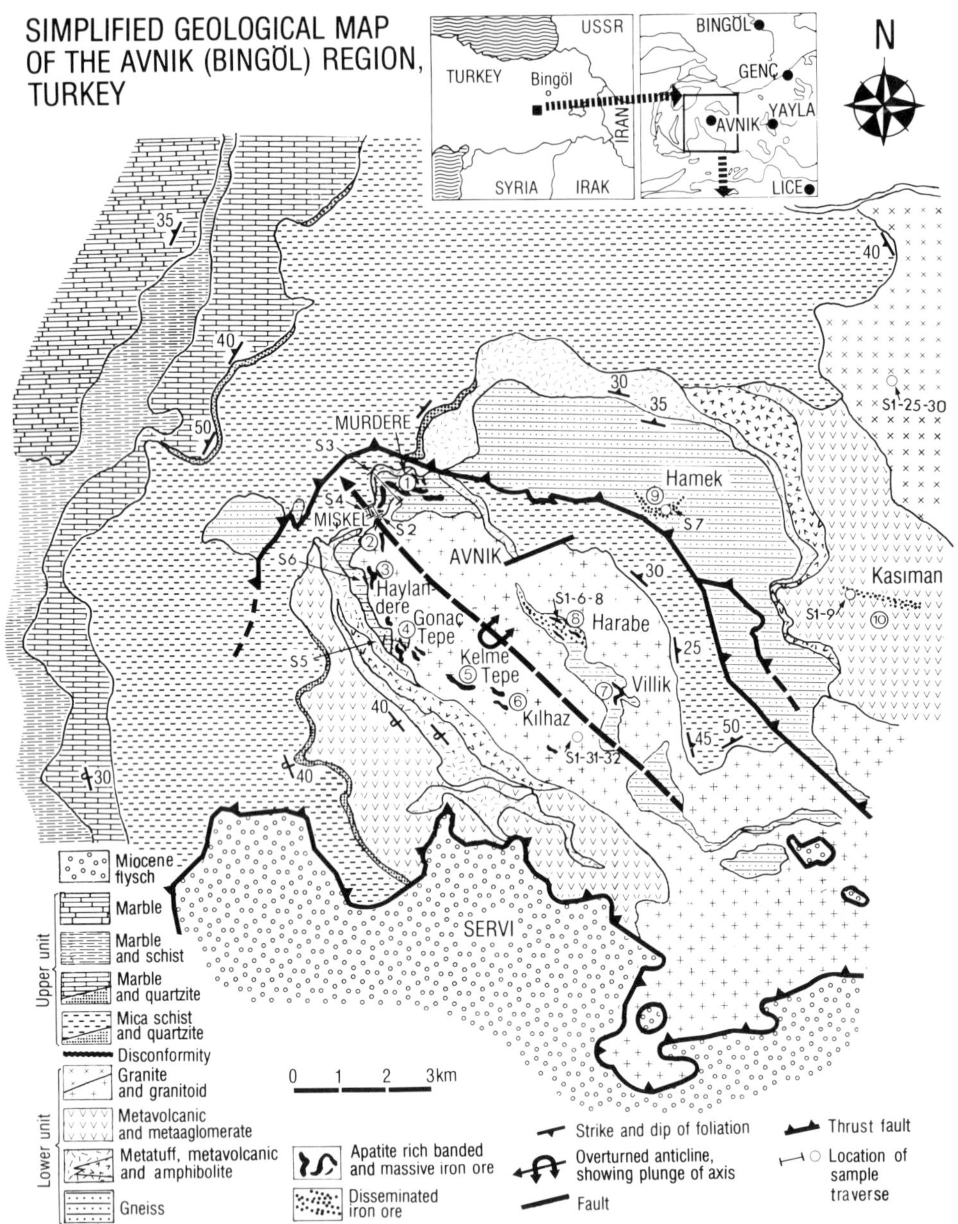

FIG. 1. Simplified geological map of the Avnik region, based on work by Erdoğan (1982).

suggest that they are basic to intermediate meta-volcanic rocks.

Meta-volcanics and *meta-tuffs* include more variable rock types and chemical compositions. They range from basic to dominantly intermediate in the composition, but the upper part of the unit is usually felsic (meta-rhyolites). Occasionally, meta-agglomerates are intercalated with these rocks. The meta-volcanics and meta-tuffs consist principally of feldspar (mainly albite and minor orthoclase), amphibole, mica, chlorite, talc, quartz, apatite, and magnetitie.

The rocks show porphyritic (porphyroblastic) and rarely spherulitic textures. Phenocrysts or porphyroblasts of quartz and feldspar, which commonly are coarsely recrystallized, polygonized and strained, are set in a fine-grained matrix. Primary feldspar and quartz phenocrysts grew outward into the groundmass, which frequently shows fluidal structure, and

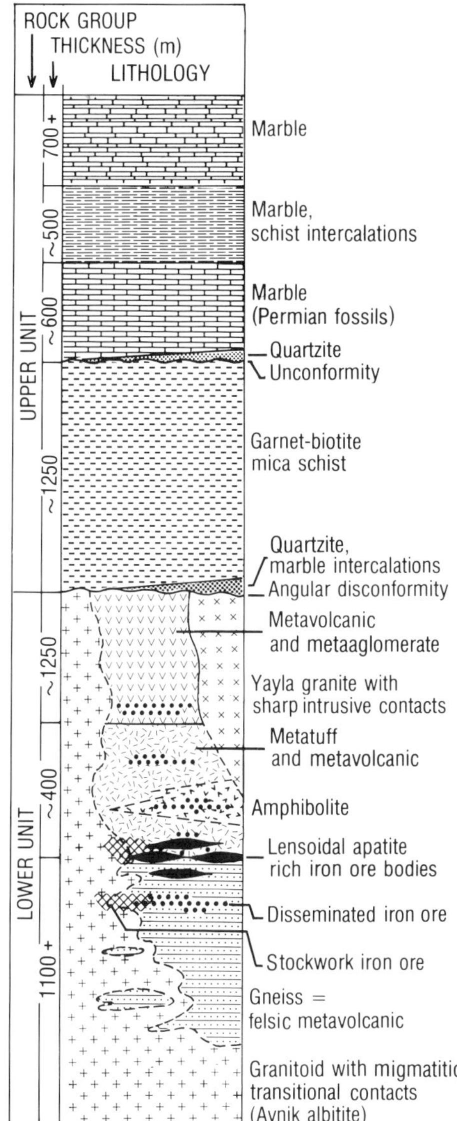

FIG. 2. Stratigraphic section of the Avnik region, modified after Erdoğan (1982).

encloses chloritized biotite, muscovite, apatite, and some larger grains of magnetitie. The groundmass textures, including relict spherulites, suggest that the rocks originally consisted largely of devitrified glass. They are strongly silicified and cut by many tiny quartz veins. Chemical analyses of the felsic metavolcanics show that their compositions have been strongly altered by post-extrusion albitization and silicification.

Meta-agglomerates are recognized in the field on the basis of their structures; they are lensoidal or form thin layers. In thin section they consist of volcanic debris (phenocrysts, groundmass material) in a fine-grained, strongly foliated micaceous matrix. Cataclastic textures are common.

The Avnik *apatite-rich iron ores* are banded, massive or disseminated in form and are stratigraphically controlled, being located in the gradational contact zone between the gneisses and the better preserved metavolcanic rocks (Fig. 2). The massive lensoidal ore zones show laminations ranging from 1–2 millimetres to a few centimetres. Where they are intruded by the Avnik granitoid the ores were remobilized to form stockwork-type veins with large apatite, amphibole and magnetite crystals.

The *Avnik Granitoid* (Avnik albitite) is heterogeneous, foliated and recrystallized at its margin, where it shows transitional, migmatitic contacts with the surrounding meta-volcanics and has assimilated the Lower Unit host rocks. Quartz and albitized K-feldspar commonly form recrystallized and strained granoblastic eyes. It becomes more homogeneous and more granitic in texture towards its central part. Field relationships, petrography and chemical analyses show that the Avnik granitoid was subjected to ablitization during post-intrusion metamorphism and that this albitization is shared by the surrounding meta-volcanic sequence.

The *Yayla Granite* is coarse grained and consists of 4–5 mm diameter crystals of quartz, orthoclase, microcline, pertite, amphibole, and biotite with equigranular textures. Chloritization and sericitization have been observed in some parts of the granite, but albitization is rare. Aplite and pegmatite dykes and veins are abundant within the body which shows sharp, intrusive contacts with the surrounding Lower Unit rocks.

Upper Unit

In the Avnik region, the Upper Unit consists of the following sequence from base to top: garnet-biotite micaschist, grey marble, marble-schist intercalations and white marble. The garnet-biotite micaschists of the Upper Unit rest with angular disconformity on the metavolcanics, the Avnik granitoid (albitite) and the Yayla granite of the Lower Unit, and are affected by the albitization event. Lensoidal quartzite-marble intercalations locally intervene between the miscaschist and both the meta-volcanics and the granitoids of the Lower Unit. These quartzites contain pebbles of the Lower Unit rocks and are interpreted as a basal conglomerate. The garnet-biotite micaschists consist principally of quartz, muscovite, biotite, garnet (almandine), albite, secondary chlorite, and minor apatite and opaques.

There is a local discordance between the micaschist and the overlying grey marble, and a lensoidal quartzite horizon is present locally between them. Marbles of Permian age rest unconformably on metamorphic rocks elsewhere within the Bitlis Massif (Tolun 1953). Regional correlations and the presence of tentatively identified Permian fossils strongly indicate a Permian age for the marbles in the Avnik area. Marble-schist intercalations and white marble lie on top of the sequence.

Metamorphism

A regional deformation and metamorphism, probably in the amphibolite facies (?) has affected the Lower Unit before the intrusion of the granitoids. The intrusive event was followed by uplift, folding and faulting that does not affect the Upper Unit. The Lower Unit has undergone a second metamorphism which has also affected the Upper Unit, and which has overprinted the previous metamorphic assemblages with greenschist-facies assemblages which are up to the epidote-amphibolite facies at some places. A similar complex evolution, including several stages of alteration and regional metamorphism, has been observed in other parts of the Bitlis Massif (Yılmaz 1975, Boray 1975, Hall 1976, Genç 1977).

In the Avnik area, the metamorphic rocks of the Bitlis massif form a regional-scale anticline overturned to the south. The Lower and the Upper Units have been affected by several stages of deformation. Several imbricated thrust planes occur within the massif. After Miocene time the Bitlis massif was carried tectonically to the south over the Miocene flysch along nearly horizontal thrust planes.

Analytical methods

All samples have been collected from fresh road-cuts, or newly opened trenches. The samples were collected in a number of traverses, each traverse or collecting area yielding a sample *series*, S2, S3 etc. (see Fig. 1). The S1 series includes sample suites from several widely separated localities (e.g. S1 25–30, S1 31–32). In all cases the samples weighed approximately 2 kg, of which half was prepared for chemical and isotopic analysis.

Rb/Sr ratios were determined in most cases by X-ray fluorescence, with a precision of ca. ± 1%. Low concentrations of Rb or Sr, and

TABLE 1. *Rb-Sr results on granitoids*

Sample	Rb, ppm	Sr, ppm	Rb/Sr	$^{87}Rb/^{86}Sr$	$^{87}Sr/^{86}Sr$
AVNIK					
S2–5	19.6	12.6	1.555	4.506	0.72401 ± 8
S2–4	4.7	18.3	0.256	0.742	0.71911 ± 12
S2–3	25.7	27.1	0.947	2.743	0.71694 ± 10
S3–1	8.2	37.7	0.218	0.630	0.71165 ± 8
S3 amph.	6.42[1]	274.1[1]	0.023	0.068	0.71117 ± 8
S3 remd.	5.28[1]	13.48[1]	0.392	1.134	0.71240 ± 18
S3–2	3.5	25.8	0.137	0.398	0.71137 ± 10
S3–3	140.0	74.9	1.868	5.414	0.72843 ± 10
S3–4	13.0	16.9	0.769	2.227	0.71547 ± 14
S4–1	12.6	19.9	0.632	1.830	0.71880 ± 8
S5–21	43.4	28.1	1.548	4.481	0.72068 ± 8
S5–22	11.8	30.1	0.392	1.135	0.71714 ± 10
S6–2	16.8	41.0	0.409	1.184	0.71496 ± 8
S6–5	25.7	63.4	0.406	1.176	0.71467 ± 10
S1–31	119.6	151.4	0.790	2.289	0.72559 ± 10
S1 bio	420.3[1]	12.33[1]	34.093	99.330	0.78188 ± 10
S1–32	108,6	92.1	1.179	3.419	0.73244 ± 12
S1 bio	263.9[1]	21.30[1]	12.392	35.969	0.74126 ± 10
YAYLA					
S1–25	117.0	35.5	3.292	9.578	0.76671 ± 12
S1–26	137.3	30.2	4.540	13.240	0.78982 ± 12
S1–27	190.7	23.4[1]	8.150	23.875	0.83716 ± 10
S1–28	12.9	14.5[1]	0.889	2.584	0.75320 ± 10
S1–29	190.0	56.5	3.359	9.760	0.75422 ± 10
S1–30	177.2	62.8	2.821	8.205	0.76316 ± 10

[1] isotope dilution analysis.

Rb and Sr contents of all mineral separates, were analyzed by isotope dilution, with a precision better than ±0.1%. $^{87}Sr/^{86}Sr$ ratios were measured on an automated VG Micromass 30 spectrometer with on-line data reduction. $^{86}Sr/^{88}Sr$ is normalized to 0.1194; $\lambda^{87}Rb$ is taken as 1.42×10^{-11} yr^{-1}. This laboratory's value for standard Sr NBS987 is 0.71030 ± 3(2SE). Between 10 and 25 sets of 10 ratios each were taken on each sample. The data are presented in Tables 1–3. Quoted uncertainties in tables and figures are ± two standard errors ($2\sigma/\sqrt{N}$).

Results

Granitoids—whole rock analyses

Four of the six analyzed samples of the Yayla granite (Fig. 3) define an age of 347 ± 52 Ma (MSWD = 25). Two samples fall so far off this line that they are not included in the regression; they are heavily sericitized and their biotite is chloritized. We tentatively interpret this Rb-Sr date as representing the age of intrusion of the granite, while the scatter about the line may reflect disturbance of Rb-Sr systems during post-intrusion metamorphism (see below). If this is the case, the high intial $^{87}Sr/^{86}Sr$ indicates that older crustal material was involved in the generation of the granite.

The 14 analyzed samples of the Avnik granitoid (albitite) give an irregular scatter of points, and no date is clearly defined. Points may be chosen to give sublinear arrays ('scatterchrons') within an envelope ranging trom 250 Ma (Fig. 3) to 425 Ma. The observed scatter is probably related to the post-intrusion albitization, which has affected all of the analyzed samples. The age of intrusion cannot be defined with our data, but *may* lie in the 250–425 Ma interval given above.

Only two of the Avnik granitoid samples have *present-day* $^{87}Sr/^{86}Sr$ as high as the *initial* $^{87}Sr/^{86}Sr$ of the Yayla granite (Fig. 3). This strongly suggests that the Yayla and Avnik bodies are not genetically related to one another. This conclusion must, however, be treated with some caution, since the albitization that has affected the Avnik body conceivably could have *lowered* the $^{87}Sr/^{86}Sr$ ratios.

Metavolcanics—whole rock samples

Nine samples were analyzed from the S3 profile of Fig. 1. The metavolcanic part of this profile extends 75 m from the contact with the Avnik albitite, to the contact with the unconformably overlying micaschist. The entire section is heavily feldspathized and silicified as described above. The lower part consists of albitites, the upper part of K-feldspar-rich rocks; the Rb/Sr ratio increases regularly upward (Table 2, Fig. 4).

Seven of the nine S3 samples define a date of 91 ± 9 Ma (MSWD = 48, IR = 0.7214 ± 34).

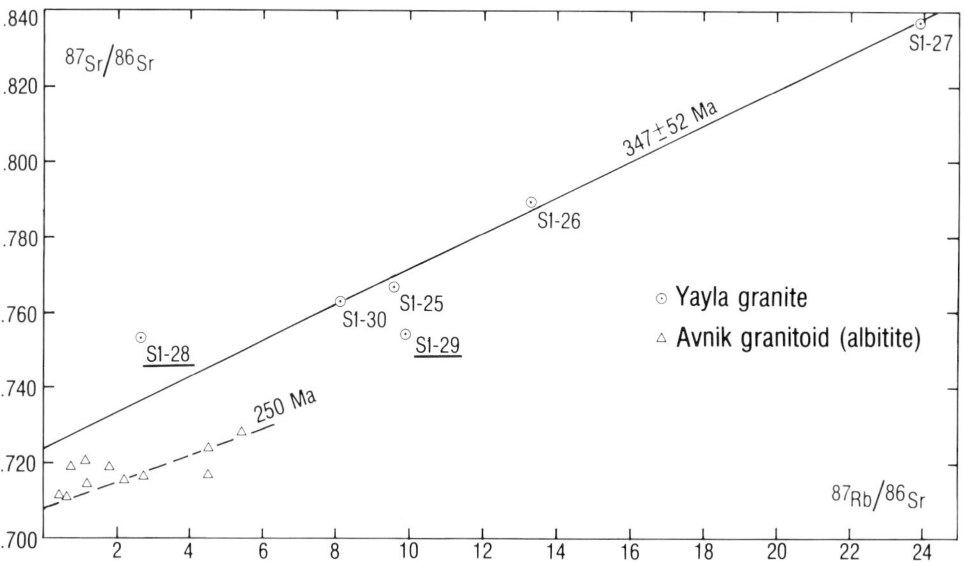

FIG. 3. Rb-Sr data for Yayla granite and Avnik albitite. Underlined Yayla samples excluded from regression. Dashed reference line through Avnik points show one possible (youngest) choice of a linear array.

TABLE 2. *Rb-Sr data on metavolcanics*

Sample	Rb, ppm	Sr, ppm	Rb/Sr	$^{87}Rb/^{86}Sr$	$^{87}Sr/^{86}Sr$
S–3 SERIES					
–25	135.1	65.3	2.068	5.995	0.72901 ± 24
–26	99.4	34.5	2.885	8.364	0.73125 ± 10
–27	87.4	23.9	3.652	10.594	0.73655 ± 14
–28	97.5	16.8	5.796	16.828	0.74431 ± 14
–29	121.8	18.2	6.703	19.447	0.73612 ± 16
–30	97.0	12.4	7.826	22.743	0.74870 ± 16
–31	217.0	6.98[1]	35.948	90.929	0.73911 ± 22
–32	225.4	4.53[1]	50.252	146.77	0.90437 ± 14
–33	215.0	3.61[1]	73.32	175.60	0.90595 ± 12
S–4 SERIES					
– 3	21.8	20.8	1.051	3.045	0.71902 ± 16
– 9	14.8	21.2	0.697	2.020	0.71742 ± 16
–16	80.0	6.56[1]	12.195	35.485	0.76391 ± 14
–18	158.5	11.03[1]	14.364	41.734	0.75547 ± 14
–21	116.2	11.26[1]	10.326	29.997	0.75317 ± 12
S–2 SERIES					
–14	135.6	6.41	21.156	61.567	0.76835 ± 16
–16	159.6	7.37	21.673	63.079	0.76955 ± 12
–17	97.7	22.2	4.411	12.790	0.73079 ± 16

1: isotope dilution analysis.

The uppermost point (S3–33) falls below the line; this sample is heavily silicified and slightly sericitized, but retains relict spherulitic textures. The position of S3–29 below the line may be due to intense sericitization and chloritization. Omission of the two points with highest Rb/Sr (S3–31, 32) from the regression does not change the age significantly. A smaller number of samples was analyzed from profiles S2 and S4, and these show more scatter. A poorly-defined line through the S4 array is subparallel to the S3 isochron (Fig. 4).

Figure 4 also includes analyses of equivalent, but less pervasively albitized, metavolcanics from the Cacas area to the east (Yılmaz 1971, Yılmaz *et al.* 1981). Yılmaz *et al.* (1981) drew several linear arrays including data from metavolcanics, paragneisses, amphibolites and granitoids, and derived 'ages' of ca. 570 Ma. They concluded from this that the metavolcanic series, at least, was Precambrian in age. We do not regard this treatment of the data as valid. Yılmaz *et al.* (1981) data for the felsic metavolcanics ('ortholeptynites') alone are shown in Fig. 5. Seven of the points define an isochron of 454 ± 13 Ma (MSWD = 0.25); the low MSWD is due to the rather large uncertainties that Yılmaz *et al.* (1981) assign to their isotopic data. Removal of Ba-5, the sample with highest Rb/Sr ratio (Fig. 5), changes the regression only slightly (465 ± 20 Ma, MSWD = 0.24); inclusion of two other points off the scale of Fig. 5 (see Fig. 4) does not change the age, but increases the uncertainty significantly. The initial $^{87}Sr/^{86}Sr$ for this line is high (0.710), but not unreasonable for continental rhyolites and dacites.

Chemical and/or modal data for a few of the samples analyzed are given by Yılmaz (1971) and Yılmaz *et al.* (1981). Where such data are available, they show that samples lying on the line in Fig. 5 are 'calc-alkaline' (i.e. not albitized) while samples lying off the line are albitized. We therefore regard the isochron in Fig. 5 as reflecting the age of extrusion of the volcanics of the Cacas area, while the scatter of some points from this line may be largely due to disturbance of Rb-Sr systems during albitiza-

TABLE 3. *Rb-Sr data on micaschist*

Sample	Rb, ppm	Sr, ppm	Rb/Sr	$^{87}Rb/^{86}Sr$	$^{87}Sr/^{86}Sr$
S2–20 rock	162[1]	37.40	4.332	12.590	0.75547 ± 12
S2–20 musc.	306.7	27.22	11.269	32.782	0.76472 ± 8
S2–20 chlor.	34.37	35.19	0.977	2.837	0.74848 ± 12
S2–20 remd.	10.47	52.48	0.200	0.579	0.74355 ± 14

1: XRF analysis, all other Rb & Sr by isotope dilution.

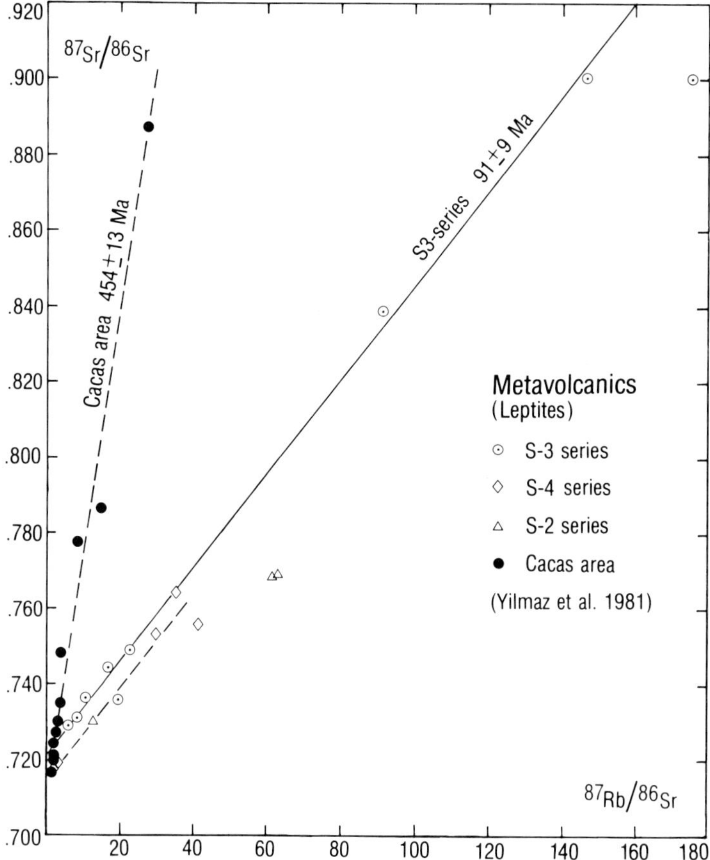

Fig. 4. Rb-Sr data for meta-volcanics. Dashed-line parallel to S3 isochron is 'scatterchron' for S4 series. Data from Cacas area from Yılmaz *et al.* (1981); see text.

tion. A similar extrusive age is inferred for the Avnik metavolcanics, but is not recorded by the Rb-Sr systems because of ubiquitous feldspathization.

Mineral analyses

Samples S3–1 is an amphibole-bearing Avnik albitite. The amphibole and 'remainder' (largely feldspar) separates give an age of 67 ± 30 Ma. The whole-rock sample falls below this line (Fig. 6), probably due to the loss of fine grained phases during crushing and sieving for mineral separation. Regression of the two mineral separates and whole rock defines an 'age' of 71 ± 28 Ma, to which no particular significance can be attached.

Sample S1–31 is a biotite-bearing Avnik albitite. The biotite-whole rock-pair gives an age of 41 ± 1 Ma. S1–32 is a similar sample in which the biotite is heavily chloritized; it was analyzed to test the meaning of the biotite age for S1–31. The biotite-whole rock pair for this sample gives an age of 19 ± 2 Ma.

A micaschist at the top of the S2 profile (S2–20) was separated into chlorite, muscovite and 'remainder' (albite, apatite, quartz) fractions (Table 3). The results (Fig. 6) suggest that the rock has not completely equilibrated during the latest metamorphic episode. The chlorite-muscovite pair defines an age of 38 ± 2 Ma.

Yılmaz *et al.* (1981) report biotite-whole rock ages of 22 Ma, on a 'paragneiss' of the Lower Unit, and 94 Ma on a granite equivalent to the Avnik and Yayla bodies. Analyses of whole rock, feldspar and biotite from a metavolcanic showed complete disequilibrium. The whole rock-feldspar pair gives an 'age' of 118 Ma, while the biotite-rock and biotite-feldspar 'ages' are negative.

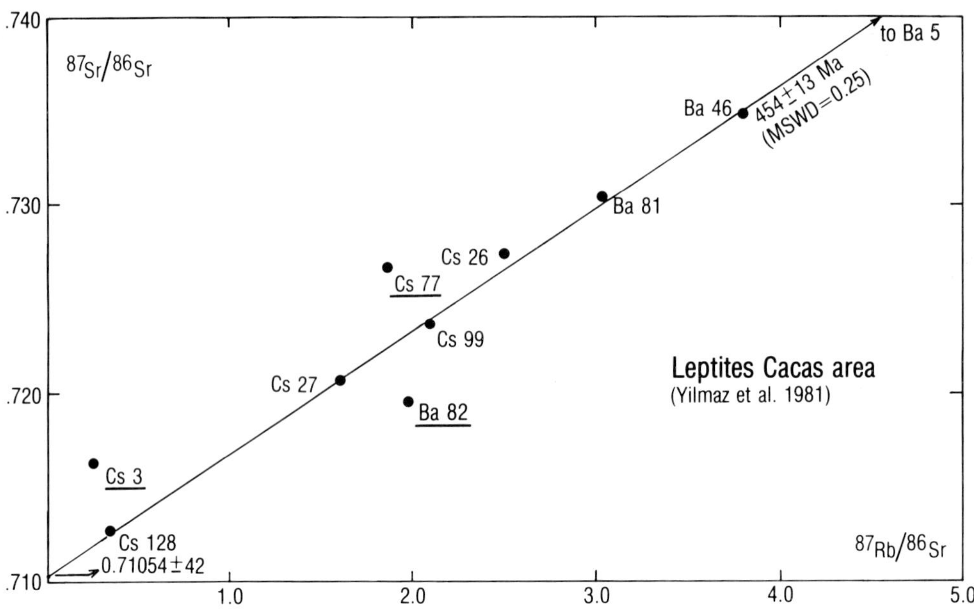

FIG. 5. Rb-Sr data for Cacas metavolcanics, from Yılmaz *et al.* (1981). Underlined samples, and 3 others outside scale, omitted from regression.

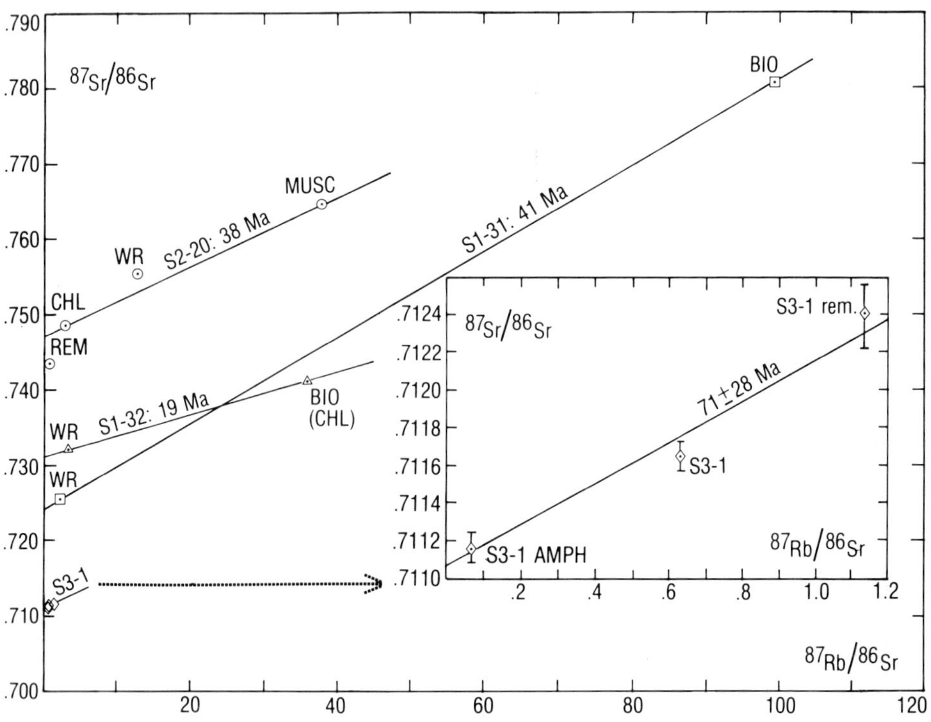

FIG. 6. Rb-Sr data for rock-mineral systems.

Discussion

The geochronological data on the Avnik area appear at first glance to contradict the field relations; the metavolcanics give a much lower age than the Yayla granite that clearly intrudes them. The data of Yılmaz *et al.* (1981), as reinterpreted by us, help to resolve this paradox. The Cacas and Avnik areas are so similar in lithological association, stratigraphy and geological history that we regard the correlation of the Cacas area's 'ortholeptynites' with the Avnik metavolcanics as reasonably certain. On the basis of this correlation, and our reinterpretation of Yılmaz *et al.*'s data, we conclude that the extrusion of the felsic volcanics in both areas took place ca. 450 Ma ago. This interpretation is consistent with the reasonably well-defined age of the Yayla granite (ca. 350 Ma).

The 90 Ma age on the pervasively feldspathitized and silicified metavolcanics of the Avnik area is interpreted as the age of the metasomatic event that caused the alteration. A similar age (70–110 Ma) was suggested by Yılmaz *et al.* (1981) on the basis of several samples with high Rb/Sr and low $^{87}Sr/^{86}Sr$, and a biotite-rock age of 94 Ma on an albitized granite.

The albitization has affected the Avnik granitoid, the metavolcanics, and the overlying micaschists. It led to massive changes in whole-rock chemistry, especially the redistribution of alkalies and silica. Exchange of material with the micaschists near the top of the section may have been responsible for extreme K-enrichment of the uppermost volcanics. The nature of the process is still not clear, but several constraints may be deduced:

(1) The process must have involved the movement of a fluid phase, to account for the redistribution of Na, K, Sr and Si, (and apparent homogenization of Sr isotopes) over distances > 100 m.

(2) The pervasive replacement of pre-existing minerals by albite suggests that the fluid was Na-rich.

(3) The albitization occurred following a period when both the metavolcanics and the Avnik granitoid served as sea floor for the deposition of the overlying micaschists. It seems reasoanble that these rocks became saturated with sea water at that time.

We therefore suggest that the albitization was caused by reaction between the rocks and sea water. Similar alteration of felsic tuffs has been described by, for example, Dickinson (1962). This might be a continuing diagenetic process, beginning in pre-Permian time, in which case the 'age' of 90 Ma is difficult to interpret. Howver, these reactions would certainly be favoured by heat and pressure and especially by directed stress, which would help to expel the fluid phase. We therefore suggest that the 90 Ma age is better interpreted as a metamorphic event, involving the final removal of the sea water responsible for the feldspathization. This is supported by the fact that some clearly metamorphic minerals appear to give similar ages (the 94 and 118 Ma ages by Yılmaz *et al.* (1981), and the amphibole-feldspar age on S3–1). This interpretation implies that the Avnik area was not affected by metamorphism between the Permian (ca. 280 Ma) and 90 Ma.

Hall & Mason (1972) and Hall (1976) have reported Cretaceous (Albian, Cenomanian and Campanian-Maastrichtian) fossils from melanges occurring as thrust slices within the Bitlis Massif. They have suggested that the subduction and high-pressure metamorphism of the melange terrane, and its thrusting together with the Bitlis massif, occurred in Upper Cretaceous time. This is consistent with our interpretation of the albitization age of ca. 90 Ma as dating a metamorphic event.

The ca. 40 Ma biotite-rock ages on the micaschist and the Avnik granitoid are interpreted as recording a late metamorphic event. This is consistent with the field evidence that these units are both involved in a major fold structure, and are cut by thrust faults. The much lower 'ages' (19 Ma) on chloritized micas show that the 40 Ma dates probably do represent metamorphic (cooling) ages rather than the later alteration and retrogression seen in some samples. No resetting of *whole-rock* Rb-Sr systems in connection with the 40 Ma metamorphism has been recognized.

The interpretation of whole-rock Rb-Sr data on metamorphic rocks is often ambiguous, and there has been considerable debate on the extent to which whole-rock Rb-Sr systems can be 'reset' by metamorphism (for example, Krill & Griffin 1981). The Avnik metavolcanics give a Rb-Sr date that is at least as well-defined as many 'isochrons' on intrusive rocks, but the geological control is such that this date can only be interpreted as a 'reset' age. This resetting of the Rb-Sr systems in the S3 profile, with its extreme variations in Rb/Sr ratio along the profile, must have required isotopic homogenization over distances of at least 100 m. The necessary conditions seem to have included (1) extremely reactive, fine-grained (locally glassy) felsic rocks, (2) expulsion of large volumes of fluid (sea water?), (3) extensive modification of initial rock compositions, including Na, K, Rb, Sr and Si contents.

In the Avnik granitoid, pervasive albitization has *not* resulted in homogenization of Sr isotopes and 'resetting' of the whole-rock system, but only in a *disturbance* of Rb-Sr systems and a scatter of points in the Rb-Sr diagram. This may be related to the originally coarser grain size, and the presence of crystalline phases such as amphibole and apatite that could serve as local traps for Sr. The Avnik area thus illustrates the difficulty of resetting coarse-grained igneous rocks to a 'metamorphic' Rb-Sr age, even under extreme metasomatic conditions.

Conclusions

The oldest rocks recognized in the Avnik area are the metavolcanics of the Lower Unit. Their extrusion cannot be dated here, because their Rb-Sr systems are reset by later albitization. Correlation with the Cacas area to the east, and reinterpretation of Yılmaz *et al.*'s (1981) Rb-Sr data, suggest that the extrusion of the largely felsic volcanics occurred ca. 450 Ma ago.

The volcanic series is intruded by the Avnik granitoid and Yayla granite. The Avnik body is albitized and yields no Rb-Sr date. The Yayla granite probably was intruded ca. 350 Ma ago, and its very radiogenic inherited Sr indicates that older crust was involved in its genesis.

The micaschists of the Upper Unit rest with unconformable depositional contact on the meta-volcanics, Avnik albitite and Yayla granite, and are affected by the albitization event. The age of the micaschists is constrained only by the 350 Ma date on the Yayla granite and by Permian fossils in correlatives of the overlying marbles.

The albitization is ascribed to reaction between the rocks and trapped sea water during an early Alpine metamorphism. Rb-Sr dating of a pervasively feldspathitized section of metavolcanics gives an age of 91 ± 9 Ma, which is regarded as a metamorphic age. Biotite-rock ages suggest a metamorphic recrystallization ca. 40 Ma ago; younger 'ages' are interpreted as the result of alteration of biotite and muscovite. The tectonic significance of the 40 Ma age is not clear, but it may be related to folding and recrystallization in connection with late overthrusting of nappes within the massif.

The Bitlis massif in the Avnik area therefore contains both Lower Palaeozoic elements (Lower Unit, Ordovician-Devonian) and Upper Palaeozoic elements (Upper Unit, Carboniferous-Permian). These have been affected by both Eo-Alpine (ca. 90 Ma, Cretaceous) and Alpine (ca. 40 Ma, Eocene) metamorphism, of which the former may have been the more important.

This work roughly doubles the geochronological data available on the Bitlis massif and the Eastern Taurides. Considerable areas of this mountain belt have now been well mapped by Turkish geologists. With the application of modern geochronological methods it should be possible to make rapid advances in our understanding of this belt, which will play an important role in interpretation of the plate tectonics in the Europe-Africa collision zone.

ACKNOWLEDGEMENTS. The geochronology laboratory at Mineralogisk-Geologisk Museum is supported by Norges Almenvitenskapelige Forskningsråd, through the Norwegian IGCP program. We thank Toril Enger for technical assistance and Magne Ranheim for considerable drafting assistance. Field work was supported by Ege University and by the M.T.A. institute in Ankara and the local division in Diyarbakır, and we thank the management and technical staff for their assistance. Author C.H. has been supported by grants from the Norwegian Government and a fellowship from Norges Teknisk Naturvitenskapelige Forskningsråd.

References

ALTINLI, I. E. 1966. Geology of eastern and south-eastern Anatolia. Part II. *Turkey Mineral Research and Explor. Inst. Bull.* **67,** 1–22.

AYKULU, A. & EVANS, A. M. 1974. Structures in the Iranides of south-eastern Turkey. *Geol. Rundschau,* **63,** 292–305.

BORAY, A. 1973. *The structure and metamorphism of the Bitlis area, south-east Turkey.* Ph.D. thesis (unpubl.) London Univ. 233 pp.

—— 1975. The structure and metamorphism of the Bitlis area. *Bull. géol. Soc. Turkey,* **18,** 81–4.

DICKINSON, W. R. 1962. Metasomatic quartz keratophyre in central Oregon. *Amer. J. Sci.* **260,** 249–66.

ERDOĞAN, B. 1982. *Bitlis Masifinin Avnik (Bingöl) yöresinde Jeolojisi ve yapısal özellikleri.* Dogentlik tezi, Ege Univ., 106 pp.

ERDOĞAN, B., HELVACI, C. & DORA, O. Ö. 1981. *Geology and genesis of the apatite-rich iron ore deposits, Avnik (Bingöl), Turkey.* Final report, Ege Univ., Yer Bilimleri Fakültesi, Izmir, 121 pp.

GENÇ, S. 1977. *Geological evolution of the southern margin of the Bitlis Massif, Lice-Kulp District, SE Turkey.* Ph.D. thesis (unpubl.) Univ. College of Wales, Aberystwyth, 281 pp.

HALL, R. 1976. Ophiolite emplacement and the evolution of the Taurus suture zone, southeast-

ern Turkey. *Bull. geol. Soc. Amer.* **87,** 1078–88.
—— & MASON, R. 1972. A tectonic mélange from the Eastern Taurus Mountains, Turkey. *J. geol. Soc. Lond.* **128,** 395–8.
KETIN, I. 1966. Tectonic units of Turkey. *Turkey Mineral Research and Explor. Inst. Bull.* **66,** 23–34.
KRILL, A. G. & GRIFFIN, W. L. 1981. Interpretation of Rb-Sr dates from the Western Gneiss Region: A cautionary note. *Norsk geol. Tidsskr.* **61,** 83–6.
TOLUN, N. 1953. Contributions á l'étude géologique des environs du sud et sud-ouest du Lac de Van. *Turkey Mineral Research and Explor. Inst. Bull.* **44/45,** 77–112.
YILMAZ, O. 1971. Etude pétrographique et géochronologique de la région de Cacas. Published Ph.D. thesis, Université Scientifique et Médicale de Grenoble, France, 230 pp.
—— 1975. Etude pétrographique et stratigraphique de la région de Cacas (Massif de Bitlis, Turquie). *Bull. geol. Soc. Turkey*, **18,** 33–40.
——, MICHEL, R., VIALETTE, Y. & BONHOMME, M. G. 1981. Réinterprétation des données isotopique Rb-Sr obtenues sur les métamorphites de la partie méridionale du massif de Bitlis (Turquie). *Sci. géol. Bull. Strasbourg*, **34,** 59–73.

CAHIT HELVACI* & WILLIAM L. GRIFFIN, Mineralogisk-Geologisk Museum, Sarsgt. 1, Oslo 5, Norway.
* Permanent address: Dokuz Eylül Universitesi, Mühendislik-Mimarlik Fakültesi, Jeoloji Mühendisligi Bölümü, Bornova, Izmir, Turkey.

The Eastern Pontide volcano-sedimentary belt and associated massive sulphide deposits

Ömer T. Akıncı

SUMMARY: The Eastern Pontide Belt is a major metallogenetic province in the Eastern Black Sea coastal region. Extensive volcanism was initiated in Liassic time. Palaeozoic granites and schists of older units form the 'basement'. The overlying volcano-sedimentary units reach 7000 m in thickness as follows; Lower Basic Series, 195–88 Ma; Dacitic Series, 88–65 Ma; Upper Basic Series, 65–37 Ma; Tertiary Granitoids, 37–22:5 Ma; Young Basic Series and Young Dykes.

The Lower Basic Series, cut by granodioritic intrusions, consists of basaltic and andesitic lavas, spilite, tuffs and agglomerates intercalated with limestones. The Dacitic Series consists mainly of andesitic, dacitic and rhyolitic lavas and pyroclastics alternating with mudstone, marly limestone and pelitic tuff. A major pillow lava flow divides the Dacitic Series into an ore-bearing lower and a less mineralized upper part. The Upper Basic Series is represented by extensive andesitic volcanism inland, while basaltic lavas and agglomerates occur along the coast. Syenite and quart-monzonites intrude this series. Granite, granodiorite, adamellite and monzonite intrusions form the Tertiary Granitoids. The youngest volcanism is represented by dykes and lavas. Radiometric ages of the intrusions show a continuity between Jurassic and Miocene, with increased activity around 80 and 40 Ma. The Lower Basic Series and the Dacitic Series are mainly tholeiitic; the Upper Basic Series and the other units are calc-alkaline.

More than 400 massive sulphide deposits are associated with tuffaceous horizons in the Dacitic Series of Upper Cretaceous age. These deposits show similarities with the Kuroko massive sulphide ores of Japan.

The Eastern Pontide Belt of Turkey shows features in common with the Carpathian Mountains, the Transylvanian Alps, the Timok mining district of Yugoslavia, the Banat-Srednogorie Belt of Bulgaria and the Caucasus. Ramovic (1966) and Popov (1981) drew attention to the distribution and similarities of mineral deposits in these volcano-sedimentary belts.

The Pontides form the northern margin of Anatolia (Ketin 1966), rising steeply from the Black Sea coast (Fig. 1). The Pontide range which extends westward for over 1200 km from the Lesser Caucasus almost to the Bulgarian border of Turkey, is divided into western and eastern segments by the Kizılırmak River west of Samsun. The Eastern Pontides are 500 km long and 75 km wide. A ca. 60000 km^2 outcrop of Cretaceous to Oligocene volcanics contain more than 400 known sulphide bodies of which 60 are of minable size. Studies of the mineralization have contributed to an understanding of the geology of the area, but the stratigraphy is less well known.

The region studied lies between 38°00″ and 42°00″ longitude and 40°30″ and 41°30″ latitude (Fig. 1). An attempt is made to correlate data from the western parts of the Pontides along the coastal strip to the Russian border. The stratigraphy, radiometric ages, mineralization and chemistry of the rocks are also considered. Extensive volcanism in the region is believed to have been initiated in Liassic time. Schists, Palaeozoic sediments and granitic intrusions form the 'basement'. The overlying volcano-sedimentary units reach 7000 m in thickness as follows: Lower Basic Series, 195–88 Ma (Lias-sic-Turonian); Dacitic Series, 88–65 Ma (Turonian-Palaeocene); Upper Basic Series, 65–37 Ma (Palaeocene-Oligocene); Tertiary Granitoids, 37–22.5 Ma (Late Eocene-Oligocene); Young Basic Series and dykes, Miocene-Pliocene.

The volcanic sequence has a gentle northerly 20–40° dip towards the Black Sea so that outcrops of older units near the coast, for example the Lower Basic Series are largely restricted to horsts. In the eastern Black Sea mountain chain, alpine structures such as major nappes or extensive overthrusts are absent.

The plate tectonic evolution of the eastern Pontides is not discussed here since the data are still incomplete. Şengör & Yılmaz (1981) reviewed the Mesozoic-Cenozoic tectonic evolution of Turkey in a plate tectonic framework and this throws some light on the evolution of the Pontides following earlier attempts at synthesis by Pejatovic (1979) and Gedikoğlu *et al.* (1979) & Ataman *et al.* 1975.

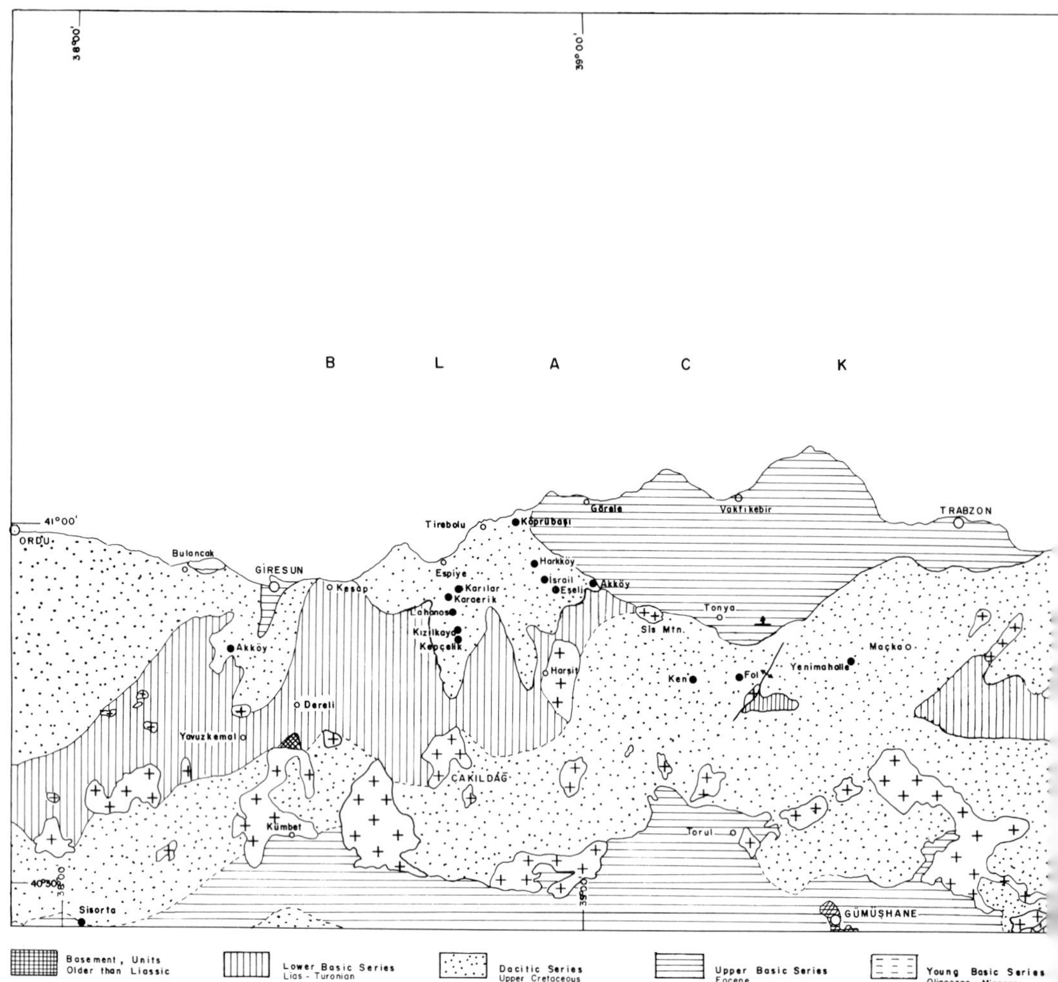

FIG. 1. Outline geological map of the Eastern Pontides showing the locations of sulphide mineralization and other places mentioned in the text.

Regional geology

Basement

Very little information is available on the metamorphic basement. The metamorphic massifs appear to form large continuous masses in the western Pontides, while in the east they are of limited extent. In some parts of the belt rocks are of low-grade greenschist facies.

The basement is exposed in the south in Gümüşhane region. An approximately 1500 m thick marine Permo-Carboniferous sequence overlies metamorphic basement, which is intruded by the Gümüşhane granite (Yılmaz 1972), the only intrusive in the region with a definite Palaeozoic age from stratigraphy and radiometric age dating. The undated complex of amphibolitic schists, gneiss, and mica-schists which forms the basement to the eastern Pontide volcanics of the Demirkent area to the south of Artvin has also been regarded as Palaeozoic by Baydar *et al.* (1969).

In the coastal strip basement rocks in the form of schistose granodiorites, are exposed in the Uzunlu area. Their age is unconfirmed but may be Variscan (TJP 1977). Similarly, pinkish intrusives in low-grade pelitic schists of the Çakıldağ area, were believed to be pre-Mesozoic by Doğan (1980). Intrusive rocks south of Dereli were also regarded as Palaeozoic by Zankl (1962).

Triassic rocks have not so far been recognised in the Eastern Pontides. However, in the Köse-Kelkit region, near Gümüşhane, Ağar (1977) states that siltstone, sandstone, conglomerate, shale and arkosic sandstone overlie Permo-Carboniferous units with a disconformity and are in turn overlain conformably by Jurassic

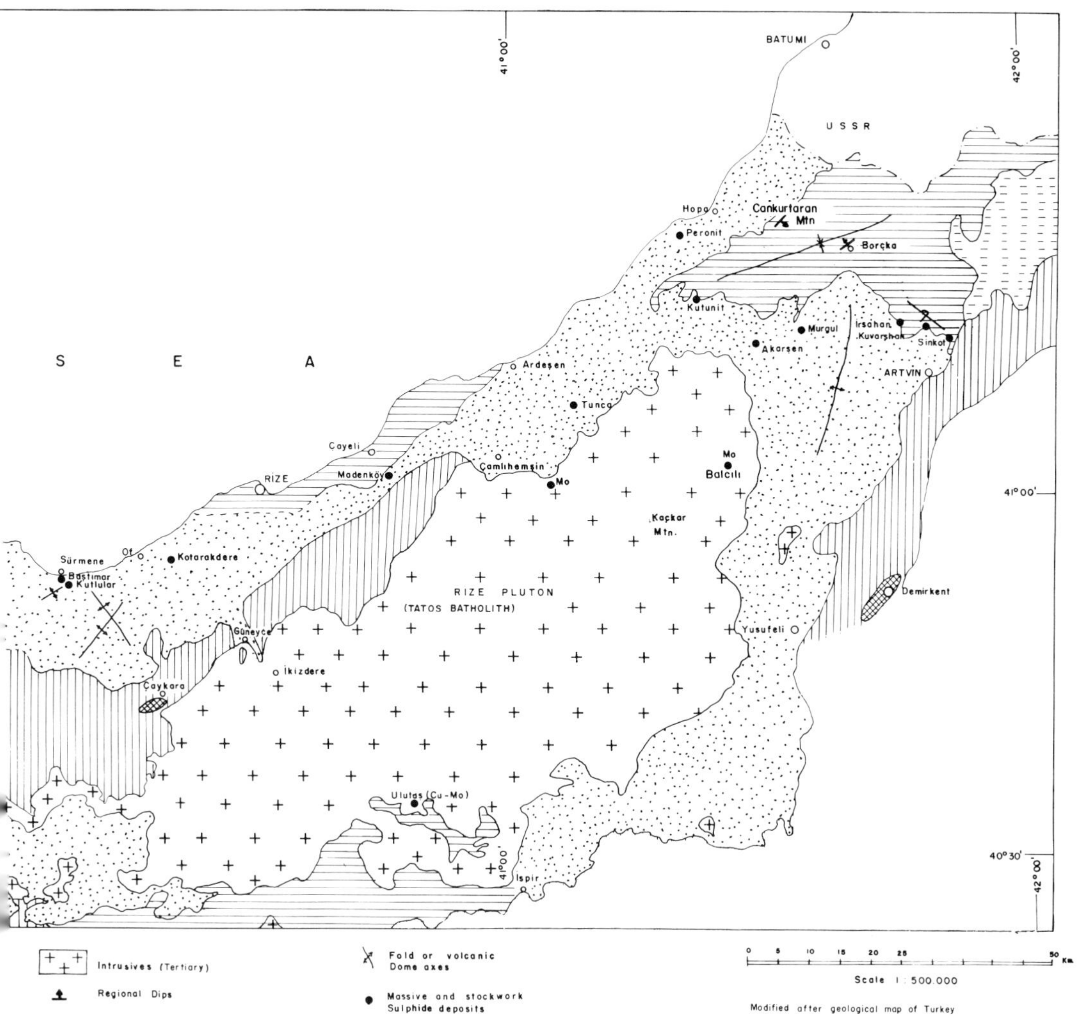

rocks which would suggest a Triassic age for the underlying sequences. On the other hand, recent works suggests a Permian rather than Triassic age for these sequences.

Lower Basic Series: Liassic to Turonian

The volcanic units are basaltic or andesitic but in places are cut by younger dacitic and rhyolitic dykes. Ignimbritic pyroclastics and, in places, relatively fresh basaltic pillow lava also occur. These rocks differ from the overlying extensively altered massive sulphide-bearing Dacitic Series and also from the fresher basaltic and andesitic Upper Basic Series, which is cut by basic dykes in most places. The Lower Basic Series is mostly exposed on the west side of the Tertiary granitoid chain (Fig. 1).

Jurassic occurrences

Liassic volcanics are well developed in the Gümüşhane region. Extensive spilitic lavas and tuffs crop out in Kırıklı area and basic dykes, dolerite plugs and tholeiitic lavas are widespread around Gümüşhane itself. The dykes cut some parts of the Jurassic sequence, but in other cases sediments overlie dykes (Yılmaz 1972; Tokel 1972). It is impossible to date the beginning and end of this magmatic activity which frequently directly overlies the Gümüşhane Granite and is in turn overlain by an Upper Cretaceous sedimentary sequence.

In the Artvin region, in the far east of the area, along the Çoruh valley, granite and gabbro bodies are cut by closely-spaced microdiorite bodies and basic dykes which intrude fossiliferous phyllitic and graphitic schists of

Jurassic age (Genye Mountains) and are then overlain by basal conglomerates below Upper Jurassic limestones (İ. Güven pers. comm., 1982).

Jurassic/Cretaceous

In the western coastal section, 15 km south of Bulancak Town, the series is predominantly dark green spilite, basalt, greenish andesite, with altered ignimbrite and breccia (Akıncı & Milutinović 1969). Thin pelitic limestone and tuff horizons separate these basic volcanics from the overlying porphyritic dacites (Akıncı 1974). Although fossils have not been found, these horizons are very similar to Upper Cretaceous reddish calcareous mudstones.

Cretaceous

In the inner parts of Harsit Valley and to the south of Dereli between Güdül and Kurtulmuş villages, Schultze-Westrum (1961) and Zankl (1962) report that the Jurassic begins with a basal conglomerate consisting of quartzite and schist overlain by recrystallized nodular and reef limestones of Upper Jurassic age. In the Kürtün area south of Sis Mountain (Fig. 1), basalt, spilite, andesite lavas, tuff and agglomerate are cut by basic dykes and intercalated with pillow lava. This whole sequence is overlain by fossiliferous tuffs of Lower Cretaceous age (Zankl *op. cit.*). These units are in places overlain by reefal Hippurite-bearing limestones of Turonian-Cenomanian age.

In the Rize-Güneyce area the lower part of the volcano-sedimentary pile contains cherty, fossiliferous neritic limestone intercalations of Albian-Aptian age (Taner & Zaninetti 1978). The higher levels of these limestones are Cenomanian neritic limestone and greywacke affected by very high temperature contact metamorphism (Taner 1977). This whole sequence is then overlain by Middle Turonian metadiabase, andesitic lavas and pyroclastics, interbedded with pyroclastics and tholeiitic pillow lavas rich in haematite.

The volcanic sequence of the Lower Basic Series, as seen in Fig. 2, is intruded in many places by granodioritic, monzonitic and tonalitic rocks. Radiometric determinations of these intrusive rocks indicate ages of 132–71 Ma (Giles 1974; Gedikoğlu 1978; Moore *et al.* 1980). This suggests that the intrusive history is complex and lasted through most of the Cretaceous.

Dacitic series

This series contains over 90 of the known sulphide reserves including all the massive- and stockwork-sulphides, and is of Upper Cretaceous age. Dacitic pyroclastics and lavas are the dominant rock types though andesitic lavas and tuffs and rhyolitic volcanics are locally important. The series is divided into two main parts by a widespread horizon of basaltic lavas of Maastrichtian age. The volcanic units are further sub-divided loosely into two units below and two above the basaltic horizon, making five in all, though the implied succession of eruptive types is not recognizable everywhere.

From base to top they are: Porphyritic Dacite Unit; Tuff/Lava/Mudstone/Pyroclastic Unit; Basaltic/Spilite Unit; Rhyodacitic-Rhyolitic Lava Unit; Biotite dacite/Rhyolite/Andesite Unit. A sixth unit of flysch-like sediments generally marks the transition at the top of the series to the overlying Eocene nummulitic limestones and marls at the base of the Upper Basic Series.

In the field the rocks of the Dacitic Series are strongly sericitized and chloritized and many are albitized. They differ markedly from the relatively fresh Upper Basic Series above. Typically, the mineralized horizons are dacitic pyroclastics or breccias with angular lava fragments set in a silicified matrix with disseminated pyrite. Volcaniclastics pass laterally and vertically into well sorted tuffs, calcareous mudstones and marls overlain by reddish cherts or haematitite-rich sediments near the sulphide deposits. The units below the Maastrichtian basaltic horizon are much more strongly altered and mineralized than those above.

In most places the base of the series is marked by a characteristic porphyritic dacite. This is generally underlain by a thin limestone or tuffaceous sedimentary horizon which is sometimes fossiliferous and gives ages always younger than Cenomanian. The age span of the series is thus established as ?Turonian to Maastrichtian/Palaeocene, the age of the uppermost flysch-like unit. Along the coast the flysch-like sediments begin in the Maastrichtian and appear to be conformable with overlying Lutetian limestones.

Porphyritic Dacite Unit

These rocks are characterized everywhere by euhedral but generally corroded quartz phenocrysts 1 mm to 1 cm long, feldspars occasionally up to 2 cm long, and altered hornblende and biotite crystals set in a light groundmass. In zones transitonal to the underlying Lower Basic Series the coarse porphyritic rocks grade to fine or medium grained dacite. In the area between the Lahanos Mine and Kızıldere, porphyritic dacite grades upwards into quartz-bearing andesites or hornblende-biotite-andesite, quartz phenocrysts disappearing gradually (Tuğal 1969).

Porphyritic dacites several tens of metres thick directly overlie the Lower Basic Series about 15 km south of Bulancak town. In an east-northeast direction to the Batlama River over 300 m occur near Akköy Mine (Metag 1972). The unit then thins out to the east to disappear completely south of the Kesap area where only pillow lavas of the Lower Basic Series are exposed. Further east in Espiye, along the Yağlıdere river, altered porphyritic dacite crops out with well developed columnar-joints. The area between the Batlama and Yağlıdere rivers was an area of uplift with the Batlama valley a narrow graben or basin filled by dacites.

Porphyritic dacites have an intrusive as well as an extrusive character locally and occur as steeply dipping dykes cutting the Lower Basic Series. These dykes are overlain by Upper Cretaceous sediments around the Görele-Akköy mine. Elsewhere Tuğal (1969) reports that Santonian limestone overlies the Lahanos orebody which represents a stratigraphically higher level in the porphyritic dacites implying a Santonian-Turonian age for them.

Tuff-Lava-Mudstone-Pyroclastic Unit

The Dacitic Series as a whole is mostly composed of alternations of lavas and tuffs, accompanied by limestone, marl and mudstone. This kind of sequence is well developed south of the Trabzon-Vakfıkebir line (Fig. 1), around Ken, and the Fol Yenimahalle Mines where the total thickness reaches almost 6000 m (TJP, 1977). The unit in this area is divided into A_1, D_1, A_2, D_2 sub-divisions. The uppermost part of D_2, the Tonya Marl, was included in the Upper Cretaceous, but recent studies in other areas have shown that it is the lowermost Eocene horizon of the Upper Basic Series.

A_1: The greater part of this sub-division is augite- and hornblende-andesite lava flows and andesitic pyroclastics of submarine origin, and subordinate biotite-dacites. The total thickness exceeds 1000 m. Intercalated limestones contain microfossils giving ages ranging from Turonian to uppermost Cretaceous. Sub-angular to rounded clasts of mudstone, granite, rhyolite, marl, siltstone and andesite or chert are found in the pyroclastic rocks, with mud-balls and green glassy fragments. Dacite, dacitic pyroclastics and marl, grade laterally into limestone in the Uzunlu area.

D_1: This consists of dacite and dacitic pyroclastics with intercalated mudstone, marl and limestone and reaches 2000 m in thickness. In places auto-brecciated dacite lavas grade into coarse porous pyroclastics. Fossils in calcareous layers indicate ages from Turonian to Campanian. Sulphide, barite and manganese mineralization occurs in various horizons.

A_2: The greater part of this unit is again andesitic pyroclastics, with augite- and two-pyroxene-andesite lavas, locally auto-brecciated and amygdaloidal. The unit is partly intercalated with green siltstone, red limestone and andesitic tuff and passes upwards into calcareous sediments. The total thickness varies from 1000 to 2000 m.

D_2: Like D_1 this sub-division is again characterized by rhyolites, dacites and dacitic pyroclastics with subordinate pyroxene andesite, basaltic andesite and biotite-dacite lavas. Sulphide mineralization is not common. The total thickness varies from 800 to 1600 m. It includes the Kaledere Rhyolite below and Tonya Marl above (TJP 1977).

Further east on the Black Sea coast in the Çayeli-Madenköy area the stratigraphic equivalents of these tuff-lava-sediment alternations are dominantly pumiceous dacitic tuffs which give way higher up the succession to purple tuffs injected by basalt sills. Similar lithologies at the eastern end of the range are host for the most important pyritic copper deposit of the Black Sea region, at Murgul.

Basalt-Spilite Unit

The basaltic lavas of this widespread unit are typically pillowed. Extensive exposures are seen along the Arhavi-Hopa-Artvin highway in the east. Pillow basalts directly overlie dacitic lavas and pyroclastics around the Kuvarshan-Irsahan mines though occasionally a thin mudstone horizon intervenes. Olivine dolerite sills assigned to this basaltic eruptive event are found intruding the overlying flysch-like sediments in several places e.g. the Çayeli-Madenköy, Arhavi-Peronit, Görele-Akköy and Harköy-Kuskunlu mines.

Rhyodacitic-Rhyolitic lavas

This unit was originally mapped as pyroxene-andesite around Lahanos-Kızılkaya Mines (Akıincı-Milutinovič, 1969) and diabase around Kutlular Mine (Gümrükçü 1974) where it blankets the mineralized horizons. A thin grey limestone horizon containing *Globotruncana* sp. of Santonian age Degirmendere and Kasapçayırı occurs immediately at the base (Tuğal, 1969). Concealed sulphide deposits are to be expected beneath this rhyodacitic-rhyolitic cover. Elsewhere this unit overlies the basaltic flows and pillow lavas of the Dacitic Series, but in places it directly overlies porphyritic dacites, as in Bulancak-Tekmezar area.

Biotite-Dacite/Rhyolite/Andesite Unit

The youngest volcanics in the Dacitic Series are biotite-bearing andesites, dacites and rhyolites occurring as volcanic domes, stocks and dykes, the most representative of this unit being the 'Lahanos Tepe Dacite' (Pollak 1961).

In the Lahanos Mine area, for example, a huge NW-SE-trending biotite-dacite dyke dated at 70 Ma cuts the Upper Cretaceous units, including the rhyodacitic lavas overlying the orebody.

Similar late biotite-bearing volcanics occur in the Gölköy region (Gedikoğlu 1970), in the Bulancak-Tekmezar and Giresun-Akköy Mine areas (Akıncı 1974), in the Keşap-Karabulduk area (A.K. Nalbantoğlu, pers. comm. 1982), Tonya area (TJP, 1977; the Kaledere Rhyolite) and finally as the Karatepe Dacite which covers ore-bearing dacitic tuffs of the Murgul Mine.

Flysch-like sediments: Akköy Formation and equivalents

Cretaceous volcanics and volcano-sedimentary units of both the western and eastern parts of the Eastern Pontides end with a flysch-like sequence which differs in the two areas but in each it grades into the overlying Eocene units with nummulitic horizons.

The Akköy Formation itself overlies the Dacitic Series and the mineralized horizons south of the Görele-Çanakçı area, and extends from Harşit-Köprübaşı Mine to Sis Mountain, through Harköy, Eseli and Görele-Akköy Mines in a southeasterly direction. From Sis Mountain eastward, the formation becomes thinner (Fig. 1). It has a maximum thickness of 400–500 m, and is composed of fossiliferous calcareous mudstones, limestone, marl and tuffs alternating with pyroclastics and lavas. Cross-bedded sandstone horizons are also reported (Akıncı 1969). The lowermost levels usually contain basic sills and sometimes dykes. Eğin's 1978 data suggest depositional depth up to 700 m in the Harsit valley. Fossiliferous horizons yield Santonian to Maastrichtian ages, with abundant *Globotruncana* sp., ammonites and casts of *Inoceramus* sp.

In the Yaslıbahçe village, south of Bulancak, further to the west, the equivalent of the Akköy Formation is composed of alternating limestone, marl and sandstone. In neighbouring areas, volcanic breccias, agglomerates and tuffaceous sandstones are the lateral equivalents. In this area fossils indicate a Santonian-Campanian age (Akıncı 1974); in addition to *Globotruncana* sp. Schultze-Westrum (1960) reports *Inoceramus balticus*. This unit extends further south-west from Bulancak to Gölköy (Ordu) as the Fatsa Formation (Terlemez & Yilmaz 1980).

In the east the equivalent of the Akköy Formation in the area between Kuvarshan and Irsahan Mines, in the Artvin region, is represented by a 50–150 m thick sequence of purplish-red, well bedded mudstones. Fossils indicate a late Upper Cretaceous age (Güven 1982). These mudstone horizons are overlain by, or intercalated with, basalts which show pillow structure and are interstratified with light grey-green, fine-grained tuffs or tuffaceous sandstones.

Upper Basic Series

The name Upper Basic Series is given to sequences of sediments and volcanics of Palaeocene to Eocene age which overlie the older units described, either conformably or unconformably. Where volcanics are present they are distinctively fresh compared to the older extrusives and dominantly basalts or basaltic andesites. The nummulitic Eocene sediments are also clearly identifiable in the field.

Fossiliferous Palaeocene rocks are only present in the western and far eastern parts of the region. In between, Eocene volcanics and sediments rest directly on Upper Cretaceous units. In the far west the Upper Basic Series is represented by the Gölköy Formation composed of Palaeocene tuffs, sandstones agglomerates, coals and clayey limestones (Terlemez & Yilmaz 1980). It rests unconformably on a syenite intrusion with a basal conglomerate and is succeeded by Ypresian pelitic sandstones and Lutetian nummulitic and marly limestones.

In the type locality of the Akköy Formation further east, the Akköy flyschoid sediments as noted above pass conformably up into Eocene nummulitic calcareous mudstones without recognizable Palaeocene beds. However, in this same area, the Hasan Dağı basalts and associated basaltic andesite pyroclastics north of the Eseli mine, which lie above the Akköy Formation and below Ypresian limestones and marls, may be Palaeocene in age. On the coast to the north, fossiliferous Palaeocene limestone blocks occur in younger basaltic agglomerates of the Upper Basic Series in the Görele-Çanakcı area.

In the eastern area between Hopa on the coast and Artvin, Palaeocene fossils are recorded from a number of localities in sediments interbedded with volcanics, as for example in biomicrites from the Cankurtaran Mountain area (Özsayer *et al.* 1981), and in volcanic conglomerates and interbedded sandstones, claystones and marly limestones around Kuvarshan (Güven 1982). This latter sequence 100–150 m thick, extends into the Eocene and similar sediments overlie a Palaeocene andesitic conglomerate further to the east (Güven, pers. comm. 1982).

The main development of the basic lavas which give their name to this series occurred in the Eocene, particularly in the coastal region where the nummulitic limestone, marl and sandy limestone alternations of Ypresian and Lutetian age are developed, in places conformably overlying the Akköy Formation as between Ordu and Arklı at Kaykara and Güveç south of Giresun (Schultze-Westrum 1961), in the Tirebolu-Kovaçpınar area (Kahraman 1981) at Gorele-Derekuşçulu (Akıncı 1969) and at Tonya (TJP 1977). These thin Eocene sequences are overlain by a thick series of fresh olivine-augite basalts, andesits, spilitic lavas, tuffs and basaltic agglomerates, with sedimentary intercalations. They are well exposed in the Tirebolu-Trabzon area and east of there in the main Upper Basic Series outcrop area between Rize and Ardeşen and further east in the Cankurtaran Mountain-Borçka syncline. These basaltic rocks reach a maximum thickness of about 1000 m.

On the southern side of the Dacitic Series belt (Fig. 1), the Upper Basic Series is characterized by extensive calc-alkaline andesitic volcanics between the Gümüşhane and Gölköy areas (Tokel 1977). Around Sis Mountain flat-lying basaltic-andesitic flows overlie basic agglomerates (Özgüneyli and Okabe 1981; TJP 1977).

In the east hornblende-dacite and andesite stocks of the Kuvarshan-Irsahan area cut Eocene flysch units; rhyolites also cut the uppermost basalts, andesites and pyroclastics, and so these intrusive members have also been included in the Upper Basic Series.

Tertiary granitoids

A number of isolated intrusions of varying size of acid to intermediate composition occupy high plateau regions. They are located roughly along the middle of an E-W trending anticlinal axis which also forms the drainage line dividing the Eastern Black Sea Mountains. These are synorogenic, shallow intrusions containing xenoliths of volcanics, metamorphic rocks and sediments. The olivine-augite basalts of Eocene age are slightly metamorphosed by these intrusions along the contacts (Akıncı 1969). Şişman (1972) reported the presence of skarns at contacts with Upper Cretaceous carbonate sediments. The intrusions cut the Eocene (Lutetian) flysch implying a Late Eocene or Oligocene age (Gattinger, *et al.* 1962; Tokel 1972).

The intrusive rocks are dominantly hornblende-biotite granodiorite and quartz-diorite, in normative composition (Moore *et al.* 1979). However, older ones are granitic (Altınlı, 1970; Gedikoğlu *op. cit.*, Gattinger *et al.* 1962; Doğan 1980).

A zonation from fine-grained tonalitic rocks at the margin to coarse porphyritic granodiorite at the centre is developed in the Rize pluton and the Tatos batholith (Taner 1977). Skarn and contact metasomatism is common along the margins. Quartz-molybdenite veins in the Çakildağı area, and Espiye are related to porphyritic adamellites and microgranites (Kamitani, *et al.* 1977; Kamitani & Akıncı 1979; Doğan *op. cit.*). Elsewhere, porphyry Cu-Mo mineralization at Ulutaş, in the Ispir area is developed in quartz-monzonite (Giles, 1974).

Giles (1974) obtained 127–132 Ma ages for the Ispir granodiorite on the southern end of Rize pluton and 41 Ma for an isolated intrusive body west of Ispir. The western edge of the Rize pluton in the Rize-Ikizdere area shows variations from 40 and 80 Ma (Taner *et al.* 1979). Similarly, Moore *et al.* (1979) obtained a 62 Ma age for the Balcılı-Balhibar (Yusifeli) granodiorite at the eastern boundary of the pluton and 37 to 84 Ma for other isolated intrusive bodies. According to Moore *et al.* (1979), plutonism started in the Mid-Cretaceous, at about 90 Ma, and continued for approximately 70 Ma until the Early Miocene, but a review of the available data suggests that intrusive activity may have started earlier.

Young Basic Series and Late Dykes

These are biotite-andesite and olivine-pyroxene-basalt dykes which cut the Upper Basic Series units and the Tertiary granitoids along the coastal strip. The Çermik rhyolite stock in the Artvin area cuts Eocene flysch and basic volcanics, exposed around the Kaçkar Mountains. Young basic volcanics to the north-east of Artvin are also included in this unit (Fig. 1). Schultze-Westrum (1961) states that some agglomerates and tuffs in the Giresun region intercalated with Oligocene-Miocene marly limestones, belong in this Series.

Chemistry of the rock units

A total of 69 representative analyses were selected from published and unpublished data, and are available as Supplementary Publication 18042. The number of analyses in the literature

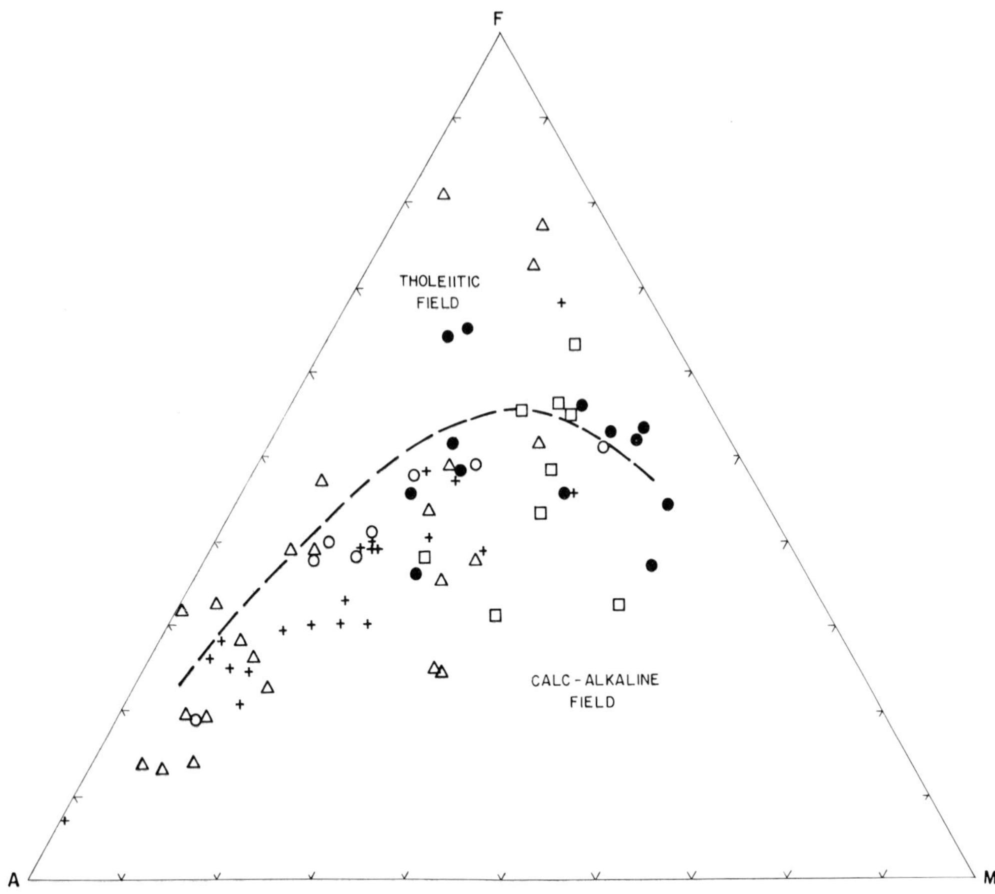

FIG. 2. AFM plot of Eastern Black Sea volcanics and intrusives. The average calc-alkaline field-boundary is taken from Kuno (1968). ●, Lower Basic Series; △, Dacitic Series; ○, Upper Basic Series; +, Tertiary Granitoids; □, Young Basic Series.

exceed 1000 (Tuğal 1969; Akıncı 1974; Gülen 1976; Taner 1977; Gedikoğlu 1978; Eğin 1978; Doğan 1980) but so far have not been considered in terms of eruption age.

Here the representative analyses have been plotted on AFM, total alkalis versus silica, and major oxides versus silica diagrams (Figs. 2, 3, 4). An immediate important conclusion can be drawn from these diagrams. In terms of their major element chemistry *all* the rocks plotted lie on a single, typically calc-alkaline trend. Much of the scatter in the trends particularly for Na_2O and K_2O is undoubtedly due to secondary alteration.

The calc-alkaline character of Eocene and younger rocks noted by Gedikoğlu (1970) is apparent but there seems no reason to distinguish them geochemically from a 'tholeiitic' Lower Basic Series or Dactic Series as did Egin *et al.* (1979) and Doğan (1980) who both worked with much smaller numbers of samples. Mineralization and spilitic alteration in the Lower Basic Series and extensive hydrothermal alteration in the Dacitic Series have undoubtedly affected bulk compositions even though the freshest rocks were selected for this compilation.

If the rocks which have been clearly depleted in alkalis are ignored the trends of the whole suite are closely comparable with those of, for example, the Cascades Volcanics (Carmichael *et al.* 1974). CaO, FeO, MgO and TiO_2 all decrease with increasing silica content. Anomalously low CaO contents in two basalts are due to spilitization. These analyses confirm the conclusions of several authors (Akıncı 1974; Tokel 1977; Pejatovic 1979; Eğin *et al.* 1979; Doğan 1980) that an active subduction zone was established under the Eastern Pontides during the Cretaceous to Oligocene period and may have been continuously active from the Liassic onwards. Further support for the calc-alkaline

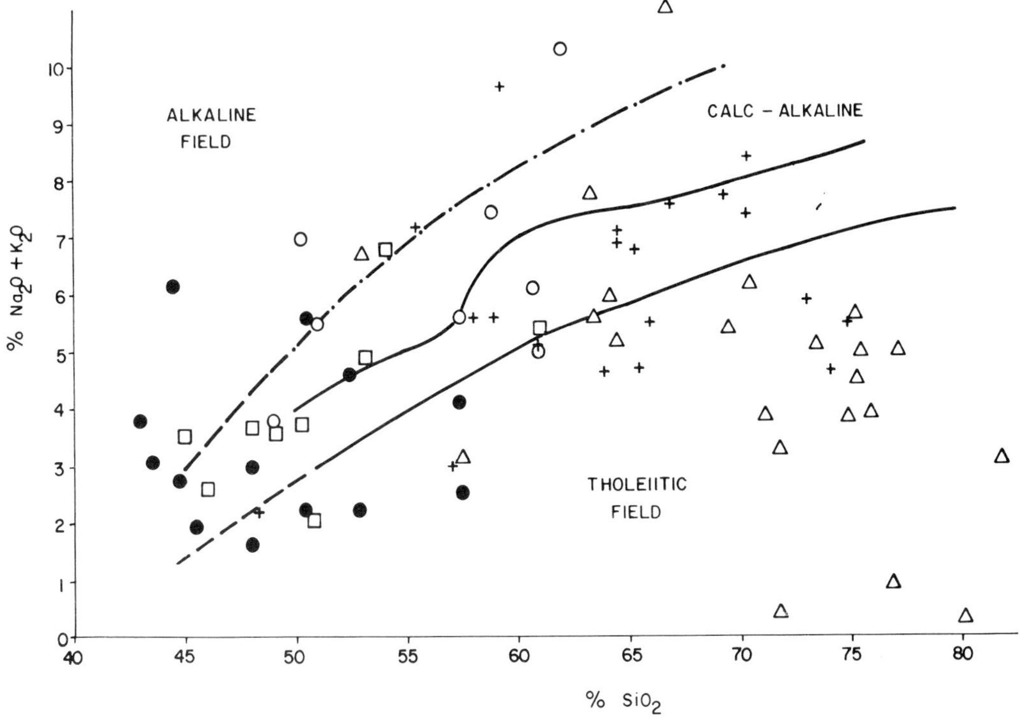

FIG. 3. Total alkalis plotted against silica for the Eastern Pontide volcanics and intrusives. Field-boundaries of Kuno (1968) are between alkaline and calc-alkaline and calc-alkaline and tholeiitic fields, while the boundary of Irvine and Baragar (1971) separates the alkaline and calc alkaline fields. For key see Fig. 2.

character of the volcanics comes also from the low TiO_2 content (<0.8%) (Pearce & Gale 1977) and from Ti/Zr relationships.

General characteristics of the massive sulphide deposits

Although opinions differ on the age and genesis of the sulphide mineralization, it is known that the most important Cu-Pb-Zn deposits of Turkey are associated with Eastern Pontide volcano-sedimentary belt (Popovic 1975). The massive and stockwork deposits which could have been root zones of the massive sulphides are confined to the different horizons of the Dacitic Series. In disagreement with the Liassic-Lower Cretaceous ages suggested by Pejatovič (1979), they are believed to have been formed within 10 Ma in the Upper Cretaceous.

The great majority of the massive and stockwork sulphide deposits are located between the Black Sea coast and the Giresun-Harşit-Çaykara-Artvin line. Numerous vein-type deposits, concentrated in Fatsa-Gölköy-Kabadüz, Bulancak-Tekmezar, Koyulhisar-Sisorta, Fol-Alacadağ, Şavşat-Artvin areas, form an arcuate trend around this massive sulphide zone (Fig. 1). Generally, mushroom-shaped orebodies consist of massive bedded and/or stockwork ore with vertical as well as horizontal mineralogical- and alteration-zoning. The ore horizons are commonly overlain by ferruginous chert, or in places by ore-breccia and brecciated wall-rock, iron and manganese-rich sediments, agglomerates, mudstone and purple tuffaceous horizons (Akıncı 1980). The uppermost ore horizon is massive, fine-grained sphalerite accompanied by chalocopyrite, barite and/or gypsum, grading down into chalcopyrite plus pyrite and finally into pyrite veinlets and disseminations. In the stockwork ores, black quartz is seen at the uppermost levels changing to white quartz in the lower levels, while sericite dominates in the lowermost zones (Sawa & Altun, 1977). Hydrothermal zoning of the orebody is characterized outwards by advanced argillic alteration, sericite alteration and silicification, intermediate argillic alteration and in the outermost zone by montmorillonite-zeolite alteration.

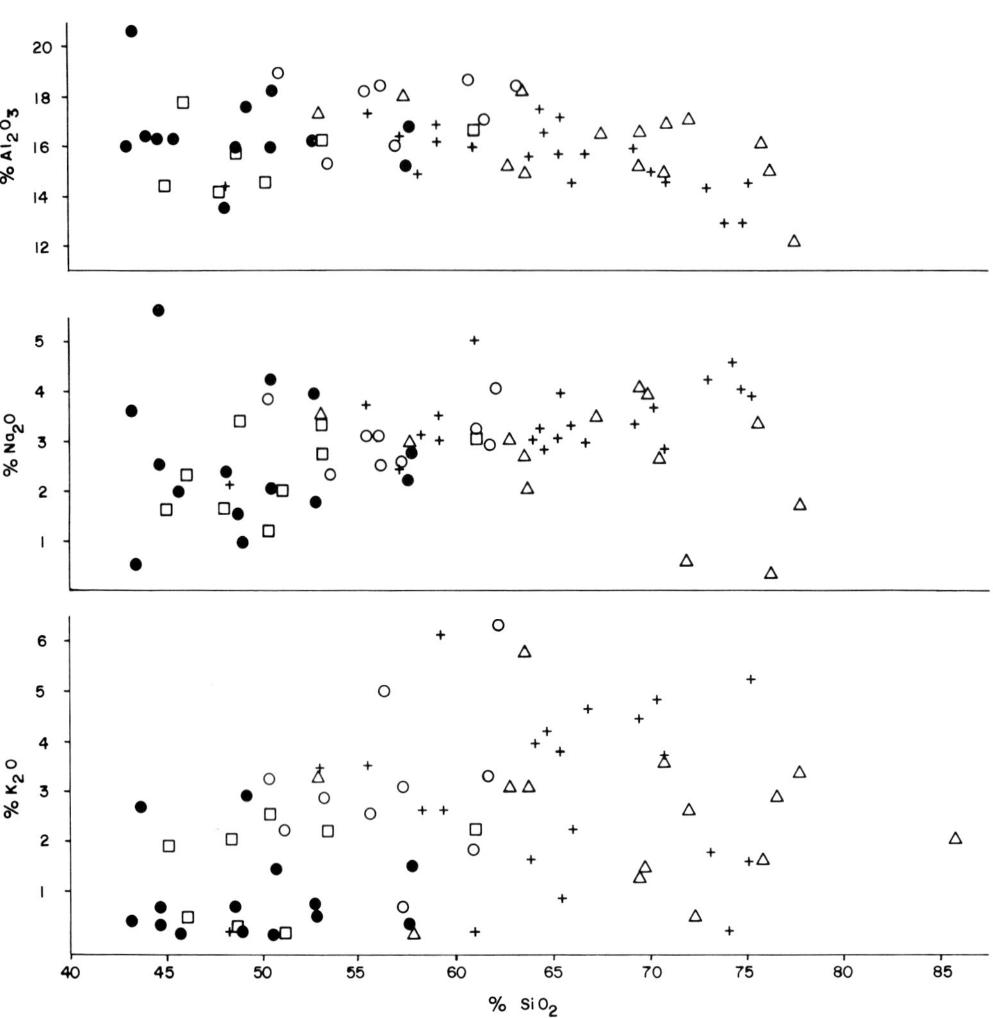

FIG. 4. Major oxides vs. SiO_2 for the Eastern Pontide volcanics and intrusives. See Fig. 2 for key.

These massive and stockwork deposits of the Eastern Black Sea metallogenetic province have various points of similarity with the Kuroko deposits of Japan: (i) a strata-bound setting of the orebody within the acidic lavas and pyroclastics; (ii) the orebody is overlain by ferruginous chert, iron- and manganese-rich sediments, agglomerate, mudstone and haematitic rocks including purple tuff with restricted permeability; (iii) a very fine grained, sometimes breccia-texture of the sulphide ores; (iv) occasional bedded structure in massive sulphide orebodies; (v) a zoned nature of the argillic alteration in and around the ore-bodies. The Kuroko deposits differ from Eastern Black Sea deposits in their Miocene age and significant content of gypsum. Gypsum is subordinate in the Turkish deposits, but barite is more common. Deposits are located at three main levels in the Dacitic Series. The first and lowermost level is represented by the Giresun-Akköy Cu-Zn deposit where mineralization occurs at the contact of quartz porphyry (porphyritic dacite) with overlying sandstone-tuff layers. Ore is localized at the brecciated upper parts of the dome-shaped dacites. Karılar-Karaerik Mines, 3–4 km south of Espiye, represent slightly higher levels of the same ore body, with 50–80 cm thick massive and lenticular pyrite orebodies with subordinate chalcopyrite. The ore is cut by 5–8 cm thick chalcopyrite veinlets associated with minor galena and sphalerite. Extensively altered host rocks are quartz-andesite and brecciated quartz-feldspar porphyry which are exposed along fault-zones.

A representative example of the second level of deposit is the Lahanos orebody, located at the base of a sequence of quartz-bearing ande-

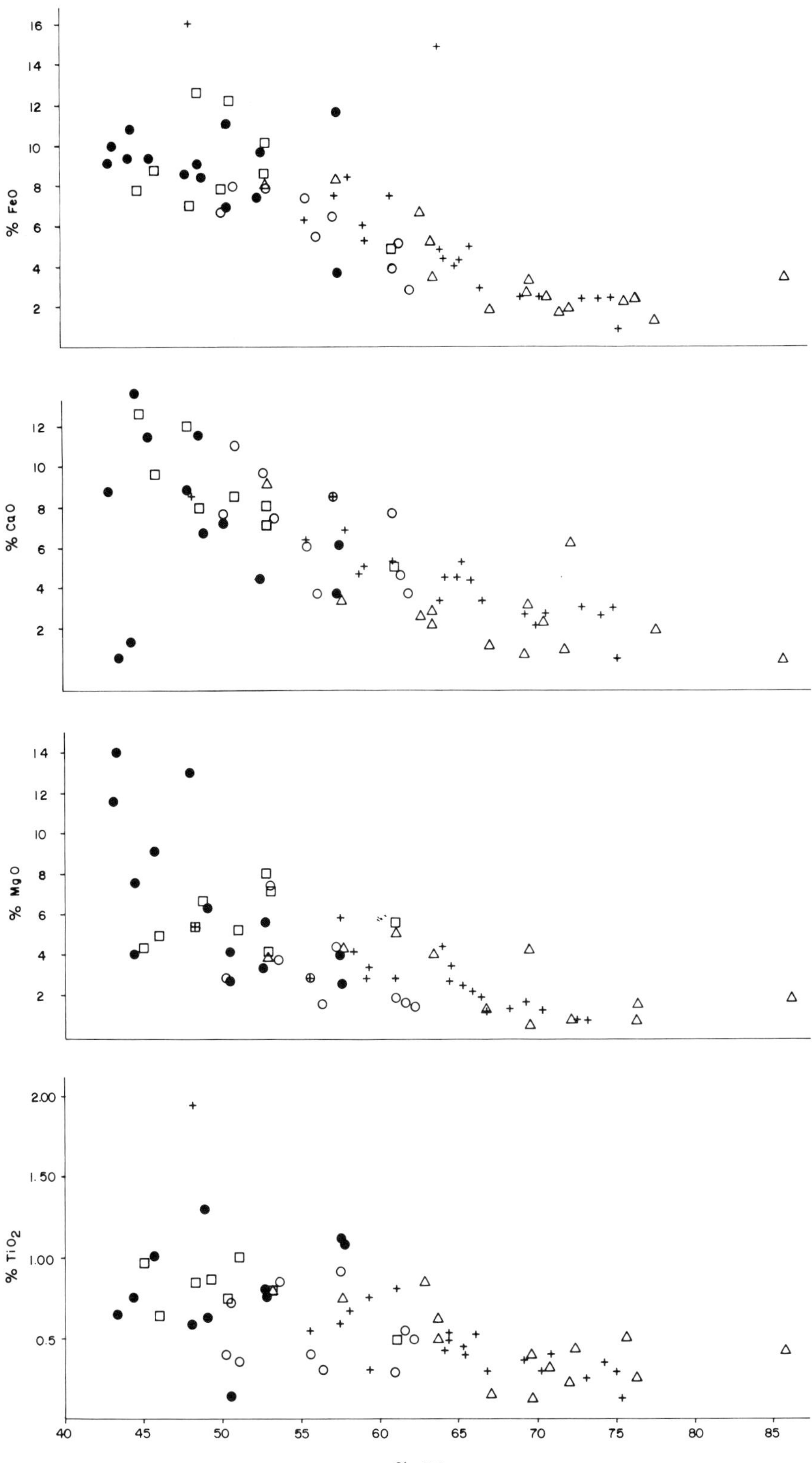

% FeO
% CaO
% MgO
% TiO2
% SiO2

sites, with calcareous mudstone and andesitic tuff intercalations which overlies a thin mudstone-agglomerate layer and is followed by fine-grained, columnar jointed dacitic-andesite lavas (Tuğal 1969).

The third ore level is represented by two major deposits: Çayeli-Madenköy and Murgul. Çayeli is a typical Kuroko-type deposit, with both massive- and stockwork-ore. The massive part of the orebody consists of Cu-Zn sulphides (black ore). The stockwork ore, composed of Cu-pyrite, shows a yellow ore character. Mineralization is in dacitic pumice and sandy tuffs overlain by purple tuff and basalts (Nebioğlu 1975; Altun 1978). The ore horizon at Murgul is overlain by dacite lava, purple and banded tuffs (Altun 1976). A 15 cm thick massive ore layer covering the stockwork ore was removed by mining.

In comparison with the third level deposits, the Kutlular, Kotarakdere and Baştımar deposits to the south of Sürmene, Trabzon, in the Gelincik Dome area are located at slightly higher levels of the dacitic tuff unit near basic flows but are not at a distinct fourth level as suggested earlier (Akıncı 1980).

ACKNOWLEDGEMENTS: The author wishes to express his gratitude to the staff of the Trabzon Branch Office of MTA and particularly to T. Y. Nebioğlu, Director, for cooperation during this study. The author also thanks Dr T. Engin for critical reading of the manuscript, also the editors. J. E. Dixon helped produce a substantially improved draft. Five tables of analyses and a stratigraphic correlation chart for the Eastern Pontides are available from the Geological Society of London as Supplementary Publication 18042 (free of charge).

References

ACAR, E. 1976. *Doğu Karadeniz Bölgesi, Tirebolu, Harşit Vadisi ve civarının jeolojisi, petrografik etüdü ve ekonomik jeolojisi hakkında.* Yük. Müh. Tezi, Ege Univ., İzmir.

AĞAR, U. 1977. *Demirözü (Bayburt) ve Köse (Kelkit) bölgesinin jeolojisi.* Ph.D. Thesis, Istanbul Univ., Istanbul.

AKINCI, Ö. T. 1969. Görele güneyindeki Koyunhamza ve Çömlekçi Dereleri arasında kalan sahanın (Çanakçı nahiyesi civarı) jeolojisi ve maden zuhurları. *MTA Derl. Rp. No.* 4875, Ankara.

—— 1974. *The Geology and mineralogy of copper lead zinc sulphide veins from Bulancak*, Turkey. Unpub. Ph.D. Thesis, Univ. Durham, 328 pp.

—— 1980. The major copper metallogenetic units and genetic igneous complexes in Turkey. In: S. JANKOVIC & R. H. SILLITOE (eds.) *European Copper Deposits*, 199–208.

—— & MILUTINOVIĆ, D. 1969. Giresun-Bulancak ve Espiye sahaları 1/10 000 ölçekli jeolojik etüd raporu: *MTA Derl. Rp. No.* 4602, Ankara.

ALTINLI, I. E. 1970. İkizdere granite complex. Istanbul Univ. *Fen Fak. Mec. Seri B*, **35,** Fasc. 3–4, Istanbul.

ALTUN, Y. 1976. Artvin-Murgul yataklarının jeolojik etüd raporu. *MTA Derl. Rp. No.* 6317, Ankara.

—— 1978. Geology of the Çayeli-Madenköy copper-zinc deposit and the problems related to mineralisation. *MTA. Bull.*, **89,** 10–23, Ankara.

ANTONOVIC, A. *et al.* 1968. Ordu-Gölköy sahası 1:25 000 ölçekli G 39–d2 ve d3 paftalarının jeoloji etüdü ve prospeksiyon raporu. *MTA Derl. Rp.* No. 4439, Ankara

ATAMAN, G., BUKET, E. & ÇAPAN, U. Z. 1975. Kuzey Anadolu fayı bir Paleo-Benioff zonu olabilir mi? *MTA Bull.*, **84,** 112–118, Ankara.

BAYDAR, O. *et al.* 1969. Yusufeli-Ögdem-Madenköy-Tortum Gölü ve Ersis arasındaki bölgenin jeolojisi. *MTA Derl. Rp. No.* 5202, Ankara.

CARMICHAEL, I. S. E., TURNER, F. J. & VERHOOGEN, J. 1974. *Igneous Petrology.* McGraw Hill, New York, N.Y. 739 pp.

ÇAĞATAY, M. N. 1977. *Development of geochemical exploration techniques for massive sulphide ore deposits Eastern Black Sea Region, Turkey.* Unpub. Ph.D. Thesis, Univ. London.

—— & BOYLE, D. R. 1980. Geology, geochemistry and hydrothermal alteration of the Madenköy massive-sulphide deposit, Eastern Black Sea Region, Turkey. *Proceedings of fifth IAGOD Symposium*, Stuttgart, Germany.

DOĞAN, R. 1980. *The granitic rocks and related molybdenite mineralisation of the Emeksan Area, Espiye, NE Turkey.* Unpubl. Ph.D. Thesis, Univ. Durham.

EĞIN, D. 1978. *Polymetallic sulphide ore deposits and associated volcanic rocks from the Harşit River area, NE Turkey.* Unpubl. Ph.D. Thesis, Univ. Durham, England.

——, HIRST, D. M. & PHILLIPS, R. 1979. The petrology and geochemistry of volcanic rocks from the northern Harşit River Area, Pontid volcanic province, northeast Turkey. *J. Volcanol. geothermal Res.* **6,** 105–123.

GEDIKOĞLU, A. 1970. *Etude Géologique de la Région De Gölköy (Province D'Ordu-Turquie).* Ph.D. Thesis, Univ. Grenoble, France.

—— 1978. *Harşit granit karmaşığı ve çevre kayaçları.* Doçentlik tezi, Karadeniz Teknik Univ., Trabzon, Turkey, 162 pp.

——, PELIN, S., ÖZSAYAR, T. 1979. The main lines of geotectonic development of the East Pontids in the Mesozoic era. *GEOCOME-1, first Geological Congress of the Middle East: MTA Publ.*, 555–580.

GATTINGER, T. E., *et al.* 1962. Explanatory text of the Geological Map of Turkey, Trabzon Sheet, 1:500 000 scale. MTA Publ., Ankara.

GILES, D. L. 1974. Geology and mineralisation of the

Ulutaş copper molybdenum prospect.: Mineral Exploration in two Areas, *UNDP Technical Report No. 6*, MTA Ankara.
GÜLEN, L. 1976. *Doğu Karadeniz volkaniklerinin petrografisi ve jeokimyası.* Dipl. Çalışması, Hacettepe Univ., Ankara.
GÜMRÜKÇÜ, A. 1974. Sürmene-Kutlular Maden sahasının jeolojisi ve bakır rezervi. *MTA Derl. Rp. No. 5455*, Ankara.
GÜVEN, I. H. 1977. Artvin-Şavşat-Madenköy ve çevresinin ayrıntılı jeolojik hartalanması hakkında rapor. *MTA Derl. Rep. No. 6342*, Ankara.
—— 1982. Artvin, Kuvarshan-Beşağıl ve Erenler çevresinin 1 : 5000 ölçekli ayrıntılı jeoloji raporu. MTA, Etüd Rp., Ankara.
IRVINE, T. N. & BARAGAR, W. R. A. 1971. A guide to the chemical classification of the common volcanic rocks. *Can. J. Earth. Sci.* **8,** 532–548.
KAHRAMAN, I. 1981. Giresun-Görele-Tirebolu yöresinin jeolojisi ve maden yatakları. *MTA Etüd Rp. No. 1785*, Ankara.
—— 1977. Molybdenum deposits of Çakıldağı area. *MTA Derl. Rp. No. 6375.*
KAMITANI, M. & AKINCI, Ö. T. 1979. Alpine granitoids and related tungsten-molybdenum deposits in Turkey. *Mining Geol.* (*Japan*), **29,** 341–350.
KETIN, I. 1966. Tectonic units Anatolia. *MTA Bull.*, **66,** 23–34, Ankara.
KRÄEFF, A. 1963. Geology and mineral deposits of the Hopa-Murgul region (western part of the province of Artvin, NE Turkey). *MTA Bull.* **60,** 37–60, Ankara.
KUNO, H. 1968. Differentiation of basaltic magmas. In: H. H. HESS & A. POLDERVAART (eds). *Basalts*, vol. 2, Interscience, New York, 623–688.
METAG 1972. Giresun-Akköy ve civarı arama raporu, Boztekke Köyü sahası arama raporu. DPT Müsteşarlığı, Ankara.
MOORE, W. J., MCKEE, E. H. & AKINCI, O. T. 1980. Chemistry and chronology of plutonic rocks in the Pontid Mountains, Northern Turkey. In: S. JANKOVIC & R. H. SILLITOE (eds.) *European Copper Deposits*, 209–216.
NEBIOĞLU, T. Y. 1975. Rize-Çayeli Madenköy-l sahasındaki bakır yatağına ait sonuç raporu. *MTA Derl. Rp. No. 5766*, Ankara.
NOCKOLDS, S. R. & ALLEN, R. 1953. The geochemistry of some igneous rock series. *Geochim. Cosmochim. Acta*, **4,** 105–142.
ÖZÜNEYLI, A. & OKABE, K. 1981. Sivas-Koyulhisar-Sisorta Kurşunluköy ve civarı kurşun-çinko-bakır madeni, ayrıntılı jeoloji ve sondaj çalışmaları raporu. MTA Etüd, Ankara.
ÖZSAYAR, T. 1971. *Palaontologie und geologie des gebiets ostlich Trabzon* (*Anatolien*). Ph.D. Thesis, Justu Liebig-Universität, Giessen, Germany.
——, PELIN, S. & GEDIKOĞLU, A. 1981. Doğu Pontidlerde Kretase. *Black Sea Techn. Univ. Earth Sci. Bull., Geology*, vol. 1, No. 2, 65–114, Trabzon.
PEARCE, J. A. & GALE, G. H. 1977. Identification of ore deposition environment from trace element geochemistry of associated igneous host rocks. In: *Volcanic Processes in Ore Genesis. Geol. Soc. London Spec. Publ.*, **7,** 14–24.
PEJATOVIC, S. 1979. Metallogeny of the Pontid-Type massive sulfide deposits. *MTA Publ. No. 177*, Ankara.
POLLAK, A. 1961. Die Lagerstatte Lahanos im Vilayet Giresun an der Türkischen Schwarmeerküste, *MTA Bull.*, **56,** 29–39, Ankara.
POPOV, P. 1981. Magmotectonic features of the Banat-Srednogorie Belt. *Geologica Balcanica*, **11,** 43–72, Sofia.
POPOVIC, R. 1975. Some of the structural and zonal distribution of nonferrous metals deposits in Eastern Pontids. *MTA Bull.* **85,** 1–16, Ankara.
RAMOVIC, M. 1966. Metalojeni ve petrolojide jeolojik zaman faktörünün önemi. *MTA Bull.* **67,** 25–37, Ankara.
SAWA, T. & ALTUN, Y. 1977. Doğu Karadeniz bölgesindeki tabakalı ve stockwork tipi bakır-kurşun-çinko yatakları. *MTA Etüd Rp. No. 1510.*
SAWAMURA, K. & YILMAZ, S. 1971. Geology and mineralisation at the Sürmene district, Trabzon, Turkey. *MTA Derl. Rp. No. 4597*, Ankara.
SCHULTZE-WESTRUM, H. H. 1960. Giresun-Trabzon vilayetlerinde yapılan harita çalışmaları hakkında rapor, MTA Archive.
—— 1961. Das geologische profil des Aksudere bei Giresun-ein beitrag zur geologie und lagerstattenkunde der ostpontischen erz und mineralprovinz, NE Anatolien. *MTA Bull.* **57,** 65–74, Ankara.
ŞATIR, F. & ERERIN, M. 1976. Artvin-Yukarımadenköy (Hot) ile yakın yöresindeki piritli Cu-Pb-Zn cevherleşmesine ait jeoloji raporu. *MTA Derl. Rp. No. 6542*, Ankara.
ŞENGÖR, A. M. C. & YILMAZ, Y. 1981. Tethyan evolution of Turkey: A plate tectonic approach. *Tectonophysics*, **75,** 181–241.
ŞIŞMAN, N. 1972. Artvin F47–a3IVa (Kokoletdere) paftası jeolojisi ve mineralizasyonu. *MTA Derl. Rp. No. 4869*, Ankara.
TANER, M. F. 1977. *Etude Géologique et pétrogaphique de la région de Güneyce-Ikizdere, située au sud de Rize* (*Pontids Orientales, Turquie*). Ph.D. Thesis, Université de Genève, Switzerland.
——, ZANINETTI, L. 1978. Etude paléontologique dans le Crétacé volcano-sédimentaire de Güneyce (Pontides Orientales, Turquie). *Riv. Itl. Paleont.* **84,** 187–198.
——, DELALOYE, M. & VUAGNAT M. 1979. On the geochronology by K-Ar method of the Rize pluton in the region of Güneyce-İkizdere, Eastern Pontids, Turkey. *Schweiz mineral. petrogr. Mitt.* **59,** 309–317.
TERLEMEZ, I. & YILMAZ, A. 1980. Ünye-Ordu-Koyulhisar-Reşadiye arasında kalan yörenin stratigrafisi. *Bull. geol. Soc. Turkey*, **23,** 179–191.
TOKEL, S. 1972. *Stratigraphical and volcanic history of the Gümüşhane region, NE Turkey.* Ph.D. Thesis, Univ. London.
—— 1977. Doğu Karadeniz bölgesinde Eosen yaşlı kalk-alkalen andezitler ve jeotektonizma. *Bull. geol. Soc. Turkey*, **20,** 49–54.
TUĞAL, H. T. 1969. *The pyritic sulphide deposits of*

the Lahanos Mine area, Eastern Black Sea region, Turkey. Ph.D. Thesis, Univ. Durham, England.

TJP. 1977. *Consolidated report on geological survey of Trabzon area, Northeastern Turkey*. MTA Inst. (Ankara)-Metal Mining Agency of Japan, 172 pp.

VICIL, M. 1975. Artvin-İrsahan madeni Cu-Pb-Zn yatağının 1:2000 lik jeoloji raporu. *MTA Etüd Rp. No. 1296*, Ankara.

VUJANOVIC, V. 1974. The basic mineralogic, paragenetic and genetic characteristics of the sulphide deposits exposed in the Eastern Black Sea coastal region (Turkey). *MTA Bull.*, **82,** 21–26, Ankara.

YILMAZ, Y. 1972. *Petrology and structure of the Gümüşhane granite and surrounding rocks, North-eastern Anatolia*. Ph.D. Thesis, Univ. London, 260 pp.

ZANKL, H. VON 1962. Magmatismus und bauplan des Ostpontischen Gebirges in querprofil des Harşit-Tales, NE Anatolien. *Geol. Rundsch.* **51,** 218–239.

This list includes references cited in *Supplementary Publication 18042*.

ÖMER T. AKINCI, MTA Enstitüsü, Ankara, Turkey.

HP/LT metamorphism and the structure of the Alanya Massif, Southern Turkey: an allochthonous composite tectonic sheet

A. I. Okay & N. Özgül

SUMMARY: The southern part of the Alanya Massif is made up of three superposed, relatively flat-lying, crystalline nappes (Alanya Nappes), which tectonically overlie the largely sedimentary lithologies of the Antalya Unit (= Antalya Complex). The predominantly Mesozoic continental margin type lithologies of the Antalya Unit outcrop beneath the Alanya Nappes in a large tectonic window.

The structurally lowest of the Alanya Nappes (Mahmutlar Nappe) consists of a heterogeneous series of shales, sandstones, dolomites, limestones and quartzites all metamorphosed under greenschist facies conditions. At least part of the sequence is Permian in age. The intermediate Nappe (Sugözü Nappe) is made up of garnet-micaschists which contain bands and lenses of eclogites and blueschist metabasites. Petrographic study has shown that rocks of the Sugözü Nappe have undergone initial HP/LT metamorphism followed by greenschist facies metamorphism. The greenschist overprint has destroyed most of the primary HP/LT mineral assemblages. The structurally highest of the Alanya Nappes (Yumrudağ Nappe) consists of a thick Permian carbonate sequence underlain by a relatively thin schist unit metamorphosed under lower greenschist facies conditions.

The absence of HP/LT metamorphism in the structurally lowest and highest of the Alanya Nappes, and the greenschist facies metamorphism which has affected all three nappes indicate that the initial HP/LT metamorphism of the Sugözü Nappe was succeeded by the tectonic stacking of the Alanya Nappes, greenschist facies metamorphism and deformation. In post-Maastrichtian times the Alanya Nappes, which were by then welded into one unit, were thrust over the sedimentary rocks of the Antalya Unit. The final thrusting of the Alanya Nappes and the underlying Antalya Unit over the Tauride carbonate platform occurred before the Middle Eocene.

The Alanya Massif is the name given to a large area of metamorphic rocks situated east of Antalya Bay in the Eastern Mediterranean (Blumenthal 1951). Here we show the Alanya Massif to be an allochthonous, composite tectonic slice overlying the predominantly Mesozoic sedimentary rocks of the Antalya Unit. We also describe for the first time eclogites and blueschists in the middle of the three crystalline nappes which make up the Alanya Massif. Interestingly this discovery follows the description of abundant detrital sodic amphibole in the beach and river sediments on the southern Turkish coast (Mange-Rajetzky 1981), some of which were certainly derived from these newly discovered HP/LT metamorphic rocks.

The Mediterranean coast of Turkey is flanked by the Tauride mountain chain which consists of superposed nappes stacked together during the Upper Cretaceous and Tertiary (Brunn *et al.* 1971; Özgül & Arpat 1973; Özgül 1976). A sedimentary sequence extending from Cambrian to Eocene, with the Mesozoic represented largely by platform carbonates, forms the Tauride autochthon in the Central Taurides. Around Antalya bay the autochthon is overthrust by the rocks of the Antalya Unit (Özgül 1976, equivalent to the Antalya Nappes of Lefèvre 1967, and the Antalya Complex of Robertson & Woodcock 1979), consisting predominantly of Mesozoic rocks of continental margin affinities. East of Antalya bay between Alanya and Anamur, rocks of the Antalya Unit are in turn tectonically overlain by the metamorphic rocks of the Alanya Massif (Fig. 1). Rocks belonging to the Antalya Unit outcrop beneath the Alanya metamorphics in a large tectonic window (Özgül & Arpat 1973; Özgül 1976), and in a narrow zone between the Alanya Massif and the autochthon (Fig. 1). In the east the Alanya Massif and the Tauride autochthon are overthrust by the rocks of the Aladağ Unit (Fig. 1, Özgül 1976, equivalent to the Hadim Nappe of Blumenthal 1944), which consist of a continuous shelf-type sedimentary sequence ranging in age from Upper Devonian to Upper Cretaceous.

The geology of the coastal part of Alanya Massif

We have mapped the south-western coastal part of the Alanya Massif between Demirtaş and west of Alanya at a scale of 1:25 000. In this area three superimposed nappes (Alanya Nappes) have been differentiated within the crystalline Alanya Massif (Figs 2 and 3). The differentiation of the nappes is essentially based

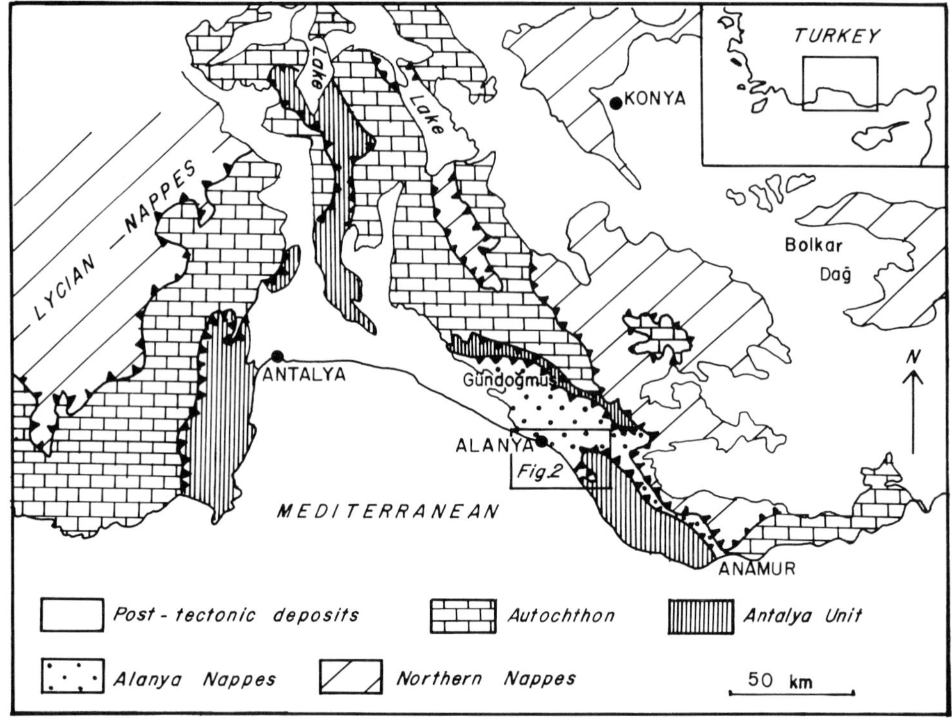

FIG. 1. Schematic tectonic map of the Western and Central Taurides showing the major tectonic units and the location of towns mentioned in the text. The location of Fig. 2 is shown by the rectangular box.

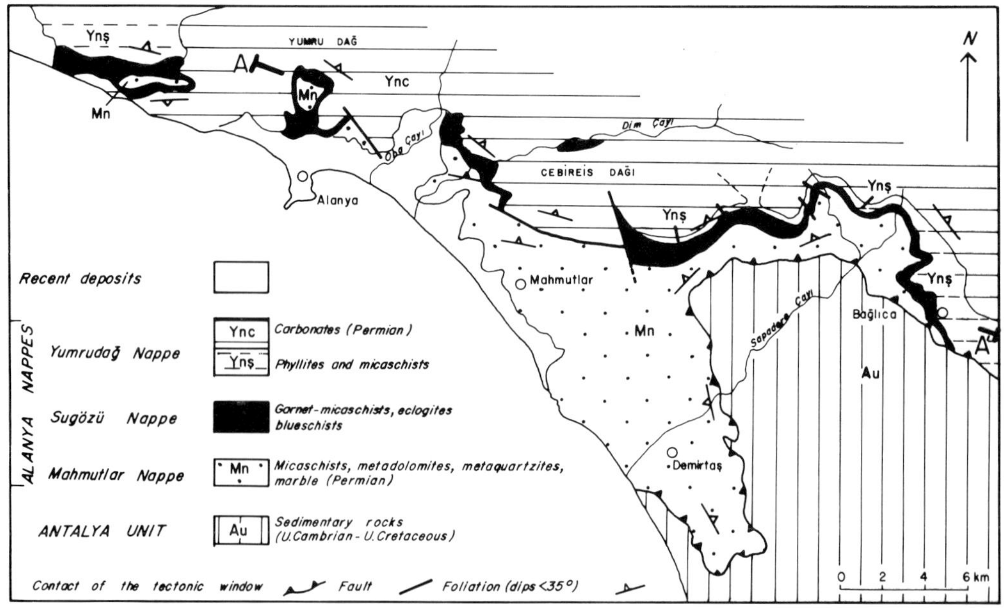

FIG. 2. Simplified tectonic map of the Alanya region showing the Alanya Nappes and part of the Alanya tectonic window.

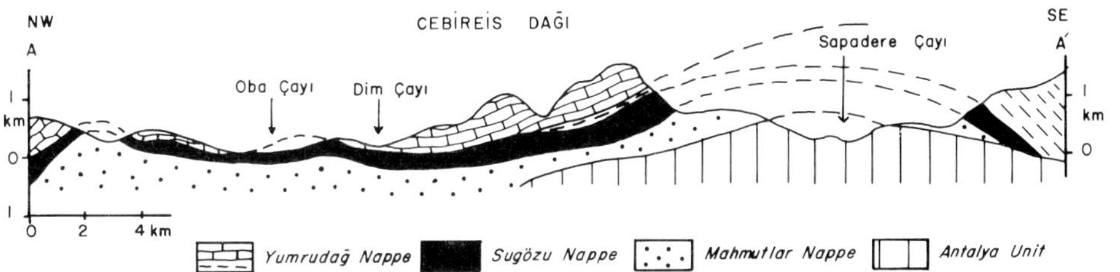

FIG. 3. Cross-section illustrating the large scale structure in the Alanya region.

on the presence of a HP/LT metamorphic slice at a structurally intermediate position in the Alanya Massif. These three Alanya Nappes tectonically overlie the predominantly sedimentary rocks belonging to the Antalya Unit, which crop out in the Alanya tectonic window.

The Alanya Nappes have a gently (<35°) undulating, but largely northerly, regional dip, such that the coastal area presents an oblique cross-section through the Alanya Nappes (Fig. 2). The formal definitions of the stratigraphy in the Alanya Nappes and especially of the Alanya tectonic window will be given elsewhere and are only summarized here.

Yumrudağ Nappe

The Yumrudağ Nappe is the structurally highest of the Alanya Nappes and constitutes the bulk of the Alanya Massif. It consists of schists overlain by a thick sequence of recrystallized limestone. The passage from the schists to the overlying carbonates is gradational with schist and carbonate bands several metres thick at the contact. Pelites, psammites, calc-schists, meta-dolomites and thin recyrstallized limestone bands are the major lithologies of the schist unit. The thickness of the schist unit is very variable; in some areas it is completely cut out and the carbonates rest directly on the garnet-mica-schists of the Sugözü Nappe. In the eastern part of the area, where the basal schist unit reaches its maximum thickness, it is 1200 m thick. The meta-clastic rocks show a well developed planar cleavage; the typical mineral assemblage in the pelites is quartz + albite + phengite + chlorite ± biotite. Some of the less well recrystallized psammites contain abundant clastic microcline.

The overlying carbonate unit forms the thick carapace of the Alanya Massif (Blumenthal 1951). Several hundred metres of generally flat-lying, grey, massively-bedded, monotonous recrystallized limestones and dolomites are the characteristic lithology. There are occasional calc-schist bands and local meta-bauxite horizons. The deceptively flat-lying structure of the carbonate unit hides strong isoclinal folding prominent in the lower levels; the real thickness of the carbonate unit must be much less than the apparent thickness (>1000 m).

Since the work of Blumenthal (1951), it is known from the well preserved *Mizzia* seen on the weathered carbonate surfaces that the carbonates are of Upper Permian age. A recently discovered exposure in the Alanya Massif shows the massive Permian carbonates to pass through a meta-bauxite horizon to thinly bedded recrystallized limestones with abundant lamellibranchs. Surprisingly similar but better preserved sequences in the Alanya tectonic window suggest that the thinly bedded limestones are Lower Triassic (Scythian) in age. Younger rocks are not found in the Yumrudağ Nappe.

Sugözü Nappe

This is a thin HP/LT metamorphic unit occupying a structurally intermediate position between the Yumrudağ Nappe above and Mahmutlar Nappe below. The dominant rock type in the Sugözü Nappe is a well-foliated, silvery-grey mica-schist with conspicuous reddish-black garnets up to 6 mm in size. These very distinctive garnet-mica-schists were also noted by Blumenthal (1951) and Peyronnet (1971). Eclogites and blueschist metabasites occur within these garnet-mica-schists either as rare intercalated bands 20–30 cm thick or as boudinaged lenses which can range up to two metres across. These basic lithologies make up less than 5% of the Sugözü Nappe. A still rarer lithology is meta-dolomite. The garnet-mica-schists were probably originally siliceous shales. Petrological studies, which are detailed in the next section, indicate that these rocks experienced a *plurifacial* metamorphism ranging from eclogite to greenschist facies.

The apparent thickness of the Sugözü Nappe ranges from 100 to 800 metres and the maximum outcrop width is about one kilometre.

Despite its relative thinness the Sugözü Nappe shows a remarkable lateral continuity. It is traceable for over 40 kilometres along the coast, beneath and south of the Yumrudağ Nappe and is cut out in only one place by a steeply dipping normal fault (Fig. 2). In the east the Sugözü Nappe is truncated by the Alanya tectonic window, whereas in the west it disappears beneath recent sediments (Fig. 2). North of the Yumrudağ Nappe it reappears in a small window in a deeply cut gorge (Dim Çayı), which gives a minimum N-S dimension of four kilometres.

Along all the observed contacts, the foliation in the garnet-mica-schists is parallel to the foliation in the overlying and underlying schists. The demarcation of the Sugözü Nappe is made in the field solely on the presence of conspicuous garnets in this unit; garnets are lacking in the Yumrudağ Nappe and in the directly underlying schists of the Mahmutlar Nappe. The conformable contacts between these three nappes, which has prevented their earlier discovery, is due to the last stage of metamorphism and deformation which affected all three nappes. However, on a larger scale there is discordance between the generally E-W trending garnet-mica-schists of the Sugözü Nappe and the NW-SE trending lithologies of the underlying Mahmutlar Nappe (Fig. 2). Near the contact, the Mahmutlar Nappe schistosity becomes concordant with that in the garnet-mica-schists of the Sugözü Nappe.

Mahmutlar Nappe

This structurally lowest of the Alanya Nappes consists predominantly of pelites, psammites, meta-dolomites, recrystallized limestones and meta-quartzites. Meta-dolerites and recrystallized radiolarian cherts occur in very minor amounts. The thickness of the Mahmutlar Nappe is difficult to estimate as its lower contact is under the sea. However, it has a minimum apparent thickness of 700–800 metres. In the east it is truncated along with the Sugözü Nappe by the Alanya tectonic window; west of Alanya it is largely covered by the overlying nappes and crops out only in small windows (Fig. 2).

The metamorphic grade in the Mahmutlar Nappe continuously increases from the north-east towards the south-west. In the north-east the clastic quartz and feldspar grains in coarse sandstones are still recognizable whereas near the coast pelites contain biotite and newly-formed small garnets along with oligoclase, quartz, phengite and chlorite. A well developed schistosity is ubiquitous in the metaclastic rocks of the Mahmutlar Nappe. The presence of *Mizzia* in some recrystallized limestones north of Demirtaş village indicates that at least part of the Mahmutlar Nappe is of Permian age.

Alanya tectonic window

Sedimentary and low-grade metasedimentary rocks belonging to the Antalya Unit outcrop in an immense tectonic window underneath the Alanya Nappes extending for 75 km along the coast from Demirtaş to Anamur (Fig. 1). Its south-western margin is under the sea whereas along its north-eastern flank the Alanya Nappes tectonically overlie the sediments of the Antalya Unit. In the north-western part of the window a thin sequence (150 m) of micaceous Upper Cambrian-Lower Ordovician sandstones and siltstones with rare, red nodular limestone intercalations is overlain by Upper Permian neritic carbonates. An ortho-quartzite horizon between the Lower Palaeozoic sandstones and Permian carbonates is interpreted as the transgressive base of the Permian. The thickly bedded, grey, neritic, fossiliferous Permian limestones, 600–700 m thick, are overlain by variegated, thinly bedded Scythian limestones and marls with abundant lamellibranchs. This 200 m thick sequence passes upwards into red radiolarian cherts and intercalated Halobia-bearing pelagic limestones. In this 30 m thick pelagic sequence there are rare intercalations of green acidic tuffs and alkali basalts. The radiolarian cherts and pelagic limestones are overlain by 600 m of greyish brown, carbonaceous Carnian sandstones with abundant plant debris. A distinctive feature of these sandstones is the presence, especially in the upper levels, of olistoliths mostly of Permian limestone, Ordovician sandstone and Cambrian red nodular limestone. These olistoliths may reach several kilometres in size.

Jurassic and Cretaceous rocks occur in only few areas in the Alanya tectonic window. They are faulted against the much more abundant Triassic rocks and comprise a 150 m sequence of multicoloured radiolarian cherts and pelagic limestones. At the top of the sequence there is a Maastrichtian shale unit with olistoliths ranging in age from Cambrian to Cretaceous.

Low-angle thrusting is common within the window so that the stratigraphic sequences are repeated several times. The thrusts trend in a north-south direction and are truncated by the overlying Alanya Nappes. On a finer scale the Triassic sandstones and shales are strongly and complexly deformed. Metamorphic recrystallization and penetrative cleavage, which is lacking in the north-eastern part of the window, develops gradually towards the coast. Along

TABLE 1. *Measured modes of typical rocks from the Sugözü Nappe*

	Metabasite			Micaschist	
	AL497B	AL497A	AL426	AL431	AL512
Sodic pyroxene	62.9	26.4	—	—	—
Garnet	11.0	14.0	9.5	5.5	9.0
Sodic amphibole	3.5	34.9	10.7	0.2i	—
Barroisite	—	1.8	17.8	—	—
Albite	—	—	23.1	22.5	25.0
Chlorite	0.5	—	9.6	9.4	1.9
Ankerite	—	2.9	9.3	—	0.2
Calcite	—	—	3.2	—	—
White mica	15.0	7.9	8.9	18.8	19.2
Biotite	—	—	1.7	4.6	tr.
Quartz	0.6	2.5	1.2	36.8	43.7
Clinozoisite	0.1	0.7	0.2	tr.i	—
Magnetite	1.6	0.2	—	—	—
Pyrite	1.6	2.1	—	—	—
Pyrrhotite	—	—	0.9	—	—
Ilmenite	—	—	—	2.0	0.9
Rutile	3.2	6.2	0.1	tri.i	tri.i
Sphene	—	—	3.2	—	—
Apatite	—	0.4	0.4	0.2	0.1
Tourmaline	—	—	0.2	tr.	—
	100.0	100.0	100.0	100.0	100.0

tr. <0.1
i inclusions in garnet

the coast the shales are converted to quartz-chlorite-sericite schists with an irregular penetrative cleavage. However, the grade of metamorphism in the window is less than in the surrounding schists of the Alanya Nappes.

Petrology of HP/LT metamorphic rocks of the Sugözü Nappe

Eclogites and blueschist metabasites occur as dense, massive, dark bands or lenses within the silvery grey, well foliated garnet-mica-schists. Petrographically there is a complete and continuous range from essentially unaltered eclogite through blueschist metabasite to barroisitic amphibolite. The earliest eclogitic and blueschist stages are best preserved in the eastern part of the Sugözü Nappe, especially around the village of Bağlica (Fig. 2). West of Alanya most of the traces of the eclogitic stage are obliterated.

Unaltered *eclogites* are rare. In hand specimen 2–3 mm large reddish-black garnets are set in a yellowish-green fine-grained matrix. Under the microscope large garnet poikiloblasts lie in a sheared groundmass composed dominantly of pale green subhedral grains of sodic pyroxene. White mica, pale blue sodic amphibole with narrow rims of bluish-green barroisite, and quartz, rutile and pyrite occur in minor amounts (Table 1). Garnet crystals are full of inclusions of sodic amphibole, clinozoisite, white mica and rutile.

Blueschist metabasites are more common than eclogites. They occur as massive, hard, banded, bluish-black rocks with conspicuous reddish garnets several millimetres in size. Rotated and deformed garnet poikiloblasts up to 5 mm in diameter are set in a groundmass of sodic amphibole, sodic pyroxene and white mica. Rutile, pyrite, quartz, altered Fe-carbonate, and clinozoisite are found in smaller amounts scattered in the groundmass (Table 1). There is a graduation from omphacite + garnet eclogites to blueschist metabasites with sodic amphibole and garnet but with no sodic pyroxene. In many blueschist metabasites sodic pyroxene is preserved in patches or in bands; there are, however, no direct replacement textures of sodic pyroxene by sodic amphibole. Pale blue sodic amphibole shows ubiquitous partial replacement by greenish blue barroisite; however, in blueschist metabasites barroisite never makes up more than 30% of the amphibole present. Garnet porphyroblasts have abundant inclusions of sodic amphibole, clinozoisite, quartz, carbonate and rutile. Quartz, calcite and white mica are concentrated in the pressure shadows and around the rims of the garnet porphyroblasts.

TABLE 2. *Stages of mineral development in metabasic rocks of the Sugözü Nappe*

Mineral \ Facies	Eclogite	Blueschist	Greenschist
Garnet	———	———	———
Na-pyroxene	———	– – –	
Na-amphibole	———	———	– – –
Barroisite			———
Albite			———
Chlorite			———
Epidote	– – –	– – –	– – –
Paragonite	———	———	———
Phengite	———	———	———
Biotite			– – –
Rutile	———	——— – –	–
Sphene		–	———
Quartz	– – –	– – –	– – –

Barroisite amphibolites represent the last stage in the retrogressive development of eclogites. They are less dense than the blueschist metabasites and eclogites, and have a pale-green colour due to the presence of abundant chlorite and albite in the rock. Red garnets are still conspicuous on weathered surfaces. These large garnet poikiloblasts are associated with sub-idioblastic barroisite, which has largely replaced the pale blue sodic amphibole, and with albite poikiloblasts, quartz, white mica and chlorite. Sphene replacing rutile, altered Fe-carbonate, clinozoisite, biotite and pyrite may occur in minor amounts (Table 1).

The mineralogy and texture of the *garnet-mica-schists* are, like their field characteristics, rather monotonous. Poikilitic, often rotated, garnet porphyroblasts up to 1 cm across are associated with helicitic albite poikiloblasts, pale green chlorite, phengite and quartz. These five minerals make up over 90% of the mode (Table 1). Small amounts of ilmenite, biotite, graphite, clinozoisite and altered Fe-carbonate may be present. Rutile (occasionally replaced by sphene), apatite and tourmaline are ubiquitous accessory minerals. The most interesting feature of the garnet-mica-schists is, however, the presence of sodic amphibole, which occurs in minor amounts along with quartz, white mica, clinozoisite and rutile as small inclusions in garnet porphyroblasts or even more rarely in large Fe-carbonate crystals.

Sodic amphibole inclusions in garnets and the interlayering of garnet-mica-schists and metabasic rocks indicate that the garnet-mica-schists have undergone an initial HP/LT metamorphism as have the basic lithologies. The initial HP/LT mineral assemblage of the garnet-mica-schists probably included jadeite instead of albite.

The evolution of mineral assemblages in the metabasic rocks is shown in Table 2. Geological evidence (cf. Table 4) indicates that HP/LT metamorphism and greenschist facies metamorphism represent two distinct events rather than a single continuous event. In this respect rocks of the Sugözü Nappe are similar to the HP/LT metamorphic rocks from the Western Alps and especially from the Sesia-Lanzo Zone (Frey *et al.* 1974). As in the Sugözü Nappe, rocks of the Sesia-Lanzo Zone are largely quartz-rich mica-schists representing metamorphosed acid magmatic rocks and pre-Alpine schists (Compagnoni *et al.* 1977). On the other hand, rocks of the Sugözü Nappe show a strong contrast in terms of the mineral assemblage, lithology and tectonic setting to the prograde blueschists of north-west Turkey, which occur 300 km north of Alanya Nappes (Okay 1982 and this volume).

Mineral chemistry

Fifty-five complete mineral microprobe analyses have been made on three samples: one eclogite (497B), one barroisite amphibolite with relict sodic amphibole (426) and one garnet-mica-schist (431). The measured modes of these specimens are given in Table 1; all three samples are kept in the Harker Collection in the Department of Earth Sciences, Cambridge.

Mineral compositions were determined for twelve elements (Na, Ca, K, Fe, Mg, Mn, V, Cr, Ti, Al, Si, P) using an electron-probe microanalyser with a Harwell Si(Li) detector and pulse processor (Statham 1976). The correction procedures are given by Sweatman & Long (1969). The method for estimating ferric ion in sodic pyroxene and in sodic amphibole is outlined in Okay (1978, 1980).

Garnet

Idioblastic garnet porphyroblasts are ubiquitous in metabasic rocks and mica-schists. Nine garnet compositions from the three samples are plotted in Fig. 4 in terms of pyrope, grossular and almandine + spessartine end-members and two of the analyses are given in Table 3. Analysed garnets have very low spessartine (<2%) and andradite (<3%) components, and show a slight zoning involving an increase in almandine component towards the rim at the expense of the grossular component. The pyrope contents range from 4% to 9%; garnets from the metabasic rocks are slightly

TABLE 3. *Representative mineral analyses*

	Garnet			Sodic pyroxene		Amphibole		Paragonite		Phengite		Epidote	Chlorite
	AL497B		AL431	AL497B		AL426		AL426	AL497B	AL426	AL431	AL497B	AL426
	core	rim											
SiO_2	37.72	38.60	38.74	57.23	56.13	57.22	50.13	46.82	46.84	48.57	51.05	39.41	27.27
Al_2O_3	21.05	21.47	21.23	14.60	12.44	11.31	10.18	38.37	38.95	27.42	27.40	29.98	20.20
TiO_2	0.17	0.00	0.17	0.00	0.00	0.00	0.20	0.00	0.13	0.33	0.29	0.00	0.00
FeO	27.26	30.56	28.26	10.06	8.87	14.06	13.62	0.68	0.72	2.57	2.83	5.98	21.59
MnO	0.48	0.28	0.91	0.11	0.00	0.00	0.00	0.00	0.00	0.00	0.00	0.19	0.00
MgO	1.46	2.09	1.07	1.69	4.40	7.03	11.63	0.00	0.00	2.82	2.46	0.00	18.48
CaO	11.05	8.51	11.05	4.32	8.39	0.45	8.31	0.42	0.14	0.22	0.00	23.61	0.15
Na_2O	0.00	0.00	0.00	12.06	9.56	6.64	3.31	6.49	7.28	0.42	0.00	0.00	0.00
K_2O	0.00	0.00	0.00	0.00	0.00	0.00	0.18	1.00	0.81	9.55	10.36	0.00	0.00
Total	99.69	101.51	101.93	100.07	99.79	96.71	97.56	93.78	94.87	91.90	94.39	99.17	87.69
Number of cations per													
	12 cations			4 cations		23 oxygens		22 oxygens		22 oxygens		12.50	28.0
Si	3.00	3.02	3.03	2.01	2.00	8.04	7.22	6.08	6.02	6.71	6.86	3.00	5.62
Al^4	0.00	0.00	0.00	0.00	0.00	0.00	0.78	1.92	1.98	1.29	1.14	0.00	2.38
Al^6	1.98	1.98	1.96	0.61	0.52	1.88	0.95	3.95	3.93	3.18	3.20	2.68	2.52
Ti	0.01	0.00	0.01	0.00	0.00	0.00	0.02	0.00	0.01	0.03	0.03	0.00	0.00
Fe^{3+}	0.01	0.02	0.03	0.21	0.14	0.00	1.64	0.07	0.08	0.30	0.32	0.38	3.72
Fe^{2+}	1.85	1.99	1.85	0.09	0.12	1.65						0.00	
Mn	0.03	0.02	0.06	0.00	0.00	0.00	0.00	0.00	0.00	0.00	0.00	0.01	0.00
Mg	0.18	0.25	0.13	0.09	0.23	1.47	2.50	0.00	0.00	0.58	0.49	0.00	5.67
				1.00	1.01	5.00	5.11	4.02	4.02	4.09	4.04		
Ca	0.94	0.72	0.93	0.17	0.32	0.07	1.28	0.06	0.02	0.03	0.00	1.92	0.03
Na	0.00	0.00	0.00	0.82	0.66	1.81	0.92	1.63	1.81	0.11	0.00	0.00	0.00
K	0.00	0.00	0.00	0.00	0.00	0.00	0.04	0.17	0.13	1.68	1.78	0.00	0.00
				0.99	0.98	1.88	2.24	1.86	1.96	1.82	1.78	7.99	19.94
Alm	61.7	66.8	62.3	Jd 61	52								
Spess	1.0	0.7	2.0	Ac 21	14								
Pyr	6.0	8.4	4.4	Au 18	34								
Gross	31.3	24.1	31.3										

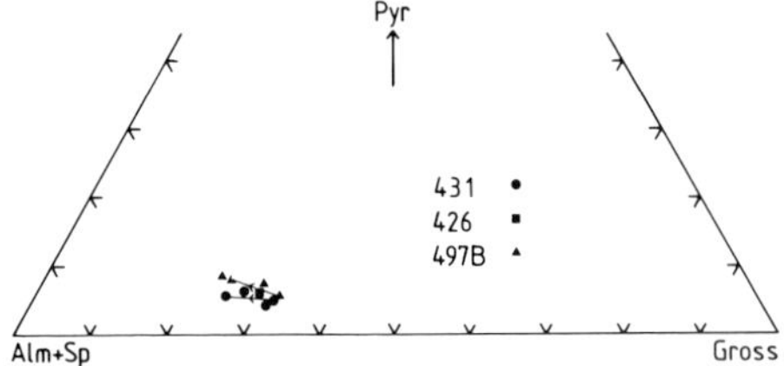

FIG. 4. Garnet compositions from the Sugözü Nappe plotted on the (almandine + spessartine)-pyrope-grossular diagram; all spessartine contents are below 2%, arrows point from core to rim compositions.

richer in pyrope component (5–9%) than garnets from the mica-schists (4–7%). Garnets from the Sugözü Nappe are similar in composition to the garnets from Type C eclogites and blueschists (Coleman *et al.* 1965).

Sodic pyroxene

Sodic pyroxenes form zoned subhedral pale yellowish-green grains or bundles of prismatic crystals up to 1 mm long. Analysed sodic pyroxenes from the specimen 497B are omphacites and aegirine-jadeites (Table 3). Their acmite contents are below 30% and their jadeite contents range up to 70%.

Amphibole

The primary amphibole is a slightly zoned colourless to pale blue sodic amphibole. Analysed amphiboles from specimen 426 plot near the ferroglaucophane/glaucophane join in the Miyashiro diagram and have very small amounts of ferric iron (Table 3, Fig. 5). They are very similar in composition to the detrital sodic amphiboles from beach sediments on the eastern part of the Alanya coast (Mange-Rajetzky 1981, Fig. 5).

In most specimens sodic amphibole has distinct and sharply bounded rims of dark bluish-green barroisite; there is no continuous compositional range between the two amphiboles. Barroisite is characterized by a small A-site occupancy (0.0–0.3 per formula unit), tetrahedral aluminium (0.6–0.9) and a Na/Na + Ca ratio of 0.3–0.5 (Fig. 5, Table 2).

Sheet silicates

Microprobe analyses have revealed that paragonite and phengite are both present in rocks of basic composition whereas only phengite is found in the garnet-mica-schists. Paragonite shows up to 12% substitution by muscovite and up to 3% by calcium mica (Table 3). Phengite coexisting with paragonite contains 6% of a paragonite component (Table 3); phengites from garnet-mica-schists on the other hand do not show any paragonite substitution.

Chlorite is largely retrogressive after garnet in garnet-mica-schists and blueschist metabasites. In barroisite amphibolites, however, it forms an important part of the primary mineral assemblage. Chlorites analysed from the barroisite amphibolite and from the garnet-mica-schist have restricted Fe/Fe + Mg ratios (0.4–0.5) and a uniform Si/Si + Al ratio (Table 3).

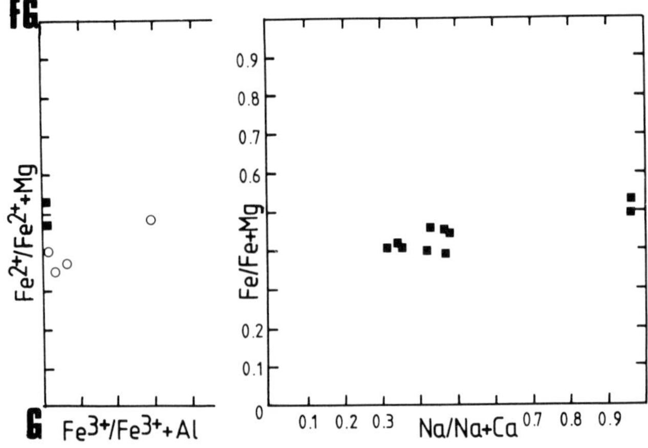

FIG. 5. Amphibole compositions from the Sugözü Nappe.
(a) Sodic amphibole compositions plotted on part of the Miyashiro diagram. Open circles indicate sodic amphibole compositions from the beach and river sediments from the eastern province, Alanya (Mange-Rajetzky 1981). G = glaucophane, FG = ferroglaucophane.
(b) Barroisite and sodic amphibole compositions from the sample AL426 plotted in terms of their Na/Na + Ca and Fe/Fe + Mg ratios. The diagram illustrates the compositional break between the sodic amphibole and barroisite.

TABLE 4. *Structural and metamorphic events in the Alanya area*

1.	Shallow carbonate platform conditions in the Alanya Nappes and Antalya Unit	Upper Permian
2.	Rifting of the carbonate platform	Anisian
3.	HP/LT metamorphism of the Sugözü Nappe	
4.	Tectonic stacking of the Alanya Nappes	
5.	Greenschist facies metamorphism of the Alanya Nappes and probably part of the Antalya Unit	pre-Maastrichtian
6.	Thrusting of the Alanya Nappes as a single sheet over the Antalya Unit	post-Maastrichtian
7.	Alanya Nappes and the underlying Antalya Unit thrust over the carbonate platform to the north	Palaeocene-Lower Eocene

Other minerals

Albite is present in barroisite amphibolites and in garnet-mica-schists. The maximum anorthite content in five analysed albites from two specimens is 3 mol %.

Epidotes, with pistacite contents of 12–17 mol % occur in minor amounts in metabasic rocks (Table 3).

Metamorphic and tectonic history of the Alanya area

The geological evolution of the Alanya area involves a complex sequence of structural and metamorphic events, which are summarised in Table 4. Petrographic and structural evidence indicates that only the Sugözü Nappe has suffered HP/LT metamorphism, whereas greenschist facies metamorphism and associated deformation have affected all three nappes.

The initial HP/LT metamorphism of the Sugözü Nappe was followed by the tectonic stacking of the Alanya Nappes, and greenschist facies metamorphism and associated deformation, which probably also resulted in the folding of the thrust surfaces of the Alanya Nappes (Fig. 2). The low-grade metamorphism of part of the Antalya Unit is also probably related to this late greenschist event. This was succeeded by the thrusting of Alanya Nappes, which were by then welded into one sheet, over the Antalya Unit. A post-Maastrichtian age for the thrusting is given by the youngest sediments of that age in the Alanya tectonic window. Similar reasoning suggests that metamorphism in the Alanya Nappes is probably pre-Maastrichtian. Metamorphosed Lower Triassic rocks of the Yumrudağ Nappe give a minimum age for the greenschist facies metamorphism. The lack of any recorded Hercynian or older metamorphism in the Taurides suggests that the metamorphism of the Alanya Nappes is Alpine in age.

In areas of similar plurifacial metamorphism, like the Western Alps or the Cyclades, Greece, an early HP/LT metamorphism was followed 20–40 Ma later by a greenschist facies metamorphism (e.g. Frey *et al.* 1974; Altherr *et al.* 1979). In the Western Alps the initial HP/LT metamorphism seems to have predated the major nappe movements. Assuming a similar pattern for the Alanya Nappes, and a Lower Maastrichtian age for the greenschist facies metamorphism, the initial HP/LT metamorphism might be Turonian in age or older.

In the north, rocks of the Antalya Unit tectonically overlie Cretaceous and Lower Eocene carbonates and clastics of the Tauride autochthon. On the other hand, Lutetian limestones are transgressive with a basal conglomerate over the Alanya Nappes in the region of Maha Yaylasi. Thus the final emplacement of the Antalya Unit and Alanya Nappes can be constrained to the Lower-Middle Eocene interval. However, emplacement had probably started in the Palaeocene as flysch-type clastics of Upper Palaeocene-Lower Eocene age in the Tauride autochthon to the north of Alanya Nappes contain abundant schist and marble fragments derived from the Alanya Nappes.

Regional implications

Lying structurally between the Tauride autochthon and the Alanya Nappes are rocks belonging to the Antalya Unit. The original place of deposition and thus the direction of tectonic transport of the Antalya Unit has been a subject of controversy. We believe that regional considerations (Robertson & Woodcock 1980; Şengör & Yilmaz 1981) and recent detailed sedimentological and structural work (e.g. Robertson & Woodcock 1981) indicate that the Mesozoic rocks of the Antalya Unit were deposited on a passive continental margin situated south of the Tauride carbonate platform as originally propounded by Dumont *et al.* (1972).

Our palaeogeographic reconstructions start from this premise.

During the Palaeozoic and the earliest Triassic, the Taurides were part of Gondwanaland. In the northern parts of Gondwanaland the Upper Permian is characterized by a far-reaching marine transgression and the establishment of shallow marine conditions over large areas with the deposition of a thick sequence of neritic limestones. At the end of the Permian part of the carbonate platform was sub-aerially exposed and subject to lateritization. Deposition in a tidal environment during the Scythian is indicated by the very widespread variegated thinly-bedded limestones. The initial rifting and continental fragmentation of the northern part of Gondwanaland began during the late Anisian and led to a horst–graben type topography (Marcoux 1978). There were narrow fault-bounded basins separted by uplifted areas. In the pelagic basins radiolarian cherts, pelagic limestones, carbonate turbidites and turbiditic sandstones were deposited. The presence of granitic rock fragments, and olistoliths of Permian limestones and Ordovician sandstones in the Triassic rocks of the Alanya tectonic window, indicate that some of the uplifted areas were subareally exposed and eroded. The structural position of the Alanya Nappes over the Antalya Unit, and the apparent absence of post-Scythian sediments in the Alanya Nappes suggest that the Yumrudağ and Mahmutlar Nappes were originally part of such an uplifted continental area separated from the major Tauride carbonate platform by a rifted pelagic basin, where rocks of the Antalya Unit were being deposited.

Although there are no ophiolitic rocks in the Alanya tectonic window, it has been suggested that during the Jurassic and Cretaceous some of these basins developed oceanic crust (e.g. Şengör and Yılmaz 1981). The well known ophiolite complexes of the Eastern Mediterranean—Troodos, Kızıldağ and Antalya ophiolites—are regarded as Late Cretaceous fragments of an oceanic basin south of the Tauride carbonate platform (Robertson & Woodcock 1980). In the Alanya region the position of this southern branch of Neotethys is not clear. At the north-western margin of the Alanya Nappes near the town of Güzelsu, there are slivers of serpentinite, spilite, chert and pelagic shales (Monod 1978). Around the town of Gündoğmuş these slivers are partly metamorphosed in the blueschist facies, and are closely associated with the rocks of Antalya Unit (Şengün *et al.* 1978, and our own observations). The presence of these poorly known ophiolitic rocks may indicate that oceanic crust was generated during the Mesozoic between an Alanya microcontinent, represented by the Mahmutlar and Yumrudağ Nappes, and the Tauride carbonate platform.

The continental basement to the Antalya Unit is not observed anywhere in the Alanya area. The Sugözü Nappe could represent part of the attenuated continental basement of the Antalya Unit, which was subducted by cover-stripping during the Cretaceous (A.M.C. Şengör, personal communication). However, it remains a perplexing question as to how rocks of the Sugözü Nappe were uplifted and then preserved as a coherent, thin metamorphic slice.

ACKNOWLEDGEMENTS: We thank S. O. Agrell and Department of Earth Sciences, Cambridge for the microprobe analyses. R. Colston & C. Hampton read and corrected the manuscript, and A. Çağatay determined the ore minerals in the analysed samples. The work was carried out while the authors were in M.T.A. (Ankara), we are grateful for the facilities provided in the field.

References

ALTHERR, R., SCHLIESTEDT, M., OKRUSCH, M., SEIDEL, E., KREUZER, H., HARRE, W., LENZ, H., WENDT, I. & WAGNER, G. A. 1979. Geochronology of high-pressure rocks on Sifnos (Cyclades, Greece). *Contrib. Mineral. Petrol.* **70**, 245–55.

BLUMENTHAL, M. M. 1944. Schichtfolge und Bau der Taurusketten in Hinterland von Dozkir (Vilayet Konya). *Revue Fac. Science Univ. Istanbul, série B*, **9**, 95–125.

—— 1951. *Recherches géologiques dans le Taurus occidental dans l'arrière-pays d'Alanya.* Publ. Miner. Res. Explor. Ins. Turkey, D5, 134 pp.

COMPAGNONI, R., DAL PIAZ, G. V., HUNZIKER, J. C., GOSSA, G., LOMBARDO, B. & WILLIAMS, P. F. 1977. The Sesia-Lanzo Zone, a slice of continental crust with Alpine high pressure-low temperature assemblages in the western Italian Alps. *Rend. Soc. Ital. Mineral. Petrol.* **33**, 281–334.

BRUNN, J. H., DUMONT, J. F., DE GRACIANSKY, P. C., GUTNIC, M., JUTEAU, T., MARCOUX, J., MONOD, O. & POISSON, A. 1971. Outline of the geology of the western Taurides. *In*: CAMPBELL, A. S. (ed.). *Geology and History of Turkey,* Petroleum Exploration Society of Libya, Tripoli, 225–55.

COLEMAN, R. G., LEE, D. E., BEATTY, L. B. & BRANNOCK, W. W. 1965. Eclogites and eclogites: their differences and similarities. *Bull. geol. Soc. Am.* **76**, 483–508.

DUMONT, J. F., GUTNIC, M., MARCOUX, J., MONOD, O. & POISSON, A. 1972. Le Trias des Taurides occidentales (Turquie). Définition du bassin pamphylien: Un nouveau domaine à ophiolites à la marge externe de la chaîne taurique. *Zeit. Deutsch. Geol. Ges.* **123,** 385–409.

FREY, M., HUNZIKER, J. C., FRANK, W., BOCQUET, J., DAL PIAZ, G. V., JÄGER, E. & NIGGLI, E. 1974. Alpine metamorphism of the Alps, a review. *Schweiz. mineral. petrogr. Mitt.* **54,** 247–90.

LEFÈVRE, R. 1967. Un nouvel élément de la géologie du Taurus Lycien: les nappes d'Antalya (Turquie). *C.r. Séances. Acad. Sci. Paris,* **D265,** 1365–8.

MANGE-RAJETZKY, M. A. 1981. Detrital blue sodic amphibole in Recent sediments, southern coast, Turkey. *J. geol. Soc. Lond.* **138,** 83–92.

MARCOUX, J. 1978. A scenario for the birth of a new oceanic realm: the Alpine Neotethys. *Abstr. Int. Congr. Sedimentology, 10th, Jersualem 1978,* **2,** 419–20.

MONOD, O. 1978. Güzelsu Akseki bölgesindeki Antalya Napları üzerıne açıklama (Orta Bati Toroslar—Türkiye). *Bull. geol. Soc. Turkey,* **21,** 27–9.

OKAY, A. I. 1978. Sodic pyroxenes from metabasites in the Eastern Mediterranean. *Contrib. Mineral. Petrol.* **68,** 7–11.

—— 1980. Sodic amphiboles as oxygen fugacity indicators in metamorphism. *J. Geol.* **88,** 225–32.

—— 1982. Incipient blueschist metamorphism and metasomatism in the Tavşanli region. Northwest Turkey. *Contrib. Mineral. Petrol.* **79,** 361–67.

ÖZGÜL, N. 1976. Torosların bazi temel jeoloji özellikleri. *Bull. geol. Soc. Turkey,* **19,** 65–78.

—— & ARPAT, E. 1973. Structural units of the Taurus orogenic belt and their continuations in neighbouring regions. *Bull. Soc. geol. Greece,* **10,** 156–164.

PEYRONNET, PH, DE. 1971. Esquisse géologique de la région d'Alanya (Taurus méridional). Origine des bauxites métamorphiques. *Bull. Miner. Explor. Inst. Turkey,* **76,** 90–116.

ROBERTSON, A. H. F. & WOODCOCK, N. H. 1980. Tectonic setting of the Troodos Massif in the East Mediterranean. *In*: PANAYIOTOU, A. (ed.). *Ophiolites, Proc. Int. Oph. Symp. Cyprus 1979,* 36–49.

—— & —— 1981. Alakır Çay Group, Antalya Complex, SW Turkey. A deformed Mesozoic carbonate margin. *Sediment. Geol.* **30,** 95–131.

ŞENGÖR, A. M. C. & YILMAZ, Y. 1981. Tethyan evolution of Turkey: a plate tectonic approach. *Tectonophysics,* **75,** 181–241.

ŞENGÜN, M., ACARLAR, M., ÇETİN, F., DOĞAN, Z. O. & GÖK, A. 1978. Alanya masifinin yapısal konumu. *Jeoloji Mühendisliği,* **6,** 39–45.

STATHAM, P. J. 1976. A comparative study of techniques for quantitative analysis of the X-ray spectra obtained with a Si(Li) detector. *X-ray Spectrom.* **5,** 16–28.

SWEATMAN, R. R. & LONG, J. V. P. 1969. Quantitative electron probe microanalysis of rock-forming minerals. *J. Petrol.* **10,** 332–79.

WOODCOCK, N. H. & ROBERTSON, A. H. F. 1977. Imbricate thrust belt tectonics and sedimentation as a guide to emplacement of part of the Antalya Complex, SW Turkey. *Abs. 6th Coll. Aegean Geol., Izmir 1977,* 98.

A. I. OKAY, Jeoloji Bölümü, ITÜ Maden Fakültesi, Teşvikiye, Istanbul, Turkey.
N. ÖZGÜL, Jeoloji Bölümü, ITÜ Maden Fakültesi. Teşvikiye, Istanbul, Turkey.

The role of the Ankara Melange in the development of Anatolia (Turkey)

Teoman N. Norman

SUMMARY: The 'Ankara Melange', as defined by Bailey & McCallien (1953), comprises several belts, sub-belts and lenses of melange units, as well as some intercalated ocean floor fragments and continental (magmatic arc?, island arc?) slivers. Two significant observations are the successive younging of melange units from north to south (or northwest to southeast locally) and mega-debris flow features in some of melange units indicating a possible flow direction from east to west (or northeast to southwest). Both observations can be explained by assuming an obliquely northward moving Tethys ocean plate, subducting under a continental mass against which successive accretion and obduction of ocean floor irregularities (such as ocean plateaux, ridges, magmatic island arcs or even continental slivers) from Early Jurassic times to Middle Oligocene, produced the present complex melange system. Flow features can be explained by the development of local high ground where a non-subducting oceanic platform transpressed obliquely against the already-formed melange material, causing it to flow successively to depressed (trench) regions. Such flows were naturally interbedded with, or accompanied by, other types of mass flows and slivers of continental and/or oceanfloor material.

The name 'Ankara Melange' was given to a group of rocks (Bailey & McCallien 1953) around Ankara in Anatolia (Turkey) which showed an unusual degree of fragmentation and mechanical mixing, bringing together material of diverse origin and age. The strict definition of the term, as well as the delineation of its boundaries was presumably considered to be of secondary importance. However, subsequent work over a period of 30 years indicates that these rocks extend over several hundred kilometres in an east to west direction, and attain an average width of about 100 km (Fig. 1). The Ankara Melange itself is made of several sub-parallel belts, sub-belts and lenses (Erol 1956; Boccaletti *et al.*, 1966; Çalgın *et al.* 1973; Norman 1975a; 1975b; Çapan & Buket 1975; Gökçen *et al.* 1978; Çapan 1981; Akyürek 1981; Ünalan 1981; Seymen 1981). Also there are several other sub-parallel melange belts

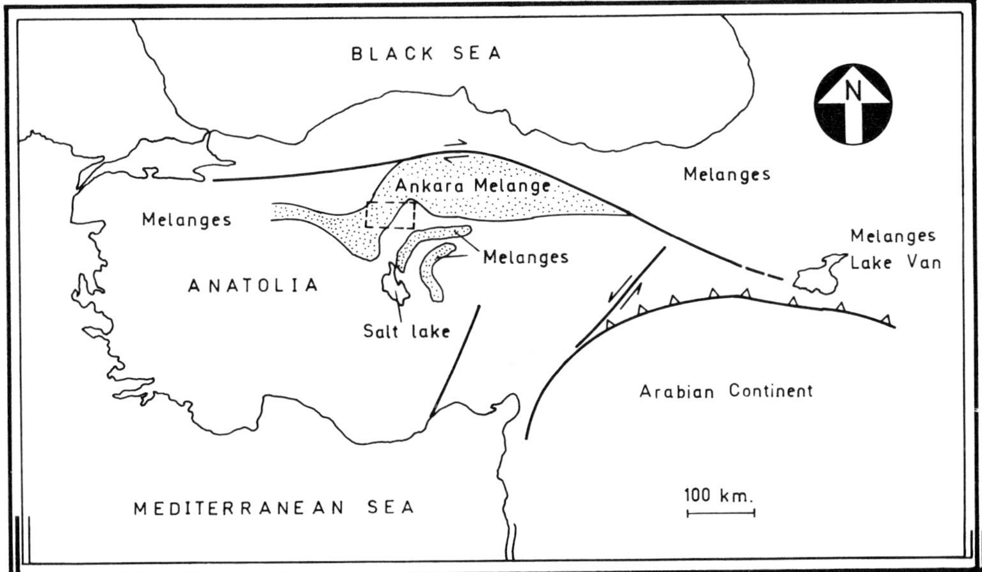

FIG. 1. Location and position of Ankara Melange and other major melange belts in Central Anatolia. Original Study area is shown by rectangle, Fig. 2 forms the southwestern quarter of this area.

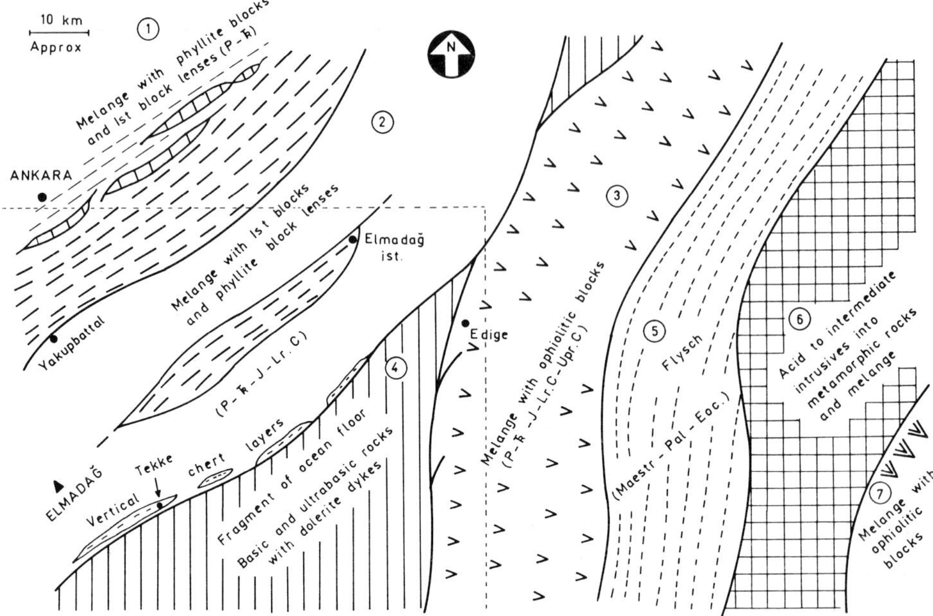

Fig. 2. Sketch map showing the major belts and sub-belts of Ankara Melange within the study area (some locality names are included for cross-reference with Fig. 3).

nearby, and other melange belts of similar character in other parts of Anatolia.

Significant features of melange units

At least 7 major belts (and several sub-belts and lenses) may be distinguished within the 'Ankara Melange' zone. In the area under study these sub-parallel belts will be described from north-west to southeast (Fig. 2), corresponding to a sequence from north to south, before local rotation by subsequent tectonic events.

Melange with phyllite blocks (metamorphic-block-melange)

This forms the outermost belt in the study area, including the foundation of the city of Ankara. The phyllite blocks of andesitic tuff origin are set in a graywacke matrix, consisting of graded turbidites (volcanic arenite), laminated mudstones and occasional channel conglomerates of a submarine fan environment. Occasional interbedded spilitic lava flows appear to have flowed over wet sediment presumably under high pressure, forming wrinkles at their soles but no pillows or vacuoles. Within this belt, there are several sub-parallel sub-belts and lenses consisting of limestone blocks (Permian), spilitic pillow lava blocks with Triassic fauna between pillows, agglomerates, tuffaceous olistostromes and radiolarite chert blocks (few), all set in a tuffaceous graywacke matrix. This part of the Ankara Melange has probably a Late Triassic-Early Jurassic age.

Melange with limestone blocks

This consists of oblong limestone blocks forming linear 'trains' which are generally sub-parallel to the general trend, but they may also curve around to form 'nose' or 'fan' (Fig. 3) shaped structures (Norman 1973, 1975). In addition to limestone blocks there are also blocks of conglomerate, agglomerate, diabase and graded turbiditic chert, set in a matrix of graded volcanic arenites and laminated black shales. The ages of limestone blocks have been palaeontologically determined as ranging from Permian through Early Cretaceous (up to Albian). A lens-shaped large sub-belt of phyllite-block melange divides this belt into two sub-belt units (Fig. 2).

A noticeable feature of this complex is the presence of mega-debris flow features, such as imbrication of blocks, concentration of larger blocks at the 'front' end, abrasion surfaces with sub-horizontal slicken-sides, curving 'noses' developed by flow lines, one-side abraded limestone blocks (Norman 1975b; Johnson 1970, 1981, personal communication). Considering

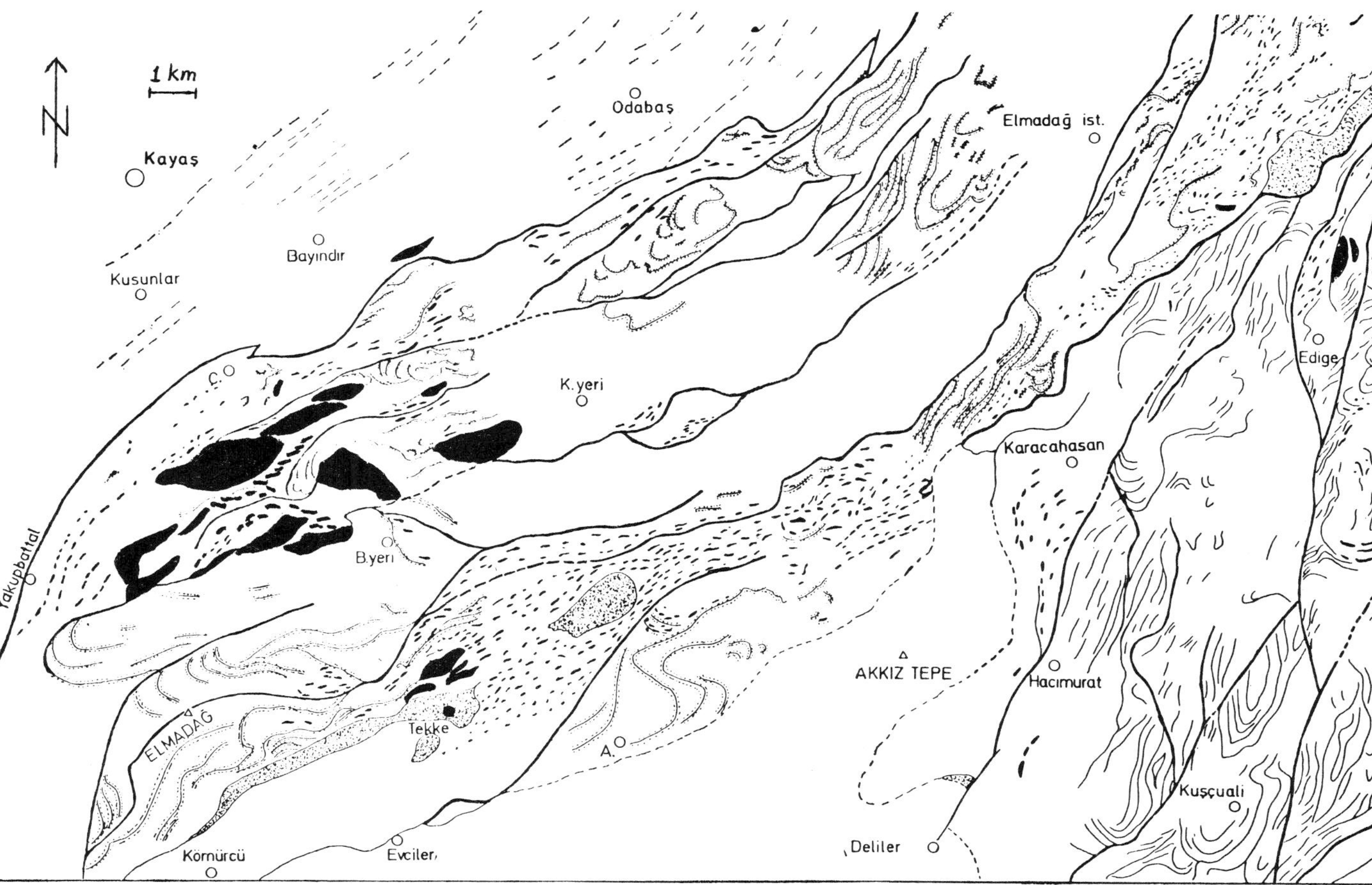

FIG. 3. Part of Ankara Melange, showing imbricated limestone blocks, cross-cutting abrasion surfaces (thick lines), 'trains' of blocks producing fan-shaped or nose-shaped flow lines. Flow movement direction from northeast to southwest (After Norman, 1975b). *Note:* Black patches indicate larger limestone blocks, stippled blocks outline radiolarite cherts (Map based on airphoto mosaic, with local ground control, at an original scale of 1:35 000).

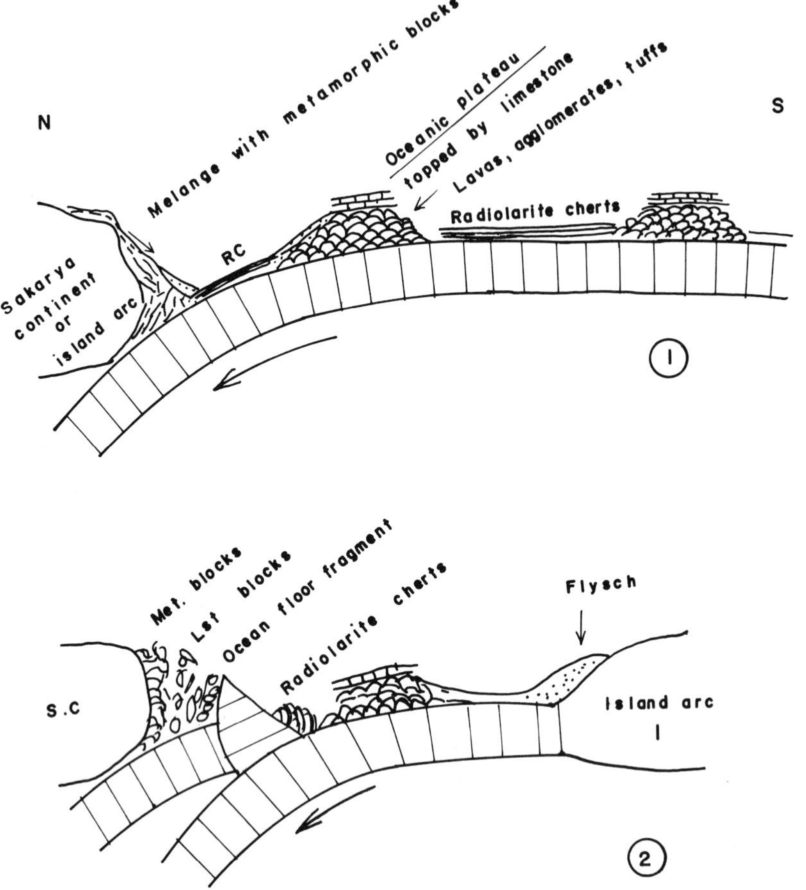

FIG. 4. Development of the various belts of Ankara Melange by accretion above a northward dipping subduction zone.

(1) Late Jurassic to Early Cretaceous: Formation of melange with metamorphic blocks as well as Permian-Triassic-Jurassic and Lower Cretaceous limestone blocks.

(2) Late Cretaceous to Paleocene: Ophiolitic melange, obduction? of ocean floor fragment.

the nearly vertical attitude of the 'layers', such flows can be interpreted as having moved from northeast to southwest or east to west in general.

Melange with ophiolitic blocks

This consists of pillow lava and agglomerate, conglomerate, sandstone and limestone blocks, red coloured radiolarite chert, clay and limestone blocks, as well as elongate lenses of serpentinite, in general all set in a tuffaceous matrix. Also noticable are serpentinite lenses which are clearly interbedded with red shales, limestones and laminar cherts, along 'normal' sedimentary contacts, while internally presenting disturbed broken-up structures of serpentinite blocks floating in essentially serpentinite matrix: a typical feature of monogenetic olistostromes. The ages of limestone blocks have been identified as ranging from Permian to Late Cretaceous (Senonian) by several workers (Gökçen 1977; Batman 1977). Since it is overlain by flysch sediments of Maastrichtian age, the formation age of this melange could be Campanian or older. Once again, typical mega-debris flow characters are observed in this melange belt, indicating a flow direction from northeast to southwest or east to west (Norman 1975b).

Ocean floor fragments

Juxtaposed with limestone-block and ophiolite-block melange belts, there are fragments of oceanic lithosphere, with sizes up to

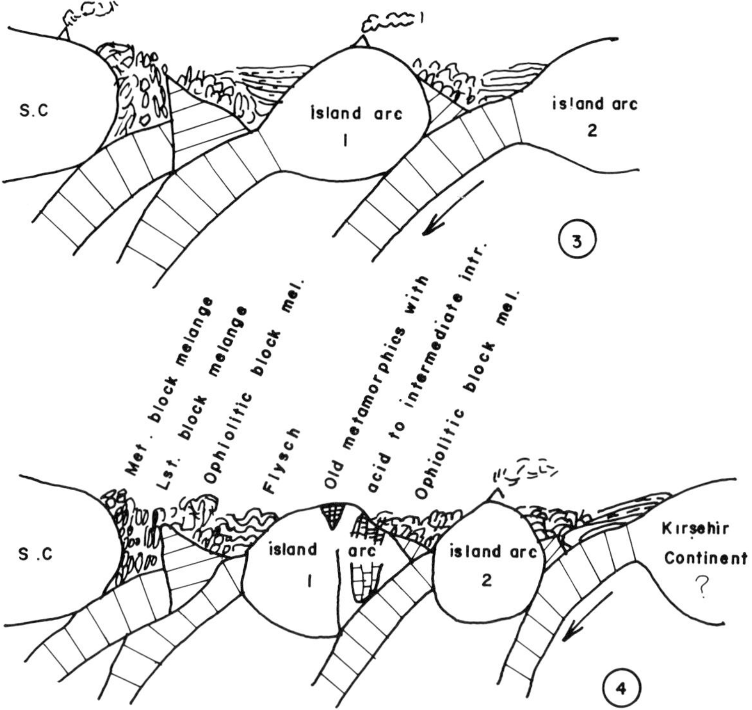

(3) Eocene: Flysch fill over melange basins.
(4) Oligocene: Further squeezing and partial uplift. No major thrusting required.

several tens of km². They have lithological compositions of basic (gabbro) and ultrabasic (harzburgite, serpentinite) rocks with occasional layered chromite. Near vertical dolerite dykes, siliceous veins and magnesite veins cut these rocks and may even become locally dominant features. These areas probably represent obducted fragments of ocean floor (Fig. 2).

Flysch

This consists of turbiditic sandstones (volcanic arenite) and shales with occasional intercalations of debris flows or olistostromes, as well as andesitic lava flows and tuff layers, and appears to rest conformably on the ophiolitic-block melange belt. Ages range from Maestrichtian to Late Eocene, or even possibly Early Oligocene. The environment of deposition is characteristically coalescing submarine fans. Palaeocurrents, though locally variable, seem to have flowed in general from northeast to southwest or east to west (Norman *et al.* 1980). When 'unfolded', the flysch basin widens to 90–100 km, indicating a great amount of shortening during the major Alpine orogeny. Petrographic studies of the clastic mineralogy indicate a steady increase in the quartz and related 'acid' minerals, going stratigraphically upwards (Norman *et al.* 1980). This suggests, for the first time in the geological history of this area, the proximity of a magmatic arc in Early Eocene time.

Acidic and intermediate magmatic rocks

These are intruded as batholiths into 'Ankara Melange'-type formations and older (Early Palaeozoic) metamorphic schists and gneisses (age of intrusion: Palaeocene). The pre-Mesozoic regional metamorphism is considered to have developed in a magmatic arc (Seymen 1982).

More 'Ankara Melange'-like belts, in juxtaposition with other ocean floor fragments and magmatic arcs, are found further southeast (Seymen 1982). New research indicates the presence of Ankara Melange-like belts, running roughly in northeast to southwest direction, within the so-called Kırşehir Massif (e.g. Belt 7, Fig. 1). They appear to include large (km size) blocks of the composition described above as ocean floor fragments. Further east and south, more and more magmatic arc rocks (acidic metamorphics and intrusions) become dominant.

Discussion

From the description above it seems clear that the present day Ankara Melange is built up of several melange belts, formed at different times and places, then brought together by gradual accretion from northwest to southeast. This process can possibly be explained by a northward moving ocean floor, carrying a variety of pillow lavas, agglomerates, tuffs, oceanic plateaux and island arcs with limestone developments, as well as flysch deposits and radiolarite cherts. This plate (Fig. 4) could have subducted under a continental sliver or island arc (Şengör & Yılmaz 1981; Tekeli 1981).

The mega-debris flow features, particularly noticeable in the limestone-block and ophiolite-block melange belts could also be explained by the same kinematic mechanism, provided an oblique shear movement is created by the northward approaching oceanic plateaux and magmatic arcs (Fig. 5). Thus the proximal end of the ridge feature (e.g. island arc) carried by the subducting oceanic plate would exert a transpressive shear, causing local squeezing up of the melange material, which in turn, could start mass flow towards depressions along the trench, parallel to the trend of the subduction zone. With the continued forward movement of the plate, this elevated area could migrate laterally, causing re-sedimentation of the melange material, particularly causing further flows from east to west, or northeast to southwest. This mechanism could also explain the interbedded or subparallel nature of the tectonic features and sedimentary units within melange belts (Norman 1975b; Norman *et al.* 1980).

Such kinematics can be observed today, through oceanographic studies, in the Eastern Pacific off South America, as well as Western Pacific, off the coast of Japan (Ben-Avraham *et al.* 1981).

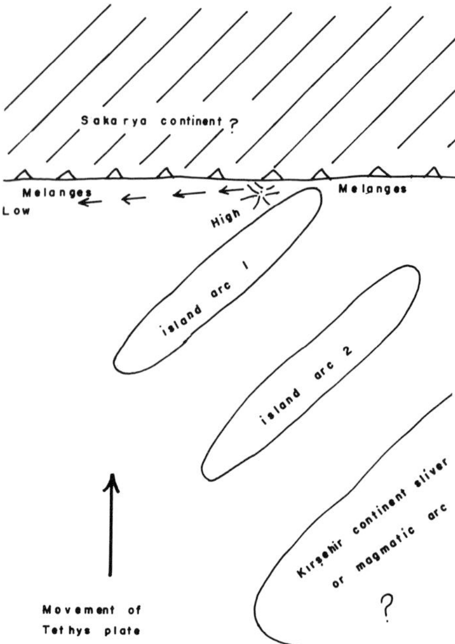

FIG. 5. Oblique approach of ocean floor topographic highs (island arcs, continental slivers, etc.) towards northward subducting zone, causing transpressive shear at a point. This squeezed and elevated site becomes a source for further westward moving melange flows. As the island arc (or any other topographic high) becomes 'jammed' against the (Sakarya?) continent, the next piece of ocean floor is subduced forming a new melange belt.

Conclusions

There appears to be an east to west trending, northward-plunging subduction zone, developed probably during Mid-Jurassic times. Obliquely northward moving Tethys (? Neo Tethys) ocean floor, successively formed the various units of the Ankara Melange. Irregularities on the oceanic plate, such as oceanic plateaux, ridges, magmatic island arcs or even possibly continental slivers, have been accreted along with melange material. Occasional fragments of ocean floor were also involved in the process. Due to oblique transpression, melange material may have flowed successively from 'high' points to depressed areas in the west or southwest.

It may be speculated that the major part of Central Anatolia has been built up by the accretion of melange material intercalated with oceanic and magmatic island arc material, without major continental microplates, *sensu stricto*.

This model suggests a 'gradual' melange formation with 'continuous' deformation and accretion, rather than one or two 'wild' periods of paroxysm and thrusting.

References

AKYÜREK, B. 1981. Fundamental charcteristics of the northern part of Ankara Melange (in Turkish). *Symposium on Central Anatolia, Geol. Soc. Turkey*, 41–52.

BAILEY, E. B. & MCCALLIEN, W. J. 1953. Serpentine lavas, the Ankara Melange and the Anatolian Thrust. *Trans. Royal Soc. Edinburgh*, **62**, 403–442.

BATMAN, B. 1977. *Geological evolution of North Haymana Region and study of the neighbouring melange.* (Unpublished dissertation, in Turkish). Hacettepe Üniv. Ankara. 172 pp.

BEN-AVRAHAM, Z., NUR, A., JONES, D. & COX, A. 1981. Continental accretion: from oceanic plateaus to allochthonous terranes. *Science*, **213**, 47–54.

BOCCALETTI, M., BARTOLETTI, V. & SAGRI, M. 1966. Ricerche sulle ofioliti delle Catene Alpine: I-Osservazioni sul'Ankara Mélange nella zona di Ankara. *Boll. Soc. geol. It.* **85**, 485–508.

ÇALGIN, R., PEHLİVANOĞLU, H., ERCAN, T. & ŞENGÜN, M. 1973. *Geology of Ankara region* (Unpublished report, in Turkish). M.T.A. Rep. No. 6487. 55 pp.

ÇAPAN, U. Z. & BUKET, E. 1975. Geology of Aktepe Gökdere region and ophiolitic melange. *Bull. Geol. Soc. Turkey*, **18**, 11–16.

—— 1981. Views on Ankara Melange and some observations on the characters of melange at Göktepe-Ankara region (in Turkish). *Symposium on Central Anatolia, Geol. Soc. Turkey*, 27–31.

EROL, O. 1956. *Research on the Geology and Geomorphology of Elmadağ region, SE of Ankara.* M.T.A. Publications, series D, No. 9, Ankara 238 pp.

GÖKÇEN, N. 1977. An investigation of the biostratigraphy of the Upper Cretaceous-Paleogenic sequences in the Irmak-Hacıbalı-Mahmutlar area (Ankara-Yahşıhan). *Yerbilimleri*, **3**, 129–144.

GÖKÇEN, S. L., NORMAN, T. N. & ŞENALP, M. 1978. Mesozoic and Cenozoic sediments in the Central Anatolian Basin. *Xth Inter. Sedim. Congress, Jerusalem, Israel, Excursion guide part III: Turkey and Cyprus*, 1–19.

JOHNSON, A. 1970. *Physical Processes in Geology.* Freeman and Co. San Francisco. 326 pp.

NORMAN, T. N. 1975a. On the structure of Ankara Melange. *Int. Geodynamics Project, Report of Turkey*, 78–92.

—— 1975b. Flow features of Ankara Melange *Proc. IXth Inter. Sedim. Congress, Nice, France*, **4**, 246–248.

——, GÖKÇEN, S. L. & ŞENALP, M. 1980. Sedimentation pattern in Central Anatolia at the Cretaceous-Tertiary boundary. *Cretaceous Research*, **1**, 61–84.

SEYMEN, I., 1981. Stratigraphy and metamorphism of the Kırşehir massif around Kaman (Kırşehir-Turkey). *Bull. Geol. Soc. Turkey*, **24**, 101–108.

—— 1982. *Geology of Kırşehir Massif in the neighbourhood of Kaman* (Unpublished dissertation, in Turkish). Technical University of İstanbul, Mining Faculty, 164 pp.

ŞENGÖR, A. M. C. & YILMAZ, Y. 1981. Tethyan evolution of Turkey: a plate tectonics approach. *Tectonophysics*, **75**, 181–241.

TEKELİ, O., 1981. Subduction complex of pre-Jurassic age, Northern Anatolia, Turkey. *Geology*, **9**, 68–72.

ÜNALAN, G. 1981. Stratigraphy of the Ankara Melange, SW of Ankara. *Symposium on Central Anatolia, Geol. Soc. Turkey*, 46–52.

TEOMAN N. NORMAN, Middle East Technical University, Institute of Marine Sciences, Erdemli-İçel, Turkey.

Basic and ultrabasic rocks from the Ankara Melange, Turkey

Ayla Tankut

SUMMARY: The Ankara Melange consists from west to east of three major parallel belts: (i) metamorphic melange, (ii) limestone melange and (iii) ophiolitic melange. A Triassic age for the metamorphic and limestone belts and a Cretaceous age for the ophiolitic belt have been suggested by several authors (Norman 1973, Çapan 1981, Yilmazer 1981).

The ophiolitic belt includes lenses of serpentinite, occasionally unserpentinized peridotite, gabbro, pillow lava, agglomerate and basic dykes. The ultrabasic rocks show tectonic fabrics and the gabbros cumulate textures. The metamorphic and limestone melange belts contain basic volcanic rocks only as blocks, clasts and dykes, and are devoid of ultrabasic rocks.

Petrochemical studies reveal that the volcanic rocks of the ophiolitic and metamorphic belts are very similar in mineralogical and chemical composition, including representatives of both the tholeiitic and calc-alkaline series.

The presence of almost all the members of an ophiolite suite may indicate an ocean floor origin for the ultrabasic and basic rocks of the ophiolitic melange belt, and the occurrence of calc-alkaline rocks in both belts may suggest at least one phase of island arc volcanism. These rocks may represent several phases of rifting and island arc generation during the evolution of the Mediterranean Tethyan domain.

The Ankara Melange crops out more or less continuously for several hundred km as a ca. 50 km wide belt (Fig. 1). The general opinion, as summarized by Erol (1981), is that the Melange should be defined to include all the chaotic formations in the area. The common lithologies in the Melange which are of Palaeozoic and Mesozoic age are metamorphosed sediments, pyroclastics, and various members of the ophiolite suite. In spite of a chaotic nature, Norman (1973 and this volume) recognises a regular alignment of blocks and lenses within the Melange; three parallel belts have been described from west to east defined by the abundance of the major constituents as follows: (i) metamorphic melange belt (the graywacke melange of Norman, 1975); (ii) limestone-block melange belt and (iii) ophiolitic melange belt (Fig. 2). Each belt may contain the components of the others in subordinate amounts. Ultrabasic and basic rocks are the major constituents of the ophiolitic melange belt, and some basic volcanics are also present in the limestone and metamorphic belts.

Although contradictory views have been put forward as regards the origin and age of the melange, many authors agree on a Cretaceous emplacement age for the ophiolitic belt and a Triassic age for the metamorphic and limestone belts (Norman 1973, Batman 1977, Erol 1981, Çapan 1981, Akyürek 1981, Ünalan 1981, Yilmazer 1981). Norman (1973) presents a stratigraphic succession indicating an age variation from west to east, formed by tectonic and

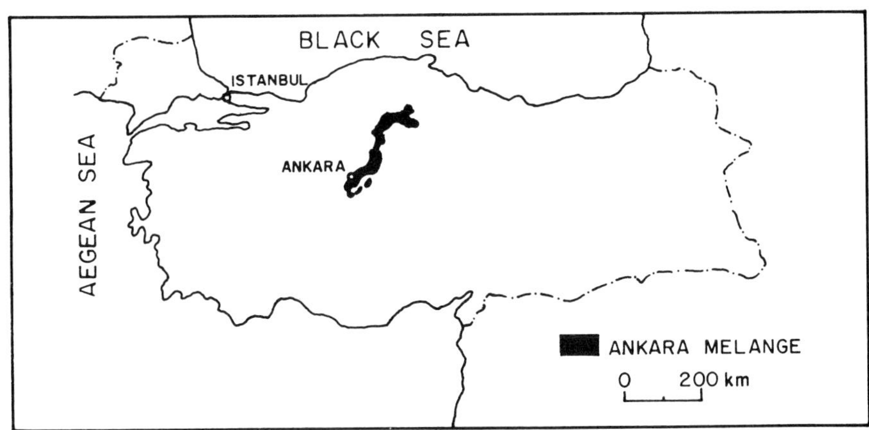

FIG. 1. The location of the Ankara Melange.

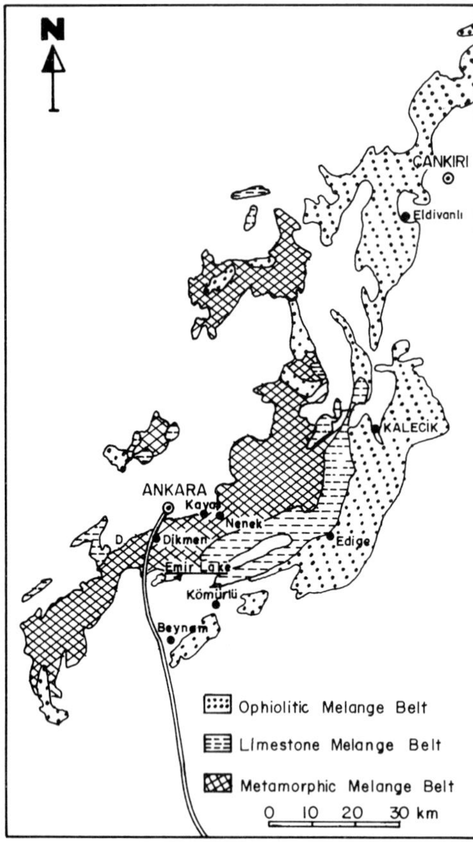

FIG. 2. Distribution of tectonic belts in the Ankara Melange.

sedimentary processes in the time interval between the end of the Carboniferous and the beginning of the Cenomanian.

The present paper reports preliminary petrological results from the ultrabasic and basic rocks of the ophiolitic melange and metamorphic melange belts.

Distribution of ultramafic and mafic rocks

The ultramafic rocks, which represent a dismembered ophiolite suite are confined to the ophiolitic belt. The metamorphic and limestone belts contain only basic volcanics: no ultramafics have been observed (Yilmazer 1981). Boccaletti *et al.* (1966) describe some ophiolitic rocks from the metamorphic zone, but these are probably basic volcanics misidentified because of surface alteration (T. Norman, pers comm. 1982). Although the members of the ophiolite suite are usually separate blocks. Akyürek (1981) reports preservation of more complete sequences around Eldivanli.

Serpentinites occur either as individual lenses 3–10 m wide and 50–100 m long, (e.g. in several road cuts along the Ankara-Kayseri highway) or are smaller bodies associated with gabbros, pillow lavas and basic dykes. One large isolated peridotite massif, of about 80 km^2, is exposed around Edige village (Norman, 1973, Yıldırım, 1974). It consists mainly of serpentinized peridotite, basic and felsic dykes and abandoned chromitite seams.

Gabbros are usually accompanied by other members of the ophiolite suite but occur in a few localities as detached cumulate gabbro bodies, up to several tens of metres wide. One of the gabbro bodies studied consists of alternating dark and light bands 4–5 cm wide marked by varying proportions of pyroxene and plagioclase. Grain-size variation has also been observed elsewhere in the same body (Fig. 3).

FIG. 3. Layering in cumulate gabbro from Kömürcü village.

The volcanics of the ophiolitic zone include agglomerates, pillow lavas and basic dykes. The metamorphic melange zone volcanics occur as blocks or clasts surrounded by a volcanic-sedimentary matrix. Some of them are basic dykes which cut the phyllitic schists.

Sampling

Serpentinites were collected from several

lenses along the Ankara-Kayseri highway, together with basic volcanics and basic dyke rocks from Beynam and Kalecik areas. Cumulate gabbros came from a body exposed near Kömürcü village. Volcanics from the metamorphic belt were collected near Nenek village and Emir Lake area (Fig. 2).

Petrography

The ultramafic rocks show tectonic fabrics and include serpentinite and highly serpentinized harzburgite. The harzburgite consists of strongly serpentinized olivine, distinguished by small relics, and orthopyroxene ranging in size from 0.5 to 1 mm. Orthopyroxene, occasionally altered to bastite, ranges modally from 10% to 25%, and commonly shows kink-banding.

The layered gabbros show primary cumulate textures. They consist of clinopyroxene and unzoned plagioclase (An_{60-70}). Plagioclase appears as an intercumulate phase enclosing small clinopyroxene grains.

The Beynam basic volcanics (Fig. 2) from the ophiolitic belt, retain their original textures in spite of intense alteration. They are holocrystalline, mostly porphyritic with low proportion of phenocrysts and are sometimes equigranular. The phenocrysts are 1–2 mm in size, composed of clinopyroxene, plagioclase and green pleochroic hornblende in which the original clinopyroxene patches are seldom preserved indicating a secondary origin for hornblende. The groundmass consists of 0.2–0.7 mm long grains of the same minerals as the phenocrysts. Secondary quartz is abundant and occupies the interstices between plagioclase grains and occasionally partly replaces them. Chlorite is a common alteration product, generally accompanied by opaque iron oxide and calcite.

The Kalecik basic volcanics, also from the ophiolitic belt (Fig. 2) are generally vesicular and amygdaloidal, and have a finer grained groundmass than those of Beynam. They also consist essentially of plagioclase and clinopyroxene. Occasionally, brown hornblende is present as a late deuteric phase after pyroxene. The groundmass phases which are strongly altered, are the same as the phenocrysts, although in some samples a little glass and palagonite are also present. Plagioclase is sericitized and occasionally saussuritized and mafic minerals are chloritized. Calcite is present filling interstices and as patches in plagioclase grains.

The metamorphic belt volcanics are similar in mineralogical composition to the opiolitic belt volcanics described above and are again strongly altered. Those from the Nenek area (Fig. 2), are predominantly porphyritic and show flow-texture. The phenocrysts are less than 1 mm in size and include clinopyroxene, plagioclase and sometimes brown hornblende of deuteric origin. The groundmass is composed of the same mineral phases, occasionally with a little glass. Plagioclase laths occur with sericite, calcite, albite, a little saussurite and occasionally free quartz; pyroxenes are partly or completely replaced by uralite, calcite and chlorite as a result of spilitization. The dolerites from Emir have again been subjected to spilitization like those of the Nenek area and contain clinopyroxene, plagioclase and green hornblende. A secondary origin of the green hornblende is indicated by indistinct clinopyroxene relics in it. Plagioclase has also been replaced by sericite, saussurite, albite and calcite, and there is abundant interstitial myrmekitic quartz.

Chemistry

Eight representative samples of volcanic rock from both the ophiolitic and the metamorphic melange zones have been analyzed for major elements. Trace element data have been obtained for 12 volcanic rock samples from both melange zones and for 4 serpentinized periodotites and 2 cumulate gabbros from the ophiolite melange zone (Tables 1, 2). All the elements except Ti and Na were determined by X-ray fluorescence. Ti was determined by spectrophotometry and Na by flame-photometry. The major element compositions of the rocks studied are given in Table 1. USGS standards AGV-1, BCR-1, DTS-1, PCC-1, W-1, G-2, a BRGM standard BR, a University of Toronto standard UTB-1 (basalt) and University of Auckland (N.Z.) standards TO (obsidian) and MEB (basalt) were used for XRF calibration.

There is little chemical difference between the bulk compositions of the metamorphic and ophiolitic belt volcanics. Most of the melange rocks contain very low K_2O, less than 0.25%. Na_2O/K_2O ratios and Al_2O_3 contents are comparable with ocean floor basalt averages: $Na_2O/K_2O = 12.50$, $Al_2O_3 = 16.01\%$ according to Cann (1971). However, Miyashiro (1975) does not accept such comparisons as reliable evidence because of the mobility of alkalis. Two rocks, one from Kalecik in the ophiolitic belt and the other one from Nenek in the metamorphic belt give very high K_2O values. Neither K-feldspar nor muscovite have been detected in these rocks and the source of the K_2O is obscure.

The bulk compositions of some rocks are

TABLE 1. *Chemical analyses of selected samples of volcanic rocks in the Ankara Melange (Sample numbers refer to the rock collection, of the Geological Engineering Department, Middle East Technical University. The letters preceding the sample numbers indicate the areas from which they were collected; B-Beynam, K-Kalecik, E-Emir Lake, N-Nenek. (FeO_T: Total iron as FeO) B1, B5-Altered porphyritic basalts-Beynam, K3-Altered vesicular Basalt-Kalecik, E1-Spilitic dolerite-Emir, N3-N15-Spilitic basalts-Nenek*

	Ophiolitic belt			Metamorphic belt				
	B1	B5	K3	E1	N3	N7	N14	N15
SiO_2	54.42	49.96	44.72	53.97	54.30	47.14	41.48	47.27
TiO_2	0.42	1.09	0.75	1.80	0.53	1.92	1.58	2.26
Al_2O_3	15.52	16.35	13.29	15.69	18.37	15.18	11.81	12.84
FeO_T	6.32	9.03	9.37	11.76	7.60	10.48	12.62	9.89
MgO	6.88	5.86	6.25	3.90	4.57	6.75	8.88	6.39
CaO	10.34	5.06	12.33	6.43	5.89	9.35	10.91	10.87
Na_2O	3.38	4.23	0.38	2.56	3.99	3.31	2.32	3.32
K_2O	<0.25	<0.25	5.76	<0.25	<0.25	0.71	<0.25	3.45
Ign. Loss	1.87	7.12	5.54	2.74	3.78	3.89	2.28	3.43
Total	99.86	99.70	99.43	100.17	99.88	99.90	99.25	100.82
FeO_T/MgO	0.92	1.54	1.50	3.02	1.66	1.55	1.42	1.55
Na_2O/K_2O	>13.52	>16.92	0.07	>10.24	>15.96	4.66	>9.28	0.61

TABLE 2. *Trace element contents (ppm) of basic volcanic rocks from the Ankara Melange. Samples as in Table 1, with additions: B2-Altered porphyritic basalt-Beynam, B6-Altered dolerite-Beynam, K2-Altered porphyritic basalt-Kalecik*

	Ophiolitic belt						Metamorphic belt					
	B1	B2	B5	B6	K2	K3	E1	N3	N7	N9	N14	N15
Nb	nil	nil	nil	1	63	4	nil	nil	11	10	7	41
Y	5	9	19	6	21	36	20	20	20	15	12	18
Zr	21	17	85	2	178	112	61	83	99	129	57	117
Ti	2500	—	6500	—	—	4500	10800	3200	11500	—	9500	11350
Rb	7	9	10	—	37	136	—	—	—	—	—	20
Sr	134	74	435	—	469	752	—	—	—	—	—	599

TABLE 3. *Average trace element contents from different worldwide tectonic settings (Pearce and Cann, 1973) OFB–Ocean floor, V.ARC–Volcanic arc, OI–Ocean island, C–Continental;*

	Mean OFB	Mean V.ARC 1*	Mean V.ARC 2*	Mean OI	Mean C
Nb	5	1.5	2.5	32	20
Y	30	19	23	29	29
Zr	92	52	106	215	215
Rb	—	—	—	—	—
Sr	131	207	375	438	460
Ti	8350	5150	5400	16250	15150

*1-Low-K tholeiite
2-Calc-alkali basalt

high in SiO_2. In thin sections these rocks contain secondary free silica in the form of interstitial quartz or granophyric intergrowths associated with spilitization.

On the alkali silica diagram of MacDonald & Katsura (1964) the samples plot both above and below the dividing line (Fig. 4). Since the rocks are intensely altered, the alkaline affinity of some of them could be explained by later metasomatism (Miyashiro 1975, Coleman 1977, Vallance 1974). The plots on the Fe O_T/MgO versus SiO_2 and FeO_T versus FeO_T/MgO diagrams (Figs 5 and 6) indicate a tholeiitic character for most of the rocks. On TiO_2 versus FeO_T/MgO diagram (Fig. 7) the tendency is towards the oceanic basaltic trend. The two doleritic rocks (B1 and N3), each from different belts, fall in the calc-alkaline field.

The content of the immobile trace elements Ti, Nb, Y and Zr in the melange rocks is given in Table 2. Table 2 also includes Rb and Sr analyses from some of the volcanic rocks. Table 3 illustrates the average trace element contents of basalts from different tectonic settings as given by Pearce & Cann (1973). The general trace element pattern of volcanic rocks from both belts is comparable with that of ocean

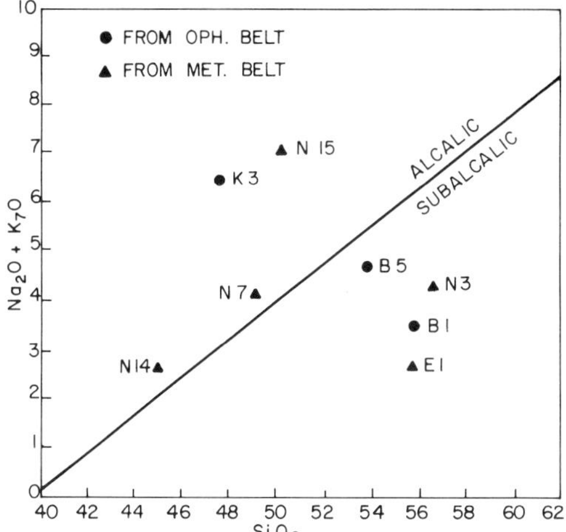

Fig. 4. Alkali SiO_2 diagram of volcanics from the Ankara Melange. Volatile-free. Boundary from MacDonald & Katsura (1964).

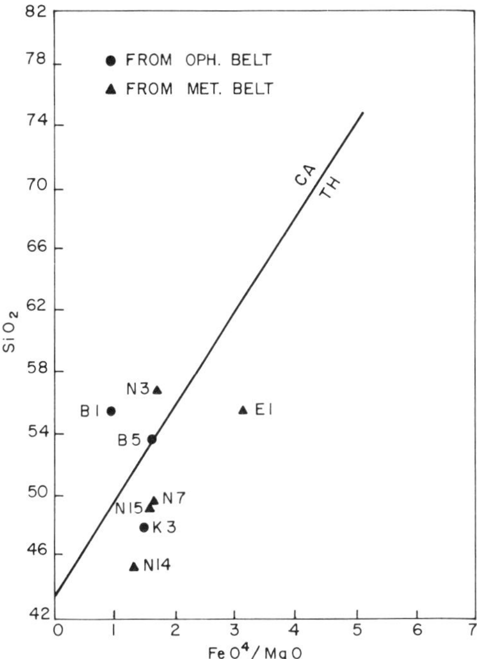

Fig. 5. SiO_2 versus FeO/MgO diagram of volcanics from the Ankara Melange. Volatile-free. Calc-alkaline/tholeiitic boundary from Miyashiro (1975).

floor basalts and island arc volcanics. Of these elements only Y could be detected in the gabbros and periodotites analysed and is at similar levels to those in the volcanic rocks of the melange.

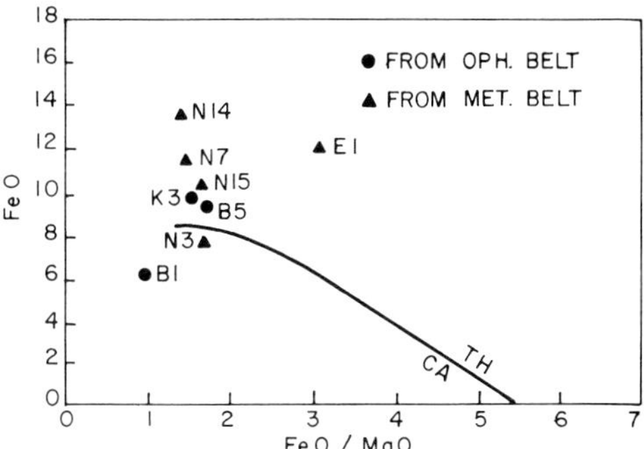

Fig. 6. FeO_T versus FeO_T/MgO diagram of volcanics from the Ankara Melange. Volatile-free. Boundary as in Fig. 5.

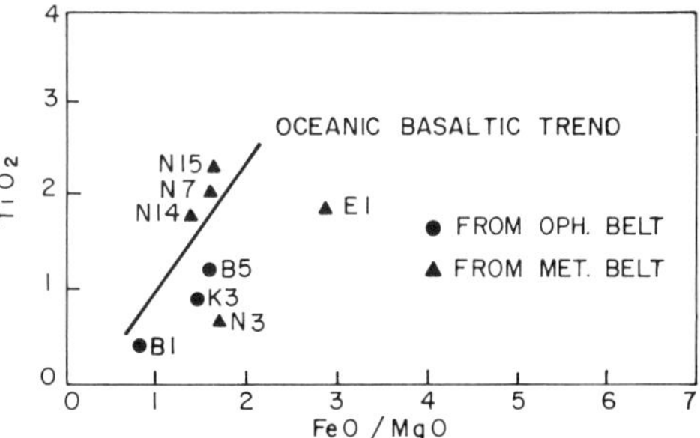

Fig. 7. TiO_2 versus FeO/MgO diagram of volcanics from the Ankara Melange. Volatile-free. Oceanic basalt trend from Miyashiro (1975).

Results and conclusions

The main findings are as follows:

(1) The ophiolitic melange belt studied includes almost all the members of an ophiolite suite. The volcanic rocks of this belt are of tholeiitic character. An ocean floor origin may thus be suggested for the ultrabasic and basic rocks of this belt.

(2) Tholeiitic rocks are also present in the metamorphic melange belt.

(3) Both belts appear to include rocks of calk-alkaline rock chemistry.

(4) The high silica content of some of the rocks is attributed to SiO_2 enrichment probably due to metasomatism associated with spilitization.

This paper is the first investigation of the petrography and chemistry of basic and ultrabasic rocks of the Ankara Melange and as yet the available data are very limited. However, the results so far suggest that the rocks studied could have been related to both rifting and island arc generation during the Late

Palaeozoic and the Mesozoic as discussed in more detail by Norman, this volume. More field and chemical data are still needed to refine understanding of tectonic significance of the mafic and ultramafic rocks of the Ankara Melange.

ACKNOWLEDGEMENTS: The study (No: 81-04-08-TH-01) is supported by the scientific Research Fund of the Middle East Technical University. The major element X-ray fluorescence analyses were carried out at the Turkish Cement Producers Association Research Centre. The help of both institutions and I. Yılmazer are appreciated.

References

AKYÜREK, B. 1981. Ankara Melanjinin kuzey bölümünün temel jeoloji özellikleri. *In: Proc. Geol. Of Central Anatolia Symp. Ankara 1981*, 41–5.

BATMAN, B. 1977. Haymana kuzeyinin jeolojik evrimi. *Yerbilimleri*, **4,** 95–124.

BOCCALETTI, M., BORTOLOTTI, V. & SAGRI, M. 1966. Ricerche sulle ofioliti delle Catene Alpine. *Boll. Soc. geol. Ital.* **85,** 485–508.

CANN, J. R. 1971. Major-element variations in ocean-floor basalts. *Phil. Trans. Roy. Soc. Lond.* **268A,** 495–505.

COLEMAN, R. G. 1977. *Ophiolites*. Springer-Verlag, New York.

ÇAPAN, U. 1981. Ankara Melanji hakkindaki görüşler ve Melanjin Gökdere Aktepe yöresindeki özelliklerine ait gözlemler. *In: Proc. Geol. of Central Anatolia Symp. Ankara 1981*, 27–30.

EROL, O. 1981. Ankara Melanji deyiminin tarihçesi. *In: Proc. Geol Central Anatolia Symp. Ankara 1981*, 32–4.

MACDONALD, G. A. & KATSURA T. 1964. Chemical composition of Hawaiian lavas. *J. Petrol.* **5,** 82–133.

MIYASHIRO, A. 1975. Classification, characteristics and origin of ophiolites. *J. Geol.* **83,** 249–81.

NORMAN, T. 1973. Ankara Yahşihan bölgesinde Üst Kretase-Alt Tersiyer sedimentasyonu. *TJK Bülteni* **16,** 41–66.

—— 1975. Flow features of Ankara Melange. Abstracts of the 9th International Congress of Sedimentology, Nice, France, 1975.

PEARCE, J. A. & CANN, J. R. 1973. Tectonic Setting of basic volcanic rocks determined using trace element analyses. *Earth planet. Sci. Lett.* **19,** 290–300.

ÜNALAN, G. 1981. Ankara güney batısındaki 'Ankara Melanji' nin stratigrafisi. *In: Proc. Geol. of Central Anatolia Symp. Ankara 1981*, 46–52.

VALLANCE, T. G. 1974. Spilitic degradation of a tholeiitic basalt. *J. Petrol.* **15,** 79–86.

YILMAZER, I. 1981. *Geology of Lalahan-Kayaş Region (Turkey)*, unpublished MSc thesis, METU, Geol. Eng. Dept. Turkey. 68 pp.

YILDIRIM, M. 1974. *Geology of Kizildağ-Edige Region, Elmadağ-East of Ankara (Turkey)*. Unpublished MSc, thesis, METU, Geol. Eng. Dept. Turkey. 80 pp.

AYLA TANKUT, Department of Geological Engineering, Middle East Technical University, Ankara-Turkey.

Distribution and characteristics of the north-west Turkish blueschists

A. I. Okay

SUMMARY: North-west Turkey encompasses two major tectonic units, the Pontides and the Anatolides, separated by the İzmir-Erzincan suture. The Anatolides, the metamorphosed northern extension of the Taurides, consist of several zones. The more southerly Afyon Zone is an Early Palaeozoic to Maastrichtian Tauride shelf-type sequence which rests on gneisses of the Menderes Massif. The Afyon Zone underwent greenschist facies metamorphism, gradually increasing in grade with depth, and overthrust by HP/LT metamorphic rocks and peridotites. A HP/LT metamorphic belt several hundred kilometres long constitutes the second Anatolide zone. There are two tectonic units in the blueschist sequence. A lower more recrystallised blueschist unit consists of a thick basal marble series overlain by meta-cherts, metabasites and meta-shales. An upper unit, with a wider distribution, shows only incipient HP/LT metamorphism, and consists of shales, radiolarian cherts, basic volcanic rocks, pelagic limestones and greywackes. The HP/LT rocks are tectonically overlain by a peridotite nappe, without HP/LT metamorphism. Steeply dipping major faults separate blueschists from the Inner Pontide sequence, which consists of a Permian-Triassic metamorphic basement overlain by carbonates and Upper Cretaceous flysch. Especially in the Ankara region, Upper Cretaceous flysch contains abundant ophiolitic olistoliths. Post-orogenic Palaeocene sediments are transgressive over the HP/LT metamorphic belt and the peridotite nappe. The blueschist sequence may have been the north-facing Anatolide/Tauride platform margin, subducted during the Late Cretaceous. Very fast uplift in the collision zone, aided by buoyancy of subducted continental crust, generated olistostromes in the Pontides and emplaced peridotite nappes over the Anatolide/Tauride platform.

The Alpine mountain chain was created by the collision of Eurasia and Gondwana during the Late Mesozoic and Tertiary. A broad zone of collision of the two megacontinents extends from the Apennines in Italy to the Taurides and Pontides in Turkey. This zone is marked by a pile of nappes, ophiolites, pelagic sediments and HP/LT metamorphic rocks. Regionally important HP/LT metamorphic provinces occur in three different parts of the Alpine chain: Western Alps, Cyclades in Greece and north-west Turkey. The aim of this paper is to give a general description of the least known of the three HP/LT metamorphic provinces. Long-standing problems of the formation and uplift of blueschists and the processes in collision zones are discussed in a regional geological framework. The description is based on the recent literature plus the author's own field-work, which together provide a general picture of much of north-west Turkey.

Geological setting

Major sutures can often be recognised by tectonic juxtaposition across narrow ophiolite zones of totally differing stratigraphies and faunal provinces. The suture representing the major oceanic area between Eurasia and Gondwana during the Mesozoic, and probably earlier, can safely be placed in Turkey in the İzmir-Ankara-Erzincan Zone (İzmir-Erzincan suture). During the Mesozoic, apart from a major İzmir-Erzincan ocean, there were also minor oceanic basins, for example separating the Apulia-Tauride microplate from Gondwana (Biju-Duval *et al.* 1977). However, these were narrow areas of oceanic crust probably of short duration, and the resulting microcontinents can be assigned to one of the megacontinents.

The İzmir-Erzincan suture divides Turkey into two major units: to the north the Pontides, and to the south the Taurides, and their metamorphosed equivalents, the Anatolides (Figs 1, 2 Ketin 1966). The İzmir-Ankara-Erzincan Zone of Brinkmann (1966) is a regional depression, where ultramafic and HP/LT metamorphic rocks, lying tectonically over the Anatolides, were preserved from erosion.

Palaeogeographically, the area south of the İzmir-Erzincan suture belongs to the Anatolide-Tauride Platform (Şengör 1979), which was linked to Gondwana. During the Mesozoic the area was a wide Bahamian-type continental platform (Bernoulli & Laubscher 1972). The Eurasian realm to the north of the İzmir-Erzincan suture was tectonically more active

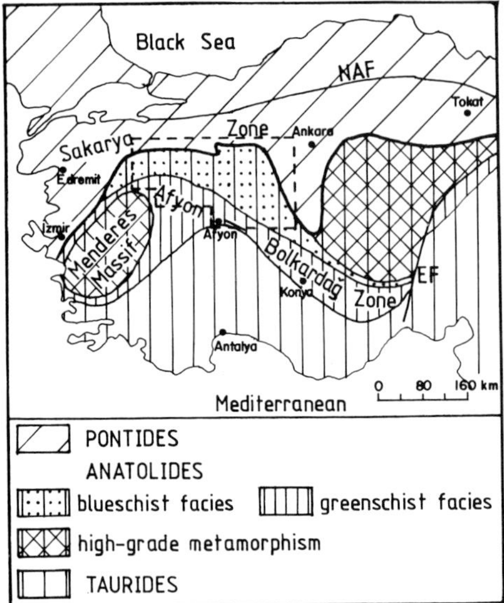

FIG. 1. Major tectonic units of western and central Turkey; only the Anatolide zones are differentiated. Area of Fig. 2 is indicated by dashed lines. NAF, North Anatolian Fault; EF, Ecemiş Fault.

during the Mesozoic, and characterized by important clastic sedimentation and basic volcanism (Fourquin 1975; Bergougnan & Fourquin 1980). North-west Turkey encompasses areas belonging to the Eurasian, the Sakarya Zone of the Pontides, and Gondwana realms, the Afyon Zone of the Anatolides, and with pelagic sediments, volcanic and ultramafic rocks recording an oceanic area separating the two continents.

Afyon Zone (Anatolides)

The Afyon Zone is a typical shelf-type Palaeozoic-Mesozoic sequence of the Taurides (Fig. 4). It covers large areas in Central Anatolia and extends as far east as the Bolkardağ region (Fig. 1); to the west it forms the cover of the gneisses of the Menderes Massif. The Afyon Zone can be correlated with the Bolkardağ Unit of Özgül (1976). Low-grade regional metamorphism, as well as its internal structural setting, places the Afyon Zone in the Anatolides (Ketin 1966).

At the base of the Afyon Zone is a thick sequence (>1500 m) of meta-sandstones, meta-siltstones and meta-quartzites with rare metabasite and recrystallised limestone horizons. Although the sequence is sparsely fossiliferous, Devonian fossils are found in the basal part of the clastic sequence in the Ilgın region

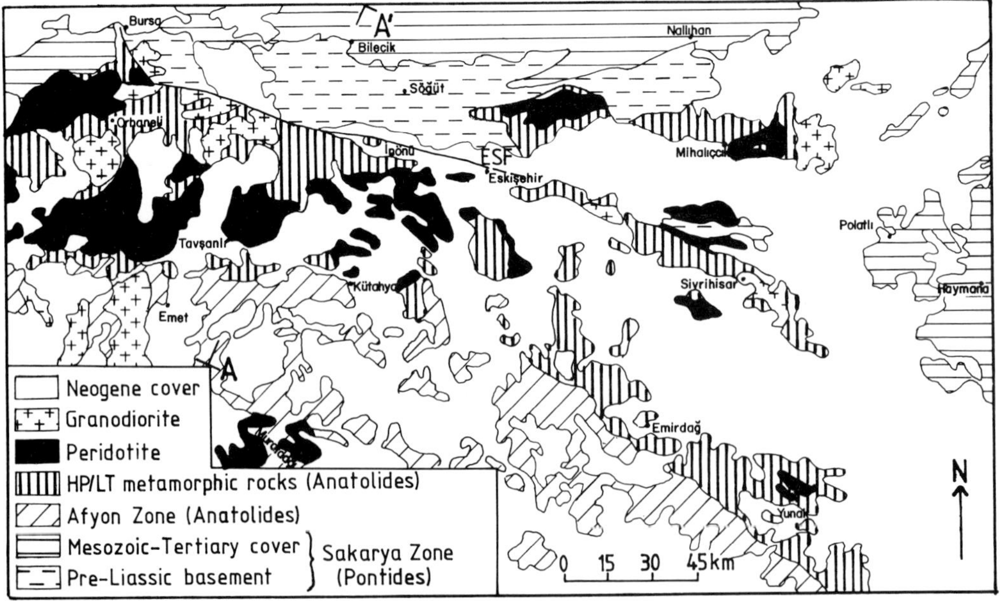

FIG. 2. Geotectonic map of north-west Turkey showing the tectonic units discussed in the text. ESF, Eskişehir Fault.

north of Konya (Niehoff 1964). The metaclastics pass up gradually into dark grey recrystallized limestones, which have yielded Carboniferous to Upper Permian fusulinids in the Afyon region (Erişen 1972). The Triassic sequence starts with a basal meta-conglomerate, consisting of meta-sandstones, meta-siltstones and intercalated algal limestones. The rest of the Mesozoic is represented by a thick sequence of platform carbonates, which extend to Maastrichtian (Fig. 4). The carbonates are locally overlain by an Upper Maastrichtian to Palaeocene 'wildflysch' comprising Permian, Jurassic and Cretaceous limestone olistoliths (Akdeniz & Konak, 1979; Konak, 1982). A strongly tectonised volcano-sedimentary complex with incipient HP/LT metamorphism and a periodotite nappe, lies tectonically over the carbonates and flysch (Fig. 3).

HP/LT metamorphic belt (Anatolides)

Rocks affected by HP/LT metamorphism occupy an elongate WNW-ESE area in north-west Turkey, extending for over 350 km from the region of Orhaneli eastward to Yunak (Figs 1 & 2). The width of the blueschist belt increases towards the east: in the Tavşanlı region blueschists are 50 km wide; in the east the width of the blueschist belt is of the order of 100–130 km. The eastward extension of the blueschist belt is abruptly terminated by a poorly defined NNW-SSE trending major fault, which juxtaposes blueschists and the Upper Mesozoic-Tertiary sedimentary sequence of the Haymana basin (Fig. 2). Farther southeast, north of Konya (Bayiç 1968) and in the Bolkardağ region (Blumenthal 1955; Çalapkulu 1980) sodic amphibole-bearing meta-volcanic rocks are intercalated with recrystallized limestones and psammitic schists. It is not known whether they represent the southeastward continuation of the north-west Turkish blueschist belt or whether they are part of the Afyon Zone, which has undergone a high-pressure greenschist facies metamorphism in these regions.

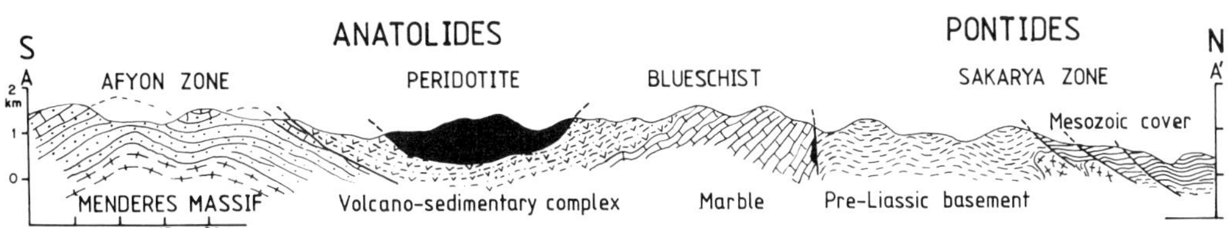

FIG. 3. Simplified structural section across north-west Turkey. The internal imbrication and faulting in the Afyon Zone and in the blueschist sequence are not shown. Line of section shown in Fig. 2 as A-A′.

Associated with the tectonic emplacement of the peridotites, the Afyon Zone was internally sliced and imbricated in post-Palaeocene time. South of İzmir the Mesozoic lithologies of the Afyon Zone rest directly on the gneisses of the Menderes Massif with a thin intervening tectonic slice of Palaeocene flysch (Başarır & Konak 1981). However, in some areas the thick basal clastic sequence of the Afyon Zone passes downward without tectonic break into the gneisses of the Menderes Massif. It seems likely that the gneisses of the Menderes Massif represent old (?Pan-African) metamorphic basement on which Palaeozoic and Mesozoic sequence of the Afyon Zone were deposited prior to the Alpine regional metamorphism (Dürr *et al.* 1978). The low-grade greenschist facies metamorphism, which has affected the Afyon Zone, shows a progressive increase in grade towards the deeper levels of the sequence. The Palaeozoic meta-clastics have the characteristic mineral assemblage: quartz + albite + phengite + chlorite ± biotite, whereas the Mesozoic carbonates and the flysch are little affected by the metamorphism and the carbonates are commonly fossiliferous.

In the south, rocks affected by the HP/LT metamorphism rest with a tectonic contact on the carbonates and flysch of the Afyon Zone. In the north, steeply-dipping faults juxtapose blueschists with the pre-Liassic basement of the Pontide sequence. The important Eskişehir fault (Fig. 2), marked by extensive brecciation and serpentinite lenses (Okay 1980c), is also characterized by sharp changes in gravity and magnetic anomalies. In a few places, such as south of Bursa (Lisenbee 1971), the basement rocks of the Pontide sequence lie with a low-angle tectonic contact over the blueschists. Blueschists are tectonically overlain by a peridotite nappe, which is now largely preserved in the regional depression between Bursa and Eskişehir.

Most of the rocks affected by the HP/LT metamorphism in north-west Turkey consist of

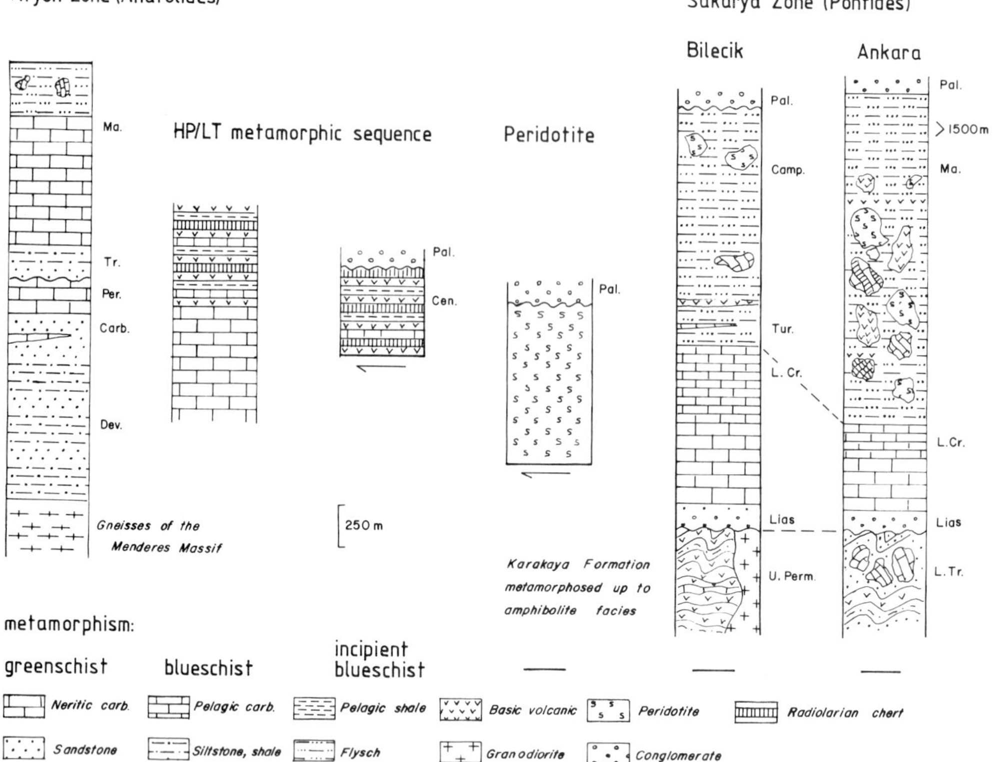

FIG. 4. Generalized synthetic stratigraphic sections of the tectonic units of north-west Turkey. The thicknesses of the HP/LT metamorphic sequences are very approximate. Dev., Devonian; Carb., Carboniferous; Per., Permian; Tr., Trias; Cr., Cretaceous; Cen., Cenomanian; Tur., Turonian; Camp., Campanian; Ma., Maastrichtian; Pal., Palaeocene; U., Upper; L., Lower.

a strongly tectonised volcano-sedimentary complex, which appears to be unmetamorphosed in the field but on careful examination reveals an incipient HP/LT metamorphism (Okay 1982). Completely recrystallized, schistose blueschists make up 30–40% of the 'blueschist' terrain. In this respect resemblance is striking to the Franciscan terrain in California, where most rocks also appear to be unmetamorphosed.

Incipiently metamorphosed volcano-sedimentary complex

These rocks occur as steeply-dipping imbricate slices underneath massive peridotites. They lie tectonically over the carbonates and the flysch of the Afyon Zone, or over the true blueschists. The rock-types in decreasing order of abundance are: pyroclastics, basic volcanic rocks, red and green pelagic shales, radiolarian cherts, pelagic limestones, greywackes, synsedimentary breccias, manganese deposits, keratophyres and neritic limestone olistoliths (Kaya 1972; Okay 1982). Penetrative deformation is absent in the sequence, and the rocks appear to be unmetamorphosed in the field. The characteristic feature of this complex is the close and repeated sedimentary intercalation of different rock types: a typical spilite-shale-chert sequence, for example, can occur 3–4 times over a 100 m wide section (Okay 1981, Fig. 4). Talc and serpentinite outcrops of various sizes mark numerous thrusts in the volcano-sedimentary complex. Tectonic slices of greenschist-facies metabasites are found in the volcano-sedimentary complex in the Tavşanlı region (Okay 1982). Much of the strong tectonism affecting this sequence was probably related to the emplacement of the peridotite nappe, and clearly post-dates HP/LT metamorphism.

Because of the strong tectonism, the thickness of the volcano-sedimentary complex is difficult to estimate. Minimum thicknesses range from 160 metres in the Orhaneli area (Lisenbee 1971) to 250 metres in the Tavşanlı area. Data on the age of deposition of the complex is sparse; pelagic limestones from the Muratdağı (Bingöl 1977) and east of Gediz (Akdeniz & Konak 1979) yield Cenomanian ages. Servais (1981) assigns a tentative Late Jurassic–Early Cretaceous age to the radiolarian cherts from the volcano-sedimentary complex south of İnönü.

A notable feature of this volcano-sedimentary complex is the presence of HP/LT minerals such as lawsonite and aragonite in the veins and amygdales in the volcanic rocks; calcite in pelagic limestones is recrystallized to coarse grains of aragonite. Associated with this incipient HP/LT metamorphism there has been extensive sodium metasomatism involving topotactic replacement of augite by sodic pyroxene in the volcanic rocks. The metamorphic conditons are estimated at 5–8 kb and 250–300°C (Okay 1982).

Blueschists

The rock types are similar to the volcano-sedimentary complex except that there is a thick basal marble unit but no ultramafic rocks.

White massive and locally cherty marble usually occurs as a several kilometre-thick unit beneath intercalated metabasites, meta-cherts and meta-shales. The contact between the marble and the meta-volcano-sedimentary sequence is primary as marble horizons and blocks are observed in the meta-volcano-sedimentary sequence near the marble contact (Çoğulu 1967; Yeniyol 1979; Okay 1980a, c). In some areas, such as in the Sivrihisar region, marble also occurs as several hundred metre-thick horizons within metabasites and meta-cherts. A foliation is generally well developed in all rock types in the glaucophane-lawsonite and glaucophane-epidote zones, where it is defined by the parallel orientation of mica and sodic amphibole grains. At least two phases of isoclinal folding are present in the glaucophane-lawsonite zone rocks (Okay 1980c). The thickness of the blueschist unit is difficult to estimate because of intense folding; apparent thicknesses of the meta-volcano-sedimentary sequence are ca. 1000–1500 m.

Lawsonite, glaucophane-lawsonite and glaucophane-epidote zones have been mapped in various parts of the Northwest Turkish blueschist belt (Çoğulu 1967; Okay 1980a, b; Kulaksız 1981). Typical mineral assemblages in the metabasites and meta-cherts from the Tavşanlı region are shown in Table 1. A characteristic feature of the north-west Turkish blueschists is the prograde nature of the mineral assemblages. The rocks generally do not exhibit the retrograde textures and minerals common in other terrains, so that prograde stages of metamorphism can easily be distinguished. The HP/LT metamorphism in north-west Turkey is characterized by an initial static metamorphic event with the development of a lawsonite + sodic pyroxene + chlorite assemblage in the basic volcanic rocks, where sodic pyroxene is pseudomorphous after augite. At this lawsonite zone stage the primary igneous texture was largely retained (Çoğulu 1967; Okay 1980b). This static recrystallization was followed by a second metamorphic event associated with the onset of penetrative deformation. Sodic amphibole formed at the expense of sodic pyroxene and chlorite during this second metamorphic phase. A late greenschist overprint, with development of epidote and calcic amphibole, is reported from the Mihaliççık (Çoğulu 1967) and Sivrihisar (Kulaksız 1981) areas.

Çoğulu and Krummenacher (1967) dated blueschists from the Mihaliççık region by the K/Ar method: phengites from two glaucophane schists gave ages of 65 and 82Ma. An upper age of blueschist metamorphism is given by Middle Palaeocene sediments unconformably overlying the volcano-sedimentary complex and peridotite in the Yunak (Yeniyol 1979) and Emirdağ (Umman & Yergök 1979) areas. Lower and Middle Eocene conglomerates and limestones are also transgressive over the peridotite and blueschist in the Tavşanlı, Yunak (Yeniyol 1979) and Simav (Akdeniz & Konak 1979) areas. Thus, the age of blueschist metamorphism is constrained to lie between Cenomanian and Middle Palaeocene. A further age constraint is given by abundant blueschist detritus in the Upper Campanian flysch of the Haymana basin (Batman 1978).

Eclogites

Eclogites are reported from the Mihaliççık and Sivrihisar areas. In the Mihaliççık region the supposed eclogites occur as tectonic inclusions in peridotite. However, chemical analysis of the eclogite (Çoğulu 1967) suggests that it may be an andradite-hedenbergite skarn rather than a true eclogite. In the Sivrihisar area, garnet-omphacite eclogites with minor glaucophane occur as small blocks along a major fault within the glaucophane-lawsonite zone blueschists (Kulaksız 1978).

Peridotite nappe

Peridotite forming the highest tectonic unit over both the blueschists and the low-grade

TABLE 1. *Progressive mineral changes in the Tavşanlı area, north-west Turkey (double lines, major (20%) mineral; single lines, minor mineral; dashed lines, sporadic occurrence)*

Metamorphic Zones	Incipient Metamorphism	Lawsonite Zone	Glaucophane-lawsonite Zone	
mineral \ rock	Basic volcanic rocks			Chert
Sodic amphibole		- - -	═══	───
Lawsonite	- - -	═══	═══	───
Sodic pyroxene	───	═══	───	
Chlorite	═══	═══	───	- - -
Albite	═══			
Quartz	- - -	- - -	- - -	═══
Phengite		───	───	───
Alm. garnet			- - -	
Spess. garnet				───
Pumpellyite	───			
Epidote			- - -	
Sphene		───	───	
Magnetite			- - -	
Hematite			- - -	───
Aragonite	- - -	- - -	- - -	
Relict augite	═══	- - -		

metamorphic rocks of the Afyon Zone is largely preserved in the regional depression between Orhaneli and Eskişehir, as a single massif over 4000 km^2 in area. In the east peridotite occurs as isolated klippen on the blueschists.

The dominant rock type is a spinel-bearing harzburgite/dunite tectonite (Çoğulu 1967; Lisenbee 1971; Okay 1980c). Serpentinization is not extensive except at the margins of the peridotite and along faults. Gabbro, occupying less than 2% of the surface area of the peridotite, occurs as small bodies a few hundred metres wide within the peridotites where it commonly shows compositional layering (Uz 1978; Tankut 1980). The common type is a pyroxene-olivine-plagioclase gabbro, with the pyroxene often uralitized and the plagioclase saussuritized.

The peridotite is cut by discontinuous microgabbroic dykes and stocks of variable width (0.2–5 m). The density of dykes is variable averaging 2–3 dykes per 500 metres horizontally. Dykes are commonly chilled against the peridotite, do not extend into the country rock, and like the gabbros, are affected by low-grade ocean floor-type metamorphism. Although the igneous texture survives, the primary plagioclase + augite assemblage in the microgabbros has been replaced by an albite + hornblende + pumpellyite assemblage (Lisenbee 1971; Yeniyol 1979; Okay 1980c). The mineral assemblages in the dykes and gabbros indicate that the peridotite unit has not undergone HP/LT metamorphism.

The upper parts of the ideal ophiolite sequence are apparently absent in north-west Turkey; this could be due to erosion, or it could be primary. It is significant that no complete ophiolite sequence has been described along the İzmir-Erzincan suture where peridotites are quite common.

There are no reliable isotopic or other data

for the age of the peridotite; its emplacement over the volcano-sedimentary complex must be post-Cenomanian but pre-Middle Palaeocene.

Sakarya Zone (Pontides)

The Sakarya Zone corresponds approximately to the Mysisch-Galatische Scholle of Brinkmann (1966), to the Inner Pontides of Bergougnan and Fourquin (1980) and to the Sakarya continent of Şengör & Yılmaz (1981). In western Turkey the Zone occupies roughly the area between the North Anatolian Fault and the İzmir-Erzincan suture (Fig. 1). The Sakarya Zone consists of a pre-Liassic metamorphic basement and a cover of Mesozoic to Tertiary clastics and carbonates. The stratigraphy, which is completely different from the Afyon Zone, is well described in the Bilecik (Altınlı 1975; Saner 1980), and Ankara (Batman 1978; Akyürek 1981; Ünalan 1981) regions, and is summarized in two generalized stratigraphic sections in Fig. 4.

The basement of the Sakarya Zone consists of a thick, deformed, variably metamorphosed sequence of greywackes interbedded with basic volcanic rocks, shales and rare limestones. The upper parts of this sequence contain abundant olistoliths of Permian and Carboniferous limestone in a matrix of greywacke and shale. Small (<100 m), rare blocks of serpentinite are also reported (Yılmaz 1977). This Karakaya Formation of Bingöl *et al.* (1975) has a very wide distribution in the Pontides (Tekeli 1981), and extends from Edremit on the Aegean coast to the region of Ankara as the Dikmen greywackes, or the Dikmen melange of Özkaya 1982); and farther east to Tokat (Özcan *et al.* 1980) and the Ağvanis Massifs. In north-west Turkey the Karakaya Formation is best exposed in a 200 km long and 20 km wide, slightly arcuate metamorphic belt between Bursa and Nallıhan (Fig. 2). Here basic volcanic rocks are the main rock type with lesser amounts of greywacke, shale, limestone and chert. Around the town of Söğüt, Yılmaz (1977) differentiated four metamorphic zones with an increase in metamorphic grade to the north-west: spilites are progressively metamorphosed to amphibolites with garnet, barroisite and albite. Sodic amphibole in iron-rich meta-cherts and in some metabasites (Yılmaz 1977; Ayaroğlu 1979), and barroisite rather than hornblende in the amphibolites are indicative of high-pressure greenschist facies metamorphism. In the Sakarya region this sequence is intruded by a granite which has been dated as Early Permian (Çoğulu *et al.* 1965).

Data on the depositional age of the Karakaya Formation are scarce. Some of the intercalated limestones have yielded Late Permian ages in the Sakarya region (Sungur 1973) and Early Triassic (Scythian) ages north of Ankara (Akyürek *et al.* 1979). As the oldest sediments unconformably overlying the metamorphosed Karakaya Formation are of Middle to Late Triassic age (Bingöl *et al.* 1975; Servais 1981), metamorphism and uplift of the Karakaya Formation must have occurred during the Triassic (Tekeli 1981). However, the major marine transgression in the Pontides was at the beginning of the Liassic. In the Sakarya Zone the deformed and metamorphosed rocks of the Karakaya Formation are overlain by widespread Lower Liassic basal conglomerates. The conglomerate consists of well-rounded granite, schist and spilite fragments in a calcareous matrix. It passes upwards into sandstones and then into neritic carbonates of Middle to Late Jurassic age. The Early Cretaceous is represented by hemipelagic cherty limestones. In the Middle Jurassic–Early Cretaceous interval carbonate deposition is ubiquitous over the whole of the Pontides.

In the Sakarya region deposition of flysch with tuffaceous intercalations starts in the Cenomanian. The first serpentinite olistoliths in the flysch occur in the Campanian (Saner 1980). In the Palaeocene, molasse deposition starts with a gradual northward marine regression.

In the Ankara region the Late Cretaceous is represented by a chaotic sequence, locally over 2500 m thick, of serpentinite, basic volcanic rocks, exotic blocks of Permian, Jurassic and Lower Cretaceous limestones, radiolarian cherts, shales and volcanogenic sandstones (Çapan & Buket 1975; Batman, 1981). This sequence passes up gradually into regularly bedded flysch of Campanian age. The chaotic ophiolitic sequence, the Dereköy Formation (Ünalan *et al.* 1976), or Kırıkkale Mélange, (Özkaya 1982) is currently interpreted as a tectonized massive olistostrome deposit (Norman 1975; Batman 1981; Norman this volume). The Ankara Melange, as originally described by Bailey & McCallien (1954), includes both the partially metamorphosed pre-Liassic Karakaya Formation with abundant Permian limestone olistoliths, and an Upper Cretaceous ophiolitic mélange (Dereköy Formation).

Following deposition of the Dereköy Formation, a deep and relatively narrow intra-continental basin developed south-west of Ankara (Haymana basin) during the Maastrichtian. This received over 5000 m of flysch-type clastics between Maastrichtian and Lutetian times (Görür *et al.* this volume). Blueschist detritus, for example as grain-flows with glaucophane-lawsonite schist fragments, first occurs in greywackes of the Upper Campanian-Maastrichtian flysch (Batman 1978) and is common along with serpentinite detritus in the Palaeocene and Eocene clastics (Norman & Rad 1971).

Calc-alkaline plutonism and Neogene deposits

During the Early Tertiary important calc-alkaline plutonism affected the whole of north-west Turkey. Several individual stocks intrude peridotite, blueschist and the low-grade metamorphic rocks of the Afyon Zone (Fig. 2). The dominant plutonic rock-type is a hornblende—biotite granodiorite (Bürküt 1966). Age determinations on these granodiorites range from Early Palaeocene to Late Oligocene (Ataman 1974; Bingöl *et al.* 1982). Miocene conglomerates, commonly containing granodiorite fragments, are transgressive over the granodiorites. The calc-alkaline plutonism of north-west Turkey seems to be closely related to the uplift of the HP/LT metamorphic rocks, and may be the result of rapid near-isothermal decompression of the thickened continental crust of north-west Turkey.

The Palaeocene and Eocene was a period of uplift for much of north-west Turkey, so that sediments of this age are restricted. At the beginning of the Miocene the area was relatively stable with numerous lakes separated by highlands of peridotite and blueschist. Terrestrial and lacustrine Neogene sediments are widely distributed in north-west Turkey (Fig. 2), and consist largely of limestone, tuff, shale and lignite of Miocene and Pliocene age (Brinkmann 1976).

Discussion

The striking contrasts of stratigraphy in the Afyon and Sakarya Zones point to the existence of a major ocean in north-west Turkey in pre-Tertiary times. Evidence for the age of opening of this İzmir-Erzincan ocean in north-west Turkey, and indeed in the 1500 km-long suture separating the Pontides and Anatolides is very sparse. Continental margin sequences, which establish fairly accurately the age of initial rifting in areas such as the Antalya region (Marcoux 1976; Robertson & Woodcock 1981), have not been described along this major suture. It is still not clear whether the İzmir-Erzincan ocean represents one of the branches of Neotethys which opened during the Liassic (Şengör & Yılmaz 1981), or whether it is the original Tethys (Palaeotethys) which was in existence from the Mid-Palaeozoic onward, as implied by Biju-Duval *et al.* (1977). Although pre-Liassic thrusting and deformation have recently been described from the Taurides (Akay 1981; Monod & Akay this volume), the restriction of the Triassic orogeny, as represented by the metamorphosed and deformed Karakaya Formation, to the Pontides suggests that important differences between the Anatolide/Taurides and Pontides existed in pre-Liassic times. Triassic basic volcanics, tuffs and pelagic sediments in the Bozkır nappes north of the Tauride autochthon (Özgül 1976), also indicate that the opening of the İzmir-Erzincan ocean dates back to pre-Triassic times.

The structural position of the blueschists between the Afyon Zone and the peridotite nappe (Fig. 3) indicates that they belong to the Anatolide rather than to the Pontide realm. It is suggested that the blueschist sequence of north-west Turkey represents part of the northward-facing continental margin of the Anatolide/Tauride platform. The main evidence for this is the sudden passage from the thick platform carbonates to the pelagic sediments and volcanics in the blueschist sequence, which is interpreted as due to rapid foundering of a carbonate platform following rifting and establishment of a continental margin. The neritic carbonate olistoliths in the volcano-sedimentary complex of the Emirdağ and Tavşanlı regions may be another indication of block-faulting and foundering of a carbonate platform. Rapid submergence of shallow water platforms by synsedimentary normal faulting and the inception of pelagic conditions over large areas is characteristic of the early Mesozoic history of the Tethys, and generally indicates rifting preceding ocean floor speading (Bernoulli & Jenkyns 1974; Laubscher & Bernoulli 1977). The predominance of basic pyroclastic rocks in the blueschist volcano-sedimentary sequence of north-west Turkey could indicate proximity to a volcanic island. The oldest available ages from the volcano-sedimentary complex indicate a pre-Late Jurassic age for the platform carbonates of the blueschist unit (Servais 1981); accurate establishment of their age will probably solve the problem of the age of opening of the İzmir-Erzincan ocean.

Any plate tectonic model attempting to explain the formation and the present day tectonic position of the blueschists in north-west Turkey must take account of the following important features of the blueschists and related rocks:

(1) HP/LT metamorphic rocks are concen-

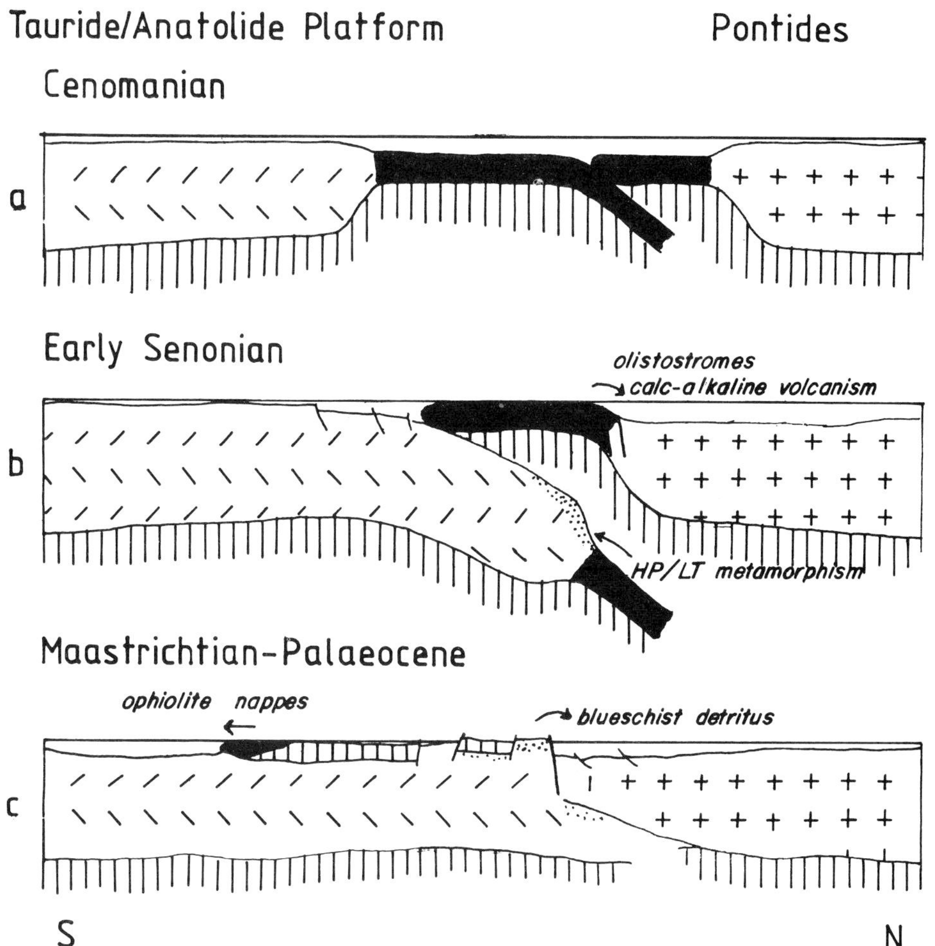

FIG. 5. Schematic sequence of N-S sections illustrating a possible model for the genesis of the north-west Turkish blueschists involving continent–continent collision. For explanations see the text.

trated in north-west Turkey, and do not occur regionally elsewhere in the İzmir-Erzincan suture.

(2) The blueschist sequences do not represent oceanic crust but may have been originally continental margin sequences.

(3) The peridotite nappe, which tectonically overlies the blueschists, has not undergone HP/LT metamorphism. A regular ophiolite sequence seems to be lacking in north-west Turkey.

It is believed that the available geological data and the inferences from these data can best be explained by invoking a northward dipping intra-oceanic subduction zone, and an irregularly shaped continental margin of the Anatolide/Tauride platform, with north-west Turkey forming a promontory during the Early Cretaceous (Fig. 5a). The consumption of the İzmir-Erzincan ocean eventually led to the subduction and HP/LT metamorphism of the north-west Turkish continental margin. The subduction of the buoyant continental crust of north-west Turkey caused uplift of the overlying oceanic crust and mantle at the same time (Fig. 5b). Deposition of flysch, calc-alkaline volcanism and huge olistostromes of ophiolitic material in the Sakarya Zone indicate that the subduction of the continental crust and uplift had already started in the Turonian. Blueschists must have been locally exposed by the Upper Campanian when detritus in the Pontide Haymana basin appears (Batman 1978). This is also an indication that the İzmir-Erzincan ocean was eliminated in north-west Turkey by the Maastrichtian (Fig. 5c). Continuing compression and strong uplift elevated the Sakarya Zone above sea level in the Palaeocene; at the same time in the south, huge ophiolite slices were detached from the uplifted oceanic crust and mantle and moved southward over the Anatolide/Tauride carbonate platform (Bozkır ophiolite nappes of Şengör & Yılmaz).

Uplift and exposure of blueschists implies several tens of kilometres of vertical upward movement relative to the rocks of Sakarya Zone. This uplift, must have been achieved along the major faults, which now juxtapose blueschists, formed at 30–35 km depth, and the Mesozoic sediments or the basement rocks of the Sakarya Zone. Another consequence is that most of the peridotites in north-west Turkey are preserved oceanic mantle rather than oceanic crust.

Uplift to the surface

A remarkable feature of the north-west Turkish blueschists is their rapid uplift to the surface. The youngest depositional ages from the HP/LT metamorphic rocks are Cenomanian. In the Upper Campanian and Maastrichtian blueschists were supplying detritus to the Haymana basin. Thus, the termination of the pelagic sediment deposition, HP/LT metamorphism and uplift to the surface must have occurred in less than 25 Ma (Cenomanian-Maastrichtian). The prograde nature of the HP/LT mineral assemblages and the general absence of a greenschist overprint also indicate very fast uplift.

The presence of aragonite, jadeite and lawsonite in the glaucophane-lawsonite schists gives a conservative pressure estimate of 10 kbar during the HP/LT metamorphism (Okay 1980c). As the blueschists occur as a coherent terrain and not as exotic blocks, a rock column 30–35 km thick must have been eroded during the Upper Cretaceous to expose the blueschists at the surface. Recent thermal modelling indicates minimum denudation rates of 1.4 mm/year or higher for the preservation of blueschists with no greenschist overprint (Draper & Bone 1981). Such a constant denudation rate gives a maximum total exhumation time of 20–25 Ma for blueschists formed at 30–35 km depth, which agrees well with the geological evidence.

As Draper & Bone (1981) point out, present day erosion rates are generally too low for the required unloading. They invoke tectonic and quasi-tectonic mechanisms, such as nappes or olistostrome flows, to achieve the necessary denudation rates. In north-west Turkey the unloading of the blueschists seems to have been achieved in two distinct ways. During the Late Cretaceous the erosion was aided by the formation of massive olistostromes which were deposited in basins in the north (Fig. 5b). Several thousand metre-thick chaotic deposits of ophiolitic material overlie carbonate platform rocks over large areas in the Central Anatolia. During the uppermost Cretaceous and Palaeocene, when the Sakarya Zone was also temporarily uplifted, the unloading was achieved by the detachment and southward movement of peridotite nappes, terminating sedimentation in the Afyon Zone (Fig. 5c).

ACKNOWLEDGEMENTS: For discussions on the geology of north-west Turkey I thank Cemal Göncüoğlu, Necdet Özgül, A. M. Celal Şengör, Okan Tekeli and Yücel Yılmaz. The manuscript was improved by suggestions from A. M. Celal Şengör.

References

AKAY, E. 1981. Beyşehir yöresinde (Orta Toroslar) olası Alt Kimmeriyen dağ oluşumu izleri. *Bull. geol. Soc. Turkey*, **24,** 105–10.

AKDENIZ, N. & KONAK, N. 1979. Simav, Emet, Tavşanlı, Dursunbey, Demirci, Kütahya dolaylarının jeolojisi. MTA Institute report no. 6547 (unpubl.), Ankara.

AKYÜREK, B. 1981. Ankara Melanjı'nın kuzey bölümünüm temel jeoloji özellikleri. *In*: *Iç Anadolu'nun jeolojisi*, Ankara, 41–5.

——, BILGINER, E., DAĞER, Z. & SUNU, O. 1979. Hacılar (Kuzey Çubuk—Ankara) bölgesinde Alt Triyasın varlığı. *Bull. geol. Soc. Turkey*, **22,** 169–74.

ALTINLI, E. 1975. Geology of the Middle Sakarya river (Turkey). *In*: *Congress of Earth Sciences on the Occasion of the Fiftieth Anniversary of the Turkish Republic*, MTA Insitute, Ankara 1973, 161–96.

ATAMAN, G. 1974. Révue géochronologique des massifs plutoniques et métamorphiques de l'Anatolie. *Hacettepe Bull. Natl. Sci. Eng.* **3,** 75–88.

AYAROĞLU, H. 1979. Bozüyük metamorfitlerinin petrokimyasal özellikleri *Bull. geol. Soc. Turkey*, **22,** 101–7.

BAILEY, E. B. & MCCALLIEN, W. J. 1954. Serpentine lavas, the Ankara melange and the Anatolian thrust. *Trans. Roy. Soc. Edinburgh*, **62,** 403–42.

BAŞARIR, E. & KONUK, Y. T. 1981. Gümüldür yöresinin kristalin temeli ve allokton birimler. *Bull. geol. Soc. Turkey*, **24,** 95–100.

BATMAN, B. 1978. Haymana kuzeyinin jeolojik evrimi ve yöredeki melanjın incelenmesi I: stratigrafi birimleri. *Yerbilimleri*, **4,** 95–124.

—— 1981. Ofiyolitli melanjın (Dereköy Formasyonu) Haymana kuzeyindeki yörede (SW Ankara) incelenmesi. *Yerbilimleri*, **8,** 61–70.

BAYIÇ, A. 1968. On metaporphyrites of the Sızma region (Konya). *Bull. Miner. Res. Explor. Ins. Turkey*, **70,** 142–56.

BERGOUGNAN, H. & FOURQUIN, C. 1980. Un ensemble d'éléments communs à une marge active alpine des Carpathes méridionales à l'Iran central: le domaine irano-balkanique. *Bull. Soc. géol. Fr.* **12,** 61–83.

BERNOULLI, D. & LAUBSCHER, H. 1972. The palinspastic problem of the Hellenides. *Eclogae Geol. Helv.* **65,** 107–18.

—— & JENKYNS, H. C. 1974. Alpine, Mediterranean and Central Atlantic Mesozoic facies in relation to the early evolution of the Tethys. *In*: DOTT, R. H. JR. & SHAVER, R. H. (eds). *Modern and Ancient Geosynclinal Sedimentation.* Spec. Publ. Soc. Econ. Paleont. Mineral. **19,** 129–60.

BIJU-DUVAL, B., DERCOURT, J. & LE PICHON, X. 1977. From the Tethys ocean to the Mediterranean seas: a plate tectonic model of the evolution of the western Alpine System. *In*: BIJU-DUVAL, B. & MONTADERT, L. (eds). *Structural History of the Mediterranean Basins.* Editions Technip, Paris, 143–64.

BINGÖL, E. 1977. Muratdağı jeolojisi ve anakayaç birimlerinin petrolojisi. *Bull. geol. Soc. Turkey*, **20,** 13–66.

——, AKYÜREK, B. & KORKMAZER, B. 1975. Geology of the Biga Peninsula and some characteristics of the Karakaya blocky series. *In*: *Congress of Earth Sciences on the Occasion of the Fiftieth Anniverary of the Turkish Republic*, MTA Institute, Ankara 1973, 71–7.

——, DELALOYE, M. & ATAMAN, G. 1982. Granitic intrusions in western Anatolia: a contribution to the geodynamic study of this area. *Eclogae Geol. Helv.* **75,** 437–46.

BLUMENTHAL, M. M. 1956. *Yüksek Bolkardağın kuzey kenar bölgelerinin ve batı uzantılarının jeolojisi.* Publ. Miner. Res. Explor. Ins. Turkey, D7, Ankara, 153 pp.

BRINKMANN, R. 1966. Geotektonische Gliederung von Westanatolian. *Neues Jahrb. Geol. Paläontol., Monatshefte*, **10,** 603–18.

—— 1976. *Geology of Turkey.* Ferdinand Enke Verlag, Stuttgart, 158 pp.

BÜRKÜT Y. 1966. *Kuzeybatı Anadolu'da yer alan plutonların mukaye seli jenetik etüdü.* İTÜ Maden Fakültesi, Istanbul, 272 pp.

ÇALAPKULU, F. 1980. Horoz granodiyoritinin jeolojik incelemesi. *Bull. geol. Soc. Turkey*, **23,** 59–68.

ÇAPAN, U. Z. & BUKET, E. 1975. Aktepe-Gökdere bölgesinin jeolojisi ve ofiyolitli melanj. *Bull. geol. Soc. Turkey*, **18,** 11–6.

ÇOĞULU, E. 1967. Etude pétrographique de la région de Mihaliççik. *Schweiz. Mineral. Petrogr. Mitt.* **47,** 683–824.

—— & KRUMMENACHER, D. 1967. Problèmes géochronométriques dans la partie N de l'Anatolie Centrale (Turquie). *Schweiz. Mineral. Petrogr. Mitt.* **47,** 825–833.

DRAPER, G. & BONE, R. 1981. Denudation rates, thermal evolution and preservation of blueschist terrains. *J. Geol.* **89,** 601–14.

DÜRR, ST., ALTHERR, R., KELLER, J., OKRUSCH, M. & SEIDEL, E. 1978. The Median Aegean Crystalline belt: stratigraphy, structure, metamorphism, magmatism. *In*: CLOSS, H., ROEDER, D. & SCHMIDT, K. (eds). *Alps, Apennines and Hellenides.* Schweizerbart'sche, Stuttgart, 455–76.

ERİŞEN, B. 1972. *Afyon-Heybeli (Kızılkilise) araştırma sahasının jeolojisi ve jeotermal enerji olanakları.* MTA Institute report no. 5490 (unpubl.), Ankara.

FOURQUIN, C. 1975. L'Anatolie du Nord-Ouest, marge méridionale du continent européen, histoire paléogéographique, tectonique et magmatique durant le Secondaire et le Tertiaire. *Bull. Soc. géol. Fr.* **17,** 1058–70.

KAADEN, G. VAN DER 1966. The significance and distribution of glaucophane rocks in Turkey. *Bull. Miner. Res. Explor. Ins. Turkey*, **67,** 37–67.

KAYA, O. 1972. Tavşanlı yöresi ofiyolit sorununun ana çizgileri. *Bull. geol. Soc. Turkey*, **15,** 26–108.

KETİN, İ. 1966. Tectonic units of Anatolia. *Bull. Miner. Res. Explor. Ins. Turkey*, **66,** 23–34.

KONAK, N. 1982. Simav dolayının jeolojisi ve metamorf kayaçların evrimi. *İstanbul Yerbilimleri*, **3,** 313–37.

KULAKSIZ, S. 1978. Sivrihisar kuzeybatı yöresi eklojitleri. *Yerbilimleri*, **4,** 89–94.

—— 1981. Sivrihisar kuzeybatı yöresinin jeolojisi. *Yerbilimleri*, **8,** 103–24.

LAUBSCHER, H. P. & BERNOULLI, D. 1977. Mediterranean and Tethys. *In*: NAIRN, A. E. M., KANES, W. H. & STEHLI, F. G. (eds). *The Ocean Basins and Margins. 4A. The Eastern Mediterranean.* Plenum, New York, 1–28.

LISENBEE, A. 1971. The Orhaneli ultramafic-gabbro thrust sheet and its surroundings. *In*: CAMPBELL, A. S. (ed.). *Geology and History of Turkey.* Pet. Explor. Soc. Libya, Tripoli, 349–60.

MARCOUX, J. 1976. Les séries triasiques des nappes à radiolarites et ophiolites d'Antalya (Turquie): homologies et signification probable. *Bull. Soc. géol. Fr.* **18,** 511–2.

NIEHOFF, W. 1964. 90/2 (Akşehir), 91/1,3,4 (Ilgın) paftaları jeolojik revizyon raporu. MTA Institute report no. 3387 (unpubl.), Ankara.

NORMAN, T. N. 1975. Flow features of Ankara Melange. *Proc. 9th Int. Sed. Cong. Nice*, **4,** 261–69.

—— & RAD, M. R. 1971. Çayraz (Haymana) cıvarındaki Harhor (Eosen) Formasyonunda alttan üste doğru doku parametrelerinde ve ağır mineral bolluk derecelerinde değişmeler. *Bull. geol. Soc. Turkey*, **14,** 205–25.

OKAY, A. I. 1980a. Mineralogy, petrology and phase relations of glaucophane-lawsonite zone blueschists from the Tavşanlı region, North-west Turkey. *Contrib. Mineral. Petrol.* **72,** 243–55.

—— 1980b. Lawsonite zone blueschists and a sodic amphibole producing reaction in the Tavşanlı region, North-west Turkey. *Contrib. Mineral. Petrol.* **75,** 179–86.

—— 1980c. *The petrology of blueschists in North-west Turkey, north-east of Tavşanlı.* Thesis. Ph.D. (unpubl.). University of Cambridge, England.

—— 1981. Kuzeybatı Anadolu'daki ofiyolitlerin

jeolojisi ve mavişist metamorfizması (Tavşanlı–Kütahya). *Bull. geol. Soc. Turkey*, **24,** 85–95.

—— 1982. Incipient blueschist metamorphism and metasomatism in the Tavşanlı region, Northwest Turkey. *Contrib. Mineral. Petrol.* **79,** 361–67.

Özcan, A., Erkan, A., Keskin, A., Oral, A., Özer, S., Sümengen, M. & Tekeli, O. 1980. *Kuzey Anadolu Fayı—Kirşehir Masifi arasının temel jeolojisi.* MTA Institute report no. 6722 (unpubl.), Ankara.

Özgül, N. 1976. Torosların bazı temel jeoloji özellikleri. *Bull. geol. Soc. Turkey*, **19,** 65–78.

Özkaya, İ. 1982. Origin and tectonic setting of some melange units in Turkey. *J. Geol.* **90,** 269–78.

Robertson, A. H. F. & Woodcock, N. H. 1981. Bilelyeri Group, Antalya Complex: deposition on a Mesozoic passive continental margin, southwest Turkey. *Sedimentology*, **28,** 381–99.

Saner, S. 1980. Mudurnu-Göynük havzasının Jura ve sonrası çökelim nitelikleriyle paleocoğrafya yorumlaması. *Bull. geol. Soc. Turkey*, **23,** 39–52.

Servais, M. 1981. Données préliminaires sur la zone de suture médio-téthysienne dans la région d'Eskişehir (NW Anatolie). *C.r. Acad. Sc. Paris*, **293,** 83–6.

Sungur, G. 1973. *Söğüt—Kızılsaray alanının jeolojisi.* Thesis. M. Sc. (unpubl.). University of İstanbul, Turkey.

Şengör, A. M. C. 1979. The North Anatolian transform fault: its age, offset and tectonic significance. *J. geol. Soc. Lond.* **136,** 269–82.

—— & Yilmaz, Y. 1981. Tethyan evolution of Turkey: a plate tectonic approach. *Tectonophysics*, **75,** 181–241.

Tankut, A. 1980. The Orhaneli Massif, Turkey. *In*: Panayiotou, A. (ed.). *Ophiolites.* Proc. Int. Oph. Symp. Cyprus 1979, 702–13.

Tekeli, O. 1981. Subduction complex of pre-Jurassic age, northern Anatolia, Turkey. *Geology*, **9,** 68–72.

Unman, Ö & Yergök, F. A. 1979. *Emirdağ (Afyon) dolayının jeolojisi.* MTA Institute report no. 6604 (unpubl.), Ankara.

Uz, B. 1978. *Sındırgı—Akhisar bölgesi ofiyolit birliğinin petrografik, petrojenetik ve jeokimyasal incelenmesi.* Thesis (unpubl.) İstanbul Technical University, İstanbul.

Ünalan, G. 1981. Ankara güneybatısındaki "Ankara Melanjı" nın stratigrafisi. *In*: *İç Anadolu'nun jeolojisi*, Ankara, 46–52.

——, Yüksel, V., Tekeli, T., Gönenç, O., Seyirt, Z. & Hüseyin. 1976. Haymana—Polatlı yöresinin (güneybatı Ankara) Ust Kretase—Alt Tersiyer stratigrafisi ve paleocoğrafik evrimi. *Bull. geol. Soc. Turkey*, **19,** 159–176.

Yeniyol, M. 1979. *Yunak (Konya) magnezitlerinin oluşum sorunları, değerlendirilmeleri ve yöre kayaçlarının petrojenezi.* Thesis. Ph.D. İstanbul University, İstanbul.

Yilmaz, Y. 1977. Bilecik—*Söğüt dolayındaki "eski temel karmaşığı" nın petrojenetik evrimi.* Thesis. İstanbul University, İstanbul.

A. I. Okay, Jeoloji Bölümü, İTÜ Maden Fakültesi, Teşvikiye, Istanbul, Turkey.

Palaeotectonic evolution of the Tuzgölü basin complex, Central Turkey: sedimentary record of a Neo-Tethyan closure

N. Görür, F. Y. Oktay, İ. Seymen & A. M. C. Şengör

SUMMARY: The Tuzgölü basin complex represents a cluster of epi-sutural depressions nested on the Ankara Knot in central Anatolia, where several sutures converge. The basin complex consists mainly of two sub-basins, the Haymana and the Tuzgölü (*s.s.*) depressions. During the Late Cretaceous to the Late Palaeocene, the Tuzgölü and the Haymana sub-basins evolved coevally, but independently, as fore-arc basins along the active margins of the Sakarya Continent and the Kırşehir block. Turbidites accumulated in the basin interiors, with shallow marine and terrestrial deposition near the basin margins. Late Palaeocene–Early Eocene collision of the Menderes-Taurus block with the Sakarya Continent and the Kırşehir block along the Inner Tauride Sture, and the coeval collision of the Kırşehir block with the Rhodope-Pontide Fragment along the Erzincan suture juxtaposed and deformed the two sub-basins. Intra-continental convergence continued during the Early to Middle Eocene with further subsidence and turbidite deposition in both sub-basins. After the Middle Eocene, a single molasse basin was characterized by extensive redbeds and evaporites. Extensive salt deposits in the Tuzgölü sub-basin probably relate to marine regression in the later Eocene. The present Tuzgölü basin is part of neotectonic Turkey and constitutes one of the large central Anatolian *ovas*.

One of the principal difficulties in the study of intracontinental basins formed in zones of complex plate convergence and continental collision is the recognition of the extremely complicated and transient tectonic processes. Spatial coincidence of pre-, syn-, and post-collisional basin-forming events frequently produces a number of superimposed basins that can mislead the geologist who sees only the product of the last event.

The Tuzgölü (Salt Lake) basin (Fig. 1) is one area where several basins are nested on top of one another as a result of a complex series of tectonic processes related to the evolution of the Anatolian branch of the Alpides since the latest Cretaceous, and which have frustrated oil company as well as academic geologists trying to formulate a satisfactory model for the structure and evolution. To date, geological and geophysical investigations including stratigraphy, sedimentology, palaeontology, structural geology, seismic reflection profiling, gravity, magnetics, and drilling have been carried out in the Tuzgölü basin because of its supposed hydrocarbon potential (e.g., Druitt & Reckamp 1959, Yüksel 1970, 1973, Norman & Rad 1970, Arikan 1975, Sirel & Gündüz 1976, Ünalan *et al.* 1980, Uygun 1981, Görür 1981, Oktay 1982). These studies reveal that the basin contains a total of ca. 10 km of sediment, of Late Maastrichtian to Recent age. The Late Maastrichtian to Late Eocene sdimentation within the basin was dominated by turbidites, grain and debris-flow deposits mainly of ophiolite, igneous, and metamorphic material, while terrestrial clastics, evaporites, and shallow marine limestones were laid down along the basin margins. Post-Eocene sedimentation was entirely terrestrial represented by widespread red-beds and evaporites.

The principal aim here is to discuss the sedimentological evolution of the Tuzgölü basin and the Kırşehir Massif area, in terms of an evolutionary tectonic model. Because of a lack of data our proposed model is somewhat speculative. Our approach is to characterize first-order tectonic environments during the evolution of the basin and to speculate about their past relations, constrained by models of the tectonic evolution of Turkey (Şengör & Yılmaz 1981, Şengör *et al.* this volume) and some independent palaeomagnetic data (Sanver & Ponat 1981).

Stratigraphy

The Tuzgölü basin *sensu lato* consists of two sub-basins that evolved coevally, but independently, during the Late Cretaceous to Eocene. These are the Haymana sub-basin and the Tuzgölü (*sensu stricto*) sub-basin. The *Tuzgölü basin complex* refers to the Tuzgölü basin *s.l.*, including the Tuzgölü and the Haymana sub-basins, whereas the *Tuzgölü basin* alone designates the Tuzgölü (*s.s.*) sub-basin. Because of their separate evolutionary histories, we present the stratigraphic data from the Tuzgölü and the Haymana sub-basins separately (Fig. 2).

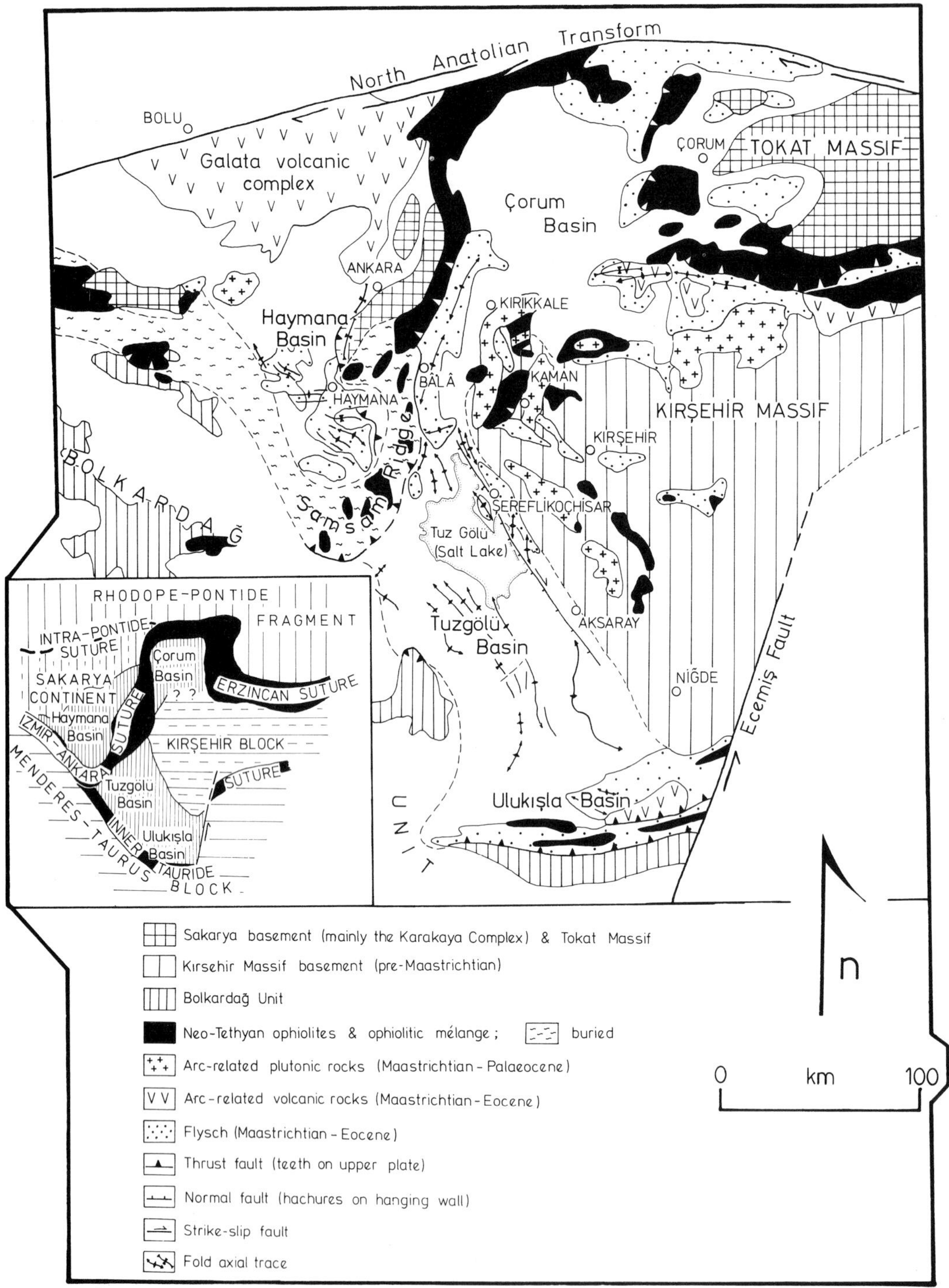

FIG. 1. Simplified geological map of the Tuzgölü basin complex and surrounding regions. Geology N of the North Anatolian transform fault and E of the Ecemiş fault is not shown. White areas are younger (Oligocene to Quaternary) cover. With the exception of the North Anatolian transform fault, the Şereflikoçhisar/Aksaray fault and the Ecemiş fault, for clarity no neotectonic structures are shown. Inset shows the first-order palaeotectonic elements, i.e. continental blocks and suture zones in the area. Vertically-ruled continental blocks are Laurasian, whereas horizontally ruled ones are of Gondwanian affinities with respect to Neo-Tethys. Closely-spaced vertical ruling shows palaeotectonic basins related to Alpide evolution. The Galata volcanic complex includes Oligocene to Pliocene post-arc volcanism.

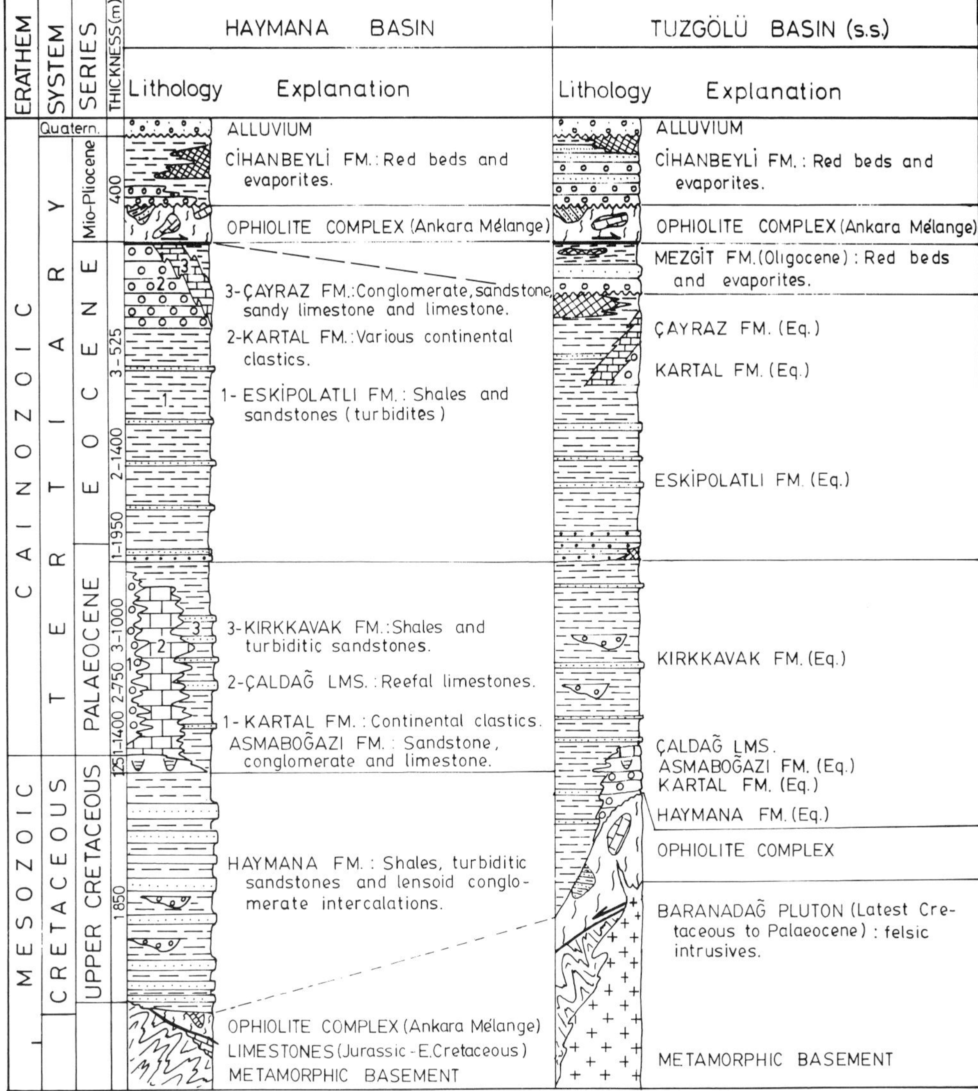

FIG. 2. Generalized stratigraphic columnar sections of the Haymana and the Tuzgölü (*s.s.*) basins.

Because the two sub-basins have always been viewed as integral parts of a single unified basin, despite the differences (e.g. Görür & Derman 1978, Turgut 1978, Görür 1981), a stratigraphic nomenclature common to both has evolved. As most of the formation names were derived from the better-exposed Haymana sub-basin, we retain this nomenclature and refer to the Tuzgölü sub-basin formations carrying the same names as those of the Haymana formations as 'equivalents' although this should not be regarded as implying strict stratigraphic equivalence. We avoid generating a new stratigraphic nomenclautre for the Tuzgölü sub-basin without having established proper type-sections (Figs 2 to 8 inclusive).

Haymana sub-basin

The basement of this sub-basin (Fig. 2), in which the cumulative thickness of the Maastrichtian-Tertiary deposits reaches thousands of metres, is composed of the Jurassic–Early Cretaceous carbonate cover of the Sakarya Continent (Şengör & Yilmaz 1981), the Karakaya Complex forming a part of the pre Jurassic basement of the Sakarya Continent (Şengör *et al.* this volume), and the Ankara Melange

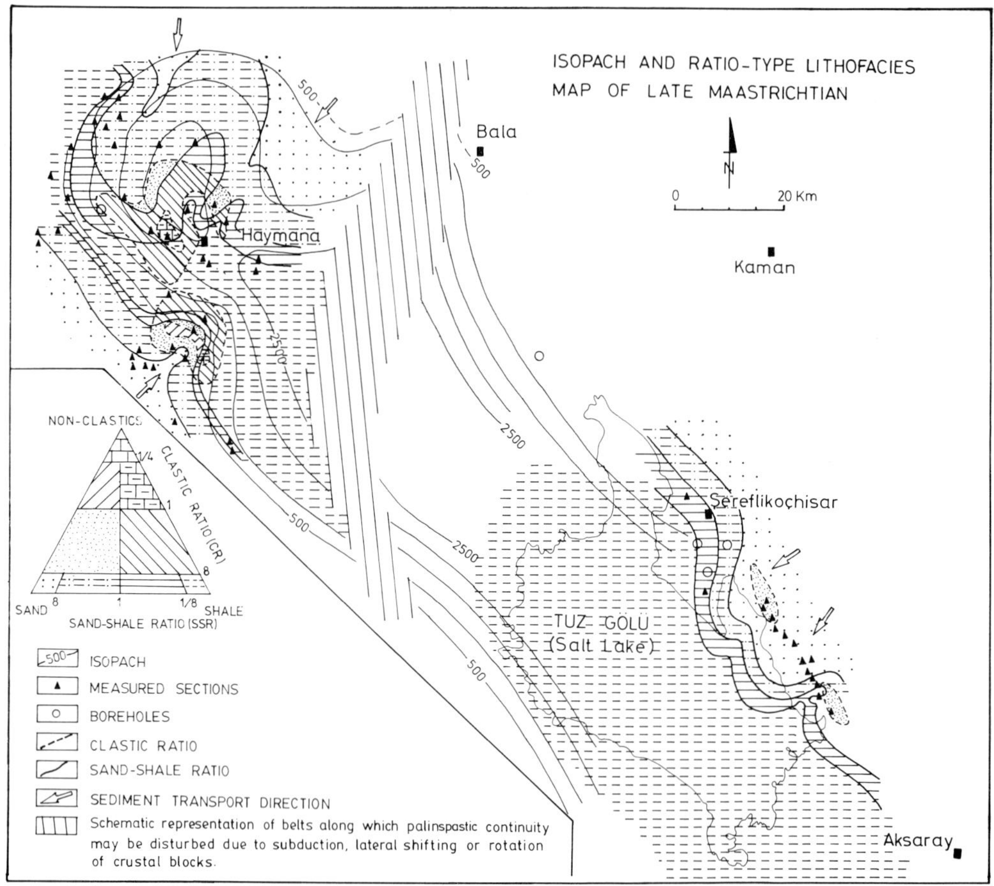

FIG. 3. Isopach and ratio-type lithofacies map of the late Maastrichtian deposits of the Tuzgölü basin complex. Note the positions of submarine fan lobes. The location of the vertically-ruled region of uncertainty is schematic.

(Bailey & McCallien 1953, Norman 1975, Ünalan *et al.*, 1976, Görür & Derman 1978, Norman *et al.*, 1980). Above this composite basement, mostly above an unconformity, are turbidites which consist dominantly of sandstone-shale intercalations with frequent conglomerates, olistostromes and debris-flow deposits forming the Haymana Formation of Late Maastrichtian age. These clastics are composed mainly of texturally and mineralogically immature mafic and ultramafic rocks of ophiolitic origin, derived most probably from the Karakaya Complex and the Ankara Mélange and metamorphic rock fragments, including composite quartz, and potassic and sodic feldspars derived from the basement of the Sakarya Continent (Ünalan *et al.*, 1976, Görür & Derman 1978, Norman *et al.*, 1980). Around the basin margins, these deep-sea deposits pass laterally and vertically into richly fossiliferous shallow marine sandstones, shales, and limestones (Fig. 3) with *Hippurites* sp., *Orbitoides* sp., *Cyclolites* sp., various gastropods and lamellibranches. This unit, the Asmaboğazı Formation, is of Late Maastrichtian age.

During the Palaeocene, deposition of terrestrial redbeds of the Kartal Formation and reefal limestones of the Çaldağ Formation characterized the basin edges, whereas the relatively deeper water shale-limestone intercalations of the Kirkkavak Formation were laid down in the interior of the basin (Fig. 4). The Palaeocene redbeds of the Kartal Formation contain clasts of ophiolites, radiolarian cherts, and limestones (Görür & Derman 1978).

The Early and Middle Eocene witnessed deposition of thick turbidite succession in the central parts of the basin, which occupied a larger area than its Palaeocene predecessor (Fig. 5). Eocene turbidites choked the depositional areas of the Çaldağ reefal limestones and the shoreline retreated away from the basin

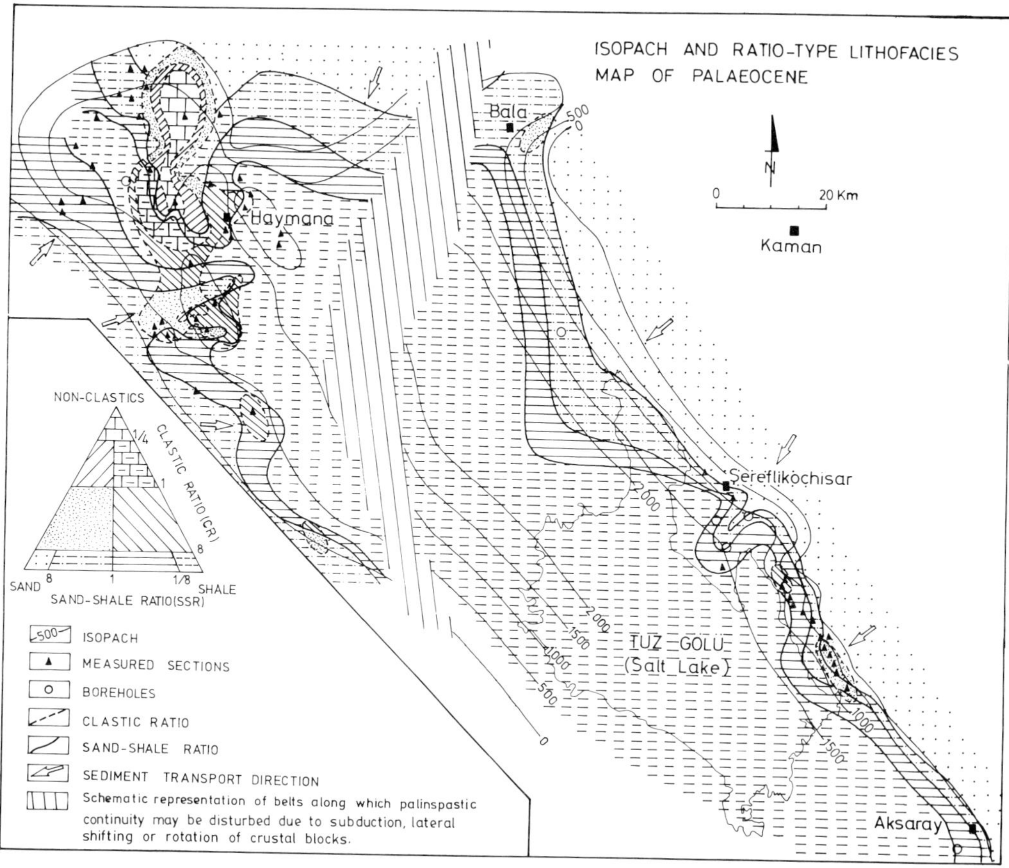

FIG. 4. Isopach and ratio-type lithofacies map of the Palaeocene deposits of the Tuzgölü basin complex. Note the positions of shallow water facies on the Samsam high and the locations of submarine fan lobes.

centre. The Eocene turbidites of the Haymana sub-basin constitute the Eski Polatli Formation and contain clasts of serpentinite, dunite, peridotite, diabase, basalt, radiolarian cherts and glaucophane schists, derived from the ophiolitic mélanges of the Karakaya Complex and the Ankara Mélange, as well as micaschists and amphibolites from the ophiolitic mélange and? the Sakarya basement, and rhyolitic lava flows from the Sakarya magmatic arc.

Towards the end of the Middle Eocene, the turbiditic depositional areas of the Haymana basin began to shrink rapidly. Eocene turbidites grade vertically and laterally into shallow marine nummulitic limestones of the Çayraz Formation and the terrestrial clastic sediments of the Kartal Formation (Fig. 2). Locally, near the eastern margin of the basin, the deposition of the Kartal and Çayraz formations was terminated by the arrival of a slice of ophiolitic mélange (Ünalan *et al.* 1976). Both the mélange nappe and the units underlying it were later covered unconformably by terrestrial conglomerates, sandstones, marls, evaporites, and the tuffs of the Mio-Pliocene Cihanbeyli Formation (Fig. 2).

Tuzgölü sub-basin

In contrast to the Haymana sub-basin, the Maastrichtian–Lower Tertiary rocks of the Tuzgölü sub-basin (Fig. 2) are largely hidden under younger cover (Fig. 1). They are exposed in a few, isolated inliers and also along a semi-continuous strip on the eastern edge of the sub-basin between Şereflikoçhisar and Aksaray (Fig. 1). A large number of boreholes drilled in the Tuzgölü sub-basin greatly supplement the surface information (Fig. 3).

The E margin of this sub-basin is formed by the pre-Maastrichtian crystalline basement of the Kırşehir Massif, structurally imbricated

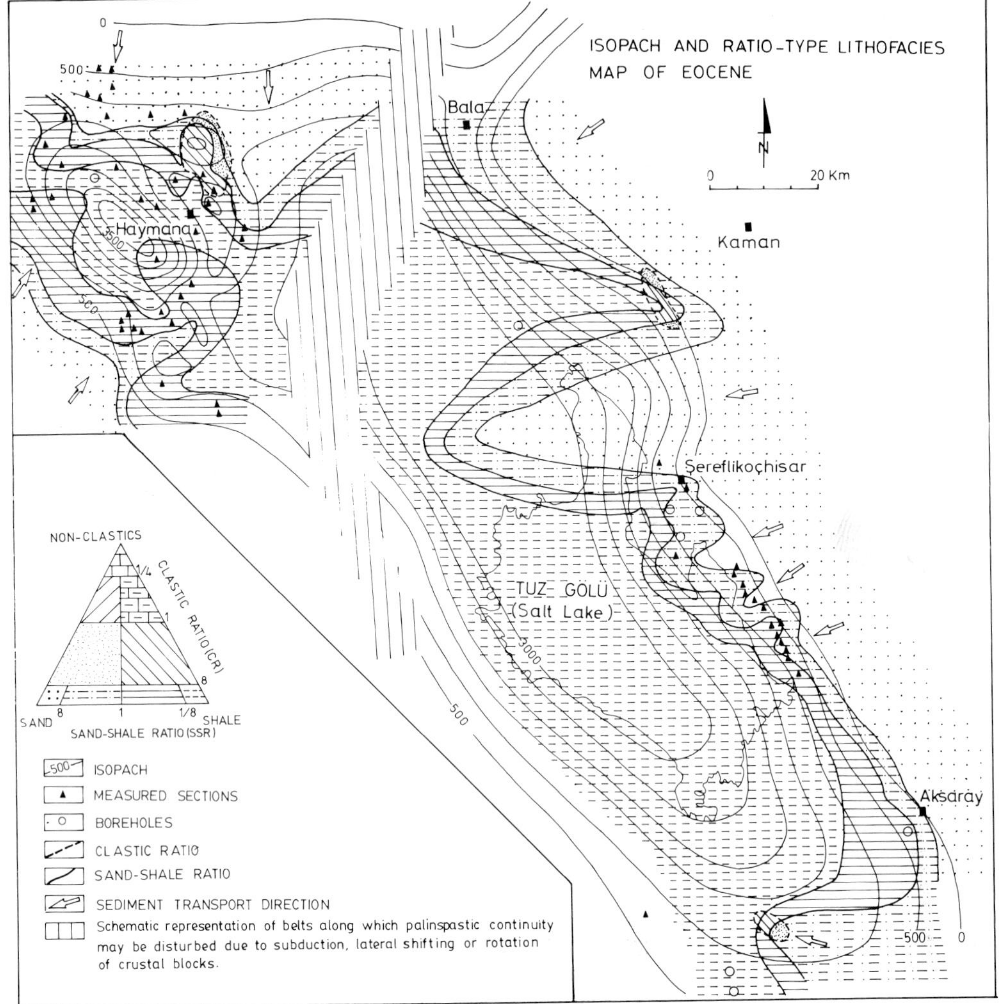

FIG. 5. Isopach and ratio-type lithofacies map of the Eocene deposits of the Tuzgölü basin complex. Notice the much enlarged turbidite fans coming off the margin of the Kırşehir Massif. Also note the pronounced NW-SE orientation of the Haymana basin axis.

with overlying ophiolitic mélange nappes (Seymen 1981, 1982). To the north is the ophiolitic Samsam high (Ankara Mélange; Uygun 1981) and to the west and south the ophiolites and sporadic blueschists (Okay, this volume) of the Inner Tauride Suture and the metamorphic Bolkardağ unit of the Menderes-Taurus block (Özgül 1976). There is no information about the nature of the basement of the Tuzgölü sub-basin. In the deep, axial part of the basin, Late Maastrichtian turbiditic deposits of the Haymana Formation equivalents form the lowest known rocks (Figs 2, 3 & 6). Along the eastern margin of the basin and also on the peripheral parts of the Kırşehir Massif bordering the Tuzgölü basin, terrestrial red clastics with coal and evaporite lenses were deposited during the same time interval (Figs 3 & 6). A small tongue of the sea adanced as far east as Kaman into the Kırşehir Massif (Fig. 6) and coincides roughly with the westerly continuation of the Eocene-Pliocene thrust zone of Savcılı (Oktay 1981) (Fig. 1) in the Kırşehir Massif, which may not be coincidental.

The Late Maastrichtian terrestrial clastics of the eastern margin of the Tuzgölü basin grade upwards and laterally into shallow marine, richly fossiliferous sandstones, siltstones, and limestones, with particularly abundant *Hippurites* sp.

The Palaeocene sediments in the Tuzgölü basin are mainly turbidites and olistostromes,

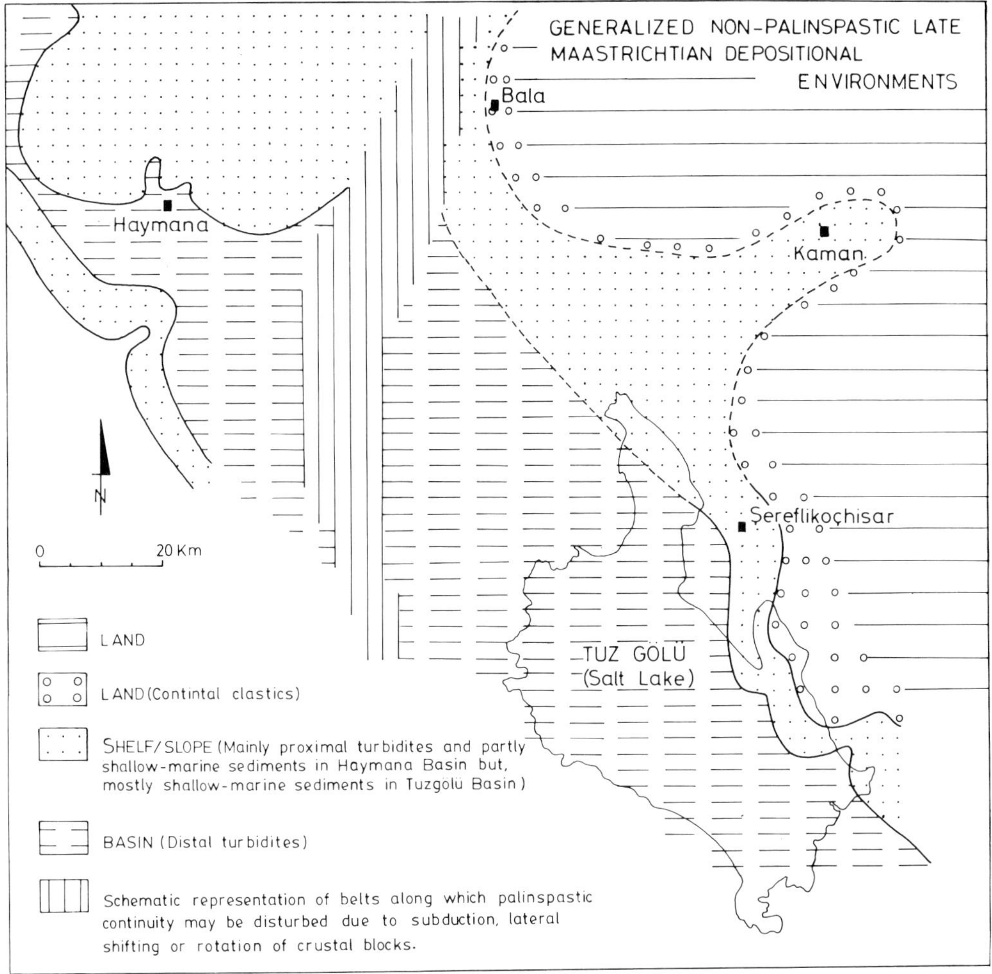

FIG. 6. Generalized non-palinspastic Late Maastrichtian depositional environments of the Tuzgölü basin complex. Notice the marine tongue extending to beyond Kaman, which may be indicative of an early initiation of the Savcılı thrust zone. See text for discussion.

which contrast sharply with coeval terrestrial and open-marine, shallower water sediments and peripheral reefal carbonates of the Haymana basin (Figs 4 & 7). In the Tuzgölü basin very localized Palaeocene reefs also developed along its eastern margin, as the Çaldağ equivalents, between Aksaray and Şereflikoçhisar (Fig. 7), but unlike those of the Çaldağ Formation, these reefs were rapidly drowned by abundant clastic influx from the east (Görür 1981), before reaching any considerable size (Fig. 7). Later during the Palaeocene, the intrusion of the Baranadağ plutonic complex (Fig. 1) uplifted the massif (Oktay 1981).

With the exception of limited shallowing along the eastern margin of the Tuzgölü basin, the Early and Middle Eocene sediments are mostly turbidites which gradually filled the basin giving way to a shallow marine-terrestrial environment during the Late Eocene, in which clastics, limestones, and evaporites were deposited (Fig. 8).

The Tuzgölü sub-basin contains major evaporite bodies, including halite, whose present structural setting and age are disputed (e.g. Görür & Derman 1978, Uygun 1981). Some view the halite as large diapirs piercing the entire sedimentary fill of the basin, and thus as of pre-Maastrichtian age. This view is based mainly on the interpretation of the Bezirci-1 well in the southern part of the basin (Turgut 1978). A recent reinterpretation of this well based on seismic reflection profiling and the better-known Karapınar-1 well has shown that

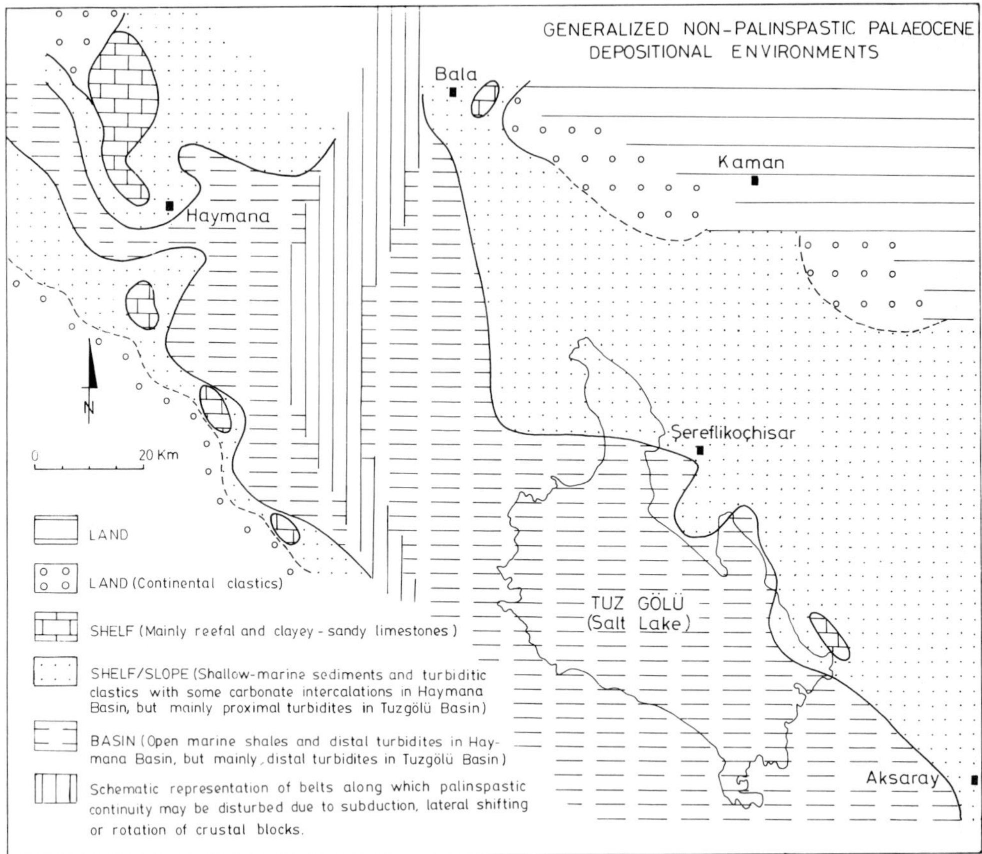

FIG. 7. Generalized non-palinspastic Palaeocene depositional environments of the Tuzgölü basin complex.

in both the major salt deposits are of Eocene age (Diker 1981). Uygun (1981) hypothesises that the major salt deposits of the Tuzgölü basin are of Late Eocene age and were deposited in a remnant marine basin restricted by the margins of the Tuzgölü basin. We follow this interpretation, because it explains the field and borehole evidence and also because the high global sea-level and the inferred location of the Tuzgölü basin in a terraced fore-arc setting during the Maastrichtian (see below), away from continental blocks would have precluded formation of a restricted marine basin without some form of topographic barrier.

Oligocene sediments in the Tuzgölü sub-basin, form the Mezgit Formation (Görür, 1981), a typical molasse deposit with red clastics and evaporites, which disconformably overlies the uppermost Eocene sediments. During (and ? after) the Oligocene, the Ankara Mélange of the Samsam ridge over-rode parts of the Mezgit Formation and the thrust contacts were later sealed by the Cihanbeyli Formation, the only formation common to both the Haymana and the Tuzgölü sub-basins.

There is a marked contrast between the basin area proper and its marginal strip along the western edge of the Kırşehir Massif. Whereas the basin itself subsided throughout its history, the edge of the Kırşehir Massif remained shallow or emergent. In a horizontal distance of only 15 km, thickness differences of at least 6.5 km came into existence between the basin and the edge of the Kırşehir Massif, during the Late Maastrichtian to the Middle Eocene. This implies an average differential motion of 0.5 mm/yr across the Aksaray/Şereflikoçhisar hinge-line, which Görür and Derman (1978) interpreted as a major dip-slip fault that originated at least as early as the Late Maastrichtian.

Structure

The structure of the Cretaceous-Eocene Tuzgölü basin complex consists of diverse axial trends

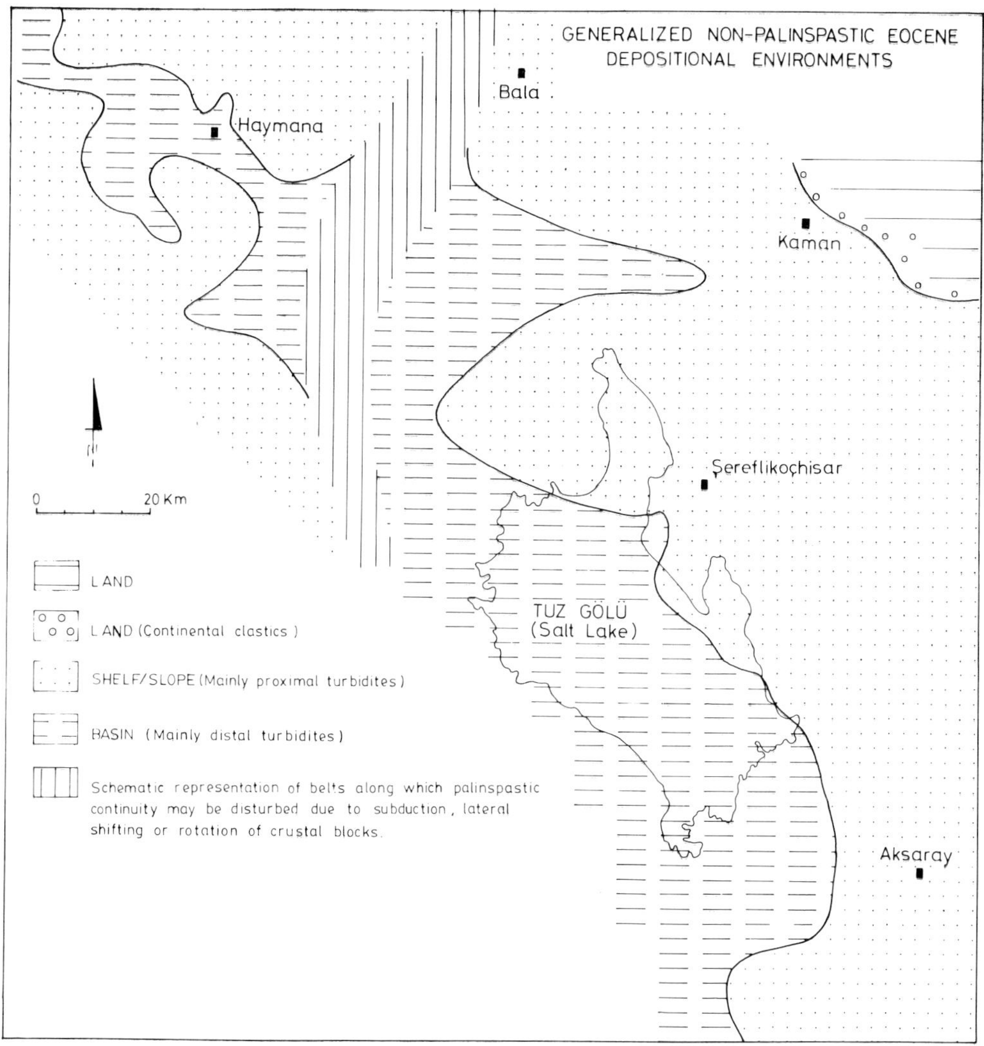

FIG. 8. Generalized non-palinspastic Eocene depositional environments of the Tuzgölü basin complex.

of major macroscopic folds and variable strike of thrust faults generated during the evolution of the basin (Fig. 1). Because most parts of the basin are covered by little-deformed Neogene-Quaternary strata, it is difficult to establish direct connections between the variously orientated structures, and the precise age of many structures cannot be established. That it is still a topic of debate whether there are any major salt diapirs in the Tuzgölü basin complex gives a fair idea of the state of structural interpretation.

Within the Haymana basin, the macroscopic fold axial traces with axial lengths on the order of 10 km, roughly follow the enclosing Samsam ophiolitic mélange ridge (Fig. 1). Along its eastern margin, the basin is over-thrust by the Ankara Mélange (Fig. 1, see also Fig. 9c). Deformation of the basin-fill must have spanned Eocene to possibly Miocene time, as the Mio-Pliocene Cihanbeyli Formation unconformably covers the basin (Görür & Derman, 1978).

The structure of the Tuzgölü sub-basin is more complicated. The dominant fold trends of the mainly pre-Miocene (? pre-Oligocene) structures follow the basin margins (Fig. 1). Along the margin of the Kırşehir Massif, the pre-Miocene fold axes closely follow the trend of the basin margin. Small deviations from parallelism between the fold axial traces and the Massif margin are believed to be late features related to the neotectonic right-lateral strike-slip motion along the Şereflikoçhisar/Aksaray Fault (Şengör & Yılmaz 1981).

In the northwestern corner of the Tuzgölü sub-basin, the Samsam ophiolitic mélange ridge overthrust the fill of the basin towards the SSE, during the Oligocene-? early Miocene. The Cihanbeyli Formation covers the thrust contacts associated with this movement. The fold axial trends in the main basinal part of the Tuzgölü sub-basin mostly follow the trend of the basin axis but mould around the southeastward projecting nose of the Samsam high (Fig. 1).

In summary, the folds in the Tuzgölü basin complex seem to have been generated during the final, syn- and post-collisional compression of the basin as they appear to have affected most of the fill. Because they are largely subsurface, and because little mesoscopic structural outcrop data are available from outcrops, a detailed kinematic picture can not be given. A number of thrusts seen on oil company seismic reflection profiles are not included in our discussion. Free access to these data will no doubt substantially modify this rather simplistic structural picture.

More, and better, structural data, are available from the western margin of the Kırşehir Massif (Seymen, 1982).

The pre-Maastrichtian structure of the Kırşehir Massif is irrelevant here. Its pre-Maastrichtian basement is covered by ophiolitic allochthons that appear to be of two provenances. Those imbricated with the basement of the Massif along moderately west-dipping thrust faults west of Kaman (Fig. 1) are parts of the Ankara Mélange that impinged upon the NW corner of the Massif. Because this imbricate structure is cut by the Baranadağ pluton of latest Cretaceous-Palaeocene age (Fig. 1) (Ayan 1959, Ataman 1972), Seymen (1982) believes this deformation to have taken place during the Late Cretaceous. The obduction age of the main ophiolitic allochthon covering major parts of the massif and rooting into the Erzincan suture (Fig. 1) (Bergougnan 1975) was also Late Cretaceous, because it too is imbricated with Cretaceous lithologies and the thrust contacts sealed by Palaeocene-Eocene sediments.

The Şereflikoçhisar/Aksaray fault is the major structure located between the Kırşehir Massif and the Tuzgölü basin. Along this the basin floor and the adjacent massif have moved in opposite directions vertically throughout evolution of the basin. The fault then seems to have functioned as a right-lateral strike-slip fault during the Neogene, in central Anatolia (Şengör 1979, Şengör & Yılmaz 1981). Whether it is still active or whether it has any relation to the right-lateral active fault of Kırşehir-Keskin (Arni 1938, see the map in Paréjas & Pamir 1938), are unknown.

In addition the large NNE-vergent thrust zone of Savcili that dissects the Kırşehir Massif (Fig. 1) may have affected the structural evolution of the Tuzgölü basin by deforming its NE margin near Kaman. Oktay (1981) has mapped the thrust zone and dated it as Eocene-Pliocene in age. Although some palaeogeographic indications exist it remains to be shown that it began its activity during the Maastrichtian.

Tectonic evolution of the Tuzgölü basin complex

Because it is located mostly within the Anatolide tectonic unit of Ketin (1966), most researchers have considered the Tuzgölü basin complex to have been 'intracratonic' since the end of the Cretaceous, when, Ketin (1966) believed, the Anatolides began to form. Oktay (1973) was the first to propose a separation of the Anatolides into two blocks, which, he postulated, had collided along a suture now underlying the Tuzgölü basin complex during the early Tertiary. However, later models of the tectonic evolution of Turkey have mostly been constructed assuming the unity of what Şengör (1979) termed the Anatolide/Tauride platform (*l'axe calcaire du Taurus* of Ricou *et al.*, 1975, also see Şengör & Yılmaz 1981) during the Mesozoic and the early Tertiary. However, later work on the Kırşehir Massif (Seymen 1981, 1982) has shown that it experienced a structural evolution different from that of the Menderes Massif and that it must have been separated from the rest of the Anatolide/Tauride platform at least since the Early Jurassic. Using Seymen's (1981, 1982) data with some other unpublished information from the Turkish Petroleum Co. (O. Sungurlu, pers. comm., 1981), Şengör *et al.* (1982) continued the Inner Tauride Suture of Şengör & Yılmaz (1981) around the Tuzgölü sub-basin to join it with the Izmir/Ankara suture near Eskişehir, thus dividing the Anatolide/Tauride platform into two blocks, *viz.* the Kırşehir block and the Menderes-Taurus block. The resulting picture is very similar to Oktay's (1973) original suggestion, and forms the basis of our model for the tectonic evolution of the Tuzgölü basin complex.

Late Maastrichtian events

As seen in Fig. 9, the three major tectonic elements that now surround the basin, the

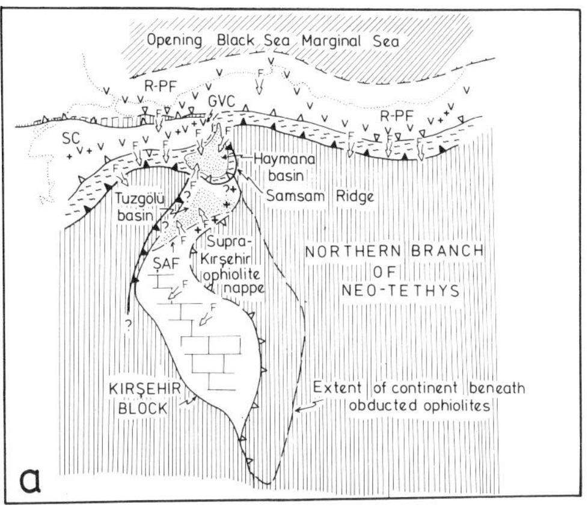

(a) Palaeotectonic map of the Late Cretaceous: R-PF = Rhodope-Pontide fragment, SC = Sakarya Continent.

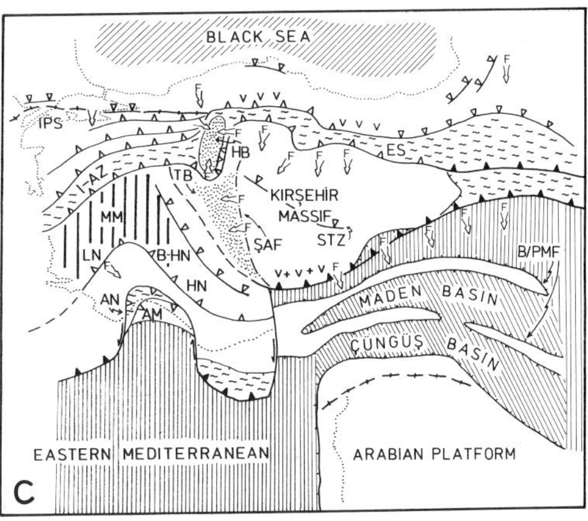

(c) Palaeotectonic map of the Early–Middle Eocene: IPS = Intra-Pontide suture, I-AZ = İzmir-Ankara Zone, MM = Menderes Massif, LN = Lycian nappes, B-HN = Beyşehir-Hoyran nappes, HN = Hadim nappes, STZ = Savcili thrust zone, B/PMF = Bitlis/Pötürge massif fragments.

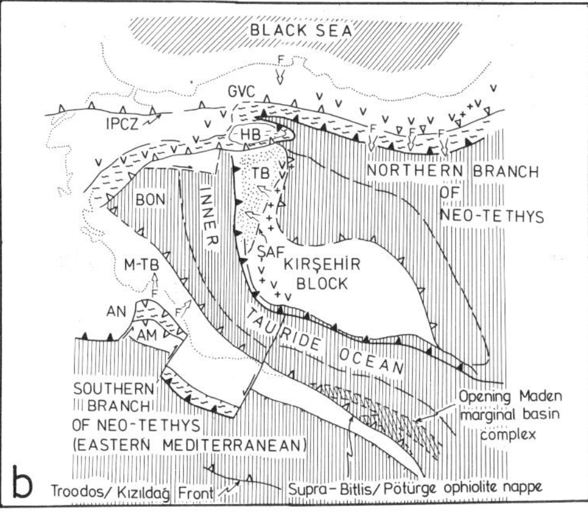

(b) Palaeotectonic map of the Palaeocene: GVC = Galata Volcanic Complex, IPCZ = Intra-Pontide Convergence Zone, HB = Haymana Basin, TB = Tuzgölü Basin, ŞAF = Şereflikoçhisar/Aksaray fault, BON = Bozkır ophiolite nappe, M-TB = Menderes–Taurus block, AN = Antalya Nappes, AM = Alanya Massif.

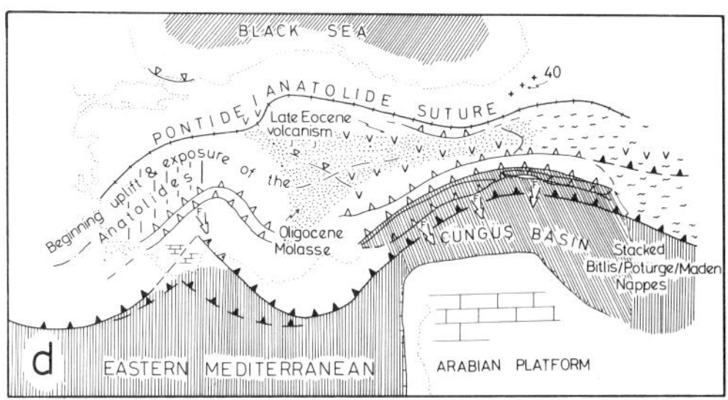

(d) Palaeotectonic map of the Oligocene.

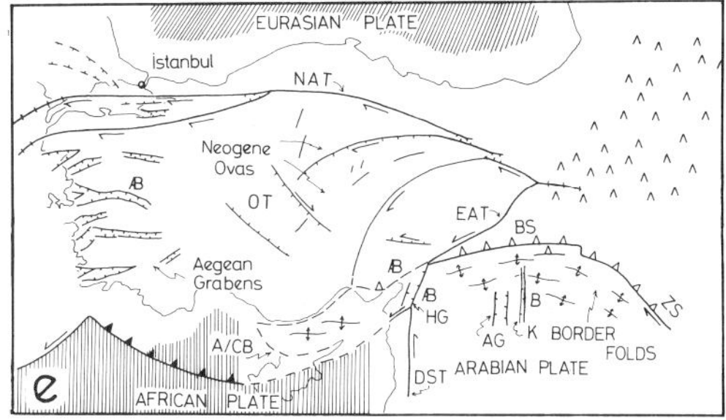

(e) Neotectonic map of Turkey. Notice that the tectonic regime here shown bears no relation to the earlier evolution. OT = Ova of Tuzgölü.

FIG. 9. Palaeotectonic maps depicting the tectonic evolution of Turkey emphasizing the tectonic development of the Tuzgölü basin complex from the late Maastrichtian to the present. Maps show oceans (widths are *not* to scale and shown only symbolically by vertical ruling for the branches of Neo-Tethys and oblique ruling for various Neo-Tethys related marginal basins), continents (white), predominant lithologic and lithofacies types (widely spaced carbonate pattern = neritic carbonates, stippled areas = turbidite depocentres, small thin discontinuous lines = non turbiditic shales, short wavy lines = ophiolitic mélanges). Open arrows show sediment dispersal directions for clastics, dominantly flysch. v = arc volcanics, ++ = arc plutonics, /\ = Tibet-type volcanics. Lines with hachures on are Atlantic-type continental margins (normal faults in figure 93), heavier lines with black triangles on them are subduction zones (triangles on the upper plate), and lines with half arrows are transform faults. Lines with open triangles are major thrust faults (triangles on the upper plate), and lines with upside-down triangles are 'rétrocharriages' (triangles on the lower plate). Dotted lines are present shorelines for reference purposes. Partly after Şengör & Yılmaz (1981).

Pontides, the Menderes–Taurus block, and the Kırşehir block were all separated from one another by various branches of Neo-Tethys during the Late Maastrichtian, when our story begins, with the exception of the contact between the toe of the Sakarya Continent (SC) and the north-west corner of the Kırşehir Massif (Fig. 9a). The latter is drawn in Fig. 9a as rotated clockwise some 90° from its present orientation using the palaeomagnetic data of Sanver & Ponat (1981). Because the Sakarya Continent had probably already been welded on to the Rhodope-Pontide Fragment (R-PF) along the Intra-Pontide suture by the Late Maastrichtian (Şengör & Yılmaz 1981), we have indicated them both simply as Pontides in Fig. 9, following Ketin's (1966) scheme.

North-dipping subduction of the floor of the northern branch of Neo-Tethys began beneath both the Rhodope-Pontide Fragment and the Sakarya Continent during the Cenomanian-Turonian interval. Local, Aptian-Albian calc-alkaline volcanics south of Bafra on the Rhodope-Pontide Fragment may indicate earlier initiation of subduction east of the Ankara Knot (M. Aydın, pers. comm. 1980). After the Maastrichtian collision of the Sakarya Continent with the Rhodope-Pontide Fragment, a single, sinuous, north-dipping subduction zone remained along the southern boundary of the Pontides, as depicted in Fig. 9a. In our area of interest (Fig. 1), its continued activity during the Maastrichtian is shown by the abundant basaltic and andesitic flows, lava blocks, breccias, and smaller trachyandesitic intrusions north of the Haymana sub-basin. Sandy-marly horizons are intercalated with pyroclastic deposits and contain *Globotruncana* sp., Radiolitidae, *Orbitoides* sp. and *Hippurites* sp. These Maastrichtian volcanics and intrusives form the *complex inférieur* of the Galata volcanic complex (Fig. 1) (Fourquin 1975), a volcanic zone which extended westwards on the Sakarya Continent, indicating its S-facing magmatic arc character (Şengör & Yılmaz 1981, Yılmaz 1981).

In the area of Fig. 1, the main depocentre of the Haymana sub-basin is located to the south of the magmatic arc axis, i.e. on its trenchward side. To the south and east, the basin was bounded by the Samsam high, made up of the ophiolitic Ankara Mélange (Fig. 9a). This high continued to grow and be uplifted throughout the Late Cretaceous-Eocene history of the Haymana basin. We interpret it as the outer non-volcanic arc associated with the north- and north-west-dipping Pontide subduction zone, delimiting the Haymana fore-arc basin. That this fore-arc basin was constructed partly on the Late Jurassic-Early Cretaceous sedimentary mantle of the Sakarya Continent and partly on the accretionary prism of the Ankara Mélange, lends strong support to the earlier contention of Yılmaz (1981) that along the southern margin of the Sakarya Continent subduction had been initiated at, or very close to, the continent/ocean interface, by converting its pre-existing, S-facing Atlantic-type continental margin to a similarly directed Pacific-type margin.

As seen in Fig. 3, coarse clastics were deposited on the Samsam outer non-volcanic arc, whereas finer grained clastics, in the form of turbidites, were laid down in the fore-arc basin itself. Fig. 3 also shows that the intersection of the sandstone-shale ratio (SSR) contours with the isopachs indicates 3 major lobes pointing in the dominant direction of sediment transport. Sediment transport directions derived independently from sedimentary structure data by Norman *et al.* (1980) are in excellent agreement with the direction of the lobes shown in Fig. 3 and document basinward influx of material from the north, north-east and south-west, i.e., both from the arc massif and the outer non-volcanic arc. As we have seen above, the provenance areas of the transported clastics also support this conclusion.

The available data from the Tuzgölü sub-basin are not sufficient to draw as complete a picture of its tectonic nature as for the Haymana sub-basin, but the following features suggest that it too was located in a fore-arc, although neither the extent of this fore-arc nor the nature of its basement can be established at present.

The Baranadağ plutonic complex, which faithfully follows the west margin of the Kırşehir Massif (Fig. 1) began its intrusion during the Maastrichtian (71 Ma, Ataman 1972) and continued through the Palaeocene (54 Ma, Ayan 1959). The Şereflikoçhisar/Aksaray fault is located 10 km to the WSW of the plutonic axis, which stretches from Aksaray all the way N to Kırıkkale (Fig. 1). The nearest subduction zone with the appropriate age and orientation that could have generated this magmatic axis was the Inner Tauride Suture (Şengör *et al.* 1982), located to the west and south-west of the Tuzgölü sub-basin. We contend that the Baranadağ plutonic axis represents an Andean-type continental arc perched along the western margin of the Kırşehir block as the latter moved westwards relative to the Menderes–Taurus block. The Tuzgölü sub-basin thus falls into a fore-arc position, whereas the Şereflkoçhisar/Aksaray fault appears as a major upper slope discontinuity separating the intermittently but consistently rising arc-massif from the subsiding fore-arc basin.

Fig. 3 shows that sedimentation in the Tuzgölü sub-basin was dominated by clastic influx from the east in the form of turbidites. The material of these initial turbidites is dominantly ophiolitic in character which at first sight is somewhat unusual for a fore-arc basin. However, during the Late Maastrichtian major parts of the Kırşehir Massif were covered by an ophiolitic nappe expelled from the northern branch of Neo-Tethys, now represented by the Erzincan suture (Bergougnan 1975). From klippen, it seems likely that this nappe covered nearly the whole of the Kırşehir Massif included in Fig. 1. Also, the toe of the Samsam melange wedge appears to have reached and surmounted the north-western extremity of the Kırşehir Massif west of Kaman. At this time the Baranadağ pluton had probably not yet reached the surface to provide detritus to the fore-arc. The only abundant material was therefore ophiolitic. Nevertheless, fragments of metamorphic rock and abundant quartz grains in the Maastrichtian turbidites do indicate that some parts of the Kırşehir basement were exposed to erosion. Significantly, no material seems to have come from the western margin of the Tuzgölü sub-basin. This shows that the Menderes–Taurus block was not where it is today, nor that an outer non-volcanic arc had emerged to supply the fore-arc basin with detritus. It seems likely that the basement of the Tuzgölü sub-basin never formed such a topographically prominent structure as that of the Haymana sub-basin during the Late Maastrichtian. It was probably a terraced fore-arc on which turbidites were ponded (see Dickinson & Seely 1979, Fig. 6).

Independent support for this hypothesis is provided by the arrival, during the Late Maastrichtian, of the toe of the Samsam accretionary wedge on the basement of the Kırşehir Massif, west of Kaman. It was probably at this point that initial contact between the Kırşehir block and the Pontides took place (Fig. 9a). Because arc magmatism both north of the Erzincan stuture and along the western margin of the Kırşehir Massif continued into the Eocene (Fig. 9a–c), neither the Erzincan nor the Inner Tauride sutures could have closed at least prior to the Palaeocene. Therefore, the Kırşehir block must have hit the accretionary wedge at the south-east corner of the Sakarya Continent with its north-west corner. Because this was a collision between two fore-arcs, much like the present collision system of the Molucca straits (Silver & Moore 1978), the topographically more prominent fore-arc of the Sakarya Continent over-rode the fore-arc of the Kırşehir block and reached as far as the basement of the Massif (Fig. 9a). This collision pinned the northwestern corner of the Kırşehir block to the Pontides, and around it the Kırşehir Massif began to rotate in an anticlockwise fashion until its entire present northern margin completely collided with the eastern Pontides along the Erzincan suture (Fig. 9c) during the Priabonian (Seymen 1975). Sanver and Ponat (1981) report palaeomagnetic evidence that the Kırşehir Massif indeed rotated for 90° in an anticlockwise fashion between the Late Cretaceous and the Lutetian, a conclusion which is in excellent agreement with the model developed here.

Palaeocene events

During the Palaeocene, arc activity continued both on the Sakarya Continent (Fourquin 1975, Şengör & Yılmaz 1981) and on the Kırşehir Massif (Seymen 1982) indicating that neither of the associated subduction zones had stopped. Although the Tuzgölü sub-basin remained much as before there were marked changes both in the structure and in the nature of sedimentation in the Haymana sub-basin (Figs 4, 7 & 9b).

The previously NW-SE orientated basin axis switched to an E-W orientation and the basin became considerably shallower. The Samsam ophiolitic ridge enlarged its width and its crest emerged over a large area on which the terrestrial Kartal Formation was deposited. Coralline algal reefs fringed both sides of the outer non-volcanic arc, whereas pelagic shale-limestone alternations were deposited in the basinal area.

The sudden rotation of the basin axis by at least 60° was probably the result of the beginning of collision between the Sakarya Continent and the Menderes part of the Menderes–Taurus block along the İzmir-Ankara suture which began squeezing the accretionary toe of the former. At this time, the area of the Çorum Basin was still underlain by oceanic crust so that escape of crustal material from the İzmir-Ankara suture in that direction was easy (Fig. 9b).

Turbidite fans in the Tuzgölü sub-basin reached considerable dimensions during the Palaeocene. Small patches of reefal limestones developed along the eastern margin of the basin, but were rapidly drowned by clastics.

Eocene events

The Eocene was the time of the assembly of the Anatolian landmass as a result of a number of collisions that welded together its constituent continental blocks. As Şengör & Yılmaz (1981) have shown, the Inner Tauride Suture, too, closed during the Late Eocene, thus completely isolating the Tuzgölü Basin (*s.l.*) between three converging continental blocks (Fig. 9c).

The basin axis of the Haymana sub-basin assumed a nearly N-S orientation during the Eocene. Large turbidite fans rapidly filled the basin from all directions by the Late Eocene (Fig. 5). The Tuzgölü sub-basin maintained its axial orientation through the Middle Eocene, but it too was choked by the Late Eocene sedimentation (Fig. 5) which overflowed from the basin that had existed since the Late Maastrichtian.

An important aspect of the Eocene sedimentation pattern in the Tuzgölü sub-basin was the continued absence of any sediment influx from the west except in the extreme south, in the Ulukışla Basin (Fig. 1) (Oktay 1973). When one considers the Eocene history of the regions now lying to the west of the Tuzgölü sub-basin in which the internal thrusting of the Menderes–Taurus block had thickened the crust to more than 60 km (Şengör 1982), and uplift of the İzmir-Ankara suture zone was still continuing until the Lutetian (Okay this volume), the lack of sediment supply to the Tuzgölü sub-basin can only be explained by assuming that the Tuzgölü depocentre was not directly accessible to the debris shed from the west.

At the end of the Eocene, the Tuzgölü Basin (*s.l.*) had become completely choked with sediment and overflowed, losing its deep-basin character and distinct outlines. Limited convergence of its bounding continental blocks constricted and deformed its contents which were un- and disconformably overlain by ? Late Eocene-Oligocene continental clastics and evaporites, which bear no relation to the earlier basin.

Post-Eocene events

Until about the late Oligocene, molasse sedimentation continued in the Tuzgölü basin area, which had become a part of the epi-Anatolide molasse basin extending from Ankara to Eastern Anatolia where its terrestrial contents interfinger with marine sediments (Lüttig & Steffens 1976). During the Early Oligocene (Lattorfian-Rupelian), small elongated basins extending in an E-W direction between the Tuzgölü basin and Erzincan, became unified into a large molasse basin in which the widespread 'Oligo-Miocene' gypsiferous series of Central Anatolia were laid down along with abundant clastics and local volcanics (Fig. 9d).

During the Early Miocene, the Tuzgölü area was exposed completely and was subject to extensive erosion. Only in the Late Miocene, from Tortonian onwards, an areally extensive, complex basin entirely unlike the earlier palaeotectonic Tuzgölü Basin formed in Central Anatolia as part of the central Anatolian '*ova*' regime (Fig. 9e) (Şengör 1979). This new Neotectonic regime (Şengör 1980) generated extensive shallow 'intra-cratonic' basins that bear no relation to the earlier orogenic structure of Anatolia as noted many years ago by Salomon-Calvi (1940).

Discussion and conclusions

The deep composite Tuzgölü Basin in Central Anatolia was produced by various convergent phenomena that accompanied the Late Cretaceous to Eocene welding together of the Anatolian landmass along the suture zones of the northern branch of Neo-Tethys (Fig. 7b, c and d in Şengör *et al.* this volume). The basin consisted of two main depocentres, the Haymana and Tuzgölü sub-basins that evolved independently during the Late Maastrichtian to latest Palaeocene when juxtaposition by continental collision began.

The Haymana sub-basin was a fore-arc basin constructed on the south-eastern edge of the Sakarya Continent and its accretionary toe was formed by the Ankara Mélange. This sub-basin is a fairly well-preserved ridged fore-arc basin (Dickinson & Seely 1979) in which a narrow, largely submarine ridge, the Late Maastrichtian Samsam ophiolitic trench slope-break, gradually widened and emerged to form a broad, largely subaerial ridge in the Palaeocene. This evolution is very similar to the Jurassic-Palaeocene evolution of the Franciscan mélange ridge/Great Valley fore-arc basin in California (Ingersoll 1979), but is here preserved in a collisional setting.

The axis of the Haymana depocentre rotated in time from a NW-SE orientation in the Late Maastrichtian to an E-W orientation during the Palaeocene and finally to a roughly N-S orientation during the Eocene. Fig. 9 shows a possible interpretation whereby the initial configuration is related to the original geometry of the trench which follows the outline of the Sakarya Continent. Continued enlargement of the Samsam ridge by accretion of oceanic material and the initial collision with the Menderes–Taurus block during the (? late) Palaeocene may have compressed the Haymana fore-arc basin in a N-S direction, giving it an E-W elongated form. Continued convergence across the Izmir-Ankara suture may have begun to drive the eastern part of the Sakarya Continent eastwards onto the Ankara Knot, where the easterly movement of the former was arrested by the Eocene tightening of the knot and then the basin began to be compressed E-W, thus switching its axis to N-S. Evidence for this later E-W compression is seen where the Samsam ridge overthrust the Haymana basin-fill with a west-vergent thrust zone during the Oligocene.

The Tuzgölü sub-basin is much less well-exposed and perhaps as a consequence, appears to have undergone a simpler evolution with a stable basin that experienced only small shifts parallel to its axis. Although its fore-arc setting seems clear, its detailed internal structure remains elusive.

The fore-arcs collided with each other and with the Menderes–Taurus block during the latest Palaeocene–Early Eocene interval, enclosing a triangular area underlain largely by the fore-arcs. This basin was rapidly filled with sediment and finally overflowed the basin margins. Continued convergence during the Late Eocene and possibly during the Oligocene deformed the basin-fill and generated macroscopic folds paralleling the basin margins.

That this post-collisional deformation was not very intense is shown by the present-day existence of the Tuzgölü basin. The reason why the basin was not totally obliterated is probably a geometry of the collisions shown in Fig. 9 which preserved a triangular 'hole', the Tuzgölü sub-basin. Thus the Tuzgölü basin appears as a cluster of remnant fore-arcs partially preserved because of their fortuitous location along an imperfectly closed suture. That it is now overlain by a more extensive and much shallower basin related to the disintegration of Turkey, as a consequence of the Arabia/Anatolia collision, is also fortuitous and irrelevant from the viewpoint of its palaeotectonic evolution. Unless future studies of the Tuzgölü basin take account of these complexities, neither the thermal evolution, nor the hydrocarbon potential of the Tuzgölü basin complex can be established.

This study also shows that the Neo-Tethyan tectonic evolution of Turkey is much more complicated than was considered only two years ago, involving major rotations of a considerable number of continental blocks whose original relations to one another remain elusive. In this context, the judicious warnings of Jackson & McKenzie (this volume) of the inherent difficulties in reconstructing continental collision zones must be taken very seriously. The reconstructions we present here should be viewed as thoughts spoken aloud, quite possibly remote from reality.

ACKNOWLEDGMENTS: We are grateful to Prof. I. Ketin for discussions on the tectonics of the Tuzgölü basin and Turkey in general. Drs N. Ozgül, O. Sungurlu, A. Okay and Y. Yılmaz contributed with unpublished data and friendly criticism. N. Görür is indebted to S. Derman for his cooperation in the study of the sedimentology of the Tuzgölü basin complex and is also grateful to the Turkish Petroleum Co. (T.P.A.O.) for support.

References

ARIKAN, Y. 1975. Tuzgölü havzasının jeolojisi ve petrol imkanları. *M.T.A. Dergisi*, **85**, 17–38.

ARNI, P. 1938. Kırşehir-Keskin ve Yerköy zelzelesi hakkinda *M.T.A. Enstitüsü yayınları*. Seri B, no. 1, Ankara.

ATAMAN, G. 1972. Ankara'nın güneydoğusundaki granitik-granodiyoritik kütlelerden Cefalık dağın radyometrik yaşı hakkında ön çalışma. *Hacettepe Fen ve Müh. Bil. Dergisi*. **2**, 44–9.

AYAN, M. 1959. *Contribution a l'étude pétrographique et géologique de la région située au Nord-Est de Kaman (Turquie)*. Thèse Fac. Sc. Univ. Nancy. **2**, 396 pp.

BAILEY, E. B. & MCCALLIEN, W. J. 1953. Ankara mélange and the Anatolian Thrust. *Trans. Roy. Soc. Edinburgh*. **62**, 403–42.

BERGOUGNAN, H. 1975. Relations entre les édifices pontique et taurique dans le Nord-Est de l'Anatolie. *Bull. Soc. géol. Fr.* **7**, 1045–57.

DEWEY, J. F. & ŞENGÖR, A. M. C. 1979. Aegean and surrounding regions: complex multi-plate and continuum tectonics in a convergent zone. *Bull. geol. Soc. Am.* **90**, 82–91.

DICKINSON, W. R. & SEELY, D. R. 1979. Structure and stratigraphy of fore-arc regions. *Bull. Am. Assoc. Petrol. Geol.* **63**, 2–31.

DİKER, S. 1981. In: Tartışma. *Türk. Jeol Kur, İç Anadolunun jeolojisi simpozyumu, Ankara*, p. 75.

DRUITT, C. E. & RECKAMP, J. U. 1959. *Çaldağ columnar section*. T.P.A.O. Arşivi, Ankara.

FOURQUIN, C. 1975. L'Anatolie du Nord-Quest, marge méridionale du continent europeén, historie paléogéographique, tectonique et magmatique durant le Secondaire et Tertiaire. *Bull. Soc. géol. Fr.* **7**, 1058–70.

GÖKÇEN, S. L. 1976. Haymana güeyinin sedimetolojik incelemesi I. stratigrafik birimler ve tektonik. *Yerbilimleri* **2**, 161–201.

——, 1977. Sedimentology and provenance of resedimented deposits in part of the Haymana basin-Central Anatolia. *Yerbilimleri* **3**, 13–23.

GÖRÜR, N. & DERMAN, A. S. 1978. *Tuzgölü-Haymana havzasının stratigrafik ve tektonik analizi*. T.P.A.O. Arşivi 1514, Ankara.

GÖRÜR, N. 1981. Tuzgölü-Haymana ḩavzasının startigrafik analizi. *Türk. Jeol. Kur, İç Anadolunun jeolojisi simpozyumu*, Ankara, 60–6.

INGERSOLL, R. V. 1979. Evolution of the Late Cretaceous forearc basin, northern and central California. *Bull. geol. Soc. Am.* **90**, 813–826.

Ketin, I. 1966. Tectonic units of Anatolia (Asia Minor). *Min. Res. Expl. Inst. Bull.* **66,** 23–34.

Luttig, G. & Steffens, P. 1976. *Explanatory notes for the paleogeographic atlas of Turkey from the Oligocene to the Pleistocene.* Bundesanstalt für Geowissenschaften und Rohstoffe, Hannover, 64 pp.

Norman, T. & Rad, M. R. 1971. Çayraz (Haymana) civarının Horhor (Eosen) formasyonunda alttan üste doğru doku parametrelerinde ve ağır mineral bolluk derecelerinde değişmeler. *Türk. Jeol. Kur, Bülteni* **14,** 205–25.

——, 1975. Paleocurrents and submarine mass-movements in the Lower Tertiary sediments of Çankırı-Çorum-Yozgat basin. *Türk. Jeol. Kur. Bülteni* **18,** 103–10.

——, Gökçen, S. L. & Şenalp, M. 1980. Sedimentation pattern in Central Anatolia at the Cretaceous-Tertiary Boundary. *Cretaceous Res.* 1, 61–84.

Oktay, F. Y. 1973. *Sedimentary and tectonic history of the Ulukışla area, southern Turkey.* Unpubl. Univ. College, London, Ph.D. Thesis, 414 pp.

——, 1981. *Savcılıbüyükoba (Kaman) çevresinde Orta Anadolu Masifi tortul örtüsünün jeolojisi ve sedimentolojisi.* I.T.Ü. Maden Fak. Doçentlik Tezi, 175 pp.

——, 1982. Ulukışla ve çevresinin stratigrafisi ve jeolojik evrimi. *Türk. Jeol. Kur. Bülteni* **25,** 15–25.

Özgül, N. 1976. Torosların bazı temel jeoloji özellikleri. *Türk. Jeol. Kur. Büteni* **19,** 65–78.

Paréjas, E. & Pamir, H. N. 1939. Le tremblement de terre du 19 avril 1938 en Anatolie centrale. *Istanbul Univ. Fen. Fak. Mec.* **4,** 183–93.

Ricou, L. E., Argyriadis, I. & Marcoux, J. 1975. L'Axe calcaire du Taurus: un alignement des fenêtres arabo-africains sous des nappes radiolaritiques et métamorphiques. *Bull. Soc. géol. Fr.* **7,** 1024–44.

Salomon-Calvi, W. 1940. Anadolunun tektonik tarzi teşekkülü hakkında kisa izahat. *M.T.A. Mecmuasi* 1/18, 35–47.

Sanver, M. & Ponat, E. 1981. Kırşehir ve dolaylarina ilişkin paleomanyetik bulgular, Kırşehir Masfinin rotasyonu. *Istanbul Yerbilimleri* **2,** 231–8.

Şenalp, M. & Gökçen, S. L. 1978. Sedimentological studies of the oil-saturated sandstones of the Haymana Region (S.W. Ankara). *Türk. Jeol. Kur. Bülteni,* **21,** 87–94.

Şengör, A. M. C. 1979. The North Anatolian transform fault: its age, offset and tectonic significance. *J. geol. Soc. London* **136,** 269–82.

——, 1980. Türkiyenin neotektoniğinin esasları. *Türk. Jeol. Kur. Konf. Ser.* **2,** 40 pp.

——, 1982. Ege'nin neotektonik evrimini yöneten etkenler. *Türk. Jeol. Kur, Batı Anadolu'nun genç tektoniği ve volkanizması paneli,* Ankara, 59–71.

——, and Yılmaz, Y. 1981. Tethyan evolution of Turkey: a plate tectonic approach. *Tectonophysics.* **75,** 181–241.

——, Yılmaz, Y. & Ketin, I. 1982. Remants of a pre-Late Jurassic ocean in northern Turkey. Fragments of Permian-Triassic Paleo-Tethys?; Reply. *Bull. geol. Soc. Am.* **93,** 932–6.

Seymen I. 1975. *Kelkit Vadisi kesiminde Kuzey Anadolu Fay Zonunun tektonik özelliği.* I.T.Ü. Maden Fak. Yayini, 198 pp.

——, 1981 Kaman (Kırşehir) dolayında, Kırşehir Masifinin metamorfizmasi. *Türk. Jeol. Kur. İç Anadolunun Jeolojisi Simpozyumu, Ankara, 12–16.*

——, 1982. *Kaman dolayında Kırşehir Masifinin Jeolojisi,* İ.T.Ü. Maden Fak. Doçentlik Tezi, 164 pp.

Silver, E. A. & Moore, J. C. 1978. The Molucca Sea collision zone, Indonesia. *J. geophys. Res.* **83,** 1681–91.

Sirel, E. & Gündüz, H. 1976. Haymana (G. Ankara) yöresindeki İlerdiyen, Kuiziyen ve Lütesiyen'deki Nummulites, Assilina ve Alveolina cinslerinin bazı türlerinin tanımlanmaları ve stratigrafik dağılımları. *Türk. Jeol. Kur. Bülteni* **19,** 31–44.

Turgut, S. 1978. Tuzgölü havzasının stratigrafik ve çökelsel gelişmesi. *In:* Eseller, G. (ed.), *Türkiye 4. Petrol Kongresi,* 115–127.

Uygun, A. 1981. Tuzgölü havzasının jeolojisi, evaporit oluşumları ve hidrokarbon olanakları. *Türk. Jeol. Kur. Iç Anadolunun jeolojisi simpozyumu,* Ankara, 66–71.

Unalan, G., Yüksel, V., Tekeli, T., Gönenç, O., Seyirt, Z. & Hüseyin, S. 1976. Haymana-Polatlı yöresinin (güney batı Ankara) Üst Kretase-Alt Tersiyer stratigrafisi ve paleocoğrafik evrimi. *Türk. Jeol. Kur. Bülteni* **19,** 159–76.

Yilmaz, Y. 1981. Atlantik tip bir kıtakenarının Pasifik tip bir kıtakenarına dönüşümüne Türkiyeden örnek. *Türk. Jeol. Kur. Konf. Ser.* **3,** 27 pp.

Yüksel, S. 1970. *Etude géologique de la région d'Haymana (Turquie centrale).* Thèse, Fac. Sc. Univ. Nancy, 179 pp.

——, 1973. Haymana yöresi tortul dizisinin düşey yönde gelişimi ve yanal fasiyes dağılışı. *M.T.A. Dergisi* **80,** 50–53.

N. Görür, F. Y. Oktay, I. Seymen & A. M. C. Şengör, I.T.Ü. Maden Fakültesi, Jeoloji Mühendisliği Bölümü, Genel Jeoloji Anabilim Dali, Teşvikiye, Istanbul, Turkey.

Geodynamic evolution of Turkey and Cyprus based on palaeomagnetic data

J. P. Lauer

SUMMARY: Using all the available palaeomagnetic results tentative reconstructions are given for the past positions of various parts of Turkey. At least two parts were located near the equator in Triassic times, presumably along the south-eastern boundary of the Arabian plate. This implies large-scale continental drift towards the north-west which must be taken into account in interpretations of the Alpine orogeny and the geology of the Eastern Mediterranean.

Palaeomagnetic studies of Turkey and Cyprus were carried out between 1973 and 1978. Around 700 orientated samples were collected from about 90 sites (Fig. 1). The ages of the units concerned were Cambrian to Miocene. It is impossible to discuss here in detail all the results, or the background to the palaeomagnetic measurements, including the separation of the stable magnetization, the ages of the remanence and the criteria for dip correction; for this we refer to Lauer (1981a). The aim here is to draw attention to the reasoning behind a new geodynamic model for the Turkish area. Separation of the various remanence directions in the rocks is generally possible, but it requires careful procedures. The uncorrected directions, determined with respect to the present-day position of units, are the only really objective data on which to base future interpretations. In particular, because of the frequent absence of information on the nature and age of tectonism it was impossible to systematically perform fold-tests to determine the absolute or relative ages of magnetization. Other criteria were therefore used, for example, for a site exhibiting two clearly different remanent directions A

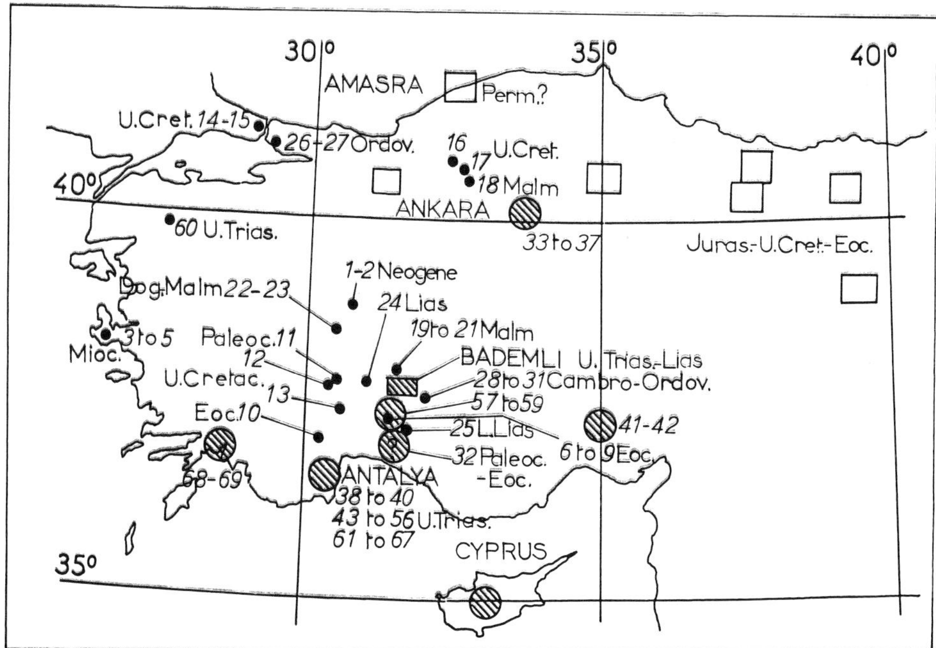

FIG. 1. Distribution of the main sites in Turkey from which palaeomagnetic results exist: the rectangles represent regions where data previously existed: 5 groups of Quaternary sites studied by Sanver (1968) should be added. The hatched areas refer to allochthonous units. Black dots are sites (or groups of sites) belonging to relatively autochthonous units. The approximate ages and serial numbers of sites are indicated.

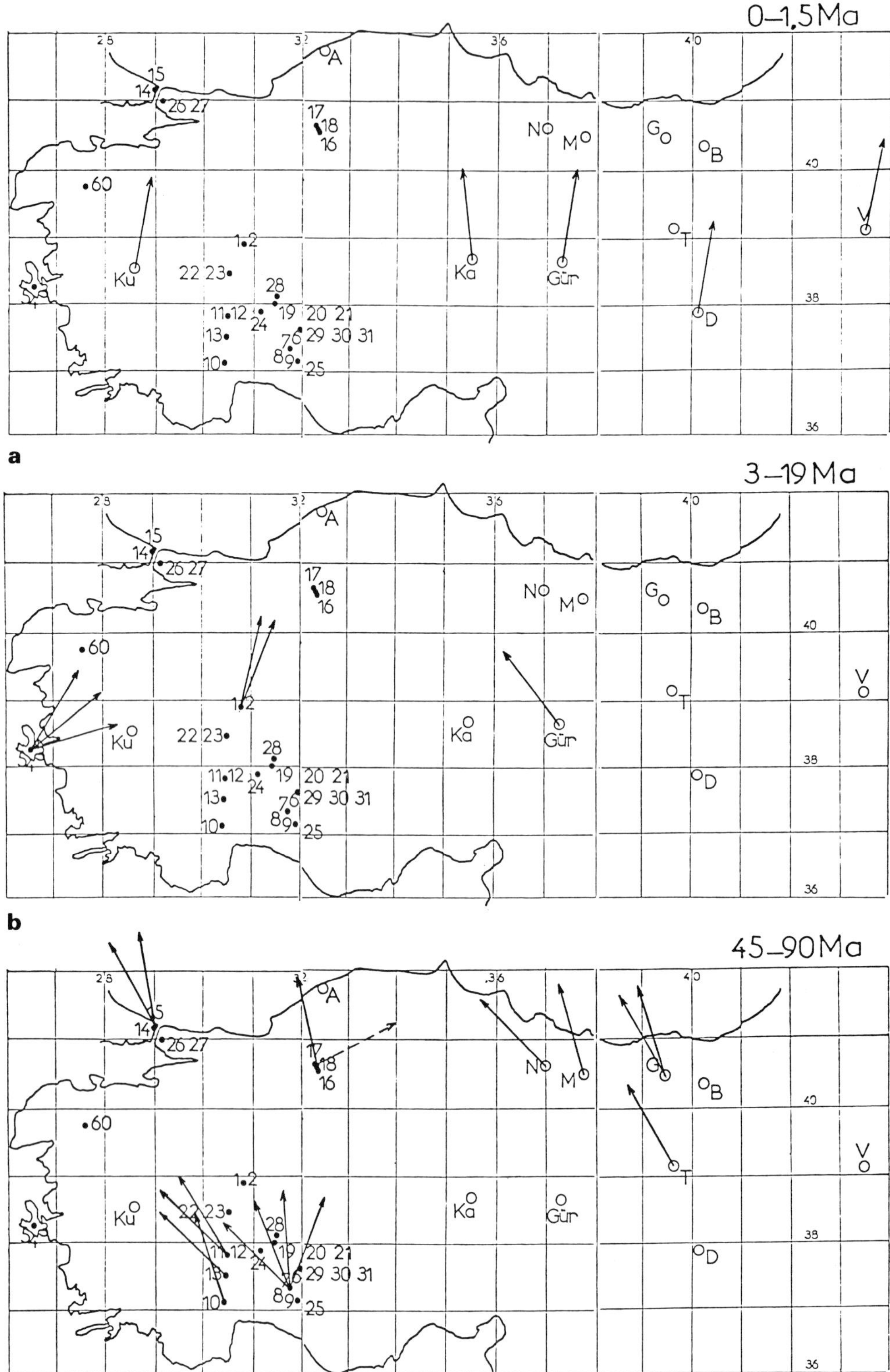

FIG. 2a–e. Directions of the horizontal components of characteristic magnetizations within the post Carboniferous relatively autochthonous units. A normal sense is assigned to each magnetization. Letters refer to sites studied by other authors.

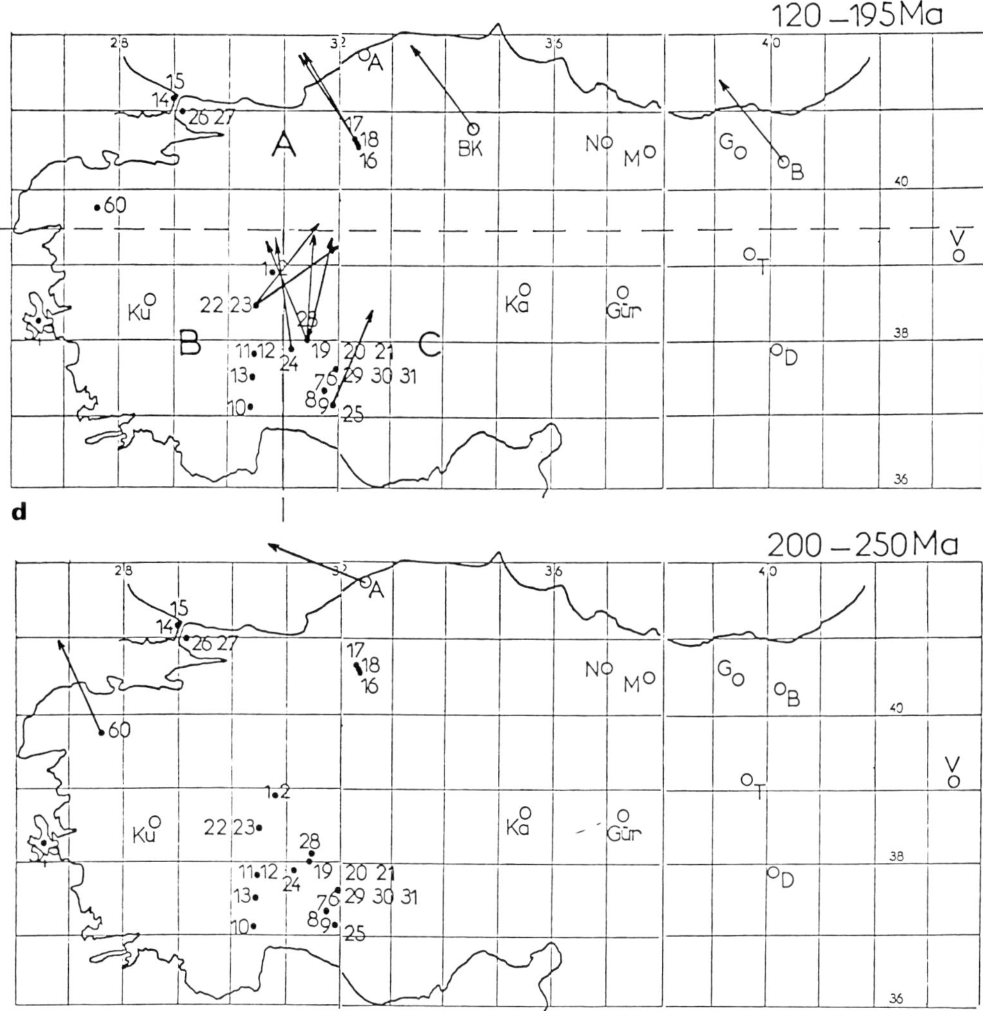

and B, if the direction A' (A corrected for dip) is near the uncorrected direction B, then it was concluded that A is probably older and B younger than the tilting. The validity of such reasoning is increased if A is an 'unusual' direction, e.g. of southern hemisphere character in a Cretaceous Turkish site.

The palaeomagnetic data

The interpretation here is based on all the presently available results (Fig. 1). The data of Gregor & Zijderveld (1964), Sanver (1968), Van der Voo (1968), Van der Voo & Van der Kleijn (1970), Orbay (1979), and Orbay & Bayburdi (1979) are also taken into account.

In Cyprus, study was concentrated on the Cretaceous Troodos Massif (Lauer & Barry 1979). Twelve sites were sampled at regular intervals around the massif, plus one site in the central gabbros. After alternating field (AF) and thermal-cleaning, the average direction for the pillow lavas is very similar to the one given by Vine *et al.* (1973) for natural-remnant-magnetism (NRM). In addition, Shelton & Gass (1980), who studied the sedimentary cover of the Troodos Massif, gave a set of NRM directions with notably westerly declinations until the Miocene. The Pliocene declinations seem to be near zero, thus the anticlockwise rotation of the Troodos would seem to have occurred in the Upper Miocene, but demagnetization data would be required to confirm this.

Palaeomagnetic interpretations are based on the assumption of an average geocentric axial magnetic dipole. In order to eliminate the dispersal of field directions due to secular variation, average permanent directions are calculated from geological units of different but closely comparable ages. These averages are only significant if all of the sites involved belonged to one rigid block. However, the studied region is part of a mobile belt between the Eurasian and Arabo–African blocks and the simple hypothesis that such large undeformed blocks exist within this belt is controversial. It would be best to regard the area as an assemblage of small blocks but this is impracticable since the palaeomagnetic data are still too scarce. The model proposed implies the division of Turkey into three blocks, which may be a simplification, but this is the best that can be achieved with the available data.

In Turkey the existence of major nappes has been clearly demonstrated by field geologists. The motion of these nappes could have differed from the associated relatively autochthonous units. It therefore seems useful to consider these clearly allochthonous units separately.

The data show very significant variations in the directions of magnetization, particularly in the declinations, recorded from different parts of Turkey, as discussed below.

Declinations of relatively autochthonous units

Figures 2a to 2e show the directions of the horizontal components of magnetization which are regarded as primary. Five time-spans are represented. In the Pontides the declinations seem to be nearly coherent for all the geological periods with the exception of one site. In the Anatolian and Tauride belts, however, contrasting trends seem to have existed between the eastern and the western parts until the most recent period. This difference is especially clear from 195 to 120 Ma, supporting the hypothesis of the separation of Turkey into three blocks.

Inclinations of relatively autochthonous units

The distribution of inclinations with time shows that the inclinations (I) progressively decrease with time (Fig. 3). This implies a correlative decrease of palaeolatitude (p) of the sites ($\tan p = 1/2 \tan I$).

In order to compare the inclinations of an area the size of Turkey a unique point has to be used as a reference. The junction of the three inferred blocks has been chosen, with coordinates 39.5°N and 31°E. Normalized inclinations I_N are thus obtained. The results of the calculations are shown in Fig. 4. If we consider the sites in blocks A, B and C separately, the scatter of I_N is reduced. Our hypothesis of separate blocks also leads to a better grouping

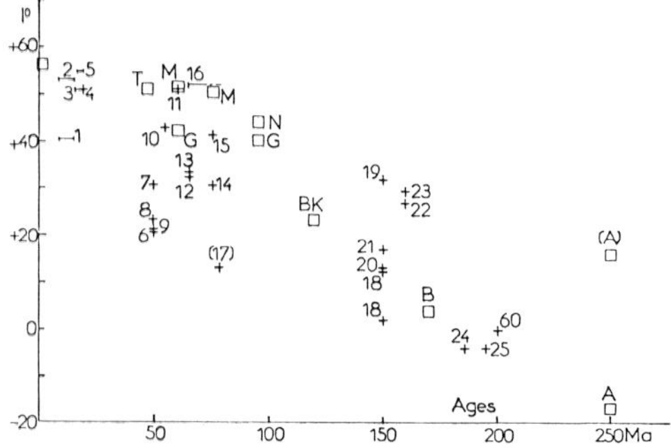

FIG. 3. Turkish relatively autochthonous sites. The distribution of the dip-corrected average inclinations with time is shown. The square (without letter) close to the I-axis gives, according to Sanver (1968), an average inclination for 5 different regions in Turkey. For Amasra (A), two possible interpretations are indicated. Solution A is preferred.

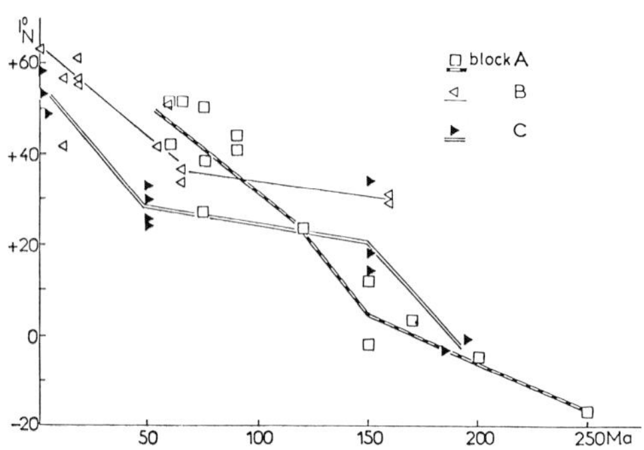

FIG. 4. Relatively authochthonous units: 'normalized' inclination, I_N, for the junction point (39.5°N, 31°E) as a function of time. Each of the three blocks has its own symbols. The evolution of inclinations with time seems to be different from one block to another. (Average curves are indicative). The very eastern sites D, V and T as well as the uncorrected site 17 (Fig. 2) are not taken into account.

of the inclinations. For each of the three blocks a mean line drawn through the clusters of points represents the time evolution of the average palaeo-inclinations, and thus of the average palaeolatitude for the inferred junction point.

Interpretation

Possible significance of separation into blocks

The three blocks A, B and C have been defined solely on palaeomagnetic criteria. We now have to take account of the geological data to see if any obvious discrepancy arises.

The southern boundary of block A separating it from blocks B and C is located between the latitudes 38.5° and 40.5°N, according to the palaeomagnetic data. We can reasonably postulate that this corresponds to the north-Anatolian ophiolitic suture zone which crosses Turkey from east to west at an average latitude of about 39.5°N. The boundary between blocks B and C is located close to the 31st meridian between 37.5° and 39°N. This corresponds to the axis of the Isparta angle which is clearly a key structure located at the junction of the Aegean and Tauride arcs.

The other block-boundaries can not be defined, but for ease of recognition we use the geographical outline of Turkey. It is possible that the extrapolation of some of the results to entire blocks is excessive as every result is valid only for the immediate site itself. However, it appears that during the Neogene (Kondopoulou & Lauer, this volume), and perhaps even earlier, a vast Aegean area from Afyon (Turkey) to the island of Corfu (Greece), thus including Block B, behaved approximately as a single unit. In this area all the known declinations are in the eastern quadrant. This is despite the fact that this area seems to have undergone important N–S extension during the same period (Le Pichon & Angelier, 1979). Thus the definition of a block in palaeomagnetic terms does not preclude important synchronous or later deformations.

Dispersal of permanent directions in relatively autochthonous units

Although some discrepancies may exist (see below) a provisional apparent polar-wander path can be drawn for the virtual geomagnetic pole (VGP) of Block A, the Pontides (Lauer 1981b). By contrast, a characteristic 'banana-shaped' dispersal is observed at any one geological time for blocks B and C, which comprise the whole of the Anatolian and the Tauride areas.

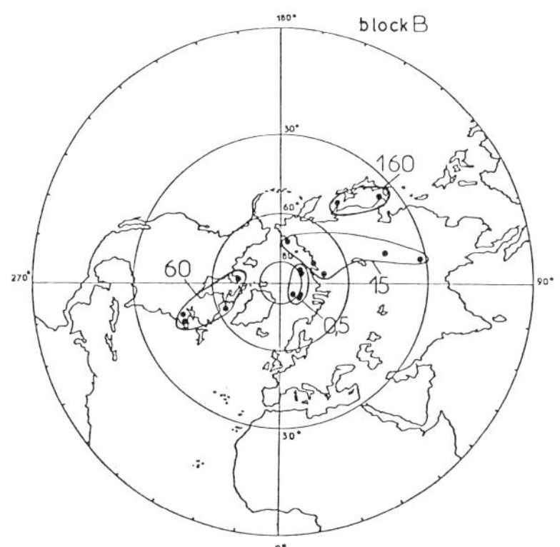

FIG. 5. Dispersal areas of virtual geomagnetic poles (VGP) for block B. Numbers stand for average ages (Ma). Stereographic projection.

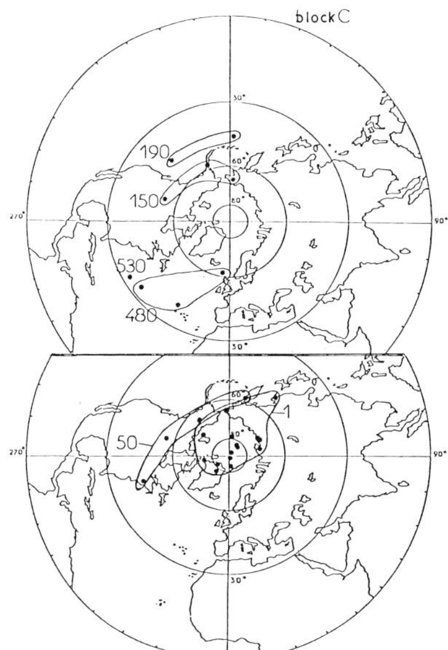

FIG. 6. Dispersal areas of VGPs for block C. As on the preceding figure, a preferential dispersal along arcs is observed, roughly centred on Turkey. Stereographic projection.

These declinations show unusual variability (Figs 5 & 6) which suggests that the inclinations and declinations should be considered separately. A particular dispersal of declinations is certainly related to a specific kind of deformation, but the cause remains unclear. The effect of a large-scale plastic deformation on the directions at individual sites, for example, is unknown.

Palaeolatitudes of the relatively autochthonous blocks A, B and C.

Using Fisher statistics, an average pole position has been calculated for each block. The average inclination of the junction point of the three blocks gives the palaeolatitude of this point (the mean curves for I_N in Fig. 4 lead to very similar results). The average declination gives the orientation of each block with respect to the palaeo-meridian. All longitudes are arbitrary. The ages are approximate for the earliest periods (±10 Ma). This leads to the reconstruction given in Fig. 7. Some uncertain-

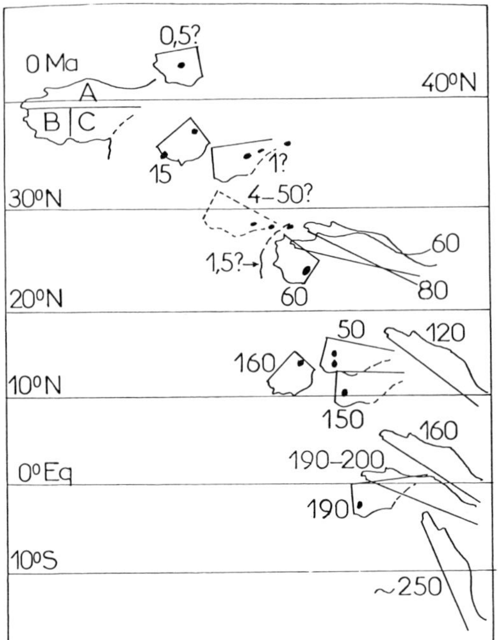

Fig. 7. Tentative latitudinal reconstruction of positions of the Turkish blocks A, B and C at various periods (average ages, Ma). The black dots on B and C represent sites for which data are used. The present position of Turkey is not deduced from the palaeomagnetism. The longitudes are arbitrary.

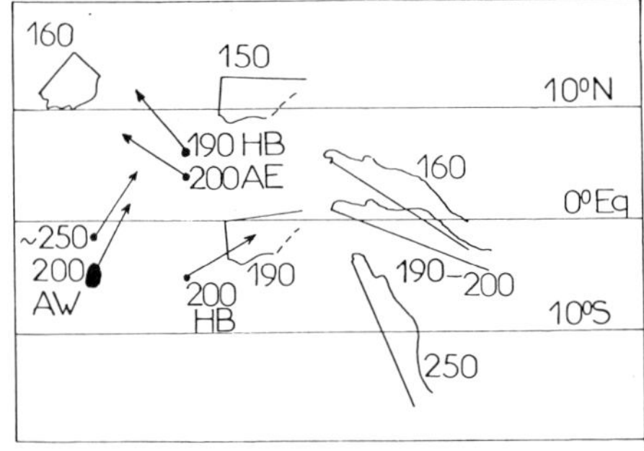

a

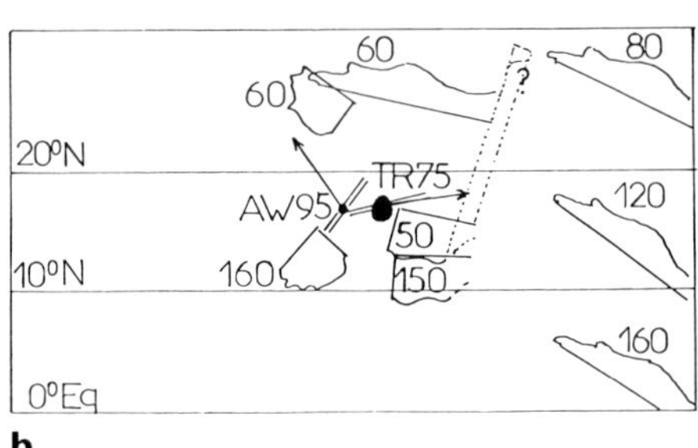

b

FIG. 8. Comparative apparent positions of blocks A, B, C and of allochthonous bodies. The numbers are average ages (Ma). The arrows represent the direction of the present geographical north of these bodies. The dimension of black spots depends on the number of sites within the nappes.

AE, = Antalya nappes, eastern part; AW, = Antalya nappes, western part; HB, = Hoyran Beyşehir nappes; (HB 200 refers to a site near, Çamlik: its magnetization is regarded as normal). TR, = Troodos.

a. Peri-Triassic period; b. Peri-Cretaceous period. The double line is the axis of expansional zones where ophiolites of AW and TR were created. (The definition of this axial direction is based on geological data). The question mark refers to the Tunçeli (T) sites.

ties remain regarding the precise position of some parts of Turkey, particularly in the most recent period (less than 5.0 Ma).

Palaeolatitudes of the allochthonous units

Here the palaeomagnetic data essentially deal with two epochs, one during the Late Cretaceous, the other in the Late Triassic. Applying the same method as before the positions of the nappes can be determined in terms of the palaeolatitudes and palaeo-orientations for each epoch. As these units usually are of smaller extent, we choose the present geographical north as the azimuth reference. It is then possible to compare the respective positions of both the relatively autochthonous units and the nappes for the same periods. The origin and emplacement mechanism of these nappes is one of the most controversial questions in Turkey. It has to be remembered that the errors in determining the locations of the in situ and transported units are additive. Nevertheless the results are reasonable for average directions.

Figures 8a and 8b give attempts to reconstruct the positions of allochthonous units of the Isparta angle, with respect to the relatively autochthonous blocks A, B and C. These are the Troodos Massif (TR, supposedly allochthonous), the eastern Antalya (AE), western Antalya (AW) and Hoyran-Beyşehir (HB) Nappes (classification of Brunn *et al.* 1970). Longitudinal positions are of course arbitrary; a position of the nappes intermediate between B and C is convenient but speculative.

The allochthonous units seem to have followed the motion of the relatively autochthonous blocks throughout. The distance between the two units never seems to have been very great. This is a very clear result.

A more detailed analysis of Figs 8a and b

leads to further conclusions, but in view of possible errors, these are only tentative. They concern first, the palaeolatitudes of the allochthonous masses with respect to blocks on which they now lie, on both sides of the Isparta angle. These respective locations seem to be preserved until the time, possibly in the Cretaceous, when the Triassic lavas of these nappes were partly re-magnetized. Secondly, an emplacement model for the Antalya nappes implies 'wedging' of oceanic material between blocks B and C. Between 60 and 15 Ma interpretation of our measurements could indicate an anticlockwise rotation of block B (Fig. 7) which might perhaps be responsible for the closure of the Isparta angle, although the data are still limited.

Discussion

The question arises as to how firm this 3-block model actually is. First we have to consider a very recent result for the western Pontides reported by Evans *et al.* (1982). They found a mean declination of 93° and a mean inclination of +53° in limestones of supposed Upper Jurassic age near Bilecik. This differs considerably from results for Jurassic relatively autochthonous units, both from near Bayburt in the eastern Pontides (D = 146°, I = + 4°) (which were previously considered as unreliable) (Van der Voo 1968) and from near Gerede, in the western Pontides in this study, (D = 328°, I = + 9°). Instead, Evans *et al*'s result (*op. cit.*) resembles the direction obtained here in the Cretaceous part of the allochthonous Ankara Melange near Gökdere (D = 77°, I = + 45°). Despite this difference there is no clear indication that the presumed Jurassic limestones of Bilecik are allochthonous. Fourquin (1975), quoting Granit & Tintant (1960), states that the underlying Bayirköy unit is of Liassic age (Pliensbachian) and is allochthonous. The Bilecik limestone commonly unconformably overlies the Bayirköy sandstones and this seems to be Lower Jurassic age (Evans *et al.* op. cit.). If confirmed, this new palaeomagnetic result would have an important bearing on the interpretation of the pre-Cretaceous evolution of Pontides. Since this would not explain other existing data the model of a southerly origin is retained.

For the southern parts of Turkey (Anatolides and Taurides), the dispersal of declinations is very organized, as described above (Fig. 8). This calls into question the use of Fisher statistics, as magnetization vectors should be distributed with an axial symmetry about their mean direction. This condition is not fulfilled and the a_{95} exaggerates the error of inclinations (Helsley & Nur 1970). All the inclinations found are low, implying a southern origin for blocks B and C. Could these 'autochthonous' blocks still have originated north of Africa? Concerning block C, our conviction stems from two different sedimentary sites of Lias age, plus several Ordovician results which must be interpreted with more caution. Concerning block B, prior to the Jurassic, there is no available paleomagnetic result. At this time, a location north of Arabia is still possible. However, the western part of the allochthonous Antalya nappes which are in tectonic contact with block B yield the best defined direction for the Late Triassic (Norian–Carnian). The average inclination of 14 separate sites of pillow lavas yields a palaeolatitude of about 5°S. It thus seems impossible that these lavas cooled north of Africa (Fig. 9). Comparable conclusions can be drawn for the location of the allochthonous Bademli redbeds (Van der Voo and Van der Kleijn 1970) of Late Triassic–Early Jurassic age which are now thrust on block C.

Comparison of the Turkish, Eurasian and Arabo–African blocks

It is also possible to restore the palaeolatitudes and orientations of these two large adjacent blocks, as shown in Fig. 9. Three periods have been chosen for which the information is reliable. The additional data used are as follows:

Arabo–African block

(1) There is a recent compilation by Van den Berg (1979) of all the available palaeomagnetic results. Africa is considered to be a single undeformed block. Arabia, for which few data exist, is positioned by closing the Red Sea.

(2) There is the apparent polar wandering (APW) path for Africa deduced from that of North America (Irving, 1977) by Van den Berg (1979) who uses poles of finite rotations for Africa with respect to North America given by Sclater *et al.* (1977).

These two data sets are independent.

Eurasian block

(3) There are the virtual geomagnetic poles (VGP) established by Irving (1977) for northern Eurasia in which Soviet results figure large.

(4) There is the compilation of exclusively Soviet palaeomagnetic results given by Khramov (1977, in Van den Berg 1979), with only cleaned data being taken into account.

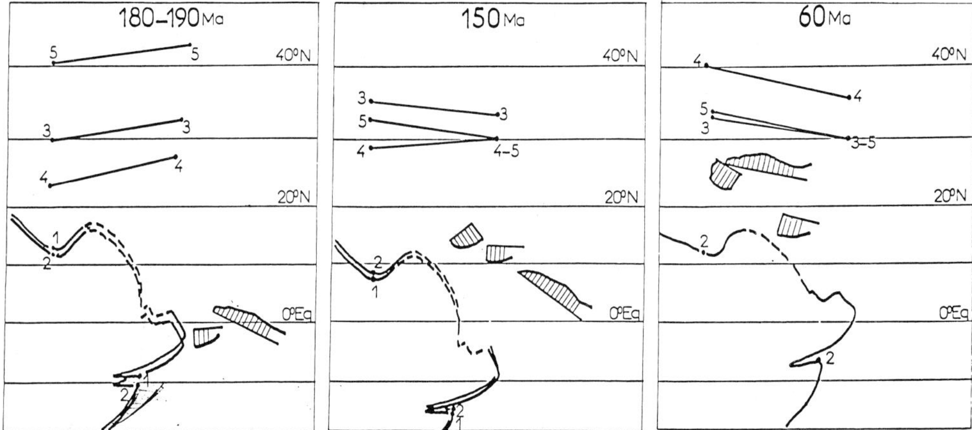

FIG. 9. Reconstruction of the positions of the Arabo–African and Eurasian blocks for three selected periods. For reference, the locations of four points have been calculated from compilations number 1 to 5 (see text). The position of Eurasia is schematically given by the segment Odessa (46.5°N, 28°E)—Baku (40°N, 47°E). The Turkish blocks A, B and C are hatched. Their northward drift seems to begin at the Triassic–Jurassic boundary. All longitudes are arbitrary.

(5) Then there is the theoretical apparent polar wandering (APW) path deduced from that of North America (Irving 1977) by Márton & Márton (1981) with the help of Europe/North America kinematic poles given by Sclater *et al.* (1977).

The data in (3) and (4) are not independent as they are partly based on identical measurements, but the poles of set 5 have been obtained separately.

The reconstruction is based upon the palaeopositions of the following reference points: for Africa 31°N, 31°E in the region of Cairo; 12°N, 51°E, near Somalia; then for Eurasia 46.5°N, 28°E, in the region of Odessa; and 40°N, 47°E in the region of Baku.

The important fact is that in the Triassic, blocks A and C of the Turkish assemblage, as well as the Tauride lavas, could not have been sited north of Africa. At this time these units must have been near the south-eastern corner of Arabia, close to the northern border of the Gondwana continent. This holds whatever the choice of palaeomagnetic data. Westphal (1976) has shown the existence, as for the Triassic–Jurassic VGPs, of a systematic difference between northwestern and southern parts of the African continent, but this does not change the conclusions.

Conclusions

The interpretation of the available palaeomagnetic data lead to the hypothesis that at least the two blocks, A and C, of the present Turkish area were located east of Arabia near the Gondwana continent until the Late Triassic. After that they underwent progressive drift to the NW ending up together as a single mass. The amplitude of the latitudinal displacement is greater than the possible maximum errors, thought to be about ±10°. Such a large movement must have affected the surrounding regions which should be taken into account in any interpretation of the Alpine orogeny and the genesis of the Eastern Mediterranean. The model cannot be regarded as definitely proved but it does point to very much greater mobility of Turkey, relative to Africa and Eurasia, than has previously been considered.

ACKNOWLEDGEMENTS: We wish to express our deep gratitude to Professor A. Roche who initiated this work. Thanks are due to Professors G. Ataman, J. H. Brunn, T. Juteau, O. Kaya, G. Rocci as well as to Drs H. Lapierre, O. Monod, M. Westphal and several other colleagues. Professors K. M. Creer and D. H. Tarling provided helpful reviews. Drs R. Montigny and A. H. F. Robertson helped with the translation into English.

References

BRUNN, J. H., GRACIANSKY, P. CH. DE, GUTNIC, M., JUTEAU, T., LEFEVRE, R., MARCOUX, J., MONOD, O. & POISSON, A. 1970. Structures majeures et corrélations stratigraphiques dans les Taurides occidentales. *Bull. Soc. géol. France*, **12,** 515–556.

EVANS, I., HALL, S. A., CARMAN, M. F, SENALP, M. & COSKUN, S. 1982. A palaeomagnetic study of the Bilecik limestone (Jurassic), northeastern Anatolia. *Earth planet. Sci. Lett.* **61,** 199–208.

FOURQUIN, C. 1975. L'Anatolie du Nord-Ouest, marge méridionale du continent européen; histoire paléogéographique, tectonique et magmatique durant le Secondaire et le Tertiaire. *Bull. Soc. géol. France*, **17,** 1058–1069.

GRANIT, Y. & TINTANT, H. 1960. Observations préliminaires sur le Jurassique de la région de Bilecik (Turquie). *C.r. Séances Acad. Sci. Paris*, **251,** 1801–1803.

GREGOR, C. B. & ZIJDERVELD, J. D. A. 1964. The magnetism of some Permian red sandstones from NW Turkey. *Tectonophysics*, **1,** 289–306.

HELSLEY, C. E. & NUR, A. 1970. The palaeomagnetism of Cretaceous rocks from Israel. *Earth planet. Sci. Lett.* **8,** 403–410.

IRVING, E. 1977. Drift of major continental blocks since the Devonian. *Nature, London*, **270,** 304–309.

LAUER, J. P. 1981a. *L'évolution géodynamique de la Turquie et de Chypre déduite de l'étude paléomagnétique.* Thesis, Dr. Sci. (limited edition), Univ. Strasbourg, I, France, 299pp.

—— 1981b. Origine mériodionale des Pontides d'après de nouveaux résultats paléomagnétiques obtenus en Turquie. *Bull. Soc. géol. France*, **23,** 619–624.

—— & BARRY, P. 1979. Etude paléomagnétique des ophiolites du Troodos (Chypre). *C.r. Séances Acad. Sci. Paris*, **D289,** 977–980.

LE PICHON, X. & ANGELIER, J. 1979. The Hellenic arc and trench system: a key to the neotectonic evolution of the Eastern Mediterranean area. *Tectonophysics*, **60,** 1–42.

MÁRTON, E. & MÁRTON, P. 1981. Mesozoic palaeomagnetism of the Transdanubian Central Mountains and its tectonic implications. *Tectonophysics*, **72,** 129–140.

ORBAY, N. 1979. The palaeomagnetic study of the North Anatolian Fault Zone. *Istanbul Univ. Fen Fak. Mec.* **C44,** 23–39.

—— & BAYBURDI, A. 1979. Palaeomagnetism of dykes and tuffs from the Mesudiye region and rotation of Turkey. *Geophys. J.R. astr. Soc.* **59,** 437–444.

SANVER, M. 1968. A palaeomagnetic study of Quaternary volcanic rocks from Turkey. *Phys. Earth Planet. Inter.* **1,** 403–421.

SCLATER, J. G., HELLINGER, S. & TAPSCOTT, CH. 1977. The palaeobathymetry of the Atlantic Ocean from the Jurassic to the Present. *J. Geol.* **85,** 509–552.

SHELTON, A. W. & GASS, I. G. 1980. Rotation of the Cyprus microplate. *In*: PANAYIOTOU, A. (ed.) *Ophiolites, Proc. Int. Oph. Symp. Cyprus 1979*, 61–65.

VAN DEN BERG, J. 1979. Palaeomagnetism and the changing configuration of the Western Mediterranean area in the Mesozoic and early Cenozoic eras. *Geol. Ultraiectina*, **20,** 1–179.

VAN DER VOO, R. 1968. Jurassic, Cretaceous and Eocene pole positions from Northeastern Turkey. *Tectonophysics*, **6,** 251–269.

—— & VAN DER KLEIJN, P. H. 1970. The complex NRM of the Permo-Carboniferous Bademli redbeds. *Geol. en Mijnbouw*, **49,** 391–395.

VINE, F. J., POSTER, C. K. & GASS, I. G. 1973. Aeromagnetic survey of the Troodos igneous massif, Cyprus. *Nature Phys. Sci.* **244,** 34–38.

WESTPHAL, M. 1976. *Contribution du paléomagnétisme à l'étude des déplacements continentaux autour de la Méditerranée Occidentale.* Thesis, Dr. Sci. (limited edition), Univ. Strasbourg I, France, 304 pp.

J. P. LAUER, Institut de Physique du Globe, 5 rue René Descartes, F-67084 Strasbourg, France.

4. NEOTETHYS: Greece and the Balkans

Stemming from a long tradition of international research the overall stratigraphy and structure of much of Greece is now quite well understood in contrast to the much larger Turkish area. Much current research now focuses on generally clearly defined regional problems or specific processes, well beyond the phase of suture-tracing still prevalent in Turkey. It follows that in any set of contributions (Fig. 1) reflecting current research, significant gaps are bound to emerge. For the general reader seeking a way into the subject we therefore have tried to integrate some of the essential background information, plus key references, into our own introductory chapter, including the Yugoslavian area which is critical to any overall Eastern Mediterranean synthesis, but otherwise receives little attention.

We order the contributions broadly working from southern (external) to northeastern (internal) zones (Fig. 2). We first give the reader a chance to sample the currently contrasting views of Cretan geology, and then turn to the mostly metamorphic geology of the southern Aegean islands. Several papers then discuss new results for the north-easterly (internal) zones, leaving to the end of the section two more wide-ranging contributions which have great potential importance in Eastern Mediterranean synthesis (Figs 1 & 2).

To what extent the well-known NS-trending isopic zones of mainland Greece can be correlated with nappes in Crete is a question taken up by **HALL *et al.***, who first highlight the differences between Crete and the Sunda arc in the SW Pacific, then summarize the lithological successions in each of the main tectonic units on the island which they divide into two major nappe groups. Their upper group is composed mostly of carbonate platform, basinal sedimentary and ophiolitic rocks, metamorphosed at the base (Phyllite nappe) to blueschists. The authors recognize a new unit, the Tripali unit, which is composed of crystalline limestone breccias considered to represent originally proximal facies located between a major carbonate platform (Tripolitza unit) and an area of basinal more pelagic accumulation (Plattenkalk unit). **HALL *et al.*** argue that their upper group of nappes underwent several phases of thrusting and deformation, with related westward flysch dispersal, followed by progressive continental collision and blueschist metamorphism in the Upper Eocene. A key difference from earlier interpretations, however, is that they believe their lower nappe, composed of 'Tripali' marginal facies and adjacent basinal pelagics (Plattenkalk), has experienced only one phase of N–S-directed deformation. The authors' radical explanation is that their lower nappe belonged to the southern passive Apulian margin, which was only finally strike-slipped into place to be in a position to be overthrust from the north by their upper group of nappes during the Oligocene.

In a comprehensive review, **BONNEAU** sets out the contrasting classical view of the southern Aegean and the Cyclades as a vast nappe pile which was transported southwards as the result of continental collision in the Upper Eocene. Summarized first are the main tectonic units in the SE ('external') area (Crete, Peloponnesus, Rhodes, Dodecanesos). A platform unit in Rhodes (Lindos neritic sequence) tectonically underlying the HP/LT metamorphosed equivalent to the Plattenkalk (Ida unit) is considered to be para-autochthonous but is not seen on Crete. **BONNEAU** considers **HALL *et al.***'s critical limestone breccia Tripali unit to be merely tectonic melange caught between two major nappes. The authors are, however, agreed that the Phyllite nappe could represent part of the basement of the Tripolitza carbonate platform, in complete contrast to **PAPANIKOLAOU** who believes this unit to be considerably more allochthonous (see below). **BONNEAU** takes folding in the overlying composite nappes (Pindos-Ethia and Arvi nappes) to indi-

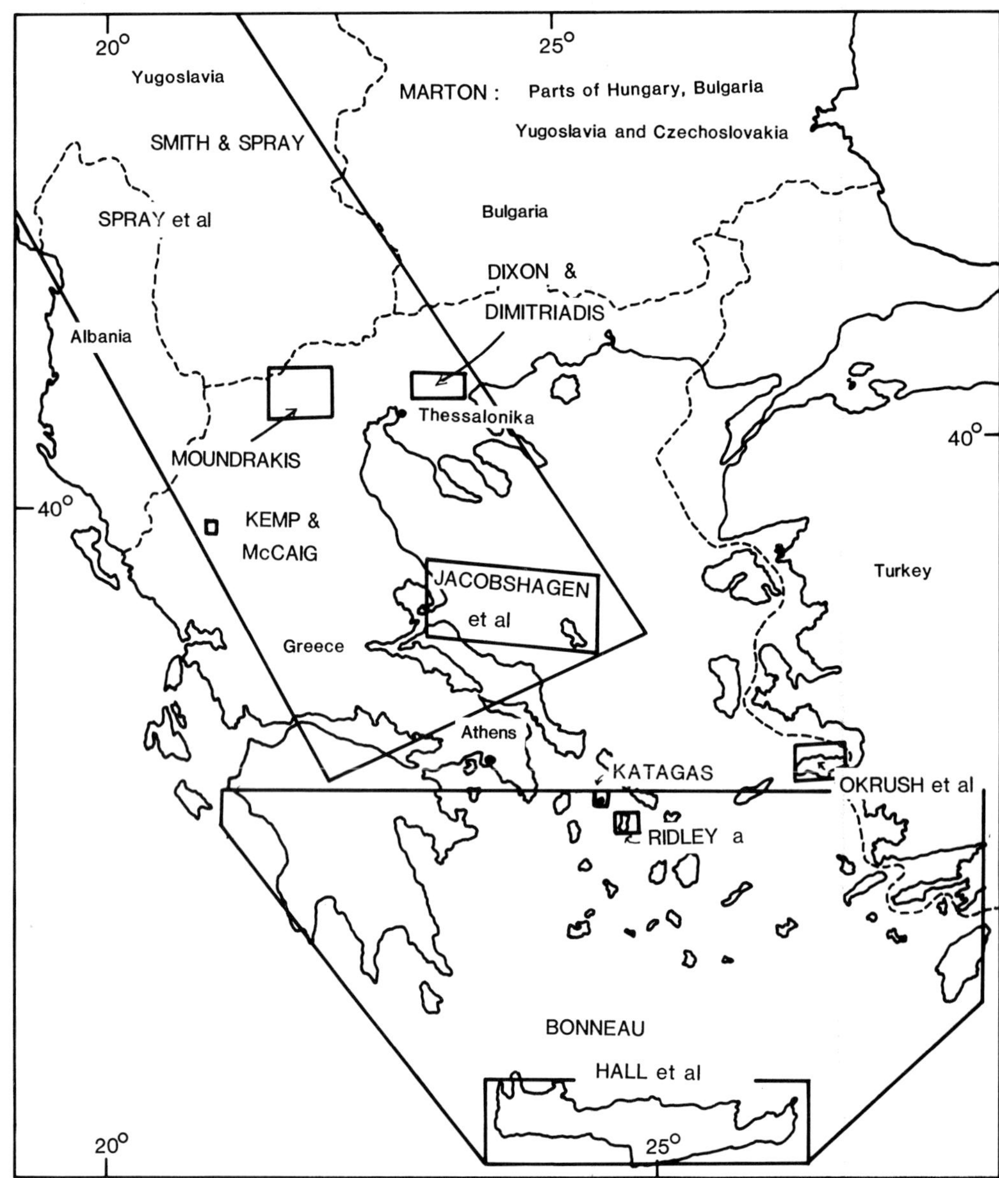

FIG. 1. The Greek and adjacent areas discussed by various authors here.

cate NE–SW compression similar to the trend of the present Aegean arc. Notably some ophiolitic melange (e.g. Anaphi) resembles that of the Lycian nappes in SW Turkey, which could suggest a link with Turkish tectonic evolution, a point also taken up in our own introductory chapter.

BONNEAU goes on to discuss the central and eastern Aegean area, including the famous blueschists, concluding that there are enough similarities with the southerly external area to suggest that both represent different structural levels in a single, vast SW-directed nappe pile. A 'Pindic ocean' in the area is thought to have closed in the Early Tertiary followed by continental collision by Late Eocene with expulsion of the vast nappe pile.

Various south Aegean units are discussed in more detail by other contributors. The HP/LT rocks of Samos (Dodecannesos) are described by **OKRUSCH *et al.*** The metamorphic complex there consist mostly of marbles and phyllites with intercalations of serpentinites, blueschists and greenschists. Notably the glaucophanites

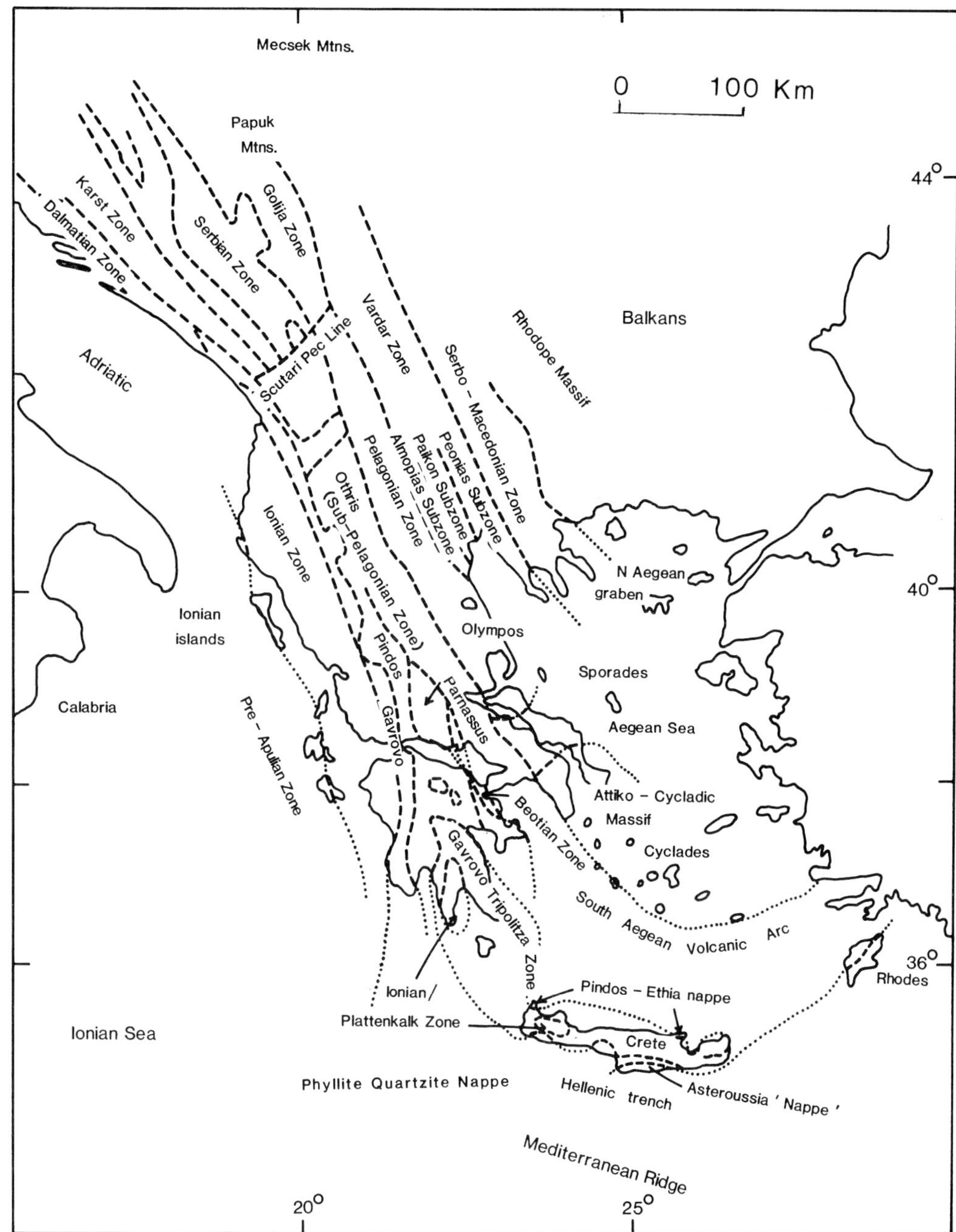

FIG. 2. The main tectonic elements of the Greek area.

possess oceanic tholeiitic protoliths. The metamorphic peak of ca. 420–470°C at 9 Kb was probably reached in Eocene time. The author concludes that the assemblage is comparable to that of the Cyclades (e.g. Syros) thus consistent with **BONNEAU**'s hypothesis which links both areas to subduction and continental collision.

The metamorphic evolution of the small Cycladic island of Ghiaros is summarized by **KATAGAS**. Meta-sedimentary and metabasic rocks underwent prograde metamorphism to upper blueschist facies at lithostatic pressures equivalent to 38 km of burial. The probable age of metamorphism was again Eocene although Palaeocene ages have come from blueschists in southern Euboea. Marked retrograde metamorphism is attributed to unroofing plus increased influx of water which took place at lower temperatures.

The structure of the island of Syros, part of the Attico-Cycladic crystalline belt, justly famous for its excellent blueschists, is described here by **RIDLEY**. Meta-sedimentary and meta-igneous rocks have again yielded Eocene metamorphic ages although these may record the time of cooling rather than primary formation. **RIDLEY** shows that the trends of the hinges of synmetamorphic recumbent folds vary from NE to SE across the island, which could have been developed, he argues, in a single deformation event with a variable wrench component, rather than during separate fold phases. Such deformation is visualized as having taken place associated with ductile thrusting at deep structural levels, by inference during subduction towards the north-west. **RIDLEY** also points out the similarities of the Cycladic HP/LT assemblages with the North Turkish blueschist belt, described by **OKAY** in Section 3.

For a third alternative view of the tectonic history of Crete and the South Aegean, based mostly on the study of the region's igneous and metamorphic rocks, we have drawn the reader's attention to recent major summaries by SEIDEL *et al.* (1982) and ALTHERR *et al.* (1982).

PAPANIKOLAOU outlines his own very personal conception of the three contrasting metamorphic belts of the Hellenides. He advocates a mode of multi-stage emplacement such that a single blueschist belt was initiated in the Cycladic area related to eastward subduction of the now vanished western part of the Pindos ocean, and then that this material was thrust westwards onto units equivalent to the Gavrovo-Tripolitza carbonate platform. The key point which distinguishes his model from all others is that later major westward rethrusting of the carbonate platform units is postulated to finally sandwich the blueschists between the Ionian zone (Plattenkalk) and more external parts of the Gavrovo platform units in the most southerly external zones in Crete and the Peloponnese.

From the SE Aegean, the Pindos zone or its equivalents, discussed by **HALL *et al.*** and by **BONNEAU**, swings NE into mainland Greece (Fig. 2). In this area **PE-PIPER & PIPER** have investigated the chemistry and tectonic significance of basic to acidic extrusives of Middle Triassic, Upper Jurassic and Lower Cretaceous age. They first summarize the evidence for the Pindos basin being originally a 200 km wide Mesozoic pelagic basin bordered westwards by a passive carbonate margin. Both the Triassic and the Upper Jurassic to Lower Cretaceous extrusives are chemically anomalous in terms of 'immobile' trace elements with both within-plate and volcanic-arc characteristics, while some of the younger extrusives also show an ocean floor character. In the light of suggestions that Late Permian alkalic volcanics in W Crete, Corsica and Italy could represent rifting to form oceanic crust, the authors see the Pindos zone as an incipient ensialic marginal basin related to subduction from an older southern ocean basin (Mesogea).

Evidence of former marginal units to the Pindos ocean basin is presented by **KEMP & McCAIG**. Structurally overlying the Pindos flysch, interpreted as a subduction-related accretionary complex, and exposed in a corridor immediately underlying the trailing edge of the Pindos ophiolite, occurs a distinctive assemblage of sub-alkalic mafic extrusives, gabbro, pelagic and hemipelagic carbonates and debris-flows of Triassic and later Mesozoic age (Perivoli Complex). This suite is considered to represent the most distal preserved part of the former Mesozoic ocean basin over which the Pindos ophiolite was thrust during its initial displacement from an ocean basin.

The previously little-known NE extension of the Pelagonian zone into northern Greece is discussed by **MOUNDRAKIS**. There are two crystalline massifs, in the east the Voras Mountains and in the west the Vernon Mountains, which show parallel crystalline basement units including granites of Carboniferous and older age. Overlying Mesozoic carbonate platforms appear to be still *in situ* in the western part (Vernon), but are allochthonous in the east (Voras). Ophiolitic units (Kastoria ophiolite) and sedimentary successions along the western margin of the Pelagonian zone in this area support the former existence of a Pindos oceanic basin to the west. The contrast in structural styles and successions between the Vernon and Voras units suggest that a basin of some significance could have been located between the two.

Units structurally above, and to the east, of the Pelagonian zone are exposed in the Sporades islands, and on the mainland in the southern part of the Pelion peninsula. **JACOBSHAGEN & WALLBRECHER** recognize four main units: at the base is the Pelagonian nappe, composed of Permian and Triassic shallow water siliciclastics and carbonates. This is overthrust by the Eohellenic nappe including meta-sediments, ophiolitic rocks, some of them metamorphosed to HP/LT and emplaced during Upper Jurassic to Early Cretaceous. After subaerial erosion both these lower nappes were overlain by the Mesoautochthonous unit of mostly neritic sediments, culminating in flysch of probable Early Tertiary age which experienced only relatively low-grade metamorphism. Isolated klippen of ophiolitic rocks, schists and melange are seen as a remnant of a major Neohellenic nappe. **JACOBSHAGEN & WALLBRECHER** view the early Mesozoic rocks as derived from a north-eastern 'Vardar' ocean which closed completely by Lower Cretaceous time, followed by subsequent re-opening in the Upper Cretaceous to Early Tertiary to accommodate the ophiolitic 'Neohellenic' nappe.

A critical area to the NE in the Serbo-Macedonian zone is discussed by **DIXON & DIMITRIADIS.** Previously regarded as behaving largely as inert metamorphic basement during the Mesozoic, an important suite of amphibolites, sheeted dykes, metamorphosed gabbros and ultramafic is described in which preserved primary igneous relationships indicate an ophiolitic origin, possibly as an *in situ* Mesozoic rift. The authors argue that the Serbo-Macedonian zone was probably very actively involved in Mesozoic and Tertiary tectonism.

Traditional palinspastic reconstructions of the Hellenides have involved the formation and destruction of either one or several Mesozoic ocean basins with varying complexities of thrusting to produce the now nearly parallel ophiolitic belts in the Greek area. A radical alternative is put forward in this volume.

First **SPRAY *et al.*** analyse available stratigraphical and radiometric age data for the formation, slicing and tectonic emplacement of ophiolites in both the parallel Greek and Yugoslavian ophiolite belts. It appears likely that both the western and the eastern ophiolite belts were formed and then sliced when still hot to form amphibolitic soles near the ridge crest within a brief period of a few Ma at most in Mid-Jurassic time. The eastern ophiolites were emplaced onto the adjacent passive margin in Upper Jurassic time, but those in the west not until the Lower Cretaceous.

Stimulated by what they consider to be a most unlikely coincidence of close parallel evolution in two genuinely separate ophiolite belts separated by a thin elongate Pelagonian micro-continent, thereby echoing **RICOU**'s belief for Turkey (Section 3), **SMITH & SPRAY** propose that the two ophiolite belts were once continuous, but that the western (Greek) belt has been translated southwards by sinistral shear to produce the present overlap. They come to the conclusion, mostly from 'immobile' trace element and petrological evidence, that the ophiolite genesis involved dehydration of a descending slab in a subduction zone. Some of the structurally underlying Triassic mafic extrusives and deep-water rift and later passive margin facies show calcalkaline tendencies suggesting they too formed above a subduction zone, in this case dipping southwards from Palaeotethys. Since the Triassic rift volcanics and sediments are thought to record a preserved marginal strip of a Triassic ocean basin, it follows that their mid-Jurassic ophiolite was involved in intra-oceanic subduction to form a narrow rift within the marginal crust stranded in the arc-trench gap. Soon after, this small basin was sliced to form amphibolite sole rocks, which can be attributed to phases of 'roll-back', then 'roll-to' which were accomplished during early North Atlantic opening, the required ca. 700 km of sinistral strike-slip faulting being completed still in the Upper Jurassic, followed by emplacement of the eastern ophiolite belt in the Upper Jurassic, and the western one in the Lower Cretaceous.

A further approach to the problem of the number and location of Mesozoic ocean basins is to use palaeomagnetic data to infer which areas were attached to either the African or Eurasian continents and thus see if major separate free-floating micro-plates are required. In Section 3 **LAUER** concluded that Turkey was split into 3 blocks in the Mesozoic, derived, in his view, well SE of the present Eastern Mediterranean areas. A very different conclusion is reached by **MÁRTON** for the (different) area further west.

MÁRTON summarizes the African and European polar wander curves (APW). From direct observation, the APW path of Europe has moved generally NE since 300 Ma, with a small 'loop' in Mesozoic time. The APW path of Africa is quite different, moving steadily north from 320 Ma to ca. 200 Ma, then taking an eastward elongate clockwise loop during the Mesozoic before resuming a northward path in the Tertiary. A large part of the Western Mediterranean and in the Eastern Mediterranean, the Istria peninsula of Yugoslavia, the Trandanubian Central Mountains, the Bükk Mountains in Hungary and the outer and inner Carpathians of Czechoslovakia are linked within unit A which, it is argued, formed part of Africa in the Mesozoic. On the other hand palaeopoles from the Mecsek and Villany Mountains in Hungary and the Papük Mountains of Yugoslavia and NW Bulgaria (unit B) are more comparable to European palaeopoles although the pattern is not yet coherent. Notably, the data are currently inadequate to determine whether the complex allochthonous terrains forming much of Greece and eastern

Yugoslavia can be as related to either of these two major units in the Mesozoic and Tertiary and this will no doubt be a major target for future palaeomagnetic investigation.

References

ALTERR, R., KREUZER, A., WENDT, I., LENZ, H., WAGNER, G. A., KELLER, J., HARRE, J. & HOHNDORF, A. 1982. A Late Oligocene/Early Miocene high temperature belt in the Attic-Cycladic crystalline complex (SE Pelagonian, Greece). *Geol. Jb.*, **E23**, 97–164.

SEIDEL, E., KREUZER, H. & HARRE, W. 1982. A Late Oligocene/Early Miocene high pressure belt in the external Hellenides. *Geol. Jb.*, **E23**, 165–206.

The significance of Crete for the evolution of the Eastern Mediterranean

R. Hall, M. G. Audley-Charles & D. J. Carter

SUMMARY: The Hellenic Arc is a convergent zone extending between Greece and Turkey associated with northward subduction of the African plate under the Aegean. The isopic zones of Greece are usually shown to extend parallel to the arc through Crete, implying that the present convergent pattern is long established. However, although Crete is situated in the centre of the Hellenic fore-arc the elongation of the island parallel to the arc is not due to convergence but to late Cenozoic regional extension. Our work in Crete indicates that Tertiary facies belts were orientated approximately N-S and that initial convergence in the region was oblique to the present arc. The Upper Nappes of Crete record two major deformational phases: D1 produced N-S folds and D2 produced E-W folds, whereas the lowest unit on Crete, the Plattenkalk Series, records only the D2 phase.

We suggest that Crete records the Mesozoic extensional, and Tertiary compressional, history of the Apulian continental margin. Eocene E-W compression (D1) caused stacking of the Upper Nappes and HP/LT metamorphism. Oligocene N-S compression (D2) caused thrusting of the Upper Nappes onto the Plattenalk Series, and initiated the present convergent pattern which is unrelated to the earlier history of the region.

Crete and the Hellenic arc

The Hellenic arc has many of the usual features of island arcs: an external trench complex, active underthrusting of the arc, seismicity indicating a slab dipping northwards beneath the arc, a large negative free air anomaly, increased heat flow behind the trench, and an internal volcanic arc about 200 km from the trench complex. These features are usually interpreted (e.g. Makris 1977; McKenzie 1978; Richter & Strobach 1978; Le Pichon & Angelier 1979) as indicating broadly northward subduction of the African plate beneath the Aegean (Fig. 1).

Crete is situated in the centre of the arc, approximately midway between mainland Greece and Turkey, and has an obvious elongation parallel to the arc. This observation suggests that the geology of Crete might in some way be related to the structure and evolution of the Hellenic arc and this is one implication of previous geological models for the region (e.g. Aubouin *et al.* 1976, Jacobshagen *et al.* 1978). The main isopic zones recognized in the Hellenides (Fig. 2) are traced by these authors into Crete and through the arc. Considerable attention has been given, not to whether the isopic zones do continue through the arc, but to which Hellenide isopic zones the Cretan geological units should be assigned. If the Mesozoic and early Tertiary isopic zones of the Hellenides can be traced through the arc from Greece to Turkey this must imply either that the arc developed in the early Mesozoic with the isopic zones parallel to the arc throughout the Mesozoic and early Tertiary, or that the isopic zones were deformed into the present arc during the Cenozoic. The persistence of the Hellenic arc through the Mesozoic to the present is explicit in many models of evolution of the region (e.g. Wunderlich 1973, Aubouin *et al.*, 1976, Jacobshagen *et al.*, 1978, Letouzey & Tremolières 1980) and implicit in others (e.g. Altherr *et al.* 1982, Bonneau 1982). In fact, at least one of the Hellenide zones, the Tripolitza zone, cannot be traced through the arc: it is not known any further east than Crete (Aubouin *et al.* 1976).

Anomalies of the Hellenic arc

The features of the Hellenic arc noted above suggest that it is comparable to other island arcs and Fig. 3 shows a comparison with the Sunda arc of Indonesia. The two arcs are remarkably similar in the dimensions of their elements and the islands of Crete and Nias occupy almost identical positions with respect to their trenches and magmatic arcs. However, when examined in more detail the differences between the two arcs are striking. Normal oceanic crust is entering the Sunda trench in contrast to the much thicker crust (regarded as thinned continental or transitional) south of the Pliny trench. Nias Island consists of tectonic melanges in the form of imbricated wedges interpreted as turbidites and other material scraped off the down-going plate (Moore *et al.* 1980). The oldest rocks exposed in these melanges are thought to be of Eocene age. The long axis of the island is

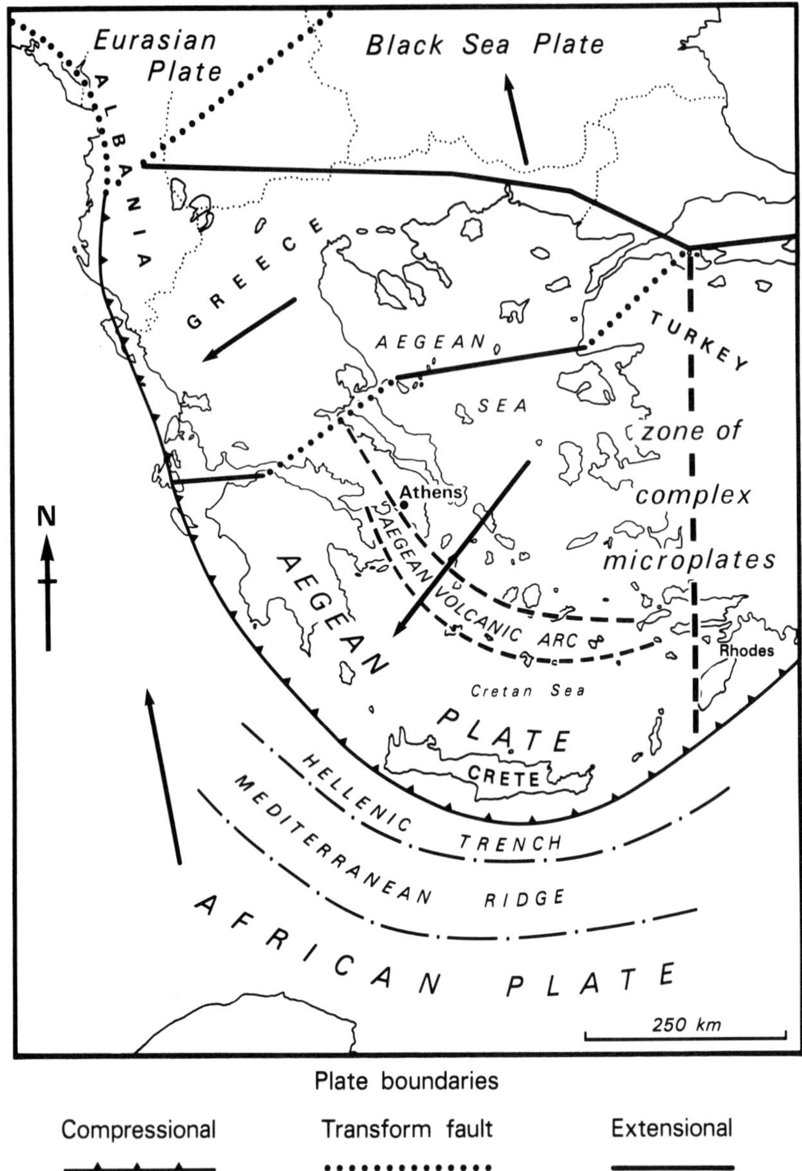

FIG. 1. Present plate boundaries and their motions after McKenzie (1978). The arrows show the directions of motion relative to Eurasia and their lengths are approximately proportional to the magnitude of the relative velocity.

parallel to the trench and is related to the imbricate structure and structural 'grain' of the island. Furthermore, volcanic rock debris from the magmatic arc makes a significant contribution to the Neogene rocks overlying the melanges of the non-volcanic arc. The Neogene rocks have sheared contacts with the underlying rocks and are folded into tight synclines between belts of melange. In contrast, Crete has an entirely different structure. It is composed of a pile of nappes with no evidence of an imbricate or accretionary wedge structure. All the exposed rocks of Crete are unrelated to the down-going slab and are not ocean plate or trench sediments but rocks deposited on continental crust. They are mainly of Mesozoic and early Tertiary age, but also include parts of their pre-Mesozoic basement and their

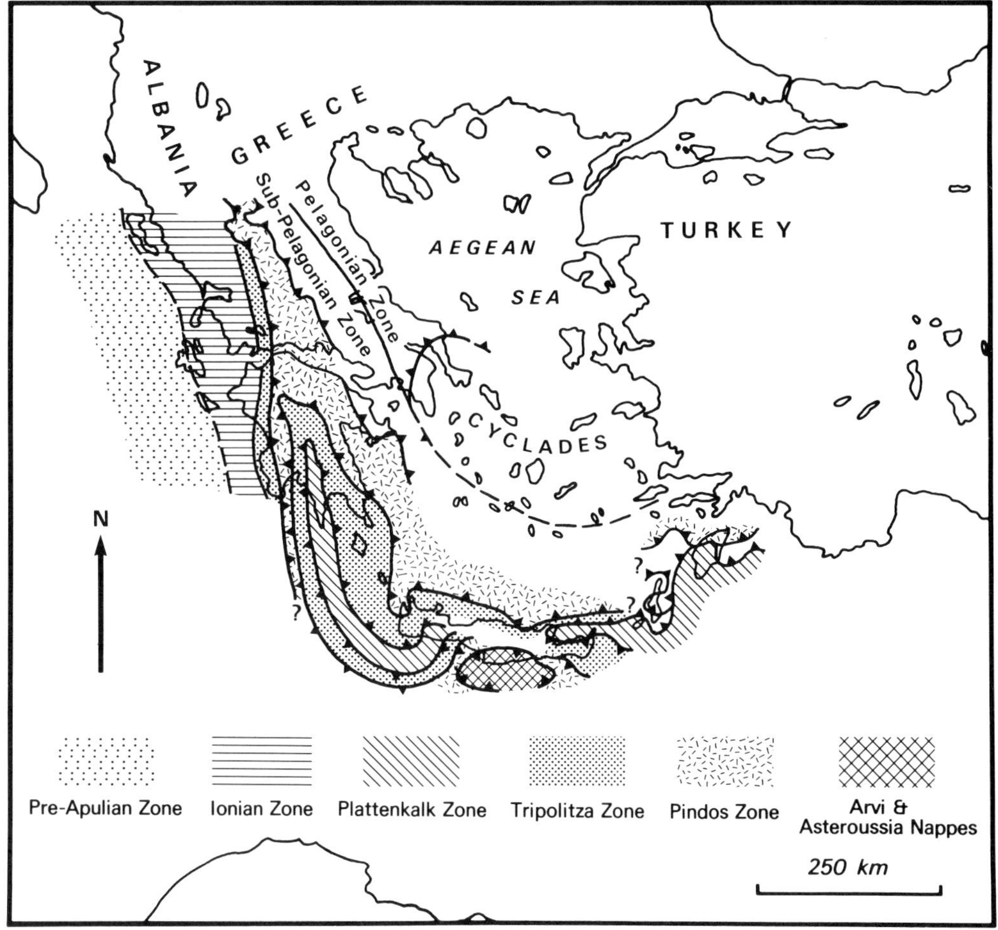

FIG. 2. Extension of the isopic zones of the Hellenides into the Hellenic arc after Aubouin *et al.* (1976).

Neogene 'molasse' cover. Debris from the present volcanic arc makes no significant contribution to the Neogene cover. Many studies of Crete have noted the importance of steep faulting on the geology and topography of the island (Fig. 4A) and the spectacular scenery for which Crete is famous is clearly a reflection of this faulting. The island has the form of a series of upfaulted horst blocks of pre-Neogene rocks separated by grabens filled by Neogene and Recent sedimentary rocks (Fig. 4B). This complicates recognition of a structural sequence and correlation between different parts of the island. The majority of contacts between different geological units are steep faults and interpretation of the structure is dependent on critical observations at a few localities, and has also depended on interpretation from the sequence of units known in the Hellenides and elsewhere in the Aegean. In contrast to Nias, the obvious elongation of Crete and its parallelism to the arc are not a result of the structure of the pre-Neogene rocks of the island but are a reflection of the late Cenozoic extension of the region (Angelier 1978, Mascle *et al.* 1982), particularly along faults parallel to the arc.

Pre-Neogene rocks

The pre-Neogene rocks of Crete form a pile of nappes which were emplaced in the Tertiary. The exact number of nappes and their names and status vary from author to author (e.g. Creutzburg & Seidel 1975, Bonneau *et al.*, 1977, Kopp 1978). Fig. 5 shows the main tectonic units we have adopted on the basis of our field-work (Tee 1982, Hall & Audley-Charles

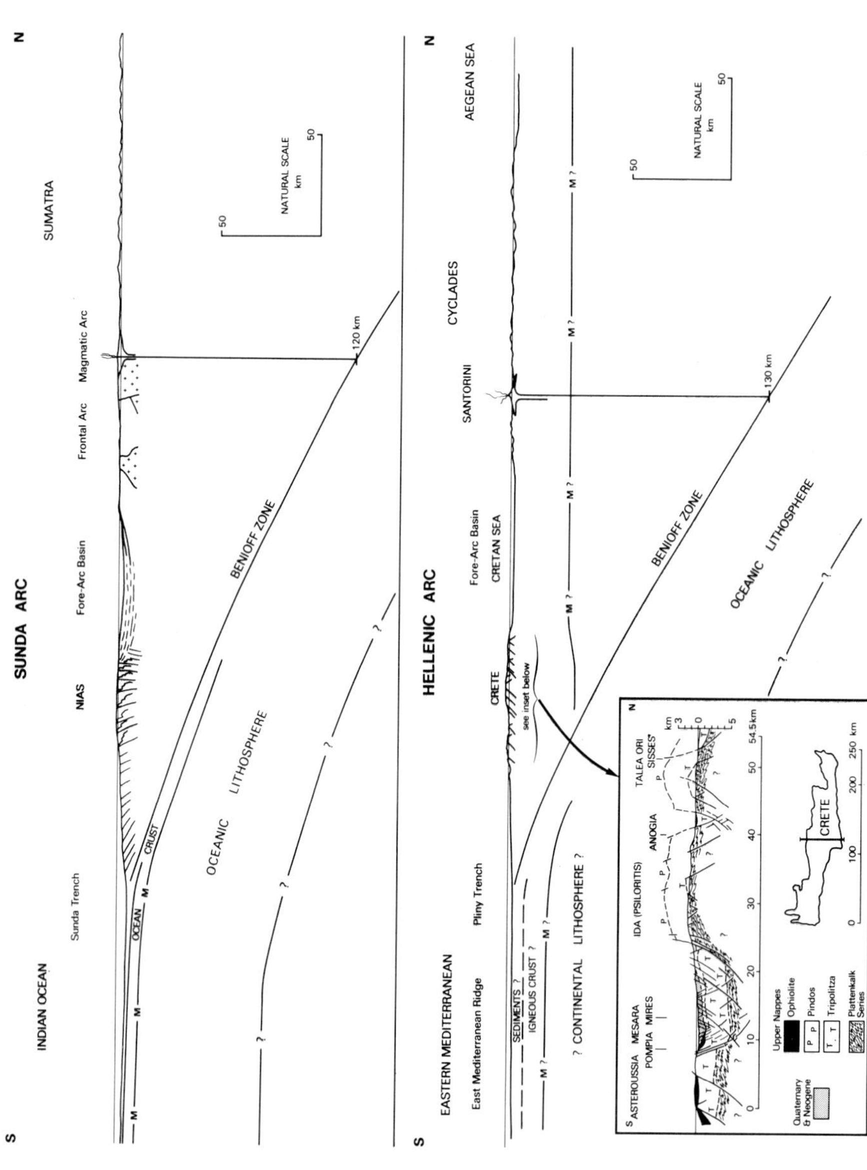

FIG. 3. A comparison of the Sunda arc through Nias Island based on Moore *et al.* (1980) with the Hellenic arc through Crete based on data in Closs *et al.* (1978). The inset diagram is modified from Le Pichon & Angelier (1979).

FIG. 4. (A) Steep faults on Crete, based largely on the 1:200000 geological map of Creutzburg (1977), and modified where we have carried out detailed mapping.

(B) Simplified map to show distribution of principal horsts and grabens of Crete. The grabens are filled by thick sequences of Neogene-Recent sediments and the horst and graben structure of the island is a reflection of considerable extension of the island since the Miocene.

1983, Hussin 1983, our unpublished observations) and their relationship to the structural units of Bonneau *et al.* (1977). The principal stratigraphic features of each of these units are summarized below and then their structural features are discussed.

Plattenkalk Series

The Plattenkalk Series begins with Norian stromatolitic dolomites indicating inter-tidal sedimentation on continental crust. Probable Early Jurassic subsidence of continental crust is indicated by coarse carbonate breccias passing up into channelled and graded calciturbidites and then into basinal conditions of a bedded chert-limestone sequence. Basinal conditions probably prevailed during most of the Mesozoic, with carbonate deposition throughout the whole period, and occasional disturbances recorded by slumped chert horizons and channelled calcisiltites. Only in the Late Eocene or Early Oligocene was there any significant siliciclastic input to the basin, with a few metres of marls or calcareous shales (Fytrolakis 1972; Bonneau 1973) sometimes referred to as Plattenkalk 'flysch'.

Epting *et al.* (1972) claim that the Norian stromatolitic dolomites rest unconformably on shallow marine carbonates and siliciclastics of Permian to Early Triassic age in the Talea Ori and interpret these rocks as the pre-Norian basement of the Plattenkalk Series. We have remapped the Talea Ori and have shown that the supposed unconformity is a high-angle fault and that the supposed basement is a downfaulted part of the overlying Phyllite-Quartzite Series nappe; this reinterpretation is discussed in detail in Hall & Audley-Charles (1983). This fault is just one of numerous late Cenozoic extensional faults that complicate interpretation of Cretan geology.

Tripali Unit

The Tripali Unit is present only in west Crete. It consists largely of recrystallized coarse carbonate breccias in which Liassic shallow-water fossils (Ott 1965, Kopp & Ott 1977) have been found. This is the only evidence for the age of these rocks, no other fossils having yet been found. The status of the Tripali Unit is somewhat controversial. Creutzburg & Seidel (1975) tentatively place it with the Phyllite-Quartzite Series on the basis of an impression of a concordant transition between them, and an obvious discordance between the Tripali Unit and the Plattenkalk Series, and consequently are forced to question the Liassic fossil dates. Kopp (1978) also considers that the Phyllite-Quartzite Series and the Tripali Unit may be in part equivalent to one another. On the other hand, Bonneau (personal communication 1982) and Karakitsios (1979) doubt the existence of a separate Tripali Unit and suggest that the breccias may be Neogene in part and elsewhere may represent mylonitized parts of an overthrust reversed limb of the folded Plattenkalk Series. The considerable tectonic thickness of these rocks of at least several hundred metres, the fossil evidence of Kopp & Ott (1977), the variety of carbonate types as clasts, the obvious sedimentary features in many of the breccia exposures, and the absence of the previously assumed large-scale inversion of the Plattenkalk Series (Hall & Audley-Charles 1983) lead us to question this interpretation. At Omalos we have observed calcarenite lenses grading up into calcisiltite horizons within these breccias, and have recognized a variety of limestone types as clasts. The similarity of the breccias to those of the lower part of the Plattenkalk Series suggests to us that they are more proximal equivalents of those breccias and both represent submarine fault-scarp deposits associated with Early Jurassic collapse of a Triassic carbonate platform.

Phyllite-Quartzite Series nappe

The Phyllite-Quartzite Series consists of continental to shallow marine siliciclastics and impure carbonates, and pyroclastic, volcanic and minor intrusive igneous rocks which range in age from Late Permian to Late Triassic (e.g. Creutzburg & Seidel 1975, Seidel 1978). In east Crete there are slices of Hercynian age high-grade metamorphic rocks included in the Phyllite-Quartzite Series. All of these rocks have been deformed and metamorphosed during the Tertiary orogenesis and the grade of metamorphism varies from very low grade zeolite/greenschist in east Crete to blueschist facies in west Crete. Part of the Phyllite-Quartzite Series may represent the pre-Carnian basement of the Tripolitza Series as there is locally a stratigraphical continuity between impure carbonates and shales of the Phyllite-Quartzite Series and basal carbonates of the Tripolitza Series, despite some tectonism (e.g. Bonneau & Karakitsios 1979).

Tripolitza Series nappe

The Tripolitza Series consists of shallow marine platform carbonates which range in age from Late Triassic to Middle Eocene. These limestones pass up into calciturbidites and then

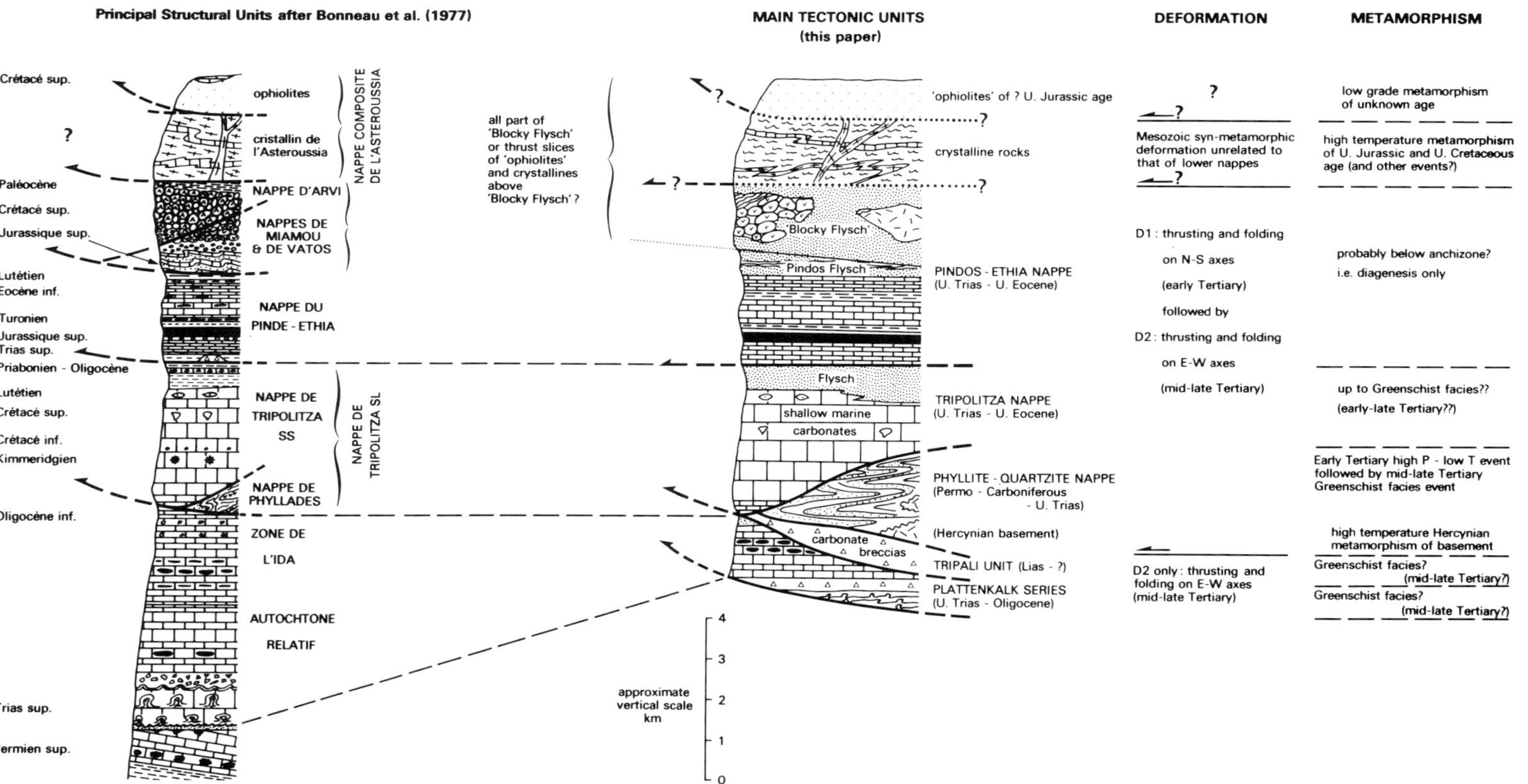

FIG. 5. Summary diagram of the main tectonic units referred to in this paper compared to those of Bonneau *et al.* (1977) with summary of the structural and metamorphic history of each of the nappes. The Zone de l'Ida of Bonneau *et al.* (1977) is equivalent to the Plattenkalk Series or Talea Ori Series of the German workers (see Epting *et al.*, 1972, Creutzburg & Seidel 1975) and is described as autochthonous or 'relative autochthon'. We have shown that the Plattenkalk Series is stratigraphically much thinner than previously estimated, and we consider that it is detached from its basement and must be at least parautochthonous and is probably a nappe. The reasons for this difference in interpretation are discussed more fully in Hall & Audley-Charles (1983).

into sandstone-shale turbidites; bottom structures in the sandstones indicate that siliciclastic transport was dominantly from east to west.

Pindos-Ethia Series nappe

Previous workers (e.g. Bonneau & Fleury 1971, Seidel 1971) have shown that the Pindos-Ethia Series is a sequence of basinal sedimentary rocks of late Triassic to Palaeocene age which include pelagic limestones, radiolarites, calciturbidites and calc-breccias, with a thin siliciclastic interval of Lower Cretaceous age ('Premier flysch'). Our field studies (Tee 1982, Hussin 1983) have shown that these pass up into turbidite sandstones and shales with intercalated carbonate conglomerate units and then into 'Blocky Flysch'. The Blocky Flysch contains a variety of exotic blocks including pillow lavas, pelagic shales, metamorphic rocks, serpentinites, limestones and turbidite sandstones. We consider that the 'nappes' of Arvi, Miamou and Vatos (Bonneau 1972, Bonneau *et al.*, 1974, Bonneau & Lys 1978) are large exotic blocks within the Blocky Flysch which we interpret as a major olistostrome.

Apparently above the Blocky Flysch are high-grade metamorphic rocks and ultramafic rocks with some subordinate microgabbros which have been assigned to a composite 'Asteroussia nappe' by Bonneau (1972) and to a 'Serpentinite-amphibolite association' by Creutzburg & Seidel (1975). These rocks are somewhat problematical. The areas over which they are exposed are usually small, often obscured by Neogene rocks, and their structural relations complicated by late Cenozoic faulting. Seidel *et al.* (1981) have shown that the 'ophiolites' of this association are of Jurassic age and are genetically unrelated to the high-grade metamorphic rocks of Late Cretaceous age with which they are now associated. These could be interpreted as exotic blocks within the Blocky Flysch or as thrust slices of ophiolites and crystalline rocks above the Blocky Flysch (Fig. 5).

Late Cretaceous-Early Tertiary facies belts

In the early Tertiary the nature of sedimentation changed from the basinal and platform carbonate deposition with calciturbidite accumulation at basin margins that had predominated throughout the Mesozoic. There was a gradual and irregular change to siliciclastic sedimentation ('flysch') in each of the main isopic zones. This change occurs in the Pindos Series in the early Eocene, in the Tripolitza Series in the late Middle Eocene and in the Plattenkalk Series in the Late Eocene to Early Oligocene. It is interesting to note that the onset of flysch sedimentation is often considered to be diagnostic of each isopic zone and for this reason the Plattenkalk Series is assigned by some authors to the Ionian zone. However, using the same criterion the Cretan Pindos Series would not be assigned to the Pindos zone since in the type area of the Hellenides flysch sedimentation begins much earlier, in the Maastrichtian.

The flysch, as is often the case in other orogenic belts, frequently resists attempts to assign it to a particular zone, being poorly exposed, difficult to date and tectonized. The steep faulting on Crete further complicates matters and many of the exposures on Crete remain as 'undifferentiated flysch'. The Asteroussia Ori of southern Crete is one of the few areas where a reasonably complete stratigraphy of the Pindos Series and its flysch can be obtained over a long and well-exposed section (70 km). The stratigraphy of this region (Hussin & Tee, 1982) is described in detail by Hussin (1983) and Tee (1982); their conclusions are summarized here only as far as they are relevant to the evolution of the region. Hussin & Tee have divided the Pindos flysch into three formations overlying calciturbidites of the Asfendilia Formation (Fig. 6). In the Asfendilia Formation thick limestone breccias are concentrated in the west and pass laterally into thick calciturbidites which thin eastwards. The most abundant material in these rocks are clasts of limestones derived from a shallow marine platform and its open sea slope, interpreted as the Tripolitza platform. The distribution of this material indicates that these were transported from the west and that facies belts of the Asfendilia Formation were orientated approximately north-south. These pass up into the Megali Kefala Formation which in the east Asteroussia is composed of thick bedded siliciclastic turbidites, amalgamated turbidite beds and channels and which pass westwards into thick shaly flysch with carbonate channel conglomerates. Sole-structures, cross-lamination and ripple-marks indicate that the main transport direction of siliciclastic sediment was from east to west with minor southwards transport. Slump-structures in the basal part of the overlying Agia Moni Olistostrome indicate a palaeoslope dipping WSW. All this evidence indicates that from at least the Late Cretaceous until the Lutetian the facies belts were orientated approximately north-south with carbonate debris derived from the Tripolitza carbonate platform and its slopes to the west, while

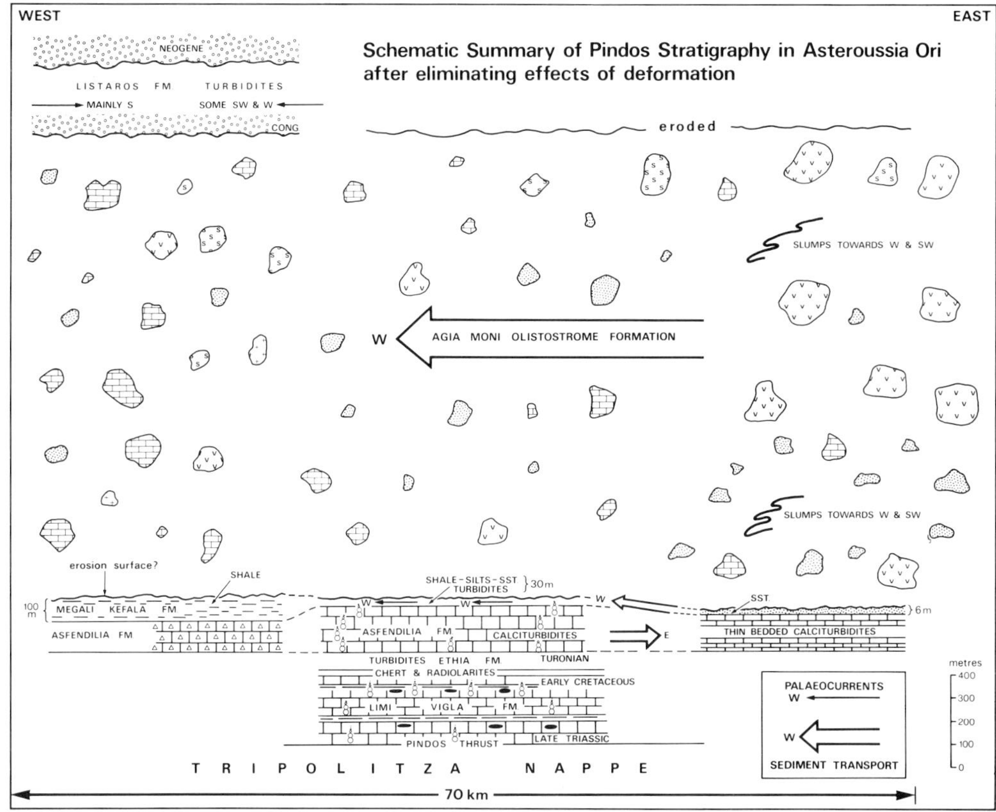

FIG. 6. Summary of Pindos stratigraphy in the Asteroussia Ori of southern Crete based on work of Hussin (1983) and Tee (1982).

siliciclastic debris, and exotic blocks in the olistostrome, were derived from a rapidly-eroding orogenic belt to the east. From the more limited exposures of Tripolitza flysch in east Crete, we have obtained a similar picture of siliciclastic sediment transport from east to west. The way in which flysch deposition migrates from the Pindos zone to the Tripolitza zone with time, the types of sediment, and their transport directions resembles the situation in the Pindos and Tripolitza zones of the Greek mainland, except that on Crete flysch sedimentation starts later. Only in the uppermost part of the Pindos flysch, the Listaros Formation, do palaeocurrent directions change to indicate that the dominant direction of siliciclastic sediment transport was from north to south.

In contrast to the overlying nappes, there is no sign in the Plattenkalk Series of the change from carbonate to siliciclastic sedimentation in the Paleocene-Eocene. At the very top of the Plattenkalk Series there are thin red and green, locally channelled, calcisiltite bands intercalated with thin bedded grey limestones overlain by a thin marl or calcareous shale with burrows and globeriginids of Upper Eocene to Lower Oligocene age. Thus, before the late Eocene to Early Oligocene the Plattenkalk Series must have been sufficiently far from the region represented by the overlying nappes to ensure its sedimentological isolation. Consideration of the regional geology indicates that the Plattenkalk sequence could only have been situated to the west, or to the south, of the Tripolitza and Pindos zones.

Structures of the pre-Neogene rocks

On the basis of major structures in the nappes we have divided the nappe pile into two parts:

(1) The Upper Nappes, which include the Phyllite-Quartzite Series and all the overlying nappes.

(2) The Lower Nappes, which include the Plattenkalk Series and the Tripali Unit.

All of the Upper Nappes are affected by two

major deformational phases (Fig. 5). The earliest deformation event (D1) has resulted in thrusting and folding on axes orientated approximately north-south. The style and degree of development of folding and thrusting is dependent on lithology and position in the nappe pile. In the Phyllite-Quartzite Series, the lowest unit of the Upper Nappes, the D1 folds are highly flattened tight to isoclinal mesoscopic folds with a penetrative axial planar cleavage. We suspect that the D1 episode may have caused very large scale isoclinal folding of the metamorphic rocks. We have so far been unable to prove it because of the absence of way-up criteria, lithological markers and the effects of later deformation. The trend of the D1 folds is not always north-south. In the Talea Ori and other parts of central Crete the folds trend almost exactly due north or just east of north (ignoring plunge), but in east and west Crete two maxima occur, one at approximately 040° and the second at approximately 320°. Greiling & Skala (1977) interpreted these two trends in west Crete as two sets of folds produced perpendicular to one another by a single deformation phase. We interpret them more simply as the result of refolding of a single axial trend by the D2 deformation phase.

The Tripolitza Series, which consists largely of massive carbonates, shows little sign of the D1 deformation phase. Only at the base of the Series, in thinly bedded dolomites, limestones and gypsum, can ductile tight to isoclinal D1 fold closures be observed. The massive carbonates have either been unaffected by the D1 phase, or have accommodated deformation by thrusting along bedding planes. We have observed listric thrust faulting in some exposures of the Tripolitza carbonates, but are unable to assess its importance.

The whole of the Pindos Series is deformed by flat-lying tight to isoclinal brittle folds with axes between N-S and NE-SW with further repetition by flat-lying thrusts.

The second major deformation phase (D2) is more variably developed in the Upper Nappes. In the Phyllite-Quartzite Series the D2 event has produced tight to isoclinal folds in the interlaminated thin limestones or quartzites and phyllites with approximately E-W axes and an axial planar strain-slip cleavage. In thick slate sequences the D2 event has often resulted only in a crenulation of the first cleavage, whereas in sequences dominated by thick quartzites no folds are observed and the D2 deformation has been accommodated by shearing and slip along lithological boundaries, particularly along thin slate interlayers. It is this phase and later brittle shearing that in places has reduced parts of the Phyllite-Quartzite Series, such as those dominated by quartzites, to a complex of sheared lenticular blocks in a sheared slaty matrix interpreted by Wachendorf *et al.* (1978) in east Crete as a tectonic melange.

The effects of the D2 phase are less well developed at higher levels in the pile of Upper Nappes. D2 folds, identified by a consistent trend approximately E-W, can be observed in all the Upper Nappes but are often restricted to parts of the pile dominated by incompetent lithologies such as parts of the flysch sequences.

In contrast to the Upper Nappes, the Plattenkalk Series shows the effects only of the D2 deformation phase. Tight to isoclinal folds with E-W axes cause considerable repetition of all parts of the Plattenkalk Series all over Crete, although frequently these folds are not obvious except where the carbonates contain thin bedded cherts. However, in the Talea Ori, and less frequently in other parts of Crete, way-up evidence is abundant in the form of stromatolitic structures, grading in calciturbidites, bottom structures, channelling, cross-lamination, and slump structures. Use of this evidence has allowed us (Hall & Audley-Charles 1983) to demonstrate that the Plattenkalk Series is repeated by mesoscopic isoclinal folds as well as thrusting. Epting *et al.* (1972) suggested inversion of the entire Plattenkalk Series sequence in the Talea Ori on the basis of observation of a limited number of way-up stuctures and this inversion was interpreted by Bonneau (1973) and Kopp (1978) as the inverted limb of a huge overturned fold. Our observations do not support this interpretation but show that although much of the Plattenkalk Series is inverted, this is due to local repetition of the sequence by isoclinal folding. Furthermore, the previous failure to recognize the significance of this isoclinal folding means that the stratigraphic thickness of the Plattenkalk Series has been considerably overestimated. We consider the entire Plattenkalk Series to be less than 1000 m thick.

Alpine metamorphism of the Cretan nappes

One of the outstanding problems of Crete has to been to explain the existence of metamorphic rocks of very high pressure origin within the pile of Cretan nappes. There are a number of aspects to this problem:

(1) Metamorphism of the Phyllite-Quartzite Series is of Tertiary age (Seidel 1978), but affects a sequence of rocks from which no rocks younger than Triassic have been reported.

(2) We have found no evidence in the Phyllite-Quartzite Series for anything other than rocks of continental and shallow marine origin. Considering the state of deformation and metamorphism of these rocks it is conceivable that rocks of deep water origin may be present. If they are, there appears to be no evidence for their existence while on the other hand there is abundant palaeontological and lithological evidence for shallow marine sedimentation at several different levels of the sequence. We consider it most probable that the Phyllite-Quartzite Series represents a series of shallow water and continental rocks deposited upon continental crust.

(3) The peak of the Alpine metamorphism of the Phyllite-Quartzite Series occurred between the D1 and D2 events. The growth of high P–low T porphyroblastic minerals (lawsonite, chloritoid, glaucophane, albite and ferrocarpholite) began during D1 (e.g. Greiling 1982) and continued during a static interval separating D1 and D2 since these minerals grow across the S1 cleavage with no preferred orientation and are deformed by the D2 folds and S2 cleavage. Ferrocarpholite, which is restricted to high pressure terrains, also occurs commonly in veins which cross-cut the D1 folds and are deformed by the D2 folds. Metamorphic conditions indicated by the mineral assemblages are temperatures of 200–300°C and pressures of 3–4 kbar in east Crete increasing to about 8 kbar in west Crete (Seidel 1978). Retrogression accompanied the D2 phase and in west Crete blueschist facies assemblages are partially replaced by greenschist facies assemblages.

(4) The Phyllite-Quartzite Series rocks are structurally overlain by the Tripolitza Series limestones and there is evidence to suggest that some stratigraphic transition originally existed between part of the Phyllite-Quartzite Series and the Tripolitza Series (Bonneau & Karakitsios 1979). The Tripolitza Series is often reported to be unmetamorphosed, or only slightly marmorized near its base (Creutzburg & Seidel 1975). In fact, the degree of metamorphism is very difficult to determine simply because virtually the entire sequence consists of very pure carbonates. We are currently examining a number of possible methods which might allow us to put some limits on the maximum temperature to which the Tripolitza sequence has been subjected. For the time being we prefer to remain cautious about metamorphic grade in view of the extensive recrystallization observed in many of the Tripolitza carbonates and the evidence suggesting that all the Upper Nappes have suffered two major deformation phases. Ductile folds at the base of the Tripolitza Series in interlaminated thin limestones, shales and gypsum suggest deformation under some considerable overburden. The problems of determining metamorphic conditions in carbonate sequences similar to the Tripolitza Series are illustrated by the recent discovery of relatively deep burial and low grade metamorphism in areas previously considered to be unmetamorphosed (for example, the external Helvetides of the Western Alps, Martini 1975) which would have been impossible had the carbonates not been associated with other lithologies to indicate grade.

(5) As in the Tripolitza Series, the grade of metamorphism of the Pindos Series is not easy to determine because of the predominance of carbonate in the sequence. However, the brittle style of folding, relative absence of widespread recrystallization, and general abundance of well-preserved fossils suggests a relatively shallow depth of burial. The flysch sequences offer more prospect of determining their burial history from their clay mineralogy and organic material and these are currently being examined.

(6) The Phyllite-Quartzite Series is structurally underlain by the Plattenkalk Series which is unaffected by the D1 deformation event which affects all the Upper Nappes. The ductile style of folding, extensive recrystallization, and mineralogy of the rare impure horizons indicate that deformation of the Plattenkalk Series occurred under a considerable overburden, probably at temperatures approximately equivalent to those of the lower greenschist facies. We are currently attempting to determine the conditions of metamorphism more precisely.

The observations summarized above suggest that although there is a general increase of metamorphic grade with depth there are important discontinuities in the nappe pile now exposed on Crete (Fig. 5). The structural and metamorphic evidence indicates that the most important of these occurs at the base of the Phyllite-Quartzite Series which is the boundary between the Upper and Lower Nappes. The metamorphism of the Phyllite-Quartzite Series is partly syn-tectonic with D1 (e.g. Greiling 1982) and our observations are that the main period of mineral growth occurred before D2 whereas metamorphism of the Plattenkalk Series, which is undeformed by D1, is syn-tectonic with D2. Other discontinuities must occur within the pile of the Upper Nappes although their location is uncertain, in part due to the predominance of pure carbonates in the nappes.

A. Late Cretaceous - Paleocene

W
E
Tripolitza Platform
?
Pindos Zone
ocean crust or thinned continental crust
Internal Hellenic Zones and Anatolia
Mesozoic sediments
Carnian evaporites
Permo - Triassic sediments
Hercynian basement

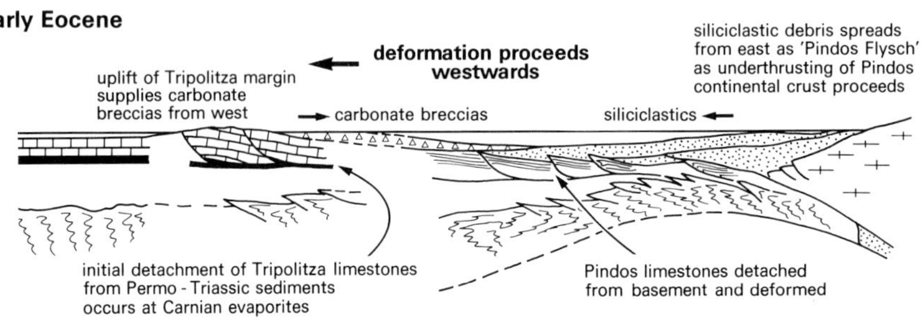

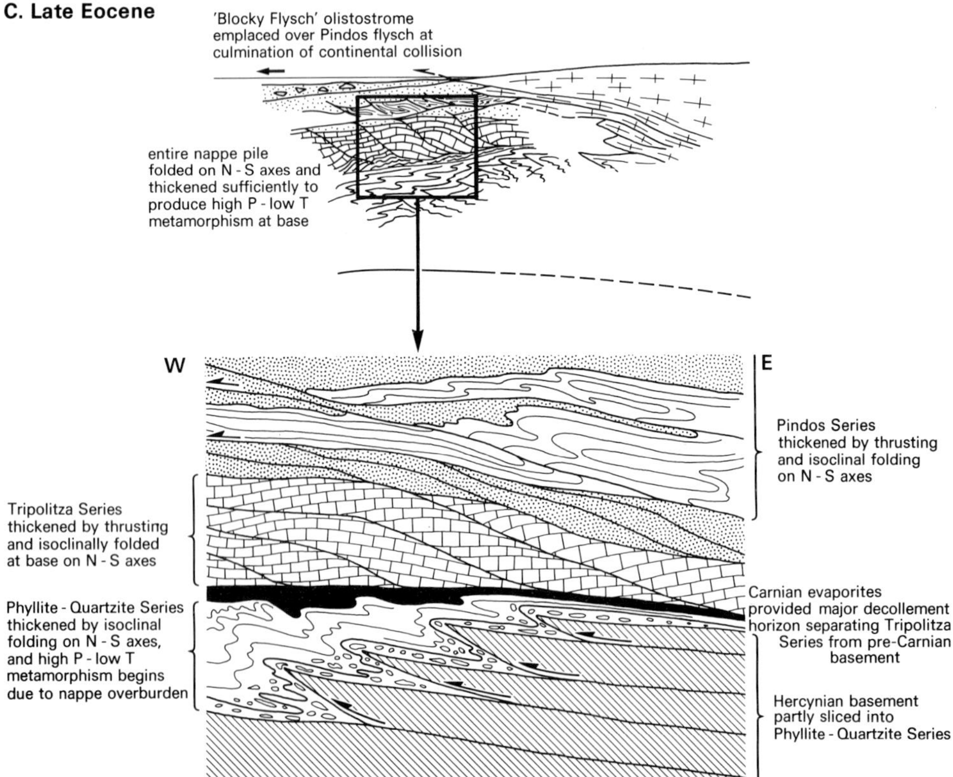

Fig. 7. Schematic summary of Late Cretaceous to Late Eocene evolution of the eastern Apulian margin now represented in the Upper Nappes of Crete. The history of the Upper Nappes, which were situated to the north of the transcurrent fault, is discussed in more detail in the text.

N
S

A. Late Mesozoic

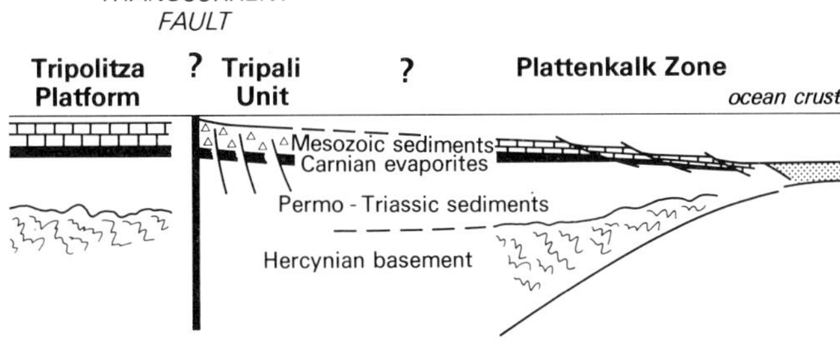

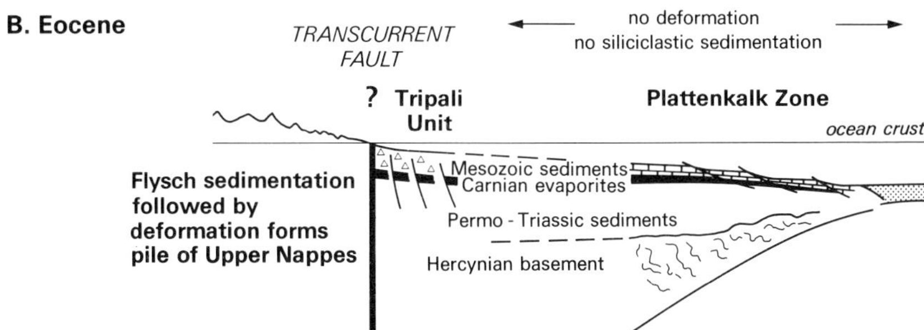

C. Oligocene - early Miocene

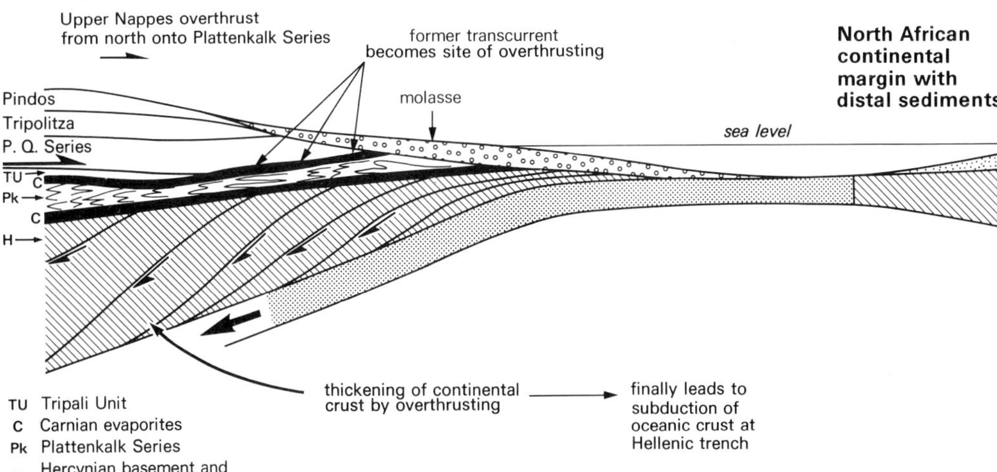

FIG. 8. Schematic summary of the evolution of the southern Apulian margin between the Late Cretaceous and the early Miocene. The Upper Nappes were deformed north of the transcurrent fault during the Eocene. The final stages of deformation and uplift of the Upper Nappes north of the fault occurred in the Late Eocene at which time the rising Upper Nappe pile began to supply clastics to the Plattenkalk Zone—Plattenkalk 'flysch' (Late Eocene to Early Oligocene). Change in motion from transcurrent to convergent on the fault at this time caused thrusting of the Upper Nappes over the Tripali Unit and Plattenkalk Series producing E-W folds (D2) in the Upper Nappes and the Plattenkalk Series.

Discussion

The volcanic arc north of Crete is of Plio-Pleistocene age (Fytikas *et al.*, 1976), and even assuming very low average rates of convergence, suggests that the Hellenic arc is a relatively recent feature of the region. This is widely accepted, and although the present tectonics and the Neogene evolution of the Aegean region remain controversial, the arc is generally considered to have been formed in the last 30 Ma. The main compressional periods recorded in the Cretan nappes, which resulted in their emplacement, occurred between the Palaeocene and the Early Miocene and thus entirely pre-date the present Hellenic arc. It is therefore not surprising that Crete has very little geological resemblance to other frontal arc regions, no matter which particular arc is chosen for comparison. The elongation and parallelism of Crete to the arc is not a feature related to the distribution or derivation of the nappes but a younger feature reflecting the late Cenozoic extension of the region along steep faults parallel to the arc.

If the Hellenic arc has developed in the last 30 Ma why should the isopic zones be parallel to the present arc or traceable through the arc in a manner shown, for example, by Aubouin *et al.* (1976), Jacobshagen *et al.* (1978), and implied by other reconstructions? We suggest that the Hellenide isopic zone model, which has proved so useful in interpreting the geology of the region, may have been taken a little too far in Crete. The evidence from Crete, which we have summarized above, indicates that the present structure of the pre-Neogene rocks developed in two stages: a first major phase in which the Upper Nappes were stacked, followed by a second major phase in which the Upper Nappes were thrust, essentially as a single unit, onto the Lower Nappes.

The evolution of the first stage of our model is shown schematically in two dimensions in Figs 7 & 8. We suggest that the Tripolitza platform and the Pindos basin extended southwards from the Greek mainland with essentially no change in their orientation, forming the eastern continental margin of Apulia (Fig. 9). For the purposes of explanation we refer to this orientation as north-south, although we accept that this is an over-simplification since the present trend of the zones on the mainland is closer to NNW-SSE, and we are neglecting the effects of the late Cenozoic deformation of the Aegean whose effects are still controversial (e.g. Le Pichon & Angelier 1979). This suggestion is consistent with the trends of the Tertiary facies belts on Crete and accounts for the transport directions of sediments in the flysch sequences. Carbonate debris was derived largely from the Tripolitza platform and its margin, situated to the west of the Pindos basin, while siliciclastic debris was derived from an orogenic belt to the east (Fig. 10; Hussin 1983, Tee 1982). This distribution explains why the onset of flysch sedimentation moves from east to west with time and why the earliest deformation phase in all the Upper Nappes produced folds with north-south axial trends. We suggest that both the Tripolitza and Pindos zones were underlain by a similar basement: Hercynian metamorphic rocks overlain by post-Hercynian molasse deposits, Permo-Triassic shallow-marine deposits, and locally Carnian evaporites. These rocks now form the Phyllite-Quartzite Series of Crete. The Eocene deformation phase on Crete (D1) caused the Pindos Series and Tripolitza Series to become detached from their basement as cover nappes, partly on the underlying Carnian evaporites. Each of the Upper Nappes appears to have been deformed in a different manner dependent on its position in the nappe pile and on its stratigraphy. The Pindos sequence, composed largely of thinly bedded limestones, cherts and shales near the top of the pile was thickened by thrusting and isoclinal folding while the Tripolitza Series, composed largely of massively bedded carbonates was probably thickened by listric thrust faults. Beneath the Carnian evaporites the Phyllite-Quartzite Series was thickened by isoclinal folding with an associated axial planar cleavage and probably detached from its more rigid Hercynian metamorphic basement. The sudden thickening of overburden by D1 folding and thrusting resulted in a rapid rise in pressure at the base of the nappe pile without a corresponding temperature increase and blueschist facies metamorphism began after the main (D1) folding. We suggest that the observed increase in metamorphic grade from east to west, indicating an increase in pressure from east to west, corresponds to deeper structural levels in the Phyllite-Quartzite Series. The D1 compressional phase culminated in the arrival of the Agia Moni olistostrome, with metamorphic rocks and ophiolites from the orogenic belt to the east. The sequence of events described so far is essentially the same as that in the Peloponnese where a similar sequence of nappes is exposed (Thiebault 1981).

As noted above, considerations of regional geology indicate that the Plattenkalk Series, and the Tripali Unit which represents its more proximal equivalent, could only have been situated to the west or south of the other zones. Other authors have considered that the Plat-

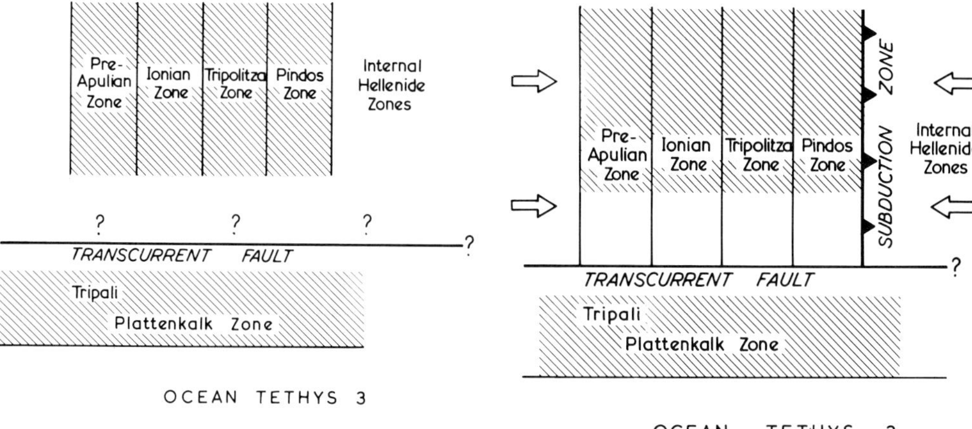

FIG. 9. Schematic and highly simplified palaeogeography of the south-eastern part of the Apulian continent in the early Cretaceous as proposed in this paper. The palaeogeographical evolution of this region is shown schematically in Figs. 9, 10 and 11 and is discussed more fully in the text. To the north of a major transcurrent fault, the External Hellenide isopic zones recognized in mainland Greece extended southwards along the *eastern* Apulian continental margin (Fig. 7A) with no change in their orientation. South of the fault, carbonates of the Tripali Unit and Plattenkalk Series, were deposited on the proximal and distal parts of the *southern* Apulian continental margin (Fig. 8A). Because of its distinct and separate stratigraphic and structural evolution we consider the Plattenkalk Series to represent part of an isopic zone not seen in mainland Greece. The amount and sense of displacement on the fault is uncertain.

FIG. 10. Middle Eocene reconstruction showing evolution of the palaeogeography from the scheme shown in Fig. 9. East to west convergence north of the transcurrent fault during the Eocene caused siliciclastic sediment to be carried westwards from an 'Internal Hellenide' orogenic belt (Fig. 7B). We are not yet certain whether oceanic crust, or only thinned continental crust, was being subducted to the east of Pindos zone during this period. South of the fault, the Plattenkalk Zone was isolated from sedimentation and deformation associated with the east to west convergence.

tenkalk Series of Crete is equivalent to the Ionian (Bonneau 1973; Kuss & Thorbecke 1974; Aubouin *et al.* 1976; Bizon *et al.* 1976), or Pre-Apulian (Jacobshagen *et al.* 1976, 1978), zones of the Hellenides and their metamorphosed equivalents in the Tagetos Mountains of the Peloponnese (Thiebault 1977). This is based on the late arrival of 'flysch' in the Plattenkalk Series. Since the Ionian and Pre-Apulian zones are west of the Tripolitza zone on the mainland it is assumed that Plattenkalk Series must have occupied a similar position in Crete or these zones must curve round the present arc. This solution not only neglects the relatively recent formation of the arc, but also the structural evidence. The D2 event recorded in the Cretan nappes deforms all the Upper Nappes *and* the Lower Nappes, and resulted in folds with east-west axial trends approximately at right angles to the D1 fold axes.

We are therefore led to propose an alternative solution: that in the late Eocene to early Oligocene the convergence direction changed from east-west to north-south, resulting in southward-directed thrusting of the Upper Nappes onto the Lower Nappes (Fig. 8C, Fig. 11). Unstacking the present sequence then puts the Plattenkalk Series as the most southerly of the Cretan nappes, the Tripali Unit as its more proximal equivalent further north, and the Upper Nappes still further north of this (Fig. 8A, Fig. 9). We therefore consider that the Plattenkalk Series and Tripali Units have no equivalents in the isopic zones of the Hellenides but represent a southern continental margin of Apulia (Fig. 9) north of oceanic crust of Mesogea (Biju-Duval *et al.* 1977) or Tethys 3 (Dewey *et al.* 1973). If this was the geographic arrangement of the main zones some mechanism is required to isolate the Plattenkalk Series from the earlier structural and sedimentological events recorded in the Upper Nappes. We propose that the southern Apulian margin was separated from the Hellenide zones by a major fault which we consider must have been a

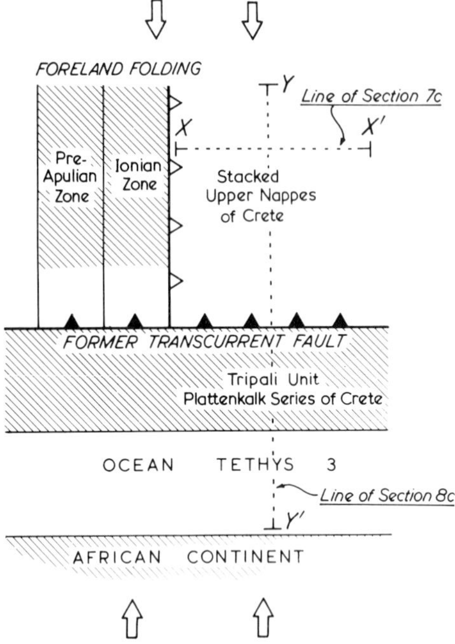

FIG. 11. Late Eocene reconstruction showing evolution of the palaeogeography from the scheme shown in Fig. 10. At the end of the Eocene east to west convergence ceased after stacking of the Upper Nappes north of the transcurrent fault. Relative motion between Africa and Apulia changed to north to south convergence and the former transcurrent fault became the site of intra-continental thrusting. Initially, the rising Upper Nappe pile became the source of siliciclastic material in the Plattenkalk 'flysch', and subsequently the stacked Upper Nappes were thrust over the Tripali Unit and Plattenkalk Series to form the present nappe pile seen on Crete.

lithospheric fracture on which there was strike-slip motion and which had an east-west orientation relative to the present geography of the region (Figs. 9 & 10). Only in the Late Eocene to Early Oligocene did the Plattenkalk Series arrive south of the Upper Nappes (Fig. 11) by transcurrent motion on this fracture at the same time that relative motion changed from transcurrent to convergent. We suggest that it was at this time that the present convergent pattern in the southern Aegean region was initiated. Overthrusting of the Upper Nappes from north to south terminated the high pressure metamorphism and produced D2 folds with east-west axes which refold D1 folds in the Upper Nappes, particularly the Phyllite-Quartzite Series, but are the only folds observed in the Plattenkalk Series.

Conclusions

(1) The Cretan nappes were emplaced in two stages.

(2) The Hellenide isopic zones do not bend round the present Hellenic arc but were terminated at a major east-west orientated fracture zone which we suggest was a fundamental transcurrent fault. This explains the absence of the Tripolitza zone to the east of Crete.

(3) Stacking of the Pindos, Tripolitza and Phyllite-Quartzite Series: the Upper Nappes of Crete, occurred during the Eocene compression of the eastern continental margin of Apulia. This compression produced the north-south (D1) structures of the Upper Nappes.

(4) The Plattenkalk Series of Crete represents distal sediments of the southern Apulian continental margin. The Tripali Unit represents part of its more proximal equivalents. They are not equivalent to any of the Hellenide isopic zones but were deposited in a separate region south of the major transcurrent fault.

(5) North-south compression began in the Late Eocene to Early Oligocene and resulted in southward overthrusting of the Upper Nappes onto the Lower Nappes and the development of east-west (D2) structures.

(6) The change from east-west convergence north of the transcurrent fault to north-south convergence across the transcurrent fault is consistent with major plate motions deduced from the Atlantic opening history. Relative motion between Africa and Europe changed from transcurrent to convergent in the Late Eocene (Dewey *et al.* 1973; Biju-Duval *et al.* 1977).

(7) High pressure-low temperature metamorphism of the Phyllite-Quartzite Series occurred in an intra-continental situation and was initiated by stacking of nappes during the D1 phase. It was terminated by north-south thrusting of the Upper Nappes during the D2 phase.

ACKNOWLEDGEMENTS. We thank G. Evans, D. H. Oswald, A. Hussin & N. Tee for help and discussion, NERC for financial support, and J. Bornovas and IGME for assistance.

References

ALTHERR, R., SEIDEL, E. & KREUZER, H. 1982. Petrological and geochronological constraints on geodynamic models for the Hellenides. *Abs. Geological Evolution of the Eastern Mediterranean, Edinburgh 1982*, 10.

ANGELIER, J. 1978. Tectonic evolution of the Hellenic Arc since the late Miocene. *Tectonophysics* **49,** 23–36.

AUBOUIN, J., BONNEAU, M., DAVIDSON, J. LEBOULENGER, P., MATESCO, S. & ZAMBETAKIS, A. 1976. Esquisse structurale de l'arc égéen externe: des Dinarides aux Taurides. *Bull. Soc. géol. Fr.* **18,** 327–336.

BIJU-DUVAL, B., DERCOURT, J. & LE PICHON, X. 1977. From the Tethys ocean to the Mediterranean seas: a plate-tectonic model of the evolution of the western Alpine system. *In:* BIJU-DUVAL, B. & MONTADERT, L. (eds.) *Structural history of the Mediterranean basins.* Editions Technip, Paris, 143–164.

BIZON, G., BONNEAU, M., LEBOULENGER, P., MATESCO, S. & THIEBAULT, F. 1976. Sur la signification et l'extension des 'massifs cristallins externes' en Péloponnèse méridional et dans l'arc égéen. *Bull. Soc. géol. Fr.* **18,** 337–345.

BONNEAU, M. 1972. La nappe métamorphique de l'Asteroussia, lambeau d'affinités pélagoniennes charrié jusque sur la zone de Tripolitza de la Crète moyenne (Grèce). *C.r. Séances Acad. Sci. Paris* **275,** 2303–2306.

—— 1973. Sur les affinités ioniennes des 'calcaires en plaquettes' épimétamorphiques de la Crète, le charriage de la série de Gavrovo-Tripolitza et la structure de l'arc égéen. *C.r. Séances Acad. Sci. Paris* **277,** 2453–2456.

—— 1982. Evolution géodynamique de l'arc égéen depuis le Jurassique supérieur jusqu'au Miocène. *Bull. Soc. géol. Fr.* **24,** 229–242.

——, ANGELIER, J. & EPTING, M. 1977. Réunion extraordinaire de la Société géologique de France en Crète. *Bull. Soc. géol. Fr.* **19,** 87–102.

——, BEAUVAIS, L. & MIDDLEMISS, F. A. 1974. L'unité de Miamou (Crète, Grèce) et sa macrofaune d'age Jurassique supérieur (Brachiopods, Madréporaires). *Ann. Soc. géol. Nord.* **94,** 71–85.

—— & FLEURY, J.-J. 1971. Précisions sur la série d'Ethia (Créte, Grèce): éxistence d'un premier flysch mésocrétacé. *C.r. Séances Acad. Sci. Paris* **272,** 1840–1842.

—— & KARAKITSIOS, V. 1979. Les niveaux inférieurs (Trias supérieur) de la nappe de Tripolitza en Crète moyenne (Grèce) et leurs relations avec la nappes des Phyllades. Problèmes stratigraphiques, tectoniques et de métamorphisme. *C.r. Séances Acad. Sci. Paris* **288,** 15–18.

—— & LYS, M. 1978. Sur la présence de Permien fossilifère dans l'unité de Vatos (Crète): sa nature interne et l'ampleur des charriages dans l'arc égéen. *C.r. Séances Acad. Sci. Paris* **287,** 423–426.

CLOSS, H., ROEDER, D. & SCHMIDT, K. (eds.) 1978 *Alps, Apennines, Hellenides.* E. Schweizerbart'sche Verlag., Stuttgart, 620 pp.

CREUTZBURG, N. 1977. *General Geological Map of Greece: Crete Island 1:200,000.* Institute of Geological and Mining Research, Athens.

—— & SEIDEL, E. 1975. Zum Stand der Geologie des Präneogens auf Kreta. *Neues Jahrb. Geol. Palaeontol. Abhandlungen* **149,** 363–383.

DEWEY, J. F., PITMAN, W. C., RYAN, W. B. F. & BONNIN, J. 1973. Plate tectonics and the evolution of the Alpine system. *Bull. geol. Soc. Am.* **84,** 3137–3180.

EPTING, M., KUDRASS, H.-R., LEPPIG, U. & SCHAFER, A. 1972. Geologie der Talea Ori/Kreta. *Neues Jahrb. Geol. Palaeontol. Abhandlungen* **141,** 259–285.

FYTIKAS, J., GIULIANI, O., INNOCENTI, F., MARINELLI, G. & MAZZUOLI, R. 1976. Geochronological data on recent magmatism of the Aegean Sea. *Tectonophysics* **31,** T29–34.

FYTROLAKIS, N. 1972. Die Einwirkung gewisser orogener Bewegungen und die Gipsbildung in Ostkreta (Prov. Sitia). *Bull. geol. Soc. Greece* **9,** 81–100.

GREILING R. 1982. The metamorphic and structural evolution of the Phyllite-Quartzite Nappe of western Crete. *J. struct. Geol.* **4,** 291–297.

—— & SKALA, W. 1977. The petrofabrics of the Phyllite-Quartzite Series of western Crete as an example for the pre-Neogenian structures of the Cretan arc. *Proc. 6th Coll. Aegean Geol., Athens 1977*, **1,** 97–102.

HALL, R. & AUDLEY-CHARLES, M. G. 1983. The structure and regional significance of the Talea Ori, Crete. *J. struct. Geol.* **5**, 167–179.

HUSSIN, A. H. 1983. *Stratigraphy and sedimentology of the Pindos Nappe in the Asteroussia Ori, Crete.* Ph.D. Thesis, Univ. London (unpubl.)

—— & TEE, A. T. 1982. Significance of sedimentary facies of the Pindos nappe of southern Crete. *Abs. Geological Evolution of the Eastern Mediterranean, Edinburgh 1982,* 53.

JACOBSHAGEN, V., MAKRIS, J., RICHTER, D., BACHMANN, G. H., DOERT, U., GIESE, P. & RISCH, H. 1976. Alpidischer Gebirgsbau und Krustenstruktur des Peleponnes. *Z. Deutsch. geol. Ges.* **127,** 337–363.

JACOBSHAGEN, V., DURR, ST., KOCKEL, F., KOPP, K.-O., KOWALCZYK, G., BERKHEMER, H. & BUTTNER, D. 1978. Structure and geodynamic evolution of the Aegean region. *In:* CLOSS, H., ROEDER, D. & SCHMIDT, K. (eds.) *Alps, Apennines, Hellenides.* E. Schweizerbart'sche Verlag., Stuttgart. 537–564.

KARAKITSIOS, V. 1979. *Contribution de l'étude géologique des Hellénides. Étude de la région de Sellia (Crète moyenne-occidentale, Grèce)* '*Les rélations lithostratigraphiques et structurales entre la série des Phyllades et la série carbonatée de Tripolitza*'. Thèse 3eme cycle, University of Paris, 167 pp.

KOPP, K.-O. 1978. Stratigraphic and tectonic sequence of Crete. *In:* CLOSS, H., ROEDER, D. & SCHMIDT, K. (eds.) *Alps, Apennines, Hellenides.*

E. Schweizerbart'sche Verlag., Stuttgart. 439–442.

—— & Ott, E. 1977. Spezialkartierungen Umkreis neuer Fossilfunde in Trypali- und Tripolitzakalken Westkretas. *Neues Jahrb. Geol. Paleontol. Monatshefte* **4,** 217–238.

Kuss, S. E. & Thorbecke, G. 1974. Die präneogenen Gesteine der Insel Kreta und ihre Korrelierbarkeit im ägaischen Raum. *Ber. naturforsch. Ges. Freiburg* **64,** 39–75.

Le Pichon, X. & Angelier, J. 1979. The Hellenic arc and trench system: a key to the neotectonic evolution of the Eastern Mediterranean area. *Tectonophysics*, **60,** 1–42.

Letouzey, J. & Tremolieres, P. 1980. Paleo-stress fields around the Mediterranean since the Mesozoic derived from microtectonics: comparisons with plate tectonic data. *Géologie des chaînes alpines issues de la Téthys, Colloque C5 26th Int. Geol. Congr. Paris 1980.* 261–273.

Makris, J. 1977. Geophysical investigations of the Hellenides. *Hamburger Geophysikalische Einzelschriften* **33,** 128 pp.

Martini, J. 1972. Le métamorphisme dans les chaînes alpines externes et ses implications dans l'orogénèse. *Schweiz. mineral. petrogr. Mitt.* **52,** 257–275.

Mascle, J., Le Quellec, P., Leite, O. & Jongsma, D. 1982. Structural sketch of the Hellenic continental margin between the western Peloponnesus and eastern Crete. *Geology* **10,** 113–116.

McKenzie, D. P. 1978. Active tectonics of the Alpine-Himalayan belt: the Aegean Sea and surrounding regions. *Geophys. J. R. astron. Soc.* **55,** 217–254.

Moore, G. F., Billman, H. G., Hehanussa, P. E. & Karig, D. E. 1980. Sedimentology and paleobathymetry of Neogene trench-slope deposits, Nias Island, Indonesia. *J. Geol.* **88,** 161–180.

Ott, E. 1965. *Dissocladella cretica*, eine neue Kalkalge (Dasycladaceae) aus dem Mesozoikum der griechischen Inselwelt und ihre phylogenetischen Beziehungen. *Neues Jahrb. Geol. Paleontol. Monatshefte* **11,** 683–693.

Richter, I. & Strobach, K. 1978. Benioff zones of the Aegean arc. *In:* Closs, H., Roeder, D. & Schmidt, K. (eds.) *Alps, Apennines, Hellenides.* E. Schweizerbart'sche Verlag., Stuttgart. 410–414.

Seidel, E. 1971. Die Pindos-Serie in West-Kreta, auf der Insel Gavdos und im Kedros-Gebiet (Mittel-Kreta). *Neues Jahrb. Geol. Paleontol. Abhandlungen* **137,** 443–460.

—— 1978. *Zur Petrologie der Phyllit-Quartzit Serie Kretas.* Habilitationsschrift, Braunschweig, 145 pp.

——, Okrusch, M., Kreuzer, H., Raschka, H. & Harre, W. 1981. Eo-alpine metamorphism in the uppermost unit of the Cretan nappe system—petrology and geochronology. *Contrib. Mineral. Petrol.* **76,** 351–361.

Tee, A. T. 1982. *Biostratigraphy of the Cretaceous-Tertiary Pindos Series of Crete.* Ph.D. Thesis, Univ. London (unpubl.)

Thiebault, F. 1977. Stratigraphie de la série des calcschistes et marbres ('plattenkalk') en fenêtre dans les massifs du Taygète et du Parnon (Péloponnèse—Grèce). *Proc. 6th Coll. Aegean Geol., Athens 1977* **2,** 691–701.

—— 1981. Modèle d'évolution géodynamique d'une portion des Hellénides externes (Peloponnèse méridional, Grèce) de L'Eocène a l'époque actuelle. *C.r. Séances Acad. Sci. Paris* **292,** 1491–1496.

Wachendorf, H., Baumann, A., Gwosdz, W. & Schneider, W. 1974. Die 'Phyllit-Serie' Ostkretas—eine Melange. *Z. Deutsch. geol. Ges.* **125,** 237–251.

Wunderlich, H. G. 1973. Gravity anomalies, shifting foredeeps, and the role of gravity in nappe transport as shown by the Minoides (Eastern Mediterranean). *In:* de Jong, K. A. & Scholten, R. (eds.) *Gravity and Tectonics*, John Wiley, 271–285.

Robert Hall, Department of Geology, University College, Gower Street, London, WCIE 6BTI.

M. G. Audley-Charles, Department of Geology, University College, Gower Street, London WC1E 6BT.

D. J. Carter, Department of Geology, Imperial College, Prince Consort Road, London SW7 2BP.

Correlation of the Hellenide nappes in the south-east Aegean and their tectonic reconstruction

Michel Bonneau

SUMMARY: The central Aegean nappe pile can be readily related to that of the mainland Hellenides and has the advantage that the deeper levels are well displayed, particularly in the Cyclades.

The following units can be distinguished, from base to top:

(1) The para-autochtonous *Ida* or *Talea Ori Sequence*, recognized in Crete, Kassos, Rhodes (Lindos sequence) and Amorgos is, in part, a neritic carbonate platform sequence. It can be correlated with the Ionian-Preapulian autochthon of the mainland and most probably with the Menderes sequence of western Anatolia, there exposed in a tectonic window.

(2) The major *Tripolitza Nappe*, a carbonate platform sequence, with the *Phyllite Nappe* at its base, is well represented in the outer arc as well as in Astypaleia and Santorini.

(3) The *Pindos* and *Arvi Nappes* of Crete, Rhodes and Tilos are the remnants of an ocean basin active from Upper Triassic to Late Cretaceous times.

(4) *The Cycladic Blueschist Unit* exposed on Syros, Siphnos, Naxos and other islands, occupies a tectonic position similar to the Pindos Nappe. It is interpreted as the subducted margin of the European plate, while the Pindos Nappe, the abyssal plain of the Apulian plate, was obducted just before continental collision in the Late Eocene-Oligocene.

(5) The *Asteroussia Nappe*, composed of high-grade metamorphic rocks yielding cooling ages around 75 Ma, is present in Crete and in the Cycladic islands of Syros, Nikouria, Denoussa and Keros. It is interpreted as the remnants of a Pelagonian basement nappe, eroded in Miocene times. In the external or outer belt it has been thrust on to the unmetamorphosed Pindos Nappe but in the Cyclades it rests on the Cycladic Blueschist unit.

Knowledge of the geology of the Central Aegean has improved considerably in the last ten years with the publication of much new petrological, radiometric, stratigraphic and structural data. Many of these studies are detailed investigations of small areas, principally because the region itself is an archipelago. It is consequently difficult to reconstruct the complex structural succession in this belt which is also completely broken up by 'neotectonic' faulting (Fig. 1).

Correlation with the Hellenides in mainland Greece and the Peloponnese is often straightforward, as for example in Crete. Important stratigraphic variations do however occur in the east, making reconstruction more speculative, as the geology of western Turkey is still poorly known. The Cycladic area, long considered to be 'ancient basement', is now accepted by all authors to have been an active part of the alpine Hellenide belt, as is clearly seen for example in southern Euboea. Any tectonic reconstruction must therefore include it along with the unmetamorphosed units.

Syntheses covering the geology of Crete and the Dodecanese have been published (see Aubouin *et al.* 1976b; Bonneau *et al.* 1977; Cadet *et al.* 1980; Creutzburg & Seidel 1975) and general works on the Cycladic area have also recently appeared (Dürr *et al*, 1978a; Blake *et al.* 1981), but so far no concentrated attempt has been made to combine these two regions in one review. That is the goal of this paper.

Any reconstruction of the complex Mesozoic and early Tertiary history of the Hellenic belt in this region must work back from a Late Miocene configuration, before the formation of the present outcrop pattern which is the result of 'neotectonics'. The Cretan Sea was then quite narrow or may even have been dry land, the South Aegean landmass of Drooger and Meulenkamp (1973), and the curvature of the Aegean Arc was then much less marked (Le Pichon & Angelier 1979).

This review begins with a description of the nappe units found in the region, first in the external or outer part of the Hellenic arc, secondly in the internal or cycladic area.

The external (southern Aegean) belt The Peloponnese, Crete and the Dodecanese

This belt contains several nappes, both metamorphic and non-metamorphic. The most com-

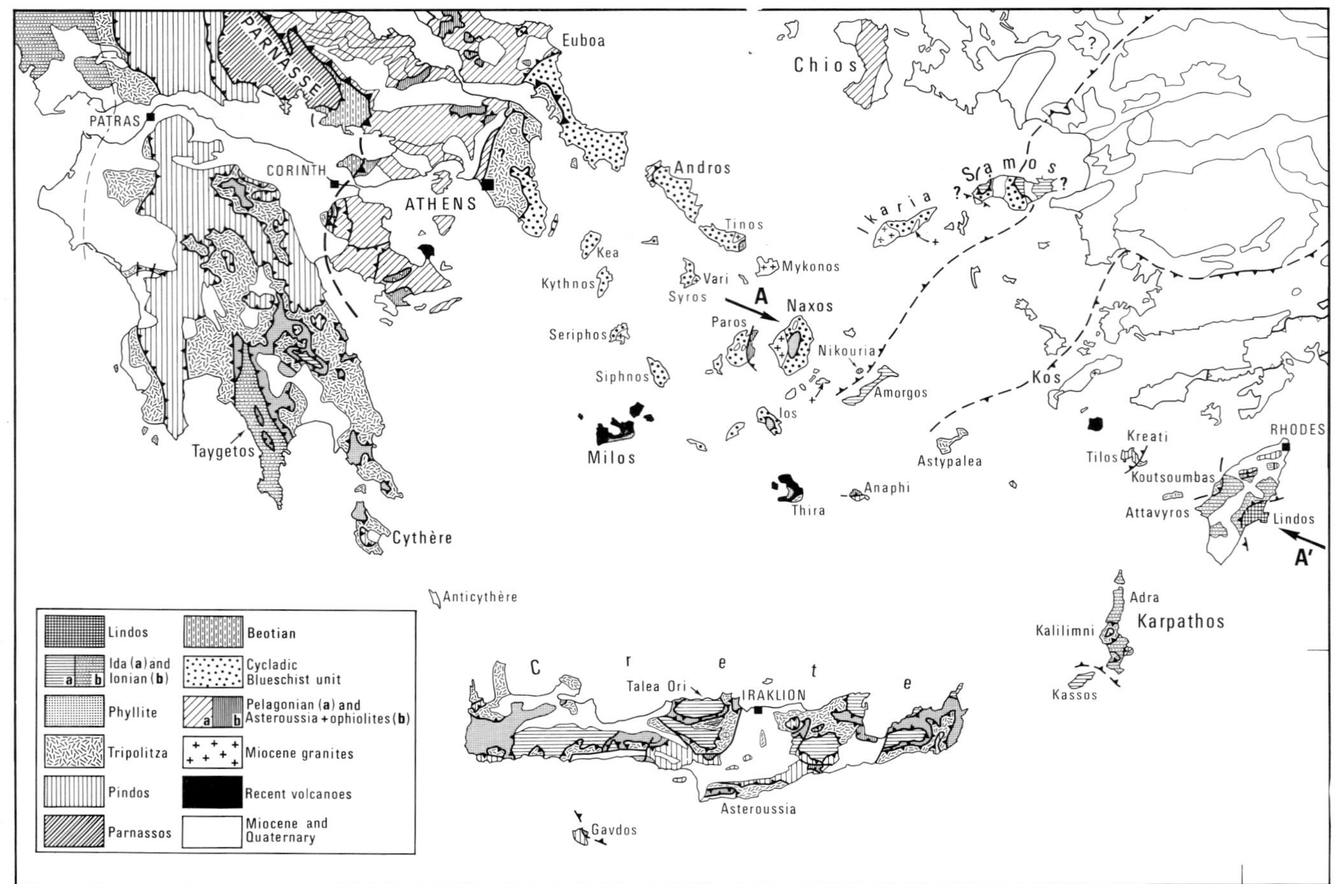

FIG. 1. Schematic geologic map of the Central Aegean. The Upper Cretaceous basalts of the Arvi Unit have been included in the Pindos Nappe. AA' indicates line of cross-section in Fig. 3.

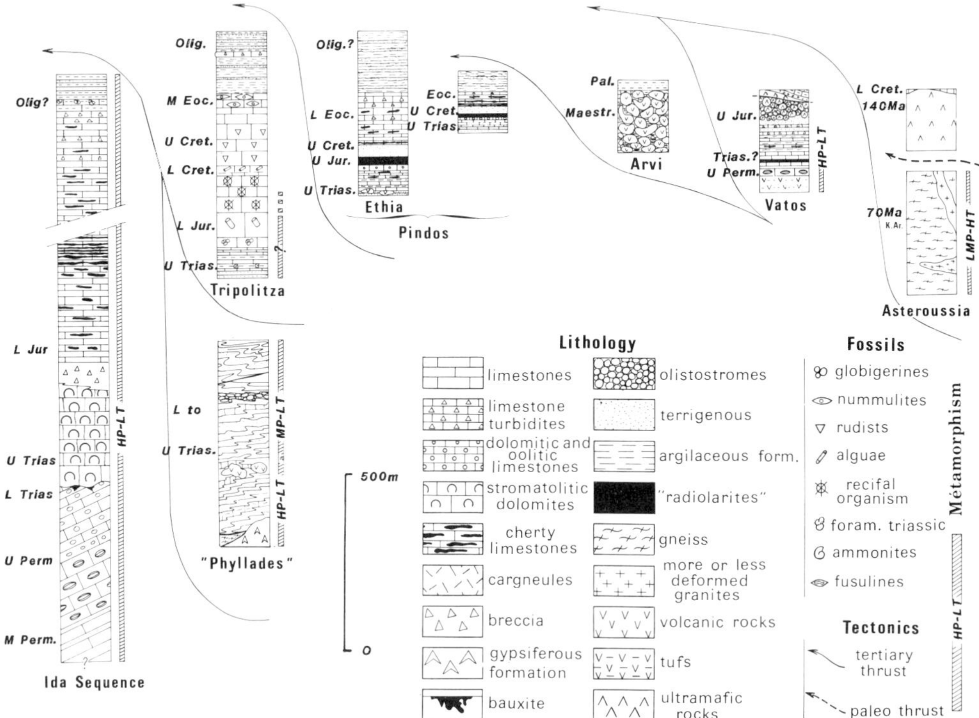

FIG. 2. Stratigraphic columns of the main units of Crete. Nature of metamorphism is indicated on the right of each column: LP, low pressure; MP, medium pressure; HP, high presure; LT, low temperature; HT, high temperature.

plete sequence is observed in Crete (Fig. 2) but useful data also come from islands to the east including Rhodes and Karpathos.

Ida Sequence (Talea Ori Sequence of Epting *et al.* 1972; Plattenkalk Series of Hall *et al.* (this volume))

The characteristic lithology in this unit is a platy limestone with secondary chert layers ('Plattenkalk'). A 2000 m thick sequence of mainly dolomitic rocks occurs in the lower part of the sequence in the Talea Ori of Crete. The unit has been metamorphosed to blueschist facies (Viswanathan & Seidel 1979) and deformed with south-verging isoclinal folds trending E-W in Crete, ENE-WNW in Kassos and NNE-SSW in the equivalent Lindos Limestone unit in Rhodes. The progressive change in fold-trend eastwards is considered to be the result of later Neogene curvature of the Hellenic Arc (Le Pichon & Angelier 1979).

Fossils are exceedingly rare. In Crete, recrystallized pelagic foraminifera in the uppermost part of the sequence give Upper Eocene to lowermost Oligocene ages (Bizon *et al.* 1976). Re-worked ?rudists are known from Crete (Renz 1932; Wachendorf *et al.* 1980) and from Lindos, Rhodes where they are associated with Cenomanian benthic foraminifera (Mutti *et al.* 1970). Triassic algae and Permian fusulinids occur in the Talea Ori carbonates.

Comparison with sequences in a similar tectonic position in the Peloponnese and mainland Greece suggests correlation with the Ionian Zone. A quartzite unit on the southern side of the Talea Ori, in Crete may be compared to the 'Schistes à Posidonies' of the typical Ionian sequence of Northern Greece. There are however some differences. The Lindos neritic sequence in Rhodes may represent a more external palaeogeographic setting, i.e. it represents a carbonate platform closer to a southern continental margin (? of Apulia) than the relatively basinal Ionian sequence. Indeed a typical unmetamorphosed Ionian sequence, the Attavyros Group is thrust over it. A hole drilled in Ionian flysch passed through gypsum, probably of Triassic age then into metamorphosed flysch before bottoming in recrystallized Eocene limestone taken to be the top of the Lindos sequence.

Further north in Amorgos in the Cyclades, a similar Ionian-type sequence, also metamor-

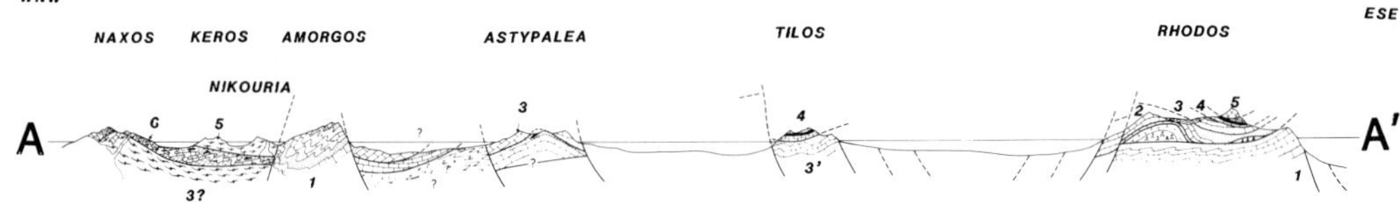

FIG. 3. Partly hypothetical cross-section from Naxos to Rhodes. Emplacement is shown on Fig. 1. 1, Lindos and Amorgos external unit (para-autochthonous), metamorphosed, at least in Amorgos under HP-LT conditions; 2, Ionian (Akramitis) sequence; 3, Tripolitza (Astypalea + Archangelos in Rhodes) sequences; 3′, Koutsoumbas sequence of Tilos (probably related to the Tripolitza Nappe); 3?, basement rocks of Ios, interpreted as a possible original basement of the Tripolitza Nappe; 4, Pindos Nappe; 5, Asteroussia and Ophiolite Nappes; and 6, Cycladic Blueschist Unit.

phosed to blueschist facies, is exposed. Here the mainly carbonate sequence contains abundant breccia indicating deposition in a transition zone between an off-shore carbonate bank, most probably the Tripolitza platform (Minoux *et al.* 1981), and a local 'swell' in the Ionian zone proper. This implies that the higher nappes in Crete, derived from north of this platform (internally), must have travelled several hundred kilometres (Fig. 3).

The complex lower unit in the central and southern Aegean can thus be correlated with the Ionian zone in the Hellenides. The structural succession in Rhodes shows that this Ionian sequence is in turn thrust on to an 'Apulian-PreApulian' continental margin carbonate platform. The frontal part of this nappe escaped metamorphism. The tectonic relationships of these units is identical to that found in the southern Peloponnese (Thiebault 1981).

The Phyllite Nappe (Phyllite-Quartzite Nappe of Hall *et al.* this volume)

This unit consists mainly of quartzites, schists and meta-tuffs with some carbonate rocks. In eastern Crete a slab of older metamorphic basement rocks is found at the base of the nappe (Seidel *et al.* 1977). The only fossils yet found are of ?Permian and Triassic age (Krahl *et al.* 1980).

The nappe, together with the metamorphic basement slab, has been metamorphosed in the blueschist facies in Oligocene-Miocene times (Seidel 1978; Seidel *et al.* 1977). The unit contains isoclinal, synmetamorphic folds, trending approximately north to south apparently parallel to the direction of tectonic transport from north to south, as indicated by a north to south stretching lineation. These folds have then been refolded by more open east to west trending folds.

This nappe is here interpreted as the Permian to Triassic sequence originally underlying the carbonates of the Tripolitza platform. A thrust now separates the two but it is considered to be a late structure as the deformation history is identical in the Phyllite and Tripolitza Nappes.

The Phyllite Nappe could be the equivalent of the low-grade metamorphic Tyros unit of the Peloponnese. It is not present in the Dodecanese but may be represented on Thira (Santorini) by the glaucophane-bearing meta-tuffs and interbedded quartzites and limestones exposed near the port of Athenion (Marinos 1947; Blake *et al.* 1981). This sequence is also overlain by marbles, similar to Tripolitza rocks. The Tripali Unit of Creutzburg & Seidel (1975) is here considered to be a tectonic melange in the thrust-zone between the Phyllite and Tripolitza Nappes rather than a separate unit. In the type locality, the 'unit' is identical to the lowest part of the Tripolitza sequence, both in lithology and fossil content. Other outcrops south-east of Rethymnon are recrystallized Tripolitza limestones, with *Cladocoropsis* near Prases, and rudists 3 km to the south. In western Crete near Kandanos, tectonically brecciated limestones assigned to this unit contain a benthic microfauna of Late Cretaceous age. Finally, north of the Omalos plain, part at least of this 'unit' is a Neogene deposit of *in-situ* brecciated Tripolitza limestones.

The Tripolitza Nappe

The thick Tripolitza sequence of the Aegean is a facies equivalent of the classical Triassic to Upper Eocene carbonate-platform sequence of continental Greece. In Crete, fossil-bearing marbles have been observed north-east of Potamies, in the Dikti Mountains, and near Prases, in the western Psiloriti Mountains. There is no information about the metamorphic grade. However, rocks near Sellia indicate temperatures of 300–400°C based on the colour index of conodonts (Epstein *et al.* 1977) from the Upper Triassic of the Tripolitza nappe (Karakitsios 1981).

In Crete this nappe is thrust on top of the Phyllite Nappe, and the bottom sections are truncated. Commonly the Phyllite Nappe is

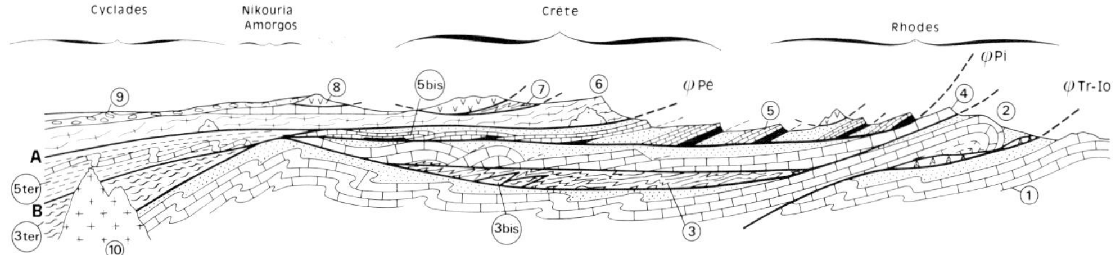

FIG. 4. Reconstructed cross-section of the South East Aegean before Middle-Upper Miocene faulting, showing the relations between the various tectonic units in the Aegean.

1, External platform (Ida sequence); 2, non-metamorphic Ionian sequence; 3, Phyllite Nappe; 3 bis, slices of basement at the base of the phyllite; 3 ter, basement rocks in Ios and Naxos, attributed to the Tripolitza basement; 4, Tripolitza carbonates and flysch; 5, Pindos-Ethia and Arvi Nappes; 5 bis, Kreati Unit of Tilos; 5 ter, Cycladic Blueschist Belt; 6, Asteroussia Nappe; 7, Vatos Unit; 8, Ophiolites; and 9, Future Marmara Nappe of the Central Aegean.

absent, such as in the eastern Lefka Ori and in most of the Psiloriti Mountains. Two distinct deformation phases have affected this nappe in Crete. The first has the same characteristics as in the Phyllite Nappe: roughly north to south trending isoclinal folds in relatively incompetent rocks; then, locally, later east to west trending folds associated with thrusting.

The Tripolitza Nappe occurs in Kassos (Thorbecke 1976). On Karpathos, the Kalilimni Nappe is mainly composed of massive neritic limestones, with intercalations of *Globotruncana*-bearing pelagic limestones. Ages range from Liassic to Lower Lutetian. This nappe could be correlated with the Tripolitza Nappe as similar sequences exist in the Peloponnese (Fleury 1980).

On Rhodes the Archangelos Nappe is composed of massive neritic limestones, cut by numerous thrust faults. Lithology and fossil content are identical with other Tripolitza sequences. Ages range from Upper Triassic to Middle Eocene. Terrigenous (?)Middle Triassic rocks with volcanic horizons similar to the Phyllite Nappe though non-metamorphic, exist locally (Leboulenger & Matesco 1975). The dismembered Archangelos or Tripolitza Nappe is thrust on top of the Ionian Nappe without any equivalent of a Phyllite Nappe between the two.

In Crete no typical non-metamorphic Ionian rocks are known, except perhaps for allochthonous gypsum lenses sometimes encountered between the Phyllite Nappe and the Ida parautochthonous sequence. By comparison with the situation in Rhodes and the Peloponnese this probably means that typical non-metamorphic Ionian sequences must be sought south of Crete, at the front edge of the Tripolitza Nappe, if they have not already been devoured in the Hellenic Arc subduction zone (Figs 3–4).

The Tripolitza Nappe is also known from the island of Astypaleia in the Northern Dodecanese (Marnelis & Bonneau 1979).

The Pindos-Ethia and Arvi Nappes

(1) The non-metamorphic Pindos Nappe of Crete and the Dodecanese comprises scattered klippen of a sequence correlative with, but not identical to, the classical Olonos-Pindos sequence of continental Greece. Notable variations are obvious from one outcrop to another. In Crete characteristic sections are present in the Kedros, Samitos and Assiderotas mountains and in Rhodes they constitute the Prophitis Ilias Nappe. Upper Cretaceous sections in the Asteroussia mountains of Crete show considerable variations from normal Pindos sequences and commonly comprise huge breccias containing neritic limestone blocks as well as abundant clasts of granite, metamorphic rocks and serpentinite (Lentas & Prinia sections). The Ethia section in Crete ends with Lower and Middle Eocene calciturbidites (Aubouin *et al.* 1965), but is typical Pindos facies in the lower part of the section. The rocks in this sequence include radiolarian cherts and pelagic carbonates typical of a deep basin (see Fig. 5). The Mangassa sequence of eastern Crete also belongs to the Pindos.

In Karpathos, the Xindothio series contains ophiolitic olistostromes in the uppermost Jurassic (Aubouin *et al.* 1976a). On the small island of Tilos, the Kreati sequence is very similar to the Pindos but has no uppermost Cretaceous limestones (Roussos & Bonneau 1979). Similarities with stratigraphic sequences in the Antalya nappes in the Taurides are also striking.

Although most of the folds in the Pindos sequences described have trends parallel to the present Aegean arc, some earlier north to south-trending folds do exist in the Pindos

Nappe of Crete in the Kedros mountains. In other places, such as in the Smarion area, stylolites and folds indicate a northwest to southeast compression.

(2) The type of Arvi Nappe is known only from central Crete but a metabasalt unit below the Asteroussia Nappe on the island of Anaphi (Reinecke *et al.* 1982) could be correlated with it. It consists of pillow-basalts, tuffs and flows, dolerite dykes and perhaps some serpentinite. The matrix comprises red pelagic limestones with Late Campanian—Early Maastrichtian *Globotruncana.* Analyses of pyroxenes suggest oceanic tholeiite affinities (Robert & Bonneau 1982) for the mafic components. Mineral assemblages are of very low grade, perhaps referable to sub-sea-floor metamorphism. Similar rocks have been described from the ophiolitic melange of the Lycian nappes in southwestern Turkey (de Graciansky 1972).

This unit might be part of an Upper Cretaceous ophiolite complex, similar to the Troodos ophiolite of Cyprus. From its present tectonic position, always closely related to the Pindos Nappe, it would appear that the Pindos pelagic basin was an oceanic trough which underwent spreading during Late Cretaceous times.

Although the Cycladic Blueschist sequence occupies a position equivalent to that of the Pindos Nappe in the area as a whole, it is appropriate to describe first the units overlying the Pindos Nappe in the southern part of the area.

The uppermost nappes (Pelagonian *sensu lato*)

The units described below have commonly been treated as several different nappes (Bonneau 1973a, 1976; Cadet *et al.* 1980). However Creutzburg & Seidel (1975) have included them together with the Arvi Nappe, in a so-called serpentinite-amphibolite association. Later Seidel *et al.* (1977) referred to this as ophiolitic melange but this term seems inappropriate as there is no recognizable matrix and the different components comprise mappable units with different geodynamic significance. In my opinion, the following individual tectonic units can be recognized.

The Vatos and Kalipso units

The Kalipso unit of Gavdos island is apparently thrust upon a Pindos-Ethia sequence. It includes meta-basalts, meta-rhyolites and meta-sediments of the glaucophanitic-greenschist facies. These rocks are overlain unconformably by unmetamorphosed Upper Cretaceous marls (Vicente 1970). K-Ar radiometric data give an Upper Jurassic age for the metamorphism.

Similar sections can be found in south-west central Crete. The Vatos unit is preserved here in a section from Moni-Preveli to Vatos (Bonneau & Lys 1978) and comprises 10 m of blueschist-facies meta-basalts and pelites, whith intercalated marbles bearing Permian fossils, and metacherts; 25 m of amphibolite; ophiolitic rocks cut by pegmatitic gabbro dykes; and a flysch-like formation with Jurassic to Early Cretaceous limestones clasts, grading up into a thick conglomerate composed entirely of ophiolitic clasts of all sizes. The whole sequence is covered unconformably by pelagic red marls and conglomerates with Campanian-Maastrichtian *Globotruncana.*

The Vatos unit is a good model in itself for the early evolution of the internal Hellenides. A Middle-Upper Jurassic event associated with the obduction of the ophiolitic slab and amphibolites produced blueschist metamorphism in the lowest part of the section. The middle part of the section shows the effects of a Cretaceous metamorphism of unknown affinity. The Uppermost Cretaceous is unmetamorphosed. From outcrops near Petrokefali and Miamou, it seems that this lies *below* the Asteroussia nappe described below.

The Asteroussia Nappe

This consists of a high-grade metamorphic complex including amphibolites, marbles, micaschists and gneisses with deformed and undeformed granitic rocks. High temperature-low pressure conditions of metamorphism were demonstrated by Seidel *et al.* (1976). Radiometric data show that the last metamorphic event took place around 75 Ma (Upper Cretaceous). This unit has been referred to as Pelagonian by Bonneau (1973b) on the basis of its tectonic position.

The thickness of the klippen ranges from 600 m in the Asteroussia Mountains, to 4–5 m in the neighbourhood of Anogia. Evidently this large unit has been severely attenuated during its Oligocene emplacement in the nappe pile. The Neogene rocks of eastern Crete, where no klippen of the Asteroussia Nappe are known, contain abundant gneissic pebbles referable to this nappe, thus indicating its much larger former extent.

The Upper Cretaceous high-temperature metamorphic event and related granitic intrusion were probably the result of the subduction of an oceanic slab below the continental crust of the Pelagonian (Bonneau 1982).

The ophiolite nappe

Ophiolitic remnants suggest that an ophiolite nappe was thrust onto the Asteroussia metamorphic complex. They are badly preserved and are mainly composed of serpentinite, except in the region of Anogia (Bonneau 1971; Thorbecke 1975) where an ophiolitic suite can be reconstructed, though without cumulate gabbros. Uppermost Jurassic to Lowermost Cretaceous neritic limestones rest on these remnants in the vicinity of Kria-Vrissi, central Crete, suggesting correlation with some of the mainland ophiolites obducted in the Upper Jurassic (Mercier 1968). This has been confirmed by radiometric determinations (Seidel *et al.* 1981), yielding ages of about 140 Ma.

The ophiolite nappe has been thrust on to the Asteroussia Nappe after the Upper Cretaceous, probably during the main Upper Eocene-Oligocene Hellenic phase. This ophiolitic nappe also occurs in Rhodes.

The Cycladic area

This region appears at first to be completely different from the external or southern belt, but some correlations are possible, most readily between units in the southern Cyclades and sequences already described:

(1) On Amorgos, only one facies sequence occurs, comprising mainly carbonates ranging from Upper Triassic to Eocene. It most probably belongs to the Ida Series, and thus represents a tectonic window in the Cycladic area.

(2) On Thira, carbonate rocks ranging from Upper Triassic (Papastamatiou 1968) to Eocene (Tataris 1964; Blake *et al.* 1981) in age most probably belong to the Tripolitza Nappe. Topographically below this unit, high pressure-low temperature metamorphic rocks crop out, probably belonging to the Phyllite Nappe. This contact is hidden by recent volcanics.

(3) Anaphi is mainly composed of metamorphic rocks (Reinecke *et al.* 1982) referrable to the Asteroussia nappe, tectonically overlain by ophiolitic klippen. The whole sequence is thrust on top of Palaeogene flysch. Greenschists of tholeiitic composition are preserved between the Asteroussia rocks and the flysch and could be the equivalent of the Arvi Nappe of Crete.

The remaining Cycladic islands further north show four kinds of tectonic unit, from the base upward (Fig. 4):

Gneissic basement rocks

This is mainly orthogneiss, with pelitic schists and gneisses, have been recognized on Ios (Henjes-Kunst 1980) and on Sikinos (Van der Maar 1981), and show a superimposed 'alpine' blueschist metamorphism. Bonneau *et al.* (1978) have shown that two units can be distinguished on Naxos, the lower one consisting of migmatites and marbles with the upper contact between the two marked by ultramafic lenses (Jansen 1973). A tectonic contact is obvious but there is no indication of the amount of displacement along the thrust-plane. The lower unit is probably correlateable with the unit seen on Ios.

It is uncertain whether this basement unit, for which I propose the name *Khora Unit*, was originally the basement of the Cycladic blueschists or the subducted basement of a sequence now represented in the more external nappes. I favour the latter interpretation and consider the unit to be the original basement of the Tripolitza Nappe.

The Cycladic Blueschist Unit

This is the main unit in the Cyclades (Durr *et al.* 1978a; Blake *et al.* 1981). It constitutes parts of Samos, Naxos, Ios, Sikinos, Folegandros, Siphnos, Syros, Giaros, Tinos, Andros, South Euboea and several other islands.

The lithologies in this unit are thick bedded marbles with neritic fossils, mainly of Triassic age, meta-tuffs, meta-basalts and meta-cherts. In the southern islands, there are also meta-bauxites (emery). At the top of the sequence is an ophiolitic olistostrome, well developed in Syros but present almost everywhere in minor amounts as in South Euboea, North Tinos, Siphnos, Sikinos and Naxos.

Metamorphism and deformation are polyphase. In the southern Cyclades the main fold trends are roughly north to south and might correspond to '*a*' folds associated with transport from north to south, as suggested locally by quartz fabrics. Further north, the early syn-metamorphic folds have NE-SW directions. They have been refolded along NNW-SSE (Dinaric) directions. Early metamorphism is typically blueschist in character with beautiful examples in Syros and Siphnos (Dixon 1976; Altherr *et al.* 1979; Bonneau *et al.* 1980). Its age is uncertain. Most published K-Ar ages are around 45 Ma, but there are indications of older ages from Syros (70–80 Ma) from ^{39}Ar-^{40}Ar dates on glaucophane (Maluski *et al.* in press).

Traces of a later (25 Ma) retrogressive event are found on almost all the islands (Andriessen *et al.* 1979): it is characteristically medium grade amphibolite facies in character but locally culminates in the formation of migmatites and associated granodioritic intrusions dated at around 10–15 Ma as on Naxos and various other islands (Dürr *et al.* 1979). I suggest that this unit represents the original Pelagonian margin of the Pindos ocean.

The Asteroussia Nappe

Dürr *et al.* (1978b) have proposed the metamorphic rocks on the islet of Nikouria, just north of Amorgos, as a probable equivalent of the Asteroussia Nappe. Indeed Upper Cretaceous radiometric ages of high grade amphibolite-facies rocks are similar in both sequences. Other islands south of Naxos show similar rocks.

On Syros, the Vari Unit, which tectonically overlies the Blueschist Unit, consists of orthogneiss giving Upper Cretaceous ages (70 Ma, Bonneau *et al.* 1980). It may also be related to the Asteroussia Nappe.

Thus a few small remnants of the Asteroussia Nappe seem to occur throughout the Cycladic area, overlying the Blueschist Unit. It is not surprising that so few remnants are preserved, since uplift and associated erosion since Upper Miocene times have evidently been extreme (Dürr *et al.* 1978a).

Marmara Nappe

This is the only unmetamorphosed unit found in the Cycladic area and is composed of Miocene sediments unconformably overlying ophiolitic remnants. On Paros, these ophiolites are overlain by Cretaceous (Barremian to Cenomanian) neritic limestones, which indicates that these ophiolites had first been obducted in ?Upper Jurassic-Lower Cretaceous times, as is the case in Crete for example.

This nappe outcrops on a number of islands (Papanikolaou 1980). It has been thrust into its present position in the Upper Miocene, as shown by Jansen in Naxos where the thrust sheet rests on a granodiorite dated at 10 Ma. However its provenance is controversial. Angelier *et al.* (1978), consider it probably came from the NE, whereas Dürr and Altherr (1979) prefer a southern provenance which seems more reasonable. It appears that this peculiar nappe, only some tens of metres thick, has been emplaced by gravitational gliding after the remaining nappe-pile in the area was already emplaced (see also Ridley, this volume).

Conclusions

Even though it is probable that lateral movements have occurred between the present Cycladic area and the external belt, the data presented above suggest beyond doubt that both areas have had a common Mesozoic and Cenozoic history (Fig. 5). More or less radially-orientated thrusts have evidently been more important that lateral movements.

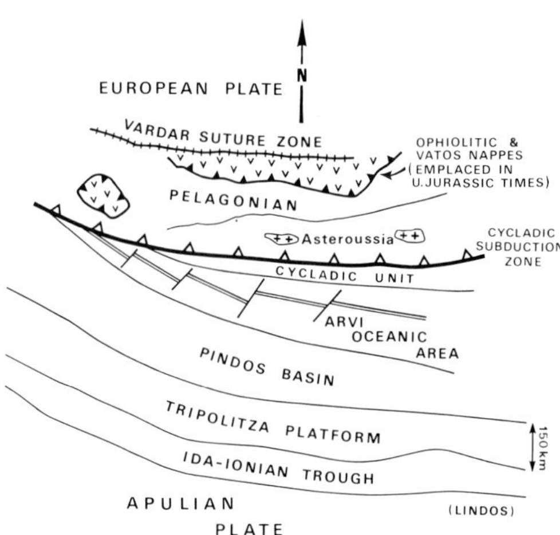

FIG. 5. Palaeogeographic cartoon of the Central Aegean in Uppermost Cretaceous times.

Two main tectonic units are preserved in both domains: the external, Ida para-autochtonous sequence, which has a much wider distribution than previously thought, and represents the complex Apulian margin, and the Asteroussia Nappe which is probably an equivalent of the Pelagonian Nappe. This Pelagonian realm was a micro-continent from Triassic until Upper Jurassic times and later became part of the European continental margin, after the closure of the Vardar (Tethys) ocean. It is proposed here that the Pindos realm was a small ocean-basin or marginal sea lying between the Apulian and European continental margins, during the Cretaceous and Palaeogene. It began being subducted during Upper Cretaceous times beneath the Pelagonian (European) plate. This subduction produced the blueschist metamorphism in the Cycladic Blueschist Unit interpreted here as the former Pelagonian margin, and amphibolite-facies metamorphism and granitic intrusion in the continental crust of the Pelagonian, the present Asteroussia Nappe. The Pindos oceanic area was closed by the end of Eocene times. Parts of its sedimentary cover and the uppermost part of the oceanic crust, the Pindos and Arvi Nappes, respectively, were obducted at this time onto the Tripolitza platform (Apulian margin).

This model is similar to that proposed by

Smith *et al.* (1980) for continental Greece, although it is based on different arguments. The Upper Eocene and Oligocene history is that of a collision. Both former margins have then suffered considerable shortening: the Tripolitza platform has been thrust onto Apulia, as well as on to the Pelagonian. Blueschist metamorphic conditions were produced at the base of the pile of nappes. From the Middle Miocene on, a completely different tectonic regime created the present outcrop pattern. This 'neotectonic' stage is related to subduction along the Hellenic Arc and resulted in the fragmentation of the Hellenic Mountain Belt.

The interesting feature of the central Aegean pre-Miocene history is the importance of Upper Cretaceous tectonism, which is a basis for comparison with the geological history of the Taurides, where ophiolite obduction took place precisely at that time.

ACKNOWLEDGEMENTS: Financial support from the CNRS (LA 215) and ATP Inag. Geodynamique I is acknowledged. J. Brouillet accomplished most of the drafting. The manuscript was reviewed by R. Hall and J. E. Dixon helped produce the final version.

References

ALTHERR, R., SCHLIESTEDT, M. OKRUSCH, M., SEIDEL, E., KREUZER, H., HARRE, W., LENZ, H., WENDT, I. & WAGNER, G. A. 1979. Geochronology of high pressure rocks on Siphnos (Cyclades, Greece). *Contr. Miner. Petrol.* **70,** 245–55.

ANDRIESSEN, P., BOELRIJK, N., HEBEDA, E., PRIEM, H., VERDURMEN, E. & VERSCHURE, R. 1979. Dating the events of metamorphism and granitiçm in the Alpine Orogen of Naxos (Cyclades, Greece). *Contr. Mineral. Petrol.* **69,** 215–25.

ANGELIER, J., GLACON, G. & MULLER, C. 1978. Sur la présence et la position tectonique du Miocène inférieur marin dans l'archipel de Naxos (Cyclades, Grèce). *C.r. Ac. Sc. Paris D*, **286,** 21–4.

AUBOUIN, J., BONNEAU, M. & DAVIDSON, J. 1976. Contribution à l'étude géologique de l'arc égéen: l'île de Karpathos. *Bull. Soc. géol. Fr.* **18,** 385–401.

——, ——, ——, LEBOULENGER, P. MATESCO, S. & ZAMBETAKIS, A. 1976. Esquisse structurale de l'Arc égéen externe: des Dinarides aux Taurides. *Bull. Soc. géol. Fr.* **18,** 327–36.

——, DERCOURT, J. NEUMANN, M. & SIGAL, J. 1965. Un élément externe de la zone du Pinde: la série d'Ethia (Crète, Grèce). *Bull. Soc. géol. Fr.* **7,** 753–7.

BIZON, G., BONNEAU, M., LEBOULENGER, P., MATESCO, S. & THIEBAULT, F. 1976. Sur la signification et l'extension des 'massifs cristallins externes' en Péloponnèse méridional et dans l'Arc Egéen. *Bull. Soc géol. Fr.* **18,** 337–45.

BLAKE, M. C., BONNEAU, M. KIENAST, J. R., LEPVRIER, C. MALUSKI, H. & PAPANIKOLAOU, D. 1981. A geologic reconnaissance of the Cycladic Blueschist belt, Greece. *Bull. geol. Soc. Amer.* **92,** 247–54.

BONNEAU, M. 1971. Les lambeaux allochtones du revers septentrional du massif des Psiloriti (Crète moyenne, Grèce), *Bull. Soc. géol. Fr.* **12,** 1124–9.

—— 1973a. Les différentes 'séries ophiolitifères' de la Crète: une mise au point. *C.r. Ac. Sc., Paris,* **276,** 1249–52.

—— 1973b. La nappe métamorphique de l'Asteroussia, lambeau d'affinités pélagoniennes charrié jusque sur la zone de Tripolitza de la Crète moyenne (Grèce). *C.R. Ac. Sc. Paris*, ? 2303–6.

—— 1982. Evolution dynamique de l'Arc Egéen depuis le Jurassique supérieur jusqu'au Miocène. *Bull. Soc. géol. Fr.* **24,** 229–42.

——, ANGELIER, J. & EPTING, M. 1977. Réunion extraordinaire de la Société géologique de France en Crète. *Bull. Soc. géol. Fr.* **19,** 87–102.

——, BLAKE, M. C., GEYSSANT, J. KIENAST, J. R. LEPVRIER, C. MALUSKI, H. & PAPANIKOLAOU, D. 1980. Sur la signification des séries métamorphiques (schistes bleus) des Cyclades (Hellénides, Grèce). L'exemple de l'île de Syros. *C.r. Ac. Sc. Paris, D*, **290,** 1463–6.

——, GEYSSANT, J. & LEPVRIER, C. 1978. Plis couchés kilométriques dans l'île de Naxos. Conséquences. *Rev. Géogr phys. Géol. dyn.* **20,** 109–22.

——, ——, KIENAST, J. R., LEPVRIER, C. & MALUSKI, H. 1980. Tectonique et métamorphisme haute pression d'âge éocène dans les Hellénides: exemple de l'île de Syros (Cyclades). *C.r. Ac. Sc. Paris, D*, **291,** 171–4.

—— & KARAKITSIOS, V. 1979. Les niveaux inférieurs (Trias supérieur) de la nappe de Tripolitza en Crète moyenne (Grèce) et leurs relations avec la nappe des phyllades. Problèmes stratigraphiques, tectoniques et de métamorphisme. *C.r. Ac. Sc. Paris, D*, **287,** 423–6.

CADET, J. P. BONNEAU, M. CHARVET, J. DÜRR, S. ELTER, P. FERRIERE, J. SCANDONE, P. & THIEBAULT, F. 1980. Les chaînes de la Méditerranée moyenne et orientale. *In: Géologie des chaînes alpines issues de la Téthys, Colloque C5, 26th I.G.C., Paris*, **98,** 118.

CREUTZBURG, N. & SEIDEL, E. 1975. Zum stand der Geologie des Präneogens auf Kreta. *N. Jb. Geol. Paläont. Abh.* **149,** 363–83.

DIXON, J. E. 1976. Glaucophane schists of Syros, Greece. *Bull. Soc. géol. France*, **18,** 280.

DROOGER, C. W. & MEULENKAMP, J. E. 1973. Stratigraphic contribution to geodynamics in the Mediterranean area: Crete as a case history. *Bull Soc. geol. Greece*, **10,** 193–200.

DÜRR, S. & ALTHERR, J. 1979. Existence d'une klippe d'une nappe composite néogène dans l'île de Mykonos, Cyclades (Grèce). *Rapp. Comm. Int. Mer Méditer.* **25/26**, 33–4.

——, ——, Keller, M. Okrush, M. & Seidel, E. 1978a. The median Aegean Crystalline Belt: stratigraphy, structure, metamorphism, magmatism. *In*: *Alps, Apennines, Hellenides* H. Closs, D. Roeder & K. Schmidt, (eds), E. Schweizerbart'sche Verlag., Stuttgart, 455–78.

——, Seidel, E. Kreuzer, H. & Harre, W. 1978b. Témoins d'un métamorphisme d'âge crétacé supérieur dans l'Egéide: datations radiométriques de minéraux provenant de l'île de Nikouria (Cyclades, Grèce) *Bull. Soc. géol. France*, **20**, 209–13.

Epstein, A. G., Epstein, J. B. & Harris, L. D. 1977. Conodont color alteration; an index to organic metamorphism. *U.S. Geol. Surv. Prof. Pap.* **995**, 27 p.

Epting, M., Kudrass, H. R. & Schaffer, A. 1972. Stratigraphie et position des séries métamorphiquest au Talea Ori. *Z. Deutsch geol. Ges.* **123**, 365–70.

Fleury, J. J. 1980. Les zones de Gavrovo-Tripolitza et du Pinde-Olonos (Grèce continentale et Péloponnèse du Nord). Evolution d'une plate-forme et d'un bassin dans leur cadre alpin. *Publ. Soc. Géol. Nord*, **4**, 648 p.

Graciansky, P. C. de 1972. *Recherches géologiques dans le Taurus Lycien.* Thèse d'Etat (unpubl.). Univ. Paris Sud-Orsay, 762 pp.

Henjes-Kunst, F. 1980. *Alpidisch Einformung des Präalpidischen Kristallins und seiner Mesozoischen Hülle auf Ios (Kykladen, Griechenland).* Dissertation (unpubl.) Tecn. Univ. Braunschweig (F.R.G.), 164 pp.

Jansen, B. H. 1973. *Geological map of Greece 1/50 000: Naxos Island.* Inst. Geol. Min. Res. (I.G.M.R.) Athens.

Karakitsios, V. 1979. *Contribution à l'étude géologique des Hellénides. Etude de la région de Sellia (Crète moyenne, Grèce).* Thèse 3e cycle (unpubl.), Univ. P. et M. Curie, Paris, 155 pp.

Krahl, J., Eberle, P., Eickhoff, J. Forster, O. & Kozur, H. 1981. *Biostratigraphical Investigations in the Phyllite-Quartzite group on Crete Island, Greece.* Abs., H.E.A.T. Symposium, Athens.

Leboulenger, P. & Matesco, S. 1975. *Contribution à l'étude géologique de l'Arc Egéen. L'île de Rhodes.* Thèse 3e cycle (unpubl.). Univ. P. et M. Curie, Paris, 217 pp.

Le Pichon, X. & Angelier, J. 1979. The Hellenic Arc and trench system: a key to the neotectonic evolution of the eastern Mediterranean area. *Tectonophysics.* **60**, 1–42.

Marinos, G. 1947. Contribution à la pétrologie du systèm cristallophyllien du SE de la Grèce. L'île de Ios. *Ann. géol. Pays Hellén.* **1**, 60–96.

Marnelis, P. & Bonneau, M. 1979. Stratigraphie et structure de l'île d'Astypalea (Dodécannèse, Grèce). *6th Coll. on the Geology of the Aegean Region*, G. Kallergis ed. Athens, 1977, 323–332.

Mercier, J. 1968. Etude géologique des zones internes des Hellénides en Macédoine centrale (Grèce). *Ann. géol. Pays Hellén.* **20**, p. 1–792.

Minoux, L. Bonneau, M. & Kienast, J. R. 1980. L'île d'Amorgos, une fenêtre des zones externes au coeur de l'Egée (Grèce), métamorphosée dans le faciès schists bleus. *C.r. Ac. Sc. Paris*, **291**, 745–748.

Mutti, E., Orombelli, G. & Pozzi, R. 1970. Geological studies on the Dodecanese Islands (Aegean Sea). Geological Map of the Island of Rhodos (Greece), Explanatory notes. *Ann. géol. Pays Hellén.* **22**, 77–226.

Papanikolaou, D. 1979. On the tectonic units of southern Aegean Sea. *Rapp. Comm. Int. Mer Méditerranée*, **25**/**26**, 51–2.

Papastamatiou, J. 1958. Sur l'âge des calcaires cristallins de l'île de Thera. *Bull. geol. Soc. Greece*, **3**, 104–13.

Reinecke, T. Altherr, R., Hartung, B., Hatzipanagiotou, K. Kreuzer, H. Harre, W., Klein, H. Keller, J., Geenen, E. Böger, H. 1982. Remnants of a late Cretaceous high temperature belt on the island of Anafi (Cyclades, Greece). *N. Jb. Miner. Abh.* **145**, 157–82.

Renz, O. 1932. Zur Geologie von Sitia, der Osthalbinsel Kretas. *Prakt. Akad. Athinon*, **7**, 105–9.

Robert, U. & Bonneau, M. 1982. Les basaltes des nappes du Pinde et d'Arvi (Crète) et leur signification dans l'évolution géodynamique de la Méditerranée orientale. *Ann. géol. Pays Hellén*, in press.

Roussos, N. & Bonneau, M. 1979. Stratigraphie et structure de l'île de Tilos (Dodécannèse, Grèce). *6th Coll. on the Geology of the Aegean Region*, G. Kallergis, ed. Athens 1977, 333–343.

Seidel, E. 1978. *Zur Petrologie der Phyllit-Quartzit Serie Kretas.* Unpub. Thesis, Braunschweig, 145 pp.

Seidel, H., Okrusch, M., Kreuzer, H., Raschka, H. & Harre, W. 1976. Eo-alpine metamorphism in the uppermost unit of the Cretan nappe system petrology and geochronology. Part I. The Lendas area (Asteroussia mountains). *Contr. Miner. Petrol.* **57**, 258–75.

—— Part II. Synopsis of high temperature metamorphics and associated ophiolites. *Ibid.* **76**, 351–61.

——, Schliestedt, M., Kreutzer, H. & Harre, W. 1977. Metamorphic rocks of late Jurassic age as components of the ophiolitic melange on Gavdos and Crete (Greece). *Geol. Jb.* **28**, 3–21.

Smith, A. G. Woodcock, N. H. & Naylor, M. A. 1979. The structural evolution of a Mesozoic continental margin, Othris Mountains, Greece. *J. geol. Soc. Lond.*, **136**, 589–603.

Tataris, A. 1964. The Eocene in the semi-metamorphosed basement of Thera island. *Bull. geol. Soc. Greece*, **6**, 232–8.

Thiebault, F. 1982. Evolution géodynamique des Hellénides externes en Péloponnèse méridional (Grèce). *Soc. Géol. Nord Publ. 6, Lille, F.*, 571 pp.

Thorbecke, G. 1973. Die Gesteine der Ophiollith-Decke von Anoja/Mittel Kreta. Berichte Naturforsch. *Gesellsch. Freiburg im Brisgau*, **63**, 81–92.

—— 1976. Nachweis von Tripolitza-Flysch auf der Insel Kasos, Griechenland. *Z. Deutsch. geol. Ges.* **127**, 125–31.

Van der Maar, P. 1981. *Metamorphism on Ios and the geological history of the Southern Cyclades (Greece)*. Unpubl. Utrecht University Thesis, 142 pp.

Vincente, J. C. 1972. Etude géologique de l'île de Gavdos (Grèce), la plus méridionale de l'Europe. *Bull. Soc. géol. France*, **12,** 481–95.

Viswanathan, H. & Seidel, E. 1979. Crystal chemistry of Fe-Mg-Carpholites. *Contr. Miner. Petrol.* **70,** 41–47.

Wachendorf, H., Gralla, P., Koll, J. & Schulze, I. 1980. Geodynamik des Mittelkrietischen Deckenstapels (nördlisches Dikti-Gebirge). *Geotekton. Forsch.* **59,** Stuttgart, 1–72.

Michel Bonneau, Département de Géologie structurale, Univ. P. et M. Curie, 75230 Paris Cedex 05 and LA 215, Tectonique, CNRS, France.

High-pressure rocks of Samos, Greece

M. Okrusch, P. Richter & G. Katsikatsos

SUMMARY: The metamorphic complex of Samos is composed mainly of marbles and phyllites, typically with intercalations of serpentinites, meta-gabbros, blueschists, and greenschists.

The meta-gabbros show typical flaser-structures, but contain frequent textural relics of the igneous protolith as well as relict clinopyroxene and brown hornblende. Primary plagioclase is totally replaced by albite, epidote or zoisite. Critical mineral assemblages of the Alpidic metamorphism are:

albite + epidote + chlorite + glaucophane ± barroisite + sphene,
albite + epidote + zoisite + tremolite + muscovite + sphene,
albite + epidote + chlorite + tremolite ± barroisite ± muscovite ± calcite ± sphene.

Associated glaucophanites show the following critical assemblages (+ sphene ± rutile):

garnet + glaucophane + epidote ± barroisite + chlorite + muscovite ± paragonite,
glaucophane + epidote + chlorite + muscovite + albite ± quartz.

The major and trace element chemistry confirms that the metabasites are derived from abyssal tholeiites and related gabbros.

The metapelites of Samos frequently contain chloritoid which, in the eastern part of the island, may be accompanied by kyanite, and rarely by Mg-carpholite. Meta-bauxites in eastern Samos contain diaspore, in western Samos diaspore + corundum (Mposkos 1978).

At the peak of metamorphism temperatures may have reached about 420° and 470°C in east and in west Samos respectively, assuming elevated pressures of about 8 kbar.

Geological review

The metamorphic complex of Samos, which is part of the median-Aegean crystalline belt (Dürr *et al.* 1978), is predominantly formed of marbles and phyllites which are repeatedly interlayered, and contain intercalations of ultramafics, metagabbros, blueschists, and greenschists. Theodoropoulos (1979) in his 1:50 000 geological map regards the metamorphic complex as a normal stratigraphic succession, which is overthrust by a nappe composed of unmetamorphosed formations in the western part of the island. In contrast, Papanikolaou (1979) interprets the metamorphic complex as a pile of nappes, a concept which we accept with slight modifications. Accordingly, the pre-Neogene succession of Samos is divided into the following tectonic units from top to bottom (Fig. 1):

(1) *The Kallithea Nappe* consists of unmetamorphosed amygdaloidal pillow basalts with minor intercalations of radiolarites, sandstones, and fossiliferous ammonitico rosso-type limestones of Upper Anisian to Lower Ladinian age. The sequence is overlain by massive limestones of Upper Triassic to Jurassic age (Theodoropoulos 1979; Papanikolaou 1979). At the base of the nappe, there are small bodies of ultramafics, in part, well preserved peridotite (Theodoropoulos 1979).

(2) *The Ampelos Nappe* consists of three tectonic units:

(a) *The Vourliotes Unit* is composed of an upper and a lower marble horizon with interbedded meta-pelites and meta-clastics containing in places, kyanite, chloritoid, and rare magnesio-carpholite (Okrusch 1981). Major intercalations are serpentinites, meta-gabbros, greenschists, and blueschists. The upper marbles in the eastern part of the island contain lenses of meta-bauxites with diaspore and chloritoid (Mposkos 1978).

(b) *The Ampelos Unit* is formed by an alternating sequence of marbles and metapelites, in part with chloritoid. Again, there are frequent intercalations of ultramafics, mainly serpentinites, meta-gabbros, blueschists, and greenschists. In the northern part of central Samos, a unit of phyllonitic garnet-mica-schists is exposed which has obviously undergone repeated metamorphism. This unit is associated with garnet-epidote glaucophanites.

(c) *The Aghios Ioannis Unit* is a small wedge of glaucophane-bearing metabasites (Papanikolaou 1979) which have not been investigated in detail.

(3) *The Kerketeas Unit* (? autochthonous) consists of thick marbles which are overlain by phyllites. A meta-bauxite occurring in the Kerketeas marbles contains corundum together with diaspore (Mposkos 1978). On the western coast of the island, below the village of Kallithea, the Kerketeas marbles are intruded by intersecting dykes of highly variable composition (Theodoropoulos 1979), including gabbros, diorites, quartz-mica diorites, tonalites,

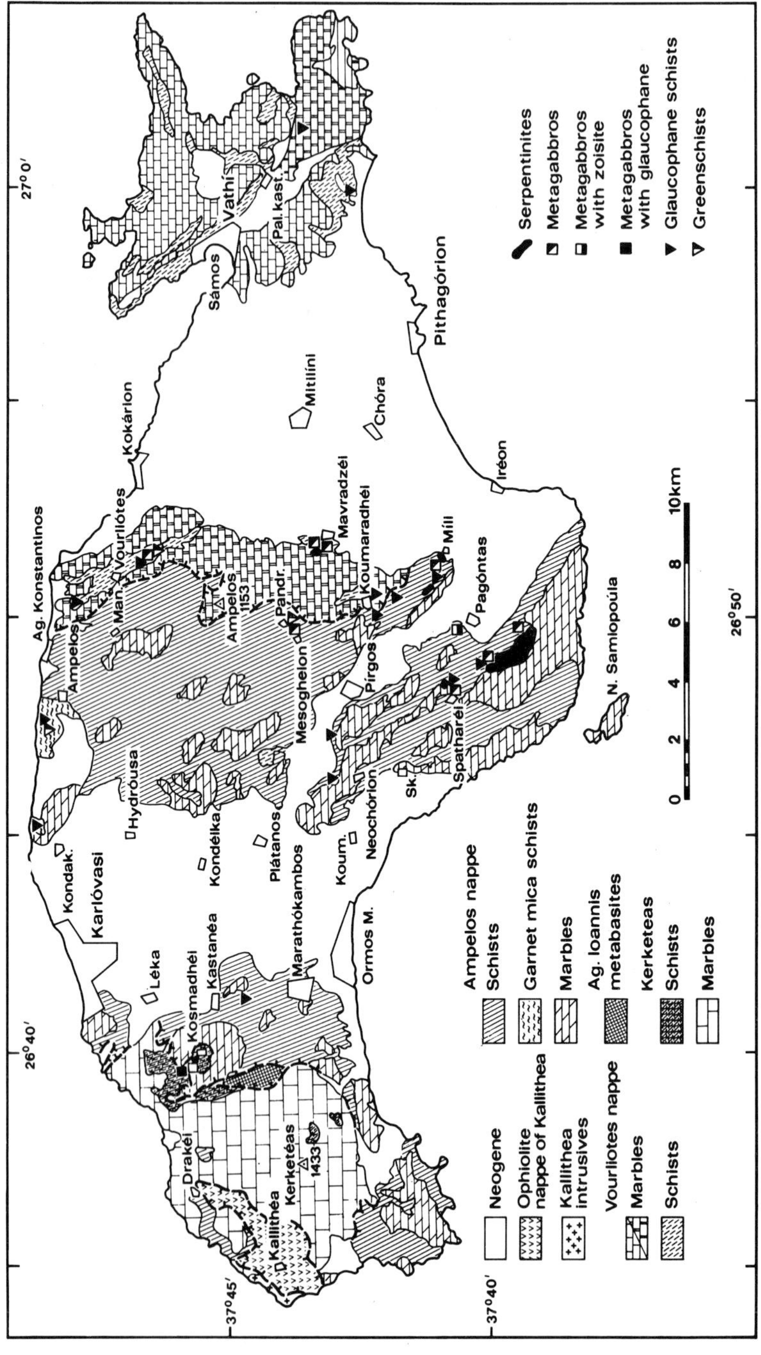

FIG. 1. Geological sketch map of Samos, simplified and modified after Theodoropoulos (1979), tectonic units after Papanikolaou (1979).

granites, and aplites. Contact metamorphism of the adjacent marble is sometimes observed. K-Ar dating of one hornblende concentrate from a diorite yielded a Miocene age (Pers. comm. H. Kreuzer, 1982), indicating that the Kallithea intrusives are the equivalents of petrographically-similar I-type granitoids in the high-temperature belt of the Attic-Cycladic crystalline complex (Altherr 1981; Altherr *et al.* 1982).

Metabasites

Meta-gabbros

The meta-gabbros of Samos, known since the work of Butz (1912), are concentrated in the central section of the island (Fig. 1). The main occurrences are at Moni Vronda near Vourliotes, near Mavradzei, near Mili, and in the Pagontas-Spatharei area. Metagabbros from different localities do not show any systematic variation in structure or mineralogical composition, although they belong to two different tectonic units. The meta-gabbros in the Kosmadhei-Nikoloudhes area, presumably belonging to the Ampelos Unit, are distinguished by the presence of glaucophane. Incidently, the names of the two villages are confused on some maps; the metagabbros are exposed in the *upper* village called *Kosmadhei* by the inhabitants.

The meta-gabbros form lens-shaped bodies up to a few hundred metres long frequently surrounded by phyllites. Direct contact with associated ultramafics, mainly serpentinites, and/or with glaucophanites is sometimes observed. The most instructive outcrop is the roadcut of the new road from Pagontas to Spatharei. The orientations of the schistosity surfaces in the gabbro bodies and in the enclosing phyllites are closely convergent indicating, that the gabbroic, as well as the associated basaltic and ultramafic rocks were (? tectonically) incorporated within the sediments before the deformation and related metamorphism took place.

The igneous origin of the meta-gabbros is established by the presence of textural relics and by relict minerals. Remnants of a primary, coarse grained, hypidiomorphic-granular structure can be observed in all major occurrences. In the bulk of the gabbro lenses, however, the original textures have been extensively modified by intense deformation and subsequent metamorphic recrystallization giving rise to a typical flaser or banded structure. Igneous relict minerals, clinopyroxene and brown hornblende, are quite common; the best examples were recorded at Mili and Mavradzei. They are not confined to meta-gabbros with relict textures, but may also be present in flasergabbros. Metamorphism has led to a partial or total replacement of clinopyroxene and brown hornblende by aggregates of tremolite and/or barroisite amphibole and/or chlorite and, at Kosmadhei, also by glaucophane. Even metagabbros exhibiting a beautiful gabbroic structure on a megascopic scale, may be totally devoid of relict minerals, a striking example being found at Kosmadhei. No original plagioclase is preserved. It is totally replaced by albite, plus epidote or zoisite, sometimes together with white mica and calcite. Frequently, albite forms large poiciloblasts full of epidote inclusions. Many meta-gabbros contain fine grained aggregates of sphene, possibly pseudomorphs after ilmenite; again, the most striking example is the glaucophane-bearing metagabbro of Kosmadhei.

Neglecting the relict minerals, the most frequent mineral assemblages in the meta-gabbros are:

alb-ep-chlor-trem and/or *barrois-sph*
alb-ep-mus-chlor-trem and/or *barrois-cc-sph*

North of Pagontas, the assemblage is:

alb-ep-zoi-mus-trem-sph

The meta-gabbros at Kosmadhei show the assemblages (± *hematite,* ± *pyrite*):

alb-ep-chlor-glauc-sph
alb-ep-chlor-glauc-barrois-sph.

An evaluation of the phase relationships must await detailed microprobe work which is planned.

Glaucophanites

Garnet-epidote glaucophanites are restricted to the central part of the northern coast of Samos between Kondakeika and Ampelos where they are intercalated with phyllonitic garnet-mica schists (Fig. 1). Typical assemblages (all ± *qz,* ± *alb,* ± *cc* & + *rut* & *hem*):

gar-glauc-ep-chlor-mus(-par)-sph
gar-glauc-barrois-ep-chlor-mus-sph-rut-hem.

Relics of brown hornblende are rare, partly replaced by glaucophane. In some of the samples, the garnets form porphyroblasts which are full of inclusions of glaucophane, white-mica, chlorite, epidote, rutile, and opaques. A quartz-rich lens in a garnet-glaucophanite contains garnet porphyroblasts full of quartz inclusions. These garnets show beautiful 'snowball' textures indicating internal rotation during growth.

Albite-epidote glaucophanites are far more

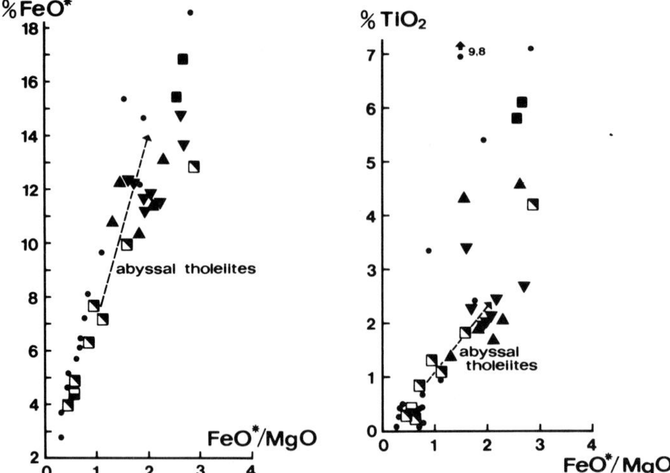

FIG. 2. Variation diagrams a, FeO^{tot} vs. DeO^{tot}/MgO and b, TiO_2 vs. FeO^{tot}/MgO, dots, other symbols see Fig. 3. Trends of abyssal tholeiites after Miyashiro (1975). Abyssal gabbros after Miyashiro and Shido (1980).

TABLE 1. *Selected Bulk Rock Analyses*

Sample #	Sa-18	Sa-230	Sa-93	Sa-196	Sa-69
Rock Type	Zoisite-bearing Metagabbro	Meta-gabbro	Glauc.-bearing Metagabbro	Garnet glauco-phanite	Epidote glauco-phanite
Locality	Pagontas	Pagontas	Kosmadhei	Agios Nikolaos	Moni Vrontas
wt.-%					
SiO_2	47.8	49.9	43.0	46.8	47.7
TiO_2	0.34	1.25	5.68	1.98	2.07
Al_2O_3	18.3	14.9	11.4	16.75	17.2
Fe_2O_3	2.47	3.87	4.97	3.89	7.08
FeO	1.96	3.97	11.57	9.15	4.93
MnO	0.08	0.15	0.21	0.21	0.14
MgO	7.53	7.89	6.04	5.49	5.52
CaO	13.8	11.25	9.66	8.64	5.47
Na_2O	2.42	3.54	2.51	3.29	3.94
K_2O	0.58	0.09	0.30	0.85	1.84
P_2O_5	0.05	0.18	0.11	0.14	0.20
CO_2	1.21	0.10	0.74	0.25	<0.05
H_2O^+	3.0	2.6	3.4	2.1	3.5
Sum	99.54	99.69	99.59	99.54	99.59
ppm					
V	94	228	1220	330	344
Cr	1100	210	44	225	258
Ni	300	129	61	51	91
Cu	23	43	57	48	42
Zn	32	162	10	125	141
Rb	10	<4	10	38	55
Sr	180	252	190	339	144
Y	<10	36	34	34	38
Zr	34	130	136	156	162
Nb	<3	5	8	4	6
Ba	15	40	265	185	215
La	<3	6	15	9	11

widespread and have the assemblages (all ± *qz*, ± *rut*, ± *cc*):
glauc-ep-alb-w.mica-chlor-sph-hem.

In many cases, these rocks show a well developed schistosity, sometimes with alternating epidote- and glaucophane-rich schlieren, a structure suggestive of volcanic tuffs or hyaloclastites as a protolith. There are also massive epidote-glaucophanites, sometimes with a texture resembling dolerite.

Greenschists

Glaucophane-free metabasites which consist of albite, epidote, chlorite, barroisite and/or actinolitic amphibole, white-mica, and sphene in various proportions, are often associated with meta-gabbros and glaucophanites. They have not been investigated in detail.

Geochemistry of meta-gabbros and glaucophanites

Geochemical investigations on meta-gabbros, glaucophanites, and related ultramafics of Samos are in progress. Selected bulk rock analyses, including trace element data, are listed in Table 1. The results so far obtained confirm the igneous parentage of the meta-gabbros as deduced from textural and mineralogical evidence. They also show that the glaucophanites are derived from an igneous protolith, i.e. basalts and basaltic tuffs. Possible exceptions are the carbonate-rich albite-epidote glaucophanites for which a mixed tuffaceous/sedimentary starting material seems to be more probable. The rather high alkali contents, relative to silica, of most metabasite samples may be due to post-igneous alterations: Minor and trace elements, e.g. Ti, P, Zr, Nb, Y, and Cr, commonly regarded as relatively immobile, indicate a sub-alkaline, tholeiitic protolith, comparable to oceanic tholeiite and related abyssal gabbro (Fig. 2, 3). The meta-gabbros show a marked compositional variation (Table 1). With increasing FeO^{tot}/MgO ratio, FeO^{tot} increases from 4.5 to 17%, TiO_2 from 0.3 to 5.7 wt.% (Fig. 2), V from 90 to 1200 ppm, Zr from 30 to 140 ppm, whereas Cr decreases from 1100 to 45 ppm (Fig. 3), and Ni from 300 to 60 ppm. A similar variation was recognized in a suite of abyssal gabbros from the Mid-Atlantic ridge by Miyashiro & Shido (1980), who classified these rocks into three groups representing an early, middle and late stage of progressive fractionation.

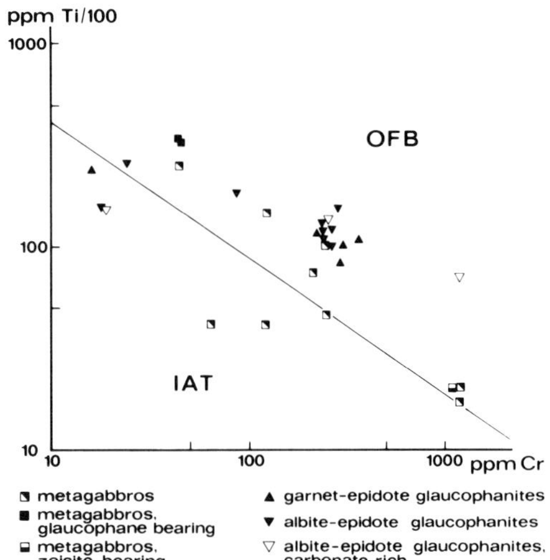

FIG. 3. Variation diagram Ti/100 vs. Cr. Limit between ocean floor basalts (OFB) and island arc tholeiites (IAT) after Pearce (1975).

Meta-sediments

Besides marbles, the dominant meta-sedimentary rocks on Samos are phyllites, quartz phyllites, albite phyllites, and calc-phyllites consisting of white-micas, chlorite, quartz, albite, calcite, and ankerite in variable proportions. We will only describe the relatively subordinate meta-sediments containing those metamorphic mineral assemblages which help to constrain P-T estimates.

Glaucophane-bearing meta-sediments

Glaucophane is not restricted to blueschists of basaltic origin, i.e. albite-epidote and garnet-epidote glaucophanites, but also occurs as a minor constituent in albite phyllites, calc-phyllites, and marbles. Occurrences are distributed over the whole island. Typical assemblages are:
qz-alb-w.mica-chlor-ep-glauc-sph;
ank ± *cc-w.mica-chlor-ep-glauc-sph;*
ank ± *cc-alb-w.mica-chlor-ep-glauc -sph* (*-qz*).

Chloritoid-bearing meta-sediments

In the marble-phyllite sequence of the Vourliotes and the Ampelos unit, chloritoid-bearing metasediments are widely distributed although generally only as minor intercalations. Chloritoid phyllites are widespread with the assemblage:
chltd-chlor-mus-par-qz-tourm-hem.

In the Ampelos area, chloritoid phyllites also contain abundant epidote. Mposkos & Perdikatzis (1981) describe chloritoid-bearing calc-phyllites in the southwestern foothills of Kerketeas mountain, containing:

chltd-chlor-mus-par-qz-cc(-ep).

Chloritoid in these assemblages contains between 58 and 77 mole % of the Fe-chloritoid molecule, and $\leqslant$3.5 mole % of Mn-chloritoid (microprobe analyses of Mposkos & Perdikatzis 1981, and our own results). Near Koumaradhei and near the town of Samos, we found chloritoid-rich ankerite marbles with the assemblage:

chltd-ank(-cc)-mus-qz.

The chloritoid porphyroblasts, up to 4 mm long, contain about 80 mole % of Fe-, and 0.5 mole % of Mn-chloritoid.

Quartzitic phyllites with chloritoid and/or kyanite are restricted to the eastern part of the island, especially to the Samos-Vathy area (Schneider 1914). They contribute the additional assemblages (± *tourm*, + *hem*):

chltd-ky (-pyroph)-mus-par-chlor-qz

ky (-pyroph)-mus-par-chlor-qz.

Similar kyanite-chloritoid-bearing meta-sediments of the schist cover of the Menderes complex near Selçuk, Turkey, about 30 km NE of Vathy, were shown to one of us (M.O.), by Dr E. Başarir (University of Izmir) in 1981. The assemblage recognized in one sample from the beach of Psili Ammos (East Samos) was:

chltd-Mg-carph-ky (-pyroph)-mus-par-chlor-qz.

The Mg-carpholite has the composition Fe-$Cph_{15.5}$Mn-Cph_3Mg-$Cph_{81.5}$, the coexisting chloritoid Fe-Ctd_{46}Mn-Ctd_9Mg-Ctd_{45}. Still richer in Mg and Mn is chloritoid which coexists with kyanite alone: Fe-Ctd_{37}Mn-Ctd_{13}Mg-Ctd_{50}.

The last group of chloritoid-bearing metasediments comprise the diasporites of eastern Samos. Typical assemblages are (Mposkos 1978):

dias-chltd-hem-mus-par-tourm-rut

dias-chltd-hem-mus-tourm-rut

dias-chltd-cc-rut.

Chlorite and spinel may be present as additional phases; kaolinite is widespread in minor amounts and is probably of secondary origin.

In contrast, the emery occurrence in the Kerketeas area contains no chloritoid. Mposkos (1978) describes the assemblage:

cor-dias-hem-ilm-rut-spin.

In one sample, he also observed kyanite.

Phyllonitic garnet-mica-schists

Garnet-mica-schists are confined to a relatively small area on the north coast of the island, between Kondakeika and Ampelos (Fig. 1). They are characterized by the occurrence of rotated garnet porphyroblasts, up to 10 mm across, some of which show beautiful 'snowball' texture. These older garnets are surrounded by smaller, idiomorphic garnets of a second generation. Some samples contain very fine grained aggregates of white-mica which are perhaps derived from primary plagioclases. The most frequent assemblage is:

qz-mus-par-chlor-gar-tourm-hem.

Albite, epidote, glaucophane, sphene, rutile, and calcite may be present as additional phases. The kyanite-bearing garnet-mica schists described by Schneider (1914) have not been re-located. Textures in the garnet-mica-schists appear to reflect two distinct episodes of metamorphism.

Petrogenetic considerations

Glaucophane is an important mineral not only in metabasites, but also in meta-sedimentary layers of the Ampelos and Vourliotes unit. In either case, it is frequently associated with albite, but not with jadeitic pyroxene. This indicates elevated, but not very high pressures of metamorphism. The possible pressure range, defined by the lower stability limit of albite (e.g. Holland 1979) and the uppermost breakdown limit of glaucophane (Maresch 1977) is relatively wide (Fig. 4). The meta-bauxites and aluminous meta-sediments of eastern Samos which are intercalated with glaucophane-schists in the Vourliotes unit establish the temperature of formation within narrow limits. Moreover, the rare occurrence of the assemblage Mg-carpholite-chloritoid-kyanite in this area is another indicator of high pressure (Seidel 1981; Chopin & Schreyer 1983, Fig. 4, 5).

Since the metabauxites contain diaspore and no corundum, the upper temperature limit is given by the univariant equilibrium curve of the reaction:

(1) diaspore = corundum + H_2O

(Haas 1972). In the kyanite-chloritoid schists, kyanite and quartz are frequently observed in mutual contact indicating that the upper stability limit of pyrophyllite according to the reaction:

(2) pyrophyllite = kyanite + quartz + H_2O

(Haas & Holdaway 1972) was crossed. On the other hand, most of the chloritoid-kyanite schists contain pyrophyllite. Although a retrograde formation of this mineral cannot be excluded, there is textural evidence that at least part of the pryophyllite was formed, together with kyanite, by the prograde reaction:

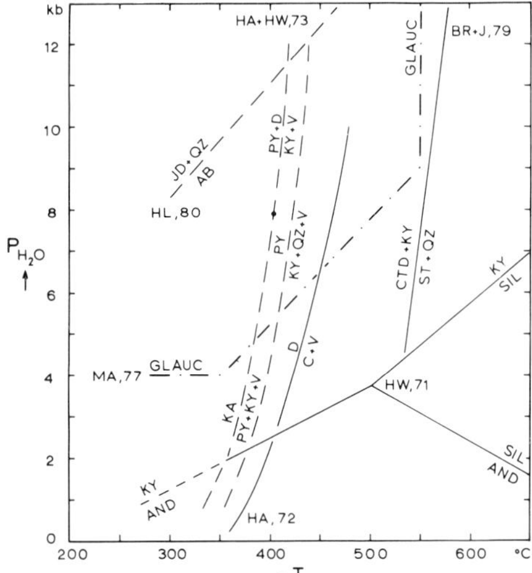

FIG. 4. P_{H_2O}-T diagram. Upper stability limits of kaolinite and pyrophyllite after Haas & Holdaway (1973), of diaspore after Haas (1972), of jadeite + quartz after Holland (1979), of chloritoid + kyanite after Bhaskar Rao & Johannes (1979); stability fields of Al silicates after Holdaway (1971), of glaucophane after Maresch (1977).

AB, albite; AND, andalusite; C, corundum; CTD, chloritoid; D diaspore; GLAUC, glaucophane; J, jadeite; KA, kaolinite; KY, kyanite; PY, pyrophyllite; QZ, quartz; SIL, sillimanite; ST, staurolite; and V, water vapour.

(3) kaolinite = 2 pyroph. + 4 ky. + 10 H_2O, which takes place at slightly lower temperatures than reaction (2), see Fig. 4. The possible stable coexistence of pyrophyllite + kyanite + quartz would correspond to the univariant equilibrium (2) and fix the P-T range of metamorphism in eastern Samos between 410°C/5.5 kbar and 440°C/12 kbar, if we assume $P_{H_2O} = P_{tot}$ and consider the stability limits of albite and glaucophane (Fig. 4).

In western Samos, temperatures were presumably higher by about 50°C judging from the (? stable) coexistence of diaspore and corundum in meta-bauxites (Mposkos 1978; Mposkos & Perdikatzis 1981); the presence of ilmenite with haematite and rutile (Mposkos 1978) also points to higher temperatures.

The meta-gabbros of Samos show a textural resemblance to the meta-gabbros of Syros which are associated with eclogites, glaucophanites, and serpentinites (Ktenas 1908; Dixon 1976). In their chemical composition, the eclogites of Syros are similar to the high FeO^{tot}/MgO meta-gabbros of Samos i.e. they have high FeO^{tot} and TiO_2, whereas the one meta-gabbro of Syros so far analysed falls in the group of the low FeO^{tot}/MgO meta-gabbros (M. Okrusch & E. Seidel, Pers. com. 1982). Altherr & Seidel (1977) interpret the high-pressure ophiolites of Syros as a suite of trench rocks which testify to the closure of the southern Tethyan ocean by continent-continent collision, during the Eocene. Such a model may also apply for the ophiolitic rocks of Samos.

ACKNOWLEDGEMENTS: Thanks are due to Deutsche Forschungsgemeinschaft, Bonn, for material financial support. We thank O. Ewald, J. Koepke, P. Meyer, E. Ockenga (Braunschweig), and Mrs R. Baur (Würzburg) for their technical assistance, and K.-P. Kelber (Würzburg) for the line drawings. The paper benefited from a critical review by John Dixon (Edinburgh) which is gratefully acknowledged.

References

ALTHERR, R. 1981. *Zur Petrologie der miozänen Granitoide der Zentralägäis (Griechenland)*. Habilitationsschrift, T. U. Braunschweig, 218 pp.

—— & SEIDEL, E. 1977. Speculations on the geodynamic evolution of the Attic-Cycladic crystalline complex during Alpidic times. *In*: KALLERGIS, G. (ed.). *Proc. VIth Coll. Geol. Aegean Region, Athens 1977*, 347–352.

——, KREUZER, H., WENDT, I., LENZ, H., WAGNER, G. A., KELLER, J., HARRE, W. & HÖHNDORF, A. 1982. A Late Oligocene/Early Miocene high temperature belt in the Attic-Cycladic crystalline complex (SE Pelagonian, Greece). *Geol. Jahrb.* **E23,** 97–164.

ANDRIESSEN, P. A. M., BOELRIJK, N. A. I. M., HEBEDA, E. A., PRIEM, H. N. A., VERDURMEN, E. A. TH. & VERSCHURE, R. H. Dating the events of metamorphism and granitic magmatism in the Alpine orogen of Naxos (Cyclades, Greece). *Contrib. Mineral. Petrol.* **69,** 215–225.

BHASKAR RAO, B. & JOHANNES, W. 1979. Further data on the stability of staurolite + quartz and related assemblages. *Neues Jahrb. Mineral. Monatsch.*, **1979,** 437–447.

BUTZ, J. 1912. Die Eruptivgesteine der Insel Samos. *Centralbl. Mineral. Geol. Paläontol.* **1912,** 609–615.

CHOPIN, C. & SCHREYER, W. 1983. Magnesiocarpholite and magnesiochloritoid: Two index minerals of pelitic blueschists and their preliminary phase relations in the model system MgO-Al_2O_3-SiO_2. *Am. J. Sci.*, Orville Vol. (in press).

DIXON, J. E. 1976. Glaucophane schists of Syros Greece. *Bull. Soc. géol. France*, **18,** 280.

Dürr, St., Altherr, R., Keller, J., Okrusch, M. & Seidel, E. 1978. The median Aegean crystalline belt: stratigraphy, structure, metamorphism, magmatism. In: Closs, H., Roeder, D. & Schmidt, K. (eds) Alps, Apennines, Hellenides. *Inter-Union Comm. Geodyn. Sci. Rep.* **38,** 455–477.

Haas, H. 1972. Equilibria in the system Al_2O_3-SiO_2-H_2O involving the stability limits of diaspore and pyrophyllite, and thermodynamic data on these minerals. *Am. Mineralogist* **57,** 1375–1385.

—— & Holdaway, M. J. 1973. Equilibria in the system Al_2O_3-SiO_2-H_2O involving the stability limits of pyrophyllite, and thermodynamic data of pyrophyllite. *Am. J. Sci.* **273,** 449–456.

Holdaway, M. J. Stability of andalusite and the aluminium silicate phase diagram. *Am. J. Sci.* **271,** 97–131.

Holland, T. J. B. 1979. Experimental determination of the reaction paragonite = jadeite + kyanite + H_2O, and internally consistent thermodynamic data for part of the system Na_2O – Al_2O_3 – SiO_2 – H_2O, with applications to eclogites and blueschists. *Contrib. Mineral. Petrol.* **68,** 293–301.

Ktenas, K. A. 1908. Die Einlagerungen im krystallinen Grundgebirge der Kycladen auf Syra und Sifnos. *Tschermaks mineral. Petrogr. Mitt.* (*N.F.*) **26,** 257–320.

Maresch, W. V. 1977. Experimental studies on glaucophane: an analysis of present knowledge. *Tectonophysics*, **43,** 109–125.

Miyashiro, A. 1975. Classification, characteristics, and origin of ophiolites. *J. Geol.* **83,** 249–281.

—— & Shido, F. 1980. Differentiation of gabbros in the Mid-Atlantic ridge near 24°N. *Geochem. J.* **14,** 145–154.

Mposkos, E. 1978. Diasporit- und Schmirgelvorkommen der Insel Samos (Griechenland). *4th Internat. Congr. for the Study of Bauxites, Alumina, and Aluminium*, **2,** 614–631.

—— & Perdikatzis, V. 1981. Die Paragonit-Chloritoid führenden Schiefer des südwestlichen Bereiches des Kerkis auf Samos (Griechenland). *Neues Jahrb. Mineral. Abhandl.* **142,** 292–308.

Okrusch, M. 1981. Chloritoid-führende Paragenesen in Hochdruckgesteinen von Samos. *Fortschr. Mineral.* **59,** Beiheft 1, 145–146.

Papanikolaou, D. 1979. Unités tectoniques et phases de déformation dans l'îsle de Samos, Mer Egée, Grèce. *Bull. Soc. géol. France*, **21,** 745–752.

Pearce, J. A. 1975. Basalt geochemistry used to investigate past tectonic environments on Cyprus. *Tectonophysics*, **25,** 41–67.

Schneider, K. 1914. *Die kristallinen Schiefer der Insel Samos.* Inaugural-Diss. Univ. Münster, 48 pp.

Seidel, E. 1981. Fe–Mg-Verteilung in koexistierenden Karpholiten und Chloritoiden. *Fortschr. Mineral.* **59,** Beiheft 1, 180–181.

Theodoropoulos. D. 1979. *Geol. Map of Greece 1:50.000, Island of Samos,* N.I.G.M.R. Athens.

High pressure metamorphism in Ghiaros Island, Cyclades, Greece

Christos G. Katagas

SUMMARY: The various rock types encountered in Ghiaros island have suffered an episode of progressive regional metamorphism which produced high pressure low temperature assemblages with Na-pyroxene and glaucophane. Later in their metamorphic history the rocks were affected by retrograde metamorphism. Glaucophane crystals became continuously zoned and sodic pyroxene crystals developed a patchy composition. Zoning in amphiboles extends from glaucophanitic cores through intermediate Na-Ca winchite compositions to actinolite rims. The compositional differences observed between the various patches in individual pyroxene crystals are interpreted to reflect contrasting P-T conditions which drive the continuous reaction jadeitic pyroxene + quartz $\leftrightharpoons$ albite + acmitic pyroxene + quartz, to progressively less jadeite-rich pyroxene.

Geological setting

Ghiaros island belongs to the Cycladic group of islands which constitutes part of the Attico-cycladic crystalline complex (Fig. 1). The latter is thought to consist of the least two main tectonic units (Dürr *et al.* 1978). The lower unit consists of a sequence of thrust sheets made up of meta-sediments and meta-volcanics which, at least in part, are Mesozoic in age. The rocks of the lower tectonic unit were affected by an Eocene high P/low T metamorphic event which was followed in the Oligocene/Miocene by a medium P/T event and the intrusion of granites (Andriessen *et al.* 1979; Altherr *et al.* 1979; Altherr *et al.* 1982).

The island of Ghiaros is covered by a series of quartzites, phyllites and quartz-mica schists with local intercalations of relatively thin marble horizons and metabasic layers. Quartz-mica schists and quartzites are the dominant lithologies but the various rock types encountered on the island show a tendency to alternate and commonly grade into each other. The metabasic rocks are also schistose and foliated and their layering is conformable with that in the meta-sediments. The whole series is intensely folded with axial planes dipping to the NE.

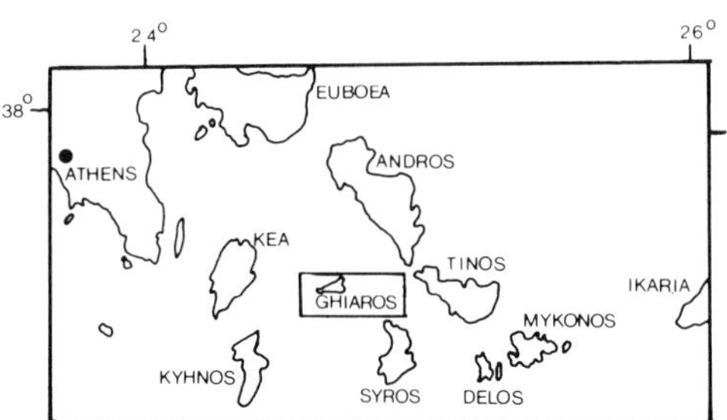

Fig. 1. Location map of Ghiaros island.

Petrography and phase relations

Semi-pelitic rocks were recrystallized to quartz-chlorite-albite-muscovite assemblages, whereas sediments in which carbonate was present recrystallized to the common assemblage quartz-chlorite-calcite. The latter is usually accompanied by minor amounts of muscovite, albite and epidote. The more quartzo-feldspathic rock types give rise to quartz-albite-sodic amphibole-bearing rocks.

The metabasic rocks show no evidence of their original texture and origin. The most commonly encountered metabasic rocks may be classified into the following three groups, on the basis of the mineral phases identified in individual thin sections. It must be noted however that the co-existing minerals listed below do not necessarily represent equilibrium assemblages.

Group A

Quartz + chlorite + albite + epidote + glaucophane + Na-pyroxene + Fe-oxides.

Quartz + chlorite + albite + epidote + zoned glaucophane + Na-pyroxene + calcite + Fe-oxides.

Quartz + chlorite + albite + epidote + actinolite + Na-pyroxene + calcite + Fe-oxides.

Group B

Quartz + chlorite + albite + epidote + glaucophane + sphene + Fe-oxides + calcite.

Quartz + chlorite + albite + epidote + zoned glaucophane + muscovite + Fe-oxides.

Group C

Quartz + chlorite + albite + pumpellyite + zoned glaucophane + stilpnomelane + epidote + Fe-oxides.

Quartz + chlorite (2) + albite + pumpellyite + actinolite + epidote + stilpnomelane + Fe-oxides.

Quartz + chlorite (2) + albite + pumpellyite + actinolite + epidote + stilpnomelane + sphene + calcite + Fe-oxides ± muscovite.

Rock textures indicate that these rocks have suffered at least two episodes of recrystallization and deformation whereas a third deformational episode produced a strain-slip cleavage. The early foliation S_1 is only rarely visible and its trend is recognized by rare muscovite flakes. The S_2 foliation is outlined by interleaved muscovite and chlorite flakes and also in the more quartzo-feldspathic and metabasic rocks by glaucophane or actinolite prisms participating in the formation of asymmetrical (F_3) microfolds. The segregation of prismatic crystals of these minerals along the limbs of the microfolds produces a strain-slip schistosity (S_3). Most of the muscovite or amphibole crystals which follow the S_2 are bent and broken in the hinges of the microfolds; they are therefore considered to have grown prior to F_3. Some of the phases which are apparently developed along the S_2 foliation (e.g. chlorite, quartz, albite, pumpellyite) may in fact be post-tectonic to F_3 but grown mimetically along the pre-existing S_2 foliation. In a number of meta-sediments in particular, bent muscovite flakes participating in the formation of the F_3 microfolds are interwoven with unstrained muscovite flakes of markedly different composition. The co-existence of two different muscovite phases has been verified by X-ray diffraction study and electron probe analyses; some of the X-rayed samples show a distinct separation of (060) muscovite peaks with b_o values around 9.054 and 9.035 Å or show a very broad (060) peak suggesting also the coexistence of muscovites differing in composition. According to the electron probe analyses presented in Table 1 the true S_2 muscovites are richer in celadonite molecule (24%) relative to those believed to have grown mimetically along S_2 (7%). Similar evidence supporting the growth in a single sample of a mineral phase showing intergrain inhomogeneity, presumably at different stages of its metamorphic evolution, is provided by the coexistence in some of the C group metabasic rocks of two different chlorite phases (see Table 1). All the analyses presented in Tables 1 and 2 were performed at the Department of Geology, University of Manchester, using a Cambridge Instrument Geoscan, fitted with a Link Systems model 290-2 kv energy-dispersive spectrometer and 2AF-4/FLS quantitative analysis software system.

TABLE 1. *Microprobe analyses of the muscovite phases coexisting in sample Y19A and of the chlorite phases coexisting in samples YB19 and Y83*

	muscovite		chlorite			
	Y19A(1)	Y19A(2)	YB19(1)	YB19(2)	Y83(1)	Y83(2)
SiO_2	45.97	52.70	27.64	26.43	26.97	25.80
TiO_2	0.63	—	—	—	—	—
Al_2O_3	35.95	23.03	17.90	18.96	18.86	19.27
FeO	1.75	2.49	22.98	25.41	24.66	26.54
MgO	0.51	4.40	16.52	14.46	15.56	14.17
MnO	—	—	0.37	0.28	0.46	0.35
CaO	0.19	0.16	—	—		
Na_2O	0.41	—	—	—		
K_2O	9.93	9.45	—	—		
Total	95.34	92.43	85.41	85.54	86.51	86.13
Si	6.11	7.19	5.90	5.72	5.74	5.59
Al^{IV}	1.89	0.81	2.10	2.28	2.26	2.41
Al^{VI}	3.74	2.89	2.40	2.55	2.47	2.51
Ti	0.06	—	—	—	—	—
Fe	0.19	0.28	4.10	4.60	4.39	4.81
Mn	—	—	0.06	0.05	0.08	0.06
Mg	0.10	0.89	5.26	4.66	4.93	4.57
Ca	0.03	0.03	—	—	—	—
Na	0.11	—	—	—	—	—
K	1.69	1.65	—	—	—	—
Cel%	7.00	24.00		—		

TABLE 2. *Electron probe analyses of coexisting phases in samples YB19 and Y83*

	Actinolite		Epidote		Pumpellyite	
	YB19	Y83	YB19	Y83	YB19	Y83
SiO_2	56.00	55.31	37.78	38.17	37.02	37.12
Al_2O_3	0.45	0.84	22.97	23.66	23.35	24.47
FeOT	11.74	14.31	—	—	6.15	5.10
Fe_2O_3T	—	—	12.28	12.74	—	—
MgO	15.22	14.72	—	—	1.96	1.98
MnO	0.32	0.29	—	0.27	0.36	0.42
CaO	12.30	12.08	24.00	24.27	22.94	23.55
Na_2O	0.32	0.82	—	—		
K_2O	—	—				
TiO_2	—	—				
	96.36	98.38	97.00	99.11	91.78	92.64

T = Total iron.

Rocks of group A are characterized by the presence of a pale green sodic pyroxene. The sodic pyroxene occurs as ragged grains which are turbid as a result of small irregular quartz and albite secondary inclusions and are inhomogeneous with patches differing in their jadeite-acmite-augite proportions. Compositions of pyroxene grains from two samples (YB1, Y81) are plotted in Fig. 2. The estimation of Fe^{3+} was made following the method of Hamm & Vieten (1971). The analysed patches range from aegirine-jadeite through chloromelanite, to aegirine-augite compositions. The YB1 plots represent analyses of patches recorded in two crystals whereas Y81 plots are

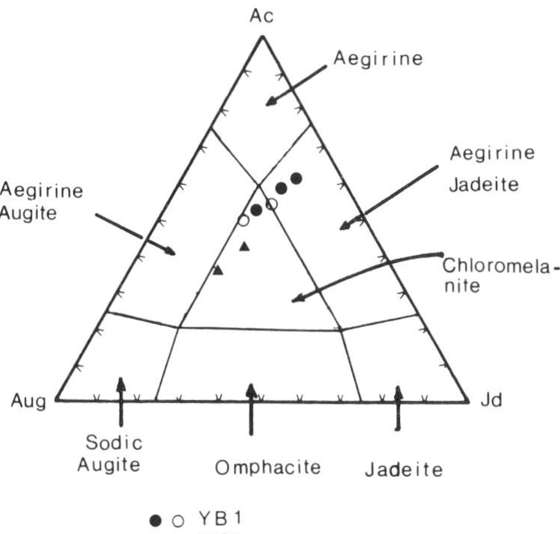

FIG. 2. Plots of sodic pyroxene compositions on the acmite-augite-jadeite diagram YB1 plots represent analyses of patches recorded in two crystals. Y81 plots are compositions of patches found in a single crystal.

compositions of patches found in a single crystal. Parts of the pyroxene crystals are often replaced by chlorite or in some cases pyroxene margins are in replacement contact with actinolite or the actinolite-rich rims of zoned amphiboles having glaucophanitic cores. Rim compositions of the zoned amphiboles approximate to the compositions of the apparently post-tectonic thin prismatic or acicular actinolitic amphiboles, although zoning in amphiboles seems to have formed prior to F_3 folds as this folding affects the zoned crystals. In some specimens however (e.g. YB1) idioblastic slightly zoned glaucophane crystals have been found growing over a fine grained sodic pyroxene-chlorite matrix, testifying to a late formation of a sodic amphibole through a reaction between these minerals. Some of the group A rocks contain polycrystalline chloritic aggregates with the external morphology of garnet. Psarianos & Charalambakis (1951) reported the occurrence in Ghiaros of a rock with zoned 'augite'-garnet-quartz-glaucophane and muscovite with subordinate plagioclase, sphene and chlorite. They note also that glaucophane and muscovite have possibly formed later than garnet and 'augite'.

The textures described for the group A rocks suggest that minerals coexisting in a single thin section do not represent stable assemblages. The early metamorphic phases present were possibly aegirine-jadeite pyroxene, garnet and glaucophane. The presence of quartz and albite inclusions which are younger phases growing in the host pyroxene suggests that later in the metamorphic history of the rocks, the aegirine-jadeite became compositionally inhomogeneous perhaps through a continuous reaction of the type:

aegirine-jadeite + quartz = albite + aegirine augite (Dixon, 1968; Okrusch *et al.* 1978).

Margins of pyroxene crystals with compositions near the boundary chloromelanite-aegirine augite and with jadeite content 22% have been found in contact with albite and quartz along sharp boundaries (Fig. 3) and are interpreted to comprise an equilibrium assemblage. Glaucophane developed actinolite rich rims during an intermediate stage, perhaps through a continuous reaction involving combinations of garnet, glaucophane, jadeitic pyroxene, epidote and quartz; garnet, if present, was finally replaced by chlorite. The random orientation of actinolite, chlorite and poiciloblastic epidote suggest that these minerals are the youngest secondary post-kinematic products.

Rocks of groups B and C differ mainly in that group C rocks contain two chlorite phases, pumpellyite, actinolite and stilpnomelane

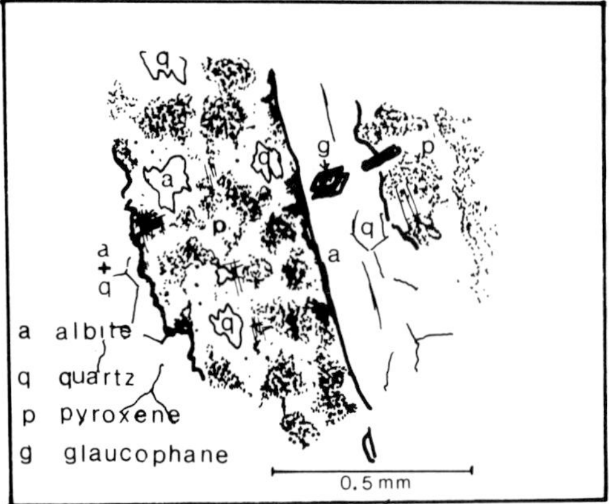

FIG. 3. Sketch drawn from a photomicrograph showing the characteristic sharp and smooth boundaries between a Na-pyroxene crystal and albite + quartz. Note also the secondary quartz and albite inclusions and the patch developed in the pyroxene crystal.

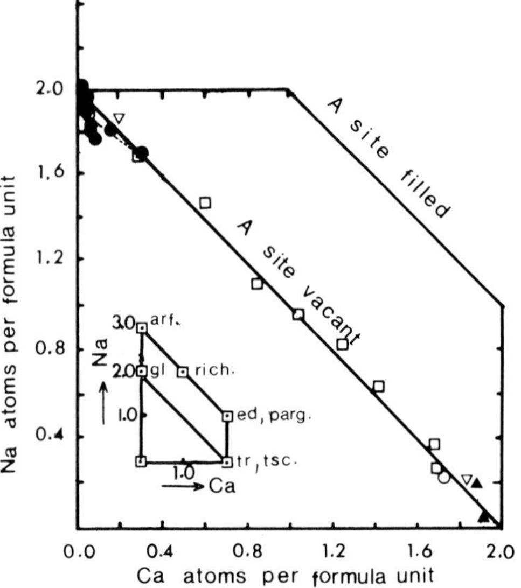

FIG. 4. Distribution of the Ca and Na cations in the M_4 site of amphiboles from Ghiaros. Y81: *quartz + albite + epidote + zoned glaucophane + Na pyroxene + chlorite + muscovite + calcite + Fe-oxides.* Y104: *quartz + albite + epidote + zoned glaucophane + chlorite + muscovite + Fe-oxides.* Y137: *quartz + chlorite + albite + epidote + zoned glaucophane + pumpellyite + chlorite + calcite + sphene + Fe-oxides.* YB1: *quartz + chlorite + albite + epidote Na-pyroxene + glaucophane + Fe-oxides.* Y83: *quartz + chlorite* (2) *+ albite + pumpellyite + actinolite + stilpnomelane + epidote + Fe-oxides.*

whereas rocks of group B contain glaucophane or a zoned amphibole with glaucophane-rich cores and actinolite-rich rims and epidote (Ps 25–30%). The latter mineral is found in small undeformed poikiloblasts showing trains of inclusions in continuity with an earlier crenulated foliation. Epidote may also be present in C-group rocks but in subordinate amounts and is never found in contact with pumpellyite. In a few of the C-group rocks rare zoned amphiboles have also been preserved.

Electron microprobe analyses on various spots of zoned amphibole crystals from A and B group of rocks reveal extremely extensive $Na \rightleftharpoons Ca$ substitution in the M_4-sites and Fe^{2+}, $Al \rightleftharpoons Mg$, Fe^{3+} substitution in the M_1, M_2 & M_3 sites, leading from glaucophanitic or crossitic cores through intermediate Na-Ca-winchite compositions to actinolitic rims. The estimation of Fe^{2+} and Fe^{3+} in the amphiboles, and calculated formulae are based on the assumption that $Si + Al + Ti + Fe^{3+} + Mn + Mg = 13$ (cf. Stout, 1972). The distribution of the Ca and Na cations in the M_4 site of core and rim amphiboles is given in Fig. 4. As shown in this figure the A sites of the analysed calcic and sodic amphiboles are vacant and the $Na \rightleftharpoons Ca$ substitution, as characteristically depicted by a series of spot analyses in amphibole Y81, is extremely extensive.

Since the chlorite group is thought to form a continuous solid solution between its various members, the presence in the same sample (e.g. YB19, Y83) of two chlorite phases differing in their compositions is taken to indicate disequilibrium. Analyses of coexisting chlorites from two group-C samples are given in Table 1. Intragrain compositional differences have not been detected and each of the analyses presented in this table is a mean of at least five spot analyses performed on the same grain. Analyses of other coexisting minerals from the same samples are given in Table 2. Pumpellyite is compositionally homogeneous and is found in contact with the chlorite having the higher Fe/Fe + Mg ratio and actinolite. The iron-rich chlorite is also found in contact with surrounding actinolite. Plots of the chemical compositions of the mineral phase observed in samples YB19 and Y83 on an ACF diagram (Fig. 5) reveals crossing of the tie lines epidote-Fe-Mg-

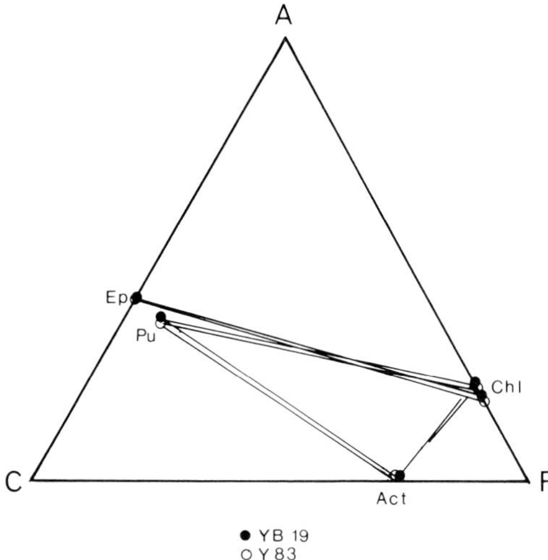

FIG. 5. Plots of coexisting minerals in samples YB19 and Y83 on an ACF diagram.

chlorite and pumpellyite-Fe-rich chlorite suggesting the reaction epidote + $chlorite_{(1)}$ = pumpellyite + $chlorite_{(2)}$, in which $chlorite_{(2)}$ is more iron-rich than $chlorite_{(1)}$.

Little can be said at this stage on the reaction which relates actinolite-pumpellyite bearing rocks to glaucophane-epidote bearing ones. Details on the chemical and physical constraints for the above reaction will be reported elsewhere. However, on the basis of textual observations on assemblages transitional between those cited for groups B and C and the available chemical data (Katagas, 1981) one may speculate that a complex reaction involving glaucophane, Fe + Mg chlorite and epidote on one side and actinolite, pumpellyite, albite and Fe-rich chlorite on the other, is involved.

Discussion

Physical conditions attending the formation and recrystallization of the Ghiaros metamorphic rocks are difficult to evaluate because of the high variance of the assemblages and the textural complexities. However, some critical data will be discussed below in an attempt to assess the probable evolution of P/T conditions during the metamorphism.

Although incompatible assemblages are common in the Ghiaros rocks, phase equilibrium was approached at least for some mineral pairs. Thus the K_D (Fe^{2+}/Mg) values for chlorite-actinolite pairs show a mean of 1.66 which conforms remarkably well with that usually reported (1.7) for pairs of the pumpellyite-actinolite facies (cf. Kawachi, 1975; Coombs *et al.* 1976; Katagas & Panagos, 1979).

X-ray diffraction study on the K-white micas from about 40 samples of suitable composition revealed that the mean value of their b_o parameter is 9.0415 Å ($\sigma = 0.006$). According to the studies of Sassi & Scolari (1974), Sassi *et al.* 1974 and Iwasaki *et al.* (1978), such a value suggests formation of their host rocks under high pressure conditions. It must be noted, however, that in the meta-sediments in which two muscovite phases are developed, the younger, less celadonite-rich muscovite phase has a significantly lower b_o value (9.035 Å). The latter value approximates to the b_o value (9.032 Å) of muscovites from similar rocks of the upper chlorite zone of the Sanbagawa terrain of Japan reported by Iwasaki *et al.* (1978). P-T estimates for the latter greenschist-blueschist facies rocks by Iwasaki *et al.* (1978) suggest pressures of about 7 kb and temperatures around 400°C. Although absolute values for the P-T conditions of metamorphism for the Ghiaros meta-sediments cannot be deduced, the presence of two muscovite phases with different b_o values is interpreted to represent growth of these phases under markedly different conditions. According to Guidotti & Sassi (1976) a change of the b_o value of this sort is taken to suggest growth of the second phase either under lower pressures or under higher temperatures.

The best pressure estimate can be made by using the compositions of Na-pyroxenes coexisting with albite and quartz in metabasic rocks. An analysis performed on the margins of the Na-pyroxene depicted in Fig. 3 shows its M_1 site to contain 0.538 Fe^{3+}, 0.253 A1, 0.24 others and no Al^{IV}. If we use the convention $100\,Fe^{3+}/Fe^{3+} + (Al^{VI} - Al^{IV}) = \%Ac$ in the acmite-jadeite part of the solution, then the pyroxene has an acmite content of 68%. On the basis of these data and Fig. 4 of Popp & Gilbert (1972), the pressure of formation of the Ghiaros rocks must have been around 7 kb at 300°C or 9 kb at 400°C. However, the compositional differences observed between the various patches of individual pyroxene crystals (Fig. 2) may be interpreted to reflect contrasting P-T conditions which drive the continuous reaction jadeitic pyroxene + quartz $\rightleftharpoons$ albite + acmitic pyroxene + quartz to progressively less jadeite-rich pyroxene. Hence the compositional differences between the patches may in fact be a result of a drop in pressure and/or as Okay (1978) argues, a decrease in temperature. Temperatures prevailing at the peak of metamorphism must have been relatively high since lawsonite has not been found so far in Ghiaros and garnet (almandine?) was certainly present. Furthermore, all the metabasic rocks are coarsely crystalline and lack textural relics of the original igneous rocks, a feature frequently reported in other metamorphic terrains of similar grade (Okay, 1980).

The effects of the retrogressive episode on the prograde metamorphic assemblages are very well preserved in some amphibole-bearing rocks. In such rocks, glaucophane tends to disappear and to be relaced by actinolite, or most commonly, the amphibole crystals become continuously zoned, with a glaucophane-rich core gradually mantled towards the rims by a more calcic actinolitic amphibole. This is taken to suggest that the crystal growth was in progress while pressure and temperature conditions were gradually changing (Laird & Albee, 1981).

According to presently accepted ideas, actinolite is considered to be formed under lower pressure conditions than glaucophane, therefore, the Ca-rich rims of the zoned crystals should have equilibrated under conditions corresponding to lower pressures or to a set of P-T conditions with a T/P ratio higher than that prevailing during the formation of the glaucophane-rich internal parts of the crystals (Brown 1974, 1977).

The lower temperature conditions for the characteristic pumpellyite-actinolite facies assemblage quartz + albite + chlorite + actinolite + pumpellyite, according to the model proposed by Nakajima *et al.* (1977), is around 275°C and its upper limit around 390°C. Since potentially useful geothermometers do not occur in the samples studied, the maximum and minimum temperatures attained are difficult to determine.

Conclusions

The data cited above suggest that the metasediments and metabasic rocks of Ghiaros have been originally metamorphosed under conditions corresponding to the upper blueschist facies. Na-pyroxene and Na-amphibole cores grew under high pressure/low temperature conditions. Later in their metamorphic history however, these rocks suffered retrogression. Amphibole zoning and patchy compositions of Na-pyroxene crystals are believed to record changes in P-T conditions during the subsequent metamorphic evolution of their host rocks. Similarly the coexistence of two chlorite phases and of two muscovite phases differing markedly in their caledonite content in the same sample and the obscure (060) reflections of K-white micas in most of the meta-sediments are considered to be due to their continuous approach to re-equilibration under gradually changing P-T conditions. This later retrograde event was either (a) at lower P and the same temperature (b) at higher temperature and the same pressure or (c) at lower pressure and temperature.

The high P assemblages formed during the peak of the prograde metamorphism in Ghiaros are commonly considered to characterize subduction zone conditions. According to the available data for the HP/LT parageneses, the rock pile must have been subducted to at least 28 km. The probable age of metamorphism for the HP/LT event of Ghiaros, as deduced from the correlation of the petrological and mineralogical character of the island with those of the adjoining areas from which isotopic dates on similar rocks are available (Wendt *et al.* 1977; Altherr *et al.* 1977; Andriessen, 1978; Andriessen *et al.* 1978; Bonneau *et al.* 1980) is around 40–45 Ma. The subduction processes, however, must have been active in the general area since Palaeocene times, as suggeted by the isotopic dating of minerals from southern Euboea (Bavay *et al.* 1980). The sequence of assemblages formed during the retrograde state of metamorphism, glaucophane/crossite–progressively less Na-rich amphiboles–actinolite and finally the appearance of pumpellyite–actinolite facies assemblages may be interpreted to represent gradual unloading at decreasing temperatures. This retrograde recrystallization may be considered as synchronous with the buoyant return of the subducted pile towards the surface after its detachment from the downgoing lithospheric slab (Ernst & Dal Piaz, 1978).

The available data suggest that extensive regional retrogression has occured not only in Ghiaros but in many other areas of the Cycladic massif; however, rarely is regional retrogression discussed in any detail (e.g. Okrusch *et al.* 1978; Kornprobst *et al.* 1979). Since a large number of the retrograde changes involve hydration reactions, a major factor controlling the retrograde metamorphism in this area is the influx of water. Thus, the incompatible assemblages which are commonly observed in adjoining rocks of similar bulk chemical compositions or disequilibrium assemblages, may in certain cases be attributed to variations in the composition and availability of the fluid phase.

ACKNOWLEDGEMENTS. Constructive criticism of this manuscript by J. E. Dixon, D. M. Laduron & J. Ridley is greatly appreciated. I also wish to thank the Department of Geology, University of Manchester where most of the analytical work was carried out.

References

ALTHERR, R., KELLER, J., HARRE, W., HOHNDORF, A., KREUZER, H., LENZ, H., RASCHKA, H. & WENDT, I. 1977. Geochronological data on granitic rocks of the Aegean Sea. Preliminary results. *In:* BIJU DIVAL, B. & MONTADERT, L. (eds.) *Structural History of the Mediterranean Basins*. Editions Technip, Paris, 317–318.

——, KREUZER, H., WENDT, I., LENZ, H., WAGNER, G. A., KELLER, J., HARRE, W. & HOHNDORF, A. 1982. A late Oligocene/early Miocene high temperature belt in the Attic–Cycladic crystalline complex (SE Pelagonian, Greece). *Geol. Jahrb.* **E23,** 97–164.

——, SCHLIESTEDT, M., OKRUSCH, M., SEIDEL, E., KREUZER, H., HARRE, W., LENZ, H., WENDT, I. & WAGNER, G. 1979. Geochronology of high pressure rocks in Sifnos (Cyclades, Greece). *Contrib. Mineral. Petrol.* **70,** 245–255.

ANDRIESSEN, P. A. M. 1978. Isotopic age relations within the polymetamorphic complex of the island of Naxos (Cyclades, Greece). *Verhandeling Nr. 3, ZWO-Laboratorium voor Isotopen-geologie, Amsterdam.*

——, BOELRIJK, N. A. I. M., HEBEDA, E. H., PRIEM, H. N. A., VERDURMEN, E. A. TH. & VERSCHURE, R. H. 1979. Dating the events of metamorphism and granitic magmatism in the Alpine orogen of Naxos (Cyclades, Greece). *Contrib. Mineral. Petrol.* **69,** 215–225.

BAVAY, D., BAVAY, P., MALUSKI, H., VERGELY, P. & KATSIKATSOS, G. 1980. Datations par la méthode $^{40}Ar/^{39}Ar$ de minéraux de métamorphisme de haute pression en Eubée du Sud (Grèce). Corrélations avec les évènements tectonométamorphiques des Hellénides internes. *C.r. Acad. Sc. Paris*, **290-D** 1051–1054.

BONNEAU M., BLAKE, M. C., GEYSSANT, J., KIENAST, J. R., LEPVRIER, C., MALUSKI, H. & PAPANIKOLAOU, D., 1980. Sur la signification de séries métamorphiques (schistes bleus) des Cyclades (Hellénides, Grèce). L'exemple de l'île de Syros. *C.r. Acad. Sc. Paris* **290-D** 1463–1466.

BROWN, E. H. 1974. Comparison of the mineralogy and phase relations of blueschists from the north Cascades, Washington and greenschists from Otago, New Zealand. *Bull geol. Soc. Am.* **85,** 333–344.

—— 1977. Phase equilibria among pumpellyite, lawsonite, epidote and associated minerals in low grade metamorphic rocks. *Contrib. Mineral. Petrol.* **64,** 123–136.

COOMBS, D. S., NAKAMURA, Y. & VUAGNAT, M. 1976. Pumpellyite-actinolite facies schists of the Taveyanne formation near Loeche, Valais, Switzerland, *J. Petrol.* **17,** 440–471.

DIXON, J. E. 1968. The glaucophane schists of Syros. Some pyroxene reactions and their possible significance. Mineral. Soc. Mtg. Cambridge [Abstract].

DÜRR, S., ALTHERR, R., KELLER, J., OKRUSCH, M. & SEIDEL, E. 1978. The Median Aegean crystalline belt: stratigraphy, structure, metamorphism, magmatism. In: CLOSS, H., ROEDER, D. & SCHMIDT, K. (eds.) Alps, Apennines, Hellenides, Inter-Union Commission on Geodynamics. Scientific report No. 38, Schweizerbart, Stuttgart, 455–477.

ERNST, W. G. & DAL PIAZ, G. V. 1978. Mineral parageneses of eclogite rocks and related mafic schists of the Piemonte ophiolite nappe, Breuill-St. Jaques area, Italian western Alps. *Am. Miner.* **63,** 621–640.

GUIDOTTI, C. V. & SASSI, F. P. 1976. Muscovites as a petrogenetic indicator mineral in pelitic schists. *N. Jb. Miner. Abh.* **127,** 97–142.

HAMM, H. M. & VIETEN, K. 1971. Zur Berechnung der kristallchemischen Formel und des Fe^{3+}-Gehaltes von Klinopyroxen aus Electronenstrahl-micro-analysen. *Neues Jb. Mineral. Monatsh.* 310–314.

IWASAKI, M., SASSI, F. P. & ZIRPOLI, G. 1978. New data on the K-white micas from the Sanbagawa metamorphic belt and their petrologic significance. *Jour. Japan. Assoc. Min. Pet. Econ. Geol.* **73,** 274–280.

KATAGAS, C., 1981. *Petrogenesis of metamorphic rocks of Ghiaros island.* Unpubl. Habil. Thesis. University of Patras, 150 pp. (in Greek with English summary).

—— & PANAGOS, A. G. 1979. Pumpellyite-actinolite and greenschist facies metamorphism in Lesvos island (Greece). *Tschermaks Min. Petr. Mitt.* **26,** 235–254.

KAWACHI, Y. 1975. Pumpellyite-actinolite and contiguous facies metamorphism in part of Upper Wakatipu district, South Island New Zealand. *N.Z.J. Geophys.* **18,** 401–441.

KORNPROBST., J., KIENAST, J. R. & VILMINOT, J. C. 1979. The high pressure assemblages at Milos, Greece. *Contrib. Mineral. Petrol.* **69,** 49–63.

LAIRD, J. & ALBEE, A. 1981. High pressure metamorphism in mafic schists from northern Vermont. *Am. J. Sci.*, **281,** 97–126.

NAKAJIMA, I., BANNO, S. & SUZUKI, T. 1977. Reactions leading to the disappearance of pumpellyite in low-grade metamorphic rocks of the Sanbagawa metamorphic belt in Central Shikoku, Japan. *J. Petrol.* **18,** 263–284.

OKAY, A. I. 1978. Sodic pyroxenes from metabasites in the Eastern Mediterranean. *Contrib. Mineral. Petrol.* **68,** 7–11.

—— 1980. Lawsonite zone blueschists and a sodic amphibole producing reaction in the Tavşanlı region, Northwest Turkey. *Contrib. Miner. Petrol.* **75,** 179–186.

OKRUSCH, M., SEIDEL, E. & DAVIS, E. N. 1978. The assemblage jadeite-quartz in the glaucophane rocks of Sifnos (Cyclades Archipelago, Greece). *N. Jb. Miner. Abh.* **132,** 284–308.

POPP, R. K. & GILBERT, M. C. 1972. Stability of acmite-jadeite pyroxenes at low pressure. *Am. Miner.* **57,** 1210–1231.

PSARIANOS, P. & CHARALAMBAKIS, S. 1951. Contribution to the geology of the island of Ghiaros. *Prakt. Akad. Athinon*, **26,** 237–258 (in Greek).

SASSI, F. P. & SCOLARI, A. 1974. The b_0 value of the potassic white micas as a barometric indicator in low-grade metamorphism of pelitic schists. *Contr. Mineral. Petrol.* **45,** 143–152.

——, ——, BOCQUET, J. & DAL PIAZ, G. V. 1974. L'utilité de la mésure de b_0 des micas blancs potassiques dans l'étude des ensembles métamorphiques. Application aux Alpes Occidentales. *Bull. Soc. géol. Fr.* **16,** 247 (Abstract).

STOUT, J. H. 1972. Phase petrology and mineral chemistry of coexisting amphiboles from Telemark, Norway. *J. Petrol.* **13,** 99–145.

WENDT, I. RASCHKA, H., LENZ, H., KREUZER, H., HOHNDORF, W., HARRE, W., WAGNER, G. A., KELLER, J. ALTHERR, R., OKRUSCH, M., SCHLIESTEDT, M. & SEIDEL, E. 1977. Radiometric dating of crystalline rocks from the Cyclades (Aegean Sea, Greece). *Proc. 5th European Coll. Geochron. Cosmochron. Isotope Geol., Pisa, Abstract.*

CHRISTOS G. KATAGAS, Department of Geology, University of Patras, Patras, Greece.

The significance of deformation associated with blueschist facies metamorphism on the Aegean island of Syros

John Ridley

SUMMARY: A thorough examination of the syn-metamorphic minor structures in the blueschists of the Aegean island of Syros allows a determination of the kinetics and nature of the deformation. The deformation is essentially related to ductile thrusting at up to 45 km depth.

The change in fold geometry with orientation, and the presence of a locally developed mineral lineation at a high-angle to the fold axes give two independent determinations of the direction of thrusting. The direction, 130 ± 10°, suggests a link between the tectonism here and that in the Izmir–Ankara zone of mainland Turkey.

The nature of the small-scale asymmetrical folding indicates shortening of the pile parallel to this thrust displacement. This, together with the suggestion from the metamorphic textures of increasing pressure and temperature during deformation suggest that the deformation is related to burial and is compressional, and hence is presumably related to a major collision event.

Syros forms part of the belt within the Attico-Cycladic crystalline massif that shows the highest grade blueschist facies metamorphism. This metamorphism has been dated at Lower Tertiary (Andriessen *et al.*, 1979, Blake *et al.*, 1981). The small and intermediate scale structures of the island have been studied, in part to give some 'pivot points' of control in any interpretation of the gross structure of the massif, and of the significance of the metamorphism and deformation within it.

Summary of the structural style and history

The observable deformation history of Syros is simple (see Ridley 1982 for a fuller description). There is a single 'high strain' deformation event synchronous with the blueschist facies metamorphism. Most of the structural features seen can be explained by deformation involving progressive, but evolving, shear within this single event. The cessation of deformation is almost exactly synchronous with the cessation of metamorphic recrystallization. Later deformation is low strain—open upright folding, and high level brittle faulting.

The high strain deformation produced an almost completely penetrative flat-lying and layer-parallel metamorphic foliation. The formation of this fabric is the earliest textural event recorded.

This metamorphic foliation is folded by scattered small to intermediate scale folds which are tight to isoclinal, and generally overturned to recumbent. Larger scale folds with the same geometry, or formed during the same interval of the deformation history, are not seen on Syros.

Folding prior to the consolidation of the continuous metamorphic fabric is seen only very rarely. This may be either an artefact of preservation or indicative of deformation without folding during that stage of the strain history.

Folds are invariably intrafolial on some scale (Fig. 2a). Rare fold interference patterns are seen, always of type three pattern (Ramsay 1967), with sensibly parallel fold hinges in outcrop (Fig. 2b).

The layer-parallel nature of the schistosity and the consistent sense of overturning and asymmetry of the small-scale folds across the whole island suggest that deformation resulted from thrust-sense shear in a low-angle crustal shear zone. Strain is concentrated along the short limbs of asymmetrical fold pairs (Fig. 2a). This is inconsistent with any model in which the folds seen are minor folds on the short limb of a megascopic, opposite-vergent, recumbent fold. The fold vergence observed on Syros is therefore regarded as the true vergence of the tectonism recorded.

Determination of the thrust displacement direction

The thrust displacement direction, or shear direction, cannot be inferred simply from the structures. The dominant lineation in outcrop is given by the hinges of the small-scale recumbent folds. The trend of this lineation varies from northeast through east to southeast across the island (Fig. 3).

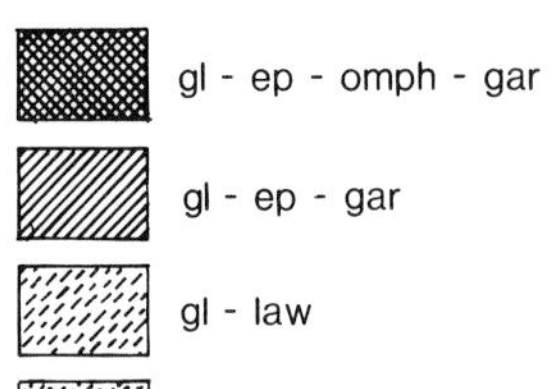

FIG. 1. The Attico-Cycladic crystalline massif emphasizing the division of the massify by blueschist metabasite assemblage—a qualitative measure of the grade of the early high pressure metamorphism

This variation is not the result of later deformation reorientating a once uni-directional fabric. The foliation is folded by broad open folds, but the trend of these, approximately northeast, is in the wrong orientation to produce patterns of the form seen (Fig. 3).

The folding loses its recumbent nature in the restricted belts where the hinge trend is close to southeasterly. Here the folding is upright and is much more pervasively developed than where recumbent, and is often symmetrical.

The variations in the fold trend and these concurrent qualitative variations in fold geometry can be simultaneously modelled if it is assumed that the average fold trend is approximately perpendicular to the thrust displacement, and that the total strain included a variable wrench shear component superimposed upon the thrust shear with the same displacement direction (Coward & Kim 1981, Sanderson 1982). All other components of the fabric are consistent with the displacement being approximately perpendicular to, rather than parallel to, the dominant lineation (Ridley, 1982).

The average fold trend of 070°, and the consistent 'z' asymmetry of the folding facing east suggests displacement approximately towards south or southeast. A more exact determination is found by analysis of details of the fold trend patterns, essentially using an analogue of the 'Hansen separation arc' method as applied to drag folds (Hansen 1971).

Where the fold trend turns sharply southeast, and the folds become symmetrical the wrench shear is presumably the dominant component in the total strain. These belts are ductile analogues of intrathrust-sheet tear faults as seen in high-level thrusting (Dahlstrom 1970).

The orientation of the most symmetrical folds therefore yields a close estimate of the tectonic displacement direction. Folds orientated to either side of this should show opposite apparent asymmetries (Fig. 4). The 'best fit' displacement direction for Syros is at $133 \pm 7°$.

An independent analysis of another component of the fabric leads to a comparable result. Locally on Syros there is a mineral lineation, predominantly in glaucophane, developed at a high-angle to the lineation given by the hinges of the small-scale folding. This fabric is only developed in rocks with isolated large glaucophane crystals in a quartz or mica matrix, and where the fold hinge trend is close to northeasterly, i.e. approximately perpendicular to the proposed displacement direction.

The relationship between this fabric and the fold hinge lineation is seen in Fig. 2c. Glaucophane prisms are aligned parallel to the

FIG. 2. Fold forms and textures in glaucophane-bearing schists on Syros.

(a) Asymmetrical fold pair with a strongly attenuated sequence and a marked strain concentration in the short limb. Note the intrafolial fold 'packet' just right of centre.

(b) Typical type 3 interference pattern as seen on Syros with two phases of asymmetrical folding with the same hinge direction and sense of asymmetry.

(c) A microscopic fold in a quartz-mica schist with isolated glaucophane crystals showing a bimodal glaucophane fabric: glaucophane prisms are aligned parallel to the fold hinge only at the hinge and are elsewhere approximately perpendicular to them.

a
b
c

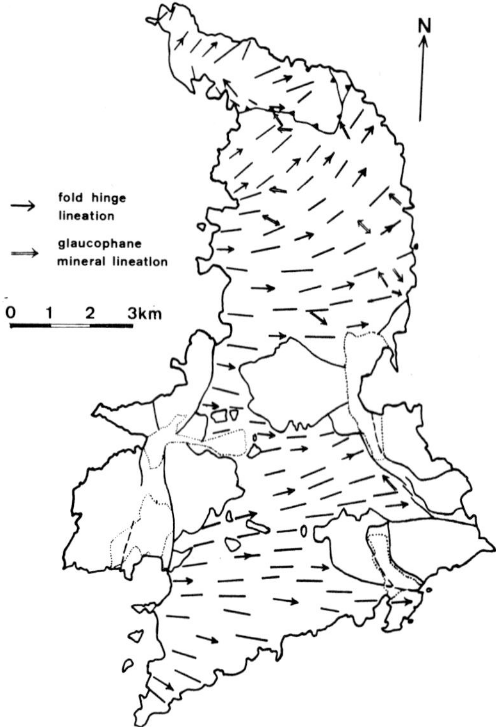

FIG. 3. Fold hinge trend of the recumbent syn-metamorphic folding in the autochthonous structural unit on Syros. Each point marked is the mean of 10–15 data points. Also shown is the trend of the approximately orthogonal glaucophane mineral lineation where developed. The 'thrust' marked in the north of the island is a semi-continuous horizon of serpentinite across which there is a discordance in the syn-metamorphic fabrics. The general foliation dip is low to moderate to the north or northeast.

fold hinge only at the hinge itself and are elsewhere approximately perpendicular to it. The glaucophane fabric in the fold limbs is presumably an elongation lineation (Schwerdtner 1970), and if strain approximates to simple shear it then also yields a close estimate of the tectonic displacement direction.

A complete plot (Fig. 5) shows a scatter from the late open and upright folding, but the best fit direction is indistinguishable from that given by the fold asymmetry analysis.

Inferences about the form of the finite strain

The deformation seen can be visualized as having taken place in a deep level ductile shear zone or thrust zone. Besides indicating thrust-sense shear the small scale structures also suggest a component of shortening parallel to the thrust displacement. Many of the asymmetrical folds could perhaps be described as asymmetrical 'kink bands'. Quartz microfabrics show that strain is concentrated along the short limbs of fold pairs. Fold tightening is essentially through thrust-sense relative translation of the two long limbs. The geometrical effect of the formation and development of such a fold pair is clearly to shorten the pile parallel to the thrust displacement direction.

This shortening component in the finite strain is also implied by the fabrics in rare meta-conglomerates. Conglomerate clasts have been deformed heterogeneously in a heterogeneous matrix and their shape fabric does not reflect exactly the form of the finite strain. On foliation surfaces the clasts are generally weakly elongate parallel to the fold hinge lineation. This is qualitatively as expected if strain included both layer-parallel shear and shear-parallel shortening (Coward & Kim 1981).

Correlations and tectonic significance of the features described

It is possible that Syros has undergone block rotation since the formation of the fabrics described. An indication of whether such is the case might be gained through similar studies on neighbouring islands. Taken at face value however, a southeasterly-directed thrust displacement in the Cycladic massif suggests that the blueschists there are genetically related to those along the northern margin of the Menderes massif in Turkey. The structures in the two areas show the same trend and the same facing direction (Çoğulu 1967, Şengör & Yilmaz 1981).

There is no suggestion from this study of any rotation of the fabrics in sympathy with the present arcuate trend of the Aegean structures. This suggests that there is no direct tectonic link between the Aegean blueschists and those of the Pelagonian zone of mainland Greece.

Fabric heterogeneities related to strain heterogeneities, as delineated for Syros, may be present throughout the Cyclades, possibly on a variety of scales. Many workers have emphasized the division of the Cycladic massif into areas where the dominant lineation trends respectively at 010° and 060° (Blake *et al.*, 1981, Papanikolaou 1980). One possible cause of this division could be a regionally varying wrench-shear component within a single tectonic belt. The division does not necessarily indicate the

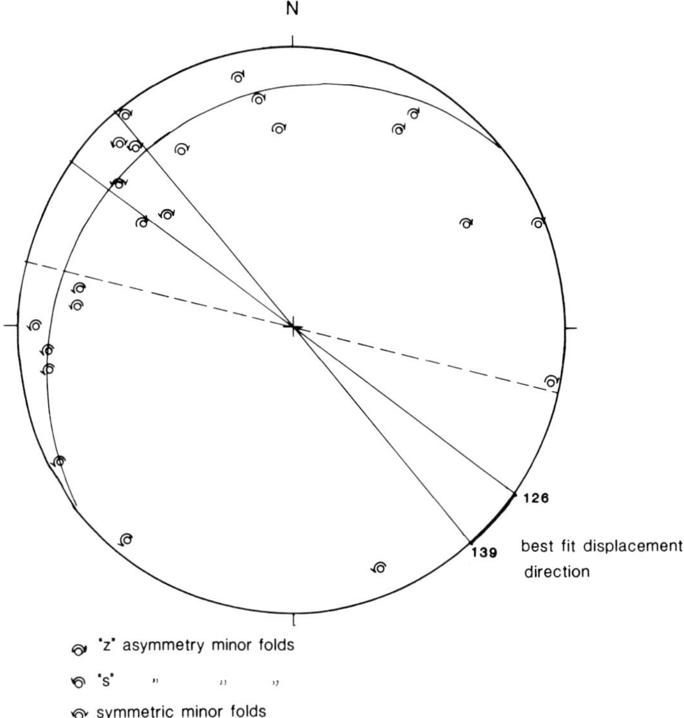

FIG. 4. Hansen 'separation arc' method for determining the shear and displacement direction from the small scale folding on Syros. The data is from the area in the north of the island above the serpentinite horizon marked on Fig. 3. For discussion see text.

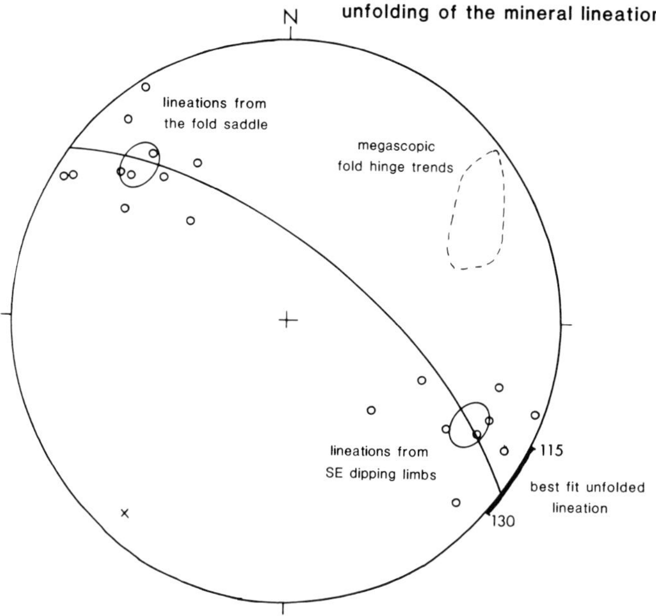

FIG. 5. Orientation data for the glaucophane mineral lineation where developed on Syros. The scatter is due to late, approximately northeast trending, open to close upright folding. The arc shown is one possible good fit 'folded lineation arc', in this case assuming buckle folding (Ramsay 1967) and giving an original lineation orientation of 127°. The range of possible folding models fitting the data gives a range for this unfolded lineation orientation of $122 \pm 8°$.

presence of two distinct tectonic belts with contrasting displacement directions.

The significance of the shortening component in the finite strain is at present unclear. It is possible that this shortening was only effective in the later stages of progressive deformation. This would be the case if the apparent absence of folding prior to the consolidation of the continuous metamorphic fabric was real and not an artefact of preservation (see above).

The shifts in the metamorphic phase equilibria during deformation clearly indicate increasing temperature, and are most consistent with increasing pressure (Ridley 1982). The deformation seen is therefore regarded as related to tectonic burial.

A strain history involving predominantly thrust-sense simple shear and a shear-parallel shortening component in the later stages of deformation seems consistent with a number of large scale tectonic regimes. It is certainly consistent with subduction-related tectonics and a process of underplating above a major thrust (Karig & Kay, 1981). It is also consistent with deformation within a wedge-shaped mass of 'sediment' immediately above a major crustal thrust.

ACKNOWLEDGMENTS: The work for this paper was carried out during the tenure of a NERC research studentship which is gratefully acknowledged. I thank IGME, Athens, for supporting the project. Dr. J. E. Dixon is thanked for useful critical discussion.

References

ANDRIESSEN, P. A. M., BOELRIJK, N. A. I. M., HEBEDA, E. M., PRIEM, H. N. A., VERDURMEN, E. A. TH. & VERSCHURE, R. H. 1979. Dating the events of metamorphism and granitic magmatism in the Alpine orogen of Naxos (Cyclades, Greece). *Contrib. Min. Pet.* **69,** 215–25.

BLAKE, M. C., BONNEAU, M., GEYSSANT, J., KIENAST, J. R., LEPVRIER, C., MALUSKI, H. & PAPANIKOLAOU, D. 1981. A geological reconnaissance of the Cycladic blueschist belt, Greece. *Bull. geol. Soc. Amer.* **92,** 247–54.

ÇOĞULU, E. 1967. Etude pétrographique de la région de Mihaliçik (Turquie). *Schweiz. Min. Pet. Mitt.* **47,** 683–824.

COWARD, M. P. & KIM, J. H. 1981. Strain within thrust sheets. *In:* MCCLAY, K. R. & PRICE, N. J. (eds) *Thrust and Nappe Tectonics.* Spec. Pub. geol. Soc. Lond. no. 9, 275–92.

DAHLSTROM, C. D. A. 1970. Structural geology in the eastern margin of the Canadian Rocky Mountains. *Bull. Can. Pet. Geol.* **18,** 332–406.

HANSEN, E. 1971. *Strain Facies.* Springer Verlag, New York, 207 pp.

KARIG, D. E. & KAY, R. W. 1981. Fate of sediments on the descending plate at convergent margins. *Phil. Trans. R. Soc. Lond. A,* **301,** 233–51.

PAPANIKOLAOU, D. 1980. Contribution to the geology of the Aegean sea: The island of Paros. *Ann. géol. Pays Hell.* **30,** 65–96.

RAMSAY, J. G. 1967. *Folding and Fracturing of Rocks.* McGraw Hill, New York, 568 pp.

RIDLEY, J. R. 1982. *Tectonic style, strain history and fabric development in a blueschist terrain, Syros, Greece.* PhD Thesis (unpubl). Edinburgh University, 283 pp.

SANDERSON, D. J. 1982. Models of strain paths and strain fields within nappes and thrust sheets. *Tectonophysics.* **88,** 201–33.

SCHWERDTNER, W. M. 1970. Hornblende lineations in Trout Lake area, Lac la Rouge map sheet, Saskatchewan. *Canadian J. Earth Sci.* **7,** 884–99.

ŞENGÖR, A. M. C. & YILMAZ, Y. 1981. Tethyan evolution of Turkey, a plate tectonic approach. *Tectonophysics.* **75,** 181–241.

J. RIDLEY, Institut für Kristallographie und Petrographie, ETH Zentrum, CH-8092, Zürich, Switzerland.

The three metamorphic belts of the Hellenides: a review and a kinematic interpretation

Dimitrios J. Papanikolaou

SUMMARY: New results obtained in the three metamorphic belts of the Hellenides suggest that each belt has a different structure, geotectonic setting and evolution. The external metamorphic belt of Peloponnesus-Crete has probably resulted from nappe movement in the vicinity of an evolving island arc system. The blueschists occurring within this external belt are allochthonous and have been transported with other nappes over the external carbonate platform of the Pre-Apulian-Plattenkalk-Ionian-Tripolitza-Almyropotamos Units. This implies that there was only one true blueschist domain, initiated in the Cycladic area during ?Early Eocene. The kinematic interpretation is based on the distinction of a- and b-structures and on the nature of the shear zones between the different domains, especially that between the metamorphic Hellenides below and the non-metamorphic Hellenides above. The existence of the probably Lower Palaeozoic meta-sedimentary sequence of Kastoria in the northern Pelagonian zone indicates that the overlying, ophiolite-bearing Almopias unit has been derived from within the Hercynian domain. The Kastoria sedimentary sequence, the ophiolites of Vertiskos and the carbonate platform of Pangeon are considered to be most likely elements of a probable Hercynian orogeny.

The metamorphic rocks of the Hellenides were for a long time considered to be pre-Alpine basement rocks on which geosynclinal sedimentation developed through the Mesozoic to early Tertiary. The distinction of external, medial and internal metamorphic massifs (Brunn 1956) was mainly based on their tectonic position during the Alpine orogeny, with the external metamorphic belt lying below the foreland platform (Gavrovo–Tripolitza zone) of the Hellenides, the medial metamorphic belt below the former Pelagonian zone and the internal metamorphic belt at the core of the arc of the Hellenides in the Rhodope Massif.

After the existence of tectonic windows in Attica, Evvia, the Cyclades, Olympus, Peloponnesus and Crete was demonstrated, a distinction was drawn between the Rhodope belt and the two others. The Rhodope massif was still considered as a 'zwischen-gebirge' whereas the two other metamorphic belts were considered as Alpine units with some pre-Alpine basement rocks involved in between, especially in Northern Greece (Aubouin 1976; Jacobshagen *et al.* 1978). The restriction of blueschists to the medial and external belts is one important difference (Papanikolaou 1979b, 1980c). The relation of the Alpine metamorphic units of the medial and external metamorphic belts, or 'metamorphic Hellenides' (Papanikolaou 1980b), to the non-metamorphic Hellenides has been the subject of much discussion.

Evolutionary models have been proposed for individual metamorphic belts but the only attempt to relate the external and the medial metamorphic belts is the proposal that during Oligocene times they constituted a paired metamorphic belt (Jacobshagen 1979, 1980; Jacobshagen *et al.*, 1978; Altherr *et al.* 1982). This is based on the existence during the Late Oligocene to Early Miocene of HP/LT metamorphism in the external belt and of LP/HT metamorphism in the medial belt (for recent reviews see: Altherr *et al.* 1982; Seidel *et al.* 1982).

This paper attempts to relate the structural style and metamorphic grade in the three belts, and their position in the overall nappe succession, to a kinematic model of Tertiary nappe emplacement. It is based on observations made in two sub-parallel transects across the Hellenides, one from Kalamata in the southern Peloponnesus to Andros in the northern Cyclades, and another through Northern Greece from Kastoria to Xanthi (Fig. 1).

The external and medial metamorphic belts

The external metamorphic belt in the Peloponnesus

The external metamorphic belt in the Hellenides is represented on the mainland of Greece by the Arna unit (Fig. 1), a sequence of HP-LT mica schists, quartzites and other meta-sediments exposed in a tectonic window in the southern part of the Peloponnesus, and by the very low grade metamorphic rocks of the Plat-

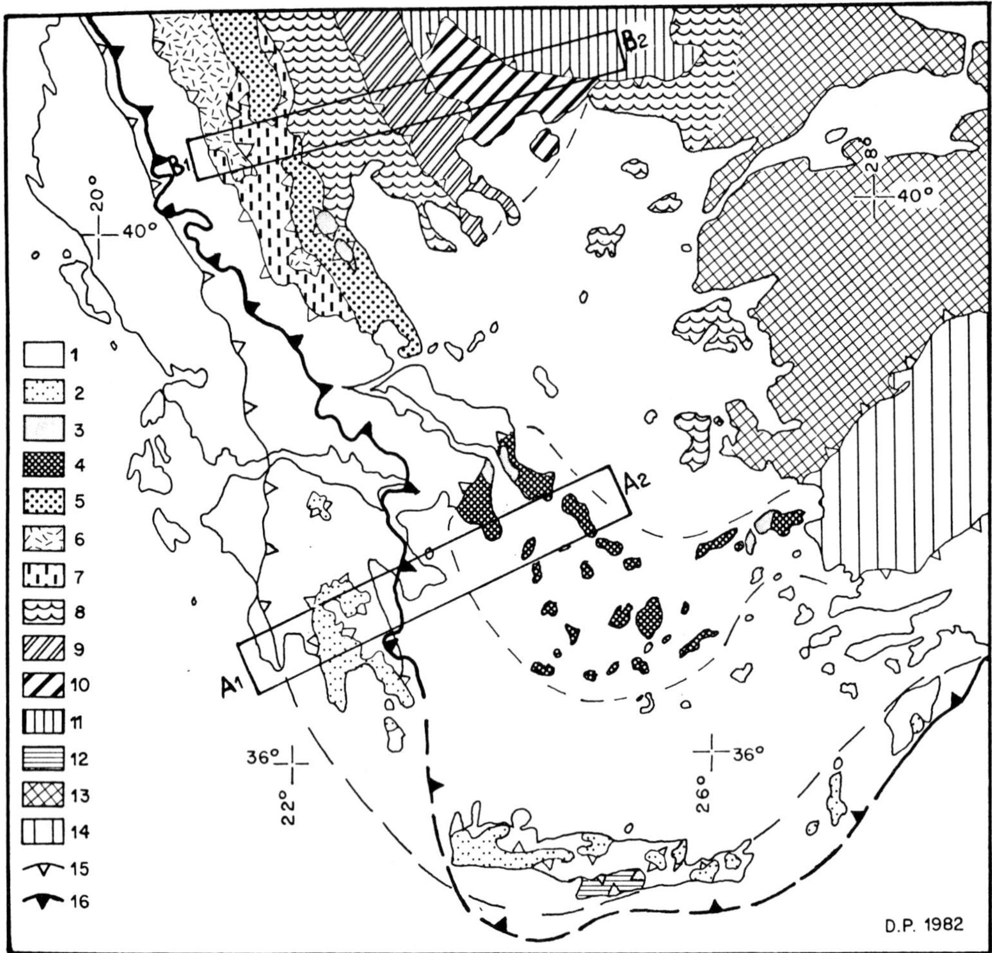

FIG. 1. Map showing the various metamorphic units of the Hellenides and the position of the two transects (A_1–A_2) (B_1–B_2) along which the kinematic interpretations are given in the text.

1, non-metamorphic Hellenides (Pre-Apulian, Ionian, Gavrovo-Tripolitza, Pindos, Parnassos and Subpelagonian) and their partly equivalent Lycian Nappes in SW Minor Asia; 2, External Metamorphic belt (Plattenkalk, Arna); 3, Lower tectonic units of the medial metamorphic belt (Olympus, Ossa, Almyropotamos, Kerketefs); 4, Cycladic units (mainly blueschists); 5, Ambelakia (blueschists) and Flambouron (crystalline basement); 6, Kastoria (Paleozoic basement); 7, Almopias (Permian–Jurassic metamorphic rocks + ophiolites + Cretaceous sediments); 8, Vardar (Axios) and circum-Rhodope units (Permian-Cretaceous); 9, Serbo-Macedonian (Kerdylia, Vertiskos); 10, Pangeon (Lower Rhodope); 11, Sidironero (Upper Rhodope); 12, Asteroussia Klippen; 13, Pontides and Izmir-Ankara zone; 14, Menderes metamorphic units; 15, Tectonic front of Pindos nappe; 16, Tectonic front of ophiolite bearing nappes ('internal' units).

tenkalk unit below it. The Arna Unit lies structurally above the Tertiary carbonate beds of the Plattenkalk unit (Fytrolakis 1972; Bizon & Thiebault 1974) and lies beneath the Mesozoic platform carbonates of the Tripolitza Nappe.

It has been interpreted in three different ways:

(1) As the original metamorphic basement of the Tripolitza nappe (Bonneau 1973; Thiebault 1975; Jacobshagen *et al.* 1976).

or

(2) As a distinct palaeogeographic zone in the Hellenides, the continuation of the 'Phyllite-Quartzite Unit' of Crete, seen as separating the Ionian zone (≡ the Plattenkalk Unit) from the Tripolitza platform (Jacobshagen *et al.* 1978; Seidel *et al.* 1982).

or

(3) As the more metamorphosed part of a flysch sequence, the Ultra-Plattenkalk flysch, originally deposited on the underlying Platten-

kalk unit (Lekkas & Papanikolaou 1978; Lekkas & Ioakim 1980; Thiebault 1982).

The first interpretation is considered to be the result of a mistaken correlation of the Arna Unit with the Permo-Triassic Tyros Beds (Ktenas 1924) which occur at the base of the Tripolitza carbonates.

It is also now clear that the unit differs both structurally and lithologically from the underlying Plattenkalk unit and so must have a separate origin. The Plattenkalk unit shows a single phase of close to isoclinal folds trending N–S, interpreted as *b*-folds, whereas the overlying Arna Unit shows the effects of two major deformation phases: an early phase of isoclinal folding with associated parallel lineations trending ENE–WSW, interpreted as a-structures, and a later phase of close to isoclinal folds trending N–S (Papanikolaou 1981b; see also Bonneau, this volume, and Hall, this volume). Lithologically, the Arna Unit comprises medium-grade metabasalts and meta-tuffs as well as the more abundant mica-schists, phyllites and quartzites, and is thus quite different from the very low-grade meta-flysch of the Plattenkalk unit (Skarpelis 1982).

Recognition of the distinct character of the Arna unit led Papanikolaou (1983) and Papanikolaou & Skarpelis (1983) to propose a Cycladic origin for it.

The first part of this paper presents a kinematic nappe-emplacement model for deriving the Arna Unit as a low-grade blueschist nappe from the Cycladic blueschist 'domain'. The main component of the *external* metamorphic belt is thus seen as a far-travelled, and probably rootless nappe derived from the *medial* metamorphic belt.

The medial metamorphic belt in the Cyclades

The metamorphic rocks of the former 'Attica–Cyclades massif' are now recognized as occurring in several distinct tectonic units (Katsikatsos 1979; Dürr *et al.* 1978; Papanikolaou 1978, 1980c, 1983). Summarizing, the main units are, from the base upwards: (1) the Attica autochthon (Triassic–Jurassic); (2) the Attica allochthon (partly Cretaceous); (3) the Almyropotamos unit in Evvia (Triassic–Eocene) and its probable equivalent, the Kerketefs unit in Samos; (4) the unit of the Northern Cyclades (partly Triassic–Jurassic) and its equivalent the Styra unit in Evvia; (5) the Makrotandalon unit of Andros Island (partly Permian) and its equivalent the Ochi unit in Evvia; (6) the unit of the Southern Cyclades (Triassic–Cretaceous) and its equivalents the Vourliotes unit in Samos and lower unit of Ikaria; (7) the topmost, non-metamorphic, Cycladic nappe, involving Permian, Triassic, Jurassic, Cretaceous and Burdigalian rocks, partly equivalent to the former 'Sub-Pelagonian zone'. Some other minor units, containing low-grade metamorphic rocks, like the Messaria unit of Ikaria, the Phourni unit of the nearby Phourni Islands and the partly Permian Dryos unit of Paros are also observed to lie below the non-metamorphic Cycladic nappe.

Kinematic interpretation

General characteristics of the model

The model proposed here for deriving the Arna unit from one of the Cycladic nappes is based on a number of stratigraphical, chronological and structural style correlations. It considers only Tertiary thrusting and for the purposes of discussion considers nappes to be of three kinds: unmetamorphosed sedimentary sequences of Mesozoic to Tertiary age; metamorphic, generally HP–LT, nappes generated in the Tertiary by subduction or nappe-loading, and thirdly, nappes derived from terrains deformed and metamorphosed prior to the Tertiary as, for example, the Pelagonian zone of Eastern Mainland Greece. The model envisages a Lower Tertiary palaeogeography with the following elements, from west to east: the Apulian platform and adjacent basinal zones (Ionian and Plattenkalk); a carbonate bank (Gavrovo-Tripolitza); a subduction zone generating blueschists near the western margin of an ocean (Pindos); the Pindos ocean itself; the Parnassos carbonate platform and furthest east, at the ocean margins, the 'internal' metamorphic units of the Pelagonian zone.

Tectonic level of nappe transport

A first point that has to be taken into consideration is the tectonic level at which nappe movement has been developed. Three depth-zones of nappe movement can be recognized. Thus, the transport of the Pindos nappe as now seen has been always at high-level, from the existence of 'wild-flysch' with olistostromes and olistoliths at the top of the underlying Gavrovo-Tripolitza flysch (Fleury 1977; Lekkas 1978), or the Ionian flysch (personal observations). The same is also true for the emplacement of the more internal, ophiolite-bearing nappes, on to the Pindos nappe, especially in northern Pindos and the Vermion Mts (personal observations). In contrast, the base of the Tripolitza nappe overlying the Plattenkalk and Arna units is characterized by marked recrystallization in the limestones and in the Permo-Triassic shales of the Tyros Beds which have become slates or phyllites, and also by the formation of 'cargneule'. The medium grade metamorphic rocks of the Arna unit have become phyllonites near the tectonic contact with the underlying Plattenkalk unit and show a strong late cleavage that obliterates all previous structures. In the Cyclades the tectonic contacts separating the various units were initiated at still deeper tectonic levels, with the exception of the topmost, unmetamorphosed 'Cycladic' and Dryos nappes. The deepest tectonic level recognized is the complex tectonic zone separating the Makrotandalon and Northern Cyclades Units in Andros (Papanikolaou 1978), where synmetamorphic folding and ductile thrust-sense shear are present.

Timing of nappe transport

As far as timing is concerned it is significant that the high-level Cycladic nappe bears an allochthonous Burdigalian molasse and its probable age of emplacement is Middle/Upper Miocene (Papanikolaou 1980a). Another important chronological constraint is that the Almyropotamos unit with its Upper Eocene, weakly metamorphosed flysch (Dubois & Bignot 1979) is overthrust onto the Attica autochthon in the Marathon area (Katsikatsos 1979). This fact when considered in relation to the structural history of Almyropotamos Unit (Katsikatsos *et al.* 1976) and of Attica autochthon (Mariolakos & Papanikolaou 1973) point to a history of repeated thrusting, with a first tectonic phase during which the Attica units were emplaced over the Almyropotamos unit, metamorphosing the flysch, and then a second tectonic phase during which the Almyropotamos unit, together with its flysch cover, was emplaced over the Attica units. The first tectonic phase must have occurred during the Late Eocene–Early Oligocene whereas the second phase probably occurred during the Late Oligocene-Early Miocene, since by Middle-Late Miocene times continental deposits were widespread in Evvia and Attica (Dermitzakis & Papanikolaou 1979). The available isotopic ages from the Cycladic area (Altherr *et al.* 1982) are in accord with this timing, pointing to a Late Eocene event around 45 Ma followed by an Early Miocene event around 20 Ma accompanied by intrusion of granitic rocks.

In the Peloponnesus the age data point to one major tectonic emplacement phase during the Late Oligocene-Early Miocene (Seidel *et al.* 1982) which coincides with the age of the youngest flysch deposits in the western Hellenides (Richter *et al.* 1978).

Structural style

The structural style of the Arna unit which is characterized by complex a- and b-structures, is similar to those of the units of Attica and Cyclades whereas the Plattenkalk unit has only b-structures with intense ductile deformation and the other, higher non-metamorphic, units have mainly b-structures and lack evidence of ductility (Papanikolaou 1981b).

Stratigraphical affinities

Stratigraphic data allow the recognition of pre-thrusting palaeogeographic domains and constrain the timing of nappe-emplacement. Thus the Pindos and Parnassos units and the intervening transitional Vardoussia thrust-sheets show sequences consistent with the evolution, through the *Mesozoic and Early Tertiary*, of a basin/carbonate platform couple (Celet 1962). Furthermore, from the *Late Cretaceous*, the Pindos and Parnassos units together show a common tectonic evolution with the more internal units to the east (Othrys or Sub-Pelagonian Zone & Pelagonian Zone) as indicated by transitional sedimentary sequences in western Thessaly (Papanikolaou & Lekkas 1979) and elsewhere in mainland Greece (Celet 1979), and by the coeval initiation of flysch sedimentation in the Maastrichtian-Danian.

In the west or external region, the Gavrovo-Tripolitza and Ionian units show a common tectonic evolution from the *Late Eocene on*, to judge from their common flysch cover in Epirus, and particularly in the Messolonghi area. The Pre-Apulian (Paxos), the Plattenkalk and the Ionian units evidently belonged in the same palaeogeographic domain right up *to the early Oligocene* as suggested by their similar stratigraphic columns (Thiebault 1977).

Metamorphic affinities

The Arna unit shares a typical blueschist metamorphic character with the Lower and Upper Attica units, and the Northern Cyclades, Makrotandalon and Southern Cyclades units (e.g. Skarpelis 1982; Marinos & Petrascheck 1956; Dixon 1976; Papanikolaou 1978; Davis 1966; Dürr *et al.* 1978; Blake *et al.* 1981). The other lower grade metamorphic units, the Plattenkalk, the Almyropotamos carbonate unit of Evvia and the Kerketefs unit of Samos show HP/LT affinities but detailed comparisons are hampered by the dominance of carbonate and the scarcity of useful lithologies (see Okrusch *et al.*, this volume). The general low-grade character does not appear compatible with the depths of subduction implied by the assemblages in the other Cycladic units.

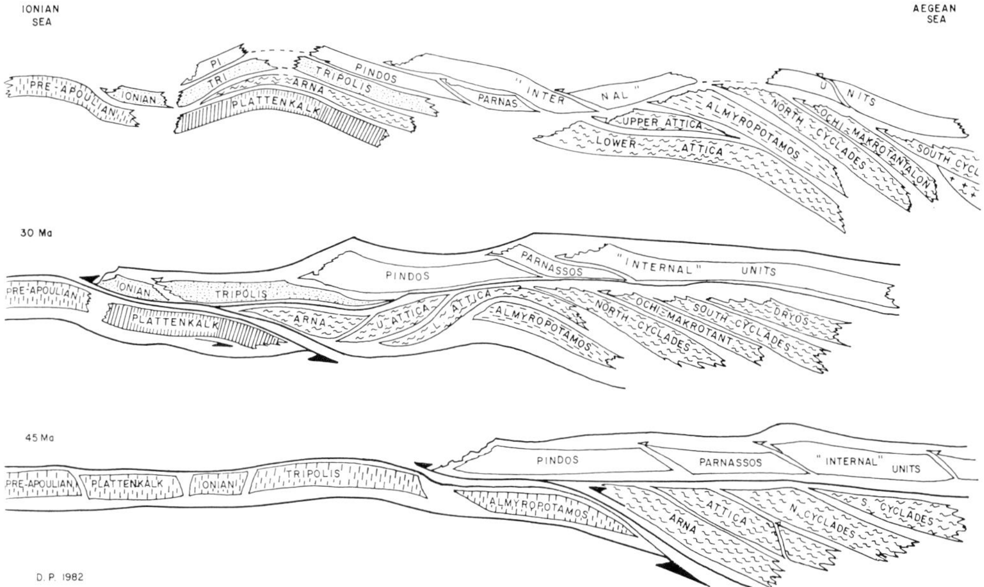

FIG. 2. Schematic profile along the A_1–A_2 traverse of the arc of the Hellenides, and Late Eocene (45 Ma) and Late Oligocene (30 Ma) reconstructions showing the development of the medial and the external metamorphic belt. Besides the resulting shear zones along the subducting or underthrusting external units and the overriding internal units there exist several other shear zones. The most important is that separating the high-level non-metamorphic units (classical Hellenides) from the underlying deep-level metamorphic units (Metamorphic Hellenides).

Status of the Almyropotamos and Kerketefs units

The palaeogeographic positions of the Almyropotamos marble unit of Evvia and the Kerketefs carbonate unit of Samos are problematic (see Katsikatsos *et al.* 1976; Dubois & Bignot 1979; Papanikolaou 1979b). The interpretation favoured here is that they represent an 'internal' (i.e. distal or eastern) margin of the Gavrovo-Tripolitza carbonate platform. This is consistent with the continuation of carbonate sedimentation in the Almyropotamos unit into the Eocene. However, a period of ?Early Cretaceous tectonism, not recorded in the Tripolitza platform proper, is indicated by the presence of discordant Upper Cretaceous marbles overlying Upper Triassic marbles (Katsikatsos 1969).

The alternative more revolutionary model is that these carbonate units, along with those exposed in the Olympus window always lay on the east side of the Pindos ocean (i.e. they are more 'internal'). This would perhaps explain some unpublished observations of the author that Cretaceous and Eocene neritic limestones have been thrust over Pindos units and also over the Northern Pindos ophiolite nappe in Western Macedonia and Western Thessaly. Schmitt (1982) has recently reported the presence of blocks of Triassic limestone in a Cretaceous carbonate unit of the Olympus sequence, analogous to the break in Almyropotamos sequence noted above. At present, the author feels that these indications of a separate tectonic history can still be accommodated if the Almyropotamos, Kerketefs and Olympus carbonate sequences were deposited on the internal margin of the Gavrovo-Tripolitza platform, west of the Pindos ocean.

Late Eocene palaeogeography

The general conclusion from the above observations is that the nappe pile of the Hellenides from the Southern Peloponnesus to the Northern Cyclades can be subdivided into three groups of units belonging to the following three palaeogeodynamic domains in Late Eocene times (Fig. 2):

(1) a complex island-arc-like domain with the Pindos, Parnassos and all the more internal non-metamorphic units. This domain has evolved since Eocene times entirely at a high structural level with successive thrusting of the internal units towards the more external, with abundant detritus being supplied to fore-arc flysch basins, which grade rapidly up to 'wildflysch' immediately before being overthrust by the advancing nappes. During the Middle Oligocene—Middle Miocene the Mesohellenic

molasse basin was created behind the tectonic front of the Pindos and related nappes (making up a 'Pindic cordillera') corresponding to a back-arc basin of that period (Papanikolaou & Dermitzakis 1981). Gravity may have played an important role in this process and in the overall internal deformation especially of the Pindos unit.

(2) a carbonate platform involving the Pre-Apulian, Plattenkalk, Ionian, Gavrovo-Tripolitza and possibly also the Almyropotamos and Kerketefs units. A relatively shallow basin involving the Plattenkalk and the Ionian zone between the Pre-Apulian and Gavrovo-Tripolitza platforms existed. This intra-platform basin was initiated through normal faulting during Middle Jurassic times within the formerly continuous shallow water platform.

(3) a very complex group of units characterized by HP/LT blueschist metamorphism and with varying lithologies comprising carbonate platforms with metabauxites, pelagic limestones alternating with siliceous layers, abundant development of meta-tuffs. Finally, the existence of ophiolitic rocks within the sequences, either in sheared tectonic bodies or as olistostromes (Blake *et al.* 1981) is remarkable.

Nappe kinematics

The tectonic evolution of these three domains is schematically illustrated in the three sections of Fig. 2, which show the present situation and the probable configurations at approximately 30 Ma and 45 Ma. The critical point is the detachment of the Almyropotamos unit from the Gavrovo-Tripolitza platform and its underthrusting below the Cycladic units, which were already metamorphosed under HP/LT conditions, and the subsequent re-thrusting of the Almyropotamos unit westwards over blueschist units that had been previously thrust right over it. The second movement phase occurred when the external part of the Plattenkalk-Ionian intra-platform basin started to be subducted below the more internal part of the basin and all the over-riding units. The Plattenkalk unit was subducted whereas the Ionian and Gavrovo-Tripolitza units continued their sedimentary evolution (with deposition of flysch) in front of the evolving island arc of the Pindos and more internal units. Thus, the external carbonate platform of the Hellenides, represented during late Eocene times by domain 2 was disrupted into several segments, each one following its own particular geodynamic evolution.

The very low grade metamorphism of the Plattenkalk unit can be easily explained by a nappe pile a few km thick being thrust onto it. On the other hand, the blueschist metamorphism of the Arna unit cannot be explained, unless it was transported from the Cycladic domain prior to Middle Eocene times and had the Tripolitza carbonates thrust over it in the shortening event that led to the subduction of the Plattenkalk.

Finally, the palaeogeographic position of at least some of the Cycladic units was probably between the external platform of the Hellenides and the Pindos basin. However, the existence within this domain of units with more internal affinities, like those of Attica which show similarities with the Almopias Unit of Northern Greece, indicates that the Pindos and Parnassos Units might have followed a similar evolution during Cretaceous times as the Ionian and Tripolitza units did during late Eocene-Oligocene times. That is, they may represent units that escaped the early Alpine (Late Jurassic-Early Cretaceous) orogenesis and the associated obduction of ophiolites and blueschist metamorphism of subducted units occurring further east.

The medial and the internal metamorphic belts

The medial metamorphic belt in Northern Greece

Deducing the possible relations between the medial and the internal metamorphic belts and arriving at a kinematic model, is possible only in Northern Greece, between the Pelagonian zone and the Rhodope massif. Thus, a complete transect of the arc of the Hellenides involving all three metamorphic belts requires an along-strike correlation of the units making up the Pelagonian zone with those of the Attico-Cycladic massif. This correlation is not easy because several different units exist in the two areas. However, the basal units, Almyropotamos in one and Olympus in the other have several common stratigraphic, tectonic and metamorphic characteristics that permit one to consider them as equivalents.

An analysis of the tectonic units of the area north of Olympus and their possible correlation with analogous units in southern Yugoslavia is presented in recent reviews by Papanikolaou & Zambetakis-Lekkas (1980), Papanikolaou (1981a, 1983), Papanikolaou *et al.* (1982) and Papanikolaou & Stojanov (1983). These results together with those of Brunn (1956), Mercier (1968), Godfriaux (1968, 1977), Yarwood & Dixon (1979), Arzovski *et al.* (1979), Katsikatsos *et al.* (1982), Schmitt (1982) and others, point to the following structural succession in

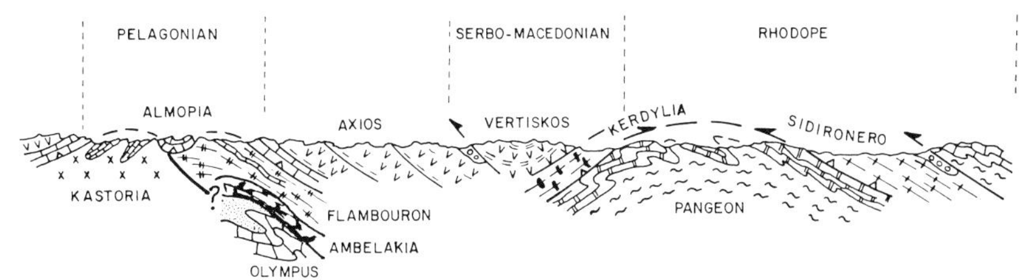

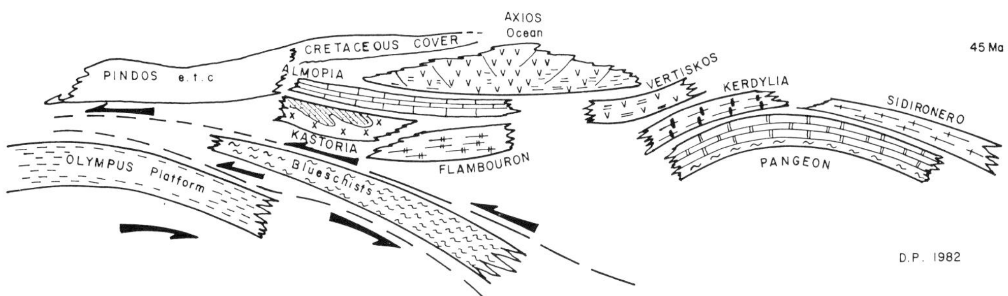

FIG. 3. Schematic profile along the B_1–B_2 traverse of the arc of the Hellenides, and Late Eocene (45 Ma) reconstruction showing the gross organization along the orogenic front and the over-riding complex 'European margin', with the ophiolites (Axios/Vardar ocean) overlying the Almopia palaeo-alpine carbonate platform and overlying the probably pre-Alpine crystalline basement units (Kastoria, Flambouron, Kerdylia, Sidironero and perhaps also Vertiskos and Pangeon).

the northern Pelagonian zone, from top to bottom and from west to east (Fig. 3):

(1) the *Almopia unit* comprising ?Permian-Middle Triassic phyllites, marbles and meta-volcanic rocks, Middle Triassic–?Middle Jurassic marbles, and ?Upper Jurassic schists, partly as a melange below the ophiolite nappe, and on top a non-metamorphic para-autochthonous Cretaceous sedimentary cover overlain by early Tertiary flysch;

(2) the *Kastoria unit* comprising two litho-stratigraphic groups: a lower group with many intrusive granitic rocks, probably of Lower and Upper Palaeozoic age creating contact metamorphic aureoles, and an upper group of meta-clastic sediments which in southern Yugoslavia are fossiliferous and are dated as Cambrian-Ordovician and Devonian (see Moundrakis, this volume);

(3) the *Kaimaktsalan-Flambouron* unit, which is made up of gneisses, mica schists, amphibolites and intercalated marbles within which a Riphean-Cambrian microflora has been reported in southern Yugoslavia. Some granitic rocks of probable Upper Palaeozoic age are also present;

(4) the *Ambelakia unit*, comprising HP/LT metamorphic rocks with obscure contacts with the overlying Flambouron nappe as well as with the underlying units of;

(5) the Olympus autochthon, comprising very low grade recrystallized carbonate rocks of Triassic to Eocene age overlain by flysch.

The internal metamorphic belt

Preliminary results of structural work in the Rhodope Massif (Papanikolaou & Panagopoulos 1981) together with the data from the literature on it and on the Serbo-Macedonian zone (e.g. Maratos & Andronopoulos 1965; Kronberg *et al.* 1970; Kronberg & Raith 1977; Kockel *et al.* 1965, 1977; Kaufmann *et al.* 1976; Kokkinakis 1977; Dimitriadis 1980) as well as on the Bulgarian part of the Rhodope (e.g. Kozhukharov *et al.* 1978; Kozhukharov & Timofeev 1980; Z. Ivanov *et al.* 1979; Ancirev *et al.* 1980; R. Ivanov 1981) point to a structural succession as follows, from top to base and from west to east: (1) the lower grade circum-Rhodope belt, which is of Triassic-Jurassic age, and partly equivalent to the Peonias Unit of Mercier (1968). This unit is usually in tectonic contact with the Serbo-Macedonian or the Rhodope units: in the west it is overthrust by Serbo-Macedonian units and in the east it is thrust on to Rhodope units. (2) the *Vertiskos unit*, which comprises gneisses, mica schists and amphibolites and a dismembered metamorphosed ophiolite (Dixon & Dimitriadis, this volume). (3) the *Kerdilion unit* which comprises gneisses, mica schists, amphibolites and marbles. (4) the *Sidironero unit*, which comprises gneisses, migmatites, mica schists, amphibolites and marbles. (5) The *Pangeon unit*, which comprises a lower group of gneisses, amphibolites and mica schists and an upper group of marbles with some mica schists at the top.

It is important to note that the few fossils found in Greece indicate a Palaeozoic age for part of Sidironero Unit and a Palaeozoic, or even a Mesozoic, age for the Pangeon Unit. In complete contradiction, the reported microflora in southern Bulgaria suggests Precambrian ages for the equivalent formations.

Kinematic interpretation

As the available data are still extremely sparse and often contradictory, the interpretation presented here is very much more speculative than that presented in the first part of the paper.

A time constraint for any interpretation of this transect of Northern Greece is that the thrusting of the Vertiskos unit over the eastern Vardar zone units took place during Late Eocene-?Early Oligocene times, judging from the tectonic involvement of molasse in the area west of Doirani Lake in south-east Yugoslavia. The same timing can be similarly deduced for the overthrust of the Kerdilion unit on the Pangeon unit, as well as for the internal imbrication in the northern Rhodope massif in southern Bulgaria. A shallow tectonic level is inferred for these late Eocene-Early Oligocene nappe movements.

The emplacement of the composite 'Pelagonian nappe' over the Olympus autochthonous units, and the intervening Ambelakia tectonic slice of blueschists, also occurred in Late Eocene times. Thus, a Late Eocene configuration along the traverse of Northern Greece involves the following palaeogeodynamic domains evolved from continuing eastward subduction of the extreme western part of the Pindos ocean:

(1) a carbonate platform of the Olympus units probably representing the internal margin of the external platform of the Hellenides and possible edge of the Apulian/African plate, partially subducted;

(2) a trench and fore-arc system further to the west made of the Pindos and the other more internal non-metamorphic units which had over-ridden the edge of the Hellenide external platform;

(3) a tectonic zone of intense shearing containing blueschists, the continuation of the subduction zone at depth, lying beneath the eastern margin of the Pindos ocean represented by domain 4;

(4) a complex tectonic domain, the margin of the European plate, with ophiolites and many metamorphic units probably belonging to various orogenic cycles. The magmatism and volcanism of Late Eocene-Early Oligocene age of the Rhodope massif corresponds very likely to the volcanic arc behind the trench and fore-arc of domain 2.

Basement units of domain 4

Following this simplified reconstruction it is appropriate to comment on the possible relations and significance of the 'basement units' of domain 4, since the lack of reliable data does not permit any extension of the kinematic interpretation to pre-Late Eocene time.

The available lithostratigraphic, tectonic, metamorphic, isotopic and other data indicate that the units of Kaimaktsalan-Flambouron, Kerdilion and Sidironero might belong in part to the Precambrian.

The Kerdilion and Sidironero units are very similar and they both overthrust the Pangeon unit which is thus a huge tectonic window. The Kastoria unit is certainly Palaeozoic in age and probably belongs to the Hercynian cycle. Thus, besides the known Mesozoic units of the Almopias and the Peri-Rhodopean Belt the only problematic units are the Vertiskos and the Pangeon. These units might well belong to the Alpine cycle but include Palaeozoic formations, or could belong only to the Hercynian cycle. In the first alternative the Vertiskos unit could include a subducted element of the Vardar ophiolites (see Dixon & Dimitriadis, this volume) with the Pangeon unit as a possible equivalent of the Almopias carbonate platform. However, in view of the resulting problems concerning the origin and significance of Flambouron, Kerdilion and Sidironero units and also in view of the lithological similarities and the overall geometry, I incline to the view that the Vertiskos and Pangeon units represent Hercynian elements.

Nevertheless, it is almost certain that the Axios ophiolites have been initiated within the Hercynian domain as indicated by the underlying lower Palaeozoic sedimentary sequence in the Kastoria area, and by the Triassic sediments transgressively overlying the Devonian formations in the less metamorphosed thrust slices near Lake Ohrid.

Concluding remarks

The comparison of the three metamorphic belts of the Hellenides and the proposed kinematic interpretations, for post-Eocene times, show that each belt has its own distinct character.

In general terms:

(1) the *external* low-grade metamorphic belt is the result of loading, by the Tertiary nappe pile which included within it a detached slice of the Medial belt blueschists—the Arna unit.

(2) the *medial* metamorphic belt has a complicated history with pre-Late Eocene subduction of a group of units originally lying between the external platform of the Hellenides and the Pindos ocean together with ophiolites and basement rocks, followed by a double westward overthrusting event during the Late Eocene-Early Oligocene and the Late Oligocene-Early Miocene, onto the internal and the medial part of the external platform of the Hellenides.

(3) the *internal* metamorphic belt has resulted from Hercynian and/or early Alpine imbrication of fragments of continental crust with ophiolitic rocks followed by a considerable intraplate shortening during Eocene-Oligocene times.

References

ALTHERR, R., KREUZER, H., WENDT, I., LENZ, H., WAGNER, G., KELLER, J., HARRE, W. & HOHNDORF, A. 1982. A late Oligocene/Early Miocene High Temperature Belt in the Attic-Cycladic Crystalline Complex (SE Pelagonian, Greece). *Geol. Jb.* **E23,** 97–164.

ANCIREV, A., GOROZANIN, O., VELICKOV, D. & BOGOYAVLENSKAYA, O. 1980. About a find of faunistic remains in the metamorphic rocks of the Western Rhodopes. *Geol. Balc.* **10,** 1, 29–32.

ARSOVSKI, M., DUMURDZANOV, N., HRISTOV, S., IVANOV, T., IVANOVA, V., PETKOVSKI, P. & STOJANOV, R. 1979. Correlation of the Precambrian complexes of the pelagonian massif, Vardar zone and Serbo-Macedonian massif. *Proc. 6th Coll. Geol. Aegean Region, Athens 1977,* **2,** 549–57.

——, ——, IVANOV, T., PETKOVSKI, P., STOJANOV, R. & TEMKOVA, V. 1979. Geologic structural characteristics of the Paleozoic complex of the southern part of the Balkan peninsula with special reference to the territory of Macedonia (Jugoslavia). *Proc. 6th Coll. Geol. Aegean Region, Athens 1977,* **2,** 559–68.

AUBOUIN, J. 1959. Contribution à l'étude géologique de la Grèce septentrionale: les confins de l'Epire et ce la Thessalie. *Ann. géol. Pays Hellén.* **19,** 1–483.

—— 1976. Alpine tectonics and plate tectonics: thoughts about the eastern Mediterranean. In: AYER, D. V. & BROOKS, M. (eds). *Europe from crust to core,* J. Wiley & Sons, London, 143–158.

——, BONNEAU, M., DAVIDSON, J., LEBOULANGER, P., MATESCO, S. & ZAMBETAKIS, A. 1976. Esquisse structurale de l'arc Egéen externe: des Dinarides aux Taurides. *Bull. Soc. géol. France,* **18,** 327–36.

BERNOULLI, D. & LAUBSCHER, H. 1972. The palinspastic problem of the Hellenides. *Eclogae geol. Helv.* **24,** 347–72.

BIZON, G., BONNEAU, M., LEBOULANGER, P., MATESCO, S. & THIEBAULT, F. 1976. Sur la signification et l'extension des 'massifs cristallins externes' en Peloponnèse méridional et dans l'arc égéen. *Bull. Soc. géol. France,* **18,** 337–45.

——, & THIEBAULT, G. 1974. Données nouvelles sur l'âge des marbres et quartzites du Taygète (Peloponnèse méridional, Grèce). *C.R. Acad. Sc. Paris,* **278,** 9–12.

BLAKE, M. C., BONNEAU, M., GEYSSANT, J. KIENAST, J. R., LEPVRIER, C., MALUSKI, H. & PAPANIKOLAOU, D. 1981. A geologic reconnaissance of the Cycladic Blueschist Belt, Greece. *Bull. geol. Soc. Amer.* **92,** 247–54.

BONNEAU, M. 1973. Sur les affinités ioniennes des 'Calcaires en plaquettes', epimétamorphiques de la Crète, le charriage de Tripolitza et la structure de l'arc égéen. *C.r. Acad. Sc. Paris,* **277,** 2453–6.

BRUNN, J. 1956. Contribution à l'étude géologique du Pinde septentrional et d'une partie de la Macédoine Occidentale. *Ann. géol. Pays Hellén.* **7,** 1–358.

CELET, P. 1962. Contribution à l'étude géologique du Parnasse-kiona et d'une partie des régions méridionales de la Grèce continentale. *Ann. géol. Pays Hellén.* **13,** 1–446.

—— 1979. Les bordures de la zone du Parnasse. Evolution paléogéographique au Mésozoique et caractères structuraux. *Proc. 6th Coll. Geol. Aegean Region, Athens 1977,* **2,** 725–40.

—— & FERRIERE, J. 1978. Les Hellénides internes: Le Pélagonien. *Eclogae geol. Helv.* **71,** 467–95.

DAVIS, E. 1966. Der geologische Bau der Insel Siphnos. *Geol. Geophys. Res. I.G.S.R. Athens,* **10,** 161–220.

DERMITZAKIS, M. & PAPANIKOLAOU, D. 1979. Paleogeography and geodynamics of the Aegean Region during the Neogene. *VII Intern. Congress Medit. Neogene, Athens 1979, Ann. géol. Pays Hellén. h. série IV, 245–289.*

DIMITRIADIS, S. 1980. A possible paleomargin evolution of the southernmost part of the Serbo-Macedonian Massif. *26th Intern. Geol. Congress, Paris 1980, Abstracts,* **1,** p. 335.

DIXON, J. E. 1976. Glaucophane schists of Syros, Greece. *Bull. Soc. géol. France,* **18,** 280.

DUBOIS, R. & BIGNOT, G. 1979. Présence d'un 'hardground' nummulitique au sommet de la série crétacée d'Almyropotamos (Eubée méridionale, Grèce). Conséquences. *C.R. Acad. Sc. Paris,* **289,** 993–5.

DÜRR, ST., ALTHERR, R., KELLER, J., OKRUSCH, M. & SEIDEL, E. 1978. The median Aegean crystalline belt: stratigraphy, structure, metamorphism, magmatism. In: CLOSS *et al.* (eds). *Alps, Apennines, Hellenides*, E. Schweiz. Barts'che. Verlag, Stuttgart. 455–477.

FLEURY, J. J. 1977. De Lamia à Messolonghi: La nappe du Pinde-Olonos et l'Unité du Megdhovas. *Bull. Soc. géol. France*, **14**, 53–60.

FYTROLAKIS, N. 1972. Die Entwinklung gewisser orogenen Bewegung un die Gipsbildung in Ostkreta (prov. Sitia). *Bull. Geol. Soc. Greece*, **11**, 81–100.

GODFRIAUX, I. 1968. Etude géologique de la région de l'Olympe (Grèce). *Ann. géol. Pays Hellén.* **19**, 1–281.

—— 1977. L'Olympe. *Bull. Soc. géol. France*, **19**, 45–9.

IVANOV, R. 1981. The deep-seated Central-Rhodope Nappe and the interference tectonics of the Rhodope crystalline basement. *Geol. Balc.* **11**, 47–66.

IVANOV, Z., MOSKOVSKI, S. & KOLCEVA, K. 1979. Basic features of the structure of the central parts of the Rhodope Massif. *Geol. Balc.* **9**, 3–50.

JACOBSHAGEN, V. 1979. Structure and geotectonic evolution of the Hellenides. *Proc. 6th Coll. Geol. Aegean Region, Athens 1977*, **3**, 1355–67.

—— 1980. *Die Eozäne Orogeneser der Helleniden.* Berl. Geow. Abh. 21–23.

——, DURR, ST. KOCKEL, F., KOPP, K. O., KOWALCZYK, G. with contr. BERCKHEMER, H. & BUTTNER, D. 1978. Structure and Geodynamic Evolution of the Aegean Region. In: CLOSS *et al.* (eds). *Alps, Apennines, Hellenides*, E. Schweiz, Bart'sche. Verlag, Stuttgart, 537–564.

——, MAKRIS, J., RICHTER, D., BACHMANN, G. H., DOERT, N., GIESE, P. & RISCH, H. 1976. Alpidischen Gebirgsbau und Krustenstruktur des Peloponnes. *Z. dt. Geol. Ges.* **127**, 337–63.

KATSIKATSOS, G. 1969. L'âge du systéme métamorphique de l'Eubée méridionale et sa subdivision stratigraphique. *Prakt. Akad. Athens*, **44**, 223–38.

—— 1979. La structure tectonique de l'Attique et de l'ile d'Eubée. *Proc. 6th Coll. Geol. Aegean Region, Athens 1977*, **1**, 211–28.

——, MERCIER, J. L. & VERGELY, P. 1976. La fenêtre d'Attique—Cyclades et les fenêtres métamorphiques des Hellénides internes. *C.r. Acad. Sc. Paris*, **283**, 1613–6.

——, MIGIROS, G. & VIDAKIS, M. 1982. Structure géologique de la région de Thessalie orientale (Grèce). *Ann. Soc. Géol. Nord*, **C1**, 177–88.

KAUFFMANN, G., KOCKEL, F. & MOLLAT, H. 1976. Notes on the stratigraphic and paleogeographic position of the Svoula Formation in the innermost zone of the Hellenides (Northern Greece). *Bull. Soc. géol. France*, **18**, 225–30.

KILIAS, A. 1980. *Geological and tectonic study of the area of east Varnous Mt., NW Macedonia.* Thesis, University of Thessaloniki.

KOBER, L. 1929. *Beiträge zur Geologie von Attica.* Sitz. Akad. Wiss. Wien, **138**, 299–327.

KOCKEL, F. & WALTHER, H. W. 1965. Die Strymonlinie als Grenze zwischen Serbo-Mazedonischen und Rila-Rhodope Massiv in Ost Mazedonien. *Geol. Jb.* **83**, 575–602.

——, MOLLAT, H. & WALTHER, H. W. 1977. Elaüterungen zur geologischen Karte der Chalkidiki und angrenzender Gebiete 1/100 000 (Nord Griechenland). *Bundesanst. für Geowiss. u. Rohstoffe*, Hannover, 119 p.

KOKKINAKIS, A. 1977. *Das Intrusivgebiet des Symvolon-Gebirges und von Kavala in Ostmakedonien, Griechenland.* Unpubl. Dokt. Diss. University of München. 255 p.

KOZHUKHAROV, D., KOZHUKHAROVA, E. & ZAGORCHEV, J. 1978. The PreCambrian in Bulgaria. *Materials to the I.G.C.P. n°22*, 65 pp.

—— & TIMOFEEV, B. 1980. First finds of microphytofossils in the PreCambrian of the Rhodope massif. In: *I.G.C.P. n°22, The PreCambrian in South Bulgaria*, 27–32.

KRONBERG, P., MEYER, W. & PILGER, A. 1970. Geologie der Rila-Rhodope Masse Zwischen Strimon und Nestos (Nord Griechenland). *Beih. Geol. Jb.* **88**, 133–80.

—— & RAITH, M. 1977. Tectonics and metamorphism of the Rhodope crystalline complex in Eastern Greek Macedonia and parts of Western Thrace. *N. Jb. Geol. Paläont. Mh.* **11**, 697–704.

MARATOS, G. & ANDRONOPOULOS, B. 1965. La faune trouvée dans les calcaires d'Aliki, Alexandroupolis (phyllites du Rhodope). *Bull. Geol. Soc. Greece*, **6**, 348–52.

KTENAS, C. 1924. Formations primaires semimétamorphiques au Péloponnèse central. *C.R. somm. Soc. géol. France*, 61–3.

LEKKAS, S. 1978. *Contribution à étude géologique de la région située au SE de Tripolis Péloponnèse central*). Thèse, Université d'Athénes, 192 p.

—— & IOAKIM, C. 1980. Données nouvelles sur l'âge des phyllades en Péloponnèse (Grèce). *Prakt. Akad. Athens*, **55**, 350–61.

—— & PAPANIKOLAOU, D. 1978. On the phyllite problem in Peloponnesus. *Ann. géol. Pays Hellén.* **29**, 395–410.

MARAKIS, G. 1970. Geochronology studies of some granites from Macedonia. *Ann. géol. Pays Hellén.* **21**, 121–52.

MARINOS, G. & PETRASCHECK, W. E. 1956. Laurium. *Geol. Geoph. Res., I.G.M.R.* **4**, 1–247.

MARIOLAKOS, I. & PAPANIKOLAOU, D. 1973. Observations on the structural geology of Western Pentelikon-Attica (Greece). *Bull. geol. Soc. Greece*, **10**, 134–79.

MERCIER, J. 1968. Etude géologique des zones internes des Hellénides en Macédoine centrale (Grèce). *Ann. géol. Pays Hellén.* **20**, 1–792.

PAPANIKOLAOU, D. 1978. Contribution to the geology of Aegean Sea. The Island of Andros. *Ann. géol. Pays Hellén.* **29**, 477–553.

—— 1979a. Unités tectoniques et phases de déformation dans l'île de Samos, Mer Egée, Grèce. *Bull. Soc. géol. France*, **21**, 744–52.

—— 1979b Stratigraphy and structure of the Paleozoic rocks in Greece: An introduction. In: SASSI, F. P. (ed.). *I.G.C.P. n°5, Newsletter*, **1**, 93–102.

—— 1980a. Contribution to the geology of Aegean Sea. The Island of Paros. *Ann. géol. Pays Hellén.* **30**, 65–96.

—— 1980b. The Metamorphic Hellenides. *26th Int. Geol. Congress, Paris 1980, Abstracts*, **1**, 371.

—— with contribution of N. Skarpelis 1980c. Geotraverse southern Rhodope-Crete. (Preliminary results). In: Sassi, F. P. (ed.), *I.G.C.P. n°25, Newsletter*, **2,** 41–8.

—— 1981a. Some problems concerning correlations within the metamorphic belts of the Pelagonian and the Rhodope. In: Sassi, F. P. (ed.), *I.G.C.P. n°5, Newsletter*, **3,** 119–23.

—— 1981b. Remarks on the kinematic interpretation of folds from some cases of the western Swiss Alps and of the Hellenides. *Ann. géol. Pays Hellén.* **30,** 741–62.

—— 1983. The PreCambrian in the Hellénides (Pelagonian, Attica-Cyclades, Peloponnesus-Crete). In: Zoubek, V. (ed.), *I.G.C.P. n°22, the Pre-Cambrian in Younger Fold Belts,* Wiley & Sons.

—— & Dermitzakis, M. 1981. The Aegean Arc during Burdigalian and Messinian: a comparison. *Riv. Ital. Paleont.* **87,** 83–92.

—— & Lekkas, E. 1979. Lateral transition between the Pindos zone and the unit of Western Thessaly in the area of Tavropos. *Bull. Geol. Soc. Greece*, **14,** 70–84.

—— & Panagopoulos, A. 1981. On the structural style of Southern Rhodope. *Geol. Balc.* **11,** 13–22.

——, Sassi, F. P. & Skarpelis, N. 1982. Outlines of the Pre-Alpine Metamorphism in Greece. In: Sassi, F. P. (ed.). *I.G.C.P. N°5, Newsletter*, **4,** 56–62.

—— & Skarpelis, N. 1983. The blueschists in the external metamorphic belt of the Hellenides: composition, structures and geotectonic significance. *Ann. géol. Pays hell.*, **31,** (in press).

—— & Stojanov, R. 1983. Geological Correlations between the Greek and the Yugoslav part of the Pelagonian Metamorphic Belt. In: Sassi, F. P. (ed.), *I.G.C.P. n°5, Newsletter* **5,** (in press).

—— & Zambetakis-Lekkas, A. 1980. Nouvelles observations et datation de la base de la série pélagonienne (s.s.) dans la région de Kastoria, Grèce. *C.r. Acad. Sc. Paris*, **291,** 155–8.

Richter, D. with contr. of Mariolakos, I. & Risch, 1978. The main Flysch stages of the Hellenides. In: Closs *et al.* (eds), *Alps, Apennines, Hellenides*, E. Schweiz. Bart'sche Verlag, Stuttgart, 434–8.

Schmitt, A. 1982. Tectonique tangentielle dans la fenêtre de l'Olympe, Grèce. *9ème Reun. An. Sc. Terre, Paris 1982*, 572.

Seidel, E., Kreuzer, H. & Harre, W. 1982. A late Oligocene/Early Miocene high pressure belt in the external Hellenides, *Geol. Jb.*, **E23,** 165–206.

Skarpelis, N. 1982. *Metallogeny of massive sulfides and petrology of the external metamorphic belt, SE Peloponnesus.* Thesis, University of Athens.

Thiebault, F. 1975. Sur l'âge alpin du métamorphisme des schistes du soubassement de Tripolitza en Péloponnèse méridional (Grèce). *C.R. Acad. Sc. Paris*, **280,** 947–50.

—— 1977. Etablissement du caractère ionien de la série des calcschistes et marbres (Plattenkalk) en fenêtre dans le massif du Taygète. Péloponnèse, (Grèce). *C.R. Somm. Soc. géol. France*, **3,** 159–61.

—— 1982. L'évolution géodynamique des Helléndes externes en Péloponnèse méridional. *Publ. Soc. géol. Nord*, **6,** 574 p.

Yarwood, G. A. & Dixon, J. E. 1979. Lower Cretaceous and younger thrusting in the Pelagonian Rocks of the High Pieria, Greece. *Proc. 6th Coll. Geol. Aegean Region, Athens 1977*, **1,** 269–80.

Dimitrios J. Papanikolaou, Department of Geology, University of Athens, Panepistimioupolis Zografou, Athens, 15771 Greece.

Tectonic setting of the Mesozoic Pindos basin of the Peloponnese, Greece

Georgia Pe-Piper & David J. W. Piper

SUMMARY: The Pindos isopic zone of western Greece comprises a 1 km thick sequence of Mesozoic and early Tertiary deep water sediments now preserved in a series of imbricate thrust slices in the Central Hellenic Nappe. This paper examines the question as to whether the basement to the Pindos Zone was oceanic or continental crust.

Mid-Triassic volcanic rocks around the margins of the Pindos basin suggest initial development as a back-arc basin by subduction of Mesogea beneath Apulia. There is then no volcanic record through the Late Triassic, but tuffs are widespread in the Jurassic. Rare lavas and pyroclastics of Late Jurassic or earliest Cretaceous age are shown by new chemical analyses to be mid-ocean ridge type basalts associated with calc-alkaline andesites and dacites characteristic of subduction. This close juxtaposition of such rock types suggests a back-arc basin setting. Thus at least part of the Pindos Zone appears to have been underlain by oceanic crust. This Late Jurassic oceanic basin is probably tectonically unrelated to the earlier mid-Triassic basin.

The Pindos isopic zone of western Greece comprises about 1 km of deep-water Triassic to Eocene sediments. It is flanked to the west by the shallow-water carbonate sequence of the Gavrovo-Tripolis Zone, and to the east by the shallow-water Pelagonian Zone. It is preserved as a series of thrust slices in the Central Hellenic nappe (Jacobshagen *et al.*, 1978). Whether the Pindos Zone sediments accumulated on subsided continental crust or on oceanic basement is unknown (Aubouin *et al.*, 1979). The distinction between these two hypotheses is important in understanding the plate tectonic history of the Hellenides, and whether the Pindos Zone is a potential source area for the major ophiolites.

In this paper, we use evidence from Mesozoic volcanic rocks of the Pindos and Gavrovo-Tripolis isopic zones (Fig. 1) to suggest a model for the Mesozoic tectonic evolution of the Pindos Zone. We briefly review recently published data on Triassic volcanic rocks (Pe-Piper, 1982) and present new data on Jurassic volcanic rocks in the 'Radiolarite Series' of the Pindos Zone.

Triassic origin of the Pindos basin

Mid-Triassic volcanic rocks are preserved in about eight localities in the nappes of the external Hellenides, including two localities at the western edge of the Pindos Zone in continental Greece (Fig. 1). Immobile trace element analyses of 35 samples show a variety of geochemical characteristics suggesting the volcanic rocks are products of subduction (Pe-Piper 1982). The most tholeiitic rocks occur in the south and west; the most shoshonitic in the north and east. The latter rocks show trace element features suggestive of both within-plate alkaline volcanism and subduction. Such 'anomalous' basalts occur in a variety of tectonic settings, including at the edges of subduction zones, and are also associated with the early stages of back-arc spreading (Weaver *et al.*, 1979). Rocks of similar age occurring to the north in Yugoslavia likewise show some evidence of subduction (Bébien *et al.*, 1978) coupled with 'within plate alkali' characteristics (Djordjević *et al.*, 1982). Seidel & Kreuzer (1982) suggest that Late Permian alkaline volcanic rocks in western Crete and similar rocks in Corsica and Italy mark the initial formation of Mesogea with the rifting of Apulia away from Africa. If this interpretation is correct, then the Pindos basin may well be a back-arc basin resulting from a phase of subduction of Mesogea beneath the Apulian plate.

(?) Late Jurassic volcanic rocks

Lavas and pyroclastics associated with the Jurassic to earliest Cretaceous 'Radiolarite Series' are known from approximately ten outcrops in the Pindos Zone (Renz 1955; Fytrolakis 1971; Richter & Lensch 1977). Some are reported as having 'ophiolitic' and others 'calc-alkali' character based on major element geochemistry (Richter & Lensch 1977). All are altered and spilitized. In addition, tuff beds occur in the 'Radiolarite Series'.

We have made petrologic and geochemical analyses (Table 1) of rocks from two localities

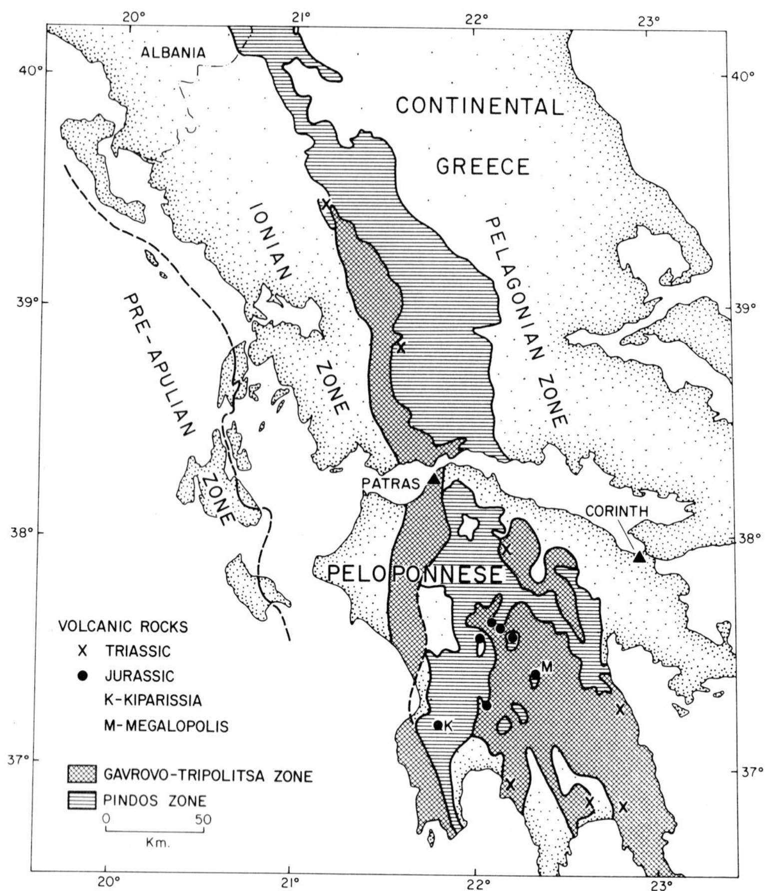

FIG. 1. Map of western Greece showing location of Pindos and Gavrovo-Tripolitsa isopic zones and volcanic rocks of (a) probable mid-Triassic age; and (b) of probable late Jurassic age.

in the central Peloponnese: altered basalts, andesites and dacites associated with pyroclastics at Paliohouni, east of Megalopolis (Richter & Lensch 1977) and a new outcrop of fresh pillow basalt at Selas, 15 km east of Kiparissia (map reference 698175). Immobile trace element abundances have been plotted on several discriminant diagrams, used to distinguish tectonic setting of ancient volcanic rocks (Table 2), which suggest that the rocks at Kiparissia are mid-ocean ridge basalt (MORB). The major element chemistry of the andesites and dacites suggests a calc-alkaline affinity at Megalopolis. The rare-earth element spectrum (Fig. 2) from Kiparissia is a typical enriched MORB or back-arc-basin basalt spectrum (Saunders & Tarney 1981) while the Megalopolis basalt shows very substantial REE enrichment more typical of subduction volcanics (Dostal *et al.*, 1977). The Kiparissia pyroxenes are those characteristic of ocean floor basalts using the criteria of Nisbet & Pearce (1977) and are similar to pyroxenes in MORB from the Phillipine Basin (Mattey *et al.*, 1980).

Such close juxtaposition of MORB and calc-alkaline rocks is known from present back-arc basins such as the Phillipine Basin, where the normal basin floor consists of MORB, while submarine calc-alkaline lavas and pyroclastics are associated with ocean-ocean subduction zones (Kroenke, Scott *et al.*, 1980). The andesites have a greater preservation potential than normal sea floor basalt because of their greater topographic elevation.

Conclusions

The trace-element interpretation of the mid-Triassic volcanic rocks of western Greece suggets that the Triassic Pindos basin was a back-arc basin, resulting from subduction of oceanic

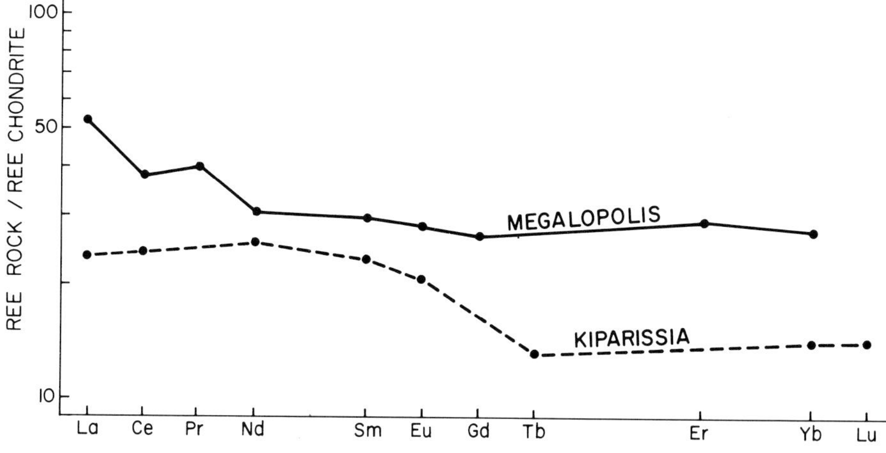

FIG. 2. REE Spectra from (?) Late Jurassic volcanic rocks.

TABLE 1. *Geochemical analyses of representative volcanic rocks from Radiolarite Series*

	Kiparissia basalt M711†	Megalopolis basalt M450*	Megalopolis andesite M453*
SiO_2	46.64	49.82	53.94
TiO_2	1.53	1.28	1.29
Al_2O_3	14.84	19.04	17.93
Fe_2O_{3t}	7.48	—	—
FeO_t	—	11.47	12.79
MnO	0.16	0.20	0.19
MgO	6.50	2.84	4.63
CaO	15.66	10.97	3.23
Na_2O	3.53	4.20	6.25
K_2O	0.96	0.05	0.07
P_2O_5	0.11	0.16	0.10
H_2O^+	2.55	n.d.	n.d.
H_2O^-	1.26	n.d.	n.d.
CO_2	3.44	n.d.	n.d.
Cr	595	0	n.d.
Ni	96	n.d.	n.d.
Rb	2	0	6
Ba	80	91	n.d.
Sr	231	40	87
Zr	119	76	101
Hf	3.23	n.d.	n.d.
Nb	n.d.	3	8
Ta	0.55	n.d.	n.d.
Zn	56	91	n.d.
La	7.68	16.7	n.d.
Ce	20.98	30.9	n.d.
Nd	15.11	18.31	n.d.
Sm	4.19	5.87	n.d.
Eu	1.43	2.04	n.d.
Tb	0.63	n.d.	n.d.
Yb	2.85	5.74	n.d.
Lu	0.49	n.d.	n.d.
Y	37	n.d.	n.d.
Th	0.73	0	n.d.

† Single analysis. Major elements by atomic adsorption; REE by INAA; trace elements by XRF.

* Two representative analyses from a total of four analysed samples. Major elements from average microprobe analysis of ten spots on fused glass disc using an ocean-floor basalt as an internal standard; REE and trace elements by XRF.

TABLE 2. *Geochemical discriminators of tectonic environment.*

Method (reference)	Kiparissia basalt	Megalopolis basalt
Ti-Zr-Y (a)	OFB	—
Th-Ta-Hf (b)	e-MORB	—
Cr-Y (c)	MORB	—
Ti-Zr (c)	MORB or WPB	(MORB or WPB)*
Ca/Yb-Ta/Yb (c)	MORB	—
Th/Yb-Ta/Yb (c)	MORB or TH	—
REE (d)	e-MORB or BAMBB	VAB
Clinopyroxenes (e)	OFB†	—

MORB = mid-ocean ridge basalt (e = enriched); OFB = ocean floor basalt; TH = tholeiite; BAMBB = back-arc marginal basin basalt; VAB = volcanic arc basalt; WPB = within plate basalt.

* Megalopolis andesite.

† Overlap with either VAB or WPB.

References: (a) Pearce & Cann, 1973; (b) Wood *et al.*, 1979; (c) Pearce, 1982; (d) Saunders & Tarney, 1981; Dostal *et al.*, 1977; (e) Nisbet & Pearce, 1977.

crust that lay to the present south-west. This oceanic crust was probably Mesogea which was subducted beneath Apulia.

The Pindos Zone is at least in part underlain by oceanic-type crust of probable Late Jurassic age, which is now preserved in only a few rare thrust slices. This resulted from complex opening of the Pindos basin, with the development both of mid-ocean-ridge-basalt and of andesites associated with ocean-ocean failed subduction zones, although we recognize that this interpretation is at present based on only a few geochemical analyses. We know of no evidence of ocean floor either older than or younger than that associated with the 'Radiolarite Series'. The age gap between this volcanic activity and that in the mid-Triassic suggests that they may have quite different origins. The lack of any Jurassic volcanism in the isopic zones to the west of the Pindos Zone suggests that subduction of Mesogea had ceased. The Late Jurassic back-arc volcanism might be related to subduction of the Vardar ocean from the (present) north-west. Alternatively, the MORB may represent an initial phase of Jurassic rifting, unrelated to subduction, followed by a phase of subduction, producing calcalkaline rocks, and perhaps obduction (Smith, 1979). Our data do not allow us to discriminate between these hypotheses.

ACKNOWLEDGMENTS: Work partly supported by NSERC grant to Piper. We thank R. M. MacKay and B. J. Fryer for assistance with analyses, and M. C. Blanchard-Williamson, L. F. Jansa and A. Kemp for review of the manuscript.

References

AUBOUIN, J., LE PICHON, X., WINTERER, E. & BONNEAU, M. 1979. Les Hellénides dans l'optique de la tectonique des plaques. *6th Colloq. Geol. Aegean Region*, **3,** 1333–54.

BÉBIEN, J., BLANCHET, R., CADET, V.-P., CHARVET, J., CHOROWICZ, J., LAPIERRE, H. & RAMPNOUX, J.-P. 1978. La volcanisme triassique des Dinarides en Yougoslavie: sa place dans l'évolution géotectonique peri-méditerranéenne. *Tectonophysics*, **47,** 159–76.

DJORDJEVIĆ, V., KARAMATA, S. & PAMIĆ, J. 1982. The Triassic rifting phase and development of oceanic basins in the Dinarides. *In:* Abstracts, *The Geological Evolution of the Eastern Mediterranean*, Edinburgh, U.K. September, 1982.

DOSTAL, J., ZENTILLI, M., CAELES, J. C. & CLARK, A. H., 1977. Geochemistry and origin of volcanic rocks of the Andes (26–28°S). *Contrib. Mineral. Petrol.* **63,** 113–28.

FYTROLAKIS, N. 1971. Geologische Untersuchungen in der Provinz von Pylias (Messenien-Peloponnes). *Ann. géol. Pays Hell.* **23,** 57–122.

JACOBSHAGEN, V., DÜRR, S., KOCKEL, F., KOPP, F.-O. & KOWALCZYK, G. 1978. Structure and geodynamic evolution of the Aegean region. *In:* H. CLOSS, D. ROEDER & K. SCHMIDT (Eds), *Alps, Apennines, Hellenides.* Schweizerbart., Stuttgart, pp. 537–64.

KROENKE, L., SCOTT, R. *et al.*, 1980. *Initial Reports of Deep Sea Drilling Project*, **59,** 820 p.

MATTEY, D. P., MARSH, N. G. & TARNEY, J. 1980. The geochemistry, mineralogy and petrology of basalts from the West Philippine and Parece Vela basins and from the Palau–Kyushu and West Mariana ridges, Deep Sea Drilling Project Leg 59. *In:* KROENKE, L., SCOTT, R. *et al.*, 1980. *Init. Repts. DSDP* **59,** 753–800.

NISBET, E. G. & PEARCE, J. A. 1977. Clinopyroxene compositions in mafic lavas in different tectonic settings. *Contrib. Mineral. Petrol.* **63,** 149–60.

PEARCE, J. A. 1982. Trace element characteristics of lavas from destructive plate boundaries. *In:* R. S. THORPE (ed), *Andesites.* Wiley, New York.

—— & CANN, J. R. 1973. Tectonic setting of basic volcanic rocks determined using trace element analyses. *Earth Planet. Sci. Lett.* **19,** 290–300.

PE-PIPER, G. 1982. Geochemistry, tectonic setting and metamorphism of mid-Triassic volcanic rocks of Greece. *Tectonophysics*, **85,** 253–72.

RENZ, C. 1955. *Die vorneogene Stratigraphie der normal sedimentären Formationen Griechenlands.* IGSR, Athens.

RICHTER, D. & LENSCH, G. 1977. Die ophiolitischen Vulkanit-Vorkommen der Olonos-Pindos-Decke im Zentralpeloponnes (Griechenland). *N. Jb. Miner. Abh.* **129,** 312–32.

SAUNDERS, A. D. & TARNEY, J. 1979. The geochemistry of basalts from a back-arc spreading centre in the East Scotia Sea. *Gerochim. Cosmochim. Acta.* **43,** 555–72.

SEIDEL, E. & KREUZER, H. 1982. Magmatism and metamorphism along the southwestern margin of The Apulian plate. *In: Abstracts, The Geological Evolution of the Eastern Mediterranean*, Edinburgh, U.K., September, 1982.

SMITH, A. G. 1979. Othris, Pindos and Vourinos ophiolites and the Pelagonian zone. *6th Colloq. Geol. Aegean Region*, **3,** 1369–74.

WEAVER, S. D., SAUNDERS, A. D., PANKHURST, R. J. & TARNEY, J. 1979. A geochemical study of magmatism associated with the initial stages of back-arc spreading. *Contrib. Mineral. Petrol.* **68,** 151–69.

WOOD, D. A., JORON, J.-L. & TREUIL, M. 1979. A re-appraisal of the use of trace elements to classify and discriminate between magma series erupted in different tectonic settings. *Earth Planet. Sci. Lett.* **45,** 326–36.

G. PE-PIPER, Department of Geology, St. Mary's University, Halifax, N.S. B3H 3C3, Canada.

D. J. W. PIPER, Atlantic Geoscience Centre, Geological Survey of Canada, Bedford Institute of Oceanography, P.O. Box 1006, Dartmouth, N.S. B2Y 4A2, Canada.

Origins and significance of rocks in an imbricate thrust zone beneath the Pindos ophiolite, northwestern Greece

Alan E. S. Kemp and Andrew M. McCaig

SUMMARY: Major zones of imbrication under the Pindos ophiolite of northwestern Greece relate to the formation of frontal and oblique ramp structures. These are marked by a northwest-trending line of tectonic windows in the ophiolite and a northeast-trending anticlinal structure which folds thrust contacts. The simplified structural sequence from top to bottom is:

(1) Ophiolite lithologies, mainly pillow basalts and basal serpentinite;

(2) Metabasites and meta-sediments of the ophiolite dynamothermal aureole ranging up to lower amphibolite facies;

(3) A complex sequence of basalt, gabbro, pelagic sediments, olistostromes, debris flows and breccias interpreted as a sea floor sequence (the Perivoli Complex);

(4) Early Cretaceous calcareous flyschoid sediments; and

(5) Tertiary Pindos Flysch and melange.

All the greenschist facies and the higher grade metamorphic rocks are interpreted as part of the metamorphic aureole and not as Palaeozoic basement. The generalized sequence of major deformational episodes is:

(A) Thrusting within the sea floor with concurrent metamorphism causing polyphase deformation and juxtaposition of the ophiolite, the dynamothermal aureole and the Perivoli Complex.

(B) The main imbrication along NNW-trending, E-dipping planes, involving all the lithologies and associated with the formation of major frontal and oblique ramp structures.

(A) occurred during the Middle Jurassic 'displacement' of the ophiolite sheet. 'Emplacement'-related deformation may have occurred in (1), (2) and (3), but is difficult to isolate due to the intensity of episode (B) which occurred during westward overthrusting in the Late-Tertiary and may have been related to an E-dipping subduction zone.

The Perivoli imbricate zone is developed along the major thrust separating the Othris Zone (Pindos ophiolite) and the Pindos zone in NW Greece (inset, Fig. 1). The Pindos ophiolite is probably continuous at depth below the Meso-Hellenic Trough molasse with the Vourinos and Othris ophiolites together forming a single ophiolite sheet (Smith *et al.* 1979). This ophiolite sheet, representing ?Upper Triassic to Middle Jurassic sea floor (Hynes *et al.* 1972; Roddick *et al.* 1979, Spray this vol.) was emplaced on to the SW margin of the Pelagonian zone in the Early Cretaceous, probably from the SW (Smith *et al.* 1979). The Othris Zone was thrust to the SW over the Pindos zone during a major regional deformation episode in the Late Tertiary (Aubouin 1965, Smith & Moores 1974). Petrological and geochemical studies of the Pindos ophiolite have been carried out by Parrot (1957) and more recently by Capedri *et al.* (1980). The structural complexity of the Perivoli corridor or 'demi-fenêtre' was noted by Brunn (1956) in the course of regional mapping at 1:50,000 scale and by Terry (1974, 1975), who also briefly described the internal structural complexity of the ophiolite.

Detailed mapping of a transect across the Perivoli corridor has shown the existence of a sub-ophiolite metamorphic aureole and a complex, tectonically underlying basalt-sediment sequence—the Perivoli Complex. These are involved with ophiolite lithologies and Cretaceous and Tertiary 'flysch' in a major imbricate thrust zone beneath the ophiolite.

Tectono-stratigraphy

The pre-imbrication stacking order is shown in Fig. 3a with estimated maximum thicknesses of the metamorphic aureole, Perivoli Complex and Early Cretaceous flysch indicated. The following are brief descriptions of the lithologies involved in the imbricate zone. The Perivoli Complex is described more fully below.

Ophiolite basalts

Dark green spilitized pillowed and massive basalts occur in tectonic contact with serpentinite in the ophiolite. The slices within the imbricate zone comprise mainly massive basalts with subordinate pillow forms. There is little intercalated sediment other than pillow debris. The amygdaloidal infill of pillow outer-margins is dominantly siliceous. Petrographically the

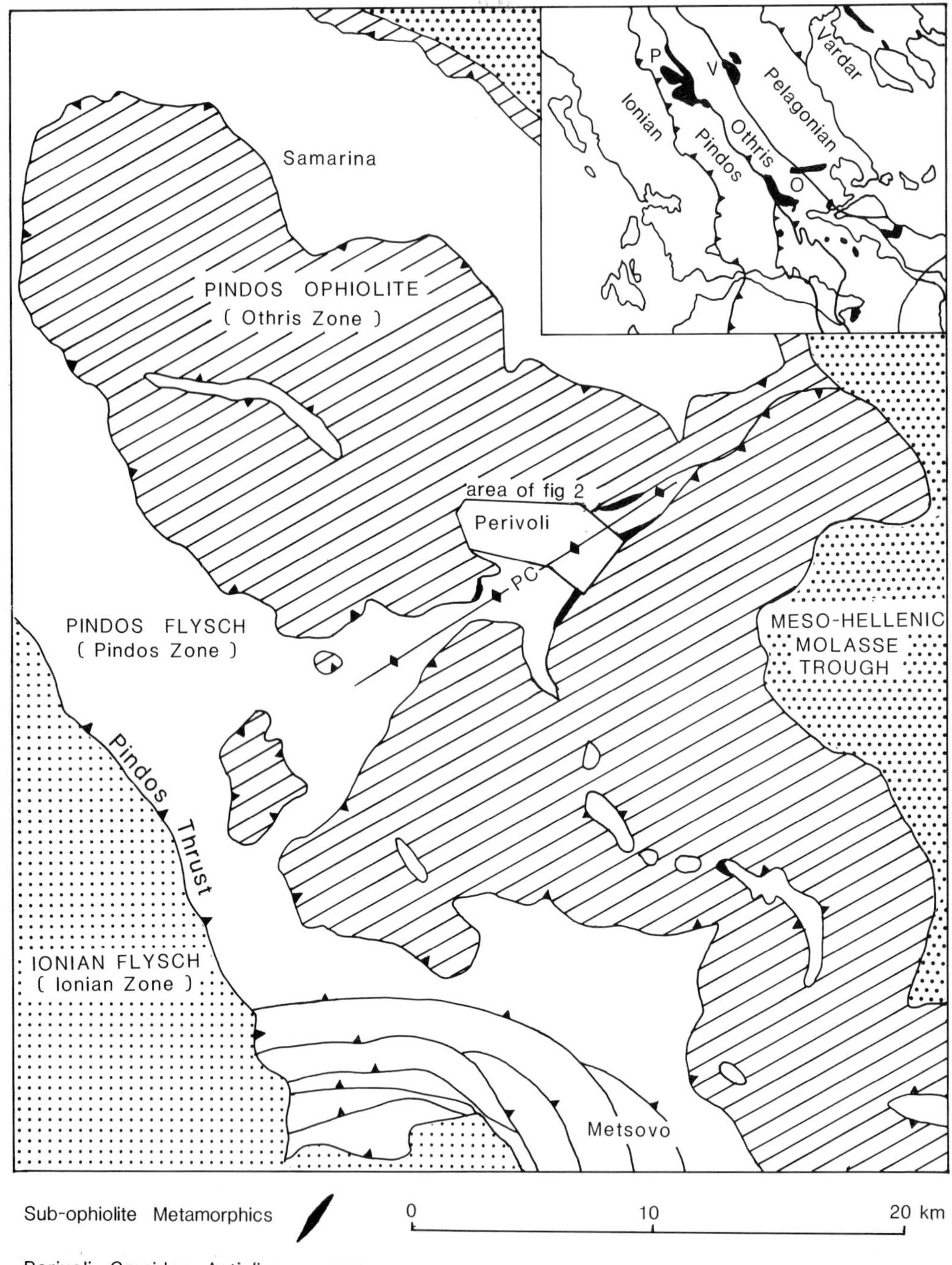

Fig. 1. Generalized map of Pindos ophiolite showing location of Perivoli corridor, northwest trending line of tectonic windows and distribution of sub-ophiolite metamorphics (modified from Brunn 1956). Inset shows location of ophiolites and isopic zones of Greece: P, Pindos; O, Othris; and V, Vourinos; after Aubouin 1965; Smith & Moores 1974.

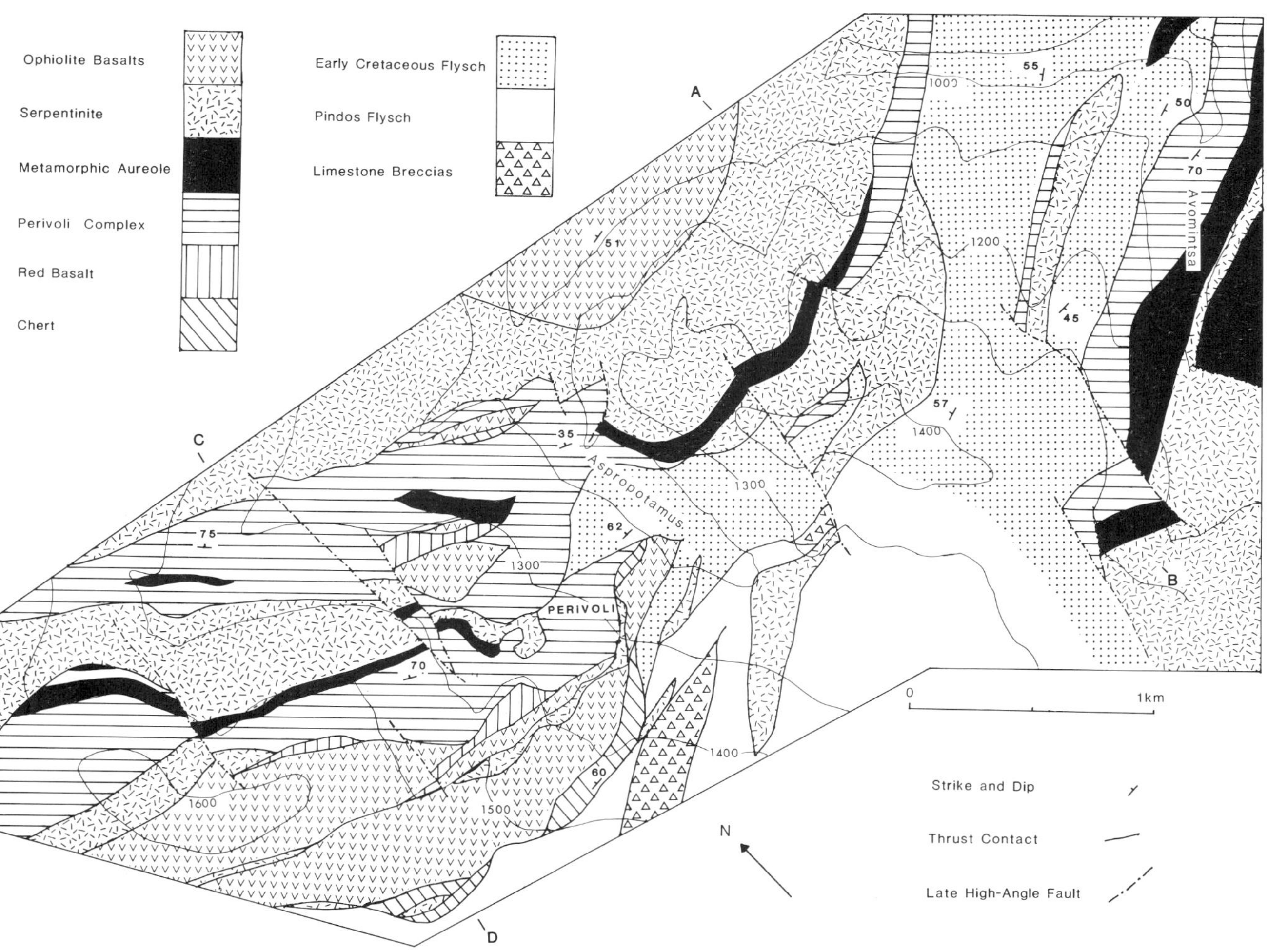

Fig. 2. Detailed map of transect across Perivoli corridor. Boundary between Pindos flysch and Early Cretaceous flysch in mid-part of corridor uncertain due to poor exposure.

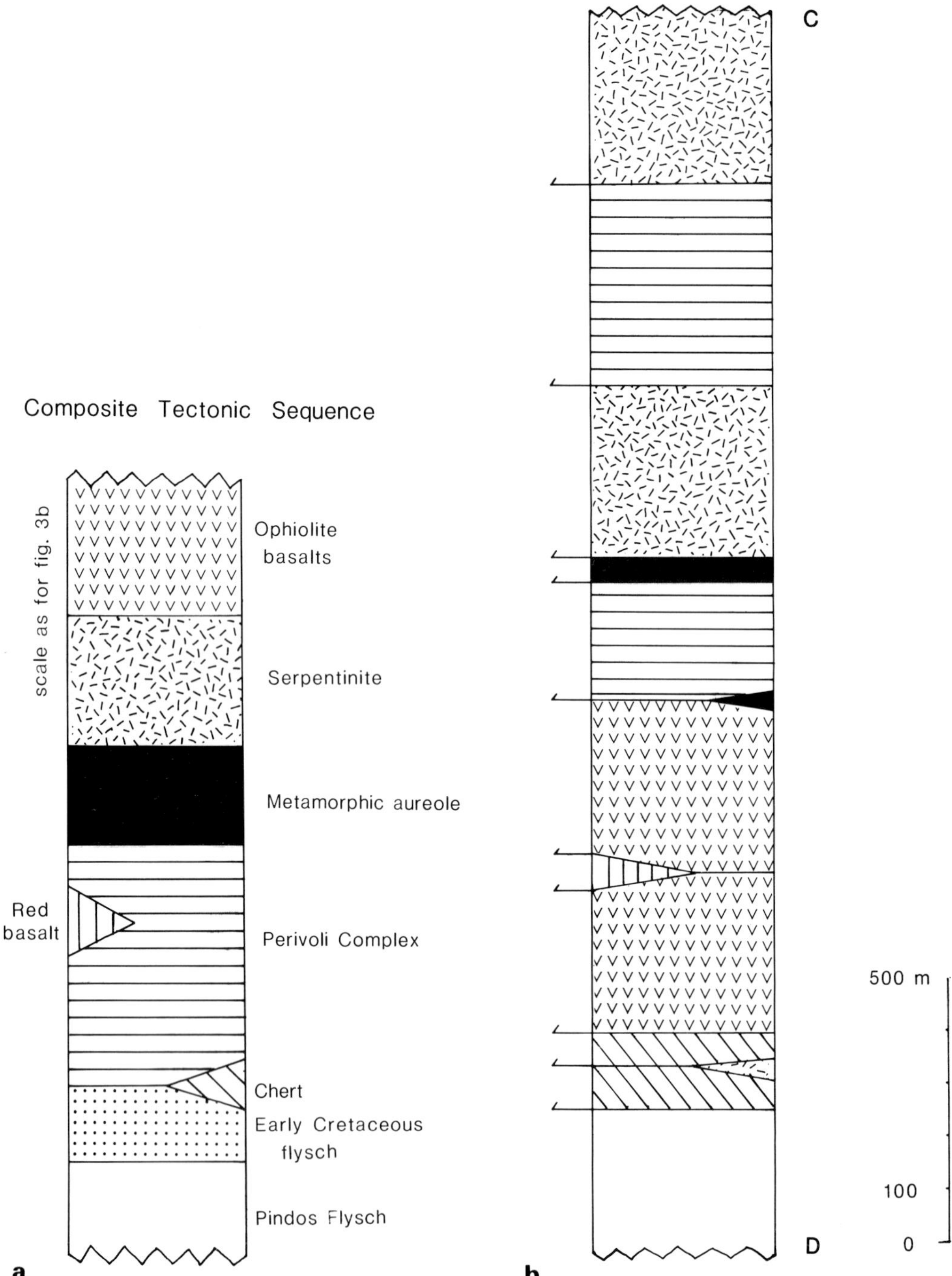

FIG. 3. (a) Generalized tectonic sequence in Perivoli area. Scale as for (b). (b) Tectonic sequence in main imbricate zone (C-D of Fig. 2).

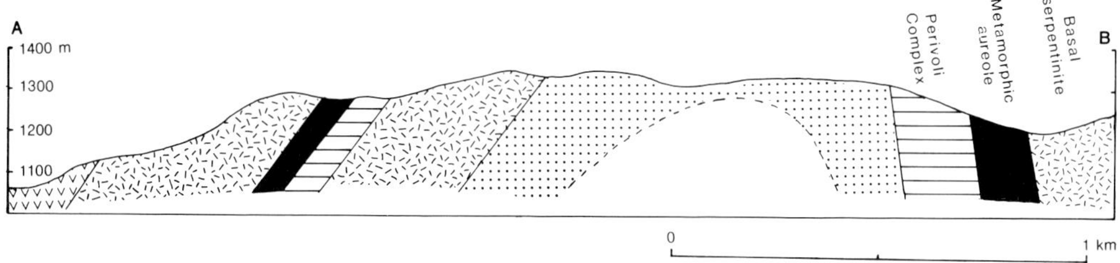

FIG. 4. Section across Perivoli corridor along line A-B of Fig. 2.

basalts commonly display an intergranular texture and resemble the Mirna group basalts of Othris. Limited major and trace element data suggests an affinity with low-K basalts of island arcs.

Serpentinite

Serpentinized ultramafics are present both in the basal part of the ophiolite and in discrete thrust slices in the imbricate zone. Cumulate textures and tectonite fabrics are rarely identifiable due to the high degree of serpentinization and intensity of shearing although they are well-developed within the ophiolite (Ross & Zimmerman 1982). In smaller thrust slices and near contacts, a characteristic lighter green serpentinite is developed, often with polished clasts in a pervasively sheared matrix.

Metamorphics

Meta-sediments and metabasites occur mainly in thin (less than 200 m), sheet-like bodies structurally below the serpentinite and above the Perivoli Complex. In the best preserved sequence in the Avomintsa River, the lower part (90 m), comprises mica-schists (becoming garnetiferous upwards), calcareous schists, marbles and rare graphitic schists intercalated with greenschist facies metabasites. The upper part (90 m) predominantly contains banded epidote-amphibolite schists of up to lower amphibolite facies. Minor folds are well developed in the amphibolites but later thrusting has substantially dismembered the sequence and orientation data are only locally consistent. Despite their disruption these metamorphic rocks closely resemble metamorphic soles occurring beneath ophiolites in the Tethyan and other belts (Woodcock & Robertson 1977; Williams & Smyth 1973, Spray *et al.* this volume). Although Brunn (1956) regarded these rocks as Palaeozoic basement, associated plagioclase amphibolites at Kiatra Fourka, 3 km SW of Perivoli and similar rocks at Othris and Vourinos have been dated as Middle Jurassic, and are thought to represent a metamorphic aureole formed during 'displacement' of the ophiolite (Spray and Roddick, 1980, Spray *et al.* this volume).

Early Cretaceous flysch

Redeposited limestones occur extensively in the core of the Perivoli corridor and in the lower part of the imbricate zone. The lithology comprises thin grey micrites and interbedded calcarenites and calcirudites with thin bedded siltstones. A coherent sequence is exposed in the slopes south-east of Perivoli showing an upwards passage from a micrite-rich (greater than 50%) facies with thin (1–3 cm) interbedded purple marls into a calcarenite/calcirudite-rich sequence with thin-bedded siltstones. The calcarenites include altered basaltic and minor chert fragments as well as redeposited limestone. The formation has been dated as Berriasian (Terry & Mercier 1971) and related to the 'Beotian Flysch' of the Peloponnese (Celet *et al.* 1976). It is lithologically and petrographically very similar to the slightly later 'first flysch' of the Pindos zone. The mafic detritus occurring in both these units is thought to have been derived from the ophiolite soon after its emplacement on to the Pelagonian zone in the earliest Cretaceous.

Pindos flysch and limestone breccias

In the south-west of the area at the base of the thrust stack, micaceous sandstones and shales typical of the Pindos flysch are present. Associated limestone breccias contain dominantly micritic fragments but also abundant Nummulitids probably indicating an Eocene age. Although coherent sequences are observed to the south-west of the area, the flysch is locally chaotic and some exotic clasts are present. It is probable therefore that elements of the Pindos flysch present below the imbricate zone may be equivalent to the melange of the

TABLE 1. *Lithologies and mode of occurrence in Perivoli Complex.*

Lithology	Nature of occurrence	Approx overall % of total outcrop area
Basalt	Pillowed and massive units forming blocks up to 50 × 30 m + with minor breccia units, hyaloclastites and tuffs	40
Gabbro	Thin (less than 3 m) breccia units providing local matrix for blocks	6
Red & Grey lime mudstones and shales and pink limestones	Blocks generally less than 20 m in stratigraphic thickness sometimes conformable with basalts. Slumped and boudinaged units merging locally with debris flows.	15
Dark grey and pale grey shales and limestones	As above	15
Red or green (5–10 cm) bedded cherts	Blocks generally less than 5 m thick	4
Matrix supported debris flow units containing clasts of the above lithologies in a shaly/calcareous mud matrix	Between and draped over blocks and forming more extensive 'Olistostrome' units with larger included blocks	15
Polygenetic breccias containing clasts of the above lithologies but including rare serpentinite	Between blocks forming local matrix	5

Anilion Complex (Lorsong 1977) occurring structurally beneath the Pindos ophiolite northeast of Metsovo.

The Perivoli Complex

The Perivoli Complex is a disordered mixture of mainly pelagic sediments and mafic igneous rocks. The best exposed and more accessible sections are in the Aspropotamos and Avomintsa rivers on the northwestern and southeastern sides of the Perivoli corridor respectively. The various lithologies recognized, together with their mode of occurrence are given in Table 1. There is no universal matrix and local matrix often appears to be derived from adjacent blocks. Stratigraphy is only discernible within individual blocks and no overall 'ghost' stratigraphy is apparent. It should be emphasized that most of the components occur throughout the outcrop area of the Complex and that as a whole it constitutes a recognizable and mappable unit.

Igneous components

Basalts

Dark red to green pillowed and massive basalts and rarer dolerite occur as blocks up to 50 m thick. Pillowed basalts frequently contain a profusion of calcite amygdales. Thin pink limestone horizons occur within some flow units and clasts of chert and limestone are also locally included. Petrographically, the basalts are highly altered spilites having undergone spilitization in a Ca-rich environment. Some contain a Ti-augite and they possess relatively high K_2O, Ti, Y and Zr, having tholeiitic to sub-alkaline characteristics and plot in the 'within-plate basalt' field (Druitt & Kemp, unpublished geochemical data).

Gabbro

Gabbro occurs in thin breccia units (less than 3 m thick) and in small blocks. It is invariably a spilitized and internally micro-cataclastic rock.

Extensive lithological units associated with the Perivoli Complex

Red basalts

Thin units of red and purple, highly amygdaloidal pillow basalts occur frequently associated with the Perivoli Complex in the imbricate zone. Most contacts with Perivoli Complex lithologies are tectonic although in some cases they may represent an original 'sedimentary' intercalation. Near Avdella, to the north, Terry (1971) described similar basalt in thin thrust slices with associated sediments in part resembling Perivoli Complex lithologies, but including

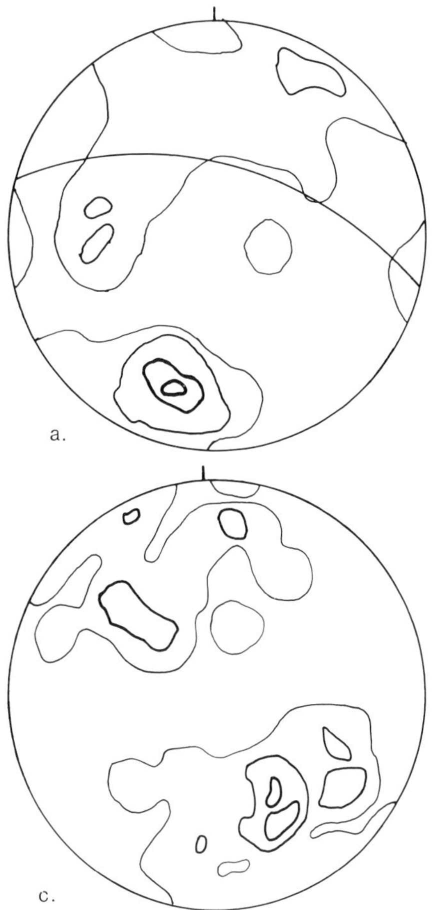

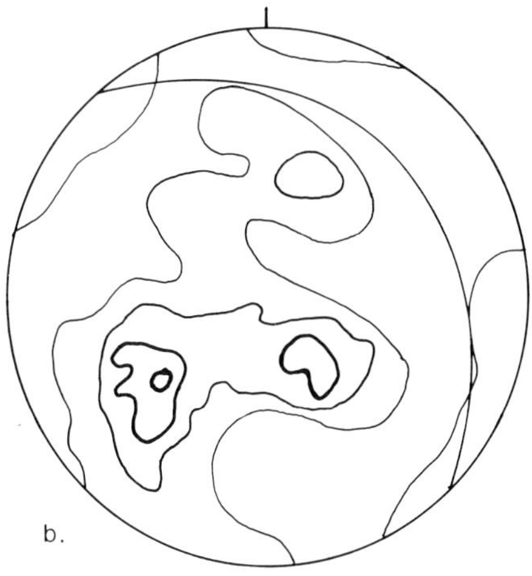

FIG. 5 (a) Poles to shear/thrust contacts in the main imbricate zone. Equal area, lower hemisphere projection contoured at 1, 3, 5 and 7 per cent intervals per 100/33 per cent area of net (33 poles). Modal plane 106/62 N.

(b) Poles to orientation of planar clasts (all sizes) in Perivoli Complex. 1, 3, 5 and 7 per cent intervals/31 poles. Modal plane 134/30 NW.

(c) Poles to bedding of Early Cretaceous flysch in NE part of Corridor: 1, 4, and 7% intervals/64 points. They define a major NE-trending fold structure parallel to the Corridor axis.

Triassic *Halobia*-limestones. In general these basalts are typical of Upper Triassic rift-associated alkali basalts developed elsewhere in the Hellenides (e.g. Hynes 1974).

Cherts

Red, thin (5–10 cm) bedded cherts occur in a number of thrust slices (maximum thickness 80 m) in the imbricate zone. They are very similar to red cherts within the Perivoli Complex and probably represent discrete thrust slices of these. However, in the main fault-bounded unit a maximum thickness of 80 m of chert is overlain by basalt-derived sandstones and rudites containing shallow water limestone clasts.

Internal structure

The sedimentary components of the Perivoli Complex contain a large variety of soft-sediment deformation structures. Some units of red/grey lime-mudstones and shales show the effects of a pervasive stretching deformation with resulting boudinage of micrite beds. Slump folds occur both in deformed bedded sequences and as isolated hinges in debris-flow units. Locally coherent sequences appear to merge laterally with slumped horizons. Some horizons of en-echelon angular micrite bed fragments in shales probably represent the *in situ* break up of partially cemented limestone in a ductile shale matrix. Coherent sediment blocks generally rest on and are draped by slumped or debris-flow units or thicker olistostrome units. The crude sedimentary fabric of slumped and olistostrome units and the bedding orientation of many of the large coherent blocks is frequently sub-parallel to thrust contacts (Fig. 5a,b) and many of the large coherent slabs of sediment probably represent slide-blocks. Local early high-angle faults cut by later thrust-related shear zones often juxtapose slumped horizons or olistostrome units with coherent sediment or

TABLE 2. *Main deformational episodes affecting Perivoli lithologies related to regional events.*

Age	Deformation in Perivoli area	Regional events
?Post-Miocene	High-angle faulting	High-angle extensional faulting
Mid-Tertiary (?Miocene)	Major imbrication and formation of frontal and oblique ramps as 'trailing' edge of ophiolite sheet (Pindos ophiolite) thrust over Pindos flysch wedge.	Westward thrusting of Othris Zone over Pindos Zone and Pindos Zone over Ionian Zone.
Lower Cretaceous	?Internal deformation and partial serpentinization of ophiolite. Folding of serpentinite-basalt contacts.	Ophiolite emplacement along W. margin of Pelagonian massif (Othris and Vourinos)
Lower-Middle Jurassic	Formation and internal deformation of sub-ophiolite metamorphic aureole and variable low grade metamorphism and thrusting in Perivoli Complex during overriding by 'displaced' ophiolite.	Displacement of ophiolite concurrent with formation of sub-ophiolite metamorphic aureole.
?Upper Triassic-Middle Jurassic	Faulting, slumping and soft-sediment deformation in Perivoli Complex	Rifting and spreading of 'Othris ocean'.

basalt blocks consistent with syn-depositional faulting. The presence of micro-cataclastic gabbro in breccias may indicate that this faulting was significant.

The Perivoli Complex was significantly modified by post-depositional thrusting. The matrix of many of the debris-flow and olistostrome units is often pervasively sheared, especially near thrust contacts. Some of the sequence is partially recrystallized, especially towards the base of the metamorphic aureole in the Avomintsa section in the southern limb of the corridor, locally a penetrative cleavage is developed in the shaly matrix of debris-flow and olistostrome units.

Structure

The main deformational episodes affecting the rocks in the Perivoli area are summarized in Table 2. Interpretation of the significance of Tertiary thrusting is made difficult by the presence of 'displacement' and 'emplacement' related structures in the Pindos ophiolite, aureole rocks and the Perivoli Complex. We have assumed that, since many of the mapped thrusts involve also Pindos Zone lithologies, the Tertiary event produced most of the imbrication observed at the base of the ophiolite.

The Perivoli area defines a broad asymmetrical anticline folding the Pindos ophiolite, the aureole and the Perivoli Complex (Fig. 4) as well as the tectonically underlying sediments and melange of the Pindos Zone (Fig. 5c).

The main imbricate zone on the northern limb of the corridor comprises a stack of steeply northward dipping thrust slices striking at 100° and swinging to about 70° towards the corridor axis. This thrust stack (Fig. 3b) represents a two-fold repetition of the basal ophiolite/Pindos Zone sequence (Fig. 3a). However, more intense imbrication is indicated by the presence of many discontinuous lenticular thrust slices (Fig. 2) and frequent major faults within the Perivoli Complex and serpentinite (not marked on Fig. 2), as well as the presence of a major thrust slice of ophiolite basalts near the base of the thrust stack (Figs 2, 3b).

Much mobilization of serpentinite occurred during thrusting so that 1–2 m thick serpentinite sheets commonly occur between thrust slices. The lubricating effect of highly ductile serpentinite may in part be responsible for the intensity of imbrication.

Imbrication is less intense on the steeply southward-dipping or overturned southern limb of the corridor (Fig. 2), where the original (?pre-Tertiary) stacking order (Fig. 3a) is preserved, although there is considerable deformation within the ophiolite itself.

Imbrication in thrust belts commonly occurs at the leading edge of major thrust sheets or by collapse of footwall ramps (Dahlstrom 1972; Elliott & Johnson 1980). Footwall imbrication frequently causes folding of higher thrusts resulting in tectonic windows in the overlying thrust sheet. The summary map (Fig. 1) shows that the Periovoli area is in line with a series of NW-SE trending tectonic windows in the ophiolite. These may have been formed over a major frontal ramp as the Pindos ophiolite 'rose' over the thick Pindos Flysch wedge during the Tertiary thrusting.

The Perivoli corridor itself contains an anomalous intensity of imbrication. Although both hanging wall and footwall are folded about the Perivoli corridor, the Pindos thrust—the basal or sole thrust of the Pindos Zone—is not affected by the folding and the Perivoli corridor passes to the NE into a major fault within the Pindos ophiolite separating higher structural levels (basalts and dolerites) to the north from lower structural levels (peridotites) to the south. It is probable, therefore, that the Perivoli corridor itself formed during thrusting in response to an oblique NE-trending ramp on which the Pindos ophiolite sheet to the south of the Perivoli corridor was partially thrust over the ophiolite sheet to the north. The structures in the Perivoli area were formed by collapse of the oblique ramp and folding over the resulting stack of imbricates. The original siting of the oblique ramp may have been due to lateral thickness variations in the footwall comprising the Pindos flysch and pelagic sediments and melange.

The structural pattern of the Perivoli corridor is similar to transverse anticlinal structures in other thrust belts which have been related to the formation of lateral or oblique ramps with associated hanging wall and footwall imbrication, e.g. the Dundonnell structure in the Scottish Moine thrust belt (Elliott & Johnson 1980).

Origins and regional significance of the Perivoli Complex

Breccias of basalt, gabbro and associated sediments have been described from various faulted, slow-spreading mid-oceanic ridges and compared with ophiolitic sequences occurring in the Apennines (Barrett & Spooner 1977). A similar origin for the Perivoli Complex is suggested from its variety of mafic igneous and almost wholly pelagic sedimentary components together with the abundant evidence of syn-depositional slumping and faulting. Although some components of the Perivoli Complex may have been shed from an over-riding or adjacent ophiolite sheet, the petrological differences with ophiolite basalts and the scarcity of plutonic components militate against this.

The consistent occurrence of the Perivoli Complex structurally below the metamorphic aureole, together with evidence of its partial recrystallization and the diversity of meta-sediments towards the base of the aureole suggest that the Perivoli Complex may have formed the protolith for the lower part of the aureole. Hence, the Perivoli Complex was probably formed before or during the Middle Jurassic 'displacement' (Spray & Roddick 1980) of the ophiolite. This age evidence is substantiated by the probable Upper Triassic age of the associated red amygdaloidal basalts and the resemblance of many of the sedimentary components of the Perivoli Complex to Upper Triassic to Middle Jurassic facies of the Pindos Zone developed further south.

The petrology and chemistry of the Perivoli Complex basalts and associated red basalts suggest formation at a marginal part of the 'Othris ocean' (Smith *et al.* 1979, Smith & Spray, this volume), or in some marginal basin, (cf. Pe-Piper & Piper, this volume) over-ridden by the Pindos ophiolite during its 'displacement'. Major decollement subsequently occurred below or within, the Perivoli Complex and it was preserved at a high structural level during later thrusting.

It is possible, then, that the Perivoli Complex might in part represent a basinward lateral correlative of the Agrilia and overlying Neokhorion Formations of Othris (Smith *et al.* 1975).

Discussion

Intermediate or alkaline basalts in the volcanic-sedimentary protolith sequences of the lower part of ophiolite dynamothermal aureoles occur elsewhere in the Tethyan region and in other orogenic belts. Their presence has been variously interpreted as recording an ophiolite sheet riding over marginal sea floor, as here, or seamounts close to the continental margin (e.g. St Anthony Complex in Newfoundland, Jamieson 1980) or both (e.g. Haybi Volcanics, in Oman, Searle *et al.* 1980, Searle & Malpas 1980) during the displacement event.

Protolith sequences have not been previously recognized associated with metamorphic aureoles in other ophiolites of the Othris zone. This may indicate that the protolith sequences were only patchily welded on to the aureole during displacement. However, many of the ophiolites were significantly deformed and eroded during emplacement, shedding blocks, particularly from their lower levels into frontal sedimentary melanges which were subsequently overridden by thrust sheets (e.g. in Othris Smith *et al.*, Celet *et al.* 1977, and in Iti and Kalidromon, Celet *et al.* 1977). Due to the thin and probably impersistent original configuration of protolith sequences at the base of ophiolites they would have been particularly susceptible to being dismembered and incorporated in subjacent melanges; perhaps these sequences are the best place to look for protolith materials.

Significantly however, most of the other Othris Zone ophiolites, e.g. Othris and Vourinos, represent the leading edge of an ophiolite sheet, i.e. the part emplaced on to the Pelagonian margin (Fig. 6). In the model of Smith *et*

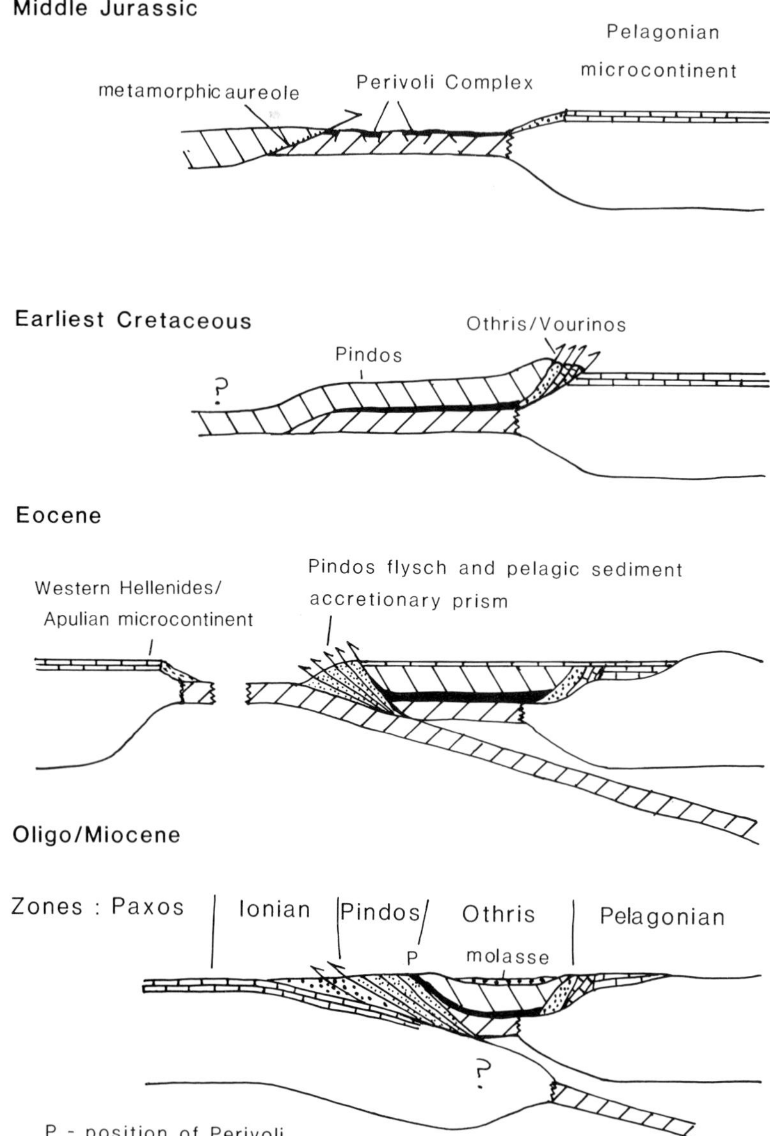

FIG. 6. Plate tectonic cartoons showing position of Perivoli area during evolution of region (modified from Smith *et al.* 1979).

al. (1979), the Pindos ophiolite would represent the trailing edge of the ophiolite sheet which was thrust over the Pindos flysch and pelagic sediment wedge (interpreted by Lorsong 1979 as an accretionary prism) during the Tertiary (Fig. 6). Thus, in the case of the NW Greek ophiolite belt the protolith for the lower part of the ophiolite dynamothermal aureole may have been left behind during emplacement and is preserved in part below the Pindos ophiolite as this represents the trailing edge of the displaced ophiolite sheet (Fig. 6).

Our geochemical data from the ophiolite (low-K tholeiite characteristics) correspond closely with the 'pillow basalts' and 'dykes' analysed by Capedri *et al.* 1980. However, the anomalous 'massive basalts' and some inconsistencies in others of Capedri *et al.*'s groups could in part be due to sampling from basalts and dolerites of the Perivoli Complex.

ACKNOWLEDGEMENTS: The fieldwork was aided by grants from the Hutchins Fund of St John's College Cambridge and from the Sheffield Church Burgesses' Educational Fund. Tim Druitt performed trace elements analyses. We are grateful to A. Gilbert Smith and J. A. Lorsong for introducing us to the Hellenides and to T. H. Green, N. H. Woodcock, J. G. Spray, M. A. Naylor and M. R. W. Johnson for many valuable discussions. Mrs M. Williams typed the manuscript.

References

AUBOUIN, J. 1959. Contribution a l'étude géologique de la Grèce septentrional: les confins de l'Epire et de la Thessalie. *Ann. géol. Pays hellén*, **10,** 1–403.

—— 1965. *Geosynclines*. Elsevier, Amsterdam. 335 pp.

BARRETT, T. J. & SPOONER, E. T. C. 1977. Ophiolitic breccias associated with allochthonous oceanic crustal rocks in the East Ligurian Apennines, Italy—a comparison with observations from rifted oceanic ridges. *Earth planet. Sci. Lett.* **35,** 75–91.

BRUNN, J. H. 1956. Contribution à l'étude géologique du Pinde septentrional et d'une partie de la Macédoine occidentale. *Ann. géol Pays hellén*, **7,** 358 pp.

ÇAPEDRI, S., VENTURELLI, G., BOCCHI, G., OSTAL, J., GARUTI, G. & ROSSI, A. 1980. The geochemistry and petrogenesis of an ophiolitic sequence from Pindos, Greece. *Cont. Mineral. Petrol.* **74,** 189–200.

CELET, P., CLEMENT, B. & FERRIERE, J. 1976. La zone béotienne en Grèce: Implications paléogéographiques et structurales. *Eclog. geol. Helv.* **69,** 577–99.

——, FERRIERE, J. & WIGNIOLLE, E. 1977. Le problème de l'origine des blocs exogènes du mélanges à éléments ophiolitiques au sud de Sperchios et dans le massif de l'Othrys (Grèce). *Bull. Soc. géol. Fr.* **19,** 935–42.

DAHLSTROM, C. D. A. 1970. Structural geology in the eastern margin of the Canadian Rocky Mountains. *Bull. Canad. Petrol. Geol.* **18,** 332–406.

ELLIOTT, D. & JOHNSON, M. R. W. 1980. Structural evolution in the northern part of the Moine thrust belt, NW Scotland. *Trans. Roy. Soc. Edinb. (Earth Sci.)* **71,** 69–96.

HYNES, A. J. 1974. Igneous activity at the birth of an ocean basin in western Greece. *Can. J. Earth Sci.* **11,** 842–853.

——, NISBET, E. G., SMITH, A. G., WELLAND, M. J. P. & REX, D. C. 1972. Spreading and emplacement ages of some ophiolites in the Othris region (eastern central Greece). *Z. Deutsch. geol. Ges.* **123,** 455–468.

JAMIESON, R. A. 1980. Ophiolite emplacement as recorded in the dynamothermal aureole of the St. Anthony complex north-western Newfoundland. *In*: A. PANAYIOTOU, (ed.), *Proceedings of the International Ophiolite Symposium, Cyprus*, 1979. Cyprus Geological Survey.

LORSONG, J. A. 1977. Stratigraphy of the Pindos Flysch in the Politses mountains, northwestern Greece. *In*: KALLERGIS, G. (ed.) *Proceedings of the 6th Colloquium on the Geology of the Aegean Region.* Inst. of Geol. and Mining Research, Athens, 1977, 703–14.

—— 1979. *Sedimentation and deformation of the Pindos and Ionian flysch, northwestern Greece* unpublished University of Cambridge Ph.D. thesis, 316 pp.

NAYLOR, M. A. & HARLE, T. J. 1976. Palaeogeographic significance of rocks and structures beneath the Vourinos ophiolite, northern Greece, *J. geol. Soc. Lond.* **132,** 667–76.

PARROT, J. F. 1957. *Le cortège ophiolithique du Pinde septentrional (Grèce)*. O.R.S.T.O.M., 114 pp.

RODDICK, J. C., CAMERON, W. E. & SMITH, A. G. 1979. Permo-Triassic and Jurassic ^{40}Ar-^{39}Ar ages from Greek ophiolites and associated rocks. *Nature*, **279,** 788–790.

ROSS, J. V. & ZIMMERMAN, J. 1983. The Pindos ophiolite complex, Northern Greece: evolution and tectonic significance of the tectonic suite. *Abstract in*: DIXON, J. E. & ROBERTSON, A. H. F. (eds), *The Geological Evolution of the Eastern Mediterranean*, Abstr., Edinburgh 1982. p. 95.

SEARLE, M. P., LIPPARD, S. J., SMEWING, J. D. & REX, D. C. 1980. Volcanic rocks beneath the Semail Ophiolite nappe and their significance in the Mesozoic evolution of Tethys. *J. geol. Soc. Lond.* **137,** 589–604.

—— & MALPAS, J. 1980. Structure and metamorphism of rocks beneath the Semail Ophiolite of Oman and their significance in ophiolite obduction. *Trans. Roy. Soc. Edinb. (Earth Sci.).* **71,** 247–262.

SPRAY, J. G. & RODDICK, J. C. 1980. Petrology of Hellenic subophiolite metamorphic rocks. *Contr. Mineral. Petrol.* **72,** 43–55.

SMITH, A. G., HYNES, A. J., MENZIES, M., NISBET, E. G., PRICE, I., WELLAND, M. J. & FERRIERE, J. 1975. The stratigraphy of the Othris Mountains, Eastern Central Greece: a deformed continental margin sequence. *Eclog. geol. Helv.* **68,** 463–81.

—— & MOORES, E. M. 1974. Hellenides. *In*: SPENCER, A. M. (ed). *Mesozoic and Cenozoic Orogenic Belts. Spec. Publ. geol. Soc. London*, **4,** 159–85.

——, WOODCOCK, N. H. & NAYLOR, M. A. 1979. The structural evolution of a Mesozoic continental margin, Othris Mountains, Greece. *J. geol. Soc. Lond.* **146,** 589–603.

TERRY, J. 1971. Sur l'âge triassique de laves associées à la nappe ophiolitique du Pinde septentrional (Épire et Macédoine, Greece). *C.r. Somm. Soc. géol. France*, 384–385.

—— 1974. Ensembles lithologiques et structures internes du cortège ophiolitique du Pinde septentrional (Grèce), construction d'un modèle petrogénétique. *Bull. Soc. géol. Fr.* **16,** 204–213.

TERRY, J. 1975. Echo d'une tectonique jurassique: les phénomènes de resédimentation dans le secteur de la nappe des ophiolites du Pinde septentrional (Grèce). *C.r. Somm. Soc. géol. France*, 49–51.

—— & MERCIER, M. 1971. Sur l'existence d'une série détritique berrasienne intercalée entre la nappe des ophiolites et le flysch éocène de la nappe du Pinde (Pinde septentrional, Grèce). *C.r. Somm. Soc. géol. France*, 71–73.

WILLIAMS, H. & SMYTH, W. R. 1973. Metamorphic aureoles beneath ophiolite suites and Alpine Peridotites: tectonic implications with Newfoundland examples. *Am. Jour. Sci.* **273,** 594–621.

WOODCOCK, N. H. & ROBERTSON, A. H. F. 1977. Origins of some ophiolite-related metamorphic rocks of the 'Tethyan' belt. *Geology*, **5,** 373–76.

ALAN E. S. KEMP, Grant Institute of Geology, West Mains Road, Edinburgh EH9 3JW.

ANDREW, M. MCCAIG, Department of Earth Sciences, University of Dundee, Dundee, DD1 4HN, Scotland.

Structural evolution of the Pelagonian Zone in Northwestern Macedonia, Greece

D. Mountrakis

SUMMARY: The pre-Upper Palaeozoic basement of the North Pelagonian zone s.l. consists of two units: the Eastern unit of the Voras Mountains and the Western unit of the Vernon Mountains. These units are two parallel crystalline sequences which have been metamorphosed under similar conditions in the greenschist facies. An Upper Carboniferous granite (302.4 ± 5/15 Ma) intrudes the metamorphic rocks of the Vernon unit. During Upper Palaeozoic time, subsidence of the western Pelagonian margin created a possible slope where a meta-clastic sequence was deposited. The clastic sediments were succeeded by a neritic carbonate cover of Middle Triassic–Jurassic age. In the eastern area of the Pelagonian zone a metamorphic carbonate mass, probably of the Triassic–Jurassic age, has been tectonically emplaced on to the crystalline basement. This carbonate cover has a possible para-autochthonous origin; thrust surfaces, folds and linear structures indicate that it was overthrust from the eastern Pelagonian margin. Two metamorphic events in the North Pelagonian zone are inferred: one pre-Upper Carboniferous which affected the crystalline basement and a second post-Jurassic which affected the Upper Carboniferous granite, the Upper Palaeozoic meta-sediments and the Triassic–Jurassic carbonates.

The concept of the Pelagonian Zone

There has been some confusion in recent years as to the meaning of the term Pelagonian Zone mainly as applied to Central Greece, where the term has been restricted by some to the neritic carbonate sediments of the Pelagonian Zone of Triassic–Jurassic age (Celet & Ferrière 1978; Papanikolaou 1981). This is inconsistent with the initial definition (Kossmat 1924) of the metamorphic rocks of the pre-Alpine basement of ancient 'Pelagonia' as the Pelagonian Massif. In order to avoid this confusion and since the present work concerns north-west Macedonia, where ancient Pelagonia existed, the term Pelagonian Zone is used in the sense of the 'Pelagonian' of Celet & Ferrière (1978) to include all the structural successions of the pre-Alpine and Alpine rocks between the Sub-pelagonian Zone and the basal thrust of the Almopias (Vardar) Zone (Fig. 1).

The purpose of this study is to provide new evidence concerning the tectonic development of the Pelagonian Zone in its northern outcrop area.

The pre-Permian basement

An initial discussion of the problems concerning the crystalline rocks of the 'Pelagonian Massif' in north-west Macedonia was given in a previous article (Mountrakis 1982), in which it was shown that the 'Pelagonian Massif' is not homogeneous, but consists of two crystalline units: the Eastern unit of the Voras (Kaimaktchalan) Mountains and the western unit of the Vernon Mountains (Fig. 1).

The pre-Alpine basement substratum of the Pelagonian Zone was shown not to be homogeneous in other areas as well. In the area of Pieria-Olympus it is divided into two thrust sheets: the Pieria Allochthon and the Livadi Complex (Barton 1976; Nance 1977; Yarwood & Dixon 1977). Further south, in Pelion, South Euboea and the Cyclades, various structural units have been recognised consisting of metamorphic rocks and Mesozoic sediments which form stacks of thrust sheets (Ferrière 1977; Katsikatsos 1977; Dürr *et al.* 1978; Jacobshagen *et al.* 1978).

For north-west Macedonia the problem is to establish the relation between the Voras and the Vernon units. Detailed petrographic observations showed that the Voras unit consists of orthogneisses, augen-gneisses and amphibolites at the base, passing up into two mica-schists, amphibolitic schists and phengite (sericite) schists. Small unmetamorphosed granodioritic bodies intrude the metamorphic rocks. The Vernon unit includes a series of metamorphic rocks which are similar to, and in analogous sequence with, those of Voras, i.e. orthogneisses, amphibolites, two mica schists, amphibolitic-schists and quartzites.

A large mass of granite, now mostly augen-gneiss (Mountrakis 1982), has intruded the Vernon metamorphic rocks of the Kastoria area. This granite extends northwards to the Florina area and into Yugoslavia (Fig. 2). Kilias (1980) examined this granite mass in the vicinity of Florina (Greece) and distinguished two main

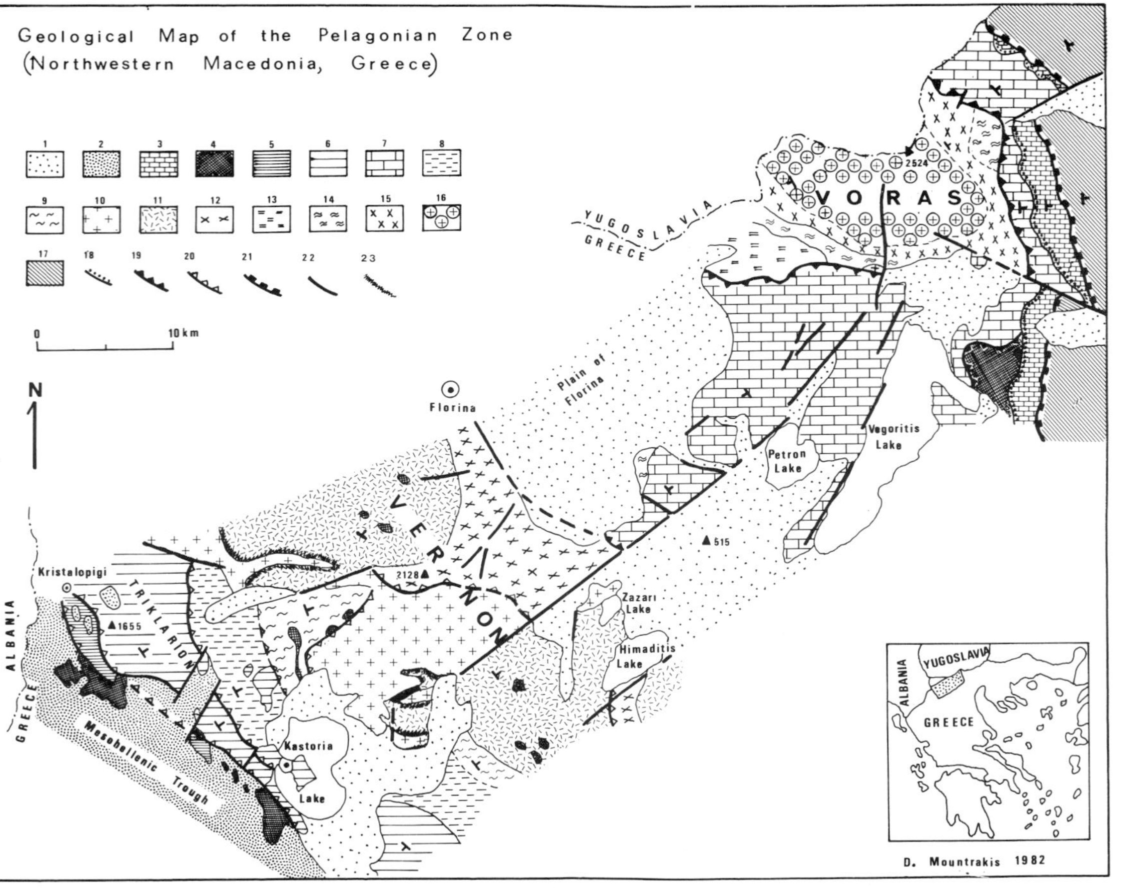

Fig. 1. Geological map of the Pelagonian Zone in Northwestern Macedonia.
1, Alluvial and Plio-quaternary deposits; 2, Molasse-type sediments of the Mesohellenic Trough (Oligocene-Miocene); 3, Upper Cretaceous sediments; 4, Ophiolites and associated sediments; 5, Subpelagonian sedimentary sequence (Upper Triasic-Lower Cretaceous); 6, carbonate cover of the western Pelagonian margin (Triassic–Jurassic); 7, carbonate cover of the eastern Pelagonian margin (Triassic–Jurassic); 8–9, metaclastic sequence; 8, Aposkepos series; 9, Sidirochori series (Permian-Lower Triassic); 10, Kastoria–Florina granite and augen-gneisses derived from it (Upper Carboniferous); 11–12, crystalline unit of Vernon; 11, Amphibolites, two mica and amphibolite-schists of Klisoura series; 12, orthogneisses, mylonites and amphibolites of Vitsi series; 13–16, crystalline unit of Voras; 13, phengite schists and phyllites; 14, amphibolites, two mica and amphibolite-schists; 15, gneisses; 16, augen-gnesisses and orthogneisses; 17, Almopias slices (Vardar zone); 18, Upper Cretaceous transgressive limestones; 19, Upper Jurassic to Lower Cretaceous thrust; 20, thrust of presumed Tertiary age; 21, thrust of the Almopias (Vardar) zone; 22, fault; and 23, intrusive contact.

granitic rock-types representing two intrusive events, with cross-cutting relations between them, which revealed their relative ages. These two granitic rock-types have very similar compositions and appear to form one complex granite body. Both granitic rock-types are partly metamorphosed to augen-gneiss, and both clearly intrude the surrounding metamorphic rocks with transformation of the schists to hornfelses containing sillimanite and relic cordierite. The older granitic rock-type extends from the Florina area southwards to Kastoria with exactly the same composition and fabric. This mass is called the Kastoria granite here. Intrusive contact phenomena, similar to those described in the Florina area, have also been observed in the Kastoria area, where the granite also cuts the earlier metamorphic fabric of the Vernon crystalline rocks and has partially assimilated schists transformed to hornfelses. Intrusive contact phenomena are also known from the continuation of the same granite body in Yugoslavia (Stojanov 1971).

K/Ar age determinations carried out on samples of the Florina granite (Fig. 2) by Marakis (1969) gave the following ages: 217 ± 6, 228 ± 6, 238 ± 6, 258 ± 7, 268 ± 7, 461 ± 12, 465 ± 12 Ma. K/Ar ages of 468 ± 14 and 526 ± 16 Ma were also obtained by Kilias (1980) for the younger granitic rock-type of the Florina area (Fig. 2). Recent U-Pb geochronology on the Kastoria granite (Fig. 2) shows that the zircon U-Pb system closed at 302.4 ± 5/15 Ma (Westphalian) during cooling of the granite magma, with a small but significant inherited earlier component from re-mobilised crustal material. The samples were collected ca. 2.5 km south of the highest peak of Mount Vitsi (2128 m). These samples represent the older rock-type of the Kastoria–Florina granitic complex and hence the Upper Carboniferous crystallisation age must be the earlier. Thus the ages of 461 ± 12, 465 ± 12, 468 ± 14 and 526 ± 16 Ma yielded by the K/Ar geochronology could be due to excess argon accumulated in parts of the granite, as suggested by Marakis (1969) and Kilias (1980). The Upper Carboniferous (302.4 ± 5/15 Ma) age of the Kastoria granite is in close agreement with the age determined for the Pieria granite by Yarwood & Aftalion (1976).

The contact metamorphism of the Vernon metamorphic rocks by intrusion of the Upper Carboniferous granite show that the pre-granitic basement of the Pelagonian zone is Lower Palaeozoic or older.

In Yugoslavia, the continuation of the Voras crystalline unit is considered by Arsovski *et al.* (1977) to be of pre-Cambrian age because the

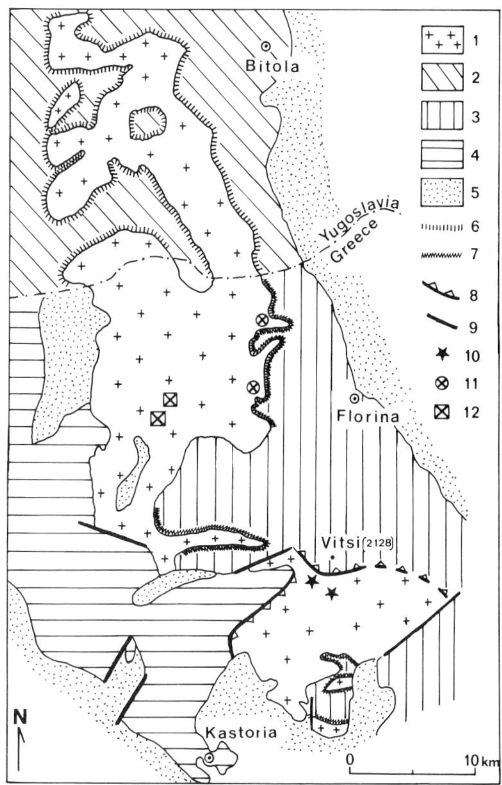

FIG. 2. Map showing the locality of the Kastoria–Florina granite in Greece and its continuation in Yugoslavia. Maps and data of Osswald 1931; Stojanov 1971 and Kilias 1980, have been taken into account.

1, granite; 2, greenschist facies rocks of Yugoslavia; 3, pre-granitic metamorphic rocks in Greece; 4, Permian and Mesozoic rocks with Alpine Metamorphism in Greece; 5, post-Alpine sediments; 6, hornfels facies of contact-metamorphism in Yugoslavia (Stojanov 1971); 7, intrusive contacts observed in Greece; 8, thrust; 9, fault; 10, sample location of U-Pb geochronology; 11, sample location of K/Ar geochronology by Kilias (1980); 12, sample location of K/Ar geochronology by Marakis (1969).

carbonate cover there has been thought to be Riphean–Cambrian. However, the same carbonate cover in the Greek Voras mountain area has been defined by Mercier (1968) as Triassic–Jurassic. Evidence which would justify an older age for the metamorphic rocks of Voras, as compared to those of Vernon, such as that reported for the Yugoslavian areas (Arsovski *et al.* 1977) has not been forthcoming in Greek Macedonia, where the pre-granitic metamorphic rocks in the two units of Voras and Veron are similar and the conditions of the meta-

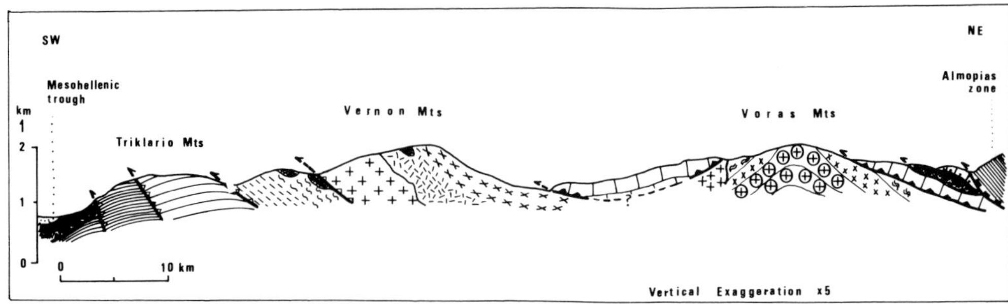

FIG. 3. Simplified cross-section through the Pelagonian Zone in Northwestern Macedonia. For key see Fig. 1.

morphism seem to be the same (upper greenschist facies). No evidence of high P/T metamorphism such as that seen in the thrust sheets of South Euboea, the Cyclades and Pelion has been observed. The metamorphic rocks in the two units seem to be similar to those observed in the metamorphic sheets of Pieria (Yarwood & Dixon 1977; Nance 1977) but no direct correlation is possible. The contact of the Voras and Vernon units is hidden by the Neogene basin of Ptolemais–Florina and hence it is not known if they are in tectonic contact, or are part of a single thrust sheet. The metamorphic rocks of the Voras unit form a megascopic anticlinal dome (Mountrakis 1982) the axis of which plunges gently towards 130° (Fig. 3). This axis direction coincides with that of the folding phase which also caused open to close mesoscopic folds with axial direction N 120–130°, refolding the primary schistosity. This phase in the Hellenides is considered by Vergely (1977) to have an Early Tertiary–Eocene age. On the other hand the metamorphic rocks of the Vernon unit are thrust sheets emplaced towards SW or NW; there is no evidence as to the age of these thrusts, but they could be Tertiary (Mountrakis 1982). The thrust of the Vernon gneisses and amphibolites onto the Kastoria granite can be seen at the foot of the Vitsi (2128 m) summit (Fig. 1 & 2).

Thus, the most likely conclusion is that the two units of Voras and Vernon represent two related crystalline sequences of probably the same pre-granitic basement which had a similar lithological development and which crystallized under similar conditions in the Palaeozoic.

The Upper Palaeozoic sequence and its palaeogeographic significance

The west side of Mt Vernon is occupied by a meta-clastic sequence, the Aposkepos and Sidirochori series, which consist of meta-arkoses, fine grained meta-pelites, phyllites, phengite-schists, coarse grained meta-sandstones, quartzose conglomerates, thin tuffaceous interbeds, lenses of recrystallized arenaceous limestones and calc-schists. The meta-arkoses contain some polymineralic clasts but are mainly composed of detrital potassium feldspar clasts. In thin section the potassium feldspar clasts appear as megacrysts which are texturally similar to those of the Kastoria granite, but rounded indicating erosion and transport. The composition and petrography of the meta-sediments, indicates a granitic source. It is therefore obvious that the meta-clastic sequence was deposited after the granite intrusion. A depositional contact of meta-clastics on granite can be observed on Yiamata hill (1198 m) at the 25 km post on the Kastoria–Florina road as well as on the same road at the 17 km post.

The age of the meta-clastics is, therefore, post-Carboniferous. The meta-clastics pass transitionally upwards into a limestone horizon, which has been defined by fossils as Middle Triassic (Mountrakis 1979, 1982, Papanikolaou & Zambetakis-Lekkas 1980), at the villages of Aposkepos and Kefalari (Fig. 4). The transitional nature of this contact is well shown by the thin lenses of recrystallized arenaceous limestones and calc-schists, which are observed within the psammitic phyllites beneath the Middle Triassic limestone. An Alpine schistosity (see below), which is developed in both underlying metaclastics and carbonate cover, intersects the bedding at an angle of about 20°. The direction N 155° of the intersection lineation coincides with the fold axes of the fold phase associated with the schistosity (Mountrakis 1982).

Consequently, the age of the meta-clastic sequence is defined as Upper Palaeozoic (Permian-Lower Triassic). As seen in Figs 1, 3 and 4 late thrusts, probably of Tertiary age caused local disruption of former depositional contacts between the Middle Triassic limestone and the

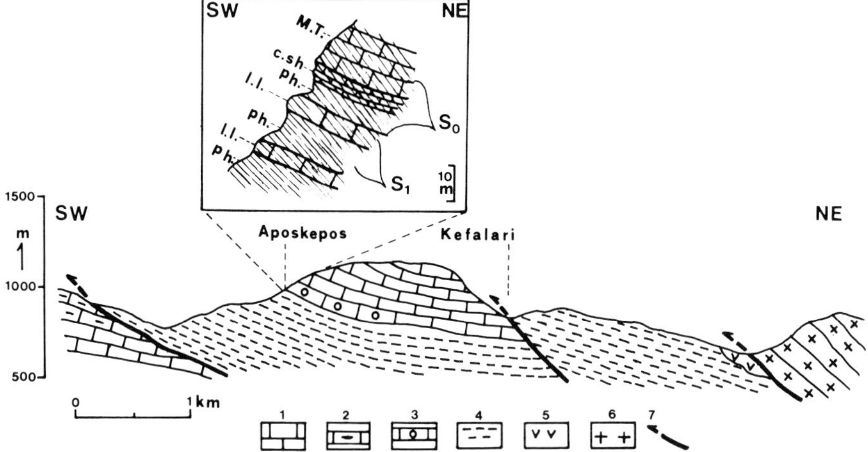

FIG. 4. Simplified cross-section showing the depositional transition between the meta-clastic sequence and the carbonate cover, and the later thrusts.

1, Mesozoic carbonate cover; 2, location of Upper Triassic fossils; 3, location of Middle Triassic fossils; 4, meta-clastics; 5, serpentinite; 6, granite; 7, thrust; S_0, bedding; S_1, schistosity; M.T., Middle Triassic limestone; c.sh., calc-schists; ph., phyllites; l.l., lenses of recrystallized limestones.

underlying meta-clastics, as well as between the meta-clastics and the granite. The meta-clastic sequence is now overlain by a stack of thrust-sheets of granite and the Vernon crystalline unit on south-east dipping thrust-planes. Serpentinite masses have been entrained along the contact planes locally (Mountrakis 1982), as can be observed along the road between the villages of Tihio and Visinia.

The meta-conglomerates contain rounded quartz, gneissic and schistose fragments, which were probably derived from the underlying basement. This confirms the occurrence of a low-grade (upper greenschist facies) metamorphic event in the Pelagonian basement prior to the Carboniferous. On the other hand very low-grade regional metamorphism, probably of the late Alpine age, is also observed in the matrix of these conglomerates, which resulted in the development of a new schistosity, involving growth of sericite and chlorite. The clasts in the meta-conglomerates are elongated parallel to the new matrix schistosity, but this later schistosity appears to cut the old (original) foliation of the gneissic and schistose fragments.

Observations relevant to the possible conditions of sedimentation of the meta-clastic sequence may be summarized as follows: Deposition occurred in close proximity to a continental margin on which continental crystalline basement was exposed to supply material for the arkosic and conglomeratic sediments. Some sedimentary structures such as grading and cross-bedding, observed at the base of the meta-clastic sequence, could indicate a fluviatile to terrestial deltaic environment. Where thickest, the meta-clastic sequence shows alternations of sediments which resemble turbidites with a great abundance of fine-grained sandstones, meta-pelites, phyllites and quartzites, which possibly indicate a continental slope environment. Towards the top of the sequence the sediments pass progressively into the neritic limestone of Middle Triassic age, though the incoming of more psammitic material.

The restriction of the meta-clastic sequence to the west side, and specifically to the immediate vicinity of the Kastoria granite, indicates that its deposition was probably related to a depositional environment specific to the western Pelagonian margin. It is possible that during Upper Palaeozoic times, subsidence of the western Pelagonian margin initially favoured deltaic conditions later developing into a slope regime.

The Mesozoic cover of the Western Pelagonian margin

A thick carbonate sequence consisting of Triassic–Jurassic neritic crystalline limestones occupies the Triklarion Mountain area west of the Vernon Mountains (Mountrakis 1982). The detailed lithostratigraphy is given in Fig. 5. The sequence represents the Mesozoic sedimentation of the Western Pelagonian margin, which was deposited on the meta-clastic sequence of

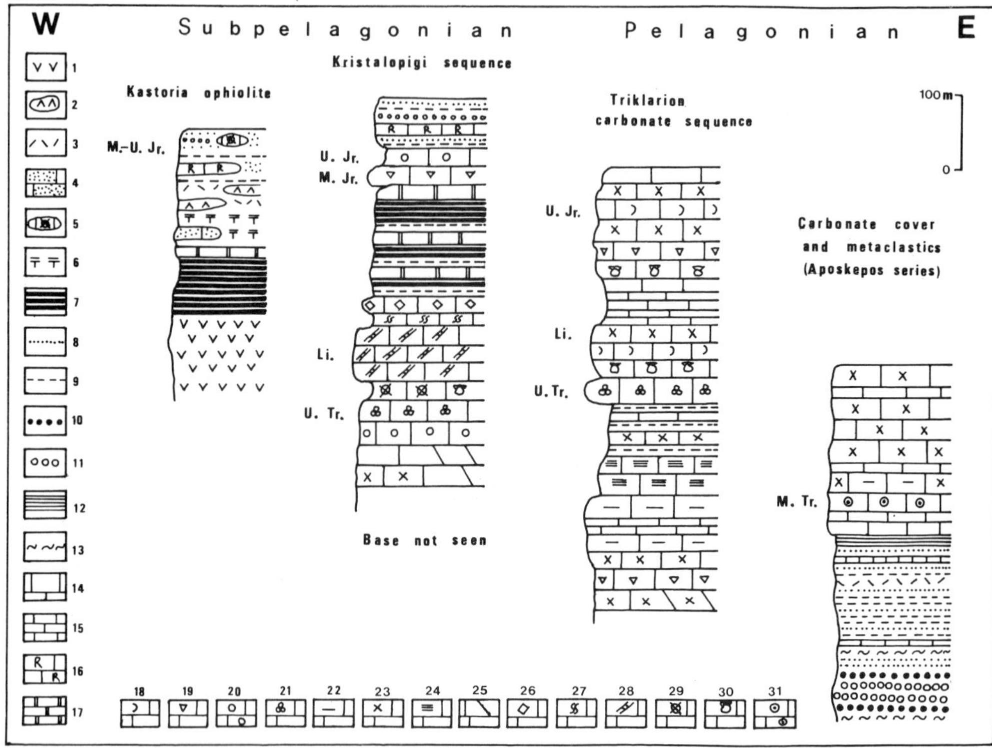

FIG. 5. Lithostratigraphic sections of the Mesozoic sediments on the western Pelagonian margin. 1, ophiolites (serpentinized harzburgite and dunites); 2, diabases and dolerites; 3, tuffs; 4, calcareous siltstones; 5, Liassic olistoliths; 6, reddish and grey siliceous shales; 7, radiolarian cherts; 8, sandstones; 9, pelites; 10, arkoses; 11, quartzose conglomerates; 12, calc-schists; 13, phyllites and phengite schists; 14–31, limestones; 14, massive; 15, platy; 16, redeposited; 17, fine grained pelagic with interbedded cherts; 18, detrital; 19, micritic; 20, oölitic; 21, light-grey gravel with foraminifera (*Involutina communis, Glomospirella friedlei*); 22, dark-grey or black argillaceous crystalline; 23, white-grey, massive, crystalline; 24, white-pink crystalline; 25, dolomitic; 26, 'bird's eye', Liassic with *Mayncina termieri*; 27, with *Litiotis*; 28, with algae (*Palaeodacycladus mediterraneous, Haurania amijii*) of Middle Liassic; 29, with corals; 30, with Jurassic *Lamellibranchs*; 31, with *Meandrospira dinarica*; M.Tr., Middle Triassic; U.Tr., Upper Triassic; Li, Liassic; M.–U.Jr., Middle–Upper Jurassic. (Other stratigraphic determinations in Mountrakis 1979, 1982; Papanikolaou & Zambetakis-Lekkas 1980).

Upper Palaeozoic age. During later thrusting the meta-clastics were emplaced westwards over these younger carbonates in the area between Mt Vernon and Mt Triklarion (Fig. 4). However, the initial transgression was confirmed by observations further south in the area of Kozani (Brunn 1956). The very-low-grade metamorphism of the carbonate sequence is analogous to that of the meta-clastics, as evidenced by thin intercalations of meta-pelites and the minor extent of calcite recrystallisation.

The neritic carbonate sequence of the Triklarion Mountains was thrust westwards over another sedimentary sequence, the Kristalopigi unit, of Upper Triassic-Lower Cretaceous age, belonging to the Subpelagonian zone (Mountrakis 1982). More precisely this unit is transitional to the Pelagonian and the Orthrys (or Maliac) units. Stratigraphical information on this sequence is given in Fig. 5, which includes for comparison the Triklarion carbonate and metaclastic successions of the western Pelagonian margin, and the sedimentary sequence associated with the Kastoria ophiolite, which lies further west in a typical Subpelagonian position. The comparison shows a change to deeper conditions of sedimentation from east to west during Mesozoic time. In the Subpelagonian sequence the sedimentation changes from neritic to pelagic after the Liassic.

The ophiolites and the associated sediments were thrust eastwards over the Subpelagonian sequence probably in Lower Cretaceous time, as evidenced by shear planes dipping SW, as well as a thrust surface marked by tectonic breccia at the base of the ophiolites, and folds with eastwards asymmetry (Mountrakis 1982). Thus, the development of an ocean basin west of the Pelagonian Zone seems very probable in this area just as already proposed for the Othrys area of Central Greece (Hynes *et al.* 1972; Smith & Woodcock 1976).

The Mesozoic cover of the Eastern Pelagonian margin

In the east Pelagonian zone the Voras unit is overlain by a metamorphosed carbonate cover which consists of white recrystallized marbles, dolomitic marbles, micaceous marbles and thin, 1 to 2 m thick schist intercalations. They are considered to be probably Triassic to Jurassic in age (Mercier 1968). Extensive research has not yielded any more information on the stratigraphy. The cover sequence was subject to low-grade metamorphism, with phengite, chlorite and epidote developed, accompanied by intense recrystallization of calcite. The carbonate cover extends westwards to cover a very small part of Vernon on its eastern side. The continuation of this contact is buried under post-Alpine formations.

The position of the carbonate cover on the pre-Alpine basement is clearly tectonic, from the following evidence:

The cover overlies different rocks in different places. Specifically, on the east side of the Voras Mountains it overlies mica-schists and gneiss, while on the west side it overlies amphibolitic-schists, mica-schists and a granodioritic intrusion. There is no progressive passage anywhere from the basement to the cover. There are no indications anywhere of conglomerates at the base of the carbonates. In all places where the contact between cover and basement is visible it is accompanied by mylonitization.

On the east side of the Voras Mountains the cover is thrust over gneiss and mica-schist, with the thrust surface dipping 30° to E or ENE. The marbles, as well as the underlying gneiss have been transformed to fine grained ultra-mylonites. The foliation in the marbles and the gneiss is usually parallel to the contact, suggesting that the foliation formed at the same time as the contact. However, there are some localities where a discordance between the two foliations is seen. The foliation in the marbles is always parallel to the contact. Recumbent, tight, mesoscopic folds with amplitudes of 50 cm to 1 m and axial direction of 150° are observed in the gneiss immediately beneath the contact, and are probably related to emplacement of the cover. These folds refold the primary foliation, and show a weak second schistosity which is axial planar. However, the intersection-lineation of the two planar fabrics is not well marked.

On the west side of the Voras Mountains the thrust surface is masked by younger faults dipping SW. However, the influence of the thrust is especially spectacular in a granodioritic body beneath the thrust, near the village of Agios Athanasios. This body has undergone mylonitisation to a thickness of 20 m from the contact with the marbles, but is not deformed internally elsewhere. Away from the contact surface the intensity of mylonitisation decreases gradually. Recumbent folds identical to those observed on the east side, with amplitudes up to 3 m, have also been observed in the marbles close to the contact. These folds are refolded by younger, almost coaxial, open folds and produce interference patterns. The younger folds belong to a Tertiary folding episode which also caused the megascopic anticlinal dome of the Voras Mountains as described above.

Further westwards, the carbonate cover has been thrust over the orthogneiss of the Vernon unit, with the thrust surface dipping 10°–20° towards 70°. Strong mylonitisation to a thickness of about 70 m is observed in the gneiss, and a 4 m thick tectonic breccia was formed at the base of the marble. Almost recumbent folds, with amplitudes of about 10 m, and with axes plunging towards 150°–160°, are found in the gneiss beneath the contact. They are close to tight, asymmetrical folds facing WSW, although the sense of asymmetry is difficult to confirm as the fold limbs cannot be traced far.

The contact of the cover on the basement is, thus, definitely tectonic. The observed structures (thrust surfaces, folds, foliations), although not considered to be strong evidence, suggest that the cover was thrust over the basement from ENE to WSW. The remaining question is its original location. It is noteworthy that the carbonate rocks contain large quantities of angular gains of quartz and feldspar which were probably derived from part of the crystalline basement which was elevated during sedimentation. In addition, marbles identical to those of the Pelagonian cover have been observed at the base of the Almopian thrust-slices (Vardar zone), near the boundary between the two zones where these slices have been thrust over the Pelagonian Zone (Mountrakis 1976). Consequently, it is possible that the carbonate cover is para-autochthonous in origin; i.e. it was deposited on the shelf of the eastern Pelagonian margin and was then thrust towards the west-south-west almost to the foot of the Vernon Mountains.

The thrusting must be of pre-Upper Cre-

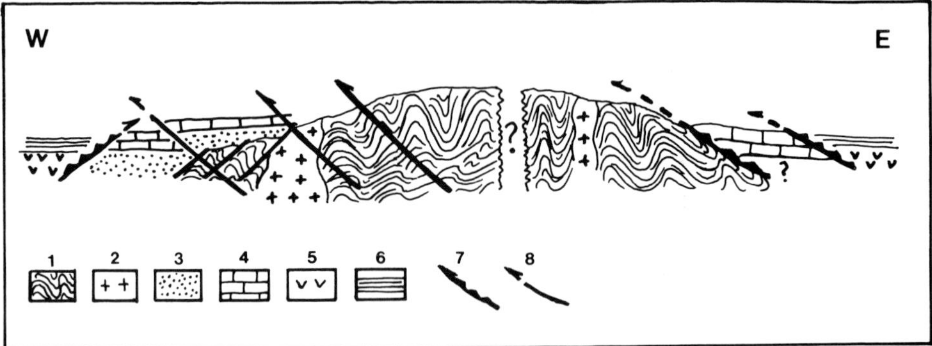

FIG. 6. Sketch of the possible palaeogeographic reconstruction of the North Pelagonian zone s.l. in pre-Upper Cretaceous time.

1, pre-granitic metamorphic rocks (pre-Upper Carboniferous); 2, granites; 3, Permian-Lower Triassic metaclastics; 4, Mesozoic carbonate covers of the Pelagonian margins; 5, ophiolites; 6, deep water sediments associated with the ophiolites.

taceous age because the mylonitisation affecting the Triassic–Jurassic marbles did not affect the Upper Cretaceous limestones, which are found as an unconformable transgressive sequence outcropping very close to the thrust. The parallelism of the marble foliation with the thrust plane suggests that thrusting and low-grade metamorphism of the marbles were synchronous and thus Upper Jurassic to Lower Cretaceous in age (Mercier 1968). This thrusting must also be related to the thrusting of the ophiolites and the associated sediments onto the Pelagonian zone, observed in the southern part of the Voras mountains (Mountrakis & Soulios 1978), and in turn related to ophiolite emplacement from the Vardar zone on to the Pelagonian Zone (Mercier *et al.* 1975).

Discussion and conclusions

The two parallel units of the pre-Permian Pelagonian basement have analogous lithological successions and can be correlated as probably originally belonging to one microcontinental block. The intrusive contacts of the Upper Carboniferous Kastoria granite and the Vernon rocks confirm the Lower Palaeozoic (or older) age of the metamorphic rocks. During Upper Palaeozoic times, subsidence of the western Pelagonian margin favoured deposition of the meta-clastic sequence of Permian-Lower Triassic age.

Comparison of the Mesozoic carbonate cover of the western margin with that of the eastern margin leads to the following conclusions. They have different lithologies and the eastern cover suffered a greater intensity of metamorphism, involving complete recrystallisation of calcite. The cover on the western margin is autochthonous with gradational contacts with the underlying clastic rocks, while on the eastern margin the cover is found in tectonic contact on the metamorphic rocks. On the western margin the meta-clastic sequence intervenes between the pre-Permian basement and the carbonate cover but is absent on the eastern margin.

There is, therefore, a case for the recognition of two separate cover successions which were deposited on two separate margins (Fig. 6) and did not originate as a single carbonate platform succession.

The thrusting of the cover of the eastern margin probably took place in Upper Jurassic to Lower Cretaceous times coeval with metamorphism and ophiolite emplacement, or nearly so. The age of the metamorphism is regarded as Alpine (133–140 Ma), for both the basement and the cover (Mercier 1968). However, although the character of the low-grade metamorphism is similar in basement and cover the metamorphic grade and the lithological succession in the Voras crystalline unit are also similar to those of the Vernon crystalline unit in which a pre-Upper Palaeozoic age of metamorphism has been proved. In addition, the recumbent folds, which are spatially related to the thrust on the eastern margin, re-fold the foliation in the gneiss. This is a clear indication that an earlier metamorphic fabric existed in the gneiss when the Upper Jurassic deformation took place. In the continuation of the Voras metamorphic belt in neighbouring Yugoslavia repeated metamorphism in at least two events is recorded (Stojanov 1967, 1971).

In conclusion, two metamorphic events can be inferred in the northern Pelagonian zone: the first pre-Upper Carboniferous with low-grade metamorphism, mainly of upper greenschist facies, which affected the Vernon crystalline unit, and perhaps the Voras unit; and the second, post-Jurassic, which on the western margin, resulted in a weak low-grade greenschist fabric in the cover meta-clastics and the development of higher-grade greenschist assemblages in the augen gneisses derived from the Kastoria–Florina granite. On the eastern margin this post-Jurassic event caused low-grade recrystallization of the carbonate cover, refolding of the underlying Voras gneiss fabric near the contact but no clear overprinting of the Voras gneiss assemblages as a whole.

Further isotopic work is clearly needed to define this sequence of events more precisely.

ACKNOWLEDGEMENTS: I wish to thank the Director of the Scottish Universities Research Reactor Centre, East Kilbride for the U-Pb determination. I also thank J. E. Dixon and P. Vergely for helpful discussion.

References

ARSOVSKI, M., DUMURDZANOV, N., HRISTOV, S., IVANOV, T., IVANOVA, V., PETKOVSKI, P. & STOJANOV, R. 1977. Correlation of the Pre-Cambrian complexes of the Pelagonian massif, Vardar Zone and Serbo-Macedonian massif. *6th Coll. Geol. Aegean region Athens*, 549–557.

——, IVANOV, T., PETKOVSKI, P., STOJANOV, R. & TEMKOVA, V. 1977. Geologic—Structural characteristics of the Paleozoic complex of the southern part of the Balkan Peninsula with special reference to the territory of Macedonia (Yugoslavia). *6th Coll. Geol. Aegean region Athens*, 559–568.

BARTON, C. M. 1976. The tectonic vector and emplacement age of an allochthonous basement slice in the Olympos area, NE Greece. *Bull. Soc. géol. France* **18,** 253–258.

BRUNN, J. H. 1956. Étude géologique du Pinde septentrional et de la Macédoine occidentale. *Annals. géol. Pays Hell.* **7,** 1–358.

CELET, P. & FERRIERE, J. 1978. Les Hellénides internes: Le Pélagonien. *Eclogae geol. Helv.* **73,** 467–495.

DÜRR, S., ALTHERR, R., KELLER, J., OKRUSCH, M. & SEIDEL, E. 1978. The median Aegean crystalline belt: stratigraphy, structure, metamorphism, magmatism. *In*: CLOSS, ROEDER & SCHMIDT (eds): *Alps, Apennines, Hellenides*. 455–477.

FERRIÈRE, J. 1977. Le secteur méridional du "massif métamorphique de Thessalie". Le massif du Pélion et ses environs. *6th Coll. Geol. Aegean region Athens*, 291–309.

HYNES, A. J., NISBET, E. G., SMITH, A. G., WELLAND, M. J. P. & REX, D. C. 1972. Spreading and emplacement ages of some ophiolites in the Othris Region, Eastern Central Greece (Proc. 4th Aegean Symposium, Hannover). *Z. dtsch. geol. Ges.* **123,** 455–468.

JACOBSHAGEN, V. 1977. Structure and geotectonic evolution of the Hellenides. *6th Coll. Geol. Aegean region Athens*, 1355–1367.

KATSIKATSOS, G. 1977. La structure tectonique d'Attique et de l'île d'Eubée. *6th Coll. Geol. Aegean region Athens*, 211–228.

KILIAS, A. 1980. *Geologische und tectonische Untersuchung des Gebietes von östlichen Varnous (NW Makedonien)*. Thesis, Univ. of Thessaloniki, Greece (unpubl.) 271 pp.

KOSSMAT, F. 1924. *Geologie der zentralen Balkanhalbinsel*. Borntraeger, Berlin. 198 pp.

MARAKIS, G. 1969. Geochronology studies of some granites from Macedonia. *Annls. géol. Pays Hell.* **21,** 121–152.

MERCIER, J. 1968. Étude géologique des zones internes des Hellénides en Macédoine centrale (Grèce). *Annals. géol. Pays Hell.* **20,** 1–792.

——, VERGELY, P. & BEBIEN, J. 1975. Les ophiolites helléniques "obductées" au Jurassique supérieur sont-elles les vestiges d'un ocean téthysien ou d'une mer marginale péri-européene? *C.r. somm. Séanc. Soc. géol. Fr.* **17,** 108–112.

MOUNTRAKIS, D. 1976. *Geological study of the Pelagonian and Vardar Zone boundary in the Almopias area (North Macedonia)*. Thesis, Univ. of Thessaloniki, Greece (unpubl.) 164 pp.

—— 1979. Résultats préliminaires de l'étude stratigraphique de la région de Kastoria (NW Macédoine, Grèce). *Sci. Annls. Fac. Phys. Mathem. Univ. Thessaloniki,* **19,** 163–173.

—— 1982. Étude géologique des terrains métamorphiques de Macédoine occidentale (Grèce). *Bull. Soc. géol. France*, **24,** 697–704.

—— 1982. Emplacement of the Kastoria ophiolite on the western edge of the Internal Hellenides (Greece). *Ofioliti*, **7,** No 2/3, Special Issue, 397–406.

—— & SOULIOS, G. 1978. Sur une formation schisto-radiolaritique à des ophiolites charriée et la présence des mélanges ophiolitiques dans la région d'Arnissa. Leur signification pour l'évolution tecto-orogénique de la zone Pélagonienne. *Bull. geol. Soc. Greece* **13,** 18–33.

NANCE, R. D. 1977. *The Livadi Mafic-Ultramafic Complex and its metamorphic basement, NE Greece*. Thesis, Ph. D. Univ. of Cambridge (unpubl.), 170 pp.

OSSWALD, K. 1931. *Geologische Ubersichskarte von Griech-Makedonien 1:300.000*. Griech. Geolog. Landesanstalt, Athen.

PAPANIKOLAOU, D. 1981. Some problems concerning correlations within the metamorphic belts of the Pelagonian and Rhodope. *In*: KARAMATA, S. & SASSI, F. P. (eds) *IGCP No 5, Newsletter* **3,** 119–123.

—— & ZAMBETAKIS-LEKKAS, A. 1980. Nouvelles observations et datation de la base de la série pélagonienne (s.s.) dans la région de Kastoria, Grèce. *C.R. Acad. Sc. Paris*, **291,** 155–158.

SMITH, A. G. & WOODCOCK, N. H. 1976. Emplacement model for some "Tethyan" ophiolites. *Geology*, **4,** 653–656.

Stojanov, R. 1967. Phengites of the Pelagonian massif. *Bull. Inst. Geol. Rep. Social. macedonienne, Skopje*, **13,** 59–73.

—— 1971. A brief review to the metamorphism of the Pelagonian massif in context with metamorphic events in west Macedonia (Yugoslavia). *Bull. Inst. Geol. Rep. Social. macedonienne, Skopje*, **15,** 21–31.

Vergely, P. 1977. Ophiolites et phases tectoniques superposées dans les Hellénides. *6th Coll. Geol. Aegean region Athens*, 1293–1302.

Yarwood, G. A. & Aftalion, M. 1976. Field relations and U-Pb geochronology of a granite from the Pelagonian zone of the Hellenides (High Pieria, Greece). *Bull. Soc. géol. France*, **18,** 259–265.

—— & Dixon, J. E. 1977. Lower Cretaceous and younger thrusting in the Pelagonian rocks of the high Pieria, Greece. *6th Coll. Geol. Aegean region Athens*, 269–280.

D. Mountrakis, Department of Geology and Palaeontology, University of Thessaloniki, Thessaloniki, Greece.

Pre-Neogene nappe structure and metamorphism of the North Sporades and the southern Pelion peninsula

Volker Jacobshagen & Eckard Wallbrecher

SUMMARY: On the North Sporades and on the southern Pelion Peninsula, four tectonic units can be distinguished. The lowermost is the Pelagonian Nappe which is a sequence of Permo-Triassic metaclastics and platform carbonates of Middle Triassic to Upper Jurassic age. It is overlain by the Eohellenic Nappe, which consists of a melange with ophiolite bodies and crystalline slices, followed by a series of calc-schists, platy marbles and spilites, and, finally, a flyschoid sequence. Subsequent to the overthrusting of the Eohellenic Nappe in the Early Cretaceous, both nappes were partly eroded and then covered by the meso-autochthonous sediments of Middle Cretaceous to Lower Tertiary age. In the Eohellenic Nappe, as well as in the higher units, blue amphiboles often appear, i.e., magnesio-riebeckite in quartzitic and phyllitic layers and crossite or glaucophane in metabasites. The pelitic layers contain phlogopite and stilpnomelane as well as chlorite and phengite. In meta-radiolarites, spessartine and piemontite have also been found. This association, indicating high pressure during very low-grade metamorphism, contrasts greatly with the albite-biotite-almandine assemblage in the underlying Pelagonian Nappe, which is typical of low-grade metamorphism at more moderate pressures. A geodynamic explanation is proposed.

The North Sporades are situated in the northwestern Aegean Sea off Eubeoa and the Pelion Peninsula (Fig. 1). Although geological observations from these islands have been published from the beginning of this century (Philippson 1901; Renz 1927; Marinos & Papastamatiou 1938, 1940; Papastamatiou 1961, 1963), a systematic investigation of the archipelago was initiated by Guernet (1971) based on small-scale geological mapping. His work was followed by Greek studies on Skiathos (Ferentinos 1973), Alonnisos and Peristera (Keleptertsis 1974) and Skyros (Melentis 1973). The southern Pelion Peninsula was studied by Ferriere (1973, 1976) and Wallbrecher (1976, 1979, 1983). Detailed mapping on the sheet 1 : 50 000 Argalasti and on Skyros is still going on and will be published soon.

During this period several opinions on the boundaries of the isopic zones of the Hellenides in the Sporades region were expressed: Renz (1940) believed that the whole archipelago was part of the Pelagonian Zone whereas according to Aubouin (1959) and Aubouin *et al.* (1963), only the western islands Skiathos and Skopelos belonged to that zone. Guernet (1971) added Skyros again, but Kelepertsis (1974), on the other hand, definitely attributed Alonnisos and Peristera to the Vardar zone.

Structural investigations by the present authors (Jacobshagen & Skala 1977, Jacobshagen *et al.* 1976, 1978) led to evidence that the structure of some of the islands is extremely complicated and, therefore, needed to be mapped and investigated in detail. Although this research is not yet complete, the main features of Skiathos (Heinitz & Richter-Heinitz 1983) and Skyros (Harder *et al.* 1983) are now known. Comparative studies of alpine metamorphism on the North Sporades and on the Pelion peninsula were also carried out by Wallbrecher (1983). These new results are briefly synthesized here; a schematic compilation is logged in Fig. 4. All the pre-Neogene rocks of the North Sporades are more or less metamorphic. They belong to four or five units, from top to bottom: Skyros Nappe; ophiolitic outliers; Meso-autochthonous complex; Eohellenic Nappe; Pelagonian Nappe.

The term 'Pelagonian Nappe' is defined in the sense of Jacobshagen *et al.* (1978) as not including the ophiolitic unit and accompanying pelagic sediments which were overthrust during the Lower Cretaceous. The latter constitute the Eohellenic Nappe (Wallbrecher 1979).

Lithology and distribution of tectonic units

Using the scheme above, we now describe the distribution of the units and their stratigraphic sequences on the main islands of the archipelago, the major structural features and the style of deformation and, finally, the character and the sequence of metamorphic events.

Pelagonian Nappe

Pelagonian rocks crop out in large areas of the Pelion Peninsula, Skiathos, Skopelos, Skantsoura and Skyros. The Pelagonian sequ-

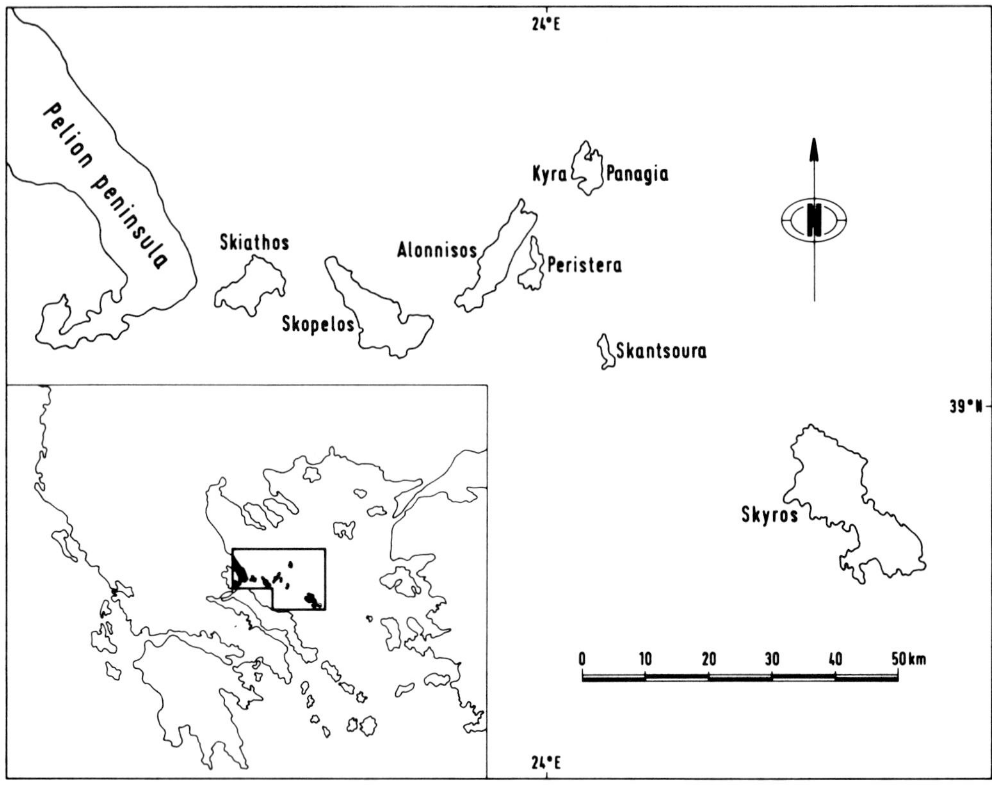

FIG. 1. Location of the North Sporades and the Pelion peninsula.

ence comprises only meta-sediments of Upper Palaeozoic to Upper Jurassic age. Underlying metamorphic rocks of Hercynian age, which are widespread in the Pelagonian complex of Thessaly, are not exposed.

Skiathos unit

The meta-clastic Skiathos unit consists mainly of meta-greywackes and schists comprising lenses of coarse quartz-conglomerates and meta-arkoses in its lower part. On Skiathos and in the Sarakiniko mountain of Pelion it is exposed to a thickness of several hundred metres. In the upper part of the Skiathos unit greyish pelites, which include lenses of dark limestones with Upper Permian faunas (Harder *et al.* 1983) become increasingly frequent. In some places these pelites pass laterally into variegated schists of the Werfen facies, which, locally, also contain volcanic rocks (Skiathos: Guernet 1971, Heinitz & Heinitz-Richter 1983; Skopelos: Papastamatiou 1963, Jacobshagen & Skala 1977; Skyros: Harder *et al.* 1983).

Pelagonian marbles

The Skiathos unit is overlain by a carbonate complex. These Pelagonian marbles consist of light-greyish meta-dolomites (Skiathos, Skopelos) or coarse-grained white calcite marbles (Skantsoura; Skyros; Southern Pelion), which exceed a thickness of 1000 m in north-western Skyros. Scarce fossils testify to a Middle Triassic to Upper Jurassic age (Papastamatiou 1963). Normally the Pelagonian marbles are sheared off the underlying schists; only in one place on Skyros are red nodular limestones and greenish tuffs exposed at this boundary. Although these limestones did not yield fossils they might be compared with the Middle Triassic Hallstatt limestones of Attika and Argolis.

Eohellenic Nappe

The Eohellenic Nappe covers large areas of northern Skyros (Fig. 2) and the southern Pelion peninsula and the overthrust contact on the underlying Pelagonian Nappe is exposed in many places. We refer mainly to Skyros and to the mainland when describing the Eohellenic Nappe and then mention the other islands later.

Although the deeper parts of the Eohellenic Nappe are extremely tectonized and neither fossils nor radiometric data are yet available,

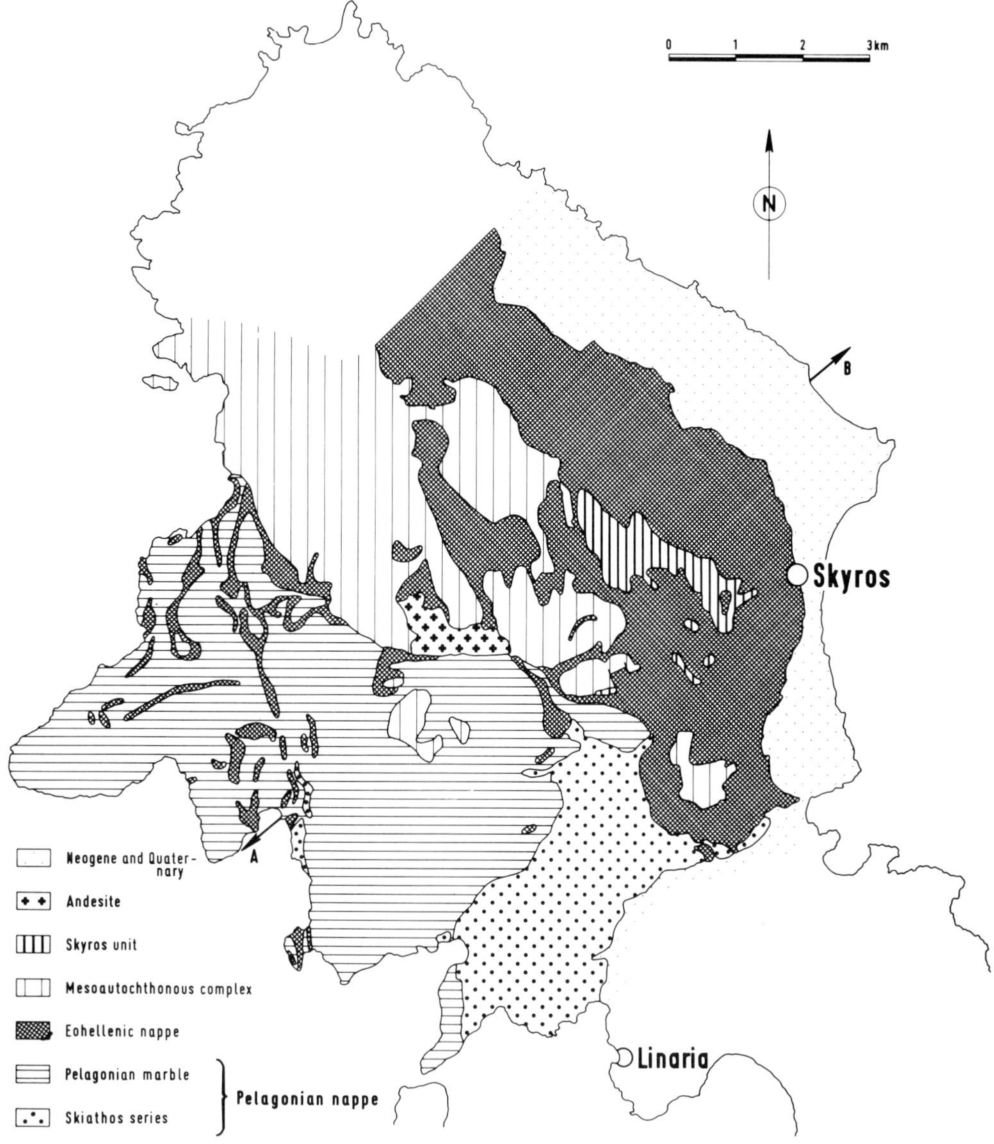

FIG. 2. Tectonic sketch-map of northern Skyros.

the following sequence was established from outcrops north of the town of Skyros. Near the basal thrust-plane, a melange made of slices of serpentinites and marbles in a phyllite or calc-schist 'matrix' is observed in many places. West of Kotounes this melange also contains mica-schists, mica-bearing marbles and breccias with volcanic debris. The serpentinite slices vary in size from centimetres to several hundred metres and locally reach a thickness of 150 m. This ophiolitic melange is overlain by platy limestones, calc-schists and meta-volcanics of spilitic character, passing into an upper series of meta-graywackes and schists. In contrast to other sections of the Eohellenic Nappe, e.g. the Vourinos, on Skyros the pelagic meta-sediments far exceed the ophiolites. A basal melange with crystalline fragments is developed in both regions.

A very similar assemblage of Eohellenic rocks was recently observed on Skopelos by D. Matarangas (pers. comm. 1982) and by one of the present authors. It is also likely that the greenschists of Skantsoura (Jacobshagen & Skala 1977) belong to the Eohellenic Nappe. On Skiathos, Eohellenic relics are missing

(Heinitz & Richter-Heinitz 1983), but the lowest unit of Alonnisos–the Mortero series with phyllites, calc-schists, micaceous marbles, amphibolitic schists, and epidotites (Kelepertsis 1974), together with the overlying Upper Jurassic marbles–are very similar to the Eohellenic sequence of Skyros, and the same is true for the ophiolite-bearing series of Kyra Panagia (Kelepertsis, in Jacobshagen & Skala 1977, Schwandt 1983).

On the Southern Pelion peninsula the Eohellenic Nappe is extremely deformed by folding and internal thrusting so that it is very difficult to reconstruct the original sequence. Here the nappe is also composed of ophiolitic melange containing meta-gabbros, mafic cumulates, serpentinites and meta-volcanics as well as calcitic and sericitic phyllites, platy marbles and thin layers of manganese- or haematite-rich quartzites. These rocks were evidently derived from oceanic crust with a thin sedimentary cover.

On the basis of fossils found on the North Sporades, the age of the Eohellenic overthrust can be determinated only as younger than Kimmeridgian and older than Cenomanian (see below). Results from Thessaly and Western Macedonia (e.g. Clement & Ferrière 1973, Pichon & Lys 1976) indicate that the thrusting probably occurred during the Early Cretaceous.

Meso-autochthonous complex

The overthrust of the Eohellenic ophiolitic nappe was followed by a period of uplift. Parts of the Eohellenic, and even parts of the Pelagonian Nappe, were eroded, and a karstic relief developed on the marbles. This relief is very often filled in by bauxites, whereas the Eohellenic serpentinites are covered with Fe-Ni-laterites or secondary cherts produced by weathering. Excellent outcrops for the study of these phenomena exist on Skyros (Harder *et al.* 1983), but the karstic relief with bauxites can also be observed on Skiathos, Skopelos and Alonnisos.

Towards the end of the Lower Cretaceous, the whole region again subsided below sea-level and was covered by breccias which are composed solely of carbonate fragments on Skiathos, Skopelos and Skyros, but also contain variegated sandstones, quartzite, chert and volcanic rocks on Alonnisos. These breccias are overlain by rudist limestone which yield Cenomanian to Santonian fossils on Alonnisos, Peristera and Skiathos (Kelepertsis 1974, Heinitz & Richter-Heinitz 1983). It seems likely, however, that the carbonate sedimentation had already started during the early Middle Cretaceous, as in Western Macedonia (Mavridis *et al.* 1979, Pichon 1979) and Argolis (Bachmann & Risch 1979). The rudist limestones pass laterally and vertically into platy limestones with cherts (Skiathos, Skyros), in places with psammitic intercalations (Palouki 'series' of Guernet 1970). From these facies changes and the fossils found by Kelepertsis (1974), the Palouki 'series' must also be Upper Cretaceous. The top of the Meso-autochthonous complex is composed of metamorphic flysch with a thickness of several hundred metres on the major islands of the North Sporades. Generally, it follows on from the Palouki 'series', but on western Alonnisos it covers the rudist limestones unconformably, and on Skiathos even sedimentary contacts with Pelagonian marbles were observed by Heinitz & Richter-Heinitz (1983). Near the base of the flysch, weakly metamorphized volcanics and tuffs of spilitic character occur in northwestern Skyros (Harder *et al.* 1983). Alonnisos and Peristera (Kelepertsis 1974) and on Skiathos and Skopelos (Heinitz & Richter-Heinitz 1983), with a maximum thickness of about 140 m, in Skyros. As fossils could not be found, the age of the latest flysch is probably Palaeocene–Eocene, as in other regions of the internal Hellenides.

On the Pelion peninsula the conglomerates are followed by platy marbles with cherts (Trikeri), platy to more massive marbles (Sarakiniko), or massive brecciated marbles (Argalasti, Metochi, Paltsi). The unit ends with flyschoid meta-greywackes and phyllites in the western-most exposures (Trikeri peninsula).

Skyros unit, ophiolitic outliers

The uppermost tectonic level on the North Sporades and on the southern Pelion peninsula is occupied by a variety of nappe relics. According to Harder *et al.* (1983), the Skyros unit is confined to some marble klippen west of the town of Skyros (Figs 2 & 3), which rest on Eohellenic rocks as well as on lateritic palaeosoils which were developed on serpentinites of this nappe and which thus confirm a post-Eohellenic age for the overthrust of the Skyros unit. This unit consists mainly of white, mica-bearing marbles, underlain by a coarse breccia with fragments of quartz veins, mica-schists and mica-bearing marbles in a phyllitic matrix. The thickness of this breccia is 10–15 m. The base of the Skyros unit is formed by a tectonic melange with large slices of micaceous quartzites and mica-schists with garnets of spessartine to almandine composition. Smaller slices consist of unmetamorphosed fine-grained sandstones and shales which show that the marbles suffered

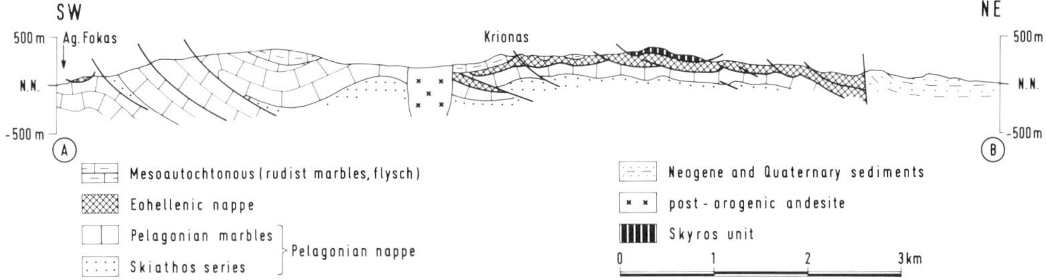

FIG. 3. Schematic cross-section through northern Skyros.

metamorphism before the Skyros unit was sheared off and overthrust, detaching metamorphic as well as sedimentary rocks, in the process.

Another type of klippe is preserved on Skiathos and at the south-western end of the Pelion peninsula, the Trikeri peninsula and the adjacent island of Palaeo Trikeri. On Skiathos, only a few small serpentinite bodies were found above the flysch (Heinitz & Richter-Heinitz 1983). The Pelion outcrops exhibit ophiolitic melange with meta-gabbros (-norites) and ultramafic cumulates, together with lenses of flyschoid sandstones and, higher up, an outlier of serpentinite (Jacobshagen *et al.* 1977). Some chemical analyses and CIPW-norms of the basic and ultrabasic rocks were given by Wallbrecher (1983).

Because of their tectonic position, both the Skyros unit and the ophiolitic outliers are believed to have been overthrust during the Eocene, i.e., in the Neohellenic event according to Katsikatsos (1979). In spite of the small dimensions of these klippen, there is no doubt that the Meso-authochthonous complex of the North Sporades and of the Pelion peninsula was covered overall by a thick nappe or nappe-pile during the Eocene orogenic events. This is the inescapable conclusion from the low-grade metamorphism of the flysch, since a thick sedimentary cover cannot be assumed. Evidence of the overthrusting event which emplaced this nappe may perhaps be seen in the large recumbent folds of the Palouki 'series' of eastern Skopelos. If this interpretation is correct, the Eocene overthrust was directed toward the west, from the vergence of the folds.

Intrusions

On the Pelion peninsula (Elafonisi peninsula, Galanos and Koukos Mountain areas) and on western Skiathos, pegmatite veins cut the Skiathos unit. They vary in thickness from centimetres to a few metres and consist of perthitic orthoclase, albite, quartz, and muscovite. The fabric has no preferred orientation, but is strongly cataclastic showing mortar structures around the quartz grains. The origin and age of these veins is still unknown.

Structure

The major tectonic features are similar throughout the whole area of investigation and may be briefly reviewed:

Nappe thrust-planes

The basal thrusts of the nappes described above are clearly exposed in many places. All of them are marked by intense tectonization of the underlying rocks and by a basal mélange; all have been subsequently folded.

The basal thrust of the Eohellenic Nappe cuts the subjacent Pelagonian Nappe at different levels. This may be the explanation for the widely varying thicknesses of the Pelagonian marbles, often over short distances, e.g., on Skyros (Fig. 3). Another explanation could, however, be the existence of isoclinal folds or imbrication within the marbles which has not been recognized. As the Eohellenic structures were intensely overprinted during the Eocene Neohellenic episode, there is no structural evidence remaining of the direction of Eohellenic nappe transport in the area studied. Facies comparisons indicate an origin in the Vardar Zone.

Direct indications of the direction of Neohellenic nappe transport are also lacking. If one accepts that the large isoclinal folds of the Palouki massif on Skopelos were caused by this event, the movements must have been directed to the NW or W. Thus the origin of the Neohellenic outliers might have been in the Vardar ocean, which is believed to have reopened in the Upper Cretaceous and early Palaeogene times.

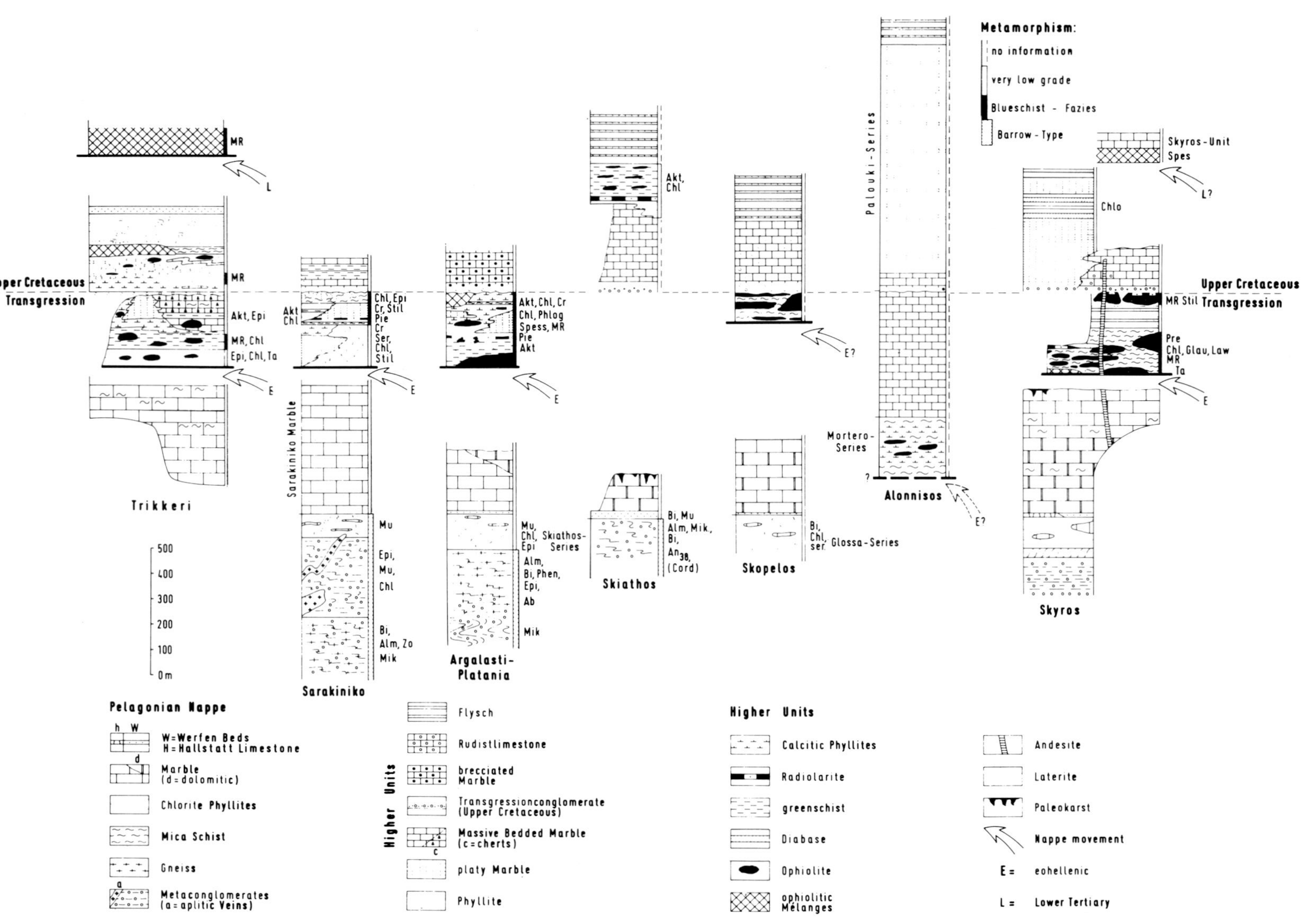

Metamorphism:
no information
very low grade
Blueschist - Fazies
Barrow - Type
MR
L
Upper Cretaceous Transgression
MR
Akt, Epi
MR, Chl
Epi, Chl, Ta
E
Trikkeri
500
400
300
200
100
0m
Sarakiniko Marble
Akt Chl
Chl, Epi
Cr, Stil
Pie
Cr
Ser, Chl, Stil
E
Mu
Epi, Mu, Chl
Bi, Alm, Zo
Mik
Sarakiniko
Akt, Chl, Cr
Chl, Phlog
Spess, MR
Pie
Akt
E
Mu, Chl, Epi
Skiathos-Series
Alm, Bi, Phen, Epi, Ab
Mik
Argalasti-Platania
Akt, Chl
Bi, Mu
Alm, Mik, Bi, An_{38}, (Cord)
Skiathos
E?
Bi, Chl, ser.
Glossa-Series
Skopelos
Palouki-Series
Mortero-Series
?
Alonnisos
E?
Skyros-Unit
Spes
L?
Chlo
Upper Cretaceous Transgression
MR Stil
Pre
Chl, Glau, Law
MR
Ta
E
Skyros
Pelagonian Nappe
h W
W=Werfen Beds
H=Hallstatt Limestone
d
Marble (d=dolomitic)
Chlorite Phyllites
Mica Schist
Gneiss
a
Metaconglomerates (a=aplitic Veins)
Higher Units
Flysch
Rudistlimestone
brecciated Marble
Transgressionconglomerate (Upper Cretaceous)
c
Massive Bedded Marble (c=cherts)
platy Marble
Phyllite
Higher Units
Calcitic Phyllites
Radiolarite
greenschist
Diabase
Ophiolite
ophiolitic Mélanges
Andesite
Laterite
Paleokarst
Nappe movement
E= eohellenic
L= Lower Tertiary

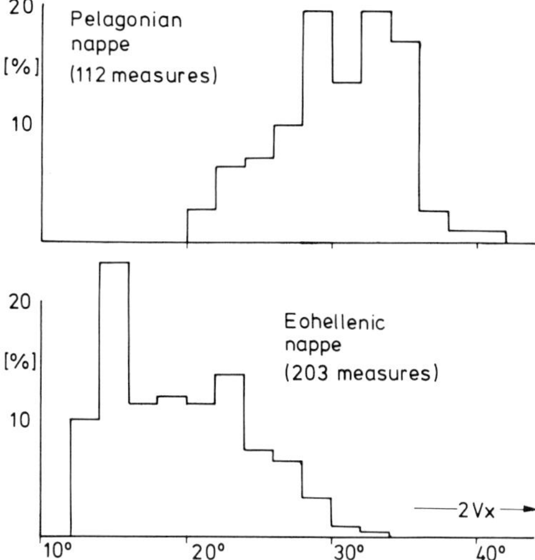

FIG. 5. Distribution of measured 2V's of white micas from samples from the Eohellenic and the Pelagonian Nappes.

Overthrust wedges

The structure of the complex nappe pile is complicated by an imbrication of overthrust wedges which cut the Pelagonian and the Eohellenic Nappes, and the Meso-autochthonous complex (Fig. 3). This probably occurred during Neohellenic thrusting.

Folds

Both the Pelagonian and Eohellenic Nappes contain large-scale isoclinal recumbent folds whose hinge zones can be observed, e.g., on the western end of the Ormos Platania and near Siki (Pelion). The Eohellenic Nappe is even more strongly folded with many drag-fold structures on the limbs of the large isoclinal folds. These are especially well developed in calcitic phyllites. Structures of metre to decimetre size can be well seen near Ag. Konstantinos on the southern Pelion peninsula.

The main structural feature is cross-folding with two axial trends, one directed to NW–NNW and one to NE–ENE (Jacobshagen *et al.* 1976, Jacobshagen & Skala 1977). This pattern can be observed on all scales. Large scale cross-folding was described from north-eastern Skiathos (Heinitz & Richter-Heinitz 1983) and from north-western Skyros (Harder *et al.*, 1983). Thin-bedded schists also show this feature well on a small scale. The intensity of cross-folding is very strong on the southern Pelion peninsula, on Skiathos, and on Skopelos. On Alonnisos and on Skyros the NW

FIG. 4. Schematic logs of the North Sporades and the southern Pelion peninsula.
Abbreviations of mineral names: Ab, = albite; Akt, = actinolite; Alm, = almandine; An, = anorthite; Bi, = biotite; Chl, = chlorite; Chlo, = chloritoid; Cord, = cordierite; Cr, = crossite; Epi, = epidote; Glau, = glaucophane; Law, = lawsonite; Mi, = microcline; Mr, = Mg-riebeckite; Mu, = muscovite; Phen, = phengite; Phlog, = phlogopite; Pie, = piemontite; Pre, = prehnite; Ser, = sericite; Spess, = spessartine; Stil, = stilpnomelane; and Ta, = talc.

trending fold set is subordinate, and on Kyra Panagia and Skantsoura only the NE trending set could be observed.

Although the northeast trending folds can be shown to be the younger ones, in general, both sets are found in the Pelagonian and the Eohellenic Nappes, as well as in the Meso-autochthonous flysch. This demonstrates clearly that the cross-folding occurred during the Neohellenic tectonic phase and also that this event was strong enough to obliterate the older Eohellenic structures.

Schistosity

Except in the hinge-zones of the large isoclinal structures, the schistosity on a mesoscopic scale is parallel to the bedding. A younger slatey cleavage, which sigmoidally deforms this schistosity, can be seen in some outcrops on the southern shore of the Pelion peninsula, east of Platania (Wallbrecher 1979).

Metamorphism

Mineral Assemblages

The Pelagonian Nappe

In the meta-clastics of the lowermost unit the metamorphic grade varies from low in Skyros to have maxima at about 29° and at 34° (Fig. 5), moderate in western Skiathos (Koukounaries and Mandraki), and on the southern shore of the Pelion peninsula. In the basement of the Sarakiniko mountain the highest metamorphic grades are observed in gneisses and mica-schists of the Elafonisi peninsula. In these gneisses quartzite, albite, microcline, white mica, biotite, epidote, and almandine co-exist. The potassium feldspars belong to a new generation of microcline porphyroblasts with rims of plagioclase, as well as to an old generation of detrital, zoned perthitic feldspars from the arkosic parent. Biotite forms porphyroblasts of up to 5 mm size, which grew parallel to the schistosity. White micas have much smaller grain-sizes. Measured 2V's of micas from different outcrops of the southern Pelion peninsula have maxima at about 29° and at 34° (Fig. 5), which shows that both phengite and muscovite occur. Almandine-rich garnet has been found in many samples on the southern shore of the Pelion peninsula as well as on West-Skiathos. It usually forms cataclastically-deformed porphyroblasts up to 5 mm in size (Mandraki). In some meta-tuff layers at Platania (Pelion) green hornblende, derived from pyroxene, has also been observed. Ferentinos (1974) even reports cordierite from west of Skiathos. The upper parts of the Skiathos unit, which consists of pelites, are less metamorphosed. Here, the following paragenesis occurs:

quartz–albite–phengite–chlorite–epidote.

The carbonate-rich sequence in the upper part of this nappe consists of strongly recrystallized calcitic or dolomitic marbles with high-angle grain borders. Aragonite has not yet been found.

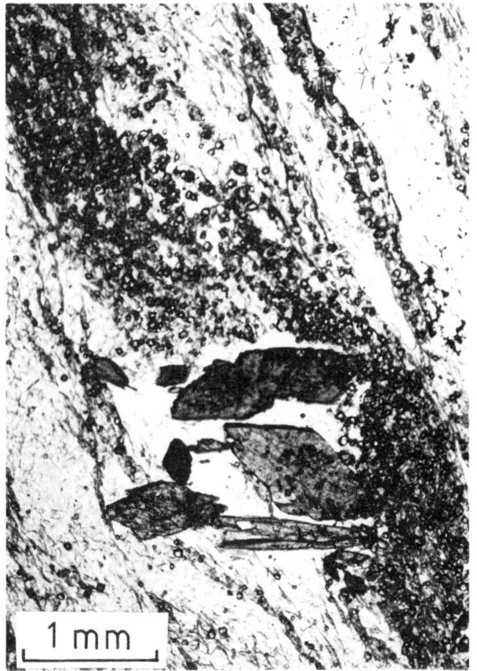

Fig. 6. Layers of spessartine porphyroblasts and idioblastic Mg-riebeckite in phyllitic quartzites of the Eohellenic Nappe (Galanos o Koukos, Pelion peninsula, sample Wa 115 ppl.).

The Eohellenic Nappe

The mineral assemblages in the pelites of this nappe differ characteristically from those of similar lithologies of the underlying unit. The typical paragenesis is:

quartz–white mica–chlorite–stilpnomelane.

The white mica is phengite and in carbonate-bearing layers there is also very often phlogopite ($2\,Vx = 14$–$16°$, Fig. 5). Garnet has only been found on the Pelion peninsula, where it occurs in quartzitic layers as very small (0.02 mm) idiomorphic spessartine crystals (Fig. 6), sometimes together with piemontite. The original sediments were probably manganese-rich radiolarites.

The most characteristic difference between the Pelagonian Nappe and the higher tectonic unit is the abundant occurrence of blue amphiboles in the latter. They have been found in all the Pelion outcrops and also on Skyros above the Pelagonian Nappe. From Skyros the following parageneses are known:

In pelites:
quartz–stilpnomelane–white mica–chlorite–(actinolite).

In cherts from the top of ophiolitic bodies:
quartz–white mica–stilpnomelane–chlorite–crossite (or Mg-riebeckite) ± lawsonite

In metabasites:
serpentine–actinolite–talc–Mg-riebeckite–prehnite.

On the Pelion peninsula the following assemblages occur:

In quartzitic layers:
quartz–stilpnomelane–chlorite–crossite/Mg-riebeckite–spessartine–hematite ± piemontite

In metabasites:
serpentine–talc–actinolite–epidote–chlorite–crossite.

Crossite or Mg-riebeckite have mostly been found in quartz and hematite-rich layers, whereas metabasites less commonly contain blue amphiboles.

Meso-autochthonous sequence

On the southern Pelion peninsula and on Skyros the pelites and metabasites of the Meso-autochthonous sequence show parageneses similar to the Eohellenic Nappe but always contain smaller amounts of extremely small blue amphiboles (only Mg-riebeckite on the Trikeri peninsula and none on Skyros). In addition to the assemblages described, post-kinematic idioblasts of chloritoid have also been found at the conglomeratic base of this sequence in Skyros. From Skiathos no blueschist parageneses are known in this unit. Here the only assemblage which has been described is:
quartz–sericite–chlorite–actinolite
(Heinitz & Richter-Heinitz 1983).

Neohellenic outliers

In the metabasites of the ophiolitic melange of the island of Palaeo Trikeri only very small needles of blue amphibole have been observed, together with actinolite. They were determined optically as Mg-riebeckite. Definite high-pressure indicators such as crossite or glaucophane have not been found. In the Skyros unit no blue amphiboles are known. Here we have observed almandine-rich garnets with spessartine rims in tectonic slices of probable Pelagonian material.

Crystallization and deformation

The mica-rich parts of the Skiathos Series show a very well-developed second crenulation cleavage. The first schistosity, formed by phengite and biotite flakes is sigmoidally deformed by the later cleavage which is oriented parallel to the bedding and thus indicates that the isoclinal folding with which it is associated, is the younger deformation. Cataclastic structures in quartzitic parts, mortar structures and sigmoidal deformation of plagioclase lamellae in the pegmatite veins also seem to follow this younger schistosity.

A crenulation cleavage is also very common in many samples of the Eohellenic Nappe. The blue amphiboles grow parallel to the earlier schistosity and show strong post-crystalline deformation. In Fig. 7 a crossite grain has grown parallel to the older sf_1 schistosity, which developed by parallel orientation of mica. A second schistosity (sf_2) developed together with strong flattening, and caused sf_1 to rotate into the direction of sf_2. The crossites were deformed and stretched in the direction of sf_2. Figure 8 also demonstrates clearly that crossite growth was synchronous with the first schistosity, now folded with axial planes parallel to sf_2. The blue amphiboles broke and were annealed

FIG. 7. Zoned crossite porphyroblasts showing post-crystallization deformation. (sample Ke 69/3, Polars at 80°).

FIG. 8. Crenulation cleavage deforming early-formed crossites into polygonal arcs (sample Ke 11, Polars at 80°).

by polygonization. The same behaviour is also found in piemontite minerals. The crossites often show a definite zonal stucture with three distinct zones (Fig. 7). In metabasites they form rims around greenish amphiboles. This observation may indicate polymetamorphic processes. Some of the Mg-riebeckite samples (Fig. 6) show a more random amphibole orientation, which might be interpreted as a later post-kinematic growth of blue amphiboles.

P–T conditions of metamorphism

All the described parageneses, with the exception of those of the Pelagonian Nappe, point to temperatures in the field of very low-grade metamorphism in the sense of Winkler (1976). According to Winkler the boundary between very-low-grade and low-grade metamorphism in pelites is the reaction forming biotite at the expense of stilpnomelane, in the presence of muscovite. Depending on the H_2O-pressures it lies between 420 and 480°C. The occurrence of prehnite in metabasites (200–400°C, Winkler 1976) and of glaucophane and lawsonite (350°C) also fits this temperature range. The exclusive occurrence of spessartine garnet in the Eohellenic Nappe is also characteristic of very-low-grade metamorphism, while almandine belongs to low-grade metamorphism. High pressures at least in the Eohellenic Nappe are indicated by the presence of minerals of the glaucophane–crossite–Mg-riebeckite group. The P–T-range in the Eohellenic Nappe was probably 200–400°C / > 6 kbar.

The Eohellenic Nappe, however, may have undergone two different phases of metamorphism (?both at high pressure), since two different generations of blue amphiboles can be observed, one grown parallel to the older schistosity and a younger generation with randomly oriented Mg-riebeckites, which have grown post-kinematically. Also, the zonal crystallization of blue amphiboles, shown in Fig. 7 may support this hypothesis. These observations are also in agreement with the findings of Maluski *et al.* (1981), who, by $^{40}Ar/^{39}Ar$-dating of glaucophanes and phengites from southern Euboea, obtained evidence of both a 100–120 Ma event. (Lower Cretaceous) and a 45–50 Ma event. (Lutetian).

The co-existence of biotite, phengite, and almandine in the Pelagonian Nappe definitely proves that here the higher temperature range of low-grade metamorphism (450°C) was reached. The lack of any blue amphiboles in this nappe also shows that here no extremely high pressures existed. We therefore assume a 'normal' P–T-gradient of Barrovian-type for the lowest tectonic unit.

Conclusions

Detailed studies of the stratigraphy, tectonics and metamorphism lead to the following conclusions about the geotectonic development of the nappe-pile in the North Sporades and the southern Pelion peninsula:

(1) The Pelagonian Nappe is of continental crustal origin. Its alpidic low-grade metamorphism was of Barrovian type.

(2) During the Early Cretaceous, the Eohellenic Nappe, which must have covered the whole of the area studied, was obducted from an oceanic basin.

(3) In a subsequent phase of uplift the Eohellenic Nappe was partly eroded.

(4) The area again subsided below sea-level in the early Middle Cretaceous. The Middle Cretaceous to Eocene sediments of the Meso-autochthonous complex cover both Pelagonian and Eohellenic rocks.

(5) The marine sequence of the Meso-autochthonous complex passes from neritic rudist limestones to pelagic limestones then to flysch with spilite intercalations. This points to a re-opening of the Vardar ocean.

(6) Oceanic material was overthrust onto the Meso-autochthonous complex in Eocene times, in a second phase of obduction.

(7) The rocks of the Skyros unit were probably deposited on continental crust. The origin and the geotectonic significance of this unit are uncertain at present.

(8) Geochemical analysis of rocks from the two oceanic complexes has not revealed significant differences which would point to different regions of origin.

(9) The lower oceanic complex suffered a high-pressure/low-temperature metamorphism, but not apparently the upper.

A geotectonic explanation which does not contradict other geodynamic models (Mercier *et al.* 1975, Ferriere & Vergely 1976, Vergely 1977, Maluski *et al.* 1981) could be as follows:

(1) During Eohellenic orogenesis, oceanic crust and associated sediments from the Vardar ocean were obducted onto its western foreland, i.e. the Pelagonian platform. This obducted material had probably experienced high-pressure metamorphism before nappe transport.

(2) After the Eohellenic event, the Vardar ocean re-opened during the Upper Cretaceous and the Palaeogene.

(3) the Neohellenic tectonism obducted material from this new ocean basin westwards onto the Meso-autochthonous complex. This tectonic cover must have been thick enough for the underlying Meso-autochthonous complex to undergo greenschist metamorphism and strong deformation.

ACKNOWLEDGEMENTS: The authors are indebted to the many colleagues who supported these investigations and the publication of the results. The studies could not have been carried out without the continuing interest and support of Drs J. Bornovas and G. Katsikatsos (I.G.M. E., Athens). Some of the minerals were determined by Dr L. J. G. Schermerhorn and M. Blümel, Berlin. Fruitful discussions and criticisms came from Drs J. Papastamatiou (†), S. S. Avgoustithis, G. Marinos, I. Mariolakos, and D. Matarangas (Athens), R. Altherr (Karlsruhe) and J. Ferrière (Villeneuve d'Ascq). Drs J. E. Dixon (Edinburgh) and A. G. Smith (Cambridge) kindly assisted with revision of the manuscript. Technical assistance was given by W. Jung, W. Michaelis, and D. Reich (Berlin).

References

AUBOUIN, J. 1959. Contribution à l'éfude géologique de la Grèce septentrionale: les confins de l'Epire et de la Thessalie. *Ann. géol. Hell.* **10,** 1–484.

——, BRUNN, J. H., CELET, P., DERCOURT, J., GODFRIAUX, J. & MERCIER, J. 1963. Esquisse de la géologie de la Grèce. *Soc. géol. France.* Livre Mém. Paul Fallot, **2,** 583–610.

BACHMANN, G. H. & RISCH, H. 1979. Die geologische Entwicklung der Argolis-Halbinsel (Peloponnes, Griechenland). *Geol. Jb.* **32,** 3–117.

CLEMENT, B. & FERRIERE, J. 1973. La phase tectonique anté-crétacé supérieur en Grèce continentale. *C.r. Acad. Sci. Paris,* (*Ser. D*), **276,** 481–484.

FERENTINOS, G. C. 1973. The geology–petrology of the island of Skiathos. *Bull. geol. Soc. Greece,* **10,** 323–358.

FERRIERE, J. 1973. Sur l'existence des terrains d'âge crétacé supérieur dans le massif du Pélion (axe de la zone pélagonienne) en Grèce continentale orientale. *C.r. Soc. géol. France,* 1973, 61–63.

——, J. 1976. Nouvelles données concernant l'âge des terrains métamorphiques de la partie méridionale de la presqu'île du Pélion (Grèce continentale orientale): présence probable du Crétacé supérieur. *C.r. Acad. Sci. Paris, Ser. D,* **282,** 1407–1410.

——, J. & Vergely, P. 1976. A propos des structures tectoniques et microtectoniques observées dans les nappes anté-crétacé supérieur d'Othrys central: conséquences. *C.r. Acad. Sci. Paris, Ser. D,* **270,** 1764–1765.

GUERNET, C. 1970. Sur l'éxistence d'un chevauchement dans les Sporades (île de Skopelos, Grèce). *C.R. Acad. Sci. Paris,* (*Ser. D*), **270,** 1764–1765.

——, C. 1971. *Études géologiques en Eubée et dans les régions voisines* (*Grèce*). Unpub. thèse Univ. Paris, 1–395.

HARDER, H., JACOBSHAGEN, V., SKALA, W., ARAFEH, M., BERNDSEN, J., HOFMANN, A., KUSSEROW, H. & SCHEDLER, W. 1983. Geologische Entwicklung und Struktur der Insel Skyros (Nord-Sporaden, Griechenland). *Berlinergeowiss. Abh.,* **A48,** 7–39.

HEINITZ, W. & RICHTER-HEINITZ, I. 1983. Geologische Untersuchungen im Nordost-Teil der Insel Skiathos (Griechenland). *Berliner geowiss. Abh.,* **A48,** 41–63.

JACOBSHAGEN, V., MÄRTZ, J. & REINHARDT, R. 1977. Eine alttertiäre Ophiolith-Decke in den inneren Helleniden NE-Griechenlands. *N.Jb. Geol. Paläont. Mh.* **1977,** 613–620.

——, V. & Skala, W. 1977. Geologie der Nord-Sporaden und die Struktur-Prägung auf der mittel-ägäischen Inselbrücke. *Ann. géol. Hell.* **28,** 233–274.

——, ——, & Wallbrecher, E. 1976. Observations sur le développement tectonique des Sporades du Nord. *Bull. Soc. géol. France,* **18,** 281–286.

——, ——, & ——, 1978. Alpine structure and development of the southern Pelion peninsula and the North Sporades. *In*: CLOSS, H., ROEDER, D. H. & SCHMIDT, K. (eds) *Alps, Apennines Hellenides,* E. Schweitzerbart'sche, Stuttgart, 484–488.

KATSIKATSOS, G. CH. 1979. La structure tectonique

d'Attique et de l'île d'Eubée. *Proc. 6th Colloq. Geol. Aegean Reg. Athens 1977*, **1**, 211–228.

Kelepertsis, A. 1974. Geological Structure of Alonnisos and Peristera islands. *Z. dt. geol. Ges.* **125**, 225–236.

Maluski, H., Vergely, P., Bavay, Ph. & Katsikatsos, G. 1981. 39Ar/40Ar dating of glaukophanes and phengites in southern Eubea (Greece): geodynamic implications. *Bull. Soc. géol. France*, **23**, 469–476.

Mavridis, A., Skourtsis-Coroneou, V. & Tsailamonopoli, St. 1979. Contribution to the geology of the Subpelagonian zone (Vourinos area, West Macedonia). *Proceed. 6th Colloq. Geol. Aegean Reg. Athens 1977*, **1**, 175–195.

Melentis, J. 1973. Die Geologie der Insel Skyros. *Bull. geol. Soc. Greece*, **10**, 298–322.

Mercier, J. L., Vergely, P. & Bebien, J. 1975. Les ophiolites helléniques, sont-elles les vestiges d'un océan ou d'une mer marginale périeuropéenne? *C.R. Soc. géol. France*, 108–112.

Papastamatiou, J. 1961. Quelques observations sur la géologie et la métallogénie de l'île de Skyros. *Bull. geol. Soc. Greece*, **4**, 219–237.

—— 1963. Les bauxites de l'île de Skopelos (Sporades du Nord). *Bull. geol. Soc. Greece*, **5**, 52–74.

—— & Marinos, G. 1938. Untersuchungen über den geologischen Bau der Nord-Sporaden. *Praktika Akad. Athen*, **13**, 45–49.

—— & —— 1940. Untersuchungen über den geologischen Bau der Nord-Sporaden. *Praktika Akad. Athen*, **15**, 344–346.

Philippson, A. 1901. Die Magnesische Inselreihe. *Petermanns Mitt., Erg.-H.* **134**, 123–142.

Pichon, J. F. 1979. Une transversale dans la zone pélagonienne, depuis les collines de Krapa (SW) jusqu'au massif du Vermion (NE): Les premières séries transgressives sur les ophiolites. *Proc. 6th Colloq. Geol. Aegean Reg. Athens 1977*, **1**, 163–171.

—— & Lys, M. 1976. Sur l'existence d'une série du Jurassique supérieur à Crétacé inférieur, surmontant les ophiolites, dans les collines de Krapa (Massif du Vourinos, Grèce). *C.R. Acad. Sci. Paris, (Ser. D)*, **282**, 523–526.

Renz, C. 1927. Beiträge zur Geologie der aegaeischen Inseln. *Praktika Akad. Athenon*, **2**, 363–369.

—— 1940. Die Tektonik der griechischen, Gebirge. *Pragm. Akad. Athen.* **8**, 1–171.

Schwandt, K. H. 1983. Das basale Kristallin der Nord-Sporaden (Griechenland). *Berliner geowiss. Abh.*, **A48**, 65–97.

Vergely, P. 1979. Ophiolites et phases superposées dans les Hellénides. *Proc. 6th Colloq. Ged. Aeglan Reg. Athens 1977*, **3**, 1293–1302.

Wallbrecher, E. 1976. Geologie und Tektonik auf dem Südteil der Magnesischen Halbinsel (Nordgriechenland). *Z. dt. geol. Ges.*, **127**, 365–371.

—— 1979. Nappe units of the southern Pelion peninsula and their origin. *Proc. 6th Colloq. Geol. Aegean Reg. Athens 1977*, **1**, 281–290.

—— 1983. Alpidischer Deckenbau und Metamorphose auf der südlichen Pelion–Halbinsel (Thessalien, Griechenland). *Berliner geowiss. Abh.*, **A48**, 99–116.

Winkler, H. G. F. 1976. New York, Heidelberg, Berlin. *Petrogenesis of Metamorphic Rocks.* Springer, 320 pp.

Volker Jacobshagen & Eckard Wallbrecher, Freie Universität Berlin, Institut für Geologie, Altensteinstrasse 34, D-1000 Berlin 33, German Federal Republic.

Metamorphosed ophiolitic rocks from the Serbo-Macedonian Massif, near Lake Volvi, North-east Greece

J. E. Dixon and S. Dimitriadis

SUMMARY: The paper describes preliminary results from the 120 sq km Volvi mafic complex within the Greek Serbo-Macedonian Massif. We conclude from published information and isotopic ages that the central and eastern part of the Massif (basement and Triassic/Jurassic cover) were affected by an amphibolite-facies regional metamorphic and deformational event in the Late Jurassic–Early Cretaceous. Delayed uplift from this same event may explain Tertiary ages from Kerdilion migmatites. The Volvi complex may thus be A: pre-Mesozoic 'basement', or B: an *in situ* Mesozoic rift complex *or* C: a Mesozoic collisional suture remnant. Field, microprobe and XRF data appear to fit model B best. Though deformed, the complex has recognisable 100% sheeted-dyke tracts and substantial areas of undeformed gabbro at lower structural levels. Deformation began before the end of magmatic activity: late sheets of pegmatic gabbro cut deformed dykes; late dykes cut sheared gabbro. Assemblages are of high-grade amphibolite facies and give garnet-hornblende temperatures of 750°C, significantly higher than inferred for staurolite-garnet schists outside the complex. Geochemically, the basaltic rocks show LIL enrichment and are intermediate between 'alkalic' and 'above-subduction-zone' in character. The complex may relate to Mesozoic rift-basins identified in E. Serbia, but is not thought to be a Palaeotethyan remnant.

In this paper we describe some preliminary field and analytical data from a complex body of ophiolitic rocks ca. 120 sq km in extent, composed mainly of variably deformed dykes and gabbros with amphibolite-facies metamorphic assemblages, lying north of Lake Volvi 40 km east of Thessaloniki, Greece (Fig. 1). The complex lies within the Serbo-Macedonian Massif (SMM), traditionally regarded as Palaeozoic or older, but we suggest that the complex and possibly much of the adjacent metasedimentary envelope may in fact be Mesozoic in age of formation and metamorphism. Şengör *et al.* (this volume) speculate that the complex may be a remnant of Palaeotethys, an idea stemming from an earlier interpretation by one of us (Dimitriadis 1980; and unpublished later manuscript). Apart from these new interpretations, the 'conventional' view is that of Kockel *et al.* (1971, 1977), that the complex is part of a pre-Mesozoic, even pre-Hercynian, basement. This view is founded on the results of detailed mapping of the Chalkidhiki peninsula and adjacent areas over several years by geologists from the BGR, Hannover, and IGME, Greece, which culminated in the publication of a 1:100 000 scale map and explanatory memoir (Kockel *et al.* 1977), followed by a series of 1:50 000 sheets published by IGME in Athens.

In the first part of this paper we discuss the basis for this interpretation of the Serbo-Macedonian Massif at some length, firstly because we feel that radically different alternatives may fit the published data and our observations better, and secondly because the reader may be better able to appreciate our own results with the alternative possibilities sketched out in advance.

We emphasise that this paper is a report of work still at an early stage. It is based on reconnaissance traverses by SD, energy-dispersive XRF analyses provided by the Open University, limited microprobe data and 3 weeks of joint field-work in 1983.

Important first results of mapping in the Volvi complex are the discovery of a disrupted sheeted (i.e. 100%) dyke-complex, and the confirmation of our provisional conclusion presented at the meeting (Dimitriadis and Dixon 1982), that magmatism and deformation overlapped in time.

Part 1: Structural and Stratigraphic Setting of the Volvi Complex

The Serbo-Macedonian Massif in Greece has been divided into two 'series' by Kockel *et al.* (1971, 1977) (Figs 1 & 2): an eastern, the *Kerdilion*, composed largely of migmatitic gneisses, but with two major marble units and minor amphibolites, and a western, the *Vertiskos* composed of schists, gneisses and amphibolites. Within the Vertiskos, close to its boundary with the Kerdilion, are three basic-ultrabasic complexes, the dominantly

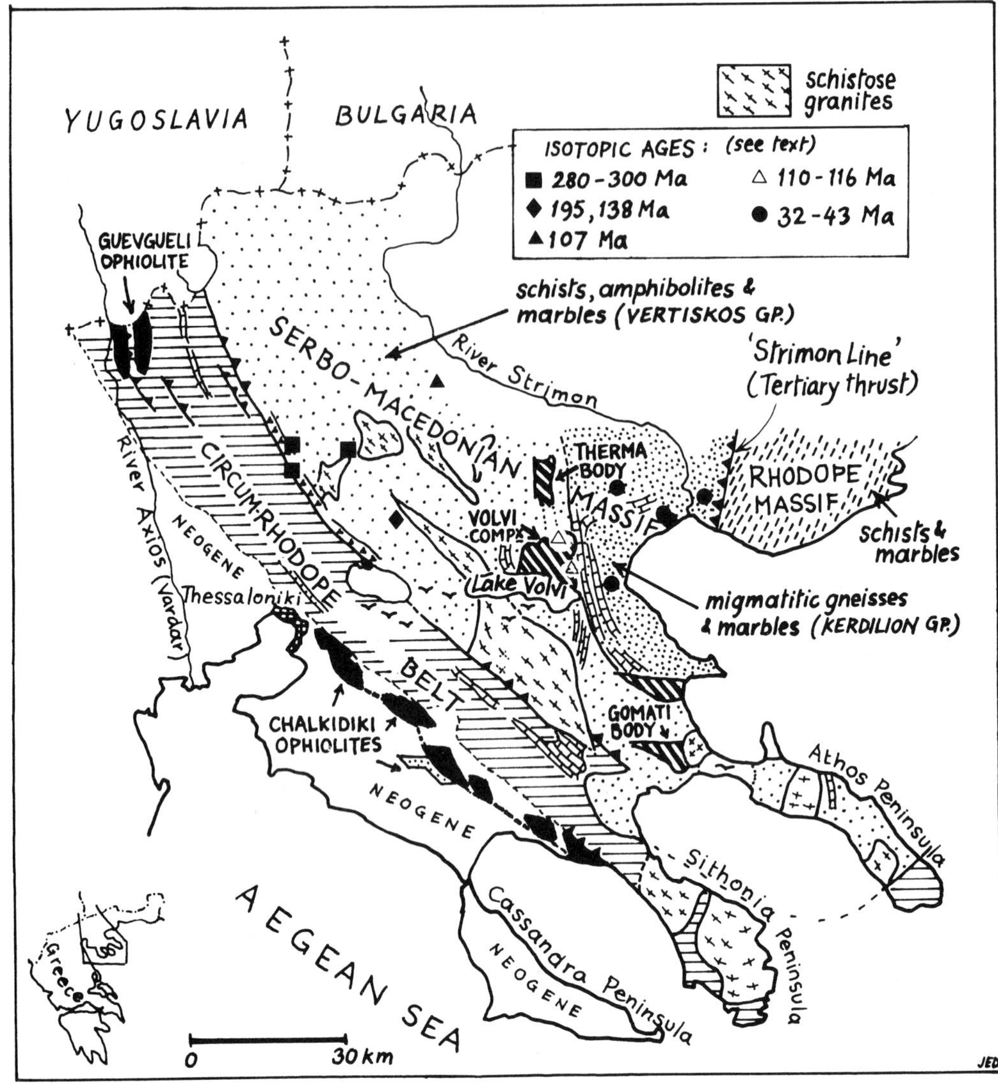

FIG. 1. Geological sketch-map of the Chalkidhiki peninsula, northern Greece, and adjacent areas, showing the main tectonic units, the Volvi complex and associated mafic/ultramafic bodies of Therma and Gomati, and isotopic age sampling points. The age data are discussed in the text.

basic Volvi complex described here, and the Therma and Gomati, dominantly ultrabasic, bodies to the north and south respectively. Dimitriadis (1980) tentatively interpreted all three as one dismembered ophiolite, the Therma–Volvi–Gomati complex. This remains a possibility but our joint work has not yet extended beyond the Volvi complex itself. Mineral assemblages within it are mostly characteristic of the high-grade amphibolite facies, whether in foliated 'equilibrated' amphibolites or in undeformed gabbros with corona textures indicating incomplete equilibrium.

The high-grade metamorphic character of the body has led hitherto to its inclusion as a component of the regional metamorphic 'basement' on the implicit assumption that the metamorphic and deformation event recorded within it was the same as that affecting the Serbo-Macedonian 'basement' surrounding it and was thus pre-Mesozoic. By implication the magmatic history of the body preceded this 'event' or perhaps was at least synchronous with it.

It is thus useful to review the evidence on which this 'conventional' view of the Serbo-Macedonian Massif is based. We will then discuss the existing radiometric data which do not readily fit this conception, and present a set of three alternative settings for the body, before describing our own results.

Stratigraphic arguments

The Serbo-Macedonian Massif in Greece is the southern extension of a large expanse of metamorphic rocks outcropping in Bulgaria and eastern Yugoslavia, bordered on the west by largely Mesozoic rocks in the Vardar Zone and on the east, in Bulgaria and Greece by the Rhodope metamorphic massif. Further north in eastern Serbia, the eastern boundary of the Serbo-Macedonian massif with the Rhodope-Moesian platform is marked by a complex series of Mesozoic sutures and associated basins and volcanic arcs (Getic Suture; Timok and Luzniča Zones: Grubić 1980). Serbo-Macedonian metamorphic rocks in Yugoslavia are overlain unconformably near Bosilgrad (E. Serbia) by fossiliferous Tremadocian strata which has led to the assignment of the entire massif to the Pre-Cambrian or earliest Palaeozoic (Dimitrievič and Cirić 1967). In Bulgaria, the Serbo-Macedonian Massif is considered to be Pre-Cambrian (Zagorchev 1974). At the western border of the Serbo-Macedonian Massif in Greece is a series of mainly NE dipping tectonically bounded slices of weakly metamorphosed sediments and igneous rocks constituting the circum-Rhodope belt of Kauffmann *et al.* (1976) and are well established as spanning in age of formation at least the Early Triassic to Early Jurassic interval. The stratigraphically lowest unit in this belt is the *Examili Formation*, composed of weakly deformed and recrystallized arkosic clastics which rest with nevertheless recognizable unconformity on schists of the Serbo-Macedonian Massif proper in a series of discontinuous, fault-bounded outcrops (Mercier 1968; Kockel *et al.* 1977 and our own observations). They are unfossiliferous but are attributed to the Permian or Lower Triassic because the overlying '*Volcano-sedimentary Series*' (rhyolitic lavas and tuffs), though itself always in tectonic contact with the Examili Formation, passes conformably up into fossiliferous Upper Skythian limestones (Kockel *et al.* 1977, p. 26). The overlying neritic carbonates of the Deve Koran–Doubia unit (Mercier 1968) contain fossils of Upper Skythian, Ladinian, and Carnian age and the highest horizons near Metaliko and Aspro Vrisi are assigned to the Rhaetian (Kockel *et al.*, *op. cit.*). Along strike to the south-east in the Chalkidhiki peninsula the neritic carbonates give way to increasingly deformed and recrystallized thin-bedded limestones, the *Svoula Marble Formation* (Fig. 2). Upper Norian conodonts and foraminifera are known from the northern, low-grade, end of the outcrop (Kauffmann *et al.* 1976). The Svoula marbles apparently pass up into phyllites and then into the thick sequence of unfossiliferous turbiditic sandstones, shales and conglomerates of the *Svoula Flysch* inferred to be of Lower Jurassic age from their apparent stratigraphic position, an attribution consistent with the occurrence of Upper Triassic limestone clasts in conglomerates within the formation (Kockel *et al.* 1977). Locally in the north-western part of the belt, rare tectonic intercalations of Lower Jurassic carbonates also occur which may be lateral equivalents. At the structurally lowest, western-most part of this belt, immediately to the SE of Thessaloniki, lies the Chortiatis magmatic suite—gabbros, diorites and more evolved granitic rocks—and the mafic and ultramafic plutonics of the Chalkidhiki ophiolites. This association lies SE along strike from the ophiolites of Guevgueli on the Yugoslavian border which are cut by the 150 Ma Fanos granite (see Spray *et al.*, this volume, for review) (Figs 1 & 2). The circum-Rhodope belt includes important thrust-bounded sheets of Serbo-Macedonian-type schists and gneisses which show a metamorphic discontinuity with the intersliced Mesozoic units. In summary, it seems clear that a pre-Triassic, possibly pre-Permian, metamorphic complex existed in this area in the Early Mesozoic on which a sequence of Permian to Lower Jurassic clastics, volcanics and limestones were deposited, followed by a thick sequence of turbiditic sandstones and shales. Whether this sequence represents the development of a rifted, passive continental margin is not clear, though parallels with the successsion on the Othris 'margin' are present (Smith *et al.* 1975) e.g. Examili Formation = Pteleon Formation of Othris?; Deve Koran Formation = Strimbes Limestone? This margin was evidently involved in at least one SW-vergent imbricate thrusting event in the mid-Jurassic which introduced ophiolitic components and slices of Serbo-Macedonian basement into the structural succession. Stratigraphic control on this event is provided by the Fanos Granite and by the *Doubcon Molasse*, a conglomeratic formation with Upper Jurassic–lowermost Cretaceous coral-bearing interbeds which unconformably overlies thrust-sheets of gabbros north of Thessaloniki (Mercier 1968).

It is apparent from the 1:100 000 map of the Chalkidhiki and from the accompanying explanatory memoir (Kockel *et al.* 1977) that later deformation must also have affected the area. For example, the Doubcon Molasse itself is

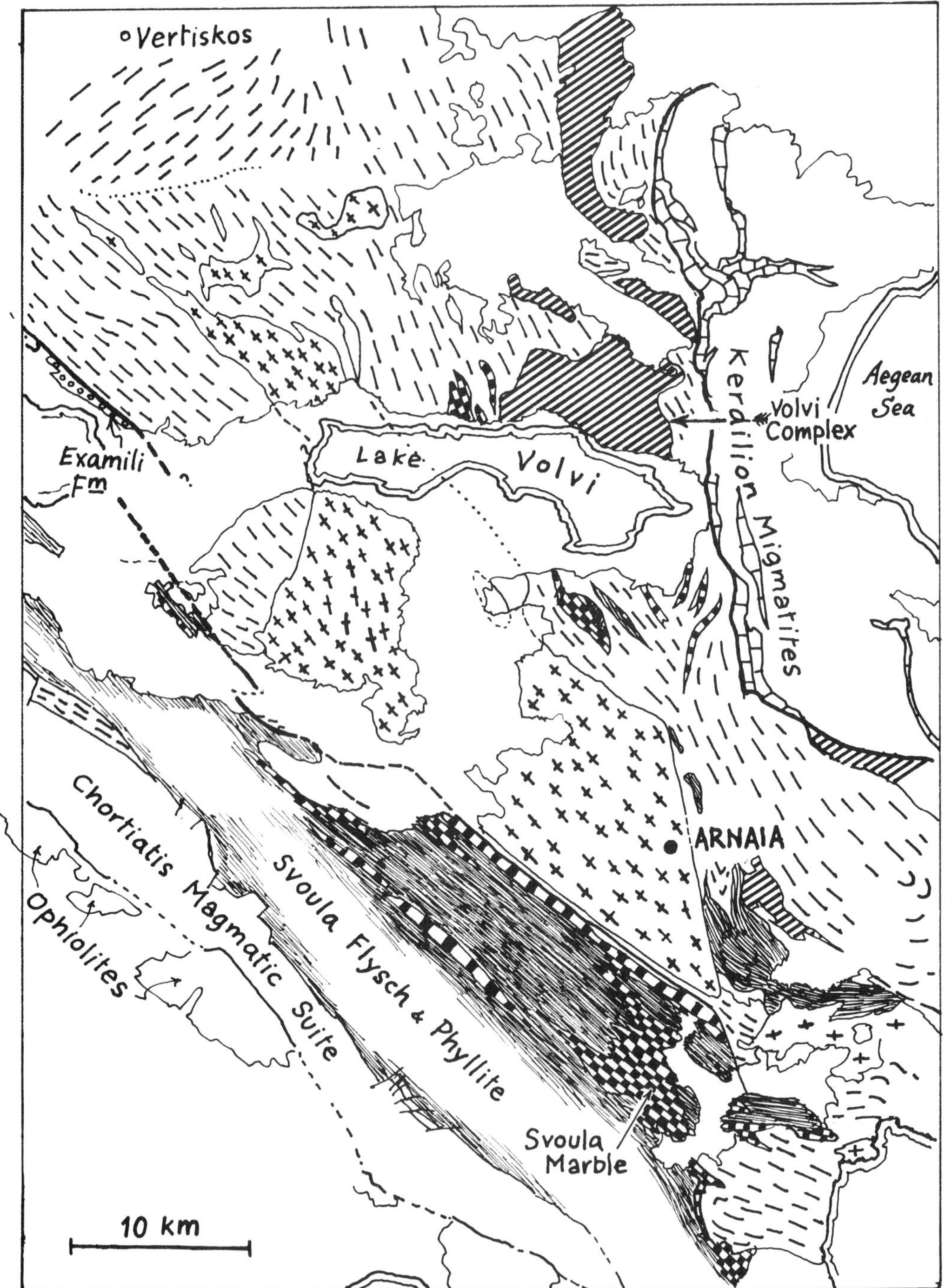

FIG. 2. Enlarged sketchmap of the central part of Fig. 1. showing structural trends in the Vertiskos schists of the Serbo-Macedonian Massif and the schistose *Arnaia Granite*. Note the inliers/intercalations of *Svoula Marbles* and *phyllites* within the S-M massif, which converge southwards on the main outcrop in the Circum-Rhodope belt, according to Kockel *et al.* (1977).

involved in SW-vergent imbricate thrusting with the Svoula flysch immediately to the SW of Deve Koran.

Correlation of units within the Serbo-Macedonian Massif

The identification by Kauffmann *et al.* (1976) of the Late Triassic age of the Svoula marble is critically important in establishing the conventional view of the internal history of the Serbo-Macedonian Massif in the vicinity of the Volvi complex. South-east of Thessaloniki the border of the Serbo-Macedonian Massif is a region of steeply dipping elongate strips of Svoula marble, metamorphosed Svoula flysch and S-M schists (Fig. 2). The Svoula marble strips appear to merge to the south with a north–south trending series of marble and phyllite horizons which strike up into the Serbo-Macedonian Massif proper, to the east of the Arnaia granite mass. These marble and relatively minor phyllite horizons within the massif are interpreted by Kockel *et al.* (1977) as infolded and/or tectonically intercalated, *outliers* of Triassic and lowermost Jurassic Svoula marble and flysch. Most significantly they are mapped as outcropping immediately adjacent to the western margin of the Volvi body. If this stratigraphic correlation is correct then the deformation and metamorphic state of these intercalations in relation to that in the Vertiskos schists and the neighbouring Volvi complex, should provide a valuable reference point in a relative and absolute chronology of events. A second potential time marker is the Arnaia granite body which lies adjacent to the Svoula marble and weakly metamorphosed flysch slices at its SW border. These sediments are cut by apophyses, veins and sheets of granitic material but they do not show the high-temperature contact aureole effects which, for example, surround the foliated Sithonia granite body (dated at 40 Ma, see Kondopoulou and Lauer, this volume) further to the south-east on the central 'finger' of the Chalkidhiki (Kockel *et al.* 1977). The Arnaia granite is thus inferred to be post-Early Jurassic, pre-Late Jurassic/Early Cretaceous on the assumption that the dominant set of structures and associated recrystallisation which clearly affect its southwestern flank are of the same pre-Doubcon-Molasse age as the main imbrication event along this margin. This granite also extends deep into the Serbo-Macedonian Massif and outcrops some 10 km west of the Volvi complex where Kockel *et al.* (1977) map it as being elongated to the north-west parallel to the dominant local fabric in the massif (including the Volvi complex) and to have within it the same SW dipping fabric as the envelope. In summary this stratigraphic and structural correlation led Kockel *et al.* (1977) to the view that the Vertiskos Series of the Serbo-Macedonian massif at the latitude of Lake Volvi is composed of pre-Mesozoic metasediments with infolded inliers of Triassic marbles and phyllites and intrusions of Jurassic granite. Inliers, granite and 'basement' are considered to have been affected by a Late Jurassic, pervasive, NE-dipping fabric-forming event linked in some way to the imbrication event affecting the circum-Rhodope belt. By implication, the original pre-Mesozoic fabric in this transect was either fortuitously NW–SE or was completely overprinted or rotated into this orientation. This overprinting is seen by Kockel *et al.* as a generally low-grade 'diaphthoritic' or retrogression phenomenon. Kockel *et al.* (1977) in fact recognize in the extensive tracts of oblique, *NE–SW* trending Vertiskos schists north of the Anaia mass, original pre-Mesozoic fabric elements.

Regional metamorphism

From the account so far it might be expected that metamorphic grade discontinuities might be preserved between the 'basement' and both the infolded Svoula marbles and the Arnaia granite, with the additional possibility that relict Hercynian or older fabrics might be present in 'basement' rocks, and absent in the cover.

Our joint observations outside the Volvi complex have been limited to the metasediments immediately adjacent to its western margin which are mapped by Kockel *et al.* as including a Svoula marble intercalation. We found no structural or metamorphic discontinuity between marble and non-calcareous rocks but instead a complex sequence of interbedded marbles and dolomites, calcareous, pelitic and graphitic schists. Assemblages in the schists contain staurolite and garnet which have clearly grown syntectonically, and the minor dolomitic interbeds have undergone syn-metamorphic reaction with adjacent carbonate-free horizons, and so were evidently juxtaposed prior to metamorphism. If our observations in the few km of across strike section examined so far, are entirely *within* a Svoula intercalation or even if they cross contacts with pre-Triassic rocks, which seems unlikely, the main fabric-forming event in this part of the massif occurred under *regional metamorphic* conditions of at least staurolite grade. This has to be Jurassic in age if the stratigraphic attribution is valid as noted above and is a very much more comprehensive overprinting event than is implied in Kockel *et al*'s (1977) interpretation.

We will argue below that the dominant fabric in the Volvi complex is also syn-metamorphic and even in part a late magmatic fabric. Likewise, the Kerdilion migmatitic gneisses to the east show high-grade mineral assemblages, syn-metamorphic fabrics and generally little evidence of retrograde reaction (Dimitriadis, 1974). The interim conclusion is that one or more post-Palaeozoic regional metamorphic events have affected the Serbo-Macedonian massif and its cover. The existing isotopic age data is now reviewed in this light.

Isotope ages

Isotopic ages from the Serbo-Macedonian Massif in Greece published to date have been K–Ar determinations on muscovites, biotites and hornblendes, and Rb–Sr model ages determined on muscovites and biotites (Borsi *et al.* 1965; Harre *et al.* 1968). Ages fall into several groups (Fig. 1). Near the western contact with the circum-Rhodope belt but well to the north of Thessaloniki, Borsi *et al.* (1965) obtained ages from white micas by both methods in the range 280–300 Ma, quite consistent with a perhaps slightly reset pre-Triassic, late Hercynian, metamorphic age. The implication is that the Jurassic deformation and recrystallization event in the north was not sufficiently intense to fully reset the isotopic systems in white mica in a metamorphic basement last re-crystallised in Late Carboniferous times. Further to the south-east, away from the contact, Harre *et al.* (1968) obtained ages of 138 ± 2 Ma and 195 ± 3 Ma (ages recalculated using the constants of Steiger and Jäger 1977) from different size fractions of white mica from a schistose pegmatite in the Vertiskos, in a region close to the boundary between 'Jurassic' fabric trends and 'pre-Mesozoic' fabric trends. Further east two amphibolites from just within the eastern and northern margins of the Volvi complex yielded late Lower Cretaceous ages in the range $116–111 \pm 3$ Ma. A schistose pegmatite at the eastern side of the Vertiskos outcrop, NE of Lachanas, yielded a K–Ar age of 107 ± 1.6 Ma from muscovite. Ages in the higher grade migmatic gneisses of the Kerdilion east of the Vertiskos are all much younger. K–Ar muscovite ages there are closely grouped from 36–43 Ma (all $\pm <1$ Ma) with biotite ages 32–38.5 Ma. A single K–Ar determination on hornblende from an amphibolite within the Kerdilion yielded 79.5 ± 1 Ma.

Harre *et al.* (1968) and Kockel *et al.* (1977) interpret the general eastward younging of isotopic ages as evidence of 're-juvenation' of an earlier pre-Hercynian basement and relegate the associated metamorphic effects to a process of retrograde 'diaphthoresis'. For them the Cretaceous ages obtained from Volvi amphibolites would presumably be mixed ages due to some arbitrary degree of argon loss.

Discussion

This conception of the evolution of the Serbo-Macedonian massif seems to us unnecessarily tentative in its reluctance to accept a direct link between the ages and the metamorphic history of the massif. We suggest that the Vertiskos schists and gneisses may well be mainly Lower Palaeozoic or even older sediments, and may well have a significant Mesozoic cover. However, we interpret the ages obtained as *cooling ages* from a *regional* metamorphic and deformational event in the Late Jurassic (?to Early Cretaceous), which was followed by uplift which was progressively later and greater towards the east, leading to the exposure of higher grade metamorphic rocks and generally younger ages eastwards. Both burial and uplift at the western edge of the belt where the ?Permian Examili formation is preserved were minimal. On this interpretation it would be inherently difficult in the higher grade parts of the Vertiskos to distinguish pre-Mesozoic metamorphic rocks on grade alone. It follows from this that more of the Vertiskos may be high grade 'cover' than solely the distinctive marble horizons recognized by Kockel *et al.* (1971, 1977). We would extend this simple equation of isotopic ages and regional metamorphic cooling ages to the Kerdilion and argue that these higher-grade 'younger' rocks represent either a deeper part of the same Late Jurassic orogen which remained deep until the early Tertiary, before being tectonically removed to higher levels, or that they are a completely separate unit which has been juxtaposed with the Vertiskos during Tertiary times.

Implications for the Volvi complex and other adjacent mafic/ultramafic bodies

What range of settings for the Volvi complex might then be envisaged in our modified concept of the evolution of at least the Vertiskos part of the Serbo-Macedonian Massif?

There would seem to be three main alternatives:

Model A

The complex belongs to the real pre-Mesozoic basement and was either tectonically or magmatically emplaced within it prior to the deposition of the Triassic cover and the last pre-Jurassic regional event. This is the view implicit in the accounts of Kockel *et al.* (*op. cit.*).

Implications: Mineral assemblages in the complex should be consistent with those in the metasedimentary envelope. A Jurassic staurolite-grade event would have been superimposed on an earlier pre-Mesozoic metamorphic assemblage. Fabrics might also be continuous with the cover and might be expected to bear little relation to intrusive contacts. K–Ar ages from hornblende might be rather older than envelope ages from muscovite or biotite because of higher closure temperatures, if recrystallization in the Jurassic was comprehensive.

Model B

The body is essentially an *in situ* syn-/or late-orogenic intrusive complex of Mesozoic age.

Implications: Assemblages might show higher temperatures of metamorphic equilibria than the envelope though structures might be continuous. No relict metamorphic fabrics would be expected. Magmatic and deformational events might be expected to show some close connections. K–Ar ages might be expected to be *younger* through locally prolonged cooling.

Model C

The complex (and the associated ultramafic bodies) were originally emplaced as a high-level ophiolite complex during the Mesozoic and subsequently metamorphosed at deeper levels during micro-continental collision. Jurassic metamorphism and deformation would be superimposed on low-pressure igneous or sub-sea-floor metamorphic assemblages. This might be considered an 'ophiolitic suture' model, see Şengör *et al.*, this volume.

Implications: Tectonic and metamorphic discontinuities would be likely between the complex and the 'basement' part of the envelope. Metamorphic discontinuities would not be expected between the Triassic 'cover' rocks and the meta-ophiolite, unless later tectonism had juxtaposed different parts of the orogen. No particular links would be expected between the magmatic and the metamorphic events.

Part 2: The Internal Structure and Evolution of the Volvi Complex

The Volvi complex lies immediately north of the eastern half of Lake Volvi, and forms a mountainous mass rising to 627 m. The relatively steep southern slope is dissected by numerous sinuous N–S ravines with excellent exposure in their floors and sides and sufficient exposure on the scrub-covered slopes between them to encourage us to attempt detailed mapping on 1:10 000 scale aided by good-quality air-photos. Field relations are none-the-less complex, because of the interaction of magmatic and deformational processes and later faulting.

Lithologies

Compositionally, the Volvi complex is dominantly basaltic in composition but has important intermediate basaltic andesite or andesitic components (see below). Very minor amounts of more evolved dioritic or granitic types are present and locally screens and angular blocks of highly metasomatized meta-sediments are found within dyke swarms. A complete range of original igneous textures is preserved undeformed, from fine-grained (1 mm) aphyric or micro-porphyritic (feldsparphyric) dykes to very coarse pegmatitic gabbros with amphibolized pyroxenes 10 cm long. Locally, feldspar-rich pegmatitic gabbros show a ramifying herring-bone feldspar texture in which individual crystals (now saussuritized) can be traced for as much as 30–40 cm (Fig. 3). The ratio of

FIG. 3. 'Herring-bone' or dendritic texture in original plagioclase feldspar, pegmatitic gabbro, Volvi Complex.

mafic to felsic components is variable, notably in the gabbros, where either component can occasionally exceed 75% of the mode. Superimposed on this range of primary grain-sizes is a wide range of strain-states which yield a correspondingly wide range of rock types from schistose fine-grained homogeneous amphibolites, amphibolites finely striped on a 1–2 mm scale, to coarse-striped amphibolites with a recognizable medium-grained gabbroic parentage. Sheared, flaser- or banded mylonitic gabbros are derived from coarser parents.

In many places these transitions can be followed in shear zones from undeformed into deformed rock and the textural progression is not in doubt. Elsewhere it is clear that a similar final texture of medium-grained striped amphibolite has been generated by more than one combination of original texture and state of strain.

All the rocks in the complex possess a metamorphic mineral assemblage to some degree and those of igneous origin should strictly be called *meta*gabbros, *meta*-granites or *amphibolites*. However, for simplicity we drop the *meta*-prefix and use the term gabbro for all rocks with a recognizable original magmatic texture of intersertal or cumulative type. Likewise, we tend to use the term *amphibolite* where the identity of the parental rock (fine-grained basalt or dolerite or gabbro) cannot be determined with certainty. Because many fine-grained 'amphibolites' are identifiable as undeformed basaltic dykes albeit with an amphibolite-facies assemblage, we use *dykes* as a more useful descriptive term for the fine-grained rocks.

Primary igneous field relations

Dykes

In the fine-grained amphibolites evidence of igneous contact relations and primary magmatic variations tends to be lost at low states of strain. However, we have observed substantial tracts, approaching 1 km length, of 100% dykes in which original chilled contacts are preserved (see Fig. 4 for a representative segment). It seems very probable that other, slightly more sheared, regions of fine-grained homogeneous amphibolite are also sheeted dyke-complex. The dykes illustrated in Fig. 4, which range from 10 cm to 2 m across, show typical variability of original texture both within and between dykes. Several are mildly feldsparphyric, with feldspars concentrated and flow-aligned in the central regions. Others show relict mafic microphenocrysts, possibly after olivine. All are now composed of a fine-grained generally isotropic-textured hornblende–garnet–plagioclase–(epidote-group-mineral)–oxide assemblage.

Clear one-way chilling is seen in the illustrated section but is scarcely usable as a statistical indication of spreading. Significantly, two of the illustrated dykes show sheared microporphyritic centres strongly suggesting that deformation occurred during the intrusive phase while the dyke centres were still hot.

Where clearly recognizable, in undeformed parts of the southern and central area, the mafic dykes are generally steeply dipping or vertical with a NE strike. Our impression is that the much rarer, isolated leucocratic aplitic sheets which are from 0.5 m to 2 m thick are more variably inclined and often flat-lying.

In some ravine sections, dykes do not comprise 100% of the outcrop and a spectacularly 'tiger-skin'-striped rock-type is preserved as screens and angular blocks between them (Fig. 5). This rock in its common variant is composed of laterally impersistent 1 cm thick plagioclase-rich bands, interlayered with similarly impersistent mafic layers made up of garnet, diopside, hornblende and plagioclase. It is provisionally interpreted as a highly modified hornfelsed and metasomatized calcareous sediment and if so is important evidence of a pre-existing 'country rock' into which the dyke components of the complex were intruded. Outcrop-scale evidence of deformation during dyke emplacement into these 'tiger-skin' rocks also exists and is discussed below.

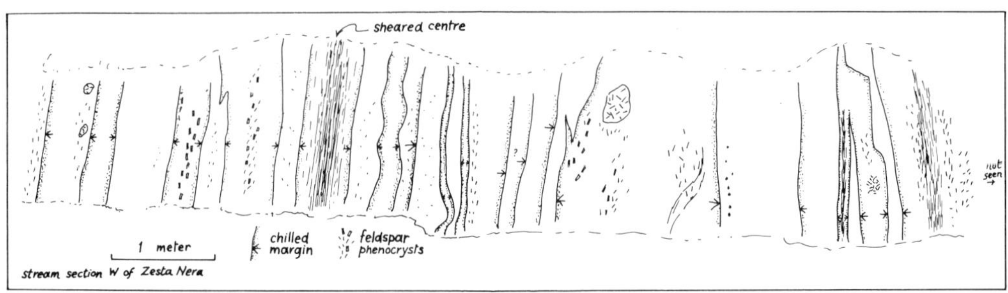

FIG. 4. Representative 10 m section in sheeted-dyke complex.

FIG. 5. 'Tiger-skin' rock. Metasomatically altered metasediment, found as screens and angular blocks between dykes.

Gabbros

Within the larger outcrop areas of gabbro the inferred primary igneous grain-size is still variable but other rock-types are generally absent. Most lithological complexity occurs in regions of transition to dyke complex, which in the few good sections examined to date occurs upwards (i.e. to higher present structural levels) over 1–200 m. Pegmatoid patches and veins in gabbro become more abundant, and the gabbro texture becomes variable on a scale of cm to m. In some localities the transition through fine-grained dolerite to homogeneous amphibolite is completely gradational over about 100 m via a nebulous 'hybrid' with impersistent diffuse boundaries between gabbroic veins and finer-grained doleritic rocks. Elsewhere the contact zone is a region where 100% gabbro gives way to decreasingly abundant, sharply bounded gabbro-pegmatite sheets apparently emanating from the upper part of the homogeneous mass and cutting fine-grained amphibolite. The sheets vary from 1–2 cm to 1–2 m in thickness but are often equally coarse-grained however thick, and show no chilled margins, textures suggestive of very high country rock temperatures at the time of intrusion.

Fine-grained dykes are also found cutting coarse-grained gabbros in these transitional regions (see Fig. 7 below). Variably resorbed, often angular, small gabbroic patches in dykes are generally quite common, implying that assimilation of earlier crystallized material was an important process.

Cumulate gabbros

At a few localities within the gabbroic parts of the body cumulates occur. These are two main types—those with plagioclase as the main cumulus phase and those with prominent rod-like clinopyroxenes lying randomly in the bedding plane, accompanying cumulus plagioclase and scattered olivine euhedra. The layering tends to have shallow dips, strengthening our view that the undeformed parts of the body, even if bounded by faults and shear zones, have not been rotated far from their original attitude.

Net veining

Angular, fracture-controlled intrusive relations occur quite widely, generally on a scale of cm to a few m for the blocks involved. Metasediment/dyke and gabbro/dyke associations have been mentioned already. More prominent are net-veining occurrences with basic dyke blocks in a granitic or ?trondjemitic vein matrix, in which 'dioritic' hybrids between block and matrix also occur (Fig. 6).

FIG. 6. Undeformed net-veining. An early high-T stage of ?diorite injection (top) in which the blocks have diffuse margins was succeeded by a later, brittle stage of leucocratic (?trondjemite) veining.

Interpretation

The parallels with structures in the so-called high-level gabbros and sheeted-dyke parts of ophiolite complexes are obviously strong (see for example Pallister and Hopson 1981; Pallister 1981). In detail it is apparent that gabbro bodies were sometimes emplaced at high levels in the Volvi complex into regions already occupied by dykes while elsewhere late basic injections appeared as fine-grained dykes in gabbro. The metasedimentary screens are intriguing but difficult to fully explain because they imply the existence of a floor on which they were deposited. If this was continental crust, it would have to have been completely displaced to both sides of the invading gabbro-dyke complex. If oceanic crust, it might suggest that pillow-lava screens, as yet unrecognized, lie at lower levels between dykes. We are following up these possibilities. What seems certain is that crustal extension approaching 100% accompanied intrusion, and thus that the complex is ophiolitic in character even though no volcanic component has yet been identified.

Deformation

The dominant structure on outcrop scale is a planar shear fabric of very variable intensity which generally dips towards the W and SW in the central and southern part of the complex. This schistosity is affected locally by close to tight folds on a scale of 100s of m with a generally consistent sense of overturn down the dip of the main fabric. Small tight or isoclinal folds on scales of 10s of cms are abundant near the large folds but generally rare overall. The body is also cut by obviously later, narrow shear zones which cut early fabrics and are associated with chlorite and actinolite formation. These are rare. The whole area is cut by several sets of late faults, some associated with intense hydrothermal alteration, which divide the body into blocks on a 1–2 km scale and hamper the process of correlation between ravine sections.

Our preliminary interpretation from the attitude of dykes and cumulate layering is that deformation has not completely destroyed an irregular, but possibly gently NE dipping, contact or transition zone between gabbro below and dykes above. A combination of a down-dip sense of shear to the SW and a probable, but not yet proved, sense of motion on E–W faults of downthrow towards the lake to the south, have the effect of exposing deeper levels of the complex towards the north.

Relationship between deformation and magmatism

We mentioned above examples of schistose fabrics developed preferentially in the central regions of dykes, implying some magmatic temperature control on the mechanical properties of the dyke. The clearest evidence of an overlap in time between intrusive activity and deformation comes from the transition zones between gabbro and dykes.

The 0.5 m scale pegmatitic gabbro sheets described earlier that emanate from a homogeneous gabbro mass into fine-grained amphibolite, cut *already deformed* (and undeformed) dykes. The dykes contain strongly flattened feldspar microphenocrysts and mm scale isoclinally folded leucocratic veinlets. The cross-cutting gabbro sheets show a very weak fabric but have an essentially unstrained coarse igneous texture quite inconsistent with the implied strain suffered by the dykes. Elsewhere in the complex the reverse relationship can be found. Figure 7 is a field illustration of a

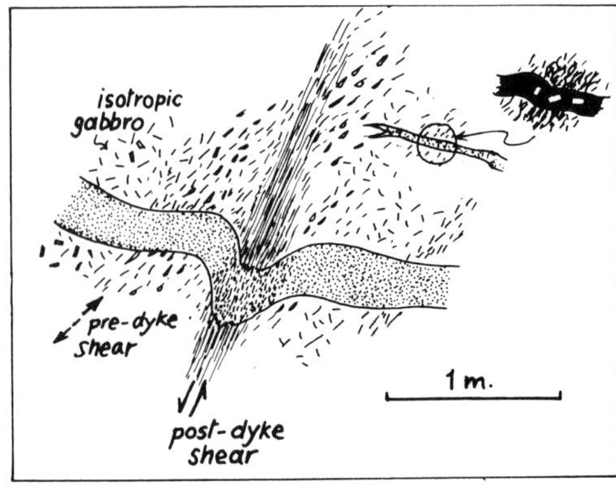

FIG. 7. Basaltic dyke cutting already sheared gabbro and subsequently affected by a later shear zone. Field sketch.

30-cm-wide, fine-grained dyke which cuts variably sheared gabbro and is itself affected by a later, more upright shear-zone which deforms the pre-dyke shear fabric. In the same place amphibolitized basalt dykelets 4 cm across cut slightly sheared gabbro but contain undistorted euhedral feldspar microphenocysts, which again imply deformation prior to dykelet intrusion.

A further clear instance of active deformation occurring between mafic and felsic dyke injection events is shown in Fig. 8, where the

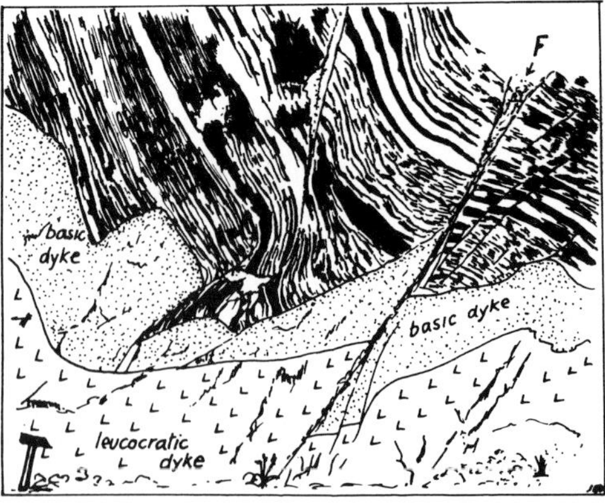

FIG. 8. Deformed and fractured 'tiger-skin' metasediment intruded by a basaltic dyke which has itself been off-set by continuing(?) fault movement. The later leucocratic sheet post-dates the faulting.

leucocratic 'meta-trondjemite' sheet is not displaced by the sinuous faults which offset the mafic dyke contact.

We seek further evidence of this overlapping of magmatism and deformation. Probably the great majority of shear deformation occurred after emplacement as it clearly deforms an igneous texture. Trondjemite net-veining associations, which are late magmatic-events are cut by shear zones which convert undeformed angular textures into striped heterogeneous amphibolites in the space of 1–2 m. We nevertheless feel from our preliminary work on the metamorphic textures and mineral assemblages that much of the deformation probably occurred *before* the present mineral assemblages had fully developed.

We cannot yet readily distinguish in thin-section between a late-syn-magmatic fabric in the few cases where this is identifiable in the field, and the 'regional' shear fabric in gabbros or amphibolites. Both may show the same kind of textural equilibrium in the same metamorphic minerals, which leads us to the working hypothesis that they may be recording the same high-temperature deformational event. This hypothesis is one of the principal targets for testing in our continuing petrography and micro-fabric study.

Complex/Envelope contact relations

We have not yet mapped the contact of the complex and our observations are limited to one ravine transect, crossing it in the south-west corner. Here amphibolites in the complex are increasingly interlayered with schistose garnet- and magnetite-bearing aplitic sheets over a distance of several hundred metres towards the contact and it is tempting to interpret these as the products of partial melting of (meta)sediments in the envelope.

Petrography and mineralogy: conditions of metamorphic equilibrium

Our petrographic work is at a very early stage. Here we summarize some preliminary observations.

In the field it is apparent that the mineral assemblage and consequently the appearance of metamorphosed gabbros is a product of two largely independent processes, a metamorphic and a metasomatic one. In completely undeformed parts of the complex, relict masses of 'fresh' gabbro occur with distinctive purplish plagioclase feldspars and a characteristic extreme toughness. These masses may be up to a few hundred metres across but more commonly are reduced to metre-sized rounded 'kernels' transected by planar alteration zones. In these zones the feldspars are milky white and pyroxenes are rimmed by, or transformed to fine-grained green amphiboles. A thin white veinlet, 1–2 mm across, frequently lies along the centre of the zone. These saussuritization zones can be traced into weakly deformed parts of the body where they are less prominent but are clearly *post-tectonic* in that they cut the schistosity and are unrelated in their orientation to shear zones. The strongly deformed gabbros are, however, also pervasively saussuritized in appearance being typically milk white/deep green in felsic and mafic areas respectively.

In the completely undeformed and unsaussuritized gabbros spectacular coronas are developed at the junctions of preserved olivine and plagioclase, involving, from olivine outwards, orthopyroxene, garnet and hornblende. Saussuritization in these rocks involves randomly oriented zoisite growth in the feldspar and a variety of other complex hydration reactions in the coronas, clinopyroxenes and remaining olivine. Deformed gabbros on the other hand typically contain hornblende, garnet, plagioclase and an epidote group mineral. Textures often look well annealed even in foliated rocks. Saussuritization can nevertheless be seen as a later random growth of a Ca–Al-silicate in the plagioclase.

The interpretation favoured at present is that the coronitic and sheared gabbros represent partial and more or less complete transitions respectively, to essentially the same high-grade amphibolite facies assemblage. Deformation clearly promoted recrystallization of gabbro to gabbroic garnet amphibolite, but it is not clear that it was necessarily synchronous with the growth of the new assemblage, which may have instead inherited an anisotropic fabric from a higher temperature stage when the igneous assemblage was still intact. Saussuritization occurred still later and is most prominent as a secondary vein-controlled process in the undeformed and unreacted gabbros because they have greater modal abundances of calcic plagioclase and no preferential schistosity—parallel paths for the entry of metasomatizing fluids.

In the finer-grained dyke rocks a similar pattern of variable, late saussuritization is recognizable but the undeformed, unsaussuritized rocks show little or no evidence of incomplete textural equilibrium. Instead they show textures similar to hornfelses with garnet-plagioclase-hornblende ± clinopyroxene-epidote (or zoisite), and even hornblende-hypersthene-plagioclase, as common assemblages. Schistose and striped amphibolites as a whole show a range of textures and mineral assemblages and although many are dominantly hornblende-garnet-plagioclase-epidote/zoisite rocks, some have evidently equilibrated at lower temperatures and contain hornblende-plagioclase-epidote-chlorite-sphene-quartz in apparent textural equilibrium. The 'tiger-skin' metasediments typically contain plagioclase-epidote in the leucocratic bands and hornblende-green diopside-garnet-epidote-plagioclase in the mafic layers. So far, our impression is one of general consistency of implied metamorphic grade for assemblages in completely undeformed rocks as well as in foliated gabbros, amphibolites and metasediments. This conclusion is backed up to a surprising degree by some preliminary geothermometry.

We have analyzed hornblende-garnet pairs in a small number of representative rocks and applied a recently calibrated Mg/Fe″ exchange thermometer (Graham and Powell, 1984).

The method has been developed by cross-calibration using garnet-hornblende-pyroxene assemblages and the experimentally calibrated garnet-pyroxene geothermometer. A correction for Ca-content in garnet is incorporated. The regression function used is

$$T(°K) = \frac{2880 + 3280X_{Ca.gt}}{\ln K_D + 2.426}$$

The results from four different rocks are summarized below.

Sample	*T°C*	*Mol% Pyr*
AAX2		
Coronitic metagabbro (*gar and hbl in contact in a corona round olivine*)	720	37
MB5		
Fine-grained 'porphyritic' amphibolite (*plag-gt-hbl-ilm*)	753	19
ZN3		
Undeformed fine-grained dyke (*plag-gt-hbl-cpx-ep*)	772	9
ANMII04		
'Tiger-skin' metasediment (*gt-hbl-plag-ep-ilm*)	768	7

Garnets are calciferous almandine-pyropes with pyrope contents ranging from 37% to 7% as noted in the table above. Hornblendes are pargasites, low in Ti. MB5 is typical, with approximate formula: (cations to 23(O))

$$Na_{0.7}K_{0.2}Ca_{1.74}Mg_{2.14}Fe^{tot}_{1.99}Ti_{0.08}Al^{VI}_{0.95}Al^{IV}_{1.79}Si_{6.21}.$$

Discussion

The consistency of implied temperature of equilibration for the deformed, undeformed and metasedimentary rocks sampled is remarkable. Taken with the evidence of an overlap in time between magmatism and deformation it is difficult to avoid the conclusion that these assemblages were acquired during the initial cooling of an actively deforming, recently intruded, mafic complex. The sheeted-dyke complex implies some form of active crustal extension. One may infer, however, that this complex was intruded at a significantly deeper level than a typical mid-ocean ridge ophiolite, from the garnet-opx coronas, and the pervasive high-grade garnet amphibolite assemblages with both cpx and opx, which contrast with the typical sub-sea floor type epidote-amphibolites found in ophiolitic sheeted-dyke complexes (Gass and Smewing 1974). Estimation of the pressure and depth of equilibration must await further microprobe-work and thermodynamic calculation.

It is also clear that the Volvi body equilibrated initially at temperatures some 200–250°C higher than the staurolite-bearing schists immediately adjacent to it, which do not appear ever to have been at that temperature from the absence of partial melting evidence. The crucial question (among many) which microprobe work in the envelope will attempt to answer is whether *pressures*, if not temperatures, are comparable, inside and outside the body. Potential geobarometers are the chlorite-biotite-white mica assemblages in the envelope (Powell and Evans, 1984) and the garnet-bearing coronas in the Volvi gabbros (Bohlen *et al.* 1983). Establishing the ages of cooling of both will be one objective of planned radiometric work.

Geochemistry

Table 1 contains some representative analyses by energy dispersive XRF of Volvi mafic and felsic rocks. On the AFM diagram (Fig. 9) the range of intermediate types is noteworthy, and the suite as a whole clearly contains rocks of broadly basaltic andesite or andesite composition as well as conventional basalts. Granitic members with SiO_2 >65% have a range of K_2O contents from 0.04% to 5.48% and do not conform to a simple 'oceanic plagiogranite' label. Figure 10 is a MORB-normalized plot of trace-element data of the type introduced by Pearce (1980). The enrichment in large-ion lithophile elements is marked though it should be remembered that all these rocks are now high-grade amphibolites which have suffered unknown additional chemical gains and losses during saussuritization. A very pronounced depletion in Nb and Zr that is taken as characteristic of island-arc character by Pearce (1982),

TABLE 1. *Representative chemical analyses of Volvi amphibolites, dykes and evolved rocks—ZN3: Fine-grained meta-basaltic dyke; Zesta Nera; Gt-Ep-Plag-Hbl-Pyrox. MB5: Fine-grained porphyritic amphibolite (after porphyritic basalt dyke); N. of Micri Volvi: Gt-Hbl-Plag-Eb-Ilm. ANMII4: Fine-grained ?meta-basaltic andesite dyke; E. of Mavro Pournaro; Gt-Hbl-Bio-Plag-Zoi. MYPIII1β: Leucocratic ?quartz-diorite matrix in a net-veining association; Mavro Pournaro; Plag-Qz-Hbl-Gt-Ep*

Wt%	ZN3	MB5	ANM*II*4	MYP*II*11β
SiO_2	47.11	43.41	54.89	67.99
TiO_2	2.68	2.59	1.32	0.54
Al_2O_3	13.75	14.60	15.67	17.97
Fe_2O_3	9.96	13.10	6.60	2.98
MnO	0.12	0.15	0.06	0.03
MgO	6.26	8.04	5.42	0.84
CaO	13.16	10.87	9.95	3.85
Na_2O	2.84	2.78	4.53	7.98
K_2O	1.44	0.69	0.34	0.23
P_2O_5	0.34	0.36	0.18	0.08
LOI	0.55	1.17	0.70	0.29
Total	98.21	97.75	99.64	102.04
ppm				
Rb	80	21	8	4
Sr	274	267	353	347
Y	51	51	23	29
Zr	211	167	369	597
Nb	10	9	10	7
Pb	4	5	6	10
Th			6	8
Cu	10	64	3	4
Zn	21	30	22	10
Ga	19	20	19	19
Ni	57	112	61	8

Analyses by energy-dispersive XRF at the Open University, Walton Hall, Milton Keynes

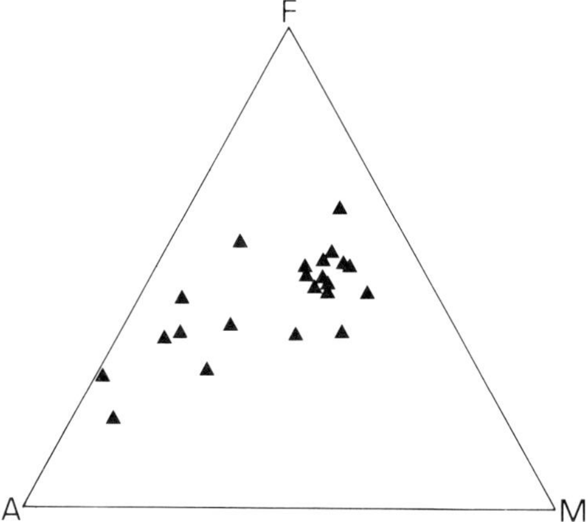

FIG. 9. AFM diagram of Volvi dykes, amphibolites, intermediate and leucocratic sheets, and net-vein matrices.

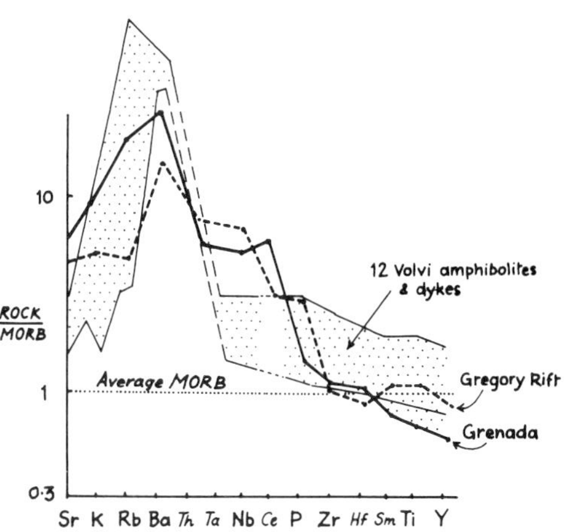

FIG. 10. MORB-normalized trace-element data 'envelope' (dotted) for Volvi basaltic dykes and amphibolites. Note: We have no data yet for the elements in italics. The Grenada and Gregory Rift patterns are from Pearce (1980 & 1982).

particularly when combined with Ba, Rb and Sr enrichment is not present nor are these rocks typically MORB-like. They show, on the basis of this limited data set, hybrid patterns of the kind that Pearce *et al.* (1984) interpret as due to the superimposition of above-subduction-zone character on an alkalic MORB 'base'. The island of Grenada is an example quoted by Pearce of hybrid tectonic setting (close to the southern transform terminating the lesser Antilles arc) and a matching hybrid trace element signature. It is included on Fig. 10 for comparison as is the intra-continental E. African Gregory Rift basalt trend. In the introductory chapter to this book, JED and co-author Robertson make the point that until the trace-element characteristics of continental rift-zone basaltic rocks are better established, particularly in areas like the Rio Grande Rift where calc-alkaline volcanics are abundant, the use of MORB-normalized plots is probably premature when formation in such a rift is a real possibility. An intra-continental rift setting above a subduction zone, active or dormant, is clearly a possibility from the regional considerations set out in the first part of this paper. Geochemically, there appear to be no obvious features that rule out such a setting. The Volvi basaltic rocks are most probably *transitional basalts* (olivine tholeiites) in geochemical character but this identification is of no help at all in establishing their tectonic setting. The affinities of the Volvi body are thus an open question on which we hope bring rare-earth and isotopic methods to bear.

Discussion and conclusions

In our earlier discussion of the setting of the Volvi complex we put forward three broad alternative 'classes' of interpretation—a 'pre-Mesozoic basement' model A; an '*in situ* Mesozoic rift' model B or an 'ophiolitic suture zone' model C, and we outlined features of field relations and petrology that might constitute tests for them. Now we briefly re-examine our alternatives.

Model A

The finding that metamorphic equilibration temperatures are markedly higher than in the envelope appears to rule out the first class of model, A. For the Volvi body to be a pre-Jurassic 'basement' complex we would have to argue that its metamorphism was synchronous with a very high-grade event in the 'true' pre-Mesozoic basement of the envelope (so far not yet seen), and that the syn-tectonic staurolite grade-event that *is* found in the envelope is represented by very localized recrystallization in shear zones within the complex which again we have so far failed to recognize as important. Our limited observations at the contact of the complex do not yet eliminate model A, but tend to favour B below. The link between magmatism and the main deformation event in the complex would not be expected if the igneous rocks were Palaeozoic or older and the deformation post-Triassic and continuous into the 'cover'.

Isotopic ages in relation to model A

We do not know the context or mineral assemblage of the amphibolite samples from the complex analyzed by Harre *et al.* (1968). However, if we were to date a random amphibolite sample we would generally expect it to give an age of initial cooling from the 750°C equilibrium temperature to a closure temperature appropriate to the method used. Uncertainty arises over the saussuritization process which certainly affects calcic plagioclase but may not disturb Ar or K in hornblende. The saussuritization 'event' is *post-tectonic* and thus not a good candidate for correlation with the staurolite-grade deformation event in the envelope. On balance, we think the Cretaceous ages obtained by Harre *et al.* (1968) are probably cooling ages and so quite inconsistent with the 'basement' model A.

Model B

Model B, that the complex is an *in situ*, deep-level extensional event occurring within the Serbo-Macedonian Massif during active deformation, fits most of the observations so far. It accounts for the higher temperatures of equilibration, for the link between magmatism and deformation and may be further substantiated if granitic melts occur in the marginal zone of the complex. The two isotopic ages are also consistent. The model suggests many lines of investigation for the future. If the original contacts are preserved, a sharp rise in regional metamorphic grade to sillimanite- or kyanite-bearing migmatites might be expected. The staurolite-garnet-schists found to date may themselves be in a syn-tectonic aureole. Systematic isotopic age progressions should occur, related to prolonged cooling in the high-temperature complex and its aureole, if uplift was uniform over the area during cooling (e.g. Dewey & Pankhurst 1970). Our planned geobarometry should provide a vital constraint on the consistency of depth inside and outside the complex. Before we comment finally on the wider implications of a model B interpretation we will consider model C briefly—the 'suture model'.

Model C

Model A would identify the complex as a fragment of *oceanic* crust emplaced between two crystalline basement blocks in some micro-continental collision. Most of the difficulties encountered by model A also apply here, particularly the metamorphic grade discontinuity and the absence so far of any lower-grade precursor assemblages. This could be explained by final rethrusting and stacking together of slices from different parts of one deeply buried suture zone, but again the *early* syn- to late-magmatic character of the deformation and mineral assemblage formation go against this. We cannot, however, yet show conclusively that the syn-magmatic fabrics are all related kinematically to the main post-magmatic fabric even though field and thin-section data point that way. Metasedimentary screens in the dyke-complex are not likely in an idealized oceanic setting but not impossible. An aureole in the envelope would rule out a simple, originally allochthonous, ophiolite model.

Models B and C are clearly not completely mutually exclusive. The Volvi complex could be a fragment of an originally continent–ocean boundary zone of Jurassic or Early Cretaceous age. The east side of the complex not yet examined could then be fundamentally different structurally from the west, a further objective for fieldwork.

Regional implications for a model B, '*in situ* rift' interpretation

At this stage a regional model would be clearly premature. Şengör *et al.* (this volume) speculate that the complex is a continuation of a missing Palaeotethyan suture of *Early* Jurassic age linking suture zones in eastern Serbia with the Rhodope-Pontide part of their Cimmerian continent. This view was partly inspired by an earlier unpublished development of the suture model C by one of us (SD). We now believe that the isotopic ages and the regional arguments rule out this interpretation on grounds of timing. However, it appears that Şengör *et al*'s interpretation of the age of the Getic suture in eastern Serbia is at variance with at least some Yugoslavian interpretations. Grubić (1980) implies that closure of the Danube basin and formation of the suture occurred in the Albian–Aptian not the Early Jurassic.

West of the Getic suture the Timok graben opened up over a west-dipping subduction zone as the Danube basin closed in the Early Cretaceous (Grubić, *op. cit.*). This intra-Serbo-Macedonian Massif extensional basin is possibly of the right age to be a northward continuation at much higher structural levels of our putative Volvi-?Therma-?Gomati rift. Grubić (1980) notes other earlier basins (e.g. the Luzniċa) subsiding in the Late Jurassic further west in the Yugoslavian Serbo-Macedonian Massif. We speculate that some of this localized subsidence may relate to Late Jurassic–Early Cretaceous *pull-apart tectonics* along and within an oblique subduction-zone margin. Such a setting would generate essentially discontinuous, intracontinental rift-basins associated with active deformation at their margins which would show a range of creation ages as the strike-slip zone evolved through tensional and compressional episodes along its length. These basins could be accompanied by complete crustal separation or only by severe attenuation of basement. In our area we would be looking presumably at a deep crustal section through such a zone. The actual depth, and the mechanism for both the burial of Triassic 'cover' rocks adjacent to the complex and the subsequent uplift, remain major questions for the future.

It seems very likely that the Serbo-Macedonian Massif, at least in Greece, will prove to have been involved in just as important a series of Mesozoic and Tertiary tectonic and metamorphic events as the Pelagonian zone further west is now known to have experienced (see Celet and Ferrière, 1978, for a review), following its many years in the rôle of inert 'basement'.

ACKNOWLEDGEMENTS: Support for field-work from the University of Edinburgh to JED is here gratefully acknowledged. The University of Thessaloniki also supported our joint field-work and we particularly thank Prof. Papazachos of the Department of Geology for his valuable help. Mrs H. Hooker processed our words.

References

Bohlen, S. R., Wall, V. J. & Boettcher, A. L. 1983. Experimental investigation and application of garnet granulite equilibria. *Contrib. Mineral. Petrol*, **83**, 52–61.

Borsi, S., Ferrara, G. & Mercier, J. 1965. Détermination de l'âge des séries métamorphiques du Massif Serbo-Macédonien au Nord-Est de Thessalonique (Grèce) par les méthode Rb/Sr et K/Ar. *Ann. Soc. géol. du Nord*, **84**, 223–225.

Celet, P. & Ferrière, J. 1978. Les Hellénides internes: Le Pélagonien. *Eclogae geol. Helv.* **73**, 467–495.

Dewey, J. F. & Pankhurst, R. J. 1970. The evolution of the Scottish Caledonides in relation to their isotopic age pattern. *Trans. R. Soc. Edinb.* **68**, 361–389.

Dimitriadis, S. 1974. *Petrological study of the migmatitic gneisses and amphibolites of Rentina-Asprovalta-Stavros-Olympias*: unpubl. Thesis, Univ. of Thessaloniki, 231pp.

—— 1980. A possible palaeomargin evolution of the southern-most part of Serbo-Macedonian Massif: *Abstr. Proc. 26th Int. Geol. Congr.* **1**, 335.

—— & Dixon, J. E. 1982. Metamorphosed ophiolitic rocks from the Serbo-Macedonian Zone, Lake Volvi, Greece. *In*: Dixon, J. E. & Robertson, A. H. F. (eds) *The Geological Evolution of the Eastern Mediterranean*, Abstracts, Edinburgh, 1982, p32.

Dimitrijević, M. D. & Cirić, B. 1967. Essai sur l'évolution de la masse Serbo-Macédonienne. *Acta Geol. Acad. Scient. Hung.* **11**, 35–47.

Gass, I. G. & Smewing, J. D. 1974. Intrusion, extrusion and metamorphism at constructive margins: evidence from the Troodos Massif, Cyprus. *Nature*, **242**, 26–29.

Graham, C. M. & Powell, R. 1984. A garnet-hornblende geothermometer: calibration, testing, and application to the Pelona Schist. *J. Metam. Geol.* **2**, 13–31.

Grubić, A. 1980. Yougoslavie. *In*: J. Dercourt and others (eds) *Géologie des pays Européens. Bordas and 26th International Geological Congress, Paris, France (Dunod)*, 291–342.

HARRE, W., KOCKEL, F., KREUZER, H., LENZ, H., MÜLLER P. & WALTHER, H. W. 1968. Uber Rejuvenationen im Serbo-Mazedonischen Massiv (Deutung radiometrischer Altersbestimmungen): *Proc. 23rd Int. Geol. Congr. Prague*, **6,** 223–236.

KAUFFMANN, G., KOCKEL, F. & MOLLAT, H. 1976. Notes on the stratigraphic and palaeogeographic position of the Svoula Formation in the Innermost Zone of the Hellenides (Northern Greece): *Bul. Soc. géol. France*, **18,** 225–230.

KOCKEL, F., MOLLAT, H. & WALTHER, H. W. 1971. Geologie des Serbo-Mazedonischen Massivs und seines mesozoischen Rahmens (Nord griechenland): *Geol. Jb.* **89,** 529–551.

—— 1977. Erläuterungen zur geologischen Karte der Chalkidhiki und agrenzender Gebiete 1:100 000 (Nord-Griechenland): Bundesanstalt für Geowissenschaften und Rohstoffe, Hannover, 119 pp.

MERCIER, J. 1968. Etude géologique des zones internes des Hellénides en Macédoine centrale (Grèce): *Ann. géol. Pays. hell.* **20,** 1–792.

PALLISTER, J. S. 1981. Structure of the sheeted dyke complex of the Samail ophiolite near Ibra, Oman. *J. geophys. Res.* **86,** 2661–2672.

—— & HOPSON, C. A. 1981. Samail ophiolitic plutonic suite: field relations, phase variation, cryptic variation and layering, and a model of a spreading ridge magma chamber. *J. geophys. Res.* **86,** 2593–2644.

PEARCE, J. A. 1980. Geochemical evidence for the genesis and eruptive setting of lavas from Tethyan ophiolites. *In*: PANAYIOTOU, A. (ed.) *Ophiolites. Proc. Int. Ophiolite Symp.*, Nicosia, Cyprus, 261–272.

—— 1982. Trace element characteristics of lavas from destructive plate boundaries. *In*: THORPE, R. S. (ed.) *Andesites*. John Wiley & Sons, London & New York, 525–48.

—— LIPPARD, S. J. & ROBERTS, S. 1984. Characteristics and tectonic significance of supra-subduction zone ophiolites. *In*: KOKELAAR, B. P. & HOWELLS, M. F. (eds) *Marginal Basin Geology*, *Spec. Publ. geol. Soc. Lond.*, No. 14.

POWELL, R. & EVANS, J. 1984. A new geobarometer for the assemblage biotite-muscovite-chlorite-quartz. *J. Metam. Geol.* **1,** 331–336.

SMITH, A. G., HYNES, A. J., MENZIES, M., NISBET, E. G., PRICE, I., WELLAND, M. J. P. & FERRIÈRE, J. 1975. The stratigraphy of the Othris mountains, eastern central Greece; a deformed Mesozoic continental margin sequence. *Eclog. geol. Helv.* **68,** 463–481.

STEIGER, R. H. & JÄGER, E. 1977. Subcommission on geochronology in geo- and cosmochronology: *Earth planet. Sci. Lett.* **36,** 359–362.

ZAGORCHEV, I. S. 1974. On the Precambrian tectonics of Bulgaria. *Precambrian Research*, **1,** 139–156.

J. E. DIXON, Grant Institute of Geology, University of Edinburgh, UK.
S. DIMITRIADIS, Department of Mineralogy and Petrology, University of Thessaloniki, Greece.

Age constraints on the igneous and metamorphic evolution of the Hellenic–Dinaric ophiolites

J. G. Spray, J. Bébien, D. C. Rex & J. C. Roddick

SUMMARY: The available age data from Greek and Yugoslavian ophiolites and related rocks are reviewed and, where necessary, have been corrected using the decay constants of Steiger & Jäger (1977). In addition, new K-Ar ages are presented from igneous mineral phases from the Guevgueli ophiolite and Fanos granitoid. Rb-Sr ages are also presented for the Furka and Štip granitoids of the Vardar Zone. Igneous ages of the dated ophiolites are approximately 160–180 Ma, although data from the Guevgueli Complex indicate that associated basic igneous activity also occurred during the Upper Jurassic. Granitoids yield ages ranging between 150 to 174 Ma. When compared with dates obtained from metamorphic soles beneath the Greek and Yugoslavian ophiolites (160–180 Ma), it is apparent that both sub-ophiolite metamorphism and granitoid intrusion closely followed ophiolite formation at a spreading centre. This indicates that the initial obduction movements occurred when the ophiolites were both young and hot and capable of forming basal dynamothermal soles mainly by the effects of their residual heat. When these ages are integrated with the field evidence for the period of final ophiolite emplacement in the Hellenides and Dinarides (Late Jurassic to Early Cretaceous), it is possible to reconstruct a simplified igneous, metamorphic and emplacement history for these displaced fragments of oceanic crust and upper mantle.

Introduction

During the last ten years considerable geochronological work has been carried out on ophiolites and related rocks from both Greece and Yugoslavia. Most age determinations have been made on mineral separates from sub-ophiolite metamorphic rocks because of their suitability for K-Ar and $^{40}Ar/^{39}Ar$ dating. Consequently, the timing of sole metamorphism is now reasonably well constrained. Limited data are also available from igneous assemblages which indicate probable crystallization ages of the ophiolites. Coupled with evaluation of the field relations of the ophiolites, the igneous and metamorphic age constraints now make it possible to attempt an integrated geological appraisal of ophiolite evolution in the Hellenides and Dinarides.

The ophiolites of Greece and Yugoslavia form two distinct NW-SE trending allochthonous belts (Nicolas & Jackson 1972; Pamić & Majer 1977; Karamata *et al.* 1980; Bébien *et al.* 1980; Fig. 1). The western ophiolite belt forms part of the Central Dinarides and Hellenides and is dominated by lherzolites, ferrogabbros and olivine tholeiites. The eastern ophiolite belt forms part of the Inner Dinarides and Hellenides and mainly comprises harzburgites and dunites, olivine gabbros, tonalites and quartz tholeiites. The petrological and chemical characteristics of the two belts reveal that the western ophiolites comprise only partly depleted upper mantle whereas the eastern ophiolites are of a more residual and highly depleted nature (Bébien *et al.* 1980; Maksimović & Majer 1981). However, more research is required to further elucidate these apparent differences. In addition, the eastern ophiolite belt is associated with and intruded by granitoids in an area stretching for 200 km from Guevgueli in the SE to Skopje in the NW; the main intrusions outcropping at Karadagh, Furka, Štip and Guevgueli (Bébien 1977). The occurrence of the ophiolites as two distinct sub-parallel belts is unusual as it is not a common feature of other Phanerozoic continental margins. Furthermore, the two Greek–Yugoslavian ophiolite belts occur along the boundary of the western lherzolitic and eastern harzburgitic divisions of Mediterranean Alpine-type peridotite (Nicolas & Jackson 1972). A possible explanation for this duality is presented elsewhere (Smith & Spray, this volume).

Field relations of the ophiolites

Apart from the geochronological evidence available, the final emplacement ages of ophiolites can also be constrained by the study of their field relations, mainly by using fossil evidence (Fig. 2). This involves evaluation of (1) the cover of the ophiolites, ranging from pelagic and related sediments formed after ophiolite inception at the speading axis, to post-emplacement continental cover. The maximum age of these sediments gives a minimum possible igneous age for the underlying ophiolites, (2) the age of the matrices of

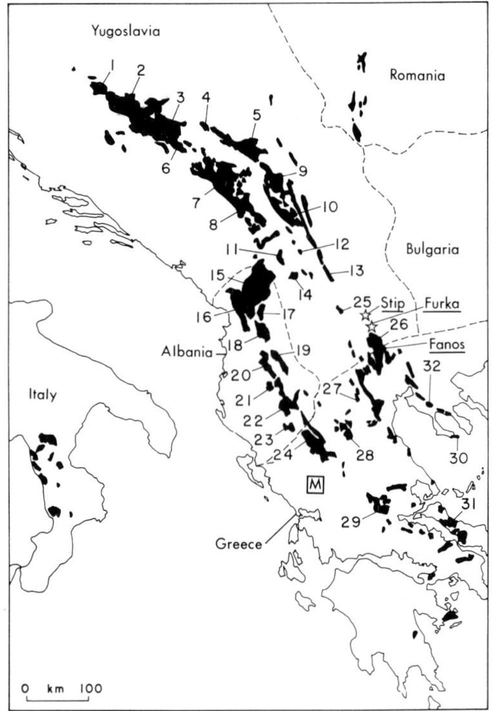

FIG. 1. Location map for the main Greek, Albanian and Yugoslavian ophiolite massifs. Also shown are locations of the Fanos, Furka and Štip granitoids of Yugoslavia and the village of Moschophyton (denoted 'M') in Greece. Numbers indicate the following massifs: 1 Banija, 2 Borja, 3 Krivaja-Konjuh, 4 Zvornik, 5 Maljen, 6 Vijaka, 7 Zlatibor, 8 Bistrica, 9 Triglav, 10 Ibar, 11 Orahovac, 12 Goles, 13 Presevo, 14 Brezovica, 15 Djakovica, 16 Mirdita (including Kam-Tropoja, Kraba, Puka, Skenderbeg and Kukēs massifs), 17 Lure, 18 Bulquize, 19 Podgradec, 20 Shpati, 21 Devolla, 22 Voskopolje, 23 Leskovica, 24 Pindos, 25 Poinja, 26 Guevgueli, 27 Vermion, 28 Vourinos, 29 Othris, 30 Kassandra, 31 Euboea, 32 Chalkidiki.

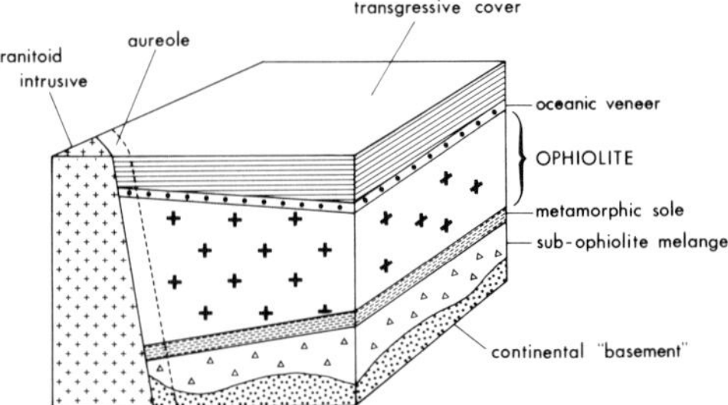

FIG. 2. Idealization of datable components (by either biostratigraphic or radiometric methods) which help to constrain the evolution of an ophiolite in time and space (see text). Not drawn to scale.

sedimentary and tectonic melanges beneath the basal thrusts of ophiolites which indicate the timing of ophiolite movement, (3) the ages of the rocks tectonically beneath the ophiolite which define the maximum possible age of their latest transport movements, (4) the ages of demonstrably autochthonous intrusive rocks and their aureoles which penetrate the ophiolites (e.g. granites) give minimum ages for final ophiolite emplacement. Table 1 summarizes some of these relations for the ophiolites considered using palaeontological control from mainly Radiolaria and rudists. The continental 'basement' commonly consists of Triassic to Upper Jurassic carbonates. Where present, the sub-ophiolite melanges, also known as the 'diabase-chert' or 'diabase-hornstein' unit, are frequently of uppermost Jurassic to Neocomian age. A cover of deep water oceanic sediments, typical of modern ocean crust, appears to be generally absent on Greek and Yugoslavian ophiolites (but, see Mavridis *et al.* 1977). Its apparent absence may be due to non-deposition, erosion or failure to identify it in the field. Non-deposition appears the most plausible. The transgressive cover ranges from Tithonian to Cretaceous clastics for the Yugoslavian examples, to Albian-Cenomanian-Coniacian carbonates in Greece. Despite the need for further studies and more precise fossil dating, the data indicate that initial emplacement onto the continental margin occurred during the uppermost Jurassic and/or Early Cretaceous (Table 1). Some ophiolites have, in addition, been subsequently transported in the Tertiary: e.g. the Pindos ophiolite now rests on a nappe of Eocene sediments via an Early Cretaceous flysch sequence (Terry & Mercier, 1971; Kemp & McCaig, this volume).

Age constraints on sub-ophiolite metamorphism

The occurrence and formation of soles

Many ophiolites possess dynamothermal aureoles (metamorphic soles) at their bases (Spray 1984). Where found intact these metamorphic sequences occur beneath the ultramafic tectonites of ophiolites, are commonly <0.5 km thick and predominantly comprise metabasic lithologies. Originally they were intepreted as static contact aureoles to

TABLE 1. *Biostratigraphic-based age data constraining the field relations of selected ophiolites*

Ophiolite	Continental 'basement'	Sub-ophiolite melange	Oceanic veneer	Transgressive cover	Data source
Krivaja	Triassic-Jurassic carbonates	'diabase-hornstein' formation of Jurassic age	?	Tithonian to Cretaceous clastics	Dimitrijević & Dimitrijević (1973) Karamata *et al.* (1980)
Zlatibor	,,	,,	?	,,	,, Bortolotti *et al.* (1971)
Brezovica	,,	,,	?	,,	,,
Pindos	Cretaceous flysch thrust onto Eocene flysch	Cretaceous ophiolitic flysch	?	?	Terry & Mercier (1971)
Vourinos	Triassic-Jurassic marbles	Ayios Nicolaos Complex: Late Jur. to Early Cret.	Tithonian-Neocomian lmst. breccia	Albian Cenomanian lmst.	Mavridis *et al.* (1977) Naylor & Harle (1976) Smith *et al.* (1975) Moores (1969)
Othris	Upper Triassic to upper Jur. carbonates	Serpentinite melange	?	Cenomanian-Coniacian limestones	Ferrière (1972) Hynes *et al.* (1972) Smith *et al.* (1975)
Euboea	Kimmeridgian-Tithonian carbonates	Berriasian Valanginian 'diabase-chert'	?	Albian Cenomanian sediments	Baumgartner & Bernoulli (1976) Parrot & Guernet (1972) Bignot & Guernet (1968)

plutonic ultramafic intrusions, or as slices of metamorphic basement terrains which had been trapped along the basal thrusts of overriding ophiolite nappes (e.g. Brunn 1956). More recent work indicates that most soles are formed during the early stages of ophiolite thrusting by the effects of residual heat from the hot ophiolite body or by frictional heat generated along the thrust plane or combinations of the two. In this respect age determinations made on metamorphic soles potentially constrain the timing of initial ophiolite displacement. The Hellenic-Dinaric ophiolites are particularly amenable to this type of investigation because they possess metamorphic soles which are discontinuously exposed along both belts. However, most data have been obtained from the western rather than the eastern belt: the metamorphic soles of the latter still await detailed investigation.

Summary of age data for soles

Table 2 summarizes age data presently available from the Greek and Yugoslavian metamorphosed ophiolite soles. All the ages have been determined using either the K-Ar or $^{40}Ar/^{39}Ar$ methods. Only results derived from amphibole or mica mineral separates have been tabulated and not whole rock or other mineral data because, for these particular radiometric methods, they are considered less reliable. The ages shown have also been corrected, where necessary, using the decay constants of Steiger & Jäger (1977). This eliminates inconsistencies due to pre-1977 calculations being made using a variety of decay constants. In most cases the modification is plus 3–5 Ma for pre-1977 data. In addition, the data of Roddick *et al.* (1979) and Spray & Roddick (1980) have been modified by recalibration of the monitor hornblendes originally used with the samples to international standards (Roddick 1983). This correction has resulted in a decrease in the ages first published (Table 2).

Metamorphic soles from the western ophiolite belt have been dated at Pindos, Othris and Euboea in Greece (Roddick *et al.* 1979; Spray & Roddick 1980; Thuizat *et al.* 1981). Soles from the eastern belt have been dated at Brezovica (Okrusch *et al.* 1978; Karamata & Lovrić 1978) and Vourinos (Spray & Roddick 1980). It is significant that the metamorphic sole ages are comparable for both the western and

TABLE 2. *Table summary of K-Ar and $^{40}Ar/^{39}Ar$ age data from ophiolite igneous phases (mantle ariegites) and sub-ophiolite metamorphic rocks*

Rock type	Sample No.	Sample location	Component dated	Data source	Published age[a]	Corrected age[b]
1 amphibolite	YUG1–71	Krivajah	pargasite	Lanphere *et al.* (1975)	157 ± 4	161 ± 4
2 —	YUG1–72	Zlatibor	pargasite	—	174 ± 14	178 ± 14
3 —	YUG1–72	—	hornblende	—	168 ± 8	172 ± 8
4 —	YUG3–72	—	pargasite	—	170 ± 11	175 ± 11
5 metapelite	B65–79	Brezovica	muscovite	Okrusch *et al.* (1978)	→	161 ± 2
6 metachert	B65–82	—	—	—	→	163 ± 2
7 amphibolite	B65–77A	—	hornblende	—	→	169 ± 3
8 —	B65–80	—	—	—	→	163 ± 3
9 —	98210	—	—	Karamata & Lovrić (1978)	172 ± 8	176 ± 8
10 —	98211	—	—	—	172 ± 9	176 ± 9
11 —	98212	—	—	—	176 ± 6	179 ± 6
12 —	98213	—	—	—	171 ± 9	176 ± 9
13 —	98231	—	—	—	→	171 ± 6
14 metapelite	98233	—	biotite	—	→	168 ± 5
15 amphibolite	98232	—	muscovite	—	→	161 ± 5
16 —	98232	—	hornblende	—	→	172 ± 8
17 —	98237	—	muscovite	—	→	159 ± 5
18 —	98237	—	hornblende	—	→	161 ± 5
19 —	123656	Pindos	—	Roddick *et al.* (1979)[c]	182 ± 4	169 ± 5
20 —	124659	—	—	Spray & Roddick (1980)[c]	172 ± 3	165 ± 3
21 —	124748	Othris	—	—	177 ± 4	169 ± 4
22 —	124765	Euboea	—	—	→	180[d]
23 —	124699	Vourinos	—	—	179 ± 4	171 ± 4
24 —	Pin II–4	Pindos	—	Thuizat *et al.* (1981)	→	176 ± 5

(a) Some previously published ages were determined using old decay constants and/or monitors. Samples 19–23 were published as total fusion ages. Arrows indicate corrections were unnecessary.
(b) corrected ages have been determined using the Steiger & Jäger (1977) decay constant and recalibrated monitors.
(c) age determinations on samples 19–23 were made by the $^{40}Ar/^{39}Ar$ method. The corrected dates were derived from hornblende plateau. All remaining ages were determined using the K-Ar method.
(d) this sample shows a complex age spectrum which rises to about 180 Ma. For this reason it has been omitted from Fig. 3.

eastern ophiolites: this indicates that sole formation occurred contemporaneously in both belts. However, it should be borne in mind that the exact boundary between the two belts is open to interpretation; for example, some workers consider Vourinos part of the western rather than the eastern belt (Moores 1969). After correcting for new decay constants the mean age of the soles lies between 155–175 Ma (Fig. 3). The differences between mica and amphibole ages are consistent with them reflecting the relatively higher Ar-blocking temperature of amphiboles rather than distinct metamorphic events (Dallmeyer 1979).

Age constraints on the igneous crystallization of the ophiolites

Previous work

The available data constraining the igneous ages of the Greek and Yugoslavian ophiolites are presently limited. Previous workers have favoured Precambrian (Kišpatić 1897), Palaeozoic (Hiessleitner 1951 and more recently Sapountzis 1980) or Mesozoic ages (Katzer 1906). Lanphere *et al.* (1975) made four K-Ar age determinations on amphibole-bearing mafic segregations found associated with the ultra-

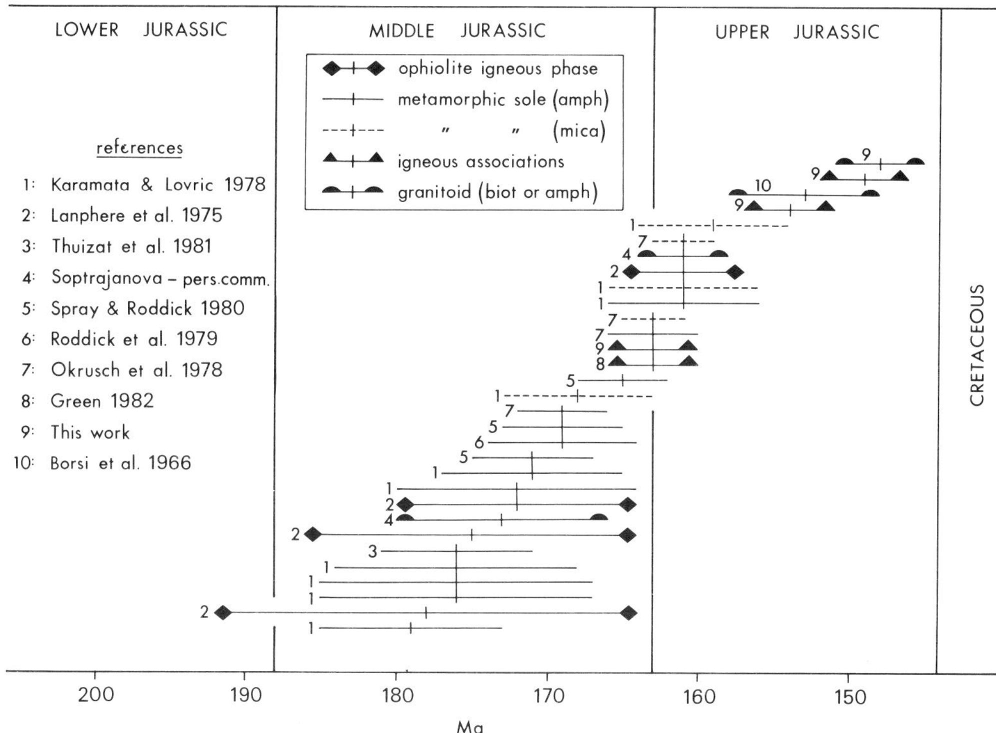

FIG. 3. Diagrammatic summary of K-Ar and $^{40}Ar/^{39}Ar$ age data from mantle ariegites, sub-ophiolite metamorphic rocks, basic igneous rocks from Guevgueli and the Fanos granitoid. Rb/Sr data for the Fanos, Furka and Štip granitoids are also shown. Period boundaries according to the time scale of Harland *et al.* (1982).

mafic tectonites of the Krivaja-Konjuh and Zlatibor ophiolites of the western belt in Yugoslavia. These mafic segregations possess mineral parageneses indicative of igneous or high pressure—high temperature metamorphic origins (i.e. they comprise corundum, pyropic garnet, spinel, aluminous pyroxene, pargasite assemblages). Comparable lithologies are commonly found in the lherzolitic bodies of the western Mediterranean (e.g. the Ronda and Lherz massifs). Many such mafic segregations have a picritic chemistry and they probably represent parental melts which would have otherwise fractionated to yield MORB-type basalts had they not remained trapped in the host peridotites (Spray 1982). It is easy to confuse these segregations with the underlying metamorphic soles, but as Lanphere *et al.* (1975 p. 274) state themselves . . . 'Amphibolites interlayered with peridotites cannot be interpreted as contact aureoles and must represent [the products of] deep crustal or upper mantle metamorphism'. Consequently, determination of their age is likely to approximate the timing of igneous crystallization of the ophiolite. Unfortunately, the analytical precision of these particular ages is poor, mainly because of the inherently low K_2O content of pargasites. Nevertheless, the ages of the samples from Zlatibor are some of the oldest obtained from the ophiolites and we interpret them as indicating a primary igneous event (Table 2; Fig. 3).

Richard & Allègre (1980) also provide a tentative constraint on the igneous ages of the ophiolites. In their Nd and Sr isotope study of ophiolites, whole rock ages of about 150 Ma were determined from gabbros and lavas from Vourinos, and about 130 Ma from gabbros from Pindos. Both these ages are low when compared with the metamorphic sole age data. However, they too support a Jurassic rather than a pre-Mesozoic age for ophiolite generation.

Sapountzis (1980) provided K-Ar ages on ophiolites from the eastern belt in Greece and dates ranging from 325 to 42 Ma were obtained. However, all but one were whole rock determinations and include samples which had undergone both alteration and cataclasis; events which render the K-Ar method unreliable. For these reasons the results are not considered to accurately reflect igneous events.

The Guevgueli Complex and the ophiolite-granitoid association

The Guevgueli Complex is situated in central Macedonia within the Vardar zone of the Inner Hellenides and Dinarides (the eastern ophiolite belt; Fig. 1). It extends 60 km N-S and attains a maximum width of about 20 km (Bébien 1977, 1981, 1982). The Complex comprises a migmatitic and gneissose basement and slivers of dismembered ophiolite thrust upon a Jurassic volcano-sedimentary sequence, all intruded by a large granitoid cupola (the Fanos granite). The ophiolite consists of two separate units: cumulate plutonic rocks overlain by sills and lavas comprise the western unit, which overall is very low in K_2O and possesses a tholeiitic differentiation trend. The eastern unit is more varied: in addition to gabbros, sheeted dykes (of calc-alkaline affinity) and lavas, there are magmatic breccias comprising doleritic inclusions in a dioritic or granitic matrix which intrude the gabbros. An Upper Jurassic age has been previously proposed for the Guevgueli Complex on the basis of stratigraphic and radiometric evidence (Borsi *et al.* 1966; Mercier 1968; Marakis 1970). New K-Ar data from samples of these rocks are shown in Table 3. An amphibole separate from an igneous amphibole-bearing gabbro associated with the magmatic breccias of the eastern unit gave an age of 149 ± 3 Ma, while an amphibole-biotite diorite (matrix to the breccias) yielded 163 ± 3 Ma (amphibole) and 154 ± 3 Ma (biotite). These results confirm an Upper Jurassic age for the plutonic components of the Complex. A third sample comprising a biotite separate from the Fanos granite gave 148 ± 3 Ma (Fig. 1; Table 3) which compares well with that of 153 Ma, corrected from 150 Ma, also determined on a biotite separate from the granite using Rb-Sr and K-Ar methods by Borsi *et al.* (1966).

The Upper Jurassic ages from the Guevgueli Complex are significant because they indicate that basic igneous activity associated with the ophiolite continued beyond the Middle Jurassic. The extent of this later phase of igneous activity must await further geochronological investigation.

Additional age data

Additional age data are presented here from a primary amphibole-bearing dolerite sill at Moschophyton in the central part of the Pindos Zone, Greece and on the Furka and Štip granitoids of the Vardar Zone, Yugoslavia (Fig. 1).

The dolerite sill was dated using the K-Ar method on a kaersutite mineral separate and it yielded an age of 163 ± 5 Ma (Table 3). The 3–4 m thick sill, which is intruded within thin-bedded Triassic *Halobia* calcisiltites and calcilutites, has a trace element content indicative of an alkalic character (Green 1982). Extrusive alkalic igneous activity has also been recognized in western Pindos, particularly as tectonic intercalations along the Pindos thrust, although these are considered to be of Triassic age, along with the many other volcanic occurrences of alkalic nature in Greece (Hynes 1974; Green 1982). The date from the dolerite sill is therefore of interest as it indicates that alkalic igneous activity spanned both Triassic and Jurassic times.

Rb/Sr dating was made on biotite separates from the Furka and Štip granitoids of Yugoslavia by G. Šoptrajanova in 1967, and she has kindly allowed us to quote the hitherto unpublished results here. 168 ± 7 Ma was obtained from Furka and 156 ± 3 from Štip. These were determined using an age constant of 1.47×10^{-16} which, after correction, yields 173 ± 7 Ma and 161 ± 3 Ma respectively. However, it should be borne in mind that many of these granitoids are intimately associated with migmatites which have clearly undergone intensive metamorphism (Bébien 1981). These two ages might therefore reflect the timing of either earlier 'basement' formation or later metamorphic overprinting. However, this complication

TABLE 3. *K-Ar age data on igneous activity associated with the Guevgueli ophiolite and from a dolerite sill at Moschophyton*

Rock type	Sample No.	Sample location	Component dated	%K	Vol. ^{40}Ar rad. scc/gm $\times 10^{-5}$	$\%^{40}Ar$ rad.	Age Ma
1 Granitoid	GV964	Guevgueli	biotite	6.23	3.7221	89.1	148 ± 3
2 Diorite	GV854	—	—	7.38	4.6105	87.0	154 ± 3
3 —	—	—	hornblende	0.881	0.5834	72.2	163 ± 3
4 Gabbro	GV801	—	—	0.302	0.1825	31.2	149 ± 3
5 Dolerite	TG389	Moschophyton	kaersutite	1.2	n.a.	n.a.	163 ± 5

does not apply to the new K-Ar age obtained from the Fanos granitoid as the mineral separate used was of igneous origin.

Synthesis

A review of available and new age data on the Greek and Yugoslavian ophiolites and related lithologies indicate that the ophiolites were (1) most probably generated at a spreading centre in the Middle Jurassic, were (2) tectonically *displaced* hot, virtually immediately after their generation, along thrust faults located within the oceanic lithosphere along which were formed the metamorphic soles and were (3) subsequently *emplaced* onto the continental margin during the Upper Jurassic/Lower Cretaceous. The obduction process therefore appears to have spanned 15–30 Ma from ophiolite inception to ensuing cold emplacement onto the continental margin. A comparable sequence of events may be applicable to many ophiolites, although the time interval between displacement and emplacement may differ. Fig. 4 presents a summary of these events as they might apply to the Greek–Yugoslavian ophiolites. Their occurrence as two distinct belts is not explained in this diagram but is elaborated upon elsewhere (Smith & Spray, this volume). The setting for ophiolite generation is a marginal basin, the exact nature of which is not considered here, although it may well have been located above a subduction zone (Fig. 4a). The second stage involves ophiolite inception due to compression of the basin. This results in overthrusting at the ridge (Fig. 4b), the most likely decoupling level being the elevated lithosphere–asthenosphere boundary beneath the spreading centre (Spray 1983). Initiation of obduction is concomitant with metamorphic sole generation along the basal thrust beneath the hot ophiolite. An alternative site of ophiolite displacement involves decoupling along the supposed limit of elastic strength within the lithosphere (Nicolas & Le Pichon 1982). This may result in overthrusting occurring at some distance away from the spreading axis (Fig. 4c). Emplacement on to the continental margin occurs during the Upper Jurassic/

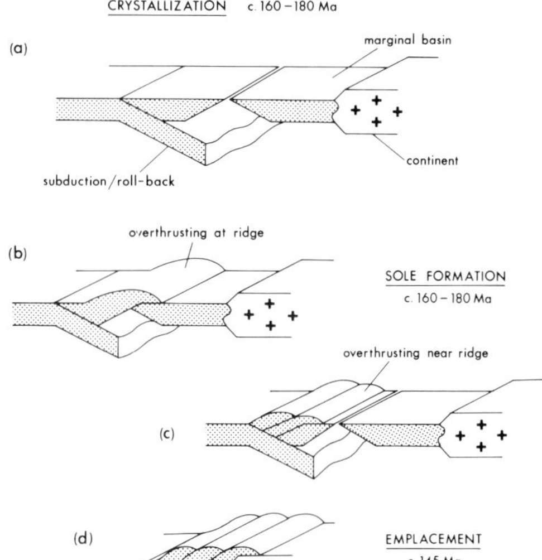

FIG. 4. Model for the evolution of the Greek–Yugoslavian ophiolites. See text for details.

Lower Cretaceous (Fig. 4d). The substantial time interval between intra-oceanic displacement and continental emplacement indicates that there was an initial Middle Jurassic compressive event which resulted in only minor shortening of the marginal basin, followed by its destruction in the Late Jurassic. Such movements are plausibly explained by changes occurring in the distance between subduction zone and continental margin throughout this period.

ACKNOWLEDGEMENTS. Tim Dempster, Alan Smith & Nigel Woodcock are thanked for reviewing the manuscript. An exchange visit to Yugoslavia was made possible through the auspices of the Royal Society, the Yugoslavian Academy of Sciences and Arts and Prof. S. Karamata, University of Beograd. We are particularly grateful to Prof. G. Šoptrajanova for allowing us to publish her Rb/Sr age determinations from the Furka and Štip granitoids and Dr. Hans Kreuzer for commenting on the age corrections. Maurice Haslop helped with the preparation of mineral separates. Cambridge Earth Sciences contribution 287.

References

BAUMGARTNER, P. O. & BERNOULLI, D. 1976. Stratigraphy and radiolarian fauna in a Late Jurassic–Early Cretaceous section near Achladi (Evvoia, Eastern Greece). *Eclogae geol. Helv.* **69**, 601–26.

BEBIEN, J. 1977. Mafic and ultramafic rocks associated with granites in the Vardar zone. *Nature, London*, **270**, 232–4.

—— 1981. A propos de l'association de certaines formations ignées basiques à caractères ophiolitiques avec des granites et des migmatites. *C.R. Acad. Sci. Paris*, **292**, 733–5.

—— 1982. *L'association ignée de Guévguéli (Macédoine grecque), expression d'un magmatisme ophiolitique dans une déchirure continentale.* Thèse d'Etat, Nancy (unpubl.) 470 pp.

——, Ohnenstetter, D., Ohnenstetter, M. & Vergely, P. 1980. Diversity of the Greek ophiolites: birth of oceanic basins in transcurrent systems. *Ofioliti* (*Spec. Iss. Tethyan Ophiolites*, **2**, 129–97.

Bignot, G. & Guernet, C. 1968. L'age de la transgression du Cretacé supérieur dans la région de Larymna, au nord des lacs Thébans (Grèce). *Rev. Micropalaeont.* **10**, 261–70.

Borsi, S., Ferrara, G., Mercier, J. & Tongiori, E. 1966. Age stratigraphique et radiométrique jurassique supérieur d'un granite des zones internes des Hellénides (granite de Fanos, Macédoine, Grèce). *Rev. Géog. phys. Géol dyn.* **8**, 279–87.

Bortolotti, V., Ficcarelli, G., Manetti, P., Passerini, P., Raddrizzani, C. P. & Torre, D. 1971. Studies on mafic and ultramafic rocks, 1. A Jurassic sequence on top of the Zlatibor ultramafic massif (Yugoslavia). *Bull. Soc. geol. Ital.* **90**, 415–28.

Brunn, J. H. 1956. Contribution à l'étude géologique du Pinde septentrionale et d'une partie de la Macédoine occidentale. *Ann. géol. Pays. Helléniques* **7**, 358 p.

Dallmeyer, R. D. 1979. $^{40}Ar/^{39}Ar$ dating: principles, techniques and applications in orogenic terranes. *In:* E. Jäger & J. C. Hunziker (eds.) *Lectures in Isotope Geology*, 77–104. Springer-Verlag, New York.

Dimitrijević, M. D. & Dimitrijević, M. N. 1973. Olistostrome melange in the Yugoslavian Dinarides and late Mesozoic plate tectonics. *J. Geol.* **81**, 328–40.

Ferriere, J. 1972. Sur l'importance du déplacement tangentiel en Othrys centrale au nord-est d'Anavra (Grèce). *C.r. Acad. Sci. Paris*, **273**, 174–6.

Green, T. J. 1982. *Structural and sedimentological studies of the Pindos Zone, Central Greece.* Thesis, PhD, Univ. Cambridge (unpubl).

Harland, W. B., Cox, A. V., Llewellyn, P. G., Pickton, C. A. G., Smith, A. G. & Walters, R. 1982. *A geologic time scale.* Cambridge University Press, London, 131 p.

Hiessleitner, G. 1951. Serpentin- und Chromerzgeologie der Balkanhalbinsel und eines Teiles von Kleinasien. *Jahrb. geol. Bundesanst. Wien* **1**, 259–683.

Hynes, A. J. 1974. Igneous activity at the birth of an ocean basin in Eastern Greece. *Can. J. Earth Sci.* **11**, 842–53.

Hynes, A. J., Nisbet, E. G., Smith, A. G., Welland, M. J. P. & Rex, D. C. 1972. Spreading and emplacement ages of some ophiolites in the Othris region, Eastern Central Greece. (Proc. 4th Aegean Symp., Hannover) *Z. Dtsch. geol. Ges.* **123**, 455–68.

Karamata, S. & Lovrić, A. 1978. The age of metamorphic rocks of Brezovica and its importance for the explanation of ophiolite emplacement. *Bull. Acad. Serbe. Sci.* **17**, 1–9.

——, Majer, V. & Pamić, J. 1980. Ophiolites of Yugoslavia. *Ofioliti (Spec. Iss. Tethyan Ophiolites)* **1**, 105–25.

Katzer, F. 1906. Historijsko rasvijanje i danasnje stanje geoloskog proucavanja Bosne i Hercegovine. *Glas. Zem. Muz. BiH. Sep. otis Sarajevo.*

Kišpatić, M. 1897. Kristalinsko kamenje serpentinske zone u Bosni. Rad. JAZU CXXXIII, Zagreb, 95–231.

Lanphere, M. A., Coleman, R. G., Karamata, S. & Pamić, J. 1975. Age of amphibolites associated with Alpine peridotites in the Dinaride ophiolite zone, Yugoslavia. *Earth planet. Sci. Lett.* **26**, 271–6.

Maksimović, Z. & Majer, V. 1981. Accessory spinels of two main zones of Alpine ultramafic rocks in Yugoslavia. *Bull. Acad. Serbe. Sci.* **21**, 47–58.

Marakis, G. 1970. Geochronology studies of some granites from Macedonia. *Ann. géol. Pays Hell.* **21**, 121–52.

Mavridis, A., Skourtsis-Coroneou, V. & Tsaila-Monopolis, St. 1977. Contribution to the geology of the Subpelagonian Zone (Vourinos area, West Macedonia). 6th Colloquium on the geology of the Aegean region, Athens 1977, 175–95.

Mercier, J. 1968. I—Etude géologique des zones internes des Hellénides en Macédoine centrale (Grèce). II—Contribution à l'étude du métamorphisme et de l'évolution magmatique des zones internes des Hellénides. *Ann. géol. Pays Hell.* **20**, 1–792.

Moores, E. 1969. Petrology and structure of the Vourinos ophiolite complex of northern Greece. *Geol. Soc. Am. spec. Paper* **118**, 74 p.

Naylor, M. A. & Harle, T. J. 1976. Palaeogeographic significance of rocks and structures beneath the Vourinos ophiolite complex, Northern Greece. *J. geol. Soc. London,* **132**, 667–75.

Nicolas, A. & Jackson, E. D. 1972. Répartition en deux provinces des péridotites des chaînes alpines longeant la méditerranée: implications géotectoniques. *Schweiz. min. petr. Mitt.* **52**, 479–95.

—— & Le Pichon, X. 1980. Thrusting of young lithosphere in subduction zones with special reference to structures in ophiolites peridotites. *Earth planet. Sci. Lett.* **46**, 397–406.

Okrusch, M., Seidel, E., Kreuzer, H. & Harre, W. 1978. Jurassic age of metamorphism at the base of the Brezovica peridotite (Yugoslavia). *Earth planet. Sci. Lett.* **39**, 291–7.

Pamić, J. & Majer, V. 1977. Ultramafic rocks of the Dinaride central ophiolite zone in Yugoslavia. *J. Geol.* **85**, 553–69.

Parrot, J. F. & Guernet, C. 1972. Le cortège ophiolitique de l'Eubée Moyenne (Grèce): étude pétrographiques des formations volcaniques et des roches métamorphiques associées dans les monts Kandilion aux radiolarites. *Cah O.R.S.T.O.M. sect. Géol.* **4**, 153–61.

Richard, P. & Allègre, C. J. 1980. Neodymium and strontium isotope study of ophiolite and orogenic lherzolite petrogenesis. *Earth planet. Sci. Lett.* **47,** 65–74.

RODDICK, J. C. 1983. High precision intercalibration of $^{40}Ar/^{39}Ar$ standards. *Geochim. Cosmochim. Acta*, **47,** 887–98.

——, CAMERON, W. A. & SMITH, A. G. 1979. Permo-Triassic and Jurassic $^{40}Ar/^{39}Ar$ ages from Greek ophiolites and associated rocks. *Nature, London*, **279,** 788–90.

SAPOUNTZIS, S. E. 1980. On the age of the ophiolitic sequence in the southeastern part of the Axios (Vardar) zone (North Greece). *N. Jb. Mineral. Abh.* **138,** 39–48.

SMITH, A. G., HYNES, A. J., MENZIES, M., NISBET, E. G., PRICE, I. WELLAND, M. J. P. & FERRIERE, J. 1975. The stratigraphy of the Othris Mountains, eastern central Greece: a deformed Mesozoic continental margin sequence. *Eclogae geol. Helv.* **68,** 463–81.

SPRAY, J. G. 1982. Mafic segregations in ophiolite mantle sequences. *Nature, London*, **299,** 524–28.

—— 1983. Lithosphere–asthenosphere decoupling at spreading centres and the initiation of obduction. *Nature, London*, **304,** 253–5.

—— 1984. Possible causes and consequences of upper mantle decoupling and ophiolite displacement. In: GASS, I. G., LIPPARD, S. J. & SHELTON, A. W. (eds). Ophiolites and oceanic lithosphere, *Spec. Publ. geol. Soc. London*, **13,** 281–94.

—— & RODDICK, J. C. 1980. Petrology and $^{40}Ar/^{39}Ar$ geochronology of some Hellenic sub-ophiolite metamorphic rocks. *Contrib. Mineral. Petrol.* **72,** 43–55.

STEIGER, R. H. & JÄGER, E. 1977. Subcommission on geochronology: convention on the use of decay constants in geo- and cosmochronology. *Earth planet. Sci. Lett.* **36,** 359–62.

TERRY, J. & MERCIER, M. 1971. Sur l'éxistence d'un série détritiques berriasienne intercalée entre la nappe des ophiolites et le flysch éocene de la nappe du Pinde. *C.r. Soc. géol. France*, **2,** 71–33.

THUIZAT, R., WHITECHURCH, H., MONTIGNY, R. & JUTEAU, T. 1981. K-Ar dating of some infra-ophiolitic metamorphic soles from the Eastern Mediterranean: new evidence for oceanic thrustings before obduction. *Earth planet. Sci. Lett.* **52,** 302–10.

J. G. SPRAY, Department of Earth Sciences, University of Cambridge, Cambridge CB2 3EQ, U.K.

J. BEBIEN, Laboratoire de Pétrologie, Faculté des Sciences, BP 239, 54506 Vandoeuvre Cedex, France.

D. C. REX, Department of Earth Sciences, University of Leeds, Leeds LS2 9JT, U.K.

J. C. RODDICK, Precambrian Division, Geological Survey of Canada, Ottawa K1A OE8, Canada.

A half-ridge transform model for the Hellenic–Dinaric ophiolites

A. G. Smith & J. G. Spray

SUMMARY: We speculate that the similarity of the E and W ophiolite belts in Greece, Albania and Yugoslavia can be plausibly explained by the disruption of a once continuous subduction-linked marginal basin and adjacent continent by a Middle to Late Jurassic sinistral transcurrent fault along the E edge of the Pelagonian Drina–Ivanjica Zones. The fault may itself have been linked to the Jurassic opening of the central Atlantic. This model is supported by the age relations between igneous crystallization of the ophiolites and the timing of metamorphic sole formation. Present isotopic dating, together with simple thermal calculations, suggests that the ages of these events differed by at most 10 Ma. Together with lateral variations in the ophiolites themselves and their soles, this apparent similarity of ages is most readily accounted for by the overthrusting at the ridge crest of one half of a narrow marginal basin by the other half.

Isotopic and geochemical data on Greek and Yugoslavian ophiolite complexes have increased dramatically over the last ten years. Coupled with Mesozoic Atlantic floor information, it is now possible to attempt preliminary space-time reconstructions for the Hellenic-Dinaric ophiolite belts. Four aspects of the ophiolites are considered in this paper:

(1) the constraints on their tectonic evolution imposed by field data;

(2) why they form two distinct belts;

(3) how their evolution may be related to the opening of the Atlantic; and

(4) an emplacement model.

The ophiolites of Greece, Albania and Yugoslavia outcrop as two belts (Bébien *et al.* 1980; Karamata *et al.* 1980; Figs 1 & 2). Data for Albanian ophiolites are sparse and they will be considered along with the Yugoslavian ophiolites for the purposes of this discussion. The W belt, outcropping in Greece in the Othris or Sub-Pelagonian Zone and in Yugoslavia as the Central Dinaric ophiolite belt, can be traced south as far as the island of Evvia, beyond which the ophiolites form much more discontinuous masses. Northwards the W ophiolites can be traced into Albania where the outcrop turns north-east, eventually merging with the E belt in central Yugoslavia (Figs 1 & 2). The E belt in Yugoslavia is known as the Inner Dinaric ophiolite belt and in Greece forms part of the Vardar Zone, subdivided by Mercier (1968) into the Almopias (W), Paikon and Peonias (E) Zones. The two belts are separated by the Pelagonian Zone in Greece and its northern continuation, the Drina–Ivanjica Zone, and possibly the Jada Zone, in Yugoslavia (Figs 1 & 2), all of which possess continental basement.

There is now substantial field evidence for the W ophiolites having been emplaced onto the E edge of the Pelagonian continent during the latest Jurassic, in pre-Tithonian time, (the JE1 phase of French geologists; Vergely 1977); and for the W ophiolites having been emplaced onto the W edge of the Pelagonian continent somewhat later during the Early Cretaceous, probably in Berriasian or later time, (the JE2 tectonic phase of French geologists; Brunn 1956; Smith *et al.* 1979; Vergely 1977). Emplacement of the E ophiolites onto the W edge of the Serbo-Macedonian continent also occurred in the latest Jurassic/earliest Cretaceous interval, but evidence is inconclusive (see Mercier 1968 for details of sequences).

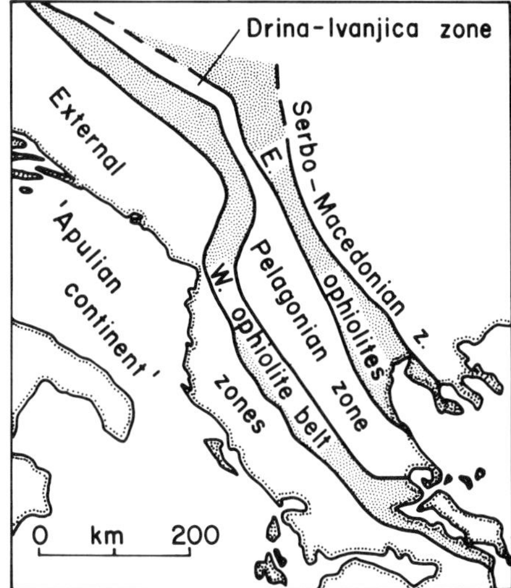

FIG. 1. Regional geological setting. The W ophiolite belt in Greece is known as the Othris or Pelagonian Zone and the Central Dinaric ophiolite belt in Yugoslavia. The E belt in Greece is known as the Vardar Zone and the Inner Dinaric ophiolite belt in Yugoslavia.

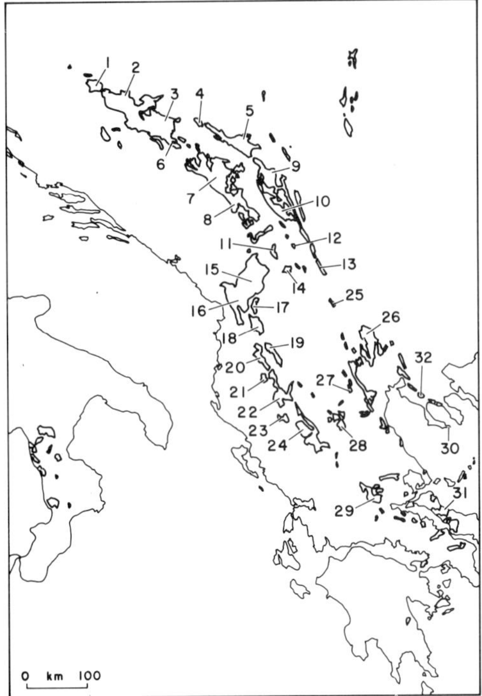

FIG. 2. Outcrop pattern and names of Greek and Yugoslavian ophiolites.

1, Banija; 2, Borja; 3, Krivaja–Konjuh; 4, Zvornik; 5, Maljen; 6, Vijaka; 7, Zlatibor; 8, Bistrica; 9, Troglav; 10, Ibar; 11, Orahovac; 12, Goles; 13, Presevo; 14, Brezovica; 15, Djakovica; 16, Mirdita (including Kam-Tropoja, Kraba, Puka, Skenderbeg and Kukës massifs); 17, Lure; 18, Bulquize; 19, Podgradec; 20, Shpati; 21, Devolla; 22, Voskopolje; 23, Leskovica; 24, Pindos; 25, Poinja; 26, Guevgueli; 27, Vermion; 28, Vourinos; 29, Othris; 30, Kassandra; 31, Evvia; and 32, Chalkidiki.

A Triassic marginal basin W of the Pelagonian/Drina–Ivanjica Zone

Prior to their emplacement the W ophiolites were bordered by 'stable' continental margins initiated during the Triassic (Bernoulli & Jenkyns 1974). The new margins consisted of the external zones of the Hellenides and Dinarides, forming the W margin of the W belt; the Pelagonian/Drina–Ivanjica Zone, separating the E and W belts, and the Serbo–Macedonian Zone forming the E margin of the E belt (Fig. 1). Though apparently stable, the presence of Triassic pyroxene cumulates (boninites) in the Othris section (Smith *et al.* 1975) suggests that water was fundamental to the creation of the ocean-floor between the W continental margins (Cameron *et al.* 1979). Middle Triassic (Anisian–Ladinian) volcanics, of the porphyrite-radiolarite unit, in most of the Dinarides contain rocks of calc-alkaline affinities resembling those of a back-arc basin (Bébien *et al.* 1978). Both the boninites and the porphyrite-radiolarite unit suggest that the Triassic marginal ocean formed as a result of Triassic subduction in a trench E of the Pelagonian/Drina–Ivanjica Zones: there is no compressional Triassic defomation to the W. We assume that these Triassic oceans were part of the 'Neotethys' (Şengör *et al.* 1980) and that at this time the Pelagonian/Drina–Ivanjica and Serbo–Macedonian Zones were part of the 'Cimmerian continent'. To the E of them lay the 'Palaeotethys' (Fig. 3). During Triassic subduction of the 'Palaeotethys', the Cimmerian continent is believed to have broken off the northern edge of Gondwanaland and migrated to a position somewhere in the Tethys, leaving the 'Neotethys' as a marginal subduction ocean behind it. Other Triassic volcanic rocks in Othris are chemically like those formed in rift zones and are not necessarily related to subduction (Hynes 1974). Thus, though adjacent to a 'back-arc basin', the new Triassic continental margins of the W belt appear to have been far enough away from any arc that may have been present to have been unaffected by it and, except at their initiation, have appeared stable and passive. That they were continental margins rather than parts of an intracontinental rift, as in the North Sea, is suggested by the amount and duration of the subsidence of the Triassic carbonate platforms–c.2 km in <35 Ma–and their area (Smith & Moores 1974).

The Pindos Zone in Greece lies between the W ophiolites and the W carbonate platform of Apulian continent. In the absence of any additional information we presume that by Middle Jurassic time, the Triassic continental margins on the W edge of the Pindos Zone and W edge of the Pelagonian Zone were separated by an ocean with Triassic ocean-floor at its edges and floor possibly as young as Jurassic in the centre. Emplacement of the W ophiolites onto the adjacent continent to the E in latest Jurassic/earliest Cretaceous time is assumed to be related to a subduction zone initiated in Jurassic time.

Evidence for ophiolite origin in a marginal ocean basin

The term 'marginal basin' is used here simply to mean part of an ocean that is marginal to a continent, or remnant arc, whatever its origin. For example, the Gulf of California is a basin marginal to the Pacific, though it does not owe

TABLE 1. *Summary of tectonic constraints on ophiolite evolution*

Observation or inference	Conclusions
Carbonate platforms to E and W; starting in Triassic time and destroyed in Late Jurassic time	Ocean bordered by passive continental margins of Triassic age, floored by Triassic (c. 240 Ma) and possibly Jurassic ocean floor
Steady lateral progression from ophiolite sheet in SW to carbonate platform in NE in Othris	Ophiolites derived from the strip of ocean adjacent to and SW of the Pelagonian continent
Trace element geochemistry suggests atypical ocean-floor possibly 'back-arc basin'	Ophiolites formed above a subduction zone
Spreading ages Middle Jurassic	Ophiolites not part of a Triassic ocean. Formed in new Middle Jurassic ocean basin during subduction of Triassic ocean-floor
Late Jurassic sediments discordant on ophiolites	Emplacement on continent in pre-Tithonian time
Metamorphic sole and spreading ages identical within statistical error of ±6 Ma	Collapse of subduction-coupled ocean started within 6 Ma of spreading probably at ridge crest
Metamorphic sole includes ocean-floor basalts and alkali basalts	Ocean-basin relatively narrow
Ophiolites 12 km thick	If shearing along 500°C isotherm ophiolites not more than 10 Ma old at time of shearing; less if T higher.
No resetting of mineral ages during Late Jurassic or Late Cretaceous emplacement	Temperatures only exceed 500°C during sole formation; ophiolites emplaced cold
Emplacement onto Pelagonian continent some 20 Ma after sole formation	Collapse of basin slow or episodic

its origin directly to subduction. The term 'marginal subduction basin' is provisionally used for all those marginal basins above and behind a subduction zone. If an arc is developed, the term will include the fore-arc region, the arc itself and the back-arc basin. If an arc is absent the term refers to the entire extensional zone between the trench and the remnant arc or continental margin from which the trench has migrated.

The Othris region of Greece comprises a series of thrust sheets interpreted as a deformed Mesozoic continental margin (Ferrière 1974, 1977; Smith *et al.* 1975; Smith *et al.* 1979). When restored palinspastically there is an orderly lateral change from the ophiolites, believed to have been emplaced from the W (Ferrière 1974; Smith & Woodcock 1976a), through a pelagic basin and submarine fan sequence to a Mesozoic carbonate platform in the E. This orderly lateral change led Smith & Woodcock (1976b) to postulate that the Othris ophiolites were once part of a marginal ocean strip. Facies variations and sedimentary structures within the sheets independently suggest that the ocean originally lay on the western side of the Pelagonian continent (Price 1976; Smith & Woodcock 1976a). The precise nature of the ocean could not be determined from these structural and stratigraphic data, though, as discussed above, some of the Triassic volcanic rocks suggest it was formed above an active subduction zone. If the Othris ophiolite was a marginal strip, then it should be the oldest part of the ocean floor. Because the passive margins are Triassic, the age of the Othris ophiolite, and by inference, all the W ophiolites might also be expected to be Triassic.

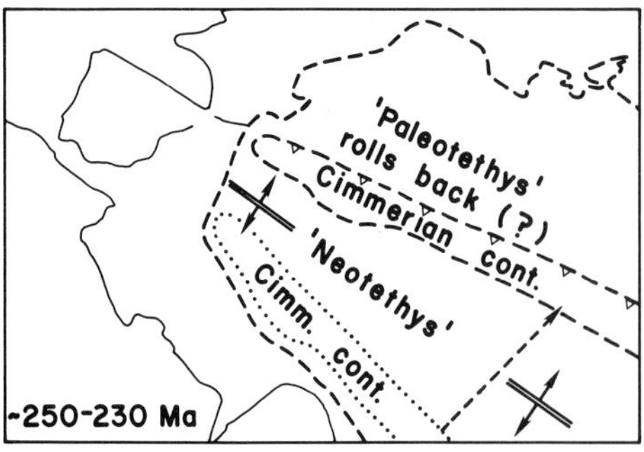

E. / M. Triassic

FIG. 3. Reconstruction of the coastline positions of Africa, Spain and W Europe taken from Livermore & Smith (in press). The Tethys has been subdivided after Şengör *et al.* (1980) into a 'Neotethys' and 'Palaeotethys' separated by a 'Cimmerian continent' whose position shape and size are not well known. The dotted line shows the possible position of the Cimmerian continent in early Triassic time. Its Middle Triassic position is shown by the darker lines. A trench is postulated along its NE edge, with the subduction-linked 'Neotethys' opening behind as the 'Palaeotethys' rolls back.

Geochemistry of ophiolites and metamorphic soles

Establishing the initial chemical signature of the Greek and Yugoslavian ophiolites is beset with difficulties. Most ophiolites exhibit the effects of alteration: essentially hydrothermal at the hypabyssal and extrusive levels, with serpentinization commonly affecting the ultramafics, particularly at ophiolite bases. Many of these retrogressive events are non-isochemical so that whole rock analyses need not reflect original igneous compositions. Consequently, relatively immobile elements have been used in attempts to 'see through' secondary overprinting. These have included Ti, Zr, Cr, Y and various rare earths. Significantly, the results for the Greek and Yugoslavian ophiolites do not indicate a uniform chemical affinity but rather a range of signatures. For example, Pindos shows a calc-alkaline trend (Capedri *et al.* 1980), Othris has a tholeiitic character (Hynes 1974), Vourinos a transitional MORB/ocean island/continental tholeiite source (Montigny 1975; Richard & Allègre 1980) and the Guevgueli magmatic complex of the Vardar Zone a tholeiitic to calc-alkaline affinity (Bébien & Mercier 1977; Bébien 1977). Coupled with this variety is the occurrence, and in some cases predominance, of lherzolite rather than harzburgite within the ophiolite mantle sequences (Spray 1982). The W zone in Yugoslavia is dominated by lherzolitic massifs with poorly developed hypabyssal-extrusive components (e.g. Pamić & Majer 1977) and in Greece the Othris ophiolite possesses both lherzolitic as well as harzburgite-dunite tectonites (Bébien *et al.* 1980). Taken together these characteristics do not define a chemical signature considered typical of mid-ocean spreading centres.

Menzies & Allen (1974) and Menzies (1976) also suggested a 'young' marginal ocean basin origin for the Othris ophiolite related to incipient rifting. A similar site has been proposed for other ophiolites in Greece and Yugoslavia on the grounds of the calc-alkaline affinities of many of the ophiolites (see Bébien *et al.* 1980 for review). A subduction-related setting is further supported by the association of granitoids with the ophiolites in the E belt (Bébien 1977; Bébien *et al.* 1980).

The chemistry of the metamorphic soles beneath the Greek–Yugoslavian ophiolites is also variable. Precise determination of the original rock compositions for the soles is inhibited by the effects of higher grades of metamorphism (locally attaining granulite facies) resulting in element mobility (Spray, in press). However, a transitional MORB to alkalic igneous character appears to predominate in the mafic lithologies (e.g. Pamić 1977 for Yugoslavia; Spray & Roddick 1980 for Greece) some of which may have been derived from Triassic volcano-sedimentary sequences.

Isotopic and sediment ages

The isotopic ages discussed by Spray *et al.* (this volume, summarized in Fig. 4 this paper) are divisible into three groups:

(1) crystallization ages, averaging 172 Ma, with a range of ±9 Ma and a standard deviation of 7 Ma;

(2) amphibole sole ages, averaging 172 Ma with a range of ±6 Ma and a standard deviation of 6 Ma; and

(3) crystallization ages of calc-alkaline granitoids and migmatites with a range of 148–174 Ma (only the Fanos granite is shown on Fig. 4).

The crystallization ages of the first group represent the igneous spreading ages of the ophiolites. These show that ophiolite spreading and metamorphic sole formation differ by at most 6 Ma, and probably less; although the igneous ages are tentative, being based on only four samples. The third group of calc-alkaline ages are presumably related to subduction and/

TABLE 2. *Provisional time-scale* (From Harland *et al.* 1982; for an alternative scale see Odin 1982)

CRETACEOUS		JURASSIC	
Hauterivian (Hau)	125–131	Oxfordian (Oxf)	156–163
Valanginian (Vlg)	131–138	Callovian (Clv)	163–169
Berriasian (Ber)	138–144	Bathonian (Bth)	169–175
JURASSIC		Bajocian (Baj)	175–181
Tithonian (Tth)	144–150	Aalenian (Aal)	181–188
Kimmeridgian (Kim)	150–156		

or collision. However, the Furka and Štip granitoids have had a complex history in that they are metamorphosed and their ages may therefore represent partial overprinting rather than the age of igneous crystallization. The relatively fresh and unmetamorphosed Fanos granite (Bébien 1977), dated at 148 ± 3 Ma (?Kimmeridgian–?Tithonian) is a biotite-rich intrusion into the tholeiitic ophiolites, roughly contemporaneous with ophiolite emplacement onto the continent. However, its relationship to the inferred Middle Jurassic subduction zone is unclear.

Detailed palaeontological studies of the fossils in successions overlying the ophiolites emplaced onto the W edge of the Pelagonian Zone show that the oldest dated sediments are probably Kimmeridgian (Mavridis *et al.* 1977) and certainly Late Jurassic in age (Pichon 1977). Tithonian sediments are discordant on both the ophiolites and the Pelagonian carbonate platform and appear to have been deposited on the ophiolites after they had been uplifted and eroded (Pichon 1977). This discordant sediment cover is itself continuous from Tithonian to Albian or early Cenomanian time. No sediments have yet been dated from the ophiolite sequence itself, though such a determination would be important. The sediments show that the Vourinos ophiolite and some of the E ophiolites emplaced onto the W edge of the Pelagonian Zone are older than at least part of the Kimmerdgian.

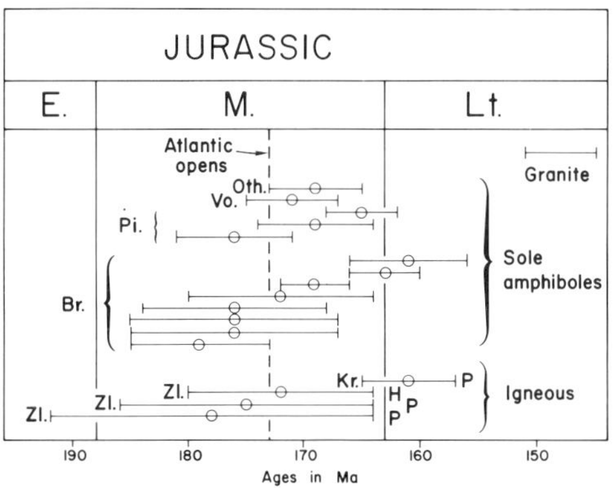

FIG. 4. Summary of ages (modified from Spray *et al.* this volume) showing K-Ar and Ar-Ar ages of the metamorphic soles, igneous amphiboles from within certain ophiolites and the Fanos granite. All data are from hornblende mineral separates (H) except for the three pargasite samples (P). The abbreviations are: Br, Brezovica; Kr, Krivaja; Oth, Othris; Pi, Pindos; Vo, Vourinos; and Zl Zlatibor. The time-scale used is from Harland *et al.* (1982). E, early; M, Middle; and Lt, Late.

The isotopic ages (Fig. 4) suggest that the ophiolites are of Middle Jurassic age, rather than the Triassic age implied by the simplest model of an emplaced marginal strip of ocean crust. This age is compatible with emplacement as a marginal strip only if there was Middle Jurassic spreading within the presumed Triassic ocean floor near the Pelagonian continent (Fig. 5). If the Neotethys was spreading throughout Triassic to Middle Jurassic time, then a second ridge or ridge jump is required to create the ophiolites next to the Pelagonian continent. A simpler model postulates that the Neotethys spread only in the early and Middle Jurassic time and that the ridge started spreading at a site next to the Pelagonian continent (Fig. 5). Although Middle Jurassic arc volcanism is absent, analogy with the remnant Pacific arcs suggests that extension could have taken place some 50–100 km above and behind the trench. The crustal strip between the embryonic ridge and the trench probably consisted of a slice of Triassic marginal volcanics or Triassic ocean floor, overlain by pelagic Triassic sediments. Similar volcanic strips have been observed adjacent to other 'Tethyan' continental margins as in Antalya, Turkey (Robertson & Woodcock 1981), Oman (Searle *et al.* 1980) and elsewhere in the region (Rocci *et al.* 1980).

Summarizing, it is concluded that the W ophiolites were generated as a marginal strip of oceanic crust; geochemical data imply that the ocean was linked to an active subduction zone; the apparent similarity of spreading and sole ages may indicate that sole formation took place at the ridge. Thus the W ophiolites were probably part of a relatively narrow, subduction-related ocean marginal to and W of the

TABLE 3. *Africa–Europe motions for 127–173 Ma* Data from Klitgord & Schouten (1982 & in press) and Livermore & Smith (in press) M = Mesozoic magnetic anomaly; BSMA = Blake Spur magnetic anomaly

	Age range	Duration	Rotation pole and angle			Anomaly	
Fossil	Ma	Ma	Lat	Long	Angle	From	To
(Hau–Vlg)	127–135	8	−50.9	203.4	1.0	M4	M11
(Vlg–Ber)	135–145	10	−63.1	201.7	2.4	M11	M16
(Tth–Kim)	145–154	9	−65.9	221.7	2.2	M16	M21
(Kim–Oxf)	154–161	7	−53.7	237.5	3.0	M21	M25
(Oxf–Cal)	161–169	8	−54.4	208.9	7.5	M25	BSMA
(Cal–Bth)	169–173	4	−54.9	212.1	3.5	BSMA	Fit

Pelagonian/Drina–Ivanjica continent. The E ophiolites also appear to be subduction-linked, but the apparent absence of soles and of exposed lateral facies changes between the ophiolites and the adjacent continental margins do not permit their setting to be so closely defined. The sedimentology of ribbon radiolarites deposited on Mesozoic ophiolites independently leads to the interpretation of these ophiolites as having originated in small ocean basins (Jenkyns & Winterer 1982).

Tectonic significance of the metamorphic soles

Metamorphic soles are best developed at the base of the ophiolites in the W zone (Karamata 1968; Pamić 1977; Pamić *et al.* 1973; Spray & Roddick 1980). They always occur at the base of the ultramafic tectonites and are therefore unlikely to represent metamorphic rocks generated in transform zones, though a relation to ultramafic diapirism cannot be excluded (Bébien *et al.* 1980). The sole rocks normally form sequences <500 m in thickness. In many cases the soles have been tectonically disrupted during later ophiolite transport such that their occurrence as blocks and slivers in sub-ophiolite melanges is commonplace. However, relatively intact examples can be found beneath the Evvia, Othris, Pindos and Vourinos ophiolites in Greece and the Zlatibor, Krivaja-Konjuh and Brezovica massifs in Yugoslavia.

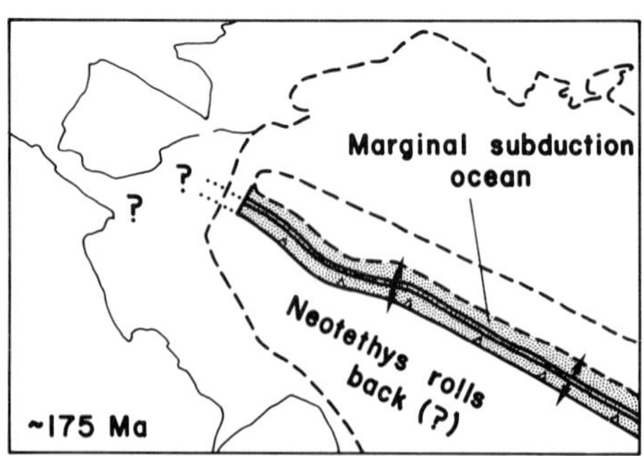

FIG. 5. 'Neotethys' rolls back in Middle Jurassic time, possibly as a result of opening of the Atlantic, creating a marginal subduction ocean—the source of the ophiolites.

Sole formation is assumed to have taken place by shortening across newly formed Middle Jurassic ridge. The alternative hypothesis that the soles represent listric faults reactivated during emplacement would not give rise to metamorphosed pelagic sediments in the sole rocks. We speculate that during the collapse of the small ocean basin, one of the Triassic borders of the marginal strip was overridden by the youngest crust and upper mantle, creating the metamorphic soles as a result (Fig. 6). We assume that the ophiolites have been tectonically detached from the upper mantle along a zone of mechanical weakness corresponding to a particular isotherm. The Vourinos complex has a thickness of about 12 km (Bébien *et al.* 1980) and the other ophiolites in Greece and Yugoslavia appear to have thicknesses less than this. The elastic thickness of oceanic lithosphere appears to lie roughly at the 500°C isotherm (McNutt & Menard 1982, their Fig. 1), below which level the lithosphere is ductile. The time taken for newly formed ocean-floor to cool to 500°C can be calculated from the infinite half-space cooling model of a ridge (Parsons & Sclater 1977). Substituting reasonable parameters into the error function describing temperature variations yields a value of about 10 Ma. However, the granulite facies grade of the sole rocks may indicate still higher temperatures, particularly if they are caused mostly by conductive heating without significant shear heating. The calculated times for ocean-floor to cool at 12 km to higher temperatures varies from about 5 Ma for 700°C to as little as 2 Ma for 1000°C. These calculations show that the time difference between spreading ages and sole formation, whether or not it occurred at a ridge, is likely to be several Ma at most, and could be so small as not to be resolvable by isotopic dating of rocks of Jurassic age. This conclusion is in agreement with the apparent similarity of sole and spreading ages (Fig. 4).

The mechanical weakness of the youngest ocean-floor would have facilitated overthrusting of one half of the ridge lithosphere and its laterally contiguous Triassic lithologies by the other half (Figs 6 & 7). It is of interest to note that the spreading and sole ages for the Semail ophiolite of Oman differ by only 5 Ma, a fact that led Coleman (1981) to propose a similar ridge displacement model. Lithosphere–asthenosphere decoupling in the vicinity of spreading centres may be an important process in the formation of ophiolites (Spray, 1983).

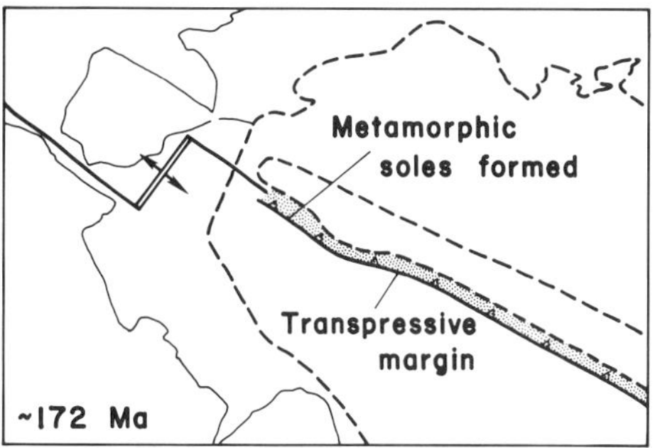

FIG. 6. Only a few Ma later (see text and Fig. 5) compression of the marginal subduction ocean occurs and the metamorphic soles are created.

The simplest half-ridge model suggests that several independent lateral gradients will exist in the preserved ophiolite (Fig. 7). Dykes in the ophiolite may show asymmetrical chilled margins, as at Troodos, and the dykes may also be preferentially inclined toward the ridge. The ultramafics may also show a systematic sense of shear in the basal tectonites. The earliest ultramafics may also be less depleted, more lherzolitic, than the younger ultramafics, which may be more harzburgitic. The mylonites in the Vourinos complex show transport to the NE, but these may result from thrusting at the ridge, rather than having been created by earlier spreading processes. After displacement, the sole rocks will become older in the direction of the overriding ridge wedge (Fig. 7). The metamorphic grade may increase away from the ridge because of the greater thickness and slower cooling rate of the ophiolites away from the ridge. Emplacement structures may also show a systematic transport sense away from the ridge wedge.

In contrast to the apparent overlap between ophiolite crystallization and sole formation, the age disparity between ophiolite displacement within oceanic lithosphere and subsequent emplacement onto continental margins in the W belt is *c.* 20–30 Ma for central Evvia and Othris (Spray *et al.* this volume) and for the Vourinos complex (Mavridis *et al.* 1977; Pichon 1977), and not well controlled elsewhere. Despite ophiolite emplacement onto the continental margin having occurred during the Late Jurassic/Early Cretaceous period, the metamorphic sole ages were not reset to register subsequent emplacement. No new mineral growth can be recognized and it is clear that continental emplacement occurred at low temperatures, well below the Ar-blocking temperature for both amphiboles and micas. This is contrary to the belief of some workers who consider the age of metamorphic soles to denote the timing of continental emplacement. At least for the Hellinic–Dinaric ophiolites, the formation of metamorphic soles occurred during the initial thrusting of ophiolites and their dating constrains only the timing of this event. Subsequent continental emplacement may ensue relatively rapidly or can take place tens of Ma later, as appears to be the case for the W Greek–Yugoslavian ophiolites.

One or two marginal basins?

Most workers regard the E and W ophiolites of Greece and Yugoslavia as parts of two distinct units. For example, distinct E and W belts are recognized in Yugoslavia (e.g. Dimitrijević & Dimitrijević 1973; Karamata 1975). Most workers in Greece also recognize two distinct belts. The original position of the Vourinos complex in Greece is unresolved. Smith (1977) argued that the present geographic separation of the Vourinos complex from the Pindos complex was the result of faulting in the Mesohellenic trough, and that the Othris, Pindos and Vourinos complexes had once been part of a single unit. Vergely (1975, 1976, 1977) used the NE to SW transport direction of presumed pre-Tithonian JE1 structures as evidence favouring an E origin for the Vourinos complex. Bébien *et al.* (1980), Brunn (1956) and Aubouin (1959) all assign an eastern origin to the Vourinos complex, but believe the Othris and Pindos ophiolites to form part of a distinct W belt.

Vergely's (1976) evidence is discussed in detail by Naylor & Harle (1977) who have pre-

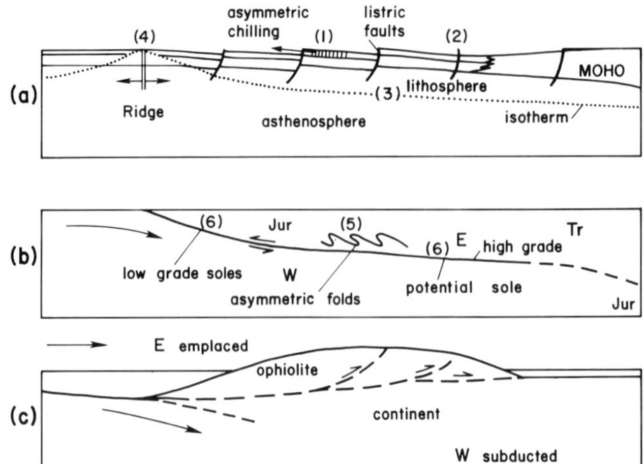

FIG. 7. Half-ridge model.

(a) Part of a ridge with an isotherm shown. Four primary features may show asymmetry: (1) one-way chilling of dykes; (2) curvature of normal faults; (3) possible shear due to relative motion of asthenosphere and lithosphere; (4) possible lateral variations in the ultramafics from lherzolitic near continent to harźburgitic at ridge.

(b) One half of the ridge overrides the other half, creating metamorphic soles. Two new asymmetric features may be created: (5) folds due to shear at base of overriding ridge; (6) low grade metamorphic soles passing laterally into higher grade rocks. E and W can be interchanged: the available data are inadequate to distinguish between them.

(c) The remnant of the overriding half is emplaced onto an adjacent continent.

sented their own evidence (Naylor & Harle 1976), in turn discussed by Vergely (1977). In essence Naylor and Harle believe that folds under the Vourinos complex showing SW-directed transport are slump folds indicating a palaeoslope to the SW, rather than tectonic transport in that direction. The second group of folds showing a similar transport direction lie some 20 km NE of the Vourinos complex and may indicate the transport direction of the Vermion ophiolite onto the E edge of the Pelagonian Zone, rather than that of the Vourinos complex itself. We also consider it possible that these folds, interpreted as JE1, may possibly be contemporaneous with the first schistosity in the Pelagonian Zone in the adjacent Olympos and Pieria regions, which has been dated by Rb–Sr methods at about 120 Ma: well into the Early Cretaceous (Barton 1976, Yarwood & Dixon 1977).

Thus the consensus is that there are two distinct ophiolite belts with only a few massifs of uncertain provenance. If this is so, then these belts are remarkable in showing a very similar evolution in space and time. Both appear to have formed adjacent to continental margins created in the Triassic. Although the available evidence is meagre, both appear to have been emplaced onto the adjacent continental margins in latest Jurassic/earliest Cretaceous time, yet both lie very close to each other (Fig. 1). As might be expected, plate tectonic interpretations of these two oceans vary: for example, in Greece compare Boillot (1977, his Fig. 6). with Katsikatsos *et al.* (1977, their Fig. 6), and in Yugoslavia contrast Dimitrijević & Dimitrijević (1973, their Fig. 7) with Karamata (1975, his Figs 3 & 4).

Despite these differences of interpretation, such a remarkable parallel evolution of two adjacent marginal ocean basins is not known from any area today, even from the complex double arcs and back-arc basins in the SW Pacific. Greece and Yugoslavian ophiolites appear to be unique. As Dimitrijević & Dimitrijević (1973, p. 336) have remarked: 'it would be difficult to envisage two synchronous subduction zones separated by only several tens of kilometers and new hypotheses should be invoked.'

There are at least two alternative explantions: the first is that the E zone is the original root zone of the ophiolites in the W. This is not a new hypothesis. For example, the Vardar Zone in Greece has been interpreted as the root zone of the W ophiolites, now detached from this root by erosion and/or tectonic processes (Bernoulli & Laubscher 1972, their Fig. 3), but as noted above, the field evidence does not appear to support these views. Dercourt (1972)

TABLE 4. *Ophiolite Evolution*

Approx. age	Event	Comment
173	Atlantic opens, subduction of Triassic ocean starts	Subduction inferred; not known from field data; Triassic ocean inferred, not preserved
172	Ophiolites spreading	'Roll-back' must be occurring, otherwise spreading impossible
172 (?Bth)	Sole formed	Ocean basin starts to collapse age not distinguishable from spreading
148 (?Tth)	Calc-alkaline granitoids formed	Uncertain of time span, subduction under-way
late Tth (?144)	Emplacement onto Pelagonian continent	Subduction probably ceases shortly afterwards
Eoc 60	Emplacement onto Apulian continent	Subduction started a few Ma earlier

has also discussed the problem. The converse hypothesis that the Vardar ophiolites were originally rooted W of the Pelagonian Zone does not appear in the literature. This would appear to be ruled out by the presence of SW-directed structures under the Vermion ophiolites (Vergely 1976), unless these formed well into Early Cretaceous time, rather than being of JE1 (=pre-Tithonian) age, in which case the evidence against a W origin for some of the Vardar ophiolites may need examination.

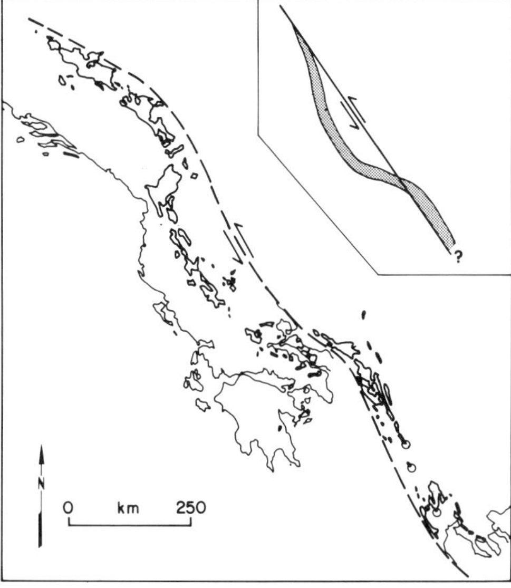

FIG. 8. The sketch shows a possible reconstruction of the Hellenic–Dinaric ophiolites as a single belt which has been cut by left-lateral transform fault of mid- to Late Jurassic age.

The second alternative, which is a new hypothesis, is to regard the two belts as having once been a single entity that has been subsequently sliced in two by large strike-slip (transcurrent) faults in the same way, for instance, that the Franciscan ophiolites of California have been dismembered by the San Andreas and other faults parallel to the North American continental margin (Fig. 8). The Drina–Ivanjica/Pelagonian sliver may well have been a slight protuberance on the SW edge of the Cimmerian continent (Figs 6 & 9).

Evidence from lherzolite–harzburgite relations within the ophiolites

The transform model can be evaluated by considering the mineralogy and bulk composition of the ultramafic rocks. Some of the ultramafic rocks in the W Yugoslavian ophiolites consist of about 85% lherzolite, with little chrome. By contrast, others in the Inner Dinaric belt in Yugoslavia average about 90% harzburgite, with an estimated 5% dunite, only 4% lherzolite and abundant chrome (Maksimović & Majer 1981). These striking differences between the W and E belts in Yugoslavia are not always so marked (Karamata, personal commun.) The Pindos ophiolite of the W zone in Greece includes harzburgite and lherzolite, as

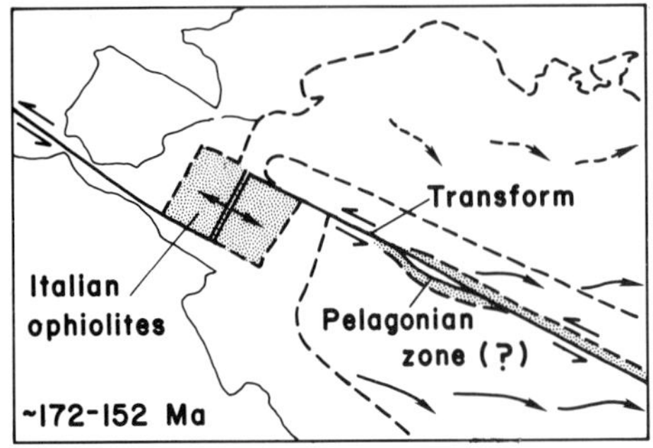

Fig. 9. During the remainder of the Middle Jurassic and much of the Late Jurassic, the Pelagonian Zone is transported SE along a transform fault and the ophiolites are emplaced onto the continent. The arrows show the calculated Africa to Europe motions in the interval 173–153 Ma during which the postulated transform of Fig. 8 is considered to have been active. Dashed arrows have displacements smaller than required. Thus, if the ophiolites formed at the S margin of the Cimmerian continent, that edge must have lain somewhere within the area of complete arrows.

does the Othris ophiolite (Bébien *et al.* 1980). The ultramafics in Pindos and Othris appear to be intermediate in character between the W and E ultramafics of Yugoslavia. If it is postulated that they were originally a single belt, with gradual changes along their length, the W belt of Yugoslavia and Greece would have joined onto the N part of the Yugoslavian belt (Fig. 8). This proposed reconstruction implies that the mantle of the original marginal subduction ocean was lherzolitic in the N, gradually passing S into more harzburgitic, chrome-rich ultramafic rocks. Depending on one's interpretation, the harzburgitic Vourinos complex could be regarded either as part of the E belt (lying S in the reconstruction), or as the lateral equivalent of the Pindos ophiolite and lying to the N on the reconstruction.

Nicolas & Jackson (1972) interpreted the generally lherzolitic ultramafic rocks of the W Mediterranean as tectonically emplaced slices of continental mantle. The predominantly harzburgitic ultramafic rocks of the E Mediterranean were interpreted as representing oceanic mantle. The dividing line between the two types separates the two ophiolite belts discussed in this paper. These views need some modification because the lherzolitic W belt seem to possess the characteristics typical of ophiolites as a whole, as well as passing laterally via tectonic slices into continental margin sequences. In other words, these ophiolites are unlikely to represent slices of continental mantle, but could represent mantle intermediate between continental and oceanic varieties (Spray 1982).

There are several alternatives to the view of Nicolas and Jackson. For example, the lherzolitic W belt in Yugoslavia and Greece, and possibly elsewhere, may represent a zone in which extension was less than it was in the harzburgitic E belt. On the proposed reconstruction the two belts are transitional, implying that the marginal subduction ocean opened southwards but narrowed or terminated in the N. Such a configuration would be compatible with the 'propagating rift' hypothesis (e.g. Vink 1982). A second possibility is that the lherzolite might represent mantle formed during the transitional period from extension to compression. Its absence in the E zone could be attributed to erosion. There is inadequate understanding and data to evaluate these possibilities.

As might be expected, the gross mineralogical changes in the ultramafic rocks are accompanied by changes in the accessory minerals, particularly spinels. For example, the spinels of the W belt in Yugoslavia (Central Dinarides) have a different composition from those of the E belt (Inner Dinarides; Maksimović & Majer 1981). Spinels in ophiolites can have several origins and can be modified after their recrystallization. If a sufficient number of spinel analyses from different levels in different ophiolites were available, systematic changes along and across the strike of each belt might be revealed. Though such data are not yet available, it is interesting to use the available analyses and to assume that the spinels may vary linearly in composition with distance along the length of two belts. Spinels from Othris (Hynes 1972) are intermediate in average composiion between the W and E spinel averages in Yugoslavia (Fig. 10). These data therefore are consistent with the postulated left-lateral movement and suggest it would be about 700–750 km. That the movement must be of this order is independently suggested by the scale of the Pelagonian/Drina–Ivanjica sliver. According to the model, this sliver has to be moved by at least the length of the ophiolite belt on the E side of the sliver, or about 700 km. Further work is in progress on spinels to further test this evidence.

Age of the hypothetical transform

The age of the speculative movements that slid the Pelagonian/Drina–Ivanjica continental sliver to the SE must obviously be younger than the Middle Jurassic spreading of the marginal sub-

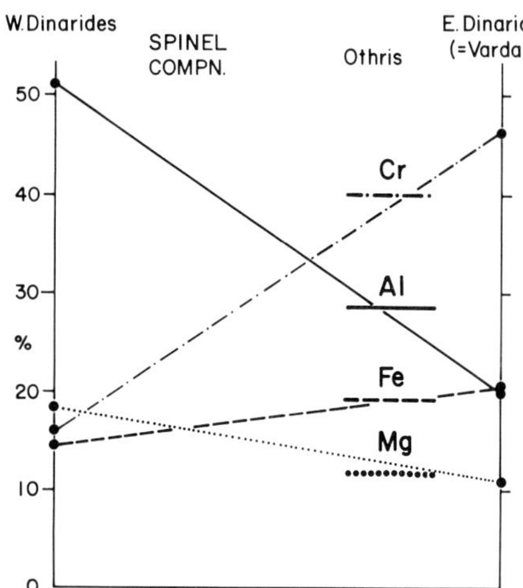

FIG. 10. Variations in the Mg, Fe, Al and Cr contents of spinels plotted at three localities: the W and E Dinarides (Maksimović & Majer 1981) and Othris (Hynes 1972). If the variations are linear, then the distance from the W Dinarides to Othris (about 550 km) fixes the original distance from the W to the E Dinarides at about 700 km.

duction ocean. They must also have begun before the E ophiolites were emplaced onto the E edge of the continental sliver because the model requires the ophiolites to be transported there prior to emplacement. In the Vermion area emplacement is of pre-Tithonian age (Pichon 1977). The faulting appears to be older than the large synmetamorphic folds in the Pelagonian discussed above. Oblique movements, possibly related to transform faults W of the Pelagonian Zone may have generated the folds trending at 57° of post-Berriasian, pre-Albian age in Othris (Smith *et al.* 1979) and the NE-trending folds beneath the Vourinos ophiolite (Naylor & Harle 1976). Uncertainties in the time-scale (Harland *et al.* 1982) preclude precise conversion of Jurassic biostratigraphical to isotopic ages. The postulated transform motion is therefore younger than 172 Ma (?post-Bathonian); largely complete by the end of Kimmeridgian time (? > 150 Ma); and had ceased by 120 Ma (?Barremian).

Late Cretaceous or younger strike-slip movements have been postulated in the Vardar Zone of Yugoslavia (Dimitrijević & Dimitrijević 1975, p. 358) and in the Vardar and Pelagonian Zones of Greece (Brunn 1960; Mercier 1968). In Greece the strike-slip zones trend at about 130°. The sense of movement is not known in Greece, but is thought to be right-lateral in Yugoslavia, in contrast to the hypothetical left-lateral earlier movements.

'Roll-back' and 'roll-to'

Elsasser (1971) discussed two limiting cases of descending slab motion at a subduction zone. In the first case the sinking slab behaves as a conveyor belt, in the second 'retrograde' motion occurs with the overriding slab moving steadily over the sinking slab. Elsasser speculates that splitting of the overriding slab might occur, but when his work was published 'back-arc' basins had only just been discovered.

For our purposes we can take the overriding plate as fixed. If the sinking slab has a finite velocity perpendicular to its dip it will move relative to the overriding plate. If the velocity is downward, the slab will move away from the overriding plate. If it is upward it will compress the overriding plate. The first condition we refer to as 'roll-back' during which stretching must occur, leading to extension of the overriding slab near the subduction zone and ultimately to formation of a subduction marginal ocean. The second condition we refer to as 'roll-to'. It involves shortening of the weakest zone in the overriding slab, probably by the collapse of an ocean basin, when it will be subducted or emplaced as ophiolites, or both.

Roll-back and roll-to are purely kinematic effects. The possible causes of these relationships have been discussed by Chase (1978) and will not be discussed further here.

Regional plate framework: constraints from the Atlantic

For brevity we adopt the interpretation of Şengör *et al.* (1980, their Fig. 1) for the Late Triassic Tethys (Figs 3 & 5). In their reconstruction, the Tethys is envisaged as comprising two oceans, a Permian or older Tethys to the NE (the 'Paleotethys') and Triassic and younger Tethys (the 'Neotethys') to the SW. A division between the two is made by the 'Cimmerian continent' (or continental fragments) that was supposedly rifted from the N edge of Gondwanaland in Triassic time. The relative sizes of the Paleotethys and Neotethys are not known. Our interpretation of the ophiolites as having belonged to a single marginal ocean basin would require the Pelagonian and Serbo–Macedonian continents to have been continuous and part of the SW edge of the Cimmerian continent prior to M Jurassic time (Figs 3 & 5).

Since the ophiolites appear to have been part

of a subduction-coupled marginal basin, then there must have been a trench W of the Pelagonian/Serbo–Macedonian continent which dipped NE under the E edge of the Triassic marginal basin (Fig. 5). This trench must be of the same age as the ophiolites, i.e. Middle Jurassic. It is the only evidence, admittedly indirect, for a compressional plate margin in a traverse from N Gondwanaland across the former Tethys to S Laurasia at this longitude in Middle Jurassic time. Ths indirect evidence is supported by an interpretation of the Svoula flysch of latest Early Jurassic to early Late Jurassic age, which may indicate an unstable continental margin and slope whose instability (Kauffmann *et al.* 1976) could have reflected the nearby existence of a subduction zone.

The initial age of the opening of the central Atlantic is fundamental to any plate tectonic interpretation of the area. Sheridan *et al.* (1982) suggest that there was a period of fast spreading from the age of the Blake Spur magnetic anomaly, probably basal Callovian (= ?169 Ma) to M25 time (Oxfordian and = ?161 Ma). Extrapolating back in time gives an initial opening age of 173 Ma, or ?Bathonian time. This age is essentially identical to the sole ages of 172 ± 6 Ma. There are therefore two distinct plate tectonic problems: the first of relating the spreading history of the ophiolites and their soles to a plate model; the second of examining the plausibility of the postulated transform fault and a plate model.

Because of the probable errors, we do not know whether the spreading of the ophiolites and the formation of the metamorphic soles should be regarded as having been essentially complete by the time the Atlantic started to open, overlapped its opening or was entirely younger than the opening. In the first case, although it is possible to attribute the spreading of the ophiolites and the inferred subduction zone to 'roll-back' of an oceanic plate, it is geometrically impossible to have 'roll-to' without creating an extensional margin somewhere else, for which there is no evidence. We therefore reject this possibiity.

If the Atlantic opening and sole ages are considered coincident, then ophiolite spreading could be attributed to 'roll-back' as before, but sole formation could be directly attributed to compression caused by the opening of the Atlantic Ocean. In the third model, the inferred trench and subduction-linked spreading are direct effects of the opening of the central Atlantic.

The second model—'roll-back' and compression caused by Atlantic opening—raises the possibility that 'roll-back' may have started a subduction zone whose pull succeeded in breaking up Pangea. We can investigate its plausibility further by trying to determine he spreading directions of the ophiolites. If 'roll-back' was the cause of extension, we might expect spreading to be perpendicular to the continental margins, but if the opening of the Atlantic was the cause of extension, then we might expect the spreading directions to be parallel to the flow lines, which are roughly E–W relative to Eurasia during this period (Fig. 9).

Aubouin & Dercourt (1975) speculated that six tranverse structures, typically offsets in the outcrop pattern of the W edge of the ophiolites—Zrmanja (N), Sarajevo, Scutari-Peč, Kastaniotikos, Sperkhios and possibly Cytheria (S)—may represent palaeo-transform faults, presumably of Middle Jurassic age. Most of the transverse structures have a trend roughly perpendicular to the ophiolite outcrops, implying that spreading was not markedly oblique to the present trend of the ophiolites.

Data on spreading directions from within these ophiolites are imprecise. The rapid lateral variations in geometry and rock type in parts of the Guevgueli, Vourinos and Othris ophiolites have been attributed to transcurrent movement contemporaneous with spreading (Bébien *et al.* 1980). Dykes in the central Vardar Zone (E Almopias) trend N (Bébien *et al.* 1980). Dykes in the Vourinos complex, when rotated back to the vertical, have an E trend, rather than an SE trend that might be expected if spreading was perpendicular to the trend of the ophiolite belt (Jackson *et al.* 1975; Ross *et al.* 1980). But spreading need not be perpendicular to dyke trends, and even if it was the dyke complex might have been rotated during emplacement.

Kinked olivines in the Vourinos ultramafic tectonites suggest that spreading was in a direction of 20°, with the ridge trending at about 110° and lying S of the outcrop (Ross *et al.* 1980, their Fig. 13), though these can be reinterpreted as initial emplacement structures (see below). In short, the available evidence suggests that spreading was not highly oblique to the present trend of the ophiolite outcrop. There is certainly no evidence suggesting that the ophiolites represent a fossil 'Gulf of California' in which ridges were at right angles to the gross trend of the ophiolite belt.

The ocean whose subduction lead to the formation of the subduction-linked marginal basin is believed to have been part of the Gonwanaland plate. We have therefore calculated the motion of Gondwanaland and Eurasia in this time interval as indicated by the central Atlantic spreading data (Fig. 6). The data show a relatively smooth arcuate sweep of Gondwanaland from west to east in the area of interest. While it is possible to set up a model in which the spreading of the subduction-linked basin is at a high angle to the SW edge of the Cimmerian continent and is then followed by transform faulting parallel to the margin, the relative motions do not suggest such a discontinuity. Several other models are possible and the data needed to discriminate between them are not yet available. The tectonic model illustrated in Fig. 9 is one in which subduction started just before the opening of the Atlantic, but where sole formation and the postulated transform faulting are directly related to Atlantic opening.

The second plate tectonic problem is whether the transform fault model can be fitted into the available plate tectonic constraints. The transform fault is believed to have occurred in the interval 172 to 150 Ma and to require at least 700 km of left-lateral (sinistral) movement. The relative motions shown on Fig. 9 are for the period 173–153 Ma. Inspection reveals that the model is compatible with the available data provided that the SW edge of the Cimmerian continent was parallel to the relative motions, and also provided that it lay within the area where the relative motions exceed 700 km. The postulated strike-slip plate boundary would account for the absence of any significant calc-alkaline igenous activity in the region in the time interval of the postulated transform.

In mid- to Late Jurassic time, the extensional margin represented by the central Atlantic opening probably passed E off N Africa (Smith 1971); and could have continued N just W of Italy, then E through the Alps (Laubscher & Bernoulli 1977, their Figs 5 & 6) and then E along the edge of the Cimmerian continent.

Conclusions

(1) In earlier Middle Jurassic time the ophiolite belts of Greece, Albania and Yugoslavia were probably part of a subduction-linked marginal ocean basin. The future ophiolites were possibly connected either across strike so that what are now the W and E belts were part of a single unit rooted in the Vardar zone, or, more probably, they were connected longitudinally so that the S end of the W belt was joined onto the N end of the E belt.

(2) In the latter case a later Middle to Late Jurassic sinistral transform fault may have broken off the Pelagonian/Drina–Ivanjica Zone as a sliver, translating it 700 km SE relative to the present-day Serbo–Macedonian zone and duplicating the ophiolite belt as a result.

(3) The emplaced ophiolites are probably one half of a narrow subduction-linked basin, which were detached at the ridge along the lithosphere–asthenosphere boundary and thrust over the other half at about 170 Ma, then transform faulted and subsequently emplaced onto the continent.

(4) Transcurrent components of motion may have been important during the spreading of the marginal basin, its initial thrusting at the ridge crust and its subsequent continental emplacement, because the associated plate margins may have been orientated roughly E–W and linked directly to the opening of the central Atlantic.

(5) Testing and refinement of the proposed half-ridge/transform model requires a better Jurassic time-scale; more precise palaeontological data from the carbonate platforms and ophiolite cover; better geochronological and geochemical data from the ophiolites and their soles, consideration of the ophiolites of Romania and Turkey and, above all, reliable palaeomagnetic data.

ACKNOWLEDGEMENTS: We thank S. Karamata for showing us the Yugoslavian ophiolites in 1981 as part of an exchange visit with the Serbian Academy of Sciences organized by The Royal Society. We are also grateful for many discussions with him, G. Šoptrajanova and J. Bébien, and for comments from J. A. Pearce, A. H. F. Robertson and N. H. Woodcock. Some of the diagrams were constructed as a result of support given by NERC grant GR3/4405. Cambridge Earth Science contribution 385.

References

AUBOUIN, J. 1959. Contributions à l'étude géologique de la Grèce septentrionale: Les confins de l'Epire et de la Thessalie. *Ann. geol. Pays Hell.* **10,** 525 p.

—— & DERCOURT, J. 1975. Les transversales dinariques derivent-elles de paléofailles transformantes? *C.r. Acad. Sci.* **218,** 347–350.

BARTON, C. M. 1976. The tectonic vector and emplacement age of an allochthonous basement slice in the Olympos area, N.E. Greece. *Bull. Soc. géol. France*, **18,** 253–258.

BEBIEN, J. 1977. Mafic and ultramafic rocks associated with granites in the Vardar zone. *Nature, London*, **270,** 232–234.

——, Blanchet, R., Cadet, J. P., Charvet, J., Chorowicz, J., Lapierre, H. & Rampnoux, J. P. 1978. Le volcanisme triasique des Dinarides en Yougoslavie: sa place dans l'évolution géotectonique. *Tectonophysics*, **47**, 159–176.

—— & Mercier, J. L. 1977. Le cadre structural de l'association ophiolites-migmatites-granites de Guévguéli (Macédoine, Grèce); une croûte de bassin interarc? *Bull. Soc. géol. France*, **19**, 927–934.

——, Ohnenstetter, D., Ohnenstetter, M. & Vergely, P. 1980. Diversity of Greek ophiolites: birth of ocean basins in transcurrent systems. *Ofioliti*, **3**, 129–197.

Bernoulli, D. & Jenkyns, H. C. 1974. Alpine, Mediterranean and North Atlantic Mesozoic facies in relation to the early evolution of the Tethys. *In*: Dott, R. H. & Shaver, R. H. (eds) Symposium on modern and ancient geosynclinal sedimentation. *Spec. Publ. Soc. Econ. Paleontol. Mineral. Tulsa*, **19**, 129–160.

—— & Laubscher, H. 1972. The palinspastic problem of the Hellenides. *Eclogae. geol Hel.* **65**, 107–118.

Boillot, G. 1977. Séance éxtraordinaire de la Société géologique de France tenue à Athènes. *Bull. Soc. géol. France*, **19**, 83.

Brunn, J. H. 1956. Etude géologique du Pinde septentrional et de la Macédoine occidentale. *Ann. géol. Pays Hell.* **7**, 358.

—— 1960. Les zones helléniques internes et leur extension. *Reflexions sur l'orogène alpine.* **2**, 470–486.

Cameron, W. E., Nisbet, E. G. & Dietrich, V. J. 1979. Boninites, komatiites and ophiolitic basalts. *Nature, London*, **280**, 550–553.

Capedri, S., Venturelli, G., Bocchi, G., Dostal, J., Garuti, G. & Rossi, A. 1980. The geochemistry and petrogenesis of an ophiolitic sequence from Pindos, Greece. *Contrib. Mineral. Petrol.* **74**, 189–200.

Chase, C. G. 1978. Extension behind island arcs and motions relative to hot spots. *J. geophys. Res.* **83**, 5385–5387.

Coleman, R. G. 1981. Tectonic setting for ophiolite obduction in Oman. *J. geophys. Res.* **86**, 2497–2508.

Dercourt, J. 1972. The Canadian cordillera, the Hellenides and the sea floor spreading theory. *Can. J. Earth Sci.* **9**, 709–743.

Dimitrijevič, M. D. & Dimitrijevič, M. N. 1973. Olistostrome melange in the Yugoslavian Dinarides and late Mesozoic plate tectonics. *J. Geol.* **81**, 328–340.

—— & Dimitrijevič, M. N. 1975. The "diabase-chert formation" of the ophiolite belt and the Vardar Zone: genetic comparison. *Acta geol. Zagreb*, **8**, 347–357.

Elsasser, W. M. 1971. Sea-floor spreading as thermal convection. *J. geophys. Res.* **76**, 1101–1112.

Ferriere, J. 1974. Etude géologique d'un secteur des zones helléniques internes subpélagoniennes (massif de l'Othrys, Grèce continentale). Importance et signification de la période orogénique anté-Cretacé supérieur. *Bull. Soc. géol. France*, **14**, 543–561.

—— 1977. Faits nouveaux concernant la zone isopique maliaque (Grèce continentale orientale). *6th geol. Coll. Aegean region, Athens*, **1**, 197–210.

Harland, W. B. Cox, A. V., Llewellyn, P. G., Pickton, C. A. G., Smith, A. G. & Walters, R. 1982. *A Geologic Time Scale.* Cambridge University Press, London, 131p.

Hynes, A. J. 1972. *The geology of the western Othris Mountains, Greece.* Thesis, Ph.D., University of Cambridge (unpubl.). pp.

—— 1974. Igneous activity at the birth of an ocean basin in Eastern Greece. *Can. J. Earth Sci.* **11**, 842–853.

Jackson, E. D., Green, H. W. & Moores, E. M. 1975. The Vourinos ophiolite, Greece: cyclic units of lineated cumulates overlying harzburgite tectonite. *Bull. Geol. Soc. Am.* **86**, 390–398.

Jenkyns, H. C. & Winterer, E. L. 1982. Paleooceanography of Mesozoic ribbon radiolarites. *Earth planet. Sci. Lett.* **60**, 351–375.

Karamata, S. 1968. Zonality in contact metamorphic rocks around the ultramafic mass of Brezovica (Serbia, Yugoslavia). *23rd Internat. geol. Congress, Beograd*, 197–207.

—— 1975. Geoloska Evolucija naseg podrucja od Trijasa do Kvartara. *Radovi Instituto za gelosko-rudarska istrazivanja i ispitivanja nuklearnih i drugih mineralnih sirovina*, **10**, 1–15.

——, Majer, V. & Pamić, J. 1980. Ophiolites of Yugoslavia. *Ofioliti*, spec. issue: Tethyan ophiolites, **1**, 105–125.

Katsikatsos, G., Mercier, J. & Vergely, P. 1977. Seance éxtraordinaire de la Societé géologique de France. *Bull. Soc. géol. France*, **19**, 83.

Kauffmann, G., Kockel, F. & Mollat, H. 1976. Notes on the stratigraphic and paleogeographic position of the Svoula Formation in the Innermost Zone of the Hellenides (Northern Greece). *Bull. Soc. géol. France*, **18**, 225–230.

Klitgord, K. D., & Schouten, H. 1982. Early Mesozoic Atlantic reconstructions from sea-floor spreading data. EOS. Trans. *Am. Geophys. Union*, **63**, 307.

—— & —— (In press). Early Mesozoic Atlantic reconstruction from sea-floor spreading data.

Laubscher, H. & Bernoulli, D. 1977. Mediterranean and Tethys. *In*: Nairn, A. E. M., Kanes, W. H. & Stehli, F. G. (eds) *The Ocean Basins and Margins.* Vol. 4A, *The Eastern Mediterranean*, Plenum Press, New York, 1–25.

Livermore, R. A. & Smith, A. G. (in press). Some boundary conditions for the evolution of the Mediterranean region. *Proc. NATO Avanced Research Institute Erice*, Sicily, Nov. 1982.

Maksimović, Z. & Majer, V. 1981. Accessory spinels of two main zones of Alpine ultramafic rocks in Yugoslavia. *Bull. Acad. Sci. Arts Serbie*, **75**, 47–58.

Mavridis, A., Skourtsis-Coroneou, V., & Tsaila-Monopolis, St. 1977. Contribution to the geology of the Subpelagonian zone (Vourinos area, west Macedonia). *6th Colloquium on the geology of the Aegean region, Athens 1977*, 175–195.

McNutt, M. K. & Menard, H. W. 1982. Constraints on the yield strength in the oceanic lithosphere derived from observations of flexure. *Geophys. J. Roy. astr. Soc.* **71,** 363–394.

MENZIES, M. A. 1976. Rare earth geochemistry of fused ophiolites and alpine lherzolites—I. Othris, Lanzo and Troodos. *Geochim. Cosmochim. Acta,* **40,** 645–56.

—— & ALLEN, C. 1974. Plagioclase lherzolite—residual mantle relationships within two Eastern Mediterranean ophiolites. *Contrib. Mineral. Petrol.* **445,** 197–213.

MERCIER, J. L. 1968. Etude géologique des zones internes des Hellénides en Macédoine Centrale (Grèce). *Ann. geol. Pays Hell.* **30,** 792 p.

MONTIGNY, R. 1975. *Géochimie comparée des cortèges de roches océaniques et ophiolitiques. Problème de leur genèse.* Thèse d'état, Paris 7, 288p.

NAYLOR, M. A. & HARLE, T. J. 1976. Paleogeographic significance of rocks and structures beneath the Vourinos ophiolite complex, Northern Greece. *J. geol. Soc. Lond.* **132,** 667–675.

—— & —— 1977. Paleogeographic significance of rocks and structures beneath the Vourinos ophiolite complex, Northern Greece: a reply. *J. geol. Soc. Lond.* **133,** 506–507.

NICOLAS, A. & JACKSON, E. D. 1972. Repartition en deux provinces des péridotites de châines alpines logeant la Mediterranée: implications géotectoniques. *Schweiz. mineral. petrogr. Mitt.* **52,** 479–495.

ODIN, G. S. 1982. *Numerical Dating in Stratigraphy.* Wiley, New York, 2 vols. 1040p.

PAMIĆ, J. 1977. Variation in geothermometry and geobarometry of peridotite intrusions in the Dinaride central ophiolite zone, Yugoslavia. *Am. Mineral.* **62,** 874–886.

—— & MAJER, V. 1977. Ultramafic rocks of the Dinaride central ophiolite zone in Yugoslavia. *J. Geol.* **85,** 553–569.

——, SČAVNIČAR, S. & MEDJIMOREĆ, S. 1973. Mineral assemblages of amphibolites associated with Alpine-type ultramafics in the Dinaride ophiolite zone (Yugoslavia). *J. Petrol.* **14,** 133–157.

PARSONS, B. & SCLATER, J. G. 1977. An analysis of the variation of ocean floor bathymetry and heat flow with age. *J. geophys. Res.* **82,** 803–827.

PICHON, J. F. 1977. Une transversale dans la zone Pélagonienne, depuis les collines de Krapa (SW) jusqu'au massif du Vermion (NE): les premiers séries transgressive sur les ophiolites. *6th Colloquium on the Geology of the Aegean region, Athens 1977,* 163–171.

PRICE, I. 1976. Carbonate sedimentology in a pre-Upper Cretaceous continental margin sequence, Othris, Greece. *Bull. Soc. géol. France,* **18,** 273–279.

RICHARD, P. & ALLEGRE, C. J. 1980. Neodymium and strontium isotope study of ophiolite and orogenic lherzolite petrogenesis. *Earth planet. Sci. Lett.* **47,** 65–74.

ROBERTSON, A. H. F. & WOODCOCK, N. H. 1981. Gödene zone, Antalya complex: volcanism and sedimentation along a Mesozoic continental margin. S.W. Turkey. *Geol. Rundschau,* **70,** 1177–1214.

ROCCI, G., BAROZ, F., BEBIEN, J., DESMET, A., LAPIERRE, H., OHNENSTETTER, D., OHNENSTETTER, M., & PARROT, J.-F. 1980. The Mediterranean ophiolites and their related Mesozoic volcano-sedimentary sequences. *Proc. Int. Ophiolite Symp. Cyprus 1979,* 273–286.

ROSS, J. V., MERCIER, J. C. C., AVE LALLEMANT, H. G., CARTER, N. L. & ZIMMERMAN, J. 1980. The Vourinos ophiolite complex, Greece: the tectonite suite. *Tectonophysics,* **70,** 63–83.

SEARLE, M. P., LIPPARD, S. J., SMEWING, J. D., & REX, D. C. 1980. Volcanic rocks beneath the Semail ophiolite nappe in the northern Oman mountains and their tectonic significance in the Mesozoic evolution of the Tethys. *J. geol. Soc. Lond.,* **137,** 589–604.

ŞENGÖR, A. M. C., YILMAZ, Y. & KETIN, I. 1980. Remnants of a pre-Late Jurassic ocean in northern Turkey: Fragments of a Permian-Triassic Paleo-Tethys? *Bull. geol. Soc. Am.* **91,** 599–609.

SHERIDAN, R. E. *et al.* 1982. Early history of the Atlantic Ocean and gas hydrates on the Blake Outer Ridge: Results of the Deep Sea Drilling Project, Leg 76, *Bull. geol. Soc. Am.* **93,** 876–885.

SMITH, A. G. 1971. Alpine deformation and the oceanic areas of the Tethys, Mediterranean and Atlantic. *Bull. geol. Soc. Am.* **82,** 2039–2070.

—— 1977. Othris, Pindos and Vourinos ophiolites and the Pelagonian Zone. *Proc. 6th geol. Coll. Aegean region, Athens,* **3,** 1369–1374.

——, HURLEY, A. M. & BRIDEN, J. C. 1981. *Phanerozoic Palaeocontinental World Maps.* Cambridge University Press, 102p.

——, HYNES, A. J., MENZIES, M., NISBET, E. G., PRICE, I., WELLAND, M. J. P. & FERRIERE, J. 1975. The stratigraphy of the Othris mountains, eastern central Greece: a deformed Mesozoic continental margin sequence. *Eclog. Geol. Hell,* **68,** 463–481.

—— & MOORES, E. M. 1974. Hellenides. *In*: SPENCER, A. (ed.) *Mesozoic and Cenzoic orogenic belts. Spec. Publ. geol. Soc. London,* **4,** 159–185.

—— & WOODCOCK, N. H. 1976a. The earliest Mesozoic structures in the Othris region, eastern central Greece. *Bull. Soc. géol. France,* **18,** 245–251.

—— & —— 1976b. Emplacement model for some 'Tethyan' ophiolites. *Geology,* **4,** 653–656.

—— & —— & NAYLOR, M. A. 1979. The structural evolution of a Mesozoic continental margin, Othris Mountains, Greece. *J. geol. Soc. Lond,* **136,** 589–603.

SPRAY, J. G. 1982. Mafic segregations in ophiolite mantle sequences. *Nature, London,* **299,** 524–528.

—— 1983 Lithosphere—asthenosphere decoupling at spreading centres and initiation of obduction. *Nature, London,* **304,** 253–255.

—— in press. Possible causes and consequences of upper mantle decoupling and ophiolite displacement. *In*: I. G. GASS, A. W. SHELTON & S. J. LIPPARD (eds) *Ophiolites and Oceanic Lithosphere. Spec. Publ. geol. Soc. Lond.*

—— & RODDICK, J. C. 1980. Petrology and $^{40}Ar/^{39}Ar$ geochronology of some Hellenic sub-ophiolite metamorphic rocks. *Contrib. Mineral. Petrol.* **72,** 43–55.

Vergely, P. 1975. Origine "vardarienne", chevauchement vers l'Ouest et rétrocharriage vers l'Est des ophiolites de Macédoine (Grèce) au cours du Jurassique supérieur-Eocrétacé. *C.r. Séances. Acad. Sci. Paris*, **D280,** 1063–1066.

—— 1976. Chevauchement vers l'Ouest et rétrocharriage vers l'Est des ophiolites: deux phases tectoniques au cours du Jurassique supérieur-Eocrétacé dans les Hellénides internes. *Bull. Soc. géol. France*, **18,** 231–244.

—— 1977. Discussion of the paleogeographic significance of the rocks beneath the Vourinos ophiolite, Northern Greece. *J. geol. Soc. Lond.* **133,** 505–507.

Vink G. E. 1982. Continental rifting and the implications for plate tectonic reconstructions. *J. geophys. Res.*, **87,** 10677–10699.

Yarwood, G. A. & Dixon, J. E. 1977. Lower Cretaceous and younger thrusting in the Pelagonian rocks of the High Pieria, Greece. *6th Coll. on the Geology of the Aegean region, Athens, 1977*, 269–280.

A. G. Smith & J. G. Spray, Dept. of Earth Sciences, Univ. of Cambridge, U.K. CB2 3EQ.

Tectonic implications of palaeomagnetic results for the Carpatho-Balkan and adjacent areas

Emö Márton

SUMMARY: From palaeomagnetic results, two different tectonic units can be recognized in the Carpatho-Balkan region and adjacent areas.

The north-western unit comprises the Italian peninsula south of the Po-basin, Sardinia, the Istria peninsula of Yugoslavia, the Transdanubian Central Mountains and the Bükk Mountains in Hungary and the outer and inner West Carpathians of Czechoslovakia. The palaeomagnetic results suggest that this unit was part of the African plate in the Mesozoic and Palaeogene, but was later decoupled from Africa.

The south-eastern unit is not very well defined palaeomagnetically. Palaeopoles are known from the Mecsek and Villány Mountains in Hungary, and the Papuk Mountains of Yugoslavia and North-West Bulgaria. They are similar to the poles of the same age from stable Europe but do not yet provide a coherent pattern. The palaeomagnetic results to date clearly testify to a different rotational history of this unit from the history of the African plate and the north-western unit, but more extensive palaeomagnetic studies especially on post-Palaeozoic rocks are needed to reveal the details of the movement history.

The geological framework in which the plate tectonic significance of the palaeomagnetic results in the Carpatho-Dinaric and adjacent areas will be discussed is illustrated in Fig. 1. This map is after Laubscher & Bernoulli (1977) but modified to include the Pannonian Basin.

According to Laubscher & Bernoulli (1977), the tectonic units between stable Europe and the present northern margin of Africa can be thought of as having belonged either to the northern or to the southern margin of the Tethys Ocean. The majority of the south-Tethyan units are relatively allochthonous. The areas marked by crosses could be relatively autochthonous, mainly from lack of evidence to the contrary. The north-western part of the Pannonian Basin represents a south-Tethyan unit, while the south-eastern part belongs to a north-Tethyan unit (Géczy 1973). These two parts are separated by the mid-Pannonian mobile belt which comprises ophiolites and associated oceanic sediments (Szepesházi 1977).

The present arrangement of the north- and south-Tethyan units implies that the tectonic history of the Carpatho-Dinaric region must have been complicated. Palaeomagnetism may help to reveal the relative rotations and latitudinal changes in the evolution of such a complicated region.

Discussion of the palaeomagnetic results

Numerous Neogene palaeomagnetic directions from the Carpatho-Dinaric and Balkan regions indicate that horizontal movements large enough to influence the palaeopole positions ended before 14–16 Ma as the average palaeopoles derived from Neogene volcanics of this age are similar to the present position of the geomagnetic pole. In contrast, Palaeogene and older poles are significantly different from the present geomagnetic pole.

In order to follow the movements of tectonic units in the time and space, Palaeogene and older palaeomagnetic poles from different parts of the Carpatho-Balkan region and adjacent areas are grouped according to their ages, assigned stratigraphically or by radiometric dating.

In Figs 2–5 the poles derived from relatively autochthonous areas are shown as circles. These poles are important for the interpretation since independent microplate movement can be seen clearly where systematic differences exist between the palaeopoles of the large plates and the poles of any relative autochthon. Nappe transport may obscure the picture completely or may have a very slight influence on the palaeopoles, as we will see in Figs 2–5.

Starting with the youngest poles, we observe that Eocene-Oligocene palaeomagnetic poles (Fig. 2) are well grouped and different from poles of similar age from both stable Europe and Africa. It is intersting to note that two of them come from Sardinia and the outer West Carpathians respectively, both classified as northern Tethyan elements by Laubscher & Bernoulli (1977). The other two are derived from Umbria and the Transdanubian Central Mountains of Hungary.

The Late Cretaceous poles are split into two

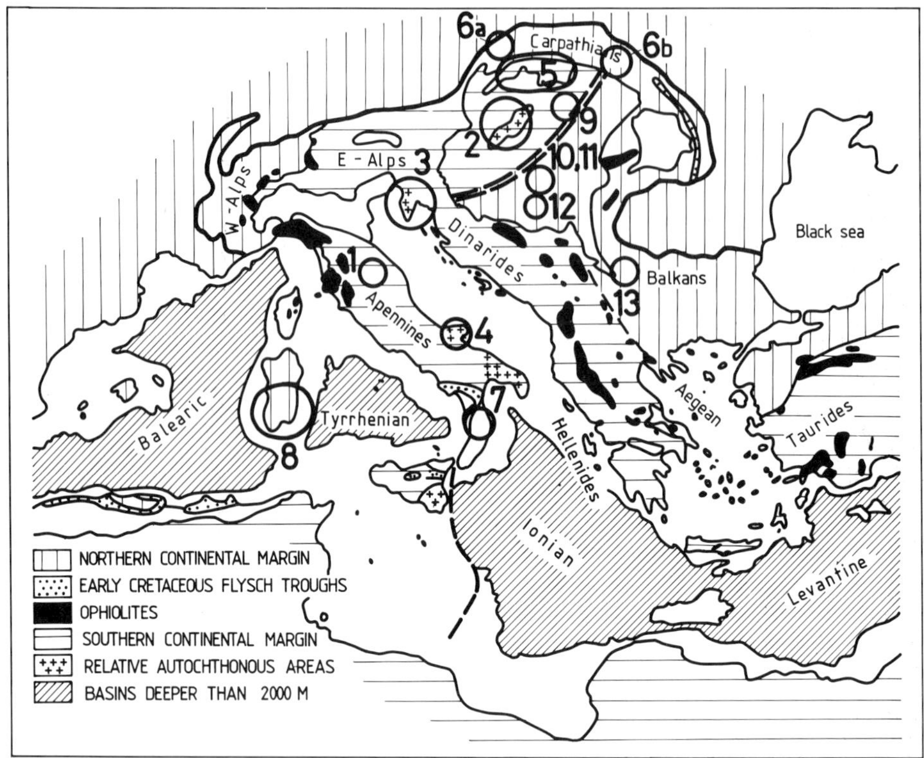

FIG. 1. The interpreted extent of the northern and southern continental margins of the Tethys ocean, modified after Laubscher & Bernoulli (1977). Palaeomagnetic results from the areas numbered on the map are discussed.

1, Umbria, Italy; 2, Transdanubian Central Mountains, Hungary; 3, Istria peninsula, Yugoslavia; 4, Gargano, Italy; 5, Inner West Carpathians, Czechoslovakia; 6a & b, Outer West Carpathians, Czechoslovakia; 7, Calabria, Italy; 8, Sardinia, Italy; 9, Bükk Mountains, Hungary; 10, Mecsek Mountains, Hungary; 11, Villány Mountains, Hungary; 12, Papuk Mountains, Yugoslavia; and 13, North-West Bulgaria.

groups. Poles from the south-Tethyan elements plus the outer West Carpathians form one group, away from the stable European and African poles. Poles from north-Tethyan units lie close to the stable European apparent polar wander curve (Fig. 3).

The Early Cretaceous poles (Fig. 4) again form one group; all the poles come from southern Tethyan units with one from the outer West Carpathians. They are different from both the stable European and African poles. This also applies to the Jurassic–Triassic poles from the south-Tethyan elements, plus Sardinia (Fig. 5). Poles from areas thought of as north-Tethyan in origin (Mecsek and Villány Mountains, Hungary) are again close to the stable European poles of corresponding age.

Palaeozoic palaeomagnetic poles are more scattered than the Mesozoic and younger poles. The scatter may be attributed to relative rotations of the Palaeozoic units within the Mesozoic mobile belt or to imperfect removal of overprinted magnetization, or to both. In spite of the poor grouping, poles from south-Tethyan elements, plus Sardinia (Figs 6, 2–9), and poles from north-Tethyan elements (Figs 6, 10–13) form different populations.

Fig. 7 summarizes the distribution of all palaeopoles from the Carpatho-Balkan and adjacent regions. Palaeogene, Mesozoic and, to a lesser extent, Palaeozoic poles, allow two distinctly different tectonic units to be recognized. Belonging to unit A (Fig. 8) are the Italian peninsula, south of the Po-basin, Sardinia, the Istria peninsula of Yugoslavia, the Transdanubian Central Mountains and Bükk Mountains of Hungary, and the outer and inner West Carpathians of Czechoslovakia. This unit is characterized by poles that are closer to the African than to the stable European palaeomagnetic poles.

Unit B is not well defined palaeomagneti-

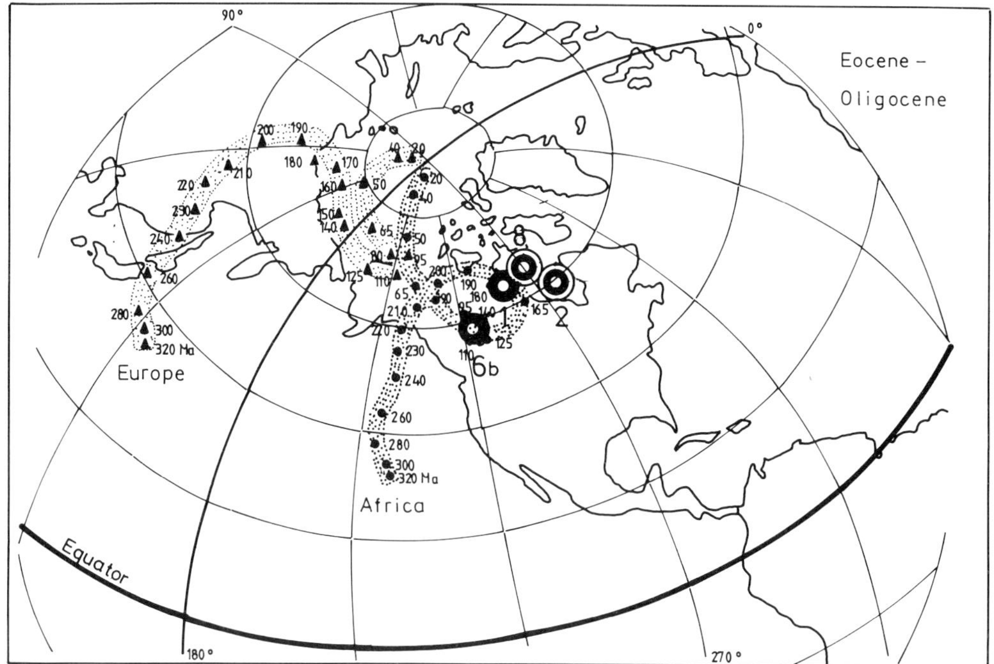

FIG. 2. Palaeogene poles from the Carpatho-Dinaric and adjacent regions. Polar path for Africa ('pseudo African') and Europe: both transferred from the North-American polar wander curve by Irving (1977) using the stage poles for the opening of the Atlantic of Sclater *et al.* (1977).

Poles from relative autochthons circled.

1, Umbria, Italy, south-Tethyan (VandenBerg, 1979); 2, Transdanubian Central Mountains, Hungary, south-Tethyan (Márton & Márton 1982); 6b, Outer West Carpathians, Czechoslovakia, north-Tethyan (Korab *et al.* 1981); and 8, Sardinia, Italy, north-Tethyan (Montigny *et al.* 1981).

cally. Further work may result in subdividing the area indicated on Fig. 8 into smaller units. Although scattered, the common feature of the poles from unit B is a stable European rather than African affinity.

The figures do not show palaeomagnetic poles from the Southern and Eastern Alps and Sicily which would unnecessarily complicate the picture.

Palaeopoles from the western and eastern parts of the Eastern Alps are very different due to the difference in declination resulting from nappe movement (Becke & Mauritsch 1982). The palaeo-directions are distorted to such a degree that they are useful only for describing relative rotations within the Eastern Alps.

Palaeopoles from Sicily and the Southern Alps do not indicate any significant relative movement with respect to Africa (VandenBerg & Wonders 1976, 1980; Schult 1973; Lowrie *et al.* 1980). Sicily can simply be assigned to the African plate. No satisfactory model for the African directions of the Southern Alps can be offered yet since unit A seems to separate the Southern Alps from Africa.

As Figs 2–5 show, contemporaneous palaeopoles from unit A are in good agreement. Moreover there are three areas within the unit where many stage poles of the Mesozoic are known: Umbria the Transdanubian Central Mts, and the Istria peninsula, the latter two being relatively autochthonous. From the stage poles detailed polar wander paths for each area can be constructed. Since coeval poles from all three areas are similar, it is possible to construct a unified curve (Fig. 9) which, as a first approximation, represents the movements of the whole unit A relative to the geomagnetic pole. The apparent polar movement of unit A is a 'loop' in the Mesozoic as a result of a counterclockwise declination rotation from the Late Triassic to the Tithonian (about 60°) and a rotation back to the present declination (about 100°). A change in palaeolatitude at the same time is less marked (Fig. 10). The 'loop' of the apparent polar path of unit A is similar to the 'loop' of the 'pseudo-African' polar wander curve of Fig. 9 (see fig. caption), but more elongated.

The analogy between the Mesozoic part of

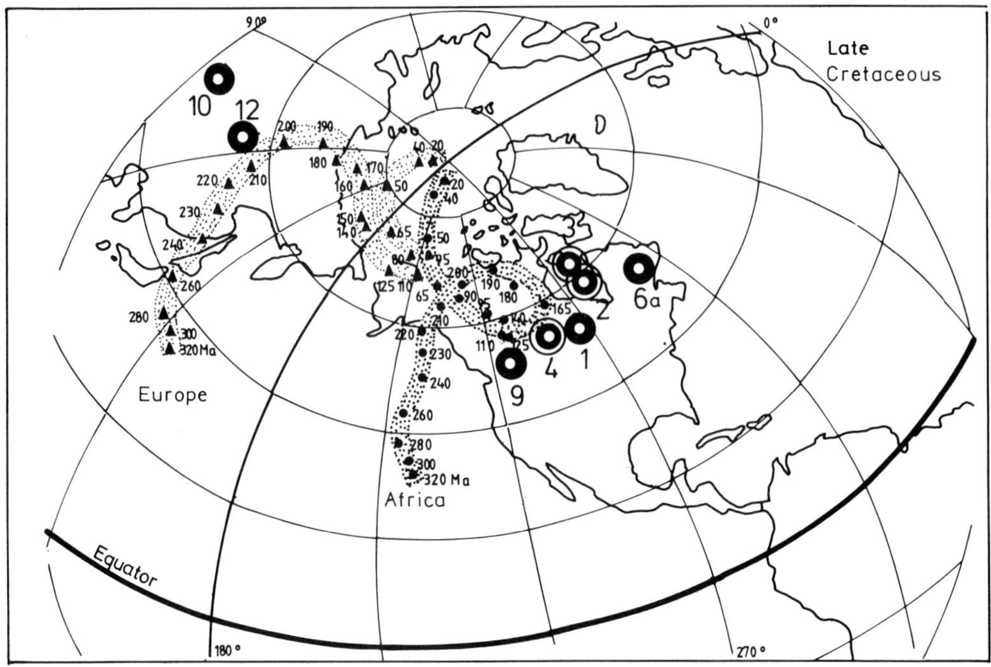

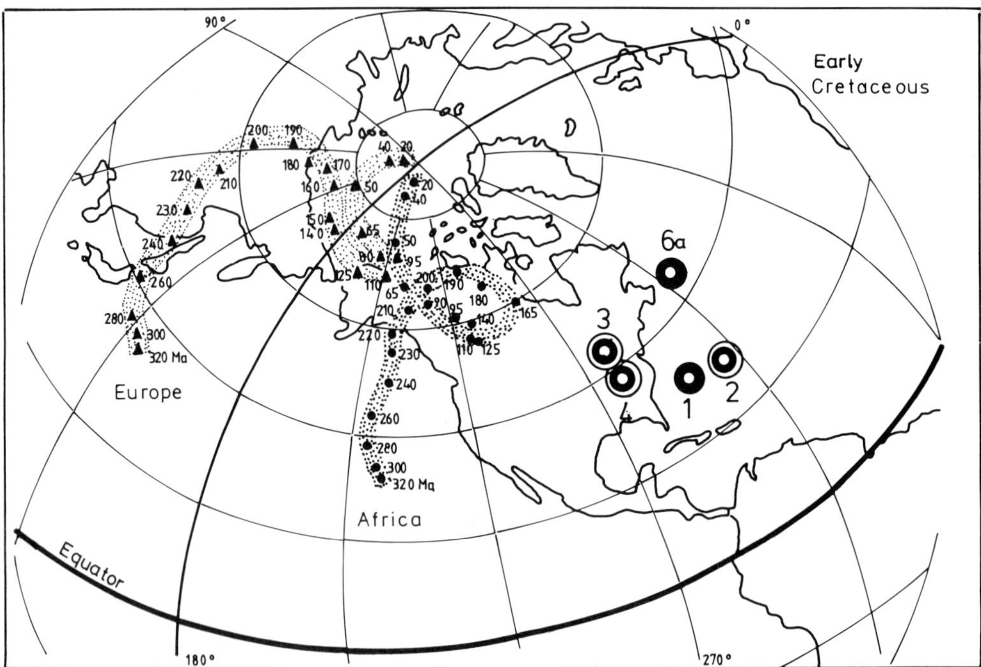

FIGS 3 & 4. Cretaceous poles from the Carpatho-Dinaric and adjacent regions. Polar path for Africa ('pseudo-African') and Europe; both transferred from the North American polar wander curve by Irving (1977) using the stage poles for the opening of the Atlantic of Scalter *et al.* (1977).

Poles from relative autochthonous areas circled.

1, Umbria, Italy, south-Tethyan (Channell *et al.* 1978; VandenBerg 1979; Lowrie *et al.* 1980); 2, Transdanubian Central Mountains, Hungary, south-Tethyan (Márton & Márton 1981, 1982); 3, Istria peninsula, Yugoslavia, south-Tethyan (Márton E. & Veljovic 1983); 4, Gargano, Italy, south-Tethyan (VandenBerg 1982); 6a, Outer West Carpathians, Czechoslovakia, north Tethyan (Krs *et al.* 1979); 9, Bükk Mountains, Hungary, south-Tethyan (Márton 1980b); 10, Mecsek Mountains, Hungary (Márton, 1980a) and 12, Papuk Mountains, Yugoslavia (unpublished, Márton & Vejorič), both north-Tethyan

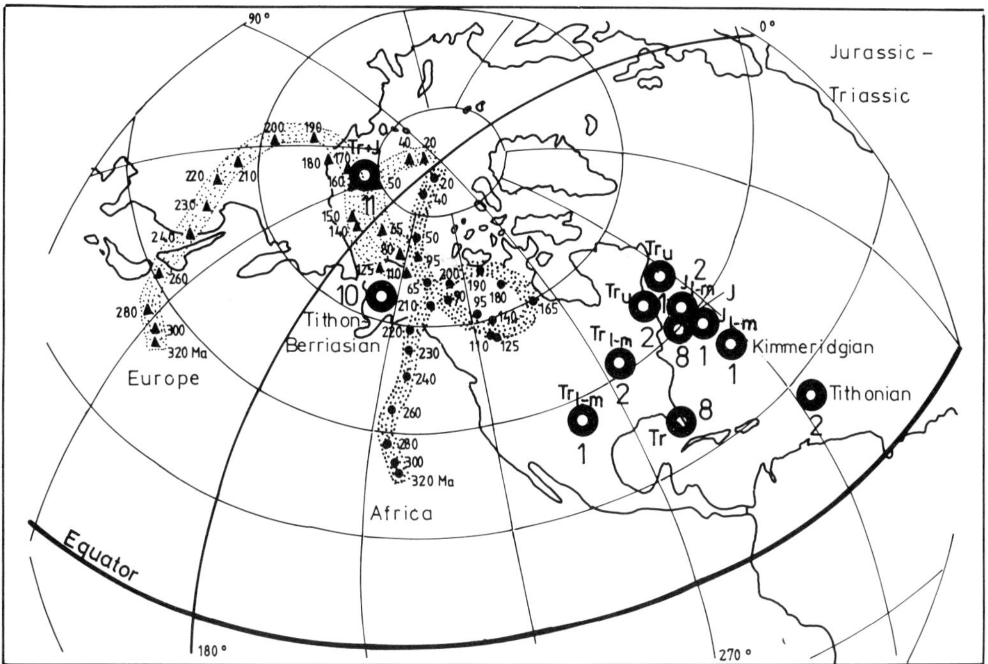

FIG. 5. Jurassic–Triassic poles from the Carpatho-Dinaric and adjacent regions. Polar path for Africa ('pseudo African') and Europe, both transferred from the North American polar wander curve by Irving (1977) using the stage poles for the opening of the Atlantic by Sclater *et al.* (1977).

Poles from relative autochthonous areas: 2, 8, 10, 11.

1, Umbria, Italy, south-Tethyan (VandenBerg 1979); 2, Transdanubian Central Mountains, Hungary, south-Tethyan (Márton & Márton 1981, 1982); 8, Sardinia, Italy, north-Tethyan (Horner & Lowrie 1981); 10, Mecsek Mountains, Hungary, north-Tethyan (Márton & Veljović unpublished); and 11, Villány Mountains, Hungary, north Tethyan (Márton & Márton 1978). Tr_{l-m}, Early–Middle Triassic; Tr_u, Late Triassic; J_{l-m}, Early–Middle Jurassic; J_u, Late Jurassic.

the African polar path and that of unit A is more significant if details from direct observations (Schult *et al.* 1981) are added to the 'pseudo-African loop' (Márton & Márton 1982).

The similarity of the polar wander curves establishes the African origin of unit A independently of any geological evidence. Although the polar wander paths are similar, a 35° clockwise rotation about a pole of rotation somewhere in the Mediterranean region is still required to bring the contemporaneous African poles and those of unit A into coincidence (Fig. 11). This means that unit A must have become decoupled from Africa and have rotated 35° counter-clockwise with respect to it. The decoupling must be definitely post-Oligocene, and also pre-mid-Miocene as it predates the 14–16 Ma volcanic activity in Hungary and Czechoslovakia. Montigny *et al.* (1981) suggest a very short duration for the rotation of Sardinia (20.5–19 Ma) and this may apply to the whole unit.

The African origin of the outer West Carpathians is less certain than that of the remainder of unit A as the units sampled are from a shorter time interval. The observed palaeopoles for this area could be equally well explained by counter-clockwise rotation of units originally from stable Europe, during nappe transport. It is interesting to note however that the poles from both the western and eastern parts of the Carpathians (Fig. 1, 6a & 6b) fit the unified polar wander curve of unit A equally well.

Reconstructing the movements of unit B from palaeomagnetic observations is more difficult and only a tentative model can be suggested. The 'stable European affinity' of the poles means simply that they lie close to the poles from stable Europe. There is an indication, however, of a post-Oligocene clockwise rotation from the Mecsek Mountains, Hungary (Márton 1980a). It is not impossible that the 'stable European affinity' of most palaeomagnetic poles from unit B is a coincidence, and the post-Jurassic rotation pattern of the unit is in

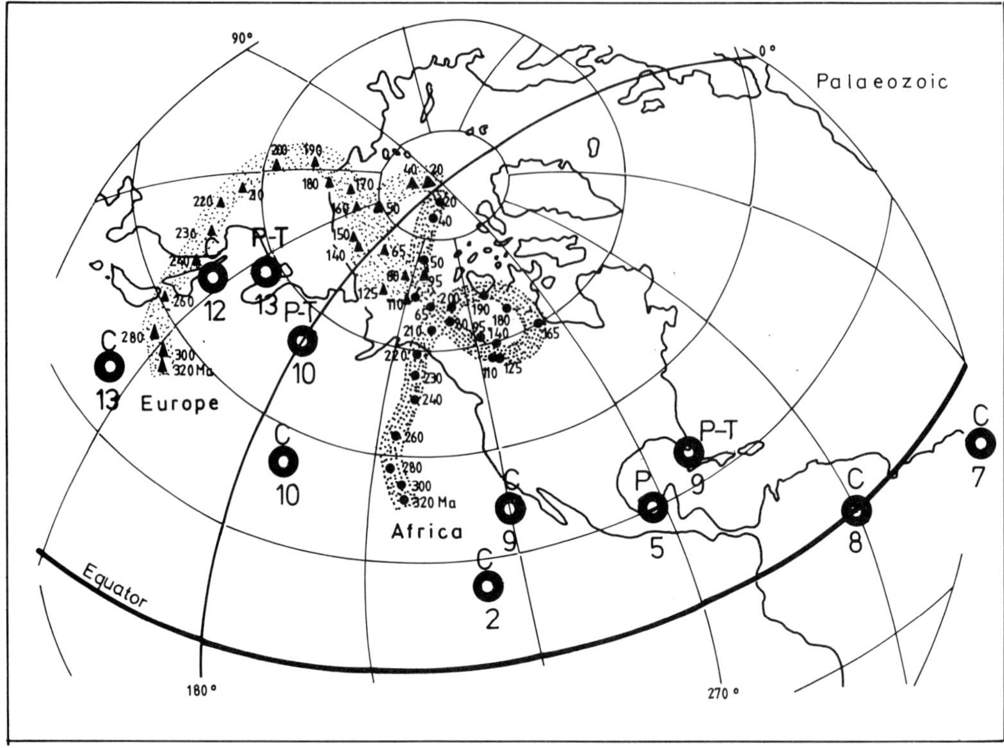

FIG. 6. Palaeozoic poles from the Carpatho-Dinaric and adjacent regions. Polar parth for Africa ('pseudo African') and Europe; both transferred from the North American polar wander curve by Irving (1977) using the stage poles for the opening of the Atlantic by Sclater *et al.* (1977).

1, Umbria, (Appenines) Italy, south-Tethyan (Zijderveld & De Jong 1969; VandenBerg 1979); 2, Transdanubian Central Mountains, Hungary, south-Tethyan (Márton unpublished); 5, Inner West Carpathians, Czechoslovakia, south-Tethyan (Kotasek *et al.* 1969); 7, Calabria, Italy, south-Tethyan (Manzoni 1979); 8, Sardinia, Italy, north-Tethyan (Edel *et al.* 1981); 9, Bükk Mountains, Hungary, south-Tethyan (Márton 1980b); 10, Mecsek Mountains, Hungary, north-Tethyan (Márton 1980a; Krs *et al.* 1969); 12, Papuk Mountains, Yugoslavia, north-Tethyan (Márton & Veljović unpublished); and 13, North-West Bulgaria, north-Tethyan (Nozarov *et al.* 1980).

fact a complicated one. Should the clockwise rotation of about 60° relative to stable Europe indicated so far by the palaeopoles of the Early Cretaceous volcanics from the Mecsek Mountains be confirmed, unit B would have to be rotated back accordingly. With this rotation the Jurassic and older poles from unit B would then be similar to the poles from unit A, and a southern Tethyan rather than northern Tethyan origin of unit B would then be supported.

Conclusions

The palaeomagnetic results accumulated rapidly over the last 3–4 years in the Carpatho-Dinaric, and to a lesser extent the Balkan, region support a unifying tectonic concept. Palaeomagnetic observations to date are sufficient to conclude that two tectonic units with basically different rotational histories exist in the region. Relative movements within either of the units can be described as superimposed on the fundamental pattern.

In order to elucidate the details, future work will have to focus on the construction of polar wander curves as complete and detailed as possible for each part of units A and B. Both units could be regarded as microplates. It is also true that they could be 'mega-nappes' on a huge scale but although palaeomagnetism cannot disprove this possibility it is unlikely that the observed consistency of the palaeomagnetic directions would have been preserved in such large-scale nappe transport.

If viewed as microplates, then the scope for *independent* movement is very limited in the case of unit A, as its polar wander path clearly indentifies it as part of the African plate until

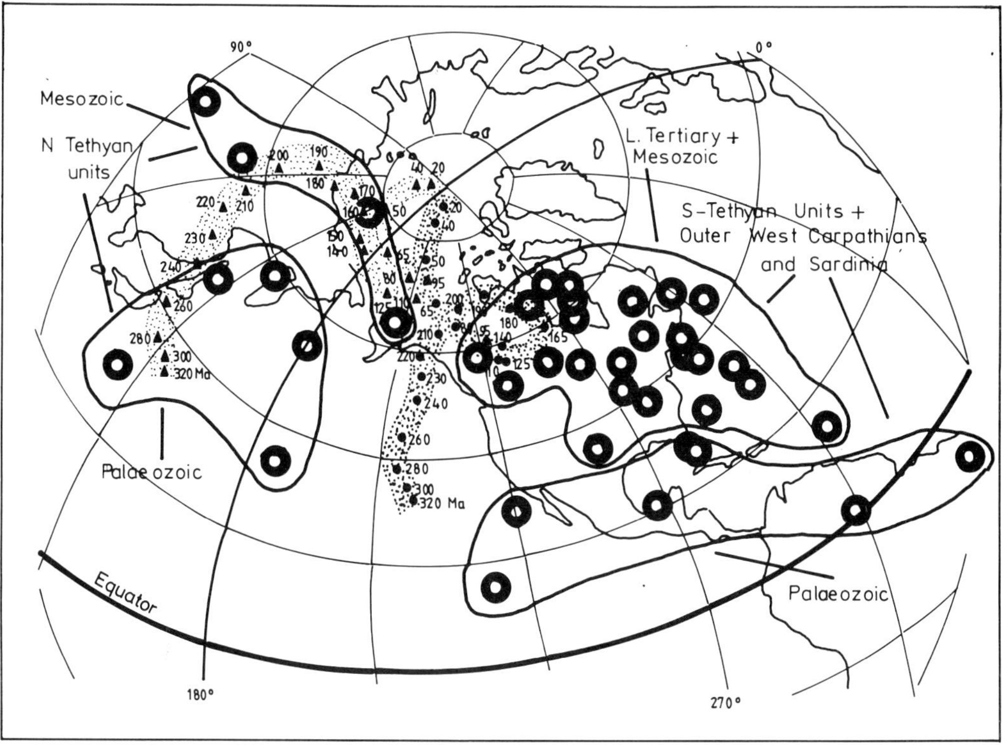

FIG. 7. Summary of the palaeopoles from the Carpatho-Dinaric and adjacent regions. Polar path for Africa ('pseudo African') and Europe: both transferred from the North American polar wander curve by Irving (1977) using the stage poles for the opening of the Atlantic of Sclater *et al*. (1977).

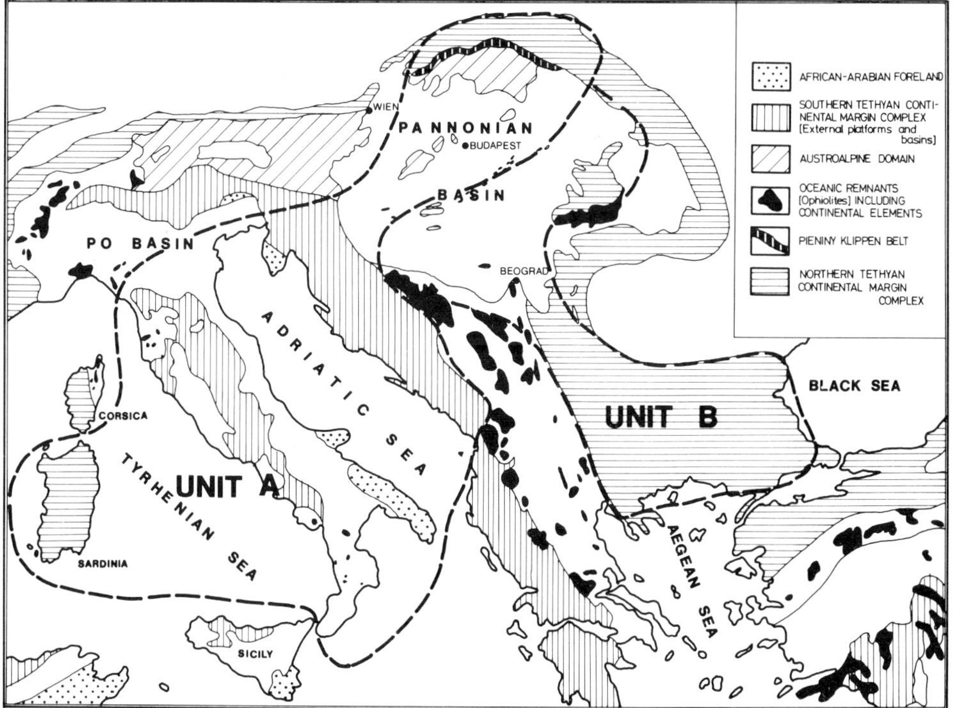

FIG. 8. The tentative outline of units with different rotational history in the Carpatho-Dinaric and adjacent regions.

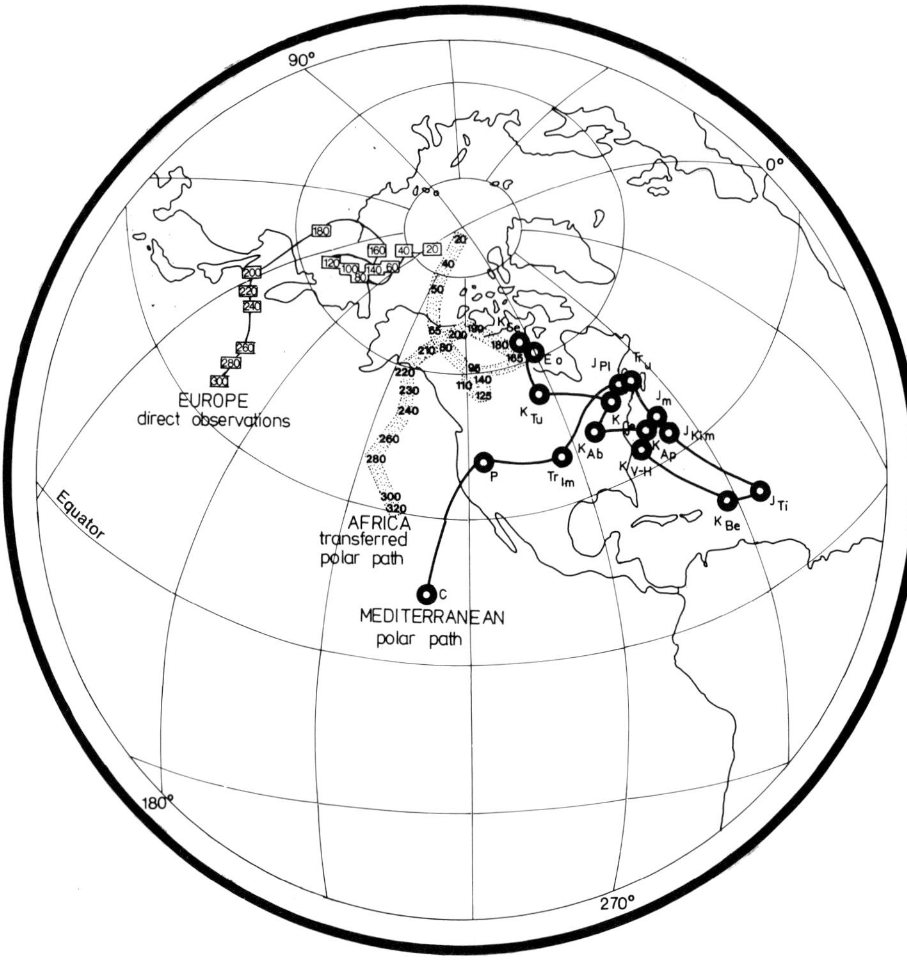

FIG. 9. Unified apparent polar wander path for unit A of the Carpatho-Dinaric and adjacent regions (Mediterranean polar path).

C, Carboniferous; P, Permian, Tr_{l-m}, Early–Middle Triassic; J_{Pl}, Pliensbachian; J_m, Middle Jurassic; J_{Kim}, Kimmeridgian; J_{Ti}, Tithonian; K_{Be}, Berriasian; K_{V-H}, Valanginian–Hauterivian; K_{Ap}, Aptian; K_{Ab}, Albian; K_{Ce}, Cenomanian; K_{Tu}, Turonian; K_{Se}, Senonian; and Eo Eocene.

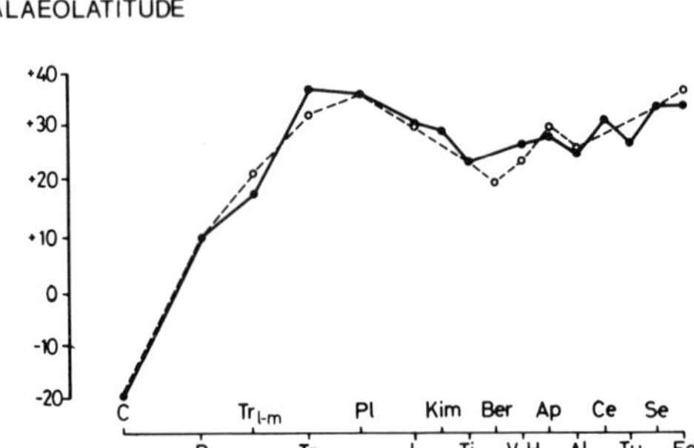

FIG. 10. The change in palaeolatitude from the Carboniferous to Eocene in the Carpatho-Adriatic region. Full circles apply to Umbria, Istria and the Transdanubian Central Mountains; open circles apply to the Transdanubian Central Mountains only. Reference point: Lat.: 47°, Long: 18°.

C, Carboniferous; P, Permian; Tr_{l-m}, Early–Middle Triassic; Tr_u, Late Triassic; Pl, Pliensbachian; J_m Middle Jurassic; Kim, Kimmeridgian; Ti, Tithonian; Ber, Berriasian; V–H, Valanginian–Hauterivian; Ap, Aptian; Al, Albian; Ce, Cenomanian; Tu, Turonian; Se, Senonian; and Eo, Eocene.

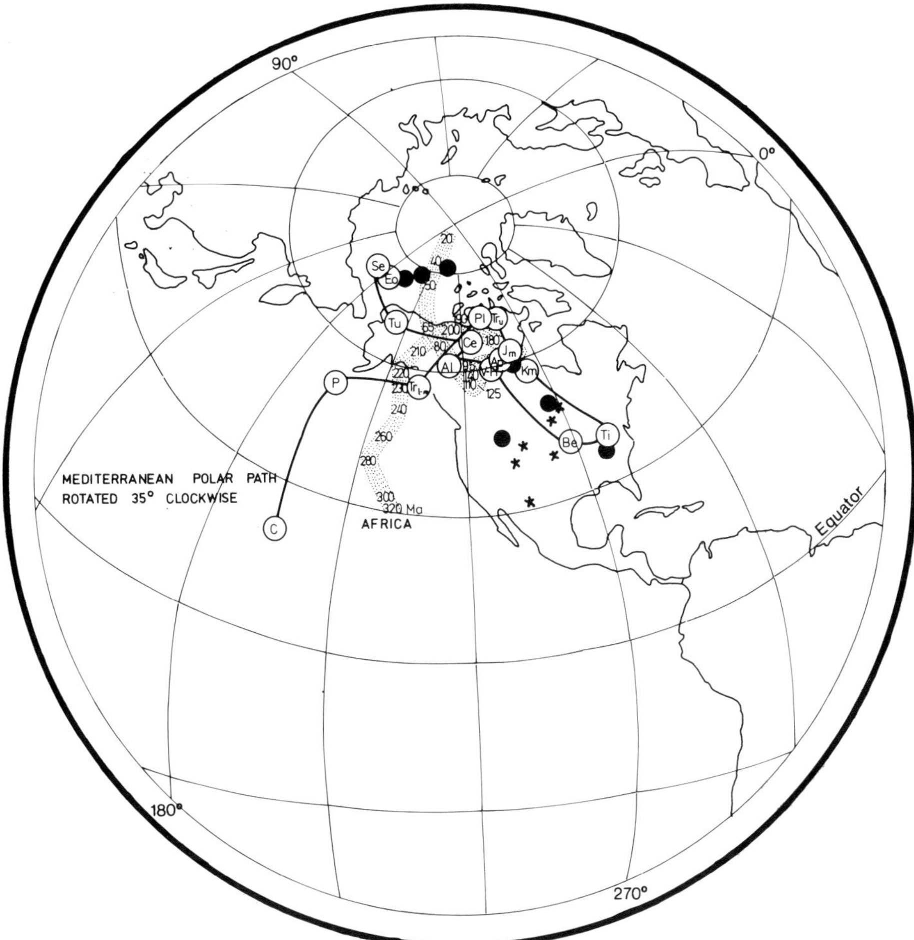

FIG. 11. The position of the unified Mediterranean polar path after rotating back unit A by 35° about a local rotation pole. Direct determination from Africa different from the 'pseudo-African' Mesozoic poles are shown by solid circles. Late Cretaceous poles are close to the 50–40 Ma part of the 'pseudo-African' polar path. Early Cretaceous poles close to the Tithonian Berriasian poles from Unit A and the Early Cretaceous South American poles (stars) transferred to Africa (Schult *et al.* 1981).

C, Carboniferous; P, Permian; Tr_{1-m}, Early–Middle Triassic; Pl, Pliensbachian; J_m, Middle Jurassic; Kim, Kimmeridgian; Ti, Tithonian; Be, Berriasian; V–H, Valanginian–Hauterivian; Ap, Aptian; Ab, Albian; Ce, Cenomanian; Tu, Turonian; Se, Senonian; and Eo, Eocene.

the end of the Oligocene. Its rotation was then completed with respect to the major plates before 14–16 Ma (mid-Miocene). Unit B is much less well constrained and it is impossible to define a time interval within which it could have moved as an independent microplate, or group of microplates.

ACKNOWLEDGEMENTS: The author is indebted to Drs S. Hall and D. H. Tarling and to the editors Drs A. H. F. Robertson and J. E. Dixon for providing critical reviews of the manuscript and helpful suggestions. Thanks are due to A. Lugosi and K. Beslényi for technical assistance.

References

BECKE, M. & MAURITSCH, H. J. 1982. Palaeomagnetic investigation in Upper Cretaceous sediments in the Northern Calcareous Alps. Abstract for EGS meeting, Leeds, 73.

CHANNELL, J. E. T., LOWRIE, W., MEDIZZA, F. & ALVAREZ, W. 1978. Palaeomagnetism and tectonics in Umbria, Italy. *Earth planet. Sci. Lett.* **39,** 199–210.

EDEL, J. B., MONTIGNY, R. & THUIZAT, R. 1981. Late Palaeozoic rotations of Corsica and Sardinia: new evidence from palaeomagnetic and K-Ar studies. *Tectonophysics*. **79,** 201–223.

GÉCZY, B. 1973. The origin of the Jurassic faunal provinces and the Mediterranean plate tectonics. *Ann. Sci. Univ. Rolando Eötvös, Sec. Geol.* **16,** 99–114.

HORNER, F. & LOWRIE, W. 1981. Palaeomagnetic evidence from Mesozoic carbonate rocks for the rotation of Sardinia. *J. Geophys.* **49,** 15–18.

IRVING, E. 1977. Drift of the major continental blocks since the Devonian. *Nature*. **270,** 304–309.

KORAB, T., KRS, M., KRSOVA, M. PAGAC, P. 1981. *Palaeomagnetic Investigations of Albian (?)—Palaeocene to Lower Oligocene sediments from the Dukla Unit, East Slovakian Flysch, Czechoslovakia.* Zapadne Karpaty. Sér. geológia 7, Geol. Ust. D. Stura, Bratislava. 127–149.

KOTASEK, J. KRS, M. & JÁMBOR, A. 1969. Palaeomagnetic studies of the Permian in the Pannonian Basin (in German). *Geofizikai Közlemények*. **18,** 43–56.

KRS, M., MUSKA, P., ORLICKY, O. & PAGÁC, P. 1979. Palaeomagnetic investigations in the West Carpathians. *Geodynamic investigations in Czechoslovakia*. Veda. Bratislava. 207–214.

LAUBSCHER, H. & BERNOULLI, D. 1977. Mediterranean and Tethys. *In*: *The Ocean Basins and their Margins: The Eastern Mediterranean*. Plenum Press, New York, 1–22.

LOWRIE, W., CHANNELL, J. E. T. & ALVAREZ, W. 1980. A review of magnetic stratigraphy investigations in Cretaceous pelagic carbonate rocks. *J. Geophys. Res.* **58,** 3597–3605.

MANZONI, M. 1979. Palaeomagnetic evidence for non-Apenninic origin of the Silia Nappes (Calabria). *Tectonophysics*. **60,** 169–188.

MÁRTON, E. & MÁRTON, P. 1978. Difference between the palaeopoles of the Transdanubian Central Mountains and the Villány Mountains (in Hungarian). *Magyar Geofizika*. **19,** 129–136.

—— 1980a. Multicomponent remanent magnetization of the migmatites, Mórágy area SW Hungary. *Earth planet Sci. Lett.* **47,** 102–112.

—— 1980b. Palaeomagnetism of Upper Palaeozoic and Triassic sediments from the Bükk Mts, Hungary. Abstract for EGS meeting, Budapest, 126.

—— & MÁRTON, P. 1981. Mesozoic palaeomagnetism of the Transdanubian Central Mountains and its tectonic implication. *Tectonophysics*. **72,** 129–140.

—— & MÁRTON, P. 1982. A refined Mesozoic polar wander path of the Transdanubian Central Mountains and its bearing on the tectonic history of the Mediterranean. Abstrct for EGS meeting, Leeds, 73.

—— & VELJOVIC, D. 1983. Palaeomagnetism of the Istria peninsula, Yugoslavia. *Tectonophysics*, **90,**

MONTIGNY, R., EDEL, J. B. & THUIZAT, R. 1981. Oligo-Miocene rotation of Sardinia: K-Ar ages and palaeomagnetic data of Tertiary volcanics. *Earth planet. Sci. Lett.* **54,** 261–271.

NOZHAROV, P., PETKOV, N., YANEV, S., KROPACEK, V., KRS, M. & PRUNER, P. 1980. A palaeomagnetic and petromagnetic study of Upper Carboniferous, Permian and Triassic sediments, NW Bulgaria. *Studia geoph. et geod.* **34,** 254–283.

SCHULT, A. 1973. Palaeomagnetism of Upper Cretaceous volcanic rocks in Sicily. *Earth planet. Sci. Lett.* **19,** 97–100.

——, HUSSAIN, A. G. & SOFFEL, H. C. 1981. Palaeomagnetism of Upper Cretaceous volcanics and Nubian Sandstones of Wadi Natash, SE-Egypt and implications on the polar wander path for Africa in the Mesozoic. *J. Geophys.* **50,** 16–22.

SCLATER, J. G., HELLINGER, S. & TAPSCOTT, CH. 1977. The palaeobathymetry of the Atlantic ocean from the Jurassic to the present. *J. Geol.* **85,** 509–552.

SZEPESHÁZY, K. 1977. Mesozoic igneous rocks of Great Hungarian Plain (in Hungarian). *Földtani Közlemények*. **107,** 384–397.

VANDENBERG, J. & WONDERS, A. A. H. 1976. Palaeomagnetic evidence of large fault displacement around the Po-basin. *Tectonophysics*. **33,** 301–320.

—— 1979. Palaeomagnetism and the changing configuration of the Western Mediterranean area in the Mesozoic and Early Cenozoic eras. *Geologica Ultraiectina, Utrecht*, **30,**

—— & WONDERS, A. A. H. 1980. Palaeomagnetism of late Mesozoic pelagic limestones from the Southern Alps. *J. Geophys. Res.* **85,** 3623–3627.

—— 1982. Reappraisal of palaeomagnetic data from Gargano (Apulia–Italy). Abstract for EGS meeting, Leeds, 73.

ZIJDERVELD, J. D. A. & DE. JONG. K. A. 1969. Palaeomagnetism of some Late Paleozoic and Triassic rocks from eastern Lombardic Alps, Italy. *Geol. Mijnbouw*. **48,** 559–564.

E. MÁRTON, Eötvös Loránd Geophysical Institute of Hungary, Columbus u. 17–23, Budapest, XIV, Hungary.

5. NEOGENE

By the Miocene most of the Eastern Mediterranean area was experiencing the increasingly pervasive effects of diachronous collision of Africa and Eurasia. Previously useful distinctions between continent and ocean break down as each individual area begins to assume its own tectonic identity (Figs 1 & 2).

For an overall introduction to the varied Mediterranean Neogene palaeoenvironments, the reader would do well to start with **STEININGER & ROEGL**'s presentation and discussion of eight maps which include the Black Sea and Caspian areas. During the Oligocene Africa and Eurasia were still separated by wide seas. Mammal migration became possible extensively for the first time in the Miocene. Around 18–17 Ma the Para-tethys was isolated as a huge land-locked brackish sea. Then, in the Langian, a major transgression again separated Africa and Eurasia and also flooded the Paratethys. Later in the Middle Miocene (15–14.5 Ma), important mammal exchanges indicate that Africa was connected to an Aegean landmass. The eastern Para-tethys was then again isolated as a low-salinity sea until cosmopolitan marine biotas were restored around 14 Ma. During the Late Miocene the well-known, but still controversial, draw-down of the Mediterranean Sea took place, while the Para-tethys was split into smaller basins showing considerable endemism. In the Pliocene the Mediterranean once again was flooded and a narrow link is inferred with the Indian Ocean through the Red Sea.

In an east–west trending orogenic belt, like the Eastern Mediterranean, palaeomagnetism really comes into its own in the collisional phases where rotation of micro-continental blocks may become a dominant feature. An excellent target for such study is the origin of the curved Aegean arc: was it an 'ancient' palaeogeographical feature or was it formed instead by tectonic 'bending' of previously linear belts, and if so when? In their study **KISSEL *et al.***, stress the need for careful and extensive sampling, noting that were their cores to be used for conventional palaeolatitude determination then errors would arise as the inclinations are typically 10–20° lower than required for Neogene Aegean sedimentary rocks. Summarizing earlier results, they go on to confirm that the SE part of the Aegan arc has not rotated significantly since the Tortonian, while the NW part from Corfu to the Peloponnesus, has undergone major anticlockwise rotation since the Mid-Tertiary. Consistent with **POISSON**'s evidence from unstretching the Lycian nappes in SW Turkey, they conclude that the Aegean arc was essentially E–W trending prior to the Oligocene. Also discussed is the timing of the inferred rotations in relation to Aegean calc-alkaline volcanism and subduction (**FYTIKAS**, see below).

From inclination and declination data from five widely spaced sites on the Greek mainland, Aegean islands and mainland Turkey, **KONDOPOULOU & LAUER** suggest that the entire area has significantly rotated clockwise during the Neogene. Notably, tilt corrections could not be made in the subaerial lava units sampled.

FYTIKAS *et al.* summarize the Tertiary, predominantly calk-alkaline, volcanism of the Aegean. From Oligocene to Miocene, after the main phase of collision to the north, northward subduction built-up a calc-alkaline volcanic arc on thick crust. Southward migration of volcanism and increase in K_2O content is then attributed to progressive steepening of the subduction zone. Renewed northward subduction in the Pliocene produced a narrow band of mostly calc-alkaline volcanism in the south Aegean, while scattered Miocene to Quaternary volcanism can be related to broad zones of extensional strain along the eastern and northern margins of an 'Aegean micro-plate'. Certain rhyolites on the island of Antiparos are by contrast attributed to sinking and anatexis of a slab left over from earlier subduction, thus

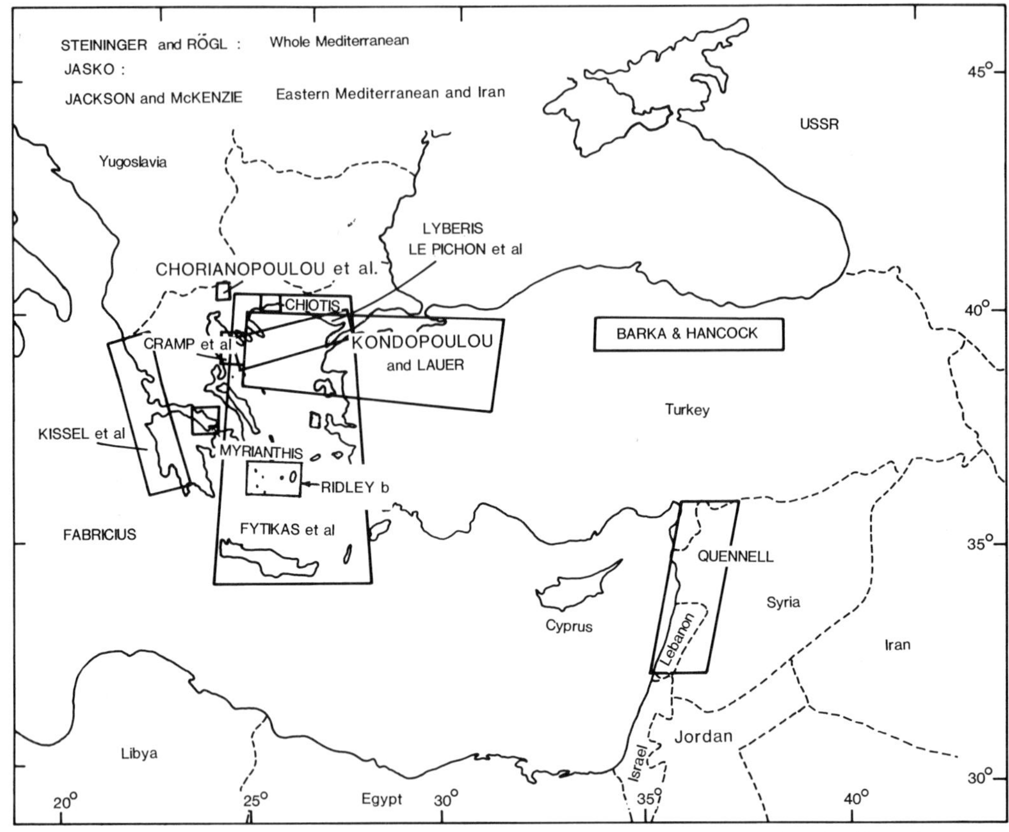

FIG. 1. The areas of the Eastern Mediterranean Neogene discussed by various authors in this section.

again stressing the problem of being sure any given calc-alkaline volcanism relates to coeval subduction.

Further evidence for Aegean graben formations is produced by **MYRIANTHIS**, who argues from two multichannel seismic reflection profiles that the Gulf of Corinth formed by normal faulting and subsidence prior to Pliocene time, accompanied uplift of the Peloponnesus block to the south. Movement continued along active normal faults which display considerable seismicity. Possible models of crustal thinning of the Corinth graben crust are also discussed.

From both regional and theoretical studies the Neogene subsidence and extension of the Aegean Sea behind the Cretan arc is well established, but here **LYBERIS** now presents a detailed analysis, based on combined geological and geophysical data, of a critical area, the North Aegean Graben system which connects eastwards with the North Anatolian Fault (see **BARKA & HANCOCK**, below). The author argues that this area has experienced crustal extension coupled with strike-slip since the late Miocene. The transition to the North Anatolian fault system is now located near the eastern part of the Marmara Sea, but earlier seems to have been further north. Notably, there is no obvious extension of the lateral motion along the North Aegean graben system into mainland Greece, which may put some limits on the total displacement along the North Anatolian fault system from Miocene onwards.

Drawing on the new evidence from the North Aegean graben specifically, in a linked contribution, **LE PICHON *et al.***, then go on to examine quantitively the subsidence history of this area with an emphasis on thermal and gravity constraints. Most of the crustal stretching, estimated at 3.5×, takes place within a surprisingly narrow, less than 20 km wide zone, which may reflect inherited basement control. An important general finding is that in such a narrow attenuated rift zone the effects of heat conduction are sufficient to suppress large-scale mantle melting and thus prevent onset of sea floor spreading.

Land-based studies too play their role in unravelling collisional orogens, but here several cautionary tales are in order. **JACKSON & McKENZIE** draw on evidence mostly from the Aegean and Iran to emphasize the importance of vertical and horizontal rotations in the evolution of geological structures. Unless the motion on master faults is known then strain fields inferred by classical field analysis of small or medium-scale structures may yield misleading results. Particularly hazardous are attempts to unravel the meaning of earlier structures in areas like much of the Eastern Mediterranean which continue to be tectonically active.

As a specific illustration of this problem, the Aegean is now known to have experienced overall Neogene crustal extension, yet a number of enigmatic thrust events had been reported which seem to imply that extension was punctuated by drastic compression for which no driving force is known. Now from evidence from the South Aegean island of Syros, **RIDLEY** (b) shows that many of the post-metamorphic faults are demonstrably of listric normal type and thus may actually signify periods of most intense crustal extension! A further implication is that peripheral units of weakly, or non-metamorphic rocks, around the edges of many of the islands can be interpreted as down-faulted remnants of higher levels of a structural pile emplaced earlier in the Tertiary, rather than as isolated remnants of a far-travelled later Tertiary nappe for which there is no other evidence.

Perhaps the expected mode in an area of head-on continental collision would be large-scale folding, thrusting and regional metamorphism, aspects which are indeed most important and have been touched on in several contributions to Sections 3 and 4: for example, **OKAY** on the NW Turkish blueschists, **HELVACI & GRIFFIN**, and **AKTAŞ & ROBERTSON** for the SE Turkish Bitlis metamorphic massif and adjacent areas, and **POISSON** for SW Turkey and possible correlations across the Aegean. Palaeogeographical irregularities exert crucial influence on structures in collisional zones. Thus as the Arabian 'salient' has ploughed into Eurasia since the Miocene, huge tracts have been expelled both east and west giving rise to major transform fault systems, especially the North Anatolian Fault, which is documented here by **BARKA & HANCOCK**. By a combination of structural and sedimentological analysis, they show that the fault originated in a broad shear zone which was later transformed to a narrow fault belt in the time interval from Late Miocene to Early Miocene. Structures in the Pontus Formation show that the main active fault strand is discontinuous with subordinate *en echelon* faults which contributed to an estimated 25 km of right-lateral slip since the end of Miocene. Westwards the North Anatolian transform fault terminates in the North Aegean graben, discussed in detail by **LYBERIS** and **LE PICHON *et al.*** (see above).

It is already appreciated that the collision of the Arabian salient with Eurasia in SE Turkey was accompanied by instigation of the major left-lateral Dead Sea transform fault in turn connected with opening of the Red Sea. Now based largely on theoretical structural considerations and geophysical data, **QUENNELL** argues that the Dead Sea transform fault system, proper, only extends to the head of the Jordan Valley. Beyond this there is first an important belt of folding, faulting and underthrusting (Palmyra fold belt), followed then by a further independent sinistral strike-slip zone connecting with the SE Turkish Border belt. Rather than treating, as traditionally, the whole of the Levant as part of the Arabian sub-plate, **QUENNELL** now argues for the existence of separate narrow Sinai–Lebanon micro-plate.

At a time of great emphasis of the formation of sedimentary basins by crustal-stretching, it is worth recalling the great diversity of sedimentary basin types in the East Mediterranean, a pointer perhaps to diverse origins. Based on a literature review, **JASCO** recognizes four main types of sedimentary basin in the area. His, 'cauldron subsidence', including the modern Mediterranean and Black Sea, forms rapidly, often independently of earlier tectonic trends. The second type 'orogenic foredeeps and intramontane troughs' form either at the margins or in the middle of young orogenic belts (e.g. Middle Danube basins). The third type is the classical tectonic graben (e.g. Dead Sea) formed by simple crustal extension; the fourth being the 'epeirogenetic depression', comprising broad shallow basins unrelated to active tectonics (e.g. North Caspian Plain).

Economically most important in sedimentary basin genesis is the accumulation of organic-rich sediments which, after burial mature to form hydrocarbons. In this context palaeobotanical studies have the considerable and, by no means widely appreciated potential, of documenting specific past climates, being one of the few means of estimating actual altitudes of sediment accumulation. In this volume **CHORIANOPOLOU *et al.***, discuss how the Pliocene Almopias calc-alkaline volcanics in NW Greece dammed a river to produce a lake into which hydrothermal solutions were released, they believe, favouring formation of chert, limestone and Fe-minerals. Study of the carbonized vegetal material then shows that the lake shore was densely wooded in a semitropical climate. Later the climate became more arid before the lake was buried by renewed volcanism.

For comprehensive discussion of the modern Mediterranean sedimentary basins, the interested reader is referred to a recent compilation edited by F. C. Wezel (1982). One finding since then, reported in this volume by **CRAMP *et al.***, is the discovery of sapropelic layers probably of Early Holocene age in the NW Aegean Sea. Preliminary micro-palaeontological data are also given. Correlation can possibly be made with the regional S1 sapropel elsewhere in the Eastern Mediterranean which is ca. 7500–9000 yrs BP.

Illustrative of the application of vitrinite reflectivity in determining temperatures of burial, **CHIOTIS** reports that coal samples from oil exploration wells in Northern Greece indicate heating to more than 200°C at inferred depths of less than 2000 m and this is attributed to nearby Miocene plutonism.

Perhaps indicating the ultimate frustration of surface geology or even seismic reflection studies as guides to deep crustal processes, it must come as a surprise to many that, despite its collisional status, much of the East Mediterranean sea area has been progressively subsiding at an average rate of 1 m/Ma since the Late Miocene. **FABRICIUS** charts this basin evolution beginning by summarizing the distribution of the Messinian evaporites in the Ionian Sea area; these he considers to have formed in shallow rather than pre-existing deep basins. Marginal areas around the subsiding zone tend to be marked by coastal terraces which occur with increasing age at heights up to 150 m, far in excess of Pleistocene eustatic effects, but which are seen as the isostatic response to localized subsidence. As the ultimate explanation **FABRICIUS** points to possible mantle diapirism, which he believes may characterize the later stages of 'Mediterranean-type' collisional orogeny. With the sheer range and complexity of the Neogene East Mediterranean collisional belt, the reader may like to reflect on the chances of disentangling and correctly interpreting the tectonic evolution of older collisional orogens, mostly metamorphosed and exposed now at deep erosional levels.

Reference

WEZEL, F. C. (ed.). *The sedimentary basins of Mediterranean margins*, Proc. C.N.S. Internat. Conf. on Sedimentary Basins of Mediterranean, Urbino University, Italy, Oct. 20–22, 1980, Technoprint, Italy.

Paleogeography and palinspastic reconstruction of the Neogene of the Mediterranean and Paratethys

Fritz F. Steininger & Fred Rögl

SUMMARY: Eight palaeogeographic-palinspastic sketches are presented for intervals from Late Oligocene to Pliocene times for the circum-Mediterranean area and the Paratethys. They are based primarily on accurate correlation of marine and continental biogeographic events and take into account the generally accepted ideas of plate tectonics. The marine and continental connections that can be inferred contribute important information on the geological evolution of the Eastern Mediterranean.

During Cenozoic time the Tethyan realm was separated into areas of contrasting sedimentation. Between the Eurasian and African plates the Mediterranean and the Paratethys sea was strongly influenced by changing connections with the Atlantic and Indo–Pacific Oceans. In Neogene time the Eastern Mediterranean and its adjacent regions became both the most important marine gateway to the Indo–Pacific and the most significant land-bridge for mammal migrations between Africa and Eurasia. Recognition of this has inspired numerous attempts at improving existing palinspastic reconstructions (e.g. Biju-Duval *et al.* 1976; Boccaletti 1979; Dewey *et al.* 1973). The approach presented here involves palinspastic reconstructions for Late Oligocene to Pliocene times, and is based on the following:

(1) A more accurate correlation of both marine and continental biostratigraphies with the stage systems in use in these regions. A decisive step in this direction taken recently by the participants in the 'Regional Committee on Mediterranean Neogene Stratigraphy' and IGCP-Project No. 25: 'Neogene Tethys-Paratethys correlations' is summarized in Fig. 1 (Seneš, 1978, 1979; Steininger & Papp 1979; Steininger *et al.* 1976). The improved correlations allow better dating and correlation of numerous major kinematic and biogeographic events in the circum-Mediterranean Neogene basins.

(2) Transgressive and regressive cycles can now be more accurately defined for the Mediterranean and Paratethys and compared with world-wide cycles. (Vail *et al.* 1977; Rögl & Steininger 1983).

(3) The facies changes of the Paratethys in space and time correlate well with the development of marine gateways to the Mediterranean and the Indo–Pacific (Rögl & Steininger 1983). These facies changes and the transgressive and regressive cycles can be readily correlated with major tectonic phases in the Alpine orogenic system (Schwan 1980).

(4) The most important biogeographic 'turnovers' in the marine and continental biotas can now be traced throughout the entire area (Steininger & Rögl 1979).

(5) The 'Correlation tables of the Neogene in Mediterranean and Paratethys' (Seneš *et al.* 1978), worked out by the IGCP-Project No. 25, enable the construction of sediment and facies distribution maps in time-slices for the entire Mediterranean area.

Late Oligocene—Early Miocene (Fig. 2)

25.0–23.0 Ma Chattian/Aquitanian–Egerian–Caucasian

A wide sea between the Afro–Arabian and Eurasian–Turkish plate systems with carbonate platforms on either side provided access from the Indo–Pacific ocean to the Eastern Mediterranean and the Eastern and Central Paratethys.

This Indo–Pacific–Atlantic seaway across the Mediterranean re-established a pronounced circum-equatorial current system, followed by major biological events: marked Late Oligocene warming trends (Haq *et al.* 1977; Hochuli, 1978); a massive appearance of larger foraminifera throughout the entire Mediterranean and Paratethys realms (Adams 1976; Drooger, 1979; McGowran 1979); the reduction of endemics within the mollusc faunas of the Paratethys (Baldi 1980). Finally, this seaway effectively hindered an African–Eurasian mammal exchange during Oligocene times (Ginsburg 1979).

The Aegean plate and the rising Pindic Cordillera were separated by deep-water troughs with flyschoid and molasse sedimentation from the mainland of modern Greece. The Pindic Cordillera and the Pelagonian zone fed vast volumes of clastic materials into the

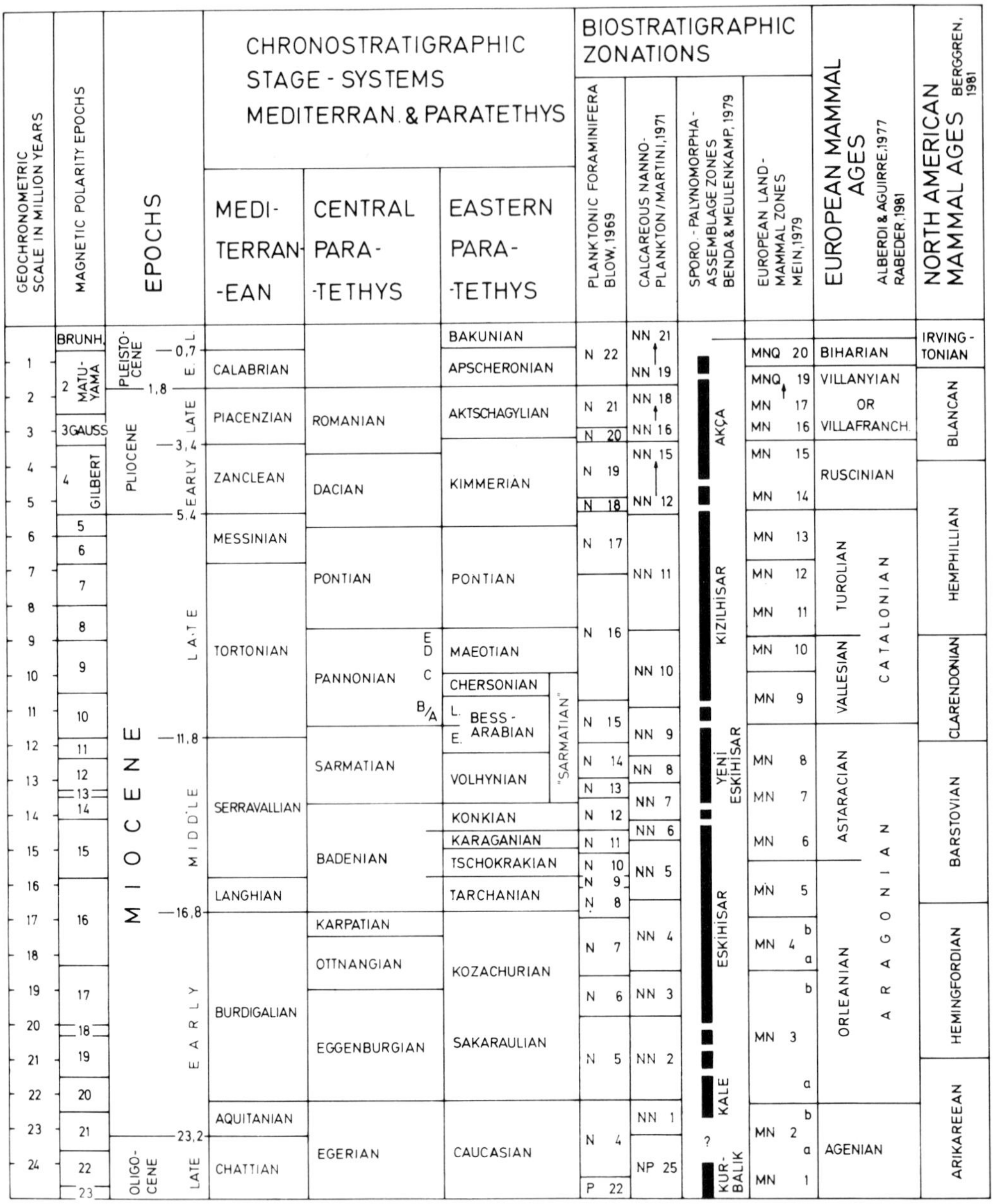

Column	Units (youngest to oldest, as printed top to bottom)
GEOCHRONOMETRIC SCALE IN MILLION YEARS	1, 2, 3, 4, 5, 6, 7, 8, 9, 10, 11, 12, 13, 14, 15, 16, 17, 18, 19, 20, 21, 22, 23, 24
MAGNETIC POLARITY EPOCHS	BRUNH.; 2 MATU-YAMA; 3 GAUSS; 4 GILBERT; 5; 6; 7; 8; 9; 10; 11; 12; 13; 14; 15; 16; 17; 18; 19; 20; 21; 22; 23
EPOCHS	PLEISTOCENE (L., E.) — 0,7 — 1,8; PLIOCENE (LATE, EARLY) — 3,4 — 5,4; MIOCENE (LATE — 11,8; MIDDLE — 16,8; EARLY) — 23,2; OLIGOCENE (LATE)
CHRONOSTRATIGRAPHIC STAGE-SYSTEMS MEDITERRAN. & PARATETHYS: MEDITERRANEAN	CALABRIAN; PIACENZIAN; ZANCLEAN; MESSINIAN; TORTONIAN; SERRAVALLIAN; LANGHIAN; BURDIGALIAN; AQUITANIAN; CHATTIAN
CENTRAL PARATETHYS	ROMANIAN; DACIAN; PONTIAN; PANNONIAN (E, D, C, B/A); SARMATIAN; BADENIAN; KARPATIAN; OTTNANGIAN; EGGENBURGIAN; EGERIAN
EASTERN PARATETHYS	BAKUNIAN; APSCHERONIAN; AKTSCHAGYLIAN; KIMMERIAN; PONTIAN; MAEOTIAN; "SARMATIAN" (CHERSONIAN; BESSARABIAN L., E.; VOLHYNIAN); KONKIAN; KARAGANIAN; TSCHOKRAKIAN; TARCHANIAN; KOZACHURIAN; SAKARAULIAN; CAUCASIAN
BIOSTRATIGRAPHIC ZONATIONS: PLANKTONIC FORAMINIFERA BLOW, 1969	N 22; N 21; N 20; N 19; N 18; N 17; N 16; N 15; N 14; N 13; N 12; N 11; N 10; N 9; N 8; N 7; N 6; N 5; N 4; P 22
CALCAREOUS NANNOPLANKTON/MARTINI, 1971	NN 21 – NN 19; NN 18 – NN 16; NN 15 – NN 12; NN 11; NN 10; NN 9; NN 8; NN 7; NN 6; NN 5; NN 4; NN 3; NN 2; NN 1; NP 25
SPORO-PALYNOMORPHA-ASSEMBLAGE ZONES BENDA & MEULENKAMP, 1979	AKÇA; KIZILHİSAR; YENİ ESKİHİSAR; ESKİHİSAR; KALE; ?; KURBALIK
EUROPEAN LAND-MAMMAL ZONES MEIN, 1979	MNQ 20; MNQ 19 – MN 17; MN 16; MN 15; MN 14; MN 13; MN 12; MN 11; MN 10; MN 9; MN 8; MN 7; MN 6; MN 5; MN 4 b, a; MN 3 b, a; MN 2 b, a; MN 1
EUROPEAN MAMMAL AGES ALBERDI & AGUIRRE, 1977 RABEDER, 1981	BIHARIAN; VILLANYIAN OR VILLAFRANCH.; RUSCINIAN; TUROLIAN; VALLESIAN; ASTARACIAN; ORLEANIAN; AGENIAN; (CATALONIAN; ARAGONIAN)
NORTH AMERICAN MAMMAL AGES BERGGREN, 1981	IRVINGTONIAN; BLANCAN; HEMPHILLIAN; CLARENDONIAN; BARSTOVIAN; HEMINGFORDIAN; ARIKAREEAN

Fig. 1. Correlation chart of Neogene biostratigraphic and chronostratigraphic units for the Mediterranean, the Central and Eastern Paratethys and the Indo-Pacific.

'Mesohellenic–Molasse-Trough' (Dermitzakis & Papanikolaou 1981). Molasse and flyschoid sedimentation continued all along the 'northern coastlines' of the Late Oligocene Mediterranean from southern Anatolia to northern Italy. Between the Eastern Mediterranean and the Paratethys we postulate a huge continental area stretching from Yugoslavia to Turkey. Temporary land-bridges must also have existed across the Yugoslavian–North Italian seaway to allow mammal exchange as correlative faunas are known from Slovenia to Turkey (Thenius 1959). East-West-trending molasse and flysch troughs were established from Bavaria, across the rising Carpathians and the Pannonian area into southern Russia.

Following a late Oligocene regression a very similar distribution of seas and land-masses

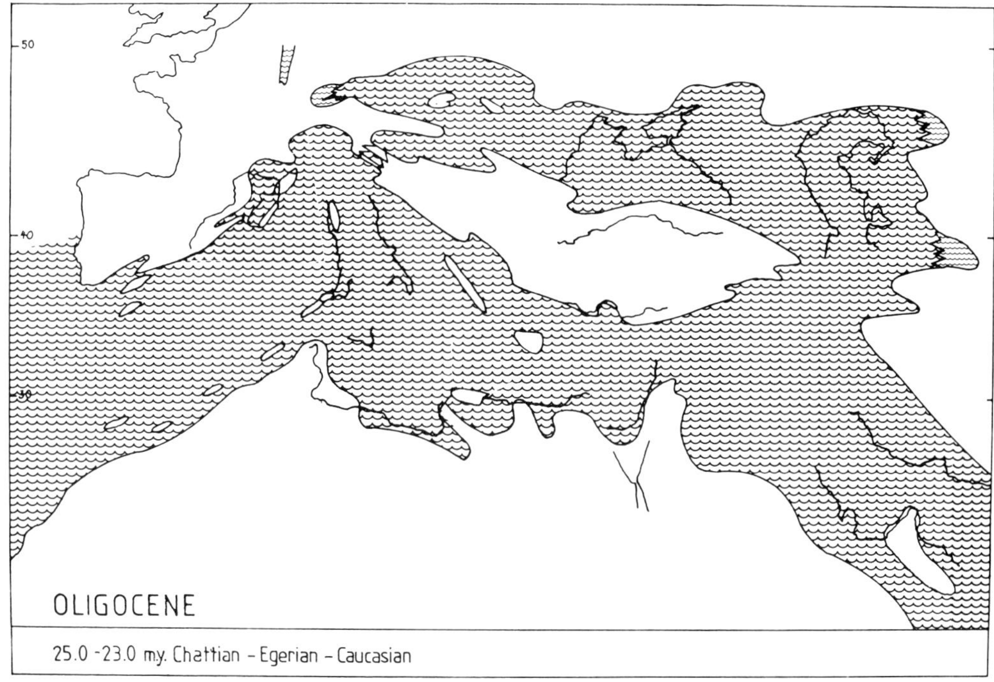

FIG. 2.

FIGS. 2–9. Explanation of ornament in sketches of the late Oligocene to Pliocene paleogeographic and palinspastic reconstructions of the Mediterranean and Paratethys shown in Figs. 2–9. 1, Marine realms; 2, reduced salinity realms; 3, endemic realms; 4, evaporitic realms; 5, continental realms.

1 2 3 4 5

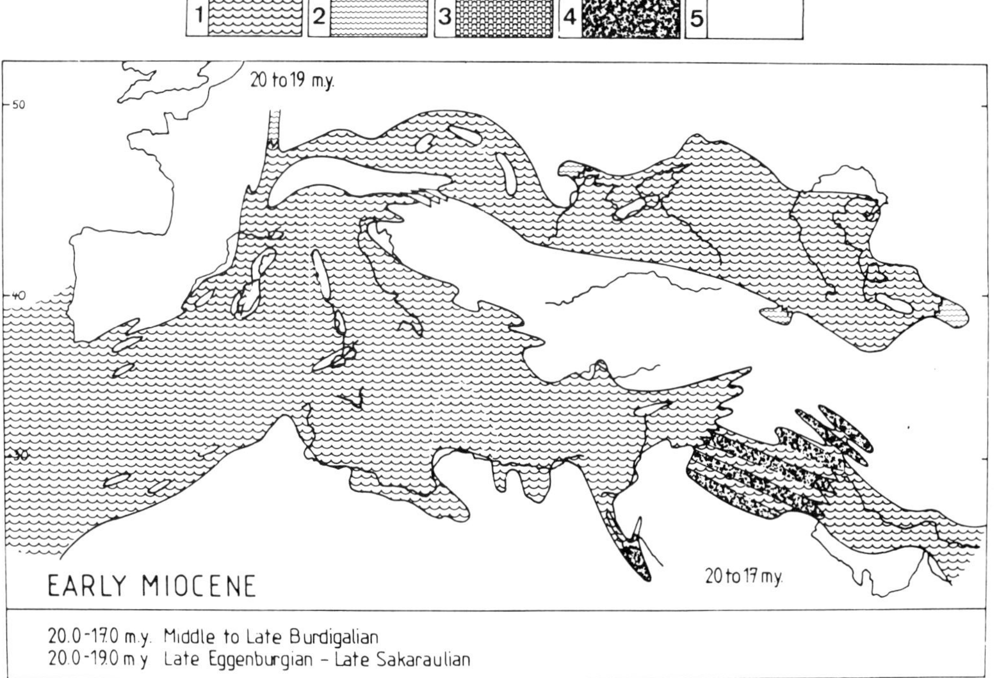

FIG. 3.

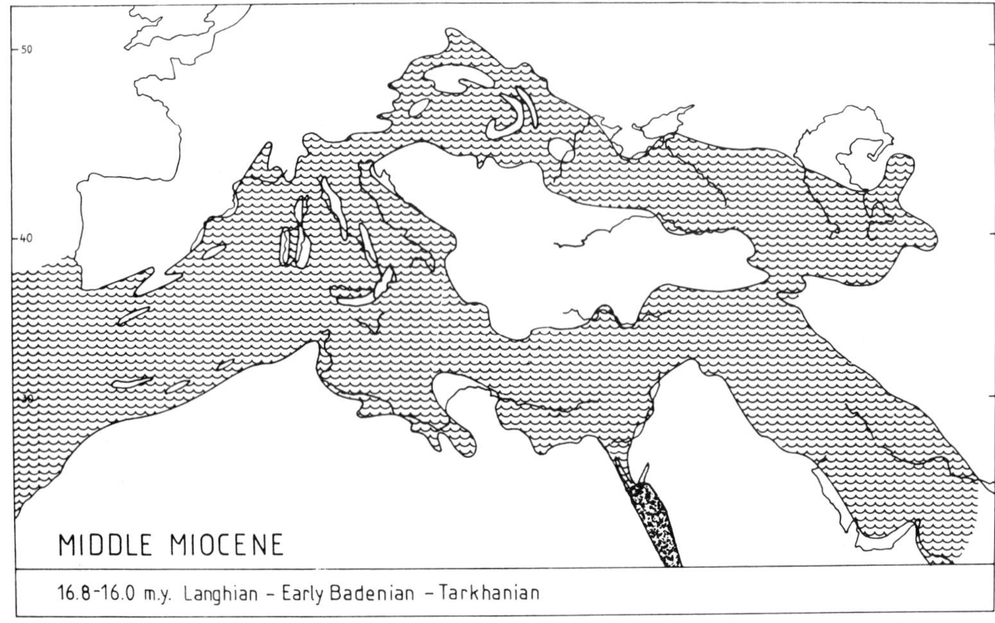

Fig. 4.

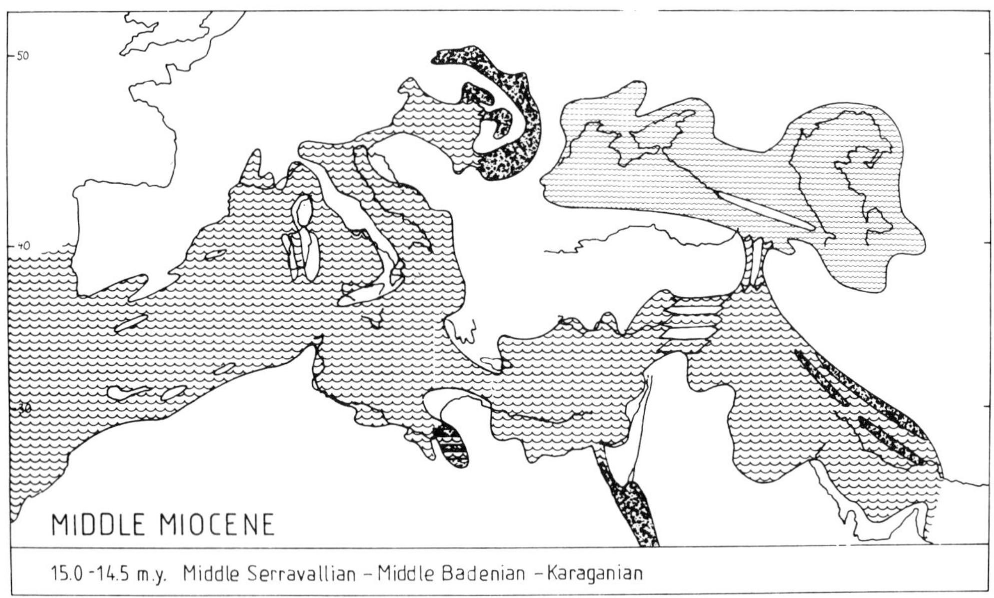

Fig. 5.

existed in the Early Miocene, which began in the entire area with facies changes and transgressive sequences.

Early Miocene (Fig. 3)

20.0–17.0 Ma Middle to Late Burdigalian
20.0–19.0 Ma Late Eggenburgian–Late Sakaraulian

Around 20 Ma it is generally accepted that collision took place between Afro–Arabia and Eurasia giving rise to a widespread regression and triggering: (1) the highly variable facies of the Lower Fars Formation between 20 and 17 Ma; and, (2) the creation of a marine embayment into the Proto-Red Sea rift zone.

Extension of the Fars Formation across the Mesopotamian Trough allowed the well-

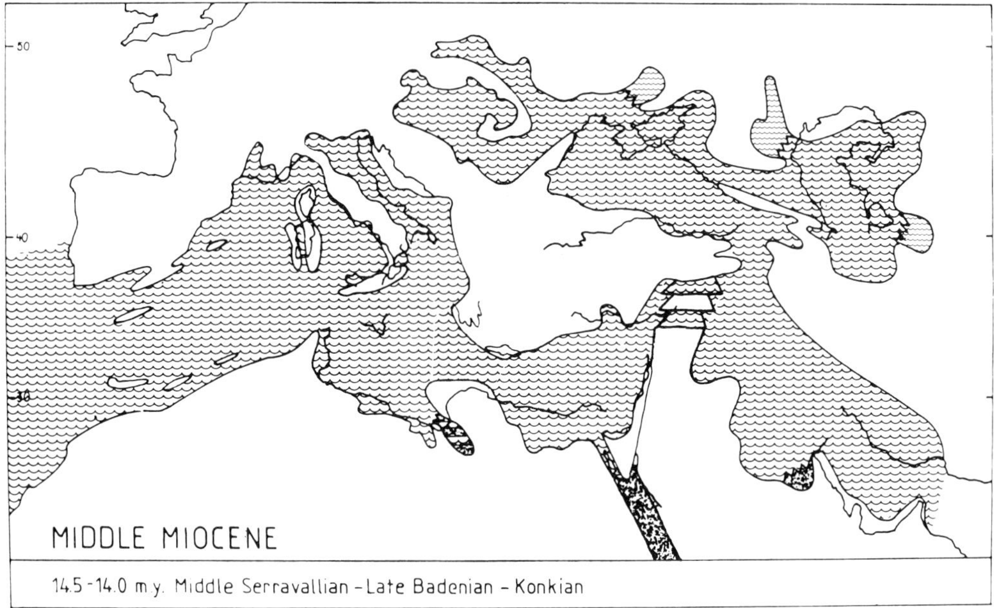

FIG. 6.

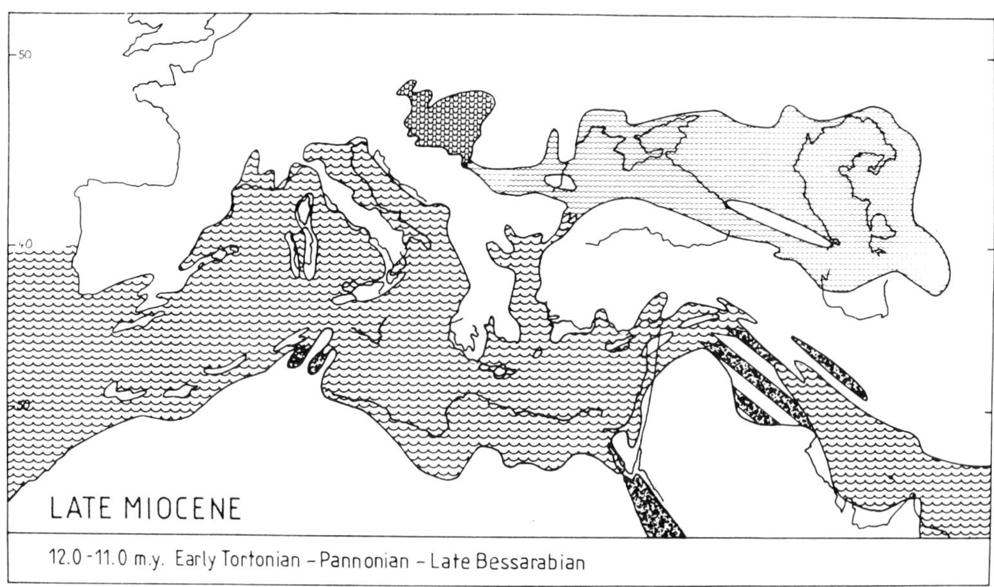

FIG. 7.

documented succession of Orleanian mammal-migrational events between Eurasia and Afro–Arabia. Mammals, such as gomphotheriids, bovids, creodont predators (e.g. *Hyaenaelurus*) and primates (*Pliopithecus*), known at this time only from Africa, migrated into Eurasia. Suids and true carnivores made their first appearance in Africa (Bernor *et al.* 1979; Bruijn & Meulen 1981; Ginsburg 1979; Mein 1979, 1981; Thenius 1979).

With the fusing of the Pindic cordillera and the European continental block by clockwise rotations (Kissel *et al.* this volume), marine sedimentation ceased around the *Globigerinoides bisphericus* level in the Mesohellenic Trough. Major tectonic re-arrangements, well known as the Styrian orogenic phase, are traceable in the entire Alpine orogen. The interruption of the open marine connection of the Eastern Paratethys to the Mesopotamian

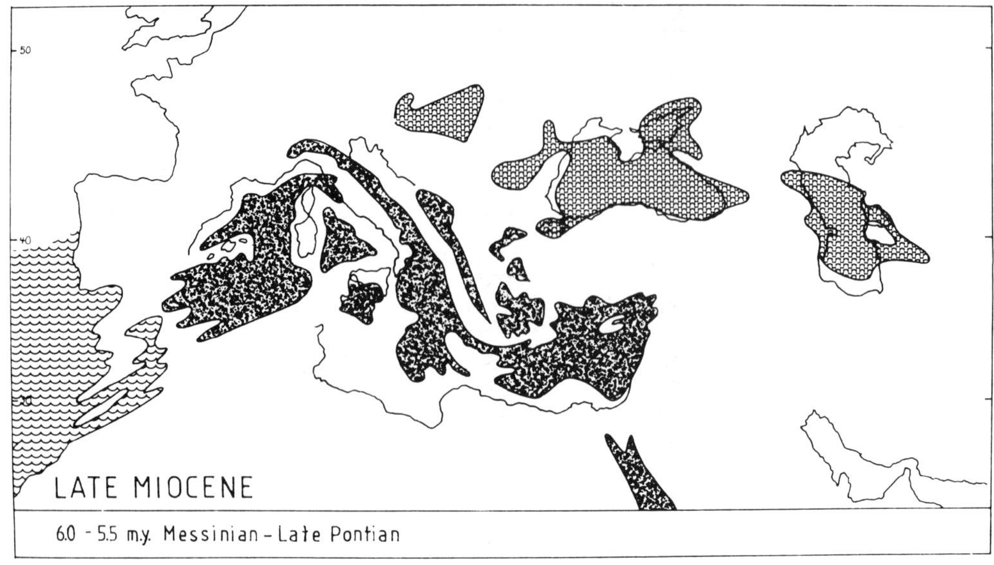

Fig. 8.

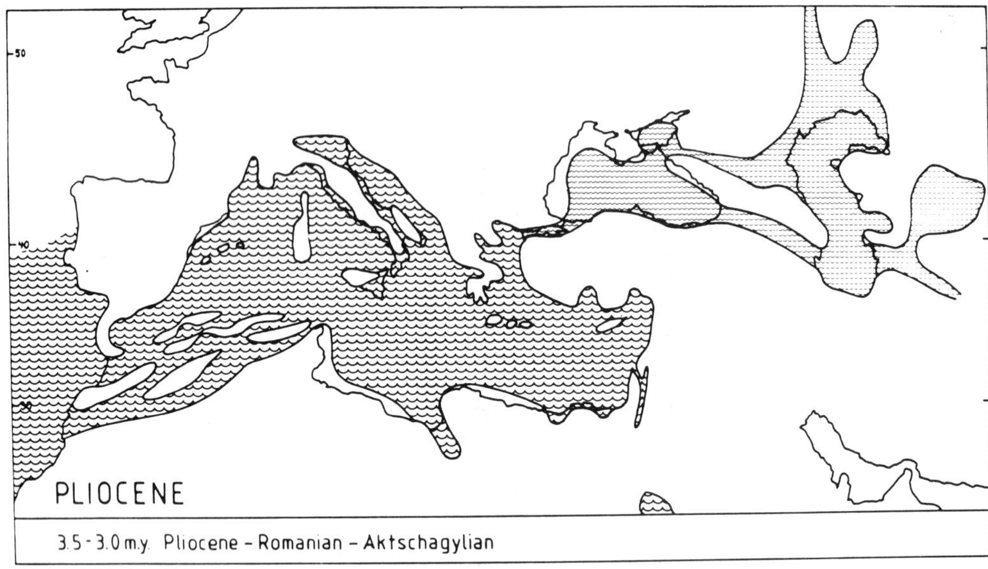

Fig. 9.

Trough around 18–17 Ma, created a huge landlocked sea with reduced salinity, characterized by the 'Kozachurian facies', which extended as far as Bavaria.

Middle Miocene (Fig. 4)

16.8–16.0 Ma Langhian–Early Badenian–Tarkhanian.

The most uniform marine faunal association known throughout the Mediterranean and Paratethys realms since Eocene times were created in the Langhian by a major transgression which once again connected the Mediterranean with the Indo–Pacific. This restored marine conditions in the entire Paratethys and gave rise to a circum-equatorial marine belt with subtropical to tropical conditions extending as far north as Poland. The deep-water trough connecting the Indo–Pacific and the Eastern Mediterranean and Paratethys is assumed to have been eliminated by underthrusting of its floor beneath the Zagros mountain belt at a later time.

No marine sediments are known on the mainland of modern Greece, or in the greater part of Yugoslavia and Turkey. An island configuration similar to that in the Late Oligocene had again emerged.

The Adriatic plate, including the Dinarides, was in a south-eastward position at this time and a deep-water connection from the Mediterranean to the Central Paratethys existed to the north of it. At this point the Carpathian arc system developed and created a series of extensional basins (e.g. Vienna Basin, Transsylvanian Basin, Pannonian Basin).

Middle Miocene (Fig. 5)

15.0–14.5 Ma Middle Serravallian–Middle Badenian–Karaganian

A facies belt of changing marine and evaporitic to continental sediments came into existence in the Middle East. This is indicative of the incipient isolation of the Levant Basin, and in turn the Eastern Mediterranean, from the Mesopotamian Trough. At the same time the marine connections from the Indo–Pacific to the Eastern Paratethys were narrowed by uplift of the Caucasus mountain ranges, turning the Eastern Paratethys into a low salinity realm.

The Serravallian regression resulted in extensive evaporitic sedimentation in the entire Carpathian foredeep as well as in the eastern Pannonian satellite basins. Marine conditions prevailed only in the Pannonian and Vienna Basins. At this time the Apennine arc began to develop into a consolidated continental area.

The marine to evaporitic/continental facies belt in the Middle East allowed the remarkable Middle Miocene–Astaracian mammal faunal exchange between Africa and Eurasia: primates (*Ramapithecus, Sivapithecus* and *Dryopithecus*) as well as smaller mammals characterized this important event (Bernor in press, Wessels *et al.* 1982).

Middle Miocene (Fig. 6)

14.5–14.0 Ma Middle Serravallian–Late Badenian–Konkian

The variable pattern of marine and evaporitic continental facies prevailed in the Middle East and allowed an ongoing mammal faunal exchange across this corridor-cum-steppingstone bridge.

Most remarkable are the uniform marine sediments, with radiolarian–diatom–pteropod marls covering both the evaporites and the hyposaline horizons of the Paratethys. Clear differences between the marine faunas of the Mediterranean and the Paratethys can be traced in the calcareous nannoplankton, the silicoflagellates, radiolarians, crustaceans and molluscs (Rögl *et al.* 1979; Dumitrica *et al.* 1978a,b; Müller 1979). This faunal evidence points to a reactivation of the Indo-Pacific seaway and at the same time to the closure of the open marine connection from the Central Paratethys across Yugoslavia towards Northern Italy.

The ongoing subduction of the floor of the Mesopotamian Trough finally began to restrict marine conditions around 14 Ma. This in turn triggered a most uniform Early Sarmatian faunal development in the Paratethys (Rögl & Steininger 1983).

Late Miocene (Fig. 7)

12.0–11.0 Ma Early Tortonian–Pannonian–Late Bessarabian/Maeotian

The accelerated movement of Arabia closed the marine connections towards the East. The continental/evaporitic upper Fars Formation covered the site of the Mesopotamian Trough resulting finally in the development of a stable continental corridor. This broad corridor permitted the simultaneous spread of the Vallesian mammal faunas, containing the North American immigrant *Hipparion* across the entire circum-Mediterranean.

The pronounced rotational and compressional events of the Late Miocene created basins and graben structures within the Aegean continental area, giving rise to the modern Aegean sea (Jacobshagen 1978; Meulenkamp 1977; Kissel *et al.* this volume). The Tortonian transgression flooded the then newly created Aegean Sea, and made short incursions into the Euxinian Basin of the Eastern Paratethys.

In early Maeotian time the extensive middle Tortonian transgression reached across the northern Aegean Sea and the Dardanelles into the Pontian-Basin. Marine conditions with corallinacean algae, calcareous nannoplankton and molluscs were re-established (Belokrys 1981; Nevesskaja *et al.* 1975; Semenenko & Ljulieva 1978). The same marine seaway allowed a number of mollusc taxa, otherwise known only from the Euxinian realms, to migrate into the Tortonian Aegean Sea as far south as Attica (Papp 1980; Papp & Steininger 1979). Iljinia & Nevesskaja (1979) postulated a marine connection to be still in existence in Maeotian time from Erevan in Armenia across the eastern Anatolian fault zone to the marine realms of Iran and Syria.

The Central Paratethys itself began to disintegrate into a number of smaller, only temporarily interconnected basins. The typical *Congeria-Melanopsis* facies of the Vienna and Pannonian Basins interfingered with the euryhaline facies in the Dacian Basin (Stevanovic 1974).

Late Miocene (Fig. 8)

6.0–5.5 Ma Messinian–Late Pontian

The well known Messinian event subdivided even the Eastern Paratethys into smaller basins, each with its own distinctive endemic faunal evolution. However, complete dessication of the Euxinian Basin, as postulated by Hsü (1978), never took place. Time-equivalent sediments with characteristic endemic faunas are very well known from land sections around the entire Euxinian Basin (Seneš 1978). Furthermore, it was shown by Kojumdgieva (1979) that the sedimentological evidence on which Hsü had based his hypothesis was founded on biostratigraphic miscorrelation.

The Late Miocene post-evaporitic 'Lago Mare' facies is characterized by low salinity faunas with congeriids, cardiids, melanopsids and specific ostracods. Because of faunal and environmental similarities this facies is generally assumed to imply connection with the Paratethys (Cita & Colombo 1979). However, the 'Lago Mare' facies is not the time-equivalent of the Pannonian stage of the Central Paratethys (see Fig. 1; Steininger & Papp 1979).

Pliocene (Fig. 9)

3.5–3.0 Ma Pliocene–Romanian–Aktschagylian

The well known Pliocene transgressive event flooded and created the modern Mediterranean Sea, extending even as far as the Euxinian Basin.

Late Pliocene marine sediments at this point covered the Late Kimmerian *Limnocardiid*-endemic facies within the Black Sea, progressing across the Caucasian foredeep into the Caspian Sea depression and perhaps even extended east as far as Lake Baikal. In our opinion several factors indicate a possible reactivation of the Turgai strait.

Conclusions

The Mediterranean functioned as an Indo–Pacific–Atlantic seaway until the beginning of the Miocene.

A first land bridge between Africa and Eurasia came into existence around 20 Ma in the Middle East.

Throughout the Miocene the tectonic instability of the Middle East gave way to recurring marine connections between the eastern Mediterranean and the Indo–Pacific, or provided gateways for mammal exchanges. At the same time the kinematic evolution of the Middle East influenced the facies development of the entire Paratethys. The Late Miocene consolidated the continental conditions in the Middle East and created the modern Aegean Sea.

The modern Mediterranean Sea and its surrounding continents were shaped at the time of the Pliocene marine flooding.

ACKNOWLEDGEMENTS: We are grateful to A. H. F. Robertson and J. F. Dewey for detailed critical comments of an earlier draft and to T. Baldi, F. Bellini, M. B. Cita, Z. Garfunkel, R. Gelati, G. Gvirtzman, V. Jacobshagen, E. Kojumdgieva, D. P. McKenzie, M. Kovacs, G. Palmieri, J. Seneš, E. Thenius and A. G. Smith for valuable discussions.

Financial support was provided by the Austrian National Committee for the International Geological Correlation Programme and the 'Hochschuljubilaeumsstiftung der Gemeinde Wien'.

References

Adams, C. G. 1976. Larger foraminifera and the Late Cenozoic history of the Mediterranean region. *Paleogeogr. Paleoclimat. Paleoecol.* **20,** 47–66.

Baldi, T. 1980. The early history of the Paratethys. *Földtani Közlöny, Bull. Hungarian Geol. Soc.* **110,** 456–472.

Belokrys, L S. 1981. Maeotian Red Algea of the Crimea (Russ.). *Paleont. J. Jg.* 1981, **2,** 117–125.

Bernor, R. L. (in press). Geochronology and zoogeographic relationships of Miocene Hominoides. *In*: Ciochon, R. L. & Corruccini, R. S. (eds). *New Interpretations of Ape and Human Ancestry*. New York, Plenum Press.

——, Andrews, P. J., Solounias, N. & Van Couvering, J. A. H. 1979. The evolution of "Pontian" mammal faunas: some zoogeographic, paleoecologic and chronostratigraphic considerations. *Ann. Géol. Pays Hellén., tome hors série*, 1979, fasc. **1,** 81–89.

Biju-Duval, B., Dercourt, J. & Le Pichon, X. 1977. From the Tethys ocean to the Mediterranean seas: a plate tectonic model of the evolution of the western Alpine System. *In:* Biju-Duval, B. & Montadert, L. (eds.) *International Symposium on the Structural History of the Mediterranean Basins.* Split (Yugoslavie) 25–29 October 1976, Editions Technip, Paris, 143–164.

Boccaletti, M. 1979. Mesogea and Mesoparatethys: their development at the Tethyan continental margins and their influence on the later evolution of the Mediterranean and Paratethys. *Ann. géol. Pays Hellénique, tome hors série.* 1979 fasc. **1**, 139–148.

Bruijn, H. de & Van Der Meulen, A. J. 1981. The distribution of land mammals in the Mediterranean through the Neogene—is there a fit with the paleogeographic and paleoclimatologic reconstructions based on data from marine basins? *Ann. Géol. Pays Hellénique, tome hors série*, 1981, fasc. **4**, 323–335.

Cita, M. B. & Colombo, L. 1979. Sedimentation in the latest Messinian at Capo Rosselo (Sicily). *Sedimentology*, **26**, 497–522.

Dermitzakis, M. D. & Papanikolaou, D. J. 1981. Paleogeography and geodynamics of the Aegean region during the Neogene. *Ann. Géol. Pays Hellén., tome hors série*, 1981, fasc. **4**, 245–289.

Dewey, J. F., Pitman, W. C. III, Ryan, W. B. F. & Bonnin, J. 1973. Plate tectonics and the evolution of the Alpine System. *Bull. Geol. Soc. Am.* **84,** 3137–3180.

Drooger, C. W. 1979. Marine Connections of the Neogene Mediterranean, deduced from the evolution and distribution of larger foraminifera. *Ann. Géol. Pays Hellén., tome hors série*, **1**, 361–369.

Dumitrica, P. 1978a. 1. Badenian Silicoflagellates form Central Paratethys. *In*: Papp, A. *et al.* (eds) M4—Badenien. *Chronostrat. & Neostrat.* **6,** 207–229.

—— 1978b. 2. Badenian Radiolaria from Central Paratethys. *In*: Papp, A. *et al.* (eds) M4—Badenien. *Chronostrat. & Neostrat.* **6,** 213–261.

Ginsburg, L. 1979. Les Migrations de mammifères carnassiers (Créodontes + Carnivores) et le problème des relations intercontinentales entre l'Europe et l'Afrique au Miocène inférieur. *Ann. Géol. Pays Hellén., tome hors série*, 1979, fasc. **1**, 461–466.

Haq, B. U. Premoli-Silva, I. & Lohmann, G. P. 1977. Calcareous plankton paleobiogeographic evidence for major climatic fluctuation in the Early Cenozoic Atlantic Ocean. *J. Geophys. Res.* **82/27,** 3861–3876.

Hochuli, P. 1978. Palynologische Untersuchungen im Oligozän und Untermiozän der Zentralen und Westlichen Paratethys. *Beitr. Paläont. Österr.* **4,** 1–132.

Hsü, K. J. 1978. Stratigraphy of the lacustrine sedimentation in the Black Sea. *In:* Ross, D. A., Neprochnov, X. P. *et al.* (eds) *Initial Reports of the Deep Sea Drilling Project*, **42,** 509–524.

Jacobshagen, V., Dürr, St., Kockel, F., Kopp, K. O. & Kowalczyk, G. 1978, with contributions of H. Berckhemer & D. Büttner. Structure and geodynamic evolution of the Aegean region. *In*: Closs, H. *et al.* (eds). *Alps, Apennines, Hellenides.* Inter-Union Comm. Geodyn. Sci. Rept. **38,** 537–564.

Kojumdgieva, E. 1979. Critical notes on the stratigraphy of Black Sea boreholes (Deep Sea Drilling Project, Leg 42 B). *Geol. Balcanica*, **9/3,** 107–110.

McGowran, B. 1979. Some Miocene configurations from an Australian standpoint. *Ann. Géol. Pays Hellén., tome hors série*, **2,** 767–779.

Mein, P. 1979a. Rapport d'activité du groupe de travail vertébrés mise au jour de la biostratigraphie du Néogène basée sur les mammifères. *Ann. Géol. Pays Hellén., tome hors série*, 1979, fasc. **3,** 1367–1372.

—— 1979b. Mammal Zonations: Introduction. *Ann. Géol. Pays Hellén., tome hors série*, 1981, fas. **4,** 83–88.

Meulenkamp, J. E. 1977. The Aegean and the Messinian salinity crisis. *Proc. 6th Coll. Geol. Aegean Region*, **3,** 1253–1263.

Müller, P. 1979. The Indo-West-Pacific character of the Badenian Decapod crustaceans of the Paratethys. *Ann. Géol. Pays Hellén., tome hors série*, **2,** 865–869.

Nevesskaja, L. A., Bagdasarjan, K. G., Nosovsky, M. F. & Paramonova, N. P. 1975. Stratigraphic distribution of bivalvia in the Eastern Paratethys. *Rep. Activ. R.C.M.N.S. Working Groups (1971–1975)*, 48–74.

Papp, A. 1980. Die Molluskenfauna von Trilophos südlich von Thessaloniki (Griechenland) und ihre paläogeographische Bedeutung. *Ann. Géol. Pays Hellén.* **30**/1 (1979), 225–247.

—— & Steininger, F. 1979. Paleogeographic implications of Late Miocene deposits in the Aegean region. *Ann. Géol. Pays Hellén., tome hors série*, **2,** 955–959.

Rabeder, G. 1981. Die Arvicoliden (Rodentia, Mammalia) aus dem Pliozän und dem älteren Pleistozän von Niederösterreich. *Beitr. Paläont. Österr.* **8,** 1–373.

Rögl, F. & Steininger, F. F. 1983. Vom Zerfall der Tethys zu Mediterran und Paratethys. Die Neogene Paläogeographie und Palinspastik des zirkum-Mediterranen Raumes. *Ann. Nat. Hist. Mus. Wien.*, **85/A,** 135–163.

——, Steininger, F. F. & Müller, C. 1978. Middle Miocene salinity crisis and paleogeography of the Paratethys (Middle and Eastern Europe). *Init. Rept. DSDP*, **42/1,** 985–990.

Schwan, W. 1980. Geodynamic peaks in alpinotype orogenies and changes in ocean-floor spreading during Late Jurassic—Late Tertiary time. *Amer. Ass. Petrol. Geol. Bull.* **64,** 359–373.

Seneš, J. *et al.* 1978. Correlation tables of the Neogene in the Mediterranean Tethys and Paratethys. First working version. XXXIV + 802. Bratislava (SAV).

—— 1978. Theoretische Erwägungen der zeitlichen Äquivalenz des Tarchanien mit den chronostratigraphischen Einheiten der Zentralen Paratethys. *Geol. Zborn. Geol. Carpat.* **29,** 177–178.

—— 1979. Correlation du Néogène de la Téthys et de la Paratéthys—Base de la reconstitution de la géodynamique récente de la région de la Méditerranée. *Geol. Zborn. Geol. Carpat.* **30,** 309–319.

Steininger, F. F. & Papp, A. 1979. Current biostratigraphic and radiometric correlations of Late Miocene Central Paratethys stages (Sarmatian s. str., Pannonian s.str. and Pontian) and Mediterranean stages (Tortonian and Messinian) and the Messinian Event in the Paratethys. *Newslett. Stratigr.* **8,** 100–110.

—— & RÖGL, F. 1979. The Paratethys history—a contribution towards the Neogene geodynamics of the Alpine orogene (an abstract). *Ann. Géol. Pays Hellén., tome hors série*, 1979, fasc. **3,** 1153–1165.

——, RÖGL, F. & MARTINI, E. 1976. Current Oligocene/Miocene biostratigraphic concept of the Central Paratethys (Middle Europe). *Newslett. Stratigr.* **4,** 174–202.

STEVANOVIĆ, P. 1974. Sur les échelles biostratigraphiques du Néogène marin et saumâtre de la Yougoslavie. *Mém. B.R.G.M.* **78**/2, 791–799.

THENIUS, E. 1959. Tertiär. 2. Teil Wirbeltierfaunen. *In*: LOTZE, F. *Handbuch der stratigraphischen Geologie.* **3**/2, XI + 328.

—— 1979. Afrikanische Elemente in der Miozaenen Saeugetierfauna Europas (African Elements in the Miocene Mammalian Fauna of Europe). *Ann. Géol. Pays Hellén., tome hors série*, 1979, fasc. **3,** 1201–1208.

VAIL, P. R., MITCHUM, R. M. Jr. & THOMPSON, S. III. 1977. Global cycles of relative changes of sea level. *In*: Seismic Stratigraphy and Global Changes of Sea Level. *Amer. Ass. Petrol. Geol. Mem.* **26,** 83–87.

WESSELS, W., BRUIJN, H. DE, HUSSAIN, S. T. & LEINDERS, J. J. M. 1982. Fossil rodents from the Chinji Formation, Banda Daud, Shah, Kohat, Pakistan. *Proc. Kon. Nederl. Akad. Wetensch., ser. B,* **85,** 337–364.

F. F. STEININGER, Institute of Paleontology, University of Vienna, 7, Universitaetsstr. A-1010 Vienna, Austria.

F. RÖGL, Department of Geology and Paleontology, Natural History Museum, 7, Burgring A-1014 Vienna, Austria.

Palaeomagnetic evidence of Miocene and Pliocene rotational deformations of the Aegean Area

Catherine Kissel, Marc Jamet & Carlo Laj

SUMMARY: Palaeomagnetic measurements have been made on the Mio-Pliocene sedimentary series along the external Hellenic Arc as well as in the Oligocene to Middle Miocene Ionian flysch of northwestern Greece.

The results show that the central and eastern parts of the arc (Crete and Rhodes) have not undergone significant rotation since the Tortonian-Messinian stage. On the other hand, the northwestern part has undergone a clockwise rotation which, in the Ionian islands, has begun 5 Ma ago and has proceeded since then at roughly 5 degrees/Ma. We suggest that the beginning of this rotation is related to the Lower Pliocene compressive phase which most probably initiated the tectonic activity of the present day Aegean Arc.

No rotational deformation occurred between 5 and 12 Ma, but the results from 16 sites in Epirus and Akarnania sampled in Ionian flysch sections indicate that another clockwise rotation, of the order of 30 degrees, occurred between 30 and 12 Ma ago.

These Oligo-Miocene and Plio-Quaternary episodes of rotational deformation may be related to the Oligo-Miocene and Plio-Quaternary arc-related volcanic activity in the Aegean Sea.

During the last 10 or 15 years many geological and geophysical data have been collected in the Aegean area, providing important information about the recent tectonic and geodynamic evolution of the Hellenic Arc.

Seismic and volcanological studies, in particular, have shown the existence in the southern part of the arc of a subduction zone with a rather well-defined Benioff zone and associated arc volcanism, while there is a progressive transition to a zone of continental collision in the northwestern part. Geological studies of both the alpine and neotectonic history of this area have demonstrated the overall structure of the nappe system and the stress patterns of the entire domain.

This paper presents some palaeomagnetic results which yield new data constraining the extent of rotational deformation in the Aegean area. Indeed, palaeomagnetism can be a quantitative technique for measuring relative displacements and rotations among different geological units. It is rather surprising that no major palaeomagnetic study of this area has been published so far. Indeed the models developed by different authors, which will not be detailed here, all come to the conclusion that the curvature of the arc has been acquired, at least partially, because of its tectonic evolution, so that rotational deformation is very likely to have occurred.

The geodynamic mechanisms involved and the timing of their activity, are subjects of considerable disagreement among different authors (Brunn 1976, Tapponnier 1977, Mercier 1977, Mercier *et al.*, 1979, Le Pichon & Angelier 1979, Angelier *et al.*, 1982). It is thus necessary, for a deeper understanding of the geodynamic evolution of this area, to distinguish among different phases of rotational deformation. For this reason, we first discuss data about rotations associated with recent neotectonic activity of the arc. We then consider 're-trotectonically' the problem of older deformation. In this way data from older formations can be corrected for eventual subsequent rotations and each phase of rotational deformation can be distinguished from the others.

Geological setting and sampling

Sampling sites were selected on geological, sedimentary and magnetic criteria. In general fine-grained sediments were chosen preferentially because their sedimentation rate is usually low enough to average out completely the secular variation of the geomagnetic field over the thickness of a sample and because this kind of sediment is usually deposited in calm water. Sections close to major faults or containing slumped beds or landslides were avoided.

Careful attention was paid to the colour of the sediments and only blue-grey marls were sampled because laboratory studies have shown that sediments of this colour have not been affected by weathering, so that they are free of any chemical remanent magnetism. Even though this colour may not be original and may

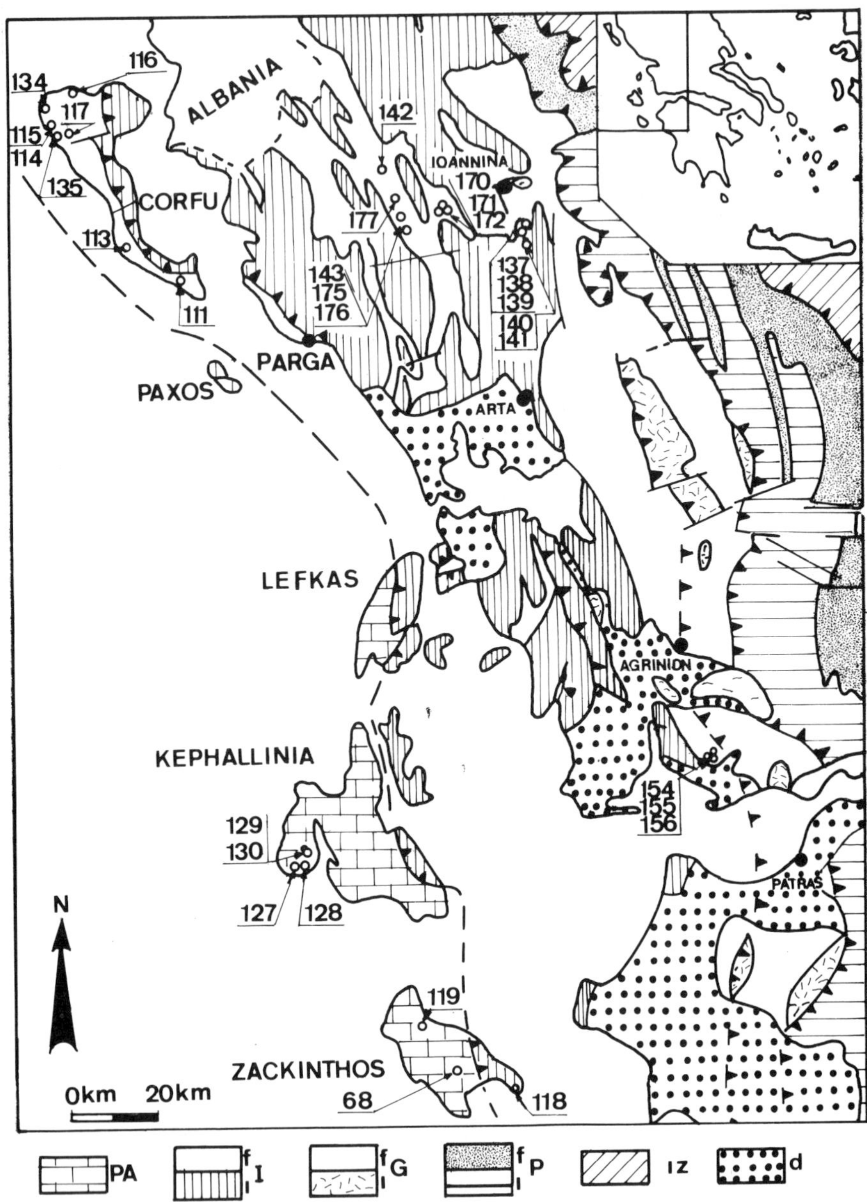

FIG. 1. Map showing the location of the sites in the Ionian islands and northwestern Greece. This map only shows the sites which have so far been studied. It should be remarked that the sampled flysch sections belong to major structures. The significance of symbols is as follows: PA Preapulian zone: I; Ionian zone; G. Gavrovo zone; P; Pindos zone; iz, Internal zone; d, post-tectonic formations; f, flysch; l, limestone.

result from reduction phenomena after deposition, we have assumed that the 'magnetic age' is not significantly different from the stratigraphic age.

Figure 1 shows the location of the sites from the Ionian islands and northwestern Greece which have so far been studied and whose results are discussed in this paper. Other sites, currently under examination, are not shown and the location of the sites in the central and southern parts of the arc can be found in a previous publication (Laj *et al.*, 1982).

The sites of Upper Miocene-Pliocene age from the Ionian islands have undergone a single compressional phase. The sites sampled in the Ionian flysch, in contrast, have undergone at least two compressive phases, so that the conventional tilt correction used in palaeomagnetism, which assumes tilting about the line of strike of bedding, could be incorrect. Therefore particular care was taken to sample sites where the available geological information could ensure that a correct tilt correction was made and could also minimize the effect of local tectonic events within the Ionian zone. For example, fourteen sites were sampled in the Botzara syncline because this is a major structure of northwestern Greece. Any data on the rotation of this structure would be highly significant for the entire region. Moreover, the axis of this syncline is very nearly horizontal all along its length, so that even if its formation is due to more that one tectonic event, folding has occurred about the same axis, and the usual tilt correction can be safely applied to the palaeomagnetic results. For the same reason many sites were sampled in another major structure, the Epirus-Akarnania syncline.

The final choice of outcrops was made directly in the field using a portable spinner magnetometer (Chiron *et al.*, 1981). Because palaeomagnetic directions can be carefully determined only using demagnetization plots, only sites with an initial magnetization intensity of at least 5–6 times the minimum measurable level could be used safely. For the Digico spinner magnetometer used to measure the samples in the laboratory, this corresponds to a NRM level of about 10^{-6} cgs emu. This is an important limitation because only 15% of the geologically suitable sites were found to meet this requirement. Samples of lower magnetization level were however collected in regions where it was impossible to locate formations with this minimum NRM intensity.

Drilling and orientation of the cores was done using usual palaeomagnetic techniques. More than 2500 cores were obtained from 142 sites of Upper Miocene to Pliocene age and more than 800 cores from 37 sites in the Ionian flysch.

Laboratory techniques

The samples were measured using standard laboratory techniques. In general a Digico spinner magnetometer was used, except for weakly magnetized samples which were studied using the cryogenic magnetometer of the palaeomagnetic laboratory at Utrecht. To distinguish between magnetite- and haematite-bearing rocks, the IRM acquisition test was applied stepwise up to 12.5 KOe, followed by AC and thermal demagnetization to determine the AC median destructive field of saturation remanence and the temperature beyond which no magnetization is left. Within the limits of this test, it was found that in 95% of the cases magnetite was the main magnetic carrier. Thermal demagnetization of the samples was used throughout the study in steps of 40–50°C to a maximum temperature ranging between 450 and 600°C depending on the samples. AC demagnetization has also been applied in some cases. Although the mean values of both declination and inclination are the same as those determined by thermal demagnetization, the overall scatter is significantly higher, so this technique has not been used systematically. Final palaeomagnetic directions were determined from demagnetization diagrams by hand fitting a straight line through at least the last 4–5 points.

Results

General

Some typical demagnetization diagrams relating to samples from Ionian flysch sections are shown in Fig. 2. Apart from a small viscous component removed at about 200°C, all the samples have a single stable component of magnetization whose direction can be determined quite precisely. Within-site scattering is very small in all cases as shown by the stereographic projections of Fig. 3. It should also be noted in this figure that whenever normal and reversed samples are present in the same site, their palaeomagnetic directions are exactly opposite, which is a good indication of magnetic stability.

A common characteristic of almost all the studied sites is that the values of the inclination are lower than the ones expected at the site latitudes, assuming a geocentric axial dipole field. This characteristic is present both in the sediments from the external arc (mean value of the anomaly = 10°) and in the flysch sections (mean value = 20°). In both cases there is no clear correlation between the inclination anomaly and the geographical or stratigraphic position of the corresponding formation. For this reason, and in spite of the fact that the inclination error is definitely greater in the older flysch sections than in the younger formations, we do not think that the low inclinations are clear evidence of a northward drift of the sampled formations. Unrealistic drift of the African plate, to which the flysch outcrops

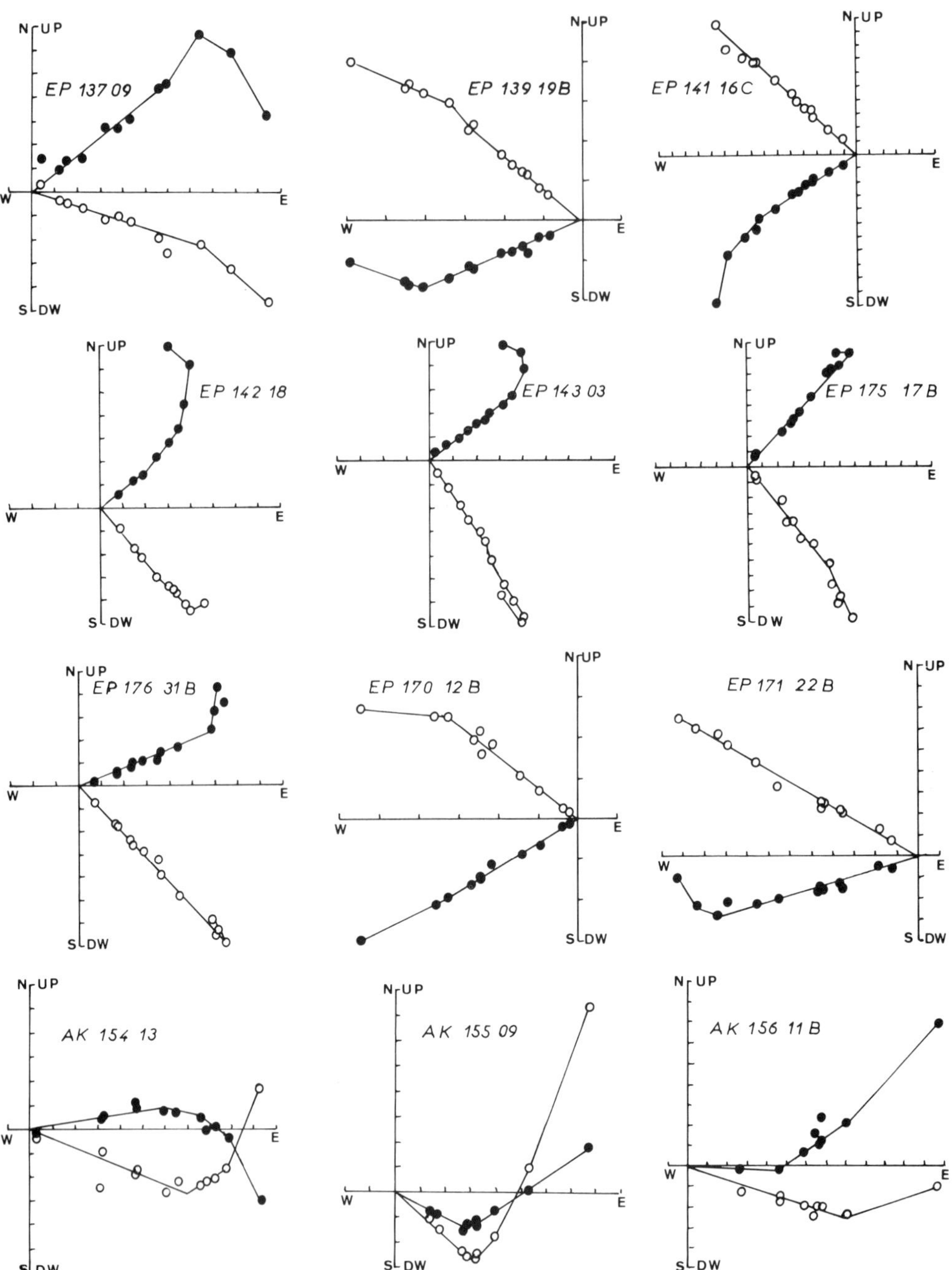

FIG. 2. Thermal demagnetization plots of various samples from sites of Ionian flysch. The temperatures corresponding to the successive steps are: 20, 120, 180, 240, 300, 350, 390, 430, 470, 510, 550 and in most cases 575°C. Open circles, inclination data; Full circles, declination data.

belonged at the time of their deposition, would moreover be needed to account for the entire inclination anomaly.

On the other hand, the presence of the inclination error could be related to the nature of the sediments and in particular to their carbonate content. In the same region, for instance, some carbonate-rich sections do not show any appreciable inclination error, while others of the same age with low carbonate

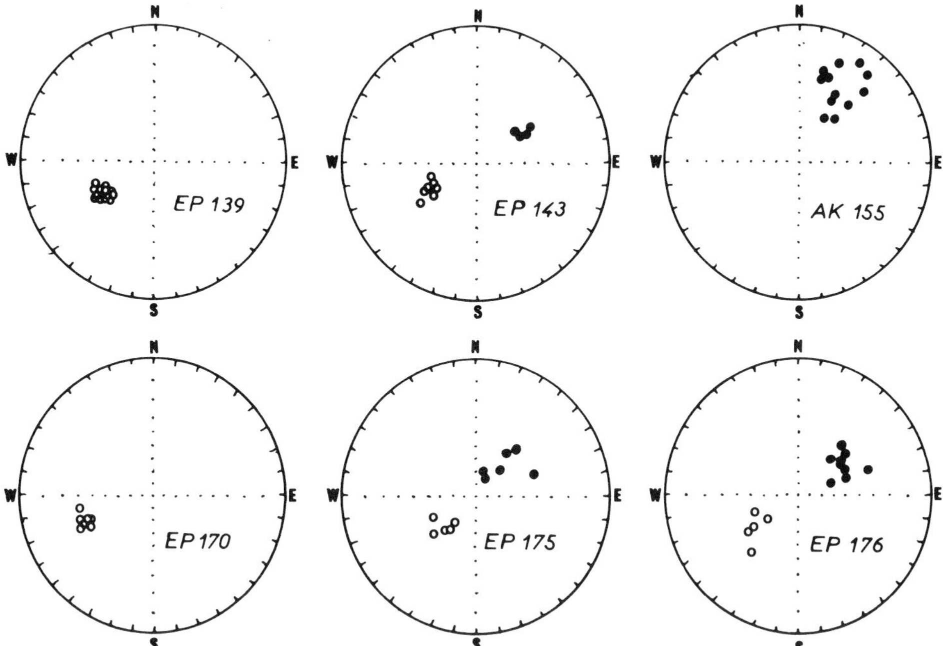

FIG. 3. Stereographic projections of results from 6 sites of Ionian flysch. The fact that normal and reversed samples are exactly opposed is a good indication of the magnetic stability of the samples.

contents have an inclination error greater than 20°.

We believe that a large fraction of the inclination error is related to compaction, which can be quite important for blue-grey marls and flysch. The flattened shape of many macrofossils in the sampled sections is evidence for the occurrence of compaction. It is known, moreover, that compaction is very limited if the amount of carbonate is high and if lithification is rapid. We cannot, however, reject the possibility of an inclination error acquired at deposition or that part of this error is due to a northward drift.

Whatever the explanation for their low value, the inclination data are certainly not particularly useful as palaeolatitude indicators for the sites which we have studied. Declination values, however, are not affected by any of the mechanisms invoked to explain the inclination error; they can thus be safely used as a measure of rotational deformation.

Results from the external sedimentary arc

Results from the external sedimentary arc have already been reported (Valente *et al.*, 1982, Laj *et al.*, 1982) so they will only be briefly discussed. It has been shown that the southeastern part of the Aegean Arc has not undergone any significant rotation since the Upper Tortonian stage, while the northwestern part, from Corfu to the southern Peloponnesos, has undergone a clockwise rotation. It thus appears necessary to postulate the existence of some sort of geological mechanism 'decoupling' Crete from Peloponnesos. There is a good agreement, in this respect, between these results, the geological observations of Lyberis *et al.* (1982) and the model suggested by Angelier *et al.* (1982).

A detailed analysis has been made of data from the Ionian islands. A recent study of the structural geology (Jamet 1982) has shown that Corfu has had, for the neotectonic period, the same structural evolution as Kephallinia and Zackinthos. For this reason all the palaeomagnetic data from these islands can be superposed on a single diagram, relating the measured angle of rotation to the age of the corresponding formation. The resulting diagram, shown in Fig. 4, differs from the one already reported (Laj *et al.*, 1982) only by the result of an additional site.

Within the limits of its accuracy, it can be seen from this diagram that data from the three Ionian islands all fall on the same curve, which is a consequence and an additional proof of their structural unity. The dashed line superimposed on the data has been obtained by the

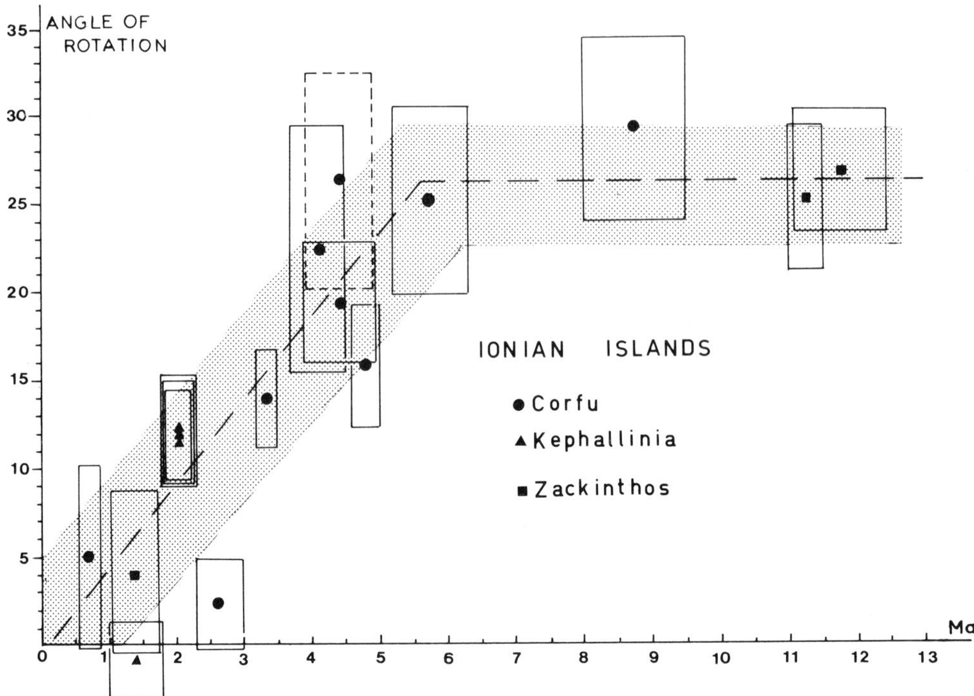

FIG. 4. Plot of the measured angle of rotation versus the age of the corresponding formation for the Ionian islands.

usual linear regression of the data performed separately in the interval 0–5.5 Ma (12 points) and 5.5–12 Ma (4 points). The shaded area is an estimate of the error.

The diagram shows that the angle of rotation increases with the age of the corresponding formation up to about 5 Ma and then remains constant up to 12 Ma. There is thus good evidence that a rotational movement began 5 Ma ago and has proceeded since then at an average rate of about 5° Ma, giving a total rotation of 26°.

The fact that the dashed line passes exactly through the origin is an indication that this rotational movement is still active today. No major rotation seems to have occurred between 5 and 12 Ma.

Results from Ionian flysch sections

Sampling of the flysch sections has been made in order to detect rotations of an earlier age. From a stratigraphical point of view these sections range from Lower Oligocene to Upper Aquitanian. Sixteen of these sites have so far been studied. The results are reported in Table 1 and are also represented by a stereographic projection in Fig. 5.

This figure shows that, in spite of significant scattering probably due to local events, palaeomagnetic directions have a declination value about 60° away from present north-south. Comparison of these data with the values of declination calculated from the available palaeomagnetic poles of the same epoch from the stable part of Africa (McElhinny & Cowley 1977, McElhinny & Cowley 1979, Besse & Pozzi oral communication 1982) indicates that this anomaly in the declination values corresponds to a clockwise rotation of about 55–60°.

Because there are no geological discontinuities between northwestern Greece and the Ionian islands (Sorel oral communication 1982), Epirus and Akarnania have certainly been affected by the same clockwise rotation which has affected these islands during the last 5 Ma. But this accounts only for about one half of the observed clockwise rotation of the flysch sections.

We thus come to the conclusion that northwestern continental Greece has undergone a clockwise rotation of at least 25–30° which took place between 30 and 12 Ma. The results are however as yet not very accurate, both because of the scattering of palaeomagnetic data and because the biostratigraphic age determinations

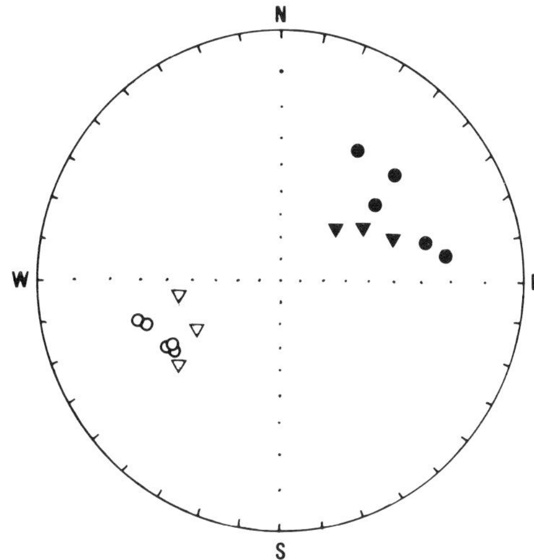

FIG. 5. Stereographic projection of the palaeomagnetic directions of Table 1. Full circles, sites containing only normal samples; open circles, sites containing only reversed samples; full triangles, sites containing samples of both polarities with a majority of normal samples; open triangles, sites containing both polarities with a majority of reversed samples.

For the sites containing samples of both polarities, the palaeomagnetic directions of samples which were not in the majority were inverted through the origin and averaged with those of the other samples.

are not completed yet, and only ages obtained from the geological maps have been used. So at present we cannot ascertain whether, within this period, the rotation has occurred gradually or alternatively whether there have been phases of rapid rotational deformation. The only speculative observation which can be made is that the results from the youngest formation studied (EP 142, Table 1) seem to indicate that a considerable amount of rotation has occurred between the Lower Burdigalian and the Serravallian-Tortonian boundary.

In summary, the present palaeomagnetic data, although incomplete, sustain the hypothesis that northwestern Greece has undergone two clockwise rotations of roughly 25° each, one during the Miocene and one during the Pliocene and Quaternary, separated by a period of at least 7 Ma during which no major rotation occurred.

Discussion

We are well aware that additional measurements are needed, both for the sediments from the Ionian islands and for the Ionian flysch sections, before any definitive interpretation can be given. Nevertheless, the present results can be discussed within the framework of the known tectonic evolution of this area.

The main tectonic events for the last 20 Ma are shown schematically in Fig. 6 for both the (Pre)apulian and Ionian zone. Also shown in this figure are the stratigraphic positions of the sites from the Ionian islands sampled in each zone and the corresponding measured angles of rotation.

It can be seen that the beginning of the rotation of the Ionian islands overlaps in time the Lower Pliocene compressive phase which, according to Mercier *et al.* (1976) marks the moment when the external Aegean Arc became an active continental margin under compression. We think that this coincidence in time shows that the beginning of the rotation of the Ionian islands is related to the formation of this margin, so that the palaeomagnetic results yield additional and independent evidence that, at least in the northwestern region, the present day Aegean Arc was initiated 5 Ma ago as suggested by Mercier *et al.* (1976).

Concerning the orogenic system which existed before then, geological data show that a carbonate platform subjected to extensional tectonics existed in the region of the Ionian islands, where there is today the compressive front of the arc. The former compressive arc lay more to the east, in the Ionian zone (Sorel 1976, Mercier *et al.*, 1979).

Back-arc tectonics and calc-alkaline volcanism in the Aegean Sea and in Turkey seem to be associated with this compressive front which is also present in southern Anatolia (Lycian overthrust nappes). In other words, there existed during the Upper Oligocene to Lower and Middle Miocene an 'Ionian-Lycian' arc whose development ended with the powerful compressive phase of the Langhian stage (16 Ma). This phase affected northwestern Greece as well as southern Anatolia, and ended with the emplacement of the Lycian nappes (Poisson 1978). No major tectonic event affected northwestern Greece after the Langhian until the Lower Pliocene compressive phase, when the Aegean arc appears at a position which represents a 'jump' from the position of the Ionian arc to a position very close to its present one (Mercier *et al.*, 1979).

As shown in Fig. 6, no significant rotation occurred during the period 12 to 5 Ma, that is immediately preceeding the 'jump' of the arc. However one can reasonably wonder whether the other phases of important tectonic activity

TABLE 1. *Palaeomagnetic results from Ionian flysch sites. The average was calculated with the usual Fisher's statistics. Sites containing samples of both polarities are indicated. The polarity assigned to thsese sites is the one of the majority of samples. The palaeomagnetic directions of samples which were not in the majority were inverted through the origin and averaged with the other samples to obtain the final direction.*

Sites	Age	Geol. position	polarity	N	D	I	K	α_{95}
EP 137			N	11	47°5	25°4	47.7	6°11
EP 138			R	9	239°	−34°7	157	3°7
EP 139	Lower	Ellinikon	R	13	238°7	−36°	355.6	2°14
			R	8	231°5	−34°	236.8	3°2
EP 140	Oligocene	Syncline	N	2	49°8	32°	1239	2°8
			total	10	231°	−33°8	286	2°6
EP 141			R	14	235°5	−36°	347	2°
EP 142	Upper Aquitanian		N	11	51°2	37°4	96.8	4°3
			R	9	242°5	−46°	155	3°7
EP 143		Botzara	N	4	58°7	44°3	190	5°1
			total	13	241°	−45°4	167.6	3°
			N	6	47°5	56°	21	12°5
EP 175	Middle	Syncline	R	5	228°3	−54°	78.5	7°1
			total	11	47°	55°	34.3	7°2
			N	9	59°	45°6	52.6	6°4
EP 176	Aquitanian		R	5	235°	−45°6	47.5	9°1
			total	14	57°5	45°6	54.1	5°1
			R	8	264°5	−44°	77.3	5°6
EP 177			N	6	75°2	44°2	62	7°25
			total	14	260°6	−44°2	66.8	4°58
EP 170	Lower	Dodoni	R	8	252°	−31°	298	2°87
EP 171			R	12	254°	−28°	86.5	4°34
EP 172	Oligocene	Syncline	N	6	62°5	35°6	61	7°3
			R	4	262°	−39°2	83.5	7°6
			total	10	70°	37°4	43	9°6
AK 154	Lower	Southern	N	7	75°3	26°4	30	9°6
AK 155			N	12	30°7	27°	27.4	7°7
AK 156	Miocene	Akarnania	N	7	81°4	20°3	25.1	10°5

have been accompanied by rotational deformation.

As already mentioned, the biostratigraphic age determinations for the sites sampled in the Ionian flysch sections are not yet complete. In addition, sites overlapping the Langhian compressive phase on both sides in age, and suitable for palaeomagnetic studies have not yet been located. So the present results are not precise enough to complete the curve of Fig. 4 for older ages with the same degree of accuracy as for the neotectonic period. In particular no details can be obtained about the effect of the Langhian compressive phase on rotational deformation.

Some conclusions can however be drawn from these results from a geodynamic point of view. Focusing attention on Epirus, where many data are available, one may note that the present day strike of the main structures is N 155 E. One can reconstruct the original orogenic system by rotating the whole area counterclockwise through an angle of about 55–60° corresponding to the sum of the two measured rotations. One then finds that the major structures were initially oriented almost perfectly east-west. This implies that the curvature of the arc was almost zero at the Upper Oligocene-Lower Miocene boundary. It is significant that the same conclusion has also been reached with entirely different arguments based on geological observations at the other termination of the arc in southern Anatolia (Poisson, this volume).

For a precise palinspastic reconstruction, the position of the eulerian pole of both rotations is necessary. A reliable determination of the pole position for the Plio-Quaternary rotation has been given by Le Pichon & Angelier (1979), but nothing is known about the Oligo-Miocene rotational deformation, which is reported here for the first time. We think, moreover, that it may be incorrect to describe this deformation in terms of rotation about a single well defined pole although this notion can certainly be used as a first approximation for a relatively small region like Epirus.

Within this approximation, geological considerations, such as the regular shape of the Bot-

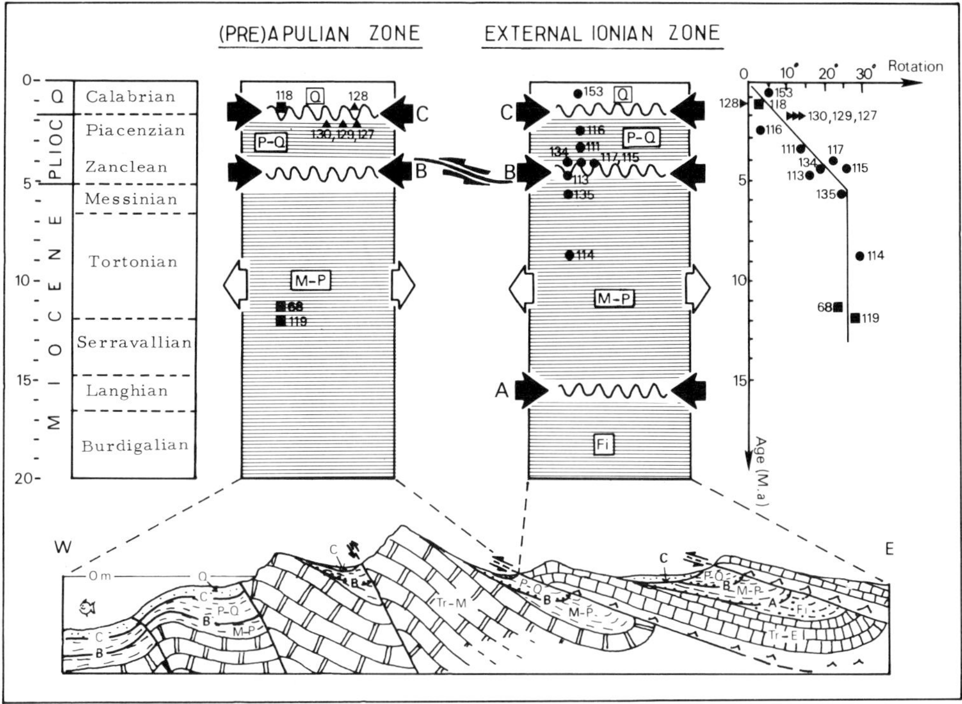

FIG. 6. Geological setting of the sampling in the Ionian islands: summary of the tectonic and sedimentary evolution and of the palaeomagnetic results.

The discordances (A, B, C) are relative to the main compressive phases. Q, Continental and marine deposits, Lower Pleistocene to present; P-Q, Pliocene to Calabrian discordant series. Preapulian zone: M-P, concordant marls, Middle Miocene to Lower Pliocene; Tr-M, Triassic-Middle Miocene carbonate series. External Ionian zone: M-P, transgressive marine molasse, Middle Miocene to Lower Pliocene; Fi, flysch ending in Burdigalian; Tr-E, Triassic-Eocene carbonate series overlying Triassic evaporites.

zara syncline which has undergone the two rotations, suggest in our opinion that the Oligo-Miocene rotation has occurred about a pole situated approximately in the same region as the Plio-Quaternary one.

Using this pole position, the displacement in Epirus corresponding to the two rotations is more than 100 km. Unfolding of the main structures only accounts for about 20 km of shortening, so these palaeomagnetic results imply the existence of large scale thrust faulting. The question then arises as to whether the sedimentary cover alone or the basement as well is affected by the thrusts.

Extensive overthrusting involving the sedimentary cover has been recognized by several authors for both the Oligo-Miocene and Plio-Quaternary stages. (IGRS-IFP 1966, Aubouin 1973, Jackson *et al.*, 1981, Sorel & Cushing 1982, Monopolis *et al.*, 1982, Jamet 1982, and many others). However, because there is no evidence of gravitational thrusts, it has also been recognized as early as 1960 that this cover deformation could be a consequence of large scale continental lithospheric shortening (Brunn 1960).

In our opinion the palaeomagnetic results, which, as far as one can judge from the presently studied sites, yield coherent results from regions separated by considerable distances and whose surface geological structures can be quite dissimilar, reflect large scale lithospheric shortening. This interpretation implies the existence of differential basement shortening increasing southward in order to account for the observed rotation. The southward increase of the displacement of allochthonous cover thrusts recognized by Aubouin (1973) could be related to this differential shortening.

As a final remark, we would like to point out the correlation which exists between the two rotational deformation phases and the occurrence of volcanic activity along the arc. Indeed it has been shown that this activity consists of two main phases one during the Oligo-Miocene and one during the Plio-Quaternary stages separated by a period of quiescence (Fytikas *et al.*, 1976, Bellon *et al*, 1979, Innocenti *et al.*, 1979, Fytikas *et al.*, this volume). It has also been suggested that the Oligo-Miocene volcanism is related to major thrusting of continental-type lithosphere (Bellon *et al.*, 1979).

Within the precision of the palaeomagnetic results, the two phases of activity and the period of quiescence are correlated in time with the two phases of rotational deformation and the period when no rotational deformation occurs.

Moreover, geochronological and volcanic data indicate that the end of the Oligo-Miocene calc-alkaline volcanism has not been synchronous over the Aegean: the oldest products can be found in Thrace, the youngest in Skiros and Evvia. A general southward migration of this volcanic cycle has been recognized. (Fytikas *et al.*, 1976, Bellon *et al.*, 1979, Innocenti *et al.*, 1979, Fytikas *et al.*, this volume). The real extent of this southward migration could however be as much as 50% smaller than might appear on the basis of the present day distribution of volcanic products because of more recent dextral strike-slip motion along the North Anatolian Fault (Fytikas *et al.*, this volume).

Nevertheless, the direction of the migration is the southward one expected assuming a clockwise rotation about a pole situated to the west of the area affected by the volcanism. Moreover, assuming as a first approximation that the migration is due to a rigid rotation of 30°, one can obtain a value of about 600 km for the distance to the pole of rotation. This is almost exactly the distance of the volcanic products to the position of the assumed 'pole' of the Oligo-Miocene rotational deformation. Of course this is only a geometrical construction, but the fact that the right order of magnitude is obtained suggests that joint volcanological and palaeomagnetic studies could yield new ideas about the geodynamic evolution of this area.

ACKNOWLEDGEMENTS: We wish to thank Prof. J. Mercier, X. Le Pichon and J. Angelier for many interesting dicussions, Dr D. Sorel for his active and invaluable participation in the field work in the south and central parts of the arc and for his help in selecting and sampling some of the sites in the Ionian flysch. J. P. Valente, J. P. Valet, E. Barrier, S. Roy and A. Mazaud helped with the sampling. Mrs G. Bizon and C. Muller supplied the biostratigraphic determinations.

Prof. J. D. A. Zijderveld made his laboratory available to us and Dr C. Langereis and J. Vandenberg helped us with discussions and in the measurements. We acknowledge the stimulating discussions with M. Fytikas, F. Innocenti, P. Manetti, R. Mazzuoli and L. Villari.

We also wish to thank Prof. K. Verosub for many discussions and for his help in improving the quality of the manuscript.

The Director of the Institute of Geology and Mining Research (IGME) at Athens kindly provided the necessary permits.

The financial support has been partially given by the CNRS-INAG ATP Geodynamique I & II.

References

ANGELIER, J., LYBERIS, N., LE PICHON, X., BARRIER, E. & HUCHON, P. 1982. The tectonic development of the Hellenic Arc and the Sea of Crete: a synthesis. *In:* LE PICHON, X., AUGUSTITHIS, S. S. & MASCLE, J. (eds). Geodynamics of the Hellenic Arc and Trench. *Tectonophysics*, **86**, 159–196.

AUBOUIN, J. 1973. Des tectoniques superposées et leur signification par rapport aux modèles géophysiques: l'exemple des Dinarides; paléotectonique, tectonique, tarditectonique, néotectonique. *Bull. Soc. géol. Fr.* **7**, 426–460.

BELLON, H., JARRIGE, J. J. & SOREL, D. 1979. Les activités magmatiques égéennes de l'Oligocène a nos jours, et leurs cadres géodynamiques, données nouvelles et synthèse. *Rev. Géogr. phys. Géol. dyn.* **21**, 41–55.

BRUNN, J. H. 1960. Les zones helléniques internes et leur extension. Réflexions sur l'orogenèse alpine. *Bull. Soc. géol. Fr.* **7**, 470–486.

—— 1976. L'arc concave zagro-taurique et les arcs convexes taurique et égéen: collision et arcs induits. *Bull. Soc. géol. Fr.* **18**, 553–567.

CHIRON, G., LAJ, C. & POCACHARD, J. 1981. A high sensitivity portable spinner magnetometer. *J. Phys. E. Sci. Instr.* **14**, 977–981.

FLEURY, J. J. 1980. *Les zones de Gavrovo-Tripolitza et du Pinde-Olonos (Grèce continentale et Péloponnèse du nord). Evolution d'une plate-forme et d'un bassin dans leur cadre alpin.* Thèse d'état, *Soc. géol. Nord, Publ.* **4**, 651 pp.

FYTIKAS, M., GIULIANI, O., INNOCENTI, F., MARINELLI, G. & MAZZUOLI, R. 1976. Geochronological data on recent magmatism of the Aegean sea. *Tectonophysics* **31**, 29–34.

IGSR-IFP, 1966. *Etude Géologique de l'Epire.* Editions Technip, Paris, 306 pp.

INNOCENTI, F., MANETTI, P., MAZZUOLI, R., PASQUARE, G. & VILLARI, L. 1981. Neogene and Quaternary volcanism in the Eastern Mediterranean. Time-space distribution and geotectonic implication. *In:* WEZEL, C. (ed.). *Sedimentary Basins of Mediterranean Margins.* Italian Project of Oceanography, Technoprint, 369–385.

JACKSON, J. A., FITCH, T. J. & MCKENZIE, D. P. 1981. Active thrusting and the evolution of the

Zagros fold belt. *In:* McClay, K. R. & Price, N. J. (eds.). *Thrust and Nappe Tectonics.* Geol. Soc. Lond. Spec. Publ. **9,** 371–379.

Jamet, M. 1982. *Etude néotectonique de Corfou et étude paléomagnetique des sédiments néogènes de îlés de Corfou, Céphalonie et Zanthe.* Thèse 3eme cycle, (unpubl.), Orsay, France, 146 pp.

Laj, C., Jamet, M., Sorel, D. & Valente, J. P. 1982. First paleomagnetic results from Mio-Pliocene series of the Hellenic sedimentary arc. *In:* Le Pichon, X. Augustithis, S. S. & Mascle, J. (eds). Geodynamics of the Hellenic Arc and Trench. *Tectonophysics,* **86,** 45–67.

Le Pichon, X. & Angelier, J. 1979. The Hellenic Arc and Trench system: a key to the neotectonic evolution of the eastern Mediterranean area. *Tectonophysics,* **60,** 1–42.

Lyberis, N., Angelier, J., Barrier, E. & Lallemant S, 1982. Active deformation of a segment of arc: the strait of Kithira, Hellenic Arc, Greece. *J. struct. Geol.* **4,** 299–311.

McElhinny, M. W. & Cowley, J. A. 1977. Paleomagnetic directions and pole positions, XIV. Pole numbers 14/1 to 14/574. *Geophys. J. R. astron. Soc.* **49,** 313–356.

—— & —— 1978. Paleomagnetic directions and pole positions, XV. Pole numbers 15/1 to 15/232. *Geophys. J. R. astron. Soc.* **52,** 259–276.

Mercier, J. 1977. L'Arc Egéen, une bordure déformée de la plaque eurasiatique, réflexions sur un exemple d'étude néotectonique. *Bull. Soc. géol. Fr.* **19,** 663–672.

——, Delibassis, M., Gauthier, A., Jarrige, J. J., Lemeille, F., Philip, H., Sebrier, M. & Sorel, D. 1979. La néotectonique de l'Arc Egéen. *Rev. Géol. dyn. Géogr. phys.* **21,** 61–72.

——, Carey, E., Philip, H. & Sorel, D. 1976. La néotectonique Plio-Quaternaire de l'Arc Egéen externe et de la mer Egée et ses relations avec la séismicité. *Bull. Soc. géol. Fr.* **19,** 355–72.

Monopolis, D. & Bruneton, A. 1982. Ionian sea (western Greece): its structural outline deduced from drilling and geophysical data. *Tectonophysics,* **83,** 227–242.

Poisson, A. 1978. *Recherches géologiques dans les Taurides occidentales (Turquie).* Thèse d'Etat, (unpubl.), Orsay, France, 1189 pp.

Sorel, D. 1976. *Etude néotectonique dans l'Arc Egéen externe occidental; les Iles Ioniennes de Kephallinia et Zackinthos et l'Elide occidentale.* Thèse 3eme cycle, (unpubl.), Orsay, France, 196 pp.

—— & Cushing, M. 1982. Mise en évidence d'un charriage de couverture dans la zone ionienne en Grèce occidentale: la nappe d'Akarnanie-Levkas. *C.r. Acad. Sc. Paris.* **294,** 675–678.

Tapponnier, P. 1977. Evolution tectonique du système alpin en Méditerranée: poinçonnement et écrasement rigido-plastique. *Bull. Soc. géol. Fr.* **19,** 437–460.

Valente, J.-P., Laj, C., Sorel, D., Roy, S. & Valet, J.-P. 1982. Paleomagnetic results from Mio-Pliocene marine sedimentary series in Crete. *Earth planet. Sci. Lett.* **57,** 159–172.

C. Kissel & C. Laj, Centre des Faibles Radioactivités, Domaine du CNRS, 91190 Gif-sur-Yvette. France.

M. Jamet, Laboratoire de Géologie Dynamique Interne, Université de Paris Sud, 91405 Orsay. France and Centre des Faibles Radioactivités, Domaine du CNRS, 91190 Gif-sur-Yvette. France.

Palaeomagnetic data from Tertiary units of the north Aegean zone

D. Kondopoulou & J. P. Lauer

SUMMARY: New palaeomagnetic data from the north Aegean area show that in this region an overall clockwise rotation is likely to have taken place in the late Tertiary.

The Aegean area is an important part of the Alpine system which stretches from Western Europe to the Himalayas. Brunn (1960), Aubouin *et al.* (1976) and others have suggested that structural features might be correlateable across the Aegean Sea.

In the southern Aegean, seismic foci are distributed along a northward dipping surface extending from the Hellenic arc (Papazachos 1973). Using focal mechanisms, Le Pichon & Angelier (1979) produced a model involving plastic deformation, north-south extension and an essentially clockwise rotation since 13 Ma. Thus, at least part of the apparent flexure of the Alpine system in this region may have occurred 'recently' and perhaps still be going on. Palaeomagnetism is an appropriate technique to check such a hypothesis.

Palaeomagnetic sampling, measurements and results

If all later overprints can be eliminated, the TRM of a lava gives the instantaneous direction of the magnetic field at the time of cooling of the flow. To average the secular variation we must use data from a number of flows of similar geological age so that the effects of small-period fluctuations away from the axial geocentric dipole can be minimized. In sediments, in contrast, acquisition of the characteristic magnetization seems to take more time. The timing of acquisition of magnetization is not well known, but the secular variation is attenuated and corrections for dip are obviously easier.

Our study deals with Neogene lavas from five groups of sites (Almopias, Lemnos, Lesbos, Karaburun, Afyon) and one granodioritic batholith (Sithonia). In almost all cases, interlayered sediments are absent so correction for tilt is impossible. The only way to minimize errors due to local tectonics is to collect samples over wide areas and average out the results. Even then the dispersal of palaeomagnetic directions could reflect regional effects, for example large scale tilting. It follows that the mean direction of magnetization of any one site is without significance in deformed areas.

Between 2 and 8 large orientated samples were collected at each site without drilling in the field. AC and/or thermal demagnetizations were carried out on all the samples. The mean directions have been computed by Fisher statistics, first as the mean site directions and then as the mean directions for groups of sites. The results are given in Table 1. More details on sampling and measuring procedures can be found in Lauer (1981) and Kondopoulou (1982).

Almopias

This region (41.1°N/21.8°E) is about 30 km north of Edessa and belongs to the Vardar zone. Sampling was of lavas and dykes in a group of 5 sites separated by a maximum distance of 10 km; the ages are between 2.4 and 4 Ma (Bellon *et al.* 1979).

Apart from one site, the directions of the characteristic magnetization of our 20 samples are rather scattered and a secondary magnetization is common. Site PE contains two stable magnetization directions. Another site is unusable. The remaining mean site directions are coherent (Fig. 1). On six different sites in the same region, Bobier (1968) obtained a comparable result with reversed polarity.

Sithonia

This is the 'middle finger' of the Chalcidiki peninsula. A large batholith of granodiorite, granite and monzo-diorite crops out over a distance of more than 20 km. A K-Ar determination on a biotite by R. Montigny and R. Thuizat (pers. comm., 1982) yields a 40 ± 1.5 Ma age. Thus this unit comes from near the Eocene-Oligocene boundary.

Our 5 sites (20 samples) are located along the eastern and western coasts of the peninsula (Fig. 2). After AC and thermal cleaning, N and R characteristic magnetizations were observed. Their directions are in good agreement at each

TABLE 1. Palaeomagnetic data from the north Aegean area. N is the number of sites, n the number of samples (sp = specimens) which have been used for a compilation of the average direction

	Age M.a.	D°	I°	k	a°95	N	n	VGP Lat N°	VGP Long E°	Reference
Almopias	2.6–4	17	+54	32	16	4		75	132	This paper
Sithonia	40	215.8	−25.4	22	16	5		49	144	This paper
Lemnos A	17	349	+23	126	22		2			This paper
C	to	9	+54	21	17		2(5sp)			This paper
E	22	31	+21	108	8.8		4			This paper
Lesbos	15–16	8.2	+56.5	47.3	13.4	4				This paper
	18	22.2	+39.9	8.3	28	5				This paper
	15–18	12.4	+48.9	11.4	14.8	10		78	146	
Karaburun	16.5–19	234	−55	34.2	21	3		48	101	This paper
Afyon 1	8 to	14.2	+41.1	612	2.2		8			This paper
2	15	202.5	−56.4	180	5.0		6			This paper
Kula	<0.7	11.6	+61.6	254	4.8	5		80.2	86.5	Sanver (1968)
Almopias	2.6–4	195	−66.5	40	10		15	77	70	Bobier (1968)
Lesbos	15–18	8.4	+40.2	8.1	7.8		47			Pe-Piper (1972)

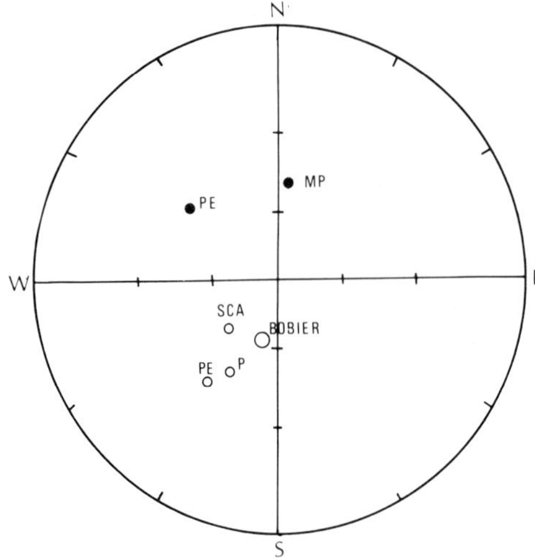

FIG. 1. Mean site directions after AC and thermal cleaning for the Almopias group (4 sites, 16 samples). Open symbols are in the upper hemisphere. The mean result by Bobier (1968) is given for comparison.

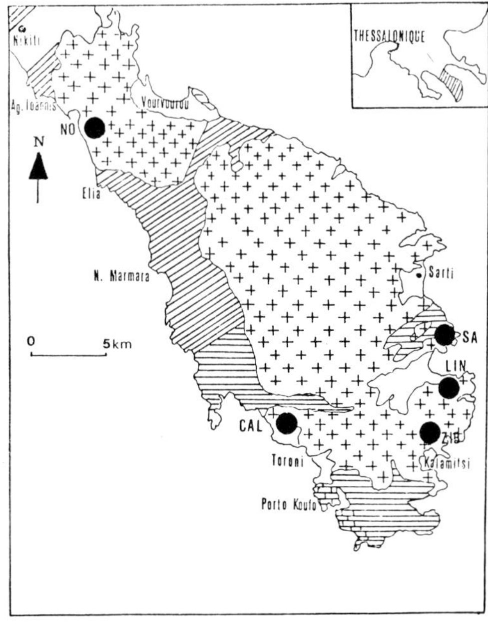

FIG. 2. Location of sampling sites on the Sithonia peninsula.

site and also from site to site over the massif (Fig. 3). The question arises as to whether this magnetization is really 40 Ma old. The mean inclination is very low.

Lemnos

The volcanic rocks of Lemnos are dated at from 17 to 22 Ma (Borsi *et al.* 1972). Our 4 sites (10 samples) are located no more than 1.5 km apart, in the south-west of the island between Myrina (Kastro) and Kondias. In this region the K-Ar ages cluster around 17 Ma.

Figure 4 shows the average site directions after AC cleaning. In Table 1, site B, with only two samples, has been eliminated because of its very disturbed results during cleaning. For the three other remaining sites the average direction corresponds to: D = 10°, I = +34°; but a_{95} is large (42°) and no firm conclusion concerning possible rotations can be drawn from these data alone; our sites are not sufficiently widely-spaced and numerous.

Lesbos

K-Ar ages on volcanics from Lesbos range from 15.5 to 18 Ma (Borsi *et al.* 1972). The

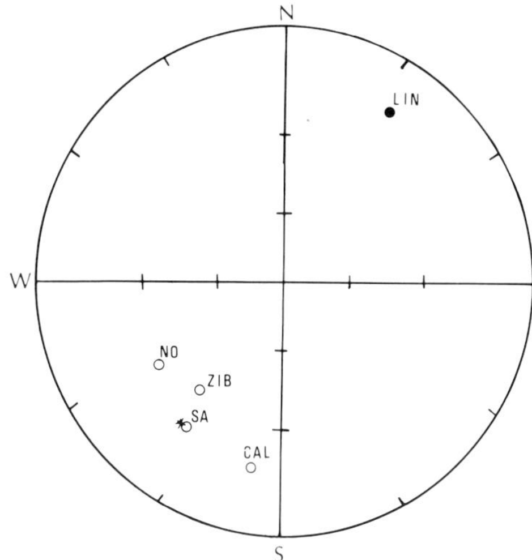

FIG. 3. Sithonia: mean site directions after AC and thermal cleaning. Open symbols: upper hemisphere.

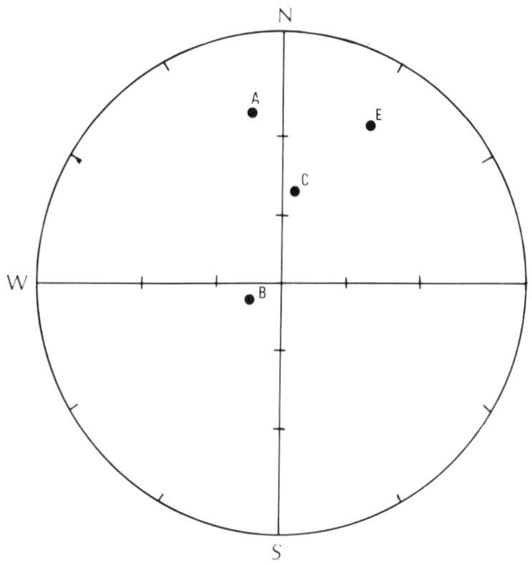

FIG. 4. Mean site directions for Lemnos island (after AC cleaning).

trend of fold axes is towards the NNE (Jacobshagen *et al.* 1976). 11 sites have been investigated (Fig. 5): 4 are near Mytilini, on the south-east coast (KATE, ML, MT and LAT); 8 others are in the north-west of the island (LEI, PHI, SCL, STY, EFT, PT and THE). During AC cleaning, the behaviour differed: KATE, LAT, ML, MT, SCL and PHI have a unique and relatively soft magnetization; the results are well grouped. For THE and STY, a complete demagnetization is impossible and one needs the help of the 'convergence method' to get a characteristic direction. The PT and LEI

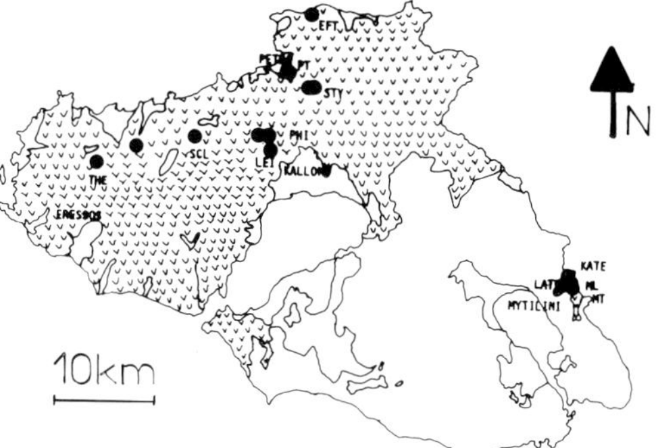

FIG. 5. Location of sampling sites on Lesbos island.

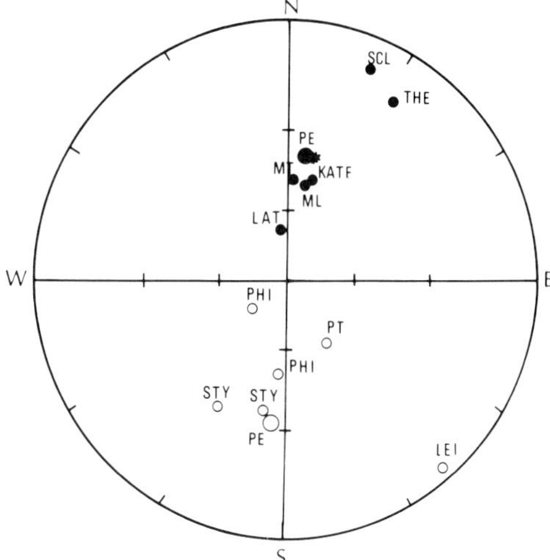

FIG. 6. Lesbos: mean site directions after AC cleaning. Open symbols are in the upper hemisphere.

sites yield intermediate directions; EFT has been eliminated because of scattered results. Fig. 6 shows the mean cleaned site directions.

In Table 1 we tentatively group our sites into older and younger units. Then, considering sites LAT, SCL, PHI 1, PT, KATE, ML, PHI 2, THE and STY we have calculated an overall average direction. Pe-Piper (1978) has presented cleaned palaeomagnetic results on 47 volcanic samples of widely scattered locations; the calculated average direction does not differ very much from ours.

Karaburun

The Karaburun peninsula is in Turkey, west of Izmir. According to Jacobshagen *et al.* (1976) the major tectonic trends are NNE-SSW.

Three sites, one of them 10 km apart, have

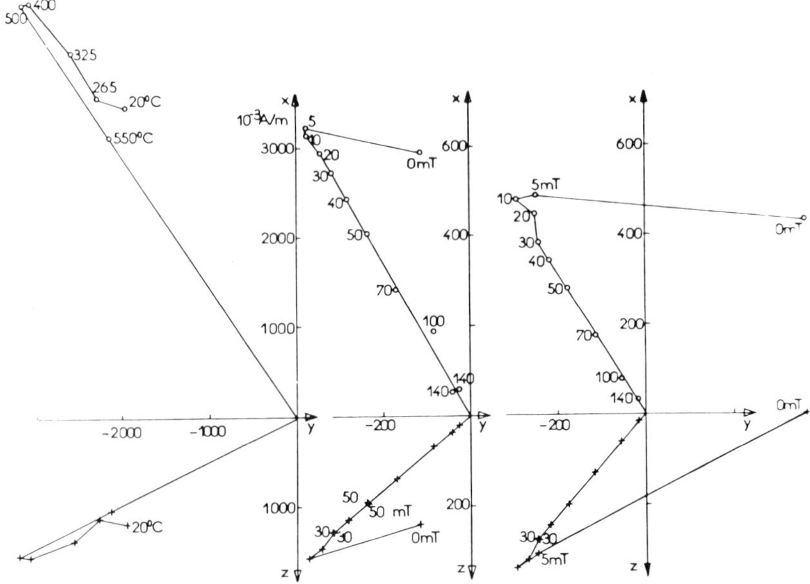

FIG. 7. Karaburun: examples of curves yielded by thermal (left) and AC demagnetization (crosses in the horizontal xy plane, open circles in the vertical yz plane).

been chosen in Miocene andesite flows. A total of 19 samples were collected. The age of these units is between 16.6 and 18.2 Ma (Borsi *et al.* 1972). A K-Ar determination by R. Montigny and R. Thuizat (pers. comm., 1982) on one of our sites yielded 18.3 ± 0.6 Ma.

Figure 7 gives examples of curves resulting from AC and thermal cleaning and these show a secondary magnetization, roughly antiparallel to the primary one which is easily destroyed during demagnetization. No sample has been omitted. A tentative correction for dip was negative and such a correction would increase the discrepancy between the sites (Fig. 8). This result is not surprising as viscous lavas like andesites can come to rest on steep slopes.

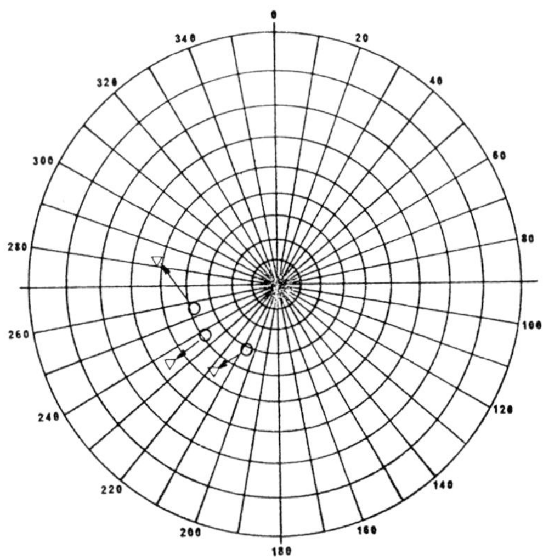

FIG. 8. Tentative correction for dip of mean characteristic directions for the three sites of Karaburun (Triangles: after correction).

Afyon

In this region Besang *et al.* (1977) obtained ages ranging from 8.6 to 14.8 Ma. Our two sites are along the road to Ankara, about 30 km NE of Afyon. One is of fine white tuffaceous agglomerate (? ignimbrite), the other includes andesitic lavas from several flows. The white unit is definitely horizontal.

Figure 9 gives examples of curves resulting from demagnetization of lavas and of the white unit. One site is normally magnetized, the other reversely magnetized. None of the 14 collected samples has been eliminated. The data yield two precise directions of the palaeo-field, but the palaeo-secular-variation is not averaged.

Conclusion

Our measurements are from 27 Tertiary sites in the north Aegean area. Taking into account results obtained by Bobier (1968), Sanver (1968) and Pe-Piper (1972), we observe that along a traverse going from the Vardar zone to Afyon, ca 750 km, all the average directions for

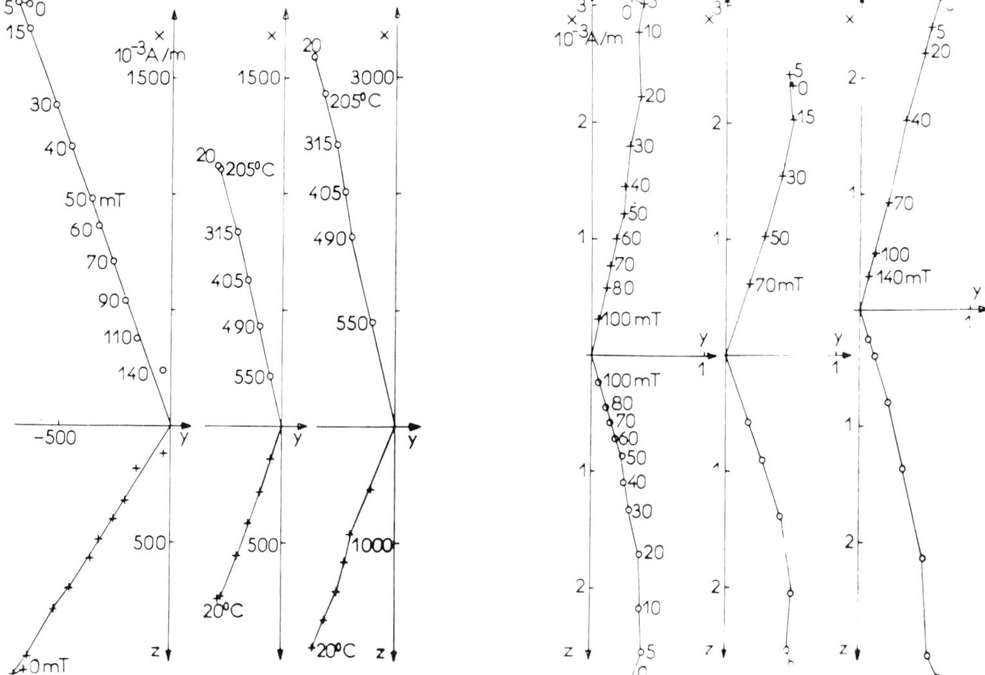

FIG. 9. AC and thermal demagnetization curves for Afyon I ('white unit', right) and Afyon 2 (lavas, left).

groups of sites are to the east and these could be the palaeo-directions of the field. A late Tertiary clockwise rotation is thus probable for this zone. Unfortunately in some places (Lemnos, Lesbos, Afyon) the semi-angle of confidence a_{95} exceeds the average declination. There the rotation although probable, can not be demonstrated. The most clearly rotated units (Sithonia, Karaburun) are among the oldest.

ACKNOWLEDGEMENTS: We thank Professor A. Roche, who was an initiator of this study, and Professors O. Kaya and B. Papazachos, Drs M. Westphal, R. Montigny, A. Gümüş and T. Chadzidimitriou for their help.

References

AUBOUIN, J., BONNEAU, M., DAVIDSON, J., LE BOULENGER, P., MATESCO, S. & ZAMBETAKIS, A. 1976. Esquisse structurale de l'arc égéen externe: des Dinarides aux Taurides. *Bull. Soc. géol. France 18,* 327–36.

BELLON, H., JARRIGE, J. J. & SOREL D. 1979. Les activités magmatiques égéennes de l'Oligocène à nos jours et leurs cadres géodynamiques: données nouvelles et synthèse. *Rev. Géol. dynam. et Géogr. phys.* **21,** 41–55.

BESANG, C., ECKHARDT, F. J., HARRE, W., KREUZER, H. & MULLER P. 1977. Radiometrische Altersbestimmungen an neogenen Eruptivgesteinen der Türkei. *Geol. Jahrb.* (*Hannover*) **B25,** 3–36.

BOBIER, C. 1968. Etude paléomagnétique de quelques formations du complèxe volcani que d'Almopias (Macédoine centrale, Grèce). *C.r. Séances Acad. Sci. Paris* **267,** 1091–4.

BORSI, S., FERRARA, G., INNOCENTI, F. MAZZUOLI, R. 1972. Geochronology and petrology of recent volcanics in the Eastern Aegean Sea (W. Anatolia and Lesvos island). *Bull. Volc.* **36,** 473–96.

BRUNN, J. H. 1960. Les zones helléniques internes et leur extension. Réflexions sur l'orogénie alpine. *Bull. Soc. géol. France* **2,** 470–86.

JACOBSHAGEN, V., SKALA, W. & WALLBRECHER, E. 1976. Observations sur le développement tectonique des Sporades du Nord. *Bull. Soc. géol. France* **18,** 281–6.

KONDOPOULOU, D. 1982. *Paléomagnétisme et déformations néogènes du Nord de la Mer Egée.* Thesis, Dr. spec. (unpublished), Univ. Strasbourg I, France, 123 pp.

LAUER, J. P. 1981. *L'évolution géodynamique de la Turquie et de Chypre déduite de l'étude paléomagnétique.* Thesis, Univ. Strasbourg I, France. Thèse Doctorat-ès-Sciences. 299 pp.

LE PICHON, X. & ANGELIER, J. 1979. The Hellenic arc and trench system: a key to the neotectonic evolution of the Eastern Mediterranean Area. *Tectonophys.* **60,** 1–42.

PAPAZACHOS, B. C. 1973. Distribution of seismic foci in the Mediterranean and surrounding area and its tectonic implication. *Geophys., J. R. astr. Soc.* **33,** 421–430.

PE-PIPER, G. G. 1978. *The Cenozoic volcanic rocks of Lesbos island* (In Greek). Thesis, Unpubl. Univ. Patras D.Sc. Thesis 379 pp.

SANVER, M. 1968. A palaeomagnetic study of Quaternary volcanic rocks from Turkey. *Earth planet. Inter.* **1,** 403–21.

D. KONDOPOULOU, Geophysical Laboratory, University of Thessaloniki, Greece.

J. P. LAUER, Inst. de Physique du Globe, 5, rue René-Descartes, 67084 Strasbourg, France.

Tertiary to Quaternary evolution of volcanism in the Aegean region

M. Fytikas, F. Innocenti, P. Manetti, R. Mazzuoli, A. Peccerillo & L. Villari

SUMMARY: Widespread orogenic volcanic activity has continued in the Aegean area from the Oligocene to present. Two main phases of activity are recognized. One developed in the North Aegean area from Oligocene to Middle Miocene times and a second started in the Pliocene, building the active South Aegean volcanic arc. Between these two phases, Upper Miocene to Quaternary volcanism of variable petrogenetic affinity occurred to a limited extent, essentially on the margins of the Aegean microplate.

The products erupted during the Oligo–Miocene phase consist mainly of calc-alkaline and shoshonitic intermediate lavas and pyroclastics with minor acidic and basic rock types. The volcanic activity started in the northernmost part of the North Aegean area with mostly calc-alkaline intermediate and acidic volcanics. The volcanism shifted successively southwards becoming progressively enriched in potassium. This evolution is interpreted as being related to an increase in the dip of the Benioff zone under the Eurasian plate, resulting from a reduction in the plate convergence rate after continental collision.

The volcanic products of the active south Aegean arc are mainly andesites with minor basalts and rhyolites which display the chemical character typical of calc-alkaline series erupted on thin continental margins. The South Aegean arc is believed to be the surface expression of active subduction of the African plate.

Scattered Upper Miocene to Quaternary activity is interpreted as occurring in zones of tensional strain along the borders of the Aegean microplate.

After the main Tertiary phase of continental collision along the contact between the Eurasian and African plates, a marked lithospheric fragmentation episode occurred, particularly in the Eastern Mediterranean (Dewey *et al.* 1973). In this sector of the Alpine chain numerous microplates were formed. From Middle to Upper Miocene time, up to the present, these have undergone independent evolution to produce a very complex neotectonic framework (McKenzie 1972, 1978).

The Aegean is an area of crucial importance as it is the only zone where active lithospheric subduction is still clearly going on with a frontal compressional region, an active volcanic arc, and behind, a vast area dominated by tensional processes with major subsided basins (Le Pichon & Angelier 1981).

In the northern part of this region a large transcurrent lineament, running approximately NE–SW, has been interpreted as the continuation of the North Anatolian Fault and appears to separate an Aegean sector from an Eurasian one (Dewey & Şengör 1979).

This work summarizes the evolution of the volcanism from the Oligocene to the present, with a view to improving understanding of the correlation between the nature of the magmas, the character of the volcanic processes and their tectonic framework. This will contribute to a clearer conception of the geodynamic evolution of the region.

The volcanism of the Aegean area

General description

From the Oligocene onwards extensive volcanism occurred in the Aegean area and western Anatolia. It has developed with varying intensity and distribution in space and time (Fig. 1).

The ages of the volcanics are now known very precisely, from stratigraphic data available on recent Greek Geological Survey (IGME) maps and particularly from K-Ar dating, summarized in Table 1.

The discussion of the petrogenesis of the volcanics and their evolution is based on 901 selected major-element analyses, including 230 new analyses, and uses the same methodology as was applied to reconstruct the evolution of volcanism in Anatolia and Northwest Iran (Innocenti *et al.* 1982a).

Two main phases of orogenic volcanism have been recognized in the Aegean area (Fytikas *et al.* 1976, 1979): one of these develops mainly from Oligocene to Middle Miocene and the other begins in Middle Pliocene and continues to the present.

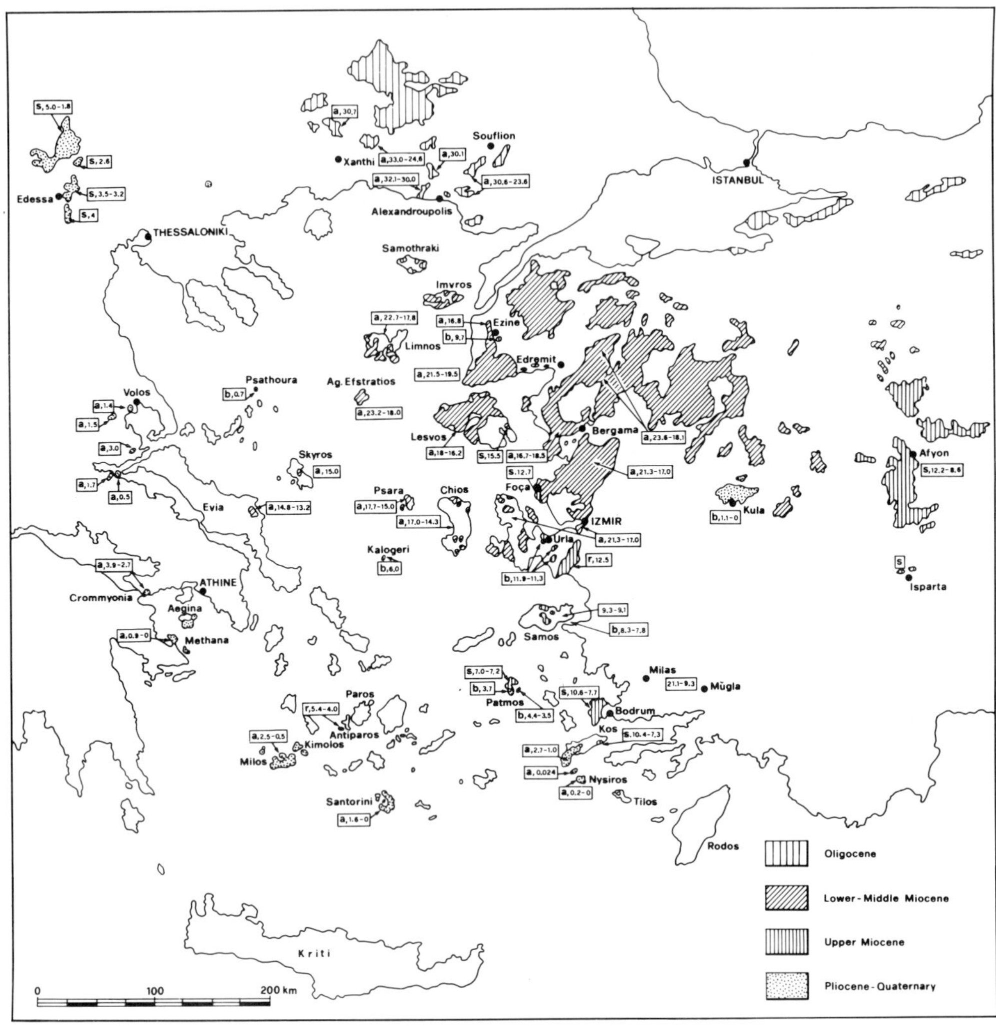

FIG. 1. Age and distribution of the volcanic products from the Oligocene to present time in the Aegean area, including Western Anatolia. Magmatic affinity is also shown within an age framework: b, basaltic; a, andesitic; s, shoshonitic; r, rhyolitic.

Between these two main phases, scattered volumetrically minor volcanism is observed with variable petrogenetic character.

Oligocene-Miocene volcanic phase: North Aegean Tertiary Activity ('NATA')

The volcanism of this phase occurred in a belt extending mostly in an E–W direction from continental Greece (Thrace) to the central Aegean, north of a line from central Evia to the Izmir region. This extensive belt continues eastwards with similar characteristics as far as Central Anatolia (Innocenti *et al.* 1982a).

The available radiometric dates (Table 1) show that the volcanic activity developed from Upper Oligocene to Middle Miocene times. However the various products are not uniformly distributed over the whole area. The oldest are confined to the northernmost part and in Greece occur on the southern margin of the Rhodope Massif and in Bulgaria extend into the Massif itself. In the Greek part of Thrace the age of these volcanics has been determined by the K-Ar method as 33.0 Ma to 23.6 Ma. However, there is stratigraphic evidence to show that the volcanic activity had already begun by the end of the Eocene from the

TABLE 1. *K-Ar ages of Tertiary and Quaternary volcanics from the Aegean area*

Locality	Ages (Ma)	References
	Oligocene	
Thrace		
N of Xanthi	30.0; 30.7; 24.6	1
W of Alexandroupolis	33.1; 32.1; 30.0	1
N of Alexandroupolis	30.1	1
S of Souflion	30.6; 29.7; 29.2; 28.5; 28.4	1
	27.4; 26.8; 26.0; 25.6; 23.6	
	Lower-Middle Miocene	
Limnos	21.6; 20.4; 19.3; 19.1; 18.7; 17.8	2
	22.7; 21.3; 19.7	Up.
Agios Evstratios	23.2; 18.0	2
Lesvos	18.0; 16.9; 16.2; 15.5	3
	17.3	4
Chios	17.0; 17.0; 15.9; 15.5; 14.6; 14.3	5
Psara-Antipsara	17.7; 14.3; 15.0	Up.
Skyros	15.0	2
Evia	14.8; 13.2	2,6
	13.2	7
	Upper Miocene	
Samos	9.3; 9.2; 9.1	8
	8.3; 7.9; 7.8	9
Patmos	7.2; 7.0	6
Kos	10.0; 7.3	7
	10.4	10
Bodrum	10.6; 10.5	Up.
	7.9; 7.8	9
Kalogeri	6.0	11
	Pliocene—Quaternary	
Edessa area (Voras etc.)	5.0; 4.6; 4.5; 4.5; 4.4; 4.3	12
	4.0; 2.5; 1.9; 1.8	12
	4.0; 3.5; 3.2; 2.7	7
Volos area	3.0; 1.6; 1.5; 1.4	13
	2.7; 1.2	14
Psathoura	0.7	Up.
Likhades	0.5	6
Kamena Vourla	1.7	7
Antiparos	5.4; 4.9; 4.7; 4.2; 4.1; 4.0	11
Patmos-Chiliomodi	4.4	6
	3.7; 3.5	9
Kos	2.7; 2.5; 1.6; 1.0	7
Yali	0.024*	15
Nisyros	0.2	6
Santorini	1.6; 1.5; 0.6	16
Milos	1.5; 1.1; 0.9; 0.5	6
	2.5; 2.3; 2.2; 1.8; 1.7; 0.9	17
Methana	0.9; 0.8; 0.6; 0.3	6
Crommyonia	3.9; 2.7	7,6

(1) Innocenti, F. *et al.* in prep.; (2) Fytikas, M. *et al.* 1979; (3) Borsi, S. *et al.* 1972; (4) Pe, G., 1980; (5) Bellon, H. *et al.* 1979b; (6) Fytikas, M. *et al.* 1976; (7) Bellon, H. *et al.* 1979a; (8) Van Couvering, J. A. & Miller, J. A. 1971; (9) Robert, U. & Cantagrel, J. M. 1977; (10) Besang, C. *et al.* 1977; (11) Innocenti, F. *et al.* 1982c; (12) Kolios, N. *et al.* 1980; (13) Innocenti, F. *et al.* 1979; (14) Pe-Piper, G. & Piper, D. J. W., 1979; (15) Wagner, G. A. *et al.* 1976; (16) Ferrara, G. *et al.* 1980; (17) Angelier, J. *et al.* 1977.

* Fission track dating.

TABLE 2. *Average chemical composition of Oligocene and Miocene calc-alkaline rocks from northern and central Aegean area*

	Basaltic Andesites		Andesites		Dacites		Rhyolites	
	(13)	(s)	(68)	(s)	(33)	(s)	(29)	(s)
SiO_2	54.87	1.09	60.17	1.93	65.45	2.04	73.09	1.88
TiO_2	0.73	0.09	0.66	0.12	0.55	0.11	0.16	0.12
Al_2O_3	17.40	1.00	16.76	0.75	15.44	1.2	13.55	1.07
Fe_2O_3	4.01	1.78	3.08	1.21	2.68	1.01	0.79	0.56
FeO	3.41	1.63	2.46	1.36	1.33	0.68	0.51	0.32
MnO	0.18	0.06	0.12	0.04	0.09	0.07	0.07	0.06
MgO	4.52	1.43	2.97	1.03	1.96	0.82	0.49	0.49
CaO	8.12	0.95	6.05	0.88	4.35	0.82	1.29	0.49
Na_2O	2.66	0.39	3.19	0.51	3.30	0.35	3.42	0.72
K_2O	1.65	0.61	2.21	0.42	2.69	0.44	4.49	0.94
P_2O_5	0.21	0.11	0.18	0.08	0.20	0.11	0.04	0.03
L.O.I.	2.29	0.97	2.17	0.79	1.90	0.84	2.11	1.39
Total	100.05		100.02		99.94		100.00	

(13) number of analyses; (s) standard deviation.

presence of volcaniclastic layers interbedded in Upper Eocene sediments, north-east of Alexandroupolis.

In the northern Aegean area volcanism is well represented on the various islands. Here it is essentially Lower Miocene in age (Limnos, Agios Evstratios, Imvros, Lesvos), except for the northernmost volcanic island of Samothraki where the eruptive products are assigned to the Oligocene on the basis of stratigraphic data (Heimann *et al.* 1972).

In the southernmost part of this belt (Central Aegean), minor volcanic products occur, dated as young as 13.2 Ma (Middle Miocene), in Evia (Fig. 1; Table 1).

From the chemical and petrographic point of view the products erupted in this volcanic phase typically display orogenic characteristics and are represented by members of the calc-alkaline and shoshonitic associations (Tables 2 and 3; Fig. 2).

The calc-alkaline members are mainly localized in Thrace where, as mentioned, they are Upper Oligocene in age. Intermediate and acid members (andesites and dacites) predominate. There is a far smaller abundance of rhyolites and less evolved products (basaltic andesites). Basalts are altogether absent (Fig. 3).

In the less evolved intermediate rocks, anhydrous phenocryst phases, plagioclase, ortho- and clinopyroxenes and ore minerals are dominant. Hydrous phenocrysts, hornblende and subordinate biotite, are sometimes present, but only in the more evolved andesites; they become common in the dacites.

The average composition of the members of this association is given in Table 2. The general chemical characteristics fit into the range of

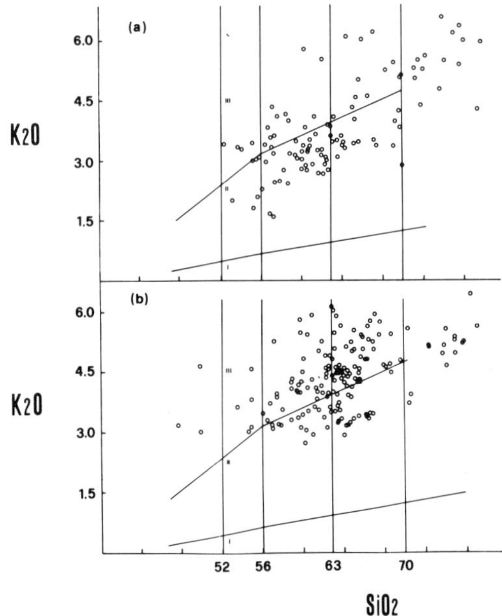

FIG. 2. K_2O vs. SiO_2 plot for the volcanic products of the Oligocene-Miocene volcanic phase (NATA). (a) Oligocene volcanics (Thrace and Samothraki); (b) Lower-Middle Miocene volcanics (Imvros, Limnos, A. Evstratios, Lesvos, Skyros, Evia, Chios, Psara-Antipsara). For the relative distribution see also Figs 1 & 6.

variation of the high-K orogenic series found in active continental margins (Ewart 1982). The relatively potassic nature of this association can also be seen in Fig. 2, which shows that the majority of samples plot near the boundary between the calc-alkaline and shoshonitic series.

TABLE 3. *Average chemical composition of Oligocene and Miocene shoshonitic rocks from northern and central Aegean area*

	Shoshonitic Basalts		Shoshonites		Latites		Trachytes	
	(4)	(s)	(13)	(s)	(70)	(s)	(44)	(s)
SiO_2	48.87	1.88	54.21	1.17	60.25	2.06	65.32	1.67
TiO_2	1.55	0.44	0.90	0.27	0.79	0.13	0.63	0.15
Al_2O_3	15.36	1.72	16.49	1.46	15.97	0.99	15.18	0.82
Fe_2O_3	4.00	1.66	4.17	1.88	3.48	1.06	2.67	0.97
FeO	4.48	1.19	2.97	1.10	2.02	0.10	1.29	0.67
MnO	0.17	0.02	0.15	0.07	0.11	0.06	0.09	0.06
MgO	7.04	2.08	3.47	1.06	2.77	0.89	1.82	0.73
CaO	9.81	1.65	8.57	1.57	5.33	1.09	3.45	0.68
Na_2O	3.12	0.78	3.04	0.54	3.33	0.44	3.21	0.42
K_2O	2.54	0.74	2.53	0.39	3.55	0.62	4.06	0.51
P_2O_5	0.57	0.26	0.42	0.35	0.38	0.14	0.30	0.14
L.O.I.	2.57	1.48	3.13	1.33	2.01	0.95	1.98	1.14
Total	100.08		100.07		9.99		100.00	

(4) number of analyses; (s) standard deviation.

In fact, a continuous variation in K_2O content is observed between the members of the two series. Rocks with the highest K_2O content are rare in the northern part of the area, but are relatively abundant in the central and southern sectors (Fig. 2).

In the shoshonitic association, intermediate rock-types predominate (Fig. 3). The rocks of this association are characterized by a greater abundance of hydrous phenocryst phases, biotite and subordinate hornblende. These minerals are also present in the less evolved members, occasionally with olivine. Shoshonitic basic rocks always contain clinopyroxene sometimes accompanied by orthopyroxene. Sanidine is rarely present in the latites but becomes dominant in the trachytes. The converse is true of plagioclase.

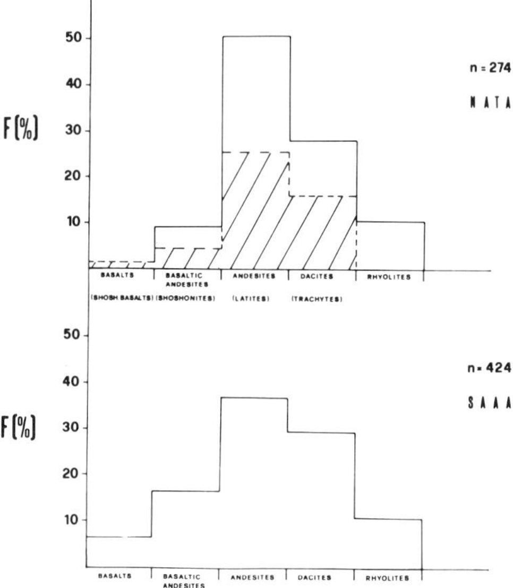

FIG. 3. Frequency distribution of members of the orogenic suites NATA (upper) and SAAA (lower). The dashed area refers to shoshonitic members.

The average chemical composition of the rocks in this association is given in Table 3. On the whole their characteristics are similar to those of shoshonitic rocks found in active continental subduction zones except that they are generally poorer in MgO and therefore have higher FeO_{tot}/MgO ratios; their K_2O content is also high (Ewart 1982).

Summarizing, the central and northern part of the Aegean area is characterized by an orogenic type of volcanism which in time has migrated towards a more southerly position. This variation is continuous both in space and in time and is accompanied by a change in the chemical character of the products erupted which tend to become richer in K_2O towards the south. The volcanics of Skyros and Evia are an exception to this. Even though they are the most recent and the most southerly of this cycle, they display practically normal calc-alkaline characteristics (Fytikas *et al.* 1979).

In the area under study the greatest extent of volcanics is in Thrace. Here activity is clearly related to the coeval volcanism on the western border of the Serbo-Macedonian Massif in Yugoslavia (Protić & Terzić 1976) and on the southern border of the Pontic Chain (Fig. 1). It can, therefore, be considered that in the Upper Oligocene the entire southern margin of the Eurasian plate in this sector represented an

TABLE 4. *Average chemical composition of calc-alkaline volcanics from the south Aegean active arc*

	Basalts		Basaltic Andesites		Andesites		Dacites		Rhyolites	
	(27)	(s)	(69)	(s)	(57)	(s)	(124)	(s)	(46)	(s)
SiO_2	50.66	0.95	54.27	1.14	59.44	1.95	65.85	1.68	73.77	2.18
TiO_2	0.85	0.10	0.89	0.25	0.73	0.25	0.61	0.19	0.17	0.11
Al_2O_3	18.62	0.83	17.34	1.15	17.13	0.99	15.48	0.73	12.87	0.90
Fe_2O_3	2.84	1.06	3.54	1.75	2.98	1.21	2.03	0.91	0.75	0.47
FeO	5.64	0.99	4.31	1.81	3.27	1.51	2.42	1.04	0.76	0.37
MnO	0.16	0.01	0.13	0.05	0.13	0.04	0.11	0.04	0.06	0.04
MgO	6.43	0.68	4.65	1.40	3.34	1.45	1.54	0.68	0.42	0.31
CaO	10.65	0.81	9.07	1.09	6.04	1.55	3.97	1.05	1.47	0.74
Na_2O	2.81	0.19	3.28	0.50	3.63	0.50	4.31	0.89	3.80	0.40
K_2O	0.60	0.21	1.33	0.49	1.87	0.57	2.44	0.58	3.75	0.66
P_2O_5	0.07	0.05	0.15	0.14	0.14	0.08	0.14	0.07	0.05	0.10
L.O.I.	0.60	0.37	0.91	0.65	1.27	0.91	1.11	0.95	2.14	1.33
Total	99.93		99.87		99.97		100.01		100.01	

(27) number of analyses; (s) standard deviation.

active continental margin characterized by a practically continuous andesite belt.

The Pliocene-Quaternary volcanic phase (South Aegean Active Arc, SAAA)

Orogenic volcanic activity died out in Middle Miocene time in the central Aegean but resumed again at the end of Lower Pliocene in a markedly more southerly position.

The volcanic centres are clustered along a restricted belt which extends in an arc from the Saronic Gulf to the west to the island of Nisyros in the east (the South Aegean Volcanic Arc). This arc is considered to be the surface expression of still-active subduction of the African plate beneath the Aegean plate (Caputo *et al.* 1970). It is thought that this began in the Middle Miocene (Le Pichon & Angelier 1981).

The products of the South Aegean Arc form a typical calc-alkaline association which displays a continuous evolution from basalts to rhyolites (Fig. 4); the general chemical characteristics (Table 4) are closely comparable with those of the volcanics of island arcs sited on thin continental margins (Barberi *et al.* 1981). Even though andesites and dacites are dominant, less evolved members, basalts and basaltic andesites, are also common and make up about 25% of the total of the products erupted (Fig. 3).

Petrographically and geochemically the arc is generally homogeneous but there are some significant variations along it, in magmatic character and mode of eruption. In the western sector, from the Saronic Gulf to the island of Milos, the volcanic centres are dominated by

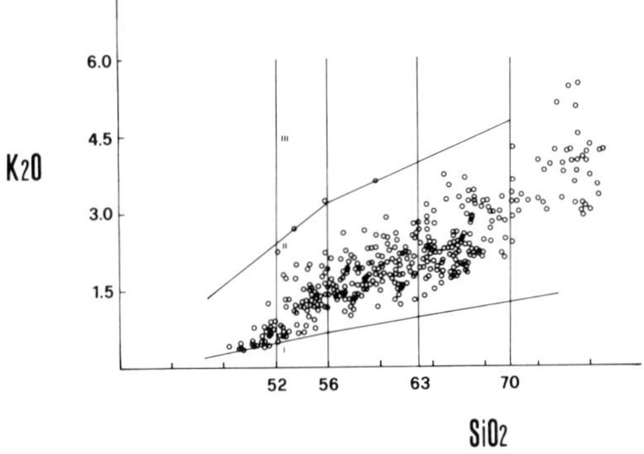

FIG. 4. K_2O vs. SiO_2 plot for the volcanics from South Aegean Active Arc (SAAA).

the presence of domes and lava flows with subordinate pyroclastics. The products erupted mainly consist of andesites and dacites which have been interpreted as having evolved from a basic parent by crystal-liquid fractionation and relatively high pressure where the plagioclase played a minor role as a separating phase (Innocenti *et al.* 1981). In the central and eastern sectors of the arc, on the other hand, there are large central volcanoes often characterized by summit calderas. This fact has been attributed to the presence of relatively shallow magma chambers, inside which fractional crystallization processes have led to highly evolved products and a markedly explosive activity.

The variations along the arc can be interpreted as an effect of the difference in the lithospheric stress-field in the various sectors of the arc. The local stress-field will be related to

TABLE 5. *Representative analyses of Upper Miocene-Pliocene Aegean volcanics. All analyses normalized to 100% including loss on ignition*

	K-4	PSI-1	PAT-12	JC-6	JC-17	BO-11	JC-87	JC-84	PAT-34	TH-99
SiO_2	47.47	50.25	50.82	54.18	56.09	58.59	60.04	64.84	65.11	76.45
TiO_2	1.74	1.71	1.58	0.71	0.79	0.75	0.57	0.45	0.61	0.11
Al_2O_3	17.00	18.86	18.80	18.30	17.92	16.74	18.28	16.77	16.40	12.27
Fe_2O_3	5.18	2.04	1.80	3.42	0.54	5.06	1.40	2.12	2.40	0.68
FeO	3.57	5.12	5.63	2.84	4.40	0.43	2.29	1.28	0.82	0.19
MnO	0.15	0.13	0.14	0.12	0.10	0.10	0.10	0.07	0.11	0.08
MgO	8.66	6.60	4.09	4.38	6.80	3.03	1.36	1.06	1.18	0.04
CaO	8.00	6.99	7.64	7.70	5.50	5.16	3.79	4.56	2.39	0.38
Na_2O	4.61	4.48	4.40	3.84	3.62	3.93	4.83	4.14	3.59	4.49
K_2O	2.07	2.64	2.03	2.60	2.79	4.11	6.40	3.73	6.11	4.37
P_2O_5	0.41	0.56	0.70	0.54	0.26	0.35	0.34	0.25	0.25	0.01
L.O.I.	1.15	0.63	2.36	1.37	1.19	1.75	0.62	0.73	1.01	0.94

K-4, Kalogeri (alkali basalt: Innocent *et al.* 1982c); PSI-l, Psathoura (alkali basalt); PAT-12, Chiliomodi-Patmos (hawaiite); JC-6, Achilleion (high-K basaltic andesites; Innocenti *et al.* 1979); JC-17, Likhades (high-K magnesian andesite; Innocenti *et al.* 1979); BO-11, Bodrum (latite); JC-87, Edessa area (latite; Kolios *et al.* 1980); JC-84, Edessa area (high-K dacite; Kolios *et al.* 1980); PAT 34, Patmos (trachyte); TH-99, Antiparos (rhyolite; Innocenti *et al.* 1982c).

the geometry of the convergent system: the main convergence direction is orthogonal in the western sector whereas it is oblique and parallel in the central and eastern sectors respectively.

The neotectonic and geophysical data (Le Pichon & Angelier 1981) lend further support to our conclusion that in the western sector there is a less marked extensional regime, making it more difficult for magmas to rise, thus favouring deeper fractionation. By contrast, in the central and eastern sectors a strongly tensional tectonic regime would enhance the tendency for basic magmas to rise and form vast magma chambers relatively near the surface.

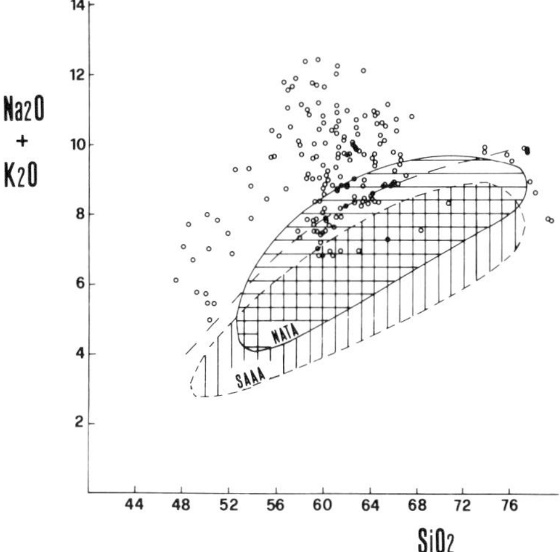

FIG. 5. $Na_2O + K_2O$ vs. SiO_2 plot for the volcanics belonging to the Upper Miocene to Quaternary scattered volcanic activity (Volos-Atlanti area, Voras Mts, Antiparos, Kalogeri, Patmos, Samos, Kos, Bodrum). Fields are also reported for NATA and SAAA respectively.

Upper Miocene to Quaternary scattered volcanic activity

From the Upper Miocene to the Quaternary there have been local eruptive events all over the Aegean area, except in the northern sector. These have generally produced small volumes of products with very different characteristics from those of the previously described orogenic association. Petrogenetic affinity, age and position all appear to be extremely variable (Fig. 5). The following associations can, however, be distinguished:

(1) Sodic alkaline products

These are represented by alkali basalts and hawaiites with clearly subordinate more evolved members (Table 5). They are mostly localized in the eastern sector of the Aegean area, and are associated with the E–W trending tensional lineaments which have affected this region and adjacent western Anatolia from the late Miocene on. These volcanics are found on Samos (8.3–7.8 Ma), Patmos (3.5–4.1 Ma), Kalogeri (6.0 Ma) and Psathoura (c. 0.5 Ma). Analogous phenomena have been described from western Anatolia at Ezine, Urla and Kula (Innocenti *et al.* 1982a).

(2) Highly potassic alkaline lavas of shoshonitic affinity

These have K_2O/Na_2O ratios always greater than 1 but have other features typical of orogenic series such as a low TiO_2 content and an absence of iron-enrichment. Chemically (Table 5) and petrographically they are assigned to the shoshonitic association and are analogous to rocks described from the regions of Afyon and Isparta in western Anatolia (Robert & Cantagrel 1977; Kolios *et al.* 1980; Innocenti *et al.* 1982). The volcanics of this association are never continuous in space or time with the volcanics of the NATA and SAAA orogenic suites.

These shoshonitic eruptives are sited exclusively on the southeastern margin of the Aegean area (Bodrum, Kos & Patmos; age ranging from 10.6–7.0 Ma) and on the northwestern margin (Voras; age ranging from 5.0–1.8 Ma). In the NW region the association is more complex and sub-alkaline products are also present (Kolios *et al.* 1980).

(3) Rhyolites formed by crustal anatexis

On the island of Antiparos on the southern margin of the Attic-Cycladic Massif volcanics of exclusively rhyolitic composition were erupted on the Miocene/Pliocene boundary. Their chemical (Table 5) and isotopic characteristics indicate an origin due to crustal melting (Innocenti *et al.* 1982c). The $^{87}Sr/^{86}Sr$ ratio is high (0.711), distinctly higher than that of the rhyolites related to the SAAA (Keller, 1982; Gale, 1981) but comparable with that of the Upper Miocene intrusives present elsewhere along the southern margin of the Attic-Cycladic Massif (Andriessen *et al.* 1979).

Volcanics with a similar composition and origin have been described from south of Izmir in western Anatolia (Borsi *et al.* 1972), and from the Afyon region (Keller & Villari 1972). However, the suggestion that this volcanism was related to a second volcanic front slightly displaced northwards with respect to the SAAA (Ninkovich & Hays 1972) is incompatible with the geochronological and geochemical data.

(4) Volos-Atalanti Group

In the northwest sector of the South Aegean Arc a group of volcanics with distinctive chemical and petrographic characteristics occurs in the Volos-Atalanti area. They consist of a series of small lava outcrops aligned approximately in a SW-NE direction along the southwest extension of the North Anatolian Fault. This volcanic activity developed between 3.4 and 0.5 Ma (Table 1) and its products consist of basaltic andesites with high K_2O and LIL elements and a high MgO content (Table 5) (Innocenti *et al.* 1979).

The geochemical characteristics of these products suggest that they had their genesis from a deep source, highly enriched with LIL elements and clearly distinct from the one that produced the calc-alkaline volcanism in the SAAA. The absence of geophysical evidence of the presence of a Benioff zone under Atalanti–Volos area (Comninakis & Papazachos 1980) also tends to exclude any relation between the active subduction processes and this volcanism. Instead this was related to ascent of magma from the mantle in an area affected by tensional strain which resulted from the southward movement of the Aegean microplate with respect to the Macedonian block (Kolios *et al.* 1980).

Discussion and conclusions

The reconstruction of the volcanic evolution in the Aegean areas has demonstrated how from Tertiary times onwards this region has been dominated by orogenic volcanism; two belts clearly separated in space and time, are respectively localized in the northern sector (NATA) and in the southern sector (SAAA) of the Aegean area (Fig. 6). These two belts display very different characteristics which can only in part be attributed to the fact that the history of the NATA can be traced until it becomes extinct, whereas the SAAA is still developing. The main differences are in the relative distribution of the members present (Fig. 3), and in the progressive and continuous increase in the potassic character of the NATA activity and its general southward migration. In particular, the NATA volcanic association is characterized by the rarity of basic products and generally by a relatively higher K_2O content in all the members of the series.

These variations are attributed to the different thickness of the crust in the two areas affected by volcanism. In the North Aegean area the NATA volcanic front has developed along the margin of the Rhodope Massif where the crust must have been relatively thick. This is also suggested by the geophysical data which show that today the thickness of the crust is around 40 km (Makris 1977), despite Neogene extension. By contrast the South Aegean volcanic front is built up on relatively thin continental crust (25–30 km) (Makris 1977). We believe that these distinctive characteristics can be explained by a 'filtering' effect dependent on the low density contrast between basic magma and continental crust (Coulon & Thorpe 1981). The occurrence of thick crust facilitates stagnation of the more basic magmas within the crust and their evolution by differentiation and possibly contamination processes prior to eruption.

The geochronological and geological data

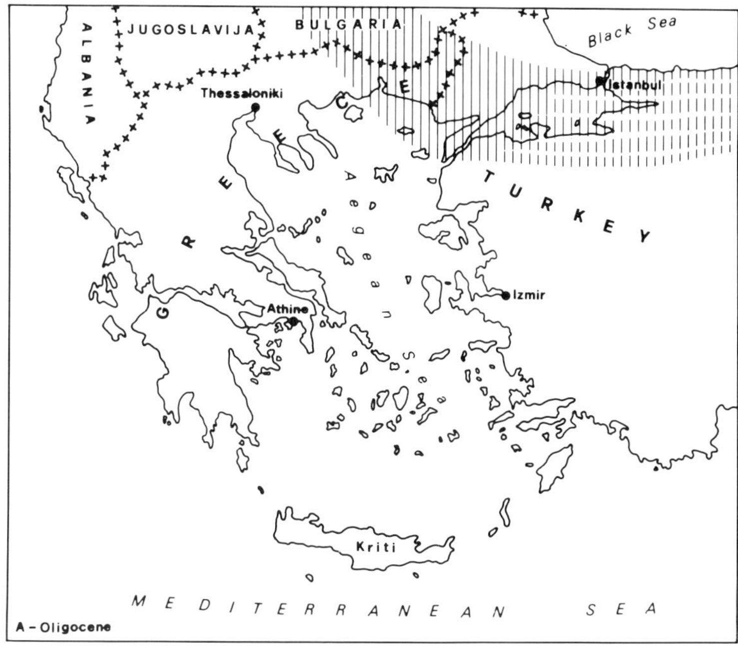

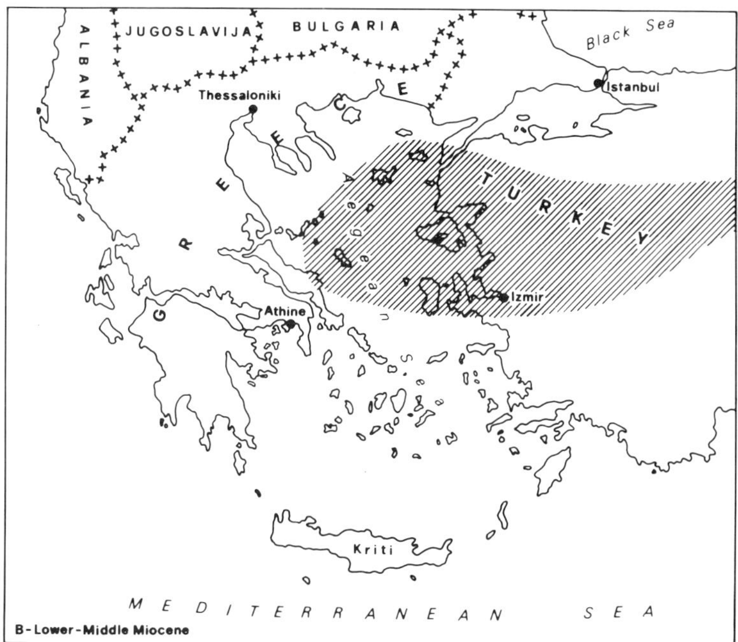

FIG. 6. Relative distribution of the volcanic activity in the Aegean area during Oligocene (A), Lower-Middle Miocene (B), Upper Miocene (C) and Pliocene-Quaternary (D).

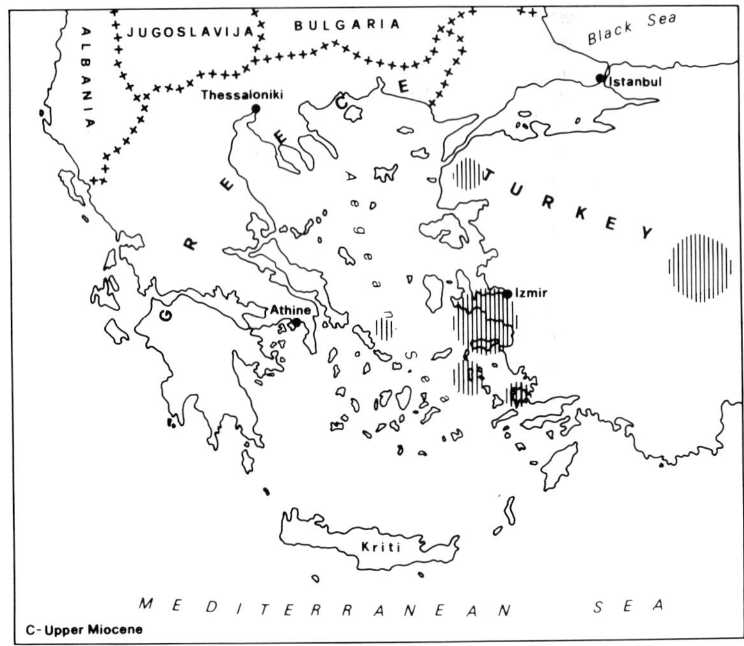
ALBANIA
JUGOSLAVIJA
BULGARIA
Black Sea
Istanbul
Thessaloniki
GREECE
TURKEY
Aegean Sea
Izmir
Athine
Kriti
MEDITERRANEAN SEA
C- Upper Miocene

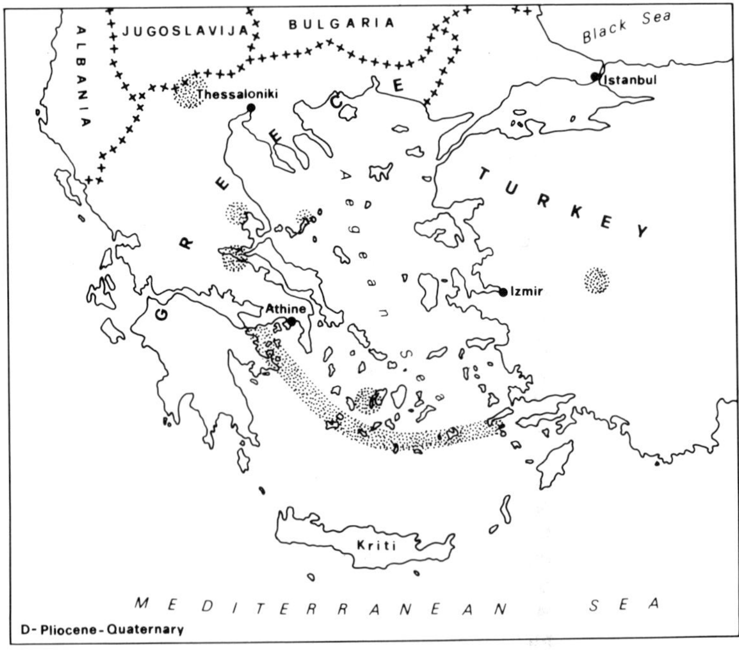
ALBANIA
JUGOSLAVIJA
BULGARIA
Black Sea
Istanbul
Thessaloniki
GREECE
TURKEY
Aegean Sea
Izmir
Athine
Kriti
MEDITERRANEAN SEA
D- Pliocene - Quaternary

FIG. 6. (*contd*)

show how the NATA volcanic activity has migrated southwards in time (Fig. 6). This migration can be well observed in the whole North Aegean area where Oligocene volcanics occur only in the northernmost part of the belts (Thrace) whereas Miocene products are confined to the southern sector of the belt. It is however difficult to estimate precisely the amount of migration, because of the important deformation by faulting which the region underwent during the Neogene. However, if one allows for the probable amount of dextral strike-slip movement along the North Anatolian Fault (Şengör 1979) and for the expansion of the area calculated by Le Pichon & Angelier (1981), the width of the belt affected by volcanism should be reduced by at last 50%. The migration of the volcanic front must therefore be less than the present-day distribution of the eruptive products might suggest.

In terms of plate-tectonics, the evolutionary characteristics of the whole North Aegean volcanic system suggest that subduction responsible for the NATA volcanism was going on from the Upper Oligocene to the Lower Miocene (Fig. 7). It is generally accepted that the volcanic activity along a convergent margin develops only where the depth (h) of the subducted lithosphere reaches suitable values, between approximately 100 and 200 km (Gill 1981). It is also generally acknowledged that, within a given convergent system, a positive correlation exists between h and the K_2O content of the erupted products (Dickinson 1975). If we assume that the gravitational force is the main force acting on the relatively dense sinking slab (Elsasser 1971) then after the onset of continental collision and reduction to zero of the convergence velocity the dip of the slab should steadily increase, and reach the vertical before the slab eventually detaches itself (Molnar & Atwater 1978). Following this approach, the southward migration of the volcanic front plus the progressive increase in K_2O in the erupted products is tentatively interpreted as the effect of an increase in the dip of the subducting slab, following the main Tertiary phase of continental collision (Fig. 7).

From the Middle Miocene the Aegean area was affected by a new structural regime, triggered by the penetration of the Arabian plate into the Eurasian mass and the consequent divergent motion of the Anatolian and Iranian blocks respectively (McKenzie 1972, 1978; Innocenti *et al.* 1982a). As a consequence of this, the Aegean microplate came into being 12–13 Ma ago (Le Pichon & Angelier 1981), with boundaries of contrasting characteristics (Dewey & Şengör 1979).

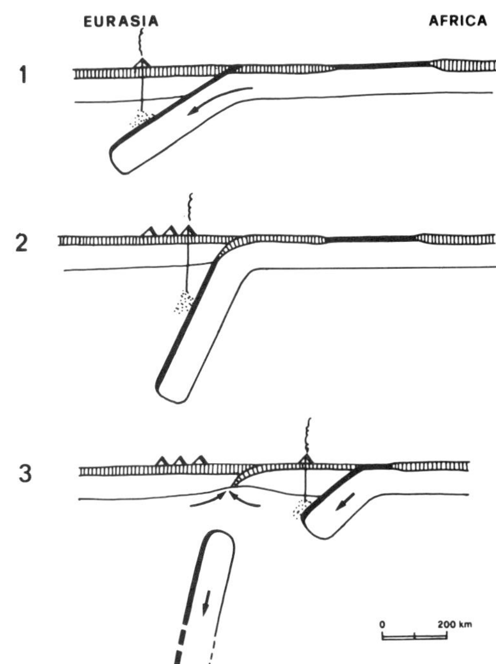

Fig. 7. Schematic cross-section depicting inferred geodynamic evolution and southward migration of orogenic volcanism in the Aegean area. During the Eocene to Oligocene the southern margin of Eurasia, represented by the Rhodope Massif, collides with the Apulian micro-plate located between Africa and Eurasia.

The southern margin of the Aegean microplate is a typical convergent plate boundary, whereas the northern and western boundaries are marked by an important transcurrent lineament, considered to be the extension into the Aegean area of the North Anatolian Fault. A clearly defined boundary is not apparent in the eastern sector, but a wide area of intra-plate extensional strain dominates the Eastern Aegean and Western Anatolia, characterized by a series of graben approximately trending E–W.

The volcanic activity following the older north-Aegean subduction phase is mainly distributed along the boundaries described above. The southern consuming margin is marked by the presence of the South Aegean active arc (SAAA), while the scattered volcanism ranging in age from Upper Miocene to Quaternary mainly occurred along the northern and western margins, i.e. along the western extension of the North Anatolian Fault, and, more diffusely, within the wide extensional area in the Eastern Aegean and Western Anatolia.

The petrogenetic affinity of the volcanic products related to the non-convergent margins is mainly alkaline, which is consistent with the tensional tectonic regime that favoured genesis and uprise of magma. The geochemical character of these products is extremely variable and suggests their derivation from heterogeneous sources in the upper mantle, which had probably undergone local modification in the earlier subduction event in the region.

Volcanic activity within the boundaries of the Aegean microplate also occurred, as in the Plio-Quaternary volcanic complex of the Voras Mountains in Macedonia, and the Mio-Pliocene rhyolitic volcanism on the Island of Antiparos in the Central Aegean Sea (Fig. 1). The volcanism of the Voras Mountains developed in a sector of Northern Macedonia characterized by a tensional tectonic setting. This area has been considered to be a wide extensional boundary between Macedonia and the Black Sea block, induced by the opposition of Eurasia to the westward motion of the Anatolian plate, which was only partially absorbed by the offset of the North Anatolian Fault (Dewey & Şengör 1979). It is therefore assumed that this area of wide continental deformation is closely comparable to that already described along the contact between the Aegean and Anatolian plates. The volcanic activity which developed in Northern Macedonia shows the same characteristics as that in the Eastern Aegean and Western Anatolia and is taken as evidence of an analogous geodynamic setting.

A completely different interpretation is suggested for the rhyolitic volcanic activity on the Island of Antiparos, which is sited far away from any plausible plate boundary. Petrological and geochemical data indicate an origin by crustal anatexis, from a source very similar to that which produced the belt of acid intrusives in the Attic-Cycladic Massif, which range in age between 8 and 15 Ma (Durr *et al.* 1978). This belt is considered to be surface expression of a well-defined thermal anomaly within the lithosphere, produced by the detachment of the subducting slab at the end of the older Tertiary convergence phase (Innocenti *et al.* 1982c) and the consequent uprising of the asthenosphere in the detachment zone. The eruption of rhyolitic magma was possible only in the central part of the belt where the maximum extension occurred, as suggested by the available geological and geophysical data (Le Pichon & Angelier 1981).

ACKNOWLEDGEMENTS: This work has been supported financially by Consiglio Nazionale delle Ricerche (C.N.R.) of Italy and the Institute of Geology and Mining Research, Greece (I.G.M.E.).

References

ANDRIESSEN, P. A. M., BOELRIJK, N. A. I. M., HEBEDA, E. H., PRIEM, H. N. A., VERDURMEN, E. A. TH. & VERSCHURE, R. H. 1979. Dating the events of metamorphism in the Alpine orogen of Naxos (Cyclades, Greece). *Contrib. Mineral. Petrol.* **69,** 215–25.

ANGELIER, J., CANTAGREL, J. M. & VILMINOT, J. C. 1977. Néotectonique cassante et volcanisme plio-quaternaire dans l'arc égéen interne: l'Ile de Milos (Grèce). *Bull. Soc. géol. France*, **19,** 119–21.

BARBERI, F., INNOCENTI, F., MANETTI, P., PECCERILLO, A., VILLARI, L. & POLI, G. 1981. The evolution of the Hellenic arc and the Aeolian arc volcanism in space and time: a comparision. IAVCEI symposium August 18–September 9, 1981. Tokio and Akone, 32–3.

BELLON, H., JARRIGE, J. J. & SOREL, D. 1979a. Les activités magmatiques égéennes de l'Oligocène à nos jours et leurs cadres géodynamiques. Données nouvelles et synthèse. *Revue Géol. dynam. Géograph. phys.* **21,** 41–55.

——, GRISSOLLET, G. & SOREL, D. 1979b. Age de l'activité volcanique néogène de l'Ile de Chios (Mer Egée, Grèce). *C.R. Acad. Sc. Paris*, *288*, ser. D, 1255–8.

BESANG, C., ECKHARDT, F. J., HARRE, W., KREUZER, G. & MULLER, P. 1977. Radiometrische Alterbestimmungen an neogenen Eruptivgesteinen der Türkei. *Geol. Jb.* **25,** 3–36.

BORSI, S., FERRARA, G., INNOCENTI, F. & MAZZUOLI, R. 1972. Geochronology and petrology of recent volcanics in Eastern Aegean Sea. *Bull. Volcanol.* **36,** 473–96.

CAPUTO, M. G., PANZA, F. & POSTPISCHL, D. 1970. Deep structure of the Mediterranean Basin. *J. Geophys. Res.* **75,** 1093–102.

COMNINAKIS, P. E. & PAPAZACHOS, B. C. 1980. Space and time distribution of the intermediate focal depth earthquakes in the Hellenic arc. *Tectonophysics*, **70,** T35–T47.

COULON, C. & THORPE, R. S 1981. Role of the continental crust in petrogenesis of orogenic volcanic associations. *Tectonophysics*, **77,** 79–93.

DEWEY, J. F., PITMAN, W. C. III, RYAN, W. B. F. & BONNIN, J. 1973. Plate tectonics and the Alpine system. *Bull. geol. Soc. Am.* **84,** 3137–80.

—— & ŞENGÖR, A. M. C. 1979. Aegean and surrounding regions: complex multiplate and continuum tectonics in convergent zone. *Bull. geol. Soc. Am.* **90,** 84–92.

DICKINSON, W. R. 1975. Potash-depth (K-h) relations in continental margin and intra-ocean magmatic arcs. *Geology*, **3,** 53–6.

DÜRR, ST., ALTHERR, R., KELLER, J., OKRUSCH, M. &

SEIDEL, E. 1978. The median Aegean crystalline belt: stratigraphy, metamorphism, magmatism. In: CLOSS, H., ROEDER, D. & SCHIMDT, K. (eds.) *Alps, Apennines, Hellenides.* IUGC Sci. Rep. **38,** 455–77.

ELSASSER, W. 1971. Sea-floor spreading as convection. *J. geophys. Res.* **76,** 1101–12.

EWART, A. 1982. The mineralogy and petrology of Tertiary-Recent orogenic volcanic rocks: with special reference to the andesitic-basaltic compositional range. In: THORPE, R. S. (ed.). *Andesites.* J. Wiley & Sons, New York, 25–87.

FERRARA, G., FYTIKAS, M., GIULIANI, O. & MARINELLI, G. 1980. Age of the formation of the Aegean active arc. In: DOUMAS, C. (ed.). *Thera and Aegean World.* **2,** 37–42.

FYTIKAS, M., GIULIANI, O., INNOCENTI, F., MARINELLI, G. & MAZZUOLI, R. 1976. Geochronological data on recent magmatism of the Aegean Sea. *Tectonophysics,* **31,** T29-T34.

——, ——, ——, MANETTI, P., MAZZUOLI, R., PECCERILLO, A. & VILLARI, L. 1979. Neogene volcanism of the Northern and Central Aegean region. *Ann. Géol. Pays Hell.* **30,** 106–29.

GALE, N. H. 1981. Mediterranean obsidian source characterisation by strontium isotope analysis. *Archeometry,* **23,** 41–51.

GILL, J. 1981. *Orogenic Andesites and Plate Tectonics.* Springer-Verlag, Berlin, 390 pp.

HEIMANN, K. O., LEBRUCHNER, H. & KRETZLER, W. 1972. *Geological map of Greece-Samothraki sheet.* IGME, Athens.

INNOCENTI, F., MANETTI, P., PECCERILLO, A. & POLI, G. 1979. Inner arc volcanism in NW Aegean Arc: geochemical and geochronological data. *N. Jb. Miner. Mh., Jg.* 1979, 145–58.

——, ——, —— & —— 1981. South Aegean volcanic arc: geochemical variations and geotectonic implications. *Bull. Volcanol.* **44,** 377–91.

——, ——, MAZZUOLI, R., PASQUARE', G. & VILLARI, L. 1982a. Anatolian and north-west Iran. In: THORPE, R. S. (ed.). *Andesites.* J. Wiley & Sons, New York, 327–49.

——, MAZZUOLI, R., PASQUARE', G., RADICATI DI BROZOLO, F. & VILLARI, L. 1982b. Tertiary and Quaternary volcanism of the Erzurum-Kars area (Eastern Turkey): geochronological data and geodynamic evolution. *Jour. Volc. and Geoth. Res.* **13,** 223–40.

——, KOLIOS, N., MANETTI, P., RITA, F. & VILLARI, L. 1982c. Acid and basic Late Neogene volcanism in Central Aegean Sea: its nature and geotectonic significance. *Bull. Volcanol.* **45,** 87–97.

KELLER, J. 1982. Mediterranean island arcs. In: THORPE, R. S. (ed.). *Andesites.* J. Wiley and Sons, New York, 307–26.

—— & VILLARI, L. 1972. Rhyolitic ignimbrites in the region of Afyon (Central Anatolia). *Bull. Volcanol.* **36,** 342–58.

KOLIOS, N., INNOCENTI, F., MANETTI, P., PECCERILLO, A. & GIULIANI, O. 1980. The Pliocene volcanism of the Voras Mts. (Central Macedonia, Greece). *Bull. Volcanol.* **43,** 553–68.

LE PICHON, X. & ANGELIER, J. 1981. The Aegean Sea. *Philos. Trans. R. Soc. Lond.* A **300,** 357–72.

MAKRIS, J. 1977. Geophysical investigations of the Hellenides. *Hamb. Geophys. Einzelschr.* **34,** 124 pp.

MCKENZIE, D. P. 1972. Active tectonics of the Mediterranean region. *Geophys. J. R. astron. Soc.* **30,** 109–85.

—— 1978. Active tectonics of the Alpine-Himalayan belt: the Aegean sea and surrounding regions. (Tectonics of Aegean region). *Geophys. J.R. astron. Soc.* **55,** 217–54.

MOLNAR, P. & ATWATER, T. 1978. Interarc speading and Cordilleran tectonics as alternates related to the age of subducted oceanic lithosphere. *Earth planet. Sci. Lett.* **41,** 330–40.

NINKOVICH, D. & HAYS, J. D. 1972. Mediterranean island arcs and origin of the high potash volcanoes. *Earth planet. Sci. Lett.* **16,** 331–45.

PE-PIPER, G. 1980. Geochemistry of Miocene Shoshonites, Lesbos, Greece. *Contrib. Mineral. Petrol.* **72,** 387–96.

—— & PIPER, D. J. W. 1979. Plio-Pleistocene age of high-potassium volcanism in the northwestern part of the Hellenic Arc. *Tschermaks Min. Petr. Mitt.* **26,** 163–5.

PROTIĆ, M. & TERZIĆ, M. 1976. Les dernières phases du volcanism Cainozoique à l'Est de la Yougoslavie. In: *Inter. Congr. on Thermal Waters, Geothermal Energy and Vulcanism of the Mediterranean area, Athens,* **3,** 195–210.

ROBERT, U. & CANTAGREL, J. M. 1977. Le volcanism basaltique dans le Sud-Est de la mer Egée. Données géochronologiques et relations avec la téctonique. In: *6th coll. on the Geol. of the Aegean Region, Athens,* **3,** 961–7.

ŞENGÖR, A. M. C. 1979. The North Anatolian transform fault: its age, offset and tectonic significance. *Jl. geol. Soc. London,* **136,** 269–82.

VAN COUVERING, J. A. & MILLER, J. A. 1971. Late Miocene marine and non-marine time scale in Europe. *Nature,* **230,** 569–73.

WAGNER, G. A., STORZER, D. & KELLER, J. 1976. Spaltspurendatierungen quartärer Gesteinsgläser aus dem Mittelmeeraum. *N. Jb. Miner. Mh. Jg. 1976,* 84–94.

M. FYTIKAS, Institute of Geological and Mineral Exploration (IGME)—Athens, Greece.

F. INNOCENTI, Dipartimento di Scienze della Terra—University of Pisa, Italy.

P. MANETTI & A. PECCERILLO, Istituto di Mineralogia, Petrografia e Geochimica—University of Florence, Italy.

R. MAZZUOLI, Dipartimento di Scienze della Terra, Università della Calabria, Cosenza, Italy.

L. VILLARI, Istituto Internazionale di Vulcanologia (CNR), Catania and Istituto di Mineralogia e Petrografia, University of Messina, Italy.

Graben formation and associated seismicity in the Gulf of Korinth (Central Greece)

M. L. Myrianthis

SUMMARY: Extensional stresses in the central Aegean region reflect isostatic subsidence and uplift of brittle crustal blocks in response to crustal thinning. Block re-arrangements along active normal faults are probably the general cause for the generation of earthquakes. Reflection seismic data for the Gulf of Korinth graben is given here. It is concluded that the main factor controlling earthquakes is continuing uplift of the Peloponnesos block relative to the Gulf of Korinth graben.

The Peloponnesos, according to Berckhemer & Kowalczyk (1978) can be regarded as a crustal block of continental type possibly underlain by pre-Apulian continental basement, and tectonically separated from the Greek mainland by a pronounced fault zone along the Gulf of Korinth, possibly with right-lateral motion. This zone extends westwards to the Gulf of Patras probably as far as the Ionian Sea, and is responsible for the creation of both the gulfs and their associated deep basins. The Gulf of Korinth (Fig. 1) has long been recognized as an asymmetrical graben formed by normal faulting and filled by Plio-Quaternary sediments. Uplift of the northern Peloponnesos occurred during Tortonian (Late Miocene) times and was followed by subsidence and marine transgression in the Lower Pliocene (Kelletat *et al.* 1978). This transgression during the Upper Pliocene affected the Northern Peloponnesos area and

FIG. 1. Map of Greece. The area of the Gulf of Korinth studies is shown in the rectangle.

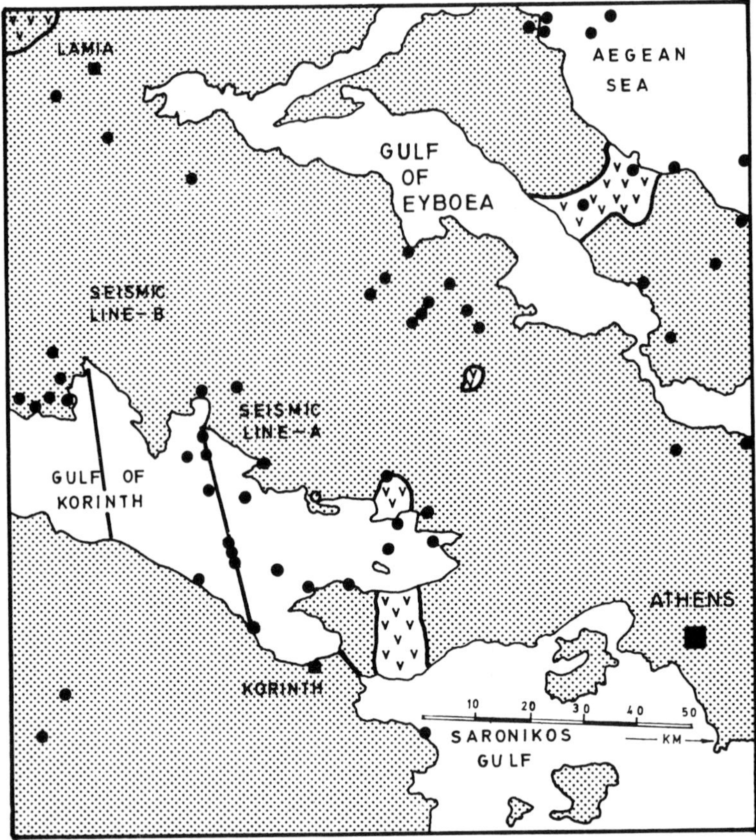

Fig. 2. Location map showing position of the two multichannel (96) seismic reflection profiles. The small circles are the epicentres of shocks with depth between 0 and 50 km taken from a conic projection published by McKenzie (1977).

led to thick sediment deposition in the southern part of the Gulf of Korinth.

Interpretation of two seismic profiles provides the basis of this study. The marine survey was conducted by the Western Geophysical Company of America using the equipment for conventional oil industry surveys, except that the source and detector were towed at a greater depth to optimize the lower frequency signal content. The source was maxipulse, operated at a depth of 12 m. Shot-point interval was 25 m providing 40 shots per kilometre. The 2375 m hydrophone streamer comprised 96 channels arranged as 20 hydrophones per channel in a linear array. The streamer was towed at a depth of 15 m. Ninety-six channels were recorded at a 4 msec sampling interval to a record length of 6.0 s. The primary positioning was by satellite navigator integrated with bottom-tracking Doppler sonar. A conventional basic processing sequence was followed to produce the final stacked sections with 48-fold subsurface coverage at a 12.5 m common depth point interval. Velocity analyses were computed at 3 km intervals.

Seismotectonic regime of the Gulf of Korinth

Two parallel seismic profiles are shown in Fig. 2 orientated NW–SE. The profiles were shot in September 1979 as part of a 96 channel digital acquisition survey. Direction of shooting was across the asymmetrical graben of the Gulf of Korinth. Line-A runs from the Antikyra Bay on the mainland to the Town of Kiato in Peloponnessos, while Line B runs from the Bay of Itea to the village of Derveni. The tectonic style of each line differs but both the seismic sections do confirm the well known NW–SE tensional regime which is to some extent responsible for the normal faulting which occurred after the main pre-Tortonian orogenic event. Line-A shows a reverse fault close to its northern end which may be the thrust contact of the Parnassos-

Ghiona zone upon the Pindos zone. In the central part of the profile a well developed graben is seen about 15 km wide.

Both flanks of the graben (Fig. 3) exhibit normal faulting with throws ranging from 1.5 to 3.5 km. Such throws are expected since Schwan (1978) pointed out that in this area major tensional faults have developed with throws of 1–3 km. A strong reflector which delineates the basement to the graben is attributed to the top of the Mesozoic sequence and is possibly of Triassic age. This horizon is heavily fractured and marked by stepped vertical normal faults. It is suggested that the faulting in the basement reflects the strong tensional field which prevailed in the area. This resulted in uplift of the south-eastern flank of the graben and subsidence of the north-western flank.

Myrianthis (1982), in a study of the earthquakes of February and March, 1981 near Alkyonides Island, Gulf of Korinth, Central Greece, used an experimental model to assess the components of the tensile triaxial state of stress and deduced that the direction of least principal stress coincided with the tensile axis T which is nearly horizontal and strikes north to south. The greatest principal stress, the vertical stress component, coincides with the compressional axis P.

The formations overlying basement are then controlled by fault tectonics which were probably synsedimentary leading to development of fault blocks with subsidence of basins and a sedimentary fill up to 1500 m thick. Movements started at the Miocene–Pliocene boundary (Kelletat, *op. cit.*).

The location of hypocentres of shocks with depths between 0 and 50 km is taken from a conic projection published by McKenzie 1977, and projected onto the vertical planes along the profiles (Fig. 2). Admittedly neither the Doppler satellite navigation of the seismic vessel, especially when it operates in water depths exceeding 500 m, nor the graphical transfer of epicentres from the conic projection allow accuracy in the location of actual azimouth positions relative to the seismic profiles. Clearly the attempted (Fig. 3) projection of hypocentres onto the seismic sections at a nominal depth of 6–8 km (i.e. 4.0 sec two-way time = 6–8 km from the sea surface) can not be precise and is aimed only at a crude qualitative understanding of the seismicity in terms of the structure of the Korinth graben.

Figures 2 & 3 illustrate a concentration of epicentres in two areas near the flanks of the graben. This is explained in Fig. 4 where as it appears from Quaternary vertical movements that the north western flank of the graben has subsided approximately more than 400 m, while the SE flank has been uplifted more than 400 m. This reverse polarity of vertical motion which amounts to a differential displacement in the order of 1000 m could to some extent be responsible for the intense seismicity. Berckhemer & Kowalczyk (1978) stressed that the rate of uplift of nearly the entire Peloponnesos increased at the end of the Pliocene.

As already discussed (Fig. 3), profile B illustrates a different tectonic style from profile A. In the middle of the section a major feature is attributed to the overthrust of the Parnassos-Ghiona zone upon the Pindos zone. This feature also delineates the northern flank of the graben. The basement of the graben can possibly be attributed to the top of the Mesozoic sequence on the basis of stratigraphical and paleogeographical evidence from outcrops in the southern part of the Parnassos, Beotian and Pelagonian zones (Celet 1977; Clement 1977; Thiebault 1977). This basement probably consists of Upper Triassic overlain by Eocene flysch. Sediments above this horizon, identified as the base of the Plio-Quaternary, appear to be intensively folded but not faulted, and create a major anticline 7 km in width. Evidence of compressional tectonics is reported by Kelletat *op. cit.* for the early Pleistocene era.

Mercier *et al.* (1979), in an analytical study of neotectonic deformation of the Aegean lithosphere, suggested recognition of four deformational periods in the history of the present Aegean Arc:

(1) During the uppermost Miocene or Lower Pliocene the back-arc basin was under compression and by Lower Pliocene the external Pre-apulian Aegean became an active continental margin under compression.

(2) During the Lower Pliocene to Upper Pliocene the internal Aegean domain was under extension and the external Pre-apulian area was only weakly active.

(3) In the Earlier Quaternary–Mid-Pleistocene the internal Aegean suffered compression and during the Lower–Mid-Pleistocene the external Pre-apulian Aegean domain was strongly compressed.

(4) From the Mid-Pleistocene to present the internal Aegean has been under extension while the external Preapulian Aegean, although only weakly deformed, has been under compression.

This chronological sequence of events could be tested by a seismic reflection stratigraphy if the data had undergone advanced processing techniques, such as wave-equation-migration, continuous-velocity-analysis and F/K filtering.

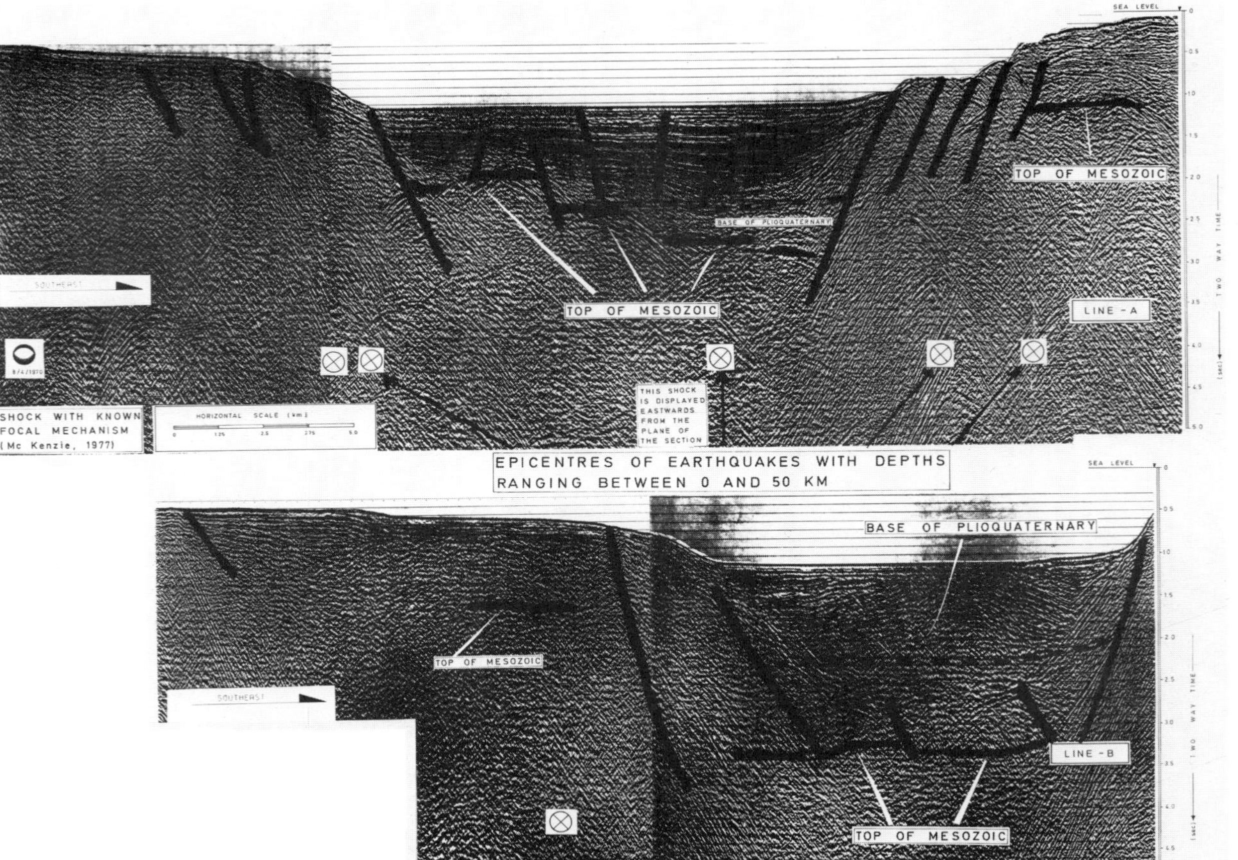

FIG. 3. Two parallel seismic profiles across the asymmetrical graben of the Gulf of Korinth. In both profiles epicentres of earthquakes with depths ranging between 0 and 50 km are included. All shocks were projected from their approximate azimouth location onto the seismic section at the same nominal depth (i.e. 4.0/sec two way time ≃6–8 km from the sea surface) for the sake of uniformity. This depth range is supported by recent detailed studies e.g. Jackson *et al.* (1982).

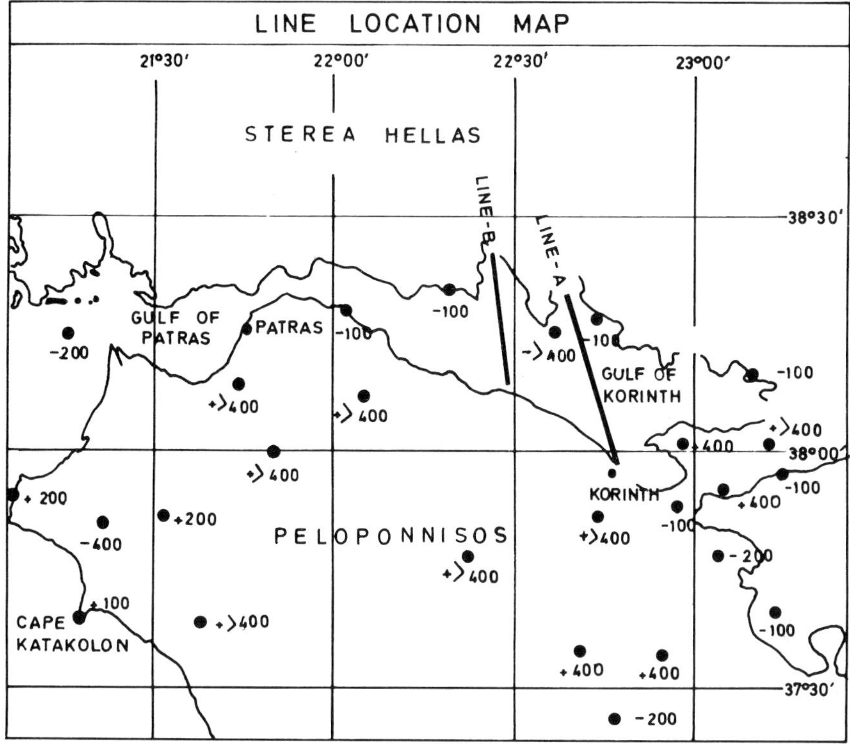

Fig. 4. Summary of vertical movements since Pliocene time. The positive (+) values refer to metres of uplift (i.e. +>400 signifies more than 400 m, while +400 is less or equal to 400 m etc.). Negative values refer to metres of subsidence. Dimensions are only relative. It is important to note that the NW part of the seismic line A has subsided more than 400 m while the SE end is uplifted more than 400 m. The differential displacement along seismic line A is in the order of 800 m. For profile B the situation is less marked. The NW part subsided 100 m, while the SE part was uplifted an unknown amount. Values for Quaternary vertical movements are taken from Kelletat *et al.* (1978).

Gulf of Korinth graben formation

The high rates of motion and the occurrence of major earthquakes indicate that brittle processes are important in the Korinth–Euboea region. Recent studies suggest that the earthquakes are commonly very shallow. Soufleris (personal communication 1982) suggested that about 80% of the hypocentres in the Gulf of Korinth are confined to a band 10–15 km below the surface. The three major earthquakes of February 24th and 25th, and March 4th, 1981 with magnitudes 6.7, 6.4 and 6.4 on the USGS scale, were estimated to have focal depths of 10, 8 and 8 km (all ± 2 km) by Jackson *et al.* (1982) using waveform modelling. The occurrence of hot-spring and solfatara activity in the vicinity of the Korinth and Euboea grabens, as at Loutraki near Korinth, Ypati and Thermopylae overlooking the Euboea graben, and Aedipsos on Euboea, suggest that heat-flow is high, with an associated high geothermal gradient. This supports the inference that the brittle-to-ductile transition zone is at a high level in the crust.

McKenzie (1978a & b) explains the high heat flow and shallow seismic activity in the Aegean region as being due to lithospheric stretching. As this is still active, subsidence is due to isostatic compensation of crustal blocks attenuated by normal faulting and not yet to thermal contraction. This model satisfactorily accounts for the formation of the Korinth and Euboea grabens as actively stretched zones lying between the relatively unaffected crust of the Peloponnesos, mainland Greece and Euboea blocks. Earthquake activity, as pointed out by McKenzie (1977) and shown on Fig. 2 is concentrated at the boundaries of the two main grabens.

The important question remaining is the origin of the continuing uplift of the Peloponnesos block, which is here considered to be responsible for enhanced seismicity on the southern side of the Korinth graben. The model preferred here (Fig. 5) is based on Bott's (1979) graben

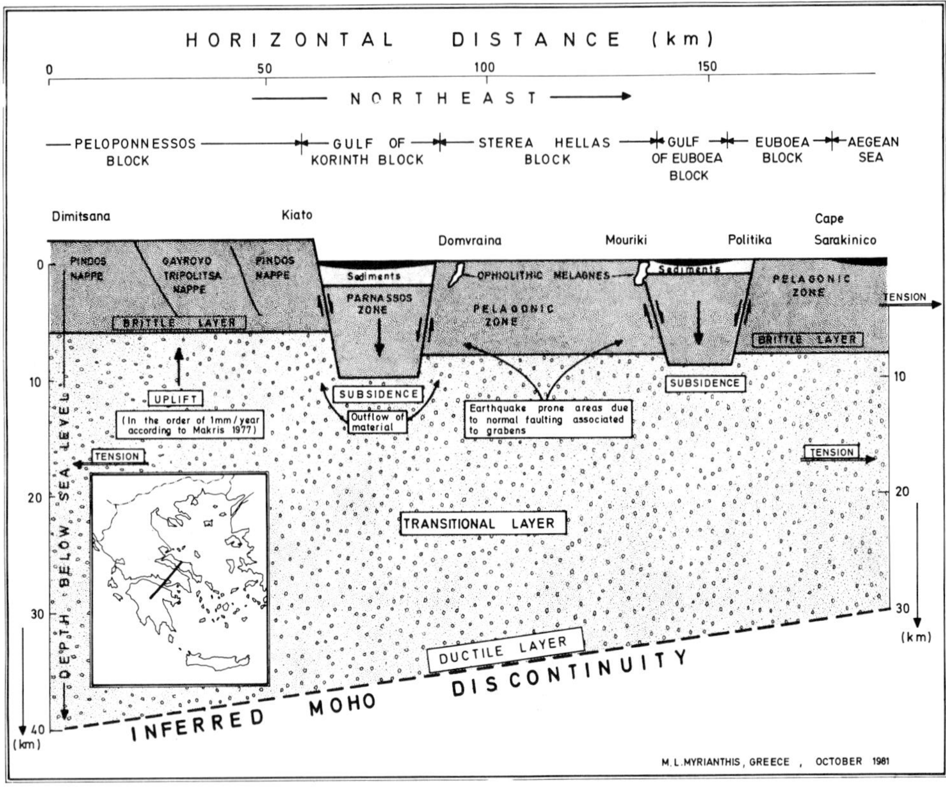

FIG. 5. A simplified model which is based on the mechanism of graben formation by wedge-subsidence affecting the upper continental crust. Subsidence compensated by horst-uplift and lateral crustal flow is probably the cause for the rapidly uplifted Peloponnesos block. Both the Korinth and Euboea grabens subsided by different amounts. Normal faults associated with these grabens are potentially active. Note that thicknesses of the various blocks are not to scale.

formation hypothesis. Isostatic compensation of the subsiding wedge of crust is considered to occur within the ductile part of the crust by lateral flow, thus thickening and uplifting the flanking blocks. An alternative model, which could also explain the regional tensional stress field is that suggested by Kusznir & Bott (1977).

This involves domal uplift, driven by thermal expansion in this case, with local rifting and associated subsidence on the crest of the dome.

Conclusions

The study of two multi-channel seismic reflection profiles lead to a better understanding of the structural setting of the Gulf of Korinth asymmetrical graben. The shallow seismicity is related to postulated vertical movements across the graben. A simple model is proposed to explain the shallow earthquakes based on graben formation by normal-fault mechanisms under the influence of horizontal deviatoric tension applied to lithosphere. Accordingly, the Gulf of Korinth and Euboea grabens are visualized as subsided, downward-narrowing wedges of brittle upper crust. Subsidence probably began earlier than the Pliocene. Outflow of material in the underlying transitional part of the crust is inferred to be the cause of Peloponnesos block-uplift, and probably began in late Pliocene times. Relative movements between the still-rising Peloponnesos block and the subsiding Gulf of Korinth block occur along active normal faults which display considerable seismicity.

The underlying cause of earthquakes in the wider area is the isostatic compensation of brittle-layer blocks, following gradual crustal stretching in the Aegean domain. Deformation related to block re-arrangements may trigger earthquakes along zones of weakness such as normal faults associated with the Korinth and Euboea grabens.

ACKNOWLEDGEMENTS: Thanks are due to Professor K. Zachos, ex-Chairman of the Public Petroleum Corporation of Greece for permission to release this paper and to Dr Jannis Samouilides, General Technical Manager of the Company, for giving the original permission. It should be noted that the opinions expressed in the paper are those of the author alone and do not necessarily reflect the views of the Public Petroleum Corporation of Greece. The manuscript was reviewed by J. A. Jackson and R. Evans.

References

BERCKHEMER, H. & KOWALCZYK, G. 1978. Postalpine geodynamics of the Peloponnesus. *In*: CLOSS, H., ROEDER, D. & SCHMIDT, K. (eds) *Alps, Apennines, Hellenides*. E. Schweizerbart'sche, Stuttgart.

BOTT, M. H. P. 1979. Subsidence mechanisms at passive continental margins. *In*: WATKINS, J. S., MONTADERT, L. & DICKERSON, P. W. (eds) *Geological and Geophysical Investigations of Continental Margins: Mem. Am. Assoc. Petrol. Geol.*, **29**, 3–9.

CELET, P. 1977. Les bordures de la zone du Parnasse (Grèce). Evolution paléogéographique au Mésozoique et caractères structuraux. *In*: KALLERGIS, G. (ed) *Proc. 6th Colloquium on the Geology of the Aegean Region*. **2**, 725–740.

CLEMENT, B. 1977. Relations structurales entre la Zone du Parnasse et la Zone Pélagonienne en Béotie (Grèce continentale). *In*: KALLERGIS, G. (ed). *Proc. 6th Colloquium on the Geology of the Aegean region*, **1**, 237–251.

JACKSON, J. A., GAGNEPAIN, J., HOUSEMAN, G., KING, G. C. P., PAPA-DIMITRIOU, P., SOUFLERIS, C. & VIRIEUX, J. 1982. Seismicity, normal faulting, and the geomorphological development of the Gulf of Korinth (Greece): the Korinth earthquakes of February and March 1981. *Earth planet. Sci. Lett.* **57**, 377–397.

KELLETAT, D., KOWALCZYK, G., SCHRÖDER, B. & WINTER, K.-P. 1978. Neotectonics in the Peloponnesian Coastal Regions. *In*: CLOSS, H. ROEDER, D. & SCHMIDT, K. (eds) *Alps. Apennines, Hellenides*. E. Schweizerbart'sche, Stuttgart, 512–518.

KUSZNIR, N. J. & BOTT, M. H. P. 1977. Stress concentration in the upper lithosphere caused by underlying visco-elastic creep: *Tectonophysics*, **43**, 247–256.

MAKRIS, J. 1977. Seismic and gravity studies in Greece and some geodynamic implications. *In*: KALLERGIS, G. (ed) *Proc. 6th Colloquium on the Geology of the Aegean Region*, **1**, 61–85.

MCKENZIE, D. 1977. Present deformation of the Aegean region. *In*: KALLERGIS, G. (ed.) *Proc. 6th Colloquium on the Geology of the Aegean Region*, **3**, 1303–1311.

—— 1978a. Some remarks on the development of sedimentary basins. *Earth planet. Sci. Lett.* **40**, 25–32.

—— 1978b. Active tectonics of the Alpine–Himalayan belt: the Aegean Sea and surrounding regions. *Geophys. J.R. astron. Soc.* **55**, 217–254.

MERCIER, J. L. DELIBASSIS, N. GAUTHIER, A. JARRIGE, J.-J. LEMEILLE, F. PHILIP, H. SEBRIER, M. & SOREL, D. La néotectonique de l'Arc Égeen. 1979. *Rev. Géol. dyn. Géogr. phys. Paris* **21**, 67–92.

MYRIANTHIS, L. M. 1982. Geophysical study of the epicentral area of Alkyonides island earthquakes,. Central Greece. *Geophysical Transactions*, Roland Eötvös Inst., **28**, 5–17.

SCHROEDER, B. 1976. Volcanism, neotectonics and postvolcanic phenomena east of Korinth/Greece. *Proc. Int. Cong. on Thermal Waters, Geothermal Energy and Vulcanism of the Mediterranean Area*. **3**, 240–248.

SCHWAN, W. 1978. Structural tectonics of the Parnassus–Ghiona Mountains in the Central Hellenides. *In*: CLOSS, H., ROEDER, D. & SCHMIDT, K. (eds) *Alps. Apennines, Hellenides*. E. Schweizerbart'sche, Stuttgart, 430–433.

THIÉBAULT, F. 1977. Stratigraphie de la série des calcschistes et marbres ("plattenkalk") en fenêtre dans les massifs du Taygète et du Parnon (Peloponnèse–Grèce) *In*: KALLERGIS, G. (ed). *Proc. 6th Colloquium on the Geology of the Aegean Region*. **1**, 691–702.

M. L. MYRIANTHIS, Public Petroleum Corporation of Greece, Department of Geophysics, 54 Academias Street, Athens (134), Greece.

Tectonic evolution of the North Aegean trough

Nicolas Lybéris

SUMMARY: Geological field data are used together with interpretation of Landsat images, results obtained during a *Jean Charcot* cruise in 1972, and other available seismic reflection profiles, to discuss the tectonic evolution of the North Aegean trough. This bathymetric trough is located over a narrow sedimentary basin with up to 6 km of sediment connected to two NNW-SSE sedimentary basins: the Thermaïkos and Kavala basins. The prevailing tectonic regime is extensional associated with strike-slip, with no evidence of compression as confirmed by seismo-tectonic observations. This regime has dominated the evolution of the North Aegean, as the prolongation of the North Anatolian fault since the Tortonian, However, the present eastern bathymetric trough does not have the same tectonic origin. The main extensional zone apparently lies to the north of it and continues eastward into the Thrace basin up to the Maritsa fault.

Geological background

The area discussed in this paper extends from the Greek peninsula to the Marmara Sea (Fig. 1). I discuss the post-Middle Miocene pattern of deformation of the North Aegean trough and the surrounding Neogene basins. The interpretation of fault patterns is based on a synthesis of land data obtained using field structural analysis in Eastern Macedonia and Western Thrace, bathymetric (Fig. 2) and available seismic reflection profiles (Biju-Duval *et al.* 1972; Needham *et al.* 1973; Lalechos & Savoyat 1979). A study of Landsat images at 1 : 500 000 scale has also been carried out to extend field analysis beyond Eastern Macedonia and Western Thrace. Finally stress trajectories are defined, taking into account of focal mechanisms obtained from recent shallow earthquakes occurring in the North Aegean and North Anatolia (McKenzie 1972 and 1978; Lybéris & Deschamps 1982; Soufleris & Stewart 1981; Jackson *et al.* 1982b).

The Neogene is characterized by numerous subsiding basins among which the most important are the North Aegean trough (Needham *et al.* 1973) and the Marmara Sea. These basins follow two main general trends (Fig. 5): the first one is NNW-SSE and the other one NE-SW to ENE-WSW, changing to E-W in the Marmara Sea. The first of these trends follows the Dinaric direction of the Hellenides (Aubouin, 1973). The structure of these basins is produced by important faulting, with up to 6–7 km of normal displacement, which is still continuing as the North Aegean trough, the Marmara Sea and the North Anatolian fault are seismically active (McKenzie 1972; Ketin 1969; Crampin & Ucer 1975; Lybéris & Deschamps 1982). The present extension of the Aegean area, which is related to the westward movement of the Turkish block has been discussed in particular by McKenzie (1972, 1978), and Le Pichon & Angelier (1979). The North Aegean trough and the surrounding basins constitute the southern margin of the European plate which is less deformed than the Hellenic arc (Le Pichon & Angelier 1979). The following sedimentary basins are distinguished from west to east (Fig. 1); the Sporades and Vardar basins; Kavala, Thasos-Samothraki; Saros trough and Marmara Sea.

The Rhodope Massif separates north-western Anatolia (Fig. 1) from the internal Hellenides. In central Macedonia, the last major orogenic phase, which created the nappe system, is of Priabonian age (Mercier 1966, 1968). According to the same author, a post-Oligocene and pre-Upper Miocene compressive phase has resulted in folding of the already deformed units and the Eocene-Oligocene molasse-type deposits. In North-West Anatolia the final compressive phase folded the Pontides prior to the Pliocene (Bergougnan 1975; Fourquin 1975) and, more precisely, in the Eastern Thrace basin, drilling has shown that the folding is of Middle Miocene age (Doust & Arikan 1974). Following the mid-Miocene compressive phase, general extension occurred in the area of the northern Aegean and Marmara Seas which is mostly responsible for the present distribution of the Neogene sedimentary basins. Information bearing upon the age of this episode is now discussed.

The first post-orogenic marine deposits in the North Aegean region are of Sarmatian age (Gillet 1961). In the Sporades basin little information is available about the Neogene sediments which occupy the deep parts of the trough. However, Lalechos & Savoyat (1979) discussed evidence of a thick Miocene interval. It should be noted that the molasse basins in the Western part of Greece stopped subsiding during the Oligo-Miocene compressive phase whereas the Vardar basin continued to subside

FIG. 1. General morphological map of the Central and Northern Aegean sea (after Needham *et al.* 1973).

throughout the Mio-Plio-Pleistocene (Mercier 1968).

Two drill-holes in the approaches to the Thermaïc Bay (Faugères & Robert 1976) (Fig. 1) show the existence of a continuous thick clastic sequence from lower Middle Miocene onwards. Evaporites are absent to the west and south of the Chalkidiki peninsula (see Fig. 5). To the east, the other basins which surround the North Aegean trough, were accumulating evaporites during the Late Miocene (Lalechos & Savoyat 1979). The marine pre-evaporitic Miocene deposits are known in Strymon and Thasos drill-holes (Gramann & Kockel 1969). The Upper Miocene marine transgression (Tortonian or Samatian) also affected the Central Aegean sea (Grekoff *et al.* 1967) and the Dardanelles (Benda & Meulenkamp 1972). In other places (Lalechos & Savoyat 1979; Faugères 1978), the Upper Miocene is represented

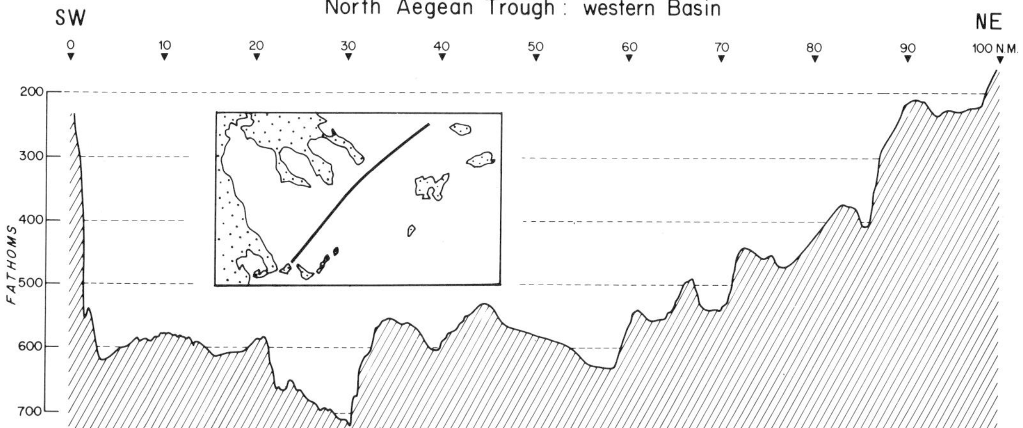

FIG. 2. Bathymetric profile over the western axis of the North Aegean trough (after Needham *et al.* 1973).

by continental, lacustrine or lagoonal sediments, resulting from subsidence followed by an important Plio-Pleistocene transgression in the North Aegean area. In north-western Anatolia, lacustrine sedimentation began in Middle Miocene and became marine during the Upper Miocene (Lütting & Steffens 1976). Thus, the main extensional phase started at the boundary of the Middle Miocene and the Upper Miocene and has continued since.

It is important to note that volcanism has been active in the proximity of the North Aegean trough since Lower-Middle Miocene. It has a high K_2O content and is mostly calc-alkaline and shoshonitic (Fytikas *et al.*, this volume). The volcanic activity in the North Aegean region dates from Upper Oligocene to Middle Miocene. On the southern edge of the Rhodope Massif, along the Kavala-Xanthi fault, volcanism dates from 33 Ma. (Fytikas *et al.* this volume). Southward, on Limnos, Ag. Evstratios and Lesbos islands, it dates from Lower to Middle Miocene (Fytikas *et al.* 1979). The Upper Miocene and Plio-Pleistocene volcanism is basaltic (Innocenti *et al.* 1979) and is situated in western Anatolia (Fourquin *et al.* 1970). In Edessa (on the Vardar basin border), the volcanism is calc-alkaline (Bellon *et al.* 1979; Kolios *et al.* 1980). In the immediate proximity of the trough axis, south of the Sporades basin, on Psathoura island (Fig. 1), basaltic alkaline volcanism occurred 0.5 Ma ago. This volcanism is similar to the Kula volcanism (Western Anatolia) (Innocenti *et al.* 1982), which occurred during the Plio-Quaternary formation of the Anatolia grabens. In the Volos area, 30 km to the west of the trough, Pliocene-Quaternary basalts with a high K_2O and MgO content erupted 3 to 0.5 Ma ago (Innocenti *et al.* 1979).

North Aegean trough

The North Aegean trough, 1000–1500 m deep, is a graben which extends from the Saros trough to the east as far as Magnesia to the west (Fig. 1). It consists of a western branch located west of Thasos island trending 045°, and an eastern one trending 075° (Fig. 1). The transition between the Macedonia-Thrace continental shelf and the North Aegean trough occurs in a series of steps. Seismic-reflection profiles demonstrate that these steps are bounded by faults with an important vertical component (Biju-Duval *et al.* 1972; Needham *et al.* 1973; Lalechos & Savoyat 1979). The one hundred kilometre-wide Sporades basin is better developed than the Saros trough, between Samothraki and Limnos, which is only 20 km wide. The morphological profile of the western part of the North Aegean trough, from the Sporades basin to the Thasos-Samothraki shelf is illustrated in Fig. 2.

The available seismic profiles (Biju-Duval *et al.* 1972; Needham *et al.* 1973; Lalechos & Savoyat 1979) exhibit the structure of the North Aegean trough which is mainly characterized by faults with an important normal component (Fig. 3). The vertical offsets on these faults are tens to hundreds of metres. The Sporades basin (Brooks & Ferentinos 1980) is not symmetrical (Fig. 3). To the northwest,there is a progressive downslope transition to the axis of the trough over a distance of 60 to 70 km. In contrast, the southeastern edge of the trough is abrupt with a transition over a distance of only about 10 km, towards the Sporades scarp (Lalechos & Savoyat 1979; Brooks & Ferentinos 1980). The basin's western margin is also quite steep with normal fault throws on a scale of kilometres along the Magnesia mountains. Normal faulting

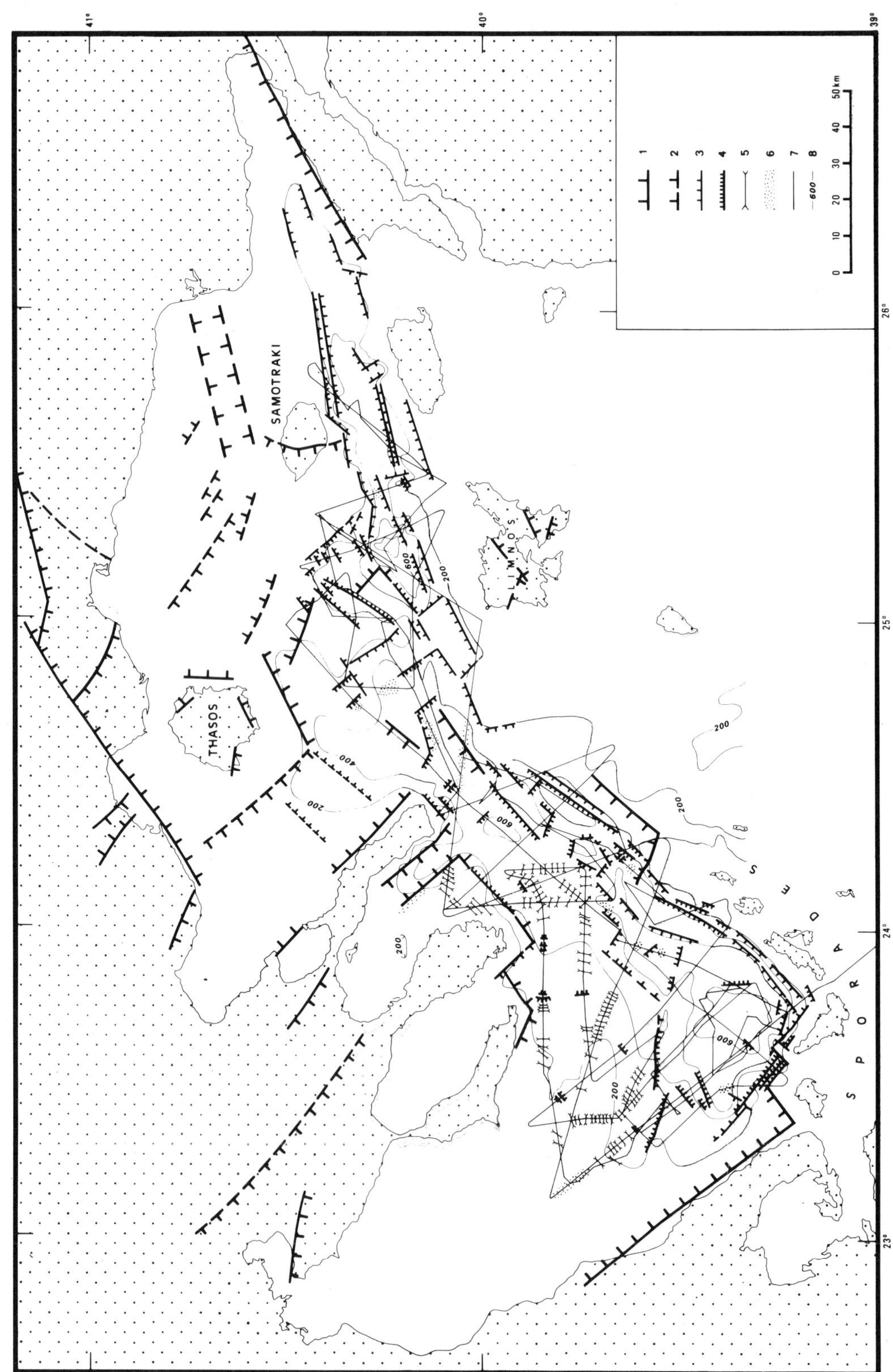
41°
40°
39°
26°
25°
24°
23°
SAMOTRAKI
THASOS
LIMNOS
SPORADES
1
2
3
4
5
6
7
8
600
0 10 20 30 40 50 km
200
400
600

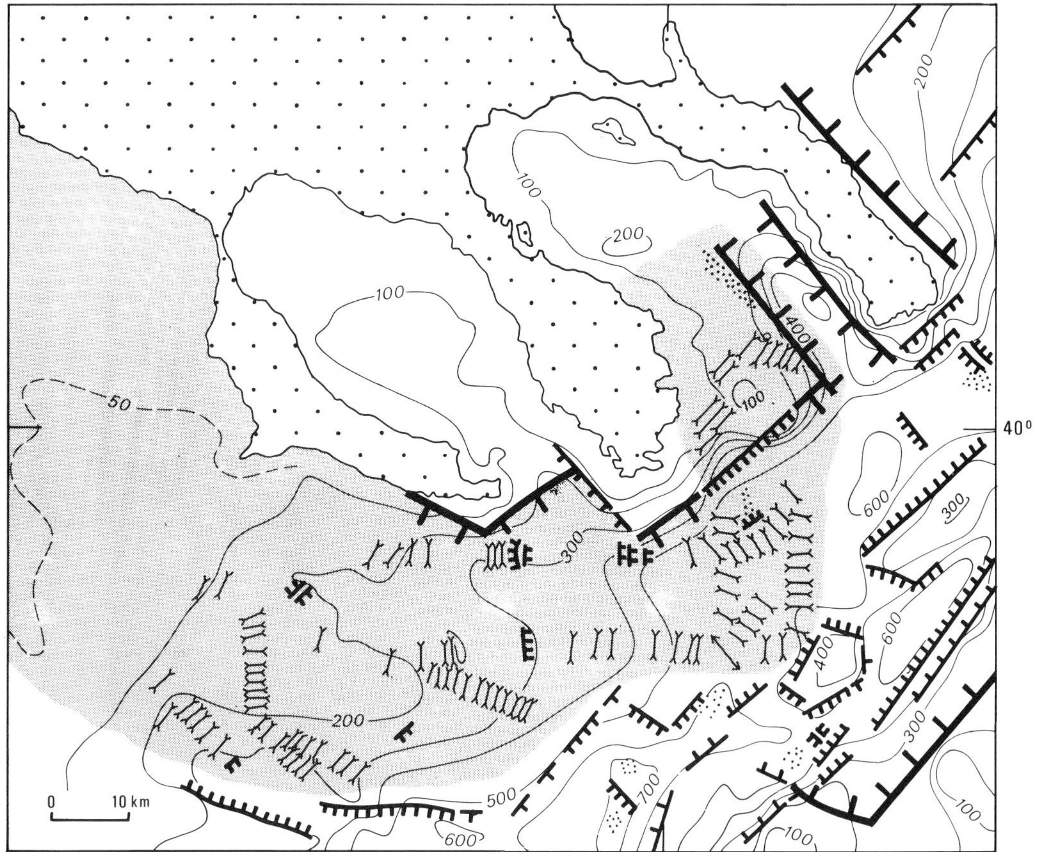

FIG. 4. Detail of Fig. 3: the gravity fault pattern. The shaded area corresponds to the extent of deltaic deposits.

affects recent sediments as shown by R.V. *Jean Charcot* 3.5 khz sounder recordings (Needham *et al.* 1973).

Air-gun and sparker profiles in the Gulf of Thermaïkos show a complex fault system with normal as well as reversed offsets to the north of the main normal faults which form the axis of the trough (Fig. 4). The fault system affects the most recent sediments which are deltaic deposits from the Vardar and neighbouring rivers. These features have been interpreted as gravity faults by Ferentinos *et al.* (1981). They are common in polygenic unconsolidated deltaic sediments because of rapid deposition. Their strike is random and does not correspond to the main direction of the deep faults. They seem to be controlled by the local morphological slope. Their spatial distribution in Fig. 4 illustrates the extent of these deltaic deposits.

In the vicinity of the Chalkidiki peninsula, the trough is divided by a horst parallel to the trough's axis. This structure is 40 km long and 700 m high. It is an outcropping basement structure (Lalechos et Savoyat, 1979) characterized by a free-air gravity anomaly maximum and a magnetic anomaly and is probably a recent volcanic structure (see Le Pichon *et al.*, this volume).

FIG. 3. Fault pattern on the North Aegean sea:
1 and 2: Faults with normal components identified on multichannels seismic reflection profiles.
2: Pre-Plio-Quaternary faults.
3 and 4: Faults with normal components identified on monotrace seismic reflection profiles.
5: gravity faults.
6: Flat basins.
7: Ship-tracks.
8: Isobaths (in fathoms).

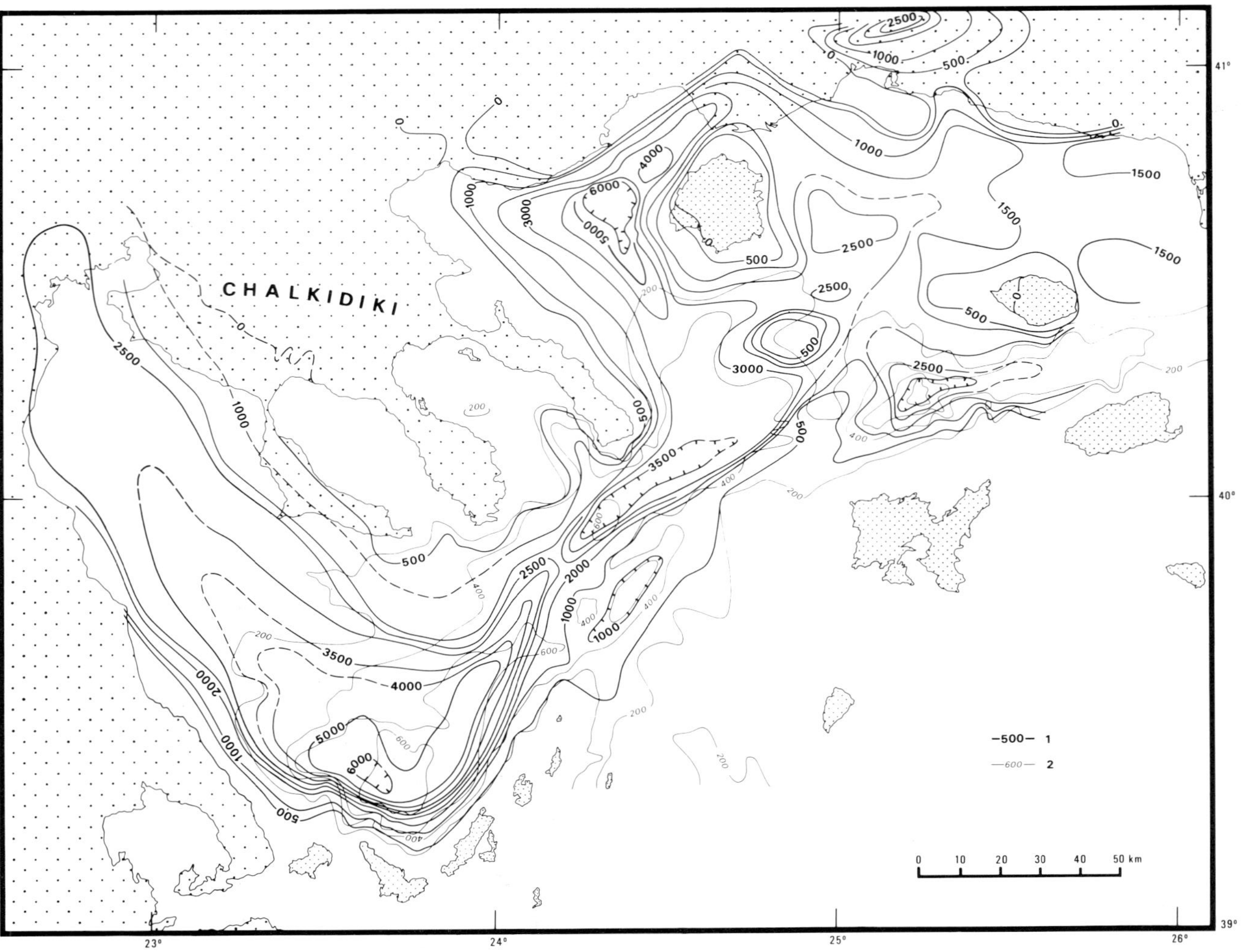

FIG. 5. Neogene (Middle Miocene to present) sediment thickness based on Lalechos and Savoyat's (1979) Neogene isopach map and on the bathymetric map given in Le Pichon *et al.* (this volume).
1: Neogene base isobaths.
2: Simplified bathymetry in fathoms.

The Strymon valley, to the east, and the Vardar valley to the west are two Palaeogene molasse basins, reactivated during Neogene times. The Strymon basin between the Serbomacedonian and the Rhodope massifs extends into the Kavala Gulf through a system of NW-SE faults (Fig. 3) which produced strong subsidence in the Neogene with deposition of more than 6 km of sediment (Fig. 5).

Biju-Duval *et al.* (1972) have demonstrated the existence of a deep, completely filled graben to the north of Saros trough, between Samothraki island and the mainland. The filling occurred during the Plio-Pleistocene within a NE-SW fault system. It is situated along the general eastward extension of the Sporades basin (Fig. 5). Possibly, this entirely buried graben connects the Sporades basin to the SW with the eastern Thrace basin to the NE. Burke & Ugurtha (1974) have shown the existence of an important Neogene basin in Thrace which is now entirely filled and emergent with 3 to 4 km of Neogene sediments. The Thrace basin terminates against the Maritsa fault (Fig. 6).

Thus, as can be seen in Fig. 6, the zone extending from Magnesia in the west to the Maritsa fault in the east along a continuous 050 to 070° trend seems to belong to the same extensional belt. It might be that the Saros trough was activated after the extension ceased along the north-eastern branch.

The submarine fault pattern is organized along the general trends: the first one along 65° and the other one along 135° (Fig. 3A). The first one corresponds to the mean direction of the scarps in the trough and the second corresponds to the trend of the Vardar and Strymon basins. The statistics in Fig. 3A take into account the main scarps of the Marmara sea. Note that scarps in the Marmara sea (bathymetric map of the Hydrographic Center) have an orientation comparable to those of the North Aegean trough described above.

Structural observations

Late Cenozoic faults, with a kilometre-scale normal component of faulting, control the morphology of the sedimentary basins. The largest ones, such as the Ganosdağ and the Kavala-Xanthi faults (Fig. 6), with vertical offsets greater than 6 km, occurred on the locations of pre-Neogene structures. The Ganosdağ fault for example, controls the Eo-Oligocene detrital deposits of the Eastern Thrace basin (Doust & Arikan 1974). Aubouin (1973) also suggested that the Kavala fault has existed since the Eo-Oligocene. Similar structures can be observed on the southern border of the Rhodope Massif: the Maritsa fault, which traverses Bulgaria, and neighbouring faults which belong to the same regime, have been active since Palaeozoic, according to Foose & Manheim (1975). These 040° to 060° faults mark one of the principal orientations of Neogene faulting. The second one, NNW-SSE to NW-SE, is the direction of the Palaeogene molasse basins, such as the Vardar basin, in which subsidence has been continuous since Lower-Middle Miocene (Faugères & Robert, 1976).

A field study of late Cenzoic faults has been carried out on the islands of Limnos, Thasos and Samothraki, as well as in the Serrai, Kavala, Xanthi and Komotini areas. Analysis of the orientation of fault planes and slickenside lineations from field measurements has enabled us to reconstruct the orientations of the principal stress axes that prevailed during the faulting event. The methods used include geometrical analysis of conjugate fault patterns, use of the method of right dihedra (Angelier & Mechler 1977) and determination of stress tensors by mathematical means (Angelier *et al.* 1982b).

The greatest fault in the North Aegean area is the Kavala-Xanthi fault (Figs 7, 8, 9) which can be clearly identified on the LANDSAT images (30209–08263 of 30.09.1978 and 213108194 of 02.06.1975) from the Strymon valley to the SW, as far as the Evros valley and the Neogene basin of Eastern Thrace to the NE. We estimate the vertical fault offset to be 7 km; the dip and motion vary with the fault's orientation. The average direction of the fault segment between Kavala and Xanthi (sites 1 and 4 on Figs 7, 8, 9) is 040° to 060° and the dip is less than 40°. The fault surface carries two principal striations: the earliest one has a normal offset (Figs 10AI and 10BI) and is cut by a later set of oblique and sub-horizontal striations. In places, one recognizes within this set of sub-horizontal lineations (Fig. 10BII), two different striation directions (Figs 10AII and 10AIII): the earliest one has a sub-horizontal pitch and the later one is oblique with a right-lateral motion sense (Fig. 10AIII). In fact, three tectonic events can be identified along this major fault. The first one corresponds to a NW-SE pure extensional motion followed by two successive right-lateral movements. The strike-slip motion was initially produced by NE-SW extension followed by a N-S extensional event (Figs 10AII and 10AIII). Thus the right-lateral component of movement results from the obliquity of these two directions of extension relative to the NE-SW to ENE-WSW trend of large pre-existing faults. The chronological order of these fault movements can be

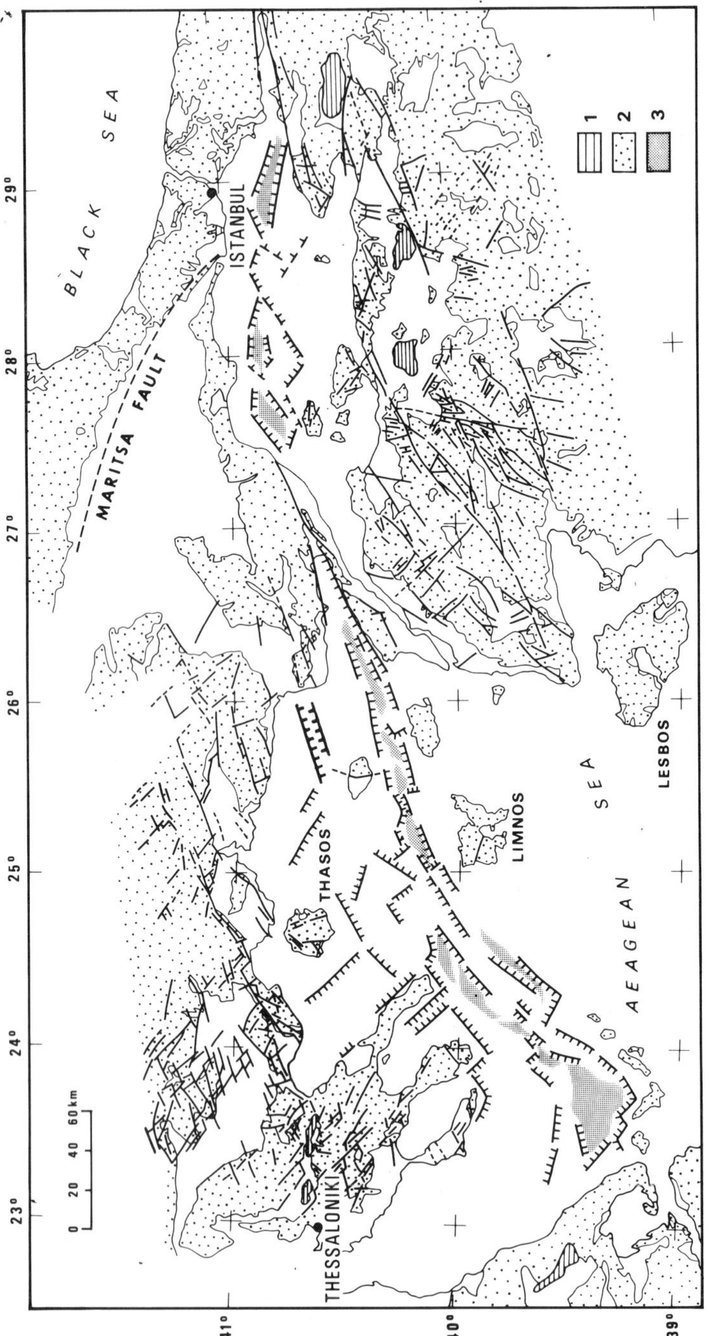

FIG. 6. Neotectonic structure of North-Western Anatolia and North-Eastern Greece based on Landsat image analysis. The sea fault pattern is based on Fig. 3: 1, Lakes; 2, pre-Middle to Upper Miocene units; 3, North Aegean trough axis.

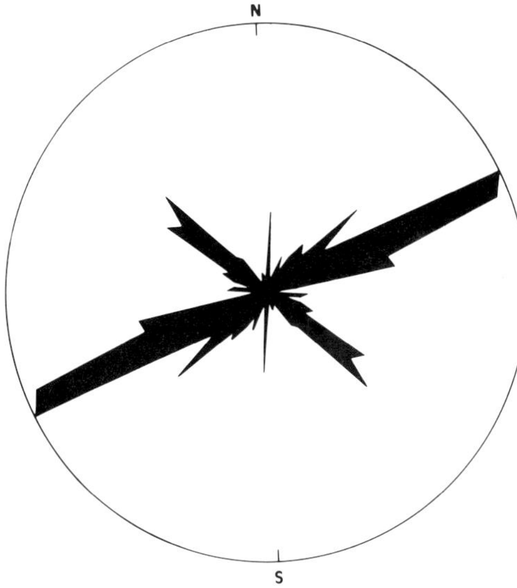

FIG. 6A. Frequency diagrams from the land fault pattern of Fig. 6 (246 measurements).

clearly determined in the field using the superimposition of successive striations.

NW-SE extension has been identified in another fault population in the Lefkon Formation of the Serrai basin. This formation is attributed to the Tortonian by Armour-Brown *et al.* (1979) (Fig. 11I and Fig. 7, site 3). The extensional pattern of these normal faults is the only one recognized so far in Late Miocene deposits. On the contrary, numerous faults with NE-SW and N-S directions of extension have been observed in the Pliocene and Quaternary sediments of the North Aegean regions (Figs 8 and 9). For example, in western Kavala, extensional faulting affects Lower Pliocene marl deposits. In this area, an initial NE-SW extension was produced by motion on NW-SE normal faults (Fig. 12I). N-S extension has occurred later (Fig. 12II). Thus, as the early NW-SE extension affects the Tortonian but does not affect the Pliocene, it must be of uppermost Miocene age.

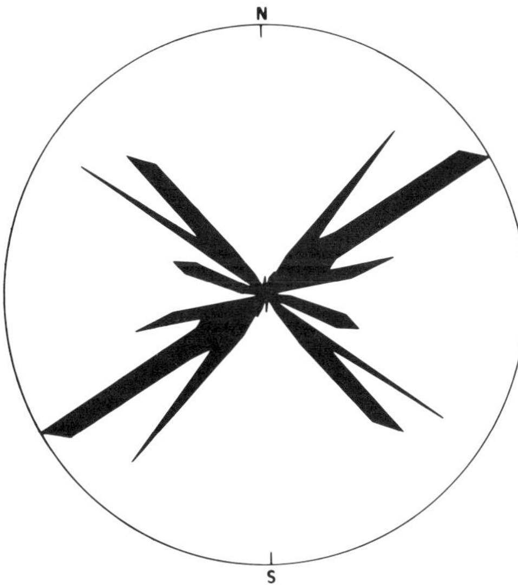

FIG. 3A. Frequency diagrams of fault strike directions in the North Aegean sea (from Fig. 3) and the Marmara Sea (from the bathymetric map N° 54020, of the Hydrographic Center, Washington).

From the fault movement chronology and the ages of the affected Neogene formations in Eastern Macedonia and Thrace, it is concluded that the major faulting events are: (1) a NW-SE extension of Upper Miocene age, (2) a NE-SW extension which resulted in a right lateral strike-slip movement on the major faults of the North Aegean region during Pliocene time and (3) a later N-S extension, which also had a right-lateral component of motion.

On Limnos, Eo-Oligocene folded clastic sediments and Lower Miocene volcanics are exposed (Fytikas *et al.* 1979). In these units, an initial compressive event has produced some disrupted folds and strike-slip faults. The computed azimuths of the sub-horizontal maximum stress axis (σ_1) are NW-SE and probably date from the Middle Miocene while this compressive event is unknown in the fault pattern which affects the Neogene deposits of the North Aegean region. The main NW-SE and NNE-SSW trend faults which cut the island are normal faults, several km long. They have offsets of hundreds of metres. The successive extensional events yield computed directions for the mimimum stress axis, σ_3, of NW-SE (Fig. 7, site 6), NE-SW (Fig. 8, sites 18 and 19) and N-S (Fig. 9, sites 16 and 17). Hot springs are common along the faults.

In Thasos, the Neogene covers the SW extremity of the island. It is represented by red proximal breccias, which are strongly fractured. Palynological analysis of samples of these breccias give a Lower Pliocene and Lower (?) Quaternary age (Lybéris & Sauvage 1983). Syn-sedimentary faulting occurred but only small normal faults have been observed near the basement. Two tectonic events have been recognized. The first one, NE-SW trending, is Lower Pliocene to (lower ?) Pleistocene in age (Fig. 8, sites 12, 13, 14) and the second one, orientated N-S, post-Lower (?) Pleistocene (Fig. 9, sites 11, 12, 13).

The eastern part of Samothraki consists of

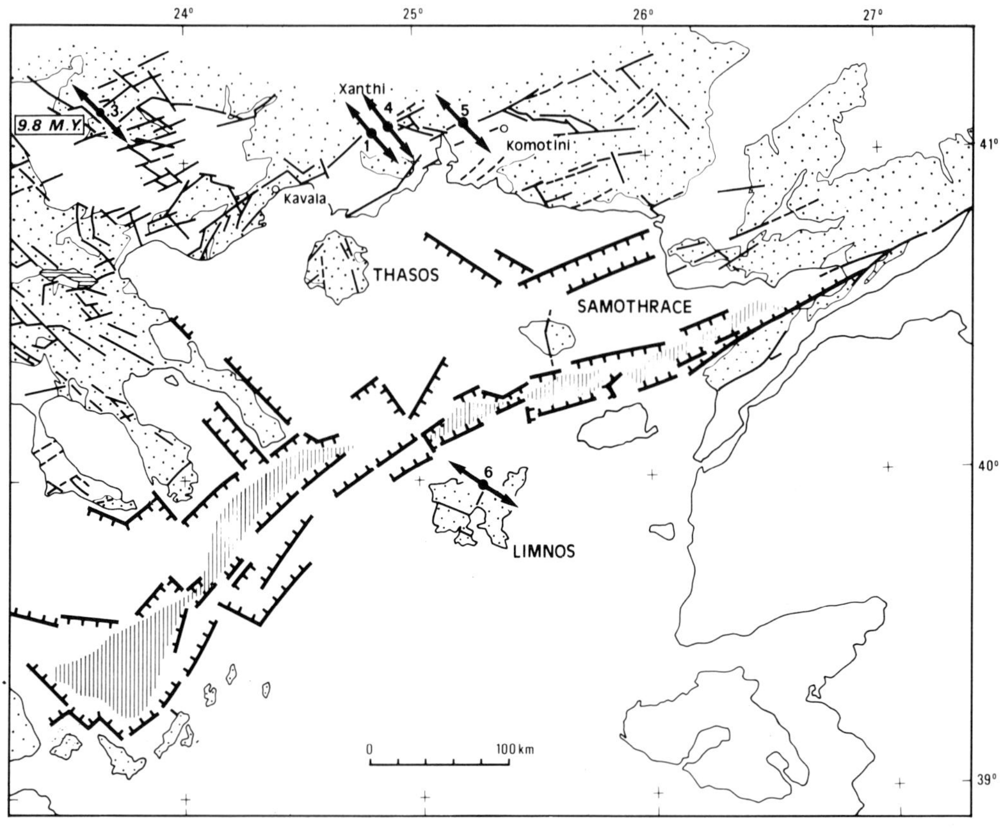

Fig. 7. Upper Miocene.

Figs 7–9. Simplified tectonic sketches of the North Aegean area from Figs 3 and 13. Black arrows indicate the main directions of extension resulting from the tectonic analysis of fault propulations. The presently known ages are indicated.

pre-Miocene volcanics (Heiman *et al.* 1972) while a Neogene basin is developed on the western side of the island. These two areas are separated by a normal NS fault with a 2 km vertical minimum offset. Most of the Neogene sedimentary units are continental cross-bedded sands and conglomerates. White marl intercalations representing local marine incursions, contain sponge spicules, and ostracods including *Cytheridea neapolitana* Kollman and *Cibicides lobatulus* Fichtel and Moll, of Lower Pliocene age (Lybéris & Sauvage, 1983). On the eastern extremity of the island, old talus was observed of probable Lower Quaternary age (Heiman *et al.* 1972). Two extensional events were recognized within the Neogene of Samothraki (Fig. 8, sites 15, 16, 17 and Fig. 9, sites 14, 15). Site 17 of Fig. 8 shows the measurements made along the large normal fault which limits the Neogene basin to the east. The mechanism of the large fault is compatible with the mechanisms of the smaller fault populations analyzed within the Lower Pliocene marls.

Neogene tectonic evolution

The north Aegean area was characterized by extensional tectonics with an important strike-slip component. The observations summarized in Figs 7–9 show coherence over the whole area, especially between the larger, kilometre offset faults (sites 1, 4, 5) and the smaller ones. Field measurements of the fault patterns are illustrated in Fig. 13. Notice that the fault trends are organized in two main directions (Fig. 13a): the first one NE-SW to ENE-SWS and the second one NW-SE. The 075° strike corresponds to the Kavala-Xanthi and the Serrai area major fault pattern. As for the dips (Fig. 13b) two maxima can be noticed: the first is at 35–40° and corresponds to the major normal faults; the second is at 50–60° and corresponds to the smaller normal faults affecting the Neogene deposits. Notice in Fig. 13c that the pitch of the striation is mostly in the range 70–90° but that sub-horizontal pitches also exist.

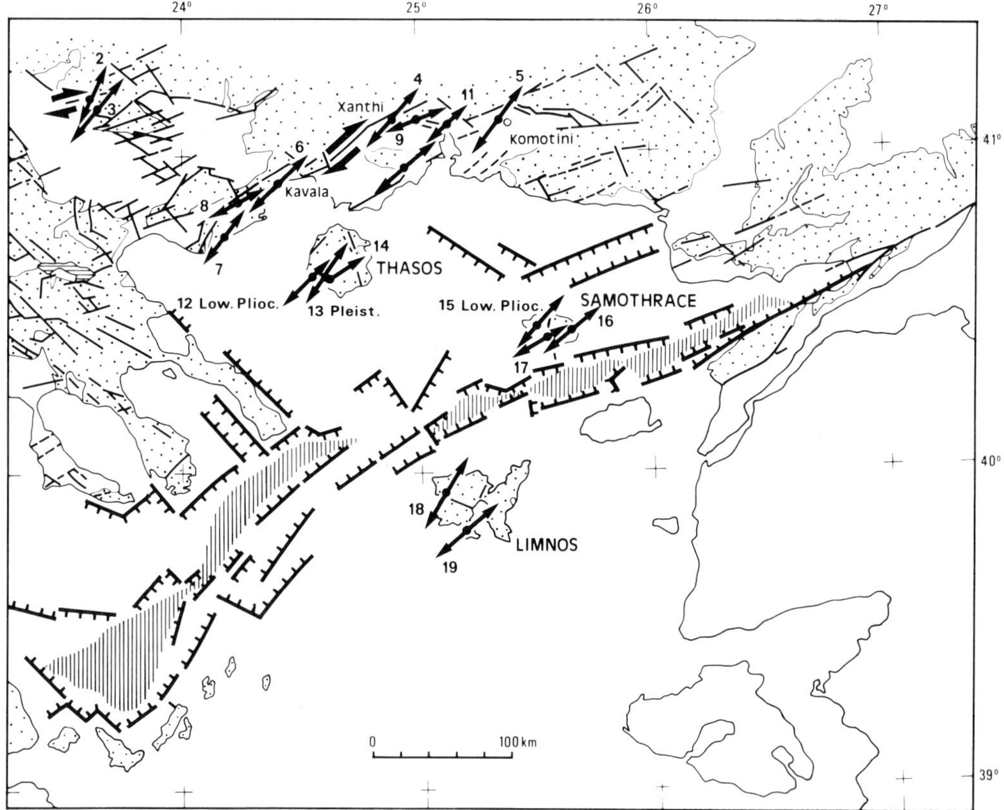

Fig. 8. Pliocene and Lower (?) Pleistocene. Double arrows indicate strike-slip faults observed in the field.

The consistency between land measurements of fault strikes (Fig. 13a), seismic reflection profiles (Fig. 3a) and orbital images (Fig. 6a) is satisfactory, although the age determination of the different tectonic events remains incomplete. The beginning of the extension in this area postdates Lower Miocene compression in the Vardar trough to the west and Middle Miocene compression in the east, in Eastern Thrace. Its average direction averages 140° prior to Lower Pliocene (Fig. 7). Important pre-middle Miocene features were reactivated during this extension. The 060° trending Kavala-Xanthi and Ganosdağ faults have a kilometre-scale normal component. They limit the main Neogene basins, as for example the North Samothraki filled basin, the Kavala basin, and the Eastern Thrace basin. It is important to notice that the NW-SE extensional event is incompatible with the North Anatolian fault's right-lateral motion. The Thrace Neogene basins are bordered to the NE by the Maritsa fault (Foose & Manheim, 1975).

A second extensional event had a 050° average orientation (Fig. 8). It occurred later than the Lower Pliocene and its affected units of definite Pleistocene age (dated in Thasos). This event is also identified in eastern Thrace (Mercier 1981) and in the Central Aegean sea. During this event the Oligocene Vardar and Strymon basins, of NNW-SSE trend, started a new cycle of subsidence, the Neogene basins being limited by right lateral strike-slip faults with a normal component. The Pliocene NE-SW extension is compatible with the right-lateral motion of the North Anatolian fault which is post-Lower Pliocene (Hancock & Barka, 1981, this volume). The North Anatolian fault seems to be independent of the older tectonic phases and has cut the orogenic belts of Northern Turkey since the Pliocene (Irrlitz 1972).

The most recent tectonic event was N-S extension (Fig. 9). This event is recognized in eastern Thrace (Mercier 1981), in Magnesia (Brooks & Williams 1982) and generally

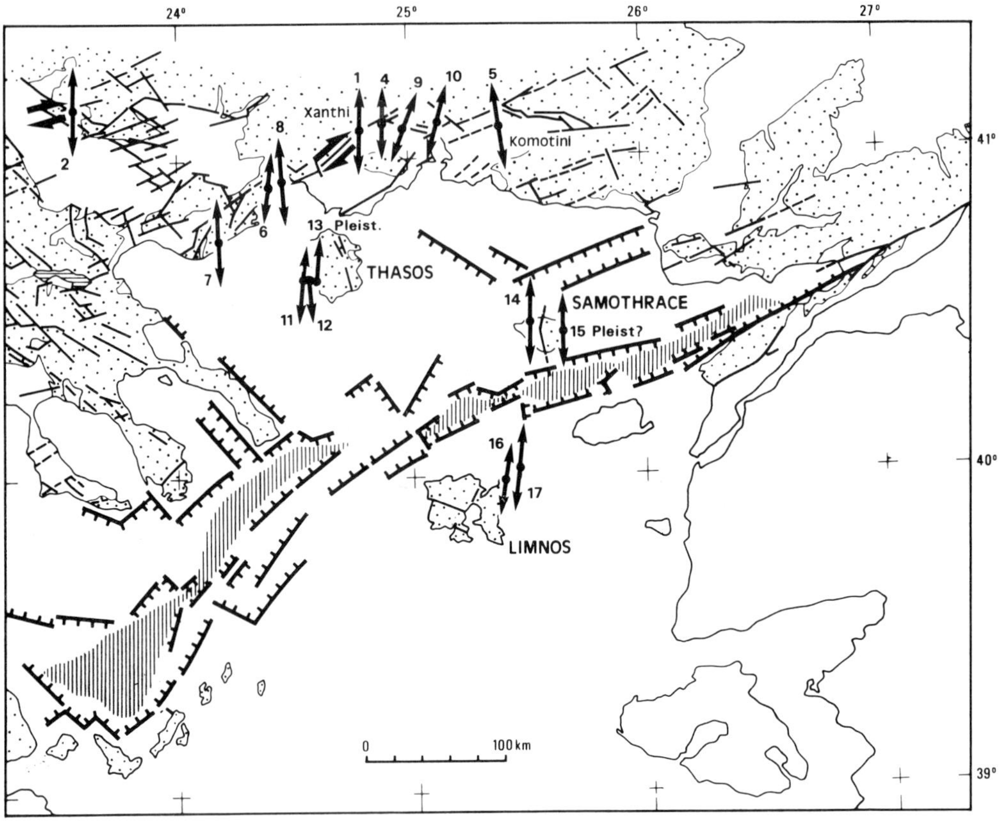

Fig. 9. Lower (?) Pleistocene to present.

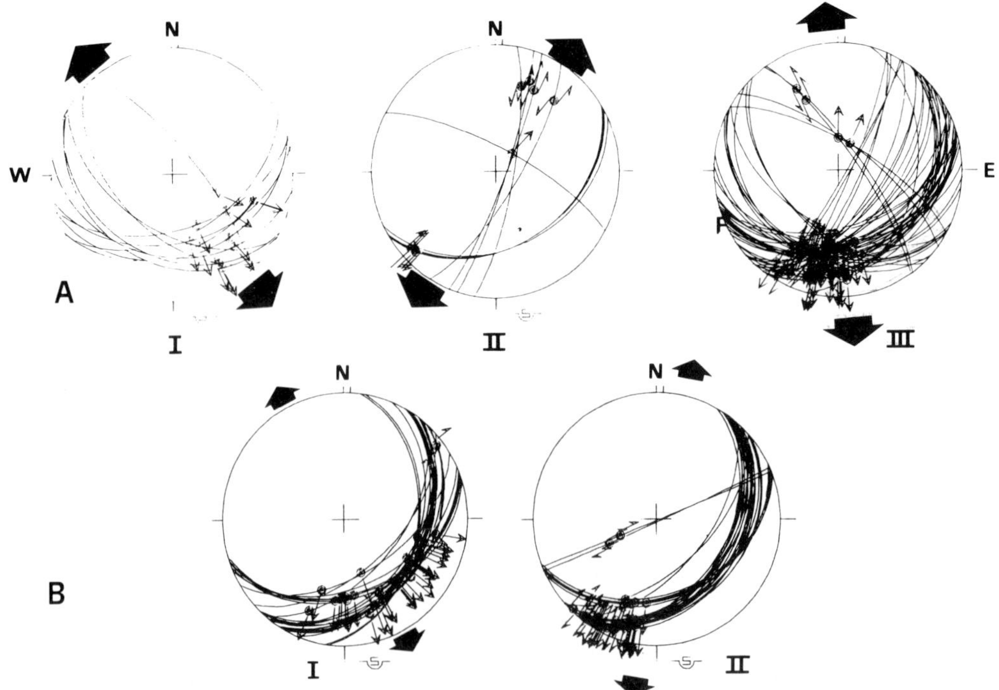

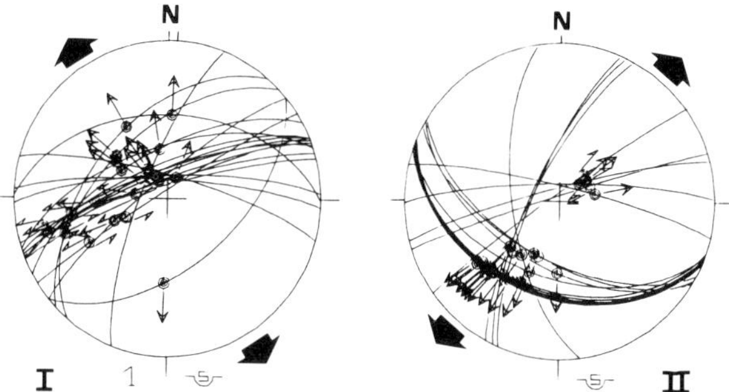

FIG. 11. Fault populations within the Lefkon formation, 9.8 Ma in age (site 3 on the Figs 7 and 8).

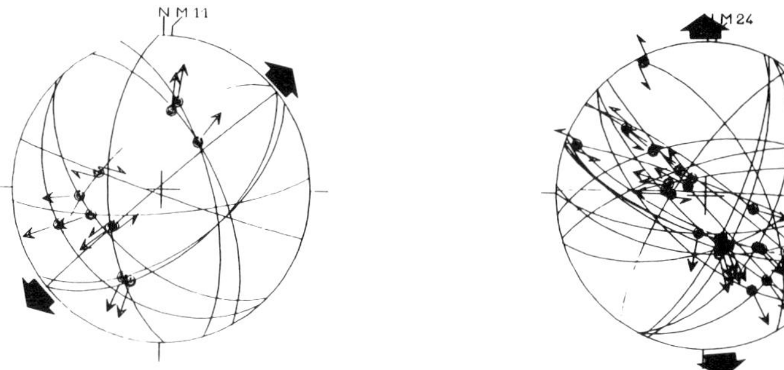

FIG. 12. Fault population analyzed within the Lower Pliocene marls of the Kavala basin border (site 7 on the Figs 8 and 9).

throughout the Central Aegean sea (Mercier *et al.* 1979a). It is later than true Pleistocene (older ?) and is still active, being accompanied by intense seismic activity (McKenzie 1972 & 1978; Papazachos & Comninakis 1978; Soufleris *et al.* 1981; Jackson *et al.* 1982b; Mercier *et al.* 1979).

The extension of the North Aegean area is related to the general Aegean post-Upper Miocene extension (Le Pichon & Angelier 1979). Numerous grabens were formed between the North Aegean trough and the Outer Hellenic arc as, for example, the sea of Crete, the Gulf of Corinth (Jackson *et al.* 1982a; Mercier *et al.* 1979a; Jongsma 1975; Lybéris *et al.* 1982a; Angelier *et al.* 1982b).

It is also concluded that Pliocene and Quaternary tectonics produced extension with a right-lateral strike-slip component around the North Aegean trough. The Rhodope Massif constitutes the southern limit of the European block, which was less deformed in the Neogene than the 'Aegean landmass'. The North Aegean trough constitutes the first important Neogene structure of this 'landmass'. The Rhodope Massif was unstable during the Neogene but its deformation is of secondary importance when compared to that of the Aegean area and of the Western Anatolia grabens (Ketin 1968).

North Aegean trough—north Anatolian fault

The Ganosdağ area of active faults (Allen, 1975) joins the North Aegean trough to the Marmara sea (Fig. 14) and establishes a tectonic connection with the North Anatolian fault. The structures of the Marmara Sea have morpho-tectonic characteristics comparble to those of the North Aegean trough. The North Anatolian fault, 1000 km long, is very active

FIG. 10. Examples of fault planes and slickenside lineations along the Kavala-Xanthi major fault. Schmidt lower-hemisphere projections. Fault planes are shown as great circles, slickenside lineations as dots with small centrifugal arrows (normal motion) or double arrows (strike-slip motion). Large black arrows indicate the direction of extension (σ_3). A: site 4 on Figs 7–9. B: site 1 on Figs 7–9.

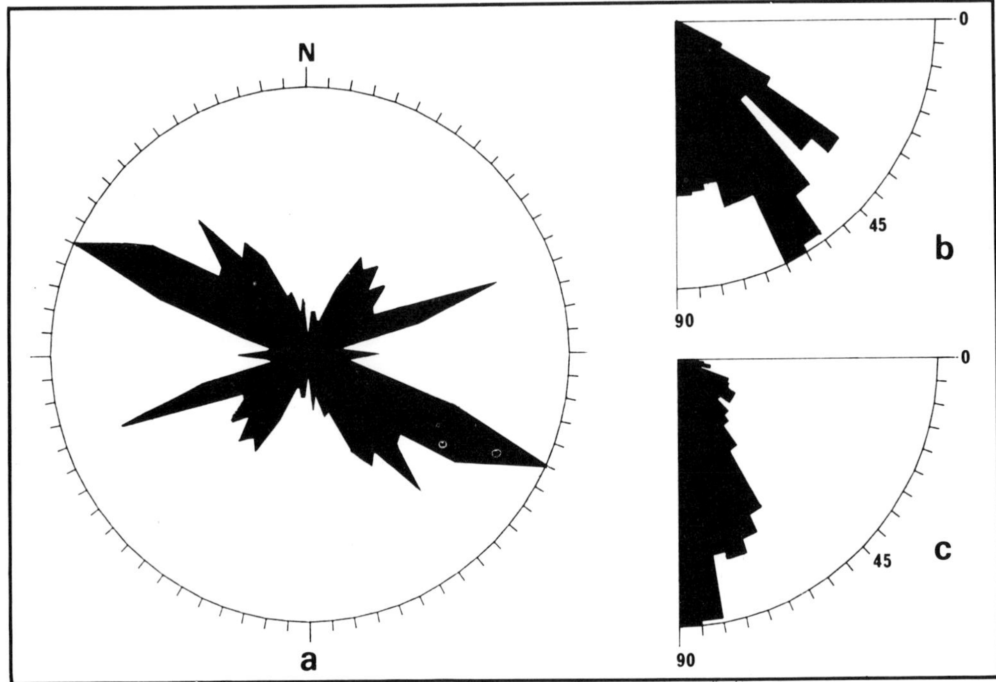

FIG. 13. Frequency diagrams from field measurements (810 measurements).
(a) Fault plane strikes,
(b) Fault dips,
(c) Pitches of striae.

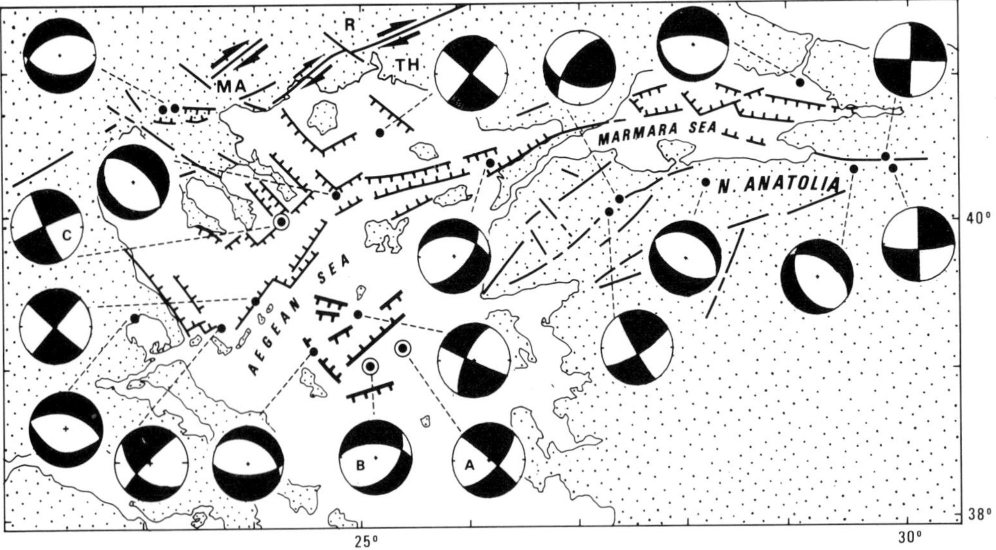

FIG. 14. Earthquake focal mechanisms in the North Aegean and north-western Anatolia, from McKenzie (1972 and 1978), Soufleris *et al.* (1982), Lybéris & Deschamps (1982) and Jackson *et al.* (1982b): the black quadrants are the extensional ones.

(Ketin 1948; Ambraseys 1970) with a right-lateral motion sense.

Focal mechanism determinations (McKenzie 1978) along the North Anatolian fault show compressive and strike-slip fault solutions. Intramontane Neogene basins (Irrlitz 1972; Biju-Duval & Letouzey 1975) occur along the fault from the Arabian-Anatolian-Eurasian plate triple-junction (McKenzie 1972) in the east to the Marmara Sea in the west. Near the Marmara Sea, the fault splays into a complex system in the Biga peninsula. To the east of the Marmara Sea basin, Toksöz *et al.* (1979) and Purkaru & Berckhemer (1982) have identified a seismic gap. In this area pure strike-slip motion changes into extensional strike-slip movement responsible for the creation of the Marmara Sea and of the North Aegean basin. Such a continuation has already been proposed by McKenzie (1970) and by Şengör (1979). Thus, Western Anatolia, south of the Marmara Sea, is affected by the Aegean extension. The exact horizontal motion along the North Aegean trough has not been determined but it is obviously much less than that along the North Anatolian fault where 80 to 100 km (Dewey & Şengör 1979) of horizontal displacement may have occurred.

The North Aegean trough to the west comes to an abrupt end in front of the Magnesia mountains (Fig. 1), on the faults bordering Magnesia (Figs 1, 3). No analogous structures are known to the west although smaller exensional basins exist in continental Greece, such as the Aliakmon basin (Lybéris *et al.* 1982b) and the Sperchios–Karpenission Gulf (Aubouin & Guernet 1963; Dercourt 1972) transverse structure. This structure is the northernmost of the major Neogene structures. The Hellenides orogenic belt has strongly subsided along it. In a sense, the Sperchios-Karpenision structure may be considered to be an attenuated westward continuation of the North Aegean trough.

Conclusions

I have shown that the North Aegean area has been affected by a prevailing tectonic regime of extension coupled with strike-slip since the uppermost Miocene, which is still active. No signs of compression were observed. The transition from this regime to the North Anatolian strike-slip and compressional regime occurs near the eastern part of the Marmara sea but an earlier connection between the two structures may have occurred farther north through the Thrace basin. At the other end of the North Anatolian trough, there is no obvious extension into continental Greece, although the Sperchios-Karpenission 'corridor' may represent an attenuated westward prolongation.

Acknowledgements. I thank D. Needham and B. Biju-Duval for access to results from their 1973 and 1972 papers which had only been presented orally. Field work was supported by CNEXO and CNRS. The numerical treatment of field measurements was supported by CNEXO. X. Le Pichon made useful comments. X. Le Pichon & J. Chorowicz critically read the manuscript.

References

Allen, C. R. 1975. Geological criteria for evaluating seismicity. *Bull. geol. Soc. Am.*, **86,** 1041–1057.

Ambraseys, N. N. 1970. Some characteristic features of the Anatolian fault zone. *Tectonophysics*, **9,** 143–165.

Angelier, J. & Mechler, P. 1977. Sur une méthode graphique de recherche des contraintes principales également utilisable en tectonique et en seismologie: la méthode des dièdres droits. *Bull. Soc. géol. Fr.*, **19,** 1309–1318.

——, J. Lybéris, N., Le Pichon, X., Barrier, E. & Huchon, P. 1982a. The tectonic development of the Hellenic Arc and the Sea of Crete: a synthesis. *Tectonophysics*, **86,** 159–196.

——, Tarantola, A., Valette, B. & Manoussis, Str. 1982b. Inversion of field data in fault tectonics to obtain the regional stress. I. Single phase fault populations: a new method of compating the stress tensor. *Geophys. J.R. astr. Soc.*, **69,** 607–621.

Armour-Brown, A., De Bruijn, H., Maniati, C., Siatos, G. & Niesen, P. 1979. The geology of the neogene sediments north of Serrai and the use of rodent faunas for biostratigraphic control. *Proc. 6th Aegean Coll., Athens*, 615–622.

Aubouin, J. 1973. Des tectoniques superposées et de leur signification par rapport aux modèles géophysiques. L'exemple des Dinarides: Paléotectonique, tectonique, tarditectonique, néotectonique. *Bull. Soc. géol. Fr.*, **15,** 426–460.

—— & Guernet, C. 1963. Sur une tectonique transversale dans le Pinde méridional au parallèle de Karpenission (province d'Euritanie, Grèce). *C.r. Somm. Soc. géol. Fr.*, 77–78.

Bellon, H., Jarrige, J. J. & Sorel, D. 1979. Les activités magmatiques égéennes de l'Oligocène à nos jours et leurs cadres géodynamiques. Données nouvelles et synthèse. *Rev. Géol. dynam. et Géog. physique*, **21,** 41–55.

Benda, L. & Meulenkamp, J. E. 1972. Discussion on biostratigraphic correlation in the Eastern Mediterranean Neogene. *Z. Dtsch. geol. Ges.*, **123,** 559–564.

Bergougnan, H. 1975. Relations entre les édifices

Pontique et Taurique dans le Nord-Est de l'Anatolie. *Bull. Soc. géol. Fr.*, **17,** 1045–1057.

Biju-Duval B. & Letouzey J. 1975. Comments about the new 'Carte géologique et structurale des bassins tertiaires du domaine méditerranéen'. *Rapp. Comm. int., Mer Médit.*, **23,** 4a 119–120.

——, Letouzey, J. & Sancho, J. 1972. Nouvelles données de sismique réflection précisant la structure du bassin Méditerranéen oriental et du Nord de la Mer Egée. Oral Comm., 4th Aegean Symposium, Hannover 1972.

Brooks, M., & Ferentinos G. 1980. Structure and evolution of the Sporadhes basin of the north Aegean trough, Northern Aegean sea. *Tectonophysics*, **68,** 15–30.

—— & Williams, G. D. 1982. Extensional tectonics in Neogene and Quaternary sequences at the western margin of the Axios basin, northern Greece. *J. geol. Soc. Lond.,* **139,** 293–297.

Burke, W. F. & Ugurtas, G. 1974. Seismic interpretation of Thrace basin. *Proc. Second Petroleum Congress of Turkey*, p. 229–249.

Crampin, S. & Ucer, S. B. 1975. The seismicity of the Marmara sea region of Turkey. *Geophys. J.R. astr. Soc.*, **40,** 269–288.

Dercourt, J. 1972. The Canadian Cordillera, the Hellenides and the sea-floor spreading theory. *Can. J. Earth Sciences*, **9,** 709–743.

Dewey, J. F. & Şengör, A. M. C. 1979. Aegean and surrounding regions: complex multiplate and continuum tectonics in a convergent zone. *Bull. geol. Soc. Am.*, **90,** 84–92.

Doust, H. & Arikan, Y. 1974. The geology of the Thrace basin. *Proc. Second Petroleum Congress of Turkey*, p. 119–134.

Faugeres, L. 1978. *Recherches gèomorphologiques en Grèce septentrionale.* Thèse, Paris, 849 p.

—— & Robert, C. 1976. Etude sédimentologique et minéralogique de deux forages du golfe Thermaïque (Mer Egée). *Ann. Univ. Provence, Géol. Méditer.*, **3,** n°4, 209–218.

Ferentinos, G., Brooks, M. & Collins, M. 1981. Gravity induced deformation on the north flank and floor of the Sporadhes basin of the North Aegean sea trough. *Marine Geology*, **44,** 289–302.

Fosse, R. M. & Manheim, F. 1975. Geology of Bulgaria: a review. *Amer. Ass. Petrol. Geol. Bull.*, **59,** 303–335.

Fourquin, C. 1975. L'Anatolie du Nord-Ouest, marge méridionale du continent européen, histoire paléogéographique, tectonique et magmatique durant le Secondaire et le Tertiaire. *Bull. Soc. géol. Fr.*, **17,** 1059–1070.

——, Paicheler, J. C. & Sauvage, J. 1970. Premières données sur la stratigraphie du 'Massif Galate d'Andésites'; étude palynologique de la base des diatomites miocènes de Beş-Konak au Nord-Est de Kızılcahamam (Anatolie-Turquie). *C.r. Acad. Sc. Paris*, **270,** D, 2253–2255.

Fytikas, M., Giuliani, O., Innocenti, F. Manetti, P., Mazzuoli, R., Peccerillo, A. & Villari, L. 1979. Neogene volcanism of the Northern and Central Aegean region. *Ann. geol. Pays Hell.*, **30,** 106–129.

Gilet, S. 1961. Esssai de paléogéographie du Néogène et du Quaternaire Inférieur d'Europe Orientale. *Rev. Géog. phys. Géol dyn.*, **4,** 218–250.

Gramann, F. & Kockel, F. 1969. Das Neogen im Strimonbecken (Grieschisch-Ostmazedonien). Teil I—Lithologie, Stratigraphie und Paläogeographie. *Geol. Jb.*, **87,** 445–484.

Grekoff, N., Guernet, C. & Lorenz, J. 1967. Existence du Miocène marin, au centre de la mer Egée, dans l'île de Skyros (Grèce). *C.r. Acad. Sc. Paris*, **265,** 1276–1277.

Hancock, P. L. & Barka, A. A. 1981. Opposed shear senses inferred from neotectonic mesofracture systems in the North Antolian fault zone. *J. struct. Geol.*, **3,** 383–392.

Heimann, K. O., Lebkuchner, H. & Kretzler, N. 1972. Geological map of Greece, 1:50 000. Samothraki. IGME, Athenes.

Innocenti, F., Manetti, P., Peccerillo, A. & Poli, G. 1979. Inner arc volcanism in NW Aegean Arc: geochemical and geochronological data. *N.Jb. Miner. Monats.*, 145–158.

——, Manetti, P., Mazzuoli, R., Pasquare, G. & Villari, L. 1982. Anatolian and North-Western Iran. In: Thorpe R. S. (ed.) *Andesites*, John Wiley & Sons, New-York, 327–349.

Irrlitz, W. 1972. Lithostratigraphie and tektonishe Entwicklung des Neogens in Nordestanatolien. *Beih. Geol. Jb.*, **120,** 111 pp.

Jackson, J. A., Gagnepain, J., Houseman, G., King, G. C. P., Papadimitriou, P., Soufleris, C. & Virieux, J. 1982a. Seismicity, normal faulting and the geomorphological development of the Gulf of Corinth (Greece): the Corinth earthquakes of February and March 1981. *Earth Planet. Sci. Lett.*, **57,** 337–397.

——, King, G. & Vita-Finzi, C. 1982b. The neotectonics of the Aegean: an alternative view. *Earth Planet. Sci. Lett.* **61,** 303–318.

Jongsma, D. 1975. *A marine geophysical study of the Hellenic Arc.* University of Cambridge, Unpubl. thesis, 69 pp.

Ketin, I. 1948. Uber die tektonisch-mechanischen Folgerungen aus den grossen anatolischen Erdbeben des letzten Dezenniums. *Geol. Rundsch.*, **36,** p. 77–83.

—— 1968. Relations between tectonic features and the main earthquake regions of Turkey. *Bull. Mineral. Res. Expl. Inst. Turkey*, **71,** 63–67.

Kolios, N., Innocenti, F., Manetti, P., Peccerillo, A. & Giuliani, O. 1980. The Pliocene volcanism of the Voras Mts (Central Macedonia, Greece). *Bull. Volcanol.*, **43,** 553–568.

Lalechos, N. & Savoyat, E. 1979. La sédimentation Néogène dans le Fossé Nord Egéen. *6th Colloquium on the Geology of the Aegean region*, **2,** 591–603.

Le Pichon, X. & Angelier, J. 1979. The Hellenic arc and trench system: a key to the neotectonic evolution of the Eastern Mediterranean area. *Tectonophysics*, **60,** 1–42.

Lüttig, G. & Steffens, P. 1976. Paleogeographic Atlas of Turkey from the Oligocene to the Pleistocene. Paleogeographic Atlas of Turkey—Explanatory Notes, 64 pp.

LYBÉRIS, N. & DESCHAMPS, A. 1982. Sismo-tectonique du fossé Nord-Egéen: relations avec la faille Nord-Anatolienne. *C.r. Acad. Sc. Paris*, **295**, Sér. **2**, 625–628.

—— & SAUVAGE, J. 1983. Sur l'âge de la tectonique extensive dans le domaine nord-égéen (Grèce): données palynologiques. *C.r. Acad. Sc. Paris*, **296**, 107–110.

——, ANGELIER, J., HUCHON, P., LE PICHON, X. & RENARD, V. 1982. La mer de Crète: extension et subsidence. *Proceedings, HEAT Symp.*, **1**, 364–382.

——, ——, BARRIER, E. & LALLEMENT, S. 1982a. Active deformation of a segment of arc: the strait of Kythira, Hellenic arc, Greece. *J. struct. Geol.* **4**, 299–311.

——, CHOROWICZ, J. & PAPAMARINOPOULOS, S. 1982b. La paléofaille transformante de Kastaniotikos (Grèce): télédétection, données de terrain et géophysique. *Bull. Soc. géol. Fr.*, **24**, 73–85.

MCKENZIE, D. 1970. Plate tectonics of the Mediterranean Region. *Nature*, **226**, 239–248.

—— 1972. Active tectonics of the Mediterranean Region. *Geophys. J.R. astr. Soc.*, **30**, 109–185.

—— 1978. Active tectonics of the Alpine-Himalayan belt: the Aegean Sea and surrounding regions. *Geophys. J.R. astr. Soc.*, **55**, 217–254.

MERCIER, J. 1966. Paléogéographie, orogenèse, métamorphisme et magmatisme des zones internes des Hellénides en Macédoine (Grèce): vue d'ensemble. *Bull. Soc. géol. Fr.*, **8**, 1020–1049.

—— 1968. I. Etude géologique des zones internes des Hellénides en Macédoine Centrale (Grèce). II. Contribution à l'étude du métamorphisme et de l'évolution magmatique des zones internes des Hellénides. *Ann. géol. Pays Hellén.*, **20**, 786 p.

MERCIER, J. L. 1981. Extensional-compressional tectonics associated with the Aegean Arc: comparison with the Andean Cordillera of South Peru–North Bolivia. *Phil. Trans. R. Soc. Lond., Ser.* **A300**, 337–355.

——, MOUYARIS, N., SIMEAKIS, C., ROUNDOYANNIS, T. & ANGELIDHIS, C. 1979. Intra-plate deformation: a quantitative study of the faults activated by the 1978 Thessaloniki earthquakes. *Nature*, **278**, 45–48.

——, DELIBASSIS, N., GAUTHIER, A., JARRIGE, J. J., LEMEILLE, F., PHILIP, H., SEBRIER, M. & SOREL, D. 1979a. La néotectonique de l'arc égéen. *Revue de Géol. dyn. Géogr. phys.*, **21**, 67–92.

NEEDHAM, D., LE PICHON, X., MELGUEN, M., PAUTOT, G., RENARD, V., AVEDIK, F. & CARRE, D. 1973. North Aegean trough: 1972 Jean Charcot cruise. *Bull. Soc. Geol. Grèce*, **10**, 152–153.

PAPAZACHOS, B. C. & COMNINAKIS, P. E. 1978. Geotectonic significance of the deep seismic zones in the Aegean Area. *Proc. Symp. Thera and the Aegean World*, **1**, London, 121–129.

PURCARU, G. & BERCKHEMER, H. 1982. Regularity patterns and zones of seismic potential for future large earthquakes in the Mediterranean region. *Tectonophysics*, **85**, 1–30.

ŞENGÖR, A. M. C. 1979. The North Anatolian transform fault: its age, offset and tectonic significance. *J. geol. Soc. Lond.*, **136**, 269–282.

SOUFLERIS, C. & STEWART, G. S. 1981. A source study of the Thessaloniki (northern Greece) 1978 earthquake sequence. *Geophys. J.R. astr. Soc.*, **67**, 343–358.

TOKSÖZ, M. N., SHAKAL, A. F. & MICHAEL, A. J. 1979. Space-time migration of earthquakes along the North Anatolian fault zone and seismic gaps. *Pageoph.*, **117**, 1258–1270.

NICOLAS LYBERIS, Laboratoire de Géodynamique, Université P. et M. Curie, 4 Pl. Jussieu 75230 Paris Cedex 05, France.

Subsidence history of the North Aegean Trough

Xavier Le Pichon, Nicolas Lybéris & Francis Alvarez

SUMMARY: The subsidence of the North-Aegean trough is examined quantitatively using available geophysical data, the emphasis being on thermal and gravity computations. The two dimensional thermal calculations incorporate conduction as well as convection with lateral variations, the effect of the sedimentary cover being treated as a perturbation. The data can be explained within the framework of the homogeneous stretching model with a maximum stretching factor of about 3.5. The stretched lithosphere has a very small flexural parameter so that local compensation prevails. The effect of conduction is sufficiently important to prevent large-scale melting of the mantle, thus making the transition to oceanic accretion rather improbable. The narrowness of the zone of extreme stretching, compared to the widths now observed on most continental margins where comparable stretching values have been observed, may be due to the presence of an earlier deep lithospheric fault which enabled the strain to concentrate along its path.

The overall tectonic evolution of the North Aegean is discussed by Lybéris in this volume. A complex of sedimentary basins (Gulf of Thermaïkos, Sporades & Kavala, (Fig. 2) has been subsiding since the Upper Miocene within an extensional tectonic regime. In one of these basins, the Sporades basin, sedimentation has not kept pace with subsidence and there is now a trough more than 1000 m deep (Fig. 1). In this paper, we examine the quantitative evolution of subsidence and show that it can be modelled reasonably well using the homogeneous stretching model of McKenzie (1978). We use available seismic reflection, refraction, gravity, magnetics, heat flow and well data. Most of the data are published (Jongsma 1974; Morelli *et al.* 1975; Makris 1977; Faugères & Robert 1976; Lalechos & Savoyat 1979). Some were previously reported (Needham *et al.* 1973; Biju-Duval *et al.* 1972) but not published. A detailed bathymetric map was made by Needham *et al.* (Fig. 1). A basement map modified after Lalechos & Savoyat is shown in Fig. 5.

The exact date of beginning of the extension which initiated this phase of subsidence in the area is unfortunately difficult to pinpoint with the data available. Following the tectonic analysis, by Lybéris (this volume) we assume that it was close to 10 Ma ago, in Upper Miocene time. Unfortunately, there are conflicting interpretations of the stratigraphy of the published drilling so that the relative importance of the subsidence in Upper Miocene and Plio–Pleistocene is not well known. The problem is different in the west from the east.

To the west, in the Gulf of Thermaïkos, the 3 km thick Neogene is known in two boreholes (Faugères & Robert 1976). At the base of the Neogene section, a continental conglomerate of probable Lower Miocene age is assumed to mark the end of the tectonic compressive phase known in the Northern Aegean. It is overlain by a thick marine clastic succession extending from Middle–Upper Miocene to the present. Two different interpretations are proposed by Faugères & Robert (1976) and Lalechos & Savoyat (1979). For the first authors, the Pliocene–Miocene boundary is at 2500 m and the total Neogene section is 3500 m thick, whereas these values are 900 and 2600 m respectively for the other authors. The first interpretation is made by comparing the successions with the adjacent land sections (Faugères & Gillet 1970, 1974; Faugères & Desprairies 1971). Brooks & Ferentinos (1980) note that an angular unconformity exists in the seismic reflection records of Lalechos & Savoyat within the Neogene section. This unconformity is at the depth of the Miocene–Pliocene transition of Lalechos & Savoyat, but at the Pliocene–Pleistocene transition of Faugères & Gillet. In Fig. 6, we show both interpretations although we favour the interpretation of Lalechos & Savoyat.

To the east, in the Kavala basin, the major difference is the presence of evaporites which belong to the Upper Miocene and probably extend into the Messinian. Consequently, the thickness of the Plio–Pleistocene is better determined. Below the evaporites, there are continental sediments of poorly known age.

Thus, a shallow marine basin appeared in the west, in the Thermaïkos basin, sometimes during or after the Middle Miocene whereas to the east, in the Kavala basin, the marine transgression started in the Upper Miocene with the evaporitic episode, although subsidence certainly started earlier from above sea-level. However, the relative amounts and rates of subsidence, in Miocene and Plio–Quaternary times, are poorly known. In this paper, we consequently reason in terms of total subsi-

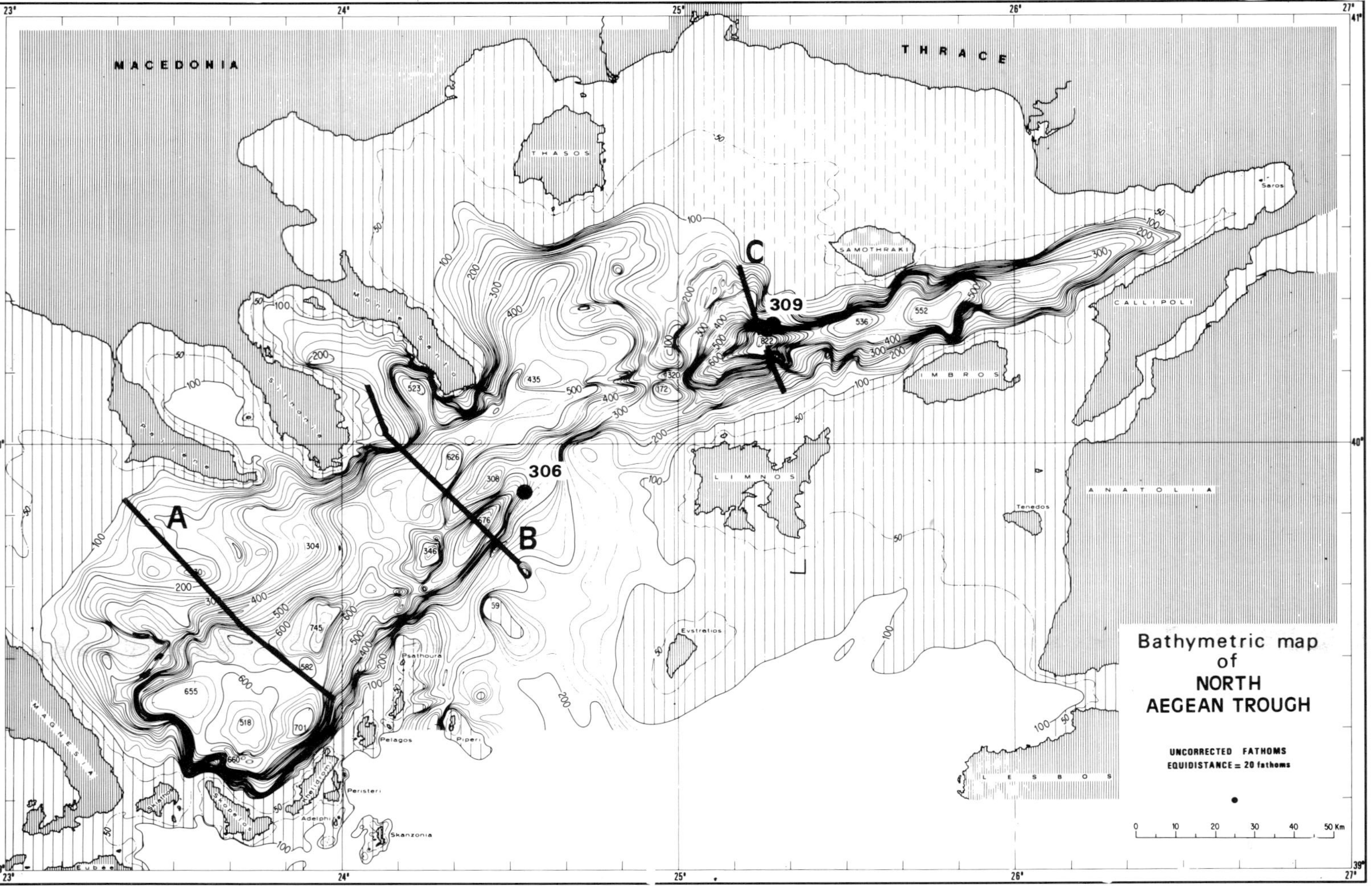

Fig. 1. Bathymetry in uncorrected fathoms (one fathom equals 1/800 second of two way travel-time), after Needham *et al.* (1973). A, B and C are the profiles of Fig. 2. 309 and 306 are the anomalous heatflow stations of Jongsma (1974).

dence. We assume further that it started roughly at 10 Ma at sea-level and that it has proceeded at a constant rate since then.

Structure from geophysical data

Free air gravity and magnetic anomaly maps for the Aegean have been published by Morelli *et al.* (1975). They indicate that the North Aegean trough area is not characterized by large free air gravity anomalies, either negative or positive and that consequently, it is in a state close to isostatic equilibrium. On the other hand, there are fairly large magnetic anomalies associated with the Sporades trough.

Three gravity and magnetic profiles were obtained during the R. V. Jean Charcot cruise in 1972 (see Fig. 1 for location) and have been previously reported by Needham *et al.* (1973). They are shown in Figs 2A to 2C. Profiles A and B cross the Sporades basin. There is a slight minimum free air anomaly over the trough and somewhat larger maxima over the shoulders. They are of the right magnitude for a topographic effect and, as will be seen later on for profile A, they are compatible with isostatic equilibrium. Thus, subsidence has occurred under a state of isostatic equilibrium.

This is not true however of profile C across the Saros trough between Imbros and Samothraki islands (Fig. 1). Here, there is a −50 mgal gravity anomaly minimum with +20 to +50 mgal level on the adjacent shoulders. Thus, the narrow 20 km wide Saros trough is not in isostatic equilibrium. Most of the relatively small subsidence of the basement (about 2 kilometres including 1 km of sediments) corresponds to a net mass deficiency. The narrow V-shaped Saros trough is still in an early stage of formation. Topography and gravity there could be interpreted as due to a 5° tilt to the north of one single 20 km wide block. We do not discuss further in this paper the formation of this trough.

We have done simple 2-D modelling of the magnetic anomalies. The magnetic anomalies in Fig. 2B can be interpreted as due to a shallow body magnetized in the direction of the present earth's field with a high susceptibility of 1.2 A m^{-1}. Part of the body crops out on the floor of the trough forming the elongate SW–NE ridge mapped in Fig. 1. The anomaly in Fig. 2A is similar although it comes from a deeper body which could be situated at the bottom of the sedimentary pile at about 5–6 km depth on the northern edge of the trough. In both cases, the simplest interpretation is that they are volcanic bodies intruded into the basement of the trough, probably during the recent extensional phase. This is corroborated by the presence of 0.5 Ma basaltic volcanism on the island of Psathoura, to the south of the Sporades basin (Fig. 1) and of 3.4 to 0.5 Ma basaltic volcanism in the region of Volos, 30 km west of the trough (Innocenti *et al.* 1979). However, the magnetic anomalies do not seem to be compatible with

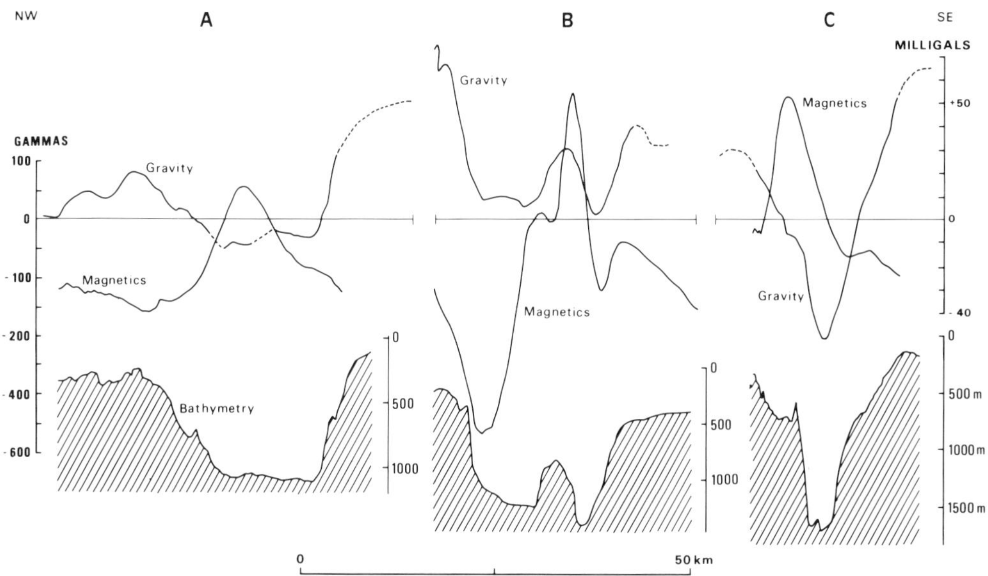

FIG. 2. Bathymetry, gravity and magnetic profiles across the North-Aegean trough after Needham *et al.* (1973), (see Fig. 1 for locations).

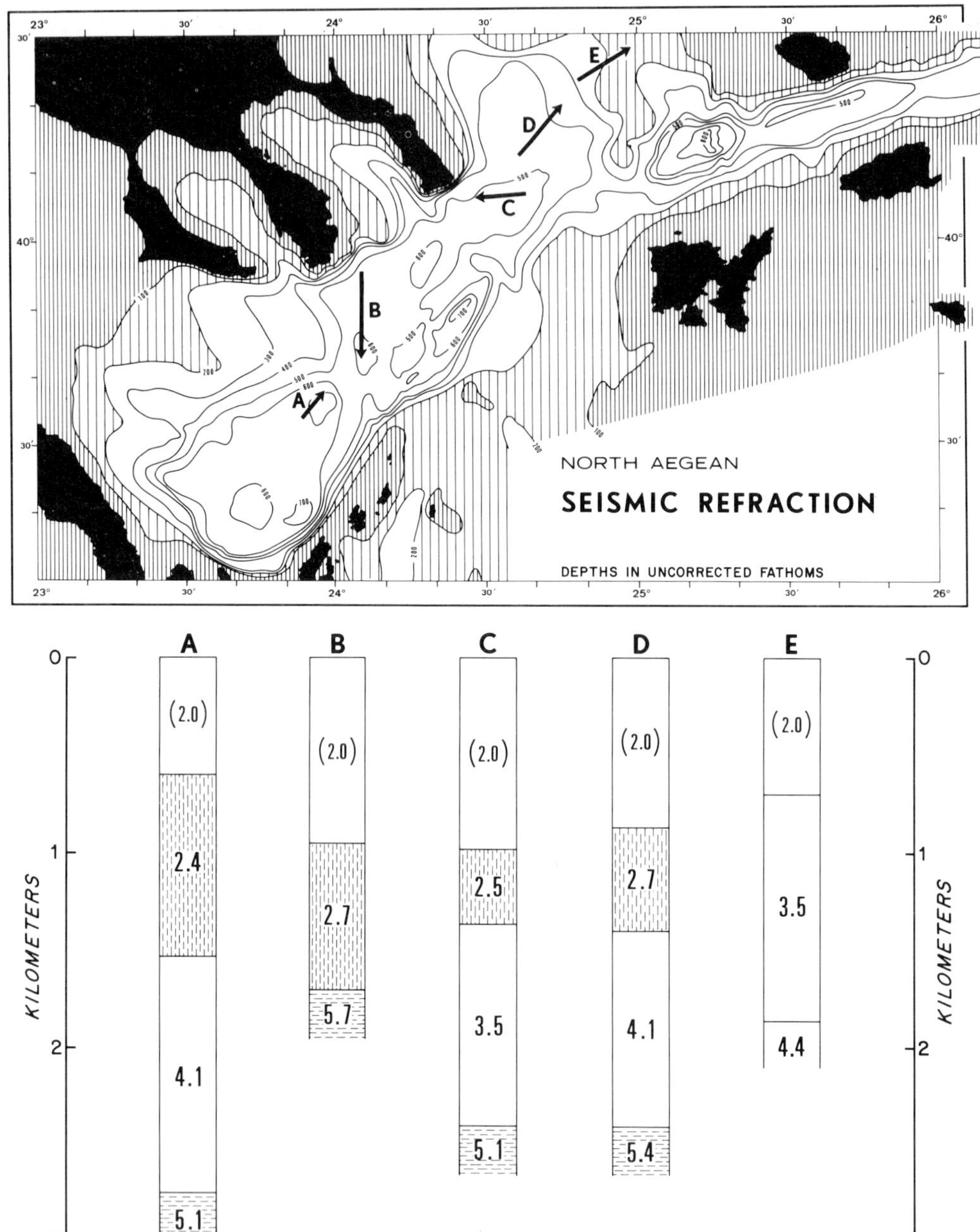

FIG. 3. Shallow seismic refraction (sonobuoy) results within the North Aegean trough after Needham *et al.* (1973). Velocities are given in km^{-1}.

the crust under the trough being entirely volcanic. In other words, they are not compatible with 100% extension and the presence of a complete gap within the continental crust under the trough.

During the R. V. Jean Charcot 1972 cruise, five small refraction profiles were made in the trough using sonobuoys (Fig. 3). These seem to record the presence of a 5+ $Km\,s^{-1}$ basement under the pile of sediment. However, these are unreversed short profiles in a region of steep sea-floor and surface basement gradients (see Fig. 5). The upper sedimentary column velocity is assumed and not measured. Finally, the sedimentary velocities measured seem high when compared to the velocity law obtained

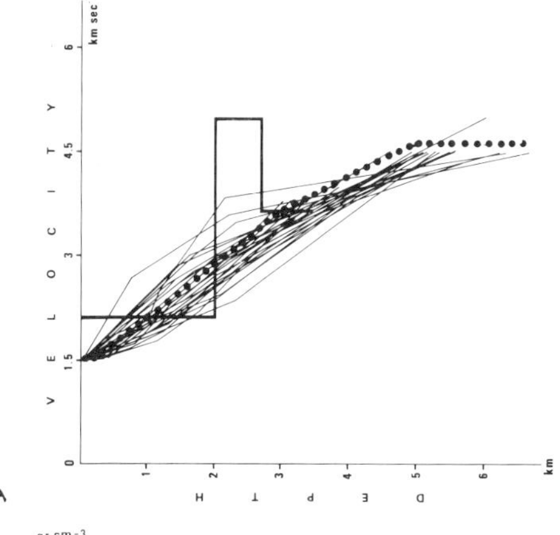

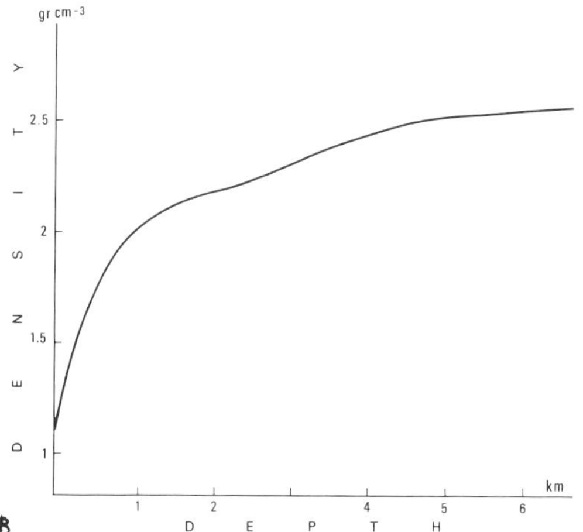

FIG. 4. Velocity curves and corresponding adopted average density curve within the North Aegean trough. A: Thin lines—velocities after Biju-Duval *et al.* (1972). Thick line—average velocity in well P1 after Lalechos & Savoyat (1979). Dotted line—adopted average velocity curve. B: The adopted average density curve is computed using the Nafe & Drake (1957) relationship.

from seismic reflection (Fig. 4A). The 3.5 to 4.1 km/sec velocities in profiles C to E could be interpreted as produced by refraction at the top of the evaporites, although the evaporites are not supposed to be present in profile A where the same velocity was measured. Thus, the velocity structure is poorly determined although the average depths to basement obtained are in fair agreement with the basement map of Fig. 5 which is based on seismic reflection results.

Fig. 4A shows different velocity profiles obtained from the multichannel seismic reflection lines reported by Biju-Duval *et al.* (1972) in the area of the North Aegean trough. The thick continuous line corresponds to average velocities in borehole P_1 above and within the evaporites (Lalechos & Savoyat 1979). The thick dotted line shows the velocity profile we have adopted in the absence of evaporites. Then, using the velocity/density relationship proposed by Nafe & Drake (1957) with slight modifications as given in Makris (1977), we obtain in Fig. 4B the density profile we have adopted in the absence of evaporites. Finally, Fig. 5 is a map of the depth to the pre-Neogene surface based on the Neogene isopach map of Lalechos & Savoyat (1979) and on the bathymetry of Fig. 1.

Evaluation of subsidence

Knowing the total Neogene sediment thickness as well as an average density vs depth law, it would be possible to obtain the subsidence history if we had good stratigraphic control and if we could assume that density only depends on depth (except for the evaporites which are not significantly compacted under pressure). The second assumption is probably a good approximation but we are far from having good stratigraphy as discussed in the introduction. Fig. 6 shows the evolution of tectonic subsidence with time for one well in the Thermaïkos basin (A_1) and four wells in the Kavala basin (P_1, AM_1, AP_1 and E.T.$_1$) based on Faugères & Robert (1976) and Lalechos & Savoyat (1979). The results depicted by this figure substantiate our decision to reason in terms of total Neogene subsidence assuming, for lack of a better choice, that it started at 10 Ma from sea-level and proceeded at a constant pace.

We now investigate whether the elasticity of the plate has played a significant role in the present structure of the basin. The relatively small widths of the basins (ca. 100 km) suggest that, in any case, the elasticity of the plate is small and that the flexural parameter does not exceed 100 km/2π which is already quite small. We choose α equal to 17 km which corresponds to a thickness of elastic crust of about 7 km. We then evaluate the total water and sedimentary load at the corners of a 20 km square grid. Finally, we compute the resulting deflection for a thin elastic plate overlying a fluid asthenosphere of density $\rho_a = 3.203\,g\,cm^{-3}$ based on a 0°C density of $3.35\,g\,cm^{-3}$ and a temperature of 1333°C using the point load approximation for each grid point and summing up the different point load contributions (Brunet & Le Pichon 1982, for details of the computation). Fig. 7 is the resulting computed deflection. The broad ellipsoidal contours can hardly be matched to

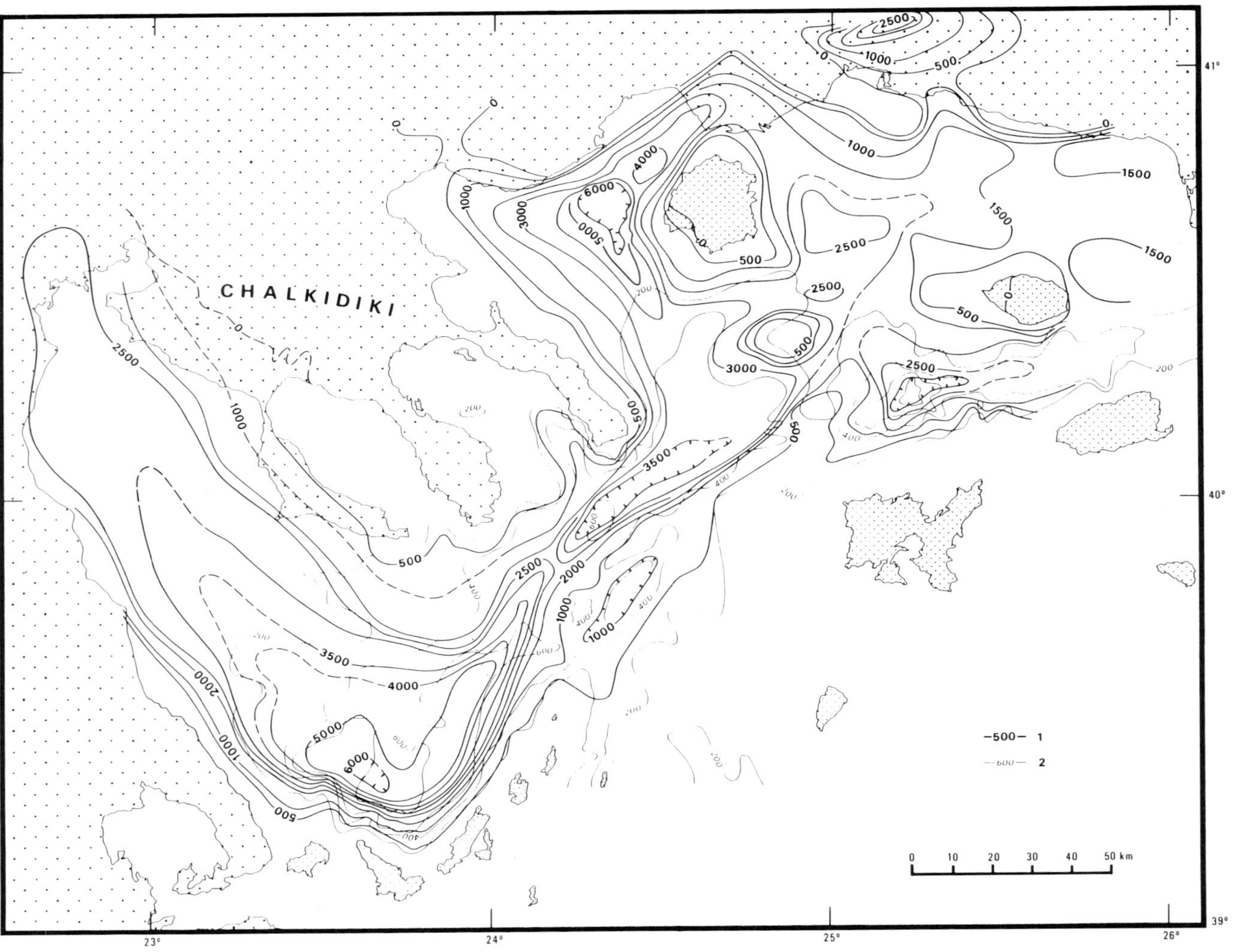

FIG. 5. Basement isobaths in metres mostly based on the Neogene isopach map of Lalechos & Savoyat (1979) and Fig. 1 with additional data from some seismic reflection profiles.

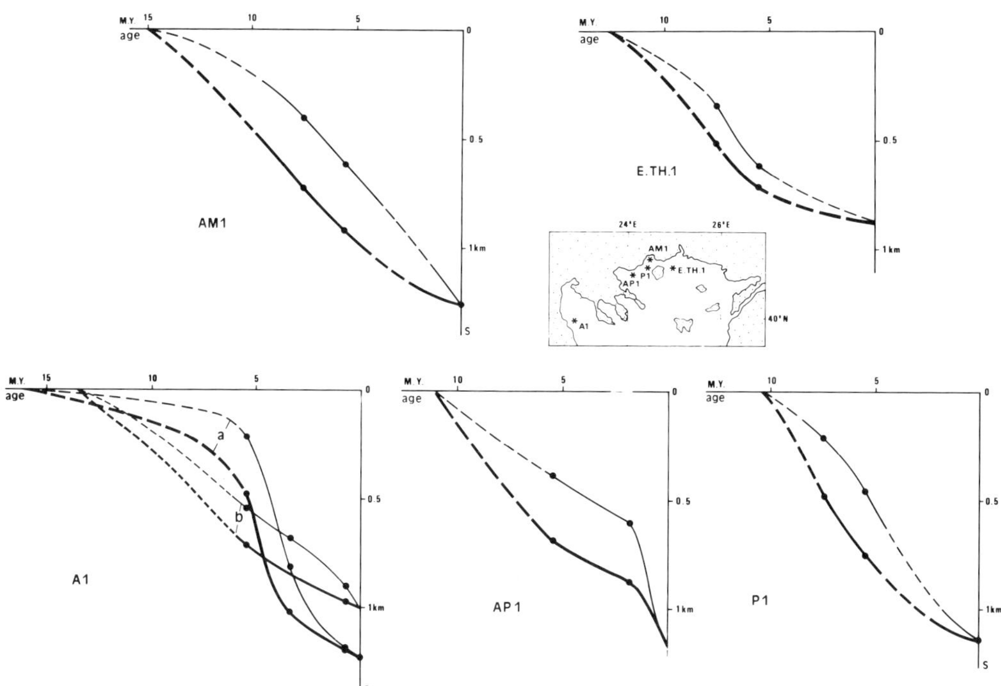

FIG. 6. Uncorrected basement tectonic subsidence in the absence of any water or sedimentary load and the same curves corrected for compaction using the average density curve of Fig. 4B (for 5 drill-holes available). It is assumed that there is no significant pre-Neogene sediment compaction during the Neogene. Dots correspond to levels which are correctly dated. A1: (a) interpretation of Faugères & Robert (1976); (b) interpretation of Lalechos & Savoyat (1979).

the present distribution of sediments as depicted in Fig. 5. We conclude that the equivalent thickness of the plate is even smaller than 7 km and that, consequently, everything happens as if the load of sediment and water were

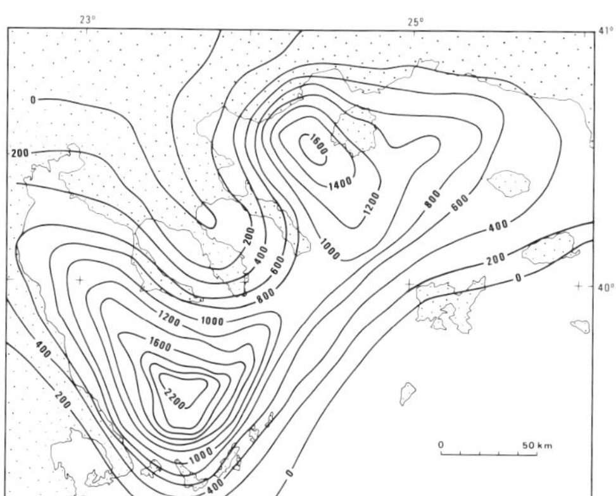

FIG. 7. Computed deflection in metres of an elastic plate with a 17 km flexural parameter under the Neogene sediment cover and water load. Note the incompatibility of the broad deflection with the actual basement deflection shown in Fig. 5.

locally compensated. Of course, it is not necessary for the local elastic thickness of the plate to be very small. Rather, it is more probable that the large faults which mark the boundaries of the basins efficiently decouple the crust under the floor of the basins from the adjacent crust. This seems reasonable at least for the time during which these faults are still active. Once the faults become inactive and frozen in, the continuity of the elastic plate is probably reestablished and any further loading is supported by the whole elastic lithosphere.

Figure 8 gives the distribution of tectonic subsidence (in the absence of water or sediment loading) with the same ρ_a density as above and assuming local isostatic equilibrium. This is a fair approximation to the depth the pre-Neogene 'basement' would have if both Neogene sediments and water were taken off. Over most of the basin, tectonic subsidence exceeds 1 km. It reaches a maximum value of 2.45 km in the south-western Sporades trough.

Using the homogeneous stretching model, Le Pichon *et al.* (1982) have shown that the instantaneous tectonic subsidence depends only on two parameters: first, the difference between the initial topography and the hydrostatic level

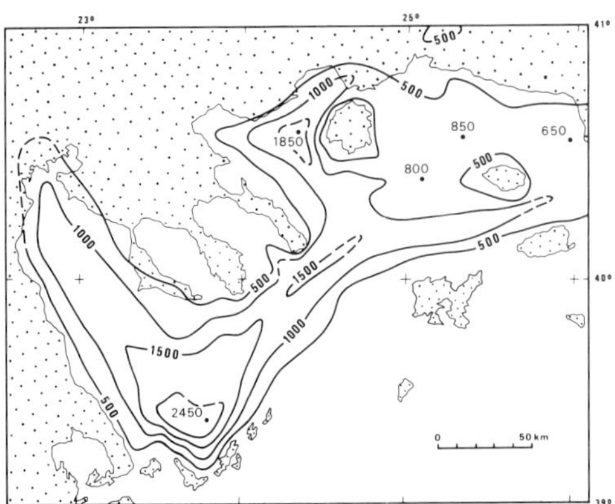

FIG. 8. Neogene tectonic subsidence in meters in the absence of water or sedimentary load.

of the asthenosphere, and second, the thinning factor γ, where $\gamma = 1 - 1/\beta$ and β is the amount of thinning. The hydrostatic level of the asthenosphere is the level to which the asthenosphere would rise if it were locally unloaded from the overlying lithosphere and if no partial fusion occurred during its rise. This level is obtained by considerations of the topography and structure of the oceanic lithosphere near mid-ocean ridges. A first estimate by Le Pichon *et al.* (1982) was 3.6 km below sea-level for the water-loaded asthenosphere (2.45 km in the absence of water). Then, if subsidence started at sea-level

$$Si = 2.45\,\gamma\,\text{km} \qquad (1)$$

where Si is the instantaneous tectonic subsidence without water-loading. The 10 Ma duration of subsidence is smaller than the 20 Ma limit considered by Jarvis & McKenzie (1980) as the time beyond which the instantaneous approximation is no longer valid. Consequently, we can use the instantaneous approximation and obtain through (1) an infinite thinning (neither crust nor lithosphere below the sediments) for a 2.45 km subsidence.

Le Pichon & Alvarez (in press) show that the constant may be closer to 4.0 or 4.2 km under water (2.7 to 2.85 km without water) giving a β of 10 and 7 respectively for the maximum subsidence. However, this 2.45 km tectonic subsidence is found at a single point and is not significant. It is better to take the average over at least 10 to 20 km. Then the maximum value of tectonic subsidence is close to 2 km which is equivalent to a β of about 5 in the south-western Sporades trough. However, these estimates ignore the finite duration of the extension (10 Ma) and the sharp lateral variations in stretching.

Thermal computations

Consequently, we now test whether such an homogenous stretching model would indeed give the required subsidence. To do so requires computing the thermal evolution of a lithosphere submitted to laterally variable amounts of stretching and simultaneously to sedimentation. This is because the width of the trough is comparable to the thickness of the lithosphere. Thus, lateral conduction of heat is significant. In addition, the high sedimentation rate and sediment cover have a shielding effect on the thermal evolution of the lithosphere (Beaumont *et al.* 1982). Alvarez *et al.* (1984) have developed a method of computation of the two dimensional evolution of a lithosphere submitted to variable rates of stretching and variable sedimentation rates. This method and the detailed computation will be published elsewhere.

In short, the finite difference numerical method used is an extension of the 'splitting-up' method (A.D.I.) of Peaceman & Racheford (1955) and Douglas (1955), taking into account a velocity field produced by a strain rate $\dot{\varepsilon}$ which does not depend either on time or on depth for a material column. This is consequently the assumption used by McKenzie (1978) in which, for a material column,

$$\beta = \exp \dot{\varepsilon}\, t$$

For the North Aegean, we use t = 10 Ma and adjust β along the profile to obtain the proper subsidence. If there were no significant conduction effect, the subsidence would be proportional to γ. Actually, subsidence is larger than predicted under the trough whereas a small uplift is produced on the edges of the trough because of the effect of lateral conduction.

Beaumont *et al.* (1982) and De Bremaecker (1983) have shown that the sedimentary layer should be taken into account in these computations. Because of its insulating effect, the lithosphere loses its heat less rapidly. Consider for example a 5 km thick sedimentary section being deposited at a constant rate over 10 Ma. Assume no compaction. Assume also an average diffusivity equal to the diffusivity of the lithosphere as justified later (Table 1). Then one can use the analytical solution of Carslaw & Jaeger (1959) for one-dimensional continuous sedimentation. Its results indicate at 10 Ma a surface heat flux which is 67% of the equilibrium pre-sedimentation flux. This is because the sedimentary column is heated and consequently the top temperature of the lithosphere continuously increases through time. Thus, part of the lithospheric heat-flow is used to heat the

TABLE 1. *Parameters for thermal computation (SI)*

Lithosphere		
Thermal conductivity	k_L	$= 3.1935$
Coefficient of thermal expansion	α	$= 3.28 \times 10^{-5}$
Thermal capacity	c_L	$= 1.172 \times 10^3$
Thermal diffusivity	κ_L	$= 8.04 \times 10^{-7}$
Thickness of crust	h_c	$= 3 \times 10^4$
Thickness of lithosphere	h_L	$= 1.25 \times 10^5$
Density of crust at 0°C	ρ_c	$= 2.78 \times 10^3$
Density of mantle at 0°C	ρ_m	$= 3.35 \times 10^3$
Temperature of asthenosphere	1333°C	
Radiogenic heat production law	$3 \times 10^{-6} \exp(-z/10^4)$	
Grid mesh	5×10^3	
Time increment	5×10^5 years	
Sediments		
Thermal conductivity $k_s = 2.10 - 0.97 \exp(-z/1210)$		
Volumetric heat capacity $\rho_s c_s = 2.09 \times 10^6 + 2.09 \times 10^6 \exp(-z/1210)$		
Average thermal diffusivity over 5 km $\bar{\kappa}_s = 8.04\ 10^{-7}$		
Density $\rho_s = 2.52 \times 10^3 - 1.39 \times 10^3 \exp(-z/1210)$		
Sea water		
Density 1.03×10^3		
Temperature 0°C		

sedimentary column and part to heat the top lithospheric column which results in a significant decrease of the surface heat flow. On the other hand, the corresponding thermal expansion of the lithospheric column is small, because the thermal anomaly due to the sedimentary cover has not penetrated sufficiently deep. After 10 Ma, it still does not exceed 40 metres for an equilibrium heat flow of 55 mWm^{-2}.

The above discussion suggests that the thermal effect of the sedimentary cover can be treated as a perturbation by adjusting the top temperature T_t of the lithosphere to the thickness of the sedimentary cover at each step of the computation. Alvarez *et al.* (1984) treat the sedimentary cover of thickness h_s as a thin insulating layer. Assume for example that it has a unit thickness, then its equivalent conductivity is k_s/h_s and its equivalent heat capacity $h_s \rho_s c_s$. At each step, the equivalent conductivity decreases by $\Delta h/h_s$ and T_t correspondingly increases. One needs, however, to make a correction for the amount of heat absorbed by the sedimentary cover and the top lithosphere. Alvarez *et al.* (1984) show that this approximation is very good at small to medium sedimentation rates when compared to the one dimensional analytical solution. The error is still less than 5% at rates of 1 km Ma^{-1}. Note that this solution assumes the absence of topography on the top surface of the lithosphere. It assumes further that heat always flows vertically through the sedimentary layer.

The constants adopted for the computations are given in Table 1. Concerning the lithosphere, they are as in Le Pichon & Sibuet (1981) following McKenzie (1978) and Parsons & Sclater (1977). The value chosen for radiogenic heat production corresponds to an equilibrium heat flow of 55 mWm^{-2} which is reasonable for the unstretched crust in the area (Fytikas & Kolios 1979). Concerning the sediments, the density law as obtained from seismic velocity measurements and is given in Fig. 4B. We approximate it by an exponential variation (Table 1).

We assume that the conductivity and volumetric capacity obey the same exponential variation, as these three parameters are dominated by the change in porosity with depth. At the surface, for $z = 0$, we adopt values in agreement with those measured by Jongsma (1974) in the North Aegean trough. At large depths, we adopt values reasonable for upper continental crust. The corresponding average thermal diffusivity over the 5 km thick sediment cover is very close to the asthenosphere one, which justifies the choice we made above when computing the analytical solution. Note that one does not need the value of thermal diffusivity within the sedimentary layer for the perturbation method but only the average conductivity k_s and the average volumetric capacity ρ_s c_s through the thickness h_s. These two parameters change continuously through time with the thickness h_s.

Discussion of the thermal profile

Figure 9 shows the results of the computation along a profile crossing the south-western portion of the North Aegean trough and passing through the point of maximum subsidence. The sea floor topography is obtained from Fig. 1 and the basement topography from Fig. 5. The vertical exaggeration is 20. Two dimensionality is assumed. Note the asymmetry of the trough, the southern flank being much steeper than the northern one. The value of γ was adjusted along the profile in order to obtain the correct sea floor topography. A perfect adjustment would of course have been possible but was not sought. We also did not try to model the topography on each side of the trough where thermal uplift and corresponding erosion has occurred. It is, however, probable that the basement uplift corresponding to the island bordering the trough to the southeast is a direct reflection of the amount of thermal uplift. Fig. 8

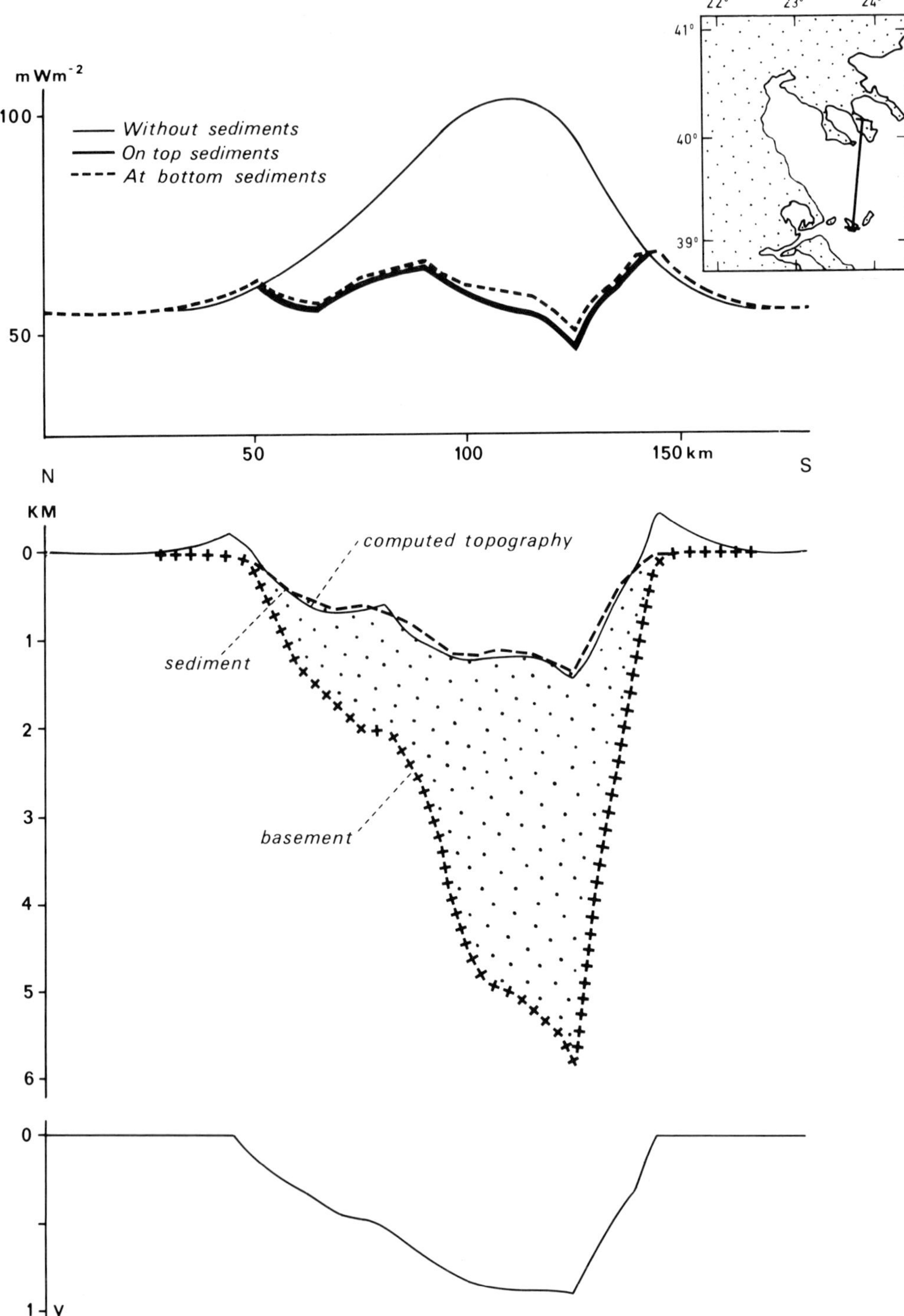

FIG. 9. Thermal computations across the North Aegean trough assuming homogeneous stretching since 10 Ma. The adopted γ is shown below, the computed heat-flow is shown at the top. Vertical exaggeration of 20 for basement and sea-floor topography.

shows that these islands occur along a zone of maximum basement slope next to the deepest section of the basin. There, γ changes from 0 to 0.7 in 10 km. This exceptional horizontal stretching gradient has produced a thermal uplift exceeding 500 m in Fig. 9.

The asymmetry in subsidence is mirrored by an asymmetry in γ. But the largest γ (0.714 which corresponds to a β of 3.5) is significantly smaller than the γ computed in the absence of conduction effect (0.95). This difference is mostly due to the lateral conduction effect. For the same γ of 0.714, the subsidence under water in the absence of conduction is 2.57 km; with only vertical conduction (unidimensional) it is 2.78 km (8% larger), and with both vertical and lateral conduction, 3.43 km (33% larger). Thus vertical conduction accounts for 25% of the difference and lateral conduction for 75%. Lateral conduction becomes very significant for such a large horizontal gradient.

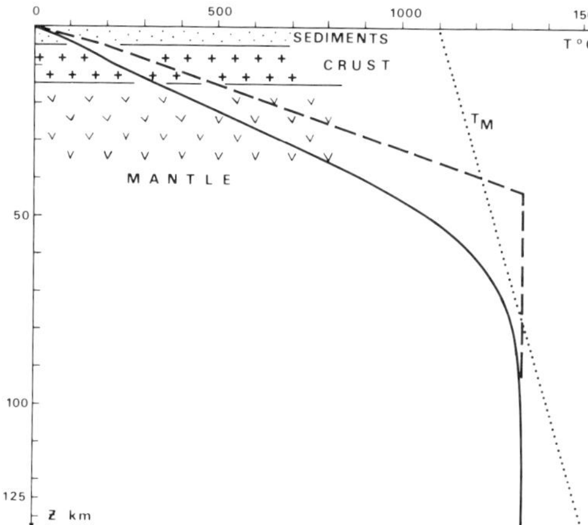

FIG. 10. Computed temperature profile at the axis of the section of Fig. 9. The stretching factor there is 3.5. The temperature profile in the absence of conduction is dashed. A simplified melting curve is dotted.

Figure 10 shows the temperature profile at the point of maximum stretching. At this point, the homogeneous stretching model predicts that the thickness of continental crust is reduced by a little more than two thirds and thus that the Moho should be 15 km deep instead of 30 km, the igneous crust being about 9 km thick. Using the simplified melting curve relationship proposed by Foucher *et al.* (1982), one would have expected large scale melting below 40 km, at the base of the thinned lithosphere. However, Fig. 9 shows that, mostly because of lateral conduction, partial fusion may only begin in a thin zone below 70 km. Of course, this result is only qualitative, considering the uncertainties in the melting relationship and in the asthenosphere temperature as well as the neglect of the adiabatic temperature gradient (see Foucher *et al.* for a more complete discussion). We have seen above that there are indications for sizeable intrusions of volcanics within the upper crust under the deeper portions of the crust. We have seen further that geologic evidence suggests a Pleistocene age for the volcanism. This volcanism may be localized along deep lithospheric fractures and probably does not represent the large scale melting that one should expect at this stretching stage, melting leading directly to the passage from continental stretching to oceanic accretion. Actually, our computations suggest that such a passage may be impossible here because the high horizontal stretching gradient produces too much lateral cooling. For oceanic accretion to occur, it is necessary not only to reach a stretching value β of the order of 3.2–3.5 as pointed out by Le Pichon & Sibuet (1981) but also that the width of the zone of stretching be sufficiently large with respect to the duration of stretching. The longer the duration of stretching, the wider the zone. Apparently, this width is not sufficient in the North Aegean trough.

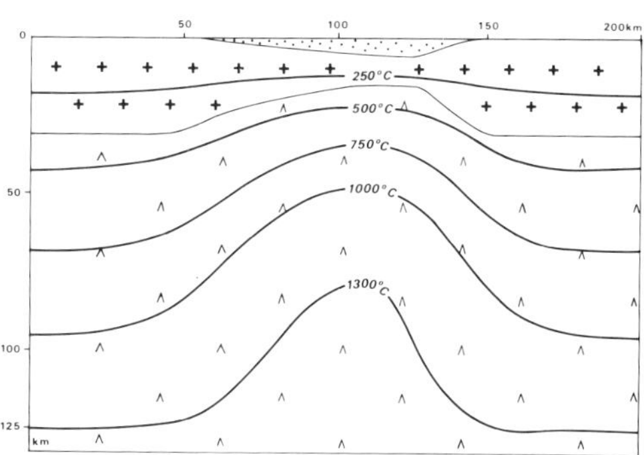

FIG. 11. Temperature section corresponding to Fig. 9. The Moho and the base of lithosphere (identified with the 1300°C isotherm) are shown as well as the sedimentary trough. No vertical exaggeration.

Figure 11 shows the computed temperature section. The base of the asthenosphere is approximated by the 1300°C isotherm whereas the base of the crust only depends on the stretching factor β. The difference in shape between these two boundaries clearly show the large effect of lateral conduction in spite of the relatively small time interval (10 Ma). Without conduction, the lithosphere would have been thinned to 36 km; instead it has only been thinned to 70 km. But, on the other hand, lateral heating is significant to 25 km on each side. As noted by England (1983), because of conduction, a given portion of the lithosphere in general tends to decrease in temperature as it is raised through time. For example, at the Moho, the temperature decreases from about 400°C to less than 300°C. Thus, if a ductile law is adopted, the strength of the uppermost mantle is considerably increased as stretching proceeds. Using the $n = 3$ exponential flow law chosen by England, the force necessary to extend the mantle portion of the lithosphere would have increased by a factor of 50 since the initiation of stretching.

However, this is an over simplification because the mantle below 13 km depth and at 300°C probably lies within the brittle field and not within the ductile one. Assume for the purpose of the discussion that the brittle-ductile transition occurs near 250°C in the crust and near 500°C in the mantle. Assume further that the deviatoric stress required for stretching is the one given by the Coulomb friction law in the brittle zone and by a power law in the ductile zone (Brace & Kohlstedt, 1980). Then, the thickness of the brittle portion of mantle stays approximately constant through time (as can be seen in Fig. 11 by comparing the temperature column at $x = 0$ and $x = 120$ km) and the corresponding average lithostatic pressure is decreased by a factor of about two. Consequently the force required to stretch this portion of lithosphere is also divided by two. The thickness of the mantle ductile zone within the lithosphere is decreased by about one half between 500 and 1000°C while the relative distribution of temperature within it does not change. Thus the force necessary to stretch this portion of lithosphere is also decreased by about one half. As for the continental crust, the thickness of the brittle zone is divided by a factor of about 2 whereas the ductile zone essentially disappears. So the force necessary to stretch the continental crust also decreases through time by a factor of more than 2.

To conclude this discussion, the force necessary to stretch the lithosphere does not increase through time when one takes the brittle field into account. It probably decreases by about one half, the role of the upper mantle zone as a layer of strength probably increasing through time. But the most remarkable fact is the nearly complete disappearance of the ductile zone of the crust as stretching increases.

Going back to Fig. 9, we see that the computed heat flow in the absence of sedimentary cover would reach a maximum of about 100 mWm^{-2}, about twice the pre-stretching value but the presence of a rapid sedimentation reduces the heat-flow to 50 to 70 mWm^{-2}. Jongsma (1974) has made six heat-flow measurements in the North Aegean trough. Four of them fall in the range 50 to 70 but one (station 306) reaches 105 mWm^{-2} and one (station 309) 88 mWm^{-2} (see Fig. 1). However, station 306 is on the flank of the steep magnetic basement ridge shown in Fig. 2B and also obvious on Fig. 5. We have pointed out that this may be a recent volcanic ridge. Thus, the topography effect, the reduced thickness of sediments and the possible recent volcanic activity might all be invoked to explain this high value. Station 309 is within the Saros trough which we have not modelled, on the northern side of profile C close to the very steep active northern fault. Thus we conclude that the average heat flow over the North Aegean trough may not exceed 60 mWm^{-2} over the deeply sedimented basin although higher values might be expected near basement ridges or more probably large active faults. Fytikas & Kolios (1979) report the presence of hot springs on Limnos, Samothraki and on the mainland, these hot springs being most probably located along active faults.

Gravity computations

We have shown above that the homogeneous stretching model can account reasonably well for the observed subsidence and heat flow provided that a much smaller continental crust thickness can be assumed under the sedimentary basin. Unfortunately, there are no seismic refraction measurements available to test this prediction. Makris (1977) has used gravity measurement to compute a crustal section through the North Aegean area where no crustal thinning is shown. In the following we show that the homogeneous stretching model is compatible with the gravity data although other quite different solutions would be possible.

Figure 12 is a computed free air gravity section along section A of Fig. 1 and 2 with a prolongation to the north west and southeast based on Morelli *et al.* (1975) gravity data. We have computed the variation of γ necessary to explain the observed subsidence in the absence of conduction. To the south of the trough, in the Aegean Sea, we assumed a β of 1.3 ($\gamma = 0.23$) following Le Pichon & Angelier (1981). We have seen that in the worst case, at the axis of the trough, close to the steep southern wall, taking conduction into account in-

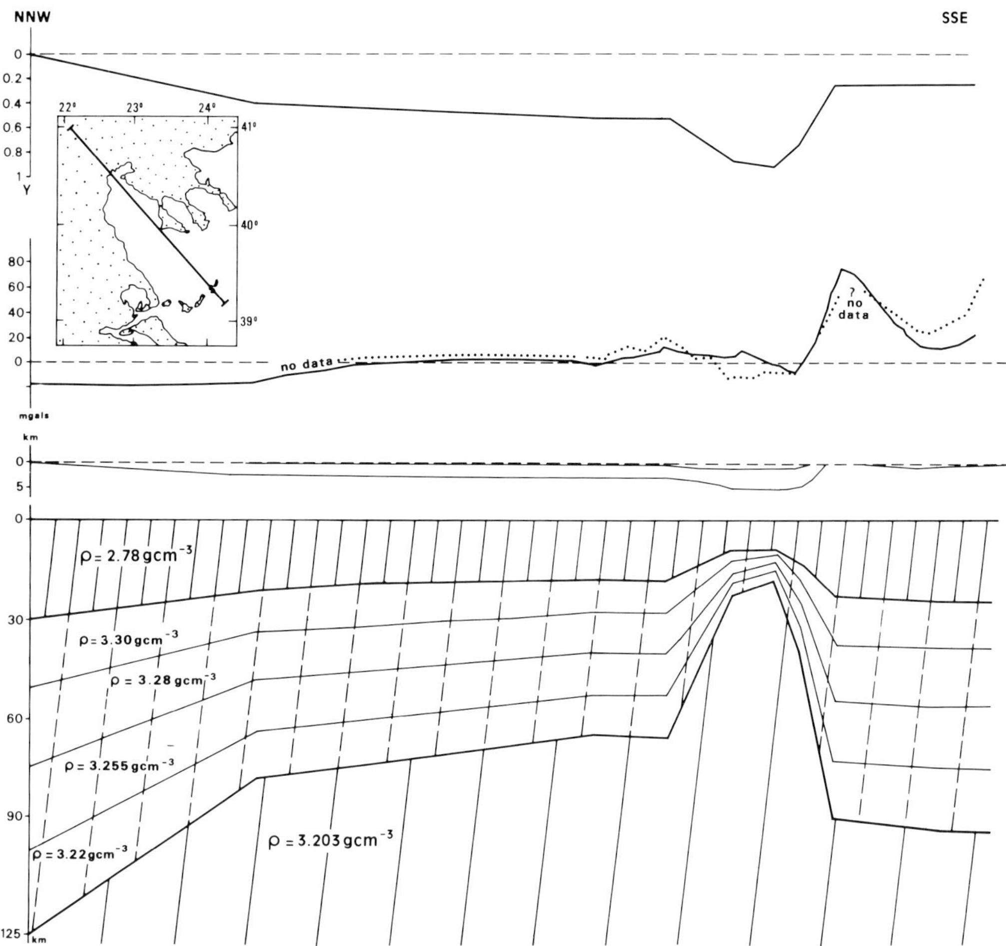

FIG. 12. Free-air gravity anomaly computations based on the homogeneous stretching model assuming negligible heat conduction. Measured anomaly: dotted line. Computed anomaly: continuous line. The adopted γ necessary to explain the tectonic subsidence is shown above. The measured water and sedimentary section on which the estimate of γ was based is shown above the corresponding density model using the densities of Table 1.

creases subsidence by 33%. But the average increase in subsidence due to conduction over the profile of Fig. 9 is closer to 10–15%. We consequently know that the values of γ in Fig. 12 are somewhat overestimated. But it would be difficult to accurately take conduction into account on the profile of Fig. 12 and it was felt that, anyway, the effect on gravity computations of such changes would be small because the computed density structure is always assumed to be in isostatic equilibrium. The densities were taken as given in Table 1 for water, sediments, crust and mantle. The computed free air anomalies in Fig. 12 match reasonably well the measured ones in view of the fact that the model is entirely theoretical and that no adjustments were made. Of course, this does not prove that the model is correct: refraction results are needed to corroborate the predicted crustal thinning.

Conclusions

We have shown that the homogeneous stretching model can account for the formation of the North Aegean trough over the last 10 Ma. Many uncertainties still exist: in particular the exact duration of the stretching phase, the possible changes in strain rates through time, the actual distribution of crustal thickness and the detailed heat-flow pattern. The tectonic analysis by Lybéris in this volume and the geophysical modelling presented here are sufficiently coherent to suggest that this model is essentially correct. If so, the implications are far-reaching because our results indicate that it is possible for extreme thinning to be concentrated in a narrow zone originally only about 50 km wide, thus resulting in very high lateral stretching gradients from a β of 3.5 to a β of about 1 in less than 20 km. Within this thinned zone, the elasticity of the plate is very small and local compensation prevails.

In these conditions, it has been shown that one should take into account lateral conduction when computing the thermal evolution. Ignoring conduction leads to a maximum underestimate of the subsidence of 33%. The effect of the high sedimentation rate is to mask the resulting heat-flow anomaly. On the other hand, the effect of the sediment cover on the computation of the lithosphere thermal expansion is small. An important consequence of conduction is to prevent large-scale melting of the mantle, thus rendering the transition to oceanic accretion rather improbable there.

An interesting byproduct of these thermal computations is that they indicate a probable disappearance of the lower crust ductile zone as stretching proceeds. Thus the highly thinned continental crust is entirely contained within the brittle field. The rheological implications will be discussed more fully elsewhere.

Finally, one can speculate about the reasons for the narrowness of the zone of extreme stretching here compared to the widths now observed on most continental margins where comparable stretching values have been reached. The tectonic analysis made by Lyberis (this volume) suggests that this may be due to the presence of earlier deep lithospheric faults which enabled the strain to concentrate along their paths.

ACKNOWLEDGEMENTS. This work was mainly supported by CNEXO and CNRS. The thermal computation program was financed by Total and Elf. We thank D. Needham and & B. Biju-Duval for access to results of their 1973 and 1972 papers which had only been presented orally and N. Lalechos for information and useful discussions. We are grateful to E. Faugères who did the elastic computations.

References

ALVAREZ, F., VIRIEUX, J. & LE PICHON, X. Thermal consequences of lithosphere extension. *Geophys. J. R. astr. Soc.*, in press.

BIJU-DUVAL, B., LETOUZEY, J. & SANCHO, J. 1972. Nouvelles données de sismique réflection précisant la structure du bassin Méditerranéen oriental et du Nord de la Mer Egée. Abstract. 4th Aegean Symposium, Hannover 1972.

BROOKS, M. & FERENTINOS, G. 1980. Structure and evolution of the Sporadhes basin of the north Aegean trough, Northern Aegean sea. *Tectonophysics*, **68**, 15–30.

BRUNET M. F. & LE PICHON, X. 1982. Subsidence of the Paris Basin. *J. Geophys. Res.* **87**, 8547–8560.

BEAUMONT, C., KEEN, C. E. & BOUTILIER, R. 1982. On the evolution of rifted continental margins: comparison of models and observations for the Nova Scotian margin. *Geophys. J.R. astr. Soc.*, **70**, 667–715.

BRACE, W. F. & KOHLSTEDT, D. L. 1980. Limits in lithospheric stress imposed by laboratory experiments. *J. Geophys. Res.*, **85**, 6248–6252.

CARSLAW, H. S. & JAEGER, J. C. 1959. *Conduction of heat in solids.* Clarendon Press, Oxford, 510 pp.

DE BREMAECKER, J. C. 1983. Temperature, subsidence and hydrocarbon maturation in extensional basins: a finite element model. *Bull. Amer. Assoc. Petrol. Geol.* in press.

DOUGLAS, J. 1955. On the numerical interpretation of $\delta^2u/\delta x^2 + \delta^2u/\delta y^2 = \delta u/\delta t$ by implicit methods. *J. Soc. Industrial App. Math.*, **3**, 42–65.

ENGLAND, P. 1983. Constraints on extension of continental lithosphere. *J. geoph. Res.*, **88**, 1145–1152.

FAUGERES, L. & DESPRAIRIES, A. 1971. Précisions stratigraphiques et sédimentologiques sur les dépôts néogènes du Bassin de Servia (Macédoine occidentale, Grèce). *Bull. Soc. géol. Fr.* **13**, 67–84.

—— & GILLET, S. 1970. Contribution à l'étude du Pontien de Macédoine: analyse géologique et sédimentologique des dépôts de Trilophos (Sud-Ouest de Salonique). *Rev. Geogr. phys. Géol. dyn.*, **12**, 9–24.

—— & GILLET, S. 1974. L'évolution morphologique des piémonts du Chortiatis (Salonique). Déformations tectoniques, mise au point sur les formations plio-quaternaires. *Ann. géol. Pays Hell.*, **25**, 407–438.

—— & ROBERT, C. 1976. Etude sédimentologique et minéralogique de deux forages du golfe Thermaïque (mer Egée). *Ann. Univ. Provence, Géol. Méditerranéene*, **3**, 209–218.

FOUCHER, J. P., LE PICHON, X. & SIBUET, J. C. 1982. The ocean-continent transition in the uniform lithospheric stretching model: role of partial melting in mantle. *Phil. Trans. R. Soc. Lond. Ser. A.* **305**, 27–43.

FYTIKAS, M. D. & KOLIOS, N. P. 1979. Preliminary heat flow map of Greece. *In:* Cermak, V. & Rybach L. (eds.) *Terrestrial heat flow in Europe*, Springer-Verlag, Berlin, Heidelberg, N.Y., 197–205.

INNOCENTI F., MANETTI, P., PECCERILLO, A. & POLI, G. 1979. Inner arc volcanism in NW Aegean Arc: geochemical and geochronological data. *N.Jb. Miner. Mn, Jg.* 145–158.

JARVIS, G. T. & MCKENZIE, D. P. 1980. Sedimentary

basin formation with finite extension rates. *Earth planet. Sci. Lett.* **48,** 42–52.

JONGSMA, D. 1974. Heat flow in the Aegean Sea. *Geophys. J.R. astr. Soc.*, **37,** 337–346.

LALECHOS, N. & SAVOYAT, E. 1979. La sédimentation néogène dans le Fossé Nord-Egéen. *6th Colloquium on the Geology of the Aegean region.* **2,** 591–603.

LE PICHON, X. & ALVAREZ, F., 1983. From stretching to subduction in back-arc regions: dynamic considerations. *Tectonophysics*, in press.

—— & ANGELIER, J. 1981. The Aegean Sea. *Proc. Roy. Soc. Lond.*, **A300,** 357–372.

—— & SIBUET, J. C. 1981. Passive margins: a model of formation. *J. geophys. Res.*, **86,** 3708–3720.

——, ANGELIER, J. & SIBUET, J. C. 1982. Plate boundaries and extensional tectonics. *Tectonophysics*, **81,** 239–256.

MAKRIS, J., 1977. Geophysical Investigations of the Hellenides. *Hamb. Geoph. Einz., R.A.*, **34,** 124.

MCKENZIE, D. P., 1978. Some remarks on the development of sedimentary basins. *Earth planet. Sci. Lett.* **40,** 25–32.

MORELLI, C., PISANI, M. & GAUTAR, G. 1975. Geophysical studies in the Aegean Sea and in the eastern Mediterranean. *Boll. Geofis. Teor. Appl.*, **18,** 127–168.

NAFE, J. E. & DRAKE, C. L. 1957. Variation with depth in shallow and deep water marine sediments of porosity, density and the velocities of compressional and shear waves. *Geophysics*, **22,** 523–552.

NEEDHAM, D., LE PICHON, X., MELGUEN, M., PAUTOT, G., RENARD, V., AVEDIK, F. & CARRE, D. 1973. North Aegean Sea trough: 1972 Jean Charcot cruise. *Bull. geol. Soc. Greece*, **10,** 152–153.

PARSONS, B. & SCLATER, J. G. 1977. An analysis of the variation of ocean floor bathymetry and heat flow with age. *J. geophys. Res.* **82,** 803–827.

PEACEMAN, D. W. & RACHEFORD, H. H. 1955. The numerical solution of parabolic and elliptic differential equations. *J. Soc. Indust. Appl. Math.*, **3,** 20–41.

X. LE PICHON, N. LYBÉRIS & F. ALVAREZ, Laboratoire de géodynamique, L.A. 215, Université P. et M. Curie, TIS, E1, 4 Place Jussieu, 75 230 Paris, CEDEX 05, France.

Rotational mechanisms of active deformation in Greece and Iran

James Jackson & Dan McKenzie

SUMMARY: The purpose of this paper is to emphasize the importance of vertical and horizontal rotations in the evolution of geological structures. In the Aegean region pervasive normal faulting, probably listric in nature, is responsible for substantial vertical rotation of footwall and hanging wall blocks as well as fault planes. The experience of the Basin and Range Province suggests that strain fields deduced from micro- or macro-structural analyses of fault planes in such an environment are unlikely to be correct unless the detailed 3-dimensional geometry of the motion on the largest faults is known. A particular hazard is the interpretation of rotated low angle gravity glides and high angle antithetic normal faults as thrusts and reverse faults respectively.

In northeast Iran, a combination of thrust and strike slip faulting has led to substantial horizontal rotations whose sense can only be deduced by a knowledge of the history of movement on major structures or by palaeomagnetism. In zones of major continental deformation present day strike azimuths, as well as horizontal and vertical directions, are of no particular significance, and the interpretation of even quite young structural data without a knowledge of the rotations involved may lead to incorrect geological reconstructions.

Old, inactive geological structures presumably formed by processes similar to those occurring now in areas of active continental deformation. The young structures forming today, which are relatively simple because they are less evolved than those in older mountain belts, can thus be helpful in understanding the origin of geological features in older orogenies. Most seismic deformation of the continents takes place during the faulting associated with large earthquakes. As these are relatively infrequent they may be investigated in some detail, and regional studies of earthquake mechanisms can be used to reveal the large scale geometry of the motions within actively deforming continents (e.g. McKenzie 1972, 1978, Molnar & Tapponnier 1975, 1978, Tapponnier & Molnar, 1977, 1979). In particular, first motion observations of teleseismic waves may be used to obtain a fault plane solution, which reveals the orientation and sense of movement where the earthquake nucleated at depth. Because the hypocentres of continental earthquakes are commonly in the range 5–20 km, these seismological studies, when combined with surface observations, allow a reconstruction of the three-dimensional geometry of fault motion that is rarely possible with studies of older, static structures. Once the geometry and sense of motion on a particular fault system is known, the origin of active geomorphological features in the area may become clear. These in turn will allow the recognition of similar structures elsewhere, both active and in the geological record. In this way, faulting that is active today can provide helpful insight into the mechanisms by which older geological structures were formed.

This paper hopes to demonstrate the use of the approach discussed above in two areas of intense present day seismicity within the Alpine—Himalayan Belt: a broad zone of active deformation resulting from the collision of Africa, Arabia and India with Eurasia (Fig. 1). In both the Aegean Sea, where normal faulting predominates, and in Eastern Iran, where crustal shortening is taking place, large and rapid vertical and horizontal rotations are evident from the seismicity and geomorphology. In such regions present day relative azimuths as well as horizontal and vertical directions may not be simply related to those in the past. It is known that the geological history of the eastern Mediterranean has involved periods of substantial crustal shortening and extension. When considering the various schemes proposed for the evolution of this region it is thus sensible to ask whether such rotations have been recognized in the geological record and accounted for in the reconstructions.

Normal faulting in the Aegean Sea

Fault plane solutions in the central and northern Aegean Sea are shown in Fig. 2, and clearly indicate that normal faulting dominates the whole region. Neotectonic studies confirm that such normal faulting has been widespread since at least the early Pliocene and probably began, in places, even earlier (e.g. Mercier *et al.*,

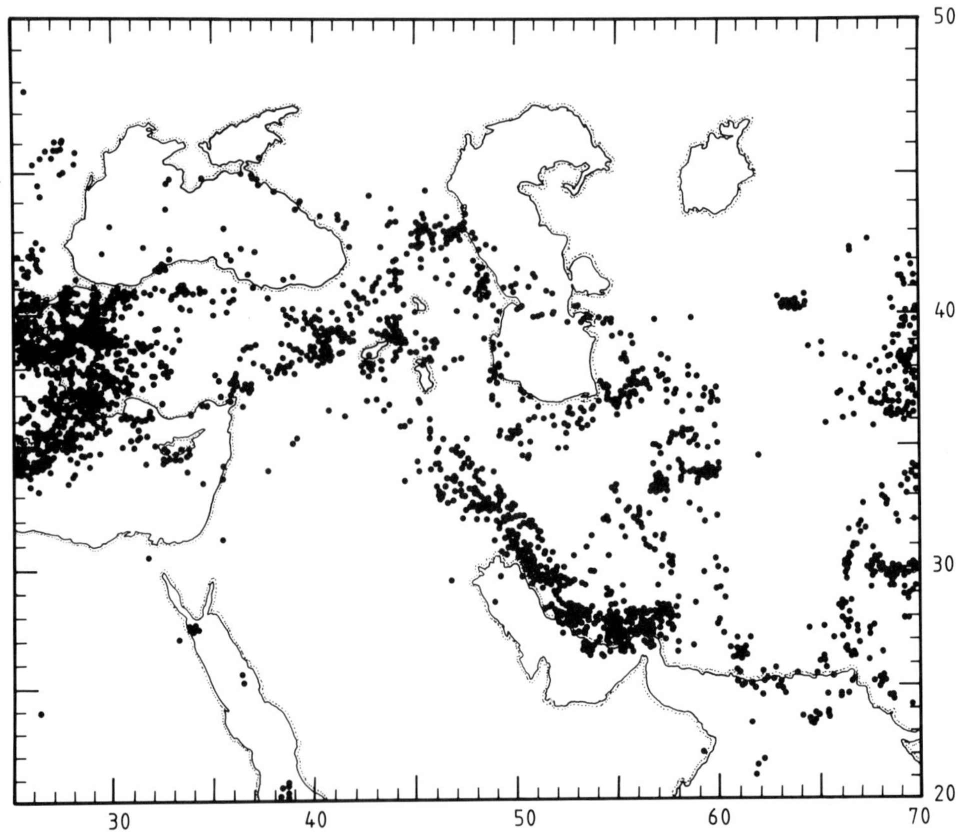

Fig. 1. Seismicity of the Middle East between Greece and Pakistan, showing epicentres reported by the U.S. Geological Survey between January 1961 and December 1980. Only earthquakes with depths reported as less than 50 km are shown.

1979a). Of all the earthquakes in Fig. 2, the best studied are those near Thessaloniki in 1978 (Mercier *et al.*, 1979b, Soufleris *et al.*, 1982) and Corinth in 1981 (Jackson *et al.*, 1982) for which observations could be made on land. Both sequences involved faulting on major topographic and bathymetric escarpments that control the local geomorphology. Many of the islands in the Aegean have large bathymetric escarpments offshore. Fault plane solutions suggest that these escarpments may be major faults, and in several places this is confirmed by seismic reflection profiling. A summary of this evidence is given by Jackson, King & Vita-Finzi (1982), who suggest that uplift in the footwall blocks of these normal faults can account for the distribution of Neogene sediments in the islands and coastal regions of the north and central Aegean.

Estimates of crustal thickness made by seismic refraction lines suggest that the crust under the Aegean has been stretched by a factor of about two compared with that under the rest of Greece and Turkey (McKenzie 1978, Le Pichon & Angelier 1979). This difference in crustal thickness is revealed by a difference in elevation as stretching leads to subsidence. Stretching of the brittle upper crust is thought to have occurred by motion on pervasive normal faults, probably with listric geometries. This belief is supported by evidence from other areas of stretched continental crust such as the Bay of Biscay, where listric normal faults can be seen in seismic reflection profiles (de Charpal *et al.*, 1978). A similar amount of stretching has been inferred for parts of the Basin and Range Province in the western U.S.A. (e.g. Proffett 1977, Wernicke 1981, Wernicke & Burchfiel 1982), where large low-angle normal faults and listric geometries are visible both in outcrop and in seismic reflection profiles. This faulting has led to a massive and widespread rotation, not only of strata but also of fault planes (Fig. 3). In some places steep antithetic faults have

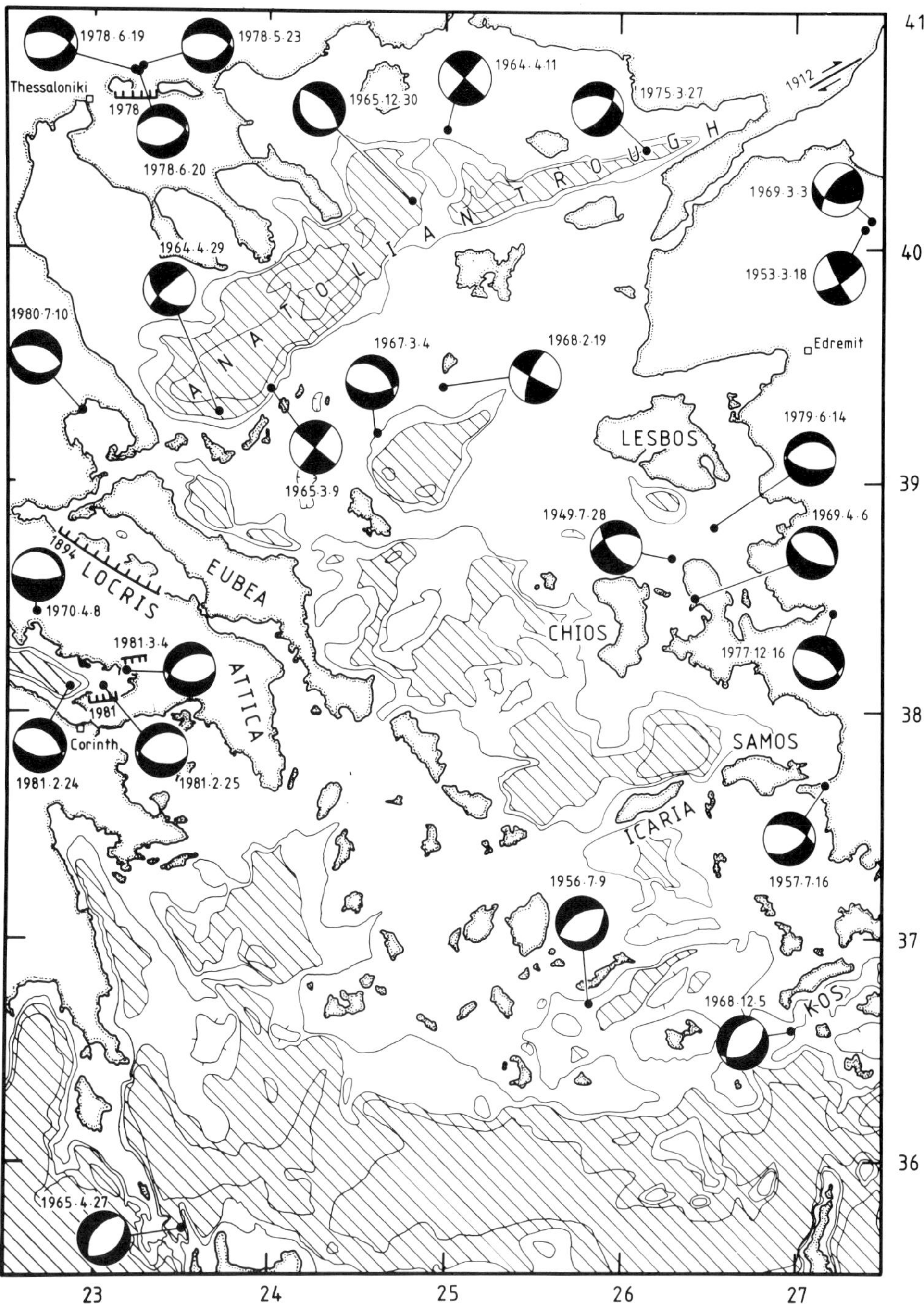

FIG. 2. Lower hemisphere fault plane solutions of shallow earthquakes in the Aegean region, taken from Jackson, King & Vita-Finzi (1982). Compressional quadrants are shaded. Bathymetric contours at 400, 600 and 1000 m are taken from Morelli, Gantar & Pisani (1975) and are shaded below 600 m. Surface fault breaks from the major earthquakes of 1894, 1912, 1978 and 1981 are shown schematically.

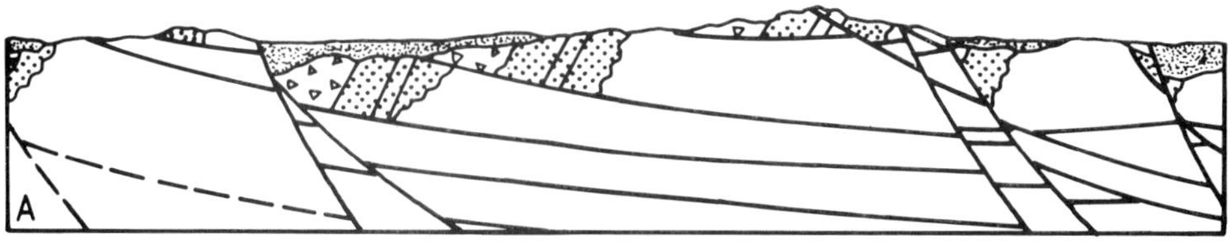

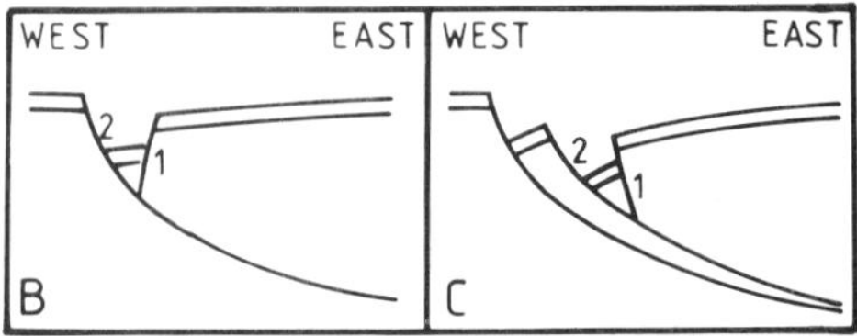

Fig. 3. (A) Cross-sections of the Yerington district, Nevada (adapted from Proffett, 1977) to show the extreme tilting associated with normal faulting and the high angle faults that cross now inactive faults which have been rotated almost flat.

(B) Diagrammatic cross-section illustrating how an antithetic normal fault (1) changes (C) to an apparent reverse fault due to tilting caused by continued movement on normal fault 2 (from Proffett, 1977). Although in these cartoons it is obvious that such apparently reverse faults are not really compressional in origin, it is often far less obvious in the field, where the main fault may be sub-horizontal and hardly exposed at all (see e.g. Fig. 6).

Fig. 4. View looking northwest near Renginion (Fig. 5) showing surfaces backtilted towards the major east-dipping fault which bounds the massif of Kallidhromon, visible as the escarpment on the left (west).

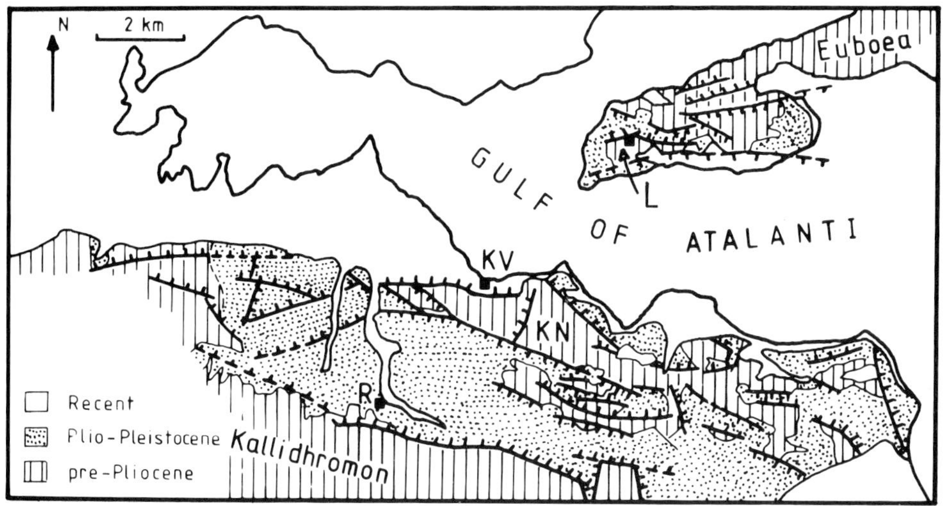

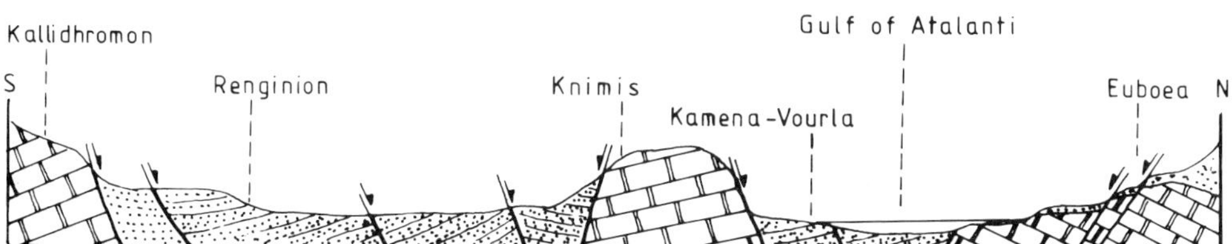

FIG. 5. Geology and cross section of the Kamena Vourla area in Locris (Fig. 2), showing the major normal faults active since the Pliocene. Simplified from Philip (1976) and Mercier *et al.* (1976). Note the antithetic relation of the faults at Lichas (L) on Euboea to those at Renginion (R) and Kamena Vourla (KV). The massif of Knimis is marked KN.

been tilted beyond the vertical by movement on neighbouring major listric faults and thus have the appearance in outcrop of minor reverse faults, though in fact their origin was extensional.

It is likely that such vertical rotations, which are a requirement of listric fault geometries, are a common feature of extensional terrains, though few areas are as well exposed, drilled and seismically explored as the Basin and Range. Although little is known of the fault geometries at depth in the Aegean, tilting is clearly visible at the surface near some of the major fault systems in coastal regions (Fig. 4). A good example is the Locris—Euboea region of the western Aegean (Fig. 2), whose evolution is discussed by Philip (1976), Mercier *et al.* (1976), and Lemeille (1977). This region is shown in Fig. 5. The major normal faults bounding pre-Pliocene formations trend WNW. Large northward dipping fault scarps, dominating the topography, outcrop near Renginion and Kamena Vourla. Southward dipping normal faults outcrop south of Knimis and on Euboea. It is likely that the Pliocene deposits in the Renginion valley have been uplifted as the footwall block by motion on the normal fault near Kamena Vourla, which is part of a large fault system on the south side of the Gulf of Atalanti that last moved in the earthquake of 1894 (Fig. 2). An emergent northeast coastline on Euboea suggests that there is a third major normal fault offshore, probably coinciding with the bathymetric escarpment in Fig. 2. At Lichas on western Euboea (Fig. 5) an almost vertical fault plane in pre-Pliocene crystalline limestone contains en echelon cracks which are taken to indicate reverse fault movement (Mercier 1976). Slickensides with different slip directions on limestone surfaces of the Kamena Vourla fault system indicate different phases of movement, one of which has a small reverse component and is taken to signify compressional motion. These sites are in very similar geometric locations to those minor faults described by Proffett (1977) in the western Basin & Range which appeared to be reverse faults as a result of rotation. Movement of the major active fault system on the south side of the Gulf of Atalanti, of which the Kamena Vourla fault system is part, will inevitably have tilted and rotated western Euboea, so that near-vertical antithetic fault planes (such as those at Lichas) could easily change their dip directions and appear as reverse faults. Similarly, minor low angle offsets in the Locris—Euboea region described as thrusts by Pechoux *et al.* (1973) may be low angle gravity driven slides, rotated so as to show apparent thrust offsets.

These small apparently reverse offset structures have been used to suggest that the large scale normal faulting, which has been the dominant mode of deformation in the north and central Aegean since at least the lower Pliocene, was interrupted by a short period of weak compression in the lower Quaternary. Not all these apparently compressional observations can easily be explained as rotated normal faults (Jackson, King & Vita-Finzi 1982), and more field data related to rotations is certainly needed. However, in a fault system as complicated as that of Locris and Euboea, where topographic differences indicate that fault motion has been large and rotations are clearly visible in the geomorphology (Fig. 4), any relations between present day horizontal and vertical directions and those at some time of faulting in the past are bound to be obscure. Similar rotations are also visible in the graben systems of western Turkey. A good example is north of Aydin in the Büyük Menderes graben, where fault planes and bedding in outwash deposits dated as early Pleistocene in age by Erinç (1955) have been rotated by up to 60° (Fig. 6). Unless the large scale rotations are known from a detailed 3-dimensional geometry of the major faults, compressions inferred from the present orientations of slip or dip directions cannot be accepted with confidence, particularly if the angle between the fault dip and the present day horizontal or vertical is small.

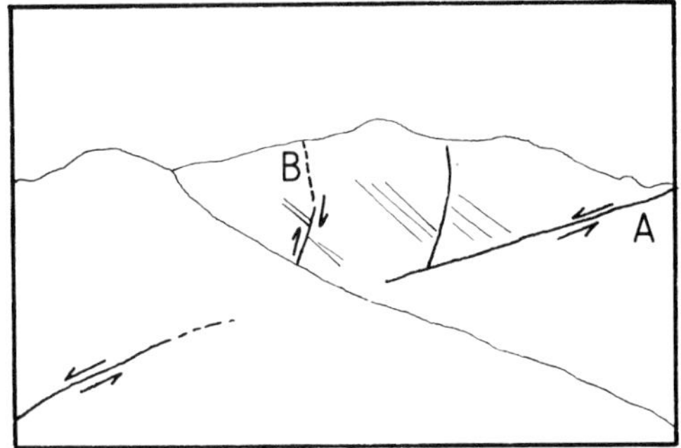

FIG. 6. Low angle normal fault near Daniş-ment, north of Aydin in the Büyük Menderes valley of western Turkey. Note the strong tilting of both the bedding and the antithetic fault, which appears to have had reverse motion as a result of bulk rotation caused by movement on a major fault zone to the north (right). View looking west. The deposits are dated as early Pleistocene in age by Erinç (1955).

Crustal shortening in northeast Iran

The seismicity of Iran shows a distinct cut-off at its northeastern and eastern boundaries, with very little activity north of the Kopet Dag (on the northeast border) and in Afghanistan (Fig. 1). Fault plane solutions (Fig. 7) show that northeast Iran is dominated by strike slip and thrust faulting and that the northeast compression seen in the active folds and thrusts of the Zagros mountains in southwest Iran persists further northeast as Iran is pushed against its stable northeastern and eastern borders (Fig. 8). The compression of northeast Iran against its borders is seen also in the topography (Fig. 8), with northwest trending ranges in the Kopet Dag and north-south ranges in eastern Iran, where faults can be seen on satellite photographs to cut the older east-west structures of western Afghanistan (Jackson & McKenzie 1984).

A closer look at the faulting in particular earthquakes (Fig. 9) shows that both east-west left-lateral and north-south right-lateral strike slip faults are present, as well as thrust faults with a northwesterly strike. This complete system of thrust and conjugate strike slip faults is consistent with a northeasterly direction of shortening. Fig. 9 shows that the northwest trending ranges are bordered by thrusts and that the dominant strike slip system is left-lateral with an easterly strike. It is not possible for two intersecting strike slip fault systems to be active simultaneously, and it appears that the north-south right-lateral set is subordinate and does not intersect the major east-west

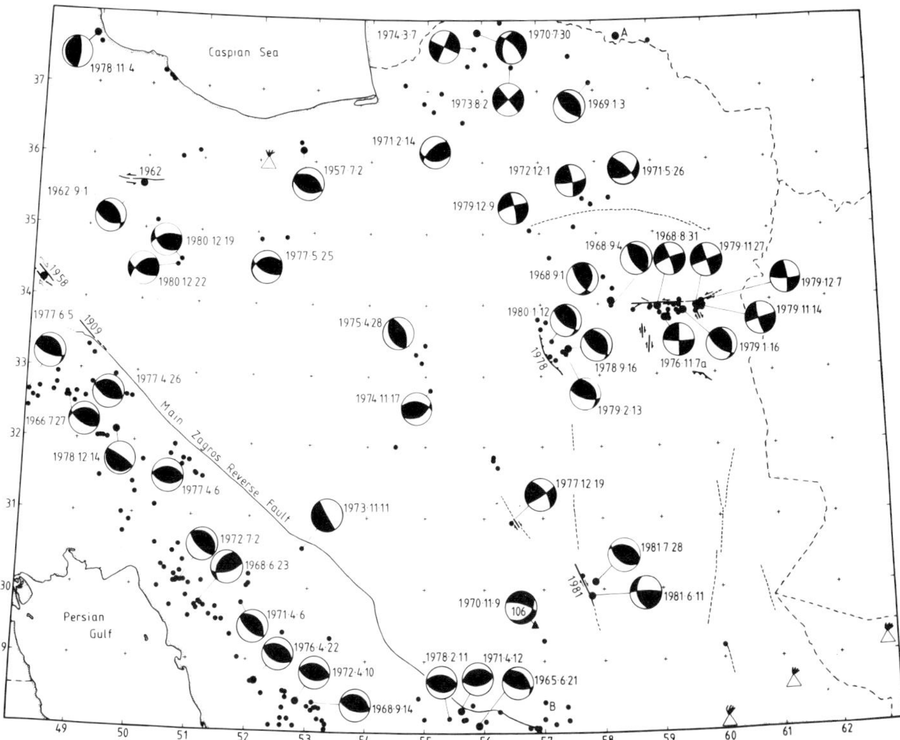

FIG. 7. Fault plane solutions in northeast Iran, from Jackson & McKenzie (1984). Other epicentres are for earthquakes of magnitude (m_b) 5.0 or greater or with more than 50 recording stations. Smoking triangles are Quaternary volcanoes. Faults which have moved historically are shown by thick lines, those inferred from satellite photographs are discontinuous lines. The Main Zagros Reverse Fault is the same as the Zagros Thrust Line of Stöcklin (1968), Falcon (1969) and others, and marks the northeastern limit of Zagros structures, rather than a fault with historic activity.

faults of Dasht-e-Bayaz (east of Ferdows) and Doruneh. This combination of one dominant strike slip fault with an accompanying mixture of subsidiary conjugate strike slip faults and thrusts will inevitably lead to major structural rotations in the areas either side of the main fault, as has been noticed in the western U.S.A. (e.g. Garfunkel 1974, Bohannan & Howell 1982). There are numerous ways in which the rotation may be accomplished, one of the easiest to visualize being when the shape of the through-going major fault remains unaltered (Fig. 10a). In practice it is not really helpful to visualize these motions in terms of blocks as the presence of thrust faulting indicates that surface area is not conserved, and some sort of continuous deformation must occur where the faults meet. The same effect may also be achieved if the magnitude of the thrusting varies along strike (Fig. 10b). Unfortunately, it is not possible to predict that the dominant east-west set in Fig. 9 will necessarily lead to an overall anticlockwise rotation in the area, as is implied in Fig. 10. If the stable western edge of Afghanistan is constrained to keep its shape the east-west left-lateral set will cause a shear in a clockwise sense, with local anticlockwise rotations caused by the subsidiary right-lateral set. Moreover, there is no particular reason why the dominant strike slip set should retain its original shape. The San Andreas Fault in the region of the Big Bend does not (Garfunkel 1974, Bohannan & Howell 1982) and east of Torbat-e-Heydarieh the Doruneh Fault curves dramatically to the south (Fig. 9), probably acquiring a component of thrust motion. This combination of strike slip and thrust faulting must eventually lead to shortening and crustal thickening in the Kopet Dag, though by that stage the various units making up the folded belt will have suffered considerable rotations and relative displacements along strike. It is likely that structures such as the Doruneh Fault are responsible for large lateral displacement during their early

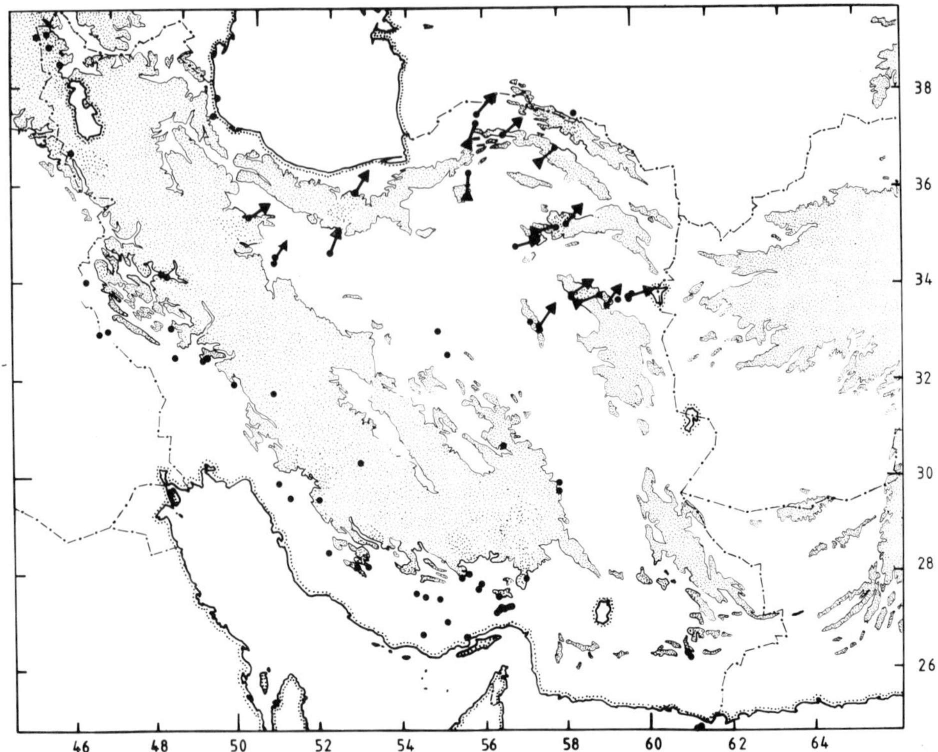

FIG. 8. Slip vectors from fault plane solutions in northeast Iran, showing the northeasterly compression of Iran against its borders. Where there was no associated surface faulting to resolve the nodal plane ambiguity, the shallow dipping thrust plane was chosen, or the plane suggested by aftershock locations (see Jackson & McKenzie 1984). Topography over 1500 m is shaded and taken from the Times Atlas (1977). Dots indicate epicentres of shocks for which fault plane solutions are given in Jackson & McKenzie (1984).

history as strike slip faults and then, as a result of progressive rotation, become thrusts. The original juxtaposition of adjacent units in a thrust belt is thus obscured and it cannot be assumed that palinspastic restoration perpendicular to the regional strike is justified. Without detailed palaeomagnetic information as well as a knowledge of the movement on the major strike slip faults involved, reconstructions of such a fold belt will be speculative at best.

Discussion

By studying active faulting and its associated geomorphology various insights may be gained into the way in which older geological structures formed. The vertical rotations apparent in extensional terranes are of great significance in structural analysis, particularly of minor faults. Because of the need to date small structures, neotectonic investigations are necessarily restricted to areas where young sediments are exposed, typically in superficial marls, gravels and sands. However, the surface faulting during individual large earthquakes, as well as the faulting at depth in aftershock sequences, strongly suggests that, at least in some circumstances, extensive minor faulting occurs in the blocks either side of a major fault as a result of internal deformation, caused by either curvature of the fault plane or non-uniform slip on it (see Jackson, King & Vita-Finzi 1982 for a review). Not only is such faulting unrepresentative of regional stresses, but substantial rotations may occur in the hanging wall blocks (Fig. 4). Interpretation of the minor faulting in the absence of information on the three-dimensional geometry of the major faults is bound to be difficult. The existence of minor reverse faults and thrusts in the Neogene outcrops of the central and northern Aegean has led to the

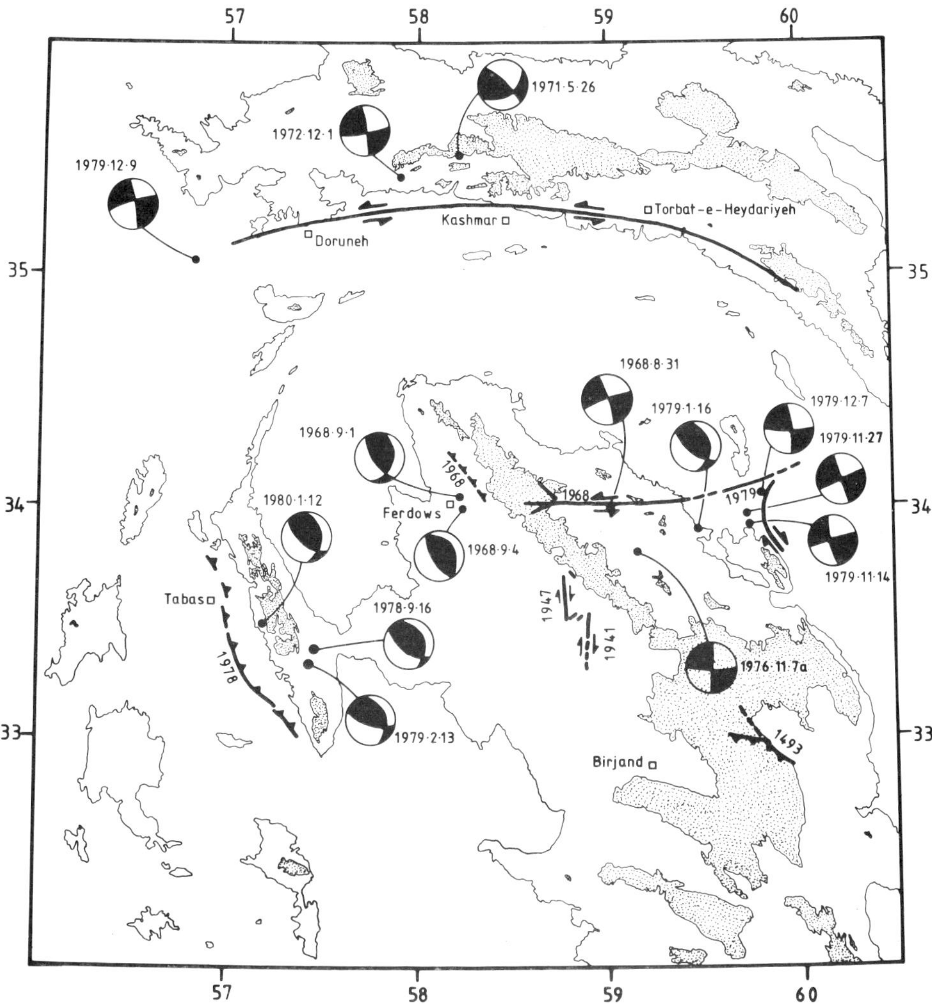

Fig. 9. Fault plane solutions, faults and topography in northeast Iran (from Jackson & McKenzie 1984). Contours are at 4000 and 6000 ft and are shaded above 6000 ft (from Operational Navigation Chart G5). Except for the Doruneh Fault, whose trace is marked from satellite photographs, all the faults shown moved during earthquakes in the year written adjacent to them, and are taken from Haghipour & Amadi (1980), Berberian (1979) and Ambraseys & Melville (1982).

suggestion that the extension which started in the early Pliocene or earlier was interrupted by short periods of compression. This view is challenged by Jackson, King & Vita-Finzi (1982) on the basis of the views expressed above, and the reader is referred to that paper for a more detailed discussion.

The northeasterly directed thrusting that pervades most of central and northeast Iran ends abruptly in the northwest striking folds of the Kopet Dag, parallel to the edge of the stable platform of Turkmenistan. It appears that the strike slip and thrust faulting of northeast Iran produces rotations that eventually align structures parallel to this stable edge. A combination of strike slip and thrust faulting will inevitably produce such rotations McKenzie & Jackson 1983) and is an alternative mechanism to the rigid block or 'ball bearing' mechanisms suggested to account for the observed palaeomagnetic rotations in the western U.S.A. (e.g. Beck 1980, Magill *et al.*, 1982). The sense of these rotations can only be deduced from a detailed knowledge of the history of movement on the major structures, or by palaeomagnetism. Observations of the large scale active

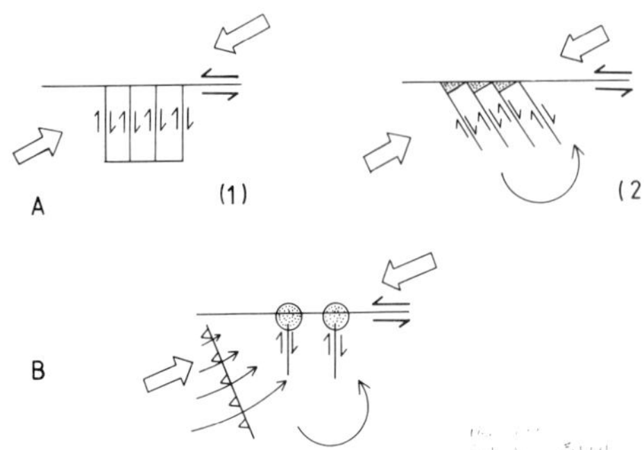

FIG. 10. Cartoons to illustrate rotation as a result of combined strike slip and thrust faulting. Plan views are shown and large arrows mark the direction of overall shortening (P axis). If the major through-going strike slip set is unchanged in shape, rotation can occur as in A (1 and 2). The stippled region of apparent extension near the main left lateral fault is a consequence of surface area being conserved from stage 1 to 2. In reality, the presence of thrusting indicates that area is not conserved, and it is likely that some form of complicated continuous deformation, on many minor faults, occurs near the junctions of the conjugate sets. An alternative visualization is shown in B, where displacement on the thrust fault(s) varies along strike (indicated by the length of the arrows on the thrust). This too will lead to a rotation in the same sense and also requires complex continuous deformation (marked by stipples) near the junctions of the strike slip sets. Note, however, that there is no particular reason why the through-going left-lateral fault should keep its geometry unchanged, and that these cartoons are, of course, gross simplifications.

tectonics of Asia, such as those of McKenzie (1972) Molnar & Tapponnier (1975, 1978) and Tapponnier & Molnar (1977, 1979) emphasize the role of stable cratons in continental collision zones. Because of their age, these cratons are likely to be colder and stronger than younger orogenic belts, which are often reactivated and compressed against them (e.g. Molnar & Tapponnier 1981). It is thus likely that the sort of rotational motions occurring in northeast Iran today have been common throughout the geological record. Structures originally oblique to the edge of the stable region may be rotated parallel to it, eventually leading to a deceptively simple appearance. However, the large amount of strike slip faulting occurring with the thrusting in the early stages of collision makes it unlikely that palinspastic restorations which assume all the shortening was perpendicular to the regional strike are correct. The present relative positions of the tectonically juxtaposed structural units in a shortened fold belt may bear no relation to their relative positions in the past, as azimuths need not be preserved.

It is interesting to consider the relationship between the vertical rotations caused by normal faulting and the horizontal rotations resulting from strike slip and thrust faulting, and in particular to compare the plan view of a strike slip system with a normal fault system in cross-section. Where a fault is embedded in a uniform elastic medium (i.e. with no free surface) there is no reference direction and consequently no meaningful distinction between normal and strike slip faulting. In such a medium there is no reason for faults to be curved. However, a distinction does exist in the presence of a free surface, which provides a reference direction and may be responsible for the curved (listric) geometry of normal faults. Where a surface separates two blocks of the elastic medium the slip vector between the blocks must lie in the plane of the surface. Only at a free surface is this condition not required. This is demonstrated in Fig. 11, where a normal fault (1) and

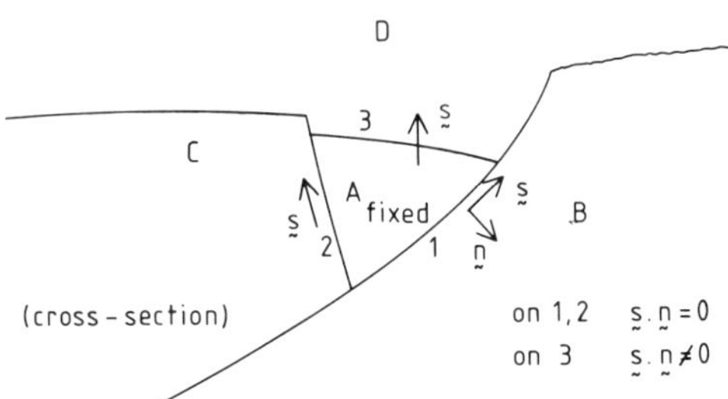

FIG. 11. Cross-section of a normal fault system to illustrate the role of the free surface, which is discussed in the text.

its antithetic (2) are shown in cross-section. Arrows show the motion of blocks B, C and the region above the Earth's surface (D) relative to block A, which is taken as fixed. On surfaces 1 and 2 the slip directions **s** must be perpendicular to the normal to the surface **n**. At the free surface (3) **s** and **n** are not perpendicular and this condition prevents the formation of voids at depth, allowing internal deformation of the hanging wall blocks. In a strike slip system **s** and **n** are always perpendicular and straight faults result.

The present day deformation of the Alpine—Himalayan collision zone can be seen to involve mechanisms, such as those discussed here, which have led to substantial structural complications, particularly on the scale at which mapping by individual geologists is feasible. There is no reason to think that older orogenies were any less complicated, and their present expressions are likely to be even more complex as these effects accumulate over millions of years. Unless detailed information on the movement history of major structures is available, the minor ones are likely to be unintelligible. In old orogenic belts the present day horizontal and vertical directions, as well as strikes and relative azimuths, are probably of no obvious significance. Without a knowledge of the rotations involved, even the interpretation of quite young structural data may lead to incorrect geological reconstructions.

ACKNOWLEDGEMENTS: This work was supported by the Natural Environment Research Council, and is contribution number 332 of the Department of Earth Sciences, Cambridge.

References

AMBRASEYS, N. N. & MELVILLE, C. 1982. *A History of Persian Earthquakes*. Cambridge Univ. Press, 219 pp.

BECK, M. E., JR., 1980. Paleomagnetic record of plate-margin tectonic processes along the western edge of North America. *J. geophys. Res.* **85,** 7115–7131.

BERBERIAN, M. 1979. Earthquake faulting and bedding-thrust associated with the Tabas-e-Golshan (Iran) earthquake of September 16, 1978. *Bull. seism. Soc. Am.* **69,** 1861–1887.

BOHANNON, R. G. & HOWELL, D. G. 1982. Kinematic evolution of the junction of the San Andreas, Garlock, and Big Pine faults, California. *Geology*, **10,** 358–363.

DE CHARPAL, O., GUENNOC, P., MONTADERT, L. & ROBERTS, D. G. 1978. Rifting, crustal attenuation, and subsidence in the Bay of Biscay. *Nature*, **275,** 706–711.

ERINÇ, S. 1955. Uber die Entstehung und morphologische Bedeutung des Tmolusschutts. *Rev. Univ. Istanbul geogr. Inst.* **2,** 57–72.

FALCON, N. L. 1969. Problems of the relationship between surface structure and deep displacements illustrated by the Zagros range. *In: Time and Place in Orogeny. Spec. Publ. Geol. Soc. Lond.*, **3,** 9–22.

GARFUNKEL, Z. 1974. Model for the late Cenozoic tectonic history of the Mojave Desert, California, and its relation to adjacent regions. *Bull. geol. Soc. Am.* **85,** 1931–1944.

HAGHIPOUR, A. & AMADI, M. 1980. The November 14 to December 25, 1979 Ghaenat earthquakes of northeast Iran and their tectonic implications. *Bull. seism. Soc. Am.* **70,** 1751–1757.

JACKSON, J. A., KING, G. & VITA-FINZI, C. 1982. The neotectonics of the Aegean: an alternative view. *Earth planet. Sci. Lett.* **16,** 303–318.

——, GAGNEPAIN, J., HOUSEMAN, G., KING, G., PAPADIMITRIOU, P., SOUFLERIS, S. & VIRIEUX, J. 1982. Seismicity, normal faulting and the geomorphological development of the Gulf of Corinth (Greece): the Corinth earthquakes of February and March 1981. *Earth planet. Sci. Lett.* **57,** 377–397.

—— & MCKENZIE, D. P. 1984. Active tectonics of the Alpine—Himalayan Belt between western Turkey and Pakistan *Geophys. J. Roy. astr. Soc.* (in press).

LEMEILLE, F. 1977. *Etudes néotectoniques en Grèce centrale nordorientale: Eubée centrale, Attique, Béotie, Locride et dans les Sporades du Nord, (Skiro)*. Thèse de 3ᵉ cycle, Univ. de Paris-Sud.

LE PICHON, X. & ANGELIER, J. 1979. The Hellenic arc and trench system: a key to the neotectonic evolution of the eastern Mediterranean area. *Tectonophysics*, **60,** 1–42.

MAGILL, J. R., WELLS, R. E., SIMPSON, R. W. & COX, A. V. 1982. Post 12 m.y. rotation of southwest Washington. *J. geophys. Res.* **87,** 3761–3776.

MCKENZIE, D. P. 1972. Active tectonics of the Mediterranean region. *Geophys. J. Roy. astr. Soc.* **30,** 109–185.

MCKENZIE, D. P. 1978. Active tectonics of the Alpine—Himalayan Belt: the Aegean Sea and surrounding regions. *Geophys. J. Roy. astr. Soc.* **55,** 217–2549

MCKENZIE, D. P. & JACKSON, J. A. 1983. The relationship between strain rates, crustal thickening, paleomagnetism, finite strain and fault movements within a deforming zone. *Earth planet. Sci. Lett.* (in press).

MERCIER, J. 1976. La néotectonique—ses methods et ses buts. Un exemple: l'arc egéen (Mediterranée orientale). *Rev. Géog. phys. Géol. dyn.* **18,** 323–346.

MERCIER, J., CAREY, E., PHILIP, H. & SOREL, D. 1976. La néotectonique plio-quaternaire de l'arc egéen externe et de la mer Egée et ses relations avec séismicité. *Bull. Soc. géol. Fr.* **18,** 159–176.

MERCIER, J., DELIBASSIS, N., GAUTHIER, A., JARRIGE, J., LEMEILLE, F., PHILIP, H., SEBRIER, M. & SOREL, D. 1979A. La neótectonique de l'arc Egéen. *Rev. Géogr. phys. Géol. dyn.* **21,** 67–92.

MERCIER, J., MOUYARIS, N., SIMEAKIS, C., ROUNDOYANNIS, T. & ANGELIDHIS, C. 1979b. Intra-plate deformation: a quantitative study of the faults activated by the 1978 Thessaloniki earthquakes. *Nature*, **278,** 45–48.

MOLNAR, P. & TAPPONNIER, P. 1975. Cenozoic tectonics of Asia: effects of a continental collision. *Science*, **189,** 419–426.

MOLNAR, P. & TAPPONNIER, P. 1978. Active tectonics of Tibet. *J. geophys. Res.* **83,** 5361–5375.

MOLNAR, P. & TAPPONNIER, P. 1981. A possible dependence of tectonic strength on the age of the crust in Asia. *Earth planet. Sci. Lett.* **52,** 107–114.

MORELLI, C., GANTAR, C. & PISANI, M. 1975. Geophysical studies in the Aegean Sea and in the eastern Mediterranean. *Boll. Geof. Teor. e appl.* **17,** 66.

PECHOUX, P. Y., PEGORARO, O., PHILIP, H. & MERCIER, J. 1973. Déformations mio-pliocène et quaternaires en extension et en compression sur les rivages du golfe Maliaque et du canal d'Atalanti (Egée, Grèce). *C.r. Acad. Sci. Paris*, **276,** 1813–1816.

PHILIP, H. 1976. Un épisode de déformation en compression à la base du Quaternaire en Grèce centrale (Locride et Eubée nord-occidentale). *Bull. Soc. géol. Fr.* **18,** 287–292.

PROFFETT, J. M. 1977. Cenozoic geology of the Yerington district, Nevada, and implications for the nature of Basin and Range faulting. *Bull. geol. Soc. Am.* **88,** 247–266.

SOUFLERIS, C., JACKSON, J. A., KING, G., SPENCER, C. & SCHOLZ, C. 1982. The 1978 earthquake sequence near Thessaloniki (northern Greece). *Geophys. J. Roy. astr. Soc.* **68,** 429–458.

STÖCKLIN, J. 1968. Structural history and tectonics of Iran: a review. *Bull. Am. Assoc. Pet. Geol.* **52,** 1229–1258.

TAPPONNIER, P. & MOLNAR, P. 1977. Active faulting and tectonics in China. *J. geophys. Res.* **82,** 2905–2930.

TAPPONNIER, P. & MOLNAR, P. 1979. Active faulting and Cenozoic tectonics of the Tien Shan, Mongolia and Baykal regions. *J. geophys. Res.* **84,** 3425–3459.

WERNICKE, B. 1981. Low angle normal faults in the Basin and Range province: nappe tectonics in an extending orogen. *Nature*, **291,** 645–648.

WERNICKE, B. & BURCHFIEL, C. 1982. Modes of extensional tectonics. *J. struct. Geol.* **4,** 105–115.

JAMES JACKSON & DAN MCKENZIE, Bullard Laboratories, Madingley Rise, Madingley Road, Cambridge, CB3 OEZ.

Listric normal faulting and the reconstruction of the synmetamorphic structural pile of the Cyclades

John Ridley

SUMMARY: There are late, post-metamorphic faults on the Aegean island of Syros that are demonstrably listric normal. Breccias associated with these faults are identical to those of late low-angle faults elsewhere on the island. It is therefore proposed that these low-angle faults are the toes of listric normal faults. This provides a basis for delineating a structural succession prior to faulting in which apparently 'allochthonous' rocks were originally at a higher level in the pile. This is consistent with the observed structural succession on Syros. The highest structural units are greenschists with similar kinematic axes to those of the blueschists below, but with a contrasting deformation style. These greenschists may have been derived from the 'roof' of the convergence zone in which the blueschist metamorphism took place.

Similar fault patterns can be inferred for many of the neighbouring islands, which are horst blocks. Many show peripheral 'upper units' of weakly- or non-metamorphic rocks. The 'upper units' are inferred to have been originally at a high level in a pile built up during early Tertiary tectonism and are not remnants of a far-travelled nappe emplaced in the Miocene.

Syros is part of the Attico-Cycladic crystalline massif—an 'Alpine' metamorphic complex covering much of the southern Aegean. The island consists largely of blueschist facies meta-sedimentary and meta-igneous rocks giving uniform Eocene radiometric dates (Bonneau *et al.* 1980, Blake *et al.* 1981). Small areas in the west and southeast of the island show a contrasting metamorphic history with a variably developed greenschist facies metamorphic imprint over earlier epidote-amphibolite facies assemblages (Ridley 1982). The metamorphism of these rocks has been tentatively dated as latest Cretaceous (Bonneau *et al.* 1980).

The island is cut by numerous late, clearly post-metamorphic, brittle faults as are most of the Cyclades. On neighbouring islands these faults in places cut dated Miocene and Pliocene 'molasse' and can be inferred to have been active from at least the Late Miocene until the Pleistocene (Angelier 1978). Angelier proposes from a study of the fault geometry a series of alternating compressional and extensional tectonic episodes in the area since the Miocene. This is somewhat at variance with the geophysical data (Makris 1978, Jongsma 1974) which suggest constant extensional tectonics (e.g. McKenzie 1977).

This paper describes the geometry and nature of the faulting seen on Syros, and suggests that all the major faults are essentially listric normal and hence consistent with the proposed extensional tectonics. A consideration of the geometrical properties of listric normal faulting allows a reconstruction of the structural pile on Syros prior to faulting, and more speculatively for the Cyclades as a whole.

Nature and pattern of faulting on Syros

Figure 1 shows the pattern of the major continuous faults on Syros. There is a complete spectrum of orientations from subvertical to subhorizontal. All, however, show very similar fault products and presumably formed under similar conditions. Within the marble units there is a characteristic breccia of even-sized, equidimensional marble fragments in a milled carbonate matrix (Fig. 2). These breccias frequently stand up above the land surface as resistant walls.

The major moderate- to high-angle faults show broadly curved traces. In the north of the island these are demonstrably both normal and listric (Figs 3 & 4). The faults at the western end of the section in Fig. 4 are such that the dip is 'opposed' to the layering. The apparent offset in the section is therefore the true offset (Redmond 1972). This example at least is an extensional fault.

Along all faults the down-faulted side is always that away from the spine of the island. The rocks to the down-thrown side either show 'reverse drag' (Hamblin 1965), or are cut by parallel secondary antithetic faults (Fig. 4) with an opposite apparent sense of throw compared to the main fault. The offset on these secondary faults is always small, generally less than 10 m. They are only seen where the trend of the

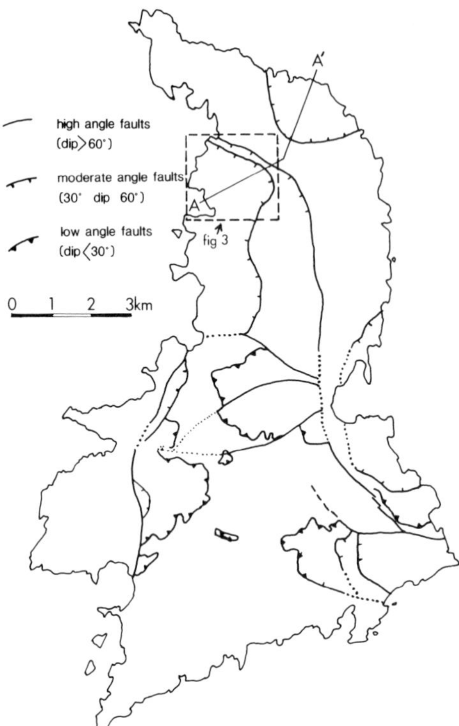

FIG. 1. Syros showing the pattern of major faults on the island (for geographical location see Fig. 5). The marking on the inclined faults indicates the upper plate. Note the curved traces of the moderate-angle faults, the constant inclination away from the 'spine' of the island, and the constant truncation relationship of higher-angle faults cutting those closer to horizontal. All these features are consistent with listric normal faulting.

major fault with which they are associated is close to northwest to southeast. It is possible that the fault kinematics, though dominated by extension, includes a minor southwest to northeast orientated strike-slip component.

In the south of the island the pattern is more complex. Individual faults may truncate earlier-formed surfaces. No clear examples of listric faults are seen. Faults are however always inclined away from the centre of the island, which effectively stands up as a horst block. There is a consistent truncation relationship: high-angle faults always truncate subhorizontal faults.

These relations are all consistent with a history of repeated listric normal faulting. The flat-lying faults are the 'toes' of early-formed faults, the high-angle faults are the higher levels of later faults.

The relationships of the rock sequences on either side of each fault is also consistent with normal faulting. A consequence of the geometry of such faulting is that rocks above any low- to moderate-angle fault were derived from higher in the pile than those below (Fig. 5). The near-horizontal faults in the south of Syros delineate a 'nappe pile'. A schematic representation of this is shown in Fig. 6.

The 'autochthon' consists of a 7 km thick succession with on average a small or moderate dip to the north. The lower half is largely pelitic; the upper half is dominated by thick repeated marble units with interbanded pelites. Towards the top of the pile there is a major variegated meta-igneous unit with serpentinite. This pile was constructed during the tectonism associated with the blueschist facies metamorphism. A static greenschist facies overprint of the blueschist assemblages becomes much more pervasive towards the base of the section.

Above the stratigraphically lower parts of the 'autochthon' (Fig. 6) there is a structural section containing four distinct allochthonous units separated by major low-angle late faults. The lowest unit contains a disrupted marble-rich sequence and shows relatively little overprinted blueschist facies assemblages. Above is a variegated meta-igneous unit, then two upper units both without a blueschist metamorphic imprint.

The rocks of the two lower allochthonous units can be correlated with the sequences towards the top of the structural pile in the autochthon, i.e. rocks higher in the pile than those they currently rest upon.

The two uppermost units are truly exotic. The lower consists of uniform chlorite–epidote–sericite–quartz meta-basic schists, which are strongly deformed with an almost mylonitic synmetamorphic fabric. Relict epidote pods suggest derivation from a slightly higher grade meta-basite antecedent.

The uppermost unit consists dominantly of coarse-grained quartzo–feldspathic rocks. The more meta-basic lithologies within it show well preserved epidote-amphibolite facies assemblages. A greenschist facies overprint of variable intensity shows assemblages similar to those in the unit below. This overprinting is synchronous with tight repeated folding that affects the whole unit.

The lower pressure paragenesis of these units is certainly consistent with derivation from higher in the syn-metamorphic structural pile. The structures and textures in these units contrast in style with those in the underlying blueschist units but are similarly orientated. It is possible that the metamorphism and deformation is

FIG. 2. Typical fault breccia as seen along all intra-marble fault lines through the whole range of attitudes from high- to low-angle. The breccia is composed of even-sized, equidimensional, randomly orientated clasts preserving the metamorphic fabric within a milled, fine grained carbonate matrix. Its exact formation mechanism is unknown.

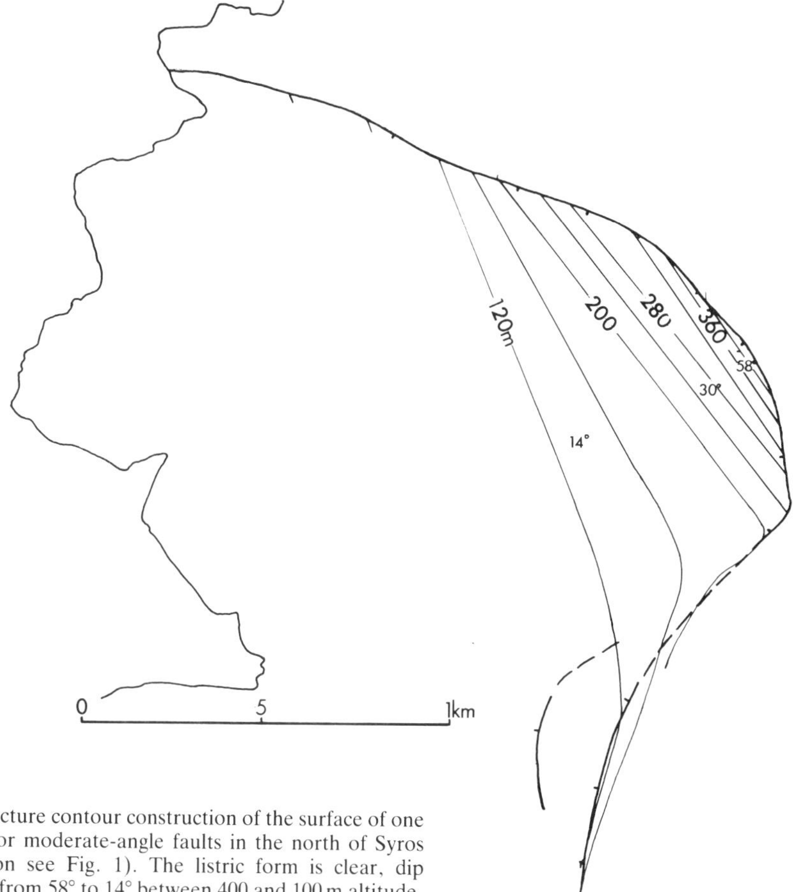

FIG. 3. Structure contour construction of the surface of one of the major moderate-angle faults in the north of Syros (for location see Fig. 1). The listric form is clear, dip decreasing from 58° to 14° between 400 and 100 m altitude. The surface need not be exactly cylindrical as shown here.

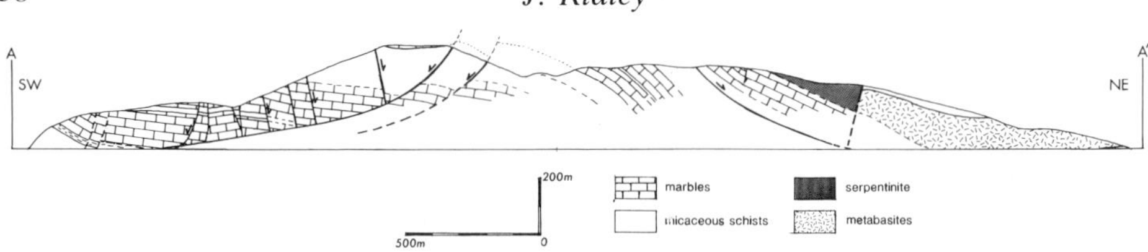

Fɪɢ. 4. Cross-section through the north of Syros (for location see Fig. 1). The listric faults are normal with downthrow away from the spine of the island. Note the localization of steep, small offset normal faults with downthrow towards the centre of the island. These are secondary, antithetic faults (Hamblin 1965) consequent on the major listric fault.

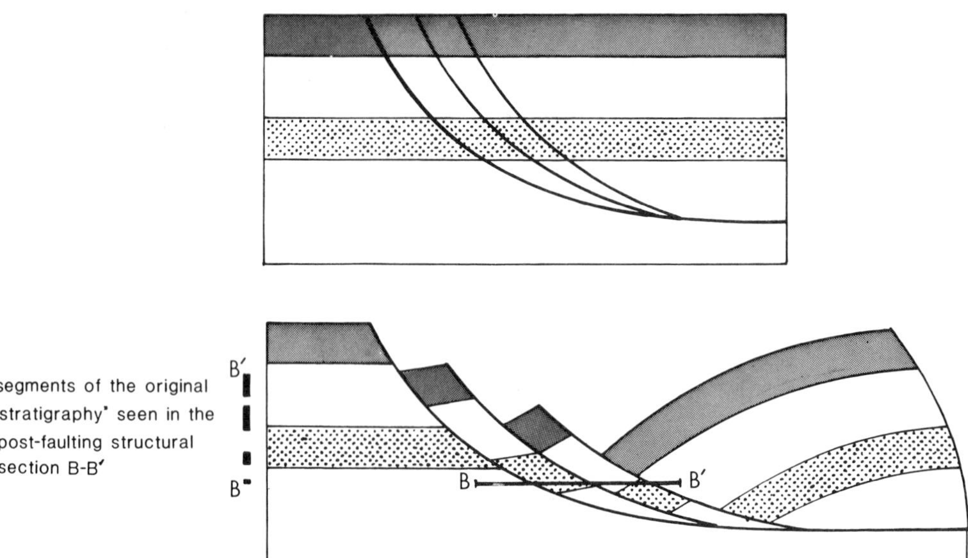

Fɪɢ. 5. Schematic gometry of imbricate listric normal faulting (after Wernicke & Burchfiel 1982). An apparent nappe pile BB′ through a series of low-angle faults will always be such that rocks above a fault derive from higher in the original pile than those below. Conceivable exceptions occur with complex patterns of intersecting faults. The sequence observed after faulting is shown on the left of the diagram. This is an incomplete, but consistently 'right-way-up' sequence through the original pile.

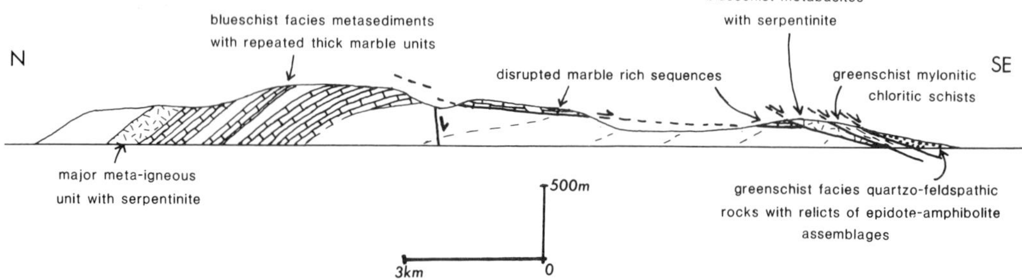

Fɪɢ. 6. Schematic section from the north coast to the southeast of Syros showing the disposition of the major structural units and the main lithological divisions within the 'autochthon'. For discussion see text.

FIG. 7. Distribution and nature of the allochthonous units in the Cyclades. These units lie in low- to moderate-angle contact over higher grade metamorphic rocks, generally blueschists or overprinted equivalents. Note that the allochthonous units are confined to the peripheries of the larger islands. Data collated from Dürr *et al.* 1978; Papanikolaou 1978, 79a, 79b, 80; Roesler 1978; Salemink 1980, and Van der Maar 1980.

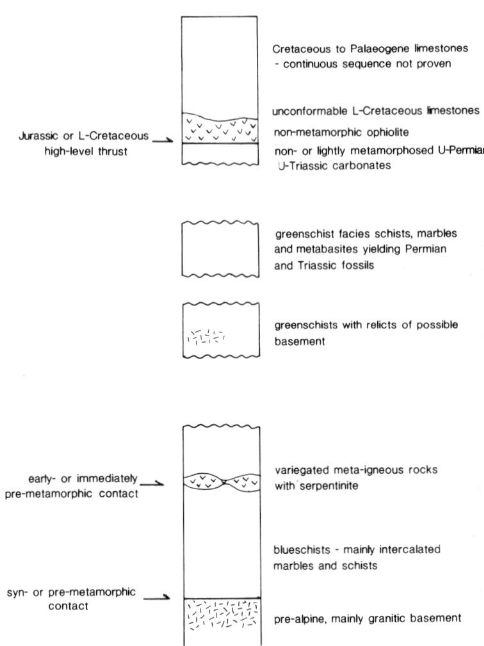

FIG. 8. Tentative suggestion for the early Tertiary and end–metamorphic structural pile in the Cyclades assuming all faulting seen is listric normal, i.e. units higher in the present structural succession have been derived from higher in the original pile. The shaded strip indicates the rock types found in the various allochthonous units. Further detail has been added by considering the clast content of the late Tertiary 'molasse' sequences (Roesler 1978).

essentially synchronous and related to the same tectonic event. The origin of the earlier epidote-amphibolite facies metamorphism is unknown.

The exact derivation and significance of these units must remain speculative until better structural and geochronological data are available from Syros and from the structural units of some of the neighbouring islands that likewise show no relics of a blueschist facies metamorphic imprint (Blake *et al.* 1981).

They may be derived from the roof of the thrust or convergence zone in which the blueschist facies metamorphism took place, so that the structures seen in them could relate to deformation at the initial formation of a crustal break.

Extension to neighbouring islands

Many of the neighbouring islands (Fig. 7) show peripheral, topographically low-lying allochthonous units of low-pressure or non-metamorphic rocks in low- to moderate-angle contact on the blueschist units. These have previously been interpreted as relics of a far-travelled high-level nappe emplaced from a distance in the Miocene. If the contacts are as on Syros, normal faults rather than thrusts, these various allochthonous units would then be locally derived, downfaulted relics of units originally at high levels in the syn-metamorphic structural pile.

On Naxos the contact metamorphism associated with a granodiorite intrusion has been dated as 11 ± 1 Ma (Andriessen *et al.*, 1979). Assemblages suggest a crystallization depth of 5–15 km (Jansen & Schuiling 1976). This granodiorite is overlain tectonically by Early and Late Miocene conglomerates, and both are overlain unconformably by Late Pliocene conglomerates (Roesler 1978). If the tectonic contact at the base of the conglomerate sequence is a thrust then some 15 km of erosion, thrusting, and renewed erosion must have taken place within 6 Ma, which seems unlikely. If the contact is a listric normal fault then the faulting itself would give the observed rapid erosion.

The conglomerate sequence on Naxos itself records a history of progressive unroofing of the structural pile (Roesler 1978) consistent with tectonic erosion by normal faulting. The Miocene strata include pebbles of rock-types, e.g. Palaeocene carbonates, which are otherwise unrecorded in the area. Pebbles of the presently exposed metamorphic rocks only appear in the Late Pliocene strata.

Figure 8 is a tentative proposal for the pre-faulting structural pile of the Cyclades collated from the published information on structural successions in the area (Dürr *et al.*, 1978, Roesler 1978). This is regarded as a representation of the structural pile at the cessation of the blueschist facies metamorphism, possibly in the Mid Eocene. No deformation appears to have taken place between the cessation of metamorphism and the onset of the faulting that could have affected the sequence.

There may be substantial gaps in the recognized pile and there may be units totally lost from the record at the present level of erosion. More work however may allow a fuller description of the structure.

ACKNOWLEDGMENTS: The work for this paper was carried out during tenure of a NERC research studentship which is gratefully acknowledged. I thank the Director of IGME for permission to pursue research in Greece. I thank Dr. J. E. Dixon for encouraging me to develop the ideas published here.

References

ANDRIESSEN, P. A. M., BOELRIJK, N. A. I. M., HEBEDA, E. M., PRIEM, H. N. A., VERDURMEN, E. A. TH. & VERSCHURE, R. H. 1979. Dating the events of metamorphism and granitic magmatism in the Alpine orogen of Naxos (Cyclades, Greece). *Contrib. Min. Pet.* **69,** 215–25.

ANGELIER, J. 1978. Tectonic evolution of the Hellenic arc since the late Miocene. *Tectonophysics*, **49,** 23–36.

BLAKE, M. C., BONNEAU, M., GEYSSANT, J., KIENAST, J. R., LEPVRIER, C., MALUSKI, H. & PAPANIKOLAOU, D. 1981. A geological reconnaissance of the Cycladic blueschist belt, Greece. *Bull. geol. Soc. Am.* **92,** 247–54.

BONNEAU, M., GEYSSANT, J., KIENAST, J.-R., LEPVRIER, C. & MALUSKI, H. 1980. Tectonique et métamorphisme haute pression d'âge éocène dans les Hellénides: exemple de l'île de Syros (Cyclades, Grèce). *C.R. Acad. Sci. Paris*, **291,** 171–4.

DÜRR, S., ALTHERR, R., KELLER, J., OKRUSCH, M. & SEIDEL, E. 1978. The Median Aegean Crystalline belt: stratigraphy, structure, metamorphism, magmatism. *In:* CLOSS, H., ROEDER, D. & SCHMIDT, K. (eds) *Alps, Apennines, Hellenides.* I.U.G.C. Scientific Report no 38, 455–76.

HAMBLIN, W. K. 1965. Origin of 'reverse drag' on the downthrown side of normal faults. *Bull. geol. Soc. Am.* **76,** 1145–64.

JANSEN, J. B. H. & SCHUILING, R. D. 1976. Metamorphism on Naxos: petrology and geothermal gradients. *Amer. J. Sci.* 276, 1225–53.

JONGSMA, D. 1974. Heat flow in the Aegean Sea. *Geophys. J. R. astron. Soc.* **37,** 337–46.

MCKENZIE, D. 1977. Present deformation of the Aegean region. *Proc 6th Coll. on the Geology Aegean Region (Athens)*, 1303–1312.

MAKRIS, J. 1978. The crust and upper mantle of the Aegean region from deep seismic soundings. *Tectonophysics*, **46,** 269–84.

PAPANIKOLAOU, D. 1978. Contributions to the geology of Ikaria Island. Aegean Sea. *Ann. géol. Pays Hell.* 29, 1–28.

—— 1979a. Contribution to the geology of the Aegean Sea: The island of Andros. *Ann. géol. Pays Hell.* **29,** 477–533.

——1979b. Unités tectoniques et phases de deformation dans l'île de Samos, Mer Égée, Grèce. *Bull. Soc. géol. France*, **21,** 747–52.

—— 1980. Contribution to the geology of the Aegean sea: The island of Paros. *Ann. géol. Pays Hell.* **30,** 65–96.

REDLAND, J. L. 1972. Null combination in fault interpretation. *Bull. Am. Assoc. Petrol. Geol.* **56,** 150–66.

RIDLEY J. R. 1982. *Tectonic style, strain history and fabric development in a blueschist terrain, Syros, Greece.* Thesis, PhD. (unpubl), Edinburgh University. 283 pp.

ROESLER, G. 1978. Relics of non-metamorphic sediments on Central Aegean Islands. *In:* CLOSS, H., ROEDER, D. & SCHMIDT, K. (eds) *Alps, Apennines, Hellenides.* I.U.G.C. Scientific Report no 38, 480–2.

SALEMINK, J. 1980. On the geology and petrology of Seriphos Island (Cyclades, Greece). *Ann. géol. Pays Hell.* **30,** 342–65.

VAN DER MAAR, P. 1980. The geology and petrology of Ios, Cyclades, Greece. *Ann. geol. Pays Hell.* **30,** 206–24.

WERNICKE, B. & BURCHFIEL, B. C. 1982. Models of extensional tectonics. *J. struct. Geol.* **4,** 105–15.

J. RIDLEY, Institut für Kristallographie und Petrographie, ETH Zentrum, CH-8092, Zürich, Switzerland.

Neotectonic deformation patterns in the convex-northwards arc of the North Anatolian fault zone

A. A. Barka & P. L. Hancock

SUMMARY: The convex-northwards arc of the North Anatolian fault zone between Çerkeş and Erbaa contains structures and landforms permitting right-lateral displacements for several time intervals to be estimated. Within the Pontus Formation (Late Miocene-Early Pleistocene) in basins along the fault zone, an unconformity, representing a time interval from the latest Tortonian to the earliest Pliocene, is interpreted as marking the transformation of the structure from a broad shear zone to a narrow fault belt. Although the amount of displacement during the shear zone phase is unknown the offset of a sedimentary facies boundary in the Lower Pontus Formation of the Havza-Ladik basin demonstrates that since the latest Tortonian there has been 25 km of right-lateral slip. The offset of valleys and ridges suggests that there has been 8 km of Quaternary displacement, about 2 km of it in the late Quaternary and at least 500 m during the Holocene.

Structures in the Pontus Formation indicate that the early history of the western Neogene basins was influenced by regional compression and that the later histories of all basins were dominated by strike-slip displacements. The main active trace is discontinuous, the 2 km-wide belt containing subordinate en échelon faults whose geometry is consistent with development during right-lateral shear.

The aim of this paper is to describe and interpret neotectonic structures within the right-lateral North Anatolian fault zone where in its 400 km central segment it curves through a convex-northwards arc of about 35° (Fig. 1). The studied segment contains an assemblage of structures deforming Neogene-Quaternary sediments and a variety of tectonic landforms which permit a displacement history to be proposed. In a regional context the proposed history is of interest because there is no agreement about when the fault zone was initiated or the amount of cumulative displacement. Comprehensive reviews of the history of investigations in the fault zone and dicussion of its tectonic significance as a transform along the northern boundary of the Anatolian *scholle* (crustal splinter) are given by Dewey & Şengör (1979), Şengör (1979) and Şengör *et al* (1982).

Stratigraphic control

Figure 2 shows the locations of five intramontane basins, each containing a Neogene-Quaternary sedimentary succession, situated astride or close to the main trace of the North Anatolian fault. The name Pontus Formation as applied to basin-fill sediments older than the late Pleistocene was introduced by Irrlitz (1971; 1972), who studied sequences in the Havza-Ladik and Taşova-Erbaa basins, provisionally correlated them with comparable successions in the three western bsins and divided the formation into upper and lower series. Hancock & Barka (1980; 1981) reported that the lower and upper series are separated by an unconformity which becomes increasingly angular as the main trace of the fault is approached.

The Lower Pontus series comprises lacustrine sands, silts, clays and marls in the centres of basins and fluvial sands and gravels around their margins. The Upper Pontus series is dominated by sands and gravels of fluvial origin.

Dating and correlating the divisions of the Pontus Formation are critical for determining the age of initiation and amount of displacement on the North Anatolian fault. Irrlitz (1971; 1972) proposed that the Lower Pontus series is of Pannonian age, which he considered to be early Pliocene, and that the Upper Pontus series is of late Pliocene-early Pleistocene age. Luttig & Steffens (1976) suggested extending the range of the lower series down into the late Miocene. Now that the Pannonian of the Paratethyan region is thought by some workers to be approximately equivalent to the Tortonian of the Tethyan region (Fahlbusch 1981), it is provisionally possible to reconcile Irrlitz's (1971; 1972) and Luttig & Steffen's (1976) dating. Correlation of the two divisions of the Pontus Formation between basins is based on lithostratigraphic similarities and faunal/floral assemblages containing ostracods, pollen and spores (Irrlitz 1971; 1972). If deposition of the Lower Pontus series continued well into the Tortonian and deposition of the upper series was estab-

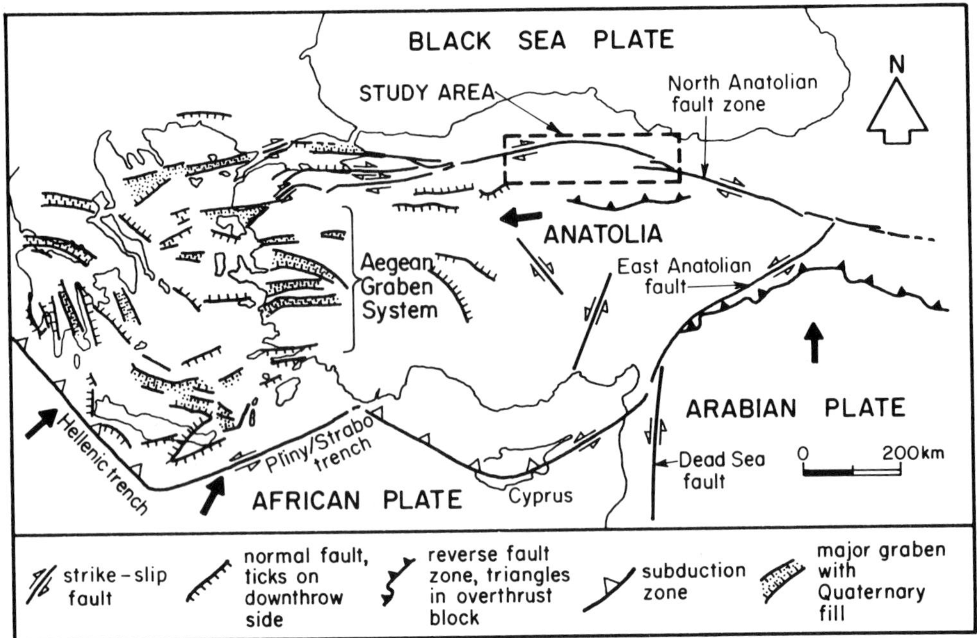

FIG. 1. Tectonic setting of the studied segment of the North Anatolian fault zone (after Şengör 1979, Fig. 1 & Hancock & Barka 1981, Fig. 1a).

lished by the early Pliocene, the time interval represented by the unconformity between them probably extended from the latest Tortonian, though the Messinian into the very earliest Pliocene.

In the three eastern basins the Pontus Formation rests with angular unconformity on pre-Neogene rocks, but in the two western basins the Sivricek Formation (volcaniclastic, pyroclastic and andesitic lavas, Barka 1981) and Devrez Formation (marls, clays and gypsum, Barka 1981) locally intervene between the Pontus Formation and the pre-Neogene basement (Fig. 2). Within each basin there is a partial and unconformable cover of late Pleistocene-Holocene terrace gravels, alluvial fan and flood-plain deposits, and colluvium.

Amounts and ages of displacements

Earthquake focal mechanism solutions and the offset of physiographic and man-made features by dated fault breaks demonstrate that present-day displacements on the North Anatolian fault are dominantly right-lateral (Allen 1969; Ambraseys 1970; McKenzie 1972; Canitez 1973). Estimates of the age of initiation include early Cenozoic (Pavoni 1961), late Oligocene (Kopp *et al.*, 1969), post-Burdigalian (Seymen 1975), between the Burdigalian and Pliocene (Şengör 1979), mid-Pliocene (Tokay 1973) and Pliocene (Irrlitz 1972). Estimates of the net right-lateral displacement are similarly varied from a maximum of 300–400 km according to Pavoni (1961), who Ketin (1969) believed made a false correlation, through intermediate values such as 100–120 km for the Erzincan region (Bergougnan 1975), 85 ± 5 km, again for the Erzincan region (Seymen 1975), 60–80 km for the Gerede-Ilgaz region (Tokay 1973), and 30 km in southern Thrace (Kopp *et al.*, 1969). It should also be borne in mind that not all parts of the fault are necessarily of the same age or have experienced the same amount of slip (Şengör *et al.*, 1982).

Figure 3 shows the distribution of coarse- and fine-grained facies within the Lower Pontus series of the Havza-Ladik basin, situated astride part of the eastern section of the main trace of the fault. If the facies boundary is restored to its inferred former position it is necessary to postulate that there has been about 25 km of right-lateral slip since deposition of the Lower Pontus series. Synsedimentary and post-depositional mesofaults related to a ENE-WSW direction of secondary extension occur within the Lower Pontus series of the Havza-Ladik and

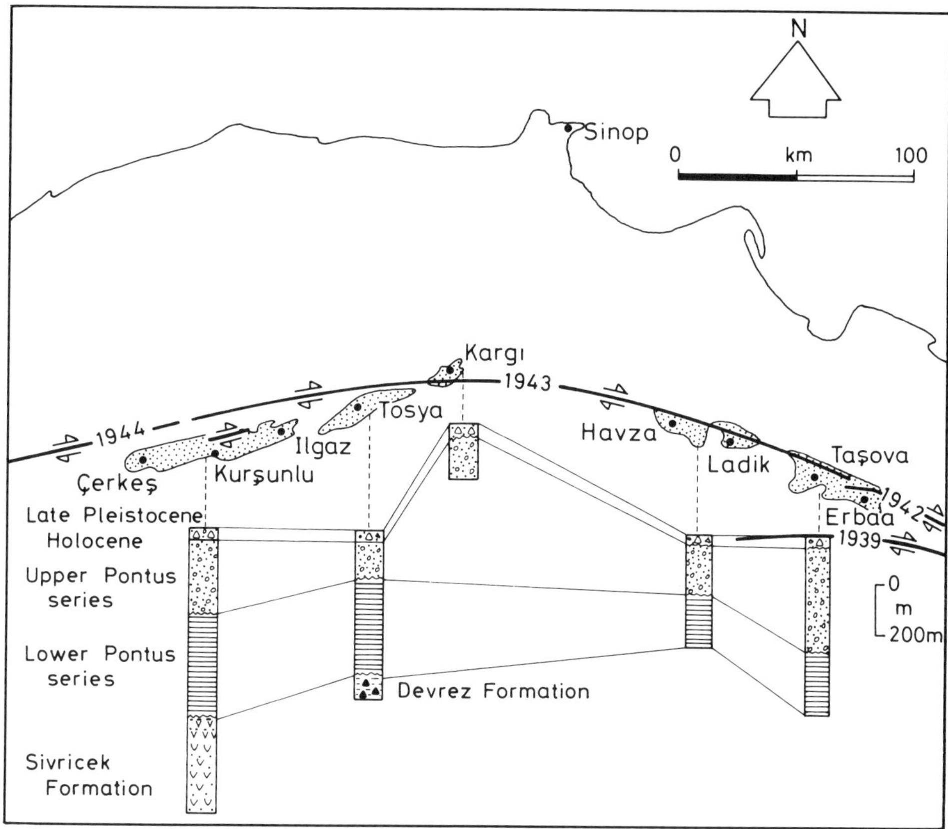

FIG. 2. Generalized lithostratigraphic successions in the five Neogene-Quaternary basins around the arc of the North Anatolian fault zone. Fault breaks formed during 20th century large-magnitude earthquakes are also shown (after Ketin, 1969, tafel I; Ambraseys 1970, Fig. 1; Hancock & Barka 1981, Fig. 1b).

other basins (Hancock & Barka 1981) and indicate that there were probably some displacements along the zone during the early stages of basin filling. Although a less reliable indicator of displacement, it is noteworthy that the present course of the Kızılırmak River at its confluence with the Devrez River is offset by 27 km (Fig. 4). It is appreciated however that where the river follows an east to west course it may in part simply be following an easily eroded zone. Evidence in favour of there having been 8 km of Quaternary displacement is provided by the arrangement of the courses of the Rivers Yeşilırmak and Kelkit where at their confluence they flow within an area of Quaternary marsh deposits (Fig. 5). About 6 km of the displacement probably occurred during the earlier Quaternary followed by an additional 2 km during the later Quaternary.

The evidence for Holocene displacements on the North Anatolian fault is well-known (e.g. Allen 1969; 1975; Ambraseys 1970). Figure 6 shows a variety of recently-mapped tectonic landforms indicating that the fault remains active and that there has been offset from a few metres for the youngest stream courses to about 500 m for older streams or small ridges transverse to the fault. It is also noteworthy that many of the physiographic indications of active faulting, such as fault scarps, linear fault valleys, side-hill ridges and sag ponds, are as equally well-developed where the fault cuts pre-Neogene basement rocks (Fig. 6a) as in the Neogene-Quaternary basins (Figs 6b & c).

Origin of basins and initiation of the fault zone

Although it is clear from the evidence just discussed that the North Anatolian fault was a shear zone before deposition of the Upper

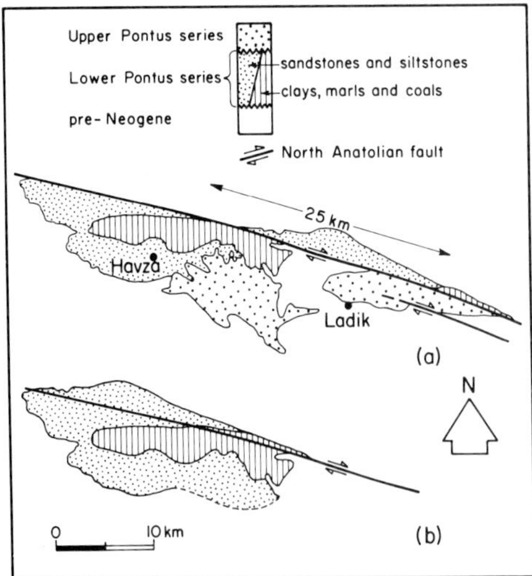

FIG. 3. Generalized distribution of coarse- and fine-grained facies in the Lower Pontus series of the Havza-Ladik basin. (a) Present-day distribution; (b) reconstructed distribution.

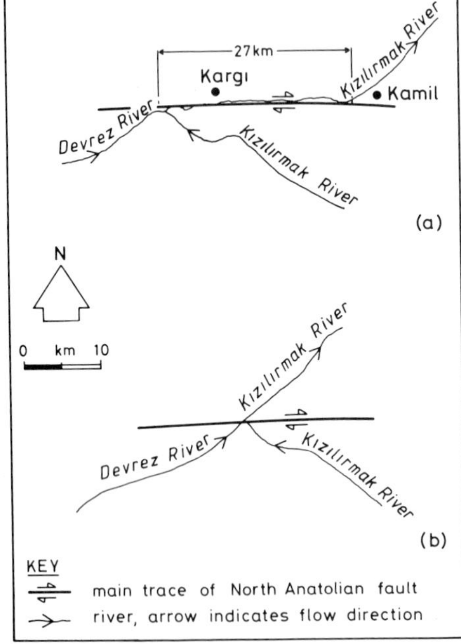

FIG. 4. Drainage system in the Kargı-Kamil area. (a) Present-day river courses; (b) reconstructed river courses.

Pontus series it is not possible from offsets of Neogene horizons to be so certain about the date of its initiation. Because the Erzincan basin (Şengör 1979; Aydin & Nur 1982) and the Niksar basin (Seymen 1975), both to the east of the studied segment, have been interpreted as pull-apart basins (rhomb-grabens), it is pertinent to consider whether any of the five studied basins are pull-aparts. Such basins should be located where the sense of offset of parallel strands permits extension parallel to the fault zone; a condition that around the studied arc is unfulfilled except by the faults bordering the Taşova-Erbaa basin (Şengör 1979) (Fig. 2). The Çerkeş-Ilgaz and Tosya basins, both of which contain sediments older than the Pontus Formation but younger than the pre-Neogene basement, are interpreted as having been regionally compressed during the earlier stages of their histories. The pre-Upper Pontus sediments of the Çerkeş-Ilgaz basin are moderately deformed by folds and thrusts trending parallel to the basin long axis (Fig. 7). Most thrust planes dip gently or moderately to the north as seen in well-exposed transverse valleys dissecting the margin of the basin between Kurşunlu and Ilgaz. Displacements generally exceed 100 m and commonly the beds beneath thrusts are overturned. The margin of the basin north of Kurşunlu and Ilgaz also displays some features which elsewhere characterize molassic sequences that have been deformed by late orogenic movements. In particular the Sivricek and Pontus Formations are overthrust by the rocks from which they were derived, and the south-facing overturned syncline north and northeast of Kurşunlu is a growth fold. In addition, some of the clasts in the Neogene conglomerates in the steep or overturned limb of the fold display solution pits and/or micro-faults of the type which have been described from other continental conglomerates that have been over-ridden by their source area (e.g. McEwen 1981).

Although it is possible to interpret the relatively deformed rocks of the Çerkeş-Ilgaz and Tosya basins as having accumulated during a modest compressional phase, it is less easy to apply the same interpretation to the Havza-Ladik basin (Fig. 8). Boundary thrusts are absent and folds are uncommon except close to the main trace of the fault where they may be related either to tilting associated with minor components of dip-slip or to secondary extension or shortening generated by lateral shear. Equally however, the Havza-Ladik basin is not readily interpreted as a pull-apart except locally in the neighbourhood of Ladik Lake where the sedimentary fill is of late Quaternary age. Perhaps the Havza-Ladik basin is an entirely post-tectonic basin not influenced by waning compressional stresses.

The widespread occurrence throughout the

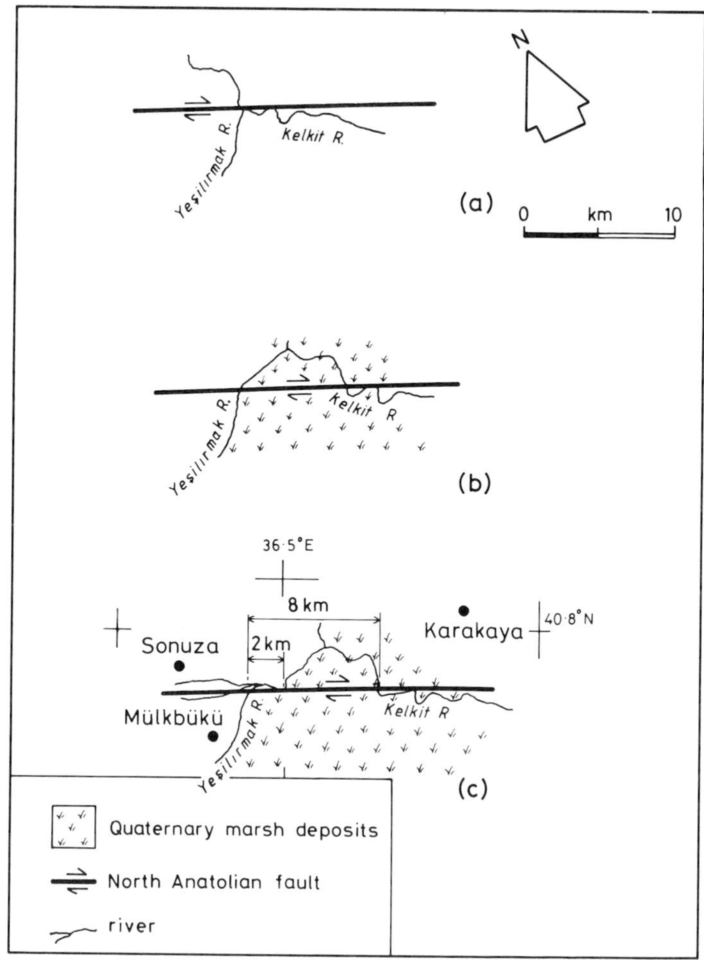

FIG. 5. Displacement of the Yeşilırmak and Kelkit Rivers. (a) Before accumulation of Quaternary deposits; (b) late Quaternary after 6 km of displacement; (c) present day after 8 km of displacement.

four basins containing the Lower Pontus series of synsedimentary and post-depositional normal and reverse mesofaults interpreted by Hancock & Barka (1981) as secondary products of right-lateral shear probably indicates that before the end of early pontus times there had been displacement along a relatively broad (10–15 km) shear zone across which strain was distributed. In the Upper Pontus series comparable mesofaults are less abundant while in post-Pontus sediments they are restricted to a narrow (2–3 km) belt astride the main trace. The change with time in the distribution of mesofaults and the observation that subordinate active macrofaults occur in the belt close to the main trace are interpreted as reflecting the transformation of an early broad shear zone into a narrow fault belt on which movement was concentrated. The initiation of the North Anatolian structure as a narrow fault belt we allocate to the approximately 3 Ma interval represented by the unconformity between the Lower and Upper Pontus series, across which there is also a marked decrease in mesofault abundance. Transformation from a shear zone to a fault belt need not have been precisely simultaneous along the length of the segment but much of it must have occurred in the Messinian.

Neotectonic deformation patterns

On small-scale maps (e.g. Fig. 1) the trace of the North Anatolian fault appears to be nearly continuous with interruptions only in the

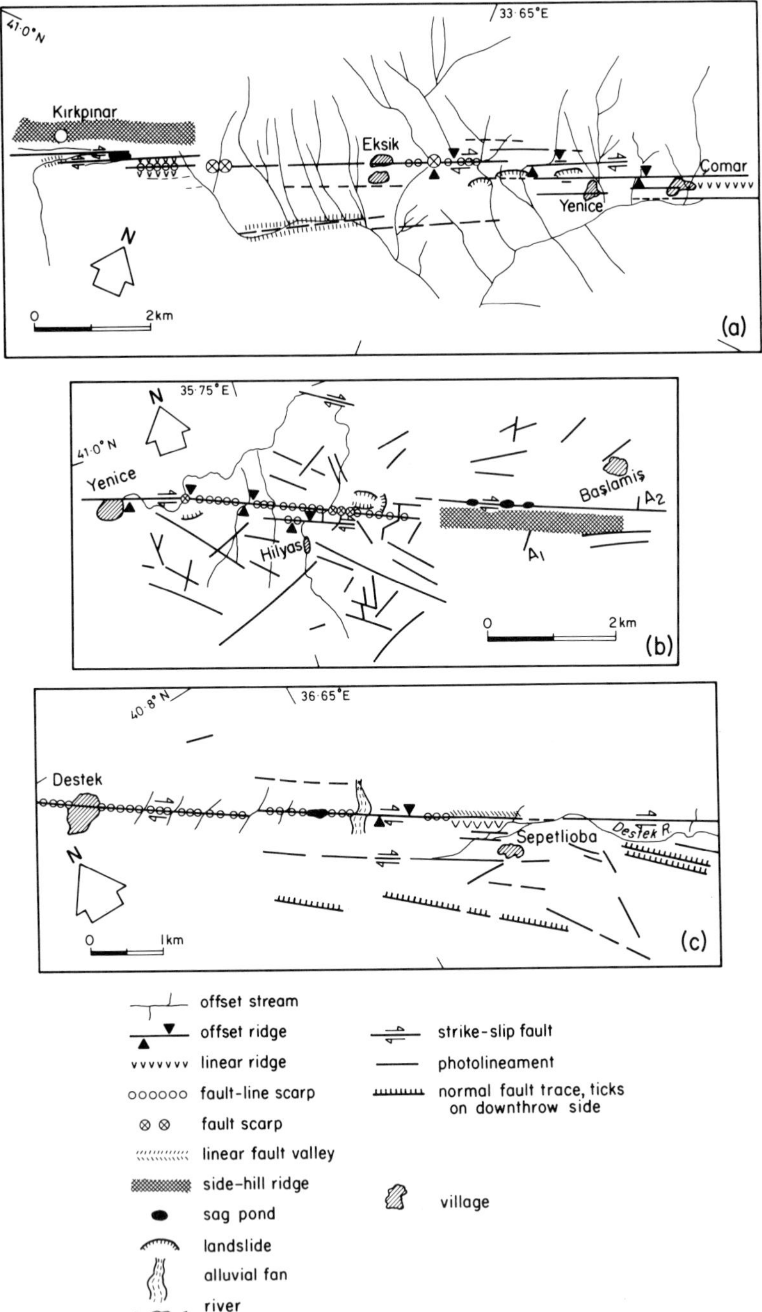

FIG. 6. Tectonic landforms associated with the active trace of the North Anatolian fault. (a) Between Kırkpınar and Çomar in pre-Neogene basement rocks north of the Çerkeş-Ilgaz basin; (b) between Yenice and Başlamiş in the Havza-Ladik basin (A1/A2 offset photolineament); (c) between Destek and Sepetlioba in the Taşova-Erbaa basin.

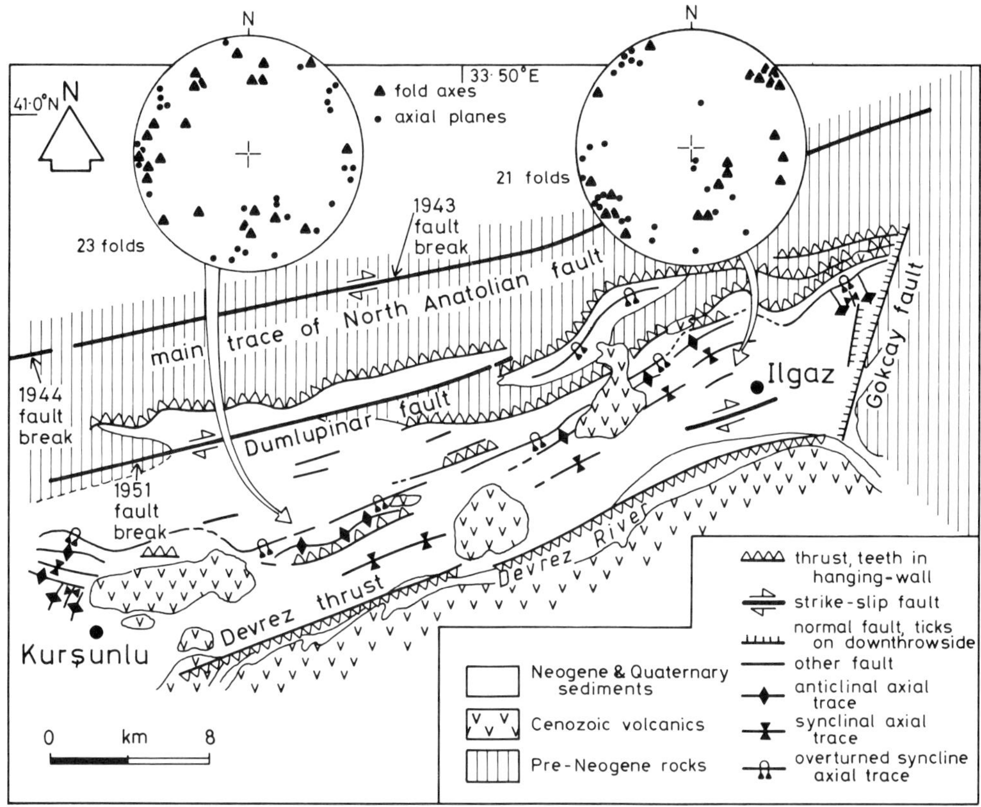

FIG. 7. Structural map of the Kurşunlu-Ilgaz area at the eastern end of the Çerkeş-Ilgaz basin. Insets are equal-area point diagrams of poles of fold axes and axial planes. Fault breaks after Ketin (1969, Figs 6 & 7).

Erbaa-Resadiye and Erzincan areas (Şengör 1979). In detail however, the main trace, as interpreted from aerial photographs, is discontinuous, comprising en échelon segments arranged in either Riedel or anti-Riedel patterns (Figs 6 & 9a). The sites of anti-Riedel offsets, commoner than Riedel offsets, are locations where sag ponds are to be expected. Individual segments of fault trace are generally straight and, as discussed by Barka & Hancock (1982), some earthquake epicentres are located where segments of different trend meet. In addition to segments being offset from each other the relatively narrow fault belt also contains abundant subordinate active faults whose en échelon organisation is compatible with their development as a consequence of right-lateral shear.

Our mechanical interpretation of the subordinate faults uses the genetic earthquake-rupture nomenclature proposed by Tchalenko & Ambraseys (1970), with the modification that normal faults, striking perpendicular to the secondary direction of stretching replace tension fissures in their scheme. Throughout Figs 9 (b-e) the main trace is taken to be a displacement shear (D). Of the two possible sets of Riedel shears only one, the R rather than the conjugate R_1 set is present, and even these right-lateral strike-slip faults subtending small angles with the main trace (Figs 9b & c) are relatively uncommon, especially in the western sector of the studied zone. P shears, displaying the same sense of motion as the principal fault and commonly regarded as forming when there has been substantial displacement, are more abundant than Riedel shears, especially between Tosya & Kargı (Figs 9 b-e). Northwest to southeast striking normal faults related to a secondary direction of horizontal stretching are likewise commonest between Tosya & Kargı (Figs 9b, c & e).

Northwest to southeast striking reverse mesofaults and northeast to southwest striking normal mesofaults, interpreted by Hancock & Barka (1980, 1981) as products of regional or

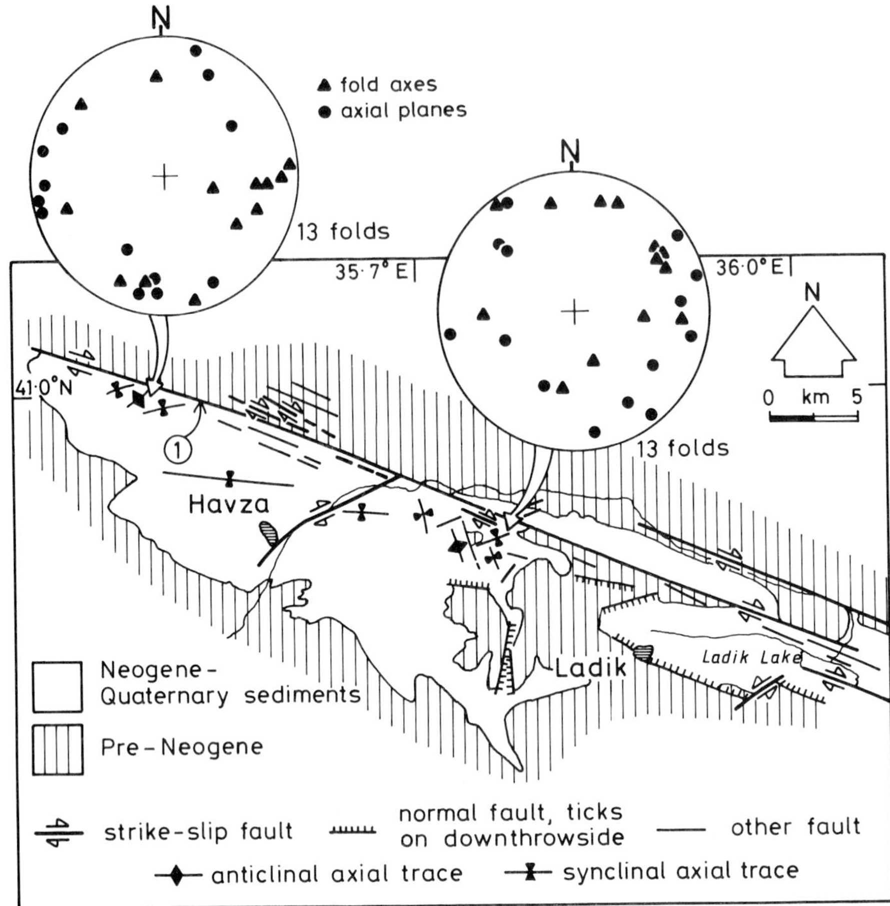

FIG. 8. Structural map of the Havza-Ladik basin. Insets are equal-area point diagrams of poles of fold axes and axial planes.

local left-lateral shear during the Pliocene-early Pleistocene do not appear to have any large-scale counterparts within the North Anatolian zone. The tectonic significance of these enigmatic structures has been the subject of further discussion by Hempton (1982), Şengör *et al.* (1983) and Hancock & Barka (1983) who conclude that they are unlikely to be related to episodes of eastwards motion of the Anatolian *scholle* but do not agree on an alternative explanation. Moreover, it should be appreciated that some of the anomalous mesofractures may be accommodation structures in the wall rocks of larger listric faults comparable to those described from the Aegean region by Jackson *et al.* (1982).

In the Havza-Ladik basin at least one sector of the 1943 earthquake fault-break is located within an approximately 1 km wide zone that has been the site of persistent deformation since the deposition of the Lower Pontus series (Fig. 10). Within the zone the angular discordance between the Lower and Upper Pontus series exceeds 15° and the maximum tilt also exceeds 15°. On either side of this zone of maximum deformation there are zones in which the angular discordance is less but the maximum tilt is still greater than 15°. At more than 2 km from the fault break both the discordance and maximum tilt are less than 15°. Tilting probably occurred as a consequence of both rotation related to subordinate dip-slip and secondary shortening during strike-slip displacements.

Near Hilyas, about 3.5 km ESE of Yenice (Fig. 10), the 20 m wide 1943 earthquake fault belt is visible in an approximately 15 m high stream section exposing a tectonic mélange comprising fault-bounded lenticles of Lower Pontus series sediments and Holocene colluvium. Figs 11 (a-c) are cartoons illustrating

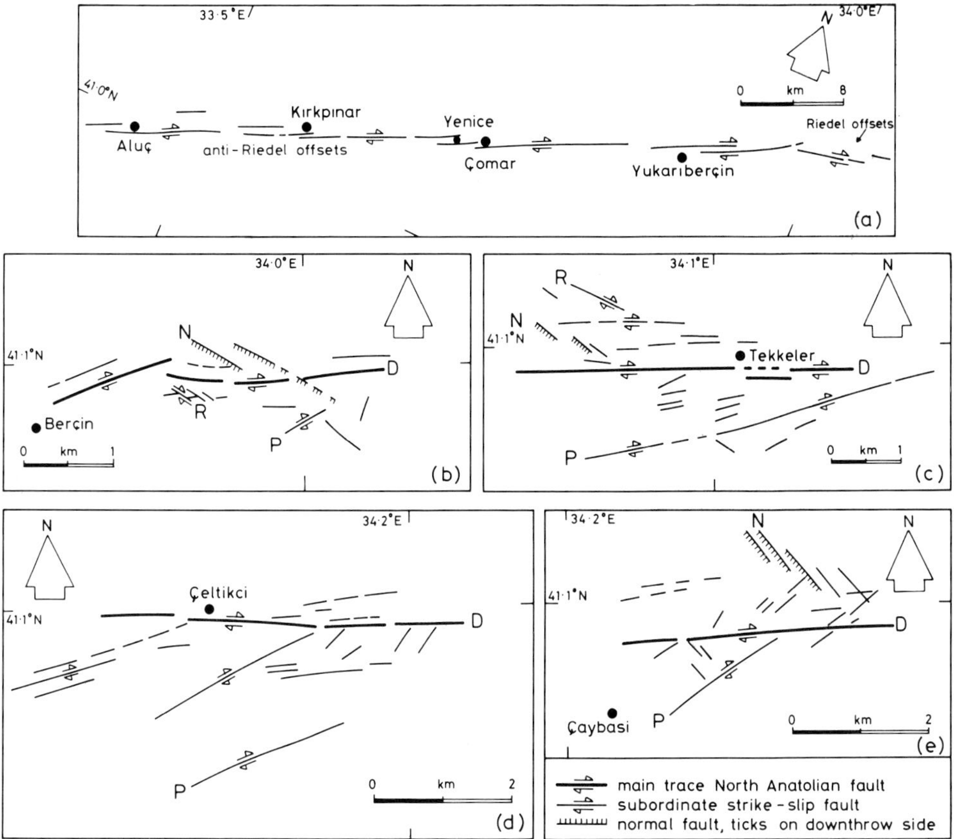

FIG. 9. Neotectonic fault patterns in pre-Neogene basement rocks; interpreted from aerial photographs of selected parts of the North Anatolian fault zone. (a) Riedel and anti-Riedel offsets of the main trace north of the Çerkeş-Ilgaz basin; (b) subordinate faults northwest of Tosya; (c) subordinate faults north of Tosya; (d) subordinate faults northeast of Tosya; (e) subordinate faults westsouthwest of Kargı. D, main trace, displacement shear; P, P shear; R, Riedel shear; N, normal fault (see text for mechanical interpretation).

how the mélange may have evolved as a result of suitably orientated shears in the anastomosing network of pre-existing faults in the Lower Pontus series having dilated obliquely during subsequent displacements along the zone. Many of the vertical or nearly vertical fault planes in the mélange are horizontally or subhorizontally striated, grooved and corrugated (Fig. 11d), providing additional evidence that recent displacements have been dominantly horizontal. At the distal ends of some grooves the unruptured pebbles or cobbles that acted as tools during tectonic erosion (scratching) are preserved.

Conclusions and discussion

(1) The evidence and arguments presented in this paper and contained in the cited literature allow the following history of the studied segment of the North Anatolian fault zone to be proposed.

Early-Middle Miocene. The Sivricek and Devrez Formations were deposited in the Çerkeş-Ilgaz and Tosya basins which were evolving in a compressional regime.

Late Miocene (*Tortonian*). The Lower Pontus series was deposited in basins within a broad shear zone responsible for widespread synsedimentary mesofaulting. The Çerkeş-Ilgaz and Tosya basins continued to be influenced by regional compression.

Latest Tortonian-earliest Pliocene. The broad shear zone developed during the previous episode was transformed into a narrow fault belt, mainly during the Messinian.

Pliocene-earliest Pleistocene. About 17 km of right-lateral displacement occurred along the

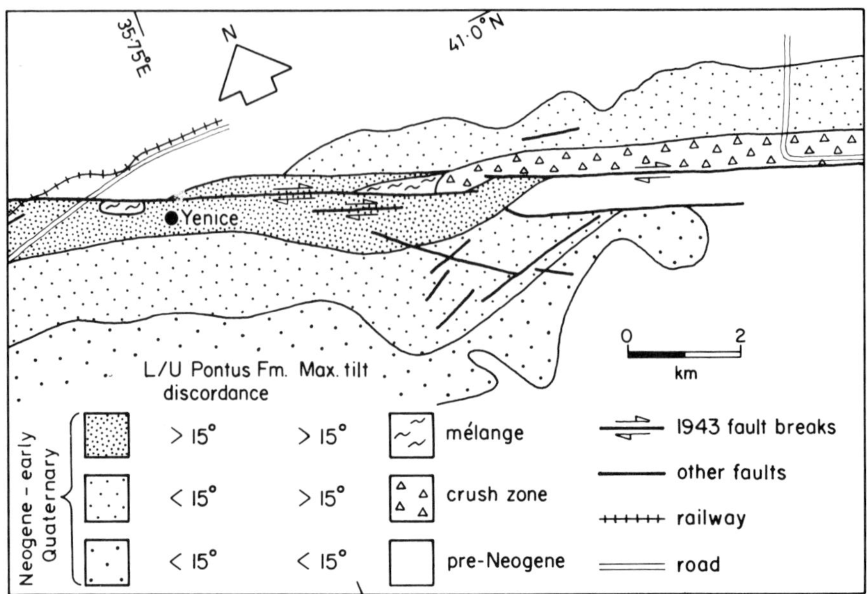

FIG. 10. Location of the 1943 Tosya earthquake fault-break (after Allen 1969, Fig. 4; Ketin 1969, Fig. 6B) in relation to relative deformation intensity within the Havza-Ladik basin.

fault belt during deposition of the Upper Pontus series.

Quaternary. During the Quaternary there has been about 8 km of displacement, possibly 2 km of it in the late Quaternary and 500 m in the Holocene. Alluvial and colluvial deposits continued to accumulate during displacements.

(2) If the net displacement since the latest Miocene is 25 km it follows that the average slip rate has been 0.38 cm/yr. The rate appears to have accelerated only slightly in the recent past being only 0.44 cm/yr since the beginning of the Quaternary. These slip rates are less than might be anticipated on the basis of evidence from elsewhere in the Eastern Mediterranean region.

(3) The organization of neotectonic subordinate faults in a narrow belt close to the main trace of the North Anatolian fault is consistent with an overall sense of right-lateral shear. The abundance of P shears in some regions provides additional support for the proposal that there has been a substantial net displacement. Although the distribution of different types of subordinate faults is not influenced by whether the main trace is contained within outcrops of pre-Neogene basement rocks or Neogene-Quaternary sediments, their distribution was controlled by location within the convex-northwards arc. P shears and en échelon normal faults are most abundant between Tosya and Kargı, that is, close to the northern 'crest' of the arc but on the side where there may be a component of convergence.

(4) Because field relationships in the Havza-Ladik basin suggest that there has been 25 km of displacement since the deposition of the Lower Pontus series it is necessary to advance some hypotheses to explain why this displacement is so much less than the 85 km offset of the Pontide-Anatolide suture near Erzincan, about 150 km ESE of Havza (Seymen 1975; Şengör 1979). Two proposals can be made: firstly, as explained earlier, a proportion of the total displacement may have occurred along a broad shear zone that was active before and during accumulation of the Lower Pontus series; and secondly some of the excess 60 km of displacement may have been accommodated on one or more convex-northwestwards faults similar to those shown by Şengör & Yilmaz (1981, Fig. 6I) as branching from the main trace and entering central Anatolia. Şengör *et al.* (1982) also report a decrease of displacement from 85 ± 5 km in the east to 20 km along the westernmost sector of the North Anatolian fault.

ACKNOWLEDGEMENTS: The M.T.A. Institute of Turkey and the University of Bristol contributed towards some of our field expenses and the M.T.A. provided a scholarship for one of us (A.A.B.). Celâl Şengör and James Jackson criticized an early draft of the paper and made numerous helpful suggestions. Alma Gregory and Jean Bees drafted the figures.

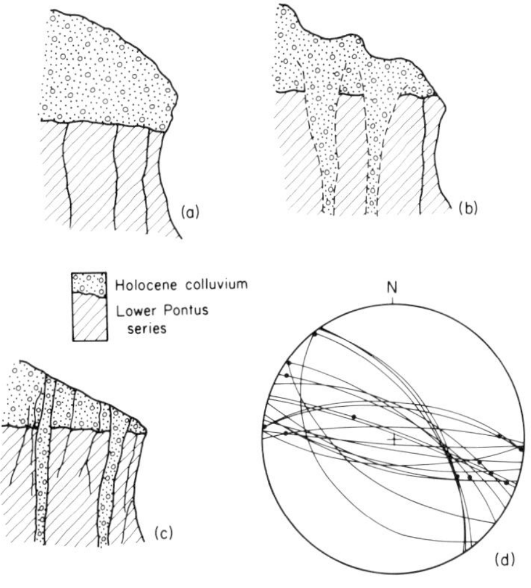

FIG. 11. Structures in the Hilyas mélange, Havza-Ladik basin. (a-c) Cartoons of mélange evolution, the sense of motion being nearly perpendicular to the page. (a) Tilted Lower Pontus rocks cut by subparallel anastomosing shears which developed before accumulation and partial erosion of colluvium. (b) Fissures formed by the dilation of those pre-existing shears which were orientated such that during a later episode of displacement there was oblique exension across them. Surfaces parallel to the slip vector, or across which there was oblique convergence, were resheared but not dilated. Dilated fissures were filled by colluvium from above. (c) Fissure margins and overlying colluvium sheared as displacements continued during the same or a subsequent event. (d) Cyclographic traces of shear planes in the Hilyas mélange; poles represent striations or grooves formed by clasts acting as tectonic tools on shear planes.

References

ALLEN, C. R. 1969. Active faulting in northern Turkey. *Contr. No.* 1577 *Div. Geol. Sci., Calif. Inst. Tech.* 1–32.

—— 1975. Geological criteria for evaluating seismicity. *Bull. geol. Soc. Am.* **86,** 1041–57.

AMBRAYSEYS, N. N. 1970. Some characteristic features of the North Anatolian fault zone. *Tectonophysics*, **9,** 143–65.

AYDIN, A. & NUR, A. 1982. Evolution of pull-apart basins and their scale independence. *Tectonics*, **1,** 91–105.

BARKA, A. A. 1981. *Seismo-tectonic aspects of the North Anatolian fault zone.* Thesis, Ph.D. (unpubl.). University of Bristol, England, 335 pp.

—— & HANCOCK, P. L. 1982. Relationship between fault geometry and some earthquake epicentres within the North Anatolian zone. *In:* IŞIKARA, A. M. & VOGEL, A. (eds.) *Multidisciplinary Approach to Earthquake Prediction.* Vieweg, Braunschweig/Wiesbaden, 137–42.

BERGOUGNAN, H. 1975. Relations entre les édifices pontique et taurique dans le nord-est de l'Anatolie. *Bull. Soc. géol. Fr.*, 7 *Ser.* **17,** 1045–57.

CANITEZ, N. 1973. Yerkabuk hareketlerine iliskin calismalar ve Kuzey Anadolu fayi problemi. *In: Symposium on the North Anatolian Fault and Earthquake Belt.* M.T.A. Enstitüsü, Ankara, 35–58.

DEWEY, J. F. & ŞENGÖR, A. M. C. 1979. Aegean and surrounding regions: complex multiplate and continuum tectonics in a convergent zone. *Bull. geol. Soc. Am.* **90,** 84–92.

FAHLBUSCH, V. 1981. Miozän und Pliozän—was ist was? *Mitt. Bayerischen Sta. Palaeön. Geol.* **21,** 121–27.

HANCOCK, P. L. & BARKA, A. A. 1980. Plio-Pleistocene reversal of displacement on the North Anatolian fault zone. *Nature, Lond.* **286,** 591–94.

—— & —— 1981. Opposed shear senses inferred from neotectonic mesofracture systems in the North Anatolian fault zone. *J. struct. Geol.* **3,** 383–92.

—— & —— 1983. Tectonic interpretations of enigmatic structures in the North Anatolian fault zone. *J. struct. Geol.* **5,** 217–20.

HEMPTON, M. R. 1982. The North Anatolian fault and complexities of continental escape. *J. struct. Geol.* **4,** 502–4.

IRRLITZ, W. 1971. Neogene and older Pleistocene of the intramontane basins in the Pontic region of Anatolia. *Newsl. Stratigr.* **1,** 33–5.

—— 1972. Lithostratigraphic und tektonische Entwicklung des Neogens in Nordostanatolien. *Beih. geol. Jb.* **120,** 1–111.

JACKSON, J. A., KING, G. & VITA-FINZI, C. 1982. The neotectonics of the Aegean: an alternative view. *Earth planet. Sci. Lett.* **61,** 303–18.

KETIN, I. 1979. Uber die Nordanatolische Horizontalverschiebung. *Bull. Miner. Res. Explor. Inst., Ankara*, **72,** 1–28.

KOPP, K.-O., PAVONI, M. & SCHINDLER, C. 1969. Geologie Thrakiens IV: das Ergene Becken. *Beih. geol. Jb.* **76,** 1–136.

LUTTIG, G. & STEFFENS, P. 1976. *Explanatory Notes for the Palaeogeographical Atlas of Turkey from the Oligocene to the Pleistocene.* Bundesanstalt für Geowissenschaften und Rohstoffe. Hannover, 164 pp.

MCEWEN, T. J. 1981. Brittle deformation in pitted pebble conglomerates. *J. struct. Geol.* **3,** 25–37.

MCKENZIE, D. 1972. Active tectonics of the Mediterranean region. *Geophys. J. R. astr. Soc.* **30,** 109–85.

PAVONI, N. 1961. Die Nordanatolische Horizontalverschiebung. *Geol. Rdsch.* **51,** 122–39.

ŞENGÖR, A. M. C. 1979. The North Anatolian fault:

its age, offset and tectonic significance. *J. geol. Soc. Lond.* **136,** 269–82.
—— & Yilmaz, Y. 1981. Tethyan evolution of Turkey: a plate tectonic approach. *Tectonophysics*, **75,** 181–241.
——, Burke, K. & Dewey, J. F. 1982. Tectonics of the North Anatolian transform fault. *In:* Işikara, A. M. & Vogel, A. (eds.) *Multidisciplinary Approach to Earthquake Prediction.* Vieweg, Braunschweig/Weisbaden, 3–22.
——, Büyukaşikoğlu, S. & Canitez, N. 1983. Neotectonics of the Pontides: implications for 'incompatible' structures along the North Anatolian fault. *J. struct. Geol.* **5,** 211–16.
Seymen, I. 1975. *Kelkit Vadisi Kesiminde Kuzey Anadolu fay zonunum tektonik ozelliği.* Doctoral thesis, Technical University of Istanbul.
Tchalenko, J. S & Ambraseys, N. N. 1970. Structural analysis of the Dasht-e Bayaz (Iran) earthquake fractures. *Bull. geol. Soc. Am.* **81,** 41–60.
Tokay, M. 1973. Kuzey Anadolu fay zonunum Gerede ile Ilgaz arasindaki kisminda jeolojik gozlemler. *In: Symposium on the North Anatolian Fault and Earthquake Belt*, M.T.A. Enstitüsü, Ankara, 19–29.

A. Aykut Barka, M.T.A. Enstitüsü, Ankara, Turkey.
Paul L. Hancock, Department of Geology, University of Bristol, Queen's Building, University Walk, Bristol BS8 1TR, England.

The Western Arabia rift system

A. M. Quennell

SUMMARY: The origin of the Dead Sea rift has generally been linked with that of the Red Sea, the widening of the latter involving left-lateral strike-slip on the rift. Its extension through the Lebanese fold belt has been denied by some because of the absence of linear and displacement continuity. Others, however, have been unable to see any other choice in spite of these difficulties, but a satisfying structural model has not yet been offered. The present author claims that the rift system is a transform plate boundary between the Arabian plate and the Sinai-Levant plate, but without uniformity along its length.

It is here suggested that during the Lower Miocene the Arabian planation surface, which before the opening of the Red Sea was the probable extension of the African mid-Tertiary surface, was ruptured by the first phase of movement on the west Arabian transform fault system, which probably reflected a geosuture zone in the continental plate and was originally a simple arc. The Arabian plate moved northward with the opening of the Red Sea and the closing of the Bitlis ocean, leaving behind the Sinai-Levantine part of the African Plate. Oblique compression on the northern Syria unstable platform across the East Anatolian transform fault caused dextral distortion on the Lebanon-Palmyra zone to create mountain ranges of two styles and a translation in the alignment of the rift segment. This inhibited further uniform strike-slip movement on the rift and displacement on the Yammouné fault was subsequently predominantly vertical. The movements on the rifts to the south and the north thus became independent. There was Arabian plate lithosphere consumption north-east of the Houlé depression to accommodate the second phase of rift movement. On the basis of this explanation, the vexed question of transmission of left lateral strike-slip across the Lebanon segment does not arise.

Lartet in 1869 recognised the Red Sea as having been formed by the separation of Arabia from Africa, accompanied by complementary shear movement along the Gulf of Aqaba and the Dead Sea rift system. The description accepted without much question has been an asymmetrical unilateral rift, a left-lateral strike-slip fault zone. The question of how far north the rift extended was not given much attention.

Dubertret (1932) accepted Lartet's model as a working hypothesis. The formation of the Red Sea presented no problem to him at this stage, and he carried the rift northward via the faults and folds of the Lebanon as far as the Taurus fold belt. However, with more extensive mapping of the Lebanon and Syria and especially after having accepted the thesis of Drake & Girdler (1964) that the Red Sea was floored by continental crust, he abandoned his view (1970) but realised that there is no simple path for the Dead Sea rift beyond the Houlé depression. However, he did propose a possible pre-Cenomanian movement related to the Roum fault. Dubertret had an intimate and detailed knowledge of the geology of the Lebanon second to none.

Quennell (1958, 1959) revived the hypothesis abandoned by Dubertret and quantified the major shift along the Dead Sea rift segment of the rift as having taken place in two stages, 62 km pre-Miocene and 45 km in Pleistocene times. He related more than ten homologous structural and stratigraphical features, and noted the displacement of the northern coast of the Red Sea and the opening of the Dead Sea as a rhomb-graben in two stages. He foresaw difficulty in the disposal of the excess crust arising from the 107 km displacement.

Dubertret's conclusion regarding extension through the Lebanese folds was ignored by Freund *et al.* (1970) who held that a solution could be found in accepting the Yammouné and Gharb faults as the northern extension of the rift system. They matched not only occurrences and structure as Quennell had done, but also columnar sections. However, their endeavour to extend this technique beyond the north Jordan valley to match Lebanese features such as Mt. Hermon and to solve the divergences in direction and anomalies in amount of displacement for the Yammouné and the Gharb faults, is not convicing. To account for the bend of the Yammouné fault they introduced oblique and lateral shift of the bordering blocks. They also proposed to use a suggestion of offset of ophiolites in the far north to account for some of the total strike-slip movement.

In his 1958 and 1959 papers, Quennell did no more than speculate about the northern extension but assumed as others had done, that the fault path would be found to be the faults and folds in the Lebanon.

The rift system

The rift system extends from the Red Sea speading zone to the plain of Antioch where it is accepted as joining the Border Zone (Fig. 1). There are three segments: Red Sea to Lebanese border; across the Lebanese fold belt; from the northern Lebanese border to Turkey.

Gulf of Aqaba, Wadi Araba and Jordan Valley

The first segment is determined by the circular arc of the boundary between the Arabian and Sinai-Levant plates. The reality of this arcuate trace of a generally concealed left-lateral strike-slip fault is well established: by photogeology and field geology in West Araba and the Jordan valley (Quennell, 1955–6, 1959); in Wadi Araba by mapping of sediments (Bender, 1974b, 1982); and from Landsat imagery (Fig. 2). There is rarely more than minor departure (1 or 2 km) from the arc which was described by trial and error originally on the scale of 1:250,000. The pole of rotation was fixed at Lat. N. 33; Long. E. 24 (Drake & Girdler, 1982).

A later contribution detailing the pattern of the Wadi Araba segment has been made by Garfunkel *et al.* (1981). This can be accepted as the definitive description. In considering irregularities in the course of the fault trace (named the er Risha fault by Quennell, 1959) it is recognised that the fault 'plane' above the plate margin was in sedimentary fill, generally unconsolidated.

The Gulf of Aqaba (Fig. 1-inset) has been described by Ben-Avraham *et al.* (1979) and by Garfunkel (1981), but without recognition that the circular arc of the plate boundary was projected to the south and that the gulf was oblique to the arc. In Fig. 1-inset the process of initiation of the gulf and its history in terms of that of the Rift system (Quennell, 1959) are set

---

Fig. 1. Geology and structure (simplified) of the West Arabian rift system from the Red Sea to the Eastern Taurus, comprising: Aqaba (Elat) Gulf; Dead Sea transform; Yammouné fault zone; Gharb fault which appears to meet the Amanos Fault of the Border Zone fault system. The Arabian plate comprises: the Arabian stable shelf; Lebanon—Palmyra fold belt; and the Syrian unstable shelf. On the west of the rift system is the Sinai—Levantine plate, an extension of the Nubian plate. The Arabian Rift system is the inter-plate boundary, while the southern Palmyra fault zone and possibly also the northern fault zone are intra-plate boundaries. Geology is after: Dubertret (1955), Ponikarov (1964), Wolfart (1967), Picard (1959), Quennell (1959), Bender (1968, 1974) and Ben Avraham *et al.* (1979). Both north and south of the Dead Sea rhomb-graben are the circular arcs of the plate boundary. The location of the pole of rotation was geometrically determined (Quennell 1959; Drake & Girdler, 1982). The Lake Tiberias rhomb-graben (Garfunkel 1981) is terminated against the Lebanon-Palmyra fold belt. The Dead Sea transform has a total sinistral displacement of 107 km. The Yammouné fault zone (Hancock & Atiya, 1979) extends across the belt, and is the only fault to do so. Sinistral displacement is of the order of 7–10 km. The el Gharb sinistral fault system containing the Gharb rhomb-graben may extend the Yammouné and is assumed to join the sinistral Amanos fault beneath the alluvial plains north of Antioch. It may have a sinistral displacement of 50 km (based on evidence of the rhomb-graben). The el Gharb fault separates the northern extension of the faulted and folded Sinai-Levantine plate from the Syrian unstable platform. The Precambrian crystalline continental crust disappears northward near Elat on the west of the rift, and on the east of the rift near the Dead Sea. Exposed cover formations, Palaeozoic to Neogene, generally marine, decrease in age northwards. The Harrat ash Shamah flood basalts, Miocene to Recent, are noteworthy.

Fig. 1 (inset). Development of the Gulf of Aqaba (Elat) and its prolongation northward as the Wadi Araba. The mechanism is the anti-clockwise rotation of the stable shelf part of the Arabian plate along the circular arc transform interplate boundary (Quennell, 1959; and Garfunkel, 1981). A zone of fracturing, stage A and a-a_1, probably extensional and normal, resulted from pre-Neogene tectonism with the formation of the Raham conglomerate (Garfunkel *et al.* 1974) and preceded the initiation of the transform interplate boundary. Rotation of the southern Arabian plate by 62 km length on the circumference was to position, stage B, b-b_1, in the early Neogene. This same movement was resumed by 45 km to reach stage C, c-c_1, and is continuing. Sedimentation probably kept pace during first phase with enlargement of the Gulf to stage B, but it has lagged behind the tectonic lengthening and widening of the gulf between stages B & C and extensional basins or rhomb-graben were formed (see figure and reference, Ben Avraham *et al.* 1979). The last phase lengthened the W. Araba to c_1. The head of the Gulf is a prograding beach. Restoration to stage A leaves the opposite Gulf margins of crystalline basement clear of each other. Impingement and overlapping of margins and submarine contours of the floor of the Gulf only involves younger sediments (see also Fig. 1b in Freund *et al.* 1970).

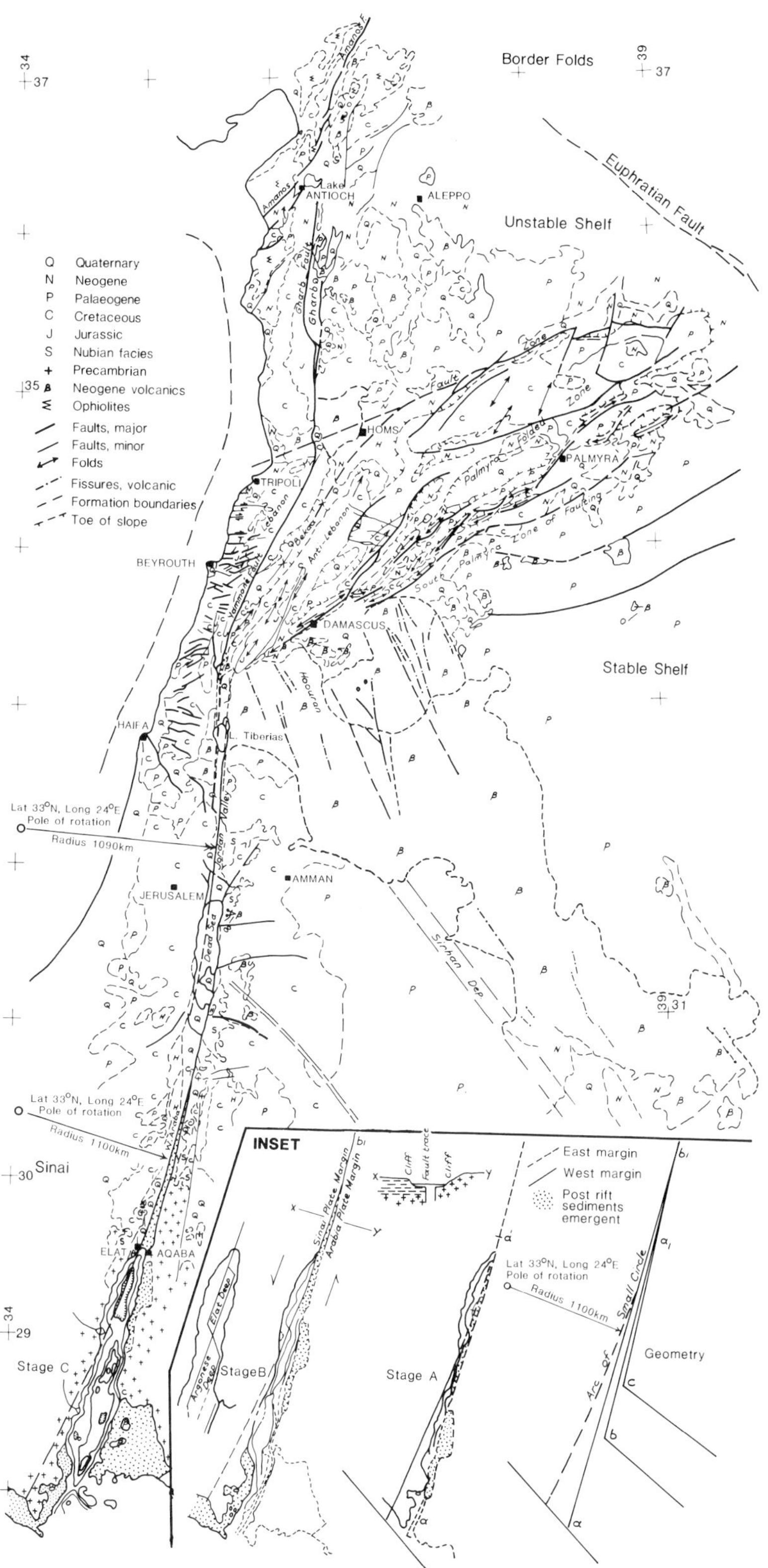
Border Folds
Euphratian Fault
Unstable Shelf
Stable Shelf
Q Quaternary
N Neogene
P Palaeogene
C Cretaceous
J Jurassic
S Nubian facies
+ Precambrian
β Neogene volcanics
Ophiolites
Faults, major
Faults, minor
Folds
Fissures, volcanic
Formation boundaries
Toe of slope
Lake ANTIOCH
ALEPPO
HOMS
TRIPOLI
PALMYRA
BEYROUTH
DAMASCUS
HAIFA
L. Tiberias
Lat 33°N, Long 24°E
Pole of rotation
Radius 1090km
JERUSALEM
AMMAN
Sirhan Dep
Lat 33°N, Long 24°E
Pole of rotation
Radius 1100km
Sinai
ELAT
AQABA
Stage C
INSET
Sinai Plate Margin
Arabia Plate Margin
Cliff
Fault trace
Cliff
East margin
West margin
Post rift sediments emergent
Stage B
Stage A
Lat 33°N, Long 24°E
Pole of rotation
Radius 1100km
Arc of Small Circle
Geometry

Fig. 2. Landsat photo covering the rift-system and Lebanon-Palmyra fold-belt. The area covered extends from the southern end of the Gharb depression in the north, to the Jordan valley south of Lake Tiberias. Compare with Fig. 1.

out. The initial form of the gulf (Stage A) has some agreement with that portrayed by Freund *et al.* (1970).

The influence of strike-slip movement on supposedly anastomozing faults in the Sinai block is important only if such faults can be seen to add materially to or subtract from total movement. This does not appear to be the case (Eyal *et al.* 1981; Bartov *et al.* 1980).

Extension of the gulf northward up the Wadi Araba involves a maximum depth of infilling sediment of perhaps 600 m between the plate margins (see also Quennell 1958). The prograding beach at the head of the gulf indicates the process by which the Wadi Araba sedimentation proceeds at the rate of approx 0.2 m per year.

The flanks of the Wadi Araba show great contrast but it should be recognized that at any one locality they were originally separated by 107 km. For example the steep but dissected eastern flank north of Aqaba was formerly the eastern coast of the gulf about half way down and opposite the western coast, while the subdued western flank of dissected Mesozoic sediments with escarpments stood opposite the dissected flank a short distance south of the Dead Sea.

The Dead Sea as a rhomb-graben has been described initially by Quennell (1959) and later by others (Freund & Garfunkel, 1981).

The northern part of the Jordan valley and its relationship to the Lebanon-Palmyra fold belt is of greatest interest. The existence of a rhomb-graben containing part of the Jordan valley and L. Tiberias as well as the Houlé depression, was suggested by Quennell (1959). Freund *et al.* (1970) appeared to accept this. The main Jordan valley rift fault, concealed beneath lava, forms its eastern margin and the probable southern extension of the Yammouné fault appears to form the western margin, also concealed beneath lava flows and a very young sedimentary cover. Movement, all strike-slip, has taken place on the eastern marginal fault which carries northward the whole flank which here is the basalt lava flow probably hundreds of metres thick, resting on Cretaceous and Eocene marine sediments. These are seen 5 km to the south. Much of the sedimentary succession from further south no doubt extends northward beneath the lava (Wolfart, 1967).

There need have been no displacement on the southward prolongation of the Yammouné fault along the west of the rhomb-graben. This borders the passive block, the continuous Sinai-Levant plate. However, on the east is the Arabian platform. This moved 62 km northward and then a further 45 km, following uplift of the Mt Hermon massif. For the movement to have continued, the geometry of the rhomb-graben would require it to be transferred to the western fault, the Yammouné or the Roum.

What then has happened to 107 km of crust? This can best be considered in the context of the marked change in tectonic style when we encounter the Lebanon-Palmyra fold belt.

The Lebanon Segment

This segment can be described as the traversing of the Lebanon-Palmyra fold belt by the rift system. It is not simply to be equated with the Yammouné fault zone.

The northern end of the Houlé depression is a focus of great significance (Figs 1, 2, 3 & 4). Radiating from this focal area are various elements. Southward is the rhomb-graben which may extend as much as 100 km southwards. It is marked by a gravity low which may indicate a great thickness of infilling sediments. The Roum fault ends 40 km to the north where it disappears beneath a mask of Cenomanian near the southern end of the Lebanon range; the Yammouné fault follows a course partially masked by young sediments to form the west flanks of the Litani and Bekaa valleys. Two faults, the Rachaya and the Serrhaya, terminate northwards within the Mt. Hermon massif and the Anti Lebanon range. Mt. Hermon is bordered by fault scarps against which lava flows and young sediments abut. The vertical displacement is many thousands of metres. An estimation based on a stratigraphic column arrived at from a number of sources (Freund *et al.* 1970; Wetzel & Morton 1959; Parker 1969) suggests that the top of the Precambrian south of Mt. Hermon lies at 3000–4000 m and the Jurassic seen in Mt. Hermon could be at say, 2000 m below. On the west there is no such displacement feature. The sandstones of Triassic and Palaeozoic age, if they extend so far north, could be at no great depth beneath Mt. Hermon which has a fault relationship with all adjacent formations to the west and south-east. It may be significant that Mt. Hermon is the site of a gravity high which may indicate a rise in the denser layers (Fig. 4).

The Yammouné fault zone is well described by Hancock and Atiya (1979) (see also Garfunkel *et al.* 1981). Evidence for any strike-slip movement is described as slickenside grooves and other features. They quantify the left-lateral slip as 7 km based on the rhomb-graben. At its northern emergence from the Lebanon range there is an offset of about 10 km. It is obvious that this fault zone, despite being the *only* such feature completely

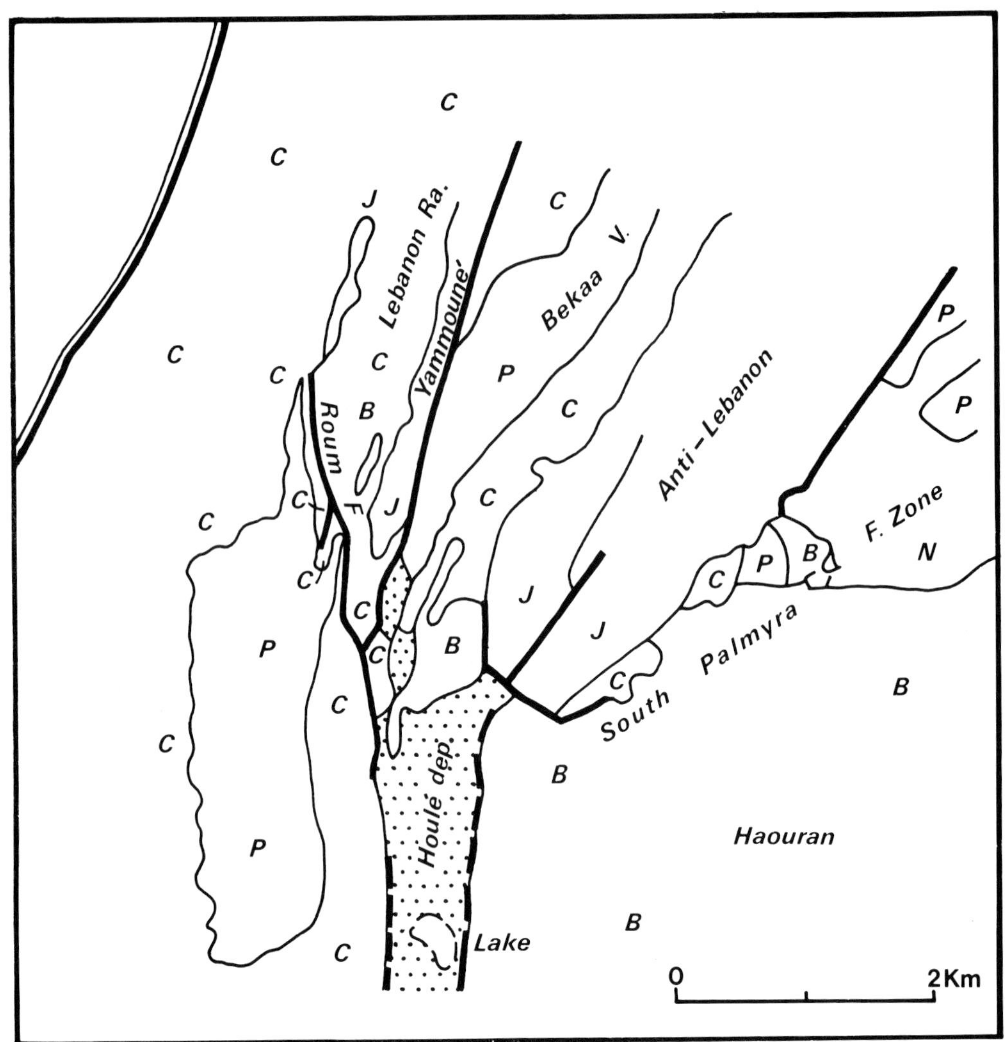

FIG. 3. Structural relations of the Dead Sea transform to faulting in Lebanon-Palmyra fold belt. Geology and faulting after Dubertret (1945; 1947; 1955), Garfunkel *et al.* (1981); Landsat image (1972). The Houlé depression is the northern part of the L. Tiberias rhomb-graben and ends beneath a basalt flow (B). The course of the Yammouné fault southward is not clear but it is accepted as being probably continuous with the west flank of the rhomb-graben. The branching Roum fault ends northward where it passes under Cretaceous marine beds (C) and there may be stream erosion simulating strike-slip movement. Faults within the Anti-Lebanon range die out, but the NE-trending fault on the east of the range continues to the NE and separates the Palmyra from the Lebanon style of folds. The South Palmyra fault-zone (Ponikarov 1964) may be an intra-plate boundary. The main Dead Sea transform fault apparently ends as the eastern flank of the Houlé rhomb-graben, but this is obscured by young basalt flows (B).
J = Jurassic sediments; P = Pleistocene deposits.

FIG. 4. Seismic and gravity data related to structure.
Gravity data from sources listed varies in reliability and value. The isogals at 10 mgal interval generally confirm disposition of crustal masses and thicknesses expected from the geology. Note: (a) the steep gradient acrosss zone west of Dead Sea transform; (b) the less steep zone across the Yammouné fault zone and Lebanon fold-belt; (c) contrast between areas of Lebanon and Palmyra styles of folding; (d) undisturbed area between Palmyra fold belt and Turkish Border Folds; (e) gravity information from Jordan could not be used but is probably as for (d). Earthquake data: Note (a) important epicentre localities are located at both ends of Dead Sea rhomb-graben; (b) virtual absence of macroseisms along Wadi Araba, but microseisms lie west of or on the plate margin of Sinai (Araba fault trace). Seven fault plane solutions (to 1976) are recorded in the Palmyra belt or close to the Dead Sea transform (Ben Menahem *et al.* 1976; Ben Menahem & Aboodi 1981). All are left-lateral strike slip.

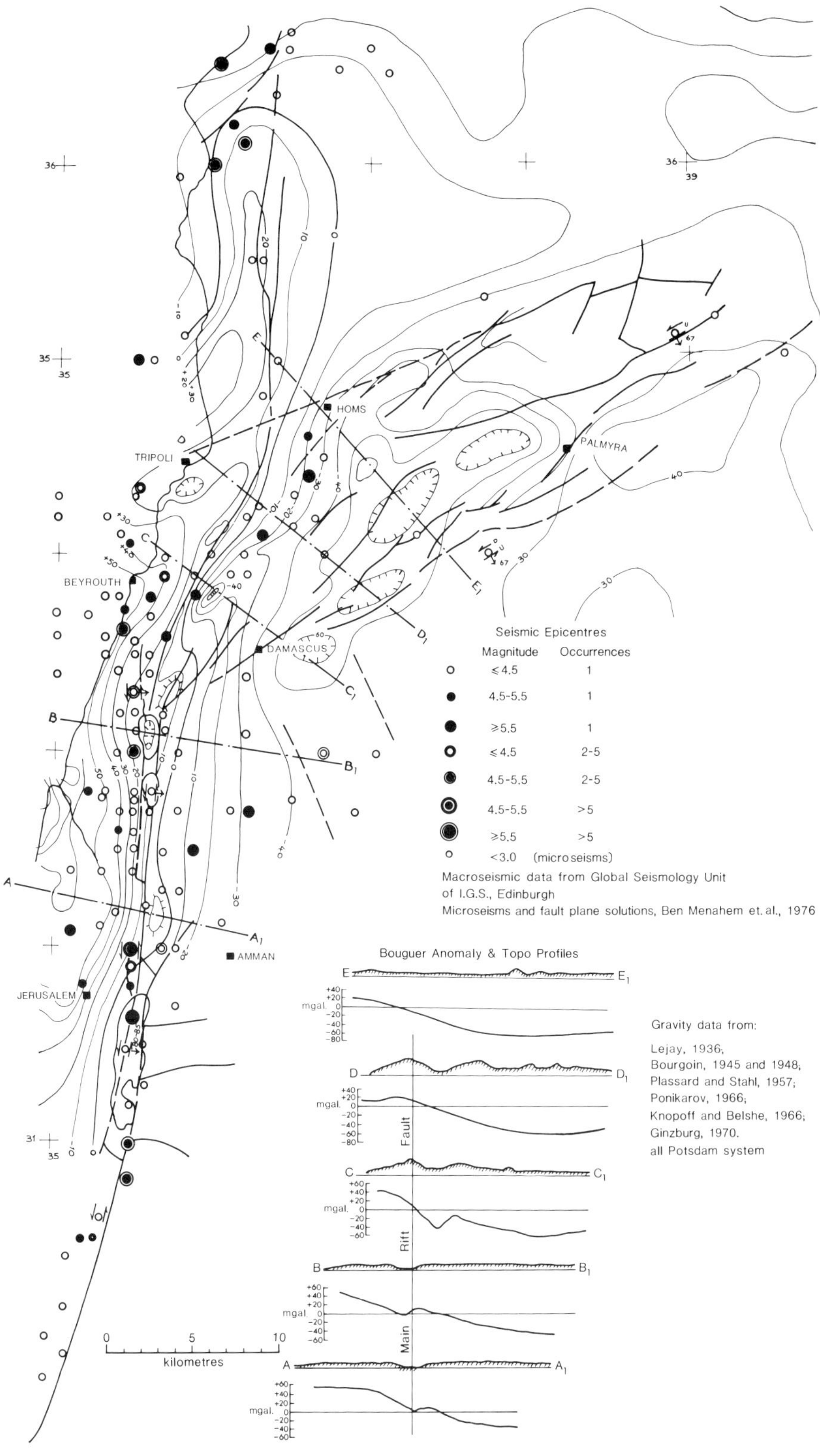
HOMS
PALMYRA
TRIPOLI
BEYROUTH
DAMASCUS
AMMAN
JERUSALEM
Seismic Epicentres
Magnitude Occurrences
≤4.5 1
4.5-5.5 1
≥5.5 1
≤4.5 2-5
4.5-5.5 2-5
4.5-5.5 >5
≥5.5 >5
<3.0 (microseisms)
Macroseismic data from Global Seismology Unit
of I.G.S., Edinburgh
Microseisms and fault plane solutions, Ben Menahem et. al., 1976
Bouguer Anomaly & Topo Profiles
mgal.
Main Rift Fault
Gravity data from:
Lejay, 1936;
Bourgoin, 1945 and 1948;
Plassard and Stahl, 1957;
Ponikarov, 1966;
Knopoff and Belshe, 1966;
Ginzburg, 1970.
all Potsdam system
kilometres

traversing the Lebanon fold belt, cannot be an adequate representative of the Dead Sea rift zone, as it has a trend of about 30° from that of the north Jordan and a totally inadequate net slip.

South Palmyra zone of faulting

Along this zone the Anti Lebanon open folds give way near Damascus to Jura-type folds trending north-east. The folds are partially overturned to the south-east and this flank is generally faulted (Fig. 1). The zone extends beyond Palmyra. The basalt flows and young masking sediments abut against the folds for 200 km along the zone where intensity of folding begins to decrease until at about 400 km the fold belt disappears.

It appears that a corner of the crustal slab, a triangle, measuring 45 km along its western side and as much as 400 km along its northern edge (Fig. 1) has been underthrust and consumed in some manner.

Lebanon-Palmyra fold belt

Probably accompanying the underthrusting by the Arabian platform, the Lebanon-Palmyra fold belt, confined between the Arabian and north Syrian platforms, was apparently subject to a clockwise rotational couple (pure shear) producing a strain rhomb (Fig. 5). There has also been compression from south to north when the movement of the Arabian plate was checked by meeting the Van plate. The axes of folds would be re-oriented clockwise and they would be lengthened and narrowed. The contrast between the Lebanon and Palmyra folds is probably related to depth to basement (Fig. 4).

Where thickness of competent formations is greatest, the folds would have greatest amplitude and underthrusting would be the manner

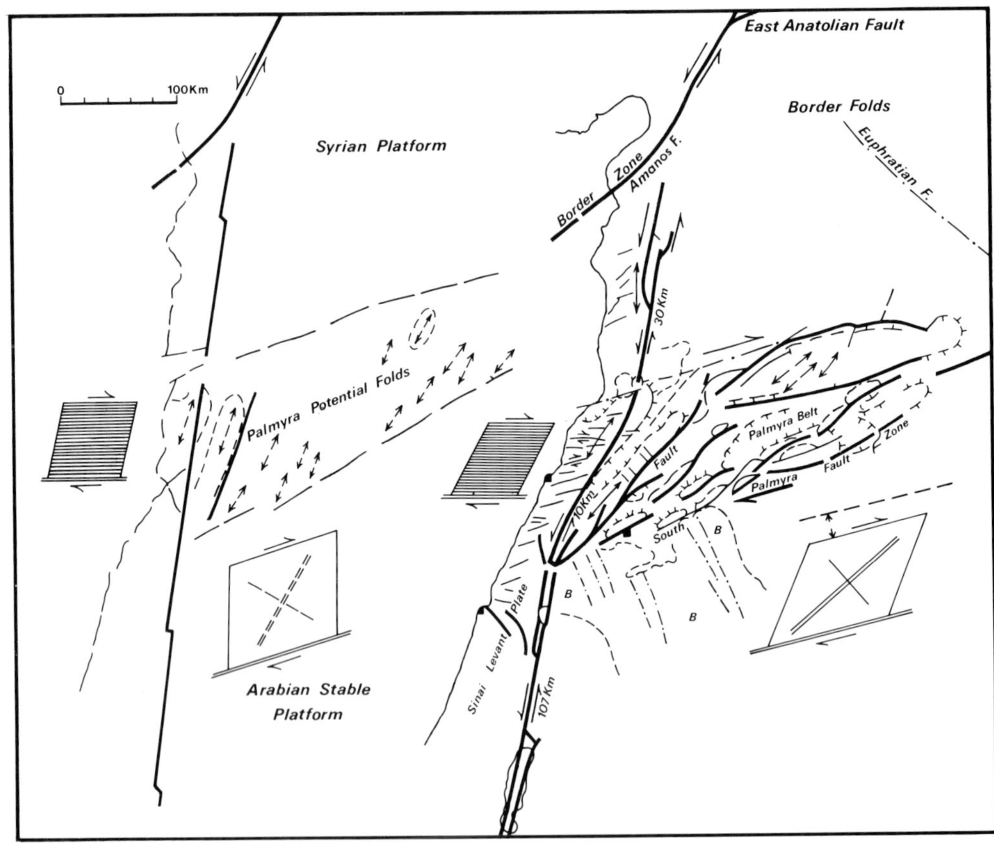

Fig. 5. Model for the generation of the Lebanon-Palmyra fold belt by oblique compression between the two platforms. Eastward movement of Syrian platform (note possible involvement of plate movement along East Anatolian transform) results in a rotational couple (pure shear) with NE-SW folds. Contrast in thickness of cover beds between Lebanon-Anti-Lebanon and Palmyra may account for differing styles and scales of folds. Superficial east-west right lateral strike-slip faults in Lebanese segment of Sinai-Levantine plate suggest simple shear west of Yammouné zone.

of disposal of the excess crust. Where the cover beds are thin-skinned and the basement shallower, detachment (décollement) would operate. Further, in this case the width of crust to be absorbed decreases to the north-east. The dividing line appears to be the fault east of the Anti Lebanon.

The segment of the Sinai-Levant plate, lying west of the Yammouné fault has apparently had imposed upon it, probably in a later phase, simple shear producing latitudinally orientated shear planes now reflected in the E-W faulting and other fold and fault patterns decribed by Hancock & Atiya (1979) (see also Palaeomagnetism below).

El Gharb segment

Where the Yammouné fault emerges from the Lebanon, it is deflected to the west. It is suggested that this may be a weak sigmoidal pattern caused by drag as the clockwise rotation (Fig. 5) of the fold belt took place. Here the fault zone disappears beneath a young sedimentary cover but it may link with the El Gharb transform fault which resumes the original trend.

The El Gharb fault described by Dubertret (1967) and by Ponikarov (1967) first traverses the Homs basalt flow of Pliocene age. The apparent strike-slip dislocation of the flow by about 20 km is post-Pliocene if the basalt is of that age (Ponikarov 1967).

The fault traverses Mesozoic formations, Jurassic on the west (the Jebel Alaouite) and Cretaceous on the east. The El Gharb depression (Dubertret 1967) commences on the northern flank of J. Alaouite. It has a length between the southern fault in Jurassic to the basalt flow and Palaeogene sediments in the north of about 50 km and a width of 10 km. The northern termination is ill-defined. Beyond this the el Gharb fault, continuing along the western flank of the depression, disappears beneath the Quaternary sediments of the plain of Antioch. There can be no simple junction with the Amanos (East Anatolian) fault.

Freund *et al.* (1970) attributed a sinistral movement of 70 km to the El Gharb fault on what they regarded as evidence from displacement of the ophiolite massifs, the Kurd Dagh on the east and the Bäer and Bassit on the west (Fig. 1).

The ophiolite masses, the 'roches vertes' of Dubertret, which are pre-Maastrichtian in age, belong to the Anatolian plate margin (see later) and lie within a zone containing these obducted bodies (Dewey *et al.* 1973). If indeed the El Gharb fault continues into this zone which is improbable, the Kurd Dagh lies well beyond its end and the relationship of the two masses cannot be used as Freund *et al.* have done to determine a displacement on the El Gharb fault of 70 km. This is therefore discounted.

Volcanics

The flood basalt volcanic activity of the Arabian stable shelf (Ponikarov, 1967; Barberi *et al.* 1980; Wolfart 1967; Burdon 1959) was chiefly centred on Jebel Druze 35 km south of Damascus (Fig. 1). The general name of the plateau is Ash Shamah while the northern part is the Haouran. The flows form a wide zone of length 360 kms extending to the southeast, with a width of 160 km reducing to 50 km. This zone lies to the northeast of the parallel Sirhan depression both features manifesting NW-SE tensional fractures.

There are NW-SE eruptive fissures marked by vents. They have been well mapped in Syria and NE Jordan. In the northwest, close to the Houlé focus of structural features (Fig. 3), detailed mapping (Ponikarov 1966) reveals a number of flows of basalt, the earliest of which is mid-Miocene in age and the youngest are Recent (ages from Barberi *et al.*). The earliest flows were erupted on the 'mid-Tertiary' Arabian planation surface (later). The fissures have a general bearing 340° nearly normal to the south Palmyra zone of faulting. All volcanic rocks surveyed south of the Syrian border (Barberi *et al.* 1980) belong to the alkalic clan, generally undersaturated. They fall within the definition of alkali basalt, basanite, nepheline basanite and hawaiite.

There is no recorded volcanic activity within the Lebanon-Palmyra fold belt. It ends abruptly at the Palmyra fault zone. However, the trend of the fissures has considerable significance in relation to the directions of shear, tension and compression of the strain ellipse for the region (Quennell 1951, 1959; Burdon 1959).

The loading on the crust is considerable, the thickness at Jebel Druze being more than 1500 m, and this may have played a part in encouraging underthrusting.

The other basalt flow of significance is that lying across the Lebanon-Syrian frontier west of Homs (also known as the Shin field). It is approximately 45 × 55 km and the basalt is chiefly Pliocene but there is an early small flow of Upper Miocene (Ponikarov 1966).

The El Gharb fault, here N-S, commences south of the flow and with left-lateral strike-slip movement displaces the flow. Fault movement of about 20 km appears to post-date the volcanics, i.e. to be post-Pliocene. Although obscured by recent sediments this fault appears to extend the Yammouné fault.

Across the northern end of the El Gharb rhomb-graben, i.e. 50 km distant from the southern cross fault, is a flow of Pliocene basalt. This is not intersected by the eastern fault nor are the Palaeogene sediments. It appears the fault had ended.

This Homs basalt could have erupted from tension fissures. It lies in the area where the N 30° E trend of the Lebanon anticlinorium changes to the N-S trend of the northern end of the Sinai-Lebanon plate (Fig. 5), the El Gharb fault making the same change. This would favour tensile stress and the formation of tension fissures.

Palaeomagnetic research

This has been carried out by Van Dongen *et al.* (1967) on Mesozoic formations of the Lebanon Mountains west of the Yammouné fault zone and northeast of Beirut; and on Pliocene basalt flows near Tartous on the Syrian coast 40 km north of the Lebanon frontier. This was followed by a survey by Gregor *et al.* (1974) on adjacent but more restricted areas in the Lebanon mountains on Mesozoic formations, and on Upper Pliocene Homs basalt north of Lebanon and west of the Yammouné zone, as well as on basalt flow remnants in the south close to Mt. Hermon and east of the Yammouné fault. Sampling was therefore from four structurally distinctive areas of contrasting tectonic histories only the last of which belongs to the Arabian plate. The others are all on the Lebanese segment of the Sinai-Levantine plate.

For our purpose it is convenient to consider only the Mesozoic sites. Gregor *et al.* (1974) state that their results '. . . indicate a progressive anticlockwise rotation of Lebanon during late Mesozoic . . . times'. As this appears on geological and tectonic evidence highly improbable, testing by a separate approach is necessary.

If the mean directions for sites in Lebanon can be made by rotation to coincide with directions at sites in the Sinai-Levant (Africa) plate, then Central Lebanon can be restored to its original orientation (Fig. 5).

Relevant work was done by Helsley & Nur (1970) on samples from the Ramon asymmetric anticline in the Negev (folded since magnetization) and from near Mt. Carmel; and by Freund & Tarling (1978) from Ramon and near Jerusalem.

Results from these formations which are comparable stratigraphically give no rational solution. On first inspection for Lower Cretaceous declinations there would have been clockwise rotation of Central Lebanon from 125° to at least 155°; and for Jurassic from 95° to 172°, an improbable situation. If the plate tectonic model proposed (Fig. 5) is accepted then the palaeomagnetic data require a different interpretation from that of Gregor *et al.*

Freund & Tarling extend the discussion beyond our present objective. They criticize the sampling plan in Central Lebanon and note that results appear to group themselves according to locality. There is not the expected consistency from sites close to each other. They suggest that these inconsistencies in palaeomagnetic results '. . . may indeed record the structural deformation of this country.' They suggest that the present orientation of declinations may be linked with Freund's model of internal rotation of faulted blocks which is integral with his model for the Lebanon segment of the rift system (Freund *et al.* 1970). The system of latitudinally-trending faults, most of them dextrally strike-slip as mapped by Dubertret (1944) and Hancock & Atiya (1979), has been stated by the latter to have been accompanied by clockwise rotation. The latter give a description of northern Lebanon, especially of the areas sampled by Van Dongen *et al.* and by Gregor *et al.* They state '. . . the Mount Lebanon anticlinorium is divided into several compartments separated by narrow zones of E-W trending, nearly vertical, dextral faults' . . . 'A conjugate set of NW-SE faults . . . resulted in a clockwise sense of rotation.' The value of palaeomagnetic results on samples from what is a unique structural and meso-structural area are limited, and cannot support the view that Lebanon west of the Yammouné fault could be a microplate which behaved as did those of the Western Mediterranean.

Palaeomagnetic results from the Ash Shamah volcanic plateau in NE Jordan and from W. el Mojib, east of the Dead Sea, are reported in Barberi *et al.* (1980). There is a marked difference between results for the two sites but this is attributed to different positions on the plate. The Ash Shamah samples give palaeomagnetic pole results close to those for the Pliocene sampling in the Lebanon and also to results for Aden (Tarling *et al.* 1967). These results confirm the comments above on the Lebansese survey regarding value of the Mesozoic sampling without reference to a structural framework.

Geomorphology and tectonics

A single planation surface of mid-Tertiary age which is reasonably well preserved and identifiable on the Arabian stable platform east of the Dead Sea rift (the Arabian surface, Quennell 1958) and on the unstable shelf of NW and SE Syria (Syrian plateau) can be recognised (Burdon 1959; Wolfart 1967). It has suffered distortion and erosion on the west and north-east. There are few residuals other than inselbergs and monadnocks, which occur mostly in the south-east. The fold mountains of the Lebanon-Palmyra fold belt and the Judean and Ajlun arches are not residual but younger tectonic landforms.

West of the rift on the Sinai-Levantine plate, the surface can be reconstructed. Although it has been folded and faulted and warped into arches, there is summit accordance and some remnants remain as in Sinai, Galilee, and the Tyre-Nabatiye Plateau.

Before the tectonic disturbance of these surfaces they were co-extensive across the site of the rift system. They are the same altitude across the Gulf of Aqaba (Quennell 1958).

An oldest age for the surface is Upper Oligocene, the cycle having been initiated in the Middle or Upper Eocene. This is based on stratigraphic evidence from Sinai. The upper age limit is that of the earliest flood basalt flows, the Haouran, of Helvetian age.

The diastrophic episode which ended the erosion cycle was probably coincident with the opening of the Red Sea, the Sinai-Levantine plate lagging behind as part of the African plate (there is not sufficient recorded field evidence that the Gulf of Suez is an inter-plate boundary). The rift was apparently initiated as the terminating cross fault of the Red Sea spreading zone, its northerly course being influenced by (a) preferred trends in the Precambrian basememt (Lenze *et al.* 1972; Bender 1982) and (b) the rotation of the Arabian platform on a

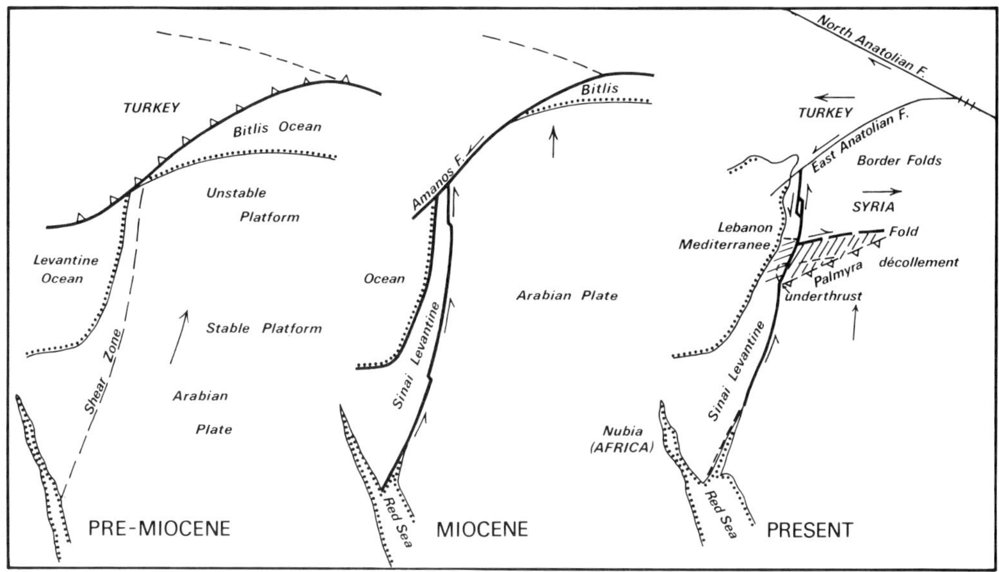

FIG. 6. Plate tectonics west Arabian rift system related to east Mediterranean region (after Dewey & Sengor 1979).

Pre- and L. Miocene: Situation prior to separation of Arabia from Africa across early tensional zone of the Red Sea. Arabian plate has freedom of movement northwards along the geosuture.

Miocene (mid): Arabian plate movement northward on circular arc transform closes Bitlis ocean and suture zone, with compression of Border Folds. Sinai-Levant (Africa) plate lags. This is the 62 km movement phase on the transform, with opening of the Dead Sea to the south by formation of a rhomb-graben. Arabian-Sinai planation surface deformed and faulted.

U. Pliocene to Present: With movement on the East Anatolian transform, westward translation of Turkey (Anatolian plate) is matched by eastward movement of north Arabia platform. Combination of latter movement and narrowing of north Arabia platform (with compression by Arabian plate on Bitlis Zone and Border folding) creates the dextral Lebanon-Palmyra fold-belt with translational kink in the geosuture (Fig. 5). Southern margin, the south Palmyra fault zone (Ponikarov 1966), is underthrust by excess lithosphere by movement of the Arabian plate from south to north by 45 km on the Dead Sea rift transform. Amount of underthrusting reduces from 45 km to nil at 400 km to NE. Gulf of Aqaba and Dead Sea lengthen. Sinai-Levant plate segment opposite fold belt suffers simple shearing and translation as does the geosuture and Yammouné fault.

small circle transform (Figs 5 & 6). The pre-existing interplate boundary between Sinai-Levant and Arabia from the Dead Sea northward, was followed.

The transform margin of the Arabian continental plate is well illustrated in the Bouguer anomaly map (Fig. 4) which also confirms the Lebanon segment as being the Yammouné fault (Fig. 5). This movement took place well before the Lebanon-Palmyra folding. Probably the whole of the 62 km first phase of movement was completed before the northward movement of Arabia closed the Bitlis ocean giving rise to a pause and the initiation of the East Anatolian transform separating the Anatolian and north Arabian plates.

The Lebanon-Palmyra fold belt appears to have formed in the weakest zone in the Arabian plate lying as it does between the stable shelf of south Arabia with its sedimentary cover on a sialic base, and the unstable shelf of Syria with its thicker sedimentary cover lying possibly on the oceanic part of the plate. It suffered a dextral distortion which apparently acted as a barrier to further northward strike-slip movement on the rift transform (Fig. 6). This must have happened after the Bitlis ocean closed (Dewey & Şengör 1979) and the northern free margin of the Arabian plate which previously was only impeded in the NW, collided with the Anatolian and Van plates. The obliquity of the East Anatolian transform could have caused the eastward migration of the unstable shelf of Syria in the same manner as the Anatolian plate was being expelled westward with dextral movement on the Lebanon-Palmyra fold belt (Fig. 6).

The Arabian platform south of the fold belt resumed its northward migration on the Dead Sea transform in Plio-Plestocene times and has moved 45 km and is continuing. It is only this excess lithosphere, and not the total of 107 km, which has to be consumed by the process of underthrusting (at Mt. Hermon) or beneath a detachment surface (below the Palmyra sedimentary cover).

Conclusion

The conclusion is that the West Arabian rift system, originally continuous from the mid-Miocene, was divided by the Upper Pliocene into: a southern transform system, Red Sea to the Lebanese frontier; and a northern transform system from the Syrian frontier to the East Anatolian transform boundary. They were both sinistral strike-slip and acted independently with differing displacements. They were separted by the belt of oblique folding, faulting and thrusting. It is probable that the Yammouné fault, a young feature, lies above the geosuture.

This explanation obviates the necessity for involved hypotheses for the passage of the faulting through the Lebanon folds and faults, with changes in strike and displacement, which Dubertret had come to deny. It also leads to reconciliation with the kinematics of the eastern Mediterranean and Aegean plates.

ACKNOWLEDGEMENTS: The author wishes to record his gratitude to the late Dr Louis Dubertret with whose classical mapping and writings, study of Levant geology has been enriched; to Dr. D. J. Burdon, late FAO; Dr. Paul Hancock of Bristol University; Prof. N. Ambraseys who assisted in the early study of the geophysics; the Global Seismology Unit of IGS; and the Jordanian Geologists' Association. The paper was reviewed by D. Neev, F. Mirsch & J. E. Dixon.

References

AMBRASEYS, N. N. 1971. Value of historical records of earthquakes. *Nature*, **232,** 375–379.

BARBERI, F., CAPALDI, G., GASPERINI, P., MARINELLI, G., SANTACROCE, R., SCANDONE, R., TREUIL, M. & VARET, J. 1980. Recent basaltic volcanism of Jordan and its implications on the Geodynamic history of the Dead Sea shear zone. *In Geodynamic Evolution of the Afro-Arabian Rift System.* Rep. International meeting, Rome, 1979, Atti dei Convegni Lincei (Rome: Accad. Natl. dei Lincei) 667–684.

BARTOV, Y., STEINITZ, G., EYAL, M. & EYAL, Y. 1980. Sinistral movement along the Gulf of Aqaba—its age and relation to the opening of the Red Sea. *Nature*, **285,** 220–222.

BEN-AVRAHAM, Z., GARFUNKEL, Z., ALMAGOR, G. & HALL, J. K. 1979. Continental breakup by a leaky transform: the Gulf of Elat (Aqaba). *Science*, **206,** 214–216.

BEN-MENAHEM, A., NUR, A. & VERED, M. 1976. Tectonics, seismicity and structure of the Afro-Eurasian junction—the breaking of an incoherent plate. *Physics of the Earth and Planetary Interiors*, **12,** 1–50.

—— & ABOODI, E. 1981. Micro- and macroseismicity of the Dead Sea rift and off-coast eastern Mediterranean. *Tectonophysics*, **80,** 199–234.

BENDER, F. 1968. Geologie von Jordanien. *Beitrage zur Regionalen Geologie der Erde*. **7,** 230 pp. Stuttgart: Borntraeger.

—— 1974a. *Geology of Jordan.* Gebruder Borntraeger, Berlin, 134 pp.

——1974b. Explanatory notes on the geological map of the Wadi Araba, Jordan. *Geol. Jb., Reihe B.* **10,** 3–62.

—— 1982. On the evolution of the Wadi Araba-Jordan Rift. *Geol. Jb.*, Reihe B, **45,** 3–20.

BOURGOIN, A. 1945–1948. Sur les anomalies de la pesanteur en Syrie et au Liban. *Notes et Mémoires, Etudes Géologiques et Géographiques sur le Liban et le Moyen-Orient*, **4,** 59–90. (Beyreuth).

BURDON, D. J. 1959. *The Geology of Jordan.* 82 pp. (Amman: Government of the Hashemite Kingdom of Jordan).

DEWEY, J. F. & PITMAN, W. C., RYAN, W. B. F. & BONNIN, J. 1973. Plate tectonics and the evolution of the Alpine system. *Bull. geol. Soc. Amer.* **84,** 3137–3180.

—— & ŞENGÖR, A. M. C. 1979. Aegean and surrounding regions. Complex multiplate and continuum tectonics in a convergent zone. *Bull. geol. Soc. Amer.* **90,** 84–92.

DRAKE, C. L. & GIRDLER, R. W. 1964. A geophysical study of the Red Sea. *Geophys. J. R. astron. Soc.* **8,** Fig. 4, 489.

—— & —— 1982. History of Rift Studies. In: PALMASON, G. (ed.) Continental and Oceanic Rifts. *Geodynamic Series. A.G.U.* **8,** 5–15.

DUBERTRET, L. 1932. Les formes structurales de la Syrie et de la Palestine: leur origine. *C. r. Seances Acad. Sci., Paris* **195,** 65–66.

—— 1943–1953. Carte géologique au 1/50 000 de la Syrie et du Liban. 21 sheets with notes. Damas et Beyreuth: Ministeres des Trav. Pub.

—— 1947. Problèmes de la Géologie du Levant: *Bull. Soc. géol. Fr.*, **17,** 3–31.

—— 1955. Carte géologique du Liban au 1/200 000e et Notice explicative. 74 pp. Beyreuth.

—— 1967. Remarques sur le fossé de la Mer Morte et ses prolonguement au Nord jusqu'au Taurus. *Rev. de Géogr. phys. Géol. dyn.* **12,** 3–16.

DUBERTRET, L. 1970. Review of structural geology of the Red Sea and surrounding areas. *Philos. Trans. R. Soc. London* **A267,** 9–20.

—— 1971–72. Sur la dislocation de l'ancienne plaque sialique Afrique–Sinai–Peninsula Arabique. *Notes et Memoires sur le Moyen-Orient. Mus. Natl. d'hist. Nat. Paris.* **7,** 227–243.

FREUND, R. & GARFUNKEL, Z. (eds). 1981. The Dead Sea Rift. I.U.C.G., Sci. Rep. 60, *Tectonophysics*, **80,** 310 pp.

——, ——, ZAK, I., GOLDBERG, M., WEISSBROD, T. & DERIN, B. 1970. The shear along the Dead Sea Rift. *Philos. Trans. Roy. Soc. London*, **A267,** 107–130.

—— & TARLING, D. H. 1979. Preliminary Mesozoic paleomagnetic results from Israel and inferences for a microplate structure in the Lebanon. *Tectonophysics*, **80,** 189–205.

GARFUNKEL, Z. 1981. Internal struture of the Dead Sea leaky transform (rift) in relation to plate kinematics. *Tectonophysics*, **80,** 81–108.

——, BARTOV, J., EYAL, Y. & STEINETZ, G. 1974. Raham conglomerate—new evidence for Neogene tectonism in the southern part of the Dead Sea Rift. *Geol. Mag.* **111,** 55–64.

——, ZAK, I. & FREUND, R. 1981. Active faulting in the Dead Sea rift. *Tectonophysics*, **80,** 1–26.

GINZBURG, A. 1970. The gravity map of Israel. *Atlas of Israel.* Amsterdam: Elsevier.

GREGOR, C. B., MERTZMAN, S., NAIRN, A. E. M. & NEGENDANK, J. 1974. The paleomagnetism of some Mesozoic and Cenozoic volcanic rocks from the Lebanon. *Tectonophysics*, **21,** 375–395.

HANCOCK, P. L. & ATIYA, M. S. 1979. Tectonic significance of meso-fracture systems associated with the Lebanese segment of the Dead Sea transform fault. *J. Struct. Geol.* **1,** 143–153.

HELSELY, C. E. & NUR, A. 1970. The paleomagnetics of Cretaceous rocks from Israel. *Earth planet. Sci. Lett.* **8,** 403–410.

KNOPOFF, L. & BELSHE, J. C. 1966. Gravity observations of the Dead Sea Rift. In: IRVINE, T. N. *The World Rift System.* I.U.M.C. Rep. 2. Ottawa: Geol. Surv. Canada, 5–21.

LARTET, L. 1869. Essay on the geology of Palestine. *Ann. Sci. Geol.* **1,** 17–18. Paris: Soc. géol. France.

LEJAY, P. 1936. Exploration gravimétrique de l'Extrême-orient. *Com. Nat. Fr. Géod. Géoph.* 75 pp. Paris.

LENZ, H., BENDER, F., BESANG, C., HARRE, W., KREUZER, H., MULLER, P. & WENDT, I. 1972. The age of early tectonic events in the zone of the Jordan geosuture, based on radiometric data. *Proc. 24th Int. Geol. Congr. 1972*, **3,** 371–379.

MCKENZIE, D. P. 1972. Active tectonics of the Mediterranean region. *Geophys. J. R. astron. Soc.* **30,** 107–185.

PARKER, D. H. 1969. *The Hydrology of the Mesozoic-Cainozoic Aquifers of the Western Highlands and Plateau of East Jordan.* Amman: UNDP/FAO, 98 pp.

PICARD, I. L. 1959. *Geological map of Israel.* 1:500 000. Geol. Surv. Israel, Jerusalem.

PLASSARD, J. & STAHL, P. 1957. Nouvelle carte gravimétrique du Liban. *Annales de l'Observatoire de Ksara, Mémoires.* **11,** 12–15.

PONIKAROV, V. P. (Editor in chief) 1966. The geological map of Syria, explan. notes. Pt. I. Stratigraphy, Igneous rocks and Tectonics. 230 pp. Pt. II. Mineral deposits and ground water. 71 pp. with geological maps scale 1:500 000, 4 sheets. *V.O. Technoexport, Moscow.* USSR contract No. 944 for Syrian Arab Republic.

PONIKAROV, V. P. & MIKHAILOV, I. (Editors) 1964. Geological map of Syria, scale 1:1 000 000 with explanatory notes. 1966, 111 pp. (with Gravity and Magnetic maps, 1:1 500 000 and Tectonic map, scale 1:1000 000). *V.O. Technoexport, Moscow*, contract for Syrian Arab Republic.

QUENNELL, A. M. 1951. Geology and mineral resources of former Transjordan. *Col. Geol. Min. Res.* **2,** 85–115.

—— 1955–6. *Geological map of Jordan (East of the rift valley)* 1:250 000 Sheets: Amman; Kerak; Ma'an. (Amman: Jordan Government).

——1958. The structure and evolution of the Dead Sea rift. *Quart. J. Geol. Soc.* **64,** 1–24.

——1959. Tectonics of the Dead Sea Rift. *Proc. 20th Inter. Geol. Congr., Mexico, 1956. Ass. de Serv. Geol. Africanos*, 385–403.

TARLING, D. H., SAUVER, N. & HUTCHINGS, A. M. J. 1967. Further paleomagnetic results from the Federation of South Arabia. *Earth planet. Sci. Lett.*, **3,** 148–154.

VAN DONGEN, P. G., VANDER VOO, R. & RAVEN, TH. 1967. Paleomagnetic research in the Central Lebanon mountains and in the Tartous area (Syria). *Tectonophysics*, **4,** 35–53.

WETZEL, R. & MORTON, D. M. 1959. Contribution à la géologie de la Transjordanie. In: DUBERTRET, L. (ed.) *Notes et Mémoires sur le Moyen-Orient.* **7,** Paris: Mus. Natl. d'Hist. Nat.

WOLFART, R. 1967. *Syrien und Libanon: Beiträge zur Regionalen Geologie der Erde.* Berlin: Borntraeger, 285 pp.

ZAK, I. & FREUND, R. 1966. Recent strike-slip movements along the Dead Sea Rift. *Israel J. Earth-Sciences*, **15,** 33–37.

A. M. QUENNELL, Department of Geology, University of Bristol, Queen's Building, University Walk, Bristol, BS8 1TR.

On the Neogene development of the Eastern Mediterranean basins

S. Jasko

SUMMARY: The cauldron-like depressions of the Eastern Mediterranean, the Black Sea and the Southern Caspian differ from the foredeep basins of young fold-mountains of Neotectonic Europe in shape and form. The formation of these collapse structures has modified the palaeogeography of the Alpine-Carpathian-Balkan-Caucasus mountain system.

The deep basins of the Mediterranean formed in the south of Neotectonic Europe and on the margins of North Africa, probably at the beginning of the Pliocene. Subsidence of the Black Sea and of the Southern Caspian took place in the north of Neotectonic Europe and on the margin of the Russian platform in the Late Pliocene and Early Quaternary.

In several recent longer works a detailed analysis has been given of the structure and development of the Neogene and Quaternary sedimentary basins of Europe (Jasko 1977, 1979, 1981). This paper summarizes the characteristics of the Eastern Mediterranean basins and those of the Black Sea and the Caspian Sea, indicating the differences with other Neogene depressions in Europe. The Neogene basins of Europe and the Middle East can be classified by orogenic structure and the amount of sediment accumulation. Figure 1 shows the locations of the basins. Sediment thickness data are given in Table 1. The basins are given the same number on the map, in Table 1 and in the text. The types of sedimentary basins with the approximate mean size and form are schematically shown in Fig. 2. This excludes the structure of the pre-Neogene basement rocks which is considered to have little bearing on the development and form of the Neogene sedimentary basins.

Basin Types

Orogenic foredeeps and intramontane troughs

Orogenic foredeeps are long narrow depressions located on the margins of young fold-mountain ranges. Foredeeps have a long axis parallel to the general strike of the mountain range and an asymmetrical cross-section (Fig. 2, section 1). Intramontane troughs are tectonic depressions developed in the middle of orogens following uplift. Intramontane troughs have irregular outlines with marginal faults directed at an angle, sometimes even perpendicularly, to the general strike of the mountain range. Intramontane troughs and orogenic foredeeps were formed more or less simultaneously but they differ in structure. Examples of intramontane troughs and orogenic foredeeps are the Middle Danube basins (1 to 6), Carpathians and Balkans (7 to 10), Caucasus (11 to 16), Elburz (17), Kopet Dag (18), Alps (19 to 22), Zagros (23 to 25), Pontus Mountains (26), Iranids (27 and the Apennines (28 to 31).

Tectonic graben

Tectonic graben are long narrow depressions caused by crustal extension. The margins are fault-planes dipping towards the graben. The strike of these faults are more or less parallel (e.g. Fig. 2, section 2). In Europe and the Middle East, tectonic graben are considerably less frequent than orogenic foredeeps. Examples of tectonic graben are the Rhine Graben (43) and the Dead Sea Graben (43a).

Epeirogenic depressions

Epeirogenic depresions are broad shallow basins which slope gently towards the depocentre (Fig. 2, section 3). No orogenic movement was involved in formation of these depressions. Locally, they are surrounded by isostatically uplifted crustal blocks, often the remains of old mountains. Examples are the North Sea basin (38 to 42) and the North Caspian Plain (47). Some authors regard the North Sea as a major tectonic graben. This is only true for the Mesozoic and Palaeozoic basin. Neogene and Pleistocene sediments were deposited unconformably on older formations and the resulting basin was thus no longer a graben bordered by steep marginal faults. Instead, the graben had become a broad bowl-shaped depression, the sides of which slope gently towards the centre from all directions. Proof for this can be found onshore in the Netherlands and Belgium

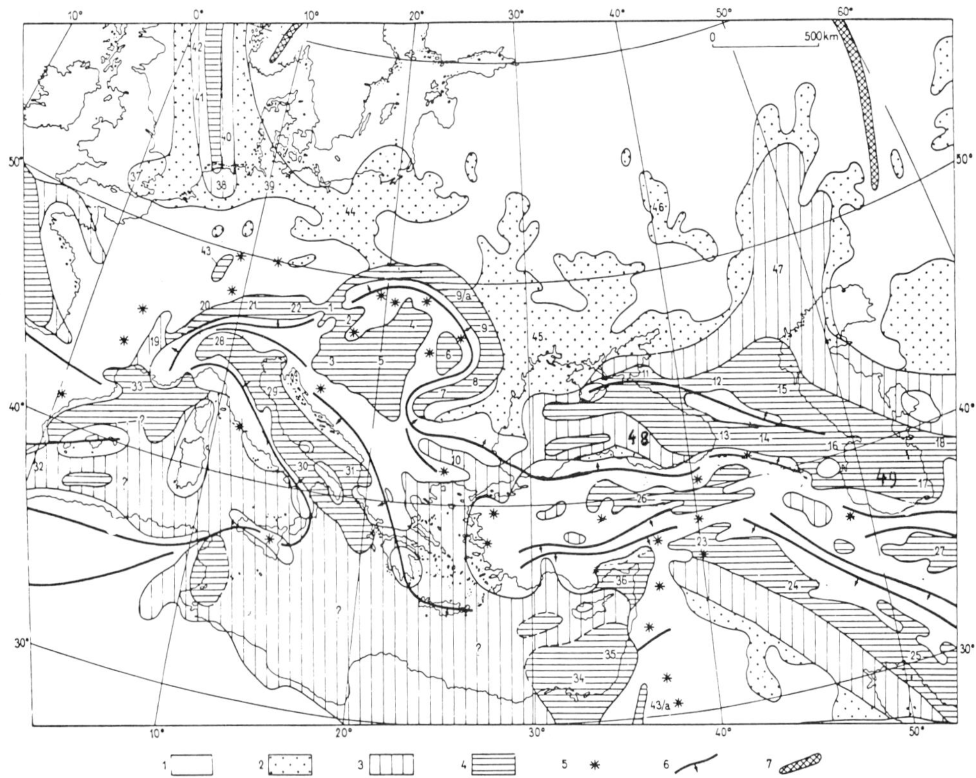

Fig. 1. Position of late Cenozoic (Neogene and Quaternary) sedimentary basins. 1, area of erosion without sediments; 2, Late Cenozoic sediments 1–300 m thick; 3, Late Cenozoic sediments 300–1000 m thick; 4, Late Cenozoic sediments over 1000 m thick; 5, Volcanic rocks; 6, Young orogenic chains; 7, Raised margins of older mountains.

(Pannekoek 1954; Zagwijn & Doppert, 1978; Voorthuysen 1954). This form of deposition can also be seen on maps compiled from offshore well data of the North Sea (Caston 1977; Kent & Walmsley 1970; Kent 1975; Rasmussen 1974, 1978). As this study is restricted to Quaternary and Neogene sedimentary basins the author considers it appropriate to classify the North Sea basin as an epeirogenic depression rather than a tectonic graben.

Cauldron-like collapse structures

Cauldron-like collapse structures evolved in the latest periods of geological history. Having an irregular oval shape they do not fit into the pre-existing structural frame. Cauldrons have steep marginal slopes and a gently undulating floor (Fig. 2, section 4). These depressions are up to 500–600 km wide with depths exceeding 2000–3000 metres at some places. Cauldrons are thus generally larger than foredeeps or tectonic graben. Examples are the Mediterranean Basin (32 to 37), Black Sea (48) and the South Caspian Depression (49).

Epicontinental sedimentary cover

The epicontinental sedimentary cover consists of formations usually less than 200–300 m thick and frequently only a few tens of metres thick. The horizontal extent of the epicontinental cover exceeds thickness by 2 to 3 orders of magnitude. The irregular outlines of the cover are determined by deposition over an uneven palaeo-relief and the subsquent development of an erosional drainage pattern (Fig. 2, section 5).

Basin evolution

Orogenic foredeeps, tectonic graben and epeirogenic depressions classified as groups (i), (ii) and (iii) vary in terms of age, tectonic structure, form and size. There is, however,

TABLE 1. *Distribution of accumulated sediments by geological age (thickness data are given in metres)*

Area	Quaternary		Pliocene		Miocene		Total
	(m)	(%)	(m)	(%)	(m)	(%)	(m)
1. Vienna basin	0	0	1250	23	4200	77	5450
PANNONIAN BASIN							
2. Northeastern part	200	3	2600	37	4200	60	7000
3. Southeastern part	50	1	2800	63	1600	36	4450
4. Northeastern part	450	7	1150	17	5000	76	6600
5. Central part	500	12	3500	88	0	0	4000
6. Transylvanian basin	0	0	1200	25	3600	75	4800
DACIAN BASIN							
7. Western extreme	0	0	1200	40	1800	60	3000
8. Central part	100	2	2800	67	1300	31	4200
9. Northeastern part	200	4	3100	55	2300	41	5600
10. Thrace	0	0	750	100	0	0	750
11. Kuban'	0	0	2100	51	2000	49	4100
12. Stavropol'	0	0	100	6	1700	94	1800
13. Kolchis	50	1	2050	48	2200	51	4300
14. Western end of Kura basin	0	0	800	22	2800	78	3600
15. Terek basin	100	2	2450	46	2800	52	5350
16. Eastern Kura basin	0	0	4050	67	1950	33	6000
17. Elburz	150	5	2800	86	300	9	3250
18. Kara Kum	450	18	900	37	1100	45	2450
19. Saone valley	0	0	250	50	250	50	500
20. Switzerland/Württemberg	0	0	0	0	3000	100	3000
21. Upper Austria	0	0	0	0	2500	100	2500
22. Lower Austria	0	0	0	0	2900	100	2900
23. Syria	1600	59			1100	41	2700
24. Mesopotamia	3000	67			1500	33	4500
25. Persian Gulf	2000	50			2000	50	4000
26. Anatolia	0	0	200	14	1200	86	1400
27. Teheran basin	200	17			1000	83	1200
28. Po basin	1600	38	1400	33	1200	29	4200
29. Rimini	1500	39	1200	32	1100	29	3800
30. Apulia	600	26	600	26	1100	48	2300
31. Albania	100	2	2000	41	2800	57	4900
32. Alboran Sea	400	45	300	33	200	22	900
33. Rhone Delta	200	7	800	30	1700	63	2700
34. Nile Delta	700	17	2100	50	1400	33	4200
35. Israel	600	25	1000	42	800	33	2400
36. Gulf of Adana	600	18	500	15	2300	67	3400
37. England	100	67	50	33	0	0	150
38. Netherlands	500	63	100	12	200	25	800
39. Northwest Germany	0	0	100	33	200	67	300
40. Well A2 (offshore Denmark)	350	24	200	14	900	62	1450
41. Argyll (offshore UK)	500	32	600	39	450	29	1550
42. Forties (offshore UK)	400	32	550	44	300	24	1250
43. Rhine Graben	300	13	500	22	1500	65	2300
44. German-Polish Plain	50	20	100	40	100	40	250
45. Southern Ukraine	0	0	50	25	150	75	200
46. Moscow/Oka region	20	40	20	40	10	20	50
47. North Caspian Plain	60	10	540	90	0	0	600

one characteristic common to all, a continuous evolution over a long geological period. Accumulation of sediments, supplied by rivers, has filled slowly sinking sedimentary basins gradually. Synorogenic sedimentation could often balance the relief created by vertical crustal movements. On the other hand, group (iv) basins, for example the Mediterranean, Black Sea and the Southern Caspian, experienced sudden subsidence to considerable depths within a short time span.

The processes leading to the formation of cauldrons are not known in detail. It is likely that the phase of rapid subsidence to consider-

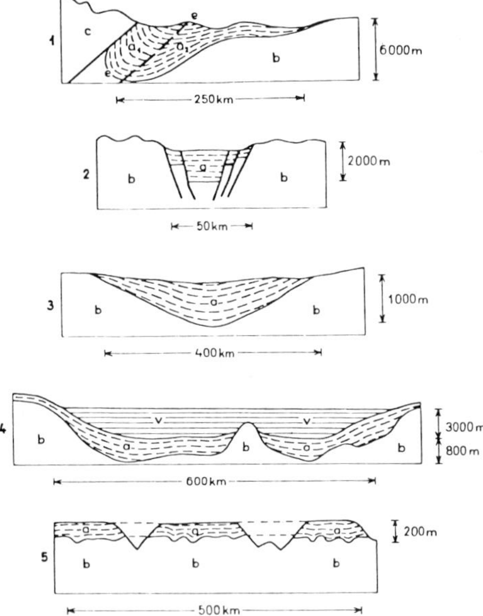

FIG. 2. Types of Neogene sedimentary basins. 1, Orogenic foredeeps (Alpine-Carpathian molasse zone); a_1, folded molasse of great thickness; a_2, non-folded thin molasse; b, basement; c, overthrust flysch zone; e-e, axis of foredeep; 2, Tectonic graben (Rhine graben and Dead Sea graben); a, Neogene and Quaternary sediments; b, basement; 3, Epeirogenic depression (North Sea); a, Neogene and Quaternary sediments; b, pre-Neogene basement; 4, Cauldron-like subsidence (Mediterranean); v, water; a, Neogene and Quaternary sediments; b, basement; 5, Epicontinental sedimentary cover (German-Polish Plain); a, Neogene and Quaternary sediments; b, basement.

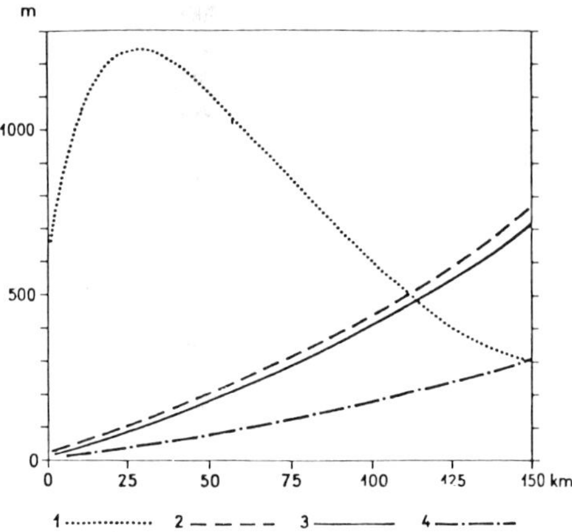

FIG. 3. Variation of sediment thickness with respect to distance from the margins towards the centres of basins. 1, Mediterranean Sea, post-Messinian (Pliocene and Quaternary); 2, North Sea (Pliocene and Quaternary); 3, Pannonian Basin (Upper Pliocene and Quaternary); 4, North Caspian (Upper Pliocene and Quaternary).

able depths did not last more than a few hundred thousand years. In a very short time interval, compared to other geological processes, 2 to 3 km deep marine depressions formed in place of land. These cauldrons subsided too fast to be filled with sediment which mostly accumulated in deltas on the margins of the basins. The volume of terrigenous deposits decreases away from the coast. The differences in sediment thickness in marginal areas of different basins are shown as curves in Fig. 3.

Until recently, it was generally accepted that the rate of sedimentation is usually an order of magnitude greater in the orogenic belts than in shelf regions. This conclusion, however, needs to be qualified. The rate of sedimentation is always greatest in places where the carrying capacity of rivers or currents undergoes a large reduction or even ceases entirely. Such circumstances are found not only in orogenic regions but also, exceptionally, in coastal areas of continents with low relief, e.g. the delta regions of the Nile, Danube and the Dnieper. An exception to this generalization is where a considerable amount of sediment is deposited by density currents at the foot of submarine canyons (Shepard 1978; Friedman & Sanders 1978).

The cauldron-like subsidences of the Eastern Mediterranean, the Black Sea and the Southern Caspian differ from the foredeep series of the young fold-mountain chains of Europe in shape and form. The formation of these collapse structures has changed the evolution of the Alpine-Carpathian-Balkan-Caucasus mountain system. The mountain chain, which was previously continuous, separated into segments; some areas sank and merged with the adjacent molasse foredeeps to form deep inland seas. The high mountain ranges of the Balkan Mountains and the Caucasus were much more extensive in the geological past than at present. These ranges continued through areas now covered by the Black Sea (Rerzanov & Sevcenko 1980). Seismic and acoustic surveys of the Bulgarian shelf have traced the eastward continuation of both the Balkan Mesozoic and Neogene units beneath the sea floor (Kuprin *et al.* 1980). This subsidence did not proceed uniformly everywhere. The deep basin of the Mediterranean formed in the south of Neotectonic Europe and on the margins of the African plate, probably at the beginning of the

Pliocene. On the other hand, the deep subsidence of the Black Sea and of the Southern Caspian Sea occurred in the north of Neotectonic Europe and on the margin of the Russian platform in the Late Pliocene and Early Quaternary.

Cauldron-like subsidence also caused changes in the lithofacies distribution. Neogene foredeeps of the Alpine-Carpathian-Caucasus mountain chains were characterized by lithofacies in zones parallel to the axis of the basin. Coarse sediments (molasse), fine grained sediments (schlier) and carbonate rocks were deposited predominantly in the coastal areas, in the middle of the basin and on the cratonic foreland. No such regular arrangement can be observed in recent sediments of the Mediterranean, Black Sea and the Caspian basin. All that can be said is that local palaeogeographic conditions controlled the great variety of deposits formed in the marginal areas. The great variety of the Quaternary and Holocene coastal units can not be analysed in terms of a single sedimentary cycle.

The biofacies distribution does not show such clear-cut differences between the foredeeps and cauldron-like subsidences. Marine, brackish and fresh-water deposits could equally well form in both types of basins, depending on oceanic connections and local palaeogeography (see Steininger & Rögl, this volume).

The Neogene sedimentary sequences of the basin floors also show certain differences from older deposits in the same basins. Messinian deposits and Pliocene foraminiferal muds, widely distributed in the Mediterranean, are entirely absent in the Neogene of the Black Sea and the Caspian. In these, the Upper Miocene, excluding Caspian oligohaline units, is represented by freshwater lacustrine chalk and diatomaceous mud. This was because the Black Sea and the Caspian belonged to the Paratethys and lacked a direct connection with the ocean. Stratigraphical cores of Deep Sea Drilling Project in the Mediterranean can be correlated with fossil-rich sediments exposures on adjacent land (Hsü *et al.* 1979). Such a correlation has not yet been established for boreholes in the Black Sea (Muratov *et al.* 1978; Hsü 1978). Thus, the central parts of the Black Sea (48) and the Southern Caspian (49) are excluded from Table 1.

The outlines of the Eastern Mediterranean have changed relatively little since the Pliocene. In contrast, major changes of coastlines occurred in the Aegean, Marmara, Black and Caspian Sea areas.

Coastal areas were submerged connecting inland seas and lakes. Subsidence took place at different rates in different areas with alternating submergence and emergence which produced a varied topography (Fedorov 1978; Keraudren 1975).

References

Biju-Duval, B. & Montadert, L. 1976. Introduction to the structural history of the Mediterranean basins. *In*: *Structural History of the Mediterranean basins*, (eds) Biju-Duval, B. & Mondadert, L. pp. 1–12. Editions Technip, Paris.

Caston, V. N. D. 1977. A new isopachyte map of the Quaternary of the North Sea. *Inst. Geol. Sci. Report.* **77,** 3–8.

Fedorov, P. V. 1978. *Pleistocene Ponto-Kaspia.* Moscow. (in Russian).

Friedman, G. M. & Sanders, E. J. 1978. *Principles of Sedimentology.* New York.

Hsü, K. L. 1978. Stratigraphy of the lacustrine sedimentation in the Black Sea. *Initial Reports of the Deep-Sea Drilling Project. Washington.* **42,** 509–24.

—— *et al.* 1978. Shipboard studies (site reports). *Initial Reports of the Deep-Sea Drilling Project. Washington.* **42,** 27–358.

Jaskó, S. 1977. Neogene subsidence and sedimentation of the Middle Danube Depression System. *Sedimentary Geology* **17,** 295–309.

—— 1979. Distribution of Miocene evaporites in the Tethys and Paratethys. *Annales géol des Pays Helléniques, Athènes. Hors. Ser.* **2,** 559–66.

—— 1981. Üledékfelhalmozódás és kőszénképződés a neogénben. *Sedimentation and Coal Formation in the Neogene.* Budapest. (in Hungarian).

Kent, P. E. 1975. Review of North Sea basin development. Tertiary. *Jl. geol. Soc. London.* **131,** 454–5.

—— & Walmsley, P. J. 1970. North Sea progress. *Amer. Assoc. Petrol. Geol. Bull.* **54,** 168–81.

Keraudren, B. 1975. Essai de stratigraphie et de paléogeographie du Plio-Pleistocene égeen. *Bull. Soc. géol. France.* **17,** 1110–20.

Kuprin, P. N. and others 1980. *Geologo geofizitseskie issledovania Bolgarskogo sektora Tsernogo Morja.* Sofia. (in Russian).

Muratov, M. *et al.* 1978. Basic features of the Black Sea Late Cenozoic history. *Initial Reports of the Deep-Sea Drilling Project (Washington)* **42,** 1141–8.

Pannekoek, A. J. 1954. Tertiary and Quaternary subsidence in the Netherlands. *Geologie en Mijnb.* **16,** 156–63.

Rasmussen, L. B. 1974. Some geological results from the first five Danish exploration wells in the North Sea. *Danmarks Geol. Undersogelse III Raekke.* **42,** 1–46.

—— 1978. Geological aspects of the Danish North Sea sector. *Danmarks Geol. Undersogelse III. Raekke.* **44,** 1–85.

REZANOV A. I. & ŠEVČENKO, V. I. 1980. Die Entwicklung des Kaukasus-Kaspigebiets im Phanerozoikum. *Zeitschrift für Geol. Wissenschaften.* **8,** 791–806.

SHEPARD, F. P. 1978. *Geological Oceanography.* London.

VAN VOORTHUYSEN, J. H. 1954. Crustal movements of the southern part of the North Sea Basin during Pliocene and early Pleistocene times. *Geologie en Mijnb.* **16,** 165–72.

ZANGWIJN, W. H. & DOPPERT, J. W. 1978. Upper Cenozoic of southern North Sea Basin. *Geologie en Mijnb.* **57,** 577–88.

S. JASKO, H-1122 Budapest, Pethenyi köz 4, Hungary.

Pliocene lacustrine sediments in the volcanic succession of Almopias, Macedonia, Greece

P. Chorianopoulou, A. Galeos & Ch. Ioakim

SUMMARY: This paper is a geological and palaeobotanical study of the lacustrine sediments occurring in the Voras mountain range NNW of Aridaea, in Western Macedonia, Greece. The sediments are interstratified with Pliocene volcanics from the Almopias volcanic centres. Alternations of chert, pyroclastics, limestone, tuff and breccia overlie a metamorphic basement or trachyandesitic lavas and tuffs.

Analysis of pollen in the lacustrine sediments indicates an Early Pliocene age. The most important species are *Pinus haplostelle, Pinus diplostelle* type, Taxodiaceae, Cupressaceae (*Taxodium, Sequoia*), cf. *Quercus* (*Scabratricolpites microhenrici, Tricolpopollenites henrici*), cf. *Castanea, Symplocos, Nyssa, Liquidambar* and *Tsuga*. Pollen analysis provides important clues to past climates and relief.

The area examined lies within the volcanics of the Voras mountain range of Western Macedonia near the border with Yugoslavia, between the mountains of Pinovo and Koziakas, and is located about 17 km NNW of Aridaea, between Strongili and Panayitsa (Fig. 1). The volcanics of this area, the Almopias volcanics, have been studied by Soldatos (1955), Mercier & Sauvage (1965), Marakis & Sideris (1973), Elefteriades (1977), Bellon *et al.* (1979), Kolios *et al.* (1980).

In this study, lacustrine sediments associated with Pliocene volcanic activity from the Almopias centre are described for the first time. Palynological study of these sediments reveals a rich microflora of spores and pollen, as well as several other micro-organisms of Lower Pliocene age.

Also abundant are carbonized vegetal remains, in the form of leaves, roots and shoots, also fragments of ostracods, gastropods and other microfossils. In some cases the vegetal material was abundant enough for thin lignite layers to form (Table 1). Intense hydrothermal activity has modified these lacustrine sediments, mainly causing pyritization and formation of haematite, goethite, limonite and siderite.

Structure

The Almopias volcanics cover an area of about 200 km^2 in the Voras and Kozuf mountain ranges on both sides of the Yugoslavian frontier with Greece. They are mainly of trachy-andesitic composition (Mercier 1968). A series of lacustrine volcaniclastic and sedimentary rocks, overlain by tuffs and lavas, occurs between Strongili and Panayitsa, on the banks of the Dassikos River, which is a tributary of the River Paroi (Fig. 2). These lacustrine sediments either overlie a metamorphic basement to the volcanics, or the volcanics themselves.

The metamorphic basement is exposed as an inlier south of the lacustrine sediments and consists of green quartz-mica-schists, calciticschists and marbles assigned to the 'Porphyroid Formation' of the Almopias Zone (Mercier 1968), and interlayered later magmatic intrusions. The 'Porphyroid' strata generally dip about 45° towards the ESE.

The trachy-andesitic lavas and tuffs preserve a pre-Lower Pliocene palaeotopography. Our present study has not extended to the volcanics of this area and only two samples were examined from the Panayitsa lavas. These are both pyroxene-biotite-hornblende andesites.

Lacustrine volcano-sedimentary unit

Stratigraphy and petrography

The lacustrine sediments exposed north of the metamorphic basement crop out on the banks and ravines of the Dassikos River below

TABLE 1. *Approximate analysis of lignite sample No. 7536*

Depth	Thickness in (m)	Moisture (%)	Ash as it is	Ash in dry	CO_2 in dry	Maximum calorific value	Minimum calorific value
Surface	0.03	5.5	34.6	36.6	2.6	3.975	3.785

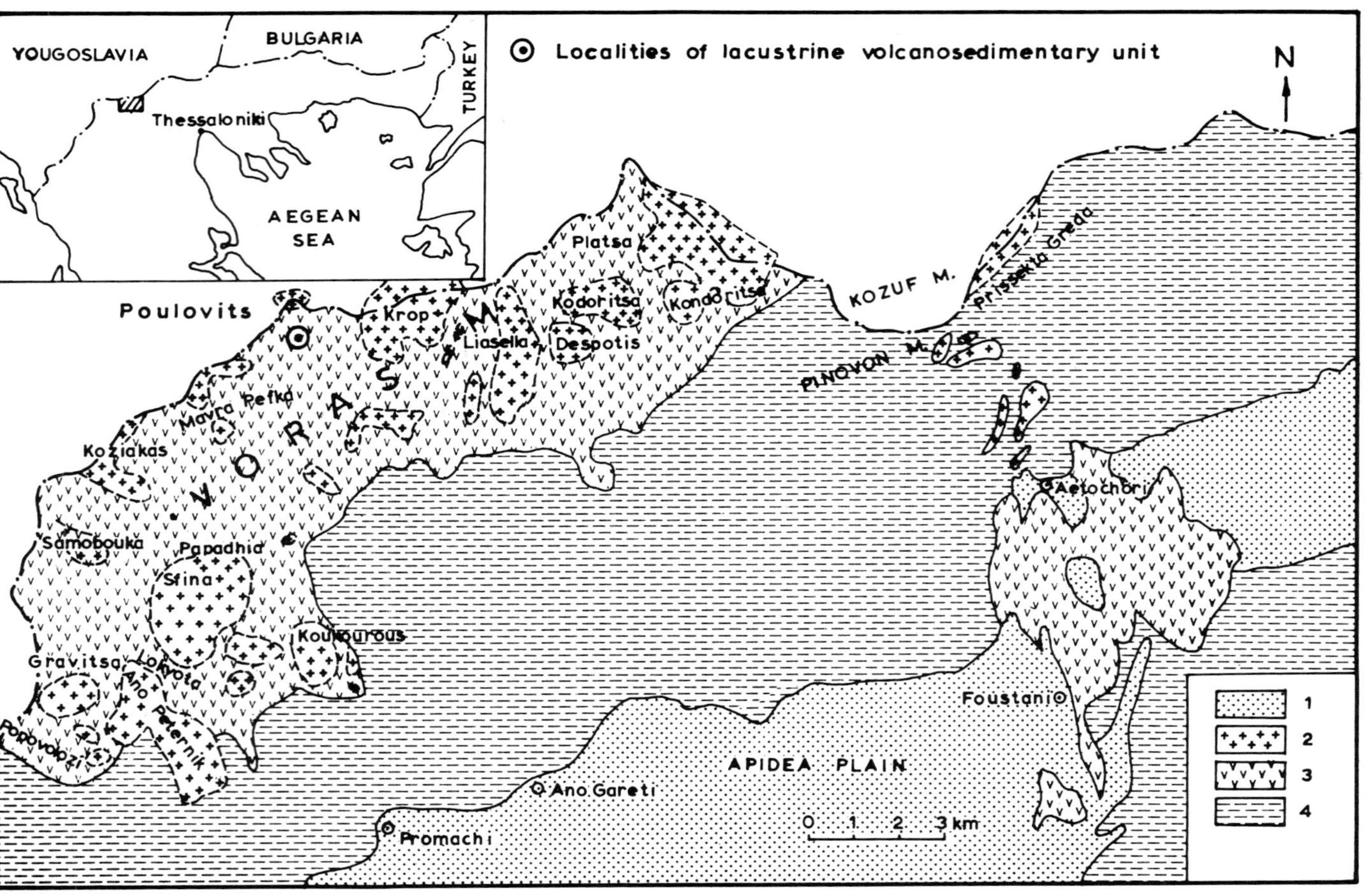

FIG. 1. Schematic geological map of the Voras Mts. area modified after Mercier 1968. 1, Quaternary deposits; 2, lava domes; 3, volcanic sequence; 4, Almopia sequence.

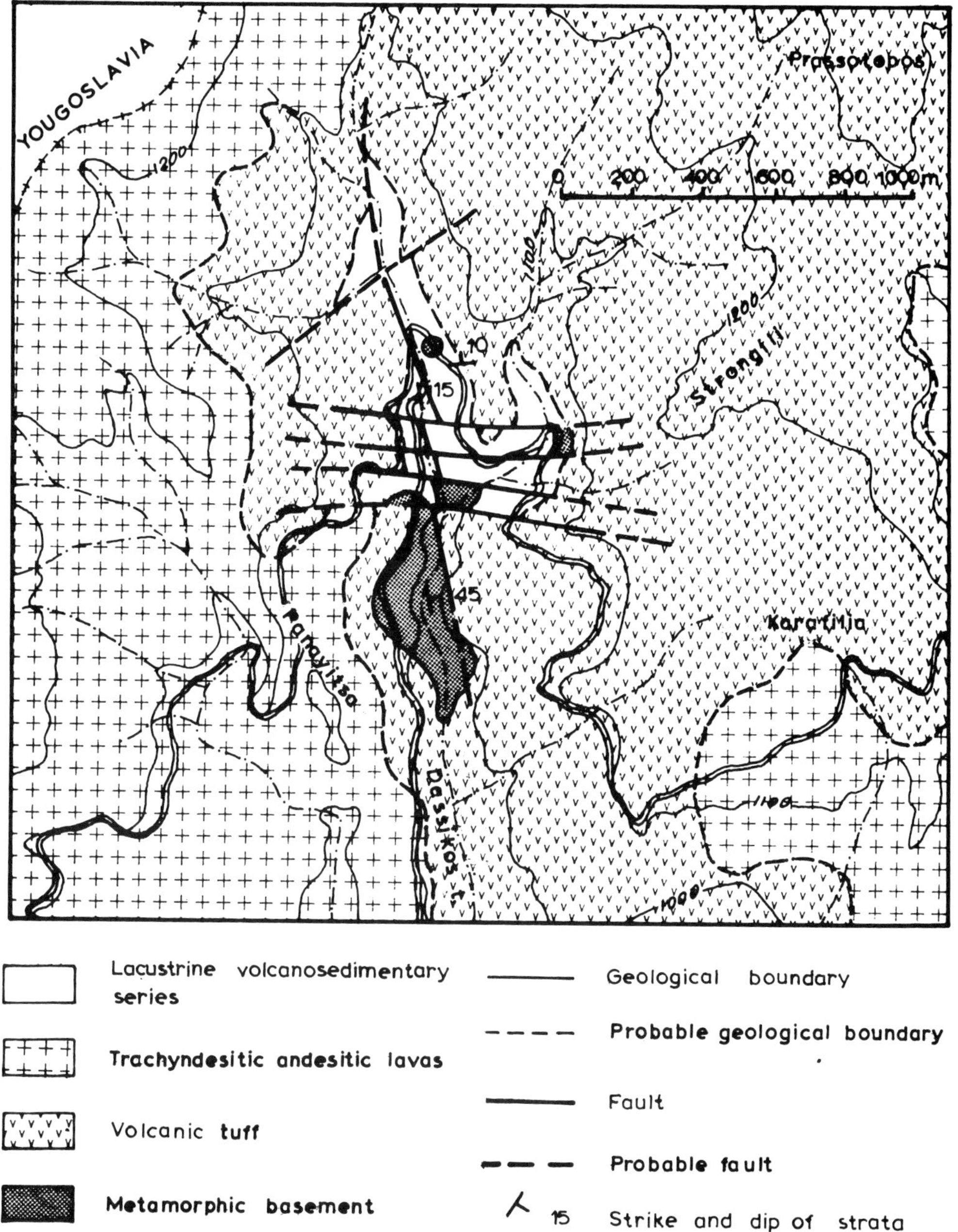

FIG. 2. Geological map of the lacustrine sedimentary basin.

the volcanics at altitudes between ca. 1000 and 1060 m. Towards higher ground the lacustrine sediments gradually disappear. A schematic cross-section is shown in Fig. 3.

The lacustrine volcano-sedimentary unit starts with a lahar deposit alternating with tuffs. The lahar consists of cobbles of metamorphic basement rocks and boulders of trachy-andesite. The cement, representing 60% of the rock, is yellow and consists of carbonate, clay and volcanic grains. The lacustrine sediments are intensely weathered along the Dassikos ravine and largely covered by hill-slope debris. The lowest visible beds consist of irregularly silicified limestones full of vegetal remains and other silicified micro-organisms including ostra-

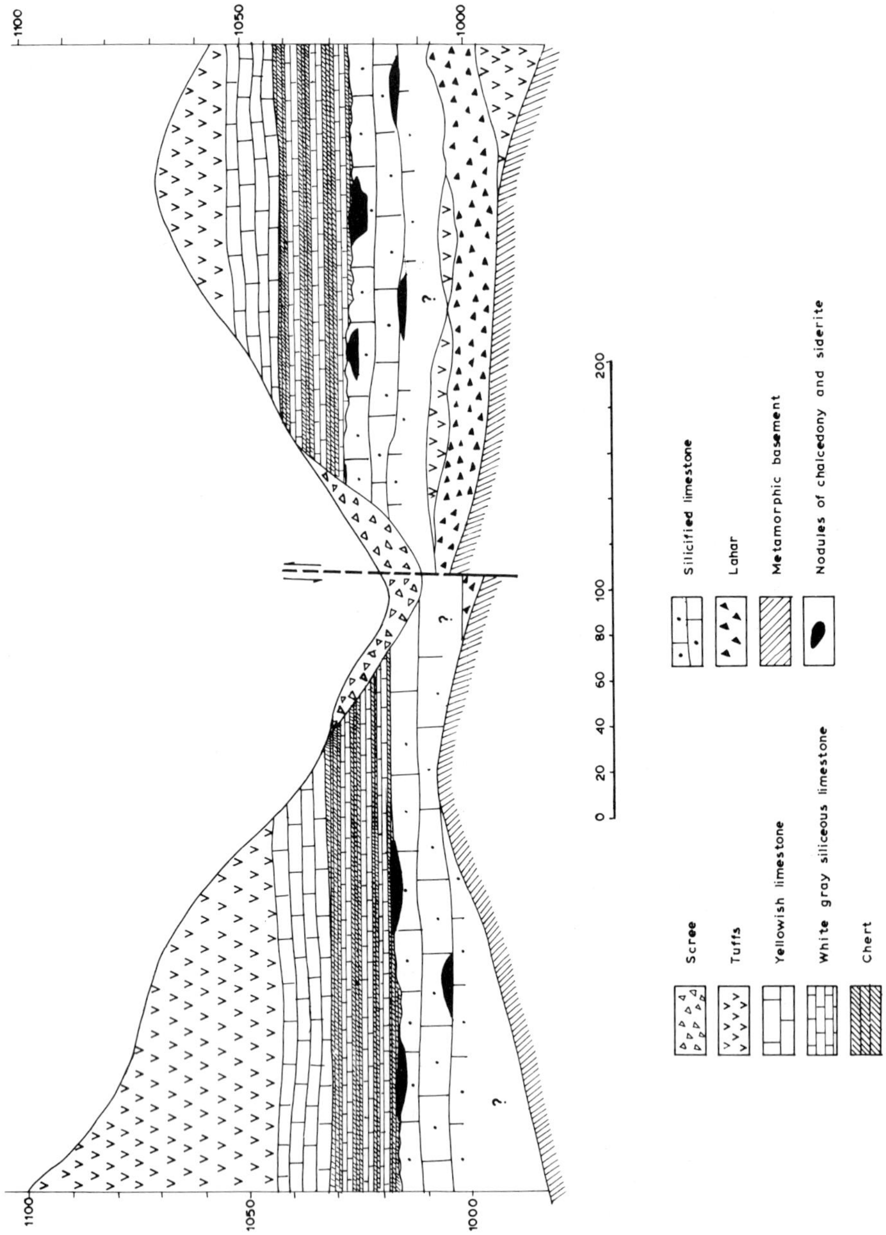

FIG. 3. Schematic cross-section through the lacustrine sediments.

cods and small gastropods. Nodules intercalated within these limestones, which are 5–30 cm in diameter, consist of granular or fibrous chalcedonic quartz with oolitic siderite concentrations. Pyrite also occurs in the nodules as veinlets, concentrations or dispersed crystals. We have also observed that vegetal remains and micro-organisms are usually silicified in nodules where siderite dominates. Concentrations of limonite also occur with total Fe up to 50% (Fig. 4).

Above the silicified limestone lies the typical lacustrine facies of the volcano-sedimentary unit. This consists of alternations of grey-white siliceous limestones, cherts and calcareous cherts. The surfaces of the layers parallel to the bedding are full of carbonized vegetal remains, leaves, roots and shoots, and occasionally insects. Layers of bituminous black limestone are observed. Generally, this facies possesses a zoned texture, alternating between white and grey according to the content of SiO_2 and organic matter. The siliceous limestones consist of sparite with abundant microcrystalline SiO_2. The cherts are thin-bedded (0.5–3 cm) and whitish-grey in colour with thin intercalations of brownish-black, waxy-looking aphanitic SiO_2. The cherts consist of slightly anisotropic SiO_2,

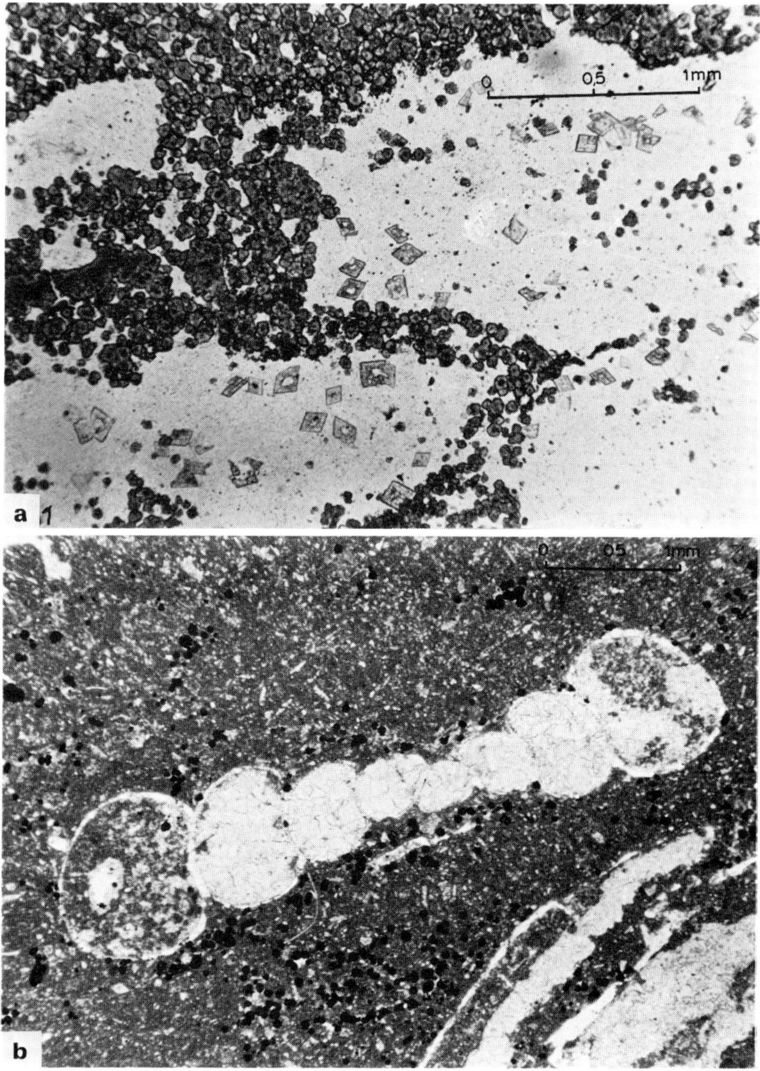

FIG. 4. (a) Nodule of chalcedony and oolitic siderite. Dispersed dolomite rhombohedra are distinguished. N II, ×36,5 (sample A_5/374 C).
(b) Fossiliferous limestone with framboids. N II, ×23 (sample A_3/460).

alternating with thin zones of sparite. X-ray diffraction shows that the SiO_2 is low-cristobalite and chalcedonic quartz.

The calcareous cherts consist of isotropic to cryptocrystalline SiO_2 with approximately 50% of small spherical sparite concentrations. The bituminous limestones possess a micritic to microsparitic texture, with abundant scattered calcite crystals of detrital origin and some quartz grains.

Clastic grains, mainly quartz, chlorite and calcite, are generally rare. Spherical concentrations of Fe-hydroxide framboids as well as dispersed pyrite grains are abundant however. Towards the upper part of the succession, bedded and unbedded yellowish fossiliferous micritic limestones become dominant with cavities full of sparite. Framboidal Fe-hydroxide and rare quartz grains are again present. At 1060 m the lacustrine sediments are covered by tuffs.

On the right bank of Dassikos River, near the contact with the metamorphic basement, a tuff with abundant thin (1 mm–2 cm) lignite layers is found in faulted contact with the lacustrine sediments. It consists of volcanic fragments and crystals of plagioclase, biotite and hornblende, cemented by microcrystalline clay minerals with Fe-Ti oxides and carbonates.

Tectonics

The area is intensely faulted. The faults were active during the Pliocene and Quaternary and are related to the formation of the volcano-sedimentary Almopias basin (Mercier 1966). In the lacustrine sediments a major fault, striking parallel to 155° with a vertical displacement of ca. 10 m, is seen in the bed of Dassikos River. Other smaller faults, trending 90°–100°, with displacements of 0.5–2 m, cut the river banks and offset the major fault (Figs 2 & 3).

Correlation between volcanism and lacustrine sediments

There is a close relation between volcanism and sedimentation. The volcanic contribution is both epiclastic and as hydrothermal solutions rich in SiO_2 which contributed to the formation of the siliceous sediments. The lake was enriched in SiO_2 from the adjacent volcanic rocks as well as from the hydrothermal solutions, and chemical chert sediments were precipitated alternating with carbonaceous sediments.

The hydrothermal solutions also contained sulphide and iron which contributed to the production of framboids and larger concentrations of pyrite and siderite ooids. The Fe-hydroxides occur almost throughout the unit in the form of thin concentrations parallel to bedding planes, and as larger limonite concentrations which mainly formed by later oxidation of pyrite. Later hydrothermal activity produced quartz, chalcedony and pyrite veins and alteration of the volcanic rocks through kaolinization and sericitization of feldspars or their replacement by calcite.

A hydrothermal origin for some of the SiO_2 and Fe-bearing minerals is favoured because:

(1) There are great quantities of SiO_2 and Fe in the lacustrine sediments which are unlikely to have been supplied by the alteration of the surrounding volcanic rocks.

(2) There is a complete lack of Radiolaria and no evidence of diatoms; i.e. the sediments are purely chemical.

(3) Clay, ferruginous clay and clastic material is generally absent from the lacustrine sediments.

(4) Hydrothermal action has been confirmed in the metamorphic basement throughout the area of outcrop of the Almopias volcanics.

Palynology

The recognition of the sediments as lacustrine prompted pollen analysis. This revealed a rich microflora, consisting of spores, pollen and various algae (Figs 5, 6 & 7). The organic material is well preserved, with the exception of one sample (378) which is highly carbonised.

The different taxa of microflora identified in these samples and their relative abundance are shown in a pollen diagram (Fig. 8). The microflora is characterized by abundant tree pollen grains of the Gymnosperms: *Pinus haplostelle, Pinus diplostelle, Abies, Tsuga, Sciadopitys* in addition to Taxodiaceae and Cupressaceae. Among the Angiosperms we see frequent pollen of *Alnus, Quercus, Ulmus-Zelkova* and *Salix*. There are also smaller amounts of the taxa *Acer, Castanea* type, *Symplocos, Platanus, Engelhardtia, Scabratricolpites fallax* and Araliaceae. Pollen grains of *Eucommia, Carpinus, Betula* and *Pterocarya* are also observed sporadically. Pollen grains of non-arboreal plants are common and are mainly represented by Gramineae, followed by Compositae, Amaranthaceae-Chenopodiaceae, Cyperaceae, Ericaceae, Typhaceae, Umbelliferae and Nymphaceae. Pteridophyte spores were observed in almost every sample, but never exceeding 10% in frequency.

The pollen analysis allows us to correlate the results with those from northern and central Europe, and from elsewhere in the Mediterra-

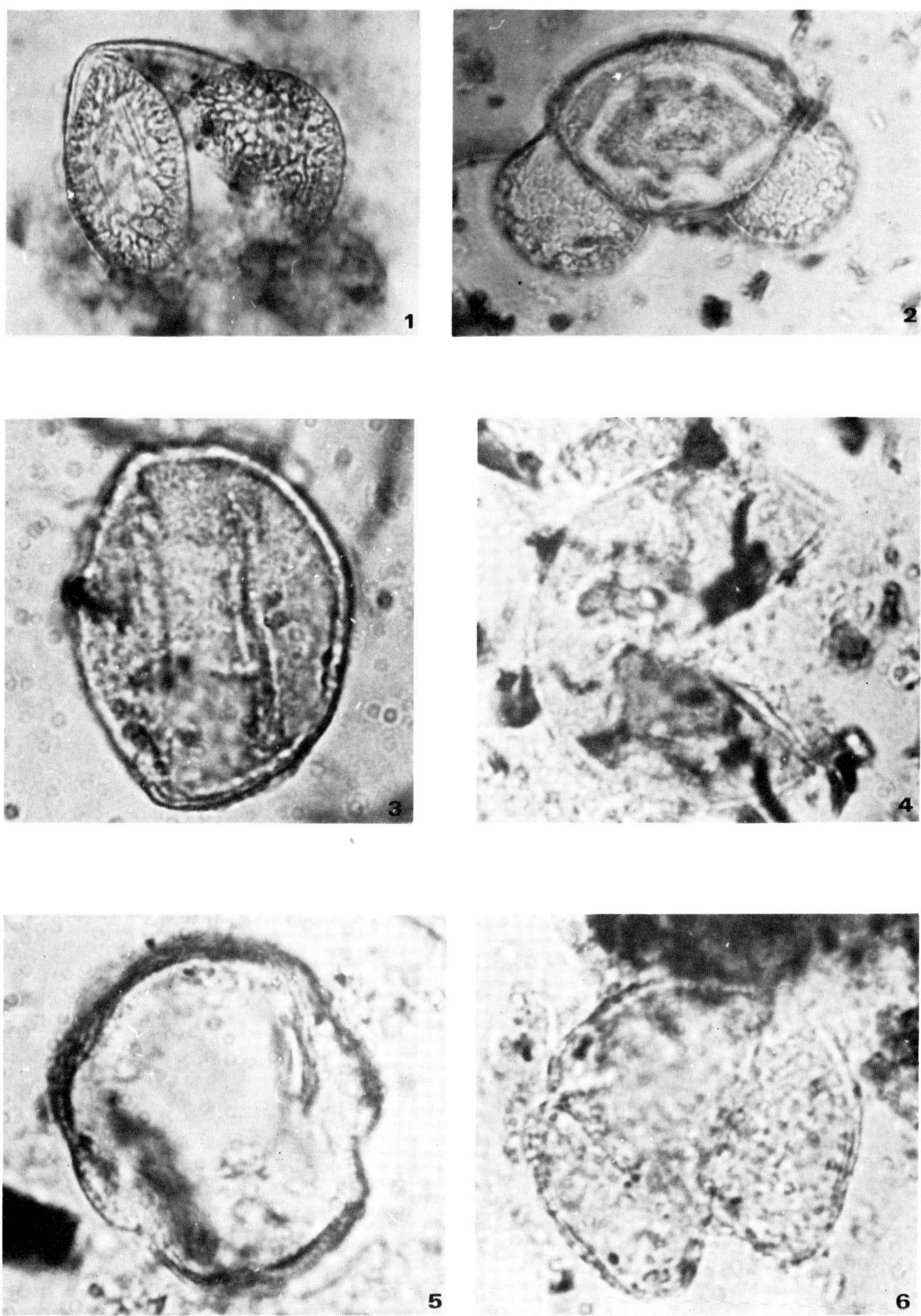

FIG. 5. (1) Pinaceae, *Pinus haplostelle* type. (2) Pinaceae, *Pinus diplostelle* type. (3) Platanaceae, *Platanus* sp. (4) Taxodiaceae, *Inaperturopollenites hiatus*. (5) Hamamelidaceae, *Liquidambar* cf. *orientalis*. (6) Aceraceae, *Acer* sp.

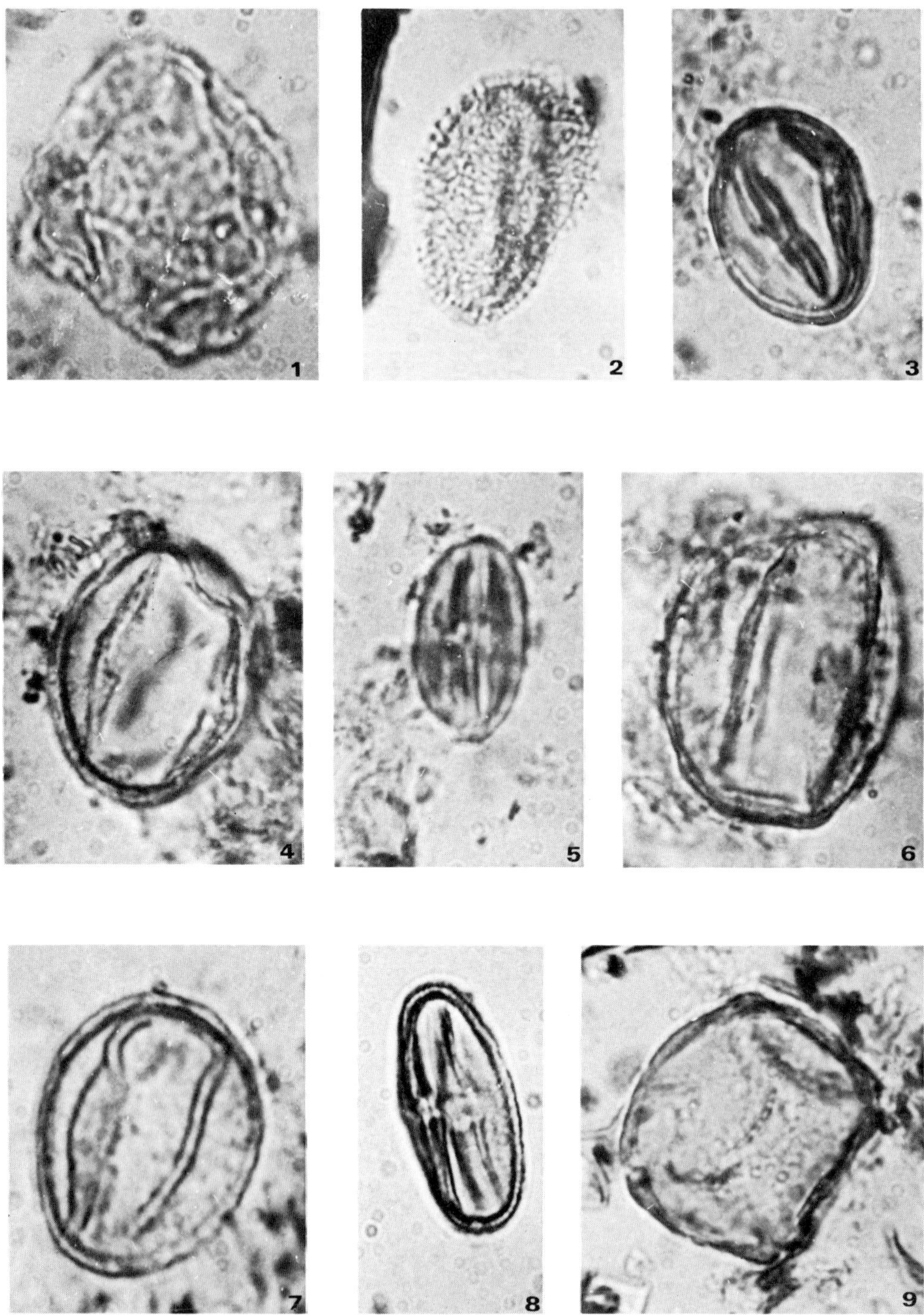

Fig. 6. (1) Ulmaceae, type-*Ulmus*. (2) Fagaceae, *Quercus* type *ilex coccifera*. (3) Fagaceae, *Tricolpopollenites henrici*. (4) Fagaceae, *Psilatricolporites megaexactus*. (5) Fagaceae, *Tricolporopollenites* gr. *cingulum*. (6 & 7) Fagaceae, *Quercus* sp. (8) Umbelliferae. (9) Betulaceae, *Polyvestibulopollenites verus*.

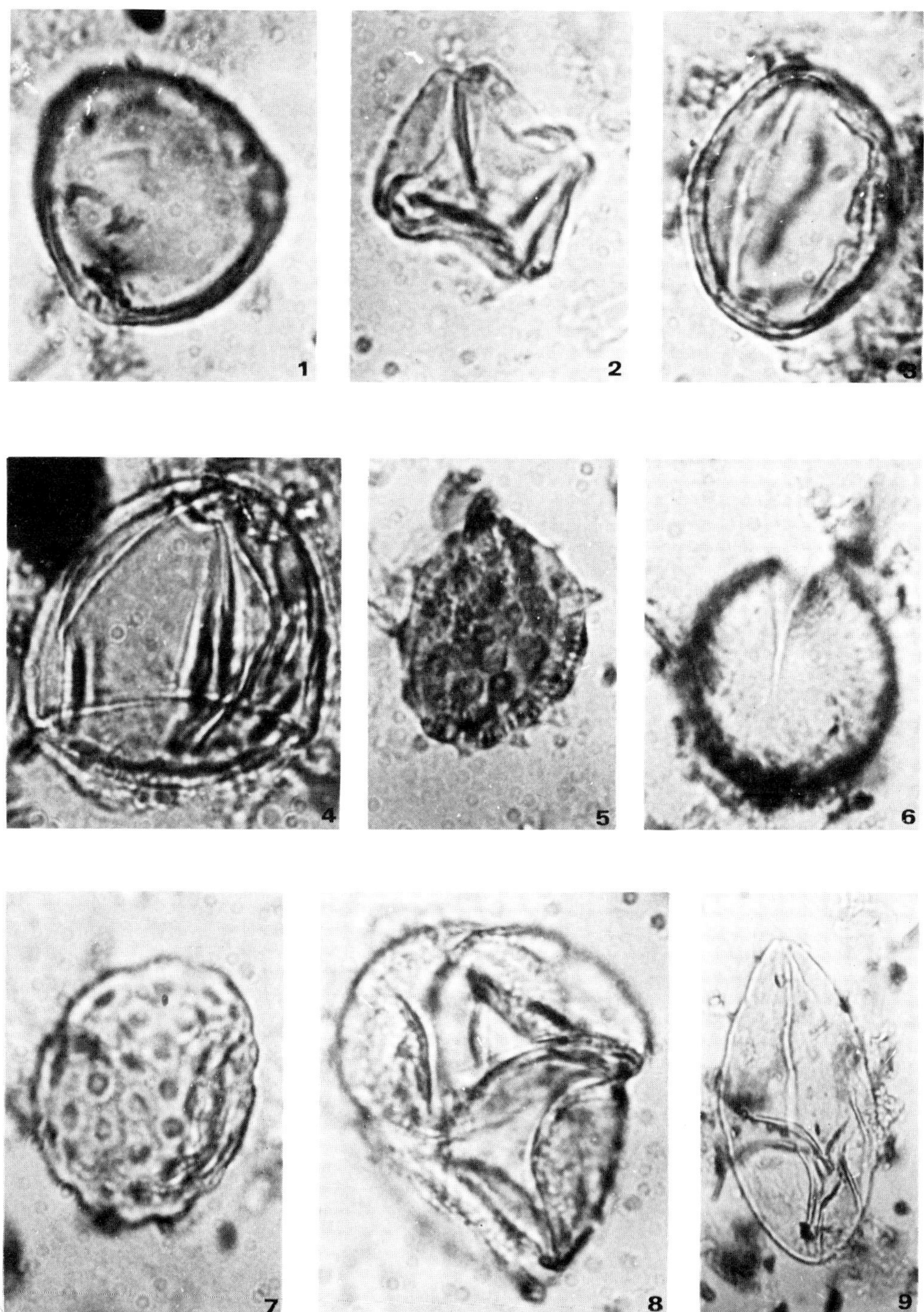

FIG. 7. (1) Betulaceae, *Trivestibulopollenites betuloides*. (2) Betulaceae, *Polyvestibulopollenites verus*. (3) Fagaceae, *Psilatricolporites megaexactus*. (4) Gramineae. (5) Compositae. (6) Nymphaceae. (7) Amaranthaceae-Chenopodiaceae. (8) Tetrads of Typhaceae. (9) Zygnemataceae, *Ovoidites elongatus*.

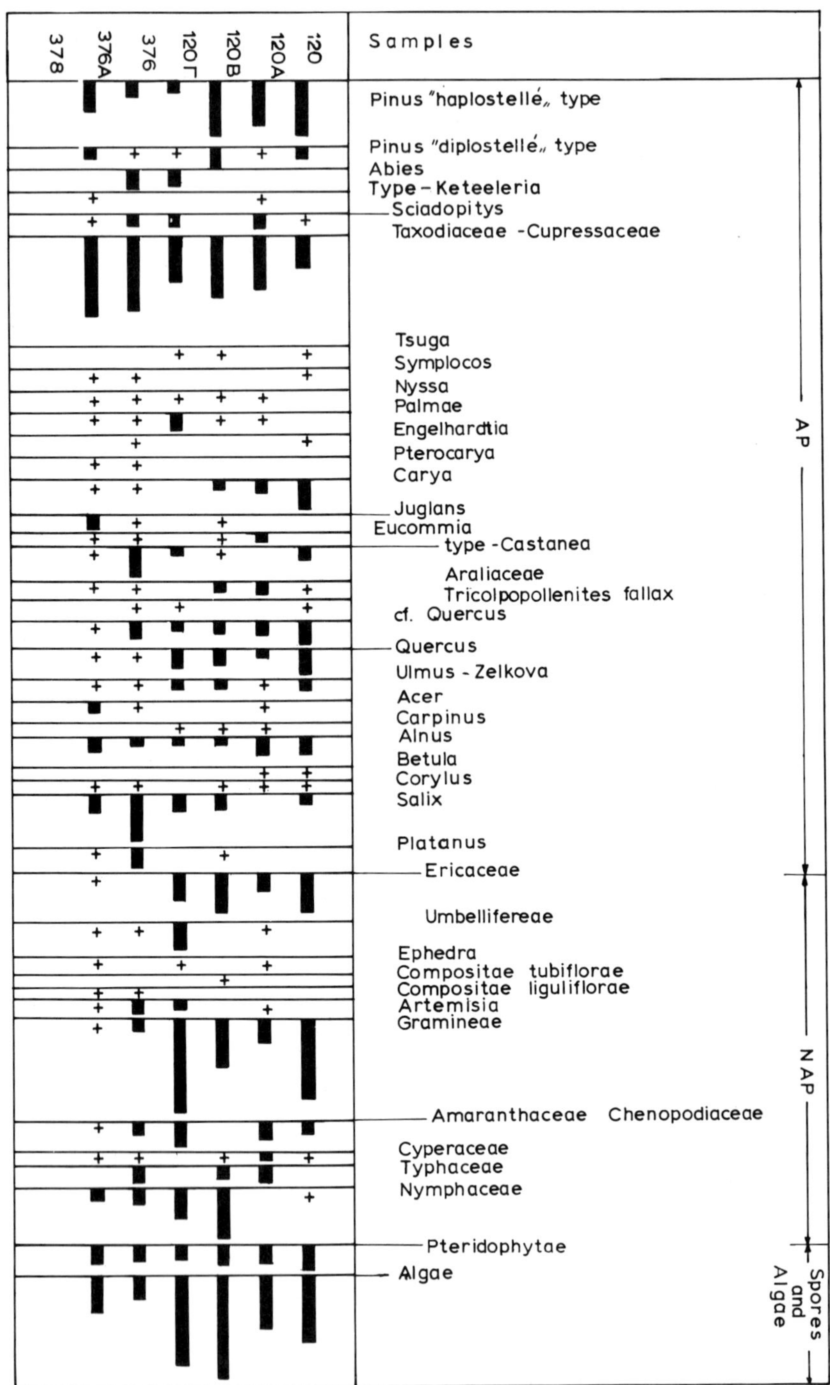

FIG. 8. Palynological diagram of Almopia lacustrine sediments.

nean area. More specifically, we can determine the age of Almopias lacustrine sediments as Lower Pliocene. The key features of the microflora contained in these sediments are: the presence of *Pinus haplostelle* type dominating over *Pinus diplostelle* type; the significant occurrence of Taxodiaceae; Cupressaceae (mainly with *Taxodium* and *Sequoia*) accompanied by the pollen grains of cf. *Quercus*, and the minor participation of *Engelhardtia, Castanea*-type, *Tricolpopollenites fallax* and Araliaceae.

A recent study of the microflora of the lignite-bearing horizons of the Komnina-Ptolemais area (Ioakim, in press) shows that they are comparable to the Almopia lacustrine deposits. They are also of Lower Pliocene age and contain a rich fauna of rodents in addition to many vegetal microfossils (e.g. leaf prints, vegetal tissues).

The microflora in the lacustrine deposits, is comparable to microfloras from Tabianien sediments in the Gulf of Catalonia, southern France and Spain (Cravatte & Suc 1981; Suc 1976, 1980) and from terrestrial deposits of Brunsumian age in Holland (Zagwijn 1960, 1974).

From pollen analysis of the tuffaceous sediments of the volcanic centres of Liasella and Papadhia, to the east and south of the area of lacustrine sediments studied here, a microflora has been identified and attributed to the Upper Pliocene according to Mercier & Sauvage (1965). In Mount Kozuf the equivalent volcanics overlie sediments of post-Pontian age. Recently, Bellon *et al.* (1979) and Kilios *et al.* (1980) have determined the ages of the volcanics to be from 2.65–4.00 Ma and 1.80–5.00 Ma respectively, using the K-Ar method.

These results support a Lower Pliocene age for the Almopias lacustrine deposits. As the deposits overlie volcanic basement we conclude that volcanic activity also began at that time. The microflora reflects dense woody vegetation where the Taxodiaceae grew accompanied by a dispersed feuillus, *Engelhardtia, Nyssa, Symplocos*, Palmae, *Carya* and *Salix*. Coexisting anemophilous pollen grains found on the basin margins such as cf. *Quercus, Ulmus-Zelkova, Castanea*-type, *Carpinus, Acer, Pterocarya* and *Pinus* make up the remainder of the vegetation suite. On the other hand, the significant presence of aquatic plant pollen grains and of various algal micro-organisms supports our view that the environment was marshy as well as the fact that the climate was hot and humid.

Palaeogeography

By the Lower Pliocene, following Tertiary folding, the morphology of the volcanic basement was similar to the present. Eruptions from the present volcanic centres began. Tuffs and lahars accumulated in the bed of Dassikos River at an altitude of ca. 1000 m, forming a natural dam and creating a lake of ca. 0.5 km^2. Volcanism in the area was accompanied by intense SiO_2-rich hydrothermal activity resulting in the formation of cherts, siliceous limestones and Fe-mineralization. At this time the climate was semitropical with dense vegetation around the lake and the lake sediments now contain abundant carbonized vegetal remains, bitumens, lignite intercalations and a rich microflora of Lower Pliocene age. With time the climate became drier and the vegetation decreased considerably. As a result carbonized vegetal remains and other organic components are rare in the sediments and their colour is yellowish. Finally, when the lake sediments reached ca. 1060 m a new volcanic eruption filled the lake, burying its sediments under tuff.

After the end of the Almopias volcanic activity, erosion exposed the lacustrine sediments and the original Lower Pliocene basement morphology.

ACKNOWLEDGEMENTS. The authors thank the General Director of the Institute of Geology and Mineral Exploration (I.G.M.E.) for permission to publish this work. Thanks also go to Professor E. Davis, for critical reading of the manuscript, and also to colleagues of the Institute for their help. Prof. M. Brooks & Dr. C. Page reviewed the manuscript.

References

BELLON, H., JARRIGE, J. J. & SOREL, D. 1979. Les activités magmatiques de l'Oligocène à nos jours et leurs cadres géodynamiques. Données nouvelles et synthèse. *Rev. Géol. dynam. Géogr. phys.*, **21**, 41–55.

CRAVATTE, J. & SUC, J. P. 1981. Climatic evolution of North Western Mediterranean area during Pliocene and Early Pleistocene by pollen analysis and forams of drill Autan 1. Chronostratigraphic correlations. *Pollen & Spores*, **23**, 247–258.

ELEFTHERIADES, G. 1977. *Contribution to the Study of the Volcanic Rocks of the Southern Almopia.* Unpubl. Univ. of Thessaloniki thesis, Greece, 173 pp.

IOAKIM, C. Etude palynologique des formations ligniteuses du Pliocène de Ptolemais (Greece). *Pollen & Spores* (in press).

KOLIOS N., INNOCENTI, F., MANETTI, P., PECCERILLO,

A. & Giuliani, O. 1980. The Pliocene volcanism of the Voras Mts (Central Macedonia, Greece). *Bull. Volcanol.* **43,** 554–568.

Marakis, G. & Sideris, C. 1972. Petrology of the Edessa area volcanic rocks, Western Macedonia. *Bull. Volcanol.* **36,** 462–472.

Mercier, J. 1968. Contribution à l'étude du metamorphisme et de l'évolution magmatiques des zones internes des Hellénides. *Ann. géol. des Pays Hellén.* **20,** 597–792.

—— & Sauvage, J. 1965. Sur la géologie de la Macedoine Centrale: les tufs volcaniques à pollens et spores d'Almopias (Grèce). *Ann. géol. des Pays Hellén.* **16,** 188–201.

Sauvage, J. 1979. Le prépleistocène en Grèce centrale et méridionale: correlations palynologiques et faunistiques, comparaisons avec l'Italie. *Quaternaria, Rome*, **21,** 1–7.

Soldatos, K. 1955. *The Volcanics of the Almopias Area.* Unpubl. Univ. of Thessaloniki thesis, Greece, 48 pp.

Suc, J. P., 1976. Rapport, de la palynologie à la connaissance du Pliocène du Rousillon (Sud de la France). *Geobios*, 741–771.

—— 1980. *Contribution à la connaissance du Pliocène et du Pleistocène inférieur des régions meditérranéennes d'Europe occidentale par l'analyse palynologique des dépôts du Langedoc-Roussilon (Sud de la France) et de la Catalogne (Nord Est de l'Espagne).* Thèse d'Etat Univ. Montpellier, 198 p.

Zagwijn, W. H. 1960. Aspects of the Pliocene and early Pleistocene vegetation in the Netherlands. *Meded. Geol. Sticht.* **5,** 78 p.

—— 1974. The Pliocene-Pleistocene boundary in Western and Southern Europe. *Boreas*, **3,** 75–97.

P. Chorianopoulou, A. Galeos & C. H. Ioakim, Institute of Geology and Mineral Exploration, 70 Messoghion Str., Athens 608, Greece.

Sapropelic layers in the NW Aegean Sea

A. Cramp, M. B. Collins, S. J. Wakefield & F. T. Banner

SUMMARY: The present study examines sapropelic layers and associated sediments in the NW Aegean Sea, based upon the analysis of cores collected during the 7/78 cruise of the R.R.S. *Shackleton*.

Preliminary sedimentological analyses of some of the cores, including the use of X-ray techniques, did not at first detect either a sapropel or a sapropelic layer (defined as >2% and 0.5 – 2.0% organic carbon, respectively—Sigl *et al.* 1978) in the sediment. Further investigations, including the analysis of organic carbon, on twelve of these cores have demonstrated the presence of horizons relatively rich in organic carbon. These horizons contain up to approximately 1.0% organic carbon and could be regarded as sapropelic layers. The horizons vary between 26 and 33 cm in thickness and are found at depths of 25 to 100 cm below the sediment-water interface, in cores which range from 101 to 240 cm in length.

Organic carbon profiles are presented, together with a discussion of the corresponding X-ray analysis and a preliminary note on the biogenic contributions to these sediments.

Cyclic variations in the organic carbon content of marine sediments have been recognized in stratigraphic sections dating back to the Jurassic (see, for example, Dean *et al.* 1977; Dean & Gardner 1980; Hay *et al.* 1982). The variations, of periods of 10^4–10^7 years, may be related to any one of a number of factors but with climatic control a favoured controlling mechanism (e.g., Fisher & Arthur 1977; Dean & Gardner 1980). Whether bottom water anoxicity is *a priori* required for these organic rich layers to form or whether their formation *causes* bottom water anoxicity will depend upon local conditions. Both situations have been recognized (e.g., Dean *et al.* 1977; Degens & Stoffers 1980). In the eastern Mediterranean, both sedimentological and micropalaeontological investigations have been directed towards explaining the periodic formation of the organic-rich layers found in the area.

The upper sapropelic layer of the Aegean Sea is tentatively correlated with the S_1 horizon identified by Stanley (1978) in the eastern Mediterranean. This sapropelic layer is an olive green/grey hemipelagic subsurface sedimentary horizon of wide extent (Fig. 1). The layer has a relatively high organic carbon level, ranging between 0.5–2.0% (Sigl *et al.* 1978). It is believed to have been formed during a warming climatic cycle in the Early Holocene (Ryan 1972) and has been dated at between 7500 and 9000 years (Van Straaten 1972). Deposition is thought to have occurred in a semi-permanently stratified water mass (e.g., Olausson 1961); this would have produced a state where vertical advective movement and thus re-oxygenation of the water column would have been limited and is believed to have been a basin-wide phenomenon (Maldonado & Stanley 1976). The factors which could have combined to produce this oceanographic feature, first recognized by Bradley in 1938, are equivocal (Kullenberg 1952; Olausson 1961; Ryan 1972; Van Straaten 1972; Huag & Stanley 1972; Nesteroff 1973; Thunell *et al.* 1977; Vergnaud-Grazzini *et al.* 1977; Sigl *et al.* 1978; Williams & Thunell 1979). A favoured theory (Olausson 1961) involves the input of relatively cold low salinity surface water, derived from glacial melt water, which issued into the Black Sea and entered the Aegean and eastern Mediterranean, over the Bosphorus sill (Fig. 1).

The Aegean Sea is an epicontinental sea situated between the Black Sea and the eastern Mediterranean; it has a highly complex bathymetry (Maley & Johnson 1971) and the basin is tectonically active (Ferentinos *et al.* 1981).

Little work has been published regarding the sedimentology of the region although data have been presented concerning volcanic ash and sapropel horizons, first recognized in the area by Ninkovich & Heezen, 1965. In particular, the NW Aegean has been studied by Collins *et al.* (1981), Lykousis *et al.* (1981) and Perissoratis *et al.* (1981).

Materials and methods

Twenty-two gravity cores were recovered from the area during cruise 7/78 of R.R.S. *Shackleton*. Previous investigations of five of these cores (Lykousis 1980) have focused on sedimentological analyses, such as grain size distribution, carbonate content, clay mineralogy and X-radiography.

The present study has concentrated upon eight cores (Table 1) forming a transect running

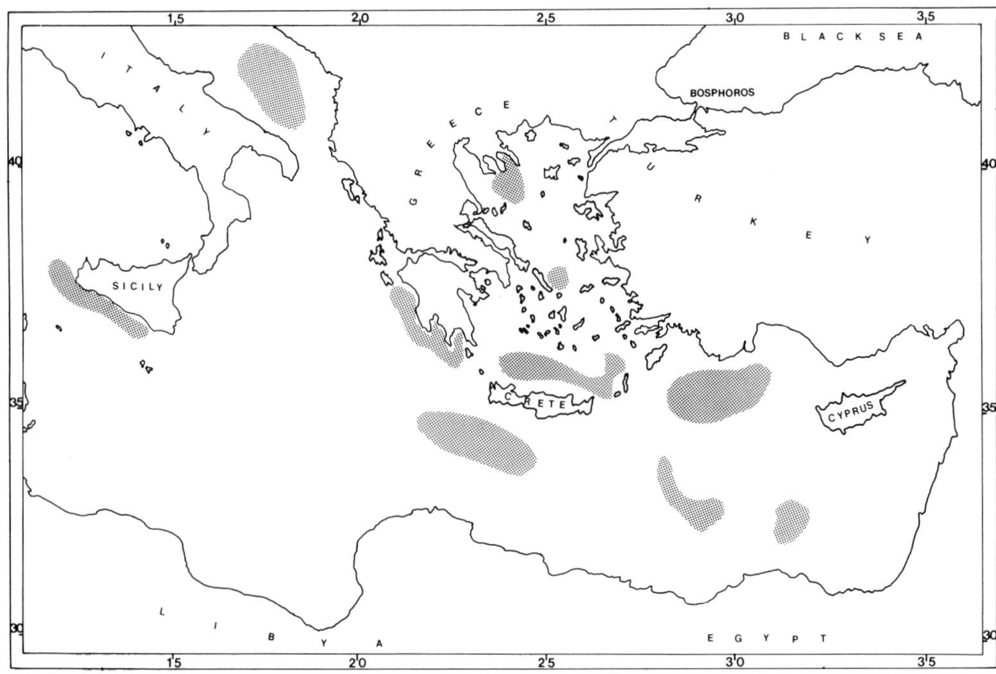

FIG. 1. Sapropelic (S_1) distribution in the eastern Mediterranean, compiled from references cited.

TABLE 1. *Details of cores used in study*

Stn. No.	Lat. (N)	Long. (E)	Water Depth (m)	Core Length (m)
NA 37	39°57.9	32°15.8	115	1.01
NA 33	39°47.6	23°20.3	240	2.00
NA 44	39°37.4	23°32.6	510	2.39
NA 42	39°37.1	23°38.2	640	1.93
NA 43	39°36.4	23°35.3	610	2.40
NA 45	39°33.2	23°31.5	830	2.30
NA 30	39°27.8	23°37.8	1150	1.93
NA 28	39°20.1	23°35.3	1055	2.30

NW/SE from Thermaicos Gulf to the Sporades Basin in the NW Aegean Sea (Fig. 2). These cores range from 101 cm to 240 cm in length and were collected in water between 115 m and 1150 m in depth.

The cores were split, with minimum disturbance, and 0.25 cm thick slices were X-radiographed. Sub-samples of the cores were analysed for organic carbon using a Leco® carbon analyser. The method adopted being similar to that used by Pedersen & Price (1980) whereby the samples are leached with phosphoric acid to eliminate $CaCO_3$, ensuring that any acid-soluble organic carbon is retained with the decalcified solid for analysis. All samples were run in duplicate and the analyses were accepted if the

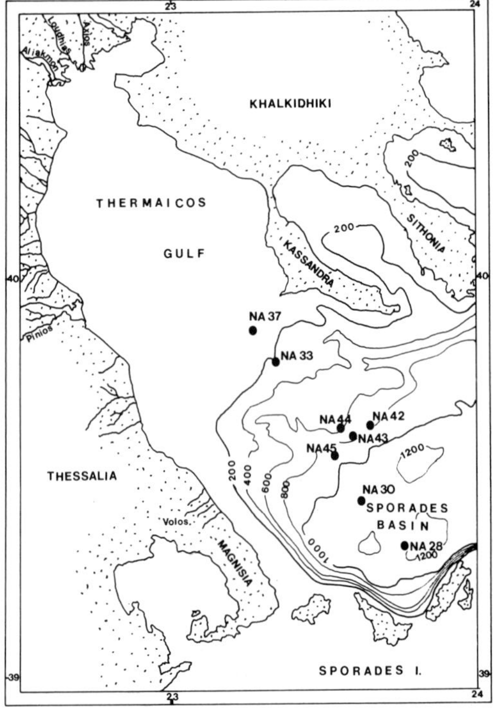

FIG. 2. Study area and core locations; bathymetry from Lykousis *et al.* 1981.

two results were within 10% of the mean; otherwise, additional replicates were run. Analytical precision was at ±9% (2σ) at the 0.5% C_{org} level.

Results

(a) Organic carbon

Vertical profiles of organic carbon levels in the cores are presented in Fig. 3. The range of measured organic carbon levels over all the samples is from 0.1 to 1.0%.

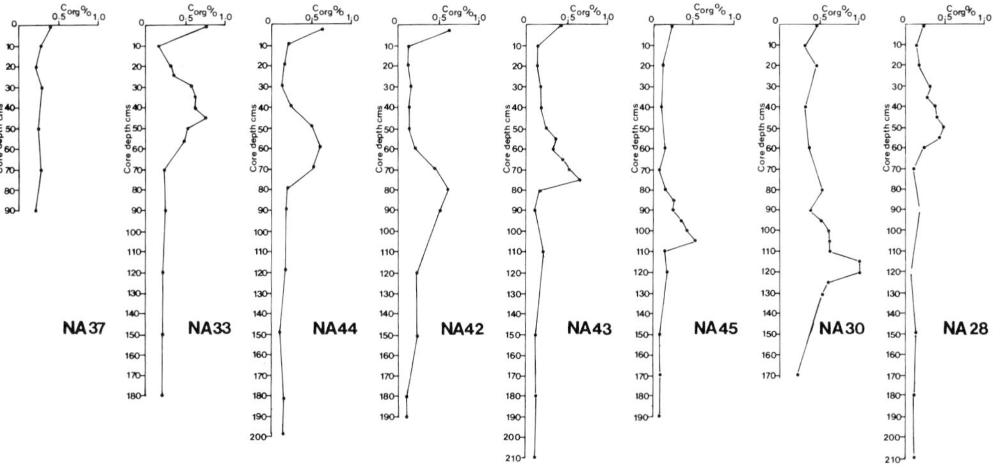

FIG. 3. Organic carbon levels for transect cores.

Seven of the eight cores show a distinct sub-surface 'peak' in organic carbon.

The level in the core at which organic carbon starts to rise is found to be progressively deeper from north–west to south–east along the sample transect, as far as core NA 30. It then shallows in the south-easternmost core, NA 28 to between 40 and 50 cm from the sediment water interface (Fig. 4). These zones of organic carbon enrichment (C_{org} > 0.2%) vary between 26 and 33 cm in thickness within the transect (NA 43: 51–77 cm; NA 30: 97–130 cm). Sections with C_{org} > 0.5% are more limited in extent (Fig. 3). Organic carbon enrichment is also found in the upper 10 cm of the cores with a maximum value at the sediment/water interface, e.g., NA 33–0.75% C_{org}.

There is a correlation between the increase in organic carbon and distinct colour changes of the sediment, identified using a Munsell colour chart. There is an abrupt change from 'a light olive gray' (5YR 6/2) to a distinctive 'olive gray' (5YR 4/2), where organic carbon levels increase. Surface enrichment also corresponds to a colour change, from a 'light olive gray' (5YR 6/2) at minimum C_{org} levels, to a 'brown' (10YR 5/3) at maximum C_{org} levels. Background levels of C_{org} are 0.1 to 0.2%.

(b) X-radiography

The majority of core material can be defined as either nearly structureless 'unifites' or 'faintly laminated muds' (Stanley & Maldonado 1981).

Cores taken from the continental slope (NA 42, 43 and 44) are more laminated in structure than cores recovered from the basin (NA 28, 30 and 45). The laminations are more distinctive below the sub-surface organic carbon enrichment units than above it.

The sub-surface organic rich horizons are 'mottled/speckled' in appearance. The base of this distinct sedimentary horizon is abrupt, whereas the upper section grades into the overlying structure. There is some evidence of limited bioturbation elsewhere in the cores. Pyritization occurs within and below the sub-surface high C_{org} layer (see also Lykousis 1980).

Visually and in the X-rays, a banded shell layer can be identified in core NA 37 at 25–30 cm, and a sand layer in NA 30 at 14–15 cm below the surface.

(c) Microfaunas

A micropalaeontological reconnaissance of cores NA 42 and 44 reveals that the high C_{org} sapropelic layer is characterized in each case by a conspicuous disappearance of *Uvigerina*

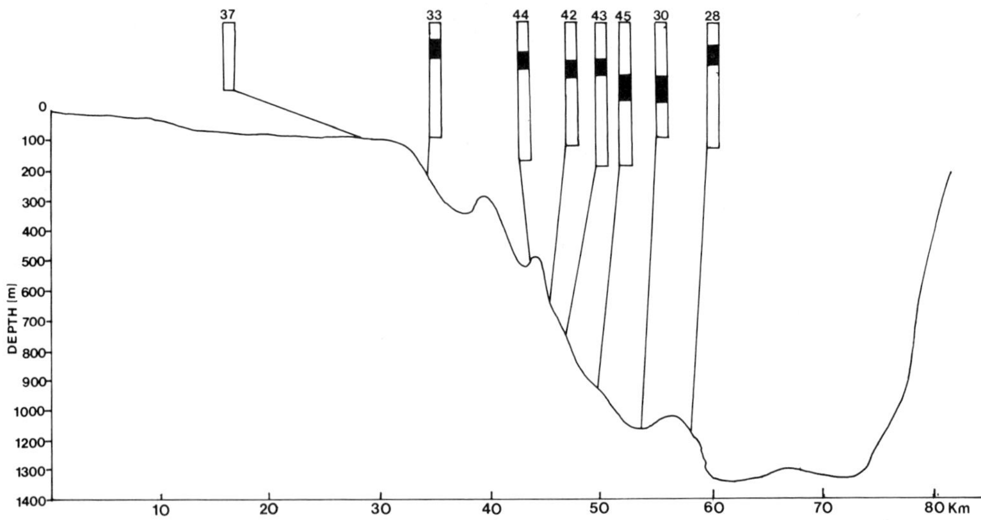

FIG. 4. Present day bathymetry of transect. Black areas indicate sapropelic horizons (derived from Lykousis *et al.* 1981). Bathymetry: vertical exaggeration ×25. Cores enlarged ×2 rel. to bathymetry.

mediterranae Hofker (present below 1300 m in the modern Mediterranean, and represented by abundant large specimens in the cores below the sapropelic layer) and a reduction in the diversity of the benthonic foraminiferal assemblage as a whole. However, the sapropelic layer noticeably contains abundant large *Globigerinoides ruber* (d'Orb.) and *Orbulina universa* (d'Orb.) with rarer *Globigerinella siphonifera* (d'Orb.) in a diverse planktonic foraminiferal assemblage, which, like those of the underlying and overlying sediment, lacks keeled *Globorotalia*). These three species prefer an upper water (top 100 m) habitat and require oceanic salinities. *Globigerinoides ruber* and *Globigerinella aequilateralis* (=*G. siphonifera*) prefer temperatures between 15°C and 24°C (e.g., Bé 1977). Additionally, the sapropelic layer contains conspicuous numbers of euthecosomatous pteropods, including *Limacina inflata* (d'Orb.), a species which migrates within the top 300 m of the water column, preferring 14–28°C temperatures and 35.5–36.7‰ salinities (Bé & Gilmer 1977); *Creseis acicula* (Rang) which is most abundant in the top 50 m but ranges through the top 500 m of the water column, with optima at temperatures of 24–27°C and salinities in the 25–45‰ range (Bé & Gilmer 1977), and *C. virgula constricta* (Chen & Bé), which is known previously from the Sargasso Sea. The microfaunal assemblages, as a whole, are very similar to those previously recorded from Mediterranean late Quaternary sapropels (e.g., Herman 1981).

Discussion

(a) Sedimentology

The data presented indicate that a sapropelic layer is present in seven of the eight cores analysed from a north–west/south–east transect in the NW Aegean. The identification of this layer is baed upon C_{org} levels, sediment colour and X-radiography. This layer was not recognized in the original sedimentological studies of Collins *et al.* (1981), but can be equated with a similar enrichment identified using the same criteria, by Perissoratis (1981) in cores taken further to the east. These two complementary data sets indicate sapropelic layer formation to be a 'basin-wide' phenomenon (Maldonado & Stanley 1976) in the NW Aegean (Fig. 1).

It is interesting to note that the sapropelic layer identified in the NW Aegean cores differs markedly in appearance from the sapropels (*sensu stricto*) identified elsewhere. The sapropelic layers in our cores are mottled rather than well-laminated in structure, olive-gray rather than black and have lower C_{org} content, compared with those analysed by Sigl *et al.* (1978).

The location of the sapropelic layer in each of the cores relative to present-day bathymetry is shown in Fig. 4. There are two main features of interest: (i) the core from the continental shelf does not contain an organic carbon-rich sapropelic layer; and (ii) the depth from the sediment/water interface to the upper boundary of the sapropelic layer increases downslope towards the basin (i.e., the sediment above that layer thickens).

The absence of the layer at the shallowest water depth is a feature which has also been noted by Perissoratis (1981) in cores collected from Toranaios Kolpos, the shallow westerly embayment of the Khalkidhiki Peninsula. This absence could be related to a number of factors, such as relative sea level changes, high sedimentation rates or subsequent erosion, acting separately or in conjunction. This is yet to be elucidated by our continuing studies.

Assuming that the sapropelic layer is contemporaneous throughout the area, the increased depth of burial of the layer towards the basin infers that the net sedimentation rate is higher within the basin than on the continental slope. This difference could be attributed to various processes: for example, (a) an increased sediment supply to the basin which is not available to the slope, e.g., input from the landmass to the west, or from the Black Sea, to the east; or (b) increased biogenic primary productivity of calcium carbonate offshore, or (c) erosional downslope processes redepositing fine-grained material from the slope to the basin. The last process would be consistent with a tectonically-active area in which 'gravitational sliding and slumping is widespread on the Thermaicos Slope and the steeper escarpment' (Ferentinos *et al.* 1981). The absence of any distinctive laminations in the sediments overlying the sapropelic layer does not preclude this explanation as similar uniform deposits from the Hellenic Trench, western Greece, have been attributed to deposition of material from low concentration turbidity flows (Blanpied & Stanley 1981). Hence, this deposition within the Sporades Basin would represent the distal end of a transport path.

(b) Micropalaeontology

Micropalaeontological data cited above and other studies (e.g., Williams *et al.* 1978) have been used to substantiate Olausson's (1961) model in which deglacial surface-water dilution increased water column stratification and reduced vertical advection, combining to produce conditions favourable for sapropel formation. It is clear, however, that regional inferences from isotopic analyses must be treated cautiously both in comparison with global events (Berger *et al.* 1978) and in consideration of the isotopic disequilibrium in the tests of certain species relative to seawater (Shackleton *et al.* 1973; Deuser *et al.* 1981). The numerically dominant species recognized in the sapropelic layers of cores NA 42 and 44 would, by themselves, suggest that upper waters were then of salinity about 35–37‰ and temperature about 24°C, which compare to the values of about 38‰ and 17–19°C of the present surface Aegean Sea.

(c) Concluding remarks

If the sapropelic layer identified here *is* correlateable with the upper sapropel (S_1) layer identified elsewhere in the eastern Mediterranean, then deposition would have occurred around 7500 to 9000 yrs BP (Van Straaten 1972). The inferred sequence of sediment transport and depositional processes, as a time-series, would then be:

(1) hemipelagic, calcareous sedimentation, possibly with low concentration turbiditic inputs (present day);

(2) hemipelagic sedimentation including the formation of the sapropelic layer (around 10,000 BP);

(3) turbiditic mud deposition, with supplementary hemipelagic sedimentation.

These three stages of deposition correspond with the lithological types identified by Collins *et al.* (1981).

ACKNOWLEDGEMENTS: The research was supported by the Natural Environment Research Council. The assistance provided by the officers and crew of R.R.S. *Shackleton* 7/78, is gratefully acknowledged.

The authors thank Dr. G. Evans and Martin Gill of the Department of Geology, Imperial College, London, for their invaluable help involving core cutting and preparation of material, and for making available the X-ray equipment.

References

BÉ, A. W. H. 1977. An ecological, zoogeographic and taxonomic review of recent planktonic foraminifera. *In:* RAMSEY, A. T. S. (ed.), *Oceanic Micropalaeontology*, **1**, 1–100, Academic Press.

—— & GILMER, R. W. 1977. A zoogeographic and taxonomic review of euthecosomatous pteropoda. *In:* RAMSEY, A. T. S. (ed.), *Oceanic Micropalaeontology*, **1**, 733–808, Academic Press.

BERGER, W. H., KILLINGLEY, J. S. & VINCENT, E. 1978. Stable isotopes in deep-sea carbonates. Boxcore ERDC-92, West Equatorial Pacific. *Oceanologia Acta*, **1**(2), 203–216.

BLANPIED, C. & STANLEY, D. J. 1981. Uniform mud

(unifite) deposition in the Hellenic Trench, Eastern Mediterranean. *Smithsonian Contrib. to the Marine Sciences*, **13,** 40 pp.

Bradley, W. M. 1938. Mediterranean sediments and Pleistocene sea levels. *Science*, **88,** 376–379.

Collins, M. B., Lykousis, V. & Ferentinos, G. 1981. Temporal variations in sedimentation patterns: NW Aegean Sea. *Marine Geology*, **43,** M39–M48.

Dean, W. E. & Gardner, J. V. 1980. Cyclic variations of organic carbon in anoxic facies of Jurassic to Eocene age, Cape Verde Basin (DSDP Site 367), continental margin of north-west Africa. *In: Abstracts, Vol. 1, 26th Int. Geol. Congress, Paris*, July 1980, 221 pp.

——, ——, Jansa, L. F., Čepek, P. & Seibold, E. 1977. Cyclic sedimentation along the continental margin of north-west Africa. *In:* Lancelot, Y., Seibold, E. *et al., Initial Reports of the Deep Sea Drilling Project*, **41,** Washington, D.C., 965–989.

Degens, E. T. & Stoffers, P. 1980. Environmental events recorded in Quaternary sediments of the Black Sea. *J. geol. Soc. London*, **137,** 131–138.

Deuser, W. G., Ross, E. M., Hemleben, C. & Spindler, M. 1981. Seasonal changes in species composition, numbers, mass, size and isotopic composition of planktonic foraminifera settling into the deep Sargasso Sea. *Palaeogeography, Palaeoecology, Palaeoclimatology*, **33,** 103–127.

Ferentinos, G., Brooks, M. & Collins, M. B. 1981. Gravity-induced deformation on the north flank and floor of the Sporades Basin of the north Aegean Sea trough. *Marine Geology*, **44,** 289–302.

Fisher, A. G. & Arthur, M. A. 1977. Secular variations in the pelagic realm *In:* Cook, H. E. & Enos, P. (eds). Deep Water Carbonate Environments, *Spec. Publ. Soc. econ. Palaeont. Miner.*, **25,** 19–50.

Hay, W. W. & Sibuet, J. C. *et al.* 1982. Sedimentation and accumulation of organic carbon in the Angola Basin and on Walvis Ridge: Preliminary Results of Deep Sea Drilling Project Leg 75. *Bull. geol. Soc. Amer.* **93,** 1038–1050.

Herman, Y. 1981. Paleoclimatic and paleohydrologic record of Mediterranean deep-sea cores based on pteropods, planktonic and benthonic foraminifera. *Revista Española de Micropaleontologia*, **13**(2), 171–200.

Huang, T. C. & Stanley, D. J. 1972. Western Alboran Sea: sediment dispersal, ponding and reversal of currents. *In:* Stanley, D. J. (ed.), *The Mediterranean Sea: A natural sedimentation laboratory*. Stroudsburg Pa., Dowden, Hutchinson & Ross, 521–559.

Kullenberg, B. 1952. On the salinity of the water contained in marine sediments. *Meddelonder fran Oceanografiska Institutet i Göteborg*, **21,** 1–38.

Lykousis, V. 1980. *Aspects of the Modern and Holocene Sedimentation of the NW Aegean.* Thesis M.Sc. (unpubl.) University of Wales, 108 pp.

——, Collins, M. B. & Ferentinos, G. 1981. Modern sedimentation in the NW Aegean Sea. *Marine Geology*, **43,** 111–130.

Maldonado, A. & Stanley, D. J. 1976. The Nile Cone: submarine fan development by cyclic sedimentation. *Marine Geology*, **20,** 29–40.

Maley, T. S. & Johnson, G. L. 1971. Morphology and structure of the Aegean Sea. *Deep-Sea Research*, **18,** 109–122.

Nesteroff, W. D. 1973. Petrography and mineralogy of sapropels. *In:* Ryan, W. B. F. *et al., Initial Reports of D.S.D.P.*, **13**(2), Washington, D.C., 713–720.

Ninkovich, D. & Heezen, B. 1965. Santorini Tephra. *XVII Symp. of the Colston Research Society*, 413–453.

Olausson, 1961. Studies of deep-sea cores. *Reports of Swedish Deep-Sea Expedition, 1947–48*, **8**(4), 353–391.

Pedersen, T. F. & Price, N. B. 1980. The geochemistry of iodine and bromine in sediments of the Panama Basin. *Journal of Marine Research*, **38,** 397–411.

Perissoratis, C. 1981. Published abstracts, International Symposium on the Hellenic Arc and Trench (H.E.A.T.), Athens.

——, Angelopoulos, J. & Mitropoulos, D. 1981. A sapropelic layer in cores taken from the Gulf of Hagion Onos, northern Aegean Sea. *CIESM, XXVII, Congr. Ass. Pl. Cagliari.*

Ryan, W. B. F. 1972. Stratigraphy of late Quaternary sediments in the eastern Mediterranean. *In:* Stanley, D. J. (ed.). *The Mediterranean Sea: A natural sedimentation laboratory*. Stroudsburg, Pa., Dowden, Hutchinson & Ross, 149–169.

Shackleton, N. J., Wiseman, J. D. H. & Buckley, H. A. 1973. Nonequilibrium isotope fractionation between seawater and planktonic foraminiferal tests. *Nature*, **242**(5394), 177–179.

Sigl, W., Chamley, H., Fabricius, S., D'Argoid, G. C. & Müller, J. 1978. Sedimentological and environmental conditions of sapropels. *In:* Hsü, K. *et al., Initial Reports D.S.D.P.*, **42**(1), Washington, D.C., 445–465.

Stanley, D. J. 1978. Ionian Sea sapropel distribution in the late Quaternary palaeogeography in the Eastern Mediterranean. *Nature*, **274**(5667), 149–152.

——, & Maldonado, A. 1981. Depositional model for fine-grained sediment in the western Hellenic Trench, eastern Mediterranean. *Sedimentology*, **28,** 273–290.

——, — & Struckenrath, R. 1975. Strait of Sicily sedimentation depositional rates and possible reversal of currents in the late Quaternary. *Palaeogeography, Palaeoecology, Palaeoclimatology*, **18,** 270–291.

Thunell, R. C., Williams, D. F. & Kennett, J. P. 1977. Late Quaternary palaeoclimatology, stratigraphy and sapropel history in the eastern Mediterranean deep-sea sediments. *Marine Micropal.* **2,** 377–388.

Van Straaten, R. M. V. U. 1972. Holocene stages of oxygen depletion in the waters of the Adriatic Sea. *In:* Stanley, D. J. (ed.). *The Mediterranean Sea: A natural sedimentation laboratory*. Stroudsberg, Pa., Dowden, Hutchinson & Ross, 631–643.

Vergnaud-Grazzini, C., Ryan, W. F. B. & Cita,

M. B. 1977. Stable isotopic fractionation, climatic change and episodic stagnation in the eastern Mediterranean during the late Quaternary. *Marine Micropal.* **2**, 353–370.

WILLIAMS, D. F. & THUNELL, R. C. 1979. Faunal and isotopic evidence for surface water salinity changes during sapropel formation in the eastern Mediterranean. *Sediment. Geol.* **23**, 81–93.

——, —— & KENNETT, J. P. 1978. Periodic freshwater flooding and stagnation of the eastern Mediterranean Sea during the late Quaternary. *Science*, **201**, 252–259.

A. CRAMP, M. B. COLLINS, S. J. WAKEFIELD & F. T. BANNER, Department of Oceanography, University College, Singleton Park, Swansea, SA2 8PP.

A Middle Miocene thermal event in northern Greece confirmed by coalification measurements

E. D. Chiotis

SUMMARY: An important thermal event is confirmed by vitrinite reflectivity measurements on coal samples from deep oil exploration wells in Kavala Bay and the Nestos delta, northern Greece. Only the lower depositional sequence of the Nestos Delta, to which a Middle Miocene age is attributed, has been affected. The coal bearing sediments have been heated to temperatures above 200°C at depths less than 2000 m. A plutonic intrusion east of the Nestos wells is postulated to have been the cause. The event corresponds in time to the well known episode of Middle Miocene local anatexis and acid plutonism in the central and southern Aegean area.

Vitrinite reflectivity measurements of Ro, the average reflectance under oil immersion, are used as a routine indicator of organic matter maturity in oil exploration. In addition they can provide valuable information about heating associated with tectonism, plutonism or depositional burial. Vitrinite is the coalification product of woody plant remains. It reacts very sensitively when heated and its optical reflectivity measured microscopically indicates the degree of coalification. The coalification process mainly depends on temperature and duration of heating. Therefore vitrinite reflectivity can not serve as a geological thermometer unless the duration of heating is also known.

Ro values on a logarithmic scale are plotted against depth in metres for the oil wells Nestos-1, -2, & -3; Peramos-1 & -2 and Ammodis-1, which penetrated Neogene sediments in Northern Greece (Figs 1 & 2). The measurements fall into two distinct groups. The first consists of the Ro values for Nestos wells from depths deeper than 2450 m. The second group comprises all the remaining measurements.

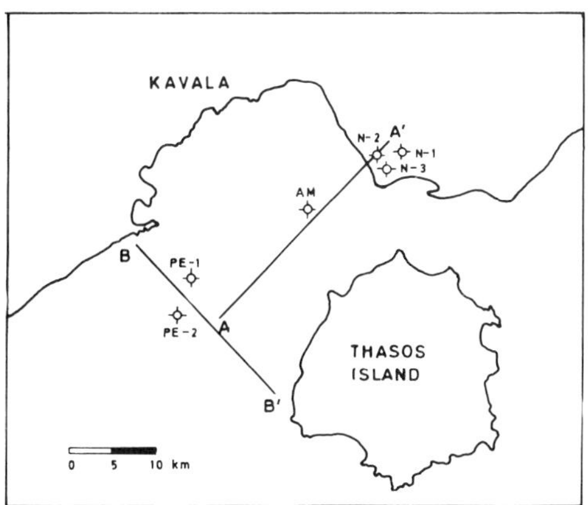

FIG. 1. Location map of wells and sections.

There is a significant discontinuity in Nestos values around a depth of about 2450 m. This Ro gap can be attributed either to erosion of some 500 m of intervening sediments or to a thermal discontinuity. The Nestos-2 sediments deeper than 2450 m have been heated intensively as evidenced by their much higher Ro values. A calorific value of about 8500 kcal/kg is estimated for the deepest coal seams (Ro = 1.86%), which rank in coalification between low volatile bituminous coal and semi-anthracite. Correlation of the data from Nestos wells and from an on-land seismic survey enabled Kousparis (1979) to distinguish four seismic sequences.

The lowest seismic sequence 1 consists of continental clastic sediments with coal layers of probable Serravallian age. Seismic sequence 2 of Tortonian age, is composed of alternating siltstone, claystone and sandstone and is referred to as the 'varve sequence'. The third seismic sequence is represented by Messinian sediments and the fourth comprises Pliocene and Pleistocene sediments.

These seismic sequences can be correlated with marine seismic sections as illustrated in Figs 3 & 4. The top of the Serravallian sequence in the well Nestos-2 was encountered at about 2450 m. It corresponds to the observed Ro-discontinuity. Thus high vitrinite reflectivity values are restricted to the Serravallian sediments. These sediments, which are 1500 m thick in the Nestos-2 well do not occur in the wells Peramos-1, -2 and Ammodis-1 where Tortonian sediments directly overlie the metamorphic basement.

Lopatin's method was applied for the estimation of temperatures developed during the Serravallian on the basis of Nestos-2 vitrinite reflectivity data. Waples's (1980) correlation of time-temperature index with Ro was used, corrected after Kettel (1981). Temperatures for isothermal heating resulting in the measured Ro values were calculated. Calculations were

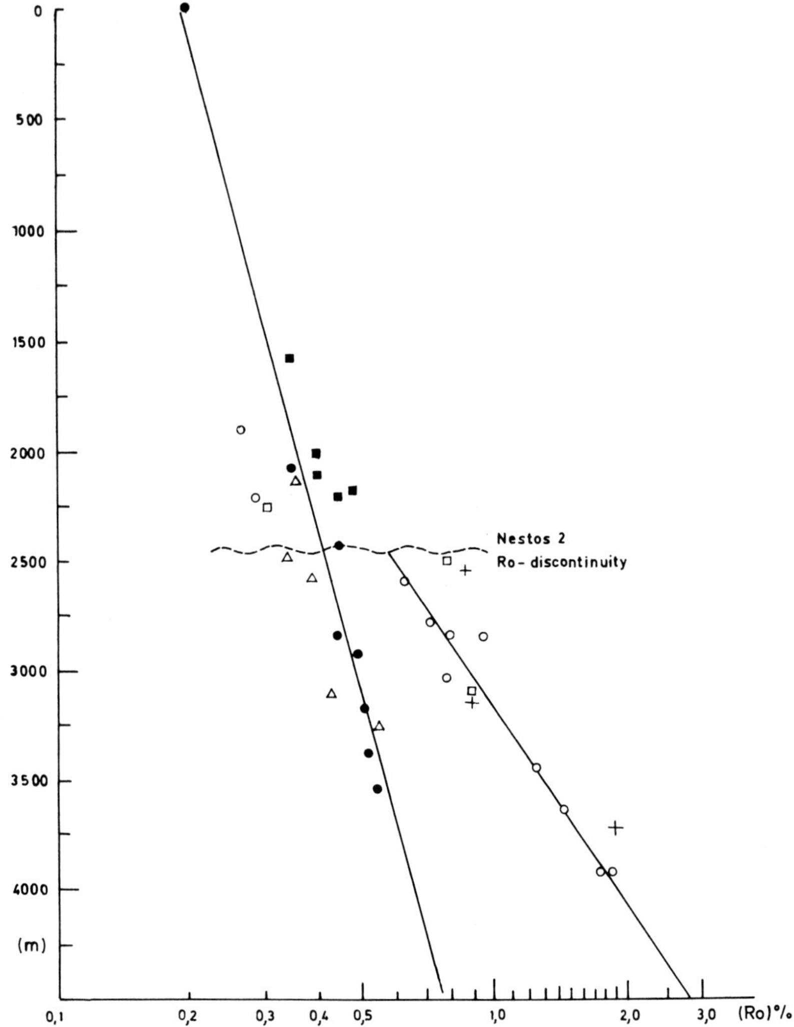

FIG. 2. Vitrinite reflectivity measurements versus depth for the wells: Nestos-1 (□), Nestos-2 (○), Nestos-3 (+), Ammodis-1 (●), Peramos-1 (△) and Peramos-2 (■). Regression lines for Ammodis-1 and Nestos-2 measurements are also shown.

based on two alternative extreme assumptions. Temperatures, T_1, (Fig. 5) were estimated for a 4 Ma duration of heating, that is throughout the whole of the Serravallian. Alternative temperatures, T_2, (Fig. 5) were estimated for a 0.5 Ma duration of heating at the end of Serravallian.

The estimated temperatures which range up to 270°C, correspond to burial depths less than 2000 m. Furthermore they are the constant temperatures needed to produce the measured degree of coalification over the same heating period.

Therefore actual maximum temperatures would have been higher than T_1 or T_2. Such high temperatures, developed at depths less than 2000 m must have been generated under magmatic influence. Therefore, a nearby plutonic intrusion is postulated to have been emplaced during the Serravallian. The small temperature difference between the upper and the lower layers (Fig. 5) further indicates an excess of lateral heat flow over vertical. The higher Ro-values for the wells Nestos-1 and -3 at depths 2500 and 3700 m suggest that these wells are closer to the magmatic intrusion than the Nestos 2 well. An aeromagnetic anomaly 7 km E-SE of well Nestos-3 might correspond to such an intrusive body.

Isotopic data on deformed, probably Palaeozoic magmatic rocks near Kavala, quoted by Kokkinakis (1980), suggest also a re-heating episode in Miocene times.

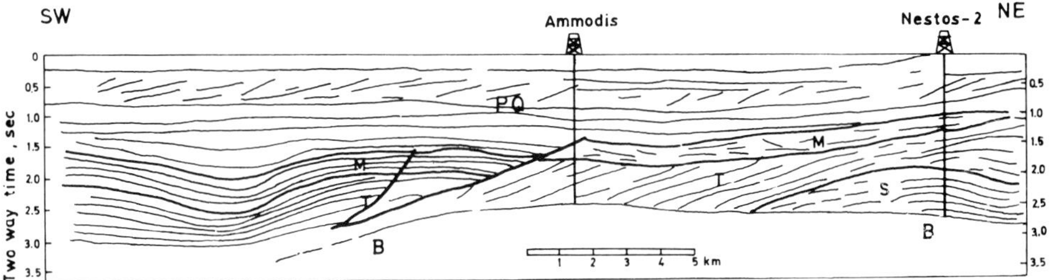

FIG. 3. Schematic seismic section AA′. PQ: Plioquaternary, M: Messinian, T: Tortonian, S: Serravallian, B: metamorphic basement.

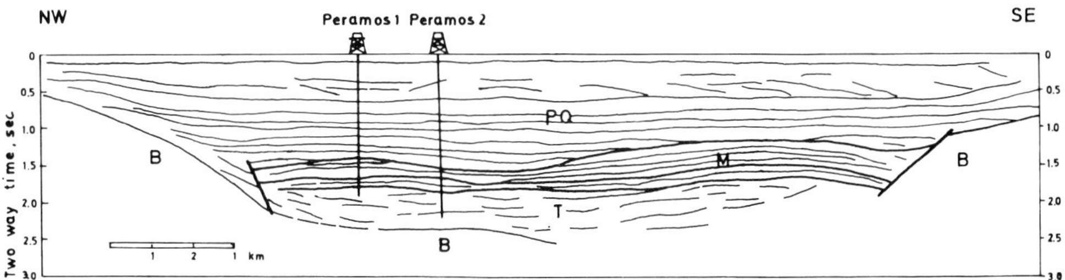

FIG. 4. Schematic seismic section BB′. Geological formations the same as in Fig. 3.

Radiometric dating indicates a minimum crystallization age of Upper Carboniferous for these magmatic rocks, whereas K-Ar data support subsequent re-heating. More specifically, an age of 15.5 ± 0.5 Ma years is reported for a granitic vein only 15 Km north-west of Nestos wells. Similarly a re-heating age of 17.8 ± 0.8 Ma is reported for magmatic rocks from the Symvolon Mountains west of Kavala. In addition, Kronberg *et al.* (1970) quoted radiometric ages between 14 and 35 Ma from various granodiorites in the area between the Strimon and Nestos rivers. The Serravallian thermal event in the Nestos Delta also corresponds in age and character to the Middle Miocene local anatexis and acid plutonism episode in the central and southern Aegean area (Dürr *et al.* 1978).

ACKNOWLEDGEMENTS: Thanks are expressed to Mr. A. Papathanassopoulos, Managing Director and to Dr. J. Samouilidis, General Technical Manager of Public Petroleum Corporation, for permission to publish this paper and for financial support to attend this conference. Thanks are also due to Dr. A. Kokkinakis and Professor C. Sideris for providing useful radiometric data.

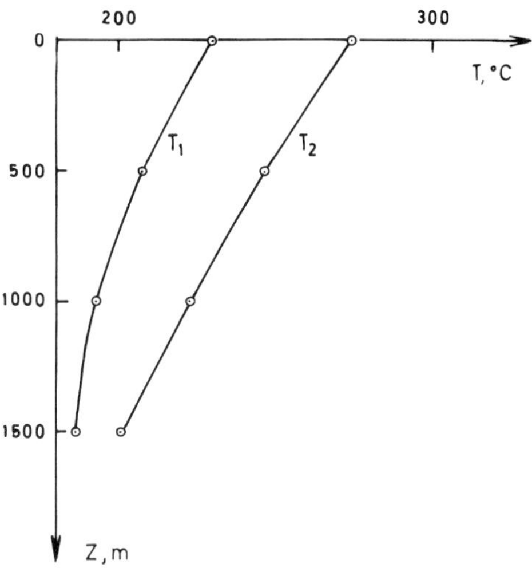

FIG. 5. Estimated temperatures of isothermal heating during the Serravallian. Z in metres is measured from the bottom of the Serravallian sediments.

References

DÜRR, S., ALTHERR, R., KELLER, J., OKRUSCH, M. & SEIDEL, E. 1978. The median Aegean crystalline belt: stratigraphy, structure, metamorphism, magmatism. *In*: CLOSS, H., ROEDER, D. & SCHMIDT, K. (eds.) *Alps, Apennines, Hellenides*. Interunion Commission on Geodynamics. Scientific Report No 38, Stuttgart, 455–501.

KOUSPARIS, D. 1979. *Seismic stratigraphy and basin*

development-Nestos Delta area, Northeastern Greece. Unpublished PhD thesis, University of Tulsa, U.S.A. 161 pp.

Kettel, D. 1981. Maturitätberechnungen für das nordwestdeutsche Oberkarbon-ein Test verschiedener Methoden. *Erdöl-Erdgas Z.* **97,** 395–404.

Kokkinakis, A. 1980. Alterbeziehungen zwischen Metamorphosen, mechanischen Deformationen und Intrusionen am Südrand des Rhodope—Massivs (Makedonien, Griechenland). *Geol. Rundschau*, **69,** 726–44.

Kronberg, P., Meyer, W. & Pilger, A. 1970. Geologie der Rila—Rhodope—Masse zwischen Strimon und Nestos (Nordgriechenland). *Beih. Geol. Jb.* **88,** 133–80.

Waples, D. W. 1980. Time and temperature in petroleum formation: application of Lopatin's method to petroleum exploration *Bull. Am. Assoc. Petrol. Geol.* **64,** 916–26.

E. D. Chiotis, Public Petroleum Corporation of Greece, 54 Academias St., Athens (143).

Neogene to Quaternary geodynamics of the area of the Ionian Sea and surrounding land masses

Frank H. Fabricius

SUMMARY: Investigations in the north-eastern and central Ionian Sea document important palaeogeographic and palaeoceanographic changes since Neogene time. These findings support the model of shallow basinal evaporation during Messinian time. On this basis the rate of subsidence of the Ionian Sea is calculated for the last 6 Ma, at about 1 m per 1000 years for basinal settings and half this for marginal settings.

Elevated terraces corresponding to high glacio-eustatic stages in the Quaternary are frequent on Mediterranean coasts at heights up to more than 150 m above present sea level. Together with elevated marine deposits of Pliocene age, they testify to general coastal uplift. These epeirogenic tectonics are interpreted as compensation for the subsidence of Mediterranean basins.

In its present general form the Mediterranean Sea has existed since Pliocene time. The first indications of the foundering of the Ionian basin are Messinian evaporites. The formation of such huge basins is interpreted as the result of the transformation of former continental crust to 'intermediate' or 'oceanized' crust. The important fast vertical tectonics of the margins of 'Mediterranean type' differ considerably from those of the Atlantic or Pacific type.

The sea floor and coastlines of the Mediterranean Sea are young features, not older than about 5 Ma. The re-flooding through the Strait of Gibraltar at the beginning of Pliocene time, after desiccation in Late Miocene time (Messinian stage), may be taken as the 'date of birth' of the modern Mediterranean. Before this we were dealing with the Tethys ocean, which in most areas lasted at least from Mesozoic to Middle Palaeogene, approximately the end of the Eocene, (ca. 40 Ma). The Oligocene to Middle Miocene may be considered as a transitional period. Although there are differences, the evolution of the Western and Eastern Mediterranean can not be totally divorced from either 'Tethyan' or later time. The area we are concerned with is the Ionian Sea, in the broad sense, including the Sirte Sea, and the Pelagian Sea.

Within the Neogene geological evolution of the western part of the Eastern Mediterranean, one has to consider two unusual periods, the Messinian and Quaternary, and a more 'normal' one, during Pliocene time. The Messinian is unusual because of desiccation of almost the entire Mediterranean area, and the Quaternary is atypical mainly because of the glacially-induced eustatic sea level changes, which affected the semi-enclosed Mediterranean more than the open oceans.

To decipher the geodynamics one has to take account of both sub-sea floor and land evidence. Information on the palaeobathymetry at different times and localities is needed. As palaeontological or biological deep-sea indicators are still controversial, we still mainly have to rely on facies analyses or indirect evidence including 'neotectonics', and features recorded by reflection seismic profiles.

Messinian time

Below the Ionian basin, evaporitic sediments more than 1000 m thick have been recorded geophysically (e.g. Finetti & Morelli 1973; Hinz 1974; Morelli 1975; Finetti 1981) and in the upper part have been provided by DSDP drilling (Site 374) (Fabricius *et al* 1978; 927; Hsü & Montadert *et al.* 1978: 175). The evaporites extend south into the area of the Sirte basin (or Sirte Rise) (Finetti & Morelli 1973; Morelli 1975). Morelli (1981) also assumes the presence of more than 1000 m of evaporites below the Mediterranean Ridge. His inferred 'salt domes' are questioned by Hieke (1982). Below the Calabrian Rise (or Messinian Cone) information is still limited (Finetti 1981). For the area from the Strait of Otranto to Corfu, Funk (in press) (Fig. 1) showed that Upper Miocene to Lower Pliocene palaeogeography and zones of evaporite deposition (Heimann 1977) do not fit with the present coast and island configuration. On the Pelagian block evaporites were mainly deposited within grabens which transect this part of the African continental plate in a NW-SE direction (Winnock 1981).

The facies patterns of Upper Miocene deposits indicate shallow-water sedimentation within a shallow basinal setting (Fabricius

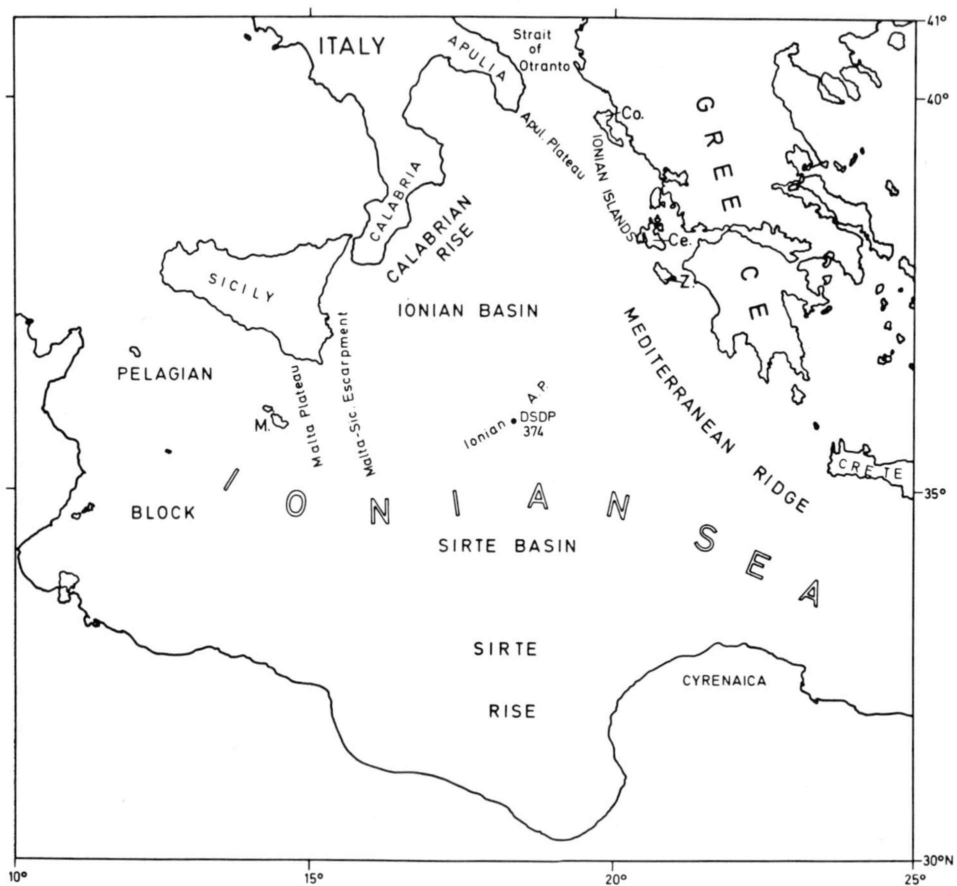

FIG. 1. Index map of the Ionian Sea.—Ce. = Kephallinia; Co. = Corfu; M. = Malta; Z. = Zakynthos.

1980). The sea-level of the Ionian basin in the broad sense was not very much below world-wide sea-level. These ideas, although not very new for Mediterranean land-geologists, were given serious consideration by relatively few participants of the two Deep Sea Drilling Project legs in the Mediterranean (c.f. Nesteroff 1973, for Leg 13; Fabricius *et al.* 1978, for Leg 42A). Two explanations exist to account for the thickness of the evaporites. Either the evaporites progressively filled pre-existing morphological basins which would have limited the thickness of evaporites to the depth of the basins; or the basins (geological, but not necessarily morphological) have been sinking at a similar rate to sedimentation.

In the first case the evaporites of the Pelagian area should have extended onto the flanks of the grabens, but this does not seem to be the case. In marginal settings in particular, the Messinian evaporites are generally found just below normal marine sediments of Lower Pliocene age (Heimann 1977, Schmolin 1981, 1983). This indicates only a small difference in sea level between uppermost Messinian and lowest Pliocene time. Therefore, the thicker deposits of Messinian evaporite, for example below the Ionian abyssal plain, seem to be good evidence for basin subsidence during Messinian time. For the Ionian Sea an average rate of subsidence can thus be calculated at about 1 m per 1000 years.

Plio-Quaternary time

For both Pliocene and Quaternary time, beginning at about 5 and 1.8 Ma respectively, basinal subsidence of the same order of magnitude of about 1 metre per 1000 years can be calculated from the depth below present sea level of the top of the Miocene (or base of Pliocene) strata, minus the assumed depth of the sea floor at the end of Messinian time. Estimating about 500 m for the latter, the average rate of subsidence would be about 0.9 m per 1000 years. This is probably too high in view of the present water depth of the Ionian abyssal plain which exceeds 4000 m. Even for an initial basinal depth of 1000 m the average rate would be about 0.7 m per 1000 years.

Vertical movement during Quaternary time

It is difficult to calculate basinal subsidence for Quaternary time in the absence of precise and reliable palaeobathymetric data for the Pliocene/Pleistocene boundary. Nevertheless, we have indirect evidence for constant basinal subsidence from coastal uplift during Quaternary time.

In many Mediterranean areas of elevated coastal terrace, coastal marine sediments and/or marine erosion cliffs ('notches') have been described (cf. Schwarzbach 1974; Woldstedt 1958, 1961, 1965; Horowitz 1979). The oldest relics of Quaternary coastlines have been taken to be those highest above the present Mediterranean sea level. Dating by ^{14}C as well as by archaeological methods has confirmed this in many areas (Horowitz 1979), but where data are inadequate many questions remain about the exact age and the nature of such uplift features.

In general, however, we can state:

(1) During Quaternary time world-wide eustatic rise was at most 50 m (Schwarzbach 1974). Nevertheless, the terraces are present at altitudes above sea level of: more than +150 m, Calabrian transgression; about +100 m, Sicilian transgression; about +60 m, Milazzian transgression; about +30 m, Tyrrhenian transgression; about +20 m, Monastirian transgression; near +5 m, Flandrian (Nizza) transgression.

The Milazzian and higher terraces could not have been caused by eustatic sea level changes alone (Fig. 2).

(2) In regions of similar tectonic style the lower terraces are always younger than the higher ones. This applies to many areas of the Mediterranean, but not in the open oceans.

A relation is thus postulated to exist between regional Mediterranean tectonics and these terraces. Many of the circum-Mediterranean coastal regions must have risen more or less steadily during Quaternary time. This general uplift trend does not exclude the possibility of local tectonic uplift or subsidence.

The coastal uplift can be compared to that generally found on the flanks of large grabens (cf. Schmidt-Thomé 1972). This is sometimes of the same order of magnitude as the subsidence, but is usually less, as for example in the Rhine graben or the East African graben system. In these cases the uplift is generally linked to the downward movement.

It would be premature to specify the amount of coastal uplift associated with basinal subsidence. Studies of Recent earthquakes show that the amount of vertical movement of the downthrown side exceeds that of the upthrown side by a factor of 10 (Jackson *et al.* 1982). Nevertheless, in many areas with high coastal terraces of Quaternary age, elevated marine deposits of Pliocene age are also known. The altitudes of the latter above present sea level may be several hundred metres or more, e.g. in Sicily, Calabria and Greece.

Another important observation is that the coastal areas of the Ionian Sea (see above) can be classified into: (i) areas of major Quaternary, and generally also Pliocene, uplift; and (ii), areas without such Plio-Quaternary uplift. The two types of area may be linked by transition zones. There is a general relationship between coastal movements and the inclination of the adjacent continental margin and size of shelf related to it. As examples of a steep continental margin and 'normal' Quaternary coastal uplift we have the eastern Peloponnese, the Ionian islands, Crete, Cyrenaica, Calabria and Sicily. Except for Cyrenaica, these areas are part of the Alpine system and have suffered Tertiary folding. Nevertheless, most of the Plio-Quaternary coastal and shelf tectonics are of block-faulting type.

The other case involves a gently inclined continental margin and no geologically important Quaternary coastal uplift as along the Pelagian Sea coast, the Sirte embayment and Apulia. In these regions the shelves extend far into the sea, as in the Pelagian Block and the southern extension of Apulia, or a distinct shelf break is missing, as in the Sirte embayment, where a distinction between 'shelf' and 'slope' is almost impossible. North-east Tunisia, which is influenced by Tyrrhenian Sea tectonics is an exception. The escarpment east of Malta is another, as a very steep 'continental margin' exists here but without coasts, except on Malta itself. Whether or not this escarpment is a true continental margin, there is, nevertheless, morphological uplift at the outermost rim of the Pelagian Block (e.g. the Maltese Platform).

Geodynamic implications

The nature of the crust beneath the central Ionian Sea, whether oceanic or 'oceanized'

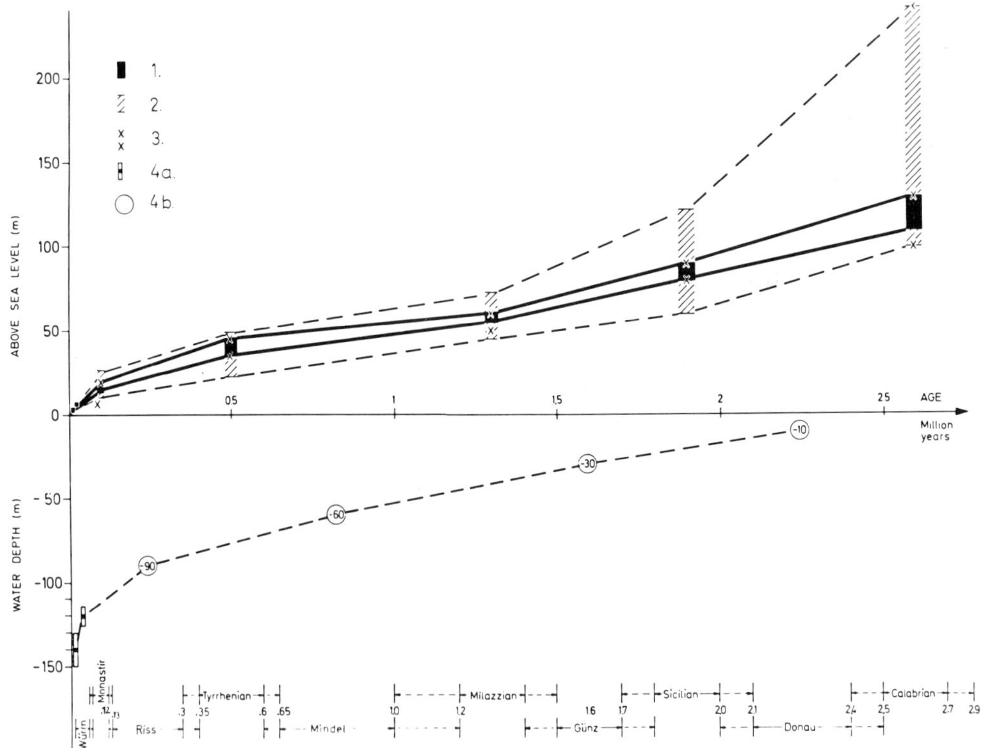

FIG. 2. Vertical distribution, relative to present Mediterranean sea level, of marine terraces and related deposits of Quaternary glacio-eustatic high and low stages (after Horowitz 1979): 1. Interval of altitude of sea level maxima; 2. range of altitudes around the Mediterranean; 3. Elevation of beach-rock in the Carmel region (Israel); 4a and 4b: minimum of Mediterranean glacio-eustatic sea level (4b: less certain). Older submarine features are less precise. Onshore and submarine 'terraces', however, show a parallel trend and are more indicative of a tectonic origin than merely eustatic changes of sea level.

continental, is still much debated (Lort 1977, Selli 1981). The rather fast vertical foundering of the Ionian basin during Neogene and Quaternary times described above, and that of other Mediterranean regions is interpreted as isostatic adjustment. Its probable causes are assumed to be a response to an increase in the average density of the northern rim of the African continental plate, perhaps due to attenuation of the sialic crust, and/or early intrusion of basaltic(?) igneous rock, during a time of extensional tectonics (Finetti 1981). This extension was still active in some regions in the Quaternary (Avedik & Hieke 1981). The foundering of huge masses beneath the Ionian basin could have produced marginal uplift as a volume compensation. Marginal uplift is especially pronounced in coastal areas of the Ionian Sea adjacent to steep slopes. In other areas, for example the Pelagian Block and the Sirte Rise, where slopes are more gentle, 'oceanization' may not have proceeded to any significant extent.

It is very likely that the Pelagian Block is a relict from the very early stages of formation of the Mediterranean Sea (Late Miocene/Early Pliocene). Since then the Pelagian Block has been tilted slightly to the east, but without much subsidence. The area of the Sirte Sea has reacted somewhat differently, but nevertheless is comparable to the Pelagian Block. In the north, near the Sirte abyssal plain, the rim of the African continental plate was forced down, with decreasing intensity to the south, thus forming a ramp without a recognizable shelf break.

The stability of the Pelagian Block contrasts strongly with the mobility of the area to the east. In the north-eastern Pelagian area (Malta Platform and parts of Sicily), the steep Sicily-Malta Escarpment must be regarded as the expression of a deep crustal or sub-crustal discontinuity, forming a still active (but possibly old, cf. Scandone *et al.* 1981) mega-lineament. To the south of the Medina (Malta) Ridge this escarpment behaved more 'scissors-like': east of the Medina Bank the fault still has a significant throw but it passes into a flexure as it approaches the coast of Africa near Missurata.

In the Tyrrhenian Sea fast vertical movements have been described by Wezel (1981), which produced roughly circular basins of a diameter exceeding 100 km in some cases. He called this intraplate deformation 'krikogenesis'. Within the Ionian area this specific mechanism has not been considered relevant because of the significant morphological, geological, and especially volcanological, contrasts with the Tyrrhenian Sea. In this respect, one might speculate whether the Ionian Sea extends over several very large basins (e.g. the Ionian basin and the Sirte basin or Rise), which could have related to earlier mantle diapirism (cf. Selli 1981) leading to oceanization and foundering. Such basin formation by foundering of large areas is also considered possible for other non-Mediterranean areas, as for instance, the Black Sea and the Caspian Sea (Jascó; this volume).

Conclusions

(1) In the Ionian Basin, and in many other Mediterranean basins there is evidence of major basinal subsidence, at least since Upper Miocene time. The average rate of subsidence is of the order of 0.5 to 1 m/1000 years for Messinian and Plio-Quaternary time. During Messinian time the formation of morphological basins was balanced by the deposition of shallow water evaporites. By contrast, during the Plio-Quaternary there was little compensation of the central basinal subsidence by sediment in-fill, resulting in the 'starved' basin of the present Eastern Mediterranean.

(2) Except for a few areas, like the Pelagian Sea, Apulian Platform, and the Sirte Embayment, the subsidence of the basinal area was probably compensated in part by uplift of the margins of adjacent landmasses. This can be demonstrated by elevated marine terraces of Quaternary age in many coastal regions. Besides these epeirogenic movements other ways of mass compensation are likely to occur to which perhaps, even the Alpine orogeny might be related.

(3) It is very probable that the palaeogeography and palaeo-oceanography of the Ionian Sea has changed drastically in the last 5 Ma. For instance, in the region to the north-west or west of the Ionian islands, from Corfu to Kephallinia, there is evidence of the existence of a former land-mass, possibly the extension of Apulia, from which clastic sediments were derived.

(4) Geophysical data (Morelli 1981, and many others) have demonstrated that the crust below the Ionian Basin differs from normal oceanic crust and this also applies to other Mediterranean basins (Selli 1981). This crust can be called 'intermediate' or 'oceanized'. Selli speaks of a 'semi-oceanic or thinned crust'. If it is modified continental crust, as most data suggest, then the Eastern Mediterranean could be regarded as an assemblage of marginal basins related to several continents. These margins are now in different stages of transformation and variable states of deformation. This dynamic situation is unlike either normal 'passive' (e.g. Atlantic) or the 'active' (e.g. Pacific) continental margins. Most margins of the Eastern Mediterranean, and probably of the Western Mediterranean also, should be regarded as 'passive margins of an active nature'; or it might be more appropriate to speak simply of a 'Mediterranean-type' continental margin.

ACKNOWLEDGEMENT: The investigations in the Ionian area were supported by the Deutsche Forschungsgemeinschaft over many years. I wish to acknowledge the kind help given by the following: my colleagues Drs K. Braune, W. Hieke, J. Schmolin and Mr S. G. Funk for many discussions on the 'genesis' of the Eastern Mediterranean Sea, which contributed much to the text; to Professor Dr. P. Sonnenfeld, Windsor/Ontario, and Professor Dr. F. C. Wezel, Urbino, for their critical reading of the manuscript and helpful comments; to Drs. J. E. Dixon and A. H. F. Robertson, Edinburgh, for correcting the English typescript; to Mrs. K. Bögel and Mr. K. Haas for technical help.

References

AVEDIK, F. & HIEKE, W. 1981. Reflection seismic profiles from the central Ionian Sea (Mediterranean) and their geodynamic interpretation. 'Meteor' *Forschungsergeb. C*, **34**, 49–64.

FABRICIUS, F. H. 1980. Messinian evaporitic facies in the Mediterranean area: Geodynamic implications. *In*: CLOSS, H., V. GEHLEN, K., ILLIES, H., KUNTZ, E., NEUMANN, J., SEIBOLD, E. (eds.): *Mobile Earth*: Internat. Geodynamics Project; final report of the Fed. Rep. of Germany/ Deutsche Forschungsgemeinschaft, 91–6.

——, HEIMANN, K. O., & BRAUNE, K. 1978. Comparison of Site 374 with circum-Mediterranean land sections: Implications for the Messinian 'Salinity Crisis' on the basis of a 'Dynamic Model'. *In*: HSÜ, K., MONTADERT, L. (eds.) *Initial Repts. DSDP*, **42**, 927–42.

FINETTI, I. 1981. Geophysical study on the evolution of the Ionian Sea. *In*: WEZEL, F. C. (ed.) Sedimentary Basins of Mediterranean margins. *Proc. C.N.R. Intern. Conf. It. Proj. Oceanogr., Bologna, Technoprint*, 465–84.

FINETTI, I. & MORELLI, C. 1973. Geophysical exploration of the Mediterranean Sea. *Boll. Geofis. Teoric. Applic.* **15** (N.60), 263–341.

FUNK, G. (in press). Paleogeographical results of Neogene sedimentation on NW-Corfu. *Rapp. Comm. Int. Mer Médit.*

HEIMANN, K. O. 1977. *Die Fazies des Messins und untersten Pliozäns auf den Ionischen Inseln (Zakynthos, Kephallinia, Korfu/Grieschenland.* Thesis, Dr. Techn. Univ. Munich, 158 pp.

HEIKE, W. 1982. 'Reflektor M' and diapiric structures in the Ionian Sea (Eastern Mediterranean). *Mar. Geol.* **46,** 235–44.

HINZ, K. 1974. Results of seismic refraction and seismic reflection measurements in the Ionian Sea. *Geol. Jahrb.* **E2,** 33–65.

HOROWITZ, A. 1979. *The Quaternary of Israel.* 394 pp. Academic Press, New York.

HSÜ, K., MONTADERT, L. *et al.* 1978. Site 374; Messina Abyssal Plain. Initial Repts. *DSDP.* **42,** 175–217.

JACKSON, J. A., GAGNEPAIN, J., HOUSMAN, G., KING, G. C. P., PAPADIMITROU, P., SOUFLERIS, C., & VIRIEUX, J., 1982. Seismicity, normal faulting, and the geomorphological development of the Gulf of Corinth (Greece): the Corinth earthquakes of February and March 1981. *Earth and Planet. Sci. Let.* **57,** 377–97.

LORT, J M. 1977. Geophysics of the Mediterranean Sea basins. *In*: NAIRN, E. M., KANES, W. H. & STEHLI, F. G. (eds.) *The Ocean Basins and Margins. 4A* The Eastern Mediterranean, pp. 151–213. Plenum Press, New York.

MORELLI, C. 1975. Geophysics of the Mediterranean. *Bull. CIESM*, **7,** 27–111.

—— 1981. Gravity anomalies and crustal structures connected with the Mediterranean margins. *In*: WEZEL, F. C. *Sedimentary Basins of Mediterranean Margins.* C.N.R. Ital. Proj. of Oceanogr. Bologna, 33–53.

NESTEROFF, W. D. 1973. Un modèl pour les évaporites messiniennes en Méditerranée des basins peu profonds avec dépôts d'évaporites lagunaires. *In*: DROOGER, C. W. (ed.) *Messinian Events in the Mediterranean.* Kon. Ned. Akad. Wetensch. Amsterdam, 68–81.

SCANDONE, P., PATACCA, E., RADOICIC, R., RYAN, W. B. F., CITA, M. B., RAWSON, M., CHEZAR, H., MILLER, E., MCKENZIE, J. & ROSSI, S. 1981. Mesozoic and Cenozoic rocks from Malta Escarpment (Central Mediterranean). *Am. Assoc. Petrol. Geol. Bull.* **65,** 1299–319.

SCHMIDT-THOMÉ, P. 1972. Tektonik. *In*: BRINKMANN, R. (ed.) *Lehrbuch der Allgemeinen Geologie. II*, 597 pp. Enke, Stuttgart.

SCHMOLIN, J. 1981. Some results about the sedimentology of Pliocene on the Ionian island Kephallinia and Zakynthos (Greece). *Rapp. Comm. int. Mer Médit.*, **27,** 87–9.

—— 1983. *Sedimentologisch-fazielle Untersuchungen des Pliozäns auf den Ionischen Inseln Kephallinia und Zakynthos* (*West-Grieschenland*). Thesis. Dr. Techn. Univ. Munich, 141 pp.

SCHWARZBACH, M. 1974. *Das Klima der Vorzeit. Einführung in die Paläoklimatologie*, 3rd ed. 380 pp, Enke, Stuttgart.

SELLI, R. 1981. Thoughts on the geology of the Mediterranean region. *In*: WEZEL, F. C. (ed.). *Sedimentary Basins of Mediterranean Margins.* C.N.R. Ital. Proj. Oceanogr. Bologna, 489–501.

WEZEL, F. C. (ed.) 1981. *Sedimentary Basins of Mediterranean Margins.* C.N.R. Ital. Proj. Oceanogr., Bologna, Technoprint, XXIII + 520 pp.

WINNOCK, E. 1981. Structure du block pélagienne. *In*: WEZEL, F. C. (ed.) *Sedimentary Basins of Mediterranean Margins.* C.N.R. Ital. Proj. of Oceanogr. Bologna, 445–64.

—— & BEA, F. 1979. Structure de la Mer Pélagienne. *In*: BUROLLET, P. F., CLAIREFOND, P., & WINNOČK, E. (eds.) *Géologie méditerranéenne; La mer Pélagienne.* Annal. Univ. de Provence, *VI*, (1), 35–40.

WOLDSTEDT, P. 1958–1965. *Das Eiszeitalter. Grundlagen einer Geologie des Quartärs. 1.* (1961) *Die allgemeinen Erscheinungen des Eiszeitalters.* (3rd ed.) 374 pp. *2.* (1958) *Europa, Vorderasien und Nordafrika.* (2nd rev. ed.), 438 pp. *3.* (1965) *Afrika, Asien, Australien und Amerika.* (2nd rev. ed.), 328 pp. Enke, Stuttgart.

FRANK H. FABRICIUS: Institute of Geology and Mineralogy; Division of Sedimentology and Marine Geology. Munich Technical University, D-8046 Garching, Lichtenbergstrasse 4, Federal Republic of Germany.